MW00450079

NUCLEAR MAGNETIC RESONANCE STUDIES OF INTERFACIAL PHENOMENA

SURFACTANT SCIENCE SERIES

FOUNDING EDITOR

MARTIN J. SCHICK

1918–1998

SERIES EDITOR

ARTHUR T. HUBBARD

Santa Barbara Science Project

Santa Barbara, California

1. Nonionic Surfactants, *edited by Martin J. Schick* (see also Volumes 19, 23, and 60)
2. Solvent Properties of Surfactant Solutions, *edited by Kozo Shinoda* (see Volume 55)
3. Surfactant Biodegradation, *R. D. Swisher* (see Volume 18)
4. Cationic Surfactants, *edited by Eric Jungermann* (see also Volumes 34, 37, and 53)
5. Detergency: Theory and Test Methods (in three parts), *edited by W. G. Cutler and R. C. Davis* (see also Volume 20)
6. Emulsions and Emulsion Technology (in three parts), *edited by Kenneth J. Lissant*
7. Anionic Surfactants (in two parts), *edited by Warner M. Linfield* (see Volume 56)
8. Anionic Surfactants: Chemical Analysis, *edited by John Cross*
9. Stabilization of Colloidal Dispersions by Polymer Adsorption, *Tatsuo Sato and Richard Ruch*
10. Anionic Surfactants: Biochemistry, Toxicology, Dermatology, *edited by Christian Gloxhuber* (see Volume 43)
11. Anionic Surfactants: Physical Chemistry of Surfactant Action, *edited by E. H. Lucassen-Reynders*
12. Amphoteric Surfactants, *edited by B. R. Bluestein and Clifford L. Hilton* (see Volume 59)
13. Demulsification: Industrial Applications, *Kenneth J. Lissant*
14. Surfactants in Textile Processing, *Arved Datyner*
15. Electrical Phenomena at Interfaces: Fundamentals, Measurements, and Applications, *edited by Ayao Kitahara and Akira Watanabe*
16. Surfactants in Cosmetics, *edited by Martin M. Rieger* (see Volume 68)
17. Interfacial Phenomena: Equilibrium and Dynamic Effects, *Clarence A. Miller and P. Neogi*
18. Surfactant Biodegradation: Second Edition, Revised and Expanded, *R. D. Swisher*
19. Nonionic Surfactants: Chemical Analysis, *edited by John Cross*
20. Detergency: Theory and Technology, *edited by W. Gale Cutler and Erik Kissa*
21. Interfacial Phenomena in Apolar Media, *edited by Hans-Friedrich Eicke and Geoffrey D. Parfitt*
22. Surfactant Solutions: New Methods of Investigation, *edited by Raoul Zana*
23. Nonionic Surfactants: Physical Chemistry, *edited by Martin J. Schick*
24. Microemulsion Systems, *edited by Henri L. Rosano and Marc Clausse*
25. Biosurfactants and Biotechnology, *edited by Naim Kosaric, W. L. Cairns, and Neil C. C. Gray*
26. Surfactants in Emerging Technologies, *edited by Milton J. Rosen*
27. Reagents in Mineral Technology, *edited by P. Somasundaran and Brij M. Moudgil*
28. Surfactants in Chemical/Process Engineering, *edited by Darsh T. Wasan, Martin E. Ginn, and Dinesh O. Shah*

29. Thin Liquid Films, *edited by I. B. Ivanov*
30. Microemulsions and Related Systems: Formulation, Solvency, and Physical Properties, *edited by Maurice Bourrel and Robert S. Schechter*
31. Crystallization and Polymorphism of Fats and Fatty Acids, *edited by Nissim Garti and Kiyotaka Sato*
32. Interfacial Phenomena in Coal Technology, *edited by Gregory D. Botsaris and Yuli M. Glazman*
33. Surfactant-Based Separation Processes, *edited by John F. Scamehorn and Jeffrey H. Harwell*
34. Cationic Surfactants: Organic Chemistry, *edited by James M. Richmond*
35. Alkylene Oxides and Their Polymers, *F. E. Bailey, Jr., and Joseph V. Koleske*
36. Interfacial Phenomena in Petroleum Recovery, *edited by Norman R. Morrow*
37. Cationic Surfactants: Physical Chemistry, *edited by Donn N. Rubingh and Paul M. Holland*
38. Kinetics and Catalysis in Microheterogeneous Systems, *edited by M. Grätzel and K. Kalyanasundaram*
39. Interfacial Phenomena in Biological Systems, *edited by Max Bender*
40. Analysis of Surfactants, *Thomas M. Schmitt* (see Volume 96)
41. Light Scattering by Liquid Surfaces and Complementary Techniques, *edited by Dominique Langevin*
42. Polymeric Surfactants, *Irja Piirma*
43. Anionic Surfactants: Biochemistry, Toxicology, Dermatology, Second Edition, Revised and Expanded, *edited by Christian Gloxhuber and Klaus Künstler*
44. Organized Solutions: Surfactants in Science and Technology, *edited by Stig E. Friberg and Björn Lindman*
45. Defoaming: Theory and Industrial Applications, *edited by P. R. Garrett*
46. Mixed Surfactant Systems, *edited by Keizo Ogino and Masahiko Abe*
47. Coagulation and Flocculation: Theory and Applications, *edited by Bohuslav Dobiás*
48. Biosurfactants: Production Properties Applications, *edited by Naim Kosaric*
49. Wettability, *edited by John C. Berg*
50. Fluorinated Surfactants: Synthesis Properties Applications, *Erik Kissa*
51. Surface and Colloid Chemistry in Advanced Ceramics Processing, *edited by Robert J. Pugh and Lennart Bergström*
52. Technological Applications of Dispersions, *edited by Robert B. McKay*
53. Cationic Surfactants: Analytical and Biological Evaluation, *edited by John Cross and Edward J. Singer*
54. Surfactants in Agrochemicals, *Tharwat F. Tadros*
55. Solubilization in Surfactant Aggregates, *edited by Sherril D. Christian and John F. Scamehorn*
56. Anionic Surfactants: Organic Chemistry, *edited by Helmut W. Stache*
57. Foams: Theory, Measurements, and Applications, *edited by Robert K. Prud'homme and Saad A. Khan*
58. The Preparation of Dispersions in Liquids, *H. N. Stein*
59. Amphoteric Surfactants: Second Edition, *edited by Eric G. Lomax*
60. Nonionic Surfactants: Polyoxyalkylene Block Copolymers, *edited by Vaughn M. Nace*
61. Emulsions and Emulsion Stability, *edited by Johan Sjöblom*
62. Vesicles, *edited by Morton Rosoff*

63. Applied Surface Thermodynamics, *edited by A. W. Neumann and Jan K. Spelt*
64. Surfactants in Solution, *edited by Arun K. Chattopadhyay and K. L. Mittal*
65. Detergents in the Environment, *edited by Milan Johann Schwuger*
66. Industrial Applications of Microemulsions, *edited by Conxita Solans and Hironobu Kunieda*
67. Liquid Detergents, *edited by Kuo-Yann Lai*
68. Surfactants in Cosmetics: Second Edition, Revised and Expanded, *edited by Martin M. Rieger and Linda D. Rhein*
69. Enzymes in Detergency, *edited by Jan H. van Ee, Onno Misset, and Erik J. Baas*
70. Structure-Performance Relationships in Surfactants, *edited by Kunio Esumi and Minoru Ueno*
71. Powdered Detergents, *edited by Michael S. Showell*
72. Nonionic Surfactants: Organic Chemistry, *edited by Nico M. van Os*
73. Anionic Surfactants: Analytical Chemistry, Second Edition, Revised and Expanded, *edited by John Cross*
74. Novel Surfactants: Preparation, Applications, and Biodegradability, *edited by Krister Holmberg*
75. Biopolymers at Interfaces, *edited by Martin Malmsten*
76. Electrical Phenomena at Interfaces: Fundamentals, Measurements, and Applications, Second Edition, Revised and Expanded, *edited by Hiroyuki Ohshima and Kunio Furusawa*
77. Polymer-Surfactant Systems, *edited by Jan C. T. Kwak*
78. Surfaces of Nanoparticles and Porous Materials, *edited by James A. Schwarz and Cristian I. Contescu*
79. Surface Chemistry and Electrochemistry of Membranes, *edited by Torben Smith Sørensen*
80. Interfacial Phenomena in Chromatography, *edited by Emile Pefferkorn*
81. Solid–Liquid Dispersions, *Bohuslav Dobiás, Xueping Qiu, and Wolfgang von Rybinski*
82. Handbook of Detergents, editor in chief: Uri Zoller Part A: Properties, *edited by Guy Broze*
83. Modern Characterization Methods of Surfactant Systems, *edited by Bernard P. Binks*
84. Dispersions: Characterization, Testing, and Measurement, *Erik Kissa*
85. Interfacial Forces and Fields: Theory and Applications, *edited by Jyh-Ping Hsu*
86. *Silicone Surfactants, edited by Randal M. Hill*
87. Surface Characterization Methods: Principles, Techniques, and Applications, *edited by Andrew J. Milling*
88. Interfacial Dynamics, *edited by Nikola Kallay*
89. Computational Methods in Surface and Colloid Science, *edited by Malgorzata Borówko*
90. Adsorption on Silica Surfaces, *edited by Eugène Papirer*
91. Nonionic Surfactants: Alkyl Polyglucosides, *edited by Dieter Balzer and Harald Lüders*
92. Fine Particles: Synthesis, Characterization, and Mechanisms of Growth, *edited by Tadao Sugimoto*
93. Thermal Behavior of Dispersed Systems, *edited by Nissim Garti*

94. Surface Characteristics of Fibers and Textiles, *edited by Christopher M. Pastore and Paul Kiekens*
95. Liquid Interfaces in Chemical, Biological, and Pharmaceutical Applications, *edited by Alexander G. Volkov*
96. Analysis of Surfactants: Second Edition, Revised and Expanded, *Thomas M. Schmitt*
97. Fluorinated Surfactants and Repellents: Second Edition, Revised and Expanded, *Erik Kissa*
98. Detergency of Specialty Surfactants, *edited by Floyd E. Friedli*
99. Physical Chemistry of Polyelectrolytes, *edited by Tsetska Radeva*
100. Reactions and Synthesis in Surfactant Systems, *edited by John Texter*
101. Protein-Based Surfactants: Synthesis, Physicochemical Properties, and Applications, *edited by Ifendu A. Nnanna and Jiding Xia*
102. Chemical Properties of Material Surfaces, *Marek Kosmulski*
103. Oxide Surfaces, *edited by James A. Wingrave*
104. Polymers in Particulate Systems: Properties and Applications, *edited by Vincent A. Hackley, P. Somasundaran, and Jennifer A. Lewis*
105. Colloid and Surface Properties of Clays and Related Minerals, *Rossman F. Giese and Carel J. van Oss*
106. Interfacial Electrokinetics and Electrophoresis, *edited by Ángel V. Delgado*
107. Adsorption: Theory, Modeling, and Analysis, *edited by József Tóth*
108. Interfacial Applications in Environmental Engineering, *edited by Mark A. Keane*
109. Adsorption and Aggregation of Surfactants in Solution, *edited by K. L. Mittal and Dinesh O. Shah*
110. Biopolymers at Interfaces: Second Edition, Revised and Expanded, *edited by Martin Malmsten*
111. Biomolecular Films: Design, Function, and Applications, *edited by James F. Rusling*
112. Structure–Performance Relationships in Surfactants: Second Edition, Revised and Expanded, *edited by Kunio Esumi and Minoru Ueno*
113. Liquid Interfacial Systems: Oscillations and Instability, *Rudolph V. Birikh, Vladimir A. Briskman, Manuel G. Velarde, and Jean-Claude Legros*
114. Novel Surfactants: Preparation, Applications, and Biodegradability: Second Edition, Revised and Expanded, *edited by Krister Holmberg*
115. Colloidal Polymers: Synthesis and Characterization, *edited by Abdelhamid Elaissari*
116. Colloidal Biomolecules, Biomaterials, and Biomedical Applications, *edited by Abdelhamid Elaissari*
117. Gemini Surfactants: Synthesis, Interfacial and Solution-Phase Behavior, and Applications, *edited by Raoul Zana and Jiding Xia*
118. Colloidal Science of Flotation, *Anh V. Nguyen and Hans Joachim Schulze*
119. Surface and Interfacial Tension: Measurement, Theory, and Applications, *edited by Stanley Hartland*
120. Microporous Media: Synthesis, Properties, and Modeling, *Freddy Romm*
121. Handbook of Detergents, editor in chief: Uri Zoller, Part B: Environmental Impact, *edited by Uri Zoller*
122. Luminous Chemical Vapor Deposition and Interface Engineering, *Hirotsugu Yasuda*
123. Handbook of Detergents, editor in chief: Uri Zoller, Part C: Analysis, *edited by Heinrich Waldhoff and Rüdiger Spilker*
124. Mixed Surfactant Systems: Second Edition, Revised and Expanded, *edited by Masahiko Abe and John F. Scamehorn*

125. Dynamics of Surfactant Self-Assemblies: Micelles, Microemulsions, Vesicles and Lyotropic Phases, *edited by Raoul Zana*
126. Coagulation and Flocculation: Second Edition, *edited by Hansjoachim Stechemesser and Bohulav Dobiás*
127. Bicontinuous Liquid Crystals, *edited by Matthew L. Lynch and Patrick T. Spicer*
128. Handbook of Detergents, editor in chief: Uri Zoller, Part D: Formulation, *edited by Michael S. Showell*
129. Liquid Detergents: Second Edition, *edited by Kuo-Yann Lai*
130. Finely Dispersed Particles: Micro-, Nano-, and Atto-Engineering, *edited by Aleksandar M. Spasic and Jyh-Ping Hsu*
131. Colloidal Silica: Fundamentals and Applications, *edited by Horacio E. Bergna and William O. Roberts*
132. Emulsions and Emulsion Stability, Second Edition, *edited by Johan Sjöblom*
133. Micellar Catalysis, *Mohammad Niyaz Khan*
134. Molecular and Colloidal Electro-Optics, *Stoyl P. Stoylov and Maria V. Stoimenova*
135. Surfactants in Personal Care Products and Decorative Cosmetics, Third Edition, *edited by Linda D. Rhein, Mitchell Schlossman, Anthony O'Lenick, and P. Somasundaran*
136. Rheology of Particulate Dispersions and Composites, *Rajinder Pal*
137. Powders and Fibers: Interfacial Science and Applications, *edited by Michel Nardin and Eugène Papirer*
138. Wetting and Spreading Dynamics, *edited by Victor Starov, Manuel G. Velarde, and Clayton Radke*
139. Interfacial Phenomena: Equilibrium and Dynamic Effects, Second Edition, *edited by Clarence A. Miller and P. Neogi*
140. Giant Micelles: Properties and Applications, *edited by Raoul Zana and Eric W. Kaler*
141. Handbook of Detergents, editor in chief: Uri Zoller, Part E: Applications, *edited by Uri Zoller*
142. Handbook of Detergents, editor in chief: Uri Zoller, Part F: Production, *edited by Uri Zoller and co-edited by Paul Sosis*
143. Sugar-Based Surfactants: Fundamentals and Applications, *edited by Cristóbal Carnero Ruiz*
144. Microemulsions: Properties and Applications, *edited by Monzer Fanun*
145. Surface Charging and Points of Zero Charge, *Marek Kosmulski*
146. Structure and Functional Properties of Colloidal Systems, *edited by Roque Hidalgo-Álvarez*
147. Nanoscience: Colloidal and Interfacial Aspects, *edited by Victor M. Starov*
148. Interfacial Chemistry of Rocks and Soils, *Noémi M. Nagy and József Kónya*
149. Electrocatalysis: Computational, Experimental, and Industrial Aspects, *edited by Carlos Fernando Zinola*
150. Colloids in Drug Delivery, *edited by Monzer Fanun*
151. Applied Surface Thermodynamics: Second Edition, *edited by A. W. Neumann, Robert David, and Yi Y. Zuo*
152. Colloids in Biotechnology, *edited by Monzer Fanun*
153. Electrokinetic Particle Transport in Micro/Nano-fluidics: Direct Numerical Simulation Analysis, *Shizhi Qian and Ye Ai*
154. Nuclear Magnetic Resonance Studies of Interfacial Phenomena, *Vladimir M. Gun'ko and Vladimir V. Turov*

NUCLEAR MAGNETIC RESONANCE STUDIES OF INTERFACIAL PHENOMENA

Vladimir M. Gun'ko

Chuiko Institute of Surface Chemistry of National Academy of Sciences of Ukraine

Vladimir V. Turov

Chuiko Institute of Surface Chemistry of National Academy of Sciences of Ukraine

CRC Press is an imprint of the
Taylor & Francis Group, an **informa** business

CRC Press
Taylor & Francis Group
6000 Broken Sound Parkway NW, Suite 300
Boca Raton, FL 33487-2742

Printed on acid-free paper
Version Date: 20130123

International Standard Book Number-13: 978-1-4665-5168-8 (Hardback)

Library of Congress Cataloging-in-Publication Data

Gun'ko, Vladimir M.
Nuclear magnetic resonance studies of interfacial phenomena / Vladimir M. Gun'ko, Vladimir V. Turov.
pages cm. -- (Surfactant science series ; 154)
Includes bibliographical references and index.
ISBN 978-1-4665-5168-8 (hardback)
1. Surface chemistry. 2. Surface tension. 3. Adsorption. 4. Absorption. 5. Nuclear magnetic resonance spectroscopy. I. Turov, Vladimir V. II. Title.

QD506.G86 2013
541'.33--dc23 2012050922

Visit the Taylor & Francis Web site at
http://www.taylorandfrancis.com

and the CRC Press Web site at
http://www.crcpress.com

Contents

Glossary of Abbreviations

A	constant
A'	normalization factor
a	adsorption (cm^3/g)
$a_1, a_2, \ldots, a_{20}$	constants in Bender equation
$a_{0.98}$	nitrogen adsorption at $p/p_0 \approx 0.98$ (cm^3 STP/g)
a_m	BET monolayer adsorption (cm^3/g)
B_0	static magnetic field (T)
B_1	radio frequency field (T)
$B, C, D, E, F, G,$ and H	constants in Bender equation
C_C	content of pyrocarbon deposits (wt%)
C_{SiO_2}	concentration of silica (wt%)
C_{uw}	amount of unfrozen water (g) per gram of adsorbent (g/g)
C_{uw}^s	concentration of unfrozen water strongly bound to surface (wt%)
C_{uw}^w	concentration of unfrozen water weakly bound to surface (wt%)
c	BET coefficient (—)
c_s	BET coefficient for the adsorption on a flat surface (—)
D	fractal dimension (—)
D	optical density (—)
D_0	diffusion coefficient (m^2/s)
D_{AJ}	fractal dimension (—)
D_c	degree of condensation (—)
D_p	average pore diameter (nm)
$D_{p,BJH(a)}$	average pore diameter determined using BJH method for adsorption data (nm)
d_{av}	average pore size (nm)
dc	direct current
dC_{uw}/dR	derivative (g/g/nm)
dC_{uw}/dT	derivative (g/g/K)
$dV_{uw}(R)/dR$	derivative (cm^3/g/nm)
E	adsorption energy (kJ/mol)
E	interaction energy (kJ/mol)
F_p	intensity of the electrostatic field (kV/m)
f	fugacity (kJ cm^3/mol/g)
$f(R_p)$	differential pore size distribution (a.u.)
$f_S(R)$	distribution function of pore size with respect to surface area (a.u.)
$f_V(R)$	distribution function of pore size with respect to pore volume (a.u.)
h	hydration (g/g)
I	intensity of ^{1}H NMR signal (—)
$I_{0,i}$	intensity of the temperature distribution curve of phase i (a.u.)
I_c	intensity of ^{1}H NMR signal of water adsorbed from the gas phase (—)
I_{uw}	intensity of ^{1}H NMR signal of unfrozen water at $T < 273$ K (—)
j	order of vector $\vec{p}^j(\vec{f})$
k_{GT}	constant in Gibbs–Thomson equation (K nm)
$k(T)$	function of temperature in GT and IGT equations based on TSDC data (K nm)
M_w	molecular weight (a.m.u.)

m	the average number of the H atoms (m) of water molecules participating in the hydrogen bonds
N	number of the grid points for f (—)
N	average coordination number of nanoparticles in aggregates (—)
p	equilibrium pressure of nitrogen (Torr)
p_0	saturation pressure of nitrogen (Torr)
$\vec{p}^j(\vec{f})$	vector in MEM
Q	heat of adsorption (kJ/mol)
Q_p	adsorption heat in pores (kJ/mol)
Q_s	adsorption heat on flat surface (kJ/mol)
q	heat of wetting (kJ/mol)
q_H	charge on H-atom (a.u.)
R	radius of pores (nm)
Re	Reynolds number (—)
R_g	gas constant (kJ/K/mol)
R_p	pore radius (nm)
R_{max}	maximal pore radius on integration (nm)
R_{min}	minimal pore radius on integration (nm)
$r_k(p)$	pore radius in Kelvin equation nm
r_k	radius of pores occupied at pressure p (nm)
r_m	meniscus radius (nm)
r_{nr}	nanoroughness (nm)
S	entropy (kJ/mol/K)
S_{BET}	specific surface area by the Brunauer–Emmett–Teller method (m^2/g)
S_C	accessible specific surface area of carbon deposits (m^2/g)
S_{DFT}	specific surface area by the DFT method (m^2/g)
S_{DS}	specific surface area of nanopores determined using DS method (m^2/g)
S_{IGT}	specific surface area determined with IGT equation (m^2/g)
S_{macro}	corrected specific surface area of macropores (m^2/g)
S^*_{macro}	noncorrected specific surface area of macropores (m^2/g)
S_{meso}	corrected specific surface area of mesopores (m^2/g)
S^*_{meso}	noncorrected specific surface area of mesopores (m^2/g)
S_{nano}	corrected specific surface area of nanopores (m^2/g)
S^*_{nano}	noncorrected specific surface area of nanopores (m^2/g)
S_{sum}	corrected total specific surface area (m^2/g)
S_ϕ	specific surface area (m^2/g)
S_{meso}	specific surface area of mesopores estimated using the α_S plot method (m^2/g)
T	absolute temperature (K)
T_1	longitudinal relaxation time (s)
T_2	transverse relaxation time (s)
T_{cr}	crystallization temperature (K)
T_f	flame temperature (°C)
T_g	glass-transition temperature (K)
T_m	melting temperature (K)
$T_{m,\infty}$	bulk melting temperature (K)
$T_m(R)$	melting temperature of a frozen liquid in pores of radius R (K)
t	thickness of an adsorbed nitrogen layer (nm)
t_m	statistical thickness of a monolayer (nm)
$t(p,R_p)$	statistical thickness of an adsorbed layer (nm)
t_T	heating time (min)

U_{LV}	total surface energy (mJ/m^2)
VAR	regularizator
V_{em}	empty volume (cm^3/g)
V_{mac}	volume of macropores (cm^3/g)
V_{mes}	volume of mesopores (cm^3/g)
V_{nano}	volume of nanopores (cm^3/g)
V_p	total pore volume (cm^3/g)
$V_{uw}(R)$	volume of unfrozen water in pores of radius R (cm^3/g)
v_f	flow velocity (m/s)
v_M	liquid molar volume (cm^3/mol)
W	adsorption (cm^3 STP/g)
w	parameter in Kelvin equation (a.u.)
w_{ef}	effective parameter in Kelvin equation (a.u.)
X	normalized inverse transition temperature (1/K)
X_{ci}	normalized inverse transition temperature of phase i (1/K)
α	regularization parameter (—)
β	heating rate (K/s)
Γ	Gibbs adsorption
γ	surface tension (N/m)
γ_S	module of total changes in the Gibbs free energy of interfacial water (mJ/m^2)
$\Delta\varepsilon$	excess of the evaporation heat (kJ/mol)
ΔE_H	energy of the hydrogen bonds (kJ/mol)
ΔG	changes in Gibbs free energy of the interfacial water (kJ/mol)
ΔG^s	changes in Gibbs free energy of the strongly bound water (kJ/mol)
ΔG^w	changes in Gibbs free energy of the weakly bound water (kJ/mol)
ΔH_f	bulk enthalpy of fusion (kJ/mol)
ΔH_{im}	immersion enthalpy (kJ/mol)
$\Delta\mu_s$	changes in the chemical potential (—)
$\Delta S_{BET}/S_{BET}$	relative changes in the BET-specific surface area (—)
ΔT_m	melting point depression (deg)
$\Delta V_p/V_p$	relative changes in the pore volume (—)
$\Delta\sigma$	changes in the surface tension (mJ/m^2)
Δw	relative deviation from the pore model (a.u.)
δ	chemical shift (ppm)
δ_H	chemical shift of protons (ppm)
π	surface pressure (mJ m^{-2})
ρ	density (g/cm^3)
ρ_b	bulk density (g/cm^3)
ρ_f	fluid density in occupied pores (g/cm^3)
ρ_m	density of the multilayered adsorbate in pores (g/cm^3)
σ_i	width of the temperature distribution curve of phase I (a.u.)
σ_s	collision diameter of surface atoms (nm)
σ_{sl}	energy of solid–liquid interaction (kJ/mol)
σ_{ss}	collision diameter of the surface atoms (nm^2)
$\phi(a)$	primary particle size distribution (a.u.)
Θ	reduced (a/a_m) adsorption (—)
Θ_C	content of the carbon phase (%)
τ	lifetime of adsorption complex (s)
τ_f	thickness of the premolten liquid layer at freezing (nm)
τ_m	thickness of the premolten liquid layer at melting (nm)
ζ	ζ-potential (mV)

A-50	fumed silica
A-100	fumed silica
A-150	fumed silica
A-200	fumed silica
A-300	fumed silica
A-380	fumed silica
A-400	fumed silica
A-500	fumed silica
AAO	anodized aluminum oxide
AC	activated carbon
AcAc	acetylacetonates
AED	adsorption energy distribution
AES	Auger electron spectroscopy
AFM	atom force microscopy
ALTADENA	adiabatic longitudinal transport after dissociation engenders nuclear alignment
AM-1	hydrophobic nanosilica
APMS	aminopropylmethylsilyl groups
APMS-MS	aminopropylmethylsilyl-modified macroporous silochrome
APMS-NS	aminopropylmethylsilyl-modified nanosilica
AST	alumina/silica/titania nanooxide
ASW	water in associates HO-H…A (A—electron donor center)
ATR	attenuated total reflection
BNNTs	boron nitride nanotubes
BSA	bovine serum albumin
BW	bound water
CASE	computer-assisted structure elucidation
CCP	cow casein powders
CFD	computational fluid dynamics
CG	collagen-glycosaminoglycan
CLSM	confocal laser scanning microscopy
CM	composite materials
CMC	critical micelle concentration
CNT	carbon nanotubes
CONTIN	regularization algorithm
COSY	correlation spectroscopy
CPACT	Center for Process Analytics and Control Technology
CPG	controlled pore glasses
CP-MAS	cross-polarization magic angle spinning
CPMG	multi-echo sequence according to Carr, Purcell, Meiboom, Gill
CRAM	carboxylic-rich acyclic molecules
CS	CarboSil
CSA	chemical shift anisotropy
CST	crystalline silicotitanate
CT	x-ray computed tomography
CTAB	cetyltrimethylammonium bromide
CTI	constant time imaging
CVD	chemical vapor deposition
Cyt *c*	cytochrome *c*
DC	diffusion coefficient
DD	dipolar dephasing
DDRS	double difference reference spectrum

DDS	dynamic dielectric spectroscopy
DE	diffusion edited ^{1}H NMR spectroscopy
DEA	diethylamine
DECRA	direct exponential curve resolution algorithm
DFT	density functional theory
DIPSI	decoupling in the presence of scalar interactions
DLPC	1,2-dilauroyl-sn-glycero-3-phosphocholine
DLS	dynamic light scattering
DMAAB	(dimethylamino) azobenzene
DMPC	1,2-dimyristoyl-sn-glycero-3-phosphatidylcholine
DMSO	dimethylsulfoxide
DNA	deoxyribonucleic acid
DNP	dynamic nuclear polarization
DON	dioxinaphthalene
DOSY	diffusion ordered spectroscopy
DP-MAS	direct polarization magic angle spinning
DPPC	dipalmitoylphosphatidylcholine
DQ	double quantum
DQF	double quantum filtered
DRS	diffuse reflectance spectroscopy
DRS	dielectric relaxation spectroscopy
DSA	4,4-dimethyl-4-silapentane-1-ammonium trifluoroacetate
DSC	differential scanning calorimetry
DSCC	DSC cryoporometry
DSS	sodium 4,4-dimethyl-4-silapentane-1-sulfonate
DTAB	dodecyl trimethyl ammonium bromide
DTG	differential TG
E.COSY	exclusive correlation spectroscopy
ECM	extracellular matrix
EDL	electric double layer
EISA	evaporation-induced self-assembly method
ERETIC	electronic reference to access in vivo concentrations
ESI	equilibrium sorption isotherm
FA	fulvic acid
FAM	fast amplitude–modulated RF pulse trains with constant
FID	free induction decay
FID-DECRA	free induction decay direct exponential curve resolution
FID-GRAM	free induction decay generalized rank annihilation method
FITC	fluorescein isothiocyanate
FLASH	fast low angle shot
FSM-16	modified mesoporous silica
FT	Fourier transformation
FT-ICR MS	Fourier transform ion cyclotron resonance–mass spectrometry
FTIR	Fourier transform infrared spectroscopy
FWHM	full width at half maximum
GC	gas chromatography
GISAXS	grazing incident small angle x-ray scattering
GLYMO	3-glycidoxypropyltrimethoxysilane
GPC	gel permeation chromatography
GPS	global protein folded state
GRAM	generalized rank annihilation method

GT	Gibbs–Thomson (equation)
H2BC	heteronuclear 2 bond correlation
HA	humic acid
HA	hyaluronic acid
HAP	hydroxyapatite
HARDSHIP	heteronuclear recoupling with dephasing by strong homonuclear interactions of protons
Hb	hemoglobin
HCA	hierarchical cluster analysis
HDW	high-density water
HEFS	2-hexyl-5-ethyl-3-furansulfonate
HEL	heterogeneous extended Langmuir theory
HEMA-AGE	poly(2-hydroxyethyl methacrylate-*co*-allyl glycidyl ether)
HETCOR	heteronuclear proton–silicon correlation spectroscopy
HF	Hartree–Fock theory
HGA	composites with nanostructured hydroxyapatite and proteins (gelatin, BSA)
hIAPP	human islet amyloid polypeptide
HIAST	heterogeneous ideal absorbed solution theory
HMBC	heteronuclear multiple bond correlation
HMDS	hexamethyldisilazane
HMQC	heteronuclear multiple quantum coherence
HPDEC	high power decoupling
HPF	human plasma fibrinogen
HPLC	high pressure liquid chromatography
HPLC-SPE	high pressure liquid chromatography–solid phase extraction
HR-MAS	high resolution magic angle spinning
HRP	horseradish peroxidase
HRSEM	high resolution scanning electron microscopy
HRTEM	high resolution transmission electron microscopy
HSA	human serum albumin
HSQC	heteronuclear single quantum coherence
H-STD	heteronuclear-saturation transfer difference
HT	high temperature
HTT	hydrothermal treatment
IAST	ideal adsorbed solution theory
IDPs	intrinsically disordered proteins
IEC	in-exchange capacity
IEP	isoelectric point
Ig	immunoglobulin
IGT	integral Gibbs–Thomson equation
Ih	hexagonal ice
IHM	indirect hard modeling
IHSS	International Humic Substances Society
IMP	inosine monophosphate
IPC	in-process control
IPSD	incremental pore size distribution
IR	infrared (spectroscopy)
JRES	J-resolved
LC	liquid chromatography
LDA	linear discriminate analysis
LDW	low-density water

LF-NMR	low-field NMR
LIA	local isotherm approximation
LJ	Lennard-Jones potential
LMM	local molecular mobility
LSC	life science
LT	low temperature
LTNA	low-temperature nitrogen adsorption
MAS	magic angle spinning
MCA	mechanochemical activated
MCC	microcrystalline cellulose
MCM-41	ordered mesoporous silica
MCM-48	ordered mesoporous silica
MCR	multivariate curve resolution
MDLT	material derived from linear terpenoids
MEM-j	maximum entropy method of the j order
MLEV	Malcolm Levitt's CPD sequence
MLG	multilayer exfoliated graphite
MLVs	multilamellar vesicles
MND	modified Nguyen–Do method
MQ	multiple quantum
MRI	magnetic resonance imaging
MR-NMR	medium resolution NMR
MS	methylated silica
MSMA	3-(trimethoxysilyl)propyl methacrylate
MSME	multi-slice multi-echo sequence (MRI)
MTBE	methyl tert-butyl ether
MTES	methyltriethoxysilane
MW	molecular weight
MWCNT	multiwall nanotubes
MWCO	molecular weight cutoff
MWS	Maxwell–Wagner–Sillars relaxation process
NAD^+	β-nicotinamide adenine dinucleotide
NAPL	nonaqueous phase liquid
NCCW	noncontact check weigher
ND	neutron diffraction
NDC	neutron diffraction cryoporometry
NIR	near infrared (spectroscopy)
NMR	nuclear magnetic resonance
NMRC	NMR cryoporometry
NMRM	NMR microscopy
NOE	nuclear Overhauser effect
NOESY	nuclear Overhauser effect spectroscopy
NOM	natural organic matter
NPCS	native phosphocaseinate suspension
NSF	National Science Foundation
OPTPD-MS	one-pass temperature-programmed desorption with mass-spectrometry control
OPU	oil pickup
OSCs	organosilicon compounds
OX-50	fumed silica
PA	polyamide
PAHs	polycyclic aromatic hydrocarbons

PASADENA	parahydrogen and synthesis allow dramatically enhanced nuclear alignment
PCA	principal component analysis
PCR	principal component regression
PCS	photon correlation spectroscopy
PD	polydispersity
PDADMAC	(1-5 bilayers) with poly(diallyl dimethylammonium chloride)
PDL	polydextrose Litesse
PDMA	poly(2-(*N*,*N*-dimethylamino)ethyl methacrylate)
PDMS	poly(dimethylsiloxane)
PEC	polyelectrolyte complex
PEEK	polyether ether ketone
PEG	poly(ethylene glycol)
PEM	polyelectrolyte multilayer
PEUA	carboxylated polyurethane
PEVBDVB	poly(ethylvinylbenzene divinylbenzene)
PFCs	perfluorinated carboxylic acids
PFG-NMR	pulsed field gradient NMR
PFOS	perfluorooctane sulfonate
PG	phloroglucinol
PHBHV	poly(3-hydroxybutyrate-3-hydroxyvalerate)
PHIP	para-hydrogen-induced polarization
PKR	RNA-dependent protein kinase
PLS	partial least squares
PLS-DA	partial least squares discriminant analysis
PLS-R	partial least squares regression
PMS	poly(methyl siloxane)
PNIPA	poly(N-isopropylacrylamide)
POA	phosphorus oxyacids
POE	poly(oxy ethylene)
PPCA	protein polymer contrast agents
PRESS	predictive residual error sum of squares
PSD	pore size distribution
PSD_{uw}	pore size distribution determined from the C_{uw} values
PSDVB	poly(styrene divinylbenzene)
PSM	protective-stimulating means
PSS	poly(sodium-4-styrenesulfonate)
PURGE	presaturation utilizing relaxation gradients and echoes
PVA	poly(vinyl alcohol)
PVD	plasma vapor deposition
PVP	poly(vinyl pyrrolidone)
PZC	point of zero charge
QC	quality control
Qc	quercetin
QCM	quartz crystal microbalance
QELS	quasi-elastic light scattering
RARE	rapid acquisition relaxation enhanced
RBC	red blood cells
RBCSC	RBC membranes (shadow corpuscles)
RCM	ring current model
REAPDOR	rotational echo adiabatic passage double resonance
REDOR	rotational echo double resonance

RFG	radial function of variation in the free energy of water in the adsorption layer
RH-STD	reverse heteronuclear saturation transfer difference
RNA	ribonucleic acid
RO	reverse osmosis
ROE	rotating frame NOE
ROESY	rotational nuclear Overhauser effect spectroscopy
RPI	recurrent ponded infiltration
RTD	residence time distribution
SA	silica/alumina nanooxides
SABRE	signal amplification by reversible exchange
SAMs	self-assembled monolayers
SANS	small-angle neutron scattering
SAPO	microporous silicoaluminophosphates
SAW	strongly associated water
SAXS	small-angle x-ray scattering
SBA-15	ordered mesoporous silica
SBW	strongly bound water
SCG	sodium hydrocitrate with glucose
SCN	type of AC
SCR	self-consistent regularization
SCV	model of a pore mixture with slit-shaped and cylindrical pores and voids between spherical nanoparticles packed in random aggregates
SDUWC	size distributions of unfrozen water clusters
SDWC	size distributions of water clusters
SE	spin echo
SEDOR	spin echo double resonance
SELTIC	side-band elimination by temporary interruption of the chemical-shift
SEM	scanning electron microscopy
SE-SPI	spin echo single point imaging method
SEVI	spin-echo-velocity imaging
SFC	solid fat content
Si-40	silica gel
Si-60	silica gel
Si-100	silica gel
SLG	single layer exfoliated graphite
S/N	signal to noise
SNIF-NMR	site-specific natural isotope fractionation-nuclear magnetic resonance
SNR	signal-to-noise ratio
SPI	single point imaging
SPIDER	saturation-pulse–induced dipolar exchange with recoupling
SPRITE	single point ramped imaging with T_1 enhancement
ST	silica/titania nanooxides
STD	saturation transfer difference
STDD	saturation transfer double difference spectroscopy
STRAFI	stray field imaging
SUVs	small unilamellar vesicles
SVE	soil vapor extraction
SW–FAM	fast amplitude-modulated RF pulse trains with incremented pulse durations
SWCNT	single wall carbon nanotubes
TCE	trichloroethylene
TD-NMR	time-domain NMR

td	recycle delay
TEA	triethylamine
TEDOR	transferred-echo double-resonance
TEM	transmission electron microscopy
TEOS	tetraethoxysilane; tetraethyl orthosilicate
TES	triethoxysilane
TFE	tetrafluorethylene
TFM	4-nitro-3-(trifluoromethyl) phenol
TG	thermogravimetry
TMS	tetramethylsilane
TMS	trimethylsilyl groups
TOCSY	total correlation spectroscopy
TOPO	trioctylphosphine oxide
TOSS	total suppression of sidebands
TPD-MS	temperature-programmed desorption with mass-spectrometry control
TRAPDOR	transfer of population in double resonance
TS	total solid
TSC	trisodium citrate
TSDC	thermally stimulated depolarization current
TSP	sodium 3-trimethylsilyltetradeuteriopropionate
TTAB	tetradecyl trimethyl ammonium bromide
UF	ultrafiltration
US	ultrasonic
UV–Vis	ultraviolet–visible (spectroscopy)
UWCSD	unfrozen water cluster size distribution
WATERGATE	water suppression by gradient tailored excitation
WAW	weakly associated water
WBW	weakly bound water
WET	water suppression enhanced through T_1 effects
WISE	wideline separation
WPC	Whey protein concentrate
XPS	x-ray photoelectron spectroscopy
XRD	x-ray diffraction

Preface

Properties of high surface area materials and their efficiency of application strongly depend on boundary effects and interfacial phenomena. These phenomena include diffusion of adsorbate and solvent molecules at the surface of the adsorbent, physisorption and chemisorption, dissolution, solvation of solutes and solid surfaces, surface reactions, heterogeneous catalysis, relaxation and segmental dynamics of adsorbed macromolecules, phase transitions, crystallization, and melting, among others. Many of the physicochemical characterization methods give very useful and complementary information regarding complex interfacial phenomena. These methods include nuclear magnetic resonance (NMR) spectroscopy, infrared (FTIR) and other optical spectroscopic methods, differential scanning calorimetry (DSC), adsorption of probe compounds, x-ray diffraction (XRD), Auger electron spectroscopy, ultrasoft x-ray emission spectroscopy, x-ray photoelectron spectroscopy (XPS), dielectric relaxation spectroscopy (DRS), and thermally stimulated depolarization current (TSDC), among others. Among these, NMR spectroscopy is the most universal, yielding very detailed structural information regarding molecules, solids, and interfaces. NMR provides information on molecular diffusion, phase transitions, melting, crystallization, relaxation, adsorption, reaction kinetics, and so on. There are a number of NMR techniques for measuring pore sizes, the two main being NMR cryoporometry, which is based on the pore-size dependence of the melting temperature of a mobile phase, and NMR relaxometry, which is based on transverse relaxation times that are sensitive to confined-space effects. The NMR cryoporometry methods can be applied to soft as well as to solid materials in their native state without the drying and degassing procedures required by some other adsorption methods. This is of importance for strongly hydrated materials, and especially biomaterials such as hydrogels and cryogels, as well as bio-objects such as cells, seeds, and tissues, the structures of which can be profoundly altered by drying.

In this book, the applications of important new NMR spectroscopic methods to a large variety of useful materials are described and compared with results from other techniques such as adsorption, DSC, TSDC, DRS, FTIR, optical microscopy, and small-angle (SAXS) and wide-angle (WAXS) x-ray scattering. A fairly comprehensive description of a large array of hard and soft materials, such as oxides, carbon, polymer–composite adsorbents, biomacromolecules, cells, tissues, seeds, and other materials, allows one to analyze the structure–property relationships and general regularities in the interfacial behavior of water that is located in confined spaces and is affected by organic solvents, adsorbed small molecules, polymers, and proteins, as well as structural changes in swollen or dried materials, phase transitions in adsorbed phases, solutions, and suspensions.

This book summarizes NMR research results collected over the past three decades that are related to very different materials, from nanomaterials and nanocomposites to biomaterials, cells, tissues, seeds, etc. It will primarily be of interest for PhD students and young scientists, although it may also prove useful to specialists and experts working on adsorption, adsorbent synthesis, surface chemistry, surface physics, biophysics, cryopreservation, and other related fields.

Authors

Vladimir M. Gun'ko, DSc, is a professor and head of the Department of Amorphous and Structurally Ordered Oxides, Chuiko Institute of Surface Chemistry of the National Academy of Sciences of Ukraine. He graduated in theoretical physics from Dnipropetrovsk State University (DSU) in 1973 and received his PhD in chemistry from the Institute of Physical-Organic Chemistry (IPOC), Kiev, in 1983 and his DSc in physics, chemistry, and technology of surface from the Institute of Surface Chemistry (ISC) in 1995. He currently serves as professor of physics and chemistry of surface at ISC, where he has been since 1985. He has been elected Marie Curie Fellow at the University of Brighton (United Kingdom), 2010–2011. He has coauthored three books, edited one book, published about 400 papers, and made about 200 presentations at conferences. He also serves on the editorial board of four journals and *Surface*, a periodic book published one to two times per year. He is a member of the American Nano Society (the United States) and an electronic member of The Royal Society of Chemistry (the United Kingdom).

Gun'ko's research interests include quantum and molecular mechanics and dynamics, development of computational algorithms in molecular physics and surface chemistry, modeling of interfacial phenomena, and synthesis and characterization of nanomaterials. His main scientific achievements include development and applications of (i) self-consistent regularization procedures; (ii) numerical algorithms for calculations of dynamics of local electronic states of atoms in molecules and clusters; (iii) algorithms to study the reaction mechanisms with consideration of liquid media effects; (iv) algorithms for calculations of changes in Gibbs free energy for adsorption, diffusion, and breakthrough of molecules; (v) algorithm and software of kinetics of competitive interactions of macromolecules with nanoparticles or solids and own mobility of microorganisms on the basis of dynamic light scattering data; (vi) thermally stimulated depolarization current (TSDC) relaxometry (porometry), integral equations for NMR cryoporometry, and TG and DSC thermoporometry; and (vii) integral equations and regularization procedures to determine the distribution functions of activation energy for dipolar and dc relaxation (TSDC), dielectric relaxation (DRS), and local mobility (NMR); synthesis and characterization of oxide and carbon nanomaterials, polymer–nanooxide, and mineral–carbon composites using the aforementioned approaches.

Vladimir V. Turov, DSc, is a professor and head of the Department of Biomedical Problems of Surface, Chuiko Institute of Surface Chemistry of the National Academy of Sciences of Ukraine. He graduated in radiophysics from Kyiv State Taras Shevchenko University (KSU) in 1973 and received his PhD in chemistry from the Institute of Physical Chemistry (IPC), Kiev, in 1985 and his DSc in physics, chemistry, and technology of surface from the Institute of Surface Chemistry (ISC) in 2002. He currently serves as professor in physics and chemistry of surface at ISC, where has been since 2007. He has coauthored three books, edited one book, published about 250 papers, and made about 100 presentations at conferences. He serves on the editorial board of the journal *Chemistry, Physics & Technology of Surface* and the periodic book *Surface*, published one to two times per year. He has been a member of the Society of Chemistry of Ukraine since 1986. Turov's research interests include molecular interactions in confined space, low-temperature NMR spectroscopy of porous nanomaterials and bio-objects, and water structure in complex mixtures. His main scientific achievements include development and applications of (i) the low-temperature NMR method for determination of thermodynamic characteristics of bound water; (ii) connection between carbon materials structure and NMR–chemical shifts of adsorbed molecules; (iii) weakly associated water at the interfaces of nanomaterials and biopolymers; and (iv) self-organization and clusterization effects of water and organics in nanoporous of solids and bio-objects.

Introduction

A variety of chemical, biochemical, physical, and other processes can occur at solid–liquid, solid–gas, liquid–gas, and liquid–liquid interfaces. To gain a deeper insight into the mechanisms and driving forces of these interfacial phenomena, a large set of methods can be applied to obtain information at nano-, micro-, and macroscales. There is a problem of harmonization of this information from different levels and different methods. Surface forces and confined space effects are most essential at the nanoscale level, especially for strongly structured nanomaterials or adsorbates, such as mixtures of adsorbates with strong competition for most active sites, which are located in nanopores. Nanomaterials are widely used in the environmental, medical, industrial, and agricultural sectors, ranging from fabrication of molecular or atomic assemblies to microbial array chips. Despite the booming applications of nanotechnology and nanomaterials, there are serious implications that have come to light in recent years within different environmental compartments, namely, air, water, and soil and their likely impact on human health (Brar et al. 2010). Health and environmental effects of common metals and materials are well known; however, when the metals, oxides, carbons, or other materials take the form of nanoparticles, consequential hazards based on shape and size are yet to be completely explored. Notice that interfacial phenomena in these systems play an important role because of the very large interfacial area of nanoparticles or strongly porous materials where the specific surface is up to 3500–4000 m^2/g. A set of nuclear magnetic resonance (NMR) techniques such as spectroscopy, imaging, relaxometry, cryoporometry, etc., can be applied to study the interfacial phenomena and to obtain information on structural, textural, kinetic, dynamic, and other aspects of the interfacial phenomena (Abragam 1989, Eckert 1992, Beysens et al. 1993, Webb 2003, 2009, *Encyclopedia of Magnetic Resonance* 2007, 2009, 2011, Petrov and Furó 2009, Simpson et al. 2011, Brown 2012, Dalitz et al. 2012).

Adsorption from gas phase or solution is the most frequently occurring interfacial phenomenon. In this aspect, NMR methods provide information on (i) chemical structure of the interfaces affecting the adsorption/desorption processes, (ii) structure of adsorption complexes and types of surface sites and adsorbate functionalities participating in the formation of these complexes, (iii) confined space effects and pore size distributions, (iv) structural and other properties of free and bound water or other solvents and solutes in an adsorption layer, and (v) state of solutes in bulk solutions and at the interfaces, among many others. NMR spectroscopy as one of the universal methods can be applied to practically any interfacial phenomena, including bulk and surface diffusion, diffusion in pores, cells, and so on; reaction kinetics; topological analysis of porous materials on the basis of changes in the temperature behavior of liquids confined in pores (within NMR cryoporometry); local mobility (relaxometry) and diffusion (diffusometry) of adsorbates; etc. Additionally, NMR as a noninvasive or weakly invasive method allows the analysis of bioobjects in their native state (in vivo, ex vivo, or in vitro) using magnetic resonance imaging (MRI), NMR microscopy, or other techniques. NMR spectroscopy is also a standard tool for structural analysis of low- and high-molecular compounds, biomacromolecules, crystalline and amorphous solids, etc. It should be noted that certain aspects of the interfacial phenomena were studied at an earlier stage of the NMR spectroscopy. For instance, the first works on investigations of porous materials with NMR relaxometry (Brown and Fatt 1956), the solvent effects (Bothner-By and Glick 1956), water adsorbed onto silica gel (Zimmerman et al. 1956), and proton relaxation in water at surfaces were published more than 50 years ago.

In this book, applications of NMR spectroscopy (as well as a set of other methods used here for deeper insight into the interfacial phenomena) to nanooxides and porous oxides, including silica, titania, alumina, and other individual, binary, and ternary oxides, carbon materials (from graphite to graphene, activated carbons, carbon blacks, carbon nanotubes, fullerenes), carbon–mineral composites, polymeric materials (both natural [cellulose, starch, chitosan, hyaluronic acid, etc.] and synthetic polymers), biomacromolecules (proteins, DNA, lipids, etc.), cells (yeast cells, red blood cells, bone marrow cells, bovine gametes), seeds, natural fibers (flax, cotton), tissues (bone, muscular), and some other bioobjects are described with increasing complexity of the objects studied from chapter to chapter. Additional features of this book are (i) wide applications of regularization procedures, including a developed self-consistent regularization procedure, to analyze the experimental data obtained by different methods (NMR, adsorption, desorption, diffusion, TSDC, DRS, PCS, rheometry, TG, TPD MS, QCM); (ii) attempts at providing a comprehensive description of the materials or the interfacial phenomena at a surface of these materials using the maximum wide set of methods, both experimental and theoretical; (iii) selection of the materials under study with varied but controlled confined space effects; (iv) results of different experimental methods published earlier recalculated using updated versions of developed computer programs.

Chapter 1, which is the largest one, is devoted to a detailed description of unmodified and modified silica materials, including nanosilicas, porous silica gels, ordered mesoporous silicas, thin silica films, as well as interactions of silicas with low- and high-molecular adsorbates and the effects of different treatments (heating, wetting, suspending, mechanochemical activation, hydrothermal treatment, chemical modification of the surfaces, etc.). The main focus is on the interfacial behavior of water adsorbed alone or in mixture with different organic and inorganic co-adsorbates.

Chapter 2 describes mixed oxides of different origin and the interfacial phenomena at their surfaces. The properties of mixed binary and ternary oxides or more complex systems and related interfacial phenomena are compared with those of such individual oxides as silica, alumina, titania, etc.

Different carbon materials (graphite, graphene, activated carbons, carbon nanotubes, etc.) are described in Chapter 3. These materials are very interesting with respect to the interfacial phenomena because they have a maximum possible specific surface area (up to 3500–4000 m^2/g), and some of them (such as graphenes and graphene oxides) represent one-two atomic layers, that is all their atoms are surface atoms.

Chapter 4 deals with hybrid carbon–mineral materials. These materials are interesting objects with respect to the interfacial phenomena because of the mosaic structure of their surfaces. This can result in strong clusterization of adsorbates and appearance of unusual properties of the materials in adsorption and other phenomena.

Chapter 5 is devoted to the interfacial phenomena at a surface of natural and synthetic polymers, including almost dry or strongly hydrated systems characterized by nano-, meso-, and macroporosity up to pores of 200–300 μm in diameter.

Chapters 6 through 8 describe a variety of biosystems, from biomacromolecules (Chapter 6), cells and tissues (Chapter 7), to seeds and other bioobjects (Chapter 8). These systems are much more complex than simple silica materials. Nevertheless, low-temperature NMR spectroscopy, cryoporometry, TSDC, DSC, FTIR, and some other methods can be successfully applied to them to obtain detailed information on their characteristics and interfacial phenomena related to their living functions.

Chapter 9 covers certain trends in the investigations of the interfacial phenomena briefly. Finally, Chapter 10 describes NMR and other methods used in this book.

1 Unmodified and Modified Silicas

Silicas (bulk SiO_2 with surface silanols) are the most popular oxide materials including porous (silica gels and ordered mesoporous silicas), nonporous highly disperse nanomaterials, microparticles, microspheres, nanotubes, etc. SiO_2 is also a component of many natural minerals. Silica materials have a relatively simple structure of bulk and surface, including siloxane bonds ≡Si–O–Si≡, terminal single ≡SiOH, and twin =$Si(OH)_2$ silanols (Iler 1979, Legrand 1998). Silanols and siloxane bonds, which are characterized by very different activity in adsorption and chemical reactions and have clear spectral difference in infrared (IR), ^{29}Si and ^{1}H nuclear magnetic resonance (NMR), and X-ray photoelectron spectroscopy (XPS), are very convenient objects in investigations of silica surfaces and interfacial phenomena there. Despite composition simplicity, highly disperse silicas (nanosilicas), as well as silica gels and other silicas, possess many useful properties for practical applications, and some of these properties can be quite unusual when nanoscaled effects appear and become essential for the interfacial behavior of adsorbates. Notice that such a simpler compound as water, both bulk-free and adsorbed (i.e., interfacial), possesses much more unusual properties than silica (Gun'ko et al. 2005d, 2009d, Chaplin 2011). Therefore, silica with bound water represents an interesting combination, properties of which play an important role in the nature and human life. These properties can be very unusual, especially after addition of another solvent (e.g., nonpolar, weakly polar, or polar organics) to water and/or surface modification of silica by polar or nonpolar compounds creating a mosaic hydrophilic/hydrophobic surface (Gun'ko et al. 2001e, 2003g, 2005d, 2007d,f,i, 2009d). Analysis of the interfacial phenomena at a silica surface could be started from nanosilicas composed of nonporous primary nanoparticles.

1.1 INTERFACIAL PHENOMENA AT A SURFACE OF NANOSILICA

Before analysis of the interfacial phenomena at a nanosilica surface pristine and modified, the textural and structural characteristics of this material should be analyzed for deeper understanding features of the interfacial phenomena. Clearly, the interfacial phenomena at a nanosilica surface can depend on confined space effects in textural pores (however, these effects are much weaker than in the case of nano/mesoporous silicas) and structural features the silica surface dependent on the type of nanosilicas (Gun'ko et al. 2005d, 2009d). To describe these phenomena and the material characteristics affecting these phenomena, we will use a set of methods. This set includes NMR spectroscopy, low-temperature nitrogen adsorption–desorption, adsorption/desorption of water, thermogravimetry (TG), temperature-programmed desorption with mass-spectrometry (TPD-MS) control, IR spectroscopy, calorimetry, thermally stimulated depolarization current (TSDC), photon correlation spectroscopy (PCS), dielectric relaxation spectroscopy (DRS), microscopic (atomic force microscopy [AFM], scanning electron microscope [SEM], and transmission electron microscopy [TEM]), and some other methods.

1.1.1 Structural and Adsorptive Characteristics of Nanosilicas

Fumed silicas (also named as nanosilicas, pyrogenic silicas, and highly disperse amorphous silicas) are synthesized by using high-temperature (HT) hydrolysis of $SiCl_4$, $SiCl_3CH_3$, or other similar precursor compounds in an oxygen–hydrogen ($O_2/H_2/N_2$) flame at high temperature $T > 1000°C$. HT flame synthesis results in the formation of amorphous nonporous primary particles forming secondary and ternary structures. Nanosilica serves as a thickening and anticaking agent (free-flow agent) in powders. It is used as a desiccant in cosmetics for its light-diffusing properties and as a light abrasive in products like toothpaste. Nanosilicas can be used as fillers in silicone elastomer and viscosity adjustment in paints, coatings, printing inks, adhesives, and unsaturated polyester resins (Iler 1979, Degussa 1997, Legrand 1998, Chuiko 2001, 2003). Major global producers of nanosilicas with different specific surface areas are Evonik (Aerosil, former Degussa, Germany), Cabot Corporation (Cab-O-Sil, USA), Wacker Chemie AG (Germany), and Wacker Chemie–Dow Corning (China). Notice that the characteristics of nanosilicas of different producers are very similar due to features of HT flame synthesis.

Nanosilicas with a spherical shape of nonporous nanoparticles, controlled particle size distribution (PaSD), and specific surface area (S_{BET}) are appropriate nanosized materials to study the interfacial phenomena (without the strong distorting effects in nanopores) in different dispersion media. Nanosilicas are fully amorphous and can possess a larger S_{BET} value from 50 up to 500 m^2/g and a narrow PaSD (which becomes narrower at a greater S_{BET} value; Iler 1979, Ulrich 1984, Degussa 1997, Barthel et al. 1998a,b, Kammler and Pratsinis 2000, Kammler et al. 2004, Cabot Corporation 2011).

Proto-particles (1–2 nm; Legrand 1998) formed in the initial zone of the flame collide, stick together, and are covered by new silica layers forming primary particles of 5–100 nm in diameter ($d \approx 6000/(S_{BET}\rho_0)$, where $\rho_0 \approx 2.2$ g/cm^3 is the true density of amorphous silica). Collision, sticking, and fusing at high temperatures of individual primary nanoparticles, which can be bonded by the ≡Si–O–Si≡ bridges with neighboring nanoparticles, result in formation of primary rigid aggregates. Subsequent attachment of individual primary particles to these initial aggregates increases their sizes. Collision and sticking together of primary aggregates lead to formation of larger secondary aggregates of 100–500 nm (mass-fractal dimension ≈ 2.5, apparent density ≈ 30% of the true density) through mainly hydrogen, electrostatic, and dispersion bonding (Iler 1979, Ulrich 1984, Auner and Weis 1996, 2003, 2005, Degussa 1997).

Bonding strength of primary particles in primary and secondary aggregates depends on temperature and their sizes (i.e., type of nanosilicas), coordination numbers, and type of bonding (chemical ≡Si–O–Si≡ or intermolecular) of primary nanoparticles. At lower temperatures and after certain hydration of the silica surfaces, aggregates form loose agglomerates (above 1 μm, fractal dimension ≈2.1–2.2, bulk density ≈2%–4% of the specific density ~0.04–0.06 g/cm^3) through hydrogen bonding, electrostatic, and dispersion interactions. The latter form visible flocks of low bulk density $\rho_b = 0.02$–0.14 g/cm^3. If $S \geq 100$ m^2/g that isolated primary particles are not practically observed without special treatment, while the smaller the particles, the stronger the bonding in the aggregates and agglomerates. For example, Aerosil 130 (Degussa) can be dispersed toward individual nanoparticles more easily than Aerosils 300 or 380. According to SEM findings, nanosilicas do not change their morphology on heating at 1000°C for 7 days, but at 1200°C, they cross-link to glass (Degussa 1997). Therefore, one can assume that changes in the temperature of the nanosilica synthesis between 1000°C and 1300°C can influence the characteristics of both primary nanoparticles and their secondary and ternary structures. These circumstances were used to synthesize a set of nanosilicas with close S_{BET} values but with very different textural organization of secondary particles, varied hydrophilicity, and adsorption properties (vide infra).

According to firm information (Cabot Corporation 2011), Figure 1.1 illustrates over-dispersion, a major concern in the use of nanosilica. The center graphic represents the silica when it is properly dispersed. After application (top right graphic), the network is able to reform throughout the coating film in the short period of time required to stop sagging. Lower left graphic represents the chain structure when the silica has been over-dispersed. After application (lower right graphic), only a partial network is able to reform in the time required to forestall sagging. This is due to the greater

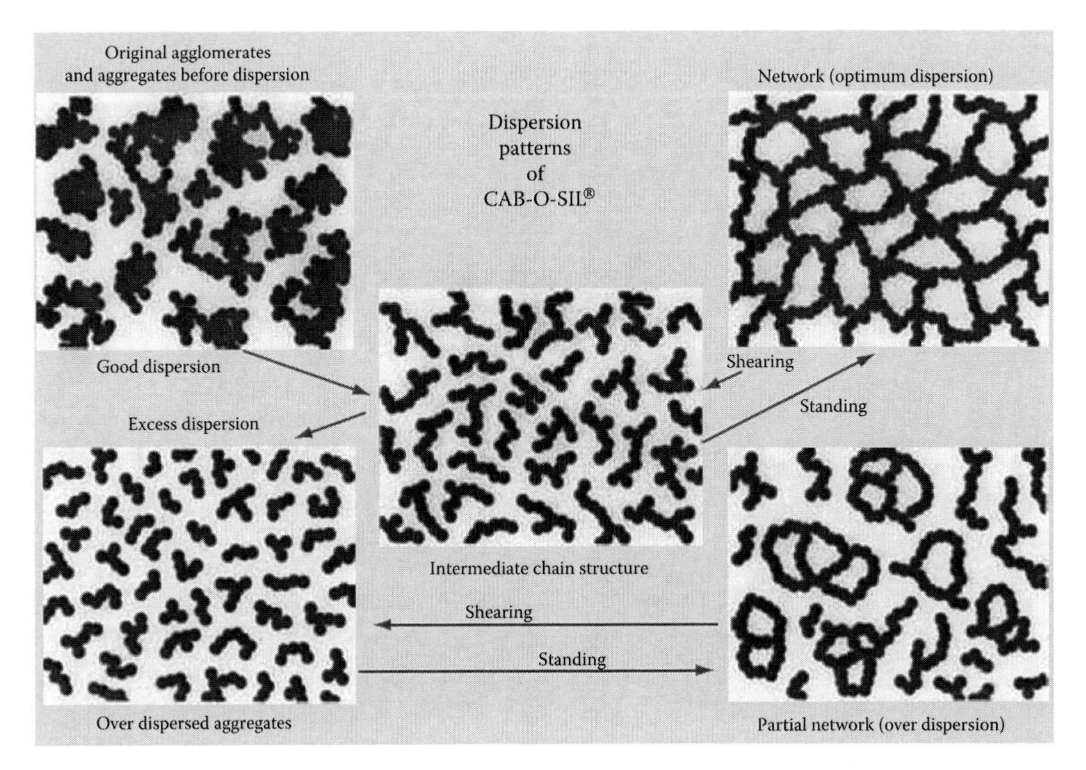

FIGURE 1.1 Dispersion patterns of nanosilica CAB-O-SIL (CAB-O-SIL, a registered trademark of Cabot Corporation, copyright Cabot Corporation 2011). (Adapted from Cabot Corp. Boston, MA, 2011, www.cabot-corp.com/fmo. With permission.)

number of inter-aggregate hydrogen bonds needed to reform the entire network when starting with a smaller agglomerate size.

Hyeon-Lee et al. (1998) compared the morphology of nanosilica and titania aggregates using small-angle x-ray scattering (SAXS). Nanostructured powders have mass-fractal morphologies, which are composed of ramified aggregates of nanosized primary particles. Sizes of primary particles and their aggregates, fractal dimension, and specific surface area were obtained from this analysis. The mass-fractal dimension varied from 2.5 to 1.6 for flame-generated silica and titania aggregates in single and double diffusion flame reactors. However, titania powders prepared in a single diffusion flame reactor appeared as non-aggregated and non-fractal. Silica powders synthesized with an imposed electric field in a laminar, premixed flame reactor corresponded to mass fractals with narrowly confined fractal dimensions from 1.5 to 1.9 regardless of aggregate size (Hyeon-Lee et al. 1998). Notice that these fractal dimension values are underestimated since, from the physical sense, it cannot be smaller than two, characteristic of a plane surface. However, this is typical result for the fractal dimension values determined from the SAXS data. The fractal dimensions of nanosilicas estimated from the adsorption data were typically between 2.2 and 2.6 (Barthel 1995, Gun'ko 2000a, Gun'ko et al. 2005d, 2009d).

Figure 1.2 shows SEM and TEM images of nanosilicas A-50 ($S_{BET}=52$ m^2/g) and A-200 ($S_{BET}\sim200$ m^2/g) from different producers. Primary particles of A-50 are of larger sizes than those of A-200 with different structures of interparticle contacts. These silica nanoparticles are amorphous (Figures 1.2 and 1.3). Calculations of the spacing distribution functions from high-resolution TEM (HRTEM) images of crystalline fumed titania and amorphous silica (Figure 1.3) clearly show the difference between crystalline and amorphous nanoparticles. Each primary nanoparticle of a nonideal spherical shape is bound to two to three neighboring particles.

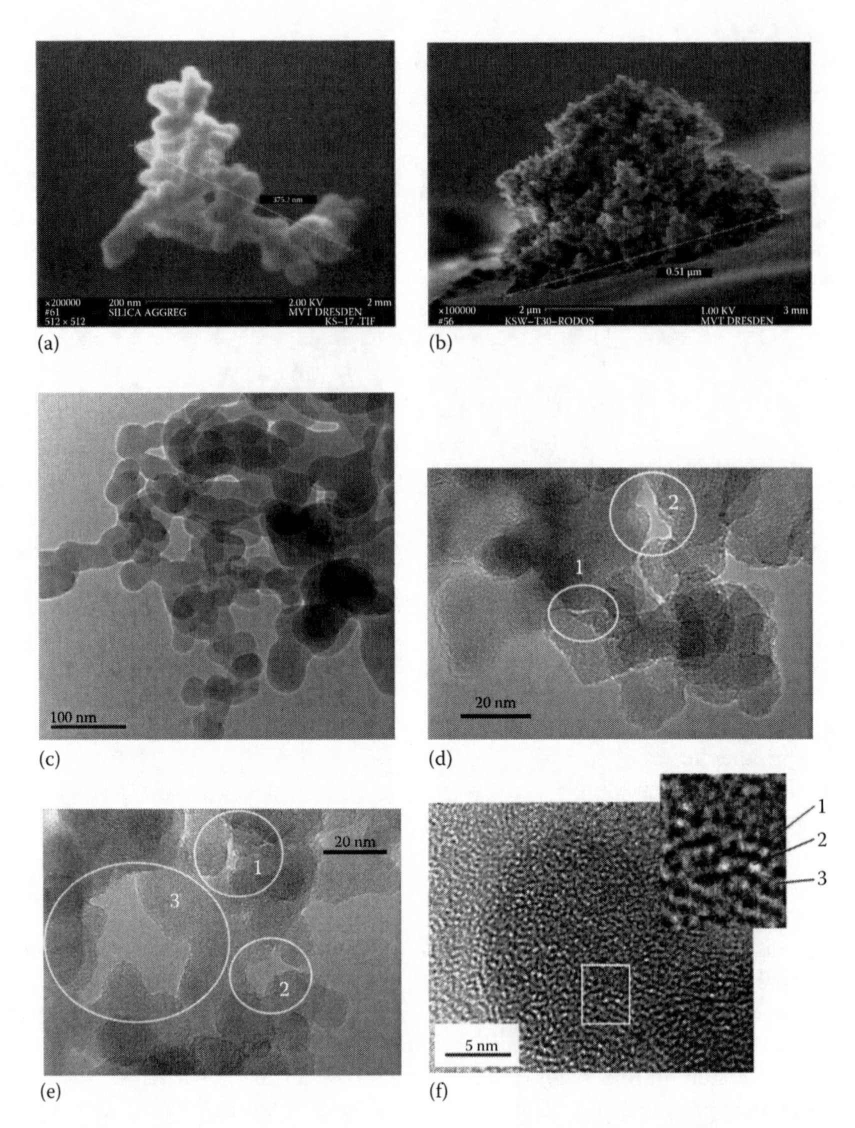

FIGURE 1.2 HRSEM (a, b) (From Barthel, H. et al., Particle sizes of fumed silica, in *An International Conference on Silica Science and Technology "Silica 98"*, Mulhouse, France, September 1–4, 1998, pp. 323–326) and HRTEM (c–f) (Adapted from Myronyuk, I.F. et al., *Phys. Chem. Solid State*, 11, 409, 2010. With permission) images of nanosilicas: (a) agglomerate (375 nm in size) of primary nanoparticles A-200, (b) aggregate of agglomerates (~6.5 μm), (c) A-50 ($S_{BET}=52$ m^2/g), (d–f) A-200 ($S_{BET}=179$ m^2/g); (d, e) 1—nanopores, 2—narrow, and 3—broad mesopores; (f) structures 1, 2, and 3 show the amorphous character of nanosilica.

The textural porosity (as voids between nonporous nanoparticles) appears in aggregates of primary particles and agglomerates of aggregates (Figure 1.2). It is mainly in the meso/macroporous range and contribution of nanopores is very low.

Nanosilica A-300 ($S_{BET}\approx 300$ m^2/g) is used as a medicinal preparation of a high sorption capability possesses certain unique properties, which make it possible to use it effectively on treatment of different diseases (Chuiko 1993, 2003, Blitz and Gun'ko 2006). These features of nanosilica are caused by several factors: (i) chemical nature of amorphous nonporous primary nanoparticles passive in redox reactions and possessing weak reactivity in acid–base reactions (with participation of the ≡SiOH groups) and a low surface charge density at pH<8; (ii) a small size

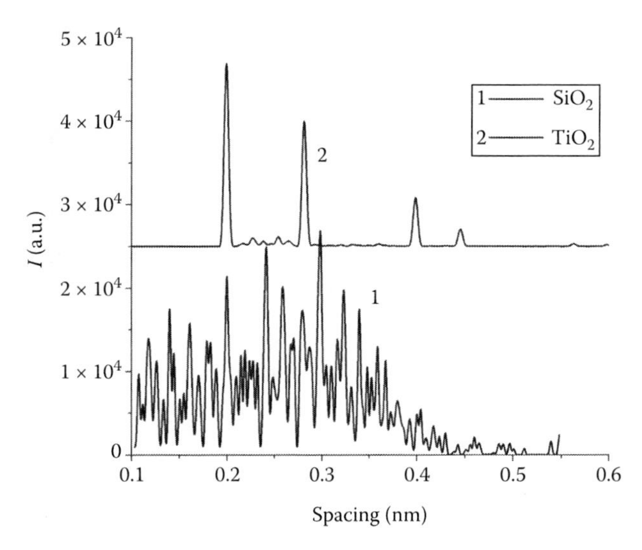

FIGURE 1.3 Spacing distribution function for nanosilica (curve 1, amorphous) and titania (2, crystalline) calculated from HRTEM images using Fiji software. (Fiji, 2009, http://pacific.mpi-cbg.de/wiki/index.php)

(average size 9.1 nm for A-300) of primary nanoparticles forming aggregates by the hydrogen bonds and partially by the siloxane bonds (bulk density of aggregates is ~30% of the true density ρ_0) and more friable agglomerates of aggregates (~0.05–0.07ρ_0), which can be easily destroyed; (iii) local interactions of nanoparticles (strongly different in comparison with micro- or macroparticles) with certain structures of membranes of cells and microorganisms (e.g., integrated proteins); (iv) good transport properties of nanoparticles in aqueous medium because of relatively fast diffusion; (v) adsorption of substances on the external surface of nonporous primary particles; and (vi) changes in the interfacial water structure in a layer of several nanometers in thickness because of chaotropic properties of a silica surface (Gun'ko et al. 2005d, 2009d, Blitz and Gun'ko 2006). Thus, good medical properties of nanosilica are caused by the morphological, structural, chemical and adsorption characteristics, and features of the interfacial phenomena play an important role here.

Now, we will analyze the structural characteristics of a variety of nanosilicas (produced at pilot plant of Chuiko Institute of Surface Chemistry, Kalush, Ukraine, during the last two decades; Tables 1.1 through 1.6) for deeper insight into the mechanisms and driving forces of the interfacial phenomena at nanosilica surfaces. Notice that comparison studies of these materials and nanosilicas produced by Degussa and Cabot Corp. demonstrated similar properties of the materials with similar S_{BET} values produced under standard conditions.

The structure of secondary particles of nanosilicas is random and corresponds to a low coordination number of particles (<3; Iler 1979, Degussa 1997, Legrand 1998). The nanosilica powder is characterized by a large total empty volume $V_{em} = 1/\rho_b - 1/\rho_0 > 7–30$ cm^3/g. Changes in synthesis conditions allow the controlled variations not only in the structure of primary nanoparticles but also in the structure of contacts between them in aggregates (Figure 1.2) that affect the properties of powders and dispersions and related interfacial phenomena (Gun'ko et al. 2005d, 2009d, Flesch et al. 2008). Additionally, different types of treatment (such as mechanical, thermal, hydrothermal ones, Gun'ko et al. 2004a,c,e,f) of the powders and the suspensions can give changes in the particle–particle interactions in the aggregates that lead to variation in the adsorption capacity with respect to different adsorbates, for example, nitrogen (Figure 1.4). On the basis of the textural characteristics (Table 1.1; Figure 1.4), we can assume that the structures of secondary particles of nanosilicas are similar for samples at different S_{BET} values. This is in agreement with microscopic images of nanosilicas.

The shape of nitrogen adsorption–desorption isotherms (Figure 1.4a) corresponding to type II (International Union of Pure and Applied Chemistry [IUPAC] classification) suggests that there is

TABLE 1.1
Structural Characteristics of Nanosilicas

Samples	S_{BET} (m²/g)	S_{nano} (m²/g)	S_{meso} (m²/g)	S_{macro} (m²/g)	V_p (cm³/g)	V_{nano} (cm³/g)	V_{meso} (cm³/g)	V_{macro} (cm³/g)	Δw_{mix}
A-50	52	30	20	2	0.13	0.01	0.06	0.06	−0.09
A-100	77	9	64	4	0.20	0.0	0.12	0.08	−0.39
A-150	144	78	53	13	0.31	0.03	0.25	0.04	0.19
A-200	226	128	77	21	0.50	0.05	0.42	0.03	0.08
A-300	337	161	139	37	0.71	0.05	0.62	0.04	0.33
A-380	378	130	201	47	0.94	0.04	0.81	0.09	0.40
A-400	409	156	209	44	0.86	0.05	0.73	0.09	0.46
A-500	492	185	258	49	0.87	0.05	0.77	0.06	0.39

Note: The values of S_{nano} and V_{nano} ($R_p \leq 1$ nm), S_{meso} and V_{meso} ($1 < R_p \leq 25$ nm), and S_{macro} and V_{macro} ($R_p > 25$ nm) were calculated using a model of complex pores including voids between spherical nonporous particles, cylindrical and slit-shaped pores with self-consistent regularization procedure (see Chapter 10 describing used methods).

TABLE 1.2
Characteristics of Nanosilica Samples (First Series) with Low Hydrophilicity

Sample Number	ρ_{ap} (g/L)	S_{BET} (m²/g)	S_α (m²/g)	S (m²/g)	V_p (cm³/g)	R_p (nm)	D_{AJ} p/p_0 <0.85	$C_{w,105}$ (wt%)	$C_{w,900}$ (wt%)
S1-1	37	144	145	118	0.261	3.7	2.618	1.0	1.1
S1-2	37	160	159	143	0.290	3.9	2.596	0.8	0.8
S1-3	49	206	219	216	0.411	4.0	2.588	1.0	0.9
S1-4	43	226	239	242	0.437	3.9	2.589	1.0	0.9
S1-5	46	337	340	328	0.608	3.6	2.608	1.2	1.0
S1-6	42	381	369	381	0.667	3.5	2.624	1.4	1.0

Note: D_{AJ} is the fractal dimension; $C_w = C_{w,105} + C_{w,900}$; $C_{w,105}$ is the amount of water desorbed on heating at $T < 105$°C, $C_{w,900}$ is the amount of water desorbed at $105 < T < 900$°C.

TABLE 1.3
Synthesis Conditions and Characteristics of Nanosilicas (Second Series)

Sample Number	d_n (mm)	$SiCl_4$ (L/h)	Ratio γ	v_f (m/s)	R_e	T_f (°C)	S_{Ar} (m²/g)
S2-1	36	65	1	27.2	51,160	1092	376
S2-2	36	70	1	30.4	55,292	1042	400
S2-3	36	80	1	33.4	62,860	1002	416
S2-4	42	70	1	21.6	47,290	1154	300
S2-5	52	120	1	24.1	67,560	1202	362
S2-6	52	120	0.8	21.2	57,440	1242	296

Note: $\gamma = 1$ corresponds to the stoichiometric ratio between $SiCl_4$ and H_2/O_2.

TABLE 1.4
Synthesis Conditions and Hydration Degree of Nanosilicas (Third Series) with Different Hydrophilicity

Sample Number	γ_{H2}	γ_{O2}	v_f (m/s)	S (m²/g)	C_{OH} (μmol/m²)	$C_{w,105}$ (wt%)	$C_{w,900}$ (wt%)	$a_{w,mono}$ (μmol/m²)
S3-1	1.0	1.0	21.6	300	3.32	1.8	1.7	8.3
S3-2	1.0[a]	1.0	20.1	267	3.67	1.6	1.4	8.2
S3-3	1.0[a]	0.8	21.2	290	3.48	1.0	0.8	9.0
S3-4	1.1[a]	0.65	19.6	260	3.81	0.6	0.3	4.9
S3-5	1.2[a]	0.8	17.0	144	5.00	0.4	0.4	2.6
S3-6	1.0[b]	0.8	21.8	308	3.30	1.3	1.5	4.4
S3-7	1.0[b]	1.0	21.6	299	3.32	1.8	2.0	8.2
S3-8	1.2[a]	0.65	22.0	319	3.30	0.5	0.6	3.0

Note: Flow velocity in the annular nozzle v_{f,H_2} = [a]2 m³/h or [b]8 m³/h; $\gamma_{H_2} = 1$ and $\gamma_{O_2} = 1$ correspond to the stoichiometric amounts of H_2 and O_2. The first sample was synthesized under standard conditions (stoichiometric ratio $H_2/O_2/SiCl_4$, laminar flow, etc.).

TABLE 1.5
Parameters of Interface Layer upon Water Adsorption onto Nanosilicas Synthesized under Different Conditions

S_{BET} (m²/g)	γ_S (mJ/m²)	$-\Delta\sigma$ (mJ/m²)	$\Delta\mu_s$ (kJ/mol)	$C_{OH,378}$ (wt%)	$C_{OH,1173}$ (wt%)	C_{OH} (mg/m²)
300	304	56	11.1	1.8	1.7	0.117
267	315	63	8.3	1.6	1.4	0.112
290	245	47	7.3	1.0	0.8	0.062
144	124	26	4.1	0.4	0.4	0.055

Note: γ_S is total changes in Gibbs free energy of interfacial water calculated from the ¹H NMR data; $C_{OH,378}$ and $C_{OH,1173}$ are the amounts of water desorbed on heating at $293 < T < 378$ K and $378 < T < 1173$ K respectively, $C_{OH} = (C_{OH,378} + C_{OH,1173})/S_{BET}$.

the similarity in the texture of nanosilicas at different S_{BET} values. However, hysteresis loops become shorter and narrower with decreasing S_{BET} value that suggests a diminution of aggregation of primary particles (Gun'ko et al. 2001a, 2009a, Hakim et al. 2005), and therefore, a decrease in the volume filled by nitrogen adsorbed even at high relative pressure $p/p_0 \rightarrow 1$. If the aggregate sizes become smaller that the bulk density of secondary particles increases, and A-50 is characterized by maximal $\rho_b \approx 0.14$ g/cm³ among studied nanosilicas. Notice that silica OX-50 (Degussa) has similar characteristics.

The incremental pore size distributions (IPSDs; calculation method is described in Chapter 10) demonstrate a certain similarity in the shape for different nanosilicas (Figure 1.4b) but intensity of a mesopores peak decreases and shifts toward larger pore radius R_p with decreasing S_{BET} value. This result suggests that compacting of primary particles in aggregates decreases with decreasing S_{BET} value. However, changes in the surface tension due to the nitrogen adsorption onto A-500 and A-50 are nearly the same since $-\Delta\sigma = 38$ and 37 mJ/m² that suggests the similarity of certain properties of both primary and secondary particles of nanosilicas despite strongly different S_{BET} and V_p values. There is tendency

TABLE 1.6
Synthesis Conditions and Characteristics of Nanosilicas (Fourth Series) with Different Specific Surface Area and Hydrophilicity

Sample Number	d_n (mm)	$SiCl_4$ (L/h)	γ_{H2}	γ_{O2}	v_f (m/s)	R_e	T_f (°C)	S_{Ar} (m²/g)	C_{OH} (μmol/m²)	$C_{w,105}$ (wt%)	$C_{w,900}$ (wt%)
S4-1	36	97.5	1	1	32.1	107,100	1082	368	2.69	1.96	1.78
S4-2	36	97.5	1	1.5	42.9	122,000	1051	398	2.38	2.40	2.20
S4-3	42	82.5	1	1	20.1	77,800	1100	325	3.48	1.88	1.80
S4-4	42	82.5	1.1	1	20.5	78,190	1105	350	3.16	1.79	1.62
S4-5	42	82.5	1.3	1	21.5	79,150	1100	376	2.82	1.32	1.26
S4-6	52	105	1	1	16.6	79,920	1190	170	4.70	1.62	1.44
S4-7	52	127.5	1	1	20.2	97,000	1202	255	3.67	1.64	1.52
S4-8	52	135	1	1	21.4	102,800	1210	275	3.64	1.72	1.60
S4-9	52	135	1	0.7	19.2	98,810	1320	320	3.28	0.66	0.72
S4-10	52	135	1	0.6	17.3	95,370	1290	262	3.67	0.42	0.46
S4-11[a]	52	135	1	1	21.4	26,690	1221	415	2.20	1.64	1.52
S4-12[a]	52	135	1	0.8	18.6	25,350	1250	512	1.88	1.40	1.20
S4-13	62	150	1	1	16.7	95,680	1264	85	5.20	0.80	0.90
S4-14	62	180	1	1	19.9	114,600	1284	196	4.10	1.60	1.72
S4-15[a]	62	180	1	1	19.9	24,350	1298	412	2.20	1.90	2.10

[a] Flow in the burner is divided by smaller ones with the effective diameter of 13.5 mm.

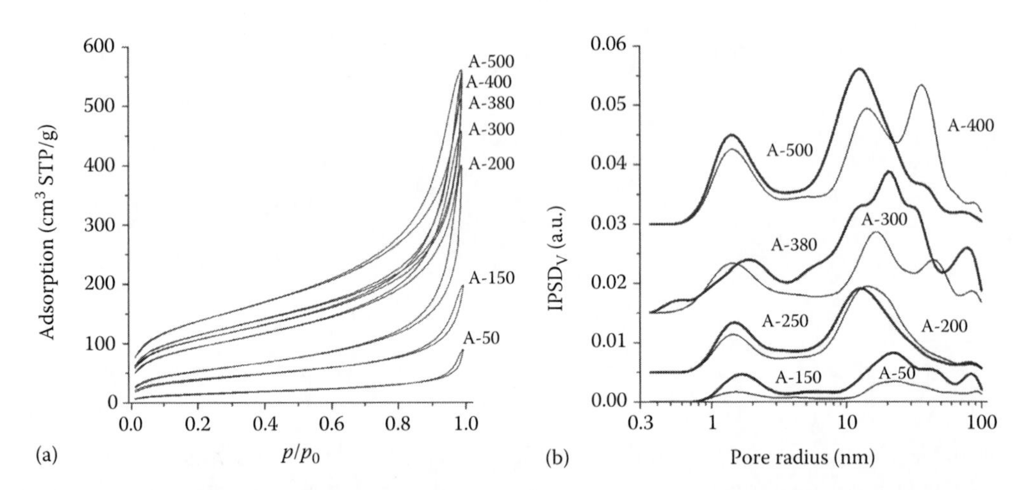

FIGURE 1.4 (a) Nitrogen adsorption–desorption isotherms and (b) incremental pore size distributions in respect to the pore volume (calculated using a mixture of slit-shaped and cylindrical pores and voids between spherical particles with self-consistent regularization procedure) of a set of nanosilicas.

of increase in the V_p value (determined from nitrogen adsorption at relative pressure p/p_0=0.98–0.99) with increasing S_{BET} value (Table 1.1). However, enhanced compacting of secondary particles up to $\rho_b \approx 0.25$ g/cm³ due to the ball-milling of the A-300 powder for 1–24 h leads to a very different relationship between S_{BET} and V_p with a sharp maximum (Gun'ko et al. 2004c, Blitz and Gun'ko 2006).

Primary nanoparticles, aggregates, and agglomerates of nanosilicas can be observed in aqueous suspensions using PCS (or quasi-elastic light scattering [QELS], dynamic light scattering [DLS]; Figure 1.5; Gun'ko et al. 2001e). The particle distributions in the aqueous suspensions depend on the

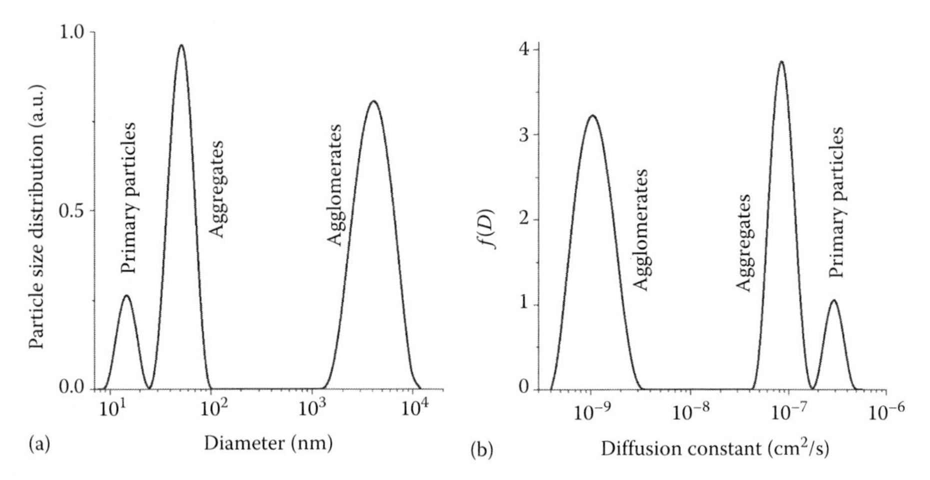

FIGURE 1.5 (a) Particle size distribution of nanosilica A-300 and (b) the corresponding diffusion constant distribution for three types of particles.

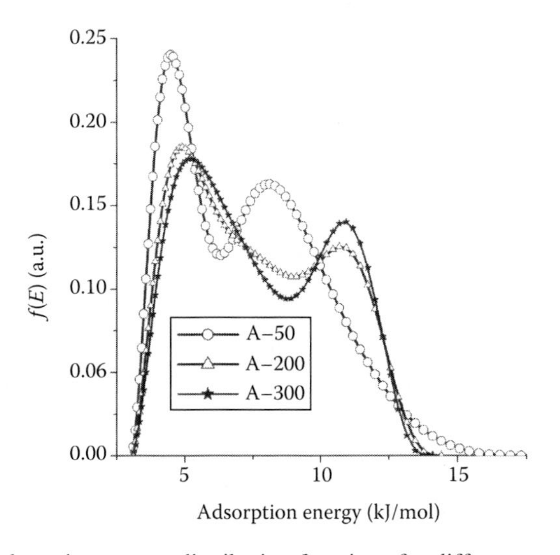

FIGURE 1.6 Nitrogen adsorption energy distribution functions for different nanosilicas calculated using integral Fowler–Guggenheim equation.

type of silica, dispersion concentration (multibody interactions), pH (i.e., surface charge density), temperature (particle mobility in the Brownian diffusion), sonication time (degree of destruction of agglomerates and aggregates), etc. (Gun'ko 2000a, Gun'ko et al. 2001e). Consequently, the interfacial phenomena in the aqueous suspensions of nanosilicas can be complex and dependent nonlinearly on many factors (vide infra).

The difference in packing of primary particles with increasing primary particle size or decreasing S_{BET} value affects the adsorption energy of nitrogen molecules because this energy increases for molecules located in narrower pores (Figure 1.6). Therefore, the high-energy peak shifts toward lower energy for A-50 characterized by much smaller aggregated larger primary nanoparticles than A-200 or A-300 (Ulrich 1984, Degussa 1997, Blitz and Gun'ko 2006, Gun'ko et al. 2009d).

Three series of nanosilica samples were synthesized (Gun'ko et al. 2001d) using $SiCl_4$ hydrolyzed in the oxygen/hydrogen flame under controlled conditions (temperature, flow velocity and turbulence, ratio of reagent amounts and their distribution in the flame, different nozzle diameter

d_n = 36, 42, or 52 mm) to produce materials possessing varied initial (native) hydrophilicity, specific surface area, and other structural and adsorption characteristics over large ranges (Tables 1.2 through 1.4). The flame temperature (T_f = 1000°C–1300°C) in the main reaction zone was measured using a Ranger II (Rayter) optical pyrometer. The specific surface area (S) for all the samples of nanosilica was also evaluated using a Jemini 2360 (SVLAB) apparatus with argon adsorption.

An increase in the reactant amounts at the stoichiometric ratio (γ = 1) between O_2/H_2 and $SiCl_4$ and at the same nozzle diameter d_n leads to a decrease in the flame temperature T_f and an increase in the flow velocity v_f of the reaction mixture leaving the nozzle (Table 1.3; S2-1, S2-2, S2-3). An increase in the nozzle diameter leads to elevating of T_f (S2-2, S2-4, S2-5) and decrease in v_f (S2-2, S2-4). Reduction of the γ value (for O_2/H_2 in respect to $SiCl_4$) to 0.8 (S2-6) gives an increase in T_f (Table 1.3) due to possible changes in the reaction mechanism (e.g., direct oxidizing Si-H or SiO to SiO_2 instead of hydrolysis of Si-Cl). There is well-seen relationship between the v_f value and the specific surface area ($S \sim \ln v_f$) or the diameter of primary particles $d \sim 1/v_f$ (Figure 1.7a). At $v_f > 25$ m/s (corresponding to $S > 380$ m²/g), changes in S (or d) are relatively small with increasing v_f. There is tendency in decreasing the specific surface area with increasing turbulence (Figure 1.7b) or temperature (Table 1.3; T_f) of the flame. However, certain scatters are observed for these relationships because of changes in the reaction conditions (e.g., reactant content ratio).

Notice that the Reynolds criterion for the studied flows (Table 1.3, *Re*) corresponds to the turbulent flame ($Re > 10^4$), whose parameters can impact the characteristics of both primary particles and their aggregates, while marked flow turbulence can promote formation of large primary aggregates in the flame and tight attachment of primary particles on sticking to these aggregates with subsequent layering of SiO_2 onto their contacts.

In the case of a laminar flow (corresponding to the standard synthetic technique), primary particles are spherical and nonporous (Figure 1.2), and their contacts in aggregates (primary aggregates can be smaller than in the turbulent flame and a large portion of aggregates is secondary ones formed mainly on the postsynthesis stage of powder treatment at 250°C–400°C with the presence of water vapor) are less tight than those in primary aggregates for nonstandard silicas formed in the turbulent flame. Additional supply of a low amount (in comparison with the main flow from a central nozzle) of hydrogen (on deficiency of O_2 in the flame) through an additional annular nozzle provides the hydrolysis of residual $SiCl_4$ and $Si\text{-}Cl_x$ on the flame periphery at 600°C–800°C that results in formation of a porous (rough) surface layer on silica particles leaving the hot zone of the flame. Thus, changes in synthesis conditions allow one to vary parameters of primary particles, primary (formed on sticking of primary particles then fused in the flame) and secondary aggregates,

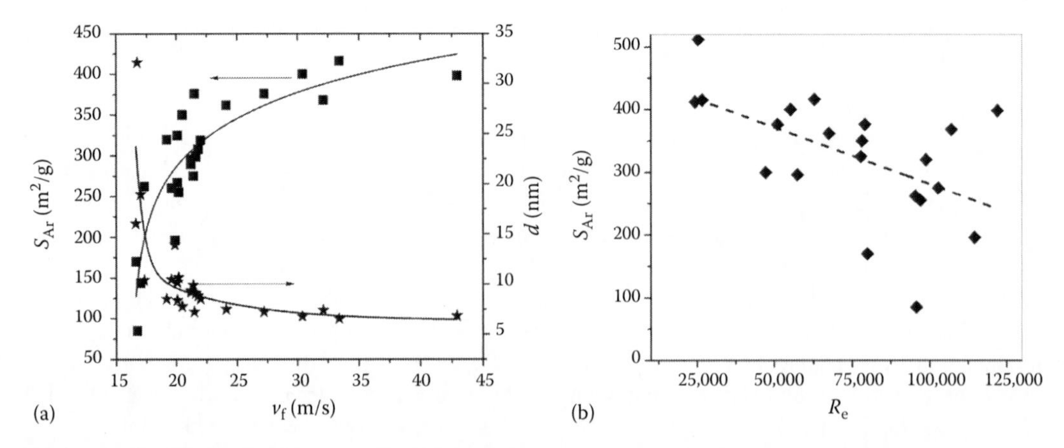

FIGURE 1.7 Relationships between (a) the flow velocity v_f and the specific surface area (S_{Ar}) or the primary particle diameter, d, and (b) the Reynolds criterion (R_e) and the S_{Ar} value for several series of nanosilica samples synthesized under different conditions that result in a certain scatter in the data.

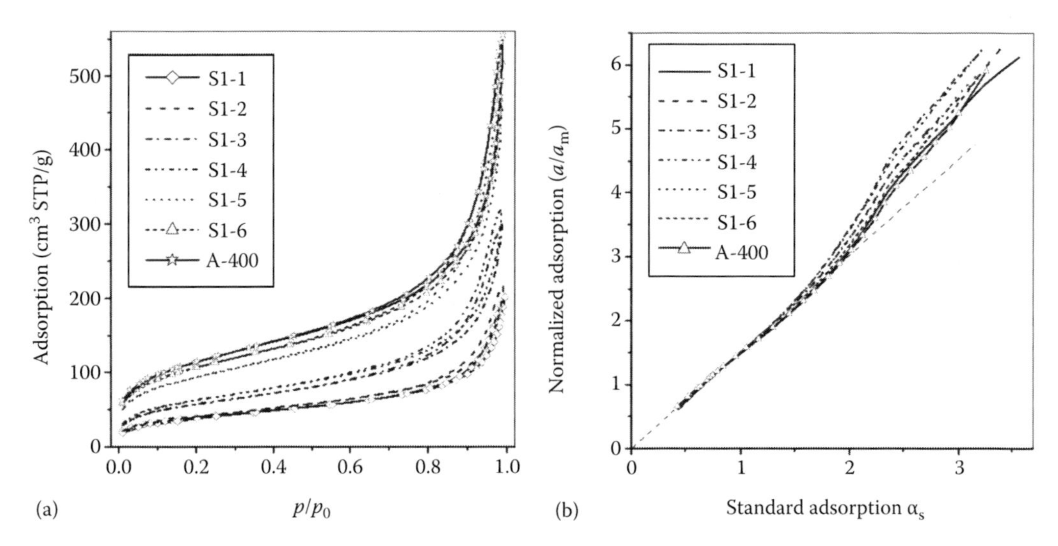

FIGURE 1.8 (a) Isotherms of nitrogen adsorption–desorption (77.4 K) on nanosilicas of the first series; the α_S plots for (b) standard adsorption, and (c) normalized by dividing by a_m. (Adapted from *J. Colloid Interface Sci.*, 242, Gun'ko, V.M., Mironyuk, I.F., Zarko, V.I. et al., Fumed silicas possessing different morphology and hydrophilicity, 90–103, 2001d. Copyright 2001, with permission from Elsevier.)

and loose agglomerates of aggregates. All these changes reflect in the interfacial phenomena in gaseous and liquid dispersion media (vide infra).

Changes in the pore volume and the specific surface area (or the primary particle size d, since $S \sim 1/d$) lead to marked alterations in the nitrogen adsorption isotherms (Figure 1.8a). However, the type of the isotherms is the same (as well as for all fumed oxides) due to textural and structural features of such materials composed of spherical nonporous primary particles (Figure 1.2) forming aggregates and agglomerates possessing only the textural porosity. The isotherms and the α_S plots (Figure 1.8) as well as average pore radius values (Table 1.2; R_p) demonstrate that nanosilicas are rather mesoporous (voids between primary particles in aggregates and between aggregates in agglomerates can be considered as mesopores of a complicated shape) independently on S or average d.

Basing on the very large values of $V_{em} = 20–25$ cm^3/g of voids, one can conclude that nanosilica is a macroporous material. However, these macropores are of very large sizes to be effectively filled by nitrogen (water or benzene vapors) even at $p/p_0 \to 1$. The normalized α_S plots for nanosilicas do not deviate from the plot for poorly porous silica gel Si-1000 ($S_{BET} \approx 26$ m^2/g, with no nanopores) at $\alpha_S < 1.5$ (Figure 1.8). Consequently, narrow nanopores are practically absent in nonstandard nanosilicas (Table 1.2) as well as in standard ones. A very low contribution of nanopores to the overall porosity of nanosilicas (provided by contact zones between adjacent primary particles) suggests that these contacts are relatively tight for nonstandard nanosilica samples resulting also in relatively low hydrophilicity of the materials (Table 1.2; C_w).

The similarity in the isotherms (Figure 1.8) corresponds to a similarity in the PSDs (Figure 1.4). Large mesopores at $R_p > 10$ nm can correspond to inter-aggregate space (voids) in agglomerates, as voids in aggregates mainly correspond to $R_p < 5–8$ nm. Changes in the apparent density of these samples (Table 1.2, ρ_{ap}) do not correlate with S and V_p, while the last parameters are mainly linked to the characteristics of primary particles and their aggregates, but the empty space in nanosilica powders is significantly larger ($V_{em} \approx 25$ cm^3/g) than V_p (<0.7 cm^3/g and close to the volume of voids in aggregates; Figure 1.2) and linked to the structure of agglomerates and visible flocks. However, an increase in ρ_{ap} and average pore radius R_p correlate with a decrease in the fractal dimension (D); that is, the increase in the apparent density of powders (typically for nanosilicas with larger primary particles and smaller S_{BET} values) corresponds to smoother particle surfaces and the increase in the dispersivity is accompanied by growing fractality (aggregation). Standard nanosilica is mainly mass fractal (for standard silica S3-1, fractal dimension $D_{AJ}=2.62$), but the samples synthesized under nonstandard conditions can possess not only mass fractality but also larger surface fractality (for silicas prepared at a greater flow of hydrogen through the annular nozzle). At the same time, the flow turbulence resulting in formation of tighter contacts between primary particles can reduce the pore (void) fractality (as void walls are smooth); therefore, D_{AJ} is higher for S3-1 than for S1-5; however, the last sample has larger S and smaller d values.

Water is much more interesting probe compound as an adsorbate than nitrogen because of formation of the strong hydrogen bonds in both bulk and bound interfacial waters. It should be noted that the adsorption energy of water can be greater in the case of formation of water clusters adsorbed near contacts between primary particles bonded to other particles by the hydrogen bonds. However, for tightly adnate primary particles (Figure 1.2), the formation of such water clusters is less probable. Therefore, the adsorbed water amounts are lower for nonstandard silicas (Tables 1.2 and 1.4; $C_{w,105}$), which are less hydrophilic than the standard sample. Additionally, the concentration of strongly bound water (SBW) (including associative desorption of hydroxyls) desorbed at T between 105°C and 900°C ($C_{w,900}$) for less hydrophilic silicas is lower by several times (Table 1.4). These structural and adsorptive features of the nanosilica surfaces cause a reduction not only of the amounts of weakly bound water (WBW) ($C_{w,105}$) adsorbed from air (desorbed at $T<105$°C) but also the monolayer capacity for water (Table 1.4; $a_{w,mono}$) adsorbed from air at room temperature.

Investigation of chemical structure and interfacial phenomena at solid surfaces, especially silicas and related materials, by NMR method has a long history, much of it motivated by the need to understand heterogeneous catalysis and water adsorption (Ernst et al. 1987, Abraham et al. 1988, Pines and Bell 1994, Brown et al. 1999). The NMR signals of solid surfaces are typically dominated, in the absence of specific line-narrowing techniques, by the same anisotropic spin interactions (magnetic dipole–dipole, chemical shift [shielding], and nuclear electric quadrupole) as govern the NMR of solids in general (Maciel 1984, Ernst et al. 1987, Brown et al. 1999). Significant efforts in solid-state NMR were devoted to the development of techniques to eliminate the line-broadening effects of these interactions, while maintaining some aspect (typically via the isotropic averages) of their information content (Ernst et al. 1987, Bell and Pines 1994, Brown et al. 1999). The most common of these techniques is magic angle spinning (MAS), which uses coherent rotational averaging to accomplish line narrowing of solid-state NMR signals (Brown et al. 1999). Other resolution-enhancing techniques include (i) straightforward "decoupling" (rapid "stirring") of one set of spins (e.g., protons of OH groups) to average to zero their dipolar effects on a set of spins being observed (e.g., ^{29}Si or ^{27}Al in their respective oxides; Maciel 1984, Ernst et al. 1987, Abraham et al. 1988); (ii) multiple-pulse decoupling (Waugh et al. 1968, Burum and Rhim 1979), for example, in combined rotation and multiple-pulse spectroscopy (CRAMPS; Gerstein et al. 1977, Warren 1990, Bell and Pines 1994) of proton–proton dipolar interactions; (iii) multiangle sample spinning (e.g., double rotation, simultaneous sample rotation about two axes [Wooten et al. 1992, Ray and Samoson 1993] or dynamic angle spinning; sequential sample spinning about two axes [Mueller et al. 1990, 1991a,b] for averaging the second-order quadrupolar effect in ^{27}Al NMR); (iv) a variety of 2D techniques (e.g., heteronuclear correlation [HETCOR] spectroscopy [Bronnimann et al. 1992] for correlating a metal nuclide with nearby protons); and (v) multiple quantum techniques, another way to simplify the spectra of quadrupolar nuclei (Frydman and Harwood 1995, Medek et al. 1995) and a strategy for estimating the number of protons in a "cluster" (Gerstein et al. 1993, Chang et al. 1996). Of course, all of these approaches have been limited largely to diamagnetic solids (such as nanooxides studied here). The role of paramagnetic centers in solid-state NMR has not yet been broadly characterized, and ferromagnetic centers are assumed to be incompatible with conventional NMR techniques.

The application of surface-selective ^{1}H → ^{29}Si cross polarization (CP) experiments for amorphous silica (Figure 1.9) gives three partially overlapping peaks: Q_2 for $(\equiv SiO)_2Si(OH)_2$ sites at about −89 ppm (relative to a liquid tetramethylsilane [TMS], peak), Q_3 for $(\equiv SiO)_3SiOH$

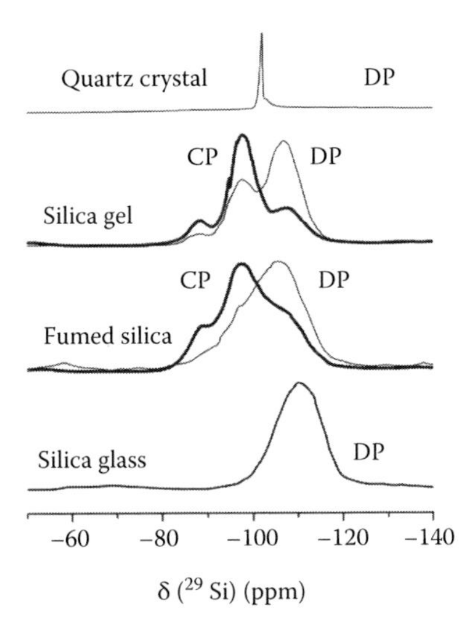

FIGURE 1.9 ^{29}Si MAS NMR spectra of silica obtained with $^{1}H \rightarrow {}^{29}Si$ cross polarization (CP) and without CP (direct polarization, DP). (Adapted with permission from *Chem. Rev.*, 99, Brown Jr., G.E., Henrich, V.E., Casey, W.H. et al., Metal oxide surfaces and their interactions with aqueous solutions and microbial organisms, 77–174, 1999. Copyright 1999, American Chemical Society.)

sites at −100 ppm, and Q_4 for $Si(OSi{\equiv})_4$ sites at −110 ppm. The positions and relative intensity of these peaks depend on both types of measurements (CP or direct polarization [DP]) and kinds of silicas (Figure 1.9).

These spectra show the dramatic difference between the narrow peak of crystalline silica (quartz) and the broad peaks of amorphous silicas (silica gel and nanosilica), caused by distributions of bond angles and/or lengths and the absence of translational order in the amorphous materials. The dominance of the Q_4 peak is observed in ^{29}Si DP MAS spectra of silicas obtained without CP. While ^{29}Si MAS NMR experiments are capable of elucidating the concentrations of Q_2, Q_3, and Q_4 sites on silica surfaces, ^{1}H MAS NMR, with or without multiple-pulse line narrowing, is not directly able to make the distinction between Q_2 and Q_3 silanols but is capable of distinguishing between hydrogen-bonded and non-hydrogen-bonded ("isolated") hydroxyl groups on the silica surface because the hydrogen bonding leads to changes in the atomic charges and shielding of protons (i.e., downfield shift increases with increasing strength of the hydrogen bonds). Compelling evidence for this distinction in ^{1}H MAS NMR can be based on the rate at which transverse proton magnetization is dephased under the influence of ^{1}H-^{1}H dipolar interactions. Since the hydrogen–hydrogen distance within a hydrogen-bonded pair of hydroxyls (silanols) is smaller than between a pair of isolated hydroxyls (silanols), this dephasing is much faster in the case of the hydrogen bonds. The ^{1}H CRAMPS spectrum of silica gel (Figure 1.10) shows the following three main contributions: a sharp peak at about 1.7 ppm due to isolated silanols (which remains even after evacuation at 500°C because residual silanols can be found after heating at 900°C [Legrand 1998, Zhuravlev 2000]), a well-defined peak at about 3.5 ppm (which is easily removed by evacuation at moderate temperatures) due to physisorbed water, and a broad underlying peak with its maximum at about 3.0 ppm (removed by evacuation at 500°C) due to a distribution of hydrogen-bonded hydroxyl groups (Brown et al. 1999).

^{1}H CRAMPS studies of silicas as a function of water content have provided valuable information on the state of water in these systems. ^{29}Si and ^{1}H NMR data of the type represented in Figures 1.9 and 1.10 are much more valuable when they can be correlated, that is, when the fraction of each type of silanol (Q_2 or Q_3, as determined by ^{29}Si NMR) that is hydrogen bonded (as determined by ^{1}H NMR) can be elucidated (Brown et al. 1999). While the most direct strategy for making such

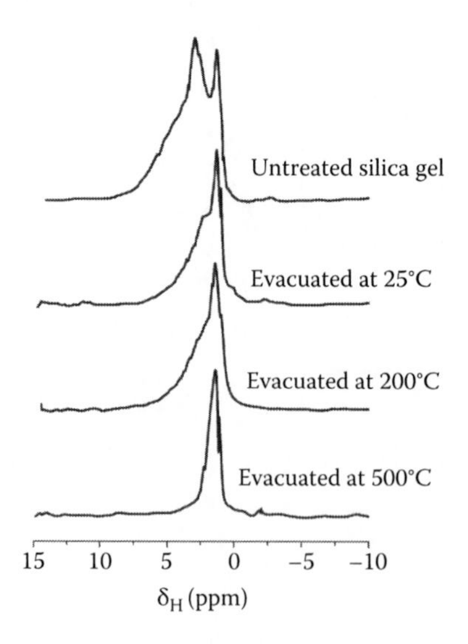

FIGURE 1.10 ^{1}H CRAMPS NMR spectra, obtained by combined MAS and multiple-pulse line narrowing, of silica gel at various degrees of water content. (Adapted with permission from *Chem. Rev.*, 99, Brown Jr., G.E., Henrich, V.E., Casey, W.H. et al., Metal oxide surfaces and their interactions with aqueous solutions and microbial organisms, 77–174, 1999. Copyright 1999, American Chemical Society.)

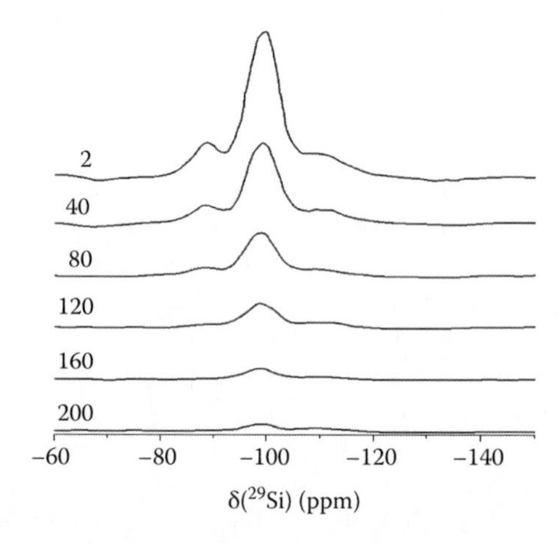

FIGURE 1.11 ^{29}Si MAS NMR spectra (absolute intensities), based on ^{1}H → ^{29}Si CP, of silica gel, obtained using various periods of ^{1}H-^{1}H dipolar dephasing prior to cross polarization. (Adapted with permission from *Chem. Rev.*, 99, Brown Jr., G.E., Henrich, V.E., Casey, W.H. et al., Metal oxide surfaces and their interactions with aqueous solutions and microbial organisms, 77–174, 1999. Copyright 1999, American Chemical Society.)

correlations by NMR would seem to be 2D HETCOR spectroscopy, to date more definitive results have been obtained by systematically manipulating the ^{1}H spins (e.g., allowing them to dephase) in an appropriate segment of a ^{29}Si CPMAS experiment. Figure 1.11 shows ^{29}Si CP/MAS spectra obtained on silica gel in experiments in which a variable ^{1}H-^{1}H dephasing period was inserted prior to the CP transfer of ^{1}H spin polarization to ^{29}Si (Chuang et al. 1992).

These spectra, and analogous results obtained by a variety of NMR techniques, show that the Q_2 silanols are more extensively hydrogen bonded than the Q_3 silanols. From results of this type in samples prepared with various water contents, it has been possible to develop a model of the silica gel (and probably nanosilica) surface in terms of surface segments that resemble specific surface faces of β-cristobalite (Iler 1979, Legrand 1998, Brown et al. 1999, Chuiko 2001, 2003).

Thus, the structural and adsorption properties of nanosilicas depend on the synthesis conditions such as the ratio of reactants ($SiCl_4$, O_2, and H_2), temperature, flow velocity, turbulence and length of the flame, and additional external actions (e.g., electrostatic field). These structural features of nanosilicas can affect the interactions with water, described later.

1.1.2 Adsorption of Water onto Nanosilica Dispersed in Gases and Organic Liquids

Interactions of nanosilicas with water adsorbed from air (typically 0.01–0.10 g H_2O per gram of dry silica) or in aqueous suspensions (up to 20 wt% of the initial nanosilica or 40–50 wt% of mechanochemically activated [MCA] silica in the suspension) or gel-like systems (C_{SiO_2} > 20 wt% of the initial nanosilica) can play an important role in practical applications of the materials (Chuiko 2001, 2003, Blitz and Gun'ko 2006). IR spectroscopy is the most frequently used method to study the adsorption of water onto silicas (Kiselev and Lygin 1975, Iler 1979, Legrand 1998, Bergna 1994, 2005). This is due to a very clear picture of the O-H stretching vibrations of surface silanols in the 2550–3750 cm^{-1} range affected by adsorbed water. Additionally, this IR band is useful to study any adsorbates at a silica surface because they interact mainly with the surface silanols. For instance, Parida et al. (2006) analyzing the adsorption behavior of various organic adsorbates on silica surface showed that most of the structural information on silica was obtained from IR spectral data and from the characteristics of water present at the silica surface. The adsorption of water from air onto nanosilicas depends on the specific surface area (i.e., sizes of primary nanoparticles), pore volume (organization of aggregates of primary particles and agglomerates of aggregates), and the number of surface-active sites (Kiselev and Lygin 1975, Iler 1979, Legrand 1998, Chuiko 2001, 2003, Gun'ko et al. 2009d). Silanol groups, single ≡SiOH and twin = $Si(OH)_2$ at $C_{SiOH}/C_{Si(OH)_2} = 0.85/0.15 - 0.6/0.4$ depending on the nanosilica type and its heating and other treatment history, are the main adsorption sites for water (as well as for other polar adsorbates) at the silica surface. Many of physicochemical properties of nanosilicas and the interfacial phenomena at their surfaces depend strongly not only on the primary particle and aggregate size distributions but also on the concentration of adsorbed water in the form of bound intact molecules (C_w) and ≡SiOH and = $Si(OH)_2$ groups, that is, dissociatively adsorbed water (C_{OH}; Figure 1.12) and treatment history of the materials (Kiselev and Lygin 1972, Iler 1979,

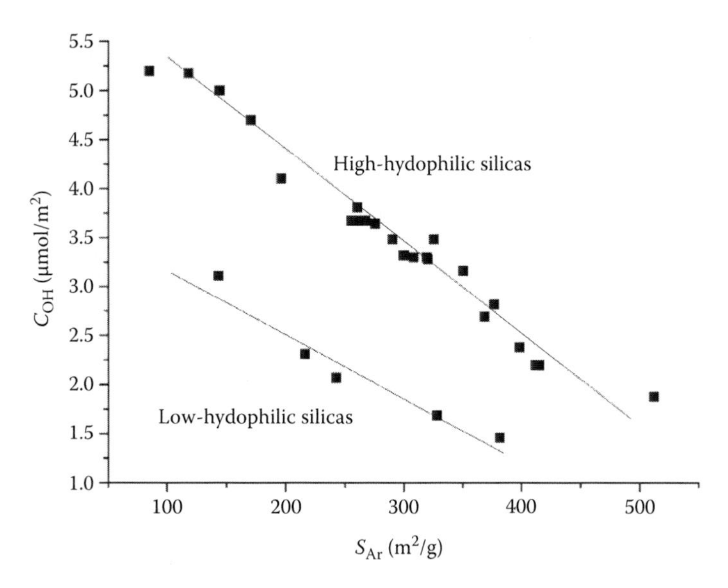

FIGURE 1.12 Relationship between the specific surface area S_{Ar} and the concentration of free silanols at $\nu_{OH} = 3750$ cm^{-1} for different nanosilicas.

Legrand et al. 1990, 1993, Bergna 1994, 2005, Vansant et al. 1995, Auner and Weis 1996, 2003, 2005, Dabrowski and Tertykh 1996, Legrand 1998, Chuiko 2001, 2003, Blitz and Gun'ko 2006, Verdaguer et al. 2006, Balard et al. 2011).

For instance, water adsorbed from air can negatively affect the characteristics of nanosilica as a filler of lyophilic media or hydrophobic polymers. In the case of hydrophilic polymers, water bound to silica provides the plasticizing effects for polymers and competitive adsorption effects for both low- and high-molecular polar compounds.

To reduce the adsorption of water (i.e., C_w value), chemical modification (hydrophobization) of the silica surfaces by organosilicon or organic compounds (Iler 1979, Leyden 1985, Leyden and Collins 1989, Legrand 1998) or heating of silica at high temperature, giving tentative diminution of the hydrophilicity, can be used. However, the first method increases the material cost and changes the nature of the silica surfaces and the size distributions of particles that can be undesirable for some applications of nanosilicas. The use of the second method can result in a decrease in the specific surface area and changes in the structure and organization of secondary particles. Therefore, production of nanosilicas possessing initially a desirable degree of the surface hydrophilicity (C_w, C_{OH}) and an appropriate specific surface area (Figure 1.12) can be of interest from both theoretical and practical points of view.

For the standard A-300 at $S \approx 300$ m²/g (standard synthesis conditions at a stoichiometric ratio $SiCl_4/O_2/H_2$, Table 1.4, sample S3-1), adsorption isotherm of water shows (Figure 1.13) that the silica affinity to water is low because the adsorption corresponds to 12.1 wt% at $p/p_0 = 0.999$. This is due to the textural porosity of nanosilica without significant contribution of nanopores (Figure 1.4b) because primary nanoparticles are nonporous (Figure 1.2). Additionally, contribution of narrow mesopores is small (Figure 1.4b). Therefore, the hysteresis loop of the nitrogen adsorption–desorption isotherm is narrow and starts at a high relative pressure (Figure 1.4a). However, some other samples (characterized by enhanced hydrophilicity) possess a higher affinity to water (Figure 1.13). For instance, sample S3-6 ($S = 308$ m²/g) adsorbs approximately 20 wt% of water at $p/p_0 = 0.999$. This difference with the standard silica A-300 can be explained by the changes in the texture of primary and secondary particles of sample S3-6. An increase in the hydrogen amount in the flame greater than stoichiometric (Table 1.4) leads to a reduction of the water adsorption (C_w) and monolayer capacity $a_{w,mono}$ (Table 1.4; samples S3-4, S3-5 and S3-8); that is, the hydrophilicity of these silicas decreases.

An increase in the hydrogen flow through the annular nozzle to 8 m³/h during the synthesis also affected the adsorbed water amounts (compare C_w for S3-2 and S3-7 or S3-3 and S3-6). However, the concentration of free silanols C_{OH} changed slightly. For S3-5 at $S_{BET} = 144$ m²/g, C_{OH} is greater due to a reciprocal dependence C_{OH} on S (Figure 1.12). Oxygen deficiency in the flame resulted in a diminution of the amount of adsorbed water, but the monolayer capacity changed slightly (S3-3) or decreased (samples 4, 5, 6, and 8 from the third series).

Changes in the hydrophilicity of the silica surfaces for samples of the third series resulted in marked differences in the water adsorption isotherms recorded (at 293 K) in mmol per gram of silica (Figure 1.13a) or per m² of the silica surface (Figure 1.13b). This difference is much larger than the normalized adsorption of nitrogen (Figure 1.8b), which is practically the same at $\alpha_S < 1.5$ ($\alpha_S = 1$ corresponds to the adsorption at $p/p_0 = 0.4$ when the capillary condensation is yet absent). However, the nitrogen adsorption differs for the nanosilicas studied at $\alpha_S > 1.5$ because of the difference in their textural organization and capillary condensation of nitrogen or water in mesopores. There is the difference in the shape of the hysteresis loops (Figure 1.13c), which are open because of the difference in the amounts of water adsorbed dissociatively. For instance, in the case of the standard silica (sample S3-1), the hysteresis loop is narrow (Figure 1.13c) in contrast to that for nonstandard silicas (samples S3-4-6).

Relative reduction of water adsorption onto nonstandard "hydrophobic" silicas (Figure 1.13; Table 1.4; S3-4, S3-5, and S3-8) is accompanied by a decrease in the adsorption potential and changes in the free energy of adsorption (Figure 1.13d) analyzed at coverage less than the water monolayer ($a < a_{w,mono}$). For less hydrophilic silicas (Table 1.4), the changes in the Gibbs free energy of water adsorption $-\Delta G < 2.5$ kJ/mol are smaller than that for more hydrophilic silicas $-\Delta G < 7$ kJ/mol (Figure 1.13d).

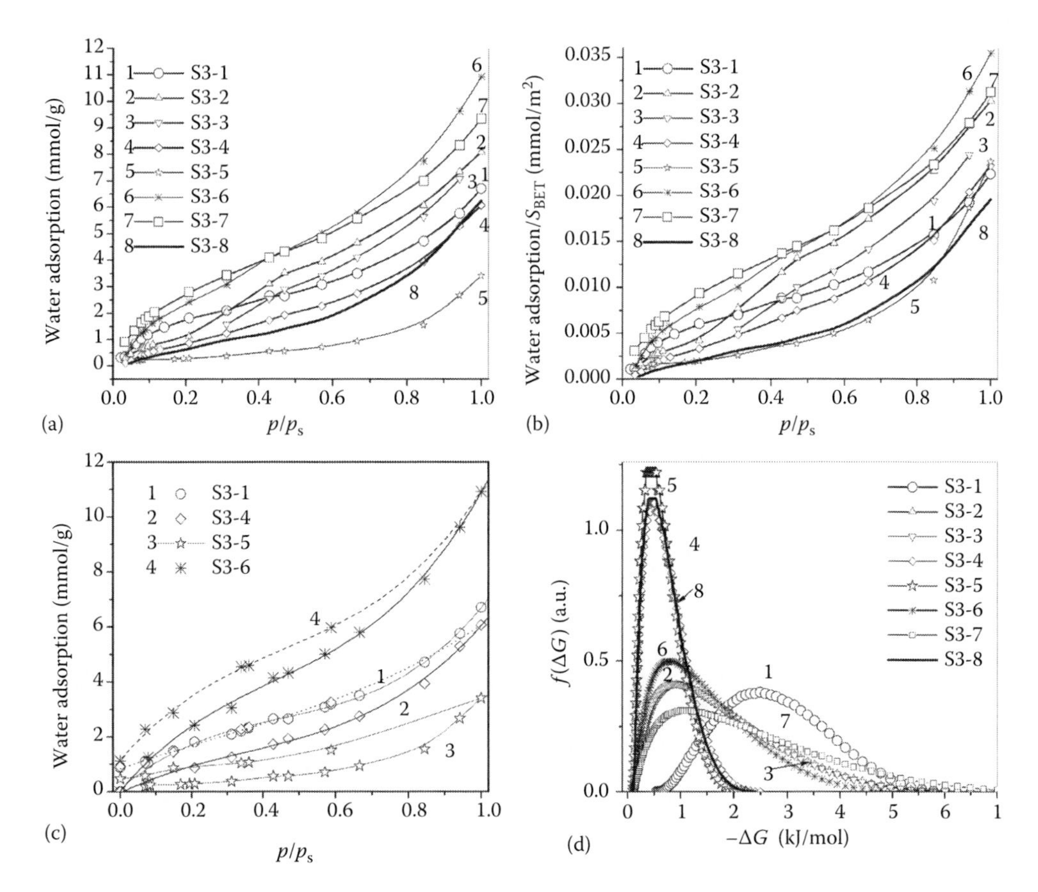

FIGURE 1.13 Isotherms of water adsorption at 293 K on silicas shown in Table 1.4 in (a, c) mmol/g and (b) mmol/m^2, (c) adsorption and desorption of water is shown, and (d) distribution functions of free energy of water adsorption onto silicas. (Adapted from *J. Colloid Interface Sci.*, 242, Gun'ko, V.M., Mironyuk, I.F., Zarko, V.I. et al., Fumed silicas possessing different morphology and hydrophilicity, 90–103, 2001d. Copyright 2001, with permission from Elsevier.)

The adsorption A of nitrogen, water, etc. can be converted to the Gibbs adsorption Γ (Frolov 1982, Adamson and Gast 1997, Dill and Bromberg 2003):

$$\Gamma = A - \rho V_a = -\frac{\rho}{R_g T}\left(\frac{\partial \sigma}{\partial \rho}\right), \tag{1.1}$$

where

V_a is the volume of the adsorption layer

ρ the density of gaseous adsorbate as a function of p/p_0

σ the surface tension

R_g the gas constant

Integrating this equation could be done with respect to σ assuming that changes in the surface tension $\Delta\sigma$ upon the adsorption correspond to changes in Gibbs free surface energy (γ_S; Table 1.5) or changes in Gibbs free energy on adsorption (ΔG_{ads}) as follows (Frolov 1982, Dill and Bromberg 2003):

$$\Delta G_{ads} = \Delta\sigma + \Gamma\Delta\mu_s, \tag{1.2}$$

where $\Delta\mu_s$ is the change in the chemical potential of adsorbate due to the adsorption (Table 1.5).

The term $\Gamma\Delta\mu_s$ determines the compaction of the adsorption layer. All the parameters shown in Table 1.5 are in correlation because a decrease in the content of silanols leads to diminution of the chemical potential and the surface becomes less active in the adsorption of polar (water) or nonpolar (nitrogen) adsorbates (the γ_S and $-\Delta\sigma$ values decrease). The difference in the γ_S and $-\Delta\sigma$ values (Table 1.5) shows that the $\Gamma\Delta\mu_s$ value in Equation 1.2 gives significant contribution to the ΔG_{ads} value. This suggests significant rearrangement of the interfacial water due to interaction with the solid surfaces in comparison with bulk water.

Heating of nanosilicas at different temperatures for different time leads to changes in the textural and structural parameters because of desorption of water and condensation of silanols to form ≡Si–O–Si≡ bridges (vide infra). This treatment can affect the reactivity and the adsorption capacity of nanosilicas.

The amounts of water adsorbed in two forms as intact water and surface hydroxyls and changes in the surface tension $\Delta\sigma$ and interfacial free energy γ_S calculated per surface unit (or changes in the chemical potential $\Delta\mu_s$) upon the water adsorption (Table 1.5) poorly correlate to changes in the specific surface area because of different conditions of the synthesis affecting not only the textural characteristics but also the chemical structure of the surface. However, there is a good correlation between γ_S, $\Delta\sigma$, and C_{OH} values (Table 1.5) that shows the dependence of the adsorption characteristics of silicas on content of silanols. For instance, an increase in the C_{OH} value leads to increase in the $\Delta\mu_s$, γ_S, and $-\Delta\sigma$ values that causes an increase in the adsorption of water (Table 1.5, $C_{OH,378}$) as well as other compounds, for example, proteins (vide infra). Thus, changes in synthesis conditions allow us to produce nanosilicas with different hydrophilicity (Tables 1.5 and 1.6; C_{OH}). It should be noted that the morphological characteristics of complex fumed oxides based on silica depend on chemical composition of primary particles much smaller than on synthesis conditions (Gun'ko et al. 2009d). Consequently, nanosilicas with different specific surface areas and prepared at different conditions can differently affect the properties of the interfacial water and its interaction with nonpolar or weakly polar coadsorbates (e.g., organic solvents).

All atoms in water molecules can form strong hydrogen bonds. Therefore, bulk or adsorbed water tends to be in a state with maximum hydrogen bonds per molecule. This leads to clustered adsorption of water at practically any surfaces. We will analyze the adsorption of water using approaches with or without consideration of the clustering effects.

Equations 10.50 and 10.51 describing clustered adsorption of water (see Chapter 10) give slightly different fitting of the water adsorption isotherm (Figure 1.14a). The best fitting over all p/p_s range gives Equation 10.51 at $m=5$ (i.e., clusters with six-member rings including $5H_2O$ and silanol group per cluster). Notice that diminution of the number of water molecules in a cluster per adsorption site from five to three results in decrease in the adsorption energy per hydrogen bond. This result shows that clustered adsorption of water occurs to form the clusters with a certain number of water molecules; however, this number increases till the Gibbs free energy decreases. In other words, this number cannot be too small (because a small number of the hydrogen bonds is energetically unfavorable) or too large (because of confined space effects in pores and diffusion of molecules toward more active surface sites). Therefore, the clustered adsorption of water depends on the PSD and the types and number of adsorption sites and their distribution in pores, as well as the amounts of adsorbed water. For instance, the character of water cluster distribution in pores of activated carbons (ACs; with O-containing functionalities at the periphery of the carbon sheets, water mainly located in mesopores), nanosilicas (water located in contact zones of adjacent nanoparticles in aggregates), and silica gels (water totally fills nanopores and mesopores) strongly differs for different adsorbents (Brennan et al. 2001, Gun'ko et al. 2009d). Equation 10.51 is more appropriate for calculations than Equation 10.50 because the former includes a smaller number of equation constants (only b). The distribution function $f(E)$ has clear physical sense and allows us to describe the nonuniformity of the adsorbent surface.

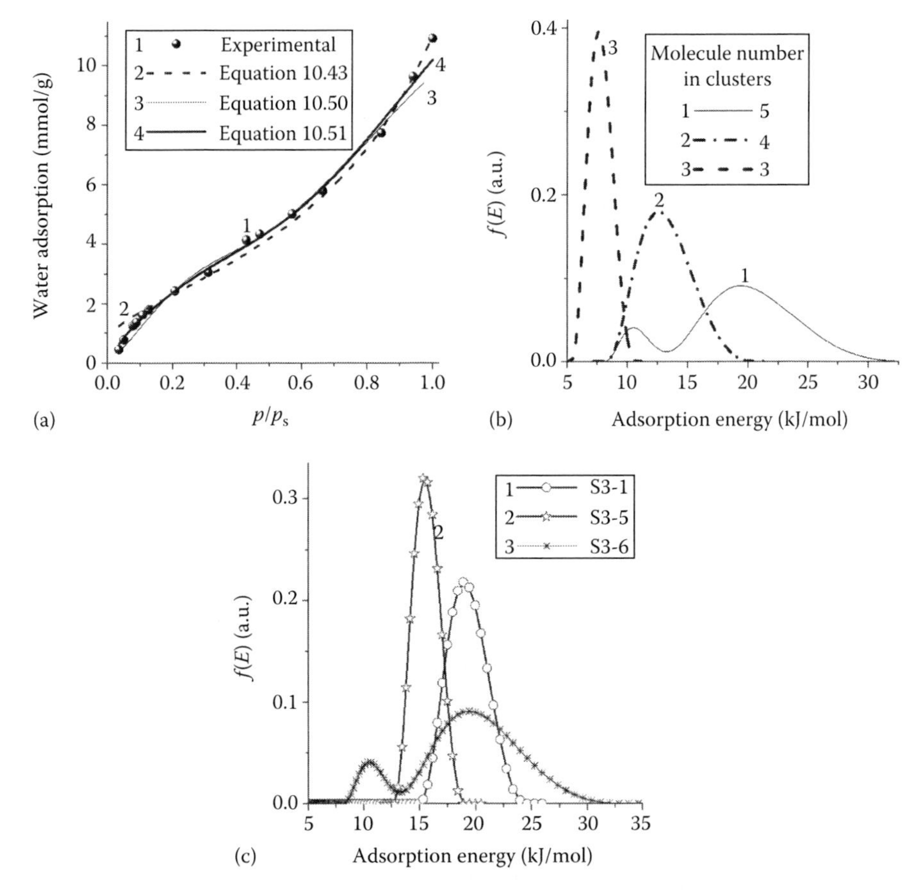

FIGURE 1.14 (a) Experimental adsorption isotherm (293 K) of water adsorbed onto A-300, sample S3-6 (Table 1.4), and fitting with standard Langmuir equation, Equation 10.43 (curve 2), Equation 10.50 (curve 3), and Equation 10.51 (curve 4) at $m=5$; (b) for sample S3-6, water adsorption energy per the hydrogen bond per molecule, $f(E)$, with Equation 10.51 and $m=5$, 4, and 3; and (c) $f(E)$ with Equation 10.51 at $m=5$ for A-300 (samples S3-1, S3-6) and A-150 (S3-5).

According to Collins et al. (2006), water at relative humidity (RH) = 51% forms monolayer coverage of a hydroxylated surface of porous silicas, and the adsorption is equal to 7.68 μmol/m^2. This allows us to calculate the specific surface area (S_w) from the water adsorption isotherms. The average deviation of the S_w values from the S_{BET} values (determined from the nitrogen adsorption) was ±12.1% (Collins et al. 2006). However, in the case of nanooxides heated at 400°C–450°C for several hours without rehydration before the measurements, the deviations of the S_w values from the S_{BET} values increase to +55.4/–29.4 (Table 1.7).

Typically, preheating of silicas at 400°C–450°C leads to the open hysteresis loops in the water adsorption isotherms because a portion of water dissociatively adsorbs and then it does not desorb. The negative deviations (Table 1.7) are due to the low hydrophilicity of silicas synthesized under special conditions described earlier in the text (Table 1.4). This results in decreased water adsorption, that is, underestimation of the S_w values and a decrease in changes in the Gibbs free energy of water adsorption (Table 1.4, γ_S). Some silicas (e.g., samples 1 and 7 in Table 1.7) were synthesized under different conditions since they have larger amounts of the surface silanols. Their preheating caused stronger distortions in the surface layer that resulted in larger amounts of water participating in rehydration of the surface, and therefore the S_w values are stronger overestimated (Table 1.7).

TABLE 1.7
Specific Surface Areas Determined for the Nitrogen or Argon (S_{BET}) and Water (S_w) Adsorption Isotherms

Silica	S_{BET} (m²/g)	S_w (m²/g)	(S_w-S_{BET})/S_{BET}	V_p (cm³/g)
A-300	299	584	0.953	0.618
A-300	260	264	0.015	0.540
A-150	144	81	−0.438	0.312
A-300	300	397	0.323	0.618
A-300	267	472	0.768	0.544
A-300	290	363	0.252	0.626
A-300	308	593	0.925	0.764
A-300	319	195	−0.389	0.563

Note: $S_w = a_{0.5} \times 130.104$, where $a_{0.5}$ is the water adsorption (mmol/g) at $p/p_0 \approx 0.5$.

The adsorption of water at a surface of porous and nonporous nanoparticles or microparticles is a complex process. Besides molecular adsorption of water, it is possible dissociative adsorption of the molecules onto strained siloxane bonds at the outer and inner surfaces of the particles. Additionally, water molecules can penetrate into the volume of nonporous particles that leads to their swelling. This process is very slow and can occur during 3–10 days depending on the size of particles and their heating history. The particle swelling results in diminution of the specific surface area of the materials. These processes are characterized by very different heat of adsorption (Q) and activation energy (E_a) for differently synthesized or pretreated silicas. For strongly dehydrated surfaces, the initial Q values can be greater than 200 kJ/mol. For hydrated surfaces, the Q values are close to the water condensation heat (45 kJ/mol). The E_a values can be small or zero during adsorption of intact water onto the outer surface of particles or $E_a > 100$ kJ/mol on dissociative adsorption of the water molecules at different sites with strained bonds at the outer or inner surfaces. The activation energy of the reactions in the inner volume of silica nanoparticles can be higher due to greater steric effects and hindered diffusion there (vide infra).

Water fills (at 20°C) 3–7 times smaller pore volume of nanosilicas than nitrogen fills (at −196°C). Therefore, the intensity of the PSD calculated using the N_2 and H_2O adsorption differs by approximately five times (Figure 1.15). Water adsorption onto silicas strongly increases with decreasing temperature toward 0°C or it strongly decreases with increasing temperature, especially at $T > 40$°C (Shim et al. 2006, Gun'ko et al. 2009d).

Water interactions with hydrated silica surface (i.e., without dissociative adsorption of the molecules) in gaseous dispersion medium can be divided into two processes such as adsorption and wetting. If liquid (subscript L) spreads at a solid surface (subscript S) in gaseous or vapor dispersion medium (subscript V) that diminution of the free surface energy F due to wetting is

$$\frac{\Delta F}{\Delta S} = \sigma_{SL} + \sigma_{LV} - \sigma_{SV}, \tag{1.3}$$

where the surface tensions are σ_{LV} at the liquid–gas (vapor) boundary, σ_{SL} at the solid–liquid boundary, and σ_{SV} at the solid–gas (vapor), and ΔS is change in the surface area. At equilibrium contact angle θ, all the forces are equilibrated that is described by the Young equation:

$$\sigma_{SV} = \sigma_{SL} + \sigma_{LV} \cos\theta. \tag{1.4}$$

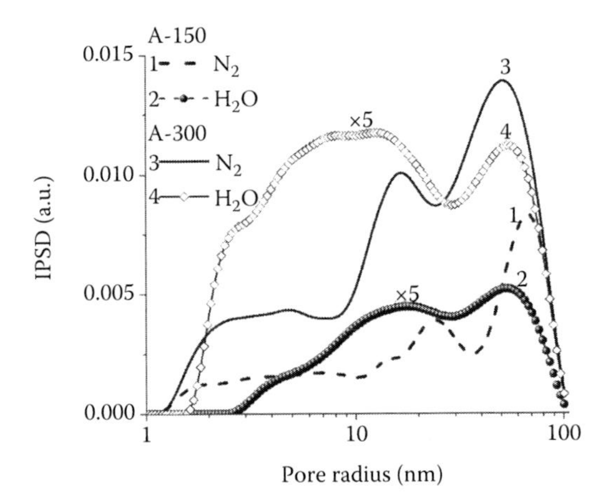

FIGURE 1.15 PSD in respect to the pore volume calculated with the model of voids between spherical particles for A-150 and A-300 using the adsorption of nitrogen (DFT method) and water (MND method, see Chapter 10).

Spreading of a liquid (totally wetting a solid surface at $\cos\theta = 1$) occurs under condition

$$\sigma_{SV} + \sigma_{LV} - \sigma_{SL} > 0. \tag{1.5}$$

The stronger the liquid interaction with the surface, the greater is the free energy diminution during the adsorption or wetting. From the condition of minimum free energy at equilibrium of contacting media, diminution of the free energy upon the formation of the interfacial boundary between a liquid and a solid surface can be used as a measure of the wetting. The heat of wetting (q) can be used as an integral characteristic of the liquid to wet the solid surface as described by the Helmholtz–Gibbs equation:

$$q = \sigma - T\frac{d\sigma}{dT}. \tag{1.6}$$

Equations 1.4 and 1.6 give the Helmholtz–Gibbs–Young equation:

$$q = \pi - T\frac{d\pi}{dT} + \sigma_{LV}\cos\theta - T\frac{d(\sigma_{LV}\cos\theta)}{dT}. \tag{1.7}$$

After differentiation (ignoring $d\sigma_{LV}/dT$), we have

$$q = \pi - T\frac{d\pi}{dT} + U_{LV}\cos\theta - T\sigma_{LV}\sin\theta\frac{d\theta}{dT}, \tag{1.8}$$

where $\pi = \sigma_S - \sigma_{SV}$ is the surface pressure of the adsorbed liquid film and U_{LV} is the total surface energy of wetting liquid at the boundary with the vapor phase (for water $U_{LV} = 118.5$ mJ/m^2 at 20°C). The q and π values (Table 1.8) can be calculated from the water adsorption isotherms:

$$\pi = \frac{R_g T}{S}\int_0^{p_s} a\, d\ln p, \tag{1.9}$$

where

- a is the water adsorption (mol/g)
- S is the specific surface area of adsorbent

TABLE 1.8
Heat of Wetting (q) and Surface Pressure (π) Calculated from the Water Adsorption Isotherms

S_{BET} (m^2/g)	q (mJ/m^2)	π (mJ/m^2)
299	207	86
260	135	40
144	113	26
300	161	56
267	170	63
290	123	33
308	195	79
319	113	26

TABLE 1.9
Heat of Wetting (Q) of Nanosilicas at Different S_{BET} Values

Parameter	A-50	A-300	A-500
S_{BET} (m^2/g)	52	232	492
C_{OH} (OH/nm^2)	6.7	4.1	1.9
Q (J/g)	11.6	51.5	69.4
Q' (J/m^2)	0.23	0.18	0.14

To calculate the q values (Table 1.8), a linear approximation of $d\pi/dT$ versus T (Tarasevich 2006) and our experimental data were used that give

$$d\pi/dT = 0.00205\pi + 0.16005 \text{ (mJ m}^{-2}\text{ K}^{-1}\text{)}. \tag{1.10}$$

The q values (Table 1.8) calculated from the water adsorption isotherms are similar to the Q values (Table 1.9) obtained using the calorimetry method. Certain differences between these values can be caused by different conditions of sample preparation and subsequent rehydration effects discussed earlier.

Changes in the structure of primary particles and their contacts influencing the water adsorption can be also analyzed using the IR spectra (Kiselev and Lygin 1975, Iler 1979, Legrand 1998, Shen and Ostroverkhov 2006). The content of free surface silanols decreases with increasing specific surface area (Figure 1.12). For instance, sample S3-6 (Table 1.4) adsorbs water significantly greater (by two times) than standard nanosilica S3-1 (Figure 1.13). This appears in great intensity of a broad band over 3000–3700 cm^{-1} (Figure 1.16; curves 1 and 4) linked to water adsorbed in different forms and disturbed surface silanols. This effect is accompanied by a substantial but nontypical reduction of the intensity of the band at 3750 cm^{-1} in comparison with the band intensity at 3400 cm^{-1} (Figure 1.16; curves 1 and 4) that is not observed for the standard silica characterized by a higher relative intensity of the band at 3750 cm^{-1} (Figure 1.16; curve 6). One can assume that this difference is connected to structural features of nonstandard silica, which can be more porous than standard nanosilica.

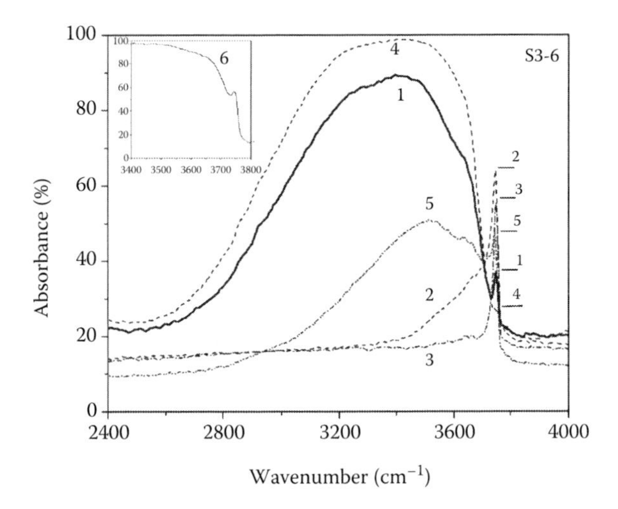

FIGURE 1.16 IR spectra of nanosilica sample S3-6 (1) in air, after degassing at (2) 450°C and (3) 650°C, (4) letting in saturated water vapor, (5) degassing at room temperature, and (6) sample S3-1 in air. (Adapted from *J. Colloid Interface Sci.*, 242, Gun'ko, V.M., Mironyuk, I.F., Zarko, V.I. et al., Fumed silicas possessing different morphology and hydrophilicity, 90–103, 2001d. Copyright 2001, with permission from Elsevier.)

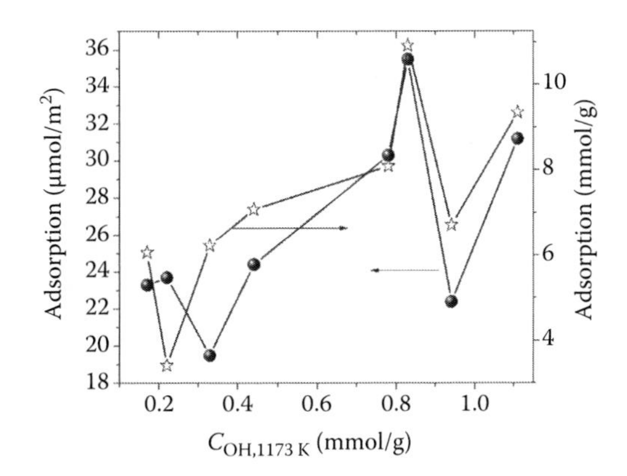

FIGURE 1.17 Water adsorption onto nanosilicas of different morphology as a function of the content of surface silanols determined from the water desorption at 105°C–900°C.

Changes in the morphology and the surface structure of nanosilicas lead to a nonlinear dependence of the maximum water adsorption versus the content of silanol groups (Figure 1.17) calculated from the thermodesorption data for the 105°C–900°C range. Notice that the IR spectra of silanols groups include several bands (Figure 1.18), which can be assigned to single undisturbed OH groups at $\nu_{OH}=3749$ cm^{-1} and slightly disturbed silanols at $\nu_{OH}=3747$ and 3742 cm^{-1}, which can be twin groups or single silanols located in shallow pores or open surface cavities (Hoffmann and Knözinger 1987, Legrand 1998). The band at $\nu_{OH}=3665$ cm^{-1} can be assigned to silanols distorted by neighboring groups in shallow cavities. The bands at lower wavenumbers (Figures 1.16 and 1.18) can be assigned to strongly disturbed silanols participating in the hydrogen bonds and bound water molecules.

The difference in the terminal and geminal silanols is well observed in the ^{29}Si CP/MAS NMR spectra of silicas (Legrand 1998, Gun'ko et al. 2007d, Murray 2010; Figure 1.19). Additionally, the ^{1}H MAS NMR spectra show the different δ_H values for single and twin OH groups because of the

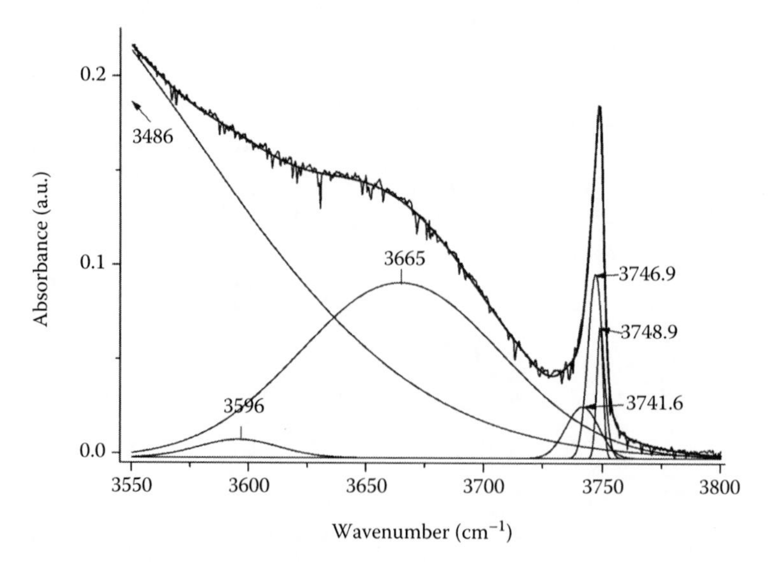

FIGURE 1.18 Decomposition of the IR spectrum of air-dried A-300 over the range of the OH-stretching vibrations.

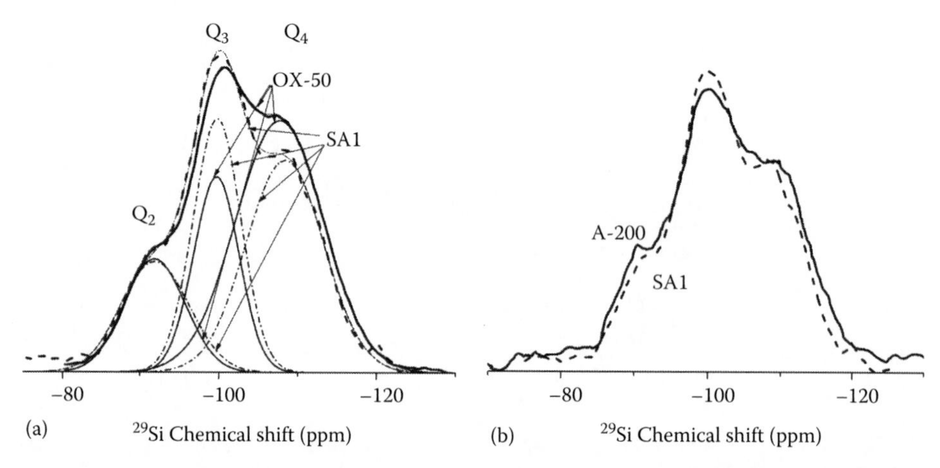

FIGURE 1.19 ^{29}Si CP/MAS spectra of (a) nanosilicas OX-50 (Degussa) and SiO_2/Al_2O_3 ($C_{Al_2O_3}$ = 1.3 wt%) decomposed into three bands: Q_4 (≡Si–O–Si≡), Q_3 (≡SiOH), and Q_2 (=Si(OH)$_2$) and (b) A-200 and SA1. (The spectra were redrawn from Murray, D.K., *J. Colloid Interface Sci.*, 352, 163, 2010 [OX-50]; *Appl. Surf. Sci.*, 253, Gun'ko, V.M. et al., *Appl. Surf. Sci.*, 253, 3215, 2007b [SA1]; Legrand, A.P. (ed.), *The Surface Properties of Silicas*, Wiley, New York, 1998 [A-200].)

difference in the surrounding of these silanols (d'Espinose de la Caillerie et al. 1997, Legrand 1998). The differences in the Fourier transform infrared (FTIR) and NMR spectra of single and geminal silanols reflect the differences in other characteristics of these silanols, for example, in interactions with water and other adsorbates, conditions of associative desorption of water molecules, and deprotonation in the aqueous media.

The content of the geminal silanols (Q_2 in Figure 1.19a) depends on the type of silica and pretreatment history of the materials. The changes in the content of geminal silanols (15%–45% of the total amounts of silanols) are one of the ways to change the hydrophilic properties of nanosilicas (Legrand 1998) due to variations in the synthesis conditions described earlier, because a surface with a larger amount of germinal silanols is more hydrophilic than a surface with a

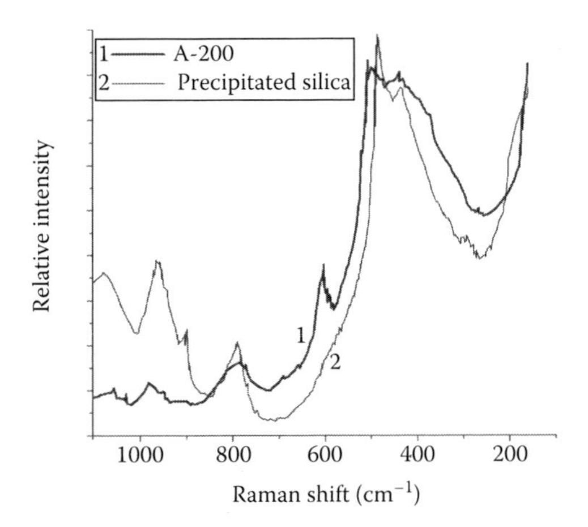

FIGURE 1.20 Raman spectra of (1) nanosilica A-200 (200 m^2/g, 3.8 OH/nm^2) and (2) precipitated silica P (175 m^2/g, 15 OH/nm^2). (Adapted from *J. Non-Crystal. Solids*, 191, Humbert, H., Estimation of hydroxyl density at the surface of pyrogenic silicas by complementary NMR and Raman experiments, 1995, 29–37. Copyright 1995, with permission from Elsevier.)

smaller number of them. Nanosilica A-200 ($S_{BET} \approx 200$ m^2/g) and silica/alumina (SA) with small content (1.3 wt%) of Al_2O_3, SA1 ($S_{BET} \approx 203$ m^2/g) have similar ^{29}Si CP/MAS NMR spectra (Figure 1.19b) because they have the same specific surface area. However, SA1 has 6.2 at% Al at the surface (practically all Al_2O_3 is at the surface). Therefore, it has slightly smaller content of Q_2 sites at the surface than A-200.

Humbert (1995) explained the overestimation of the Q_3 content in nanosilicas using their Raman spectra. The Raman spectrum of A-200 exhibits a symmetrical peak at 607 cm^{-1} (Figure 1.20) assigned to cyclic trisiloxane rings (with ∠SiOSi = 142°). Notice that precipitated silica with close S_{BET} value does not have this Raman line. In these rings in nanosilicas, the Q_4 sites are characterized by $\delta(^{29}Si) = -106$ ppm instead of −110 ppm. Consideration for its contribution allows one to estimate the Q_3 contribution as 4 OH/nm^2 (this value is close to that estimated from spectral and other data; Zhuravlev 2000). However, there is an additional explanation of the large intensity of the Q_3 silicon states based on strong coupling of Si atoms in silanols with protons in the bonded hydroxyls (Li and Ba 2008). Notice that according to Legrand et al. (1999), precipitated silica is characterized by a large Q_3 value (43%).

The NMR spectral characteristics of water adsorbed to a silica surface are affected by the proton exchange reactions between silanols and water molecules. A large difference (two to three orders of magnitude) in transverse relaxation time of protons of the silanols and adsorbed water molecules can lead to a significant broadening of signal of adsorbed water even in the case of a small rate of the proton exchange (cross relaxation effect). To analyze this effect, the temperature dependences of the ^{1}H NMR band shape and transverse relaxation time (T_2) were studied for nanosilica A-300 hydrated in air (Figure 1.21; Gun'ko et al. 2009d).

Signal width decreases with increasing temperature from 195 to 265 K (Figure 1.21a) and then it decreases and signal is not observed at $T > 355$ K in the high-resolution spectra. The $T_2(1/T)$ function has an inflection point at 265 K. Signal broadening (i.e., a decrease in the T_2 value) with decreasing temperature to 265 K is due to a decrease in the proton exchange rate and the molecular mobility of adsorbed water (O'Reilly 1974, Callahan et al. 1983, Mank and Lebovka 1988). In a first approximation, the slope ratio of the ln $T_2(1/T)$ curve determines the activation energy of self-diffusion of adsorbed water molecules (to the right of the maximum, $E_a \approx 7$ kJ/mol) and the activation energy of the proton exchange reaction between water molecules and silanols (to the left of the maximum, $E_a = 13.3$ kJ/mol). We can assume on the basis of these results (Figure 1.21) that the cross relaxation becomes especially significant for water adsorbed to nanosilica at $T > 270$ K. In the case

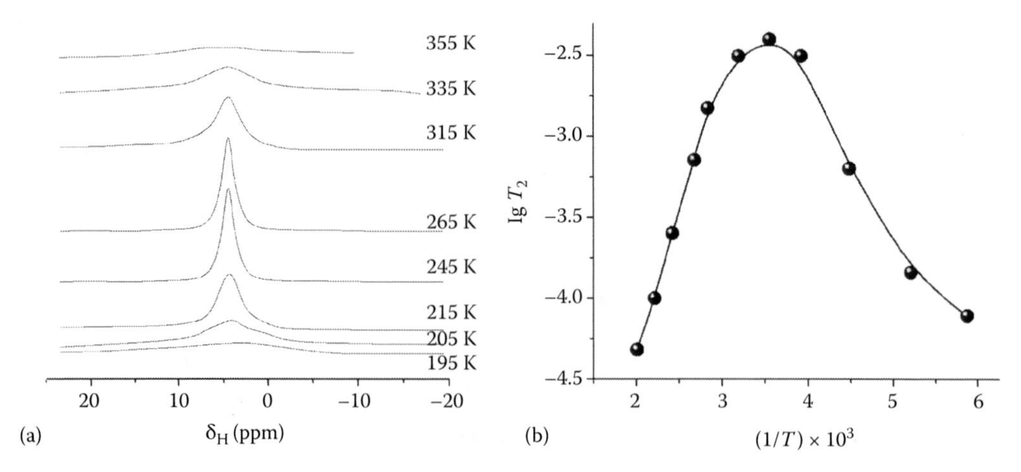

FIGURE 1.21 (a) ^{1}H NMR spectra of water adsorbed onto A-300 from air and (b) its temperature dependence (lg T_2 vs. $1/T$) of the transverse relaxation time.

of the aqueous suspensions of nanosilicas, the amounts of structured water in the interfacial layer but unbound directly to the silica surface are much larger (1–2 orders of magnitude) than in the case of weakly hydrated silica powder (<5 wt%). Therefore, contribution of water molecules participating in the proton exchange reactions with silanols to the total width of signal is low in the suspensions even at $T < 273$ K because of a significant amount of bound unfrozen water in a thick adsorption layer. Thus, the cross relaxation effect can be ignored for the frozen aqueous suspensions of nanosilicas because of two mentioned effects at low temperatures and at T close to 270–273 K.

Silicas with different specific surface areas can differently affect the properties of the interfacial water and its interaction with nonpolar or weakly polar solvents (coadsorbates). Aggregates of primary silica nanoparticles (Figure 1.22) can remain after adsorption of water. However, aging of nanosilicas can lead to changes in their textural and other characteristics (Morel et al. 2009,

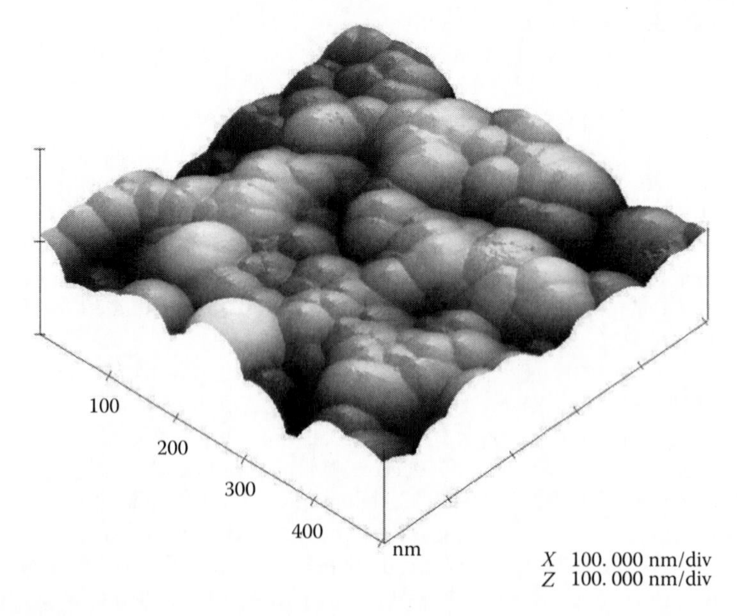

FIGURE 1.22 AFM image of nanosilica A-300 (500 × 500 nm^2) with visible aggregates (50–100 nm in size) of primary nanoparticles.

TABLE 1.10
Immersion Enthalpy (ΔH_{im}) of Fumed Oxides (with Different Specific Surface Area S_{BET} and Bulk Density ρ_b) in Water

Sample	S_{BET} (m^2/g)	ΔH_{im}[a] (J/g)	ΔH_{im} (mJ/m^2)	ρ_b (g/cm^3)
OX-50	54	6.7	124	0.142
A-100	143	22.4	157	0.043
A-150	162	26.6	164	0.041
A-200	206	31.6	153	0.053
A-250	266	34.1	128	0.070
A-300	337	48.9	145	0.066
A-380	381	55.9	147	0.041
A-400	409	57.5	141	0.085

[a] Errors are between ±0.3 and ±8.0 J/g (293 K).

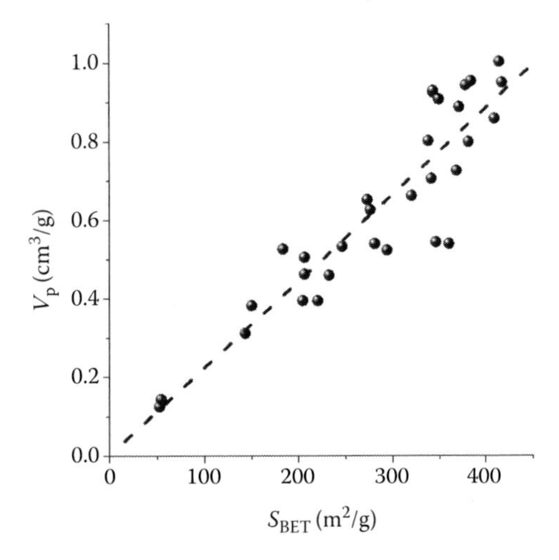

FIGURE 1.23 Relationship between specific surface area S_{BET} and pore volume V_p of nanosilicas.

Balard et al. 2011). Clearly, the binding strength of primary particles in aggregates formed in the flame and upon subsequent treatments depends on temperature, size, coordination numbers, and bonding type (≡Si–O–Si≡, ≡SiOH···O(H)Si≡ or ≡SiOH···$(OH_2)_n$···O(H)Si≡, etc.) of primary particles in secondary structures. After synthesis at lower temperatures and after hydration, aggregates form loose agglomerates (>1 μm, $D_m \approx 2.1$–2.2, bulk density $\rho_b \approx 2\%$–6% [Table 1.10] of the true density of 2.2 g/cm^3 for amorphous silica) through the hydrogen and electrostatic binding.

If $S_{BET} \geq 100$ m^2/g that isolated primary particles are not observed separately without a special treatment, while the smaller the particles, the stronger the bonding in the aggregates and the adsorption capacity (i.e., V_p) increases with S_{BET} (Figure 1.23).

The state of water at surfaces of unmodified and modified (e.g., partially hydrophobized) silicas differs because any complete modification displaces water from the surface in the region with lower electrostatic field, or partial modification results in the formation of a clustered adsorption layer. The latter is due to surface hydrophobic functionalities playing a role of barriers for separated water clusters located between them. Additionally, kosmotropic or chaotropic effects, the type of which depends on the structure of surface groups (Chaplin 2011, Gun'ko et al. 2005d), lead to changes in the interfacial water structure in a relatively thick layer (Gun'ko et al. 2009d). Polar heads of hydrophobic groups (e.g., NH_2) add an additional disturbing factor for the interfacial water layer.

Therefore, several series of nonporous and porous silicas unmodified and modified, differently hydrated, and placed in different dispersion media (air, nonpolar, weakly, or strongly polar) are analyzed here to elucidate their influence on the structure and properties of interfacial water. These investigations start from initial nanosilicas consisting with amorphous nonporous spherical primary particles forming secondary structures (aggregates and agglomerates).

Figure 1.24a shows temperature dependences of the chemical shift of the proton resonance δ_H for hydrogen atoms in water molecules adsorbed onto a surface of nanosilicas A-150, A-200, and A-380 having different specific surface area and adsorbing different amounts of water and placed in dispersion medium with a hydrophobic solvent (chloroform-*d*). The δ_H values for water bound to silicas are smaller than that for the bulk water because water adsorbs in the form of clusters in which the molecular associativity is lower than that in the bulk water. The δ_H value increases

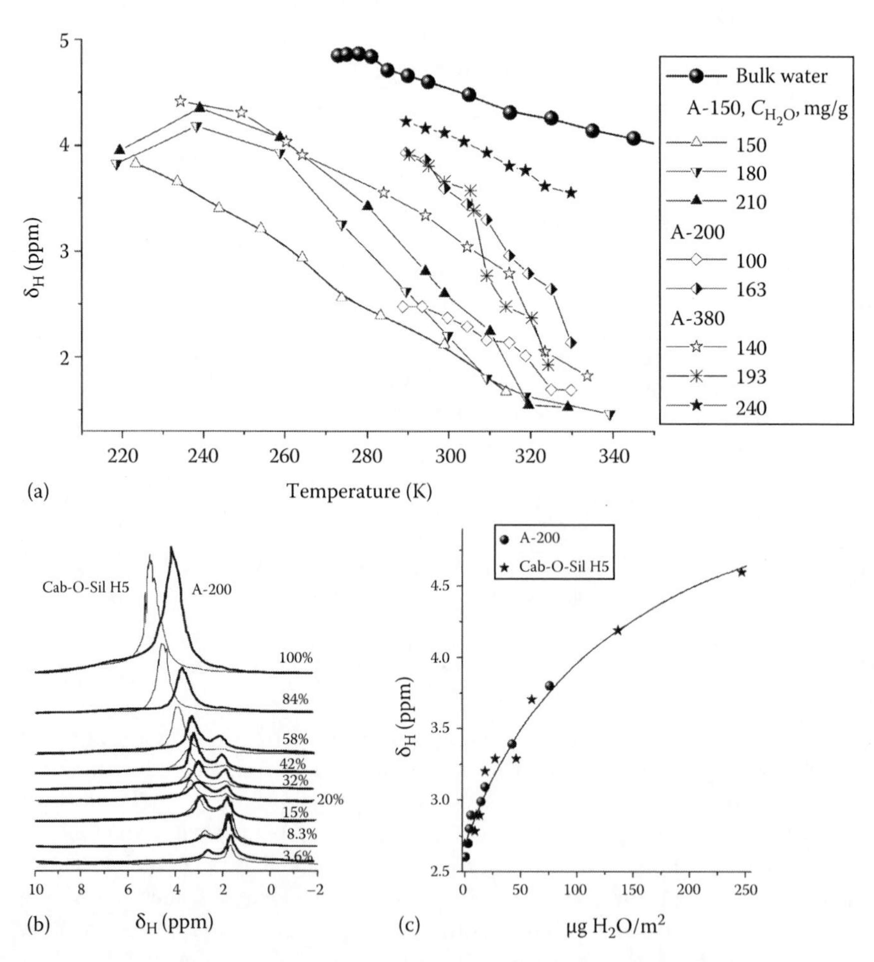

FIGURE 1.24 (a) Temperature dependence of δ_H for bulk water and water adsorbed onto nanosilicas at different S_{BET} values and different water content C_{H_2O} (static samples); (b) one-pulse ^1H MAS NMR spectra of nanosilicas A-200 and Cab-O-Sil H5 ($S_{BET}=326\ m^2/g$) equilibrated at different RH values; and (c) the δ_H value (2.6–4.6 ppm ^1H NMR peaks in (b)) as a function of the amounts of adsorbed water at room temperature. (a: Adapted from *Adv. Colloid Interface Sci.* 118, Gun'ko, V.M., Turov, V.V., Bogatyrev, V.M. et al., Unusual properties of water at hydrophilic/hydrophobic interfaces, 125–172, 2005d. Copyright 2005, with permission from Elsevier; b and c: redrawn from Legrand, A.P. (ed.), *The Surface Properties of Silicas*, Wiley, New York, 1998 and d'Espinose de la Caillerie, J.-B. et al., *J. Colloid Interface Sci.*, 194, 434, 1997.)

with lowering temperature (frequently nonlinearly) and with increasing amount of adsorbed water (C_{H_2O}) at a fixed temperature. For instance, the δ_H value increases from 1.4–2.0 ppm to 4.5–5.0 ppm with lowering T from 350 to 220 K. The δ_H value depends weakly on the S_{BET} value at relatively high temperatures (330–340 K) but it increases to 3.5 ppm with increasing C_{H_2O} value to 240 mg/g for A-380.

Interfacial water is layer-by-layer frozen with lowering temperature $T<273$ K that leads to diminution of the δ_H values for certain silica samples, for example, A-150 (characterized by minimal δ_H values) at different amounts of adsorbed water, because of freezing-out of low-density water (LDW) (with expanded structure [ES], at $\rho<1$ g/cm^3), or strongly associated water (SAW)/WBW (which is distant from the surface and characterized by higher δ_H values than high-density water [HDW] with collapsed structure [CS], at $\rho\geq 1$ g/cm^3 under standard conditions [SBW]) and then of HDW (in proximity to the surface) at lower temperatures. The δ_H values of interfacial water (HDW) are smaller than those of bulk water (HDW/LDW), but the $\delta_H(T)$ slope is larger for the interfacial water because of another ratio between HDW and LDW in the interfacial layer (electric double layer [EDL]), which also depends on temperature due to a decrease in a portion of mobile water with decreasing temperature. These results suggest that the structure of interfacial water depends on both the amounts of adsorbed water and surface structure. The lower the content of the adsorbed water, the higher is the content of CS/HDW with decreased associativity of molecules (i.e., clustered) at the hydrophilic surface and without nanopores. Morphological features of nanosilicas affect the interfacial water structure because packing of primary particles in aggregates and agglomerates of aggregates typically intensifies with decreasing size of primary particles; that is, the adsorption capacity changes as well as the structure of water clusters in interparticle space (Gun'ko et al. 2005d, 2007f, 2009d). Therefore, the associativity of water molecules differently depends on C_{H_2O} at different S_{BET} values. Additionally, silanols can be assigned to chaotropic groups. Therefore, the surface force effects on the EDL give relatively small changes in the Gibbs free energy of the interfacial layer, and the thickness of this layer is much smaller than that at a surface of mixed oxides (vide infra). The adsorption of water (from air) occurs close to contacts of adjacent primary particles in aggregates and is characterized by lower Gibbs free energy because the adsorption potential between two or several particles is much stronger than that at the outer surface of aggregates. The shape of voids between primary particles depends on their size (i.e., S_{BET}). An increase in aggregation of primary particles with decreasing size (i.e., increasing S_{BET} value) results in the increase in water adsorption. Additionally, the curvature of a surface of smaller particles is higher that provides greater water condensation effects between adjacent smaller particles of silicas at higher S_{BET} values. An increase in the intra-aggregate volume with lowering size of primary particles (or with increasing S_{BET} that leads to enhancing particle aggregation) provides an increase in the V_p value up to 1 cm^3/g for nanosilicas at large S_{BET} values (300–500 m^2/g) in comparison with A-50 or OX-50 ($V_p\approx 0.13$ cm^3/g; Figure 1.23). This changes conditions for water adsorption (i.e., its associativity) onto nanosilicas with increasing S_{BET} value.

Temperature lowering leads to a decrease in the molecular mobility and the average length of the hydrogen bonds in water, and average number of the hydrogen bonds per molecule increases (similar to ice). However, the coordination number of water molecules decreases to 4; that is, water (HDW → LDW) becomes more ordered with higher associativity that causes an increase in the average δ_H value for interfacial and bulk water. This value can be used to estimate the average number of the hydrogen bonds per molecule. The effect of the hydrogen bonding on the δ_H value is substantially smaller in a complex with a water molecule as an electron donor, for example, ≡SiO–H···**OH**$_2$ compared to ≡SiO(H)···**H–**OH with a molecule as a proton donor. Therefore, the average number of the H atoms (m) of water molecules participating in the hydrogen bonds at the interfaces is useful to analyze the ^{1}H NMR spectra. The m value can be estimated from the experimental value of the chemical shift $\delta_{H,obs}$:

$$m = (\delta_{H,obs} - 1.4)/2.4. \quad (1.11)$$

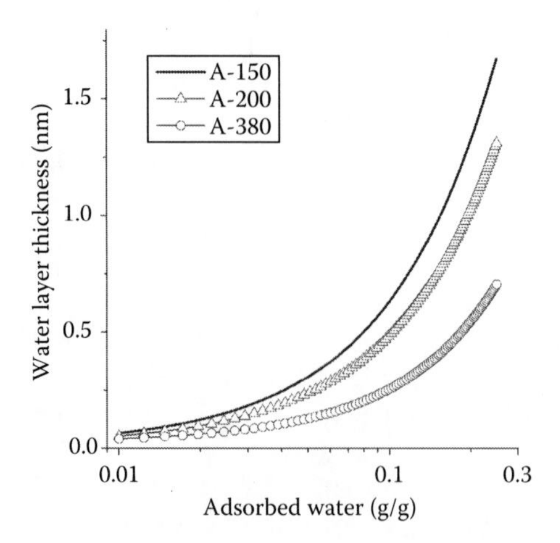

FIGURE 1.25 Average thickness of an interfacial water layer as a function of the amounts of adsorbed water for different nanosilicas assuming that the coordination number of primary particles in aggregates is 2.5 (A-150), 2.8 (A-200), and 3.5 (A-380) and the layer is uniform.

where constants are determined by the δ_H values for ice Ih (7 ppm) and a molecule without the hydrogen bonds, that gives $m = 0.0$–1.5 for interfacial water at $\delta_{H,obs} = 1.4$–5.0 ppm.

A stronger decrease in the chemical shift δ_H of water adsorbed on to nanosilica is observed with elevating temperature (Figure 1.24a) or due to decrease in the water content (Figure 1.24b and c) in comparison with bulk water. This can be explained by a smaller size of adsorbed water clusters at a lower average number of the hydrogen bonds per molecule and higher molecular mobility of water molecules in open space between nonporous silica particles forming loose aggregates and agglomerates (bulk density of the nanosilica powder is 0.04–0.14 g/cm^3; Table 1.10); that is, the associativity of the molecules decreases. The amounts of water adsorbed onto nanosilicas (Figure 1.24) provide the formation of a layer at the thickness larger (Figure 1.25) than the statistical monolayer (0.3 nm) and the number of water molecules per silanol ($C_{OH} \approx 2$ OH/nm^2) is close or larger than $10H_2O$.

However, adsorbed water does not form a continuous layer (this will be discussed in the subsequent sections) and the water clusters form near silanols between adjacent primary particles. The absence of pores in primary nanoparticles and their loose packing in aggregates and agglomerates (the ratio between the length l and the diameter D of voids between primary particles in aggregates is $l/D = 1$–10) result in higher molecular mobility in comparison with mesoporous silica gels, where contribution of LDW can be, therefore, higher than that at the nanosilica surface. Therefore, the associativity of the molecules is greater in narrower pores of silica gels. Structural features of nanosilicas and the temperature effects (decrease in the hydrogen bonding of water molecules with elevating temperature due to bending of the O–H···H bonds and increase in the coordination number of water molecules) can lead to boost of mixing of water clusters with chloroform clusters at the interfaces of nanosilica. Water molecules can penetrate in the interfacial chloroform layer but not vice versa, because according to quantum chemical calculations of the solvation effects $\Delta G_{s,H_2O \rightarrow CHCl_3} < \Delta G_{s,CHCl_3 \rightarrow H_2O}$ (Tables 1.11 and 1.12); that is, Gibbs free energy is lower on solution of water in chloroform than on solution of chloroform in water. However, the activity of water as a solvent decreases in the interfacial layer (Gun'ko et al. 2009d, Chaplin 2011).

The effects of confined space in pores of silica gels (l/D is up to 10^5) lead to stronger hydrogen bonding (greater associativity) of adsorbed water molecules, diminution of their mobility, and lower mixing of water and chloroform because of the formation of thicker interfacial HDW/LDW layers. This creates a barrier for motion of water molecules into $CDCl_3$ and vice versa. Thus, water in the case of the absence of strong confined space effects in narrow pores and in the presence of

TABLE 1.11
Electronic Structure of Water Clusters in the Gas and $CHCl_3$ Media (SM5.42/6-31G(d)) and Experimental and Estimated Chemical Shifts of Protons of Water Molecules in Chloroform

Cluster	$q_{H\ (free\ H)}$	$q_{H\ (H\ bond)}$	$\delta_{H\ (free\ H)}$ (ppm)	$\delta_{H\ (H\ bond)}$ (ppm)	ΔG_s, (kJ/mol)
H_2O^g	0.434	—	−0.06[b]	—	
H_2O^l	0.467		1.7[c]	—	−9.8
$H_2O...HOH^g$	0.450	0.475	0.55[b]	2.42[b]	
	0.450		0.55		
	0.424		−0.37		
$H_2O...HOH^l$	0.480	0.502	2.92[d]	4.44[d]	−15.4
	0.480				
	0.449				
$4H_2O$ (cycle)[g]	0.440	0.507	−0.1[a]	5.02[a]	
$4H_2O$ (cycle)[l]	0.465	0.508	1.88[d]	4.85[d]	−4.4
$4H_2O$ (cycle)[h]	0.449	0.510	0.78[d]	4.99[d]	15.8
$6H_2O$ (cycle)[g]	0.436	0.523	−0.3[a]	6.01[a]	
$6H_2O$ (cycle)[l]	0.466	0.518	1.95[d]	5.55[d]	−3.8
$6H_2O$ (prism)[g]	0.462	0.519	2.11[a]	8.37[a]	
	0.462	0.508	2.11	5.72	
	0.450	0.500	1.47	4.83	
	0.449	0.496	1.29	4.82	
	0.445	0.488	0.60	3.78	
	0.444	0.474	0.45	2.35	
$6H_2O$ (prism)[l]	0.487	0.510	3.40[d]	4.99[d]	−0.9
	0.482	0.510	3.06	4.99	
	0.480	0.508	2.92	4.85	
	0.475	0.493	2.57	3.82	
	0.475	0.493	2.57	3.82	
	0.471	0.489	2.30	3.54	

Note: The basis sets B3LYP/6-311G(d,p)[a] and B3LYP/6-311++G(3df,2p)//6-311(d,p)[b] were used for calculation of the δ_H values for complexes in the gas phase; [c] experimental value for chloroform medium; and [d] δ_H values estimated using a linear relationship $\delta_{H,iso} = a + bq_H$, where a and b are constants determined from experimental ^1H NMR data and q_H charges calculated for complexes in the gas phase; [l] in $CHCl_3$ media, [g] in gas phase; and [h] in cyclohexene.

hydrophobic chloroform can be in a weakly associated state forming small clusters, interstitial water, or due to nonassociated state (individual molecules as in complexes ≡SiO–H···OH_2 surrounded by chloroform molecules or interstitial water) that corresponds to the unusual interfacial water. Thus, a significant diminution of the δ_H values with elevating temperature (Figure 1.24) is caused by several reasons: (i) weakening of the hydrogen bonds (diminution of the associativity of molecules, increasing clustering) because of the increased mobility of molecules (larger average length of the bonds and deviation of the ∠OHO value from optimal 170°–180° characteristic of LDW with ES) and an increase in the amount of interstitial water and an enhanced HDW effects, (ii) increased mixing of small water clusters with chloroform in mesopores, and (iii) transfer of a portion of adsorbed liquid water into the vapor phase (in the NMR ampoule) that causes diminution of the interfacial water layer thickness and, consequently, reduction of the average m value.

TABLE 1.12
Calculated and Experimental Free Energy of Solvation ΔG_s (kJ/mol)

		Solute					
Solvent	**ε**	**H_2O[a]**	**H_2O**	**H_2O[d]**	**$CHCl_3$**	**$CHCl_3$**	**CCl_4**
H_2O	78.36	−25.4	−24.6[c]	−26.4	−2.8[a]	−2.7[c]	−0.3[a]
C_7H_{16}	1.91	0.8					
CCl_4	2.23	−0.8	−4.1[c]				
$CHCl_3$	4.71	−9.8	−8.2[b]	−4.6			
C_6H_6	2.27	−2.7		−3.8			
CH_3CN	35.69	−20.7		−16.3			
$(CH_3)_2CO$	20.49	−20.9					
$(CH_3)_2SO$	46.83	−25.0					

Note: ε is the permittivity of a solvent; used methods: [a]SM5.42/6-31G(d) at the solute geometry optimized with consideration for the solvent effects; [b]IEFPCM/B3LYP/6-31G(d,p); [c]IEFPCM/B3LYP/6-311++G(3df,2p); and [d]experimental data.

Fast molecular exchange (in the NMR timescale) is possible between water structures being in these states that give relatively broad averaged ^1H NMR signal of the interfacial water. Change in temperature leads to change in contribution of water structures being in different states (i.e., contributions of two signals at $\delta_H=3–5$ [SAW] and 1–2 ppm [WAW]), and the signal shape changes. The existence of pores at large *l/D* values (e.g., in silica gels) causes greater associativity and smaller mobility of water molecules in the confined space of long and relatively narrow pores. Similar pores are absent in aggregates of primary particles of nanosilicas that results in different temperature dependence of the δ_H value in comparison with that for silica gels (Figure 1.24 and Section 1.2). Additionally, structural and morphological features of nanosilica and silica gels cause the difference in changes in the Gibbs free energy of interfacial water in the aqueous suspensions of these oxides as functions of the pore size and volume (vide infra) because of the difference in the HDW and LDW balance and clusterization of bound water confined in long and narrow (silica gel) and short and broad (voids in nanosilica aggregates; Figure 1.2) pores. However, the main changes in the ΔG value for water adsorbed onto A-300 and different silica gels occur in the R range of 1.0–1.5 nm connected with a layer of SBW. This suggests that the surface of both types of silicas can strongly change the structure of the hydrogen bond network in the interfacial layer at the thickness of 3–5 statistical monolayers of water. The texture of silochrome, which can be considered as an intermediate material between nanosilica and silica gels, provides clear observation of the unusual interfacial water (Gun'ko et al. 2005d).

The behavior of water adsorbed onto a nanosilica surface depends on the water content because the continuous layer of the interfacial water forms at much higher C_{H_2O} value than the formal content necessary to form a monolayer. This can be easily checked using the FTIR method. A tablet of pressed nanosilica A-300 (25.7 mg) was saturated by water vapor up to hydration $h=1.2$ g/g (average error ± 5%) by adsorbed water (with respect to the weight of the sample heated at 120°C for 2 h and cooled to room temperature for 1 h) for 24 h. The FTIR spectra (Figure 1.26) were recorded at the time step of approximately 5 min. After the measurement, the excess amount of water was approximately 30 mg/g. The rapid loss of the humidity is nonuniform in the sample because of local heating of the tablet by the IR light beam. The tablet at $h=1.2$ g/g is completely transparent because voids between primary particles in aggregates are completely filled by adsorbed water. The loss of moisture is accompanied by increased turbidity of the tablet up to the complete opacity (milk-white)

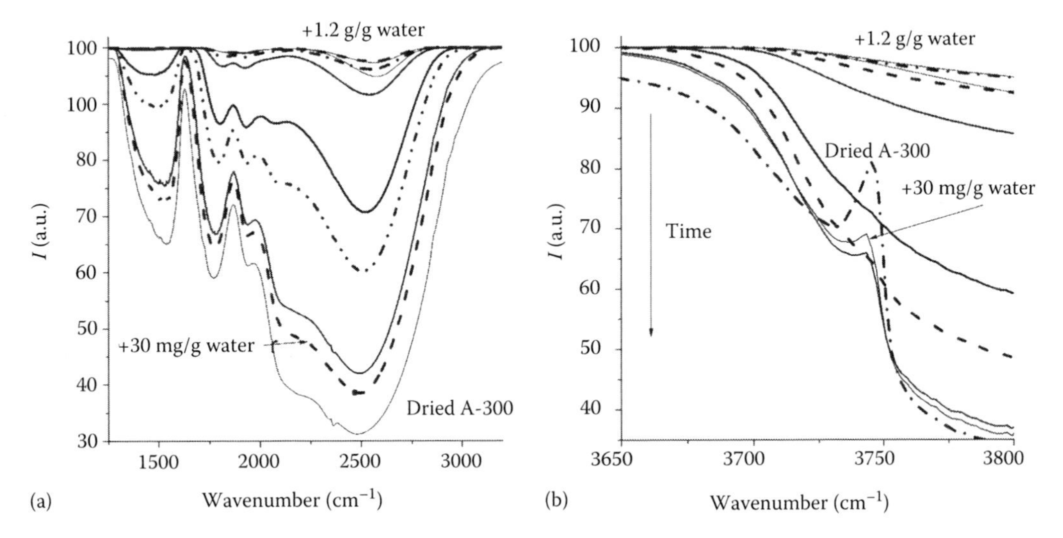

FIGURE 1.26 FTIR spectra over (a, b) different wavenumber ranges for nanosilica A-300 at different amounts of adsorbed water (excess water from 1.2 g/g to 30 mg/g) desorbed in air for approximately 40 min. (Adapted from *Adv. Colloid Interface Sci.*, 118, Gun'ko, V.M., Turov, V.V., Bogatyrev, V.M. et al., Unusual properties of water at hydrophilic/hydrophobic interfaces, 125–172, 2005d. Copyright 2005, with permission from Elsevier.)

for the visible light. Further loss of moisture reduces the turbidity to a transparent–opalescent form akin to that for the initial tablet after pressing.

The FTIR spectra (Figure 1.26) of a nanosilica sample with different amounts of adsorbed water reveal certain features of the interfacial water. Complete disappearance of the band of free silanols is observed only at $C_{H_2O}>25$ wt% ($h>0.3$ g/g) that suggests that a continue water layer is absent at $C_{H_2O}<25$ wt%. This result is in agreement with a linear dependence of the dielectric permittivity ε' on the water content only at $C_{H_2O}<25$ wt% in contrast to binary titania/silica nanooxides (Gun'ko et al. 2009d). This difference in the behavior of the interfacial water at the silica and titania/silica surfaces is caused by the difference in the content and the acid/base properties of surface sites of these oxides. These structural features of titania/silica provide the formation of a thicker EDL (in the aqueous suspensions), higher amount of unfrozen water per gram of oxide and γ_S greater by two to three times, and larger amounts of water adsorbed from air (vide infra). However, the ^{1}H NMR spectra demonstrate the presence of weakly associated water (WAW; water adsorbed on mixed oxides in air), which gives a shoulder at $\delta_H=1$–2 ppm. This can be explained by the formation of a thin layer of water that can result in the formation of the structures with WAW. Therefore, one can assume that the HDW/LDW balance and clustering at the surface of mixed fumed oxides strongly differs from those at the nanosilica surface.

The IR spectra of a weakly wetted powder of nanosilica A-380 ($h\approx50$ mg/g) in the air and CCl_4 media (nonpolar in contrast to slightly polar $CDCl_3$) are shown in Figure 1.27. According to these spectra, it is possible to expect that a significant portion of water molecules will be in a weakly associated state in the hydrophobic environment at 293 K. In air (Figure 1.28a), a band at 3750 cm^{-1} of free silanols is observed. Besides, there are the O-H stretching vibrations at 3680 cm^{-1} (poorly accessible or inaccessible silanols), 3500 cm^{-1} (hydrogen-bonded silanols and HDW), and 3200 cm^{-1} (adsorbed water LDW and LDW/HDW). In the hydrophobic CCl_4 medium (Figure 1.28b), the band of free silanols at 3750 cm^{-1} disappears because of the interaction of silanols with the CCl_4 molecules that result in the displacement of this band by 85 cm^{-1} toward lower wavenumber values.

However, the shape of the broad band of adsorbed water remains near the same but the intensity reduces, especially for adsorbed water (3200 cm^{-1}) and silanols disturbed by the hydrogen bonding with water and contribution of HDW (3500 cm^{-1}; Figure 1.28). Thus, water (adsorbed in amount

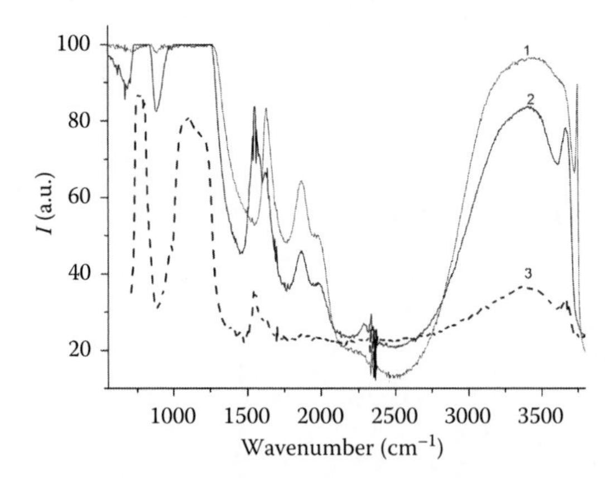

FIGURE 1.27 IR spectra of hydrated silica powder ($h \approx 50$ mg/g) in (1) air and (2, 3) CCl_4 media at the layer thickness in the IR dish of (2) 200 μm and (3) 20 μm. (Adapted from *Adv. Colloid Interface Sci.*, 118, Gun'ko, V.M., Turov, V.V., Bogatyrev, V.M. et al., Unusual properties of water at hydrophilic/hydrophobic interfaces, 125–172, 2005d. Copyright 2005, with permission from Elsevier.)

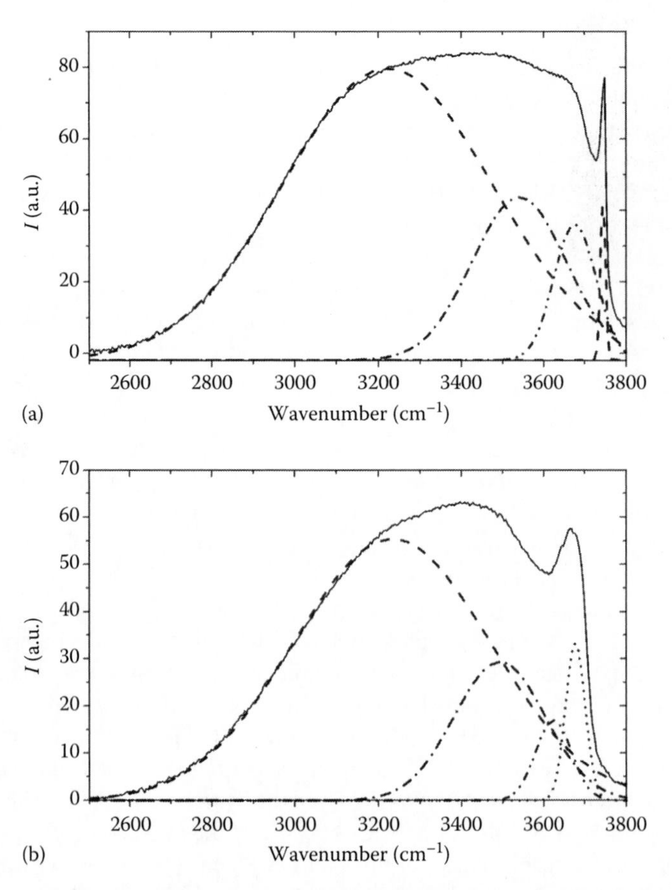

FIGURE 1.28 Deconvolution of the IR spectra of silica A-380 in (a) air and (b) CCl_4 environment. (Adapted from *Adv. Colloid Interface Sci.*, 118, Gun'ko, V.M., Turov, V.V., Bogatyrev, V.M. et al., Unusual properties of water at hydrophilic/hydrophobic interfaces, 125–172, 2005d. Copyright 2005, with permission from Elsevier.)

higher than that corresponding to a statistical monolayer coverage) remains in the associated state, for example, located in the form of clusters between adjacent primary silica particles in aggregates with voids between nanoparticles of 0.3–20 nm in radius (Gun'ko et al. 2005d, 2007f, 2009d). However, a portion of water is displaced from the silanols by CCl_4 molecules because the band intensity at 3500 cm^{-1} (HDW) decreases. These water molecules can transform to LDW or it can be dissolved in the CCl_4 medium but in a very low amount because $\Delta G_{s,H_2O \rightarrow CCl_4}$ is close to zero (Table 1.12). This amount can be significantly lower than that dissolved in chloroform because $\Delta G_{s,H_2O \rightarrow CHCl_3} \ll \Delta G_{s,H_2O \rightarrow CCl_4}$. Now initial and MCA silicas are compared to elucidate the textural and confined space effects on adsorbed water.

Primary nonporous nanoparticles of A-300 form aggregates (Figure 1.29a) responsible for the textural porosity of the powder. These nanoparticles have a relative narrow size distribution (Figure 1.29b; curve 3) calculated from the nitrogen adsorption data using a self-consistent regularization procedure (Gun'ko 2000b) based on the CONTIN algorithm (Provencher 1982). Treatment of the SEM image (Figure 1.29a) using specific software (ImageJ with granulometry plugin or Fiji with local thickness plugin) allows us to calculate the size distribution functions of primary particles and their aggregates (Figure 1.29b). The textural porosity of the silica powder is due to voids between nanoparticles in their aggregates. This porosity causes the appearance of hysteresis loops in the

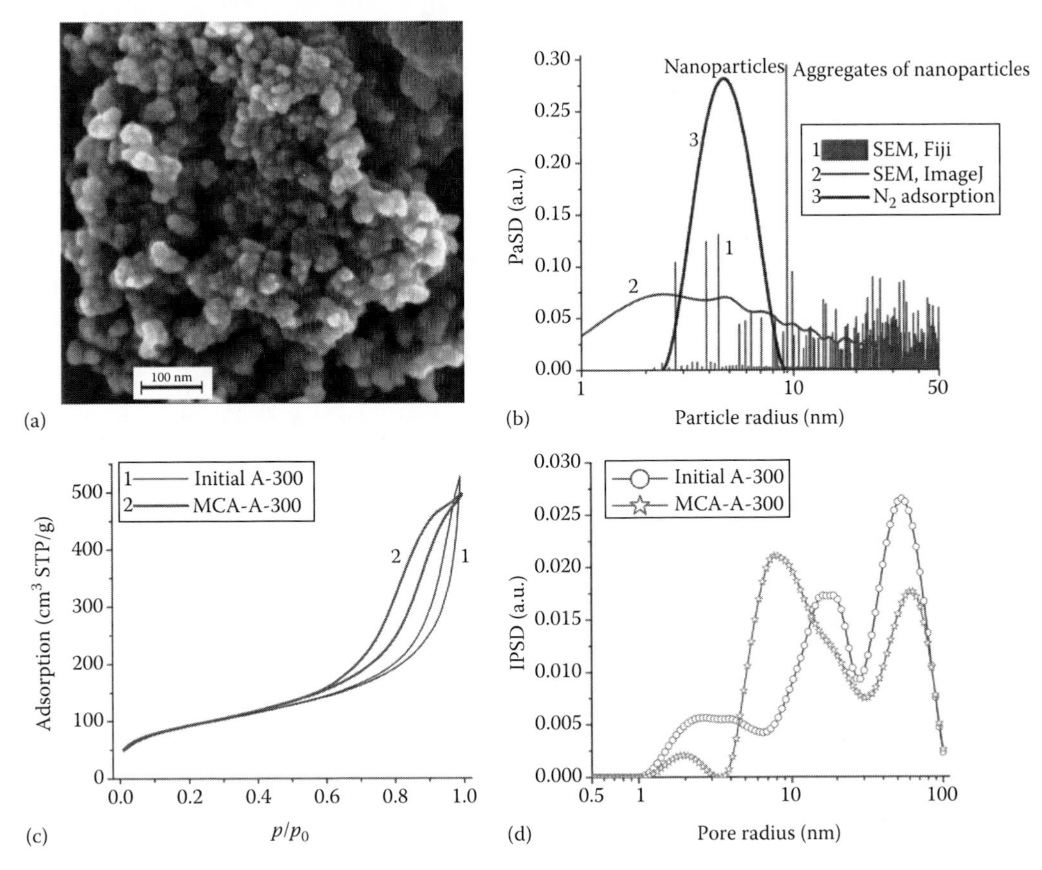

FIGURE 1.29 (a) SEM image (JSM-6310, Japan Electron Optics Ltd) of nanosilica A-300, (b) size distribution functions of primary particles and their aggregates of A-300, (c) nitrogen adsorption–desorption isotherms (recorded using a Micromeritics ASAP 2405N adsorption analyzer), and (d) DFT pore size distributions calculated using a model of voids between spherical silica nanoparticles for initial and MCA-silica. (Adapted from *Colloids Surf. A: Physicochem. Eng.* Aspects., 390, Turov, V.V., Gun'ko, V.M., Turova, A.A., Morozova, L., and Voronin, E.F., 48–55, 2011e. Copyright 2011, with permission from Elsevier.)

nitrogen adsorption–desorption isotherms (Figure 1.29c) typical for mesoporous materials. The pore (void) size distributions (Figure 1.29d) show that MCA-A-300 is characterized by the PSD shifted toward narrower pore sizes in comparison with the initial material. The difference in the textural characteristics of initial and MCA silica samples can affect their interactions with water.

Water adsorbed on the initial silica is characterized by two ^{1}H NMR signals at $\delta_H = 2.8$–4.8 and 1.5 ppm (Figure 1.30a). Similar δ_H values were observed for water adsorbed onto porous silicas and assigned to the hydrogen-bound and non-bound protons in the water molecules (Grünberg et al. 2004, Vyalikh et al. 2007).

Signals at $\delta_H = 0$ and 7.16 ppm are due to tetramethylsilane and $CHCl_3$ (admixture in $CDCl_3$), respectively. According to the ^{1}H NMR signal classification (Gun'ko et al. 2005d, 2009d), observed water signals (Figure 1.30a) can be assigned to SAW and WAW, respectively. The SAW signal demonstrates the upfield shift with increasing temperature. This can be explained by breaking of a portion of the hydrogen bonds, decomposition of large polyassociates of unfrozen water, and stabilization of smaller structures of unfrozen water in narrow pores. At $T < 250$ K, the SAW signal intensity decreases due to freezing of this water with lowering temperature. This process is accompanied by signal broadening because of decreased mobility of adsorbed water at lower temperatures. A certain broadening of the SAW signal at $T > 250$ K is due to such factors as accelerating exchange processes between SAW and WAW and an increase in nonuniform broadening due to appearance of solvent vapor bubbles formed with elevating temperature. The latter can lead to a decrease in the SAW signal intensity.

The WAW signal decreased with lowering temperature is narrower than SAW signal (Figure 1.30a). Two split signals of different intensity are observed for WAW because of nonuniformity of this bound water. At $T > 250$ K, the WAW signal intensity increases due to decreased content of SAW. Two signals of WAW (Figure 1.30a) can be attributed to water dissolved in chloroform (the right signal of lower intensity) and water bound to silica nanoparticles (the left signal of greater intensity).

TMS and $CHCl_3$ signals (Figure 1.30a) grow with decreasing temperature due to an increase in the population of the nuclear levels (Curie law; Abragam 1961). This effect is larger for protons in structures with decreased efficiency of the relaxation mechanisms of nuclear magnetization. The TMS and chloroform molecules interact with surroundings by the dispersion forces, which do not provide high exchange efficiency. Therefore, the Curie law occurring is stronger for them than for water.

Replacement of weakly polar $CDCl_3$ medium by nonpolar CCl_4 (Figure 1.30b) results in a significant decrease in the WAW signal intensity and certain changes in the SAW structure. Therefore, the upfield shift of the SAW signal with increasing temperature is smaller than that in the $CDCl_3$ medium. At $T < 240$ K, the signal width of TMS increases (Figure 1.30b) because of freezing of CCl_4. TMS is frozen at much lower temperature than CCl_4. Therefore, liquid TMS is concentrated in voids in frozen CCl_4 in which the freezing temperature of both components is lowering due to the colligative properties of the mixtures and the freezing point depression for liquids confined in narrow pores (voids; Gun'ko et al. 2005d, Mitchell et al. 2005, 2008, Petrov and Furó 2009).

The ^{1}H NMR spectra shape strongly change in the case of MCA silica (Figure 1.30c). Denser packing of primary particles in the MCA silica than in the initial A-300 enhances the water adsorption (and retention). Therefore, water adsorbed in narrower voids (Figure 1.29d) is more strongly bound to the silica surface, as well as CCl_4. The ^{1}H NMR signal intensity of bound water increases in respect to the TMS signal. A significant portion of CCl_4 is unfrozen at $T = 230$ K. Therefore, the signal width of TMS does not change over the total temperature range (Figure 1.30c) in contrast to that for the initial silica (Figure 1.30b). The freezing temperature of water bound to the MCA silica surface is lower than that for the initial silica. This is due to the adsorption of water in narrower voids between nonporous silica nanoparticles.

NMR cryoporometry shows the difference in the organization of water clusters adsorbed onto the initial and MCA-A-300 (Figure 1.31) because of the difference in the PSD of these silicas (Figure 1.29d).

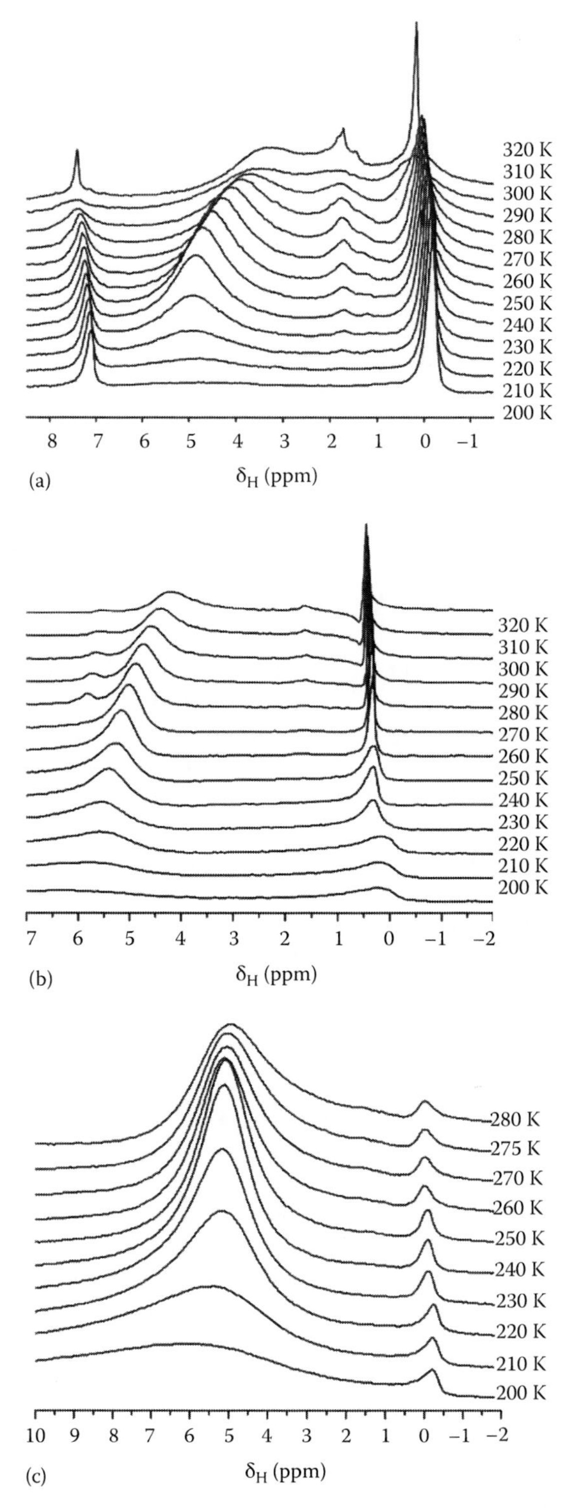

FIGURE 1.30 ^{1}H NMR spectra recorded at different temperatures of water (5 wt%) adsorbed on A-300 (a, b) initial and (c) MCA in (a) $CDCl_3$ and (b, c) CCl_4 media. (Adapted from *Colloids Surf. A: Physicochem. Eng.* Aspects., 390, Turov, V.V., Gun'ko, V.M., Turova, A.A., Morozova, L., and Voronin, E.F., 48–55, 2011a. Copyright 2011, with permission from Elsevier.)

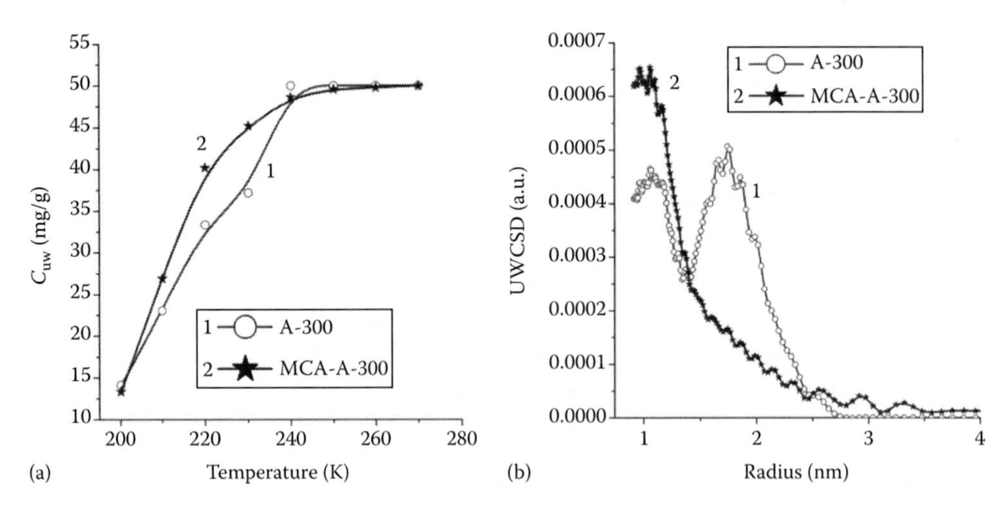

FIGURE 1.31 (a) Temperature dependence of content of unfrozen water, C_{uw} (total water content 50 mg/g) adsorbed to initial A-300 and MCA-A-300 and (b) unfrozen water cluster size distributions, UWCSD (CCl_4 medium). (Adapted from *Colloids Surf. A: Physicochem. Eng.* Aspects., 390, Turov, V.V., Gun'ko, V.M., Turova, A.A., Morozova, L., and Voronin, E.F., 48–55, 2011e. Copyright 2011, with permission from Elsevier.)

Despite an increase in contribution of smaller water clusters in MCA-A-300, contact area between water and silica (determined by integration of the corresponding distribution functions) is larger for the initial silica (S_{uw} = 78 m^2/g) than for MCA-A-300 (56 m^2/g). Notice that 50 mg/g H_2O is too low a content to form a continuous water layer at the silica surface and to fill all textural pores (Figure 1.29d). Therefore, water is strongly bound (freezing starts at T < 250 K; Figure 1.31a) and can cover only a portion of the silica surface and fill only a small portion of pores, mainly narrow mesopores and nanopores, since water forms clusters at R < 3 nm in radius (Figure 1.31b). Additionally, CCl_4 can reduce the contact area between the silica surface and adsorbed water, which tends to reduce the contact area with the hydrophobic solvent (Gun'ko et al. 2009d). However, both $CDCl_3$ and CCl_4 cannot remove water from the surface because there are no significant changes in the intensity at subzero temperatures 273–250 K.

Figure 1.32 shows the temperature dependences of the amounts of SAW (signal 1) and the δ_H values for weakly hydrated A-300 (50 mg/g H_2O) in different media. The SAW amount (Figure 1.32a) increases to a certain value with elevating temperature and then it decreases. This can be explained by changes in contributions of water located in different clusters and nanodomains. In the CCl_4 and $CDCl_3$ media, the WAW amounts (signal at δ_H = 1.3 ppm; Figure 1.30) increase with elevating temperature. In the mixtures of $CDCl_3$ with electron-donor solvents (CD_3CN and dimethyl sulfoxide [DMSO]), rearrangement of the water/organics mixtures occurs that result in a certain diminution of the δ_H values with increasing temperature. This is due to decomposition of a part of the hydrogen bonds between water molecules under action of polar organic solvents.

The most hydrophobic medium with CCl_4 enhances the associativity of bound water (Figures 1.30 and 1.32) to reduce the contact area between two immiscible liquid phases. In the $CDCl_3$ medium, the water associativity decreases, and WAW appears because of water clustering. The process of decreasing associativity becomes stronger with addition of electron-donor solvents into the $CDCl_3$ medium since the δ_H values decrease (Figure 1.32b). The relationships between the ΔG and C_{SAW} values (Figure 1.32c) show that the major portion of water is strongly bound to the silica surface, and this value is maximal in the CCl_4 medium. However, this water fraction decreases with addition of more polar solvents, which can strongly interact with both water, silica surface, and nonpolar solvent. This allows these solvents to dissolve a portion of water and to displace SAW from the silica surface. The sizes of SAW clusters and domains are mainly in the range of 0.8–5 nm in radius (Figure 1.32d).

Despite close chemical structure of H_2O and H_2O_2, their physicochemical properties strongly differ (Buckingham 1958, March 1992, Hess 1995, Jones 1999). First of all, this difference is caused by the difference in their hydrogen bond network resulting in a clear difference in the ^1H NMR spectra of water and hydrogen peroxide (Stephenson and Bell 2005, Zhang 2007). Low temperature ^1H NMR spectroscopy is a very informative method to study changes in the hydrogen network of water depending on the presence of H_2O_2 and organic co-solvents and silica nanoparticles. Interactions of H_2O_2/H_2O mixture with nanosilica are studied in different media ($CDCl_3$, CD_3CN, $(CD_3)_2SO$, and CCl_4) in comparison with liquid samples without silica.

Nanosilica A-400 ($S_{BET}=409$ m²/g and bulk density $\rho_b=0.061$ g/cm³) was selected as a material with small nanoparticles ($d_{av}=6.7$ nm) relatively strongly aggregated (pore volume $V_p=0.86$ cm³/g) and having a broad PSD calculated using the nitrogen adsorption isotherm with the model of a

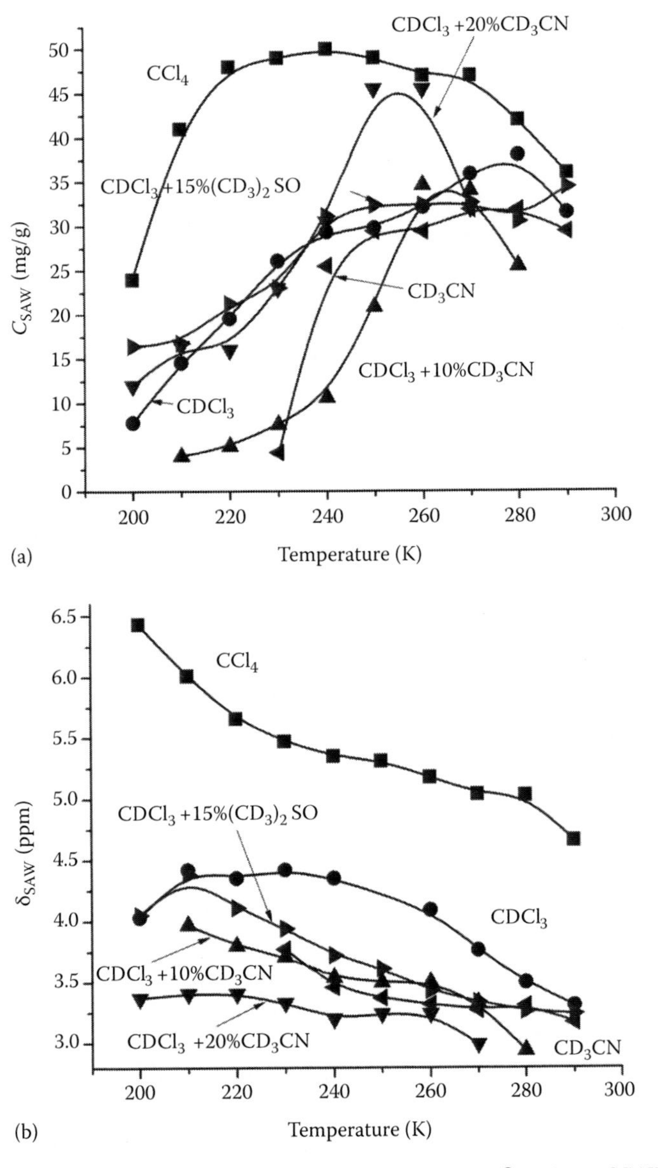

FIGURE 1.32 Temperature dependences of (a) the amounts and (b) the δ_H values of SAW.

(continued)

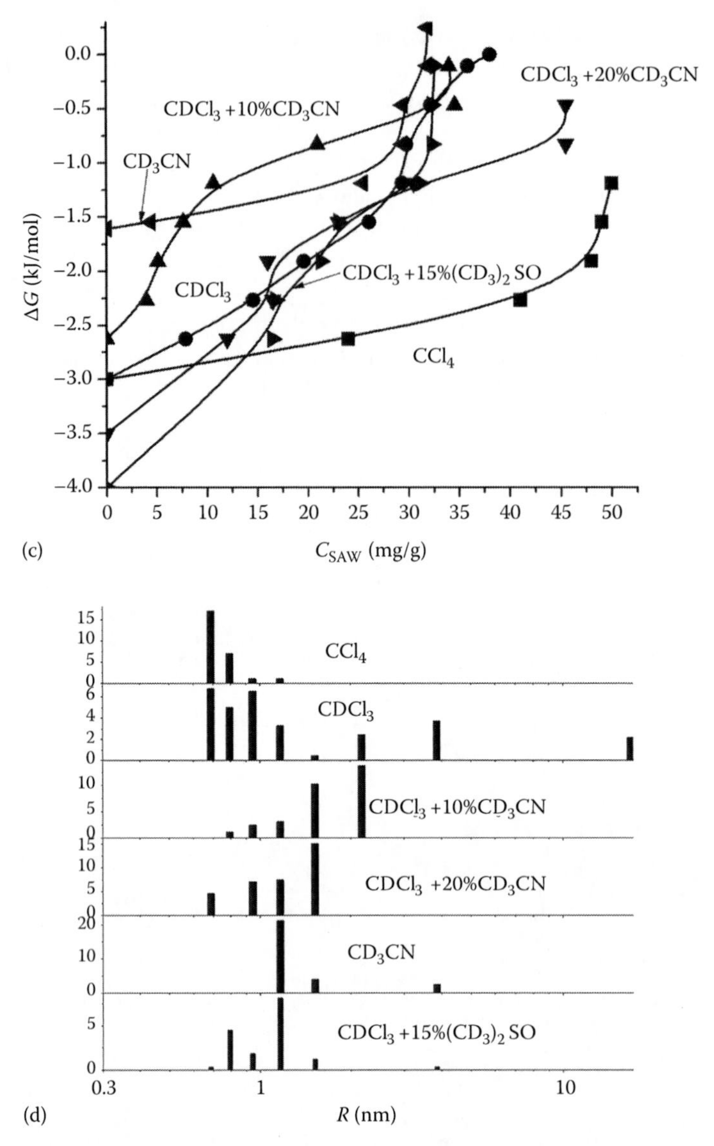

FIGURE 1.32 (continued) Temperature dependences of (c) changes in the Gibbs free energy of SAW; and (d) distribution functions of clusters and domains of SAW for hydrated A-300 at $h = 50$ mg/g placed in different media.

mixture of slit-shaped and cylindrical pores and voids between spherical particles using self-consistent regularization procedure. A total of 30 wt% aqueous solution of H_2O_2 and solvents $CDCl_3$, CD_3CN, $(CD_3)_2SO$, and CCl_4 (for NMR spectroscopy) was used in low-temperature (LT) ^1H NMR measurements of static samples (to suppress signals of immobile phases; Gun'ko et al. 2012c).

The ^1H NMR spectra were recorded at 200–290 K using a Varian 400 Mercury spectrometer (working frequency 400.4475 MHz, magnetic field 9.4 T) with the probing 90° pulses at duration of 3 μs with eight scans and 1.5 s delay between them. The ^1H NMR spectra of H_2O_2/H_2O published in the literature (Deng et al. 2003, Stephenson and Bell 2005, Zhang 2007) show that the δ_H values for mixtures depend on component concentrations. To prevent supercooling of water in the systems studied, the spectra of bound H_2O_2/H_2O (unfrozen at $T < 273$ K) were recorded for samples precooled from room temperature to 200 K for 10 min and then heated to 290 K at a step $\Delta T = 10$ K, a heating

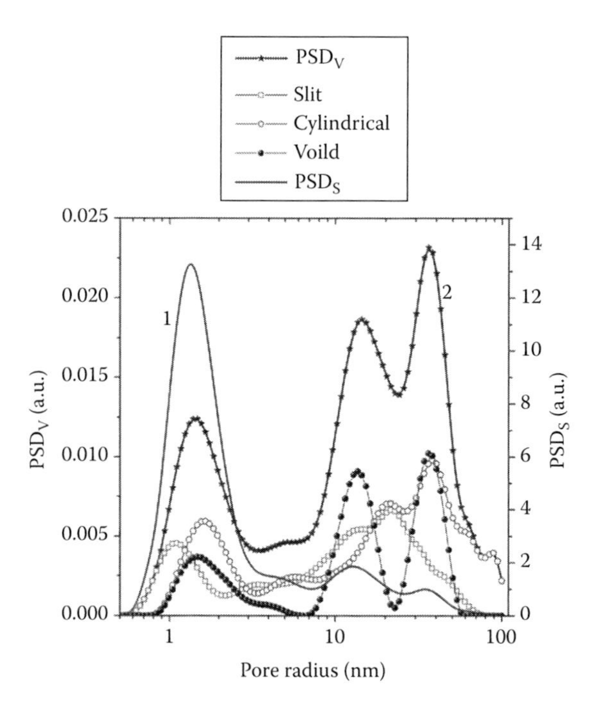

FIGURE 1.33 Incremental pore size distributions for A-400 in respect to the specific surface area (PSD$_S$, curve 1) and pore volume (PSD$_V$, curve 2). Contributions of slit-shaped and cylindrical pores and voids between spherical nanoparticles to PSD$_V$ were calculated using self-consistent regularization procedure. (Adapted from *Chem. Phys. Lett.*, 631, Gun'ko, V.M., Turov, V.V., and Turov, A.V., Hydrogen peroxide—water mixture bound to nanostructured silica, 132–137, 2012c. Copyright 2012, with permission from Elsevier.)

rate of 5 K/min for 2 min, and a fixed temperature for 8 min. The melting curves were used in the NMR cryoporometry calculations.

Quantum chemical calculations were carried out by ab initio and density functional theory (DFT) methods with the 6-31G(d,p) basis set using Gaussian 03 and GAMESS Firefly 7.1G program suites to full geometry optimization. Large models were calculated using the PM6 method (MOPAC 2009 package, Stewart 2008). To calculate the $f(\delta_H)$ functions using the PM6 results, two calibration functions for H_2O and H_2O_2 were used to describe the dependencies between atomic charges q_H (PM6) and the δ_H values (GIAO/B3LYP/6-31G(d,p)) for certain clusters.

Aggregates and agglomerates of silica nanoparticles of A-400 are characterized by the textural porosity with a broad PSD (Figure 1.33).

Contribution of nanopores at radius $R<1$ nm is relatively small ($V_{nano}=0.033$ cm^3/g and $S_{nano}=35$ m^2/g). However, contribution of narrow mesopores at $1<R<3$ nm is much higher than that of nanopores (Figure 1.33). In the NMR measurements, 100 mg/g of the aqueous (30%) solution of H_2O_2 with respect to the weight of dry silica was added to dry A-400 powder and then an organic solvent was added. On the basis of the PSD$_V$ curve, one can assume that practically total amount of the H_2O_2/H_2O solution can be located in nanopores and narrow mesopores of the aggregates of silica nanoparticles because the adsorption potential is higher in narrow pores than in broader ones. Additionally, water tends to be adsorbed in the clustered state, and location of clusters in pores of appropriate size is more favorable than their adsorption in broader pores or onto the outer surface of nanoparticles far from contacts with adjacent nanoparticles in their aggregates.

The aqueous solution of hydrogen peroxide is characterized by two ^{1}H NMR signals at $\delta_H\approx 6$ (H_2O) and 12 (H_2O_2) ppm (Figure 1.34a).

The signal width increases with increasing temperature. This effect is stronger for H_2O_2 due to enhancement of the proton exchange $H_2O \leftrightarrow H_2O_2$. A weak signal at $\delta_H\approx 2$ ppm can be

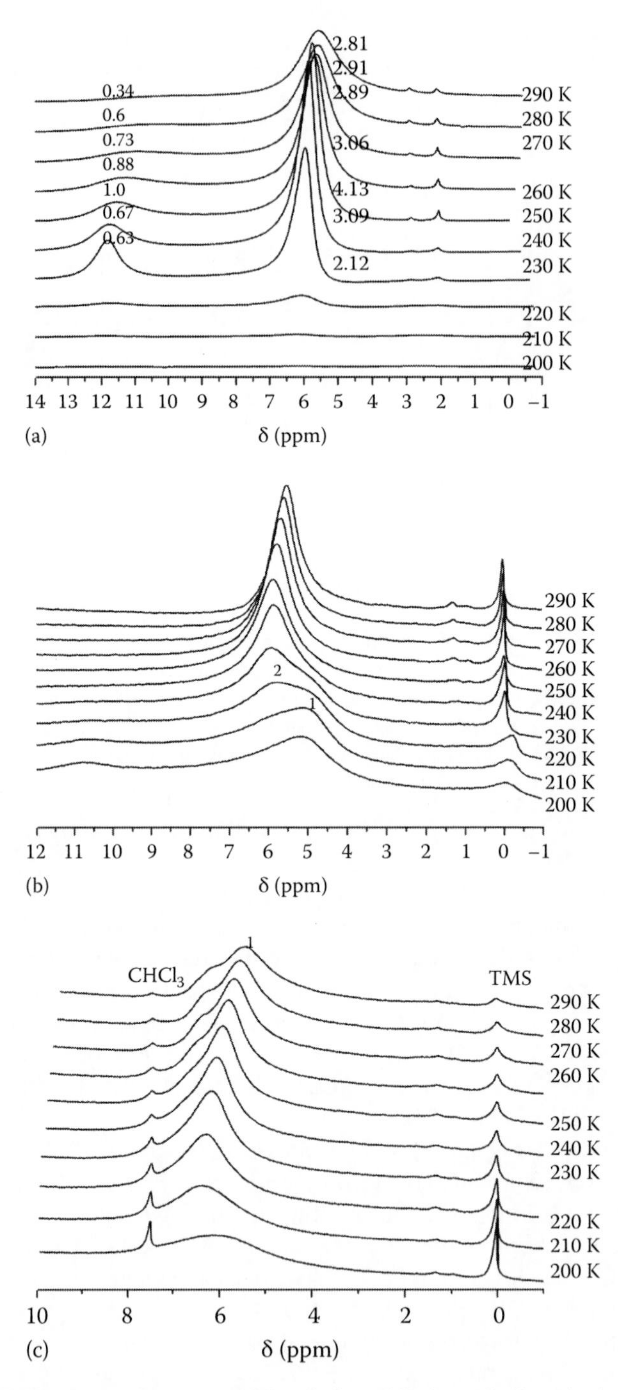

FIGURE 1.34 ^{1}H NMR spectra of aqueous 30% solution of H_2O_2 (a) bulk and adsorbed (100 mg/g) onto A-400 being in (b) CCl_4 and (c) $CDCl_3$ media. (Adapted from *Chem. Phys. Lett.*, 631, Gun'ko, V.M., Turov, V.V., and Turov, A.V., Hydrogen peroxide—water mixture bound to nanostructured silica, 132–137, 2012c. Copyright 2012, with permission from Elsevier.)

assigned to methyl groups of acetone added as a standard to determine the chemical shift. The chemical shift of water interacting with H_2O_2 is 1–1.5 ppm larger than that for bulk liquid water. This downfield shift for water and the upfield shift for H_2O_2 are due to their mixing. Symbols at the peaks (Figure 1.34a) show the integral intensity of these peaks. At $T<230$ K, the intensity drops down because of freezing of water and hydrogen peroxide. Complete melting of the frozen components occurs at $T_m = 250$ K. The ratio of the integral intensity of signals is 4:1 that is close to that of their molar fractions in the sample. A decrease in the intensity at $T>T_m$ is due to changes in nuclear level population (Curie law; Pople et al. 1959). Additionally, H_2O_2 can decompose with elimination of oxygen, which solving in water can provide additional channel (paramagnetic centers) of relaxation of the nuclear magnetization. It is possible that the presence of a certain amount of dissolved oxygen in the H_2O_2 solution can be a reason of a stronger manifestation of the Curie law in comparison with other similar heterogeneous systems. The intensity of the H_2O_2 signal (Figure 1.34a) decreases with increasing temperature much stronger than that of water. This can be explained by changes in the component clusterization (due to cryoconcentration), which decreases with increasing temperature. In other words, there are two states of H_2O_2 characterized by slow and fast (in NMR timescale) proton exchange with water and the contribution of the latter grows with increasing temperature.

According to clustered structure of water (Chaplin 2011), liquid water is composed of clusters and nanodomains. In the case of miscible liquids such as water and hydrogen peroxide, more clustered structure can be formed at lower temperatures (Figure 1.35a; left picture). This clustered structure can be destroyed with increasing temperature due to enhanced motion, diffusion of the molecules forming random structures (Figure 1.35a; right picture). The latter is characterized by broadened ^1H NMR spectrum theoretically calculated (Figure 1.35). In the case of the solution, there is a downfield shift of the water peak (curves 2w and 3w) due to interaction of water molecules with hydrogen peroxide molecules.

At $T<250$ K, water (SAW) mixed with H_2O_2 is partially unfrozen but characterized by a lower activity as a solvent worse dissolving H_2O_2. Therefore, the latter tends to be more clustered (as shown in Figure 1.35, right picture). Greater δ_H values for water in mixture with H_2O_2 than that for pure liquid water are due to the hydrogen bonding of water molecules with H_2O_2 molecules. With increasing temperature, the solubility of H_2O_2 in SAW increases that results in enhancement of the proton exchange between water and hydrogen peroxide. In this case, one can expect the appearance of broad but non-split signal of the H_2O–H_2O_2 mixture with the downfield shift in comparison with pure water. However, this effect is absent since the separated signals of water and hydrogen peroxide are observed at $T>250$ K (Figure 1.34a). This effect can be explained by the existence of the clustered H_2O_2 structure even at 280–290 K.

In the case of interaction of the aqueous solution of H_2O_2 with nanosilica (dehydrated before mixing with the solution) being in nonpolar (CCl_4) or weakly polar ($CDCl_3$) solvents, the temperature behavior of the ^1H NMR signals differs from that for the pure solution (Figure 1.34). Several signals are in the spectra: WAW ($\delta_H \approx 1$ ppm), two signals of SAW with different amounts of dissolved H_2O_2 ($\delta_H \approx 5$–6.5 ppm). Additionally, in the CCl_4 medium, a separated signal of hydrogen peroxide is observed at $\delta_H \approx 11$ ppm. Freezing temperature of CCl_4 is 240 K. The dispersion medium freezing (Figure 1.34b) results in broadening of TMS signal at $\delta_H = 0$ ppm. Near freezing temperature of CCl_4 transition of SAW from state 1 at $\delta_H \approx 5$ ppm into state 2 at $\delta_H \approx 6$ ppm is observed. Both signals are observed at 210–240 K. The H_2O_2 signal decreases with temperature and it disappears at $T>240$ K.

In weakly polar $CDCl_3$ medium (Figure 1.34), the δ_H value of SAW increases. Splitting of the SAW peak into 1 and 2 states occurs at $T>250$ K. Signal 2 has the upfield shift with increasing temperature because of decreasing clusterization of the solution. This effect for signal 1 is much smaller. Silica A-400 as a nanostructured material affects the clusterization of the solution located in narrow nano- and mesopores (voids between silica nanoparticles). This process is also dependent on cryoconcentration, that is, relative concentration of H_2O_2 in certain structures increases due to partial freezing of water in the form of pure ice.

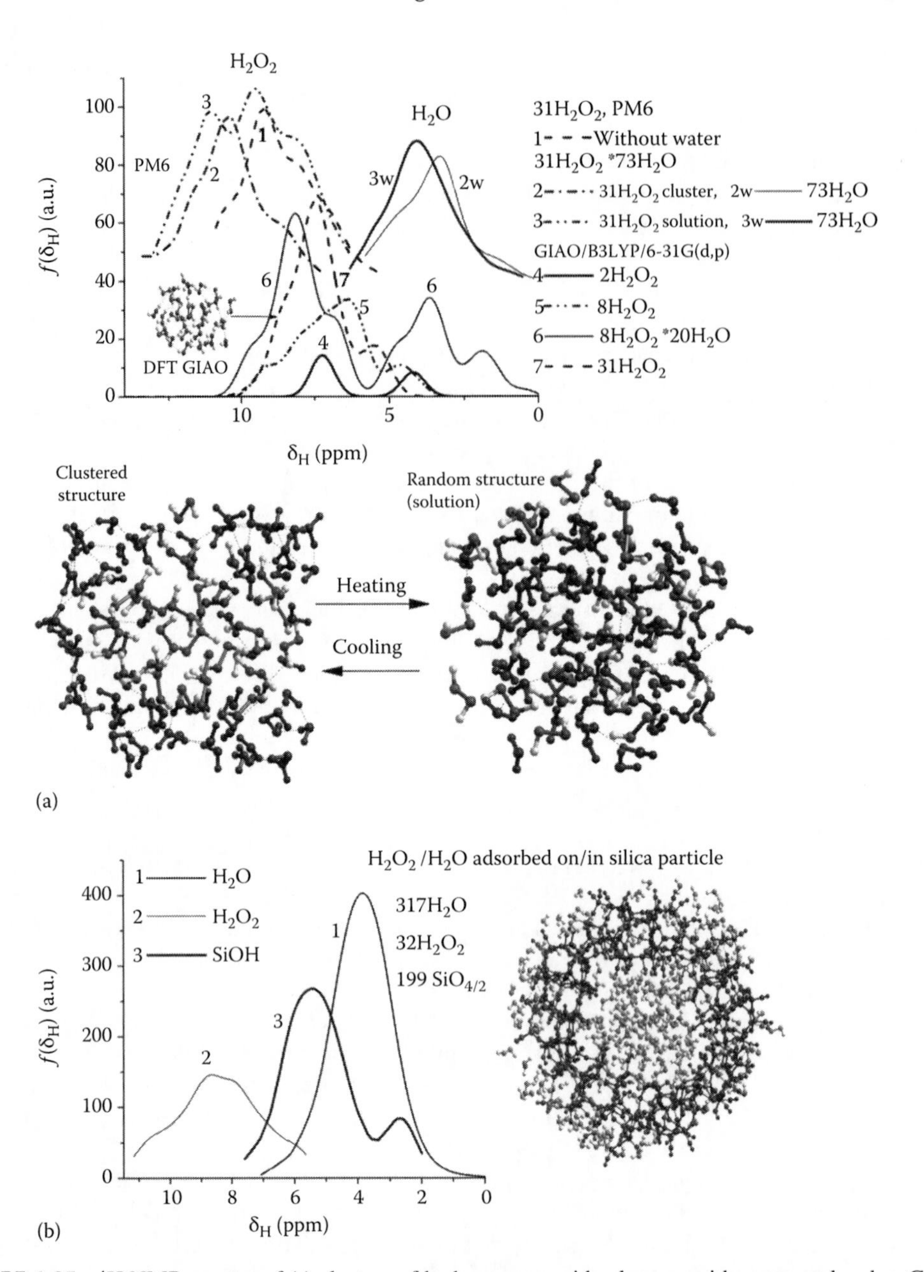

FIGURE 1.35 ^{1}H NMR spectra of (a) clusters of hydrogen peroxide alone or with water molecules, GIAO/IEFPCM/B3LYP/6-31G(d,p) (curve 7), and two types of the mixtures were shown with pure solution and clustered structure; (b) H_2O_2/H_2O mixture adsorbed onto/in silica particle. (Adapted from *Chem. Phys. Lett.*, 631, Gun'ko, V.M., Turov, V.V., and Turov, A.V., Hydrogen peroxide—water mixture bound to nanostructured silica, 132–137, 2012c. Copyright 2012, with permission from Elsevier.)

The ^{1}H NMR spectra of the H_2O/H_2O_2 mixture adsorbed onto nanosilica being in different mixtures of nonpolar (CCl_4) or weakly ($CDCl_3$) solvents with addition of polar solvents (CD_3CN and $(CD_3)_2SO$; Figure 1.36) demonstrate a complex temperature behavior. Besides signals of water, H_2O_2, and TMS, signals of methyl groups of CH_3CN (as admixture in CD_3CN) at $\delta_H = 2.2$ ppm and $(CH_3)_2SO$ (admixture in $(CD_3)_2SO$) at $\delta_H = 2.5$ ppm are observed. In frozen CCl_4 medium ($T < 240$ K), signals of SAW and H_2O_2 are observed separately (Figure 1.36a). The SAW signal demonstrates the upfield shift with increasing temperature and minimal value $\delta_H = 3.5$ ppm is observed at 230 K and

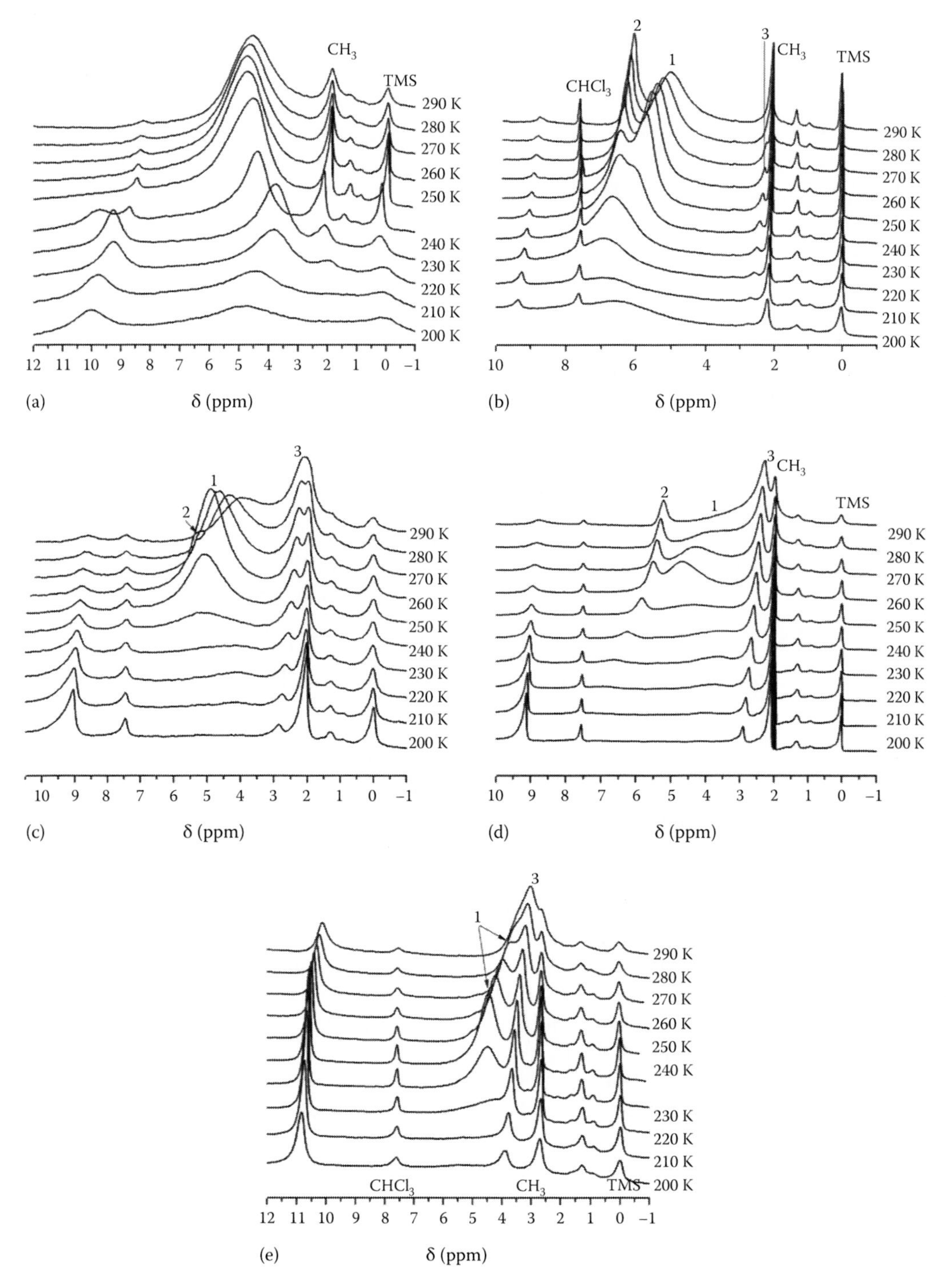

FIGURE 1.36 ^{1}H NMR spectra of aqueous 30% solution of H_2O_2 adsorbed (10 wt%) onto A-400 being in (a) CCl_4 + 8% CD_3CN, (b) $CDCl_3$ + 8% CD_3CN, (c) $CDCl_3$ + 15% CD_3CN, (d) $CDCl_3$ + 22% CD_3CN, and (e) $CDCl_3$ + 8% DMSO media. (Adapted from *Chem. Phys. Lett.*, 631, Gun'ko, V.M., Turov, V.V., and Turov, A.V., Hydrogen peroxide—water mixture bound to nanostructured silica, 132–137, 2012c. Copyright 2012, with permission from Elsevier.)

then it has the downfield shift. At 240 K, two signals of H_2O_2 in the adsorbed solution are observed at $\delta_H = 10$ and 8.5 ppm. However, at higher temperatures, only one weak signal is observed and its δ_H value decreases with temperature. Since the δ_H value of this signal is lower than that of pure H_2O_2, one can assume that the H_2O_2 molecules form clusters with water and acetonitrile molecules.

After melting of CCl_4 ($T > 240$ K) a major portion of H_2O_2 dissolves in the SAW nanodomains (that result in increasing δ_H value of SAW) or forms the adsorption complexes with surface silanols groups. In the last case, their signal can be absent in the spectra due to the proton exchange SiOH↔HOH causing significant shortening of the relaxation time (signal becomes too broad to be registered).

In the $CDCl_3$ with addition of CD_3CN (Figure 1.36b), a weak signal of H_2O_2 is observed at $\delta_H \approx 9$ ppm. It can be due to the formation of clusters with the participation of H_2O_2, H_2O, and $CDCl_3$ or CD_3CN molecules (as well as in the system shown in Figure 1.36a). SAW is observed as one ($T < 240$ K) or two signals (signals 1 and 2), which demonstrate the upfield shift with increasing temperature. Signal 1 shifts stronger than signal 2. The maximal δ_H value for SAW is 7 ppm at 220 K. At lower temperatures, the signal intensity and its δ_H value decrease due to partial freezing of water and H_2O_2. Weak signal of water (signal 3) is also observed at $\delta_H = 2.2$–2.4 ppm (to the left from signal of methyl groups of acetonitrile). This δ_H value is intermediate between the values of WAW (1–1.5 ppm) and water bound to acetonitrile molecules (3 ppm). This signal intensity increases with increasing concentration of acetonitrile in the mixture with CD_3Cl (Figure 1.36b through d). It is possible that it can be caused by clusters with CD_3Cl-CD_3CN-H_2O bound to the A-400 surface in narrow voids (Figure 1.33). The H_2O_2 signal intensity at $\delta_H \approx 9$ ppm increases with increasing CD_3CN concentration (Figure 1.36b through d). The minimal δ_H value of SAW decreases to 4 ppm. Signals 1 and 2 are observed separately only at $T > 270$ K. Signal 3 is the most intensive at 22% CD_3CN in a mixture with CD_3Cl. Its intensity increases with increasing temperature but intensity of signal 1 decreases. In the case of a $CDCl_3$ mixture with DMSO (Figure 1.36e), the δ_H value of hydrogen peroxide increases to 11 ppm. Significant changes in the intensity of signals 3 and 1 (similar to that at 22% CD_3CN) are already observed at 5% DMSO added to CD_3Cl.

Thus, observed patterns of the relationships of the ^1H NMR signals of aqueous 30% solution of hydrogen peroxide bound to nanosilica A-400 being in the dispersion media with solvents of different polarity and hydrophobicity can be interpreted in terms of the formation of clusters and nanodomains in voids between spherical silica nanoparticles, and these clusters contain different amounts of H_2O_2. At low temperatures, the confined space effects increase due to partial freezing of components forming crystallites in voids between silica nanoparticles that can add nanopores and narrow mesopores to the textural porosity of A-400. This leads to cryoconcentration of the dissolved components remained in liquid state. In the hydrophobic nonpolar medium, the solution of H_2O_2 becomes inhomogeneous and forms two types of structures with SAW with different contents of hydrogen peroxide. The content of WAW is much smaller than in the case of similar samples with water without H_2O_2. In the case of the H_2O/H_2O_2 mixtures in weakly polar CD_3Cl with addition of polar CD_3CN or strongly polar DMSO, three types of clusters and domains are formed at the silica surface: (i) clusters with concentrated H_2O_2 at $\delta_H = 9$–11 ppm, (ii) clusters with concentrated water and dissolved certain amounts of co-solvents, and (iii) clusters with organic solvents containing certain amounts of water and H_2O_2. The activity of solvents increases with increasing temperature, but the cryoconcentration effects decrease that result in the formation of more homogeneous interfacial solutions including organic solvents. Hydrogen peroxide forms clusters with water molecules, which can slowly exchange by protons with the clustered SAW and WAW. Significant amounts of WAW are observed only in the case of the CD_3Cl/DMSO medium.

Thus, structure of water bound to a surface of nanosilicas depends on (i) the specific surface area of nanosilica; (ii) pretreatments (heating, MCA, hydrothermal treatment [HTT], etc.); (iii) the presence of coadsorbates, their type (polar or nonpolar, low- or high [vide infra] -molecular compounds), and content; (iv) dispersion medium (nonpolar, weakly polar, or polar) and size of co-solvent molecules; and (v) temperature.

1.1.3 Adsorption of CH_4 in the Presence of Preadsorbed Water

Silica matrices are frequently used as a substrate for metallic catalysts of organic reactions with hydrogen transfer (Fackler 1990, Narayan and King 1998, VanderWiel et al. 1999, Deleplanque et al. 2009 Walaszek et al. 2009). In these reactions, the interactions of hydrogen with silica can play a certain role. Information on the H_2 adsorption at low temperatures and interactions with oxide materials can be obtained using diffuse reflectance infrared fourier transform (DRIFT) spectroscopy (Kazansky 2003). The spectra show changes in the OH-stretching vibrations of the surface hydroxyls and adsorbed H_2 molecules (stretching frequencies in the 4400–3800 cm^{-1} range). The strongest interactions of the H_2 molecules occur with metal cations located in pores (channels, cavities) of zeolites and other oxides. We tried to study the adsorption of hydrogen onto fumed oxides using low-temperature ^1H NMR spectroscopy. However, nanooxides do not have nanopores; therefore, the hydrogen adsorption was too low (with or without preadsorbed water) to be studied by the ^1H NMR spectroscopy in contrast to the adsorption of hydrogen onto ACs (vide infra). Therefore, the methane adsorption onto nanosilicas is analyzed here.

There is a problem related to storage of combustible gases such as hydrogen and methane onto adsorbents that require low temperature and/or high pressures for transformation into the liquid state or clathrate formation (Wergzyn and Gurevich 1996, Kuochih 2001, Lozano-Castello et al. 2002, Østergaard et al. 2002, Celzard et al. 2007). Adsorption methods are attractive alternatives since they require less energy input (Gregg et al. 1979, Fenelonov 1995, Salame and Bandosz 1999a,b, Subramanian and Sloan 1999, Ballard and Sloan 2004). Therefore, the development of adsorption methods for light hydrocarbons storage is of considerable interest. These gases can effectively adsorb onto different adsorbents only in narrow pores, the diameter of which must be $\leq 2d$ with respect to the adsorbate size (d). Coadsorbed water influences the adsorption of light hydrocarbons on different adsorbents, and in the case of high pressures, optimal adsorbent hydration should be relatively high ($h \approx 1$ gram of water per gram of adsorbent; Celzard and Marêché 2006, Zhou et al. 2006). However, for the adsorption of methane at atmospheric pressure and low temperatures, optimal hydration of different adsorbents is much lower ($h \approx 0.1$ g/g) (Turov et al. 2008b, Turov and Gun'ko 2011). At these conditions, the adsorption of methane on wetted adsorbents can be several times greater than on dry adsorbents. This effect for adsorbents composed of mesoporous particles can be explained by an increase in effective nanoporosity, due to the formation of water clusters and ice nanocrystallites at polar surface sites in narrow mesopores. However, in the case of adsorbents (such as fumed oxides composed of nonporous primary nanoparticles) possessing textural porosity practically without nanoporosity, adsorbed water structures do not undergo strong confined space effects as in narrow intraparticle pores in porous adsorbents. Therefore, the effects of coadsorbed water on the adsorption of methane are less predictable for the nanooxide systems. The effects of preadsorbed water on the adsorption of methane onto nanosilica possessing only textural porosity were analyzed in comparison with the adsorption onto a nano/mesoporous silica gel 200DF (Crosfield; Gun'ko et al. 2011c).

Standard nanosilicas A-300 and A-380 (S_{BET}=337 and 378 m^2/g, pore volume V_p=0.714 and 0.943 cm^3/g, respectively, estimated from LT [77.4 K] adsorption isotherms of nitrogen) were heated at 673 K for several hours to remove adsorbed organics and residual HCl and rehydrated in air. Silica 200DF (Crosfield, S_{BET}=540 m^2/g, V_p=0.34 cm^3/g) possessing nanopores and narrow mesopores was used to compare intraparticle (200DF) and textural (nanosilica) porosity effects on the behavior of bound water and coadsorption of water and methane. Bidistilled water and commercial methane (99% purity) were used (Gun'ko et al. 2011c).

The PSD was calculated using the DFT method using the Micromeritics DFT applied to silica gel 200DF with model of cylindrical pores and another DFT version developed for the model of voids between spherical particles randomly packed in agglomerates (Gun'ko 2007, Gun'ko et al. 2007f).

The FTIR spectra of nanosilica were recorded using a ThermoNicolet FTIR spectrometer at 288 K. To study the adsorption of methane, it was fed into a special cell (blown by a methane flow at 1 bar) with a nanosilica pellet. The spectra were recorded for initial nanosilica (hydration $h \approx 0.06$ g/g), methane, and methane and water coadsorbed on nanosilica in the cell. Additionally, the spectra were recorded for samples in air. A diffuse reflectance FTIR spectrum of silica 200DF was obtained with a Digilab FTS3000 FTIR spectrometer (Gun'ko et al. 2011c).

The ^{1}H NMR signals of immobile water molecules from ice and OH groups from silica were not observed in the spectra recorded here for static samples because of features of the measurement technique (Gun'ko et al. 2005d, 2009d) and a significant difference in the spectral width of mobile and immobile structures without the use of the MAS technique. Investigations of water adsorbed on a silica surface and surface silanols using MAS and CRAMPS ^{1}H NMR spectroscopy (Kinney et al. 1993, Humbert 1995, Chuang and Maciel 1996, Liu and Maciel 1996, Hu et al. 2005) give temperature-dependent spectra different from that characteristic of mobile water (Gun'ko et al. 2005d, 2009d,e) studied here. To compare the MAS and non-MAS (static samples) ^{1}H NMR spectra of initial silica A-300, they were recorded at room temperature and a rotor frequency of 0 and 9 kHz.

Before the static ^{1}H NMR measurements of coadsorbed mobile water and methane, samples placed in closed NMR ampoules (5 mm in diameter) with preadsorbed water were equilibrated for 5–10 min (for each temperature) at a methane pressure of 1.1 bar. Water loss was insignificant during the preparation of wetted samples and measurements.

Changes in the Gibbs free energy (ΔG) and other characteristics of coadsorbed water and methane, due to their interaction with the silica surface, were determined from the temperature dependence of the amounts of unfrozen water (C_{uw}) at $200 < T < 280$ at different prehydration of silica ($h = 0.005$–1.0 g/g) using the melting curves. Water in the studied samples can be divided into several types such as (i) mobile (liquid) and immobile (crystalline and amorphous ice), (ii) bound (strongly bound and unfrozen at $T < 260$ K and $-\Delta G > 0.5$ kJ/mol) and weakly bound (frozen at $T > 260$ K and $-\Delta G < 0.5$ kJ/mol), and (iii) unbound (bulk, $h = 1$ g/g), strongly associated (chemical shift of the proton resonance $\delta_H = 3$–5 ppm) and weakly associated ($\delta_H = 1$–2 ppm). The amounts of adsorbed water were estimated using integral ^{1}H NMR intensity and a calibrated function obtained by the weight method.

The FTIR spectra of nanosilicas A-300 and A-380 with adsorbed water (from air) and water/methane coadsorbed in the spectral cell (Figure 1.37) are characterized by a broad band at 2500–3700 cm^{-1}, which can be assigned to the OH-stretching vibrations of adsorbed water and disturbed surface silanols (Kiselev and Lygin 1972, Legrand 1998). A narrow band at 3747 cm^{-1} linked to free silanols (Kiselev and Lygin 1972, Legrand 1998) is observed for nanosilicas (more readily observed for A-300 than A-380) but not for silica 200DF (Figure 1.37a). This result can be explained by clustering of water adsorbed on nanosilicas (i.e., a continuous layer of adsorbed water is absent), in contrast to porous silica 200DF whose narrow pores (Figure 1.38) are practically entirely filled by water adsorbed from air. Gaseous methane alone and adsorbed onto nanosilica is characterized by an intensive narrow band at $\nu_{CH} = 3016$ cm^{-1} and a set of low-intensity satellite bands split due to vibrational–rotational motions of methane molecules in the vapor phase in the FTIR cell. A small quantity of methane adsorbed on nanosilica is observed after purging the cell with air (Figure 1.37b; curve 2).

The adsorption of methane does not lead to significant changes in the band intensity of free silanols because of the weak interaction of nonpolar CH_4 molecules with the silica surface. However, calculations of the free surface area (S_{IR}) on the basis of the integral intensity of free silanols at $\nu_{OH} = 3747$ cm^{-1} (as a surface characteristic) and a band at 1870 cm^{-1} as a bulk Si–O combination mode (used as an internal standard related to the nanoparticle volume), according to a method described by McCool et al. (2006), give $S_{IR} = 263.3$ m^2/g for A-300 alone (with adsorbed water from air) and 266.9 and 265.9 m^2/g for silica with methane adsorbed for 1 and 7 min, respectively. An increase in the S_{IR} value is due to partial desorption of water resulting from methane flow the cell. It should be noted that the S_{IR} value can be considered as a portion of the surface area free from adsorbates, since it is determined from the band of free silanols at 3747 cm^{-1}, and it is close

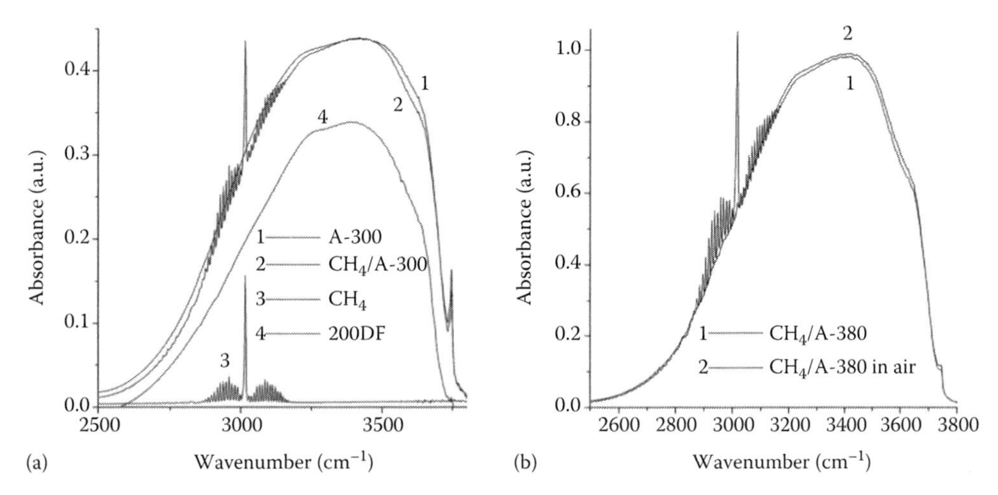

FIGURE 1.37 FTIR spectra over the OH-stretching vibrations range for (a) silica A-300 with adsorbed water (from air) (curve 1), after blowing by methane (2), methane alone (3), and silica 200DF with water adsorbed from air (4) and (b) A-380 with coadsorbed water and methane (1) and after blowing by air (2). (Adapted from *Appl. Surf. Sci.*, 258, Gun'ko, V.M., Turov, V.V., Bogatyrev, V.M. et al., The influence of pre-adsorbed water on adsorption of methane on fumed and nanoporous silicas, 1306–1316, 2011c. Copyright 2011, with permission from Elsevier.)

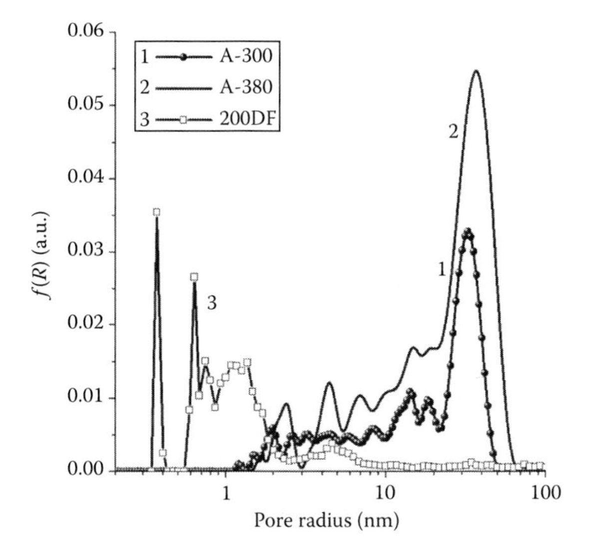

FIGURE 1.38 Incremental pore size distributions of silicas A-300 (1), A-380 (2), and 200DF (3) calculated from the nitrogen adsorption/desorption isotherms (77.4 K) using the DFT method. A model of cylindrical pores was used for 200DF, and a model of pores as voids between spherical nonporous primary particles forming aggregates of random structures for nanosilicas was employed. (Adapted from *Appl. Surf. Sci.*, 258, Gun'ko, V.M., Turov, V.V., Bogatyrev, V.M. et al., The influence of pre-adsorbed water on adsorption of methane on fumed and nanoporous silicas, 1306–1316, 2011c. Copyright 2011, with permission from Elsevier.)

to the S_{BET} value for dehydrated silicas (McCool et al. 2006). According to the S_{IR} and S_{BET} values, adsorbed water ($h \approx 0.1$ g/g) disturbs about 22% of surface silanols. According to the S_{IR} values after exposure to methane for 1 and 7 min, methane disturbs about 0.4% of surface silanols free from adsorbed water. Comparing this result with the changes caused by known amounts of water, it is possible to estimate the amount of adsorbed methane that gives 1.8% CH_4 with respect to adsorbed water (i.e., ~0.2 wt% with respect to silica since $h \approx 0.1$ g/g) at 288 K. This value is smaller than that obtained from ^{1}H NMR spectral measurements (discussed in the subsequent sections), giving a greater ratio of adsorbed methane and water as 0.116: 1, but at a slightly lower temperature (280 K).

Nanosilica A-380 is more aggregated and has a larger textural porosity than A-300 (Figure 1.38), since the pore volume determined from the nitrogen adsorption isotherm is $V_p = 0.943$ and 0.714 cm^3/g, respectively. Therefore, nanosilica A-380 can adsorb larger amounts of water from air. This results in a much lower intensity of the band at 3747 cm^{-1} (Figure 1.37) and much smaller free surface area $S_{IR} = 186$ m^2/g at $S_{BET} = 378$ m^2/g (50.8% reduction) than that for A-300. Purging this sample with methane gives $S_{IR} = 194$ m^2/g because of partial desorption of water, as seen with A-300. Subsequent purging with air leads to diminution of the S_{IR} value to 183 m^2/g, because of coadsorption of water and residual methane. Assuming that water displaces adsorbed methane, one can estimate the value of free surface area occupied by methane as 1.8%, which is higher than that observed for A-300, due to stronger aggregation of silica nanoparticles in A-380 (Figure 1.38). Notice that these estimations are not exact due to features of the used calculation methods but clearly show tendency of the influence of preadsorbed water on the adsorption of methane and vice versa. Silica 200DF includes both nano- and narrow mesopores in contrast to nanosilicas characterized by broader textural pores (Figure 1.38). Therefore, silica 200DF can adsorb larger amounts of water from air than nanosilicas. Thus, the confined space effects in pores of different types, and the influence of preadsorbed water on the adsorption of methane, can significantly differ for nanosilicas and 200DF.

The ^{1}H NMR spectra of mobile water and water/methane at $T < 273$ K were recorded here without the use of the MAS technique. Therefore, the spectral bandwidth (non-MAS) of surface silanols and OH groups from ice is much broader than mobile water. However, immobile OH structures can affect the baseline of the mobile water spectra. The chemical shift of the proton resonance of free unbound silanols is $\delta_H = 1.5$–2.0 ppm (Figure 1.39; rotor frequency $f_r = 9$ kHz; curve 2; Kinney et al. 1993, Humbert 1995, Chuang and Maciel 1996, Liu and Maciel 1996, Hu et al. 2005). This band is absent for A-300 at $h = 0.05$ (Figure 1.59; curve 1, $f_r = 0$ kHz) and 0.08 g/g (Figure 1.40a). However, the O-H stretching vibrations of free silanols are observed at 3747 cm^{-1} in the FTIR spectra of A-300 at a similar h value (Figure 1.37), because free silanols can be observed at the nanosilica surface at $h < 0.2$ g/g due to clustered adsorption of water. The bands at $\delta_H = 1$–2 ppm (i.e., in the range of free silanols observed in the ^{1}H MAS NMR spectra of silica [Figure 1.39]) are observed

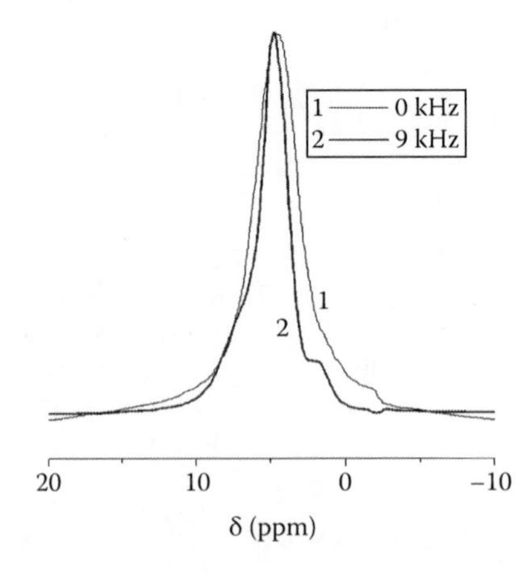

FIGURE 1.39 ^{1}H NMR spectra of initial A-300 ($h \approx 50$ mg/g) recorded at 293 K using the MAS technique (rotor frequency $f_r = 0$ [1] and 9 [2] kHz). (Adapted from *Appl. Surf. Sci.*, 258, Gun'ko, V.M., Turov, V.V., Bogatyrev, V.M. et al., The influence of pre-adsorbed water on adsorption of methane on fumed and nanoporous silicas, 1306–1316, 2011c. Copyright 2011, with permission from Elsevier.)

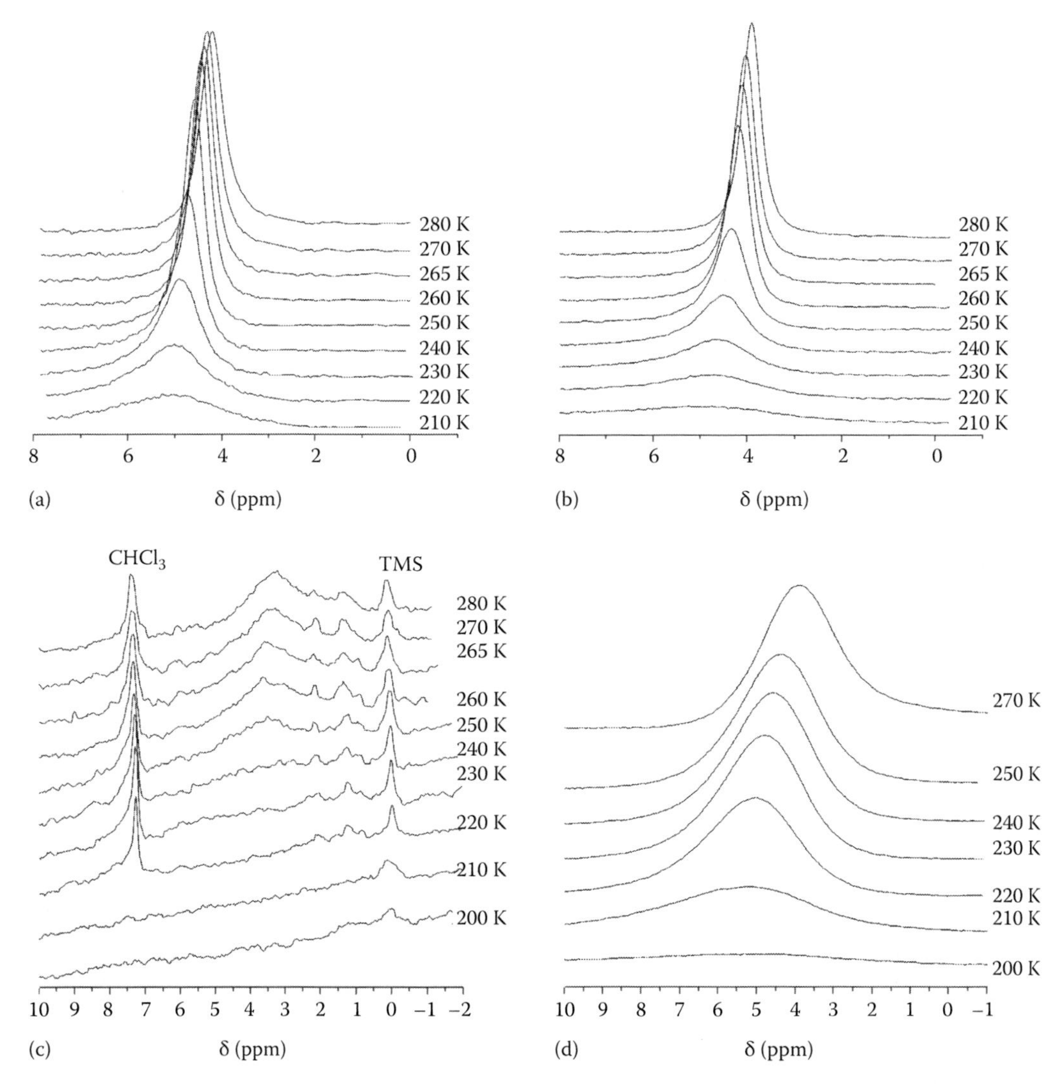

FIGURE 1.40 ^{1}H NMR spectra of mobile water adsorbed on nanosilica A-300 (a, b) and silica 200DF (c, d) and h = (a) 0.08, (b, d) 0.15, and (c) 0.005 g/g, the spectra shown in (c) correspond to weakly hydrated 200DF being in chloroform-d containing a portion of $CHCl_3$ (<1%) ($\delta_{H,CHCl_3}$ = 7.26 ppm) and tetramethylsilane (TMS) as a standard ($\delta_{H,TMS}$ = 0 ppm). (Adapted from *Appl. Surf. Sci.*, 258, Gun'ko, V.M., Turov, V.V., Bogatyrev, V.M. et al., The influence of pre-adsorbed water on adsorption of methane on fumed and nanoporous silicas, 1306–1316, 2011c. Copyright 2011, with permission from Elsevier.)

for silica 200DF at $h \approx 0.005$ g/g (Figure 1.40c). However, intensity of these bands decreases with lowering temperature and they are not observed at 210 K. This behavior is more characteristic of weakly associated but SBW than for free silanols. Additionally, the O-H stretching vibrations of free silanols are not observed for silica 200DF (Figure 1.37a), that is, all the SiOH groups are hydrogen bonded. Notice that a small portion of water can be dissolved in chloroform and contribute to the spectra at $\delta_H \approx 1.2$ ppm. The band at $\delta_H = 3–4$ ppm, which is not observed at $T < 240$ K, can be assigned to SAW with a major fraction of WBW (frozen at $T > 260$ K). As a whole, clustering of water adsorbed on nanosilica diminishes with increasing amounts of water, that is, it becomes more strongly associated. Therefore, ^{1}H NMR signal intensity decreases faster with lowering temperature for A-300 than 200DF with a greater amount of adsorbed water, which is readily seen at $T \leq 240$ K (Figure 1.40). Thus, this NMR technique allows us to observe the ^{1}H NMR spectra of mobile water, in contrast to the ^{1}H MAS NMR technique, giving the spectra of all OH structures (silanols, bulk and adsorbed liquid, and frozen water). This technique, with layer-by-layer freezing-out of bound water at 200–280 K, can give detailed information on coadsorption of methane and water on the studied silicas.

The ^{1}H NMR spectra of water and methane (Figure 1.41) coadsorbed on nanosilica A-300 at different hydration degrees were recorded over the 200–280 K range. These spectra include a single or weakly split signal of methane at $\delta_H = 0.0–0.3$ ppm. This signal slightly increases with lower temperature due to increasing adsorption of methane and an enhanced interaction with the silica surface. Water is responsible for two signals at $\delta_H \approx 1$ and 4–5 ppm corresponding to WAW with a significant portion of unbound H atoms, and SAW, with dipolar coupled molecules due to O–H···O hydrogen bonds, respectively. The presence of several ^{1}H NMR signals of an adsorbate suggests slow (on the NMR timescale) molecular exchange between the corresponding states (Pople et al. 1959). The signal intensity of SAW at $\delta_H \approx 4–5$ ppm decreases, but the δ_H value increases, with lower temperature because of layer-by-layer freezing-out of fractions of more weakly bound SAW and WAW (Figure 1.41). However, complete freezing of WAW is observed only at $h = 0.005$ g/g and $T < 220$ K (Figure 1.41a), and at $h > 0.03$ g/g, only a certain portion of WAW is frozen at 200 K.

Equilibration of the water/A-300 system for a week leads to several important effects illustrated in Figure 1.42: (i) the adsorption of methane decreases for the equilibrated compared with the non-equilibrated system at the same h values; (ii) water becomes more strongly associated due to equilibration, that is, it becomes less clustered with a greater number of hydrogen bonds per molecule; (iii) this latter effect is stronger with increasing hydration of nanosilica as well as silica 200DF (Figures 1.43 and 1.44). For the latter, clustered adsorption of water is observed only at minimal hydration $h = 0.005$ g/g (Figure 1.43a; a shoulder at $\delta_H = 1–2$ ppm). At greater amounts of water adsorbed onto silica gel 200DF, only SAW is observed at δ_H 4–6 ppm (Figure 1.43b and c). This leads to strong diminution of the adsorption of methane with increasing amounts of preadsorbed water.

When $h < 0.3$ g/g, all water is strongly bound to the nanosilica surface (Gun'ko et al. 2007a) because it freezes at $T < 260$ K and changes in the Gibbs free energy $\Delta G < -0.5$ kJ/mol (Figures 1.44 and 1.45). Four of six non-equilibrated A-300/H_2O/CH_4 samples (with the exception of samples at $h = 0.005$ and 0.1 g/g) show an increase of WAW with lower temperature (Figure 1.44b). The most methane adsorption is observed at $h = 0.1$ g/g (Figure 1.44c) when the amount of WAW is largest and the system is non-equilibrated (Figure 1.44b). Equilibration of this system for 7 days causes significant diminution of the adsorption of methane. However, this effect is much smaller at lower hydration (Figure 1.44c) because adsorbed water is more clustered at smaller amounts despite the long equilibration time. Notice that changes in the ΔG and γ_S values after equilibration are relatively small (Figure 1.45); however, changes in contributions of SAW (Figure 1.44a) and WAW (Figure 1.44b) are significant. Consequently, changes in the adsorption of methane after preequilibration of the A-300/water system are more influenced by structural changes in adsorbed water (e.g., increase in the associativity and the size of adsorbed water structures) than because of changes in its energetic characteristics.

There is a correlation between the amount of adsorbed methane and WAW (Figure 1.44d). Lower amounts of adsorbed methane is observed for samples at minimal ($h = 0.005$ g/g) and maximal (1 g/g) hydration of silica. The former includes too little water to form effective nanoporosity (between the surface of adjacent silica nanoparticles, clusters, and domains of unfrozen water and ice nanocrystallites), and the latter includes too much water, which can form a nearly continuous surface SAW film (Figure 1.44a), and bound water is less clustered. In other words, strong clustering of bound water (Figure 1.46; Table 1.13) is a necessary condition for the maximal adsorption of methane onto nanosilica.

The value of free surface energy, γ_s, determined as the modulus of the total decrease in the Gibbs free energy of the adsorbent–adsorbate system, increases with increasing water content (Table 1.13). However, normalized free surface energy ($\gamma_{s,SAW}/C_{SAW}$ for SAW and γ_s/h for all bound water)

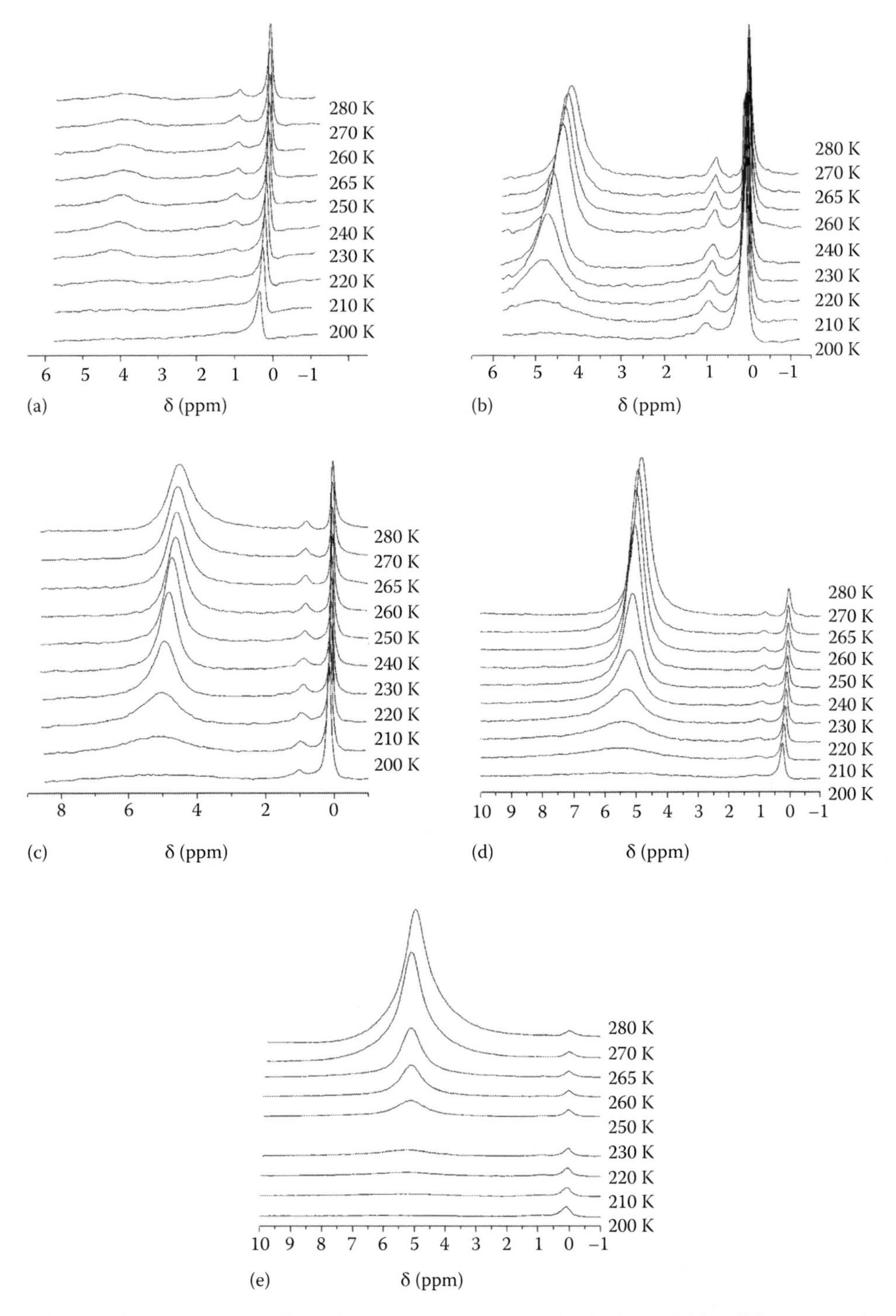

FIGURE 1.41 ^{1}H NMR spectra of mobile water and methane coadsorbed on A-300 at different temperatures and h = (a) 0.005, (b) 0.035, (c) 0.065, (d) 0.3, and (e) 1.0 g/g. (Adapted from *Appl. Surf. Sci.*, 258, Gun'ko, V.M., Turov, V.V., Bogatyrev, V.M. et al., The influence of pre-adsorbed water on adsorption of methane on fumed and nanoporous silicas, 1306–1316, 2011c. Copyright 2011, with permission from Elsevier.)

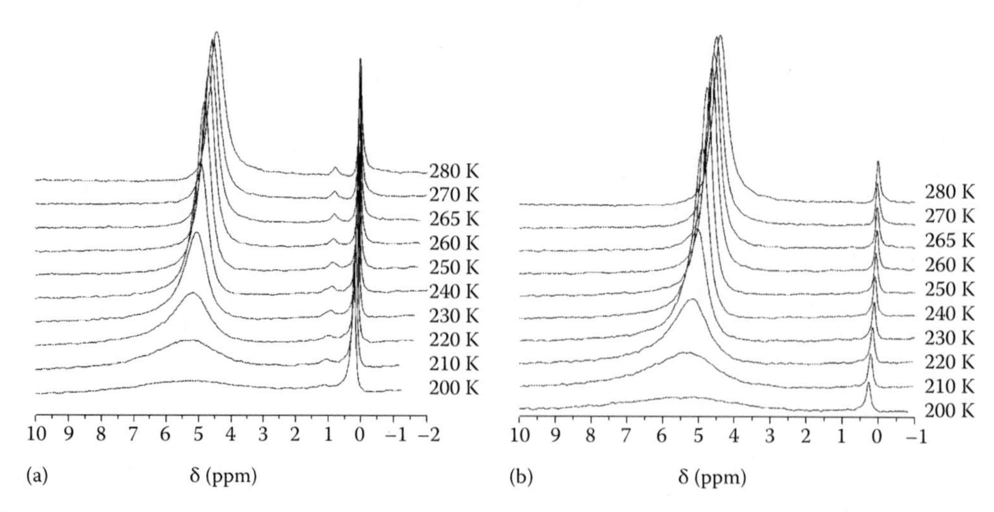

FIGURE 1.42 ^{1}H NMR spectra of mobile water and methane coadsorbed on A-300 at different temperatures and $h=0.1$ g/g 15 min after sample preparation and 5–10 min equilibration for each temperature (a) and after equilibration for 7 days (b). (Adapted from *Appl. Surf. Sci.*, 258, Gun'ko, V.M., Turov, V.V., Bogatyrev, V.M. et al., The influence of pre-adsorbed water on adsorption of methane on fumed and nanoporous silicas, 1306–1316, 2011c. Copyright 2011, with permission from Elsevier.)

decreases at $h=0.3$ and 1.0 g/g (Figure 1.45b), because the influence of the silica surface decreases for distant layers of bound water. In the case of a sample at $h=0.1$ g/g, the γ_s/h and $\gamma_{s,SAW}/C_{SAW}$ curves have a maximum. This is due to the equilibrium state of intact sample (with water adsorbed from air and equilibrated for long periods) compared to the incomplete equilibration of heated (at 150°C to $h=0.005$ g/g) and subsequently wetted samples. Complete equilibration of treated and wetted nanosilica requires several days due to the slow penetration of water molecules into the particle volume (Collins et al. 2006). Notice that the γ_s/h and $\gamma_{s,SAW}/C_{SAW}$ maxima at $h=0.1$ g/g corresponds to the largest contribution of WAW and greatest adsorption of methane.

The relative contribution of WAW of total bound water is maximal (up to 20%) at the smallest hydration (0.005 g/g) of silica, and minimal (<1%) at the largest hydration (1.0 g/g) (Figure 1.47). Consequently, the characteristics of bound water strongly correspond to the characteristics of SAW (Table 1.13), the major contributor of bound water. There are four-three types of SAW structures (Figures 1.48 and 1.49), for example, clusters, nanodomains, and microdomains. These SAW structures freeze at different temperatures depending on the strength of their interactions with the silica surface, reflecting changes in the relationship between ΔG and C_{SAW} (Figure 1.45a). A displacement of the dC_{SAW}/dT maxima toward higher temperatures with increasing h value (Figure 1.48) is seen because of the reduced interaction of SAW fractions with the silica surface with growing hydration. This results in the appearance of WBW (characterized by $\Delta G > -0.5$ kJ/mol and frozen at $T > 260$ K) at $h \geq 0.3$ g/g (Figures 1.44 and 1.45) due to the formation of relatively large water structures. These water structure sizes are determined by the Gibbs–Thomson relation for the freezing point depression of confined liquids (Aksnes and Kimtys 2004, Mitchell et al. 2008), are relatively large (Figure 1.49; Table 1.13), and are frozen at $T > 260$ K (Figure 1.48). Notice that the relative contribution of nanopores (Table 1.13, S_{nano}, V_{nano}) is largest at $h=0.1$ g/g; however, absolute values are larger at $h=0.3$ g/g. At $h=1$ g/g; the S_{nano} and V_{nano} values strongly decrease but the S_{meso} and V_{meso} values strongly increase. These results correspond to significant enhancement of water associativity with increasing h value.

For weakly hydrated samples at $h \leq 0.1$ g/g, the maximal contributions of dC/dT (Figure 1.48) and $f(R)$ (Figure 1.49) are for structures with SBW frozen at $T < 230$ K with sizes $R < 3$ nm. These water structures, attributed to clusters at $R < 1$ nm and nanodomains at $R = 1$–3 nm, correspond to nanopores and narrow mesopores, respectively. Water partially filling narrow pores at $R < 10$ nm

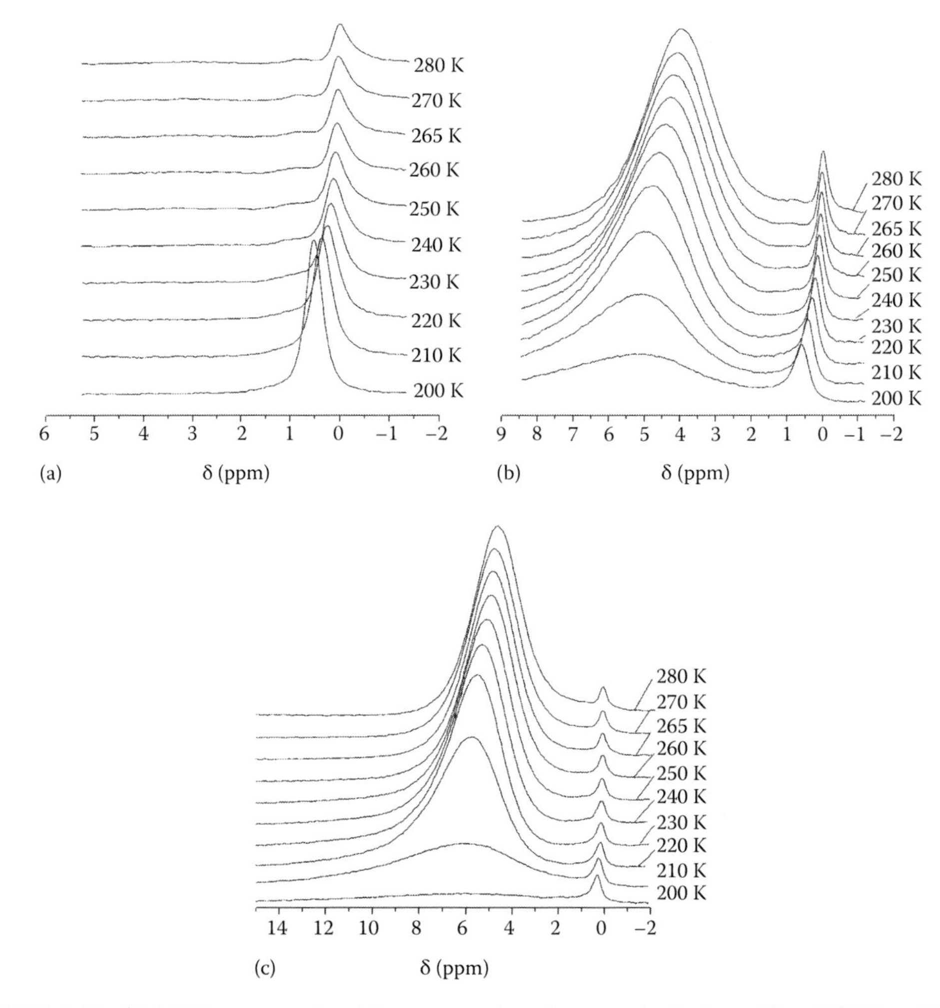

FIGURE 1.43 ^{1}H NMR spectra of mobile water and methane coadsorbed on silica 200DF at different temperatures and h=(a) 0.005, (b) 0.045, and (c) 0.15 g/g. (Adapted from *Appl. Surf. Sci.*, 258, Gun'ko, V.M., Turov, V.V., Bogatyrev, V.M. et al., The influence of pre-adsorbed water on adsorption of methane on fumed and nanoporous silicas, 1306–1316, 2011c. Copyright 2011, with permission from Elsevier.)

between primary silica nanoparticles (Figures 1.46, 1.49, and 1.50) provides enhancement of effective nanoporosity at $R<1$ nm. This results in an increase in the adsorption of methane (Figure 1.44c), especially at $h=0.1$ g/g, when the contribution of small water structures at $R\leq 1$ nm is maximal (Figure 1.49; Table 1.13) as the relative contribution of nanopores filled by bound unfrozen water.

The molecular mobility of water can appear in different structures where water molecules have different numbers of hydrogen bonds (per molecule) of different strength. Therefore, the distribution functions of the activation energy ($f(E)$) of the molecular mobility (calculated using the temperature dependences of $dC_{SAW}(T)/dT$ assuming that this process obeys the Arrhenius law; Gun'ko et al. 2009e) are relatively broad over the 15–85 kJ/mol range (Figure 1.51a). This range of E values corresponds to the motion of water molecules having from one to four hydrogen bonds per molecule (Gun'ko et al. 2007i).

The activation energy of the molecular motion of methane (determined from the temperature dependences of dC_{CH_4}/dT) is 1–10 kJ/mol for a minimally hydrated sample (Figure 1.51b). However, for more strongly hydrated silica ($h=0.1$ and 1.0 g/g), the E values increase by several times. This can be explained by several reasons. First, enhanced nanoporosity formed by water clusters,

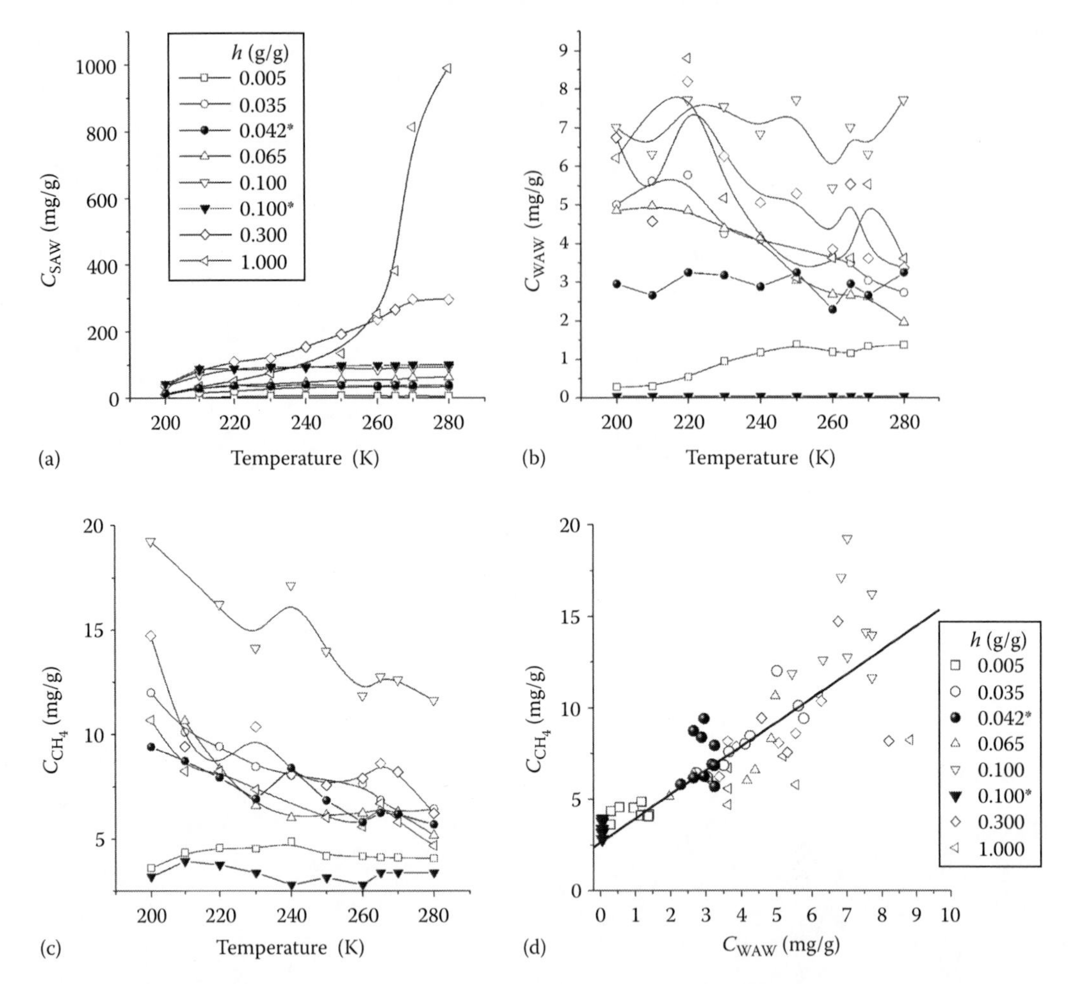

FIGURE 1.44 Temperature dependences of concentration of (a) SAW, (b) WAW, and (c) methane coadsorbed on nanosilica A-300 and (d) relationship between concentrations of methane and WAW. (Adapted from *Appl. Surf. Sci.*, 258, Gun'ko, V.M., Turov, V.V., Bogatyrev, V.M. et al., The influence of pre-adsorbed water on adsorption of methane on fumed and nanoporous silicas, 1306–1316, 2011c. Copyright 2011, with permission from Elsevier.)

nanodomains and ice nanocrystallites, partly or completely frozen with lowering temperature, enhances the barriers for methane motion. Second, mixed water/methane structures can demonstrate cooperative motion, therefore, the activation energy of the methane motion depends on the water motion. Thus, coadsorbed water and methane are not individual phases independent of one another. This is also confirmed by the correlation between the amounts of adsorbed methane and WAW (Figure 1.44d).

Thus, the adsorption of methane onto nanosilica A-300, composed of nonporous primary nanoparticles (average diameter 8.1 nm), at standard pressure is a function of temperature and silica hydration. The silica hydration dependence is nonlinear, and maximal adsorption of methane (1.9–1.2 wt% at 200–280 K) is observed at hydration $h=0.1$ g/g for intact silica. Decrease (on heating) and increase (on wetting) of the silica hydration both lead to a reduction of methane adsorption. Coadsorption of methane and water leads to the appearance of a ^{1}H NMR signal from WAW at $\delta_H \approx 1$ ppm. The amount of this water correlates to concentration of adsorbed methane, because weakly associated bound water is most clustered at the surface of nanosilica composed of nonporous primary nanoparticles. The adsorption of methane on nano/mesoporous

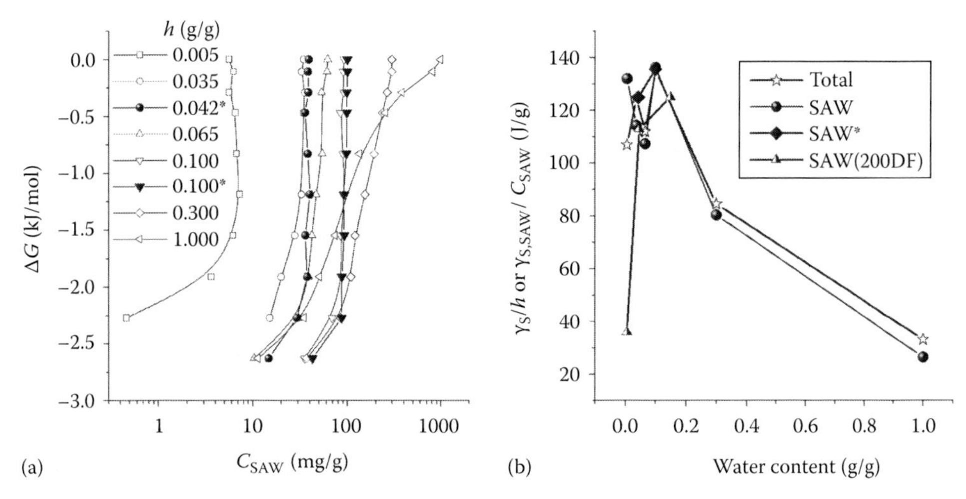

FIGURE 1.45 (a) Relationship between changes in the Gibbs free energy and the SAW amounts on coadsorption of water and methane onto nanosilica A-300, and (b) normalized free surface energy (γ_s/h or $\gamma_{s,SAW}/C_{SAW}$) as a function of water content (SAW* corresponds to samples equilibrated for a week) for A-300 and 200DF. (Adapted from *Appl. Surf. Sci.*, 258, Gun'ko, V.M., Turov, V.V., Bogatyrev, V.M. et al., The influence of pre-adsorbed water on adsorption of methane on fumed and nanoporous silicas, 1306–1316, 2011c. Copyright 2011, with permission from Elsevier.)

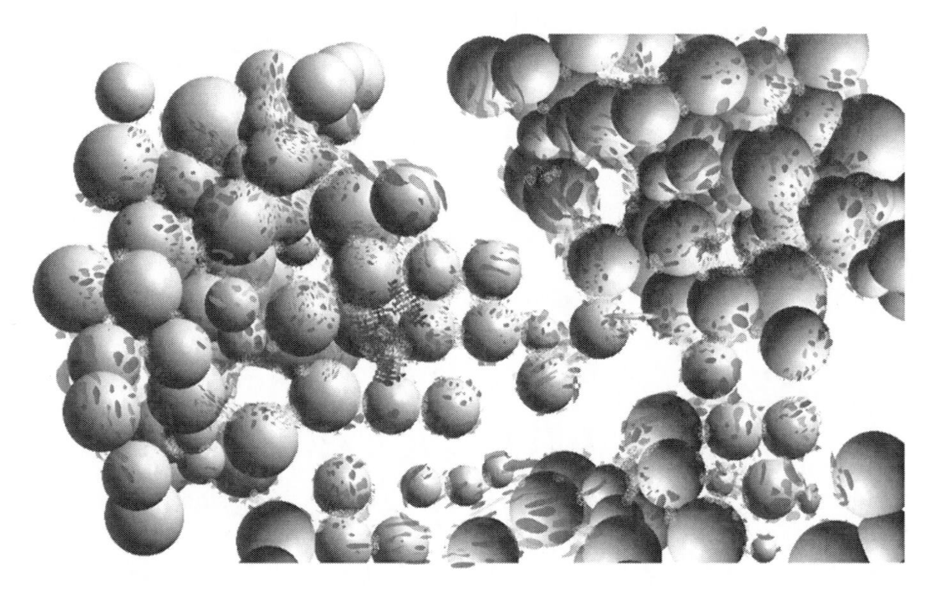

FIGURE 1.46 Model of clustered adsorption of water increasing nanoporosity and enhancing the adsorption of methane onto nanosilica composed of aggregates of nanoparticles. (Adapted from *Appl. Surf. Sci.*, 258, Gun'ko, V.M., Turov, V.V., Bogatyrev, V.M. et al., The influence of pre-adsorbed water on adsorption of methane on fumed and nanoporous silicas, 1306–1316, 2011c. Copyright 2011, with permission from Elsevier.)

silica 200DF decreases with increasing amount of preadsorbed water characterized by significant associativity ($\delta_H \approx 5$ ppm) at $h \geq 0.005$ g/g.

The thermodynamic characteristics of water bound by nanosilica OX-50 (Degussa, $S_{BET} \approx 52$ m^2/g) were determined for different dispersion media (air, liquid water, gaseous methane, liquid chloroform) using LT ^{1}H NMR spectroscopy. It was established that freezing-out of water in aqueous suspension or concentrated gel of OX-50 occurs at relatively high

TABLE 1.13
Structural and Energetic Characteristics of All Bound Water and the SAW Portion (Shown in Brackets) on Coadsorption with Methane on A-300 (S_{BET} = 337 m²/g, V_p = 0.714 cm³/g)

h (g/g)	γ_s (J/g)	$-\Delta G_s$ (kJ/mol)	S_{nano} (m²/g)	S_{meso} (m²/g)	V_{nano} (cm³/g)	V_{meso} (cm³/g)
0.005	0.92 (0.74)	2.85 (2.51)	0.7 (0.0)	2.3 (2.3)	0.001 (0)	0.008 (0.006)
0.035	4.87 (3.91)	3.09 (2.65)	50.4 (21.8)	7.8 (7.0)	0.022 (0.010)	0.017 (0.016)
0.065	7.17 (6.62)	2.84 (2.78)	58.3 (48.4)	5.9 (3.9)	0.027 (0.022)	0.035 (0.035)
0.100	13.51 (12.60)	2.98 (2.80)	136.2 (128.3)	8.0 (1.9)	0.061 (0.059)	0.038 (0.038)
0.300	24.90 (23.80)	2.91 (2.80)	154.0 (147.0)	42.8 (42.2)	0.070 (0.068)	0.222 (0.222)
1.000	27.16 (26.10)	2.84 (2.78)	70.4 (54.5)	101.7 (69.8)	0.032 (0.025)	0.631 (0.631)

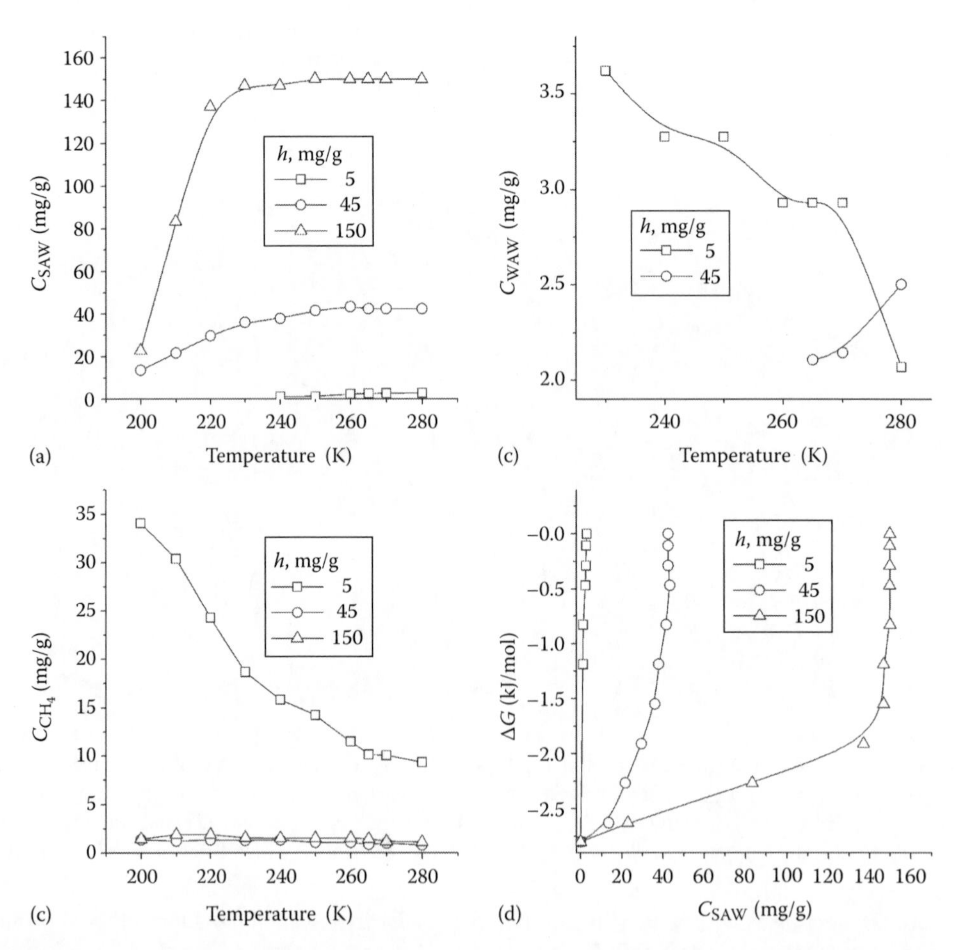

FIGURE 1.47 Temperature dependence of the concentration of (a) SAW, (b) WAW, and (c) methane coadsorbed on silica 200DF and (d) relationship between concentration of SAW and changes in Gibbs free energy of bound water. (Adapted from *Appl. Surf. Sci.*, 258, Gun'ko, V.M., Turov, V.V., Bogatyrev, V.M. et al., The influence of pre-adsorbed water on adsorption of methane on fumed and nanoporous silicas, 1306–1316, 2011c. Copyright 2011, with permission from Elsevier.)

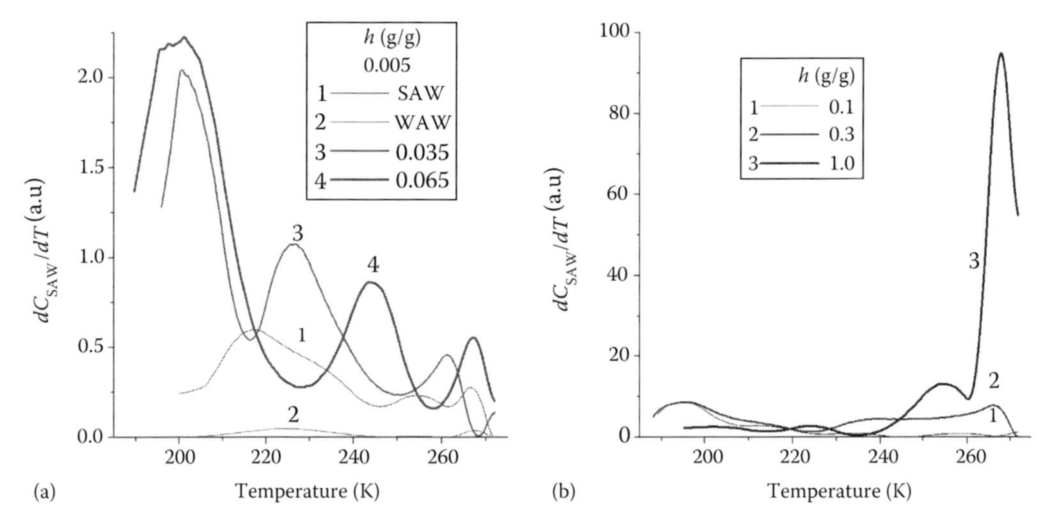

FIGURE 1.48 Temperature derivatives of the SAW content for (a) weakly and (b) strongly hydrated silica samples, dC_{WAW}/dT is shown for a sample at $h = 0.005$ g/g (a, curve 2). (Adapted from *Appl. Surf. Sci.*, 258, Gun'ko, V.M., Turov, V.V., Bogatyrev, V.M. et al., The influence of pre-adsorbed water on adsorption of methane on fumed and nanoporous silicas, 1306–1316, 2011c. Copyright 2011, with permission from Elsevier.)

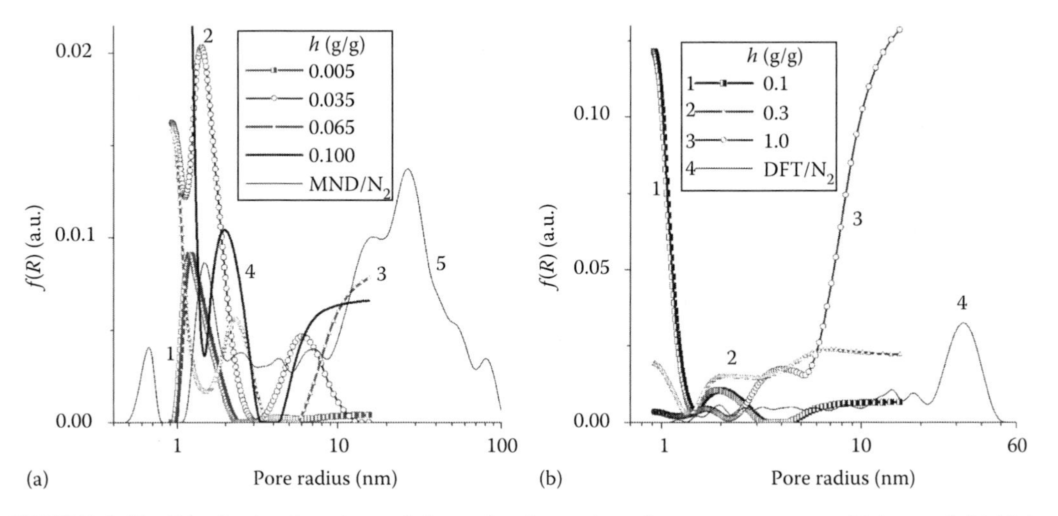

FIGURE 1.49 Distribution functions of sizes of unfrozen bound water structures at (a) low and (b) high hydration and the pore size distribution of A-300 calculated using modified Nguyen–Do equation with the model of cylindrical pores (a, curve 5) and DFT method with the model of voids between spherical particles (b, curve 4). (Adapted from *Appl. Surf. Sci.*, 258, Gun'ko, V.M., Turov, V.V., Bogatyrev, V.M. et al., The influence of pre-adsorbed water on adsorption of methane on fumed and nanoporous silicas, 1306–1316, 2011c. Copyright 2011, with permission from Elsevier.)

temperature $T > 240$ K but this occurs in the hydrated powder at lower temperature $T > 200$ K. Diminution of the Gibbs free energy (ΔG^s) of strongly bound first layer of water is greater for the hydrated powder than that for silica in the aqueous medium (Table 1.14). In the methane medium, the adsorption of methane causes certain diminution of the amounts of SBW (Table 1.14, C^s_{uw}). A similar effect is observed for weakly hydrated silica in liquid $CDCl_3$. The difference of the powder and the concentrated suspension of silica is due to the character of particle–particle interactions since their aggregation is much stronger in the powder (tight contacts between

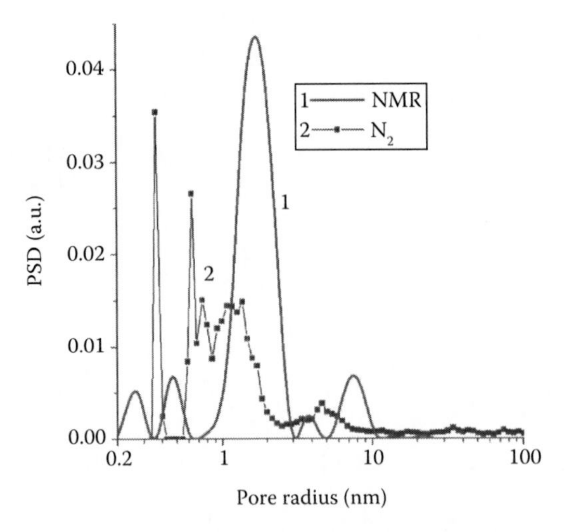

FIGURE 1.50 Distribution function of sizes of unfrozen water structures (curve 1, h=0.15 g/g) and pore size distribution of silica 200DF (2). (Adapted from *Appl. Surf. Sci.*, 258, Gun'ko, V.M., Turov, V.V., Bogatyrev, V.M. et al., The influence of pre-adsorbed water on adsorption of methane on fumed and nanoporous silicas, 1306–1316, 2011c. Copyright 2011, with permission from Elsevier.)

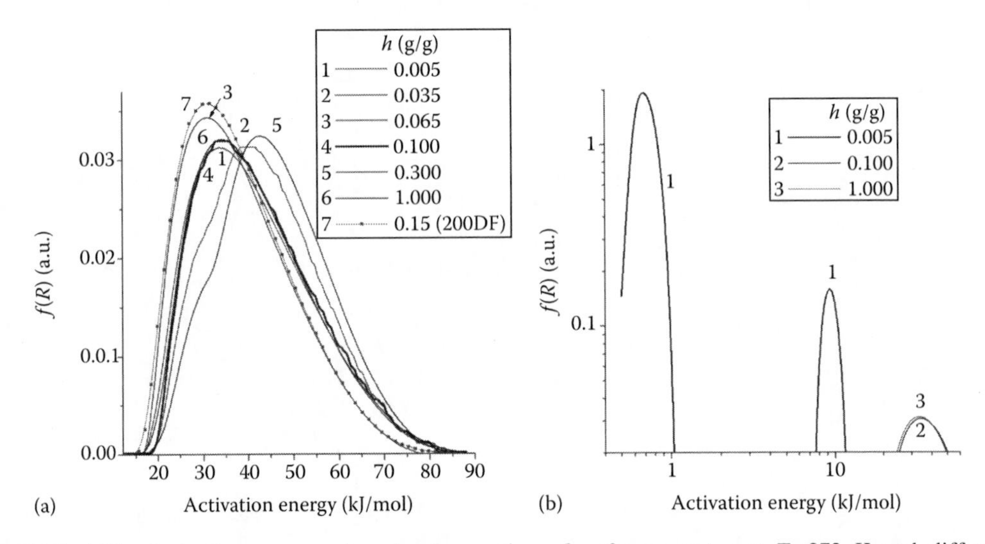

FIGURE 1.51 Activation energy of molecular motion of unfrozen water at T<273 K and different hydration of A-300 and 200DF with coadsorbed methane. (Adapted from *Appl. Surf. Sci.*, 258, Gun'ko, V.M., Turov, V.V., Bogatyrev, V.M. et al., The influence of pre-adsorbed water on adsorption of methane on fumed and nanoporous silicas, 1306–1316, 2011c. Copyright 2011, with permission from Elsevier.)

adjacent particles up to siloxane bonding) than in the suspension (particles are distant one from another due to electrostatic repulsive interactions; Figure 1.52a).

The size distributions of water clusters (SDWCs) and domains (Figure 1.52b), calculated with the Gibbs–Thomson equation for the freezing point depression, show that their average size is much larger for the suspension (gel) than that for the hydrated powder. The minimal size is 0.6 and 1.6 nm for the powder and the suspension, respectively. All surface silanols interact with water molecules in the suspension (gel). Therefore, low amounts of unfrozen water at low temperatures suggest that bound water molecules participate in the formation of ice (amorphous ice; Gun'ko et al. 2009d, Chaplin 2011). Unfrozen water registered as SBW can locate in voids between both silica particles

TABLE 1.14
Thermodynamic Characteristics of Water Bound to Silica OX-50 Particles in Different Media

Medium	h (g/g)	C^{s}_{uw} (mg/g)	C^{w}_{uw} (mg/g)	ΔG^{s} (kJ/mol)	ΔG^{w} (kJ/mol)	γ_S (J/g)
Water (suspension)	0.82	17	387	−1.6	−0.6	4.4
Water (gel)	0.67	17	287	−1.6	−0.6	3.2
Air	0.02	10.5	9.5	−3.2	−1.2	1.2
$CDCl_3$	0.02	8	12	−2.7	−0.7	0.8
CH_4	0.10	9	91	−3.2	−0.4	1.8

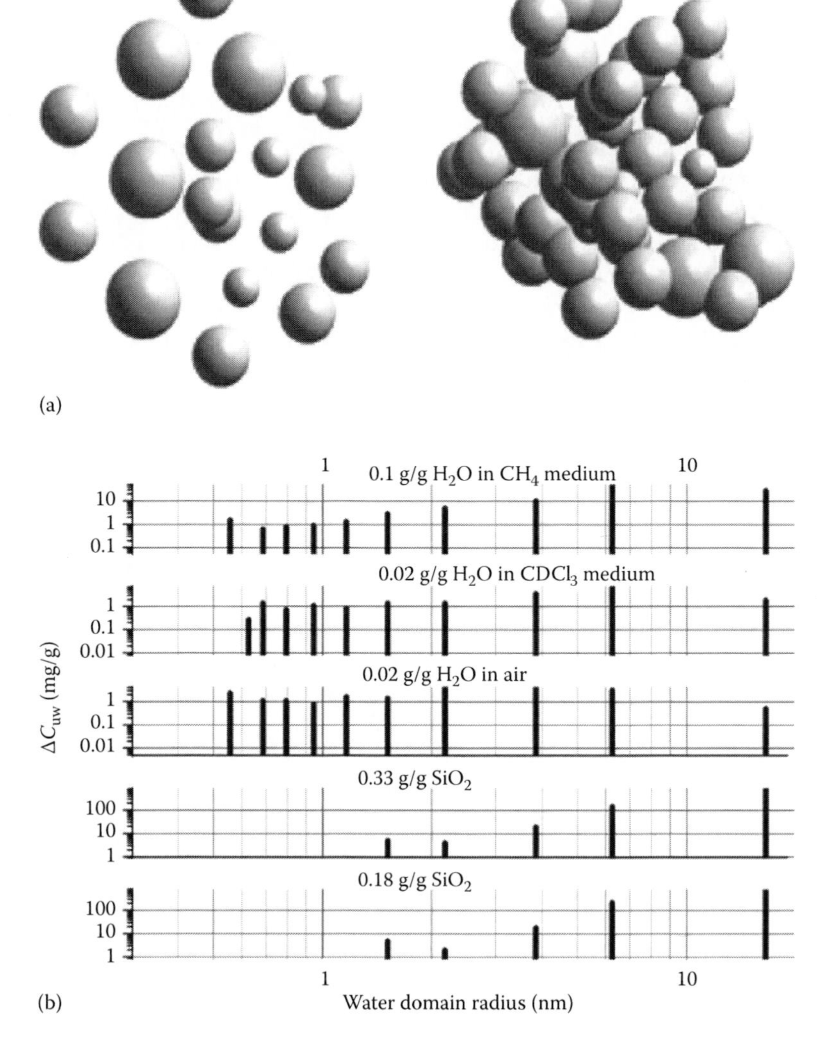

FIGURE 1.52 (a) Scheme of silica nanoparticle location in the suspension and the hydrated powder of nanosilica OX-50 and (b) size distributions of clusters and domains of bound water.

and ice crystallites. In the case of gel, the average distance between particles diminishes and, therefore, the amounts of weakly bound water decrease (Table 1.14, C^{w}_{uw}). The difference in the values of free surface energy (γ_S) for the suspension and the gel is determined by the difference in the particle–particle interactions in these systems. The γ_S value for the OX-50 suspension is small in comparison with that for nanosilicas with much larger S_{BET} values because a small S_{BET} value causes low amounts of bound water per gram of silica.

Thus, the difference in the hydration characteristics of silica OX-50 in the aqueous suspension and hydrated powder in different media depends on the particle–particle interactions (distances between particles) and the textural porosity in the powders. Nonpolar (methane) or weakly polar (chloroform) coadsorbates can reduce interaction of water with the silica surface because water–water and water–silica interactions are much stronger than that of water–organics. Therefore, rearrangement of the system organization leads to an increase in the size of water domains with parallel diminution of the contact area between water–organics and water–silica but with increasing silica–organics contact area.

Thus, to enhance the adsorption of methane onto adsorbents with preadsorbed water at standard pressure and $T < 280$ K, this water should exhibit maximal clustering. The amount of clustered water should be equal to a value characteristic of a given adsorbent, to maximize the contribution of narrow nanopores appropriate for the adsorption of methane.

1.1.4 Competitive Adsorption of Water and Low-Molecular-Weight Organics

Typically, different adsorbates present at the interfaces of the real systems that cause competitive adsorption at the most active surface sites to reduce free energy (G) of the system; however, the $T\Delta S$ term in ΔG can be destabilizing (Adamson and Gast 1997, Dill and Bromberg 2003). The adsorption of polar and nonpolar adsorbates can occur onto different sites, especially in the case of a mosaic hydrophilic/hydrophobic (i.e., polar/nonpolar) surface. Therefore, at low coverage values, the competitive adsorption effects at such a surface can be negligible. Investigations of the competitive adsorption are of importance because the adsorbent selectivity, concentration, separation, filtration, heterogeneous catalysis, biocompatibility, etc. depend on occurring competitive adsorption at the interfaces (McGuire and Suffet 1983, Adamson and Gast 1997, Do 1998, Schwarz and Contescu 1999, Ding 2002).

One of the important trends in surface science deals with features of chemical and physical processes in nanoconfined space (Pool and Owens 2003). In such space, adsorbed substances can lose homogeneity due to clusterization and strong dependence of the properties on clusters localization and solid surface characteristics. Structural heterogeneity of adsorbates can be especially great for strongly associated liquids such as water. The thickness of an interfacial water layer disturbed by surface forces can reach 10 or more molecular layers (Derjaguin et al. 1987, Israelachvili 1994). Water in nanoconfined space, especially with mosaic hydrophilic/hydrophobic walls, has the hydrogen bond network strongly different from that of bulk water. Clusters of hexagonal ice with four hydrogen bonds per molecule can be considered as a certain model of clustered liquid water (Chaplin 2011). However, thermal motions of molecules at ambient conditions or proton tunneling at low temperatures can destroy a portion of the hydrogen bonds, and certain water molecules can be in structural voids without the hydrogen bonds with neighboring molecules (interstitial water). This results in greater density of liquid water than that of ice and in decrease of the average number of the hydrogenous bonds per molecule (<4), despite their larger coordination number (~4.5) (Chaplin 2011). Clustered structure of water confined in narrow pores undergoes strong changes because of inconsistency in the positions of adsorption sites (e.g., surface hydroxyls) and hydroxyls of water molecules in the network characteristic of ice Ih. This results in additional structural and energetic heterogeneity of bound water and the formation of amorphous ice during freezing-out of bound water. Coadsorbates such as polar and nonpolar organics are additional disturbing factors at the interfaces.

Narrow voids (i.e., nanopores and narrow mesopores) between adjacent primary silica nanoparticles are the main adsorption places for water molecules. However, the adsorption of water onto nanosilicas from air is relatively low (1–7 wt%) and depends on features of aggregation of nanoparticles (Degussa 1997, Gun'ko et al. 2001a, 2009a, Mueller et al. 2004). An increase in water amounts by wetting nanosilica can lead to strong changes in the structure and characteristics of wetted powder materials or aqueous suspensions (Gun'ko et al. 2005d, 2009d). Additionally, the characteristics of water bound to nanosilica can be changed on coadsorption with organic solvents, low-molecular organics, polymers, etc. (Gun'ko et al. 2005d, 2009d). These effects depend on many factors; therefore, sometimes it is difficult to interpret the observed results. In this chapter, the influence of the hydration degree and coadsorption of different organic solvents on the characteristics of water bound to nanosilica, characterized by significant aggregation of primary particles, will be analyzed using LT ^{1}H NMR spectroscopy with layer-by-layer freezing-out of bound water in static samples and NMR cryoporometry.

Standard nanosilica A-380 (specific surface area $S_{BET}=378$ m^2/g) was heated at 673 K for several hours before the measurements to remove organic adsorbates and residual HCl. Polar and weakly nonpolar deuterated (D) solvents, CD_3CN, $(CD_3)_2SO$, and $CDCl_3$ (Aldrich, qualification "for NMR spectroscopy" at content of main isotope of 99.5%), soluble or practically insoluble in water, were used in the deuterium form to avoid their contribution to the ^{1}H NMR signal intensity of unfrozen interfacial water at $T<273$ K. Bidistilled water was used here.

Samples (Table 1.15) at various hydration, $h=0.07$–1 g of water per gram of dry silica and coadsorbed solvents, $CDCl_3$ (0.5–9 g/g), CD_3CN (1 g/g), and DMSO (0.5–1 g/g), were studied using the ^{1}H NMR method. Total amounts of water and organic solvents were much smaller than the empty volume (V_{em}) in the dry silica powder because bulk density $\rho_b \approx 0.05$ g/cm^3 provides $V_{em} \approx 20$ cm^3/g. Therefore, a liquid phase was absent in all studied samples representing wetted powders. Before the measurements, samples were equilibrated at room temperature for 20–60 min.

Dry powder of A-380 can be characterized by certain structural hierarchy of particles starting from primary particles (5–15 nm in size, average diameter 7.2 nm), aggregates of primary particles, and agglomerates of aggregates (see Section 1.1.1). Adsorbed water at low amounts (<20 wt%) localizes mainly in the zone of contacts between adjacent nanoparticles (Section 1.1.2). At higher amounts, adsorbed water can form continuous layers covering whole nanoparticles. Consequently, for two samples at $h=0.5$ and 1 g/g, continuous water layers should be formed, but for other samples at $h=0.07$, 0.08, and 0.15 g/g, island adsorption of water could only form partial coverage of the surface.

Despite this structural difference, water bound to nanosilica gives a single ^{1}H NMR signal at $\delta_H=4$–5 ppm (Figure 1.53a), that is, it is SAW, whose intensity decreases with lowering temperature at $T<250$ K due to partial freezing of water. The δ_H value decreases with elevating temperature because of the influence of thermal motion of water molecules on the hydrogen bond network structure. There are fractions of SBW and WBW in SAW but WAW is absent. Addition of organic solvents to A-380/water (Figure 1.53) slightly affects the shape and the position of the water signal. A weak signal of the CHD_2 groups of admixtures in CD_3CN and DMSO is observed at $\delta_H=2$–2.5 ppm. Coadsorbed DMSO (in a mixture with $CDCl_3$) more strongly affects freezing-out of water since even at 210 K, an intensive signal of unfrozen water is observed (Figure 1.53d).

To study features of the influence of organic solvents on the characteristics of bound water located in voids between adjacent silica nanoparticles in their aggregates on the basis of the temperature dependences of changes in the integral intensities of the ^{1}H NMR signals of unfrozen water, the relationships between changes in Gibbs free energy on water adsorption and the amounts of unfrozen bound water were determined (Figure 1.54). At relatively high degree of hydration of A-380 (Figure 1.54a and b), the influence of organic solvents is weaker than that at low hydration (Figure 1.54c through e). For the former, there are almost vertical (related to SBW) and almost horizontal (WBW) sections in the $\Delta G(C_{uw})$ graphs. The maximal amount of SBW is 250 mg/g, which corresponds to approximately a third of the volume of pores (voids between adjacent primary

TABLE 1.15
Characteristics of Bound Water in Hydrated Powders of Nanosilica A-380 with the Presence of Organic Solvents

h (g/g)	Organics (g/g)	C_{SBW} (mg/g)	C_{WBW} (mg/g)	$-\Delta G^S$ (kJ/mol)	γ_S (J/g)	S (m²/g)	S_{nano} (m²/g)	S_{meso} (m²/g)	S_{macro} (m²/g)	V_{nano} (cm³/g)	V_{meso} (cm³/g)	V_{macro} (cm³/g)
1	—	250	750	2.52	30.2	116	14	102	0.4	0.007	0.976	0.007
	1.0 $CDCl_3$	150	850	2.87	23.0	150	35	115	0.5	0.016	0.966	0.008
	2.0 $CDCl_3$	170	830	2.37	27.0	130	0	130	0.1	0	0.996	0.002
	6.0 $CDCl_3$	30	970	2.57	21.6	121	6	115	0.3	0.003	0.985	0.005
0.5	—	125	375	2.84	16.5	87	43	43	0.5	0.020	0.451	0.007
	2.0 $CDCl_3$	100	400	2.59	15.8	56	23	32	0.8	0.011	0.449	0.014
	8.0 $CDCl_3$	75	425	3.16	15.2	53	19	34	0.7	0.009	0.458	0.011
0.15	—	100	50	2.78	11.1	78	67	11	0.1	0.029	0.117	0.001
	3 $CDCl_3$	100	50	2.43	9.3	40	28	12	0.1	0.013	0.134	0.001
	6 $CDCl_3$	100	50	2.70	8.8	10	0	10	0.1	0	0.145	0.002
	0.5 DMSO	100	50	3.37	14.7	181	177	4	0.4	0.073	0.055	0.006
	0.5 DMSO + 1.5 $CDCl_3$	145	5	3.35	21.4	227	221	6	0	0.094	0.056	0
0.08	—	60	20	3.48	7.8	75	71	4	0.2	0.029	0.043	0.002
	2.0 $CDCl_3$	55	25	2.47	4.6	5	1	4	0.1	0	0.074	0.002
	1.0 DMSO	70	10	3.92	12.3	178	177	1	0	0.065	0.013	0
	1.0 DMSO + 2.0 $CDCl_3$	70	10	3.76	10.7	116	114	2	0	0.046	0.029	0
0.07	1.0 CD_3CN + 1.0 $CDCl_3$	35	35	2.42	3.3	17	0	17	0	0	0.059	0
	1.0 CD_3CN + 0.5 $CDCl_3$	35	35	2.32	3.1	15	0	15	0	0	0.054	0
	1.0 CD_3CN + 2.0 $CDCl_3$	30	40	2.14	2.9	15	0	15	0	0	0.056	0

Note: S, S_{nano}, S_{meso} and S_{macro} are the surface area total, nano- (radius $R < 1$ nm), meso- ($1 < R < 25$ nm) and macropores ($R > 25$ nm), respectively, of silica in contact with unfrozen water, V_{nano}, V_{meso}, and V_{macro} are the pore volume of nano-, meso-, and macropores, respectively.

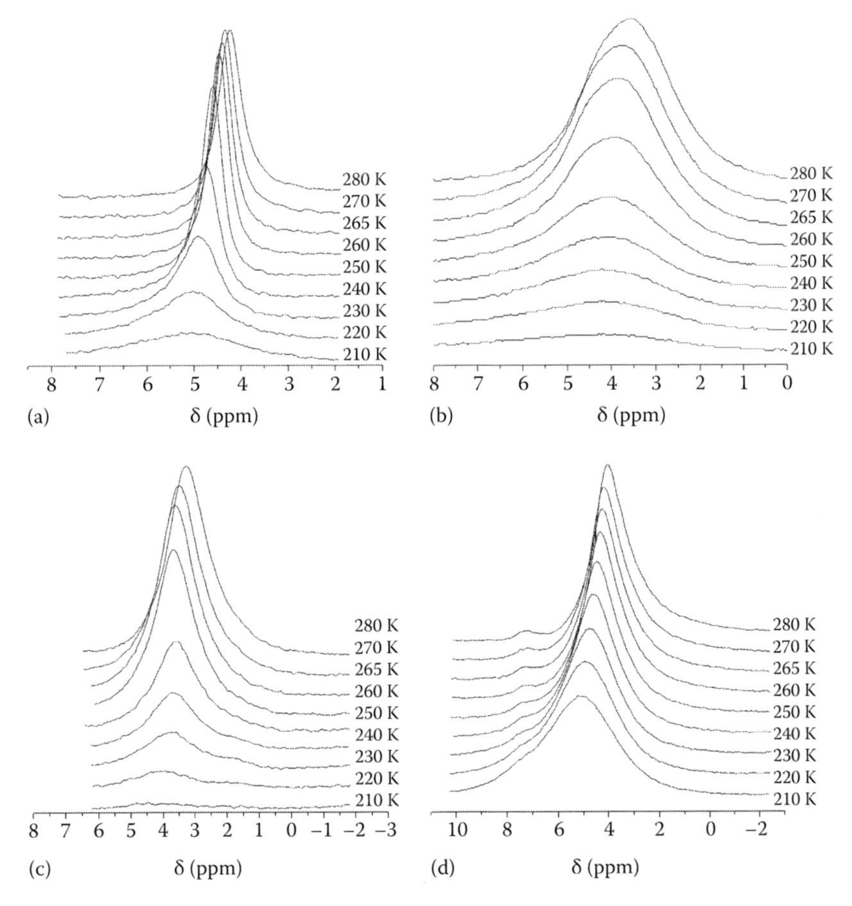

FIGURE 1.53 ^{1}H NMR spectra of water bound to nanosilica A-380 (a) alone at $h=0.08$ g/g and with addition of (b) $CDCl_3$ (3 g/g) at $h=0.15$ g/g, (c) CD_3CN (1 g/g)+$CDCl_3$ (1 g/g), $h=0.07$ g/g and (d) DMSO (0.5 g/g) +$CDCl_3$ (1 g/g), $h=0.15$ g/g recorded at different temperatures.

particles in aggregates) filled by nitrogen at 77.4 K. Addition of chloroform alone or in a mixture with DMSO leads to stronger changes in the characteristics of bound water at lower hydration (Table 1.15; Figure 1.54). The effects of chloroform in a mixture with acetonitrile are smaller than that of $CDCl_3$+DMSO. These effects could be explained by the difference in the solubility of water in these organics individual and in mixtures. Coadsorbed chloroform decreases the free surface energy (Table 1.15; γ_S) of bound water at different hydration. The free surface energy normalized to $h=1$ g/g as $\gamma_s^* = \gamma_s/h$ increases with decreasing h value (Figure 1.55) because of a decrease in the amounts of WBW. There is a tendency of a decrease in the γ_s^* value with increasing content of chloroform (Figure 1.55a).

This can be explained by the displacement of a portion of water from narrow voids to broader ones, and this effect is stronger at lower hydration (Figure 1.56) when bound water does not form continuous adsorption layers and coadsorbed organics can easily contact the silica surface. Such electron-donor solvents as CD_3CN and DMSO are well dissolved in water and chloroform. However, chloroform–DMSO (CD_3CN) mixtures can stratify on addition of a small quantity of water to form layers enriched by chloroform and an electron donor. To avoid this effect, such quantities of mixtures were added to the silica powder to prevent the formation of a liquid phase. CD_3CN and DMSO differently affect the interaction of water with the silica surface (Table 1.15, Figures 1.54 through 1.56) since CD_3CN enhances but DMSO diminishes the free surface energy value of bound water.

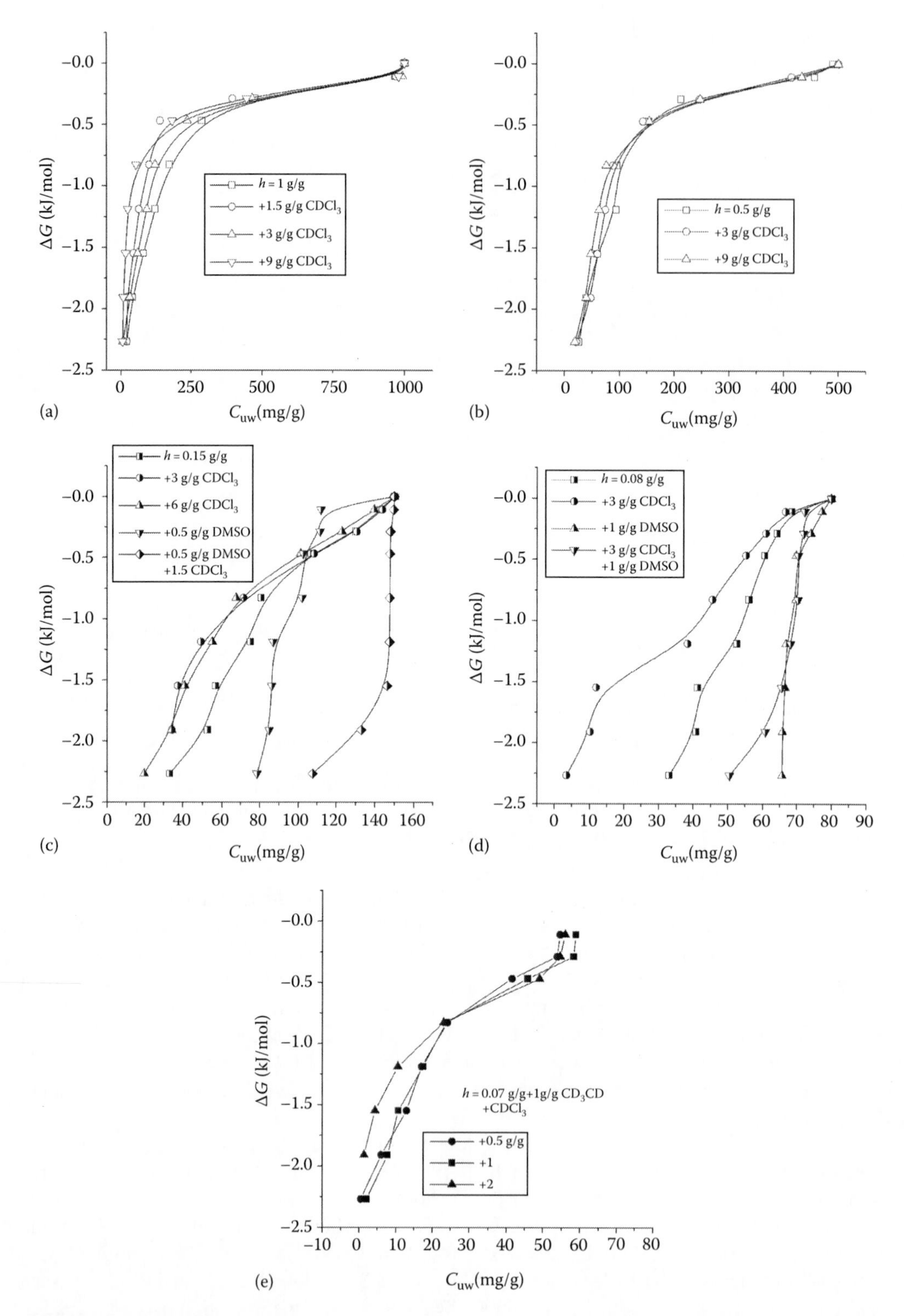

FIGURE 1.54 Relationships between content of unfrozen water and changes in the Gibbs free energy of unfrozen water in differently hydrated powders of nanosilica A-380 at h=(a) 1, (b) 0.5, (c) 0.15, (d) 0.08, and (c) 0.07 g/g and different amounts of organic solvents.

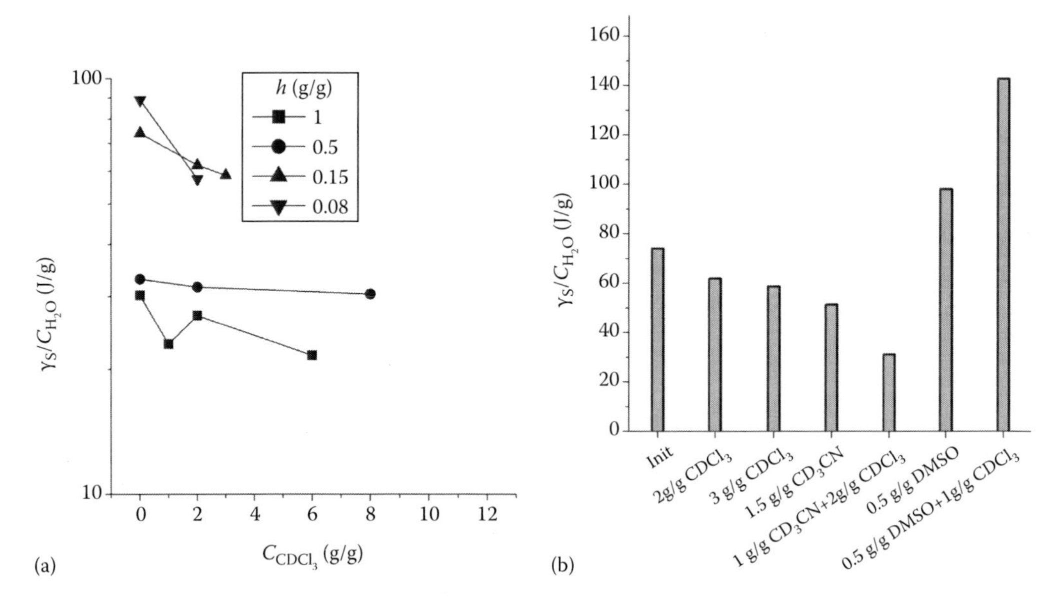

FIGURE 1.55 Normalized free surface energy (a) as a function of the content of chloroform at different hydration of silica A-380 and (b) after addition of $CDCl_3$, CD_3CN, DMSO, and their mixtures at a fixed hydration $h = 0.15$ g/g.

During coadsorption of water and studied organics, there are several factors affecting changes in the structure (clusterization) of bound water. If only weakly polar chloroform is coadsorbed with water, then the phase boundary between liquids tends to be minimal. Therefore, if preadsorbed water forms a continuous coverage of the silica surface, the effects of chloroform are minimal. These effects increase if the island coverage with preadsorbed water is realized (Figure 1.54). Chloroform in a mixture with polar organics can be mixed with bound water (Figure 1.57) because DMSO effectively interacting with both polar water and nonpolar chloroform can separate water and chloroform clusters. In other words, there is no tendency to decrease the contact area between water and chloroform at the silica surface because they contact only with DMSO and the silica surface. This is well seen from the structural characteristics of bound water (Table 1.15) since maximal specific surface area (total and of nanopores) of unfrozen water structures is observed for the DMSO + $CDCl_3$ (or DMSO alone). However, even for these cases, these values are much smaller than the S_{BET} value of silica A-380 (378 m^2/g). Notice that $S < S_{BET}$ on the adsorption of water alone at $h = 1$ g/g. This can be explained by structural nonuniformity of bound water (i.e., its clusterization, formation of nanodomains, and absence uniform continuous water layers covering silica nanoparticles) and the absence of contacts between total surface area of silica particles and water.

Thus, water bound to nanosilica does not form uniform layers even at relatively high hydration (1 g of water per 1 g of dry silica). Coadsorption of water with weakly polar chloroform alone or in a mixture with polar DMSO or acetonitrile more strongly affects the structure of water bound to nanosilica A-380 at lower hydration when significant portion of silica surface remains uncovered by water. Maximal clusterization of bound water is observed on the coadsorption with chloroform/DMSO mixture because DMSO can effectively interact with both nonpolar (through dispersion interactions of chloroform with CH_3 groups) and polar (water interaction with the S=O bonds) structures.

Interactions of standard nanosilica A-300 ($S_{BET} \approx 300$ m^2/g, $\rho_b = 0.045$ g/cm^3, which is less aggregated than A-380 and having a smaller pore volume; see Section 1.1.1) with water in the presence of $CDCl_3$, CD_3CN, $(CD_3)_2SO$, DCl (36 wt% in D_2O), or CCl_4 were studied. TMS (0.2 wt%) was added to $CDCl_3$ or CCl_4 to determine the δ_H values using a Varian 400 Mercury spectrometer.

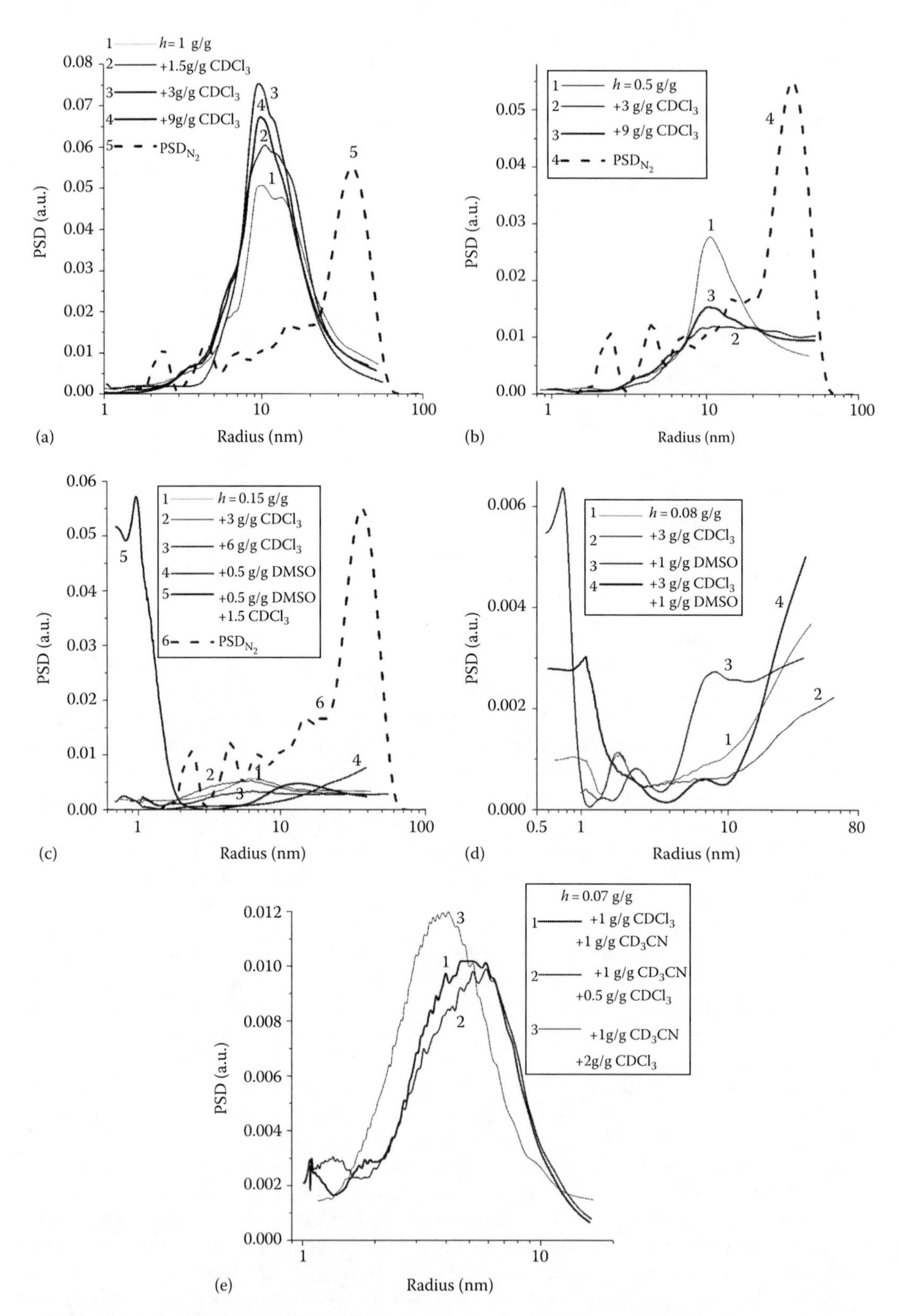

FIGURE 1.56 Size distribution of unfrozen water structures for hydrated A-380 powder at h = (a) 1, (b) 0.5, (c) 0.15, (d) 0.08, and (e) 0.07 g/g and different amounts of organic solvents.

FIGURE 1.57 Displacement of water molecules by chloroform and DMSO from the silica surface (force field optimized geometry).

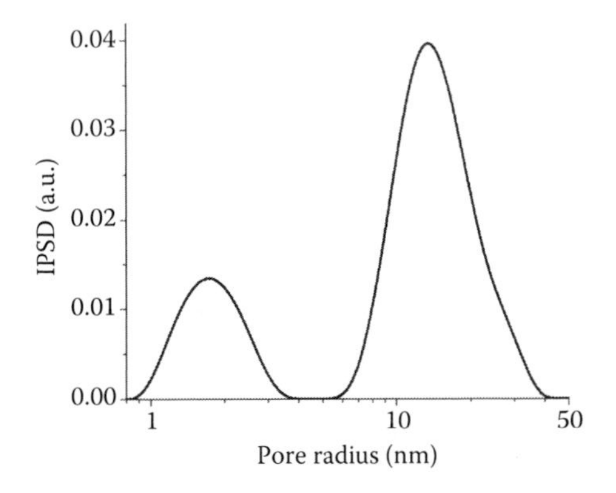

FIGURE 1.58 Incremental PSD of voids between silica nanoparticles in aggregates.

Nanosilica is composed of primary nanoparticles of a spherical shape, which form aggregates (Figure 1.29a) characterized by the textural porosity (Figure 1.58).

WAW (δ_H = 1.5 ppm) and SAW (2.8–4.8 ppm) are observed for weakly (h = 50 mg/g) hydrated nanosilica A-300 in both nonpolar CCl_4 and weakly polar $CDCl_3$ medium (Figure 1.59). SAW and WAW demonstrate typical dependence of δ_H on temperature since it increases or remains constant, respectively.

Addition of 10% or 20% acetonitrile to chloroform enhances contribution of WAW (1.3 ppm) and a portion of water dissolved in CD_3CN (2.5–3 ppm), as well as SAW (3.5–4 ppm; Figure 1.60). CH_3CN as an admixture in CD_3CN gives signal at ~2 ppm. At 290 K, only one broad signal is observed due to increased efficiency of exchange processes with increasing temperature. In pure acetonitrile (Figure 1.60c), two signals of water (2.5–3 ppm) and CH_3CN (2 ppm) are observed. At T < 240 K, intensity of both signals strongly decreases due to freezing of acetonitrile.

Addition of DCl transformed into HCl due to proton exchange reaction results in the downfield shift (Figure 1.60d) that is a typical effect for strong acids. This effect is stronger at lower temperatures because of concentrating acid since ice is much pure than the solution. However, even at 200 K,

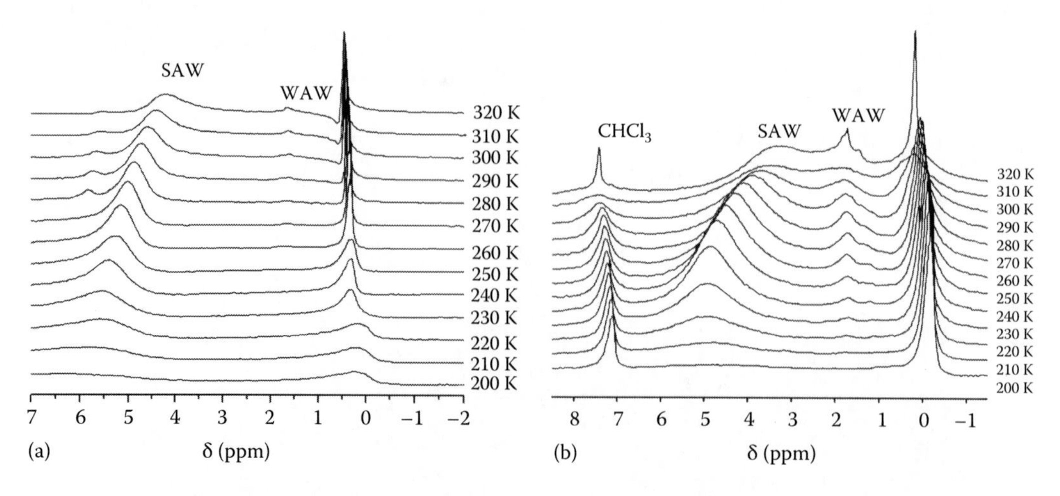

FIGURE 1.59 ^{1}H NMR spectra of water (h = 50 mg/g) bound to A-300 in (a) CCl_4 or (b) $CDCl_3$ medium.

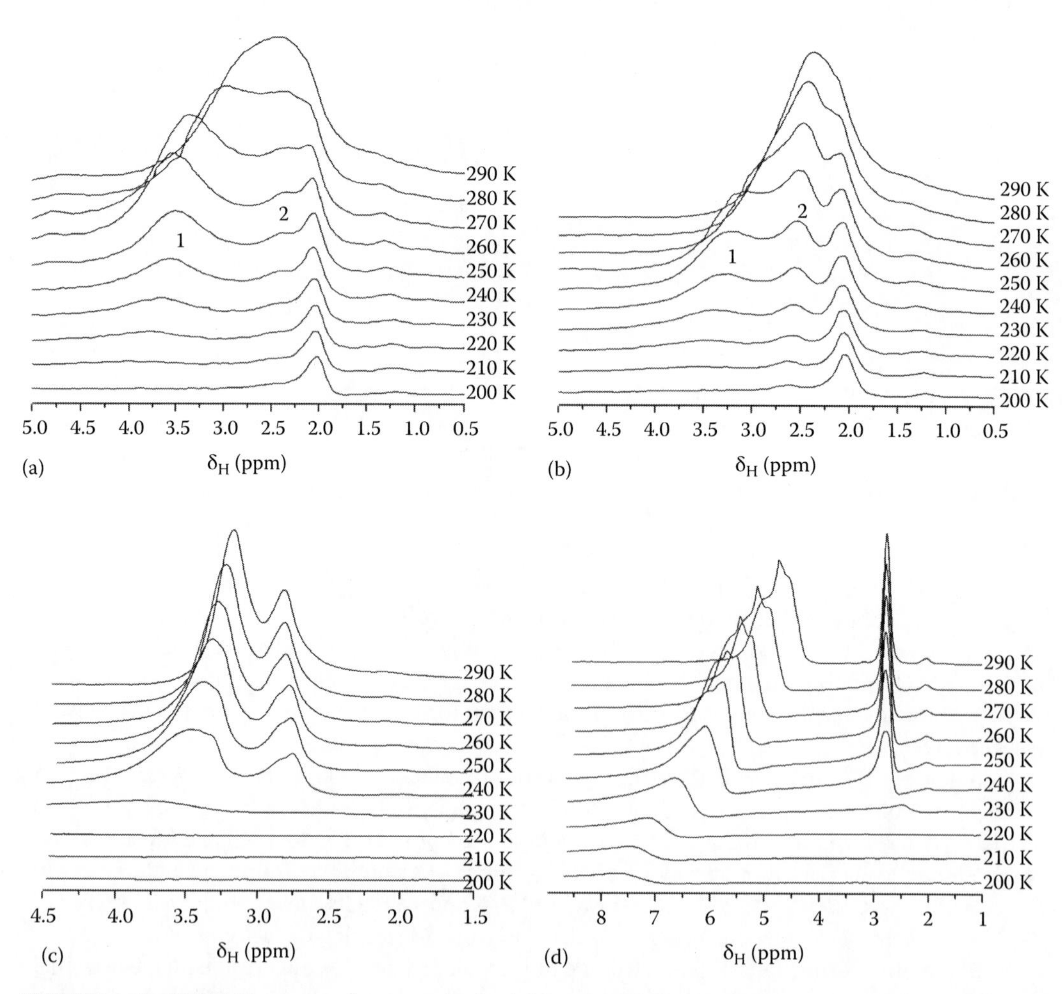

FIGURE 1.60 ^{1}H NMR spectra of water at h = (a–c) 50 mg/g and (d) 88 mg/g bound to A-300 in $CDCl_3$ + (a) 10% or (b) 20% CD_3CN; (c, d) CD_3CN with addition of (d) DCl (44 mg/g).

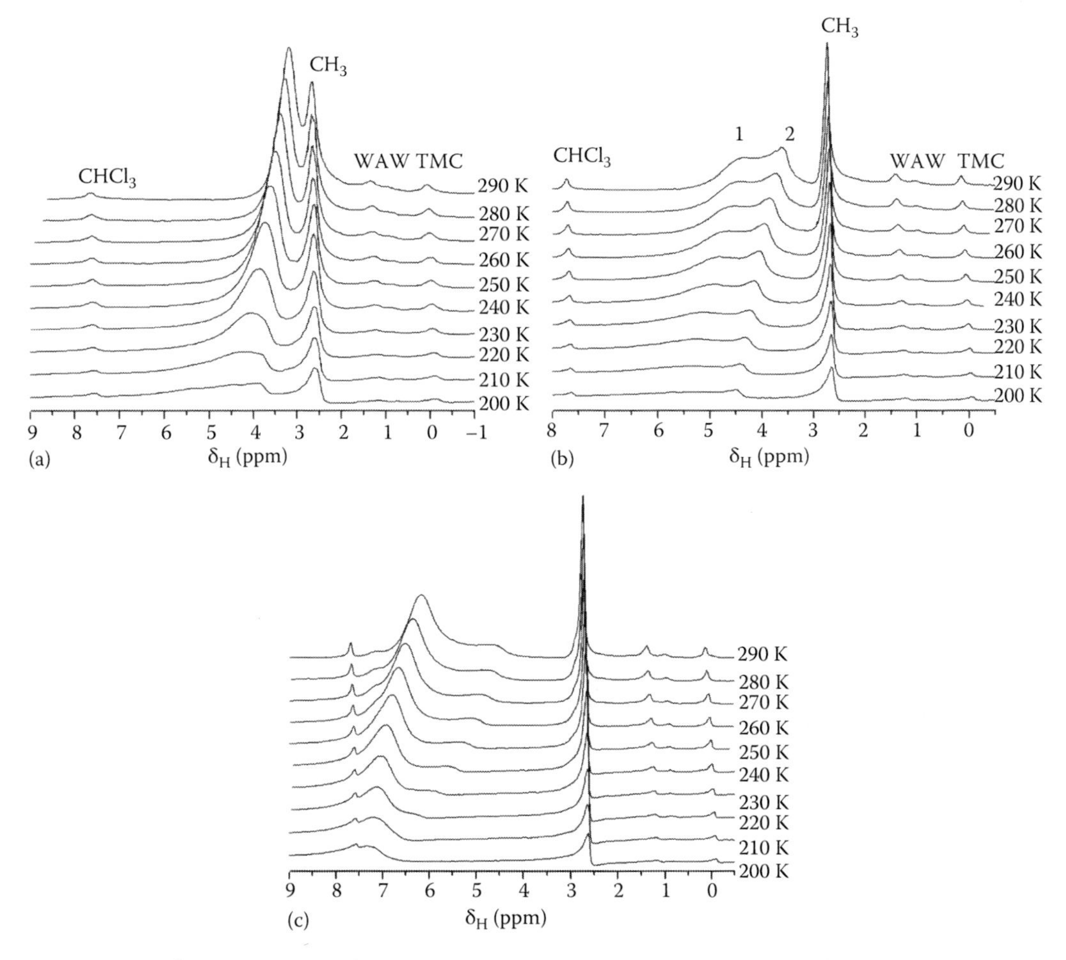

FIGURE 1.61 ^{1}H NMR spectra of water at h = (a) 50, (b) 88, and (c) 110 mg/g with addition of HCl (b) 44 and (c) 60 mg/g bound to A-300 in $CDCl_3$ + 15% $(CD_3)_2SO$ medium.

the δ_H values are lower (6–7 ppm) than that for concentrated HCl solution (9.5 ppm). This result can be explained by lower activity of bound water (there is a low amount of water in the system) than bulk water. In other words, HCl/DCl forms clusters with low content of water and water dissolves a small amount of the acid (see Section 1.1.7).

Addition of DMSO, a stronger electron donor than acetonitrile, to chloroform (Figure 1.61) gives the ^{1}H NMR spectra shape (Figure 1.61a) similar to that with acetonitrile (Figure 1.60c).

However, the δ_H values for water are larger (3.5–4.5 ppm, SAW) in the $CDCl_3$–DMSO medium than in CD_3CN. A weak signal of WAW is observed at 1.5 ppm. Addition of DCl/D_2O (Figure 1.61b) causes splitting of the water signal into two signals with a downfield shift with lowering temperature. With increasing contents of water and acid (Figure 1.61c), the δ_H values increase because the activity of water as a solvent increases and it can dissolve a greater amount of HCl/DCl.

The amounts of SAW are nonlinear functions of temperature (Figure 1.32a) because of enhanced thawing of ice and changes in the content ratio for different mobile unfrozen components. In CCl_4 and $CDCl_3$ media, the amounts of WAW (δ_H = 1.3 ppm) increase with increasing temperature. In more polar solvents (acetonitrile, DMSO), contributions of SAW and water dissolved in organic solvents change with increasing temperature.

Strongly acidic or basic compounds can affect interaction of water with nanosilica (Figure 1.62). This occurs due to several processes: (i) an increase in surface charge density of silica with

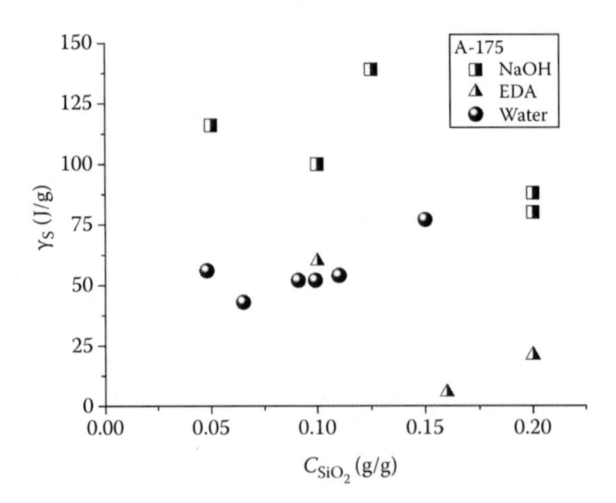

FIGURE 1.62 Changes in the free surface energy in the aqueous suspensions of nanosilica A-175 alone or with addition of NaOH or ethylenediamine (EDA) as a function of nanosilica concentration (according to ^{1}H NMR data).

increasing pH value, (ii) changes in the structure of the EDL, (iii) competitive adsorption of water molecules and solutes, and (iv) changes in the organization of secondary particles.

To describe the competitive adsorption of two compounds (A and B) under isothermal conditions, an additive Langmuir model can be used. The adsorption rate of A molecules (r_A) depends on partial pressure p_A and a number of unoccupied adsorption sites $(1-\theta_A-\theta_B)$:

$$r_{A,a} \propto (1-\theta_A-\theta_B)p_A \tag{1.12}$$

Desorption rate $r_{A,d} \propto \theta_A$ is equal to the adsorption rate upon equilibrium conditions:

$$(1-\theta_A-\theta_B)b_A p_A = \theta_A, \tag{1.13}$$

where b_A is the equation parameter dependent on free energy of adsorption of compound A. A similar equation is for the adsorption of compound B. From these two equations, we can obtain equation for the adsorption of A:

$$\theta_A = \frac{b_A p_A}{1+b_A p_A + b_B p_B}. \tag{1.14}$$

and a similar equation for B. This equation can be generalized for N adsorbates:

$$\theta_i = \frac{b_i p_i}{1+\sum_{j=1}^{N} b_j p_j}. \tag{1.15}$$

The b_i parameters are functions of the structural and adsorption characteristics of adsorbent. For known PSD $f(R)$, Equation 1.15 can be transformed into integral equation for the ith compound:

$$\langle\theta_i\rangle = \int_{R_{min}}^{R_{max}} \frac{b_i(R)p_i}{1+\sum_{j=1}^{N} b_j(R)p_j} f(R)dR. \tag{1.16}$$

Since b_i is a function of the free energy of adsorption ΔG dependent on the PSD and lateral interactions (z) dependent of the partial pressure (p_i), and $\phi_i(\Delta G)$ describes the distribution function for the ith adsorbate, we can write the generalized equation

$$\langle \theta_i \rangle = \int_{R_{\min}}^{R_{\max}} \int_{\varepsilon_{\min}}^{\varepsilon_{\max}} \frac{p_i \exp\left(\frac{-(\Delta G_i(R) + z_i(R))}{R_g T} \right)}{1 + \sum_{j=1}^{N} p_j \exp\left(\frac{-(\Delta G_j(R) + z_j(R))}{R_g T} \right)} \phi_i(\Delta G) f(R) d(\Delta G) dR. \quad (1.17)$$

Equation 1.16 can be written as a set of equations for adsorbates with their increasing molecular sizes (Wang and Do 1999):

$$\langle \theta_i \rangle = \int_{a_i}^{a_{i+1}} \theta_i^{(i)}(p_1, p_2, \ldots, p_i; R) f(R) dR + \int_{a_{i+1}}^{a_{i+2}} \theta_i^{(i+1)}(p_1, p_2, \ldots, p_{i+1}; R) f(R) dR + \cdots$$
$$\int_{a_N}^{r_{\max}} \theta_i^{(i)}(p_1, p_2, \ldots, p_i, \ldots p_N; R) f(R) dR \quad (1.18)$$

where $i = 1, 2, \ldots, N$, a_i is the minimal pore half-width in which the ith adsorbate can penetrate, and

$$\theta_i^M(p_1, p_2, \ldots, p_M; R) = \frac{b_i(R) p_i}{1 + \sum_{j=1}^{M} b_j(R) p_j}, \quad (1.19)$$

where $M \leq N$, $M = k, k+1, \ldots, N$. The first term in Equation 1.18 corresponds to contribution of the ith compound in the adsorption into pores accessible from the first to the ith compounds, and the last term corresponds to the adsorption into pores accessible for all adsorbates.

Excess free energy of binary adsorption is (Myers and Siperstein 2001, Siperstein and Myers 2001)

$$g^e = \sum_{i=1}^{N_C} \sum_{j=1}^{N_C} (A_{ij} + B_{ij} T) x_i x_j (1 - \exp(-C_{ij} \Psi)), \quad (1.20)$$

where $j > i$, A_{ij}, B_{ij}, C_{ij} are equation parameters, x_i and x_j are the molar fractions of adsorbates,

$$\Psi = -\frac{\Phi}{R_g T} = \int_0^p a \, d \ln f = \int_0^a \left(\frac{\partial \ln f}{\partial \ln a} \right)_T da \quad (1.21)$$

Φ is the adsorption potential, f is the fugacity, a is the adsorption value, and differential enthalpy of desorption (isosteric heat) for the ith component for multicomponent adsorption of real gas and nonideal adsorbed mixture

$$\Delta\bar{h}_i = \Delta h_i^0 + R_g T^2 \left(\frac{\partial \ln \gamma_i}{\partial T}\right)_{\Psi,x} + \left[\frac{1}{a_i^0} + \left(\frac{\partial \ln \gamma_i}{\partial \Psi}\right)_{T,x}\right] \times \left[\frac{\sum_j x_j G_j^0 a_j^0 (\Delta\bar{h}_j^0 - \Delta h_j^0) + R_g T^2 \left(\frac{\partial a^{-e}}{\partial T}\right)_{\Psi,x}}{\sum_j x_j G_j^0 - \left(\frac{\partial a^{-e}}{\partial \Psi}\right)_{T,x}}\right] \tag{1.22}$$

where Δh_j^0 is the differential enthalpy for individual adsorbates and γ_i is the activity parameter

$$\left(\frac{\partial \ln \gamma_i}{\partial T}\right)_{\Psi,x} = -\frac{A}{R_g T^2}(1 - \exp(-C\Psi))x_j^2 \text{ at } i \neq j \tag{1.23}$$

$$G_j^0 \equiv \frac{1}{(a_j^0)^2}\left(\frac{\partial \ln(a_j^0)^{-e}}{\partial \ln f_j^0}\right)_T . \tag{1.24}$$

For binary mixture,

$$\Delta\bar{h}_1 = \Delta h_1^0 + \frac{1}{a_1^0}\left[\frac{x_1 G_1^0 a_1^0 (\Delta\bar{h}_1^0 - \Delta h_1^0) + x_2 G_2^0 a_2^0 (\Delta\bar{h}_2^0 - \Delta h_2^0)}{x_1 G_1^0 + x_2 G_2^0}\right] \tag{1.25}$$

Comparison (Wang et al. 2000) of calculations of multicomponent adsorption using three models (ideal adsorbed solution theory [IAST], heterogeneous ideal absorbed solution theory [HIAST], and heterogeneous extended Langmuir [HEL]) showed that HIAST can give worse results than IAST depending on the used distribution of adsorption energy. In the HIAST and HEL models, the concentration of adsorbed phase corresponds to the integral

$$C_\mu(k) = \int_0^\infty C_\mu[k, E(k)] F[k, E(k)] dE(k) \tag{1.26}$$

where $F[k, E(k)] = \frac{1}{E_{max}(k) - E_{min}(k)}$ at $E_{min} < E < E_{max}$ and $F = 0$ at other values, $E(k)$ is the energy of local interactions of the kth compound, and $C_\mu[k,E(k)]$ is the local concentration of adsorbed phase on sites with given energy, calculated from IAST for individual component using the Langmuir isotherm (model 1) or expanded multicomponent Langmuir equation (model 2). Local Langmuir equation for individual k-component is

$$C_\mu[k, E(k)] = C_{\mu s}(k) \frac{b_0(k) C(k) \exp\left(\frac{E(k)}{R_g T}\right)}{1 + b_0(k) C(k) \exp\left(\frac{E(k)}{R_g T}\right)}, \tag{1.27}$$

where

b_0 is affinity constant at zero energy level
C is the concentration of the gaseous phase

A similar equation was used to describe clustered adsorption of water (Chapter 10).

The dynamic adsorption of gas/vapor mixtures, for example, toxic gases or vapors from wet air, is of importance from a practical point of view. Total adsorption of an organic compound onto an adsorbent or a filter for time t can be described by equation (Gun'ko et al. 2006c)

$$C_\Sigma(t) = A\int_0^{-\Delta G_{max}}\int_0^t f(\Delta G)\exp\left(-\frac{\Delta G}{RT}\right)\exp\left[\frac{v_E(t-t_b)}{\alpha L V_p}\right]dtd(-\Delta G), \qquad (1.28)$$

where

A and α are equation constants
$f(\Delta G)$ is the distribution function of free energy of adsorption ΔG
v_L if the flow velocity
L is the filter (bed) thickness
V_p is pore volume
t_b is the breakthrough time at minimal critical concentration

Overall adsorption can be written with consideration for the distribution function of the adsorption rate constant $f(\beta_e)$:

$$C_\Sigma(t) = A\int_{\beta_{min}}^{\beta_{max}}\int_0^t f(\beta_e)\exp\left[-\frac{\beta_e}{v_L}\left(L-\frac{c_0 v_L t}{\alpha[c_0-c(t)]}\right)\right]dtd\beta_e, \qquad (1.29)$$

where

β_e is the adsorption rate constant
c_0 and c are the initial and current (after bed) concentration of organic compound

Equation constants depend on preadsorption of water and the RH value, and the competitive adsorption of water and organics results in worse efficiency of the filter. It should be noted that preadsorbed water provides stronger negative effects of subsequent adsorption of organics than water vapor present in the gas flow in mixture with organics (Gun'ko et al. 2006c).

Notice that several aspects of competitive adsorption of water and organics onto different adsorbents are analyzed in this book. Here, we could note that the main results of this competition is (i) displacement of water from narrow pores into larger or narrower pores; (ii) changes in contributions of SAW and WAW, as well as SBW and WBW that strongly depend on the water amounts, adsorbent structure, and polarity of a coadsorbate; and (iii) changes in clusterization of bound water. These results allow us to explain cryopreservation (DMSO, glucose, etc.) or ice-forming effects of different organics.

1.1.5 Adsorption of Ice Nucleators

The behavior of interfacial water at a solid surface is affected by the morphology of solid particles, their porosity, and surface area and chemistry as well as by the properties of coadsorbates (DeMott 1995, Harrison and van Grieken 1998, Hass et al. 1998, Richet and Polian 1998, Scatena et al. 2001, Raviv and Klein 2002, Kuo and Mundy 2004, Chaplin 2011). If the interfacial water is studied at subzero temperatures using ^{1}H NMR method with freezing-out of bulk water (Gun'ko et al. 2005d, 2007f, 2009d), that information about interfacial ice different from bulk ice could be of importance. There are several mechanisms of the formation of interfacial ice that are governed by the effects of deposition, immersion, and condensation freezing, as well as by the topology, the hydrophilicity, and the ice-nucleating ability of a surface (DeMott 1995, Harrison and van Grieken 1998, Gun'ko et al. 2006e). Coadsorbed organics can affect the interaction of solid particles with water at both low and high temperatures. Polar organic compounds possessing ice-nucleating ability can enhance the adsorption of water and the formation of ice nanocrystallites or amorphous ice in the interfacial layer at subzero temperatures. On the other hand, coadsorbed organics can be responsible for changes in desorption of water on heating because of the ordering effect, stimulated changes in the intermolecular interactions, and the displacement of a portion of water from a solid surface with enhanced contribution of SAW.

Organic and inorganic compounds capable of forming dispersion by sublimation (Novakov and Penner 1993, Reid 1997) or liquefied gases, such as CO_2, propane, and nitrogen (Xuecheng and Weimin 1996), have been investigated as ice nucleators. The mechanism of the heterogeneous nucleation of supercooled water has been studied in detail (Masuda and Takahashi 1990, Wagner and Vali 1988). However, the influence of organic compounds coadsorbed onto a surface of micro(nano)particles on the ice nucleation is complex and many details of this phenomenon are not clear. This effect depends on the relationships between the hydrophilicity and the hydrophobicity of a solid surface (or even patches of a mosaic hydrophilic/hydrophobic surface) and an adsorbate, ice-nucleating ability of an adsorbent and an adsorbate, and content and type of distribution (individual molecules, oligomers, nanocrystallites) of the latter at a surface of porous or nonporous solid particles. It was found that the effect of a fraction of nanoparticles participating in the formation of ice crystals was influenced by the mean radius of the particles, the degree of oxidizing of their surfaces and temperature (Gorbunov et al. 2001). Fumed oxides with adsorbed ice nucleators can be used as a model of complex particles to study the mentioned effects important for the environment because of the effects of aerosol microparticles on the atmospheric phenomena. For instance, adsorbed 1,3,5-trihydroxybenzene (phloroglucinol [PG]) as an active ice nucleator (e.g., the yield of ice nanocrystallites on dispersion of fine PG particles in supercooled fog corresponds to 10^{12}–10^{13} g^{-1}) can strongly affect the structure and the temperature behavior of interfacial water. For deeper insight into the problem of changes in the ice-nucleating ability of aromatic compounds, uni- (PG), bi- (1,5-dioxinaphthalene [DON]), and tri-ring (phenazine $C_6H_4N_2C_6H_4$ possessing ice-nucleating ability and anthraquinone $C_{14}H_8O_2$ passive compound) compounds were studied using LT ^1H NMR spectroscopy applied to static samples.

As mentioned earlier, primary nanoparticles of nanosilicas (with or without adsorbed PG) can form aggregates (50–500 nm) and agglomerates (>1 μm; Gun'ko et al. 2005d, 2007d,f,i, 2009d). One can assume that the properties of the water/PG/nanosilica system can depend on structural features of the mentioned secondary particles because the structural features of aggregates and agglomerates differ, and their changes can affect the structure of water filling free space (voids between silica nanoparticles) in these secondary particles. The ^1H NMR spectroscopy with layer-by-layer freezing-out of both bulk and interfacial waters is one of methods that allow one to study nonfreezable structured water at $T<273$ K with consideration for the confined space effects on interfacial water. It has been found that the thickness of a layer of water strongly bound to a surface of particles with polar compounds having the ice-nucleating ability (e.g., AgI, PG, and 1,5-dioxynaphthalene) is small at $T<273$ K and lower than that for polar but inactive analogues (Gun'ko et al. 2006e, 2009d). However, a thick layer of amorphous ice can form near this surface. The mobility of molecules in amorphous ice is greater (i.e., transverse relaxation time T_2 is longer) than that in bulk crystalline ice. The ^1H NMR signal (proton resonance band) width of the molecules in amorphous ice is larger by several times than that of water adsorbed onto a surface of mineral particles with a low ice-nucleating ability. However, this width is significantly smaller than that of bulk ice. As a whole, the structure of bound water at various interfaces (e.g., water/solids, vapor/liquid water, water/organics, and organics/water/solids) has certain peculiarities. For instance, an ab initio molecular dynamics simulation of the aqueous liquid–vapor interface showed stabilized region of bulk water in the center of a water slab and confirmed the experimentally observed abundance of surface "acceptor-only" (19%) and "single-donor" (66%) moieties as well as substantial surface relaxation approaching the liquid–vapor interface (Kuo and Mundy 2004). The average value of the dipole decreases and the average value of the highest occupied molecular orbital for each water molecule increases approaching the liquid–vapor interface. Examination of the orientational dynamics showed a faster relaxation in the interfacial region. This is in agreement with enhanced mobility of interfacial water in comparison with bulk water observed using the ^{1}H NMR method at $T<273$ K.

Pristine nanosilica surface or modified silicas (MSs) with such functionalities as $\equiv$Si–O–Si(CH_3)$_3$, $\equiv$Si–O–Si((CH_3)$_2$)(CH_2)$_3$$NH_2$ or other similar surface structures studied (Gun'ko et al. 2005d, 2009d) are incapable of strongly ordering the structure of interfacial water close to that of ice because they rather provide disordering (chaotropic; Chaplin 2011) effect that result in increased mobility of the interfacial water unfrozen at $T<273$ K. Therefore, it is of interest to study the effects of very active ice nucleators in the adsorbed state on the interfacial water. In this chapter, the interaction of water with nanosilica and different amounts of adsorbed PG (which as an active ice nucleator can order the structure of interfacial water close to ice) will be analyzed at different temperatures using the ^{1}H NMR spectroscopy with layer-by-layer freezing-out of bulk and interfacial waters ($200<T<273$ K), IR and ultraviolet (UV) spectroscopies, TG and TPD-MS at $293<T<1073$ K, SEM, and quantum chemistry (Gun'ko et al. 2006e).

PG (1,3,5-trihydroxybenzene, PG, Sigma) samples (121, 253, 379, 507, 632, and 947 mg) were dissolved in 100 mL of 96% ethanol. The PG solution was added to 5 g of standard nanosilica A-380. This mixture was stirred and aged at room temperature for 24 h. Then ethanol was removed from the suspension on heating at 333 K for 3 h. PG/silica powder samples contained from 0.024 to 0.189 g of PG per gram of dry silica. To record the IR spectra using a Specord M-80 (Carl Zeiss) spectrophotometer, PG/silica samples were pressed in plates (8×26 mm, 15–18 mg) at 160 MPa. Individual PG was stirred with KBr powder. PG solution ($C_{PG}=0.001$ M) in ethanol/water (96: 4) was utilized in the UV/vis measurements using a Specord M-40 (Carl Zeiss) spectrophotometer with the absorption mode for the solutions and the reflectance mode for the PG/A-380 powders. The pH value was changed by addition of the HCl or NaOH solution and controlled using a pH meter with a glass electrode. The PG adsorption on nanosilica was carried out from the ethanol solution at pH 8.3. To study the stability of the PG/A-380 systems, the UV spectra were recorded over a wide pH range at a constant PG content. TG measurements of PG and water desorption were carried out using a V2.5H TA (Philips, USA) apparatus. Samples (2.5–4.0 mg) of nanosilica/PG/water were heated at a rate of 2 K/min at $T<450$ K and 5 K/min at $T>450$ K at an air flow rate of 105 cm^3/min. Calculation of a distribution function of activation energy $f(E_a)$ of desorption of water or water/PG was carried out using integral equation (Gun'ko et al. 2005d, 2009d) described in Chapter 10.

The TPD-MS measurements of water desorption from nanosilica A-380 (treated by the same way that the samples upon the adsorption of PG) and PG/A-380 were carried out using a MX-7304A ("Electron," Sumy, Ukraine) mass spectrometer. Water and PG desorption from nanosilica was also studied using one-pass TPD time-of-flight (TOF) MS (pressure in the chamber $\sim10^{-7}$ Torr, sample weight 20–40 mg, heating rate 2 K/s, with a short distance [~0.5 cm] between sample and an MS detector) with an MSC-3 ("Electron", Sumy, Ukraine) TOF mass spectrometer (sensitivity 2.2×10^{-5} A Torr^{-1}, accelerating voltage 0.5 kV, pulse frequency 3 kHz). The distribution functions of the activation energy of water desorption were determined from the TPD-MS data as in the case of the TG experiments with an integral equation solved using the regularization method for each TPD peak.

SEM images of PG/silica were recorded using a JEOL 6100 SEM (JEOL Optical Laboratory, Japan) with acceleration voltage of 18 kV.

A silica cluster with eight tetrahedra and eight $\equiv$SiOH groups as a model of silica surface and the corresponding complexes with PG and water molecules were calculated by using Hartree–Fock (HF) and DFT with the GAMESS (Schmidt et al. 1993, Granovsky 2009) and Gaussian 03 (Frisch et al. 2004) program packages. All these calculations were carried out to total optimization of the geometry. DFT with a hybrid functional including a mixture of HF exchange with DFT exchange-correlation B3LYP using the 6-31G(d,p) basis set was applied in calculations of the NMR spectra using gauge-independent atomic orbital (GIAO) method (Gaussian 03). Kitaura–Morokuma method (GAMESS) with consideration for the basis set superposition error was used to analyze components of the interaction energy between the PG molecule and the ≡SiO–H group of the silica cluster with the relaxed geometry (HF/6-31G(d,p)). The silica clusters with 8 (open surface) and 65 (nanopore at diameter ≈ 0.8 nm) tetrahedra were used in PM3 calculations of the dynamic behavior of adsorbed PG/water complexes. Their initial kinetic energy E_k corresponds to $\frac{3}{2}kT$ for each molecule that has a random velocity vector. The initial geometry of these complexes was entirely optimized. The number of water molecules in the droplets was up to 20 in ab initio calculations and 302 in PM3 calculations (Gun'ko et al. 2006e). The distribution functions $f(r_{\alpha\beta})$ of distances $r_{\alpha\beta}$ between atoms of α and β kinds (e.g., α = O and β = O and H), that is, the number of atom pairs at the distance in the $r_{\alpha\beta} \pm \delta$ range, were calculated:

$$f(r_{\alpha\beta}) = \sum_{r_{\alpha\beta}-\delta \le \Delta x_{i,j} \le r_{\alpha\beta}+\delta} \sqrt{\sum_{l} (x_{l,i,\alpha} - x_{l,j,\beta})^2}, \tag{1.30}$$

where

$$\Delta x_{i,j} = \sqrt{\sum_{l} (x_{l,i,\alpha} - x_{l,j,\beta})^2} \tag{1.31}$$

is the distance between atoms with $x_{l,i,\alpha}$ and $x_{l,j,\beta}$ Cartesian coordinates (l = 1, 2, 3) of all possible atom i,j-pairs of α and β kinds in the system, and δ is a small distance increment between 0.01 and 0.0001 nm dependent on the size of the system.

The IR spectra of PG (Figure 1.63a; curve 1) and nanosilica individual (curve 2) and with adsorbed PG (0.25 mmol/g) (curve 3) demonstrate an influence of PG on a broad IR band at ν_{OH} = 3300 cm^{-1} assigned to adsorbed water.

Typically, there are four bands of the O–H stretching vibrations (ν_{OH}) in the IR spectra of nanosilica with adsorbed water (Legrand 1998). A band at 3746 cm^{-1} corresponds to free silanols (Figure 1.63b). A band at 3640–3680 cm^{-1} is assigned to silanols that are poorly accessible for practically all adsorbates, despite weak disturbance of these silanols. A band at 3450–3550 cm^{-1} is linked to silanols that are disturbed by hydrogen bonds with adsorbed water (interaction energy is 35–40 kJ/mol). Adsorbed water is characterized by a broad band at 3200–3300 cm^{-1}. Very strongly disturbed silanols can contribute the last band. Notice that the O–H stretching vibrations of non-hydrogen-bonded OH groups of water molecules at vapor/water and CCl_4 (or hexane)/water interfaces correspond to ~3700 cm^{-1} and 3669 cm^{-1}, respectively, that is, nonpolar compounds can slightly disturb the OH groups because this displacement corresponds to the interaction energy in 10–15 kJ/mol. A band at 3600–3700 cm^{-1} (Figure 1.63a; curve 1) corresponds to hydroxyls of PG. Several bands at 1460, 1510, and 1620 cm^{-1} are linked to vibrations of the aromatic ring of PG. The IR spectra of PG/silica demonstrate broadening of the band related to adsorbed water because of the PG effect (Figure 1.63a; curves 2 and 3). This effect is due to an increase in the amount of adsorbed water and strengthening of the hydrogen bonding in the presence of adsorbed PG as a strong ice nucleator. Consequently, the adsorbed PG molecules can strongly affect the structure of interfacial water. However, the properties of adsorbed PG as an ice nucleator can change at the silica surface (this effect will be analyzed in the subsequent sections).

The adsorption of PG onto a surface of nanosilica through the hydrogen bonds can be evaluated from the optical density of the IR band at ν_{OH} = 3746 cm^{-1} or of the bending vibrations at δ_{COH} = 1420 cm^{-1} in PG. In the case of nanosilica, the ν_{OH} band of free silanols is intensive, therefore, the optical density $D = \lg(I_0/I)$, where I_0 and I are the intensities of free silanols of the initial silica and silica/PG samples, can be easily calculated at relatively low errors. The bending vibrations δ_{COH} in PG have a middle intensity, therefore, an internal standard should be used for estimation of the PG adsorption characteristics from the δ_{COH} values. Considering the amounts of adsorbed PG,

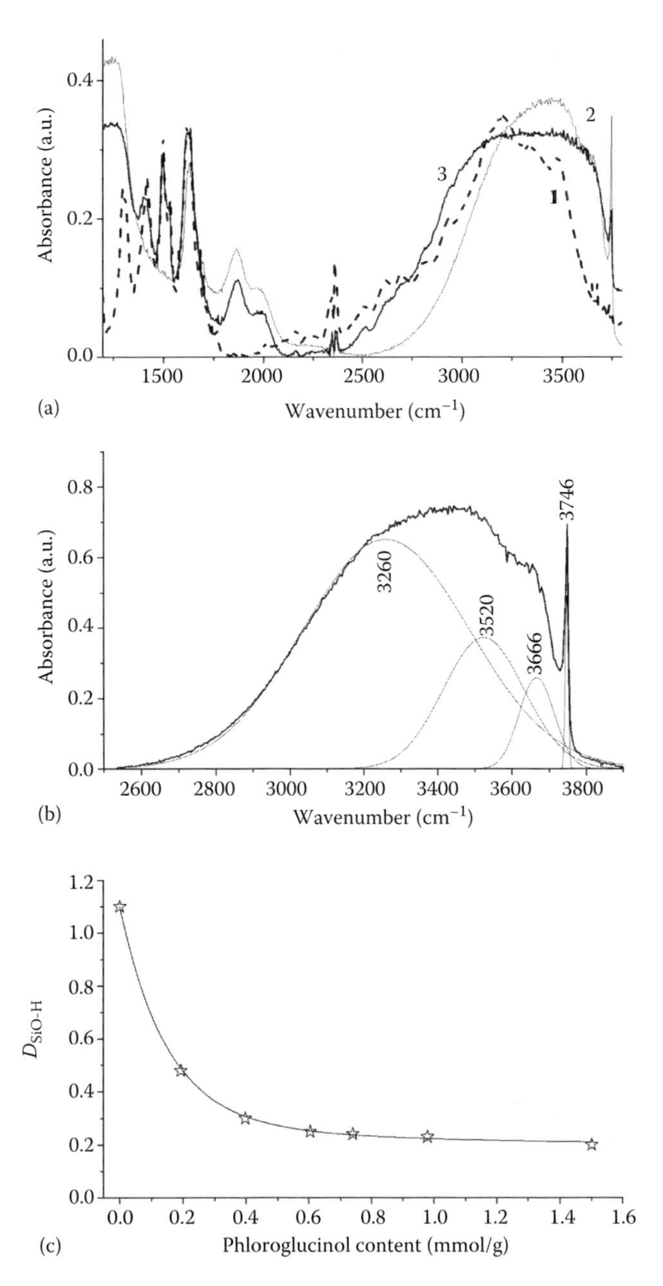

FIGURE 1.63 IR spectra of (a) phloroglucinol (curve 1), nanosilica (2), and the sample 2 of PG/silica (3), (b) decomposition of the IR spectrum of initial nanosilica for the O-H stretching vibrations, and (c) changes in the optical density of the band of free silanols at 3746 cm^{-1} depending on the PG amount adsorbed onto nanosilica. (Adapted from *Colloids Surf. A: Physicochem. Eng. Aspects*, 278, Gun'ko, V.M., Turov, V.V., Barvinchenko, V.N. et al., Characteristics of interfacial water at nanosilica surface with adsorbed 1,3,5-trihydroxybenzene over wide temperature range, 106–122, 2006e. Copyright 2006, with permission from Elsevier.)

a band of the CC stretching vibrations at 1460–1550 cm^{-1} in the aromatic ring can be used for this purpose, assuming that changes in the intensity of this band are significantly smaller than that of δ_{COH} upon the hydrogen bonding of the PG molecules. However, according to the IR spectra, the shape of the ν_{CC} band changes upon the PG adsorption (Figure 1.63a), but these changes are much smaller than that of the δ_{COH} value. Therefore, the PG adsorption interaction can be evaluated using

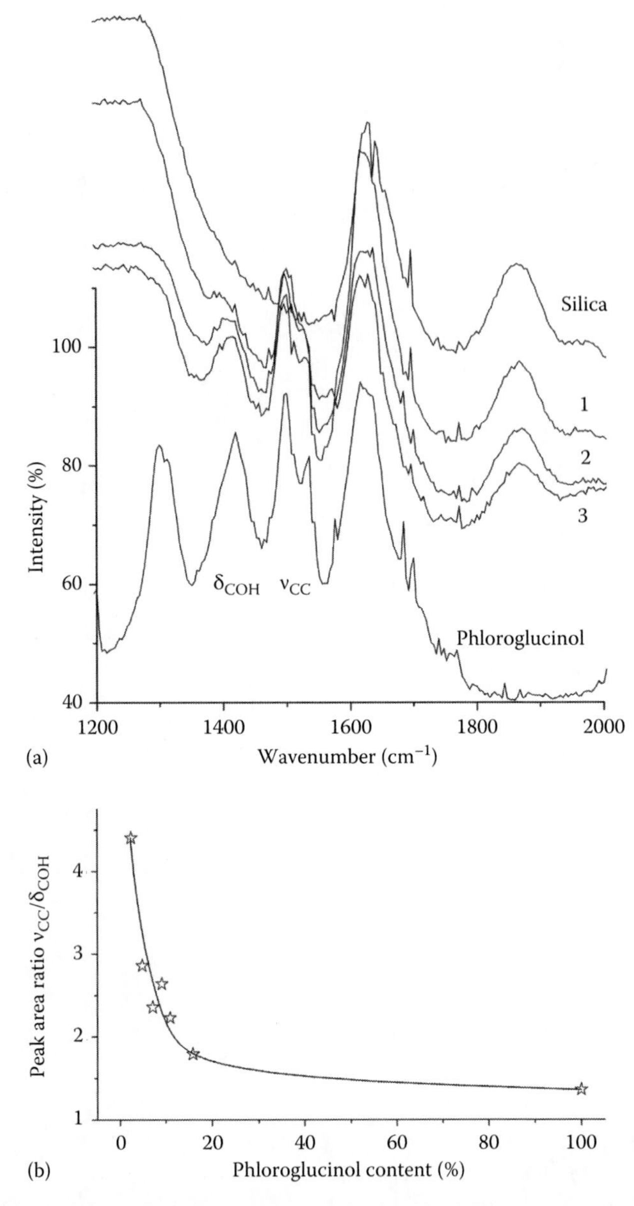

FIGURE 1.64 (a) IR spectra of phloroglucinol, nanosilica and samples with PG adsorbed onto nanosilica at $C_{PG}=0.193$ (1), 0.397 (2), and 0.980 (3) mmol/g. (b) Changes in the ratio (ν_{CC}/δ_{COH}) of integral intensities of bands ν_{CC} and δ_{COH} of PG depending on C_{PG}. (Adapted from *Colloids Surf. A: Physicochem. Eng. Aspects*, 278, Gun'ko, V.M., Turov, V.V., Barvinchenko, V.N. et al., Characteristics of interfacial water at nanosilica surface with adsorbed 1,3,5-trihydroxybenzene over wide temperature range, 106–122, 2006e. Copyright 2006, with permission from Elsevier.)

a relationship ν_{CC}/δ_{COH} for peak area in the $\lg(I_0/I)$—wavenumber (cm^{-1}) coordinates. At a smaller amount of adsorbed PG, the δ_{COH} maximum shifts from 1420 cm^{-1} in individual PG toward 1408 cm^{-1} in adsorbed PG. The optical density of the band of free silanols decreases with increasing amount of adsorbed PG (Figure 1.63c). The opposite picture is observed for the δ_{COH} band (Figure 1.64a).

Features of the IR spectra of adsorbed PG suggest that the hydrogen bonding of the surface silanols with the adsorbed PG molecules occurs through ≡SiO–H···O(H)R but the formation of the bonds ≡SiO(H)···H–OR is less probably. A statistical PG monolayer corresponds to $C_{PG} \approx 10$ wt%.

However, an increase in the amount of adsorbed PG that is higher than the monolayer is accompanied by small changes in the shape of the IR spectra. This effect can be caused by aggregation of PG molecules in oligomers (nanocrystallites).

Degassing (10^{-2} Torr) of PG/water/nanosilica (sample 6) at 293 K leads to desorption of a portion of water and PG (Figure 1.65; curves 1 and 2).

Despite the C_{PG} value of the sample 6 is near twice larger than that of the monolayer coverage, the band of free silanols is observed. Consequently, PG forms small particles (nanocrystallites) at a silica surface, and only a portion of PG molecules is in contact with the silica surface. Notice that the formation of PG particles is confirmed by the distribution function of size of pores (voids between primary particles) filled by unfrozen water in the aqueous suspensions that is analyzed in the subsequent sections. An increase in the degassing temperature to 373 K results in desorption of a major portion of water because the intensity of the broad IR band at 3100–3400 cm^{-1} strongly diminishes. Rehydration of the sample in air during 20 h does not lead to the spectrum shape characteristic of the initial sample 6 because of desorption of a portion of PG on previous degassing. The intensity of the OH vibrations of the free silanols at 3748 cm^{-1} as well as of the ν_{OH} band of the COH groups significantly decreases after rehydration because of the hydrogen bonding with adsorbed water, which can strongly interact with hydroxyls of both adsorbed PG molecules and surface silanols.

PG is not a very stable compound, for example, it can be transformed into the oxidized form due to interaction with oxygen from air; therefore, this aspect was analyzed. PG in the ethanol solution at pH 2.45–9.35 is stable and characterized by a UV band at 269 nm (Figure 1.66) assigned to PG in the molecular form. The spectra do not change after aging of these samples for 48 h in air. At high pH values of 10.4–13.5, the characteristic band shifts toward higher λ values, and a new band at $\lambda_{max}=352$ nm appears. This band is due to deprotonation of the PG hydroxyls. The intensity of the band at 269 nm that is characteristic of the molecular form of PG increases with increasing content of PG adsorbed onto nanosilica (Figure 1.67).

Consequently, the PG molecules form the adsorption complexes by the hydrogen bonds, and this result is in agreement with the IR spectra of the PG/silica samples (Figures 1.63 through 1.65). However, a new band appears at $\lambda_{max}=445$ nm (Figure 1.67b; spectra normalized at $\lambda_{max}=269$ nm), which could be assigned to the oxidized form of PG with the quinone groups that are typically characterized by the UV/vis bands at $\lambda=400$–650 nm caused by electron transition $S(\pi\rightarrow\pi^*)$ and $S(n\rightarrow\pi^*)$. The intensity of the band at $\lambda_{max}=445$ nm decreases with increasing C_{PG} value because of diminution of the interaction of the adsorbed PG molecules with oxygen present in the adsorption layer since the amount of this oxygen is low. Therefore, a relative amount of oxidized PG molecules decreases with increasing C_{PG} value. Additionally, adsorbed oligomers of PG can weaker interact with the oxygen than individual adsorbed molecules.

Morphological changes of nanosilica are possible on wetting (suspending) in the aqueous media and subsequent drying. To estimate these changes, the SEM method was used. The SEM images of treated powders (Figure 1.68) demonstrate the morphology of secondary particles typical for fumed oxide powders.

Typically, the powders treated in aqueous suspensions consist of tightened microscale agglomerates with the bulk density (ρ_b) up to 0.25 g/cm^3 instead of 0.05–0.07 g/cm^3 for the initial nanosilica powder. Such increased ρ_b values are characteristic of the fumed oxide powders that were wetted or suspended in the aqueous or water/alcohol medium and then dried under mild conditions. However, treatment in nearly pure (96%) alcohol medium with dissolved PG can cause smaller morphological changes in the nanosilica in comparison with treatment in pure water. Thus, treated PG/nanosilica powders remain in the free non-bound state with easily rearranged structure of secondary particles.

The porosity of secondary silica particles (voids between primary particles in their aggregates and agglomerates) can affect the adsorption behavior of water and PG/water. Notice that the empty space ($V_{em} = \rho_b^{-1}$, where $\rho_0=2.2$ g/cm^3 is the true density of nanosilica) in the powder of wetted and then dried nanosilica is equal to 5–10 cm^3/g that is lower than $V_{em}=15$–25 cm^3/g of the initial powder. The TG data demonstrate that adsorbed water and PG can fill only a small portion of V_{em} (Figures 1.69 and 1.70).

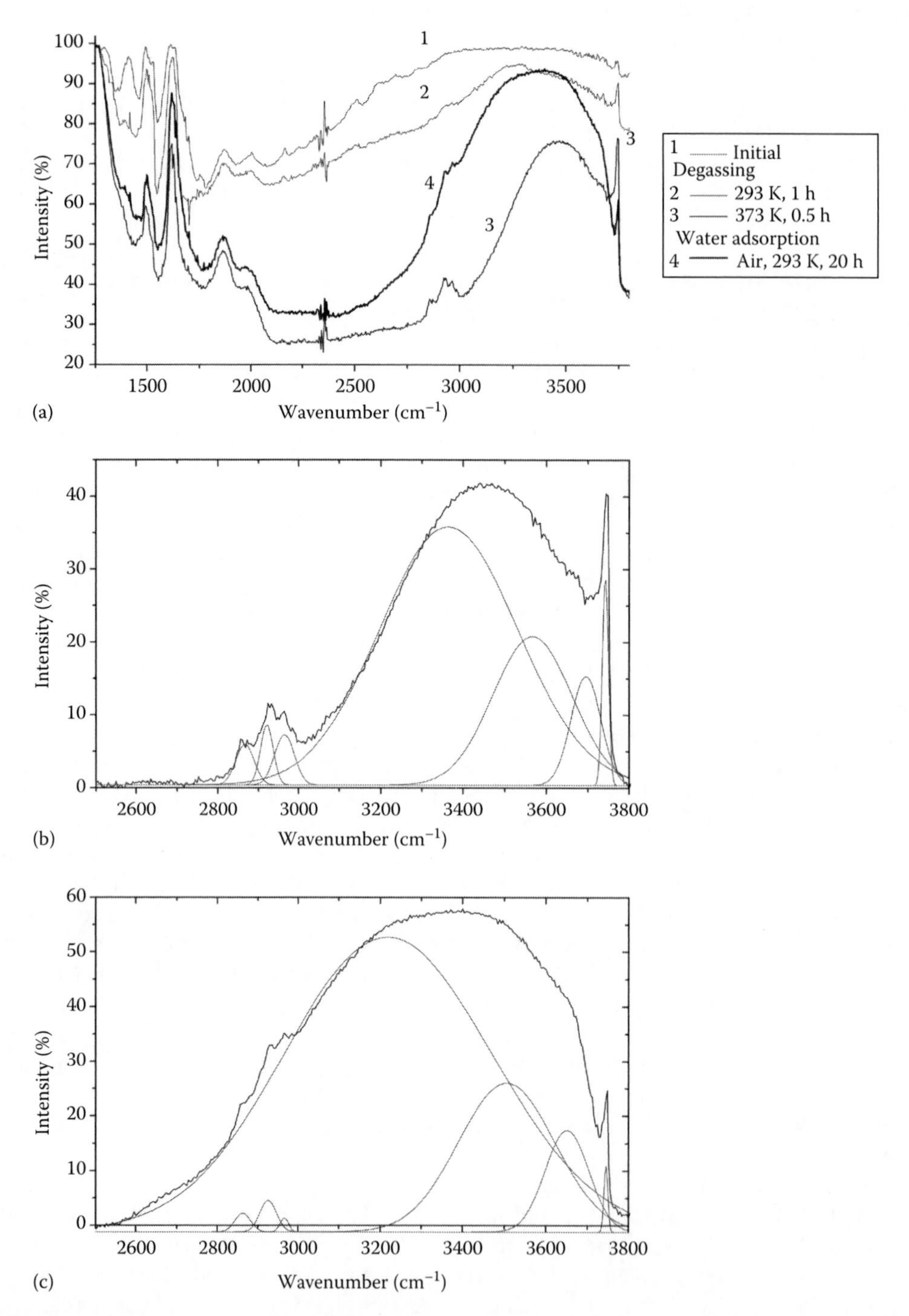

FIGURE 1.65 IR spectra of PG/nanosilica (sample 6, curve 1) differently treated (a) degasses at room temperature for 1 h (curve 2) and at 373 K for 0.5 h (curve 3, b), rehydration of a sample due to water adsorption from air at room temperature for 20 h (curve 4, c), decomposition of the IR spectra (curves 3 and 4) for the O-H stretching vibrations is shown (b, c), $C_{PG} \approx 1$ mmol/g. (Adapted from *Colloids Surf. A: Physicochem. Eng. Aspects*, 278, Gun'ko, V.M., Turov, V.V., Barvinchenko, V.N. et al., Characteristics of interfacial water at nanosilica surface with adsorbed 1,3,5-trihydroxybenzene over wide temperature range, 106–122, 2006e. Copyright 2006, with permission from Elsevier.)

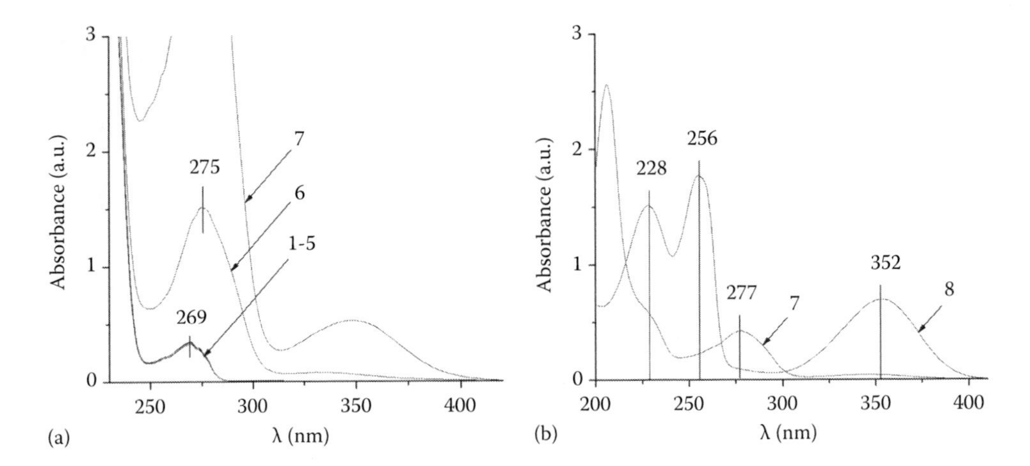

FIGURE 1.66 UV spectra of the PG solution (0.001 M) in ethanol at different pH values at the thickness of (a) 1 cm and (b) 0.1 cm at pH (1) 2.45, (2) 3.25, (3) 3.75, (4) 8.25, (5) 9.35, (6) 10.4, (7)11.2, and (8) 13.5. (Adapted from *Colloids Surf. A: Physicochem. Eng. Aspects*, 278, Gun'ko, V.M., Turov, V.V., Barvinchenko, V.N. et al., Characteristics of interfacial water at nanosilica surface with adsorbed 1,3,5-trihydroxybenzene over wide temperature range, 106–122, 2006e. Copyright 2006, with permission from Elsevier.)

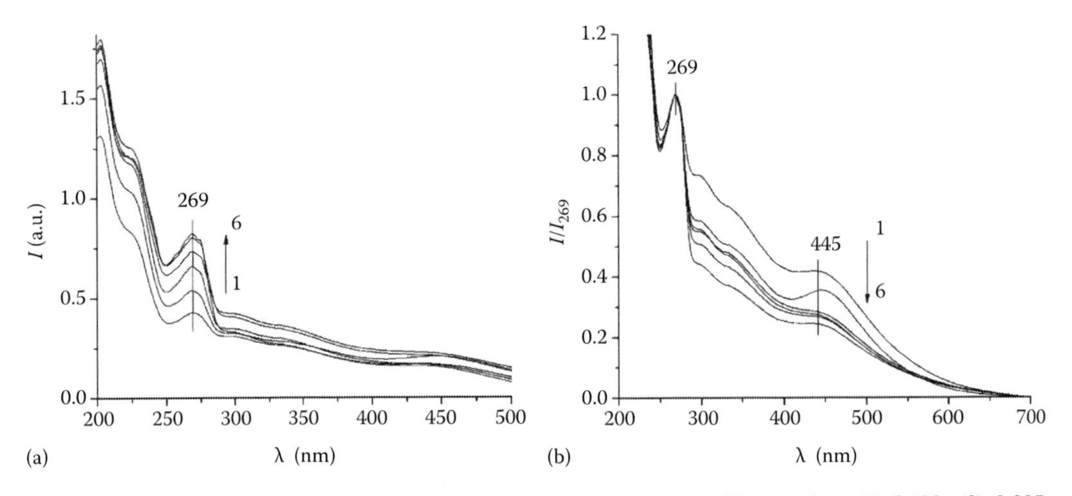

FIGURE 1.67 (a) UV spectra of PG adsorbed (at pH=8.3) on nanosilica at C_{PG}=(1) 0.193, (2) 0.397, (3) 0.604, (4) 0.793, (5) 0.98, and (6) 1.5 mmol/g and (b) the same normalized spectra. (Adapted from *Colloids Surf. A: Physicochem. Eng. Aspects*, 278, Gun'ko, V.M., Turov, V.V., Barvinchenko, V.N. et al., Characteristics of interfacial water at nanosilica surface with adsorbed 1,3,5-trihydroxybenzene over wide temperature range, 106–122, 2006e. Copyright 2006, with permission from Elsevier.)

This result can be caused by the PG/water adsorption mainly in the interior volume of aggregates (0.3–0.8 cm^3/g) of primary silica particles which are substantially smaller than the V_{em} value. Additionally, these relatively low amounts of adsorbed water can be evidence of insignificant compacting of the silica on the sample preparation because significant compacting of the powder causes substantial increase in the adsorption capacity with respect to water or other adsorbates. In the aqueous suspensions, total "empty" space (V_{em} is not filled by silica) is filled by water and PG molecules and features of this filling can be elucidated using the LT ^1H NMR data (analyzed in the subsequent sections).

The TG curves (Figure 1.69) have two portions of relatively fast changes that correspond to the minima of the differential thermogravimetry (DTG) curves. The first one at $T<50°C$ can be

(a) (b)

FIGURE 1.68 SEM images of the sample 5 of the PG/nanosilica powder with (a, b) different magnification. (Adapted from *Colloids Surf. A: Physicochem. Eng. Aspects*, 278, Gun'ko, V.M., Turov, V.V., Barvinchenko, V.N. et al., Characteristics of interfacial water at nanosilica surface with adsorbed 1,3,5-trihydroxybenzene over wide temperature range, 106–122, 2006b. Copyright 2006, with permission from Elsevier.)

connected with desorption of a portion of PG and physically adsorbed water. Notice that in the case of desorption from degassed samples in vacuum, this peak shifts toward higher temperature (Figure 1.71a). The second DTG minimum at $150 < T < 225°C$ (Figure 1.69) is close to the melting temperature of PG (218°C–221°C for anhydrous PG) and caused by PG desorption because the TG/DTG curve shapes depend on the PG loading. Notice that the TPD-MS thermograms (Figure 1.71a) give the peaks at the same temperatures. A subsequent decrease in the sample weight at $T > 250°C$ (Figure 1.69) is caused by desorption of residual PG molecules, which can be strongly adsorbed between primary silica particles in aggregates, and by associative desorption of water dissociatively adsorbed on the silica surface (surface hydroxyls). Certain broad peaks in DTG between 400°C and 700°C (Figure 1.69) correspond to continuation of these desorption processes. At $T > 700°C$, alone silanols can associatively desorb to form water molecules at high activation energy (250 kJ/mol or higher); therefore, this portion of the TG and DTG curves is not shown here. An increase in the PG loading enhances the weight loss at $T = 150°C–225°C$ due to PG desorption because the amount of adsorbed water (desorbed at $25°C < T < 700°C$) is not higher than 5.6 wt% (it is maximum for the sample 1). In the case of individual nanosilica, this value (2.2 wt%) is even smaller (Figure 1.70a). Consequently, despite filling of voids between primary particles in aggregates and agglomerates by the PG molecules, the water adsorption (from air) increases for the samples 1–5 compared to the initial silica. In the case of the maximum amount of adsorbed PG (sample 6), a portion of PG can be desorbed on sample storage because the weight loss (desorption of PG + water) on heating up to 600°C is close to the initial amount of loaded PG. Thus, adsorbed PG strongly affects the adsorption/desorption of interfacial water.

Calculations of the distribution function ($f(E_a)$) of the activation energy of desorption of water (pure A-380) or water/PG (sample 6), which is related to the main DTG minima, show the $f(E_a)$ displacement toward higher E_a values for the sample 6 (Figure 1.70b). This effect is caused by two factors. The first one is connected to desorption of PG molecules with greater molecular weight (MW) and larger number of intermolecular bonds with the neighbors than water molecules. The second one is linked to appearance of more strongly bound interfacial water in the presence of PG because of its interaction with PG molecules and surface silanols simultaneously and blocking

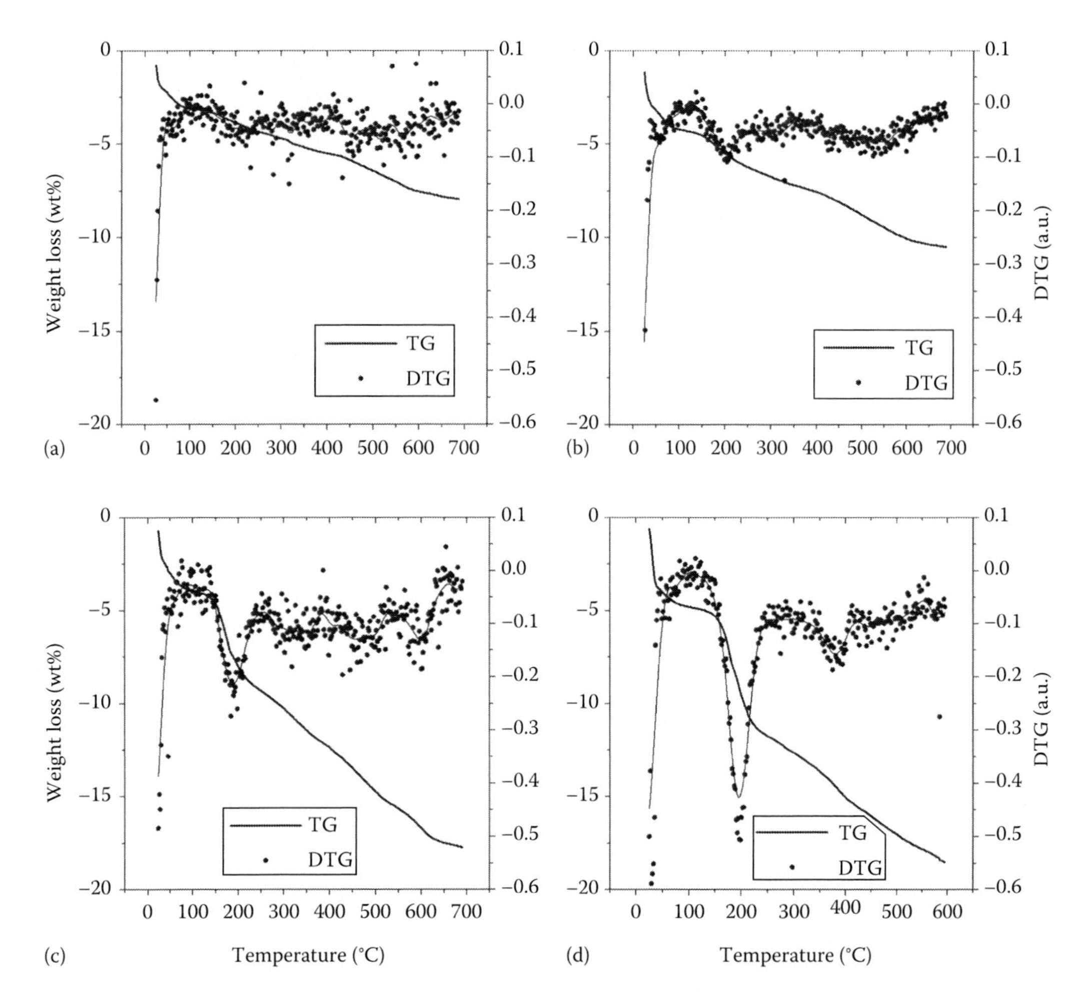

FIGURE 1.69 Weight loss (TG) and derivative of weight loss (DTG) of adsorbed PG/water/nanosilica samples on heating: (a) sample 1, (b) 2, (c) 5, and (d) 6. (Adapted from *Colloids Surf. A: Physicochem. Eng. Aspects*, 278, Gun'ko, V.M., Turov, V.V., Barvinchenko, V.N. et al., Characteristics of interfacial water at nanosilica surface with adsorbed 1,3,5-trihydroxybenzene over wide temperature range, 106–122, 2006b. Copyright 2006, with permission from Elsevier.)

of water molecules in narrow voids of aggregates by PG molecules. However, in the case of the sample 6, a portion of water desorbs at lower temperatures than that of pure silica (Figures 1.69 and 1.70). This effect can be caused by the location of the PG molecules in places with the maximum adsorption potential. In other words, the PG molecules can displace a portion of water from narrow voids between adjacent primary particles in aggregates into larger voids, where water is weaker bound to the surface and can be removed at lower temperatures than water adsorbed in narrow voids of pure nanosilica.

The TPD-MS thermograms and the corresponding distribution functions of the activation energy $f(E)$ of water desorption from pure silica and PG/silica have a complex shape (Figures 1.71 and 1.72) because of many possible processes contributing the desorption curves.

The TPD-MS graphs (Figures 1.71a and 1.72a) differ significantly over the range 100°C–700°C related to associative desorption of water molecules (because of several possible reactions with the participation of OH-containing PG molecules and surface silanols) and desorption of intact water molecules from the voids with the maximal adsorption potential and the volume of nanoparticles (up to 250°C or even higher). An additional broad peak at 550°C (Figure 1.71a, curve 2 and

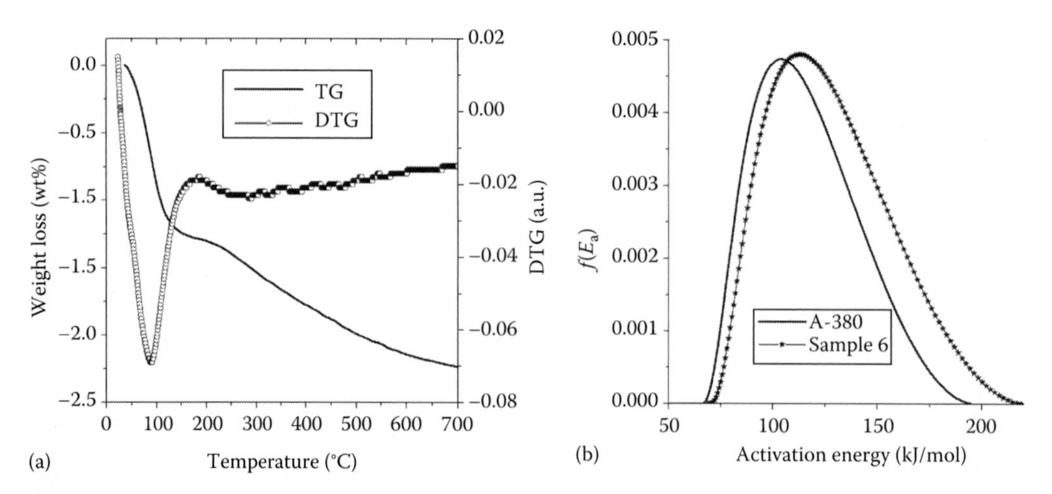

FIGURE 1.70 (a) Weight loss (TG) and derivative of weight loss (DTG) of water/nanosilica, and (b) distribution functions of activation energy of desorption of water from A-380 and of water/PG from the sample 6. (Adapted from *Colloids Surf. A: Physicochem. Eng. Aspects*, 278, Gun'ko, V.M., Turov, V.V., Barvinchenko, V.N. et al., Characteristics of interfacial water at nanosilica surface with adsorbed 1,3,5-trihydroxybenzene over wide temperature range, 106–122, 2006e. Copyright 2006, with permission from Elsevier.)

Figure 1.72a, curve 3) is in the range of decomposition of the PG molecules, which contributes desorbed water. As a whole, PG adsorbed on silica has a noticeable effect on the interfacial water because the shape of the TPD-MS curve changes over the total temperature range. This effect results in certain changes in the distribution functions $f(E)$ calculated separately for four (Figure 1.71) or three (Figure 1.72; silica) and five (Figure 1.71) and three (Figure 1.72; PG/silica) portions of the TPD-MS curves. An additional $f(E)$ peak appears at $E = 100$ kJ/mol (Figure 1.71c), which can be linked to water interacting with PG molecules in narrow voids of secondary silica particles. A peak at $E = 155$ kJ/mol is broadened due to the PG effects on the adsorbed water and elimination of water from destroyed PG molecules. OPTOF-TPD-MS thermograms recorded at higher vacuum show (Figure 1.72) that adsorbed PG enhances desorption of water at lower temperatures (because of water clustering) and decreases desorption at higher temperatures because of displacement of water from places with maximal adsorption potentials. The desorption curves recorded after different degassing using different equipments differ as well as the $f(E)$ functions (Figures 1.71 and 1.72) because of changes in the interfacial layer state on degassing to higher vacuum (up to 10^{-7} Torr). Thus, adsorbed PG molecules can (i) strengthen the binding of a portion of water in narrow voids of secondary silica particles and (ii) weaken the binding of another portion of water displaced into large voids, and these two effects depend on the content of adsorbed PG and pretreatment conditions. Certain changes in the structure of the interfacial water in the presence of adsorbed PG can be analyzed on the basis of the ^{1}H NMR spectra and quantum chemical calculations of the corresponding models.

The ^{1}H NMR spectra of the frozen aqueous suspension of the sample 2 were recorded at the bandwidth of 50 kHz (Figure 1.73a with amorphous ice/unfrozen water) or 10 kHz (Figure 1.73b with unfrozen water only) and $220 < T < 268$ K.

Notice that other PG/silica samples have the ^{1}H NMR spectra (not shown here) of a similar shape. Broad (proton resonance line width > 30 ppm) and narrow (width < 3 ppm) signal components are observed at $250 < T < 268$ K (Figure 1.73a). The narrow component is caused by protons of interfacial mobile (so-called nonfreezable) water at the boundary of ice/water/surface of nanoparticles. The broad component is connected to the amorphous ice, which strongly differs from the bulk ice. The mobility of water molecules in such amorphous ice is higher than in bulk ice but it is slower than that in mobile (unfrozen at $T < 273$ K) water adsorbed on a surface of mineral particles. It is known

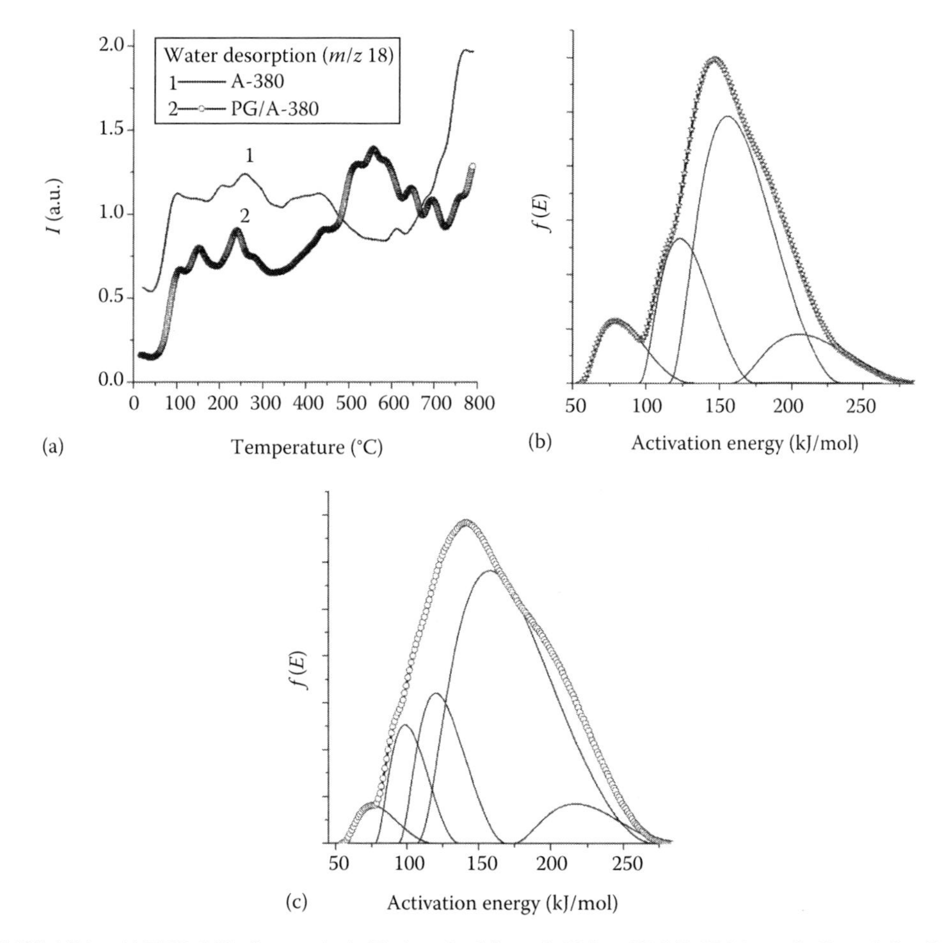

FIGURE 1.71 (a) TPD-MS of water (*m/z* 18) desorbed from A-380 and PG/A-380 (sample 5), and distribution function of activation energy of water desorption from (b) A-380 and (c) PG/A-380 (sample 5). (Adapted from *Colloids Surf. A: Physicochem. Eng. Aspects*, 278, Gun'ko, V.M., Turov, V.V., Barvinchenko, V.N. et al., Characteristics of interfacial water at nanosilica surface with adsorbed 1,3,5-trihydroxybenzene over wide temperature range, 106–122, 2006e. Copyright 2006, with permission from Elsevier.)

that the ^{1}H NMR signal of amorphous ice is observed in the aqueous suspensions ($T<273$ K) of substances with the ice-nucleating ability. The mentioned broad component is not observed at the 10 kHz bandwidth (Figure 1.73b).

An inflection observed in the $\Delta G(C_{uw})$ curves separates two portions corresponding to strongly and WBWs (Figure 1.74a). The Gibbs free energy characterizing WBW (distant from the surface) is slightly lower than that of the bulk water because of a weak influence of the solid surface and the PG molecules (which remain in the adsorption state because its solubility in water is low) on distant layers of water. This water is frozen at T close to 273 K. The Gibbs free energy of water strongly bound to the solid surface or the PG molecules is significantly lower than that of the WBW. The SBW (close located to the surface) becomes frozen at T significantly lower than 273 K. For most heterogeneous systems, the freezing temperature of SBW is significantly lower than the temperature of heterogeneous nucleation of supercooled water. Table 1.16 shows the characteristics of the interfacial unfrozen water layers and the values of the surface free energy (γ_S). The initial point of the $\gamma_S(C_{PG})$ graph (Figure 1.74b) corresponds to the aqueous suspension with the initial silica without PG. The characteristics of the unfrozen water layers strongly change at $C_{PG}>0.1$ g/g. At lower PG amounts, tendency of C^{s}_{uw} diminution is observed with increasing C_{PG} value (Table 1.16) because a portion

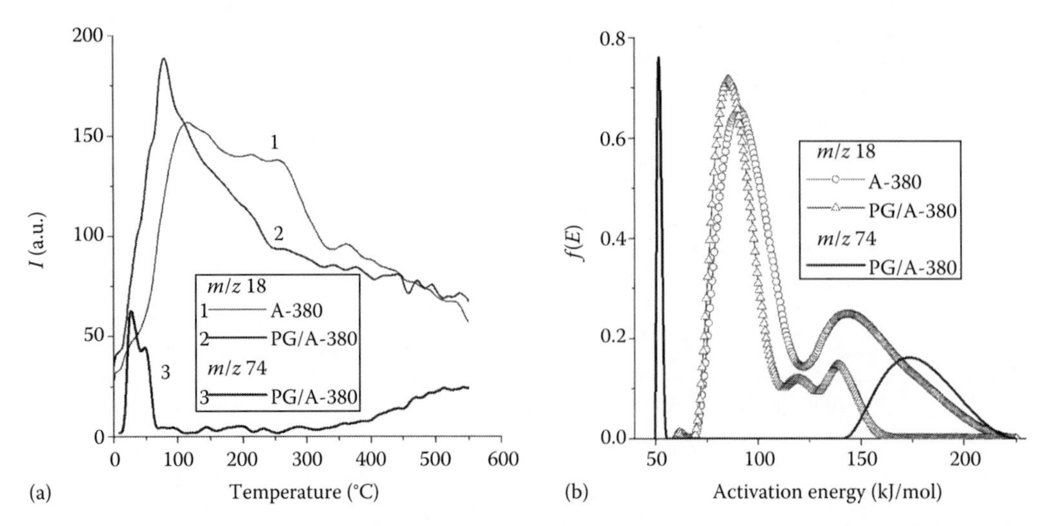

FIGURE 1.72 (a) OPTOF-TPD-MS thermograms of water (*m/z* 18) desorbed from initial silica and water/PG/A-380 (sample 5) and line at *m/z* 74 corresponding to a fragment of PG (sample 5) and (b) corresponding distribution functions of the activation energy of desorption. (Adapted from *Colloids Surf. A: Physicochem. Eng. Aspects*, 278, Gun'ko, V.M., Turov, V.V., Barvinchenko, V.N. et al., Characteristics of interfacial water at nanosilica surface with adsorbed 1,3,5-trihydroxybenzene over wide temperature range, 106–122, 2006e. Copyright 2006, with permission from Elsevier.)

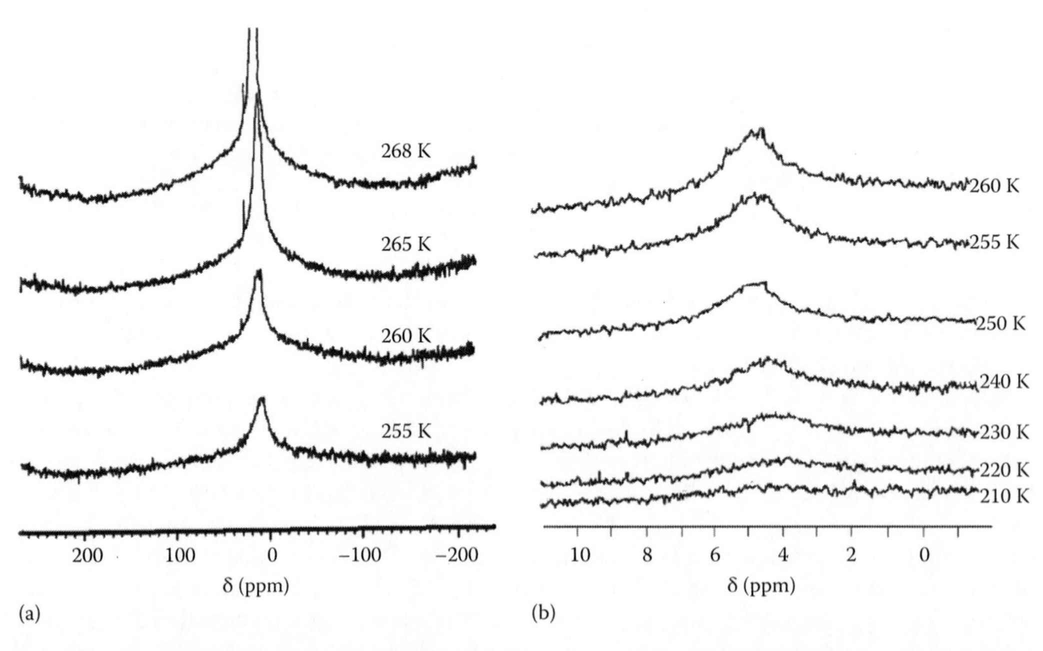

FIGURE 1.73 ^{1}H NMR spectra of the interfacial water in the sample 2 with PG/nanosilica at different width of (a) 50 kHz and (b) 10 kHz. (Adapted from *Colloids Surf. A: Physicochem. Eng. Aspects*, 278, Gun'ko, V.M., Turov, V.V., Barvinchenko, V.N. et al., Characteristics of interfacial water at nanosilica surface with adsorbed 1,3,5-trihydroxybenzene over wide temperature range, 106–122, 2006e. Copyright 2006, with permission from Elsevier.)

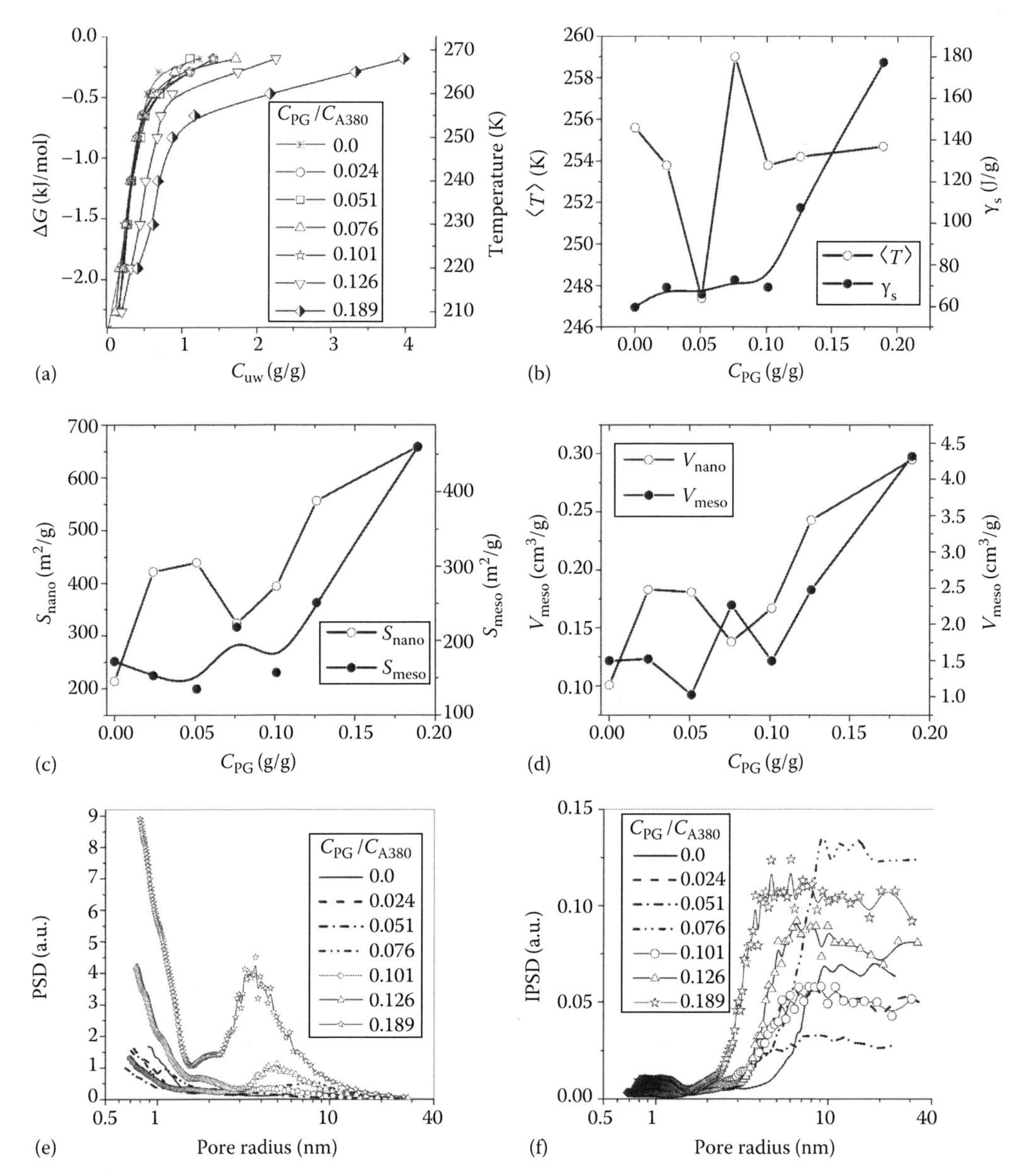

FIGURE 1.74 (a) Amounts of unfrozen water, C_{uw} (bandwidth 10 kHz) as a function of temperature at different C_{PG} values and the corresponding changes in differential Gibbs free energy of interfacial water, relationships between concentrations of PG and (b) modulus overall changes in the Gibbs free energy of the interfacial water (γ_S) and average melting temperature ⟨T⟩, (c) specific surface area and (d) volume of nanopores and mesopores filled by unfrozen water, and PSD (e) differential and (f) incremental (k_{GT}=58.4 K nm). (a: Adapted from *Colloids Surf. A: Physicochem. Eng. Aspects*, 278, Gun'ko, V.M., Turov, V.V., Barvinchenko, V.N. et al., Characteristics of interfacial water at nanosilica surface with adsorbed 1,3,5-trihydroxybenzene over wide temperature range, 106–122, 2006e. Copyright 2006, with permission from Elsevier.)

of interfacial water is displaced by PG molecules from the places with the maximum adsorption potential. However, the maximal C^{s}_{uw} value, as well as C^{w}_{uw}, is found at maximal C_{PG} values (samples 5 and 6). A clear picture of the influence of the adsorbed PG molecules on the characteristics of bound water is observed in the $\gamma_S(C_{PG})$ graph (Figure 1.74b) including two portions. The first one at $C_{PG} \leq 0.1$ g/g (samples 1–4) shows an increase in γ_S with increasing C_{PG} value. However, at higher

TABLE 1.16
Characteristics of Bound Unfrozen Water (Bandwidth 10 kHz) Layers in Aqueous Suspensions of Nanosilica with Various Amounts of Adsorbed PG (C_{PG} Is Given With Respect to the Silica Weight)

C_{PG} (g/g)	γ_S (J/g)	ΔG_w (kJ/mol)	ΔG_s (kJ/mol)	C^w_{uw} (g/g)	C^s_{uw} (g/g)	V_{uw} (cm³/g)	V_{nano} (cm³/g)	V_{meso} (cm³/g)	V_{macro} (cm³/g)	S_{uw} (m²/g)	S_{nano} (m²/g)	S_{meso} (m²/g)	S_{macro} (m²/g)	$\langle T \rangle$ (K)
0	59.7	−0.4	−2.42	1.20	0.71	1.91	0.101	1.497	0	386	214	172	0	255.6
0.024	69.3	−0.7	−2.87	1.19	0.61	1.80	0.183	1.519	0.050	579	422	153	4	253.8
0.051	66.0	−0.8	−3.19	0.66	0.65	1.31	0.181	1.029	0	575	439	135	0	247.4
0.076	72.8	−0.4	−3.04	1.99	0.62	2.61	0.138	2.265	0.124	551	324	218	10	259.0
0.101	69.4	−0.6	−3.08	1.11	0.69	1.80	0.167	1.496	0	551	394	157	0	253.8
0.126	107.4	−0.6	−2.75	2.04	0.91	2.95	0.243	2.476	0.081	813	556	251	6	254.2
0.189	177.3	−0.8	−2.65	3.62	1.26	4.88	0.295	4.315	0	1119	659	460	0	254.7

amounts of PG (samples 5 and 6), $\gamma_S(C_{PG})$ sharply increases>100 J/g, which is significantly higher than that for the initial nanosilica. These effects can be caused by a difference in the influence of PG on the interfacial water at the PG loading lower or larger than the monolayer. The hydrogen bonding of the PG molecules (from the monolayer) to surface silanols can cause diminution of the effect of PG on the interfacial water; therefore, the PG/silica influence on interfacial water is comparable to that of the initial silica.

At the PG loading that is higher than the monolayer, the PG molecules (which are not hydrogen bonded to silanols) can more strongly affect the interfacial water structure because a portion of PG molecules does not directly interact with the silica surface and, therefore, possess the ice-nucleating ability close to that of individual PG. Additionally, another reason of the sharp increase in the signal intensity of the samples 5 and 6 is total disturbance of water in the volume of PG/silica aggregates that also results in increase in the amount of amorphous ice (high intensity of the broad component in Figure 1.73a). As a whole, the intensity of the broad component correlates to the ice-nucleating ability of suspended particles, and PG in the microcrystalline state is the most active ice-nucleating agent. Therefore, an increase in the PG amount higher than the monolayer (samples 5 and 6) can be accompanied by an increase in the ice-nucleating ability of adsorbed PG because of aggregation of the PG molecules in nanocrystallites. This leads to the increase in the amount of the amorphous ice responsible for the broad ^{1}H NMR band and the strong changes in the γ_S value.

As stated earlier, the empty volume of voids in the initial dry nanosilica powder is equal to 17–25 cm^3/g; however, after suspending and drying, it typically reduces to V_{em}=5–10 cm^3/g or smaller because of compacting of secondary particles. The latter value is in agreement with $C_{uw} = C^{S}_{uw} + C^{W}_{uw}$ for the sample 6 as $V_{uw} \approx 4.9$ cm^3/g, which is much larger than that for other samples (Table 1.16). Consequently, one can assume that for the samples at low C_{PG} values, a portion of water located inside agglomerates is akin to the bulk water (as $V_{em} \gg V_{uw}$) because the distances between primary particles in agglomerates are much larger than the thickness of a layer of disturbed interfacial water unfrozen at $T<273$ K. The distribution functions of voids filled by unfrozen water (Figure 1.74e and f) demonstrate not only rearrangement of silica nanoparticles occurred in the aqueous suspension compared to the powder but also the formation of new pores between ice crystallites. This effect is observed even at low PG content since S_{uw}, S_{nano}, V_{nano}, and V_{meso} increase at C_{PG}=0.024 g/g (Table 1.16). Ice nano- and microcrystallites most strongly increase the specific surface area and volume filled by unfrozen water that is observed for samples 5 and 6. The S_{uw} and V values increase by approximately three times for sample 6 in comparison with A-300 alone. Notice that structural rearrangement of the interfacial layer is not a linear function of the PG content that is observed for S, V, and $\langle T \rangle$ values (Figure 1.74). These effects can be caused by changes in the structure of PG nanoparticles (nanocrystallites).

Ab initio quantum chemical calculations of a hydrated PG molecule in the free (Figure 1.75a) and adsorbed states (Figure 1.75b) show that the hydration shell changes upon the PG adsorption onto a silica cluster.

This shell becomes slightly disordered and certain water molecules change their orientation with respect to the free (without silica cluster) complex with PG. The ice-nucleating ability of PG is caused by the same distance between oxygen atoms in PG (0.471 nm by HF/6-31G(d,p)) and hexagonal ice (r_{OO}=0.470 nm for molecules in the second coordination sphere). Additionally, in the case of the water complexes with PG molecules, the oxygen atoms from PG have a smaller number of degrees of freedom compared to the water molecules in the same positions in a water droplet. Therefore, distortion of the water shell of the PG molecule on the adsorption can cause diminution of its ice-nucleating ability. This assumption is confirmed by other calculated values. For instance, the average value δ_H=4.33 ppm (calculated using GIAO method with B3LYP/6-31G(d,p)//6-31G(d,p)) for the H atoms from water molecules participating in the hydrogen bonding in the PG/water complex (average energy of a hydrogen bond is equal to E_H=−41.7 kJ/mol) is higher than that of a water droplet with $18H_2O$ at δ_H=3.22 ppm and E_H=−38.5 kJ/mol and for the PG/$9H_2O$ complex adsorbed onto the silica

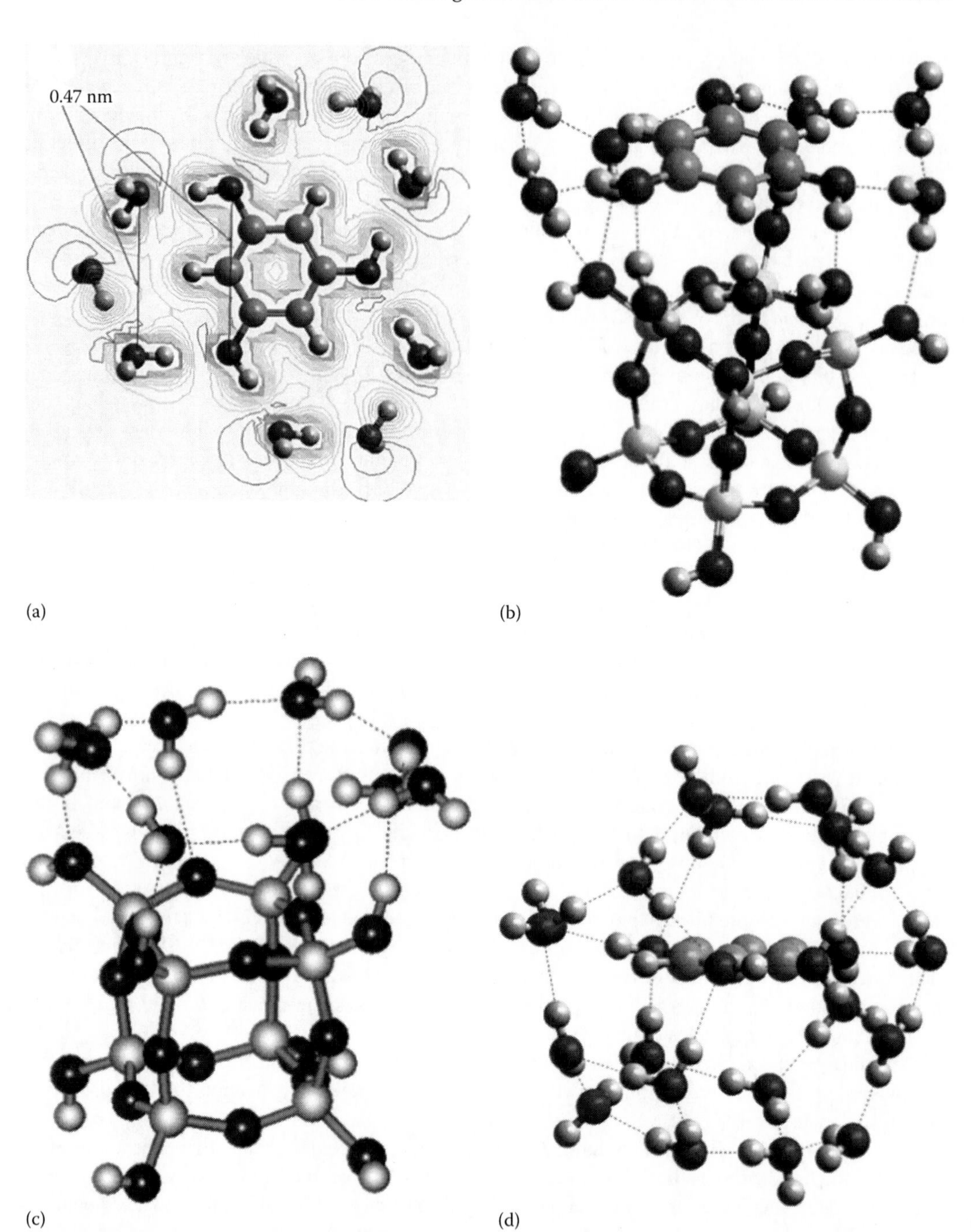

FIGURE 1.75 Hydrogen bonding of PG with nine water molecules in (a) the free complex (average $\delta_H = 4.33$ ppm for H from water molecules participating in the hydrogen bonds, B3LYP/6-31G(d,p)//HF/6-31G(d,p)), (b) on adsorption onto a silica cluster (average $\delta_H = 3.97$ ppm only for H from water molecules participating in the hydrogen bonds), electrostatic potential is shown in the plane of PG (increment 0.05 a.u.), (c) nine water molecules adsorbed onto the silica cluster (average $\delta_H = 3.51$ ppm), the geometry was optimized using the HF/6-31(d,p) basis set, and (d) hydrogen bonding of PG molecules with $20H_2O$. (Adapted from *Colloids Surf. A: Physicochem. Eng. Aspects*, 278, Gun'ko, V.M., Turov, V.V., Barvinchenko, V.N. et al., Characteristics of interfacial water at nanosilica surface with adsorbed 1,3,5-trihydroxybenzene over wide temperature range, 106–122, 2006b. Copyright 2006, with permission from Elsevier.)

cluster at $\delta_H = 3.97$ ppm (only for H from water molecules) and $E_H = -33.9$ kJ/mol. However, in the case of the adsorption of nine H_2O molecules (the solvate shell of the PG molecule in the complex shown in Figure 1.75b with the geometry optimized after removal of the PG molecule) average $\delta_H = 3.51$ ppm. These results show certain loosening of adsorbed water cluster around the PG molecule due to its adsorption accompanied by changes in the ordered effect of PG molecules on water. Calculations of the distribution functions of the distances between atoms O–O and O–H in water droplets individual or with a PG molecule or on the adsorption onto the silica cluster (Figure 1.76) show that a portion of the hydrogen bonds becomes stronger but another portion becomes weaker on interaction of water/PG with silica (see $f(r_{OO})$ at $0.2 < r_{OO} < 0.5$ nm and $f(r_{OH})$ at $0.15 < r_{OH} < 0.3$ nm with the displacement of certain peaks toward shorter r values but other peaks toward larger r values).

On the other hand, a PG molecule has a weaker effect on the water structure than silica because the corresponding displacements of the $f(r)$ peaks are smaller (however, this is a concentration dependent factor). The $f(r)$ distribution function calculated on the basis of the x-ray diffraction (XRD) data of wetted silica using a regularization procedure is in agreement with the function, which was calculated from PM3 geometry, at $r > 0.4$ nm (Figure 1.76c). Observed difference at $r < 0.4$ nm is caused by silica (its Si–O and O–O strongly contribute $f(r)$ at $r < 0.35$ nm) present in larger amount than water in the corresponding XRD sample. Notice that the initial geometry of the relaxed water droplet with $18H_2O$ corresponds to hexagonal ice (Figure 1.77c).

This geometry significantly changes upon the relaxation (Figure 1.77a and b) to diminish the free energy at the droplet surface. In the case of the $302H_2O$ droplet, the water structure corresponds to liquid state with ordered small clusters. The initial structure of the outer layer of this droplet was close to that of the **icosahedral** nanodomain with $280H_2O$ with $120H_2O$ at a surface (Gun'ko et al. 2005d, Chaplin 2011). Thus, obtained results (Figures 1.75 through 1.77) suggest that the hydrogen bonds between PG and water molecules are slightly weakened near a silica surface because of the disordering effect of the silica surface. Therefore, one may expect certain diminution of the ice-nucleating ability of the PG molecules from the interfacial layer (in the absence of PG nanocrystallites) at the nanosilica surface. This effect can lead to changes in the amounts of amorphous interfacial ice depending on the PG concentration, that is, at the C_{PG} values smaller and higher than that of the monolayer coverage.

An electrostatic component is one of the main contributors to the attractive interaction in the hydrogen bonds. For instance, according to the Kitaura–Morokuma analysis (using HF/6-31G(d,p)) of the complex shown in Figure 1.75a, the average energy of the hydrogen bonding is equal to –33 kJ/mol with the electrostatic and charge transfer energies of –27 and –31 kJ/mol, respectively, and the electron-exchange repulsion energy is equal to 29 kJ/mol. The corresponding energies for water dimer are smaller and equal to –21, –17, –19, and 16 kJ/mol, respectively. This result is in agreement with the IR spectra (Figures 1.63 through 1.65) because the intensity of the IR band of SBW at 3300 cm^{-1} grows with increasing amount of adsorbed PG. Ab initio calculations of the IR spectra of PG (Figure 1.78a), $PG + H_2O$ (Figure 1.78b), and $PG + 9H_2O$ (Figure 1.78c) show that the intensity of the mentioned band increases with increasing amount of water molecules interacting with the PG molecule. In the case of a larger droplet formed around the PG molecule (Figure 1.75d), the average value $\delta_H = 3.54$ ppm for H atoms (from water molecules) participating in the hydrogen bonding is slightly smaller than that in the complex of the PG molecule with nine water molecules (Figure 1.75a). This is due to the formation of the hydrogen bonds close to the aromatic ring of the PG molecule (Figure 1.75d), whose π-electrons can increase the proton shielding of closely located water molecules, despite the formation of an empty cavity near the aromatic ring.

Modeling of the dynamic behavior of the PG/water adsorption complex at 150 K and 300 K (Figure 1.79) using PM3 method reveals that similar complexes can be decomposed at room temperature (or slightly higher) if the PG adsorption occurs onto the outer (open) surface of silica aggregates (desorption of a portion of PG at room temperature was observed for the stored sample 6 with the maximal C_{PG} value when the PG adsorption can occur on the outer surfaces of aggregates).

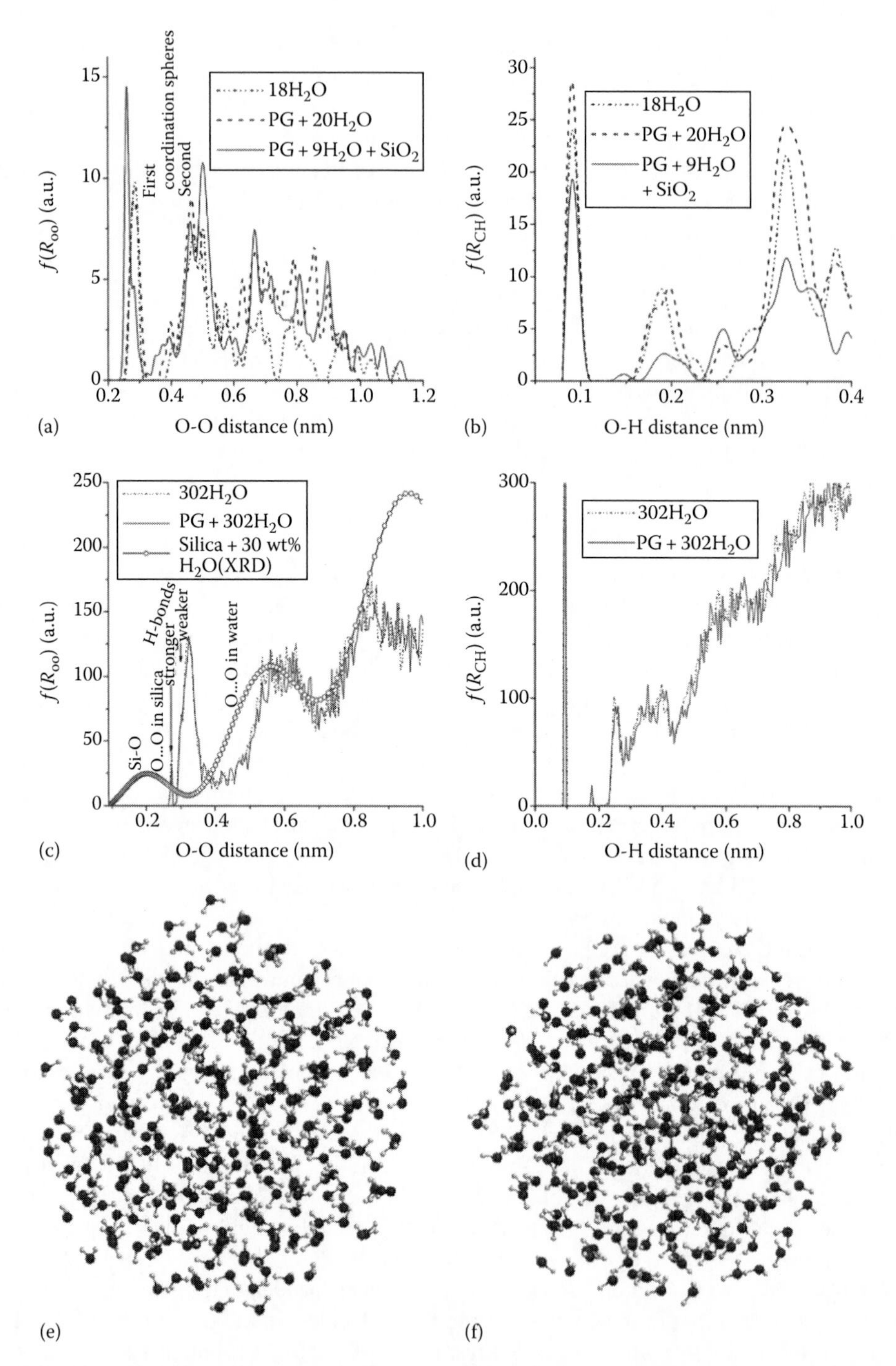

FIGURE 1.76 Distribution functions of distances between atoms (a, c) O-O and (b, d) O-H in water droplets with (a, b) 9, 18 (shown in Figure 1.77a), and 20 (shown in Figure 1.75d) and (c, d) 302 molecules individual or with a PG molecules or on adsorption onto a silica cluster (Figure 1.75b) calculated using (a, b) HF/6-31G(d,p) and (c, d) PM3 methods, water droplet with 302 molecules (e) without and (f) with PG molecule and the geometry optimized using the PM3 method, (c) $f(r)$ calculated on the basis of the XRD data was shown for wetted silica with 30 wt% water. (Adapted from *Colloids Surf. A: Physicochem. Eng. Aspects*, 278, Gun'ko, V.M., Turov, V.V., Barvinchenko, V.N. et al., Characteristics of interfacial water at nanosilica surface with adsorbed 1,3,5-trihydroxybenzene over wide temperature range, 106–122, 2006e. Copyright 2006, with permission from Elsevier.)

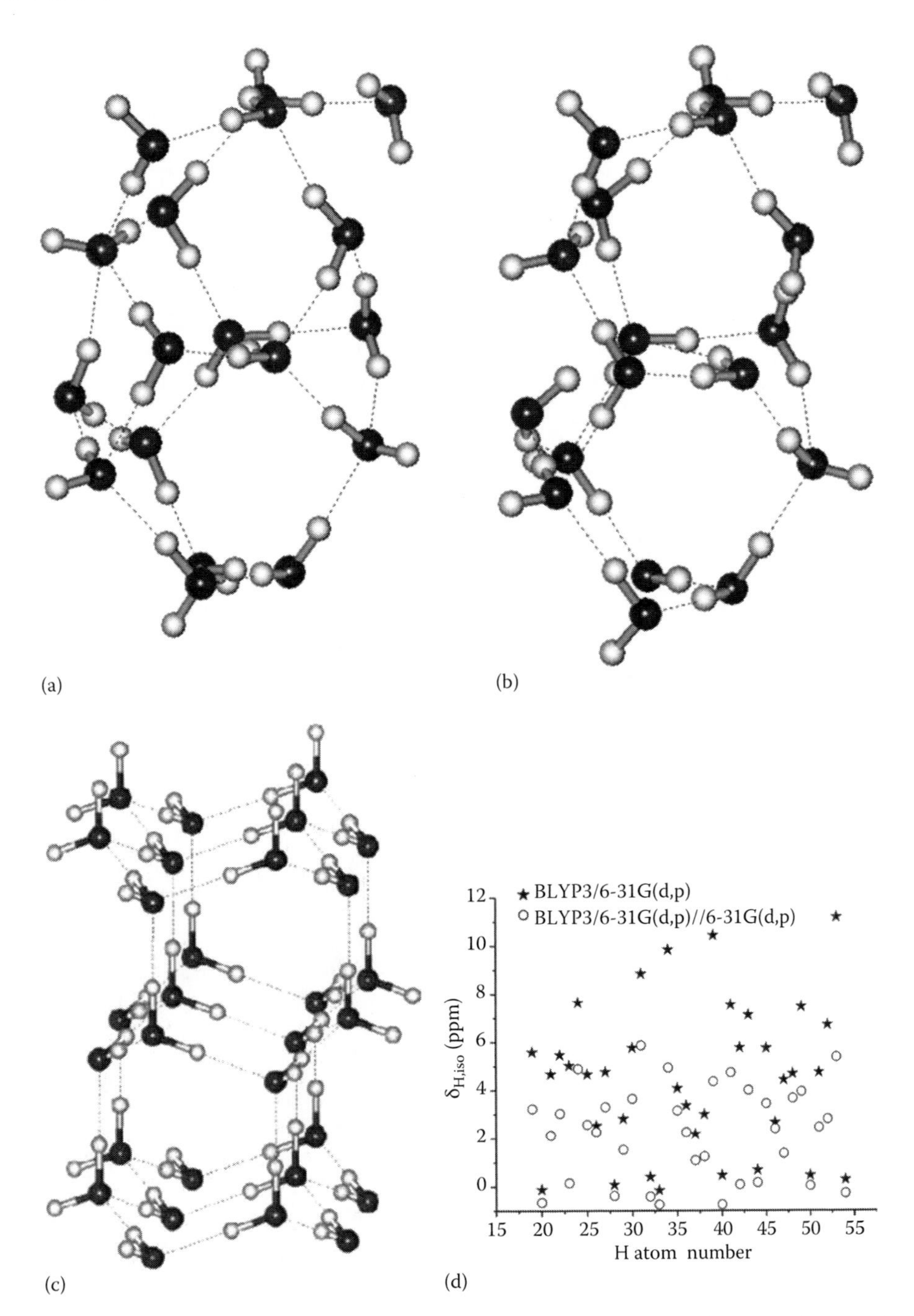

FIGURE 1.77 Water droplet ($18H_2O$) geometry optimized by using (a) B3LYP/6-31G(d,p) (accessible area 2.734 nm^2) and (b) HF/6-31G(d,p) (accessible area 3.021 nm^2) basis sets, average $\delta_H = 3.22$ ppm (4.47 ppm at B3LYP/6-31G(d,p) for optimization of the geometry and NMR spectrum calculations) for H atoms participating in the hydrogen bonds at B3LYP/6-31G(d,p)//HF/6-31G(d,p), (c) structure of hexagonal ice, and (d) δ_H values (GIAO method) for all water molecules in the droplet calculated by using B3LYP/6-31G(d,p) and B3LYP/6-31G(d,p)//HF/6-31G(d,p). (Adapted from *Colloids Surf. A: Physicochem. Eng. Aspects*, 278, Gun'ko, V.M., Turov, V.V., Barvinchenko, V.N. et al., Characteristics of interfacial water at nanosilica surface with adsorbed 1,3,5-trihydroxybenzene over wide temperature range, 106–122, 2006e. Copyright 2006, with permission from Elsevier.)

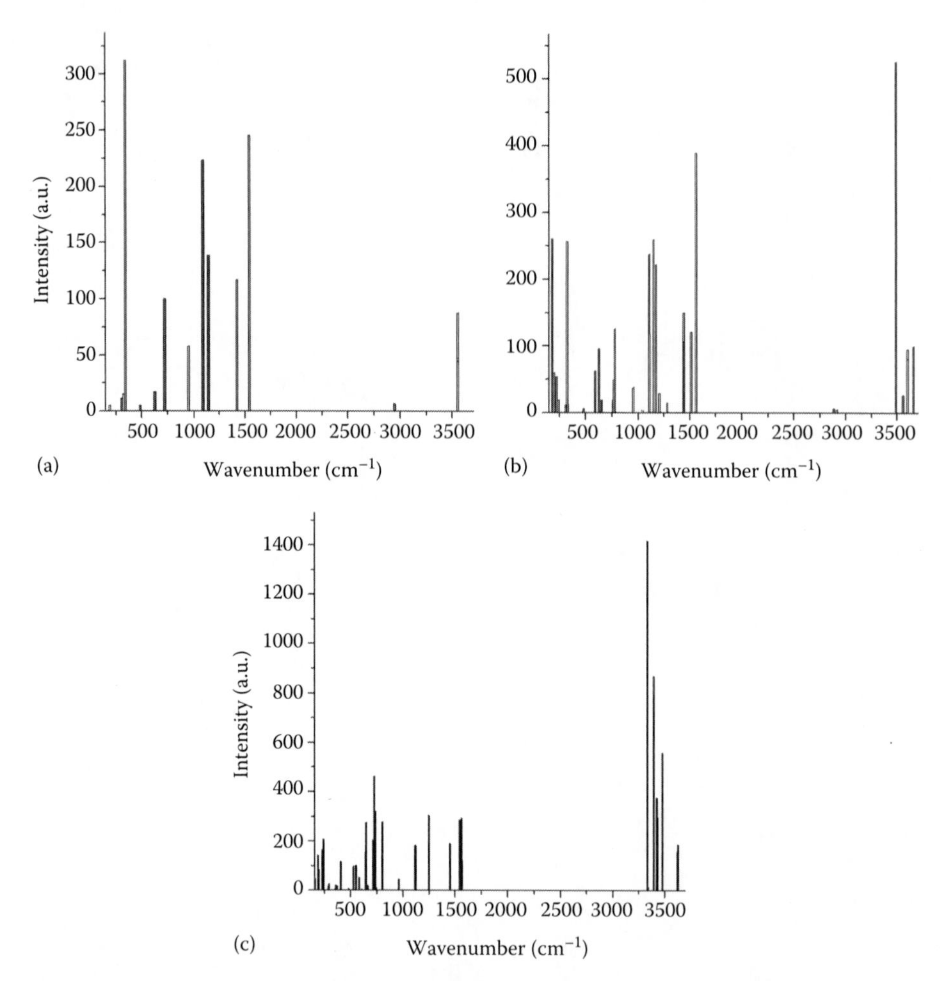

FIGURE 1.78 IR spectra calculated using (a) B3LYP/6-311(d,p) (normalizing factor 0.926), (b, c) HF/6-31(d,p) (normalizing factor 0.8585) for (a) individual PG molecule, (b) PG+H_2O with the hydrogen bond CO-H···OH_2, and (c) PG+9H_2O (complex shown in Figure 1.74a).

However, decomposition of similar complexes in pores (or voids between adjacent primary particles in their aggregates) occurs more slowly (Figure 1.80) because of the steric effect caused by the pore walls and stronger adsorption potentials in nanopores in comparison with that of the outer (open) surfaces of aggregates.

The complex at the lower temperature is more stable, but the diffusion of water molecules at room temperature can lead to changes in its structure (Figures 1.75b and 1.79). It should be noted that changes in the potential energy at the mentioned temperatures significantly differ (Figure 1.79b and d) due to total decomposition of the complex at 300 K (Figure 1.79d). Results of modeling of the dynamic behavior of PG/water/nanosilica interface are in agreement with the NMR data, which show the mobility of adsorbed water molecules at temperatures significantly lower than 273 K because of lower number of the hydrogen bonds than that in ice and their significant bending that lead to diminution of the activation energy of the diffusion of interfacial water molecules in comparison with bulk ice.

Thus, the structure and the temperature behavior of interfacial water in the systems with water/PG/nanosilica depend strongly on the concentration of PG possessing the ice-nucleating ability. This effect is maximal at the PG content higher than that required to form the monolayer coverage. The presence of PG causes the formation of a thick layer of interfacial amorphous ice at $T<273$ K.

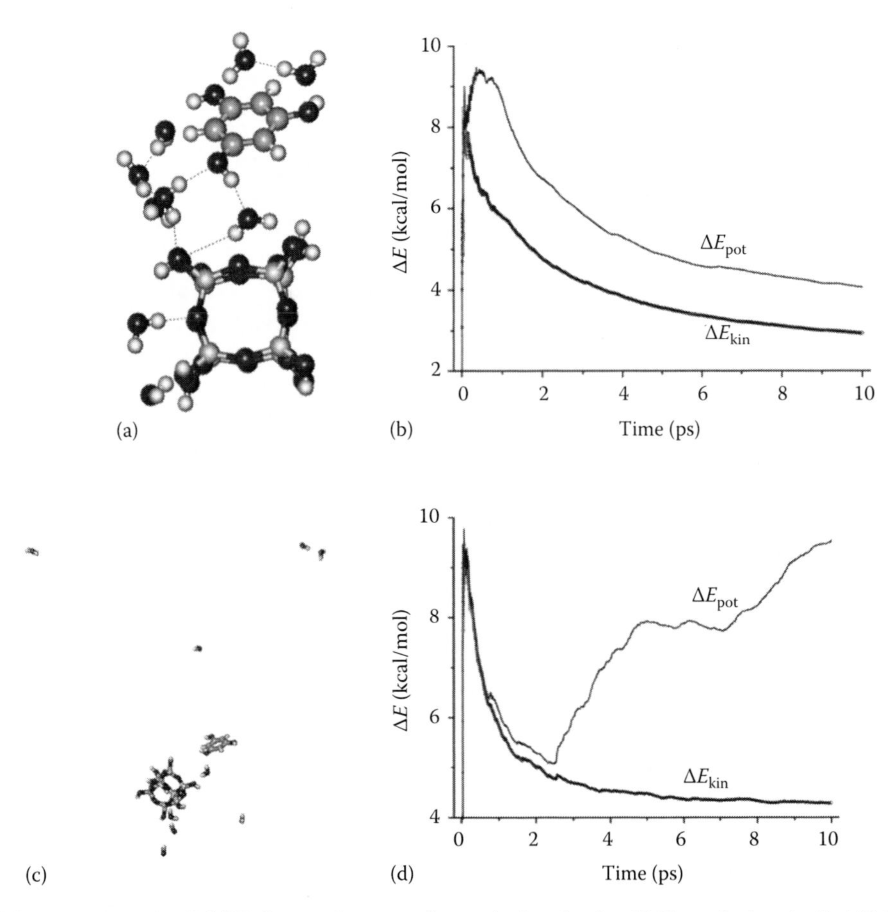

FIGURE 1.79 Complex PG/9H_2O at a cluster surface calculated using PM3 method at (a, b) 150 K (c, d) 300 K in 10 ps (starting point corresponds to the equilibrium geometry calculated using HF/6-31G(d,p)). (Adapted from *Colloids Surf. A: Physicochem. Eng. Aspects*, 278, Gun'ko, V.M., Turov, V.V., Barvinchenko, V.N. et al., Characteristics of interfacial water at nanosilica surface with adsorbed 1,3,5-trihydroxybenzene over wide temperature range, 106–122, 2006e. Copyright 2006, with permission from Elsevier.)

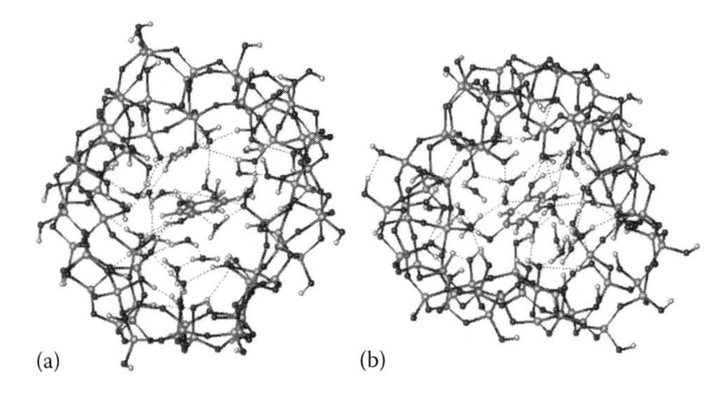

FIGURE 1.80 Molecule PG with a solvate shell (20H_2O) in a silica nanopore (diameter ≈ 0.8 nm) (PM3 geometry): (a) initial and (b) after heating at 300 K for 1 ps. (Adapted from *Colloids Surf. A: Physicochem. Eng. Aspects*, 278, Gun'ko, V.M., Turov, V.V., Barvinchenko, V.N. et al., Characteristics of interfacial water at nanosilica surface with adsorbed 1,3,5-trihydroxybenzene over wide temperature range, 106–122, 2006e. Copyright 2006, with permission from Elsevier.)

The mobility of water molecules in this layer is slower than that of the molecules adsorbed onto nanosilica without PG but it is higher than that of the molecules in bulk ice. The overall changes in the Gibbs free energy of the interfacial water layer in the water/PG/nanosilica system is a nonlinear function of the PG concentration due to a stronger influence of PG on the structure of interfacial water with increasing C_{PG} value. Nonlinear changes in the weight loss are observed in the TG curves for PG/silica samples at various C_{PG} values. The ice-nucleating ability of adsorbed PG can be a nonlinear function of its content due to changes in the structures of adsorbed complexes with increasing PG loading because of formation of PG oligomers at high C_{PG} values. These investigations show the ways of control of the ordering effect (including ice-nucleating ability) of adsorbed PG (as well as other compounds) on interfacial water. This effect depends also on treatment conditions of water/PG/solid nanoparticles because the structure of secondary particles of nanosilica, as well as oligomers of adsorbed PG molecules, is not stable and can change on treatment.

Standard nanosilica A-300 was impregnated by 0.2–1.0 mmol/g of 1,5-DON from water–ethanol solution, dried, and studied in the form of tablets with a mixture with KBr (1:5) using a ThermoNicolet Nexus FTIR spectrophotometer. The interaction of DON with a surface of nanosilica was estimated by changes in the FTIR spectra related to OH-stretching vibrations of free surface silanols (ν_{OH} at 3748 cm^{-1}) depending on the DON amounts of 0.2–1.0 mmol/g (Figure 1.81a). The IR bands of stretching and deformation vibrations of bonds in adsorbed DON molecules as well as the ν_{OH} vibrations of surface silanols can be identified in these spectra. The intensity of the ν_{OH} band of free silanols decreases with increasing amounts of adsorbed DON (C_{DON}) because of the hydrogen bonding of these molecules to the silanols. A sharp increase in the degree of distortion of free silanols by the DON molecules is observed at $C_{DON}<0.5$ mmol/g (Figure 1.81b). Notice that the S_{IR} values for DON/A-300 are relatively high (300–338 m^2/g) and are in agreement with a relatively low degree of distortion of surface silanols (Figure 1.81b). This result can be explained by the adsorption of DON in the form of clusters but not in the form of individual molecules. This is characteristic of polyaromatic compounds because of the great energy of dispersion interactions upon their cluster formation.

The initial ethanol solution of 1,5-DON ($C=0.001$ M) was prepared using DON purified by sublimation. This solution was diluted to $C=0.0001$ M for UV measurements. The pH value was changed by addition of HCl or NaOH and controlled using a pH meter with a glass electrode. The absorption electron spectra of the solutions and the reflectance spectra of solid samples were recorded using a

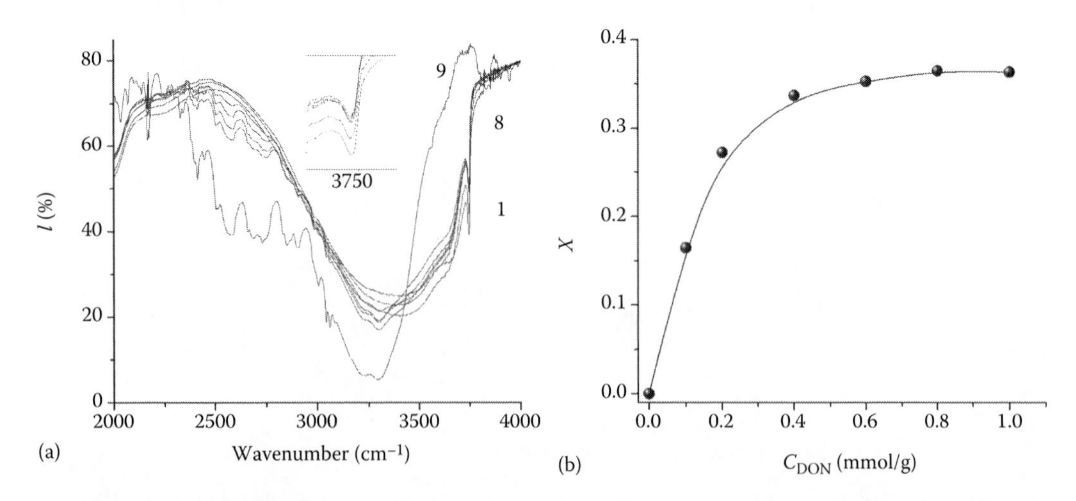

FIGURE 1.81 (a) FTIR spectra (in a transmission mode) of nanosilica with adsorbed DON at $C_{DON}=0$ (1), 0.1 (2), 0.2 (3), 0.4 (4), 0.6 (5), 0.8 (6), 1.0 (7), and 1.5 mmol/g (8) and (b) degree of distortion of surface silanols ($\nu_{OH}=3748$ cm^{-1}) as a function of the adsorbed amount of DON.

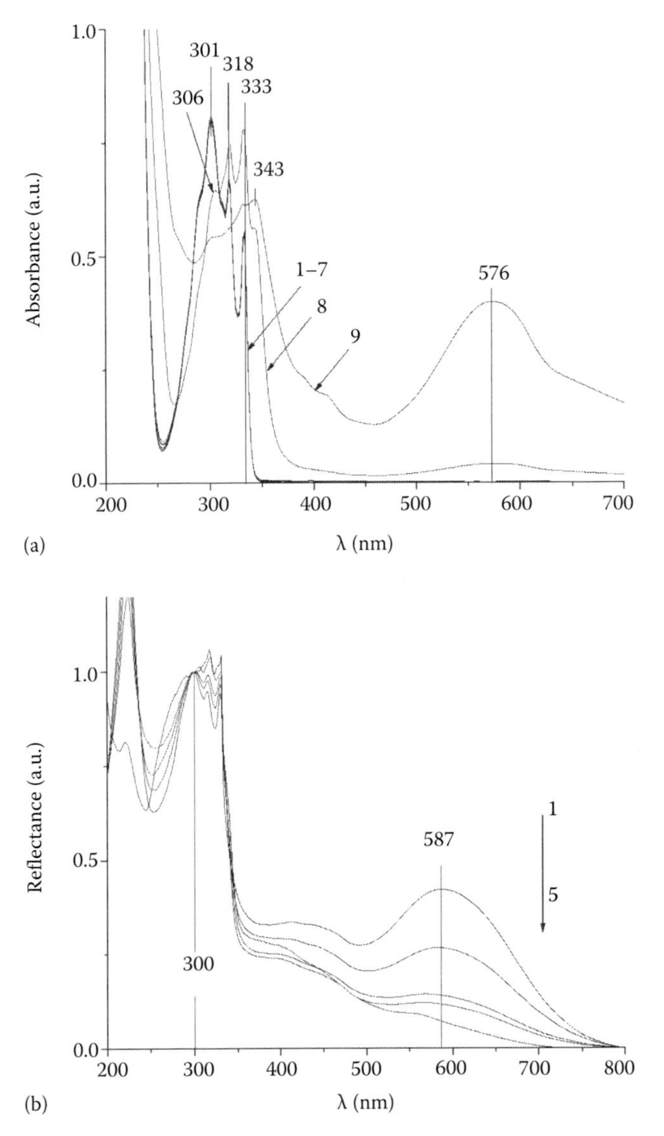

FIGURE 1.82 (a) The absorbance spectra (l = 1 cm) of the ethanol solution of DON at C_{DON} = 0.0001 M at different pH values 2.7 (1), 3.05 (2), 3.3 (3), 4.05 (4), 8.3 (5), 8.5 (6), 9 (7), 11.6 (8), and 12.65 (9) and (b) reflectance spectra of DON adsorbed onto nanosilica at C_{DON} = 0.1 (1), 0.4 (2), 0.6 (3), and 0.8 mmol/g (4) and pure 1,5-DON (5).

Specord M-40 (Carl Zeiss). The impregnation of nanosilica A-300 was carried out by the ethanol solution of DON at pH = 8.3. To estimate the stability of the organic-mineral system, changes in the electronic spectra of adsorbed DON at different pH values at a constant C_{DON} value (Figure 1.82a).

The ethanol solutions of DON are stable at pH = 2.7–9.0 (Figure 1.82a) as they are characterized by the same bands at λ_{max} = 301, 318, and 333 nm that correspond to the molecular form of 1,5-DON. These spectra do not change after aging for 27 h. Consequently, DON does not transform during preparation procedures over this pH range. However, certain changes in the spectra are observed at high pH values because there are a displacement of λ_{max}, an increase in the intensity at larger wavelength, and an appearance of the band at λ_{max} = 576 nm. These effects are caused by deprotonation of hydroxyl groups of DON as well as oxidation of certain impurities. Figure 1.82b

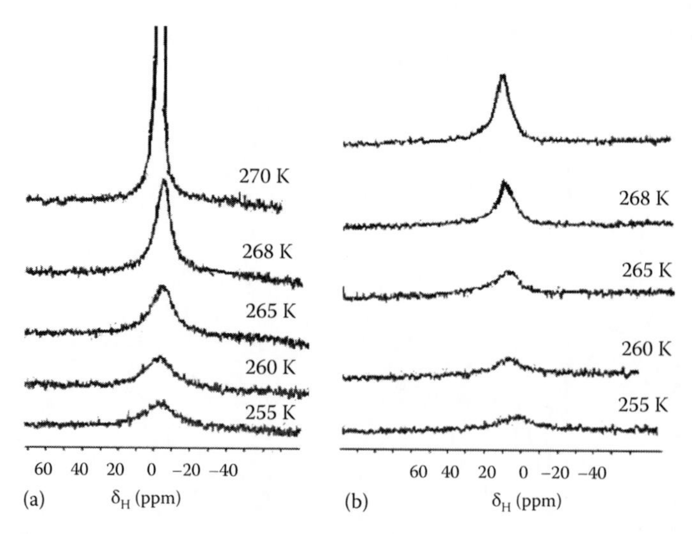

FIGURE 1.83 Temperature dependence of the ^{1}H NMR spectra of water at the interfaces of nanosilica particles/ice in the presence of adsorbed 1,5-DON at C_{DON} = (a) 1.6 wt% and (b) 14 wt%.

shows the reflectance spectra of DON adsorbed onto nanosilica at different C_{DON} values and pH 8.3. The intensity of the band at $\lambda_{max} = 576$ nm decreases with increasing content of DON.

The characteristics of the water layers at the interfaces of nanosilica particles with grafted 1,5-DON (C_{DON} = 1.6–14 wt%) were investigated by using LT ^{1}H NMR spectroscopy according to the techniques described in previous chapters. The ^{1}H NMR spectra of interfacial water recorded at T = 255–270 K for two samples with minimal (1.6 wt%) and maximal (14 wt%) C_{DON} values are shown in Figure 1.83.

The spectra can be assigned to adsorbed unfrozen water, that is, signal of "mobile" ice (observed in the case of adsorbed PG) is absent. Broad component of the unfrozen water is absent even in the case of temperatures close to 273 K, characteristic of freezing of the bulk water. Temperature dependence of the content of unfrozen water (C_{uw}) and changes in Gibbs free energy of this water with respect to the C_{uw} values are shown in Figure 1.84. The characteristics of bound water layers and Gibbs free surface energy that are calculated on the basis of these graphs are shown in Table 1.17.

The Gibbs free surface energy (γ_S) of the initial nanosilica in the aqueous medium is close to 50 J/g. Therefore, obtained results reveal that the adsorption modification of a surface of nanosilica by 1,5-DON does not lead to significant changes in its hydration characteristics. There are two portions in the $\Delta G(C_{uw})$ graphs (Figure 1.84), as well as in the case of the initial silica, related to both strongly and WBWs. The thickness of the layer of SBW is near the same for all the studied samples and corresponds to 500–600 mg/g. There are larger changes in the C_{uw}^{w} value. There is tendency of a decrease in the amounts of WBW with increasing content of adsorbed DON. However, even in the case of samples 4 and 5 with minimal amounts of bound water, the thickness of the bound water layer corresponds to several statistical monomolecular water layers. This tendency is also observed for concentration dependence of γ_S (Table 1.17). This value changes not more than by 20% with changing amounts of adsorbed DON. One can assume that nanosilica with adsorbed DON does not have the ice-nucleating ability because this ability of highly disperse particles is observed when the thickness of bound water layer is not larger than 1–2 statistical monolayers.

Quantum chemical calculations of the complexes with 1,5 DON molecules with water molecules and silica clusters have been carried out by using ab initio (HF/6-31G(d,p), DFT/B3LYP/6-31G(d,p)) and semiempirical (PM3) methods. Comparison of the structure of the complex of 1,5 DON molecule with eight water molecules (Figure 1.85a) with that of hexagonal ice or clusters of

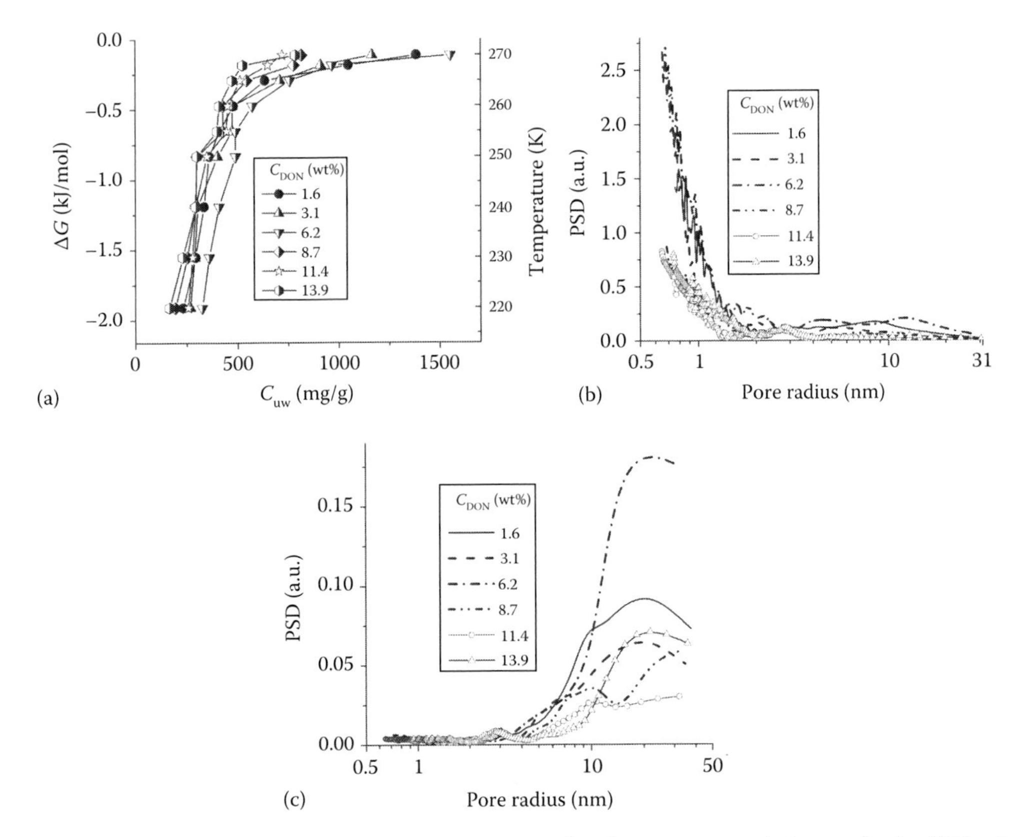

FIGURE 1.84 (a) Temperature dependence of content of unfrozen water and changes in the Gibbs free energy of interfacial water in the aqueous suspensions of nanosilica A-300 at different amounts of adsorbed 1,5-DON; PSD (b) differential and (c) incremental of voids filled by unfrozen water ($k_{GT}=60$ K nm).

this ice after the geometry optimization (Figure 1.77) shows that these structures strongly differ. Additionally, the distance between oxygen atoms in water molecules of the second coordination sphere in ice and in the corresponding clusters or in the clusters with PG (0.470–0.476 nm) is shorter than that in cluster with 1,5-DON (0.5–0.515 nm). Consequently, the ice structure around of a DON molecule can be distorted. Therefore, average value of the chemical shift δ_H of hydrogen atoms (in water molecules) participating in the hydrogen binding is 3.42 ppm that is lower than that in the water clusters without DON (Figure 1.85). In the case of water molecules interacting with a PG molecule (Figure 1.72a), $\delta_H=4.33$ ppm. Calculations of different complexes with 1–2 DON molecules, 1–2 silica clusters, and 6–39H_2O (Figure 1.85) reveal that the ice geometry can be strongly distorted. Thus, 1,5-DON in form of a molecule or two-molecule cluster free or adsorbed onto silica has lower ice-nucleating ability than PG molecules. However, in the case of the slab structure with certain position of neighboring molecules, the distance between oxygen atoms is 0.472 nm (Figure 1.85e) that corresponds to marked ice-nucleating ability of DON microcrystallites because this distance corresponds to the distance between oxygen atoms in hexagonal ice. Consequently, the results of theoretical modeling explain diminution of the ice-nucleating ability of DON adsorbed onto nanosilica because the DON molecules can be distributed as individual molecules or clusters with the geometry inappropriate to form ice crystallites; that is, they do not have the ice-nucleating ability.

Crystallites of compounds, which do not have the ice-nucleating ability, can affect the ice formation because this process at a surface of organic particles depends on many factors such as the geometry of hydrophilic and hydrophobic sites, the thickness and the structure of adsorbed

TABLE 1.17
Characteristics of Interfacial Water for Nanosilica/DON Samples

C_{DON} (wt%)	γ_S (J/g)	C_{max} (mg/g)	ΔG_S (kJ/mol)	C^s_{uw} (mg/g)	C^w_{uw} (mg/g)	V_{nano} (cm³/g)	V_{meso} (cm³/g)	V_{macro} (cm³/g)	S_{uw} (m²/g)	S_{nano} (m²/g)	S_{meso} (m²/g)	S_{macro} (m²/g)	$\langle T \rangle$ (K)
1.6	59.4	1736	−2.83	500	1236	0.162	1.427	0.073	501	370	127	5	255.3
3.1	61.4	1382	−2.96	600	782	0.230	1.046	0.050	652	550	99	4	250.0
6.2	75.8	2251	−3.21	500	1751	0.255	1.579	0	764	621	143	0	253.6
8.7	49.9	1089	−3.11	500	589	0.152	0.792	0.059	437	362	70	5	248.4
11.4	52.5	843	−3.24	500	343	0.201	0.566	0	546	492	54	0	240.2
13.9	44.3	1077	−2.88	600	477	0.123	0.812	0.064	356	282	69	5	250.9

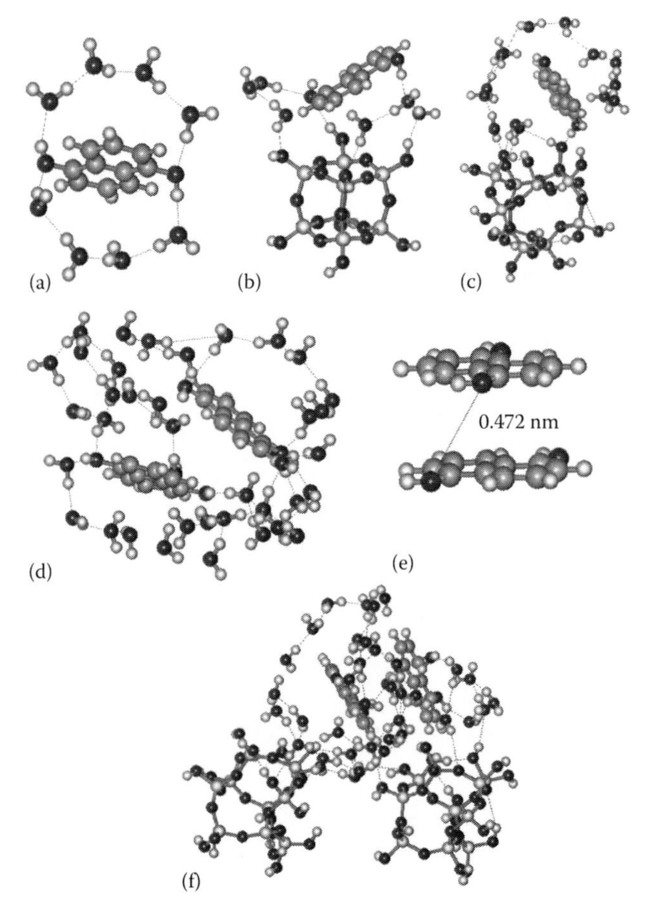

FIGURE 1.85 Structures of a cluster of eight water molecules upon interaction with 1,5-DON (a) alone or (b) with a silica cluster (6-31G(d,p)); or (c) a larger cluster (PM3); (d) two molecules of 1,5-DON interacting with water molecules (PM3) and (e) with no water; (f) complex with two silica clusters, two molecules of 1,5-DON and $39H_2O$ (PM3).

water layer, the size of crystallites, and the morphology of solid particles. This phenomenon was studied by the example of a mechanical blend of such tri-ring compounds as phenazine $C_6H_4N_2C_6H_4$ (possessing ice-nucleating ability, the distance between N atoms in the molecule is 0.28 nm, which is close to the OO distance in ice) and anthraquinone $C_{14}H_8O_2$ (passive compound, the distance between O atoms in the molecule is about 0.53 nm, which does not match to any OO distance in ice). Seven samples at concentration of phenazine $C_{phen}=0\%–100\%$ were prepared and studied by the ^{1}H NMR spectroscopy. Microcrystalline powders were prepared by stirring of the corresponding blends using a ball mill. The size of microcrystallites was not larger than 1 μm. A sample (100–200 mg) was placed in 4 mm ampoule (length 15–25 cm), which then was filled by water so that the powder was wetted. The experiments are the same with what has been described earlier.

Temperature dependence of the ^{1}H NMR spectra of frozen aqueous suspensions of pure phenazine is shown in Figure 1.86.

Broad and narrow components of the signal are observed in Figure 1.86. The narrow component at the width of 3–4 ppm is caused by water adsorbed onto a surface of organic crystallites. The broad band (width > 100 ppm) is linked to an ice layer located near the surface of crystallites. This ice is characterized by middle mobility of the water molecules (between the mobility of molecules

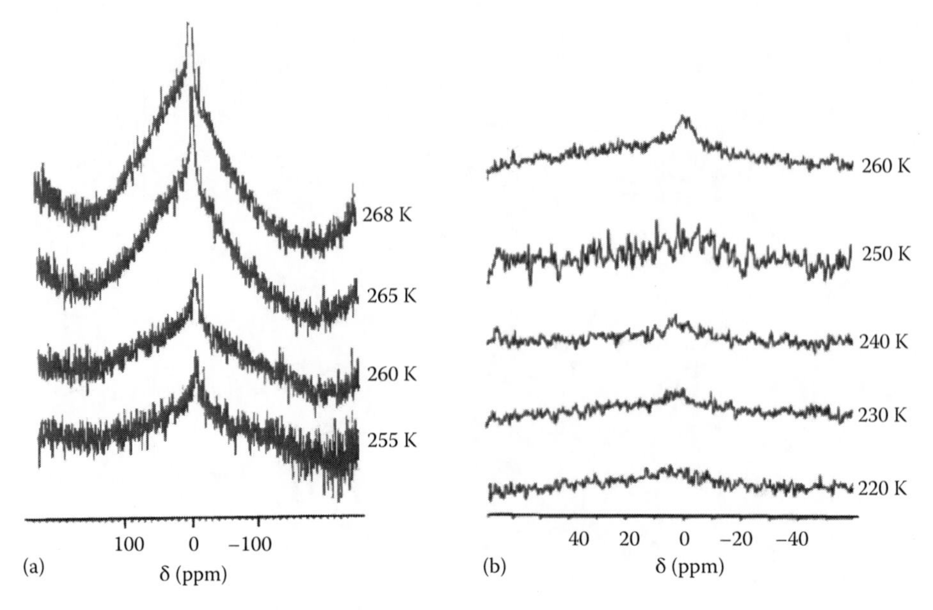

FIGURE 1.86 Temperature dependence of the ^{1}H NMR spectra of the aqueous suspension of microcrystalline phenazine.

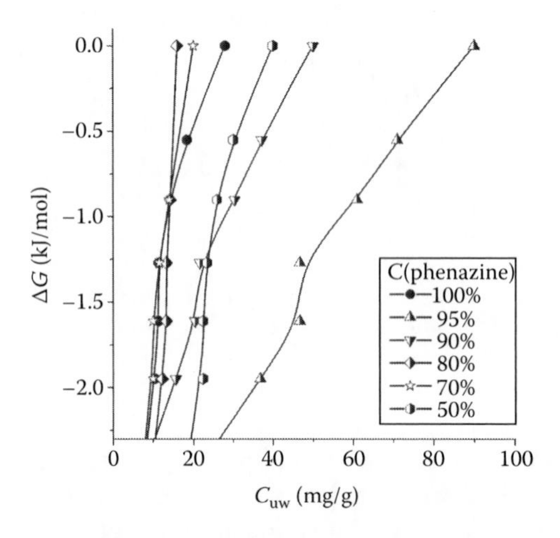

FIGURE 1.87 Dependences of changes in the Gibbs free energy versus the concentration of unfrozen water at different ratio between the amounts of phenazine and anthraquinone.

in unfrozen adsorbed water and bulk ice). Temperature dependences of signals for other samples have practically the same shape. The intensity of both signals decreases with lowering temperature. However, the narrow component is observed over the total temperature range of 220–273 K in contrast to the broad component, which can be observed only at $T > 250$ K.

Figure 1.87 shows dependences of changes in the Gibbs free energy ΔG versus concentration of unfrozen water (calculated from the integral intensity of the narrow component of signal) for all the studied mixtures with phenazine and anthraquinone using the temperature range characterized by the absence of the broad component. These dependences are close to linear and show changes in ΔG for SBW. Integration of the $\Delta G(C_{uw})$ graphs extrapolated to the axes allows us to estimate the interfacial Gibbs free energy linked to the layer of SBW (Table 1.18).

TABLE 1.18
Characteristics of Water Layers Adsorbed onto a Surface of Particles with Phenazine and Anthraquinone

C_{phen} (wt%)	ΔG_s (kJ/mol)	C_s (mg/g)	γ_S (J/g)
100	−4	28	2.3
95	−3.2	90	7.8
90	−3	50	3.7
80	−4.5	16	2
70	−4	20	2.1
50	−5	40	4.9

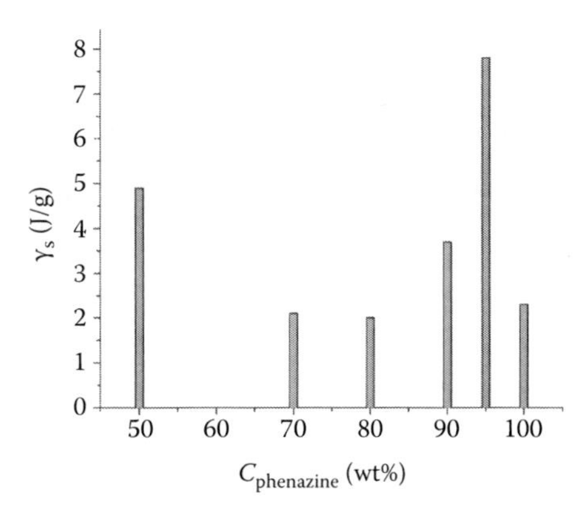

FIGURE 1.88 Interfacial free energy as function of the phenazine content.

The ΔG^s value characterizing maximum changes in the Gibbs free energy of adsorbed water layer has the maximum value (by modulus) for a mixture with 50% of phenazine. However, the maximum thickness of the bound water layer, as well as γ_S (Figure 1.88), is for a mixture at $C_{phen}=95\%$.

The maximum ice-nucleating ability of dustlike compounds is in inverse proportion to the thickness of SBW (Turov et al. 1998a). Consequently, one can expect that the maximum ice-nucleating ability of the mixtures with phenazine and anthraquinone corresponds to the samples at $C_{phen}=70\%–80\%$. This assumption is in agreement with direct measurements of the ice-nucleating ability of these compounds carried out using a climatic chamber. Notice that the ice-nucleating ability of phenazine can have the same origin that 1,5-DON because the distance between N atoms can correspond to the distance in oxygen atoms in water molecules in ice only in the case of the N atoms from neighboring phenazine molecules in slabs. However, their packing method should be different because the N atoms in phenazine and O atoms in 1,5-DON are in different positions.

Thus, the dependence of the ice-nucleating ability of the mixtures with two organics with strongly different the ice-nucleating ability is nonlinear. Addition of 5%–20% of passive anthraquinone to active phenazine reduces the activity of the latter. The same effect is observed at $C_{phen}=50\%$. However, the activity is close to that of pure phenazine if $C_{phen}=70\%–80\%$.

The structure and the behavior of interfacial water adsorbed onto PG/nanosilica depends strongly on concentration of an organic compound as an ice-nucleating agent. The presence of PG causes the formation of thick layer of "mobile" ice at the interfaces of solid particles/ice. The mobility of water molecules in this layer is smaller in comparison with that of water molecules adsorbed on nanosilica but it is higher than that of water molecules in bulk ice. Free surface energy (i.e., entire changes in the Gibbs free energy of the interfacial water) of the PG/nanosilica/water systems depends nonlinearly on the PG content because of the effect of PG on the structure of the bound water. Nonlinear changes in the weight loss are observed in the TG graphs for samples of PG/nanosilica/water. Therefore, the ice-nucleating ability of adsorbed PG should be a nonlinear function of its content at a surface. Thus, the ice-nucleating ability of PG in the adsorption state can be controlled by means of the spectral characteristics of the ^{1}H NMR signal of the interfacial water as well as the concentration dependence of the free surface energy.

1,5-DON possessing certain ice-nucleating ability in the microcrystalline state loses this ability upon the adsorption at a surface of nanosilica. A broad band of "mobile" ice, which is typically observed in frozen suspensions of active compounds, is absent in the ^{1}H NMR spectra of 1,5-DON/water adsorbed onto nanosilica independent of the DON amounts. Additionally, the thickness of an unfrozen water layer in the aqueous suspensions of silica/DON corresponds to several molecular diameter of water that is much higher than that in the suspensions of the active compounds. It is possible that this layer of the interfacial water (with the structure that differs from that of ice) prevents contacts between DON and supercooled water.

A mixture of phenazine possessing ice-nucleating ability with inactive anthraquinone has the ice-nucleating ability according to the ^{1}H NMR spectra. Concentration dependence of the ice-nucleating ability of this mixture is nonlinear. Addition of 5%–20% of anthraquinone to phenazine reduces the ice-nucleating ability of the latter. A similar effect is observed at $C_{phen} = 50\%$. At the same time, the activity of the mixture at $C_{phen} = 20\%–30\%$ is close to that of pure phenazine.

1.1.6 Aqueous Suspensions

Light scattering occurs on polarizable solid and liquid particles (or molecules) bathed in the electromagnetic field of the light beam because of the difference in the dielectric properties of the material and the surrounding media (e.g., aqueous), and the varying field induces oscillating dipoles in the particles radiating light in all directions. The intensity of scattered electromagnetic field depends on the ratio between the particle size and the incident light wavelength (λ), and the shorter the λ value, the smaller the particles, which can be effectively investigated due to their light scattering. Therefore, not only the visible light but also x-ray is used in the PCS techniques to increase the resolution power. Similar physical effects (e.g., scattering of phonons) are utilized in the acoustic and electroacoustic techniques applied for investigations of the PaSDs and related phenomena (Hunter 1993, Hubbard 2002, Birdi 2009).

The short-term intensity fluctuations (dynamics) of the scattered light arise from the fact that the scattering particles are in the motion, for example, diffusive Brownian motion (PCS of particles of 10 μm diameter and smaller), own motion of living flagellar microorganisms, or the particle motion under external force (e.g., particle sedimentation because of gravitation, electrophoretic mobility of charged particles in the external electrostatic field). These motions cause short-term fluctuations in the measured intensity of the scattered light. Various terms have been used for this phenomenon: PCS, DLS, QELS, spectroscopy of optical displacement, laser correlation spectroscopy, and others. The pace of the movement is inversely proportional to particle size (the smaller the particles, the faster is their motion or diffusion), and the velocity can be detected by analyzing the time dependency of the light intensity fluctuations scattered from the particles when they are illuminated with a laser beam.

PCS is a powerful and fruitful physical method successfully applied to solve a variety of problems in many technological and scientific branches such as colloidal chemistry, biochemistry, biophysics, molecular biology, and food technology. The utilization of this method in exploration not only of solid nanoparticles but also of biopolymers and microorganisms is effective, since they are appropriate objects for PCS because their characteristic sizes are in the 1 nm–10 μm range (Hunter 1993). The PCS information is of importance to understand certain interfacial phenomena, especially for the systems with nanoparticles and macromolecules studied here using NMR and other methods.

A random motion of nanoparticles in aqueous suspension changes the time intensity of the scattered light and the fluctuating signal is processed by forming the autocorrelation function $G(t)$. For a monodisperse suspension of globular particles in Brownian motion (Dahneke 1983, Chu 1991, Brown 1993, Hunter 1993), it can be written as follows:

$$G(t) = A + B\,exp(-2\Gamma t) \quad (1.32)$$

where

A and B are the constants
Γ is the decay rate
t is the time

This function can be used to determine the decay rate, which is linked to the translational diffusion coefficient D:

$$\Gamma = Dq^2 \quad (1.33)$$

where q denotes the scattering vector. Then the particle diameter d can be determined using the Stokes–Einstein expression for D_0:

$$D_0 = kT/(3\pi\eta(t)d) \quad (1.34)$$

where

k is the Boltzmann's constant
T is the temperature
η denotes the viscosity of the dispersing liquid

Notice that Equation 1.34 can be used only for the dispersions of the noninteracting spherical particles in lowly concentrated suspensions. Due to averaging, the diffusion coefficient distribution is given by

$$D = \frac{\sum Nm^2F(q,d)D}{\sum Nm^2F(q,d)} \quad (1.35)$$

where

N is the number of particles
m is the particle mass
$F(q,d)$ is the particle form factor dependent on the particle size d and the scattering vector q

Also, d_{PCS} can be deduced from decay time τ:

$$d = kT\tau q^2/(3\pi\eta) \quad (1.36)$$

For strongly anisotropic particles, the decay rate of the intensity autocorrelation function depends also on the rotational diffusion coefficient D_R (Pecora 1985) and Equation 1.33 is replaced:

$$\Gamma = Dq^2 + 6D_R \quad (1.37)$$

To characterize the PaSD uniformity, a value of polydispersity (PD; Chen et al. 1981) can be used as a measure of the nonuniformity, and PD can be written as follows:

$$PD = \mu_2/\Gamma^2 \tag{1.38}$$

where

$$\mu_2 = (D^{2*} - D^{*2})q^4 \tag{1.39}$$

$$q = (4\pi n/\lambda)\sin(\Theta/2) \tag{1.40}$$

D^* is the average value
λ is the light wavelength
Θ is the scattering angle (typically 90°–152°)
n is the index of refraction of the suspending liquid

Monodisperse particles correspond to $PD<0.02$, and PD between 0.02 and 0.08 represents a narrow distribution. Notice that features of aggregation (aggregation rates) of the particles and aggregate fractality depend strongly on initial PD of particles, and preparation of monodisperse uniform inorganic colloid dispersions has a great practical importance.

There are such two views on the type of bonding of primary nanoparticles in aggregates as the hydrogen bonds and ≡Si–O–Si≡ linkages due their collision and sticking at high temperatures at the end of the pyrogenic synthesis (Iler 1979, Auner and Weis 1996, 2003, 2005, Degussa 1997, Legrand 1998). We can assume that the number of the siloxane bonds between primary particles in the aggregates can be very small because the number of particles of the size corresponding to that of primary particles can be 90%–95% (Figure 1.89; Gun'ko et al. 2001e).

However, frequently the average size of particles in the aqueous suspensions is in the range of aggregate sizes (Figures 1.90 and 1.91).

By and large, the concentrated aqueous suspensions of nanosilicas are relatively stable (Figure 1.91) due to many-body electrostatic interactions between particles and hydration forces. However, even the concentrated suspensions of nanosilicas are sensitive to the changes in the concentration

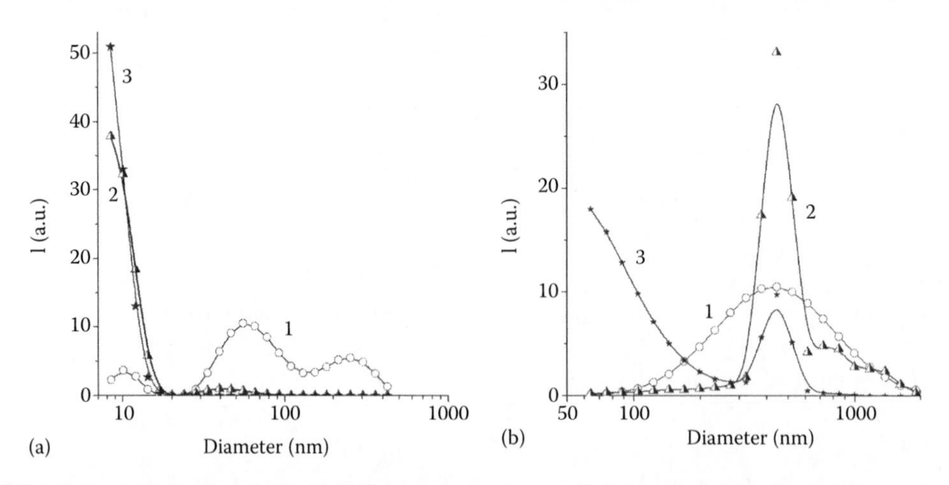

FIGURE 1.89 Particle size distributions in respect to (curves 1) light scattering, (2) particle volume and (3) particle numbers in the aqueous suspensions of (a) standard nanosilicas A-300 ($C_{A\text{-}300}=3$ wt%, $t_{us}=25$ min) and (b) A-50 ($C_{A\text{-}50}=1$ wt%, 0.9 wt% NaCl, $t_{us}=10$ min). (Adapted from *J. Colloid Interface Sci.*, 300, Gun'ko, V.M., Zarko, V.I., Voronin, E.F. et al., Successive interaction of pairs of soluble organics with nanosilica in aqueous media, 20–32, 2006i. Copyright 2006, with permission from Elsevier.)

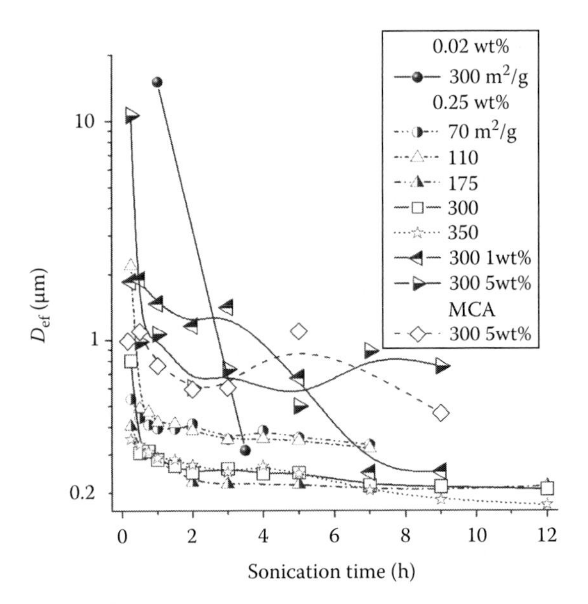

FIGURE 1.90 Effective diameter as a function of the sonication time of the aqueous suspensions of nanosilicas (C_{SiO_2}=0.02, 0.25, 1.0, and 5.0 wt%) at the S_{BET} values from 70 to 350 m²/g, $D_{ef}(t)$ is also shown for MCA suspension (5 wt%). (Adapted from *Adv. Colloid Interface Sci.*, 91, Gun'ko, V.M., Zarko, V.I., Leboda, R., and Chibowski, E., Aqueous suspensions of fumed oxides: particle size distribution and zeta potential, 1–112, 2001e. Copyright 2001, with permission from Elsevier.)

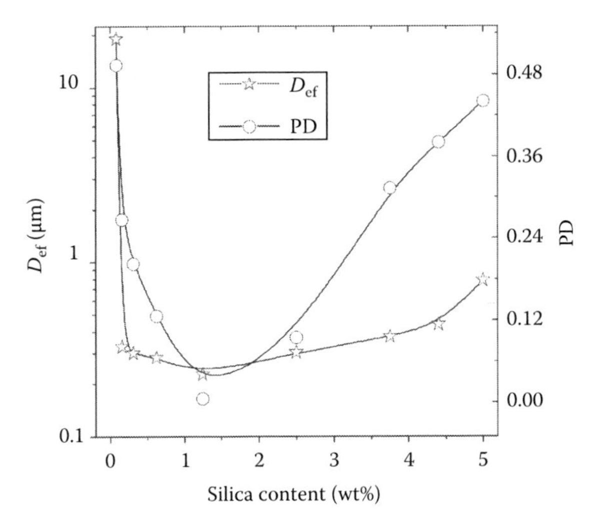

FIGURE 1.91 Effective diameter and polydispersity (PD) as a function of the concentration of A-300 in the aqueous suspension ball-milled for 5 h. (Adapted from *Adv. Colloid Interface Sci.*, 91, Gun'ko, V.M., Zarko, V.I., Leboda, R., and Chibowski, E., Aqueous suspensions of fumed oxides: particle size distribution and zeta potential, 1–112, 2001e. Copyright 2001, with permission from Elsevier.)

of oxide, for example, addition of water to the suspension at C_{SiO_2}=5 wt% reduces the D_{ef} and *PD* values but shifts the first PSD peak toward larger particles, as C_{SiO_2} reduces to 1 wt%. Then, subsequent reduction of C_{SiO_2} causes a magnification of the D_{ef} and *PD* values (Gun'ko et al. 2001a). Typically, the smallest particles give the main contributions to the total number of particles and to the total surface area. Dilution of the suspensions determines not only reduction of the size of agglomerates but also magnification of the size of aggregates at $d \approx 0.1–0.2$ μm, and the peak at $d \approx 0.01$ μm (size of primary particles), which gives a maximal contribution to the surface area (Figure 1.89), disappears totally at any dilution of the initial suspension. At a low C_{SiO_2} value, large agglomerates at $d \approx 10$ μm form due to a diminution of the stability of diluted suspensions.

The aggregation effects are very important in NMR investigations of bound water because the thickness of this water layer depends on the organization of secondary particles in the systems. For instance, silica gels are composed with nanoparticles forming much stronger aggregated structures than nanoparticles of fumed silicas. The effects of confined space in pores of silica gels (l/D is up to 10^5) leads to stronger hydrogen bonding (greater associativity) of adsorbed water molecules, diminution of their mobility and lower mixing of water with organic co-solvents because of thicker interfacial HDW/LDW layers in comparison with nanosilicas. Water in the case of the absence of strong effects of pores and in the presence of coadsorbates (e.g., hydrophobic chloroform) can be in a weakly associated state in small clusters and interstitial water or nonassociated state (individual molecules as in complexes $\equiv$SiO–H···**OH$_2$** surrounded by chloroform molecules or interstitial water) that corresponds to the unusual interfacial water. Notice that the specific surface area S of fumed oxides is mainly determined by the size of primary particles ($\sim 1/d$) and slightly by its packing type. The PaSD observed in the aqueous suspensions of nanosilica after a given pretreatment can be connected with transformation of agglomerates or aggregates, their decomposition during pretreatment and subsequent rearrangement due to different aggregation processes (Figure 1.91), and, also, governed by the characteristics of suspending liquids and media as a whole (polarity, salinity, pH, etc.), the types of oxide pretreatment before the suspension preparation, such as heating, wetting, immersion, drying, and chemical modification of the surface (Gun'ko et al. 2001e, 2009d).

The shape of radial dependence of the adhesion forces shown in Figure 1.92 for nanosilica in the aqueous medium differs significantly from similar dependences measured directly as forces between two crossed cylinders (Claesson et al. 1995, Atkins et al. 1997, Spalla and Kékicheff 1997). Clearly from these results (Figure 1.92), the adhesion force value decreases practically linearly with the distance (not far from the surface). At the same time, the interaction force between two cylinders changes in the sign at a short distance between them due to too much decrease in the liquid layer thickness and can be described as follows $F = x^{-n}$. When x is smaller than the adsorbed layer, the work is made against the adsorption forces, and the structure of this

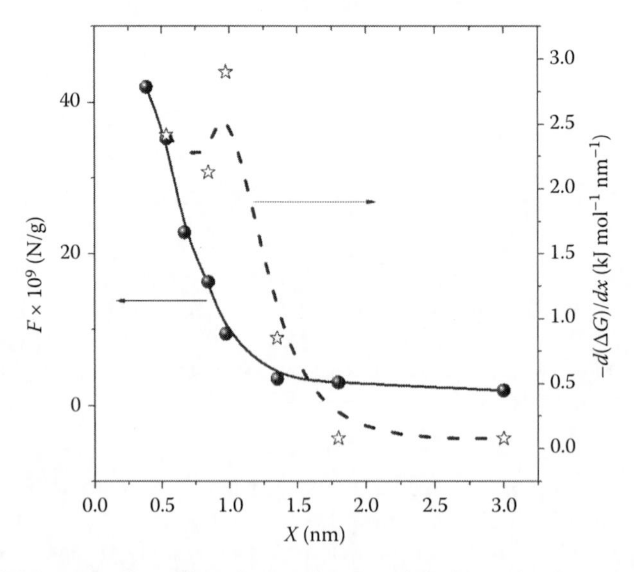

FIGURE 1.92 Radial dependencies of (a) adhesion forces and (b) $\theta(x)$ value for water bound to nanosilica A-300 (pressed at 150 atm, $\rho_b = 0.25$ g/cm^3 and labeled A_p) in aqueous suspension.

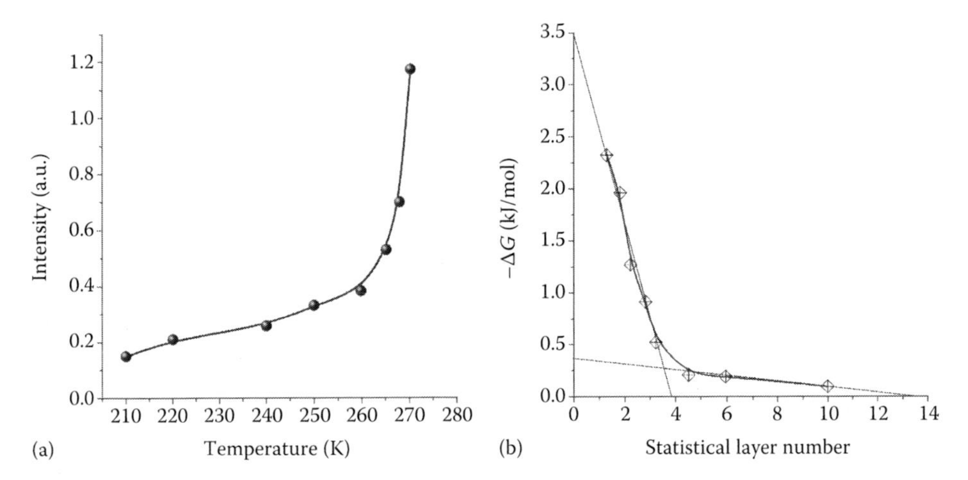

FIGURE 1.93 (a) Dependence of the ^{1}H NMR signal intensity of unfrozen water versus temperature and (b) dependence of variation in the free energy of water at the silica/water interface as a function of distance between the surface and the frozen/unfrozen water boundary in the aqueous suspension of pressed nanosilica A-300 (A_p).

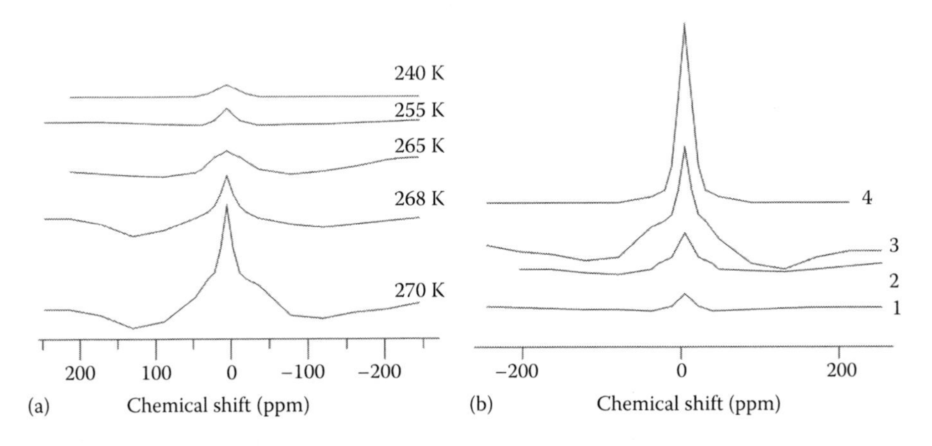

FIGURE 1.94 Dependencies of ^{1}H NMR spectra of water in frozen aqueous suspensions of initial nanosilica A-300 (A_{nt}) on (a) temperature and (b) silica concentration: 1% w/w (1), 2% w/w (2), 6% w/w (3), and 20% w/w (4).

layer is destroyed. In the NMR experiments on freezing-out of the bulk phase, the adhesion force $F = dG/dx$ can be determined as $\Delta G/x$ (where x is the distance between the surface and the outer boundary of the unfrozen water layer).

While the energy of interaction between the oxide surface and adsorbed water molecules typically increases with decreasing number of molecules in the interfacial layer, change in the sign of F is not observed. One can assume that such a method (however, as an indirect one) to estimate the work of forces of adhesion and $F(x)$ is more adequate than that based on direct measurements between two cylinders, as a cylinder can change the structure of the interfacial layer near the surface of second one producing additional errors in the measurements.

A method of determination of the Gibbs free surface energy and the radial dependence of adhesion forces of adsorbents in respect to the interfacial water was developed on the basis of measurements of the dependence of ^{1}H NMR signal intensity of unfrozen water on temperature at $T < 273$K (Figures 1.93 and 1.94). The free surface energy values are computed for a variety of oxide, carbon, carbon-mineral, polymeric, and composite materials (Turov and Leboda 1999, 2000, Gun'ko et al. 2005d).

Clearly, the availability of the four-stepwise hierarchy in the structure of the nanosilica powder could cause some dependencies of the structural features and other characteristics of the colloidal dispersions on the pretreatment techniques and other conditions, which can influence the particles differently from different hierarchic levels. Therefore, several types of the sample preparation technique were applied. There were (a) ultrasonic (US) treatment of the freshly prepared suspensions of nanosilica (at different concentrations) at t_{us} = 1–12 h (Figures 1.89, 1.90, and 1.95); (b) pretreatment of a dry silica powder in the ball mill for t_{MCA} = 1–24 h, then preparation of the aqueous suspensions sonicated for 1–9 h; (c) MCA of the aqueous suspensions of silica in the ball mill for 0.1–24 h; (d) MCA of the mixtures of the aqueous suspensions of silica and other compounds for 5–7 h; and (e) addition of some compounds (polymers, biomacromolecules, surfactants, drugs) to the ball-milled suspensions of silica (vide infra).

There is a problem of the stability of diluted and concentrated aqueous suspensions of nanosilica (Gun'ko et al. 2001e, 2005b). We studied concentrated suspensions of A-380, A-300, A-200, and A-50 at C_{SiO_2} = 1.0–20.0 wt% sonicated (labeled as US) for 5 min (power 500 W and frequency 22 kHz) or ball-milled (labeled MCA) for 3 h. Notice that the suspensions at C_{SiO_2} > 10 wt% can transform to a gel-like bound state during several days (C_{SiO_2} = 5–8 wt% corresponds to the bulk density of the nanosilica powder). Subsequent increase in C_{SiO_2} is accompanied by rearrangement of the silica particles network in the suspension and the γ_S value increases (Figure 1.95f). The main changes in the Gibbs free energy of the interfacial water occur in one-two statistical layers ($h \approx 0.3$–0.6 nm; Figure 1.95b). This thickness can be also estimated with consideration for the structure of secondary particles, that is, voids in aggregates filled by unfrozen water clusters (Figure 1.95c and d). This shows that SBW locates in narrow voids $R_p < 1$ nm between adjacent primary particles in aggregates and a portion of occupied pore volume is less than 0.5 cm^3/g. The long-range forces (of electrostatic and hydrogen-binding nature and gradually decreasing with the distance from the solid surface) provide distortion of a thicker layer of interfacial water up to $h \approx 10$ nm that corresponds to the distance in 20 nm between particles. This value corresponds to a cubic packing of primary particles of A-300 at C_{SiO_2} = 17–18 wt% and corresponds to the main PSD peak (Figures 1.95c,d and 1.96). However, the corresponding changes in the ΔG value are observed at smaller C_{SiO_2} values (Figure 1.95). This can be caused by different types of packing of particles depending on C_{SiO_2} and keeping of a portion of aggregates after sonication. The distances between adjacent primary particles can be shorter with increasing C_{SiO_2} value, despite changes in the type of packing due to variation in the coordination number (n_c). This value can be relatively low n_c = 2–4 in the initial aggregates of primary particles. An increase in n_c in the concentrated suspensions, as well as shortening of distances between primary particles with increasing C_{SiO_2} value, dissolution (analyzed in the subsequent sections), and transferring of a portion of silica, affects the structurization of the dispersion. Therefore, gelation of the concentrated dispersion (at room temperature) occurs faster, for example, at C_{SiO_2} = 16.7 wt% in 2 days, than that in less concentrated systems at C_{SiO_2} < 8 wt%, in which this process is not practically observed for both ball-milled and sonicated suspensions. However, less concentrated suspensions can form the sediment. Notice that sonication leads to lower turbidity of the suspension compared to ball-milled one because of a different size of residual aggregates possessing substantially higher light-scattering ability in comparison with individual primary particles. This difference reflects in the $\gamma_S(C_{SiO_2})$ graphs (Figure 1.95f) as the MCA suspension demonstrates higher γ_S values. Calculations of the PSD as pore volume increment versus pore radius on the basis of the $C_{uw}(T)$ curves (Figure 1.95a) for concentrated aqueous suspensions of nanosilica A-200 give the PSD_{uw} curves akin to that calculated on the basis of the LT nitrogen adsorption/desorption isotherms, PSD_{N2} (Figure 1.96).

The dependence of the PSD_{uw} shape on the C_{SiO_2} value can be caused by rearrangement of secondary particles on sonication and short aging of the suspensions that can also affect the structure of the HDW/LDW layers. It should be mentioned that sonication reduces the density of both bulk water and aqueous suspensions by 1.5%–1.7% because of partial decomposition of the clustered HDW structure and strengthening of the effect of absorption of gases by water. The graphs

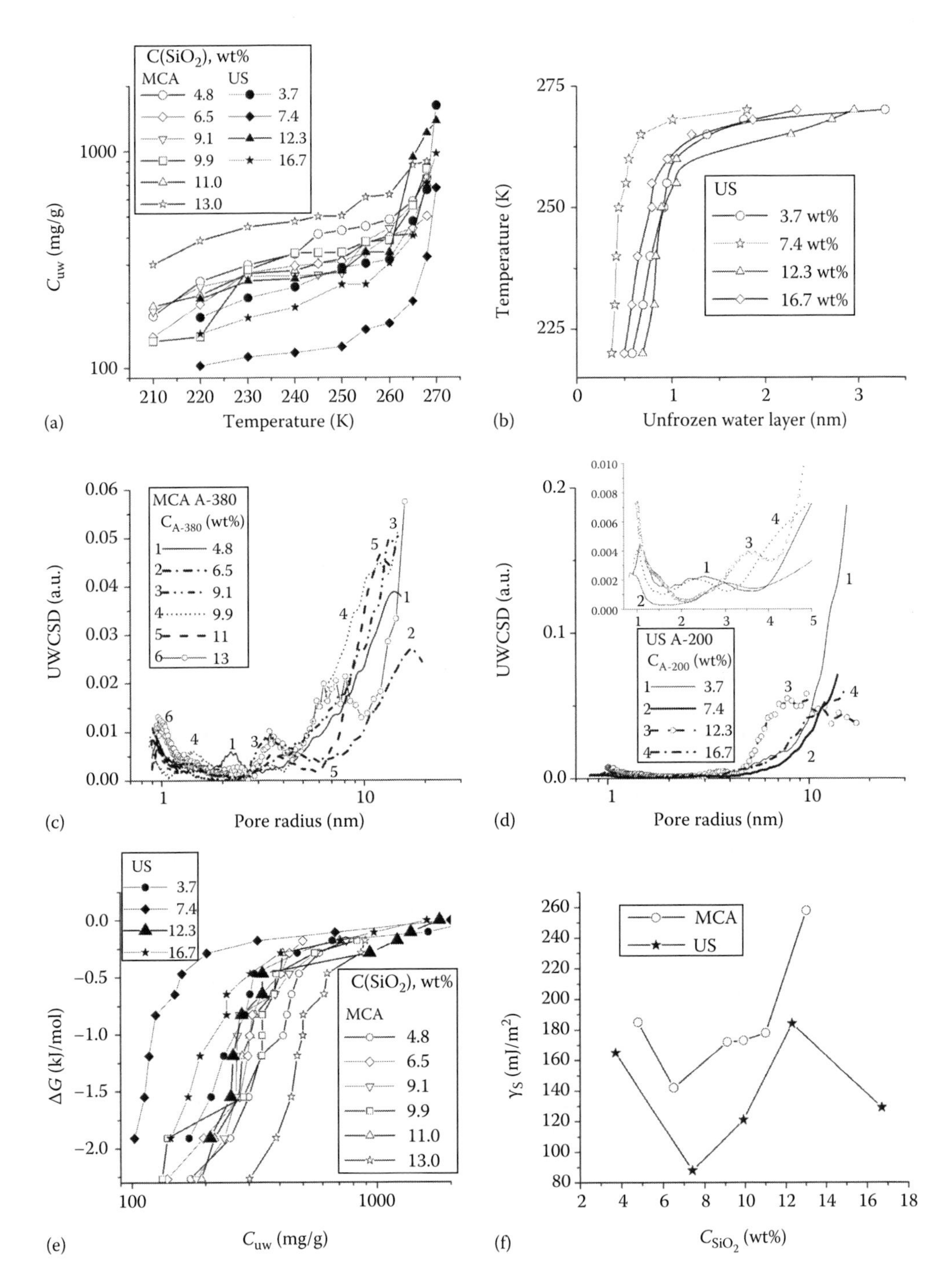

FIGURE 1.95 Characteristics of interfacial water in aqueous suspensions of A-200 sonicated (US) or A-380 treated in a ball mill (MCA) at different concentrations of silica: (a) amounts of unfrozen water as a function of temperature at $T < 273$ K, (b) relationship between the thickness of unfrozen water layer and temperature, and unfrozen water cluster size distributions for (c) MCA A-380 and (d) US A-200, (e) relationship between the amounts of water unfrozen and changes its Gibbs free energy, and (f) interfacial Gibbs free energy as a function of silica concentration in the suspensions differently treated. (Adapted from *J. Colloid Interface Sci.*, 289, Gun'ko, V.M., Mironyuk, I.F., Zarko, V.I. et al., Morphology and surface properties of fumed silicas, 427–445, 2005b. Copyright 2005, with permission from Elsevier.)

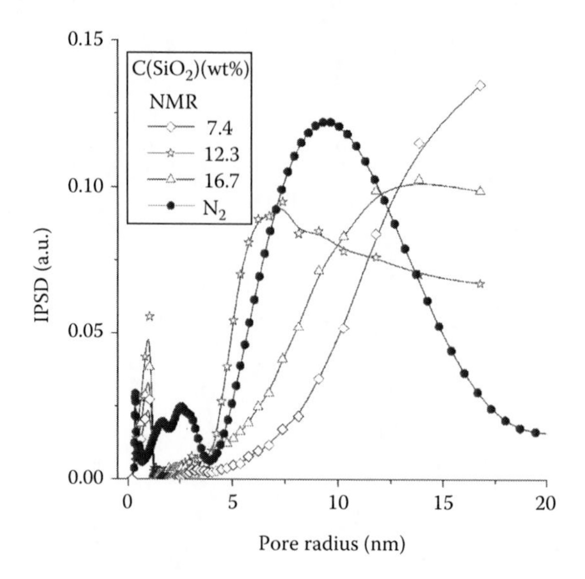

FIGURE 1.96 Pore size distributions of standard nanosilica A-200 calculated on the basis of the low-temperature nitrogen adsorption isotherm for the dry powder and the ^{1}H NMR spectra of water in the concentrated aqueous suspension of A-200 at $T<273$ K. (Adapted from *Adv. Colloid Interface Sci.*, 118, Gun'ko, V.M., Turov, V.V., Bogatyrev, V.M. et al., Unusual properties of water at hydrophilic/hydrophobic interfaces, 125–172, 2005d. Copyright 2005, with permission from Elsevier.)

(Figure 1.96) show that mobile unfrozen water can be found relatively far from the silica surface. It should be pointed out that overextended curves at $R>5$ nm can be connected to water locating at the outer surface of aggregates where the distances between primary particles from neighboring aggregates are larger than 10 nm. Consequently, this water is closer to the surface of silica than it is calculated because the effect of the outer surface of aggregates on the PSD_{uw} functions.

The influence of the specific surface area (i.e., the sizes of primary nanoparticles) on the characteristics of the interfacial water was analyzed using 5 wt% aqueous suspensions of nanosilicas at $S_{BET}=144–378$ m^2g (Figure 1.95 through 1.98 and Tables 1.19 and 1.20). The characteristics of bound water layers nonlinearly change with the S_{BET} value because of the difference in the

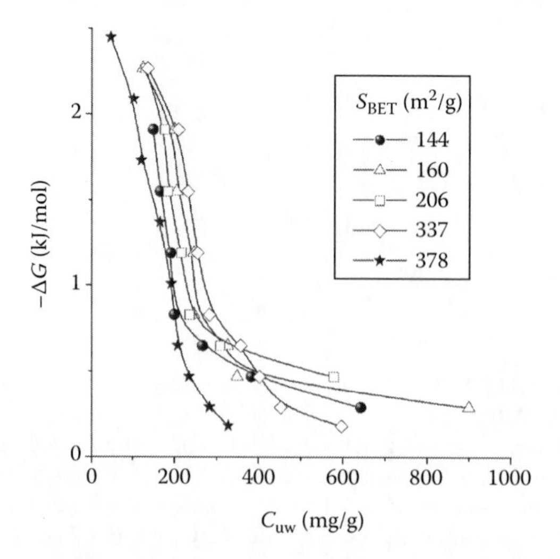

FIGURE 1.97 Relationship between the amounts of unfrozen water and changes in its Gibbs free energy for nanosilicas at different S_{BET} value.

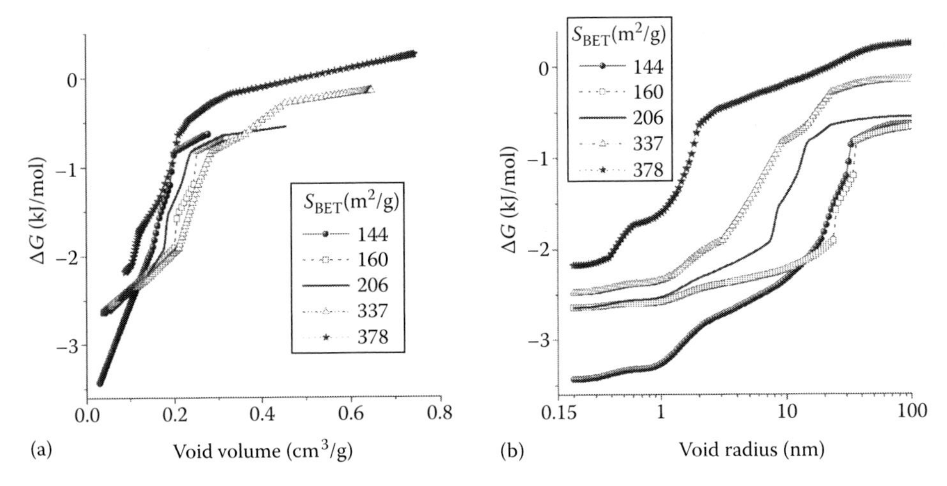

FIGURE 1.98 Relationships between the changes in the Gibbs free energy of water bound to different silicas and (a) volume of voids filled by unfrozen water and (b) void radius. (Adapted from *Adv. Colloid Interface Sci.*, 118, Gun'ko, V.M., Turov, V.V., Bogatyrev, V.M. et al., Unusual properties of water at hydrophilic/hydrophobic interfaces, 125–172, 2005d. Copyright 2005, with permission from Elsevier.)

TABLE 1.19
Characteristics of Water Bound to MCA A-380 (S_{BET} = 378 m²/g) and US A-200 (S_{BET} = 232 m²/g)

C_{SiO_2} wt%	$-\Delta G_s$ (kJ/mol)	$-\Delta G_w$ (kJ/mol)	C^s_{uw} (mg/g)	C^w_{uw} (mg/g)	S_{uw} (m²/g)	S_{nano} (m²/g)	S_{meso} (m²/g)
4.8	2.7	0.5	500	800	363	254	109
6.5	2.7	0.7	450	250	255	190	65
9.1	2.68	0.5	500	800	395	283	112
9.9	2.78	0.5	400	1000	370	246	123
11	2.89	0.4	400	900	371	280	90
13	2.78	0.7	650	550	358	231	128
3.7	2.72	0.3	320	2240	236	100	136
7.4	2.65	0.3	280	740	212	151	61
12.3	2.74	0.7	320	1230	225	30	195
16.7	2.62	0.3	290	850	216	101	115

TABLE 1.20
Characteristics of Water Bound to Nanosilicas at Different S_{BET} Values

Sample	S_{BET} (m²/g)	$-\Delta G_s$ (kJ/mol)	$-\Delta G_w$ (kJ/mol)	C^s_{uw} (mg/g)	C^w_{uw} (mg/g)	γ_S (mJ/m²)
A-150	144	3.8	0.8	200	900	304
A-150	160	2.8	0.6	400	1200	315
A-200	206	2.9	0.8	250	1150	245
A-300	337	2.75	0.6	300	600	124
A-380	378	2.75	0.7	230	270	67

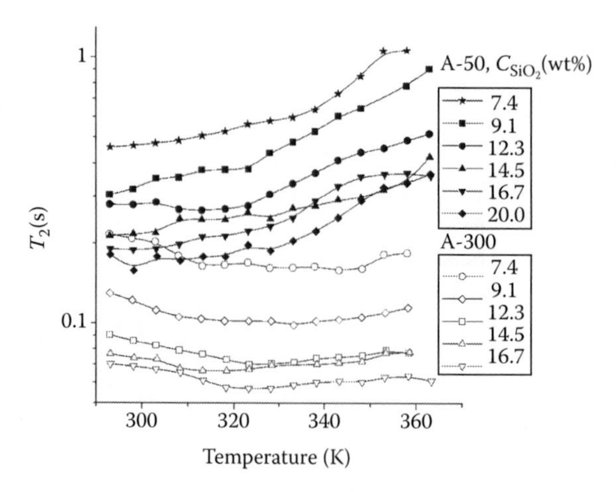

FIGURE 1.99 Transverse relaxation time T_2 as a function of temperature for water in concentrated aqueous suspensions of A-50 and A-300. (Adapted from *Adv. Colloid Interface Sci.*, 118, Gun'ko, V.M., Turov, V.V., Bogatyrev, V.M. et al., Unusual properties of water at hydrophilic/hydrophobic interfaces, 125–172, 2005d. Copyright 2005, with permission from Elsevier.)

nanoparticle organization in the suspensions and certain difference in the surface properties of nanosilicas synthesized under varied conditions (see Sections 1.1.1 and 1.1.2).

The mobility of water molecules, particularly their rotational characteristics in liquid water at the interfaces, is influenced by the number of the hydrogen bonds per molecule. The proton exchange between these clusters can be represented by scheme described in Section 10.1 (Equations 10.24 through 10.27). The concentrated aqueous suspensions of nanosilicas A-50 and A-300 are characterized by different temperature dependences of transverse relaxation time T_2 (Figure 1.99).

This effect corresponds to diminution of the diffusion rate of water molecules with increasing content of silica in the suspension because of additional structuring and clustering of silica particle aggregates and the interfacial HDW/LDW layers, and these effects are stronger for A-300 as more aggregated nanosilica and composed of smaller nanoparticles than A-50. In other words, the viscosity of interfacial water is higher and contribution of this water increases with increasing C_{SiO_2} and S_{BET} values. An increase in the S_{BET} value (i.e., a decrease in the size of primary particles) leads to tighter packing of primary particles, and therefore voids between them in secondary particles become narrow. Additionally, relative contribution of interfacial water increases with decreasing size of nanoparticles and increasing their specific surface area. In the case of nanosilica A-500 (S_{BET}=492 m^2/g), these voids are so narrow that water can be frozen over a relatively narrow temperature range, and a peak at small R values (see Section 1.1.1) is absent for A-500, in contrast to A-50 and A-300. Narrowing of voids between particles of nanosilica with a larger S_{BET} value causes a diminution of the transverse relaxation time T_2 (Figure 1.100).

This corresponds to a decrease in the diffusion of interfacial water molecules more strongly for A-300 (average diameter of primary particles of 9.2 nm) than for A-50 (average size of primary particles 40–50 nm), because the amount of interfacial water is greater for silica suspensions with greater S_{BET} values. An increase in the C_{ox} value leads to a similar effect because of an increase in interaction of interfacial water with particle surfaces.

Calculation of the distribution function $f(\Delta G_a)$ according to Equations 10.25 through 10.27 reveals the existence of two states of water characterized by different free energy of activation of relaxation depending on temperature (Figure 1.101), which can be assigned to water of different associativity.

This effect can be explained in the terms of two-state mixture with HDW and LDW because of nonlinear changes in bulk and interfacial waters with elevating temperature (Chaplin 2011). The low-energy peak is connected to WAW, and high-energy peak is linked to strongly associated

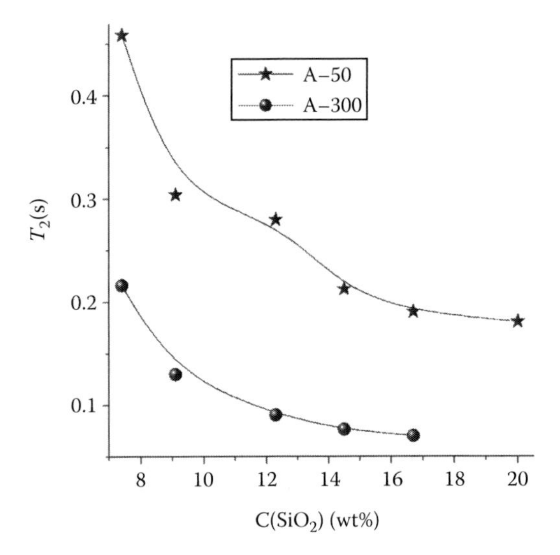

FIGURE 1.100 Transverse relaxation time T_2 as a function of concentration of nanosilicas A-50 and A-300 in the aqueous suspension at 293 K.

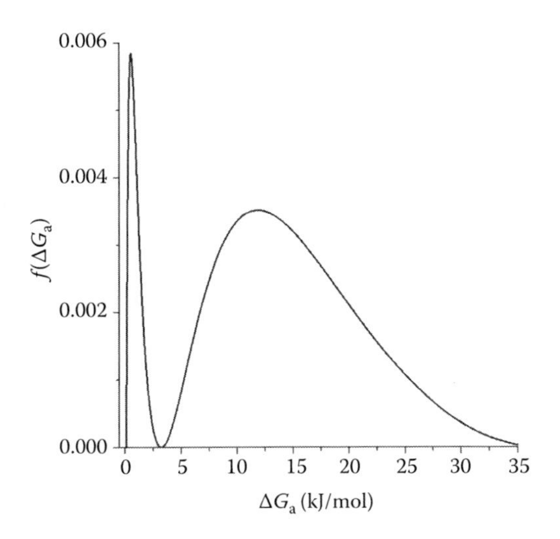

FIGURE 1.101 Distribution function of the free activation energy of water relaxation (ΔG_a) calculated on the basis of the $T_2(T)$ dependence for the aqueous suspension of A-300 at $C_{SiO_2}=7.4$ wt%.

interfacial and bulk water. Contribution of several types of water to this peak causes its relatively large width. A low area of the first peak depicts that the amount of WAW is low in these systems because of the too high content of water. Even in the case of hydrated powders of nanosilica in the hydrophobic medium ($CDCl_3$), water appears in the ^1H NMR spectra in the form the single signal, which is the averaged signal for WAW and SAW. Existence of rapid molecular exchange (only one peak is observed) testifies the absence of barriers between structures with the different forms of water. A decrease in concentration of adsorbed water enhances contribution of WAW at $T>273$ K. However, the ^1H NMR signal becomes too weak at $C_{H_2O}<8$ wt% for reliable determination of the spectral parameters because bulk density of nanosilica is very low (i.e., total amount of water under investigation becomes too low).

To study the effects of different pretreatments, standard nanosilica A-300 ($S_{BET} \approx 300$ m²/g, bulk density $\rho_b \approx 0.04$ g/cm³) was investigated as non-treated (A_{nt}) (Figure 1.94), preheated up to 520 K for 3 h (A_h), wetted with water (A_w) or hexane (supplied by Aldrich) (A_{hex}) followed by drying at 293 K ($\rho_b = 0.09$ g/cm³), and a sample (A_p) pressed at 15 MPa, $\rho_b \approx 0.25$ g/cm³.

The ^{1}H NMR spectrum of water adsorbed on non-treated silica has a single signal, whose width increases while the temperature goes down due to a decline in the molecular mobility of adsorbed water and the intensity falls as the thickness of region aqueous close to the silica surface decreases through freezing similar to other silica samples (Turov and Barvinchenko 1997, Gun'ko et al. 2005d). The same type of temperature-dependent changes is observed in the spectra registered for A_w and A_{hex}. The signal of adsorbed water for A_h and A_{nt} is observed on the background of a considerably wider signal having a shape different from the Gaussian one (because of their similarity only the spectra for water on A_h are given). The intensity of a broad component quickly falls as a function of decrease in temperature, and it is not observed in the spectra at $T < 255$ K (Figure 1.94). The broad component of the signal for water diminishes with decrease in the distance between the silica particles. It has been suggested that it is due to the formation of amorphous modification of ice near the surface.

Figure 1.102 shows the $\Delta G(d)$ and $F(x)$ graphs for silicas differently pretreated. These graphs reflect the type of the radial functions relating to changes in the free energy of water in the adsorption layer. Two segments may be revealed on the dependencies obtained: the segment of a quick decrease in the thickness of the unfrozen water layer in a narrow range of the ΔG value changes (temperature is about 273 K) and the one, where d decreases relatively weakly in a broad range of changes in the ΔG values. As was shown earlier, there are WBW and SBWs.

The data given in Table 1.21 show that the similar values of ΔG_s and thickness of the SBW layer have been obtained for all the adsorbents with the exception of A_w. For WBW, the thickness of the layer of structurally ordered liquid is minimal for A_p and maximal for A_w. For A_{nt} and A_h samples, the n_s, ΔG_w, and γ_S parameters have not been determined taking into account superposition of the narrow and wide components of the signals. The wide component of the bonded water signal disappears as a function of rise in the bulk density of silica. However, the maximal thickness of the water layer structurally ordered by the surface is observed on slight compaction of the material. Thus, for practically equal C_{uw} values for A_w and A_{hex}, the thickness of the bound water layer varies more than twice. The dependencies obtained may be explained by the fact that the characteristics of the water layers structurally ordered by the surface depend on interparticle interactions. Nanosilica particles in

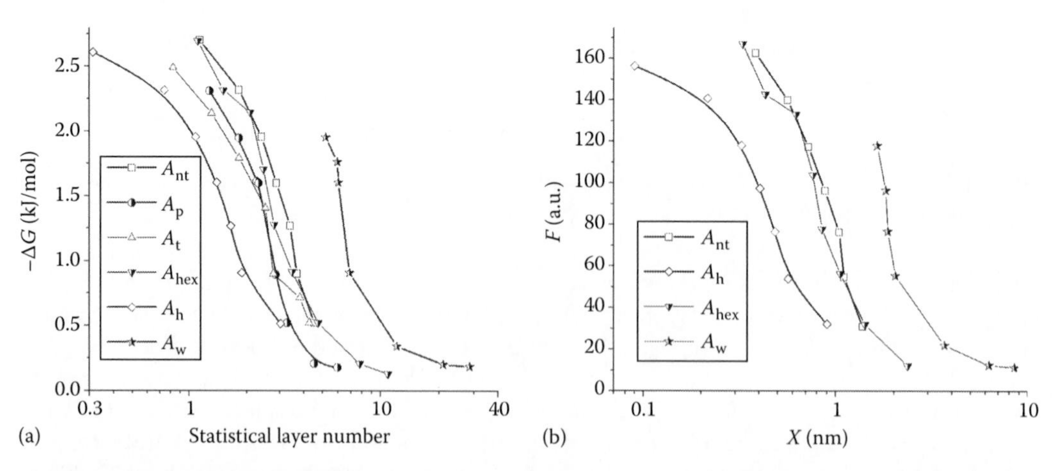

FIGURE 1.102 Effect of the pretreatment procedure of nanosilica A-300 on the shape of (a) radial function of variation in the free energy of adsorbed water and (b) adhesion forces for aqueous suspensions of nanosilica. (Adapted from *Colloids Surf. B: Biointerfaces*, 8, Turov and Barvinchenko, Structurally ordered surface layers of water at the SiO_2/ice interface and influence of adsorbed molecules of protein hydrolysate on them, 125–132, 1997. Copyright 1997, with permission from Elsevier.)

TABLE 1.21
Influence of Silica Pretreatment on the Characteristics of the Interfacial Water

Sample	$-\Delta G_s$ (kJ/mol)	$-\Delta G_w$ (kJ/mol)	n_s	n_w	γ_S (mJ/m^2)
A_{nt}	3.4	—	5	—	—
A_{hex}	3.5	0.35	4.5	17	195
A_p	3.8	0.35	3.2	13	160
A_h	3.3	—	4	—	—
A_t	2.9	—	2.7	—	—
A_w	5.3	0.5	8	35	540

Note: Unfrozen water layer thickness is shown as the number statistical layers n.

the aqueous suspensions are present in the form of aggregates (100–500 nm) and agglomerates (>1 μm). These swarms are not stable and even exposure to negligible external factors (e.g., shaking of suspensions) gives rise to dispersion as well as to amalgamation of particles. The amorphous form of frozen water is likely to be formed only at the appropriate distances between the particles and favorable dimensions of the secondary particles of the adsorbent. Therefore, the influence of the adsorbent pretreatment on its adsorption characteristics should be taken into account while studying adsorption of organics on highly disperse materials. By means of modifying the pretreatment conditions for the adsorbent, one may vary its adsorption capacity in a wide range of values. This will have an effect on the adsorption value of biomacromolecules (vide infra).

Additional information on the temperature dependence of the interfacial behavior of water and other adsorbates can be obtained with the TSDC technique. Typically when TSDC investigations of wetted powders and aqueous suspensions of oxides and bio-objects were carried out, the systems were cooled to 265 K, polarized by an electrostatic field at an intensity $F_p=0.1$–1 MV/m at 263–265 K for several minutes, cooled to 90–100 K with the field still applied, and then heated without the field at a heating rate $b=0.05$ K/s to 265 K (Gun'ko et al. 2007i). The TSDC method can be effectively used to analyze the associativity of interfacial water and compared to the results of ^{1}H NMR investigations with layer-by-layer freezing-out of bulk and interfacial liquids at a surface of porous or disperse materials. Comparison of the TSDC data not only with NMR results but also with the DRS, differential scanning calorimeter (DSC), and adsorption data can give much better understanding of the observed interfacial phenomena.

The interfacial phenomena at the surface of fine oxides in aqueous suspensions depend on concentration of oxide (C_{ox}) because of changes in the amounts of interfacial and bulk water and in interparticle interactions. For instance, interfacial Gibbs free energy depends strongly on the concentration of nanosilica in the aqueous suspension (Figures 1.95 and 1.103e).

Bulk water (ice) gives an intensive TSDC peak at 120 K (Figure 1.103a). An increase in silica concentration in the aqueous suspension causes relative diminution of the intensity at this temperature, and the LT band peak shifts toward higher temperature from 120 to 150 K (Figures 1.103a and 1.104).

This effect is linked to decreased molecular mobility of interfacial water (ice) in comparison with the bulk and strong effects of surface forces near the particles. In the case of $C_{ox}=30$ (Figure 1.103a) or 33.4 wt% (Figure 1.104), the intensity of the LT band is minimal because of maximal structurization of water. Certain scatter in the displacement and rearrangement of the LT band at 100–160 K is due to several factors: (i) certain characteristics, for example, free surface energy γ_s (Figure 1.103e), of the aqueous suspension are nonlinear functions of C_{ox} because of the rearrangement of the dispersion structure with changes in oxide concentration and (ii) an increase in

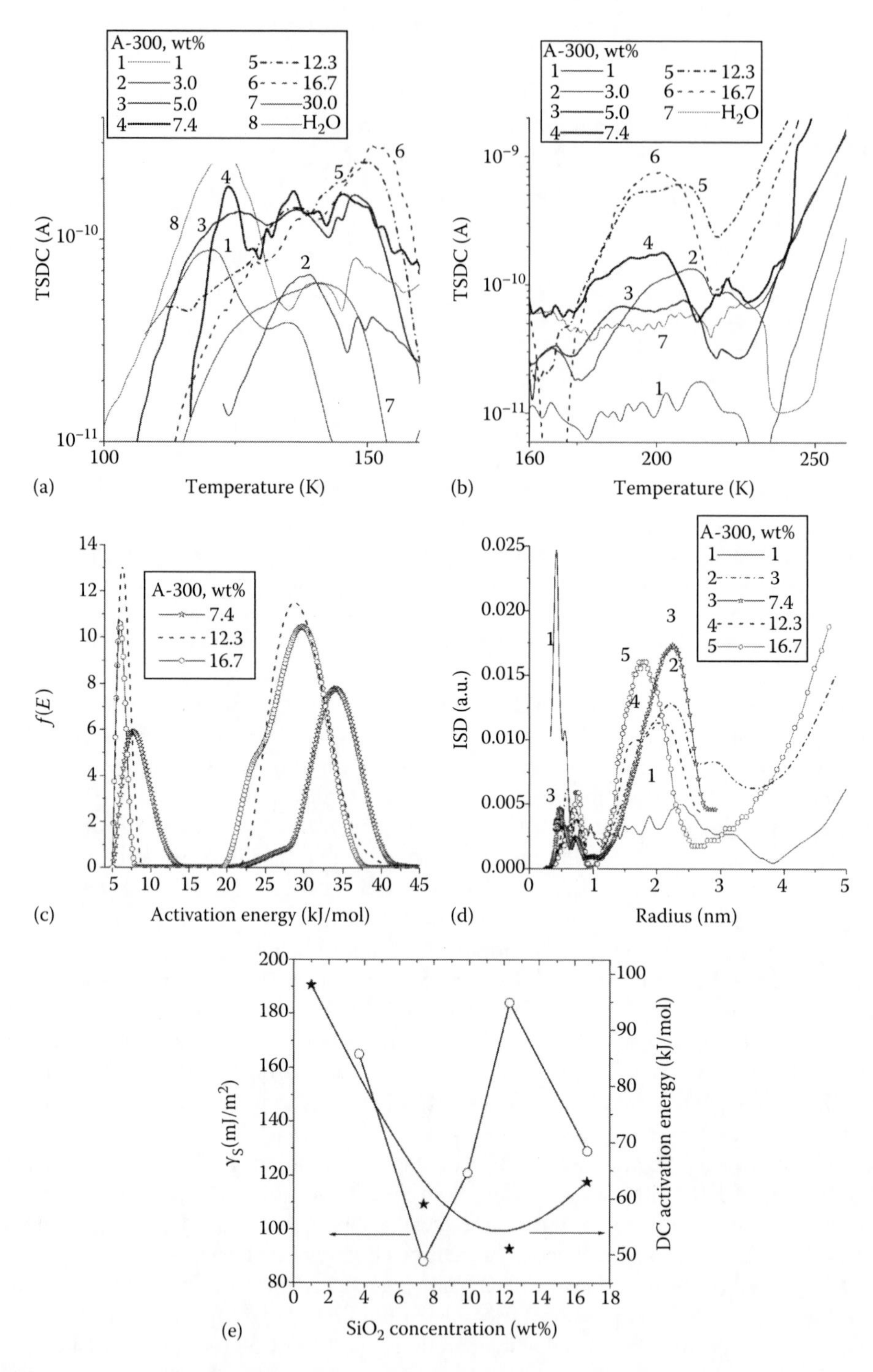

FIGURE 1.103 (a, b) TSDC thermograms of the aqueous suspensions of standard nanosilica A-300 at different concentration (intensity normalized to $F_p = 100$ kV/m), (c) distribution function of activation energy of dipolar relaxation of water, (d) distribution function of the size of relaxing water structures, and (e) relationship between concentration of nanosilica A-300 in the aqueous suspension (sonicated) and Gibbs free energy of interfacial water (γ_S) and activation energy of the dc relaxation. (Adapted from *Adv. Colloid Interface Sci.*, 131, Gun'ko, V.M., Zarko, V.I., Goncharuk, E.V. et al., TSDC spectroscopy of relaxational and interfacial phenomena, 1–89, 2007i. Copyright 2007, with permission from Elsevier.)

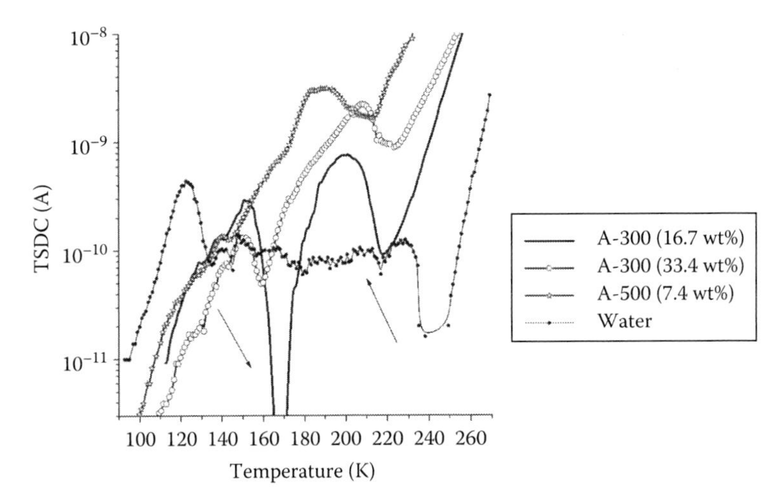

FIGURE 1.104 TSDC thermograms of concentrated aqueous suspensions of standard nanosilicas A-300 and A-500 and pure water. (Adapted from *Adv. Colloid Interface Sci.*, 131, Gun'ko, V.M., Zarko, V.I., Goncharuk, E.V. et al., TSDC spectroscopy of relaxational and interfacial phenomena, 1–89, 2007i. Copyright 2007, with permission from Elsevier.)

the C_{ox} value leads to changes in the structure of secondary particles and the coordination number of primary particles can increase. The HT band at 160–270 K is characterized by increased TSDC intensity with increasing C_{ox} value, and the peak shifts toward lower temperature from 220 K at C_{ox} = 1–3 wt% to 200 K at C_{ox} = 12–17 wt% because of enhanced water clusterization. The effective dipole moment of water increases at the silica surface, which also affects the LT and HT band intensity and position. An increase in the C_{ox} value leads to enhancement of formation of interfacial water clusters similar to HDW nanodomains (Gun'ko et al. 2005d, 2009d), and the concentration of charge carriers (e.g., protons) increases because of the hydrolysis of surface silanols. These factors affect the starting point of the dc relaxation as the through conductivity (Figure 1.103b), and the activation energy of the dc relaxation ($E_{a,dc}$) decreases with increasing C_{ox} value (Figure 1.103e) as well as the activation energy (Figure 1.103c) of dipolar relaxation at $T > 180$ K. It is possible that dc relaxation can give certain contributions to the TSDC at these temperatures because, according to ^{1}H NMR data (Figure 1.95), mobile water molecules appear in aqueous suspensions of fumed oxides at $T \approx 200$ K (Gun'ko et al. 2007i). The TSDC cryoporometry shows (Figure 1.103d) that the size of relaxing water structures exhibit two tendencies: the size of the smallest structures ($R < 1$ nm) increases, but larger structures at $R > 1$ nm decrease because interparticle distances become shorter with increasing C_{ox} value and water becomes more strongly clustered.

According to Velikov et al. (2001), there are two T_g values for hyperquenched water at 136 and 165 K. Interfacial water in the TSDC measurements can be similar in structure to hyperquenched water. For instance, in the case of the TSDC measurements of pure water, pressed snow, and fast frozen water (Figure 1.105), the corresponding shoulders or peaks are observed at the same temperatures. Notice that fast freezing of pure water (Figure 1.105a, curve 4) or the aqueous suspension of nanosilica A-300 (Figure 1.105b; curve 1) results in much lower TSDC intensity because the field is not enough strong to change the orientation of water dipoles fast frozen at low temperature (by liquid nitrogen at 77 K) despite $T_p = 260$ K.

As a whole, an increase in the silica concentration strongly affects the TSDC spectra (Figures 1.103 and 1.104) over the temperature ranges related to all the relaxation mechanisms, and the shape of all bands change. An LT peak at 120–125 K is absent at high C_{ox} values (12.3 and 16.7 wt%). A peak at 134 K at C_{ox} = 1 wt% shifts toward 152 K at C_{ox} = 16.7 wt%. Thus, the peaks in the LT band shift toward higher temperatures; however, the peaks in the HT band located at $T > 160$ K shift in the opposite

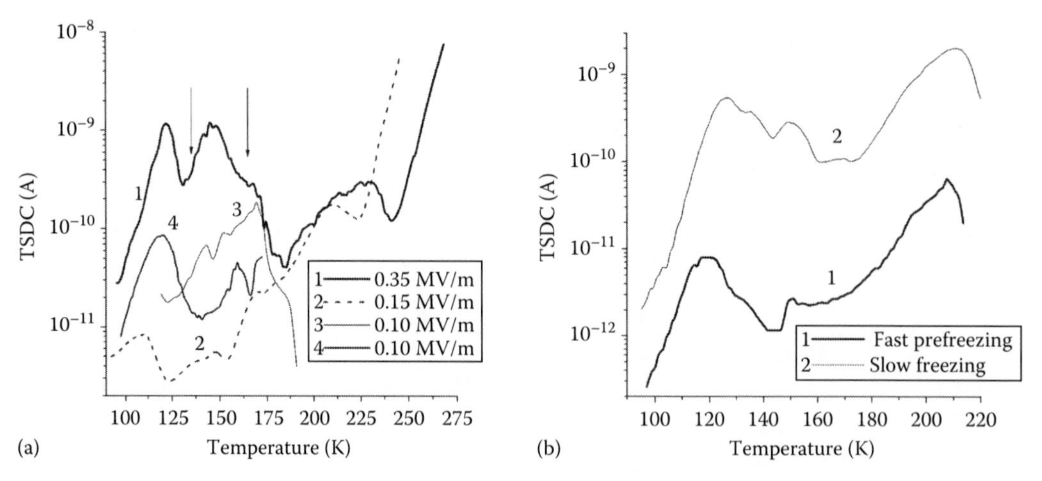

FIGURE 1.105 TSDC thermograms of (a) water (curves 1 and 2), pressed snow (curve 3) polarized at different F_p values (arrows show T of 136 and 165 K), (curve 4) ice prepared by fast freezing of water by liquid nitrogen, (b) aqueous suspension of nanosilica A-300 (7.4 wt%) prefreezing by liquid nitrogen without field for 0.5 min (curve 1) and freezing with applied field for 5 min (curve 2), T_p=260 K (1–4). (Adapted from *Adv. Colloid Interface Sci.*, 131, Gun'ko, V.M., Zarko, V.I., Goncharuk, E.V. et al., TSDC spectroscopy of relaxational and interfacial phenomena, 1–89, 2007i. Copyright 2007, with permission from Elsevier.)

direction. These displacements and changes in the spectral shape lead to the corresponding changes in the distribution function of the activation energy of dipolar relaxation (Figure 1.103c). Notice that piecewise linearization of the TSDC spectrum for the aqueous suspension of A-300 at C_{ox} = 16.7 wt% in the coordinates log $I - 1/(k_B T)$ gives an activation energy E_a=9–42 kJ/mol (in agreement with the distribution function in Figure 1.103c) for dipolar relaxation and ~60 kJ/mol for the dc relaxation.

An increase in the C_{ox} value is accompanied by a displacement of temperature onset of the dc relaxation by approximately 30 K. The main reason for this change is a reduction of bulk water content, and an increase in the amounts of structured water and mobile charges (i.e., protons), because an increase in the C_{ox} value results in an increase in the negative charge density on silica particles (Gun'ko et al. 2001e). The size of relaxing water structures decreases with increasing C_{ox} value (Figure 1.103d) because of the smaller distances between neighboring particles.

An increase in the F_p value gives higher TSDC intensity (Figures 1.105a and 1.106a). However, the peak position and shape change slightly (Figure 1.106a), resulting in close shapes of the SDWCs (Figure 1.106b).

The similarity in the TSDC or SDWC shapes is also observed for different nanosilicas even at different F_p values (Figure 1.107) because of similar morphology, causing similar confined space effects (spatial restrictions) for water clusters located in voids between primary particles in their aggregates. However, an increase in the S_{BET} value is accompanied by increased aggregation of primary particles, and voids between them can slightly decrease (comp. SDWCs for A-300 and A-400 at S_{BET}=294 and 409 m^2/g, respectively; Figure 1.107b).

Thus, an increase in the C_{ox} value leads to the following effects: (i) contribution of relaxation of small water (ice) structures characterized by minimal activation energy (i.e., minimal deviation from the equilibrium state characteristic of ice) and relaxing at the lowest temperatures decreases and (ii) relaxation of larger structures appearing at higher temperatures (T> 160 K) and at larger TSDC intensity shifts toward lower temperatures with increasing C_{ox} value, that is, the microdomain structures in water undergo certain differentiation because of the distance dependence of the surface effects and a decrease in the average distance between silica primary particles in the dispersion. This results in a decrease and then an increase in the activation energy of the dc relaxation with C_{ox} (Figure 1.103d). Similar changes are observed for the γ_S value; however, the system at C_{ox} = 12.3 wt%

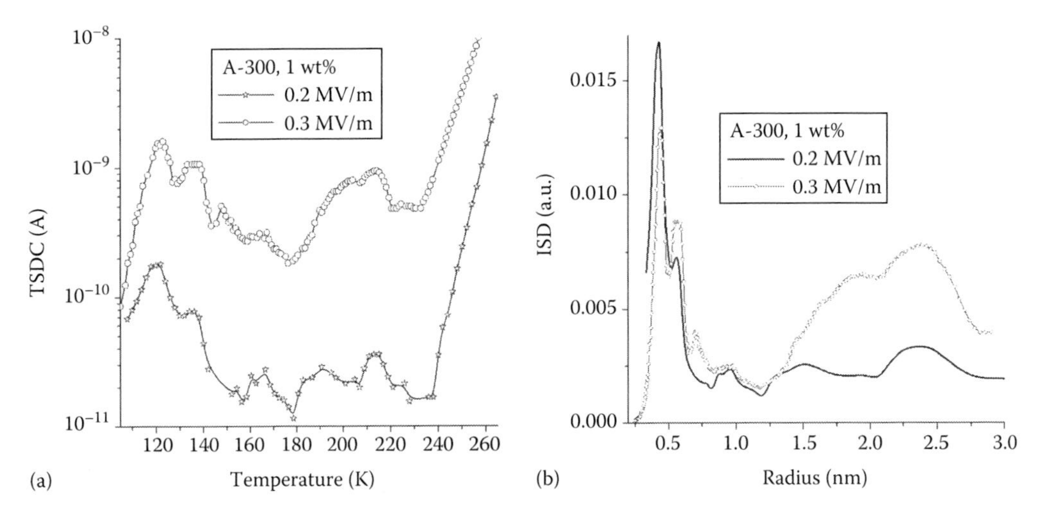

FIGURE 1.106 Influence of the field intensity F_p on (a) TSDC spectra and (b) size distribution functions of water structures relaxing at different temperatures in frozen aqueous suspension of nanosilica A-300 (C_{ox} = 1 wt%). (Adapted from *Adv. Colloid Interface Sci.*, 131, Gun'ko, V.M., Zarko, V.I., Goncharuk, E.V. et al., TSDC spectroscopy of relaxational and interfacial phenomena, 1–89, 2007i. Copyright 2007, with permission from Elsevier.)

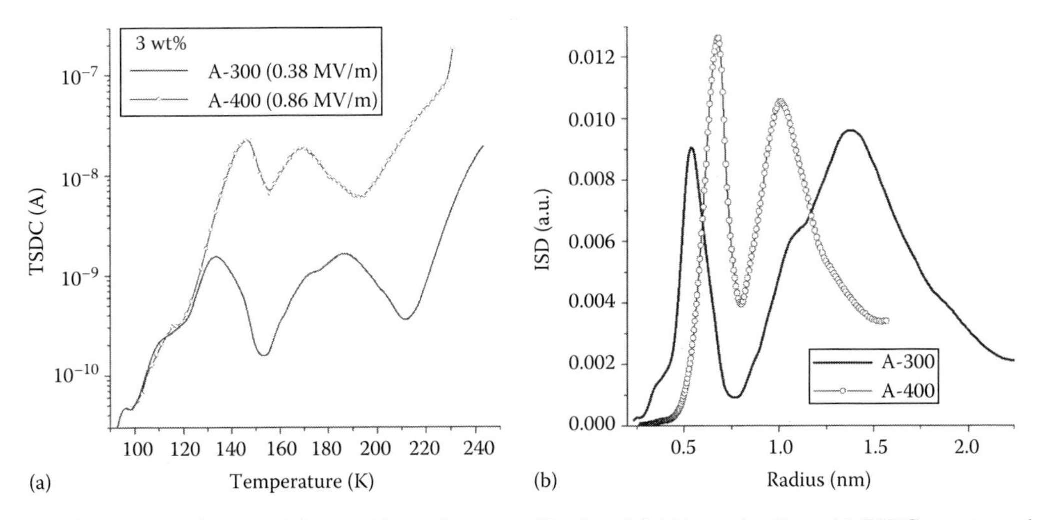

FIGURE 1.107 Influence of the specific surface area (S_{BET}) and field intensity F_p on (a) TSDC spectra and (b) size distribution functions of water structures relaxing at different temperatures of frozen aqueous suspension of nanosilica A-300 (standard nanosilica) and A-400 (non-standard nanosilica) (C_{ox} = 3 wt%). (Adapted from *Adv. Colloid Interface Sci.*, 131, Gun'ko, V.M., Zarko, V.I., Goncharuk, E.V. et al., TSDC spectroscopy of relaxational and interfacial phenomena, 1–89, 2007i. Copyright 2007, with permission from Elsevier.)

gives an increase in the γ_S value but the activation energy of the dc relaxation ($E_{a,dc}$) decreases because the physical sense of these two values differs.

Additional factors affecting TSDC intensity are variations in the F_p value and changes in the effective dipole values of interfacial water molecules. The size distributions of relaxing water structures can be similar despite the difference in the F_p value (Figure 1.106b). Changes in the S_{BET} value and the primary particle packing density cause a displacement of the TSDC peaks (Figures 1.106a and 1.108a) and the corresponding changes in the size of relaxing water structures (Figures 1.106b and 1.108b).

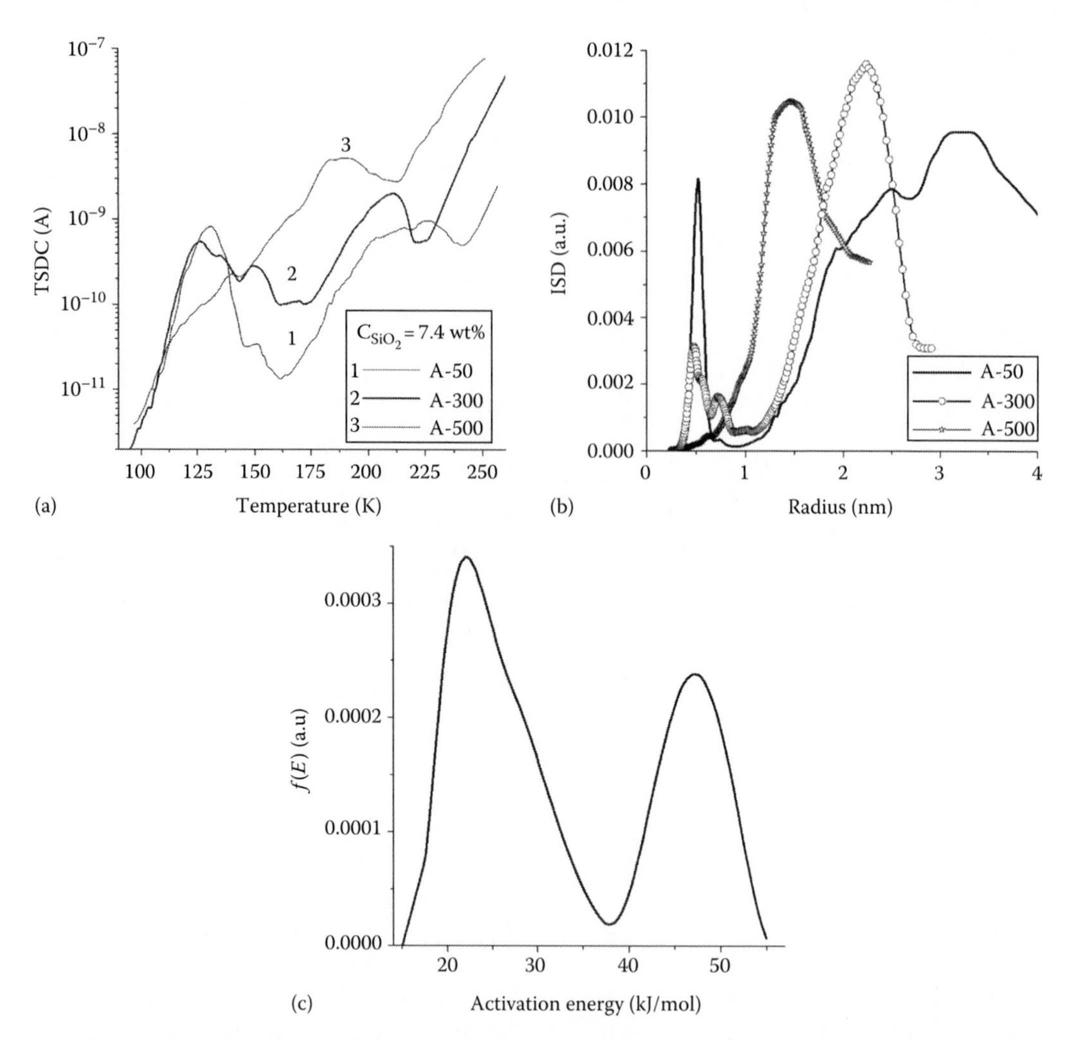

FIGURE 1.108 (a) TSDC for aqueous suspension of nanosilicas at different specific surface area and C_{ox} = 7.4 wt%, (b) corresponding size distribution functions of relaxing water structures, and (c) distribution function of activation energy of dipolar relaxation in the aqueous suspension of A-50 (7.4 wt%). (Adapted from *Adv. Colloid Interface Sci.*, 131, Gun'ko, V.M., Zarko, V.I., Goncharuk, E.V. et al., TSDC spectroscopy of relaxational and interfacial phenomena, 1–89, 2007i. Copyright 2007, with permission from Elsevier.)

Changes in the C_{ox} values affect the distances between particles. However, these distances are larger in suspension than after drying of solid residua of these dispersions (Figure 1.109; peaks at R between 1 and 4 nm). A certain similarity in the PSD of dry and dried powders and suspensions is observed.

Zaporozhets et al. (2000) studied the adsorption of different salts, $(C_{10}H_{21})_2N(CH_2)_2N(C_{10}H_{21})_3I$ (I) (Figure 1.110a and b), $(CH_3)(C_{10}H_{21})_2N(CH_2)_2N(C_{10}H_{21})_3I_2$ (II), $(C_{10}H_{21})3N(CH_2)_6N(C_{10}H_{21})_3I_2$ (III), and $(C_2H_5)_4NBr$ (IV) (Figure 1.100c and d), on standard nanosilica A-300 and silica gel (SG) using LT ^{1}H NMR spectroscopy applied to frozen aqueous suspensions. Besides electrostatic interactions of cations, intramolecular dispersive interaction plays an essential role in the interfacial phenomena studied.

To compare the molecular size and silica type effects, salts I and IV (Figure 1.110) were used. An increase in the content of salt I of the larger size adsorbed onto silica gel results in strong diminution of the amounts of bound water (Table 1.22 and Figure 1.111). This can be explained by filling

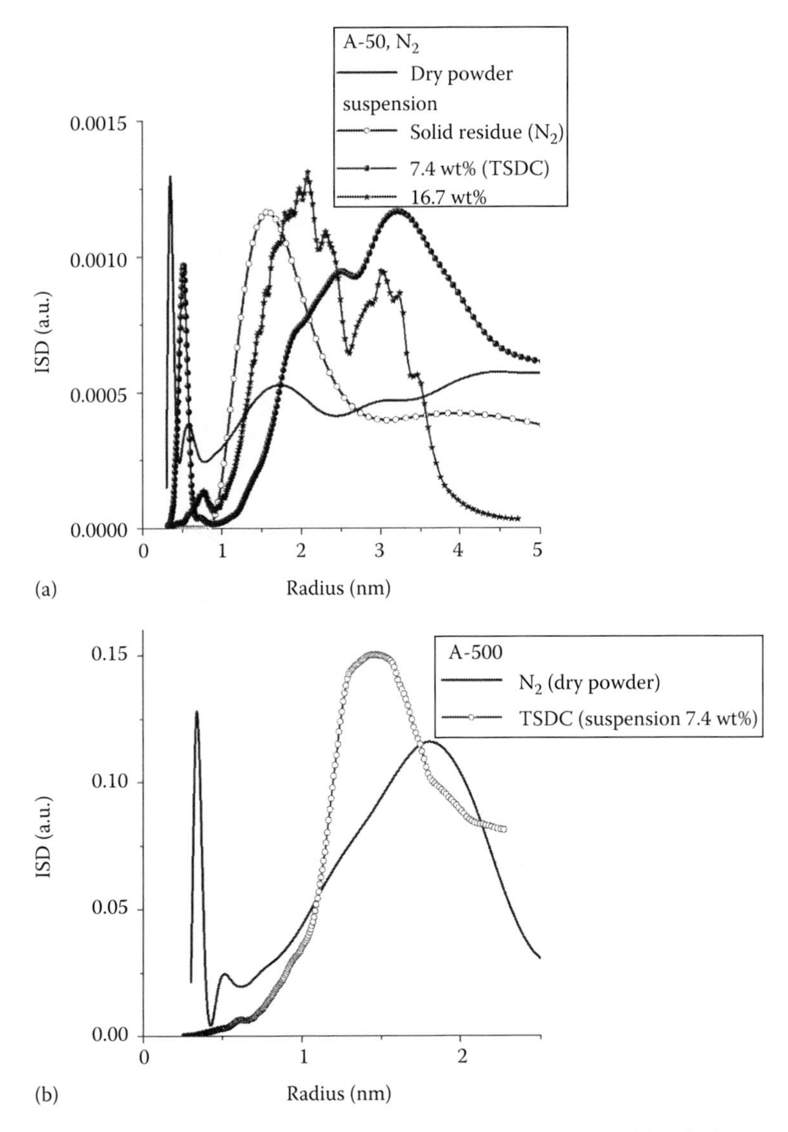

FIGURE 1.109 Pore size distributions for nanosilicas (a) A-50 and (b) A-500 calculated on the basis of nitrogen desorption isotherms for dry powders and dried solid residue and size distribution functions of water structures relaxing on the TSDC measurements of aqueous suspensions at $C_{ox} = 7.4$ and 16.7 wt%. (Adapted from *Adv. Colloid Interface Sci.*, 131, Gun'ko, V.M., Zarko, V.I., Goncharuk, E.V. et al., TSDC spectroscopy of relaxational and interfacial phenomena, 1–89, 2007i. Copyright 2007, with permission from Elsevier.)

of SG pores with adsorbed cations with large hydrophobic chains, which could partially block pores from water. Therefore, the volume of bound water (Table 1.22, C_{max}) is minimal, as well as γ_S, at a maximal content of salt I. In the case of small cations IV without long hydrophobic chains adsorbed onto A-300, aggregation of silica nanoparticles with adsorbed $(C_2H_5)_4NBr$ occurs (since V_{meso} increases but V_{nano} and S_{nano} are zero) but without strong screening of voids (γ_S decreases but slightly). Therefore, the C_{max} is maximal at maximal content of salt IV. These effects are well seen as the opposite displacement of the PSDs with increasing salt content for salts I (toward smaller sizes because of filling of SG pores) and IV (toward larger sizes because of aggregation of A-300 nanoparticles; Figure 1.111).

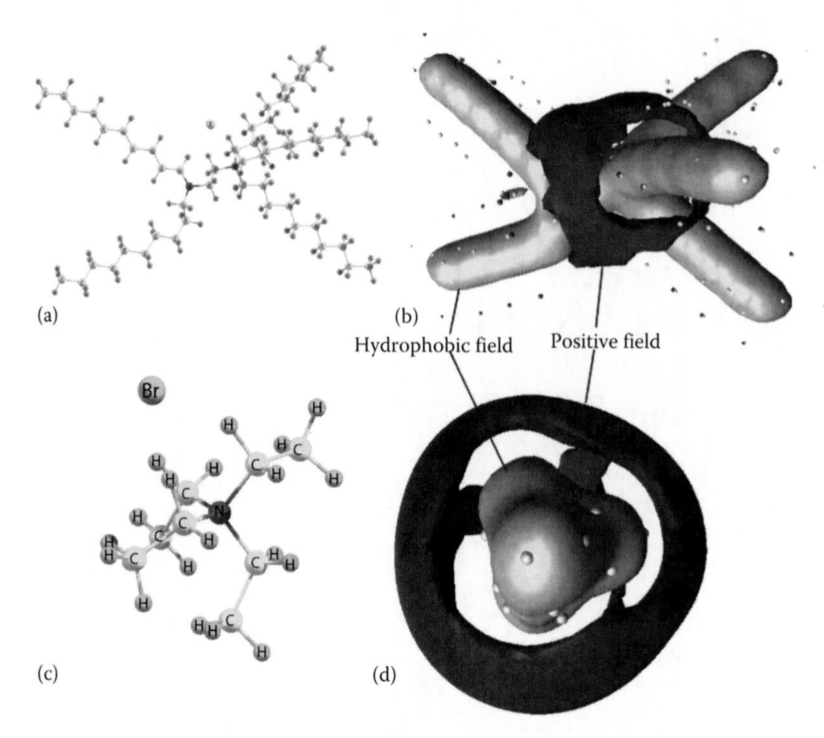

FIGURE 1.110 Molecular structures (geometry was optimized using the PM6 method) of (a) $(C_{10}H_{21})_2N(CH_2)_2N(C_{10}H_{21})_3I$, (b) $(C_{10}H_{21})_2N(CH_2)_2N(C_{10}H_{21})_3^+$, (c) $(C_2H_5)_4NBr$, and (d) $(C_2H_5)_4N^+$, main fields calculated with FieldView 2.0.2 (2011): positive field was around (b) charged cationic fragment of a large molecule or (d) whole small cation and hydrophobic field was around weakly polar CH chains.

TABLE 1.22
Characteristics of Water Bound to A-300 Alone or with Adsorbed Salts $(C_{10}H_{21})_2N(CH_2)_2N(C_{10}H_{21})_3I$ (I) or $(C_2H_5)_4NBr$ (IV)

Salt	C_{salt} (mg/g)	S_{uw} (m²/g)	S_{nano} (m²/g)	S_{meso} (m²/g)	V_{nano} (cm³/g)	V_{meso} (cm³/g)	V_{macro} (cm³/g)	C_{max} (mg/g)	$\langle T \rangle$ (K)	$-\Delta G_s$ (kJ/mol)	γ_S (J/g)
I	0	410	280	130	0.124	1.127	0	1382	244.9	3.38	73.8
I	8.9	283	213	69	0.099	0.750	0	890	244.0	3.09	52.3
I	142	262	188	74	0.082	0.702	0	849	248.2	3.55	42.1
IV	0	312	176	133	0.084	1.358	0.028	1567	240.4	3.66	99.1
IV	100	122	0	120	0	1.278	0.023	1409	244.5	2.98	76.9
IV	1000	162	0	161	0	1.693	0.011	1796	248.8	2.28	87.0

1.1.7 Adsorption of Dissolved Compounds

Cations and anions of dissolved salts can be both chaotropic (water-structure breakers) and kosmotropic (structure-makers) agents (Chaplin 2011). If ions belong to different groups and have similar activity that their total effect on water ordering can be relatively small and they are good solutes. However, in the case of ions from the same group, the structuring effect is much stronger and the solubility of these compounds can be low. The distortion of the hydrogenous bond network in the interface layer at a solid surface is maximal; therefore, the effect of solutes (e.g., salts) in this layer can be maximal too (Zhao et al. 2006, Chaplin 2011).

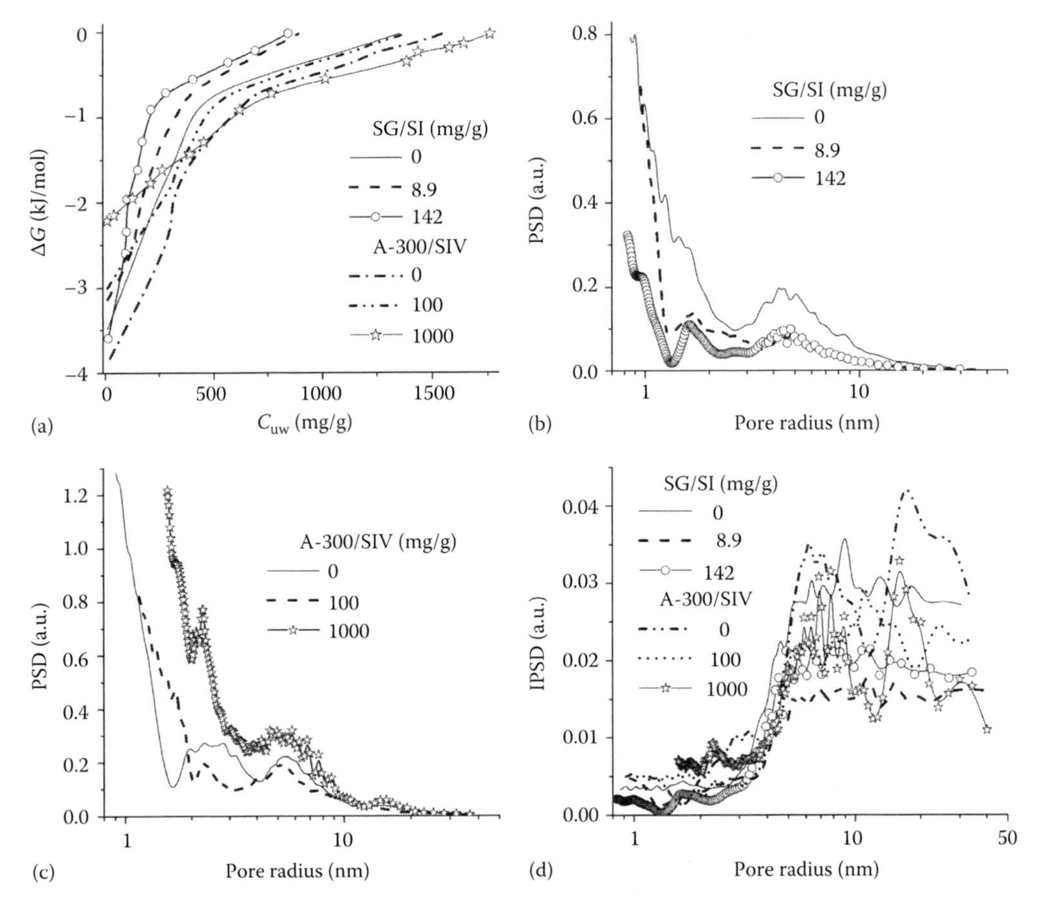

FIGURE 1.111 (a) Relationship between changes in Gibbs free energy and amounts of unfrozen water in aqueous suspensions of SG or A-300 alone or with addition of salt I (SI) on SG and IV (SIV) on A-300, and size distributions of unfrozen water clusters and domains (in SG pores or voids between A-300 nanoparticles): (b, c) differential and (d) incremental.

Roose et al. (1996) studied the magnetic-field dependence of the proton spin–lattice relaxation time T_1 (referred to as nuclear-magnetic-relaxation dispersion) in aqueous colloidal silica containing paramagnetic Mn^{2+} ions (Figure 1.112). The experimental relaxation rate of solvent protons in aqueous colloidal silica suspensions containing Mn^{2+} ions can be expressed as a weighted mean of several contributions:

$$T_1^{-1} = (1 - P_{M,ads} - P_{M,free})T_{1D}^{-1} + T_{1P,ads}^{-1} + T_{1P,free}^{-1} \tag{1.41}$$

where each term takes into account a different fraction of solvent protons sensing a particular relaxation mechanism. The variables $1/T_{1P,ads}$ and $1/T_{1P,free}$ are given by equation describing the paramagnetic spin–lattice relaxation enhancement:

$$T_{1P}^{-1} = P_M(T_{1M} + \tau_M)^{-1} \tag{1.42}$$

where P_M is a weight factor corresponding to the fraction of water protons bound to the paramagnetic ions and should be equal to $n[M]/[H_2O]$, n is the number of water molecules coordinated to the metal ion, [M] is the ion concentration expressed in molarity, τ_M is the residence time of a proton on the

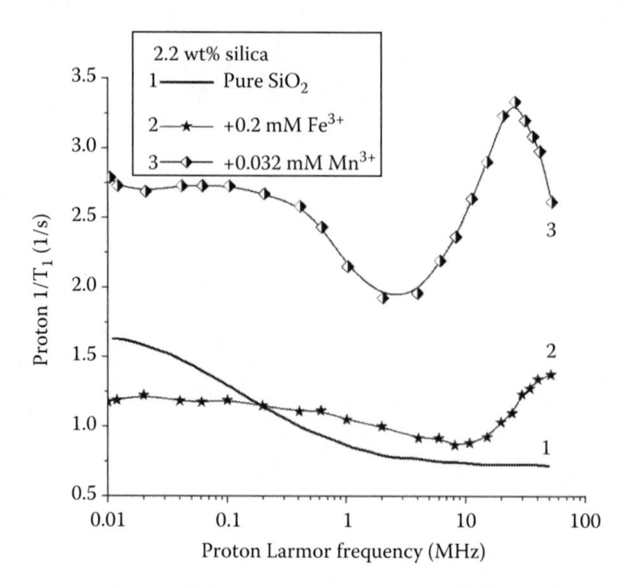

FIGURE 1.112 Water-proton $1/T_1$ NMRD profiles for colloid silica ($S_{BET}=265$ m²/g) sols (2.2 wt%) at 4°C alone and with metal ions. (Adapted from *J. Magn. Reson. A.*, 120, Roose, P., Van Craen, J., Andriessens, G., and Eisendrath, H., NMR study of spin–lattice relaxation of water protons by Mn^{2+} adsorbed onto colloidal silica, 206–213, 1996. Copyright 1996, with permission from Elsevier.)

paramagnetic complex, and $1/T_{1M}$ is the relaxation rate of a single nuclear spin interacting with a paramagnetic ion. According to Roose et al. (1996), the n value for Mn^{2+} is equal to six.

To study the chaotropic effects of cations and anions of dissolved salts, KCl was selected because both ions are chaotropic but not maximum strong. It is known that interfacial water at a surface of nanosilica is in the strongly associated state at $C_{H_2O}>200$ mg/g (Gun'ko et al. 2005d, 2009d). Therefore, nanosilica A-380 at $C_{H_2O}=275$ mg/g was selected as an initial sample. The ^{1}H NMR spectra of the sample in deuterochloroform are given in Figure 1.113 at $210<T<320$ K.

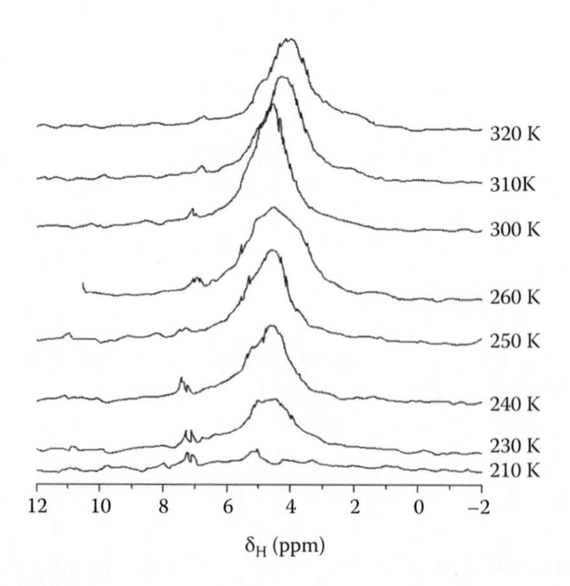

FIGURE 1.113 The ^{1}H NMR spectra of water bound to standard nanosilica A-380 at $C_{H_2O}=275$ mg/g in deuterochloroform medium at different temperatures.

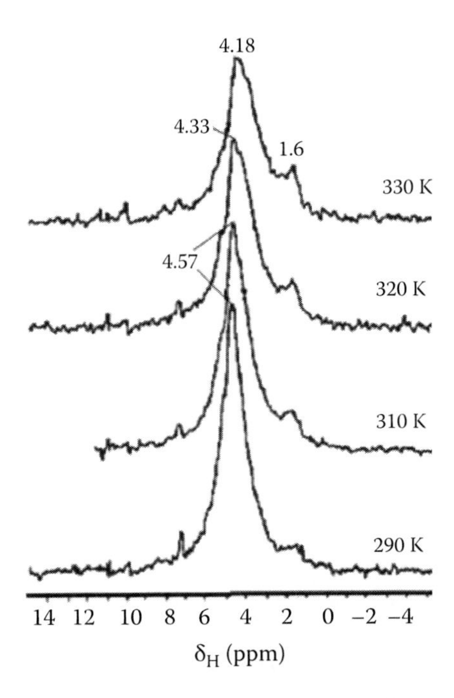

FIGURE 1.114 The ^{1}H NMR spectra of water bound to nanosilica A-380 at C_{H_2O}=440 mg/g and C_{KCl} = 16 mg/g in deuterochloroform medium at different temperatures. (Adapted from *Adv. Colloid Interface Sci.*, 118, Gun'ko, V.M., Turov, V.V., Bogatyrev, V.M. et al., Unusual properties of water at hydrophilic/hydrophobic interfaces, 125–172, 2005d. Copyright 2005, with permission from Elsevier.)

The chemical shift of the adsorbed water weakly depends on temperature both on heating and freezing. After addition of 16 mg KCl per gram of silica into the $CDCl_3$ suspension of the hydrated powder of nanosilica and intensive mixing besides signal 1 of SAW, a new signal appears at δ_H = 1.6 ppm related to WAW (Figure 1.114). Notice that similar unusual water is not observed in the ^{1}H NMR spectra of bulk water with dissolved salts; consequently, this is the property of only interfacial water.

An increase in C_{KCl} to 100 mg/g (Figure 1.115; KCl was adsorbed onto silica by its moistening with the salt solution and desiccation at 333 K) affects the ^{1}H NMR spectra depending on C_{H_2O}. In the case of greater amount of water C_{H_2O}=2 g/g (Figure 1.115a), the adsorbed water (about of 10% of the empty volume in the powder of nanosilica) fills voids between primary particles in aggregates (<1 cm^3/g) and partially in agglomerates. Therefore, the chemical shift δ_H depends weakly on temperature and water is in the strongly associated state. However, in the case of decreasing concentration of adsorbed water, signal at δ_H = 1.4 ppm linked to WAW appears in the spectra (Figure 1.115b). Correlating data shown in Figures 1.113 and 1.115b, it is possible to conclude that KCl at the interfaces of water/nanosilica has a strong influence.

It should be noted that water can be in weakly associated state at the interfaces of the wetted powders of nanosilica only when hydrophobic medium it is located in proximity to the silica surface and actively participates in structuring (clustering) of interfacial water. Probably, under the action of the surface forces and hydrophobic medium, water and chloroform are capable to form a mixture in which separate molecules or small clusters of water are divided by chloroform clusters. This state can be stabilized by chaotropic ions (K^+, and Cl^-), which can be localized in a layer slightly distant from the surface. For instance, chaotropic Cs^+ is weakly adsorbed onto nanosilica in comparison with kosmotropic cations (Gun'ko et al. 2004e). These conditions cause the formation of very small HDW structures in the vicinity of the surface and with strongly distorted network of the hydrogen bonds that appears as the WAW at small δ_H values. The results obtained

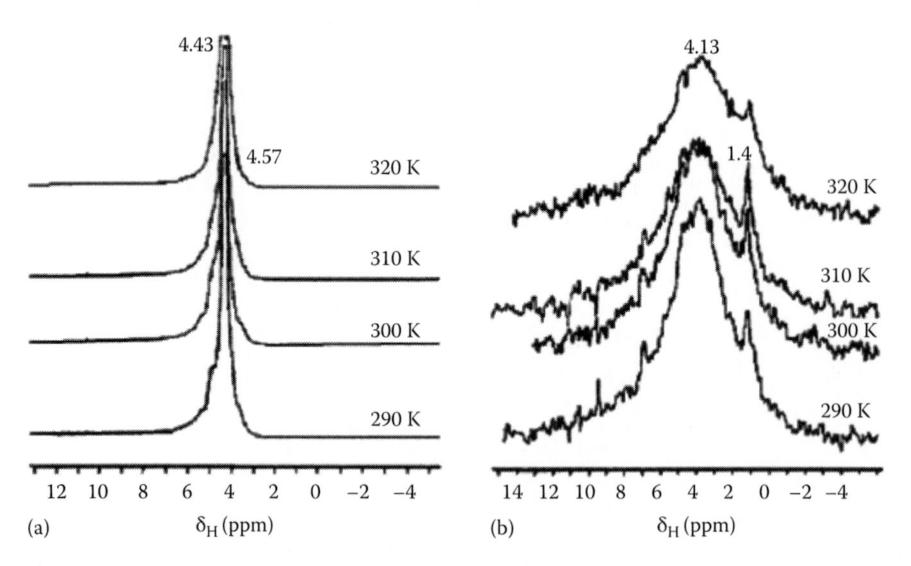

FIGURE 1.115 The ^{1}H NMR spectra of water bound to nanosilica A-380 at C_{H_2O} = (a) 2 g/g and (b) 0.2 g/g at C_{KCl} = 0.1 g/g in deuterochloroform medium at different temperatures. (Adapted from *Adv. Colloid Interface Sci.*, 118, Gun'ko, V.M., Turov, V.V., Bogatyrev, V.M. et al., Unusual properties of water at hydrophilic/hydrophobic interfaces, 125–172, 2005d. Copyright 2005, with permission from Elsevier.)

on the investigations of relatively simple systems with solid adsorbents can be useful on the study of much more complex biosystems (vide infra).

To elucidate the bioactivity mechanism of nanosilica used as an enterosorbent (Chuiko 1993, 2003, Blitz and Gun'ko 2006), it is needed to understand the influence of the clustered state of bound water on occurring of chemical reactions in this water. Dissociation and solvation of acids (e.g., HCl), which are relatively simple but important processes, can be used as model reactions. Mineral acids, such as hydrochloric acid, are strong acids, which are in completely dissociated state in diluted aqueous solutions (Drago 1977). This dissociation results in the formation of hydrated protons in clusters $H_3O^+(H_2O)_n$ and acidic residua (e.g., Cl^- $(H_2O)_m$) both strongly affecting water structure (Jieli and Aida 2009). In concentrated HCl solutions, the dissociation degree is low (about several percents). This degree and corresponding changes in the water structure can be studied using the LT ^{1}H NMR spectroscopy giving changes in the δ_H values in comparison with pure and maximum concentrated acids and pure water. "Acidic" H atoms in the most numbers of acids are characterized by low electron density. Therefore, their δ_H values can be of 9–13 ppm that are, however, smaller than δ_H = 16–21 ppm of protons in the symmetric strong hydrogen bonds according to experimental and theoretical results (Tolstoy et al. 2004, Murakhtina et al. 2006).

The interfacial behavior of water and HCl solution interacting with initial and MCA nanosilica A-300 at the bulk density of 0.045 and 0.394 g/cm^3, respectively, was studied in nonpolar (CCl_4) and weakly polar ($CDCl_3$) media used to enhance structurization of bound water and water/HCl (Turov et al. 2011c).

Initial standard nanosilica A-300 (S_{BET} = 330 m^2/g and bulk density ρ_b = 0.045 g/cm^3) was MCA in a ball mill for 4 h with the presence of water (50 wt% with respect to dry silica). MCA silica has S_{BET} = 332 m^2/g and ρ_b = 0.394 g/cm^3 (Gun'ko et al. 2011e).

Distilled water, chloroform-d, DCl (36% solution in D_2O), and CCl_4 (grade for NMR spectroscopy) were used in NMR spectroscopy measurements. The use of the DCl solution (including approximately 3% of HCl), in which the D-H exchange occurs during contact with silica, provides relatively low intensity of the ^{1}H NMR signals to avoid the phase distortions.

To prepare hydrated powders, certain amounts of distilled water and DCl (or HCl) were added to air-dry silica samples. Before the measurements, samples (placed in closed NMR ampoules) were shaken for 10 min and then equilibrated for 1 h. To determine the chemical shift of the proton resonance, δ_H, tetramethylsilane (TMS) was used as an internal standard (δ_H=0) added to $CDCl_3$ and CCl_4 in amount of 0.2 wt%.

Quantum chemical calculations of HCl and HCl/water clusters were carried out with consideration for solvation effects in the aqueous medium using the polarizable continuum model (PCM) with integral equation formalism (IEFPCM) method (Frisch et al. 2004) with the Gaussian 03 program package and the DFT (B3LYP/6-31G(d,p) technique (Frisch et al. 2004). The NMR spectra were calculated using the GIAO method (Frisch et al. 2004). The δ_H values were calculated as the difference between isotropic values (average value of diagonal values of the magnetic shielding tensor of protons, $\sigma_{H,iso}$) of TMS ($\sigma_{H,iso}$=31.76 ppm by GIAO/B3LYP/6-31G(d,p)) and studied compounds $\delta_H = \frac{1}{3}Tr\sigma_{TMS} - \frac{1}{3}Tr\sigma_H$. The distribution functions of the δ_H values were calculated using a simple Equation 10.92 described in Chapter 10.

Clusterization of adsorbed HCl solution can occur similar to that for water adsorbed alone onto nanosilicas. Pure liquid hydrochloric acid (36 wt%) has signal at δ_H=8.8–9.4 ppm increasing with lowering temperature (Figure 1.116a). Hydrated complexes of HCl (Figure 1.116b; signal 1) and SAW (signal 2) bound by initial silica differ in their temperature dependence. The δ_H value and the intensity of signal 1 depend weakly on temperature in contrast to signal 2, which increases and demonstrates the upfield shift with increasing temperature. Signal 2 is not observed at T<240 K. Addition of hydrochloric acid (Figure 1.116b) leads to diminution of signal of WAW interacting with silica (δ_H= 1.5–2 ppm) in comparison with the A-300-water system described earlier in the text. However signal of water dissolved in chloroform (δ_H= 1 ppm, very low amounts) does not depend on the presence of hydrochloric acid.

The spectra shape of SAW bound in samples in the CCl_4 medium changes weakly (Figure 1.116c). However, the difference in the chemical shift of two signals of SAW decreases. The difference between signals 1 and 2 decreases with increasing HCl concentration since relative intensity of signal 1 increases. A decrease in the intensity of signal 2 is observed at T<240 K (Figure 1.116).

In 36% hydrochloric acid used, approximately six water molecules are per HCl molecule. This solution is frozen at T<200 K (Figure 1.116a) with simultaneous freezing of water and HCl (Leninsky et al. 1985). Freezing of the HCl solutions is possible at higher temperatures if the HCl concentration decreases. Consequently, the results shown in Figure 1.116b through d can be interpreted as the formation of two types of HCl/water structures with different concentrations of HCl (C_{HCl} is smaller in structures corresponding to signal 2 at a lower δ_H value and these structures include non-dissociated HCl). The chemical shift of signal 1 is close to that of individual HCl solution, that is, the structures corresponding to signal 1 have the properties similar to that of the bulk HCl solution with significant contribution of dissociated HCl molecules.

In the case of water/HCl adsorbed onto MCA-A-300, the ^{1}H NMR spectra (Figure 1.116e) are similar to that of water adsorbed on MCA-A-300. Freezing of the adsorbates in these systems occurs at similar temperatures. Water in confined space of smaller voids in MCA-A-300 has lower activity as a solvent. Therefore, water clusters adsorbed on MCA-A-300 include a lower amount of dissolved HCl molecules. Quantum chemical calculations of the ^{1}H NMR spectra (Figures 1.117 and 1.118) show that non-dissociated HCl molecules dissolved in water have δ_H<6 ppm that corresponds to signal observed in Figure 1.116e.

The cluster with 16 HCl calculated with consideration of the solvation effects (GIAO/IEFPCM/B3LYP/6-31G(d,p)) is characterized by a narrow $f(\delta_H)$ distribution function (Figure 1.118) because all the HCl molecules are non-dissociated. In the case of dissociated HCl (Figure 1.117), dissolved HCl and non-dissolved HCl show that the ^{1}H NMR spectra of these systems strongly differ (Figure 1.118). ^{1}H NMR spectrum of the system with dissociated HCl is much broader than that of non-dissociated HCl molecules (Figure 1.118). Dissociated HCl being in the $16H_2O$ cluster (Figure 1.117) is characterized by greater δ_H values, and this cluster is characterized by a broad $f(\delta_H)$ distribution function (Figure 1.118). These calculations show that signal 2 (Figure 1.117b) can include a signal

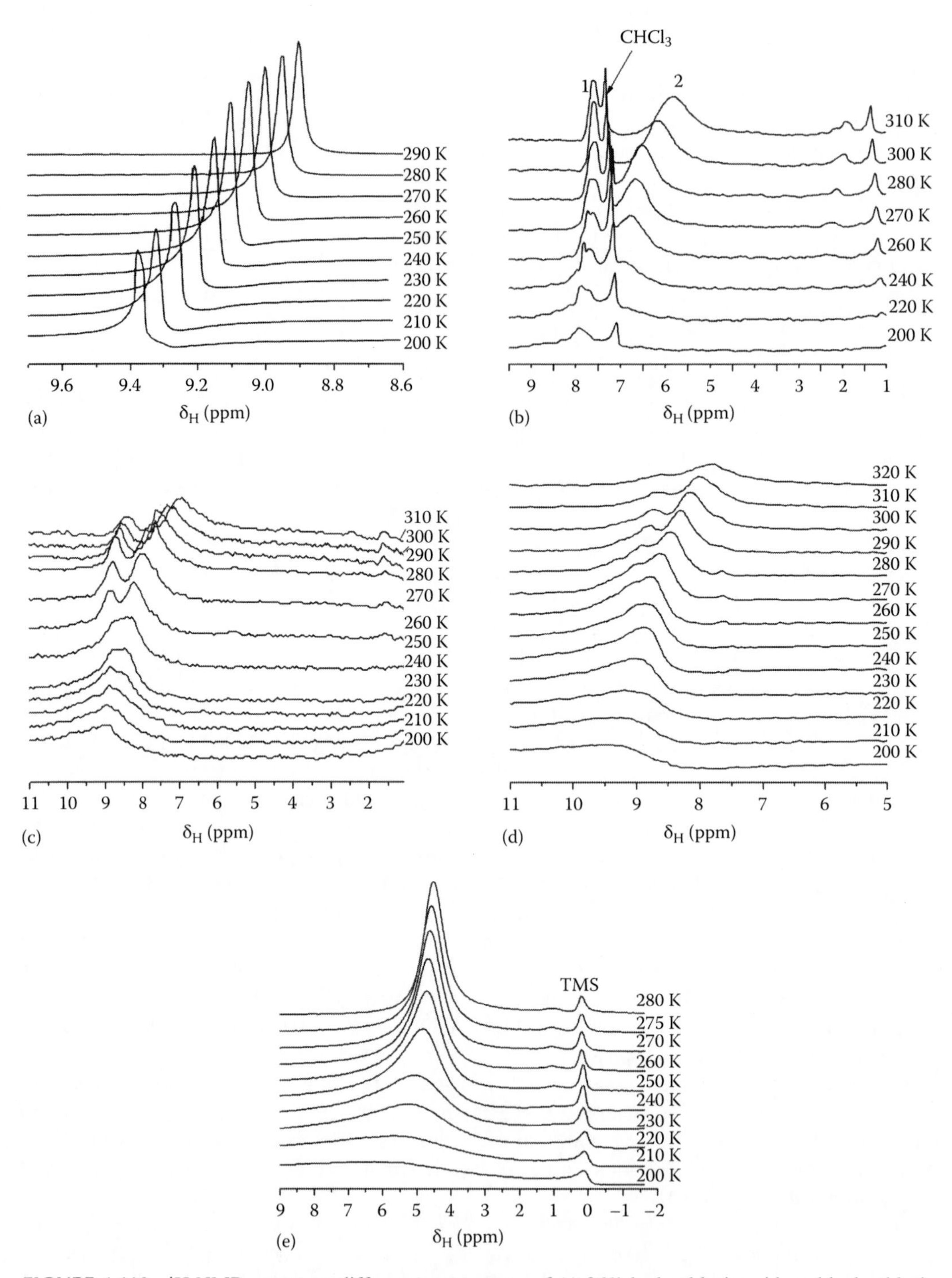

FIGURE 1.116 ^{1}H NMR spectra at different temperatures of (a) 36% hydrochloric acid, and hydrochloric acid and water adsorbed to (b)–(d) initial and (e) MCA A-300 in (b) $CDCl_3$ or (c)-(e) CCl_4; (b, c, e) 38 mg/g H_2O and 44 mg/g HCl, (d) 140 mg/g H_2O and 120 mg/g HCl. (Adapted from *Colloids Surf. A: Physicochem. Eng.* Aspects., 390, Turov, V.V., Gun'ko, V.M., Turova, A.A., Morozova, L., and Voronin, E.F., 48–55, 2011e. Copyright 2011, with permission from Elsevier.)

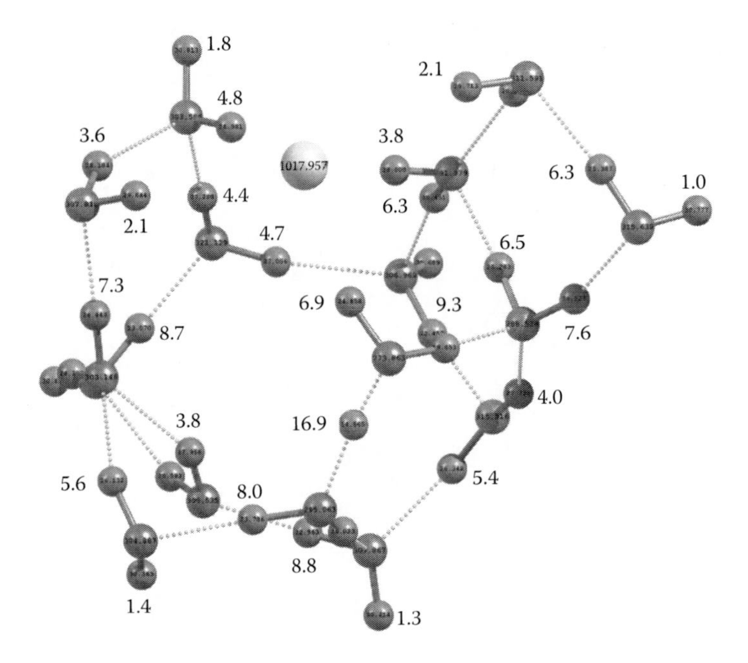

FIGURE 1.117 The δ_H values in a cluster with dissociated HCl and $16H_2O$. (Adapted from *Colloids Surf. A: Physicochem. Eng.* Aspects., 390, Turov, V.V., Gun'ko, V.M., Turova, A.A., Morozova, L., and Voronin, E.F., 48–55, 2011e. Copyright 2011, with permission from Elsevier.)

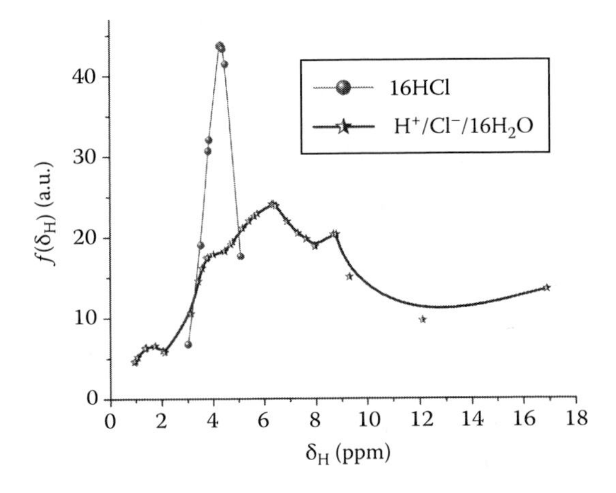

FIGURE 1.118 The distribution function of the δ_H values for clusters with 16HCl and dissociated HCl interacting with $16H_2O$ (geometry was optimized using IEFPMC/B3LYP/6-31G(d,p) and the NMR spectra were calculated using GIAO with consideration of solvation effects). (Adapted from *Colloids Surf. A: Physicochem. Eng.* Aspects., 390, Turov, V.V., Gun'ko, V.M., Turova, A.A., Morozova, L., and Voronin, E.F., 48–55, 2011e. Copyright 2011, with permission from Elsevier.)

of non-dissociated HCl molecules. Eigen and Zundel cations located far from Cl^- ions (Murakhtina et al. 2006) provide signals at high δ_H values (signal 1; Figure 1.116b).

Schematic distribution of nanodroplets of water (several percents) adsorbed in textural pores of nanosilica being in the CCl_4 (or $CDCl_3$) media is shown in Figure 1.119. Water structures (with or without dissolved HCl) tend to have a minimal contact area with the nonpolar solvent. At the boundary of water/solvent, the δ_H values are in the 1–2 ppm range corresponding to WAW. The presence of HCl dissolved in water can weakly affect the interaction of water with hydrophobic solvents.

FIGURE 1.119 Scheme of distribution of water nanodroplets (2–5 nm in size) in voids between silica nanoparticles (~10 nm in size) in the case of low content (several percents) of water adsorbed to nanosilica being in CCl_4 or $CDCl_3$ media (adapted from *Colloids Surf. A: Physicochem. Eng.* Aspects., 390, Turov, V.V., Gun'ko, V.M., Turova, A.A., Morozova, L., and Voronin, E.F., 48–55, 2011e. Copyright 2011, with permission from Elsevier.)

Adsorption of a homogeneous HCl/water mixture onto nanosilica A-300 initial and MCA results in diminution of HCl dissolution in water (at low content of water in the system), especially in the case of MCA silica. This effect appears as changes in the ^1H NMR spectra of the HCl solution characterized by signals at the δ_H values corresponding to both dissolved and non-dissolved HCl molecules. One can assume that the use of nanosilica as an enterosorbent for oral administration can change the characteristics of gastric juice with the HCl solution. These changes can depend on the type of nanosilica, for example, initial or MCA silica.

Anhydrous phosphorus oxyacids (POA) are characterized by high proton conductivity because of a dense hydrogen bond network (Zhao et al. 2007, Schuster et al. 2008). Typically, POA are used in aqueous solutions, in which POA molecules are strongly hydrated but only partially deprotonated. POA are well water soluble but relatively weak acids with a low degree of dissociation of PO-H bonds in the aqueous solution (Nekrasov 1973). Chemical shift of the proton resonance (δ_H) in the ^1H NMR spectra of POA in concentrated aqueous solutions can reach 10 ppm or more (Emsley et al. 1965) that suggests the presence of hydrated protons appearing due to dissociation of a portion of the PO-H bonds. In phosphonic acid, doublet signals of protons from the P-H bonds are also observed due to large J-coupling to the P nuclei but of a lower intensity than that of the PO-H bonds (Schuster et al. 2008).

The NMR spectroscopy is informative in investigations of POA and their derivatives because the δ_H values are sensitive to the hydrogen bond network structure (Chruszcz et al. 2003, Holland et al. 2007, Schuster et al. 2008, Faßbender 2009). Conductivity and diffusion effects playing an important role in the properties and applications of POA and their derivatives (Dippel and Kreuer 1991, Ruiz-Bevia et al. 1995, Chakrabarti 1996, Kreuer 1996, Chung at al. 2000, Aihara et al. 2006, Schuster et al. 2008) depend also on features of this network. In pure phosphoric acid, the proton conduction mechanism corresponds to the Grotthuss mechanism (Dippel et al. 1993, Schuster et al. 2008) with fast proton transfer between OH groups and some structural reorganization of the hydrogen bond network. These effects appear in the dependences of the ^1H NMR spectra shape on acid concentration, temperature, presence of co-solvents, dissolved compounds, or solid particles with polar surface functionalities. Analysis of states of water bound by POA alone and adsorbed onto nanosilica depending on concentration, temperature, and POA and silica types was carried out using LT NMR and FTIR spectroscopies, NMR cryoporometry, and quantum chemical calculations.

Phosphoric $O{=}P(OH)_3$ and phosphonic $O{=}P(H)(OH)_2$ acids, nanosilicas OX-50 (Degussa; specific surface area $S_{BET}=52$ m^2/g), and A-300 (pilot plant of Chuiko Institute of Surface Chemistry, Kalush, Ukraine; $S_{BET}=297$ m^2/g) were used as initial materials (Gun'ko et al. 2012b). Concentrated aqueous suspension of OX-50 (16 wt%) stabilized by addition of 4 wt% H_3PO_4 was dried at 293 K for 8 days then at 330 K for 10 min. The residual content of water was 0.5 wt%. To prepare more hydrated systems, certain amounts of water were added to the powder and equilibrated for 1 h. A-300/POA was prepared using A-300 powder mixed with a solution of H_3PO_4 in isopropanol, stirred and heated at 353 and 383 K to complete removal of the solvent. After drying, samples containing 14 and 28 wt% of POA were milled and used in a FTIR study. Initial solid H_3PO_3 included 50 mg of water per gram of acid. A concentrated aqueous solution of POA was with 30 wt% of acid. The aqueous suspension of silica A-300/H_3PO_3 (as 4/1) was stirred and dried at 293 K for 14 days to obtain the powder at a low content of adsorbed water. A-300/POA powder containing 14 and 28 wt% of H_3PO_4 was used for recording of FTIR spectra (diffusive reflectance mode) with a FTIR ThermoNicolet spectrophotometer (Gun'ko et al. 2012b).

The ^{1}H NMR spectra of fresh 30% solution of phosphonic acid in D_2O (Figure 1.120a) demonstrate the main signal of the protons in the PO-H groups (and H_2O or HDO due to H-D exchange reactions between PO-H and D_2O) with $\delta_H=6$ ppm (290 K) increased to 9.8 ppm (200 K) with decreasing temperature.

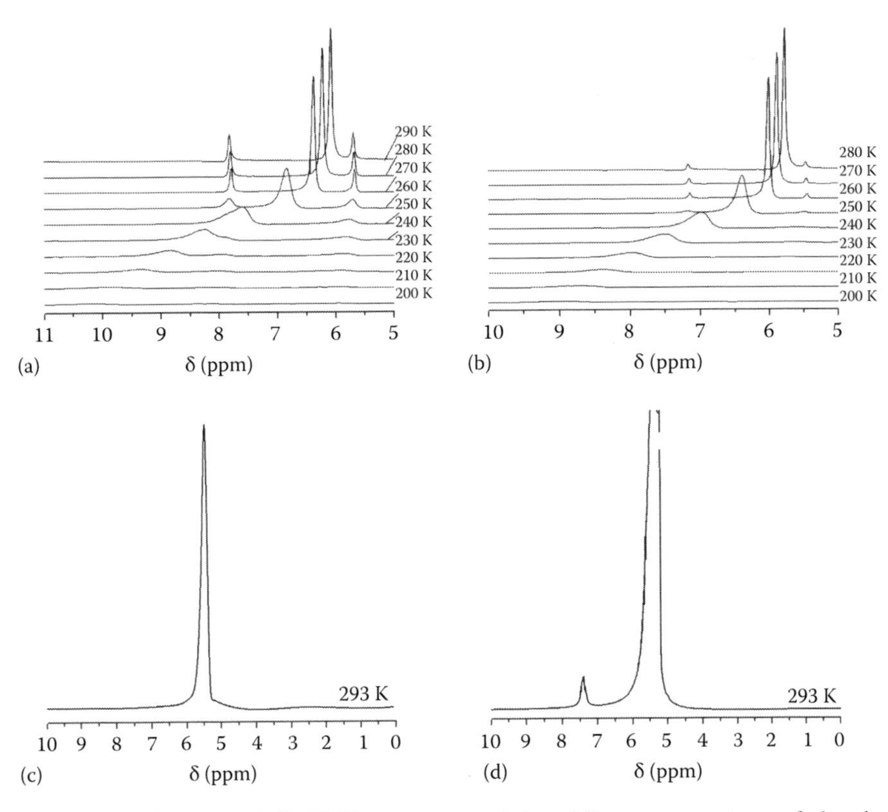

FIGURE 1.120 (a,b) ^{1}H and (c,d) ^{2}D NMR spectra recorded at different temperatures of phosphonic acid (30%) in D_2O (a,c) fresh and (b,d) boiled samples, and (b) sample was also stored for 7 days after boiling. (Adapted from *J. Colloid Interface Sci.*, 368, Gun'ko, V.M., Morozova, L.P., Turova, A.A. et al., Hydrated phosphorus oxyacids alone and adsorbed on nanosilica, 263–272, 2012b. Copyright 2012, with permission from Elsevier.)

The signal intensity decreases with lowering temperature and its width increases because of freezing of a portion of water and acid as well as concentrating of the acid solution with decreasing temperature. Notice that in the case of a similarly concentrated HCl solution, $\delta_H = 9.0–9.4$ ppm at 290–200 K (Turov et al. 2011b) because the dissociation degree of HCl is much higher than POA. Two signals of lower intensity are observed at weaker and stronger fields than the main signal (Figure 1.120a). The difference between their δ_H values is approximately 2.1 ppm. These δ_H values depend weakly on temperature; however, the intensity sharply increases at $T \geq 260$ K but the width increases at $T < 260$ K. This occurs due to freezing of a portion of POA. After boiling of the sample and its storage at 290 K for a week, the intensity of the H_P doublet signals decreases (Figure 1.120b) because of the H_P-D_{D2O} exchange reaction.

The H-D exchange reaction for H-P is slow for the H_3PO_3 solution in D_2O since the exchange P-H $\leftrightarrow$ PO-H is slow. The slow rate of the H-D exchange reaction is confirmed by the ^{2}D NMR spectra (Figure 1.120c and d). For a fresh sample, only one ^{2}D NMR signal is observed (Figure 1.120c) related to D_2O and PO-D. After boiling of the sample for 10 min, a weak signal of the P-D groups appears (Figure 1.120d). The absence of splitting of this signal into doublet (as for P-H) is due to much smaller spin–orbit coupling constant of P-D in comparison with P-H (Emsley et al. 1965).

An increase in the signal width at $T \leq 250$ K (Figure 1.120b) is due to partial freezing of the solution. An increase in the δ_H value of the main signal is due to an increase in contribution of hydrated protons (Hindman 1966, Kinney et al. 1993, Turov et al. 1996a, Gun'ko and Turov 1999) of dissociated acid molecules concentrated with lowering temperature. The ^{1}H NMR spectra of solid phosphonic acid with a low content of water (50 mg per gram of solid POA) in CCl_4 medium include three signals at $\delta_H = 6$ (signal 1), 5 (weak signal as a shoulder of signal 1), and 3.5 (signal 2) ppm at 290 K (Figure 1.121a). The shape of signal 1 is asymmetrical. Its intensity decreases with decreasing temperature (Figure 1.121a), as well as the δ_H value decreases to 5 ppm at 200 K (in contrast to the behavior of signal in the concentrated POA solution; Figure 1.120a). The δ_H values for other two signals depend weaker on temperature than signal 1. Addition of water to 70 mg/g results in an increase in the signal intensity and width (Figure 1.121b). However, the temperature behavior of the spectra is similar to that at $C_{H_2O} = 50$ mg/g.

The downfield shift of signal 1 with increasing temperature suggests that contribution of protons in water with dissolved acid increases with temperature (Figure 1.121a and b) because of increased solubility of H_3PO_3. Contribution of hydrated protons (i.e., dissociated POA) is small at a small content of water (Figure 1.121a) because the δ_H value decreases with decreasing temperature in contrast to that for the concentrated POA solution (Figure 1.120a). The spectra shape shows that structures with water/dissolved H_3PO_3 are nonuniform since several signals can be found. Signal 2 at 3.5 ppm can be attributed to water molecules with one hydrogen bond as a proton donor and PO-H without the hydrogen bonding (see quantum chemical calculation results).

Dried suspension with 16 wt% of silica and 4 wt% H_3PO_3 as a powder with approximately 10 wt% of water is characterized by the ^{1}H NMR spectra with only one relatively broad signal at 4–4.5 ppm (Figure 1.121c). Similarly, hydrated nanosilica A-300 modified by phosphonic acid obtained by hydrolysis of adsorbed PCl_3 was characterized by larger values $\delta_H = 6.5–10.5$ ppm (Turov et al. 1996a,b). During drying of the concentrated suspension with silica and POA, solution supersaturation results in formation of acid crystallites. Additionally, residual water bound by nanosilica has low activity as a solvent. Therefore, according to the ^{1}H NMR spectra (Figure 1.121c), bound water can dissolve a low amount of H_3PO_3 since the δ_H value depends weakly on temperature as well as the signal shape.

Water unfrozen at $T < 273$ K due to bonding to solid POA or dried A-300/POA powder (Figure 1.122a) can be attributed to SBW (unfrozen at $T < 250$ K) and WBW (frozen at $T < 250$ K) (Gun'ko et al. 2005d, 2009d). However, there is not a clear boundary between SBW and WBW because of changes in the concentration of dissolved POA.

A major fraction of water responsible for signal 1 ($C_{uw1}(T)$; Figure 1.122a) corresponds to SBW but a major fraction of water giving signal 2 corresponds to WBW (Figure 1.122b).

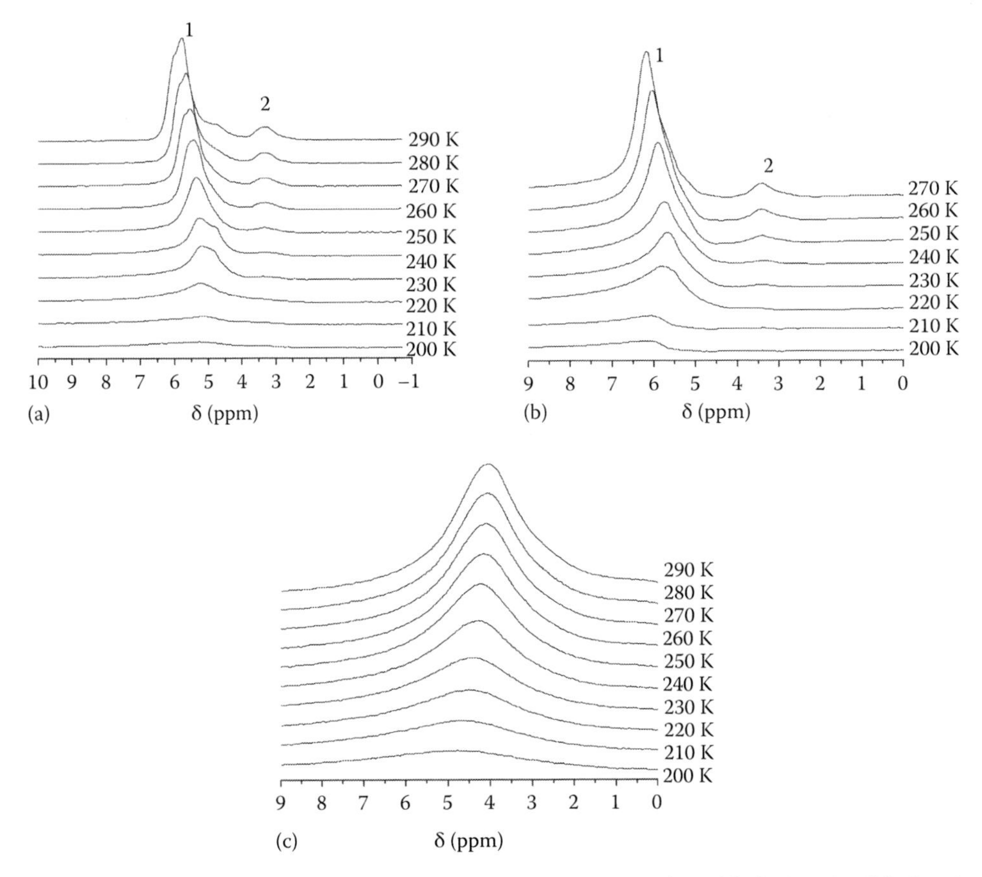

FIGURE 1.121 ^{1}H NMR spectra recorded at different temperatures of weakly hydrated solid phosphonic acid (a,b) alone and (c) adsorbed onto nanosilica at A-300/POA = 4/1 at C_{H_2O} = (a) 50, (b) 70 and (c) 100 mg/g in CCl_4 medium. (Adapted from *J. Colloid Interface Sci.*, 368, Gun'ko, V.M., Morozova, L.P., Turova, A.A. et al., Hydrated phosphorus oxyacids alone and adsorbed on nanosilica, 263–272, 2012b. Copyright 2012, with permission from Elsevier.)

The δ_H value of sample with 50 mg H_2O per gram of acid depends weakly on temperature (Figure 1.122c). It slightly increases at $T > 230$ K. An increase in the water content to 70 mg/g gives an increase in the δ_H value with a minimum at 240 K. For the dried A-300/H_3PO_3 powder, the δ_H value decreases with increasing temperature (Figure 1.122c). These results are caused by two competitive processes such as diminution of associativity of water molecules (δ_H decreases) and increase of concentration of dissolved acid (δ_H increases) with temperature. During freezing of the solution, pure ice and acid crystallites (which do not contribute the ^{1}H NMR spectra here) form separately. However, interfacial water can be frozen at lower temperature than acid that affects the concentration of acid dissolved in the interfacial water, and the chemical shift δ_H decreases with lowering temperature. However, signal of the concentrated acid solution is absent in weakly hydrated A-300/H_3PO_3 powder because water bound to the silica surface does not dissolve the acid, which forms crystallites weakly bound to silica surface, according to the FTIR spectra (vide infra).

Solid POA has a relatively low surface area in contact with unfrozen water at low C_{H_2O} values (Table 1.23, S_{uw}) at a major contribution of nanoclusters of water at $R < 1$ nm to the S_{uw} value (Table 1.23, S_{nano}). However, the volume of water nanoclusters (V_{nano}) is smaller than the volume of larger domains of unfrozen water at $1 < R < 25$ nm. Addition of water to 70 mg/g increases mainly domain structures (S_{dom} and V_{dom}). Signal 2 corresponds water forming domain structures at $R > 1$ nm. In the case of dried A-300/POA powder, $S_{uw} < S_{BET,A\text{-}300}$ but $S_{nano,uw} > S_{nano,N2} = 22$ m^2/g

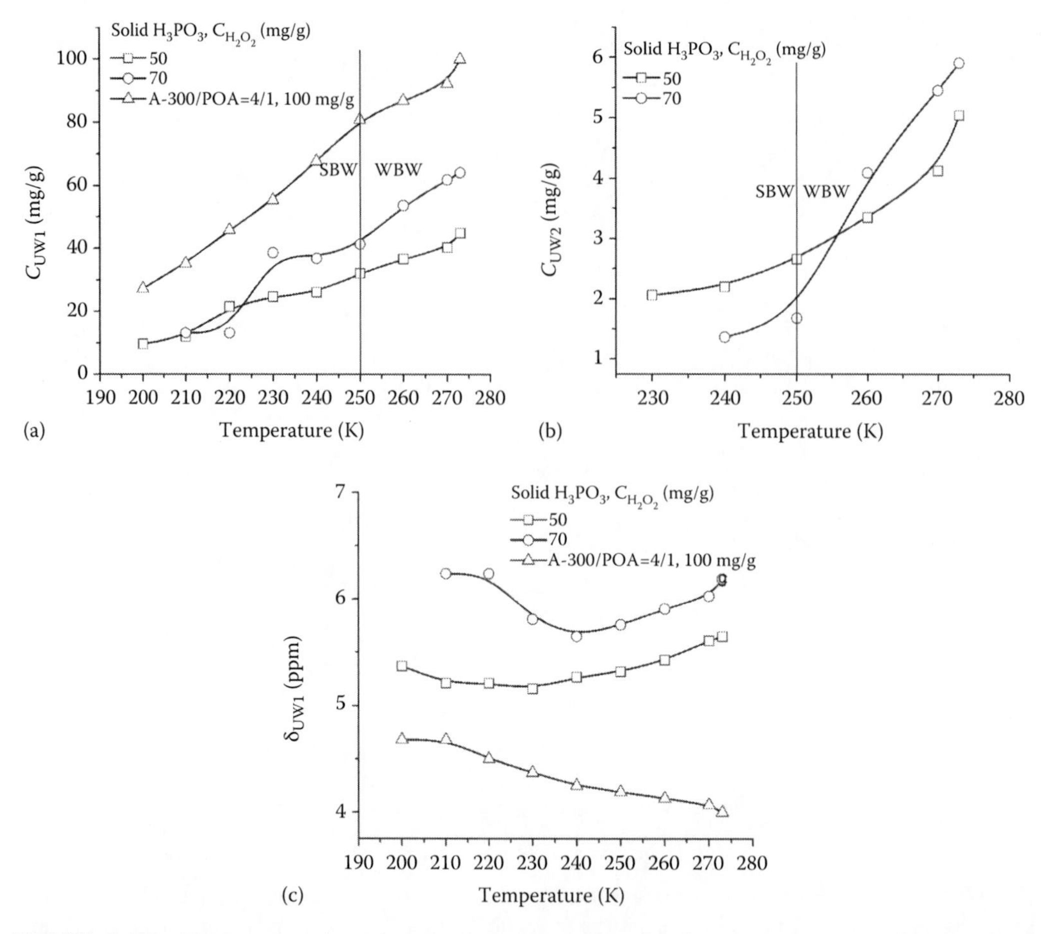

FIGURE 1.122 Temperature dependences of concentration of unfrozen water responsible for signal (a) 1 and (b) 2 and (c) chemical shift of signal 1 for weakly hydrated solid H_3PO_3 (C_{H_2O}=50 or 70 mg/g) and A-300/H_3PO_3=4/1 (C_{H_2O}=100 mg/g) in CCl_4 medium. (Adapted from *J. Colloid Interface Sci.*, 368, Gun'ko, V.M., Morozova, L.P., Turova, A.A. et al., Hydrated phosphorus oxyacids alone and adsorbed on nanosilica, 263–272, 2012b. Copyright 2012, with permission from Elsevier.)

TABLE 1.23
Characteristics of Water Bound in POA or POA/Silica Powders

Sample	C_{H_2O} (mg/g)	S_{uw} (m²/g)	S_{nano} (m²/g)	S_{dom} (m²/g)	V_{nano} (cm³/g)	V_{dom} (cm³/g)	γ_S (J/g)
H_3PO_3 (signal 1)	50	30	26	4	0.011	0.030	4.1
H_3PO_3 (signal 2)	50	0.4	0	0.4	0	0.004	0.6
H_3PO_3 (signal 1)	70	32	24	8	0.010	0.051	5.3
H_3PO_3 (signal 2)	70	0.6	0	0.6	0	0.006	0.3
H_3PO_3/A-300	100	82	74	8	0.032	0.062	9.8
H_3PO_4/OX-50	5	1	0	1	0	0.005	0.5
H_3PO_4/OX-50	25	12	10	2	0.005	0.010	1.7
H_3PO_4/OX-50	45	16	12	4	0.006	0.034	3.3
H_3PO_4/OX-50[a]	4000	651	374	277	0.169	1.624	110

[a] Concentrated aqueous suspension. Other samples are weakly hydrated powders.

because POA forms nanocrystallites smaller than the size of silica nanoparticles and unfrozen water locates in narrow pores (voids between silica nanoparticles and POA crystallites).

Water bound to solid POA or silica/POA powders is energetically and structurally nonuniform (Figure 1.123) because of spatial restrictions dependent on void size distribution between nanoparticles and the difference in the surface effects on water layers nearest to the surface and located in the next layers. As a whole, at low content of adsorbed water (50–100 mg/g), it tends to form small clusters (Figure 1.123d) characterized by relatively large changes in the Gibbs free energy (Figure 1.123a through c) of bound water with the water cluster size distributions (WCSDs) over 0.6–10 nm in radius (Figure 1.123d).

Water responsible for signal 2 is nonuniform too but it does not form nanoclusters at $R<1$ nm (Figure 1.123g). Its nanodomains weaker interact with solid POA (Figure 1.123e and f) than water responsible for signal 1. To analyze the H_3PO_4 distribution on a silica A-300 surface, the FITR spectra were recorded at two contents of POA (Figure 1.124). The PO-H stretching vibrations at 3672 cm^{-1} are observed for both samples. However, at 28 wt% of POA, the intensity of this band (as well as the SiO-H stretching vibrations at 3741 cm^{-1}) decreases because the formation of larger acid crystallites with lower contribution of non-bound surface PO-H groups strongly covered the silica surface. The POA amount of 28 wt% corresponds to practically total coverage of the silica surface.

Despite melting of anhydrous H_3PO_4 occurs at 290 K, the aqueous solution of the acid can be frozen at much lower temperature because of colligative properties of the aqueous solutions (Chaplin 2011). To elucidate the effects of different nanosilicas on the interfacial behavior of strong acids, nanosilica OX-50 (Degussa) with small specific surface area (52 m^2/g) was used to compare with A-300 studied earlier.

Initial sample OX-50/H_3PO_4 (Turov et al. 2010a, Gun'ko et al. 2012a) at $C_{H_2O}=5$ mg/g has only a weak ^{1}H NMR signal at $\delta_H=4$ ppm (Figure 1.125a).

Its intensity decreases with lowering temperature because of freezing of water, which is weakly bound. Signal at 0 ppm is due to tetramethylsilane (0.2 wt%) used as a reference compound. In the case of dried OX-50/POA powder at $C_{H_2O}=45$ mg/g, the values of all the structural characteristics are much lower than those of A-300/POA at $C_{H_2O}=50$ mg/g (Table 1.23) because OX-50 has the S_{BET} value six times smaller than that of A-300. However, the difference in the S_{uw} values is smaller than that in the S_{BET} values because of contributions of POA nanocrystallites. Very large values of the structural characteristics of unfrozen water are observed for the concentrated suspension of OX-50/POA (Table 1.23). In other words, water in this suspension is strongly clustered (vide infra) due to interactions with POA molecules and oligomers. To increase the amounts of water, it was introduced in the suspension in CCl_4 stirred, shaken for 10 min, and equilibrated for 2 h at 293 K. The spectrum of OX-50/H_3PO_4 at $C_{H_2O}=25$ mg/g at 290 K (Figure 1.125b) includes two signals of SAW at 5.5 and 3.7 ppm and a signal of WAW at 0.8 ppm. The SAW signal at 3.7 ppm disappears at 260 K. Therefore, this water is WBW with the size of clusters larger than 10 nm in radius. Addition of water ($C_{H_2O}=45$ mg/g) leads to disappearance of this signal (Figure 1.125c). In the case of totally unfrozen concentrated aqueous suspension of OX-50 (16 wt%) and POA (4 wt%; Figure 1.125d), the chemical shift increases from 4.5 ppm at 280 K to 6.5 ppm at 210 K with a small fraction of unfrozen water (Figure 1.126b). A major amount of water is frozen at $T\leq 240$ K. Concentrating of POA in residual liquid water, whose amounts decrease with decreasing temperature, results in increasing δ_H value.

In the case of a low content of water (5 and 25 mg/g), it is SBW frozen at $T<260$ K (Figure 1.126a). A small amount of WBW is at $C_{H_2O}=45$ mg/g. The δ_H value increases with decreasing temperature if $C_{H_2O}=25$ or 45 mg/g in contrast to that at 5 mg/g (Figure 1.126b). Thus, in the last case, water does not dissolve POA. The POA solubility increases with increasing C_{H_2O} and the δ_H value increases. Concentrating of POA in liquid water is observed at lowering temperature because of partial freezing of bound water.

In the aqueous suspension of OX-50/H_3PO_4 (Figure 1.127a), the amount of bound water is larger than 3 g per gram of solids including 1 g/g of SBW.

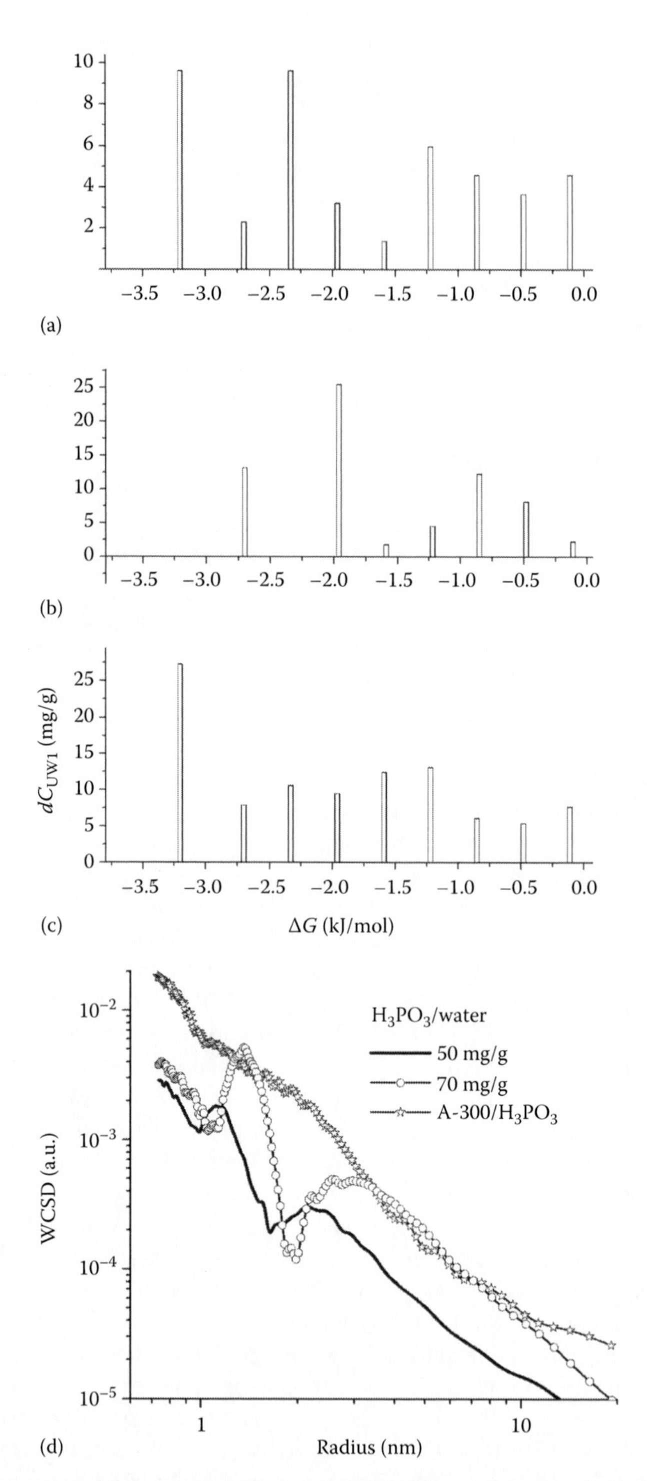

FIGURE 1.123 Changes in the Gibbs free energy of bound water in weakly hydrated solid H_3PO_3 in CCl_4 medium at C_{H_2O}=(a, e) 50 or (b, f) 70 mg/g and (c) A-300/POA at C_{H_2O}=100 mg/g and (d, g) water cluster size distributions (WCSD) with respect to signals (d) 1 and (g) 2. (Adapted from *J. Colloid Interface Sci.*, 368, Gun'ko, V.M., Morozova, L.P., Turova, A.A. et al., Hydrated phosphorus oxyacids alone and adsorbed on nanosilica, 263–272, 2012b. Copyright 2012, with permission from Elsevier.)

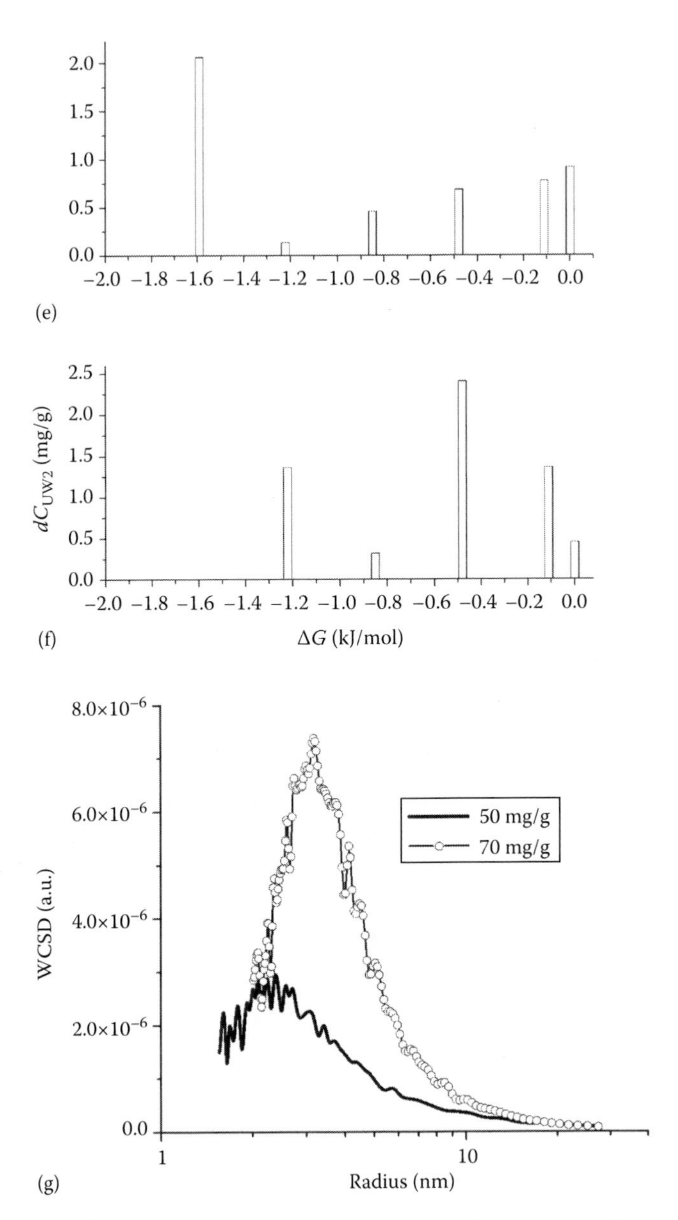

FIGURE 1.123 (continued) Changes in the Gibbs free energy of bound water in weakly hydrated solid H_3PO_3 in CCl_4 medium at C_{H_2O}=(a, e) 50 or (b, f) 70 mg/g and (c) A-300/POA at C_{H_2O}= 100 mg/g and (d, g) water cluster size distributions (WCSD) with respect to signals (d) 1 and (g) 2. (Adapted from *J. Colloid Interface Sci.*, 368, Gun'ko, V.M., Morozova, L.P., Turova, A.A. et al., Hydrated phosphorus oxyacids alone and adsorbed on nanosilica, 263–272, 2012b. Copyright 2012, with permission from Elsevier.)

The δ_H value (related to water and PO-H groups) linearly increases with decreasing temperature (Figure 1.127b) to 6.7 ppm at 200 K because of dissolution of POA. This temperature behavior of the δ_H value is similar to that of H_3PO_3 (Figure 1.120). Smaller δ_H values (Figure 1.127b) are due to smaller content of POA.

Theoretical calculations of the ^{1}H NMR spectra of phosphonic and phosphoric acids in different states (without water, hydrated, and deprotonated) in CCl_4 medium give relatively broad distribution functions (Figure 1.128) because a portion of protons does not participate in the hydrogen

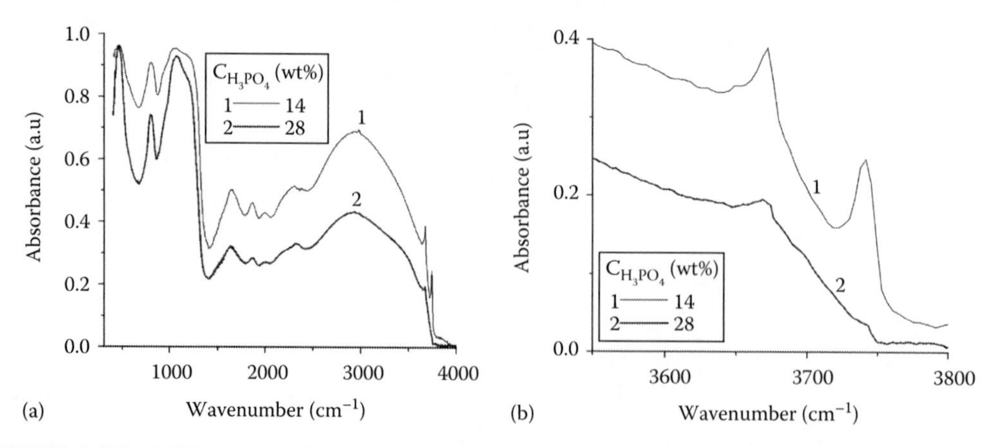

FIGURE 1.124 FTIR spectra (diffusive reflectance mode) of nanosilica A-300 with adsorbed H_3PO_4 at 14 wt% [96°C] and 28 wt% 115°C) and low amounts of adsorbed water. (Adapted from *J. Colloid Interface Sci.*, 368, Gun'ko, V.M., Morozova, L.P., Turova, A.A. et al., Hydrated phosphorus oxyacids alone and adsorbed on nanosilica, 263–272, 2012b. Copyright 2012, with permission from Elsevier.)

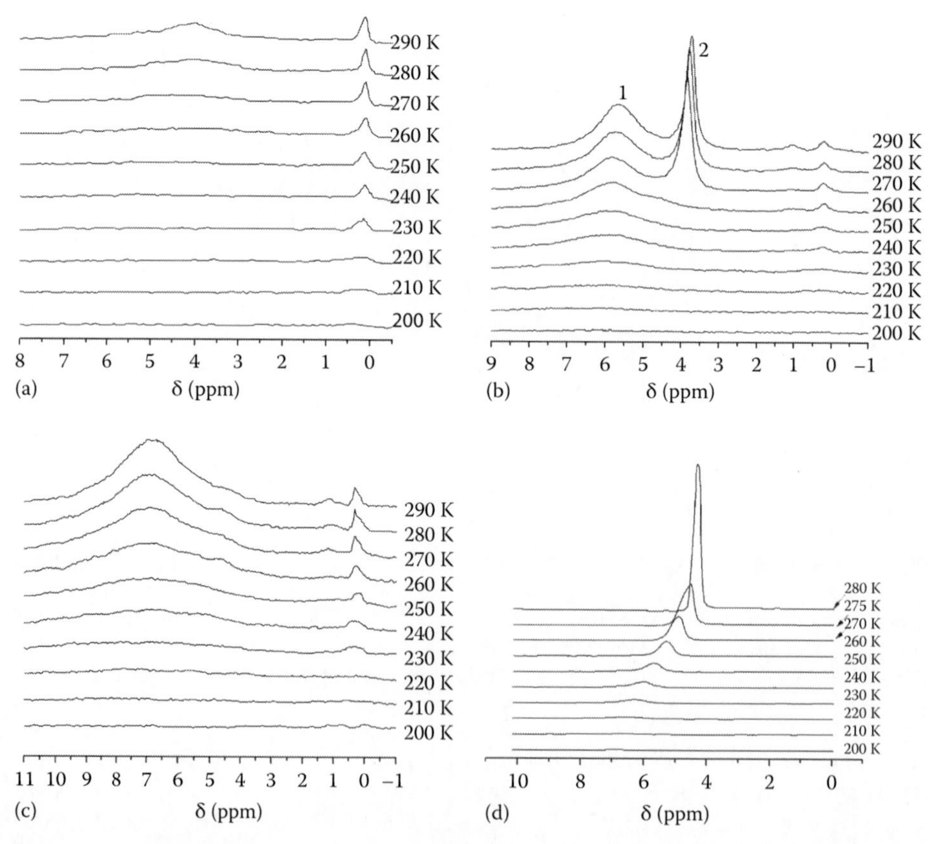

FIGURE 1.125 ^{1}H NMR spectra recorded at different temperatures of dried OX-50/H_3PO_4 powder at C_{H_2O} = (a) 5, (b) 25 and (c) 45 mg/g in CCl_4 medium and (d) frozen aqueous suspension with OX-50 (16 wt%) and H_3PO_4 (4 wt%). (Adapted from *J. Colloid Interface Sci.*, 368, Gun'ko, V.M., Morozova, L.P., Turova, A.A. et al., Hydrated phosphorus oxyacids alone and adsorbed on nanosilica, 263–272, 2012b. Copyright 2012, with permission from Elsevier.)

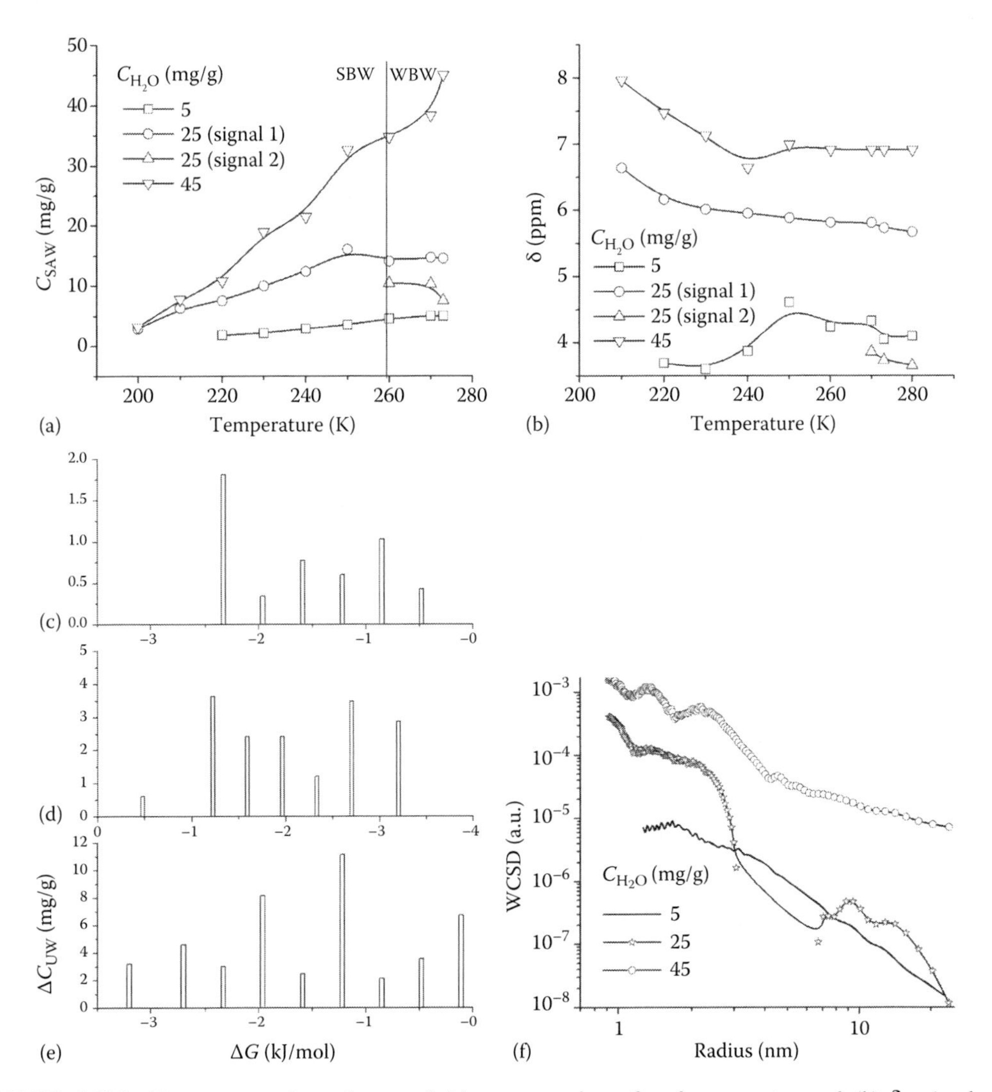

FIGURE 1.126 Temperature dependences of (a) concentration of unfrozen water and (b) δ_H, (c, d, e) relationships between changes in the Gibbs free energy and the amounts of unfrozen water and (f) unfrozen water cluster size distributions for OX-50/H_3PO_4 at C_{H_2O}=(c) 5, (d) 25 and (e) 45 mg/g. (Adapted from *J. Colloid Interface Sci.*, 368, Gun'ko, V.M., Morozova, L.P., Turova, A.A. et al., Hydrated phosphorus oxyacids alone and adsorbed on nanosilica, 263–272, 2012b. Copyright 2012, with permission from Elsevier.)

bonds (i.e., WAW at δ_H=1–2 ppm) but other OH groups form the hydrogen bonds of different strength from typical bonds at δ_H=4–6 ppm to very strong H bonds or bonds with H^+ transfer at δ_H=8–17 ppm. The energy of the formation of acid dimers is high (Figure 1.129). However, these dimers can be destroyed due to interaction with water (inserts in Figure 1.128). The distribution functions (Figure 1.128) include all δ_H values observed in the experimental spectra (Figures 1.120 and 1.121) and confirm the aforementioned interpretations.

There is a significant difference in the solubility of POA in bulk and bound water because of decreased activity of the interfacial water. LT ^{1}H NMR spectroscopy investigations of water bound by phosphoric and phosphonic acids solid alone or adsorbed onto nanosilicas OX-50 or A-300 show that concentrated solutions or weakly hydrated solid POA or dried silica/POA powders being in CCl_4 medium are characterized by different temperature dependences of the δ_H values because of only partial dissociation of the PO-H bonds.

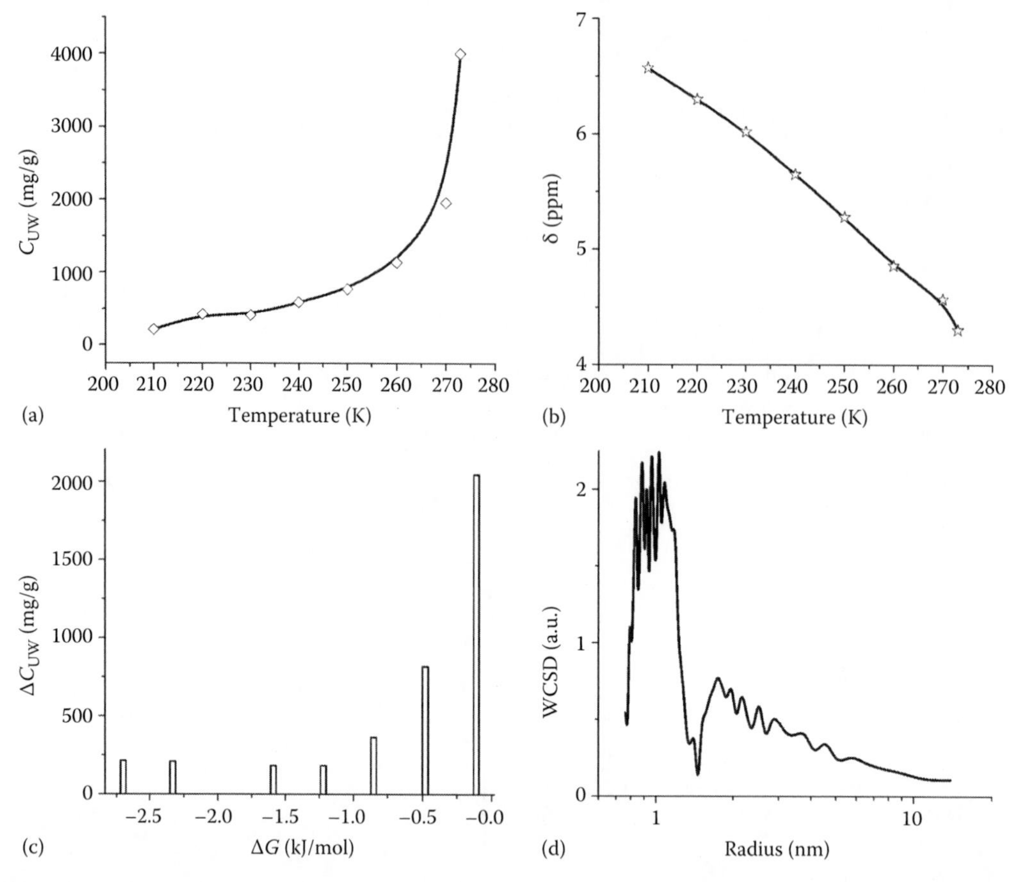

FIGURE 1.127 Temperature dependences of (a) concentration of unfrozen water in concentrated suspension of OX-50 (16 wt%) and H_3PO_4 (4 wt%); (b) chemical shift δ_H, (c) relationship between changes in the Gibbs free energy and changes in the C_{uw} values, and (d) unfrozen water cluster size distribution. (Adapted from *J. Colloid Interface Sci.*, 368, Gun'ko, V.M., Morozova, L.P., Turova, A.A. et al., Hydrated phosphorus oxyacids alone and adsorbed on nanosilica, 263–272, 2012b. Copyright 2012, with permission from Elsevier.)

The δ_H values depend strongly on water amounts, silica type, and temperature. NMR cryoporometry shows that small water clusters (<1 nm) and nanodomains (up to 20 nm in size) are present at the interfaces of hydrated solid POA and silica/POA powders. Quantum chemical calculations of the ^{1}H NMR spectra demonstrate the influence of POA/water cluster structure and dissociation of the PO-H bonds on the δ_H values.

Water adsorbed on nanosilica modified by phosphorus chlorides, which were then hydrolyzed, was studied by ^{1}H NMR, dielectric, and IR spectroscopies and quantum chemical ab initio and semiempirical methods (Turov et al. 1996a). Hydrophosphinic groups (HPA) bound to nanosilica surfaces are stable at the adsorbed water content $h \leq 0.12$ g/g, which corresponds to water adsorption from air at standard conditions. Increase of the adsorbed water content can lead to hydrolysis of the ≡Si–O–P bonds via nucleophilic substitution S_Ni(Si) mechanism. Upon heating, the adsorbed phosphinic acid can cleave the ≡Si–O–Si≡ bridges and form new sites responsible for changes of structure of the adsorbed water clusters and chemical shift values for the ^{1}H NMR spectra. Large clusters of the adsorbed water molecules, which form at $h \geq 0.12$ g/g, are localized near ionized bound hydrophosphinic groups or adsorbed acid molecules. Therefore, the δ_H values are in the 6–12 ppm range (Figure 1.130) that are greater than that of bulk water (~5 ppm) and close to the δ_H values observed for POA/nanosilica system analyzed earlier.

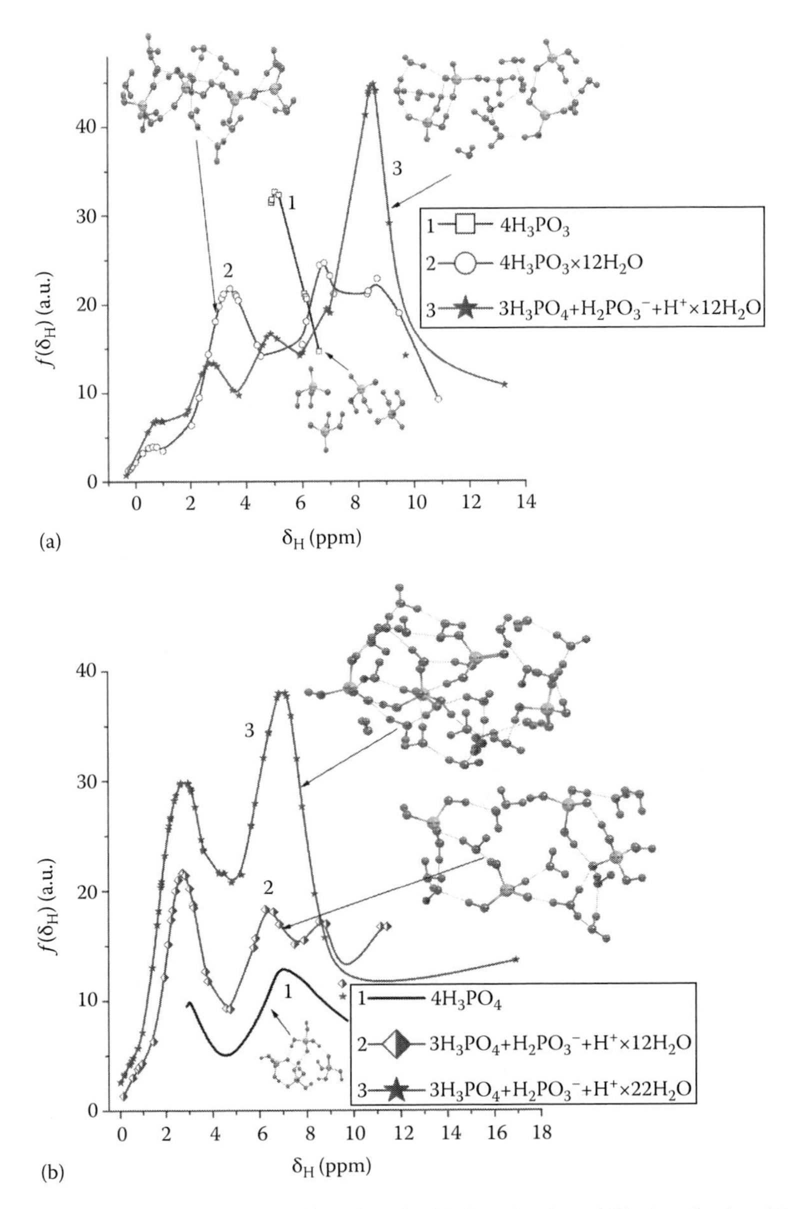

FIGURE 1.128 Chemical shift distribution functions for (a) phosphonic and (b) phosphoric acids in molecular and deprotonated forms alone and hydrated in CCl_4 medium (GIAO/IEFPCM/B3LYP/6-31G(d,p)). (Adapted from *J. Colloid Interface Sci.*, 368, Gun'ko, V.M., Morozova, L.P., Turova, A.A. et al., Hydrated phosphorus oxyacids alone and adsorbed on nanosilica, 263–272, 2012b. Copyright 2012, with permission from Elsevier.)

The ^1H NMR spectra of $H_2O/HPA/SiO_2$ have significant distinctions in comparison with those for the parent nanosilica, and these differences depend not only on the adsorbed water content but also on reaction temperature, dispersion medium (Figure 1.131 and Table 1.24), and other parameters. Notice that the $\delta_H(h)$ function becomes more nonlinear with increasing temperature and this is accompanied by diminution of the δ_H values (Figure 1.130b).

The effects of $CDCl_3$ medium on the interfacial behavior of bound water depend on the water content. For instance, changes in the $\Delta G(C_{uw})$ graphs are opposite at $h=0.12$ g/g (more WBW in $CDCl_3$) and 0.26 g/g (more SBW in $CDCl_3$) (Figure 1.131 and Table 1.24; γ_S, ΔG_s, and $\langle T \rangle$). The amounts of bound water reduce in $CDCl_3$ medium (Table 1.24; V and C_{max}).

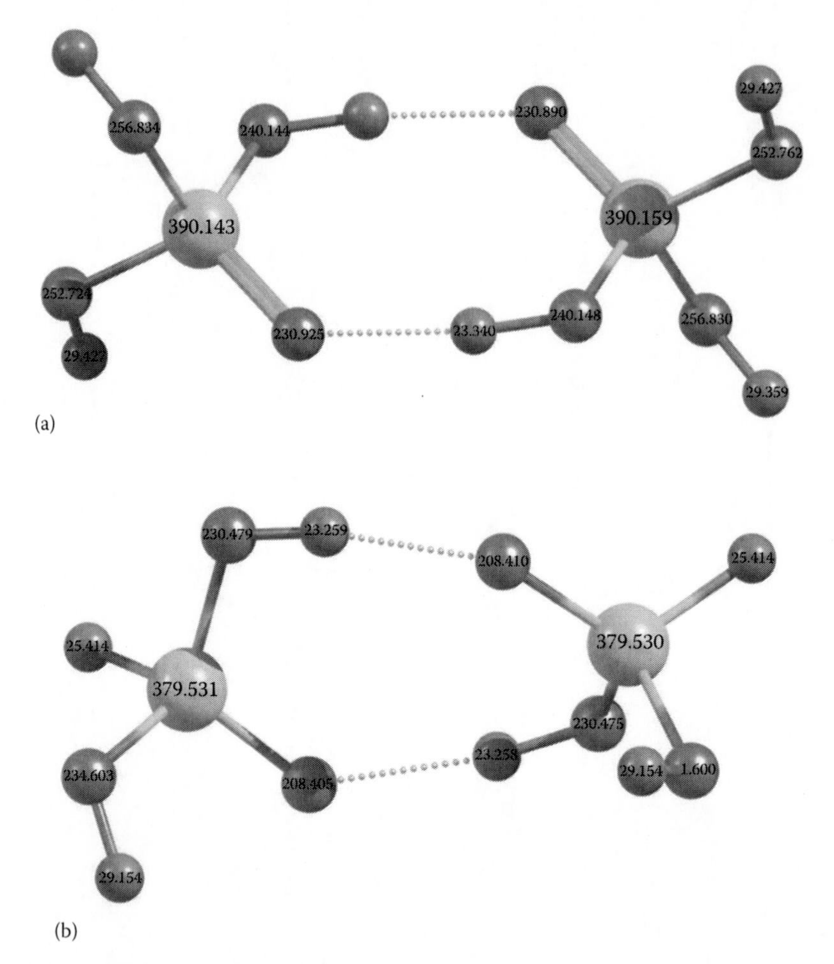

FIGURE 1.129 Dimers of (a) phosphoric and (b) phosphonic acids at changes in the total energy $\Delta E_t = -114.1$ and -107.5 kJ/mol, respectively, and $\delta_H \approx 8.5$ ppm. (Adapted from *J. Colloid Interface Sci.*, 368, Gun'ko, V.M., Morozova, L.P., Turova, A.A. et al., Hydrated phosphorus oxyacids alone and adsorbed on nanosilica, 263–272, 2012b. Copyright 2012, with permission from Elsevier.)

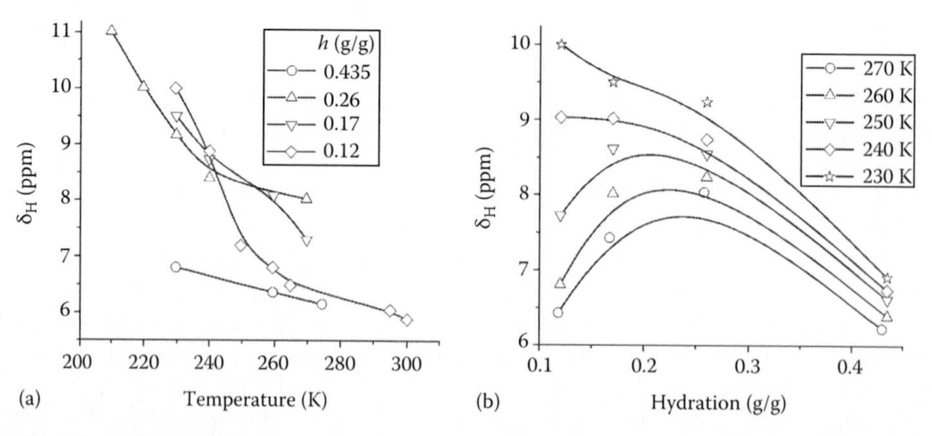

FIGURE 1.130 Chemical shift δ_H as a function of (a) temperature and (b) hydration of nanosilica A-300 with bound hydrophosphinic acid.

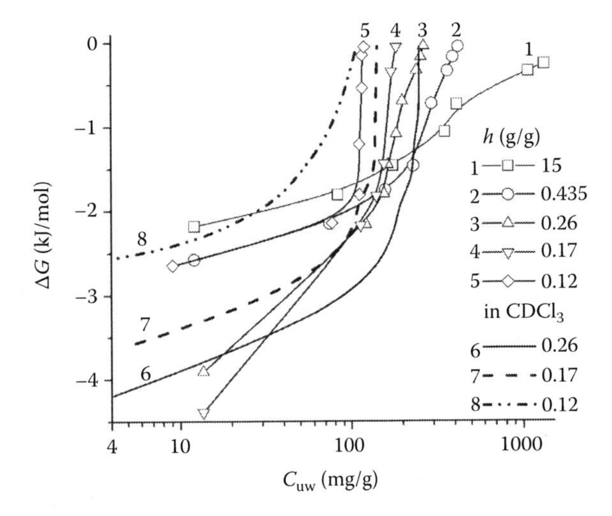

FIGURE 1.131 Relationship between changes in Gibbs free energy and amounts of unfrozen water bound to nanosilica A-300 with attached hydrophosphinic groups at different hydration in air or $CDCl_3$ medium.

TABLE 1.24
Characteristics of Water Bound to Modified A-300 at Different Hydration and in Different Dispersion Medium (Water, Air, or Chloroform)

Medium	h (g/g)	S_{uw} (m²/g)	S_{nano} (m²/g)	S_{meso} (m²/g)	V_{nano} (cm³/g)	V_{meso} (cm³/g)	C_{max} (mg/g)	⟨T⟩ (K)	$-\Delta G_s$ (kJ/mol)	γ_S (J/g)
Water	15	245	59	186	0.029	1.574	1701	254.1	2.26	64.2
Air	0.435	238	192	47	0.087	0.320	429	234.8	2.65	33.1
Air	0.26	189	160	29	0.069	0.187	260	216.3	3.81	29.2
Air	0.17	200	184	16	0.076	0.104	170	199.0	4.20	26.3
Air	0.12	12	0	12	0	0.118	119	213.1	2.70	14.7
$CDCl_3$	0.26	288	232	56	0.101	0.148	250	195.8	3.88	37.8
$CDCl_3$	0.17	94	67	27	0.032	0.110	143	202.5	3.46	20.2
$CDCl_3$	0.12	12	0	12	0	0.106	108	232.9	2.65	9.0

1.1.8 Effects of Surface Modification on Adsorption Phenomena

Chemical modification of silica by alkylsilanes or other organosilicon compounds (OSCs) is widely used to change the surface properties important in interactions with polar and nonpolar substances. This modification (hydrophobization or other specific functionalization) results in diminution of surface affinity to water or other polar compounds, but for nonpolar organics, the opposite effect may be observed. However, nitrogen (which is nonpolar) adsorption energy and adsorption potentials of the nanosilica (or silica gel) surfaces modified by different OSCs decrease with increasing amounts of grafted functionalities or growing their CH chains. The heat of immersion of silylated nanosilica A-380 in water significantly reduces from approximately 50 J/g at low Θ values to 10 J/g at high Θ values (Tsutsumi and Takahashi 1985). Nevertheless, the IR and ^{1}H NMR spectroscopy data showed that the significant amounts of adsorbed water were observed on similar materials (Gun'ko et al. 2003e, 2005d, 2009d, Turova et al. 2007).

Initially, the reaction of hexamethyldisilazane (HMDS) or other OSCs with silica surface results in a random distribution of well-separated TMS or other functional groups. One can assume that a portion of silanols remaining after partial modification of the silica surface by OSC in contact zones between adjacent primary particles or in narrow pores is difficult to be accessible for OSC molecules. However, these surface patches can be much more easily accessible for water molecules with the size significantly smaller than that of the OSC molecules. Therefore, the ^{1}H NMR spectroscopy investigations showed that a similar surface can disturb a relatively thick interfacial water layer (unfrozen at $T<273$ K) and its thickness can exceed (especially for a small content of grafted functionalities) that of a similar layer at a hydrophilic surface. One of possible reasons of the observed effects is polarization of the heterogeneous surface with different charging of hydrophobic and hydrophilic patches, and changes in the structure of the water layers near totally hydrophobic and hydrophilic surfaces. Besides, nonuniform electrostatic field of the oxide surface can reorient dipoles of water molecules that results in enhancement of their rotational mobility and disordering of the hydrogen bond network in the interfacial water, especially near mosaic hydrophilic–hydrophobic surface patches. Therefore, the interfacial water remains unfrozen at $T<273$ K. In the case of totally hydrophobic surfaces, the number of hydrogen bonds between water molecules in the first interfacial layers becomes greater. Consequently, one can expect essential dependence of the structure of the interfacial water on amounts of grafted hydrophobic TMS groups. The use of the ^{1}H NMR spectroscopy in combination with freezing-out of bulk water and layer-by-layer freezing-out of the interfacial water allows one to determine the thickness of the water layers differently disturbed by unmodified and modified patches of a silica surface.

The results shown in previous chapters were obtained on investigations of unmodified silicas. Clearly, partial or complete hydrophobization of the silica surface can cause significant changes in the structure of the interfacial water and change contributions of different forms of interfacial water (weakly, WAW at $\delta_H = 1–2$ ppm, and strongly, SAW at $\delta_H > 3$ ppm, associated and weakly, WBW at $-\Delta G < 0.5–0.8$ kJ/mol, and strongly, SBW $-\Delta G > 0.8$ kJ/mol, bound waters). There are several types of surface modification of nanosilicas. First, functionalized silanes with 1–3 reactive groups (such as Si-Cl or Si-OR) are frequently used to modify the silica surface (Iler 1979, Leyden 1985, Leyden and Collins 1989, Bergna 1994, 2005, Gun'ko et al. 1997c, Legrand 1998). The chemical modification of solid surfaces by OSCs is commonly employed for a wide variety of applications. Performing these reactions is a well-established method for altering many surface characteristics such as free surface energy, topography, PSD, and specific surface area. The adsorptive properties of surface-modified materials can strongly differ from unmodified materials and can, to some extent, be tailored. Some factors that can contribute to the properties of surface-modified materials include the characteristics of the unmodified solid, the type of OSC used for surface modification, and the extent to which the surface modification reaction occurs. The second type of nanosilica modification is grafting of nanoparticles (oxide, carbon, metal) or a thin film (polymer) with a new phase. The third type of the modification can be assigned to geometrical modification using HTT or strong mechanical activation or MCA. The results of the influence of all these modifications of nanosilicas on the interfacial phenomena are present in this section (chemical modification) and other Sections 1.2–1.4.

Scholten et al. (1994) studied silicas modified by different OSCs (Figure 1.132).

They showed, using ^{29}Si CP/MAS NMR, that isobutyl side groups of ligand silane chains of a silica-based reversed phase high-performance liquid chromatography (RP-HPLC) phase provide a significant increased steric protection of the ligand siloxane bond compared to their methyl analogues. Hydrogen bonding of residual silanols to the ligand siloxane oxygen atoms is shown to occur much less in the diisobutyl-n-octadecylsilane MS. It is likely that this also explains the increased chromatographic stability at low pH of this stable bond phase, as the siloxane bond is protected from contact with aggressive, hydrolyzing eluents. Arce et al. (2009) studied butoxylated silica nanoparticles (BSN) prepared by esterification of the silanol groups of nanosilica nanoparticles with butanol and characterized by ^{13}C and ^{29}Si NMR and TG. The molecular probes benzophenone and safranine-T were used to investigate the BSN suspensions in water/acetonitrile.

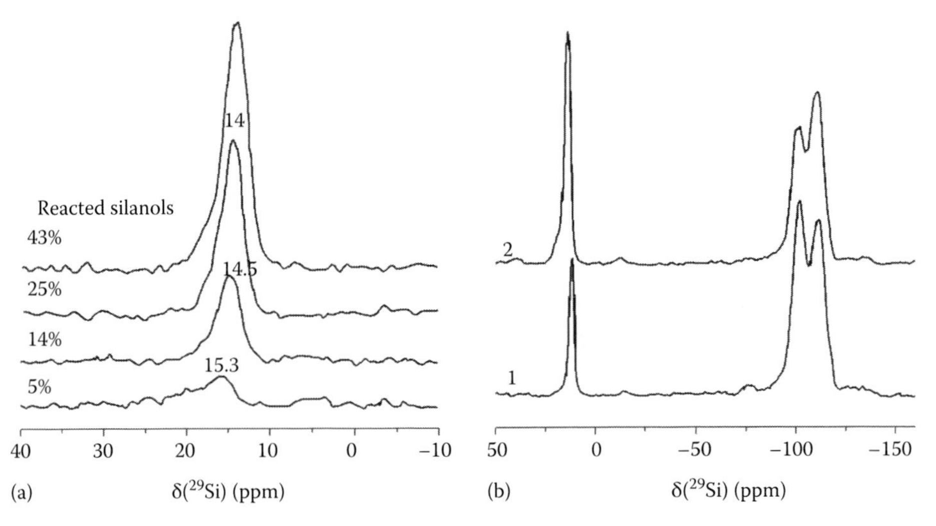

FIGURE 1.132 ^{29}Si CP/MAS NMR spectra of (a) modified nanosilica A-200 with different amounts of grafted trimethylsiloxane groups and (b) Zorbax octadecyl RP-HPLC phases with (1) diisobutyl-n-octadecylsiloxane and (2) dimethyl-n-octadecylsiloxane. (Adapted from *J. Chromatography* A., 688, Scholten, A.B., de Haan, J.W., Claessens, H.A., van der Ven, L.J.M., and Cramers, C.A., 29-Silicon NMR evidence for the improved chromatographic siloxane bond stability of bulk alkylsilane ligands on a silica surface, 25–29, 1994. Copyright 1994, with permission from Elsevier.)

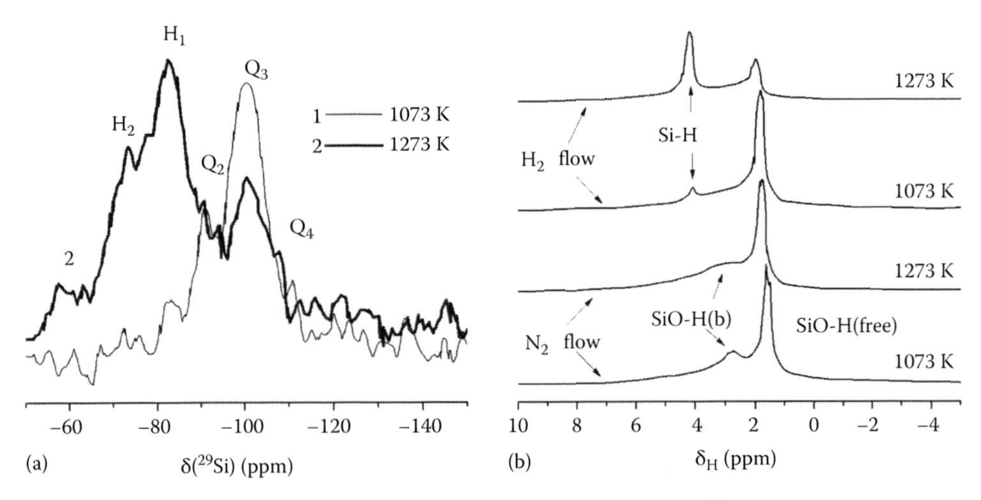

FIGURE 1.133 (a) ^{29}Si CP/MAS NMR spectra of hydrogen treated silica A-300 treated at 1073 and 1273 K and (b) ^{1}H MAS NMR spectra of A-300 treated in the nitrogen or hydrogen flow. (Adapted from *J. Colloid Interface Sci.*, 215, Heeribout, L., d'Espinose de la Caillerie, J.B., Legrand, A.P., and Mignani, G., A new straightforward approach to generate Si–H groups on silica, 296–299, 1999. Copyright 1999, with permission from Elsevier.)

Heeribout et al. (1999) studied the reaction of silica surface with hydrogen at high temperatures. Reaction occurs more strongly at 1273 K than 1073 K (Figure 1.133). Silicon states H_1 (−82 ppm) and H_2 (−72 ppm) correspond to $(\equiv SiO)_3SiH$ and $(\equiv SiO)_2Si(OH)H$, respectively. The chemical shift of protons at 1.7 ppm and 2.9 ppm corresponds to free and hydrogen-bonded silanols. The band at $\delta_H = 4$ ppm corresponds to Si-H groups.

Li et al. (2009b) studied nanosilica modified by poly(acrylamide-co-2-(dimethylamino) ethyl methacrylate, methyl chloride quaternized (poly(AM-co-DMC). The ^{1}H NMR signals (Figure 1.134)

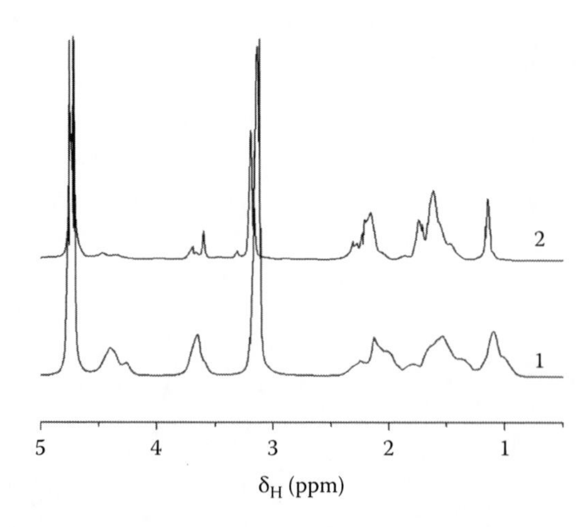

FIGURE 1.134 ^{1}H NMR spectra (1) obtained for the original linear cationic poly(acrylamide-co-2-(dimethylamino) ethyl methacrylate, methyl chloride quaternized) (poly(AM-co-DMC)) with 30 mol% of DMC and (2) obtained for the hybrid copolymer–silica nanoparticles. (Adapted from *J. Colloid Interface Sci.*, 338, Li, X., Yang, T., Gao, Q., Yuan, J., and Cheng, S., Biomimetic synthesis of copolymer–silica nanoparticles with tunable compositions and surface property, 99–104, 2009b. Copyright 2009, with permission from Elsevier.)

attributed to copolymer (3.5–4.6 ppm: $-O-CH_2-CH_2-$; 3.14 ppm: $-N(CH_3)_3$) were observed for the hybrid copolymer–silica nanoparticles, indicating that a significant proportion of the cationic poly(AM-co-DMC) chains were not part of the silica framework.

Mijatovic et al. (2000) studied unmodified and silane-modified nanosilica ($S_{BET}=200$ m^2/g) using ^{29}Si solid-state NMR spectroscopy. The mode of bonding onto the silanol surface (mono- or divalent, seven different surface functionalities) was specified by comparison with chemical shifts from solution. A detailed analysis was performed with respect to products formed via hydrosilylation reactions using a rough quantification of the surface loading with a signal deconvolution process of the silanol resonances on the Aerosil surface. Thus, the NMR techniques were mainly used to analyze the chemical structure of whole silica materials or their surface.

Here, we will study the relationships between the characteristics of nanosilica surface partially modified by HMDS at different TMS loading (C_{TMS}) and the properties of the interfacial water layers in different media (air, liquid and frozen water, and weakly polar chloroform) at different temperatures by means of NMR, adsorption, desorption, spectroscopic, and theoretical methods (Gun'ko et al. 2003e). Detailed investigations of the influence of the hydrophobic environment on the ^{1}H NMR spectra of adsorbed water and its thermodynamic characteristics in wetted powders of nanosilica partially modified by grafted TMS groups showed that signal of water in the absence of chloroform is observed in the singlet form. This water can be assigned to HDW in narrow pores (voids between primary nanoparticles in the same aggregate) and HDW/LDW in larger pores (between primary particles from different aggregates in agglomerates). The signal width increases and intensity decreases with lowering temperature because of reduction of the mobility of water molecules and partial freezing-out of interfacial water with moving boundary of immobile ice/mobile water from larger pores toward narrower ones. Such a behavior of the signals of adsorbed water at $T<273$ K is characteristic of the majority of disperse and porous materials. The chemical shift of the proton resonance for adsorbed water is close to chemical shift of liquid water (5 ppm) because several mutually compensating factors such as (i) existence only HDW in pores that should reduce δ_H in comparison with that for LDW, (ii) bending of the hydrogen bonds near active surface sites in very narrow pores (δ_H decreases), (iii) surface electrostatic field results in deshielding of protons (δ_H increases), (iv) overlapping of EDL of opposite pore walls (complex effect on δ_H), and (vi) difference in concentrations of counterions in different pores (complex effect on δ_H).

Controlled hydrophobization of nanosilica A-380 was carried out by HMDS reacting with silanols (Gun'ko et al. 2003e):

$$2(\equiv SiOH) + ((CH_3)_3Si)_2NH \rightarrow 2(\equiv SiOSi(CH_3)_3) + NH_3.$$

HMDS chemisorption was occurred in a reactor with IR spectroscopy monitoring of the synthesis. Before HMDS chemisorption, silica powder and plate were heated in the reactor in air at 673–693 K for 1.5 h. Then the sample was degassed for 10–15 min and cooled to ambient temperature. HMDS vapor (dosed in dependence on a desirable degree of surface modification) interacted with the silica for 15–18 h (so long time of adsorption was caused by the necessity to diminution of the difference in HMDS reaction with silica powder and pressed silica samples) at ambient temperature, then the sample was heated and degassed at 473 K for 2 h. Carbon content (C_C) in modified samples SM1-SM6 was determined by chemical analysis. The C_C values were used to estimate the amounts of grafted TMS groups (Table 1.25, C_{TMS}).

Increase in the concentration of TMS groups (Table 1.25; C_{TMS}) leads to changes in the interaction with water. These changes are clearly seen in the TG and DTG curves at $T<473$ K (Figure 1.135). A fast decrease in the sample weight is due to desorption of intact water at $T<150°C$ and at 400°C–450°C (marked only for SM4-SM6) is linked to destruction of TMS groups.

Increase in C_{TMS} leads to diminution of desorbed water, since surface modification reduces the amounts of silanols, which are adsorption centers for water. However for SM1-SM4, diminution of amounts of water adsorbed in air in comparison with A-380 is not observed (Figure 1.135). In the case of water adsorption on additionally heated samples at a fixed water vapor pressure $p/p_0=0.8$ for 24–72 h, amounts of adsorbed water depend stronger (nonlinearly for SM1-SM3) on C_{TMS} (Gun'ko et al. 2003e). These changes can be connected not only with increased hydrophobicity but also with changes in the spatial structure of the silica surface per se and nanoparticle arrangement in aggregates and agglomerates.

Increase in C_{TMS} leads to reduction of the specific surface area (Table 1.25; S_{BET}) but differently affects contributions on nano-, meso-, and macropores in degassed (determined from LT nitrogen adsorption) and suspended (estimated using cryoporometry) samples (Table 1.25). These results are connected to features of the morphology of nonporous primary and "porous" secondary particles of nanosilica with mosaic hydrophilic/hydrophobic surfaces of partially modified A-380. This is also seen in the PSD graphs (Figure 1.136b). Only a minor portion of large voids between primary particles in secondary ones can be entirely filled by nitrogen even at $p/p_0 \rightarrow 1$ (Figure 1.136a), since $V_p \ll V_{em}$. However, filling of large pores gives a marked contribution to total adsorption. Silylation of silica leads to reduction of the nitrogen adsorption potential, since a high-energy adsorption energy distribution (AED) peak at 11–12 kJ/mol decreases (Figure 1.136c). Contribution of the first monolayer (providing the second [at 7–8 kJ/mol] and third [at 11–12 kJ/mol] AED peaks due to adsorption on different sites placed in broad and narrow pores) to the adsorbed amounts of nitrogen is less 10%. Notice that according to calculations of adsorption complexes by using B3LYP/6-31G(d,p), the difference in interaction energy of a nitrogen molecule with silanol and siloxane bonds is equal to approximately 3 kJ/mol, which can affect AED for the first monolayer.

The samples show a complex textural character (Figure 1.136b) of secondary particles of nanosilica. Therefore, the differences in AEDs or PSDs for the initial silica and maximum modified sample SM6 are relatively small. Thus, basing on analysis of results of nitrogen adsorption, one can assume that water interaction with MSs should be weaker than that with unmodified A-380. However, the difference in binding of water molecules to the initial and MSs may be greater than that for nitrogen molecules because of the interaction of water with surface silanols by the hydrogen bonds and with $\equiv SiOSi(CH_3)_3$ by weaker dispersive forces. Results obtained on water desorption (Figure 1.135) are confirmed by the ^1H NMR and microcalorimetric data (Figures 1.137 and 1.138) showing nonlinear changes in the Gibbs free energy of the interfacial water in the heat of immersion in water with increasing C_{TMS} and with their significant diminution of ΔH_{im} observed only for SM6.

TABLE 1.25
Characteristics of Interfacial Water Layer in Aqueous Suspensions of Initial and HMDS (SM1–SM6) Modified Nanosilica and Textural Parameters Determined from Nitrogen Adsorption Data

Sample (Bound Water)	γ_S (J/g)	$-\Delta G_s$ (kJ/mol)	C_{uw}^{w} (g/g)	C_{uw}^{s} (mg/g)	V_{nano} (cm³/g)	V_{meso} (cm³/g)	V_{macro} (cm³/g)	S_{uw} (m²/g)	S_{nano} (m²/g)	S_{meso} (m²/g)	S_{macro} (m²/g)	⟨T⟩ (K)	k_{GT} (K nm)
A-380	65.0	2.43	2.21	0.70	0.087	2.299	0.177	379	183	184	12	260.4	59.640
SM1	67.1	2.80	1.47	0.60	0.110	1.590	0.116	383	237	138	9	254.5	66.753
SM2	49.7	2.93	1.25	0.50	0.114	1.320	0.124	366	248	109	9	256.8	61.311
SM3	52.8	2.71	0.67	0.45	0.134	0.852	0.049	373	295	74	4	247.9	61.0
SM4	48.3	2.80	1.08	0.40	0.115	1.070	0.106	346	251	87	8	254.1	64.9
SM5	51.8	2.54	0.47	0.40	0.131	0.674	0.021	338	275	61	2	242.5	62.781
SM6	36.9	2.39	0.94	0.30	0.090	0.967	0.064	284	195	84	5	256.8	55.183
APMS/A-300[a]	32.3	3.63	0	0.24	0.126	0.109	0	333	310	24	0	204.1	67
APMS/A-300[b]	15.3	2.86	0.12	0.12	0.037	0.180	0.006	95	81	13	0.4	239.6	67
APMS/A-300[c]	20.4	2.31	0.20	0.19	0	0.337	0.006	23	0	23	0.3	244.4	67
APMS/A-300[d]	35.9	3.43	0.16	0.23	0.152	0.215	0.011	368	351	16	0.7	224.0	78
Sample (N_2 adsorption)	**C_{TMS} (mmol/g)**	**S_{BET} (m²/g)**	**V_p (cm³/g)**	**V_{nano} (cm³/g)**	**V_{meso} (cm³/g)**	**V_{macro} (cm³/g)**	**S_{nano} (m²/g)**	**S_{meso} (m²/g)**	**S_{macro} (m²/g)**	**c_{slit}**	**c_{cyl}**	**c_{void}**	**Δw**
A-380*	0	378	0.778	0.030	0.550	0.362	127	232	20	0.051	0.939	0.009	−0.055
SM1	0.09	379	0.776	0.044	0.672	0.229	93	273	14	0.630	0.158	0.212	−0.254
SM2	0.14	362	0.850	0.039	0.700	0.633	80	247	35	0.701	0.093	0.205	−0.237
SM3	0.12	372	0.775	0.038	0.701	0.204	80	280	12	0.585	0.188	0.227	−0.137
SM4	0.23	345	0.726	0.025	0.690	0.162	50	285	10	0.560	0.186	0.253	−0.114
SM5	0.42	330	0.771	0.018	0.680	0.270	32	281	17	0.515	0.206	0.280	−0.154
SM6	0.79	285	0.678	0.011	0.607	0.171	17	257	11	0.382	0.256	0.362	−0.115

Note: k_{GT} values were determined from condition $S_{uw} \approx S_{BET}$. *A-380 was treated in conditions corresponding to modification reactions but without silane. APMS/A-300 (silica with grafted aminopropylmethylsilyl at 30% substitution of silanols) in [a,c]air and [b,d]chloroform at $h =$ [a,b]0.236 and [c,d]0.387 g/g.

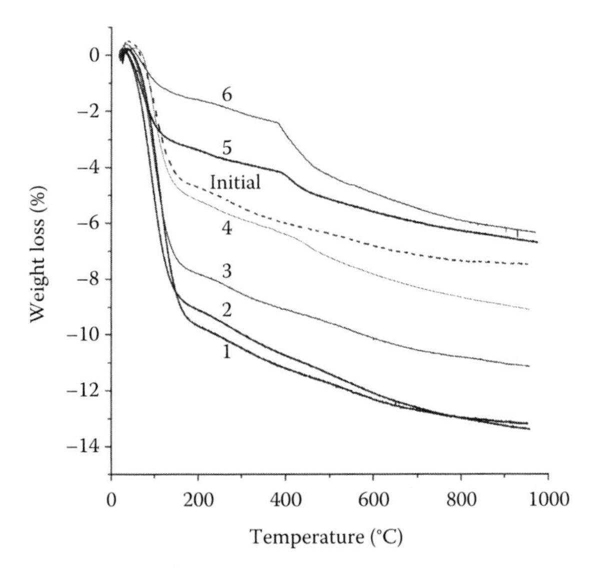

FIGURE 1.135 TG graphs for initial and modified silicas (curve numbers correspond to numbers of SM samples in Table 1.25).

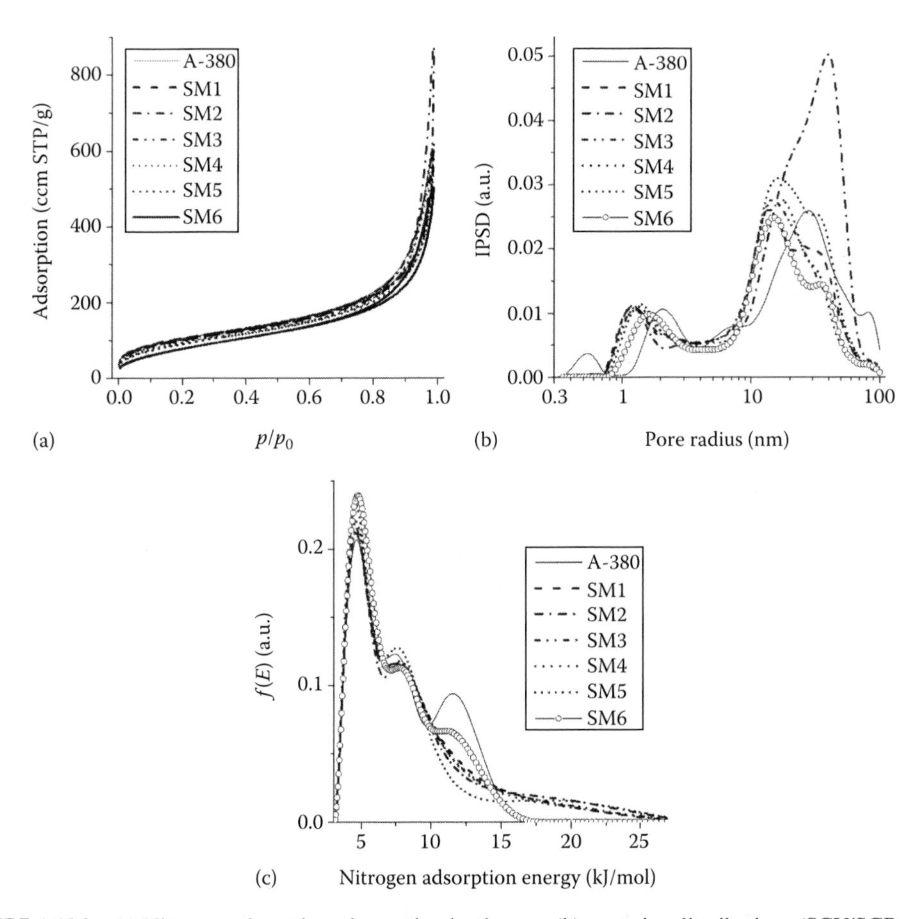

FIGURE 1.136 (a) Nitrogen adsorption–desorption isotherms, (b) pore size distributions (SCV/SCR model), and (c) nitrogen adsorption energy distributions (AED) for initial and modified nanosilicas.

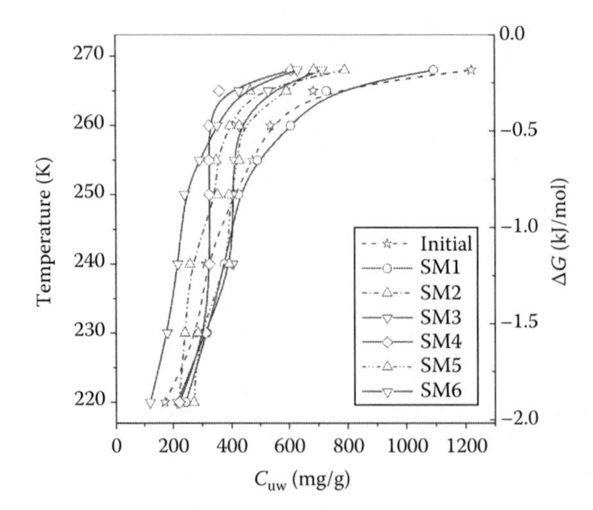

FIGURE 1.137 Temperature dependence of amounts of unfrozen interfacial water (C_{uw}) in the aqueous suspensions of initial and modified silicas and changes in the Gibbs free energy versus C_{uw}. (Adapted with permission from *Langmuir,* 19, Gun'ko, V.M., Turov, V.V., Bogatyrev, V.M. et al., Influence of partial hydrophobization of fumed silica by hexamethyldisilazane on interaction with water, 10816–10828, 2003e. Copyright 2003, American Chemical Society.)

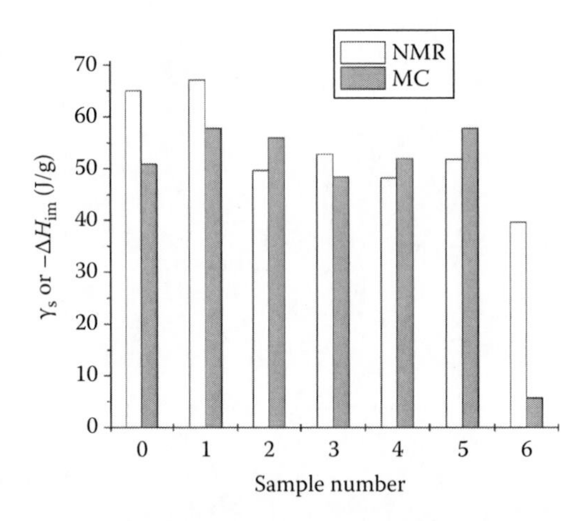

FIGURE 1.138 Comparison of the heat of immersion of silicas in liquid water at room temperature (ΔH_{im}) and changes in the Gibbs free energy of interfacial water (in Joule per gram of oxide) at $T \rightarrow 273$ K. (Adapted with permission from *Langmuir,* 19, Gun'ko, V.M., Turov, V.V., Bogatyrev, V.M. et al., Influence of partial hydrophobization of fumed silica by hexamethyldisilazane on interaction with water, 10816–10828, 2003e. Copyright 2003, American Chemical Society.)

One can assume that interaction of partially hydrophobic silicas with water in the aqueous suspensions (in contrast to interaction with water adsorbed from air and desorbed on TG measurements or on interaction with nonpolar nitrogen) can lead to significant rearrangement of secondary particles of nanosilica. These particles are instable in the aqueous suspension and can be decomposed on different treatment (mixing, stirring, ball-milling, or sonication). For instance, a major portion of particles observed in the MCA aqueous suspension of initial nanosilica A-300 (5 wt%) corresponds to individual primary particles (Figure 1.89) or small aggregates (Figure 1.139).

Notice that the hydrodynamic diameter of particles estimated from the PCS measurements (Figure 1.139) is larger by 10%–15% than the geometrical diameter of these particles (Gun'ko et al. 2001e).

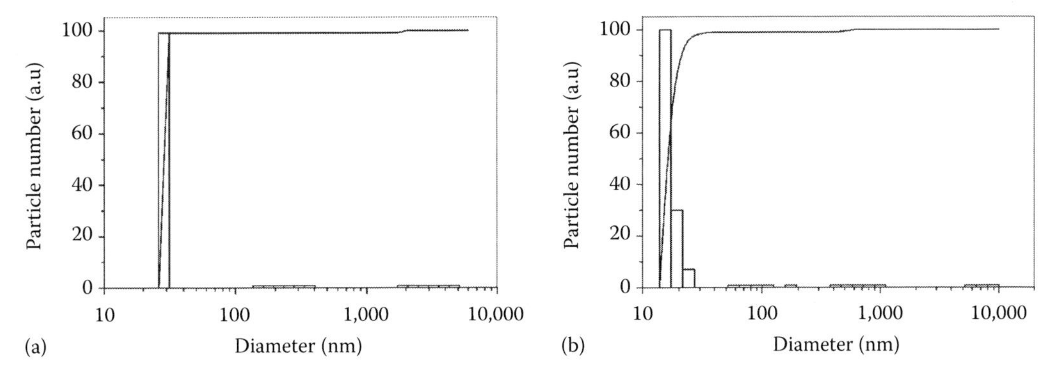

FIGURE 1.139 Particle size distributions in MCA (at t_{MCA} = (a) 10 min and (b) 5 h) aqueous suspension of initial nanosilica A-300 at concentration of 5 wt% with respect to the particle number measured by the photon correlation spectroscopy method. (*Adv. Colloid Interface Sci.*, 91, Gun'ko, V.M., Zarko, V.I., Leboda, R., and Chibowski, E., Aqueous suspensions of fumed oxides: particle size distribution and zeta potential, 1–112, 2001e. Copyright 2001, with permission from Elsevier.)

Hydrophobic patches on a surface of primary particles formed due to chemical modification of the surface by HMDS can cause rearrangement of secondary particles (aggregates and agglomerates) in the suspension to provide maximal shielding of patches of hydrophobic TMS groups from liquid water that can also lead to increase in the size of secondary particles. Similar effects were observed for blends of hydrophilic and hydrophobic nanosilicas (see Chapter 2). Therefore, the interaction energy between unmodified silica patches (which have been inaccessible [between adjacent primary particles] for HMDS molecules on surface modification but are accessible for water molecules because of their smaller size and the mentioned rearrangement of secondary particles in the suspension) and water can be great that gives a significant contribution to ΔH_{im} or γ_S for modified samples (Figure 1.138 and Table 1.25). Magnitudes of the Gibbs free energy of the interfacial water γ_s (calculated from the data shown in Table 1.25) per mass unit of adsorbent and the values of the heat of immersion ΔH_{im} measured for the same samples by the microcalorimetry method show consistent changes with increasing C_{TMS} value for practically all samples excepting SM6 because of a greater contribution of changes in enthalpy than entropy on water interaction with these solids. A small ΔH_{im} value for SM6 is caused by its poor wetting on calorimetric measurements, but in the NMR experiments, it has been suspended using alcohol/water with subsequent decanting. Notice that maximal γ_s values were typically observed for complex adsorbents with a low amount (0.5–3 wt%) of the second oxide or pyrocarbon grafted on the oxide matrix (Gun'ko et al. 1999c). Consequently, a maximum of the ΔH_{im} and γ_S values for SM1 corresponds to a general regularity observed for the interfacial water at the surfaces of heterogeneous adsorbents, and one can assume that these materials possess a maximal surface nonuniformity at low amounts of grafted (second) material.

To elucidate changes in the properties of the interfacial water versus the C_{TMS} value, we have calculated the free energy of solvation (ΔG_s) for the silica cluster with varied amounts of OH and CH_3 groups ($n_{OH}+m_{CH_3}=4$ at n_{OH} and m_{CH_3} varied from 4 to 0) using different methods (Figure 1.140).

There is a difference in $-\Delta G_s$ (Figure 1.140) and ΔH_{im} or γ_S (Table 1.25 and Figure 1.138), which may be caused by too great ratio between the "surface" and the "volume" of the used cluster in comparison with primary particles, since their average diameter for A-380 is approximately 7.2 nm, and the first water monolayer (which gives the lion's share to ΔH_{im}) contacts to approximately 12 wt% (a surface layer with the Si-O-Si and SiOH bonds) of silica. Consequently, ΔH_{im} (or γ_S) can be lower by order (since it was calculated per gram of silica) than $-\Delta G_s$ calculated for the small cluster having only "surface" atoms. With consideration for this difference, the $-\Delta G_s$ value is in agreement with the ΔH_{im} and γ_S values (recalculated to kJ per mole of the interfacial water).

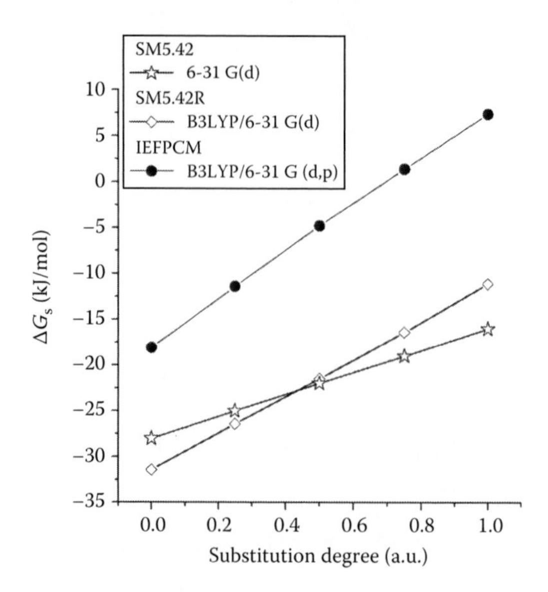

FIGURE 1.140 Relationship between free energy of solvation ΔG_s (calculated using the SM5.42R method with 6-31G(d) and B3LYP/6-31G(d) or IEFPCM with B3LYP/6-31G(d,p)) and the substitution degree for $\equiv$SiOH to $\equiv SiCH_3$.

Calculations of ΔG_s with consideration for the electron correlation effects (SM5.42R/B3LYP/6-31G(d) and IEFPCM/B3LYP/6-31G(d,p)) or without that (SM5.42R/HF/6-31G(d)) show that ΔG_s is a linear function of the substitution degree (calculated as $m_{CH_3}/4$; Figure 1.140). Consideration for the electron correlation effects leads to slightly steeper lowering $-\Delta G_s$ with increasing concentration of CH_3 groups. Thus, for the same texture of an adsorbent, one can anticipate a linear decrease in the amounts of disturbed adsorbed (bound) water (unfrozen at $T<273$ K) with increasing C_{TMS} value. Deviation from a similar linear dependence can be caused by the textural nonuniformity of MSs (Gun'ko et al. 2003e).

In the case of contact of hydrated silica surface with air (water vapor), the structure of adsorbed water is determined not only by the adsorbent surface (i.e., phase boundary of silica/water) but also by the phase boundary of water/air. Appearance of these phase boundaries leads to reduction of the free energy of the interfacial water accompanied by lowering of its freezing temperature. The information on a structure of the adsorption complexes at a surface of oxide adsorbents can be obtained from temperature dependences of chemical shift of protons of interfacial water molecules. It is necessary to take into account that δ_H is defined by strength of the hydrogen bonds between water molecule and active surface sites and depends on the amounts of hydrogen bonds per water molecule.

The ^{1}H NMR spectra of hydrated powders of SM4-SM6 (in chloroform) are shown at different temperatures (Figure 1.141). For SM1-SM3 (which are not shown in Figure 1.141), the $\delta_H(T)$ graphs are akin to that for SM4. Since chloroform is a poor proton donor, its molecules are not capable to compete with water molecules on interaction with residual silanols on modified samples to form complexes $\equiv$SiO(H)$\cdots$**H**–X. Therefore, one can assume that chloroform weakly affects the structure of water clusters adsorbed on the hydrophilic surfaces.

On the other hand, chloroform hampers the proton exchanging between the water molecules interacting with different active surface sites and allows us to specify the δ_H values in adsorbed water. In the case of partially hydrophobic surfaces, chloroform can promote the formation of larger water droplets weakly interacting with hydrophobic groups but strongly interacting with remained hydrophilic patches. This circumstance causes enhancement of contribution of the ^{1}H NMR signal intensity corresponding to fluid water at $\delta_H \approx 5$ ppm (Figure 1.141) with increasing C_{TMS}; that is, the size of water droplets located between aggregates in agglomerates increases with C_{TMS}.

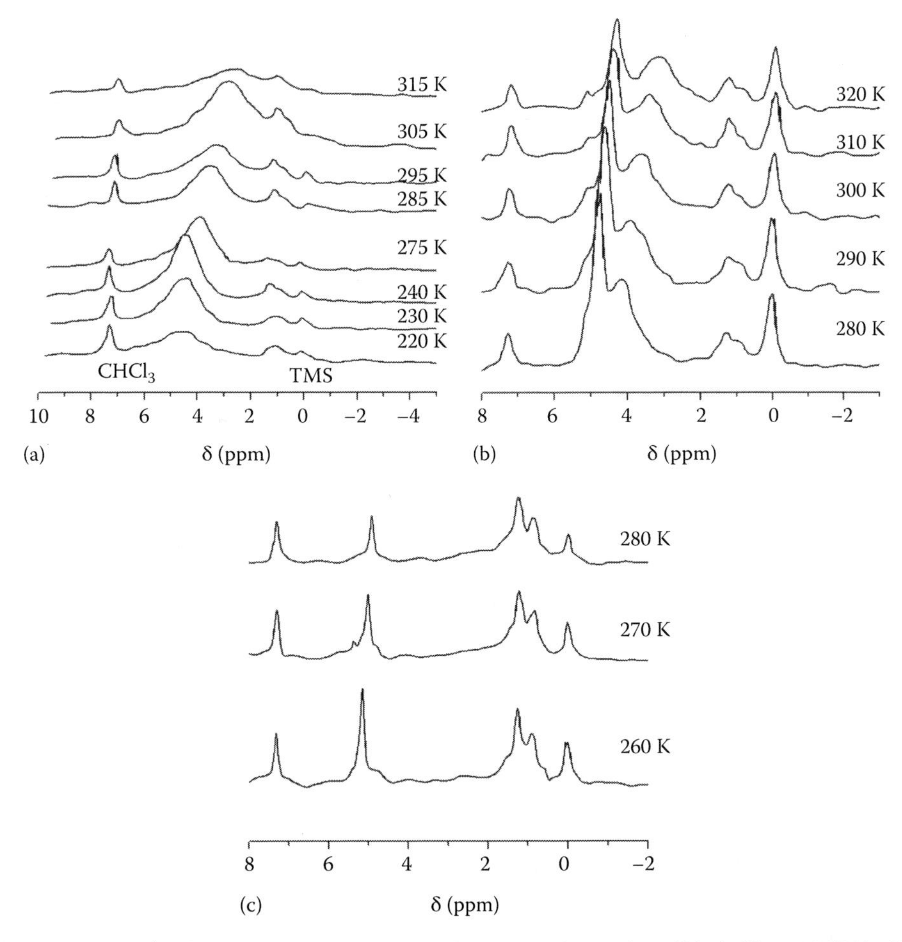

FIGURE 1.141 The ^{1}H NMR spectra of water adsorbed on surfaces of modified silicas (a) SM4, (b) SM5, and (c) SM6 at different temperatures (^{1}H signals of $CHCl_3$ and tetramethylsilane [TMS] are shown). (Adapted with permission from *Langmuir*, 19, Gun'ko, V.M., Turov, V.V., Bogatyrev, V.M. et al., Influence of partial hydrophobization of fumed silica by hexamethyldisilazane on interaction with water, 10816–10828, 2003e. Copyright 2003, American Chemical Society.)

Chemical shift of protons in water molecules, which are not participating in hydrogen binding, is 1.5–1.7 ppm. If water molecules form ice-like polyassociates, in the volume of which each molecule forms ideal (as in hexagonal ice) four hydrogen bonds (two of them are formed by hydrogen atoms and two remaining are formed with the participation of the alone electron pairs of oxygen atom), that $\delta_H = 7$ ppm. Additionally, it is necessary to take into consideration that binding $\mathbf{H_2}$O···H–OSi≡ weaker influences δ_H of water molecule than that for HO–**H**···OX. The formation of the later hydrogen bonds enhances δ_H value by 2.7 ppm.

For samples SM1-SM4, adsorbed water gives signal with a minimal width at $T = 240$ K. The signal intensity decreases at low temperatures because of layer-by-layer freezing out of the interfacial water. An increase of the signal width with lowering temperature is caused by reduction of the molecular motility of adsorbed water, and the HT effect is due to acceleration of proton exchanging between water molecules and surface hydroxyls. Chemical shift δ_H of adsorbed water is observed over the 3–4.5 ppm range (Figures 1.141 and 1.142), and δ_H decreases with elevating temperature.

Additionally, the signals of H atoms from chloroform **C**$\mathbf{HCl_3}$ are observed at $\delta_H = 7.26$ ppm, and $\mathbf{CH_3}$ groups of tetramethylsilane (used as a standard) correspond to $\delta_H = 0$ ppm, and low-intensity signal of water dissolved in chloroform is observed at $\delta_H = 1.7$ ppm. The main signal of adsorbed

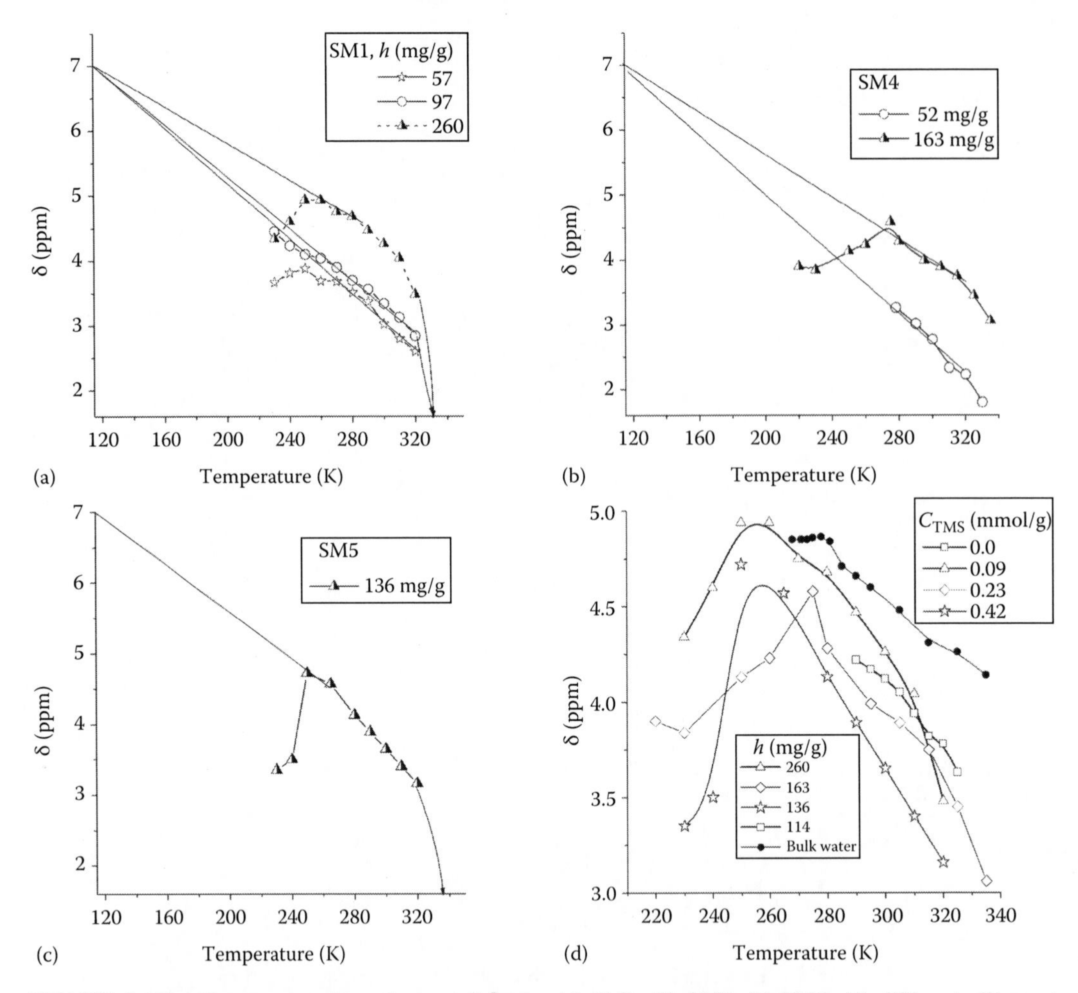

FIGURE 1.142 Temperature dependences of δ_H for (a) SM1, (b) SM4, (c) SM6, (d) different silicas at different amounts of adsorbed water. (Adapted with permission from *Langmuir*, 19, Gun'ko, V.M., Turov, V.V., Bogatyrev, V.M. et al., Influence of partial hydrophobization of fumed silica by hexamethyldisilazane on interaction with water, 10816–10828, 2003e. Copyright 2003, American Chemical Society.)

water splits with increasing C_{TMS} (Figure 1.141) and a narrow line at 5 ppm corresponding to fluid water becomes intensive. Practically, only this signal is observed for water adsorbed on SM6 (Figure 1.141c). This can also correspond to increase in the average number of the hydrogen bonds per molecule in the first interfacial water layers near the hydrophobic surfaces. Additionally, the intensity of an IR band at 3660 cm^{-1} (corresponding to silanols in sites difficult of access even for water molecules) increases with C_{TMS} (Gun'ko et al. 2003e). Most probably that for modified nanosilica, adsorbed water forms clusters and SAW nanodomains (Figure 1.143) in loosely packed agglomerates between aggregates of primary particles.

There are certain regularities in the $\delta_H(T)$ graphs at $T > T_m$ of adsorbed water, as δ_H decreases with elevating temperature and diminution of water concentration in samples (Figure 1.142). A minimal value $\delta_H = 1.8$ ppm is observed for SM4 at $h = 52$ mg/g and $T = 330$ K (Figure 1.142b). A maximal value $\delta_H \approx 4.8$ ppm is observed at a high content of adsorbed water and temperature close to 270 K, and at lower temperatures, the δ_H value decreases because of freezing of water. A linear portion of the $\delta_H(T)$ plots can be found for all samples at any hydration. Prolongation of these lines to lower temperatures gives their crossing at 115 K at a point corresponding to $\delta_H = 7$ ppm characteristic of molecules possessing four hydrogen bonds, that is, in the ice-like structure. It is possible that this

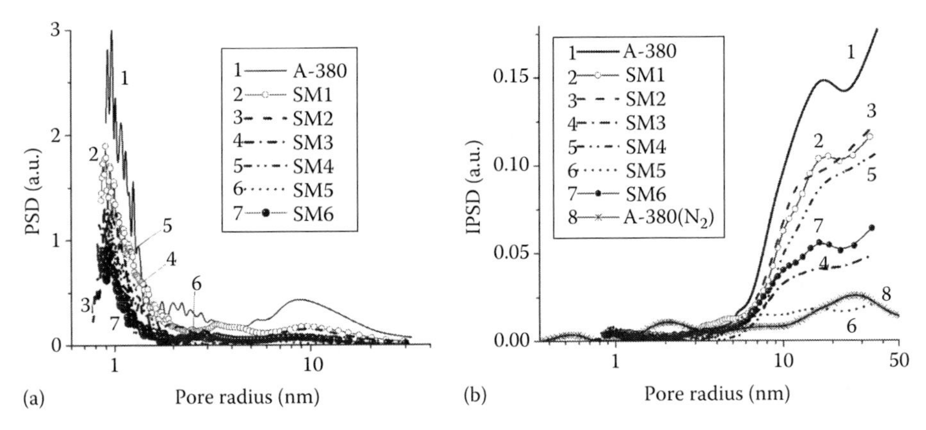

FIGURE 1.143 Size distributions of voids filled by structured unfrozen water: (a) differential and (b) incremental for initial and modified silicas.

temperature corresponds to an ordered structure of adsorbed water akin to that of free ice, which, however, is not observed in our experiments because of features of the used measurement technique.

A small δ_H value at T close to 320 K testifies changes in the structure of water clusters adsorbed on the surface of MS in comparison with that observed at lower temperatures in consequence of diminution of the average number of the hydrogen bonds per molecule with elevating temperature, because of enlargement of the hydrogen bonds with T and increased contribution of WAW that results in reduction of δ_H. A similar picture was observed earlier for the initial silica. So small δ_H values for water molecules taking part in the hydrogen bonds with the surface hydroxyls can be also explained by their participation in complexes ≡SiO–H···**OH**$_2$ as an electron donor, that is, a marked portion of protons of adsorbed water molecules (forming small clusters of WAW with only several molecules) do not participate in the hydrogen bonds. Additionally, water interaction with siloxane bonds (≡Si)$_2$O···**H**–OH results in lower δ_H values than that for ≡SiO(H)···**H**–OH (vide infra). Therefore, small water clusters located under "umbrellas" of TMS groups near siloxane bonds can be characterized by relatively small δ_H values. Additionally, the δ_H values for water molecules and silanols demonstrate different dependences on the atomic charges (q_H) affected by different surroundings (Figure 1.144).

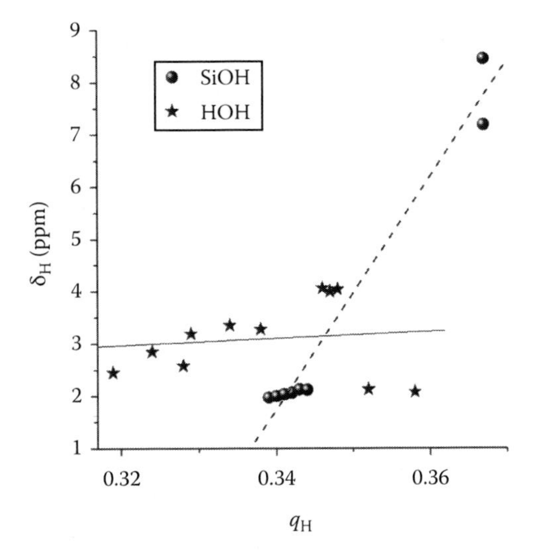

FIGURE 1.144 Relationships between chemical shift $\delta_{H,iso}$ (calculated by the GIAO method) and charge of H atoms (BLYP3/6-31G(d,p)//6-31G(d,p)).

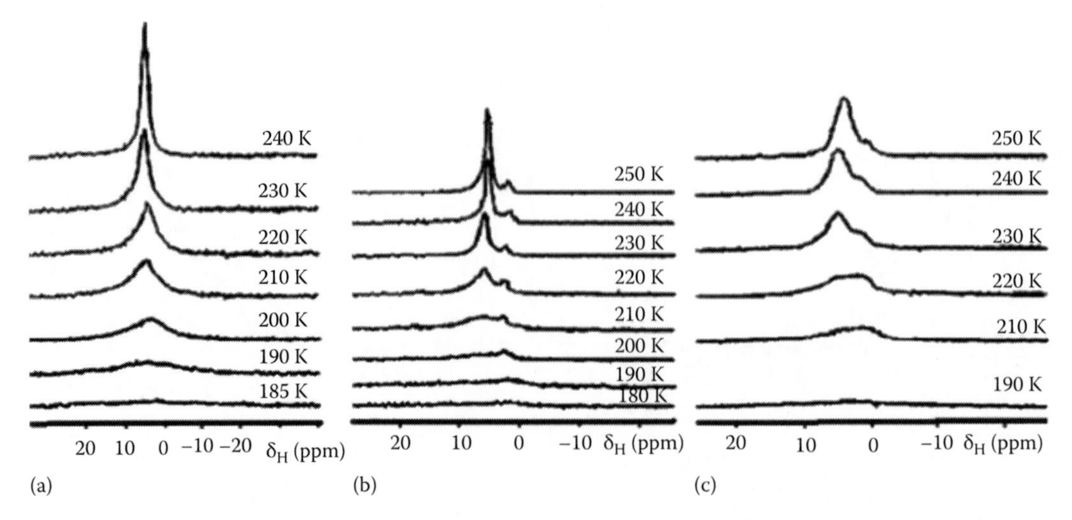

FIGURE 1.145 ^{1}H NMR spectra of water adsorbed on SM3 (h = 240 mg/g) in the presence of chloroform-d at (a) C_{CDCl3} = 0; (b) 0.6 g/g; and (c) 4.8 g/g at different temperatures. (Adapted from *Adv. Colloid Interface Sci.*, 118, Gun'ko, V.M., Turov, V.V., Bogatyrev, V.M. et al., Unusual properties of water at hydrophilic/hydrophobic interfaces, 125–172, 2005a. Copyright 2005, with permission from Elsevier.)

On simultaneous adsorption of water and chloroform-*d* on a surface of partially hydrophobic modified nanosilica samples (Figures 1.145 through 1.147), a signal of smaller intensity at 1.7 ppm (signal 2 linked to WAW) is observed in addition to the main signal (signal 1).

The intensity of the second signal increases with lowering temperature, whereas the first signal diminishes and only the second signal is observed at T < 210 K. Changes in the ratio of the first and second signals with lowering temperature are especially evidently shown in the case of the maximum amount of chloroform (Figures 1.145 through 1.147). The fact that water responsible for the second signal does not freeze with lowering temperature to 190 K allows one to conclude that this water is strongly bound (i.e., with a strongly distorted structure of the hydrogen bonds in HDW/interstitial water) to a surface or it is under special conditions steadier than water responsible for the first signal. The water unfrozen at $T \rightarrow 200$ K and having chemical shift close to 1.7 ppm was also registered in a mechanical mixture of unmodified and totally silylated nanosilicas. It should be noted that this mixture (16 wt% of hydrophilic nanosilica A-300, 4 wt% of hydrophobic silica, and 80 wt% of water) prepared at a high mixing rate (1.2×10^3 rpm) includes a great portion of air because the initial bulk density of the mixture is close to 0.6 g/cm^3, which increases with time; however, residual amounts of air remains during long storage period. Therefore, one can assume that remaining air nanobubbles at the boundary of hydrophobic silica and water, as well as TMS groups, strongly affect the structure of the interfacial water, and WAW appears at δ_H = 1.7 ppm.

The γ_S value (as total changes in the Gibbs free energy of interfacial water) decreases with increasing hydrophobicity of silylated silicas and amount of hydrophobic chloroform-*d* (Figure 1.148); that is, contribution of interfacial WBW/LDW can increase. Consequently, the hydrophobic environment reduces the interaction energy of interfacial water with remained hydrophilic sites of the surface of partially silylated silicas. This process is accompanied by an increase in contribution of the ^{1}H NMR signal of WAW.

Despite diminution of the concentration of silanols (main surface centers for water adsorption) for MSs by several times (e.g., if a maximal concentration of ≡SiOH groups on a nanosilica surface is approximately 2.5 μmol/m^2; i.e., 9.4 mmol/g, that for SM6, the residual amount of ≡SiOH groups is approximately 16% of their initial content and the IR band ν_{OH} at 3750 cm^{-1} is yet observed [Gun'ko et al. 2003e]), changes in the Gibbs free energy of the interfacial water layer (γ_S) in the aqueous medium are less 20%. A great nonuniformity of partially modified surfaces

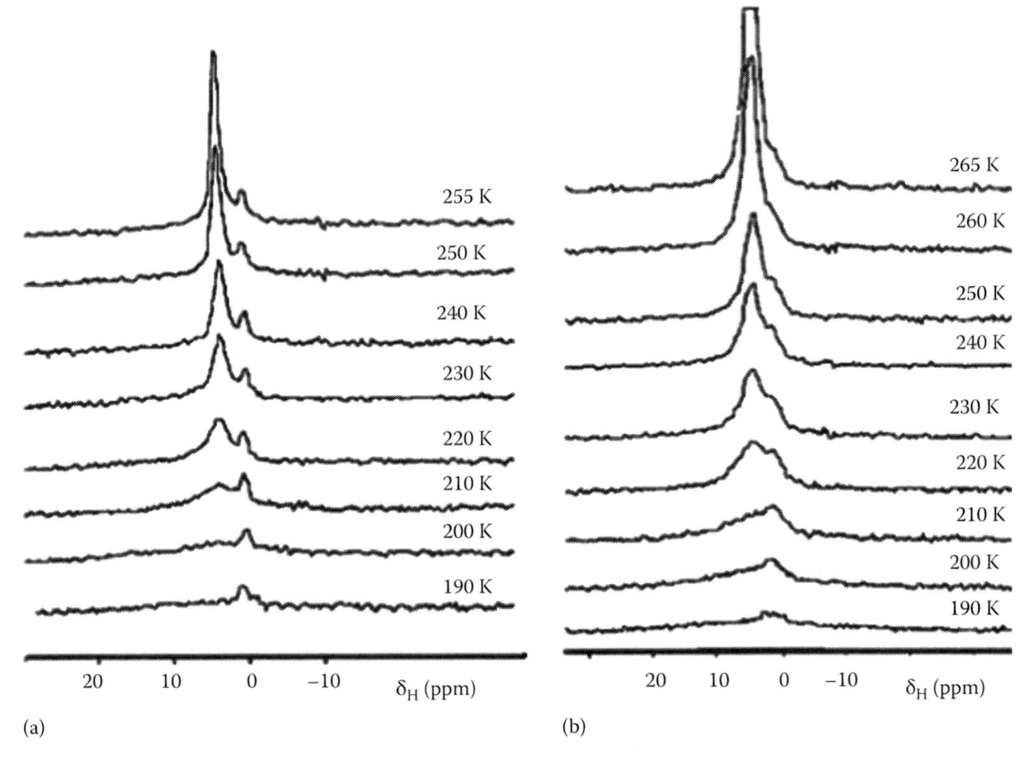

FIGURE 1.146 ^{1}H NMR spectra of water adsorbed on SM4 (h = 220 mg/g) in the presence of chloroform-*d* at C_{CDCl_3} = (a) 0.6 g/g and (b) 3.3 g/g at different temperatures. (Adapted from *Adv. Colloid Interface Sci.*, 118, Gun'ko, V.M., Turov, V.V., Bogatyrev, V.M. et al., Unusual properties of water at hydrophilic/hydrophobic interfaces, 125–172, 2005d. Copyright 2005, with permission from Elsevier.)

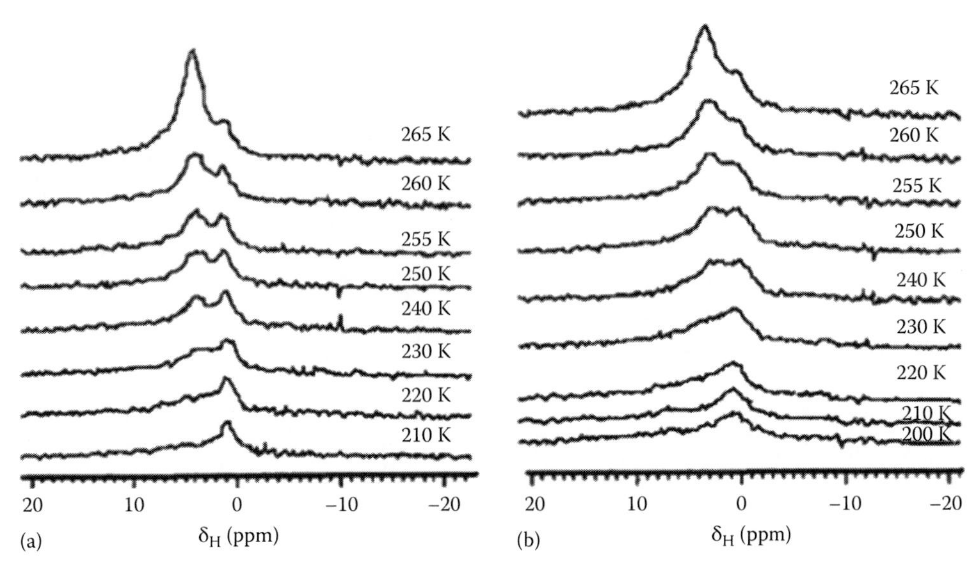

FIGURE 1.147 ^{1}H NMR spectra of water adsorbed on SM5 (h = 200 mg/g) in the presence of chloroform-*d* at C_{CDCl_3} = (a) 1.4 g/g and (b) 2.7 g/g at different temperatures. (Adapted from *Adv. Colloid Interface Sci.*, 118, Gun'ko, V.M., Turov, V.V., Bogatyrev, V.M. et al., Unusual properties of water at hydrophilic/hydrophobic interfaces, 125–172, 2005d. Copyright 2005, with permission from Elsevier.)

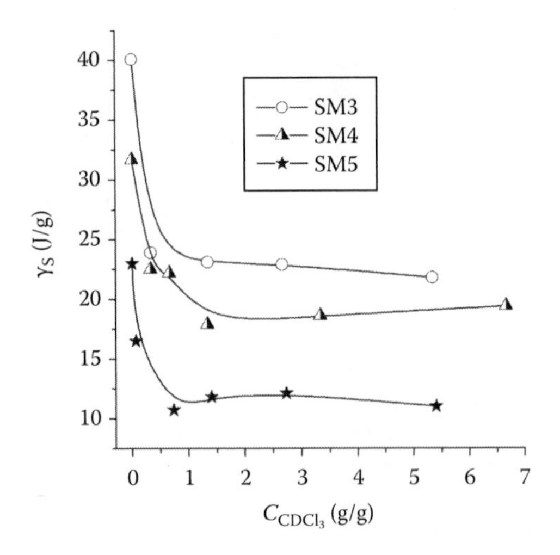

FIGURE 1.148 Dependence of interfacial free energy on concentration of chloroform-*d* added to in hydrated powders of silylated silicas. (Adapted from *Adv. Colloid Interface Sci.*, 118, Gun'ko, V.M., Turov, V.V., Bogatyrev, V.M. et al., Unusual properties of water at hydrophilic/hydrophobic interfaces, 125–172, 2005a. Copyright 2005, with permission from Elsevier.)

causes disorder of several monolayers of interfacial water and enhances the rotational mobility of water molecules that leads to reduction of the freezing temperature of bound water. Notice that an increase in the electrical conductivity for alternating current (increase in bound charges, e.g., dipoles of molecules) and its decrease for constant current (mobile charges, e.g., protons and other ions) were observed for the suspensions of a blend of hydrophilic and hydrophobic (silylated) nanosilicas. Consequently, changes in polarization of interfacial water (enhancement of bound charges) can affect its free energy because of the influence of the electrostatic field of the surface on the rotational mobility of water molecules (i.e., bound charges—dipoles of molecules) and the hydrogen bond network as a whole.

The TSDC spectrum of A-300 with grafted TMS groups and adsorbed water in the CCl_4 medium (Figure 1.149) shows changes in the structure of the interfacial water at the mosaic hydrophobic/hydrophilic surface in comparison with its structure at the surface of hydrophilic silica A-300 or in bulk water.

Water forms small clusters (TSDC peak at 135 K) and relatively large clusters and nanodomains (peak at 210 K). However, the structures with individual water molecules are practically not observed because only discharging current is measured at $T < 120$ K. The free energy of solvation of water in CCl_4 and $CHCl_3$ (Gun'ko et al. 2005d, 2009d) reveals that the interfacial solubility of water in $CHCl_3$ should be higher than that in CCl_4 (the solubility of CCl_4 and $CDCl_3$ in the interfacial water can be neglected). Consequently, the TSDC spectra (Figure 1.149) and quantum chemical calculations suggest that the use of $CDCl_3$ (as a weakly polar compound) in the NMR investigations of the interfacial behavior of water at the mosaic surfaces is more preferable than that of nonpolar CCl_4. One can expect that benzene-d_6 used instead of $CDCl_3$ can give a similar effect but more polar CH_3CN, $(CH_3)_2CO$, or $(CH_3)_2SO$ possessing chaotropic and kosmotropic moieties can give very different pictures because of changes in the HDW/LDW balance at the interfaces and larger molecular sizes (affecting their penetration into narrow pores), as well as because of changes in the solubility of water in these solvents. Similar effects were observed on the adsorption of water on ACs in the presence of different solvents (vide infra).

Modified nanosilica A-300 with grafted aminopropylmethylsilyl (APMS) groups (about 30% of the initial content of the surface silanols were reacted with the silane molecules) was used to study the interaction of a mosaic surface with water in a hydrophobic medium. It should be noted

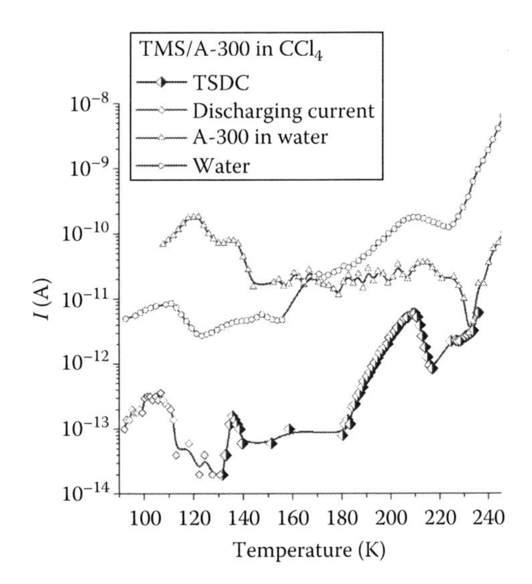

FIGURE 1.149 TSDC thermograms of hydrated nanosilica TMS/A-300 at $C_{TMS} \approx 0.64$ mmol/g in the CCl_4 medium ($C_{SiO_2} = 1$ wt%; polarization at $F = 4.0 \times 10^6$ V/m and $T = 240$ K), aqueous suspension of A-300 ($C_{SiO_2} = 1$ wt%), and bulk water. (Adapted from *Adv. Colloid Interface Sci.*, 118, Gun'ko, V.M., Turov, V.V., Bogatyrev, V.M. et al., Unusual properties of water at hydrophilic/hydrophobic interfaces, 125–172, 2005a. Copyright 2005, with permission from Elsevier.)

that grafting of the APMS groups can cause a more complex behavior of the interfacial water in comparison with TMS/silica because of the presence not only of residual silanols and $SiCH_3$ groups but also of a polar NH_2 head and a nonpolar CH_2 chain in the APMS. Like initial silica, APMS/A-300 is capable to bind a considerable amount of water $C_{H_2O} > 200$ mg/g. Furthermore, it easily forms stable suspensions both in water and in weakly polar or nonpolar solvents such as chloroform or carbon tetrachloride. Therefore, it is possible to study the dependence of the characteristics of interfacial water on concentration of a disperse phase over a wide concentration range of a hydrophobic dispersion medium.

The ^1H NMR spectrum of water bound to APMS/A-300 in air has a singlet shape at the width more than 1 kHz, and the average value of chemical shift is close to 5 ppm. In the $CDCl_3$ medium, the signal width of adsorbed water is sharply reduced, and two signals of water become distinguished in the spectra. One of them (signal 1) is linked with the chemical shift $\delta_H = 4.8$ ppm close to that of bulk water. The second signal at $\delta_H = 0.8$–1.1 ppm (Figure 1.150) corresponds to WAW.

The spectra of the adsorbed water change insignificantly with elevating temperature $T > 273$ K. A decrease in the signal intensity is observed at $T < 273$ K due to partial freezing-out of the interfacial water. Freezing-out of water starts from bulk water at $T = 273$ K then (at $T < 273$ K) of water characterized by signal 1. Freezing-out of WAW (signal 2) occurs at lower temperatures $T < 220$ K.

The relationship between the intensity of signals 1 and 2 (Figure 1.150) depends on the ratio of concentrations of the disperse phase and the dispersion medium. At a constant concentration of adsorbed water, an increase in deuterochloroform concentration is accompanied by an increase in the intensity of signal 1 and by the corresponding decrease in the intensity of signal 2. Besides signals of adsorbed water, there are signals of the CH groups of chloroform ($\delta_H = 7.26$ ppm) and methylene chloride ($\delta_H = 4$ ppm) present in deuterochloroform as admixtures. Furthermore, signal 2 of WAWs has a doublet structure, which is probably caused by the difference in chaotropic/kosmotropic effects of NH_2, CH_2, CH_3, and silanols differently structuring HDW and LDW at the interfaces. With increasing concentration of adsorbed water to 387 mg/g (Figure 1.151), the intensity of signal 1 grows to a greater extent than signal 2, while the chemical shifts of both signals remain without changes. Consequently, the amount of larger HDW structures increases but the content of

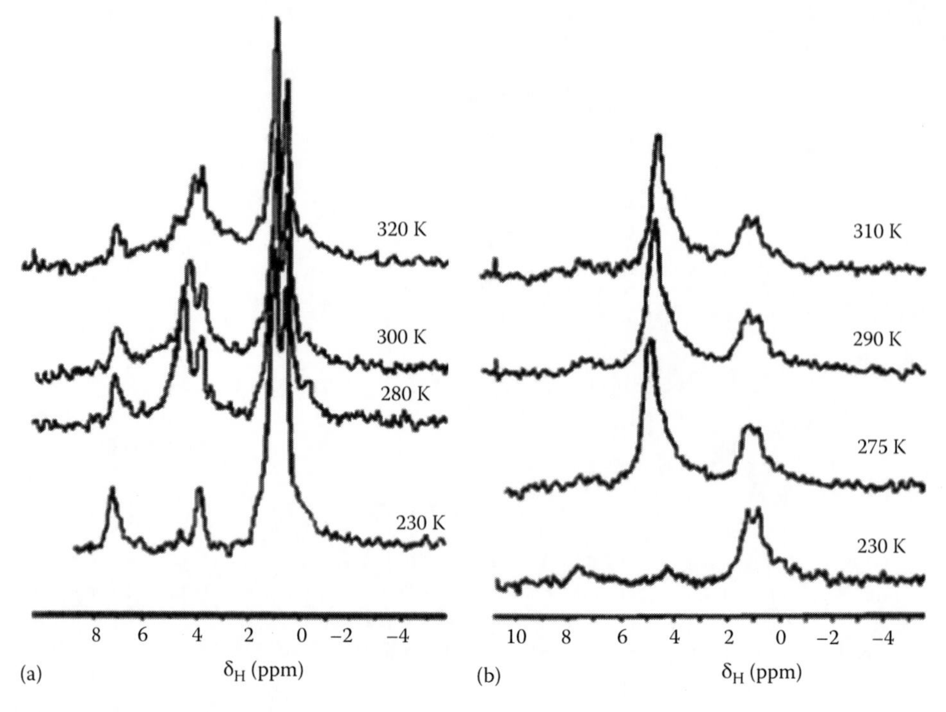

FIGURE 1.150 ^{1}H NMR spectra of water in the suspensions of modified nanosilica APMS/A-300 at h = 236 mg/g in the deuterochloroform medium at C_{SiO_2} = (a) 21.6 wt% and (b) 11.5 wt%.

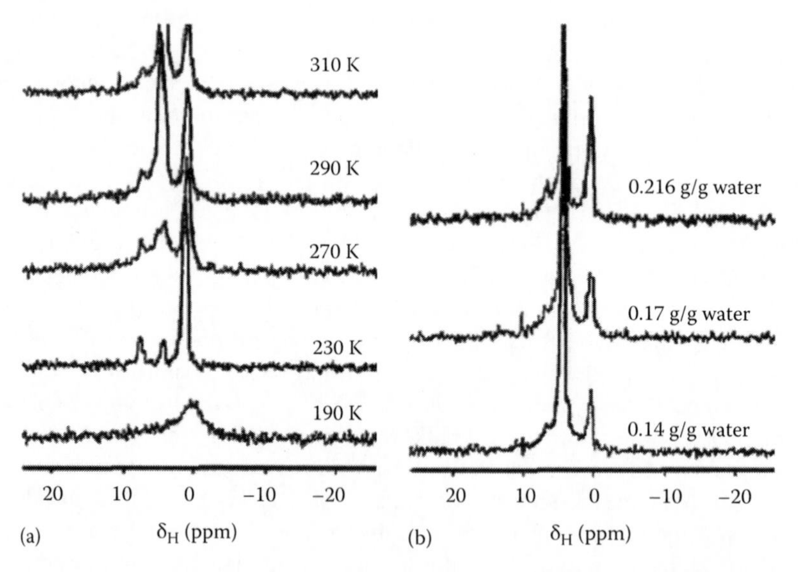

FIGURE 1.151 ^{1}H NMR spectra of water in the suspension of APMS/A-300 at h = 387 mg/g in $CDCl_3$ (a) at different temperatures and (b) after its dilution in the deuterochloroform (room temperature).

water responsible for signal 2 changes much smaller. The doublet structure of signal 2 becomes less distinguished possibly because of an increase in the signal width. The suspension dilution leads to a relative decrease in the intensity of signal 2 (Figure 1.151b). Figure 1.152 shows the temperature dependence of the content of unfrozen water in wetted APMS/A-300 powders in air and $CDCl_3$ and the relationship between the differential Gibbs free energy and the concentration of unfrozen

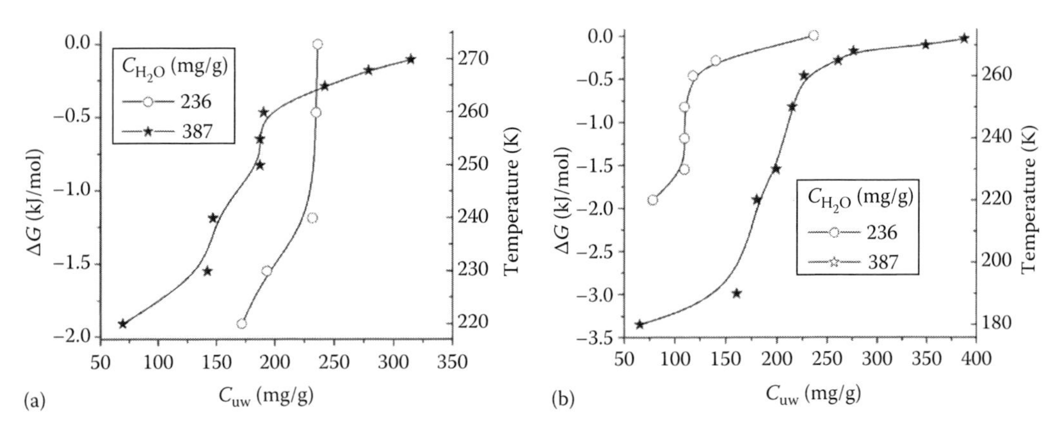

FIGURE 1.152 The temperature dependence of the concentration of the unfrozen water (C_{uw}) and relationship between differential Gibbs free energy and C_{uw} for wetted APMS/A-300 powder in (a) air and (b) $CDCl_3$ medium.

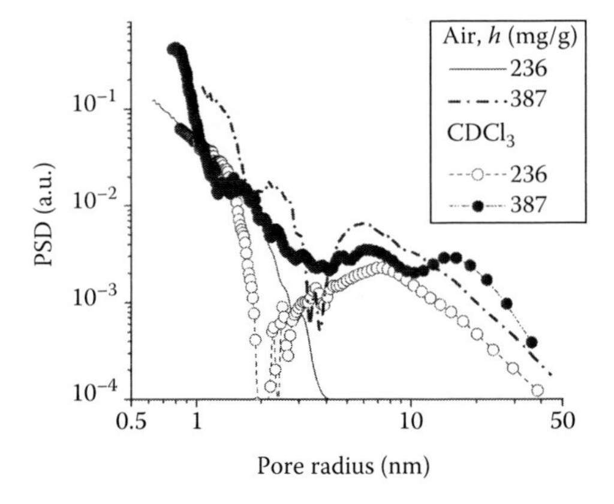

FIGURE 1.153 Size distributions of pores filled by structured water for APMS/A-300 powder (at hydration $h=236$ and 387 g/g) in air and the $CDCl_3$ medium.

water ($\Delta G(C_{uw})$). The $\Delta G(C_{uw})$ graph for APMS/A-300 at $C_{H_2O}=236$ mg/g has a vertical section caused by the absence of freezing-out of adsorbed water over a relatively wide temperature range ($240<T<273$ K) that corresponds to a weak dependence of the corresponding chemical shifts on temperature. With increasing concentration of adsorbed water to 387 mg/g (sample 2), a section related to WBW is observed if $C_{uw}>175$ mg/g.

The replacement of air by chloroform-*d* for the sample 1 at $h=236$ mg/g leads to a decrease in the content of SBW to 120 mg/g (Table 1.25). Another portion of adsorbed water is weakly bound. In accordance with the data shown in Figure 1.150, it is possible to assume that this process is caused by differentiation of adsorbed water to strongly (signal 1 of HDW) and weakly (signal 2 of HDW/interstitial water and individual molecules) associated water molecules. However, it should be noted that transferring of sample 2 from air to $CDCl_3$ leads to smaller changes in the relationship of contents of SBWs and WBWs. Consequently, water adsorbed onto sample 2 and characterized by signal 1 is strongly bound, and it is characterized by different changes in the structure under action of the $CDCl_3$ environment (Figure 1.153 and Table 1.25). The displacement of water by $CDCl_3$ occurs toward both smaller and larger pores.

Figure 1.154 shows the dependences of concentrations of SBW (signal 1) and WAW (signal 2) on the content of the disperse phase (C_{SiO_2}) for the suspensions of APMS/A-300 in $CDCl_3$.

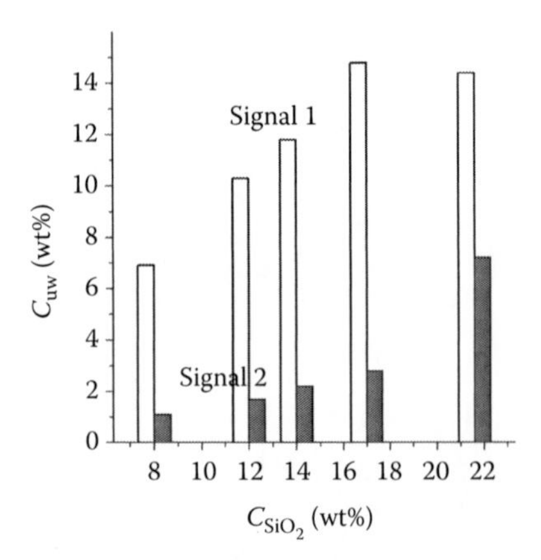

FIGURE 1.154 Concentration dependence of two types of water on content of the solid phase in the suspensions of hydrated APMS/A-300 powder ($h=216$ mg/g) in the deuterochloroform medium. (Adapted from *Adv. Colloid Interface Sci.*, 118, Gun'ko, V.M., Turov, V.V., Bogatyrev, V.M. et al., Unusual properties of water at hydrophilic/hydrophobic interfaces, 125–172, 2005d. Copyright 2005, with permission from Elsevier.)

The relationship of concentrations of both types of water remains near constant at $C_{SiO_2} < 12$ wt%. The redistribution of the intensities of signals 1 and 2 begins at $C_{SiO_2} \geq 14$ wt% and most actively at $C_{SiO_2} > 20$ wt%. This is the region of the concentrations of the organic/inorganic phase and water in, for example, biosystems when high intensity of signal of WAW is observed (vide infra). Notice that APMS/A-300 capable to form homogeneous suspensions in chloroform at a high concentration of the adsorbed water and the solid phase can be considered as a simple system for modeling of the behavior of interfacial water in partially dehydrated biosystems.

Signal 2 (Figures 1.150 and 1.151) coincides with signal of water dissolved in chloroform at $\delta_H = 1.7$ ppm. Therefore, it is possible to assume that the appearance of this signal is caused by enhanced interfacial mixing of water and chloroform stimulated by the electrostatic and dispersive surface forces at solid nanoparticles with mosaic surfaces with hydrophilic and hydrophobic functionalities. The solubility of chloroform (as well as organic solvents) in interfacial water decreases in comparison with the bulk solubility because the activity of adsorbed water decreases. Therefore rather individual (according to quantum chemical calculations [Gun'ko et al. 2005d]) water molecules can penetrate into an interfacial chloroform layer that provides signal of WAW. Notice that there are other factors affecting appearance of this unusual water such as an increase in the amount of interstitial water in HDW, bending of the hydrogen bonds in small water clusters being in proximity to mosaic hydrophilic/hydrophobic patches, and displacement of water molecules by chloroform into much stronger confined space of very narrow pores and structural channels in solid particles where the hydrogen bonds are strongly bent or absent. In accordance with the data shown in Figure 1.154, the content of WAW in APMS/A-300–$CDCl_3$ can reach 7 wt%, which is greater by order than the solubility of water in bulk chloroform at room temperature. Consequently, alternation of hydrophobic and hydrophilic surface patches affects mixing of water with weakly polar solvents (e.g., $CDCl_3$) at the interfaces that can result in an increase in "interfacial" solubility of water in the mentioned solvents but not them in water. It should be noted that this interfacial solubility depends more weakly on temperature in comparison with the bulk solubility because of the effects of the mosaic hydrophilic/hydrophobic surface.

Quantum chemical calculations of the NMR spectra of water molecules located between different hydrophobic groups show that the δ_H values for the H atoms without the hydrogen bonds are between 0.8 and 1.6 ppm (Figures 1.155 and 1.156, the first basis set shown was used for calculations of the NMR spectra, and the second one was applied on the geometry optimization). The H atoms participating in the hydrogen bonds are characterized by the δ_H values between 3.0 and 5.6 ppm. These δ_H values for the hydrogen atoms (participating or nonparticipating in the hydrogen bonds) are in agreement with the experimental values and the earlier interpretation of the ^1H NMR spectra.

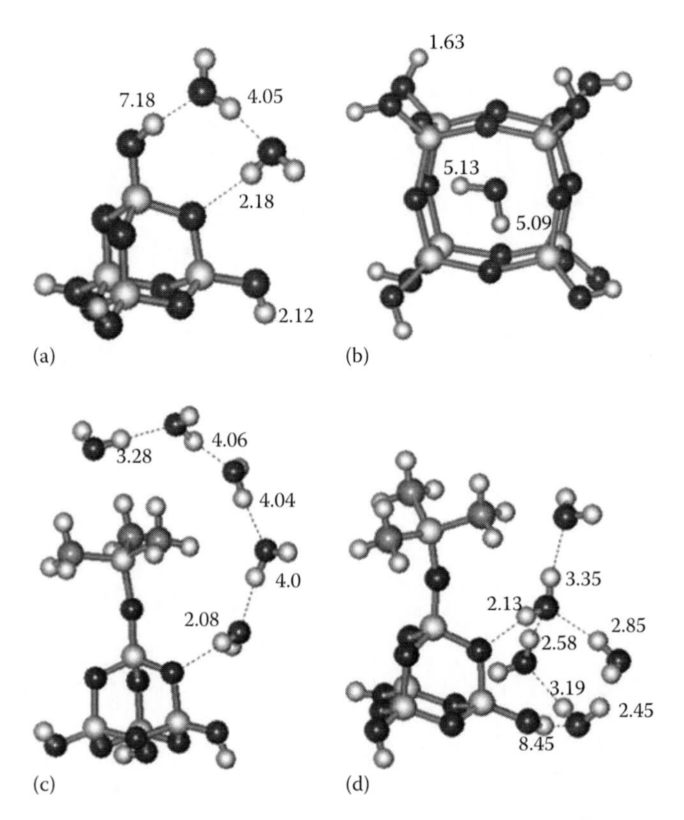

FIGURE 1.155 Chemical shift $\delta_{H,iso}$ for adsorbed water molecules and silanols calculated by the GIAO method with BLYP3/6-31G(d,p)//6-31G(d,p) for (a,b) unmodified and (c,d) TMS-modified silica clusters. (Adapted from *Adv. Colloid Interface Sci*., 118, Gun'ko, V.M., Turov, V.V., Bogatyrev, V.M. et al., Unusual properties of water at hydrophilic/hydrophobic interfaces, 125–172, 2005d. Copyright 2005, with permission from Elsevier.)

From the experimental data and quantum chemical calculations of the water clusters in the gas phase and with consideration for the solvation effects (Table 1.26), one can confirm the aforementioned reasons of appearance of ^{1}H NMR signal at $\delta_H = 1.1$–1.7 ppm. Notice that a significant decrease in $\Delta\sigma_{H,aniso}$ is observed for a water molecule being between two TMS groups in comparison with a molecule interacting with unmodified surface because of weakening of the polarization effect of the silylated surface onto adsorbed molecules, despite the effect of CCl_4 that reduces the total charge $q_{H_2O} = 0.044$ in comparison with $q_{H_2O} = 0.061$ for the molecule adsorbed between two TMS groups. This change in the anisotropy of the shielding tensor is an additional factor inducing appearance of unusual water with $\delta_H = 1.1$–1.7 ppm. In small water clusters (2–6 molecules akin to that found on the adsorption of water on nanoporous hydrophobic adsorbents), at least half of the hydrogen atoms do not form the hydrogen bonds and are in contact with $CDCl_3$ molecules and nonpolar or weakly polar functionalities or a hydrophobic surface. According to ab initio calculations of the chemical shifts in the gas phase (Figures 1.155 and 1.156) and consideration for their correction (using linear approximation $\delta_H = a + bq_H$ where a and b are constants, and q_H is the atom charge determined on the basis of the ab initio calculations of the δ_H values for a variety of associates of water molecules) due to the solvent effects, the δ_H values for "free" hydrogen atoms (in water molecules) may be between 0.5 and 2.5 ppm depending on polarity of the surroundings and the number of water molecules in a cluster, as well as on bending of the hydrogen bonds. Notice that the free energy of solvation of the water clusters in chloroform becomes higher with increasing their size. Consequently, there are two tendencies on the interaction of water with partly hydrophobized silica surface in the chloroform-*d* medium. The first is the formation of larger and larger water clusters and nanodomains ($\delta_H = 4$–5 ppm) with

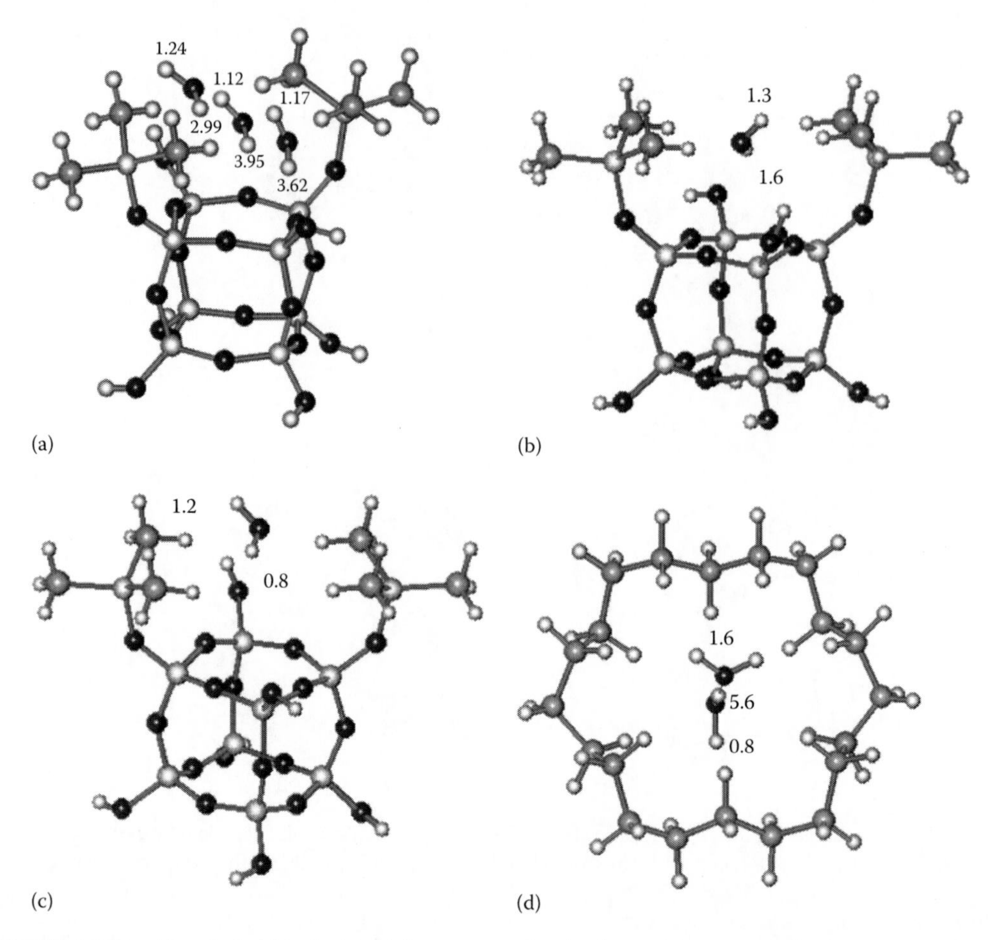

FIGURE 1.156 Chemical shifts δ_H (ppm) for water molecules located between (a, b, c) two trimethylsilyl groups attached to a silica cluster (B3LYP/6-31G(d,p)//HF/6-31G(d,p)), and (d) nonpolar molecule $C_{20}H_{40}$ (B3LYP/6-31G(d,p)//HF/3-21G). (Adapted from *Adv. Colloid Interface Sci.*, 118, Gun'ko, V.M., Turov, V.V., Bogatyrev, V.M. et al., Unusual properties of water at hydrophilic/hydrophobic interfaces, 125–172, 2005d. Copyright 2005, with permission from Elsevier.)

increasing hydration degree of the materials (compare the signal intensity at 4–5 and 1.1–1.7 ppm in Figures 1.150 and 1.151). The second is the water distribution in the form of small clusters (with considerable bending of the hydrogen bonds), dimers, and alone molecules (δ_H = 1.1–1.7 ppm) in the MS/chloroform-*d* medium, and the size of these clusters decreases with decreasing hydration degree, since signal at 4–5 ppm is much lower than that at δ_H = 1.1–1.7 ppm.

Thus, a substantial portion of water bound to the MS surface can be in the weakly associated state, which can exist in the limited range of the concentrations of bound water (typically when the continuous layer of water is absent). With decreasing temperature, typical interfacial water (SAW) partially freezes, while WAW can remain in the mobile state up to 200 K. Similar effects are observed for weakly hydrated bio-objects such as seeds and cells (vide infra).

Water bound at the hydrophilic/hydrophobic interfaces can be assigned to several structural types as WAWs and SAWs, which can be also weakly or strongly bound to the solid surface or macromolecules. These types of water can be analyzed in the terms of HDW and LDW or CS and ES waters (Gun'ko et al. 2005d). The molecular mobility of weakly associated interfacial water (unusual water because atypical value δ_H = 1.1–1.7 ppm) depends weaker on temperature, and it is higher than that of SAW because of smaller number of the hydrogen bonds per molecule. Additionally, this water

TABLE 1.26
Changes in Free Energy of Solvation (kJ/mol) Due to Formation of Adsorption Complexes of Water and CCl_4 Molecules on Clusters of the Initial and Partially Hydrophobized Silica in the Aqueous, Carbon Tetrachloride, and $CHCl_3$ Media (SM5.42R/6-31G(d)) and the $\delta_{H,iso}$ and $\Delta\sigma_{H,aniso}$ Values Calculated with No Consideration for the Solvent Effects

Complex		$\Delta\Delta G_s$ (in Water)	$\Delta\Delta G_s$ (in CCl_4)	$\delta_{H,iso}$ (ppm)	$\Delta\sigma_{H,aniso}$ (ppm)
$H_2O^*CCl_4{}^*Si_8O_{12}(OH)_8$		−24.5	−40.4	2.17 0.72	22.02 22.42
$H_2O^*Si_8O_{12}(OH)_6(OSi(CH_3)_3)_2$		−21.5	−28.9 in $CHCl_3$ −26.5	1.17 0.64	18.53 13.13

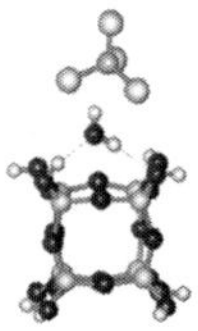

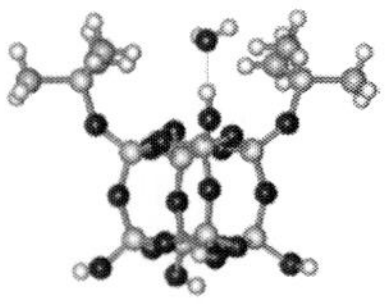

can be characterized by considerable bending of the hydrogen bonds or it corresponds to interstitial water in the interfacial HDW layer. These differences in typical and unusual interfacial waters result in the different temperature dependences of their ^{1}H NMR spectra at $T<273$ K. Immersion of the mosaic hydrophobic/hydrophilic systems (e.g., partially silylated nanosilica or silica gel, proteins, yeast cells, wheat seeds, and bone or muscular tissues containing certain amounts of adsorbed water, Gun'ko et al. 2005d, 2009d) in a weakly polar solvent ($CDCl_3$) allows increasing contribution of weakly associated bound water. This water gives a peak in the ^{1}H NMR spectra at $\delta_H = 1.1$–1.7 ppm, which demonstrates the atypical behavior with lowering temperature.

The unusual properties of the weakly associated interfacial water can be explained by its augmented mixing their clusters with clusters of hydrophobic weakly polar solvent at the mosaic hydrophobic/hydrophilic interfaces in comparison with its low solubility in the bulk of this solvent because of very local chaotropic and kosmotropic effects of adjacent different surface and macromolecular hydrophobic/hydrophilic functionalities in confined space of pockets (voids) filled by water or water/organic solvent. The weakly polar solvent can shield nonpolar surface functionalities from interfacial water because this can be more energetically favorable than contact of water directly with hydrophobic surface functionalities (e.g., the hydrophobicity of TMS groups can be higher than that of $CDCl_3$). The Gibbs free energy of water solved in $CHCl_3$ is negative (experimental $\Delta G_{s,exp} \approx -5$ kJ/mol) in contrast to its solution in nonpolar solvents (e.g., $\Delta G_{s,exp} \approx 11$ kJ/mol for water solved in heptane), and ΔG_s is positive for water solved other aliphatic solvents and vice versa; however, it is negative for aromatic solvents (e.g., $\Delta G_{s,exp} \approx -4$ kJ/mol for water solved in benzene or toluene), which can be considered for explanation of separation of water in tiny clusters in vicinity to both polar silanols or other polar groups and nonpolar CH functionalities of hydrophobized solid surfaces (or intracellular biostructures, vide infra). The polarity both of water and weakly polar solvent molecules can increase near polar patches of the surface that can result in increase in their mixing at the mosaic interfaces, despite decreased activity of the adsorbed water; that is, dissolution of nonpolar or weakly polar solvents in the interfacial water (as well in the bulk water) is lower than that for water in these solvents. The use of solvents of lower polarity than chloroform cannot provide the appearance of unusual interfacial water at $\delta_H = 1.1$–1.7 ppm if $\Delta G_s \rightarrow 0$ or $\Delta G_s > 0$. This can be explained by much lower probability of shielding of hydrophobic surface functionalities by hydrophobic molecules from water clusters ($\Delta G_s > 0$). However, in the case of a low content of the interfacial water, similar effects could be observed even without a weakly polar solvent because water is distributed in the form of tiny clusters with strong bending of the hydrogen bonds and individual molecules interacting with nonpolar or weakly polar organic functionalities. This results in the appearance of the ^{1}H NMR signal at $\delta_H = 1.2–1.7$ ppm.

Thus, partially MS surfaces can bind a certain quantity of WAW. One can assume that this unusual water is typical for mosaic hydrophilic/hydrophobic systems of different. This makes it possible to assume that WAW can fulfill an important function in biosystems, especially when these systems are near dry such as dry cells or seeds (vide infra).

Thus, nanosilicas partially modified by HMDS and containing different amounts of trimethylsilyl groups (from 0.09 to 0.79 mmol/g) are characterized by nonlinear dependences of changes in the structural, adsorptive, and other parameters such as the specific surface area, the pore volume (total, nano- and mesoporous), PSDs, concentrations of WBWs and SBWs, Gibbs free energy of these waters, the heat of immersion in liquid water, and chemical shift $\delta_H(T)$ of water adsorbed from air. On water adsorption on MSs from air, the size of water droplets (located between aggregates of primary particles in agglomerates) increases with amounts of grafted TMS groups, and for SM6 maximum hydrophobic, the main ^{1}H NMR signal of adsorbed water becomes close to that for fluid water. Enhancement of the intensity of the IR band ν_{OH} at 3660 cm^{-1} (corresponding to silanols difficult of access for water and other adsorbates) occurs with increasing C_{TMS}. The formation of small water clusters between grafted TMS groups causes appearance of the ^{1}H NMR signal with low chemical shift of ^{1}H (in adsorbed water molecules) because of the formation of complexes ($\equiv$Si)$_2$O···**H**–OH (characterized by smaller δ_H in comparison with that for HO–**H**···OH$_2$ or $\equiv$SiO(H)···**H**–OH) and due to smaller number of the hydrogen bonds per water molecule in the small water clusters in comparison with that in larger water droplets on maximum hydrophobic silica.

Now we will analyze the interfacial phenomena in mixtures with totally hydrophilic and totally hydrophobic nanosilicas differently hydrated.

1.1.9 Mixtures of Hydrophilic and Hydrophobic Silicas

Specifically treated mixtures (mixed at 5000–18000 rpm) with hydrophilic and hydrophobic nanosilicas, water, and air can form two-shell microspheres. These microparticles can include both air and water, for example, water on the outside and air inside microspheres (a hydrophobic part is inside and a hydrophilic shell is outside) or vice versa depending on the preparation technique.

The water-rich powders containing up to 98% (by weight) of water and a small amount of hydrophobic nanosilica and characterized by the same flow properties as dry silica powder ("dry water") can be prepared by a mixing process at a high rpm value (Berthod et al. 1988, Hamer et al. 2001, Lahanas et al. 2001, Yoichiro et al. 2002, Koga et al. 2004, Dampeirou 2005, Hasenzahl et al. 2005, Binks and Murakami 2006, Forny et al. 2007, 2009, Saleh et al. 2011). "Dry water" materials with water droplets covered by the hydrophobic shells or related dried materials without water are of interest for applications in medicine, cosmetics, biotechnology, etc.

If in a hydrophilic—hydrophobic silicas—water mixture (prepared at >10,000 rpm), the inner layer is composed of hydrophobic silica and the outer layer is of hydrophilic silica that air bubbles can be entrapped in the microspheres. The bulk density of this fresh gel-like mixture can be relatively low, ~0.5–0.6 g/cm^3, due to entrapped air (Mironyuk et al. 1994, 1999). In the case of the opposite structure of the composition, water is located inside microparticles with the hydrophobic outer shell that give the dry water material mentioned earlier. In the dry water materials, instead of hydrophilic silica, starch or other superhydrophilic polymers or other hydrophilic compounds can be used to provide stronger and longer retention of water inside microparticles.

Several techniques can be used to produce composite or hybrid materials based on the hydrophobic–hydrophilic mixtures with water or other liquids (de Gennes et al. 2004, Eshtiaghi et al. 2009, Boissiere et al. 2011). For instance, a mixture of hydrophilic nanosilica A-300 and hydrophobic nanosilica R812 was dispersed in a Schiff-base-type liquid crystal (LC), *p*-ethoxy(ben zylidene)-*p*-*n*-butylaniline, and analyzed using deuterium NMR spectroscopy (Milette et al. 2008) showing concentration changes in the silica particles-LC network structure. Here, we will analyze the structural characteristics and certain physicochemical properties of mixtures with hydrophilic and hydrophobic nanosilicas and water prepared using an agitator with a high rotating velocity >10,000 rpm (Mironyuk et al. 1994, 1999, Turov and Mironyuk 1998, Ding et al. 2009).

Hydrophobization of silica by grafting of tri- (TMS) or dimethylsilyl groups (DMS) $Si(CH_3)_n$ ($n=2$ or 3) is accompanied by a decrease in a concentration of the surface hydroxyls and formation of extended hydrophobic areas on the surface characterized by poor sorption of water. Therefore, the particles of DMS silica (MS) are hydrophobic, and a contact angle with water drops can be large up to 120°. The mixed material (MIX) consisting of hydrophilic silica A-300 ($S_{BET} \approx 300$ m^2/g) and hydrophobic MS ($S_{BET}=285$ m^2/g) (4:1 and 20 wt% in total in the gel-like material) and water (80 wt%) is composed of nanoparticles and their aggregates interacting with each other. It can be prepared by wetting the mixture of parent powders with an organic solvent or by intense mixing of the A-300 aqueous suspension with addition of the MS powder. The resultant material does not break down into its initial components either in water or in organic solvents. This gives an argument in favor of strong interparticle or inter-aggregate interactions in MIX. Despite the hydrophobic component, the MIX particles may, however, retain water in an amount which several times exceeds the weight of the parent silica powder itself (Mironyuk et al. 1994, 1999, Turov and Mironyuk 1998).

The reasons for the strong interparticle interactions in complex MIX system, as well as for the formation of thick layers of adsorbed water at the adsorbent–water interface, relate to the existence of the complex surface forces at the mosaic interfaces. Direct methods for the determination of the magnitude of surface forces have been developed (Derjaguin et al. 1987, Ralston et al. 2002, de Gennes et al. 2004). These methods provide the opportunity to obtain the value of surface forces as a function of the distance between the particles for a variety of substances. The processes of stabilization or destabilization (coagulation) of colloid dispersion of nanoparticles can be described by forces with electrostatic and non-electrostatic components with Derjaguin and Landau, Verwey and Overbeek (DLVO) and -DLVO theories (Lyklema 1991, 1995). During mixing at a high rate, agglomerates of hydrophilic and hydrophobic silicas are destroyed (Ding et al. 2009), and these effects depend on energy density (rotating velocity), pH, and solid concentration.

Structure of interfacial water in hydrated powders and aqueous suspensions of mixed adsorbents can be analyzed using the temperature dependences of the intensity of the ^{1}H NMR signal originating from bound water within a temperature range lower than 273 K. Water at the interface freezes if the value of its free energy is equal to that of ice. The temperature dependence graph of the content of unfrozen water may be transformed into one of the variations in the free energy of bound water as a function of the distance of a water layer toward the surface. The hydration properties of a mixture of four parts hydrophilic silica and one part MS at different content of water corresponding to wetted powders (partially dried MIX), initial MIX, and aqueous suspension of MIX, as well as individual A-300 and MS, were studied using the low temperature ^{1}H NMR spectroscopy (Turov and Mironyuk 1998).

MS was synthesized by treating the surface of the dehydrated silica A-300 ($S_{BET} \approx 300$ m^2/g) with dimethylchlorosilane ($S_{BET}=285$ m^2/g) to maximum substitution of silanols by the DMS groups. The reaction temperature was 280°C–300°C, which provided total coverage of the silica surface by the $=Si(CH_3)_2$ groups. However, the presence of two reactive Si-Cl groups in the silane can cause cross-linking of the surface functionalities with the formation of strained Si-O-Si bonds (Gun'ko et al. 2000g). These strained bonds can be partially hydrolyzed during suspending and mixing of the MS/A-300-water system at 12000 rpm to prepare the MIX system.

The MIX sample with MS/A-300-water was prepared by mixing hydrophobic and hydrophilic silicas in 1:4 ratio using aqueous suspension of hydrophilic silica and dry powder of hydrophobic silica (80 wt% of water in the final composition) mixed at a high velocity (12000 rpm; Mironyuk et al. 1994, 1999). ^{1}H NMR spectra were recorded using a high-resolution Bruker WP-100 SY NMR spectrometer using a transmission band of 50 kHz. An aqueous suspension of individual hydrophobic MS was prepared by wetting the MS powder with ethyl alcohol followed by subsequent washing with distilled water by means of 30-fold decanting.

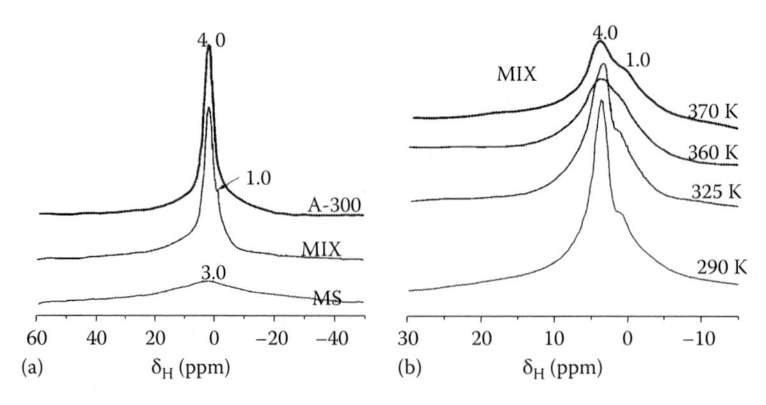

FIGURE 1.157 ^{1}H NMR spectra of water (a) adsorbed from air at 293 K and on individual A-300 (h=0.2 g/g), MS and their dried mixture, (b) bound to the MIX sample during free evaporation of water from the measuring ampoule at different temperatures. (Adapted from *Colloids Surf. A: Physicochem. Eng. Aspects*, 134, Turov and Mironyuk, Adsorption layers of water on the surface of hydrophilic, hydrophobic and mixed silicas, 257–263, 1998. Copyright 1998, with permission from Elsevier.)

Figure 1.157a shows the ^{1}H NMR spectra of water adsorbed on the MIX, MS, and A-300 samples at 293 K and 0.2 g/g moisture content.

The ratio of the signal intensities reflects the real differences in adsorptive capacity toward the water vapors (Figure 1.157a). The intensities of signals caused by water adsorbed on the MIX and A-300 samples are similar, while the amount of water adsorbed on the MS sample is four times lower. The spectra of the MIX sample have a peculiarity which reflects itself in the appearance of two signals of adsorbed water characterized by 4 ppm (signal 1) and 1 ppm (signal 2) chemical shifts. The chemical shift of the downfield signal coincides with that caused by water adsorbed on A-300. Water adsorbed on MS gives one wide signal. It is characterized by a chemical shift of 3 ppm. When a portion of water is eliminated from the initial MIX sample (80 wt% water and 20 wt% of oxides), a ratio of intensities undergoes transformations. Figure 1.157b shows changes in the shape of the spectra of bound water when the MIX sample is subjected to heating in an open ampoule. A certain amount of water freely evaporates from the surface of the adsorbent. The intensity of the first signal in a course of evaporation decreases, while that of the second remains practically unchanged. Therefore, the second signal relates to the water molecules, which are more strongly bound to the adsorbent surface. While forming hydrogen-bound associates (clusters, nanodomains), water molecules may act either as proton donors or acceptors. One water molecule can participate in the formation of several hydrogen bonds (three to four ones in liquid water and four ones in ice).

The chemical shift of water molecules is determined by the strength of the hydrogen bonds and the average number (m) of hydrogen bonds per water molecule as a proton donor. The chemical shift of the molecules increases from 1.7 ppm for water dissolved in chloroform or benzene to δ_H=2.5 ppm for water dissolved in dimethylsulfoxide; however, in ice δ_H=7 ppm and δ_H=4–5 ppm for bulk liquid water. The chemical shift for water adsorbed on the silica surface ranges from 3 ppm in the case when the concentration of water is considerably lower than the number of the surface hydroxyl groups and 4.5–4.7 ppm for strongly hydrated nanosilicas. It may be concluded that the chemical shift of small water clusters in the course of their adsorption on the primary adsorption sites of the unmodified silica surface is close to 3 ppm. The sizes of the surface associates of water molecules increase with increasing hydration degree where m goes up to a value close to that of bulk water. On the basis on the aforementioned arguments, the data in Figure 1.157 may be interpreted as follows: the water molecules on the A-300 surface are the components of polyassociates (clusters, nanodomains) structured by the hydrogen bonds with the surface, where the m value is, however, lower than in liquid water. Taking into account that A-300 is characterized by $C_{H_2O} \approx 75$ mg/g (approximately four water molecules per silanol group), it may be concluded that the m value is in a range of $1<m<2$ (for water molecules as a proton donor). This result can be explained by the formation of cyclic clusters with water molecules and silanols when half protons are without the hydrogen bonds.

The chemical shift of the water molecules adsorbed on MS is $\delta_H \approx 3$ ppm but for the MIX sample, it is characterized by two signals $\delta_H = 1$ and 4 ppm. The first one can be attributed to single water molecules localized under cross-linked DMS "umbrella." This δ_H value is lower than those for nonassociated water molecules in inert chloroform or benzene because the DMS groups are very hydrophobic (the siloxane bonds Si–O–Si are rather hydrophobic too) and create the barriers for the formation of larger water clusters in contact with residual silanols on MS. More intensive signal at $\delta_H = 4$ ppm is due to SAW forming large clusters and nanodomains.

The changes in the free energy of adsorbed water are the functions of the concentration of unfrozen water for the MIX samples differing in water content in powders (curves 1–3, dried MIX), gel (4, initial MIX), and aqueous suspensions (suspended MIX [curve 5] and individual A-300 and MS, Figure 1.158). Points where the plots intersect with the x-axis determine the thickness of the layers of bound water (WBW and SBW). On the one hand, this value for hydrated powders, which are characterized by a relatively low water content (curves 1–3), coincide with the initial concentration of water in the sample as almost all the water in the sample is in a bound state. On the other hand, it reflects the maximum amount of water, which is subjected to perturbation caused by the surface ($C_{uw} = C_s + C_w$, Table 1.27). However, for the initial MIX at $h = 4$ g/g (curve 4) and MIX in the aqueous suspension (curve 5, as well as A-300 and MS) the amounts of SBW and WBW (typically <1 g/g) are smaller than the total amount of water. An extrapolation of the plots up to their intersection with an ordinate axis makes it possible to determine the maximum change in the free energy of adsorbed water caused by adsorption (Table 1.27, ΔG_s), as well as the ΔG value for WBW layer (ΔG_w). If we assume a uniform distribution of water on the surface, the shape of the $\Delta G(C_{uw})$ dependence coincides with that of the curve defining the radial function of variation in the Gibbs free energy, RFG, of water in the adsorption layer (Gun'ko et al. 2009d).

The highest degree of surface hydration is observed in aqueous suspensions. Two segments may be revealed on the dependencies $\Delta G(C_{uw})$ obtained: the segment of the quick decrease in the thickness of the unfrozen water layer in the narrow range of the ΔG value changes (the temperature is about 273 K) and that where C_{uw} decreases relatively weakly in a broad range of changes in the ΔG values. The portion of water adsorbed on the surface, which is responsible for the appearance

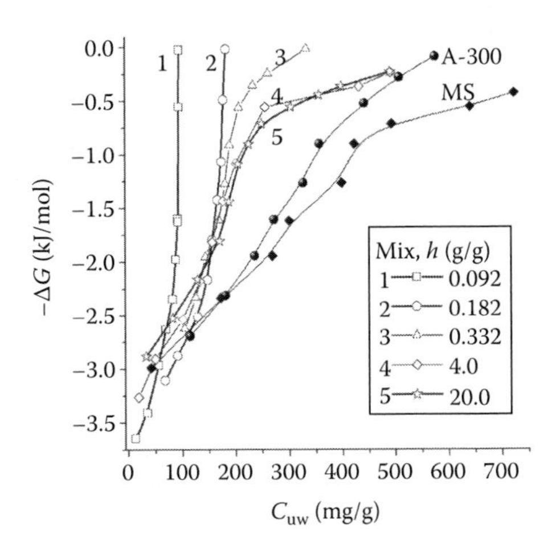

FIGURE 1.158 Change in free energy of bound water versus concentration of unfrozen water in the MIX sample at hydration h=(1) 0.092, (2) 0.182, (3) 0.332, (4) 4, and (5) 20 g/g, and A-300 and MS in the aqueous suspension. (Adapted from *Colloids Surf. A: Physicochem. Eng. Aspects*, 134, Turov and Mironyuk, Adsorption layers of water on the surface of hydrophilic, hydrophobic and mixed silicas, 257–263, 1998. Copyright 1998, with permission from Elsevier.)

TABLE 1.27
Characteristics of Water Bound to the Surface of A-300, MS, and MIX

Sample	h (g/g)	S_{uw} (m²/g)	$S_{nano,uw}$ (m²/g)	$S_{meso,uw}$ (m²/g)	$-\Delta G_s$ (kJ/mol)	$-\Delta G_w$ (kJ/mol)	C_s (g/g)	C_w (g/g)	γ_S (J/g)
A-300	20	317	245	73	3.26	0.8	0.353	0.250	51.2
MS	20	301	51	250	3.19	0.7	0.426	0.650	69.4
MIX	0.092	205	196	8	3.60	-	0.092	-	15.0
MIX	0.182	309	299	10	3.51	-	0.182	-	26.8
MIX	0.332	262	246	16	4.24	0.5	0.227	0.105	31.7
MIX	4	360	251	109	3.33	0.5	0.260	0.547	36.5
MIX	20	308	208	100	3.04	0.7	0.251	0.391	36.1

of the first of the previously mentioned areas, corresponds to the signs of WBW and is due to the long-distance component of RFG value. The portion of water responsible for the appearance of the second area corresponds to the signs of SBW. This type of bound water relates to a short-distance component of the RFG value. The thickness of the WBW and SBW layers (or their amounts C_w and C_s) and the corresponding changes in the free energy (ΔG_w and ΔG_s) may be determined by extrapolation of the corresponding segments of the $\Delta G(C_{uw})$ plot to the x- and y-axes.

The presence of a vertical segments on the $\Delta G(C_{uw})$ dependence (Figure 1.158, curves 1 and 2) reflects the fact that almost all the adsorbed water is bound to the surface having a low degree of hydration. Water starts to freeze only in the case when a drop in the free energy of water, which is caused by decrease in a temperature, becomes equal to that in the free energy due to adsorption at the surface. The characteristics of the layers of water adsorbed on the surface of the A-300, MS, and MIX samples and the corresponding values of the free surface energy in an aqueous medium (γ_S) are set out in Table 1.27. It may be concluded on the basis of the data given in Figure 1.158 and Table 1.27 that the maximum thickness of the layers of water (as well as the γ_S value), which are subjected to perturbation arising from the surface, is observed for the hydrophobic MS sample. The lowest values of the thickness of the layer of SBW as well as the γ_S values are observed for the MIX sample. It should be emphasized that the low thickness of the layers of adsorbed water may be expected for the MS sample if we focus our attention on the data given in Figures 1.157 through 1.159. Therefore, the long-distance acting component of the surface forces appears on the surface of particles in contrast to the hydrated powders in aqueous suspensions.

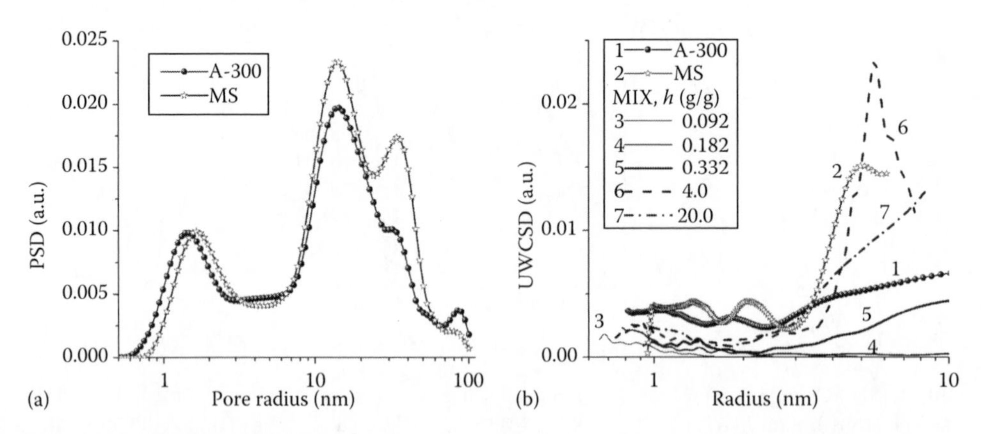

FIGURE 1.159 The distribution functions of (a) pore sizes of A-300 and MS calculated using the nitrogen adsorption–desorption isotherms with the SCV/SCR model, and (b) unfrozen water clusters filled voids in hydrated powders and aqueous suspensions of A-300, MS and MIX samples.

One of the main reasons responsible for the appearance of the long-distance acting component of the surface forces relates to charge inhomogeneity of the particles. The aforementioned type of inhomogeneity determines, in particular, such important processes in colloid systems as electric osmosis, electrophoresis, and surface conductivity. Surface charge density inhomogeneity of the MS surface may appear due to dissociation of residual hydroxyl groups followed by adsorption of a proton on an area remote from the dissociated site, as well as hydrolysis of strained siloxane bonds formed during cross-linking neighboring DMS groups. At the same time, an area where electrostatic forces are operating may be represented as a function of the distance among the charges. Homogeneity of the silica surface in the process of its methylation decreases. Hydroxyl coverage of the MS particles consists not only of the residual hydroxyl groups but also of the hydroxyl groups formed during the hydrolysis of the Si–Cl bonds, cross-linked but then hydrolyzed again during sample treatment. Differences in the dissociation constants of protons and adsorption of a proton solvated by several water molecules on different types of adsorption sites may result in the strong polarization of hydrophobic particles in an aqueous medium. As has already been mentioned, the MIX sample is distinguished by its ability to form a gel with the water content four times exceeding the weight of dry silicas. The data presented in Figure 1.159b and Table 1.27 show that the thickness of the layer of bound water is thinner than 10 nm. In the case of dried in air MIX samples at $h=0.332$ g/g, water tends to form relatively large clusters (Figure 1.159b, curve 5). However at $h=0.092$ and 0.182 g/g, the cluster size is smaller 1 nm in radius. Notice than the positions of the main PSD peak of dry A-300 and MS (Figure 1.159a) and the unfrozen WCSD (UWCSD) peak for the initial MIX and the suspensions of A-300 and MS (Figure 1.159b) are close. This can suggest that the structures of aggregates of primary particles in dry powders, suspended A-300 and MS samples, and the initial MIX are similar. This is also confirmed by the values of the surface area in contact with unfrozen water (Table 1.27, S_{uw}), which are close to the specific surface area of A-300 ($S_{BET}=300$ m^2/g) and MS ($S_{BET}=285$ m^2/g) for strongly hydrated systems. However, this value is much smaller for the MIX at $h=0.092$ g/g because this water amount is not enough to form a continuous layer at the silica surface.

This study gives sufficient grounds to make a conclusion that the thickness of the layer of water in aqueous suspensions, which is subjected to a perturbation arising from the surface, depends on polarization of the dispersed particles rather than hydrophobic properties of the surface of MS. A strong interaction among hydrophilic and hydrophobic particles is observed in the adsorbent comprising a stable mixture of A-300 and MS silicas with water (80 wt%). It reveals itself in the appearance of signal for water in anomalously strong magnetic fields in the ^{1}H NMR spectra of adsorbed water (~1 ppm). Apparently, the earlier signal is caused by the water molecules that are located under umbrellas of the cross-linked DMS groups. The free surface energy of adsorbents in the aqueous medium, calculated on the basis of changes in the concentration of unfrozen water in the process of heating frozen aqueous suspensions, increases in accordance with all MIX samples < A-300 < MS series.

Stable mixtures of hydrophilic and hydrophobic silicas with water at different ratio between the concentrations of these oxides and a constant total amount of oxides were also studied using electrophysical methods (Mironyuk et al. 1999). The conductivity (σ) of the oxide blend at different frequencies of the applied electromagnetic field increases relative to that for hydrophilic silica. However, the σ value at the constant electrical current decreases as the concentration of hydrophobic silica increases. An increase in the frequency from 0.1 to 10 kHz gives the exponential growth of the ratio between dielectric loss (ε'') and dielectric permittivity (ε') due to a strong decrease in the ε' value. These effects are caused by additional ordering in the system to reduce the contact area between hydrophobic silica particles and water that influences formation of micelles with a high polarizability.

The nature of nanooxide surfaces affects the structure of interfacial water layers and their electrophysical characteristics (Gun'ko et al. 1995, 1997a,b, 2007f,i, 2009d). For instance, a continuous layer with adsorbed water at $C_{H_2O} < 20$ wt% does not form at a nanosilica surface in contrast to fumed titania. Hydrophobic DMS A-300 can adsorb a certain but small amount (~1 wt%) of water from air (Figure 1.157); however, it cannot be wetted by liquid water due to a floating effect. In the case of the MIX sample with 20 wt% of A-300/MS (4:1) in a mixture with water (80 wt%, i.e., $h=4$ g/g), a high viscosity of the system and hydrophobic/hydrophilic interactions prevent the stratification and removal of air bubbles from MIX microparticles (Mironyuk et al. 1994, 1999).

Mixtures with A-300 and MS were studied at $\beta=C_{MS}/(C_{A\text{-}300}+C_{MS})$ from 0 to 0.3. The bulk density of fresh MIX samples depends on the β value. It is equal to $\rho_b=0.6$ g/cm³ at $\beta=0.2$ (Flotosorb-1, Mironyuk et al. 1994) and it is approximately 1.2 g/cm³ at $\beta=0$. During long storage (several months), it increases to 1.05–1.09 g/cm³ at $\beta=0.2$ due to removal of a portion of air from the bubbles.

The dielectric measurements were performed at room temperature and frequency (f) from 0.1 kHz to 8 MHz and strength $U=40$ V/m. At $U=2$ kV/m, the registered parameters were close to those at $U=40$ V/m (Mironyuk et al. 1999).

The resistivity value (ρ_x) depends on the β value (at a fixed f value) stronger (Figure 1.160a) than on frequency at a fixed β value (Figure 1.160b).

At $\beta=0$, the ρ_x value is 25 times smaller than that of distilled water at pH 5.8 and depends weakly on frequency. Unexpectedly, the ρ_x value decreases with increasing content of hydrophobic silica (Figure 1.160). The conductivity value $\sigma_x=\rho^{-1}$ characterizes the density of "free" charge carriers at a given frequency. The behavior of the function $tg\delta(f,\beta)=\varepsilon''/\varepsilon'$ (where ε'' and ε' are imaginary [dielectric loss] and real parts of complex permittivity dependent on the ratio of the density of free [σ_f] and bound [σ_b] charges or conduction current and displacement current ratio [Von Hippel 1954]) can be used for evaluation of the charge density assuming that changes in the σ_f value can be estimated from the behavior of the ρ_x value. A relative value of σ_f (from the ρ_x dependence on f) weakly changes with increasing frequency at a constant β value (Figure 1.160). Thus, it is close to active component; however, the tgδ increases significantly (Figure 1.161a). This can be explained by greater changes in the displacement current, that is, ε', than in the conduction current (ε''). The tgδ value slightly decreases at $\beta=0.05$ and 0.2 (Figure 1.161b).

The dependence of tgδ on β is weaker at $f=8$ MHz than that at 10 kHz (Figure 1.161b). At f=8 MHz $\varepsilon'=15$–17 and $\varepsilon''=3$–4 at different β values. The tgδ values increase exponentially with increasing frequency from 0.1 to 10 kHz (Figure 1.161a). Thus, the ε' and ε'' values are much greater

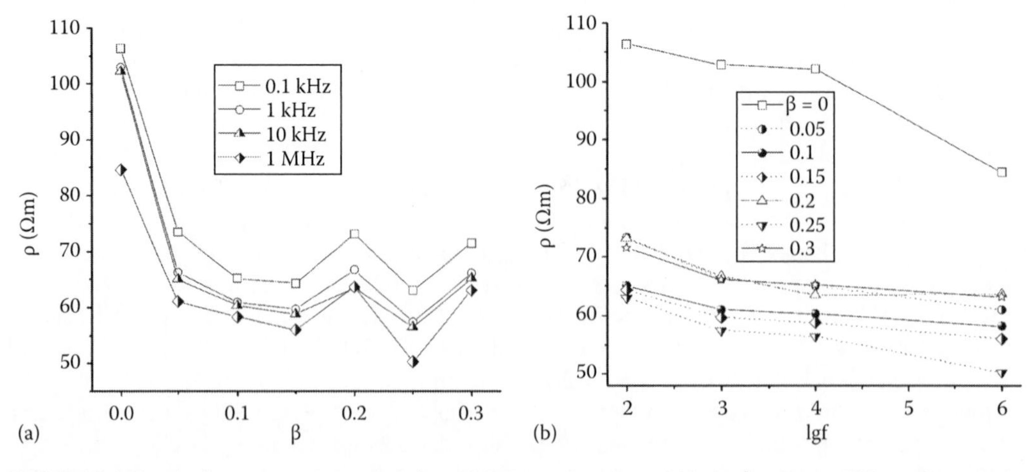

FIGURE 1.160 Active current (ac) resistivity of MIX as a function of (a) the β value at different frequencies and (b) of frequency at different β values and $\beta=0.2$ corresponds to initial MIX sample or Flotosorb-1. (Redrawn according to Mironyuk, I.F. et al., *Rep. NAS Ukr.*, N3, 149, 1999.)

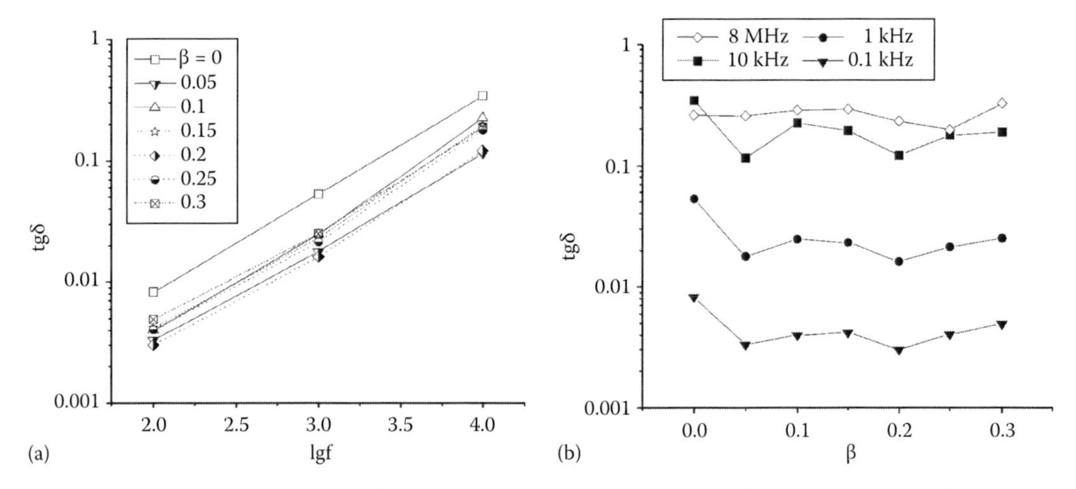

FIGURE 1.161 Dependences of $tg\delta = \varepsilon''/\varepsilon'$ on (a) frequency at different β values and (b) the β values at different frequencies. (Redrawn according to Mironyuk, I.F. et al., *Rep. NAS Ukr.*, N3, 149, 1999.)

at lower frequencies; however, the ε' value increases faster since $tg\ \delta \ll 1$. This behavior different from that of bulk water (Zatsepina 1998) shows that interfacial structures are characterized by a greater polarizability and the σ_f values are greater than that for bulk water especially at low frequencies. Consequence, dielectric (dipolar) relaxation times for these interfacial structures are greater than that for bulk water. This can be caused by interfacial polarization or Maxwell–Wagner–Sillars (MWS) relaxation process (Von Hippel 1954, Ngai et al. 1994). The ε' value can be lower at $\beta = 0$ (i.e., A-300) than that for MIX because the conductivity for the A-300-water system is lower (Figure 1.160) but $tg\ \delta$ is greater (Figure 1.161). However, this does not arise from the surface structure of the oxides since MS has the $=Si(CH_3)_2$ groups, which are not sources of mobile protons in aqueous media in contrast to the $\equiv SiOH$ groups at the pristine A-300 surface.

The direct current (dc) resistance (measured in 20 min after voltage application) of composite systems increases with increasing β value (Figure 1.162). However, recalculations of the dc conductivity as $(\rho')^{-1} = \rho^{-1}/(1-\beta)$ shows an increase in $(\rho')^{-1}$ with increasing β value. This result can be explained by changes in the conductivity conditions at the interfaces with changing content of the MS component, which was observed using the ^1H NMR spectroscopy.

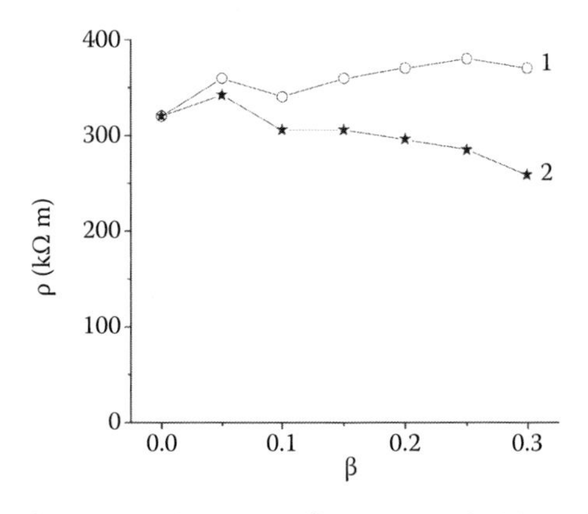

FIGURE 1.162 Dependence of the dc resistivity on the β value (1) and with consideration for only hydrophilic component (2). (Redrawn according to Mironyuk, I.F. et al., *Rep. NAS Ukr.*, N3, 149, 1999.)

Thus, to decrease the free energy of MIX contact between water and hydrophobic silica should be minimal. A high viscosity of the MIX system and absence of bulk water prevent floating of MS particles that can occur at small amounts of both oxides in a diluted suspension. Therefore, primary particles of A-300 and their aggregates form dense shells (micelles) around MS particles with entrapped air that create a barrier preventing contact of water with MS and the formation of separated MS phase. The LT ^{1}H NMR and NMR cryoporometry methods give useful structural information on the hydrophilic/hydrophobic silica–water mixtures and can be used to characterize the properties of "dry water" materials.

1.1.10 Adsorption of Low- and High-Molecular Organic Compounds

Nanosilica can be used in medicine as an enerosorbent or a drug carrier (Chuiko 1993, 2003, Blitz and Gun'ko 2006). Drug transport and release from carriers, drug solubility, and bioaccessibility are of importance as key problems in medicine. Features of drug interactions with carriers (determining release rate) depend strongly on the porosity type of carriers. The narrower and longer the pore, the longer is the release of a drug from this pore that, however, depends on the drug solubility (Dubin 2006). Typically, the smaller the drug particles, the faster is its dissolution, that is, drugs in a nanostructured state are characterized by better and faster solubility and, therefore, by a better bioaccessiblity. Organic systems such as liposomes, drug lipidic complexes, virosomes, and polymer-based conjugates can be used to increase the solubility of hydrophobic drugs, to minimize side effects of the direct applications of the drug and to achieve specific targeting (Wagner et al. 2006, Zhang et al. 2007a, Arruebo 2012). Drug carriers with porous and nonporous silicas, many of them commercially available, increase the solubility of poorly soluble drugs by hosting them within their pores or voids in nanoparticle aggregates (Chuiko 1993, 2003, Rigby et al. 2008, Arruebo 2012). Detailed investigations of the use of such drug carriers as porous silica aerogels (Smirnova et al. 2003), silica zeolite (Fukahori et al. 2011), solgel silica nanospheres (Cardoso et al. 2010), ultrastructural silica nanoparticles (Leirose and Cardoso 2011), nanostructured mesoporous silica (Vallet-Regi 2010), nonordered mesoporous silica (Limnell et al. 2011), nanosilicas (Chuiko 1993, 2003, Blitz and Gun'ko 2006), and other porous and micro-/nanostructured materials (Xue et al. 2006, Bladgen et al. 2007, Cao et al. 2008, Jannin et al. 2008, Shegokar and Müller 2010, Heimink et al. 2011, Qian and Zhang 2011, Vassiliou et al. 2011) showed that the drug release can be varied from several minutes to several months. This huge difference is mainly caused by the textural characteristics of drug carriers. Nanosilica provides the fastest drug release and the highest bioaccessibility of drugs being in a nanostructured state at the open surface of nonporous silica nanoparticles forming loose aggregates (Chuiko 1993, 2003, Blitz and Gun'ko 2006, Gun'ko et al. 2009d, Turov and Gun'ko 2011) and strongly interacts with polar polymers, such as poly(ethylene oxide) (PEO), poly(ethylene glycol) (PEG) (Tajouri and Bouchriha 2002), poly(vinyl alcohol) (PVA), and poly(vinyl pyrrolidone) (PVP; Gun'ko et al. 2009d), which can be used in a variety of compositions to better control drug release.

Nanosilica as a medicinal preparation (enterosorbent) of a high sorption capability possesses certain unique properties, which make it possible to use it effectively on treatment of different diseases (Chuiko 1993, 2003, Blitz and Gun'ko 2006). These features of nanosilica are caused by several factors: (i) chemical nature of amorphous nonporous primary nanoparticles passive in redox reactions and possessing weak reactivity in acid–base reactions (with ≡SiOH) and a low surface charge density at pH<8 (isoelectric point [IEP], pH ~2.2, point of zero charge [PZC], pH ~ 3); (ii) a small size of primary nanoparticles forming aggregates by the hydrogen bonds and partially by the siloxane bonds (density of aggregates is ~30% of the true density ρ_0) and more loose agglomerates of aggregates (~0.05–0.07ρ_0), which can be easily destroyed; (iii) local interactions of nanoparticles with certain structures of membranes of cells and microorganisms (e.g., integrated proteins); (iv) good transport properties of nanoparticles in aqueous medium because of faster diffusion in comparison with microparticles; (v) adsorption of substances on the external surface of nonporous primary particles; and (vi) changes in the interfacial water structure in a layer of several nanometers in thickness because of chaotropic properties of a silica surface and surface potentials generated by charged Si (Mulliken $q_{Si} \approx 1.5$ a.u. at HF/6-31G(d,p) basis set), O ($q_O = -0.65 - 0.8$ a.u.), and H ($q_H = 0.35 - 0.4$ a.u.) atoms. The uniqueness of nanosilica suggests that its medical applications can change treatment techniques of many diseases. Therefore, the applications of nanosilica in medicine become wide during last decade. In this chapter, to understand the mentioned features, the adsorption of certain drugs, polymers, proteins, and DNA onto nanosilicas will be analyzed using LT ^{1}H NMR spectroscopy and other methods.

A major portion of drugs represents low-molecular organic or inorganic compounds of different polarity and water solubility. In the case of adsorption of such compounds onto a nanosilica surface from aqueous media, there are certain regularities depending on pH value, size and polarity of a drug molecule, its charge state (neutral, anionic, or cationic), and concentration. Polar and charged compounds (acids, salts) are poorly adsorbed onto nanosilica due to great desolvation energy and absence of nanopores in silica primary particles. Figure 1.163a shows that such a salt as propranolol hydrochloride (a nonselective adrenaline blocking compound with two functionalized aromatic rings) adsorbs worse onto nanosilica A-300 at the same pH values than larger hydrochlorides Ethmosine (antiarrhythmic drug with functionalized four-ring structure) or Verapamil (calcium antagonist with functionalized tri-ring structure; Gun'ko et al. 2000f).

The cationic nature of these drugs results in a significant increase in their adsorption onto nanosilica with increasing pH value because the silica surface charge density becomes more negative. At lower pH values, the largest Ethmosine cations adsorb better than smaller ones (Figure 1.163a).

Upon coadsorption of human serum albumin (HSA) and Verapamil onto the silica surface from aqueous solution, the influence of protein is relatively small (Figure 1.163b). However, the concentration of the drug in this experiment was lower than upon its adsorption without HSA and the curve slope slightly increases. This can be explained by adsorption of the drug onto both the silica

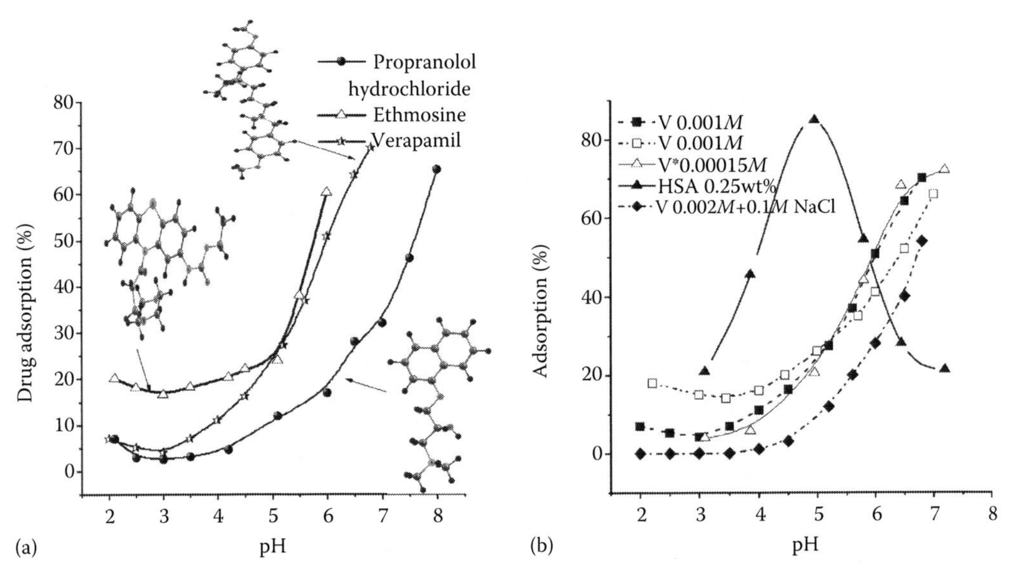

FIGURE 1.163 Adsorption of (a) Propranolol hydrochloride, Ethmosine, and Verapamil (in per cent of the initial concentration) onto silica A-300 at $C_{A\text{-}300} = 1$ wt% and drugs content of 0.001 M (b) adsorption of Verapamil onto silica in aqueous suspension stirred during a week (■) and 48 h (□) and at 0.1 M NaCl (υ) and mixed adsorption of Verapamil (△) and HSA (▲). (Adapted from *Colloids Surf. A: Physicochem. Eng. Aspects*, 167, Gun'ko, V.M., Vlasova, N.N., Golovkova, L.P. et al., Interaction of proteins and substituted aromatic drugs with highly disperse oxides in aqueous suspensions, 229–243, 2000f. Copyright 2000, with permission from Elsevier.)

surface and albumin, since HSA is the main transportation protein in blood and drug transportation in the human body occurs with its participation. Additionally, maxima of adsorption of Verapamil and HSA are not coincident. The pH_{PZC} values of these compounds differ. Typically, adsorption of charged particles or molecules strongly increases at a pH near IEP due to disappearance of the electrostatic repulsive interactions. At $PZC_{protein} > pH > PZC_{SiO_2}$, protein can effectively interact with the oxide surface through amino groups with attached protons. At $pH > PZC_{protein}$, the -COO^- groups of HSA turned into the bulk solution can strongly interact with protons attached to amino groups of the drugs, which are dissolved poorly in the molecular state and used only as hydrochlorides. Consequently, adsorption of cationic drugs or proteins onto the oxide surface depends mainly on the difference between a given pH value of the suspension and pH_{PZC} of these compounds as well as solubility and free energy of dissolution and adsorption of a drug. Additionally, interactions of globular proteins (such as HSA) with silica nanoparticles of the same size ranges leave some voids (Figure 1.164), which can be occupied by small drug molecules (ions).

The solubility of drugs can be improved using the corresponding salt compounds (e.g., hydrochlorides), as well as using preparations with drugs transformed into a nanostructured state at a surface of nanostructured matrices.

The solubility of antibiotic chloramphenicol (2,2-dichloro-N-[1,3-dihydroxy-1-(4-nitrophenyl) propan-2-yl]acetamide) in water is relatively low (~2.5 g/L at 25°C). Therefore, to prepare chloramphenicol/silica composites, the impregnation method was used (Krupska et al. 2006). The LT ^{1}H NMR spectroscopy study of these composites showed that the free surface energy γ_S of interfacial water decreases with increasing amount of chloramphenicol in the composition (Figure 1.165, curve 1). This can be interpreted as displacement of water from the silica surface by the drug molecules. However, despite the maximal amount of chloramphenicol (1 mmol/g) greater than the content of surface silanols (C_{OH} ~0.6–0.7 mmol/g for A-300 samples), the perturbation

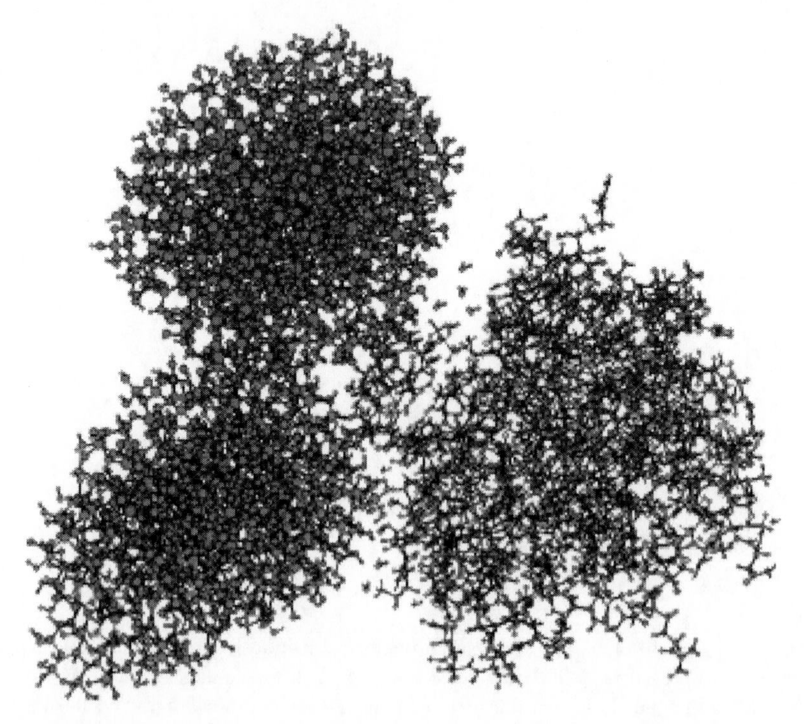

FIGURE 1.164 Interaction of to oxide nanoparticles with a globular protein and a certain number of water molecules (CharMM calculations). (Adapted from *Adv. Colloid Interface Sci.*, 105, Gun'ko, V.M., Klyueva, A.V., Levchuk, Yu.N., and Leboda, R., Photon correlation spectroscopy investigations of proteins, 201–328, 2003b. Copyright 2003, with permission from Elsevier.)

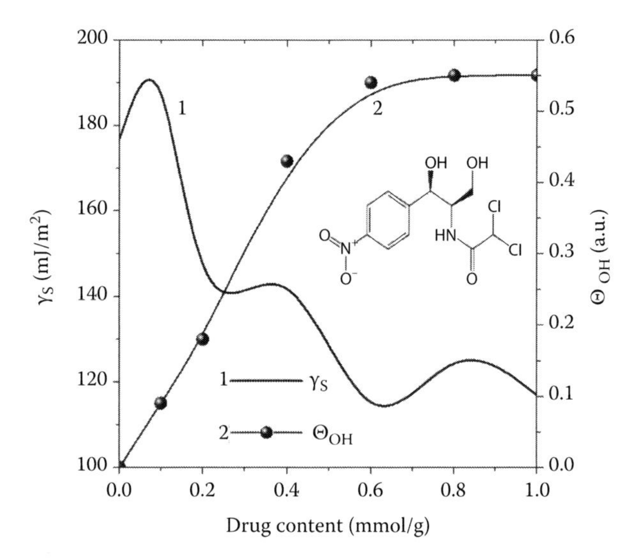

FIGURE 1.165 Changes in (a) the free surface energy of nanosilica A-300 impregned by chloramphenicol (^{1}H NMR), and (b) perturbation degree of surface silanols due to the chloramphenicol adsorption (FTIR).

degree of the silanols (measured using the FTIR spectroscopy of dried samples) interacting with chloramphenicol corresponds to approximately $\Theta_{OH} \approx 55\%$ (Figure 1.165, curve 2).

This value is relatively low because the molecular size of chloramphenicol allows a molecule to interact simultaneously with two to three surface SiOH groups and it also can shield neighboring silanols, which remain undisturbed, from other molecules. The results with respect to the concentration behavior of the γ_S and Θ_{OH} values can be explained by a clustered adsorption of chloramphenicol. This type of the adsorption occurs due to a specific characteristic of the impregnation method and the presence of both electron-donor and proton-donor sites in the chloramphenicol molecule causing strong attractive intermolecular interactions between neighboring molecules. However, from the Θ_{OH} value and the molecular size of chloramphenicol, one can estimate that the adsorbed clusters can include only several chloramphenicol molecules. In other words, the adsorbed drug is in a nanostructured state. The transition of chloramphenicol into this state in the composite with A-300 causes enhancement of its antimicrobial activity in respect to tested microorganisms (*E. coli*; Figure 1.166) because of increased bioaccessibility and solubility of the drug.

Natural bioactive flavonoid quercetin (Qc) (2-(3,4-dihydroxyphenyl)-3,5,7-trihydroxy-4H-chromen-4-one) is poorly dissolved in water (<0.1%), slightly better in ethanol (~0.2%), and much

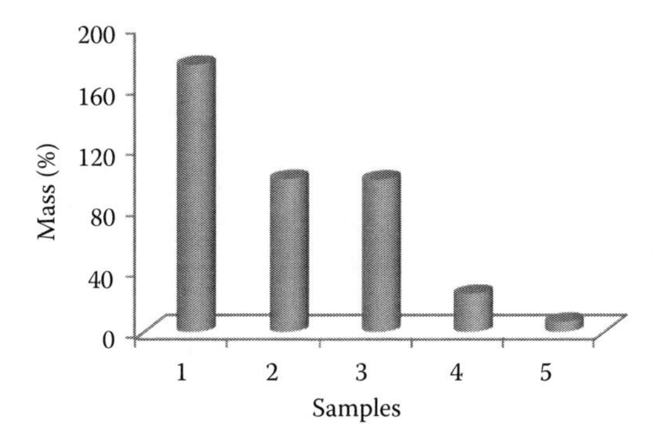

FIGURE 1.166 Changes in the mass of *E. coli* due to cell proliferation in the presence of (1) A-300 treated in alcohol, (2) control sample, (3) initial silica, (4) pure chloramphenicol (0.06 g), and (5) 1 mmol chloramphenicol per gram of A-300 (all other conditions were identical).

better in DMSO or dimethyl formamide (DMF) (~3%). Recently Hangzhou New Asia International Co., Ltd. (http://www.herb-extract.com/rutin/569615.html) informed that water-soluble quercetin has been produced. Quercetin is widely distributed in the nature as a component of some herbs. It is used in medicinal (phyto) preparations since it has antioxidant and enzyme modulator properties.

Starch, cellulose, and related substances are widely used in pharmaceutical, food, and other industries (Whistler et al. 1984, Young 1986, Frazier et al. 1997, Stenius 2000, Eliasson 2004), and their extraction from natural materials is carried out by water or water/organic mixtures (Minina and Kaukhova 2004) that can influence the product structure and other characteristics. Composites with starch or cellulose, drugs and highly disperse adsorbents can be used in medicine. These nanocomposites are characterized by enhanced activity of biopolymers and drugs because of their transformation into a nanostructured state during interactions with solid nanoparticles that enhance their bioaccessibility.

Quercetin adsorption onto nanosilica is low from both aqueous or water/alcohol solutions (Kazakova et al. 2002). Therefore, the Qc adsorption onto nanosilica was carried out using starch impregnated by water–alcohol solution of Qc, then dried at 60°C, and Qc content (C_{Qc}) was 4.8 wt% with respect to dry starch (Gun'ko et al. 2010c). A-300/starch or A-300/starch/Qc composites were prepared by homogenization of the aqueous suspension of starch (or starch/Qc) and silica (starch/silica = 20:1) then dried at 70°C for 6 h. Notice that peak temperature of gelatinization of potato starch used is $T_{gel} \approx 62°C$, that is, the Qc/starch mixture was prepared at temperature close to T_{gel} and Qc/starch/A-300 was prepared at T slightly higher than T_{gel}. The corresponding composites were also prepared as mechanical mixtures at room temperature (to prevent starch gelatinization).

Starches of different origin are widely used in food industry as thickening agents and their swelling and solubilization properties depend on temperature. Nanosilica (especially modified one such as studied here) can also be used as a thickening agent. Therefore, a mixture of starch with modified nanosilica can be of interest as a complex thickening agent. Double-helix structure of starch consisting of essentially linear poly(1,4-α-D-glucan) (amylose, $M_w \approx 0.1$–1.0 MDa) and branched amylopectin ($M_w \approx 0.01$–1.0 GDa) with 1,4-α-D-glucan chains connected through 1,6-α linkages depends on biological origin, amounts of water and temperature. In the case of potato starch (studied here) having hexagonal unit cell ("B" type) of the space group $P6_1$, content of amylose is 17%–20% (apparent amylose ~ 23%; Galliard 1987, Butler and Cameron 2000). Gelatinization of starch is accompanied by several processes such as swelling, leaching of amylose, loss of crystallinity, increase in viscosity, and solubilization of macromolecules (Atwell et al. 1988). Some of these effects lead to a decrease in the mobility of macromolecules and water molecules and influence the dielectric and other properties of the starch–water systems (Jaska 1971, Ryynänen et al. 1996) dependent on water content, starch gelatinization, and temperature (Bircan and Barringer 1998, Ndife et al. 1998a,b, Piyasena et al. 2003). Generally, the dielectric constant and the loss factor increase with increasing water content in the starch systems (Tsoubelli et al. 1995).

The starch structure is well studied and documented; however, detailed changes in its structure at different hydration degrees are complex, especially with consideration for the role of amylopectin and amylose structures forming granules (1–100 μm depending on the natural source of starch), which change (unfold) the structure on hydration/gelatinization or/and on interaction with solid nanoparticles. Clearly, the spatial structure of composites with polymers (such as starch, cellulose) and solid nanoparticles (e.g., nanosilica) is much more complex than that of individual polymers or nanosilica. Additionally, the properties of nanocomposites depend on features of interaction between nanoparticles and polymer molecules and individual components itself affected by the morphology, chemical structure, unfolding of polymer molecules, concentration and distribution of components in the system, temperature, pH, and salinity. Therefore, the structural and relaxational characteristics of starch and starch/nanosilica powders and hydrogels were compared at different hydration and temperatures using the ^{1}H NMR spectroscopy, cryoporometry, and DRS methods (Gun'ko et al. 2008c).

According to the TG (TG/DTG/DTA) data (Figure 1.167), interaction of water with starch/Qc (weight loss 10.4 wt% at 200°C) is stronger than with starch/Qc/A-300 (weight loss 11.3 wt%

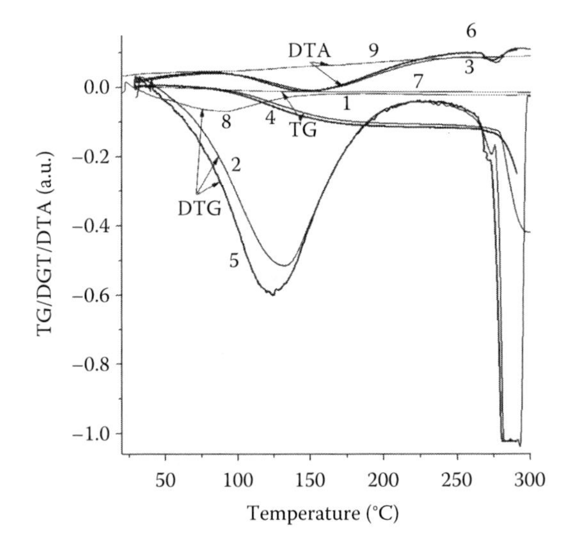

FIGURE 1.167 Thermogravimetry (TG) (curves 1, 4, and 7), differential TG (2, 5, 8), and DTA (3, 6, 9) curves of starch with Qc (1–3) (4.8 wt%) and with addition of A-300 (4–6) (4.8 wt%), and initial A-300 (7–9). (Adapted from *Appl. Surf. Sci.*, 256, Gun'ko, V.M., Turov, V.V., Barvinchenko, V.N. et al., Nonuniformity of starch/nanosilica composites and interfacial behaviour of water and organic compounds, 5275–5280, 2010c. Copyright 2010, with permission from Elsevier.)

at 200°C). This corresponds to a typical tendency; in the case of the same materials, the smaller the amounts of adsorbed water, the higher the adsorption energy. Additionally, slightly higher temperature of preparation of the second system (70°C) than the first one (60°C) can cause stronger unfolding of macromolecules with increasing sizes of pores (voids) and decreasing interaction with water located there. Water desorption at $T > 100$°C can be also accompanied by Qc desorption. Notice that water desorption from initial A-300 (weight loss 1.3 wt% at 200°C) occurs at lower temperatures (minimum DTG is at 95°C, Figure 1.167, curve 8) than that from composites (124 and 132°C, curves 5 and 2, respectively). Consequently, addition of silica leads to changes in bound water structure and possibly promotes desorption of Qc. A significant weight loss at $T > 275$°C (endothermic process) is due to destruction of starch (dehydration and then oxidation since heating was in air).

The influence of A-300 on the starch/Qc/water state is also observed in the FTIR spectra (Figure 1.168). The intensity of a broad band at 3320 cm^{-1} related to WBW increases with the presence of A-300; and in this case, the intensity of C=O and C=C bands of Qc (1600–1670 cm^{-1}) is higher than that without silica.

This result can be explained by diminution of Qc interaction with starch ($\nu_{C=O}$ of Qc at 1662 cm^{-1} shifts toward 1651 cm^{-1}) and water in the presence of silica ($\nu_{C=O}$ of Qc at 1659 cm^{-1}) because of strong interactions of A-300 with starch and water that is in agreement with the TG data (Figure 1.167). Upon Qc interaction with A-300 or starch a band at 3590 cm^{-1} (CO–H stretching vibrations of Qc alone) is absent (Figure 1.168) and intensity of a band of free silanols at 3750 cm^{-1} decreases because of the hydrogen bond formation. Additionally, fine structure of the Qc bands at 1000–1700 cm^{-1} is less visible, and the C–O and C–C stretching vibrations of starch at 1000–1200 cm^{-1} have higher intensity (due to unfolding of starch macromolecules) for Qc/starch prepared by the impregnation method than that for the mechanical mixtures.

Features of interactions of composite components depend on preparation technique (starch impregnated by alcohol solution of Qc with subsequent heating [causing starch gelatinization] or a mechanical mixture of components at room temperature) that lead to changes in the UV/vis spectra (Figure 1.169). The observed difference in the spectra is due to the difference in the form of Qc, which is better distributed on starch, and A-300 on the use of the impregnation method

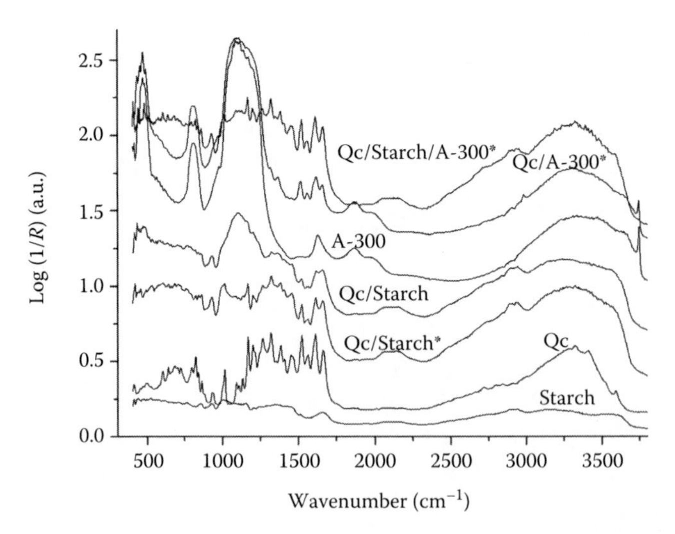

FIGURE 1.168 FTIR spectra of starch, Qc, A-300, Qc(4.85 wt%)/starch, Qc/A-300 (1:1) and Qc(4.85 wt%)/starch/A-300(4.85 wt%) (mechanical mixtures were marked by asterisk). (Adapted from *Appl. Surf. Sci.*, 256, Gun'ko, V.M., Turov, V.V., Barvinchenko, V.N. et al., Nonuniformity of starch/nanosilica composites and interfacial behaviour of water and organic compounds, 5275–5280, 2010c. Copyright 2010, with permission from Elsevier.)

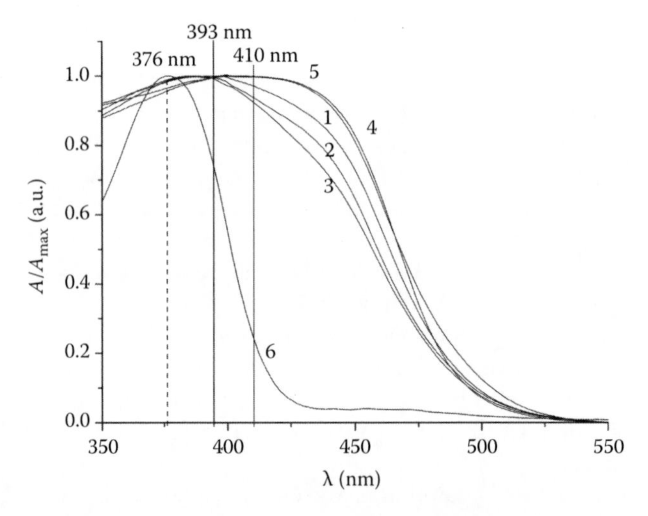

FIGURE 1.169 UV/visible spectra of starch with Qc (C_{Qc} = 1.15 (curve 2) and 4.85 wt% (1, 3, 4, and 5) and A-300 (4.85 wt% (3, 4)), impregnation (1–3) and mechanical blend (4, 5), Qc solution (1.6×10^{-4} mol/L) in ethanol (6). (Adapted from *Appl. Surf. Sci.*, 256, Gun'ko, V.M., Turov, V.V., Barvinchenko, V.N. et al., Nonuniformity of starch/nanosilica composites and interfacial behaviour of water and organic compounds, 5275–5280, 2010c. Copyright 2010, with permission from Elsevier.)

(providing nanostate of adsorbates) than in the case of mechanical mixture where a portion of Qc can remain in the form of microparticles. Thus, Qc interaction with starch gel is stronger than that for macromolecules or silica nanoparticles in the mechanical blend (Figure 1.169, curves 4 and 5).

Features of the interactions between starch gel and A-300 reflect also in the TSDC thermograms (Figure 1.170). An LT band (T < 170 K) shifts toward higher temperatures and its intensity for water/starch/A-300 is smaller than that for water alone or A-300 suspension. An HT band (T > 170 K) shifts toward lower temperatures and its intensity grows. The first effect is due to stronger hydrogen

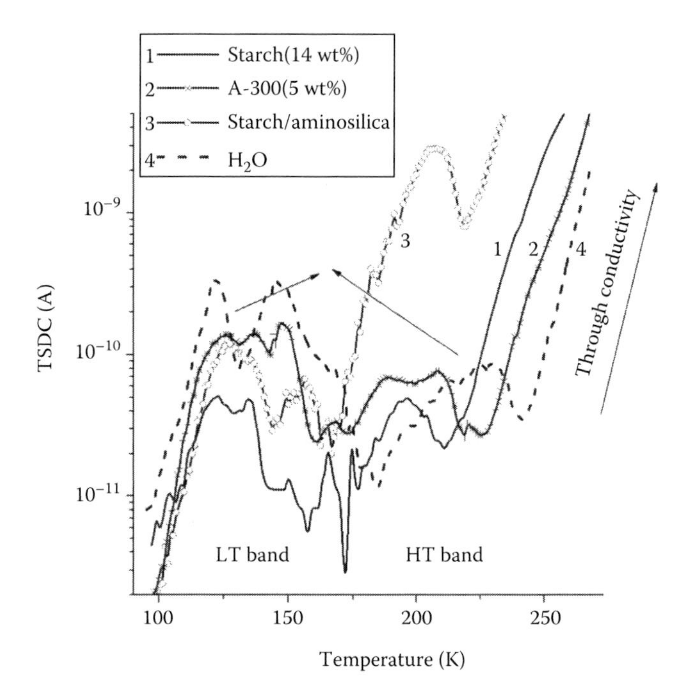

FIGURE 1.170 TSDC thermograms of hydrogel of gelatinized starch (14.2 wt%) (1), aqueous suspension of A-300 (5 wt%) (2), gelatinized starch/nanosilica (7.7:1) at $h = 16.9$ g/g (3), and water (4), arrows show directions of the displacement of LT and HT bands on addition of silica and starch to water. (Adapted from *Appl. Surf. Sci.*, 256, Gun'ko, V.M., Turov, V.V., Barvinchenko, V.N. et al., Nonuniformity of starch/nanosilica composites and interfacial behaviour of water and organic compounds, 5275–5280, 2010a. Copyright 2010, with permission from Elsevier.)

bonding of interfacial water molecules in composite than in starch (Gun'ko et al. 2007i). The second effect is due to enhancement of plasticizing of starch by water (Gun'ko et al. 2008c) in the presence of A-300 because of transition of adsorbate (starch macromolecules) into nanostate (i.e., diminution of supramolecular interactions between macromolecules) and an increase in the amounts of more mobile relaxing dipoles in glycoside structures characterized by greater polarity due to interaction with neighboring water molecules or surface silanols of silica nanoparticles.

The ^{1}H NMR spectra of water bound in the starch powder (Figure 1.171a) and gel (Gun'ko et al. 2008c) are characterized by a broadened single signal at chemical shift of the proton resonance δ_H in the 4.5–6.5 ppm range similar to that for liquid water. The signal intensity decreases with lowering temperature as a result of partial freezing of bound water, and the δ_H value increases due to a stronger influence of macromolecules on closer located water molecules characterized by stronger polarization causing reduction of the proton shielding and, therefore, resulting in the downfield shift. Chloroform medium weakly influences the spectrum shape (Figure 1.171a). However, hydrated Qc/starch/A-300 (treated at 70°C) in $CDCl_3$ (Figure 1.171b) is characterized by a weak additional signal at $\delta_H = 1.3$ ppm, which grows after addition of DMSO.

Another signal of water appears at $\delta_H = 3$–4 ppm (split into two signals at $T < 260$ K due to slowing of the molecular exchange) in the spectra of hydrated Qc/starch/A-300 (treated at 70°C) placed in $CDCl_3$/DMSO mixture (Figures 1.171c) that is due to the formation of the $HO–H\cdots O=S(CH_3)_2$ complexes. These complexes are solvated by $CDCl_3$/DMSO that causes a certain downfield shift of signal with lowering temperature. The intensity of this signal is determined by the ratio of the free energy of water binding with starch and its dissolution in the solvent mixture.

The theoretically calculated ^{1}H NMR spectra (Figure 1.172a) are broadened for water interacting with DMSO. The presence of the quercetin molecule or two starch (amylose) fragments in water

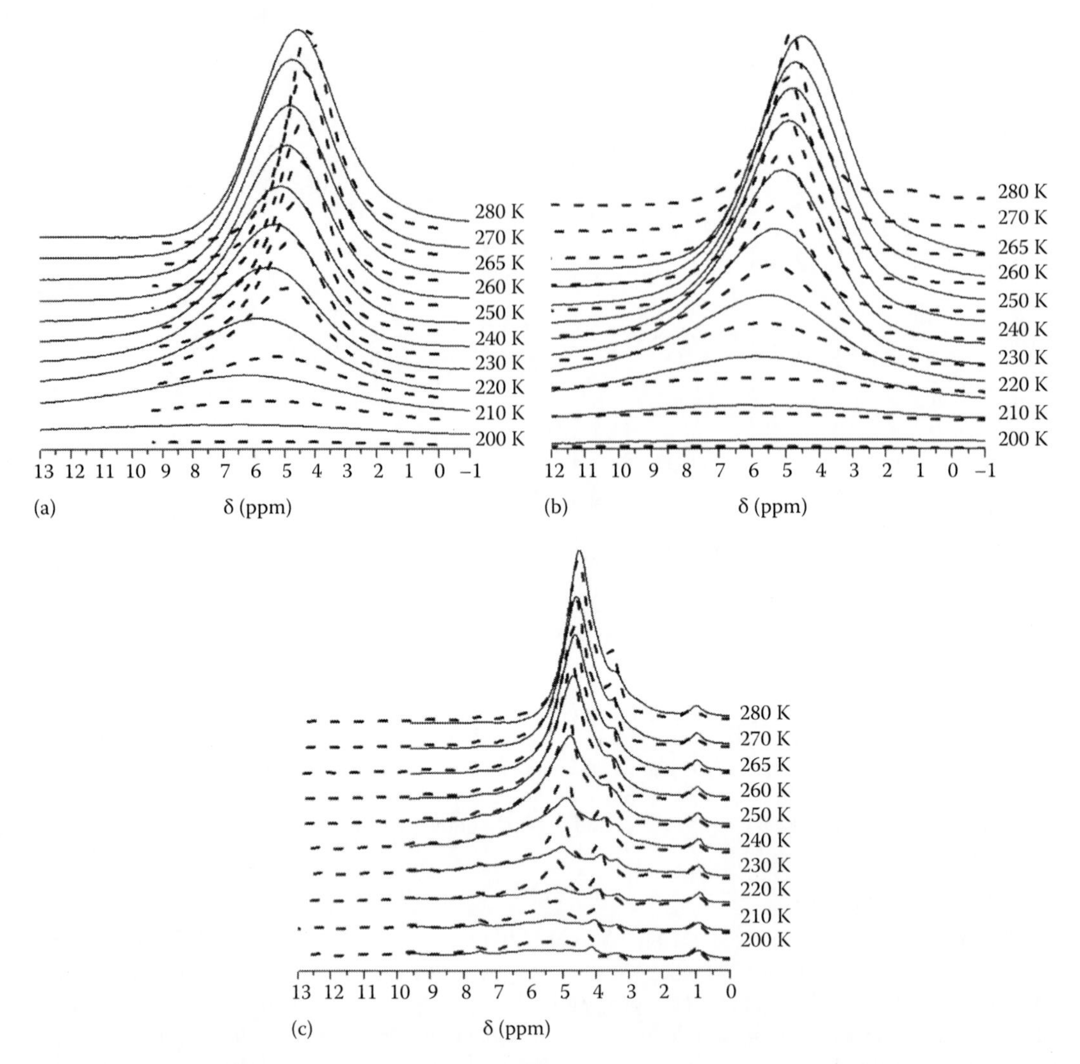

FIGURE 1.171 ^{1}H NMR spectra of unfrozen water in hydrated starch powder at (a) $h = 0.3$ g/g (solid lines) and in $CDCl_3$ medium (dashed lines), (b) on addition of Qc (4.8 wt%) + A-300 (4.8 wt%) at $h = 0.2$ g/g (pretreated at 70°C) (solid lines) and then placed in $CDCl_3$ (dashed lines), and (c) with addition of 0.5 g/g (solid lines) or 1 g/g DMSO (dashed lines). (Adapted from *Appl. Surf. Sci.*, 256, Gun'ko, V.M., Turov, V.V., Barvinchenko, V.N. et al., Nonuniformity of starch/nanosilica composites and interfacial behaviour of water and organic compounds, 5275–5280, 2010c. Copyright 2010, with permission from Elsevier.)

nanodomain results in the upfield shift (Figure 1.172b). This suggests weakening of the hydrogen bonding or changes in orientation of the O–H bonds of water molecules with respect to lone-electron pairs of the O atoms of neighboring molecules. As a whole, changes in the short-range order (OH distances) in a water nanodomain disturbed by the quercetin molecule are small (Figure 1.173) since the first three maxima (0.099 (**O–H**···O), 0.172 (O–**H**···**O**), and 0.293 nm [second coordination sphere]) in the $f(R_{OH})$ function are practically identical. However, certain changes in the whole nanodomain structure are observed due to addition of Qc or starch/Qc into the water droplet that reflect in the differences for the $f(R_{OH})$ peaks at $R_{OH} > 0.3$ nm. Notice that the density of the water nanodrop alone is $\rho = 1.0744$ g/cm^3 and 1.0346 g/cm^3 with Qc (the right insert in Figure 1.172b). However, the water nanodrop with two amylose fragments has $\rho = 1.046$ g/cm^3 and 1.048 g/cm^3 after addition of the Qc molecule (PM6 optimization of the geometry; insert in Figure 1.172b). In other words, Qc stronger interacts with hydrated starch than with water alone since Qc is poorly

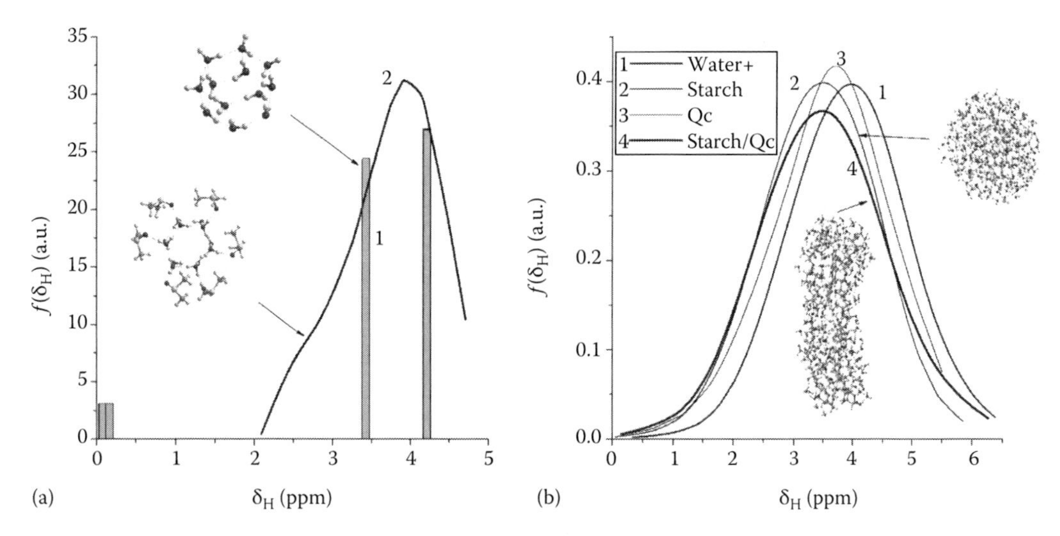

FIGURE 1.172 Theoretically calculated ${}^{1}H$ NMR spectra for (a) water cluster with $12H_2O$ (1) and distribution function $f(\delta_H)$ for $12H_2O^{*}6((CH_3)_2SO)$ (only for water) (2) (method GIAO/B3LYP/6-31G (d,p)//6-31G (d,p)), (b) distribution functions $f(\delta_H)$ for water nanodomain (1), hydrated two starch fragments (2), quercetin molecule in water nanodrop (3), and hydrated starch/Qc (4) calculated using correlation equation (geometry was optimized using PM6). (Adapted from *Appl. Surf. Sci.*, 256, Gun'ko, V.M., Turov, V.V., Barvinchenko, V.N. et al., Nonuniformity of starch/nanosilica composites and interfacial behaviour of water and organic compounds, 5275–5280, 2010c. Copyright 2010, with permission from Elsevier.)

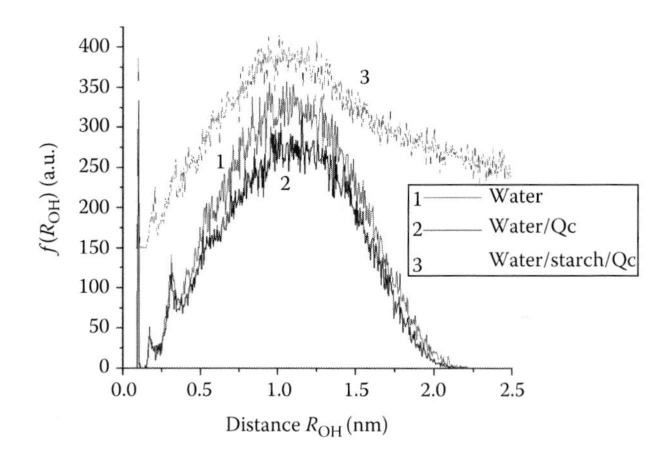

FIGURE 1.173 Distribution functions of OH distances in water nanodomain alone (curve 1), with Qc (2) and starch/Qc (3) (PM6 geometry). (Adapted from *Appl. Surf. Sci.*, 256, Gun'ko, V.M., Turov, V.V., Barvinchenko, V.N. et al., Nonuniformity of starch/nanosilica composites and interfacial behaviour of water and organic compounds, 5275–5280, 2010a. Copyright 2010, with permission from Elsevier.)

dissolved in water. For hydrated two fragments of amylose alone or with Qc, changes are observed for both $f(\delta_H)$ (Figure 1.172b) and $f(R_{OH})$ (Figure 1.173) distribution functions. Consequently, the water structure is sensitive even to small concentrations of dissolved organic substances.

Comparison of changes in the dC_{uw}/dT and $f_V(R)$ functions for unfrozen water (Figure 1.174) show that SBW ($\Delta G < -(0.5 \div 0.8)$ kJ/mol) corresponds to mainly water structures at $R < 2$–3 nm (WAW and SAW), and WBW (($\Delta G > -(0.5 \div 0.8)$ kJ/mol) is linked to structures at $R > 2$–3 nm (SAW). The boundary ΔG and R values between SBW and WBW depend on system composition. For more complex systems with starch/Qc/A-300/water, this boundary is nonabrupt (Figure 1.174e) in contrast

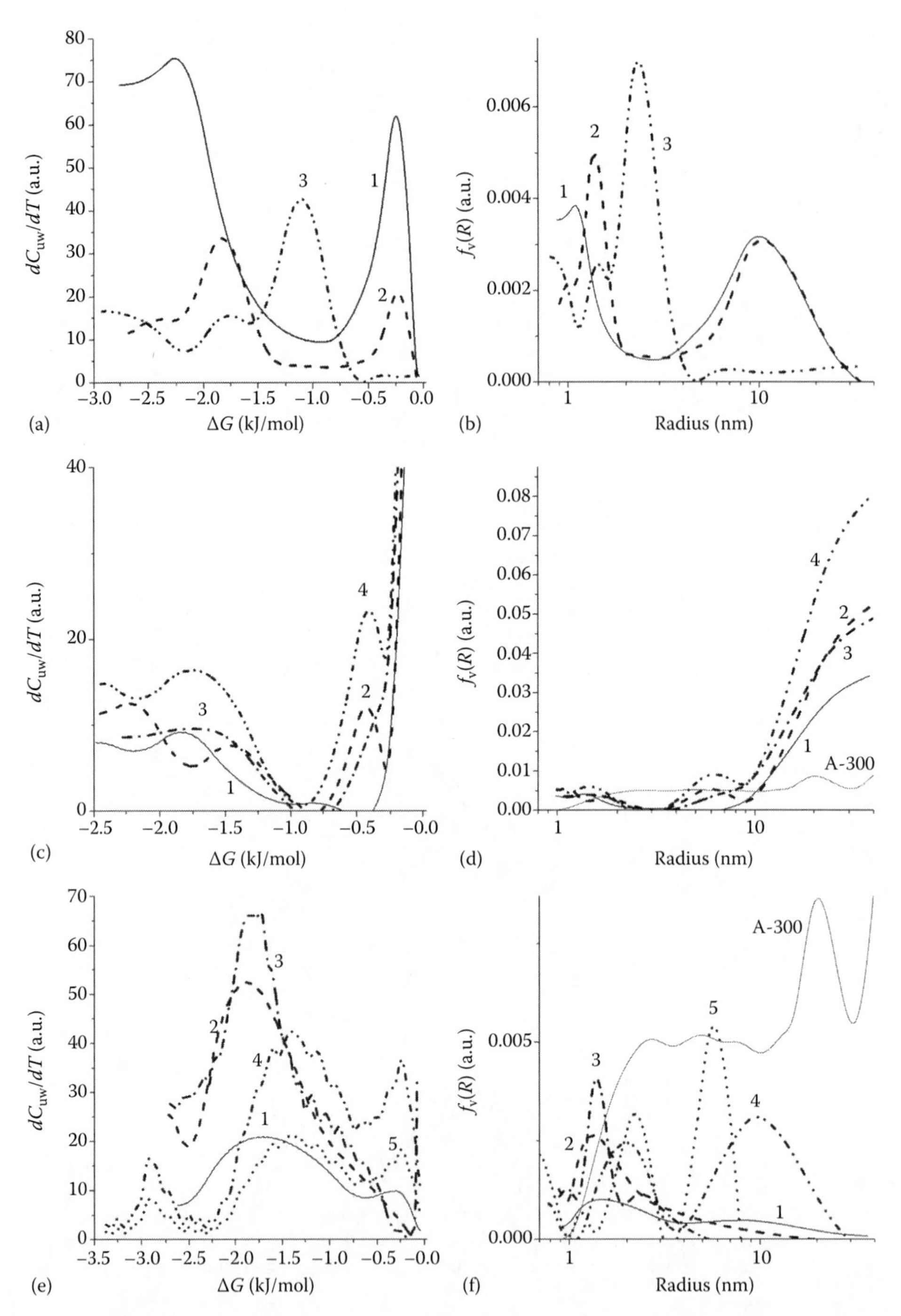

FIGURE 1.174 Changes in (a, c, e) the amounts of unfrozen water dC_{uw}/dT as a function of changes in the Gibbs free energy, and (b, d, f) size of unfrozen water structures in starch at h = (a, b) 0.3 g/g, (c, d) 1 (curve 1), 1.5 (2, 3) and 2.3 g/g (4), (e, f) 0.1 (1), 0.2 (2–5) g/g, on addition of $CDCl_3$ at C_{CDCl_3} = (a, b) 2 g/g (2, 3), (e, f) 2 g/g (3–5), and C_{DMSO} = (a, b) 1 g/g (3), (e, f) 0.5 (4) and 1 (5) g/g, C_{Qc} = 4.8 wt% (c, d, curves 2–4) and (e, f, 1–5), $C_{A\text{-}300}$ = 4.8 wt% (c, d, curve 4) and (e, f curves 1–5), PSD for A-300 calculated on the basis of the nitrogen adsorption is shown (d, f). (Adapted from *Appl. Surf. Sci.*, 256, Gun'ko, V.M., Turov, V.V., Barvinchenko, V.N. et al., Nonuniformity of starch/nanosilica composites and interfacial behaviour of water and organic compounds, 5275–5280, 2010c. Copyright 2010, with permission from Elsevier.)

to starch/water or starch/water/$CDCl_3$ (Figure 1.174a) because there are the differences in the freezing point depression (i.e., the corresponding size [*R*] of pores where water is unfrozen at this temperature) and the ΔG values for water interacting with organics and silica nanoparticles (Gun'ko et al. 2009d). Addition of DMSO to starch/water/$CDCl_3$ causes significant changes in the dC_{uw}/dT (Figure 1.174a, curve 3) and $f_V(R)$ (Figure 1.174b, curve 3) functions. Addition of Qc and A-300 (Figure 1.174) to hydrated starch leads to more complicated shapes of both dependences because of different tendencies in the DMSO, $CDCl_3$, Qc, starch, and A-300 effects on the interfacial water structure. For instance, the bound water displacement to both broader and narrower pores (in starch and/or silica) occurs due to its interactions with organic solvents.

Thus, the interfacial behavior, energetic characteristics, and structure of both SBW and WBW depend strongly on the amounts of water and the type and amounts of coadsorbates (organic solvents) in hydrated starch with addition of quercetin, nanosilica, and weakly polar ($CDCl_3$) and polar (DMSO) organic solvents. The addition of these solvents to hydrated quercetin/starch/nanosilica causes the appearance of three additional water signals at $\delta_H = 1.3$ (WAW), 3 (ASW), and 4 (SAW) ppm differently dependent on temperature. Thus, the composite systems with hydrated starch/quercetin/nanosilica are characterized by structural and energetic nonuniformities, which can be varied due to changes in the amounts of water and treatment temperature.

PVP (used in different medical preparations) in contrast to starch or PVA, does not have hydroxyl groups that can strongly change its behavior at the interfaces with nanosilica. Interaction of PVP with nanosilica is of interest because of possible medicinal applications of their appropriate mixtures (Chuiko 2003, Blitz and Gun'ko 2006). The PVP/silica blends were studied in the form of dry and wet powders and aqueous suspensions by means of the ^{1}H NMR, adsorption, IR spectroscopy, TSDC, electrophoresis, PCS, DSC, rheometry, and other methods (Gun'ko et al. 2001f, 2004a, 2006g,h, 2007i, Bershtein et al. 2009, 2010a–c). Here, we show some results obtained using LT ^{1}H NMR spectroscopy and TSDC methods.

Two typical regions in the graph of ΔG versus the amounts of unfrozen interfacial water (C_{uw}) can be found for pure silica, PVP/silica, and ball-milled PVP/silica suspensions (Figure 1.175). The first corresponds to a significant decline in the amount of unfrozen water in a narrow range of ΔG

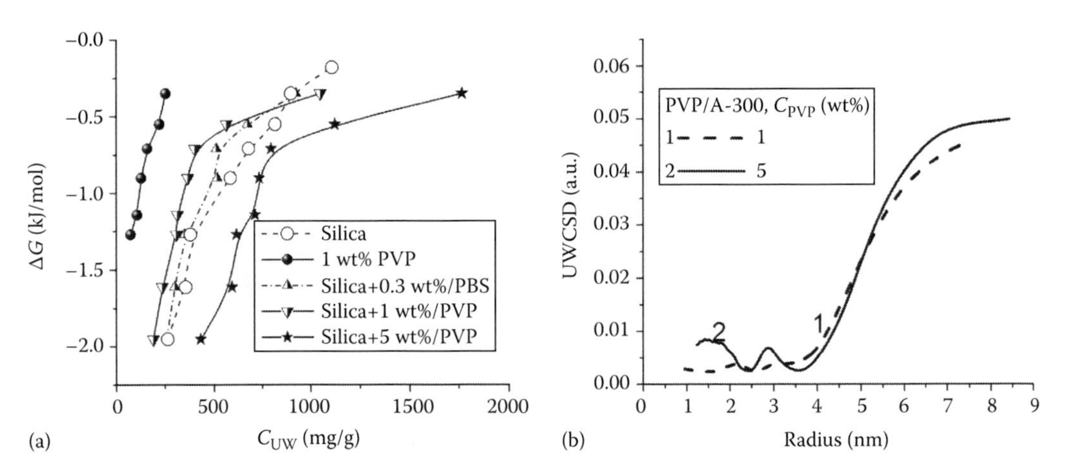

FIGURE 1.175 Gibbs free energy changes of the interfacial water as a function of the amounts of unfrozen water (mg) per g of oxide for month-stored ball-milled suspensions of pure silica ($C_{SiO_2} = 6$ wt%), pure solution of PVP ($C_{PVP} = 1$ wt%), ball-milled suspension of PVP/silica (storage period of 1 month) at $C_{SiO_2} = 6$ wt% and $C_{PVP} = 0.3$ wt%, and silica suspension at $C_{SiO_2} = 6$ wt%, $C_{PVP} = 1$ or 5 wt% of pure PVP, and (b) UWCSD for samples at 1 and 5 wt% PVP (^{1}H NMR study). (Adapted from *Colloids Surf. A: Physicochem. Eng. Aspects*, 233, Gun'ko, V.M., Voronin, E.F., Zarko, V.I. et al., Interaction of poly(vinyl pyrrolidone) with fumed silica in dry and wet powders and aqueous suspensions, 63–78, 2004e. Copyright 2004, with permission from Elsevier.)

at T close to 273 K. For another region, the amount of unfrozen water goes down slightly but ΔG changes stronger. The interfacial water responsible for the first region corresponds to a thick layer of water weakly bound to the surface and perturbed by long-range intermolecular forces (mainly electrostatic). This water can be also assigned to SAW.

The water strongly bound to the surface responsible for the second region in the ΔG graph is linked to a thin layer adjacent to the silica surface. Its characteristics are affected by short-range forces (polar, Lifshitz–van der Waals, and electrostatic) between the surface and adsorbate molecules. The properties of the SBW and its interaction with the surface determine the immersion enthalpy and related characteristics of oxides. The amounts of each type of water and maximal values of the corresponding changes in the Gibbs free energy of the interfacial water layers (ΔG_s and ΔG_w) can be estimated by extrapolation of the corresponding dependencies to the x- and y-axes on the $\Delta G(C_{uw})$ graphs. The γ_S (overall reduction of the Gibbs free surface energy), ΔG_s, ΔG_w, C_{uw}^s, and C_{uw}^w values are given in Table 1.28 for the ball-milled suspensions of pure silica and PVP/silica.

The interaction between the PVP molecules and the silica surface reduces ΔG and C_{uw}^s but C_{uw}^w increases. This result is in agreement with reduction of the zeta potential of particles in the PVP/silica suspension (Gun'ko et al. 2004a). An enhancement in C_{uw}^w may be due to nonuniform electrostatic field in a relatively thick interfacial layer containing tails and segments of adsorbed polymer molecules. The amount of WBW in this layer is larger than that around pure silica particles (Table 1.28), as well as the thickness of the interfacial layer perturbed by PVP+dissolved ions and silica particles in comparison with that for the initial silica suspension. The polymer molecules set a significant portion (all PVP molecules are bound to the silica surface) in this layer. However, the sum $C_{uw}^s + C_{uw}^w$ is the same. It should be noted that long-period storage of the ball-milled suspension of silica has a weak effect on ΔG, and the difference between the corresponding γ_S values is only 12 mJ/m^2 that is in agreement with a high adsorptive ability of nanosilica in the aged ball-milled suspensions tested by the adsorption of proteins. A total of 0.3 wt% of PVP impacts the strongly bound interfacial water similar to that for 1 wt% of pure PVP (Table 1.28, C_{uw}^s and ΔG_s). WBW is perturbed by 0.3 wt% of PVP in the physiological buffer solution weaker than by 1 wt% of pure PVP; however, changes in the Gibbs free energy of this layer (ΔG_w) are similar. An increase in C_{PVP} to 5 wt% leads to growth of the thickness of the unfrozen water layer (Figure 1.175), $C_{uw}^s + C_{uw}^w$ and γ_S (Table 1.28). Notice that the value of the sum $\Delta G_s + \Delta G_w$ (Table 1.28) is in agreement with the results based on the PVP adsorption data (Gun'ko et al. 2004a). These changes in the interfacial water (i.e., EDL) suggest an increase in binding of the polymer molecules by the silica particles in the PVP/silica system that results in the enhancement of their aggregation (Gun'ko et al. 2004a, 2009d).

TABLE 1.28
Parameters of Interfacial Water for Pure Silica and Silica/PVP Suspensions ($C_{SiO2} = 6$ wt%)

System	$-\Delta G_s$ (kJ/mol)	$-\Delta Gw$ (kJ/mol)	C_{uw}^s (mg/g)	C_{uw}^w (mg/g)	γ_S (mJ/m^2)
SiO_2(I)	3.2	1.3	700	700	253
SiO_2(I)[a]+0.3 wt% PVP	2.5	1.0	500	900	186
SiO_2(II)	3.0	1.4	730	680	279
1 wt% PVP	1.6		300		
SiO_2(II)+1 wt% PVP	3.0	0.9	520	1600	220
SiO_2(II)+5 wt% PVP	3.0	1.0	1100	1400	403

Note: $S_{BET} \approx 300$ and 190 m^2 g^{-1} for SiO_2(I) and SiO_2(II), respectively.

[a] In the physiological buffer.

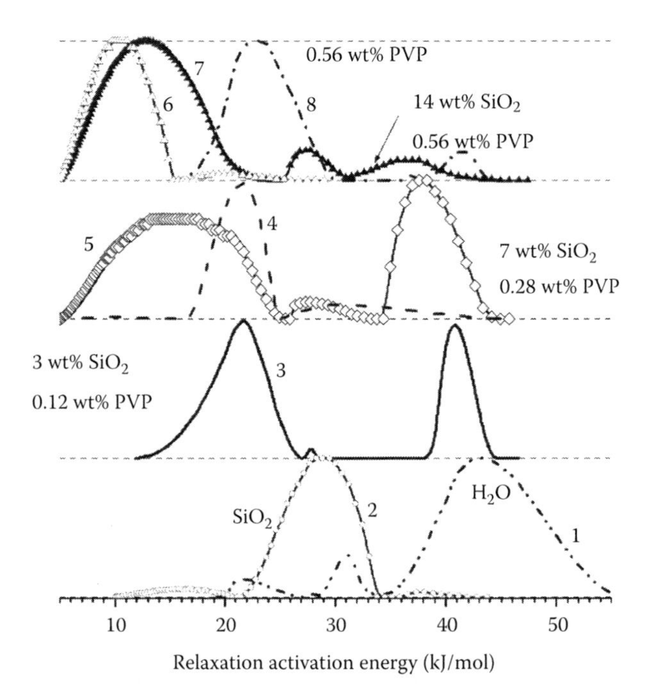

FIGURE 1.176 Distributions of the activation energy of TSDC for pure water (curve 1), water (7 wt%) adsorbed on silica in the gas phase (2), silica suspension at C_{SiO_2}=(4) 7 and (6) 14 wt%, PVP/silica at (3) C_{SiO_2}=3 wt% and C_{PVP}=0.12 wt%, (5) C_{SiO_2}=7 wt% and C_{PVP}=0.28 wt%, (7) C_{SiO_2}= 14 wt% and C_{PVP}=0.56 wt%, and pure PVP solution at C_{PVP}=0.56 wt% (8). (Adapted from *Colloids Surf. A: Physicochem. Eng. Aspects*, 233, Gun'ko, V.M., Voronin, E.F., Zarko, V.I. et al., Interaction of poly(vinyl pyrrolidone) with fumed silica in dry and wet powders and aqueous suspensions, 63–78, 2004e. Copyright 2004, with permission from Elsevier.)

Calculations of the structural characteristics of two last samples in Table 1.28 (Figure 1.175b) show that the PVP layer is not dense in the aqueous media because the specific surface area corresponds to 430 m²/g at C_{PVP}=5 wt%. The UWCSD for this sample shows two peaks of narrow mesopores.

The TSDC study of the PVP/silica suspensions shows changes in the state of water and polymer molecules with increasing C_{PVP} and C_{SiO_2} (Figure 1.176). Silica particles in the concentrated aqueous suspension (C_{SiO_2}=7 wt% [curve 4] or 14 wt% [curve 6]) disrupt a significant portion of the hydrogen bond network in water, as the bulk free water ($f(E_a)$ peak at 44 kJ/mol in curve 1) is practically absent at such a concentration of silica (Gun'ko et al. 2007i). In the case of PVP/silica suspension at the same C_{SiO_2} (curves 5 and 7) or lower C_{SiO_2} (curve 3), the amounts of bulk water are larger than that in the pure silica suspensions. However, the displacement of this peak toward a lower energy with C_{SiO_2} is observed (i.e., the average number of the hydrogen bonds per molecule tends to diminish). Consequently, the shielding of the silica surface by polymer molecules results in appearance of nearly free bulk water disappearing in the PVP/silica suspensions at a large C_{SiO_2}= 14 wt% (curve 7) or in the pure silica suspension (curves 4 and 6) or the pure PVP solution at C_{PVP}=0.56 wt% (curve 8). Additionally, an increase in the concentration of silica in the PVP/silica dispersion results in appearance of water molecules with a low number of the hydrogen bonds per molecule corresponding to the $f(E_a)$ peak at low E_a= 10–15 kJ/mol. This peak shifts toward lower E_a values with increasing $C_{SiO_2}+C_{PVP}$.

At a low concentration of PVP, the polymer molecules adsorb strongly (great changes in the Gibbs free energy) and the number of free pyrrolidone groups (unbound by the silica surface) can be relatively low. Consequently, the adsorbed PVP layer is relatively dense and the interaction between water molecules and oxide particles or PVP molecules disturbs a minor portion of the water. At greater C_{PVP} values, a significant portion of PVP molecules has free tails (at $\gamma \approx 0.2$, the relative number of free C=O groups in PVP adsorbed on silica from solution and then dried is approximately 2/3, according to the IR spectra), which do not interact with the silica surface but effectively interact with water molecules. Therefore a large portion of the water is disturbed. The $\Delta G(C_{uw})$ graph shifts toward larger C_{uw} values (Figure 1.175) and $f(E_a)$ shifts toward lower E_a values (Figure 1.176) due to a reduction of the number of the hydrogen bonds per molecule. These features of the adsorbed PVP layer and the reduction of the number of the PVP molecules interacting simultaneously with several oxide particles cause the diminution of the viscosity with $C_{PVP} > 0.3$ wt% (Gun'ko et al. 2001f, 2004a, 2009d).

The presence of weak acidic sites SiOH leads to appearance of the negative surface charge at pH>7 due to transferring of protons from silanols into the solution or because of the adsorption of OH^-. The surface charge density of nanosilica is smaller than that of other nanooxides. Silica particles can relatively strongly interact with positive charged structures, for example, protonated amino groups of proteins. However, these interactions are weaker than those of mixed oxides (vide infra), which have higher surface charge density. Silanols interact as both proton donor and electron donor (proton acceptor) with protein molecules, water, and other polar substances. As a result, nanosilica sorbs proteins strongly (because of multicentered adsorption complexes) and in large amounts, 300–800 mg per gram of silica (Chuiko 2003).

Comparison of equilibrium adsorption of proteins onto silica A-300 (Figures 1.177 and 1.178) and minute protein adsorption/flocculation with nanosilica as a function of protein concentration C_p demonstrates strong but different effects of pH and salinity (Gun'ko et al. 2003a,c). The equilibrium adsorption of proteins is up to 1 mg/m^2 (or ≈300 mg/g) at pH 3.5 (i.e., between pH [IEP] of silica and proteins) for bovine serum albumin (BSA) with 0.9 wt% NaCl and gelatin without NaCl or at pH (IEP) of protein for ovalbumin without NaCl. The lowest equilibrium adsorption 0.1–0.2 mg/m^2 is typically observed at pH=2, which is close to pH (IEP_{SiO_2})≈2.2, and without NaCl (Figure 1.177). It should be noted that the maximal adsorption values for BSA preparations with marked amounts of lipids and fatty acids can be up to 600 mg/g (at C_{SiO_2}=3–5 wt%) and significantly lower (280–300 mg/g) for pure BSA that can be caused by the effect of hydrophobic admixtures on the structure of adsorbed complexes, as polar and charged proteins should shield hydrophobic compounds from water.

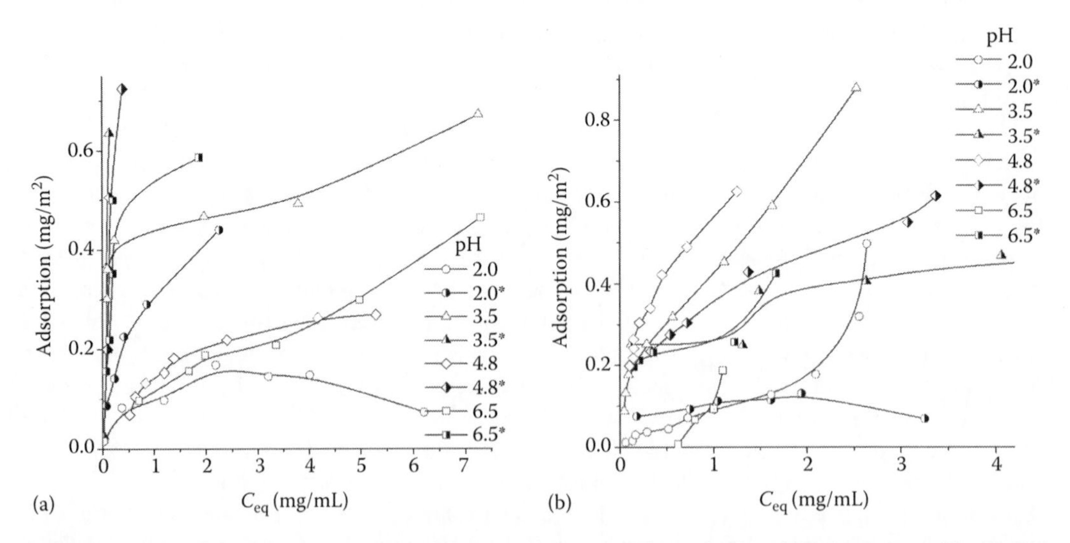

FIGURE 1.177 Adsorption isotherms of different proteins: (a) BSA and (b) ovalbumin on nanosilica A-300 (3.6 wt%) at different pH values, which with asterisk show the systems with addition of 0.9 wt% NaCl. (Adapted from *J. Colloid Interface Sci.*, 260, Gun'ko, V.M., Mikhailova, I.V., Zarko, V.I. et al., Study of interaction of proteins with fumed silica in aqueous suspensions by adsorption and photon correlation spectroscopy methods, 56–69, 2003c. Copyright 2003, with permission from Elsevier.)

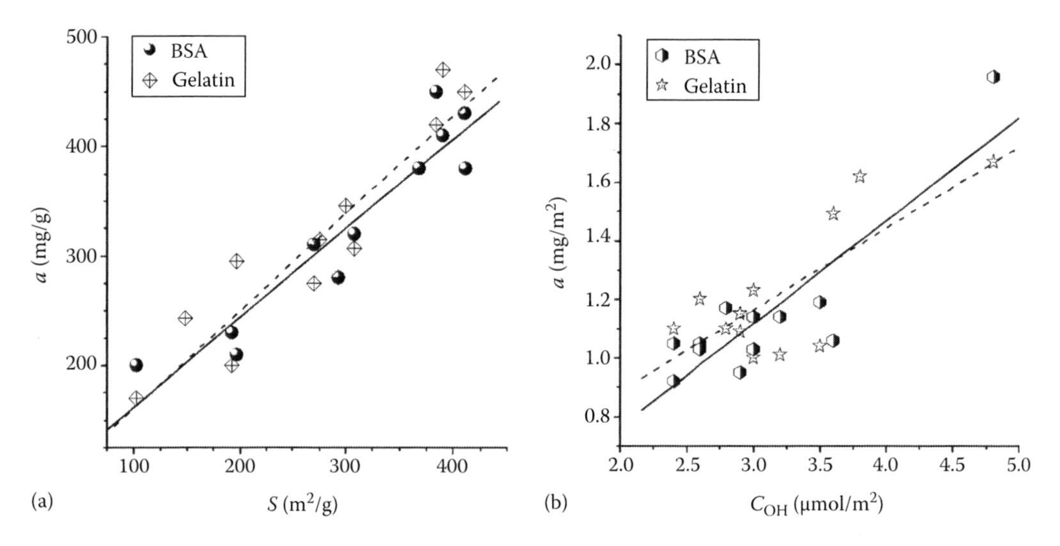

FIGURE 1.178 Protein adsorption in (a) mg/g and (b) mg/m^2 as a function of (a) S_{BET} and (b) content of silanols. (Adapted from *Colloids Surf. A: Physicochem. Eng. Aspects*, 180, Mironyuk, I.F., Gun'ko, V.M., Turov, V.V., Zarko, V.I., Leboda, R., and Skubiszewska-Zięba, J. et al., Characterization of fumed silicas and their interaction with water and dissolved proteins, 87–101, 2001. Copyright 2001, with permission from Elsevier.)

In the case of diluted suspensions (C_{SiO_2}=0.1–0.2 wt%), the adsorption increases up to 1.5–4.5 g/g due to intensive flocculation of proteins with silica aggregates by formation of protein bridges between them. The adsorption of proteins increases with increasing S_{BET} and C_{OH} values (Figure 1.178).

To elucidate some of these effects, the PCS measurements of the protein/A-300 systems were performed. The $D_{ef}(pH)$ graphs have a maximum close to pH (IEP) of proteins (Figure 1.179).

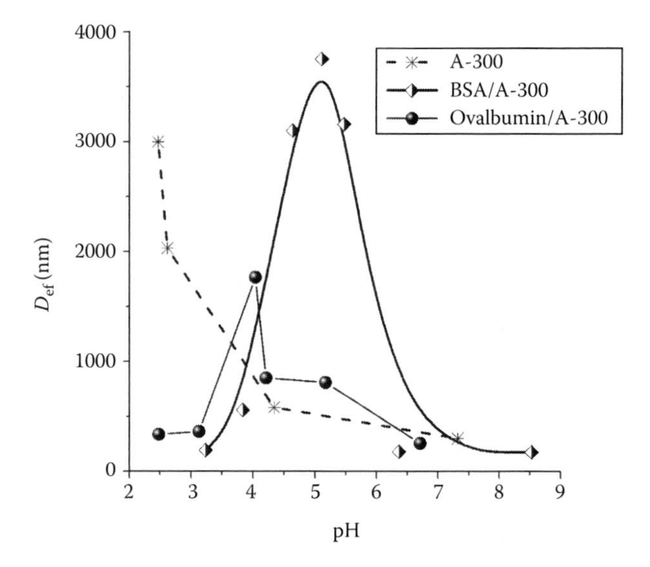

FIGURE 1.179 Effective diameter of particles as a function of pH in aqueous suspensions of nanosilica A-300 and protein/A-300. (Adapted from *J. Colloid Interface Sci.*, 260, Gun'ko, V.M., Mikhailova, I.V., Zarko, V.I. et al., Study of interaction of proteins with fumed silica in aqueous suspensions by adsorption and photon correlation spectroscopy methods, 56–69, 2003c. Copyright 2003, with permission from Elsevier.)

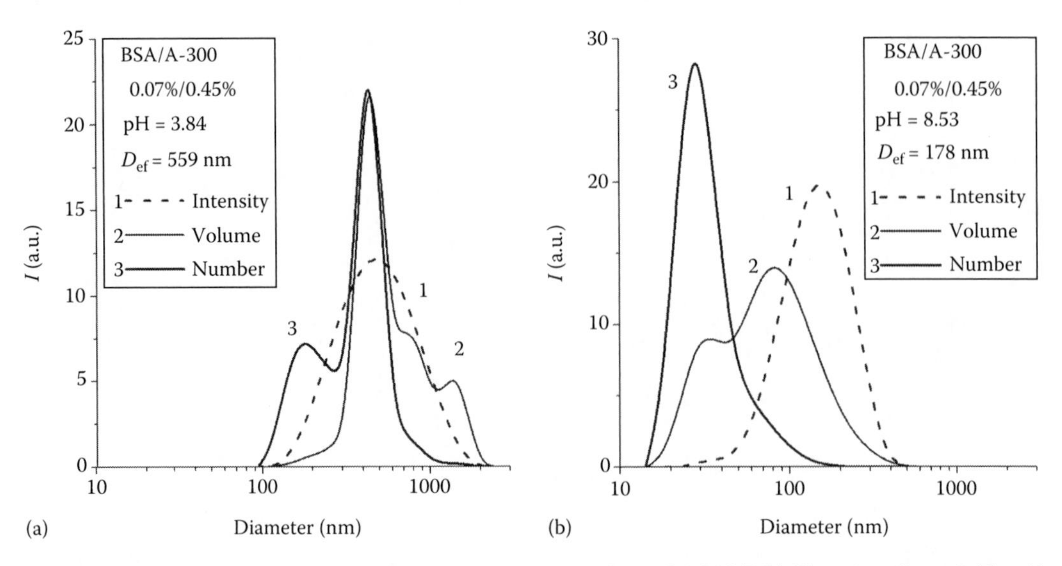

FIGURE 1.180 Particle size distributions in aqueous suspensions of A-300/BSA(Sigma) at C_{SiO_2}=0.45 wt% and C_{BSA}=0.07 wt% and pH: (a) 3.84 and (b) 8.53 with respect to the light-scattering intensity, particle volume and number. (Adapted from *J. Colloid Interface Sci.*, 260, Gun'ko, V.M., Mikhailova, I.V., Zarko, V.I. et al., Study of interaction of proteins with fumed silica in aqueous suspensions by adsorption and photon correlation spectroscopy methods, 56–69, 2003c. Copyright 2003, with permission from Elsevier.)

However, at pH far from pH (IEP) of proteins the PSDs of protein/A-300 are akin to those for pure silica suspension but D_{ef} is smaller than that for pure silica due to decomposition of silica agglomerates and aggregates under action of protein molecules. This effect is independent of the protein type, as interaction of protein molecules with silica particles can be stronger than that between silica particles especially in loose agglomerates. The PSDs of A-300 in the aqueous suspensions at C_{SiO_2}=0.1–3.0 wt% (without NaCl) and pH far from IEP_{SiO_2} are typically bimodal and correspond to small aggregates of 20–60 nm (including from several to dozens of primary particles with the size of 5–12 nm) and larger ones with the size up to 500 nm (up to dozens of thousands of primary particles; Figure 1.180).

Agglomerates (which are typical for untreated or weakly treated suspensions) with the size d_{PCS}>1 μm are not observed in the strongly sonicated suspensions at pH>5. Addition of 0.15 M NaCl to 3% suspension results in disappearance of the first PSD peak at d_{PCS}= 10–30 nm observed without NaCl. As a whole, the bimodal PSD slightly shifts toward larger d_{PCS} values. Interaction of BSA with nanosilica after preadsorption of polymers (PEG, MW ≈ 2 kDa or PVA, MW = 43 ± 7 kDa) depends on time (Figure 1.181) because of the rearrangement of secondary particles including silica and organic molecules. This effect can play an important role on the use of nanosilica as a component of complex medicinal preparations including polymers. The second and perhaps most important and unique factor, which causes a high efficiency of nanosilica as a medicinal preparation, is the nano-dimensionality of primary particles, which can form stable concentrated suspensions with major contribution of individual primary particles (Figures 1.89, 1.139, and 1.181). This nano-dimensionality provides effective local interaction of silica particles with small and positively charged structures of cell membranes, which have total negative charge.

These local structures can be fragments of integrated transport proteins, products of the metabolism, or cell decay and others. If the size of negatively charged colloidal particles considerably exceeds the size of positively charged structures (fragments) of the membranes, then negatively charged cells can repulse these particles due to the long-range electrostatic forces. Therefore, straight contact between large sorbent particles and cell membranes can be absent. This can considerably decrease the effectiveness of sorbents consisting with particles of micron or millimeter

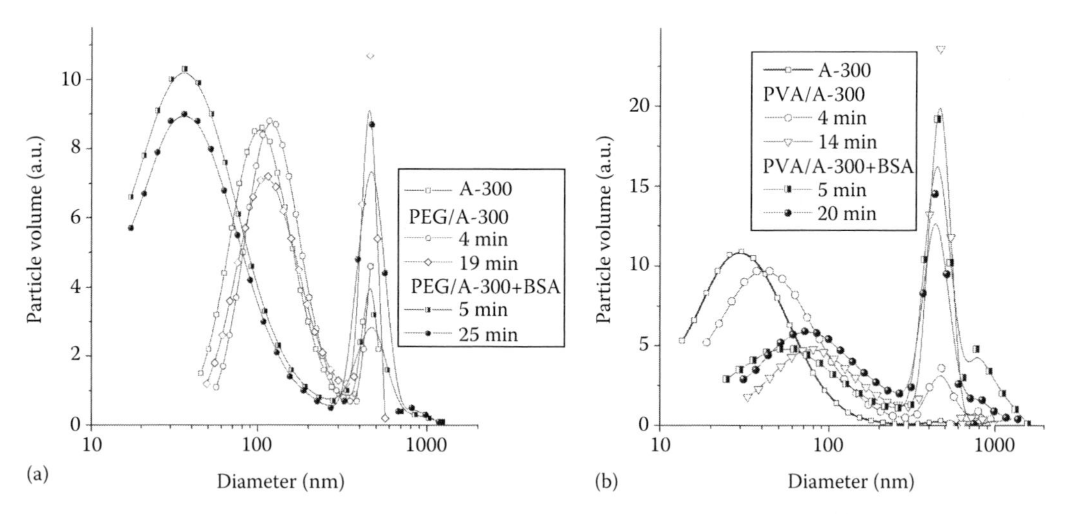

FIGURE 1.181 Particle size distributions with respect to particle volume in the aqueous suspension of A-300, (a) PEG/A-300, BSA(II)/PEG(I)/A-300 and (b) PVA/A-300, BSA(II)/PVA(I)/A-300, I and II is the first and second adsorbates respectively. (Adapted from *J. Colloid Interface Sci.*, 300, Gun'ko, V.M., Zarko, V.I., Voronin, E.F. et al., Successive interaction of pairs of soluble organics with nanosilica in aqueous media, 20–32, 2006i. Copyright 2006, with permission from Elsevier.)

sizes. Furthermore, low-mobility large particles of a roughly disperse sorbent are less effective as a carrier of drugs in contrast to silica nanoparticles, which can be excellent carriers for medicinal substances immobilized on their surface because they are very mobile and can provide close contact with the cell membranes. Nanoparticles and their aggregates adsorbing macromolecules near the cell membranes can change the state of interfacial water, the diffusion of molecules both to the membrane and from it. Nanosilica particles can adsorb substances (e.g., toxins), which should be removed from the organism and they can simultaneously fulfill a carrier function for drugs transported to the cell membranes. Desorption of medicines occurs more easily from nonporous primary particles than from porous ones. It can be also governed by changes in the morphology of silica particles and the modification of their surface. Furthermore, because of the nanodimensionality, nanosilica particles diffuse rapidly, and their mobility ensures the fast absorption of endo- and exotoxins (Chuiko 2003). This is important on poisoning or elapsing diseases, when the life depends on the detoxication speed. In this situation, no other sorbent can act so rapidly as nanosilica. Practically all other sorbents consist of particles of considerably larger sizes than nanosilica and possessing much smaller mobility that cannot provide very rapid treatment. Time of the adsorption of macromolecules (toxins) in pores of porous sorbents is long. This reduces a quantity of sorbed substance in comparison with nonporous nanosilica, which sorbs, for example, proteins much faster and effectively (~50% of proteins for 1 min and ~90% for 10 min) than porous adsorbents. Biomacromolecules are rapidly sorbed only on the external surface of porous adsorbents. However, they practically do not penetrate into nanopores ($d<2$ nm) and the adsorption in mesopores ($d=2$–50 nm) and macropores ($d>50$ nm) requires certain time. This is a very crucial point, since a substantial portion of harmful substances, which must be moved away from the organism, wounds, and damaged organs, are of protein origin and their molecules are of high MW. This again emphasizes the uniqueness of nanosilica. Notice that interaction of polymers (PVP, PEG, and others) or proteins with nanosilica leads to partial destruction of secondary structures (sometimes up to primary particles), but new secondary structures (50–1000 nm) form with oxide particles and macromolecules. This causes certain diminution of the diffusion of modified nanosilica particles; however, this can provide a higher biocompatibility because of the formation of an outer adsorption layer with biomacromolecules (e.g., hemolysis of red blood cell [RBC] strongly decreases, Blitz and Gun'ko 2006, Gun'ko et al. 2009d). Therefore, it is possible to expect that the rapid adsorption of proteins can reduce the negative influence of nanosilica on the membranes of erythrocytes and other cells. For example, strong hemolysis is observed in the case of nanosilica interaction with erythrocytes in the absence of plasma proteins. However, PVA immobilized on A-300 causes reduction of the hemolysis (Chuiko 2003, Gun'ko et al. 2009d).

The aggregate structure can impact the adsorption efficiency of nanosilica in respect to proteins since the molecular size of BSA and gelatin is close to the size of primary particles of nanosilica, and a major portion of voids in aggregates has the corresponding sizes. Therefore, the protein adsorption can occur on the major portion of the surface of primary nanoparticles. Adsorption of proteins of the different nature on nanosilica in the aqueous suspensions can depend not only on the particle morphology but also on a pH value of the suspension, as a maximal adsorption is typically observed at pH close to the PZC of proteins when electrostatic repulsive interactions are weak or absent and the volume of compacted protein molecules is minimal. For simplicity of comparison of different nanosilica samples, we did not adjust pH values of the suspensions during protein adsorption (Mironyuk et al. 2001), which can, however, result in only slight changes in pH of the suspensions, as the concentration of silica was only 0.8 wt% and the nature of the oxide surface was the same and an increase in C_{OH} (influencing pH) was factually compensated due to diminution of the specific surface area.

An increase in the S value (i.e., decrease in the primary particle size) results in an enhancement of adsorption (in mg/g) of both BSA and gelatin (Table 1.29; Mironyuk et al. 2001).

However, adsorption in mg/m^2 goes down with growing S due to reduction of C_{OH} (as silanols represent the main adsorption sites for proteins) and a possible increase in the stability of aggregates (interacting with proteins) with decreasing primary particle size. The relationship between protein adsorption and the bulk density (ρ_b) of nanosilica powders (Table 1.29) is relatively complicated, as the aggregate and agglomerate structures (determining ρ_b) in the gas and aqueous media can

TABLE 1.29

Characteristics of Nanosilicas and Plateau Adsorption of BSA and Gelatin

No	S^a m^2/g	d^b nm	ρ_b g l^{-1}	$C_{w,105}$^{c} wt%	$C_{w,900}$^{d} wt%	C_{OH} μmol/m^2	BSA mg g^{-1}	BSA mg m^{-2}	Gelatin mg g^{-1}	Gelatin mg m^{-2}
1	102	26.7	75	0.6	0.6	4.8	200	1.96	170	1.67
2	148	18.4	60	0.6	0.5	3.8			243	1.62
3	197	13.8	54	0.7	0.5	3.6	210	1.06	295	1.49
4	192	14.2	38	3.0	3.8	3.5	230	1.19	200	1.04
5	270	10.1	25	2.8	3.1	3.2	310	1.14	275	1.01
6	293	9.3	28	1.5	1.7	2.9	280	0.95	320	1.09
7	300	9.1	24	0.7	0.6	2.9			346	1.15
8	275	9.9	24	0.9	1.0	3.0			315	1.14
9	308	8.9	24	1.6	1.5	3.0	320	1.03	307	1.0
10	308	8.9	23	1.0	0.9	3.0	350	1.14	380	1.23
11	368	7.4	47	1.1	0.8	2.6	380	1.03	250	0.68
12	384	7.1	24	2.2	3.8	2.8	450	1.17	420	1.09
13	390	7.0	24	0.7	0.6	2.6	410	1.05	470	1.21
14	410	6.7	24	1.9	1.8	2.4	430	1.05	450	1.10
15	411	6.6	61	1.4	1.2	2.4	380	0.92	200	0.48

[a] The specific surface area S was evaluated using argon adsorption data.
[b] Average diameter of primary particles.
[c] Amount of water desorbed at $T < 105$°C.
[d] amount of water desorbed at $105 < T < 900$°C.

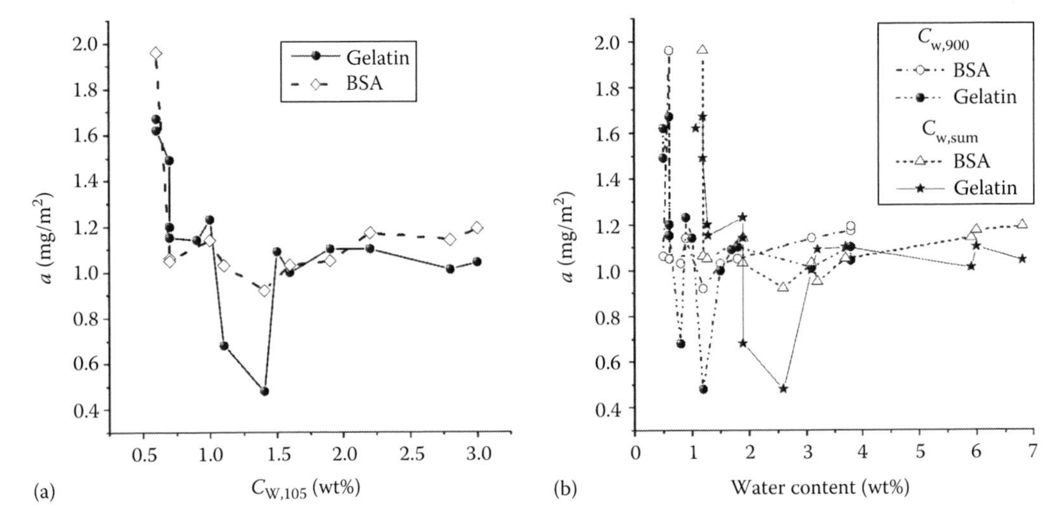

FIGURE 1.182 Relationships between adsorption of proteins (in mg per m^2 of surface area of oxide) and concentration of adsorbed water (a) weakly bound (desorbed at $T < 105°C$) and (b) summary amount desorbed from room temperature up to 900°C. (Adapted from *Colloids Surf. A: Physicochem. Eng. Aspects*, 180, Mironyuk, I.F., Gun'ko, V.M., Turov, V.V., Zarko, V.I., Leboda, R., and Skubiszewska-Zięba, J. et al., Characterization of fumed silicas and their interaction with water and dissolved proteins, 87–101, 2001. Copyright 2001, with permission from Elsevier.)

be different. However, tendency of reduction of adsorption (in mg/g) with growing ρ_b seems quite explicable. For instance, samples 11 and 15 with a large apparent density (Table 1.29) are characterized by a strong decrease in gelatin adsorption, but for BSA, such an effect is smaller maybe due to stronger impact of albumin on the aggregate rearrangement during protein adsorption.

Protein adsorption rises with C_{OH} independently on the protein nature (Table 1.29 and Figure 1.182), as the main adsorption sites for proteins are silanols.

The C_{OH} value can influence the amounts of differently adsorbed water; however, the relationship between them can be complicated due to the impact of other structural factors changed due to variations in synthesis conditions. For example, in the case of samples 7, 8, and 9 possessing close C_{OH} and S values, the amounts of weakly and strongly adsorbed water (Table 1.29) significantly differ and gelatin adsorption correlates with changes in the concentrations of these waters, that is, polarity and polarizability of the silica surface. The relationship between the BSA adsorption in the liquid medium and the amounts of waters differently adsorbed from the gas phase is simpler than that of gelatin (Figure 1.182 and Table 1.29). It should be noted that the existence of correlation between the parameters linked to adsorption of different compounds on nanosilica in different media can be caused by a global impact of the nature of the silica surface on these processes depending on structural and energetic features of differently prepared samples.

For samples with large S values (No 11–15), protein adsorption (in mg/m^2) is smaller than that for samples with lower S values (Table 1.29). Clearly, the size of oxide aggregates can influence the protein adsorption and there is such a tendency: the smaller the D_{ef} value (or the effective diameter D_{min} corresponding to the first peak in PSD), the larger the adsorption in mg per gram of oxide (Figure 1.183a).

However, the scatter for this correlation is large, as the relationships shown in Figure 1.183 can be really different due to the impact of dissolved protein molecules on the structure of oxide particle aggregates and dispersion stability, as attractive polymer–particle interaction can be stronger than that between oxide particles (also electrostatic repulsive forces between charged silica particles can promote decomposition of aggregates and stabilizing dispersion, and these effects depend on primary PSD). The stability of aggregates increases with decreasing size of primary particles; therefore, decomposition

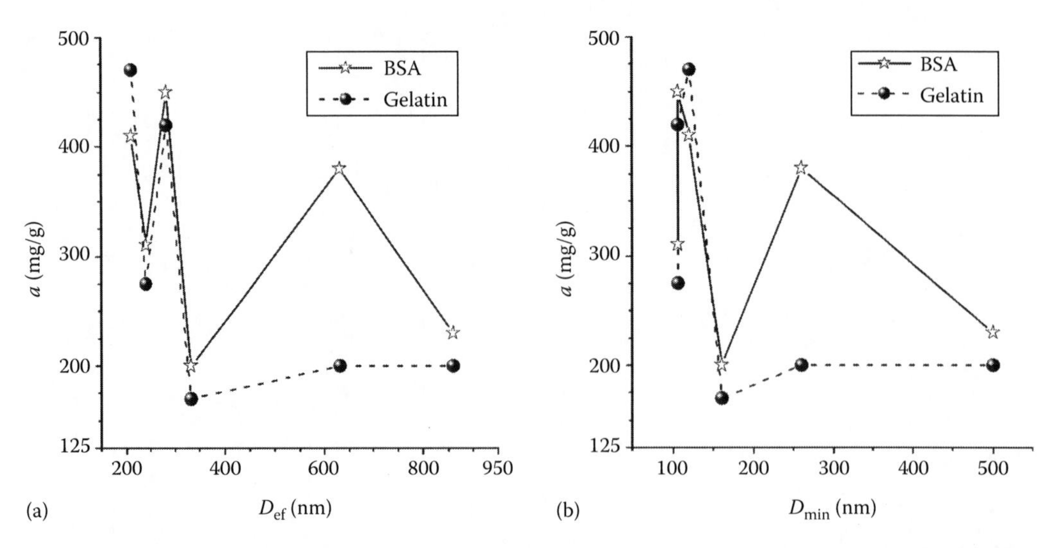

FIGURE 1.183 Relationship between protein adsorption (in (a) mg per gram of oxide powder and (b) mg per m^2 of oxide surface) and effective diameter of nanosilica aggregates in the aqueous suspensions (with no proteins) average D_{ef} and minimal D_{min} observed in PSD.

of agglomerates and aggregates of larger primary particles (i.e., samples with low *S* values) due to interaction with protein molecules can cause an increase in the protein adsorption (in mg/m^2), as well as parallel enhancement of the C_{OH} value (Table 1.29). However, it should be noted that an increase in average *d* is accompanied by broadening of primary PSD, which can markedly influence the rearrangement of aggregates in the suspensions due to interaction with dissolved protein molecules.

Changes in the oxide surface properties accompanied by a reduction of the adsorbed water amounts can cause an increase in gelatin adsorption, but the opposite result is observed for BSA. Consequently, controlled changes in synthetic conditions allow one to prepare nanosilicas possessing close values of the specific surface area and concentration of silanols, but strongly different in their hydrophilic characteristics (amounts of water adsorbed from air or bound to the surfaces in the aqueous suspension) and adsorption ability in respect to proteins.

Samples of suspended–dried polymer/oxides were prepared with the solid residual of the centrifuged dispersions dried at room temperature, then heated at 350 K for 3 h. The perturbation degree was calculated as $\Phi = 1 - \frac{I}{I_0}$, where I_0 and I are intensities of a band of free silanols at 3750 cm^{-1} of initial silica and silica/polymer samples, respectively. Φ was used as a measure of interaction of polymers with the silica surface, because free silanols are the main adsorption sites for polar compounds adsorbed on the silica surface.

The adsorption of several types of polymers onto nanosilica A-300 gives different interfacial layer structures (Gun'ko et al. 2008b). PEG and PEO having the same structure of the segments are characterized by a close $\Phi(C_{pol})$ shape (Figure 1.184a). Despite weak polymer–polymer interactions, polydimethylsiloxane (PDMS) gives the lowest $\Phi(C_{pol})$ values (as a function of C_{pol} in mg/g) because of (i) a large segment (m_{seg}) weight –$(CH_3)_2Si$–O–$Si(CH_3)_2$–, (ii) its weak interaction with silanols due to relatively poor electron-donor properties of oxygen atoms in the siloxane bonds, (iii) steric effects of the CH_3 groups, and (iv) the helical structure of the PDMS chain. PVA molecules can form strong hydrogen bonds with both silanols and OH groups of neighboring molecules, leading to low Φ values as a function of C_{pol} and the lowest Φ values as a function of C_{pol}/m_{seg} (Figure 1.184c). PVP has a larger segment weight than PEG, PEO, and PVA by a factor of 2.4. Therefore, $\Phi(C_{pol})$ for PVP is lower than PEG or PEO, but close to that of PVA (Figure 1.184a). PVP molecules can more effectively interact with surface silanols than other linear polymer molecules because of the rotational

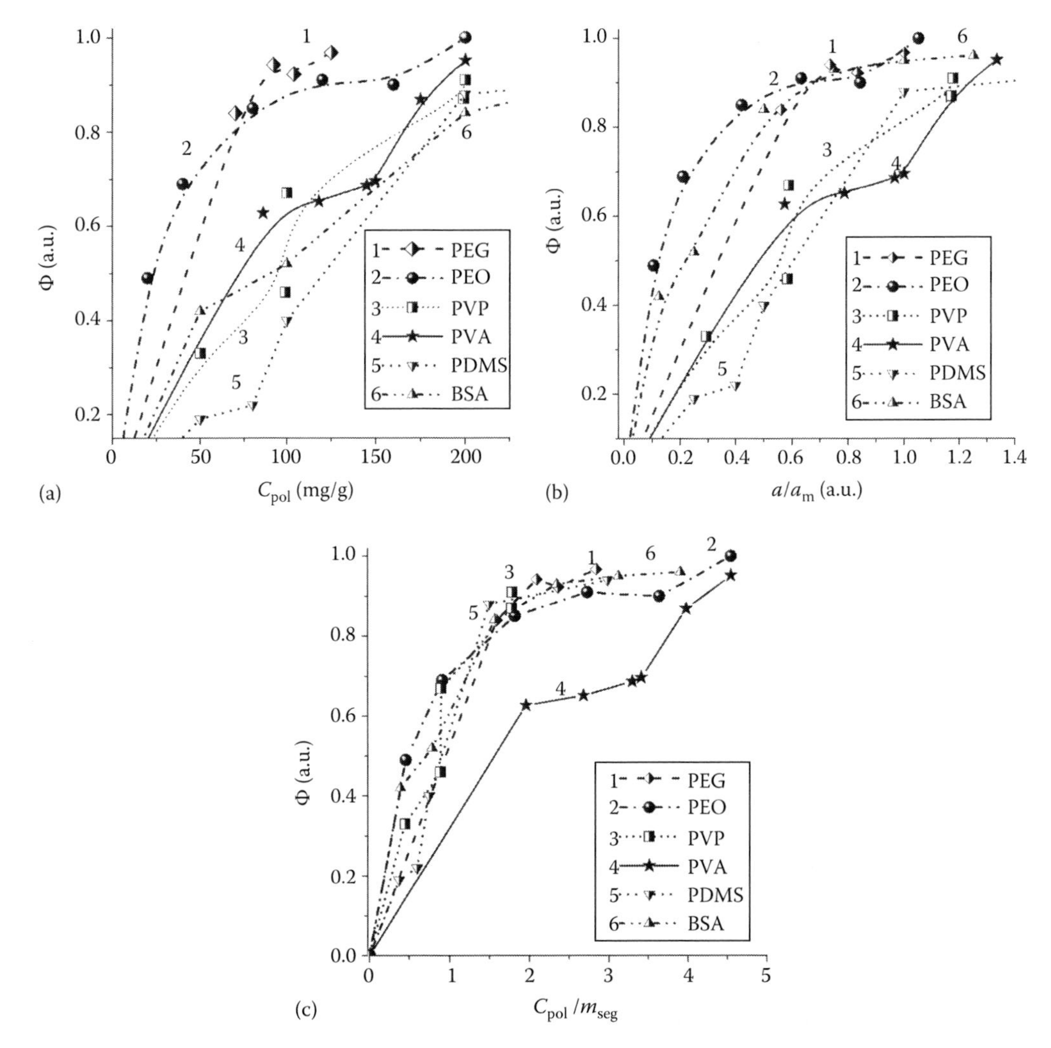

FIGURE 1.184 Perturbation degree (Φ) of free surface silanols as a function of the polymer loading (a) C_{pol} in mg per gram of silica, (b) $\Theta = C_{pol}/C_{pol,monolayer}$ and (c) C_{pol} normalized by dividing by the molecular weight of a segment (m_{seg}) for PEG (35 kDa), POE (600 kDa), PVP (12.7 kDa), PVA (43 kDa), PDMS (8 kDa) and BSA (67 kDa) onto A-300. (Adapted from *Powder Technology*, 187, Gun'ko, V.M., Leboda, R., Skubiszewska-Zięba, J. et al., Influence of different treatments on characteristics of nanooxide powders alone or with adsorbed polar polymers or proteins, 146–158, 2008b. Copyright 2008, with permission from Elsevier.)

mobility of the side groups responsible for the formation of hydrogen bonds (Figure 1.184c). BSA molecules demonstrate relatively low $\Phi(C_{pol})$ values because of the high average m_{seg} value, and the globular protein shape. Therefore, a significant portion of the molecules cannot be in contact with the silica surface. However, protein molecules have several types of polar side groups in addition to the polypeptide chain, which can form strong hydrogen bonds with silanols. That can partially compensate for the effect of the globular shape of the molecules on the $\Phi(C_{pol})$ values. Normalization of C_{pol} by dividing by the segment m_{seg} MW (Figure 1.184c) results in similar $\Phi(C_{pol}/m_{seg})$ graphs for all A-300/polymer systems except A-300/PVA, probably because of the effects of strong hydrogen bonds between PVA molecules.

The globular structure of BSA molecules prevents strong lateral interactions, therefore $\Phi(C_{pol}/m_{seg})$ for BSA is similar to other systems characterized by relatively weak polymer–polymer interactions. The effects of residual interfacial water, which can disturb silanols and form hydrogen bond bridges

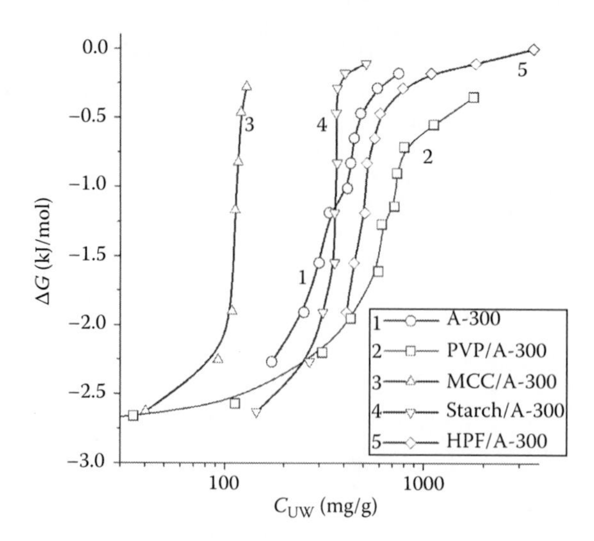

FIGURE 1.185 Comparison of the relationships between amounts of unfrozen water and changes in its Gibbs free energy for individual aqueous suspension of A-300 ($C_{A\text{-}300}$=4.8 wt%, curve 1), PVP(5wt%)/A-300(5wt%)inPBS(curve2),microcrystallinecellulose,MCC(24.6wt%)/A-300(4.4wt%)(curve3), starch(26.5 wt%)/A-300(3.4 wt%) (curve 4), and human plasma fibrinogen, HPF(2.6 wt%)/A-300(5.2 wt%) in PBS (curve 5).

between polar groups of polymer molecules and surface silanols, can provide a small difference in the $\Phi(C_{pol}/m_{seg})$ graphs for different polymers (especially at $C_{pol}/m_{seg}>1$) (Figure 1.184c). Thus a minimal loss of the specific surface area for the nanooxide/polymer powders results from a monolayer coating of nanoparticles by PEO or PEG, because of the formation of strong hydrogen bonds with silanols and weak polymer–polymer interactions, result in a dense coverage with more intensive interaction between polymer molecules and surface silanols. Notice that the shape of the $\Phi(\Theta)$ graphs (Figure 1.184b) slightly differs from that of $\Phi(C_{pol})$ (Figure 1.184a) due to certain differences in the monolayer capacity for the studied polymers: 125 (PEG), 150 (PVA), 170 (PVP), 190 (PEO), 200 (PDMS), and 400 mg/g (BSA).

Comparison of the ^{1}H NMR spectroscopy data for concentrated aqueous suspensions of A-300 alone or with different macromolecules (Figure 1.185) shows that the amount of water bound by microcrystalline cellulose (MCC)/A-300 is much smaller than in the suspension of A-300 alone. In the latter, this amount is close to that bound by the starch/A-300, despite a great amount of starch. Both results can be explained by strong interactions between macromolecules and silica nanoparticles and macromolecule–macromolecule that result in the displacement of water from the interfacial layer. In the systems with PVP/A-300 and HPF/A-300 in PBS, the content of bound water is larger because of less denser interfacial layer than in the case of MCC or starch. These results qualitatively differ from the earlier FTIR results (Figure 1.184) because the latter were obtained for dried systems in contrast to highly hydrated systems studied by the ^{1}H NMR spectroscopy. In other words, the interfacial layer structure depends not only on the type of an adsorbate but also on the content of water. Thus, the properties of the interfacial layer at the nanosilica surface with different adsorbates hydrated, air-dried, or degassed can be very different. Therefore, the results obtained using different methods for similar systems different only in the hydration degree should be compared very carefully.

1.1.11 Influences of Thermal, Hydrothermal, and Mechanical Treatments

The morphology of nanooxides is sensitive to any treatment such as thermal, hydrothermal, and mechanical because of nonrigid binding of nanoparticles in aggregates, aggregates in agglomerates

TABLE 1.30
Characteristics of Two Samples of Nanosilica A-300 Differently Hydrated Due to Synthesis Conditions

Parameter	No 1	No 2
[a] S_{Ar}, m²/g	299	317
[b] ρ_b, g l⁻¹ (fresh)	28	25
[c] $C_{w,105}$, wt%	1.2	0.5
[d] $C_{w,900}$, wt%	2.0	0.6
S_{BET}, m²/g	309	358
S_{nano}, m²/g	62	51
S_{meso}, m²/g	235	282
S_{macro}, m²/g	12	25
V_p, cm³/g	0.659	1.011

[a] The specific surface area S_{Ar} was evaluated using argon adsorption
[b] ρ_b is the bulk density of dry powder.
[c] Amount of water desorbed at $T < 105°C$.
[d] Amount of water desorbed at $105 < T < 900°C$, S_{BET}, and V_p were determined using nitrogen adsorption, ρ_b is the bulk density of dry powder.

and agglomerates in visible powder particles. The analysis of these factors can be started from heating of nanosilicas at different temperatures.

For one of nanosilica A-300 samples shown in Table 1.30, synthesis conditions corresponded to 30% excess of oxygen and hydrogen in comparison with their stoichiometric contents in respect to $SiCl_4$ to provide a high degree of hydration of particles (No 1, Table 1.30).

To produce lowly hydrated sample A-300 (No 2, Table 1.30), the flame temperature was elevated, and the amounts of oxygen and hydrogen were lower than their stoichiometric concentrations. Prepared samples were subjected to heating at different temperatures (200°C–900°C) for $t_T = 10$–160 min and studied right away and after aging during several months by means of the adsorption methods.

It is known that heating of nanosilica up to 1000°C for several days does not practically change the morphology of silica particles, which remain amorphous (Degussa 1997). However, such a rigid treatment can change the specific surface area and the pore volume in aggregates due to dehydration with removal of both surface and bulk hydroxyls (Iler 1979, Legrand 1998, Gun'ko et al. 2009d). Heating of nanosilica samples over a wide temperature range (200°C–900°C) for varied time (t_T) affects their structural properties related to both nanopores and mesopores (Figure 1.186), since changes in the nitrogen adsorption observed over the total p/p_0 range show changes in the PSD of the materials.

The structural changes in heated nanosilica are ambiguous and nonlinear in respect to the S values (Figures 1.186 and 1.187), since heating causes several processes such as desorption of intact water from the surface and the bulk of particles, associative desorption of surface and inner hydroxyls, formation of strained siloxane bonds on the surface and in the bulk of particles, bonding adjacent primary particles due to formed siloxane linkages, rearrangement of the silica lattice, etc. whose impact on the structural parameters of nanosilica depends differently on the heating time and temperature.

For instance, removal of interior intact and structural waters and surface hydroxyls can reduce the size of primary particles enhancing S. However, rearrangement of aggregates of primary particles

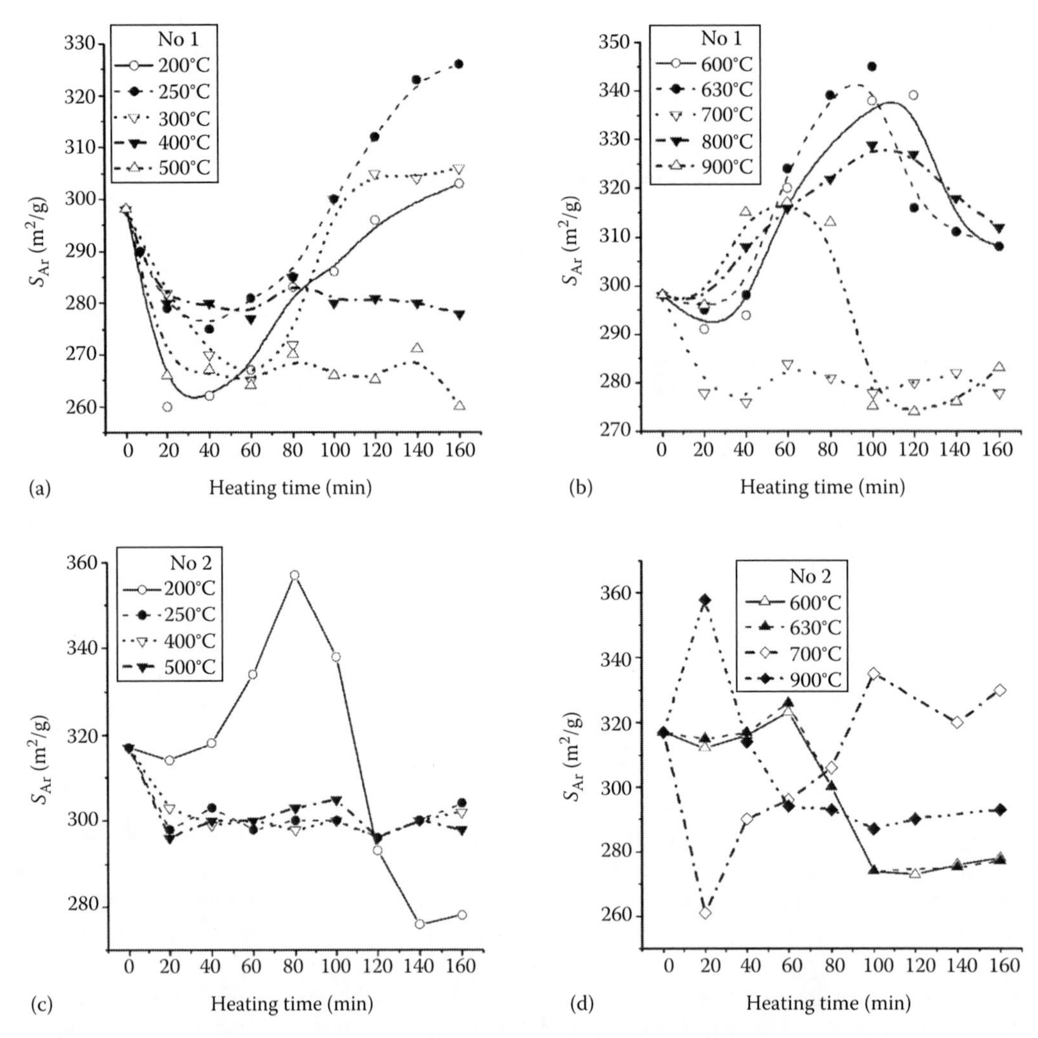

FIGURE 1.186 Specific surface area S_{Ar} as a function of the heating time at different temperatures for fresh nanosilica A-300 samples (a, b) No 1 and (c, d) No 2 calculated using the argon adsorption data. (Adapted from *Colloids Surf. A: Physicochem. Eng. Aspects,* 218, Gun'ko, V.M., Voronin, E.F., Mironyuk, I.F. et al., The effect of heat, adsorption and mechanochemical treatments on stuck structure and adsorption properties of fumed silicas, 125–135, 2003g. Copyright 2003, with permission from Elsevier.)

(e.g., sticking of contacts with changing accessibility of the particle surfaces) and agglomerates, which are accompanied by changes in the structure of their channels (pores), affects nanopore and mesopore volumes and the corresponding specific surface areas otherwise. The shape of the S_{Ar} functions of T and t_T depends on the origin of nanosilica (its hydration degree; Figure 1.186). Analysis in details of the PSD computed using the model of slitshaped and cylindrical pores and voids between spherical particles (SCV) with self-consistent regularization (SCR) method shows a certain dependence on the calcination temperature (Figure 1.187). However, the PSD profiles remain similar with four peaks of mesopores and macropores.

Now, we will analyze the influence of compacting of nanosilica during MCA in a ball mill in the presence of low amounts of water, alcohol, or water/alcohol mixture (Gun'ko et al. 2011e).

One of important textural features of fumed oxides is their low bulk density $\rho_b = 0.02–0.14$ g/cm³. This characteristic depends on the specific surface area (S_{BET}), size distribution, chemical composition and true density (ρ_0) of primary particles, and treatment conditions. Nanosilicas A-300, A-380, A-400,

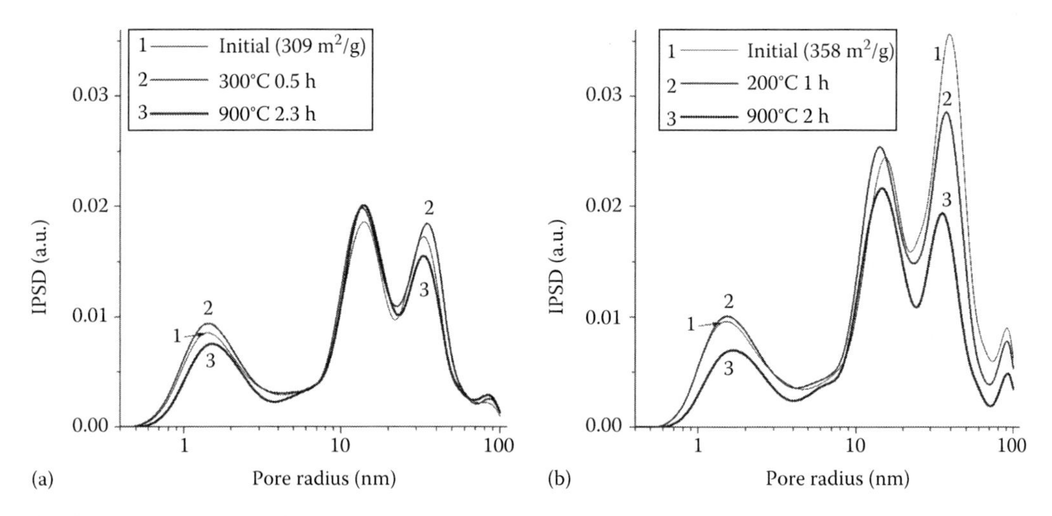

FIGURE 1.187 Pore size distribution for initial and heated A-300 samples No (a) 1 and (b) 2 computed with the SCV/SCR method.

and A-500 composed of primary nanoparticles at $d<10$ nm in average size and $S_{BET} \geq 300$ m^2/g are characterized by $\rho_b = 0.02$–0.06 g/cm^3, which is smaller than that of other nanooxides. For instance, initial silica OX-50 (Degussa, $S_{BET} \approx 50$ m^2/g) has $\rho_b = 0.13$ g/cm^3, which is much greater than that of nanosilica A-300 ($S_{BET} \approx 300$ m^2/g, $\rho_b = 0.045$ g/cm^3); however, $S_{BET,2}/S_{BET,1} > \rho_{b,1}/\rho_{b,2}$. Fumed oxides with alumina, titania, etc. (see Chapter 2) have the ρ_b values greater than that of nanosilicas with the same S_{BET} values because of a larger ρ_0 value. This morphological feature of nanosilicas, as well as the physical binding (due to electrostatic, dispersion and hydrogen bonds) of a particle to other particles at a small coordination number of them in aggregates, can cause significant dust formation. Mechanical compaction of the nanosilica powder can be used to prevent this effect (rather negative). Such compacted nanosilicas can be of interest for both medical and industrial applications.

MCA of nanosilicas can cause significant changes at all levels of their structural hierarchy, as well as changes in other properties (Abe et al. 2008, Buyanov et al. 2009, Sydorchuk et al. 2010, Gun'ko et al. 2011e). The MCA influence on the silica properties depends on a material type (nanosilica, silica gel, quartz, etc.), medium type, pH, temperature, mechanical loading and time, etc. Changes in the specific surface area, particle shape, pore volume, and surface structure on the MCA are especially important for disperse and porous materials since their textural, structural, and morphological characteristics determine the application efficiency of unmodified and modified materials. Changes in the interfacial behavior of water bound in MCA-treated nanooxides, dry or wet powders, nanooxide-filled polymers, or aqueous suspensions are additional and important aspects sensitive to the textural and structural changes in the treated materials used in medical and biotechnological applications. All the mentioned effects can be varied due to changes, first of all, in MCA time (t_{MCA}) and a medium type, as well as in post-MCA treatment conditions. Therefore, here we will analyze the influence of different factors (MCA time, medium type, post-MCA treatment temperature) on the textural, structural, and adsorption characteristics of MCA-treated nanosilica A-300.

Initial standard nanosilica A-300 ($S_{BET} = 330$ m^2/g, $\rho_b = 0.045$ g/cm^3, empty volume in the powder $V_{em} = 1/\rho_b - 1/\rho_0 = 21.8$ cm^3/g, textural porosity $\chi = 100V_{em}/(V_{em} + 1/\rho_0) = 98\%$) was heated at 450°C for 4 h to remove residual HCl and adsorbed compounds. MCA was performed in a ceramic ball mill (volume 1 dm^3 2/3 filled by balls of 2–3 cm in diameter, 60 rpm) using dry or ethanol, water/ethanol (1 vol/1 vol), or water-wetted powders (0.5 g of a solvent per gram of silica) for 1, 2, 4, and 6 h (Gun'ko et al. 2011e). Approximately 75% of water was in the adsorbed state and 25% as vapor in the ball mill. The use of a tight-head drum provided the absence of the water or alcohol vapor loss on the MCA treatment. The MCA-treated samples were dried in air for 24 h. The maximal bulk density $\rho_b = 0.394$ g/cm^3 ($V_{em} = 2.08$ cm^3/g, $\chi = 82\%$) was reached on MCA treatment of the water-wetted powder for 6 h (labeled 6h-MCA). Treatments of dry silica or ethanol or water/ethanol-wetted samples give a lower bulk density of the powders.

To analyze the effects of structural changes during MCA, poly(vinyl) alcohol (PVA; Kuraray Ltd, Germany, ®Mowiol grade 13–88, viscosity 11.5–14.5 mPa·s, hydrolysis 86.7–88.7 mol.%), films were filled with initial and MCA (6h-MCA) nanosilicas. Aqueous (10 wt%) PVA solution was prepared by dissolving dry PVA pellets in deionized water and then refluxing at 80°C. The nanosilica powder (initial or 6h-MCA) was suspended in deionized water. The PVA and nanosilica solutions were mixed and sonicated at room temperature for 20 min. Silica content was between 0.1 and 20 wt% with respect to dry PVA. The solutions were poured onto a glass plate, and casting using a stainless steel bar with a fixed distance to the glass plate. The wet films of 500 μm in thickness were dried at 70°C for 30 min. A portion of the dried PVA/silica composites was left in air for overnight to study the effect of water adsorption from air. The second dry portion of the materials studied was with no contact with air.

To analyze the textural characteristics of the initial and MCA-treated nanosilicas, LT (77.4 K) nitrogen adsorption–desorption isotherms (Figure 1.188a) were recorded using a Micromeritics ASAP 2405N adsorption analyzer. Before measurements, the samples were heated at 200°C for 2 h to remove adsorbed water.

An overall adsorption equation derived within the framework of DFT (Do et al. 2001) improved to be used for different pore models (Gun'ko et al. 2007f) and nonlocal DFT (NLDFT; Quantachrome Instruments software, version 2.02) with the model of cylindrical pores in silica was applied to calculate the PSDs (PoSD). The differential distribution functions $f_V(R_p)$ ($\int f_V(R_p)dR_p \sim V_p$) were converted to incremental PoSDs (IPSDs, $\Phi(R_p)$ at $\Sigma\Phi_i(R_p) \sim V_p$) (Figure 1.189).

Calculations of the specific surface area S_ϕ (at $N = 5$, 4, and 3 at $t_{MCA} = 6$, 1, and 0 h, respectively, t and r_m at $0.05 < p/p_0 < 0.2$) were carried out with equation described in Chapter 10 (Gun'ko et al. 2007f).

Thermal analysis of silica samples (sample weight ≈ 200 mg) was carried out in air at 293–1173 K using a Q-1500D (Paulik, Paulik & Erdey, MOM, Budapest) apparatus at a heating rate of 10 K/min.

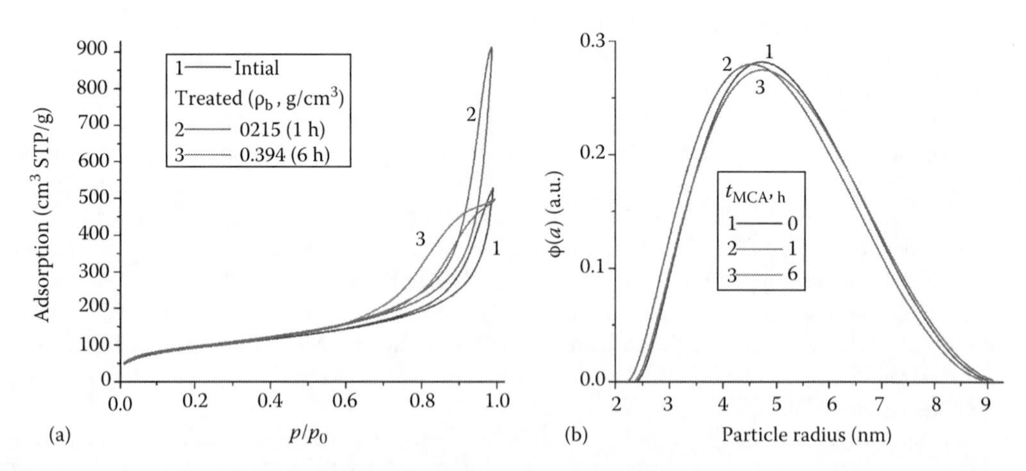

FIGURE 1.188 (a) Nitrogen adsorption–desorption isotherms, and (b) primary particle size distributions for silica initial and MCA-treated wetted powder for 1 and 6 h calculated using the adsorption data with the model of voids between spherical particles and the SCR procedure. (Adapted from *J. Colloid Interface Sci.*, 355, Gun'ko, V.M., Voronin, E.F., Nosach, L.V. et al., Structural, textural and adsorption characteristics of nanosilica mechanochemically activated in different media, 300–311, 2011e. Copyright 2011, with permission from Elsevier.)

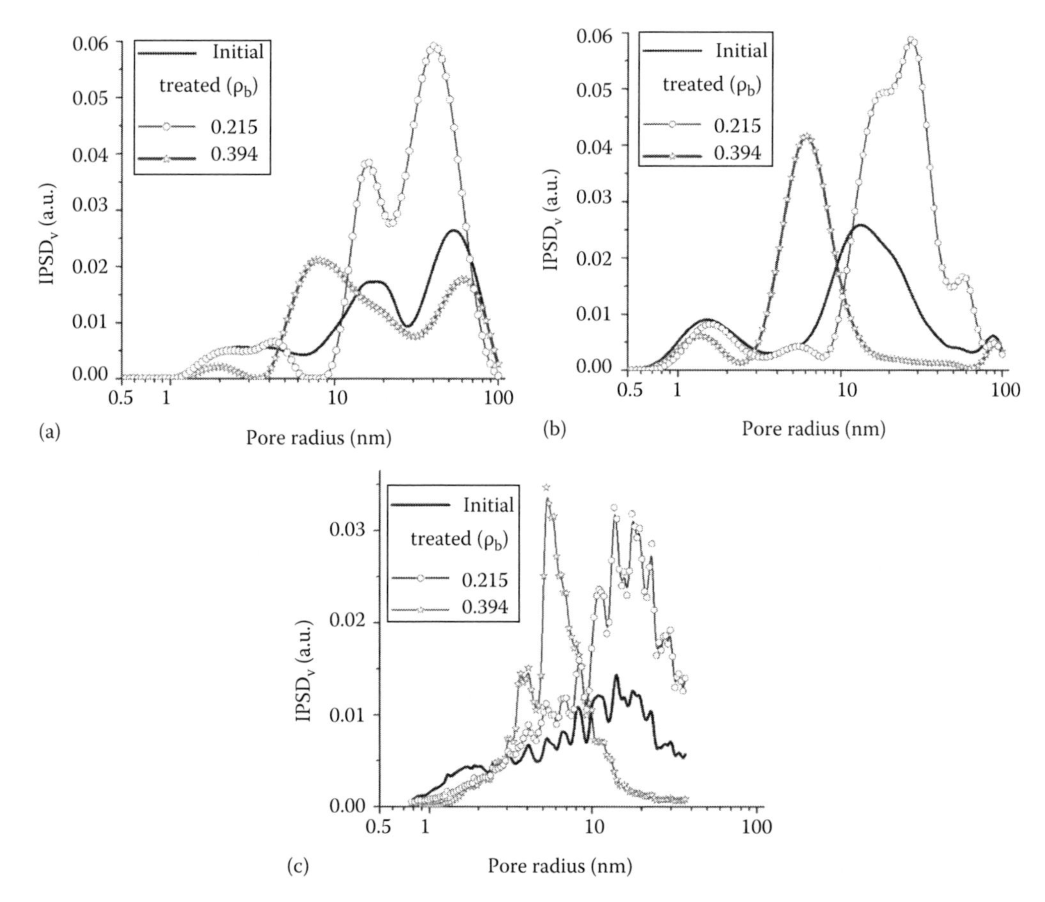

FIGURE 1.189 IPSD for initial and MCA-treated wetted powders for 1 and 6 h with respect to the pore volume calculated using (a) DFT/PaSD for the model of voids between spherical particles, (b) the self-consistent model of mixture of pores (slit-shaped and cylindrical pores and voids), and (c) NLDFT (cylindrical pores). (Adapted from *J. Colloid Interface Sci.*, 355, Gun'ko, V.M., Voronin, E.F., Nosach, L.V. et al., Structural, textural and adsorption characteristics of nanosilica mechanochemically activated in different media, 300–311, 2011e. Copyright 2011, with permission from Elsevier.)

The FTIR spectra of nanosilica were recorded in the mid-IR region using a ThermoNicolet FTIR spectrometer in diffuse reflectance mode. The FTIR spectra of PVA/silica films were recorded using a FT/IR-4100 (JASCO Benelux B.V.) spectrometer in transmittance mode.

DLS investigations were performed using a Zetasizer 3000 (Malvern Instruments) apparatus (λ=633 nm, Θ=90°) at 298 K. Deionized distilled water or 0.1 M NaCl solution was used for preparation of nanosilica suspensions at C_{SiO2}=3 wt% sonicated for 5–10 min using an US disperser (Sonicator Misonix Inc., 22 kHz, 500 W).

The DSC measurements were performed using a DSC (DSC 822e, Mettler Toledo) equipped with an intracooler. Samples (5.0±0.01 mg) were cooled at a cooling rate V=5, 10 or 20°C/min to T=−60°C and then heated at the same rate to 150°C or 300°C. Activation energy (E) values were determined with the accuracy of 10% using ln V versus T^{-1} (K^{-1}) linear dependencies by the formula (Bershtein and Egorov 1994):

$$E = -R\frac{d\ln V}{d(T^{-1})} \quad (1.43)$$

Additionally, calculation of a distribution function of activation energy of water evaporation $f(E)$ was carried out using the integral equation (Gun'ko et al. 2009d) for the corresponding endotherm around 100°C:

$$F(T) = \frac{c_1 T[\Theta(T)]}{\sqrt{2\pi\sigma^2}} \exp\left(-\frac{(T-T_m)^2}{2\sigma^2}\right) \int_{E_{min}}^{E_{max}} f(E) \exp\left(-\frac{E}{R_g T}\right) dE \tag{1.44}$$

where c_1 is a constant dependent on the heating rate, σ is the half-width of an endotherm with the peak temperature T_m, $\Theta(T)$ is the temperature dependence of the amounts of water bound by PVA or PVA/silica, R_g is the gas constant, and E_{min} and E_{max} are the minimal and maximal values of the activation energy E. Equation 1.44 was solved using a regularization procedure with unfixed regularization parameter. The equation parameters (c_1, σ) and the $\Theta(T)$ function were determined before and after regularization using additional minimization and self-consistent procedures with condition of the best fitting to the experimental endotherm of water evaporation.

MCA of nanosilica leads to both small (Table 1.31, S_{BET}) and large (V_p, V_{meso}, and V_{macro}) changes in the textural characteristics. This different influence of MCA on the S and V values can be explained by two factors. Firstly, there are very small changes in the primary nanoparticles (5–18 nm in size determining S values; Figure 1.188b). Secondly, there is significant rearrangement of aggregates of primary particles (<1 μm) and especially agglomerates of aggregates (>1 μm) both responsible of the textural porosity (Table 1.31, V) and PoSD of the powders (Figure 1.189). These results are also confirmed by the same shape of the nitrogen adsorption isotherms at $p/p_0 < 0.4$ (dependent on the specific surface area) and its significant changes at $p/p_0 > 0.7$ (Figure 1.188a) dependent on the pore volume (Table 1.31) and the PoSD, especially the shape of meso-/macropores (Figure 1.189). The PoSD calculated using different models are characterized by relatively close positions of the main peaks, and, therefore, close contributions of nano-, meso-, and macropores into the total porosity and surface area (Table 1.31). However, the detailed shape of the distribution functions differs more strongly.

The NLDFT PoSD calculated with the model of cylindrical pores (Figure 1.189c) have a sawtooth shape because of too small default value of the regularization parameter (α) used in the Quantachrome Instruments software.

The DFT PoSD calculated using the model of voids and the $\phi(a)$ function (Figure 1.189a and b), as well as the PoSD calculated using the mixture of voids and cylindrical and slit-shaped pores with the SCR procedure (Figure 1.189c and d), have a smoother shape than NLDFT PoSD since a relative large $\alpha = 0.01$ was used. However, all the PoSDs are broad and show that the initial and MCA-treated nanosilicas are meso-/macroporous materials with the textural porosity (i.e., without pores in nanoparticles but with voids [pores] between nanoparticles in aggregates and agglomerates). The PoSD peaks shift toward smaller pore sizes with increasing MCA time due to compaction of secondary particles. Notice that the fractal dimension (Table 1.31, D_{FHH}) decreases with increasing MCA time that can be explained by increased uniformity of both treated silica nanoparticles and secondary particles (aggregates and agglomerates).

Rearrangement of nanoparticles packing on the MCA (i.e., powder compaction) is affected by a solvent (water, ethanol, water/ethanol = 1 vol/1 vol) present in the ball mill in the amounts of 0.5 g per gram of dry silica remaining, however, in the powder form. This rearrangement is confirmed by a significant increase in the bulk density of the powders dried after MCA at room temperature for 24 h (Figure 1.190a).

The solvents in the amounts used can form a multilayer coverage of all primary particles in the wetted powders. The water or ethanol molecules can penetrate in the contact area between adjacent nanoparticles in aggregates. The water molecules (half smaller in volume than ethanol molecules) can penetrate in narrower voids and strongly decompose the particle–particle linkages in the aggregates. Contrariwise, water clusters can be bridges between nanoparticles from neighboring

TABLE 1.31
Textural and Adsorption Characteristics of Initial and MCA-Treated Wetted Silica Samples

Sample	t_{MCA} (h)	ρ_b (g/cm³)	V_{em} (cm³/g)	S_{BET} (m²/g)	S_{nano} (m²/g)	S_{meso} (m²/g)	S_{macro} (m²/g)	V_p (cm³/g)	V_{nano} (cm³/g)	V_{meso} (cm³/g)	V_{macro} (cm³/g)	Δw	D_{FHH}	A_{gel} (mg/g)
1*	0	0.045	21.8	330.1	0.1	294.3	35.7	0.826	0.0	0.422	0.404	−0.273	2.5846	320
1**					51.4	270.0	8.7		0.026	0.655	0.145	−0.021		
1***				329.3[a]	26.0	301.0	3.2		0.011	0.774	0.040	0.002[b]		
2*	1	0.215	4.20	345.3	0	271.1	74.1	1.419	0	0.592	0.827	0.031	2.5516	186
2**					24.9	285.0	35.4		0.010	0.884	0.525	0.184		
2***				345.6[a]	19.3	300.1	26.0		0.008	1.073	0.338	−0.00065[b]		
3*	6	0.394	2.08	332.2	0	305.1	27.1	0.771	0	0.494	0.277	−0.086	2.5127	140
3**					29.3	301.2	1.6		0.017	0.714	0.040	0.034		
3***				332.2[a]	9.8	321.3	1.1		0.005	0.721	0.045	0.00003 [b]		

Note: * DFT/PaSD with void model, ** self-consistent model of a mixture of voids, cylindrical and slit-shaped pores, *** self-consistent regularization with respect to both PoSD ($f_V(R_p)$) and PaSD ($\phi(a)$) with the model of voids, [a] $\langle S_\phi \rangle$, [b] $\Delta w = S_{BET}/\langle S_\phi \rangle - 1$, D_{FHH} is the fractal dimension with Frenkel–Halsey–Hill equation accounting for adsorbate surface tension effects (Quantachrome Instruments software), A_{gel} is the gelatin adsorption in mg per gram of silica.

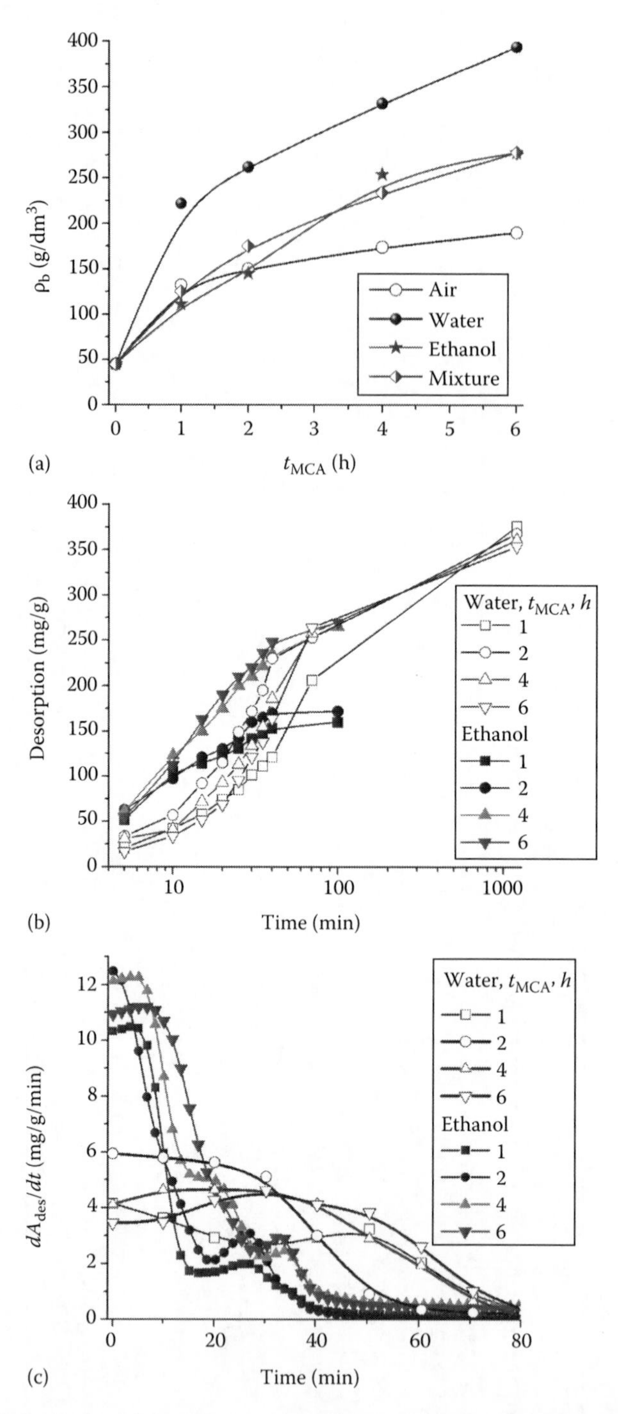

FIGURE 1.190 (a) Bulk density of the silica powders (dried in air for 24 h) treated in different media (air and water, ethanol or water/ethanol ~0.5 g per gram of dry silica) as a function of MCA time, (b) desorption (in air at room temperature) of water or ethanol from MCA-treated silicas, and (c) corresponding desorption rate as a function of time at the initial fast stage of the desorption. (Adapted from *J. Colloid Interface Sci.*, 355, Gun'ko, V.M., Voronin, E.F., Nosach, L.V. et al., Structural, textural and adsorption characteristics of nanosilica mechanochemically activated in different media, 300–311, 2011e. Copyright 2011, with permission from Elsevier.)

aggregates that cause diminishing voids between these aggregates in the wetted powders. Primary particles in the initial silica practically do not have the siloxane bonds with neighboring particles. They are attached one to others by the hydrogen bonds and electrostatic and dispersion interactions. The solvent molecules (especially water) can break these interparticle bonds and increase the mobility of nanoparticles on the MCA treatment. In other words, the solvents in the used amounts provide essential plasticizing effects on the wetted silica powders. The MCA treatment of water-wetted silica gives the powder with a larger bulk density (~0.4 g/cm^3, Figure 1.190a) due to better packing of primary particles than that after the MCA in the dry or ethanol atmosphere. However, this maximal ρ_b value corresponds to approximately 35% of dense and ordered-packed silica spheres of the same size ($d=6000/(\rho_0 S_{BET})=8.26$ nm at $\rho_0=2.2$ g/cm^3 and $S_{BET}=330$ m^2/g) that in a cubic lattice give $\rho_b \approx 1.15$ g/cm^3. This difference can be explained by, at least, two factors: (i) primary particles of nanosilica have different sizes (Figure 1.188b) and (ii) certain amounts of non-decomposed aggregates can remain after the MCA (vide infra). Both factors, as well as a random structure of initial aggregates, prevent the ordered packing of silica nanoparticles in secondary particles on MCA treatment. However, compaction of primary particles becomes stronger with increasing MCA time (t_{MCA}) under the plasticizing effect of a solvent because of much deeper rearrangement of agglomerates and aggregates.

The particle compaction affects the desorption rate of water or ethanol on evaporation at room temperature in air (Figure 1.190b and c). However, this dependence is complex, as well as changes in the PoSDs (Figure 1.189) affecting this process. The pores (voids between spherical nanoparticles can be considered as textural pores) size distributions (Figure 1.189) depend on t_{MCA}. For instance, at $t_{MCA}=1$ h (water-wetted powder) contribution of large voids at $R>10$ nm increases, and contribution of cylindrical pores increases to 83%, but slit-shaped pores and voids give 2% and 15%, respectively, according to calculations with the SCR procedure. At $t_{MCA}=6$ h (water-wetted powder) the PoSD at $R<10$ nm gives the main contribution to the porosity, and contribution of cylindrical pores decreases to 49% (slit-shaped pores and voids give 30% and 21%, respectively). Closer packing of nanoparticles provides an increase in the adsorption of water from air (Figure 1.191) in narrower

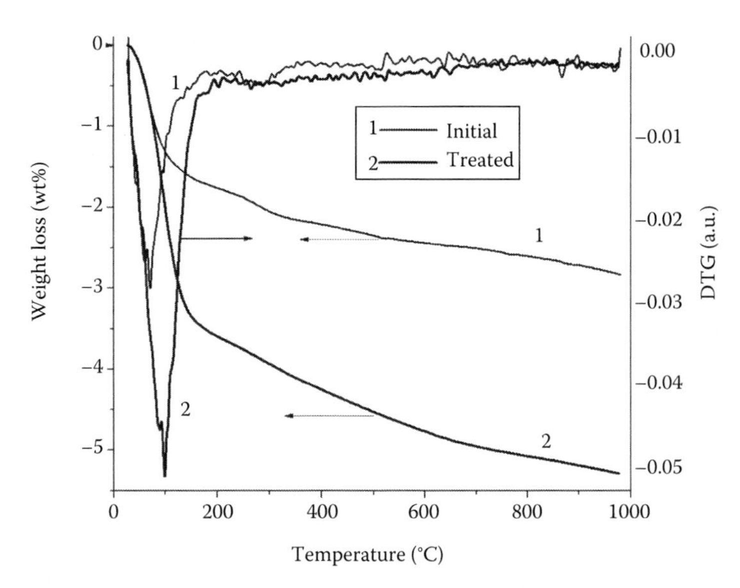

FIGURE 1.191 Thermogravimetric (TG) and differential TG (DTG) curves of weight loss on heating of initial silica (curves 1) and water-wetted MCA-sample treated for 6 h (curves 2). (Adapted from *J. Colloid Interface Sci.*, 355, Gun'ko, V.M., Voronin, E.F., Nosach, L.V. et al., Structural, textural and adsorption characteristics of nanosilica mechanochemically activated in different media, 300–311, 2011e. Copyright 2011, with permission from Elsevier.)

pores (Figure 1.189) and simultaneously reduces the desorption rate (Figure 1.190c). Therefore, the desorption rate peaks shift toward longer time with increasing MCA time for both ethanol and water desorbed (Figure 1.190c).

On heating, desorption of both intact water ($T < 150°C$) and associatively desorbed water (or water desorbed from the inner volume of primary particles at $T > 150°C$) increases after the MCA treatment (Figure 1.191). The difference in the amounts of water desorbed from initial and MCA-treated silica slightly increases with heating temperature to 900°C. This can be explained by larger amounts of water molecules and OH groups inside silica nanoparticles treated with 0.5 g of water per gram of silica (added into the ball mill) for 6 h, as well as retardation of water desorption from narrower pores in denser secondary particles.

Desorption of water from the treated powder is a relatively slow process at room temperature (Figure 1.190b). Desorption of ethanol occurs much faster (for 2 h) with a higher rate than water (>20 h; Figure 1.190b and c). The desorption rate strongly decreases after 40 and 80 min on ethanol and water desorption, respectively. The total amounts of water desorbed from the 6 h-MCA-treated sample (dried at room temperature for 24 h) on heating to 900°C are about 5.5 wt% (and a DTG minimum shifts toward higher temperature, 3.4 wt% water is desorbed at $T < 150°C$) and 2.8 wt% (1.6 wt% desorbed at $T < 150°C$) from the initial sample (Figure 1.191). Thus, the main portion of adsorbed water (~0.4 g/g from 0.5 g/g used, since a portion of water remained in the vapor phase) was desorbed from MCA-treated silica in air for 24 h (Figure 1.190b). This can be explained by a small contribution of nanopores (Table 1.31, V_{mic}) and narrow mesopores at $R < 3$ nm (Figure 1.189) to the total porosity where water can be strongly adsorbed.

The band intensity of the O-H stretching vibrations at 3750 cm^{-1} of free surface silanols (Legrand 1998) decreases with increasing t_{MCA} value (Figure 1.192a). This is due to an increase in the amounts of adsorbed water disturbing silanols, stronger contacts between adjacent nanoparticles in aggregates, and enhanced scattering of the IR beam by denser packed aggregates (Figure 1.193a). Heating at 80°C–600°C and then cooling of MCA-treated samples to room temperature leads to nonlinear changes in the optical density of this band (D_{3750}). Heating at 80°C enhances D_{3750} only for the initial silica, but for MCA-treated samples, it decreases. The latter can be due to associative desorption of OH groups close located on the surfaces of denser packed primary particles (simultaneously with desorbed intact water) or certain rearrangement of aggregates, and enhanced IR beam scattering (notice that the FTIR spectra were recorded in diffuse reflectance mode). Then the D_{3750} value increases for samples heated at $T > 120°C$ due to desorption of water bound to the surface silanols. Minimal D_{3750} values are observed for denser silica MCA-treated for 4–6 h because of larger amounts of adsorbed water and enhanced IR beam scattering effects by denser packed aggregates.

The specific area of free surface (S_{IR}) with free silanols can be estimated from the FTIR spectra (McCool et al. 2006) using a linear relationship between the ratio of integral intensity of the bands of free silanols at 3750 cm^{-1} (integrating at 3710–3760 cm^{-1} on subtraction of the baseline) and Si-O overtone at 1870 cm^{-1} (integrating at 1760–1940 cm^{-1}) I_{sb} = SiOH/overtone Si-O as follows $S_{IR} = I_{sb} \times 299.48874 - 18.29504$ (m^2/g). The S_{IR} values show that the sample treated for 1 h in dry air has the S_{IR} value smaller than that for sample after the MCA for 2 h (Figure 1.192b and c). However, this difference decreases with increasing heating temperature (Figure 1.192c). Heating of the initial sample practically does not change the S_{IR} value since the adsorbed water amounts are small (~1.5 wt%, Figure 1.191). In the case of the MCA-treated wetted powder, the mentioned effect at $t_{MCA} = 1$ and 2 h is small, and it is absent for MCA-treated water/ethanol-wetted powder (Figure 1.192b).

Nonlinear changes in the textural characteristics with t_{MCA} can be caused by MCA-time-dependent rearrangement of agglomerates and then aggregates since the latter are more stable than the former (Gun'ko et al. 2001e, 2009d). This difference trace is well seen in the PaSDs observed in the aqueous media (Figure 1.193). Notice that the PaSD$_S$ for the scattering light intensity (Figure 1.193a) and the PaSD$_N$ for particle number (Figure 1.193c) for 6-h-MCA-treated sample in the aqueous medium

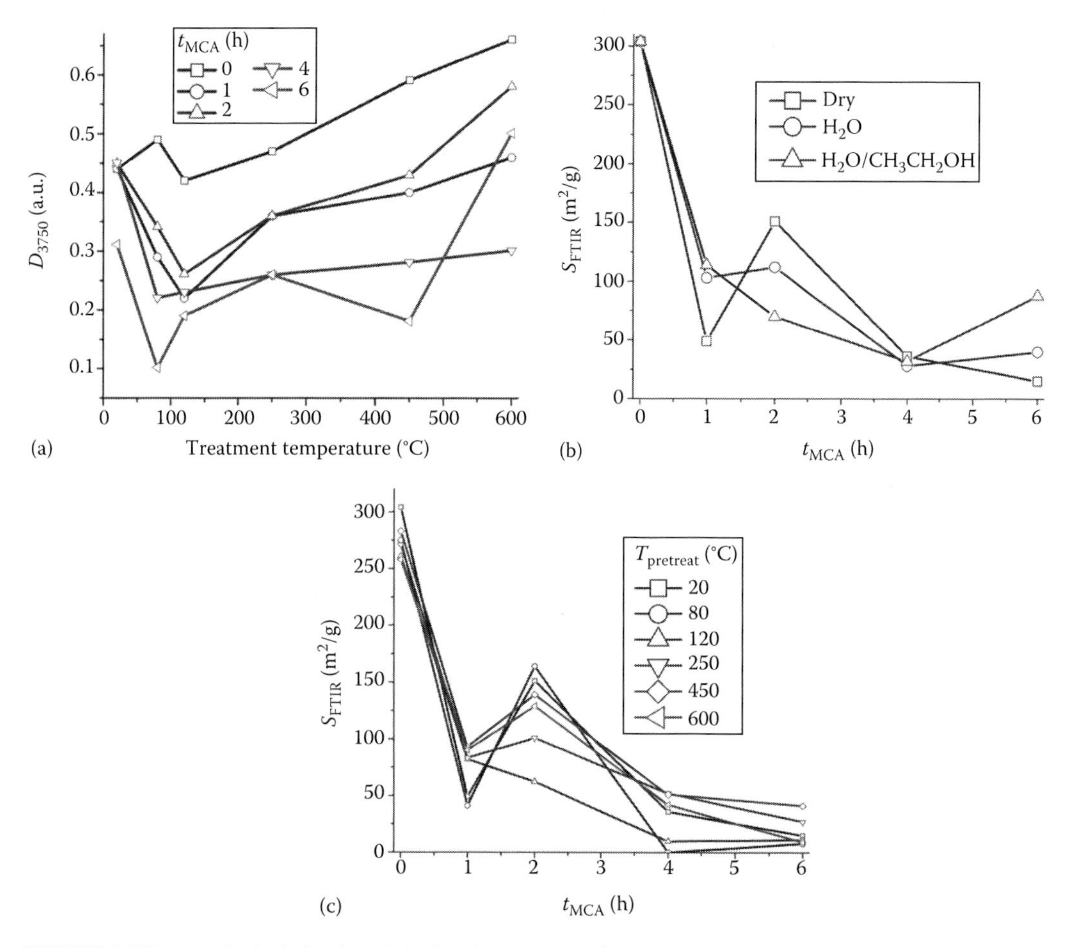

FIGURE 1.192 (a) Optical density of the band at 3750 cm^{-1} as a function of heating temperature of initial and MCA-treated samples then cooled to room temperature in air, and the surface area free of adsorbed water and contacts with adjacent particles as a function of MCA time for samples MCA treated (b) in different media (c) air and heated and cooled after MCA to room temperature in air. (Adapted from *J. Colloid Interface Sci.*, 355, Gun'ko, V.M., Voronin, E.F., Nosach, L.V. et al., Structural, textural and adsorption characteristics of nanosilica mechanochemically activated in different media, 300–311, 2011e. Copyright 2011, with permission from Elsevier.)

or in the 0.1 M NaCl solution are the same but the $PaSD_V$ functions for the particle volume differ because of an increased number of agglomerates with increasing ionic strength of the suspension (Figure 1.193b). Agglomerates are absent in the sonicated suspension of the initial silica in contrast to MCA-treated samples keeping large particles at $d > 1$ μm after sonication and characterized by their increased amounts with increasing MCA time.

The calculation of the specific surface area averaged from the $PaSD_N$ (S_{DLS}) for the initial silica (3 wt% suspension, 10 min sonication) gives $S_{DLS} = 325$ m^2/g. For MCA-treated sample at $t_{MCA} = 6$ h, this value is 72 m^2/g (water) and 76 m^2/g (NaCl solution) calculated assuming that all observed particles are nonporous and spherical. Consideration for the porosity of aggregates can significantly enhance the S_{DLS} value, which becomes close to that for the initial silica because of keeping the sizes of primary particles on MCA (Figure 1.188b).

The influence of MCA-treated silica as a filler of PVA films differs from that of initial A-300 (Figures 1.194 and 1.195) because of more compacted secondary particles in the treated silica powder. The activation energy of water evaporation from the films (Figure 1.194g and h) depends

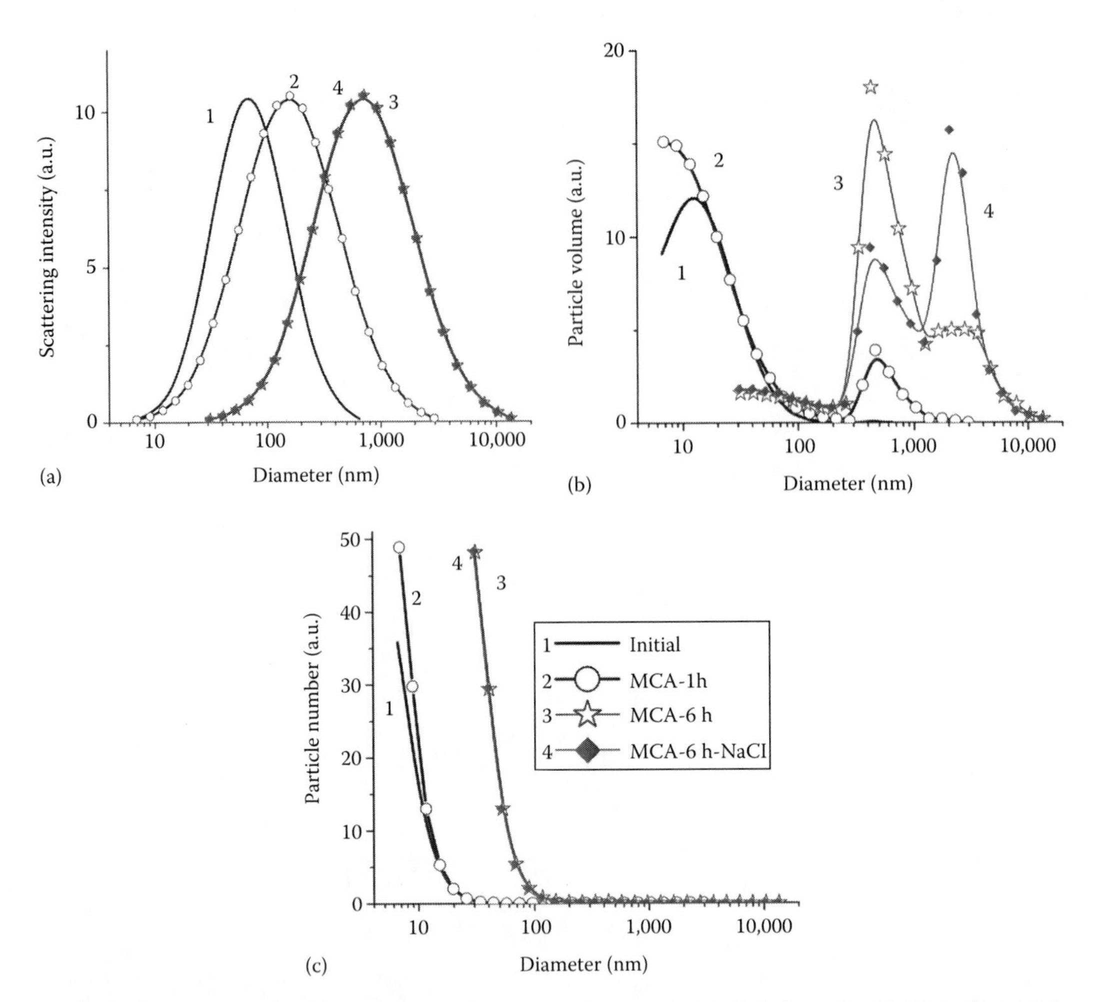

FIGURE 1.193 Particle size distributions with respect to (a) scattering light intensity ($PaSD_S$), (b) particle volume ($PaSD_V$), and (c) particle number ($PaSD_N$) for the initial silica (sonication 10 min) and MCA-samples treated for 1 and 6 h (silica concentration 3 wt%, sonication 5 min, without and with 0.1 M NaCl). (Adapted from *J. Colloid Interface Sci.*, 355, Gun'ko, V.M., Voronin, E.F., Nosach, L.V. et al., Structural, textural and adsorption characteristics of nanosilica mechanochemically activated in different media, 300–311, 2011e. Copyright 2011, with permission from Elsevier.)

more strongly on silica content for the MCA-treated filler. The SiO-H stretching vibrations of free surface silanols at 3747 cm^{-1} are not observed (or at the noise level) in FTIR spectra of PVA/silica composites (Figure 1.195b) because of interactions of silanols with PVA or bound water. In the case of MCA-treated silica, the activation energy increases with increasing silica content.

This is due to the changes in the textural characteristics of MCA-treated silica (Figures 1.188 and 1.189) in which aggregates with denser packed nanoparticles give the effects similar to inner porosity in silica gels. These changes can be rather attributed to changes in the texture of the secondary silica particles than in the PVA film itself. For instance, endotherms located around 200°C (melting of PVA) and 300°C (associatively desorbed water molecules, decomposition of cross-linked portion of PVA) more weakly depend on the type of silica (Figure 1.195a) than the endotherm around 100°C related to evaporation of molecularly adsorbed water (Figures 1.194 and 1.195a). Consequently, water organization in the PVA/silica films depends on the type of silica. The mobility of water is higher in the case of PVA/MCA-treated silica, and it is minimal for PVA/initial silica composite with tighter contacts between primary particles and PVA molecules.

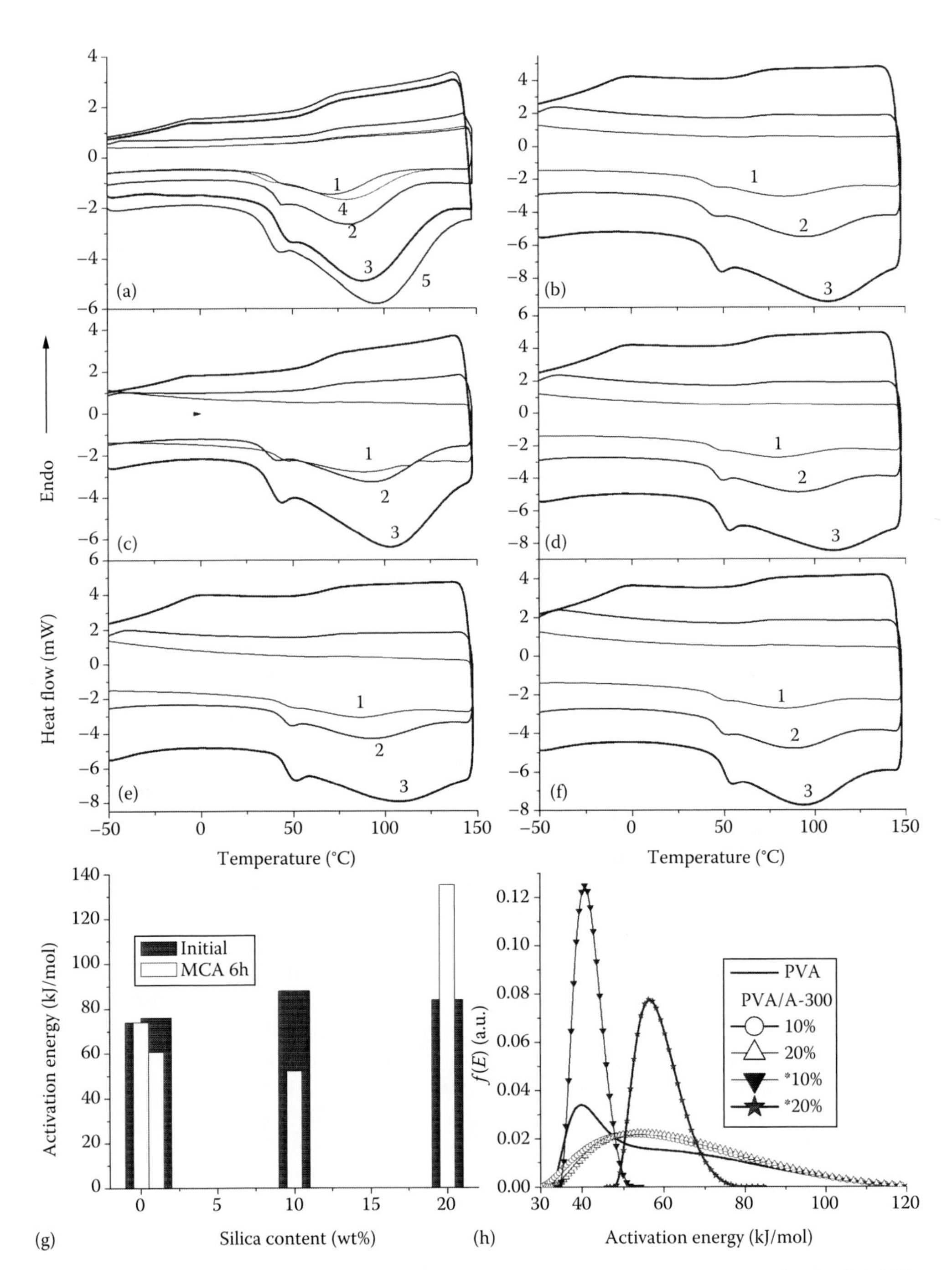

FIGURE 1.194 DSC thermograms for PVA/A-300 (a, c, e) initial and (b, d, f) 6h-MCA at $C_{A\text{-}300}$ = (a, b) 1, (c, d) 10 and (e, f) 20 wt% at a heating rate V = 5 (curves 1), 10 (2) and 20 (3) °C/min, pure PVA (a) at V = 5 (4) and 20 (5) °C/min. Activation energy of for water evaporation (endotherms around 100°C) calculated using Equation 1.43 (g) and Equation 1.44 (h) (*6h-MCA A-300 as a filler). (Adapted from *J. Colloid Interface Sci.*, 355, Gun'ko, V.M., Voronin, E.F., Nosach, L.V. et al., Structural, textural and adsorption characteristics of nanosilica mechanochemically activated in different media, 300–311, 2011a. Copyright 2011, with permission from Elsevier.)

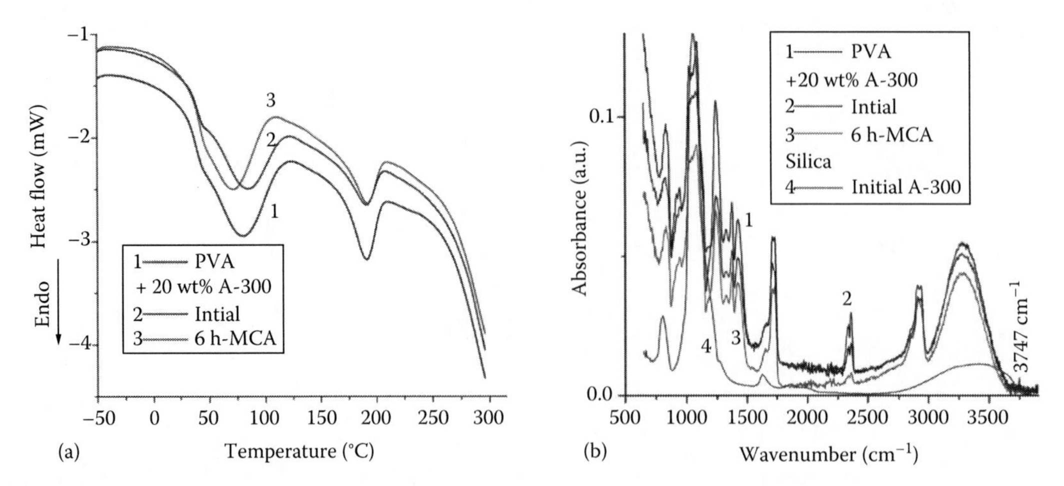

FIGURE 1.195 (a) DSC thermograms (heating rate 20°C/min) and (b) FTIR spectra of PVA (curve 1) and PVA/A-300 (2) initial and (3) 6h-MCA, and initial silica alone (4). (Adapted from *J. Colloid Interface Sci.*, 355, Gun'ko, V.M., Voronin, E.F., Nosach, L.V. et al., Structural, textural and adsorption characteristics of nanosilica mechanochemically activated in different media, 300–311, 2011e. Copyright 2011, with permission from Elsevier.)

The behavior of water bound by silica initial and after the MCA of the water-wetted powder for 1 and 6 h was studied by the ^{1}H NMR spectroscopy with layer-by-layer freezing-out of bulk and bound water at $200<T<273$ K in different media (air, water, chloroform; Gun'ko et al. 2011e). The NMR study shows that the longer the MCA leading to narrowing voids between silica nanoparticles (Figure 1.189), the stronger bound the interfacial water and the stronger the $CDCl_3$ effects on this water (Figure 1.196). In the case of low hydration $h=0.11$ g/g, the displacement of bound water by chloroform into larger voids becomes more visible at $T<260$ K since the amounts of SBW frozen at $T<260$ K decrease differently (Figure 1.196c). For strongly wetted powders (Figure 1.196a and b), the ^{1}H NMR spectra of unfrozen water are observed at lower temperatures to 200 K with the single signal at chemical shift of the proton resonance $\delta_H=4.5$–5.5 ppm close to that of bulk water (Gun'ko et al. 2005d). For the aqueous suspension of the initial silica (Figure 1.196a), the ^{1}H NMR signal is observed at $T\geq 220$ K. Water in both cases is strongly associated (SAW; Gun'ko et al. 2005d) since the δ_H value is close to that of bulk water. A portion of adsorbed water frozen close to 273 K can be assigned to WBW at $-\Delta G<0.5$ kJ/mol. At $h=0.11$ g/g (Figure 1.196c through e), the spectra shape depends on the t_{MCA} value and medium (air or $CDCl_3$). Water adsorbed on all samples being in the air medium is strongly bound since signal decreases only at $T<250$ K. In the case of the chloroform medium, a portion of adsorbed water is weakly bound since signal strongly decreases at $T<265$ K. The downfield shifts with lowering temperature is due to a stronger influence of the silica surface on closer located water layers frozen at lower temperatures. Additional signals at $\delta_H\approx 7$ and 0 ppm are due to H in chloroform (as an admixture in $CDCl_3$) and trimethylsilane (added to chloroform as a reference compound), respectively.

A signal of WAW appears at $\delta_H=1.0$–1.5 ppm (Gun'ko et al. 2005d, 2009d, 2011e) for the sample in $CDCl_3$ medium (Figure 1.196c). This signal can be assigned to water dissolved in chloroform or forming small clusters in contact with both chloroform and the silica surface in narrow voids. The amounts of this water decrease for MCA-treated samples because of changes in the textural porosity (Figure 1.189) and associativity of water molecules adsorbed in different voids. At temperatures close to 273 K, the WAW signal splits into two signals due to structural nonuniformity of the silica–water–chloroform system.

SBW (frozen at $T<260$ K, Figure 1.197a) at $h=0.11$ g/g is located in narrow voids at $R<5$ nm between neighboring primary particles (comp. Figures 1.189c and 1.197c). The pores filled by

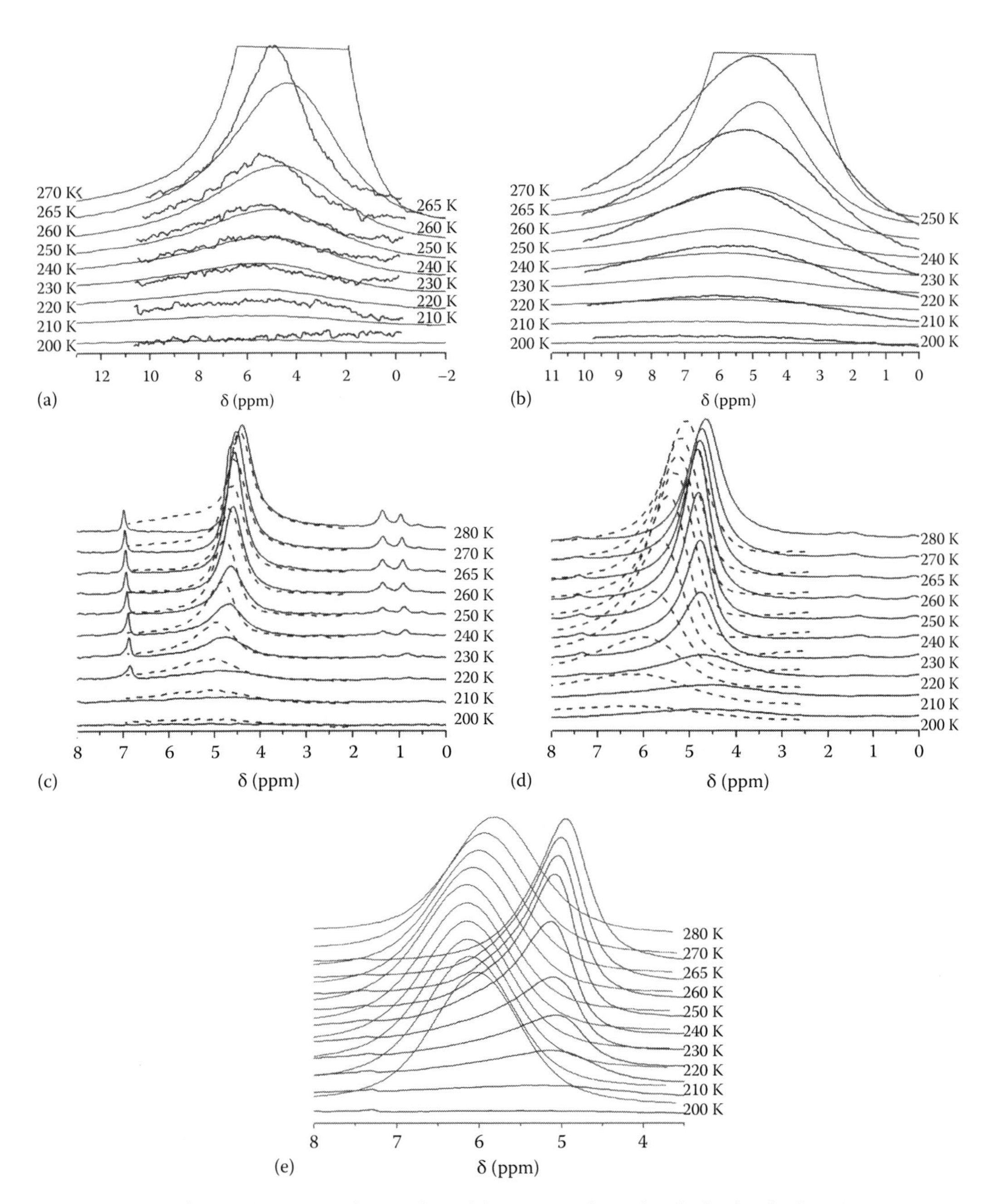

FIGURE 1.196 ^{1}H NMR spectra of water bound in (a) wetted powder (hydration $h=3$ g water per gram of silica) and aqueous suspension (2.4 wt% of silica) of initial silica, (b) wetted MCA-samples at $t_{MCA}=1$ ($h=1.2$ g/g) and 6 h ($h=1$ g/g), and weakly wetted samples ($h=0.1$ g/g) (c) initial and treated for (d) 1 h and (e) 6 h without (left peaks) and with $CDCl_3$ medium (right peaks). (Adapted from *J. Colloid Interface Sci.*, 355, Gun'ko, V.M., Voronin, E.F., Nosach, L.V. et al., Structural, textural and adsorption characteristics of nanosilica mechanochemically activated in different media, 300–311, 2011e. Copyright 2011, with permission from Elsevier.)

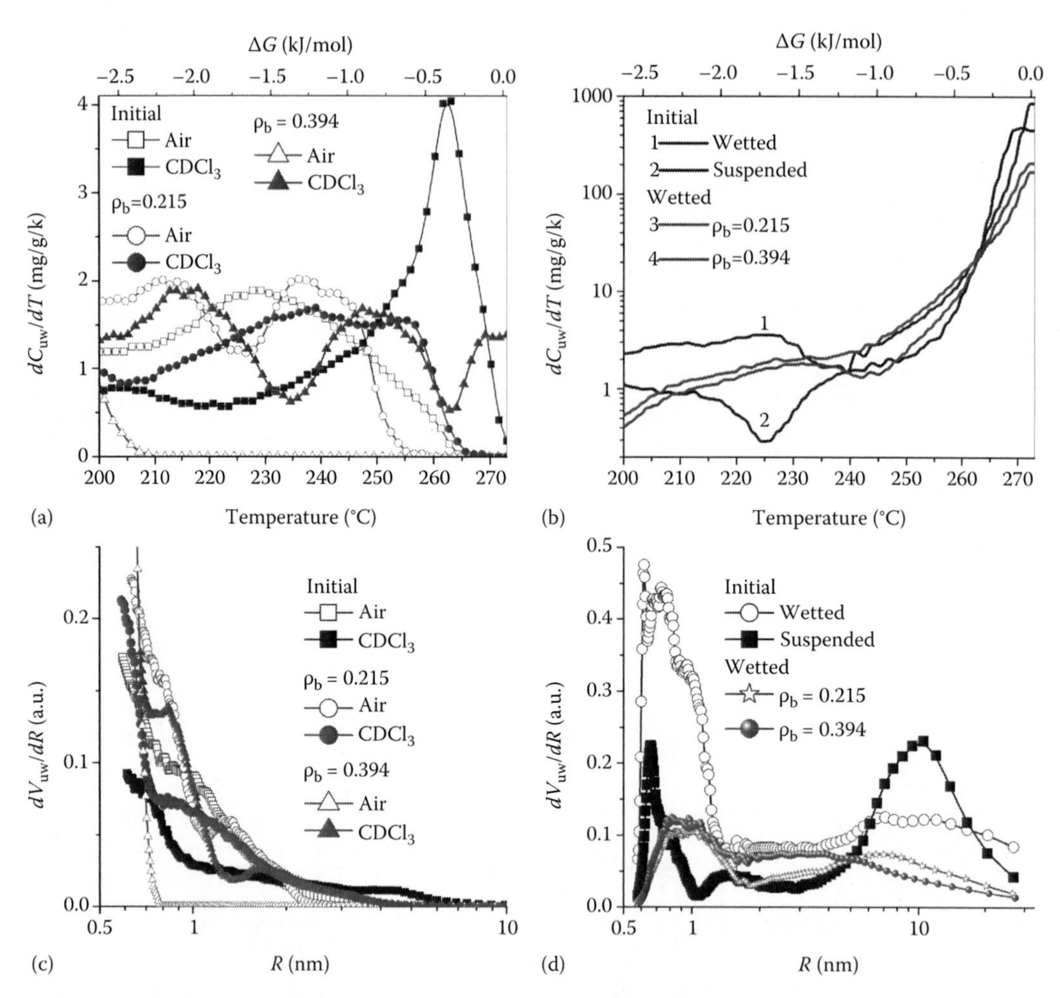

FIGURE 1.197 Differential changes in the amounts of unfrozen water (dC_{uw}/dT) as a function of temperature (bottom x-axis) and changes in the Gibbs free energy (ΔG, top x-axis) in (a) weakly wetted samples ($h=0.11$ g/g) initial ($\rho_b=0.045$ g/cm^3) and treated for 1 h ($\rho_b=0.215$ g/cm^3) and 6 h ($\rho_b=0.394$ g/cm^3) without and with $CDCl_3$ medium and (b) strongly hydrated ($h=3$, 40.7, 1.2 and 1 g/g, respectively) silicas, (c, d) differential functions with respect to the sizes of unfrozen water structures for (c) weakly and (d) strongly hydrated silicas. (Adapted from *J. Colloid Interface Sci.*, 355, Gun'ko, V.M., Voronin, E.F., Nosach, L.V. et al., Structural, textural and adsorption characteristics of nanosilica mechanochemically activated in different media, 300–311, 2011e. Copyright 2011, with permission from Elsevier.)

unfrozen water are narrower in MCA-treated samples. Chloroform can displace bound water into larger voids (Figure 1.197c and Table 1.32) since this provides a smaller boundary area between larger water structures and weakly polar chloroform located in voids. This displacement depends on t_{MCA} since the larger the bulk density of the powder (longer MCA time), the smaller the displacement of the water (Figure 1.197c). For weakly hydrated initial silica at $h=0.11$ g/g, the specific surface area determined with NMR cryoporometry $S_{cryo}<S_{BET}$ (Tables 1.31 and 1.32) because water can cover only a portion of the particle surface at $h=0.11$ g/g.

This portion decreases in the $CDCl_3$ medium because of the displacement of bound water (Figure 1.197c) and the S_{cryo} value decreases (Table 1.32). For MCA-treated powders at $h=1.2$ and 1 g/g, $S_{cryo}<S_{BET}$ is due to the same reason and possible remaining of air bubbles in narrow voids. For the initial silica at $h=40.7$ g/g, $S_{c,meso}>S_{c,nano}$ but at $h=3$ g/g, $S_{c,meso}<S_{c,nano}$ (Table 1.32) because of attractive interactions between neighboring nanoparticles through water bridges in the wetted

TABLE 1.32
Characteristics of Water Unfrozen at $T<273$ K and Bound to Initial and Wet-MCA-Treated Nanosilica Estimated by NMR Cryoporometry

t_{MCA} h	ρ_b g/cm^3	h g/g	S_{cryo} m^2/g	$S_{c,nano}$ m^2/g	$S_{c,meso}$ m^2/g	$V_{c,nano}$ cm^3/g	$V_{c,meso}$ cm^3/g	γ_S J/g	$-\Delta G$ kJ/mol	Medium
0	0.045	0.11	159	111	48	0.045	0.055	10.3	3.03	Air
0	0.045	0.11	69	56	13	0.022	0.077	6.3	2.96	$CDCl_3$
0	0.045	3.0	322	207	115	0.085	1.367	32.2	3.08	H_2O
0	0.045	40.7	281	80	201	0.031	2.136	26.6	3.12	H_2O
1	0.215	0.11	188	131	57	0.053	0.076	10.8	2.87	Air
1	0.215	0.11	142	101	41	0.024	0.076	9.4	3.00	$CDCl_3$
1	0.215	1.2	123	52	71	0.022	0.772	15.6	3.21	H_2O
6	0.394	0.11	332	332	0	0.100	0	16.7	3.21	Air
6	0.394	0.11	100	94	6	0.040	0.058	8.3	2.71	$CDCl_3$
6	0.394	1.0	124	65	59	0.028	0.676	17.9	3.09	H_2O

Note: ΔG corresponds to the changes in the Gibbs free energy of a water layer closely located to the silica surface, γ_S is the modulus of overall changes in the Gibbs free energy of all bound water.

powder and repulsive interactions between charged particles in the suspension. Therefore, PCS shows mainly individual nanoparticles for the initial silica (Figure 1.193c). The $S_{c,meso}$ value is small at $h=0.11$ g/g since water in the form of clusters locates only in narrow voids (Figure 1.197c) and does not form coverage of the total surface of particles. In the suspension or strongly hydrated powder, the $S_{c,meso}$ value strongly increases, especially for the initial silica.

The γ_s values varied between 32.2 J/g (98 mJ/m^2) and 6.3 J/g (19 mJ/m^2) for initial silica at $h=3$ and 0.11 g/g (in $CDCl_3$) are relatively small (Gun'ko et al. 2005d). This is due to relatively small amounts of the surface silanols on nanosilica synthesized at high temperature (Degussa 1997, Chuiko 2001, 2003). Therefore, this silica adsorbs ~1.6 wt% of water from air (desorbed at $T<150$°C, Figure 1.191), and only 7.9% (from the FTIR data and S_{IR} value) of the surface area is occupied by the adsorbed water molecules. More hydrophilic nanosilicas can adsorb 5–7 wt% of water from air and have $\gamma_s=150$–200 mJ/m^2 (Gun'ko et al. 2005d). The γ_s values for MCA-treated silica are smaller than that of suspended or strongly hydrated initial silica (Table 1.32). This result suggests that the number of surface silanols can insignificantly change on MCA but particles are strongly aggregated that reduce total amounts of bound water since $C_{uw}<h$ (1.2 or 1 g/g) and the number of silanols accessible for adsorbed water.

The TG data (Figure 1.191) confirm the small changes in the content of the surface silanols since the difference (Δ) in the TG curves for the initial and MCA-treated silicas not strongly increases with elevating T from 150°C to 900°C ($\Delta=1.63$ and 2.35 wt%, respectively). The specific surface area of the initial silica is 304, 322, 281, 325, and 330 m^2/g calculated from the FTIR spectrum (2.5 wt% of adsorbed water), NMR cryoporometry ($h=3$ and 40.7 g/g for wetted powder and suspension, respectively), PCS (3 wt% suspension), and nitrogen adsorption data, respectively. As a whole the specific surface area is determined by the PaSD, a surface portion accessible for probe molecules (adsorption methods) or residual amounts of water (FTIR). Consequently, in all the mentioned measurements, the primary particle aggregation (in air), disaggregation (in suspension), and adsorption of water from air weakly affect the accessible surface area for water or nitrogen molecules. The 6h-MCA treatment affects the results of these measurements. In the case of the nitrogen adsorption, the S_{BET} value is practically the same for the initial and treated samples (Table 1.31). The NMR cryoporometry gives 332 m^2/g for 6h-MCA-treated sample with low amounts of adsorbed water ($h=0.11$ g/g) and 124 m^2/g for the wet sample, PCS (3 wt% suspension) gives

76 m²/g, S_{IR} = 15 m²/g (MCA of dry powder) and 40 m²/g (MCA of wet powder). Thus, in contrast to the initial silica, all the measurements for 6 h-MCA-treated silica give very different S values. Additionally, the critical concentration of gel formation is 12 and 50 wt% and the osmotic activity with respect to water is 500 and 135 wt% for the initial and 6 h-MCA-treated silicas, respectively. All these results show that aggregation of primary particles strongly enhanced by the MCA is not easily reversible as for the initial silica (Figure 1.193). Since temperature of the silica powder increases to 50°C on MCA, one can assume that local heating of samples (due to ball blows on the mill rotation) promotes cross-linking of a portion of neighboring nanoparticles by the siloxane bonds. These aggregates cannot be decomposed on sonication. Additionally, the coordination number of primary particles increases in the aggregates. This affects many characteristics of MCA-treated silica.

The compaction of primary particles on the MCA leads to diminution of the adsorption of gelatin from the solution (Figure 1.198) used as a probe protein.

The MCA treatment of the wetted nanosilica powder reduces the adsorption of gelatin stronger than the MCA of the dry powder (Figure 1.198a) because stronger compaction of the wetted powder (Figure 1.190a). As a whole, the adsorption of gelatin is a smooth function of the ρ_b value

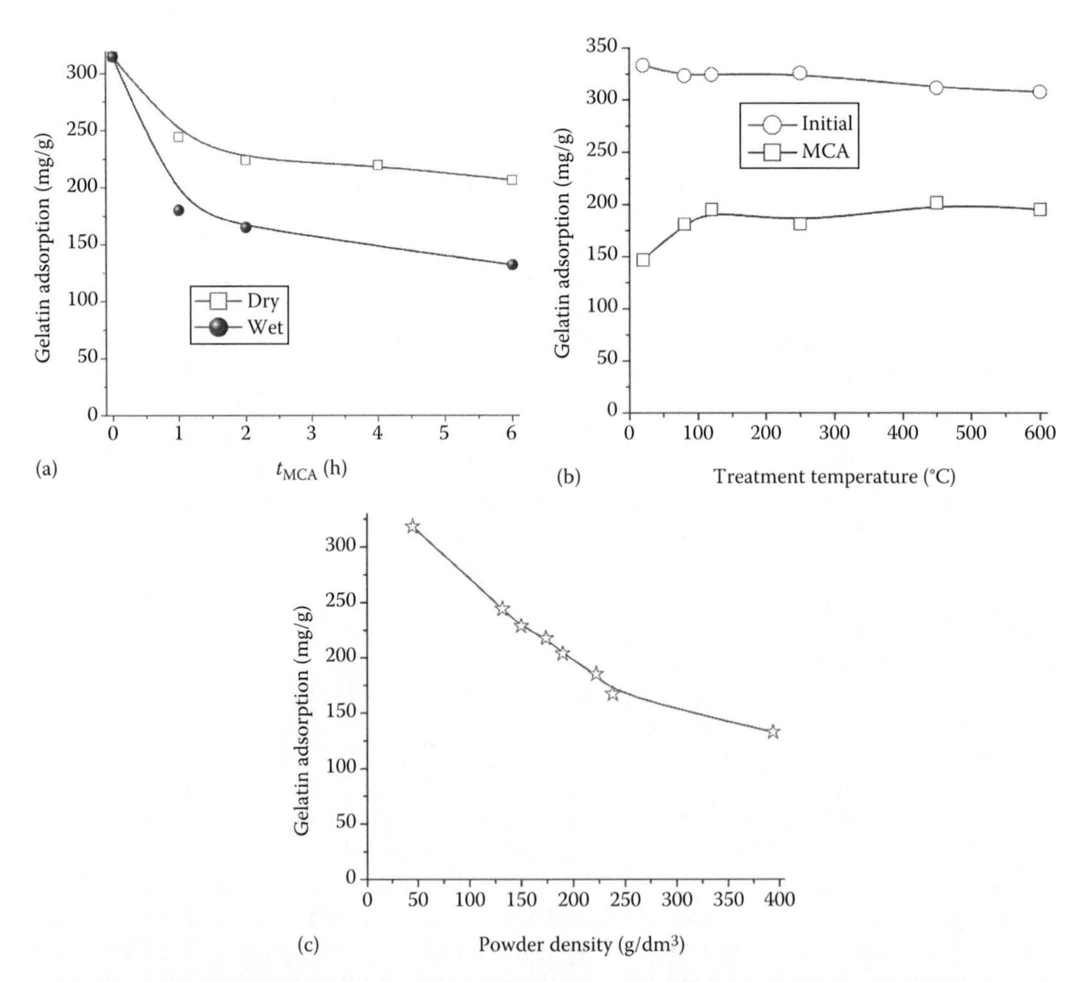

FIGURE 1.198 Gelatin adsorption on silica as a function of (a) MCA time for dry and wetted samples, (b) preheating temperature of initial silica and sample MCA-treated in air, and (c) bulk density of both samples. (Adapted from *J. Colloid Interface Sci.*, 355, Gun'ko, V.M., Voronin, E.F., Nosach, L.V. et al., Structural, textural and adsorption characteristics of nanosilica mechanochemically activated in different media, 300–311, 2011e. Copyright 2011, with permission from Elsevier.)

(i.e., total textural porosity of nanosilica) independently of the medium type on the MCA (Figure 1.198c). Heating–cooling of the initial and MCA-treated silicas gives small changes in the adsorption of gelatin (Figure 1.198b). These features of the protein adsorption are due to worse decomposition of secondary silica particles after longer MCA (Figure 1.193). A stronger compacted powder is characterized by narrower voids (Figure 1.189) in which the adsorption of gelatin is smaller than on the outer surfaces of non-compacted initial material. Thus, a smaller outer surface and narrower voids of more strongly compacted secondary particles cause reduction of the gelatin adsorption. One can assume that similar effects can be observed on interaction of other macromolecules with treated nanosilicas.

Thus, the investigations of nanosilica MCA in the form of dry and wetted (by water, water/ethanol, or ethanol at 0.5 g of a solvent per gram of silica) powders for 1–6 h show both significant (porosity, PoSD, aggregation, adsorption) and insignificant (S_{BET}, PaSD) changes in the textural and adsorption characteristics. The MCA treatment enhances the bulk density (ρ_b) of the powder from 0.045 g/cm^3 for the initial silica to 0.4 g/cm^3 for 6 h-MCA-treated wetted silica. The ρ value depends on MCA time and the medium type, since it is smaller for MCA-treated ethanol- or water/ethanol-wetted powders. The specific surface area of MCA-treated samples does not practically change. However, PSD, porosity, particle aggregation and size distribution in aqueous suspension, the behavior of interfacial water, and adsorption of gelatin are the t_{MCA} and ρ_b dependent functions. Some of the changes in the characteristics (e.g., gelatin adsorption) depend on the ρ_b value but are independent of the medium type. Practically all the characteristics studied demonstrate a nonlinear dependence on both t_{MCA} and ρ_b values. MCA-treated silica is characterized by a third lower osmotic activity and four times as higher critical concentration of gel formation than the initial silica A-300. Obtained results show that the MCA treatment allows one to keep the specific surface area but strongly change textural porosity and some other characteristics of nanosilica that can be of importance on application of compacted fumed oxides.

Much stronger actions than MCA can be caused by HTT of nanosilicas. HTT can be used for significant changes of the particle morphology and structural and adsorptive properties of the silica materials in consequence of continuous dissolution–condensation and mass transfer. These effects depend on HTT conditions (temperature, pressure, and ratio between concentrations of water and solids) and a medium type (aqueous or steam; Gun'ko et al. 2004c). During HTT, nanosilica can transform its state from free-dispersion (with relatively weak interparticle interactions in aggregates of primary particles and agglomerates of aggregates with hydrogen and electrostatic binding) to porous bonded-dispersion (e.g., Silochrome). Dissolution of the smallest particles and deposition of $Si(OH)_4$, its oligomers, clusters, and fragments of small particles on larger ones accompany this process in the aqueous media. Similar processes in overheated water vapor do not lead to complete disappearance of the free-dispersion state, despite transferring of $Si(OH)_4$ complexes with water molecules, recondensation,

$$\equiv \mathrm{SiOH} + \mathrm{HOSi} \equiv \rightarrow \equiv \mathrm{Si} - \mathrm{O} - \mathrm{Si} \equiv + \mathrm{H_2O},$$

and growth of the particle size. In other words, continuous and tight linkage of primary particles by a large number of the ≡Si–O–Si≡ bridges (characteristic, e.g., for silica gel) does not occur. It is known (Iler 1979, Leboda et al. 1995) that HTT of porous silicas enhances the average pore size and reduces the specific surface area and a pore number because of disruption of pore walls upon intensive hydrolysis of the ≡Si–O–Si≡ bonds depending on the treatment temperature, and the surface of silica gel particles (0.1–0.5 mm in size) becomes sponge-like. For nanosilicas (Leboda et al. 1998a), structural changes affected by HTT may strongly differ from those observed for porous silica treated under the same conditions in consequence of the difference in the particle morphology and pore structure (Legrand 1998). Typically, the higher the specific surface area of modified nanosilica, the lower is the treatment temperature required for appearance of textural changes, for example, this temperature corresponds to 100, 150, and 200°C for A-380, A-300, and A-175,

TABLE 1.33
Structural Characteristics of Nanosilica Samples before and after HTT

Sample	T_{HTT} (°C)	S_{BET} (m²/g)	V_p (cm³/g)
OX-50	—	54	0.108
	250	40	0.065
	350	24	0.030
A-150	—	144	0.261
	150	129	0.377
A-200	—	192	0.452
	250	124	0.250
	350	86	0.157
A-210	—	206	0.411
	150	143	0.355
A-230	—	232	0.483
	250	89	0.170
	350	70	0.123
A-250	—	264	0.499
	250	131	0.325
	350	57	0.101
A-300	—	294	0.442
	250	89	0.105
	350	70	0.084
A-340	—	337	0.608
	150	240	1.436
A-380	—	381	0.667
	150	192	1.292
A-400	—	409	0.696
	250	118	0.331
	350	94	0.218

respectively (Leboda et al. 1998a). HTT of nanosilicas practically doubles the surface hydroxylation degree to 4.8–4.9 μmol/m² close to that of precipitated silicas (Iler 1979, Legrand 1998).

Nanosilicas (Table 1.33) and silica OX-50 (Degussa) were used as the initial materials to study the HTT effects (Gun'ko et al. 2004a). Nanosilicas were hydrothermally treated in the steam phase at T_{HTT} = 150°C, 250°C, and 350°C. Notice that three samples of A-200 and three samples of A-300 were labeled as A-200, A-210, and A-230 and A-250, A-300, and A-340, respectively, to distinguish them.

HTT was realized in a stainless steel autoclave (0.3 dm³). Sample (2 g) placed in a vessel in the autoclave containing 20 mL of water was treated at 250°C or 350°C for 6 h. After modification, samples were dried at 200°C for 6 h. Certain samples of nanosilica were hydrothermally treated at 150°C for 4 h using a larger ratio of silica/water (10.5 g/40 mL) to reduce mass transferring and structural changes.

The FTIR spectra of certain oxide samples (initial and after HTT at 350°C; Figure 1.201) were recorded over the 4000–400 cm⁻¹ range by means of a FTIR 1725× (Perkin-Elmer) spectrophotometer. Before FTIR measurements, samples were dried at 200°C for 2 h. They (0.33 wt%) were stirred with KBr (Merck, spectroscopy grade) and then pressed to form appropriate tablets. The baselines were subtracted from all the FTIR spectra shown in Figure 1.201.

Structural changes of fumed oxides occurring on HTT in the steam phase may be caused by such reasons (dependent, of course, on HTT conditions) as (i) mass transferring from smaller primary particles to larger ones and merger of these particles due to coverage of their contacts by new silica

TABLE 1.34
Bulk Density (g/cm³) of Certain Oxides after Synthesis, Stored Initial and Heated at 150°C, and Additionally Heated after Hydrothermal Treatment at 350°C

Sample	Initial (Just after Synthesis)	Heated Initial	Heated HTT
OX-50	0.130	0.127	0.139
A-200	0.054	0.143	0.211
A-400	0.061	0.083	0.164

layers; (ii) enhancement of the packing density of particles in aggregates and agglomerates and, consequently, the bulk density (ρ_b) of samples (Table 1.34), which is accompanied by reduction of the accessible surface area and changes in the pore (void) volume (calculated per gram of the adsorbent) in aggregates and empty volume ($V_{em} = 1/\rho_b - 1/\rho_0$, where ρ_0 denotes the true density of material) in a sample as a whole; (iii) impact of the "porosity" of secondary particles on mass transferring through the diffusion and kinetics factors; (iv) topographical reconstruction of the surfaces of primary particles becoming less smooth or even porous (sometimes crystalline such as treated fumed alumina); and (v) difference in reactivity of silica and other nanooxides (e.g., Al_2O_3 and TiO_2 in mixed oxides) in chemical reactions taking place on HTT (Gun'ko et al. 2004a).

To elucidate some aspects of the phenomena caused by HTT of fumed oxides (possessing different specific surface area, degree of aggregation of primary particles, chemical composition, and reactivity) in the steam phase, their structural changes occurring at different treatment temperatures are analyzed in the subsequent sections on the basis of the adsorption data, the FTIR spectra, the AFM images of initial and treated samples, and theoretical calculations (Gun'ko et al. 2004a).

From analysis of the nitrogen adsorption–desorption isotherms (attributed to type II of the conventional classification) for fumed oxides (Figure 1.199), one can conclude that the morphology of primary particles does not change cardinally during HTT in the steam phase, because the isotherm type is practically the same, despite decrease in the adsorption for many samples and certain changes in the isotherm shape. It means that the structure of fumed oxides after HTT under relatively mild conditions is not yet close to that of silica gel or xerogel, that is, oxides remain highly disperse. Primary particles grow but remain roughly spherical (however, their shape can be distorted because of nonuniform distribution of transferred fragments with preferred coverage of contacts between adjacent primary particles that is seen in the AFM images of treated samples) and practically nonporous and form secondary particles, whose structure corresponds to the disperse state, despite its significant alterations on HTT. The last assumption is based on the fact that main changes in the isotherm shapes are connected to a portion at high pressures corresponding to secondary (volume) filling of mesopores and macropores, that is, voids between primary particles in secondary structures. A minimal decrease in the nitrogen adsorption at $p/p_0 < 0.8$ is observed for silica samples treated at 150°C (Figure 1.199), despite increase in their bulk density (Table 1.34). Increase in the specific surface area S_{BET} of different initial nanosilicas as well as elevating T_{HTT} leads to greater changes in the structural and adsorptive properties of treated samples. Decrease in the nitrogen adsorption before the hysteresis loop onset for treated oxides is a result of reduction of S_{BET}, as the adsorptive capacity (i.e., effective "porosity" [Table 1.33, V_p] of secondary particles, as primary nanoparticles are typically nonporous) increases for many treated samples. For instance, HTT of nanosilica at 150°C leads to considerable enhancement of the effective porosity, which is maximal for treated A-340 and A-380 characterized by large hysteresis loops (Figure 1.199) and strongly enhanced mesoporosity (main $f(R_p)$ peak of mesopores is narrower than that for other samples; Table 1.33, V_p, Figure 1.200), despite diminution of S_{BET}. However, one

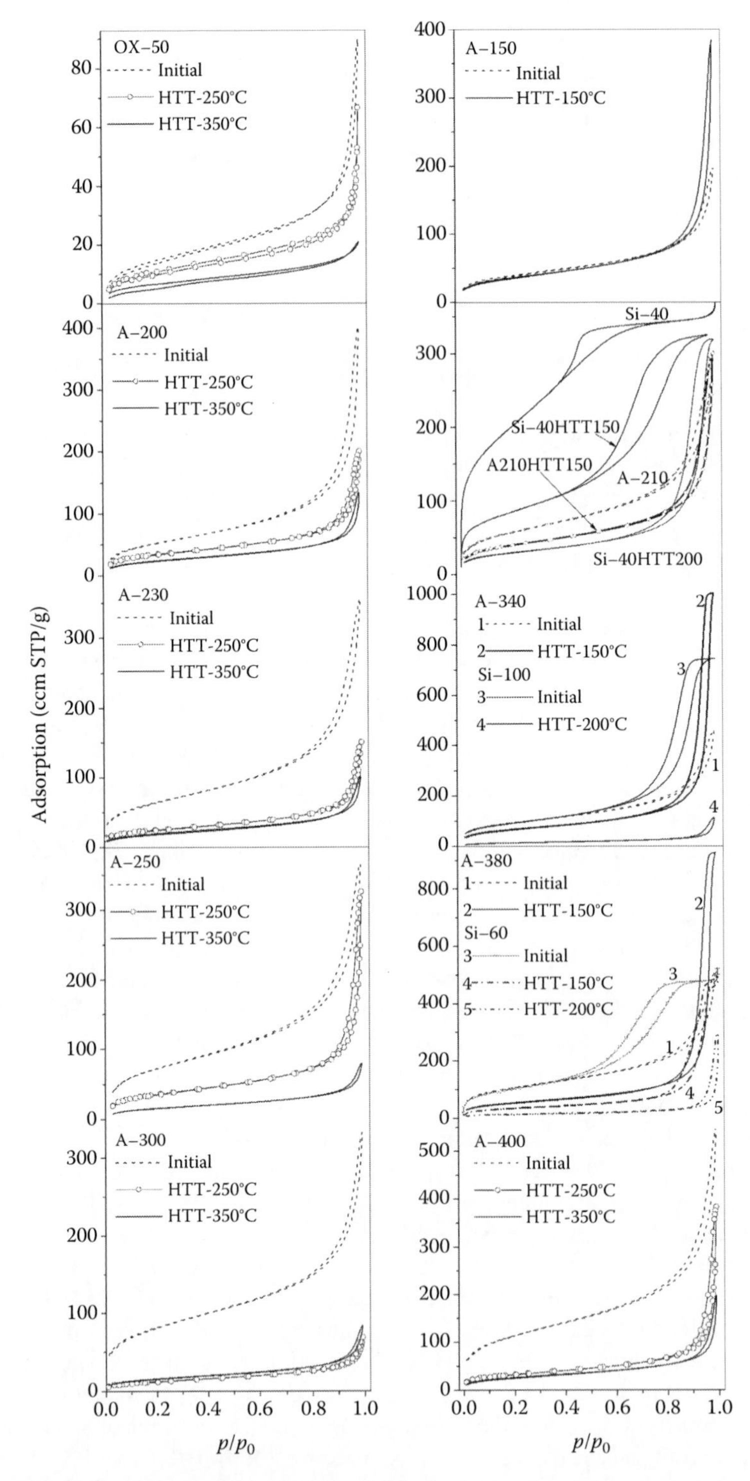

FIGURE 1.199 Nitrogen adsorption–desorption isotherms for initial and hydrothermally treated nanosilica and silica gel Si-40, Si-60, and Si-100 samples. (Adapted from *J. Colloid Interface Sci.*, 269, Gun'ko, V.M., Skubiszewska-Zięba, J., Leboda, R. et al., Influence of morphology and composition of fumed oxides on changes in their structural and adsorptive characteristics on hydrothermal treatment at different temperatures, 403–424, 2004a. Copyright 2004, with permission from Elsevier.)

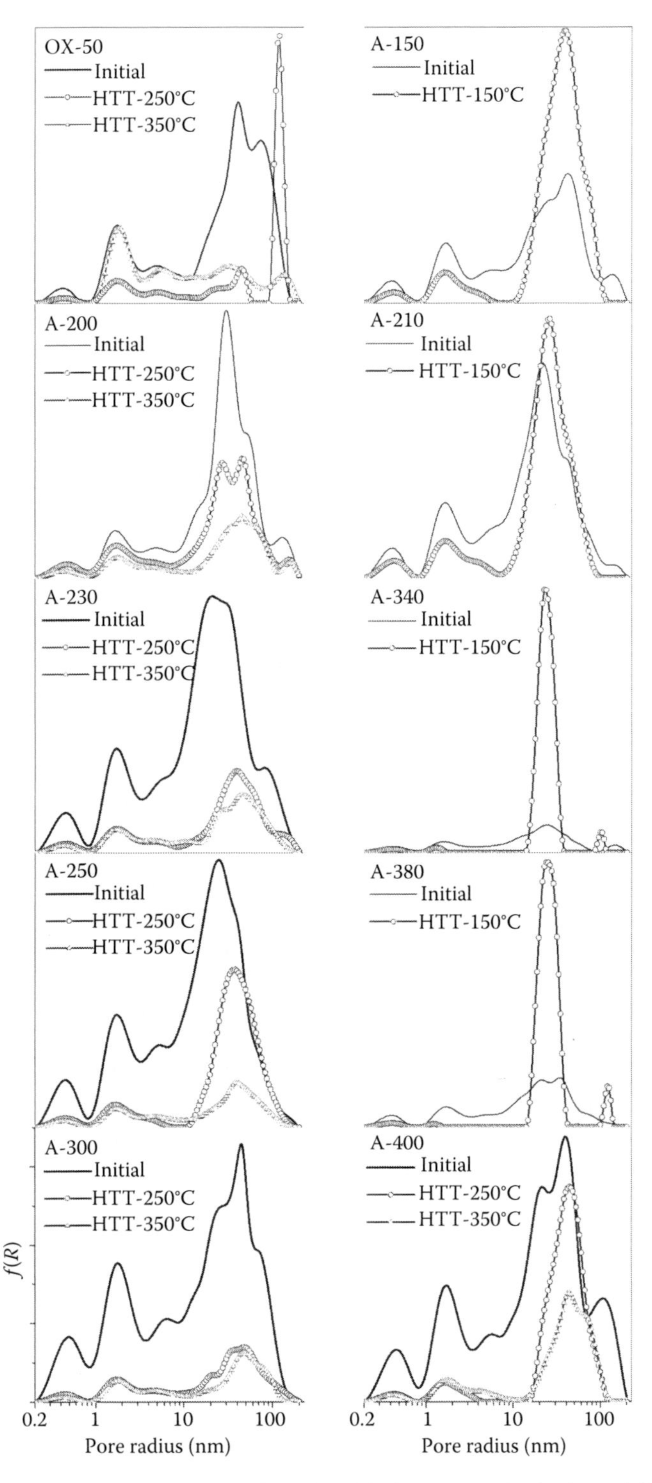

FIGURE 1.200 Pore size distributions (model of voids between spherical particles) for initial and hydrothermally treated nanosilicas. (Adapted from *J. Colloid Interface Sci.*, 269, Gun'ko, V.M., Skubiszewska-Zięba, J., Leboda, R. et al., Influence of morphology and composition of fumed oxides on changes in their structural and adsorptive characteristics on hydrothermal treatment at different temperatures, 403–424, 2004a. Copyright 2004, with permission from Elsevier.)

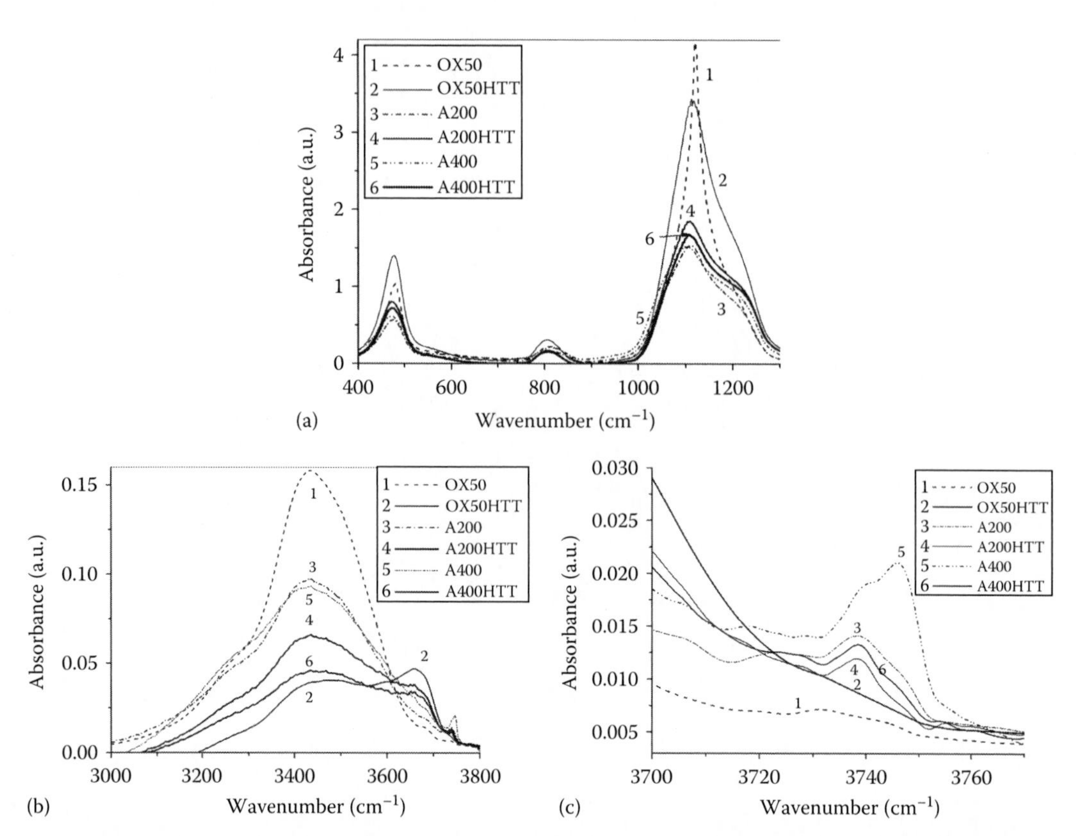

FIGURE 1.201 FTIR spectra of initial and HTT-350°C nanosilicas OX-50, A-200, and A-400 over different ranges: (a) 400–1300 cm^{-1}, (b) 3000–3800 cm^{-1}, and (c) 3700–3770 cm^{-1}. (Adapted from *J. Colloid Interface Sci.*, 269, Gun'ko, V.M., Skubiszewska-Zięba, J., Leboda, R. et al., Influence of morphology and composition of fumed oxides on changes in their structural and adsorptive characteristics on hydrothermal treatment at different temperatures, 403–424, 2004a. Copyright 2004, with permission from Elsevier.)

should remember that this porosity corresponds to only a portion of the empty volume in the fumed oxide powder. The volume of nanopores (R_p < 1 nm) corresponds to an insignificant portion of the total porosity for all samples, which is clearly seen from the PSDs and decreases with growing size of primary particles and reduction of their aggregation. Thus changes in the adsorption caused by alterations of the oxide texture on HTT can be explained by changes in the structure not only of primary particles (changes in their size and surface topography) but also of their aggregates and agglomerates (Figures 1.202 through 1.204).

The FTIR spectra of silica samples after HTT at 350°C demonstrate appearance of a band at 3650–3680 cm^{-1} (Figure 1.201b) connected to groups ≡SiO-H in hard-to-reach places (i.e., inner OH groups [Legrand 1998]) in shallow but narrow pores, or between adjacent particles, or shielded by deposited silica fragments. Therefore, these silanols are undisturbed by adsorbed water molecules, as $\Delta\nu_{OH}$ < 100 cm^{-1} is too small for adsorption complexes of water molecules bound by the hydrogen bonds. This band intensity is maximal for OX-50HTT-350°C (Figure 1.201b). A broad band at 3500 cm^{-1} (corresponding to OH groups of adsorbed water molecules and silanols disturbed by these molecules) is maximal for OX-50, which can be caused by some differences of its surface structure in comparison with other nanosilicas (it should be noted that OX-50 behavior on HTT differs from that for other nanosilicas). Observed regularities in the FTIR spectra can be explained by a relatively low temperature of HTT (in comparison with temperature of the pyrogenic synthesis), that is, the silica particle skeleton is not distinct but the particle surfaces (certain surface layer) change in result of transferring of $Si(OH)_4$ and related compounds from one particle to another

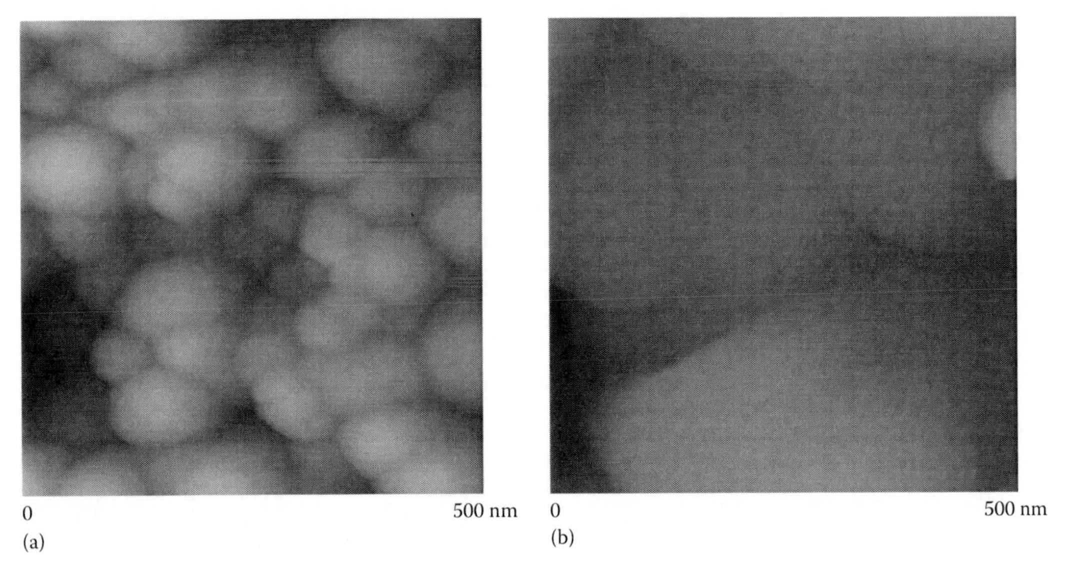

FIGURE 1.202 AFM images (500×500 nm^2) of primary particles of OX-50 packed in aggregates (a) initial and (b) after HTT at 350°C (Nanoscope III (Digital Instruments, USA) apparatus with a tapping mode AFM measurement technique). (Adapted from *J. Colloid Interface Sci.*, 269, Gun'ko, V.M., Skubiszewska-Zięba, J., Leboda, R. et al., Influence of morphology and composition of fumed oxides on changes in their structural and adsorptive characteristics on hydrothermal treatment at different temperatures, 403–424, 2004c. Copyright 2004, with permission from Elsevier.)

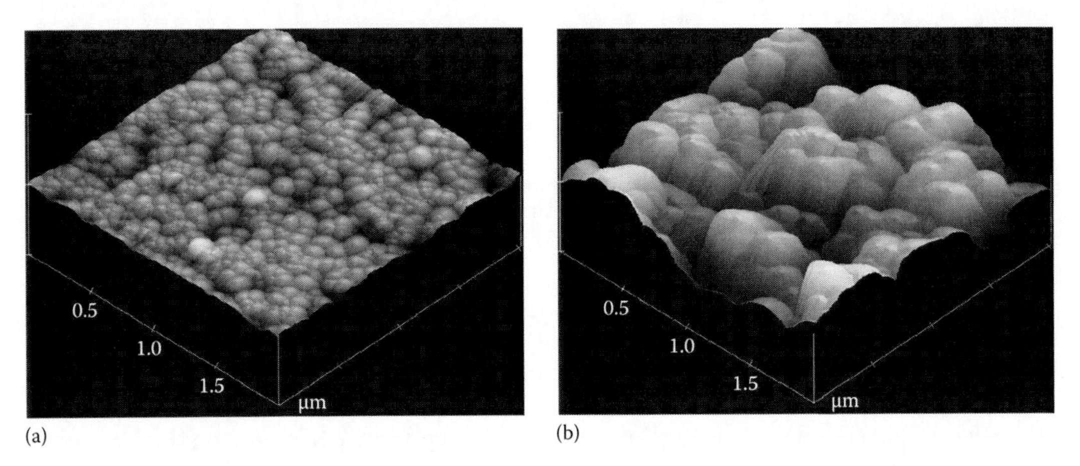

FIGURE 1.203 3D AFM images of primary particles of OX-50 (a) initial and (b) after HTT at 350°C (2×2 μm^2), x and y scale division is 500 nm, and 1000 nm/div for z. (Adapted from *J. Colloid Interface Sci.*, 269, Gun'ko, V.M., Skubiszewska-Zięba, J., Leboda, R. et al., Influence of morphology and composition of fumed oxides on changes in their structural and adsorptive characteristics on hydrothermal treatment at different temperatures, 403–424, 2004c. Copyright 2004, with permission from Elsevier.)

on HTT, and not all silanols of transferred fragments and target particle surfaces react to form the ≡Si–O–Si≡ bonds. Subsequent silica deposits shield these remained ≡SiO-H groups (proving in the band ν_{OH} at 3650–3680 cm^{-1}), which become inaccessible (after completion of HTT) for adsorbed water molecules.

HTT of A-400 leads to changes in the ν_{OH} band intensity at 3747 and 3740 cm^{-1} (free single and twin silanols) and the ratio between the intensity of these bands (Figure 1.201c). In the case of A-200, similar changes are smaller and the band shapes for initial and treated samples are close.

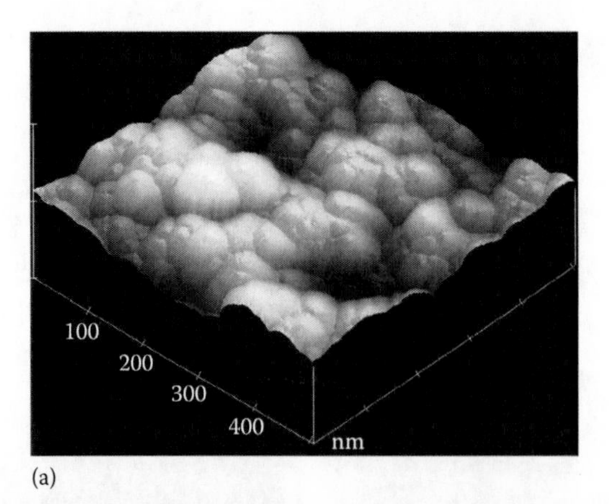

(a)

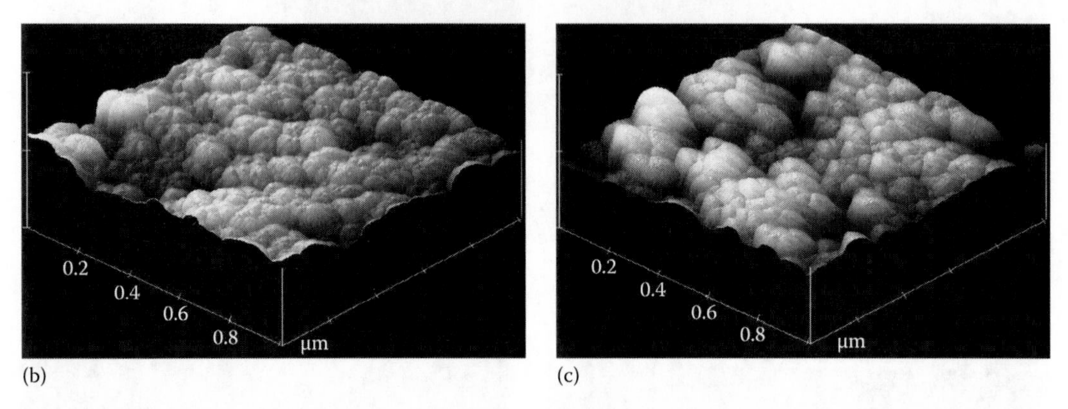

(b) (c)

FIGURE 1.204 3D AFM images of A-400 (a) initial (500 × 500 nm²), x and y scale division is 100 nm, and 200 nm for z, and (b) (1 × 1 μm²), and (c) after HTT at 350°C (1 × 1 μm²) (primary and secondary particles are visible), (b, c) x and y scale division is 200 nm, and 500 nm/div for z. (Adapted from *J. Colloid Interface Sci.*, 269, Gun'ko, V.M., Skubiszewska-Zięba, J., Leboda, R. et al., Influence of morphology and composition of fumed oxides on changes in their structural and adsorptive characteristics on hydrothermal treatment at different temperatures, 403–424, 2004a. Copyright 2004, with permission from Elsevier.)

This difference for A-400 and A-200 can be caused by larger structural changes of A-400 on HTT (Table 1.33 and Figures 1.200 and 1.204) also reflected in the isotherm shapes (Figure 1.199).

Maximal changes in the Si-O stretching vibrations over the $\nu_{SiO} = 1000–1250$ cm^{-1} range are observed for OX-50 after HTT-350°C, as the band becomes more broadened (Figure 1.201a). The ν_{SiO} band at 1100 cm^{-1} becomes more intensive for all samples after HTT as a result of growth of the silica particle size.

According to the AFM images of OX-50 (Figures 1.202 and 1.203) with the minimal initial S_{BET} value (Table 1.33), and A-400 (Figure 1.204) with maximal initial S_{BET} value (Table 1.33), HTT at 350°C results in the formation of denser aggregates and agglomerates. However, changes in the structure of primary particles depend on their sizes and the chemical composition of oxides (Gun'ko et al. 2004a). For instance, the size of primary particles of treated OX-50 and A-400 increases from 50 to 114 nm and from 7 to 29 nm, respectively, according to their S_{BET} values.

Thus, HTT of pure nanosilicas in the steam phase leads to significant structural changes in the primary and secondary particles depending not only on their specific surface area, the treatment temperature and other conditions but also on the density of packing of primary particles in aggregates that is clearly seen from comparison of structural changes of fumed oxides and silica gels (Gun'ko et al. 2004a). During HTT, the average size of primary silica particles increases (greater for samples with primary particles denser packed in secondary particles) because of mass transferring from smaller particles to larger ones and their tighter merger due to coverage of their contacts by transferred silica. Clearly, the latter process can be more effective in denser aggregates (as the distances of mass transferring are shorter), which appears in greater structural changes of silicas with a large specific surface area characterized by greater aggregation of primary particles revealed from the nitrogen adsorption–desorption isotherms.

Thus, the structural and adsorption properties of nanosilicas depend not only on the synthesis conditions such as the ratio of reactants ($SiCl_4$, O_2, and H_2), temperature, flow velocity, turbulence and length of the flame, and additional external actions (e.g., electrostatic field) but also on heating, hydrothermal, or mechanochemical treatment that affect the structural, textural, and morphological characteristics, and therefore, the interfacial phenomena at a surface of treated nanosilicas.

1.1.12 Cryoporometry and Relaxometry

NMR cryoporometry method was developed (Overloop and Van Gerven 1993, Strange et al. 1993, Akporiaye et al. 1994) on the basis of the thermodynamic theory of the phase transition in materials in confined space pioneered by J.W. Gibbs, J. Thomson, and W. Thomson. According to Gibbs–Thomson equation for the melting point depression, ΔT_m, for a small isolated spherical crystal of diameter x in its own liquid can be written as (Jackson and McKenna 1990)

$$\Delta T_m = T_m(x) - T_{m,\infty} = \frac{4\sigma_{sl} T_{m,\infty}}{\Delta H_f \rho x} \tag{1.45}$$

where

$T_m(x)$ is the melting temperature of ice crystallite of diameter x
$T_{m,\infty}$ is the bulk melting temperature
ρ is the density of solid
σ_{sl} is the surface energy of solid–liquid interface
ΔH_f is the bulk enthalpy of fusion

If we assume that the interactions between the pore walls and crystal and liquid located there have a weak influence on the phase transition that Equation 1.45 can be interpreted as the dependence between the pore size (x) and melting temperature of ice in this pore. However, a more accurate equation should include the contact angle φ (which is commonly assumed to be 0° in the liquid–vapor [Kelvin] case and 180° in the solid–liquid [Gibbs–Thomson] case), that is, cos(φ) as a parameter describing interactions in the system vapor–liquid–solid in pores (Mitchell et al. 2008):

$$\Delta T_m = T_m(x) - T_{m,\infty} = \frac{4\sigma_{sl} T_{m,\infty}}{\Delta H_f \rho x}\cos(\phi) \tag{1.46}$$

where

$$\cos(\phi) = W_{ps}/\sigma_{sl} - 1 = (\sigma_{sl} + \sigma_{pl} - \sigma_{ps})/\sigma_{sl} - 1 \tag{1.47}$$

s, l, and p are related to solid (ice), liquid (water), and pore wall. The effects of the cos(φ) values on the melting effects were observed for different adsorbents (Mitchell et al. 2008). This aspect could be more important on the use of organic liquids as probe compounds because the contact angle φ in the this case can significantly differ from zero. However, in this book, we mainly use water as a probe compound; therefore, condition $\varphi = 0$ is quite appropriate for hydrophilic materials.

A simple equation derived from Equation 1.45

$$\Delta T_m = \frac{k_{GT}}{x} \tag{1.48}$$

where k_{GT} is a certain constant, which can be written as (Webber et al. 2007a,b, Mitchell et al. 2008)

$$\Delta T_m = \frac{k_g k_s k_i}{x} \quad (1.49)$$

where k_g is a geometric interface-dependent constant, $k_s = T_{m,\infty}/\Delta H_f \rho$ is a constant dependent on type of the system, and k_i is a constant dependent on free surface energy term $k_i = \sigma_{sl} - W_{ps}$ or $k_i = \sigma_{ps} - \sigma_{pl}$ (Mitchell et al. 2008). Equation 1.49 is more appropriate to describe the effects in pores of different geometry (spherical, cylindrical, slit-shaped). Equation 1.48 can be transformed into the Strange–Rahman–Smith (Strange et al. 1993) appropriate to calculate the PSD:

$$\frac{dv}{dx} = \frac{dv}{dT_m(x)} \frac{k_{GT}}{x^2}, \quad (1.50)$$

where v is the volume of a particular pore size x. Additionally, Equation 1.48 can be transformed into differential equation assuming that $\frac{dV_{uw}(x)}{dx} = \frac{dC_{uw}}{dx}$ and using $\frac{dT_m(x)}{dx} = \frac{k_{GT}}{x^2} = \frac{(T_m(x) - T_{m,\infty})^2}{k_{GT}}$ that give

$$\frac{dV_{uw}(x)}{dx} = \frac{A}{k_{GT}} (T_m(x) - T_{m,\infty})^2 \frac{dC_{uw}(T)}{dT}, \quad (1.51)$$

where

$V_{uw}(x)$ is the volume of unfrozen water in pores of radius x

C_{uw} the amount of unfrozen water per gram of adsorbent as a function of temperature

A is a constant

Equations 1.48 and 1.51 can be transformed into integral equation on replacing $dV_{uw}(x)/dx$ by $f_V(x)$, converting dC_{uw}/dT to dC_{uw}/dx as $dT = \frac{dT}{dx} dx$ and integrating by x,

$$C_{uw}(T_m) = A' \int_{x_{min}}^{x_{max}} \left(\frac{k_{IGT}}{(T_{m,\infty} - T_m(x))x} \right)^2 f_V(x)\, dx, \quad (1.52)$$

where

x_{max} and x_{min} are the maximal and minimal pore radii (or sizes of unfrozen liquid structures), respectively

A' is a normalization factor depending on the values of units used in Equation 1.52

This equation describes both exact Gibbs–Thomson relationship and random deviation in the x value at a certain T value because of nonuniformity of the adsorbent studied. Additionally, the formation of a solid (ice)—liquid—pore wall "sandwich" structure can lead to deviation from the Gibbs–Thomson equation (Petrov and Furó 2009). In the latter review, they considered the effects of the formation of the mentioned sandwich on the dependence of the melting temperature on the pore size expressed via the surface area and the volume characteristics of the sandwich model. Some results discussed in this book could be interpreted in the term of this sandwich model because of broad PSDs, which cause a significant difference in the values of the force field near the pore walls and in the center of broad mesopores of macropores.

Notice that not only water can be used as a probe compound in NMR cryoporometry but also organic, organosilicon, and other liquids (Petrov and Furó 2006, 2009, 2010, 2011, Vargas-Florencia et al. 2006, 2007, Petrov et al. 2007, Mitchell et al. 2008) because the Gibbs–Thomson effect is observed for any liquid confined in pores. The temperature dependences of the NMR signal intensity for freezing and melting of confined liquid are characterized by the hysteresis effect. Therefore, the PSDs calculated from the melting and freezing curves demonstrate a certain shift, which is informative on the pore shape.

Petrov and Furó (2006) analyzed freezing–melting hysteresis in cryoporometry and attributed it to a free energy barrier between metastable and stable states of pore-filling material. Freezing point depression can be given by (Petrov and Furó 2006)

$$\Delta T_{\mathrm{f}} \cong -\frac{\nu\gamma_{\mathrm{sl}}T^{0}}{\Delta H}\frac{S}{V}, \tag{1.53}$$

while the melting point depression is given by

$$\Delta T_{\mathrm{m}} \cong -\frac{\nu\gamma_{\mathrm{sl}}T^{0}}{\Delta H}\frac{\partial S}{\partial V}, \tag{1.54}$$

where

ν is the molar volume
γ_{sl} is the surface free energy
T^0 is the bulk melting point
ΔH is the latent heat of melting
S is the surface area of the pore
V is the volume of the pore

Using Steiner's formula for equidistant surfaces, Petrov and Furó (2006) showed that Equation 1.54 can be rewritten as

$$\Delta T_{\mathrm{m}} \cong -\frac{\nu\gamma_{\mathrm{sl}}T^{0}}{\Delta H}2\kappa = \Delta T_{\mathrm{f}}\frac{2\kappa V}{S}, \tag{1.55}$$

where κ is the integral mean curvature of the pore surface. For an open-ended cylindrical pore $2\kappa V/S = 0.5$, and, hence, the numerical difference in ΔT_{m} and ΔT_{f} can be used to determine whether a pore has a cylindrical geometry. It is then also implicit in Equation 1.55 that, upon freezing, liquid solidifies along the pore from the end where the liquid is in contact with bulk solid, while melting is initiated at the liquid film at the pore surface and propagates from the surface toward the pore bulk (Perkins et al. 2008).

Notice that Equation 1.52 was obtained from Equation 1.48 with simple mathematical procedures and only one assumption that the volume of pores corresponds to the volume of unfrozen water and the density of this water is equal to 1 g/cm^3 without any additional assumption or simplification. However, solution of Equation 1.52 corresponds to well-known ill-posed problem due to the impact of noise on measured data (i.e., the NMR signal versus temperature), which does not allow one to utilize exact inversion formulas or iterative algorithms. Therefore, Equation 1.52 (as well as other integral equations used in this book) can be solved using a regularization procedure based on the CONTIN algorithm (Provencher 1982) under nonnegativity condition ($f(x) \geq 0$ at any x) and an fixed or unfixed value of the regularization parameter (α) determined on the basis of the F-test and confidence regions using the parsimony principle.

To more accurately calculate the distribution function on the basis of the IGT equation, an additional regularizer was derived using the maximum entropy principle (Muniz et al. 2000) applied to $f_V(R)$ written as N-dimension vector (N is the number of the grid points for f):

$$\mathrm{VAR} + \alpha^2 \left(1 - \frac{S(\vec{p}(\vec{f}))}{S_{\max}} \right) \rightarrow \min, \tag{1.56}$$

where
VAR is the regularizer
α the regularization parameter
S is the entropy

$$\vec{p}^{\,0}(\vec{f}) = \vec{f},\ p_i^1(\vec{f}) = f_{i+1} - f_i + (f_{\max} - f_{\min});\ i = 1, \ldots N-1,$$

$$p_i^2(\vec{f}) = f_{i+1} - 2f_i + f_{i-1} + 2(f_{\max} - f_{\min}),\ i = 2, \ldots N-1, \tag{1.57}$$

$$S(\vec{f}) = -\sum_{k=1}^{N} s_k \ln(s_k),\ s_k = \frac{f_k}{\sum_{k=1}^{N} f_k},\ \text{and } S_{\max} = -\ln\left(\frac{1}{N}\right).$$

The $\vec{p}^{\,j}(\vec{f})$ vector corresponds to the maximum entropy principle of the j-order. This procedure was used to modify the CONTIN algorithm (CONTIN/MEM-j, where j denotes the order of $\vec{p}^{\,j}(\vec{f})$). A self-consistent regularization procedure (starting calculations were done without application of MEM) with an unfixed regularization parameter (for better fitting) was used on CONTIN/MEM-j calculations, which were applied for mesoporous ordered silicas (vide infra).

The cryoporometry results for the PSDs can be used to estimate contributions of pores of different sizes to the total porosity. The $f_V(R)$ function can be converted into the distribution function $f_S(R)$ with respect to the specific surface area:

$$f_S(x) = \frac{w}{x}\left(f_V(x) - \frac{V(x)}{x} \right), \tag{1.58}$$

where $w = 1$, 2, and 1.36 for slit-shaped, cylindrical pores, and voids between spherical particles packed in the cubic lattice, respectively. Integration of the $f_S(x)$ function determined with Equation 1.58 on the basis of the IGT equation gives the specific surface area (S_{IGT}) of the studied materials in contact with structured water (S_{IGT} values are shown in Tables 1.35 and 1.36). Integration of the $f_V(x)$ and $f_S(x)$ functions at $x < 1$ nm, $1 < x < 25$ nm, and $x > 25$ nm gives the volume and the specific surface area of nano-, meso-, and macropores, respectively. Notice that the cryoporometry was used in the this book mainly with the model of cylindrical pores.

In the case of TSDC cryoporometry, modified Equations 1.48 and 1.52 with k as a linear function of temperature

$$k(T) = 40 + \frac{5}{6}(T - 90)\ (\mathrm{K\ nm}) \tag{1.59}$$

at T between 90 and 270 K obtained on the basis of the calibration curves for silica gels (Si-40 and Si-60) were used to estimate the PSDs.

TABLE 1.35
Structural Characteristics of Wetted Powders and Suspensions with Nanosilica A-300, Silylated A-300 (DMS/A-300), and a Mixture with 80 wt% of A-300 and 20 wt% of DMS/A-300 (Mix)

Parameter	A-300	DMS/A-300	Mix	Mix	Mix	Mix	Mix
h (g/g)	17.0	17.0	0.092	0.182	0.332	4.0	17.0
S_{IGT} (m^2/g)	240	322	195	304	262	348	310
V_{IGT} (cm^3/g)	0.434	0.615	0.091	0.181	0.324	0.471	0.465
γ_S (J/g)	50.2	67.6	15.0	26.5	30.8	35.2	36.2

TABLE 1.36
Structural Characteristics of PMS Hydrogels and Dried/Wetted Xerogel

	C_{PMS} (wt%)				
Parameter	**1.25**	**2.5**	**5.0**	**10.0**	**28.0**
S_{IGT} (m^2/g)	177	243	201	197	225
$S_{IGT,nano}$ (m^2/g)	62	125	132	131	116
$S_{IGT,meso}$ (m^2/g)	115	108	64	65	104
$S_{IGT,macro}$ (m^2/g)	—	10	5	1	5
$V_{IGT,nano}$ (cm^3/g)	0.014	0.023	0.020	0.052	0.036
$V_{IGT,meso}$ (cm^3/g)	0.265	0.144	0.102	0.060	0.416
$V_{IGT,macro}$ (cm^3/g)	—	0.286	0.131	0.041	0.073
γ_S (J/g)	19.3	23.2	12.6	10.1	30.4
ΔG_s (kJ/mol)	−2.92	−3.71	−2.87	−3.44	−4.05
C_s (mg/g)	140	130	105	70	150
C_{uw} (mg/g)	280	454	259	153	527

Note: C_s and ΔG_s corresponds to strongly bound water.

Multidimensional NMR spectroscopy provides more information than the 1D one since it allows us to establish correlations between variables and distinguish between physical or chemical environments (Elyashberg et al. 2008, Mitchell et al. 2012). Data having two or more dimensions are encountered commonly in spectroscopy (Keeler 2005, Levitt 2008) and imaging (Haacke 1999, Hertrich 2008), where the conversion from time domain to frequency domain is achieved by the application of a Fourier transform (FT) along each dimension. The FT, and its inverse, provides essentially exact solutions that are stable in the presence of noise. Such computations are achieved efficiently using the fast FT algorithm. Mitchell et al. (2012) analyzed multidimensional experiments, which probe relaxation times (longitudinal T_1 or transverse T_2) and self-diffusion coefficients (D); these NMR parameters are sensitive to the local physical environment in addition to providing chemical information. Since, in the general case, there may be a range of T_1, T_2, or D values characterizing a given system, to interpret these data, it is necessary to establish the distributions of these parameters (Mitchell et al. 2012). In this book, we used mainly the CONTIN algorithm (Provencher 1982) improved with maximum entropy method and self-consistent regularization (Gun'ko 2000b, Gun'ko et al. 2007f).

Low-field NMR (LF-NMR) spectroscopy is a developing field having a significant potential as an important analytical tool for reaction monitoring, in hyphenated techniques and for systematic investigations of complex mixtures (Dalitz et al. 2012) and interfacial phenomena in these liquids. Solid and soft materials can be investigated using LF-NMR for deeper insights into structures and dynamics. Developments in electronics and magnet technology should allow the development of new dedicated pulses and NMR sequences which, together with gradient assisted spectroscopy could open up a large field for improving performance and consequently for the application of LF-NMR instrumentation. In both time-domain NMR and high-field NMR, a huge amount of experience is available, which can be explored and adapted to the needs of medium resolution (MR) NMR, thus enabling methodic development in MR-NMR, which could be a valuable tool for reliable and reproducible process and reaction monitoring even in industrial environments. Apart from convenient fluidic reactions (Figure 1.205), a wide application range is possible for MR-NMR, starting from quasi-solid reactions, followed by the measurement of relaxation weighted spectra or relaxation rates up to gas phase reactions using hyperpolarization techniques as dynamic nuclear polarization or para-hydrogen induced polarization (Dalitz et al. 2012).

Most often the samples under investigation do not contain just one species with ^1H so that a contrast parameter is needed for discrimination of different ^1H signals (Dalitz et al. 2012). Relaxation properties, especially transverse relaxation T_2, are suitable for such a contrast creation, mainly for highly viscous fluids and solids. The main relaxation path is the homonuclear dipolar interaction,

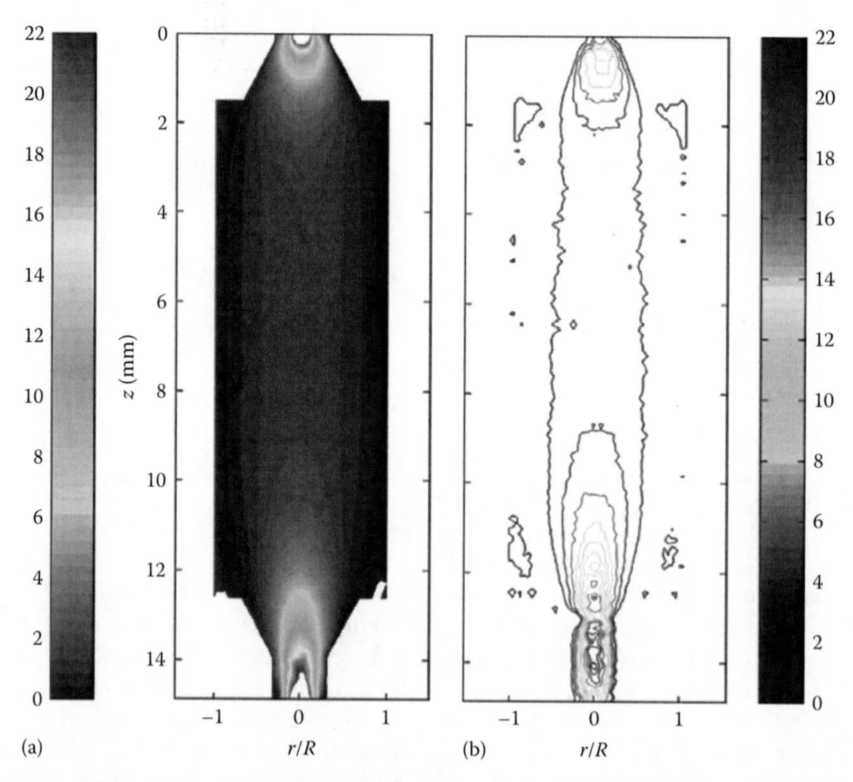

FIGURE 1.205 Characterization of a MR-NMR fluid cell: comparison of (a) computational fluid dynamics (CFD) simulation and (b) NMR velocity imaging measurement. The volume flow rate was 1 mL/min, the axis represents the velocity at each spatial point. A qualitatively good agreement is found, dead zones and deviations from laminar flow occur especially at the inlet. (Adapted from *Prog. Nuclear Magn. Reson. Spectr.*, 60, Dalitz, F., Cudaj, M., Maiwald, M., and Guthausen, G., Process and reaction monitoring by low-field NMR spectroscopy, 52–70, 2012. Copyright 2012, with permission from Elsevier.)

which is given by the squared coupling strength and the spectral densities at defined frequencies (Abragam 1989, Kimmich 1997):

$$\frac{1}{T_1(\tau_c)} = 0.3\left(\frac{\mu_0}{4\pi}\right)^2 \frac{\gamma^4 h^2}{4\pi^2 r^6}\left(\frac{\tau_c}{1+\omega_0^2\tau_c^2} + \frac{4\tau_c}{1+4\omega_0^2\tau_c^2}\right) \quad (1.60)$$

$$\frac{1}{T_2(\tau_c)} = 0.15\left(\frac{\mu_0}{4\pi}\right)^2 \frac{\gamma^4 h^2}{4\pi^2 r^6}\left(\frac{3\tau_c}{1+\Delta\omega^2\tau_c^2} + \frac{5\tau_c}{1+\omega_0^2\tau_c^2} + \frac{2\tau_c}{1+4\omega_0^2\tau_c^2}\right) \quad (1.61)$$

where
- T_1 is the longitudinal relaxation time
- τ_c defines the motional correlation time
- μ_0 is the magnetic permeability
- γ is the gyromagnetic ratio
- h is the Planck constant
- r is the distance of nearby ^1H
- ω_0 is the Larmor frequency
- $\Delta\omega$ is the line width for rigid molecules

In the case of macromolecules with restricted molecular motion, these material properties can be exploited in order to discriminate among different signal contributions and facilitate the assignment and quantitative determination of the sample properties. Apart from sophisticated analytical approaches to describe the molecular mobility and its impact on ^1H NMR relaxation in detail, more often, empirical relations are found and exploited. For example, the transverse relaxation of elastomers can be related to the elasticity modulus; in other examples a relation to the viscosity can be established. A third approach is the phenomenological approach, which does not consider the origin of relaxation at all, but relates the signal directly with the material property searched for. The optimization of pulse sequence parameters is done by statistical means, which provide the best correlation and the lowest statistical deviation between NMR signal and material properties (Dalitz et al. 2012).

In liquids, simple functions are applied like Lorentzian, Gaussian, or Voigt lines, taking instrumental and physical facts into account. Best results are currently achieved using a Lorentz–Gauss function for representing NMR peaks (Dalitz et al. 2012):

$$I(\delta) = I_{max}\frac{\exp(-b^2(\delta-\delta_{max})^2}{1+a^2(\delta-\delta_{max})^2} \quad (1.62)$$

The function includes four adjustable parameters: the maximum intensity of the peak, I_{max}, the chemical shift at maximum intensity of the peak, δ_{max}, and the Lorentz and Gaussian parameters a and b, respectively.

The ability of NMR imaging, spectroscopy, and relaxometry to probe noninvasively fluids within porous media provides many opportunities for characterizing fluids and flow in porous media (Watson and Chang 1997). The sensitivity of fluid relaxation to solid surfaces and pore topology provides a number of opportunities for characterizing pore structures and surface properties. Suitable analysis of experimental data, particularly of pulsed-field-gradient (PFG) experiments, allows to estimate a number of different properties at both the microscopic and macroscopic scale. The transverse relaxation times can be measured with Carr–Purcell–Meiboom–Gill (CPMG; Elyashberg et al. 2008) pulse sequence: $90x$-(τ-$180y$-τ-echo)$_n$, here with 90° pulse of 3 μs and total echo time of 500 μs. A total of eight scans were used for each measurement. The relaxation time between scans was 0.5 s. Equilibration time of each temperature was 5 min. To calculate a distribution function of transverse relaxation time $f(T_2)$, CPMG echo decay envelopes ($I(t)$) were used as the left term of integral equation:

$$I(t) = A \int_{T_{2,\min}}^{T_{2,\max}} \exp\left(-\frac{t}{T_2}\right) f(T_2) dT_2, \tag{1.63}$$

where

t is the time

$T_{2,\min}$ and $T_{2,\max}$ are the minimal (10^{-6} s) and maximal (1 s) T_2 values on integration, respectively

A is a constant

Equation 1.63 can be solved using a regularization procedure based on the CONTIN algorithm under nonnegativity condition ($f(T_2) \geq 0$ at any T_2) and an unfixed value of the regularization parameter (α) determined on the basis of the F-test and confidence regions using the parsimony principle. It should be noted that the solution of Equation 1.80 is insensitive to the A value, which can affect only relative intensity of $f(T_2)$.

In the case of the use of the NMR relaxometry for estimation of the sizes of the water and benzene (or other liquids) structures (layers, clusters, etc.) filling pores, Equation 1.63 can be transformed for consideration for the dependence of the CPMG echo decay envelopes (i.e., transverse relaxation time) on the pore size:

$$I_i(t) = B \int_{R_{\min}}^{R_{\max}} \exp\left(-\frac{t}{T_{2,i,\mathrm{m}}}\right) \frac{(T_{\mathrm{m},\infty,i} - T_{\mathrm{m},i}(x))}{k_i} f_i(x)\, dx, \tag{1.64}$$

where

B is a normalization factor

k_i is a constant analogous to that in Equations 1.48 and (1.52)

Figure 1.206 shows the PSD (or UWCSD) calculated using the NMR and TSDC cryoporometry methods, NMR relaxometry (aqueous suspensions), and nitrogen adsorption method.

The concentrations of A-300 (3–7 wt%) in the aqueous suspensions were smaller than the value corresponding (10–13 wt%) to the bulk density of the A-300 powder. Additionally, during preparation of the suspensions (sonication or MCA), agglomerate of aggregates can be destroyed and aggregates of primary particles can be rearranged (Gun'ko et al. 2001e, 2005d, 2007f, 2009d). Therefore, the PSDs of A-300 being in the aqueous suspensions or as degassed powder differ. However, there are certain regularities: (i) small contribution of nanopores ($R<1$ nm) and narrow mesopores ($1<R<3$ nm); (ii) mesopores at $R>5$ nm give the main contributions to the pore volume; and (iii) the cryoporometry methods are weakly sensitive to large pores if water is used as a probe compound.

Comparison of the results obtained using nonintegral and integral Gibbs–Thomson equations (Figure 1.207) shows that the PSD functions are similar but with certain differences caused by the aforementioned factors. The solution of the integral equation using the regularization procedure gives smoother PSD and it can be extended toward larger pore sizes due to method power (similar to image reconstruction from a small part).

Thus, different versions of NMR and TSDC cryoporometry (as well as DSC thermoporometry (Landry 2005), vide infra) and relaxometry allow one to obtain the structural characteristics not only for powders but also for strongly hydrated systems that is of importance for biomaterials, cells, and other bio-objects, which can be studied in the native state. In other words, these methods are nondestructive methods.

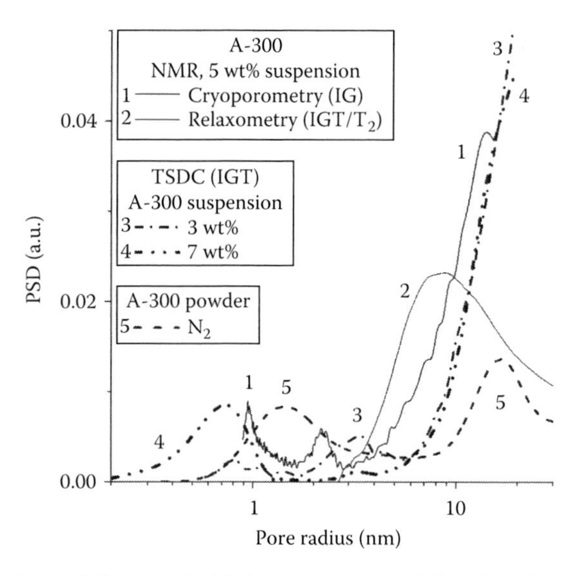

FIGURE 1.206 Comparison of the pore (voids between nanoparticles) size distributions calculated using four methods: (1) NMR cryoporometry and (2) NMR relaxometry, (3) and (4) TSDC cryoporometry (aqueous suspensions at $C_{A\text{-}300}=3$–7 wt%), and (5) PSD calculated using the nitrogen adsorption isotherm (SCV/SCR model, see Section 1.1.1).

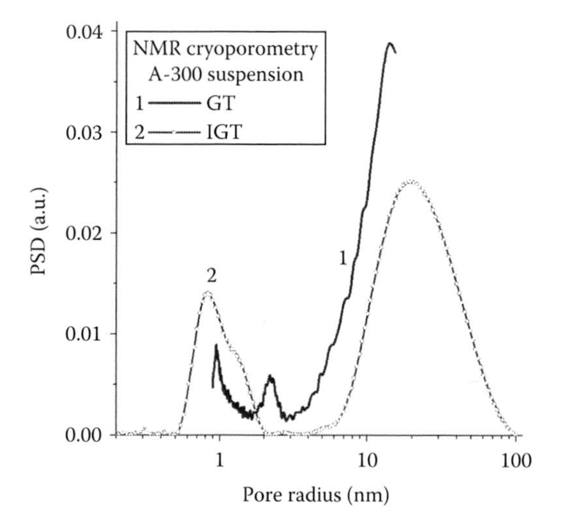

FIGURE 1.207 Comparison of the PSD calculated for 5 wt% suspension of A-300 using nonintegral (1) and integral (2) Gibbs–Thomson equations.

1.1.13 Comparison of NMR, DSC, TSDC, Adsorption Data, and Quantum Chemical Models

It was shown in the previous chapters that the behavior of bound water is affected by many factors, which could be elucidated from comparison of the results obtained such methods as low temperature ^{1}H NMR spectroscopy, TSDC, DSC, adsorption, and quantum chemical modeling. The NMR, TSDC, and DSC are sensitive to transition of phase water↔ice and can distinguish different phases in terms of quantity. These methods allow obtaining structural information on interfacial water (UWCSD), pore structure (PSD, surface area, and volume of pores in different ranges), and the thermodynamic characteristics of bound water or other liquids (such as changes in Gibbs free energy on adsorption [ΔG], free surface energy [γ_s], temperature of transition of phase [e.g., glass transition, T_g, melting, T_m, crystallization, T_{cr}, α, β, and γ relaxation, and throughout conductivity temperatures], heat of liquid adsorption [Q], or immersion in liquid [ΔH_{im}]). According to investigations carried out using the ^{1}H NMR spectroscopy with layer-by-layer freezing-out of water at $T<273$ K described earlier in the text (Gun'ko et al. 2005d, 2007f, 2009d), several types of interfacial water can be distinguished such as SBWs ($\Delta G<-0.5 - -1$ kJ/mol) and WBWs ($\Delta G>-0.5 - -1$ kJ/mol depending on the interface characteristics) (i.e., energetic-dependent parameter of water interaction with a surface), and SAWs (chemical shift of the proton resonance $\delta_H>3$ ppm) and WAWs ($\delta_H=1$–2 ppm) (structure-dependent parameter). SBW locates close to a surface but WBW is more distant from the surface. SAW represents nano- and microdomains but WAW is strongly clustered water, for example, at a mosaic hydrophobic/hydrophilic surface with surface barriers preventing formation of large water clusters and domains.

The ^{1}H NMR spectra described here and obtained with layer-by-layer freezing-out of bulk and interfacial waters include information on both local and full mobility of water molecules. However, dc relaxation observed using the TSDC method is linked to throughout conductivity (i.e., the full mobility of molecules at the interfaces providing percolation effects for ions) in contrast to dipolar relaxation linked to the mobility of polar bonds, molecular segments or fragments and whole molecules.

During TSDC or DRS investigations, several types of secondary (γ and β) and primary (α) relaxations can be observed in glass-forming materials (Ngai et al. 1994, Murakami et al. 1998) appearing with increasing temperature. Smaller structures, such as polar bonds, are responsible for the γ-relaxation due to their rotations (or proton tunneling effects), and larger structures, such as polymer fragments with bound water, give the β-relaxation. Therefore, the γ-relaxation is faster (with a timescale on the order of picoseconds at room temperature) than β-relaxation and occurs at lower temperatures in isochronal (constant frequency) experiments. The main α-relaxation (related to dynamic glass transition) occurring at temperatures higher than the glass transition temperature (T_g) is characterized by the nonexponential (e.g., stretched-exponential) time dependence and the non-Arrhenius temperature dependence of relaxation time expressed by, for example, the Vogel–Tammann–Fulcher equation (Tamman and Hesse 1926). The Johari-Goldstein process (β_{JG}; Johari and Goldstein 1970) decoupled from the α-process appears roughly below T_g, and the relaxation time shows Arrhenius temperature dependence, that is, without the cooperative effects. An interfacial polarization or MWS relaxation process (Ngai et al. 1994) was identified for temperatures above the glass transition and the α-relaxation. The effects of confined space for polymers penetrating into pores result in changes in both temperature and frequency dependences. Notice that adsorbed water can strongly affect these phenomena because of plasticization of polymers and changes in polymer–oxide interactions. Consequently, comparative analysis of the temperature dependences of the ^{1}H NMR and TSDC signals linked to different types of the mobility of confined water can give additional information on the behavior of this water and the characteristics of confined space where this water locates. Therefore, to analyze certain regularities observed in the characteristics of confined water (bond, local, and throughout molecular mobility of water contributing SAW, WAW, SBW, and WBW) unfrozen at $T<273$ K in different adsorbents and bio-objects, the ^{1}H NMR spectroscopy (190–$200<T<280$ K) and the TSDC method ($90<T<270$ K) were used.

Solid adsorbents (Table 1.37) (nanosilicas, alumina, titania, SA, silica/titania [ST], and alumina/silica/titania [AST] assigned to nanooxides); silica gels Si-40, Si-60, and Si-100 (Merck); ordered mesoporous silicas Mobil Crystalline Material (MCM)-41, MCM-48, and Santa Barbara Amorphous (SBA)-15, AC with 47% burn-off degree, C-47 (Gun'ko et al. 2008a,d); and polymer adsorbent LiChrolut EN (Merck; Gun'ko et al. 2008e) were used in comparative investigations carried out using the ^{1}H NMR and TSDC methods applied to dry and wetted powders ($C_{H_2O}=1$–30 wt%) and aqueous suspensions (solid phase concentration $C=1$–20 wt%). Bio-objects such as human plasma fibrinogen (HPF; obtained from the plasma of donor blood by fractional salting out with sodium sulfate), collagen, RBCs at 1×10^7 cell/mL in a buffer

TABLE 1.37
Composition, Specific Surface Area (S_{BET}) and Pore Volume (V_p) of Adsorbents

Sample	C_{SiO_2} (wt%)	C_{TiO_2} (wt%)	$C_{Al_2O_3}$ (wt%)	S_{BET} (m^2/g)	V_p (cm^3/g)
A-50	100			52	0.13
A-100	100			89	0.20
A-150	100			144	0.31
A-200	100			206	0.46
A-300	100			294	0.52
A-380	100			378	0.94
A-400	100			409	0.86
A-500	100			492	0.87
SA1	98.7		1.3	207	0.42
SA3	97		3	188	0.41
SA8	92		8	308	0.69
SA23	77		23	353	0.79
SA30	70		30	180	0.64
SA75	25		75	85	0.32
Al_2O_3			100	140	0.37
ST2	98	2		77	0.26
ST9	91	9		238	0.58
ST14	86	14		137	0.39
ST20	80	20		86	0.17
ST40	60	40		109	0.33
ST65	35	65		34	0.08
ST94	6	94		30	0.10
TiO_2		100		50	0.19
AST50	28	50	22	38	0.10
AST71	8	71	21	74	0.13
AST82	6	82	12	39	0.15
AST87	4	87	9	42	0.15
AST88	8	88	4	39	0.12
Si-40	100			732	0.54
Si-60	100			447	0.82
Si-100	100			349	1.23
MCM-41	100			997	0.70
MCM-48	100			1104	1.17
SBA-15	100			809	1.33
C-47				1648	1.88
LiChrolut EN				1512	0.83

Note: The S_{BET} and V_p values were determined using the nitrogen adsorption isotherms (77.4 K).

with sodium hydrocitrate (2%) and glucose (3%), dry and wetted yeast *Saccharomyces cerevisiae* cells, and wheat and rape seeds were studied by the mentioned methods at different hydration ($h = 0.07$–99.0 g of water per gram of dry materials) of samples (Gun'ko et al. 2005d, 2009d,e). Note that some of the mentioned materials and bio-object will be described in detail in next chapters of this book.

The temperature range in the TSDC measurements was much broader (90–270 K) than that in the ^{1}H NMR measurements (190–280 K) because the dipolar relaxation (TSDC) observed at $T<200$–220 K is linked to the polar bond rotations or molecular fragment relaxation, which cannot be registered in the high-resolution ^{1}H NMR spectra of static samples attributed only to mobile water, which appears at higher temperatures. However, the temperature ranges of the throughout conductivity (dc relaxation in TSDC) and the molecular mobility (^{1}H NMR) are relatively close but not the same because the dc relaxation requires the throughout percolation of ions but the molecular mobility reflected in the ^{1}H NMR spectra can be due to local mobility of bound and bulk water (from individual molecules, clusters, nanodomains to bulk water with elevating temperature).

Typically, water confined in narrower pores is characterized by a smaller average number of the hydrogen bonds per molecule. Therefore, the chemical shift of the proton resonance δ_H of confined water decreases in comparison with bulk water (Figure 1.208). Notice that $\delta_H \approx 1$ ppm is observed for individual water molecules in the gas phase or dissolved in nonpolar or weakly polar organic solvents, $\delta_H \approx 4$–5 ppm is for bulk water and $\delta_H \approx 7$ ppm is for Ih ice. The hydrogen bond network structure of bound water differs from that of bulk water or ice and depends on the topology of pores and chemical composition of the pore walls affecting the energy of the hydrogen bonds with water molecules (ΔE_H). The hydrogen bonds between water molecules and active surface sites (hydroxyls) of silica or other oxides are stronger ($-\Delta E_H=40$–50 kJ/mol) than the hydrogen bonds between water molecules ($-\Delta E_H=25$–28 kJ/mol; Brennan et al. 2001, Gun'ko et al. 2005d, 2007f, 2009d). Therefore, the heat of water adsorption on hydrophilic adsorbents is greater than the latent heat of bulk water condensation ($Q=45$ kJ/mol; Brennan et al. 2001). This leads to reduction of the Gibbs free energy ($\Delta G<0$) of interfacial water in comparison with the bulk depending on the pore size and chemical structure of the pore walls.

The δ_H values of bound water characterize both topological and surface nature effects on this water (Figure 1.208). For instance, the δ_H values are larger for water adsorbed on mesoporous silica gels than on nanosilica, alumina and SA composed of nonporous primary nanoparticles forming aggregates

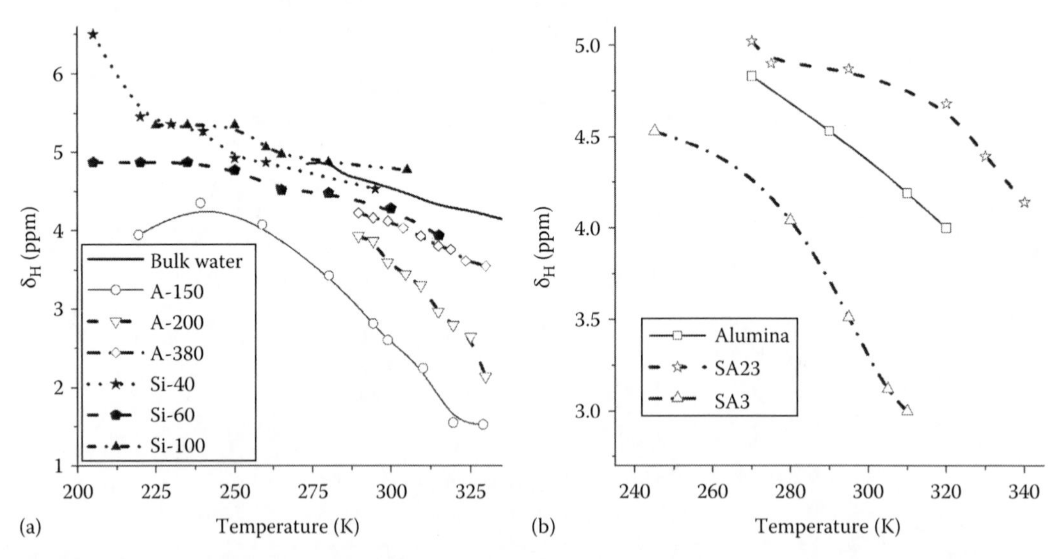

FIGURE 1.208 Temperature dependence of the δ_H value for bulk water (line without symbols) and water adsorbed on (a) nanosilicas and silica gels at $h=0.21$ (A-150), 0.163 (A-200), 0.24 (A-380), 0.12 (Si-40), 0.115 (Si-60), and 0.114 (Si-100) g/g, and (b) fumed alumina and SA at $h=0.2$ (Al_2O_3), 0.37 (SA3) and 0.223 (SA23) g/g. (Adapted from *Colloids Surf. A: Physicochem. Eng. Aspects*, 336, Gun'ko, V.M., Turov, V.V., Zarko, V.I., Goncharuk, E.V., and Turova, A.A., Regularities in the behaviour of water confined in adsorbents and bioobjects studied by ^{1}H NMR spectroscopy and TSDC methods at low temperatures, 147–158, 2009a. Copyright 2009, with permission from Elsevier.)

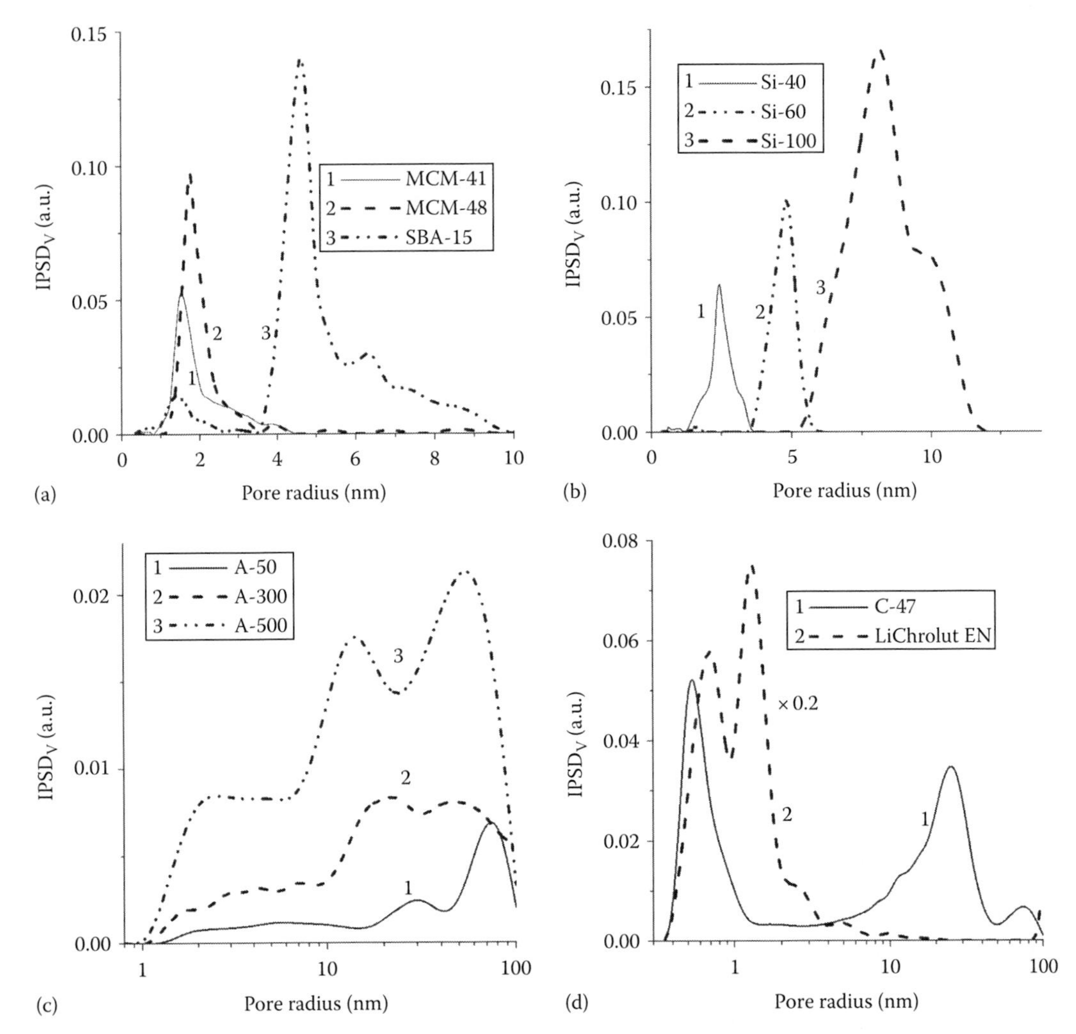

FIGURE 1.209 Incremental PSDs (in arbitrary units) for (a) ordered mesoporous silicas; (b) silica gels; (c) nanosilicas; and (d) activated carbon C-47 and LiChrolut EN calculated using the nitrogen desorption data and DFT method with the models of cylindrical pores (porous silicas and LiChrolut EN), voids between spherical particles (nanooxides) and slit-shaped pores (C-47). (Adapted from *Colloids Surf. A: Physicochem. Eng. Aspects*, 336, Gun'ko, V.M., Turov, V.V., Zarko, V.I., Goncharuk, E.V., and Turova, A.A., Regularities in the behaviour of water confined in adsorbents and bioobjects studied by ^{1}H NMR spectroscopy and TSDC methods at low temperatures, 147–158, 2009e. Copyright 2009, with permission from Elsevier.)

and agglomerates of aggregates. Nanooxides possess only textural porosity caused by voids between spherical nanoparticles with broad PSD with main contributions of mesopores ($1<R<25$ nm) and macropores ($R>25$ nm) but practically without nanopores ($R<1$ nm; Figure 1.209).

Therefore, the confined space effects are weaker and the associativity is lower for water adsorbed on nanooxides than on porous silicas. Aggregation of primary particles of nanooxides decreases with their increasing sizes (i.e., decreasing S_{BET} value). Therefore, the PSD (normalized to the pore volume) of A-50 is much lower than that of A-300 or A-500 (Figure 1.209c). These structural, morphological, and adsorption features result in smaller δ_H values for water adsorbed on nanosilica A-150 than that for A-200 or A-380, first of all, because of a decrease in the water associativity with decreasing porosity. The PSD effects are also characteristic of silica gels studied. Therefore, water adsorbed on mesoporous silica gel Si-100 is characterized by larger δ_H values than that for silica gels Si-60 and Si-40 (Figure 1.208) having narrower pores (Figure 1.209b). In other words, water adsorbed in broader pores of Si-100 has a higher associativity in larger nanodomains located in broader pores. However, stronger interactions of water molecules with the silica surface than with other water molecules lead to larger δ_H values for water adsorbed on Si-100 than that for bulk water, despite the lower associativity of water confined in pores than that in the bulk.

Features of the primary PaSD ($\phi(a)$) and their aggregation in secondary particles (i.e., textural porosity characteristics) influence the diffusion of water molecules in concentrated aqueous suspensions of nanosilicas. Therefore, transverse relaxation time (T_2) is two to three times longer (i.e., the molecular mobility is higher) for water in the aqueous suspensions of A-50 (average diameter of primary particles $d \approx 52$ nm) than that for A-300 ($d \approx 9.3$ nm) at the same silica concentrations (Figures 1.99 and 1.100). Additionally, the temperature effects are stronger for the A-50 suspension. In other words, at the same weight concentration of nanosilicas, water is more strongly bound in the suspension with smaller particles (A-300) because of larger amounts of the interfacial water in the suspension of A-300 (by several times) because this silica has much larger S_{BET} value (Table 1.37) and more strongly aggregated primary particles than A-50. Notice that the ^{1}H NMR spectroscopy investigations of a variety of nanosilicas and other oxides showed that concentrations of surface hydroxyls and their acidity can strongly affect the amounts and the characteristics of bound water in the aqueous suspensions. Therefore, not only the morphology, the $\phi(a)$ distributions, and the S_{BET} values but also the surface structure of nanoparticles can influence the molecular mobility, the dipolar and dc relaxations, and other dynamic processes in the interfacial water layer.

As mentioned earlier, the high-resolution ^{1}H NMR spectra recorded here for static samples are linked to mobile water molecules but the dc relaxation (TSDC) is caused by mobile protons and other ions when condition of throughout percolation of ions (between two electrodes in a TSDC cell) is achieved on heating of a frozen system. Both dynamic phenomena are temperature dependent (Figure 1.210).

For all samples, a linear dependence of $\ln(I_{TSDC})$ versus $1/T$ is observed for the dc relaxation; that is, it obeys the Arrhenius law. The mobility of water molecules (NMR) demonstrates a more complex character caused by non-Arrhenius or several Arrhenius processes since two to three linear portions are observed for the curves of $\ln(C_{uw}(T))$ versus $1/T$ (where C_{uw} is the amount of unfrozen water as a function of temperature in the melting curves). This effect can be explained by the presence of several types of interfacial water (i.e., SAW, WAW, SBW, and WBW) located in different pores at different distances from the oxide surfaces and responsible for different local molecular mobility (LMM) of interfacial water observed at different temperatures (due to layer-by-layer freezing-out of confined water with lowering temperature and a similar process of thawing but with a hysteresis effect described earlier). Additionally, certain cooperative effects can be observed in water bound in nanopores or interacting with surface functionalities because of a clustered structure of bound water characterized by a dense hydrogen bond network that lead to cooperative relaxations of water clusters or nanodomains, especially in the case of relaxation of hydrated hydrophilic polymers or polymer/oxide composites. Typically, a more complex and broader PSD of a material (Figure 1.209) leads to a more complex $\ln C_{uw}(1/T)$ function (Figure 1.210).

The complexity of the behavior of interfacial water reflecting in complex temperature dependences of the molecular mobility and related characteristics reflects in a broad distribution of the activation energy of the LMM at $E_{LMM} = 5$–83 kJ/mol (NMR) and the dc relaxation at $E_{dc} = 46$–98 kJ/mol (TSDC) calculated assuming that the processes obey the Arrhenius law, and typically $E_{LMM} < E_{dc}$ for the same systems. The E_{LMM} values are greater for the suspensions with A-300 (maximal values $E_{LMM} = 46$ and 52 kJ/mol at hydration $h = 7.1$ and 5 g/g, respectively) than E_{LMM} for porous silicas: 40 (MCM-48), 25 (SBA-15), 34 (Si-100), 18 (Si-60), and 21 kJ/mol (Si-40). These results can be due to larger S_{BET} values of porous silicas than that of nanosilicas and stronger confined space effects in narrower pores of silica gels. These effects lead to diminution of the average number of the hydrogen bonds per molecule that diminish the E_{LMM} values (Gun'ko et al. 2007f,i, 2009d,e). Calculations of the distribution functions of activation energy, $f(E)$ of the local mobility (NMR),

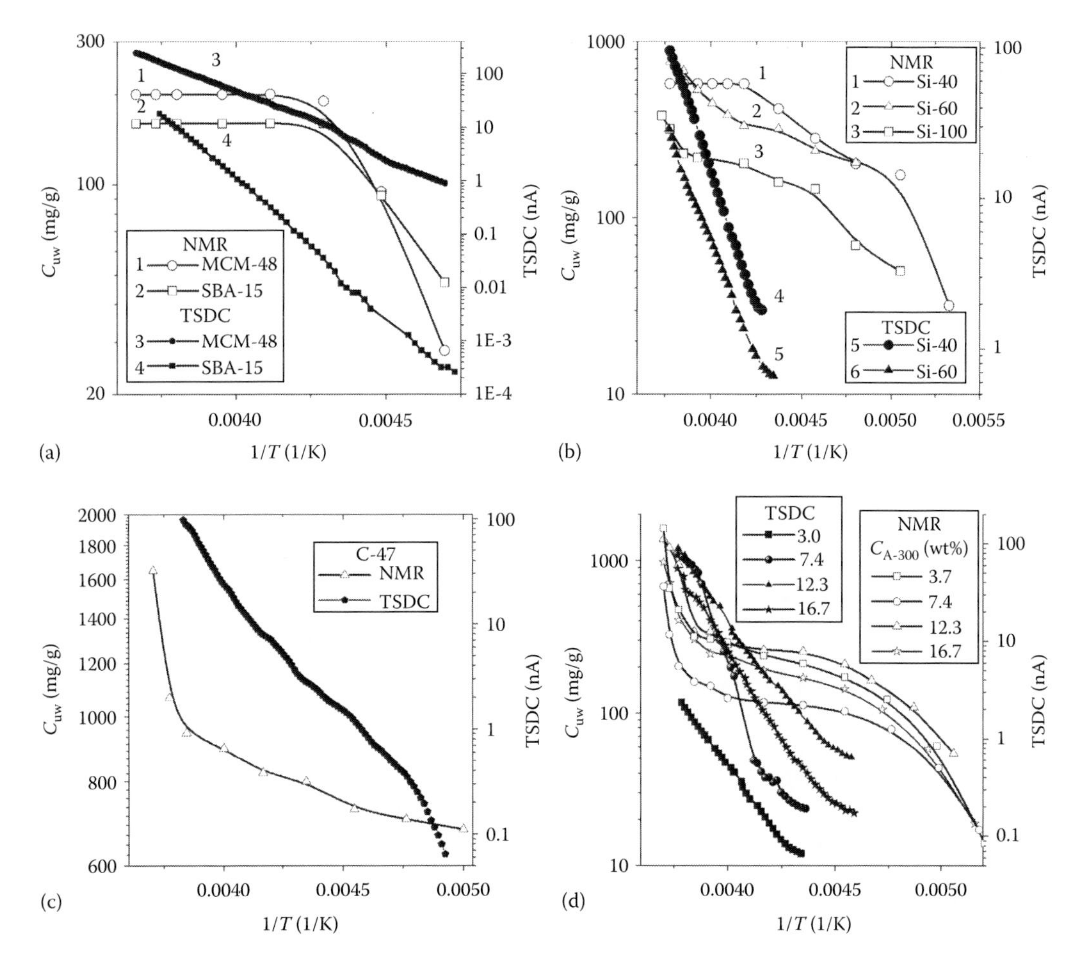

FIGURE 1.210 Temperature dependences of the amounts of unfrozen water (C_{uw}) and TSD current (dc) on the adsorption of water on (a) ordered mesoporous silicas MCM-48 (hydration $h=0.25$ [NMR] and 5 g/g [TSDC]) and SBA-15 (0.19 [NMR] and 5 g/g [TSDC], (b) silica gels Si-40, Si-60 and Si-100 ($h=19$ [NMR] and 5 g/g [TSDC]), (c) activated carbon C-47 ($h=4.71$ [NMR] and 13.3 g/g [TSDC]); and (d) nanosilica A-300 at $C_{A\text{-}300}=3.0$ (TSDC, $h\approx32.3$ g/g), 3.7 (NMR, $h\approx26.0$), 7.4 ($h\approx12.5$), 12.3 ($h\approx7.1$) and 16.7 wt% ($h\approx5.0$ g/g). (Adapted from *Colloids Surf. A: Physicochem. Eng. Aspects*, 336, Gun'ko, V.M., Turov, V.V., Zarko, V.I., Goncharuk, E.V., and Turova, A.A., Regularities in the behaviour of water confined in adsorbents and bioobjects studied by ^{1}H NMR spectroscopy and TSDC methods at low temperatures, 147–158, 2009e. Copyright 2009, with permission from Elsevier.)

and the dipolar and dc relaxations (TSDC) in the interfacial water at a surface of polymer adsorbent LiChrolut EN (Figure 1.211) and nanosilica A-300 (Figure 1.212c; there are representative samples here with high nanoporosity (AC C-47) or textural porosity (nanosilica A-300)) reveal the presence of several dynamic processes in bound water studied by both methods.

The dipolar relaxation (TSDC) is observed at temperatures ($90<T<210$–220 K) lower than that characteristic of the appearance of the LMM (NMR). However, the throughout molecular mobility causing the dc relaxation (TSDC) appears at higher temperatures ($T>210$–230 K) than LMM (NMR) and both processes depend (but differently) on the confined space effects. These results are due to the influence of surface electrostatic fields on the structure of the hydrogen bond network in interfacial water (i.e., the average number of these bonds per molecule and their strength and geometry) and due to additional conditions necessary for the dc relaxation (ion percolation) in comparison with the LMM. An increase in the content of an adsorbent (and, therefore, contribution of interfacial

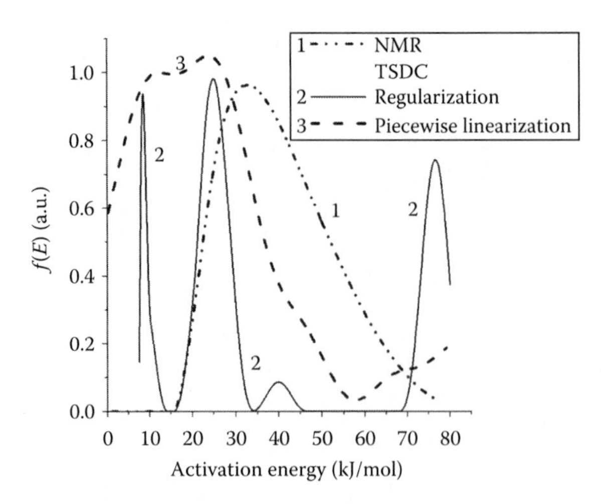

FIGURE 1.211 Distribution functions of the activation energy of the mobility of water molecules at a surface of polymer LiChrolut EN adsorbent calculated on the basis of the ^{1}H NMR data (curve 1, hydration h=2 g/g) and dipolar and dc relaxation (TSDC, curves 2 (regularization) and 3 (piecewise linearization), h=5.4 g/g). (Adapted from *Colloids Surf. A: Physicochem. Eng. Aspects*, 336, Gun'ko, V.M., Turov, V.V., Zarko, V.I., Goncharuk, E.V., and Turova, A.A., Regularities in the behaviour of water confined in adsorbents and bioobjects studied by ^{1}H NMR spectroscopy and TSDC methods at low temperatures, 147–158, 2009e. Copyright 2009, with permission from Elsevier.)

water) leads to diminution of the LT peak of the dipolar relaxation (TSDC) at 120 K characteristic of bulk water and related to the dipolar relaxations of OH groups in small clusters (without the cooperative effects) and the relaxation of interstitial water molecules (without the hydrogen bonds and $\delta_H \approx 1$ ppm). The displacement of the LT band ($T<160$ K) toward higher temperatures and the opposite displacement of HT band at $160<T<220$–240 K (Figure 1.212) is due to the interaction of water with solid surfaces leading to enhancement of clusterization of bound water and diminution of the activation energy for the relaxation processes responsible for the HT band and the opposite effect for the LT band. This also affects temperature of the dc relaxation beginning. The stronger the confined space effects (in narrower pores), the greater is the current (I_{TSDC}) in the HT band (Figure 1.212) and the lower is the temperature of the dc relaxation. This is due to additional polarization of bound water molecules in the electrostatic field at a surface (the dipole moment values of bound water significantly increase; Tischenko and Gun'ko 1995), water clusterization, and the difference in the freezing point depression for water being in pores of different sizes (Gun'ko et al. 2005d, 2007i, 2009d,e). As a whole, the appearance of the dc relaxation (TSDC) corresponds to the appearance of the ^{1}H NMR signals of mobile water (Figure 1.210). However, the ^{1}H NMR signals of bound water typically appear at lower temperatures than the dc relaxation (TSDC) starts in the same systems (Figure 1.210). This difference is due to features of the measurement techniques because the mobility of water molecules registered in the NMR measurements is related to the dynamic processes in any local structures in pores near the pore walls (i.e., LMM). However, this mobility could not provide the dc relaxation since the ion percolation (throughout conductivity) between two electrodes is yet absent and it appears only at higher temperatures when the throughout molecular mobility appears. A smaller size of solid particles provides the formation of overlapping layers of mobile interfacial water (necessary for the appearance of the throughout conductivity in the suspensions) at lower temperatures because of tighter and better contacts between smaller particles more strongly aggregated and their electrical double layers, where bound water is located, more strongly overlap. Therefore, in the case of relatively large silica gel granules (0.15–0.5 mm in diameter, d), the ion percolation is observed at higher temperatures than that for nanosilica (primary particle diameter < 15 nm for A-300) or MCM-48 and SBA-15 (d<0.1 mm), despite the LMM, which

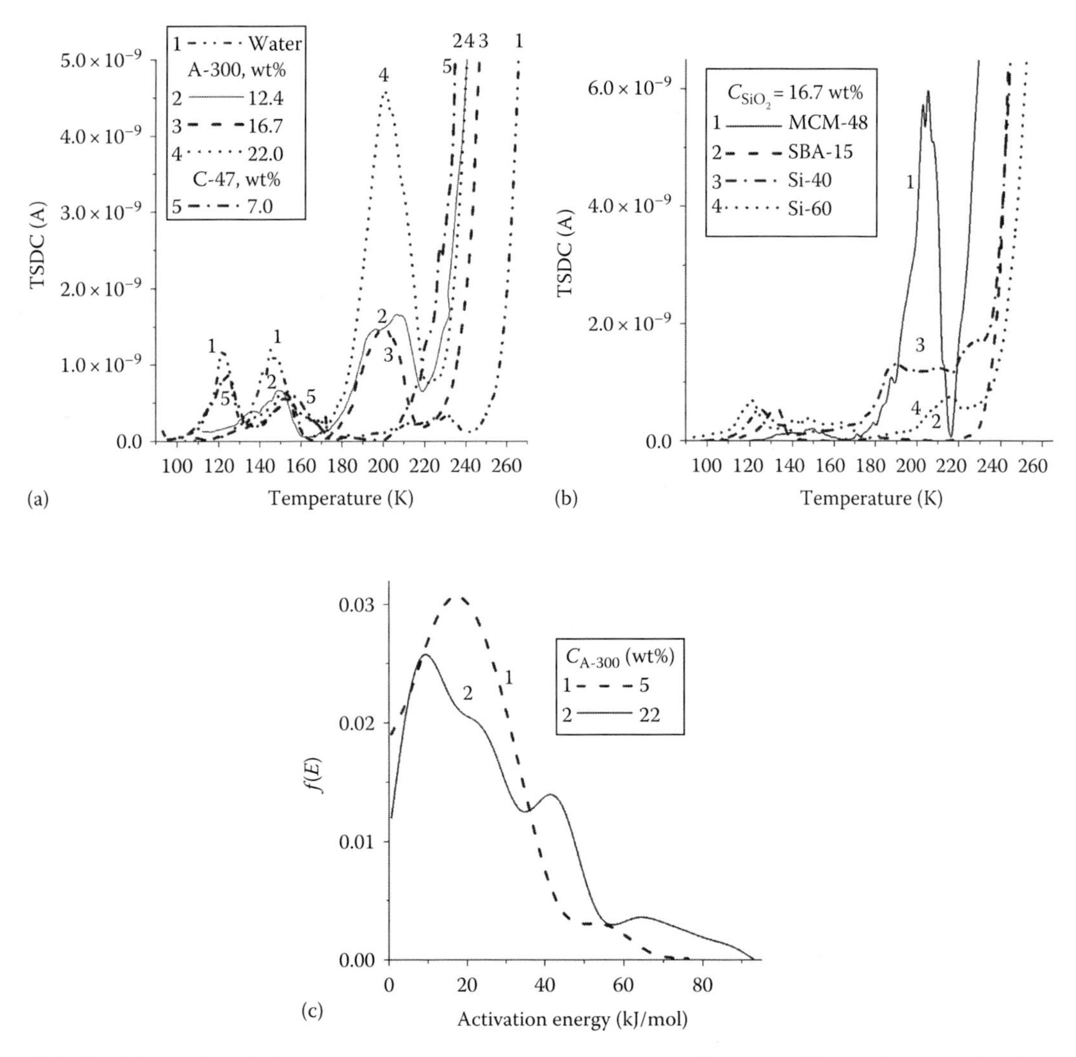

FIGURE 1.212 TSDC thermograms for the aqueous suspensions of (a) nanosilica and activated carbon C-47, and (b) porous silicas; and (c) the distribution functions of activation energy of dipolar and dc relaxation for the aqueous suspensions of A-300 at $90<T<265$ K (piecewise linearization assuming Arrhenius-type processes over small temperature ranges). (Adapted from *Colloids Surf. A: Physicochem. Eng. Aspects*, 336, Gun'ko, V.M., Turov, V.V., Zarko, V.I., Goncharuk, E.V., and Turova, A.A., Regularities in the behaviour of water confined in adsorbents and bioobjects studied by ^{1}H NMR spectroscopy and TSDC methods at low temperatures, 147–158, 2009e. Copyright 2009, with permission from Elsevier.)

is observed for porous silicas at low temperatures. Notice that an increase in A-300 concentration in the aqueous suspension leads to enhanced clusterization of bound water that reflects in a more complex shape of the *f*(*E*) function (Figure 1.212c). It should be noted that the silica concentration effects are nonlinear (Figures 1.210 and 1.212) because of a strong rearrangement of secondary particles with increasing concentration at C_{SiO_2} = 12–13 wt%. The behavior of confined water can be changed due to changes in the surface structure of nanoparticles or the pore walls, for example, for mixed oxides such as SA, ST, and AST in comparison with individual silica (vide infra).

Surface chemistry (i.e., the number and the type of surface acid/base sites or grafted functionalities of different hydrophilicity) influences the hydrogen bond network structure in a relatively thick layer of interfacial water (3–10 nm or larger). Partial hydrophobization of a silica surface (e.g., by grafting of trimethylsilyl, TMS groups) affects the amounts of unfrozen water and the temperature dependence of the amounts of unfrozen water (Figure 1.137). The amounts of structured water decrease with increasing content of the TMS groups as well as the Gibbs free surface energy (surface energy γ_s estimated as the modulus of total changes in the Gibbs free energy of all layers of bound water) and the heat of immersion in water (ΔH_{im}). However, changes in the amounts of unfrozen water ($C_{uw}(T)$) and the γ_s and ΔH_{im} values depend nonlinearly on content of the TMC groups (C_{TMS}; Figure 1.138) because contributions of four types of bound water and their clusterization depend differently on the C_{TMS} value. Therefore, partial functionalization of adsorbent surface can be used to control the clusterization and other characteristics of confined water or other polar liquids.

Mixed nanooxides (SA and ST) have strong Brønsted acid sites (≡SiO(H)M≡ where M = Al or Ti) and a set of weaker sites (bridging (≡MO(H)M≡) and terminal (≡MOH, ≡SiOH) hydroxyls, Lewis acid sites, etc.) that affect the behavior of interfacial water (Figure 1.208). The chemical nonuniformity of the SA, ST or AST surfaces can cause a stronger non-Arrhenius character of the dc relaxation and the molecular mobility in the aqueous suspension (Figures 1.213 and 1.214) in comparison with nanosilica (Figure 1.212). The molecular mobility of interfacial water appears in the aqueous suspensions of SA at lower temperatures than that for nanosilica (Figure 1.213). However, the dc relaxation begins at temperatures close to that for nanosilica (Figures 1.212 and 1.214). Consequently, changes in the local structures in interfacial water and the temperature behavior of this water (NMR) more strongly depend on the oxide surface structure than the ion percolation effects (TSDC) depend on the composition of oxide particles and their aggregation in the suspensions.

The oxide surface structure and the type of adsorbent pretreatment influence the thermodynamic characteristics of interfacial water (Gibbs free energy, heat of immersion in water, activation energy of the molecular mobility, etc.). The analyzed structural and other effects cause a significant scatter in the relationship between the specific surface area and the γ_s and ΔH_{im} values (Figure 1.215).

The γ_s and ΔH_{im} values depend also on sample history, especially heating temperature and subsequent relaxation on adsorption of water from air or wetting. For instance, the average difference in the γ_s and ΔH_{im} values is ~34% if oxide samples were degassed at 473 K for 6 h before the microcalorimetric measurements and equilibrated in air before the NMR measurements. However,

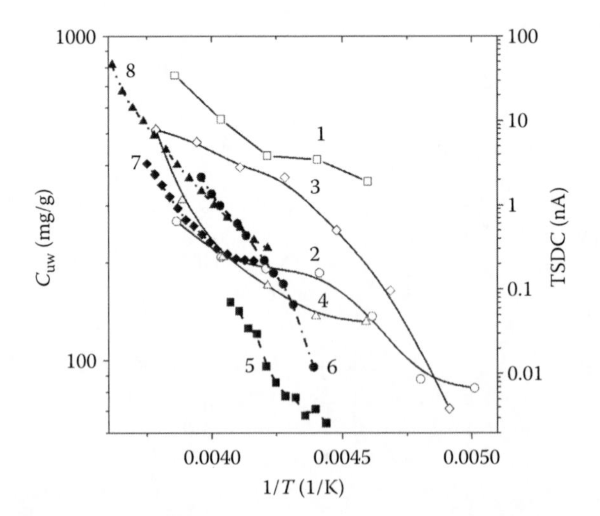

FIGURE 1.213 Amounts of unfrozen water (NMR, curves 1–4) and dc relaxation (TSDC, curves 5–8) as functions of reciprocal temperature for fumed Al_2O_3 (curves 1 and 5), SA1 (2 and 6), SA3 (3 and 7) and SA23 (4 and 8) at C_{ox} = 5 wt% (1–4) and 3 wt% (5–8). (Adapted from *Colloids Surf. A: Physicochem. Eng. Aspects*, 336, Gun'ko, V.M., Turov, V.V., Zarko, V.I., Goncharuk, E.V., and Turova, A.A., Regularities in the behaviour of water confined in adsorbents and bioobjects studied by ^{1}H NMR spectroscopy and TSDC methods at low temperatures, 147–158, 2009e. Copyright 2009, with permission from Elsevier.)

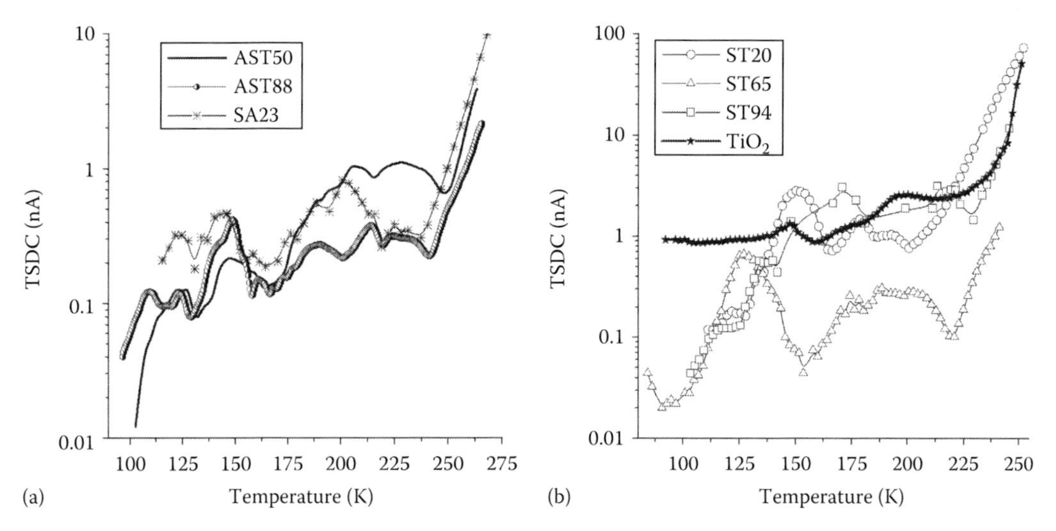

FIGURE 1.214 TSDC thermograms of aqueous suspensions of (a) AST and SA23 and (b) ST and TiO_2 at concentration of oxides $C_{ox}=3$ wt% and $F_p=0.5$–2.0 MV/m. (Adapted from *Colloids Surf. A: Physicochem. Eng. Aspects*, 336, Gun'ko, V.M., Turov, V.V., Zarko, V.I., Goncharuk, E.V., and Turova, A.A., Regularities in the behaviour of water confined in adsorbents and bioobjects studied by ^{1}H NMR spectroscopy and TSDC methods at low temperatures, 147–158, 2009e. Copyright 2009, with permission from Elsevier.)

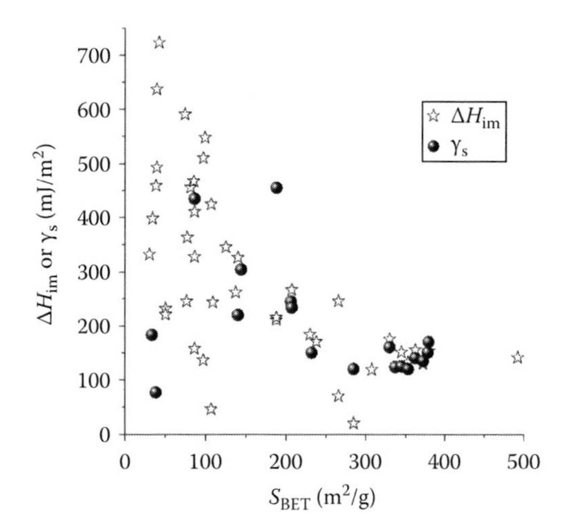

FIGURE 1.215 Relationships between the specific surface area S_{BET} and the heat of immersion in water (ΔH_{im}) of individual and complex nanooxides at hydration $h=60$ g/g or the surface energy (γ_s) in their aqueous suspensions ($h=16$–19 g/g); the average difference between the ΔH_{im} and γ_s values is 34.3% (83 mJ/m^2). (Adapted from *Colloids Surf. A: Physicochem. Eng. Aspects*, 336, Gun'ko, V.M., Turov, V.V., Zarko, V.I., Goncharuk, E.V., and Turova, A.A., Regularities in the behaviour of water confined in adsorbents and bioobjects studied by ^{1}H NMR spectroscopy and TSDC methods at low temperatures, 147–158, 2009e. Copyright 2009, with permission from Elsevier.)

degassing of MS at 353 K for 6 h gives a smaller difference (~9%), which is, however, larger than the difference in the ΔG and ΔH values calculated on the basis of the ^{1}H NMR data (~5%). The average difference in the S_{BET} values calculated using the adsorption isotherms of nitrogen and water is ~12% for equilibrated silicas and ~46% if the water adsorption was measured for samples degassed at 473 K. The strong effects of heating and subsequent relaxation on the ΔH_{im} and S_{BET}

values are caused by dissociative adsorption of water molecules on dehydrated surface and in the bulk of oxide particles preheated, which are characterized by very different rate values.

The behavior of confined water depends not only on the presence of different surface functionalities but also on coadsorption of organic compounds. The effects of different adsorbents on coadsorption of water and organics were studied earlier. Here, this is done only for a water/methane pair adsorbed onto different adsorbents (Figure 1.216). Two ^{1}H NMR signals at $\delta_H = 0.7$ ppm (methane) and 5 ppm (water) are observed on the coadsorption of water and methane onto porous silica Silochrome-120 modified by 3-APMS groups substituting a third of surface hydroxyls.

This adsorbed water is strongly associated ($\delta_H \approx 5$ ppm) but weakly bound because the ^{1}H NMR signal is observed only at $T > 260$ K ($\Delta G > -0.6$ kJ/mol). The signal of methane at $\delta_H = 0.7$ ppm does not practically change with temperature on the adsorption onto adsorbents without nanopores

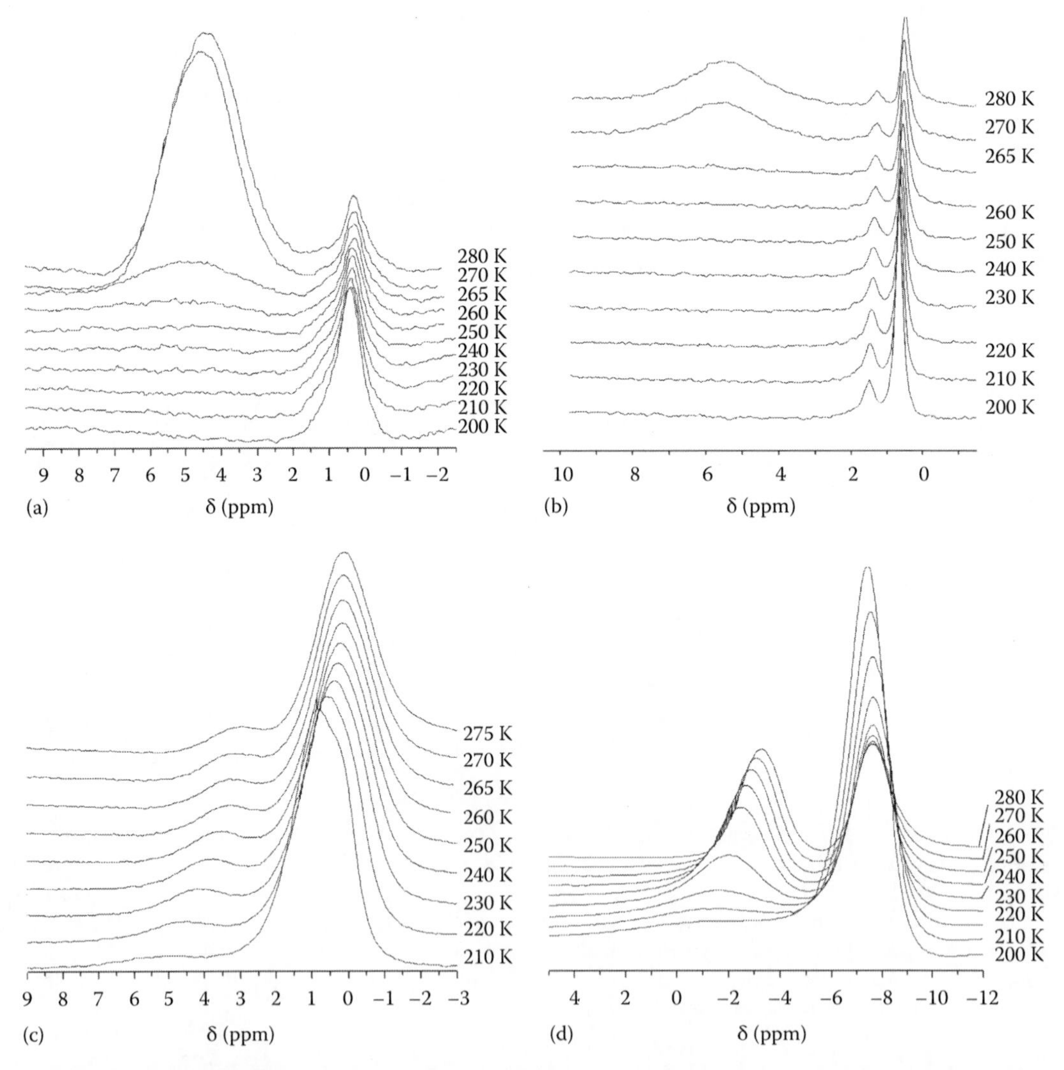

FIGURE 1.216 ^{1}H NMR spectra of water and methane coadsorbed onto (a) APMS-Silochrome-120, (b) APMS-alumina, (c) LiChrolut EN, and (d) activated carbon C-47 at h = (a) 50 and (b, c, d) 100 mg/g, and C_{CH4} = (a) 10, (b) 50, (c) 10, and (d) 100 mg/g. (Adapted from *Colloids Surf. A: Physicochem. Eng. Aspects*, 336, Gun'ko, V.M., Turov, V.V., Zarko, V.I., Goncharuk, E.V., and Turova, A.A., Regularities in the behaviour of water confined in adsorbents and bioobjects studied by ^{1}H NMR spectroscopy and TSDC methods at low temperatures, 147–158, 2009e. Copyright 2009, with permission from Elsevier.)

(modified Silochrome and nanoalumina). If adsorbents have nanopores (e.g., LiChrolut EN, carbon C-47) that adsorbed water includes both WBW and SBW fractions. In this case, the ^{1}H NMR signal of methane coadsorbed with water increases with lowering temperature because of an increase in the adsorption of methane and enhancement of the structural effects of confined and clustered water and ice nanocrystallites on the adsorption of methane. Water adsorbed on the polymeric adsorbent (LiChrolut EN) includes both WAW and SAW fractions since the ^{1}H NMR signals at $\delta_H = 4–5$ (SAW) and 1–2 (WAW) ppm are observed at $T < 273$ K. The effects of π-electrons at the basal planes of carbon adsorbent C-47 lead to enhanced shielding of protons and, therefore, the ^{1}H NMR signals demonstrate upfield shift since $\delta_{H,H_2O} < -2$ ppm and $\delta_{H,CH_4} < -6$ ppm. Adsorbed methane can displace a portion of unfrozen water from narrow pores to larger ones.

Thus, comparative investigations of water confined in pores of solid materials using the ^{1}H NMR, TSDC, and microcalorimetry methods show that the difference in temperature of transition from the LMM of water bound at the pore walls to ion percolation effects (throughout conductivity) can be 10–40 K. This value depends on several factors such as the morphology of particles, textural, and structural characteristics of adsorbents, chemical structure of the surfaces, concentrations of components, and the presence of dissolved salts (e.g., NaCl) and other compounds. There are four types of water confined in pores solid adsorbents such as strongly ($\Delta G < -1$ kJ/mol) and weakly ($\Delta G > -1$ kJ/mol) bound and strongly ($\delta_H = 4–5$ ppm) and weakly ($\delta_H = 1–2$ ppm) associated waters. At low amounts of adsorbed water (much smaller than the pore volume) typically all water is strongly bound. A portion of this water can be weakly associated if the pore walls are mosaic and composed with hydrophilic and hydrophobic patches, for example, in the case of partial silylation of silica surface. The LMM of bound water depicts a non-Arrhenius character or includes several Arrhenius-type processes characterized by different activation energies because the corresponding water is differently clustered and bound to different structures at the interfaces. Therefore, the dynamic characteristics of these types of bound water can significantly differ, as well as its temperature behavior. The temperature dependence of ion percolation in bound water has the Arrhenius character for relatively uniform systems such as the aqueous suspensions of nanosilica or mesoporous silicas. However, in the case of complex materials (e.g., mixed nanooxides) or adsorbents possessing both broad and very narrow pores (e.g., ACs) certain deviations from the Arrhenius behavior are observed on the dc relaxation. The behavior of interfacial water depends strongly on the presence of nonpolar or polar organic solvents, which can displace a portion of water from narrow pores into larger ones, change the associativity of bound water and its interaction energy with solid surfaces. However, confined water and polar organics are much poorly mixed in narrow pores than in the bulk that leads to enhanced clusterization of water bound to the mosaic surfaces with the presence of organic coadsorbates.

Water structure in confined space of nanopores in microparticles, in surface layers at nonporous nanoparticles or macromolecular pockets depends strongly on size of water clusters or domains, water content, features of surroundings, composition of the pore walls or particle surfaces, and the presence of small polar or nonpolar molecules or ions. During analysis of the experimental data of, for example, ^{1}H NMR spectroscopy, TSDC, calorimetry (DSC), TG, and IR spectroscopy, it is needed to understand what changes occur in the bound water structure on changes in component concentrations, temperature, media composition, exterior forces, etc. This is important not only from the theoretical point of view but also from practical one since these effects definitely influence bioactivity of compounds adsorbed in the surface layer of particles or bound by macromolecules, on transport, pharmacokinetics, and pharmacodynamics of medicinal matters and allow us to change the doses of medications and to promote efficiency of their action and bioavailability (Chuiko 1993, 2003, Blitz and Gun'ko 2006). These effects are complex and multivariable; therefore, they need careful, comprehensive, experimental and theoretical analysis for finding out features of driving forces, correlations "structure–property," etc. Theoretical modeling of the proper structures and direct calculations of such characteristics as the magnetic (electronic) shielding tensors, chemical shifts, polarizability, interaction energy, free energy of solvation,

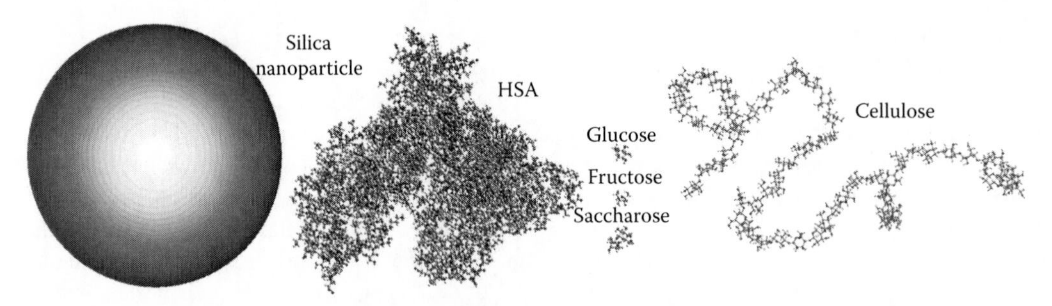

FIGURE 1.217 Silica A-300 nanoparticle (diameter 8.5 nm) and HSA (8×9.2 nm), sugar molecules and cellulose molecule (molecular weight 10/4 kDa, length 28 nm, surface area 98 nm^2, polarizability 859 cm^3 and hydration energy 663 kcal/mol).

dipole moments, atomic charges, geometry of molecules, and complexes allow us more detailed analysis of the experimental data and proper effects. The second important aspect of the applied theoretical analysis is the use of experimental information to obtain additional information due to the use of analytical models and proper functional or integral equations, which describe certain experimental dependences. The solution of these equations allows us to obtain the distribution functions of some important parameters, for example, size distributions of bound water structures or pores, adsorption or activation energy or free energy distributions of reactions, and time and activation energy distributions of relaxations.

Theoretical calculations of the NMR spectra using quantum chemical ab initio or DFT methods need the use of adequate bases sets (not worse than 6-31G(d,p)) and proper molecular or cluster models. In the case of bioactive molecules, which interact with water and adsorbent surface, the model systems include several hundreds of atoms, and in the case of biomacromolecules they include dozens of thousands of atoms. For the calculations of NMR spectra, it is needed fully to optimize geometry of these systems, which is a complex problem already for the first type of the systems, and in the case of biomacromolecules (Figure 1.217) calculations are more difficult.

Thus, a task is very difficult for the "frontal" calculations of the NMR spectra. It can be solved by two ways: with the use of supercomputers (for "frontal" calculations) or with the use of simplified approaches and less powerful computers. For the last case, such approach was developed (Gun'ko et al. 2005d, 2007i, 2009d)

1. Calculations of the NMR spectra for small systems (similar to ones shown in Figure 1.218) were carried out using GIAO/B3LYP/6-31G(d,p) method (Gaussian 03; Frisch et al. 2004).
2. Calculations of those systems were also performed using PM3 or PM6 semiempirical methods (which well reproduce the structure of hydrogen bonds).
3. Then correlation dependences of chemical shifts of protons (GIAO; Frisch et al. 2004) on charges on atoms (PM3, PM6) were determined as $\delta_H = a + bq_H$ (values of a and b were constant only for certain set of the same types systems), for example, $\delta_{H,w} = -18.81135 + q_H \times 92.83742$ (ppm) for water and $\delta_H = -2.33089 + q_H \times 26.24548$ (ppm) for methane on adsorption of water–methane mixtures on polymeric adsorbent (PM3) and $\delta_{H,w} = -27.97889 + q_H \times 87.56668$ (ppm) for water (PM6).
4. Calculations of large systems using PM3 or PM6 methods.
5. Calculations of the δ_H values using correlation equations.
6. Calculations of model spectra (distribution functions $f(\delta_H)$ of chemical shifts of protons) using Gaussian functions and parameters of dispersion of peaks from the experimental NMR spectra (or theoretical estimations) as described in Chapter 10. Such approach allows us to calculate appropriate ^{1}H NMR spectra of large systems. This information can be used for more reliable and detailed analysis of the experimental NMR spectra.

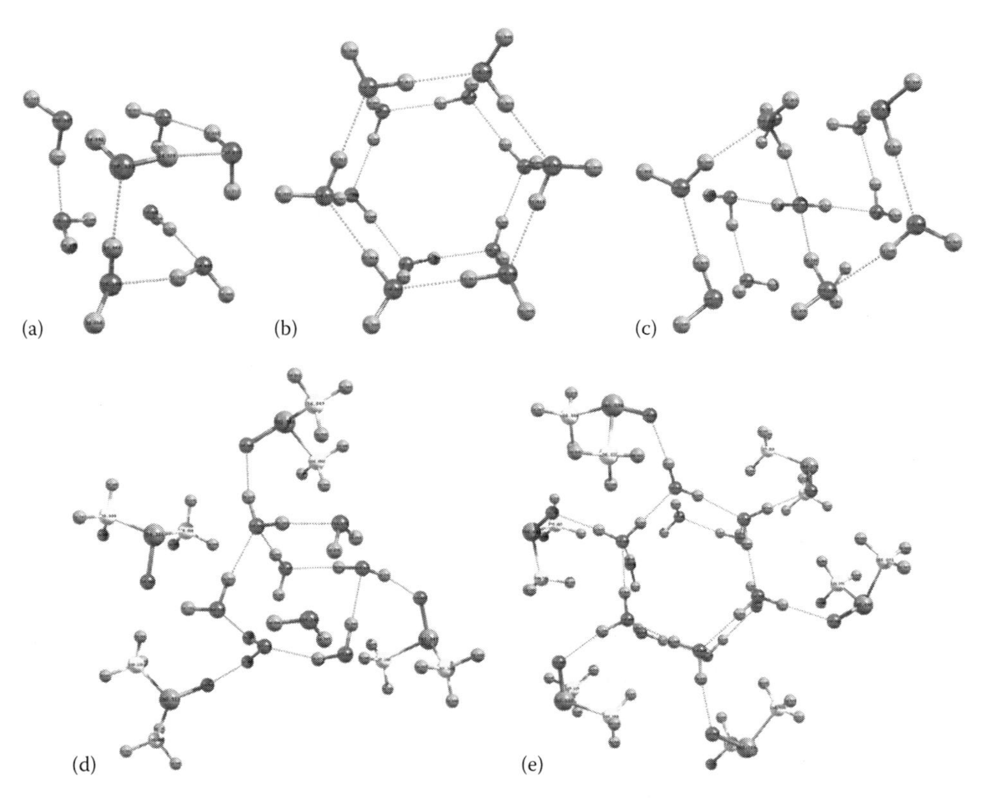

FIGURE 1.218 Clusters of water molecules (a–c) free and (d,e) surrounded by DMSO molecules (average diagonal elements to the tensor of the magnetic shielding are shown), which were used for calculations of correlation equations with GIAO/B3LYP/6-31G(d,p) and PM6.

> Within the framework of this approach, it is possible to explore the influence of liquid environment (water or organic) on the NMR spectra calculating the electronic structure of the systems using IEFPCM (Frisch et al. 2004) or SM5/GAMESOL (Xidos et al. 2002) methods, which give the parameter values (q_H, etc.) taking into account the solvation effects. The last corrections improve agreement of theoretical and experimental information.

Quantum chemical calculations were performed using Gaussian 03 (Frisch et al. 2004) and GAMESS (version 7.1G) (Granovsky 2009) with ab initio, DFT and semiempirical (PM3, PM6) methods. Several models were used: (1) water nanodroplets (250–300 molecules) containing molecules of rutin or quercetin and (2) model fragments of cellulose with one or two chains, surrounded by molecules of (a) water, (b) water and chloroform, and (c) water and DMSO. Calculations of clusters of silica (to 100 atoms in the case of ab initio or DFT calculations and 500 atoms [PM3] and to 6000 atoms [PM6, MOZYME algorithm], associates of water molecules (2–20 molecules on ab initio or DFT calculations, 300 molecules [PM3] and 2000 molecules [PM6/MOZYME]). The calculations of small clusters (size$<$1 nm; Figure 1.218) and nanodomains (Figure 1.219) of water were performed by quantum chemical nonempirical (6-31G(d,p)), DFT (B3LYP/6-31G(d,p)) (Figure 1.218), semiempirical (PM3, PM6; Figure 1.219) methods using Gaussian 03, GAMESS PC 7.1G, WINGAMESS 2009, 2010, 2011 (Schmidt et al. 1993) and MOPAC2009 (Stewart 2008). The initial geometry of nanodroplets and domains (Figure 1.219) of water was determined using molecular mechanics (MM) with VEGA ZZ (Pedretti et al. 2004), Jmol (Jmol 2011), and Chemcraft (Zhurko and Zhurko 2011) programs.

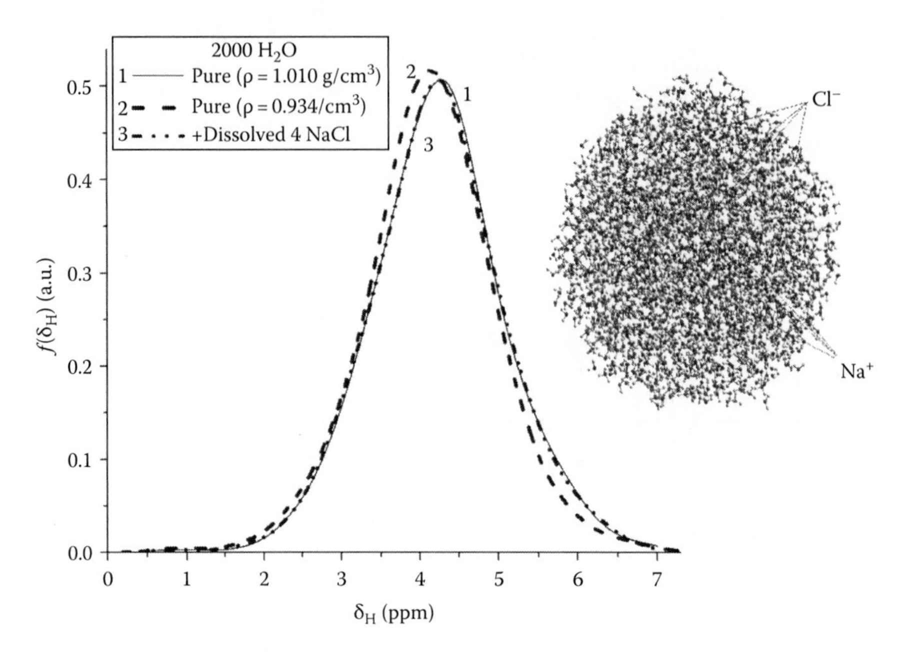

FIGURE 1.219 ^{1}H NMR spectra of water nanodrop with 2000 molecules (pure water and with dissolved 4 NaCl) using the correlation between the H atom charges and the electron shielding effects of protons and the calibrated function obtained using GIAO/DFT/6-31G(d,p) and PM6 calculations.

For simple complexes HO–**H**···O=S$(CH_3)_2$ and HO–**H**···OH_2 (calculated using GIAO/B3LYP/6-31G(d,p)) $\delta_{H,iso}=3.2$ and 3.1 ppm, respectively, and $\delta_{H,iso}=3.3$ ppm for $(CH_3)_2S$=O···**H**–O–**H**···O=S$(CH_3)_2$. In larger water domains (for water alone), the δ_H value grows because the number of molecules having four hydrogen bonds (per molecule) increases in dense structures with reduced shielding of protons similar to that in ice (Gun'ko et al. 2005d, 2009d). Therefore, the formation of larger water clusters and domains in the mixtures with Me_2SO with decreasing C_{Me2SO} leads to an increase in the δ_H value to 5–5.5 ppm. A relative small water/Me_2SO droplet with the geometry optimized using the PM3 method is characterized by a smaller average value $\delta_H=2.1$ ppm (the δ_H values change over the 0.5–5.5 ppm range) for water molecules because some H atoms (from the H_2O molecules) do not participate in the hydrogen bonds and have $\delta_H<2$ ppm. The average value $\delta_H=3.6$ ppm is for the H atoms participating in the hydrogen bonds (whose structure differs from that for ice). These δ_H values were calculated on the basis of the atom charges q_H (PM3) in the Me_2SO/water nanodroplet and a linear approximation $\delta_{H(H_2O)}=-11.72921+q_{H(H_2O)}\times 62.30708$ obtained using B3LYP/6-31G(d,p) (to calculate δ_H) and PM3 (q_H) calculations of small complexes with one or two Me_2SO molecules and H_2O.

The calculations of the ^{1}H NMR spectra on the basis of the electronic structure of large systems calculated using the PM6 method (Stewart 2008) and calibration function obtained from the results of calculations of set of water clusters with GIAO/DFT/6-31G(d,p) (δ_H) and PM6 (q_H) (Gun'ko et al. 2005d, 2007f, 2009d) show (Figure 1.219) that the spectra are close to typical spectra of SAW. This approach is useful for both quantitative and qualitative description of complex systems, which are too large to be calculated using ab initio or DFT methods.

The results of calculations (Figures 1.219 and 1.220) are in agreement with the experimental ^{1}H NMR spectra regarding both the chemical shift values for pure water ($\delta_H=4$–5 ppm) and the temperature dependence of δ_H (δ_H diminishes with elevating T value) and weak dependence of δ_H on the presence of NaCl ($C_{NaCl}=0.65$ wt%). It should be noted that diminishing of the water cluster size results in the displacement of the spectrum toward smaller δ_H values and contribution of WAW ($\delta_H=1$–2 ppm; Figure 1.220) increases due to increase in the relative number of water molecules

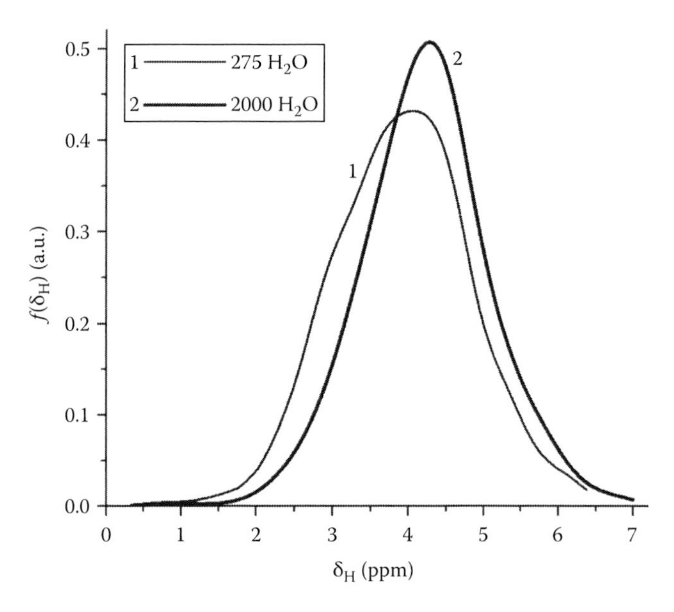

FIGURE 1.220 Theoretical ^{1}H NMR spectra of water nanodrops of different sizes: 275 (1) and 2000 (2) molecules.

being at the surface of a droplet. These results are in agreement with the experimental data in relation to the influence of water clusterization on the δ_H value. Thus, developed methodology of theoretical calculations of the NMR spectra of large systems, which cannot be calculated using nonempirical methods, can be used for the analysis of interactions of medicinal matters (e.g., quercetin, rutin, doxorubicin, salinomycin), biopolymers (starch, cellulose, proteins), and nanooxide particles with water.

Theoretical calculations show that the presence of molecules of quercetin or rutin in the water droplet (Figure 1.221) results in growth of clusterization of water (δ_H in 1–2 ppm range grows, Figure 1.222), as these molecules have both polar and nonpolar fragments, which differently

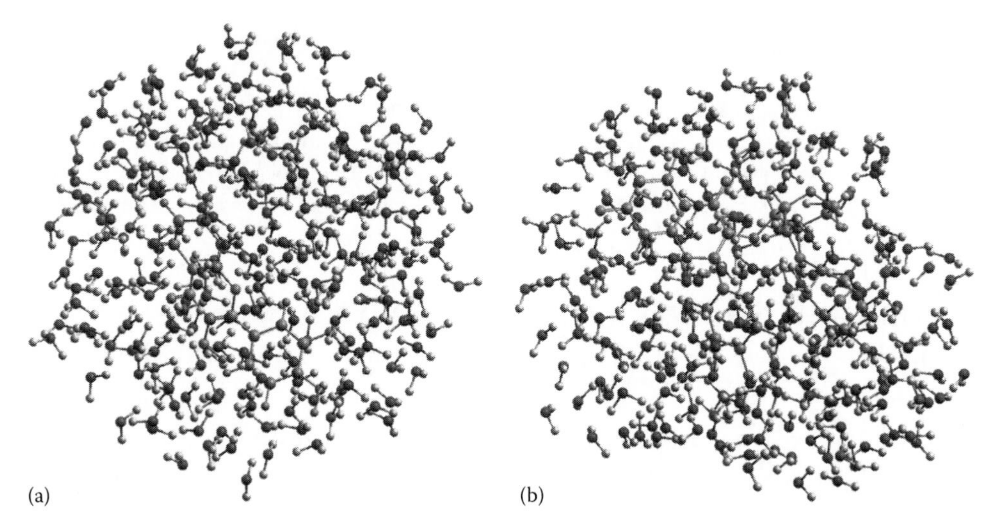

FIGURE 1.221 Clusters of water molecules with (a) quercetin and (b) rutin (geometry was optimized by PM6 method).

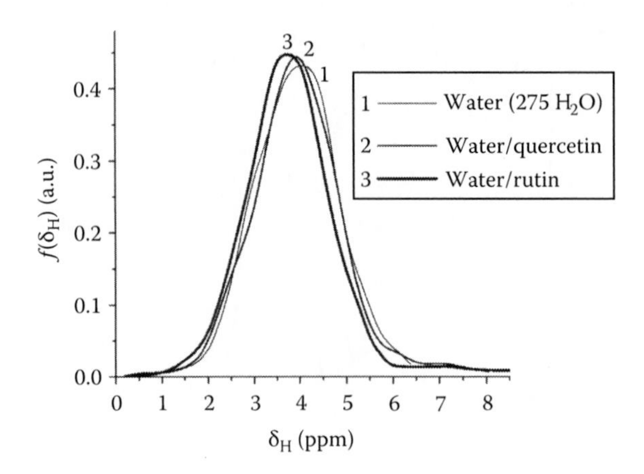

FIGURE 1.222 ^{1}H NMR spectra to the water clusters free (1) and with molecules of quercetin (2) and rutin (3) (only for water molecules).

influence the structure of hydrogen bond network in the structured water. Molecule of rutin is almost twice greater than the molecule of quercetin, therefore, the effect in relation to the displacement of the spectrum toward the smaller δ values is greater for rutin (Figure 1.222). The presence of significant number of O-containing groups in these molecules (which are antioxidants) stipulates appearance of small peak at $\delta = 7$ ppm (Figure 1.222).

To model adsorption of different compounds (N_2, H_2O, CH_4/H_2O, CH_3CN/H_2O, and $CHCl_3/H_2O$), the geometry of a model porous silica nanoparticle (3.8 nm in size with the pore diameter of 1.3 nm, Figure 1.223a) was optimized by PM6 method. Subsequent calculations were carried out by the PM6 method to full optimization (Figure 1.223). The pore surface was with silanols (Figure 1.223a) or TMS (Figure 1.223e). Adsorption of nitrogen (Figure 1.223b) gives different orientation of the N_2 molecules at different sites. Therefore, the average area of silica occupied by a nitrogen molecule is 0.137 nm^2, which is smaller than that (0.162 nm^2) a plane adsorption (e.g., on carbon). For nitrogen, adsorption is rather non-clustered (because nitrogen condensation heat $Q_L = 5.6$ kJ/mol at 77.4 K) but for water $Q_L = 45$ kJ/mol and interaction energy with silica is high (Table 1.38). This leads to clustered adsorption of water (Figures 1.223c through g and 1.224) at any adsorbents even at low amounts of water. Clustered water structures are characterized by the spectra with upfield shift in comparison with that of an individual water nanodomain with 2000H_2O. The appearance of hydrophobic TMS groups or methane (as a coadsorbate) leads to a certain diminution of the δ_H values, that is, water becomes WAW.

Thus, the theoretical calculations of the different water systems in comparison with the experimental data allow considerably deeper understanding of features of changes in the structure and parameters of interfacial water layers and to forecast the proper changes in the properties of composite materials at changes in concentrations of components, a dispersion medium type, coadsorption of water and organics, etc.

1.2 SILICA GELS, AEROGELS, SILOCHROME, AND POLY(METHYLSILOXANE): STRUCTURAL, INTERFACIAL AND ADSORPTION CHARACTERISTICS, AND STRUCTURE–PROPERTY RELATIONSHIPS

Silica gels with porous amorphous beads, the specific surface area in the 25–800 m^2/g range and the pore volume of 0.05–1.5 cm^3/g, are typically produced from sodium silicate (tetraethyl orthosilicate [TEOS]). Silica gels are widely used in adsorption applications as a desiccant (to control local humidity), chromatographic adsorbents (unmodified and functionalized), drug and catalysts carriers, etc. (Iler 1979, Bergna 1994, 2005, Legrand 1998,). Surface chemistry of

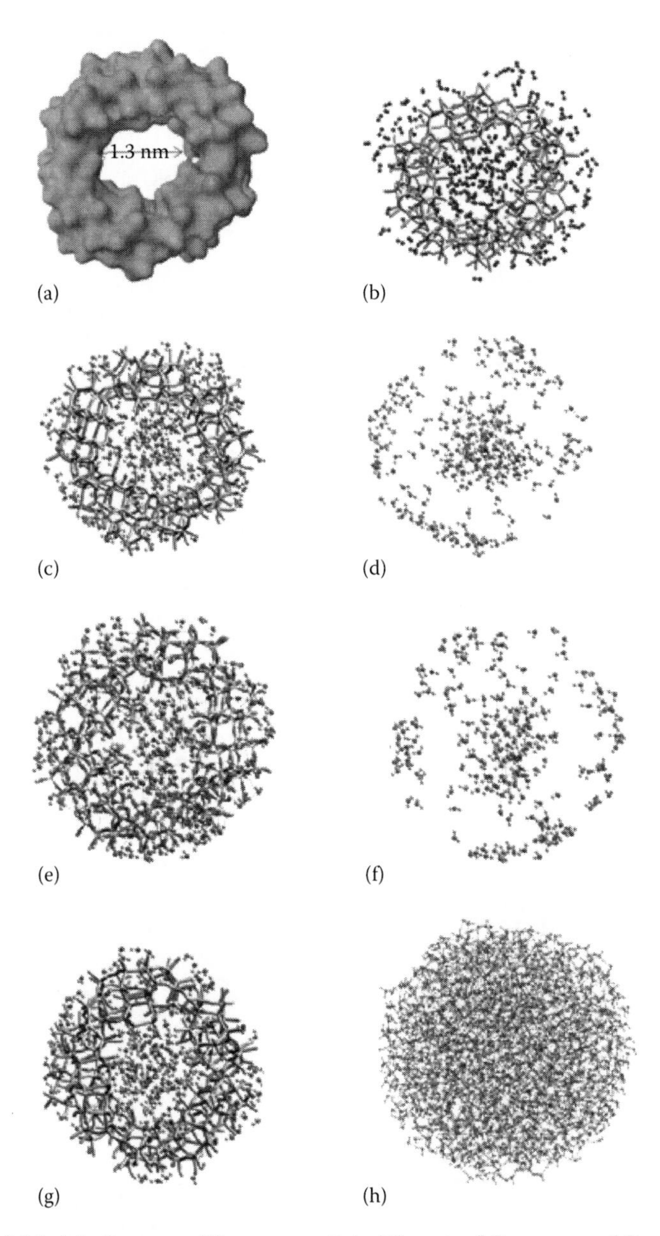

FIGURE 1.223 (a) Model of porous silica nanoparticle (diameter 3.8 nm, pore 1.3 nm in diameter) with different adsorbates: (b) nitrogen; water at the (c, d) initial surface and (e, f) with trimethylsilyl groups and (g) in a mixture with methane; (d, f) cluster structure of adsorbed water; (h) individual nanodomain of water of 4.5 nm in diameter (density 1.01 g/cm^3); calculations were carried out by the PM6 method (MOPAC2009).

silica gels, as well as other pure silicas, is mainly determined by silanols, which differ in their accessibility because of a non-smooth relief of a silica surface that appear in features of the IR spectra of amorphous silica (Table 1.39), as well as significant differences in their adsorption and other characteristics (Iler 1979, Legrand 1998). The IR spectroscopy gives rather qualitative than quantitative results with respect to the content of different silanols (Legrand 1998, Kiselev and Lygin 1975); however, TG, ^{1}H MAS NMR (Legrand 1998, Ek et al. 2001), and TPD-MS (Gun'ko et al. 1998a, Zhuravlev 2000) give more accurate quantitative information on the content of surface silanols on different silicas.

TABLE 1.38
$-\Delta E$ Values (kJ/mol) for H-Complexes of Different Molecules with Alone SiOH (First Row) and H-Bonded Vicinal SiO(H)-O-SiOH (Second Row) Groups

H_2O		CH_3OH		NH_3		$(CH_3CH_2)_2O$		$((CH_3)_3Si)_2O$	
45*	57**	36*	47**	46*	63**	36*	49**	21*	34**
54*	90**	51*	67**	53*	70**	43*	53**	18*	31**

Note: * Ab initio calculations with HF/6-31G(d,p); ** DFT calculations with B3LYP/6-31G(d,p) (HF/6-31G(d,p) geometry for all cases).

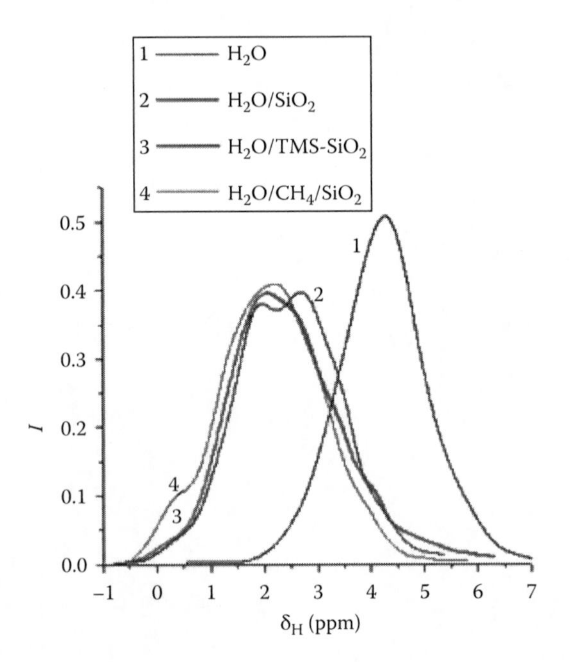

FIGURE 1.224 ^{1}H NMR spectra calculated for structures shown in previous Figure using B3LYP/6-31G9d,p)//6-31G(d,p) (GIAO) and PM6 method.

Despite the same chemical composition of silica gels, precipitated silica, aerogel, and nanosilicas, comparison of their ^{29}Si CP MAS NMR spectra (Figure 1.225) shows the difference not only in the content of Q_2 (=Si(OH)$_2$) and Q_3 (≡SiOH) sites but also in the δ(^{29}Si) shift values for Q_2, Q_3, and Q_4 (≡Si-O-Si≡) sites. This differences can be explained by features of particle morphology (from individual nanoparticles of nanosilicas to millimeter-sized or larger particles of other silicas), surface structure (content of the ≡SiOH and =Si(OH)$_2$ groups), heating history (affecting both surface and volume of the particles), and some admixtures. The textural differences of silicas (especially the presence of nanopores at radius $R<1$nm and narrow mesopores at $1<R<5$ nm) add the differences in their adsorption properties (Iler 1979, Bergna 1994, 2005, Legrand 1998, Chuiko 2001, 2003, Blitz and Gun'ko 2006).

Silica gels have cylindrical pores with relatively small deviations (~5%–20%) of their shape from the model of ideal cylindrical pores with relatively narrow PSDs (Figure 1.226). Typically, silica gels have broader PSDs than templated ordered mesoporous silicas (M41S, SBA, Folded sheet mesoporous (FSM), etc., vide infra). Many silica materials are well structurally

TABLE 1.39
Infrared Bands of Amorphous Silica at Room Temperature

Band Types	Wavenumber (cm^{-1})
$\nu_{c(SiOSi)}$, δ_{OSiO}	457
	507
$\nu'_{c(SiOSi)}$	810
	820
$\nu_{a(SiOSi)}$	1076
	1256
$\nu'_{a(SiOSi)}$	1200
	1160
$\delta_{SiOH,iso}$	770/(820)
$\delta_{SiOH,I,II}$	(775)/830 – 850
$\delta_{SiOH,b}$	870–900
$\delta_{SiOH\cdots[OHH]}$	950
$\nu_{OH\cdots[OHH]}$	<3450
$\nu_{OH,b}$	3520
$\nu_{OH,int}$	3670
$\nu_{OH,I}$	3690–3735
$\nu_{OH,II}$	3735–3742
$\nu_{OH,iso}$	3747
$(\nu+\tau)_{OH}$	≈3850
$(\nu+\delta)_{SiOH\cdots[OHH]}$	<4400
$(\nu+\delta)_{SiOH,b}$	4420
$(\nu+\delta)_{SiOH,iso}$	4515/(4570)
$(\nu+\delta)_{SiOH,I}$	(4490)/4540
$(\nu+\delta)_{SiOH,II}$	4565
$2\nu_{OH,int}$	7220
$2\nu_{OH,I}$	7220–7290
$2\nu_{OH,II}$	7290–7315
$2\nu_{OH,}$	7329
$(2\nu+\tau)_{OH}$	≈7410

Note: Bands: iso, isolated; int, internal; disturbed: b, hydrogen bonded; I are silanols of primary hydration sites; II are silanols at outer surface (according to Legrand 1998, Kiselev and Lygin 1972).

characterized. Therefore, both silica gels and ordered mesoporous silicas are frequently used as model adsorbents for tailoring of nonstandard methods such as different versions of NMR and TSDC cryoporometry and relaxometry, and DSC thermoporometry (Strange et al. 1996, Aksnes and Kimtys 2004, Gun'ko et al. 2005d, 2007f,i, 2009a, Webber et al. 2007a,b, Mitchell et al. 2008, Petrov and Furó 2009).

Bredereck et al. (2011) synthesized aquasols containing silica nanoparticles with diameters of 75–95 nm obtained directly by hydrolysis of 2 wt% TEOS in water in the presence of a nonionic surfactant and used ^{29}Si MAS NMR spectroscopy of the freeze-dried particles to obtain information about the degree of condensation and the ratio of free hydroxyl groups. This is a typical example of the use of NMR spectroscopy as a tool in analytical chemistry of solids.

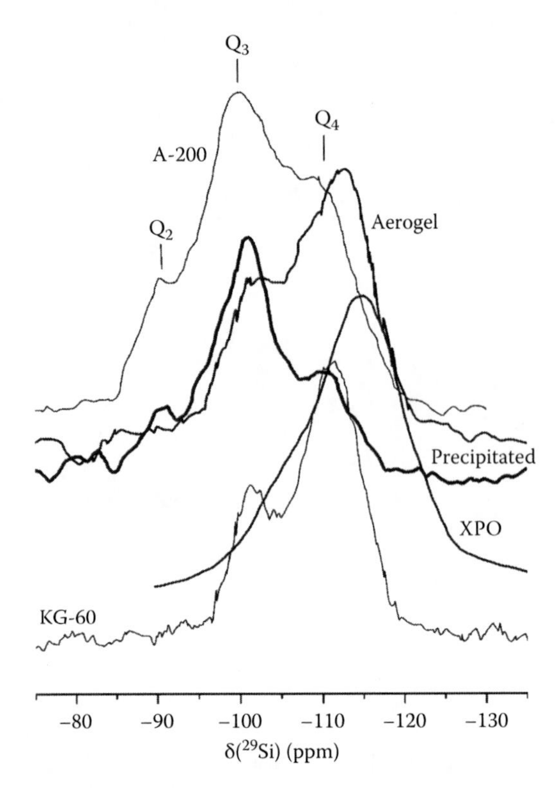

FIGURE 1.225 ^{29}Si CP MAS NMR spectra (redrawn according to different publications) of amorphous silicas: silica gel Kieselgel 60 (KG-60, Oepen and Günther 1996), silica gel Grace Davison XPO2407 (Capel-Sanchez et al. 2004), precipitated silica (Jesionowski et al. 2003), aerogel (Marzouk et al. 2004), and nanosilica A-200 (Legrand 1998).

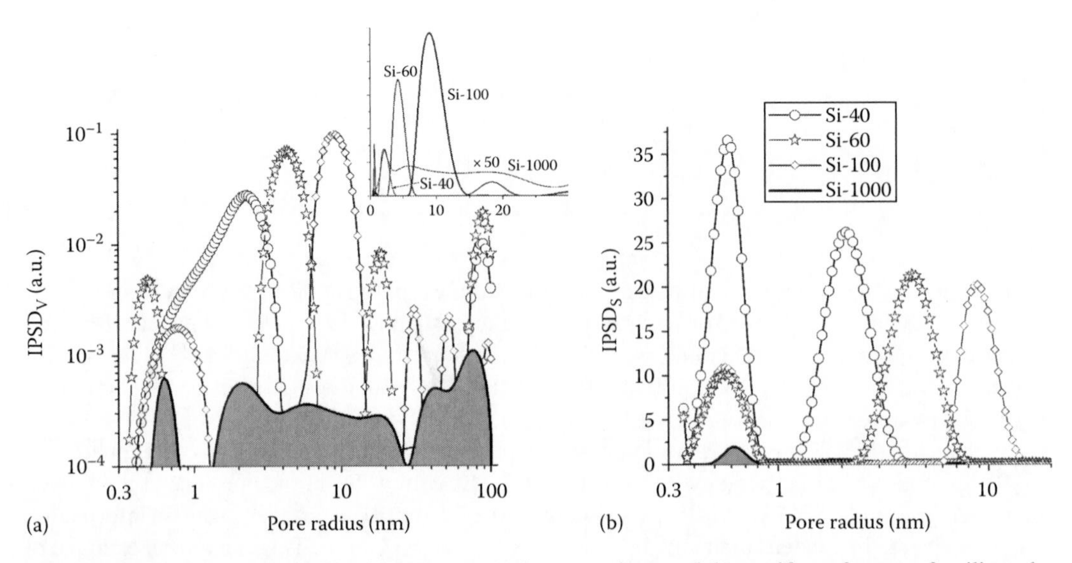

FIGURE 1.226 Incremental PSD with respect to (a) the pore volume and (b) specific surface area for silica gels.

The nitrogen adsorption isotherms for silica gels (Figure 1.227a, type IV of IUPAC pore classification) are characterized by broad hysteresis loops because of strong capillary condensation in mesopores. However, the water adsorption isotherms have broader hysteresis loops because water can more strongly interact (by hydrogen bonding) with the silica surface than nitrogen (dispersion interactions). The adsorption potential for water bound to silica gels is much greater than that for nitrogen (Figure 1.228) because of the stronger bonding of water to surface silanols by the hydrogen bonds, which play a very important role in the interfacial phenomena at the silica surfaces (Iler 1979, Legrand 1998, Gun'ko et al. 2005d, 2007i, 2009d).

Typically, the smaller the specific surface area of a silica gel (e.g., Si-1000 < Si-100 < Si-60 < Si-40) and the larger the pore diameter, the broader is the main PSD peak in the mesopore range, especially for low-porosity Si-1000 at $S_{BET}=26$ m^2/g and $V_p=0.05$ cm^3/g (Figure 1.226). For instance, silica gel Si-40 (Merck, $S_{BET}=732$ m^2/g, average pore diameter of approximately 4 nm) has a narrower main PSD peak than Si-60 (447 m^2/g) or Si-100 (349 m^2/g) (Merck).

Silica gels adsorb water vapor from air more strongly (by order of magnitude) than nanosilicas because inner pores in silica gel granules have much greater adsorption potentials than nonporous nanoparticles of nanosilicas. ^{1}H NMR, TSDC, and TG studies of water bound to silica gels show (Figures 1.226 and 1.227) that water is more strongly adsorbed in narrower pores and desorbed from them at higher temperatures. The analysis of the activation energies of water mobility from rotation of OH bonds and H^+ tunnel transfer at 100–130 K (TSDC) at minimal activation energy $E \approx 10$ kJ/mol to molecular mobility at T close to 273 K (NMR) or evaporation at 350–390 K (TG) and $E=50$–90 kJ/mol (Figure 1.227) reveals certain regularities observed in the data obtained by different methods. These regularities are caused by the nature of intermolecular bonds in water located in confined space of nanopores and mesopores and changes in the Gibbs free energy of bound water depending on size and volume of pores in silica gels (Figure 1.229).

There are two tendencies: (i) water is more strongly associated (δ_H grows) in silica gels with larger pores despite partial filling of pores by water and (ii) water more strongly interacts with silica in narrower pores (Figure 1.230). The first effect is due to clustered adsorption of water. The second one is due to that these clusters tend to be adsorbed in narrower pores where the adsorption potential is higher. Overlapping of the force fields produced by opposite pore walls and a higher ratio of a number of water molecules directly bonded to the silica surface to the total number of molecules in the clusters bound in narrower pores result in a decrease in the electron shielding of protons in the water molecules (downfield shift is observed). For Si-100, the formation of larger clusters results in a stronger effect than filling of narrower pores (Si-40 or Si-60; Figure 1.230). However, water located in narrower pores of Si-40 is characterized by greater δ_H values than Si-60. The corresponding changes in the Gibbs free energy are observed for the aqueous suspensions of silica gels with totally filing of pores by water. A significant diminution of the δ_H values with elevating temperature (Figure 1.230) is caused by several reasons: (i) weakening of the hydrogen bonds (diminution of associativity of molecules) because of the increased mobility of molecules (larger average length of the bonds and deviation of the ∠OHO value from optimal 170°–180° characteristic of ice with a low density, LDW) and an increase in the amount of interstitial water and enhanced HDW effects; (ii) increased mixing of water with chloroform in mesopores; and (iii) transfer of a portion of adsorbed liquid water into the vapor phase (in the NMR ampoule) that causes diminution of the interfacial water layer thickness and, consequently, reduction of the average value of the hydrogen bond number per molecule. Fast molecular exchange is possible between these water states (in the NMR timescale) that gives relatively broad averaged signal (Pople et al. 1959) of the interfacial water. Change in temperature leads to change in contribution of different states of water (i.e., contributions of two signals at $\delta_H=3$–5 and 1–2 ppm), and the signal shape changes.

The existence of pores at large l/D values in silica gels causes greater associativity and smaller mobility of water molecules in the confined space of long and relatively narrow pores. Similar pores are absent in aggregates of primary particles of nanosilicas (as described earlier) that result in different temperature dependence of the δ_H value in comparison with silica gels. Additionally,

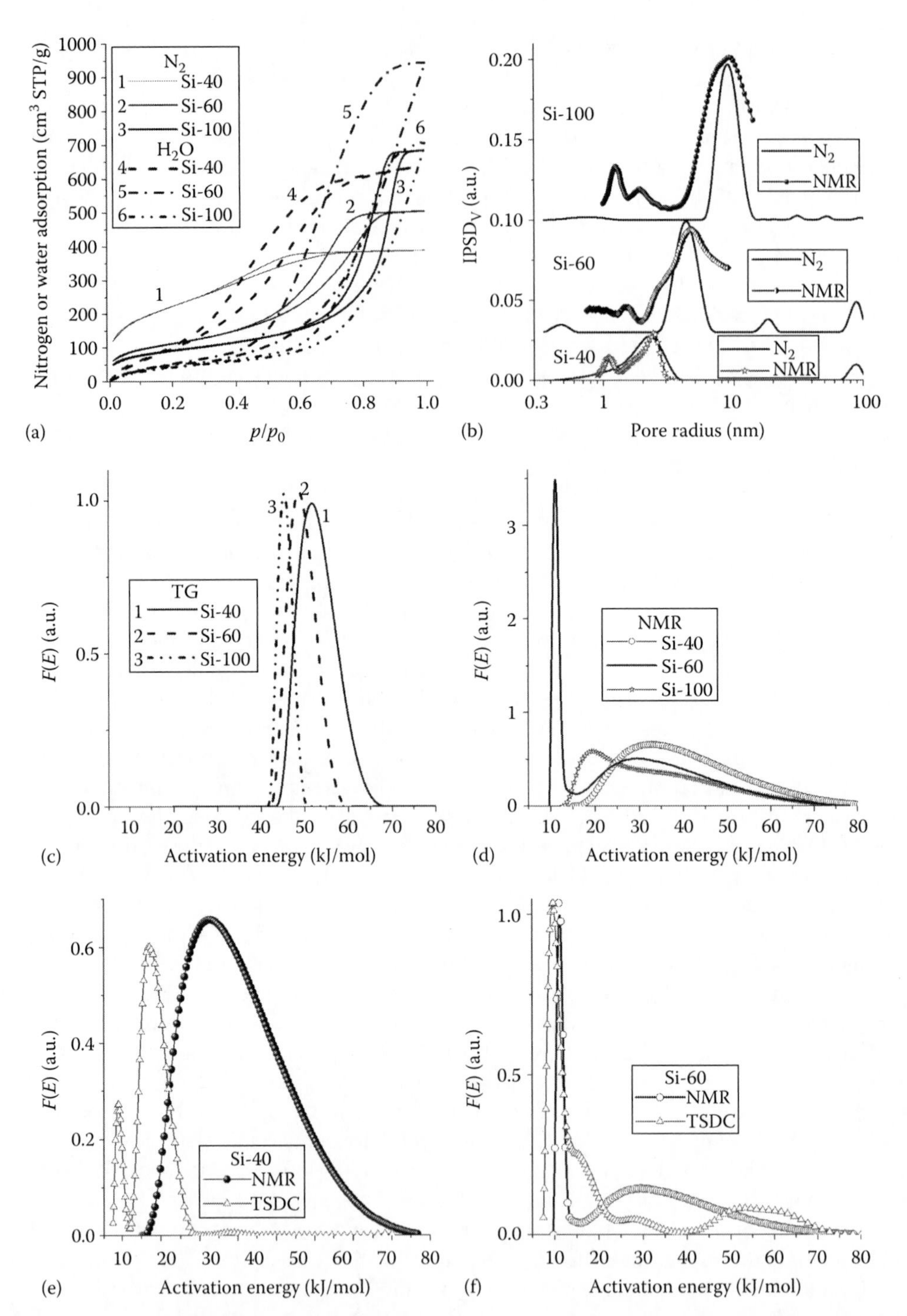

FIGURE 1.227 (a) Nitrogen (77.4 K) and water vapor (293 K) adsorption isotherms for silica gels Si-40, Si-60 and Si-100; (b) pore size distributions of silica gels calculated on the basis of the low-temperature nitrogen adsorption isotherms for dry adsorbents and the ^{1}H NMR spectra of water in the concentrated aqueous suspensions at $T<273$ K; (c) activation energy of molecular water desorption during TG experiments ($T<140$°C); (d) activation energy of molecular mobility in NMR measurements at $T<0$°C; and comparison of the activation energy distribution functions of molecular (NMR) and bond/molecules (TSDC dipolar relaxation) for (e) Si-40 and (f) Si-60.

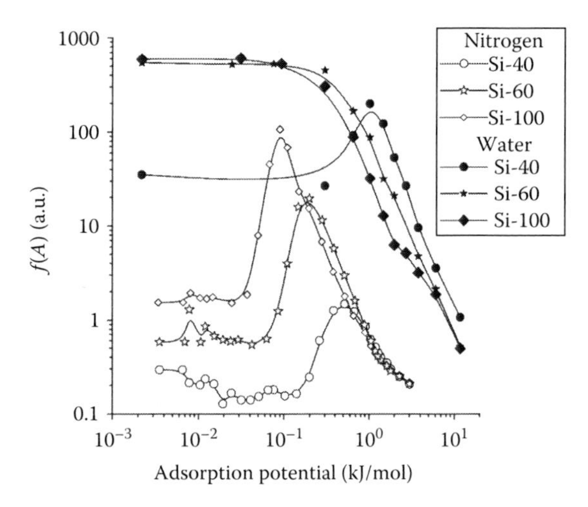

FIGURE 1.228 Adsorption potential ($f(A) = -d[spline(a(p/p_0))]/dA$ (where a refers to the adsorbed amount of nitrogen or water; $A = -\Delta G = R_g T \ln(p_0/p)$ represents the differential molar work equal to the negative change in the Gibbs free energy) for nitrogen and water adsorbed onto silica gels.

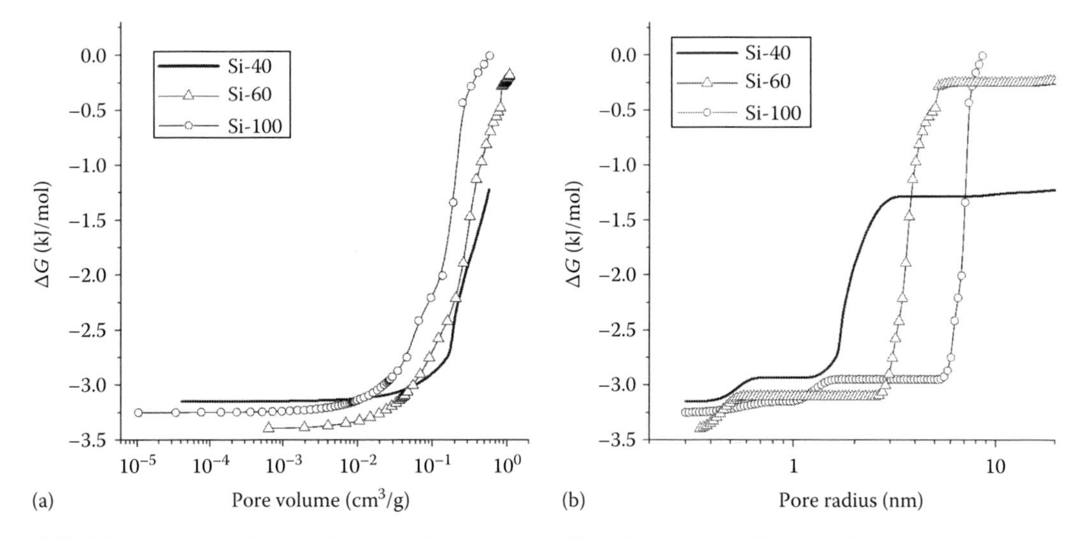

FIGURE 1.229 Relationships between changes in the Gibbs free energy of interfacial water in the aqueous suspension (C_{SiO_2} = 5 wt%) of silica gels Si-40, Si-60 and Si-100 and (a) pore volume and (b) radius of cylindrical pores.

structural and morphological features of nanosilica and silica gels cause the difference in changes in the Gibbs free energy of the interfacial water in the aqueous suspensions of these oxides as functions of the pore size and volume because of the difference in the HDW and LDW balance, amounts of SAW, SBW, and WBW. However, the main changes in the ΔG value for water adsorbed onto A-300 and different silica gels occur at $R < 6$ nm connected with SBW. This suggests that the surface of both types of silicas can strongly change the structure of the hydrogen bond network in the interfacial layer at the thickness of 3–10 statistical monolayers of water.

Notice that the texture of silica Silochrome (prepared by HTT of nanosilica), which can be considered as an intermediate material between nanosilica and silica gels, provides clear

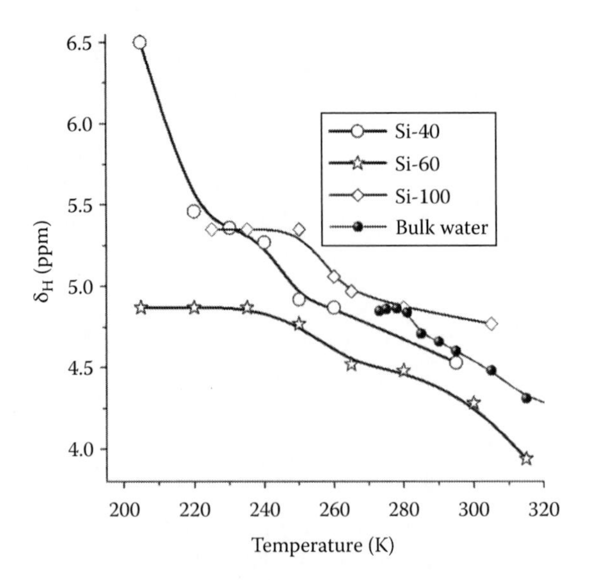

FIGURE 1.230 Temperature dependence of the chemical shift of bulk water and water bound to silica gels at low hydration $h=0.131$ (Si-40), 0.115 (Si-60) and 0.114 g/g (Si-100).

observation of such unusual interfacial water as WAW even in the case of the pristine material (Gun'ko et al. 2005a). It should be noted that the PSD_{uw} for Si-40 and Si-60 are in agreement with the PSD_{N2} (Figure 1.227b). Observed deviations between two kinds of the PSDs can be caused by the dependence of σ_{sl} on R (i.e., the interaction energy between a liquid and a solid depends on the pore size), the pore topology effects, the presence of both ice and water in the same pore at a certain temperature, and other factors affecting the results of applications of the Gibbs–Thomson equation (Webber et al. 2007a,b, Mitchell et al. 2008, Petrov and Furó 2009, Webber 2010). The Gibbs–Thomson equation applied to bound water to pores at $R>10$ nm can give incorrect results because more distant water does not "sense" the pore surface, that is, it is akin to bulk water. In other words, LDW can appear in these broad pores whereas mainly HDW can be in narrow pores of unmodified silica gels. Therefore, other liquids, for example, organics, can be used in cryoporometry as probe compounds to characterize large pores.

A series of NMR cryoporometry investigations of mesoporous silica gels were carried out using cyclohexane (Figures 1.231 through 1.233), mixtures cyclohexane with water (Figure 1.234), or decane with water (Figure 1.235; Strange et al. 1993, Allen et al. 1998, Alnaimi et al. 2004, Mitchell et al. 2005, 2008). The NMR or DSC PSDs are in good agreement with the gas phase (N_2) PSD (Figures 1.231 and 1.232). Water adsorbed onto a mixture of three different silica gels demonstrates a clear dependence of the melting temperature on the pore size of these materials (Figure 1.231b). This result shows the NMR cryoporometry ability to resolve multimodal PSD in complex materials. Comparison of the PSDs (Figure 1.232) is obtained using three different methods (NMR cryoporometry with naphthalene, DSC thermoporometry with cyclohexane, and nitrogen adsorption). Certain differences between the PSDs can be caused by a lower resolution of the "stepped" NMR cryoporometry for larger pores.

For 10 nm silica, the NMR and DSC results showed good agreement although the gas adsorption PSD has larger median pore diameter. For 6 nm silica, the NMR cryoporometry gives larger minimum pore diameter because of a large unfrozen layer of naphthalene preventing the smallest pores from being measured (Mitchell and Strange 2004, Mitchell et al. 2008). Notice that benzene used as an adsorbate in NMR cryoporometry gives good results for the PSD of porous glasses; however, certain deviations of the NMR PSDs in comparison with N_2 PSD are observed (Aksnes and Kimtys 2004).

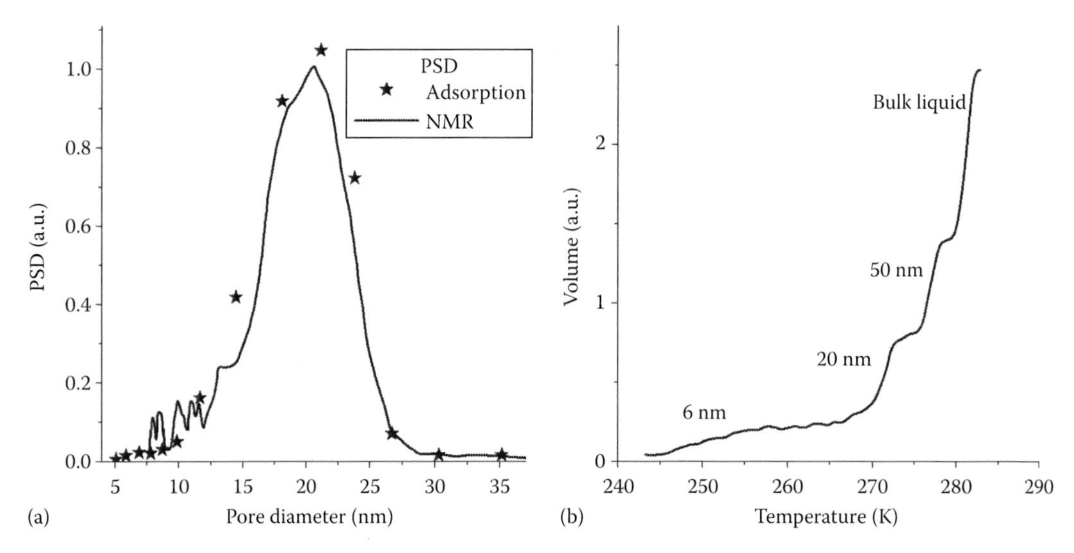

FIGURE 1.231 (a) Pore size distribution for silica gel at the median pore diameter of 20 nm; (b) melting curve measured from a sample containing silica gels with median pore diameters of 6, 20, and 50 nm (cyclohexane was used as an adsorbate) (according to Strange et al. 1993). (Adapted from *Physics Reports*, 461, Mitchell, J., Webber, J.B.W., and Strange, J.H., Nuclear magnetic resonance cryoporometry, 1–36, 2008. Copyright 2008, with permission from Elsevier.)

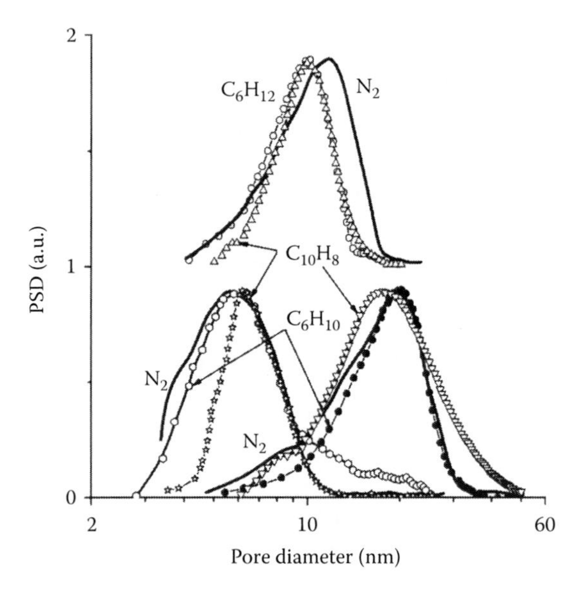

FIGURE 1.232 Pore size distributions from silica gels with different median pore diameters of 6, 10, and 20 nm measured using NMR cryoporometry with naphthalene, DSC thermoporosimetry with cyclohexane, and nitrogen gas adsorption (according to Strange et al. 1993, 2003). (Adapted from *Physics Reports*, 461, Mitchell, J., Webber, J.B.W., and Strange, J.H., Nuclear magnetic resonance cryoporometry, 1–36, 2008. Copyright 2008, with permission from Elsevier.)

Mixtures with water and organics (cyclohexane, Figure 1.234, or decane, Figure 1.235) demonstrate the behavior dependent on both pore sizes of silica gel and the mixture type.

The water/cyclohexane mixture shows that both water and organic molecules can occupy narrower pores formed between the silica gel pore walls and coadsorbate crystallites or unfrozen structures. At lower amounts of water, it tends to form clusters at the pore surfaces characterized

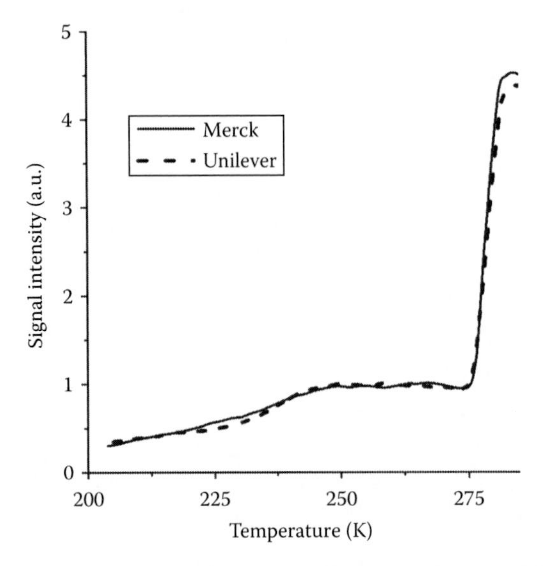

FIGURE 1.233 The temperature behavior of cyclohexane in silica gels at a median pore diameter of 6 nm from different producers. (Adapted from *Physics Reports*, 461, Mitchell, J., Webber, J.B.W., and Strange, J.H., Nuclear magnetic resonance cryoporometry, 1–36, 2008. Copyright 2008, with permission from Elsevier.)

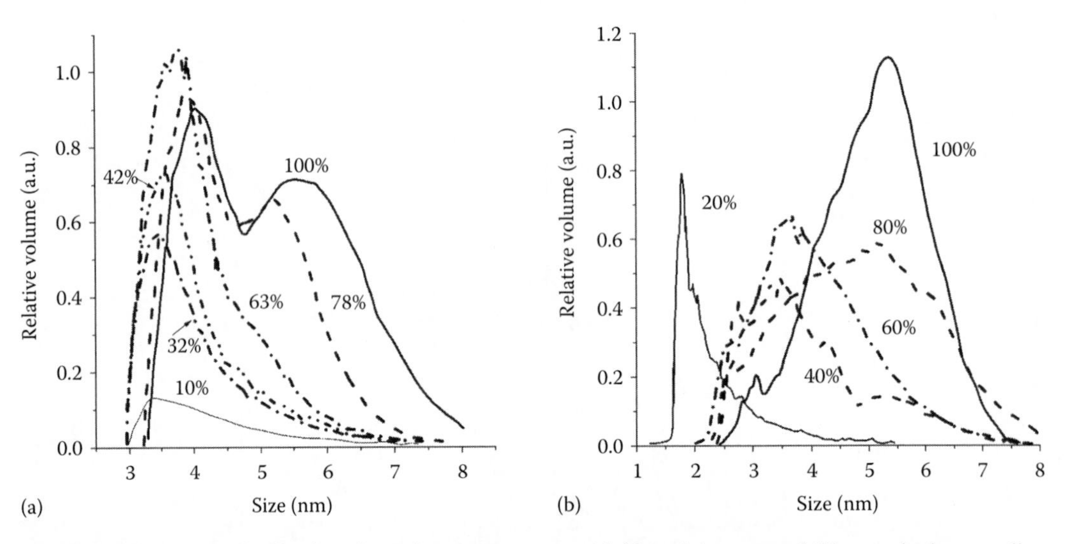

FIGURE 1.234 Size distributions for unfrozen structures with (a) cyclohexane and (b) water in 6 nm median pore diameter silica gel. (Adapted from *Physics Reports*, 461, Mitchell, J., Webber, J.B.W., and Strange, J.H., Nuclear magnetic resonance cryoporometry, 1–36, 2008. Copyright 2008, with permission from Elsevier.)

by smaller sizes than the smallest pores in the sample (~3 nm). At larger content of water, the clusters (of a puddle type) grew until a uniform coverage was achieved. Cyclohexane was seen to preferentially fill the smallest pores first.

These different mechanisms were associated with the different surface interactions of water (hydrogen bonding) and cyclohexane (dispersion interactions). The analysis incorporating magnetic susceptibility variations can be a potential method to provide information on the textural characteristics (Allen et al. 2001, Strange et al. 2003, Mitchell et al. 2008). Water and decane were distinguishable in the melting curves (Figure 1.235a) due to their different properties. If the water content increases, the decane is displaced from the narrowest pores first. Water can displace cyclohexane (Mitchell et al.

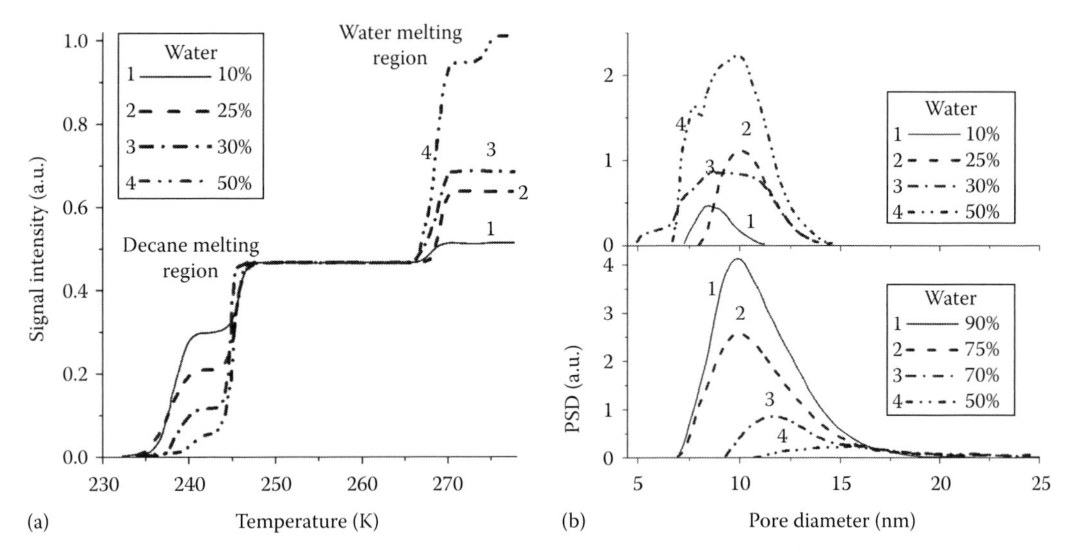

FIGURE 1.235 NMR cryoporometry measurements of water and decane at various ratios in 10 nm median pore diameter silica: (a) melting curves showing separated decane and water fractions, and (b) PSD for the decane and water, which displaces decane from the narrowest pores. (Adapted from *J. Chem. Phys.*, 120, Alnaimi, S.M., Mitchell, J., Strange, J.H., and Webber, J.B.W., Binary liquid mixtures in porous solids, 2075–2077, 2004. Copyright 2004, with permission from Elsevier.)

2008) and a miscible alkane mixture of heptadecane and decane in silica gels when the organics absorbed preferentially (Mitchell et al. 2005). A similar effect was observed for binary mixtures of nitrobenzene and n-hexane in porous glass (Valiullin and Furo 2002a–c) interpreted as nitrobenzene droplets formed within a pore wetting layer of n-hexane. The droplet sizes were analyzed using PFG techniques (Valiullin and Furo 2002a). The displacement effects of water by organics were also observed for different oxide, polymer and carbon adsorbents (Gun'ko et al. 2005d, 2009d).

Nanoscaled phase separation of liquids (Figure 1.236) occurs in confined space of mesoporous and macroporous glasses (Valiullin et al. 2004).

There are domains with distinct molecular composition (e.g., nitrobenzene-rich domains) and thereby distinct freezing point depression. ^{1}H PGSE NMR diffusion data indicated that the domains are small on the microscale (Valiullin and Furó 2002a). The phase separation is not a consequence of freezing: below crystallization temperature T_{cr}, the ^{15}N NMR spectrum is split into two lines, one small (from nitrobenzene molecules in the hexane-rich domains), and one large (from the nitrobenzene-rich phase). Moreover, ^{1}H cross relaxation NMR data show a decrease in contacts between nitrobenzene and hexane upon cooling at $T<T_{cr}$ but keeping it unfrozen. As the average pore size increases toward the macropore range, the relative size of nitrobenzene domains increases. While at the smallest pore diameter, apparently no nitrobenzene domain fills a pore completely; at the largest pore size, nitrobenzene seems to fill completely those pores that contain nitrobenzene-rich domains. At intermediate pore sizes (broad mesopores and narrow macropores), a mixture of these two processes occurs (Valiullin et al. 2004).

Petrov and Furó (2010) compared water, benzene, cyclohexane, and cyclooctane as NMR cryoporometry adsorbates used for structural characterization of porous glasses. They found that the freezing (ΔT_f) and melting (ΔT_m) temperature shifts correlate linearly, giving a constant ratio $\Delta T_m/\Delta T_f$ for the mentioned adsorbates. This ratio can be used as a measure of the pore morphology. According to the theoretical analysis, ΔT_f is controlled by the surface-to-volume ratio (S/V) and ΔT_m is controlled by the pore curvature (dS/dV). Therefore, the ratio $\Delta T_m/\Delta T_f$ provides additional information on the pore shape, unavailable from ΔT_m or ΔT_f values alone. For glasses samples, this ratio decreases from 0.67 in 7.5 nm pores, which is expected for spherical geometry, to 0.57 in 27.3

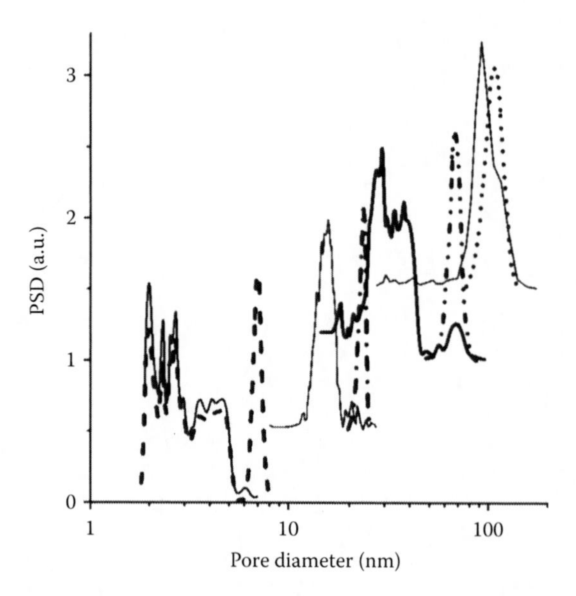

FIGURE 1.236 The size distributions of the nitrobenzene-rich domains after nitrobenzene–hexane phase separation in porous glasses of 7.5 nm, 24 nm, 73 nm and 127 nm average pore size as obtained by ^{1}H NMR cryoporometry. The dashed lines show the PSDs obtained for the adsorption of pure nitrobenzene. (Adapted from *Current Appl. Phys.*, 4, Valiullin, R., Vargas-Kruså, D., and Furó, I., 370–372, 2004. Copyright 2004, with permission from Elsevier.)

and 72.9 nm pores of cylindrical or tubular shape. The PSD functions obtained from the freezing and melting data are similar in shape and width that melting and freezing processes occur in pores of glasses on the same length scale of the pore structure.

Petrov and Furó (2011) compared two models of pores (cylindrical and spherical) with respect to the ^{1}H NMR PSD for unfrozen water bound to glasses adsorbents (Figure 1.237) calculated using freezing and melting branches of the hysteresis loop. To separate the signal of unfrozen water from that of a solid, a T_2-filter based on the CPMG sequence was exploited with a relaxation delay 2.4 ms. The coincidence of the peaks suggests that the cylindrical pore model is adequate for Vycor. For controlled-pore glass (CPG), unlike for Vycor, the cylindrical model gives distributions that essentially shifted one from another, whereas PSDs computed within spherical model are consistent. However, the average pore size estimated is larger than the firm d_{av} value (Figure 1.237). These results suggest that having both freezing and melting data, one can ascertain the interconnectivity of the pore structure and estimate the goodness of a chosen model of the pore geometry by comparing PSD's computed from freezing and melting data.

However, additional information on the PSD (e.g., from the gas adsorption, SAXS, or HRTEM) is needed to estimated the accuracy of the NMR cryoporometry results, especially for complex materials. According to Hitchcock et al. (2011), to interpret scanning loop experiments in NMR cryoporometry properly, it is necessary to use PFG-NMR and relaxometry data to probe the configuration of the molten phase to eliminate ambiguities. They showed how different scenarios for the form of the pore network and the freezing/melting mechanisms can be tested using these techniques. They demonstrated that incorporation of advanced melting effects allows one to interpret scanning loop data properly, without invoking unproven dead-end pores. The freeze–thaw hysteresis in cryoporometry can be purged from scanning loops to give a reversible process, as expected if advanced melting was occurring.

Dvoyashkin et al. (2009) studied surface diffusion of n-heptane in two mesoporous adsorbents with different morphologies of the pore network (Vycor random porous glass and porous silicon with linear pores) using PFG-NMR (Figure 1.238). The obtained diffusivities revealed increasing

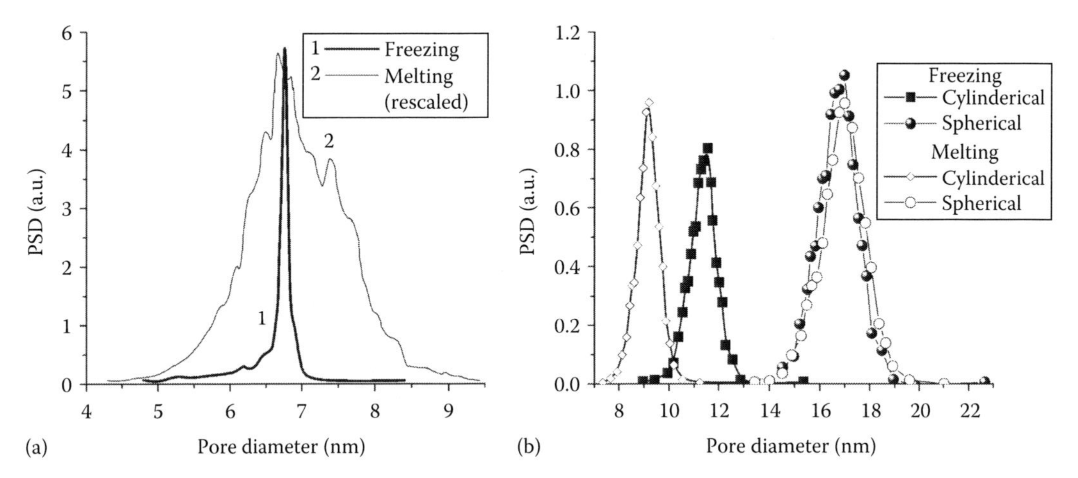

FIGURE 1.237 The pore size distributions in (a) Vycor (average pore diameter $d_{av}=7$ nm) assuming cylindrical pore geometry and (b) CPG ($d_{av}=7.5$ nm) with cylindrical and spherical pore models from freezing and melting. (Adapted from *Micropor. Mesopor. Mater.*, 136., Petrov and Furó, A joint use of melting and freezing data in NMR cryoporometry, 83–91, 2010. Copyright 2010, with permission from Elsevier.)

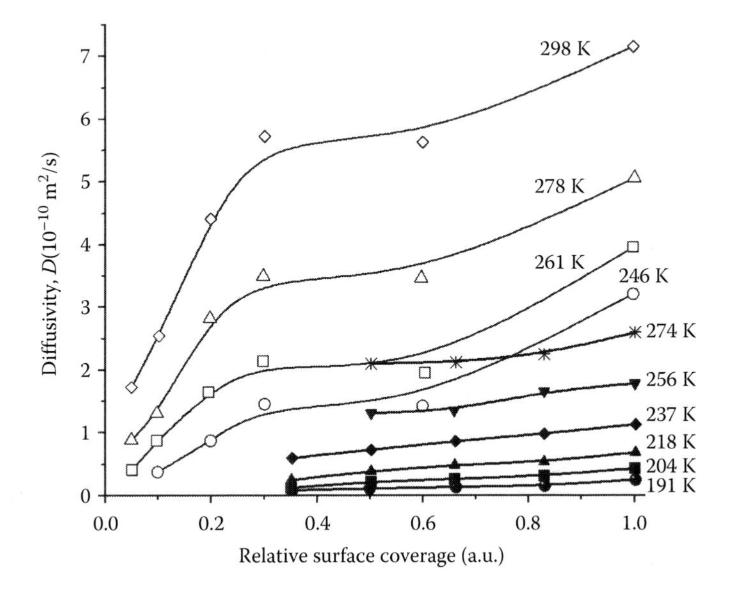

FIGURE 1.238 Diffusivities of n-heptane in Vycor porous glass (open symbols) and in porous silicon (solid symbols) as a function of surface coverage measured at different temperatures. (Adapted from *Micropor. Mesopor. Mater.*, 125, Dvoyashkin, M., Khokhlov, A., Naumov, S., and Valiullin, R., Pulsed field gradient NMR study of surface diffusion in mesoporous adsorbents, 58–62, 2009. Copyright 2009, with permission from Elsevier.)

mobility of adsorbed molecules with increasing surface coverage, indicating not only surface heterogeneity but also the dependence of interaction of adsorbed molecules located in the first and next layers in pores. The activation energy (E_a) for surface diffusion turned out to be a function of surface coverage. The E_a values were 10–25 and 12–17 kJ/mol for the heptane diffusion in porous glass Vycor and silicon, respectively (Dvoyashkin et al. 2009).

Troyer et al. (2005) studied differently hydrated CPG at the pore diameter of 23.7 nm using analysis of water proton spin–spin relaxation decay curves modeled as two-component exponential decays as a function of hydration. The results were consistent with a geometric model involving a surface water layer and a bulk-like liquid fraction in the form of a plug. The amount of surface water increases as the sample hydrates, until hydration reached approximately a monolayer, at which point a water plug starts to form in the pore, and grow in length at the expense of the surface layer. The results were also analyzed in terms of a puddle pore-filling model.

Valiullin et al. (2005) studied self-diffusion of acetone and cyclohexane in CPG (4 nm) and mesoporous silicon (average pore size 3.5 and 9.6 nm) as a function of pore loading using the PFG-NMR method. It was found that the effective diffusivities of adsorbate molecules in mesopores at partial loadings are related to two mechanisms, the Knudsen diffusion through the gaseous phase in the pore space and the diffusion within the layer of molecules (fluid) adsorbed on the pore walls. The relative contributions of these modes, which were determined by the details of the interphase equilibrium, change with variation of the pore loading, leading to a complex behavior of the effective self-diffusion coefficients. These coefficients of the adsorbate molecules have been found to depend on the molecular concentration in pores, and in most cases, pass through a maximum for cyclohexane at the relative coverage $\Theta \approx 0.2$ and larger values for acetone dependent on the average pore size and the material kind. Thus, the position and the magnitude of the maximum were well correlated with the pore size, details of the adsorbate–surface interactions, and physical properties of acetone and cyclohexane. The formation of the maximum in diffusivities was explained within the frame of the modified two-phase exchange model where a particular shape of the adsorption isotherm was taken into account. This made it possible to include the effects of the shifted interphase equilibrium with respect to that in bulk liquids in the model. The latter is of special importance in nanoporous materials. The molecular self-diffusion coefficient, which is quite sensitive to the liquid configuration within the pore network, was found to depend on the saturation direction when plotted via the external vapor pressure. When diffusivities were compared at equal pore concentrations, the occurrence of the hysteresis was observed only in pores with smaller pore diameters (about 4 nm), whereas in bigger pores (9.6 nm) did not (Valiullin et al. 2005). Notice that the geometric criteria predict the absence of a hysteresis loop upon adsorption–desorption of probe gases in pores of mesoporous silicas at $d \leq 4$ nm (Tovbin et al. 2006). Thus, the effects observed by Valiullin et al. (2005) and Tovbin et al. (2006) are governed by the surface forces, which strongly overlap in narrow mesopores or nanopores at $d<4$ nm. Consequently, in pores at $d>4$ nm, the organization of adsorbates can be less energetically uniform.

Jobic et al. (1995) studied the mobility of cyclohexane in a nanoporous silica powder using neutron scattering and NMR techniques at different temperatures and loadings. The silica powder had the same characteristics as the corresponding supported membrane. Self-diffusion coefficients of the order of 10^{-10} m^2/s were obtained by both techniques at 300 K, the activation energy being ca. 11 kJ/mol. The same value was also found for benzene diffusing in the same silica sample. This implies that there were no specific interactions with silica. The dimension of the pores between the voids of the nonporous silica particles was of the order of 1 nm. Unfortunately, the textural characteristics of the materials studied by Jobic et al. (1995) were not obtained, despite the presence of the corresponding raw data obtained by several methods.

The ^{1}H NMR cryoporometry method can be used to measure broad pore sizes (Figure 1.239) if octamethylcyclotetrasiloxane is used as an adsorbate (Ono et al. 2009). Three very different methods such as the nitrogen adsorption, the ^{1}H NMR cryoporometry, and the mercury porosimetry give the PSDs, which are mutually complementary. These results confirm that organic molecules are better probe compounds than water for the structural characterization of macroporous materials with NMR cryoporometry or relaxometry.

The characterization of porous materials using the NMR relaxometry is based on the fact that the interactions between the fluid molecules undergoing confined space effects in pores and interacting with the pore walls change the NMR relaxation times (Kimmich 1997, Stallmach and Kärger 1999). The longitudinal (T_1) and transverse (T_2) relaxation times of the fluid molecules are proportional to the V_p/S_p ratio for pores according to equation

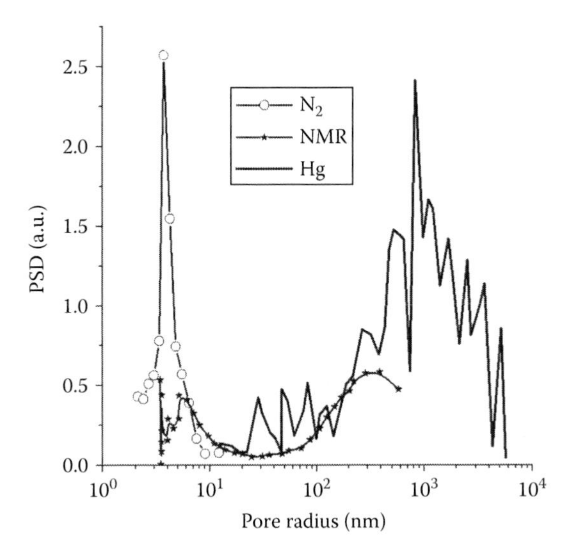

FIGURE 1.239 Pore size distributions from $R=1$ nm to 10 μm for silica calculated from the nitrogen adsorption, ^{1}H NMR spectroscopy data for octamethylcyclotetrasiloxane and the mercury porosimetry. (Adapted from *J. Colloid Interface Sci.*, 336, Ono, Y., Mayama, H., Furó, I. et al., Characterization and structural investigation of fractal porous-silica over an extremely wide scale range of pore size, 215–225, 2009. Copyright 2009, with permission from Elsevier.)

$$T_{1,2} = \frac{1}{\rho_{1,2}}\frac{V_p}{S_p} \sim \frac{1}{\rho_{1,2}}R_p \quad (1.65)$$

Here, $\rho_{1,2}$ is the surface relaxivity for longitudinal and transverse relaxations. In addition, it has been assumed that the V_p/S_p ratio is directly proportional to the pore radius R_p (as for cylindrical pores). Therefore, the distribution of the relaxation times observed from the porous sample reflects the PSD. The prerequisite for using this equation is that the pore is so small that the fluid molecules collide many times with the pore wall during their contribution to the NMR signal, and this imposes the upper size limit of the method. Surface relaxivity is material dependent, and it must be determined separately for each material before pore sizes can be determined. The chemical shift of xenon depends on the surface-to-volume ratio of the pocket. On the basis of the geometry of the pocket, it can be shown that the surface-to-volume ratio is proportional to pore size, and therefore the following relationship exists between the chemical shift, δ, and the pore radius, R_p:

$$\delta = \delta_s/(1 + bR_p/a) \quad (1.66)$$

Here, δ_s is the chemical shift of a xenon atom adsorbed on the surface of the pocket, $a=KR_gT$, where K is Henry's constant and R_g is the universal gas constant. Parameter b depends on the geometry of the pocket. Xenon porometry/relaxometry was successfully used to study different adsorbents (Telkki et al. 2005a,b, 2006a,b).

The relaxation of hydrogen nuclei in a fluid confined in pores is determined by the self-diffusion of the molecules within the pores, the bulk relaxation of the liquid, and the surface relaxation at the pore walls (Valckenborg et al. 2000, 2001). For NMR spin-echo times of a few ms, the diffusion length of the water molecules in broad pores is about 1 μm, and hence the surface relaxation is the dominating process at pore sizes below 0.1 μm (fast diffusion limit; Brownstein and Tarr 1979). In this case, a one to one correspondence exists between the observed distribution of relaxation times and the PSD (Halperin et al. 1989, Valckenborg et al. 2000). Since the transverse relaxation time of

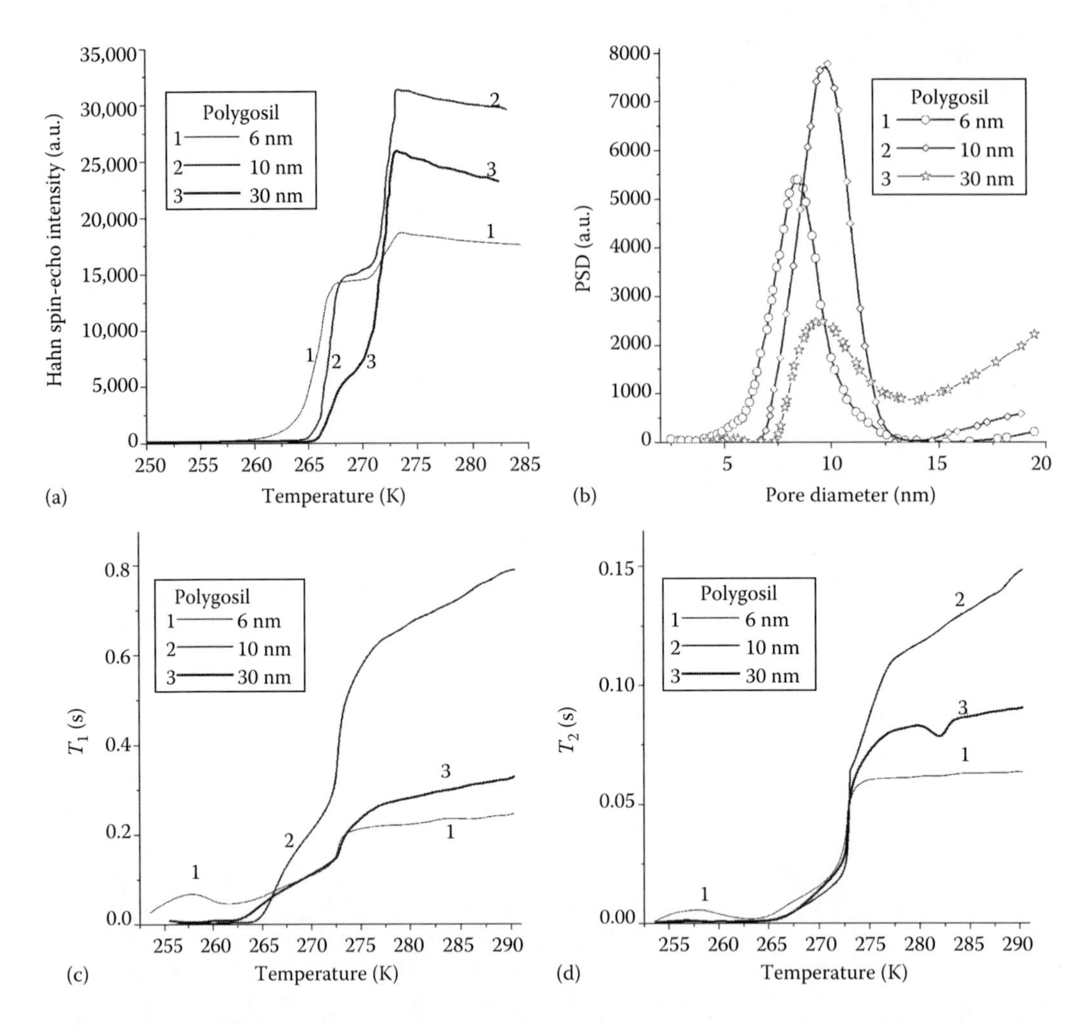

FIGURE 1.240 Temperature dependences of (a) the Hahn spin-echo intensity, (c) longitudinal and (d) transverse relaxation times, and (b) PSD calculated with the NMR cryoporometry (k_{GT}=62 K nm) for such irregular silicas for analytical applications as Polygosils. (According to Valckenborg R. et al., Cryoporometry and relaxation of water in porous materials, in *Proceedings of the 15th European Experimental NMR Conference (EENC 2000)*, June 12–17, 2000, University of Leipzig, Leipzig, Germany.)

the frozen fraction of the pore fluid is usually very short ($T_2 \sim 6$ μs; Overloop and Van Gerven 1993), only the liquid fraction will be observed in pulsed NMR experiments with echo times larger than 100 μs. This makes NMR an ideal tool for cryoporometry (Strange et al. 1993). The melting process of ice located in pores of silica particles starts in narrower pores at $T > 250$ K (Figure 1.240). Both longitudinal and transverse relaxation times monotonically increase with temperature. In this temperature region, the T_1 value is proportional to the NMR signal intensity, that is, the liquid water fraction in the pores. These results suggest a rather homogeneous melting process of water in mesopores of silicas (Valckenborg et al. 2000). However, as mentioned earlier, the homogeneous melting process can be expected only for relatively narrow pores.

The transverse relaxation time $T_{2,avg}$ for adsorbate occupying the total porous structure is the weighted average of all individual pores and can be both measured and calculated for every pore size a (Valckenborg et al. 2001):

$$\frac{V_0}{T_{2,\text{avg}}(a)} = \int_0^a \frac{\partial V_i(a')}{\partial a'} \frac{1}{T_{2,i}(a')} da' \tag{1.67}$$

where

V_0 is the total volume of pores smaller than a

$\partial V_i(a')/\partial a'$ is the volume of pores with size between a' and $a' + \partial a'$

The PSD can be measured by NMR cryoporometry, SAXS, small-angle neutron scattering (SANS), nitrogen or argon adsorption, etc. The transverse relaxivity as a function of pore size $T_{2,i}(a')$ is also known from the classical one pore situation, as shown earlier. Thus, the NMR cryoporometry pore diameters for silica gels are in agreement with the pore diameters as calculated from the gas adsorption. Additionally, the pore diameters of silicas estimated from the neutron scattering patterns show good agreement with other measurements (Webber et al. 2001).

Perkins et al. (2008) showed that the hysteresis observed in NMR cryoporometry for the freezing–melting curves can be deconvoluted into separate single pore and pore-blocking contributions using a model of a disordered mesoporous solid with a spatial arrangement of pore sizes analogous to a macroscopic "ink-bottle." The suggestion of pore-blocking effects was confirmed using PFG-NMR. The percolation theory was used to analyze the pore-blocking contribution to the hysteresis to determine the accessibility function and the pore connectivity of the void space of mesoporous silica. The NMR cryoporometry can be used to obtain a full range of void space descriptors, and thus offers a full, and potentially better understood, complementary pore structure characterization method to gas sorption and mercury porosimetry.

Mitra et al. (1993) have shown that a perturbation expansion of the measured diffusivity can be deviated from the macroscopically unrestricted intraparticle diffusion coefficient D_0:

$$D(\Delta) = D_0 - \frac{4D_0^{1.5}S}{9\pi^{0.5}V}\Delta^{0.5} \tag{1.68}$$

where

S is the particle external surface area

V is the volume of the powder particle

A straight-line fit of PFG-NMR data for measured diffusivity (or reciprocal tortuosity) against $\Delta^{0.5}$ should thus yield an intercept equal to the unrestricted intraparticle diffusion coefficient D_0 (or $1/\tau_0$), and $D_{\text{PFG}} = D_w/\tau_p$ (where D_w is the free diffusion coefficient for bulk water at the temperature at which the PFG-NMR experiment is conducted, and τ_p is the tortuosity of the pore space occupied by the fluid).

Wikberg et al. (2011) studied porous silicas (average pore diameter 6, 20, and 30 nm [Purospher® STAR] with 5 μm particles, and silica 10 nm [Merck] with 3 μm particles used as hydrophilic stationary phases in liquid chromatography) using the ^{2}H NMR spectra of bound water at temperatures between 193 and 277 K (Figures 1.241 and 1.242).

Freezing/thawing of bound water is characterized by the hysteresis loop (Figure 1.241). The used amounts of water were 1.5–2 times larger than the pore volume of silicas. Therefore, a certain portion of water corresponds to bulk water. Silicas are characterized by certain PSDs and water is unfrozen in narrower pores at lower temperatures. It can be assigned to bound freezable (located far from the pore walls in broader pores) and nonfreezable water located in the nearest surface layers or being in narrowest pores (Figures 1.241 and 1.242). Calculations of the textural characteristics of silicas with the Gibbs–Thomson equation ($k_{GT} = 50$ K nm) give the values close to that calculated from the adsorption data (Table 1.40). In the case of silica 30 nm, a certain overestimation is observed and the free surface energy γ_S (in mJ/m^2) is higher for this silica than that for others.

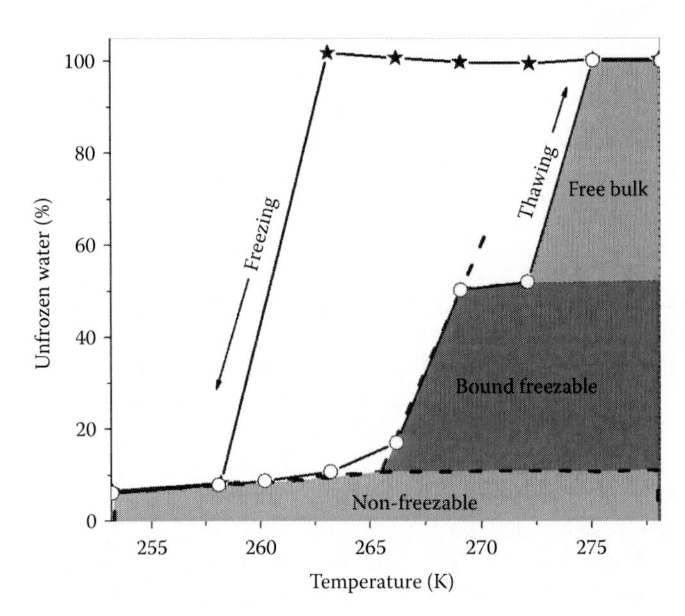

FIGURE 1.241 Temperature-corrected peak integrals from the ^{2}H NMR spectra acquired during the freezing and thawing cycle for water bound to porous silica (average pore size 10 nm) showing the hysteresis due to supercooling. (Adapted from *J. Chromatography A.*, 1218, Wikberg, E., Sparrman, T., Viklund, C., Jonsson, T., and Irgum, K., A ^{2}H nuclear magnetic resonance study of the state of water in neat silica and zwitterionic stationary phases and its influence on the chromatographic retention characteristics in hydrophilic interaction high-performance liquid chromatography, 6630–6638, 2011. Copyright 2011, with permission from Elsevier.)

Tombari et al. (2009) studied the thermodynamic features of water confined in 4 nm pores of glass Vycor (Figure 1.243). They showed that water behavior depends strongly on the pore-filling degree. For a given amount of bound water, the distribution of water is envisaged (i) as uncrystallizable layer of H_2O bonded to the silica surface, (ii) as H_2O clusters only partially filling the pores, and (iii) as regions of hydrogen-bonded and packed H_2O molecules in the pores. When the total amount of water is comparable to the amount needed to form an uncrystallizable monolayer attached to the silica wall, some H_2O still form clusters that may crystallize to an ice or else to a low entropy structure on cooling, with an exothermic peak at ~233 K. As the filling is increased, a larger population of H_2O molecules aggregate to form clusters at random sites inside the pores and the crystallization peak at ~233 K grows. On heating, an endothermic peak due to melting appears at ~255 K. For 46% filling of pores, some of the water forms the clusters and some fills certain regions of the pores. A study of 100% water-filled pores of MCM has shown two melting endotherms, which may indicate lack of radially symmetrical layer of water in these pores. Freezing of water in the completely filled pores produces a prominent exothermic peak at 250 K and indicates crystallization of the packed regions in the mesopores. Heating melts the state of solid formed in both stages of crystallization and broadens the endotherm. For 100% pore filling, there are no H_2O clusters and therefore only the prominent exothermic peak due to crystallization in the filled pores is observed. The assumption of a radially symmetrical layer of free H_2O molecules seems to be inconsistent with the observed crystallization of water in partially filled pores. Rather, H_2O molecules would likely be clustered in the pores in addition to the partial covering by an uncrystallizable layer on the pore wall. A separate study of cooling to different temperatures and thereafter heating confirms the occurrence of a two-stage crystallization and two-stage melting in the partially filled pores, but the two melting endotherms overlap, thereby broadening the endothermic peak on its LT side. These results are in agreement with the clustered adsorption of water onto other silicas described earlier.

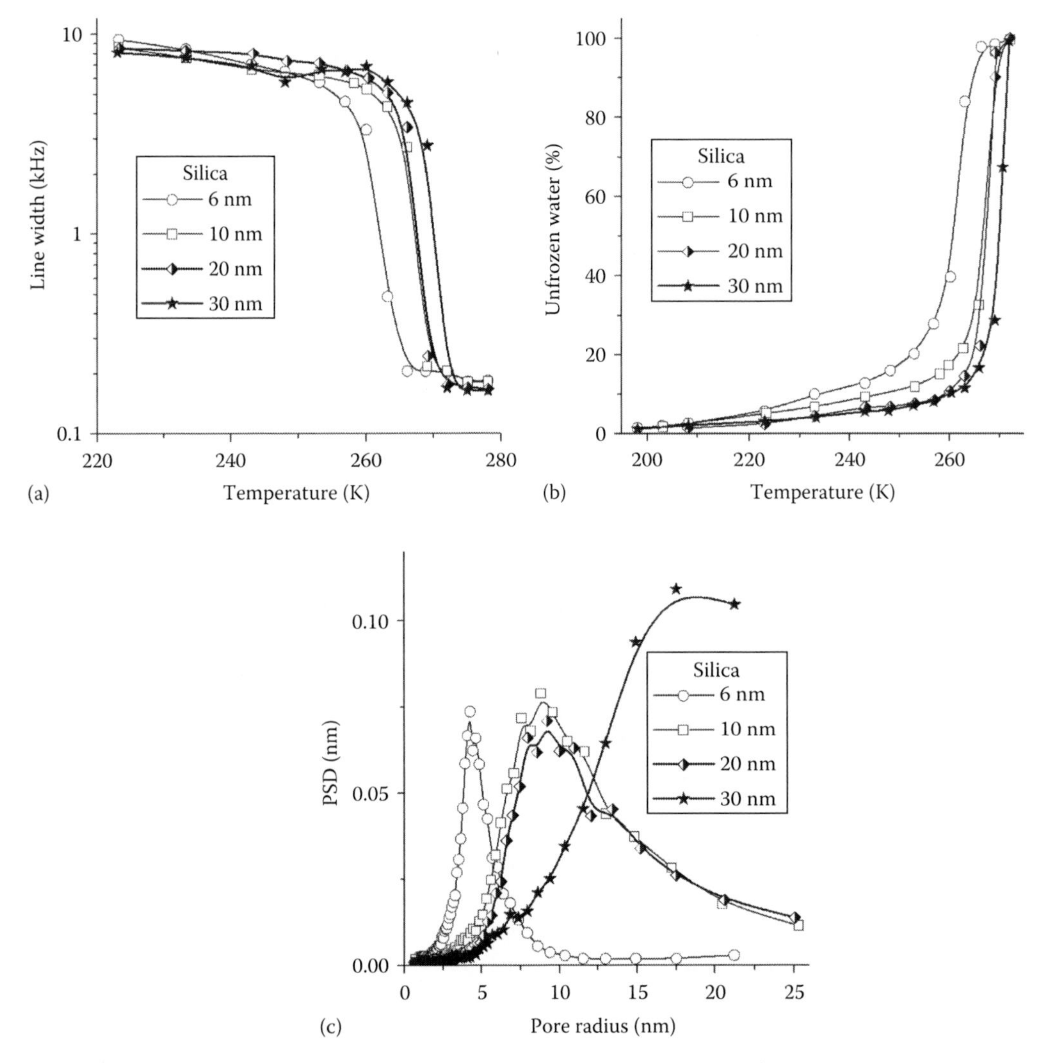

FIGURE 1.242 (a) ^{2}H NMR line width observed during the thawing part of the measurement cycle for silica samples; (b) relative content of unfrozen water confined in silicas; and (c) PSD calculated here using IG equation. (Adapted from *J. Chromatogr. A.*, 1218, Wikberg, E., Sparrman, T., Viklund, C., Jonsson, T., and Irgum, K., A ^{2}H nuclear magnetic resonance study of the state of water in neat silica and zwitterionic stationary phases and its influence on the chromatographic retention characteristics in hydrophilic interaction high-performance liquid chromatography, 6630–6638, 2011. Copyright 2011, with permission from Elsevier.)

TABLE 1.40
Textural Characteristics of Silicas from Adsorption and Cryoporometry Data

Silica (Pore Diameter) (nm)	S_{BET} (m^2/g)	S_{GT} (m^2/g)	$S_{GT,nano}$ (m^2/g)	$S_{GT,meso}$ (m^2/g)	V_p (cm^3/g)	γ_S (J/g)	γ_S (mJ/m^2)
6	477	444	156	288	1.15	43.1	90
10	296	271	121	150	1.06	25.0	84
20	190	166	49	117	0.87	16.2	85
30	113	134	67	67	0.92	12.9	114

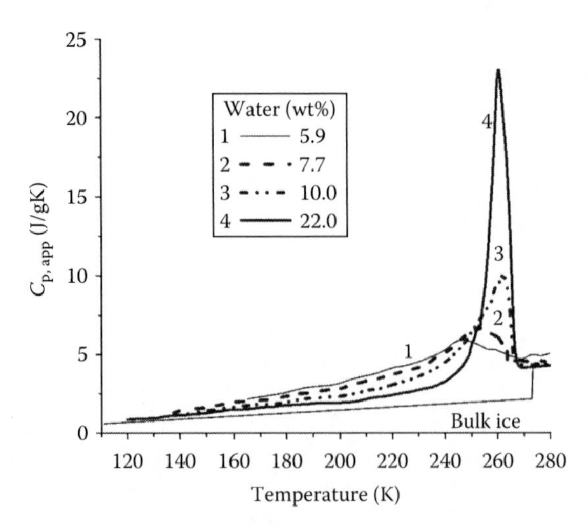

FIGURE 1.243 The plots of apparent heat capacity $C_{p,app}$ for water in 4 nm pores of Vycor glass containing 5.9, 7.7, 10, and 22 wt% water (27, 35, 46 and 100% pore filling) as measured during the heating from 110 to 280K at 12 K/h. The samples were precooled to 110 K at 12 K/h and kept for 1 h before heating. (Adapted from *Thermochim. Acta*, 492, Tombari, E., Ferrari, C., Salvetti, G., and Johari, G.P., Dynamic and apparent specific heats during transformation of water in partly filled nanopores during slow cooling to 110 K and heating, 37–44, 2009. Copyright 2009, with permission from Elsevier.)

The NMR techniques were also used to analyze different structural, surface reactions, reaction mechanisms, and other aspects related to the surface chemistry of silicas. For instance, Brunet et al. (2008) studied the electron irradiation effects on controlled-pore borosilicate glasses (CPGs at 8 and 50 nm in pore size, 96% SiO_2 and 3% B_2O_3) using multinuclear solid-state NMR technique. ^{1}H MAS NMR was used to study the surface proton sites. They showed that the irradiation leads to a dehydration of the material. The observed variation of the Q_4, Q_3, and Q_2 species from ^{1}H–^{29}Si CPMAS spectra showed an increase of the surface polymerization under irradiation, implying in majority a Q_2 to Q_3/Q_4 conversion mechanism, and ^{1}H–^{17}O CPMAS measurements exhibited an increase in contribution of Si–O–Si groups at the expenses of Si–OH groups. Modifications of the environment of the residual boron atoms were also put in evidence from ^{11}B MAS and MQMAS NMR. These data showed that MAS NMR can be used as sensitive tools to characterize CPG modifications occurring under electron irradiation. Notice that Cataldo et al. (2008) observed similar effects of the γ irradiation of amorphous silica Ultrasil 7000 (Degussa).

Carroll et al. (2002) studied silica glass and silica gel using batch and flow-through dissolution experiments, multinuclear NMR spectroscopy, and surface complexation modeling to reevaluate amorphous silica reactivity as a function of solution pH and reaction affinity in NaCl and CsCl solutions. The NMR data suggested that changes in surface speciation were driven by solution pH and to a lesser extent alkali concentrations, and not by reaction time or saturation state. The ^{29}Si CP NMR results showed that the concentration of silanol surface complexes decreased with increasing pH, suggesting that silanol sites polymerized to form siloxane bonds with increasing pH. Increases in silica surface charge were offset by sorption of alkali cations to ionized sites with increasing pH. It was the increase in these ionized sites that appeared to control silica polymorph dissolution rates as a function of pH. The ^{23}Na and ^{133}Cs NMR spectra showed that the alkali cations form outer sphere surface complexes and that the concentration of these complexes increased with increasing pH because of increasing negative charge density. Changes in surface chemistry could not explain decreases in dissolution rates as amorphous silica saturation was approached. There was no evidence for repolymerization of the silanol surface complexes to siloxane complexes at longer reaction times and constant pH (Carroll et al. 2002).

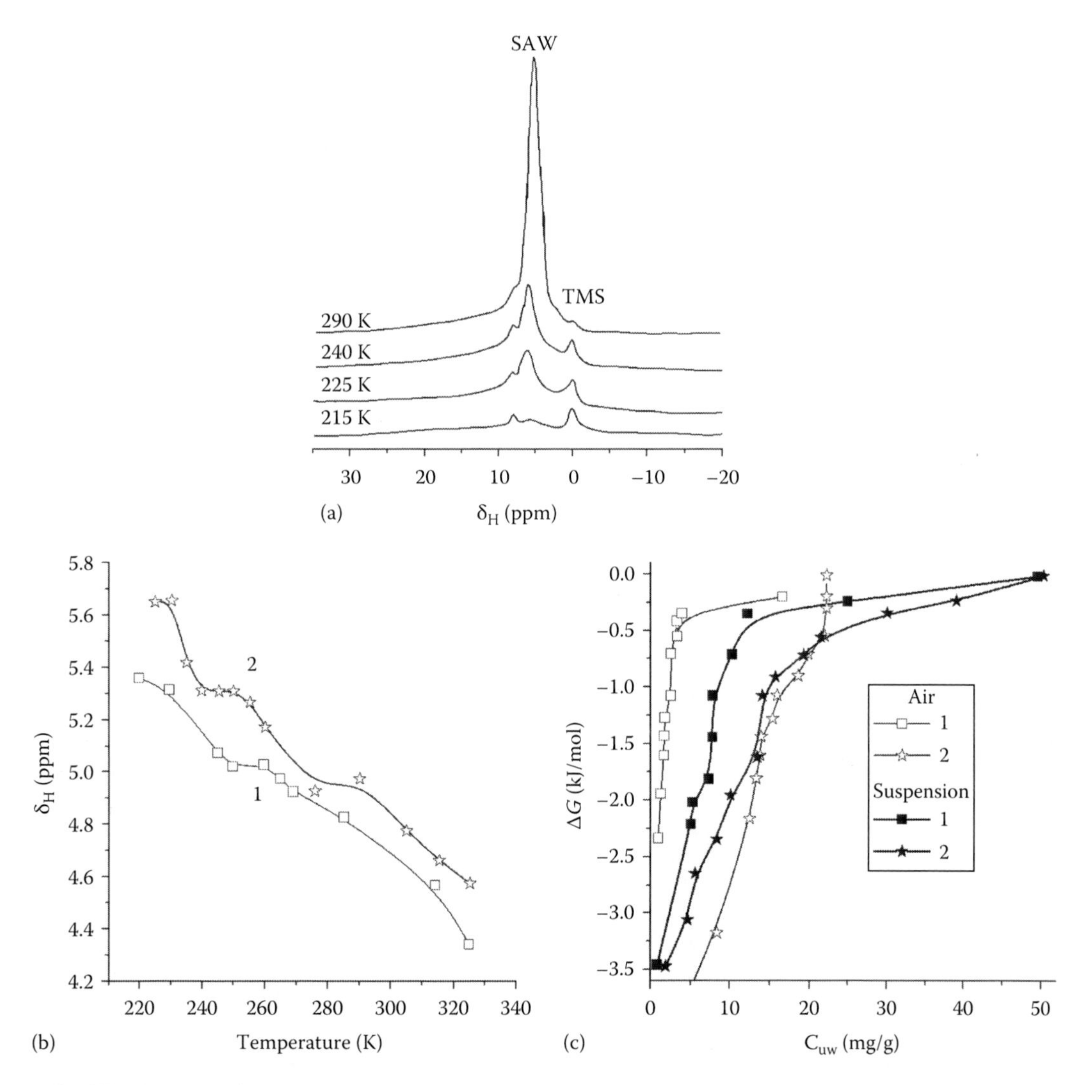

FIGURE 1.244 (a) ^{1}H NMR spectra of water (h=0.022 g/g) bound to sample 2 in $CDCl_3$ medium, (b) temperature dependence of the δ_H value for sample 1 of silicalite (h=0.017 g/g) and (h=0.022 g/g), and (c) relationships between the amounts of unfrozen water adsorbed from vapor phase and aqueous suspensions.

Two silicalite samples with S_{BET}=302 (sample 1) and 241 (sample 2) m^2/g synthesized without aluminum have a ZSM-5-like structure with 100 and 66% of crystallinity, respectively (Turov et al. 1998a, 1999). It has been found that in the case of both water vapor adsorption and water adsorption from aqueous suspensions, the volume of silicalite pores occupied by adsorbed water molecules is considerably smaller (0.02–0.05 cm^3/g, Figure 1.244c) than the total volume of pores (0.229 and 0.153 cm^3/g) determined from the nitrogen adsorption. This due to a low hydrophilicity of the siloxane bond network and long narrow pores in the particles. The chemical shift of water molecules on the silicalite surface depends on temperature and varies from 4.5 ppm at 330 K to 5.6 ppm at 220 K (Figure 1.244a and b). Thus, despite relatively low content of water, it represents only SAW including both SBW and WBW (Figure 1.244c).

Features of water organization in pores are well seen from the PSD (Figure 1.245) since IPSD (Figure 1.245c) show that the main volume of water is located in broad pores but not in narrow channels of silicalite. This is in agreement with partial filling of pores by water.

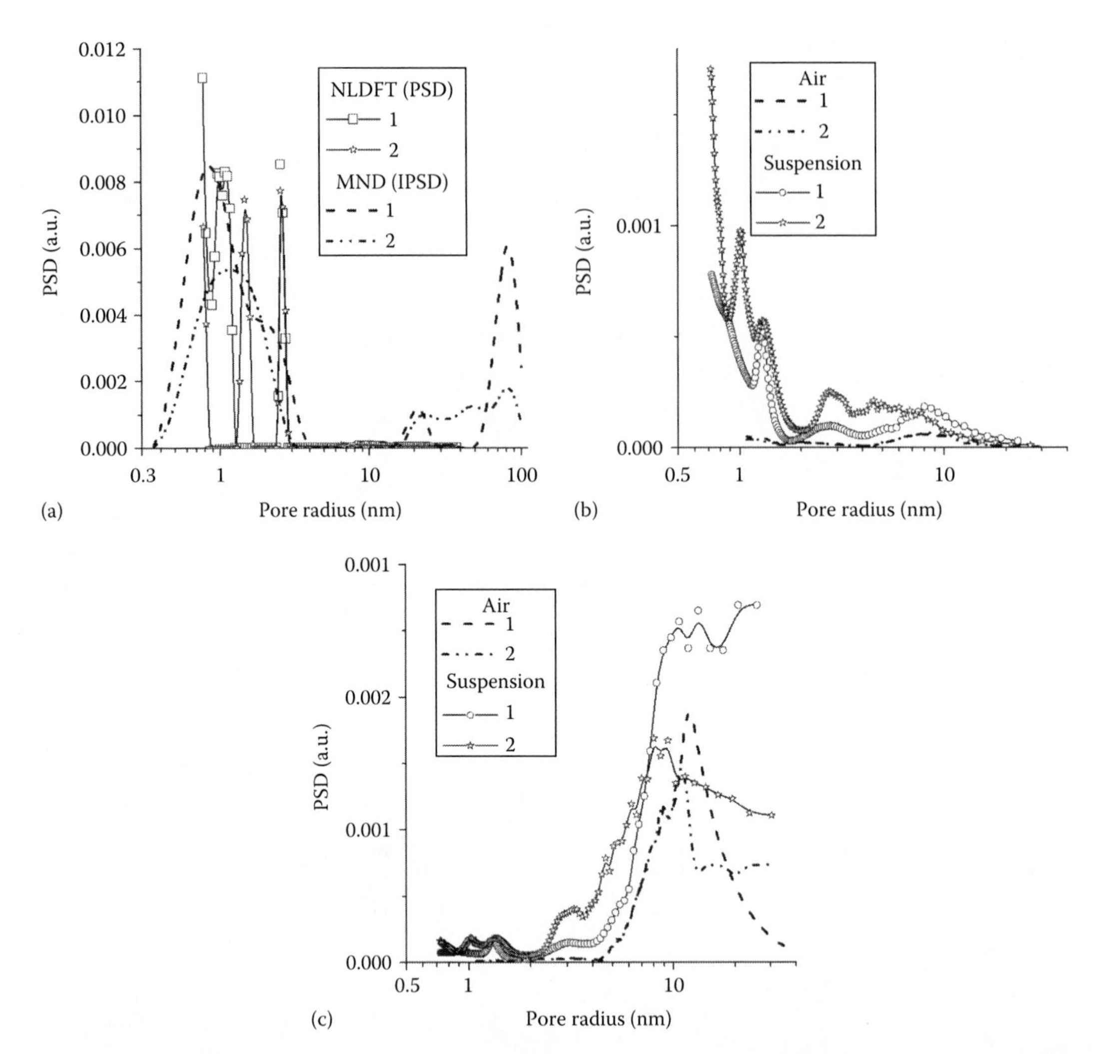

FIGURE 1.245 (a) Pore size distributions of silicalite samples calculated from the nitrogen adsorption data with NLDFT and MND methods (cylindrical pore model); and NMR cryoporometry results with GT equation (k_{GT}=70 K nm) (b) differential and (c) incremental PSDs.

A certain portion of water can penetrate in narrow channels (Figure 1.245a and b). However, even in the aqueous suspensions, this volume is 0.004–0.007 cm^3/g and the contact area in nanopores is 9 and 16 m^2/g but the S_{nano} values from the nitrogen adsorption are equal to 223 and 152 m^2/g, respectively.

Hybrid organic–inorganic materials can exhibit unique electro-optical properties and improved thermal stability. In the case of hybrid glasses, their inherently mechanically fragile nature, however, which derives from the oxide component of the hybrid glass network together with the presence of terminal groups that reduce network connectivity, remains a fundamental challenge for their integration in nanoscience and energy technologies (Oliver et al. 2010). Silica gels as well as other silicas are used as unmodified and modified materials, for example, partially or completely hydrophobized by silanes (Iler 1979, Bergna 1994, 2005, Legrand 1998, Chuiko 2001, 2003). The effects of hydrophobization can be elucidated to study the behavior of interfacial water at a silica surface partially modified by grafted trimethylsilyl (TMS) groups (approximately 30% ≡SiOH groups were substituted for $\equiv SiOSi(CH_3)_3$) (Gun'ko et al. 2005d). Such a modification of silica gel Si-60 leads to changes in the shape and the temperature dependence of the ^1H NMR spectra of

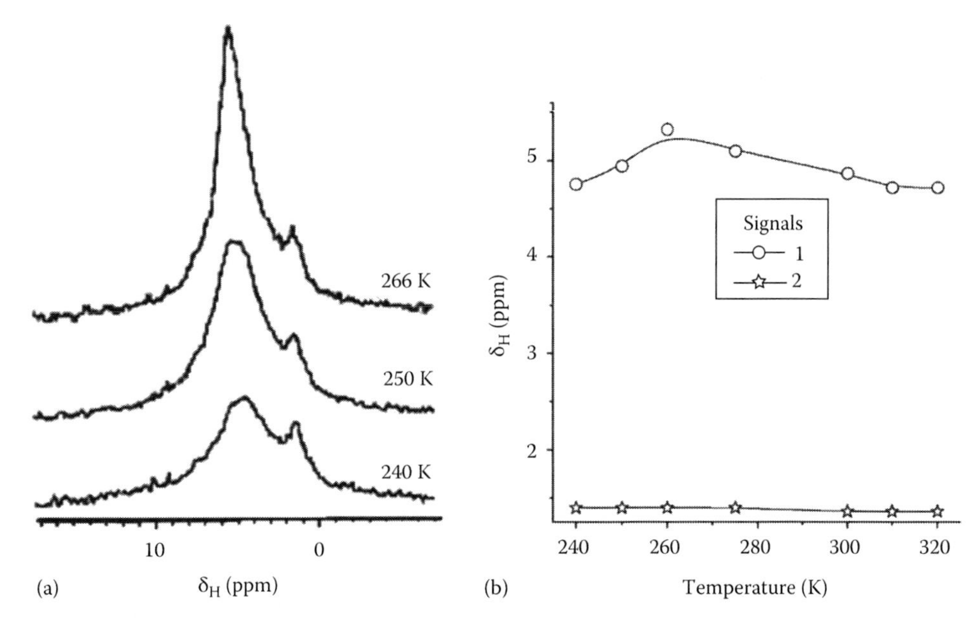

FIGURE 1.246 (a) ^{1}H NMR spectra at different temperatures and (b) $\delta_H(T)$ for two peaks for water (C_{H_2O} = 244 mg/g) adsorbed on silylated Si-60 (in chloroform-d medium). (Adapted from *Adv. Colloid Interface Sci.*, 118, Gun'ko, V.M., Turov, V.V., Bogatyrev, V.M. et al., Unusual properties of water at hydrophilic/hydrophobic interfaces, 125–172, 2005d. Copyright 2005, with permission from Elsevier.)

adsorbed water (Figures 1.244 and 1.246) because ≡SiOH is chaotropic group but ≡SiOSi(CH_3)$_3$ is kosmotropic group. There are two signals at 5 ppm (first signal, SAW) and 1.4 ppm (second signal, WAW) (Figure 1.246a). The peak positions of these signals are practically independent of temperature (Figure 1.246b) (in contrast to Silochrome of a different texture, vide infra); that is, structural features (e.g., contributions of nanodomains of SAW and clusters of WAW) of two types of mobile water, its clustering in pores, depend on temperature only slightly for this adsorbent. Consequently, the effect of relatively narrow pores of partially hydrophobized silica gel is strong and contribution of WAW is enough to be registered. The intensities of signals depend differently on temperature because freezing-out of bound HDW and LDW layers or SAW and WAW structures differs.

The first signal decreases with lowering temperature because of freezing-out of a portion of water distant from the surface and form relatively large domains (i.e., boundary zone between HDW and LDW or LDW). The intensity of the second signal is practically the same over a wide range of temperature; that is, this water is out from freezing under these conditions and keeps the structure of WAW. Consequently, the appearance of hydrophobic patches at the silica surface, which are more appropriate for contacts with molecules of chloroform, provides the appearance of a noticeable amount of interfacial water in an anomalous weakly associated state at small δ_H = 1.4 ppm. Taking into account relatively low intensity of the second signal one can assume that WAW locates only near hydrophobic TMS groups (predominantly between them) to form certain mixture with the $CDCl_3$ clusters contacting with these TMS groups. The molecular exchange between the mentioned two water states is slow (in the NMR timescale) because there are different dependences of their intensity on temperature and practical independence of the peak position on temperature. This effect suggests that there are space barriers between waters possessing different states; that is, direct contact between these waters is absent or weak. The adsorption complexes of chloroform molecules on the TMS groups can play a role of these barriers. Similar WAW is observed at the interfaces of hydrophilic solid surfaces in the presence of hydrophobic solvents in the suspensions of mixed oxides and partially silylated nanosilicas as described earlier.

Detailed investigations of the influence of the hydrophobic environment on the ^{1}H NMR spectra of adsorbed water and its thermodynamic characteristics for wetted powders of nanosilica partially modified by grafted TMS groups showed (Gun'ko et al. 2005d, 2009d) that signal of water in the absence of chloroform is observed in the singlet form. This water can be assigned to HDW in narrow pores (voids between primary nanoparticles in the same aggregate) and HDW/LDW in larger pores (between primary particles from different aggregates in agglomerates). The signal width increases and intensity decreases with lowering temperature because of reduction of the mobility of water molecules and partial freezing-out of interfacial water with moving boundary of immobile ice/mobile water from larger pores toward narrower ones. Such a behavior of signals of adsorbed water at $T<273$ K is characteristic of the majority of disperse and porous materials. The chemical shift of the proton resonance for adsorbed water is close to chemical shift of liquid water (5 ppm) because of several mutually compensating factors such as (i) existence only HDW in pores that should reduces δ_H in comparison with that for LDW; (ii) bending of the hydrogen bonds near active surface sites in very narrow pores (δ_H decreases); (iii) surface electrostatic field results in deshielding of protons (δ_H increases); (iv) overlapping of EDL of opposite pore walls (complex effect on δ_H); and (vi) difference in concentrations of counterions in different pores (complex effect on δ_H) (Gun'ko et al. 2005d).

Surface modification of mesoporous silica gel (average pore diameter 15 nm) by β-nicotinamide adenine dinucleotide (NAD^+) (Vartzouma et al. 2004) leads to changes in the ^{29}Si CP MAS NMR spectrum (Figure 1.247). The intensity of Q_3 and Q_2 sites decreases because of replacing of the surface H atoms by NAD^+. This changes can be used to estimate the adsorption degree of a modifier that is of importance to elucidate the influence of modified surfaces on the interfacial behavior of water or other adsorbates.

Silochrome is a silica macroporous material with broad pores and much higher bulk density than that of the nanosilica powder. However, HTT (during Silochrome synthesis) can lead to formation of not only broad pores but also shallow and narrow pores, which can provide appropriate conditions for appearance of WAW (Gun'ko et al. 2005d). Additionally, low amounts of adsorbed water can provide the ^{1}H NMR spectra with visible signal of WAW. The temperature effect on the ^{1}H NMR spectra of water adsorbed on the surface of Silochrome ($S_{BET} \approx 120$ m^2/g) at $C_{H_2O}=31$ mg/g is depicted in Figure 1.248.

Signals of strongly and WAWs are well distinguished over a wide temperature range. The total intensity of signal decreases at $T<230$ K due to a decrease in the intensity of signal 1 (HDW adsorbed

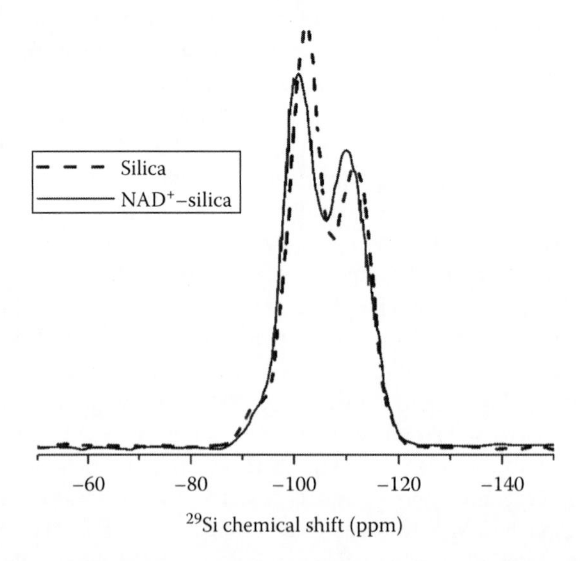

FIGURE 1.247 ^{29}Si CP MAS NMR spectra of unmodified and NAD^+-modified silica gel. (Adapted from *Mater. Sci. Eng. C.*, 24, Vartzouma, Ch., Louloudi, M., Prodromidis, M., and Hadjiliadis, N., Synthesis and characterization of NAD^+-modified silica: a convenient immobilization of biomolecule via its phosphate group, 473–477, 2004. Copyright 2004, with permission from Elsevier.)

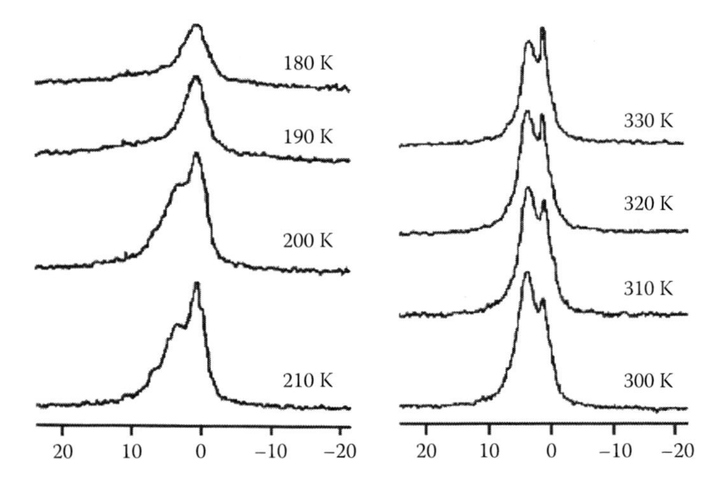

FIGURE 1.248 ^{1}H NMR spectra of water (C_{H_2O} = 31 mg/g) adsorbed on Silochrome at different temperatures. (Adapted from *Adv. Colloid Interface Sci.*, 118, Gun'ko, V.M., Turov, V.V., Bogatyrev, V.M. et al., Unusual properties of water at hydrophilic/hydrophobic interfaces, 125–172, 2005d. Copyright 2005, with permission from Elsevier.)

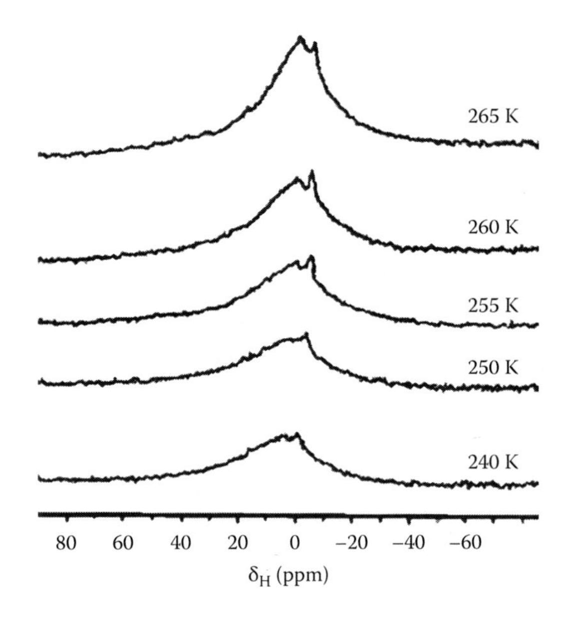

FIGURE 1.249 ^{1}H NMR spectra of unfrozen water adsorbed on Silochrome ($C_{SiO_2} \approx 35$ wt%). (Adapted from *Adv. Colloid Interface Sci.*, 118, Gun'ko, V.M., Turov, V.V., Bogatyrev, V.M. et al., Unusual properties of water at hydrophilic/hydrophobic interfaces, 125–172, 2005d. Copyright 2005, with permission from Elsevier.)

in the form of relative large CS clusters). Only signal of WAW (HDW in the form of small clusters with strongly distorted structure of the hydrogen bonds, interstitial water, or even individual adsorbed water molecules mixed with interfacial chloroform) is observed at $T < 200$ K (Figure 1.248a). The intensity of the second signal increases more strongly with elevating temperature (Figure 1.248b) because of faster melting of the corresponding ice structures on heating. An increase in the water content leads to significant broadening of signal; however, signal 2 is yet observed (Figure 1.249). This result can be provided by a specific texture of Silochrome (see its PSD_{uw} in the subsequent sections).

Figure 1.250 shows the temperature dependences of the chemical shifts of signals 1 and 2 at $180<T<330$ K, and Figure 1.251 depicts the temperature dependence of C_{uw} and the graph of ΔG versus C_{uw} for the suspension of Silochrome in $CDCl_3$ and water.

The maximum of the δ_H value of signal 1 is observed at $T \approx 290$ K, but the chemical shift linked to signal 2 increases with temperature nearly monotonically from 0.9 to 1.5 ppm. Entire adsorbed water is strongly bound (HDW) in the suspension of Silochrome in the chloroform-*d* medium that could be expected because of low concentration of bound water. A section caused by WBW appears in the $\Delta G(C_{uw})$ graph at $C_{uw}>60$ mg/g for silica in the aqueous medium but it is absent in the $CDCl_3$ medium. Signal 2 is present in the spectra at lower temperatures than signal 1; that is, molecules

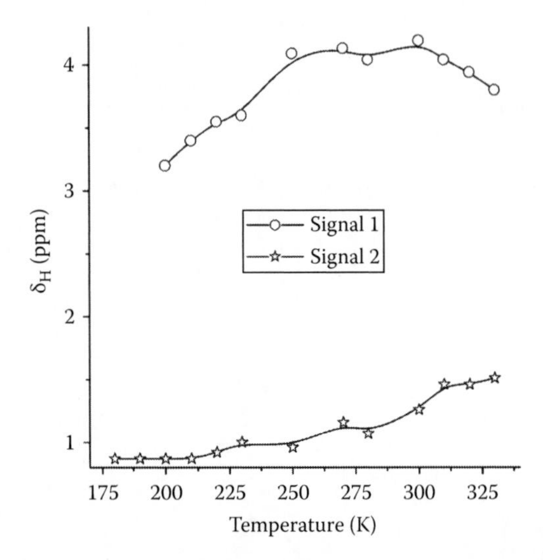

FIGURE 1.250 The temperature dependence of the chemical shift of signals 1 (δ_H=3.2–4.2 ppm) and 2 (δ_H=0.9–1.5 ppm) for the suspension of Silochrome (C_{H_2O}=31 mg/g) in $CDCl_3$. (Adapted from *Adv. Colloid Interface Sci.*, 118, Gun'ko, V.M., Turov, V.V., Bogatyrev, V.M. et al., Unusual properties of water at hydrophilic/hydrophobic interfaces, 125–172, 2005a. Copyright 2005, with permission from Elsevier.)

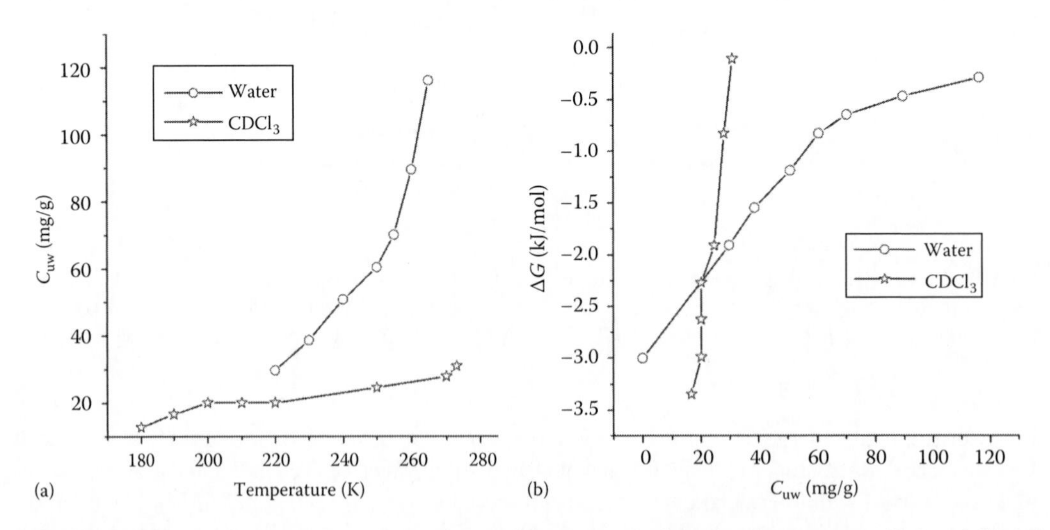

FIGURE 1.251 (a) The temperature dependence of the concentration of unfrozen water (C_{uw}) and (b) ΔG versus C_{uw} in the aqueous suspension of Silochrome and hydrated silica in $CDCl_3$. (Adapted from *Adv. Colloid Interface Sci.*, 118, Gun'ko, V.M., Turov, V.V., Bogatyrev, V.M. et al., Unusual properties of water at hydrophilic/hydrophobic interfaces, 125–172, 2005a. Copyright 2005, with permission from Elsevier.)

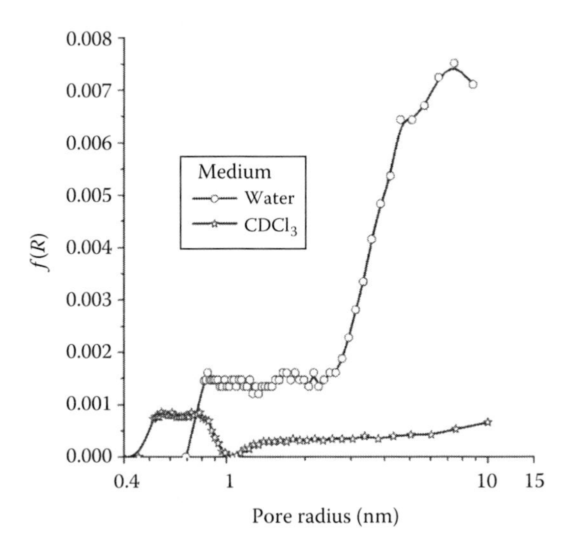

FIGURE 1.252 Distribution functions of pores filled by unfrozen water, PSD_{uw} (only for a portion of pores at $R \leq 10$ nm) of Silochrome in various environment. (Adapted from *Adv. Colloid Interface Sci.*, 118, Gun'ko, V.M., Turov, V.V., Bogatyrev, V.M. et al., Unusual properties of water at hydrophilic/hydrophobic interfaces, 125–172, 2005d. Copyright 2005, with permission from Elsevier.)

of WAW remain mobile even after freezing of bulk chloroform. One can assume that clusters of interfacial water (WAW) and chloroform are capable to form a certain mixture in pores, and their nucleation on freezing differs from that in the bulk liquids.

The distribution functions of pores filled by unfrozen water (Figure 1.252), that are calculated on the basis of the $C_{uw}(T)$ graphs for interfacial water in the different environments, depict that water in the presence of chloroform fills narrow pores at $R<0.7$ nm or form certain structures in larger pores that are mostly filled by chloroform.

Therefore, according to changes in the shape of the PSD_{uw}, one can assume that WAW (assigned to signal 2) forms a mixture with chloroform in pores at $R>0.5$ nm. It should be noted that the pore size calculated on the basis of the $C_{uw}(T)$ graphs for macroporous Silochrome can underestimate this size because the adsorbed water mainly "senses" the pore walls at distances <10–15 nm.

Polymethylsiloxane (PMS) materials with 1D (PDMS) or 2D/3D (PMS) siloxane backbone have drawn considerable fundamental and technological interest because of their applications as components of nanocomposites, copolymers for synthesis of ion conducting polymeric materials, chromatographic adsorbent, a component of medicinal preparations (e.g., Cleocin, Universal Washaid, United States), implants, adjuvant Capsil (Aquatrols, USA; Scotts, USA), a vaccine adjuvant, etc. Additionally, PMS in the form of hydrogel ($C_{PMS} \approx 10$ wt%) is utilized as a medicinal enterosorbent Enterosgel (Kreoma-Pharm, Ukraine). Functionalized PMSs are used for modification and functionalization of solid surfaces. They are also used as supports for catalysts, or as polymer backbones for preparation of liquid crystalline polymers.

The NMR investigations and dielectric relaxation (DRS, TSDC) measurements showed a strong dependence of the mobility of polysiloxanes depending on the structure of side groups (Gun'ko et al. 2007b,h, 2009d, Sulym et al. 2011). This structure determines the shear elasticity of polymers and other characteristics. The behavior of interfacial water in xerogels, hydrogels, and suspensions of branched PMS was studied in comparison with silylated nanosilica possessing close specific surface area, the same surface functionalities but differently composed, and different particle morphology (Gun'ko et al. 2007h, 2009d). The properties of PMS materials in the forms of hydrogels, xerogels, or dry powders depend on their cross-linking degree (maximal in dry xerogel with all Si atoms bonded by three siloxane bonds $(\equiv Si{-}O)_3SiCH_3$) and hydration degree. Heating/drying of PMS hydrogel can lead to loss of the hydrophilic properties because of enhancement of cross-linking by the Si–O–Si bonds on condensation of hydrophilic silanols. Heated PMS xerogel with the maximal cross-linking degree is hydrophobic and practically not wetted by water (contact angle>95°). However, PMS hydrogel can retain significant amounts of water and its aqueous suspension is stable for long time because of incomplete cross-linking degree and the presence of residual hydrophilic silanols.

The hydrophilic–hydrophobic properties of the PMS hydrogel and xerogel surfaces (Gun'ko et al. 2007h) can be close to that of partially and completely silylated nanosilica, respectively, because these materials have close specific surface area (200–300 m^2/g) and the same surface functionalities ($SiCH_3$, SiOH and Si–O–Si) and are characterized by the textural porosity (i.e., voids between particles non-densely packed in secondary structures). The hydrophobic functionalities (trimethylsilyl, TMS, or dimethylsilyl, DMS) on a silylated silica surface inhibit formation of a continuous layer of adsorbed water. Therefore, one can expect clusterization of interfacial water in PMS hydrogel or water adsorbed on PMS xerogel. Comparison of the adsorption characteristics of unmodified and differently MSs shows that even partial silylation of the silica surface reduces the adsorption of water by several times. Nevertheless, partially or completely silylated nanosilica placed in the aqueous medium can disturb a relatively thick layer of interfacial water. A similar effect can be expected for PMS hydrogel and suspension. However, the differences in the structure of primary particles of PMS and silylated nanosilica as well as in the composition of surface sites ($(\equiv Si–O)_3SiCH_3$ and $(\equiv Si–O)_2Si(OH)CH_3$ for PMS and $\equiv Si–OSi(CH_3)_3$ and $(\equiv Si–O)_2Si(CH_3)_2$ for TMS- and DMS-nanosilica, respectively) can lead to certain differences in the behavior of bound water as well as in the physicochemical properties of concentrated and diluted aqueous suspensions, hydrogels and dry powders with these materials. Wetting/drying of nanosilica, which is composed of rigid primary particles forming nonrigid aggregates and agglomerates of aggregates, changes both structural and morphological characteristics. Resembling effects caused by transformation of PMS hydrogel (composed of soft primary and secondary particles) into xerogel can be stronger than that for nanosilica because of the enhancement of the cross-linking degree in dried PMS particles.

To analyze the effects of the liquid media on rigid and soft materials, such techniques as NMR cryoporometry, TSDC, and QELS can be used without removal of the liquids (i.e., as nondestructive methods) in combination with standard adsorption and other methods applied to both aqueous dispersions and dry powders (Gun'ko et al. 2005d, 2007f,h,i, 2009d). Here, we will analyze regularities in the structural and adsorption characteristics of PMS in the forms of suspension, hydrogel and xerogel compared with those of aqueous suspensions and dry powders of unmodified and silylated nanosilica and silica gel using ^{1}H NMR and TSDC with layer-by-layer freezing-out of bulk and interfacial water, QELS, rheometry, and adsorption of proteins.

Commercial PMS hydrogel (Kreoma-Pharm, Kiev, Ukraine) including 10 wt% of PMS and 90 wt% of water was used as the initial material (Gun'ko et al. 2007h). Aqueous suspensions (C_{PMS} = 1.25, 2.5 and 5 wt%) were prepared by dilution of the initial PMS hydrogel by distilled water. Dry PMS xerogel ($C_{PMS} \approx 97$ wt%) was obtained from the initial material dried in air at 300 K for five days. PMS xerogel wetted by ethanol and then ethanol/water and repeatedly washed off by water contains approximately 72 wt% of water. Nanosilicas A-300 and A-380 at $S_{BET} \approx 337$ and 378 m^2/g, respectively, were silylated by dimethyldichlorosilane (C_{DMS}=0.24, 0.57, 0.84, 1.12 and 1.21 mmol/g) and trimethylchlorosilane (C_{TMS}=0.09, 0.14, 0.12, 0.23, 0.42, and 0.79 mmol/g), respectively. The structural and other characteristics of silylated A-380 samples were described in previous chapters. The mentioned materials are used here for comparative analysis of the interfacial phenomena at a surfaces of different silicas.

The rheological behavior of aqueous suspensions of PMS (C_{PMS}= 1, 3, and 5 wt%) was studied with increasing shear rate from 0.1 to 1312 s^{-1} and then its decreasing to 0.1 s^{-1} at 293 K using a Rheotest 2.1 (VEB MLW Prufgerate-Werk Medingen Sitz Ftreital, Germany) rotational viscometer with a cylindrical measuring system. The distribution function of activation energy $f(E)$ of the shear viscosity was determined using integral equation shown in Chapter 10 (Gun'ko et al. 2006b).

TABLE 1.41
Structural Characteristics of Dry Powder with Nanosilica A-300 and Differently Silylated A-300

Parameter	A-300	S1	S2	S3	S4	S5
C_{DMS} (mmol/g)	—	0.24	0.57	0.84	1.12	1.21
C_{OH} (mmol/g)	0.53	0.55	0.42	0.31	0.11	—
ρ_{bulk} (g/cm^3)	0.041	0.028	0.031	0.035	0.034	0.36
S_{BET} (m^2/g)	337	343	273	214	183	150
S_{ϕ}[a] (m^2/g)	334	319	272	220	178	137
S_{meso} (m^2/g)	303	295	243	178	156	126
S_{macro} (m^2/g)	34	48	30	36	27	24
V_p (cm^3/g)	0.714	0.925	0.652	0.626	0.527	0.429
V_{meso} (cm^3/g)	0.379	0.420	0.314	0.244	0.217	0.170
V_{macro} (cm^3/g)	0.335	0.505	0.338	0.382	0.310	0.259

[a] $N=3$.

Aqueous solution (0.6 wt%) of a protein (ovalbumin, gelatin) and 0.2 g of PMS ($C_{PMS}=10$ wt%) or 0.02 g of nanosilica A-300 was mixed for 1 h and then centrifuged. The equilibrium concentration of protein in the liquid was determined by the Biuret method. A distribution function $f(-\Delta G)$ of Gibbs free energy (ΔG) of the protein adsorption was calculated using the kernel of the integral equation in the form of the Langmuir equation.

The cryoporometry methods were used for the structural characterization of different materials dried or being in the aqueous media (Tables 1.35, 1.36 and 1.41). This is of importance because removal of water can damage the structure of the systems such PMS hydrogel.

Hydrophobic functionalities at a silica gel surface change the amounts and the temperature behavior of bound water (Figure 1.253a), which undergoes clusterization (Figure 1.253b) leading to formation of small clusters with weakly associated interfacial water characterized by $\delta_H=1.2$–1.7 ppm and large clusters (SAW) with $\delta_H\approx 5$ ppm (Figure 1.253c). Thus, even partial (approximately 30%) silylation of silica gel Si-60 causes the formation of relatively small water clusters on incomplete filling of pores by water (Figure 1.253b). This structured water is unfrozen at very low temperatures (Figure 1.253a) and characterized by a low δ_H value and similar effects were observed for other materials with hydrophilic–hydrophobic structures. Replacing of the air medium by $CDCl_3$ leads to formation of larger water clusters frozen at higher temperature because of the displacement of water by chloroform from the surface and from the narrowest pores (Figure 1.253b). Thus, application of the IGT equation allows us to analyze rearrangement of water bound in silica gels depending on the pore structure or the surface modification and the presence of a weakly polar solvent (Figure 1.253).

Silylation of nanosilica characterized by the textural porosity caused by voids between nanoparticles forming relatively soft secondary particles (aggregates and agglomerates) affects not only the behavior of structured water but also the structural characteristics of the material as a whole (Tables 1.35, 1.36 and 1.41 and Figures 1.254 through 1.256).

Calculations of the S_{ϕ} values (Table 1.41) with Equation 10.37 described in Chapter 10 show that these values are close to the S_{BET} values at a low coordination number of primary particles in aggregates $N=3$. This result and low values of the bulk density of the powders (Table 1.41, ρ_{bulk}) suggests non-dense packing of particles. Estimation of the characteristics of nanopores (S_{nano}, V_{nano}), mesopores (S_{meso}, V_{meso}), and macropores (S_{macro}, V_{macro}) (Table 1.41) from the nitrogen adsorption shows predominant contribution of mesopores and macropores. However, in the case of the aqueous suspensions, contribution of nanopores (that is close to zero for the powders) strongly increases for

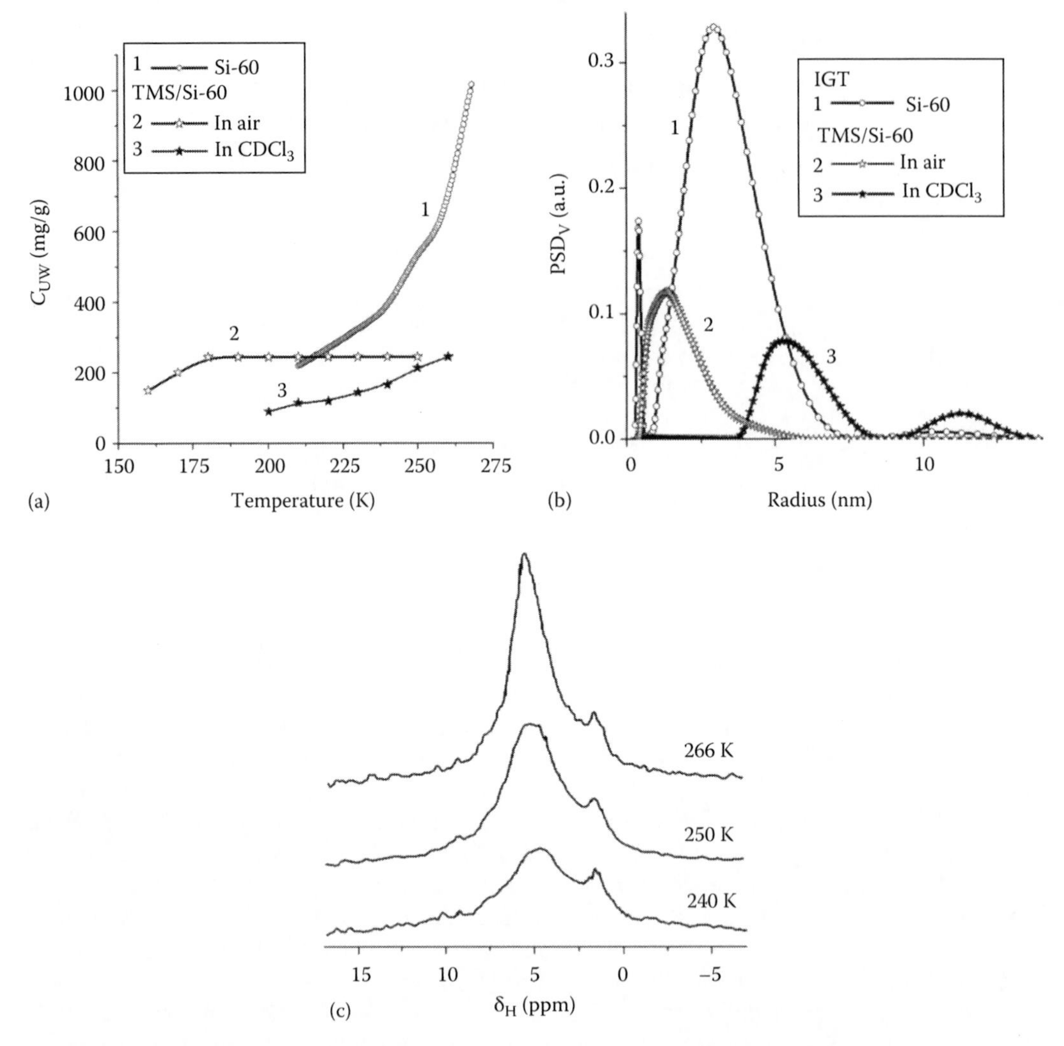

FIGURE 1.253 (a) Temperature dependence of the content of unfrozen water for unmodified and silylated (30%) Si-60; (b) the corresponding distribution functions calculated using NMR cryoporometry with IGT eq., and (c) ^{1}H NMR spectra of water (hydration $h=0.244$ g/g) adsorbed on silylated Si-60 (in chloroform-d medium) at different temperatures. (Adapted from *J. Colloid Interface Sci.*, 308, Gun'ko, V.M., Turov, V.V., Zarko, V.I. et al., Comparative characterization of polymethylsiloxane hydrogel and silylated fumed silica and silica gel, 142–156, 2007h. Copyright 2007, with permission from Elsevier.)

silylated samples (Table 1.25, S_{nano} and V_{nano}). This can be due to tendency of the formation of maximal number of contacts between silylated surface patches of neighboring particles to reduce their interaction with liquid water. Partial silylation of nanosilica leads to enhancement of clusterization of interfacial water (small structures in Figures 1.254 and 1.255), which freezes at lower temperatures (Figure 1.253) similar to that for silylated silica gel.

The amounts of structured water (Table 1.25) decrease for silylated silicas despite nearly the same adsorption capacity for adsorbed nitrogen because of reduction of the adsorption potential for water. Enhancement of the clusterization of interfacial water is clearly observed for silylated A-300 (Figure 1.255) and a mechanical mixture of unmodified nanosilica and silylated A-300 (with DMS functionalities) (Figure 1.256). The PSDs of unmodified A-300 and A-300-DMS/A-300 mixture are close. A decrease in hydration (h) leads to appearance of small water clusters at $R<1$ nm (Figure 1.256a).

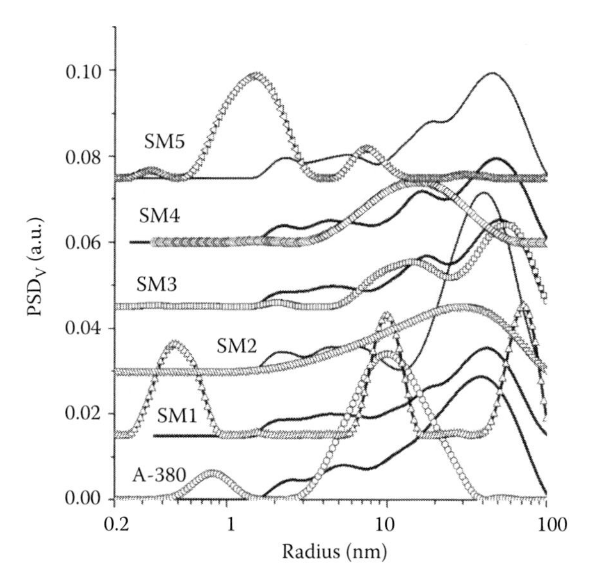

FIGURE 1.254 Pore size distributions of unmodified and modified nanosilica A-380 (Table 1. 3) calculated on the basis of the nitrogen adsorption/desorption isotherms using DFT method with the model of pores as voids between spherical particles and NMR cryoporometry with IGT equation. (Adapted from *J. Colloid Interface Sci.*, 308, Gun'ko, V.M., Turov, V.V., Zarko, V.I. et al., Comparative characterization of polymethylsiloxane hydrogel and silylated fumed silica and silica gel, 142–156, 2007h. Copyright 2007, with permission from Elsevier.)

Unmodified and silylated nanosilicas (Tables 1.25 and 1.41 and Figures 1.254 and 1.255) and the A-300-DMS/A-300 mixture (Figure 1.256 and Table 1.35) are mesoporous–macroporous materials; that is, the size of voids between particles are close or larger than the size of primary particles. Notice that wetting/drying of unmodified nanosilica leads to certain changes not only in the structure of secondary particles but also in the primary ones, which slightly grow (Figure 1.255d). Clearly similar processes for PMS hydrogel can more strongly affect the structure of PMS particles.

Results obtained for unmodified and MSs by using NMR and TSDC cryoporometry suggest that these techniques can be used to characterize the PMS hydrogel structure and the behavior of the interfacial water there. The PMS macromolecules in hydrogel are characterized by incomplete condensation of silanols because this material remains hydrophilic. Complete condensation of silanols leads to appearance of the hydrophobic properties of dried PMS xerogel because of the presence of CH_3 groups attached to each Si atom and very low content of residual hydrophilic silanols. Figure 1.257 shows the PMS models: a small primary particle (4.3 nm in size) at $C_{OH}/C_{CH_3}=0.546$ and small secondary particle (11.5 nm) with voids in the range of nanopores and mesopores.

MM calculation of a small particle in water (Figure 1.257b) gives a decrease in the particle size because of water pressure onto hydrophobic functionalities to reduce the contact area between water and the hydrophobic structures. Therefore PMS particles in hydrogel ($C_{PMS}=10$ wt%) or in the aqueous suspensions ($C_{PMS} \leq 5$ wt%) can tend to form a continuous network to reduce the interaction of hydrophobic functionalities with water similar to observed for modified nanosilicas.

PaSDs in the PMS suspensions can be relatively broad (Figure 1.258) or very narrow (Figure 1.259) depending on the concentration, pH value and sonication time.

It should be noted that the QELS intensity for PMS was less by order than that for the suspensions of nanosilica with the same concentration. This is due to the differences in the morphology of secondary PMS particles and in the nature of primary particles because 3D network in 2D/3D PMS particles is looser and close to the 2D network due to CH_3 groups attached to each Si atom and residual silanols. For maximum diluted PMS suspensions, the presence of monodisperse particle

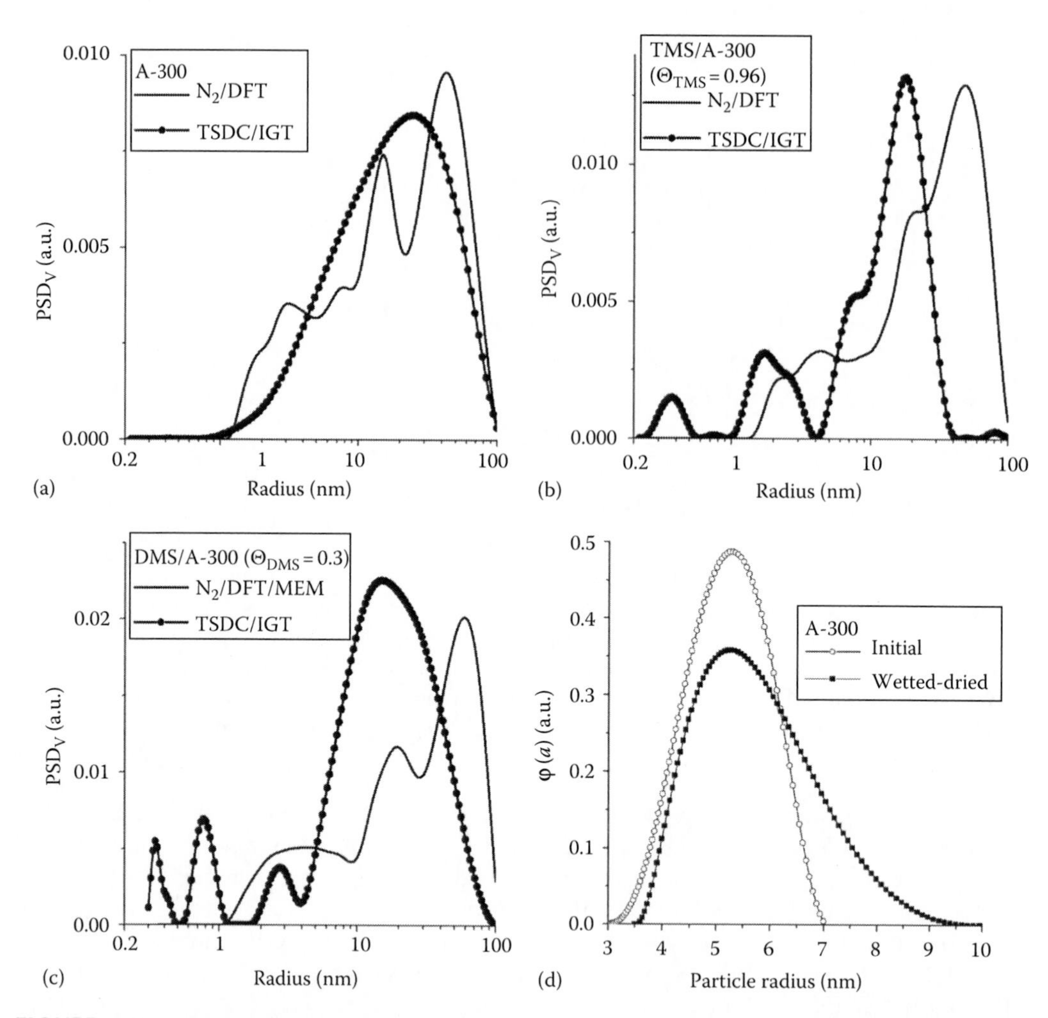

FIGURE 1.255 Pore size distributions of unmodified and silylated nanosilica A-300 calculated on the basis of the nitrogen adsorption/desorption isotherm using DFT method with the model of pores as voids between spherical particles and TSDC cryoporometry with IGT equation: (a) initial A-300; (b) TMS/A-300 and (c) DMS/A-300; and (d) particle size distribution of initial and wetted/dried A-300. (Adapted from *J. Colloid Interface Sci.*, 308, Gun'ko, V.M., Turov, V.V., Zarko, V.I. et al., Comparative characterization of polymethylsiloxane hydrogel and silylated fumed silica and silica gel, 142–156, 2007h. Copyright 2007, with permission from Elsevier.)

distribution is characteristic of PMS because this concentration is much smaller than that enough to form a continuous network with primary and secondary particles. The D_{ef} value does not change with decreasing PMS concentration at $C_{PMS} < 0.06$ wt% and it is equal to 7–8 nm. This size can be assumed as that of primary PMS particles, which are similar to those of A-380 studied.

The rheological behavior of PMS suspensions was investigated to elucidate the influence of PMS concentration on interaction between particles. Diluted PMS hydrogel (i.e., suspension or colloidal solution) is characterized by the viscosity increasing with concentration (Figure 1.260a).

The distribution function of the activation energy of the shear viscosity shifts toward higher energy with increasing C_{PMS} value (Figure 1.260b). The structure of the dispersion becomes more complex since the $f(E)$ function is multimodal at $C_{PMS} = 5$ wt%. The viscosity of the aqueous suspension of nanosilica even at higher concentration (8 wt%) is lower and the $f(E)$ peak is at lower E values. Features of different interactions between colloidal particles of PMS or nanosilica

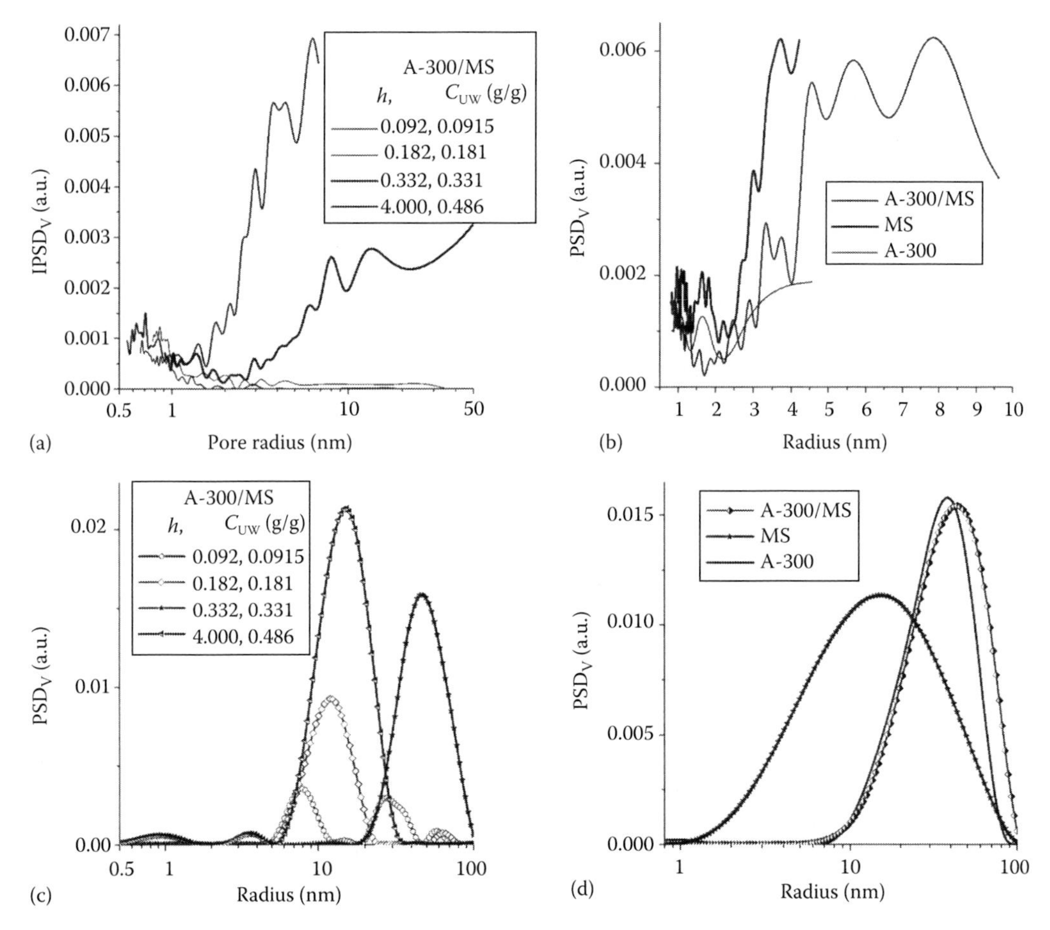

FIGURE 1.256 PSDs calculated with the NMR cryoporometry with (a, b) GT and (c, d) IGT equations for A-300, modified silica (MS) and their mixture. (Adapted from *J. Colloid Interface Sci.*, 308, Gun'ko, V.M., Turov, V.V., Zarko, V.I. et al., Comparative characterization of polymethylsiloxane hydrogel and silylated fumed silica and silica gel, 142–156, 2007h. Copyright 2007, with permission from Elsevier.)

determining the formation of the 3D network with primary particles forming secondary ones in the dispersions can be responsible for this difference in the $f(E)$ functions. On rheological studies, destruction of secondary structures occurs that is manifested in changes in the rheological parameters of the dispersions during their measurement. This destruction progressive with increasing shearing rate is manifested in deviations from that characteristic of the Newtonian flow. If the recovery rate of destroyed interparticle bonds does not exceed the rate of their destruction then the viscosity decreases until the rates of both processes become equal one to another. This equilibrium can be reached for a short time at $C_{PMS} = 1$ wt%. (Figure 1.260a, curve 1) that indicates a relatively low thixotropy of the system. An increase in the C_{PMS} value leads to an increase in the thixotropy (curves 2 and 3). The viscosity and the shear stress practically linearly depend on PMS concentration at the maximal shear rate $\dot{\gamma} = 1312$ s^{-1}. The effect of an increase in the viscosity value on the measurements due to a higher rate of bond formation than a rate of their destruction is not observed at different concentrations of PMS. Although the viscosity is noticeably higher at smaller $\dot{\gamma}$ values and depends on C_{PMS}. These results suggest the absence of stable aggregates in the PMS suspensions that is in agreement with the QELS data showing the monodispersity of the diluted suspensions and the rapid destruction of aggregates on sonication of the suspensions. Although fast structurization can occur in the concentrated suspensions at $C_{PMS} \geq 3$ wt% because of aggregation of

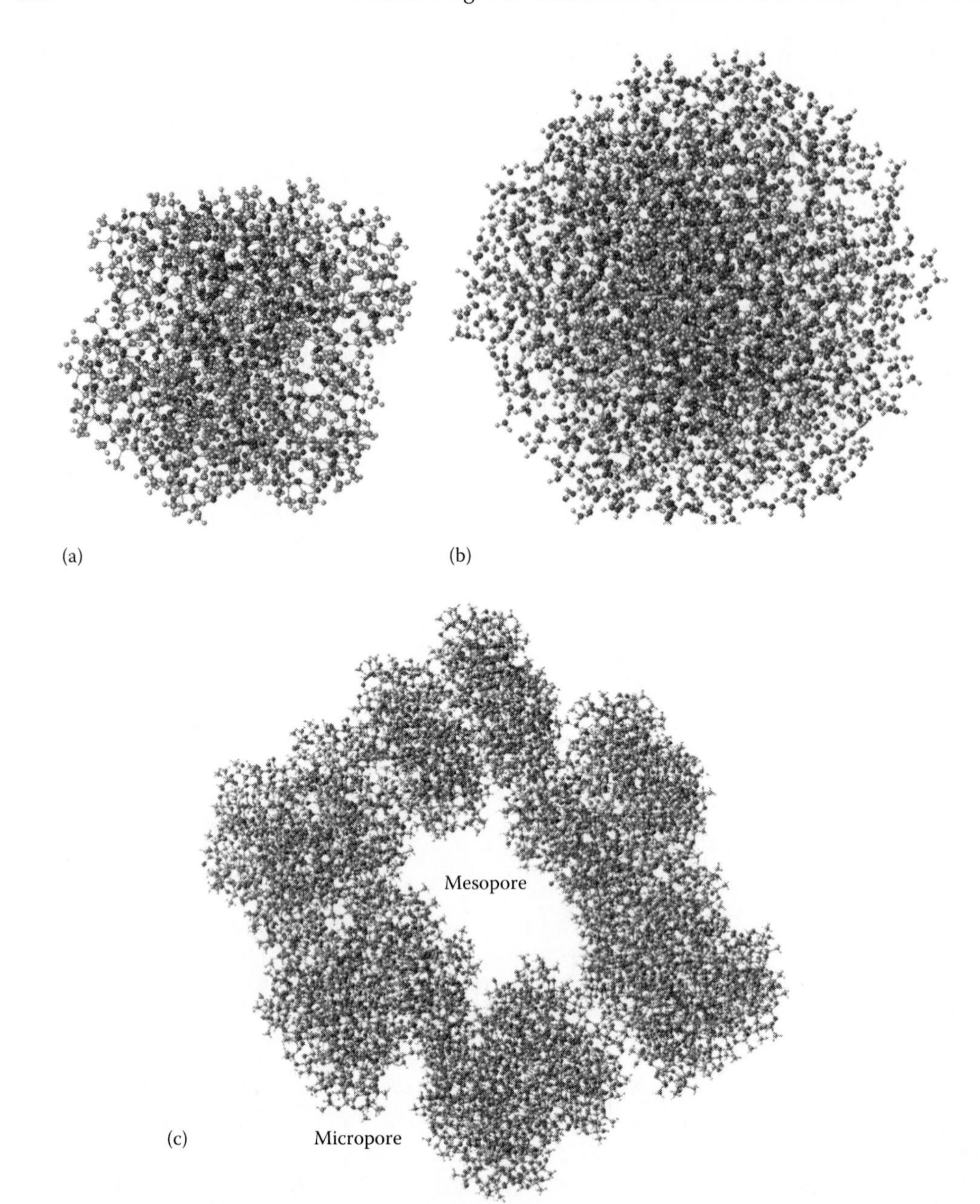

FIGURE 1.257 Models of (a) dry and (b) hydrated primary PMS particle and (c) dry secondary particle. (Adapted from *J. Colloid Interface Sci.*, 308, Gun'ko, V.M., Turov, V.V., Zarko, V.I. et al., Comparative characterization of polymethylsiloxane hydrogel and silylated fumed silica and silica gel, 142–156, 2007h. Copyright 2007, with permission from Elsevier.)

primary particles. The presence of aggregates and agglomerates at $C_{PMS} \geq 3$ wt% causes significant enhancement of the viscosity especially at small $\dot{\gamma}$ values.

Thus, the diluted PMS suspensions have practically monodisperse distribution of primary nanoparticles forming aggregates and agglomerates (which can be easily destroyed) with increasing PMS concentration. Clearly, the absence of stable aggregates and agglomerates plays an important role on applications of PMS as an enterosorbent providing a relatively high rate of diffusion of primary particles in the liquid medium.

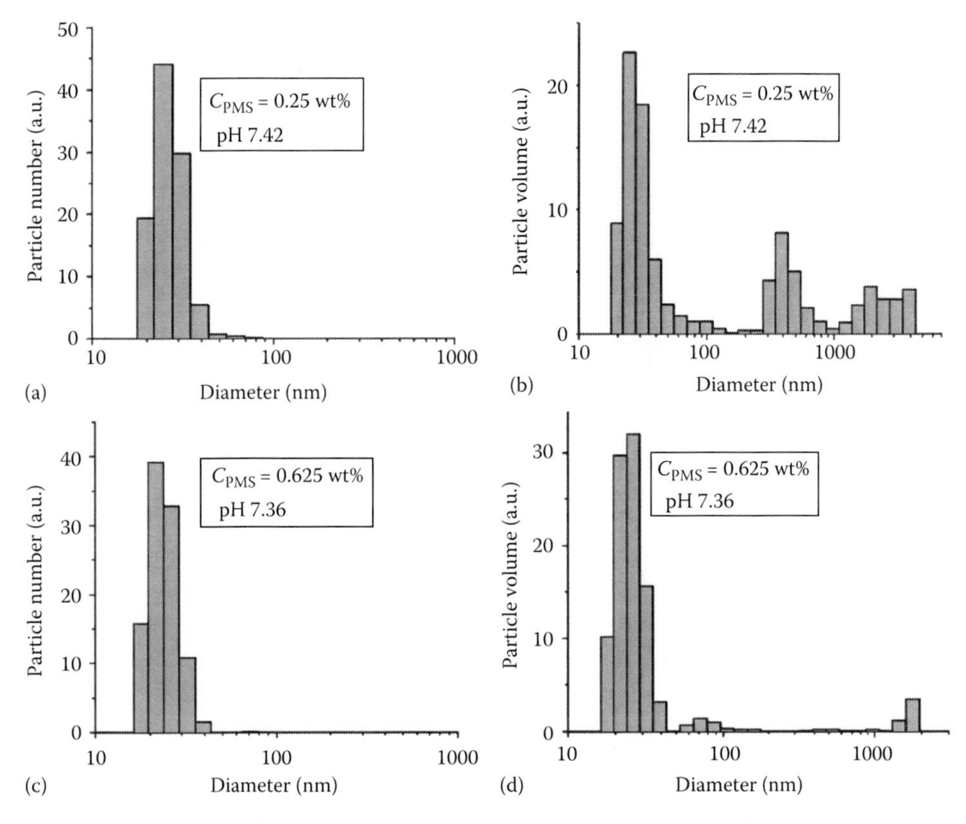

FIGURE 1.258 Particle size distributions in the diluted aqueous suspensions of PMS at C_{PMS} = (a, b) 0.25 and (c,d) 0.0625 wt% shown with respect to (a, c) the particle number and (b, d) particle volume. (Adapted from *J. Colloid Interface Sci.*, 308, Gun'ko, V.M., Turov, V.V., Zarko, V.I. et al., Comparative characterization of polymethylsiloxane hydrogel and silylated fumed silica and silica gel, 142–156, 2007h. Copyright 2007, with permission from Elsevier.)

Additional information about the properties of the PMS hydrogel and suspensions can be obtained on a study of interfacial water by the ^{1}H NMR spectroscopy with layer-by-layer freezing-out of bulk and structured bound water. The measurements of the ^{1}H NMR spectra were carried out in the deuterochloroform medium to accurately determine the value of the chemical shift of the proton resonance of interfacial water and to decrease the signal width for static samples. Since deuterochloroform contained a small amount of tetramethylsilane (as an additive) used as the standard for the measurement of chemical shifts, the ^{1}H NMR spectra include not only signal of unfrozen bound water but also signals of tetramethylsilane ($\delta = 0$ ppm) and CH group of chloroform ($\delta = 7.26$ ppm) (Figure 1.261).

A signal observed at $\delta_H = 1.7$ ppm corresponds to WAW, that is, small water clusters, which appear at a PMS surface with residual hydrophilic silanols surrounded by hydrophobic $\equiv SiCH_3$ groups. The intensity of this signal decreases with lowering temperature slower than that for other forms of interfacial water. The main signal at $\delta_H > 5$ ppm corresponds to SAW, which freezes at $T < 255$ K. Water with signal $\delta_H = 3.7$ ppm at 285 K can be attributed to WBW, which freezes at T close to 273 K. The amount of this water is lower than that with $\delta_H > 5$ ppm because of low hydration ($h \approx 0.03$ g/g) of dry PMS xerogel. PMS particles do not have dense 3D bond network (Figure 1.257), therefore the presence of intraparticle water is possible in the form of relatively small clusters between residual silanols and $\equiv SiCH_3$ groups (Figure 1.257b). This water can be responsible for signal at $\delta_H = 1.7$ ppm. Water characterized by signal at 5 ppm can locate in relatively narrow mesopores (Figure 1.257c). Water giving signal $\delta_H = 3.7$ ppm can be located in macropores since it is frozen at T close to 273 K, and the main portion of water in macropores can be frozen at

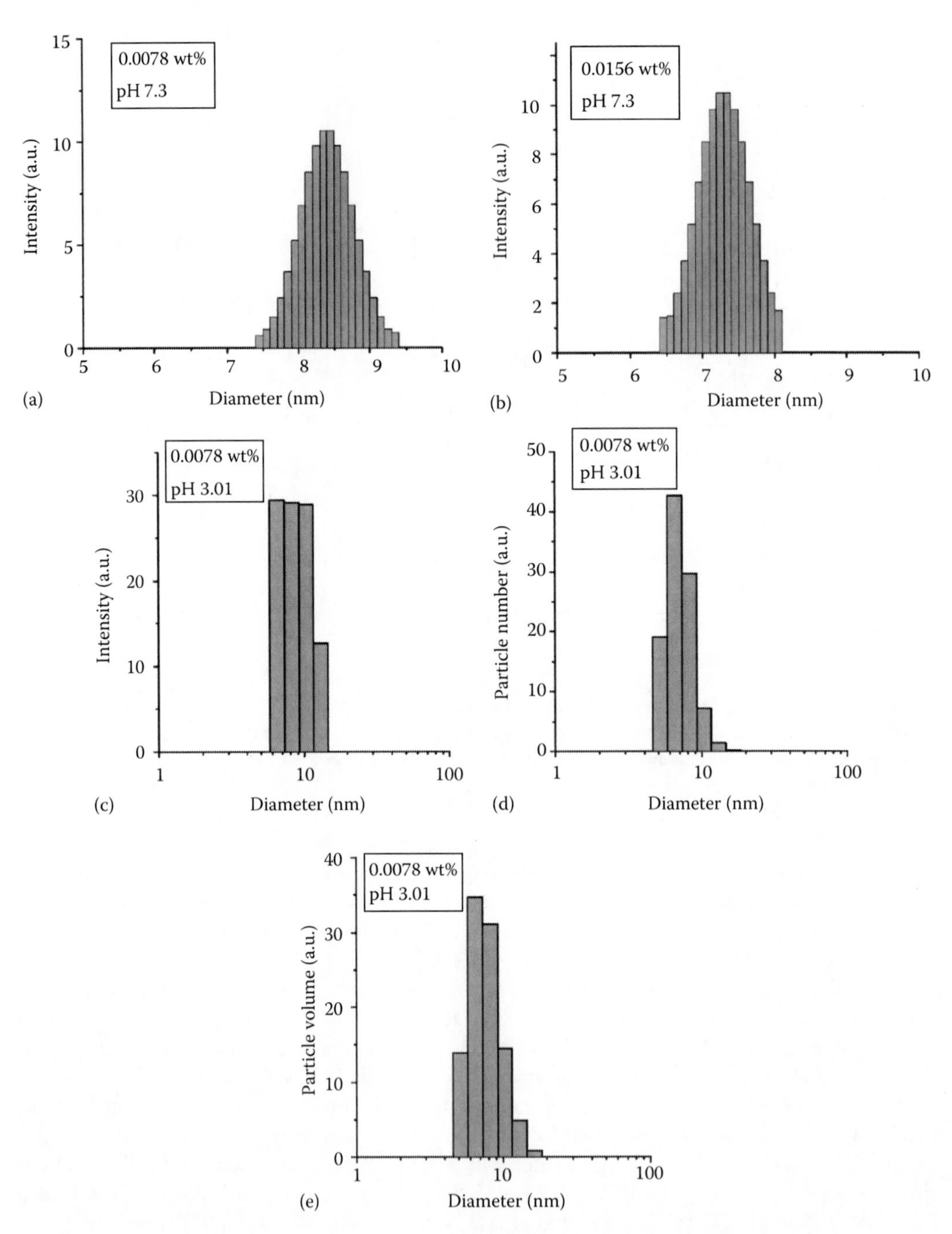

FIGURE 1.259 Particle size distributions in the diluted aqueous suspensions of PMS at C_{PMS} = (a, c, d, e) 0.0078 and (b) 0.0156 wt% shown with respect to (a, b, c) intensity of scattered light, (d) particle number and (e) particle volume at pH (a, b) 7.3 and (c, d, e) 3.01. (Adapted from *J. Colloid Interface Sci.*, 308, Gun'ko, V.M., Turov, V.V., Zarko, V.I. et al., Comparative characterization of polymethylsiloxane hydrogel and silylated fumed silica and silica gel, 142–156, 2007h. Copyright 2007, with permission from Elsevier.)

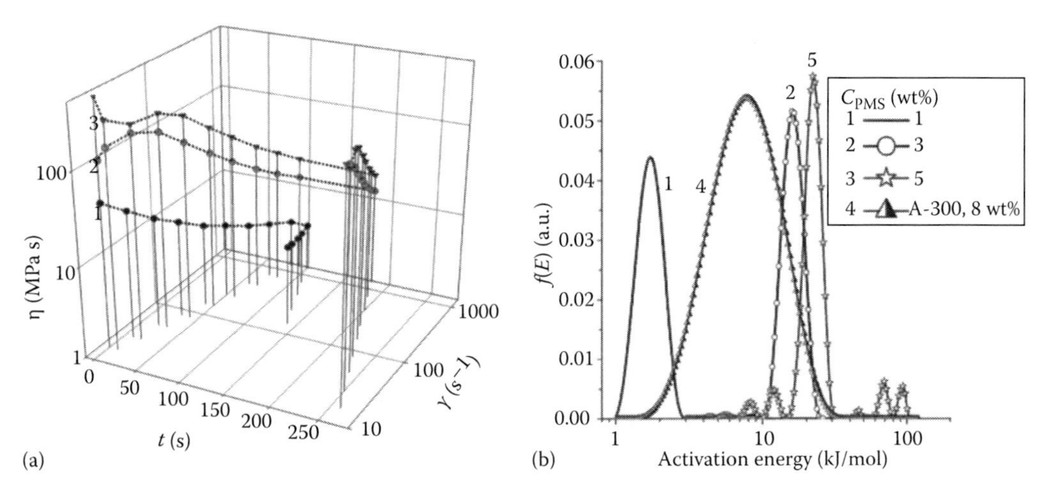

FIGURE 1.260 (a) Viscosity as a function of the shear rate and measurement time for aqueous suspensions of PMS at C_{PMS}=(1) 1, (2) 3 and (3) 5 wt%, and (b) corresponding distribution functions of activation energy of the shear viscosity ($f(E)$ is also shown for the suspension of A-300 at C_{ox}=8 wt%). (Adapted from *J. Colloid Interface Sci.*, 308, Gun'ko, V.M., Turov, V.V., Zarko, V.I. et al., Comparative characterization of polymethylsiloxane hydrogel and silylated fumed silica and silica gel, 142–156, 2007h. Copyright 2007, with permission from Elsevier.)

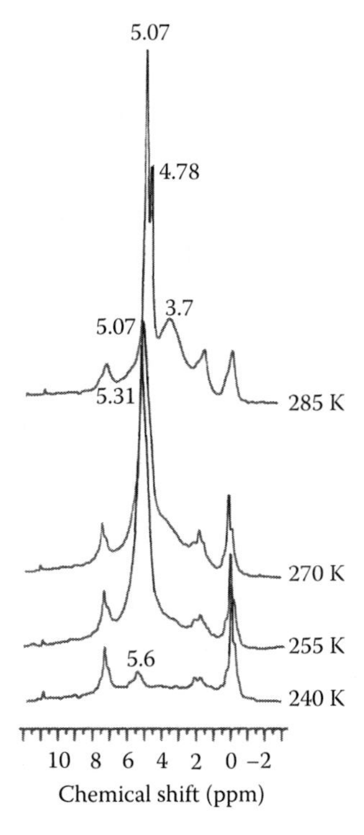

FIGURE 1.261 ^{1}H NMR spectra of dry PMS xerogel ($h \approx 0.03$ g/g) recorded at different temperatures. (Adapted from *J. Colloid Interface Sci.*, 308, Gun'ko, V.M., Turov, V.V., Zarko, V.I. et al., Comparative characterization of polymethylsiloxane hydrogel and silylated fumed silica and silica gel, 142–156, 2007h. Copyright 2007, with permission from Elsevier.)

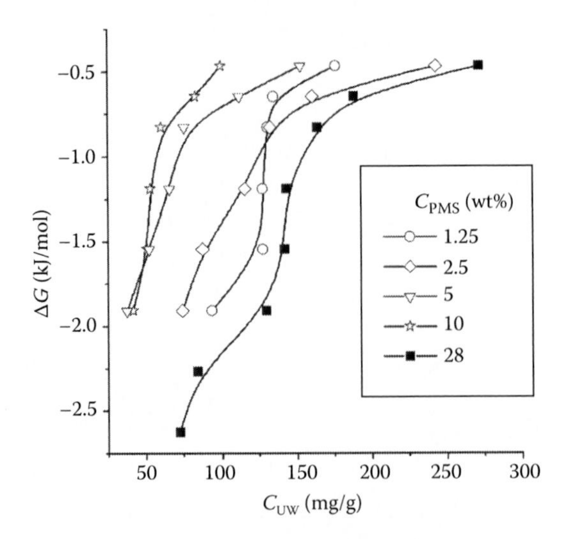

FIGURE 1.262 Amounts of unfrozen water in differently hydrated PMS samples. (Adapted from *J. Colloid Interface Sci.*, 308, Gun'ko, V.M., Turov, V.V., Zarko, V.I. et al., Comparative characterization of polymethylsiloxane hydrogel and silylated fumed silica and silica gel, 142–156, 2007h. Copyright 2007, with permission from Elsevier.)

273 K as bulk water. As a whole, the amounts of water unfrozen at $T<273$ are relatively low in dry PMS xerogel because small $h\approx 0.03$ g/g.

Addition of water to the initial PMS hydrogel or its drying/wetting leads to rearrangement of particles since both treatments lead to an increase in the amounts of structured interfacial water (Figure 1.262) and the structural characteristics noticeable change (Table 1.36).

The maximal amount of structured water is observed at a lower hydration of dried/wetted PMS xerogel ($C_{PMS}=28$ wt%). The amounts of unfrozen water are relatively small for all PMS samples $C_{uw}<530$ mg per gram of dry material (because of the mentioned mesoporous–macroporous structure of PMS and, therefore, freezing the main amounts of water at $T\approx 273$ K). The amounts of water structured by silylated nanosilica (Table 1.25) are larger than that for PMS hydrogels. This can be due to a smaller surface area of PMS particles aggregated in secondary particles (which can be destroyed on sonication). Additionally, silylation of any silica results in diminution of the adsorption potential for both nonpolar (e.g., nitrogen) and polar (water, polymers, proteins) compounds. Therefore, PMS, which is similar to partially silylated nanosilica, is characterized by low adsorption potential with respect to water. Therefore, the amounts of SBW (Table 1.36, C_s) are smaller than that of WBW (i.e., the difference $C_{uw}-C_s$). The corresponding changes in the surface energy (Figure 1.263) show that the initial PMS hydrogel ($C_{PMS}=10$ wt%) is characterized by the smallest γ_S value.

In other words, the state of the initial PMS hydrogel is metastable since changes in the water content in both directions lead to reduction of the Gibbs free energy of the system. The γ_S values for the PMS systems (Figure 1.263) are smaller than that for unmodified silica and silylated A-380 (Table 1.25) or A-300 (Table 1.35) but close to that for the A-300–DMS/A-300 mixtures (Table 1.35). Consequently, the metastable state of the PMS systems and relatively high Gibbs free energy of the interface are caused by a mosaic hydrophobic–hydrophilic PMS structure.

Calculations of the size distribution of unfrozen water structures (Figure 1.264) show that the initial material is the most compacted (V is minimal, Table 1.36) but the pore size filled by this water is nearly maximal (peak maximum at $R=22$ nm). It is also possible that the initial material is characterized by smaller number of residual silanols than the diluted suspensions.

Therefore, its capability in the structurization of water is minimal, and the γ_S value is minimal. This capability increases on decomposition of the PMS particles leading to an increase in

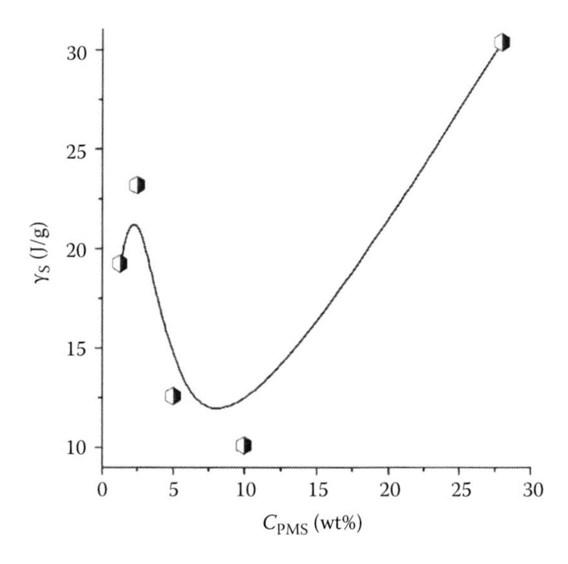

FIGURE 1.263 Free surface energy as a function of PMS concentration in the PMS/water systems. (Adapted from *J. Colloid Interface Sci.*, 308, Gun'ko, V.M., Turov, V.V., Zarko, V.I. et al., Comparative characterization of polymethylsiloxane hydrogel and silylated fumed silica and silica gel, 142–156, 2007h. Copyright 2007, with permission from Elsevier.)

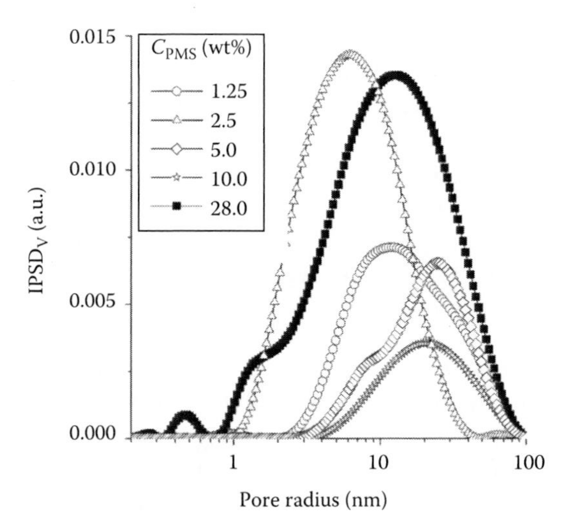

FIGURE 1.264 PSDs of differently hydrated PMS samples. (Adapted from *J. Colloid Interface Sci.*, 308, Gun'ko, V.M., Turov, V.V., Zarko, V.I. et al., Comparative characterization of polymethylsiloxane hydrogel and silylated fumed silica and silica gel, 142–156, 2007h. Copyright 2007, with permission from Elsevier.)

the number of silanols. The first diluted system ($C_{PMS}=5$ wt%) has both broader (maximum at $R=24$ nm) and narrower ($R=8$ nm) pores. Subsequent dilution leads to formation of smaller water structures. This can be due to rearrangement of particles, which can be decomposed and form more compacted structures since the suspension was stirred upon dilution. The dilution of the PMS hydrogel enhances the mesoporosity. Maybe nanopores transform into mesopores since an increase in the $S_{IGT,mes}$ and $V_{IGT,mes}$ values is accompanied by a decrease in the $S_{IGT,nano}$ and $V_{IGT,nano}$ values (Table 1.36) and total surface area grows. These results can be explained by decomposition of PMS particles into smaller ones during dilution of the initial PMS hydrogel and rearrangement of

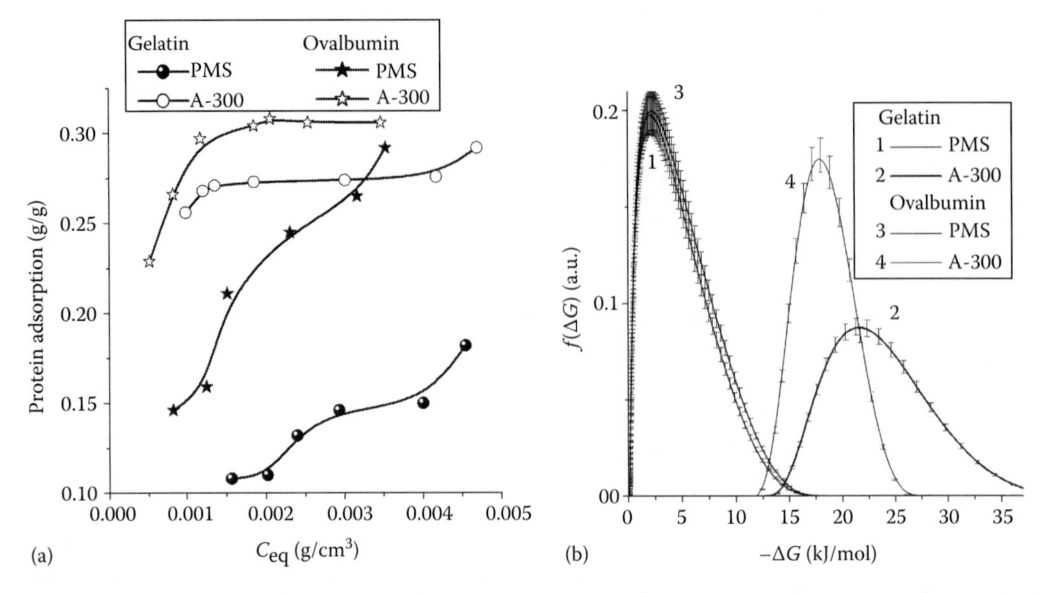

FIGURE 1.265 (a) Isotherms of adsorption of gelatin and ovalbumin onto PMS and nanosilica and (b) the corresponding distribution functions of the Gibbs free energy of protein adsorption. (Adapted from *J. Colloid Interface Sci.*, 308, Gun'ko, V.M., Turov, V.V., Zarko, V.I. et al., Comparative characterization of polymethylsiloxane hydrogel and silylated fumed silica and silica gel, 142–156, 2007h. Copyright 2007, with permission from Elsevier.)

them into new secondary structures. Another type of rearrangement of particles occurs on drying of hydrogel to dry xerogel ($C_{PMS} \approx 97$ wt% at $h \approx 3$ wt%) then wetted ($C_{PMS} = 28$ wt%) because of condensation of residual silanols. The narrowest pores at $R < 1$ nm are characteristic of this material; however, broad pores remain (Figure 1.264). Consequently, the collapse of the textural pores does not occur during drying/wetting of PMS.

The comparative analysis of the adsorption of gelatin and ovalbumin on PMS and nanosilica A-300 shows that the adsorption capacity of PMS is lower than that of A-300 (Figure 1.265a).

The presence of $\equiv SiCH_3$ groups in PMS can be responsible for this effect. Notice that the difference in the adsorption of ovalbumin is higher than that for gelatin. However, the distribution functions of the Gibbs free energy of adsorption are similar for both proteins adsorbed onto PMS (Figure 1.265b) since the difference in the peak position is smaller than 0.3 kJ/mol.

Diminution of the interaction of PMS with proteins in comparison with nanosilica can provide a higher biocompatibility of PMS particles. One can assume that, for example, hemolysis of RBCs caused by PMS can be much smaller than that by unmodified nanosilica since nanosilica modified by immobilized polymers causes lower hemolysis of RBCs (Blitz and Gun'ko 2006).

According to the results of NMR cryoporometry and QELS, PMS hydrogel can be assigned to a mesoporous–macroporous material with mainly textural porosity caused by voids between nanosized primary particles, which are non-dense and nonrigid because of the presence of CH_3 groups attached to each Si atoms and residual silanols. Consequently, PMS particles can be modeled by rather strongly crumpled 2D sheets than 3D solid particles because the Si-O-Si linkages do not form solid 3D network characteristic of silica. This structural feature of PMS causes the formation of three types of the water structures: (i) very small intraparticle water clusters located inside primary PMS particles and characterized by the chemical shift of the proton resonance at $\delta_H \approx 1.7$ ppm corresponding to weakly associated but SBW frozen at $T < 240$ K; (ii) interparticle water locating close to the particle surface and forming larger clusters in mesopores (strongly associated and both SBWs and WBWs at $\delta_H \geq 5$ ppm) frozen at $T > 240$ K; and (iii) WBW at $\delta_H \approx 3.7$ ppm located in macropores and frozen at T close to 273 K. Despite a soft character of secondary particles, the viscosity of the aqueous suspensions of PMS at $C_{PMS}=3–5$ wt% is higher than that of the aqueous suspension of nanosilica A-300 at $C_{A\text{-}300}=8$ wt% because PMS particles tend to form a continuous network in the suspensions to reduce the interaction of hydrophobic $SiCH_3$ groups with water. Additionally, the initial PMS hydrogel at $C_{PMS}=10$ wt% is not liquid since it looks like dry, soft, and very thixotropic gel (i.e., there is a continuous structure there) in contrast to the 10 wt% suspension of fumed nanosilica, which remains liquid during a very long time. The adsorption capacity of PMS with respect to proteins (gelatin, ovalbumin) is smaller than that of nanosilica despite close values of the specific surface area because of the effects of $SiCH_3$ functionalities in PMS reducing the adsorption potentials with respect to any adsorbate.

Now, we will consider the interfacial phenomena at silica gel surfaces with the participation of very different coadsorbates such as water, methane and organic solvents. One of possible methods to increase the capacity of methane accumulators is preadsorption of certain substances capable to form secondary nanoporosity in adsorbent mesopores. This causes more effective van der Waals adsorption of methane in "new" pores smaller than four molecular diameters of the adsorbate. As a whole, it is difficult to investigate these effects because of their dependence on many factors related to structure of adsorbents, features, and amounts of coadsorbates, temperature, and pressure. The analysis of the van der Waals adsorption of methane with preadsorption or coadsorption of water, benzene, and dichloromethane onto silica Si-40 undergoing HTT (Si-40HTT) and characterized by relatively narrow mesopores or other silicas (Gun'ko et al. 2004d, 2009d, 2011c,e, Petin et al. 2010) shows the strong effects of preadsorbed water or organics according to the measurements of the amounts of adsorbed methane under isobaric conditions at 200–280 K using LT ^{1}H NMR spectroscopy method (Figure 1.266).

The temperature dependence of quantity of adsorbed methane ($C_{CH_4}(T)$) at fixed concentration of preadsorbed dichloromethane (Figure 1.266a) or benzene (Figure 1.266b) shows that the maximum adsorption of CH_4 is observed at minimal amounts of coadsorbates (Petin et al. 2010). If the amounts of coadsorbates, which interact with silica surface stronger than methane, are larger than the volume of nanopores of Si-40HTT, the amounts of adsorbed methane strongly decrease. The $C_{CH4}(T)$ dependences are nonlinear, especially at the minimal amounts of coadsorbates. At high (close to 273K) temperatures, the essential increase in adsorption of methane is observed with increasing concentration of CH_2Cl_2; and at 280 K, the adsorption of methane in the presence of CH_2Cl_2 exceeds the adsorption of individually adsorbed methane by 20%–30%. For the sample

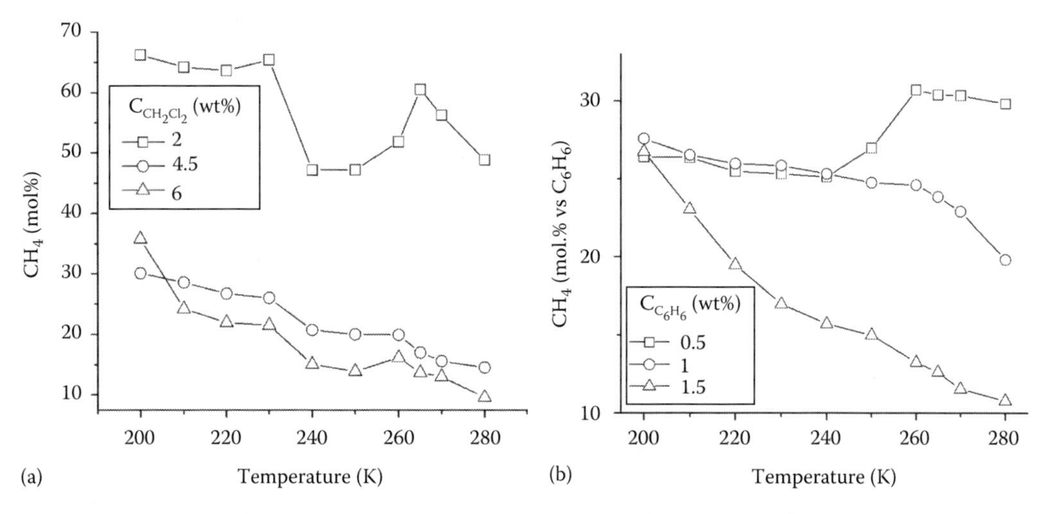

FIGURE 1.266 Temperature dependence of quantity of methane adsorbed onto Si-40HTT in the presence of a small content of (a) CH_2Cl_2 or (b) C_6H_6. (According to Petin, A.Yu. et al., *Chem. Phys. Technol. Surf.*, 1, 138, 2010.)

with 2 wt% of CH_2Cl_2, there is a broad temperature interval with increasing adsorption of methane with elevating temperature. This can be explained by the kinetic and diffusion effects since the methane adsorption can be limited by the diffusion in narrow pores partially filled by coadsorbates, clustered structure of which depends on temperature too. The complex shape of the $C_{CH_4}(T)$ curves (Figure 1.266) can be also caused by the possibility of the adsorption of methane in the form of individual molecules (in very narrow pores) and clusters with dichloromethane (in supernanopores and narrow mesopores). At about 273 K, dichloromethane can locate not only in pores of silica with maximal adsorption potential (nanopores) but also in mesopores because of the diffusion effects (energy of activation of the diffusion is comparable with the difference in the adsorption energy in pores of different sizes) that provide a portion of empty nanopores and narrow mesopores. The adsorption of CH_4/CH_2Cl_2 in the narrowest pores as well as the quantity of adsorbed methane in the form of the clusters increases at $T<240$ K (Figure 1.266a). However, subsequent lowering of temperature does not lead to significant changes in the adsorption of methane because the structure of adsorption complexes of coadsorbates becomes practically fixed, that is, dichloromethane molecules (as a host coadsorbate and methane as a guest adsorbate in formed clusters) are localized in narrow pores with the high adsorption potential and empty from coadsorbates. Thus, there is an optimum ratio of the amounts of host and guest coadsorbates necessary to form the clustered structures in mesopores (i.e., producing effective nanopores with the host coadsorbate in mesopores) without filling of the total volume of nanopores by the host coadsorbate. This aspect is clear from the introduction of additional portions of dichloromethane that sharply reduce the adsorption of methane and the formation of the clusters with CH_4–CH_2Cl_2 (Petin et al. 2010).

Similar effects are observed upon the adsorption of methane with preadsorbed benzene (Figure 1.266b). Unlike dichloromethane, benzene easily freezes especially with increasing its concentration. Optimum conditions of clustered coadsorption are at $C_{C_6H_6}=0.5$ wt%. For this sample, the methane adsorption grows with elevating temperature over relatively broad temperature range. At very low temperature (200 K), the adsorption of methane is practically the same for all samples independent of the amounts of benzene. It is possible that benzene is frozen in mesopores and partially in nanopores that restrict the clustered coadsorption with methane. At 280 K, the quantity of adsorbed methane decreases by three times with increasing benzene concentration from 0.5 to 1.5 wt%. The methane adsorption in the presence of CH_2Cl_2 is higher than that at the preadsorption of benzene since benzene interacts with the silica surface more strongly than dichloromethane.

Note that HTT of Si-40 results in pore enlargement (Gun'ko et al. 2004c); therefore, the associativity of water adsorbed onto Si-40HTT is greater than that adsorbed onto the initial silica gel (Figure 1.267). This pore morphology can affect the coadsorption of methane with benzene or dichloromethane (Figure 1.266).

Youngman and Sen (2004) studied the composition and temperature dependence of the coordination environments of F atoms in fluorinated silica glasses using high-resolution ^{19}F and ^{29}Si NMR spectroscopies (Figure 1.268). ^{19}F MAS and wideline NMR spectra have revealed the presence of two distinct types of fluorine environments in glasses containing 1–3.3 wt% fluorine at $\delta(^{19}F)=-137$ and -146 ppm at contributions 16% and 84%, respectively. The majority of the fluorine environments are formed by replacing one of the bridging oxygens around a silicon atom with a non-bridging fluorine atom, forming $SiO_{3/2}F$ polyhedral (with two ^{29}Si lines at -103 and -125 ppm). The less abundant species is found to be highly unusual in that it involves bonding of a non-bridging fluorine to a silicon atom that is already coordinated to four bridging oxygens, yielding a fivefold coordinated silicon of the type $SiO_{4/2}F$. The relative concentration of the $SiO_{4/2}F$ species in these glasses is found to increase with both increasing fluorine content.

Juvaste et al. (1999) used ^{1}H, ^{13}C, and ^{29}Si solid-state NMR spectroscopy to characterize silicas (EP 10, Crosfield Ltd., 300 m^2/g, $V_p=1.8$ cm^3/g, pore $d=24$ nm, 0.1 mm particles) modified with aminopropyldimethylethoxysilane (APDMES), $Me_2Si(C_5Me_4H)Cl$, and $MeHSi(C_5Me_4H)Cl$ (immobilized on the APDMES-functionalized silica by utilizing n-butyllithium). The amine groups react with n-BuLi by producing lithiated amine groups. Furthermore, the lithiated amine reacts with

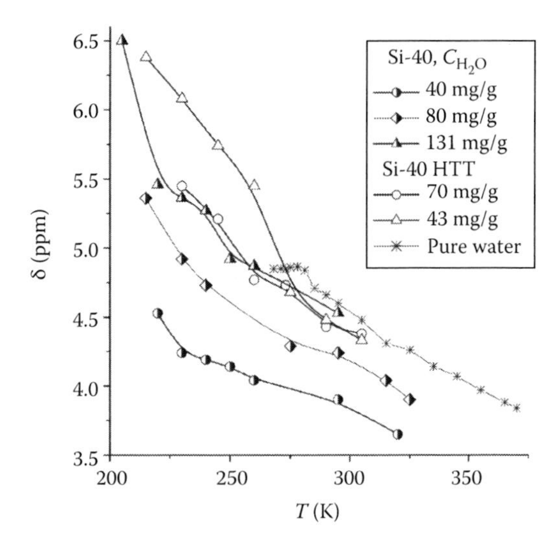

FIGURE 1.267 The ^{1}H chemical shift as a function of temperature at different amounts of water adsorbed onto initial and hydrothermally treated (HTT) silica gel Si-40. (Adapted from *Colloids Surf. A: Physicochem. Eng. Aspects.*, 235., Gun'ko, V.M., Skubiszewska- Zięba, J., Leboda, R., and Turov, V.V., Impact of thermal and hydrothermal treatments on structural characteristics of silica gel Si-40 and carbon/silica gel adsorbents, 101–111, 2004c. Copyright 2004, with permission from Elsevier.)

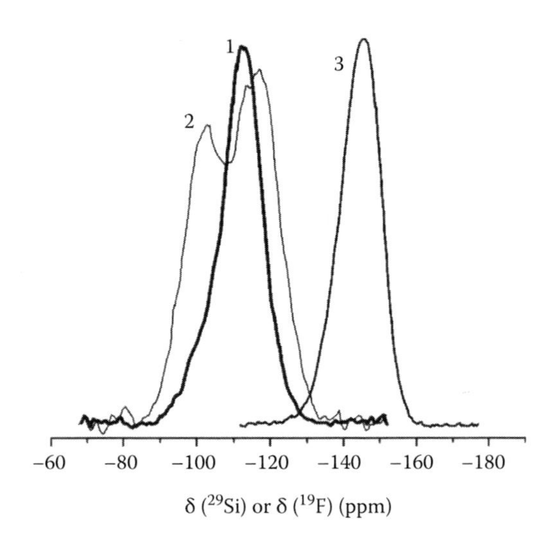

FIGURE 1.268 NMR spectra of 3wt% F-doped silica (1) ^{29}Si MAS NMR spectrum with ^{19}F decoupling, (2) ^{19}F → ^{29}Si CP/MAS NMR spectrum using a short contact time of 0.5 ms, and (3) ^{19}F MAS NMR spectrum. (Adapted from *J. Non-Crystal. Solid.*, 349, Youngman and Sen, Structural role of fluorine in amorphous silica, 10–15, 2004. Copyright 2004, with permission from Elsevier.)

chloride of the silane and a chemical bond between nitrogen and silicon of the silane is formed. These stepwise prepared surfaces were characterized with NMR and FTIR spectroscopies. This is typical example of the use of the NMR spectroscopy in parallel with the FTIR spectroscopy as an analytical tool for chemical analysis of a modified surface of silicas and related materials (Bergna 1994, 2005, Luhmer et al. 1996, Legrand 1998, Juvaste et al. 1999, Mijatovic et al. 2000, Vartzouma et al. 2004, Jerman et al. 2010).

Thus, detailed investigations of textural and adsorption properties of materials with 1D (PDMS), 2D (PMS), and 3D (silica gels, aerogels, Silochrome) siloxane backbones with unmodified (pure silanols) or functionalized ($Si(CH_3)_n$ groups) surfaces show that the interfacial phenomena depend strongly on many factors including the S_{BET} and V_p values, PSD, surface functionalization degree, type, and amounts of adsorbates and coadsorbates, as well as temperature and dispersion medium. The ^1H, ^{13}C, ^{19}F, and ^{29}Si NMR spectroscopy, NMR cryoporometry, and relaxometry methods are very informative with respect to porous silicas characterization and describing interfacial phenomena in confined space of nano- and mesopores. The parallel analysis of the data of TSDC, DSC, QELS, FTIR, adsorption, and desorption methods allows deeper insight into these problems and better understanding of the NMR results.

1.3 INTERFACIAL PHENOMENA AT SURFACES OF STRUCTURALLY ORDERED SILICAS

Mesoporous silicas with ordered structure of pores (such as M41S, MCM, SBA, FSM) are interesting as adsorbents, catalyst carriers, drug carriers, etc. (Beck et al. 1992, Bonneviot et al. 1998, Davidson 2002, He and Shi 2011). Morphological, textural, and adsorptive characteristics of these relatively new (with 20-year history) materials have been studied and well documented (Brinker and Scherer 1990, Huo et al. 1996, Legrand 1998, Zhou and Klinowski 1998, Ciesla and Schüth 1999, Jaroniec et al. 2001, Kumar et al. 2001, Ribeiro Carrott et al. 2001, Guliants et al. 2004, He and Shi 2011). Series of studies of mesoporous silicas was carried out by Wiench, Pruski et al. (Huh et al. 2003, 2004, 2005, Chen et al. 2005, Trebosc et al. 2005a,b, Kumar et al. 2006, Wiench et al. 2007a,b, 2008, Mao et al. 2009, 2010, Mao and Pruski 2009, 2010, Kobayashi et al. 2011) by combining 1D and 2D NMR techniques (MAS, CRAMPS, HETCOR, insensitive nuclei enhanced by polarization transfer [INEPT], CPMG, 2D double-quantum [DQ], etc.) with several decoupling schemes. Typically, ordered mesoporous silicas have a narrow size distribution of pores of a cylindrical shape that is of importance for NMR investigations of the structural features of them and the interfacial behavior of adsorbed compounds.

The structural characteristics of ordered mesoporous silicas depend on the molecular size and the structure of a template, composition of a reaction solution (e.g., the presence of alcohol) and reaction conditions (temperature, time, pH, etc.). Their adsorption properties depend on the PSD and the structure of pore walls. Surface silanols (2–4 $\mu mol/m^2$ for MCM-41 [Kumar et al. 2001], 4–5OH/nm^2 for other silicas) are main adsorption sites for polar adsorbates (Legrand 1998, Bergna 2005). Braga et al. (2011) studied silica MCM-41 using solid-state ^{29}Si MAS NMR measurements showed a high amount of silanol groups (74.7% before heating and 52% after heating at 550°C) in the structure (Q_2 and Q_3 species). The oxygen atoms in the ≡Si–O–Si≡ siloxane bonds possess weaker electron-donor properties than that in silanols. Therefore, the ≡Si–O–Si≡ bonds demonstrate much weaker hydrophilic properties than silanols. Dispersion forces are predominant upon interactions of nonpolar and weakly polar adsorbates with the siloxane bonds, as well as with the whole silica surface. These structural features of silicas allow polar and nonpolar adsorbates to interact with different surface sites (even in the case of unmodified silicas) due to the hydrogen and/or dispersion binding. Therefore, the structure of the interfacial layers of mixtures of different immiscible or poorly miscible polar and nonpolar liquids or fluids in pores of silicas can have an inhomogeneous structure. Additionally, competitive adsorption of different adsorbates can lead to the displacement of one by another from pores and pore walls. The behavior of water/organic mixtures in the form of solutions, emulsions or separated phases, which is of importance from practical point of view, can be strongly different at a plane open surface, in narrow pores and in the bulk liquids. This is due to lower solubility of organics in the interfacial water (Chaplin 2011) and lower molecular mobility in the interfacial layers than that in the bulk. Numerous papers with very detailed description of the synthesis and the characterization of ordered mesoporous silicas were published but the synthesis aspects are out of this book. Therefore, these aspects are shown here only with respect to the effects of confined space and the surface chemistry on bound water and its mixtures with other liquids (benzene and chloroform) adsorbed onto ordered mesoporous silicas.

The behavior of bound water in mesoporous silicas was studied by ^{1}H NMR spectroscopy (giving temperature dependence of transverse relaxation time and chemical shift of the proton resonance), XRD (freezing/melting behavior and ice crystalline structure), TSDC (relaxation phenomena for bound water depending on temperature and pore structure), FTIR and Raman spectroscopies (characteristics of adsorption sites and water binding), and other methods (see previous chapters and Sklari et al. 2001, Sliwinska-Bartkowiak et al. 2001).

Webber and Dore (2008) compared the textural characteristics of silica SBA-15 calculated using neutron diffraction (NDC), NMR (NMRC), and DSC (DSCC) cryoporometry methods. The information from a set of neutron diffraction measurements of liquids and their crystalline forms in mesopores, as a function of temperature, was displayed as a cryoporometry graph, which was then conveniently interpreted using the Gibbs–Thomson relationship by analogy with NMRC. The method described by Webber and Dore (2008) uses global pattern matching (a 1D morphing algorithm inside a linear least-squares fitting algorithm) applied to the full range of the diffraction data. This is a rapid method by comparison with the conventional method of fitting individual (overlapping) peaks, and has already led to NMR observations indicating plastic (rotator phase) ice in the same system. These cryoporometry methods gave close results for the pore diameter $d = 8.4 \pm 0.15$, 8.56 ± 0.06, and 9.33 ± 0.6 nm (median) and 8.3 ± 0.3, 8.54 ± 0.06, 9.19 ± 0.6 nm (peak) with the DSCC, NMRC and NDC methods, respectively (Webber and Dore 2008).

Morishige and Iwasaki (2003) studied the freezing and melting behavior of water in partially or totally filled pores of SBA-15 (average pore diameter $d = 7.8$ nm) using XRD method at different temperatures (Figure 1.269). They showed that the freezing temperature increased continuously with increasing pore filling even in the region of capillary condensation because of diminution of the influence of the surface forces on distant water layers. The results are related to the different states of the pore-confined water depending on the degree of pore filling. This result can be interpreted according to our classification as an increase in the amounts of WBW and SAW and a decrease in

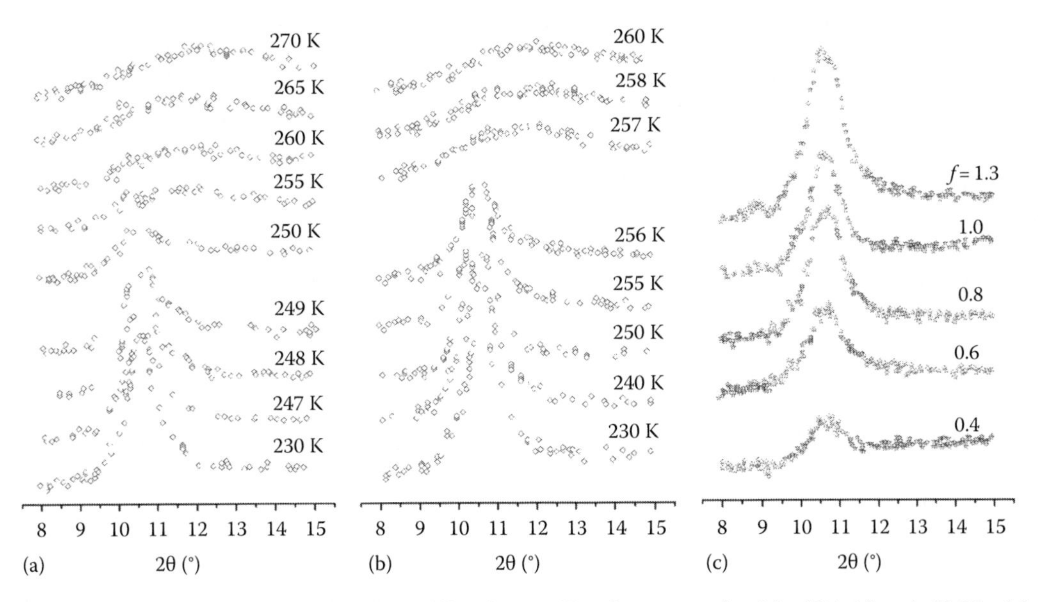

FIGURE 1.269 The change of the x-ray diffraction profile of water confined in SBA-15 at $f = V_w/V_p = 1.3$ upon (a) cooling and (b) heating, and as a function of pore filling by water at 230 K. (Adapted with permission from Langmuir, 19, Morishige and Iwasaki, X-ray study of freezing and melting of water confined within SBA-15, 2808–2811, 2003. Copyright 2003, American Chemical Society.)

relative contributions of SBW and WAW (Gun'ko et al. 2009d). The XRD patterns show that the freezing of the confined water results in formation of ice microcrystallites with almost the same structure and size, irrespective of the different states of the bound water. Therefore, the melting of the frozen confined water took place at a well-defined temperature of 256 K (Figure 1.269), independent of the degree of pore filling. This result can be typical for narrow mesopores in which the surface fields of the opposite pore walls strongly overlap.

There are several methods used for estimation of the PSDs of ordered mesoporous silicas on the basis of the specific behavior of liquids confined in nano- and mesopores. This behavior changes in the case of the adsorption of a mixture of water and organic coadsorbates. In general, the structurization (clustering) of these mixtures occurs in confined space of pores to minimize the Gibbs free energy of the system. Clearly, the structural characteristics of these mixtures depend on the structure and the chemistry of both adsorbates and adsorbents. Additionally, the order of the loading of liquids (as well as treatments of the mixtures) can affect this structurization because the molecular mobility of liquids in narrow pores is much lower than in the bulk (Franks 1982, Brovchenko and Oleinikova 2008) and observed picture can be far from the equilibrium state. These circumstances restrict or slow down the rearrangement of the structure of confined mixtures affected by temperature and a ratio of their content. Confinement in narrow pores has a strong influence on gas/liquid and liquid/solid phase transitions. Liquid water can be thermodynamically stable in pores at temperatures well below the temperature of homogeneous nucleation of ice in undercooled bulk water ($T_H \approx 235$ K) (Findenegg et al. 2008, Chaplin 2011). There are well-known thermal anomalies of water, which include the sharp increase in the isothermal compressibility, the isobaric heat capacity, the magnitude of the thermal expansion coefficient, and other effects can appear upon supercooling (Chaplin 2011). These anomalies can be represented by power laws $Y = A(T - T_s)^x$, where A and x are positive parameters chosen to fit the variation of the property Y with temperature, and T_s is a singularity temperature, for which values near 228 K were obtained for all the anomalous quantities. These observations can be rationalized by the existence of a critical point at which two distinct forms of supercooled water, low-density liquid, and high-density liquid become identical as the temperature is increased. The singularity temperature T_s can be seen as the critical temperature of the liquid–liquid coexistence. The fact that this temperature (228 K) can be attained with water in nanopores has motivated several recent studies of static and dynamic properties of water in nanoporous solids (Findenegg et al. 2008). Liquid water can be thermodynamically stable in nanopores of MCM-41 at $d < 4$ nm (Figure 1.270) at temperatures below the limit of homogeneous nucleation of bulk water (~235 K) that correspond to the anomalous behavior of deeply supercooled water (Findenegg et al. 2008).

According to Findenegg et al. (2008), the chemical shift of water and silanols of silicas depends on the water content (Figure 1.271). Free silanols in both MCM-41 and SBA-15 are characterized by $\delta_H \approx 1.7$ ppm similar to that for other silicas described earlier. For MCM-41, the δ_H value of bound water increases from 2.5 to 4.7 ppm with increasing water content. The band of free silanols at 1.74 ppm disappears with increasing water content. The appearance of two distinct lines for hydrogen-bonded water (3–3.5 and 4.7 ppm) is due to the difference in the associativity of the molecules located in pores of different sizes. Water bound in broader pores of SBA-15 give only a single line, which shifts from 2.5 ppm at the lowest water loading toward 5 ppm for completely filled pores. This result can be explained by fast exchange of the protons between the molecules in different structures of the total hydrogen-bonded network. This finding is consistent with the fact that in SBA-15, because of the greater pore width, condensation of water starts at a significantly larger film thickness than in MCM-41 where the condensation occurs after completion of the first monolayer, while pore filling in SBA-15 proceeds in a layer-by-layer mode before pore condensation commences at a higher pore filling (Schreiber et al. 2001, Findenegg et al. 2008). These results are similar to those observed for silica gels and discussed earlier.

The pore size effects on the water freezing/melting processes depend on the pore size (Figure 1.272). The hysteresis effect decreases with decreasing pore size (compare positions of open

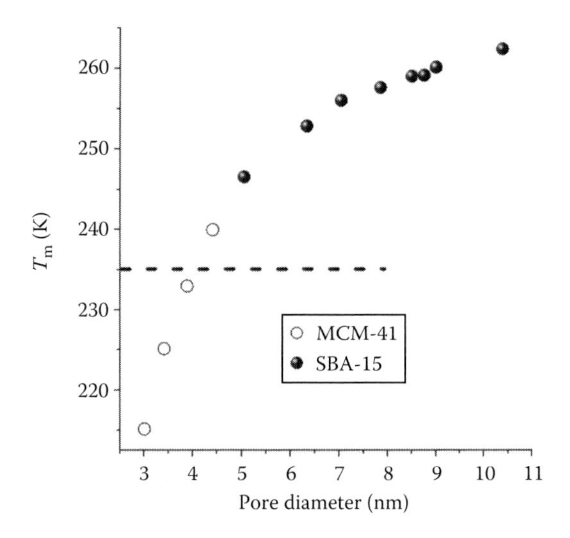

FIGURE 1.270 Effect of the pore diameter d on the melting temperature T_m of water in cylindrical pores of silicas MCM-41 and SBA-15; the dashed line at 235 K marks the limit of homogeneous nucleation of bulk water. (From Findenegg, G.H., Jähnert, S., Akcakayiran, D., and Schreiber, A.: Freezing and melting of water confined in silica nanopores. *Chem. Phys. Chem*.. 2008. 9. 2651–2569. Copyright Wiley-VCH Verlag GmbH & Co. KGaA. Adapted with permission.)

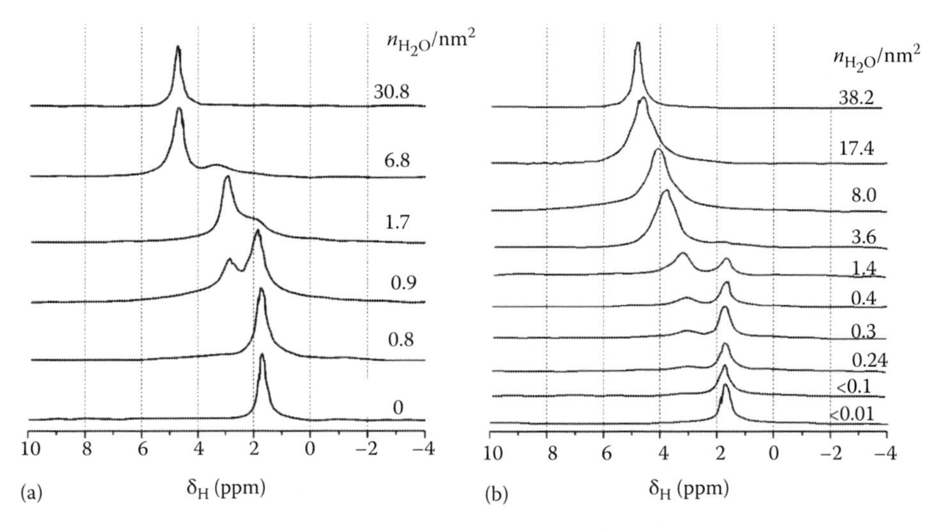

FIGURE 1.271 Room-temperature ^{1}H solid-state NMR spectra of water adsorbed onto (a) MCM-41 (d=4.0 nm) and (b) SBA-15 (d=8.9 nm) at different number of water molecules (n_{H_2O}) per nm^2 of the pore walls. (From Findenegg, G.H., Jähnert, S., Akcakayiran, D., and Schreiber, A.: Freezing and melting of water confined in silica nanopores. *Chem. Phys. Chem*.. 2008. 9. 2651–2569. Copyright Wiley-VCH Verlag GmbH & Co. KGaA. Adapted with permission.)

symbols for freezing and full symbols for melting for MCM-41 and SBA-15). These dependences can be approximated as $\Delta T = k_{GT}/(R - t)$ (t=0.6 nm) and k_{GT}=52.4 K nm for melting of ice (Findenegg et al. 2008).

Melting and freezing of water in completely filled pores of SBA-15 gives a single DSC peak at pore-filling factor $f<1$ (Findenegg et al. 2008). At $f<1$, the melting scans exhibit only a single peak, but two or three peaks appear in the freezing scans, depending on the degree of pore filling (Figure 1.272). This behavior indicates the existence of two or more distinct arrangements of solid

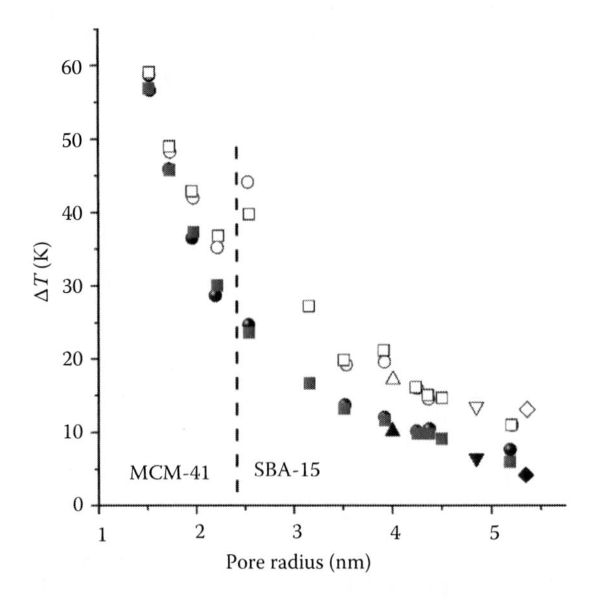

FIGURE 1.272 Melting point depression ΔT_m and freezing point depression ΔT_f as a function of the pore radius R for H_2O (□) and D_2O (O) in the pores of MCM-41 and SBA-15. Full symbols indicate ΔT_m, and open symbols correspond to ΔT_f. Results for H_2O in acid-functionalized SBA-15 materials: (Δ), carboxylic acid; (∇), sulfonic acid; (◇), phosphonic acid. (Redrawn according to Findenegg, G.H. et al., *Chem. Phys. Chem.*, 9, 2651, 2008.)

ice and liquid water in the pores. The sample with excess water ($f = V_w/V_p = 1.3$, where V_p is the pore volume and V_w the volume of liquid added) exhibits two prominent peaks, designated as I (258 K) and II (248 K). Peak I with the sharp onset at the HT side is caused by supercooled freezing of external water, while the large, nearly symmetric peak II is due to freezing of water located in pores (Figure 1.273). A related peak (designated as II*) also appears in a sample at $f = 0.6$, but now shifted toward lower temperatures (245 K). Its sharp peak onset indicates that the freezing of bound water is supercooling in partially filled pores. This sample ($f = 0.6$) exhibits two further peaks, designated as III (236 K) and IV (233 K). These peaks, which are found at all pore fillings $f < 1$ (though with different relative peak areas), are attributed to phase transitions of the water film at the pore walls. At even lower pore fillings ($f = 0.4$), peak II* has disappeared, indicating that pores or pore sections completely filled with water are now absent (Findenegg et al. 2008).

The peaks for silanol protons in the high-resolution ^1H NMR spectrum obtained on SBA-15 are broadened and shifted downfield by hydrogen bonding with adsorbed water molecules (Hu et al. 2005). Overlapping of the resonance for hydrogen-bonded silanol with the corresponding broad peak of hydrogen-bonded water complicates the spectrum that hamper a quantitative analysis of the spectra. Hu et al. (2005) demonstrated that adsorbed water can be removed by exposing the sample to dry nitrogen during MAS. This results in significant line narrowing for the silanol protons in the ^1H MAS spectrum. The enhanced spectral resolution makes it possible to quantify the various hydroxyl groups (Hu et al. 2005).

Wouters et al. (2001) used the ^{29}Si CP MAS NMR spectra of unmodified and silylated MCM-41 to estimate the content of surface hydroxyls and TMS groups with equations

$$[\mathrm{SiOH}] = \frac{I_{Q_3} + 2I_{Q_2}}{60I_{Q_4} + 69I_{Q_3} + 78I_{Q_2}} \quad \text{and} \quad [\mathrm{TMS}] = \frac{I_{\mathrm{TMS}}}{60I_{Q_4} + 69I_{Q_3} + 78I_{Q_2}} \tag{1.69}$$

This approach in combination with FTIR spectra, TG, and TPD-MS data can be useful for quantitative characterization of unmodified and MSs with respect to the surface structure.

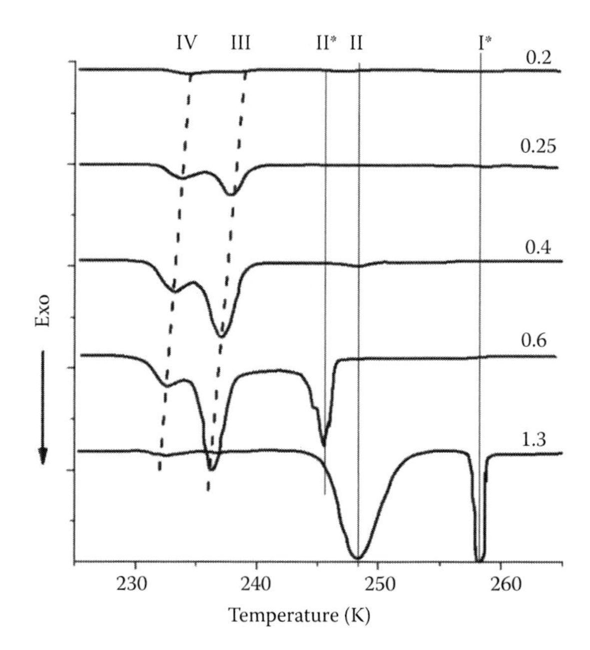

FIGURE 1.273 DSC freezing scans for water adsorbed to SBA-15 (d=8.7 nm) at five different pore fillings $f = V_w/V_p$ as indicated by the numbers. (From Findenegg, G.H., Jähnert, S., Akcakayiran, D., and Schreiber, A.: Freezing and melting of water confined in silica nanopores. *Chem. Phys. Chem.*. 2008. 9. 2651–2569. Copyright Wiley-VCH Verlag GmbH & Co. KGaA. Adapted with permission.)

Johari (2009) studied in detail the LT (90–230 K) behavior of water bound in MCM-41. The results on the enthalpy relaxation rate, dH_m/dt, and specific heat, $C_{p,m}$, of water bound in 1.8 nm pores of MCM-41 were interpreted by considering surface interactions and the number of H_2O molecules available to crystallize or vitrify. Of a maximum of five H_2O molecules in a close-packing along the 1.8 nm diameter pore, at least two would form an uncrystallizable shell bonded to the silica wall and three would remain as a 1.1 nm diameter nanocore, which is too small for nucleation of the usual ice crystals. These results are in agreement with the data (Findenegg et al. 2008) discussed earlier (Figure 1.273) despite the different temperature ranges used. The dH_m/dt features observed in the temperature range of 90–130 K and 125–175 K showed kinetic unfreezing or glass-softening characteristics of confined water. The first was attributed by Johari (2009) to the reorientation of H_2O in the nanoshell with little change in their center of mass position and the second to the change in the population of bonds between water and silica surface. The third dH_m/dt feature observed in the 180–230 K range is inconsistent with kinetic unfreezing or glass-softening and was attributed to the formation and melting of distorted ice-like unit cells with or without the growth- and stacking-faults that remain at equilibrium with the melt. The large increase in the $C_{p,m}$ value at T near 210 K is attributed to the latent heat of their "melting," as occurs on premelting of fine-grained ice and other solids. Data on the pore size dependence qualitatively support this interpretation. Structure of the water in nanopores depends on the pore size, and its properties differ from those of bulk water (Johari 2009).

Dore et al. (2004) compared the characteristics of several amorphous silicas studied by SANS and NMR techniques. The measurements reveal different characteristics for sol-gel silicas and ordered mesoporous silicas (MCM and SBA type). They obtained good agreement between gas adsorption, NMR and SANS results for the pore sizes. The neutron diffraction results for the water/ice phase transformation in SBA silicas indicated a complex relationship that can be due to effects of nanopores and certain complexity of the water behavior in SBA mesopores which are broader than MCM mesopores.

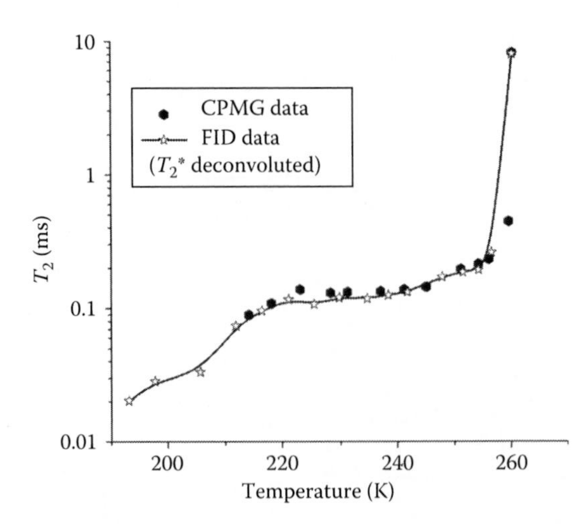

FIGURE 1.274 Temperature dependence of T_2 relaxation time (Carr–Purcell–Meiboom–Gill and $T_2^* d$-corrected FID) component of plastic ice and water located in pores of SBA-15. (Adapted from *Magn. Reson. Imaging*, 25, Webber, J.B.W., Anderson, R., Strange, J.H., and Tohidi, B., Clathrate formation and dissociation in vapour/water/ice/hydrate systems in SBA-15, Sol-Gel and CPG porous media, as probed by NMR relaxation, novel protocol NMR cryoporometry, neutron scattering and ab-initio quantum-mechanical molecular dynamics simulation, 533–536, 2007a. Copyright 2007, with permission from Elsevier.)

As shown earlier, the NMR relaxation behavior of water/ice in SBA-15 depends on temperature, and two "water-like" and "ice-like" phases and a third phase with intermediate properties were observed by Webber et al. (2007a). Bulk water has T_2 of a few seconds at a 0.5-T $\boldsymbol{B}_0$ static magnetic field as used for this relaxation work. Water located in pores of SBA-15 has T_2 about 8 ms (surface relaxivity reduces T_2) (Figure 1.274). The bulk phase of brittle ice has approximately Gaussian free induction decay (FID) with a relaxation time on the order of 10 μs. The FID for this sample at ~190 K is very similar to that of bulk ice, but as the temperature is raised by 20–210 K, some of the brittle ice transforms to a component with a longer relaxation time and with a form that becomes increasingly exponential. Above 220 K, the shorter brittle ice component reduces in amplitude as it converts to the longer form, while the longer component of FID remains fairly constant in relaxation time (between 100 and 200 μs), and its amplitude progressively increases with temperature. Just below the melting temperature of ice in these pores, at the Gibbs–Thomson-lowered temperature of about 260 K, there is a rapid increase in the relaxation time of the longer component. Neutron diffraction measurements showed a similar component as a broad peak characteristic of defective ice/water system (Webber et al. 2007a).

Baccile and Babonneau (2008) synthesized of various organo-modified mesoporous silicas using the template synthesis and studied the adsorption properties toward a methoxy-modified chlorophenol in water. Phenyl, propyl, hexyl, and hexadecyl groups were selected to study the possible interactions that the pollutant could preferentially develop with the surface sites. The best performing sample was the hexyl-modified porous silica. The physical state of the adsorbed 2-chloro-4-methoxyphenol (CMP) confined in mesopores as well as the pollutant/surface interactions were characterized by a combination of solid-state NMR techniques. The ^{1}H and ^{13}C NMR responses showed a high mobility of the molecules in both the modified and unmodified silicas, suggesting the absence of strong interactions between the pollutant and the surface, despite the large difference in the chemical nature of the silica surface sites. Figure 1.275 shows ^{1}H MAS NMR spectra of SiO_2 and hexyl-SiO_2 recorded at different temperatures and with a moderate MAS rate (5 kHz).

At 298 K, for both samples, the two signals at 6.6 and 3.6 ppm were attributed, respectively, to phenyl ring and methoxy groups of CMP molecules (Baccile and Babonneau 2008). Their narrow

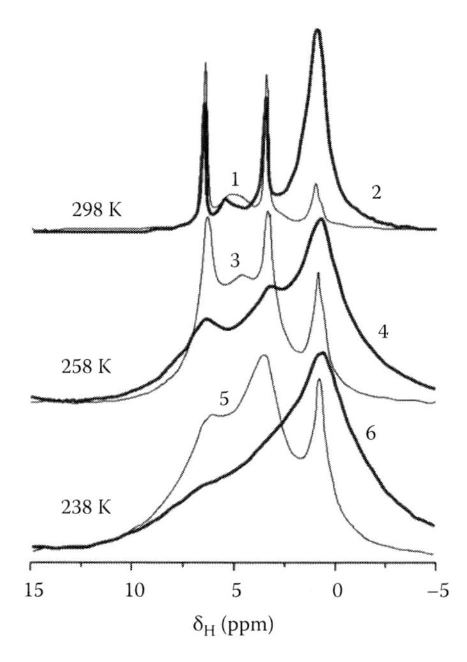

FIGURE 1.275 ^1H MAS NMR (ν_{MAS} = 5 kHz) spectra of (1, 3, 5) SiO_2 and (2, 4, 6) hexyl-SiO_2 after CMP adsorption performed at 298, 258 K and 238 K. (Adapted from *Micropor. Mesopor. Mater.*, 110, Baccile and Babonneau, Organo-modified mesoporous silicas for organic pollutant removal in water: Solid-state NMR study of the organic/silica interactions, 534–542, 2008. Copyright 2008, with permission from Elsevier.)

full width at half maximum (FWHM), especially the one of the phenyl ring (50 Hz), indicate clearly high molecular mobility. The broader peak present around 5.5 ppm for both samples is due to silanol species involved in hydrogen bonding with water molecules and possibly ethoxy groups (^{13}C NMR results in Figure 1.276). The broad peak at 1.2 ppm corresponds to the grafted hexyl chains in hexyl-SiO_2, while in SiO_2, the small peaks around 1 ppm is due to residual surfactant, incompletely removed by the solvent extraction procedure. When temperature is lowered down to 238 K, all peaks are homogeneously broadened due to a reintroduction of the ^1H–^1H homonuclear dipolar coupling as a consequence of a reduced mobility of CMP molecules within the mesopores of silicas. Temperature effect over molecular mobility of CMP in confined space of mesopores was used in similar systems to study, for instance, liquid-to-solid transitions (Xia et al. 2006). However, from the ^1H line width evolution, which is governed primarily by the extent of the homonuclear dipolar coupling, it is not possible to discriminate between the formation of a glassy and a crystalline state at lower temperature. The evolution of the ^{13}C line width gives more information. Also the broadening of the CMP peaks seems less important at 258 K for SiO_2 than for hexyl-SiO_2. However, the extensive overlap with other peaks, especially the peak due to the hexyl chains, makes the analysis of line-widths difficult. In both samples, the transition between a quasi-liquid-like state to a solid-like state occurs over a broad range of temperature characteristic of confined species (Baccile and Babonneau 2008).

In the one-pulse ^{13}C MAS NMR spectrum (Figure 1.276, curve 1), all carbon sites grafted onto silica surface can be detected (Baccile and Babonneau 2008) and all the aromatic C peaks from CMP are observed between δ_C = 100 and 160 ppm. Peaks at δ_C = 125.6, 144.8, and 148.3 ppm are due to quaternary carbons while peaks at δ_C = 112.6, 116.2, and 121.7 ppm are due to the aromatic CH sites. The peak at δ_C = 56.5 ppm is assigned to OCH_3 group of CMP. The three intense peaks at δ_C = 32.4, 23.3, and 14.3 ppm are due to the pendant hexyl chains grafted to silica surface. As for the remaining observed peaks at δ_C = 60.4 and 17.9 ppm, they are due to ethoxy groups that have been grafted to the surface during the surfactant removal procedure. Signal intensity integration gives a

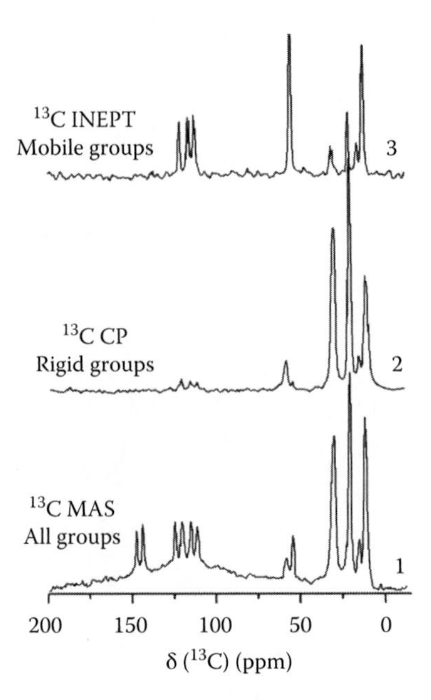

FIGURE 1.276 Comparison between ^{13}C MAS NMR spectra of hexyl-SiO_2 sample after CMP adsorption (ν_{MAS} = 5 kHz): (1) one-pulse experiment; (2) CP experiment (t^{CP} = 3 ms); (3) INEPT spectrum. (Adapted from *Micropor.Mesopor.Mater.*, 110, Baccile and Babonneau, Organo-modified mesoporous silicas for organic pollutant removal in water: Solid-state NMR study of the organic/silica interactions, 534–542, 2008. Copyright 2008, with permission from Elsevier.)

ratio of roughly 1 CMP molecule per 3.5 hexyl chain after impregnation. The ^{13}C CP MAS NMR spectrum based on through-space magnetization transfer via dipolar coupling favors detection of rigid protonated moieties. Thus, CH groups in totally rigid systems can be easily detected for very short contact times less than 100 μs. In this system, peaks belonging to CMP are barely detected for t_{CP} = 500 μs and t_{CP} has to be increased to 3 ms (Figure 1.276, curve 2) for the peaks due to protonated C (aromatic CH and methoxy) to appear. One can, however, notice slight differences in the relative intensities of the hexyl-related peaks, especially the peak at 14 ppm that corresponds to the overlap of the CH_3 and Si–CH_2 signals, is underestimated, because of the high mobility of the CH_3 end group, which prevents a good polarization transfer efficiency. The ^{13}C INEPT MAS NMR spectrum, based on through-bond magnetization transfer, was also recorded following an approach developed by Alonso and Massiot (2003) (Figure 1.276, curve 3). This experiment favors detection of mobile carbon sites directly bonded to protons, and brings thus complementary information with respect to CP experiment. Indeed, all the C sites of CMP (CH and O–CH_3), but the quaternary carbons (due to the lack of directly bonded H), are now very well detected, confirming attribution given in the one-pulse spectrum (Figure 1.276, curve 1). The OCH_2 peak due to the ethoxy groups has disappeared, which suggests that these groups are truly grafted to the surface, and not simply adsorbed. Interestingly, three sharp peaks are detected for the hexyl chains with relative intensities and line-widths different from what could be seen in the previous spectra. They can be assigned to the sites exhibiting the highest mobility, for example, –CH_2–CH_2–CH_3. These experiments clearly demonstrate the rather high mobility that CMP molecules at room temperature can experience within the mesoporous hexyl-modified silica network (Baccile and Babonneau 2008).

Jähnert et al. (2008) studied freezing and melting of H_2O and D_2O in the cylindrical pores of well-characterized MCM-41 silica materials with pore diameter, *d* from 2.5 to 4.4 nm using DSC and ^{1}H NMR cryoporometry. Well-resolved DSC melting and freezing peaks were obtained for pore diameters down to 3.0 nm, but not in 2.5 nm pores. The pore size dependence of the melting

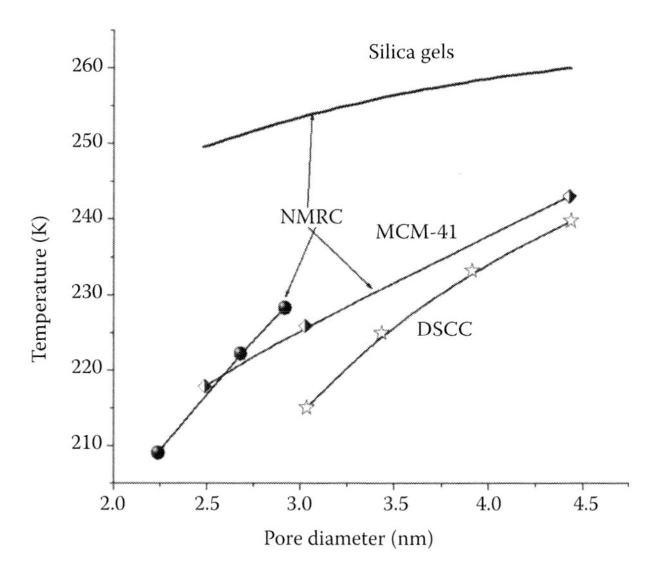

FIGURE 1.277 Comparison of the DSC melting temperatures with the NMR transition temperature for water in a different set of MCM-41 materials and silica gels. (Drawn according to *Phys. Chem. Chem. Phys.*, 10, Jähnert, S., Vaca Chávez, F., Schaumann, G.E., Schreiber, A., Schönhoff, M., and Findenegg, G.H., Melting and freezing of water in cylindrical silica nanopores, 6039–6051, 2008.)

point depression ΔT_m was represented by the Gibbs–Thomson equation (Jähnert et al. (2008) used $k_{GT}=51.9$ K nm) when the existence of a layer of nonfreezing water at the pore walls is taken into account. The DSC measurements also showed that the hysteresis connected with the phase transition, and the melting enthalpy of water in the pores, both vanish near a pore diameter $d^* \approx 2.8$ nm. They concluded that d^* represents a lower limit for first order melting/freezing in the pores. The NMR spin-echo measurements showed that a transition from low to high mobility of water molecules takes place in all MCM-41 materials, including the one with 2.5 nm pores, but the transition revealed by NMR occurs at a higher temperature than indicated by the DSC melting peaks. The disagreement between the NMR and DSC transition temperatures (Figure 1.277) becomes more pronounced as the pore size decreases. This is attributed to the fact that with decreasing pore size an increasing fraction of the water molecules is situated in the first and second molecular layers next to the pore walls, and these molecules have slower dynamics than the molecules in the core of the pore.

Jähnert et al. (2008) showed that there are the pronounced differences in the magnitude and pore size dependence of the transition temperature $T_c(d)$ of water obtained by the spin-echo method due in part to the chosen delay time τ (Figure 1.277). For instance, the τ value (2–20 ms) plays an important role as the pore size of the silica samples decreases. The melting point depression of water in silica pores of nominal pore size 4 nm was found to be 14 K when measured with $\tau=2$ ms, but only 6 K when measured with $\tau=20$ ms. This was attributed to a pore size dependence of the relaxation time T_2 of the confined liquid (Jähnert et al. 2008).

The interfacial behavior of water/organics (Franks 2000, Chaplin 2011), which can be very important on practical applications of porous materials as adsorbents, catalysts, etc., can be investigated by using both NMR cryoporometry and relaxometry techniques and TSDC method for deeper understanding of structurization of liquids confined in pores of ordered mesoporous silicas. Here, we will analyze results of a study of (i) the structural characteristics of synthesized ordered mesoporous silicas MCM-41, MCM-48, and SBA-15 by XRD, adsorption, and FTIR methods to explain the difference in their adsorption characteristics with respect to polar (water), weakly polar (chloroform), and nonpolar (benzene) adsorbates; (ii) the relationships between these characteristics and the behavior of adsorbed pure water and its mixtures with benzene or chloroform-*d* over a wide temperature range and on variation of the mixture composition by using the ^1H NMR spectroscopy (with both cryoporometry and relaxometry) with layer-by-layer freezing-out of bulk and pore liquids at 170–283 K; and (iii) concentrated aqueous suspensions of MCM-48 and SBA-15 by the TSDC method at 90–265 K.

Ordered mesoporous silica MCM-41 with spherical particles was synthesized using cetyltrimethyl ammonium bromide (CTAB) as a template and tetraethoxysilane (TEOS) as a precursor of silica (Gun'ko et al. 2007g). The samples were treated at 823 K for 5 h to remove the template. The structural characteristics of the synthesized two samples of MCM-41 are shown in Table 1.42. The pore size of MCM-41, as well as SBA-15 and MCM-48, was calculated on the basis of the XRD and adsorption data (according to Wang and Kabe 1999, Schumacher et al. 2000). Synthesis of MCM-48 was carried out using a reaction mixture with CTAB, TEOS, and NaOH at a molar ratio TEOS:CTAB:NaOH:H_2O = 1:0.5:0.54:110. The organic compounds were removed by calcination of the samples in air flow (9 L/h) initially heated at a heating rate 24 K/h and then at 823 K for 6 h. SBA-15 was synthesized by using a nonionic oligomeric alkyl-ethylene oxide surfactant (Pluronic P123) as a structure-directing agent dissolved in water and HCl at 313 K, and TEOS. The organic compounds were removed on calcination as in the case of MCM-48 preparation (Gun'ko et al. 2007g).

Benzene (melting point $T_{m,\infty,b} = 278$ K) and chloroform-*d* ($T_{m,\infty,c} = 209$ K) used in the NMR measurements are poorly soluble in water and vice versa because benzene is nonpolar (dielectric constant $\varepsilon = 2.27$) and chloroform is weakly polar ($\varepsilon = 4.71$). Chloroform-*d* was used to avoid contribution of protons of an organic solvent to the ^1H NMR signal.

Benzene was used in a mixture with water to measure the relationship between the amounts of unfrozen adsorbates dependent on temperature and PSD of silica.

Small-angle XRD patterns (Figure 1.278a) were recorded over $2\theta = 0.5°–10°$ range using DRON4-07 and DRON-3M diffractometers (LOMO and Bourevestnik, Inc., St. Petersburg) with Cu K_α ($\lambda = 0.15418$ nm) radiation and a Ni filter. A MCM-41 sample initial and treated at 473 K for 2 h was stirred with dry KBr (1: 20) to record the FTIR spectra by using a FTIR ThermoNicolet spectrophotometer over the 400–4000 cm^{-1} range. The FTIR spectra of MCM-48 and SBA stirred with dry KBr powder (1: 100) were recorded using a FTIR Nicolet Nexus spectrophotometer without thermal pretreatment of silicas (Gun'ko et al. 2007g).

The ^1H NMR spectroscopy with layer-by-layer freezing-out of bulk and interfacial liquids was used to determine the characteristics of interaction of water with silica surface affected by benzene or chloroform-*d*. The ^1H NMR spectra were recorded using a Varian 400 Mercury spectrometer and a Bruker WP-100 SY spectrometer of high resolution with the probing 90 or 40° pulses at duration of 4–5 μs. The transverse relaxation times were measured with CPMG pulse sequence: $90x$-(τ-$180y$-τ-echo)$_n$ with 90° pulse of 3 μs and total echo time of 500 μs. A total of eight scans were used for each measurement. The relaxation time between scans was 0.5 s. Equilibration time of each temperature was 5 min (Gun'ko et al. 2007g).

SBW can be unfrozen even on significant cooling of the system and corresponds to the maximum disturbed water layer at the interfaces. The thickness of the layers of each type of water or water + benzene (C_u^s and C_u^w denoting amounts of strongly and weakly bound liquids) and the maximum values for lowering of the Gibbs free energy of these liquids (ΔG_s and ΔG_w) can be estimated from a linear extrapolation of appropriate linear portions of the $\Delta G(C_u)$ graphs to the corresponding axis.

Notice that there are two types of water (parameter subscript "w") and benzene (subscript "b") or chloroform-*d* amounts: (i) unfrozen fraction ($C_u = C_{uw} + C_{ub}$) and (ii) total amounts of water (h), benzene (b), and chloroform-*d* (c) in mL per gram of dry silica. Derivative dC_u/dT (calculated using spline approximation of $C_u(T)$) as a function of temperature can be used to show characteristic changes in the amounts of structured unfrozen water and water/benzene dependent on temperature. A portion of unfrozen water or water/benzene corresponding to $\Delta G > -0.5$ kJ/mol may be considered as a liquid weakly bound in pores. According to Aksnes and Kimtys (2004), $k_{GT,b} = 44$ K nm for benzene freezing in porous silicas. For pure water adsorbed in pores of

TABLE 1.42
Structural Characteristics of MCM-41 (Two Samples), MCM-48, and SBA-15 Calculated from the Nitrogen Adsorption Isotherms

Sample	S_{BET}[a] (m²/g)	S_{BET}[b] (m²/g)	V_p (cm³/g)	S_{nano} (m²/g)	S_{meso} (m²/g)	S_{macro} (m²/g)	V_{nano} (cm³/g)	V_{meso} (cm³/g)	V_{macro} (cm³/g)	Δw_{cyl}[a]	χ[a] (%)	Δw_{cyl}[b]	χ[b] (%)	d_{N2} (nm)	d_{XRD} (nm)	d_{hkl} (nm)	d_{PSD} (nm)	t_{wall} (nm)
MCM-41(1)	969	788	0.604	13	956	1	0.003	0.583	0.018	0.38	27.5	0.17	14.5	2.45	3.11	3.40	3.36	0.82
MCM-41(2)	1196	997	0.702	26	1170	1	0.009	0.683	0.010	0.43	29.9	0.19	16.0	2.34	3.09	3.27	3.14	0.69
MCM-48	1324	1104	1.174	340	975	7	0.094	0.956	0.130	0.14	12.3	0.07	6.5	3.92	3.25*	3.53	3.76	0.69*
SBA-15	974	809	1.333	165	807	2	0.058	1.231	0.045	0.18	15.3	0.02	2.0	6.10	11.43	10.91	9.55	1.17

Note: Diameter mesopores $d_{N2}=4V_{meso}/S_{meso}$; d_{PSD} is the diameter of pores corresponding to the main maximum of the IPSD.

$d_{XRD} = 1.213 d_{100}\sqrt{\varepsilon}$ is the pore diameter for MCM-41 and SBA-15, $\varepsilon = \dfrac{\rho_0 V_p}{1+\rho_0 V_p}$; $\rho_0 = 2.2$ g/cm³ is the true density of amorphous silicas.

$d_{hkl} = \dfrac{\lambda}{2\sin\Theta_m}$ is the spacing value and Θ_m is the angle corresponding to (hkl) reflection peak.

$t_{wall} = a_0 - d_{XRD}$ is the thickness of pore walls and $a_0 = \dfrac{2d_{100}}{\sqrt{3}}$ is the distance between pore centers for MCM-41 and SBA-15.

$a_0 = d_{211}(h^2+k^2+l^2)^{0.5}$ for MCM-48.

[a] $s_{N2}=0.162$ nm²; [b] $s_{N2}=0.135$ nm²; *Parameters for MCM-48 were calculated from the d_{211} spacing.

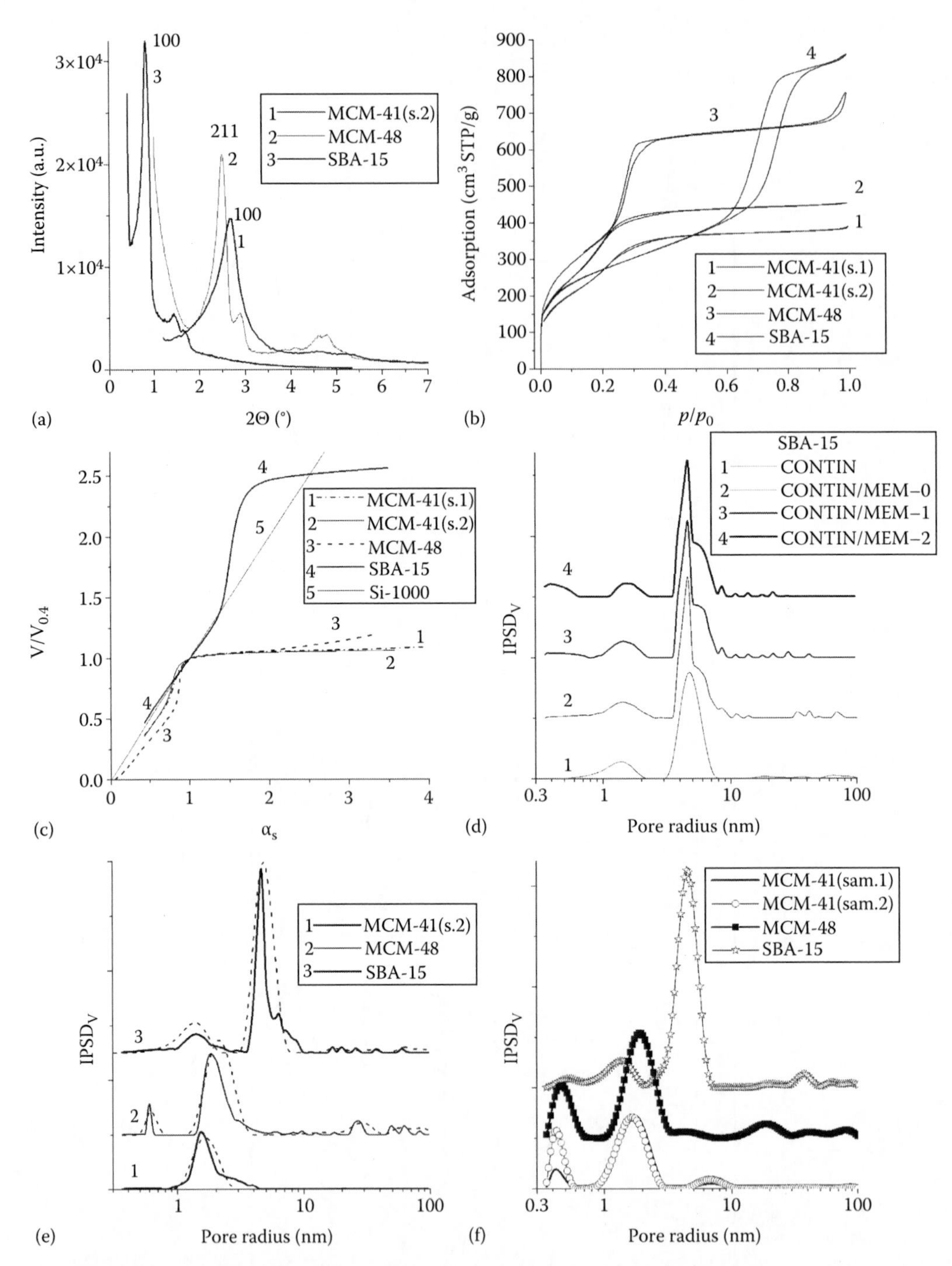

FIGURE 1.278 (a) XRD patterns of ordered mesoporous silicas; (b) nitrogen adsorption isotherms; (c) α_S plots (silica gel Si-1000 as a reference) corresponding to the nitrogen adsorption isotherms; and incremental pore size distributions for (d) SBA-15 with CONTIN and CONTIN/MEM-j and the model of cylindrical pores (d); and (e, f) MCM-41 ((e) sample 2 and (f) samples 1 and 2), MCM-48 and SBA-15 with respect of the pore volume with the model of (e) cylindrical pores (CONTIN/MEM-0 (solid line) and CONTIN (dashed line)) and (f) cylindrical pores plus voids between spherical particles (CONTIN) calculated with self-consistent regularization procedure. (Adapted from *Central Eur. J. Chem.*, 5, Gun'ko, V.M., Turov, V.V., Turov, A.V. et al., Behaviour of pure water and water mixture with benzene or chloroform adsorbed onto ordered mesoporous silicas, 420–454, 2007g. Copyright 2007, with permission from Springer.)

mesoporous silica, $k_{GT,w}=67$ K nm. The GT equation can be utilized to calculate the IPSD on the basis of the temperature dependence $C_u(T)$ estimated from the ^{1}H NMR spectra recorded at $T<T_{m,\infty}$ for individually adsorbed liquids. This approach was employed for different materials and gave the IPSDs close to those calculated on the basis of nitrogen adsorption/desorption isotherms (Gun'ko et al. 2005d). In the case of a mixture of immiscible liquids (e.g., water and benzene or chloroform-*d*), GT equation can be used for both liquids with consideration for their amounts in a mixture and their temperature-dependent signal intensity (obtained on deconvolution of the total signal):

$$\frac{dV_{u,i}}{dR} = \frac{A_i}{k_i}(T - T_{m,\infty,i})^2 \frac{dC_{u,i}}{dT}, \tag{1.70}$$

where i denotes an adsorbate number, $C_{u,i}(T)$ is the integral intensity of a δ_H band for the ith adsorbate as a function of temperature, and A_i is a weight constant dependent on the molecular volume (v_i), the number of protons (n_i) in a molecule of the ith adsorbate, and the used units. Notice that $\frac{v_w}{n_w} \approx \frac{v_b}{n_b}$ because of $v_b/v_w \approx 2.83$ and $n_b/n_w = 3$. In the case of the water/benzene mixtures, the $\sigma_{sl,w}$ value determined for the water/silica system was multiplied by $\Delta G_{s,h,b}/\Delta G_{s,h}$ (where $\Delta G_{s,h}$ and $\Delta G_{s,h,b}$ are changes in the free energy of SBW without and with the presence of benzene respectively) for consideration for the effect of benzene. Clearly, this effect leads to a certain diminution of the k_w value varied between 67 and 40 K nm for different water/benzene mixtures. For simplicity, the k_b value was assumed as a constant (44 K nm) for all the studied systems because benzene is being in contact with the pore walls since it can displace water from pores and the pore walls. Equation 1.70 allows us to calculate the distribution functions of the sizes of pores filled by an individual adsorbate (e.g., water) and an incremental size distribution (ISD) of structures in a liquid mixture (e.g., water/benzene) filling pores. In the last case, the pore size can correspond to a sum of the sizes (thickness) of the pore liquid layers if liquids simultaneously locate in the same pores. In the case of a mixture of water with chloroform-*d*, the latter does not contribute the ^{1}H NMR spectra. However, chloroform-*d* can fill a portion of pores and affect their occupation by water, that is, change the $C_{uw}(T)$ shape. Equation 1.70 can be transformed into integral equation (IGT), replacing dV/dR by $f(R)$, converting dC/dT to dC/dR with GT equation and integrating by R,

$$C_{u,i}(T) = A \int_{R_{min}}^{R_{max}} \left(\frac{k_i}{(T_{m,\infty,i} - T_{m,i}(R))R} \right)^2 f_i(R)dR \tag{1.71}$$

where

R_{max} and R_{min} are the maximal and minimal pore radii (or sizes of unfrozen liquid structures), respectively
i is the index corresponding to ith pore liquid
A is a normalization factor

In the case of the use of the relaxometry method for estimation of the sizes of the water and benzene structures (layers, clusters, etc.) filling pores, Equation 10.9 can be transformed for consideration for the dependence of the CPMG echo decay envelopes (i.e., transverse relaxation time) on the pore size:

$$I_i(t) = B \int_{R_{min}}^{R_{max}} \exp\left(-\frac{t}{T_{2,i,m}} \right) \frac{(T_{m,\infty,i} - T_{m,i}(R))}{k_i} f_i(R)dR, \tag{1.72}$$

where

B is a normalization factor

k_i is a constant analogous to that in GT equation

Equations 1.71 and 1.72 can be solved using the regularization procedure based on the CONTIN algorithm (Provencher 1982).

Pores of MCM-41 (sample 2 was chosen for the NMR measurements because it has larger V_p and S_{BET} values than sample 1, Table 1.42) were partially filled by water at hydration $h=0.186$ or 0.317 mL of water per gram of dry MCM-41, then a portion of benzene ($b=0.196$ mL per gram of dry MCM-41) or chloroform-*d* ($c=5.0$ mL/g) was added, then a portion of water was added that gives $h=0.37$ mL/g, and then a portion of benzene was added that gives $b=0.4$ mL/g. Sample at $h=0.37$ mL/g and $b=5.8$ mL/g was prepared on the first addition of water ($h=0.37$ mL/g) and then benzene ($b=5.8$ mL/g). In the case of MCM-48 ($h=0.2$ mL/g and $b=0.2$ mL/g) and SBA-15 ($h=0.16$ mL/g and $b=0.13$ mL/g), water was adsorbed before benzene.

LT (77.4 K) adsorption/desorption isotherms of nitrogen (Figure 1.278b), which is type IV in the IUPAC classification (Gregg and Sing 1982) (however, for MCM-41 it tends to type I), were measured using a Micromeritics ASAP 2405N adsorption analyzer. The specific surface area S_{BET} (Table 1.42) was calculated according to the standard BET method but at different pressure ranges, $p/p_0=0.05$–0.17 (MCM-41), 0.05–0.2 (MCM-48), and 0.05–0.23 (SBA-15), to avoid overestimation of the Brunauer-Emmett-Teller (BET) S_{BET} values because of the beginning of the formation of the second layer with adsorbed nitrogen in narrow mesopores at lower pressures (Thommes et al. 2002). The dependence of this effect on the used pressure range results in different dependences of the S_{BET} values on the high boundary value $p/p_{0,max}$ for the studied silicas (Figure 1.279). The pore volume V_p was estimated at $p/p_0 \approx 0.98$–0.99 (where p and p_0 denote the equilibrium and saturation pressures of nitrogen, respectively), converting the adsorbed amount $a_{0.98}$ (in cm^3 STP of gaseous nitrogen per gram of the adsorbent) to the liquid adsorbate $V_p \approx 0.0015468a_{0.98}$.

PSDs ($f_V(R)$ and $f_S(R)$ with respect to the pore volume and the specific surface area, respectively) were calculated using overall isotherm equation based on equation proposed by Nguyen and Do for

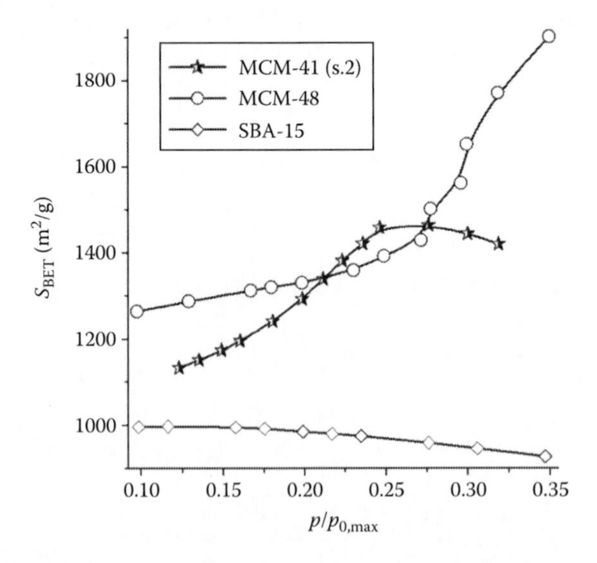

FIGURE 1.279 S_{BET} as a function of the maximal p/p_0 value of the pressure range (0.05 – $p/p_{0,max}$) used on calculation of the S_{BET} values for MCM-41 (sample 2), MCM-48 and SBA-15. (With kind permission from Springer Science+Business Media: *Central Eur. J. Chem.*, Behaviour of pure water and water mixture with benzene or chloroform adsorbed onto ordered mesoporous silicas, 5, 2007g, 420–454, Gun'ko, V.M., Turov, V.V., Turov, A.V. et al., Copyright 2007.)

carbon adsorbents with slit-like pores (Nguyen and Do 1999) and modified for cylindrical pores and voids between spherical particles (Gun'ko and Mikhalovsky 2004). This equation was solved by means of a regularization procedure based on the CONTIN algorithm. The differential PSDs $f_V(R)$ and $f_S(R)$ were converted to incremental PSDs (IPSD) (Figure 1.278e and f). The $f_V(R)$ and $f_S(R)$ functions were used to calculate contributions of nanopores (S^*_{nano}, V_{nano}) at $R<1$ nm, mesopores (S^*_{meso}, V_{meso}) at $1<R<25$ nm, and macropores (S^*_{macro}, V_{macro}) at $25<R<100$ nm to the specific surface area and the total porosity. To compute the α_S plots (Figure 1.278c), silica gel Si-1000 was used as a standard adsorbent (Jaroniec et al. 1999).

The tablets (diameter 30 mm, thickness ~1 mm) with the frozen aqueous suspension of silicas (16.7 wt%, i.e., $h=5$ mL/g) were polarized by the electrostatic field at the intensity $F_p=200$–300 kV/m at 260 K then cooled to 90 K with the field still applied and heated without the field at a heating rate $\beta=0.05$ K/s. The current evolving due to sample depolarization was recorded by an electrometer over the 10^{-15}–10^{-7} A range. Relative mean errors for measured TSD current were ±5% for the current, ±2 K for temperature, and ±5% for the temperature change rate. Modified GT equation with k_{GT} as a linear function of temperature ($k_{GT}(T)=40+\frac{5}{6}(T-90)$ K nm at T between 90 and 270 K) obtained on the basis of the calibration curves for silica gels Si-40 and Si-60 was used for estimations of the IPSDs on the basis of the TSDC data.

Figure 1.278a shows XRD patterns for MCM-41 (observed reflections suggest that it can be assigned to 2D hexagonal structure), SBA-15 (hexagonal structure), and MCM-48 (3D cubic structure). For hexagonal structure the lattice constant $a_0=\frac{2d_{100}}{\sqrt{3}}$ corresponding to the distance between pore centers, the pore size $d_{XRD}=1.213d_{100}\sqrt{\frac{\rho_0 V_p}{1+\rho_0 V_p}}$ ($\rho_0=2.2$ g/cm^3 is the true density of amorphous silica), and the pore wall thickness $t_{wall}=a_0-d_{XRD}$ can be estimated from the d_{100} spacing value. The d_{XRD} and t_{wall} values for MCM-48 (Table 1.42) were calculated using the d_{211} spacing value.

The average pore wall thickness is between two (0.62 nm) and four (1.24 nm) silica layers for the studied samples (Table 1.42, t_{wall}). The pore sizes estimated from the XRD (Table 1.42, d_{XRD}) and adsorption data (d_{N2} and d_{PSD}) are close (with one exception of SBA-15). A significant difference between the d_{N2} and d_{XRD} values for SBA-15 is due to contribution of narrow mesopores at $1<R<2$ nm (Figure 1.278e and f) (these pores can be attributed to inner-wall pores), which are out of the main mesopore peak, on calculation of d_{N2}. The d_{PSD} value for the main mesopore peak for SBA-15 is closer to d_{XRD} than d_{N2}. Calculation of the d_{N2} value only for this peak gives 9.92 nm, which is closer to the d_{XRD} value. Consequently, on the calculation of the size of the main mesopores from the adsorption data, not only external surface and the corresponding pore volume but also the surface and the volume of narrow pores should be taken into account. Notice that a certain difference between the pore sizes determined from the XRD and adsorption data can be caused by neglect or accounting of the size of surface atoms.

The shapes of the nitrogen adsorption/desorption isotherms and hysteresis loops (Figure 1.278b) reveal that MCM-48 and SBA-15 have the porosity of two types (Tanev and Pinnavaia 1996): (i) framework-confined porosity caused by relatively uniform channels of the templated framework (tested by the presence of a significant adsorption step at p/p_0 between 0.1 and 0.5) and (ii) textural porosity caused by intra-aggregate voids and spaces formed by interparticle contacts (tested by the presence of the hysteresis loop at $p/p_0>0.5$). The presence of the textural porosity is also confirmed by the IPSD for MCM-48 and SBA-15 characterized by certain contribution of large mesopores and macropores (Figure 1.278d through f). The porosity of the second type for MCM-41 is very small because the isotherm does not include the hysteresis loop and the adsorption step at $p/p_0>0.5$ (Figure 1.278b). The presence of two types of the porosity can affect the interfacial behavior of water and adsorbate mixtures.

The Δw_{cyl} values (Table 1.42) suggest that a pore wall surface is not smooth because the specific surface area (S^*) calculated using the model of cylindrical pores is smaller than the S_{BET} value, especially in the case of MCM-41 samples. Notice that the use of the S^*_{meso} value which was not multiplied by $(\Delta w_{cyl}+1)$ on calculation of the diameter of the mesopores gives 3.33 nm for MCM-41 (sample 2). This value is close to d_{XRD} and d_{PSD}. There are, at least, two reasons resulting in $S^*<S_{BET}$: (i) certain roughness of the pore walls, which cannot be described in the framework of the model of ideal cylindrical pores, and (ii) overestimation of the S_{BET} value because of the nitrogen adsorption in the second and next layers (i.e., pore filling) at $p/p_0<0.3$ and $d<4$ nm and because of overestimation of the cross-sectional area for a nitrogen molecule ($s_{N2}=0.162$ nm^2) for hydroxylated silica surfaces. For instance, this area of 0.135 nm^2 gives better agreement with the S_{BET} values obtained from the Ar adsorption data (Thommes et al. 2002). Diminution of the $p/p_{0,max}$ values on the calculations of the S_{BET} values allows us to reduce the second effect. The use of reduced $s_{N2}=0.135$ nm^2 gives much smaller S_{BET} and Δw_{cyl} values (Table 1.42); however, inequality $S^*<S_{BET}$ remains. Therefore, one can assume that this residual inequality $S^*<S_{BET}$ is due to a nonideal shape of the pores, for example, the roughness of the pore walls. Consequently, consideration for the roughness of the pore walls, which can be estimated as $\chi=100(S_{BET}-S^*)/S_{BET}=100\Delta w_{cyl}/(\Delta w_{cyl}+1)$ (%) (Table 1.42), is important on estimation of the pore size from the adsorption data. For instance, the χ value is maximal for MCM-41 samples; therefore underestimation of the d_{N2} values is maximal for these silicas. In other words, the pore walls of MCM-41 samples are rougher than that of MCM-48 or SBA-15. However, according to the IPSDs obtained with the model of cylindrical pores contribution of nanopores (which can be also linked to the roughness of the pore walls) for MCM-41 is lower than that for other silicas. This roughness can be modeled by tiny spherical-like hills and the corresponding shallow hollows on the surface of cylindrical mesopores; that is, the pore walls are composed of spherical globules (~1 nm in diameter). Additionally, the textural porosity can be modeled as voids between spherical globules of larger sizes (10–100 nm or more). This allows us to use a complex model of pores of a cylindrical shape and voids between spherical particles with self-consistent regularization procedure used to obtain independent $f(R)$ functions for both pore shapes. The complex model of pores adds intensity of the IPSD of nanopores for all samples (Figure 1.278f); however, the main peaks of mesopores remain nearly the same. This model gives the Δw values much lower (e.g., 0.01 for MCM-48 and 0.12 for sample 2 of MCM-41) than that obtained with the model of cylindrical pores with different s_{N2} values.

Thus, the studied samples of MCM-41 (especially sample 2) do not have the ideal conventional structure. For instance, the nitrogen adsorption/desorption isotherm does not have a large adsorption/desorption step (Figure 1.278b) characteristic of MCM-41 synthesized using $C_{16}TAB$ but without alcohol. However, this step is more clearly observed for sample 1 than for sample 2 of MCM-41 (Figure 1.278b). Alcohol in the reaction solution interacting with the template molecules can affect their micellar/tubular supramolecular structure and, therefore, the finished mesostructure of silica, and the roughness of the pore walls increases (especially on additional aging of sample 2). Notice that similar XRD patterns and nitrogen adsorption/desorption isotherms were previously observed for MCM-41 synthesized with alcohol in the reaction solution (Lind et al. 2003). The synthesis of MCM-41 without alcohol gives materials characterized by a typical shape of the nitrogen adsorption isotherms with a typical step shape. The reaction time effect is well observed on comparison of the structural characteristics of two MCM-41 samples (Table 1.42 and Figure 1.278) synthesized under nearly the same conditions only at different time of the filtration stage after slurry stirring.

The IPSDs (Figure 1.278d through f) demonstrate the presence of large and narrow pores in addition to the main narrow mesopores for all the studied silicas. The IPSDs of MCM-48 and SBA-15 shift toward larger pore sizes in comparison with MCM-41. The α_S plots (Figure 1.278c) and the IPSDs (Figure 1.278d through f) reveal certain (however, relatively small) contribution of nanopores at $R<1$ nm. Notice that the model of cylindrical pores (Figure 1.278e) (with both CONTIN and CONTIN/MEM-0 regularizations) does not practically give nanopores for MCM-41 in contrast to the complex model of pores (Figure 1.278f). Calculations of the IPSDs for SBA-15 with CONTIN/MEM-j at $j=0$, 1, and 2 (Figure 1.278d) show that the main differences are observed at $R<0.7$ nm and $R>30$ nm. Therefore, calculations for other silicas were carried out with

CONTIN/MEM-0 (Figure 1.278e). Thus the studied silicas are characterized by different IPSDs; therefore the behavior of confined liquids in these adsorbents can depend differently on temperature at $T<273$ K. To study the influence of benzene and chloroform-*d* on the structure of bound water, the LT NMR measurements of static samples were started from the amounts of adsorbed liquids much smaller than the pore volume of silicas.

In the case of partial filling of pores by water and benzene, the latter tends to displace a portion of water from narrow pores to larger ones or from the pore walls to the center of pores. This is well seen on comparison of the ^{1}H NMR spectra of adsorbed pure water and water/benzene mixture (Figure 1.280). The ^{1}H NMR spectra show a stronger decrease in the signal intensity of the proton resonance of adsorbed water at $\delta_H \approx 5$ ppm (SAW) at $T<230$–240 K with the presence of benzene.

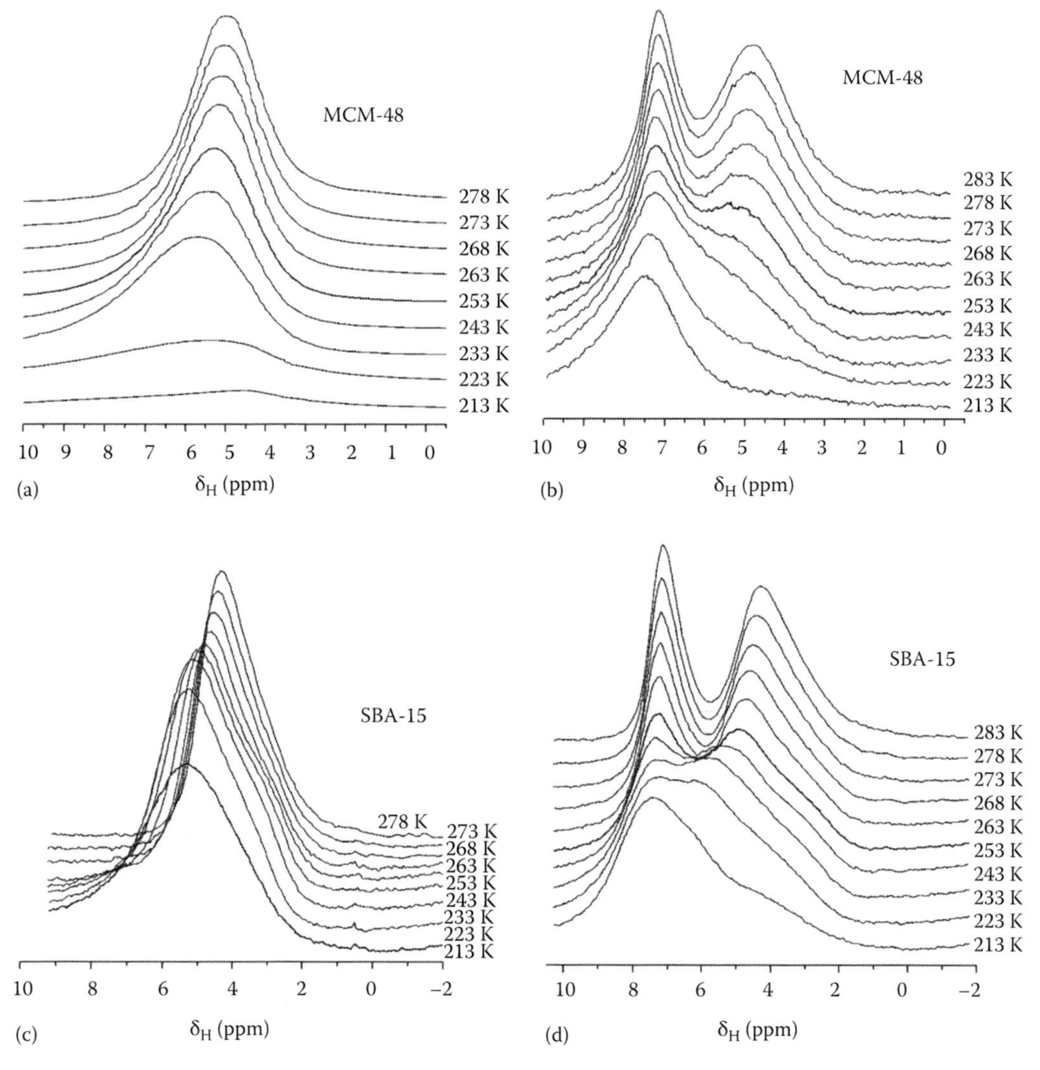

FIGURE 1.280 Chemical shit of proton resonance of (a) water (h=0.2 mL/g) and (b) water/benzene mixture (h=0.2 mL/g, b=0.2 mL/g) adsorbed onto MCM-48; (c) water (h=0.16 mL/g) and (d) water/benzene mixture (h=0.16 mL/g, b=0.13 mL/g) adsorbed onto SBA-15. (With kind permission from Springer Science+Business Media: *Central Eur. J. Chem.*, Behaviour of pure water and water mixture with benzene or chloroform adsorbed onto ordered mesoporous silicas, 5, 2007g, 420–454, Gun'ko, V.M., Turov, V.V., Turov, A.V. et al., Copyright 2007.)

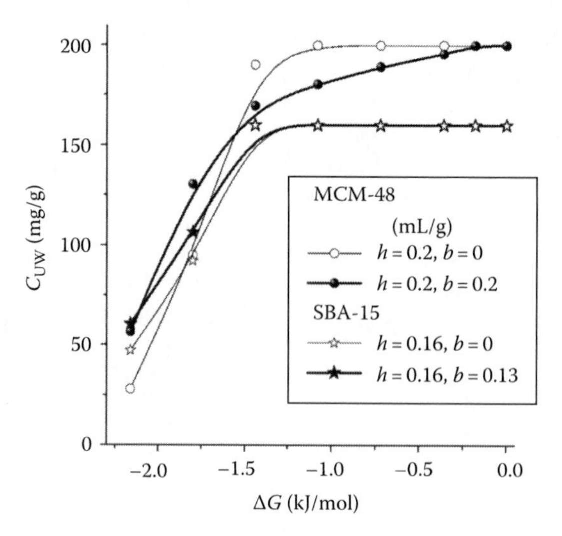

FIGURE 1.281 Relationships between amounts of unfrozen water and changes in the Gibbs free energy for water and water/benzene adsorbed onto MCM-48 and SBA-15. (With kind permission from Springer Science+Business Media: *Central Eur. J. Chem.*, Behaviour of pure water and water mixture with benzene or chloroform adsorbed onto ordered mesoporous silicas, 5, 2007g, 420–454, Gun'ko, V.M., Turov, V.V., Turov, A.V. et al., Copyright 2007.)

Therefore, the relationship between the ΔG and C_{uw} values (Figure 1.281) demonstrates diminution of the C_{uw} value for the water/benzene mixture in comparison with pure water at the same ΔG and temperature values. These effects can be explained by the displacement of water by benzene from the pore walls or narrow pores into larger ones (e.g., textural pores discussed earlier). These effects can seem unusual since water can form strong hydrogen bonds with a silica surface in contrast to benzene. However, dispersive interaction of benzene with siloxane bonds can be characterized much higher energy than that for water. Additionally, there is a phase boundary between water and benzene with high Gibbs free energy. To reduce this energy and the area of contact between water and benzene, the latter can displace water into large pores where water forms larger clusters and domains than in narrow pores. Typically water tends to form clustered coverage of a silica surface and a continuous layer is not formed at relative high hydration $h < 0.3$ g/g (Gun'ko et al. 2005d). Such clustered coverage of a silica surface by water leads to a great area of contact between water clusters and benzene. Therefore, formation of much larger water structures (in larger pores) with smaller contact area with the benzene phase is energetically favorable.

The $f(T_2)$ distribution functions (Figure 1.282) demonstrate that the proton transverse relaxation of confined benzene is more complex (as characterized by bi- or trimodal $f(T_2)$ distributions) in comparison with bound water (mainly monomodal $f(T_2)$ distributions). Additionally, there are both faster and slower relaxation processes of benzene in comparison with water. These results can be caused by the difference in the spatial distribution of benzene and water in pores of different sizes and the effects of the pore size and the structure of the pore walls on the T_2 values because of the corresponding changes in the molecular mobility of liquids undergoing different confined space effects. This conclusion is confirmed by size calculations of the bound water and benzene structures (Figures 1.283 through 1.285).

The analysis of the $f(T_2)$ and ISD graphs for pore liquids reveals that these liquids can (i) fill narrow pores (e.g., nanopores at $R < 1$ nm), (ii) adsorb onto the pore walls in the form of adsorbed films in larger pores ($R > 1$ nm) in which certain free space remains far from the pore walls because of partial filling of pores, and (iii) form clusters, nanodomains, droplets, films, or continuous condensate in large mesopores ($R > 10$ nm) or macropores ($R > 25$ nm). The thickness (size) of the layers (clusters) with unfrozen liquids decreases with lowering temperature (Figures 1.283 and

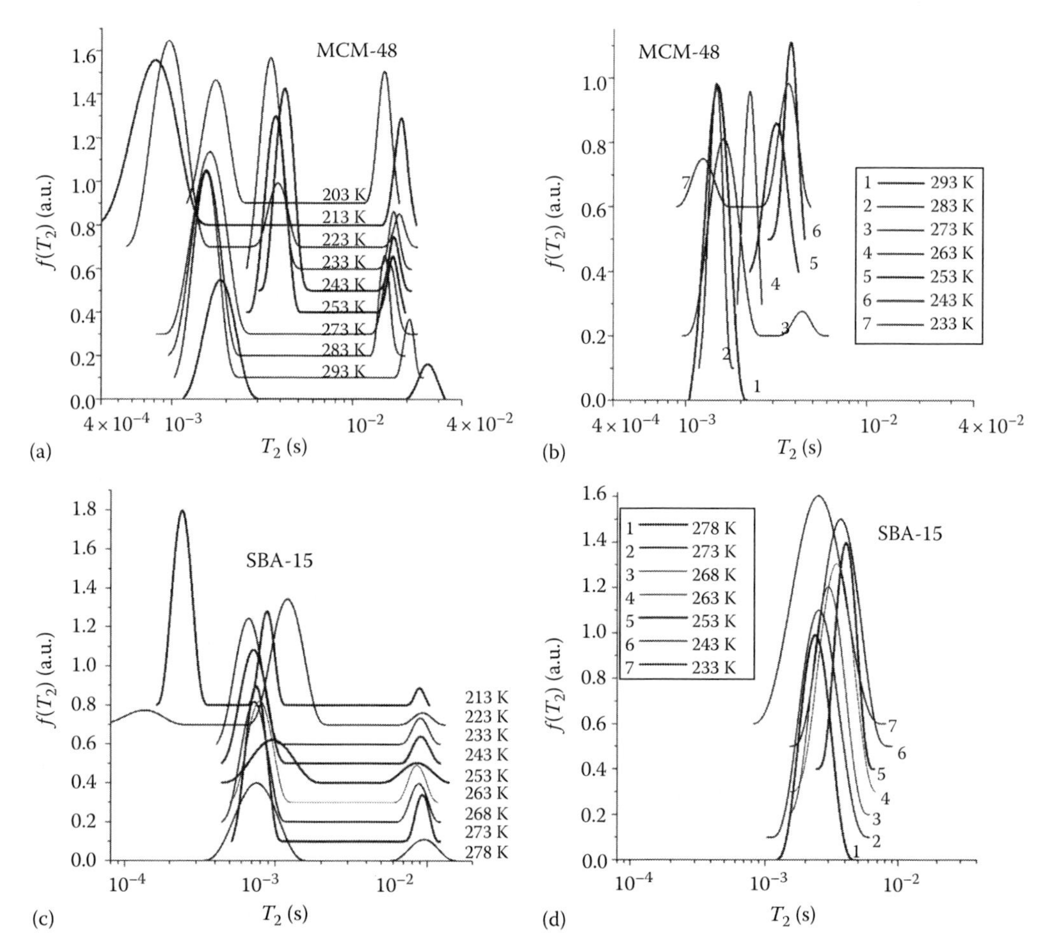

FIGURE 1.282 Distribution functions of proton transverse relaxation time (T_2) of (a) benzene and (b) water adsorbed onto MCM-48 as a mixture with C_6H_6 and H_2O (h=0.2 mL/g, b=0.2 mL/g); (c) benzene and (d) water adsorbed onto SBA-15 as a mixture at h=0.16 mL/g and b=0.13 mL/g. (With kind permission from Springer Science+Business Media: *Central Eur. J. Chem.*, Behaviour of pure water and water mixture with benzene or chloroform adsorbed onto ordered mesoporous silicas, 5, 2007g, 420–454, Gun'ko, V.M., Turov, V.V., Turov, A.V. et al., Copyright 2007.)

1.284) because of partial (layer-by-layer) freezing-out of liquids confined in different pores. Pore filling by the liquids is larger for MCM-48 than for SBA-15 because of the difference in the pore volume of silicas (Table 1.42, V_p) and in the amounts of adsorbates. The structures formed in MCM-48 and SBA-15 have the sizes close to that of the main mesopores (Figure 1.285); however, the ISD curves (normalized to the volume of adsorbates) are below the IPSDs (normalized to the pore volume) because of partial filling of pores by adsorbates.

Similar results with respect to various structurization of adsorbed water in pores of different sizes were obtained at much higher hydration of these samples ($h \approx 5$ mL/g) by using the TSDC method (Figure 1.286). The relaxing water structures correspond to mesopore sizes of MCM-48 but they are slightly smaller than the pore size of SBA-15. The dipolar and dc relaxations of water adsorbed on different mesoporous silicas are characterized by two typical LT and HT bands in the TSDC thermograms (Figure 1.286a). The effect of pore walls is minimal for a fraction of water located in large pores, for example, of SBA-15 (Figure 1.286c) or silica gel Si-60 (discussed in Section 1.2). MCM-48 and SBA-15 (with a more ordered pore structure than that of silica gels) have a stronger effect on the adsorbed water than silica gel and the LT peak shifts toward higher

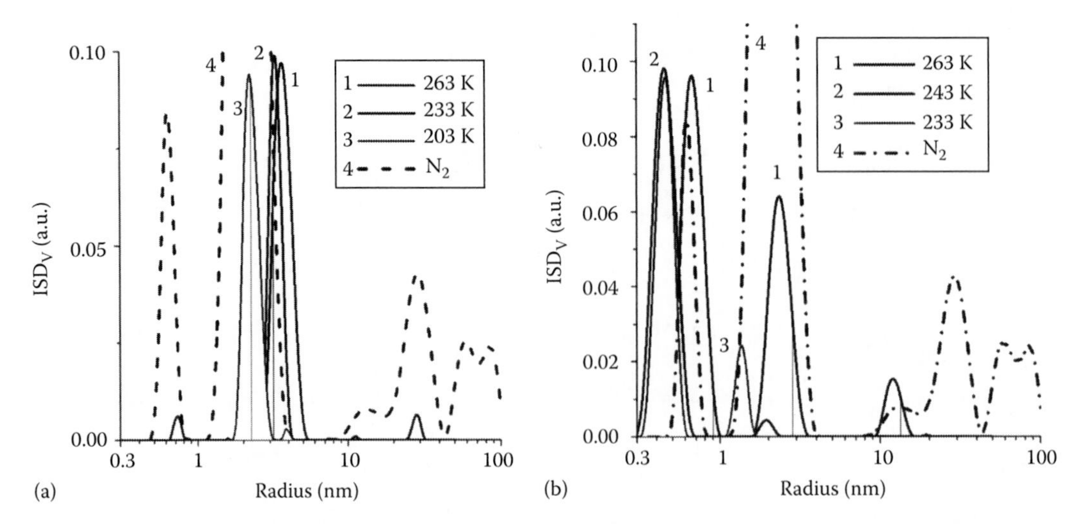

FIGURE 1.283 Pore size distributions calculated on the basis of the nitrogen adsorption isotherm and ISDs based on CPMG echo decay envelopes for (a) benzene and (b) water adsorbed onto MCM-48 as a mixture with 200 mg C_6H_6 and 200 mg H_2O per gram of dry silica. (With kind permission from Springer Science+Business Media: *Central Eur. J. Chem.*, Behaviour of pure water and water mixture with benzene or chloroform adsorbed onto ordered mesoporous silicas, 5, 2007g, 420–454, Gun'ko, V.M., Turov, V.V., Turov, A.V. et al., Copyright 2007.)

temperatures (Figure 1.286a). The dipolar relaxation of water frozen in pores of different sizes occurs step-by-step if the pore size is larger than the size of relaxing water structures. For instance, in the case of SBA-15 with relatively large pores, the size of water clusters relaxing at certain temperatures is smaller than the size of pores (Figure 1.286c) because water film freezes at the pore walls at lower temperatures (it is in the liquid state under the applied field for a longer time) than layers far from the pore walls. A similar effect was observed on the ^{1}H NMR investigations of water in silica gels.

In the case of MCM-48 (Figure 1.286b) with narrower pores than SBA-15, water can freeze in pores over a narrow temperature range (a similar effect was observed for silica gel Si-40) but at lower temperatures. A small amount of water (small clusters) can be frozen at the pore walls at lower temperatures. According to the TSDC data, mobile water molecules (participating in the dc relaxation on proton transferring) appear at $T \approx 210$ K (Figure 1.286a) that is in agreement with the ^{1}H NMR spectra (Figure 1.280) showing that mobile water molecules appear at $T \geq 213$ K.

Thus freezing/thawing of pore liquids in relative large mesopores of SBA-15, especially on total filling of pores (on the TSDC measurements), can occur step-by-step. This can result in very nonuniform frozen/unfrozen structures of liquids mixed in pores. To diminish these effects, the pore liquids can be studied on the adsorption onto MCM-41 possessing narrower pores than MCM-48 and SBA-15 (Figure 1.278). For the NMR measurements, sample 2 of MCM-41 was used because it has a larger adsorption capacity, narrower pores, and a larger roughness of the pore walls in comparison with other samples (Table 1.42) that can provide the appearance of unusual WAW. Additionally, water adsorbed alone in narrower pores at partial filling of them can form smaller clusters that cause a larger area of boundary contacts between water and subsequently adsorbed benzene and stronger effects of rearrangement of water clusters to reduce the boundary contact area with benzene.

The ^{1}H NMR spectra of the water/benzene mixtures at different h and b values include signals of water (~5 ppm) and benzene (~7 ppm) differently dependent on temperature (Figure 1.287) similarly to those for MCM-48 and SBA-15 (Figure 1.280).

In the case of the sample at $h = 0.37$ mL/g and $b = 5.8$ mL/g, a low-intensity signal is also observed at $\delta_H = 1.2$–1.4 ppm (Figure 1.287d), which corresponds to WAW. Notice that the roughness of the

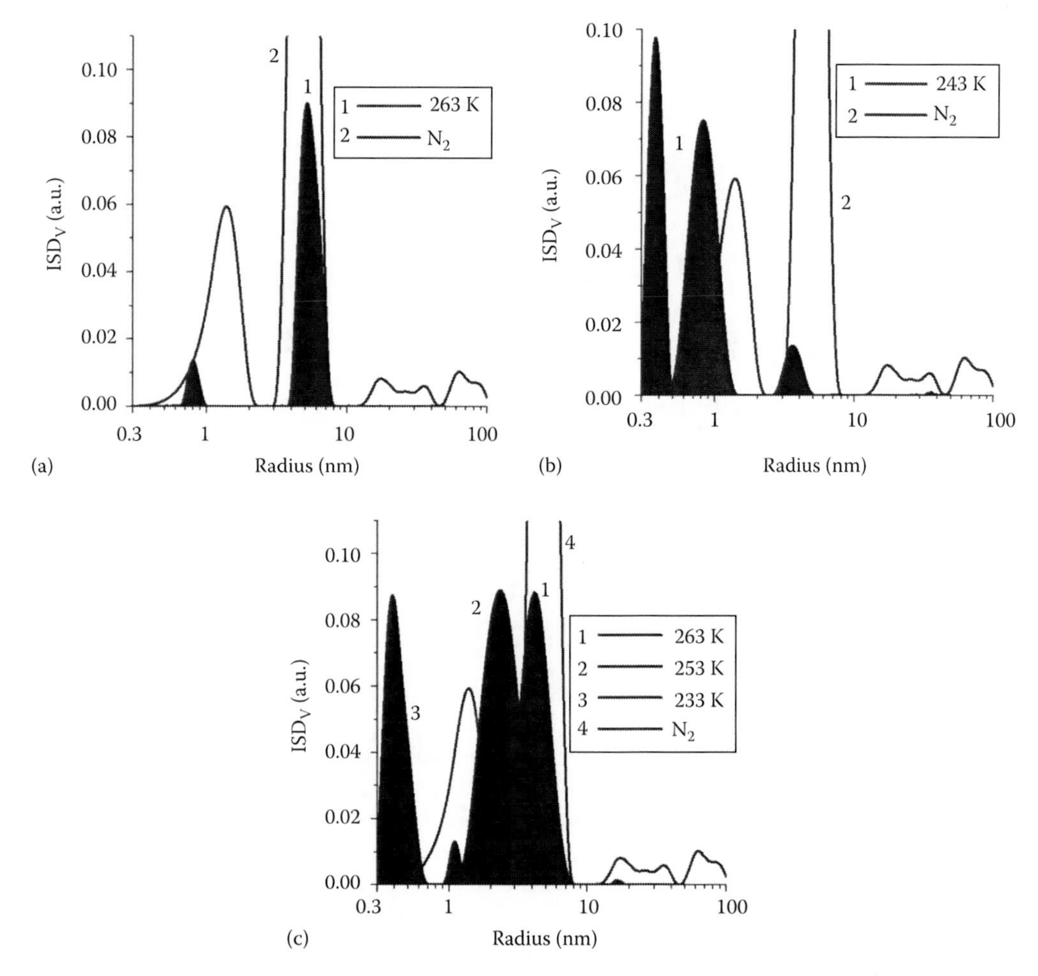

FIGURE 1.284 Incremental size distributions of unfrozen structures calculated on the basis of CPMG echo decay envelopes for (a and b) benzene and (c) water adsorbed onto SBA-15 as a mixture at $h=0.16$ mL/g and $b=0.13$ mL/g; $IPSD_V$ based on the nitrogen adsorption isotherm is also shown. (With kind permission from Springer Science+Business Media: *Central Eur. J. Chem.*, Behaviour of pure water and water mixture with benzene or chloroform adsorbed onto ordered mesoporous silicas, 5, 2007g, 420–454, Gun'ko, V.M., Turov, V.V., Turov, A.V. et al., Copyright 2007.)

pore walls (providing formation of very small clusters of WAW) can be one of necessary conditions for observation of similar water. The main peak of adsorbed water at 5 ppm (SAW) is close to that of liquid water characterized by the average number of the hydrogen bonds per molecule close to four. Consequently, the major portion of water in the pores of MCM-41 is strongly associated because it keeps the hydrogen bond network, causing the mentioned magnitude of δ_H. If the summary volume of the adsorbates (at $h=0.186$ mL/g and $b=0.196$ mL/g) is substantially lower than the pore volume V_p, the relationship of the signal intensities of water and benzene slightly decreases with lowering temperature to 223 K. This is due to filling of the narrowest pores that provides the maximal freezing point depression. The intensity of the water and benzene signals weakly changes at 220–280 K (Figure 1.288c, $h=0.186$ mL/g and $b=0.196$ mL/g).

In this case, the temperature dependence of signal is nearly the same as for pure water adsorbed on MCM-41 in air at $h=0.186$ mL/g. Notice that $C_u(T)$ in Figure 1.288c and d corresponds to the sum $C_u=C_{uw}+C_{ub}$ for unfrozen water (subscript "uw") and benzene (subscript "ub"). Figure 1.288e shows only the $C_{uw}(T)$ graphs calculated by deconvolution of the ^1H NMR spectra.

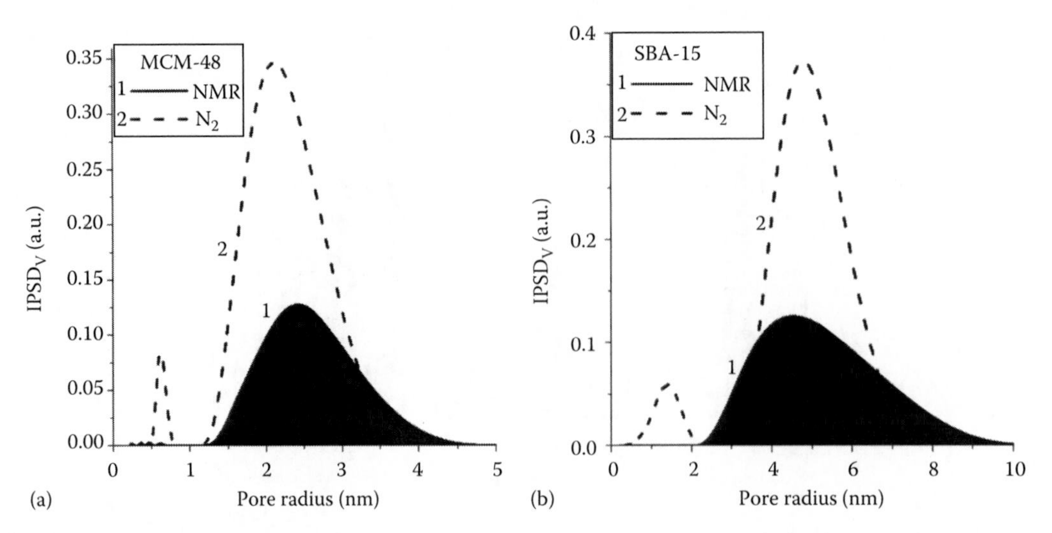

FIGURE 1.285 IPSDs for (a) MCM-48 and (b) SBA-15 calculated on the basis of the nitrogen adsorption isotherms and the ISDs based on the ^{1}H NMR spectra of water and water/benzene calculated using the IGT/MEM-0 method. (With kind permission from Springer Science+Business Media: *Central Eur. J. Chem.*, Behaviour of pure water and water mixture with benzene or chloroform adsorbed onto ordered mesoporous silicas, 5, 2007g, 420–454, Gun'ko, V.M., Turov, V.V., Turov, A.V. et al., Copyright 2007.)

If the volume of the adsorbates is greater than the total pore volume (e.g., $h = 0.37$ mL/g and $b = 5.8$ mL/g), a portion of benzene locating out of pores can freeze at T between $T_{m,\infty,b} \approx 278$ K and $T_{m,\infty,w} \approx 273$ K because of a certain freezing point depression for benzene mixed with water even out of pores since $T_{m,\infty,b} > T_{m,\infty,w}$. Composition of the frozen portions of liquids at $T < T_{m,\infty,i}$ depends on the affinity of the adsorbates to the pore walls, a clustered structure of preadsorbed water, order of the loading of liquids into pores, and a period of time before measurements (if the equilibrium state is not reached before freezing). According to the data shown in Figure 1.287e, the relationship of the signal intensities of water and benzene is practically identical at 243 K independent of the initial relationships of the amounts of components ($h = 0.186$ mL/g and $b = 0.196$ or 0.4 mL/g). In the case of a small excess of the liquids ($h = 0.37$ mL/g and $b = 0.4$ mL/g) and the same amount of water but a great excess of benzene ($h = 0.37$ mL/g and $b = 5.8$ mL/g), the main signals at ~5 ppm (shoulder related to SAW) and ~7 ppm (peak related to benzene) have a similar shape; however, an additional signal of WAW appears at 1.2–1.5 ppm. This suggests that a portion of water is strongly clustered, that is, it is distributed in the form of very small clusters (locating in shallow nanopores in rough walls of mesopores) or individual molecules dissolved in benzene in silica pores. It is possible that both the roughness of the pore walls of MCM-41 and residual organic functionalities promote appearance of WAW (Gun'ko et al. 2005d). However, the chemical shift of the last signal does not practically change with temperature (Figure 1.287d) in contrast to that of water dissolved in benzene and characterized by decreased signal intensity with lowering temperature. Consequently, there are certain effects of the silica surface (nanopores and residual organic groups), that is, its structure and nature, on WAW. Thus, residual organic groups at the silica surface, the roughness of the pore walls of MCM-41 (sample 2) (Table 1.42, χ), and significant excess of the organic solvent can facilitate the appearance of WAW characterized by signal at $\delta_H = 1.2$–1.5 ppm.

In all the case of the water/benzene mixtures, a portion of water is displaced from the pore walls and narrow pores into larger ones (Figures 1.288 through 1.290) to reduce the contact area between these immiscible liquids.

Adsorbed pure water forms several types of structures (e.g., clusters, thin surface films, and structures more completely filling certain narrow pores) because the ISDs have two overlapping

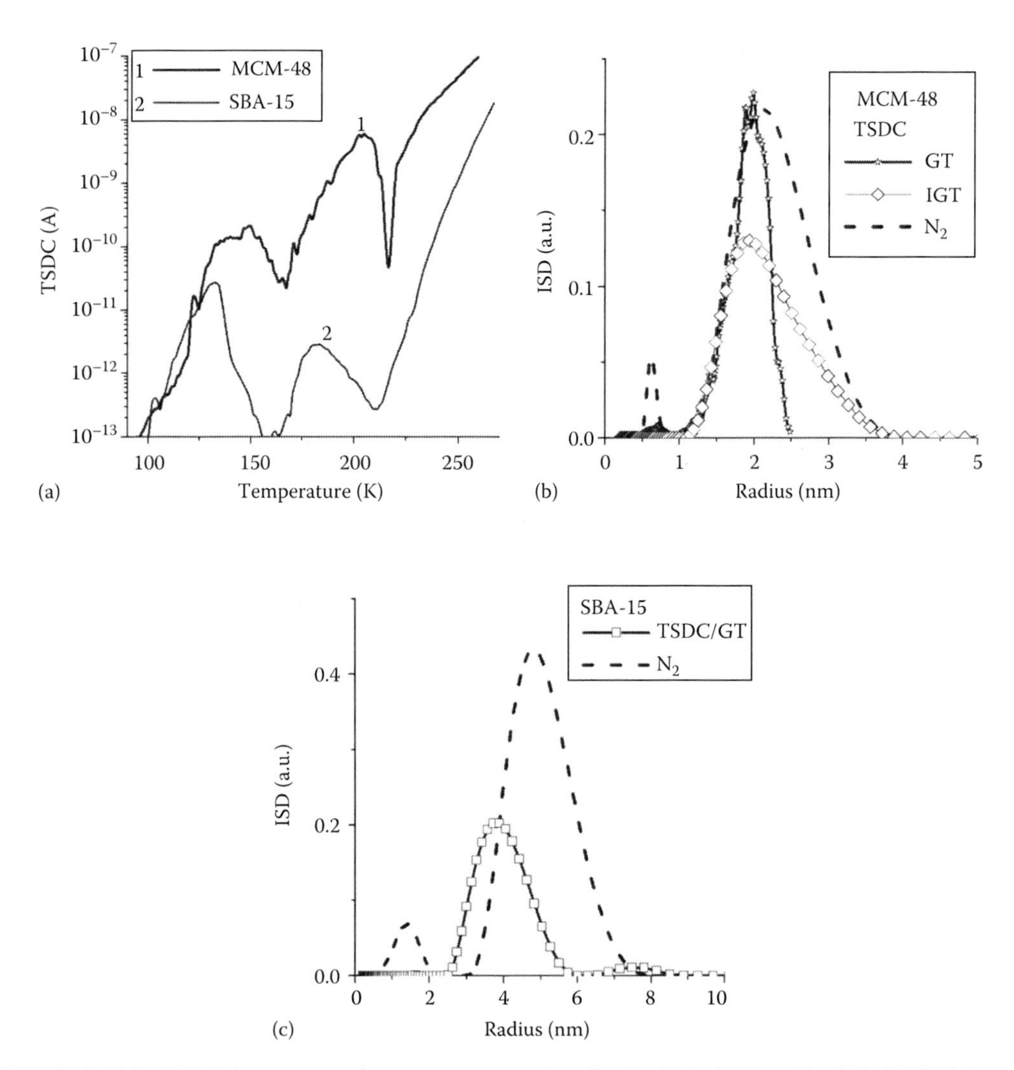

FIGURE 1.286 TSDC thermograms for aqueous suspensions (h=5 mL/g) of silicas (F_p=200–300 kV/m); and incremental pore size distributions calculated on the basis of nitrogen desorption isotherms for dry powders and the size distributions of water clusters relaxing on the TSDC measurements of aqueous suspensions (h=5 mL/g) for (b) MCM-48 (using modified GT and IGT equations), and (c) SBA-15 (modified IGT). (With kind permission from Springer Science+Business Media: *Central Eur. J. Chem.*, Behaviour of pure water and water mixture with benzene or chloroform adsorbed onto ordered mesoporous silicas, 5, 2007g, 420–454, Gun'ko, V.M., Turov, V.V., Turov, A.V. et al., Copyright 2007.)

peaks (Figure 1.290a), despite the IPSD has only one main peak of mesopores (Figure 1.278). The formation of more distinct water (as well as benzene) structures is observed on the adsorption of water/benzene mixtures because of the effects of benzene (Figure 1.290). In the case of the water/chloroform-*d* mixture (Figure 1.290e), water is mainly located in narrow mesopores of MCM-41 (only a small portion of adsorbed water is in pores at $R>2$ nm) similar to that for adsorbed pure water (Figure 1.290a). Consequently, the effect of nonpolar benzene on the bound water structure is stronger than that of weakly polar chloroform-*d*. This is due to a smaller excess of the Gibbs free energy on interaction of chloroform with clustered water than that on interaction of benzene with this water. It should be noted that at $T \leq 213$ K only benzene (signal at 7.2 ppm) remains unfrozen in the water/benzene mixture filling pores because adsorbed water completely freezes at $T<233$ K. However, in the case of adsorbed pure water, its portion freezes at much lower temperatures.

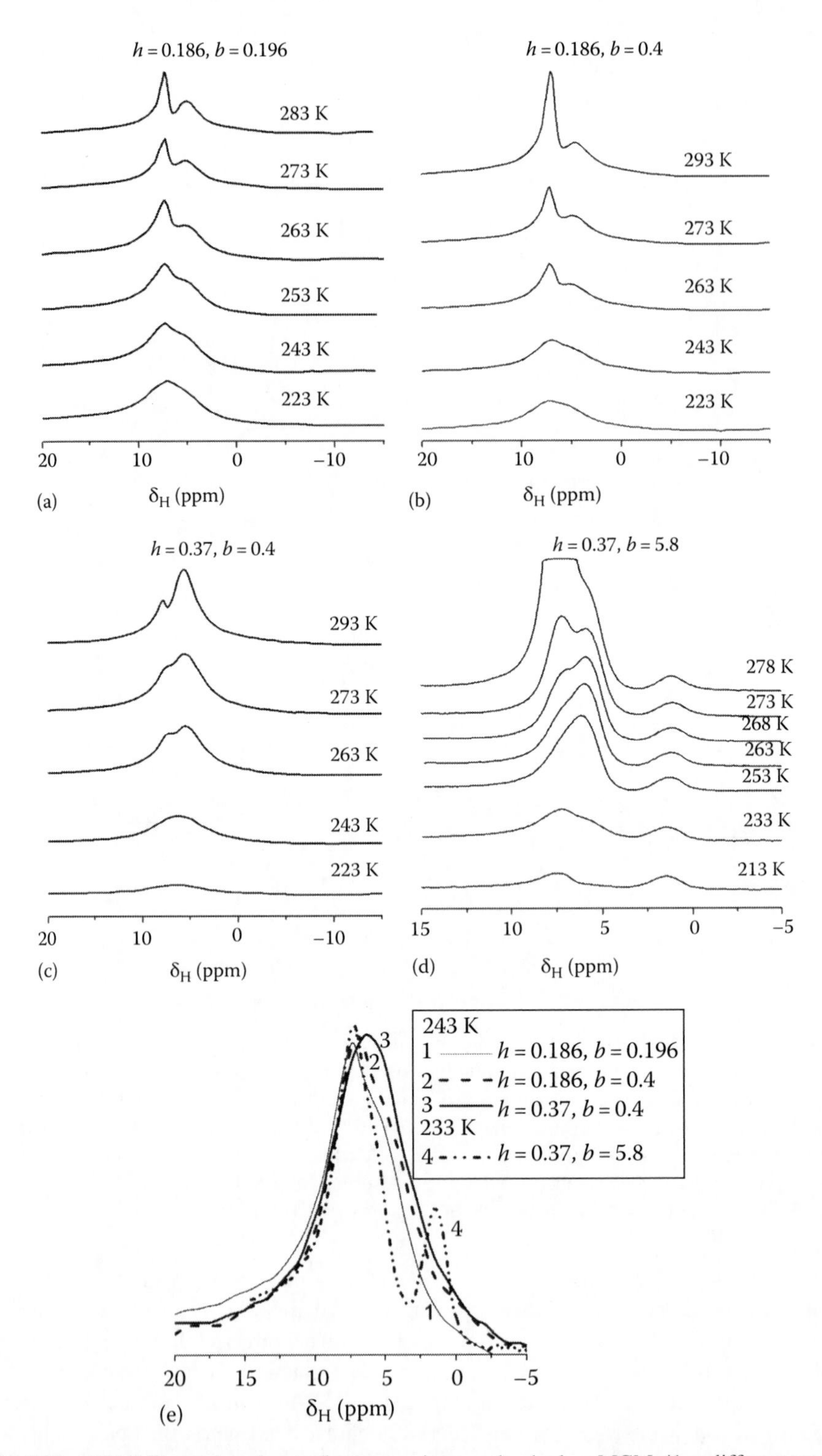

FIGURE 1.287 1H NMR spectra of water/benzene mixture adsorbed on MCM-41 at different temperatures and different amounts of water and benzene (b): h=(a, b, e) 0.186, (c, d, e) 0.37 mL/g and b=(a, e) 0.196, (b, c, e) 0.4 and (d, e) 5.8 mL/g (sample 2). (With kind permission from Springer Science+Business Media: *Central Eur. J. Chem.*, Behaviour of pure water and water mixture with benzene or chloroform adsorbed onto ordered mesoporous silicas, 5, 2007g, 420–454, Gun'ko, V.M., Turov, V.V., Turov, A.V. et al., Copyright 2007.)

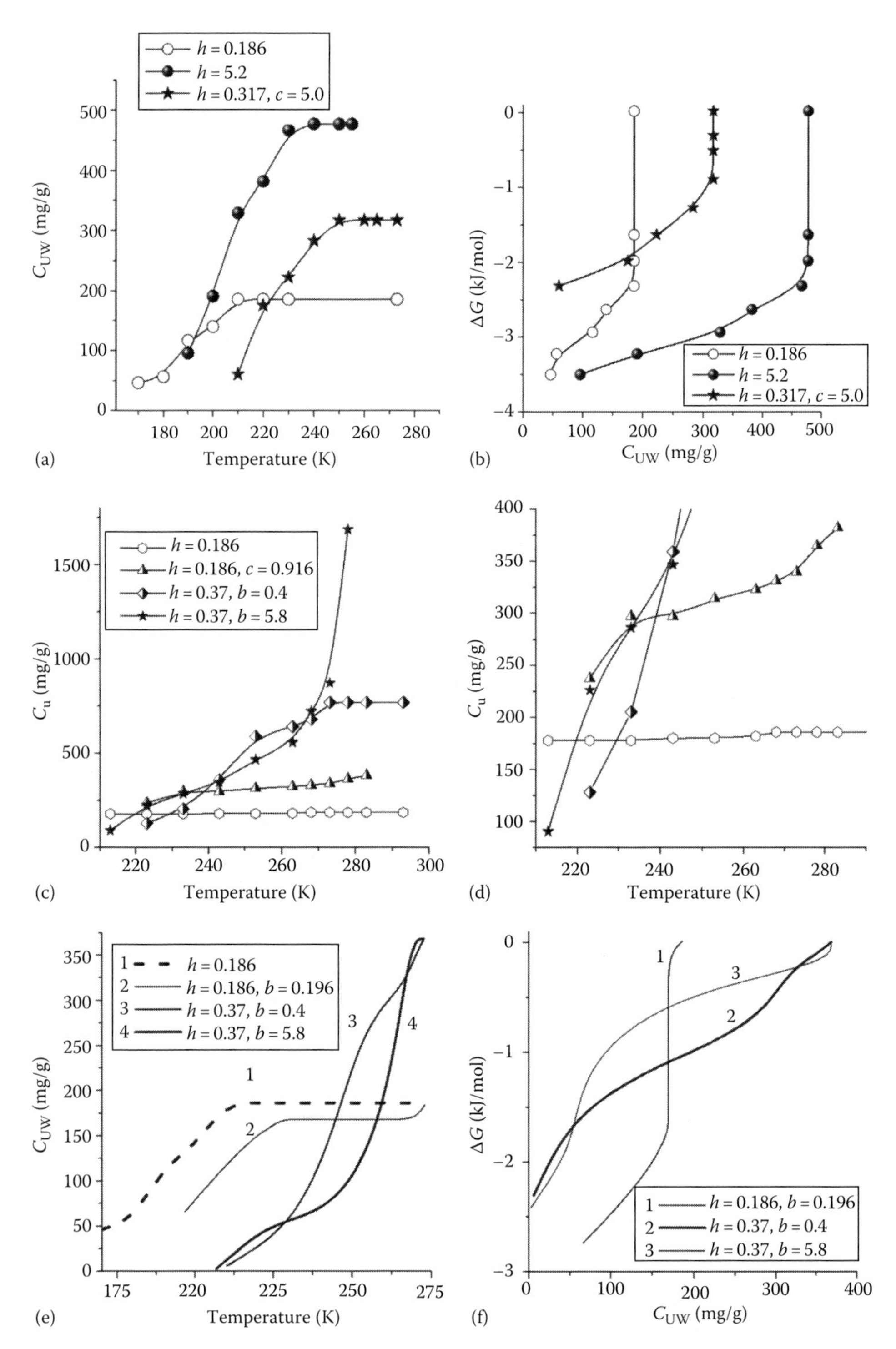

FIGURE 1.288 Amounts of unfrozen (a, e) water and (c, d) water and benzene, and (b, f) the relationship between changes in the Gibbs free energy of adsorbed water and unfrozen water amounts on MCM-41 (sample 2); (d) shows an initial portion of the graphs shown in (c). (With kind permission from Springer Science+Business Media: *Central Eur. J. Chem.*, Behaviour of pure water and water mixture with benzene or chloroform adsorbed onto ordered mesoporous silicas, 5, 2007g, 420–454, Gun'ko, V.M., Turov, V.V., Turov, A.V. et al., Copyright 2007.)

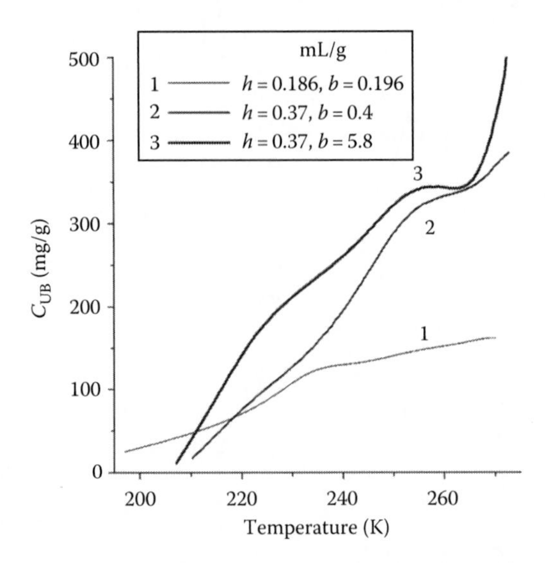

FIGURE 1.289 Amounts of unfrozen benzene for different water/benzene mixtures adsorbed onto MCM-41 (sample 2). (With kind permission from Springer Science+Business Media: *Central Eur. J. Chem.*, Behaviour of pure water and water mixture with benzene or chloroform adsorbed onto ordered mesoporous silicas, 5, 2007g, 420–454, Gun'ko, V.M., Turov, V.V., Turov, A.V. et al., Copyright 2007.)

In other words, benzene weakens the interaction of water with the pore walls because water displaced from narrow pores forms larger structures.

The amounts of water adsorbed onto silica in air (h=0.186 mL/g) or on addition of a small amount of benzene (h=0.186 mL/g and b=0.196 mL/g) provide filling of approximately quarter of the total pore volume by water and high clusterization of adsorbed water, and approximately half volume (h=0.317 and 0.37 mL/g respectively) in the case of addition of chloroform-*d* (c=5.0 mL/g) and benzene (b=0.4 and 5.8 mL/g). A vertical section observed on the $\Delta G(C_{uw})$ graphs (Figure 1.288b) corresponds to nonfreezing of bound water over a wide temperature range (Figure 1.288a). This bound water is strongly bound to the silica surface ($\Delta G<-1$ kJ/mol, Figure 1.288b), and the smaller the adsorbed water amount, the lower is the freezing temperature of water. This effect is caused by predominant occupation of the narrowest pores (characterized by the maximal adsorption potential) on their partial filling (Figure 1.290a). At a low hydration, the adsorbed water can form clusters or thin films smaller in size (thickness) than pores. An increase in the hydration up to h=5.2 mL/g leads to filling of larger pores because the ISD of the structures with unfrozen bound water shifts toward larger R values (Figure 1.290a). However, the amount of unfrozen water located in pores at h=5.2 mL/g remains smaller than the total pore volume. This effect can be caused by several reasons: (i) air bubbles remaining in pores inhibit their filling by water (as it was discussed earlier for silicalites); (ii) a small portion of pyrocarbon and organics remains (Figure 1.291) in pores of silica after its calcination during the template removal; and (iii) a portion of water (far from the pore walls) filling broader mesopores at R>4 nm (Figure 1.278) can be frozen at temperature near 273 K.

However, the first two factors can be predominant because contribution of broad mesopores in MCM-41 is very low (Figure 1.278) and WBW (which can be located in broad pores and satisfies condition $\Delta G>-0.5$ kJ/mol) is not observed (Figure 1.288b). A slightly decreased hydrophilicity of the pore walls of MCM-41 due to small amounts of pyrocarbon and trace organic functionalities can strengthen bubble plugs in narrow mesopores at R<2 nm in which the capillary condensation of water is absent. The FTIR spectra (Figure 1.291, MCM-41 [sample 1], MCM-48, and SBA-15) show that small amounts of the template remain in all the samples because there are low three peaks at 2925 cm^{-1} ($\nu_{CH,as}$), 2850 cm^{-1} ($\nu_{CH,s}$), and 1465 cm^{-1} ($\delta_{CH2,as}$) corresponding to vibrations of the CH_2 groups. Additional heating of MCM-41 in air at 473 K for 2 h affects only adsorbed water because the

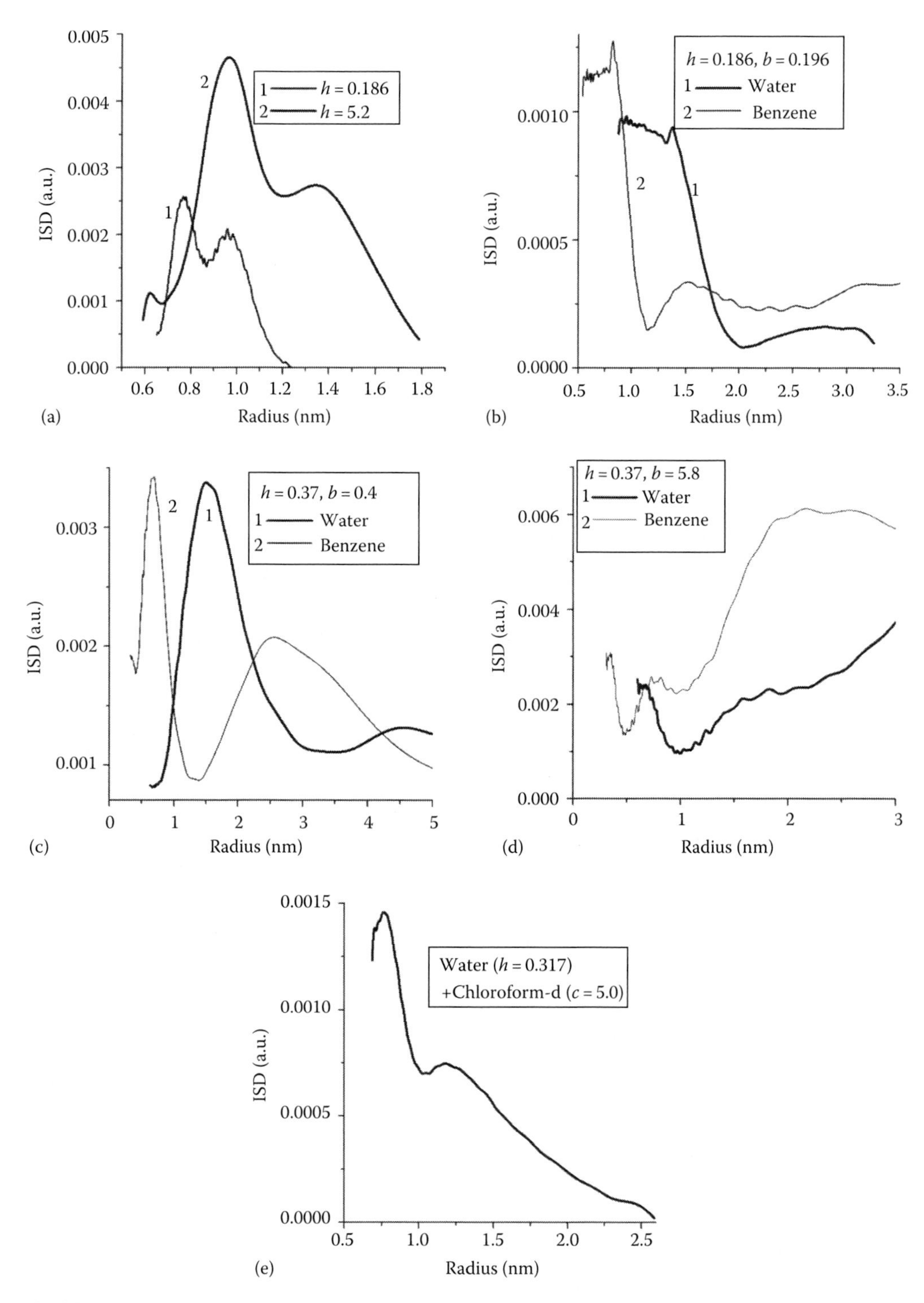

FIGURE 1.290 Distribution function of sizes of structures unfrozen liquids in MCM-41 (sample 2) pores: (a) pure water, (b, c, d) water and benzene, and (e) water (with the presence of chloroform-*d*). (With kind permission from Springer Science+Business Media: *Central Eur. J. Chem.*, Behaviour of pure water and water mixture with benzene or chloroform adsorbed onto ordered mesoporous silicas, 5, 2007g, 420–454, Gun'ko, V.M., Turov, V.V., Turov, A.V. et al., Copyright 2007.)

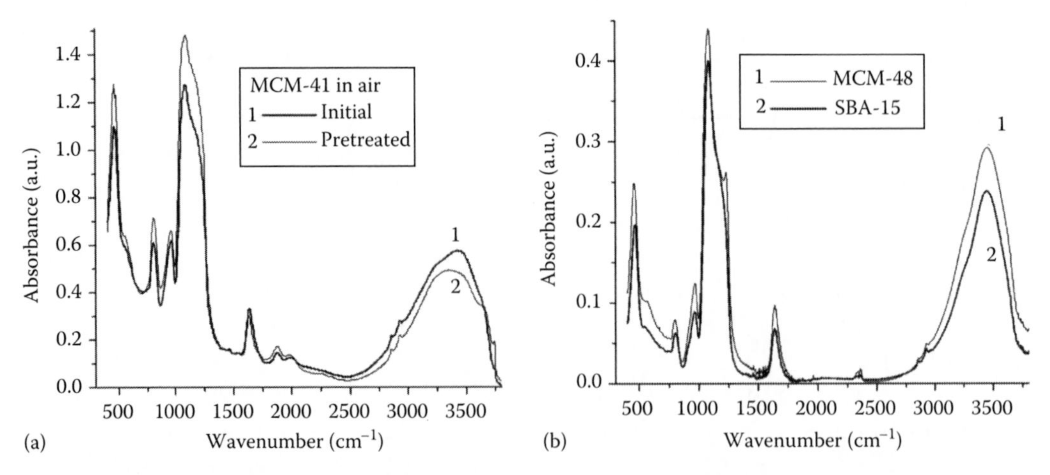

FIGURE 1.291 FTIR spectra of (a) MCM-41 (sample 1) initial and treated in air at 473 K for 2 h, and (b) MCM-48 and SBA-15. (With kind permission from Springer Science+Business Media: *Central Eur. J. Chem.*, Behaviour of pure water and water mixture with benzene or chloroform adsorbed onto ordered mesoporous silicas, 5, 2007g, 420–454, Gun'ko, V.M., Turov, V.V., Turov, A.V. et al., Copyright 2007.)

intensity of broad band of the O-H stretching vibrations of water at 3200 cm^{-1} decreases after heating and the intensity of free silanols at 3740 cm^{-1} increases and a shoulder at 3650 cm^{-1} appears. The presence of low amounts of residual organics is typical for silicas synthesized using organic templates because it is difficult to provide their complete removal from pores without damage of the silica structure during HT calcination. Residual hydrophobic CH_2 groups can affect the amounts of bound water (i.e., pore filling by water) and its structure (WAW can appear). However, the water/benzene mixture can fill the total pore volume that may confirm the previously mentioned assumption.

The enthalpy of immersion of MCM-41 in water $\Delta_{im}H = -102$ mJ/m^2 and of the enthalpy of its wetting in water $\Delta_w H = -78$ mJ/m^2 (Meziani et al. 2001). Recalculation of the γ_S values (Table 1.43) with consideration for a portion of filled pores (multiplying γ_S by $z = 10^3\rho_{uw}V_p/C_{uw}|_{T=273K}$ because C_{uw} is given in mg/g and assuming $\rho_{uw} = 1$ g/cm^3) gives the values (Table 1.43, $z\gamma_S$) close to $\Delta_{im}H$ at different h values for the air and $CDCl_3$ media. However, in the case of the water/benzene mixtures, the $z\gamma_S$ values related to the total changes in the Gibbs free energy of the bound water is much smaller (by a factor of 3–4) because of the displacement of adsorbed water by benzene from narrow pores and the pore walls. This effect is caused by the fact that direct interaction of benzene with the silica surface instead of interaction with water can cause a reduction of the Gibbs free energy of the system on simultaneous diminution of the contact interfacial area between poorly immiscible benzene and water.

The $z\gamma_S$ (as analogous to $\Delta_{im}H$) values for MCM-48 and SBA-15 are close to that of MCM-41 (Table 1.43). An increase in these values with decreasing h value can be caused by higher energy of the adsorption of water in narrower pores that can give a slight overestimation of the $z\gamma_S$ values.

Chloroform can fill a portion of pores and displace a fraction of water from the pore walls and the narrowest pores into larger pores (comp. Figures 1.278 and 1.290a,e). Therefore, a starting freezing temperature depression (corresponding to $dC_w/dT = 0$) decreases by approximately 20 K (Figures 1.288a and 1.292, curves 2 and 6).

The corresponding maximum temperature below which bound water begins to freeze determines minimal reduction of the Gibbs free energy caused by adsorption interactions with the pore walls. These changes are similar for the air and aqueous media (bending of the $\Delta G(C_{uw})$ curves at −2 kJ/mol characteristic of SBW). However, in the case of the $CDCl_3$ medium, it is equal to approximately −1 kJ/mol (Figure 1.288). The extrapolation of the $\Delta G(C_{uw})$ curves to the y-axis determines the maximum reduction in the Gibbs free energy of the first surface layer of water at $\Delta G_s \approx -4$ kJ/mol. This ΔG_s value is lower (by −1.5 kJ/mol) for the air and aqueous media in comparison with $CDCl_3$ ($\Delta G_s \approx -2.5$ kJ/mol). Consequently, pore chloroform-*d* causes certain diminution of bound water

TABLE 1.43
Energetic Characteristics of Interfacial Water Adsorbed Individually or in Mixture with Benzene or Chloroform-*d* onto MCM-41 (Sample 2), MCM-48, and SBA-15

Parameter	MCM-41
γ_S, mJ/m^2	68.0 (h=5.2 mL/g)
γ_S, mJ/m^2	26.0 (h=0.186 mL/g)
$\gamma_{S,w}$, mJ/m^2	25.0 (h=0.186 mL/g, c=5.0 mL/g)
$\gamma_{S,w}$, mJ/m^2	10.4 (h=0.186 mL/g, b=0.196 mL/g)
$\gamma_{S,w}$, mJ/m^2	15.6 (h=0.37 mL/g, b=0.4 mL/g)
$\gamma_{S,w}$, mJ/m^2	12.0 (h=0.37 mL/g, b=5.8 mL/g)
$z\gamma_S$, mJ/m^2	100.1 (h=5.2 mL/g)
$z\gamma_S$, mJ/m^2	98.1 (h=0.186 mL/g)
$z\gamma_{S,w}$, mJ/m^2	94.3 (h=0.186 mL/g, c=5.0 mL/g)
$z\gamma_{S,w}$, mJ/m^2	39.3 (h=0.186 mL/g, b=0.196 mL/g)
$z\gamma_{S,w}$, mJ/m^2	29.6 (h=0.37 mL/g, b=0.4 mL/g)
$z\gamma_{S,w}$, mJ/m^2	22.8 (h=0.37 mL/g, b=5.8 mL/g)
$z\gamma_{S,w}$, mJ/m^2 (MCM-48)	101 (h=0.2 mL/g)
$z\gamma_{S,w}$, mJ/m^2 (SBA-15)	116 (h=0.16 mL/g)

interaction with the MCM-41 surface; however, WBW at $\Delta G > -0.5$ kJ/mol is absent with the presence of $CDCl_3$ as in the case of adsorbed pure water (Figure 1.288b). Freezing of pore fluid is sensitive to its states dependent on the degree of pore filling, and the adsorbed film (closest to the pore walls) and fluid at the middle of pores freeze at different temperatures. Therefore, the effect of chloroform-*d* on the bound water can be explained by the displacement of a portion of water from the narrowest pores into larger ones or from the pore walls toward the middle of pores or out of pores. For instance, water fills pores at $R > 1$ nm (Figure 1.290e), and there are two maxima of the water structures (ISD), as well as in the case of the adsorption of pure water (Figure 1.290a); however, the left portion of the ISD is "cut" by the pore chloroform-*d*. This suggests nonuniform freezing of bound water or water/chloroform-*d* because thin liquid layers or small clusters locating close to the pore walls freeze at lower temperatures than liquid domains locating far from the pore surface that causes appearance of two ISD peaks of water.

Benzene affects the characteristics of the bound water more strongly than chloroform-*d* (Figures 1.288 through 1.291) because it can more strongly displace water from the pore walls and pores (Figure 1.290), especially at its excess at b=5.8 mL/g (Figure 1.290c and d). An ISD peak at $R < 1$ nm of unfrozen benzene (Figure 1.290b through d) can correspond to a thin layer locating at the pore walls when water is displaced from the pore walls or out of pores. This effect causes diminution of the freezing point depression of water (comp. Figure 1.288a and c), and changes in the Gibbs free energy of adsorbed water are smaller (Figure 1.288f) than in the case of the adsorption of pure water or water/chloroform-*d* (Figure 1.288b). The benzene/water mixture can fill total pore volume because the C_u value corresponds to the V_p value (Figure 1.288c). Probably, they form the clustered structure (especially in the case of sequential addition of water and benzene to MCM-41) in which contacts of benzene with the pore walls can be predominant because the ISDs related to benzene shift toward smaller R values in comparison with the ISDs linked to water (Figure 1.290b through d). There is certain broadening of these ISDs because of incomplete compensation of the interference of water and benzene on calculations. This rearrangement of pore liquids diminishes the interaction energy of water with the silica surface (Figure 1.290f).

However, a major fraction of water locating in pores and keeping the hydrogen bond network forms relatively large structures, which can be attributed to SAW with $\delta_H \approx 5$ ppm (Figure 1.287). On the other hand, a minor portion of water at large excess of benzene ($b = 5.8$ mL/g) can be assigned to weakly associated ($\delta_H = 1.2$–1.4 ppm, Figure 1.287d) but strongly bound (as it freezes at very low temperatures) water. Another portion of the bound water is weakly bound ($\Delta G > -0.5$ kJ/mol, Figure 1.288f) because its interaction with benzene occurs without formation of strong hydrogen bonds characteristic of the bulk water ($\delta_H \approx 5$ ppm) or water interacting with the silica surface (at the same δ_H values). The effects of the surface forces and intermolecular interaction between benzene and water molecules resulting in the structurization (clusterization) of the mixture in pores that ensures the minimum value of Gibbs free energy for this complex system. However, interaction energy of the water–water, water–silica, benzene–benzene, and benzene–silica types is greater than that of water–benzene. Therefore, the $\Delta G(C_{uw})$ curves for the bound water in the benzene/water mixtures (Figure 1.288f) locate over the curves of the pure water (Figure 1.288b) and the benzene/water/MCM-41 system should tend to reach a state with a minimal area of the water/benzene interfaces in pores. For instance, maxima of the ISDs at $R = 1.0$–1.5 nm related to adsorbed unfrozen water (Figure 1.290b through d) correspond to minima of the ISD related to adsorbed benzene. Additionally, benzene fills a fraction of the narrowest pores (Figure 1.290) and changes in the ΔG_b values on benzene interaction with silica (Figure 1.289) are larger than that of water (Figure 1.288). According to Aksnes and Kimtys (2004), all benzene freezes in silica gel pores (average $R \approx 2$ nm) at $T > 227$ K that is in agreement with the dC_u/dT peak at 230–250 K of benzene adsorbed on MCM-41 (Figure 1.292). Notice that MCM-41 has narrower pores than the mentioned silica gel, that is, the freezing temperatures of pore benzene adsorbed onto MCM-41 can be lower. The dC_u/dT peak at 270 K corresponds to thawing of benzene in larger pores and far from the pore walls because its melting point in the bulk is only slightly higher (~278 K). A minor portion of adsorbed water, which corresponds to weakly bound one, can thaw at temperatures close to 270 K (Figure 1.288e and f).

A simple additive model of water/benzene mixture completely filling pores ($h = 0.37$ mL/g and $b = 5.8$ mL/g) based on Equation 1.71 and assumption of the additivity of water and benzene contributions

$$\frac{dV_u(R_{uw} + R_{ub})}{dR} = \frac{dV_{uw}(R_{uw})}{dR} + \frac{dV_{ub}(R_{ub})}{dR} \tag{1.73}$$

gives the distribution function close in R values to $IPSD_{N2}$ but with a more complex shape (Figure 1.292d). This is due to the presence of several structures with benzene and water reflecting in the $f(T_2)$ distributions (Figure 1.282) and the ISDs (Figures 1.283 through 1.285, 1.290, and 1.292d). This complexity causes certain broadening of the ISD. Calculation of the ISD with the IGT equation gives a complex curve (because of the bound benzene effects), which, however, has a maximum close to that of IPSD.

Thus, the study of the structural characteristics of ordered mesoporous silicas using improved procedure (CONTIN/MEM-j) or a complex model of cylindrical pores and voids between spherical particles suggests that the use of alcohol in the reaction solution and additional aging of the reaction product causes increased roughness of the pore walls. However, the textural porosity of MCM-41 samples synthesized with the presence of ethanol is lower than that of MCM-48 and SBA-15 synthesized without alcohol. The roughness of the pore walls and residual organics provide the appearance of WAW adsorbed in a mixture with an organic solvent.

The investigations of pure water and water/chloroform-*d* or water/benzene mixtures adsorbed onto MCM-41 by using the ^{1}H NMR spectroscopy with layer-by-layer freezing-out of bulk and pore liquids show that both chloroform-*d* and benzene can displace a portion of water from narrow pores and/or from the pore walls to reduce the contact area between immiscible liquids. However, the effects of nonpolar benzene are stronger than that of weakly polar chloroform-*d*. This effect

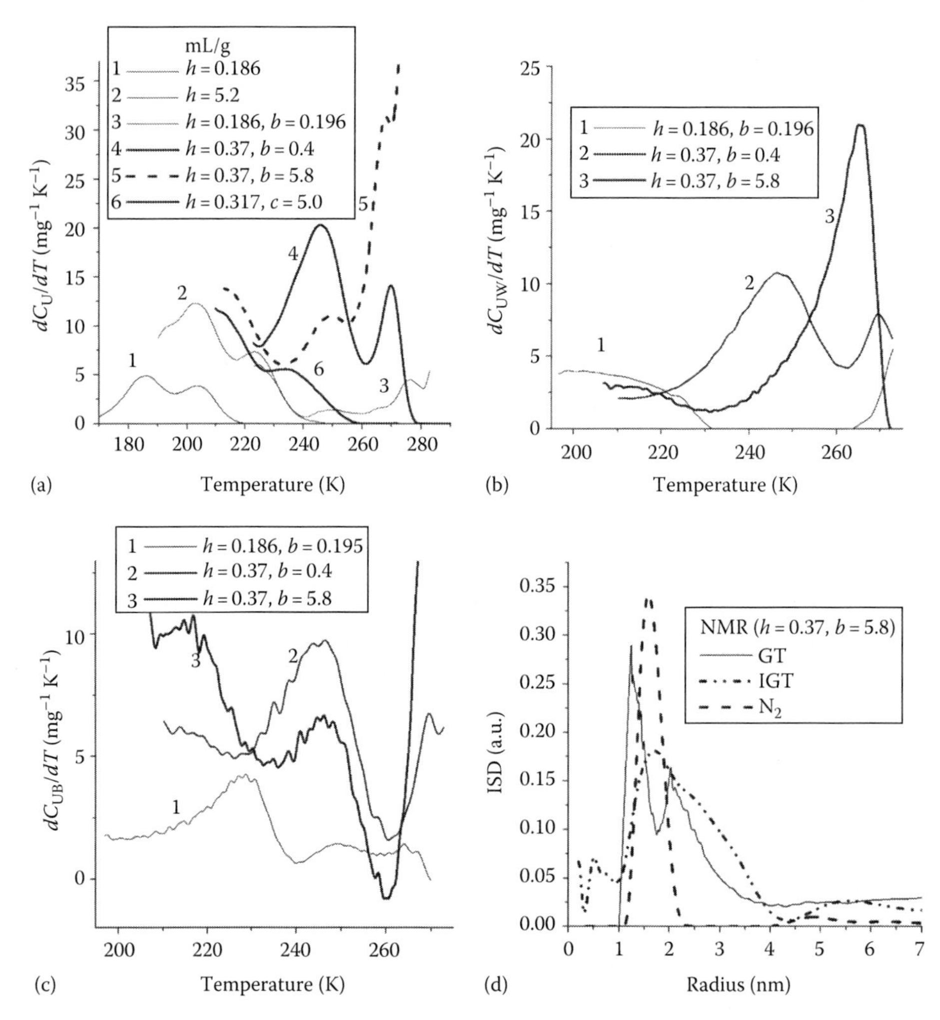

FIGURE 1.292 Derivative dC_u/dT as a function of temperature at different amounts of water, benzene and chloroform-*d* (sample 2) for (a) total amounts of water or water/benzene; (b) water and (c) benzene adsorbed onto MCM-41; and (d) IPSD calculated on the basis of the nitrogen adsorption isotherm and ISD for water/benzene adsorbed onto MCM-41 (using GT and IGT equations). (With kind permission from Springer Science+Business Media: *Central Eur. J. Chem.*, Behaviour of pure water and water mixture with benzene or chloroform adsorbed onto ordered mesoporous silicas, 5, 2007g, 420–454, Gun'ko, V.M., Turov, V.V., Turov, A.V. et al., Copyright 2007.)

of benzene leads to diminution of freezing temperature depression of adsorbed water and its interaction energy with the pore walls. Water fills only a portion of MCM-41 pores at high hydration (h=5.2 mL/g); however, the water/benzene mixture fills total pore volume. The modified Gibbs–Thomson relation for the freezing point depression and equations related to transverse relation processes in adsorbed water/benzene mixtures allow to elucidate the difference in structurization and rearrangement of pure bound water and water/benzene or water/chloroform-*d* mixtures. This structurization depends on the PSD, amounts and type of coadsorbates, their clusterization, the order of their adsorption, the pore wall structure, and residual amounts of organic templates. One could expect that preadsorption of water in narrow pores with metal oxides having more hydrophilic surface than silica (e.g., mesoporous mixed alumina/silica, titania/silica), on which a continuous water film can be formed, the effects of subsequent adsorption of benzene or other nonpolar solvents on the water organization can be smaller.

Freezing of a pore fluid is sensitive to its different states dependent on the degree of pore filling: the adsorbed surface film and capillary condensate (far from the pore walls) freeze at lower and higher temperatures, respectively. As shown earlier, Morishige and Iwasaki (2003) studied the freezing/melting behavior of water in partially filled pores of mesoporous silica SBA-15 (average pore radius 3.9 nm) as a function of temperature and pore filling. The freezing temperature increased continuously with increasing pore filling even in the region of capillary condensation. The results are related to the different states of the bound water depending on the degree of pore filling. On the other hand, the melting of the frozen bound water took place at a well-defined temperature of 256 K, independent of the degree of pore filling. The XRD patterns show that the freezing of the bound water results in formation of ice microcrystallites with almost the same structure and size, regardless of the different states of the bound water.

Now, we compare some results obtained for ordered mesoporous silicas and silica gels. The dipolar and dc relaxations of water adsorbed in different mesoporous silicas are characterized by two typical LT and HT bands in the TSDC thermograms (Figure 1.293). Silica gels Si-60 and Si-40 with less ordered porous structure than MCM-48 or SBA-15 have LT peaks at 120–125 K, which are close to that of pure water. Consequently, the effect of pore walls is minimal for a fraction of water located in the largest pores of silica gels. MCM-48 and SBA-15, with a more ordered pore structure, have a larger effect on the bound water and the LT peak shifts significantly toward higher temperatures (Figure 1.293), even though the PSDs of mesopores are close for SBA-15 and Si-60 (Figure 1.294).

The dipolar relaxation of water frozen in pores of different sizes occurs step-by-step if the pore size is larger than the size of relaxing water structures. For instance, in the case of SBA-15 with relatively large pores, the size of water clusters relaxing at certain temperatures is smaller than the size of pores (Figure 1.294b) because water freezes at the pore walls at lower temperatures (i.e., it is in the liquid state under the applied field for a longer time) than far from the walls. This effect was observed on the ^{1}H NMR investigations of water in silica gels. In the case of MCM-48 (Figure 1.294a), with narrower pores than SBA-15, water can freeze in pores over a narrow temperature range (a similar effect was observed for silica gel Si-40). Therefore, relaxing water clusters have a size close to that of pores (Figure 1.294a) as well as in the case of Si-40 (Figure 1.294c). However,

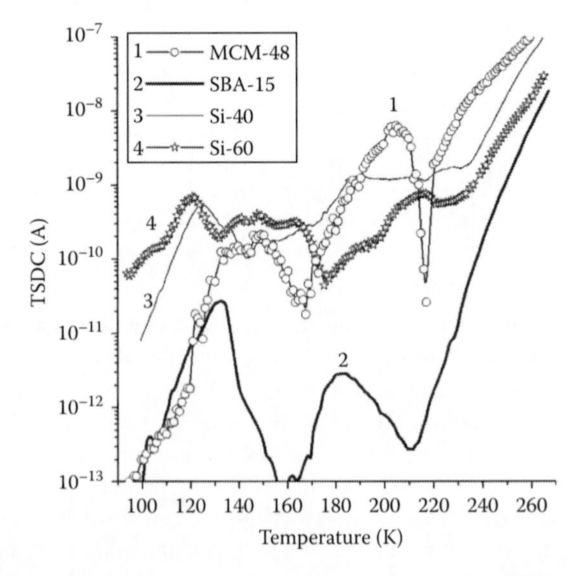

FIGURE 1.293 TSDC thermograms for aqueous suspensions (C_{ox} = 16.7 wt%) of porous silicas (F_p = 200–300 kV/m). (Adapted from *Adv. Colloid Interface Sci.*, 131, Gun'ko, V.M., Zarko, V.I., Goncharuk, E.V. et al., TSDC spectroscopy of relaxational and interfacial phenomena, 1–89, 2007i. Copyright 2007, with permission from Elsevier.)

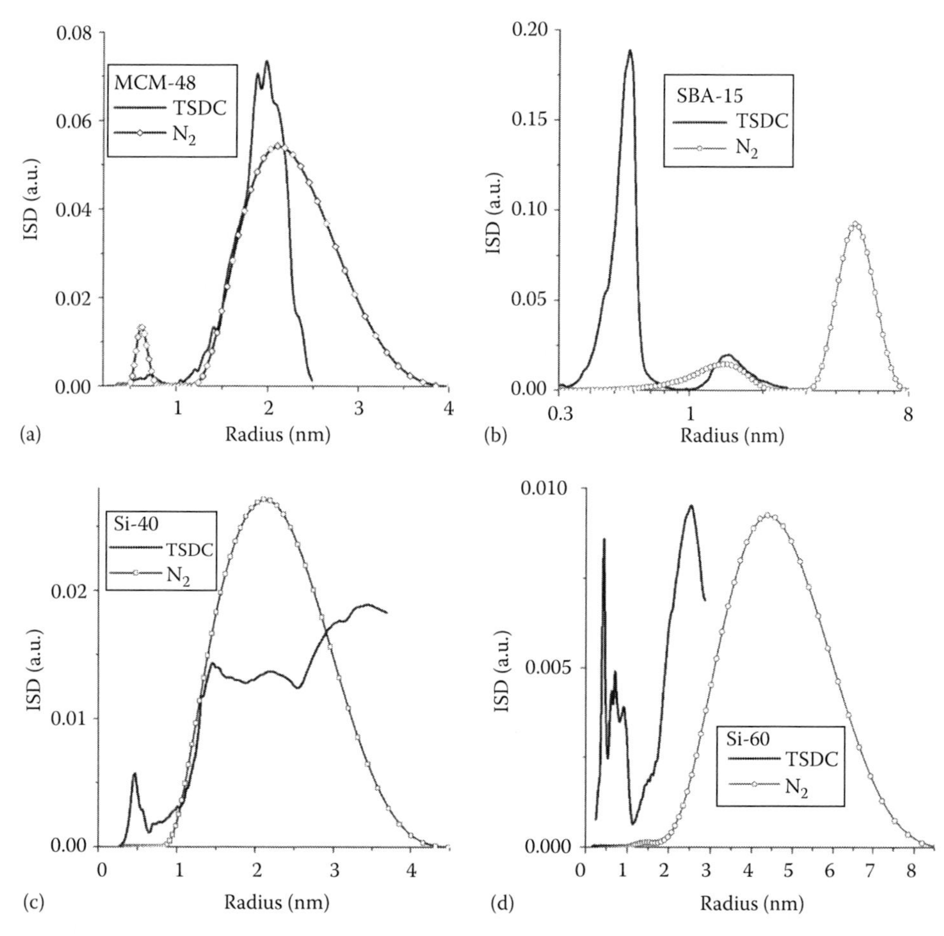

FIGURE 1.294 Incremental pore size distributions calculated on the basis of nitrogen desorption isotherms for dry powders and the size distributions of water clusters relaxing on the TSDC measurements of aqueous suspensions (16.7 wt%) for (a) MCM-48, (b) SBA-15, (c) Si-40, and (d) Si-60. (Adapted from *Adv. Colloid Interface Sci.*, 131, Gun'ko, V.M., Zarko, V.I., Goncharuk, E.V. et al., TSDC spectroscopy of relaxational and interfacial phenomena, 1–89, 2007i. Copyright 2007, with permission from Elsevier.)

a small amount of water (small clusters) can be frozen at the pore walls at lower temperatures. These small clusters give a low-intensity peak at $R < 1$ nm.

1.4 THIN FILMS AND OTHER MOIETIES ON SILICA SUPPORTS

To prepare mesoporous materials as thin films, the template synthesis technique can be applied. Appropriate improvement of these methods yielded stable films with well-defined symmetries, controlled-pore orientation, continuity, and film thickness (Pevzner et al. 2000). The ability to tailor mesoporous film properties is important to use the materials in applications ranging from catalysis to microelectronics, where morphological control in the meso-domain is important. For instance, the fluoride-mediated synthesis of pure silica zeolite thin films with different topologies on surface-modified (100) Si wafers was reported by Hunt et al. (2010) for low dielectric constant (low k) material applications. Some of the films were polycrystalline, intergrown, continuous, and well adhered to their substrates. The films were characterized by XRD and field emission scanning electron microscopy. Zeolites investigated gave k-values lower than those predicted from their structures using the Bruggeman effective medium model that has been commonly employed and found able to predict dielectric constants for amorphous silicas (Hunt et al. 2010). Grunenwald et al. (2012) synthesized hydrophobic mesostructured organosilica-based thin films with tunable mesoporosity as a material for ultra low *k* applications. They used spin-on solutions and evaporation-induced self-assembly (EISA) method with polystyrene-b-polyethylene oxide block copolymers as structure-directing agents and methyltriethoxysilane as organosilica precursor. Lee et al. (2009) prepared polyimide/mesoporous silica hybrid nanocomposites having relatively good phase mixing behavior by utilizing polyimide synthesized from a water-soluble poly(amic acid) ammonium salt, which lead to low *k* up to 2.45. The NMR spectroscopy was applied in this work only as a method for chemical/structural analysis. Yang et al. (1997) prepared mesoporous silica films on graphite substrate. Alberius et al. (2002) synthesized cubic, hexagonal, and lamellar silica and titania mesostructured thin films. Agoudjila et al. (2008) prepared inorganic membrane by solgel method. Cui et al. (2008) synthesized second-order nonlinear optical chromophores (Zyss 1994), containing two reactive sites and covalently bonded to a silica film (~1 μm in thickness), and characterized by ^{1}H NMR and other methods. The silica films, as well as other oxides, for example, alumina (Bäumer and Freund 1999) or titania (Chen and Mao 2007, Ismail and Bahnemann 2011), can be used as a matrix to prepare metal nanoparticles (Epifani et al. 2000, Krylova et al. 2009). Socha and Fransaer (2005) studied the mechanism of the formation of silica–silicate thin films on zinc from a colloidal solution (or sol) of silica. Thus, unmodified and modified or functionalized thin silica and other oxide films can be used as protective coatings of metals and other materials, catalysts and catalyst carriers, optical coatings with self-purification capacity, etc. The morphological and textural characteristics of the films play an important role in these applications and for deeper understanding interfacial phenomena occurring at the film surfaces. The properties of thin films change with changing porosity; therefore, it is important to know and understand these effects; however, it is difficult to measure them. Taylor et al. (1998) studied several thin films of different porosity (deposited by solgel methods onto quartz substrates) using surface acoustic wave (SAW) mass sensor with standard BET technique and ellipsometry. Nitrogen adsorption was measured by a SAW device, and porosity was calculated using BET equation. Porosity data were compared and correlated with measurements made by ellipsometry (film refractive index and thickness) and SEM (surface topography). As these techniques do not yield PSD, they are complimentary to the SAW technique, which directly measures the porosity. Notice that obtained nitrogen adsorption isotherms have a distorted shape at p/p_0>0.7–0.8 because of a small weight of the deposited films. Thus, the structural characterization of thin films is a more difficult task than the characterization of powders or monolith particles. Additionally, the film structure is strongly affected by the substrate (Figure 1.295; Krylova et al. 2009).

These films can be also used to synthesize metal nanoparticles with Ag (Figure 1.296) (Krylova et al. 2009) or Au, for example, for catalytic applications. Both metal particle sizes and distributions through the films can be varied depending on reaction and treatment conditions.

Dourdain et al. (2006) determined the structural characteristics of mesoporous silica thin films using x-ray reflectivity (XR). The total porosity was estimated from the total reflection plateau. The calculation of the XR curves by the matrix technique allowed to distinguish between nano and meso porosities of the films. In combination with grazing incident small-angle x-ray scattering (GISAXS), the surface area of the mesopores was ascertained, thereby providing a complete analysis of the porosity in thin films by x-ray scattering methods. We used the GISAXS data published by Dourdain et al. (2006) to compute the PSD *f*(*r*) (Figure 1.297a) using integral equation for the scattering intensity *I*(*q*) (Pujari et al. 2007):

$$I(q) = C\int_{R_{\min}}^{R_{\max}} \frac{\left(\sin qr - qr\cos qr\right)^2}{(qr)^2} v(r)f(r)dr, \tag{1.74}$$

where $q = 4\pi\sin(\theta)/\lambda$ the scattering vector value, 2θ is the scattering angle, λ is the wavelength of incident x-ray, *v*(*r*) is the volume of a pore with radius *r* (proportional to r^3), and *f*(*r*)d*r* represents

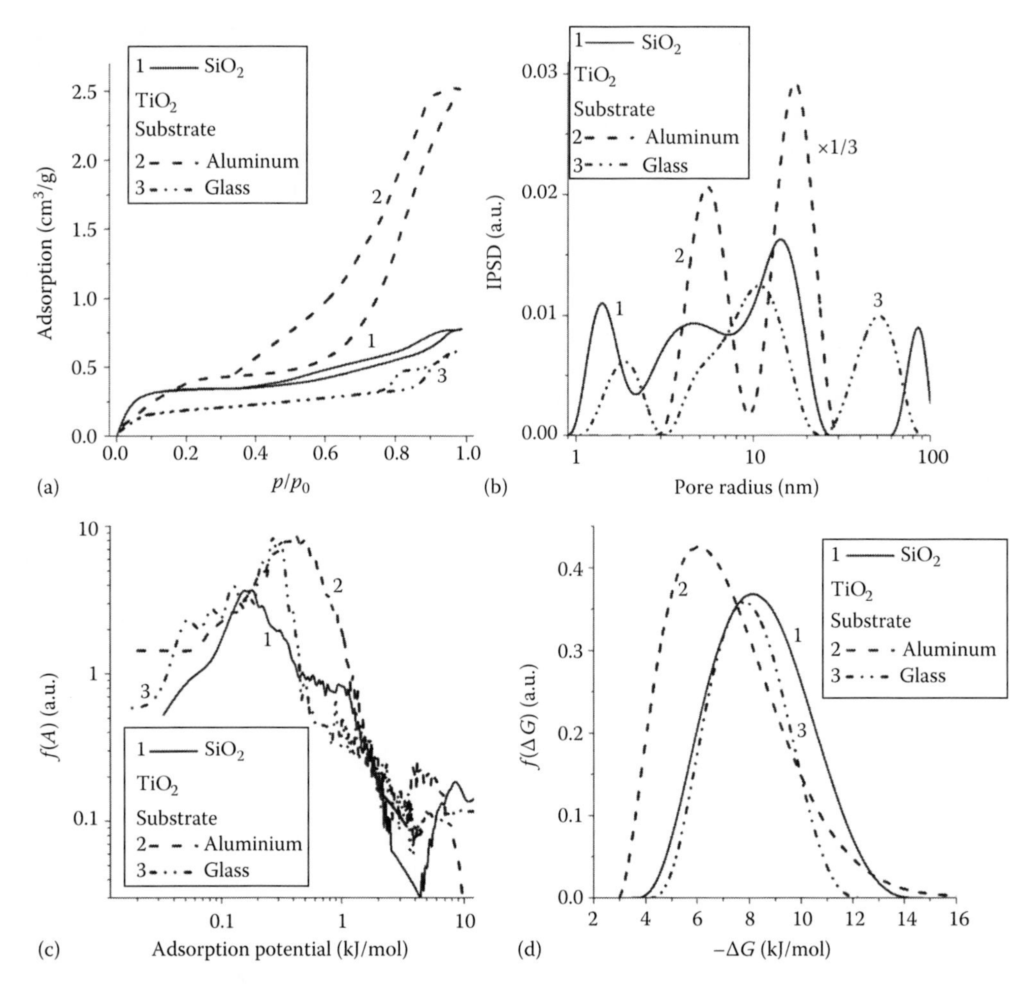

FIGURE 1.295 (a) Hexane adsorption isotherms; and distribution functions of (b) pore size of SiO_2 (1, glass substrate) and TiO_2 (curves 2 (aluminum substrate) and 3 (glass substrate)) films sintered at 773K, (c) adsorption potential, and (d) Gibbs free energy of hexane adsorption. (With kind permission from Springer Science+Business Media: *J. Sol-Gel Sci. Technol.*, Ag nanoparticles deposited onto silica, titania and zirconia mesoporous films synthesized by sol-gel template method, 50, 2009, 216–228, Krylova, G.V., Gnatyuk, Yu.I., Smirnova, N.P., Eremenko, A.M., and Gun'ko, V.M., Copyright 2009.)

the probability of having pores with radius r to $r+dr$. $R_{\min}$ ($=\pi/q_{\max}$) and $R_{\max}$ ($=\pi/q_{\min}$) correspond to lower and upper limit of the resolvable real space due to instrument resolution. The chord size distribution, $G(r)$ as a geometrical statistic description of a multiphase medium, was calculated (Figure 1.297b) from the GISAXS data using integral equation (Dieudonné et al. 2000)

$$G(r) = C\int_0^{\infty}\left[K - q^4 I(q)\right]\frac{d^2}{dr^2}\left(-4\frac{\sin qr}{qr}\right)dq, \tag{1.75}$$

where K is the Porod constant (scattering intensity $I(q) \sim Kq^{-4}$ in the Porod range).

The results shown in Figure 1.297 are in agreement with the structural results given by Dourdain et al. (2006) and add useful information on the texture and morphology of the silica films. The PSD (Figure 1.297a) demonstrate trimodal functions. However, the chord length distribution

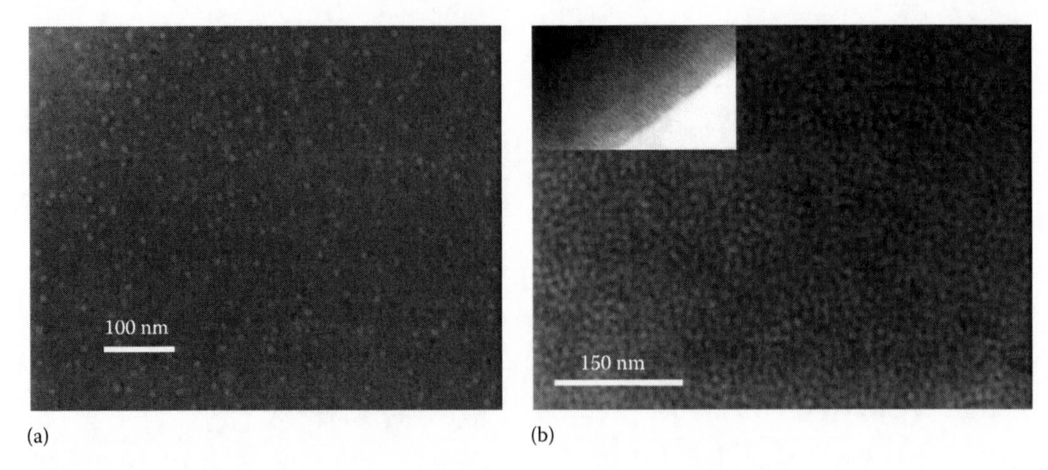

FIGURE 1.296 SEM image of mesoporous silica film (a) with Ag nanoparticles or (b) alone (inset shows normal cross-section of the film). (With kind permission from Springer Science+Business Media: *J. Sol-Gel Sci. Technol.*, Ag nanoparticles deposited onto silica, titania and zirconia mesoporous films synthesized by sol-gel template method, 50, 2009, 216–228, Krylova, G.V., Gnatyuk, Yu.I., Smirnova, N.P., Eremenko, A.M., and Gun'ko, V.M., Copyright 2009.)

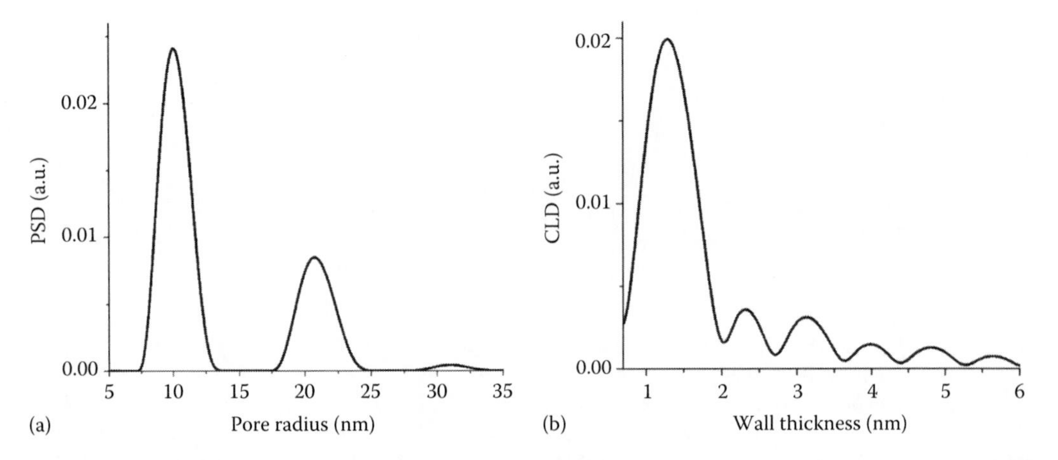

FIGURE 1.297 Distribution functions of (a) pore sizes and (b) chord length of a silica film calculated here from the GISAXS data published by Dourdain et al. (2006).

(Figure 1.297b) shows that the pore wall thickness is mainly between 1 and 2 nm. These results are in agreement with the structural characteristics presented by Dourdain et al. (2006).

Haddad et al. (2004) used continuously circulating flow of laser hyperpolarized (HP) ^{129}Xe NMR technique to study the porous structure of MCM-41, SBA-15 and mesoporous SiO_2 and TiO_2 thin films. The use of HP xenon allowed them to measure spectra at very low concentration of xenon where xenon reflects mainly interaction between the adsorbed xenon atoms and the surface. Variable temperature measurements allowed Haddad et al. (2004) to obtain information on the heat of adsorption of xenon on the surface and to evaluate the chemical shift of xenon in interaction with the surface.

Borsacchi et al. (2008) studied $BaSO_4$ submicronic (0.20–0.25 μm) particles (sample A), modified by a thin silica film (1 wt%) (sample B) and then treated with stearic acid (sample C), characterized by means of ^{29}Si, ^{13}C, and ^{1}H MAS high-resolution techniques, and low-resolution ^{1}H-FID analysis.

Despite the fact that work is very interesting, some questions arose because the authors did not use other methods, which could add structural and chemical information for deeper understanding of obtained NMR results. Borsacchi et al. (2008) recorded ^{29}Si-CP/MAS spectra for samples B and C at 25°C at a MAS frequency of 5 kHz with a recycle delay of 3 s, a contact time of 3 ms, and 26,000 transients. ^{29}Si-delayed CP/MAS spectra were recorded, introducing between the ^{1}H 90° pulse and the contact pulse a delay, τ, of 200 μs with a refocusing 180° pulse in the middle of it. The experimental conditions were the same as for the ^{29}Si-CP/MAS spectra, with the only exception of the number of transients, which in the ^{29}Si-delayed CP/MAS spectra, was 100000. ^{29}Si-single pulse excitation (SPE)/MAS spectra were recorded for samples B and C with a recycle delay of 120 s and a MAS frequency of 5 kHz. The ^{29}Si-SPE/MAS spectra recorded with a long-recycle delay (Figure 1.298a, dashed lines) show that the content of Q_2 sites ($Si(OH)_2$) is very low. This can be explained not only by features of the used measurement technique but also by the synthesis and treatment conditions because germinal groups are more sensitive to heating than single silanols (Legrand 1998, Gun'ko et al. 2009d).

Observed small changes in the spectra after modification with stearic acid (samples B and C) suggest only physical adsorption of the fatty acid onto the surface. The ^{29}Si-CP/MAS spectra of samples B and C (Figure 1.298a, solid lines) exhibited significant differences with respect to the SPE/MAS spectra. For instance, the relative intensity ratio between Q_4 and Q_3 Si signals was inverted because Q_3 silicon atoms in SiOH are usually spatially closer, and therefore more strongly dipolarly coupled, to ^{1}H nuclei with respect to Q_4 silicon nuclei in $(\equiv Si{-}O{-})_4Si$. The CP spectra of samples B and C were similar with peaks of Q_3 and Q_4 sites but with an additional signal as a shoulder of the Q_3 peak at about −94 ppm, which can be ascribed to Q_2 silicon nuclei in germinal silanols of a very small percentage in both the samples, as indicated by the impossibility of detecting their signals in the SPE spectra. In the CP spectra, the intensity of both Q_2 and Q_3 sites is significantly enhanced due to ^{1}H–^{29}Si dipolar interactions. In sample A, the only possible source of protons is water adsorbed onto a surface of $BaSO_4$ particles. Borsacchi et al. (2008) fitted the experimental ^{1}H-FID by a linear combination of a Gaussian and an exponential

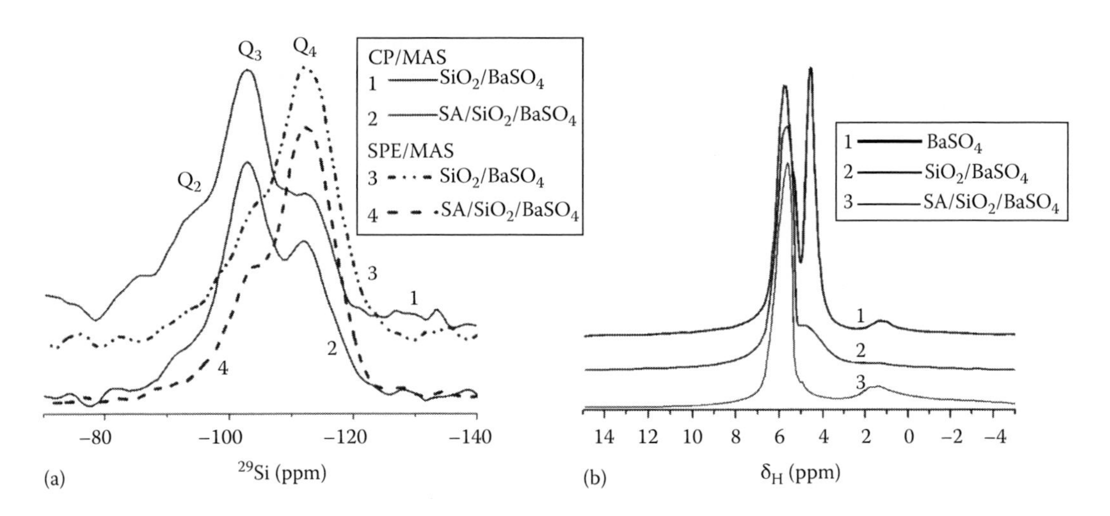

FIGURE 1.298 (a) ^{29}Si CP or SPE MAS spectra of (1, 3) $BaSO_4$ microparticles coated by silica film (1 wt%) and (2, 4) with deposited stearic acid (2.5 wt%) recorded at a MAS frequency of 5 kHz. (b) ^{1}H-MAS spectra of: (1) initial $BaSO_4$, (2) covered by silica and (3) stearic acid (by removing the first 50 μs of the FID prior to Fourier transform), (a, b) MAS frequency of 5 kHz. (From Borsacchi, S., Geppi, M., Veracini, C.A., Lazzeri, A., Di Cuia, F., and Geloni, C.: A multinuclear solid-state magnetic resonance study of the interactions between the inorganic and organic coatings of $BaSO_4$ submicronic particles. *Magn. Reson. Chem.* 20058. 46. 52–57. Copyright Wiley-VCH Verlag GmbH & Co. KGaA. Adapted with permission.)

function, characterized by T_2 of 50.7 μs and 2.0 ms, respectively. These components can indicate the presence of two motionally distinct domains in the sample, probably corresponding to two kinds of water molecules (i.e., differently bound to the solid surface) exhibiting very different dynamic behavior. The ^{1}H-FID analysis of sample B gave similar results to the FID (at 65.1 μs and 1.6 ms). Thus, the addition of a silica film covering $BaSO_4$ particles did not dramatically change the motional distribution of protons in the sample. Treatment with stearic acid strongly modified the ^{1}H-FID: the FID of sample C was well reproduced by a linear combination of two Gaussian functions and one exponential. The latter is characterized by a T_2 value very similar to that found for the exponential components of samples A and B, while its weight percentage is considerably smaller (26.5% against 66%–68%). Even though a detailed interpretation of these results in terms of dynamic behavior is not straightforward, a few hypotheses can be proposed. As far as the Gaussian components are concerned, they can be associated with two distinct rigid sample fractions: the most populated (59.7%, T_2 = 16.1 μs) in which motional processes with characteristic frequencies in the range of kHz and faster are substantially absent (25.6%, T_2 = 1.9 ms), and another, involving 13.8% of protons in the sample, experiencing a restricted mobility (T_2 = 97 μs). Even though it is very difficult to evaluate the percentage contribution of stearic acid to the whole proton content of the sample, it is possible to hypothesize that most of the protons described by the short-T_2 Gaussian component belong to the fatty acid and to water or silanols close to the silica surface. This is indeed in agreement with the results of the ^{29}Si-delayed CP/MAS experiment, which indicated that only protons with T_2 much shorter than 100 μs were significantly dipolarly coupled to ^{29}Si nuclei, and therefore spatially close to the silica surface. On the other hand, the FID components characterized by longer T_2 values can be ascribed to protons spatially far from silica, for instance, water in $BaSO_4$ particles or mobile stearic acid chains (Borsacchi et al. 2008).

Borsacchi et al. (2008) obtained useful information for a more detailed assignment of at least the ^{1}H nuclei present in the more mobile environments, that is, contributing to the slow-decaying FID components, from the ^{1}H-MAS spectra of the three samples studied (Figure 1.298b). It must be pointed out that all the three spectra contained a broad component, substantially unaffected by MAS, ascribable to both probe background and sample protons in rigid environments, whose dipolar interactions were therefore not sufficiently averaged by the molecular motions. For the sake of clarity, in the spectra shown here, such broad component has been suppressed by removing the first 50 μs of the FIDs prior to FT. The ^{1}H-MAS spectrum of sample A (Figure 1.298b, curve 1) showed two very intense and sharp peaks at 5.6 and 4.7 ppm and one, broader and weaker, centered at about 1.4 ppm. While the latter must be probably assigned to impurities present in the $BaSO_4$ powder, the two intense peaks are both due to water molecules, located either inside or on the surface of the nanoparticles, characterized by a remarkable mobility, and presumably associated with the slow-decaying exponential FID component. Comparing this spectrum with those of samples B and C, an assignment of the two different water signals can be attempted. In fact, the peak resonating at 5.6 ppm remained substantially unchanged in all the three samples, while signal at 4.7 ppm considerably decreased passing from $BaSO_4$ to the silica-coated powder and it almost disappeared in the sample treated with stearic acid. Therefore, the peak at 5.6 ppm can be assigned to water located inside the $BaSO_4$ particles, not affected by the various sample treatments. On the contrary, signal at 4.7 ppm can be ascribed to water physisorbed on the surface of the $BaSO_4$ nanoparticles, which is clearly strongly affected by silica and stearic acid treatments that partially remove this surface water and reduce its mobility. No signals arising from silica protons can be recognized in the ^{1}H-MAS spectra: their contribution is probably small and distributed between the two motionally distinct domains highlighted by the ^{1}H-FID analysis. It is reasonable that silica protons in the more mobile environment give signal in the ^{1}H-MAS spectrum probably undistinguishable from signal centered at 4.7 ppm, as expected for silanols located on the surface of non-dehydrated silica, involved in hydrogen bonds with water molecules. In the ^{1}H-MAS spectrum of sample C, a quite broad signal is also present at about 1.4 ppm, similar to that ascribed to impurities in the spectrum of sample A; in this case it is slightly more intense, reasonably because of the contribution arising from the more mobile chains of the stearic acid molecules. Therefore, the influence of the organic acid on the ^{1}H-MAS spectrum is scarce, indicating that most of its molecules experience a quite rigid environment, in agreement with the results obtained from ^{29}Si-delayed CP/MAS and ^{1}H-FID analysis (Borsacchi et al. 2008).

Zhang et al. (2007b) synthesized mesoporous silica thin films using cetyltrimethylammonium bromide $CH_3(CH_2)_{15}N^+(CH_3)_3Br^-$ (CTAB) as a template on glass slides by EISA process with a dip-coating method. The effects of sol aging on the mesophase structure of the thin films organization were investigated (Figure 1.299).

The films were studied using XRD, TEM, ^{29}Si NMR, and UV–Vis methods showing that sol aging has great effects on the mesophase structure. To obtain a better understanding of the effects of sol aging on the mesophase structure, the theories of apparent mass fractal dimension and charge density matching were introduced to explain the self-assembly process. The ^{29}Si NMR spectra show that Q_1 sites remain in a small amount even in four days aging of the sol. In a continuation of the previous work, Zhang et al. (2008) synthesized a series of continuous, crack-free, highly ordered amino-functionalized mesoporous silica thin films by cocondensation of TEOS and 3-aminopropyltriethoxysilane in the presence of CTAB, nonionic $C_{16}H_{33}(OCH_2CH_2)_{10}OH$ (Brij-56), or triblock copolymer $H(OCH_2CH_2)_{20}(OCH(CH_3)CH_2)_{70}(OCH_2CH_2)_{20})OH$ (P123) surfactants under acidic conditions by solgel dip-coating. The effect of the sol aging on the mesostructure of thin films was studied to determine the optimal sol aging time for different templates. The amino-functionalized mesoporous silica thin films exhibited long-range ordering of 2D hexagonal (p6mm) and 3D cubic (Fm3m) pore arrays of 2.2–8.3 nm in size (Zhang et al. 2008). Huang et al. (2006) used Brij-56 and P123 templates to synthesize mesostructured silica films supported on glass and silica wafer substrates. They used ^{29}Si NMR spectra to analyze the aging and composition effects.

Li and Ba (2008) prepared silica sol-gel coating of aluminum for corrosion protection using triethoxysilane (TES) as a precursor and characterized the materials using spectroscopic (NMR, FTIR, and Raman) and other methods. Electrochemical data have shown that the sol-gel coating significantly improves the corrosion protection properties of aluminum. The ^{29}Si NMR spectra of

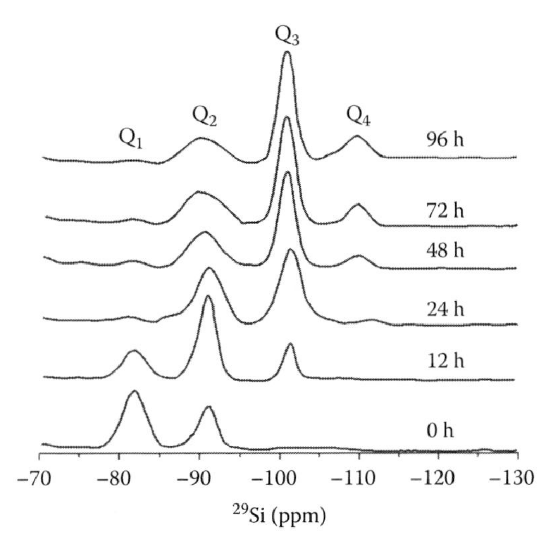

FIGURE 1.299 ^{29}Si NMR spectra from the precursor sol with different aging time under continuous stirring. (Adapted from *Thin Solid Films*, 515, Zhang, X., Wu, W., Wang, J., and Liu, C., Effects of sol aging on mesoporous silica thin films organization, 8376–8380, 2007b. Copyright 2007, with permission from Elsevier.)

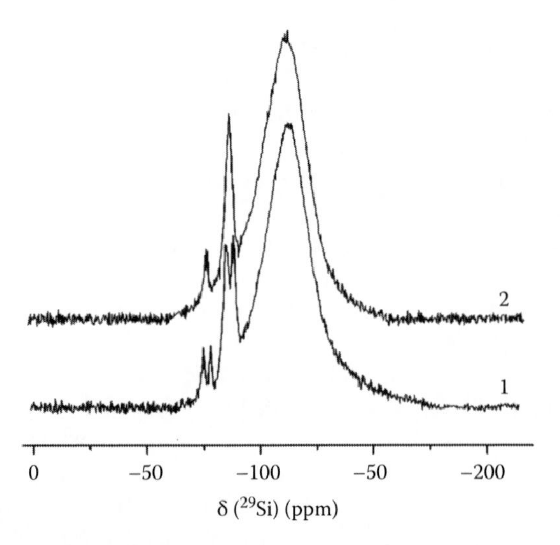

FIGURE 1.300 ^{29}Si NMR spectrum of TES sol–gel (curve 1) and spectrum with proton decoupling (curve 2). (Adapted from *Spectrochim. Acta Part A.*, 70, Li and Ba, Spectroscopic studies of triethoxysilane sol–gel and coating process, 1013–1019, 2008. Copyright 2008, with permission from Elsevier.)

TES sol–gel (broad strong line at −112.5 ppm is due to the glass sample tube) include two doublets at mean values of chemical shifts at −86.5 and −76.5 ppm (Figure 1.300, curve 1). For each of these doublets, the splitting is contributed by coupling of the hydrogen bonded directly to Si in the sol–gel. From the splitting, the coupling constant between H and Si was estimated as 300.1 Hz. As the proton was decoupled, each doublet changed into a singlet (Figure 1.300, curve 2). The presence of two Si lines indicates that all the silicon atoms in the sol-gel solution are not equivalent. This is not the expected result if the solgel process is completed and the Si–O–Si network is perfect. This result provides additional evidence that the hydrolysis involved in the sol-gel formation is incomplete. Such incomplete hydrolysis would cause the presence of nonequivalent silicon atoms in the solgel structure, confirming the result derived from vibrational spectroscopy (Li and Ba 2008). Similar results were obtained for Enterosgel, hydrated PMS (Gun'ko et al. 2007h).

Metroke et al. (2002) studied the corrosion resistance properties of spray- and dip-coated 3-glycidoxypropyltrimethoxysilane (GLYMO)–tetraethoxysilane (TEOS) Ormosil films using salt spray analysis and ^{1}H–^{13}C and ^{1}H–^{29}Si CP/MAS NMR (Figure 1.301). The latter indicated that organic content and hydrolysis water ratio affect the film structure and its corrosion resistance properties. These films comprised a dense network structure with organic groups dispersed throughout the film, providing a hydrophobic barrier coating capable of repelling water and corrosion initiators.

Silicon atoms in states T_1, T_2, T_3, Q_2, Q_3, and Q_4 correspond to R–Si(OR)$_2$(OH), R–Si(OR)(OH)$_2$, R–Si(OH)$_3$, Si(OSi)$_2$(OH)$_2$, Si(OSi)$_3$(OH), and Si(OSi)$_4$, respectively. Distinct peaks for silica network units in the GLYMO–TEOS Ormosils prepared are observed at $\delta(^{29}\mathrm{Si}) = -49.4$ to −50.7 ppm (T_1), −54.6 to −59.9 ppm (T_2), −65.1 to −68.8 ppm (T_3), −92.8 to −94.0 ppm (Q_2), −102.7 to −103.4 ppm (Q_3), and −111.1 to −113.0 ppm (Q_4) (Figure 1.301). These ^{29}Si CP/MAS NMR spectra show a decrease in the intensities of the T_n and Q_n peaks as hydrolysis water content increases due to changes in the film structure. At low water content, the Si atoms are in structural environments, which allow them to cross polarize efficiently with surface water and hydroxyl groups, presumably through a hydrated network structure. At water content increase, the relaxation time for the Si nuclei has increased, presumably due to the formation of dense SiO_2 particles. The Si nuclei within these particles appear to not be in close contact with surface water or hydroxyl groups, making effective CP more difficult and, as a result, decreased peak intensities are observed in the ^{29}Si CP/MAS NMR spectra. A similar decrease in the ^{29}Si CP/MAS NMR spectral intensities of Q_n peaks was observed by Chuang et al. (1993) in a comparative study

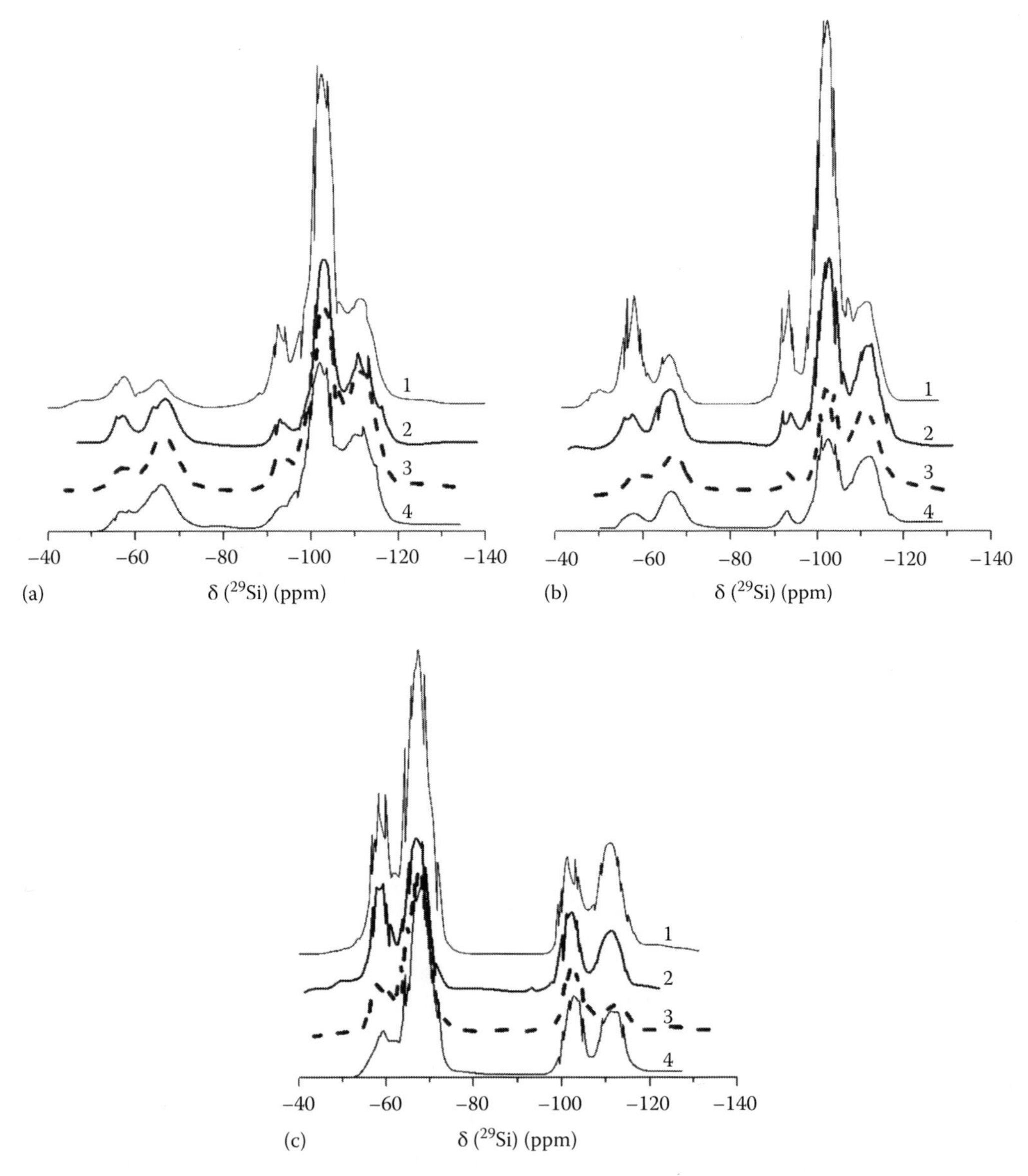

FIGURE 1.301 ^{29}Si CP/MAS NMR spectra for 11, 25, and 67 mol% GLYMO Ormosils prepared using different water content $h = (1)$ 1, (2) 2, (3) 4, and (4) 6. (Adapted from *Prog. Organ. Coating.*, 44, Metroke, T.L., Kachurina, O., and Knobbe, E.T., Spectroscopic and corrosion resistance characterization of GLYMO–TEOS Ormosil coatings for aluminium alloy corrosion inhibition, 295–305, 2002. Copyright 2002, with permission from Elsevier.)

of high and low surface area silicas. Metroke et al. (2002) assumed that as hydrolysis water content increases, the structure of the GLYMO–TEOS Ormosils changes from a high surface area, network structure, composed primarily of linear polymers, throughout which epoxide functionalities are dispersed to a lower surface area, particulate structure in which the surfaces of dense silica particles are covered with epoxide functionalities.

Szekeres et al. (2003) prepared monolayers of Stöber silica particles on the surface of water and deposited onto glass substrates by the Langmuir–Blodgett method. Prior to film formation the surface of the silica particles was methoxylated by washing with methanol at room temperature. This reaction

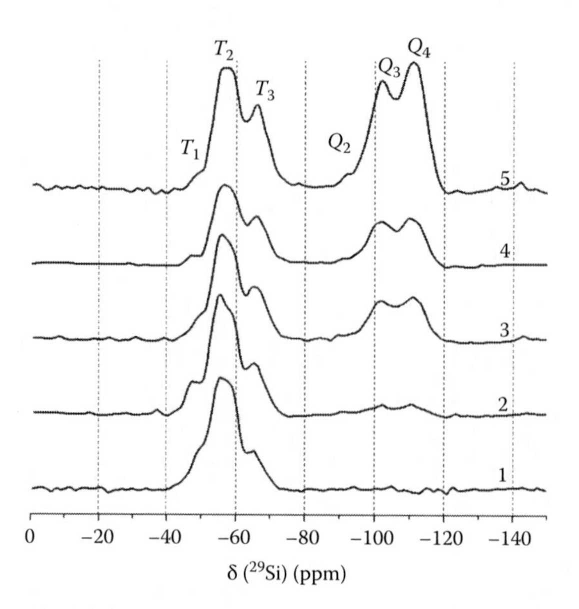

FIGURE 1.302 ^{29}Si CP/MAS NMR spectra of the hybrid films at the silica content C_{SiO_2}=(1) 0, (2) 10, (3) 33, (4) 40 and (5) 50 wt%. (Adapted from *Polymer*, 44, Yu, Y.-Y., Chen, C.-Y., and Chen, W.-C., Synthesis and characterization of organic–inorganic hybrid thin films from poly(acrylic) and monodispersed colloidal silica, 593–601, 2003. Copyright 2002, with permission from Elsevier.)

was controlled using ^{13}C CP MAS NMR and ^{29}Si MAS NMR spectroscopy in parallel to FTIR study. It was shown that hydrophilic silica particles with surface silanol groups are less suitable for 2D crystalline ordering using the Langmuir–Blodgett (LB) technique than methoxylated particles.

Yu et al. (2003) synthesized hybrid thin films containing nanosized inorganic domains from poly(acrylic) and monodispersed colloidal silica with bound 3-(trimethoxysilyl)propyl methacrylate (MSMA) for potential applications as passive films in optical devices. The latter was polymerized with acrylic monomer to form a precursor solution, which was spin-coated and cured to form hybrid films. The silica content in the hybrid thin films was varied from 0 to 50 wt% (Figure 1.302). The coverage area of silica particles by the MSMA decreased with increasing silica content and resulted in the aggregation of silica particles in the hybrid films, and the silica domains were 20–35 nm in size.

The spectra show six peaks at −49.5 to −50.9, −55.4 to −56.6, −65.1 to −66.1, −90.2 to −93.4, −101.3 to −102.2, and −110.3 to −111.4 ppm, which are assigned to T_1, T_2, T_3, Q_2, Q_3, and Q_4, respectively. The splitting of the T_2 signal is interpreted by the spin–spin splitting of alkyl group. The proportion of the T_i, Q_i, and the degree of condensation D_c in the hybrid materials was determined by a quantitative analysis based on the peak areas of species. The D_c value

$$D_c(\%) = 100\left(\frac{T_1 + 2T_2 + 3T_3}{3} + \frac{Q_1 + 2Q_2 + 3Q_3 + 4Q_4}{4}\right) \quad (1.76)$$

was changed between 72.0 (C_{SiO_2}=0) and 83.6% (C_{SiO_2}=50 wt%).

Thus, the ^{1}H, ^{13}C, and ^{29}Si NMR spectroscopy was frequently used to study silica thin films, hybrid organosilica, and other films. However, these applications deal with chemical structure of the films. These are no publication with NMR cryoporometry applied to thin silica films.

Oxide materials based on silica but including other oxides such as titania, alumina, germania, and zirconia are next objects under investigations.

2 Interfacial Phenomena at Surfaces of Mixed Oxides

2.1 MIXED NANOOXIDES

Binary and ternary fumed oxides (or nanooxides) as more complex materials than individual oxides can have properties different from that of the latter. This corresponds to a more complex interfacial behavior of adsorbates including water, low and high molecular organics, proteins, cells, etc. (Gun'ko et al. 2001e, 2005d, 2007d,f,i, 2009d, Blitz and Gun'ko 2006) at the surface of mixed oxides. The surface of mixed oxides can more strongly influence a thicker adsorption layer than silicas studied in Chapter 1. The main reason of this difference is due to appearance of novel sites at the surface of complex nanoparticles and a specific role of bridging hydroxyls. The concentration and types of hydroxyl groups such as terminal ≡MOH and bridging ≡MO(H)M≡ or $(\equiv M)_3OH$ at the surface of individual oxide fragments (which can be also characterized by hypervalent surroundings of Al, Ti, Zr, Ge, or other metal atoms), $\equiv M_1O(H)M_2\equiv$ present only at the surface of mixed oxides (where M, M_1, and M_2 are metal atoms and $M_1 \neq M_2$), and other structures with hydroxyls and neighboring incomplete O-coordinated metal atoms (Lewes [L] acid sites) determine the physicochemical properties of the oxide materials used as adsorbents, catalysts, fillers, pigments, carriers, etc. (Tanabe 1970, Iler 1979, Dabrowski and Tertykh 1996, Busca 1998, Brown et al. 1999, Gun'ko et al. 2001e, Tsantilis et al. 2002, Zhou and Fedkiw 2004, Teleki et al. 2008, Buesser and Pratsinis 2010). Typically, bridging hydroxyls $\equiv M_1O(H)M_2\equiv$ (e.g., M_1 = Si and M_2 = Al or Ti) corresponding to Brønsted (B) acid sites (i.e., proton-donor sites) possess higher acidity than ≡MO(H)M≡ (M = Al, Ti, etc.) (Tanabe 1970). Terminal hydroxyls ≡MOH can be amphoteric (on silica, titania, etc.), i.e., a proton-donor or proton-acceptor depending on the kind of an adsorbate, or basic electron-donor sites (on alumina, magnesia, etc.). These features of surface hydroxyls (B sites) and metal atoms (L sites) determine the catalytic properties of oxides in acid–base reactions, the structure of electrical double layer (EDL) in aqueous media, the adsorption phenomena in different dispersion media, etc. (Tanabe 1970, Kiselev and Lygin 1972, Iler 1979, Dabrowski and Tertykh 1996, Brown et al. 1999, Gun'ko et al. 2005b,d, 2007a,d,f,i, 2009d). Therefore, it is evident that investigations of the properties of surface hydroxyls as the main adsorption and reaction sites at the oxide surfaces are of importance for detailed characterization of complex oxide materials and better understanding of the interfacial phenomena occurring at their surfaces. Different methods such as nuclear magnetic resonance (NMR), Fourier transform infrared (FTIR), Raman spectroscopy, x-ray photoelectron spectroscopy (XPS), Auger electron spectroscopy (AES), adsorption/desorption of probe compounds from gaseous or liquid media, thermogravimetry with differential thermal analysis (TG/DTA), differential scanning calorimetry (DSC), temperature-programmed desorption with mass-spectrometry (TPD-MS) control, chemical reactions, theoretical modeling, etc. can be applied to study metal oxide hydroxyls, whole surface, adsorption layer structure, and interfacial behavior of adsorbates. Combinations of several mentioned methods provide a deeper insight into the relationships between the structure of the hydroxyl cover and the total surface layer, properties of the materials, and features of the interfacial phenomena (Gun'ko et al. 2009d). Therefore, not only the NMR spectroscopy results for mixed nanooxides would be analyzed here but also results obtained using other methods mentioned earlier as it was done for silicas discussed earlier. Notice that combination of different investigation methods can be more important for mixed oxides than silica because of their greater complexity. Frequently, we will compare the properties of complex oxide materials with those of the corresponding individual oxides such as silica, alumina, titania, etc.

Complex nanooxides, such as SiO_2/Al_2O_3 (SA), SiO_2/TiO_2 (ST), $Al_2O_3/SiO_2/TiO_2$ (AST) (Table 2.1), SiO_2/ZrO_2, SiO_2/B_2O_3, and others can be synthesized using a mixture of metal chlorides MCl_n as precursors and a flame synthesis technique similar to that used to synthesize fumed silica (Degussa 1996, 1997, Chuiko 2001, 2003, Blitz and Gun'ko 2006, Gun'ko et al. 2007a, 2009a,b,g, 2010a,d,e). However, there are several essential differences between fumed silica and mixed oxides (Table 2.1).

This is due to a set of the effects. First, there is different number and types of surface hydroxyls, (M_iOH, $M_i(OH)_2$, $M_iO(H)M_j$, and $(MO)_3H$), which can be both stronger and weaker Brønsted acid sites than ≡SiOH (and can affect the formation of nanoparticles during the synthesis). Second, Lewes acid sites are absent at the silica surface as well as Lewes basic sites (O with high negative charge). Third, different oxide phases can be nonuniformly distributed at the surface and volume of nanoparticles. Additionally, mixed oxides have larger true and bulk densities of MO_x than silica, and typically $S_{BET}<200$ m²/g if TiO_2 is a component of the nanooxides (Table 2.1). However, the shape of nanoparticles of mixed nanooxides and nanosilica is practically the same, roughly spherical due to high temperatures of the flame synthesis.

TABLE 2.1
Structural Characteristics and the Heat of Immersion of Nanooxides in Water

Sample	$C(SiO_2)$ (wt%)	$C(TiO_2)$ (wt%)	$C(Al_2O_3)$ (wt%)	S_{BET} (m²/g)	V_p (cm³/g)	ΔH_{im} (J/m²)
A-50	100	0	0	52	0.126	0.23
A-200	100	0	0	230	0.463	0.18
A-500	100	0	0	492	0.874	0.14
SA1	98.7	0	1.3	203	0.416	0.28
SA3	97	0	3	185	0.405	0.21
SA5	95	0	5	266	0.719	0.25
SA8	92	0	8	303	0.688	0.12
SA23	77	0	23	347	0.788	0.13
SA30	70	0	30	238	0.643	
SA75	25	0	75	118	0.320	0.34
SA96	4	0	96	81	0.163	0.46
Al_2O_3		0	100	125	0.262	0.37
ST2	98	2	0	77	0.263	0.36
ST9	91	9	0	235	0.580	0.18
ST14	86	14	0	156	0.386	0.26
ST20	80	20	0	84	0.174	0.34
ST40	60	40	0	148	0.333	0.24
ST63	33	63	0	84	0.215	
ST65	35	65	0	34	0.080	0.40
ST94	6	94	0	30	0.100	0.33
TiO_2	0	100	0	42	0.117	0.26
AST03	1.7	0.3	98	125	0.308	0.35
AST06	20	0.6	79.4	97	0.234	
AST1	10	1.0	89	99	0.253	0.55
AST50	28	50	22	37	0.095	0.47
AST71	8	71	21	74	0.127	0.59
AST82	6	82	12	39	0.150	0.49
AST87	4	87	9	42	0.148	0.72
AST88	8	88	4	39	0.123	0.64

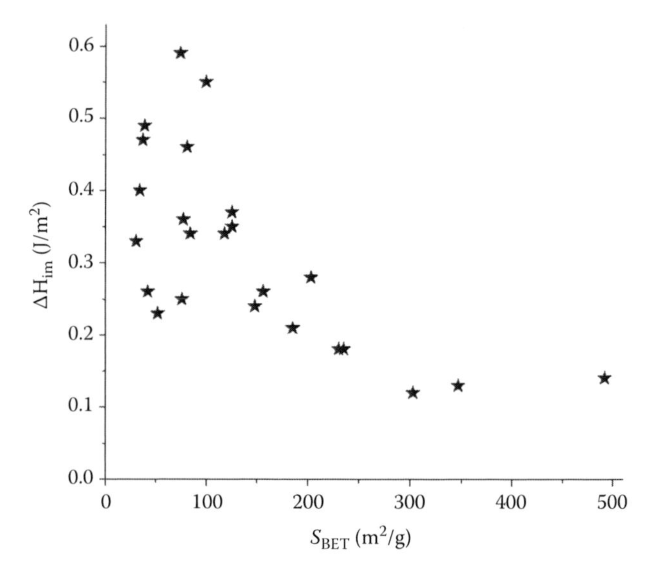

FIGURE 2.1 Relationship between the specific surface area of nanooxides and the enthalpy of immersion of them in water.

The structural features of mixed oxides with a mosaic surface of nanoparticles, including patches of different oxide phases or a solid solution of a lower concentrated oxide in a more concentrated oxide, can strongly affect the interfacial phenomena in any media. For instance, the enthalpy of immersion in water (Table 2.1; ΔH_{im}) is greater for mixed nanooxides than nanosilica. However, surface nonuniformity and the differences in the properties of a variety of surface sites result in a significant scatter in the relationship between the ΔH_{im} and S_{BET} values (Figure 2.1), despite a tendency of a decrease in the ΔH_{im} value (calculated per surface area unit) with increasing S_{BET} value.

Greater differences in the properties of silica and non-silica phases in mixed oxides can be elucidated in detail if the properties of mixed nanooxides are analyzed with consideration of the surface content of all oxide phases (Gun'ko et al. 2005b, 2007d,i, 2009d). For complex nanooxides, fumed or prepared using chemical vapor deposition (CVD) method, practically all textural or physicochemical characteristics are nonlinear functions of the total content (C_X) of a more active phase (e.g., alumina, titania, germania, and zirconia) in fumed SA and ST or CVD-TiO_2/SiO_2 and CVD-GeO_2/SiO_2 (Figure 2.2) and other materials (Gun'ko et al. 2009d).

The curve courses of the specific surface area are opposite for SA and ST but for CVD oxides, it decreases with C_X (Figure 2.2a) because of the formation of larger X particles (of a greater true density than silica) with increasing C_X value. These effects depend also on different hydrolytic stability of the Si-O-Ti, Si-O-Al, Si-O-Ge, and Si-O-Si bonds (affecting changes in the particle morphology during synthesis), different true density of these oxides, larger or smaller differences in their lattice parameters, etc. (Gun'ko et al. 2005d, 2007f, 2009d).

The effective diameter of particles in the aqueous suspensions determined using photon correlation spectroscopy, photon correlation spectroscopy (PCS) is maximal at low C_X values (Figure 2.2b) because of the difference in the isoelectric points (IEPs) of silica ($pH_{IEP} \approx 2.2$), alumina (9.8), and titania (~6.0), as well as in point of zero charge (PZC), that can cause attractive interactions between particles (or surface patches) with different phases with the charges of the opposite signs (Gun'ko et al. 2001e, Kang et al. 2011).

The free surface energy, γ_S, value determined using low-temperature (LT) ^{1}H NMR spectroscopy is a measure of both free surface energy of a solid surface (in vacuum) and changes in the free energy of water due to interactions with this surface. It is maximal at small C_X values (Figure 2.2c) when the surface of complex materials is maximum nonuniform and guest oxide phase represents

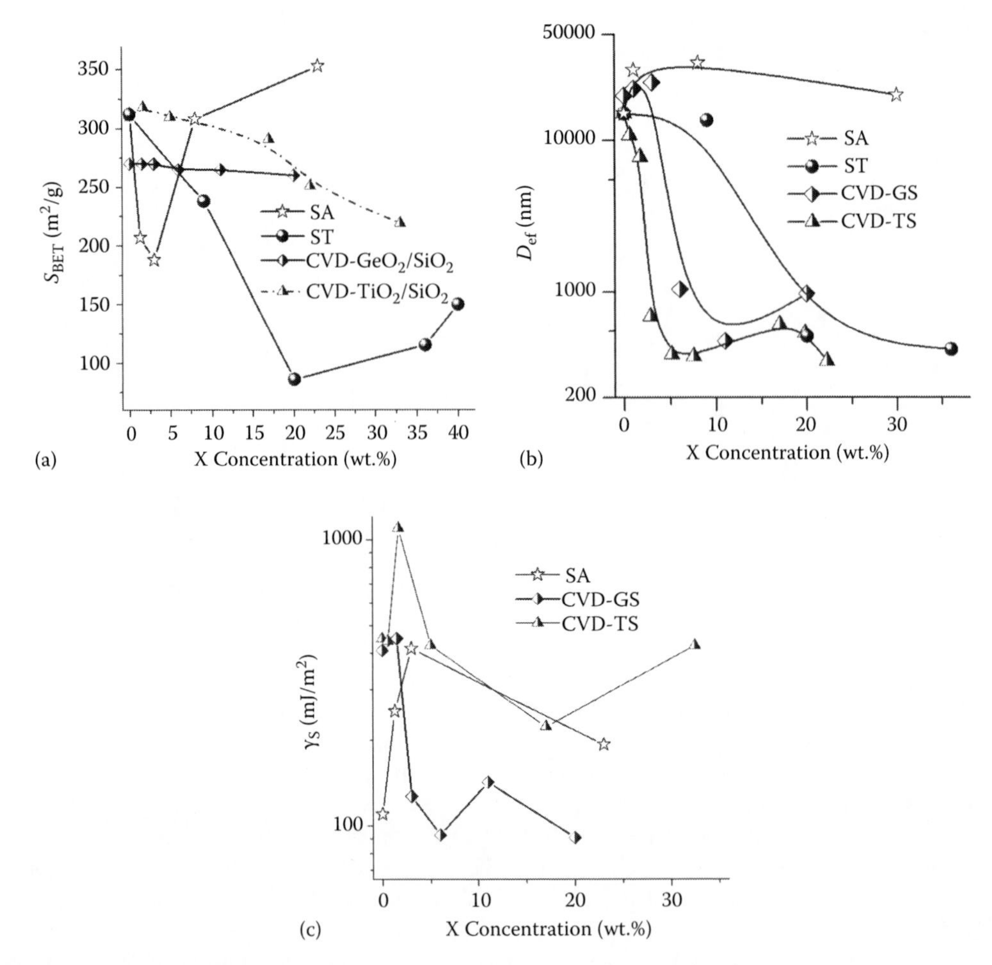

FIGURE 2.2 Changes in (a) S_{BET} (LTNA); (b) effective diameter of X/SiO_2 particles at pH between 2.9 and 6.8 (PCS); and (c) free surface energy of oxides in aqueous media (^1H NMR) as a function of concentration of X phase in X/SiO_2. (Adapted with permission from Gun'ko, V.M., Zarko, V.I., Turov, V.V., Leboda, R., and Chibowski, E., The effect of second phase distribution in disperse X/silica (X = Al_2O_3, TiO_2, and GeO_2) on its surface properties, *Langmuir*, 15, 5694–5702, 1999c, Copyright 1999, American Chemical Society.)

of the smallest nanoparticles (e.g., in the CVD synthesis) or surface patches in fumed mixed oxides (Gun'ko et al. 1999b). To deeper insight into the nonlinear dependences of different characteristics of mixed nanooxides, the surface structure should be analyzed in detail using a set of methods (AES, NMR, x-ray diffraction [XRD], XPS, Raman spectroscopy, etc.), and first of all, the surface content of alumina and titania should be studied in such fumed mixed oxides as SA, ST, and AST (Table 2.1).

2.1.1 Surface Structure of Nanoparticles

Surface content of aluminum (at%) in SA and AST and titanium in ST and AST was determined from the Auger electron spectra (AES) recorded using a JAMP-10S (JEOL) spectrometer (Gun'ko et al. 2007a and Chapter 10). Complex nanooxides, SA, ST, and AST, are characterized by nonlinear dependences of surface content of alumina $\left(C^{s}_{Al_2O_3}\right)$ and titania $\left(C^{s}_{TiO_2}\right)$ on their total content (C_{Al_2O} and C_{TiO_2}) in the materials (Figure 2.3). These nonlinear dependences are due to differences in (i) reactivity of $SiCl_4$ (vapor) and $AlCl_3$ (forming dimers Al_2Cl_6 and being in solid state, therefore it should be sublimated to supply to the flame) in the H_2/O_2 flame; (ii) reactive species concentrations

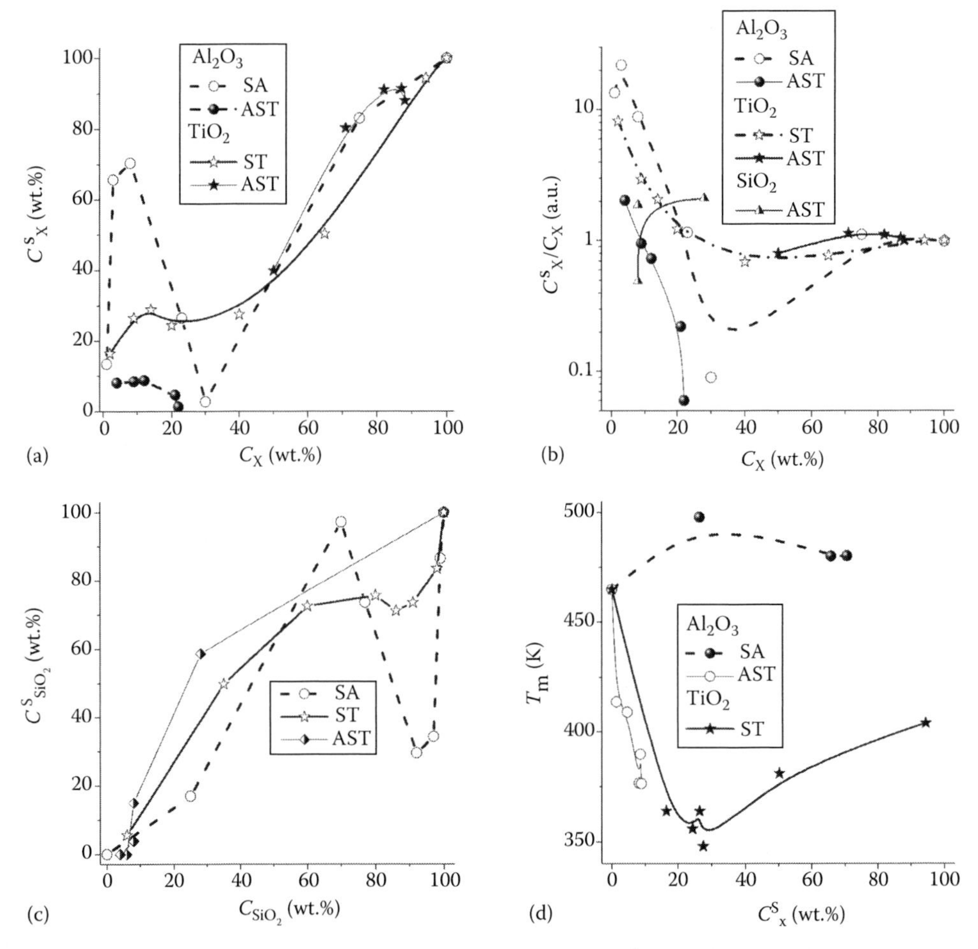

FIGURE 2.3 Relationships between total C_X and surface content C^S_X of (a) alumina and titania, (c) silica in SA, ST and AST samples; (b) relationship between ratio C^S_X/C_X and the total content of the second phases, and (d) relationship between the surface content of alumina or titania and the peak temperature of TPD-MS thermograms of desorbed water. (Adapted from *J. Colloid Interface Sci.*, 314, Gun'ko, V.M., Blitz, J.P., Gude, K. et al., Surface structure and properties of mixed fumed oxides, 119–130, 2007a, Copyright 2007, with permission from Elsevier.)

in different zones of the flame; and (iii) synthesis conditions for different mixed oxides (average temperature, temperature gradient in the flame, flame length, turbulence, velocity, etc.). The $TiCl_4$ (vapor) reactivity is similar to that of $SiCl_4$. Therefore, surface titania content exhibits a more linear dependence as a function of the total titania content than that for alumina (Figure 2.3). Small deviations from linearity for titania can be due to variations in the synthesis conditions.

Surface hydroxyl properties (Figures 2.3d and 2.4) depend on the surface content of alumina or titania in nanooxides characterized by the AES method (Figure 2.3). For instance, one-pass temperature-programmed desorption with mass-spectrometry control (OPTPD-MS; Figure 2.3d) shows a reduction in H_2O desorption peak temperature with increasing surface alumina content in AST and titania content <30% in ST. However, an increase in the alumina content in SA gives the opposite result. This can be caused by differences in the nanooxide surface composition with respect to the surface content of alumina or titania (Figure 2.3) that affect the properties of surface hydroxyls (both terminal and bridging). In addition, the distribution of Al atoms in silica (SA) and titania (AST) phases and Ti atoms in silica (ST) and alumina (AST) phases can also differ that

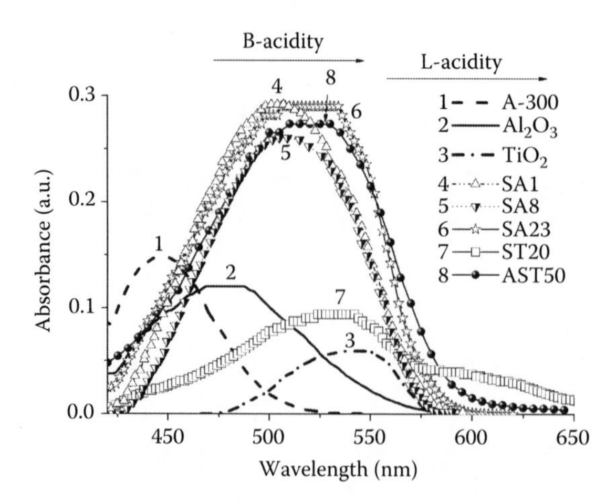

FIGURE 2.4 Optical spectra of DMAAB adsorbed onto different nanooxides (sample weight was selected to provide close surface area of all samples). (Adapted from *Appl. Surf. Sci.*, 253, Gun'ko, V.M., Nychiporuk, Yu.M., Zarko, V.I. et al., Relationships between surface compositions and properties of surfaces of mixed fumed oxides, 3215–3230, 2007d, Copyright 2007, with permission from Elsevier.)

result in varying surface properties such as Brønsted acidity of surface hydroxyls (Figure 2.4); i.e., the proton mobility appearing in low-temperature NMR and thermally stimulated depolarization current (TSDC) data (vide infra). This also leads to different dependences in the peak temperature of associative desorption of water from SA, ST, and AST samples (Figure 2.3d).

To study acid–base features of the surface of mixed oxides, (dimethylamino)azobenzene (DMAAB; $pK_a = 3.3$; see Chapter 10) was chosen as a color indicator to estimate the content of different active sites on the oxide surfaces by the optical spectroscopy (Figure 2.4). Various surface centers such as strong B and L acid sites are characteristic for mixed oxides with SA, ST, and AST in contrast to silica, having only weak amphoteric sites with surface silanols (Figure 2.4). Changes in the surface content of titania in ST, or alumina in SA, affect the content of acidic bridges ≡SiO(H)Ti≡ or ≡SiO(H)Al≡, and these changes are nonlinear functions of the total content of these oxides. Therefore, the spectra of DMAAB adsorbed onto mixed oxides demonstrate much higher intensity than for individual oxides (Figure 2.4) at a given surface area. Silica has minimal B acidity among the studied oxides since the corresponding spectrum shifts toward shorter wavelengths. An increase in the number of the types of surface sites (mainly B and L acid sites) for mixed oxides causes significant broadening of the spectra of DMAAB adsorbed onto SA, ST, and AST, in comparison with that for individual silica, alumina, and titania.

It should be noted that the local properties of surface sites depend not only on the nearest surroundings but also on the structure of several coordination spheres. For instance, the B acidity of bridging hydroxyls increases with decreasing number of Al atoms in alumina/silica materials (Tanabe 1970). This influence reflects in such features of Si atoms in mixed oxides as the ^{29}Si chemical shift in different structures with $Si(OM-)_4$ (Q^4_{si} in Table 2.2; Figure 2.5a and b), $Si(OM-)_3(OH)$ $\left(Q^3_{si}\right)$, and $Si(OM-)_2(OH)_2$ $\left(Q^2_{si}\right)$ (where M = Si, Al or Ti) and their contributions to the total spectra (Humbert 1995, Peeters and Kentgens 1997, Müller et al. 2000a,b, Babonneau and Maquet 2000, Ganapathy et al. 2003, Isobe et al. 2003, MacKenzie 2004, Al-27 2011, Ti-47,49 2011).

The ^{29}Si (resonance frequency 59.595 MHz) cross polarization with magic angle spinning (CP/MAS) NMR spectra and the ^{27}Al (resonance frequency 78.172 MHz) MAS-NMR spectra (program ZG) of mixed oxides were recorded using a Bruker Avance™ 300 NMR spectrometer (magnetic field of 7.046 T, a spinning rate of 8 kHz of 4 mm zirconia rotor) (Gun'ko et al. 2007d). Chemical shifts of ^{29}Si and ^{27}Al were referenced to tetramethylsilane (TMS) and an $Al(NO_3)_3$ aqueous solution, respectively; i.e., the resonance of $Si(CH_3)_4$ and $Al(H_2O_6)_6^{3+}$ was set to 0 ppm.

TABLE 2.2
Contributions (%) of Different Centers with Si and Al in SA, ST, and AST Samples Determined as Relative Integral Intensity of the Bands Obtained by Deconvolution of the ^{29}Si CP/MAS and ^{27}Al MAS NMR Spectra

Sample	Q^4_{si}	Q^3_{si}	Q^2_{si}	$Q_{Al(VI)}$	$Q_{Al(V)}$	$Q_{Al(IV)}$
SA1	45.7	33.3	21.0	60.3		39.7
SA3	31.1	52.6	14.2	77.9	1	21.1
SA8	48.0	45.0	7.0	73.4		26.6
SA23	45.5	33.4	21.1	55.9	3.5	40.6
SA30	28.8	50.7	20.5	51.0	4.2	44.8
ST9	33.6	44.7	21.7			
ST14	44.0	31.8	24.2			
ST20	29.1	54.6	16.3			
ST63	57.3	16.6	26.1			
ST65	64.9	19.5	15.6			
AST50	68.1	23.6	8.3	86.2		13.8
AST82				94.1	3.4	2.5

The chemical shifts are affected by the number and types of surface and bulk structures with different O-coordination numbers of aluminum atoms changing with increasing $C_{Al_2O_3}$ value as well as contributions of six-, five-, and fourfold O-coordinated Al atoms. In the case of a great titania content in ST94 or AST88 ($C_{SiO_2}=6$ and 8 wt% respectively), silica does not represent a separate phase because it forms only a solid solution in titania. However, titania in ST9 or ST with larger C_{TiO_2} values forms its own phase, observed in XRD patterns (vide infra).

Additionally, the distribution of alumina in SA and titania in ST at the particle surface as a function of the total content of alumina or titania differs (a smoother curve is for ST than SA). Therefore, a small amount of Si atoms at the surface of ST samples at high C_{TiO_2} values causes low intensity of the Q^3_{si} and Q^2_{si} states and high intensity of the Q^4_{si} state (Figure 2.5) corresponding to the Si atoms in the bulk. This is in agreement with the FTIR spectra of the materials (Figure 2.6) showing decreased intensity of the O–H-stretching vibrations of free silanols at 3740–3750 cm^{-1} with increasing surface content of alumina or titania in mixed oxides. A similar tendency for SA samples is more complex than that for ST samples because of more strongly nonlinear changes in the surface content of alumina (Figure 2.3).

An increase in the content of alumina in SA samples and titania in ST samples leads to the opposite results with respect to contribution of Q^4_{si} at −110 ppm (Figure 2.5; Table 2.2) because of the difference in the electronic properties of titania (semiconductor) and alumina (dielectric) that change the shielding of the Si cores in the similar Q sites but with different surroundings, i.e., the position of the Q^4_{si} peak can change in mixed oxides in comparison with pure silica. Changes in Q^3_{si} at −101 ppm and Q^2_{si} at −91 ppm do not have clear tendency to a decrease or an increase depending on the $C_{Al_2O_3}$ and C_{TiO_2} values. The peak positions of all the Si sites can change in the ^{29}Si CP/MAS spectra of complex oxides in comparison with pure silica. Deviation from a linear dependence of contribution of these sites on the $C_{Al_2O_3}$ and C_{TiO_2} values, is caused by nonlinear changes in the surface content of these oxides (Figure 2.3) because the Q^3_{si} and Q^2_{si} sites correspond to surface silanols whose content directly depends on the surface content of alumina and titania (a greater content of them causes the decrease in surface content of silica, Figure 2.3). Additionally, the effects of titania on the shielding of the Si cores can more strongly affect the position of the corresponding components in the ^{29}Si

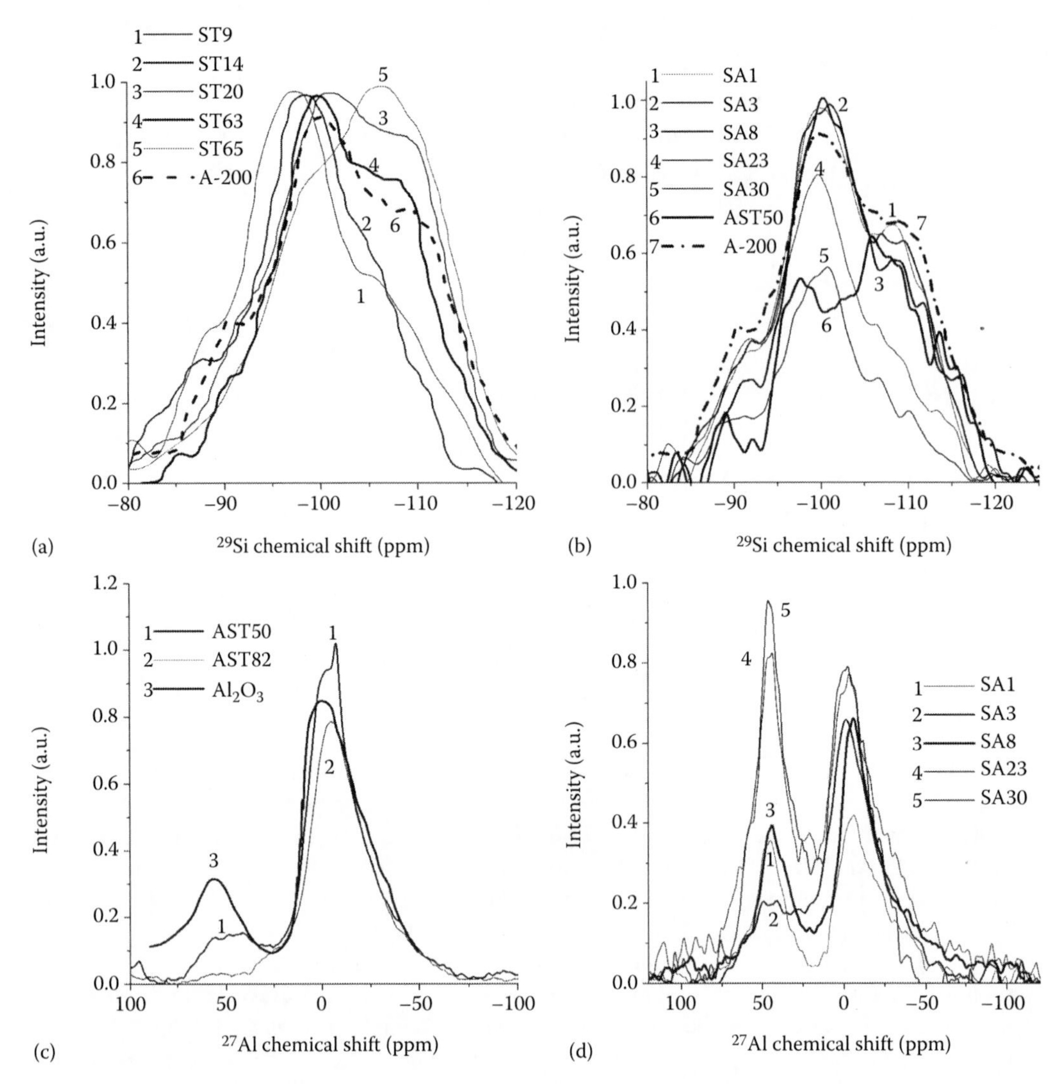

FIGURE 2.5 (a and b) ^{29}Si CP/MAS and (c and d) ^{27}Al MAS NMR spectra of nanooxides (a) ST and A-200, (b and d) SA and A-200, and (c) AST and Al_2O_3 samples. (Adapted from *Appl. Surf. Sci.*, 253, Gun'ko, V.M., Nychiporuk, Yu.M., Zarko, V.I. et al., Relationships between surface compositions and properties of surfaces of mixed fumed oxides, 3215–3230, 2007d, Copyright 2007, with permission from Elsevier.)

CP/MAS NMR spectra than alumina does (Figure 2.5) because of the difference in the electronic structure of these oxides.

A semiconducting titania phase as an electron-donor for the silica phase can change shielding of the Si nuclei located close to the titania fragments. Additionally, the Si(OM≡)$_4$ structure (i.e., bond lengths and valence angles) changes if M ≠ Si. Therefore, the ^{29}Si NMR spectra slightly shift and the Q^4_{si} intensity increases with increasing C_{TiO_2} value in ST samples (Figure 2.5a). The electronic properties of dielectric alumina and silica (as well as the sizes of Si and Al atoms) are closer than those of titania and silica. Therefore, a similar displacement of the ^{29}Si NMR spectra is not observed for SA samples with increasing $C_{Al_2O_3}$ value (Figure 2.5b). The number of the Q^2_{si} sites decreases with increasing surface content of alumina or titania in mixed oxides (Table 2.2). For instance, it is minimal for SA8 (Figure 2.5b) with the maximal content of surface alumina (Figure 2.3). The number of the Q^3_{si} sites is maximal for ST20 and SA3 because of several reasons related to the

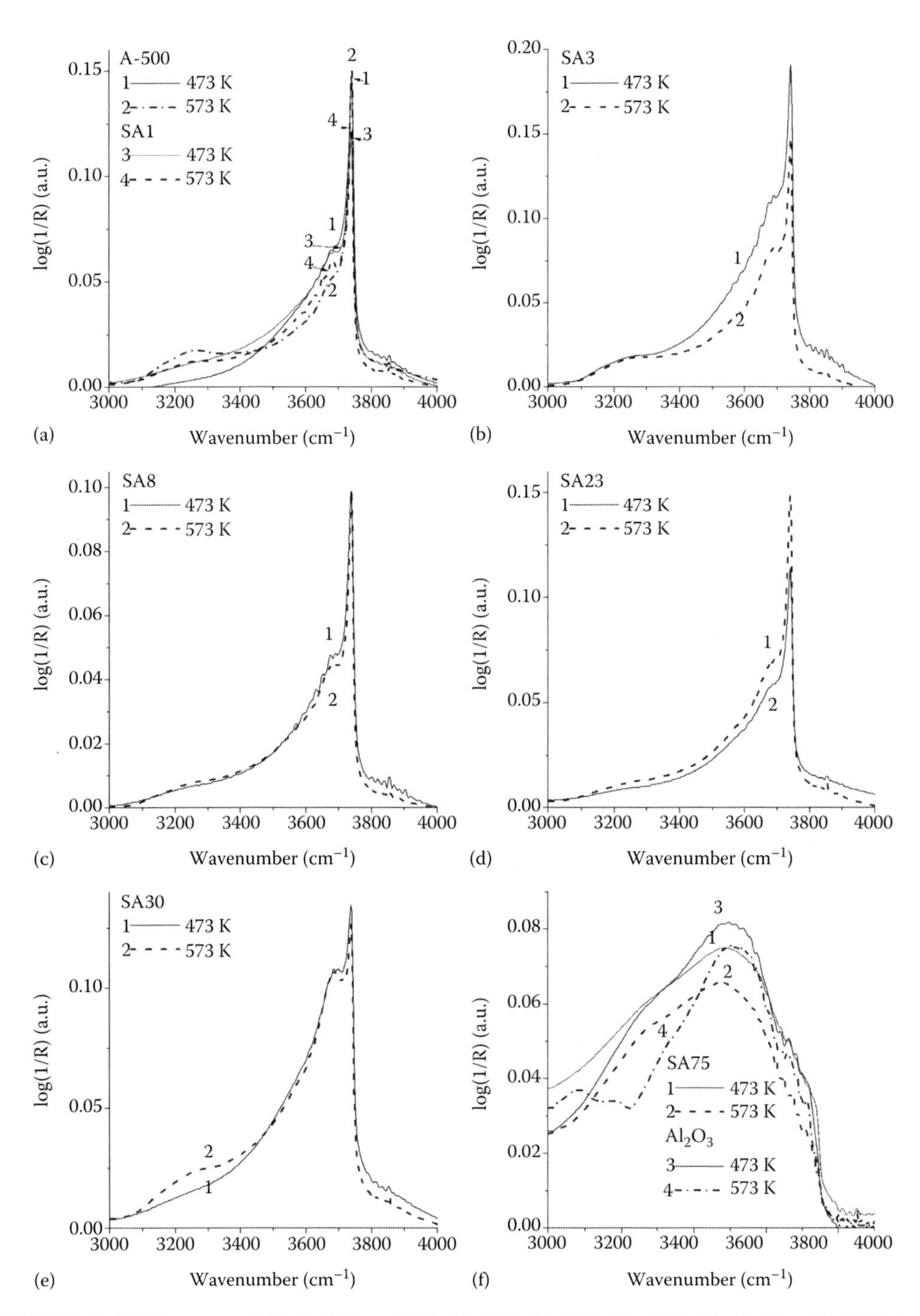

FIGURE 2.6 FTIR spectra of (a) A-500 and SA1, (b) SA3, (c) SA8, (d) SA23, (e) SA30, and (f) SA75 and Al_2O_3 recorded in the diffuse reflectance mode at 473 and 573 K. (Adapted from *J. Colloid Interface Sci.*, 314, Gun'ko, V.M., Blitz, J.P., Gude, K. et al., Surface structure and properties of mixed fumed oxides, 119–130, 2007a, Copyright 2007, with permission from Elsevier.)

surface content of the second oxide, its distribution form in the bulk, the size of its particles and surface patches, the number of contacts of these patches with silica, and the number of centers with isomorphic substitution of Si for Al or Ti atoms.

According to the ^{27}Al MAS-NMR spectra (Figure 2.5c and d), the content of sixfold O-coordinated Al(VI) (^{27}Al resonance at ~0 ppm) in the alumina phase is larger than that of fourfold O-coordinated Al(IV) (^{27}Al resonance at ~50 ppm) in SA samples (Table 2.2). This suggests that alumina forms rather individual phase (e.g., γ-Al_2O_3 [which represents ~20% in individual fumed alumina] includes approximately 80% Al(VI) and 20% Al(IV) or individual amorphous phase) than a solid solution in the silica matrix with substitution of fourfold O-coordinated Si and keeping this coordination state. Notice that a peak intensity of the Al(IV) resonance is higher than that of Al(VI) for SA23 and SA30; however, the Al(VI) band is broader (Figure 2.5), and a larger full-width at half-maximum value cause a greater integral contribution of Al(VI) sites (Table 2.2). In the case of AST82 synthesized at a high temperature (>1700 K) than SA samples, the alumina phase includes mainly Al(VI) because (i) patches with individual alumina can have a structure close to that of a high-temperature (HT) modification of alumina (e.g., α-Al_2O_3 with only Al(VI)) and (ii) a dense solid solution of alumina in titania corresponds to substitution of sixfold O-coordinated Ti(VI) by Al(VI). In the case of AST50 including a larger content of silica but a smaller content of titania (Table 2.1) and synthesized at a lower temperature than AST82, an individual alumina phase can include patches with the structure close to that of γ-Al_2O_3 or amorphous alumina including both Al(VI) and Al(IV). Contribution of Al(VI) in AST50 is greater than that in γ-Al_2O_3 because of the reasons mentioned earlier with respect to AST82. However, lower synthesis temperature and possible incorporation of Al(IV) into the silica lattice (as AST50 includes 28% of silica) give a larger amount of Al(IV) in AST50 than that in AST82. Notice that the ^{29}Si CP/MAS NMR spectrum of AST82 (which is not shown here) in contrast to AST50 does not have clear bands, and the Si–O(Si) stretching vibrations at 1100 cm^{-1} are not observed in the FTIR spectrum of AST82 as well as AST88. These results suggest that silica is absent as an individual phase in AST82, AST87, and AST88 because it forms only the solid solution in titania and alumina. The opposite effects of titania and alumina on the ^{29}Si chemical shift provide low signal intensity of ^{29}Si (at a noise level) in the AST samples with the exception of AST50. A low amount of Al(V) (^{27}Al chemical shift at 30 ppm) is observed for certain samples (Table 2.2); however, the clear Al(V) band is observed only for SA23 (Figure 2.5d) characterized by a low amount of surface alumina (Figure 2.3).

If the ratio $Q_{Al(VI)}/Q_{Al(IV)} \approx 4$ corresponds to an individual Al_2O_3 phase (γ-Al_2O_3 or amorphous) in mixed oxides, the value $\beta = Q_{Al(IV)}/(Q_{Al(IV)} + Q_{Al(IV)}) - 0.2$ can be used as a measure of a number of Al atoms isomorphically substituting Si atoms in the silica lattice. This estimation for SA3 ($\beta = 0.011$) shows that practically all alumina is at the surface that is in agreement with the AES data. However, in the case of other SA samples, a portion of the Al atoms can be embedded in the silica matrix, and a maximal number of these Al atoms ($\beta > 0.2$) is observed for SA30 and SA23 (Table 2.2), which are characterized by relatively low surface content of alumina according to the AES data (Figure 2.3).

The dependence of Brønsted acidity on the $C^{s}_{Al_2O_3}$ value and the nonlinear dependence of surface alumina concentration on the total content of alumina in SA and AST samples (Figure 2.3), as well as differences in the morphology of particles and the size distribution of voids between these particles, can cause nonlinearity of the interfacial characteristics of both mixed oxides and adsorbed water or water layers structured by the oxide surface in aqueous suspensions. For analysis of the interfacial phenomena in the aqueous suspensions, the data related to dry powders and dried solid residues obtained from the suspensions of nanooxides can be used. This comparison allows the elucidation of some changes in the characteristics of oxides occurring on their transfer from the gaseous medium to liquid and then into air.

The FTIR spectra of SA samples at $C_{Al_2O_3} \leq 30$ wt% (Figure 2.6a through e) demonstrate a sharp peak of free silanols at ~3740 cm^{-1} as pure fumed silica.

At $C_{Al_2O_3} = 75$ wt% (Figure 2.6f), the intensity of this silanol peak is very low. The surface alumina content of SA75 is >80 wt%, higher than its total Al_2O_3 content, consistent with the reflectance FTIR spectra. One or more peaks are observed in the 3700–3600 cm^{-1} region, but it is difficult to

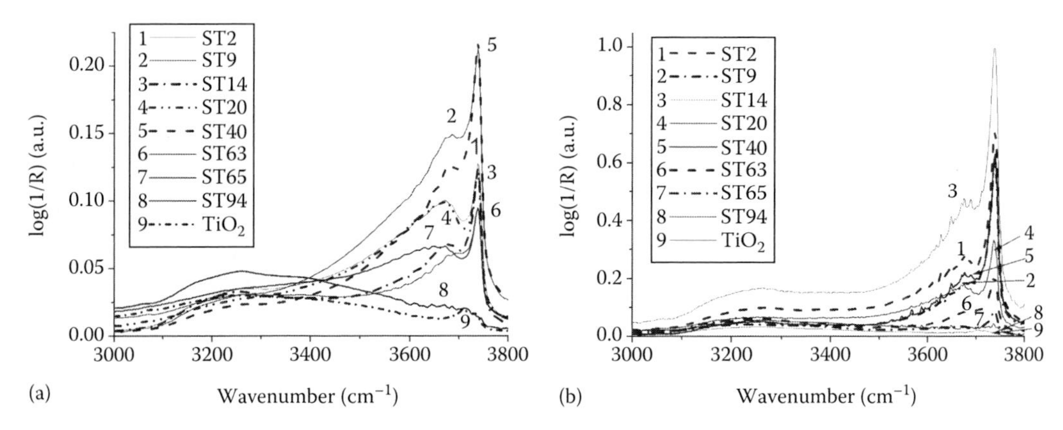

FIGURE 2.7 FTIR spectra of TiO_2 and ST samples recorded in the diffuse reflectance mode at (a) 473 K and (b) 573 K. (Adapted from *J. Colloid Interface Sci.*, 314, Gun'ko, V.M., Blitz, J.P., Gude, K. et al., Surface structure and properties of mixed fumed oxides, 119–130, 2007a, Copyright 2007, with permission from Elsevier.)

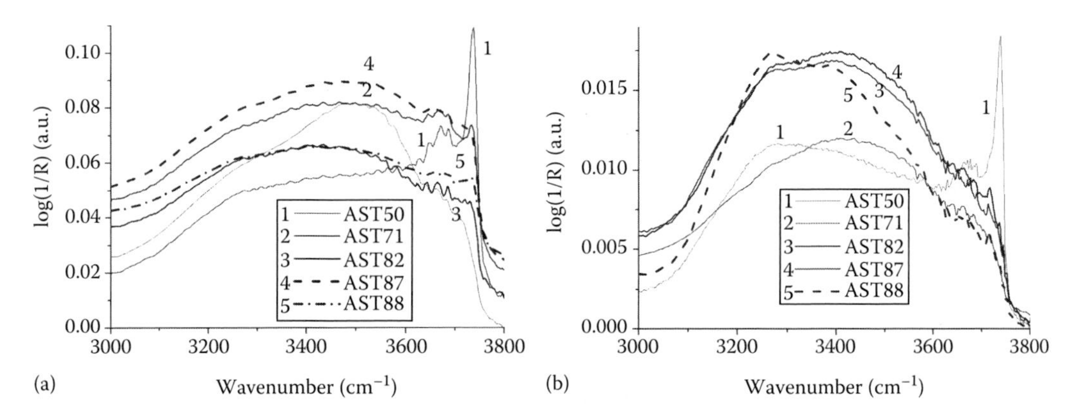

FIGURE 2.8 FTIR spectra of AST samples recorded in the diffuse reflectance mode at (a) 473 K and (b) 573 K. (Adapted from *J. Colloid Interface Sci.*, 314, Gun'ko, V.M., Blitz, J.P., Gude, K. et al., Surface structure and properties of mixed fumed oxides, 119–130, 2007a, Copyright 2007, with permission from Elsevier.)

assign these to bridging mixed oxide groups as assigned in zeolites or weakly hydrogen-bonded silanols. The OH-stretching vibrations at ν_{OH} between 3650 and 3400 cm^{-1} correspond to surface hydrogen-bonded hydroxyls with water molecules of neighboring hydroxyls of the same particle or an adjacent particle since the FTIR spectra were recorded at 473 and 573 K, and heating could cause the formation of denser contacts between adjacent particles.

FTIR spectra in the OH-stretching region are shown for ST samples in Figure 2.7 and AST samples in Figure 2.8.

Bands are seen in the 3700–3600 cm^{-1} region where one would expect bridging hydroxyls, and it is interesting to note that these bands are more visible at higher temperature, possibly because of less perturbation due to surface adsorbed water. Although these spectra possess bands, which may be due to bridging OH groups, given the low concentration of these groups and possible interference from vapor phase water rotation bands, it is not possible to assign specific bands to bridging surface OH groups. This is also the case for the spectra of SA samples in Figure 2.6. A band at 3300–3200 cm^{-1} (Figures 2.6 through 2.8) is due to surface hydroxyls and strongly adsorbed water molecules located in very narrow nanopores (or inside primary particles), which undergo strong hydrogen bonding interactions. These hydroxyls readily undergo thermal condensation reactions, as this band disappears after sample heating at higher temperatures.

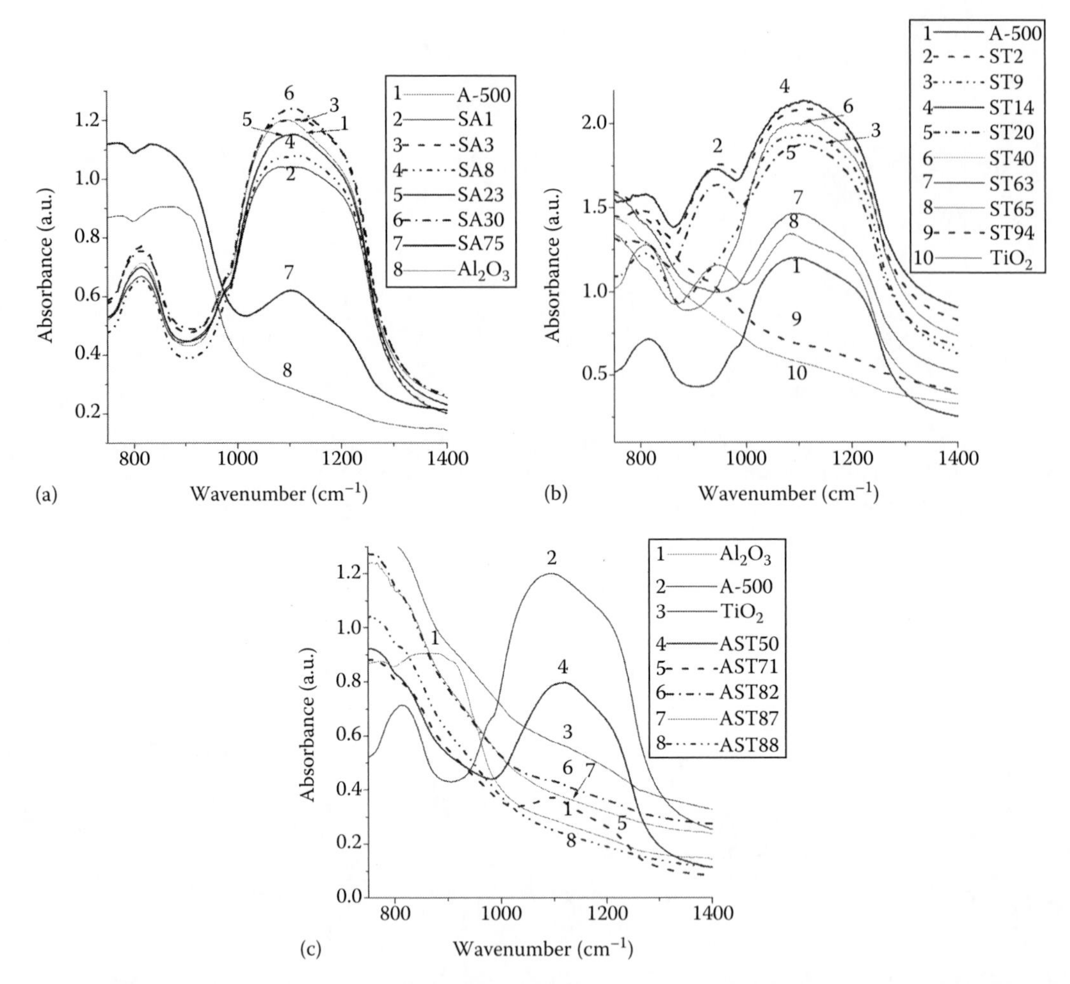

FIGURE 2.9 FTIR spectra over the SiO-stretching vibration range for (a) A-500, Al_2O_3, and SA; (b) A-500, TiO_2, and ST; and (c) A-500, Al_2O_3, TiO_2, and AST samples recorded in the diffuse reflectance mode at 573 K. (Adapted from *J. Colloid Interface Sci.*, 314, Gun'ko, V.M., Blitz, J.P., Gude, K. et al., Surface structure and properties of mixed fumed oxides, 119–130, 2007a, Copyright 2007, with permission from Elsevier.)

Additionally, small changes are observed for the ν_{SiO} band of the Si–O asymmetric stretching vibration at 1100 cm^{-1} (Figure 2.9a) and the combination vibration at 1860 cm^{-1} (not shown here). Only SA75 demonstrates a significant difference compared with A-500 (pure SiO_2) and approaches that of pure alumina. Ti-containing nanooxides spectral changes in these wavenumber regions are larger even at low titania content (Figure 2.9b). A new band appears at 960 cm^{-1} corresponding to the SiO-stretching vibrations from asymmetrical bridged ≡Si–O–Ti≡ bonds. Among AST samples, only AST50 has a well-defined ν_{SiO} (Figure 2.9c) and overtone bands because of the low silica content in other AST samples (Table 2.1).

Correlations can be made between integrated infrared band intensity values (calculated using spectral deconvolution with baseline subtraction) and results from other characterization methods. The content of ≡Si–O–Ti≡ bonds in ST samples (Figure 2.10a, 960/800 cm^{-1} band area ratio) is found from integrated absorbance values from transmission spectra of thin films at room temperature, since surface adsorbed water does not interfere with these measurements. Additionally, diffuse reflectance spectra recorded at 573 K (Figure 2.9a, 960/1860 cm^{-1} band area ratio) give a similar

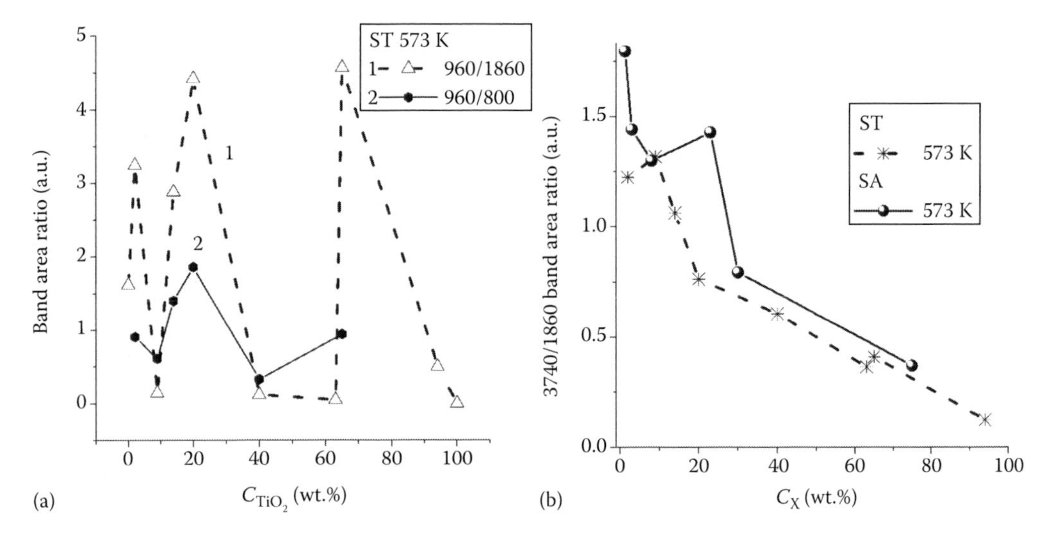

FIGURE 2.10 (a) Relative amounts of Si–O–Ti bonds as a function of the total content of titania in ST samples estimated from the spectra recorded in (1) reflectance and (2) transmittance (recorded at 293 K) modes; and (b) relative amounts of silanols as a function of the total content of alumina in SA or titania in ST samples from the diffuse reflectance spectra (all diffuse reflectance spectra were recorded at 573 K). (Adapted from *J. Colloid Interface Sci*., 314, Gun'ko, V.M., Blitz, J.P., Gude, K. et al., Surface structure and properties of mixed fumed oxides, 119–130, 2007a, Copyright 2007, with permission from Elsevier.)

curve showing the maximal amount of the ≡Si–O–Ti≡ bonds at $C_{TiO_2}=20$ wt%. Notice that use of the 960/800 cm^{-1} or 960/1860 cm^{-1} band area ratio (i.e., the bands at 800 and 1860 cm^{-1} function as an internal standard) allows us to avoid the effects of different sample weights and, therefore, differences in spectral intensity.

The Si–O vibrational intensity is used as an internal standard to normalize for the amount of sample in the infrared (IR) beam, taking into account the percent silica in each sample (Table 2.1). The silanol content of SA and ST samples (Figure 2.10b) is performed similarly from integrated log(1/R) values from diffuse reflectance spectra obtained at 300°C to eliminate surface adsorbed water.

The ~3740 cm^{-1} band intensity is used for silanol content. Band position does not vary with titania or alumina content, indicating that this band does not contain significant contributions from other possible OH species in these materials. While these band positions do not vary with second phase content, this band position is consistently 3742 cm^{-1} for SA samples and 3738 cm^{-1} for ST samples.

The presence of several structural factors results in complex correlation graphs related to the relative amount of Si–O–Ti bonds shown as the ratio between the peak area (1000–870 cm^{-1}) at $\nu_{SiO}=960$ cm^{-1} and the symmetric Si–O stretch (870–770 cm^{-1}) at 800 cm^{-1}. Changes in the distribution of titania in the bulk and on the surface of nanoparticles (Figure 2.3) has a very large effect on the number of Si–O–Ti bonds. For instance, ST65 clearly has a large Si–O–Ti band area, whereas the Si–O–Ti bond is barely if at all visible on ST63. Note that ST20 has the largest contribution from Si–O–Ti bonds, in agreement with previous work that shows ST20 has the highest Brønsted acidity of all ST samples, which is caused by the bridging hydroxyls ≡Si–O(H)–Ti≡.

The silanol content decreases with increasing $C_{Al_2O_3}$ or C_{TiO_2}; however, this trend is not smooth at low alumina and titania contents (Figure 2.10b). This can be partially explained by the fact that there is a nonlinear dependence of surface alumina and titania contents as the total content of these phases increases (Figure 2.3).

Thus, investigations of a variety of mixed nanooxides using several methods reveal that the surface properties of SA, ST, and AST materials (such as content of SiOH, $Si(OH)_2$, $(-O)_4Si$, Al(VI), Al(V), and Al(IV), intensity of the FTIR bands of surface hydroxyls, heat of immersion in water, desorption temperature, and content of desorbed water) are complex functions of the specific surface area, total and surface content of all oxide components, composition of the surface (both oxide phase distributions and distributions of the Ti and Al atoms in the silica phase and vice versa), and treatment temperature (differently affecting SA, ST, and AST samples because ternary oxides can lose hydroxyl coverage at lower temperatures than binary oxides). Many of the mentioned properties very clearly correlate with the surface content of alumina and/or titania in mixed oxides. Despite these differences, nanooxides are morphologically similar even at different specific surface areas because of features of HT synthesis of these oxides.

2.1.2 Morphology, Structural Hierarchy, and Textural Characteristics of Powders

Many physicochemical properties of mixed nanooxides depend not only on the surface composition but also on the morphological and structural characteristics of whole nanoparticles and their aggregates (Figure 2.11c and d), their crystallinity, size distribution, shape, etc. The XRD patterns (Figures 2.11 and 2.12) show that the crystallinity degree of titania and alumina depends on the

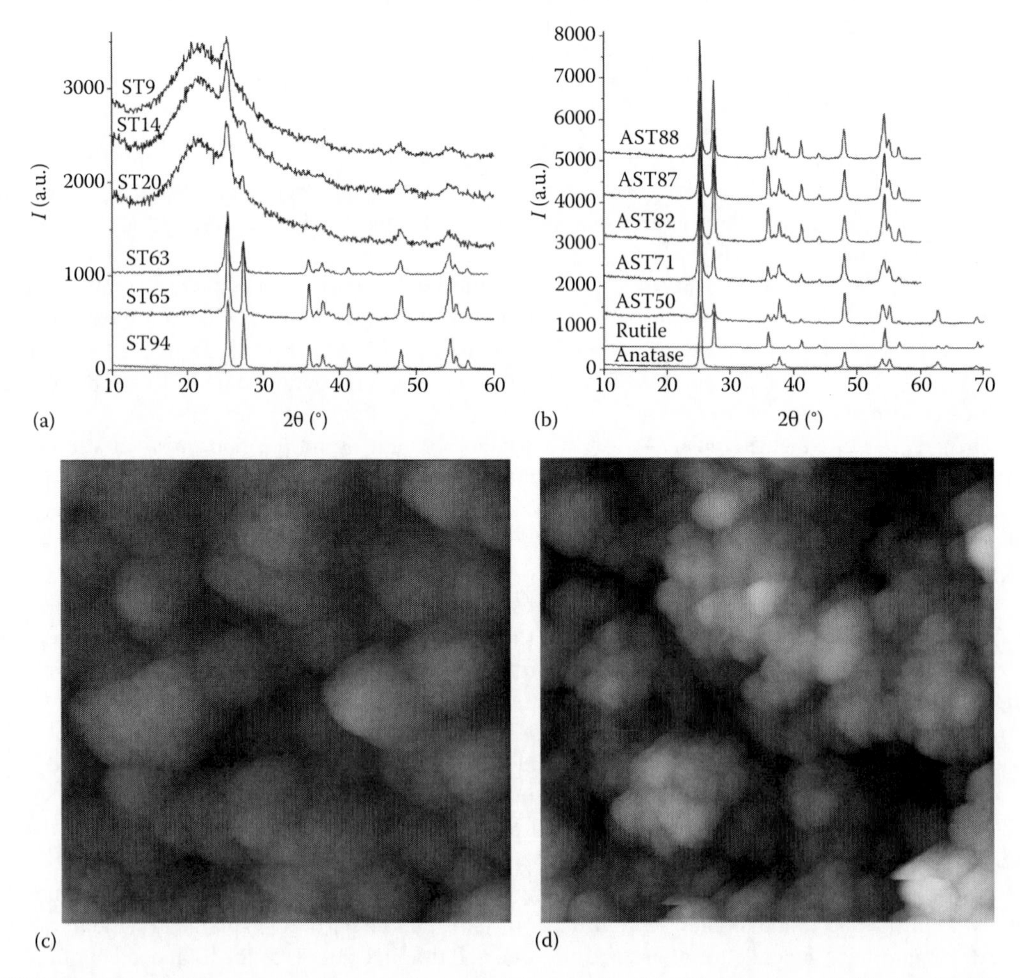

FIGURE 2.11 XRD patterns of nanooxides containing titania: (a) silica/titania and (b) alumina/silica/titania, and AFM images of (c) ST9 (500 × 500 nm^2) and (d) AST50 (1 × 1 μm^2).

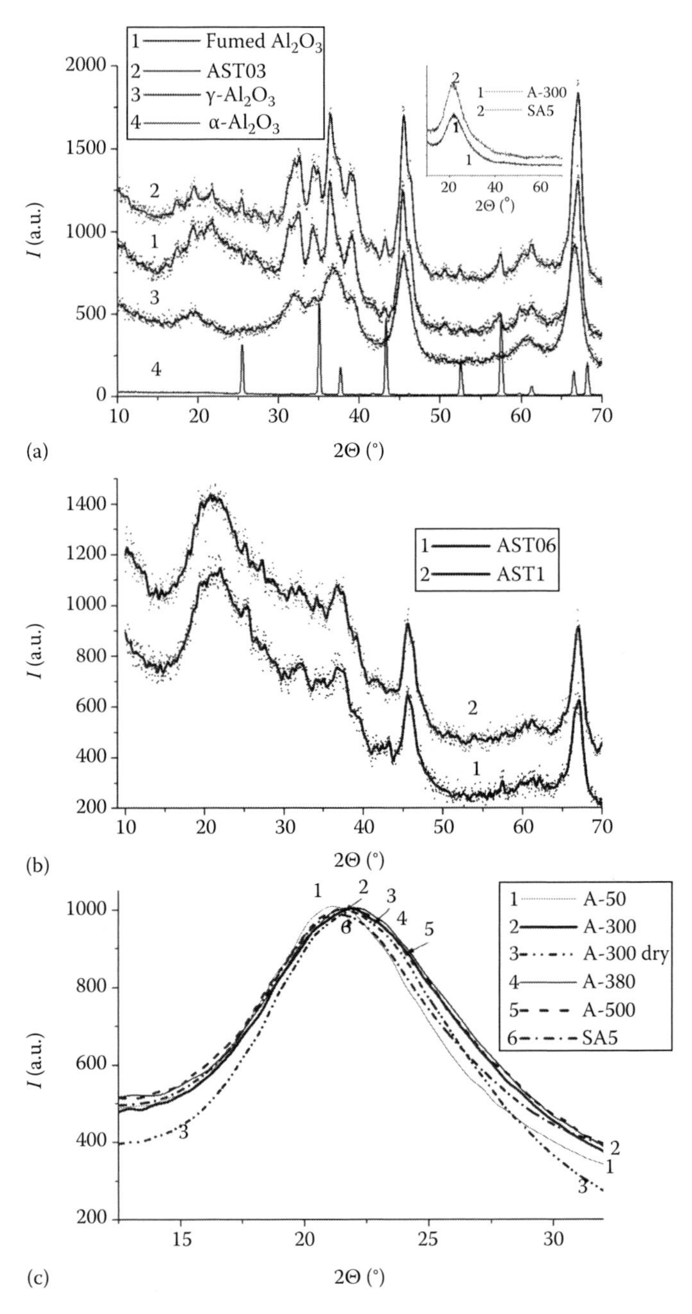

FIGURE 2.12 XRD patterns of nanooxides of different compositions: (a) fumed alumina (curve 1), AST03 (2), crystalline alumina such as γ-Al_2O_3 (3), and α-Al_2O_3 (4); (b) A-300 (1) and SA5 (2); (c) AST06 (1) and AST1 (2); and (d) different nanosilicas (curves 1–5) and SA5 (6). (Adapted from *Powder Technol.*, 195, Gun'ko, V.M., Zarko, V.I., Turov, V.V. et al., Morphological and structural features of individual and composite nanooxides with alumina, silica, and titania in powders and aqueous suspensions, 245–258, 2009g, Copyright 2009, with permission from Elsevier.)

TABLE 2.3
Structural Characteristics of Alumina, SA5, and AST from the XRD Data

Sample	Phase Composition	d_{cr} (Al_2O_3) (nm)	χ_{cr} (%)
Al_2O_3[a]	γ-Al_2O_3	9–10	45
	α-Al_2O_3	≥30	
	amorphous Al_2O_3		
AST03	γ-Al_2O_3	9–10	45
	α-Al_2O_3	≥30	
	amorphous Al_2O_3		
AST1	γ-Al_2O_3	9–10	~30
	α-Al_2O_3		
	Amorphous $Al_2O_3+SiO_2$		
AST06	γ-Al_2O_3	9–10	~30
	Amorphous $Al_2O_3+SiO_2$		
SA5	Totally amorphous		

[a] Alumina at $S_{BET,473K}=86$ m^2/g; d_{cr} is the average size of alumina crystallites; and χ_{cr} is the relative crystallinity.

composition of mixed oxides (as well as on reaction and treatment temperatures, Gun'ko et al. 1997d, 2009c,d). Silica is amorphous in both individual nanosilicas (as described earlier) and mixed oxides. Titania appears in the form of crystalline phases with anatase and rutile. However, the rutile content is low if $C_{TiO_2} \leq 20$ wt% (Figure 2.11a). Relative contribution of rutile increases in AST samples with increasing C_{TiO_2} value (Figure 2.11b) because of higher synthesis temperatures. The $\gamma = C_{anatase}/C_{rutile}$ value changes from 7.33 (AST50) to 0.83 (AST82).

A significant portion of alumina in both individual and composite samples at $C_{Al_2O_3} \geq 79.4$ wt% has a crystalline structure with mainly γ-phase and a smaller content of α-phase (Figure 2.12a, Table 2.3). Diffraction peaks appearing at $2\theta = 67.24$, 46.23, 37.46, 39.73, 31.37, and 61.70° can be assigned to the (440), (400), (311), (222), (220), and (511) planes of γ-alumina (JCPDS Database 2001, Ma et al. 2008).

The γ-alumina phase as a defect spinel includes both fourfold (~20%) and sixfold (~80%) O-coordinated Al atoms (i.e., it has a more complex structure than the α-phase with only sixfold O-coordinated Al atoms in the bulk). Therefore, γ-alumina has broader lines in the XRD patterns than corundum (Figure 2.12). The content of the latter is slightly larger in AST03 than in nanoalumina alone because of the higher temperature synthesis conditions, since γ-to-α phase transition is effective at $T > 1300$ K (Tkalcec et al. 2007, Chary et al. 2008, Gocmez and Özcan 2008). Samples AST06 and AST1 have a larger portion of silica than AST03 (Table 2.4), possess lower crystallinity (Table 2.3) and a higher intensity of a broad line from the amorphous phase at $2\theta = 21$–$22°$ (Figure 2.12). The γ-alumina phase has smaller crystallites than α-alumina (Table 2.3). Notice that a similar relationship between crystallite sizes of low- and HT phases was observed for fumed titania (anatase/rutile; Gun'ko et al. 1997d). This is due to the large temperature gradient (1000–2000 K) in the flame during the synthesis of nanooxides and the formation of core-shell particles (e.g., rutile-core/anatase-shell) or particles of different sizes and phase composition in different temperature zones in the flame (Degussa AG 1996, 1997). Fumed silica alone or in composition with other oxides is amorphous (Figure 2.12b). Titania at low content ($C_{TiO_2} \leq 1$ wt%), in AST samples studied here, does not appear as a crystalline phase in the XRD patterns. The low alumina content material SA5 is amorphous and similar to fumed

TABLE 2.4
Composition and Structural Characteristics of Nanooxides Preheated at Different Temperatures

Oxide	C_{SiO_2} (wt%)	$C_{Al_2O_3}$ (wt%)	C_{TiO_2} (wt%)	$S_{BET,373K}$ (m²/g)	$S_{BET,473K}$ (m²/g)	$S_{BET,623K}$ (m²/g)	$V_{p,373K}$ (cm³/g)	$V_{p,473K}$ (cm³/g)	$V_{p,623K}$ (cm³/g)
A-300	99.9			318	331	195	0.695	0.689	0.371
A-150	99.9			128	152	206	0.514	0.408	0.294
A-100	99.9			80	92	63	0.158	0.155	0.161
Al_2O_3		99.8		100	107	103	0.224	0.281	0.194
Al_2O_3		99.8		91	95	92	0.197	0.220	0.248
Al_2O_3		99.8		79	89	73	0.134	0.167	0.177
Al_2O_3		99.8		81	86	81	0.206	0.188	0.156
SA5	94.6	5.3	0.1	261	266	234	0.623	0.719	0.637
SA96	3.8	96.1	0.1	80	81	75	0.149	0.163	0.138
AST1	10.0	89.0	1.0	81	99	108	0.199	0.253	0.253
AST06	20.0	79.4	0.6	78	97	96	0.208	0.234	0.153
AST03	2.75	97.0	0.25	80	125	121	0.307	0.308	0.224

silica (Figure 2.12), with similar shapes of the broad line at $2\theta = 21°–22°$. However, the half-width of this band slightly differs for a set of amorphous nanooxides: from 9 (A-500 and A-380), 8 (A-300 and SA5), and 7.4 (A-50) to 7.2° (dry A-300) (Figure 2.12d). For nanosilicas with smaller primary particle size (A-500 and A-380), characterized by stronger strain and bond deformation in primary particles (which are more strongly aggregated, and therefore, adsorption of water from air is greater), this line is broader than for A-50 or dried A-300. Similar results are observed for the asymmetric Si–O stretching vibrations at $\nu_{SiO} = 1100$ cm^{-1} in the FTIR spectra of these silicas, since this band is broader for A-500 than for A-50 or A-150. The presence of silica (SA), titania (AST), and the γ, α, and amorphous phases of alumina in complex nanooxides causes the appearance of several types of surface hydroxyls (especially ≡Si–O(H)–M≡ at M = Al or Ti strongly interacting with adsorbed water and other polar compounds) affecting the behavior of the adsorption phase (Gun'ko et al. 1995, 1997a,d, 1998a–c, 1999b,c, 2000f,g,h, 2001e, 2004f, 2005b,d, 2007a,d,f,i, 2009d,g, 2010d).

The textural characteristics of mixed oxides were derived from LT N_2 adsorption data (Tables 2.1 and 2.4; Figures 2.13 through 2.15). Certain regularities exist because of several factors. First, spherical primary particles form aggregates and agglomerates of aggregates, and both types of secondary particles are responsible for the textural porosity of the powders. Second, diminution of the S_{BET} value is accompanied by a decrease in the V_p value due to reduced aggregation of primary particles. Third, pore (voids between primary particles) size distributions are very broad (Figures 2.14 and 2.15) and include two main modes at $R < 20$ nm (voids mainly in aggregates) and $R > 20$ nm (voids mainly in agglomerates, Figure 2.11c and d). These structural features can affect the properties of the hydroxyl coverage of primary particles and the interfacial behavior of molecularly adsorbed water (vide infra).

Heating of nanosilica at different temperatures (over the 473–1173 K range) for different times (from 10 min to 5 h) weakly affects the morphology of primary particles (Gun'ko et al. 2003g). Several processes occur on heating of nanooxides from room temperature to 1100 K or higher because of molecular and associative desorption of water from the particle surface and volume. Changes in the pore size distributions (PSDs) (Figure 2.15) and the primary particle size distributions, $\phi(a)$ (Figure 2.16) reflect these processes, depending on the chemical composition of nanooxides. In contrast to the nitrogen adsorption on different oxides when their difference appears

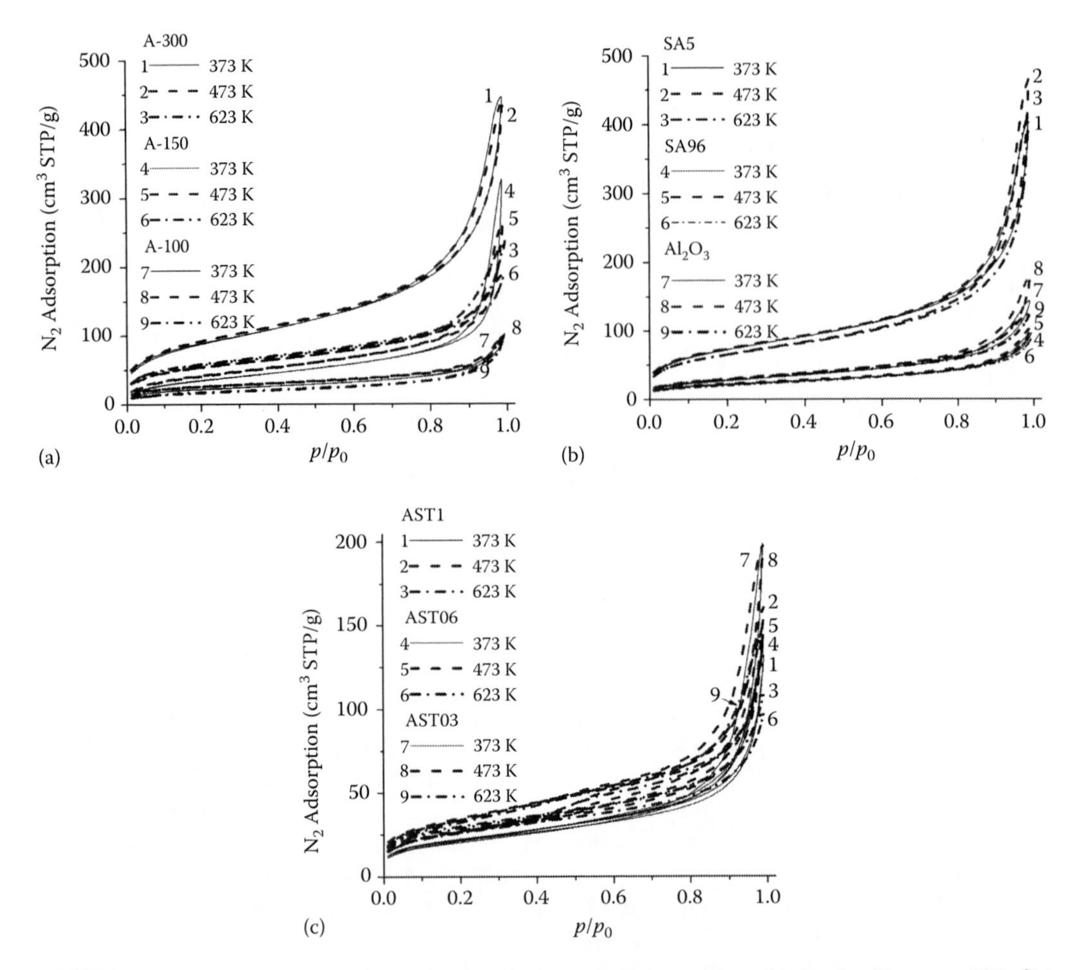

FIGURE 2.13 Adsorption–desorption isotherms of nitrogen for (a) nanosilicas; (b) alumina ($S_{BET,473K} = 107\ m^2/g$) and SA; and (c) AST samples preheated at 373, 473, and 623 K. (Adapted from *Powder Technol.*, 195, Gun'ko, V.M., Zarko, V.I., Turov, V.V. et al., Morphological and structural features of individual and composite nanooxides with alumina, silica, and titania in powders and aqueous suspensions, 245–258, 2009g, Copyright 2009, with permission from Elsevier.)

at low pressures (Figure 2.13), heating of nanosilicas leads to a larger difference in the adsorption at higher pressures than in the case of mixed oxides.

This is due to changes in the characteristics of both primary and secondary particles that are readily seen from the PSDs (Figure 2.15). These changes are over the total range of the pore (void) sizes caused by associative desorption of water from both the surface and bulk of primary particles. Therefore, the specific surface area depends on preheating (Table 2.4). Heating of nanooxides at different temperatures (for 2 h) and subsequent measurements without relaxation of nanoparticles in air show changes in the temperature dependences of the nitrogen adsorption (Figure 2.13) and organization of secondary and higher level particles. This results in changes in the relationships between the S and V values (Figure 2.17) and the shape of the PSDs (Figure 2.15). The textural characteristics of nanooxide powders (especially macropores) depend on the heating temperature, because stronger changes occur in the structure of agglomerates of aggregates than aggregates of primary particles. There is a nearly linear relationship between the specific surface area and the pore volume of meso- and macropores (Figure 2.17).

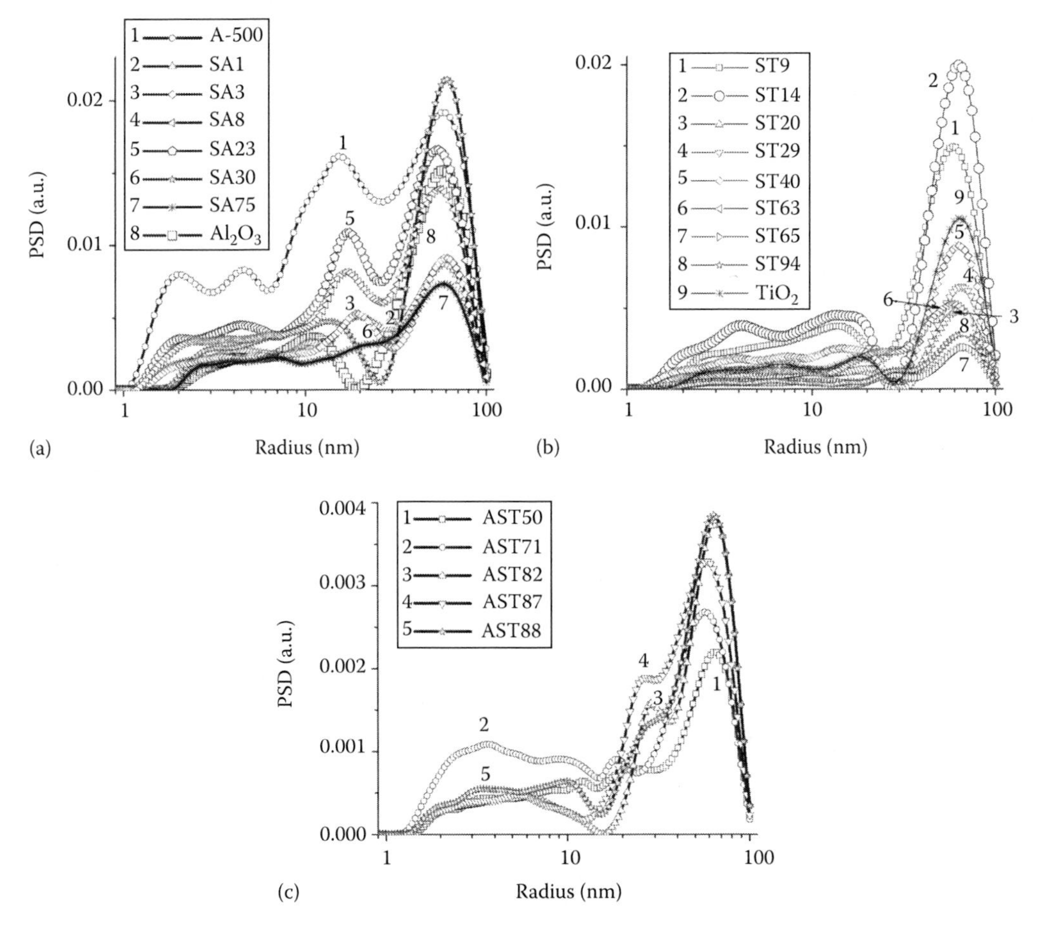

FIGURE 2.14 Incremental distributions of pore (void) size for (a) A-500, SA, and Al_2O_3; (b) ST and TiO_2; and (c) AST samples calculated using the DFT method with a model of voids between spherical particles. (Adapted from *Appl. Surf. Sci.*, 253, Gun'ko, V.M., Nychiporuk, Yu.M., Zarko, V.I. et al., Relationships between surface compositions and properties of surfaces of mixed fumed oxides, 3215–3230, 2007d, Copyright 2007, with permission from Elsevier; *J. Colloid Interface Sci.*, 314, Gun'ko, V.M., Blitz, J.P., Gude, K. et al., Surface structure and properties of mixed fumed oxides, 119–130, 2007a, Copyright 2007, with permission from Elsevier.)

However, significant scatter is observed for the relationship between the S_{BET} value and S_{macro} or V_{macro} due to random structures of aggregates and agglomerates, and different sensitivity of the organization of different oxides to heating and other treatments. Aggregates are more stable than agglomerates to any mechanical or thermal treatments. Therefore, heating of samples at different temperatures more strongly affects the characteristics of macropores (Table 2.4; Figures 2.13 and 2.17), i.e., voids between secondary particles in ternary particles. Samples with alumina and alumina/silica (SA5 and SA96) are more stable than other nanooxides shown here, and silica possesses a minimal structural and morphological stability on heating. This is due to the totally amorphous character of silica, in contrast to alumina (γ and α phases are observed in the XRD patterns of these samples, Figure 2.12) and titania (crystalline phase is observed in ST and AST samples at $C_{TiO_2} > 5$ wt%, Figure 2.11). Additionally, associative desorption of water from silica results in larger changes in the structure of particles than that for alumina or titania, because the Si atoms in silica do not have hyperstoichiometric surroundings, in contrast to Al and Ti in the Al_2O_3 and TiO_2 phases, respectively. The morphological stability of SA5 and SA96 on heating can

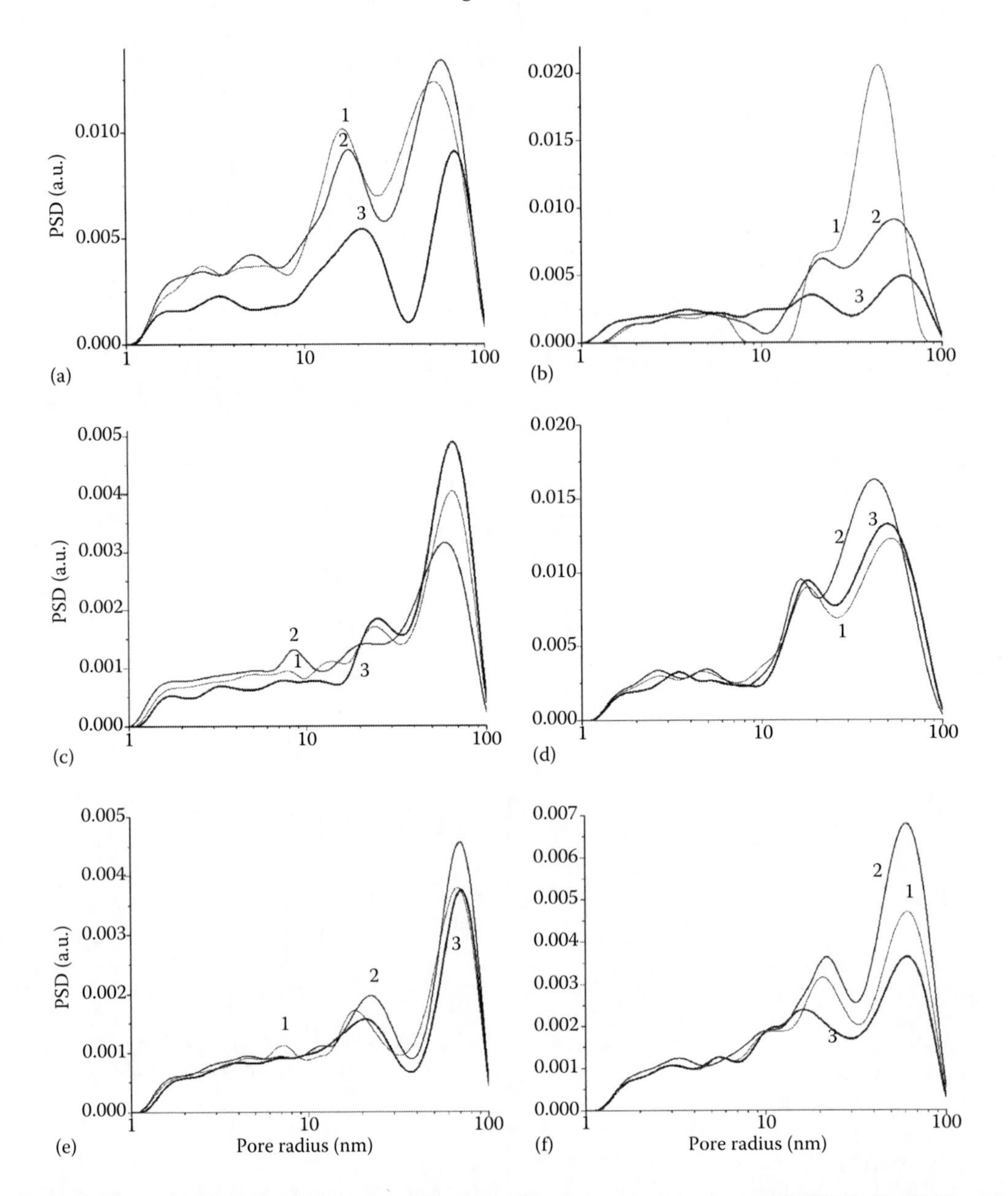

FIGURE 2.15 DFT $IPSD_V$ for nanooxides heated at different temperatures (curves 1) 373 K, (2) 473 K, and (3) 623 K: (a) A-300; (b) A-150; (c) A-100; (d) SA5; (e) SA96; (f) Al_2O_3. (Adapted from *Powder Technol.*, 195, Gun'ko, V.M., Zarko, V.I., Turov, V.V. et al., Morphological and structural features of individual and composite nanooxides with alumina, silica, and titania in powders and aqueous suspensions, 245–258, 2009g, Copyright 2009, with permission from Elsevier.)

be also caused by lower amounts of adsorbed water (from air), smaller than that for other samples (Figures 2.18 and 2.19).

Results in Figure 2.20 show that the FTIR determined silanol content nearly quantitatively correlates to the samples' surface area. For pure silicas, a correlation of surface silanol content and surface area has been described earlier. For mixed nanooxides, at least two factors (S_{BET} and C_X^S) affect the silanol content (Figure 2.20). The S_{BET} value (similar to that for pure nanosilica) is the

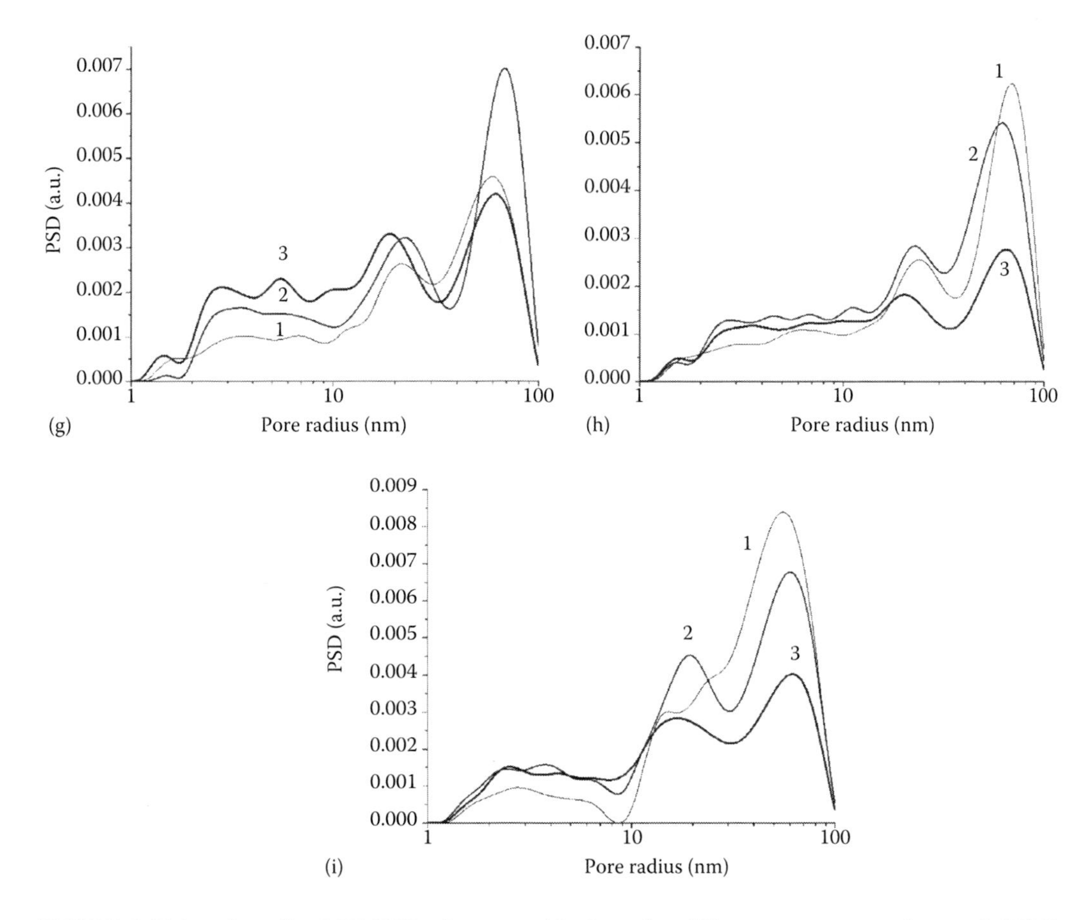

FIGURE 2.15 (continued) DFT $IPSD_V$ for nanooxides heated at different temperatures (curves 1) 373 K, (2) 473 K, and (3) 623 K: (g) AST1; (h) AST06; and (i) AST03. (Adapted from *Powder Technol.*, 195, Gun'ko, V.M., Zarko, V.I., Turov, V.V. et al., Morphological and structural features of individual and composite nanooxides with alumina, silica, and titania in powders and aqueous suspensions, 245–258, 2009g, Copyright 2009, with permission from Elsevier.)

predominant factor at concentrations (C_X) of the second oxide, but at high C_X values, the effect of C_X^S predominates.

The differences mentioned in the surface structure of mixed oxides in comparison with silica lead to certain peculiarities in the behavior of adsorbed water and desorption of water upon heating and degassing (Gun'ko et al. 1998d). The oxide samples (compared here using OPTPD with MS control method) contained from 5 wt% (silica A-300) to 13 wt% (ST20) adsorbed water at room temperature (Table 2.5). The lion's share of this water is molecularly adsorbed (the corresponding first TPD or differential thermogravimetry (DTG) peak is the largest and its integral intensity contributes more than 70% of the total signal) and it can be desorbed on heating to 450–500 K (Gun'ko et al. 1998d).

As mentioned earlier, the nanooxide surfaces are heterogeneous, and several types of surface sites can influence the water desorption peaks and the corresponding centers ($E_{0,i}$) in the desorption energy distribution for these peaks. If the TPD spectrum is decomposed into a few peaks without any restriction, the activation energy calculated for the total temperature intervals for each of the peaks can be underestimated due to the overestimation of the peak width. Therefore, we used some modification of the activation energy ($E^{\neq}$) calculations (Gun'ko et al. 1998d). We assumed that the distribution of the effective desorption energy for each of the TPD peaks has a Gaussian shape

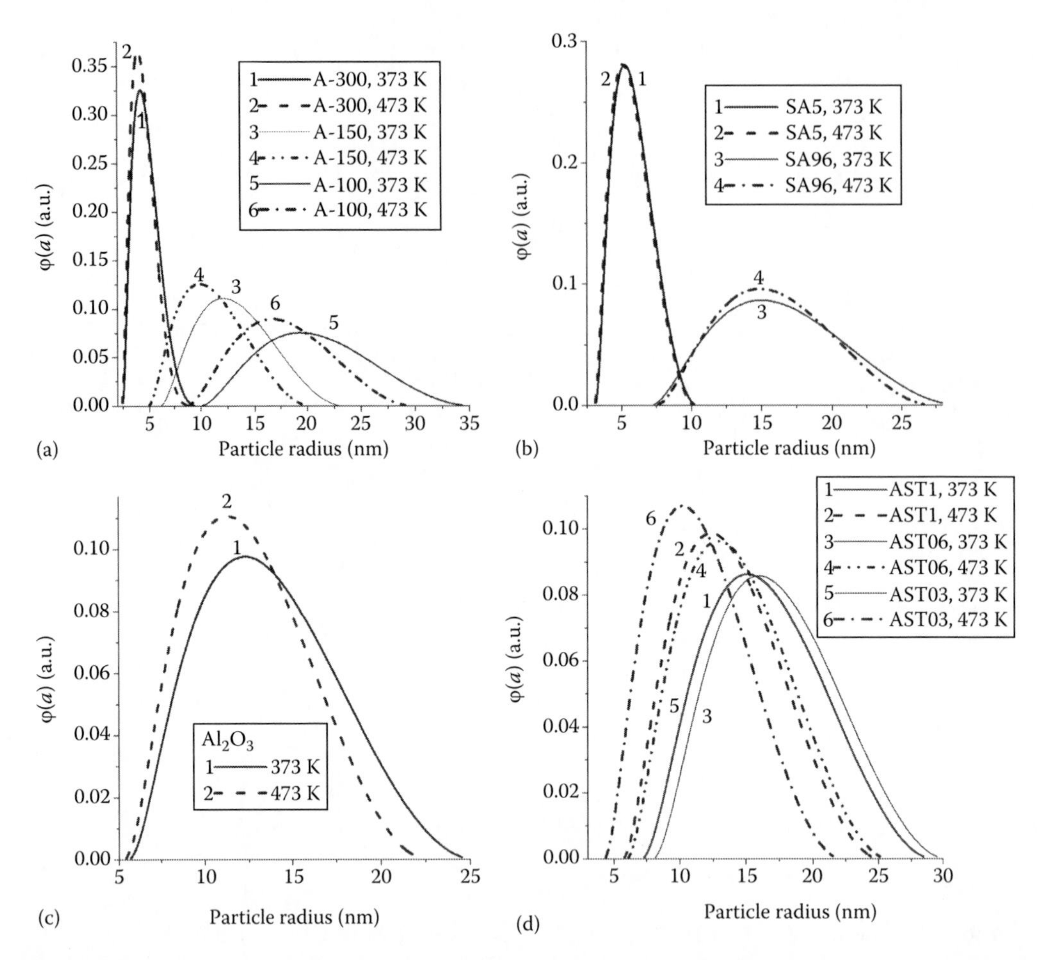

FIGURE 2.16 Primary particle size distributions for nanooxides heated at different temperatures: (a) different silicas; (b) SA5 and SA96; (c) Al_2O_3; and (d) AST. (Adapted from *Powder Technol.*, 195, Gun'ko, V.M., Zarko, V.I., Turov, V.V. et al., Morphological and structural features of individual and composite nanooxides with alumina, silica, and titania in powders and aqueous suspensions, 245–258, 2009e, Copyright 2009, with permission from Elsevier.)

$A\exp(-(E_{0,i} - E)^2/2\sigma^2)$, where A is a constant, i is the peak number, and σ^2 is the energy distribution dispersion. Therefore, a TPD peak can be represented as follows

$$I(T) = CI_0T\int_0^{\infty}\exp\left(-\frac{(E-E_0)^2}{2\sigma^2}\right)\exp\left(-\frac{E}{RT}\right)f(T,E,k_0)dE \tag{2.1}$$

where

$f(T,E,k_0) = \exp(-\Phi(T,E,k_0))$ or $1/(1+\Phi(T,E,k_0))$ for reactions of first and second order, respectively, for initial $\Theta_{OH} = 1$, $\Phi(T,E,k_0) = \frac{k_0}{\beta}\int_{T_{min}}^{T_{max}}\exp(-E/RT)dT$

C is the normalizing constant

β is the heating rate

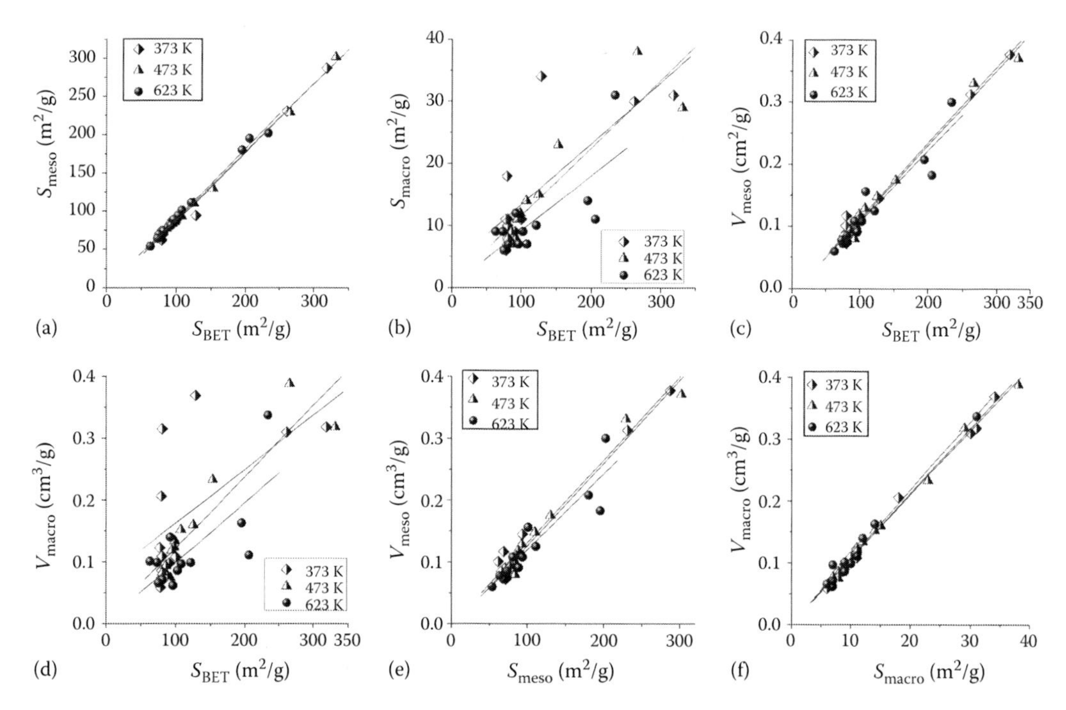

FIGURE 2.17 Relationships between the structural characteristics of nanooxides heated at different temperatures: S_{BET} and (a) S_{meso}; (b) S_{macro}; (c) V_{meso}; (d) V_{macro}; and (e) S_{meso} and V_{meso}; (f) S_{macro} and V_{macro} (samples shown in Table 2.4). (Adapted from *Powder Technol.*, 195, Gun'ko, V.M., Zarko, V.I., Turov, V.V. et al., Morphological and structural features of individual and composite nanooxides with alumina, silica, and titania in powders and aqueous suspensions, 245–258, 2009g, Copyright 2009, with permission from Elsevier.)

Equation 2.1 can be rewritten as

$$I(T) = CI_0T\exp\left(-\frac{E_0}{RT}\left(1-\frac{\sigma^2}{2E_0RT}\right)\right)\int_0^{\infty}\exp\left(-\frac{(E-E_0(1-\sigma^2/E_0RT))^2}{2\sigma^2}\right)f(T,E,k_0)dE \quad (2.2)$$

Inasmuch as $f(T,E,k_0)$ is a monotone function of E and the peaks are assumed to be narrow, Equation 2.2 can be transformed to

$$I(T) = I_0T\exp\left(-\frac{E_0}{RT}\left(1-\frac{\sigma^2}{2E_0RT}\right)\right)f\left(T,E_0\left(1-\frac{\sigma^2}{E_0RT}\right),k_0\right) \quad (2.3)$$

or

$$I(T) = I_0T\exp\left(-\frac{E_0}{RT}\left(1-\frac{\phi^2E_0}{2RT}\right)\right)f\left(T,E_0\left(1-\phi^2\frac{E_0}{RT}\right),k_0\right) \quad (2.4)$$

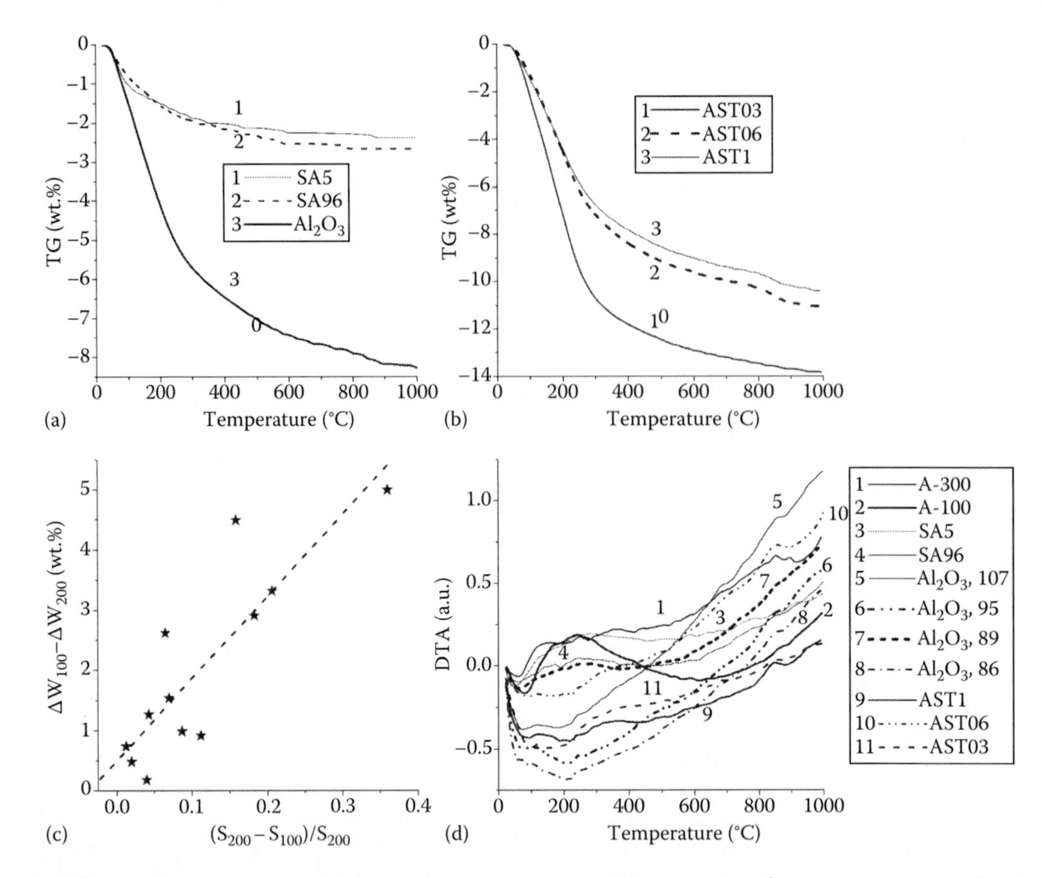

FIGURE 2.18 TG curves for nanooxides (a) SA and Al_2O_3 ($S_{BET,473K}$ = 107 m^2/g); (b) AST; (c) relationship between changes in the weight at 100°C and 200°C and relative changes in the specific surface area measured at 100 (S_{100}) and 200°C (S_{200}) for silica, alumina, and AST samples shown in Table 2.4; and (d) DTA curves for oxides ($S_{BET,473K}$ is shown for four Al_2O_3 samples). (Adapted from *Powder Technol.*, 195, Gun'ko, V.M., Zarko, V.I., Turov, V.V. et al., Morphological and structural features of individual and composite nanooxides with alumina, silica, and titania in powders and aqueous suspensions, 245–258, 2009g, Copyright 2009, with permission from Elsevier.)

where $\sigma = \varphi E_0$, $\varphi^2 \ll 1$ for narrow peaks. We used Equation 2.4 at $\varphi \leq 0.2$ for decomposition of the TPD spectra into 2–4 peaks; i.e., the equation for total ion current can be written as follows

$$I_{\Sigma}(T) = \sum_{i=1}^{L} I_i(T) \tag{2.5}$$

where
 $L = 2 - 4$
 $I_i(T)$ is given by Equation 2.4

The k_0 values for bimolecular processes were calculated as formal preexponential factors with dimensions of s^{-1}, but it can easily be converted to k_0 dimension of cm^2/s with consideration of the Θ_{OH} value (surface coverage by hydroxyls) for each TPD peak. We assumed that initial dimensionless Θ_{OH} equals 1 for the first peak for associative desorption of water and $\Theta_{OH,i}$ for the next i-peaks was calculated as follows

$$\Theta_{OH,i} = 1 - \sum_{j}^{i-1} \Theta_{OH,j} \tag{2.6}$$

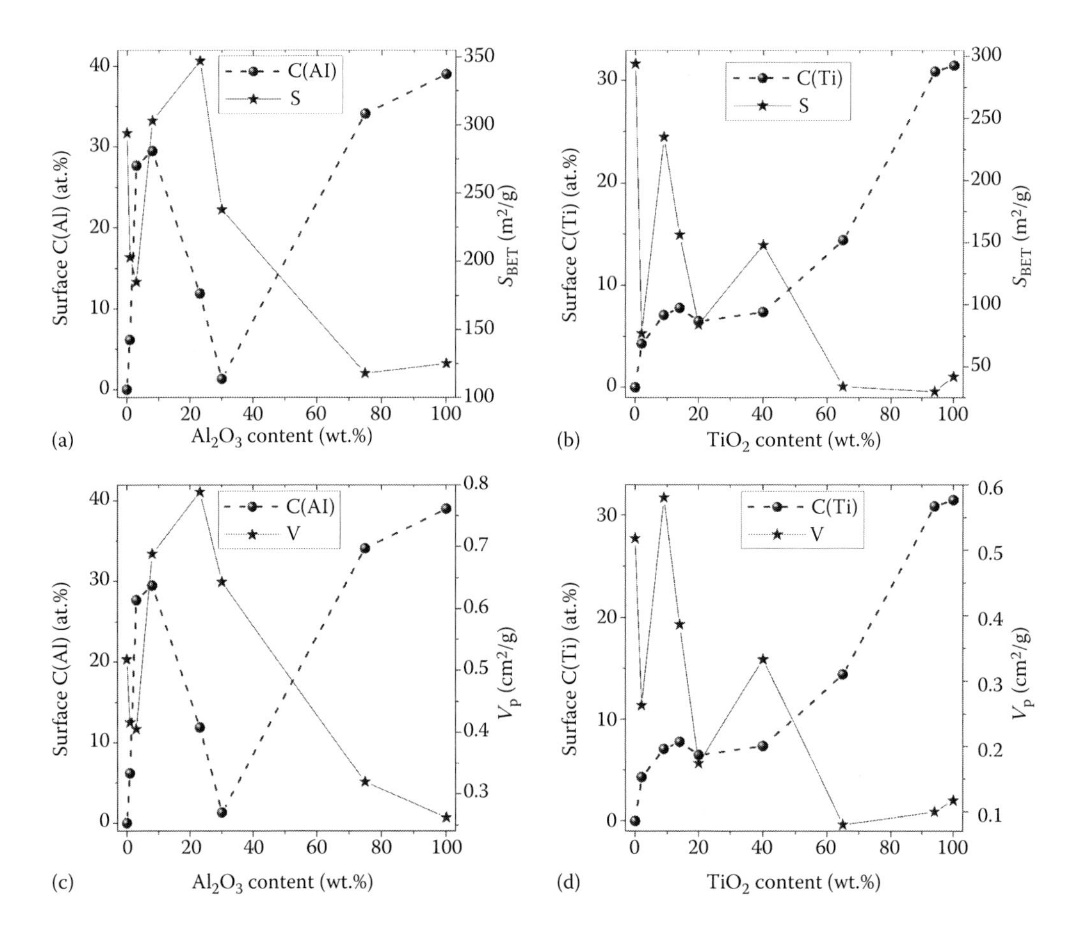

FIGURE 2.19 Relationships between the total content of (a and c) alumina in SA samples and (b and d) titania in TS samples and surface content of (a and c) aluminum and (b and d) titanium and (a and b) the specific surface area S_{BET} and (c and d) pore volume V_p. (Adapted from *Appl. Surf. Sci.*, 253, Gun'ko, V.M., Nychiporuk, Yu.M., Zarko, V.I. et al., Relationships between surface compositions and properties of surfaces of mixed fumed oxides, 3215–3230, 2007d, Copyright 2007, with permission from Elsevier.)

Θ_{OH} can be above zero after heating to 1000 K, e.g., for amorphous silica ($S \approx 300$ m²/g) heated to 1000 K $\Theta_{OH} \approx 0.15$ as an initial coverage is 4.7 OH per nm², and after heating to 1000 K, it equals 0.7 OH per nm² (Zhuravlev 2000).

The first OPTPD (*m/z* 18) or DTG peak for all studied oxides is caused by unimolecular desorption of intact H_2O molecules; i.e., the reaction (desorption) equation is of first order. All other peaks at higher temperatures could be assigned to associative desorption of water molecules in a second order reaction (as bimolecular process). Maybe some part of water observed at $T > 550$–600 K can be due to water desorbed via the unimolecular process (e.g., water desorption from nanopores, defects, and bulk of nonporous particles), but this contribution is not dominant. The amount of intact water is minor in the OPTPD measurements performed at 10^{-9} Torr as the amount of adsorbed molecules depends almost linearly on the pressure. The DTG data (Table 2.5) correspond to desorption of ca. 7 H_2O/nm² for silica (less than the number of molecules from one molecular layer as the area for one adsorbed H_2O is about 0.12 nm²), near to 40 H_2O/nm² for TiO_2 (i.e., the titania surface is covered by a few molecular layers), and about 60 H_2O/nm² for ST20. Consequently, OPTPD (Table 2.6; N_Σ) gives only a small part of this value: 33% for silica, 13% for titania, and 3% for ST. Water desorbed during degassing can be assigned to SAW/WBW and water remained at the oxide surfaces can be

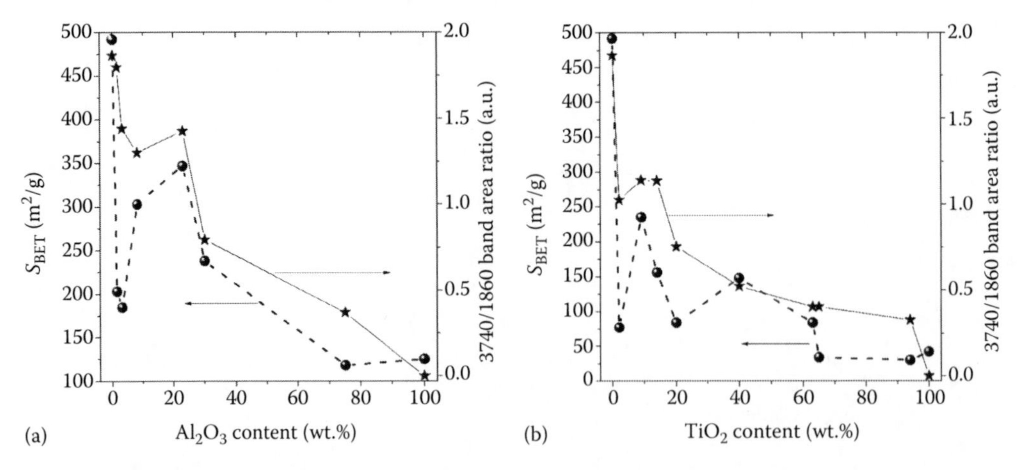

FIGURE 2.20 Relationships between total content of (a) alumina in SA and (b) titania in ST and the specific surface area and relative amounts of silanols in binary oxides (diffuse reflectance FTIR spectra were recorded at 473 K). (Adapted from *J. Colloid Interface Sci.*, 314, Gun'ko, V.M., Blitz, J.P., Gude, K. et al., Surface structure and properties of mixed fumed oxides, 119–130, 2007a, Copyright 2007, with permission from Elsevier.)

TABLE 2.5
DTG Data: the Temperature of the Peaks (T_{max}), the Amount of Desorbed Water (a_i) Corresponding to Every Approximating *i*-Curve and Total Content (a_Σ) of Desorbed Water

Sample	T_{max} (K)	a_i (wt%)	a_Σ (wt%)
A-300	346	2.79	4.91
	596	1.28	
	926	0.84	
Fumed titania	337	3.76	9.06
	405	2.54	
	628	2.76	
ST20	343	4.17	12.91
	397	4.83	
	414	2.02	
	789	1.89	
ST21(CVD)	385	4.99	8.98
	603	3.31	
	806	0.68	
SA23	371	4.99	7.37
	535	0.81	
	858	1.57	

weakly associated water (WAW or SAW)/SBW. Water adsorbed on ST is more strongly associated and therefore more weakly bound to the surface than in the case of silica characterized by a smaller amount of water adsorbed from air than that for titania or mixed oxides (Table 2.5).

The OPTPD spectrum of water desorbed from fumed silica has three well-defined maxima (Figure 2.21), but our calculation gives four peaks (relatively narrow peaks; see Table 2.6; φ); a shoulder

TABLE 2.6
Parameters of the OPTPD Spectra

Sample	l	x	T_{max} (K)	$E^{\neq}$ (kJ/mol)	k_0 (s^{-1})	$n_{mol} \times 10^{17}$	N_{H_2O} (n/m²)	N_{Σ} (n/m²)	φ
SiO_2	1	1	451	66	1.6×10^6	0.75	0.25	2.34	0.150
	2	2	580–596	87–106	2.8×10^6–9.5×10^7	1.10	0.37		0.175
	3	2	725–747	114–171	2.2×10^6–4.9×10^{10}	2.76	0.92		0.194
	4	2	879–883	187–252	9.4×10^{10}–5.2×10^{14}	2.40	0.80		0.190
SiO_2(r)	1	1	474	69	4.3×10^6	0.49	0.16	1.03	0.158
	2	2	607–631	91–112	1.3×10^7–1.2×10^8	0.81	0.27		0.142
	3	2	707–709	109–160	5.1×10^8–4.8×10^{10}	1.26	0.42		0.198
	4	2	812–818	174–209	8.8×10^9–2.7×10^{12}	0.55	0.18		0.200
TiO_2	1	1	480	71	3.6×10^6	0.85	1.70	5.06	0.158
	2	2	549–566	85–96	9.8×10^6–4.0×10^7	1.26	2.52		0.163
	3	2	638–645	100–138	1.5×10^7–1.9×10^{10}	0.43	0.86		0.173
TiO_2(r)	1	1	438	63	2.3×10^6	1.25	2.50	3.38	0.154
	2	2	536–544	84–87	9.6×10^6–1.2×10^7	0.44	0.88		0.164
ST20	1	1	438	63	2.7×10^6	1.81	0.72	1.94	0.153
	2	2	511–576	84–88	1.4×10^7	0.34	0.14		0.166
	3	2	678–694	108–154	1.3×10^7–3.1×10^{10}	1.51	0.60		0.178
	4	2	834–835	179–221	1.1×10^{10}–1.1×10^{13}	1.21	0.48		0.200
ST20(r)	1	1	398	55	1.9×10^6	0.77	0.31	1.23	0.147
	2	2	537–545	84–91	1.7×10^7–3.2×10^7	1.03	0.41		0.160
	3	2	657–679	100–144	1.8×10^6–1.2×10^{10}	0.52	0.21		0.145
	4	2	944–950	195–278	1.7×10^{10}–1.5×10^{15}	0.74	0.30		0.200
SA23	1	1	458	67	3.0×10^6	2.09	1.23	4.18	0.157
	2	2	557–566	86–102	3.6×10^6–6.7×10^7	1.18	0.69		0.173
	3	2	703–718	108–160	1.8×10^6–3.7×10^{10}	2.77	1.63		0.183
	4	2	858–867	179–232	5.3×10^9–3.2×10^{13}	1.06	0.62		0.200
SA23(r)	1	1	429	61	2.0×10^6	1.44	0.85	2.49	0.152
	2	2	532–551	83–91	7.3×10^6–2.3×10^7	0.96	0.56		0.167
	3	2	669–693	102–152	1.8×10^6–1.5×10^{10}	0.98	0.58		0.186
	4	2	892	191–254	5.3×10^{10}–5.7×10^{14}	0.85	0.50		0.200
ST21(CVD)	1	1	430	60	1.3×10^6	1.84	0.88	2.51	0.153
	2	2	524–549	83–92	1.4×10^7–2.9×10^7	0.91	0.43		0.163
	3	2	664–676	100–145	2.4×10^6–1.5×10^{10}	1.72	0.82		0.180
	4	2	787	166–196	3.6×10^9–5.7×10^{11}	0.79	0.38		0.200
ST21(CVD) (r)	1	1	411	61	2.3×10^7	0.50	0.24	0.63	0.153
	2	2	472–485	72–75	3.1×10^7–9.5×10^6	0.36	0.17		0.155
	3	2	606–615	95–109	1.6×10^7–1.3×10^8	0.22	0.22		0.174

l is the number of peak, x is the reaction order, T_{max} is the peak temperature, k_0 is the preexponential factor, n_{mol} and N_{H_2O} are the number of water molecules per mg and per nm², respectively, N_{Σ} is the sum of N_{H_2O}, φ is the parameter for Equation 2.4.

Note: The intervals of the T_{max}, $E^{\neq}$, and k_0 values are given for the reactions of associative desorption of water molecules.

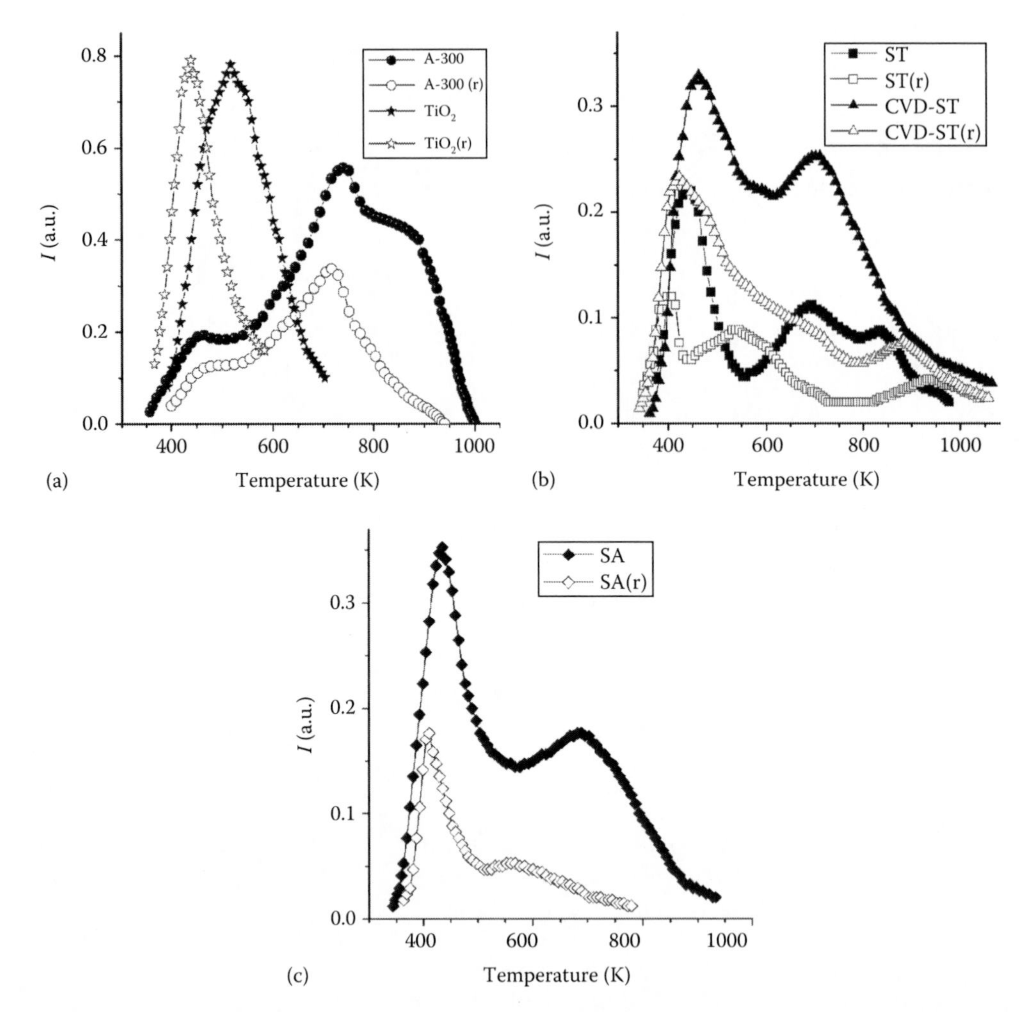

FIGURE 2.21 OPTPD spectra of water (*m/z* 18) for initial and heated and rehydrated (r) nanooxides: (a) silica and titania, (b) fumed ST20 ($C_{TiO_2}=20$ wt%) and CVD-TiO_2/A-300 at $C_{TiO_2}=21$ wt%, and (c) SA23. (Adapted from *Int. J. Mass Spectrom. Ion Proces.*, 172, Gun'ko, V.M., Zarko, V.I., Chuikov, B.A. et al., Temperature-programmed desorption of water from fumed silica, titania, silica/titania, and silica/alumina, 161–179, 1998a, Copyright 1998, with permission from Elsevier.)

at 580 K (as the second peak) is discernible. The lower limits of the $E^{\neq}$ and k_0 values for associative desorption (Table 2.6; the first values) are close to those for amorphous silica with the same S value.

The number of water molecules desorbed per one nm^2 (Table 2.6) corresponds to four OH groups (plus 0.25 of molecules of molecular adsorbed H_2O per nm^2). The amount of OH groups (α_{OH}) is consistent with the data published by Zhuravlev (2000). In addition, the α_{OH} value estimation for fumed silica heated to 700–900 K gives 1.5–1.7 OH/nm^2 (Dabrowski and Tertykh 1996) and that corresponds to the N_{H_2O} ($N_{H_2O}=2\alpha_{OH}$) value for the last peak (Table 2.6).

We calculated $E^{\neq}$ and k_0 (Table 2.6; the first values for $x=2$ and $l=2-4$) according to Zhuravlev (2000), but obtained $E^{\neq}$ values are lower than experimental heats of dissociative adsorption of water ($Q=100-280$ kJ/mol) on different oxides. Therefore, we recalculated the $E^{\neq}$ and k_0 values for bimolecular processes (Table 2.6; the second values for $x=2$ and $l=2$–4). An increase of $E^{\neq}$ leads to the growth of k_0 values, which corresponds better to the defined range for k_0 (Zhdanov 1991).

Dense islands of a few (3–10) OH groups can be formed on the silica surface. In such islands, the distances between some OH groups are relatively short and this can provide relatively easy dehydroxylation so that the islands disappear after evacuation at T about 770 K. Therefore, we can assume that the second OPTPD peak for A-300 (Figure 2.21a) corresponds to water desorbed from such dense islands, which contain not only vicinal OH groups but also twin $>Si(OH)_2$ groups. Such islands can be formed near contacts between primary particles in their aggregates.

The third peak (Figure 2.21; Table 2.6) can be linked to water desorbed via the interaction of vicinal or twin OH groups but not from the dense clusters of OH groups. Dissociative adsorption of water occurs easily for those sites, at which each metal atom already bonds one OH group or more; i.e., the products of such adsorption correspond to the structures as $M_1(OH)_2$-O(H)-M_2(OH) or $M_1(OH)_2$-O-$M_2(OH)_2$. Therefore, the $E\neq$ value (Table 2.6) for the reaction

$$\underset{A}{-Si(OH)_2\text{-}O\text{-}Si(OH)_2-} \rightarrow \underset{B}{-(HO)Si{<_O^O}{>}Si(OH)(\leftarrow OH_2)-} \rightarrow \underset{C}{-(HO)Si{<_O^O}{>}Si(OH)\text{-}} + H_2O \tag{2.7}$$

is lower than that for $>Si(OH)\text{-}O\text{-}Si(OH)< \rightarrow >Si{<_O^O}{>}Si< + H_2O$ as bond strain is lower for Equation 2.7 due to the lesser number of the lattice Si-O bonds.

The fourth OPTPD peak (Table 2.6) can be caused by the reaction between isolated and distant OH groups, when reaction is limited by $H^{\bullet}$ or $OH^{\bullet}$ diffusion from one OH group to another or by strong deformation of the surface, e.g., near contact between the primary particles in aggregates and agglomerates at small Θ_{OH} values.

The OPTPD spectrum of water desorbed from silica after rehydration of the heated sample displays strong changes relative to the initial spectrum (Figure 2.21). The maximal intensity decreases by half and the last peak is reduced by 80%, but the relative contribution of molecularly adsorbed water increases. The third and fourth peaks shift to lower temperatures and the $E\neq$ values decrease (Table 2.6). These effects are caused by close localization of hydroxyl groups formed upon dissociative adsorption of water during rehydration; i.e., the amount of isolated OH groups spaced at a large distance is smaller than that for the initial oxide, as such isolated groups make the main contribution to the highest temperature peak. The intensity of the second peak decreases slightly after rehydration that suggests our assumption that the second peak is due to water desorbed from the dense islands of OH groups with the participation in the reaction of twin groups. Additionally, after heating and rehydration, the $E\neq$ and T_{max} values for the first and the second peaks do not decrease (Table 2.6). This is due to smaller amounts of readsorbed water giving a higher enthalpy of adsorption than for original samples and to the occurrence of the associative reactions between OH groups in the dense islands (for the second peak). The third peak can be connected with the condensation of OH groups linked by hydrogen bonds (vicinal hydroxyls) or single but closely placed OH groups (not involved in dense clusters of OH groups). In this case, the contribution of the deformation energy to $E\neq$ increases and k_0 decreases as the number of active neighboring sites (i.e., possible pathways of associative reactions) reduces. The HT peak strongly decreases after rehydration as the number of isolated groups is small (Table 2.6; N_{H_2O}) due to the relatively short time of rehydration. It should be noted that for low Θ_{OH} (i.e., for HT peaks), the k_0 and $E\neq$ values can strongly depend on diffusion of $H^{\bullet}$ that leads to a reduction of k_0, but increasing the reaction temperature can compensate this effect (Table 2.6).

The number of hydroxyl groups and adsorbed water molecules per nm^2 for fumed titania is higher than that for silica (Table 2.6) but the OPTPD spectrum shape is more simple for titania (Figure 2.21). Water desorption from titania occurs under milder conditions than for silica due to a lower value of $E\neq$ for associative desorption of water from neighboring terminal (TiOH) and bridging (TiO(H)Ti) groups than that for two terminal SiOH groups (the energy in the corresponding process $Ti^{VI} \rightarrow Ti^{V}$ is lower than for $Si^{IV} \rightarrow Si^{III}$ [Gun'ko et al. 1997d]). Additionally, α_{OH} (or N_{Σ}) for titania is higher than that for silica (Table 2.6), i.e., the number of dense islands of OH groups can be greater for titania, and this can lead to lowering $E\neq$ value and the desorption temperature.

After heating and rehydration, the HT low peak disappears (Figure 2.21). This effect can be explained from the absence of isolated ≡TiOH or ≡TiO(H)Ti≡ groups; i.e., all dissociatively adsorbed water molecules form only adjacent ≡TiOH and ≡TiO(H)Ti≡ groups (dense clusters) as H or OH can diffuse relatively slow at 300 K (due to a high energy of the bond cleavage of M-OH or MO-H [Gun'ko et al 1997d]). Water desorption with the participation of such pairs makes a contribution only to the second OPTPD peak, which is lower after rehydration. Consequently, the third low peak for original titania is caused by water desorption with the participation of residual hydroxyl groups (spaced at a larger distance) in a bimolecular process. The first peak shifts to lower temperatures and grows, i.e., the number of water molecules, which do not dissociate on the surface, is relatively high.

The amount of water desorbed from ST in OPTPD measurements is lower than that for silica or titania (Table 2.6), but under standard conditions, ST contains a greater amount of water than silica or titania. The main part of molecularly adsorbed water on ST can locate at the interfaces of TiO_2-SiO_2 in the form of large clusters or nanodomains (SAW). Because of this and the fast desorption of physisorbed water (SAW) upon pumping to high vacuum, the amount of residual water (SBW) on ST after degassing is lower than that on silica or titania as the heterogeneity of the individual oxides is lower and water molecules distribution on them is more uniform than on ST. This leads to stronger interactions between these molecules and the surfaces of individual oxides, as molecules nearby the surface having a lower free energy (Gun'ko et al. 1995). These proposals are supported by a maximum intensity of the first peak for mixed oxides (Figure 2.21). In addition for all peaks, T_{max} is lower for original ST20 than that for original silica (Figure 2.21) (for ST, two HT peaks lie lower than that for silica, but their shape is close to those for silica) or titania (Figure 2.21; Table 2.6). A HT shoulder above 900 K can be due not only to isolated ≡SiOH groups but also to bulk hydroxyl groups at the phase boundary TiO_2-SiO_2, as these phases strongly differ in the crystalline structures determining the appearance of the strained bonds relaxed via hydrolysis upon dissociative adsorption of water. However, for pure silica, this region of the TPD spectrum is linked only with the interaction between isolated OH groups.

After heating and rehydration (and degassing) of ST20, the OPTPD spectrum changes greatly especially in the region of the peaks 2 and 3 (Figure 2.21b) and a decrease in the $E\neq$ and T_{max} values is observed for all the peaks with one exception of the fourth peak (Table 2.6). However, the HT shoulder is more intensive for ST20(r) (Figure 2.21b) than for ST20. These changes in the OPTPD spectrum can be caused by the alterations in the phase boundary TiO_2-SiO_2 induced by heating–cooling and dehydration–hydration cycles. In addition, according to the intensity of the corresponding peaks, the rehydration of the silica phase is lower (desorption at 700–900 K) than that for the interfaces and the titania phase (desorption at 400–600 K). The interfaces of ST in the bulk can influence these processes as well as in the case of ST(CVD). However, heating and rehydration do not give such dramatic changes in the OPTPD spectrum in the region corresponding to desorption from the titania phase or ST(CVD) phase boundary at the surface (Figure 2.21b). At the same time, rehydration of the silica phase is not observed as the peaks corresponding to water desorbed from silica are absent, i.e., a part of CVD-titania is deposed on the silica particles as small clusters (other part of titania presents as large individual titania particles), which are the centers of water adsorption and desorption from ST(CVD). However, for ST(CVD), molecularly adsorbed water has a desorption peak close to that for fumed ST (Figure 2.21b, the first peak) but water desorption from the last oxide occurs at lower temperatures.

In the case of fumed SA, molecularly adsorbed water desorbs at higher temperatures than from ST (Figure 2.21c; Table 2.6) due to higher acidic properties of SA leading to stronger interaction of water molecules with the surface. It should be noted that according to the NMR data, the changes in the free energy of adsorbed water is greater for Al_2O_3 and Al_2O_3/SiO_2 than for silica (vide infra). Because of these facts, the SA surface interacts more strongly with water molecules, and the OPTPD spectra have only two separated maxima (Figure 2.21c). The total amount of water adsorbed on SA after rehydration is higher than that on ST (Table 2.6; N_Σ) as well as the intensity of the first peak. However, the spectrum shapes for SA and ST are similar at $T > 550$ K (Figure 2.21).

Thus, the OPTPD time-of-flight mass-spectrometry (TOFMS) technique allowed the direct measurement of the amounts of associatively desorbed water per each type of the sites and per nm^2 and enabled to elucidate the details of dehydration and dehydroxylation of fumed silica, titania, ST, SA, and ST(CVD). Quantitative results were obtained regarding water desorption via various mechanisms with the participation of different hydroxyl groups contributing to different TPD peaks. The maximum activation energy of water desorption corresponds to the silica phase in ST and SA. The initial amounts of water adsorbed on ST or SA are higher than those for individual oxides, but the main body of adsorbed water is localized at the interfaces in the large clusters of water molecules, which can easily be desorbed by pumping to high vacuum or heating to ca. 400 K. Therefore, only a small part of initial amounts of adsorbed water (observed in TPD-DTG measurements) remains on the oxide surfaces in OPTPD measurements at low pressure. Dehydration of the oxides studied (after heating and rehydration) occurs with lower values of T_{max} and $E\neq$ probably due to slow diffusion of H and OH along the oxide surfaces; i.e., hydroxyl groups formed upon dissociative adsorption of water on heated samples remain adjacent in dense islands oriented in a favorable way for the associative desorption due to a lower deformation contribution to the activation energy. Additionally, a significant fraction of adsorbed water does not dissociate upon re-adsorption since the first OPTPD peak relatively increases and shifts to lower temperature. Rehydration of the silica phase in fumed ST or SA occurs more slowly than for the titania or alumina phases or at the interfaces. For ST(CVD), the silica phase remains practically dry upon rehydration.

2.1.3 Effects of Dispersion Media: Gaseous Phase and Aqueous Suspensions

Typically, nanooxides demonstrate multimodal particle size distributions (Figure 2.22) in aqueous suspensions because aggregation binding is stronger than repulsive electrostatic interactions, caused by surface charges appearing at pHs different from the pH at the PZC (Gun'ko et al. 2001e, 2009d,g). The shape of the particle size distributions depends on pH, salinity, oxide concentration, pretreatments (sonication and heating), and the chemical structure of the particle surface (Gun'ko et al. 2001e). This structure has a strong influence on the particle size distributions in the case of SA samples because $pH_{PZC} \approx 2.2$ and 9.8 (or pH of the IEP, pH_{IEP}) for fumed silica and alumina, respectively. The surface charges of silica and alumina between these pH values are of opposite signs, resulting in strong electrostatic attractive interactions of silica and alumina patches of SA particles.

Titania ($pH_{PZC} \approx 6$) does not exhibit strong effects of aggregation for ST samples, because of the low surface charge density (σ_0) of silica at pH < 7. Notice that for AST materials at higher content of titania, the σ_0(pH) and κ(pH) graphs (κ^{-1} is the Debye screening length) have an unusual shape, and the pH_{PZC} and pH_{IEP} values strongly differ. Individual primary particles were observed for high titania content AST samples at low C_{ox} = 0.04 wt%. However, for AST studied here at low titania content, primary particles are not observed (Figure 2.22). All particle size distributions shown demonstrate the presence of aggregates of primary particles at $d < 1$ μm, and only Al_2O_3 and SA96 do not have agglomerates at $d > 1$ μm. For Al_2O_3 (Figure 2.22c), a small contribution of individual primary particles is observed at the left wing of aggregates because of strong primary particle–particle repulsive interactions. For other samples, individual primary particles are practically absent because of weaker repulsive and stronger attractive interactions between primary particles.

The structure of interfacial water can be also characterized by the Debye screening length (κ^{-1}) as a function of pH (Figure 2.23) and ζ potential as a function of pH and the surface charge density (Figure 2.24). The κ^{-1} value determines the EDL thickness (it is much denser far from IEP or PZC) and the amounts of bound water determined by LT ^{1}H NMR method (Gun'ko et al. 2009d). Notice that the ζ potential and the surface charge density deal with different planes (i.e., shear and surface planes, respectively). In the case of individual oxides (e.g., silica, titania, and alumina), the κ^{-1}(pH) function has only one maximum observed in the pH range close to the IEP of the materials (Figure 2.23a) since IEP(SiO_2) ≈ 2.2, IEP(TiO_2) ≈ 6, and IEP(Al_2O_3) ≈ 9.8. Notice that the κ^{-1}(pH)

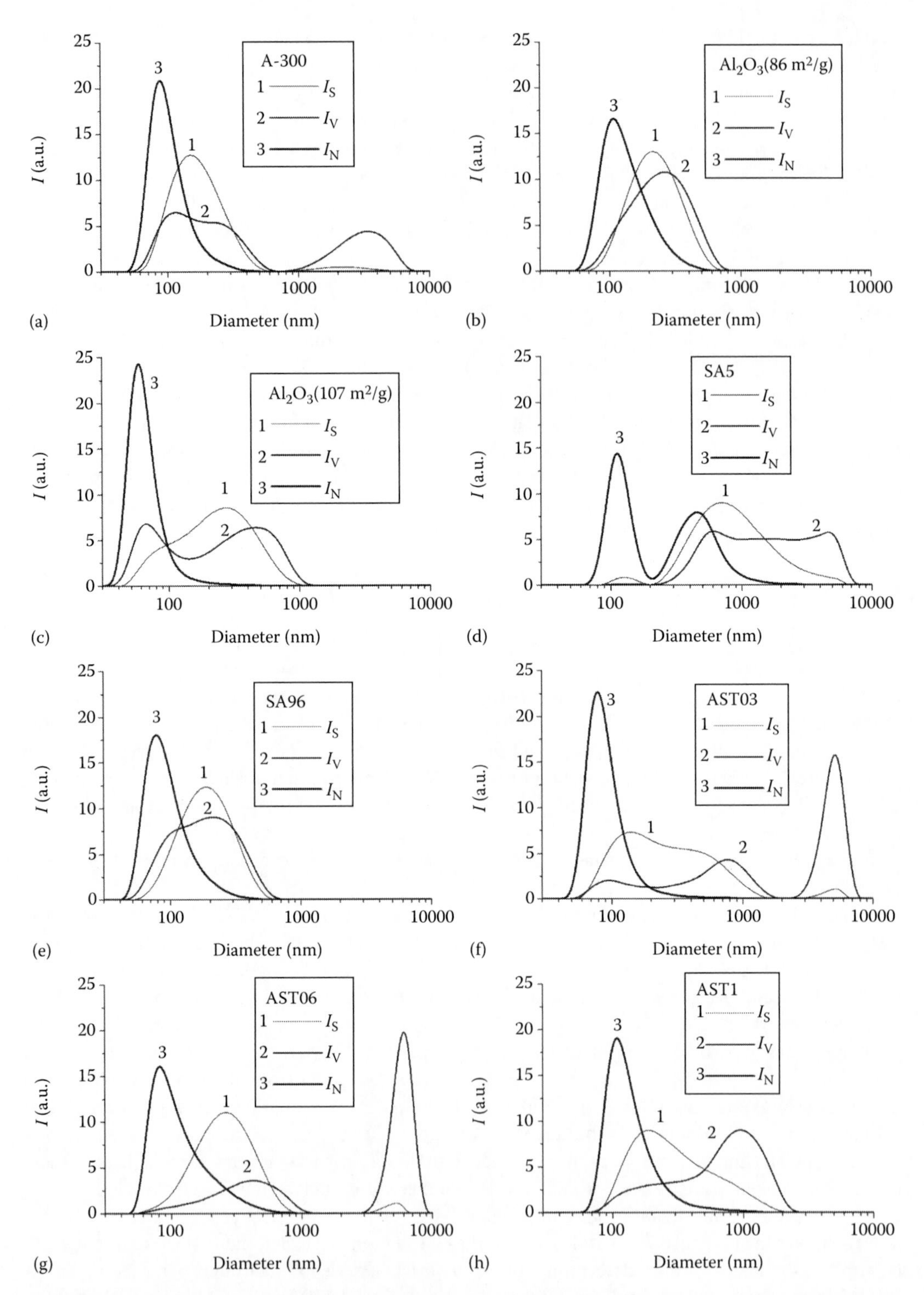

FIGURE 2.22 Particle size distributions with respect to the light scattering (I_S), particle volume (I_V) and particle number (I_N) in the aqueous suspensions of nanooxides at their concentration of 3 wt%. (Adapted from *Powder Technol.*, 195, Gun'ko, V.M., Zarko, V.I., Turov, V.V. et al., Morphological and structural features of individual and composite nanooxides with alumina, silica, and titania in powders and aqueous suspensions, 245–258, 2009g, Copyright 2009, with permission from Elsevier.)

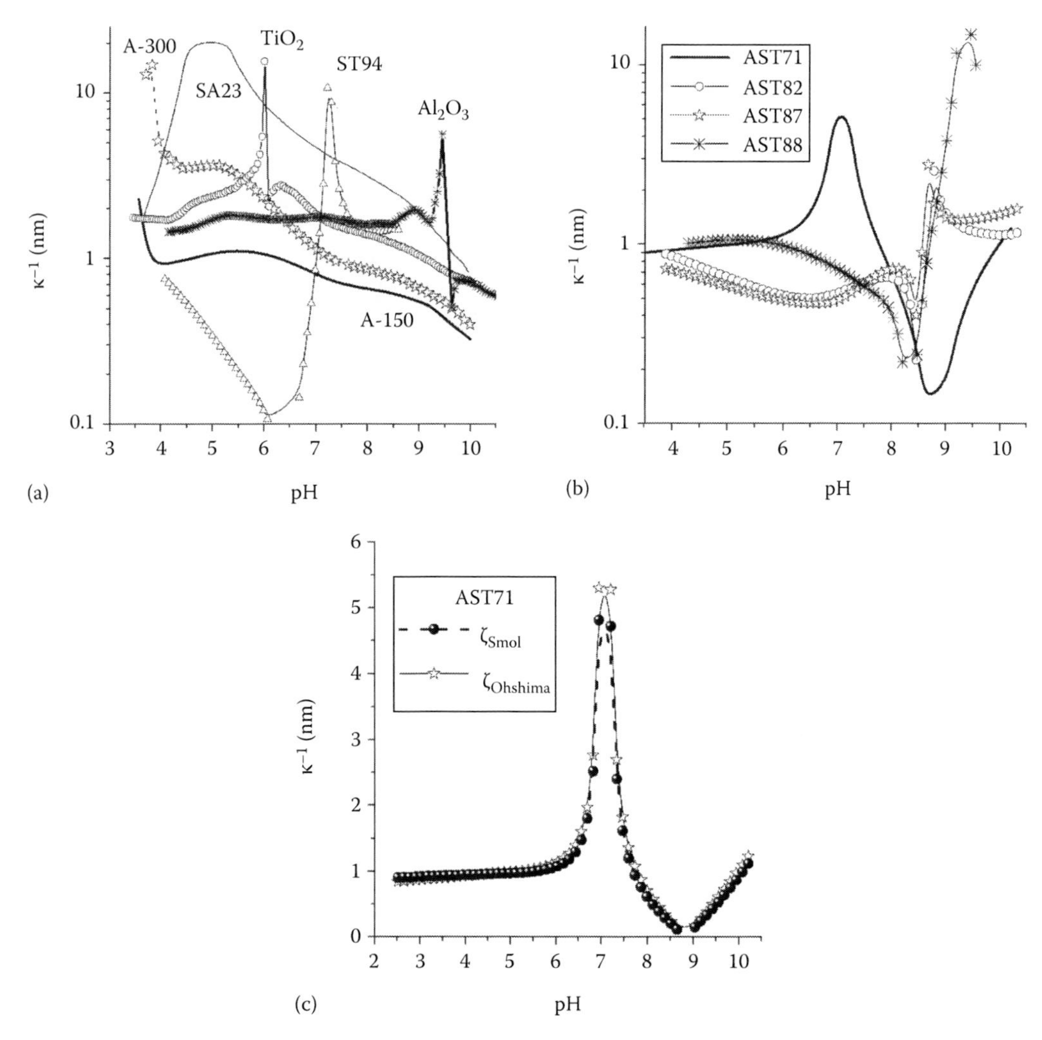

FIGURE 2.23 Debye screening length (κ^{-1}) for dispersions with (a) fumed silica A-150 and A-300, SA23, ST94, TiO_2, and Al_2O_3 (b) AST samples, and (c) comparison of the results obtained with Smoluchowski and Ohshima equations for ζ potential (Gun'ko et al. 2001a). (Adapted from *Appl. Surf. Sci.*, 253, Gun'ko, V.M., Nychiporuk, Yu.M., Zarko, V.I. et al., Relationships between surface compositions and properties of surfaces of mixed fumed oxides, 3215–3230, 2007e, Copyright 2007, with permission from Elsevier.)

function is determined on the basis of three dependences ζ(pH), σ_0(pH), and effective diameter of particles D_{ef}(pH) (with consideration for the porosity of aggregates), and these dependences are nonlinear functions of pH (Figure 2.24).

However, for mixed oxides, the κ^{-1}(pH) function can have an unexpected shape, e.g., for SA23 (at a low amount of surface alumina and maximal Brønsted acidity) with a maximum far from IEP(SiO_2) and IEP(Al_2O_3) (Figure 2.23a). There is a difference in the peak position of the κ^{-1}(pH) function for TiO_2 and titania-containing ST and AST oxides because of complex structure of their surface. The improved equation for the ζ potential and the Smoluchowski theory used to determine the ζ potential from particle mobility (Hunter 1981) give practically the same κ^{-1}(pH) curves (Figure 2.23c). The difference in the surface composition of AST samples described earlier causes noticeable differences in their behavior in the aqueous media (Figures 2.23 and 2.24), especially for AST71 in comparison with other AST samples because of maximal content of alumina and specific surface area among AST sample (Table 2.1). This results in the differences (discussed earlier

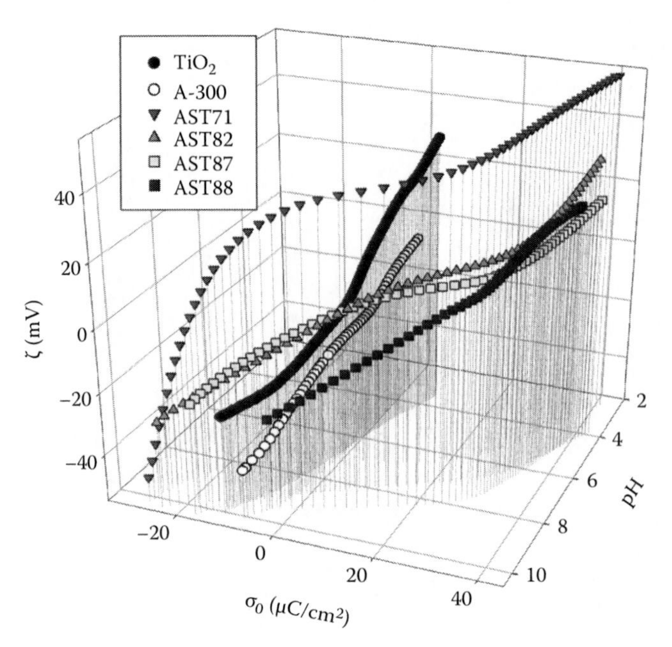

FIGURE 2.24 ζ Potential as a function of pH and surface charge density of A-300, TiO_2 and AST samples. (Adapted from *Appl. Surf. Sci.*, 253, Gun'ko, V.M., Nychiporuk, Yu.M., Zarko, V.I. et al., Relationships between surface compositions and properties of surfaces of mixed fumed oxides, 3215–3230, 2007d, Copyright 2007, with permission from Elsevier.)

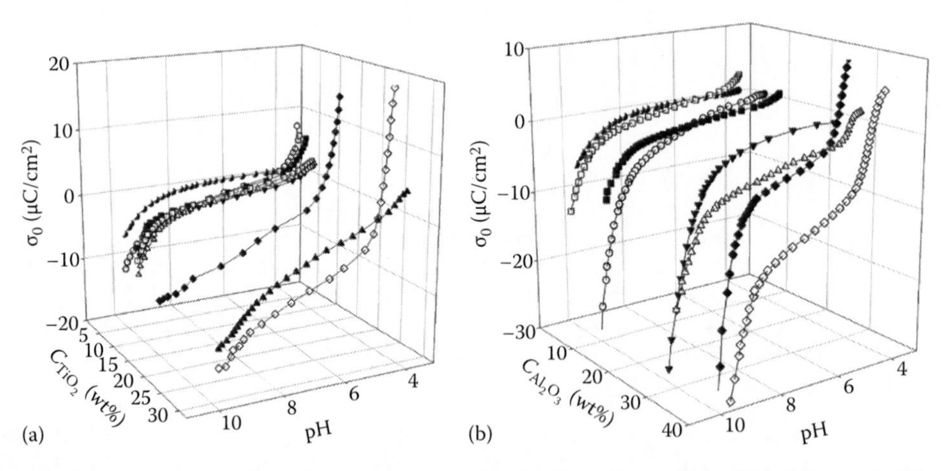

FIGURE 2.25 Dependence of surface charge density for mixed oxides on pH for fumed (a) titania/silica and (b) alumina/silica samples.

and subsequently in the text) in many of the properties of AST71 and other samples. Complex shapes of the κ^{-1}(pH) function as well as of the ζ(pH) potential and σ_0(pH) can lead to complex pH dependences of other properties of the nanooxide dispersions such as aggregation of particles (i.e., D_{ef}(pH)), the adsorption of dissolved compounds and metal ions, the suspension viscosity, the thickening ability, etc.

The pH dependences of the surface charge density are similar for SA and ST samples (Figure 2.25) because both type materials are acidic due to the B type acidity of bridging hydroxyls in the SiO(H)Al and SiO(H)Ti bonds.

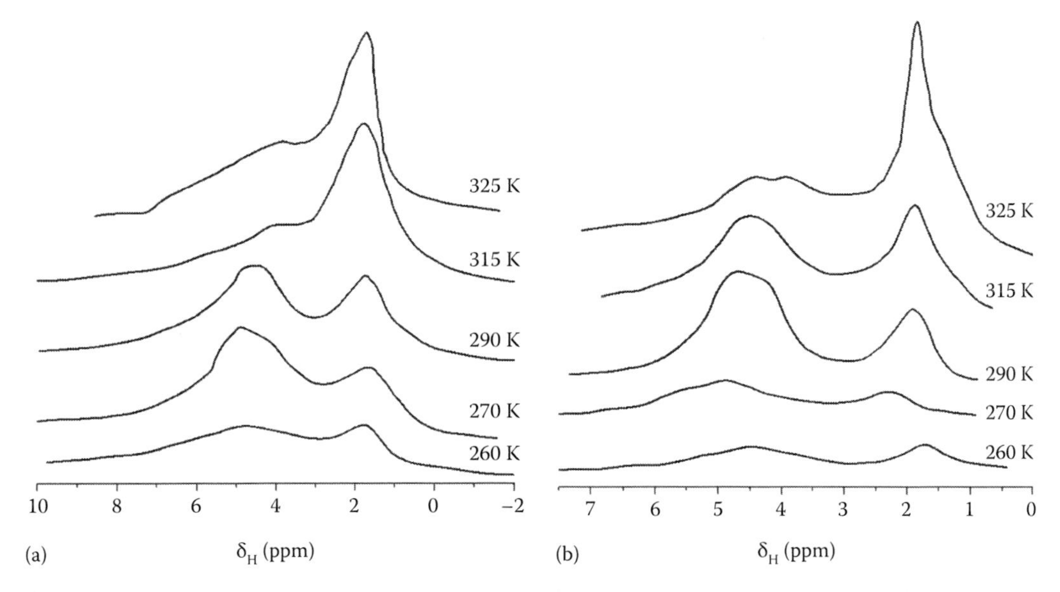

FIGURE 2.26 The ^{1}H NMR spectra of hydrated (a) ST20 and (b) CVD-TS (C_{TiO_2} = 22 wt%) at different temperatures. (Adapted with permission from Gun'ko, V.M., Zarko, V.I., Turov, V.V., Voronin, E.F., Tischenko, V.A., and Chuiko, A.A., Dielectric properties and dynamic simulation of water bound to titania/silica surfaces, *Langmuir*, 11, 2115–2122, 1995, Copyright 1995 American Chemical Society.)

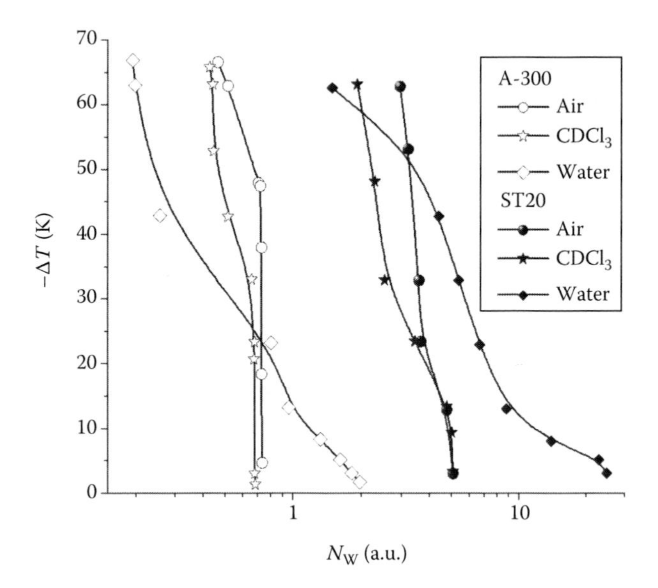

FIGURE 2.27 Relationships between the thicknesses of unfrozen water layer (measured in the numbers of water monolayer of 0.3 nm in thickness) and changes in the melting temperature for A-300 and ST20 in different dispersion media (air, $CDCl_3$, and water). (Adapted with permission from Gun'ko, V.M., Zarko, V.I., Turov, V.V., Voronin, E.F., Tischenko, V.A., and Chuiko, A.A., Dielectric properties and dynamic simulation of water bound to titania/silica surfaces, *Langmuir*, 11, 2115–2122, 1995, Copyright 1995 American Chemical Society.)

Mixed fumed and CVD oxides differ in their properties even at the same chemical composition because of the difference in the surface structure of the materials. This difference affects the behavior of water bound to fumed ST20 (Figure 2.26a) and CVD-TiO_2/A-300 at C_{TiO_2}=22 wt% (Figure 2.26b). The medium type and amounts of bound water affect the freezing/melting processes (Figure 2.27). However, the amounts of water disturbed by the ST surface are much greater than

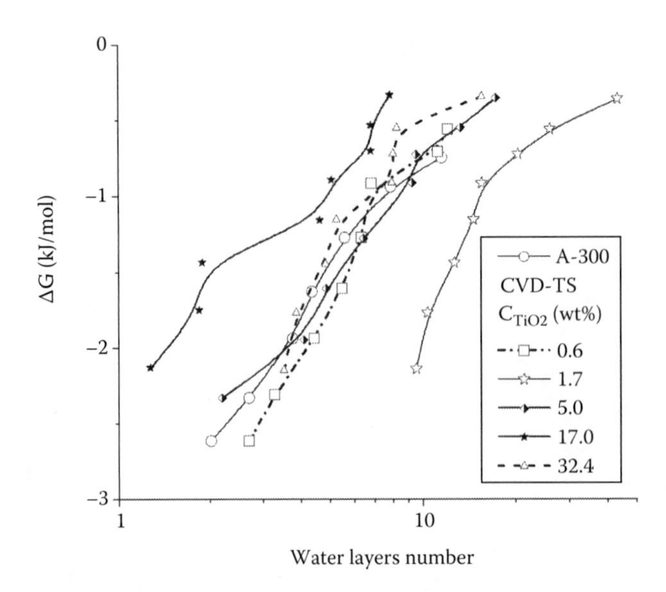

FIGURE 2.28 Relationships between the thicknesses of unfrozen water layer (the numbers of water monolayer) and changes in the Gibbs free energy of bound water in the aqueous suspensions (~5 wt%) of nanosilica A-300 and CVD-TS at different contents of grafted titania. (Adapted from *J. Colloid Interface Sci.*, 198, Gun'ko, V.M., Zarko, V.I., Turov, V.V. et al., CVD-titania on fumed silica substrate, 141–156, 1998b, Copyright 1998, with permission from Elsevier.)

that in the aqueous suspension of nanosilica (Figures 2.27 and 2.28). This result is due to the acidic properties of the ST surface discussed earlier.

Exchange of the air dispersion medium by weakly polar chloroform (Figure 2.27) diminishes the interactions of water with the oxide surface and this diminution is larger for ST. In the aqueous suspension, ST20 (it was selected as containing maximal amount of SiO(H)Ti bridges among ST samples) particles can disturb the water layer thicker by order of magnitude than A-300 does. In the case of ST(CVD) (Figure 2.28), the effect in the aqueous suspensions depends nonlinearly on the titania content since it is maximal at $C_{TiO_2} = 1.7$ wt%.

Titania and titania/silica possess stronger B and L sites than germania or GeO_2/SiO_2. Therefore, the bound water layer is thicker in the case of the aqueous suspension of CVD-TS than CVD-GeO_2/SiO_2 (Figure 2.29). Less uniform surface with a smaller content of the grafted phase with TiO_2 or GeO_2 can disturb greater amounts of water and this effect is stronger for CVD-TS (Figures 2.28 and 2.29).

The difference in the interaction strength of water with the oxide nanoparticles appears in the temperature dependences of the chemical shift of the proton resonance of bound water molecules (Figure 2.30).

The δ_H values increase with lowering temperature and increasing total water content (Figure 2.30). These effects are similar to the effects observed for nanosilicas and discussed in Chapter 1. Briefly, the first effect is due to decreased mobility of the molecules and the formation of more ordered hydrogen bond network (for SAW structures, it is similar to that in amorphous or crystalline ice). The second effect is due to the increased number of the hydrogen bonds per a molecule with increasing total content of water, i.e., contribution of SAW increases. However, there is an additional effect caused by the type of nanooxide. Titania, alumina, and mixed oxides give stronger effects than nanosilica that result in larger δ_H values. This is in agreement with the difference in the spectra of DMAAB adsorbed onto different nanooxides (Figure 2.4) and caused by stronger acidic and basic sites at non-silica oxides than those on silica. This difference is also well seen in the difference in the thickness of the disturbed water layers interacting with nanosilica and SA samples (Figure 2.31).

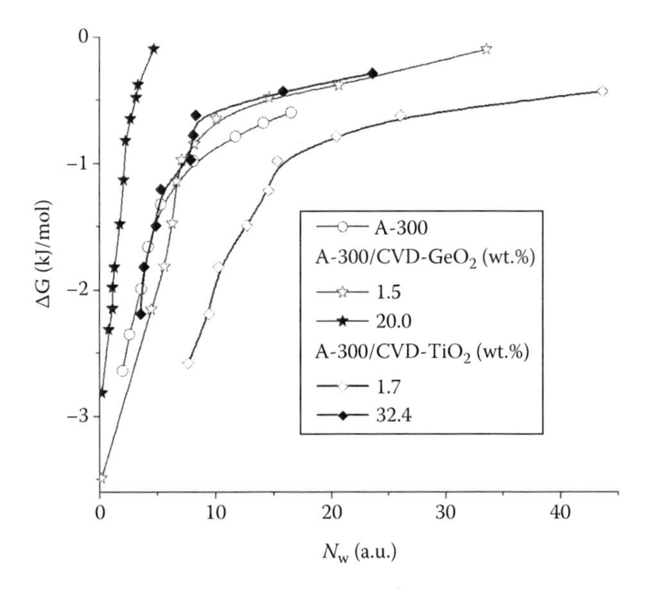

FIGURE 2.29 Relationships between the thicknesses of unfrozen water layer (the numbers of water monolayer) and changes in the Gibbs free energy of bound water in the aqueous suspensions (~5 wt%) of nanosilica alone and with grafted CVD oxides with germania and titania at small and great C_X values. (Adapted from *J. Colloid Interface Sci.*, 198, Gun'ko, V.M., Zarko, V.I., Turov, V.V. et al., CVD-titania on fumed silica substrate, 141–156, 1998b, Copyright 1998, with permission from Elsevier; *J. Colloid Interface Sci.*, 205, Gun'ko, V.M., Zarko, V.I., Turov, V.V., Leboda, R., Chibowski, E., and Gun'ko, V.V., Aqueous suspensions of highly disperse silica and germania/silica, 106–120, 1998c, Copyright 1998, with permission from Elsevier.)

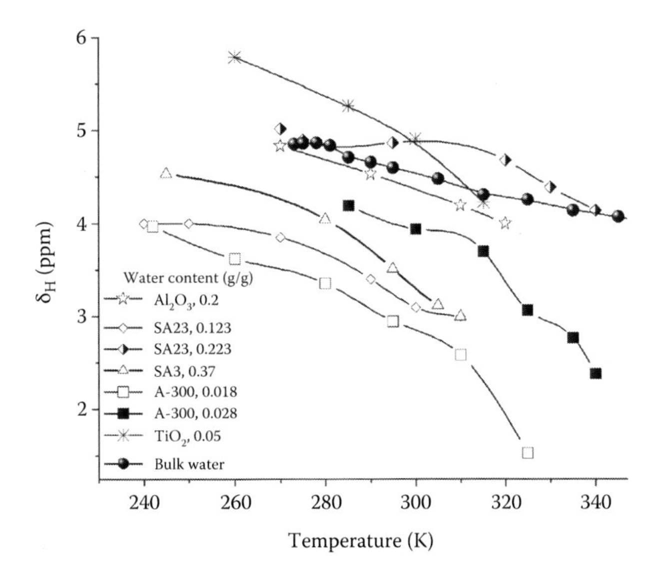

FIGURE 2.30 Chemical shift of protons in water bound to different oxides as a function of temperature compared to bulk water. (Adapted with permission from Gun'ko, V.M., and Turov, V.V., Structure of hydrogen bonds and ^{1}H NMR spectra of water at the interface of oxides, *Langmuir*, 15, 6405–6415, 1999, Copyright 1999 American Chemical Society.)

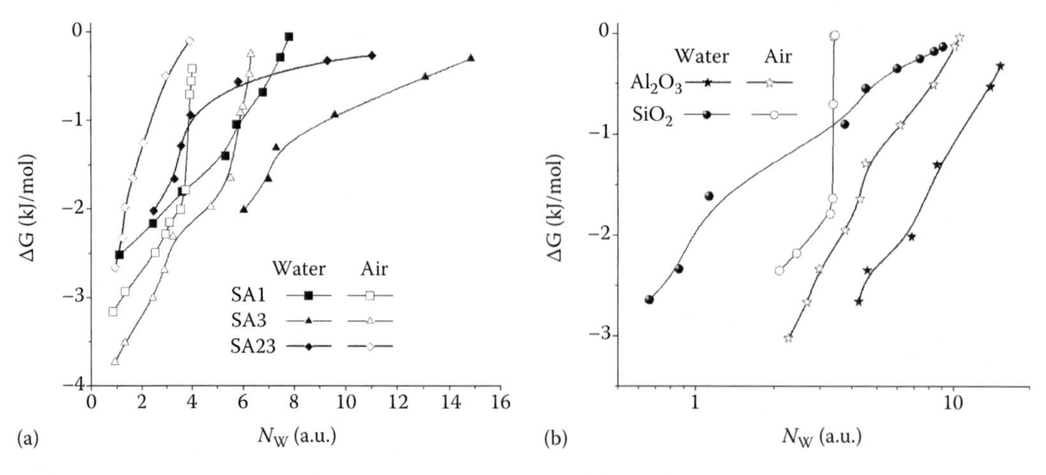

FIGURE 2.31 Relationships between the thicknesses of unfrozen water layer (the number of water monolayers) and changes in the Gibbs free energy for (a) SA and (b) A-300 and Al_2O_3. (Adapted with permission from Gun'ko, V.M., Turov, V.V., Zarko, V.I. et al., Active site nature of pyrogenic alumina/silica and water bound to surfaces, *Langmuir*, 13, 1529–1544, 1997a, Copyright 1997 American Chemical Society.)

TABLE 2.7
Characteristics of Nanooxides and Bound Water Layers

Sample	S_{BET} (m²/g)	$N_{w,max}$ (a.u.)	$N_{w,max}$ (nm)	γ_S (mJ/m²)
A-300	300	3.2	1.0	150
Al_2O_3	60	13.5	4.1	513
SA1	219	10.0	3.0	234
SA3	181	12.5	3.8	455
SA23	171	5.2	1.6	248

A nonlinear dependence is for the free surface energy for SA samples (Table 2.7) since it is larger for SA3 than for SA1 or SA23. However, the γ_S value (per unit of surface area) is larger for pure Al_2O_3 having minimal surface area.

Certain additional information on features of interactions of water with different nanooxides can be obtained from the direct adsorption/desorption data described in the subsequent sections.

2.1.4 Adsorption Phenomena

Organization of secondary particles and textural porosity of the nanooxide powders (Figures 2.11 through 2.14), dependent on the distribution function of size of primary particles, $\phi(a)$, and S_{BET}, influence adsorption of water and nitrogen (Figure 2.32), as well as other compounds. Notice that the adsorption of nitrogen onto nanooxides is greater than that of water at the same relative pressures because of several factors. First, temperature of water adsorption is higher (293 K) than that of nitrogen (77.4 K) that causes a higher kinetic energy of H_2O molecules (approximately by four times) favorable for transferring molecules from the adsorbed liquid (fluid) phase to the gas phase. Second, saturation pressure of water is much lower (~17 Torr at ambient conditions and ~760 Torr for nitrogen). Third, preferable adsorption of water occurs in the form of clusters in voids at surface hydroxyls near contacts between adjacent primary particles (Iler 1979, Legrand 1998, Gun'ko et al. 2009d). In the case of adsorption of a small amount of water onto solid surfaces at very low temperatures (~100–200 K), the molecules are weakly sensitive to neighbors (Henderson 2002).

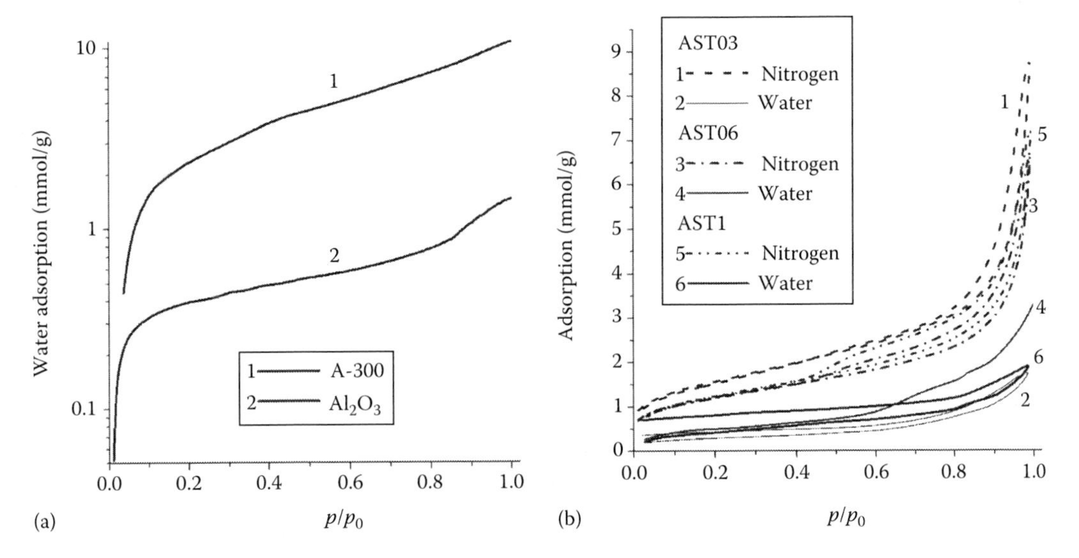

FIGURE 2.32 Adsorption-desorption isotherms of (a) water and (b) water or nitrogen on nanooxides. (Adapted from *Powder Technol.*, 195, Gun'ko, V.M., Zarko, V.I., Turov, V.V. et al., Morphological and structural features of individual and composite nanooxides with alumina, silica, and titania in powders and aqueous suspensions, 245–258, 2009g, Copyright 2009, with permission from Elsevier.)

However, the adsorption of water at ambient conditions results in the formation of clusters at solid surfaces, especially in pores and at the surface containing hydroxyl groups (Brennan et al. 2001, Gun'ko et al. 2005d, 2007d,f,i, 2009a, Chaplin 2011).

The morphology of primary nanoparticles of nanooxides, as well as particles of higher hierarchic levels such as aggregates of primary particles and agglomerates of aggregates, is typical and nearly the same for all individual and complex nanooxides because of the flame synthesis conditions. Therefore, the nitrogen adsorption–desorption isotherms are of the same type for different oxides (Figure 2.32).

However, the interaction energy is higher for nitrogen molecules adsorbed onto titania than silica or alumina due to stronger dispersive, electrostatic (−13.3, −9.5, and −5.4 kJ/mol), and charge transfer (−5.2, −2.9, and −1.6 kJ/mol) interactions in complexes –O–H···N≡N with TiO(H)Ti, SiO(H)Al, and SiOH groups, respectively. These values were estimated using the Kitaura–Morokuma method applied to the complexes calculated with the general atomic and molecular electronic structure system (GAMESS) program suit (Schmidt et al. 1993) and the 6–31G(d,p) basis set.

Both structural and energetic factors of nanooxides cause small differences in the shape of the nitrogen adsorption isotherms at low pressures when the adsorption monolayer forms. This is due to the absence of nanopores in nanooxide powders and a nonpolar character of the nitrogen molecule. The aforementioned factors can more strongly affect the adsorption/desorption of water because of its more specific interactions with oxide surfaces by the hydrogen bonding. Open hysteresis loops are characteristic for the water adsorption/desorption isotherms because of preheating of samples during degassing before the adsorption. This pretreatment causes irreversible dissociative adsorption of water (to form surface hydroxyls) as well as its adsorption in the volume of primary particles (where desorption is more difficult than from the surface). However, the later process is very slow occurring during days. In the case of totally hydrated oxides, this effect is absent and the specific surface area estimated from the water adsorption is very close to the S_{BET} value estimated from the nitrogen adsorption (Gun'ko et al. 2009a). Water adsorbed onto nanooxides forms certain clustered structures. However, the adsorption of water is relatively low because of the absence of nanopores and weak aggregation of nanoparticles, especially at $S_{BET} < 100$ m^2/g, causing a low adsorption capacity of the powders with respect to water. Therefore, the adsorption of water from air onto nanooxides is relatively low (<13 wt%; Table 2.5) in comparison with mesoporous silica gels (up to 50–70 wt%).

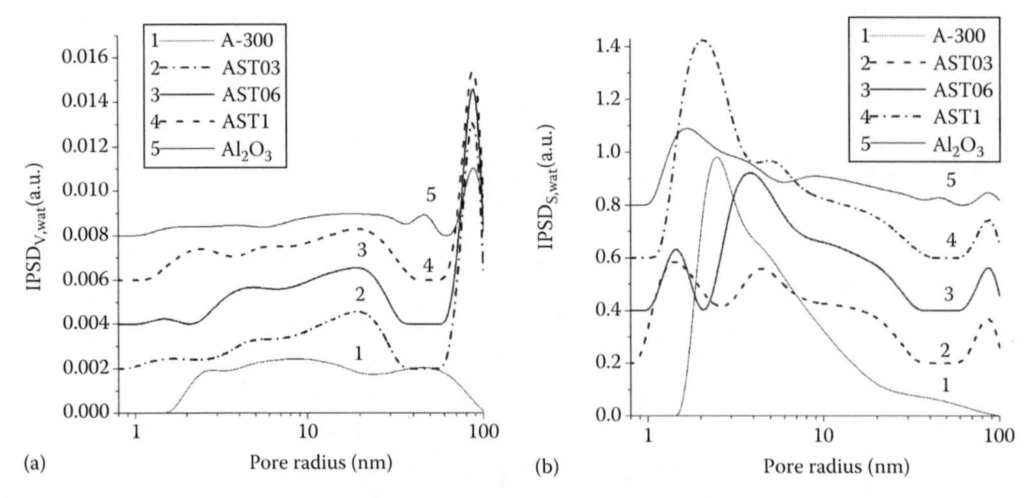

FIGURE 2.33 DTF IPSDs with respect to (a) the pore volume and (b) the specific surface area calculated on the basis of the water adsorption isotherms for nanooxides. (Adapted from *Powder Technol.*, 195, Gun'ko, V.M., Zarko, V.I., Turov, V.V. et al., Morphological and structural features of individual and composite nanooxides with alumina, silica, and titania in powders and aqueous suspensions, 245–258, 2009g, Copyright 2009, with permission from Elsevier.)

The adsorbed water structures (specific adsorption) differ from those formed on the adsorption of nitrogen (non-specific adsorption) because of the hydrogen bonding of water molecules to other molecules and surface hydroxyls (resulting in clustered adsorption of water). However, the adsorption isotherms of water and nitrogen are of the same type II because of the structural, textural, and morphological features of nanooxides mentioned earlier. Water does not form a continuous film at the silica surface since it forms clusters and nanodomains in narrow voids between adjacent nanoparticles (Figure 2.33), and a significant portion of the surface of nanoparticles remains free of adsorbed water even at $C_{H_2O} > 10$ wt%. This is readily seen from the FTIR spectra of nanooxides demonstrating the presence of free silanol groups (Figures 2.6 through 2.8) even with the adsorption of relatively large amounts of water from air (up to 10 wt% or more; Gun'ko et al. 2005d). In the case of mixed nanooxides, a continuous surface water layer forms at smaller amounts of adsorbed water because of a larger content of surface hydroxyls (Chuiko 2001, Gun'ko et al. 2009d). Adsorbed water appears in the FTIR spectra as a broad band with a maximum at 3200–3500 cm^{-1}. Therefore, it is difficult to estimate the different forms of water only from the IR spectra.

The adsorption (Figures 2.32 and 2.34a) and thermodesorption (Figures 2.34b and 2.35) of water from nanooxides depend on the concentration and kind of surface hydroxyls, specific surface area (Table 2.1; S_{BET}), and type of packing of primary particles in aggregates.

For instance, the adsorption/desorption of water is maximal for SA23 among SA samples because SA23 has the maximal S_{BET} value (Table 2.1) and maximal Brønsted acidity (Figure 2.4) because of a low $C^{s}_{Al_2O_3}$ value (Figure 2.3). Typically, the lower the $C^{s}_{Al_2O_3}$ value in amorphous (e.g., fumed) or crystalline (e.g., zeolites) SA, the higher is the Brønsted acidity. The sample SA8 has larger S_{BET} and $C^{s}_{Al_2O_3}$ values than SA3. Therefore, desorption of water (Figure 2.34b) from SA8 is greater at $T < 550$ K (effect of surface alumina and Al_2O_3/SiO_2 interfaces) and lower at $T > 550$ K (effect of surface silica) because associative desorption of water with condensation of surface hydroxyls occurs more easily from SA than silica (Figures 2.34 and 2.35). However, the adsorption of water is higher on SA3 than SA8 (Figure 2.34a). Notice that water fills (adsorbed water volume $V_w < 0.25$ cm^3/g) only a portion of the pore volume filled by nitrogen because the V_p values (Table 2.1) are much larger than the maximal adsorption of water (in cm^3/g). Consequently, the lower adsorption of water onto SA8 than SA3 (despite larger S_{BET} and V_p values for SA8) can be caused by a difference in packing of primary particles in aggregates. This difference appears in the incremental pore size

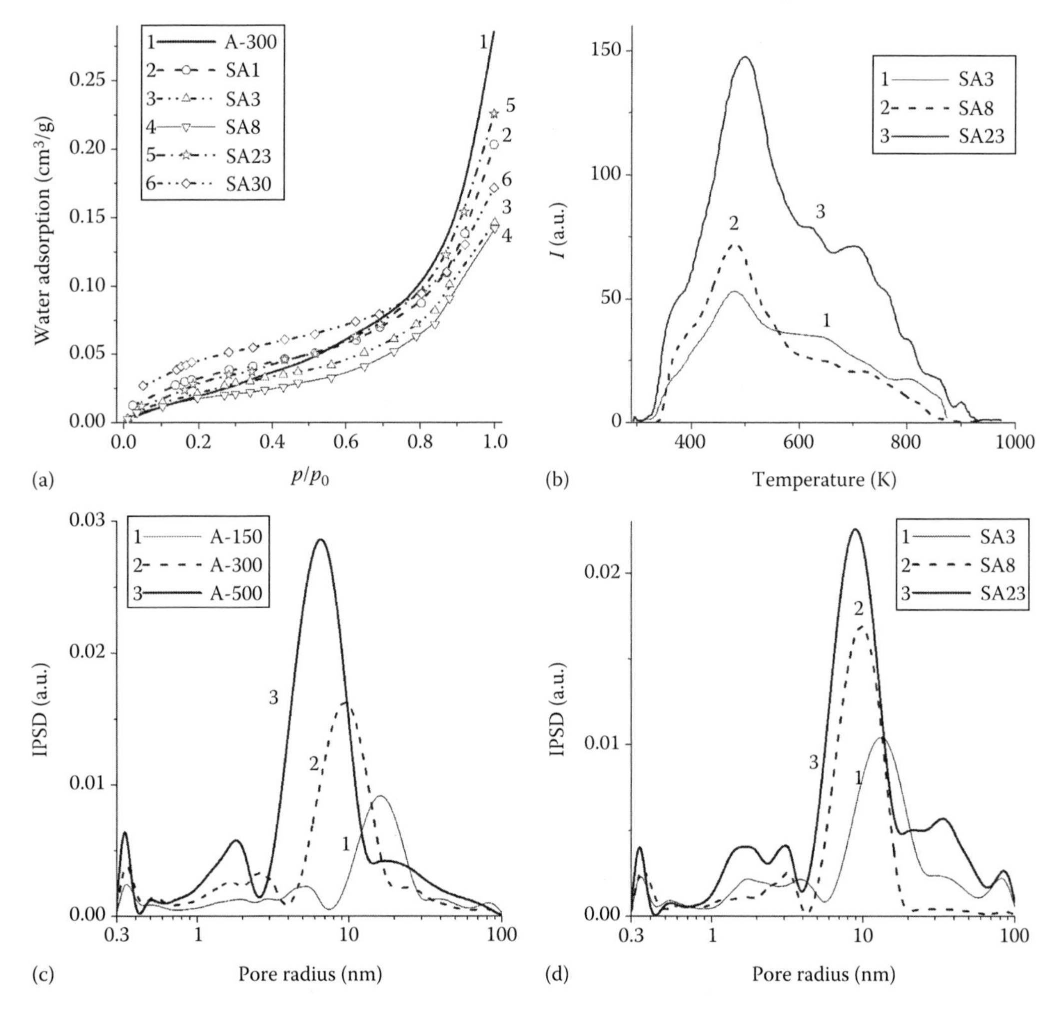

FIGURE 2.34 (a) Adsorption isotherms of water, (b) TPD-MS spectra of water (*m/z* 18); and IPSD calculated on the basis of nitrogen desorption data for (c) fumed silica and (d) SA samples. (Adapted from *Appl. Surf. Sci.*, 253, Gun'ko, V.M., Nychiporuk, Yu.M., Zarko, V.I. et al., Relationships between surface compositions and properties of surfaces of mixed fumed oxides, 3215–3230, 2007d, Copyright 2007, with permission from Elsevier.)

distribution (IPSDs) (Figure 2.34d) for both narrow pores at R between 1 and 5 nm and for broader pores at $R>20$ nm. Water more effectively fills narrow pores ($V_w<V_p$), and the IPSD of SA3 is higher at $1<R<5$ nm than that of SA8 (Figure 2.34d). Therefore the adsorption of water is higher for SA3 than for SA8 (Figure 2.34a). Notice that there is a certain difference in the shape of the IPSDs of SA (Figure 2.34d) and fumed silica (Figure 2.34c) samples over the ranges of narrow mesopores and macropores. This can be caused by the presence of surface alumina and silica in SA samples that leads to changes in the particle–particle interaction on the formation of aggregates and agglomerates, whose sizes are larger for SA than that for fumed silica samples (Gun'ko et al. 2001c).

The major portion of water desorbs from ST at $T=330$–450 K and from AST at $T=330$–550 K (Figure 2.35). However, the tail of this desorption is longer for the ST samples, as well as for the SA samples (Figure 2.34b), than for AST since in the last case the desorption stops at $T=650$–700 K because of a small content of silica in these samples. The role of silica in mixed nanooxides is much lower than for CVD-TiO_2/A-300 (Figure 2.35c; HT portions of the water desorption curves for A-300 and CVD-TiO_2/A-300 are similar), because in the case of grafted CVD-TiO_2, separated titania particles can form and their contacts with silica particles are weak. Therefore, a significant

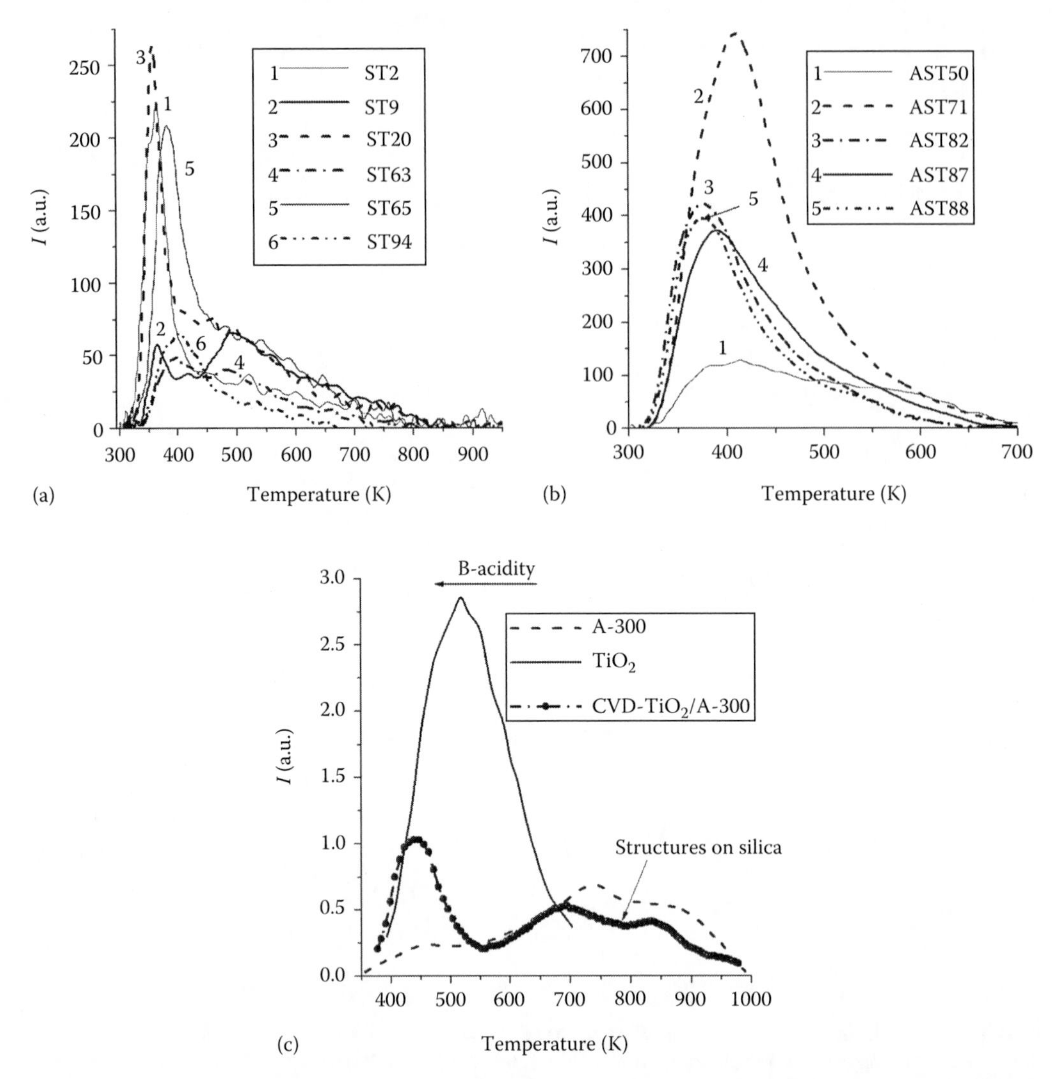

FIGURE 2.35 TPD-MS spectra of water (m/z 18) desorbed from degassed (a) ST, (b) AST, and (c) silica, titania, and CVD-TiO_2/A-300 ($C_{TiO_2}=22$ wt%) (spectra in (c) were normalized to the amounts of desorbed water per a surface area unit). (Adapted from *Appl. Surf. Sci.*, 253, Gun'ko, V.M., Nychiporuk, Yu.M., Zarko, V.I. et al., Relationships between surface compositions and properties of surfaces of mixed fumed oxides, 3215–3230, 2007d, Copyright 2007, with permission from Elsevier.)

portion of the silica surface remains accessible. This result confirms the importance of the formation of bridges ≡SiO(H)M≡ and ≡SiOM≡ and tight contacts between different phases in mixed nanooxides for different surface processes. These structural features can play a specific role on the relaxations of interfacial water in aqueous suspensions of these oxides shown next.

The larger number of different surface sites at the surface of mixed oxides leads to broadening of the TPD-MS spectra for ST samples in comparison with individual titania (Figure 2.35). In the case of AST samples, desorption of water from AST82, AST87, and AST88 is close, but for AST50 and AST71, it differs (Figure 2.35) because of the effect of larger S_{BET} (AST71) and lower $C^{s}_{TiO_2}$ values (AST50). The TPD-MS spectra of desorbed water (Figure 2.35) are dependent on the oxide composition, illustrated in the graphs of the dependence of the peak temperature on the surface content of alumina and titania (Figure 2.36). Notice that in the case of SA and ST, the T_m value (temperature of the main TPD maximum) increases with increasing C^{s}_{Al} or C^{s}_{Ti} value (Figure 2.36a),

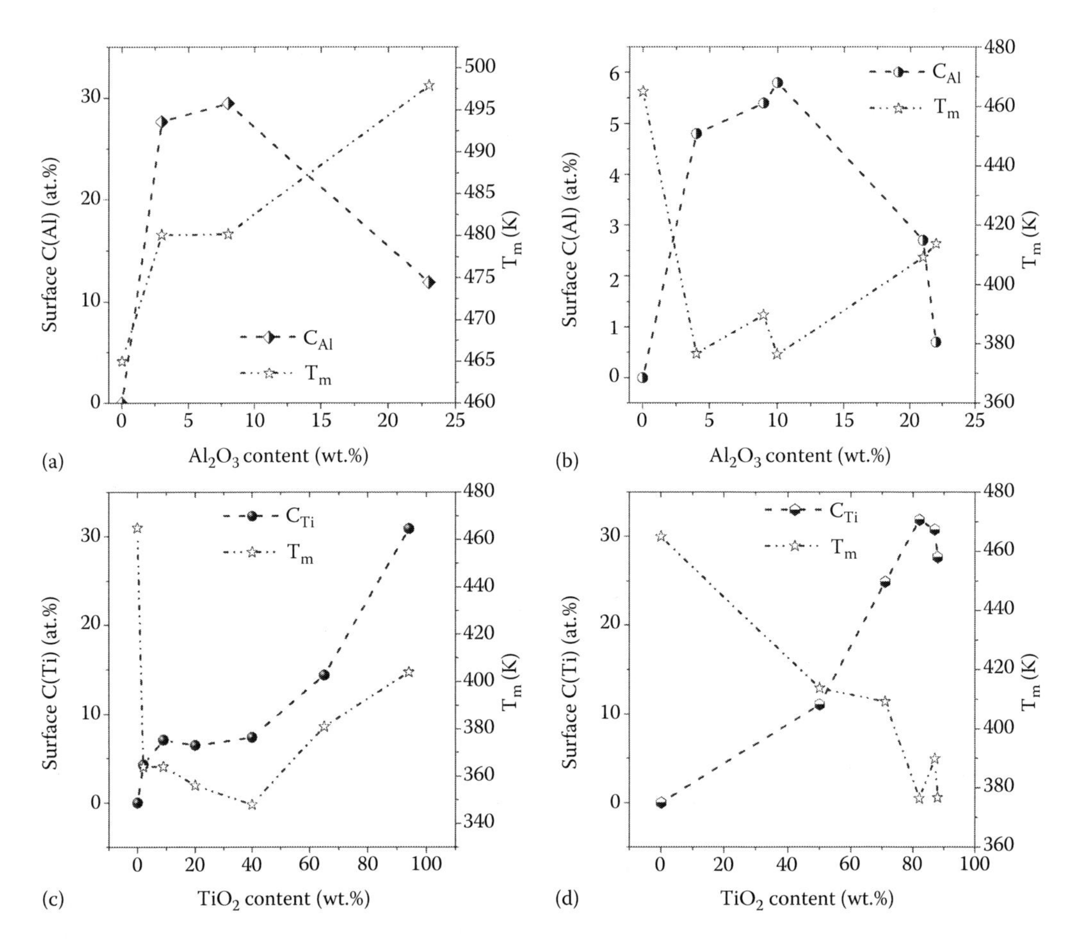

FIGURE 2.36 Relationships between the total content of (a and b) alumina and (c and d) titania and the surface content of Al in (a) SA and (b) AST and Ti in (c) ST and (d) AST and the temperature of the main peak in the OPTPD TOFMS spectra. (Adapted from *Appl. Surf. Sci.*, 253, Gun'ko, V.M., Nychiporuk, Yu.M., Zarko, V.I. et al., Relationships between surface compositions and properties of surfaces of mixed fumed oxides, 3215–3230, 2007d, Copyright 2007, with permission from Elsevier.)

and in AST (Figure 2.36b, $C_{Al_2O_3} > 10$ wt%) and C^s_{Ti} value (Figure 2.36c, $C_{TiO_2} \geq 40$ wt%). In the case of AST samples, the opposite dependence is observed (Figure 2.36b, $C_{Al_2O_3} < 10$ wt%, and Figure 2.36d). Several factors, in addition to the surface content of the second oxide, can be responsible for this effect: (i) different contents of terminal and bridging hydroxyls for different samples; (ii) different topology and morphology of primary and secondary particles; (iii) different S_{BET} values; and (iv) different amounts of adsorbed intact water. These factors also affect the distribution functions of the activation energy of water desorption (Figure 2.37) calculated from the OPTPD-MS data.

The difference in the *f*(*E*) shapes for SA, ST, and AST is much smaller than that for A-300 because silica has only terminal hydroxyls (single ≡SiOH and twin = Si(OH)$_2$). An increase in the surface B acidity at a higher S_{BET} value for SA23 results in higher intensity of *f*(*E*) at $E > 150$ kJ/mol for SA3 and SA8 than that for SA23 (Figure 2.37a). A smaller S_{BET} value (Table 2.1), i.e., larger size of primary particles, corresponds to lower aggregation of these particles in secondary ones. The V_p value, which can be considered as a measure of this aggregation typically decreases with decreasing S_{BET} value (Table 2.1; Figure 2.17). Notice that hydrothermal treatment (HTT) of fumed silicas resulting in a decrease in the S_{BET} value gives strong enhancement of the V_p value because of denser packing of primary particles in secondary ones, but the MCA remaining the S_{BET} values

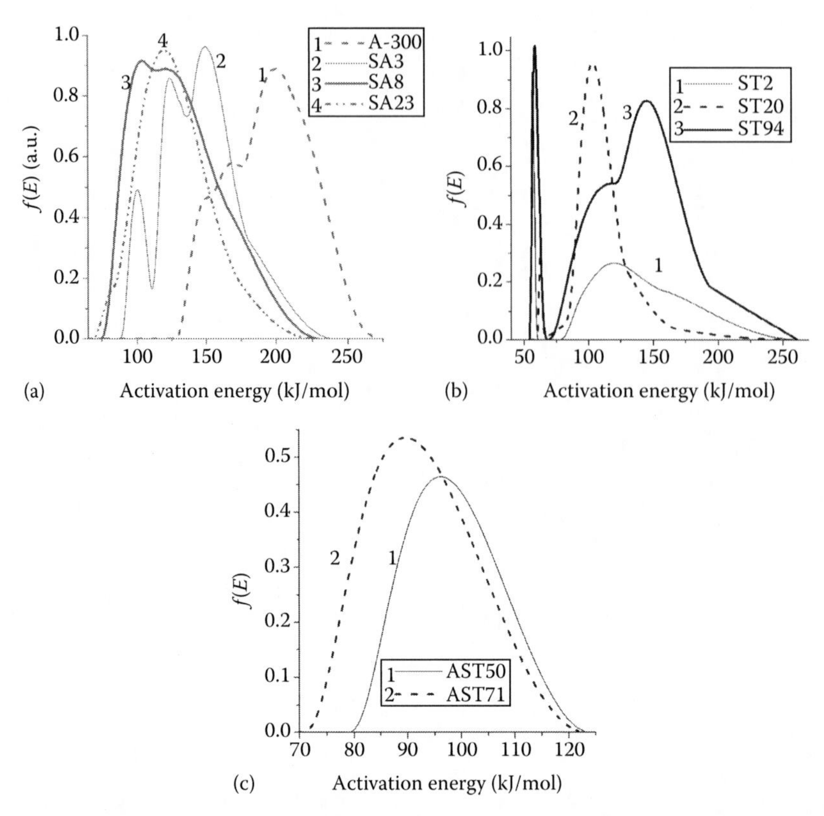

FIGURE 2.37 Distribution functions of the activation energy of water desorption (*m/z* 18) from (a) A-300 and SA, (b) ST, and (c) AST samples. (Adapted from *Appl. Surf. Sci.*, 253, Gun'ko, V.M., Nychiporuk, Yu.M., Zarko, V.I. et al., Relationships between surface compositions and properties of surfaces of mixed fumed oxides, 3215–3230, 2007d, Copyright 2007, with permission from Elsevier.)

without changes firstly increases but then decreases the V_p values. The aforementioned reduction of interparticle interactions (aggregation) for the initial nanooxides with decreasing S_{BET} value causes diminution of interparticle transfer of mass during water desorption. This effect can add the E values in addition to decreasing B-acidity of SA3 and SA8 in comparison with SA23. A similar effect related to B acidity is observed for TS samples (Figure 2.37b). ST20 has a maximal number of bridging hydroxyls ≡SiO(H)Ti≡ with maximal B acidity among ST samples. Therefore, the high-energy band of $f(E)$ at $E > 130$ kJ/mol is lower for ST20 in comparison with ST2 and ST94 (shown as boundary ST samples).

The activation energy of water desorption (by both molecular and associative desorption) is lower for AST71 than for AST50 (Figure 2.37c) because of lower silica content (characterized by maximal E values on water desorption) in AST71 (Table 2.1). Desorption of water from AST samples (Figure 2.35) stops at lower temperatures than that for SA and ST, and especially for silica (Figure 2.35). Therefore, the E values for AST samples are much smaller than those for SA and especially for fumed silica because of a higher content of surface bridging hydroxyls on the AST surface than on other oxides studied. Notice that the TPD activation energy of 70–80 kJ/mol for desorption of water from titania corresponds to the initial portion of the $f(E)$ function for mixed oxides (Figure 2.37) related to desorption of intact water molecules.

A minimal $C^s_{Al_2O_3}$ value for SA23 (Figure 2.3a) results in the smallest amounts (C_{uw}) of structured water unfrozen at $T < 273$ K (Figure 2.38) determined from the ^{1}H NMR spectra. Larger C_{uw} values are for the suspensions of alumina and SA3 (high $C^s_{Al_2O_3}$ value). Changes in the Gibbs free energy of bound water are between −0.2 and −2.7 kJ/mol (Figure 2.38b). Derivatives dC_{uw}/dT show the curve

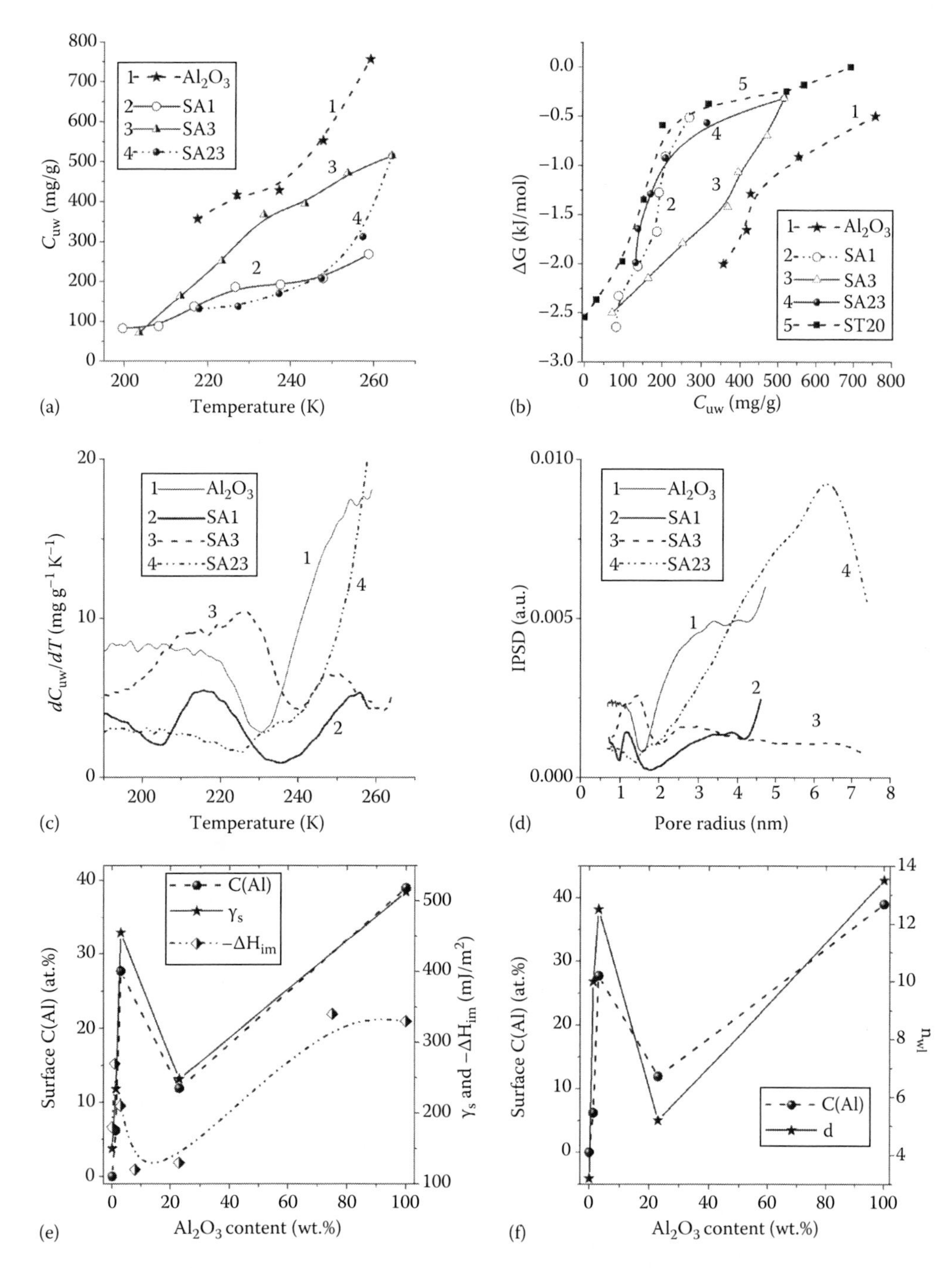

FIGURE 2.38 (a) Concentration of unfrozen water (C_{uw}) as a function of temperature for aqueous suspension (5 wt%) of alumina and SA samples determined from the ^{1}H NMR spectra; (b) relationship between Gibbs free energy ΔG and C_{uw} (curve for ST20 is also shown); (c) derivatives $dC_{uw}(T)/dT$; (d) pore size distribution calculated by using the NMR cryoporometry; (e and f) relationships between total content of alumina is SA samples and surface content of aluminum and (e) interfacial free energy γ_S and the enthalpy of immersion in water (ΔH_{im}); and (f) the number of statistical water layers (n_{wl}) at a thickness of each of them of 0.3 nm. (Adapted from *Appl. Surf. Sci.*, 253, Gun'ko, V.M., Nychiporuk, Yu.M., Zarko, V.I. et al., Relationships between surface compositions and properties of surfaces of mixed fumed oxides, 3215–3230, 2007d, Copyright 2007, with permission from Elsevier.)

shapes (Figure 2.38c), which correspond to the TSDC peaks in the HT band at 210–215 K for SA1 and SA23 (Gun'ko et al. 2007i). Calculations of the size distributions of unfrozen water clusters (SDUWC) using the NMR cryoporometry show that structured water fills voids between primary particles corresponding to narrow mesopores at $R<7$ nm (Figure 2.38d). Notice that the TSDC cryoporometry gives similar SDUWC with slightly smaller sizes. The $IPSD_{NMR}$ (Figure 2.38d) and $IPSD_{N2}$ (Figures 2.14 and 2.35d) have similar shapes. Their differences can be caused by several reasons. First, water fills a smaller pore volume than nitrogen does. Second, the ^1H NMR spectra were not recorded at $T \rightarrow 273$ K (when water thaws/freezes in large mesopores and macropores). Third, water can thaw/freeze differently near the pore walls (surface of oxide nanoparticles) than at the center of broad pores.

Figure 2.39 shows the ^1H NMR spectra of aqueous suspensions of AST03 as a representative sample, since the spectra of other AST samples (Table 2.8) have practically the same shapes at the same temperatures.

All bound water unfrozen at $T<273$ K can be assigned to strongly associated water at $\delta_H \approx 5$–5.5 ppm. WAW at $\delta_H = 1$–2 ppm is not observed in any of the studied suspensions of the initial nanooxides. The total amounts of water bound by oxide nanoparticles in these suspensions (Table 2.8; $C_{uw} = C^{s}_{uw} + C^{w}_{uw} = 1 \quad 2.5 \text{ g/g}$ at a major contribution of weakly bound water frozen at $T>260$ K; Figure 2.40a) are much larger than that observed from the measurements of the water

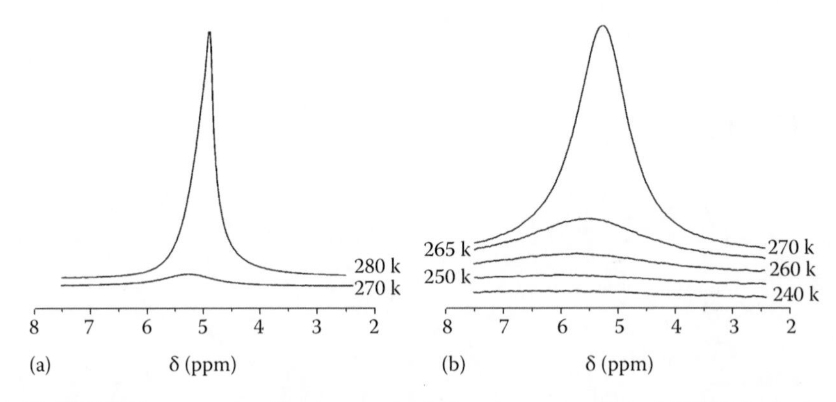

FIGURE 2.39 ^1H NMR spectra of unfrozen water in the aqueous suspension of AST03 (hydration $h=13$ g/g) recorded at different temperatures. (Adapted from *Powder Technol.*, 195, Gun'ko, V.M., Zarko, V.I., Turov, V.V. et al., Morphological and structural features of individual and composite nanooxides with alumina, silica, and titania in powders and aqueous suspensions, 245–258, 2009g, Copyright 2009, with permission from Elsevier.)

TABLE 2.8
Characteristics of Bound Water in the Suspensions of SA and AST Oxides

Sample	h (g/g)	C^{s}_{uw} (g/g)	C^{w}_{uw} (g/g)	$-\Delta G^{s}$ (kJ/mol)	$-\Delta G^{w}$ (kJ/mol)	γ_S (J/g)
SA5	31.8	0.07	1.03	2.54	0.32	14.5
SA96	13.9	0.05	1.15	2.06	0.50	12.3
AST1	13.0	0.12	1.18	1.75	0.32	16.1
AST06	11.5	0.15	0.85	1.30	0.32	9.9
AST03	13.0	0.12	2.38	2.15	0.30	30.0

Note: C^{s}_{uw} and C^{w}_{uw} are the amounts of strongly and weakly bound waters, respectively; ΔG^{s} and ΔG^{w} are the corresponding changes in the Gibbs free energy; and γ_S is the free surface energy.

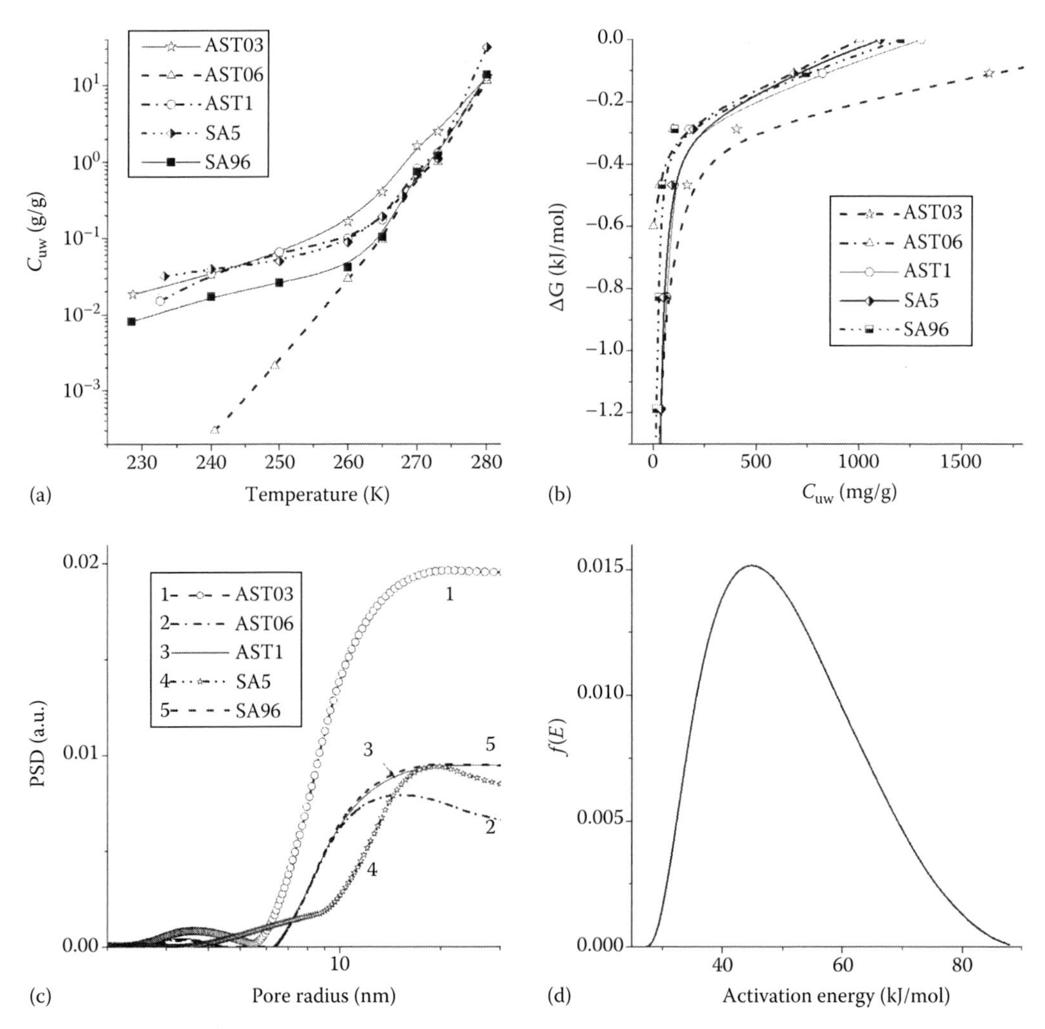

FIGURE 2.40 (a) Amounts of unfrozen water (C_{uw}) in the aqueous suspensions of nanooxides (h values are shown in Table 2.8) as a function of temperature; (b) relationships between the C_{uw} and ΔG values; (c) distribution functions of pore (void) size filled by unfrozen water; and (d) distribution function of the activation energy of the molecular mobility of unfrozen water for the aqueous suspension of AST03. (Adapted from *Appl. Surf. Sci.*, 253, Gun'ko, V.M., Nychiporuk, Yu.M., Zarko, V.I. et al., Relationships between surface compositions and properties of surfaces of mixed fumed oxides, 3215–3230, 2007b, Copyright 2007, with permission from Elsevier.)

adsorption isotherms (Figure 2.32). This confirms the explanation of the difference in the ΔH_{im} ($h=60$ g/g) and Q_{ads} ($h<1$ g/g) values (Table 2.9).

However, the γ_S values (Table 2.8) are less than the ΔH_{im} and Q_{ads} values, because samples were not heated and degassed before the NMR measurements but they were degassed before the water adsorption measurements. In other words, the γ_S values characterize the interaction of water with the preequilibrated oxide surfaces. The ΔG^s and ΔG^w values (Table 2.8) and the $\Delta G(C_{uw})$ graphs (Figure 2.40b) reveal that the studied SA and AST samples do not have a high affinity for water because primary particles are nonporous, and the textural porosity of secondary particles corresponds to meso- and macroporosity as described earlier. At $T<240$ K, the amounts of unfrozen water are less than 3.5 wt% (with respect to the weight of dry oxides; Figure 2.40a). Notice that a similar weight corresponds to the amounts of water adsorbed from air and desorbed during heating at $T<420$ K (Figure 2.36). Consequently, water adsorbed from air onto nanooxide at room temperature is mainly strongly bound.

TABLE 2.9
Specific Surface Area Determined from Water Adsorption, Heats of Immersion, and Adsorption of Water

Oxide	$S_{BET,w}$ (m²/g)	ΔH_{im} (J/g)	Q_{ads} (J/g)
A-300	255	51.5	62.3
A-150	26		16.2
A-50		11.6	
Al_2O_3	97		21.0
AST1	97	54.3	26.6
AST06	107	43.2	32.7
AST03	65		20.8

The PSD_{NMR} and PSD_{TSDC} (Figure 2.40c) are estimated from the analysis of different processes, local molecular mobility (NMR), and local mobility of polar bonds (TSDC), which can demonstrate a cooperative character on different scales. The PSD_{TSDC} functions correspond to narrower pores than PSD_{NMR}. The latter is close to the PSD_{N2} (Figure 2.15). This is due to the appearance of the local molecular mobility of water on thawing of ice in voids of certain size at a certain temperature (according to the Gibbs–Thomson (GT) relation).

An assumption that the molecular mobility of bound unfrozen water obeys the Arrhenius law allows us to calculate the distribution function of the activation energy ($f(E)$) of this mobility using the corresponding integral equation for dC_{uw}/dT. Since the dC_{uw}/dT graphs for all the studied suspensions are close, the $f(E)$ function was calculated only for the suspension with AST03 (Figure 2.40d). The position of this function maximum is in agreement with the E_{dc} value (44 kJ/mol). This is due to the mechanism of the direct current, dc relaxation linked to proton mobility, which depends on the molecular mobility of water, because protons are attached to water molecules and the mobility of the H_3O^+ ions is registered as the dc.

The TSDC thermograms (Figure 2.41a) are shown only for selected AST samples at different C_{TiO_2} values. The dc relaxation can start in frozen suspensions at very low temperature, e.g., at ~200 K for AST71 (activation energy $E_{dc}=19$ kJ/mol). This oxide possesses the most negative σ_0 values from −20 (pH 5) to −100 μC/cm² (pH 10) among AST at high C_{TiO_2} values, providing large amounts of mobile protons (as counter ions in the EDL) causing the effective dc relaxation at lower temperatures. The suspensions with AST1 and AST50 demonstrate much lower dc at higher activation energy $E_{dc}=126$ and 74 kJ/mol, respectively, and this dc relaxation appears at higher temperatures (~250 K). This difference between AST samples can be explained by smaller σ_0 values and lower amounts of mobile protons and bound water unfrozen over this temperature range (this water is the necessary medium for the mobile protons). Notice that for dc relaxation, the condition of ion percolation between two electrodes should be satisfied, i.e., the dc relaxation corresponding to the through current. For AST03 and AST06, the dc relaxation starts at 210–220 K at $E_{dc}=44$ and 36 kJ/mol, respectively. Besides dc relaxation, the dipolar relaxation is observed at lower temperatures (Figure 2.41a). This relaxation is caused by rotational mobility of polar bonds (e.g., O-H), tunnel proton transferring, rotation of whole water molecules, and then small and larger clusters and nanodomains with increasing temperature. For aqueous suspensions of nanooxides, mainly water molecules are responsible for this (dipolar) type of TSDC relaxation, i.e., contribution of surface hydroxyls is much smaller due to the difference in the contents of structured water and surface hydroxyls. Comparison of the TSDC thermograms for water alone and aqueous suspensions reveals the opposite displacements of a LT band toward higher temperatures, and a HT band toward lower temperatures, as well as changes in the intensity of these bands. This is a typical effect for

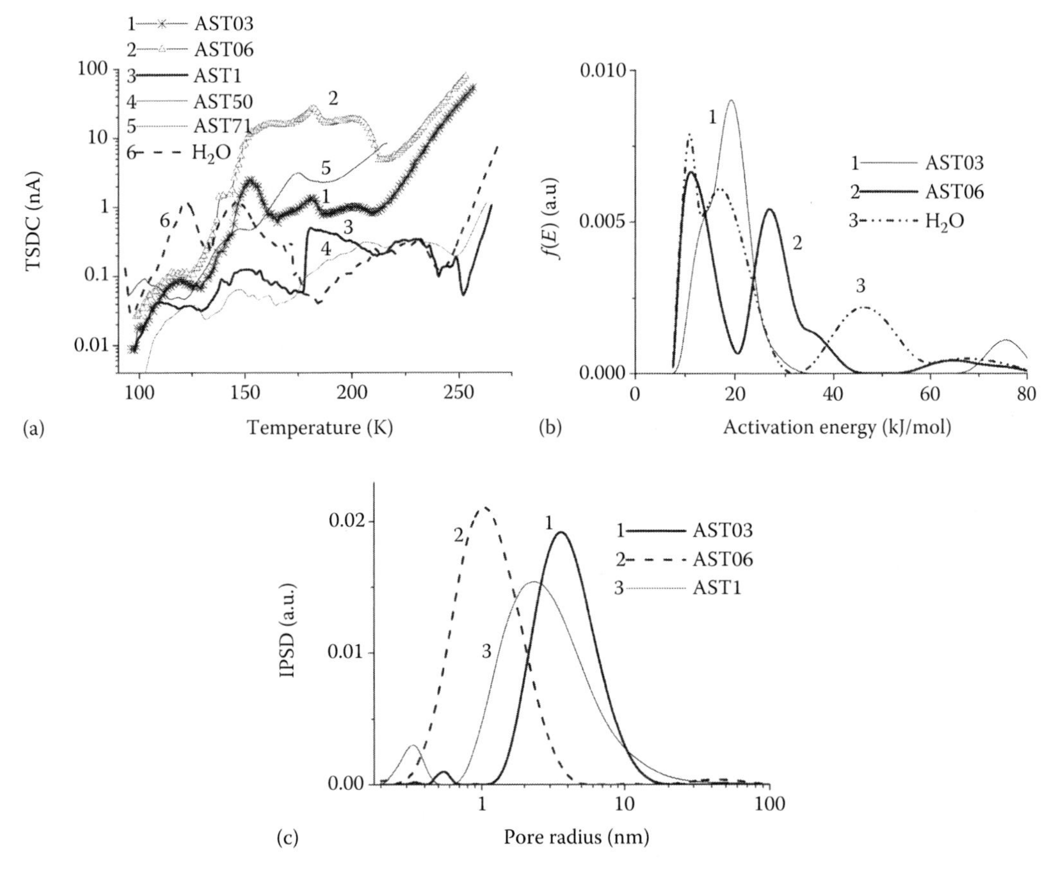

FIGURE 2.41 (a) TSDC thermograms of aqueous suspensions of AST samples (3 wt%) recorded at T=90–265 K; (b) corresponding distribution functions of activation energy of dipolar relaxation for the frozen suspensions with AST03 and AST06 and frozen pure water; and (c) size distribution of water structures on dipolar relaxation of frozen AST suspensions calculated using the TSDC cryoporometry. (Adapted from *Powder Technol.*, 195, Gun'ko, V.M., Zarko, V.I., Turov, V.V. et al., Morphological and structural features of individual and composite nanooxides with alumina, silica, and titania in powders and aqueous suspensions, 245–258, 2009g, Copyright 2009, with permission from Elsevier.)

aqueous suspensions of nanooxides, which increases with increasing concentration of oxides (Gun'ko et al. 2007i). The displacement of the LT band (related to maximum clustered water molecules, i.e., WAW molecules) accompanied by an increase in the activation energy (Figure 2.41b) occurs due to (i) stronger interactions of water molecules with the oxide surfaces than with other H_2O molecules and (ii) additional polarization of interfacial water molecules in the surface electrostatic field, especially in narrow voids (pores). The displacement of the HT band (related to relaxations of clusters and nanodomains of bound water) is due to disordering effects of nanoparticles on bound water, forming clustered structures at $R < 10$ nm (Figure 2.41c). Nanoparticles at $C_{ox} = 3$ wt% can disturb a significant fraction of water in the suspension because of the long-range interactions. Therefore, the TSDC thermograms of these suspensions strongly differ from that of water alone (Figure 2.41).

The spectra of DMAAB adsorbed onto SA23 and ST20 are similar in their width (Figure 2.4) because of a variety of surface sites. However, ST20 has a larger amount of L sites. This similarity in the types of surface sites results in a similar shape of the $\Delta G(C_{uw})$ graphs for interfacial water in the aqueous suspensions of these oxides (Figure 2.38b). However, the relationships between the enthalpy of immersion in water (ΔH_{im}) and surface content of Al (Figure 2.38e) and Ti (Figure 2.42)

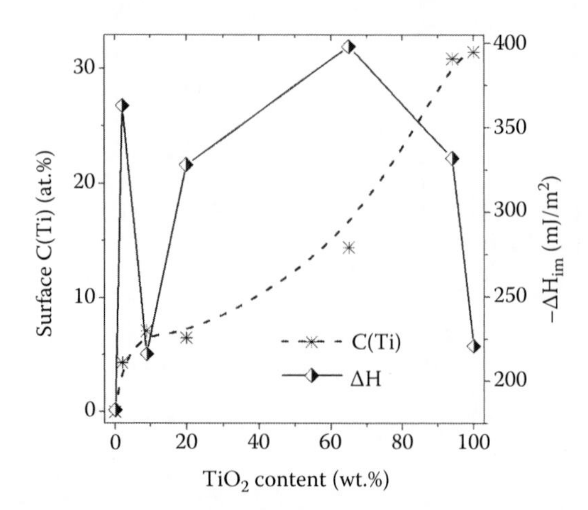

FIGURE 2.42 Relationship between the total titania content in TS samples and surface content of titanium and the enthalpy of immersion of TS samples in water (boundary points correspond to A-300 and pure titania). (Adapted from *Appl. Surf. Sci.*, 253, Gun'ko, V.M., Nychiporuk, Yu.M., Zarko, V.I. et al., Relationships between surface compositions and properties of surfaces of mixed fumed oxides, 3215–3230, 2007d, Copyright 2007, with permission from Elsevier.)

versus the total content of the second oxides differ for SA and ST. This can be caused by differences in the distribution of alumina and titania in the bulk and at the surfaces and the active sites at the surface of mixed oxides.

The γ_s and ΔH_{im} values as functions of $C_{Al_2O_3}$ in SA correlate with the surface content of aluminum C^s_{Al} (Figure 2.38e), as well as the thickness of the interfacial structured water (Figure 2.38f). These correlations are caused by several factors: (i) number and acidity of surface sites, (ii) nonuniformity of the surface, and (iii) specific surface area (γ_s and ΔH_{im} are calculated per surface area unit), aggregation of primary particles (i.e., V_p value). The influence not only of the number of B- sites but also other factors on the surface properties of mixed oxides is clearly depicted from comparison of the data shown in Figures 2.38 and 2.43.

One can expect that the activation energy of the dc relaxation ($E_{a,dc}$) caused by the mobile charge carriers (mainly protons for the studied suspensions) and observed by the TSDC method at $T > 210$–220 K for the studied oxides should decrease with increasing C^s_{Al}. However, the opposite result is observed since the $E_{a,dc}$ value increases with C^s_{Al} (Figure 2.43). This effect can be caused by diminution of the B acidity with increasing C^s_{Al} value. Consequently, the $E_{a,dc}$ value correlates with B acidity of SA; however, the γ_s and ΔH_{im} values do not correlate with the B acidity of SA but correlate with the surface content of alumina.

In the case of ST and AST samples, an increase in the C^s_{Ti} value exhibits the opposite effect (Figure 2.43) compared with that of C^s_{Al} changes in SA, because titania possesses much higher electronic conductivity (as a semiconductor) than alumina and silica (dielectric materials). The electronic conductivity, dependent on the surface (providing percolation effects) and total (determining the total number of mobile electrons) titania contents, affects the $E_{a,dc}$ values in addition to the ion conductivity (determined mainly by the number of B sites). A combination of several factors gives complex effects and causing an increase in the $E_{a,dc}$ value for pure titania (Figure 2.43). This result can be caused by significant diminution of B acidity of TiO_2 in comparison with ST and AST samples (Figure 2.4).

A multiple-factor influence of the surface composition of mixed nanooxides is observed for such interfacial phenomenon as release of heat of immersion in water (Figures 2.44 and 2.45). For SA samples, two maxima are observed on the $\Delta H_{im}(C_X)$ graphs, at low alumina content and for pure alumina. Both maxima correspond to maximal amounts of hydroxyls (see relative desorption of

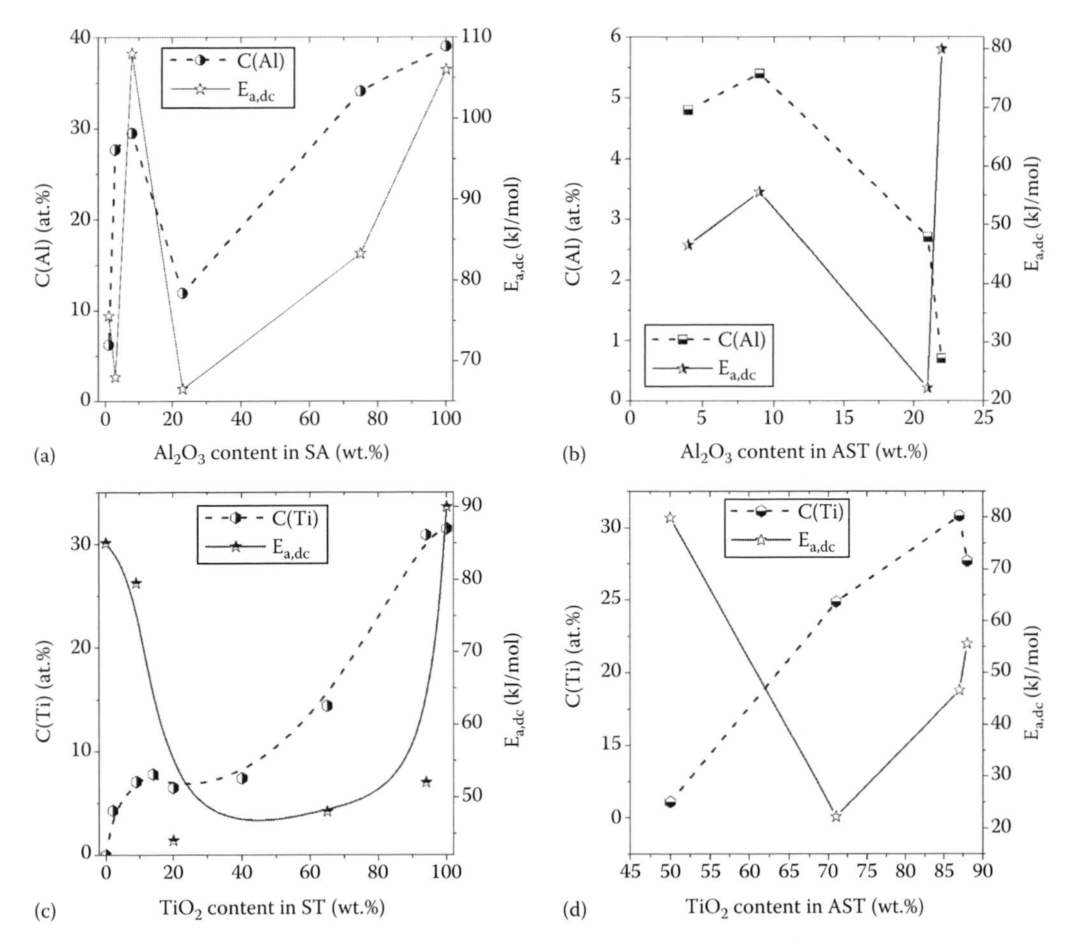

FIGURE 2.43 Relationship between the total content of (a and b) alumina in (a) SA and (b) AST and surface content of Al and the activation energy of the dc relaxation for aqueous suspension of SA; (c and d) titania in (c) ST and (d) AST and surface content of titanium and activation energy of dc relaxation for aqueous suspensions at $C_{ox}=3$ wt%. (Adapted from *Appl. Surf. Sci.*, 253, Gun'ko, V.M., Nychiporuk, Yu.M., Zarko, V.I. et al., Relationships between surface compositions and properties of surfaces of mixed fumed oxides, 3215–3230, 2007d, Copyright 2007, with permission from Elsevier.)

water in Figure 2.45). A similar picture is observed for ST samples at low C_{TiO_2} values. However, at high C_{TiO_2} values the $\Delta H_{im}C_{TiO_2}$ maximum is observed at $C_{TiO_2}=65$ wt%.

Thus, adsorption/desorption, relaxation, and diffusion processes at the surfaces of mixed oxides are governed by the structure of their surfaces, i.e., surface content and distribution of alumina and titania as more active phases in comparison with silica, rather than the specific surface area and the total content of the second oxide, because practically, all the studied parameters of mixed oxides correlate with the surface content of alumina in SA and AST and titania in ST and AST samples. On the basis of these results, specific recommendations related to "structure-property" relationships can be derived. Depending on the type of tasks, which should be solved by using mixed oxides (such as redox or acid–base catalysis, adsorption of ions or molecules, delivery of adsorbates, filling of polymers, etc.), certain properties of the materials (dc relaxation, Gibbs free surface energy, and the heat of immersion in water) depend on some surface characteristics, e.g., the amounts of acid sites, but other properties (e.g., adsorption of metal ions) depend on the general surface composition more than on the number of given sites. Thus, during the synthesis of mixed nanooxides, it is possible to create appropriate surface compositions and to design required surface structures; however, the

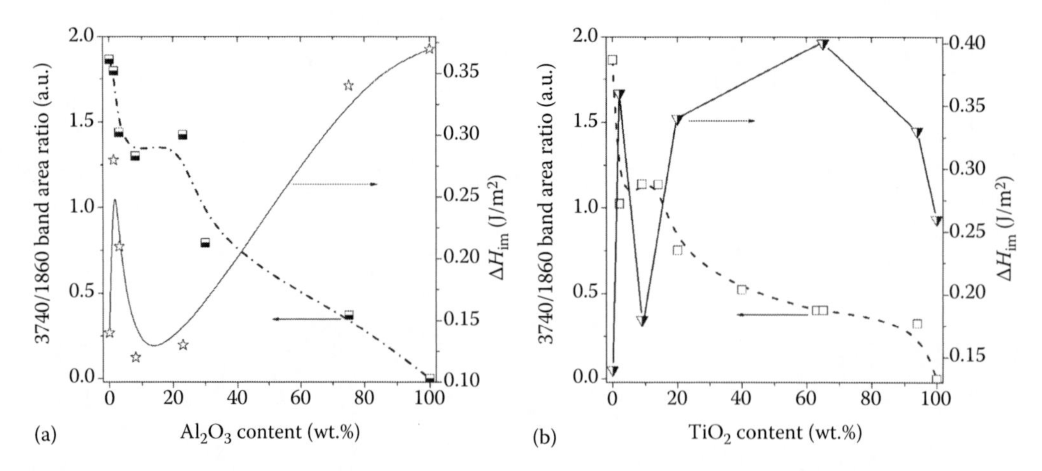

FIGURE 2.44 Relationships between the total content of (a) alumina in SA and (b) titania in ST and the amounts of silanols (spectra were recorded in the diffuse reflectance mode at 473 K) and the heat of immersion of oxides in water. (Adapted from *J. Colloid Interface Sci.*, 314, Gun'ko, V.M., Blitz, J.P., Gude, K. et al., Surface structure and properties of mixed fumed oxides, 119–130, 2007a, Copyright 2007, with permission from Elsevier.)

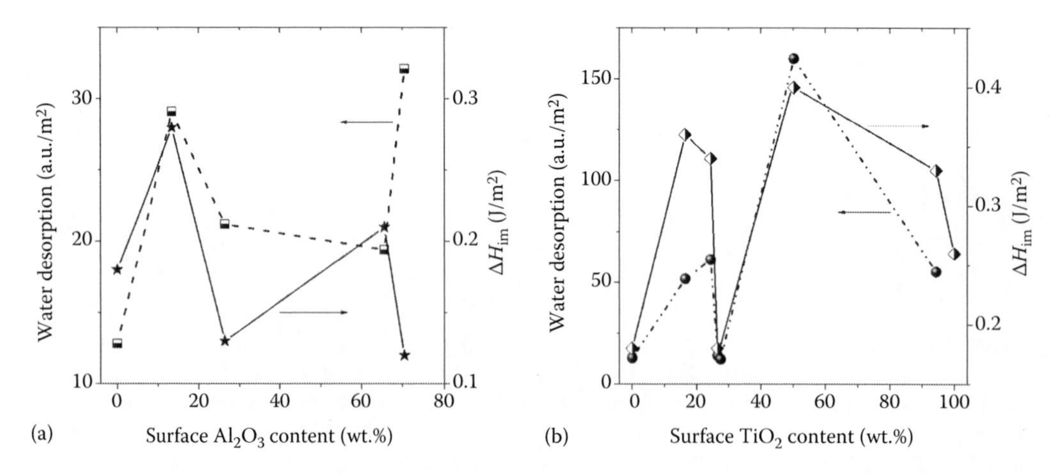

FIGURE 2.45 Relationships between the surface content of (a) alumina in SA and (b) titania in ST and total water desorption (TPDMS thermograms) and the heat of immersion of oxides in water. (Adapted from *J. Colloid Interface Sci.*, 314, Gun'ko, V.M., Blitz, J.P., Gude, K. et al., Surface structure and properties of mixed fumed oxides, 119–130, 2007g, Copyright 2007, with permission from Elsevier.)

cores of primary particles can be composed of mainly individual silica, titania, or other oxides depending on the kind of practical applications of the materials.

The effects of not only B sites but also other surface structures (and several adsorption mechanisms) on the adsorption of metal ions are clearly observed in Figure 2.46 because the shapes of the plateau adsorption curves of Pb(II) onto SA samples and Ni(II) onto ST samples correlate with the surface content of alumina and titania, respectively.

A larger deviation in the curve shapes for the adsorption of Pb(II) onto the SA samples corresponds to a similar deviation in the relationship between the S_{BET} and C_{Al}^{s} curves versus the total $C_{Al_2O_3}$ values (Figure 2.3) because the ion adsorption is calculated per surface area unit. The observed correlations show that the number of B sites is not limiting (as in the case of dc relaxation and $E_{a,dc}(C_{Al_2O_3})$ shape) for the plateau adsorption of Pb(II) and Ni(II) because this maximal adsorption is observed at relatively high pH values (close to pH 7–8) when the

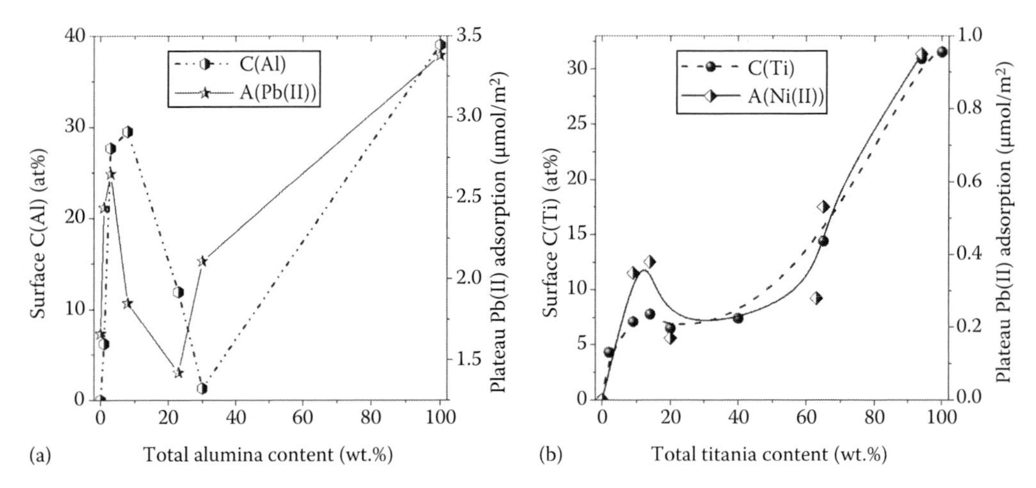

FIGURE 2.46 Surface content of (a) Al in fumed silica/alumina and (b) Ti in titania/silica and the plateau adsorption (A) of (a) Pb(II) and (b) Ni(II) as a function of the total (a) alumina and (b) titania content in mixed nanooxides. (Adapted from *Appl. Surf. Sci.*, 253, Gun'ko, V.M., Nychiporuk, Yu.M., Zarko, V.I. et al., Relationships between surface compositions and properties of surfaces of mixed fumed oxides, 3215–3230, 2007b, Copyright 2007, with permission from Elsevier.)

contribution of hydroxy species of Pb(II) and Ni(II) increases. These hydroxy species can adsorb by the formation of hydrogen bonds rather than due to ion-exchange reactions responsible for adsorption of cations at lower pH values (when the adsorption is lower than the plateau by several times). Consequently, parallel analysis of the surface composition and the adsorption of metal ions allows us deeper insight into their adsorption mechanisms.

To analyze the adsorption characteristics of nanooxides, adsorption of nonpolar hexane ("hex" was used as a subscript), weakly polar acetonitrile (ac), and polar diethylamine (DEA), triethylamine (TEA), and water (w) were studied using an adsorption apparatus with a McBain–Bark quartz scale at 293 K (Table 2.10). Oxide samples were evacuated at 10^{-3} Torr and 473 K for several hours to a constant weight, then cooled to 293 ± 0.2 K. The measurement accuracy was 1 ± 10^{-3} mg with a relative mean error of ±5%.

For studied nanooxides, the nitrogen adsorption–desorption isotherms are characterized by narrow hysteresis loops (Figures 2.47 and 2.48) and can be assigned to the type II of International Union of Pure and Applied Chemistry (IUPAC) classification. This regularity is due to a spherical (but non-ideal) shape of nonporous primary particles (average diameter $d_{av}=6.9$–52.5 nm for studied oxides; Table 2.10) forming aggregates (secondary particles of 50–1000 nm in size) and looser agglomerates of aggregates (ternary particles >1 μm in size). Voids in secondary and ternary particles provide the textural porosity of the materials but with virtually no contribution from nanopores (Figure 2.49). Notice that the adsorption isotherms of other studied adsorbates (Figures 2.47 and 2.48) can also be assigned to the type II because of the textural porosity type of nanooxides. The empty volume in loose nanooxide powders is very great $V_{em}=1/\rho_b - 1/\rho_0 = 10$–30 cm³/g (where $\rho_b=0.03$–0.10 g/cm³ is the bulk density of the powders).

However, a major contribution of broad voids (macropores; Figure 2.49) in secondary and ternary particles leads to thermodynamically unfavorable condensation of gases or vapors in such pores and results in ineffectively filling of them by any gaseous or vapor adsorbates even at $p/p_0=0.98$–0.99. For instance, the maximal $V_{p,N2}$ value among studied oxides is 0.82 cm³/g for SA23 (Table 2.10; possessing the maximal $S_{BET,N2}$ value), which corresponds to about 3% of the V_{em} value ($\rho_b=0.036$ g/cm³). Another effect appearing due to the textural porosity is that the nitrogen adsorption is proportional to the $S_{BET,N2}$ value at p/p_0 up to 0.5–0.6 because of weak capillary condensation and insignificant volume filling of broad voids.

TABLE 2.10
Structural Characteristics of Nanooxide Powders

Sample	$C^s_{SiO_2}$ (wt%)	$C^s_{TiO_2}$ (wt%)	$C^s_{Al_2O_3}$ (wt%)	d_{av} (nm)	Δw	S_{BET,N_2} (m²/g)	V_{p,N_2} (cm³/g)	$S_{BET,ac}$ (m²/g)	$S_{BET,hex}$ (m²/g)	$S_{BET,DEA}$ (m²/g)	$S_{BET,TEA}$ (m²/g)
SA8	29.5		70.5	8.6	0.33	303	0.68	223	207	271	265
SA23	73.6		26.4	6.9	0.23	347	0.82	362	247	434	384
ST20	75.6	24.4		27.9	0.40	84	0.18	116	80	97	115
AST03				13.9	0.30	125	0.31	162	101	114	126
AST1				18.0	0.33	99	0.25	180	99	113	117
AST50	58.7	40.0	1.3	47.9	0.31	37	0.08	44	31	44	42
AST71	5.0	80.4	4.6	21.6	0.43	74	0.13	83	63	60	70
AST82	0	91.2	8.8	40.2	0.27	39	0.11	43	48	46	58
A-50	>99.8			52.5	0.38	52	0.13	42	77		
A-200	>99.8			11.9	0.31	230	0.48	232	187		
A-300	>99.8			9.2	0.17	295	0.52		261		
Al_2O_3			>99.8	12.9	0.16	133	0.45	246	187	205	238
Al_2O_3			>99.8	19.3	0.35	89	0.13	95	90		
TiO_2		>99.8		35.7	0.32	42	0.09	58	43		

Note: C^s is the surface content of oxides, and Δw is relative error of the model of voids between spherical particles packed in random aggregates.

However, the $S_{BET,N2}$ values were calculated here using the standard range of adsorption at $0.05<p/p_0<0.3$. This does not give large errors in estimation of the $S_{BET,N2}$ values because of similar results of adsorption linearization the coordinates in the Brunauer-Emmett-Teller (BET) method for narrow and broad p/p_0 ranges. For instance, $S_{BET,N2}=318$ and 35 m²/g for SA23 and AST50 (with the largest and smallest $S_{BET,N2}$ values) at $0.05<p/p_0<0.5$, respectively, that are close to the $S_{BET,N2}$ values calculated at $0.05<p/p_0<0.3$ (Table 2.10). Thus, despite a certain difference in the normalized (divided by $S_{BET,N2}$) nitrogen adsorption at $p/p_0>0.5$ caused by variations in the primary nanoparticle size distribution (NPSD) and packing of primary particles in secondary and ternary structures, the nanooxides morphology can be assumed as practically the same for all the oxide samples. This is due to the flame synthesis of them at high temperatures $T>1200$°C; i.e., primary particles can be in state close to liquid one in the flame. However, the studied oxide powders strongly differ in the specific surface area (Table 2.10; S_{BET}), NPSD, and the aggregation degree of particles (V_p) that, as well as the difference in the nature of oxide surfaces, can affect the adsorption of any adsorbates. The analysis of these aspects can be undertaken by comparison of the structural characteristics (S, V) calculated from the adsorption isotherms of different adsorbates studied here.

Greater differences in the adsorption on the same oxide are observed for more complex and polar adsorbates (DEA, TEA, CH_3CN, and H_2O) than nitrogen (Figure 2.47). These effects are caused by stronger bonding of stronger electron-donor and polar adsorbates to Brønsted (bridging hydroxyls) and Lewis (incompletely O-coordinated Al or Ti atoms) acid sites and other centers (e.g., terminal hydroxyls). However, there is a clear tendency toward a decrease in the adsorption of all adsorbates with decreasing S_{BET} value, despite the difference in their polarity, electron-donor properties, and molecular size. Clearly, the specific surface area and textural porosity features of adsorbents play an important role on adsorption of various adsorbates. The strong influence of the S_{BET} value on the adsorption of any adsorbates onto nanooxides is explained by the fact that a significant contribution to the adsorbed amounts is due to the first monolayer with insignificant capillary condensation of adsorbates in broad but short voids between neighboring nanoparticles in their aggregates and agglomerates. Notice that some isotherms were recorded at the maximal value $p/p_0=0.62$–0.75 due to certain difficulties in the measurements caused by the behavior of the adsorbates studied at higher p/p_0 values. Additional effects are due to the difference in the nature and content of active surface sites (terminal [≡MOH] and bridging [≡M′O(H)M″≡] hydroxyls and others) responsible for the formation of strong adsorption complexes with electron-donor and proton-donor molecules.

The temperature at which the nanooxides are degassed plays an important role on measurements of the adsorption and, therefore, estimation of the structural characteristics. However, morphological changes of fumed alumina and titania are much smaller on heating than that of silica. This effect is due to the difference in the types of hydroxyls (only terminal for silica and terminal and bridging for alumina and titania) and the O-coordinating numbers of Si (only fourfold O-coordinated) and Al (from fourfold to sixfold) or Ti (fivefold and sixfold O-coordinated) atoms. Therefore, associative desorption of water from silica leads to stronger changes in the lattice structure than for other studied oxides.

Alumina ($S_{BET,N2}$ = 133 m²/g) degassed at 600°C or 900°C and cooled to room temperature (without contact to air) adsorbs greater amounts of acetonitrile and especially TEA than alumina degassed at 200°C (Figure 2.48b). This result, as well as the maximal $S_{BET,X}/S_{BET,N2}$ values (Figure 2.50; Table 2.10), can be caused by enhancement of donor–acceptor interactions between the N atoms of adsorbates and surface Lewis acid sites (effectively formed during strong dehydration of alumina at high temperatures) and certain diminution of the size of primary particles during thermoevacuation.

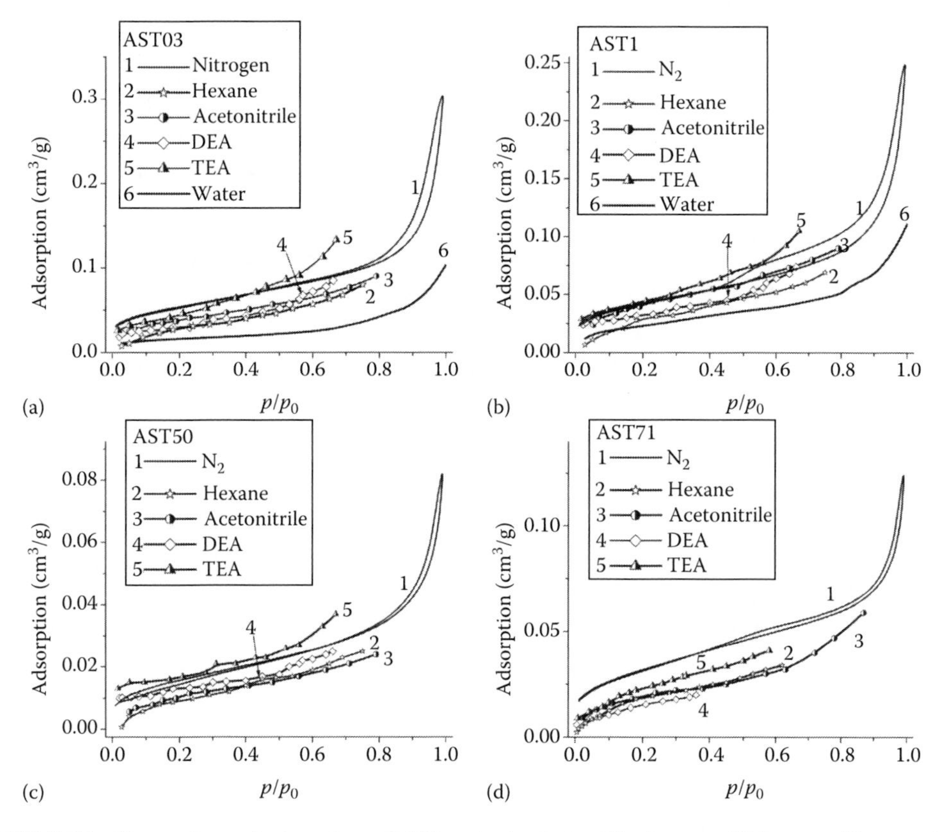

FIGURE 2.47 Comparison of adsorption of different adsorbates (1) nitrogen, (2) hexane, (3) acetonitrile, (4) DEA, (5) TEA, and (6) water onto (a) AST03, (b) AST1, (c) AST50, (d) AST71. (Adapted from *J. Colloid Interface Sci.*, 348, Gun'ko, V.M., G.R. Yurchenko, Turov, V.V. et al., Adsorption of polar and nonpolar compounds onto complex nanooxides with silica, alumina, and titania, 546–558, 2010d, Copyright 2010, with permission from Elsevier.)

(continued)

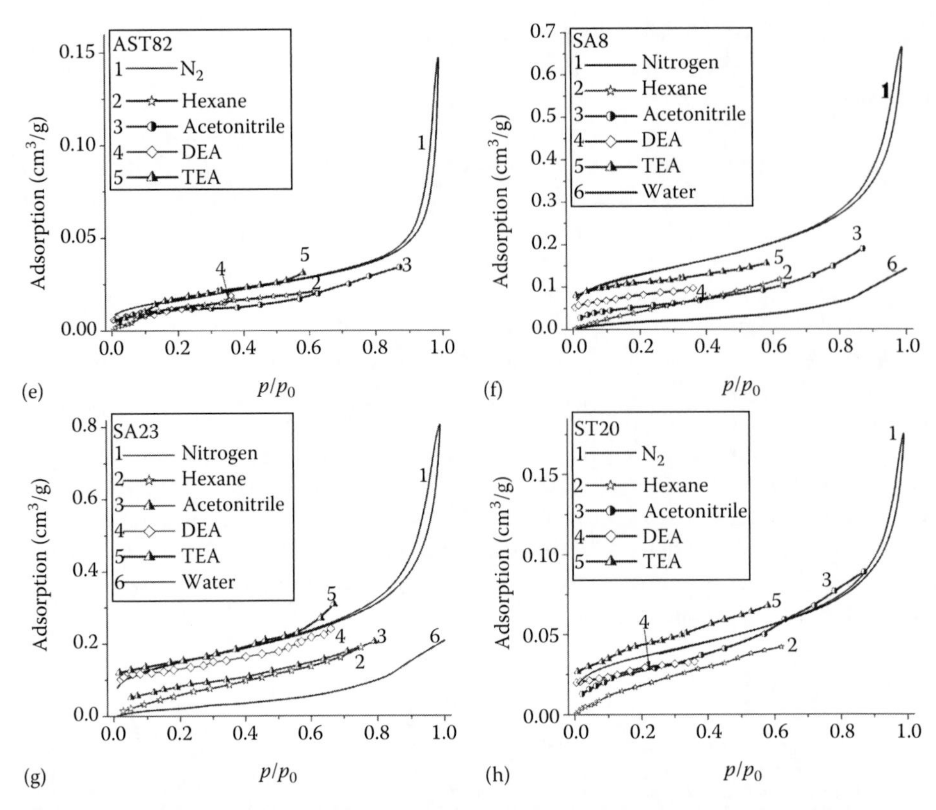

FIGURE 2.47 (continued) Comparison of adsorption of different adsorbates (1) nitrogen, (2) hexane, (3) acetonitrile, (4) DEA, (5) TEA, and (6) water onto (e) AST82, (f) SA8, (g) SA23, and (h) ST20 degassed at 200°C. (Adapted from *J. Colloid Interface Sci.*, 348, Gun'ko, V.M., G.R. Yurchenko, Turov, V.V. et al., Adsorption of polar and nonpolar compounds onto complex nanooxides with silica, alumina, and titania, 546–558, 2010d, Copyright 2010, with permission from Elsevier.)

Additionally, calcination at 900°C can increase the crystallinity degree of alumina that can also influence the adsorption of electron-donor compounds. Notice that the first portion of adsorbates (especially water) can dissociatively adsorb onto strained bonds appearing at the oxide surface degassed at high temperatures.

The degree of aggregation of primary particles (i.e., V_p value), which depends on treatment conditions, plays a certain role in the adsorption and changes in the $S_{BET,X}/S_{BET,N2}$ values. For instance, for two samples of alumina, the ratio $S_{BET,X}/S_{BET,N2}$ is greater for a sample with a larger V_p value (Table 2.10) because the adsorption is greater for nanooxides with more aggregated nanoparticles. Additionally, variations in orientation of adsorbed molecules at a surface affects the surface area (σ) occupied by each molecule and, therefore, the $S_{BET,X}/S_{BET,N2}$ ratio.

A minimal adsorption onto oxides degassed at 200°C is observed for water (Figures 2.47 and 2.48), despite the fact that it can form strong hydrogen bonds with surface hydroxyls. This low adsorption is due to several reasons. First, saturated vapor of water has low pressure (17.5 mmHg at 293 K). Second, water adsorption is clustered rather than monolayered, e.g., the surface is incompletely covered by water even at $C_w = 15–20$ wt% corresponding to several statistical monolayers. Third, nanopores are absent and the contribution of narrow mesopores is low in the powders of nanooxides (Figures 2.13 through 2.15 and 2.49). Among the adsorbates studied, only water adsorbs in the form of clusters. Even during single molecule adsorption, it tends to form two hydrogen bonds with neighboring surface hydroxyls. Therefore, nanopores and narrow mesopores are more appropriate for effective clustered or nanodomain adsorption of water than broad mesopores and macropores.

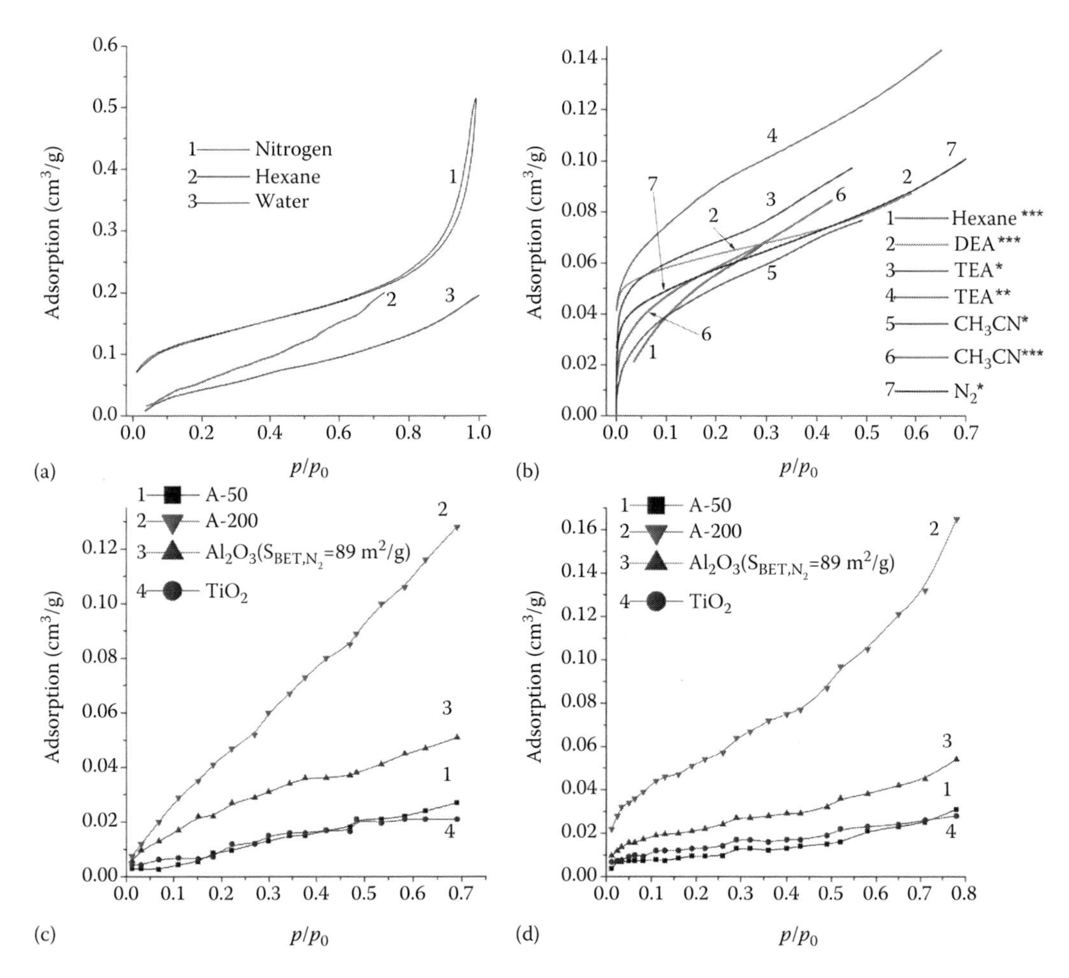

FIGURE 2.48 Adsorption of different adsorbates onto (a) A-300 degassed at 200°C; (b) fumed alumina ($S_{BET,N2}$ = 133 m²/g) degassed at *200, **600, and ***900°C; and adsorption of (c) hexane and (d) acetonitrile onto A-50, A-200, alumina ($S_{BET,N2}$ = 89 m²/g), and titania. (Adapted from *J. Colloid Interface Sci.*, 348, Gun'ko, V.M., G.R. Yurchenko, Turov, V.V. et al., Adsorption of polar and nonpolar compounds onto complex nanooxides with silica, alumina, and titania, 546–558, 2010d, Copyright 2010, with permission from Elsevier.)

In the latter, the formation of 3D structures totally filling voids needs very large amounts of adsorbed water. However, the adsorption of great amounts of water in broad pores at $p/p_0 < 0.98$ is thermodynamically unfavorable, as well as for other studied adsorbates. All the mentioned factors can play a certain role in the adsorption of organic adsorbates. However, thermodynamic conditions of their adsorption differ from that of water. This leads to greater adsorption of organic adsorbates onto nanooxides in comparison with water (Figures 2.47 and 2.48) at the same temperatures (~20°C).

Nonpolar hexane and weakly polar acetonitrile adsorb better than water but worse than polar DEA and TEA. This is due to strong hydrogen bonding of the amines to the surface hydroxyls and Lewis acid sites. The adsorption of TEA is typically greater or close to that of nitrogen (at the same p/p_0 values), despite a large difference in the adsorption temperatures (293–297 and 77.4 K, respectively). Temperature determines the average kinetic energy of molecules. Elevating temperature leads to diminution of the adsorption in voids between nonporous spherical nanoparticles where the steric effects are absent. For AST71, the nitrogen adsorption is much higher than that of TEA. This result can be caused by dense packing of primary particles of AST71 because the hysteresis loop in the nitrogen adsorption isotherm is long. This loop starts at $p/p_0 = 0.4$ for AST71 and AST1, but for other

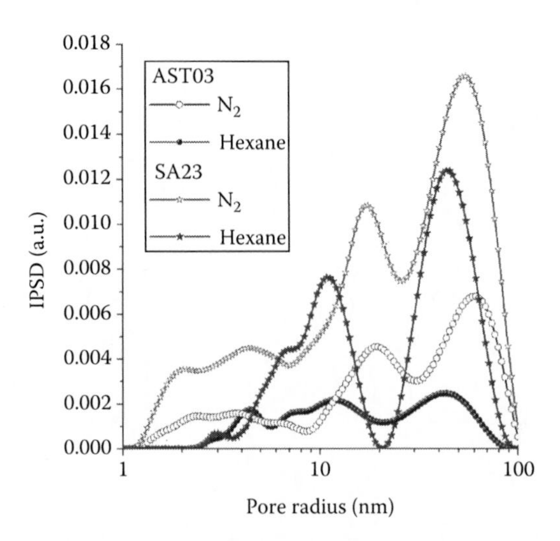

FIGURE 2.49 Pore size distributions for selected oxides calculated on the basis of nitrogen desorption and hexane adsorption with the model of voids between spherical particles. (Adapted from *J. Colloid Interface Sci.*, 348, Gun'ko, V.M., G.R. Yurchenko, Turov, V.V. et al., Adsorption of polar and nonpolar compounds onto complex nanooxides with silica, alumina, and titania, 546–558, 2010d, Copyright 2010, with permission from Elsevier.)

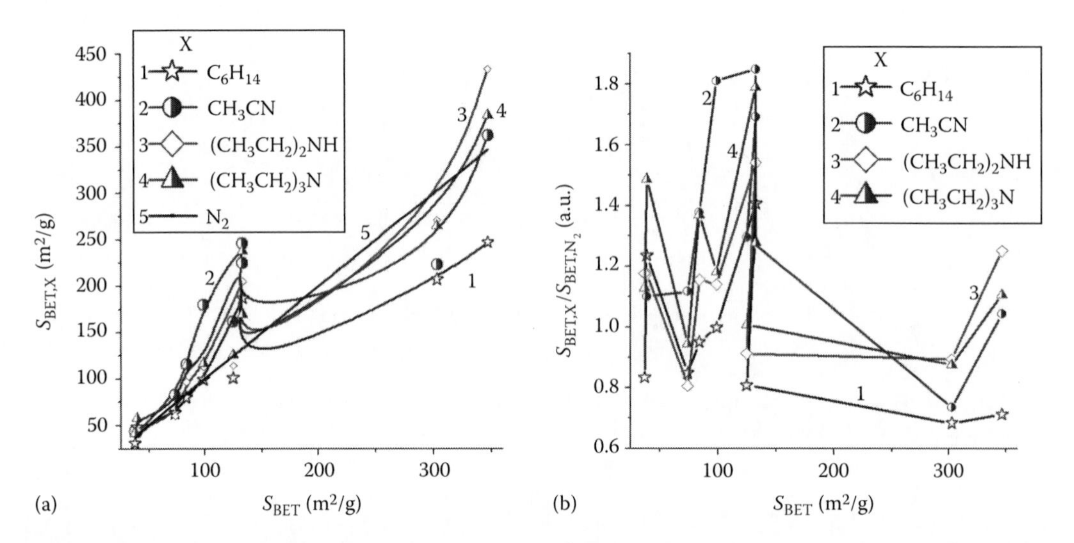

FIGURE 2.50 Relationships between the $S_{BET,N2}$ and (a) $S_{BET,X}$ or (b) $S_{BET,X}/S_{BET,N2}$ values. (Adapted from *J. Colloid Interface Sci.*, 348, Gun'ko, V.M., G.R. Yurchenko, Turov, V.V. et al., Adsorption of polar and nonpolar compounds onto complex nanooxides with silica, alumina, and titania, 546–558, 2010d, Copyright 2010, with permission from Elsevier.)

samples, it begins at $p/p_0 \geq 0.7$. Consequently, capillary condensation of nitrogen begins on AST71 (or AST1) at lower pressures than for other oxides that enhances the adsorption.

In contrast to dispersion interactions of hexane and nitrogen, specific interactions of acetonitrile and amines with surface active sites can cause an overestimation of the $S_{BET,X}$ value in comparison with the $S_{BET,N2}$ or $S_{BET,hex}$ values (Figure 2.50; Table 2.10). This is due to several effects linked to features of orientation and packing of molecules in the first adsorbed layer. Orientation of adsorbed molecules interacting with surface hydroxyls, especially bridging ones, can correspond to their non-maximal projection onto the surface. Therefore, the effective σ value can be smaller than the value estimated from the molecular geometry of the adsorbates. This overestimation is maximal for acetonitrile.

Its adsorption complexes mainly correspond to non-lengthwise orientation of the molecules that causes a significant diminution of the effective σ value. However, the σ value for acetonitrile used for estimation of $S_{BET,ac}$ (Table 2.10) corresponds to the average projection area of a molecule to a surface.

The $S_{BET,X}$ overestimation for TEA is greater than that for DEA for samples at $S_{BET,N2} \leq 125$ m²/g, significant crystallinity (Figure 2.51) and surface roughness (Table 2.10; Δw). However, for SA8 and SA23 with large $S_{BET,N2}$ values and amorphous silica and alumina phases (Figure 2.51), this effect is greater for DEA (Figure 2.50). These results can be caused by different changes in the packing of adsorbed molecules of DEA and TEA in the first monolayer at the surface of smaller nanoparticles more strongly aggregated in secondary ones because the steric factor is lower for smaller DEA molecules than for larger TEA.

This factor is greater for samples with a larger S_{BET} value because they are characterized by enhanced aggregation of nanoparticles. For hexane, the $S_{BET,hex}$ overestimation is observed for A-50, AST82, and alumina degassed at 200°C and 900°C, respectively, that can be explained by, at least, two factors. First, a low S_{BET} value for both AST82 and A-50 with large primary nanoparticles (Table 2.10; d_{av}) weakly aggregated (Figure 2.47g, loop beginning at $p/p_0 > 0.8$) provides better accessibility of the surface for relatively large C_6H_{14} molecules in comparison with other oxides with larger S_{BET} values and more strongly aggregated nanoparticles (Figure 2.49). Second, a significant surface content of titania in AST82 (Table 2.10) enhances dispersive interactions of hexane molecules with nanoparticles. In the case of alumina degassed at 900°C, there are additional factors such

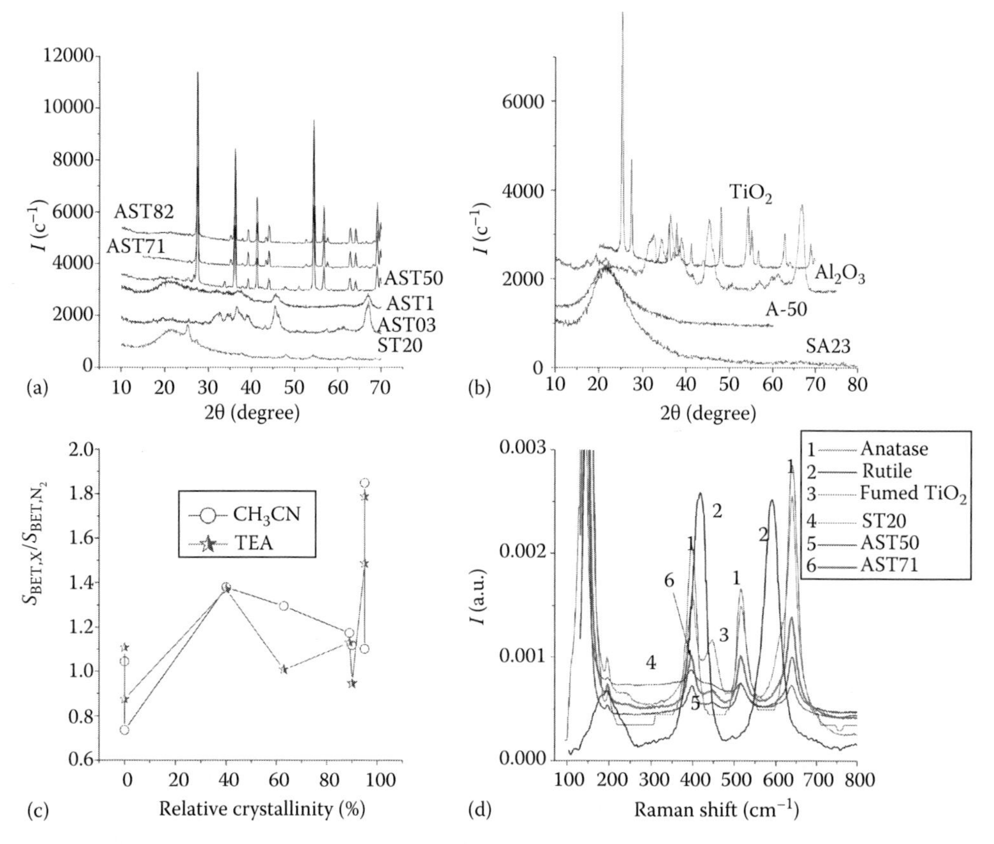

FIGURE 2.51 (a and b) XRD patterns of nanooxides; (c) changes in the $S_{BET,X}/S_{BET,N2}$ value as function of relative crystallinity of nanooxides determined from the XRD data; and (d) Raman spectra of some oxides containing titania. (Adapted from *J. Colloid Interface Sci.*, 348, Gun'ko, V.M., G.R. Yurchenko, Turov, V.V. et al., Adsorption of polar and nonpolar compounds onto complex nanooxides with silica, alumina, and titania, 546–558, 2010d, Copyright 2010, with permission from Elsevier.)

as diminution of sizes of nanoparticles (due to dehydration and crystallization without sintering), increased crystallinity, and dissociative adsorption of a portion of the molecules.

The sample crystallinity depends mainly on the crystallinity of titania if alumina is not the main phase (Table 2.10) since silica is always amorphous (Figure 2.51). Individual titania includes both anatase and rutile (Figure 2.51b and d). For other titania-containing samples, the titania phase is composed mainly of anatase. This is of importance because the crystallinity and the type of titania phases can influence the adsorption, catalytic, and other important characteristics of the materials. Additionally, an increase in the Δw values (Table 2.10), i.e., enhancement of the surface roughness of nanoparticles and deviation of their shape from spherical, corresponds to increasing $S_{BET,X}/S_{BET,N2}$ ratio for, e.g., X = acetonitrile. This correlation for acetonitrile is clear for 10 oxide samples from 14 ones studied. In other words, it is not strong because many structural and other effects overlap here.

Notice that quantum chemical calculations of the adsorption of a CO_2 cluster onto a silicon surface using periodic boundary conditions showed that the density of the monolayer can be higher than the critical density of relatively rigid CO_2 molecules. Consequently, the adsorption layer density of molecules such as TEA and DEA "softer" than CO_2 can be relatively high. Thus, any strong interactions of adsorbates with the adsorbent surface cause compaction of, at least, the first adsorption monolayer (due to both orientation effects, e.g., for acetonitrile, and conformation changes, e.g., for DEA and TEA) that result in diminution of the effective surface area occupied by a molecule and an increase in the $S_{BET,X}/S_{BET,N2}$ value.

Brønsted acid sites ≡M-O(H)-M≡ and other surface hydroxyls on ST, AST, silica, alumina, and titania are responsible for reactions (Tanabe 1970, Hunter 1981, 1993, Lyklema 1991, 1995)

$$\equiv \text{M-O(H)-M} \equiv \rightarrow \equiv \text{M-(O}^-\text{)-M} \equiv + \text{H}^+ \qquad (2.8)$$

$$\equiv \text{SiOH} \rightarrow \equiv \text{SiO}^- + \text{H}^+ \qquad (2.9)$$

$$\equiv \text{MOH} + \text{H}^+ \rightarrow \equiv \text{MOH}_2^+ \qquad (2.10)$$

$$\equiv \text{MOH} + \text{A}^- \rightarrow \equiv \text{(OH)M} \leftarrow \text{A}^- \qquad (2.11)$$

where $A^- = OH^-$, Cl^-, etc. (they are also the main adsorption sites for metal ions). These reactions can cause non-additive changes in the electrophoretic potential (ζ) as a function of pH depending on the composition of complex oxide particles (Figure 2.52) and the interfacial phenomena at their surfaces (Gun'ko et al. 2004f, 2005d, 2009d). Estimation of the Debye–Huckel parameter κ^{-1}(pH) from the electrical conductivity of the aqueous suspensions of ST and AST (C_{ox} = 0.04 wt%, 0.001 M NaCl) according to Xu (1998) gives the κ^{-1} values over the 2–7 nm range. One can assume that the shear layer thickness between an oxide surface and a shear plane is close to κ^{-1}. Notice that the thickness of a disturbed interfacial water layer (unfrozen at $T < 273$ K) estimated from the ^{1}H NMR measurements with freezing-out of bulk water in the aqueous suspensions of nanooxides is close to these κ^{-1} values. Water flows in voids between primary particles inside their moving aggregates affect the mobility, U_e of oxide particles (that may give overestimated values of ζ(pH) calculated with the Smoluchowski theory; Hunter 1981). The difference in the ζ(pH) graphs calculated according to the Smoluchowski theory and equation corrected for porous aggregates of primary spherical particles (see Section 10.15) increases at pH far from the IEP (Figure 2.52a), since κ^{-1} is a function of pH and the flows (as well as an immobile portion of the EDL) inside aggregates can depend on κ^{-1} and pH.

Consideration of the primary particle size distribution $f(a)$ in the ζ(pH) calculations using Equation 10.80 results in a slight decrease in the modulus of ζ(pH) in comparison with that obtained with Equation 10.77 (Figure 2.52b) due to an asymmetrical shape of the $f(a)$ distributions (particles

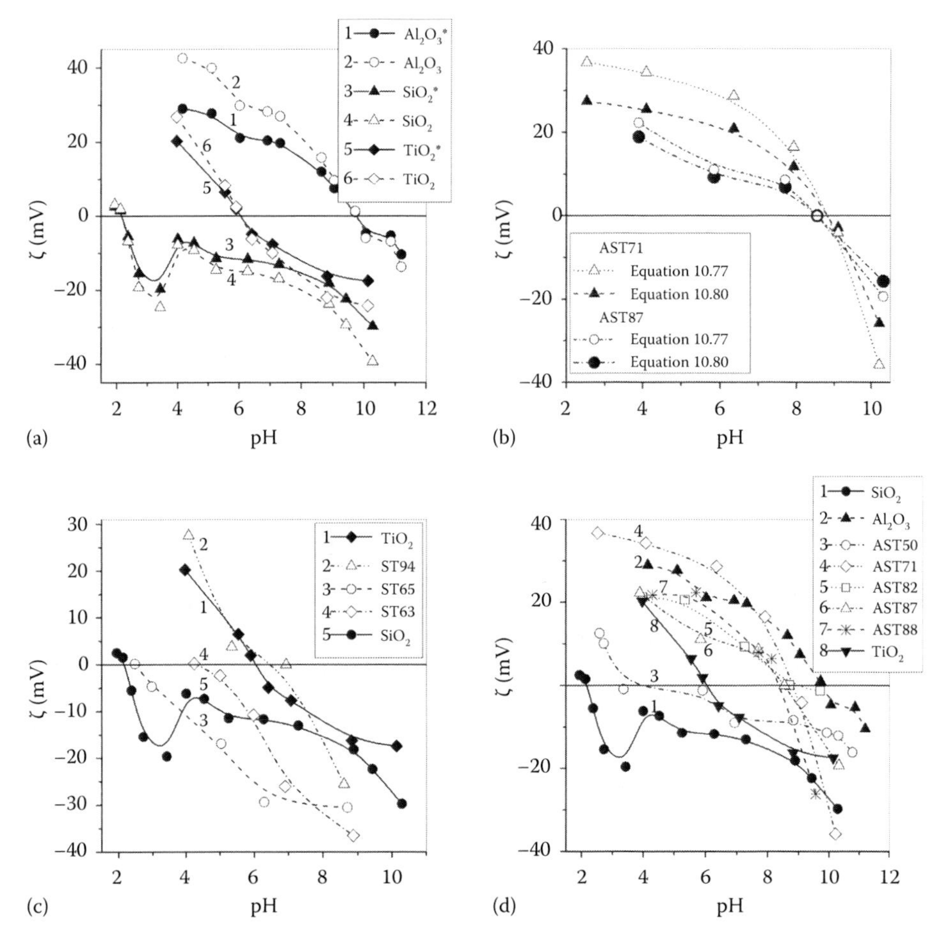

FIGURE 2.52 Electrokinetic potential as a function of pH for (a) fumed individual oxides computed according to Smoluchowski's theory (open symbols) and corrected with consideration for the porosity of aggregates (see Chapter 10); corrected ζ potential (b) calculated using Equation 10.77 or 10.80; and Equation 10.77 for (c) silica, titania and ST samples; and (d) individual and ternary oxides in the aqueous suspensions at C_{ox}=0.02–0.04 wt%, 0.001 M NaCl. (Adapted from *Colloids Surf. A: Physicochem. Eng. Aspects*, 240, Gun'ko, V.M., Zarko, V.I., Mironyuk, I.F. et al., Surface electric and titration behaviour of fumed oxides, 9–25, 2004f, Copyright 2004, with permission from Elsevier.)

of different sizes have different electrokinetic characteristics; Gun'ko et al. 2001e). The ζ(pH) graphs shown in Figure 2.52c and d were calculated using Equation 10.77 because the difference between the ζ(pH) curves determined with Equations 10.77 and 10.80 is not too great.

In the case of a low concentration of silica in ST94, ζ(pH) is similar to that for titania (Figure 2.52c), but it shifts toward negative ζ values with increasing C_{SiO_2} and its shape changes. This difference is caused by contribution of asymmetrical Si-O(H)-Ti bridges in ST (having higher B acidic properties than the sites on individual silica or titania) giving more negative ζ values at pH>6 for ST in comparison with that of silica and titania. Notice that the FTIR spectra show that the Si-O-Si bonds are absent in ST94 and AST88; i.e., silica forms a solid solution in titania. The ζ(pH) curves for ST65 and ST63 lie below than those for other mixed oxides at pH>4 (Figure 2.52c). These binary oxides have a greater charge density than silica, titania, ST94, or ternary oxides with increasing pH. However, the IEPs of ST samples lie to the right in comparison with IEP_{SiO_2}, which can be explained by availability (in ST) not only of acidic bridging groups ≡Si-O(H)-Ti≡ but also basic terminal groups such as ≡TiOH (IEP_{TiO_2}≈6).

In the case of AST at C_{TiO_2}>70 wt% (Table 2.1; $C_{Al_2O_3}$ between 4 and 21 wt%), ζ(pH) is positive over a large pH range (Figure 2.52d) due to surface richness by alumina ($IEP_{Al_2O_3}$≈9.8) and titania (IEP_{TiO_2}≈6). However, at pH close to IEP of alumina, the ζ(pH) curves have a stronger decline than that of pure alumina due to the influence of titania and silica (charged negatively at these pH values). At pH between 3 and 7, ζ(pH) for AST50 lies below than that for other AST samples and close to ζ(pH) for silica or titania. AST50 and AST71 characterized by different concentrations and distributions of oxide phases (however, $C_{Al_2O_3}$ is near the same) (Table 2.1) show different shapes of the ζ(pH) curves (especially at pH<8), as well as the dependence of the effective diameter D_{ef} (hydrodynamic diameter corresponding to the geometrical diameter plus the shear layer thickness) on pH (Figure 2.53b).

One can assume that the AST50 surface is more uniform (with a low surface concentration of alumina, since ζ(pH) is closer to that of titania or silica than alumina) than other AST samples. Therefore, D_{ef} (as well as ζ) of AST50 depends on pH weaker than for other AST samples. For the latter, D_{ef} increases by approximately 50 times with pH and becomes maximal when titania and alumina phases have opposite charges at pH between their IEP values (Figure 2.53b), which provides strong attractive interaction between these alumina and titania patches on neighboring primary particles. Thus, the surface alumina on AST samples (with one exception for AST50) strongly affects the dispersion and electrophoretic properties.

Pyrocarbon (C_C≈10 wt%) grafted onto nanooxides changes ζ(pH) differently for C/ST and C/AST (Figure 2.54c and d; carbon–mineral adsorbents will be analyzed in detail in Chapter 4).

This effect depends on the surface alumina distribution on AST nanoparticles. For AST71 characterized by a maximal influence of the surface alumina on ζ(pH) (Figures 2.52 and 2.54), pyrocarbon grafting leads to a displacement of ζ(pH) toward negative ζ values (at pH<10) (Figure 2.54c). For carbonized ST94, ST65, AST87, and AST88, pyrocarbon causes the opposite displacement of the ζ(pH) curves in comparison with C/AST71, since their surface composition differs strongly. Position of ζ(pH) for initial and carbonized oxides differs because of changes in the concentrations of Brønsted and Lewis sites and other polar functionalities on oxides after grafting of pyrocarbon. The effect of partial hydrophobization of oxide particles by pyrocarbon (which is more hydrophobic than oxides) results in large D_{ef} (>10 μm) in consequence of strong particle aggregation (Gun'ko et al. 2000d,h, 2001c,e).

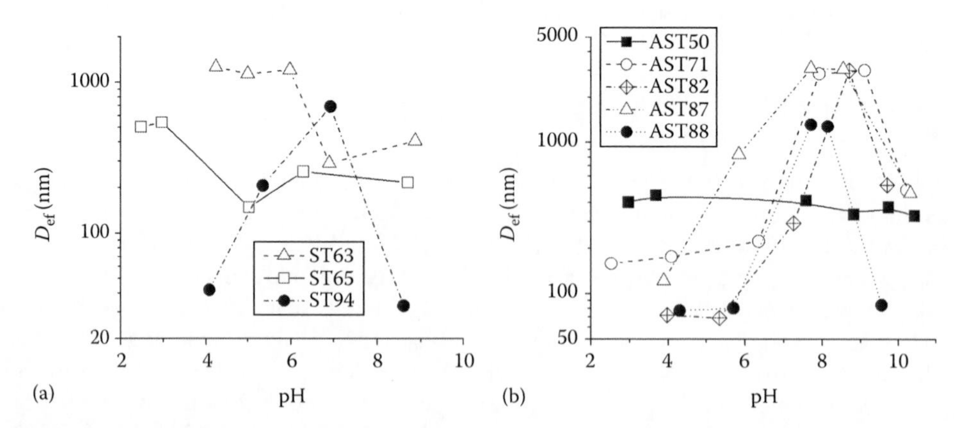

FIGURE 2.53 Effective diameter D_{ef} as a function of pH for (a) binary and (b) ternary oxides at C_{ox}=0.04 wt%, 0.001 M NaCl. (Adapted from *Colloids Surf. A: Physicochem. Eng. Aspects*, 240, Gun'ko, V.M., Zarko, V.I., Mironyuk, I.F. et al., Surface electric and titration behaviour of fumed oxides, 9–25, 2004f, Copyright 2004, with permission from Elsevier.)

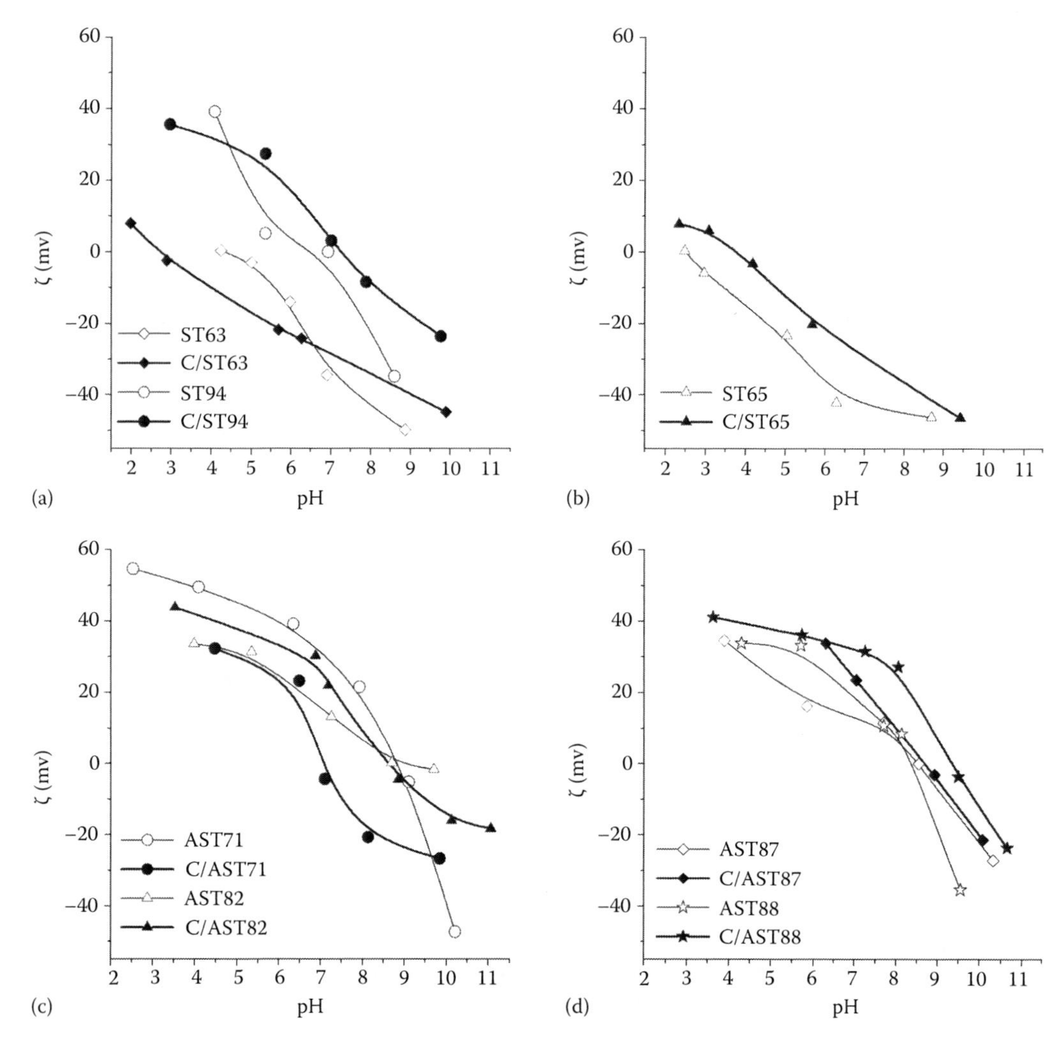

FIGURE 2.54 Electrokinetic potential as a function of pH for initial and carbonized (a and b) binary and (c and d) ternary nanooxides calculated according to the Smoluchowski theory. (Adapted from *Colloids Surf. A: Physicochem. Eng. Aspects*, 240, Gun'ko, V.M., Zarko, V.I., Mironyuk, I.F. et al., Surface electric and titration behaviour of fumed oxides, 9–25, 2004f, Copyright 2004, with permission from Elsevier.)

The SPSDs of ST (Figure 2.55) depend on pH stronger than those of AST (Figure 2.56) (e.g., AST71 and AST88) with the exception of pH region corresponding to maximal attractive interactions between alumina and titania (or silica) patches on neighboring particles.

D_{ef}(pH) changes over the total pH range stronger for AST than that for ST (Figure 2.53) due to interactions between three different surface phases. For ST94, D_{ef}(pH) is maximal at pH close to IEP_{TiO_2} due to reduction of repulsive interaction between titania patches and attractive forces between silica and titania at pH between their IEPs. Similar maxima are observed for AST samples at pH between 7 and 9.5, which, however, shift slightly toward higher pH values due to the influence of the surface alumina. Thus, one can assume that primary AST particles are characterized by a relatively greater surface content of alumina in comparison with the total $C_{Al_2O_3}$ values for these samples, since their ζ(pH) lies closer to that for pure alumina than that for titania and AST50 (which does not reveal such a ζ(pH) dependence).

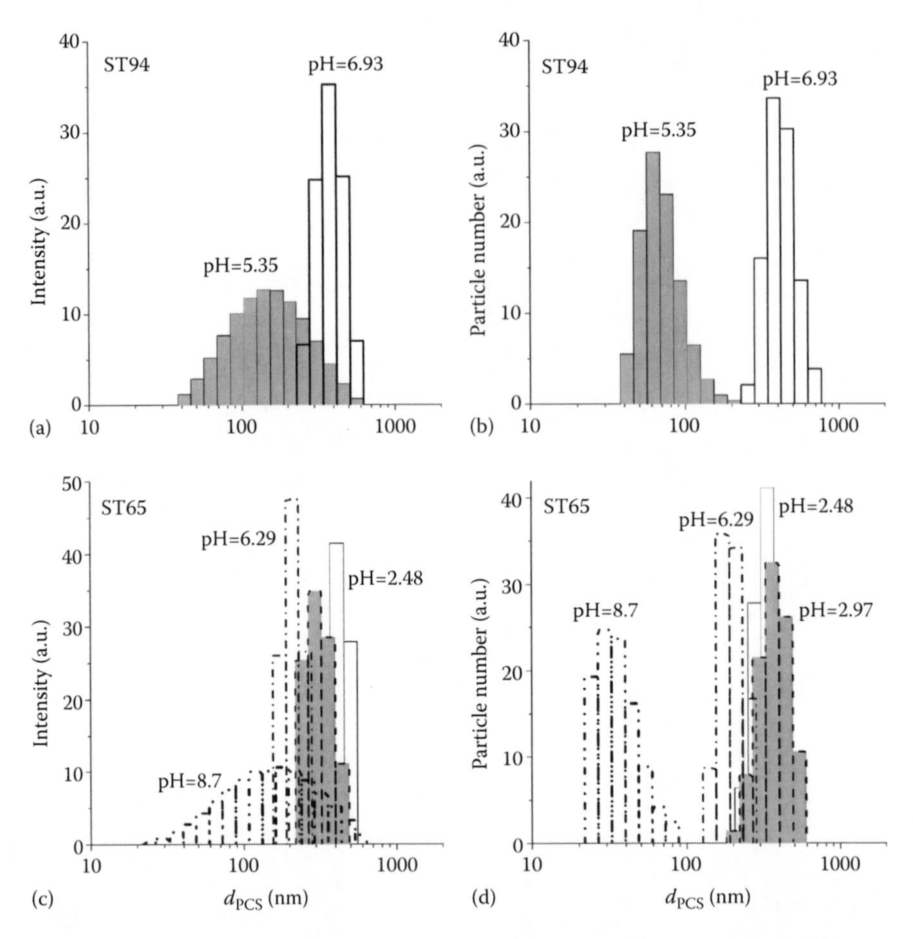

FIGURE 2.55 Particle size distributions in respect to (a and c) the light scattering, $SPDS_I$ and (b and d) particle number, $SPDS_N$ for ST samples at different pH values (shown in Figure), C_{ox}=0.04 wt%, 0.001 M NaCl. (Adapted from *Colloids Surf. A: Physicochem. Eng. Aspects*, 240, Gun'ko, V.M., Zarko, V.I., Mironyuk, I.F. et al., Surface electric and titration behaviour of fumed oxides, 9–25, 2004f, Copyright 2004, with permission from Elsevier.)

The observation of individual AST particles with the hydrodynamic diameter of 10–20 nm (Figure 2.56), i.e., smaller than the average geometrical diameter of primary particles, is untypical for the aqueous suspensions of fumed silica, alumina, titania, or binary oxides (Gun'ko et al. 2001e). This result can be caused by a very broad size distribution of primary particles of AST (broad primary particle size distribution is characteristic for nanooxides with a low specific surface area [vide supra; Degussa 1997]). Therefore, one can assume that primary AST particles of strongly different sizes are characterized by different contributions of titania, alumina, and silica, since they can be formed in different zones of the flame during the synthesis.

The SPSDs of mixed oxides (initial and carbonized) depend strongly on pH (Figures 2.53, 2.55, and 2.56) since the electrostatic interactions between silica, titania, alumina, and pyrocarbon phases characterized by different points of zero charge, PZC are differently affected by pH (Figures 2.52 and 2.54). The studied characteristics of nanooxides can correlate with the adsorption of metal ions because of the common basis (structure of oxide particle surfaces and their behavior in the aqueous medium) of these phenomena.

Active surface sites on the AST surfaces are responsible for noticeable adsorptive ability of ternary oxides with respect to such ions as Pb(II), Cs(I), and Sr(II) (Figures 2.57 through 2.60).

The shapes of the Γ(pH) curves for the studied ions adsorbed onto different nanooxides are quite typical. The main sites providing these effects are different hydroxyls especially in asymmetrical bridges $\equiv M_1\text{-}O(H)\text{-}M_2\equiv$, which can be easily deprotonated (at $pH > pH_{IEP}$) in the aqueous media and take part in the ion-exchange reactions with metal ions and related species or bind them by the electrostatic and hydrogen bonds. The adsorption of ions depends on the surface properties of solids; on the other hand, the interfaces can be modified by adsorbed ions in respect to the value and the sign of the surface charge and the electrokinetic potential (Gun'ko et al. 2004f). The density of metal ions adsorbed at the solid/liquid interfaces depends not only on the types of an adsorbent and a solvent but also on the kind and the form of metal species, especially for ions undergoing hydrolysis (such as Pb(II) and Sr(II)) to form $M_n(OH)_m^{k+}$. Typically, a dramatic increase in the adsorption of metal ions is observed over a narrow pH range, usually no more than two pH units. Specifically adsorbed metal ions (without attached OH^-) are localized in the inner Helmholtz plane close to the surface charge plane. Therefore, they can affect the surface charge density, PZC, and potential distribution within the electrical interfacial layer (Brown et al. 1999).

A number of the surface hydroxyls taking part in the adsorption (2.2–4.7 OH/nm^2 for silica, 3.9–8.0 OH/nm^2 for $\gamma\text{-}Al_2O_3$, 4–6 OH/nm^2 for titania) of ions are less than that determined from the crystallographic data. In the case of porous or highly disperse oxides, a portion of the surface hydroxyls does not take part in adsorption and chemical reactions due to steric effects. Contribution of the

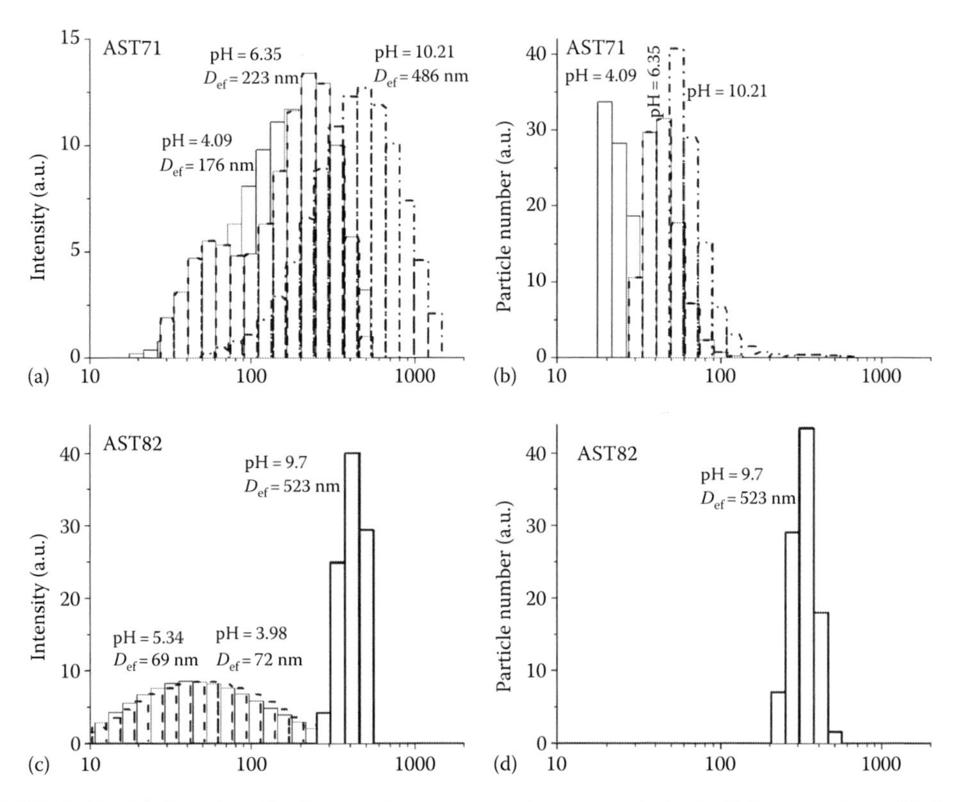

FIGURE 2.56 Particle size distributions in respect to (a, c, e, and g) the light scattering, $SPDS_I$ and (b, d, f, and h) particle number, $SPDS_N$ for AST samples at different pH values, C_{ox} = 0.04 wt%, 0.001 M NaCl. (Adapted from *Colloids Surf. A: Physicochem. Eng. Aspects*, 240, Gun'ko, V.M., Zarko, V.I., Mironyuk, I.F. et al., Surface electric and titration behaviour of fumed oxides, 9–25, 2004f, Copyright 2004, with permission from Elsevier.)

(*continued*)

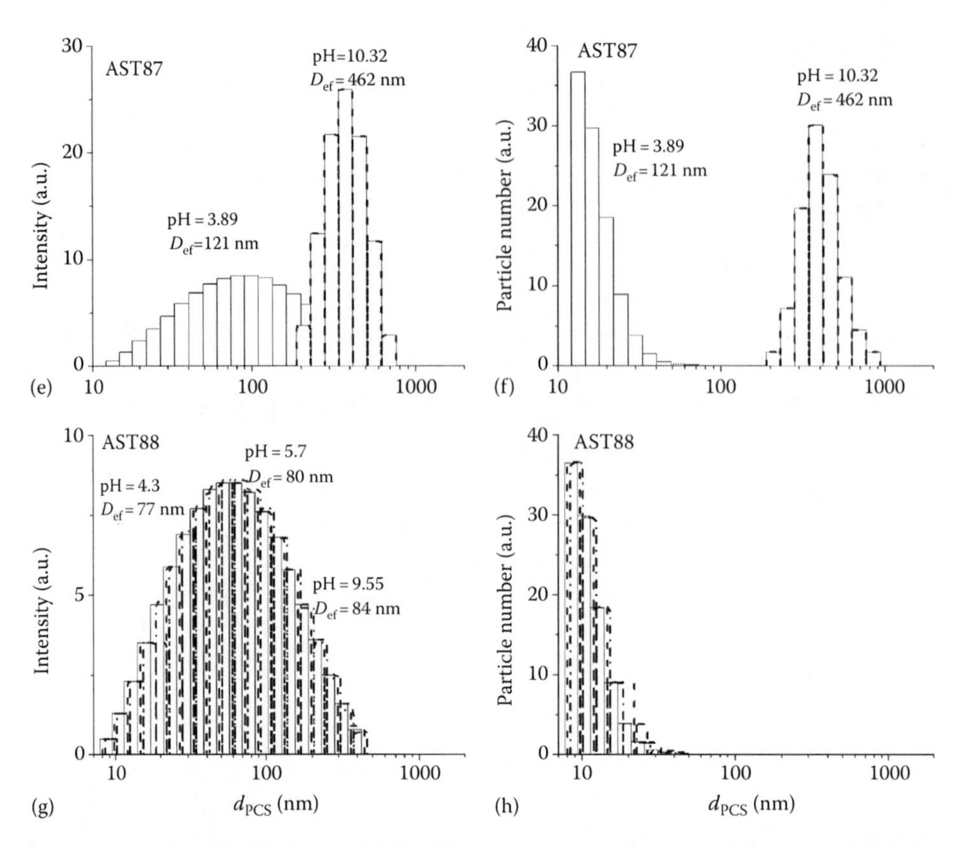

FIGURE 2.56 (continued) Particle size distributions in respect to (a, c, e, and g) the light scattering, $SPDS_I$ and (b, d, f, and h) particle number, $SPDS_N$ for AST samples at different pH values, C_{ox}=0.04 wt%, 0.001 M NaCl. (Adapted from *Colloids Surf. A: Physicochem. Eng. Aspects*, 240, Gun'ko, V.M., Zarko, V.I., Mironyuk, I.F. et al., Surface electric and titration behaviour of fumed oxides, 9–25, 2004f, Copyright 2004, with permission from Elsevier.)

acidic surface hydroxyls (B sites) on ST and SA at $C_{SiO_2}>50$ wt% give $pH_{IEP}<pH_{IEP}(SiO_2)\approx 2.2$. However, at $C_{SiO2}<50$ wt%, pH_{IEP} shifts toward greater pH values with increasing C_{TiO_2} or $C_{Al_2O_3}$ (Figure 2.52) because the number of basic hydroxyls increases.

On the Pb(II) uptake, an increase in the C_{Pb} concentration leads to the displacement of the plateau adsorption toward greater pH values in consequence of (i) a lack of the surface negative charge for compensation of the positive charge of adsorbed cations and (ii) the electrostatic repulsive interactions between adsorbed cations. Therefore, the uptake of Pb(II) and related species (Figures 2.57 and 2.58) grows on negative charging of titania and the interfaces, as the PZC of fumed silica lies at pH=2–3 (pH_{PZC}=9–10 for fumed alumina) but the surface charge density of silica at pH<7 is very small (Figure 2.61). At a low C_{Pb} value (Figure 2.57a), the lead adsorption graphs for AST50 ($pH_{IEP}\approx 3$) and ST20 ($pH_{IEP}<IEP_{SiO_2}$) are close and lie significantly higher than that for silica or alumina. This result is caused by the fact that the silica and alumina surfaces do not have strong Brønsted acid sites (which can effectively interact with cations through ion-exchange reactions) in contrast to mixed oxides. Besides, the entire concentration of the surface hydroxyls on fumed silica is lower than that on titania or mixed oxides.

It should be noted that concentration of oxide in the aqueous suspension affects the surface charge density, which is shown for A-300 (Figure 2.61a), and the PZC slightly shifts toward lower pH values with C_{SiO_2} (Figure 2.61b). The state of the oxide surface is affected by the oxide concentration because bulk concentrations of H^+ and OH^- ions as well as other ions

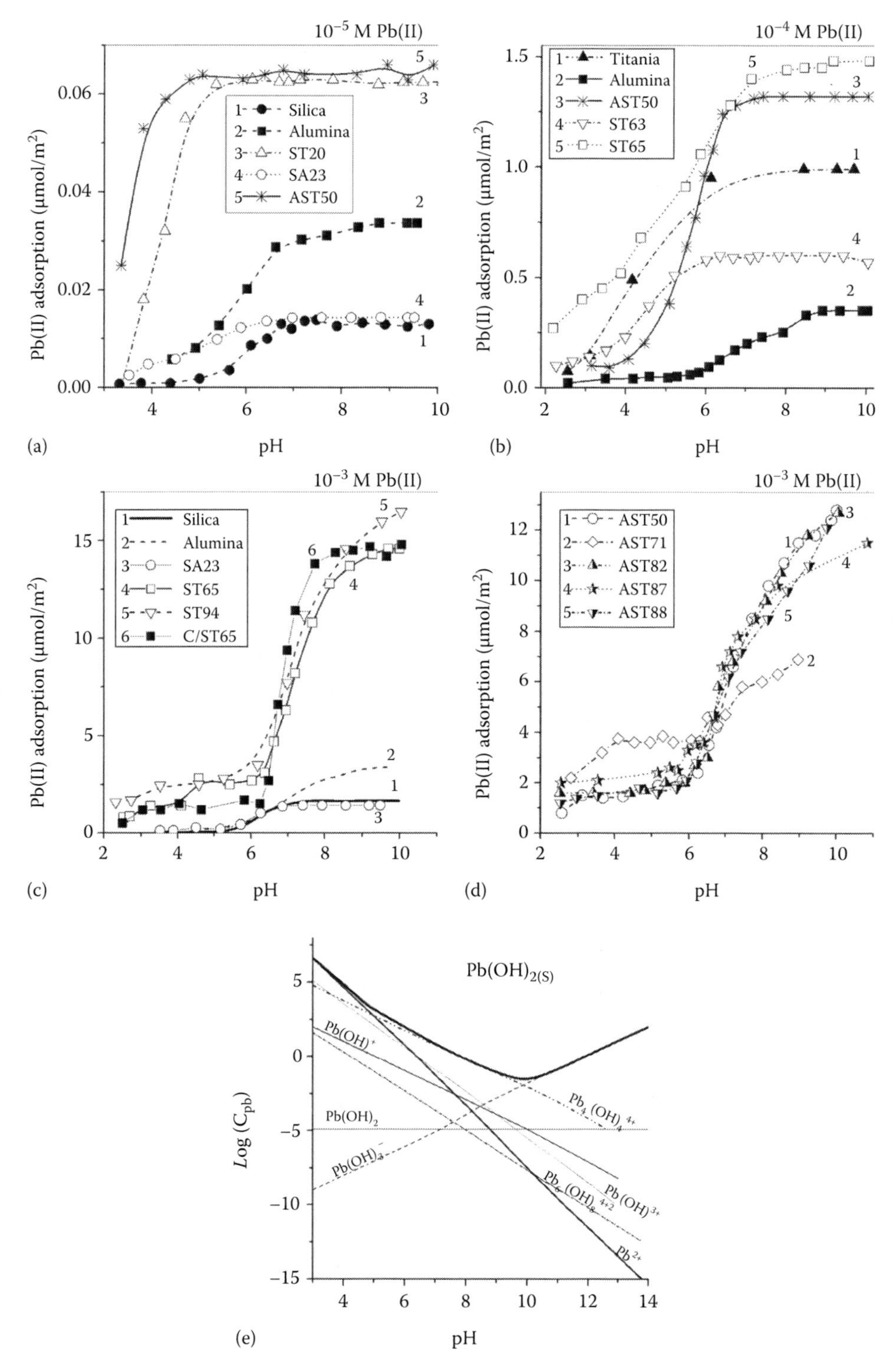

FIGURE 2.57 Adsorption of Pb(II) on individual (silica, alumina, and titania) and mixed nanooxides as a function of pH at $C_{ox} = 0.2$ wt% and $C_{Pb} =$ (a) 10^{-5} M; (b) 10^{-4} M; (c) and (d) 10^{-3} M; and (e) solubility diagram of $Pb(OH)_2$. (Adapted from *Colloids Surf. A: Physicochem. Eng. Aspects*, 240, Gun'ko, V.M., Zarko, V.I., Mironyuk, I.F. et al., Surface electric and titration behaviour of fumed oxides, 9–25, 2004f, Copyright 2004, with permission from Elsevier.)

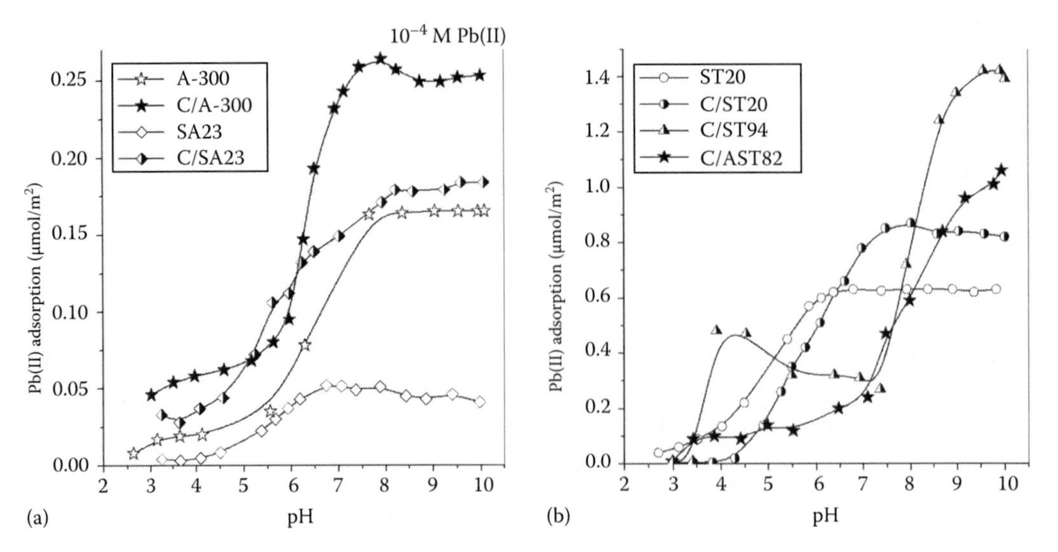

FIGURE 2.58 Adsorption of Pb(II) (10^{-4} M) on initial and carbonized oxides. (Adapted from *Colloids Surf. A: Physicochem. Eng. Aspects*, 240, Gun'ko, V.M., Zarko, V.I., Mironyuk, I.F. et al., Surface electric and titration behaviour of fumed oxides, 9–25, 2004f, Copyright 2004, with permission from Elsevier.)

(e.g., Na^+ and Cl^-) differ from their surface concentrations (determined on the titration or influencing results of the titration). The surface concentration of ions changes with C_{ox}; therefore, reactions 2.8–2.11 depend on C_{ox} and the ratio C_{ion}/C_{ox}. Additionally, the PZC and IEP values and the dependencies of σ_0 (Figures 2.61 through 2.63), ζ (Figures 2.52 and 2.54), and mobility (Figure 2.64) on pH demonstrate the influence of concentrations of oxides and inert electrolytes on the states of the particle surfaces and the shear plane. For instance, the PZC and IEP values for AST samples differ by several units (that is atypical for oxides).

At a maximal value of C_{Pb}, a minimal Pb(II) adsorption per sq. m is observed for A-300 and SA23 (Figure 2.57c) having large S_{BET} values. However, the Pb(II) adsorption per gram of oxide is relatively higher for A-300. A similar result is for AST71 (at pH>6; Figure 2.57d) having a maximal S_{BET} value among the AST samples. The Pb(II) adsorption at pH<6 is maximal for AST71 that is in agreement with a maximal negative charge density on its surface (Figure 2.62b), despite maximum positive values of ζ at pH<7 (Figure 2.52d). This difference in the ζ(pH) and σ_0(pH) dependences may be explained by the difference in the state of the shear plane (related to ζ(pH)) and the oxide surfaces titrated on the σ_0(pH) measurements. Notice that these properties of AST71 affect its SPSD in comparison with that for other AST samples (light scattering intensity $SPSD_I$ is shown in Figure 2.56a, c, e, and g and the particle number $SPSD_N$ in Figure 2.56b, d, f, and h). The SPSDs shift toward larger particle sizes at higher pH values and pictures for $SPSD_N$ and $SPSD_I$ differ. These effects can be caused by a specific role of the surface alumina on AST71 particles. Additionally, SA_{23} possessing stronger Brønsted acid sites than those on silica or alumina shows similar adsorptive capability (per square meter) as these individual oxides (Figure 2.57) due to a low surface content of alumina in SA23.

Pyrocarbon grafted on ST65 (Figure 2.57c) gives a weaker effect than that on ST20 (Figure 2.58b), A-300, and SA23 (Figure 2.58a). Clearly, the influence of pyrocarbon deposits is greater in the case of the oxide matrices having a lower adsorptive capability, since pyrocarbon per se is characterized by a significant absorptive capability with respect to metal ions due to a variety of oxygen-containing surface sites effectively interacting with cations by the ion-exchange mechanism or electrostatic intermolecular interactions with polar or charged metal species and strong dispersive interactions with uncharged metal species adsorbed in carbon nanopores.

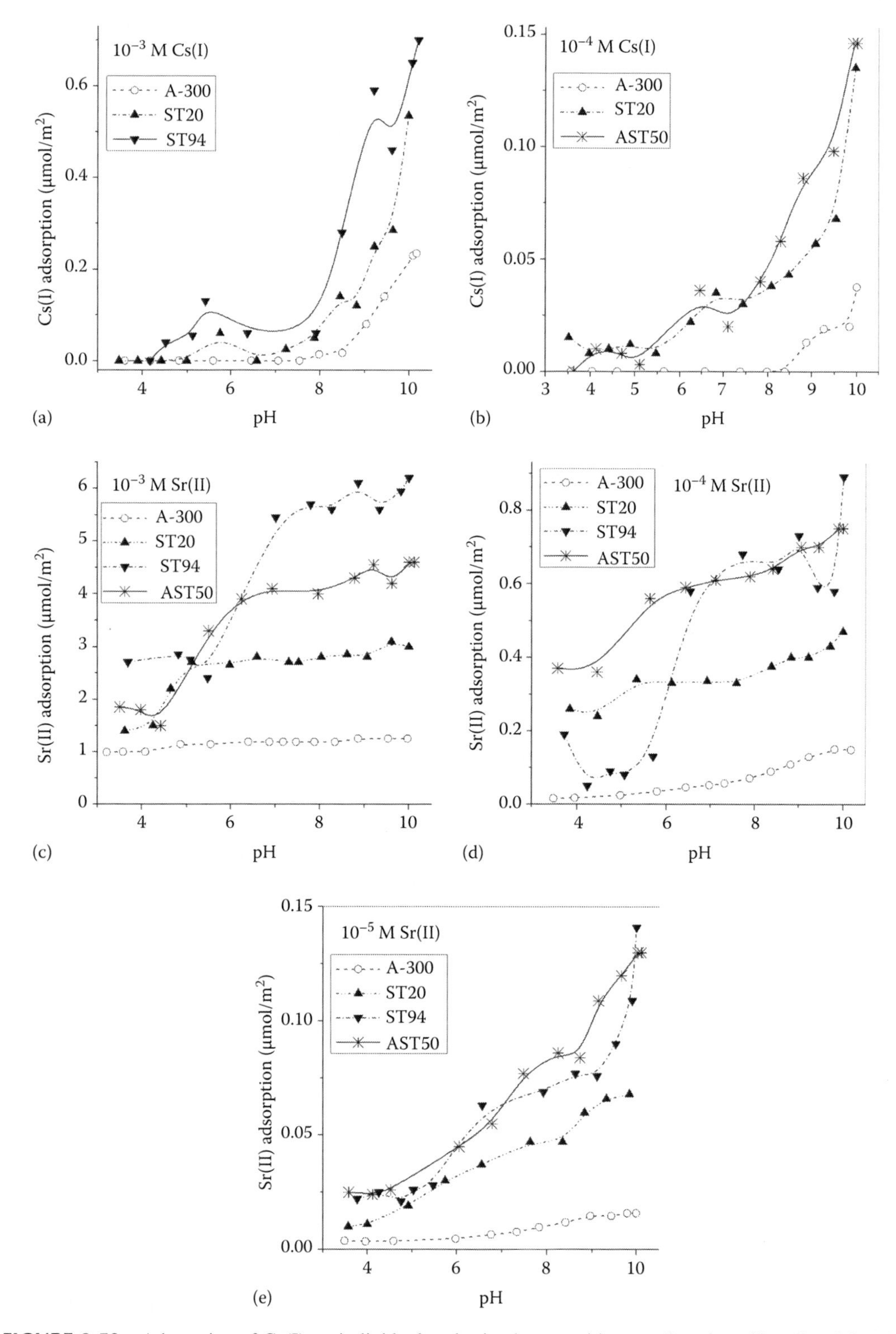

FIGURE 2.59 Adsorption of Cs(I) on individual and mixed nanooxides as a function pH at $C_{ox}=0.2$ wt% and $C_{Cs}=$(a) 10^{-3} M and (b) 10^{-4} M; and Sr(II) at $C_{Sr}=$(c) 10^{-3} M, (d) 10^{-4} M, and (e) 10^{-5} M. (Adapted from *Colloids Surf. A: Physicochem. Eng. Aspects*, 240, Gun'ko, V.M., Zarko, V.I., Mironyuk, I.F. et al., Surface electric and titration behaviour of fumed oxides, 9–25, 2004f, Copyright 2004, with permission from Elsevier.)

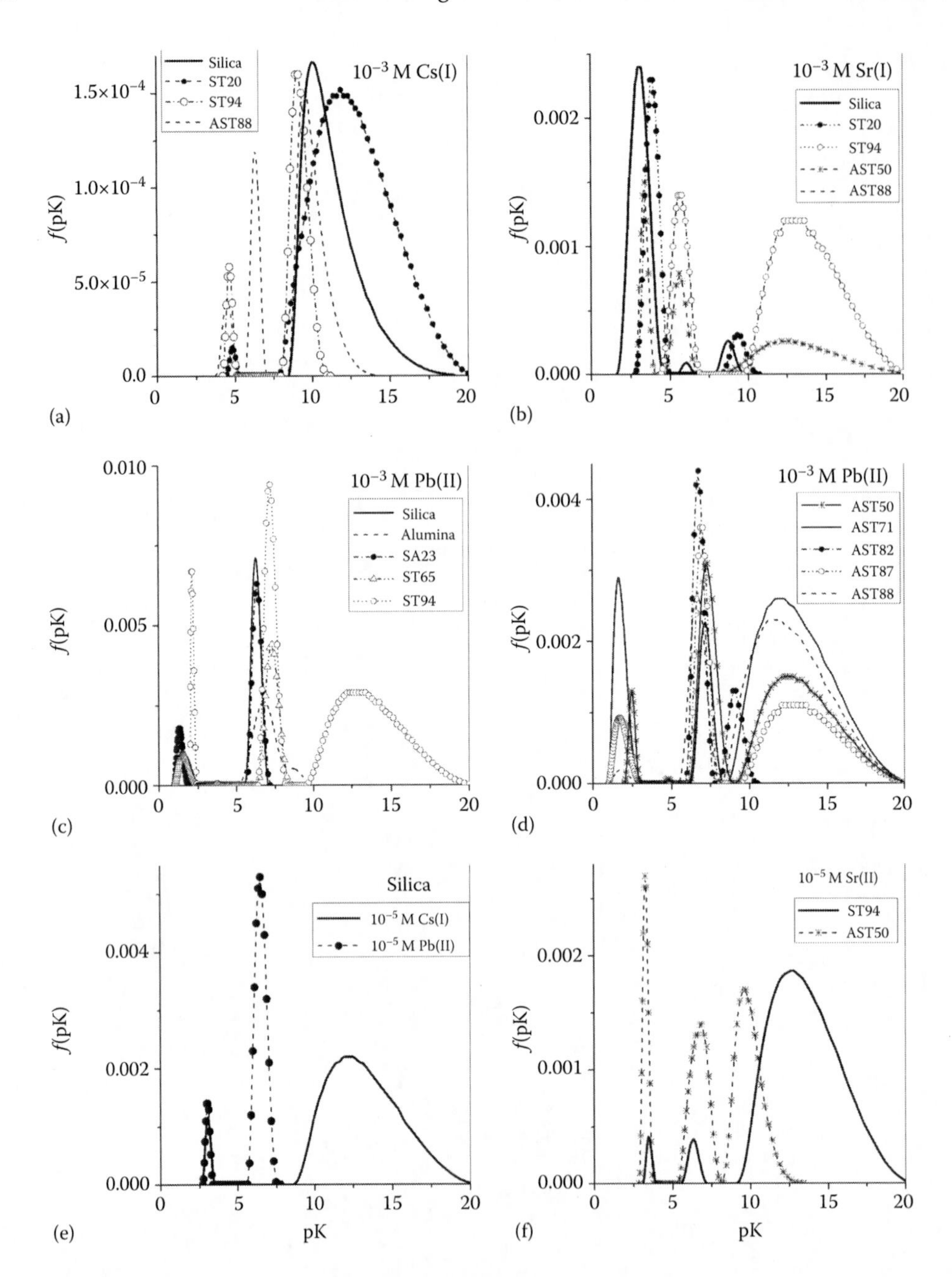

FIGURE 2.60 Adsorption constant distributions of metal ions Cs(I) at (a) 10^{-3} M and (e) 10^{-5} M; Pb(II) at (c and d) 10^{-3} M and (e) 10^{-5} M; and Sr(II) (b) at 10^{-3} M and (f) 10^{-5} M on individual, binary and ternary oxides at $C_{ox} = 0.2$ wt%. (Adapted from *Colloids Surf. A: Physicochem. Eng. Aspects*, 240, Gun'ko, V.M., Zarko, V.I., Mironyuk, I.F. et al., Surface electric and titration behaviour of fumed oxides, 9–25, 2004f, Copyright 2004, with permission from Elsevier.)

The Cs(I) adsorption is approximately 10 times as lower than that of Pb(II) or Sr(II) (Figures 2.57 through 2.59) due to differences in adsorption and solvation energies of these species, as Cs(I) in contrast to Pb(II) and Sr(II) is only in the cation form Cs^+. The adsorption (in μmol per m^2 of adsorbent surface) of different metal cations (Ctn+) on mixed oxides (ST20, ST94, and AST50) is significantly greater than that on silica A-300 because of the availability of

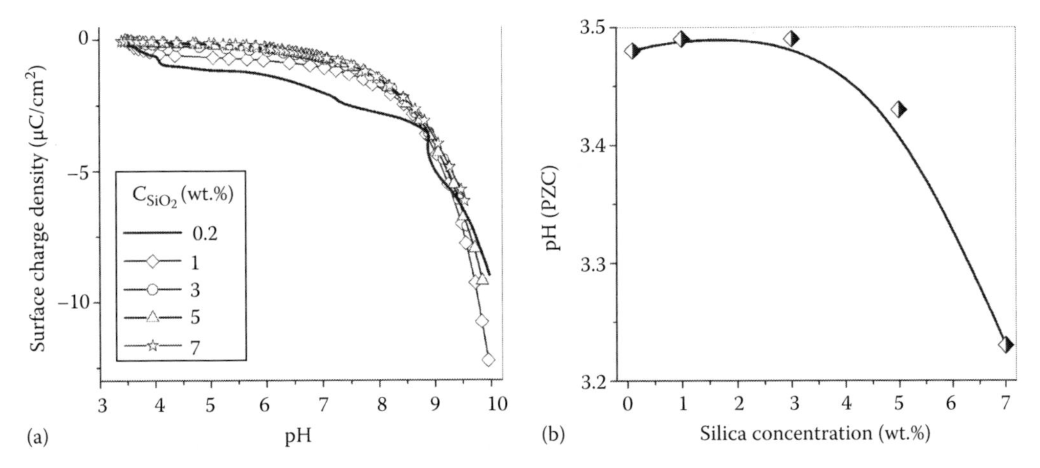

FIGURE 2.61 (a) Surface charge density as a function of pH for fumed silica A-300 at different C_{ox} values and (b) pH(PZC) as a function of C_{ox}. (Adapted from *Colloids Surf. A: Physicochem. Eng. Aspects*, 240, Gun'ko, V.M., Zarko, V.I., Mironyuk, I.F. et al., Surface electric and titration behaviour of fumed oxides, 9–25, 2004a, Copyright 2004, with permission from Elsevier.)

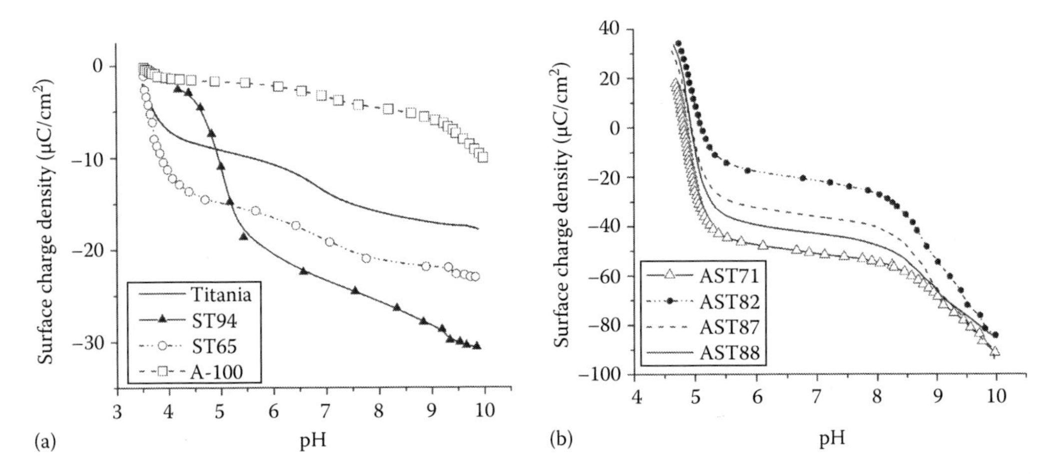

FIGURE 2.62 Surface charge density as a function of pH for (a) individual and binary oxides and (b) AST. (Adapted from *Colloids Surf. A: Physicochem. Eng. Aspects*, 240, Gun'ko, V.M., Zarko, V.I., Mironyuk, I.F. et al., Surface electric and titration behaviour of fumed oxides, 9–25, 2004f, Copyright 2004, with permission from Elsevier.)

stronger acidic (≡Si-O(H)-Al≡ or ≡Si-O(H)-Ti≡) and basic (≡AlOH and ≡TiOH) sites reacting as follows:

$$\equiv \text{Si-O(H)-Al} \equiv +\text{Ct}^{n+} \rightarrow \equiv \text{Si-O(Ct}^{(n+)-1}\text{)Al} \equiv +\text{H}^{+} \quad (2.12)$$

$$\equiv \text{AlOH} + \text{Ct}^{n+} \rightarrow \equiv \text{AlO(H)Ct}^{n+} \quad (2.13)$$

or due to interaction of Ct*n*+ (and related species) with other sites such as ≡TiO(H)Ti≡, ≡TiOH, $(\equiv AlO)_2OH$, $(\equiv AlO)_3OH$, ≡TiO(H)Al≡, etc. Notice that in the case of Pb(II) and Sr(II), different

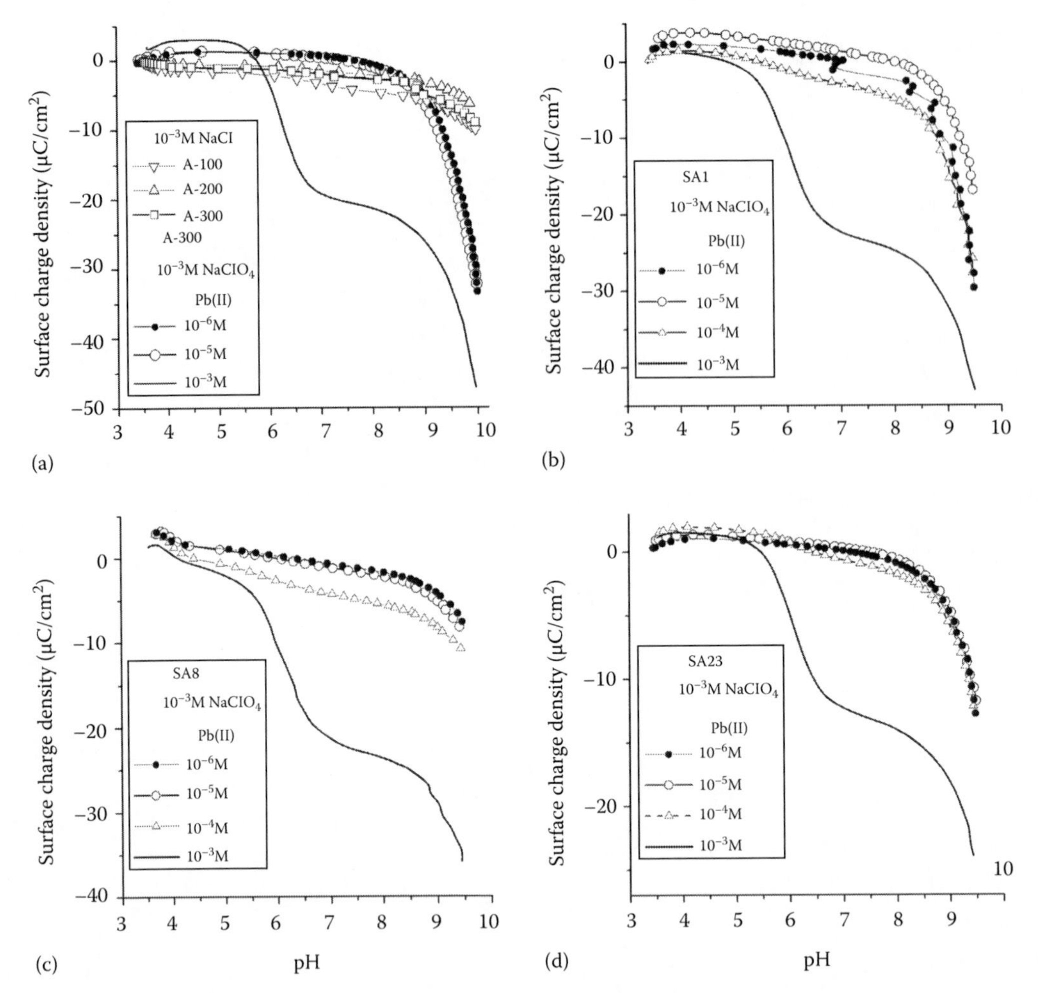

FIGURE 2.63 Surface charge density as a function of pH for the aqueous suspensions of (a) silica, (b) SA1, (c) SA8, and (d) SA23 at different concentration of Pb(II). (Adapted from *Colloids Surf. A: Physicochem. Eng. Aspects*, 240, Gun'ko, V.M., Zarko, V.I., Mironyuk, I.F. et al., Surface electric and titration behaviour of fumed oxides, 9–25, 2004f, Copyright 2004, with permission from Elsevier.)

hydrated species can be formed and adsorbed not only by the ion-exchange mechanism but also due to the electrostatic, polar, and hydrogen bonding. The complexation equilibrium constant for the reaction (2.12) can be written as follows:

$$K_{\mathrm{Ct}} = \frac{[\equiv \mathrm{SO^-Ct^+}][\mathrm{H^+}]}{[\equiv \mathrm{SOH}][\mathrm{Ct^+}]} \frac{f_{\mp\mathrm{Ct}}\gamma_{\mathrm{H}}}{\gamma_0\gamma_{\mathrm{Ct}}} \exp\left[\frac{e(\psi_{\mathrm{Ct}} - \psi_{\mathrm{H}})}{kT}\right] \quad (2.14)$$

where

$f_{\mp\mathrm{Ct}}$ is the surface activity coefficient
γ_i is the mean activity coefficient of i ion in solution
ψ_{H} and ψ_{Ct} are the mean potential at the planes of the H^+ and cation adsorption, respectively

Calculations based on these assumptions give, e.g., $pK_{\mathrm{Ct}} = 9.4$ on the Cs(I) adsorption on anatase and $pK_{\mathrm{Ct}} = 5.5$ (≡SiOCs) and −2.05 (≡SiO(H)Cs^+) on its adsorption on silica gel in the presence of Na^+. The Pb(II) adsorption on iron hydroxide corresponds to pK_{Ct} in the 6.9–13 range. The pK_{Ct} value

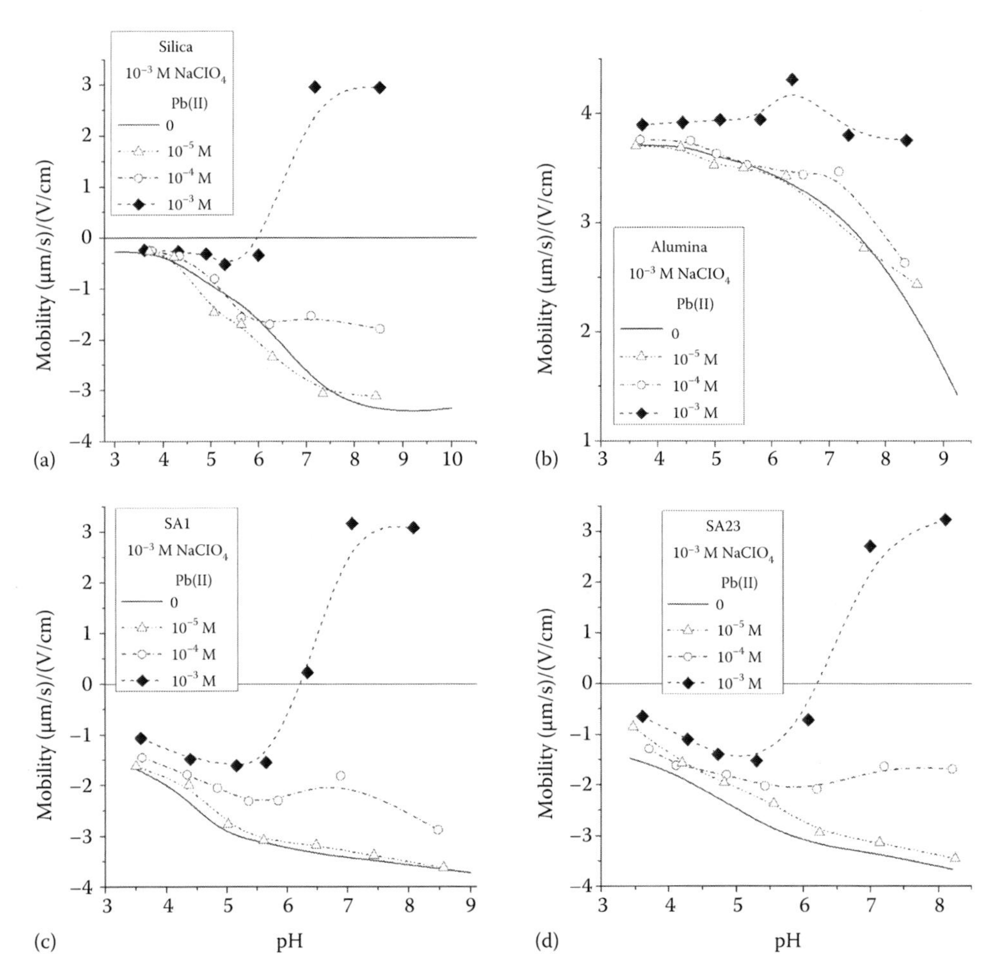

FIGURE 2.64 Electrophoretic mobility as a function of pH for the aqueous suspensions of (a) silica, (b) alumina, (c) SA1, and (d) SA23 at different concentration of Pb(II). (Adapted from *Colloids Surf. A: Physicochem. Eng. Aspects*, 240, Gun'ko, V.M., Zarko, V.I., Mironyuk, I.F. et al., Surface electric and titration behaviour of fumed oxides, 9–25, 2004f, Copyright 2004, with permission from Elsevier.)

equals to 7.75 or 5.09 for the Pb(II) adsorption on a ≡SiOH group on silica and it is equal to 17.2 or 10.68 on the Pb(II) binding to two ≡SiOH groups simultaneously (Gun'ko et al. 2004f).

Calculations of the adsorption constant distributions (Figure 2.60) give bi- or trimodal $f(pK_n)$ functions. The $f(pK_n)$ peaks corresponding to a strong interaction ($pK_n < 5$) appear due to a perceptible adsorption of cations at low pH values (Figures 2.57 through 2.59). A sharp enhancement of the adsorption up to 100% of Pb(II) at higher pH values gives a $f(pK_n)$ peak at pK_n between 5 and 10. These pK_n values are in agreement with results obtained using triple layer model (TLM) (Table 2.11).

Broadened $f(pK_n)$ peaks at $pK_n > 10$ are caused by certain changes in the adsorption after the "edge" one and can be attributed to the cation interaction with weaker sites or with two sites simultaneously.

A similar broad $f(pK_n)$ peak is absent in the case of the Pb(II) uptake up to 100% by oxides with a high specific surface area (e.g., SA23 at $S_{BET} = 353$ m²/g; Figure 2.60c), a relatively low density of adsorbed cations, and a smooth adsorption Γ(pH) curve at pH > 7 (Figure 2.57). Notice the smoothness of the adsorption curve (as demonstration of the surface uniformity) affects the distribution function determined by the regularization method: typically the smoother the experimental curves (low noise) the narrower the peaks of the distribution functions.

TABLE 2.11
Complexation Equilibrium Constants for Metal Ions Adsorbed on Different Oxides Calculated Using TLM (GRFIT)

Oxide	Cation	C_{Ct} (M)	Complexation Reaction	–log$K$
A-300	Pb(II)	10^{-4}	$\equiv SiOH + Pb^{2+} \rightarrow \equiv SiO^{-}Pb^{2+} + H^{+}$	5.6
			$\equiv SiOH + PbOH^{+} \rightarrow \equiv SiO^{-}Pb(OH)^{+} + H^{+}$	6.1
			$\equiv SiOH + Pb(OH)_2 \rightarrow \equiv SiO(H)Pb(OH)_2$	1.1
A-300	Sr(II)	10^{-4}	$\equiv SiOH + Sr^{2+} \rightarrow \equiv SiO^{-}Sr^{2+} + H^{+}$	14.3
			$\equiv SiOH + SrOH^{+} \rightarrow \equiv SiO(H)SrOH^{+}$	3.8
			$\equiv SiOH + SrOH^{+} \rightarrow \equiv SiO^{-}SrOH^{+} + H^{+}$	6.1
			$\equiv SiOH + Sr(OH)_2 \rightarrow \equiv SiOHSr(OH)_2$	1.3
A-300	Cs(I)	10^{-4}	$\equiv SiOH + Cs^{+} \rightarrow \equiv SiO^{-}Cs^{+} + H^{+}$	16.4
			$\equiv SiOH + Cs^{+} \rightarrow \equiv SiO(H)Cs^{+}$	3.3
ST94	Sr(II)	10^{-4}	$(\equiv M)_2OH + Sr^{2+} \rightarrow (\equiv S)_2O^{-}Sr^{2+} + H^{+}$	5.2
			$(\equiv M)_2OH + SrOH^{+} \rightarrow (\equiv S)_2O(H)SrOH^{+}$	11.5
			$(\equiv M)_2OH + SrOH^{+} \rightarrow (\equiv S)_2O^{-}SrOH^{+} + H^{+}$	4.1
			$(\equiv M)_2OH + Sr(OH)_2 \rightarrow (\equiv S)_2O(H)Sr(OH)_2$	7.1
ST94	Cs(I)	10^{-3}	$(\equiv M)_2OH + Cs^{+} \rightarrow (\equiv S)_2O^{-}Cs^{+} + H^{+}$	9.7
			$(\equiv M)_2OH + Cs^{+} \rightarrow (\equiv S)_2O(H)Cs^{+}$	1.4
AST50	Pb(II)	10^{-4}	$(\equiv M)_2OH + Pb^{2+} \rightarrow (\equiv S)_2O^{-}Pb^{2+} + H^{+}$	3.3
			$(\equiv M)_2OH + PbOH^{+} \rightarrow (\equiv S)_2O^{-}PbOH^{+} + H^{+}$	4.2
			$(\equiv M)_2OH + Pb(OH)_2 \rightarrow (\equiv S)_2O(H)Pb(OH)_2$	6.8
AST50	Sr(II)	10^{-4}	$(\equiv M)_2OH + Sr^{2+} \rightarrow (\equiv S)_2O^{-}Sr^{2+} + H^{+}$	5.7
			$(\equiv M)_2OH + SrOH^{+} \rightarrow (\equiv S)_2O^{-}SrOH^{+} + H^{+}$	2.6
			$(\equiv M)_2OH + Sr(OH)_2 \rightarrow (\equiv S)_2O(H)Sr(OH)_2$	9.3
AST50	Cs(I)	10^{-4}	$(\equiv M)_2OH + Cs^{+} \rightarrow (\equiv S)_2O^{-}Cs^{+} + H^{+}$	16.8
			$(\equiv M)_2OH + Cs^{+} \rightarrow (\equiv S)_2O(H)Cs^{+}$	3.8

Note: M = Si, Ti, or Al for mixed oxides.

In the case of the Cs(I) adsorption, the $f(pK_n)$ distribution functions (Figure 2.60a and e) reveal relatively low adsorption constants since a maximal uptake of Cs(I) on the studied samples is not more than 14% (typically 4%–6%). A maximal uptake of Sr(II) depends on its concentration and it can reach up to 100% at $C_{Sr(II)} = 10^{-5}$ M, but at 10^{-3} M, it is between 35%–70% depending on the specific surface area of oxides and their kind. Additionally, the Sr(II) adsorption is characterized by relatively great initial values of Γ(pH) at pH < 6 and by its non-steep enhancement at higher pH values (Figure 2.59). This effect gives the $f(pK_n)$ distribution with a large peak at $pK_n < 7$ and a broad peak at $pK_n > 10$ (only for certain samples; Figure 2.60b and f). These differences in $f(pK_n)$ for Cs(I), Sr(II), and Pb(II) are linked to the difference in the form of metal species, composition of their solvation shells, and the distributions of active surface sites on oxides most appropriate for the adsorption of given species. As a whole, the complexation equilibrium constants (Table 2.11) calculated within the scope of the TLM or the $f(pK_n)$ distributions (Figure 2.60) are in agreement. Observed differences are caused by features of the used models and the corresponding equations. In the case of Pb(II)/SiO_2, the first two complexes in Table 2.11 give the main contribution to the total adsorption. This result is in agreement with the main contribution of the $f(pK_n)$ peak at $pK_n = 6$ (Figure 2.60c). Similar picture is observed for Sr(II)/SiO_2 (Table 2.11; Figure 2.60b). The difference between the Pb(II) and Cs(I) adsorption on silica is much greater, as Cs(I) adsorbs in a minor portion that gives broadened $f(pK_n)$ peaks at $pK_n > 8$ (Figure 2.60a and e). According to calculations of pK_{Ct} (Table 2.11), probability of the adsorption of $M(OH)_n^{(2-n)+}$ ($n = 1, 2$; M = Pb and Sr) is greater than that for the adsorption of cations Pb^{2+} and Sr^{2+} not only on weakly acid silica but also on mixed oxides having stronger B sites. Cs(I) rather adsorbs by electrostatic binding to oxygen atoms of the surface hydroxyls with a low probability of the exchange for H^+. These results can be explained by the fact that the studied nanooxides are not very strong acidic oxides (such as zeolites, etc.).

The nonuniformity (heterogeneity) of the surface of primary particles of binary and ternary nanooxides (caused by the presence of individual oxide patches on the surfaces, solid solution of one oxide in another one with formation of surface asymmetrical bridges $\equiv M_1O(H)M_2\equiv$, etc.) results in large differences in the surface charge density σ_0(pH) for the studied oxides (Figure 2.62). Additionally, the σ_0(pH) curve shapes significantly differ from that for ζ(pH) (Figure 2.52). Notice that the difference in the oxide concentration (by a factor of 5 at the same salinity of 0.001 M NaCl) used on the measurements of ζ(pH) (C_{ox} = 0.04 wt%) and σ_0(pH) (0.2 wt%) can play a significant role.

A large difference (more than five pH units) between the pH_{IEP} (Figure 2.52d) and pH_{PZC} (Figure 2.62b) values for AST is noteworthy. Notice that σ_0(pH) for AST does not strongly correlate to the chemical composition as a whole (similar effects are also seen for other parameters) due to the influence of three oxide phases differently distributed in the volume and on the surface of primary particles and differently affected by pH. The σ_0(pH) curves have a simple shape for fumed silica (Figure 2.61) because this oxide is characterized by only one type of surface sites such SiOH groups (since single, twin, and vicinal silanols possess close properties). Mixed oxides, titania, and alumina are characterized by a variety of active surface sites including terminal and bridging hydroxyls and L-sites. Therefore, the shape of σ_0(pH) is characterized by, at least, two portions with fast changes in σ_0 with pH and one–two fragments with its slow changes. A decrease in the rate of changes in σ_0(pH) is observed near regions of PZC of certain components (e.g., near pH at 6 for titania).

Concentration of adsorbed ions can play a substantial role in changes in the σ_0(pH) and ζ(pH) values if this concentration is higher than a certain critical value corresponding to a dramatic change in the surface charge density (Figure 2.63) and the electrophoretic mobility (Figure 2.64). The latter parameter was used instead of the electrophoretic potential, since for similar systems, the conductivity of the surface layer should be also considered in addition to the porosity of aggregates on transform of U_e to ζ. Corrections in ζ considering the solution conductivity and the porosity of aggregates of primary particles of fumed silica give close results even at a large salinity (0.01 M NaCl). Therefore, correction related to the conductivity was not considered in our calculations (Figure 2.52). The electrophoretic mobility directly measured was analyzed on the Pb(II) adsorption at different C_{Pb} values (Figure 2.64). Notice that on the adsorption of tetraalkylammonium or alkali ions, changes in the surface charge density on fumed silica OX-50 with nonporous primary particles (average size 40 nm) is significantly lower than that for porous silica (Stöber silica). This result is in agreement with a relatively weak effect of the conductivity on the ζ values of fumed silica. It should be noted that the surface charge density increases for fumed silica with increasing mean size of primary particles (Figure 2.63a). This is in agreement with an enhancement of amounts of free surface $\equiv$SiOH groups with growing primary particle diameter, as deprotonated silanols are responsible for negative charging of the silica surface.

Character of changes in σ_0 and U_e on the Pb(II) adsorption differs, since one can see an enhancement of negative σ_0(pH) value (Figure 2.63) and appearance of positive U_e(pH) (Figure 2.64) at $C_{Pb} = 10^{-3}$ M. Relative changes in U_e(pH) for Pb(II)/alumina with C_{Pb} are significantly lower (Figure 2.64b) than that for oxides with acidic properties such as silica (Figure 2.64a) and SA (Figure 2.64c and d). However, amounts of Pb(II) adsorbed on alumina is greater than that for silica or SA23 at the critical Pb(II) concentration of 10^{-3} M (Figure 2.57c).

This concentration corresponds approximately one–two Pb(II) ions per nm^2 of A-300 (this value is close to the concentration of the surface hydroxyls 2.5 OH/nm^2 for A-300). There is another effect contributing change in the sign of U_e at $C_{Pb} = 10^{-3}$ M due to a difference in the mobility of oxide aggregates and metal cations under applied external electrostatic field, which results in assemblage of cations near or inside porous aggregates more than the equilibrium concentration of adsorbed ions. These cations can speed up oxide aggregates in the applied field. Clearly, this effect can increase with increasing size and porosity of oxide aggregates and pH due to reactions 2.8 and 2.9, but reaction 2.10 can inhibit this process (Figure 2.64). The ion current near the oxide surfaces cannot entirely compensate the effect of this nonequilibrium "adsorption" of cations. The Pb(II) concentration of 10^{-4} M (i.e., smaller than the critical concentration) or less provides too low surface coverage to change the sing of σ_0(pH) (Figure 2.63) and U_e(pH) (Figure 2.64) as at $C_{Pb} = 10^{-3}$ M.

The influence of the surface composition of SA samples of the Pb(II) uptake is well observed in Figure 2.46 showing the relationships between the surface alumina content ($C_{Al_2O_3,surf}$ was determined on the basis of the Auger electron spectra of a thin surface layer of SA particles), the plateau Pb(II) adsorption (at $C_{Pb} = 10^{-3}$ M) and the total content of alumina. The shape of $\Gamma_{Pb}(C_{Al_2O_3})$ is akin to that of $C_{Al_2O_3,surf}(C_{Al_2O_3})$ due to the difference in the interaction between Pb(II) species and the surfaces of silica (with no B sites), alumina/silica (with B sites), and certain sites of the alumina phase (AlOH, AlO(H)Al, etc.) depending on the specific surface area (compare $\Gamma_{Pb}(C_{Al_2O_3})$ in μmol/m^2 [Figure 2.46a] and mmol/g [Figure 2.46b]), the number of different surface hydroxyls and their acidity. Notice that the acidity of B sites increases with decreasing content of alumina, which can affect the Pb(II) adsorption (Figure 2.46), which is maximal (in mmol/g) at $C_{Al_2O_3} = 8$ wt%. Notice that Γ_{Pb} in μmol/m^2 depends on the specific surface area of oxides, since the plateau Pb(II) ($C_{Pb} = 0.001$ M) adsorption (shown in Figure 2.46) corresponds to its 100% uptake.

Thus, the electrophoretic potential of AST and C/AST samples as a function of pH (especially at pH < 8) demonstrates a great effect of the surface alumina. Titania in ST samples causes smaller changes in ζ(pH) in comparison with alumina in AST. The pyrocarbon effect on ζ(pH) (its displacement and sign) depends on the surface alumina distribution in C/AST and the titania concentration in C/ST, since pyrocarbon deposits can block the most active surface sites, both acidic and basic. The secondary particle size distributions in the diluted aqueous suspensions of ST and AST are strongly affected by pH due to the influence of attractive interaction between oxide phases (patches) having different charges and placed on neighboring primary particles. The effective diameter of particle aggregates depends on pH nonlinearly and stronger for AST than for ST due to the effect of the surface alumina since its patches cause additional attractive interaction between ternary oxide particles at pH between IEP of silica and alumina. AST and ST (as well as C/AST or C/ST in respect to Pb(II)) reveal a greater adsorptive ability in respect to metal ions Cs(I), Sr(II), and Pb(II) in comparison with individual (alumina, silica, and titania) and binary (SA) nanooxides. The dependencies of the surface concentration of alumina and the Pb(II) uptake versus the total content of alumina in SA samples have close curve shapes that suggest strong influence of the surface composition on the adsorption of metal ions and related species. The impact of adsorbed Pb(II) on the surface charge density and the electrophoretical mobility essentially differs at the critical concentration (10^{-3} M) when the state of the shear plane sharply changes. The parameters mentioned change in opposite directions: U_e becomes positive with increasing adsorption of Pb(II) at pH > 6, but σ_0 becomes more negative at the same pH values.

The hydrophilicity of nanooxides, which plays a very important role in their applications and affects many of their properties, was analyzed using calorimetry (oxides were degassed at 473 K at low pressures for several hours) and ^{1}H NMR spectroscopy (oxides were equilibrated in air) methods applied to samples after different pretreatments. This characteristic is linked to the possibility of the formation of strong hydrogen and donor–acceptor bonds or/and dissociative adsorption of water. The treatments before the calorimetric measurements resulted in desorption of intact water and a portion of dissociatively adsorbed water (≡MOH, ≡M′O(H)M″≡, where M = Si, Al, or Ti) from both surface and volume of oxide nanoparticles. However, in the case of the NMR measurements, surface and volume water was readsorbed from air. Therefore, one could expect that the heat effects on the adsorption of water on the calorimetric measurements should be stronger than that on the NMR measurements. This is typically observed for the samples studied with the exception of SA8 and ST20 (Table 2.12).

There is a tendency for an increase in the difference between the Q_w and γ_S values with increasing size of primary particles. For instance, it is minimal (and $K < 1$) for amorphous SA23 (average particle diameter $d_{av} = 6.9$ nm, deviation of the pore shape from the random voids between spherical particles $\Delta w = 0.23$) and maximal for AST50 ($d_{av} = 49.7$ nm, $\Delta w = 0.31$, i.e., roughness AST50 > SA23) with crystallinity > 90% ($K = 1.83$). However, this ratio depends on the particle composition, e.g., AST50 (crystalline) and A-50 (amorphous, $\Delta w = 0.38$) have close d_{av} values but the reverse relationships between the Q_w and γ_S values; however, their γ_S values are close. These results can be caused by stronger heat effects on the water adsorption onto degassed samples composed of larger primary particles but depending on silica content and surface structure. Larger particles can adsorb greater amounts of water in the volume because the *V/S* ratio increases (especially for silica) as well as the ratio of volume and surface amounts of adsorbed water. Similar effects observed for different silicas result in a significant overestimation of the S_{BET,H_2O} value for heated and degassed samples and $S_{BET,H_2O} \approx S_{BET,N_2}$ for samples equilibrated in air (Gun'ko et al. 2009d). Notice that samples studied with maximal S_{BET,N_2} values are characterized by minimal hydrophilicity. For instance, the K value, determined as the ratio of the heats of immersion in water and decane $K = Q_w/Q_d$, is low (<1) for SA23 (Table 2.12), and this nanooxide possesses the maximal S_{BET,N_2} value among SA samples. Titania-containing samples (typically with small S_{BET,N_2} values) have larger K values than samples with alumina/silica, alumina or silica (Table 2.12).

Figure 2.65 shows typical ^{1}H NMR spectra of unfrozen water bound in wetted powder (total amount of water $h = 0.11$ g/g) and aqueous suspension [$C_w = 90$ wt%] of SA8 (as a representative sample) recorded at different temperatures.

TABLE 2.12
Adsorption Characteristics Determined from Calorimetrical (Q_w, Q_d, and K) and ^{1}H NMR Spectroscopy (γ_S) Measurements

Sample	Q_w (J/m^2)	Q_d (J/m^2)	$K = Q_w/Q_d$	γ_S (J/m^2)
SA8	0.119	0.087	1.37	0.288
SA23	0.132	0.151	0.87	0.122
ST20	0.328	0.082	4.0	0.435
AST03	0.345			0.240
AST1	0.548	0.136	4.03	0.163
AST50	0.459	0.251	1.83	0.077
AST71	0.591	0.161	3.67	
AST82	0.493	0.148	3.33	0.479
Al_2O_3[a]	0.411	0.295	1.39	0.231
Al_2O_3[b]	0.400	0.160	2.50	
A-200[c]	0.183	0.149	1.23	0.124
A-50	0.231			0.177
TiO_2	0.262	0.083	3.16	

Source: Adapted from *J. Colloid Interface Sci.*, 348, Gun'ko, V.M., G.R. Yurchenko, Turov, V.V. et al., Adsorption of polar and nonpolar compounds onto complex nanooxides with silica, alumina, and titania, 546–558, 2010e, Copyright 2010, with permission from Elsevier.

[a] $S_{BET,N_2} = 133\,m^2/g$.

[b] $S_{BET,N_2} = 86\,m^2/g$.

[c] $S_{BET,N_2} = 230\,m^2/g$.

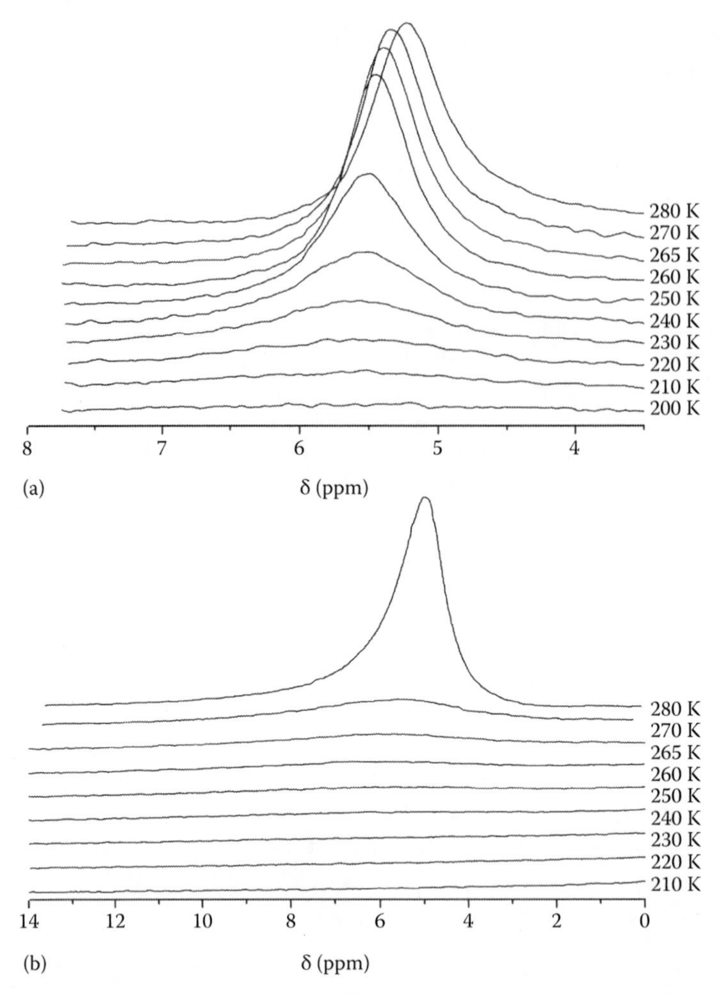

FIGURE 2.65 ^{1}H NMR spectra of unfrozen water in (a) hydrated powder of SA8 (h=0.11 g/g) and (b) aqueous suspension of SA8 (C_w=90 wt%). (Adapted from *J. Colloid Interface Sci.*, 348, Gun'ko, V.M., G.R. Yurchenko, Turov, V.V. et al., Adsorption of polar and nonpolar compounds onto complex nanooxides with silica, alumina, and titania, 546–558, 2010d, Copyright 2010, with permission from Elsevier.)

Relative amounts of bound unfrozen water ($C_{uw,270K}/C_w$) in the suspension are much smaller (< 30%) than that in the hydrated powder (100%) in which practically all water ($C_{uw,270K}$) is unfrozen at 270 K. In the suspensions of AST82 and SA8, the $C_{uw,270K}$ value corresponds to 17 and 27% of C_w, respectively. The S_{BET,N_2} value of SA8 is seven and four times larger than that of AST82 and AST71, respectively (Table 2.1). Therefore, the amounts of interfacial bound unfrozen water are larger for both hydrated powder and aqueous suspension of SA8 with lowering temperature than that for AST71 or AST82 (Figure 2.66a). This corresponds to stronger binding of interfacial water to smaller nanoparticles of SA8 (i.e., with larger specific surface area) than to larger particles of AST samples, and the Gibbs free energy is lower (at the same C_{uw} values) for the systems with SA8 (Figure 2.66b).

The secondary and ternary particles of nanooxides can be assigned to relatively soft systems, which can be rearranged on interactions with adsorbates, especially water, other small polar molecules or polar polymers such as poly(vinyl alcohol), poly(ethylene glycol), etc. These rearrangements are revealed on comparison of the PSDs (Figure 2.67) calculated on the basis of the nitrogen adsorption onto dry samples and using NMR cryoporometry (solving the integral Gibbs-Thomson equation; see Section 10.1) applied to the $C_{uw}(T)$ functions for wetted samples or aqueous suspensions.

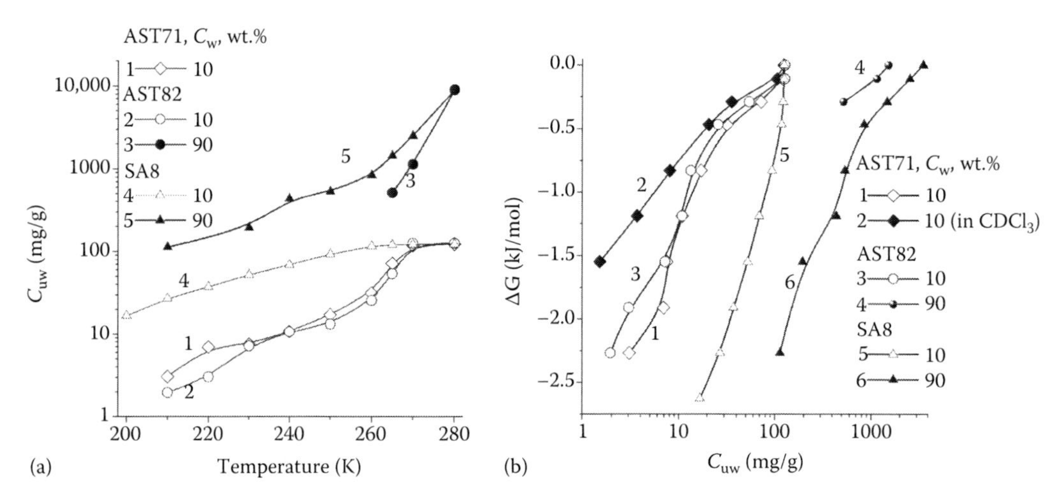

FIGURE 2.66 (a) Temperature dependence of unfrozen water content (C_{uw}) and (b) relationships between C_{uw} and changes in the Gibbs free energy of bound water for hydrated powders ($C_w = 10$ wt%) and aqueous suspensions ($C_w = 90$ wt%) of different oxides. (Adapted from *J. Colloid Interface Sci.*, 348, Gun'ko, V.M., G.R. Yurchenko, Turov, V.V. et al., Adsorption of polar and nonpolar compounds onto complex nanooxides with silica, alumina, and titania, 546–558, 2010e, Copyright 2010, with permission from Elsevier.)

Contribution of narrow voids at $R < 3$ nm filled by structured water in wetted powder at $C_w = 10$ wt% is greater than that of voids filled by N_2 fluid in dry powder. The changes in the wetted powder can be due to several effects. Water molecules penetrating between adjacent nanoparticles (nanopore contribution grows) plasticize their contacts and contract neighboring particles in denser aggregated structures (contribution of narrow mesopores grows). Notice that these effects clearly appeared on wetting and drying of the nanooxide powders. On suspending, the rearrangements of secondary and ternary particles are enhanced. This causes stronger textural changes in nanooxides after suspendingdrying than after wetting–drying or interaction of the powders with organic solvents (e.g., ethanol). In the suspension of SA8 (Figure 2.66), $C_{uw} = 2.5$ g/g at 270 K, i.e., the volume of bound water ($V = 2.5$ cm^3/g) is approximately two and four times larger than the pore volume $\left(V_{p,N_2}\right)$ of suspended-dried SA8 (1.33 cm^3/g) and the initial powder of SA8 (0.68 cm^3/g), respectively. These effects are due to loosing of contacts between nanoparticles on wetting or suspending and then strong binding (up to formation of siloxane bonds) of nanoparticles one to others on drying. These aspects of the behavior of nanooxides are of importance on treatment or use of these materials in liquid and polymer/solvents media. Co-adsorption of water and organic solvents onto nanooxides leads to diminution of the water interaction with the oxide surfaces. This effect is clearly observed on suspending of wetted powder of AST71 in the $CDCl_3$ medium since the $-\Delta G$ value of bound water becomes smaller as well as C_{uw} (Figure 2.66b, curves 1 and 2).

Thus, nitrogen adsorption isotherms can be used to determine not only the S_{BET} and V_p values, and PSDs, but also the primary particle size distributions using pair integral equations related to PSD and $\phi(a)$, and solved with the self-consistent regularization procedure. Heating of nanooxides caused several processes (such as dehydration of the surface and the bulk of primary particles, binding of adjacent nanoparticles due to condensation of hydroxyls, and rearrangement of secondary and ternary particles), which result in structural changes of nanooxides dependent on the heating temperature. As a whole, the morphology of primary nanoparticles of nanooxides is stable to different treatments, in contrast to the structure of secondary particles and especially particles from higher structural levels.

The amounts of water adsorbed onto nanooxides from air correspond to the amounts of water strongly bound (unfrozen at $T < 273$ K) to the surface of nanooxides in aqueous suspensions. In these suspensions, multimodal particle size distributions are observed due to aggregation of primary

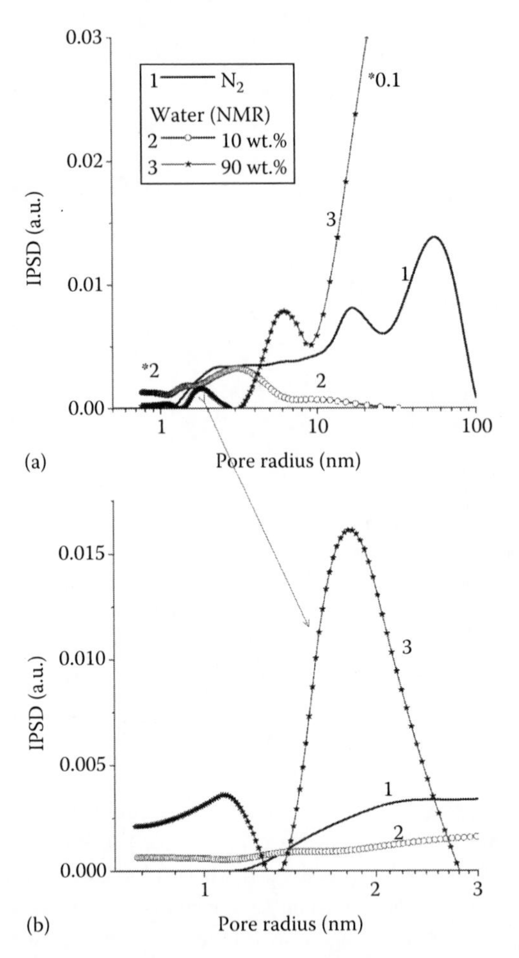

FIGURE 2.67 PSDs for SA8 calculated from the nitrogen desorption data (1) and the $C_{uw}(T)$ function (NMR) for hydrated powder (2) and aqueous suspension (3) (a) with normalized curves 2 and 3, and (b) initial range of the PSDs at $R < 3$ nm without the normalization of the curves. (Adapted from *J. Colloid Interface Sci.*, 348, Gun'ko, V.M., G.R. Yurchenko, Turov, V.V. et al., Adsorption of polar and nonpolar compounds onto complex nanooxides with silica, alumina, and titania, 546–558, 2010d, Copyright 2010, with permission from Elsevier.)

particles and agglomeration of secondary particles. The absence of nanopores in nanooxides, which are characterized by the textural porosity, causes relatively low values of the free surface energy γ_S. The γ_S values are smaller than the heat of immersion of nanooxides in water, and the heat of water adsorption from the vapor phase. This difference is due to the difference in pretreatment (heating) of samples. Thus, heating history can strongly affect many physicochemical properties of nanooxides, because of changes in the amounts of intact and dissociatively adsorbed water at the surface of nanoparticles and in their volume.

2.2 POROUS OXIDES AS A FUNCTION OF MORPHOLOGY

Mixed porous oxides can be divided into two main classes (similar to nanooxides) with mixed oxides synthesized simultaneously or matrix oxides (host oxides) modified by grafting of another oxide (guest oxide) using CVD, PVD, precipitation, or other methods. Clearly, the properties of the materials with the same gross composition but from different classes can be strongly different. Many of regularities found for mixed nanooxides can be true for porous oxides. However, the difference in the textural organization of these materials can play an important role, especially for mixed oxides, prepared on the basis of silica gels (or ordered mesoporous silicas) and fumed silicas used as substrates for grafted oxides because the latter have open surface of nonporous primary nanoparticles, but for porous oxides, the main porosity is the inner porosity of particles. Therefore, the guest phases grafted on these different matrices can be strongly different in the morphology of their particles and their influence on the textural characteristics of the whole materials. Porous mixed materials can be in the form of porous microparticles (i.e., in the 1–1000 μm range of sizes) and macroparticles (>1 mm), monoliths, films, layered structures, membranes, nano- and microtubes, hollow spheres, as well as in the form of composites with polymers or carbons (Iler 1979, Brinker and Scherer 1990, Ohring 1992, Ertl et al. 1997, Hüsing and Schubert 1998, Brown et al. 1999, Wright and Sommerdijk 2001, Al-Abadleh and Grassian 2003, Diebold 2003, Smith et al. 2004, Ma and Zaera 2006, Chen and Mao 2007, Ma et al. 2007a, Kanan et al. 2009, D'Alessandro et al. 2010). Porous mixed materials can be of both synthetic and natural origin.

Before analysis of mixed porous oxides, some individual non-silica materials such as titania, alumina, etc., which are components of mixed oxides, should be analyzed. Ryu et al. (2010) synthesized mesoporous titania microspheres (Figure 2.68) using Pluronic F108 as a template agent and titanium(IV) isopropoxide as a precursor of titania. The textural characteristics of the spheres were studied using ^{1}H NMR cryoporometry (water and cyclohexane as probe liquids), XRD, and nitrogen adsorption. It was shown that the specific surface area decreases (S_{BET} = 337, 225, 177, 128, and 89 m^2/g for MT300, MT400, MT500, MT600, and MT700 calcined at 300, 400, 500, 600, and 700°C, respectively), but the anatase crystallite sizes grow from 6, 6.7, 7.5, 10, to 19.2 nm with increasing calcination temperature. The NMR spin-echo intensities from the liquid phase confined in pores were continuously measured until the probe molecules had completely melted. At each new temperature, the sample was equilibrated for at least 10 min before any measurements. All measurements were performed using a spin-echo pulse (90°-τ-180°-τ-echo) with a 90° pulse length of 2.14 μs and a 180° pulse length of 4.28 μs. For cyclohexane, used k_{GT} = 178 K nm. The NMR and BJH PSDs are in semi-qualitative agreement (Figure 2.68b). Notice that Barrett-Joyner-Halenda (BJH) method (Barrett et al. 1951) for titania can give certain errors because the own errors of the BJH method and the use of silica parameters (used in the firm software) instead of titania parameters.

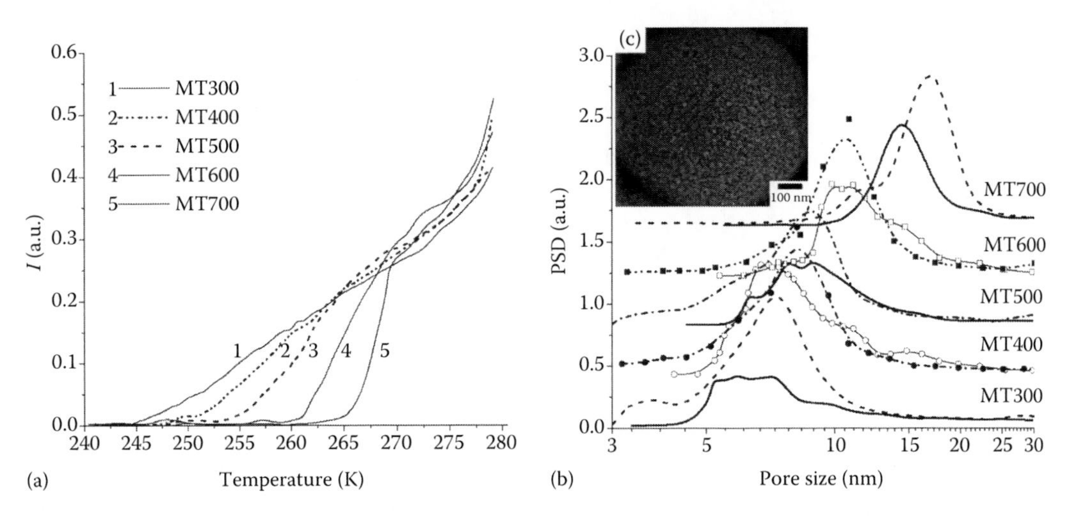

FIGURE 2.68 (a) Relative NMR spin-echo intensities versus temperature for cyclohexane confined within the pores of titania spherical particles; (b) pore size distribution curves of the MTx samples determined by ^{1}H NMR cryoporometry and BJH method (dashed lines); (c) insert with field-emission (FE) SEM image of MT400. (Adapted with permission from Ryu, S.-Y., Kim, D.S., Jeon, J.-D., and Kwak, S.-Y., Pore size distribution analysis of mesoporous TiO_2 spheres by ^{1}H nuclear magnetic resonance (NMR) cryoporometry, *J. Phys. Chem. C*, 114, 17440–17445, 2010, Copyright 2010 American Chemical Society.)

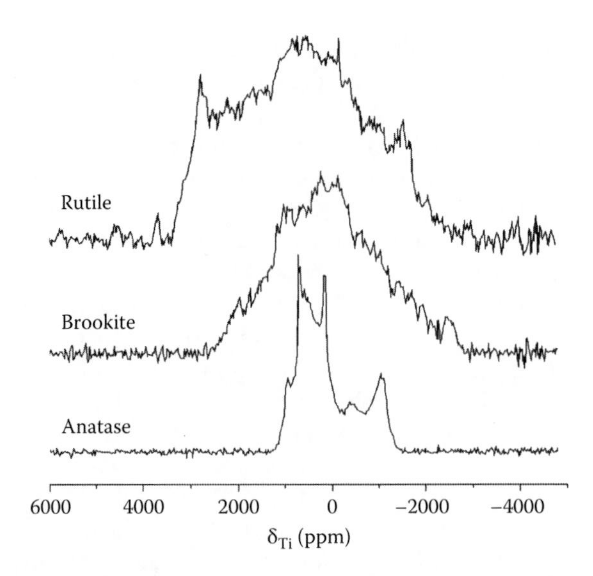

FIGURE 2.69 ^{47}Ti and ^{49}Ti MAS NMR spectra of anatase, brookite and rutile (chemical shift was estimated with respect to TiC_4). (Adapted from *Chem. Phys. Lett.*, 270, Labouriau, A. and Earl, W.L., Titanium solid-state NMR in anatase, brookite and rutile, 278–284, 1997, Copyright 1997, with permission from Elsevier.)

Labouriau and Earl (1997) studied anatase, brookite, and rutile using the ^{47}Ti and ^{49}Ti MAS NMR spectroscopy (9.4 T, 5×10^5 scans). The ^{47}Ti and ^{49}Ti isotopes having spin 5/2 and 7/2, respectively, possess nuclear quadrupole moments (therefore, there is a small number of published papers based on similar effortful investigations). The ^{47}Ti and ^{49}Ti MAS NMR spectra of anatase, brookite, and rutile (Figure 2.69) demonstrate very broad resonance lines.

This due to lattice features, which are not of cubic symmetry, therefore second-order quadrupolar effects dominate, and line broadening is greater for rutile than for anatase (Figure 2.69).

Navrotsky (2007) analyzed water adsorption onto different titania materials (Figures 2.70 and 2.71) using the DSC method. The adsorbed water amounts increased with increasing specific surface area (typical result for all hydrophilic oxides) but it was stronger for rutile than for anatase (Figure 2.70). Notice that this difference can be caused by the textural differences of the materials but not only by the phase composition. The partial molar enthalpy of adsorption levels off at a value that is indeed the heat of condensation of gaseous water (about −44 kJ/mol) for high water content (Figure 2.71a). At low water content, the heat of adsorption becomes much more exothermic. The partial pressure dependence of the water content gives the chemical potential as well. Therefore, one has the entropies of adsorption, which become strongly negative (Figure 2.71b) as the enthalpies of adsorption become strongly exothermic. These strongly negative enthalpy and entropy values provide evidence that a significant fraction of the first portion of water adsorbed onto titania (as well as on other oxides) is dissociated into OH groups. Those OH groups are strongly bound, so they are not mobile on the surface. This is not the same case as dissociating water into ions in solution. So, in fact, the entropy change is significantly negative because of the formation of strong bonds (Navrotsky 2007).

Hagaman et al. (2012) studied interaction of benzoic acid with metal oxides using solid-state ^{17}O NMR spectroscopy. Complexes formed by dry benzoic acid with mesoporous silica and nonporous titania and alumina were analyzed. Chemical reactions with silica were not observed, but the behavior of benzoic acid on silica was a function of the water content. The acid was characterized by high mobility as evidenced by a liquid-like, Lorentzian NMR resonance. Excess benzoic acid remained as the crystalline hydrogen-bonded dimer. Benzoic acid reacted with titania and alumina surfaces in equilibrium with air to form the corresponding titanium and aluminum benzoates. In both materials, the oxygen of the ^{17}O-labeled acid was bound to the metal, showing the bond

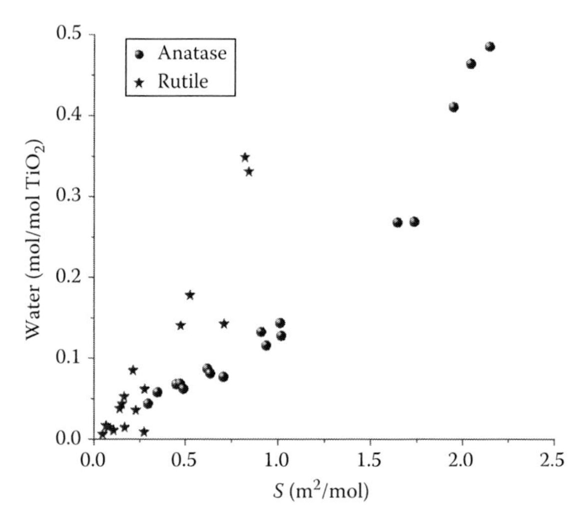

FIGURE 2.70 Dependence of bound water versus surface area of titania samples. (Adapted from *J. Chem. Thermodynamics*, 39, Navrotsky, A., Calorimetry of nanoparticles, surfaces, interfaces, thin films, and multilayers, 2–9, 2007, Copyright 2007, with permission from Elsevier.)

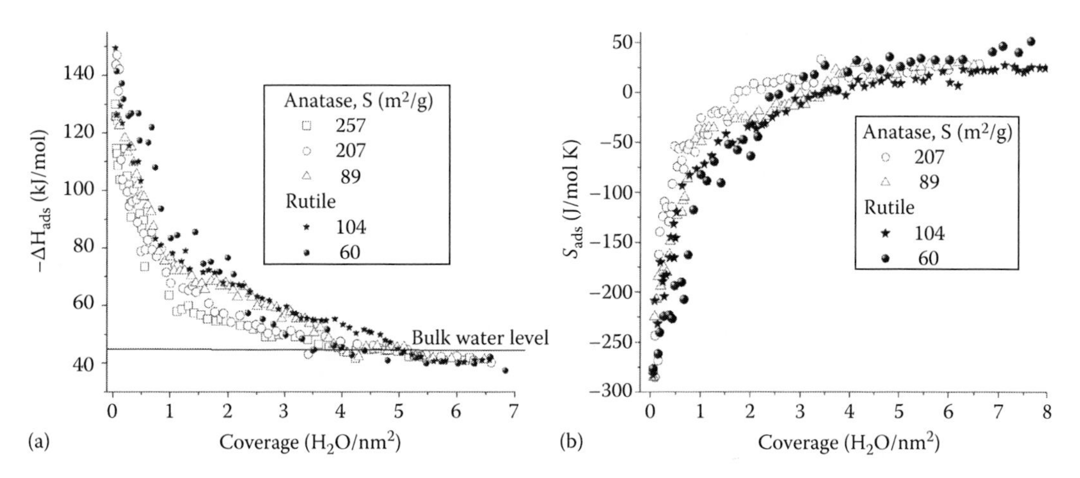

FIGURE 2.71 Dependence of partial molar (a) enthalpy and (b) entropy of adsorption of water against surface coverage of anatase and rutile of different specific surface area. (Adapted from *J. Chem. Thermodynamics*, 39, Navrotsky, A., Calorimetry of nanoparticles, surfaces, interfaces, thin films, and multilayers, 2–9, 2007, Copyright 2007, with permission from Elsevier.)

formation between oxygen deficient metal sites and the oxygen of the carboxylic acid. The ^{27}Al MAS NMR spectra confirmed this mechanism for the reaction with the alumina surface. Dry mixing of benzoic acid with alumina rapidly quenched pentacoordinate aluminum sites, excellent evidence that these sites were confined to the surface of the alumina particles (Hagaman et al. 2012). Anderson et al. (1996) studied the processes of isomorphous substitution of the Si atoms by Al and Ga in the nanoporous titanosilicate using the ^{29}Si, ^{27}Al, and ^{71}Ga MAS NMR spectroscopy. It was found that both Al and Ga substitute exclusively in tetrahedral Si sites in a manner that avoids Ti-O-Al and Ti-O-Ga linkages.

Huittinen et al. (2011) studied the specific sorption of Eu(III) and Y(III) on γ-Al_2O_3 with solid-state ^{1}H NMR spectroscopy (Figure 2.72). Solution pH affected the ^{1}H MAS NMR spectra (800.13 MHz, 18.8 T, 18 kHz rotor) of γ-alumina (63 m^2/g, IEP ~ 8.3) and the ion adsorption; therefore, controlled pH was 8.00 ± 0.05. The metal ion concentration was varied between 6.58×10^{-7}–3.95×10^{-4} M Y(III)

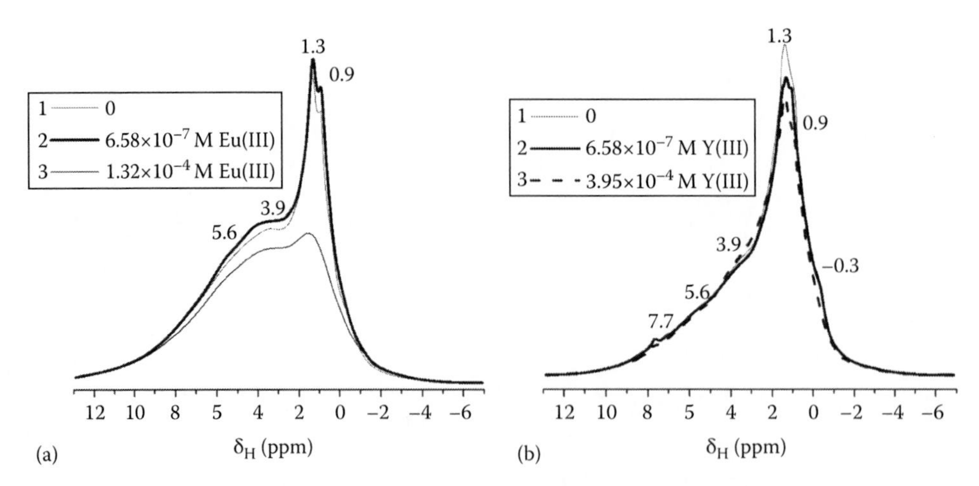

FIGURE 2.72 ^{1}H NMR spectra of (a) Eu(III) and (b) Y(III)-containing γ-alumina samples (normalized to same mass, 1 mg). (Adapted from *J. Colloid Interface Sci.*, 361, Huittinen, N., Sarv, P., and Lehto, J., A proton NMR study on the specific sorption of yttrium(III) and europium(III) on gamma-alumina [γ-Al_2O_3], 252–258, 2011, Copyright 2011, with permission from Elsevier.)

and 6.58×10^{-7}–1.32×10^{-4} M Eu(III). After separation of the liquid phase, the samples were dried under vacuum to remove physisorbed water. However, even after 48 h of drying at 150°C and 20 mTorr, water was still detected in the NMR spectra as two distinct peaks at $\delta_H = 1.3$ and 0.9 ppm (Figure 2.72). These peaks can be assigned to WAW according to our classification of bound water (Gun'ko et al. 2005d, 2009d) or non-hydrogen-bonded OH groups.

The Eu addition to the γ-alumina samples demonstrated significant spectral changes in comparison with Y-containing samples. These changes were attributed to the paramagnetism of europium rather than to complexation reactions occurring on the mineral surface. The ^{1}H NMR spectra of yttrium samples were therefore used to detect the spectral changes induced by the sorption reaction itself. The results revealed a large distribution of protons being removed from the mineral surface upon yttrium complexation. Removed protons were assigned to both bridging surface hydroxyls such as $(Al_{VI})_2$–OH as well as to terminal hydroxyls, e.g., of type Al_{VI}–OH. Acidic protons belonging to $(Al_{VI})_3$–OH groups were not observed to participate in the surface reaction. The shoulder at −0.3 ppm can be assigned to the most basic OH groups Al_{IV}OH. The most acidic bridging groups have $\delta_H = 2.4$ ppm $(Al_{VI})_3$OH and three additional types of the OH groups have $\delta_H = 1.4$–1.5 $(Al_{VI})_2$OH, 0.9 (Al_{Vi}O(H)Al_{IV}), and 0.0–0.1 (Al_{VI}OH) ppm. The 5.6 and 3.9 ppm features decreased as a function of increasing pH, while the opposite effect was for the −0.3 ppm feature, and 7.7 ppm was observed only at higher drying temperature (Figure 2.72). Huittinen et al. (2011) assigned the structures at high δ_H values to charged sites with attached protons. However, these structures can be also assigned to surface hydroxyls with strongly H-bonded water molecules. Notice that static LT ^{1}H NMR spectra could add needed information on the presence of mobile water there.

Metal-semiconductor phase transition (MSPT) in VO_2 reflected in the interfacial behavior of bound water because of changes in the adsorptivity of the materials with respect to water and efficiency of nuclear magnetic relaxation (Turov et al. 1995a). Therefore, the signal width and intensity of bound water strongly changed. However, these materials were not structurally characterized using adsorption or/and cryoporometry methods.

Leboda et al. (1999b) synthesized titania onto a silica gel matrix of a relatively large porosity ($V_p = 0.91$ cm^3/g) but not large S_{BET} value (278 m^2/g). To avoid strong blocking of silica gel pores by CVD-titania nanoparticles, the TiO_2 content was relatively low (3.2, 4.5 and 7.7 wt%). Water bound in pores of the initial silica gel was characterized by greater $\gamma_S = 160$ mJ/m^2 and lower free energy in the first layers (Figure 2.73). Both γ_S (90 and 130 mJ/m^2 for SGT2 and SGT3, respectively) and

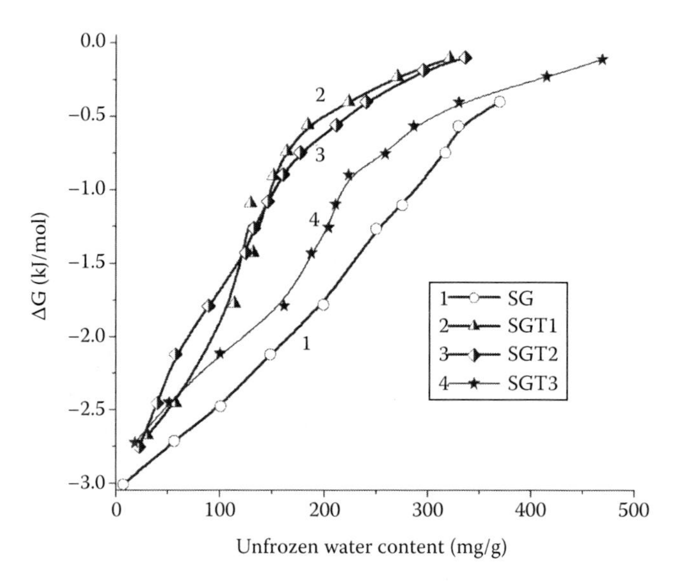

FIGURE 2.73 Relationships between unfrozen water amounts and changes in the Gibbs free energy of water bound to (1) silica gel and CVD-TiO_2/SiO_2 at C_{TiO_2}=(2) 3.2, (3) 4.5 and (4) 7.7 wt%.

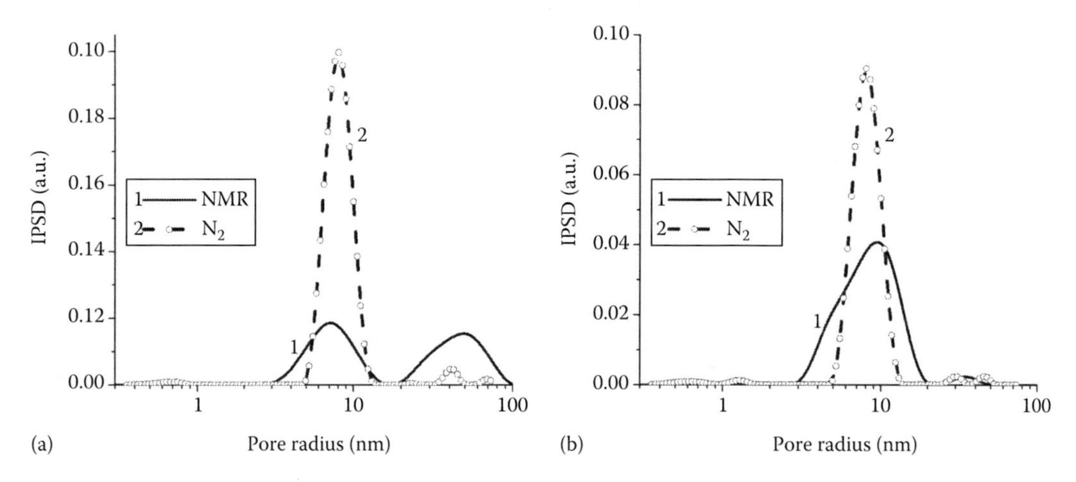

FIGURE 2.74 Pore size distributions calculated from the nitrogen adsorption isotherms and the ^{1}H NMR data (IGT) for unfrozen water for (a) silica gel and (b) CVD-TiO_2/silica gel (C_{TiO_2}=7.7 wt%).

ΔG values are nonlinear functions of the titania content because of several effects caused by textural changes of the materials and the formation of new phases (amorphous titania, anatase, and rutile) with particles of different sizes formed in pores and at the outer surface of the silica gel particles. Water becomes more strongly bound with increasing titania content (Figure 2.73). However, the $\Delta G(C_{uw})$ curves for all the SGT samples are above the curve for silica gel. The secondary porosity of silica gel decreases due to grafting of titania, and the peak intensity at $R=40$ nm of textural pores decreases (Figure 2.74), but a peak at $R=30$ nm appears because of partial filling of these broad pores by titania nanoparticles.

The shape of the main mesopore peak does not change but its intensity slightly decreases, however, the position remains the same (both PSDs maxima are at $R=8.04$ nm). This suggests that the main amount of titania is located out of these mesopores. Partial filling of broad pores at $R>20$ nm by titania particles reduces the amounts of water located there (Figure 2.74). The NMR PSD shows narrower pores than N_2 PSD because of non-simultaneous melting of ice in broad pores; i.e., in these pores, both unfrozen water and ice crystallites can be present at certain temperatures. Despite this, there is appropriate agreement between the PSDs calculated using two methods, and parallel analysis of them gives additional information on complex titania–silica materials (Figures 2.73 and 2.74). Notice that the formation of a CVD-phase with titania (or other materials) can faster occur out of narrow mesopores, e.g., in macropores. This is typical situation for CVD-MO_x formed on different silicas resulting in the formation of relatively large particles (much larger than the average size of the main mesopores) on the outer surface of silica gel granules (Leboda et al. 1999a,b, 2000, Gun'ko et al. 2000d,e, 2002a, Borysenko et al. 2005).

Babonneau and Maquet (2000) described structural features of siloxane–oxide hybrid systems prepared by sol-gel processes using solid-state NMR techniques. The ^{29}Si and ^{17}O NMR spectra can give unique information on the formation and stability of co-condensing species. For the final gels, solid-state ^{29}Si MAS, used in conjunction with CP and related two-dimensional ^{29}Si–^{1}H heteronuclear NMR correlation techniques, provided detailed information on dipole–dipole coupled ^{29}Si and ^{1}H species among different molecular moieties. The ^{17}O MAS NMR with MQ MAS sequence provided a direct identification of the various oxo bridges (Figure 2.75) in the gels (Babonneau and Maquet 2000). Notice that the ^{29}Si NMR spectra of ST gels can be characterized by broad lines over Q_2, Q_3, and Q_4 sites dependent on both reaction conditions and precursor composition (Jung 2001).

Cherry et al. (2004) studied crystalline silicotitanate (CST), $HNa_3Ti_4Si_2O_{14} \cdot 4H_2O$, and the Nb-substituted CST (Nb-CST), $HNa_2Ti_3NbSi_2O_{14} \cdot 4H_2O$, as highly selective Cs^+ sorbents appropriate for the selective removal of radioactive species from nuclear waste solutions. The structural basis for the improved Cs^+ selectivity in the niobium analogs was investigated using MAS NMR spectroscopy. Changes in the local environment of the Na^+ and Cs^+ cations in both CST and Nb-CST materials as a function of weight percent cesium exchange were investigated using ^{23}Na and ^{133}Cs MAS NMR spectroscopy. Framework changes, induced by Cs^+ loading and hydration state, were investigated with ^{29}Si MAS NMR. The relative population of different Cs^+ environments varied with the extent of Cs^+ loading. Marked changes in the framework Si environment were noted with the initial incorporation of Cs^+; however, with increased Cs^+ loading the impact to the Si environment was less pronounced. The Cs^+ environment and Si framework structure were influenced

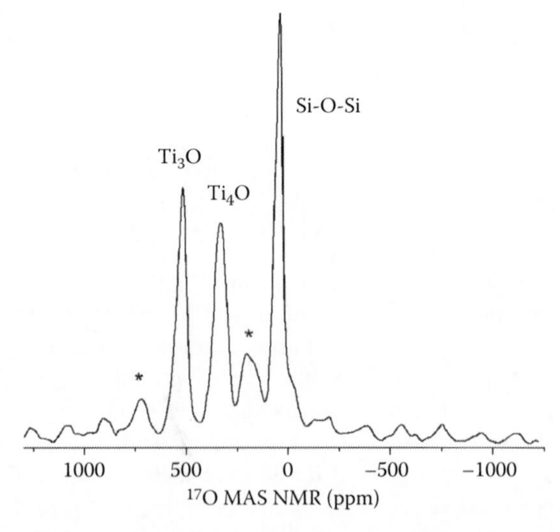

FIGURE 2.75 ^{17}O MAS NMR spectrum of a dried silica/titania gel prepared at Si/Ti molar ratio 70/30, a rotor frequency of 12.5 kHz; *spinning side bands. (Adapted from *Polyhedron*, 19, Babonneau, F. and Maquet, J., Nuclear magnetic resonance techniques for the structural characterization of siloxane–oxide hybrid materials, 315–322, 2000, Copyright 2000, with permission from Elsevier.)

by the Nb-substitution and were greatly affected by the amount of water present in the materials. The increased Cs^+ selectivity of the Nb-CST materials arises from both the chemistry and geometry of the tunnels and pores (Cherry et al. 2004).

Pilkenton and Raftery (2003) investigated photocatalytic oxidation of ethanol over a series of SnO_2-based photocatalysts using solid-state NMR spectroscopy. The adsorption of ethanol on SnO_2 powder was studied using both cross-polarization ^{13}C NMR and REDOR techniques and showed the formation of two surface ethanol species, hydrogen-bonded ethanol at surface hydroxyl groups, and ethanol chemisorbed to the SnO_2 surface. The ^{13}C NMR spectroscopy of the adsorbed ethanol was used to characterize the surface of monolayer SnO_2–TiO_2 coupled photocatalysts supported on porous Vycor glass. These studies showed that the photooxidation of ethanol over the monolayer photocatalysts was slower than that over a supported TiO_2 monolayer photocatalyst due to the build-up of reaction intermediates such as acetic acid on the catalyst surface. The ^{119}Sn NMR spectra were used to characterize the tin species on the porous Vycor glass support (Pilkenton and Raftery 2003).

Hensen et al. (2010) used two different silica matrices such as precipitated Sipernat-50 (Degussa) and fumed silica (VWR) at S_{BET}=400 and 390 m^2/g, respectively, to synthesize alumina/silica at 5–20 wt% of precipitated alumina. The ^{27}Al MAS NMR spectra were recorded at different stages of preparation, treatment and adsorption measurements. Calcination results in such competing process as the diffusion of aluminum into the silica matrix and sintering of alumina into separate patches mainly consisting with sixfold O-coordinated Al. The materials exhibited Brønsted acidity similar to amorphous SA prepared by grafting of alumina on silica gel. The acidity was not varied systematically with the Al content but it increased with the calcination temperature. The active sites form due to the diffusion of aluminum into the silica network at high temperatures, leading to Al substitutions of Si atoms and formation of SiO(H)Al structures responsible for the Brønsted acidity. In this paper, Hensen et al. (2010) did not study the textural characteristics of the materials, interfacial phenomena, crystallinity, particle size distributions, etc. Therefore, despite detailed analysis of the ^{27}Al MAS NMR spectra of the materials, a lot of unclear aspects remained. Notice that LT MAS NMR experiments (150–160 K) with wet gels frozen directly in the NMR rotor can give more appropriate results than the measurements at ambient conditions (König et al. 2007).

Fitzgerald et al. (1997) studied the dealumination of kaolinite by the solid/solution "interfacial" reaction with HCl(aq) at 98°C using solid-state ^1H CRAMPS and MAS ^{29}Si NMR spectroscopy. The single-pulse (SP)/MAS ^{29}Si NMR spectra of kaolinite-derived solids (2–83% dealuminated) showed the dependence on the degree of dealumination for resonances at −89, −100, and −109 ppm. The −89 ppm resonance corresponds to Q_3-type sites with $Si(OSi)_3(OAl_2)$ in unreacted kaolinite, while the latter two resonances were assigned to Q_3-type silica–alumina (Si–O(H^+)–ype silica–a4-type amorphous silica ($Si(OSi)_4$) sites contained in dealuminated kaolinite solids following partial dealumination. The ^1H CRAMPS spectra of these dealuminated solids were dependent upon the degree of removal of Al^{3+}ions from the kaolinite Al–O(H)–Al layer. In addition to a broad proton peak at 4.0 ppm due to structural protons of the intact Al–O(H)–Al layer in unreacted kaolinite, two new resonances at 0.4 and 7.0 ppm were assigned to protons of the silanol (≡Si–OH) and alumina–silica (≡Si–O(H)–Al≡) sites. The spectral intensities of the SP/MAS ^{29}Si and ^1H CRAMPS NMR results were correlated with increasing dealumination, permitting the development of a structural model for partially dealuminated kaolinite. This structural model is consistent with a kinetic model involving a chemically controlled "heterogeneous" reaction process whereby H^+ ion attack of the Si–O–Al linkages at the edges of the mineral surface leads to liberation of aluminum ions into the solution medium (Fitzgerald et al. 1997).

The reactivity of the silanol groups on the zeolite surfaces was studied by Kawai and Tsutsumi (1998), Tanabe (1970), Kazansky (2003), and other authors. However, in the case of natural minerals characterized by a low specific surface area, there is a problem of a low content of surface functionalities (e.g., hydroxyls). Fry et al. (2003b) overcame the problem of low analyte concentration through NMR observation of the highly sensitive ^{19}F nuclei contained in a fluorinated

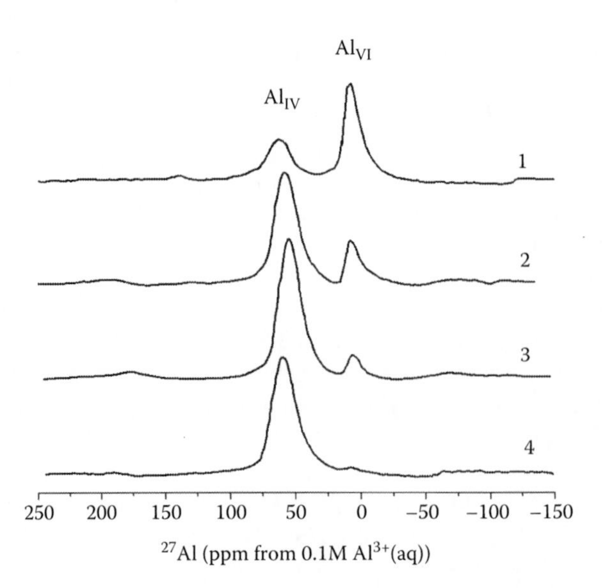

FIGURE 2.76 (1) ^{27}Al direct polarization MAS NMR spectra of clay weathering products and fines from the first sonication (2 min) of clay/rhyolite sample and subsequent sonication for (2) 10 min, (3) 2 h, and (4) 10 h. (Adapted from *Geochim. Cosmochim. Acta*, 72, Washton, N.M., Brantley, S.L., and Mueller, K.T., Probing the molecular-level control of aluminosilicate dissolution: A sensitive solid-state NMR proxy for reactive surface area, 5949–5961, 2008, Copyright 2008, with permission from Elsevier.)

organosilane used for surface modification of the materials. Using 3,3,3-(trifluoropropyl) dimethylchlorosilane (TFS) probe molecules they demonstrated the viability of quantifying the number of reactive hydroxyls on low surface area silicate and alumino[boro]silicate glass fibers with specific surface areas of <0.25 m^2/g (Fry et al. 2003a,b). Washton et al. (2008) studied a set of natural aluminosilicates using the ^{19}F and ^{27}Al NMR spectroscopy (Figure 2.76). The parent rhyolite contains only tetrahedrally coordinated aluminum (Al_{IV}) species, whereas the clay predominantly contains octahedrally coordinated aluminum (Al_{VI}) species, facilitating a quantitative analysis using ^{27}Al direct polarization solid-state NMR experiments (Figure 2.76). An increase in sonication time results in a decrease in the Al_{VI} signal at approximately 3 ppm arising from aluminum within clays contained in the sample (Washton 2007, Washton et al. 2008). Thus, TFS chemisorption, coupled with quantitative ^{27}Al MAS NMR spectroscopy, can be used to determine the extent of contamination of primary minerals with secondary clay minerals for weathered samples.

Isobe et al. (2003) studied different aluminum hydroxides using the NMR technique. Free waters and OH groups coupled with aluminum for amorphous samples were observed at ~5 and ~4.5 ppm (Figure 2.77), respectively, the latter peak being broader. This was consistent with the differential spectra between spin echo and transfer of populations in double resonance. They concluded that the subunits of AlO_4, AlO_5, and AlO_6 in amorphous aluminum hydroxides were bound through hydrogen bonds with a wide distribution of bonding strength.

Gunawidjaja et al. (2003) studied the effects of different heat treatments (i.e., successively or directly heated to particular temperatures) and atmospheres (air or nitrogen) on the solid-state ^{1}H, ^{13}C, ^{17}O, and ^{29}Si NMR spectra obtained from $(TiO_2)_{0.15}(ZrO_2)_{0.05}(SiO_2)_{0.80}$ sol–gel materials. The ^{29}Si MAS NMR spectra showed that the extent of condensation of the silica-based network strongly depends on the maximum temperature for a sample, but the condensation is largely independent of the details of the heat treatments and atmosphere used. The ^{17}O NMR results confirmed that a nitrogen atmosphere did significantly reduce loss of ^{17}O from the structure, but care must be taken since there could be differential loss of ^{17}O from the regions having different local structural characteristics. Stronger changes were observed for the ^{1}H NMR spectra due to loss of different forms of organics and bound water during heating (Figure 2.78).

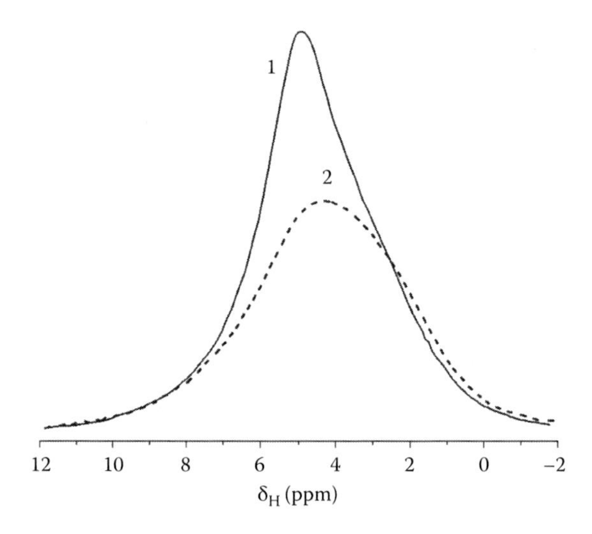

FIGURE 2.77 ^{1}H MAS NMR spectra measured at a spinning rate of 35 kHz and a magnetic field of 14.09 T for amorphous aluminum hydroxide (1) initial and (2) dried at 90°C for 90 min. (Redrawn from *J. Colloid Interface Sci.*, 261, Isobe, T., Watanabe, T., d'Espinose de la Caillerie, J.B., Legrand, A.P., Massiot, D., Solid-state ^{1}H and ^{27}Al NMR studies of amorphous aluminum hydroxides, 320–324, 2003, Copyright 2003, with permission from Elsevier.)

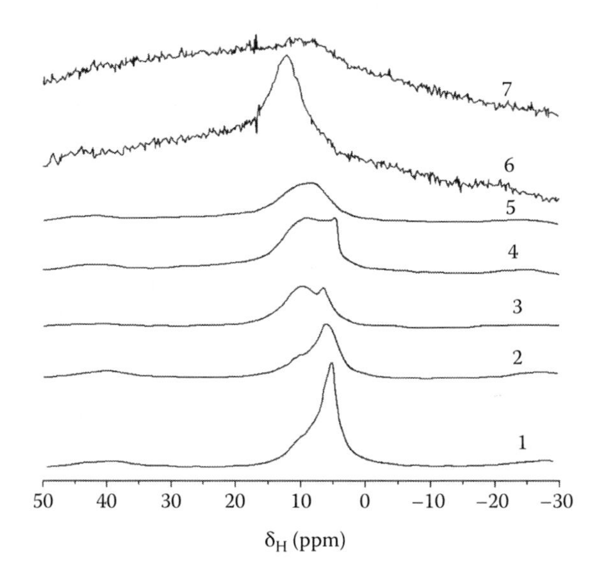

FIGURE 2.78 ^{1}H MAS NMR spectra of the $(TiO_2)_{0.15}(ZrO_2)_{0.05}(SiO_2)_{0.80}$ xerogels after various heat treatment: (1) unheated, direct 250°C (2) N_2 and (3) air, step 500°C (4) N_2 (×8) and (5) air (×8), and step750°C (6) N_2 (×64) and (7) air (×64). (Adapted from *Solid State Nucl. Magn. Reson.*, 23, Gunawidjaja, P.N., Holland, M.A., Mountjoy, G., Pickup, D.M., Newport, R.J., and Smith, M.E, The effects of different heat treatment and atmospheres on the NMR signal and structure of TiO_2–ZrO_2–SiO_2 sol–gel materials, 88–106, 2003, Copyright 2003, with permission from Elsevier.)

According to Gunawidjaja et al. (2003), in the dried but unheated xerogels, the ^{1}H NMR spectra are characterized by a resonance close to 6 ppm (Figure 2.78, curve 1). The authors assigned this band to protons in C–H bonds, along with a second, lower intensity peak at higher shift value. After heating to 250°C there is a difference between the samples step-heated under nitrogen (Figure 2.78; curve 2) and air (curve 3). Under the nitrogen atmosphere, the form of the spectrum is very similar to that at room temperature but with the total signal intensity reduced by 32% ± 3%. However, under air there is a 49% ± 3% loss of total signal and this difference in loss of intensity is largely due to the greatly reduced intensity of the 6 ppm peak in the air-heated sample. The remaining resonance observed at 10 ppm can be assigned to O–H bonds participating in the hydrogen bonds. Further heating causes even greater signal loss with only the –OH signal eventually remaining. A small signal of between 1% and 6% of the original signal intensity from this species exists even after heating to 750°C. For both atmospheres, stepped heat treatment (i.e., greater total heating time) leads to the removal of more OH groups than direct heating. For both regimes, heating under air leads to more OH loss than heating under nitrogen (Gunawidjaja et al. 2003). Notice that the ^{17}O NMR spectroscopy is frequently used to characterize mixed oxides because the oxygen atoms are sensitive to different surroundings (Bastow et al. 1993, Delattre and Babonneau 1997, van Eck et al. 1999, Babonneau and Maquet 2000, Chadwick et al. 2001, Gervais et al. 2001, Massiot et al. 2002, Gunawidjaja et al. 2003).

Buchholz et al. (2002) studied the thermal stability and dehydroxylation of nanoporous silicoaluminophosphates H-SAPO-11, H-SAPO-18, H-SAPO-31, and H-SAPO-34 heated at 773–1173 K. The crystallinity upon thermal treatment at 1173 K was not changed according to the XRD patterns. However, ^{1}H MAS NMR spectroscopy indicated a dehydroxylation of 40%–50% of the initially existing bridging OH groups (Figure 2.79).

The ^{27}Al and ^{29}Si MAS NMR spectra showed that the dehydroxylation was not accompanied by a dealumination but rather by the removal of silicon (desilication) in the local structures of the former bridging OH groups. The dehydroxylation and desilication of the silicoaluminophosphate framework did not lead to the formation of defect OH groups as evidenced by ^{1}H MAS NMR spectroscopy. To explain the high stability of the silicoaluminophosphates after dehydroxylation and desilication, a healing process was proposed, which is based on the migration of phosphorus atoms to framework vacancies and their transformation to $P(OAl)_4$ species. The ^{1}H MAS NMR spectra of dehydrated silicoaluminophosphates included signals at 1.7–1.8, 2.4, 3.7, and 4.7 ppm. The signals at 1.7–2.4 ppm are due to SiOH, POH, and AlOH groups at framework defects and on

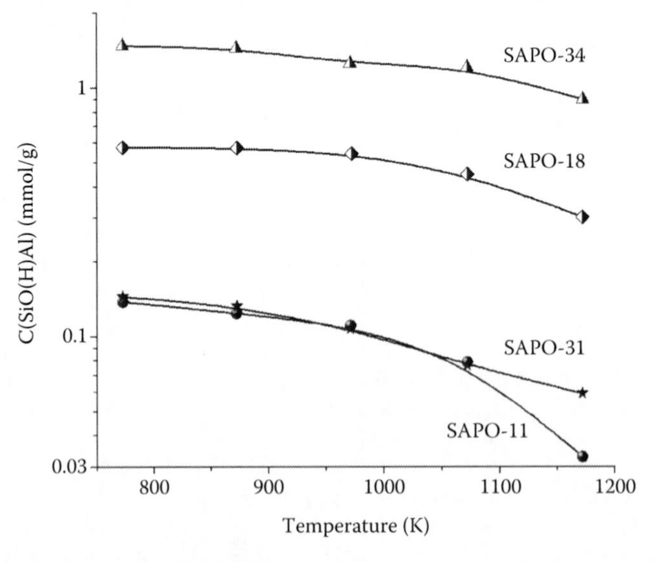

FIGURE 2.79 Concentration of bridging OH groups a function of the treatment temperature of silicoaluminophosphates H-SAPO-11, H-SAPO-18, H-SAPO-31, and H-SAPO-34 obtained by ^{1}H MAS NMR spectroscopy. (Adapted from *Micropor. Mesopor. Mater.*, 56, Buchholz, A., Wang, W., Xu, M., Arnold, A., and Hunger, M., Thermal stability and dehydroxylation of Brønsted acid sites in silicoaluminophosphates H-SAPO-11, H-SAPO-18, H-SAPO-31, and H-SAPO-34 investigated by multi-nuclear solid-state NMR spectroscopy, 267–278, 2002, Copyright 2002, with permission from Elsevier.)

the outer surface of the silicoaluminophosphate particles. Depending on the location of bridging OH groups in large pores and cages or in small six ring windows, the ^{1}H MAS NMR signals of these OH groups appear at 3.7 and 4.7 ppm, respectively (Buchholz et al. 2002).

Buchholz et al. (2003) studied the hydration and dehydration of silicoaluminophosphates H-SAPO-34 and H-SAPO-37 applying ^{1}H and ^{27}Al CF (continuous flow) MAS NMR spectroscopy under in situ conditions in a flow of nitrogen loaded with water vapor or in a flow of dry nitrogen. It was found that the hydration of H-SAPO-34 and H-SAPO-37 is composed of two successive steps: (i) hydration of Brønsted acidic bridging OH groups (SiO(H)Al) followed by (ii) coordination of water to aluminum atoms. In hydrated H-SAPO-34, the hydration-induced transformation of tetrahedrally coordinated framework aluminum atoms into octahedrally coordinated aluminum atoms was a reversible process at 298 K. In the case of hydrated H-SAPO-37, the dehydration of octahedrally coordinated aluminum atoms requires temperatures >373 K corresponding to a stronger bonding of the water molecules to the aluminum atoms than in H-SAPO-34. This indicates an immediate hydrolysis of the H-SAPO-37 framework. Since no defect SiOH groups were observed upon hydration and dehydration of H-SAPO-37 at 298 K, the hydrolysis of this materials started at ≡P–O–Al≡ bonds in the aluminophosphate regions of the framework. Upon hydration of H-SAPO-37 at $T > 353$ K, only a hydration of bridging OH groups occurs, and no hydrolysis of BP–O–AlB bonds responsible for a damage of the framework was found (Buchholz et al. 2003).

Nakayama et al. (2002) studied nanoporous silicoaluminophosphate using solid-state NMR spectroscopy. They found correlation between pore size (d) and the chemical shift of adsorbed ^{129}Xe ($d \sim 1/\delta_{iso}$), as well as the relationship between adsorption amount and isotropic chemical shift, δ_{iso}, and ^{129}Xe chemical shift anisotropy.

Thus, both NMR cryoporometry and solid-state NMR spectroscopy give useful textural and structural information on mixed porous oxides, and this information can be broadened and deepened with the use of DSC, TG, adsorption, FTIR, Raman, XPS, XRD, and other methods. The authors not always used obtained information for maximum comprehensive description of the materials in terms "structure—properties." Frequently, this is due to a small number of used techniques, as well as incomplete analysis of obtained experimental results.

2.3 STRUCTURALLY ORDERED OXIDES

Mixed structurally ordered mesoporous oxides can be interesting as catalysts (because ordered texture can positively influence the selectivity of catalytic processes) and catalyst (e.g., metal or metal oxide nanoparticles) supports. Meynen et al. (2009) verified synthesis methods for a large number of mesoporous silica and titania materials and some super-microporous (according to Dubinin's classification of pores, but it corresponds to nanopore range) materials (pore diameter 1.5–2 nm). They analyzed a set of material characteristics based on the most commonly applied characterization techniques (nitrogen sorption, XRD, transmission electron microscopy [TEM], scanning electron microscopy, NMR, etc.) for mesoporous materials. An aluminum-containing hydrated MCM-41 was characterized by more intensive ^{27}Al signal of fourfold O-coordinated Al_{IV} than Al_{VI} (Figure 2.80). These signals have broader (Al_{VI} at −5.22 and Al_{IV} at 51.5 ppm) and narrower (at −0.31 and 52.9 ppm) components, which can be due to less and more ordered alumina and alumina/silica structures.

Sixfold O-coordinated Al in hydrated aluminosilicates studied by Meynen et al. (2009) can be due to extra-framework aluminum compounds (i.e., alumina nanoparticles). If Al_{VI} is in polymeric aluminum oxides or oxide hydrates, a strong quadrupolar line broadening can occur owing to distortions of the octahedral symmetry. Figure 2.80 shows decomposed ^{27}Al MAS NMR spectrum of an aluminum-containing hydrated MCM-41. The broad background signal is due to the presence of polymeric aluminum oxides or oxide hydrates. Čejka (2003) analyzed the textural, structural, and catalytic properties of templated aluminas at S_{BET} values up to 800 m^2/g and pore size of

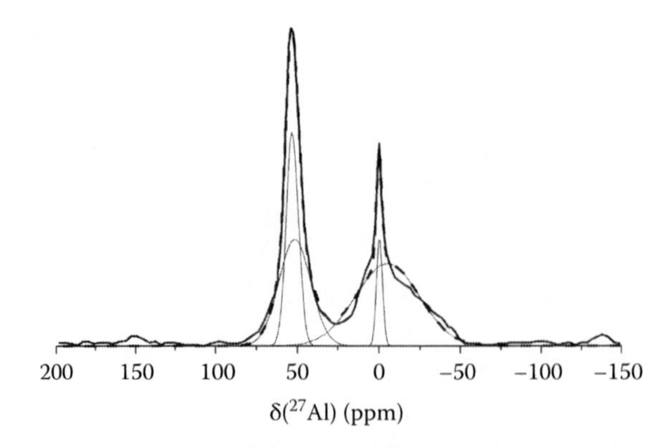

FIGURE 2.80 ^{27}Al MAS NMR spectrum of an aluminum-containing hydrated MCM-41 with signals at 53 and 0 ppm due to Al_{IV} and Al_{VI} species, respectively. (Taken from *Micropor. Mesopor. Mater.*, 125, Meynen, V., Cool, P., and Vansant, E.F., Verified syntheses of mesoporous materials, 170–223, 2009, Copyright 2009, with permission from Elsevier, and decomposed here.)

2–10 nm. The ^{27}Al MAS NMR spectra of initial and calcined mesoporous alumina (synthesized using "anionic" route) included besides the signal at 0 ppm a signal at 70 ppm and a weak signal at 35 ppm, which are temperature dependent. Schmücker et al. (1997) observed the most intensive ^{27}Al MAS NMR line at 30–32 ppm in the spectra of the SiO_2-Al_2O_3 and SiO_2-Al_2O_3-Na_2O glasses. They assigned this peak to distorted, tricluster-forming AlO_4 tetrahedra rather than to pentahedrally O-coordinated Al atoms. Peeters and Kentgens (1997) observed the temperature-dependent ^{27}Al MAS NMR signals at 0 (Al_{VI}), 30 (Al_V and Al_{IV} in strained polyhedrons), and 55 (Al_{IV}) ppm in Al-containing silica-based sol-gel. The content of Al_{VI} and Al_{IV} depend oppositely on the water amounts.

Ganapathy et al. (2004) studied molecular sieve ETAS-10 and showed that the silicon sites tetrahedrally connected to aluminum in framework positions of a molecular sieve can be identified by a selective reintroduction of the heteronuclear ^{27}Al–^{29}Si dipolar interaction through Rotational Echo Adiabatic Passage DOuble Resonance (REAPDOR) NMR. They used an effective dipolar dephasing of the Si–O–Al, over Si–O–Si, environments to identify silicon sites in the immediate vicinity of aluminum. The direct NMR estimation of Al–Si distance gave $r_{Al-Si} \approx 0.323$ nm. REAPDOR experiments in the aluminotitano-silicate ETAS-10 showed that a preferential aluminum substitution occurs at the siliceous Q_4 sites in the ETS-10 structure and confirms the Al–Ti avoidance.

Chen et al. (1999a) studied the coordination state of Al atoms in molecular sieve SAPO-37 using ^{27}Al multiple quantum (MQ) MAS NMR spectroscopy. The 1D ^{27}Al MAS NMR spectrum of the aluminum sites showed a complex pattern, and the 3Q MAS NMR spectra of SAPO-37 differently treated showed the presence of four different Al species: (i) tetrahedral Al sites with 4P; (ii) tetrahedral Al sites with one (or more) Si as nearest neighbors interacting with tetrapropylammonium; and framework Al coordinated with (iii) tetramethylammonium; or (iv) water. The ^{27}Al MAS NMR spectrum of the as-synthesized sample contains two peaks at 34 and 9 ppm, respectively, as well as a broad hump around 0 ppm. The peak at 34 ppm has been assigned to tetrahedral coordinated aluminum. By the transferred-echo double-resonance method, it was attributed to Al(4P) and Al(4Si) framework configurations. The 9 ppm peak was regarded by some authors as framework Al coordinated to TMA, while on the contrary, this line was also assigned to octahedral extra-framework Al (Chen et al. 1999). Fyfe et al. (1995) studying SAPO-37 molecular sieve by coherence-transfer and dipolar-dephasing solid-state NMR method showed that Al_{IV} with the symmetrical surroundings is characterized by a sharp peak at 37.4 ppm. In the case of asymmetrical environments, a broad resonance with a large anisotropy (over 300 ppm) is at 7.4 ppm.

Chen et al. (2007) modified Si-MCM-41 by depositing a heteropolyacid $H_3PW_{12}O_{40}$ on the surface and by introducing Zr^{4+} ions into the silica framework (Figure 2.81).

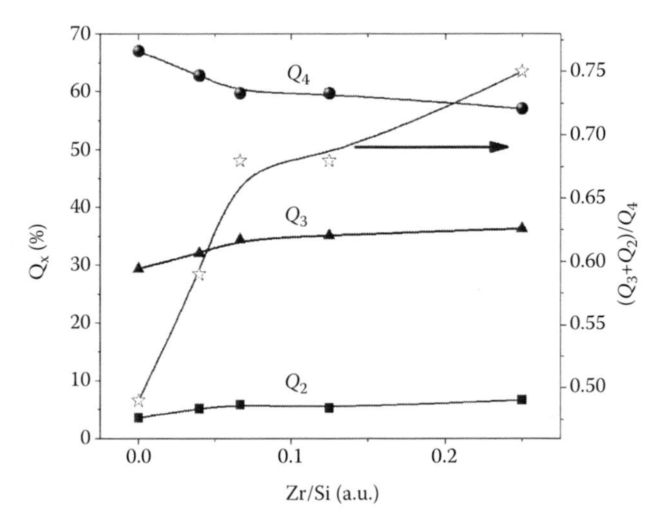

FIGURE 2.81 Changes in the content of Si states: Q_4 (−115 ppm), Q_3 (−103 ppm), and Q_2 (−92 ppm) in WSZ_n materials calculated here on the basis of the ^{29}Si MAS NMR spectra. (Published by *J. Solid-State Chem.*, 180, Chen, L.F., Wang, J.A., Noreña, L.E. et al., Synthesis and physicochemical properties of Zr-MCM-41 mesoporous molecular sieves and Pt/$H_3PW_{12}O_{40}$/Zr-MCM-41 catalysts, 2958–2972, 2007, Copyright 2007, with permission from Elsevier.)

The Zr-modified Si-MCM-41 mesoporous materials (referred as WSZ_n, n = Si/Zr = 25, 15, 8, and 4) were prepared using the template synthesis and fumed silica as a precursor. After impregnation with 25 wt% of $H_3PW_{12}O_{40}$, the surface Brønsted acidity of the Pt/$H_3PW_{12}O_{40}$/WSZ_n catalysts was enhanced by 2–10 times relative to the bare WSZ_n support. Two kinds of supported heteropolyacids were formed: (i) bulk-like heteropolyacid crystals with unchanged Keggin structures and (ii) highly dispersed heteropolyacid with distorted Keggin units (Chen et al. 2007). The formation of various kinds of heteropolyacid structures is closely related to the interaction between the heteropolyanions and the hydroxyl groups in the host support. Solid-state ^{31}P and ^{29}Si MAS-NMR spectra (9.4 T, 7.5 kHz rotor) were used for analysis of the complex materials (Figure 2.81).

Shouro et al. (2000) modified mesoporous silica FSM-16 with various oxides. The BET areas of oxides (8 wt%) supported on FSM-16 were smaller than that of silica alone (1065 m^2/g): Fe_2O_3/FSM-16, CdO/FSM-16, and Al_2O_3/FSM-16 have S_{BET} = 979, 796, and 546 m^2/g, respectively. Intensities of all the XRD peaks of the catalysts were less than those of FSM-16. Thus, the pore structure of the catalysts was slightly destroyed by the impregnation process.

Oxide nanotubes are a new class of ordered materials. Significant efforts in titania research have been even more intensified by the finding of self-organized nanotubular oxide architectures that can be prepared by a simple but optimized anodization of Ti metal surfaces (Ou and Lo 2007, Nah et al. 2010). The nanotubular geometry provides large potential for enhanced and novel functional features. Jung et al. (2004) used ^{11}B NMR spectra and the spin lattice relaxation times, T_1, to analyze the structure of boron nitride nanotubes (BNNTs), grown by CVD, and found two components arising from the hexagonal and rhombohedral phases in the BNNTs.

Confined space effects in nanotubes were studied for lipid bilayers using solid-state NMR spectroscopy (Gaede et al. 2004, Lorigan et al. 2004, Chekmenev et al. 2005, Karp et al. 2007). Aligning lipid bilayers in nanoporous anodized aluminum oxide (AAO) is a new method, which can be used to study membrane proteins by NMR spectroscopy and other methods. This method allows to maintain hydration, sample stability, and compartmentalization over long periods of time and to easily change solvent composition. Such phospholipid as 1,2-dimyristoyl-sn-glycero-3-phosphatidylcholine was used in these studies with AAO. ^2H and ^{31}P solid-state NMR spectroscopy was used to analyze the alignment of lipid molecules and compare the efficiency of alignment.

Three lipids were studied in three different AAO at 200, 100, and 20 nm in pore diameter. The ^{31}P spectrum becomes narrower with decreasing size of the pores. In the case of 20 nm AAO, the spectrum demonstrates down-field shift by 2.4 ppm (Karp et al. 2007). Unfortunately, NMR cryoporometry was not applied for these systems; however, one can assume that results could be very interesting.

2.4 NANOCRYSTALLINE AND MICROCRYSTALLINE MATERIALS

Metal, nonmetal, and oxide nanoparticles are attracting increasing attention with the growth of interest in nanotechnological disciplines, especially in medicine (drug, adsorbents, drug delivery, and MRI), optics, data storage, catalysis, magnetic fluids, etc. (Shipway et al. 2000, Seayad and Antonelli 2004, Evanoff and Chumanov 2005, Klingshirn 2007, Kinge et al. 2008, Su et al. 2010). The physical (electronic and optical) and chemical (reactivity) properties of nanoparticles can strongly differ from those of bulk solids because of size-dependent changes in the electronic structure, free surface energy, Laplace pressure, surface bond length, and other characteristics. These changes can be stronger in the materials with delocalized electrons (metals and carbons) and a narrow or zero bandgap (i.e., conductors or semiconductors) than in the materials with strongly localized electrons (oxides and polymers) and a broad bandgap (dielectrics) because of the strong sensitivity of the electrons to the dimensional and boundary effects. Nanoparticles tend to form such secondary structures as ordered assemblies or random aggregates. 1D, 2D, and 3D assemblies of nanoparticles have a great potential of applications (Eberhardt 2002, Gutsch et al. 2002, Lu et al. 2006, Kinge et al. 2008, Narayanan and Sakthivel 2011).

O'Donnell et al. (2008) characterized the chain conformation and dynamics of hydrocarbon $CH_3(CH_2)_nCOOH$ and perfluorocarbon ($CF_3(CF_2)_nCOOH$) fatty acids (n=6–18) adsorbed on zirconia nanoparticles ($d \approx 4$ nm in size) using solid-state ^{13}C chemical shift and ^{19}F NMR relaxation measurements, respectively (Figure 2.82). They compared obtained results for nanozirconia (140 m^2/g) and lower surface area (40 m^2/g) fumed metal oxide powders. Notice that from the

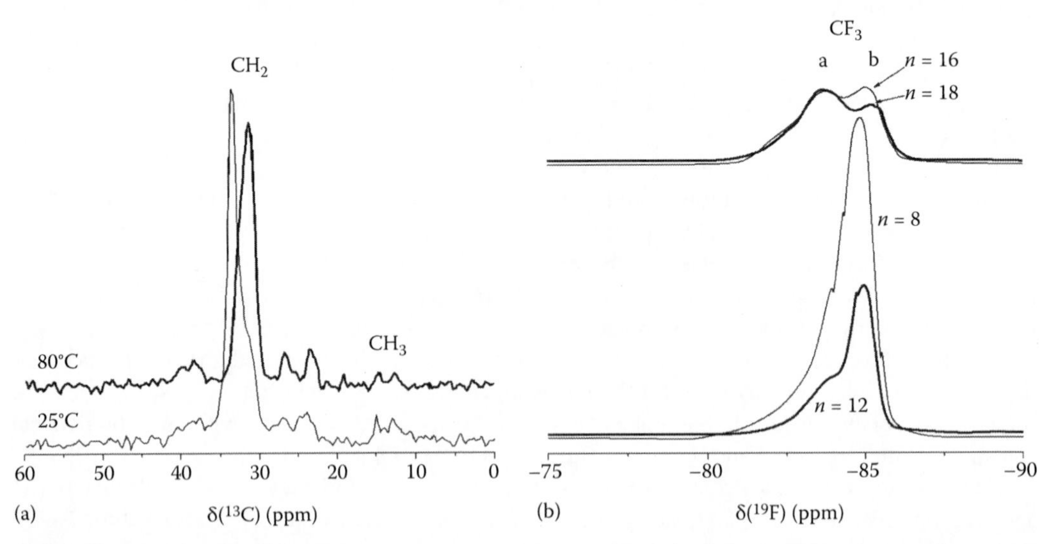

FIGURE 2.82 (a) ^{13}C MAS NMR spectra of the CH_2 and CH_3 groups of fatty acid $CH_3(CH_2)_{17}COOH$ adsorbed on ZrO_2, where the chain disordering transition temperature is at~55°C; (b) 19F MAS NMR spectra of adsorbed perfluoro acids $CF_3(CF_2)_nCOOH$ at different n values (a and b symbols correspond to adsorbed and bulk acids). (Adapted with permission from O'Donnell, A., Yach, K., and Reven, L., Particle–particle interactions and chain dynamics of fluorocarbon and hydrocarbon functionalized ZrO_2 nanoparticles, *Langmuir* 24, 2465–2471, 2008, Copyright 2008 American Chemical Society.)

specific surface area, the average diameter of zirconia nanoparticles should be 7.5 nm (estimated using simple equation $d=6000/(S_{BET}\rho)$ at $S_{BET}=140$ m^2/g and the true density $\rho=5.68$ g/cm^3) but not 4 nm (given by the authors).

This larger d value is confirmed by asymmetry of the particle size distribution function based on the TEM image analysis (O'Donnell et al. 2008). Additionally, before the adsorption measurements, nanozirconia should be heated/degassed that could result in strong particle aggregation. The interdigitation of fatty acid chains between neighboring particles, which increases with chain length, was analyzed from the splitting of the ^{13}C NMR and ^{19}F NMR signals of the CH_3 and CF_3 groups, respectively. This interdigitation allowed for efficient chain packing despite the high surface curvature similar to the case of alkanethiol self-assembled monolayers on gold nanoparticles. The hydrocarbon chains on nanozirconia were more ordered, and the reversible chain length dependent order–disorder transition temperatures were elevated relative to those of the same fatty acids adsorbed on fumed ZrO_2 powder (O'Donnell et al. 2008).

Parneix et al. (2009) published a very interesting paper in respect to the interfacial phenomena at the surface of colloidal nanosilica (two colloidal silicas were synthesized at the average diameter of particles of 13 and 16.6 nm). Aggregation of these dispersions was triggered by addition of an aqueous "flocculant" solution. This solution was prepared by partial neutralization of an aluminum chloride solution ($[AlCl_3]=0.25$ M) with sodium hydroxide ([NaOH]$=0.25$ M) at 70°C. The final ratio $r_{OH/Al}=[OH]/[Al]$ was 2.4, and the final pH was 4.5. Under these conditions, it is known that the hydrolysis of the aluminum salt leads to the formation of soluble polynuclear species, among which the polycation $[Al^{IV}O_4Al_{12}{}^{VI}(OH)_{24}(H_2O)_{12}]^{7+}$ (abbreviated Al_{13}) was the predominant one. The structure of this polycation consists of 12 sixfold O-coordinated aluminum ions, surrounding a central aluminum ion with tetrahedral O-coordination. At the hydrolysis ratio $r_{OH/Al}=2.4$, the Al_{13} units can be either isolated or slightly aggregated, yielding linear-shaped clusters. Parneix et al. (2009) studied interactions of Al_{13} polycations with silica in aqueous suspensions at different pH values and different content of the cations (Figure 2.83). The silica nanoparticles were aggregated through the addition of the Al_{13} polycations and then submitted to osmotic compression. The structures of these dispersions have been determined using SANS, before and after compression. Some dispersions consisted of a mixture with aggregated and nonaggregated nanoparticles, the resistance to osmotic compression was due to the ionic repulsions of the nonaggregated particles (Figure 2.83a and b).

The compression law that related the applied osmotic pressure Π to the silica volume fraction Φ was $\Pi\sim[\Phi/(1-\Phi)]^2$. Other dispersions fully aggregated the resistance to compression was due to surface–surface interparticle bonds. The application of low osmotic pressures (<50 kPa) resulted in compression at macroscopic scales only (>300 nm), while the structure of the network at local and mesoscopic scales was unchanged. Accordingly, few interparticle bonds were broken, and the deformation was primarily elastic. The compression law for this elastic deformation was in agreement with the predicted scaling law $\Pi\sim\Phi^4$. The application of higher osmotic pressures (>50 kPa) resulted in compression at macroscopic and mesoscopic scales (30–300 nm), while the local structure was still retained. Accordingly, many more interparticle bonds were broken. The compression law for this plastic deformation was in agreement with a scaling prediction of $\Pi\sim\Phi^{1.7}$. The location of the elastic–plastic transition indicated that the strength of the interparticle bonds was on the order of five times the thermal energies at ambient temperature (Parneix et al. 2009). They performed NMR experiments to determine the chemical environment of the adsorbed Al atoms. For this purpose, silica dispersions were equilibrated with different concentrations of the aqueous solution containing Al_{13} as for the determination of adsorption isotherms. They were then centrifuged, separated from the supernatant, freeze-dried, and examined through ^{27}Al MAS NMR (Figure 2.83d). Contribution of Al^{VI} species increases with increasing concentration of Al_{13} in the dispersions resulting in stronger aggregation of the particles (Figure 2.83b).

The interactions of Al_{13} with the surface of colloidal silica can occur in two stages. First, the Al_{13} polycations react with the surface silanols, and the Al atoms change their coordination to form

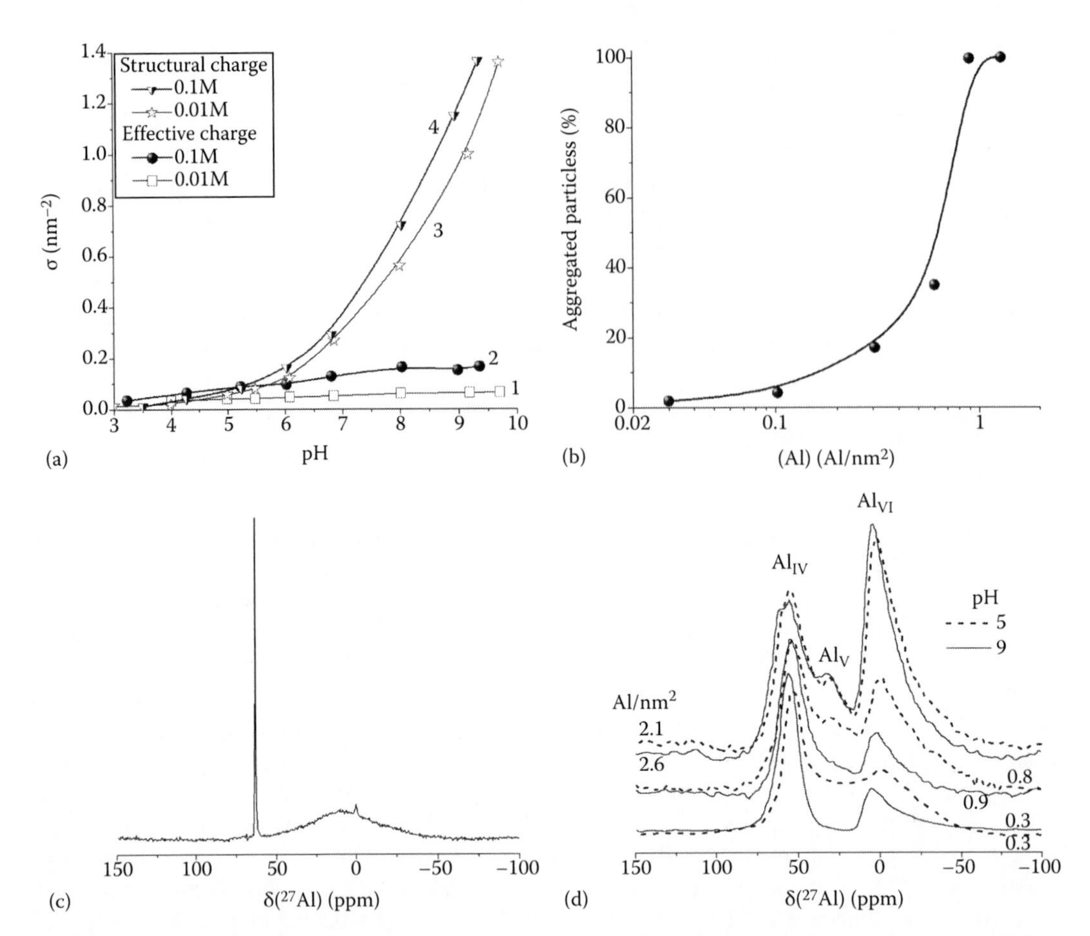

FIGURE 2.83 (a) Comparison of the effective charge number density (number of counterions in the diffuse layer, per nm^2 of silica surface) calculated from electrophoretic mobility measurements at ionic strengths 0.1 and 0.01 M with the structural charge number density determined from potentiometric titrations performed at the same ionic strengths (4 and 2); (b) Fractions of aggregated particles, according to the fit of the experimental structure factors by the calculated ones; (c) ^{27}Al liquid-state NMR spectrum of the flocculant solution ([Al] = 0.072 M as measured by inductively coupled plasma spectroscopy) showing the intensive peak of tetrahedral Al states at 63 ppm; (d) ^{27}Al MAS NMR spectra of powders obtained from the reaction of silica particles with Al_{13} polycations at pH 5 or 9 (the amounts of reacted aluminum atoms were calculated from the adsorption isotherms, and the spectra have been scaled to a constant height for the peak at 53–55 ppm, corresponding to Al atoms in a tetrahedral coordination in aluminosilicates). (Adapted with permission from Parneix, C., Persello, J., Schweins, R., and Cabane, B., How do colloidal aggregates yield to compressive stress? *Langmuir*, 25, 4692–4707, 2009, Copyright 2009 American Chemical Society, curves in (d) were smoothed with FFT.)

a layer of negatively charged tetrahedral aluminosilicate sites on the silica surface corresponding to a broad peak at 53 ppm (Figure 2.83d). Additional Al_{13} polycations adsorbing on this negative aluminosilicate surface layer compensate its charge (increase of the 2 ppm resonance and appearance of the 63 ppm peak that is characteristic of the central tetrahedral Al atom of Al_{13} polycation). The 32 ppm resonance observed on the upper spectra probably reveals the presence in the Al_{13} polycations of Al atoms with a tetrahedral configuration that was distorted due to the removal of water by freeze-drying (Parneix et al. 2009).

Le Feunteun et al. (2011) showed that coupling NMR 1D-imaging with the measure of NMR relaxation times and self-diffusion coefficients can be a powerful tool to investigate fluid infiltration into porous media. They studied very slow seeping of pure water into hydrophobic materials with three model samples of nuclear waste conditioning matrices, which consist in a dispersion of $NaNO_3$ (highly soluble) and/or $BaSO_4$ (poorly soluble) salt grains embedded in a bitumen matrix.

Bräuniger et al. (2004) demonstrated that the use of fast amplitude-modulated radio-frequency (RF) pulse trains with constant (FAM-I) and incremented pulse durations (SW–FAM) can provide considerable sensitivity enhancement for the central-transition signal (via spin population transfer from the satellite transitions) for solid-state NMR spectra of titanium, ^{47}Ti (5/2) and ^{49}Ti (7/2). For the MAS spectra of TiO_2 and $BaTiO_3$; the intensity of the ^{49}Ti central-transition line was more than doubled compared to simple Hahn-echo acquisition, while for the static case, enhancement factors of 1.6 (TiO_2) and 1.8 ($BaTiO_3$) were obtained. No lineshape distortions were observed in either MAS or static spectra of both compounds. Employment of the FAM and SW–FAM sequences should be useful in the routine acquisition of ^{47,49}Ti spectra, as the NMR signal can be detected much faster (Bräuniger et al. 2004). Notice that the ^{47,49}Ti band width for anatase obtained by Bräuniger et al. (2004) was much narrower than that published by Labouriau and Earl (1997).

Disperse vanadium dioxide was studied in a matrix of polyethylene glycol (PEG; 1500 Da) doped with tetraethylammonium bromide (TEABr) using ^{1}H NMR spectroscopy (Turov et al. 2001). It has been established that under the influence of the dopant, the heating of a sample up to the temperature of the metal-semiconductor phase transition (MSPT) results in a phase inhomogeneity of the sample: one portion of a disperse VO_2 particle passes to the metallic state, while the other retains its semiconducting state. On the basis of the results of this study, it is possible to conclude that small concentrations of TEA cations in a PEG matrix can exert a strong effect on the electronic structure of disperse VO_2 particles. This is accompanied by the appearance of phase heterogeneity of VO_2, which manifests itself in the fact that with the onset of MSPT with increasing temperature one portion of the substance passes to the metallic state, and the other remains in the semiconducting state. The heterogeneity of a sample is observed in the wide temperature interval $305 < T < 390$ K. One of the probable causes of the observed phenomenon is a strong bonding of TEABr cations with the polymeric matrix, while bromine anions diffuse to the VO_2 particle surface and discharge on it in the process of chemisorption. As a result, an electric field sets up, and the field lines are directed from cations localized in the polymer matrix to the VO_2 particle surface. The ^{1}H NMR signal intensity demonstrate a hysteresis loops during cooling–heating (Figure 2.84), which also depends on the content of TEABr and water in the PEG matrix.

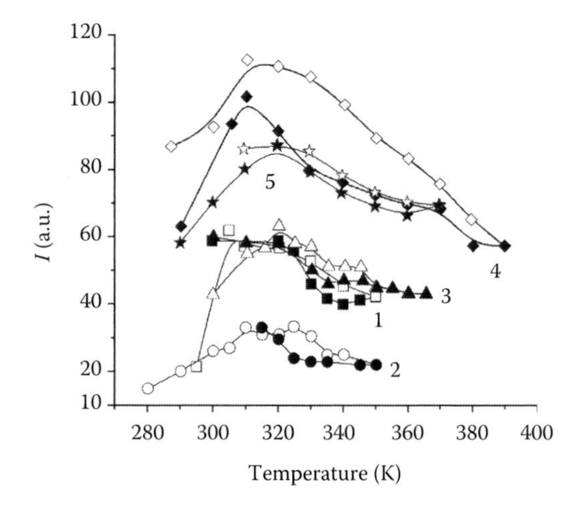

FIGURE 2.84 Temperature-induced variations of the ^{1}H NMR signal intensity for protons in VO_2 powders in a PEG matrix (curves 1), with addition of 100 mg/g of water (2), TEABr initial (3) and heated at 295 K for 20 days (4) and TEABr with 50 mg/g H_2O (5) (cooling—solid symbols and heating—open symbols). (Redrawn according to Turov, V.V. et al., *Colloids Surf. A: Physicochem. Eng. Aspects*, 178, 105, 2001.)

2.5 CLAYS, ZEOLITES, AND OTHER NATURAL MINERALS

Stable nano/mesoporous materials with mixed oxides such as zeolites, clays, and other minerals are widely used in various fields: catalysis, adsorption, ion-exchange, separation, etc. because of their catalytic activity in acid/base and redox (e.g., materials with titania phase) reactions, ability to sorb selective molecules of diverse types, participate in ion-exchange reactions, providing sieve effects, etc. (Tanabe 1970, Grandjean and László 1989, Rocha and Anderson 2000, Cundy and Cox 2005, Tao et al. 2006). The most important processes that utilize the selective properties of these materials are alkylation and isomerization of aromatic hydrocarbons as well as conversion of methanol to hydrocarbons, and some other reactions. Silicalite is an extreme type of the materials with the ZSM-5 zeolite structure but whose aluminum content is negligible. Therefore, unlike conventional zeolites, silicalite does not possess ion-exchange properties, and its surface has a weak affinity to water.

2.5.1 Silicalite

The porous structure of silicalite is formed by longitudinal and transversal channels bounded by ten 10- and 12-member rings of alternating atoms of silicon and oxygen. Structure of silicalite crystals obtained by ESM and TEM methods (Hunger et al. 1987, Burton et al. 2007) corresponds to the channel width ca. 0.6 nm, due to which the material has properties of a molecular sieve. The researches into adsorption of inorganic molecules, hydrocarbons, and binary mixtures of linear and cyclic hydrocarbons by the techniques of the neutron diffraction (LeVan 1996) and NMR spectroscopy (Hunger et al. 1987, Portsmouth et al. 1995) made it possible to establish that small molecules of hydrocarbons are predominantly adsorbed in longitudinal channels of silicalite. At the same time, molecules of longer alkanes and benzene are located at the position of intersection of channels. The interaction between adsorbed molecules is one of the important characteristics that determine adsorption properties of silicalite.

Interesting aspects of the properties of silicalite are revealed by studies of its interaction with water molecules. The size of water molecules is half as large as the silicalite channel diameter, and so, water can find its way into pores of the material. Since the silicalite surface is made up of siloxane bridges that do not practically form hydrogen-bonded associates with water molecules (Iler 1979), one can expect that at a small surface coverage the water in silicalite pores can be in the form of nonassociated molecules, dimers or larger 1D clusters. Under the conditions of a high surface coverage, water molecules in silicalite pores can form polyassociates whose structure differs from that of the hydration sheath of other silicas, especially with a great number of silanols.

Chodorowski et al. (1998) synthesized samples of silicalite using the methodology described by Grose and Flanigen (1977) with a reaction mixture of the following molar composition: $0.08Na_2O$: 0.02TPABr: $1.0SiO_2$: $13H_2O$. In order to prepare the reaction mixture, 2.01 g of NaOH were dissolved in 50 ml of distilled water, then 18.4 g of nanosilica A-380 was added to the solution obtained, and subsequently 3.26 g of tetrapropylammonium bromide (TPABr) dissolved in 26.3 g of water was added to the aforementioned solution. The hydrothermal crystallization was carried out in a stainless steel autoclave at autogeneous pressure at a temperature of 473 K for 72 h (sample 1) and 144 h (sample 2). The reaction product was washed up to the neutral reaction of washing waters, dried in the air at 383 K, and calcined at 813 K. In order to remove residual sodium ions, samples were boiled in 2 N solution of NH_4Cl, following which the samples were dried and calcined at 813 K in the air.

The XRD studies of the samples prepared were conducted with an automated diffractometer DRON-UM1 using Ni-filtered CuKα radiation. The comparison of the XRD patterns of the samples prepared with the literature data on ZSM-5 (Argauer and Landolt 1972) shows that the x-ray spectrum lines for these samples correspond to the spectral lines of ZSM-5. The degrees of crystallinity of samples 1 and 2 were 100% and 66%, respectively. The specific surface area of the silicalite samples measured by the method of LT adsorption of nitrogen was 312 and 240 m^2/g,

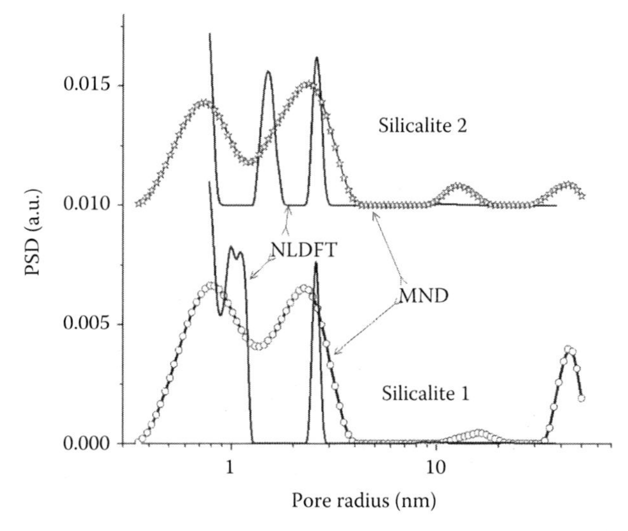

FIGURE 2.85 Pore size distributions of silicalites calculated using NLDFT (Quantachrome software, version 2.02) and modified Nguyen-Do (MND) methods with the model of cylindrical pores.

and $V_p=0.23$ and 0.15 cm^3/g, respectively. Though the conditions of the synthesis were similar, the samples were somewhat different in their structural characteristics and PSD (Figure 2.85). This results in their different ability to sorb water (adsorption capacity of sample 2 was higher), which made it necessary to analyze both samples. The PSDs calculated with NLDFT and MND methods have different shapes because of the difference in the regularization parameter values (it was larger in MND calculations, $\alpha=0.01$). Both methods show the presence of nanopores at $R<1$ nm and narrow mesopores at $1<R<4$ nm. Larger meso- (10–20 nm) and macropores at $R>30$ nm can be due to defects in particles and interparticle voids. Cleary, this nonuniformity of pores can strongly affect the behavior of bound water.

Water suspensions of silicalite were prepared in a glass vacuum installation where silicalite was subjected to a thermal vacuum treatment (600 K, 2 h), following which the water vapor was condensed on the cooled silicalite surface. It was assumed that such a method for preparation of samples facilitates the permeation of water molecules into adsorbent pores free from molecules of adsorbed nitrogen and oxygen of the air. However, check measurements showed that the adsorbed water concentration in pores of the adsorbent samples prepared by the method described earlier is only two times greater than that bound to the adsorbents from air. Water filling of pores is about 20% in the suspensions. This can be explained by the residual air bubbles and plugs in long and narrow pores.

Chemical shifts were measured with reference to TMS with a powdered sample of the hydrated adsorbent being introduced into the deuterochloroform medium containing a small additive of TMS. This procedure did not lead to any variations in the intensity and shape of the adsorbed water signal, which gave ground to consider the values of the chemical shift of adsorbed water in the air and in the $CDCl_3$ medium to be identical. The accuracy of chemical shift determination was ±0.05 ppm.

Figure 2.86 shows ^{1}H NMR spectra of water in a frozen aqueous suspension of silicalite (sample 2). The spectra were recorded at two transmission bands of the spectrometer: 20 kHz (Figure 2.86a) and 50 kHz (Figure 2.86b). In the spectra with the more wide transmission band, there are broad and narrow components of the signal for non-freezing water. With lowering temperature, the width of the broad component of the signal decreases sharply, and at $T<250$ K, the component ceased to be registered in the spectra. Turov and Leboda (1999) showed that the broad component of the signal with the frozen aqueous suspension of nanosilica (A-300) was attributed to the formation of an ice interlayer near the adsorbent surface, where the mobility of water molecules

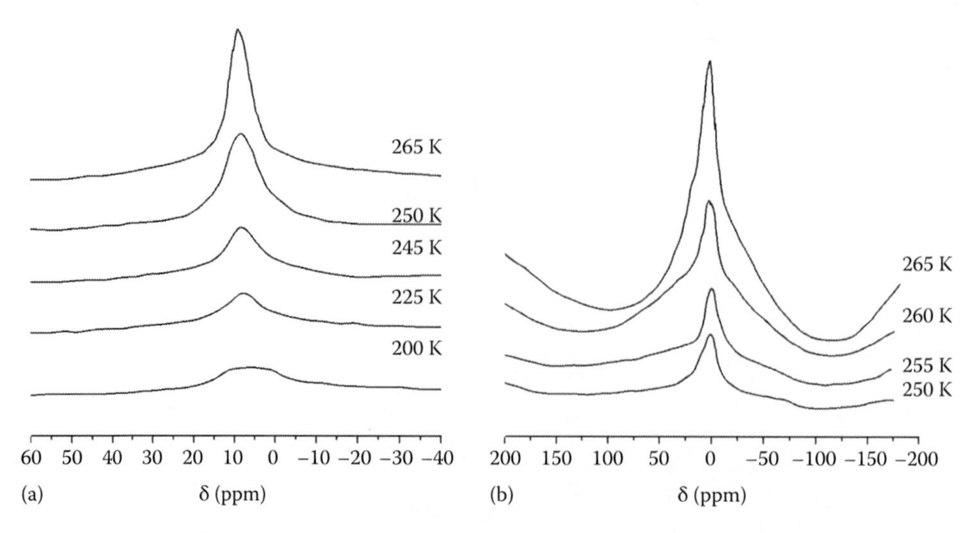

FIGURE 2.86 ^1H NMR spectra of water in an aqueous suspension of silicalite at temperature $T<273$ K at various transmission bands: (a) 20 kHz and (b) 50 kHz. (Adapted from *Micropor. Mesopor. Mater.*, 23, Turov, V.V., Brei, V.V., Khomenko, K.N., and Leboda, R., ^1H NMR studies of the adsorption of water on silicalite, 189–196, 1998a, Copyright 1998, with permission from Elsevier.)

takes a mid-position between the mobility of molecules of ice and that of the adsorbed phase. The narrow component of the signal is assigned to molecules of water adsorbed on the surface.

As temperature decreases, the signal intensity for adsorbed water diminishes owing to partial freezing-out of water in an adsorption layer, with the width of the signal being increased because of the decreasing of the mobility of adsorbed water molecules (Figure 2.86a). In the ^1H NMR spectra, the signal for ice molecules is not observed, which is due to a small time of the transverse relaxation of water molecules in the crystal lattice of ice. Despite the fact that in the spectra, there is a broad signal component whose intensity sometimes exceeds the intensity of the signal for water adsorbed on the surface, in view of the great difference in the signal width at temperatures $T<265$ K, it is possible to determine precisely enough the narrow component intensity by means of the compensatory linearization of the zero line. The shape of the ^1H NMR spectra for water in a frozen aqueous suspension of sample 1 varies in much the same way.

Figure 2.87 shows the temperature variations of spectra for water adsorbed on sample 2 at $h=22$ mg/g in the $CDCl_3$ medium. Besides the signal of water in these spectra, there are signals of TMS ($\delta_H=0$ ppm) and chloroform ($\delta=7.26$ ppm). At 290 K on the shoulder of the adsorbed water signal ($\delta_H=4.97$ ppm), there appears a weak signal of water dissolved in chloroform (WAW, $\delta_H=1.7$ ppm). The appearance of this signal is caused by the increase in the solubility of water in chloroform (in solution and interface) with increasing temperature. With decreasing temperature and partial freezing-out of adsorbed water, one can observe a shift of the adsorbed water signal to lower magnetic fields.

Figure 2.88 displays the temperature dependencies of the chemical shift of adsorbed water for both synthesized samples of silicalite with water contents h of 17 and 22 mg/g for sample 1 and sample 2, respectively. As is seen from Figure 2.88, the dependencies are complex in character. The common features observed for both samples include the down-field shift of signal by about 1 ppm with decreasing temperature from 325 to 220 K and the presence (with the dependencies $\delta_H(C_{uw})$) of a horizontal section in the temperature interval $230<T<250$ K. The complexity of the dependencies displayed is partly caused by the variation of the concentration of unfrozen water in the freezing-out process. The dependencies of the chemical shift and free energy of adsorbed water on concentration of unfrozen water in the case of hydrated powders of silicalite is shown in Figure 2.88. Figure 2.89 displays the plots of the dependence $\Delta G(C_{uw})$ for frozen aqueous suspensions of silicalite.

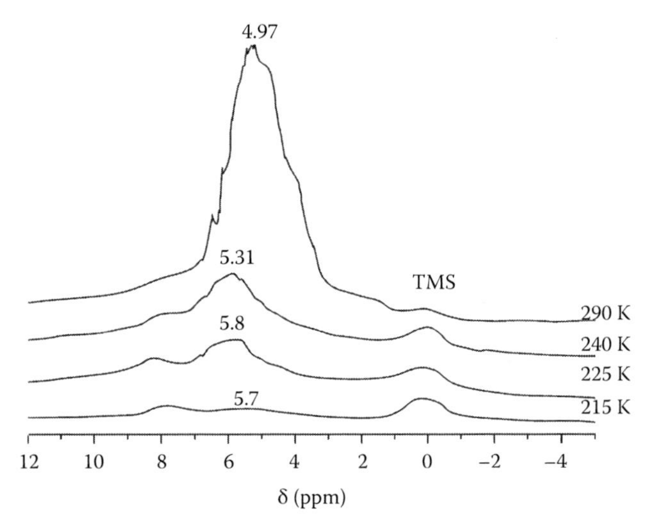

FIGURE 2.87 Temperature variation of the ^{1}H NMR spectra of water adsorbed on silicalite surface in the deuterochloroform medium. (Adapted from *Micropor. Mesopor. Mater.*, 23, Turov, V.V., Brei, V.V., Khomenko, K.N., and Leboda, R., ^{1}H NMR studies of the adsorption of water on silicalite, 189–196, 1998a, Copyright 1998, with permission from Elsevier.)

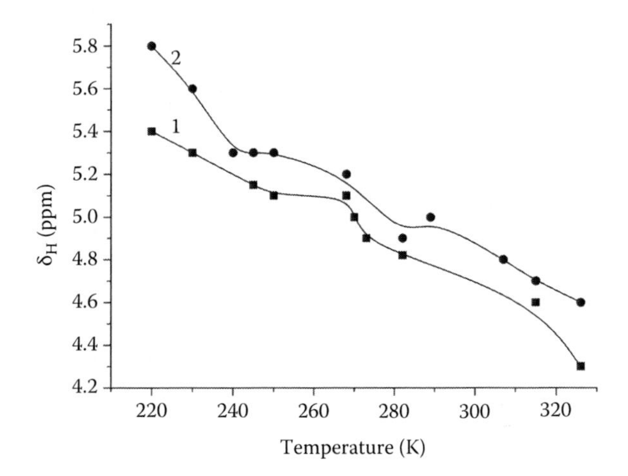

FIGURE 2.88 Temperature dependence of the chemical shift of water in pores of silicalite samples 1 and 2. (Adapted from *Micropor. Mesopor. Mater.*, 23, Turov, V.V., Brei, V.V., Khomenko, K.N., and Leboda, R., ^{1}H NMR studies of the adsorption of water on silicalite, 189–196, 1998a, Copyright 1998, with permission from Elsevier.)

Water adsorbed on the surface of silicalite powders freezes when the decrease in temperature compensates for the decrease in free energy of molecules caused by adsorption. In this case, on the curves representing the dependence $\Delta G(C_{uw})$, there appears a section where the ΔG value does not change. It corresponds to the range of temperatures ($T<273$ K) at which the freezing of water adsorbed on the surface does not occur (Figure 2.89b, curve 2). The absence of such a section for sample 1 gives evidence that the sample contains water that is weakly bound to surface, and this water freezes as a result of even a slight lowering of temperature. As follows from the data of Figure 2.89, the freezing of the main part of water is not accompanied by any significant changes of the chemical shift of molecules of adsorbed water. Only at temperatures $T<240$ K on the curves representing the dependence $\delta_H(C_{uw})$, there appears a section of a sharp increase in the δ_H value (the reasons of a similar down-field shift was discussed in detail in Chapter 1).

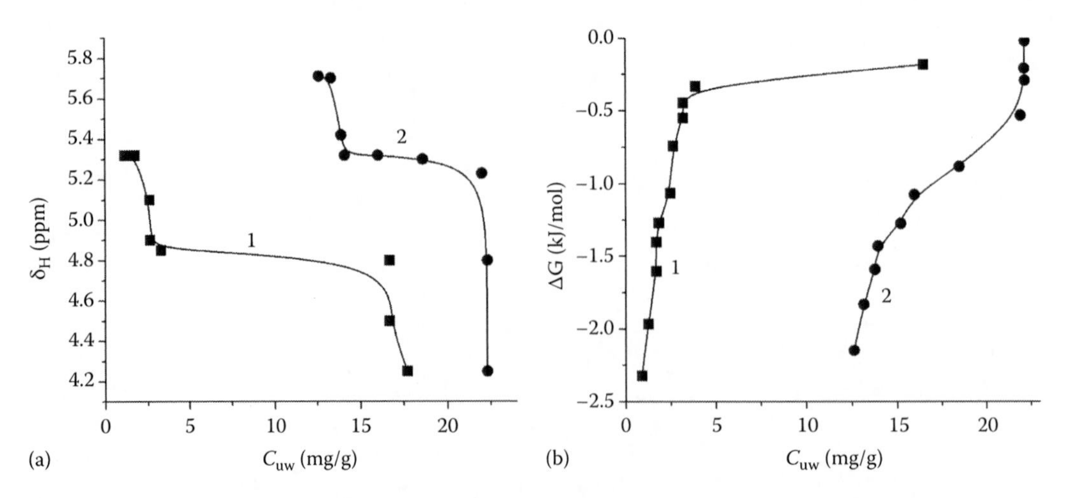

FIGURE 2.89 Dependence of the chemical shift (a) and variation of free energy of adsorbed water (b) on concentration of unfrozen water in samples 1 and 2. (Adapted from *Micropor. Mesopor. Mater.*, 23, Turov, V.V., Brei, V.V., Khomenko, K.N., and Leboda, R., ^{1}H NMR studies of the adsorption of water on silicalite, 189–196, 1998c, Copyright 1998, with permission from Elsevier.)

From the data of Figure 2.89, one can calculate characteristics of water layers adsorbed on silicalite.

On the curves of the function $\Delta G(C_{uw})$ for both samples, it is possible to discern two sections. One of them is the section of a rapid variation of the thickness of the unfrozen water layer at small changes of the ΔG value. This section is related to that part of adsorbed water whose free energy undergoes only a slight decrease resulting from the adsorption interactions with the silicalite surface (weakly bound water). The second section reflects relatively slight variations of the thickness of the adsorbed water layer at a wide range of ΔG. This section of the dependence is due to molecules of strongly bound water.

The concentration of strongly and weakly bound water (C^s and C^w) as well as the maximum decrease in the free energy of water due to adsorption for both types of bound water (ΔG^s and ΔG^w, respectively) can be determined from the data of Figure 2.90a by means of extrapolation of

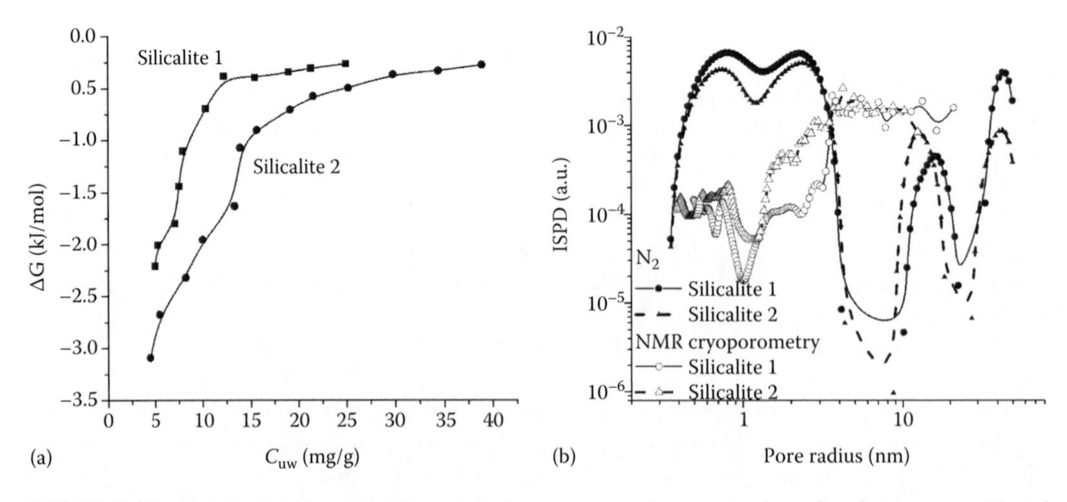

FIGURE 2.90 Relationship between changes in free energy and concentration of unfrozen water (a) and IPSD (bound cluster) size distributions (b) in frozen aqueous suspensions of silicalites calculated with the NMR cryoporometry (bound water) and MND (nitrogen) methods.

TABLE 2.13
Characteristics of Layers of Water Adsorbed on Silicalite and A-300 Surfaces

Material	C_{max} (mg/g)	C^w (mg/g)	C^s (mg/g)	$-\Delta G^w$ (kJ/mol)	$-\Delta G^s$ (kJ/mol)	γ_S (mJ/m^2)
Silicalite 1	55	43	12	0.5	3.2	5.0
Silicalite 2	55	35	20	1.0	3.6	12.0
Nanosilica A-300	1500	1200	300	0.35	3.8	150

corresponding sections of the dependence $\Delta G(C_{uw})$. In this case, $C^w = C_{max} - C^s$, where $C_{max} = C$ as $\Delta G \to 0$. The characteristics of layers of water adsorbed on the surface of both silicalite samples determined in this way and values of free surface energy are presented in the Table 2.13. For comparison, Table 2.13 lists also the characteristics of layers of water adsorbed onto nanosilica A-300 (Gun'ko et al. 2005d).

As follows from the data of the Table 2.13 and Figure 2.90a, the concentration of strongly bound water onto silicalite does not exceed 20 mg/g, which is by an order of magnitude smaller than that for the nanosilica sample. The results obtained are in good agreement with the concepts on hydrophobic properties of the silicalite surface. Taking into account that the silicalite has a relatively high specific surface ($S_{BET} = 312$ and 240 m^2/g) and $V_p = 0.23$ and 0.15 cm^3/g, one can come to the conclusion that water adsorbed in its pores occupies only a small part of the total pore volume. The specific surface area in contact with unfrozen water (calculated with the NMR cryoporometry) is 27 and 48 m^2/g for silicalites 1 and 2, respectively, which are much smaller than the S_{BET} values.

Using the value of the chemical shift of adsorbed water molecules, it is possible to make a judgment on the structure of adsorption complexes of water on the silicalite surface. As follows from Figure 2.88, at $T < 273$ K, the value of the chemical shift δ_H is more than 5 ppm. Since on the silicalite surface, there are no acid or base sites that could form strong hydrogen-bonded complexes with adsorbed water molecules, the high δ_H value for water must be due to the formation of surface polyassociates of water whose average number of neighbors at each molecule exceeds 3.5. In accordance with the Gibbs-Thomson equation, the size of clusters and domains of unfrozen bound water (Figure 2.90b) may be in a broad range, which corresponds or significantly exceeds the size of the main channels in the crystal structure of the zeolite. This implies that most of strongly associated water attributed not to the water in narrow pores of the zeolite structure, but to the interparticles water, and water adsorbed on structural defects of the crystal lattice. The bonding with the surface is affected through a relatively small number of surface hydroxyl groups whose nature is that of defects of the crystal structure of silicalite, which are able to act as primary adsorption sites where clusters of adsorbed water begin to nucleate and grow. The difference in hydration properties of samples 1 and 2 can be consistently explained by the difference in the concentration of OH-groups in these samples, it being known that the structure of sample 2 is more imperfect. With increasing temperature, one can observe disintegration of hydrogen-bonded complexes, and the balance is shifted toward the formation of more simple aqueous associates, and the chemical shift of molecules of adsorbed water decreases (Figure 2.88).

The values of free surface energy obtained for both samples of silicalite are very small. However, in view of the fact that the main part of the pore volume is not occupied by water, γ_S values listed in the Table 2.13 should be related to that part of the adsorbent surface where clusters of adsorbed water are formed and not to the whole adsorbent surface. In this case, the ΔG value for the hydrated part of the surface should be multiplied by the coefficient that is a ratio of the specific surface of adsorbent to that part of the surface that is in contact with clusters of adsorbed water.

2.5.2 Influence of Al Content on Hydration of Zeolites of a ZSM-5 Type in Aqueous Suspension

Turov et al. (2000) determined the characteristics of water bound to H-ZSM-5 zeolites differing in their aluminum content under conditions of a high degree of hydration and possibility of a competing sorption of ambient air and water vapors. Samples of zeolites were synthesized according to the methodology described by Grose and Flanigen (1977) on the basis of highly disperse aluminosilicas prepared using a simultaneous hydrolysis of $SiCl_4$ and $TiCl_4$ in the flame of a hydrogen-air burner. Such a procedure allowed us to simplify synthesis of zeolites since aluminosilica served as a source for formation of the both oxides in the zeolite crystal lattice.

The zeolites were synthesized using aluminoaerosils with molar ratios SA equal to 131, 100, 26, and 19.5 for preparation of samples 1–4 (Table 2.14), respectively. The template was tetrapropylammonium bromide. In order to prepare a reaction mixture, sodium hydroxide and template were dissolved in distilled water, and then a suitable amount of aluminosilica was added to this solution under conditions of mixing. In terms of oxides, the component ratios in the reaction mixture were as follows: $Na_2/SiO_2=0.05$, $[(C_3H_7)_4N]_2O/SiO_2=0.04$, $H_2O/Na_2O=300$.

Hydrothermal crystallization was effected in a stainless steel autoclave with a fluoroplastic beaker 30 cm^3 in volume at 433 K for 8 h. The reaction product was washed up to a neutral reaction of the washing water, dried at 383 K in air, and calcined at 813 K. The H-form was imparted to the prepared zeolites through ionic exchange in 2 N solution of NH_4Cl, following which the samples were dried and calcined in air at 813 K.

The XRD study of the samples synthesized was made with an automated DRON-UMI (LOMO) diffractometer using Ni-filtered Cu Kα radiation. The comparison of the XRD patterns of the prepared samples with the literature data on ZSM-5 shows that the x-ray spectra lines for these samples correspond to the lines of ZSM-5. The degrees of crystallinity of these samples were 94%–96%. The specific surface of the zeolite samples was measured by the method of thermal desorption of argon; the pore volume was determined in terms of methanol isotherms. The values obtained are listed in Table 2.14.

Figure 2.91 illustrates the dependencies of variation in the free energy of adsorbed water on its concentration for aqueous suspensions of zeolites (1), hydrated powders in air (2), and in deuterochloroform media (3). Figure 2.92 shows temperature dependencies of the chemical shift for water adsorbed in pores of zeolites established for hydrated powders in the $CDCl_3$ medium. The characteristics of layers of water bound in hydrated powders and in aqueous suspensions of adsorbents are listed in Table 2.15. As distinct from the analogous dependencies established for mesoporous materials (Turov and Leboda 1999, Gun'ko et al. 2005d, 2009d), the dependencies $\Delta G(C_{uw})$ for all the studied zeolites do not contain portions corresponding to water weakly bound to the adsorbent surface. At the same time, there are portions where the value of the unfrozen water layer remains practically constant over a wide range of variation of the ΔG value. The appearance of such portions on the dependence $\Delta G(C_{uw})$ is due to the fact that the free energy of the absolutely all

TABLE 2.14
Structural Characteristics of the Synthesized Zeolites

Sample	Al_2O_3 Content (wt%)	S (m^2/g)	V_p (cm^3/g)
1	1.96	515	0.26
2	3.15	482	0.23
3	7.38	475	0.26
4	8.95	450	—

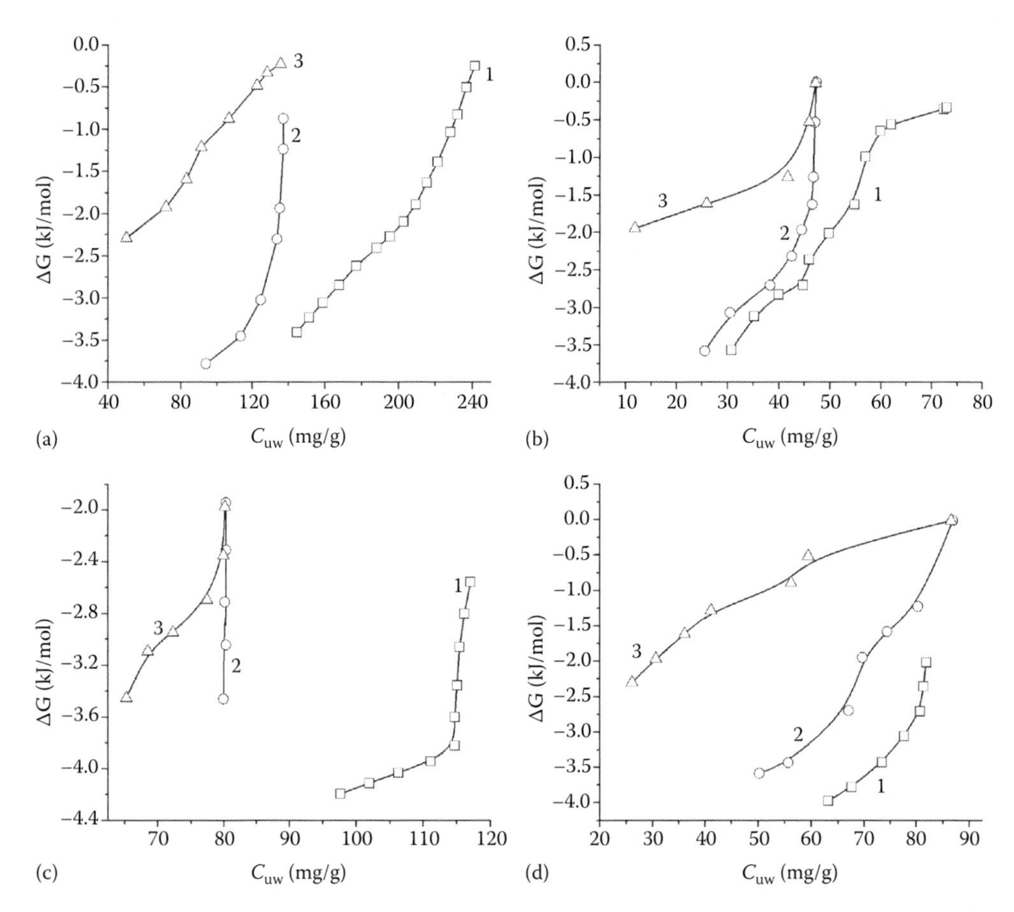

FIGURE 2.91 Dependence of variation of free energy of unfrozen water on its concentration for a suspension of zeolite in water (1), hydrated powder in air (2), and a suspension of zeolite in deuterochloroform (3) for samples at different content of alumina (a) 1.96, (b) 3.15, (c) 7.38, and (d) 8.95 wt%. (Redrawn according to Turov, V.V. et al., *Adsorption Sci. Technol.*, 18, 75, 2000.)

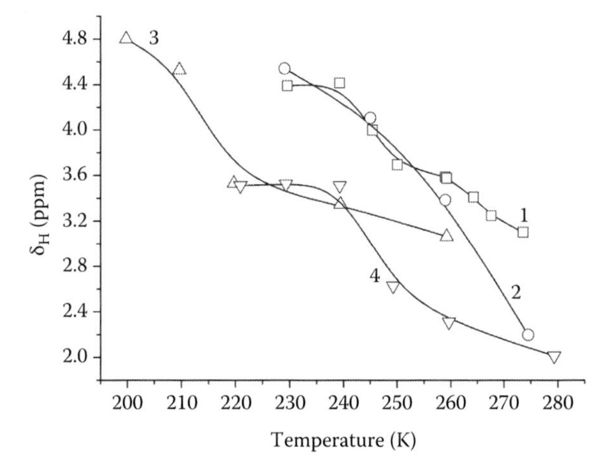

FIGURE 2.92 Temperature dependence of the chemical shift for water adsorbed on zeolites containing various concentrations of Al_2O_3 (wt%) in a deuterochloroform medium: (1) 1.96, (2) 3.15, (3) 7.38, and (4) 8.95. (Redrawn according to Turov, V.V. et al., *Adsorption Sci. Technol.*, 18, 75, 2000; Turov, V.V. et al., *Colloids Surf. A: Physicochem. Eng. Aspects*, 178, 105, 2001.)

TABLE 2.15
Characteristics of Layers of Water Bound in Aqueous Suspensions of Zeolites

Al_2O_3 Content		ΔG^{max} (kJ/mol)			
(wt%)	C_{max} (mg/g)	Water	Air	$CDCl_3$	γ_S (mJ/m²)
1.96	250	6.1	4.8	3.0	37
3.15	80	6.1	5.5	2.3	6
7.38	120	6.0	—	6.0	26
8.95	85	6.2	5.0	4.0	19

molecules of water adsorbed in pores is lowered by adsorption interactions with the surface so that it is necessary to cool the sample substantially in order to attain equality between the free energies for adsorbed water and ice. This is typical for nanoporous adsorbents.

One more feature of the studied type of nanoporous materials manifests itself in that the curves of the dependencies $\Delta G(C_{uw})$ for the aqueous suspensions of adsorbents in the whole range of variation in the concentration of unfrozen water are displaced toward large value of C_{uw} in comparative of the corresponding curves for hydrated powders. Additionally, in the presence of chloroform, water freezes at higher temperatures and the $-\Delta G$ value decreases (notice that $\Delta G = -4$ and -2 kJ/mol corresponds to melting temperatures $T = 162$ and 217 K, respectively, of confined ice crystallites). The observed regularities can be caused by the adsorption of water not only in the conventional zeolite pores but also in voids between micro- and nanoparticles, crystal defects with larger content of silanols than in the main pores. Water can form SAW structures in these voids and defects. However, the structure of water bound in the zeolite powder placed in the chloroform medium changes. Weakly polar chloroform can easily penetrate into narrow pores of low hydrophilicity and displace water from them into pores of larger sizes or into interparticles voids. This results in increase in the freezing temperature of bound water.

The range of variation in chemical shifts for water adsorbed in zeolite pores (Figure 2.92) makes up 2.5–5.6 ppm. The complex shape of the plot for the temperature dependence of the chemical shift is attributed to the fact that with decreasing temperature, a partial freezing-out of adsorbed water takes place. This freezing results in decreasing sizes of clusters of liquid water adsorbed in the pores. The chemical shift value for water in polyassociates is determined by the number of hydrogen bonds (n), in whose formation every molecule of water participates simultaneously. With decreasing temperature, the number of such bonds increases at the expense of retardation of molecular motion, and, as a result, one can observe a down-field shift of adsorbed water signal (an increase in the chemical shift). However, as some part of water in pores freezes, the average sizes of clusters of adsorbed water become smaller, which must manifests itself in a decrease of n and, therefore, in a reduction of the chemical shift value. The relatively small range of the temperature-induced variation of the chemical shift is probably due to simultaneous contributions made by both the aforementioned factors.

As follows from the data listed in Table 2.15, the concentration of water adsorbed in zeolite pores and the value of the free surface energy of adsorbents (with the exception of the sample containing 3.15 wt% of Al_2O_3) decrease as the aluminum content in a sample increases. At the same time, the value of the maximum decrease in the free energy in a layer of bound water is practically the same for all the studied zeolites and about twice as large as the corresponding value for silicalite. Under these conditions, all the adsorbent pore volume is filled with water only in the case of the sample whose Al_2O_3 content is 1.96. The observed regularity disagrees with the data presented by (Olson et al. 1980) for adsorption of water vapors in zeolites differing in the concentration of Al in their crystal lattice.

The cause of the observed features may be two main factors, namely the change in the structure of water polyassociates with increasing concentration of Al and variation of conditions for the competing sorption of water and gases that are present in air. Really, in the situation of a high degree of hydration of the surface, water not only interacts with primary adsorption sites but also becomes part of molecular aggregates that contain Brønsted acid sites and several molecules of water bound to such a site (Resing and Davidson 1975, Buzzoni et al., 1995, Pazé et al. 1997, Zecchina et al., 2005). However, high acidity is characteristic only of a relatively small portion of surface hydroxyl groups, which is evidenced for by the data of calorimetric titration (Drago et al. 1970) or NMR-spectroscopy (Brunner and Pfeifer 2008). With increasing concentration of aluminum in the crystal lattice, the percentage of such active OH groups decreases, and, as a consequence, one may also observe a decrease in the sizes of clusters of water molecules adsorbed on the surface. Thus, despite the increase in the concentration of primary sites for adsorption of water, the increase in the concentration of aluminum can be accompanied with a reduction of the degree of hydration of the zeolite surface.

If adsorption of water is affected in air, there may appear a second factor that affects the dependence of the degree of hydration of the surface on the aluminum content in a sample. Owing to the presence of air, the zeolite surface can adsorb simultaneously not only water but also molecules of nitrogen and oxygen that are not able to form hydrogen-bonded complexes with surface hydroxyl groups and are adsorbed predominantly on hydrophobic portions of the surface. With increasing sizes of clusters of water adsorbed on the surface, some portion of the adsorbed gases must be removed from the surface. Then the value of adsorption of water should depend on the relationship between the free energy of formation of water clusters and that of desorption of gas molecules from the surface. In this situation, the formation of water clusters on hydrophilic sites of the surface may be energetically less favorable than adsorption of gas molecules on hydrophobic portions of this surface. As the content of aluminum in the zeolite increases, the number of primary sites for adsorption of water increases too, but the water clusters formed nearby these sites make contact predominantly with hydrophobic portions of the surface created by the siloxane bridges. As a result, one can observe an effect that manifests itself in a decrease in the free surface energy of zeolites with increasing content of aluminum in their lattice.

2.5.3 Zeolites and Other Aluminosilicates

Dorémieux-Morin et al. (1991) using ^{1}H MAS NMR spectroscopy showed that partly dehydrated silica–alumina and strongly dealuminated HY zeolite have similar properties with respect to surface hydroxyls. However, there is a fundamental difference between the dehydration processes of crystalline and amorphous samples due to the differences in the surface structure and OH group location.

Beutel et al. (2001) studied the interactions of phenol with zeolite NaX using solid-state ^{1}H and ^{29}Si MAS and ^{29}Si CP-MAS NMR spectroscopy. The ^{1}H MAS NMR spectrum of NaX degassed at 400°C included peaks at $\delta_H = 1.24$, 2.05, 4.05, and 5.11 ppm. The signal at 2.05 ppm is characteristic of silanol protons at the external surface of the zeolite. On NaY zeolite (Si/Al = 2.4), a peak at $\delta_H = 1.6$ ppm was found after pretreatment in air and in vacuum at 400°C and assigned to traces of water adsorbed on cations. Therefore, the peak at $\delta_H = 1.24$ ppm was attributed to free hydroxyl protons of water molecules strongly adsorbed on Na^+ ions. The peaks at great δ_H values could be attributed to water bound to different active sites.

The bands of neat phenol (Figure 2.93, curve 1) at $\delta_H = 6.4$ and 0.7 ppm correspond to the hydrogen atoms of the aromatic ring and the hydroxyl proton of phenol. After phenol adsorption, a relatively sharp signal is observed at $\delta_H = 1.2$ ppm, which was attributed to hydroxyl protons of a condensed phenol phase outside the zeolite pores. The position of the chemical shift of the hydrogen atoms of the aromatic ring was unchanged after adsorption, but a broad feature appeared at $\delta_H = 9$ ppm. Additionally, two shoulders at $\delta_H = 5.3$ and 4.2 ppm were detected and assigned to zeolitic hydroxyl protons as generated by proton transfer from phenolic hydroxyl groups to basic framework oxygen.

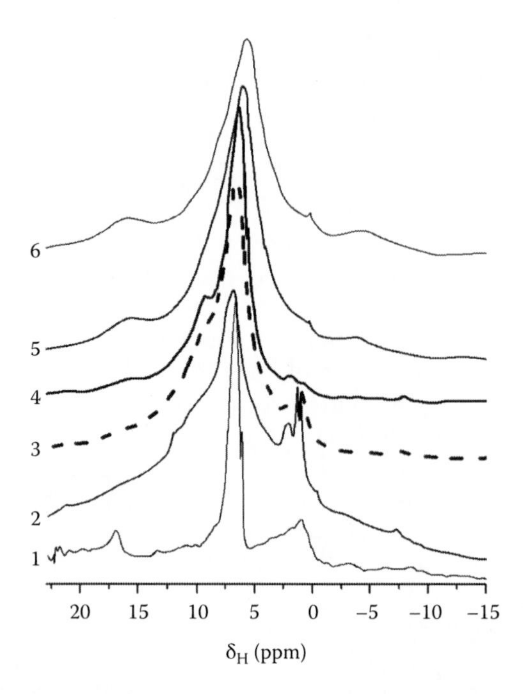

FIGURE 2.93 ^{1}H MAS NMR spectra of (1) pure phenol and of calcined NaX after adsorption of phenol at room temperature (2), after heating at 50°C for 1 h (3) and 15 h (4), and subsequent heating at 220°C (5) and 320°C for 1 h (6), respectively. (Adapted from *Colloids Surf. A: Physicochem. Eng. Aspects*, 187–188, Beutel, T., Peltre, M.-J., and Su, B.L., Interaction of phenol with NaX zeolite as studied by ^{1}H MAS NMR, ^{29}Si MAS NMR and ^{29}Si CP MAS NMR spectroscopy, 319–325, 2001, Copyright 2001, with permission from Elsevier.)

Spectra 3 and 4 showed that heating at 50°C for 1 and 15 h resulted in a decrease in the peak at $\delta_H = 1.2$ ppm, while the broad feature at $\delta_H = 9$ ppm increased. After heating at 50°C for 15 h, the peak at $\delta_H = 1.2$ ppm has almost vanished. Apparently, phenol crystals characterized by the peak at $\delta_H = 1.2$ ppm decomposed, and phenol migrated into the zeolite pores. The depletion of the signal at $\delta_H = 2.06$ ppm with increasing time of heat treatment at 50°C proved an interaction of phenol with silanol groups by hydrogen bonding.

An enhancement of the ^{29}Si CP MAS signal was observed for the phenol loaded NaX zeolite and interpreted in terms of an interaction of the aromatic ring with either Na^+ cations or oxygen atoms of the 12-member ring of the supercage. A small fraction of framework Si atoms in fourfold aluminum coordination (through O atoms), most likely Si atoms in four member ring positions of S_{III} or S'_{III} sites, was electronically affected by interaction of phenol with Na^+ ions located in these sites. After thermal treatment to 320°C, a part of the phenol is deprotonated and rearranged to linear Na+…−OPh complexes (Beutel et al. 2001).

The solid-state ^{29}Si MAS NMR spectra are informative with respect to structural features of mixed oxides because the peak position of the Si atoms depends on the number of hydroxyls bonded to these atoms (as discussed earlier in Chapter 1) and the structure of the second coordination sphere (Figure 2.94). For instance, an increase in the number of Al atoms in the second coordination sphere of the Si atoms in zeolite ZK-5 (Nakata et al. 1998) results in the down-field shift of the signal (curve 6); i.e., electron shielding of the Si atoms decreases. Additionally, the position of the Q_4 site (Si(OSi)$_4$) band depends on the composition of the materials (compare curves for ETAS-10 and ZK-5). Triple-quantum ^{23}Na MAS-NMR spectra of ETS-10 as-prepared and dehydrated showed different heating effects for different sites (Rocha and Anderson 2000).

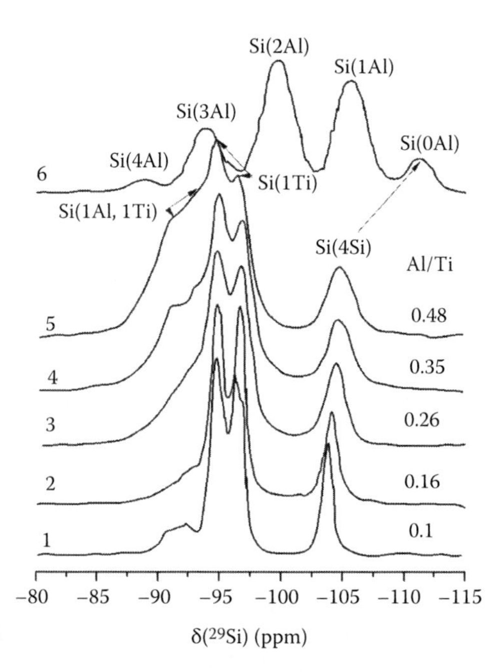

FIGURE 2.94 ^{29}Si MAS NMR spectra of titanoaluminosilicate ETAS-10 with different Al/Ti ratios (curves 1–5). (Adapted from Rocha and Anderson, *Eur. J. Inorg. Chem.*, 801, 2000) and aluminosilicate zeolite ZK-5 (curve 6) at Si/Al = 2.2 and from *J. Mol. Struct.*, 441, Nakata, S., Tanaka, Y., Asaoka, S., and Nakamura, M., Recent advances in applications of multinuclear solid-state NMR to heterogeneous catalysis and inorganic materials, 267–281, 1998, Copyright 1998, with permission from Elsevier.)

Müller et al. (2000a,b) studied zeolite beta samples PB1 and PB3, mordenite, ZSM-5, and ferrierite using solid-state ^{1}H, ^{27}Al, and ^{29}Si MAS NMR spectroscopy, and showed the presence of several types of the surface hydroxyls. They assigned the bands at $\delta_H = 3.8–4.2$ ppm to free Brønsted acid sites (i.e., bridging hydroxyls), and $\delta_H = 5.1–5.9$ ppm was attributed to the bridging hydroxyl groups with the zeolite framework. Notice that there is nearly linear ($R^2 = 0.98$) correlation between the frequency of the OH-stretching vibrations and the δ_H value (Figure 2.95).

Zhang et al. (2001) studied the HTT effects on the structure of zeolite HZSM-5 (with micro- and nanosized particles) using ^{1}H (Figure 2.96), ^{27}Al, and ^{29}Si solid-state MAS NMR. The thermal stability of nanosized HZSM-5 (~70 nm) is lower than that of microsized particles (~1 μm) due to dealumination and desilicification. After HTT at 700°C, the Brønsted acid sites ($\delta_H = 3.9$ ppm) disappear (as well as nonframework AlOH at $\delta_H = 2.7$ ppm) in nanosized HZSM-5 in contrast to microsized HZSM-5 (Figure 2.96). However, the silanol peak at 1.7 ppm increases for nanosized particles. For microsized particles, the dealumination process is dominant. Upon HTT, the amorphous silica can "heal" dealuminated fragments. Therefore, the hydrothermal stability of nanosized HZSM-5 particles is similar to that of the microsized HZSM-5 (Zhang et al. 2001).

Kneller et al. (2003) studied dealuminated zeolites NaY and mordenite (MOR) characterized by ^{27}Al and ^{29}Si MAS NMR and temperature dependent ^{129}Xe NMR. However, they did not use the ^{129}Xe NMR spectra for quantitative structural characterization of the zeolites because very low xenon loadings were used to analyze the surface structure but not the PSD.

Kazansky (2003) presented results on interactions of H_2 with different zeolites using the FTIR spectroscopy, which could be more interesting if the ^{1}H NMR spectra were recorded in parallel for the same samples. The same conclusion is for the paper (Prasanth et al. 2008) describing the hydrogen adsorption onto zeolite Z. Ganapathy et al. (2003) studied titanium

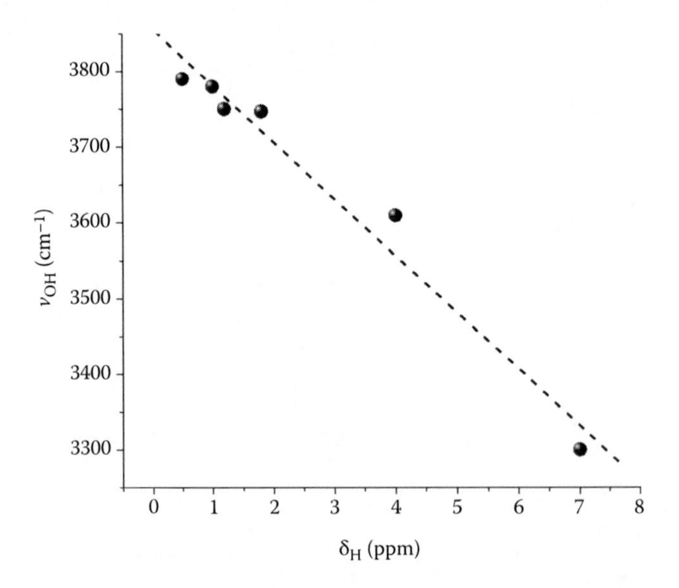

FIGURE 2.95 Relationship between the chemical shift of the H atoms in SiOH and AlOH groups alone and with adsorbed water. (Drawn using the data from Müller et al., *Micropor. Mesopor. Mater.*, 34, 135, 2000a; Müller et al., *Micropor. Mesopor. Mater.*, 34, 281, 2000b.)

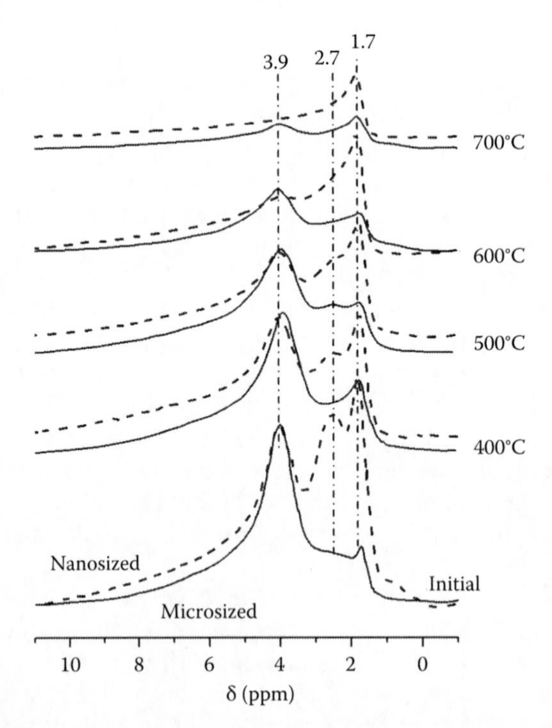

FIGURE 2.96 ^{1}H MAS NMR spectra of microsized (solid lines) and nanosized (dashed lines) HZSM-5 zeolites initial and after hydrothermal treatment at different temperatures. (Adapted from *Micropor. Mesopor. Mater.*, 50, Zhang, W., Han, X., Liu, X., and Bao, X., The stability of nanosized HZSM-5 zeolite: A high-resolution solid-state NMR study, 13–23, 2001, Copyright 2001, with permission from Elsevier.)

substituted ultrastable Y (USY) zeolite using multinuclear (^{27}Al, ^{29}Si, and ^{47,49}Ti) solid-state NMR spectroscopy. The absence of the ^{1}H NMR spectra in this work did not allow one to analyze the aspects important for the interfacial behavior of the materials. Similar to the aforementioned papers, Jacobsen et al. (2000) described interaction of ammonia with ZSM-5 without the use of ^{1}H NMR spectroscopy, as well as Andersen et al. (2004) studied the effects of hydrating a white Portland cement, Wang and Scrivener (2003) analyzed alkali-activated slag, He et al. (2004) investigated mullites from different kaolinites. Kennedy et al. (2002) analyzed crystallographically different tetrahedral Al sites in zeolite merlinoite using the ultra-high-field (18.8 T) ^{27}Al and ^{29}Si MAS NMR spectroscopy. The chemical shifts and peak areas indicated a preferential siting of Al in site T_{II}. Datka et al. (2003) studied heterogeneity of OH groups in zeolite HY using IR and ^{29}Si MAS NMR spectroscopies. This is not optimal set of methods to study the zeolite surface because ^{1}H NMR could give additional and useful information. Freude et al. (2001) used multiple-quantum magic-angle spinning and double-rotation NMR techniques in the high field of 17.6 T to the study of oxygen-17-enriched zeolites A and LSX with the ratio Si/Al = 1. A monotonic correlation between the isotropic value of the chemical shift and the Si–O–Al bond angle α (taken from x-ray data) could be found. Hydration of the zeolites causes a downfield ^{17}O NMR chemical shift of about 8 ppm with respect to the dehydrated zeolites. However, the ^{1}H NMR spectra (were not recorded) could give additional information, as well as in the case of a series of aluminum-containing kanemite (Al–kanemite) samples with several Si/Al molar ratios synthesized by Toriya et al. (2003). They used the ^{27}Al and ^{29}Si MAS NMR, and FTIR spectra to characterize the materials.

The ^{1}H NMR spectroscopy with T_1 and T_2 measurements of bound water was used by Totland et al. (2011) to study the properties of sodium dodecyl sulfate (SDS), tetradecyl trimethyl ammonium bromide (TTAB), and dodecyl trimethyl ammonium bromide (DTAB) adsorbed onto kaolin. They showed that adsorbed surfactants form a barrier between water and the paramagnetic species present on the clay surface that lead to a significant increase in the proton T_1 values of water. The total surface area covered by the cationic (DTAB and TTAB) and anionic (SDS) surfactants could be estimated from the water T_1 data and found to correspond to the fractions of negatively and positively charged surface area, respectively. T_1 measurements demonstrated that the total kaolin surface area covered by TTAB or DTAB was larger than the corresponding area covered by SDS, indicated by larger water T_1 values in the samples containing TTAB or DTAB (Totland et al. 2011).

Fitzgerald et al. (1996) studied the thermal transformations of the 2:1 phyllosilicate mineral, pyrophyllite, at 150°C–1350°C using solid-state multinuclear MAS NMR spectroscopy. The ^{27}Al and ^{29}Si NMR and ^{1}H CRAMPS techniques were used to analyze mineral dehydroxylation at 150°C–550°C. At 550°C, pyrophyllite was completely dehydroxylated in 7 days to pyrophyllite dehydroxylate, an aluminosilicate intermediate containing fivefold O-coordinate Al. MAS ^{27}Al and CP/MAS (cross-polarization) and SP/MAS ^{29}Si NMR spectra showed that the dehydroxylate was formed prior to the separation of the silica-alumina layer but this occurred at 950°C. A transition-alumina-type phase, containing four- and sixfold O-coordinate Al, was formed between 950°C and 1050°C. A high content of amorphous silica glass and a small amount of a poorly ordered Si/Al-containing mullite phase formed between 950°C and 1050°C. At 1250°C–1350°C, the ^{29}Si NMR showed that this glassy silica was converted to cristobalite, while the ^{27}Al NMR indicated that this process was accompanied by conversion of octahedral Al to tetrahedral Al, possibly by incorporation of aluminums into an amorphous Si/Al-containing phase.

Timonen and Pakkanen (1999) used solution state ^{1}H NMR to study the mode and strength of chloroform adsorption on conjugated acid–base pairs of cation-exchanged Y-zeolites. The ^{1}HNMR results indicated that there are two different surface bonding orientations of the adsorbed molecules. An acidic hydrogen site of zeolite (SiO(H)Al) favors orientation where the chloroform hydrogen points toward the zeolite cage center. An exchanged electropositive cation increased the basicity of

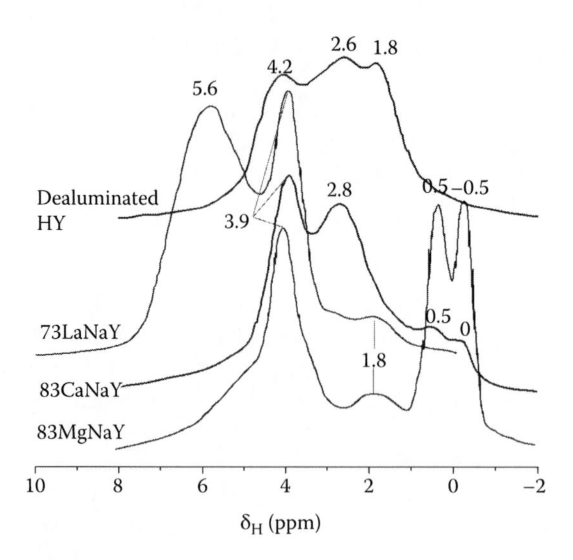

FIGURE 2.97 ^{1}H MAS NMR spectra of zeolites calcined at 433 K and dealuminated HY calcined at 673 K, rotor frequency was 10 kHz for 73LaNaY and 3 kHz for others. (Adapted from *Solid State Nuclear Magn. Reson.*, 6, Hunger, M., Multinuclear solid-state NMR studies of acidic and non-acidic hydroxyl protons in zeolites, 1–29, 1996, Copyright 1996, with permission from Elsevier.)

framework oxygens of zeolite and directed the chloroform to adsorb on framework oxygens with a hydrogen bond. In this adsorption site, the chemical shift of the chloroform hydrogen has a linear dependence on the electronegativity of the cation. The ^{1}H NMR spectra of chloroform adsorbed onto zeolites was decomposed into three to four lines around a narrow peak at $\delta_H = 7.23$ ppm (free bulk chloroform), because these spectra were relatively broad: $\delta_H = 7.5$–8 ppm for HY, 6.0–9.5 ppm for KY, and 5.7–8.7 ppm for BaY. The δ_H value of $CHCl_3$ adsorbed onto zeolite with different cations can be described as a linear function of the Pauling electronegativity (χ): $\delta_H = 9.9905 - 2.48488\chi$. Only Li^+ falls out from this dependence for K, Ba, Na, Sr, Ca, and La. Based on the chemical shift the strength of adsorption sites for $CHCl_3$ decreases in the order K > Ba > Na > Sr > Ca > Li > La (Timonen and Pakkanen 1999).

Hunger (1996) reviewed applications of the solid-state NMR spectroscopy to study Brønsted acid sites in zeolites depending on interactions with probe molecules, dehydrohylation, and structural features of the materials. Silanols are characterized by $\delta_H = 1.3$–2.2 ppm, AlOH can have δ_H from −0.2 or lower to 2.6.3.6 for extra-framework aluminum species. Acidic groups can have $\delta_H = 3.8$–7 ppm (Figure 2.97). In the case of their interactions with strongly basic compounds, proton transfer is possible that results in very large δ_H values. For instance, pyridine adsorption onto HZSM-5 gives $\delta_H = 15.5$–19.5 ppm. The δ_H value or its changes linearly correlate with the hydrogen bond length (the shorter the bond, the larger is the δ_H value) or with the OH-stretching vibration (the smaller the wavenumber ν_{OH} the greater is the δ_H value).

Thus, the NMR spectroscopy as a powerful and complementary technique for acquiring insights on the molecular processes and structures under crystallization is frequently used to study the bulk structure of solids and their surface sites (Epping and Chmelka 2006). However, the NMR technique is also a very powerful tool to study the interfacial phenomena in complex systems (Gun'ko et al. 2005d, 2009d, Mitchell et al. 2008, Petrov and Furó 2009). This is of importance not only for such fine materials as zeolites but also for such "prosaic" matters as cement, concrete, mortar, limewater, slag, etc. (Valckenborg 2001, Wang and Scrivener 2003, Andersen et al. 2004, Korb et al. 2007, Schönfelder et al. 2007, Sun and Scherer 2010, Koptyug 2011, Stark 2011) or such unusual objects as meteorites studied by Bland et al. (2009) with the cryoporometry to characterize the texture of

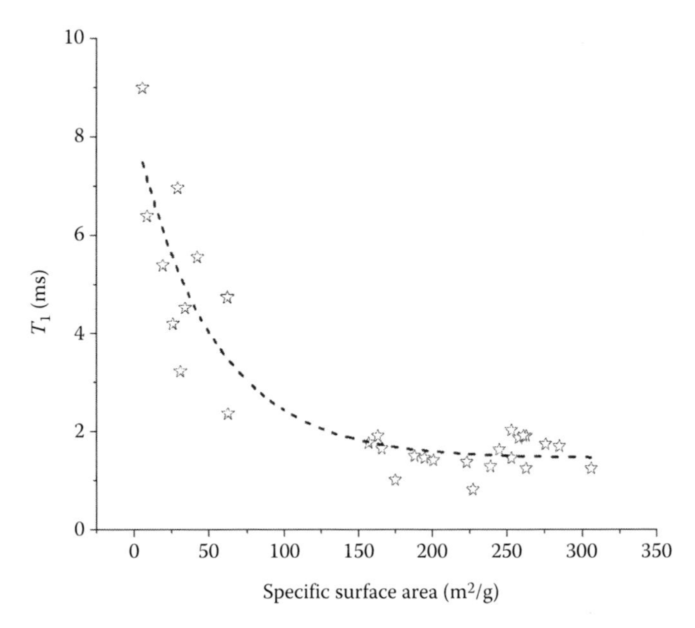

FIGURE 2.98 Spin lattice relaxation time, T_1, as a function of the specific surface area of a set of cement pastes. (Drawn from the data shown by Gran and Hansen, *Cement Concrete Res*., 27, 1319, 1997.)

chondrite meteorites, and showed that these materials are meso/macroporous. Mitchell et al. (2008) and Petrov and Furó (2009) analyzed the NMR cryoporometry applications to different industrial materials including cement, concrete, mortar, rocks, marine sediment, etc. They showed that the number of potential cryoporometric compounds for which there exist experimental estimates of the k_{GB} value is about 25. The k_{GT} value is in the 20–220 K nm range depending on the types of adsorbates and adsorbents. For instance, k_{GT}=21–70 K nm for water adsorbed onto different materials (Gun'ko et al. 2005d, 2009d, Petrov and Furó 2009). Additionally, the k_{GT} values differ for melting and freezing processes.

^{1}H NMR spectroscopy (Figure 2.98) was applied by Gran and Hansen (1997) to study a set of cement pastes of different structural and hydration characteristics, the freeze/thaw cycling effects on the pore structure.

There is relation $1/T_{1,av} = 1/T_{1,bulk} + (S/V)(\lambda/T_{1,S})$ giving the average $T_{1,av}$, where S and V are the specific surface area and pore volume, $T_{1,S}$ is the relaxation time of water at the inner surface of pores, which explains the dependence shown in Figure 2.98. This equation is valid in the case of fast exchange (on an NMR time scale) between fluid structures located in different places of the particles. The $T_{1,bulk}$ is in the order of one second, but $T_{1,S}$ is two–three orders shorter (1–10 ms). Therefore, the narrower the pores and the larger the S value, the shorter is the T_1 time (Figure 2.98).

Sun and Scherer (2010) described in detail the thermoporometry technique based on DSC measurements of the temperature behavior of bound water over a broad temperature range for freezing and melting cycles. They analyzed the pore structure of mortar using thermoporometry in comparison with the results of nitrogen adsorption/desorption and mercury intrusion porosimetry. Typically, DSC thermoporomertry gives the results similar to that of the NMR cryoporometry because both methods are based on the same Gibbs–Thomson relation for freezing point depression for liquids confined in pores.

In the case of cylindrical pores, there are relationships between the temperature shifts and the pore radius with the corrections related to the thickness of the premolten liquid layer at freezing (τ_f) and at melting (τ_m) conditions (Figure 2.99)

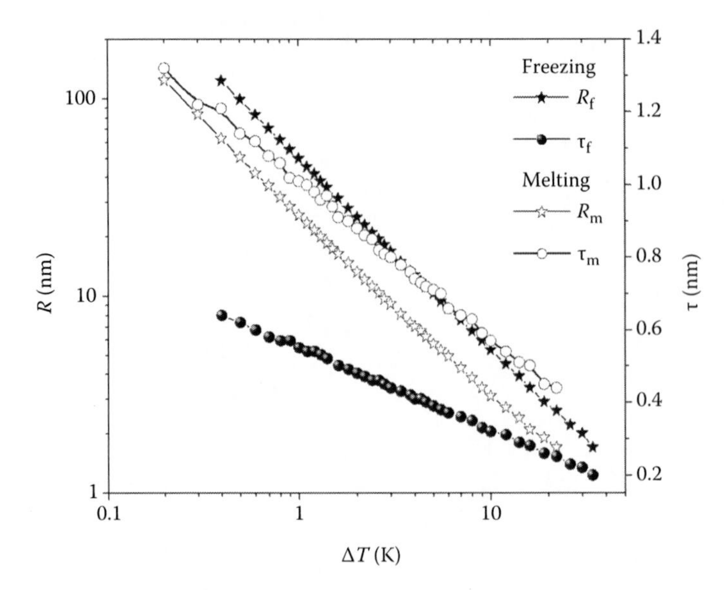

FIGURE 2.99 Relationships between the radius R of cylindrical pores or the thickness of the premolten liquid layer at freezing (τ_f) and at melting (τ_m) conditions and the freezing (ΔT_f) and melting (ΔT_m) temperature shifts for confined water. (Adapted from *Prog. Nuclear Magn. Reson. Spectr.*, 54, Petrov, O.V. and Furó, I., NMR cryoporometry: Principles, applications and potential, 97–122, 2009, Copyright 2009, with permission from Elsevier.)

$$\Delta T_m = -\frac{K_c}{R - (\tau_m - \xi)} \tag{2.15}$$

$$\Delta T_f = -\frac{2K_c}{R - \tau_f} \frac{R - (\tau_f + \xi)}{R - (\tau_f + 2\xi)} \tag{2.16}$$

where

K_c is the cryoporometry constant (similar to k_{GT})
ξ is the potential parameter (Petrov and Furó 2009)

Kopinga and co-workers (vanderHeijden et al. 1997, Valckenborg et al. 2000, 2001, Schönfelder et al. 2007) published a series of papers on NMR cryoporometry and relaxometry. A portion of their results was collected by Valckenborg in the thesis (Valckenborg 2001; Figure 2.100) devoted to NMR investigations of technological porous materials. In general, water in mortar can be divided into non-evaporable and evaporable water (Hansen 1986). Water can be chemically bound to the cement gel. It has a very fast transverse relaxation time ($T_2 \approx 20$ μs) and cannot be detected according to experimental conditions. The evaporable water can be divided into two different states. The physically adsorbed water is bound to the surface of gel particles and occupies the gel pores. The free water, which occupies the capillary pores, is the remainder of the water in the mortar. Therefore, the double exponential behavior of the relaxation of the water reflects the pore geometry (Valckenborg 2001). During drying water is removed from mesopores (Figure 2.100).

Notice that the PSD calculated from the NMR data overlaps with the PSDs calculated from water adsorption and mercury intrusion porosimetry (Valckenborg 2001).

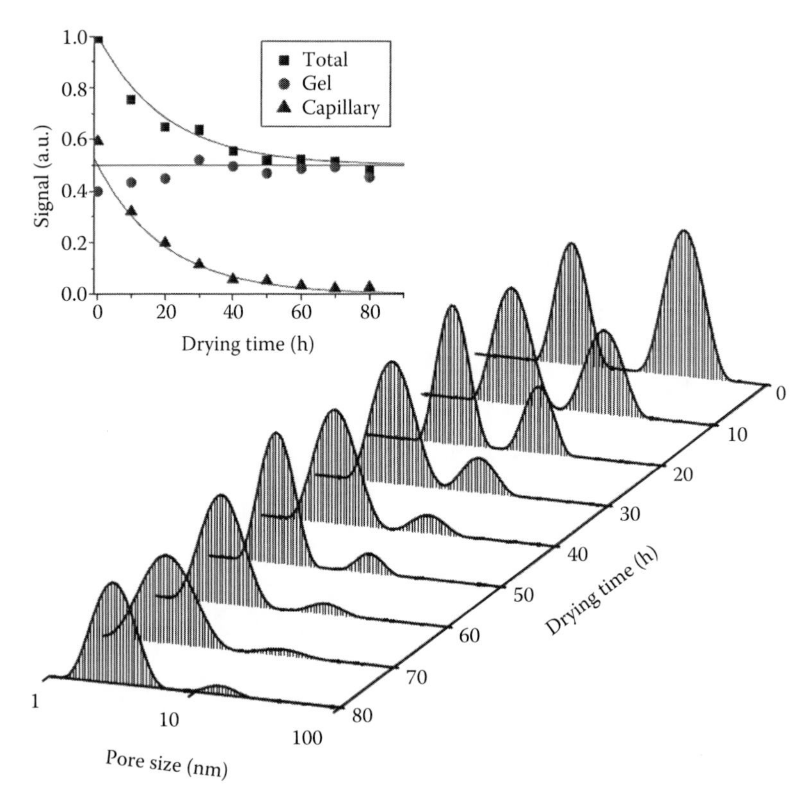

FIGURE 2.100 Bound water distribution in the middle of the mortar sample for 10 h steps as determined from relaxation (T_2) measurements; in the inset, the total amount of water, the water in the capillary pores, and the water in the gel pores is given as a function of time. (According to Valckenborg, R.M.E., NMR on technological porous materials, PhD thesis, Eindhoven, the Netherlands, Eindhoven University of Technology, 2001.)

Llewellyn and Maurin (2005) demonstrated that gas (nitrogen and argon) adsorption microcalorimetry can be used a powerful technique for depth examination of the surface state of adsorbents and a minute following of adsorption mechanisms such as phase changes and transitions. The use of this technique in parallel to the measurements (and appropriate analysis) of the adsorption isotherms of the same gases, and DSC and/or NMR cryoporometry measurements can provide deeper insight into the interfacial phenomena over a broad temperature range.

3 Interfacial Phenomena at Surfaces of Carbon Materials

Carbon materials are the most effective adsorbents to uptake low-molecular compounds and ions from different dispersion media or flows because of a very large specific surface area (1000–3500 m^2/g), large pore volume (1–3 cm^3/g), controlled pore size distribution (PSD) with a great nanoporosity contribution, and a variety of H- and O-containing functionalities (Smisek and Cerny 1970, Gregg and Sing 1982, Bansal et al. 1988, Fenelonov 1995, Adamson and Gast 1997, Bansal and Goyal 2005, Bandosz 2006). Some carbons can also include functionalities with N and other heteroatoms. Nanoporous activated carbons (ACs) are inappropriate to uptake high-molecular compounds because of a major contribution of narrow pores, which are inaccessible for large molecules, especially of a globular shape. However, macroporous or meso/macroporous carbons can be appropriate for adsorption of large molecules, e.g., proteins. Thus, the accessibility of inner pores of particles of carbon adsorbents for different adsorbents can be varied due to controlled changes in contributions from nanopores to macropores to the PSDs and variations in AC granule sizes.

In the case of carbon adsorbents stored or applied in air, water adsorbed from air can partially fill pores and reduce the pore volume accessible for other adsorbates (e.g., toxic gases). In aqueous media, water can be a strong competitor for dissolved polar or charged adsorbates, especially adsorbed in broad pores. Therefore, the interfacial phenomena in the systems with carbon adsorbent—water—adsorbate should be studied with consideration for the textural characteristics as very important for the performance of carbons.

Application of ^{13}C NMR spectroscopy to carbon materials and interfacial phenomena at their surfaces is complicated due to the presence of C atoms in a large number of structures of various types with sp^2- and sp^3-hybridized orbitals (Gregg et al. 1979, Gregg and Sing 1982, Adamson and Gast 1997, Gomes and Mallion 2001, Stein et al. 2009) affecting the shielding of the C nuclei. The ^{13}C chemical shifts can be in a broad range of 20 to 200 ppm with a large number of lines in the NMR spectra (Freitas et al. 2001). Additionally, graphite-like and graphene clusters and other structures with carbon sheets as main structural elements of carbon materials are characterized by anisotropic broadening of NMR signals caused by a strong temperature dependence of their diamagnetic susceptibility (Tabony et al. 1980). Therefore, when recording ^{13}C MAS NMR spectra for carbonaceous materials, it often turns out that we can succeed in observing only one broad signal whose maximum position is determined by the concentration ratio of sp^2- and sp^3-hybridized C atoms.

The overwhelming majority of experimental results that have been achieved by the NMR method to date are related to research into organic substances and a variety of adsorbates bound to different carbon adsorbents. Here, we can distinguish three main directions of studies: measurements of chemical shifts for adsorbed molecules, determination of their dynamic characteristics (mobility, diffusion coefficient, etc.), and investigations of thick layers of a liquid (fluid) adsorbed in pores and on the outer surface of carbon microparticles. Each of these directions permits one to characterize a certain property of carbon materials. Thus, the chemical shift value makes it possible to form judgments on coordination positions of molecules on a carbon surface as well as the enhanced shielding effects in narrow slit-shaped pores (Gradstajn et al. 1970, Boddenberg and Moreno 1977, Voloshchuk et al. 1986, Grosse and Boddenberg 1987, Boddenberg and Grosse 1988, Boddenberg and Neue 1989, Maniwa et al. 1996). The first experiments with recording

^1H and ^{19}F NMR spectra of toluene and benzene adsorbed on graphitized carbon blacks (Spheron, Graphon, and Carbopack; Gradstain et al. 1970) have already shown that such a spectrum may have two signals with up-field shift. Therefore, it was assumed that the main cause of the phenomenon observed was the influence of the local magnetic anisotropy of carbon sheets (especially in narrow slit-shaped pores) and adsorption of a substance on adsorption sites of various types (e.g., O-containing) and in pores of different sizes and shapes. The chemical shift value calculated from local magnetic field strengths proved to be close to that acquired by experiment. Thus, it has been shown that spectral characteristics of signals of adsorbed molecules can be used for identification of the adsorption site and adsorption complex structures (Tabony et al. 1980, Tabony 1980, Harris et al. 1996). The range of the studied carbon adsorbents includes not only ACs, graphitized carbon blacks, and exfoliated graphite (EG) but also carbon nanotubes (CNTs; single wall, SWCNTs and multiwall, MWCNTs), single (SLG) and multilayer (MLG) graphenes, SLG and MLG oxides (SLGO and MLGO), carbon fibers, and carbon-containing nanocomposites. All these materials are characterized by different texture, morphology, and surface composition that result in, e.g., strong, weak, or zero up-field shifts of NMR signals (e.g., ^1H) of adsorbates in comparison with non-carbon adsorbents with similar pore sizes.

Dynamic characteristics of adsorbed molecules can be determined in terms of temperature dependences of relaxation times (Grosse and Boddenberg 1987, Boddenberg and Grosse 1988, Boddenberg and Neue 1989) and by measurements of self-diffusion coefficients applying the pulsed-gradient, spin-echo method (Karger and Ruthven 1992, Heink et al. 1993, Nicholson and Cracknell 1996). Both of these two methods enable one to estimate the mobility of molecules in adsorbent pores and the rotary mobility of molecular groups. The methods are based on the fact that the nuclear spin relaxation time of a molecule depends on the feasibility for adsorbed molecules to move in adsorbent pores. The lower the molecular mobility, the more effective is the interaction between nuclear magnetic dipoles of adsorbed molecules and the shorter is the nuclear spin relaxation time. The results of measuring relaxation times at various temperatures may form the basis for calculations of activation characteristics of molecular motions of adsorbed molecules in an adsorption layer. These characteristics are of utmost importance for application of adsorbents in a capacity of catalyst carriers. They determine the diffusion of reagent molecules toward active sites of a catalyst and removal of the reaction products. Sometimes, the data on the temperature dependence of a diffusion coefficient allows one to ascertain subtle mechanisms of filling in nanopores in ACs.

Of great importance for colloid chemistry and processes of adsorption and adhesion is formation of thick layers of a substance situated in the region of the action of the adsorbent surface forces on the adsorbent–liquid interface. There are not many techniques (within the scope of the NMR method) for research into the structure of thick layers of a liquid adsorbed on the surface of solids, with one of them being the technique of freezing a liquid phase (Strange et al. 1993, 1996, Gun'ko et al. 1995, 1997a,b, 2005d, 2007i, Turov et al. 1996a,b, 1997a,b, 1998b,c, Turov and Leboda 2000). The basis of the technique was formed by the results of investigations on nonfreezing water in aqueous suspensions of adsorbents (Ramseir 1967, Tabony 1980, Bogdan et al. 1996, Hansen et al. 1996). The technique consists in lowering the temperature of a suspension containing an adsorbent that was immersed into the liquid under study at a temperature below its freezing point. Then, in the absence of impurities dissolved in this liquid, its main part freezes. The unfrozen part of the liquid is the one that interacts with the adsorbent surface. The freezing point of the liquid bound to the surface is lowered due to the adsorption interactions. The stronger the adsorption interactions, the lower are the temperatures at which molecules go from the adsorbed state to solid state. Since the relaxation times for adsorbed molecules and solid bodies may differ by several orders of magnitude, at a low value of the spectrometer transmission bandwidth, no NMR signal of the frozen substance is observed, it is possible to discern only the signal of adsorbed mobile molecules of the substance. At any temperature below the freezing point of the liquid, its molecules go from the adsorbed phase to solid phase in the case of the equality between the free energy of adsorbed molecules and that of the frozen substance. In this situation for suspensions of adsorbents, it is possible to determine such important characteristics as the decrease in the free energy caused by adsorption and the thickness of the substance layer that experiences the perturbing influence of the surface. Besides, it is also possible to calculate values of the free surface energy of an adsorbent that determine the total decrease in the free energy of a substance caused by the presence of a phase interface.

3.1 TEXTURE OF CARBONACEOUS MATERIALS AND CHEMICAL SHIFT OF ADSORBED MOLECULES

Adsorption properties of carbonaceous materials are determined by their textural and structural (surface chemistry) characteristics. Depending on the starting raw materials, temperature, and method for carbonization, it is possible to produce carbonaceous materials differing both in their macroscopic parameters (grain size, density, and mechanical properties) and characteristics of the pore structure at the nanoscale level. In conformity to the classification introduced by Dubinin (1960), all the porous materials can be subdivided into three groups, namely macroporous materials with the pore size $d > 50$ nm, mesoporous materials ($d = 2$–50 nm), and nanoporous (obsolete "microporous") materials at $d < 2$ nm. Using various starting raw materials and procedures of activation of carbonaceous adsorbents, it is possible to produce adsorbents with preassigned adsorption properties and a wide range of porosity (from nanoporous to macroporous adsorbents). In some cases, carbonization can result in the formation of carbonaceous materials with an undeveloped pore structure that in a first approximation can be regarded as nonporous. The constitution of carbonaceous adsorbents of various types and methods for researches into their characteristics were expounded in many monographs (Smisek and Cerny 1970, Gregg and Sing 1982, Bansal et al. 1988, Fenelonov 1995, Adamson and Gast 1997, Marsh et al. 1997, Marsh and Rodríguez-Reinoso 2006).

The carbon lattice formation is a very complex multistage process involving thermolysis of simple organic substances, polymers, or natural precursors (fruit stones, shells, wood, etc.) and results in the development of polyaromatic structures. These structures are graphene clusters (containing 15–20 fused benzene rings in each of several sheets aggregated in a graphene cluster) on whose boundaries there are hydrogen atoms and oxidized carbon groups. Such structures can be identified in some types of carbonaceous materials (Donnet and Voet 1976). When primary carbonization products (carbonizates) are exposed to temperatures high enough for chemical reactions to proceed, these graphene clusters interact with each other forming complex carbon aggregates. With increasing temperature and time, the carbonization leads to formation of a regular carbon lattice whose main building blocks are primary graphene clusters.

There are substantial differences in the structural characteristics of materials synthesized under various conditions of pyrolysis of gases (carbon black and carbon-containing mineral adsorbents), carbonization of liquids (cokes), and solids (chars and ACs). It is caused by the fact that materials of the first type are formed under conditions of a free motion of graphene clusters relative to each other, which provides the possibility of a comparatively easy formation of ordered carbon structures. With increasing temperature and time of carbonization, the structural order of such materials increases, and, in the limit, they may be converted to graphitized forms of carbon. In the case of solid carbonized materials, their carbonization proceeds in the presence of a rigid skeleton, and the graphene clusters formed as a result of heat treatment remained to be chemically bound to non-carbonized part of the material. Under such conditions, a regular carbon lattice cannot be formed. The process of its formation can be affected only in very severe conditions involving application of high pressures and temperatures. Such materials are referred to as non-graphitizable (Ruland 1965, Walker 1990, Darmstadt et al. 1994, El Horr et al. 1994, Rannou et al. 1994).

3.1.1 Graphitized Materials

The major successes in studying the structure of carbon black particles are related to application of transmission electron microscopy and x-ray diffraction (XRD) methods. As early as 1934, XRD structure investigations (Warren 1934) showed that carbon black particles were not amorphous (as it was believed earlier); rather, they consisted of ordered carbon layers whose structure resembles that of carbon layers in graphite, though these layers have regular reiteration breaks with respect to a perpendicular to the graphite layers. In view of this, a term "turbostratic structure" was introduced (Warren 1934, Biscoc and Warren 1942), which was used by many authors to describe constitution of graphite and some graphitized materials (Fenelonov 1995). The turbostratic layers of carbon do not exist independently of each other. On the contrary, they are parts of tightly packed crystallites consisting of several parallel or almost parallel layers. The structure of such a crystallite is represented schematically in Figure 3.1 (Bourrat 1993). Positions of fractures of graphite planes are defects of the crystal lattice, which contain sp^3-hybridized carbon atoms. The sizes of a crystallite are its true diameter (L_1) and thickness (L_2). The angle γ characterizes the intrinsic defects of graphite planes. In the case of a turbostratic structure, the interplanar spacing value in a package of graphite planes is higher than in graphite and is equal to ca. 0.344 nm (Biscoc and Warren 1942).

On the basis of the results of investigations on a large number of commercial carbon blacks synthesized under various conditions, it was established (Bourrat 1993) that the texture of all the materials could be described in terms of the characteristics represented in Figure 3.1. Depending on the method for production of a material, its parameters L_1, L_2, and γ may vary. A decrease in γ and increase in L_1 give evidence for ordering of the particle structure. Individual crystallites can be found in carbon black particles, but the surface of each of the crystallites has the turbostratic structure (Bourrat 1993).

The scanning tunnel microscopy revealed a somewhat different structure of carbon black particles (Donnet and Custodero 1992, Donnet 1994, Donnet et al. 1995). The crystallites forming carbon black particles have differently extended structures and exhibit several clearly visible turbostratic bends.

The particle surface has the shape of a curved staircase formed by overlapped crystallites. In the case of both of the types of carbon black particles on the boundaries of crystallites, there can be a great number of hydrogen atoms, oxidized carbon atoms, and broken carbon bonds. Atom force microscopy (AFM) images of carbon black (Tanahashi et al. 1990, Donnet 1994) showed that crystallites are presented on the surface and they have the form of rectangles, but their arrangement can be random. It is evident that the differences in the carbon black particle structure observed by various techniques for diverse materials reflect the diversity of synthesized carbonaceous materials

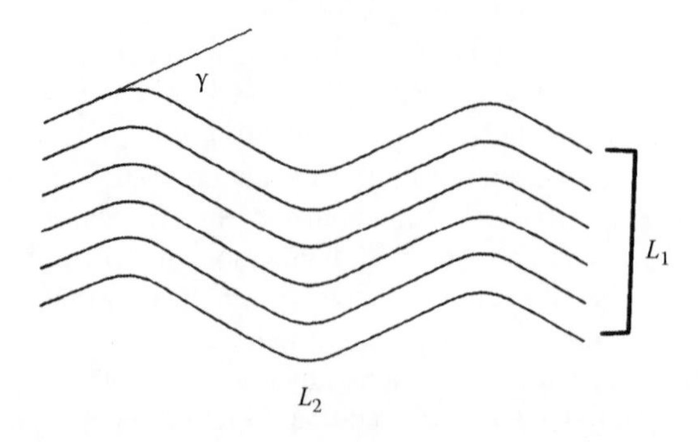

FIGURE 3.1 Schematic representation of the constitution of a carbon crystallite where L_1 and L_2 is the thickness and sizes a crystallite, γ is the angle characterizes concentration of intrinsic defects of graphite planes. (Taken from *Carbon*, 31, Bourrat, X., Electrically conductive grades of carbon black: Structure and properties, 287–302, 1993. Copyright 1993, with permission from Elsevier.)

and strong dependence of their structure on carbonization conditions. The pore structure of carbon blacks is practically undeveloped. Nevertheless, it is possible to expect the presence of a small number of slit-shaped voids between graphene clusters.

3.1.2 Nongraphitized Carbonaceous Materials

ACs belong to the main type of carbonaceous adsorbents that have been used for the past several decades. To a great extent, it is due to a high specific surface of such materials that can amount to 1000 m^2/g and more (Matsumoto et al. 1997, Kruk et al. 1999, Salame and Bandosz 1999). Such high specific surface values provide evidence for the presence in these materials of a substantial number of slit-shaped pores formed by graphene planes. In contrast to interplanar gaps in carbon black particles, the majority of these pores are accessible to adsorbed molecules, especially if $S_{BET} > 2000$ m^2/g. This high accessibility of slit-shaped gaps in ACs is attributed to large values of interplanar spacing in comparison with graphene clusters in carbon blacks. The available transmission electron microscopy (TEM) data (e.g., Figure 3.2a) give ground for construction of two models of the constitution of ACs, namely the model of disordered arrangement of graphene planes (Bradley and Rand 1995) and that of "crumpled paper sheets" (CPS; Huttepain and Oberlin 1990, Stoeckli 1990).

In conformity to both of the models, the material pore structure is formed by gaps between graphene planes of different sizes, as it is shown in Figure 3.2b. This model is in agreement with recent theoretical models (Furmaniak et al. 2011, Palmer and Gubbins 2012). However, in the case of the CPS model, the degree of the material surface order is considerably higher. This model is most commonly employed for ACs with a high specific surface produced by carbonization of polymeric products and a subsequent high-temperature activation of the material. The second model is associated with a relatively low degree of the carbon lattice order and can be used when describing the structure of carbons produced by carbonization of natural materials without special procedures of activation.

The textural features of the constitution of nanoporous ACs, synthesized as a result of pyrolysis of cellulose, were studied by the TEM method (Kaneko et al. 1992), with the method being applied to carbons differing in the temperature of their heat treatment in the argon atmosphere over

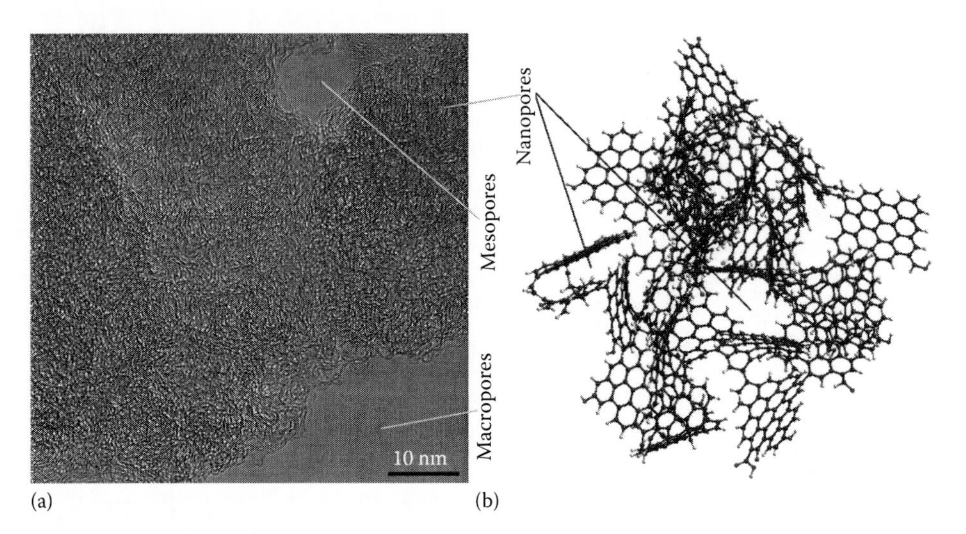

FIGURE 3.2 (a) HRTEM image of AC (S_{BET} = 1664 m^2/g) showing the turbostratic structure and (b) an AC model with the geometry optimized by MM (CharMM force field) and then by semiempirical method PM6. (A: Taken from *Carbon*, 50, Gun'ko, V.M., Kozynchenko, O.P., Tennison, S.R., Leboda, R., Skubiszewska-Zięba, J., and Mikhalovsky, S.V., Comparative study of nanopores in activated carbons by HRTEM and adsorption methods, 3146–3153, 2012a. Copyright 2012, with permission from Elsevier.)

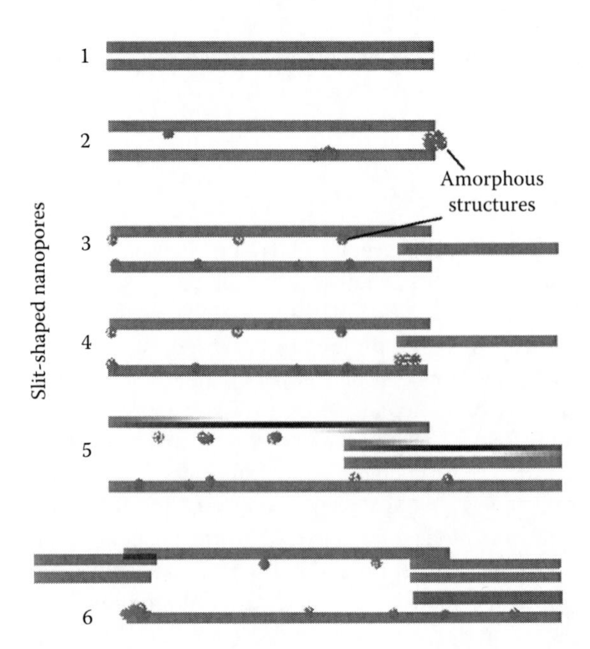

FIGURE 3.3 (1) Nonporous fragment ($x<0.2$ nm with consideration of sizes of C atoms), pores at $x=(2)$ 0.3, (3) 0.4, (4) 0.45, (5) 0.6, and (6) 0.8 nm (small particles shown are carbon clusters with several atoms as structural defect or surface functionalities). (Taken from *Carbon*, 50, Gun'ko, V.M., Kozynchenko, O.P., Tennison, S.R., Leboda, R., Skubiszewska-Zięba, J., and Mikhalovsky, S.V., Comparative study of nanopores in activated carbons by HRTEM and adsorption methods, 3146–3153, 2012a. Copyright 2012, with permission from Elsevier.)

temperature range 1000–3000 K. At low temperatures of activation of an adsorbent, its texture can be represented by a system of bound graphene planes forming a layered structure. Figure 3.3 shows possible schemes of the formation of slit-shaped nanopores at "fixed" sizes (slightly varied due to the presence of disordered small clustered structures) corresponding to the high resolution transmission electron microscopy (HRTEM) PSD with narrow peaks (Gun'ko et al. 2012a). However, non-planarity of the sheets (Figure 3.2) can lead to broadening of these sizes with the appearance of relatively smooth PSDs.

The interlayer distance is rather large owing to the presence of sp^3-hybridized carbon atoms forming chemical bonds with atoms belonging to different carbon layers. In the carbon lattice, there are many defects caused by a nonplanar arrangement of some graphene clusters. In such a material, there are clearly visible portions of a disordered arrangement of graphene clusters (Figure 3.2). At high temperatures of heat treatment, the material order increases. Individual graphene clusters gather into dense graphite-like packages consisting of several graphene layers. Graphene clusters interact with each other forming quite extended layers. The adsorbent pore structure is now formed not by individual graphene layers but by gaps between ordered graphene clusters. The major types of pores in such materials are slit- and wedge-shaped pores that may be both open and partially closed (with one pore orifice closed). The same types of pores are regarded as main types of pores in nanoporous carbonaceous adsorbents in the case of a generalized model of the nanoporous structure of carbonaceous materials, which was constructed on the basis of measurements of adsorption isotherms for a wide range of substances differing in their molecular mass and heats of adsorption (Wickens 1990).

3.1.3 Polar Sites on a Carbon Surface

Oxygen-containing sites on the surface of carbonaceous materials are situated predominantly on boundaries of graphene layers. It is attributed to a great difference in the chemical activity of basal and prismatic faces of graphite. Thus, on the basis of the data collected when cleaving crystals of graphite in a high vacuum (Abrahamson 1973), it was established that the free surface energies of its faces differ by a factor of 13 and more (0.14 and 4.80 mJ/m^2 for prismatic and basal faces, respectively). The same ratio was also observed for reactivity of the crystallographic planes of graphene clusters (Abrahamson 1973). Therefore, even in the cases when carbonization is effected without access for oxygen, on the boundaries of graphene clusters (formed as a result of pyrolysis of the starting material), there remains a large number of reactive groups that react readily with atmospheric oxygen after reestablishment of contact with atmosphere.

The constitution of oxygen-containing sites on carbonaceous materials of various types (Figure 3.4) was a subject of investigations by many authors. These sites strongly affect the adsorption properties of carbons because of significant changes in the fields around the O-containing sheets (Figure 3.4a and c) in comparison with sheets without O-containing sites (Figure 3.4b and d). The first systematized results on O-containing functionalities in carbons were outlined by Boehm (1966). Numerous studies were carried out applying the diffuse reflectance infrared Fourier transform spectrometry technique (Ko et al. 1977, Ito et al. 1988, Meldrum and Rochester 1990), and their results were summarized by Fanning and Vannice (1993). The analogous results concerning the constitution of oxygen-containing sites were achieved from the data by the x-ray photoelectron spectroscopy (Nakayama et al. 1990). The aforementioned results made the foundation for identification of the several types of sites on carbon surface (Brennan et al. 2001).

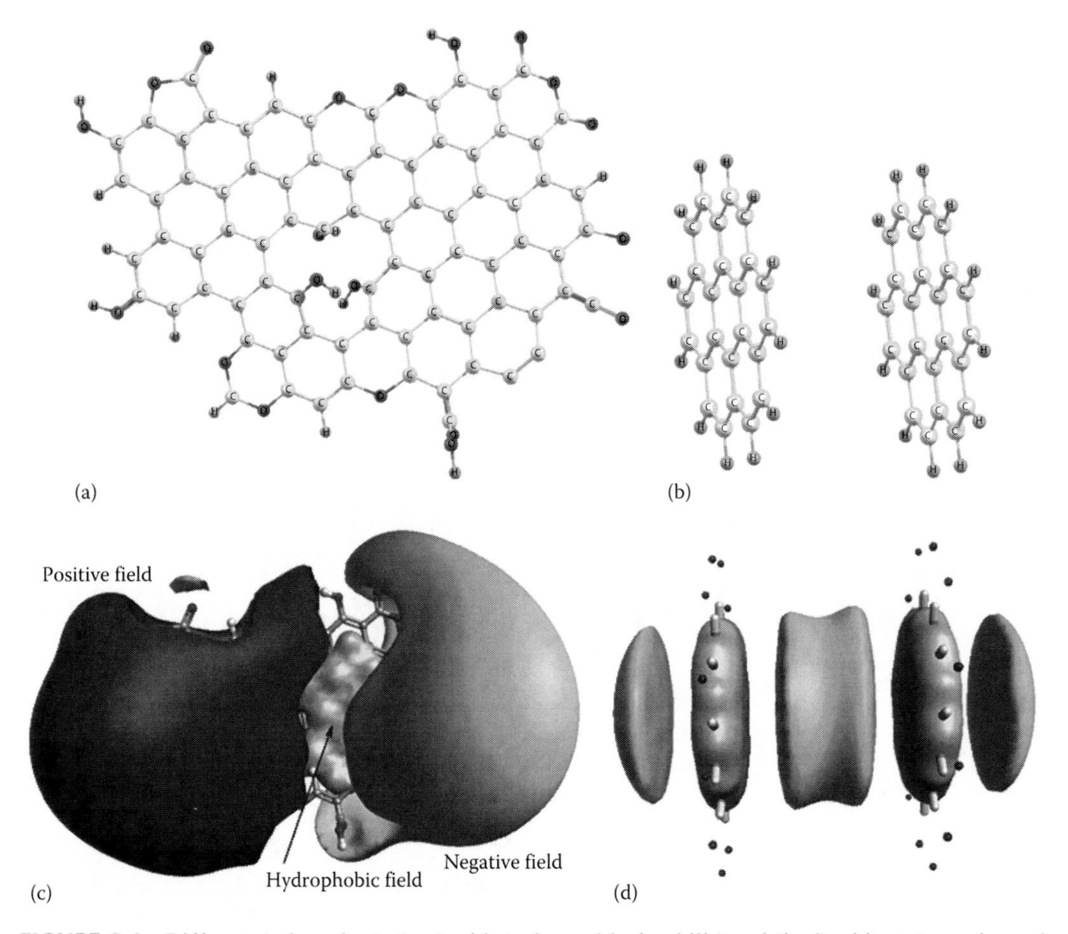

FIGURE 3.4 Different carbon sheets (a, c) with (polar and hydrophilic) and (b, d) without (nonpolar and hydrophobic) O-containing sites (c, d) producing different surface fields (calculated using FieldView 2.0.2 2011); distance between sheets (b, d) is 0.91 nm.

These sites seem to be present to a greater or lesser extent in the overwhelming majority of carbonaceous materials. Oxygen-containing groups can play a significant role during adsorption of substances capable of forming hydrogen-bonded complexes on the surface of carbonaceous adsorbents. Of special importance, their influence may be in the case of nonporous materials whose formation involved overlapping of graphene layers. In such carbonaceous particles, the surface portion built up by prismatic faces may be substantial.

On a thorough analysis of the data on the carbonaceous material surface morphology, it can be concluded that in the case of graphitized adsorbents (carbon blacks and carbon-containing mineral adsorbents), the main types of surface adsorption sites can be graphene layers, fractures of turbostratic structures, formed mainly at the expense of sp^3-hybridized carbon atoms, oxygen-containing groups, and slit-shaped gaps between graphene clusters. In non-graphitized materials, the main types of adsorption sites are slit- and wedge-shaped pores whose interplanar spacing may vary in a wide range, disordered carbon lattice, and oxidized carbon atoms. Depending on the method for production of carbonaceous adsorbents, it is possible to increase concentration of either type of adsorption sites.

3.1.4 Peculiarities of NMR Spectroscopy for Molecules Adsorbed on Carbon Surface

The ring current model (RCM) proposed in 1936 can be used to describe the magnetic properties of aromatic molecules (Lazzeretti 2000), especially to explain the fact that the diamagnetic susceptibility of aromatics is much greater in a direction normal to the plane of the molecule than in the directions parallel to the molecular plane (Lonsdale 1937). According to quantum mechanical perturbation theory (van Vleck 1932, Ramsey 1952), magnetic susceptibility and nuclear magnetic shielding were expressed as sum of diamagnetic and paramagnetic contributions. For the out-of-plane tensor components, symmetry considerations showed that in aromatic molecules such as benzene (Figure 3.5a), the paramagnetic terms should almost vanish, if the origin of the gauge of the vector potential is taken in the center of the ring, and only the diamagnetic contribution to magnetic properties remains. The RCM concept uses delocalization of π-electrons as the fact determining their special mobility in the presence of a magnetic field perpendicular to the molecular

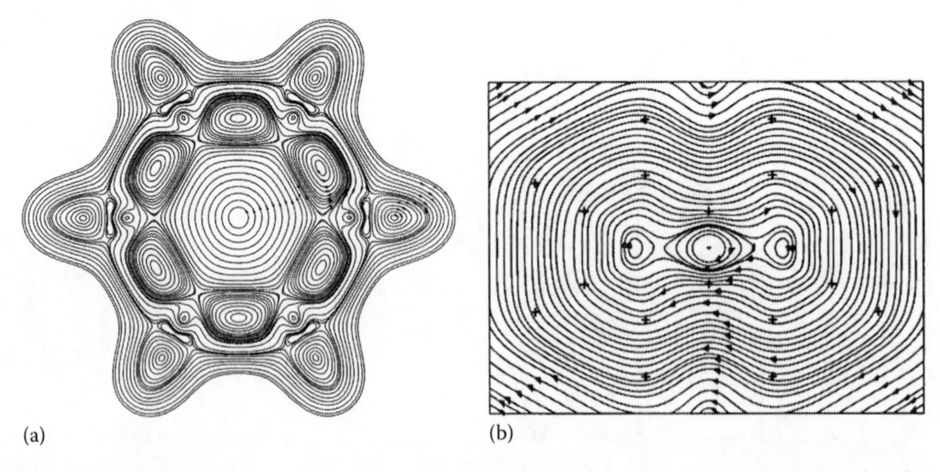

FIGURE 3.5 Projected direction of the current density at points in a plane parallel to the molecular plane of (a) benzene or (b) naphthalene molecule displaced above the molecular plane. (Taken from *Prog. Nucl. Magn. Reson. Spectr.*, 36, Lazzeretti, P., Ring currents, 1–88, 2000. Copyright 2000, with permission from Elsevier.)

plane, and, consequently, the existence of intense stationary currents, which are responsible for enhancement of diamagnetic contributions to the perpendicular component of magnetic tensors.

In fact, the historical RCM can alternatively be referred to Pauling, Lonsdale, and London model an acronym that would avoid possible misunderstanding, in that a magnetic field induces "ring currents" in any charge distribution (Corcoran and Hirschfelder 1980). The chemical shift of benzene observed in proton magnetic resonance was interpreted in terms of ring currents (Pople 1956). He calculated the secondary magnetic field, acting upon a test magnetic dipole at the proton positions, arising from ring currents induced by the primary magnetic field perpendicular to the molecular plane. A simple model with the assumption of equal currents in each ring (Bernstein et al. 1956) was used to interpret the chemical shifts in aromatic systems.

Large basis sets used in ab initio calculations of magnetic susceptibility and nuclear magnetic shielding of aromatics became available in the 1980s. The current density maps (Lazzeretti and Zanasi 1983) provided the first nonempirical RCM for benzene via an all-electron calculations. The results of calculated current density of π-electrons gave clear picture at points in a plane parallel to the molecular plane of (a) benzene or (b) naphthalene molecule displaced above the molecular plane (Figure 3.5). The magnetic field was perpendicular to the plane of figures.

The region of local magnetic anisotropy appears at the plane of two adjacent rings that can provide shielding effect on nuclei localized there. An increase in the number of the rings leads to enlargement of this region. Therefore, the chemical shift of adsorbed molecules should be smaller at the carbon surface than in the bulk. This effect increases in narrow slit-shaped pores because of the effects of both pore walls. The magnetic field of ring currents of a polyaromatic system can be equivalent to the field caused by current occurring through the circuit boundary for this system (as a consequence of the second Kirchhoff law). The RCM method is used for calculations of magnetic anisotropy in different aromatic systems (Coulson et al. 1975, Gomes and Mallion 2001, Compernolle et al. 2006, Mallion 2008, Rai et al. 2010).

The ability of the carbon material surface to influence the chemical shift value of the molecules adsorbed in pores can be used to determine the area of localization of the water clusters (or other substances) in the pores. The local magnetic anisotropy of the system of condensed aromatic rings (grapheme clusters) forms a surface of slit-like pores in ACs. The result is a shielding effect, and the up-field shift of the signal of adsorbed molecules is observed (vide infra). Its value depends strongly on the distance from the carbon surface and the structure of the graphene clusters. The greater the value, the narrower the pore width (due to the rise of contributions from the second wall) and the closer to the wall the adsorbed molecules or molecular aggregate (cluster) is located.

3.2 ACTIVATED CARBONS

ACs can be produced using both natural (coal, wood, fruit stones and shells, etc.) and synthetic (polymers) precursors. The first stage is carbonization of the precursors at high temperature in a certain atmosphere, and the second stage is activation of carbonizates by CO_2, CO, H_2O, etc. at high temperature to open closed pores and to enhance the specific surface area and pore volume. Besides carbon atoms, clusters, sheets, and graphenes, H, O, N, S, and other atoms can be included. Clearly, ACs prepared from natural precursors are more chemically and structurally nonuniform than the materials based on uniform polymers. Theoretically, the specific surface area of ACs can be about 4200 m^2/g if the pore walls are composed of single layer carbon sheets. However, commercial ACs have the S_{BET} value in the 1000–2700 m^2/g range that correspond to multilayered structure of the pore walls. A very large specific surface area, great porosity, chemical nonuniformity of the surface with the presence of different active polar sites, and hydrophobicity of pure carbon sheets cause more complicated interfacial phenomena at a surface of ACs than in the case of silica or other oxides described earlier in Chapters 1 and 2.

3.2.1 ACs Obtained from Natural Precursors

ACs produced from natural precursors are cheap materials that allows deep activation with a greater burn-off degree (i.e., with a great loss of the mass) to obtain the materials with developed porosity (1–2 cm^3/g) and great specific surface area (>1000 m^2/g). However, deep activation results in removal of active functionalities present in the precursors. Therefore, certain ACs produced with natural precursors can undergo weaker activation than results in the materials with lower V_p (<1 cm^3/g) and S_{BET} (<1000 m^2/g) values. From both practical and theoretical points of view, interactions of ACs with water adsorbed from air is important for adsorbents used in air (e.g., in gas masks, gas flow filters, etc.) and solutions (adsorbents working in liquid media, e.g., for purification of drinking water, hemoperfusion, hemofiltration, etc.).

Harris et al. (1996) studied adsorption of water (D_2O) by different ACs using ^2H MAS NMR spectroscopy. The effects of activation to different extents of nutshell and coal-based ACs (Table 3.1) on the adsorption of water were analyzed in respect to the structural characteristics.

Notice that the ACs studied (Harris et al. 1996) were of a low porosity and specific surface area. Separate signals were observed for water adsorbed in nanopores and larger mesopores or at the external surfaces of carbon particles. There were two characteristic ^2H peaks in the MAS NMR spectra. One, the sharper of the two peaks occurred at 0 ppm relative to the external standard and the second broader peak was in –3 to –9 ppm range. The quoted chemical shifts were relative to a sample of liquid D_2O set at 0 ppm. In the case when each sample was loaded with 40% w/w deuterium oxide, without preconditioning under vacuum, and the MAS NMR spectra are shown in Figure 3.6a. Nutshell A exhibits a single narrow peak at 0 ppm (this is SAW), while nutshell B shows a large peak at 0 ppm (SAW) with, in addition, a smaller, broader peak at –9.0 ppm (water confined in narrow pores, WAW). Nutshell C retains a dominant peak at 0 ppm (SAW), but the low-frequency peak (or up-field shift peak) is more pronounced and has moved to –7.9 ppm (water in narrow pores but broader than in the B sample). In nutshell D, the peak at 0 ppm has become the minor of the two peaks, while the major peak has shifted to –7.2 ppm, i.e., contribution of water confined in narrow pores increased but the up-field shift decreases since nanopores become broader (or less slit-shaped). Thus, the observed changes are due to increased porosity and the surface area of AC from A to D samples.

At lower hydration (10% w/w loading of D_2O), the NMR spectra are shown in Figure 3.6b.

Nutshell A shows a small, broad peak at –3.0 ppm and a dominating narrow peak at 0 ppm. Nutshell B has a broad peak centered at –9.5 ppm, and a slight peak at 0 ppm is just evident. The broad peak for nutshell C is at –7.9 ppm, and there is a very small peak at 0 ppm, whilst nutshell D shows a single peak at –6.3 ppm with similar general trends to those in experiment 1 mentioned earlier were observed.

Activated nutshell carbons with S_{BET} of 1200–1300 m^2/g can adsorb about 40 wt% of water (Hall and Holmes 1992) that, however, could be smaller than the V_p value determined from N_2

TABLE 3.1
Structural Characteristics of Nutshell Carbons

Carbon	S_{BET} (m^2/g)	V_p (cm^3/g)	V_{nano} (cm^3/g)
A	310	0.06[a]	0.15
B	400	0.22	0.21
C	480	0.27	0.26
D	570	0.43	0.32

[a] Underestimated value.

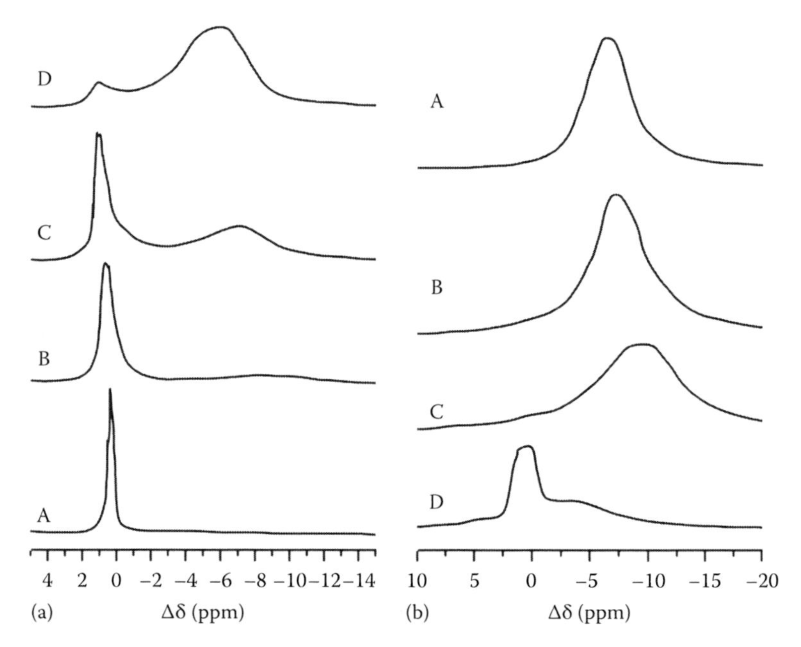

FIGURE 3.6 30.72 MHz ^{2}H NMR spectra of nutshell carbons A to D with a 40% w/w (a) and 10 (b) loading of D_2O adapted. (Taken from *Carbon*, 34, Harris, R.K., Thompson, T.V., Forshaw, P. et al., A magic angle spinning NMR study into the adsorption of deuterated water by activated carbon, 1275–1279, 1996. Copyright 1996, with permission from Elsevier.)

adsorption. This means that the most AC nutshell D could be effectively fully saturated with adsorbate at a 40% loading; it naturally follows that nutshell carbons (A–D), with lower pore volume, will also be fully saturated with D_2O at a 40% w/w. However, Harris et al. (1996) did not consider that water can incompletely fill the AC pores. With this in mind, the ^{2}H NMR spectra (Figure 3.6) can be used to analyze the mechanisms of pore filling by water. Nutshell A has a very low pore volume; therefore, adsorbed water is mainly located on the external surface of AC particles. The NMR spectrum does indeed show a ^{2}H signal at an identical chemical shift and similar peak width to that found for water in the liquid state. Thus, water adsorbed onto A is typical SAW/WBW. It might be expected that a small proportion of the D_2O would be adsorbed in the limited nanopore volume of the material, but it is likely that these pores may be blocked or filled by air or non-removed water, which will prevent ingress of the D_2O adsorbate. Pretreatment of the material under vacuum clearly shows this (Figure 3.7), indicating that additional peak(s) are observed in the spectrum at a lower frequency than that for pure water.

The inference is that vacuum treatment has removed blocking molecules from the limited number of nanopores in the unactivated carbon and allowed ingress of deuterated water. Shifts to lower frequencies of adsorbates have been noted for alcohols adsorbed on carbon and have been assigned to the shielding effects of the graphene planes (Harris et al. 1995). These shifts to lower frequency are thus indicative of adsorption processes occurring in the pore system within the bulk of the adsorbent material.

Yao and Liu (2012) compared results obtained with mercury intrusion porosimetry, constant-rate-controlled mercury porosimetry, low-field NMR spectral analysis (LFNMR), and micro focus computerized tomography for structural characterization of coals. They showed that LFNMR (Figure 3.8) is an efficient tool for nondestructively quantifying the PSD of coals.

However, the LFNMR data (Figure 3.8) were not converted into the PSD for direct comparison with the mercury porosimetry results.

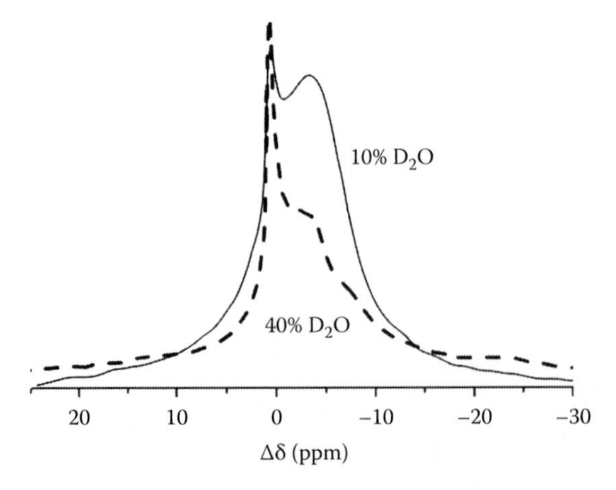

FIGURE 3.7 30.72 MHz ^{2}H NMR spectra for loadings of D_2O on nutshell A, which was subjected to vacuum prior to loading. (According to Harris, R.K. et al., *J. Chem. Soc. Faraday Trans.*, 91, 1795, 1995.)

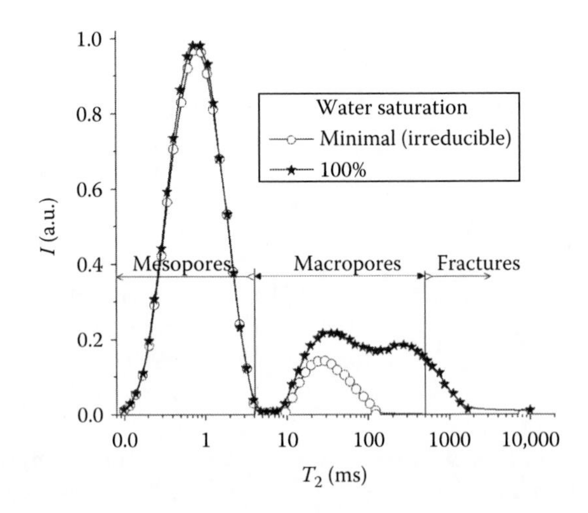

FIGURE 3.8 NMR relaxation of water bound to a coal sample. (Taken from *Fuel*, 95, Yao, Y. and Liu, D., Comparison of low-field NMR and mercury intrusion porosimetry in characterizing pore size distributions of coals, 152–158, 2012. Copyright 2012, with permission from Elsevier.)

Fruit stones are very attractive as AC precursors, and they are frequently used for the production of active carbons (Marsh et al. 1975, Lopez-Gonzalez et al. 1980, Lussier et al. 1994, Leboda et al. 1997) having a wide range of applications for concentration of organic substances present in air or water and separation of light gases (Leboda et al. 1997, 1998).

The starting *a*ctive carbon (APS0) produced from *p*lum *s*tones (PS-*i*; Wood Dry Distillation Works, Hajnówka, Poland) was with the grain fraction of 0.045–0.063 mm. Surface contamination and poorly carbonized parts of the particles were removed from the commercial APS0 by washing with a mixture of mineral acid (HCl) and methanol. This sample was subjected to hydrothermal modification (HTM) in a stainless autoclave of 0.3 dm^3 in volume at 523 K (sample APS1) or 623 K (APS2) for 6 h. For preparation of the oxidized adsorbents (APS1 and APS2), 5 g of the initial APS0 was placed in the quartz thimble. On the bottom of the autoclave was placed 20 cm^3 of 30% H_2O_2. In this way, the HTM occurred in the steam phase. Reduced AC (APSH) was obtained by heating APS0 in a quartz flow reactor in a stream of H_2 at 1073 K for 8 h.

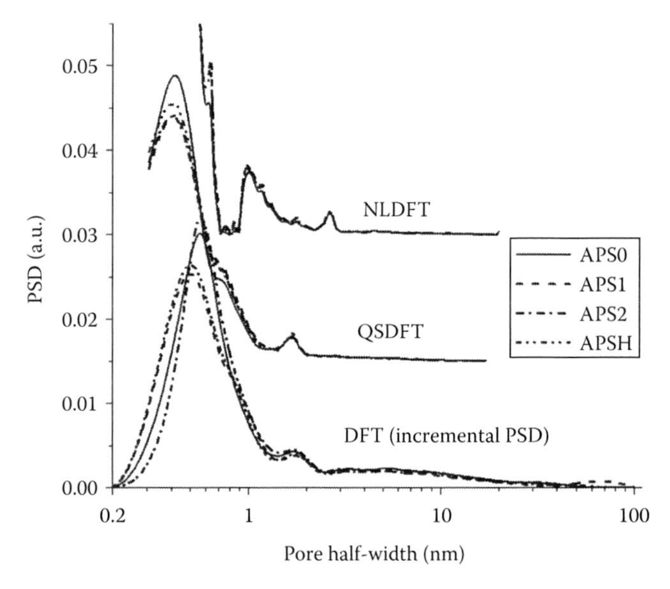

FIGURE 3.9 Pore size distributions for carbons APS0, APS1, APS2, and APSH calculated with three versions of DFT method using slit-shaped pore (DFT and QSDFT) or slit-shaped/cylindrical pore (NLDFT) models.

TABLE 3.2
Structural Parameters of APS-*i* Carbons

AC	S_{BET} (m^2/g)	S_{nano} (m^2/g)	S_{meso} (m^2/g)	V_p (cm^3/g)	V_{nano} (cm^3/g)	V_{meso} (cm^3/g)	V_{macro} (cm^3/g)	Δw	γ_S (in water) (mJ/m^2)
APS0	1088	1018	70	0.72	0.542	0.171	0.007	−0.031	64
APS1	1149	1083	66	0.72	0.552	0.157	0.014	−0.023	66
APS2	1162	1072	90	0.71	0.523	0.182	0.008	0.161	112
APSH	1201	1137	64	0.73	0.570	0.156	0.007	−0.010	32

Nitrogen adsorption isotherms at 77.4 K were measured using a Micrometrics ASAP 2010 adsorption analyzer. The pore size distributions for the materials were evaluated from the desorption data using the density functional theory (DFT) (Do et al. 2001, Gun'ko et al. 2007f), quenched solid DFT (QSDFT), and nonlocal DFT (NLDFT) (Quantachrome software, version 2.02) methods. Additional activation of APS0 results in certain changes mainly in pores at $x<2$ nm (Figure 3.9; Table 3.2). The deviation of the shape from the model of slit-shaped pores is not great (Table 3.2; Δw). It is maximal for APS2 (16%) and 1%–3% for other AC. Therefore, the difference in the behavior of adsorbates can be assigned to changes in the nature of the AC surface after oxidation or reduction.

The textural characteristics (Table 3.2; Figure 3.9) show that APS-*i* samples are not pure nanoporous but contributions of nanopores into the specific surface area and pore volume are predominant. Therefore, one could anticipate strong shielding effects for adsorbates located in narrow pores.

The values of chemical shifts (δ_H) for the substances adsorbed on the surface of the starting, reduced, and oxidized AC are summarized in Table 3.3 also showing differences in chemical shifts for the adsorbed molecules and liquid substance (as regards methane its solution in chloroform was used). The $\Delta\delta = \delta - \delta_0$ value determines the measure of the adsorbent surface effect on the chemical shifts of adsorbed substances.

TABLE 3.3
Chemical Shifts for Molecules Adsorbed on Carbonaceous Adsorbents Based On Stones of Fruits

	Chemical Shift (ppm)									
	Water		Benzene		Acetone		Methane		DMSO	
Adsorbent	$\delta_{T=260}$	$\Delta\delta$	$\delta_{T=260}$	$\Delta\delta$	$\delta_{T=260}$	$\Delta\delta$	$\delta_{T=260}$	$\Delta\delta$	$\delta_{T=260}$	$\Delta\delta$
APS0	0.2	−4.3	−1.6	−8.9	−6.0	−8.2	−6.0	−5.0	3.0	0.5
APS1	−2.8	−7.3	0.2	−7.0	−4.7	−6.9	−6.0	−5.0	—	—
APS2	0.7	−3.8	0.0	−7.3	−5.3	−7.5	−5.5	−4.5	2.8	0.3
APSH	−4.4	−8.9	−2.0	−9.3	−5.3	−7.5	—	—	2.4	−0.1

As follows from the data in Table 3.3, the $\Delta\delta$ values for various adsorbates are different but mainly negative. The maximum effect of the surface on the chemical shift value is observed for benzene (the $\Delta\delta$ value reaches to −9.3 ppm). The chemical shifts for water, acetone, and methane were somewhat lower, namely $\Delta\delta = -(4.0/8.2)$ ppm. In the case of dimethylsulfoxide (DMSO), the $\Delta\delta$ value is close to zero. This can be explained by DMSO adsorption in mesopores where the shielding effect is much smaller than in narrow nanopores.

The $\Delta\delta$ value for various adsorbates is also differently affected by the surface modification of the starting carbon. The most marked changes are observed for water. In this case, the treatment of the surface with hydrogen leads to a substantial increase in the $\Delta\delta$ value (−4.3 ppm for the starting carbon and −8.9 ppm for the reduced one). Oxidation of the carbon with a solution of hydrogen peroxide results in a decrease in the $\Delta\delta$ value, namely for the APS-2 sample, it becomes close to the corresponding $\Delta\delta$ value for the starting adsorbent. In contrast to water in the case of benzene, the $\Delta\delta$ values for carbons oxidized with hydrogen peroxide are lower than for the starting carbon. For the rest of adsorbates, the surface modification exerts a weak effect on the chemical shift value for adsorbed molecules. As previously shown (Turov and Leboda 2000), there are up-field shifts of signals of molecules adsorbed on carbons. This is due to the existence of more or less sizes of π-electron systems of carbon sheets and graphenes strongly affecting adsorbates in narrow slit-shaped pores.

The results, summarized in Table 3.3, can be interpreted in the following way. Molecules of water are adsorbed in slit-like pores that are created by graphene planes. They also adsorb through formation of hydrogen-bonded complexes with O-containing surface functionalities. The ^{1}H NMR signal observed for adsorbed water molecules is an averaged signal for molecules of water adsorbed on these two types of sites. The treatment of the surface with molecular hydrogen results in a decrease in concentration of oxidized carbon atoms. Therefore, the contribution of water molecules in slit-like pores to the averaged signal increases, which manifests itself by the up-field shift of the signal. Oxidation of the surface with hydrogen peroxide leads to an increase in the surface concentration of oxygen-containing groups, which results in an increase in the contribution of water molecules bound to these groups to the total signal.

Methane molecules interact with the surface by the dispersion mechanism. If doing so, they can adsorb both in nanopores and on oxidized carbon atoms and positions of junction of graphene planes. The maximum value of the $\Delta\delta$ parameter for methane is equal to −6 ppm, which is by 3 ppm smaller than for water. On the assumption that graphene plane nanopores affect the chemical shifts for any adsorbed molecules in the same way, it is possible to draw a conclusion that the nanopores contain about two-thirds of all the adsorbed methane molecules.

The fact that for adsorbed benzene molecules the value of the $\Delta\delta$ parameter is maximal seems to be caused by the specific interactions of π-electron systems of benzene and graphenes. Acetone molecules having electron-donor properties are also readily adsorbed in nanopores of the adsorbents studied. In this case, the modification of the carbon surface does not practically affect the proportion of molecules adsorbed in nanopores and on the other types of active sites (the $\Delta\delta$ value varies by approximately 1 ppm). In contrast to other types of adsorbates, DMSO molecules having high electron-donor characteristics practically do not adsorb in nanopores of ACs. This is evidenced for by the absence of any influence of the surface on the shift for the adsorbed DMSO molecules. The major type of active sites interacting with DMSO molecules is likely to be hydroxyl, carbonyl, and carboxyl groups of the AC surface.

Low-temperature ^{1}H NMR spectroscopy measurements were performed with freezing-out of water. Therefore, the influence of this process on the structural characteristics of carbons should be considered (Turov et al. 2002). Additionally, freezing of air-dry samples and samples in aqueous suspensions can differently affect the AC structure because of different filling of pores by water (and ice) and different external conditions.

Fast freezing of AC, produced from PS-*i* with water adsorbed from air by using liquid nitrogen (with subsequent heating and degassing of the carbons), results in an increase in the porosity and specific surface area of the treated samples (Table 3.4) because of enlargement of slit-like pores by ice crystallites possessing larger volume than water droplets. Changes in the PSD are observed for both nanopores and mesopores. They are similar for all PS-*i* samples; however, the amounts of water adsorbed on them in air differ.

Treatments of the same AC in the aqueous suspension more strongly increase the specific surface area and pore volume than treatment of the samples with water adsorbed from air (Table 3.4). In the case of less activated PS1, this effect causes the changes of mainly nanopores at $x < 1$ nm (Figure 3.10). For more strongly activated PS2 and PS3, this pore enlargement occurs for both nano- and mesopores (Table 3.4).

Notice that the suspending effect without freezing of the suspension gives a stronger effect (compare PS** and PS3*** in Table 3.4).

The amounts of unfrozen water (Figure 3.11; Table 3.5, C_{uw}) were evaluated from the ^{1}H NMR signal corrected in according with Curie's law.

Changes in the free energy of water unfrozen in carbon pores dependent on its concentration (on temperature) are shown for PS-*i* samples in different media (Figures 3.12 and 3.13; Table 3.5).

Deuterated compounds (C_6D_6 and $CDCl_3$) used to prevent contribution of the ^{1}H signals of the second liquid phase are not practically mixed with water and they exist in pores as a separated liquid phase, which is in contact with the water phase. The trends in $\Delta G(C_{uw})$ show the vertical sections for PS1 samples in all media (Figure 3.12). Their appearance is linked to the fact that water is unfrozen

TABLE 3.4
Structural Characteristic of PS-*i* Carbons (DFT Calculations)

Parameter	PS1	PS1*	PS1**	PS2	PS2*	PS2**	PS3	PS3*	PS3**	PS3***
S_{BET}	868	929	958	1351	1449	1483	1873	1955	2070	2233
S_{nano}	845	904	931	1283	1372	1406	1737	1810	1923	2058
S_{meso}	23	25	27	68	77	77	136	145	147	175
V_p	0.445	0.471	0.488	0.749	0.812	0.825	1.037	1.091	1.145	1.255
V_{nano}	0.400	0.423	0.437	0.611	0.658	0.667	0.819	0.849	0.904	0.972
V_{meso}	0.044	0.047	0.050	0.133	0.151	0.149	0.215	0.231	0.233	0.280
V_{macro}	0.001	0.001	0.001	0.005	0.003	0.010	0.003	0.010	0.007	0.003
Δw	0.025	0.047	0.045	0.083	0.080	0.089	0.158	0.160	0.157	0.149

Note: The *S* and *V* values are in m^2/g and cm^3/g, respectively. Structural characteristic of carbons after freezing with water adsorbed from *air or **suspension and subsequent drying and degassing to 10^{-3} Torr at 200°C for 2 h, and ***suspending–drying–degassing of AC.

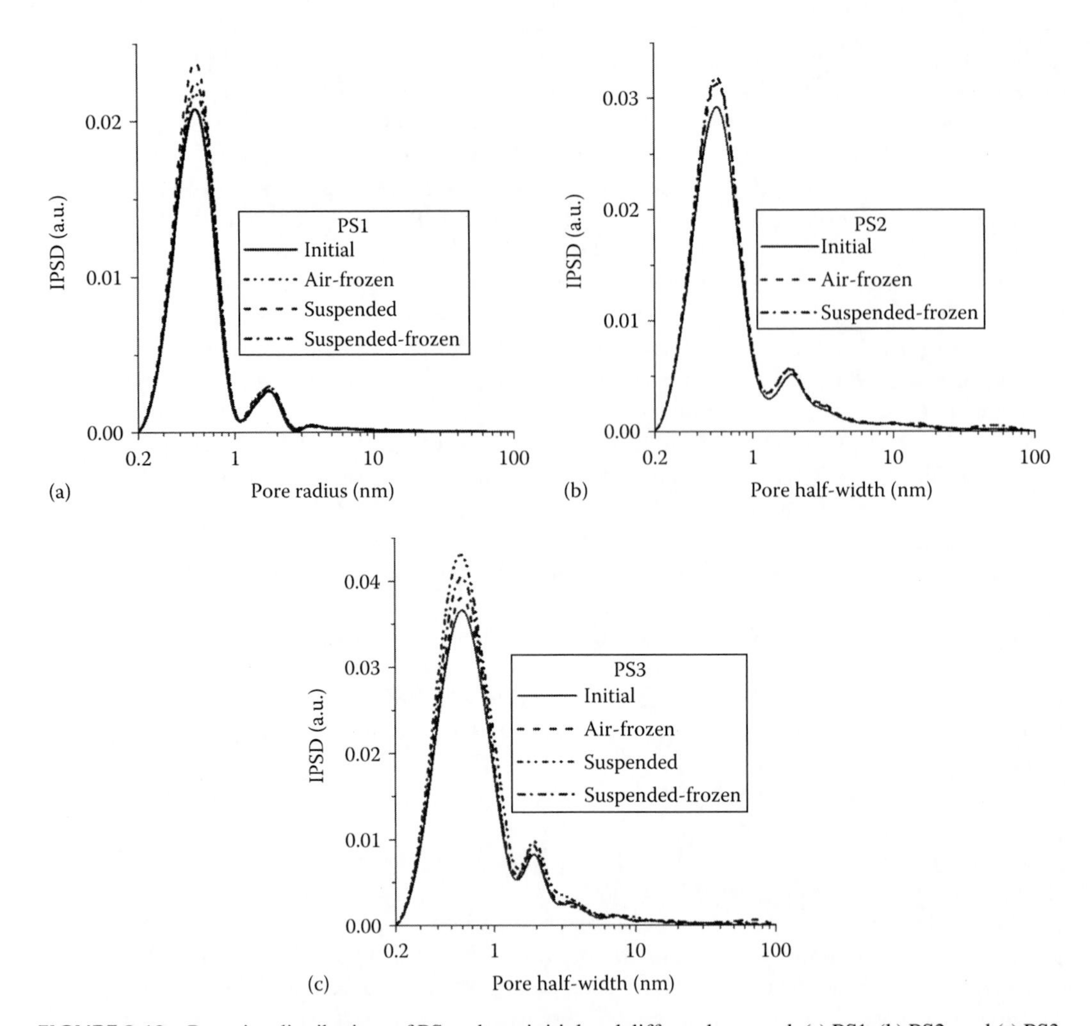

FIGURE 3.10 Pore size distributions of PS carbons initial and differently treated: (a) PS1, (b) PS2, and (c) PS3.

in nanopores even at significant lowering temperature and this water is strongly bound to the pore walls. In contrast to nonporous and mesoporous adsorbents (Turov and Leboda 1999, Gun'ko et al. 2005d, 2009d), the availability of the boundary between water and air for nanoporous carbons wetted in air does not lead to a noticeable increase in $\Delta G(C_{uw})$ in comparison with carbons suspended in liquid water.

This effect can be due to a small surface area in contact with water adsorbed in pores and being in contact with air as a consequence of textural features of nanoporous carbon grains (i.e., water forms large clusters and nanodomains). Replacement of air in pores by benzene-d_6 or especially chloroform-d leads to a significant reduction of the ΔG modulus (i.e., water is practically removed from pores). These changes are demonstrated on the shape of $\Delta G(C_{uw})$ where the sections with a small alteration in ΔG over a large C_{uw} range appear at ΔG values close to zero. This can be assigned to weakly bound water. Thus, a portion of water in pores was transformed from strongly to weakly bound one. This can occur because of two factors: (i) relocation of water to larger pores and (ii) removal of a portion of water from the first interfacial layer (which is possible in pores of the sizes larger than the thickness of two layers of organics plus at least one layer of water). It should be noted that low amounts (several percents) of water frozen in pores provide significant reduction of the pore volume accessible for nitrogen molecules ($\Delta V_p/V_p \sim -0.5$); i.e., water can locate in broad pores. The water clusters or nanodomains removed from the pore walls into organic media can be considered

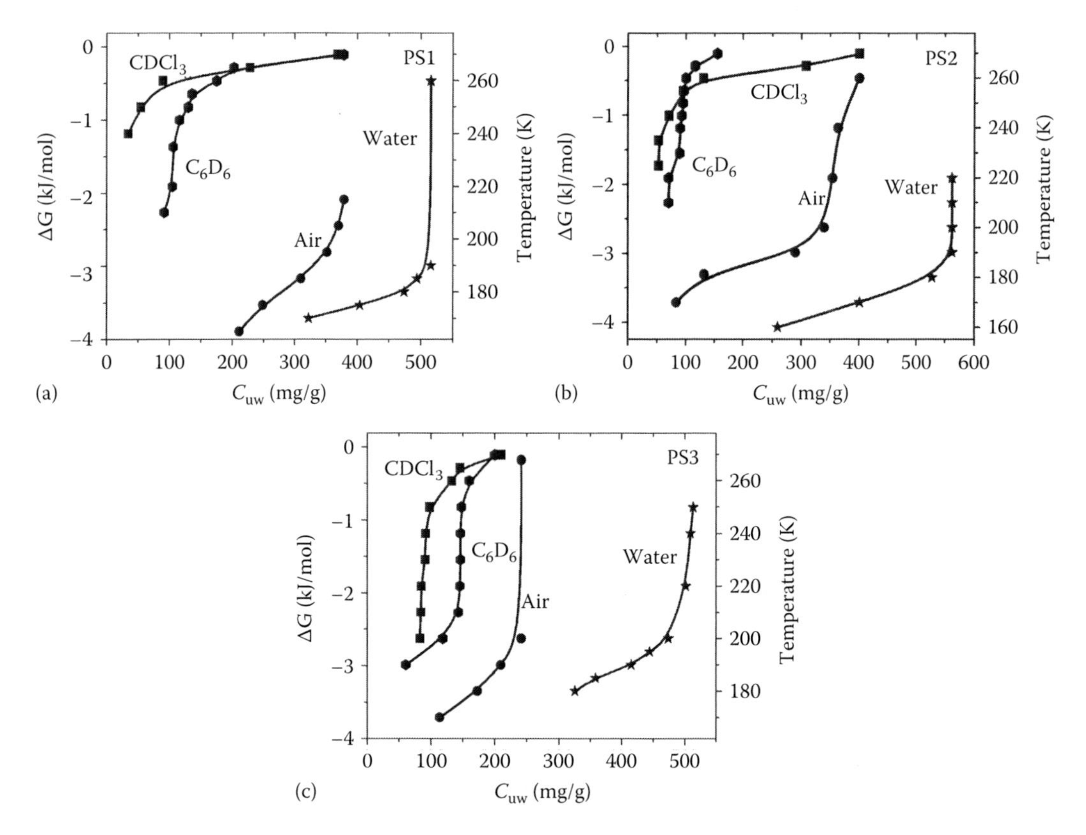

FIGURE 3.11 Relationships between concentration of unfrozen water and changes in Gibbs free energy and melting temperature for water bound to carbons (a) PS1, (b) PS2, and (c) PS3 in different media (samples were placed into chloroform-*d* or benzene-d_6 after their stay in air at room temperature).

as emulsion, which fills carbon pores. The free energy of the interfacial layer depends on the distribution of water droplets in this emulsion.

Volume filling of pores by water begins from narrow pores. Then water occupies larger pores with increasing vapor pressure (capillary condensation). For simplicity, one can assume that with an increase in the amount of adsorbed water, the boundary between water and air shifts toward pores of larger sizes. Since the carbon surface is rather hydrophobic, a major portion of adsorbed water is in the condensed phase locating in pores as nanodomains, which do not spread as a continuous water layer. If the pore volume filled by water is V_p^f that equation for the γ_s value deals with a part of the surface, which is in contact with adsorbed water. Therefore, one can write equation for the effective surface free energy,

$$\gamma_S^* = K\frac{V_p}{V_p^f}\int_0^{C_{uw}^{max}} \Delta G dC_{uw} \qquad (3.1)$$

The shape of $\Delta G(C_{uw})$, shown as an example for PS1 (Figure 3.12a), changes due to filling of pores by different amounts of water in the chloroform medium. Effective value of the surface free energy γ_S^*, which was calculated with Equation 3.1 depending on the degree of pore filling by water, strongly decreases with pore filling by water for PS-*i* carbons in the chloroform-*d* medium (Figures 3.12b and 3.13b). The $\Delta G(C_{uw})$ plots for PS3 at a fixed concentration of water and varied amounts of chloroform-*d* (Figure 3.13a) and dependence of γ_S^* on a relative portion of pores filled by chloroform-*d* (Figure 3.13b) demonstrate a complexity of transformation of the water phase under of chloroform.

TABLE 3.5
Characteristics of Water Bound in Pores of Carbons Being in Different Media

Carbon	Medium	h (mg/g)	C^{u}_{uw} (mg/g)	$-\Delta G_{max}$ (kJ/mol)	γ^{*}_{S} (mJ/m²)
PS1	Air	370	0	5	112
	$CDCl_3$	16	0	5	94
	$CDCl_3$	309	209	5	21
	$CDCl_3$	330	190	5	25
	$CDCl_3$	370	230	5	14
	Water	515	0	4.25	107
PS2	Air	400	0	4	93
	$CDCl_3$	400	220	4	23
	Water	561	0	4.75	123
PS3	Air	240	0	4.3	112
	$H_2O:CDCl_3 = 1:1$	240	0	3.75	100
	$H_2O:CDCl_3 = 1:2$	240	0	3.4	86
	$H_2O:CDCl_3 = 1:3$	240	0	3.5	62
	$H_2O:CDCl_3 = 1:4$	240	110	3.3	41
	$CDCl_3$	240	140	5	48
	Water	501	0	4.5	109

Note: h refers to the total amount of adsorbed water, in the case of the aqueous medium h corresponds to SBW; C^{u}_{uw} is the concentration of weakly bound water; ΔG_{max} is the changes in the Gibbs free energy of the first water monolayer being in direct contact with the carbon surfaces.

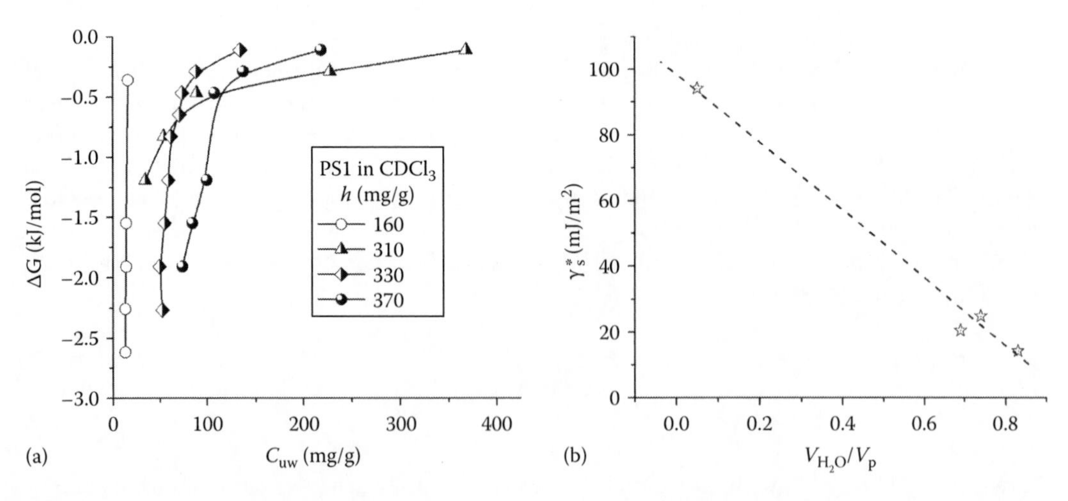

FIGURE 3.12 Influence of the total concentration of water adsorbed on PS1 placed in chloroform-d on (a) changes in the Gibbs free energy versus amounts of unfrozen water and (b) surface free energy as a function of pore fraction filled by water. (Taken from *J. Colloid. Interface Sci.*, 253, Turov, V.V., Gun'ko, V.M., Leboda, R. et al., Influence of organics on structure of water adsorbed on activated carbons, 23–34, 2002c. Copyright 2002, with permission from Elsevier.)

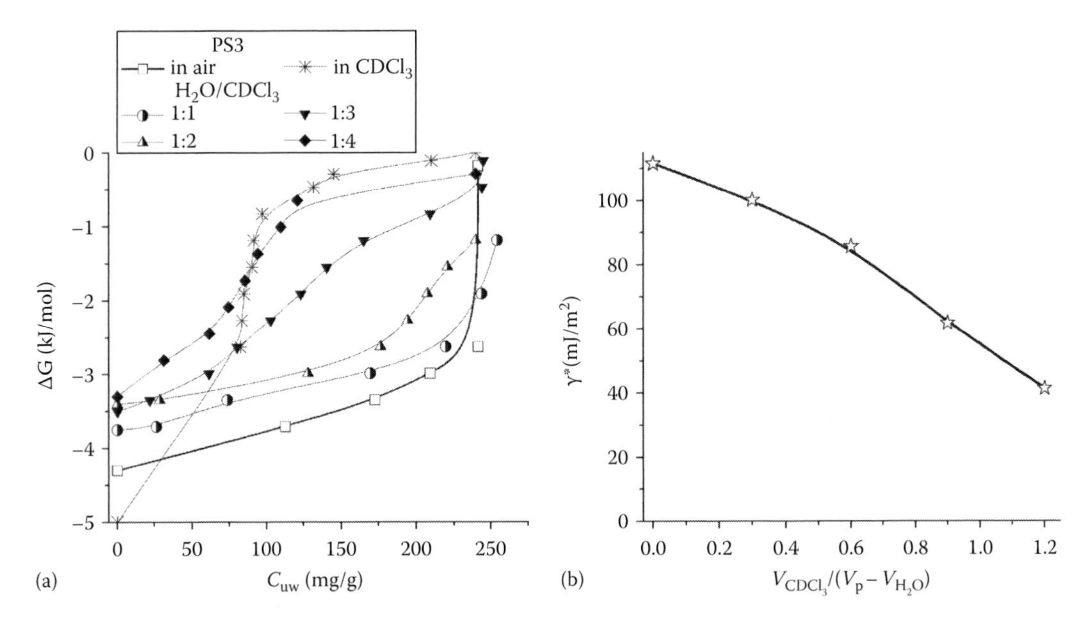

FIGURE 3.13 (a) Changes in the Gibbs free energy of water adsorbed on PS3 being in air or chloroform-*d* versus the concentration of unfrozen water (C_{uw}) at the varied ratio between total concentration of water (h) and CDC13; (b) surface free energy as a function of pore fraction filled by chloroform-*d*. (Taken from *J. Colloid. Interface Sci.*, 253, Turov, V.V., Gun'ko, V.M., Leboda, R. et al., Influence of organics on structure of water adsorbed on activated carbons, 23–34, 2002c. Copyright 2002, with permission from Elsevier.)

This complexity is demonstrated in the shape of $\Delta G(C_{uw})$ function, which at a varied ratio between H_2O and $CDCl_3$ intersect the boundary $\Delta G(C_{uw})$ curves related to PS3 in air and $CDCl_3$ (Figure 3.13a) and in the characteristics of bound water, which are nonlinear functions of the total amounts of water adsorbed (Table 3.5; h).

In the case of APS-*i* carbons, the trends in $\Delta G(C_{uw}$; Figure 3.14) are in agreement with the degree of surface oxidation (Table 3.6). A shift toward larger amounts of unfrozen water is observed, and these curves are similar to those (at $\Delta G < -0.5$ kJ/mol) obtained for water adsorption on PS-*i* (Figure 3.11). An initial portion of the curves (slope at $\Delta G < -2.5$ kJ/mol) and their vertical parts correspond to water adsorbed in narrow pores. The amounts are related to the pore volumes (assuming specific density of water in pores equals to approximately 1 g/cm^3) and they are similar or smaller than the volume of nanopores (Tables 3.2 and 3.4). Despite similar values of V_p and V_{nano} for APS-*i* samples (Table 3.2), the portions of strongly bound water differ significantly depending on the degree of surface oxidation (Figure 3.14; Table 3.6). This is the result of cluster (from dimers to pentamers or larger) and droplet (dozens of molecules) formation in the vicinity of oxygen-containing groups. This phenomenon does not occur on nonpolar graphite-like sheets. The differences in the textural and chemical properties of carbons result in noticeable differences in the plots of NMR cryoporometry PSD (Figure 3.15). These results were also conditioned by incomplete filling of APS0 and APSH pores by water. The filling degree is higher for oxidized carbons APS1 and APS2. The average ice melting temperature $\langle T \rangle$ is maximal for the reduced carbon APSH (226.9 K; this water is located in narrow nanopores, see $\Delta\delta_H$ in Table 3.3) and it decreases from 211.7 K (APS1) to 206.5 K (APS2) with increasing oxidation degree.

However, minimal $\langle T \rangle = 168.7$ K is for APS0 because of minimal pore filling (~50%) by water located in nanopores ($V_{water} < V_{nano}$ for APS0 and APSH). Incomplete filling of pores by water affects the NMR cryoporometry PSDs, which are lower for APS0, APS1, and APSH than N_2 PSD (Figure 3.15b).

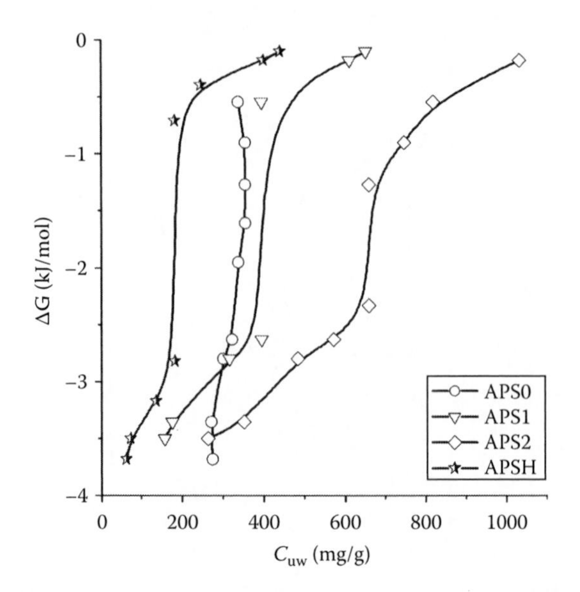

FIGURE 3.14 Changes in the Gibbs free energy of water adsorbed on APS carbons versus the concentration of unfrozen water. (Taken from *J. Colloid. Interface Sci.*, 253, Turov, V.V., Gun'ko, V.M., Leboda, R. et al., Influence of organics on structure of water adsorbed on activated carbons, 23–34, 2002c. Copyright 2002, with permission from Elsevier.)

TABLE 3.6
Concentration (meq/g) of Surface Functional Groups Calculated Using Boehm Titration Method

Adsorbent	pH	–COOH–	–COO–	>C–OH	>C–O	Basic
APS0	5.4	0.208	0.191	0.203	0.201	0.226
APSH	10.43	0	0	0.076	0.041	0.512
APS1	4.82	0.309	0.145	0.316	0.209	0.154
APS2	6.11	0.250	0.163	0.322	0.215	0.257

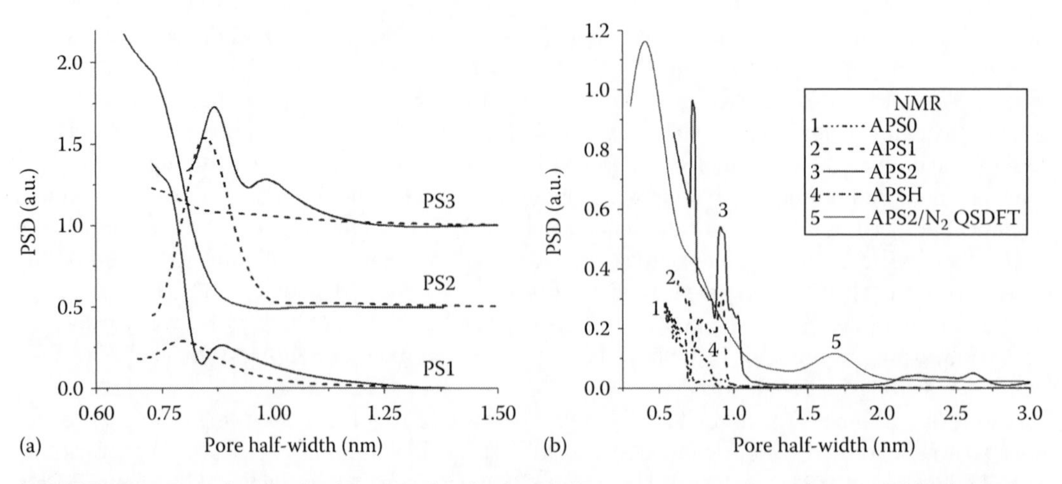

FIGURE 3.15 Pore size distributions calculated using NMR cryoporometry for carbons (a) PS in air (dashed lines) or suspended in water (solid lines) and (b) APS-*i* suspended in water.

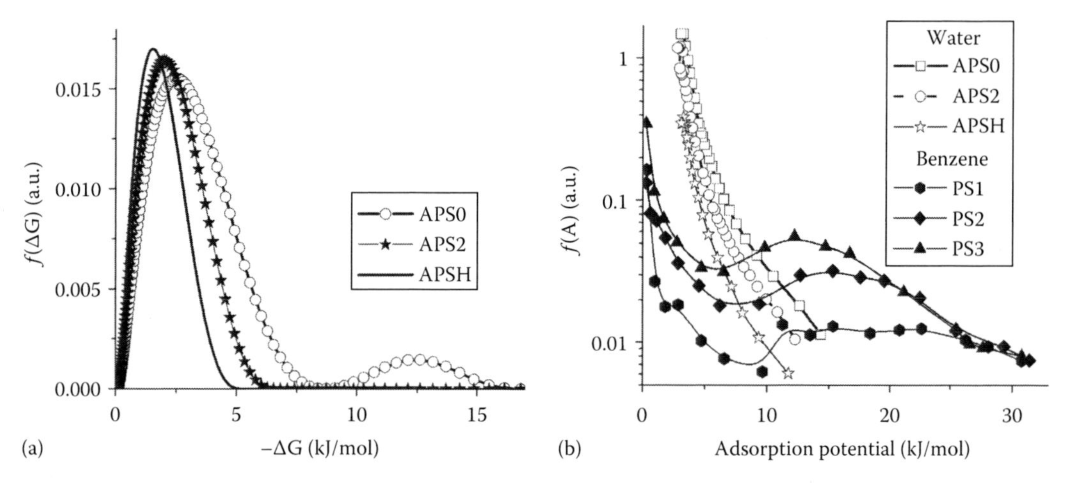

FIGURE 3.16 Distribution functions of (a) free energy of water adsorption and (b) adsorption potential for water and benzene adsorbed onto APS-*i* and PS-*i* samples.

In the case of APS2, the NMR cryoporometry PSD intensity is much higher and the volume of bound water (1.03 cm^3/g) is larger than $V_p = 0.71$ cm^3/g. An increase in the pore volume can be caused by a larger ice volume than that of water completely filling pores of APS2. Similar freezing effects on the textural characteristics were observed for different materials (Gun'ko et al. 2003f, 2005a).

The differences in the degree of oxidation of the APS-*i* samples (Table 3.6) result also in the differences in the energetic characteristics of water adsorption (Figure 3.16). Additional oxidation of the surfaces (APS2) leads to an increase in the ΔG modulus (Figure 3.16a) and an increase in the water adsorption potential (Figure 3.16b) in comparison with reduced APSH. It is worth mentioning that the adsorption potential for benzene is significantly larger in narrow pores ($A > 15$ kJ/mol) than that for water (Figure 3.16b), which is in agreement with a decrease in the affinity of the carbon surfaces for water when benzene is present (Figure 3.11). This happens because of the replacement of water with benzene from low-polar and hydrophobic patches of the carbon surfaces.

This process can occur due to an increase in the contact angle between adsorbed water droplets and carbon surfaces leading to thinning of droplet "foots" bound to the surfaces their breaking down and formation of unbound droplets in the organic medium (but with a minimal contact area between them).

Based on the obtained data, one could find some regularity in the properties of water in pores of ACs prepared from natural precursors. Firstly, weakly bound water exists in pores if a portion of the pore volume filled by this water is larger than some minimal value (corresponding to the volume of narrow pores) or a certain amount of organics (chloroform-*d*) is also present in the pore system. Secondly, the amount of weakly bound water increases with increasing fraction of the pore volume filled by water in the chloroform medium and with increasing a portion of the pore volume occupied by chloroform at a fixed concentration of water. The value of effective surface free energy is close to 100 mJ/m^2 for PS-*i* carbons in both air and aqueous media. This occurs also at a low degree of pore volume filling by water in the chloroform-*d* medium. The benzene-d_6 medium also reduces the free energy of water interaction with the carbon surfaces. This effect is smaller than that for chloroform-*d* (Figure 3.11). These changes in the behavior of water adsorbed on carbons in the presence of poorly water-soluble organics having stronger dispersive interactions with graphene sheets than water are likely due to relocation of adsorbed water and its adsorption in larger pores or on the external surfaces of carbon grains.

A portion of water nonremoved from narrow pores (characterized by a maximum adsorption potential [Figure 3.16b]) by organics corresponds to approximately 10% of the total pore volume. Based on the main contribution of dispersive interactions to adsorption on hydrophobic patches of the carbon surfaces, one could assume that adsorbed water should be removed simultaneously from all the pores. This should occur due to the fact that a portion of the surfaces containing oxygen groups (mainly on the edges of graphite-like sheets at the narrow pore entrances) is smaller than the surface area of unoxidized graphite sheets. However, even in the case of absolute excess of chloroform-*d*, a small amount of water adsorbed in nanopores remains strongly bound to the carbon surfaces, and it is practically insensitive to the presence of $CDCl_3$. The explanation of this phenomenon can be in a significant affinity for association of water adsorbed on oxygen-containing groups and forming of dense droplets or clusters there. This is possible because the interaction energy between water molecules and carbon surfaces is lower (15–32 kJ/mol; Brennan et al. 2001) than that between water molecules in a 3D space (up to 44–45 kJ/mol for bulk condensation) or between organics (e.g., benzene; Figure 3.16b) and carbons.

Notice that freezing of carbons with adsorbed water leads not only to the increase in the porosity of dried carbons (Table 3.4) but also to enhancement of the adsorption potential with respect to nitrogen. Therefore, one can assume that the estimation of changes in the Gibbs free energy of adsorbed water by using the ^{1}H NMR technique with freezing-out of water gives slightly overestimated values. However, relative changes for different samples remain the same, as the profiles of the PSDs and $f(A)$ for initial and frozen samples can be similar (Gun'ko et al. 2003f, 2005a).

Thus, the interactions of water with carbon can be considered as interactions of clusters or nanodomains localized near oxygen-containing groups. In those entities, water molecules interact with each other stronger than with hydrophobic patches of the carbon surface. It is important to mention that for removal of water molecule from a dense water cluster or a nanodomain formed near oxygen-containing groups, several hydrogen bonds should be broken. It follows that the water droplets could be substituted by organics from broad pores as a whole entity but such a process in some narrow pores is likely suppressed because of the steric factor and interaction of water molecules with oxygen-containing groups there.

Thus, interactions of water with ACs produced from PS-*i* depend on the structural characteristics and the chemical properties of the adsorbents as shown for PS-*i* and APS-*i* carbons. The values of the surface free energy are similar to each other for some samples (PS-*i* and APS2). The minimum value was found for reduced carbon APSH. Chloroform-*d* and benzene-d_6 immiscible with water can replace a significant portion of water from nanopores of ACs. This process leads to a noticeable decrease of the surface free energy and localization of water in mesopores or macropores or on the outer surfaces of carbon grains near oxygen containing groups. Water remaining in the pores of carbons exposed to adsorption of organic species occupies approximately 10% of the total pore volume and likely adsorbs in narrow pores where interactions with surface functional groups are possible. Probably, more intensive "dehydration" of ACs can be imposed by adsorption of organics on adsorbents possessing developed mesopores and macropores. Such carbons seem to be more appropriate for purification of air from organics.

Several ACs differently pretreated (deashed, oxidized, reduced, wetted, frozen, and dried) were investigated using static (equilibrium) and dynamic adsorption of probe compounds (nitrogen, benzene, and *tert*-butylbenzene [TBB]; Gun'ko et al. 2006c). Treatments of carbons at relatively mild conditions leading to not great changes in their textural characteristics affect the TBB breakthrough concentration because of changes in the chemistry of the surfaces (oxidized or reduced) and the presence of water in airstream or pre-adsorbed on carbon beds. Calculations of the distribution functions of adsorption potential (A), energy (E), Gibbs free energy (ΔG) of adsorption of benzene and TBB, and effective adsorption (first-order) rate pseudoconstant β_e over different ranges of relative exit TBB concentration $c(t)/c_0$ (from $\approx 10^{-5}$ to 0.01–0.4) revealed nonlinear effects caused by the size of carbon granules, the pore size distributions, the presence of water, oxidation or reduction of the surfaces, and other treatments resulting in distribution functions $f(y)$ with nonzero intensity in relatively broad ranges of $y = A$, E, ΔG, and β_e (Gun'ko et al. 2006c).

Krutyeva et al. (2009) used combined NMR cryoporometry, relaxometry, and diffusometry to characterize porous carbon materials. They studied two carbon molecular sieves (CMS; Takeda 4A and Takeda 5A, Takeda Chemical Industries Ltd.) and three AC (PK13 [Norit], RGC30 [Nuchar], and MAC-LMA12; produced from olive stones, Laboratorio de Materiales Avanzados). CMS and PK13 were mainly nanoporous, but RGC30 and MAC-LMA12 were with both nano- and mesopores. Pore characterization with NMR cryoporometry (melting point depression of the confined liquids), NMR relaxometry, and diffusometry (confined space effects on the random motion of the molecules) gave compatible results with sorption experiments. It should be noted that NMR relaxometry and diffusometry (acetone, ethanol, toluene, *n*-decane, and nitrobenzene were used as adsorbates) can be applied to broader pores (up to 20 μm) than NMR cryoporometry (nitrobenzene as an adsorbate). However, the PSD were not calculated from the nitrogen adsorption data (despite N_2 isotherms were given) that could be used for direct comparison with the PSD obtained with the NMR techniques.

3.2.2 ACs Obtained from Synthetic Precursors

3.2.2.1 Location of Adsorbates in Porous Media

Structural and energetic differentiation of water and water/organic mixtures adsorbed onto ACs with narrow intraparticle pores and broad pores between primary particles was studied using ^1H NMR spectroscopy with layer-by-layer freezing-out of adsorbed liquids and adsorption methods (Gun'ko et al. 2008a,d, 2011d). Spherical porous particles of phenol–formaldehyde resin (MAST Carbon Technology Ltd., Guildford, United Kingdom; Tennison et al. 2004) were carbonized in the CO_2 atmosphere at elevating temperature from ambient to 1073 K at a heating rate of 3 K/min (sample C-0) and additionally activated by CO_2 at 1173 K at 47% burn-off (sample C-47). Carbon particle size was 0.15–0.50 mm. The textural and adsorption characteristics of carbon adsorbents C-0, C-47, and similar materials were studied in detail (Gun'ko and Mikhalovsky 2004, Melillo et al. 2004, Mikhalovsky et al. 2005a, Gun'ko et al. 2006g, 2008a,d, 2011a, 2012a). Several solvents (acetonitrile, acetone, and DMSO), miscible (or soluble) or immiscible (chloroform) with water, were used in the deuterated form to avoid their contribution to the ^1H NMR signal of unfrozen adsorbed water at $T<273$ K. Additionally, several solvents in the H-form (acetone, DMSO, and benzene) and methane adsorbed after preadsorption of water were used to study their mutual effects on the interfacial behavior of water/organics mixtures. Two ACs (PS3 described earlier and CS2 produced from coconut shells [CS], Gryfskand Hajnowka, Poland), having nanopores similar to that of carbon C-47 (Figure 3.17a) were used to study the adsorption of methane affected by preadsorbed water.

Carbon adsorbents studied differ in origin but being nongraphitizing ones should follow the model of crumpled sheets of paper (Huttepain and Oberlin 1990). However, we believe that in phenolic carbons (C-0, C-30, C-47, C-86, and others), there is strong similarity between the morphology of precursor and final carbon particles. Thus nanoporosity in phenolic carbons is created not only by slit-shaped nanopores inside nanosized (20–40 nm) spheroid carbon grains but also by voids between these particles derived from domains in resin precursor. These grains are aggregated into network of randomly shaped clusters separated by interconnected channels and voids that are meso- or macropores (Tennison 1998, Gun'ko 2008a,d, 2011a,d).

The studied carbons possess a significant nanoporosity giving a major contribution to the S_{BET} (all samples) and V_p (CS2 and PS3) values (Figure 3.17; Table 3.7; S_{nano} and V_{nano}). However, the total nanoporosity of CS2 and PS3 (Table 3.7; S_{nano} and V_{nano}) is higher than that of C-0 and C-47 because of larger contribution of supernanopores (PSD peak at $x=0.7$–0.8 nm). Carbon C-47 (as well as C-0) has broader PSD.

The adsorption of nitrogen onto C-47 at low pressures $p/p_0<0.05$ was close to that for CS2 (Figure 3.18a). This similarity is due to the similarity in the nanoporosity of these samples reflected in similar PSD functions of nanopores (Figure 3.17a; peak at 0.5 nm). However, the difference in the total porosity of the studied carbons leads to a significant difference in the adsorption potential distributions $f(A)$ (Figure 3.18c), especially at $A<3$ kJ/mol corresponding to filling

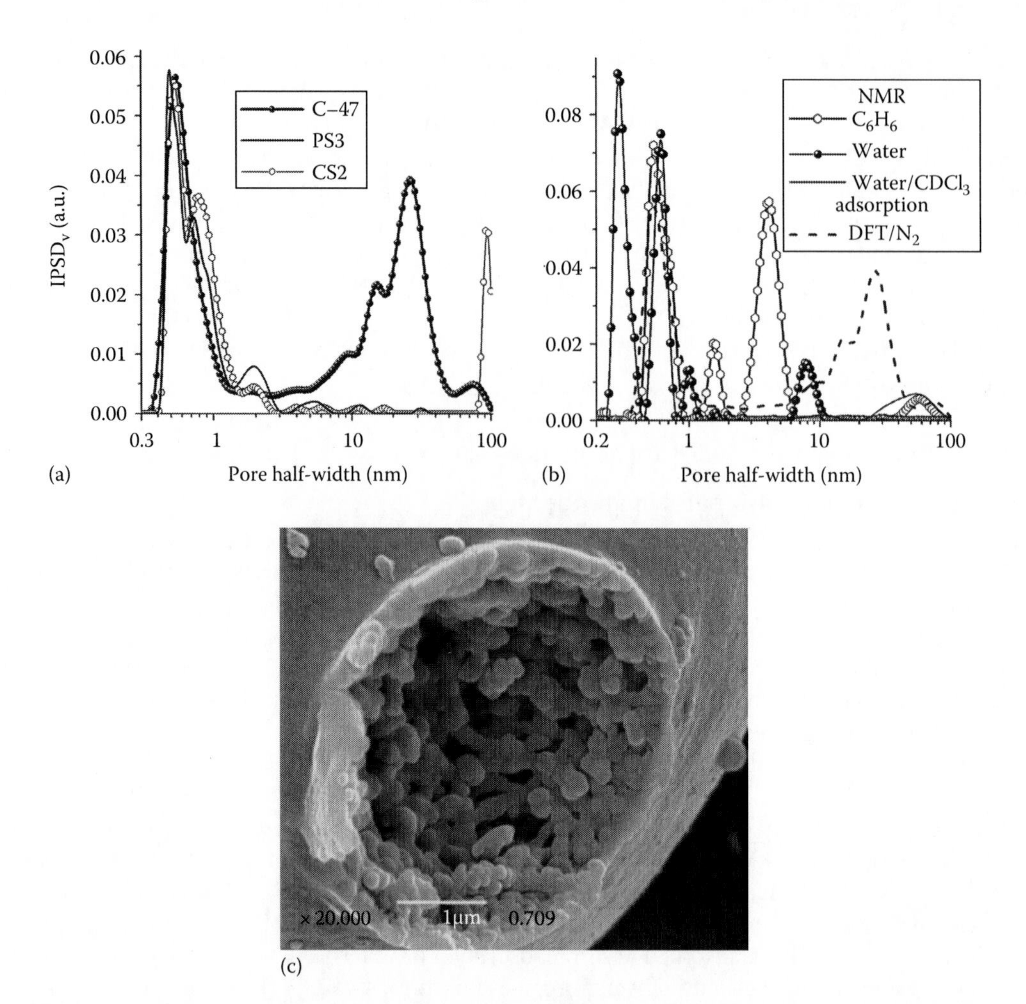

FIGURE 3.17 DFT incremental pore size distributions for activated carbons (a) C-47, PS3, and CS2 calculated on the basis of the N_2 adsorption data and (b) size distribution functions of structures with unfrozen liquids calculated with IGT equation and NMR cryoporometry for adsorbed water (alone and with the presence of chloroform-*d*) and benzene, and DFT PSD on the basis of nitrogen adsorption on carbon C-47, and (c) SEM image of a granule of C-0. (a, b: Taken from *Appl. Surf. Sci.*, 254, Gun'ko, V.M., Turov, V.V., Kozynchenko, O.P. et al., Characteristics of adsorption phase with water/organic mixtures at a surface of activated carbons possessing intraparticle and textural porosities, 3220–3231, 2008d. Copyright 2008, with permission from Elsevier.)

TABLE 3.7
Structural Characteristic of Carbons

Parameter	C-0	C-47	C-86	CS2
S_{BET} (m²/g)	579	1648	3463	2164
S_{nano} (m²/g)	520	1493	2181	2040
S_{meso} (m²/g)	52	142	1279	124
S_{macro} (m²/g)	7	13	3	0.1
V_p (cm³/g)	0.956	1.876	2.320	1.048
V_{nano} (cm³/g)	0.270	0.742	1.314	0.902
V_{meso} (cm³/g)	0.425	0.753	0.887	0.143
V_{macro} (cm³/g)	0.261	0.382	0.119	0.003

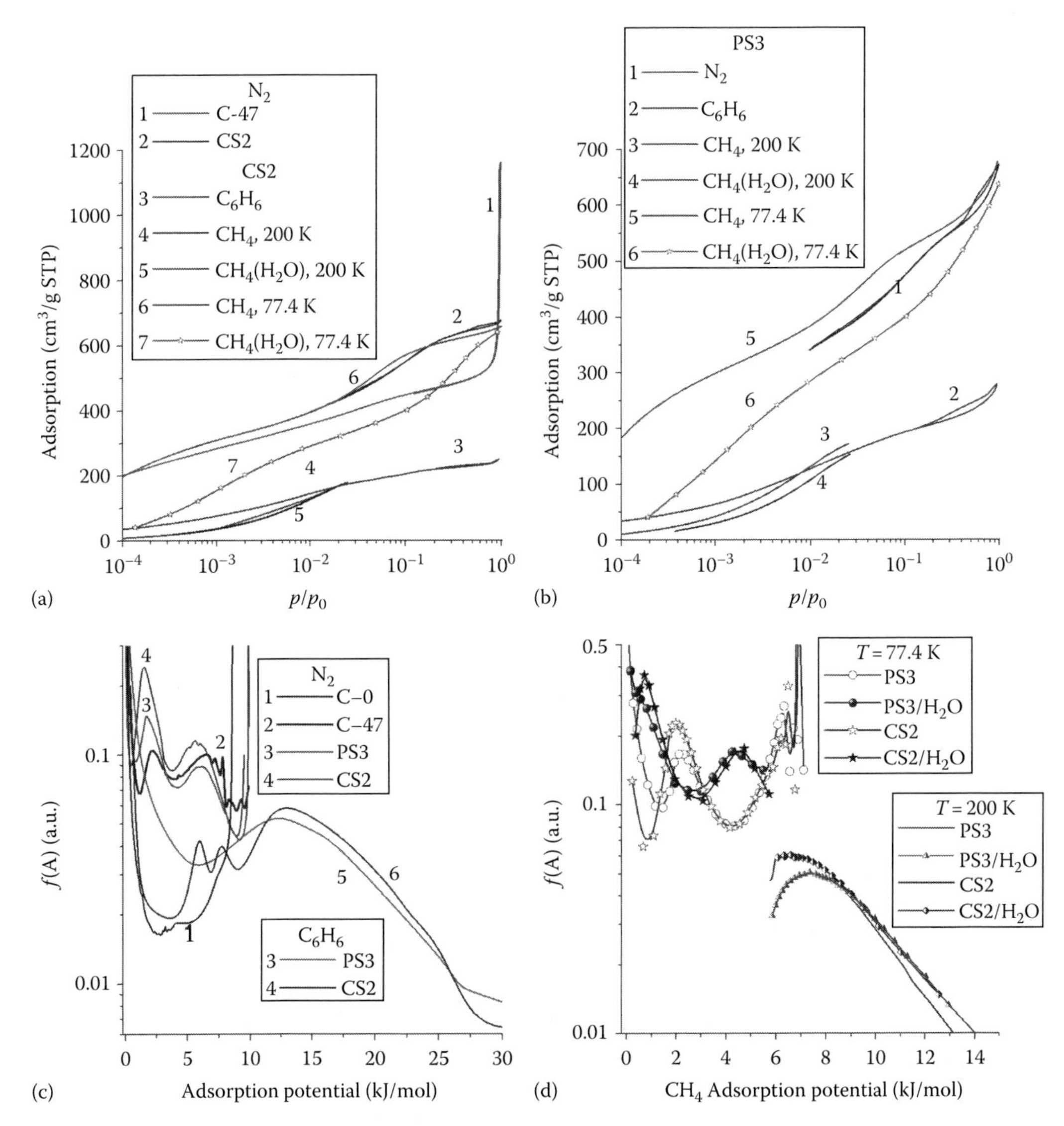

FIGURE 3.18 Adsorption-desorption isotherms of nitrogen (at 77.4 K), benzene (293 K), and CH_4 (77.4 and 200 K) on activated carbons (a) C-47 and CS2 and (b) PS3; adsorption potential distributions of (c) nitrogen adsorbed on C-47, PS3, and CS2 and benzene adsorbed on PS3 and CS2; and (d) methane adsorbed on PS3 and CS2 without and with preadsorbed water (50 mg/g) at 77.4 and 200 K. (Taken from *Appl. Surf. Sci.*, 254, Gun'ko, V.M., Turov, V.V., Kozynchenko, O.P. et al., Characteristics of adsorption phase with water/organic mixtures at a surface of activated carbons possessing intraparticle and textural porosities, 3220–3231, 2008d. Copyright 2008, with permission from Elsevier.)

of mesopores. The mentioned structural features of carbons affect the behavior of adsorbed water and/or organics.

According to the TPD mass-spectrometry data (Gun'ko et al. 2008a), concentration of oxygen-containing groups (eliminated on heating in vacuum as CO and CO_2) at a surface of C-0 and C-47 is not high. This is in agreement with low intensity of infrared (IR) bands of functional groups on these carbons (Figure 3.19). After converting spectra into Kubelka–Munk units and subtraction of the baseline, it is possible to observe the IR bands related to adsorbed water (3000–3600 cm^{-1}) and surface functionalities with C=O (in carboxylic, anhydride, lactone, and ketene at 1750–1630 cm^{-1}),

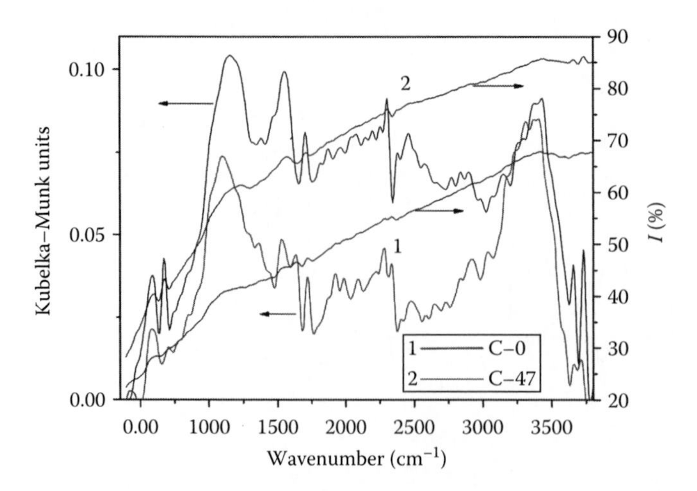

FIGURE 3.19 Infrared spectra of carbons C-0 (1) and C-47 (2) in two scales. (Taken from *Appl. Surf. Sci.*, 254, Gun'ko, V.M., Turov, V.V., Kozynchenko, O.P. et al., Characteristics of adsorption phase with water/organic mixtures at a surface of activated carbons possessing intraparticle and textural porosities, 3220–3231, 2008d. Copyright 2008, with permission from Elsevier.)

C–O (lactonic, ether, phenol, etc. at 1300–1000 cm^{-1}), C–H (3100–2800 cm^{-1}), CC (1600–1450 cm^{-1}), and CH (3070–3030 cm^{-1}) in aromatic groups, and twinned bonds C=C=C and C=C=O (2070–2040 cm^{-1}; Morterra and Low 1983, Ito et al. 1988, Zawadzki 1988, Fanning and Vannice 1993, Simms and Yang 1994).

The amount of water adsorbed onto studied carbons from air is small, i.e., they are rather hydrophobic (due to low amounts of oxygen-containing groups and the hydrophobicity of the basal planes) than hydrophilic adsorbents since adsorbed water fills only 3%–6% of the pore volume. Notice that the adsorption of water from air onto hydrophilic porous silica gels is much higher (water fills a major portion of the pore volume; Chapter 1) than on the studied carbons.

The NMR spectra were recorded by means of a Varian 400 Mercury spectrometer at 190–280 K with the probing 90° pulses at duration of 2 μs. For prevention of supercooling of the studied systems, the measurements of amounts of unfrozen adsorbates were carried out on heating of samples preliminary cooled to 190 K. The signals of water molecules from ice and frozen organics did not contribute the ^{1}H NMR spectra because of features of the measurement technique and the short time of transverse relaxation of protons in solids. During ^{1}H NMR spectroscopy measurements, a given amount of a liquid solvent was added to a NMR ampoule with a carbon adsorbent after the adsorption of water (samples were shaken for 2–3 min and equilibrated for 15–20 min). Feeding of methane from a rubber reservoir to an ampoule with a hydrated adsorbent was carried out at pressure slightly higher than atmospheric one. Therefore, the adsorbed amounts of methane were higher on the NMR study than on the adsorption measurements. The adsorption of light gases on ACs and other carbon nanomaterials is of interest from both practical and theoretical points of view (Morris and Wheatley 2008, Suyetin and Vakhrushev 2009, Vakhrushev and Suyetin 2009, Oriňáková and Oriňák 2011, Peng and Morris 2012).

On wetting of C-47 at hydration $h = 1.33$ g of water per gram of carbon, only a minor portion of the pore volume (approximately $V_p/4$) was filled by water (Figure 3.20a). On preparation of this system, a significant quantity of liquid water was observed out of carbon granules; consequently, air bubbles remained in pores. Boiling of C-47 in water (hydration $h > 10$ g/g) for 10 min leaded to complete filling of pores by water (Figure 3.20). Scheme of pore filling by unfrozen water (Figure 3.20c) was based on the $C_{uw}(T)$ dependence and the Gibbs–Thomson (GT) relation for the freezing point depression; i.e., we assumed that water was frozen in narrower pores at lower temperatures. Four bands (chemical shift of the proton resonance, δ_H) in the ^{1}H NMR spectra are observed for

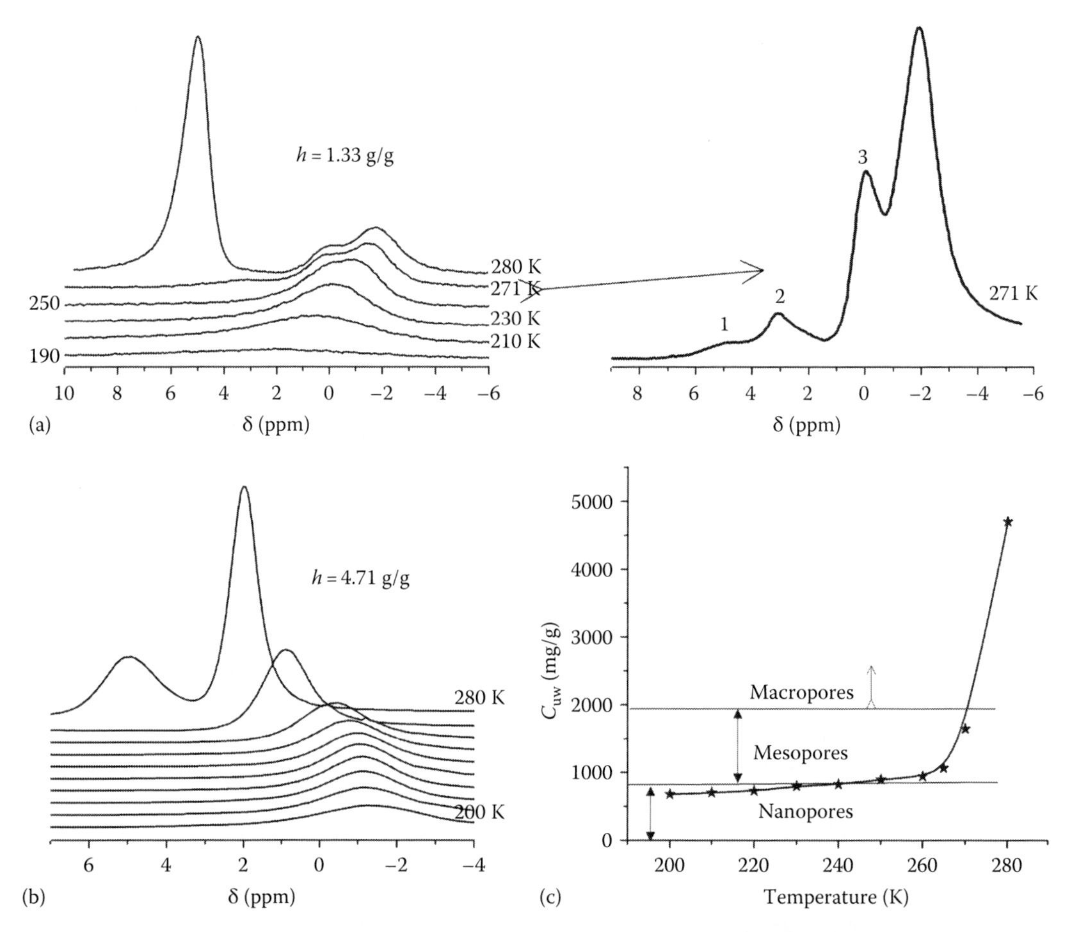

FIGURE 3.20 ^{1}H NMR spectra of water at h=(a) 1.33 g/g and (b) 4.71 g/g (carbon boiled in water) adsorbed on carbon C-47 at different temperatures and (c) the dependence of content of unfrozen water on temperature at h=4.71 g/g. (Adapted with permission from Gun'ko, V.M., et al., *Colloids Surf. A Physicochem. Eng. Aspects*, 317, 377, 2008a; Gun'ko, V.M., et al., *Appl. Surf. Sci.*, 254, 3220, 2008d.)

water adsorbed onto C-47 at T>270 K (Figure 3.20). The most intense signal 1 at $\delta_H \approx 5$ ppm (SAW) can be attributed to bulk water located out of carbon granules or/and in broad macropores because this signal of mobile water is observed only at T>273 K; i.e., this water is not influenced by the carbon surface. Another signal at $\delta_H \approx 3$ ppm (signal 2) observed at T>270 K and linked to water structures of a smaller size (because the δ_H value is smaller) than that characterized by the first signal of SAW can be assigned to water adsorbed in meso- and macropores (corresponding to the main second peak in the PSD of C-47 at x> 10 nm) where the pore walls weakly influence the freezing point depression of adsorbed water. Water located in narrow pores (the first main peak PSD at $x \approx 0.6$ nm) gives two signals ($\delta_H \approx 0$ ppm (signal 3) and δ_H between −1 and −2 ppm (signal 4)) observed at T>250 K.

At lower temperatures, only one signal (signal 4) has down-field shift with decreasing temperature (Figure 3.20a). The shielding effects of circular currents of π-electrons of basal planes at a carbon surface on the protons in mobile adsorbed molecules are maximal in narrow nanopores. Therefore, signals 3 and 4 can be attributed to water locating in narrow mesopores or/and super-nanopores and nanopores, respectively, i.e., in pores corresponding to the first peak at $x \approx 0.5$ nm and its right wing at x=0.9–2.0 nm in the PSD (Figure 3.17). On the basis of the down-field shift of the NMR signal with lowering temperature (i.e. deshielding increases), one can assume that

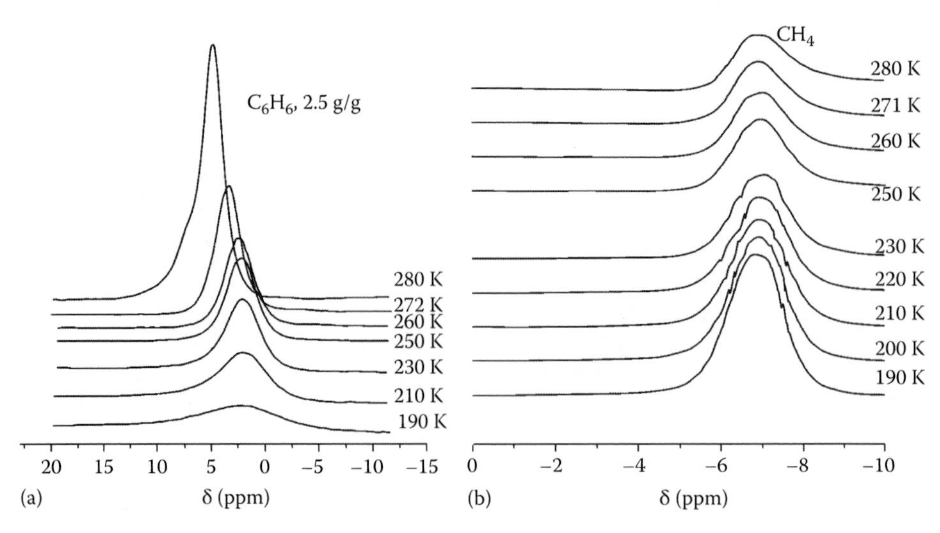

FIGURE 3.21 ^{1}H NMR spectra of (a) benzene (2.5 g/g of dry carbon) and (b) methane adsorbed on carbon C-47 at different temperatures. (Taken from *Appl. Surf. Sci.*, 254, Gun'ko, V.M., Turov, V.V., Kozynchenko, O.P. et al., Characteristics of adsorption phase with water/organic mixtures at a surface of activated carbons possessing intraparticle and textural porosities, 3220–3231, 2008d. Copyright 2008, with permission from Elsevier.)

the corresponding water clusters rather interact with oxygen-containing functionalities (in narrow pores or at the entrances into narrow nanopores (Brennan et al. 2001) than only with hydrophobic basal sheets in narrow nanopores because in the last case the shielding effect should increase and the displacement of the signal should be in the opposite direction (as in the case of the boiled carbon sample). Total filling of pores by water after boiling of the sample results in a different temperature dependence of the signals (compare Figure 3.20a and b) since the up-field shift of the signal is with lowering temperature. Consequently, in the last case, water more completely fills narrow nanopores.

The temperature-dependent up-field shift of the ^{1}H NMR signal is observed for nonpolar benzene adsorbed on C-47 (Figure 3.21a) because of its preferable interaction (by dispersive forces) with basal planes than oxygen-containing functionalities. Benzene can easily penetrate in all pores and fill the total pore volume in contrast to water (Figure 3.20a). The chemical shift δ_H of benzene is close to 0 ppm at $T<260$ K that indicate a strong shielding effect of the surface (because $\Delta\delta_H \approx -7$ ppm) characteristic of adsorbates located in narrow slit-shaped nanopores of carbons. The high adsorption potential in nanopores leads to strong disturbance of the structure of adsorbed benzene that causes nonfreezing of the liquid over a wide temperature range. The adsorption potential in broad mesopores is much lower than that in nanopores; therefore, the freezing point depression is small for benzene adsorbed in mesopores and macropores. There are several overlapping bands in the ^{1}H NMR spectra of benzene (Figure 3.21a) when the total amount of benzene is larger than the pore volume. A weak signal of benzene at $\delta_H = 7.26$ ppm (at $T \geq T_{m(b)} = 278$ K) corresponds to bulk liquid (i.e., benzene located out of pores). A signal at $\delta_H = 3.0$–4.5 ppm rapidly reduces with lowering temperature and it can be attributed to benzene filling macro and mesopores. A signal at $\delta_H \approx 2$ ppm depends weakly on temperature and it can be assigned to benzene adsorbed in nanopores as this portion of benzene is frozen at low temperatures $T<210$ K.

Methane adsorbed onto C-47 (Figure 3.21b) is characterized by a single signal at $\delta_H \approx -7$ ppm practically independent of temperature because methane is unfrozen over the total temperature range used here. On the NMR study, a measuring ampoule was in contact with a vessel with methane that allowing an increase (by a factor of 2.5 at 190 K) in the adsorbed amounts of CH_4 with lowering temperature. The observed signal position reveals a great shielding effect of the

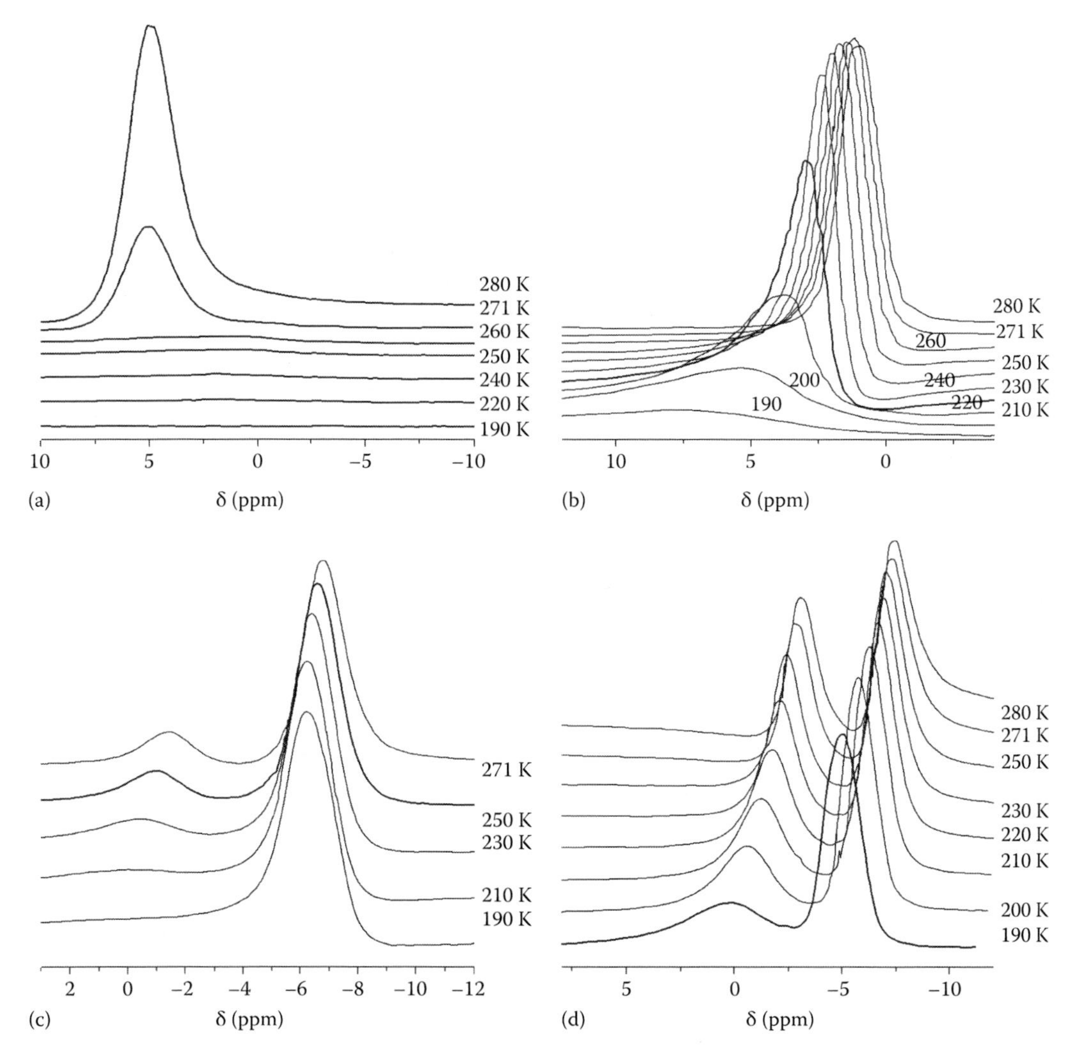

FIGURE 3.22 ^{1}HNMR spectra of (a) water (h=0.36 g/g) at chloroform-d amount of 0.8 g/g; (b) benzene (0.05 g/g) and water (0.05 g/g); and methane with water h=(c) 0.05 g/g and (d) 0.1 g/g adsorbed on carbon C-47 at different temperatures. (Taken from *Appl. Surf. Sci.*, 254, Gun'ko, V.M., Turov, V.V., Kozynchenko, O.P. et al., Characteristics of adsorption phase with water/organic mixtures at a surface of activated carbons possessing intraparticle and textural porosities, 3220–3231, 2008d. Copyright 2008, with permission from Elsevier.)

π-electron current at the carbon surfaces on the protons in adsorbed CH_4 molecules. This corresponds to the fact that with elevating temperature, the lion's share of methane is accumulated in narrow pores of carbons.

Chloroform, methane, and benzene differently affect the behavior of water adsorbed on C-47 (Figure 3.22).

The shape of the ^{1}H NMR spectra of water essentially changes in the presence of chloroform (Figure 3.22a), and water gives the signal at $\delta_H \approx 5$ ppm (SAW) characteristic of bulk water or water adsorbed in macropores. Its intensity strongly decreases at $T < 273$ K and it is not observed at $T \leq 271$ K; i.e., this water is located only in broad pores. A very weak and broad signal of water is also observed at $\delta_H \approx 0$ ppm and $T > 250$ K. Consequently, adsorbed chloroform strongly reduces the interaction of water with the carbon surface because it drives a substantial portion of water from nanopores into meso- and macropores (Figure 3.23). Similar effects are characteristic for different

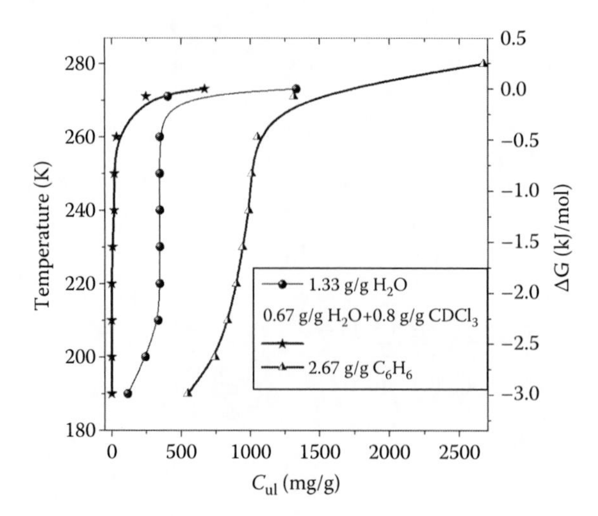

FIGURE 3.23 Amounts of unfrozen water (individual at $h = 1.33$ g/g and in a mixture with chloroform-*d* (0.8 g/g) at $h = 0.67$ g/g) and individual benzene ($C_{ul} = C_{uw}$ or C_{ub}) as a function of temperature on the adsorption on carbon C-47, and corresponding relationships between the amounts of unfrozen water and changes in the Gibbs free energy of the interfacial water. (Taken from *Appl. Surf. Sci.*, 254, Gun'ko, V.M., Turov, V.V., Kozynchenko, O.P. et al., Characteristics of adsorption phase with water/organic mixtures at a surface of activated carbons possessing intraparticle and textural porosities, 3220–3231, 2008d. Copyright 2008, with permission from Elsevier.)

carbon materials because of a greater affinity of weakly polar or nonpolar organic molecules to hydrophobic carbon surface in comparison with the affinity of polar water molecules (Gun'ko et al. 2005d, 2008a,d, 2009d, 2011d).

On simultaneous adsorption of small amounts of water and benzene (Figure 3.22b), only a single signal is observed because both adsorbates locating in narrow pores have close δ_H values (Figures 3.20 and 3.21). Lowering temperature leads to a considerable down-field shift of the signal ($\Delta\delta \approx 6$ ppm) that is especially pronounced when freezing of a substantial fraction of adsorbates occurs. This signal (Figure 3.22b) drifts in to the opposite direction in comparison with benzene adsorbed alone (Figure 3.21a). This difference can be explained by the different dependencies of the freezing point depression for adsorbed water and benzene on the presence of the second liquid because of changes in their location in narrower and broader pores that affect the $T_m(x)$ and δ_H values. Consequently, a portion of benzene is in broad pores (Figure 3.22b). On simultaneous adsorption of methane and water, their signals are well-separated (Figure 3.22). Lowering temperature leads to the down-field shifts of signals of water and methane. This effect is greater at a larger amount of adsorbed water. Consequently, adsorbed water (even at small amounts of 5–10 wt%) is responsible for a significant decrease in the shielding effect of π-electrons at the carbon surface for protons in methane molecules. This can be caused by changes in the location of water after thawing that affects the location of adsorbed methane (e.g., displaced into larger pores), and the amounts of adsorbed methane decrease with elevating temperature.

The temperature dependences of concentrations of unfrozen water and benzene and the relationships between changes in the Gibbs free energy and concentrations of the unfrozen liquids (Figure 3.23) reveal that benzene can fill practically the total pore volume in contrast to water, which occupies only a portion of the pore volume (Figure 3.20a), since another portion of the pore volume is occupied by remained air bubbles in contact with hydrophobic basal planes of the adsorbent. Despite the total amount of unfrozen adsorbed water is ≈0.5 cm^3/g and a surface area in contact with unfrozen adsorbed water is 703 m^2/g that are smaller than the V_{nano} and S_{nano} values (Table 3.7), a portion of water remains out of pores since hydration is $h = 1.33$ g/g. Added chloroform displaces adsorbed water (and air) from pores because the amount of unfrozen water decreases as well as its freezing point depression.

The NMR-PSDs show (Figure 3.17b) the structural effects for both water clusters/droplets and carbon pores. For instance, a peak of very narrow nanopores at $x=0.3$ nm is observed on the adsorption of water, and the PSD_{uw} has several peaks, which coincide (at $x\approx0.6$ nm) or do not coincide (8 nm) with the PSD_{N2}. Adsorbed benzene (PSD_{ub}) does not give a peak at $x=0.3$ nm; however, the PSD_{ub} coincides with the PSD_{N2} and PSD_{uw} at $x\approx0.6$ nm but the PSD_{ub} at $x=1.5$ and 4 nm does not coincide with the PSD_{N2}. These differences in the PSD_{N2}, PSD_{ub}, and PSD_{uw} functions are due to (i) incompletely filling of pores by water and (ii) layer-by-layer freezing of adsorbed liquids in meso- and macropores at different temperatures. The PSD_{uw} corresponding to unfrozen adsorbed water with the presence of chloroform demonstrates that water locates only in broad pores at $x>20$ nm. The main peak of narrow pores at $x\approx0.6$ nm is similar for all types of the PSDs (nitrogen, benzene, CH_4, and water; Gun'ko et al. 2008a,d, 2011d). However, benzene and water frozen in broad pores demonstrate the PSDs different from the PSD_{N2} because the freezing in broad pores occurs through the layer-by-layer process. Therefore, the NMR cryoporometry results with water as a probe liquid are most reliable for pores at the half-width $x<10$ nm (Gun'ko et al. 2005d). It is known than the correlation length for liquids located in confined space at temperature close to phase transition temperature can be 1–10 nm (Beysens et al. 1993, Chalyi 2000). These values correspond to our estimations of the maximal x values appropriate to be determined by using the Gibbs–Thomson relation for the freezing point depression for adsorbed liquids.

To elucidate the interfacial behavior of adsorbed water, organic solvents were used in the deuterated form. The ^{1}H NMR spectra of water/CD_3CN were recorded on partial filling of pores (Figure 3.24a) or for the hydrated adsorbent in liquid acetonitrile (2 g/g), i.e., at $V_{acetonitrile}>V_p$ (Figure 3.24b). In the first case, water is characterized by a single signal at δ_H increasing from 1 to 4 ppm ($\delta_H\approx3$ ppm for water dissolved in acetonitrile) and simultaneous broadening of the signal with lowering temperature, which is due to a decrease in the molecular mobility of adsorbed water. The signal intensity remains practically without changes that indicate strong adsorption interaction of water with the pore walls. The obtained δ_H values correspond to transition from a weak proton shielding effect at room temperature to a small deshielding effect at $T=230$–200 K. Consequently, the observed signal relates to water adsorbed in broader pores because acetonitrile substitutes water in

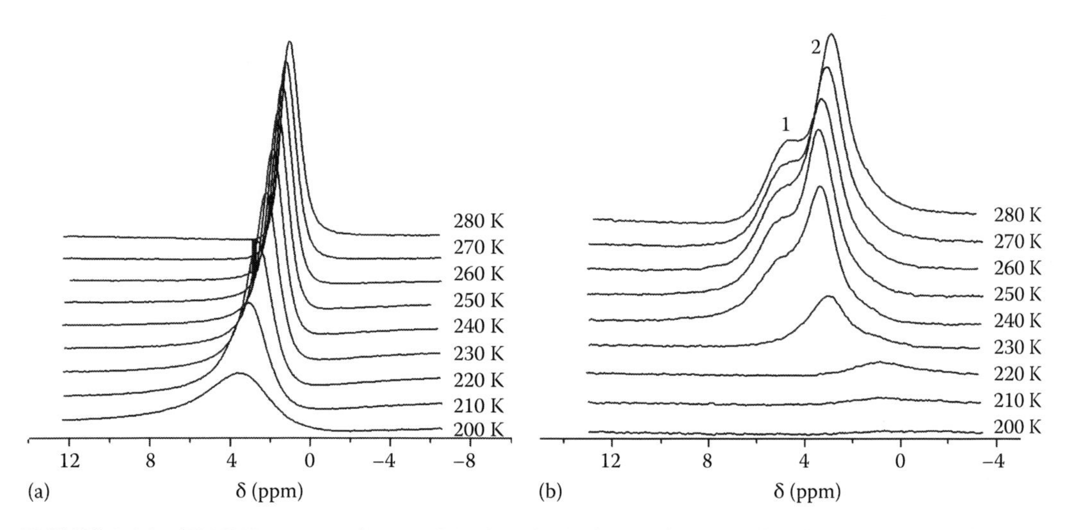

FIGURE 3.24 ^{1}H NMR spectra of water (0.1 g/g) adsorbed on carbon C-47 in mixtures with CD_3CN at (a) 0.1 g/g and (b) 2 g/g. (Taken from *Appl. Surf. Sci.*, 254, Gun'ko, V.M., Turov, V.V., Kozynchenko, O.P. et al., Characteristics of adsorption phase with water/organic mixtures at a surface of activated carbons possessing intraparticle and textural porosities, 3220–3231, 2008d. Copyright 2008, with permission from Elsevier.)

narrower pores. For hydrated C-47 being in acetonitrile-d_3 (2 g/g), the spectra shape substantially changes (Figure 3.24b) since two signals are observed at $T>240$ K.

A signal with down-field shift (signal 1) corresponds to δ_H close to that of water dissolved in acetonitrile. Its intensity rapidly decreases with lowering temperature, and it is not observed at temperature close to the freezing point of bulk acetonitrile (245 K). Probably, this is a signal of water dissolved in bulk acetonitrile out of C-47 granules. The second signal (signal 2) can be linked to the protons in the water molecules under a weak shielding effect. Its spectral characteristics are close to that of water shown in Figure 3.24a. However, its intensity sharply decreases at $T<240$ K and it has up-field shift; i.e., a portion of water located in narrow nanopores. At $T<230$ K, a weak signal of water is also observed at $\delta_H=-2$ ppm corresponding to the chemical shift of water adsorbed in narrow slit-shaped nanopores. To explain features of signal 2 over the total temperature range, one can speculate that it is the averaged signal of water molecules located not only in nanopores but also in mesopores in which water freezes at higher temperatures.

Upon adsorption of water/acetone (acetone is a stronger electron-donor at $DN=71$ kJ/mol than acetonitrile at $DN=59$ kJ/mol; Beysens et al. 1993; Figure 3.25), the spectra shapes and the temperature behavior are similar for two types of acetone; however, contribution of the protons in $(CH_3)_2CO$ results in enhancement of the signal with up-field shift (Figure 3.25b). There are two signals of water (Figure 3.25a) adsorbed in nano- and mesopores and characterized by the up-field and down-field shifts, respectively, with lowering temperature. On freezing-out of water in mesopores ($T<200$ K), single signal is observed at $\delta_H=0$ ppm (Figure 3.25a). Therefore, one can assume that a major portion of adsorbates is located in mesopores, where oxygen-containing groups attached to edges of basal carbon planes are the main adsorption sites for polar adsorbates (Brennan et al. 2001).

The spectra of water/DMSO (Figure 3.26; DMSO is the strongest electron-donor at DN = 124.4 kJ/mol (Coetzee and Chang 1986) among the studied solvents) on partial filling of pores show that the behavior of signals is close to that for the water/acetone (Figure 3.25a).

There are two signals of water adsorbed in nanopores and mesopores, and its major portion locates in mesopores. The spectra of water/DMSO (Figure 3.26b) are more complicated because of contribution of the protons from the CH_3 groups of DMSO. The signal of adsorbates locating in nanopores is bifurcated, moreover the distance between the maxima of these signal components

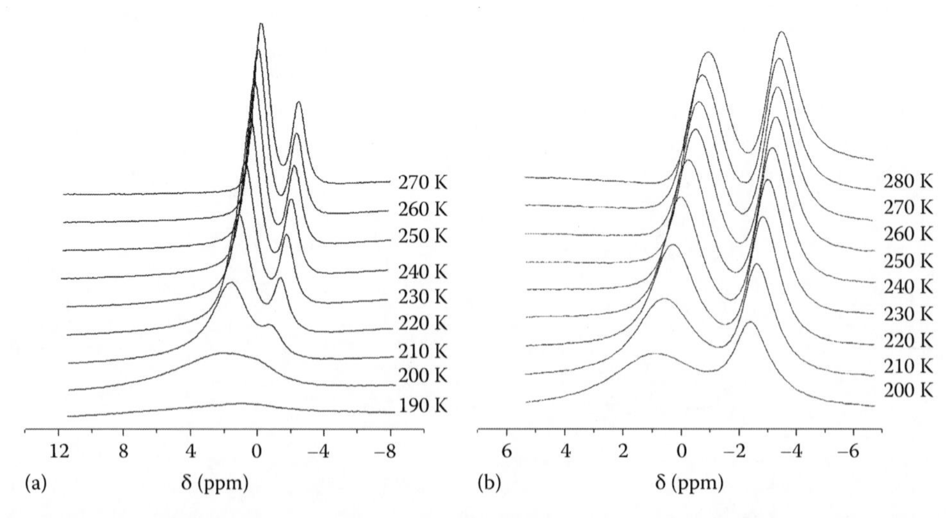

FIGURE 3.25 ^{1}H NMR spectra of water (0.1 g/g) adsorbed in mixtures with (a) $(CD_3)_2CO$ and (b) $(CH_3)_2CO$ (0.1 g/g) on carbon C-47 at different temperatures. (Taken from *Appl. Surf. Sci.*, 254, Gun'ko, V.M., Turov, V.V., Kozynchenko, O.P. et al., Characteristics of adsorption phase with water/organic mixtures at a surface of activated carbons possessing intraparticle and textural porosities, 3220–3231, 2008d. Copyright 2008, with permission from Elsevier.)

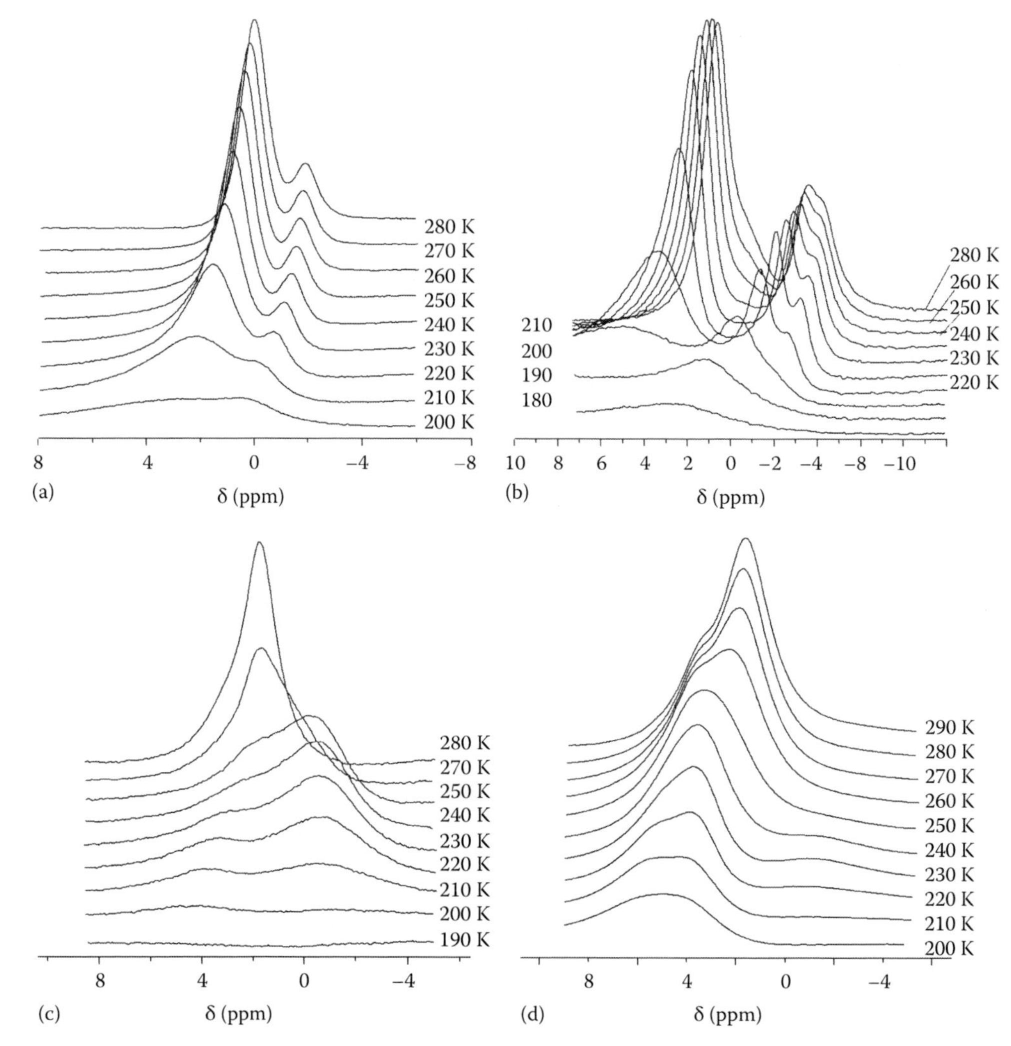

FIGURE 3.26 ^{1}H NMR spectra of water (0.1 g/g (a–d)) adsorbed in a mixture with $(CD_3)_2SO$ (0.1 g/g (a) and DMSO-d_6 medium (2 g/g) (c)) and $(CH_3)_2SO$ (0.1 g/g (b) and 2 g/g of DMSO (d)) adsorbed on carbon C-47. (Taken from *Appl. Surf. Sci.*, 254, Gun'ko, V.M., Turov, V.V., Kozynchenko, O.P. et al., Characteristics of adsorption phase with water/organic mixtures at a surface of activated carbons possessing intraparticle and textural porosities, 3220–3231, 2008d. Copyright 2008, with permission from Elsevier.)

increases with lowering temperature, and at T between 280 and 190 K, it demonstrates the down-field shift from −3 to 3 ppm. A smaller signal drift is observed for adsorbates located in mesopores since δ_H changes from 1 to 5 ppm. The signal splitting in the strong magnetic field is caused by separate observation of the signals of the protons of water (from 0 to −3 ppm) and methyl groups of DMSO (from −2 to −4 ppm).

On the adsorption of water/DMSO-d_6 at a great excess of the latter (Figure 3.26c), the observed temperature changes in the water spectra suggest that at least there are two states. At $T > 273$ K there is rapid (in the NMR timescale) molecular exchange between these states but a slow molecular exchange is at lower temperatures. The intensity and the δ_H value of the up-field-shifted signal change with temperature more weakly than that of the signal with down-field shift. The main changes in the signal intensity occur at $T > 250$ K that indicate weak interaction of this type of water with the surface. Probably, this water is located in broad mesopores or macropores in a mixture with DMSO. After its freezing at $T<270$ K, two signals are in the spectra with down-field and up-field shifts. The chemical shift δ_H of the latter is approximately −1 ppm and the position of which depends weakly on temperature, but its intensity decreases with lowering temperature and it is not observed at $T<210$ K. The intensity of the signal with the down-field shift is considerably smaller. Its δ_H value increases from 3 ppm at 250 K to 6 ppm at 200 K. These signals can be attributed to water located in nanopores (signal with up-field shift) and broad mesopores (signal with down-field shift). Consequently, a quantity of water in mesopores is considerably lower than that in nanopores because of preferred adsorption of DMSO-d_6 in mesopores.

On combined adsorption of water/DMSO on C-47 at $C_{DMSO}=2$ g/g, the signals of water and methyl groups of DMSO are observed (Figure 3.26d). Two signals at $\delta_H\approx 2$ and 3–4 ppm are observed at $T=280$ K. On the basis of the relationship of component concentrations, one can conclude that the first signal was caused by the protons of DMSO and the second one is linked to water. A low-intensity signal at $\delta_H\approx -2$ ppm observed at $T>210$ K is linked to water and DMSO located in nanopores.

The relationships between the ΔG and C_{uw} values are rather nonlinear for different types of water characterized by different δ_H values (Figure 3.27).

This is due to the possibility of the transformation of one type of water into another type with elevating temperature because thawing of a portion of adsorbed water leads to changes in its structure (Figure 3.27a), associativity, mobility, and mixing with organics. This is also dependent on the content of organics (Figure 3.27).

Thus, investigations of the interfacial behavior of water/organic mixtures in ACs by using the ^{1}H NMR and adsorption methods allow us to analyze the rearrangement effects of adsorbed water and organics depending on the character of their interaction, concentrations, and temperature. Water and organics adsorbed in pores of carbon adsorbents give several ^{1}H NMR signals with up-field shift compared to the signals of bulk liquids alone, and the maximum shielding effect ($-\Delta\delta=5$–6 ppm) for the protons in adsorbed molecules is observed for adsorbates locating in narrow nanopores at $x<0.7$ nm. On simultaneous adsorption of water and methane, the observed shielding effect of circular currents of π-electrons at the carbon surface strongly reduces with formation of ice nanocrystallites in narrow pores with increasing amounts of adsorbed water because location

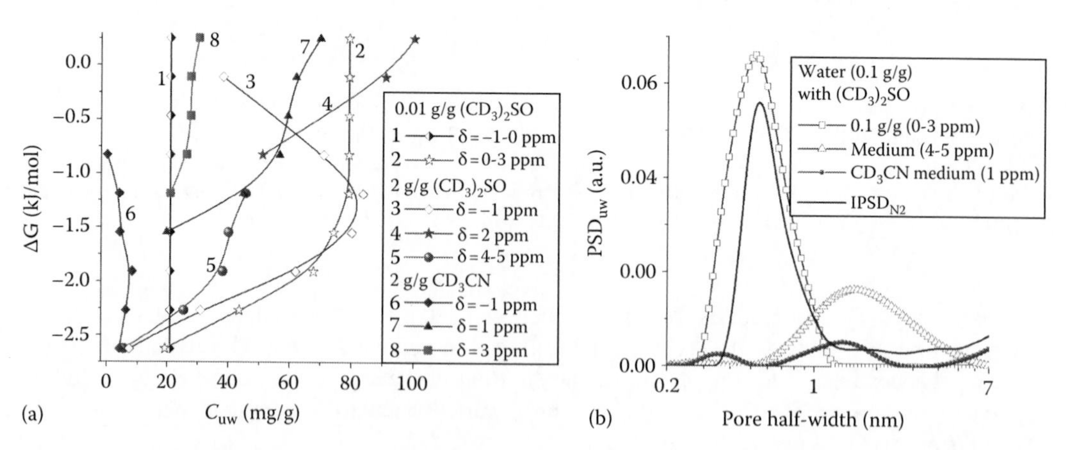

FIGURE 3.27 Relationships between changes in the Gibbs free energy and concentration of unfrozen water (0.1 g/g) characterized by different chemical shifts of proton resonance on co-adsorption of (a) 0.1 g/g of $(CD_3)_2SO$ and in organic media, $(CD_3)_2SO$ (2 g/g) and CD_3CN (2 g/g) and (b) distribution functions of sizes of unfrozen water structures on binary adsorption calculated for different signals and DFT IPSD calculated using the nitrogen desorption data for carbon C-47. (Taken from *Appl. Surf. Sci.*, 254, Gun'ko, V.M., Turov, V.V., Kozynchenko, O.P. et al., Characteristics of adsorption phase with water/organic mixtures at a surface of activated carbons possessing intraparticle and textural porosities, 3220–3231, 2008d. Copyright 2008, with permission from Elsevier.)

of methane occurs in pores of larger sizes on the preadsorption of water. Weakly polar chloroform can easily displace adsorbed water from nanopores because of the difference in the nature of their interactions with nonpolar and polar structures at the surface of carbon adsorbents. For the majority of the systems upon binary adsorption of water/organics, there is down-field shift of the signals of the protons of adsorbates with lowering temperature. In certain cases, this shift is so great that the δ_H values become even greater than that in the liquid state; i.e., the shielding effect is replaced by the deshielding one. The main reasons of this phenomenon can be migration of adsorbate molecules from slit-shaped nanopores into mesopores.

3.2.2.2 Effect of Weakly Polar Organics on Adsorbate Localization in Pores of ACs

The question of the localization of adsorbed molecules in AC pores is central to adsorption theory (Gregg et al. 1979, Gregg and Sing 1982, Dubinin 1983, 1960, Fenelonov 1995). In terms of thermodynamics, the adsorbed substance is localized in the region of maximum adsorption potential in the narrowest pores or at adsorption sites specific to the adsorbate molecules. However, this is true only under conditions of low surface coverage by the adsorbed substance. Upon greater coverage, adsorbate molecules can form in the pores the supramolecular cluster structures stabilized by adsorbate–adsorbate molecular interactions (i.e., there is volume filling of pores). This is especially important for associated liquids, one of which is water. For water, the energy gain as a result of formation of clusters can considerably exceed the differences in energy during adsorption of the individual molecules in pores of various sizes. Cluster adsorption on almost all adsorbents is thus characteristic for water. In recent years, the question of cluster formation in the nanopores of ACs has been actively discussed in the literature (Vartapetyan et al. 1989, Brennan et al. 2001, Fomkin 2005, Gun'ko 2005a, 2009a, Shkolin and Fomkin 2009, Chaplin 2011).

It should be noted that the possibility of forming supramolecular cluster systems consisting of water and organic molecules in the pores of AC can be very sensitive to the pore size; sterically, such complexing can occur more efficiently in wide pores even when an appreciable part of nanopores space remains free. For our investigations, we chose a carbon adsorbent with high specific area and broad pore size distribution. Although such materials can undergo substantial structural deformation at the high pressures of gaseous adsorbate (Shkolin and Fomkin 2009), we may expect that these effects are negligible under conditions of the low filling of pores.

Localization of water and co-adsorption of water with organic co-adsorbates in confined space of narrow pores were studied for highly activated AC ($S_{BET}=3463$ m^2/g, $V_p=2.04$ cm^3/g; Turov et al. 2011a–c). To synthesize this AC, phenol–formaldehyde resin was carbonized in a CO_2 flow, and additional activation was performed at 1183 K (Tennison 1998, Tennison et al. 2004). The burn-off degree was 86% (C-86). The adsorption isotherm of nitrogen (Figure 3.28a) and the pore size distributions (Figure 3.28b) suggest a complex texture of C-86. Several peaks of nanopores (pore half-width $x<1$ nm) and narrow mesopores ($x<10$ nm) are observed. The peak in the region of 90–100 nm is due to the textural porosity, i.e., to voids between the nanoparticles forming the aggregates and globules of AC.

^{1}H NMR spectra of water ($h=50$ mg/g) adsorbed on C-86 in the presence of small and fixed (~0.5 wt%) amount of methane under conditions, when the most of the volume of pores remained free from adsorbed substance, are shown in Figure 3.29. From Figure 3.29 it is visible that both water and methane are observed in the spectra as two main signals with up-field shift relatively to liquid water (methane) signals, at that the value of the chemical shift of all signals significantly increases with decreasing temperature. Taking into account that the liquid water and methane signals have the chemical shifts $\delta_H=5$ and 0 ppm, respectively, it can be concluded that the maximal shielding effect ($\Delta\delta$; at $T=280$ K) for the adsorbed water is $\Delta\delta=-1.8$ and -3.2 ppm for weak field and strong field signals, respectively. Slightly greater values of the maximal shielding effect ($\Delta\delta=-3.5$ and -5 ppm) are registered for methane. Since for the separate registration of NMR signals the time of the molecular exchange must be more than the relaxation time of nucleus spins (Abragam 1961), it can be supposed that the signals of the adsorbates with various value of the

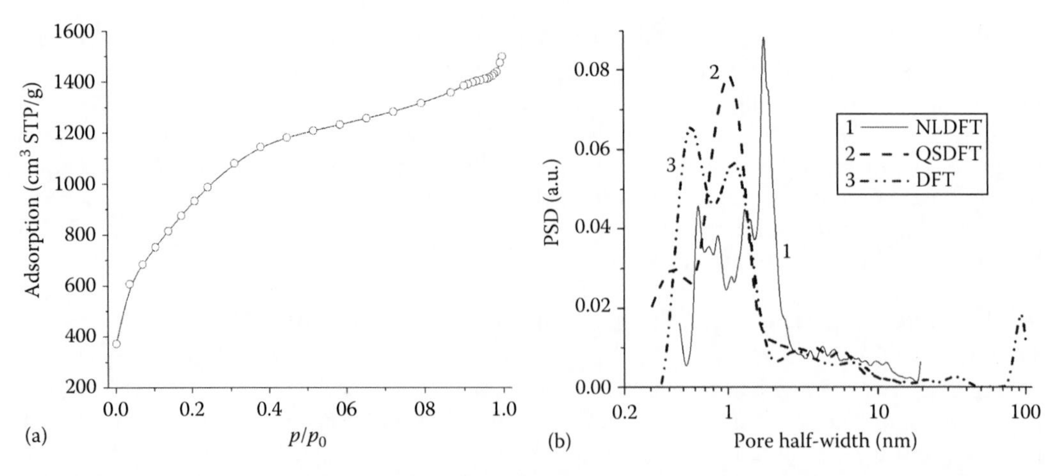

FIGURE 3.28 Adsorption isotherm of nitrogen on activated carbon C-86 (a) and incremental DFT pore size distributions (b) within the model of (2, 3) slit-like pores and (1) a mixed model of slit-shaped and cylindrical pores.

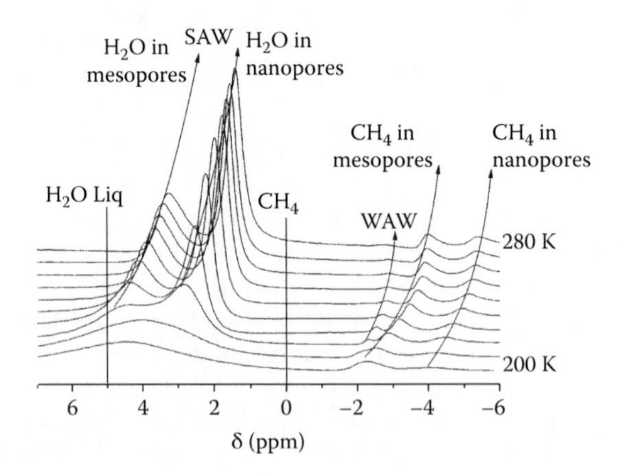

FIGURE 3.29 Spectra of water adsorbed on C-86 ($h = 0.05$ g/g) taken at various temperatures in the presence of small amounts of methane.

chemical shift corresponds to the substance localized in various parts of the pore space, most likely in nano- and mesopores (pores with half-width $x < 1$ nm and $x > 1$ nm, Figure 3.28). At that the signal in strongest magnetic fields (with lesser value of the chemical shift) corresponds to the substance in nanopores. Consequently, at the chosen amount of adsorbate both water and methane can simultaneously adsorb both in nano- and mesopores despite the fact that the greatest part of the nanopore volume remains free.

Since the value of the shielding effect for water was slightly lower than for methane, it can be concluded that degradation of the water polyassociates in the AC pores does not occur, and the adsorbed both in nano- and mesopores water is in strongly associated state, i.e., in the form of large clusters (SAW domains). For weakly associated form of water, for which δ_H does not exceed 2 ppm (Gun'ko et al. 2009d), the appearance of signal in the region of −2 ppm was to be expected. Such signal is really registered on the spectra; however, its intensity is significantly lower than that of the main signals. The significant down-field shift of all signals of the adsorbed substances observed in the experiments with decreasing temperature (i.e., decrease in the shielding effect) can be explained by changes in location of the mobile molecules with decreasing temperature and the formation of

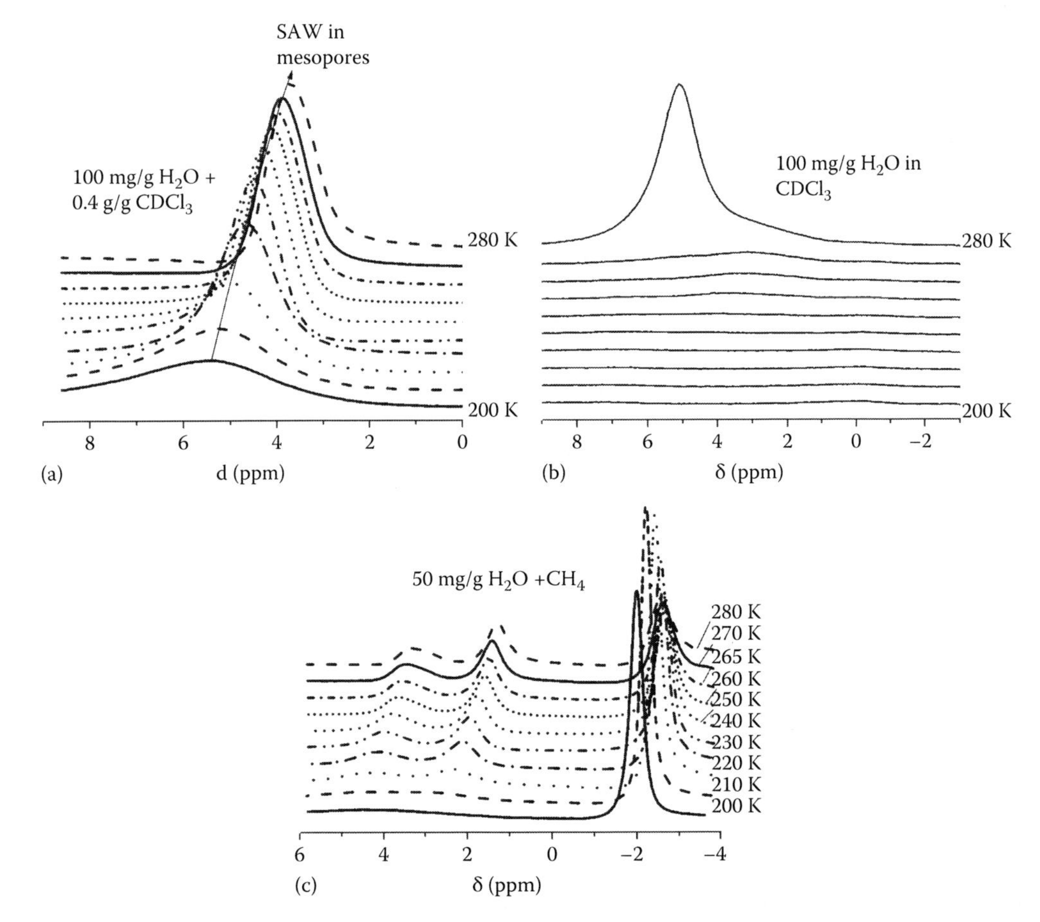

FIGURE 3.30 Effect of weakly polar substances, chloroform (a) 0.1 g/g H_2O + 0.4 g/g $CDCl_3$; (b) 0.1 g/g H_2O in a $CDCl_3$ medium; and methane (c) 0.05 g/g $H_2O + CH_4$ on ^1H NMR spectra of water adsorbed by activated carbon C-86.

ice crystallites, which cannot provide ring current effects for the mobile molecules. The greater value of the shielding effect for methane, for the molecules of which the efficiency of the dispersion interactions with the carbon surface can be higher than for the water molecules, which take part in the formation of hydrogen bonds, is also explained by more close location to the walls in narrow pores. It can be also supposed the presence of the change of the nanopore sizes during water freezing in them (ice volume is higher than that of liquid water). Thus, there are two stable states for adsorbate molecules close to walls and in the pore center, which is stabilized by the temperature decrease, in the slit-like pores of the AC.

In the case of co-adsorption of water and weakly polar organic substances, the area of cluster localization of strongly associated water can change. Figure 3.30a shows the results of the co-adsorption of water and deuterated chloroform (which is not registered in ^1H NMR spectra).

With the addition of 0.4 g/g of $CDCl_3$ (relative to the AC weight), the water signal in the nanopores disappears, while the signal of strongly associated water in mesopores rises respectively; i.e., the chloroform displaces the water from nanopores into larger pores. A subsequent increase in the chloroform amount (Figure 3.30b) is accompanied by almost complete water removal from both nano- and mesopores of the AC since the major portion of water is frozen at temperatures close to 273 K. A similar effect has been observed for a great number of carbon and silica materials (Turov et al. 2002a–c, 2004, Gun'ko et al. 2005d, 2007i, 2009d).

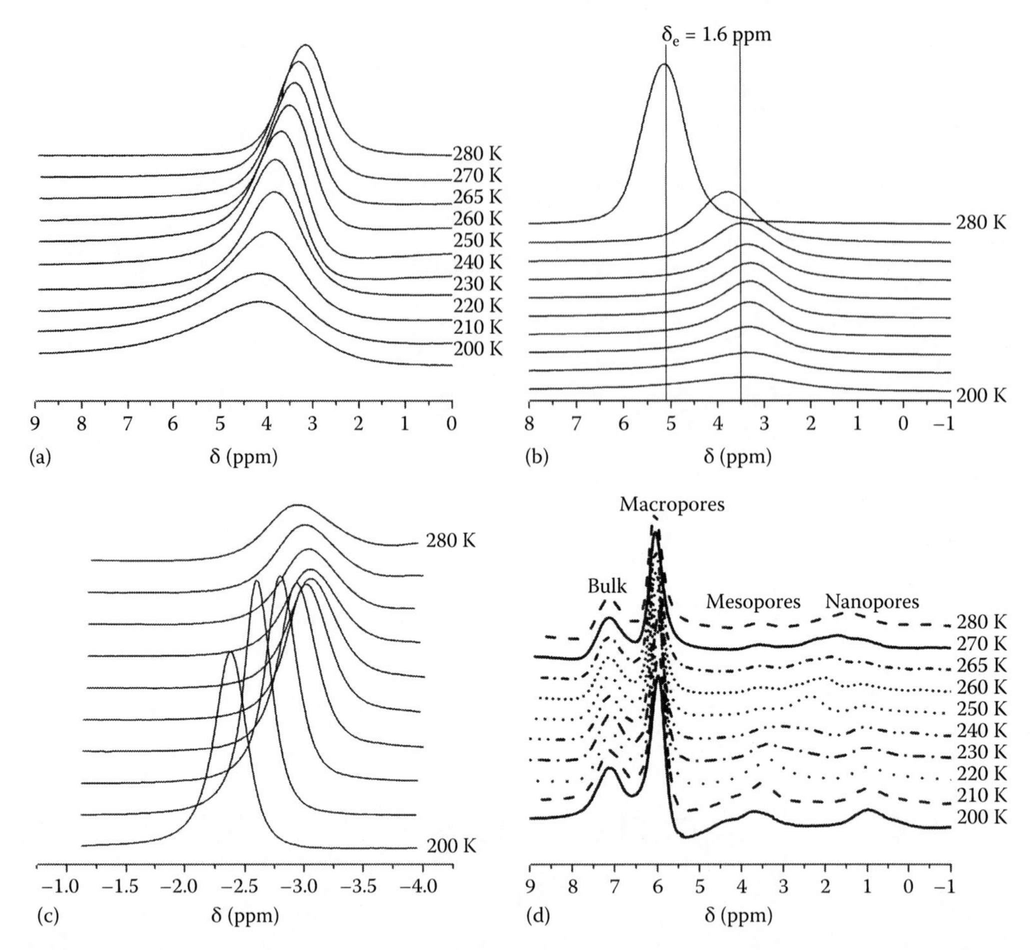

FIGURE 3.31 ^{1}H NMR spectra of water ((a) 1 and (b) 10 g/g H_2O), methane (c), and chloroform ((d) 10 mL/g $CHCl_3$) adsorbed on C-86.

The complete displacement of water from AC mesopores does not occur in the case of co-adsorption of 0.05 g/g of water and methane from the container (Figure 3.30c). Only a small redistribution of the signal intensities toward a relative increase in the water signal intensity in mesopores is observed. Since the maximum value of methane adsorption does not exceed 5 wt%, we may assume that the phases of the adsorbed methane and water coexist in narrow pores independently of one another. The adsorption of weakly polar organic substances in narrow pores is thus preferable to clusters of strongly associated water.

With complete and substantial filling of the AC pores by adsorbate, the form of the spectra and the temperature dependence of the chemical shift of the adsorbed substances significantly change (Figure 3.31). At $h = 1$ g/g, one signal of water, the chemical shift of which increases from 3.5 to 4.5 ppm with a drop in temperature from 280 to 200 K, is registered in the spectra (Figure 3.31a). At $h = 10$ g/g, the water in AC pores is observed as a signal whose intensity is determined by the total volume of nano- and mesopores. The chemical shift is virtually independent of temperature and has a value averaged for nano- and mesopores, and for water localized at the pore walls and in the pore center (or in contact with ice crystallites, as in Figure 3.31b). It is likely that during the filling of all pore spaces with water, there is spatial confinement for molecular exchanges between water molecules in various states and localized in the pores of various size disappears. A comparison of the intensities of the water signals before and after the freezing of the bulk water gives grounds for concluding that the total amount of unfrozen water is considerably higher than the total volume of pores determined from the adsorption isotherms of nitrogen (Figure 3.28a).

A similar conclusion can be drawn using $CHCl_3$ as the liquid medium (Figure 3.31d). Since the chloroform molecules are considerably larger than molecules of water or methane, the molecular exchange between water localized in the different pores and voids of the sample slows down, and bulk $CHCl_3$ signals ($\Delta\delta = 0$ ppm), in the internal granule space ($\Delta\delta = -1$ ppm), and in the slit-like pores ($-\Delta\delta = 2$–6 ppm) can be observed. The total amount of $CHCl_3$ exposed to the considerable shielding effect (Figure 3.31d) is close to the pore volume measured by the adsorption isotherms of nitrogen (Figure 3.28a). The low value of the apparent density of the material (≈0.1 g/mL) leads to a great number of voids inside the AC granules, exceeding the volume of the slit-like pores by a factor of 2.7. The position and intensity of $CHCl_3$ signals in the pores are strongly dependent on temperature, testifying to a change in the areas of localization of the adsorbate molecules in the pores. Because of the complexity of these processes, however, it is difficult to analyze within the limits of the performed experiment.

For methane adsorbed from the container, the maximum adsorption does not exceed 5 wt%. A single signal averaged for meso- and nanopores is observed on the spectra (Figure 3.31c). With decreasing temperature, filling of the nano- and then mesopores occurs. The movement of the adsorbate inside the pores, which determines the rather complicated form of the chemical shift's temperature dependence, can be superimposed on this effect (Figures 3.30 and 3.31b).

In the ^{1}H NMR spectra (Figure 3.32, 200–280 K) of hydrogen adsorbed onto C-86, single signal is observed at $\delta_H \approx 4$ ppm (Turov et al. 2011c) similar to that observed for H_2 adsorbed onto other adsorbent (Turov et al. 2010f,g). The signal intensity increases due the adsorption of additional portions from a H_2 reservoir with lowering temperature. Water gives a broad signal with decreasing intensity with lowering temperature because of partially freezing of bound water. The δ_H value of water decreases from 3 to −1 ppm with increasing total amount of water from 0.005 to 0.04 g/g. Additionally, the δ_H value for samples at $h = 0.01$ and 0.04 g/g increases by 2 ppm with lowering temperature (similar to the adsorption of water alone).

The amounts of water and hydrogen were much smaller than the volume of nanopores. Therefore, one could expect that both adsorbates are located in nanopores. However, small shielding effects do not support this assumption. C-86 has well-developed structure of narrow mesopores, where water can be adsorbed in the form of clusters and nanodomains. These structures (with unfrozen water and ice crystallites) can form effective nanopores where hydrogen can be adsorbed. The adsorption of water in mesopores is also confirmed by the signal shape suggesting fast (in NMR timescale) exchange processes in water.

Comparison of the $C_{SAW}(T)$ and $C_{H2}(T)$ curves (Figure 3.33) shows that the H_2 adsorption increases with adsorption of water. This confirms the aforementioned assumption on the δ_H value because it shows the increasing confined space effects on H_2 due to adsorbed water without strong shielding effects characteristic for the adsorption in nanopores of pure carbons. The size distributions of water clusters (Figure 3.33c) correspond to main PSD peak of C-86 (Figure 3.28b). However, the displacement toward smaller sizes is observed with increasing water content.

In the case of co-adsorption of water and hydrogen with the presence of small amounts of weakly polar methylene chloride and benzene (Figure 3.34), the $\Delta\delta_H$ values (−6 and −7 ppm) for the latter show their adsorption in nanopores. Therefore, benzene ($T_m \approx 280$ K for bulk benzene) is unfrozen in the total temperature range.

Co-adsorption of hydrogen with benzene (Figure 3.35) demonstrates the opposite results in comparison with co-adsorption with water because the hydrogen adsorption decreases with increasing content of adsorbed benzene. Thus, benzene completely filling nanopores of C-86 does not form additional nanopores for adsorbed hydrogen. This is due to different localization of water and benzene in pores of AC. The study of co-adsorption of benzene and hydrogen at great $C_{C_6H_6}$ values is difficult due to a large difference in their signal intensities. Therefore, nonpolar CCl_4 was used for completely filling of nanopores (Figure 3.35d) at low content of water (2 mg/g) and TMS (1 mg/g).

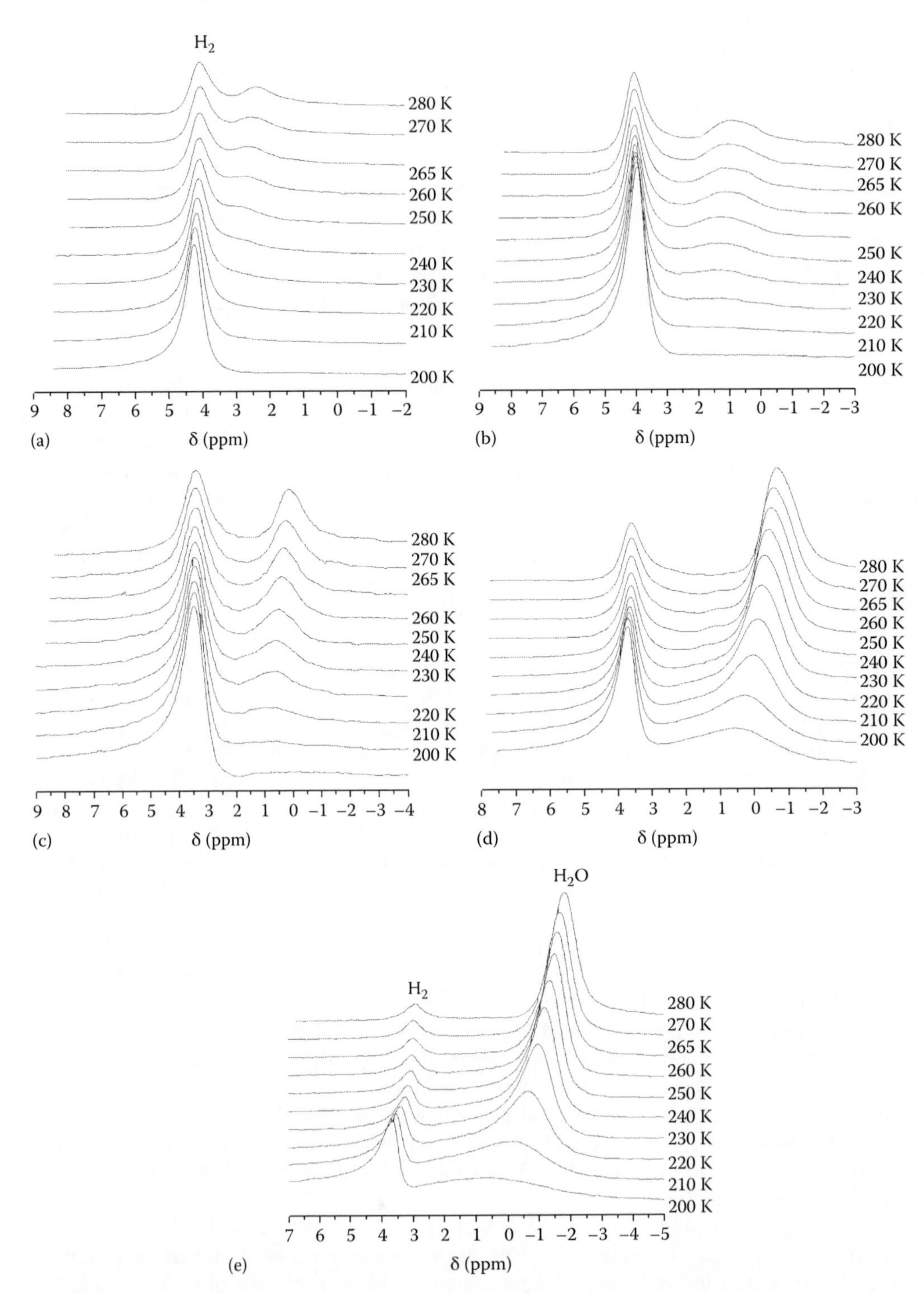

FIGURE 3.32 ^{1}H NMR spectra of water and hydrogen co-adsorbed onto C-86 recorded at different temperatures at different water content: (a) 0.75, (b) 1.5, (c) 2, (d) 10, and (e) 40 mg/g.

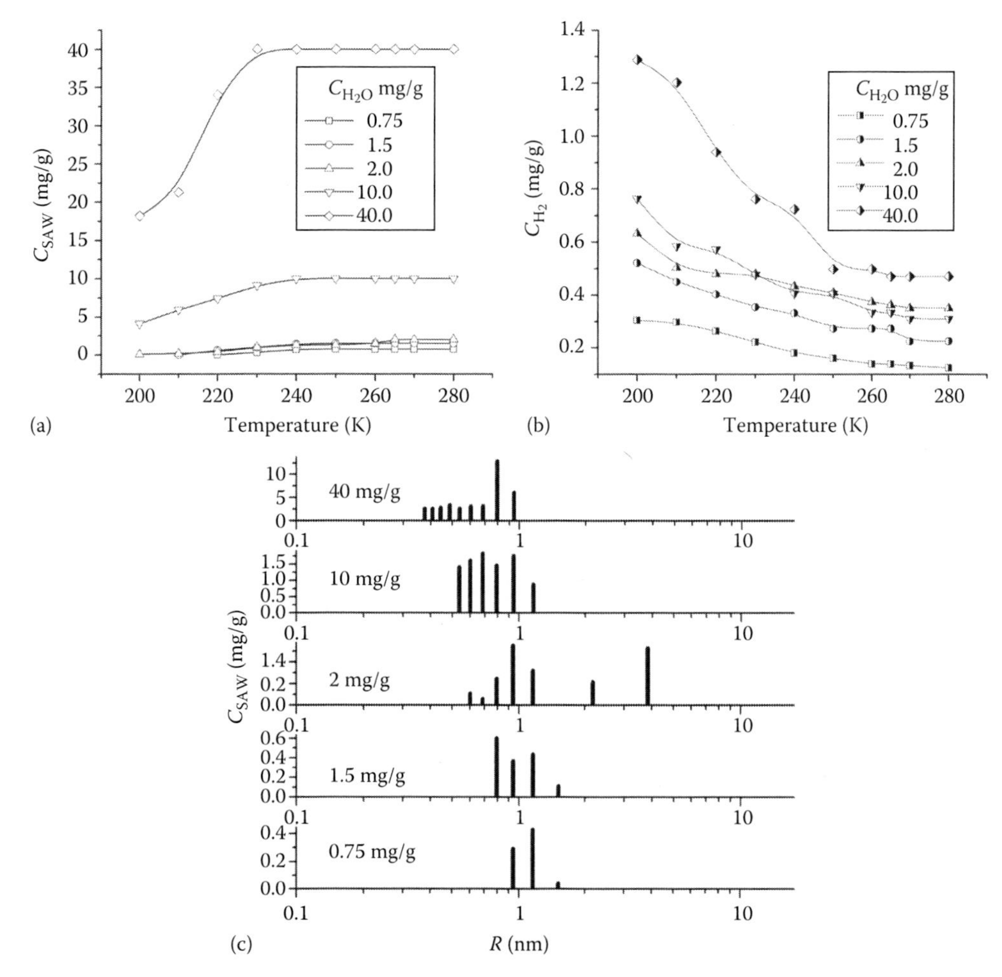

FIGURE 3.33 Temperature dependences of the content of (a) unfrozen water (SAW) and (b) adsorbed hydrogen onto C-86 and (c) corresponding size distribution functions for SAW.

In this case, the hydrogen adsorption was very low. Thus, weakly polar or nonpolar adsorbates (especially with plane molecules as benzene), which tend to be adsorbed in slit-shaped nanopores, are rather negative co-adsorbates for hydrogen in contrast to water which locates at the edges of carbon sheets around O-containing functionalities in narrow mesopores that results in the formation of secondary nanoporosity. Thus, clustered adsorption of water provides enhancement of effective nanoporosity appropriate for hydrogen adsorption.

AC C-86 is characterized by relatively decreased contribution of broad mesopores and macropores because of very strong burn-off activation. Therefore, a lower activated C-47 (47% burn-off degree) was used in similar experiment to analyze the textural effects (Figure 3.36). Co-adsorption of 10 mg/g water leads to small adsorption of hydrogen. This can be explained by smaller size of narrow mesopores ($\Delta\delta_H = -6$ ppm for water) where water is bound. Water fills these pores, blocks entrances into nanopores, and, therefore, the adsorption of hydrogen decreases in comparison with C-86.

In the case of co-adsorption of water and organic solvents, localization of bound water can change in comparison with the adsorption of water alone. Addition of 0.4 g $CDCl_3$ per gram of AC leads to disappearance of the signal of water adsorbed in nanopores but the signal of water (SAW) located in mesopores increases (Figure 3.37a). Thus, chloroform displaces water from narrow pores.

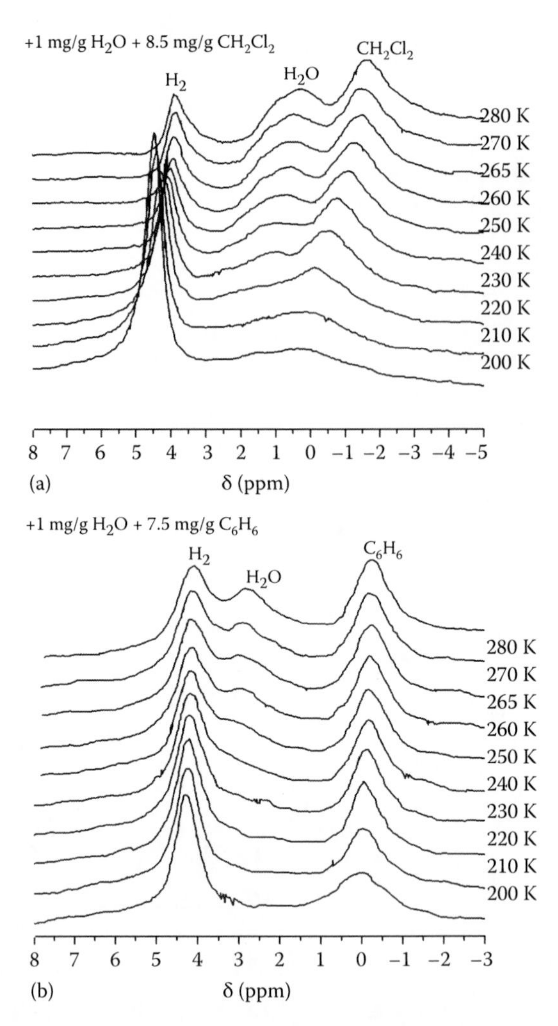

FIGURE 3.34 ^{1}H NMR spectra of H_2 and H_2O co-adsorbed with (a) CH_2Cl_2 or (b) C_6H_6 onto C-86.

An increase in the chloroform amount (Figure 3.37b) leads to almost complete removal of water from nano- and mesopores. Similar effects were observed for different adsorbents (Fenelonov 1995, Turov et al. 2002a–c, 2004, Gun'ko et al. 2003f, 2005d). It can be explained in the terms of free energy. The interaction energy of water with the pore walls in AC nanopores (where there is a low content of O-containing sites) can be much smaller (15–20 kJ/mol) than the liquefaction heat of water (45 kJ/mol) (Brennan et al. 2001). Interaction energy of water with such hydrophobic solvents as chloroform, benzene, etc. is three to five times lower (IEFPCM/B3LYP/6–311++G(3df,3pd) calculations give $\Delta G_s = -12.6$ (chloroform) and -7.9 (benzene) kJ/mol for water solvated molecule). Therefore, water tends to reduce the surface area of contact with nonpolar solvents. This is possible if water forms larger SAW structures in larger pores.

In the case of co-adsorption of 5 mg H_2O per gram of AC and methane (from a reservoir at slightly increased pressure of 1.1 bar), a significant displacement of water does not occur (Figure 3.37c). A certain increase in the signal of water located in mesopores occurs. The adsorption of methane was not high (<50 mg/g). Therefore, water and methane can co-adsorb in different pores without strong contacts between them (as in the case of liquid chloroform or benzene media).

The δ_H values of methane adsorbed onto C-47 is between −7 and −8 ppm (≈0 ppm for free methane) and $\delta_H = -2$ to -4 ppm for bound water (Figure 3.38). These shielding effects suggest that the

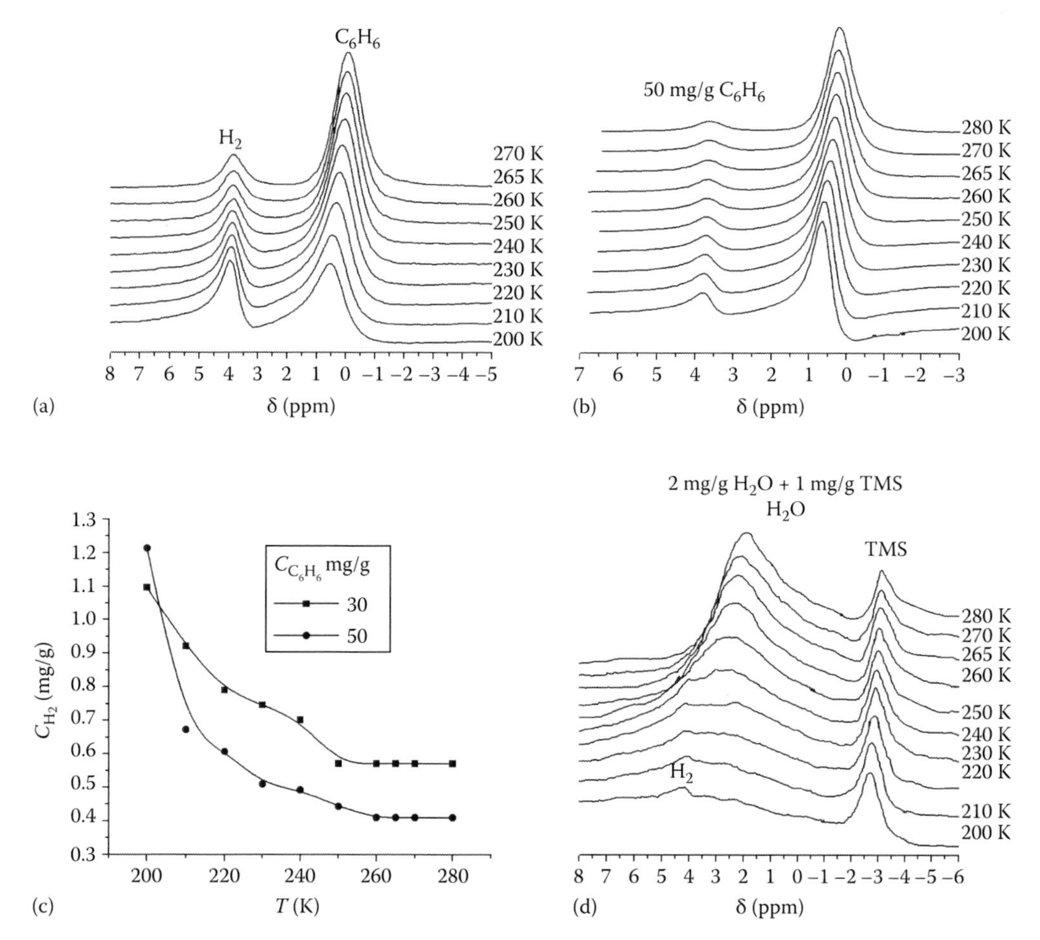

FIGURE 3.35 ^{1}H NMR spectra of H_2 co-adsorbed with (a and b) benzene and (d) CCl_4, and the temperature dependence of the content of adsorbed H_2 at 30 or 50 mg/g of adsorbed benzene.

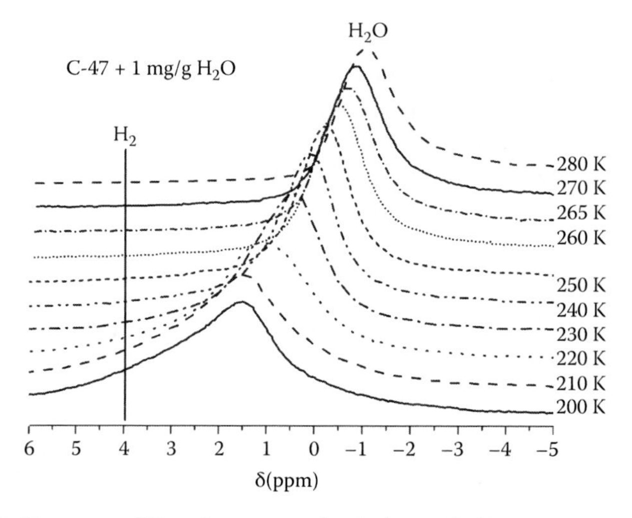

FIGURE 3.36 ^{1}H NMR spectra of H_2 and water co-adsorbed onto C-47.

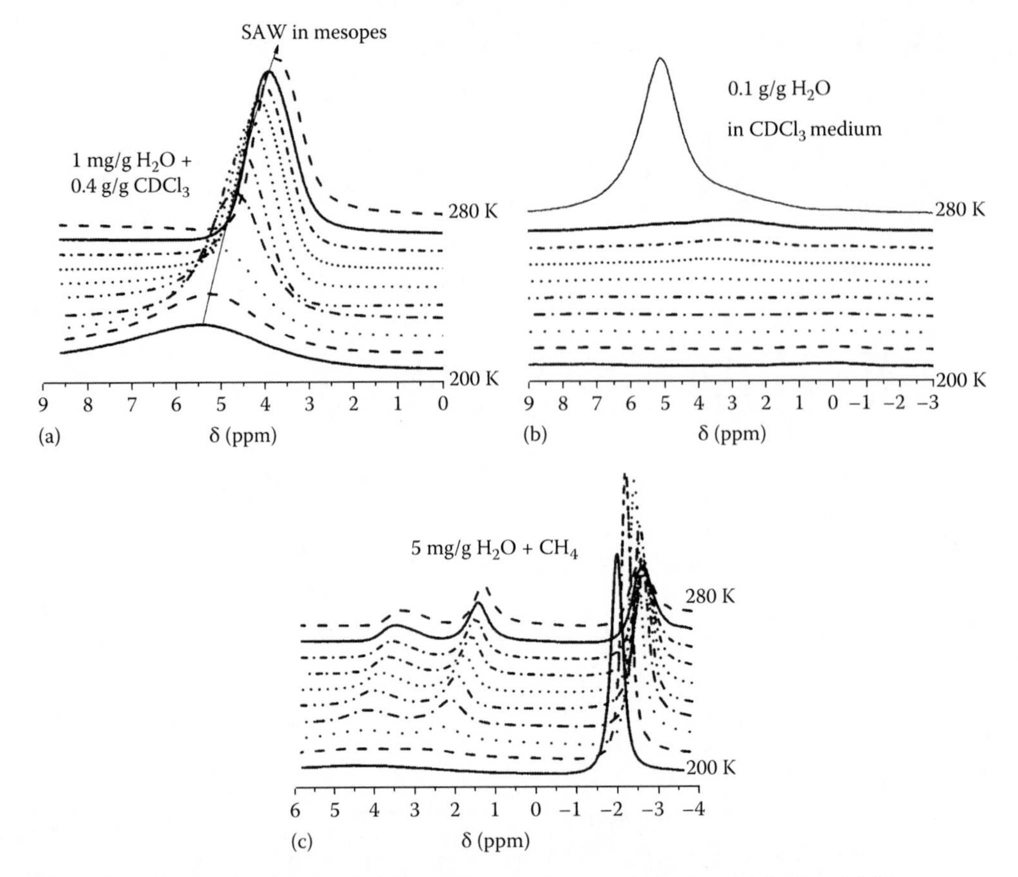

FIGURE 3.37 Influence of (a, b) chloroform and (c) methane on the water adsorbed to C-86.

both adsorbates are located in nanopores. Co-adsorption of water can increase the adsorption of methane (Figure 3.39). The CH_4 adsorption as a function of water content has a maximum (at 50–70 mg/g water), which shifts toward lower water values with decreasing temperature (Figure 3.39b). The influence of water on the methane adsorption, as mentioned earlier for adsorbed hydrogen, is due to changes in the effective textural characteristics.

In the case of 1.33 g/g water adsorbed onto C-47, the relationship $\Delta G(C_{uw})$ has a vertical portion (Figure 3.39c), which corresponds to the absence of water freezing over relatively broad temperature range (Gun'ko et al. 2005d, 2009d). The position of this curve portion (i.e., the C_{uw} value) determines the content of SBW ($C^{s}_{uw} = 0.35\,g/g$ for C-47).

The γ_s value for water bound to C-47 is relatively large (40 J/g) for weakly polar adsorbent. One can assume that water will be SBW with decreasing content. However, water co-adsorbed (in low amounts) with methane is SBW if $h = 0.1$ g/g (Figure 3.39c). At $h = 0.05$ g/g, the vertical portion is absent and a significant portion of water can be attributed to WBW at $-\Delta G < 0.5$ kJ/mol. Thus, even methane can be a relatively effective competitor displacing water into larger pores during co-adsorption to carbon adsorbents. Similar results were observed for other, e.g., polymeric, adsorbent (Turov et al. 2004, Gun'ko et al. 2005d, 2007d, 2008a,d,e 2009d).

The NMR cryoporometry results suggest that water adsorbs predominantly in mesopores of C-47 (Figure 3.40) forming nanodomains affected by methane co-adsorption.

Quantum chemical calculations of water/methane structures clathrate (Figure 3.41a and b) and non-clathrate (Figure 3.41c and d) showed that clusters with water (WAW) and methane molecules can form mosaic structures in confined space of pores with hydrophilic (O-containing) and

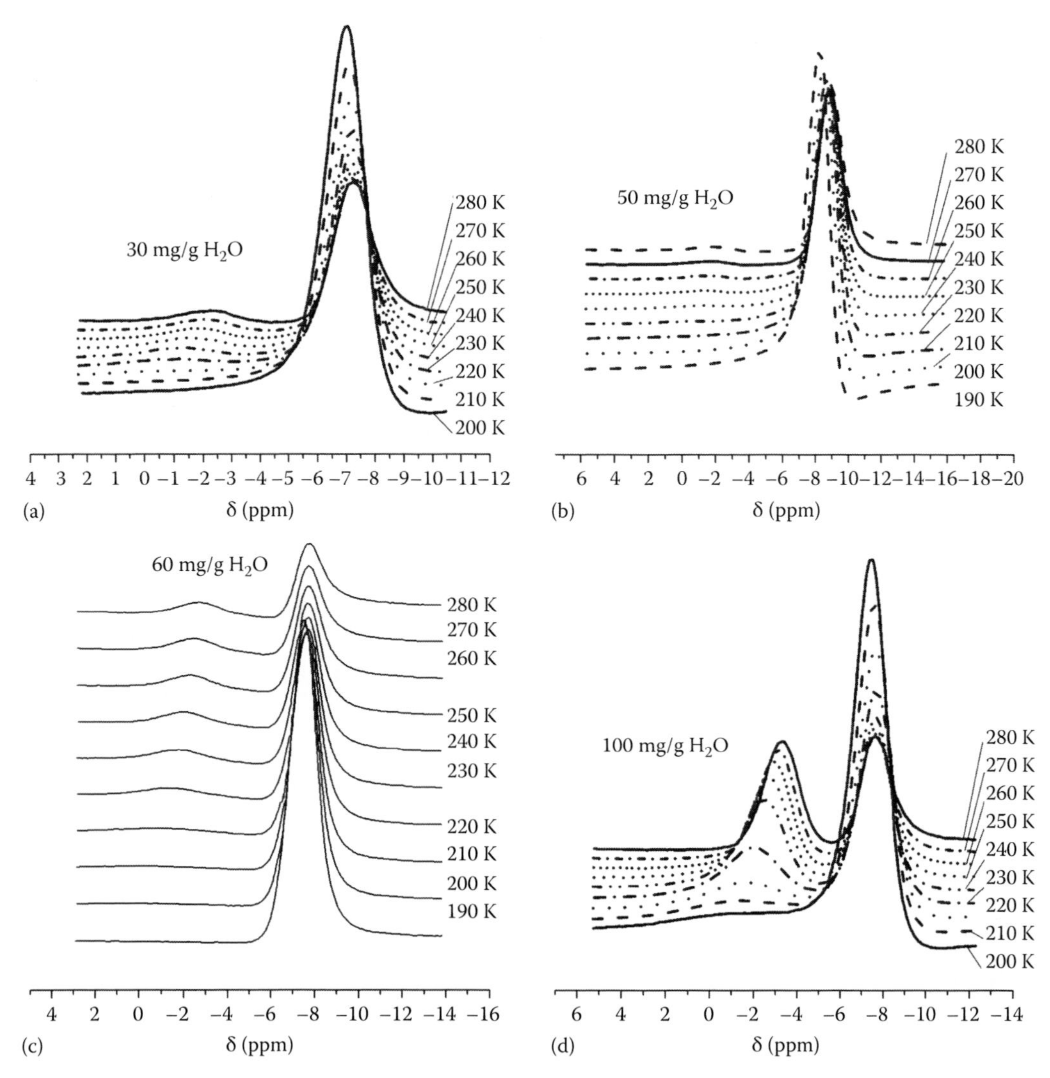

FIGURE 3.38 ^{1}H NMR spectra of water and methane co-adsorbed onto C-47 at different water amounts: (a) 30, (b) 50, (c) 60, and (d) 100 mg of water per gram of dry AC. (Taken from *Appl. Surf. Sci.*, 255, Turov, V.V., Turova, A.A., Goncharuk, E.V., and Gun'ko, V.M., Adsorption of methane with the presence of water on oxide, polymer, carbon adsorbents studied using ^{1}H NMR spectroscopy at low temperatures, 3310–3317, 2008a. Copyright 2008, with permission from Elsevier.)

hydrophobic (aromatic) patches. In the case of clathrates 5^{12} and $5^{12}6^2$, SAW is dominant (Figure 3.42, peak at $\delta_H = 3–6$ ppm).

Changes in the energy due to the formation of methane clathrate $5^{12}/CH_4$ and $5^{12}6^2/CH_4$ is equal to −37 and −29 kJ/mol (PM3) or −18 and −20 kJ/mol (B3LYP/6-31G(d,p)//6-31G(d,p)), respectively. Calculations of water/methane clusters in hydrophobic nanopores (Figure 3.41d and e) show the formation of WAW structures (Figure 3.44, peak at $\delta_H = 0–1$ ppm). In the case of isolated (without adsorbent) clusters (Figure 3.41c), contribution of SAW is greater than that for adsorbed systems (Figure 3.42, peak at $\delta_H = 3$ ppm). Methane in different structures (Figure 3.41) has $\delta_H = -2$ to 0 ppm (Figure 3.42).

Thus, co-adsorbed water and methane molecules can form clustered (mosaic) structures in nanopores of different adsorbents without the formation of clathrate structures. One of

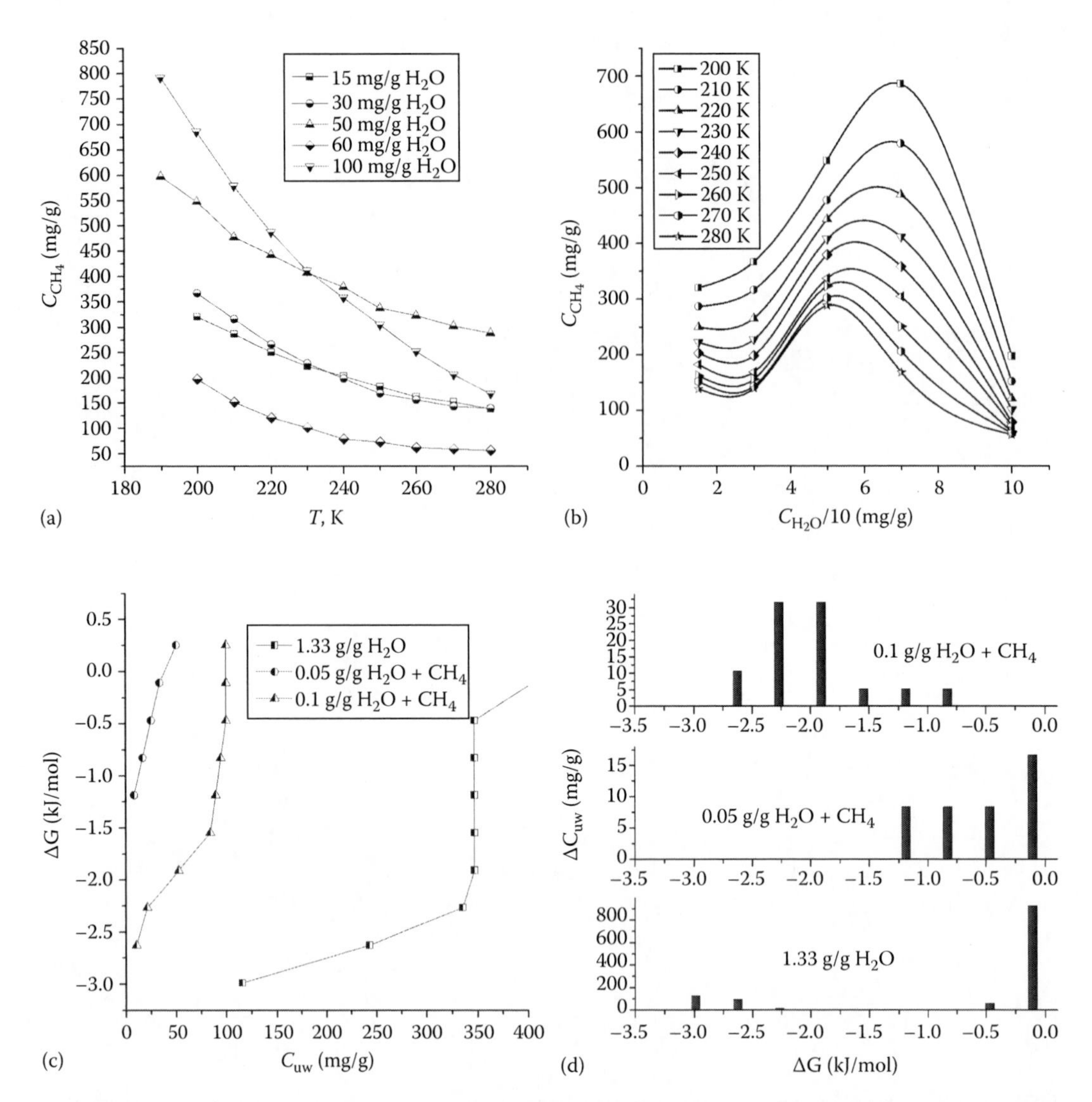

FIGURE 3.39 (a) Temperature dependences of methane adsorption onto C-47 at different content of co-adsorbed water; (b) relationships between the contents of co-adsorbed water and methane at different temperatures; (c) changes in the Gibbs free energy versus amounts of unfrozen water for water alone and for the mixtures; and (d) corresponding size distributions of unfrozen water structures AC. (Taken from *Appl. Surf. Sci.*, 255, Turov, V.V., Turova, A.A., Goncharuk, E.V., and Gun'ko, V.M., Adsorption of methane with the presence of water on oxide, polymer, carbon adsorbents studied using ^{1}H NMR spectroscopy at low temperatures, 3310–3317, 2008b. Copyright 2008, with permission from Elsevier.)

favorable conditions is a mosaic hydrophilic/hydrophobic structure of pores similar to hydrophilic O-containing functionalities on the edges of carbon sheets in narrow mesopores of AC at the entrances into nanopores and hydrophobic graphene structures in nanopores. The modeling results explain the experimentally observed different (opposite) effects of the co-adsorbed water on the co-adsorbed methane.

To analyze changes in the interfacial behavior of different adsorbates with a small size of the molecules (e.g., H_2, CH_4, etc.), carbon C-30 (S_{BET} 1145 m^2/g, V_p = 1.187 cm^3/g) was chosen (Gun'ko et al. 2011d) as a representative sample with relatively small contribution of broad pores to the specific surface area since $(S_{meso} + S_{macro})/S_{BET} = 0.081$, broad PSD, low contribution of narrow mesopores and

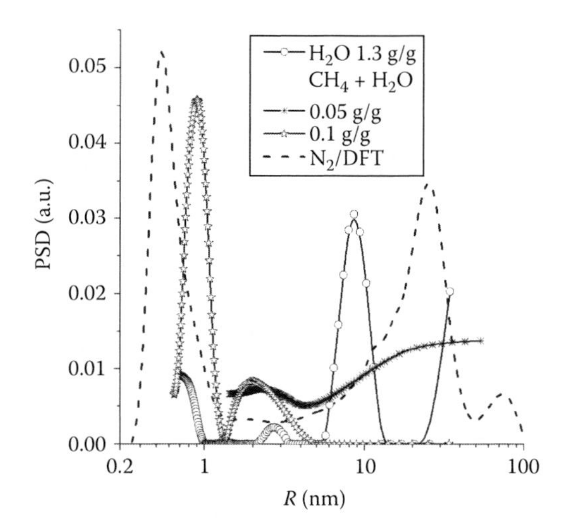

FIGURE 3.40 Size distribution functions of unfrozen water bound to C-47 at different hydration and without or with co-adsorption of methane, PSD of C-47 (dashed line) was calculated from the nitrogen adsorption/desorption isotherm AC. (Taken from *Appl. Surf. Sci.*, 255, Turov, V.V., Turova, A.A., Goncharuk, E.V., and Gun'ko, V.M., Adsorption of methane with the presence of water on oxide, polymer, carbon adsorbents studied using ^{1}H NMR spectroscopy at low temperatures, 3310–3317, 2008b. Copyright 2008, with permission from Elsevier.)

$V_{nano}=0.554$ cm^3/g close to $V_{meso}+V_{macro}=0.633$ cm^3/g. Notice that broad pores are not appropriate for adsorption of small molecules of the probes studied. Water adsorbed in textural broad voids and narrow pores in C-30 is characterized by two signals in the ^{1}H NMR spectra (Figure 3.43a).

However, these signals do not split because of the large amount of adsorbed water ($h=0.75$ g/g) and the existence of fast molecular exchange between water structures located in different pores. Water in broad pores is weakly bound, and therefore it is completely frozen at $T<265$ K (Figure 3.43a) since signal 2 disappears at $T<265$ K. Water in narrow slit-shaped pores is unfrozen even at 200 K (Figure 3.43a; signal 1) and characterized by the signal with up-field shift because of the shielding effects of the π-electron current at the basal planes that increase in narrower pores. Additionally, the ^{1}H NMR signals of adsorbed water are characterized by up-field shift with lowering temperature because water freezes in narrower pores (where the shielding effect of the π-electrons is stronger) at lower temperatures due to stronger freezing point depression for liquids confined in narrower pores.

Co-adsorption of water and chloroform-*d* results in the displacement of water by weakly polar (hydrophobic) chloroform from narrow pores. Therefore, all water is completely frozen at $T<260$ K (Figure 3.43b), which suggests that all displaced water is weakly bound and located in broad pores (i.e., in textural voids between carbon nanoparticles) or on the outside of carbon granules. Notice that the signal of displaced water is at $\delta_H=5$ ppm (typical value for bulk water with high associativity of the molecules), which is greater than $\delta_H=3$ ppm without chloroform-*d*, i.e., water associativity increases due to interaction with chloroform. This can be caused by displacement of water into larger pores or out of them where the associativity of water molecules can increase (similar to the effects in C-86 and C-47 described earlier).

Upon the adsorption of methane and hydrogen, a rubber reservoir with a gas (pressure 1.1 atm) was connected to a NMR ampoule. Therefore, the adsorbed amounts of CH_4 or H_2 can increase with lowering temperature (Figure 3.44a and b). For both methane and hydrogen, the up-field shift of the ^{1}H NMR signal is observed since for free gases $\delta_H=0$ and 4 ppm, respectively. In contrast to CH_4 (Figure 3.44a), the H_2 adsorption is very small (Figure 3.44b) because C-30 is not a pure microporous carbon and has a broad PSD (Gun'ko et al. 2011d). In contrast to water (Figure 3.43a), adsorbed methane is characterized by the ^{1}H NMR signal with down-field shift with lowering temperature

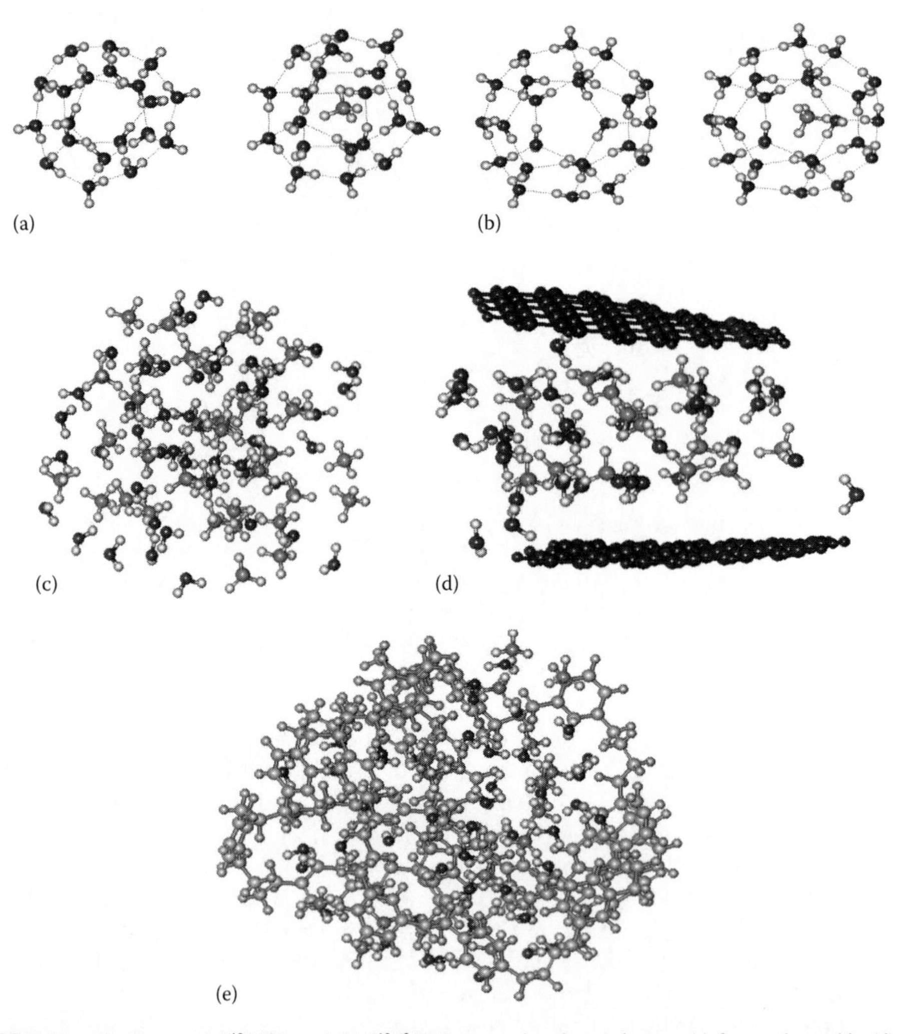

FIGURE 3.41 Clathrates (a) $5^{12}/CH_4$ and (b) $5^{12}6^2/CH_4$, water/methane clusters (c) free or bound in (d) carbon or polymeric nanopores AC. (Taken from *Appl. Surf. Sci.*, 255, Turov, V.V., Turova, A.A., Goncharuk, E.V., and Gun'ko, V.M., Adsorption of methane with the presence of water on oxide, polymer, carbon adsorbents studied using ^{1}H NMR spectroscopy at low temperatures, 3310–3317, 2008b. Copyright 2008, with permission from Elsevier.)

(Figure 3.44a) because it is unfrozen (freezing temperature $T_f=90.7$ K) over the total range of used temperatures and its adsorption increases ($p=1.1$ atm in the CH_4 reservoir) in larger pores with the reduced shielding effect with decreasing temperature.

On the adsorption of chloroform-H, four ^{1}H NMR signals are observed at $\delta_H=0.5$, 3.5–3, 4.5–5, and 7.2 ppm at $T>210$ K (Figure 3.44c). The intensity of signals 3 and 4 strongly decreases at $T<210$ K since freezing temperature of $CHCl_3$ is $T_f=209.7$ K. Therefore, these signals can be assigned to chloroform-H located in broad pores or out of pores. Signal 4 at 7.2 ppm corresponds to practically free liquid (e.g., adsorbed in broad pores) but signal 3 at 4.5–5 ppm can be due to chloroform located in mesopores. Signals 1 and 2 are due to the adsorption of chloroform in narrow mesopores and nanopores. Thus, the lower the δ_H value, the narrower are the pores where the $CHCl_3$ molecules are located. The signal of chloroform located in the narrowest pores at $x<1$ nm does not diminish at lower temperatures in contrast to that of molecules located in broader pores because of stronger freezing point depression for adsorbates confined in narrower pores.

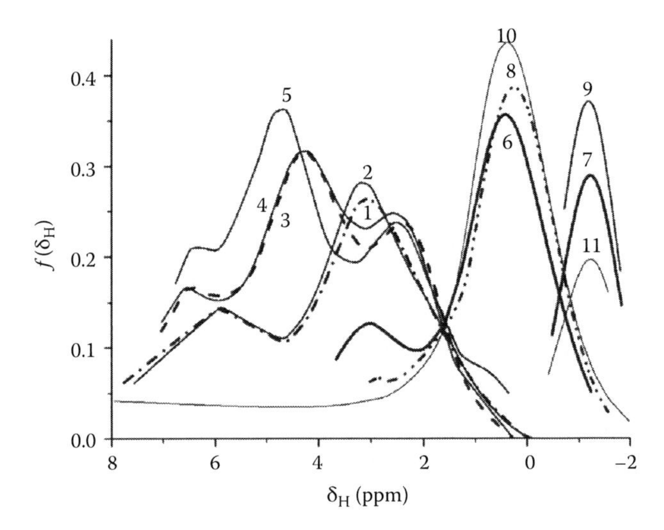

FIGURE 3.42 ^{1}H NMR spectra calculated for water and water/methane clusters: clathrates 5^{12} (1), $5^{12}/CH_4$ (2), $5^{12}6^2$ (3), $5^{12}6^2/CH_4$ (4) (GIAO/B3LYP/6-31G(d,p)//6-31G(d,p)) and $5^{12}6^2/CH_4$ (PM3 + calibration function $\delta_H = aq_H + b$), (5) for H in water (6, 8, and 10) and methane (7, 9, and 11), and non-clathrate structures, free $53H_2O/35CH_4$ (6, 7), and $46H_2O/22CH_4$ bound to polymeric adsorbent (8 and 9) and $37H_2O/14CH_4$/carbon adsorbent (10 and 11) AC. (Taken from *Appl. Surf. Sci.*, 255, Turov, V.V., Turova, A.A., Goncharuk, E.V., and Gun'ko, V.M., Adsorption of methane with the presence of water on oxide, polymer, carbon adsorbents studied using ^{1}H NMR spectroscopy at low temperatures, 3310–3317, 2008b. Copyright 2008, with permission from Elsevier.)

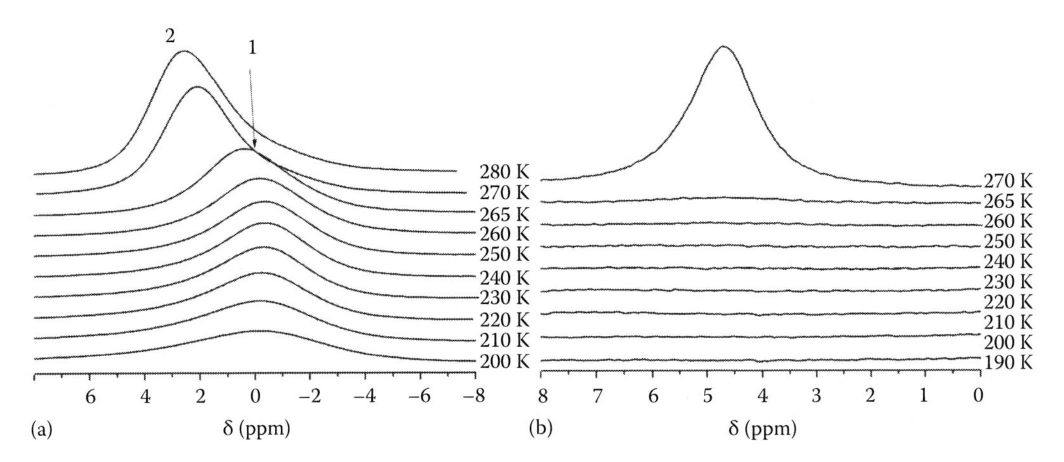

FIGURE 3.43 ^{1}H NMR spectra of water (0.75 g per gram of dry carbon) adsorbed on C-30 at different temperatures and different media: (a) air and (b) chloroform-*d*. (With kind permission from Springer Science+Business Media: *Adsorption*, Activation and structural and adsorption features of activated carbons with highly developed micro-, meso- and macroporosity, 17, 2011d, 453–460, Gun'ko, V.M., Turov, V.V., Kozynchenko, O.P. et al. Copyright 2011.)

The freezing effect is not observed for CH_2Cl_2 (Figure 3.44d) because its $T_f = 176.2$ K is lower than the temperature range used. A decrease in the signal at 2 ppm with lowering temperature can be due to the formation of larger Van der Waals clusters in larger pores with decreasing temperature. A similar but stronger effect is also observed for the CH_2Cl_2-DMSO-d_6 mixture (Figure 3.44e) due to stronger rearrangement of this mixture with lowering temperature. This can be due to large differences in the T_f values (~115 K), polarity (donor number ~0 and 125 kJ/mol,

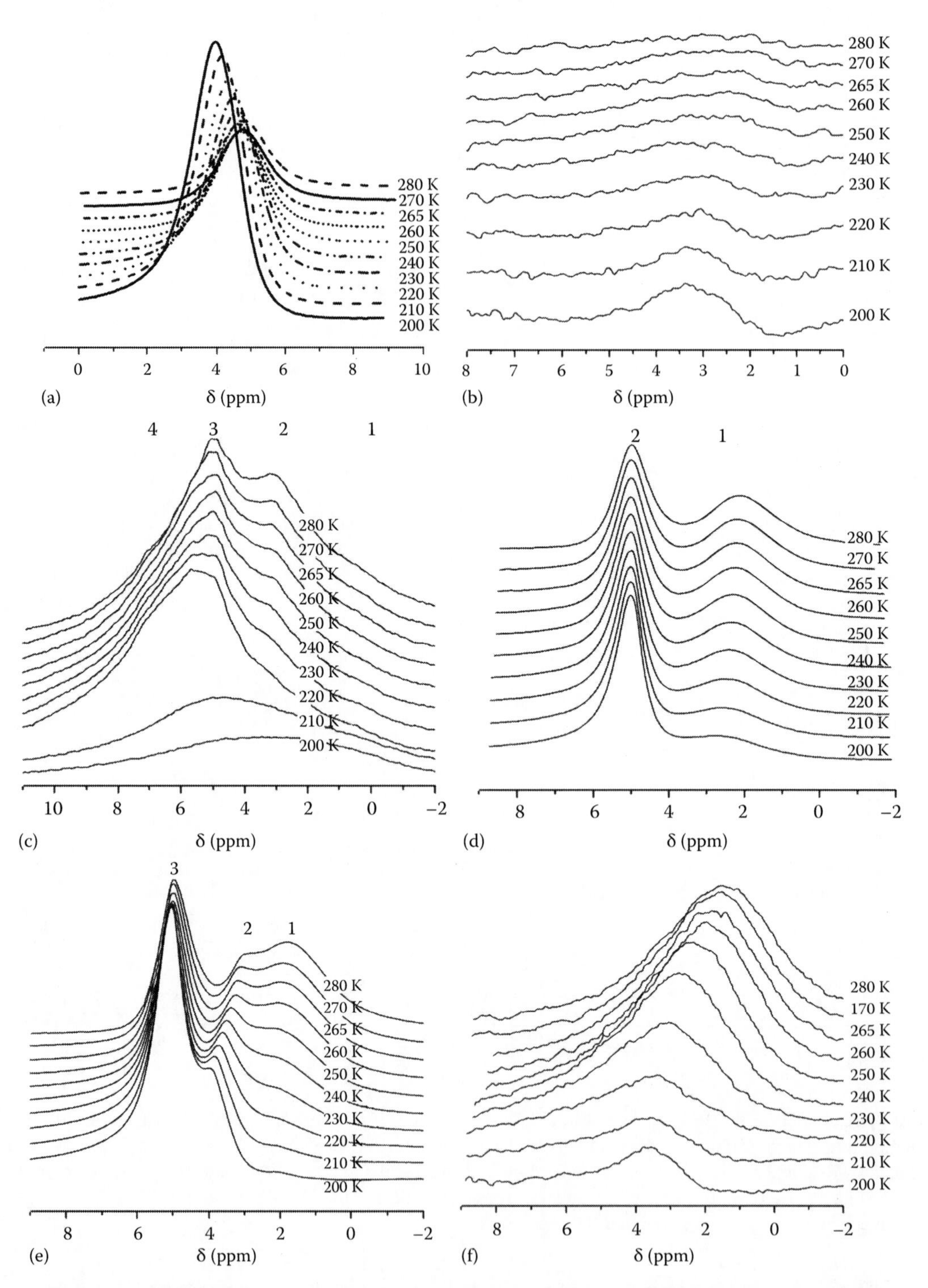

FIGURE 3.44 ^{1}H NMR spectra of (a) CH_4 (freezing point $T_f = 90.7$ K), (b) H_2 ($T_f = 14$ K), (c) $CHCl_3$ ($T_f = 209.7$ K), (d) CH_2Cl_2 ($T_f = 176.2$ K), (e) CH_2Cl_2 + DMSO-d_6 (0.4 g/g pre-adsorbed, $T_f = 291.6$ K), and (f) $H_2 + H_2O$ (0.02 g/g preadsorbed). (With kind permission from Springer Science+Business Media: *Adsorption*, Activation and structural and adsorption features of activated carbons with highly developed micro-, meso- and macroporosity, 17, 2011d, 453–460, Gun'ko, V.M., Turov, V.V., Kozynchenko, O.P. et al. Copyright 2011.)

respectively), interaction energy with AC and molecular sizes of co-adsorbates. Similar effects were observed for different adsorbents analyzed in this book, and a similar effect of "freezing" of the up-field-shifted signal is observed for the H_2-H_2O mixture (Figure 3.44f). Thus, this phenomenon reflects certain regularities in the interfacial behavior of the mixtures with compounds unfrozen and frozen in a given temperature range due to, at least, three reasons: (i) variations in the freezing point depression for different liquids confined in different pores, (ii) diminution of the mobility of small molecules with decreasing temperature that allows them to be adsorbed in larger pores, and (iii) a frozen phase (or an immobilized phase) can change the topology of pores filled by an unfrozen phase. Therefore, the adsorption of H_2 or CH_4 onto different adsorbents can increase due to preadsorption of water or other adsorbates with relatively high T_f values (Gun'ko et al. 2009d). Another important factor is the affinity of the AC surface with respect to one of adsorbates that can be changed due to surface modification (e.g., oxidation, grafting functionalities with H, N, S, P, and other atoms). Uniform or nonuniform modification can affect adsorption of a given adsorbate in the form of films (completely filling narrow pores) or clusters (e.g., at the edges of carbon sheets).

Many of the mentioned effects depend on the textural characteristics of carbons. Therefore, we compare here three sets of ACs. An increase in the carbon activation by CO_2 to the 86% burn-off degree (C-0 to C-86) leads to parallel enhancement of contribution of micro- and mesopores to the pore volume (Figure 3.45). However, contribution of mesopores to the specific surface area is significant only for C-86 and W-4 with the maximal activation by CO_2 and water vapor, respectively. A reason of these changes in the characteristics is clearly visible in Figure 3.46 showing the appearance of a peak of narrow mesopores at $1 < x < 2$ nm for both C-86 and W-4, which are absent for all other samples. However, an increased activation degree gives the displacement of the PSD peaks of broad mesopores and macropores toward smaller x values (Figure 3.46). This result can be explained by the origin of these pores, which have a textural character as voids between porous nanoparticles (30–50 nm in size according to AFM and SEM images (Gun'ko et al. 2011d)). The increased activation leads to diminution of the sizes of these nanoparticles, and voids between smaller particles become narrower. Water molecules are smaller than CO_2 molecules; therefore the former penetrating into smaller pores can more strongly add the nanoporosity with respect to both pore volume and specific surface area (Figures 3.45 and 3.46, C-86 and W-4). An increase in the activation from 47% to 86% burn-off degree results in a diminution of contribution of the narrowest nanopores due to partial disruption of the pore walls and increasing contribution of narrow mesopores at $x < 2$ nm (Figure 3.46).

Notice that the relative contribution of nanopores decreases (Figure 3.45). The maximal activation with water vapor more strongly increases contribution of large mesopores than the activation with CO_2. However, the character of structural changes depends on the type of a reactor used and other conditions on activation by water vapor. In the case of the fluidized bed reactor, contribution of narrow mesopores is larger than that for samples prepared in the fixed bed reactor. Notice that further activation of carbons at burn-off >86%–88% was not carried out because of a low yield of AC.

Thus, ACs prepared with the same precursor but differently activated by CO_2 or water vapor demonstrate increased deviation of the pore shape from the slit-shaped model with increased contribution from pores over a broad range with increasing burn-off degree. This occurs due to enhanced crumpling of single sheets or stacks with a small number of sheets and reduction of the sizes on nanoparticles (formed with these crumpled sheets) packed in carbon globules with increasing activation. Activation by both CO_2 and water vapor results in close changes in contributions of nano-, meso-, and macropores depending on burn-off degree, and contribution of macropores decreases in contract to mesoporosity. Relative contribution of nanopores to the specific surface area decreases with increasing burn-off degree. Changes in the mesoporosity have a more complex character because tendency in changes of narrow mesopores (close in size to nanopores) can differ from that for broad mesopores (close in size to macropores). The interfacial behavior of different adsorbates adsorbed alone or in various mixtures on ACs depends strongly on temperature, especially when

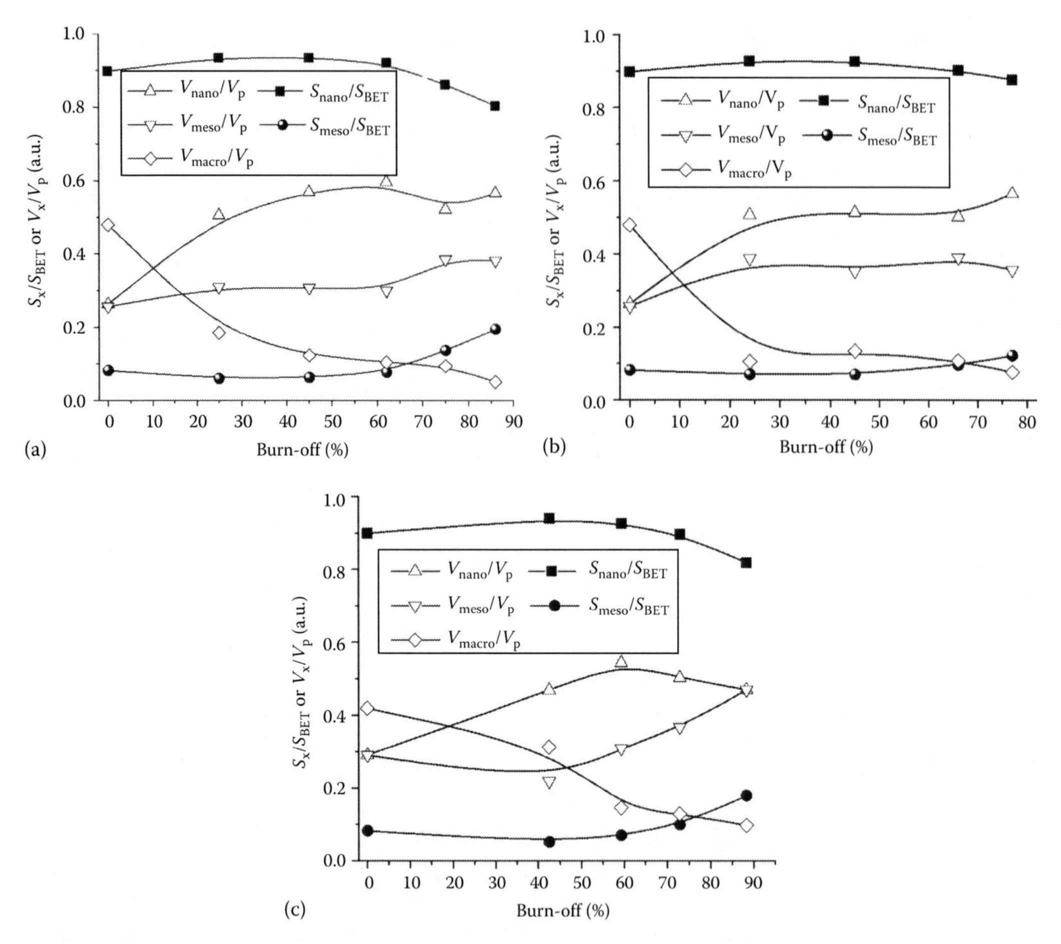

FIGURE 3.45 Relationships between S_{nano}, S_{meso}, and S_{BET} and V_{nano}, V_{meso}, V_{macro}, and V_p for ACs activated by (a) CO_2 and water vapor in (b) fixed and (c) fluidized bed reactors. (With kind permission from Springer Science+Business Media: *Adsorption*, Activation and structural and adsorption features of activated carbons with highly developed micro-, meso- and macroporosity, 17, 2011d, 453–460, Gun'ko, V.M., Turov, V.V., Kozynchenko, O.P. et al. Copyright 2011.)

current temperature becomes lower than freezing temperature of one of the adsorbed compounds. Significant re-arrangement of adsorption complexes, especially of the Van der Waals–type characteristic for nonpolar or weakly polar adsorbates, occurs in both nanopores and mesopores with decreasing temperature. The temperature behavior of the mixtures can strongly differ from that of individual adsorbates. These effects should be considered in practical applications of ACs for adsorption of various mixtures at different, especially low, temperatures.

3.3 GRAPHITIZED CARBONS AND GRAPHITE

3.3.1 Exfoliated Graphite

Graphite is an interesting subject of inquiry while investigating spectral characteristics of adsorbed molecules since the main type of adsorption sites on its surface is basal graphite planes. However, in this case the recording of ^{1}H NMR spectra for adsorbed molecules is complicated on account of small values of specific surface of graphite powders and large widths of NMR signals due to paramagnetic relaxation of nuclear spins of protons affected by conduction electrons. Even in

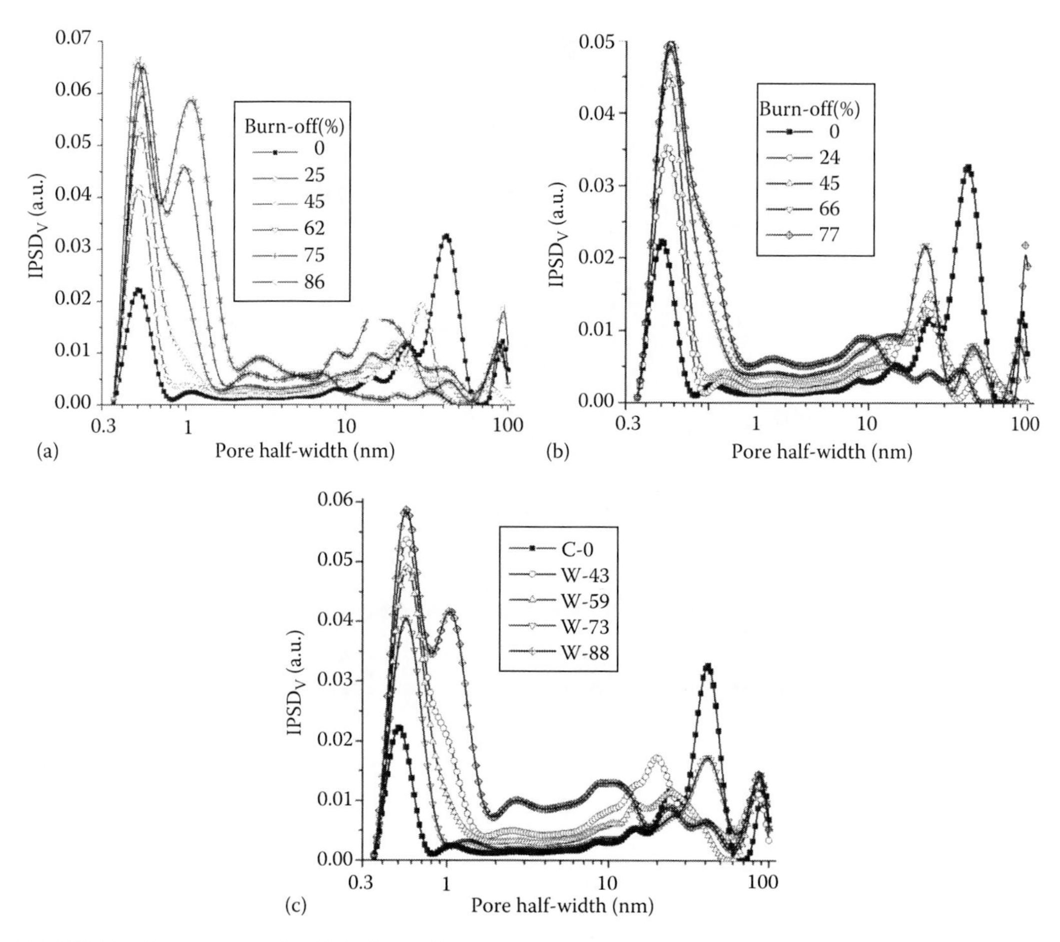

FIGURE 3.46 Incremental PSDV for carbon samples activated by (a) CO_2 and water vapor in (b) fixed and (c) fluidized bed reactors, DFT method with the model of slit-shaped pores. (With kind permission from Springer Science+Business Media: *Adsorption*, Activation and structural and adsorption features of activated carbons with highly developed micro-, meso- and macroporosity, 17, 2011d, 453–460, Gun'ko, V.M., Turov, V.V., Kozynchenko, O.P. et al. Copyright 2011.)

the situation of adsorption of alkanes on the surface of graphitized carbon blacks (Grosse and Boddenberg 1987, Boddenberg and Grosse 1988), the NMR signal width can amount to several kHz, which does not allow one to make exact measurements of chemical shifts.

To decrease the peak widths of ^{1}H NMR spectra of substances adsorbed on graphite surface (Turov et al. 1991a–c), water, benzene, and acetonitrile were adsorbed on the surface of EG (Figure 3.47) produced by fast heat treatment of a crystalline graphite powder intercalated with sulfuric acid.

As is known, the specific surface of such an EG sample is higher than that of pristine graphite, and, in addition, at the expense of reduction of the concentration of free electrons in the sample the intensity of relaxation effects decreases. The measurements were carried out for suspensions of the substances under study, with the suspensions being frozen. Applying the ring current method (Abragam 1961, Emsley 1965), Pogorelyi et al. (1991) theoretically estimated chemical shift values for some molecules adsorbed on graphite clusters.

In Table 3.8 for a number of compounds adsorbed on EG, chemical shift values observed in experiments and those calculated by the method of ring currents are compared. The parameters for adsorption complexes were obtained on the basis of simulating their interactions with the graphite

FIGURE 3.47 SEM image of exfoliated graphite (bar 50 μm).

TABLE 3.8
Measured and Calculated Values of ¹H NMR Chemical Shifts for Water, Acetonitrile, and Nitrobenzene Adsorbed on the EG Surface

	δ_H (ppm)	
Substance	**Condensed State**	**Adsorbed State**
H_2O	5	2.5
CH_3CN	2	0
$C_6H_5NO_2$	7.5	5

surface by the method of interatomic Scheraga's potentials. For all the investigated compounds, chemical shifts for the adsorbed state are smaller than those for the condensed state. A strong shielding effect for compounds susceptible to various molecular interactions enables one to conclude that it is the surface of basal planes that exerts a magnetic shielding effect on adsorbed molecules. As can be seen from Table 3.8, the calculated chemical shifts for acetonitrile and nitrobenzene are less than those observed in liquid phase freezing experiments. This is due to the fact that on recording ¹H NMR spectra under freezing conditions, the resonance signal is an average for molecules adsorbed on the surface and for molecules that are at some distance from it but are also subjected to the perturbing action of the surface. In this case, due to a contribution of molecules, which do not interact directly with areas of local magnetic anisotropy, the recorded signal has down-field shift as compared with a signal obtained under conditions of gas phase adsorption. For this reason, the chemical shift for water adsorbed on the graphite surface upon freezing is greater than that for water adsorbed from a gas phase. In the calculations performed, the feasibility of multilayer adsorption was not considered, so that the calculated values of chemical shifts for acetonitrile and nitrobenzene are related to ^1H NMR signals in the case of gas phase adsorption.

It should be noted that positions of adsorbed molecules with respect to the magnetic axis of ring currents for a ring may vary depending on the chemical nature of an adsorbed molecule. Here, one would also expect changes in the chemical shift of the adsorbed molecules similar to those observed for protons of molecules interacting with π-electron systems of benzene. However, for carbon sheets or graphene clusters the integrated ring current generated by the system of conjugated rings is equal to the current flowing along the outer contour of this system. Then, a molecule adsorbed at any point inside the contour will be subjected to the shielding effect similar to that near the axes of single benzene ring. It is precisely this fact that explains why the up-field shifts of protons are observed for any molecules irrespective of their coordination position on surface.

3.3.2 Oxidized Graphite and Graphene

Graphite surface (in particular, EG surface) contains a lot of structural imperfections, such as oxidized groups and residual molecules of oxidizers (Morimoto and Miura 1985). They can be involved in formation of H-bonded complexes with water-type molecules. Active protons in such associations should be subjected to a deshielding effect of electron-donating atoms, and, besides, in this case, one could expect the appearance of ^1H NMR signals with down-field shift relative to the corresponding resonance lines for a condensed phase. As no such peaks were found in the spectra, and, in addition, experimentally determined chemical shifts for water are close to calculated values, it may be concluded that the concentration of these sites in EG is low in comparison with hydrophobic adsorption sites on basal graphite planes of a carbon surface.

To obtain ^1H NMR spectra for water molecules interacting with surface oxidized groups (carbonyls, carboxyls, and hydroxyls), NMR investigations have been made on frozen aqueous suspensions of graphite oxide (GO; Karpenko et al. 1990a,b). GO particles are the colloidal particles consisting of strongly oxidized graphite sheets (Hennig 1959) similar to SLGO (vide infra). In gaps among these planes, there is a great deal of surface-bonded water. A ^1H NMR spectrum for water in such a suspension at a solid phase concentration C_{sol} = 1.92 wt% is shown in Figure 3.48. As evident from this figure, the spectrum is composed of three signals, two of which (at δ_H = 7 and 0 ppm) make

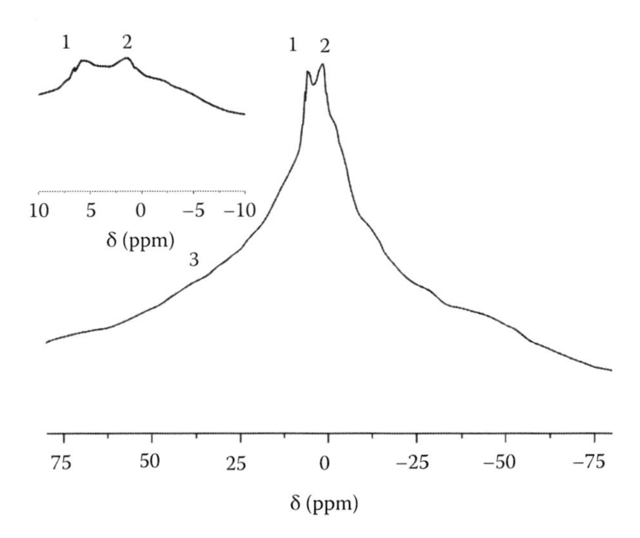

FIGURE 3.48 ^1H NMR spectra for water in a frozen aqueous suspension of oxidized graphite. (According to Karpenko, G.A. et al., *Teoret. Eksperim. Khim.*, 26, 102, 1990b.)

up a narrow component observed on the background of a very broad signal (3) with the half-width of 5 kHz. The former two narrow lines may be assigned to water molecules attached to oxygen-containing and free-radical surface sites while the third signal may be assigned to structurized water in interplanar gaps. The oxygen-containing surface sites were carbonyl, carboxyl, and hydroxyl groups. Free-radical sites are mainly located at terminal carbon atoms on graphite planes. The presence of a free radical excess in a GO suspension was corroborated by observation of an intensive electron paramagnetic resonance (EPR) signal with a width of 0.2 mT and *g*-factor of 2.003. Since signal 1 (Figure 3.48) has the chemical shift greater (~7 ppm) than that for liquid water (~5 ppm), it can be ascribed to water molecules incorporated into complexes of the C=O···HOH and COOH···(H)OH (cyclic complex with two H bonds) types, where hydrogen bonds are stronger than in pure water. Signal 2 was related (Karpenko et al. 1990b) to water bonded to free-radical sites or water located in narrow hydrophobic pores in the shape of small clusters. However, this assignment is not unambiguous as GO particles may contain cluster-like structures of nonoxidized graphite.

When interacting with such inclusions, water molecules are subjected to a shielding effect of ring currents in aromatic systems similar to the effect observed in EG. The value of *g*-factor that is close to *g*-factor for free electrons provides evidence for the validity of our argument.

The fact that in the spectra for frozen suspensions of GO there is a signal with a width of more than 5 kHz gives ground to believe that water molecules responsible for this signal have a very low molecular mobility, which is substantially lower than that for water molecules bound to the surface active sites. In all likelihood these slow-moving molecules are molecules situated in gaps among planes of disperse particles of GO in the field of surface forces. As judged by the intensity of signal 3, this part of water in the suspensions studied is a great part of water contained in such a suspension. Karpenko et al. (1990a,b) investigated diluted aqueous suspensions of GO whose concentrations were not higher than 1 wt%. Therefore, in conformity to the data of Figure 3.49, the thickness of a layer of water bound to the surface must be equal to several dozens or even hundreds of molecular diameters. In the case of such a great thickness of a water film, the energy of interaction of water molecules with the surface should be low. It can be assumed that signal 3 is attributed to solid water (amorphous ice) whose constitution differs from that of ordinary crystalline ice. The mobility of water in such layers takes up an intermediate position between the mobilities of water adsorbed on surface and water in the bulk of ice.

The effect of electron-donor agents on the hydration of GO colloidal particles was studied using ^{1}H NMR (Turov et al. 1991a–c). The water/GO dispersion (1.4 wt% of solid phase) was examined

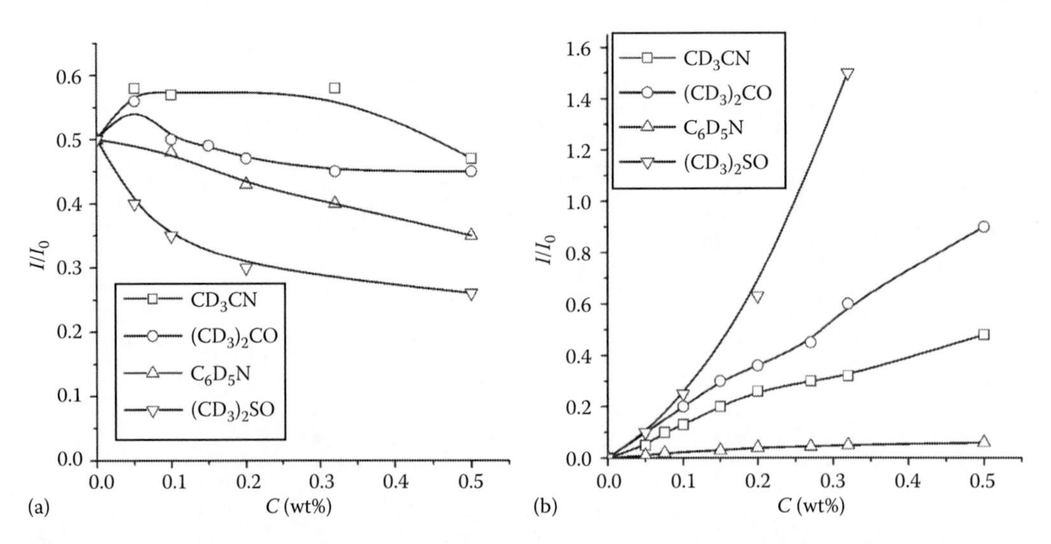

FIGURE 3.49 Broad (a) and narrow (b) component intensities of ^{1}H NMR signal of water bound in 1.4 wt% GO dispersion as a function of concentration of organic additives.

for investigation. The observed spectral parameters of bound water for GO dispersion were found to vary substantially according to the electron-donor properties of the introduced organic agents (Figure 3.49). The intensity curves for the broad signal component (BSC; Figure 3.49a, $\delta_H \approx 5$ ppm) display the effect of electron-donors nature and its concentration on state of water polyassociates.

For acetonitrile, as the weakest electron-donor in a set used $CD_3CN < (CH_3)_2CO < (CH_3)_2SO < C_6H_6N$, a certain increase in the signal is observed at low content. Then the intensity of a narrow signal component (NSC) decreases. Addition of acetone did not practically change the BSC intensity. Stronger electron-donors (DMSO, pyridine) cause a significant decrease in the BSC intensity.

Changes in the NSC intensity (splitting of the narrow signal is not observed with the presence of organic solvents) dependent on the type of additives (Figure 3.49b) differ from the BSC behavior (Figure 3.49a). Addition of pyridine does not affect the NSC. However, addition of acetonitrile or acetone (linear increase), and especially DMSO (nonlinear growth) leads to enhancement of NSC. The BSC can be attributed to amorphous ice, which is characterized by a higher molecular mobility than the molecules in crystalline hexagonal ice. This frozen water (SAW/WBW) is characterized by the hydrogen bond network strongly disturbed by GO nano- and microparticles. The decrease in the BSC intensity can be explained by competitive interactions of DMSO or Py molecules with O-containing sites of GO, i.e., they displace water molecules from the GO surface. Therefore, a larger portion of water becomes nonbound but remains SAW, and it can form more ordered ice crystallites, which were not observed in the ^{1}H NMR spectra because of strong shortening of the relaxation time. A similar displacement of water by organic solvents was observed for porous AC as discussed earlier.

For NSC of unfrozen water more strongly interacting with GO particles (i.e., this water is SBW), one could anticipate a linear increase in the intensity with increasing content of organic co-adsorbate because of the colligative properties of the mixtures. However, this linear effect is observed for the mixtures with acetone and acetonitrile, but for Py and DMSO the pictures are opposite (Figure 3.49b). This suggests that the structure of clusters of co-adsorbates depends strongly on the type of organics.

2D carbon sheets, also known as SLG, have demonstrated superior electronic and thermal properties over other carbon materials including CNTs and have therefore become the latest focus in the nanoscience field (Novoselov et al. 2004, Subrahmanyam et al. 2008, Soldano et al. 2010). The effect of downsizing physical dimensions of carbons on their chemical reactivity and physicochemical properties has been extensively studied for CNTs (Tasis et al. 2006) whilst publications on chemical properties of SLG have started to appear only recently (Hu et al. 2010, Zhu et al. 2010). Thus, SLG and MLG and related oxides (SLGO and MLGO, Figure 3.50) are interesting materials due to their unusual physical properties.

HRTEM images (Figure 3.50) show that large 2D sheets of SLGO include small ordered fragments with a graphite motif (Figure 3.50b). In the case of MLGO, similar ordered structures were not observed (Figure 3.50d). Therefore, one can expect relatively weak shielding effects (up-field shift of ^{1}H NMR signal) for water bound to SLGO and MLGO.

Water bound to GO and SLGO (Bockel and Thomy 1981, Liu et al. 2008, Erickson et al. 2010, Voitko et al. 2011, Whitby et al. 2011, 2012) can have similar structure. As whole, there are certain common features for water bound in the suspensions of GO (Figure 3.48) and SLGO (Figure 3.51b, $\delta_H \approx 5$ ppm at 271 K, but the signal is not split). However, in the case of lower content of water (~160 mg/g), the ^{1}H NMR spectra are complex (Figure 3.51a). There are SAW at $\delta_H = 5.5$ (290 K) and 6 (200 K) ppm and WAW with three signals at $\delta_H = 0.8$, 1.2, and 1.8 ppm. The signal intensity decreases with lowering temperature due to freezing of a portion of bound water. There are boundary signals at $\delta_H = 0$ (TMS) and 7.2 (chloroform) ppm of the added standard compounds.

In the case of GO, the carbon sheet stacks can include many layers in contrast to SLGO (1-2-3 layered structures). Therefore, spatial hindrances for the exchange processes between different water clusters, domains, or layers are stronger for GO particles than for SLGO in the suspensions. In the suspensions, SLGO can be present in the form of single layers or more complex structures

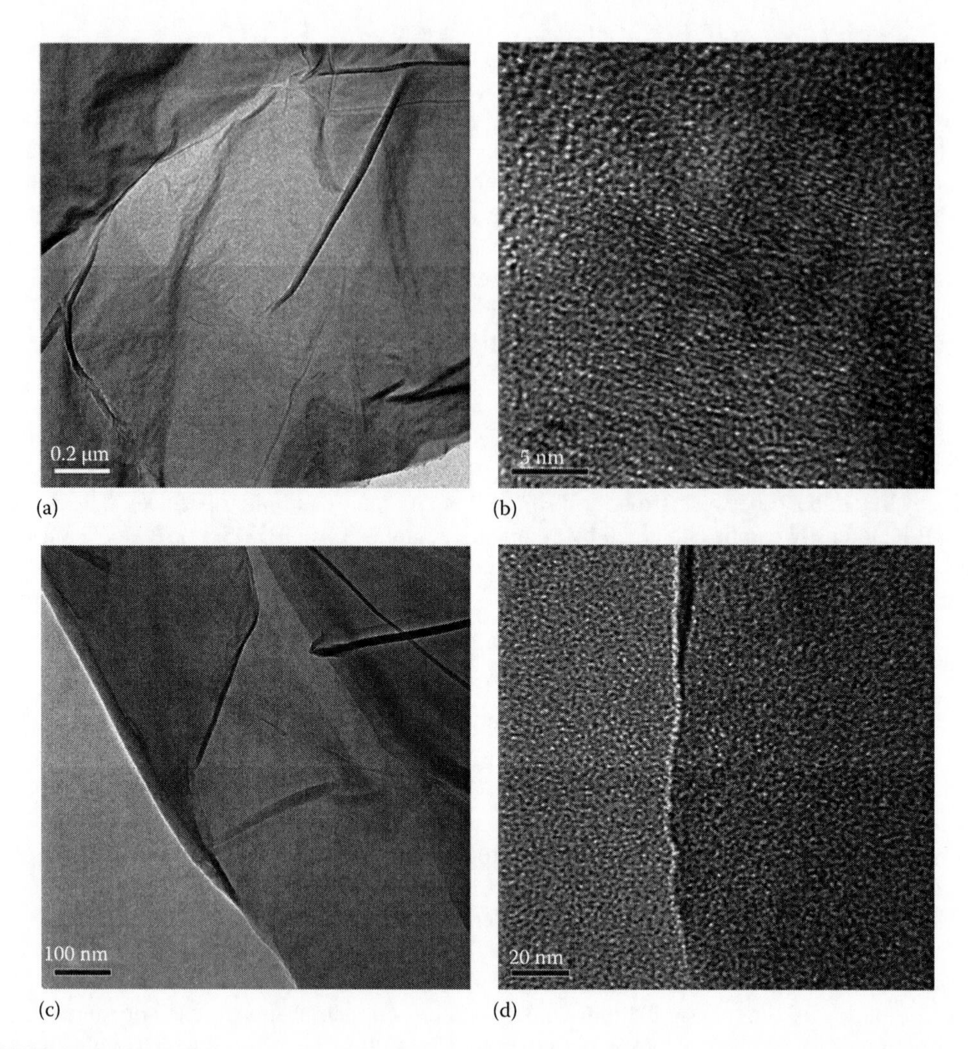

FIGURE 3.50 HRTEM images of (a, b) SLGO (bar 200 nm and 5 nm) and (c, d) MLGO (bar 100 and 20 nm).

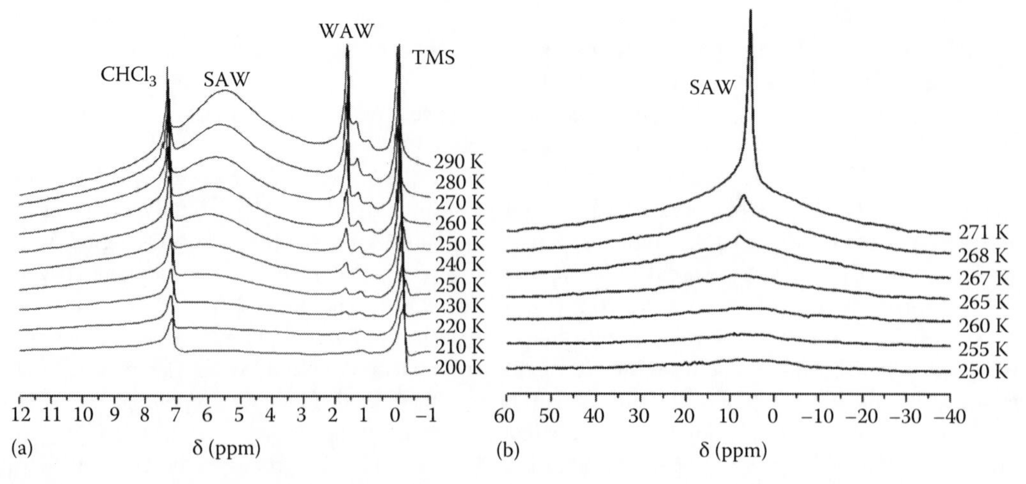

FIGURE 3.51 ^{1}H NMR spectra of water bound to SLGO at (a) 160 mg/g of water (in $CDCl_3$ medium) and (b) 1.6 wt% of SLGO in the aqueous suspension.

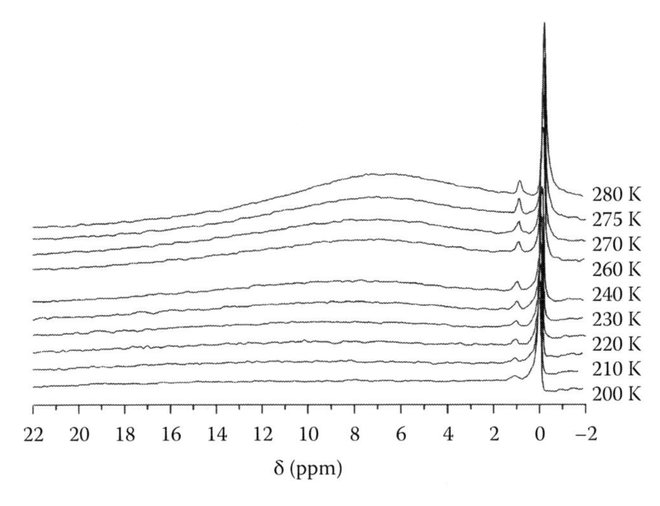

FIGURE 3.52 ^{1}H NMR spectra of water and methane bound to SLGO.

depending on the pH value (Whitby et al. 2011, 2012). In the case of wetted SLGO (160 mg/g water), small water clusters (WAW) can be located at a surface of a single layer at the O-containing functionalities at the edges or inside the sheet (minimal δ_H values). Additionally, dry or wetted SLGO sheets can form complex 3D structures. Therefore, small water clusters can form hydrogen bonds with two sheets (WAW with larger δ_H values). An increase in the pore size and water cluster there results in the formation of SAW structures.

Water (h=0.17 g/g) and methane adsorbed onto SLGO give three signals (Figure 3.52) at δ_H=6–8 ppm (SAW), 1 ppm (WAW), and 0 ppm (CH_4). The methane signal does not practically depend on temperature but its intensity slightly increases with increasing temperature that can be due to reorganization of water structures with melting a portion of ice. A great δ_H value of SAW is due to interacting with numerous O-containing functionalities of SLGO (O atoms represent about 10 wt% in SLGO). A portion of SAW (distant from a surface) is frozen at lowering temperature but other portion interacting with the surface remains unfrozen and it is characterized by greater δ_H values. A narrow WAW signal at δ_H = 1 ppm is observed in the presence of methane (or weakly polar $CDCl_3$, Figure 3.51a).

The interfacial behavior of water bound to SLGO differs in a mixture (1.5/1) with nonpolar CCl_4 and polar DMSO-d_6 (Figure 3.53). Besides two signals of water (3–3.5 and 4.5–6.0 ppm), intensive signal of CH_3 groups of $(CH_3)_2SO$ as an admixture in DMSO-d_6 is observed at 2.5 ppm. The CCl_4/DMSO mixture is frozen at T<260 K, which allows us to study the behavior of bound water when the dispersion medium changes its phase state. Complete freezing of the medium occurs at $T \leq$ 240 K and water has δ_H=5–6 ppm. At higher temperatures $T \geq$ 265 K, water gives signal at 3.5 ppm characteristic for complexes HO-H···O=S$(CD_3)_2$ or HO-H···O=S$(CH_3)_2$. Consequently, a major fraction of water transforms into dissolution state in the liquid organic solvent and loses direct contact with the SLGO surface.

3.3.3 Intercalated Graphite

The nuclear magnetic resonance spectroscopy is one of the techniques that are widely used for researches into intercalated graphite (Estrade-Szwarckopf 1985). Here, the major part of experimental results has been achieved for graphite intercalated with alkali metals since nuclei of isotopes ^{7}Li, ^{89}Ru, and ^{133}Cs are known to have a magnetic moment differing from zero, which allows one to make a record of them in NMR spectra. In the case of graphite intercalated with potassium, it proved possible to record well-resolved ^{13}C spectra for samples with various concentrations of

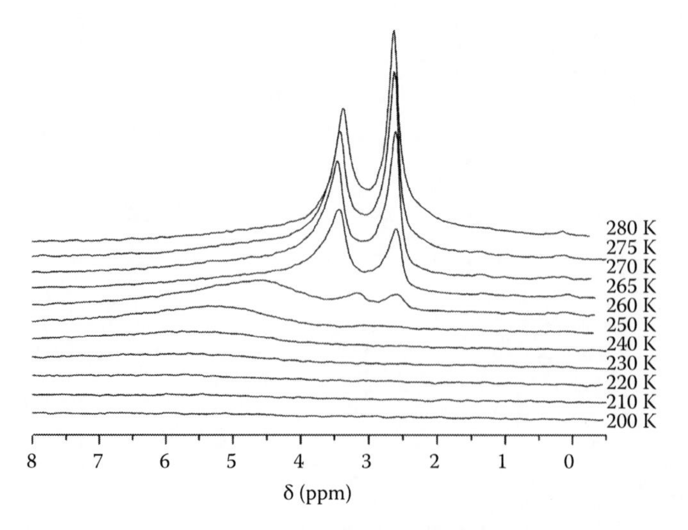

FIGURE 3.53 ^{1}H NMR spectra of water bound to SLGO in CCl_4/DMSO-d_6 (1.5/1) medium.

the intercalant and to measure chemical shifts of intercalated compounds (intercalates) for various stages of intercalation. For alkali metal nuclei, the chemical shift is dependent first of all on the interaction of an intercalant nucleus with conduction electrons and is predominantly determined by the value of the electric charge on the nucleus. Since in the case of the second and higher stages of graphite intercalation with alkali metals, the metal atoms are in the ionization state, the NMR signals are more often with down-field shift.

Graphite intercalation compounds (GIC) can form ternary GICs whose interplanar gaps contain also organic molecules (Gómez-de-Salazar et al. 2000). In this case, they can be studied by the ^{1}H or ^{13}C NMR spectroscopy (Quinton et al. 1982, 1986, 1988, Tsang and Resing 1985, Davidov and Selig 1986). Such studies show that NMR signals of organic molecules situated in interplanar gaps of graphite have up-field shift, which is due to the screening effect of π-electron systems of carbon sheets. The characteristics of water molecules situated in interplanar gaps of ternary GICs ascertained by the ^{1}H NMR spectroscopy were described in detail (Turov et al. 1991a–c). When conducting their experiments, the authors used graphite samples intercalated by the persulfate method (Bockel and Thomy 1981). In the course of the reaction, sulfuric acid molecules were inserted into interplanar gaps of crystalline graphite and some of them were covalently bound to structural elements of the surface. Unreacted acid molecules remained sufficiently labile. The intercalation reaction was arrested by addition of excess water to the system. Under these conditions, labile acid molecules were substituted by water molecules. By prolonged washing, almost all the acid can be removed from interplanar gaps.

Figure 3.54 displays ^{1}H NMR spectra for samples treated with washing waters at pH of 0, 2, 5, and 7 (spectra 1–4, respectively). It should be mentioned that the instrument amplification factor was adjusted to suit conditions for recording spectra, thus signal intensities in Figure 3.54 do not reflect the actual concentration ratios for adsorbed molecules. As seen in Figure 3.54, chemical shifts for water molecules trapped in interplanar gaps of washed intercalated graphite decreased from 7 ppm at pH 0 to −4 ppm for a neutral sample. Such a shift with decreasing pH values was due to proton exchange between acid and water molecules. As protons in acids give signals with significant down-field shifts (ca 15–20 ppm), the presence of acids even in small amounts results in a down-field shift of the signal. After the complete removal of acid molecules from interplanar gaps in intercalated graphite, the chemical shift for water protons proved to be an up-field shift. In this case, the surface screening effect for water in the interplanar gaps causes the up-field shift of the water signal by about 7.5 ppm.

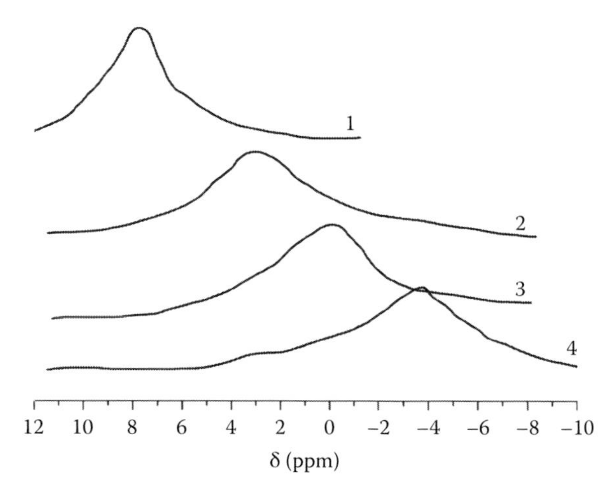

FIGURE 3.54 ^{1}H NMR spectra of water in interplanar gaps of intercalated graphite at various pH values of aqueous suspensions at pH (curve 1) 0, (2) 2, (3) 5, and (4) 7.

3.3.4 Partly Graphitized AC

Despite the fact that ACs are usually regarded as nongraphitized materials, many of them include such structural elements as more or less extended graphite planes forming a nanoporous structure of the material. The extent of graphitization of such materials depends on regularity in arrangement of the graphite planes, size and thickness of graphite clusters, and spacing between planes in a nanoporous structure. A slit-shaped gap between two graphite planes is the most commonly encountered type of adsorption sites in ACs subjected to a high-temperature treatment in the absence of oxygen. When a carbon material is subjected to a high-temperature treatment, graphene clusters can nucleate and grow, and the first to undergo local graphitization are nanoporous areas of the forming adsorbent surface. As a consequence, the local extent of graphitization in pores of different sizes may vary. The ^{1}H NMR spectra recorded for compounds adsorbed on such surfaces are rather complex (Figure 3.55).

Figure 3.55 illustrates the temperature dependence of ^{1}H NMR spectra for (a) water, (b) acetonitrile, and (c) benzene in pores of synthetic carbon adsorbents (SCAs) prepared by carbonization of porous poly(divinylbenzene) co-polymers.

Samples of SCA 2000 treated in vacuum at 2273 K were used in the experiments. The specific surface area of this adsorbent was 360 m^2/g, and pore volume amounted to 0.56 cm^3/g. The measurements were carried out under liquid phase freezing conditions. At $T<273$ K, water (SAW) in intergranular space and in the most large pores froze, and in the ^{1}H NMR spectrum it was possible to detect signals only for the water molecules bound to the surface.

The spectrum recorded was composed of three signals, with signal 1 being related to bulk water, signals 2 and 3 being assigned to water attached to the adsorbent surface. With lowering temperature, signal 1 disappeared, while the intensity of signal 2 decreased and the peak was shifted toward stronger magnetic field. The intensity of signal 3, however, monotonously increased at a constant chemical shift.

It was shown that the appearance of signal 3 depended on the thermal treatment of the material, i.e., on the extent of its local graphitization, rather than on the presence of molecularly adsorbed and chemisorbed oxygen in the adsorbent (Turov et al. 1990). This led to a conclusion that signal 3 was caused by water adsorbed on graphitized areas of the surface. A similar type of the temperature dependence was also found for the following organic compounds frozen in pores of SCA 2000: acetonitrile and 1,2-dichloroethane (Kolychev et al. 1992, Turov et al. 1992).

To ascertain the relationship between a pore size and a chemical shift of an adsorbate, the following approach was used. Benzene vapors were adsorbed at a low relative pressure $p/p_0=0.1$.

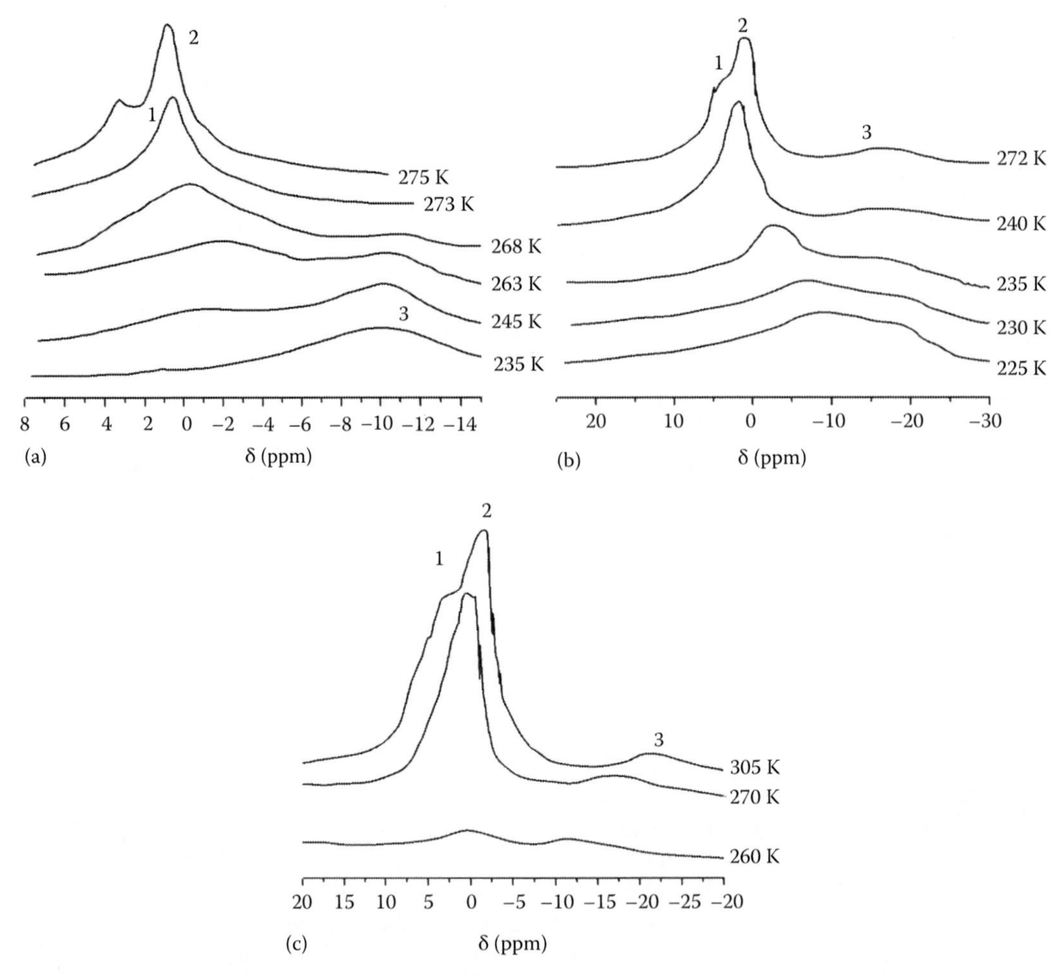

FIGURE 3.55 Temperature-dependent variations of the shape of ^{1}H NMR spectra for (a) water, (b) acetonitrile, and (c) benzene adsorbed onto partly graphitized activated carbon SCA 2000.

Under these conditions, it was adsorbed in nanopores of an adsorbent. The measurements performed for a number of compounds (Figure 3.56) showed that at low values of relative vapor pressure the ^{1}H NMR spectroscopy detected only signal 3′ with up-field shift. Thus, the graphitized areas are predominantly localized in nanopores of the adsorbent.

Spectral characteristics of various compounds measured by methods of liquid phase freezing and adsorption from a gas phase are summarized in Table 3.9. On the basis of the data obtained, the following trends can be noted for adsorbed compounds irrespective of their capability to interact with various molecules. (1) Upon a decrease in temperature by 30 K relative to the freezing point, signal 2 has up-field shift by 4–7 ppm. (2) Signal 3 was temperature independent and was positioned in somewhat stronger fields as compared with signal 3′ of molecules adsorbed at low p/p_0 values. The latter fact implied that in liquid phase freezing experiments signal 3 resulted not only from the presence of a compound in nanopores, rather it was an averaged signal caused by the presence of the compound in nanopores and pores of larger sizes. To verify the validity of this assumption for benzene, by way of example, spin-spin and spin-lattice relaxation times were determined for signals 2 and 3 after freezing of benzene in pores of SCA 2000 and for signal 3′ after adsorption at $p/p_0=0.1$ (Turov et al. 1992). The obtained temperature dependence of the transverse relaxation time, T_2 is shown in Figure 3.57.

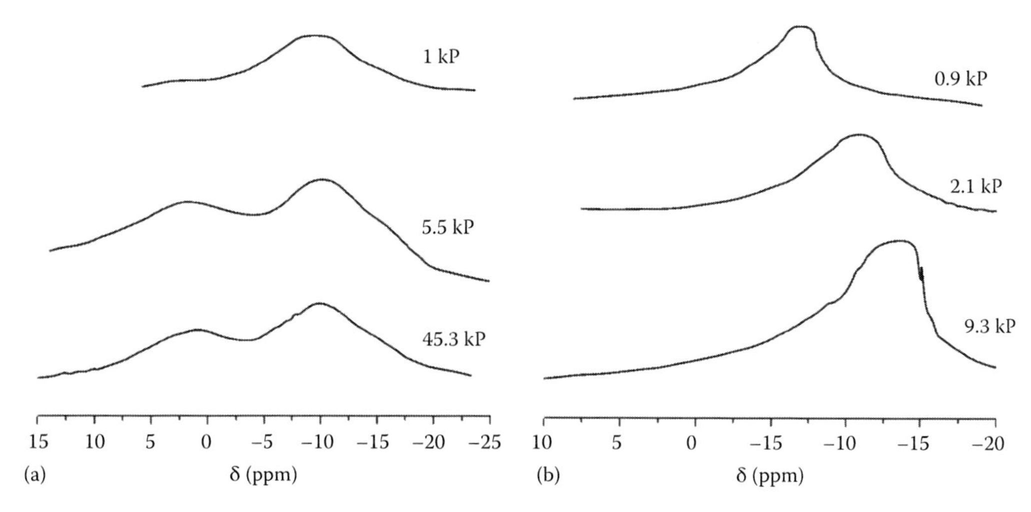

FIGURE 3.56 Pressure-dependent variations of the shape of ^1H NMR spectra for (a) methane and (b) CH_2Cl_2 adsorbed onto partly graphitized activated carbon SCA 2000.

TABLE 3.9
Chemical Shifts for Simple Organic Molecules and Water Adsorbed on SCA

	Chemical Shift (ppm)					
		Liquid Phase		Adsorption		Extent of
Compound	Pure Compound	Signal 2	Signal 3	at $p/p_0 = 0.1$ Signal 3′	Calculation	Graphitiz α
C_6H_6	7.20	4.0 to 0.9	−11	−6.2	−6.7	0.060
$(CH_3)_2CO$	2.07	—	—	−15.3	−11.4	0.115
CH_4	0.30	—	—	−13.6	−12.7	0.084
CH_2Cl_2	5.28	—	—	−5.1	−9.1	0.020
$C_2H_2Cl_2$	3.70	2.5 to −0.5	−14.8	−12	−12.1	0.074
CH_3CN	1.96	−0.76 to −8.20	−18	−13.2	−13.0	0.071
H_2O	4.50	3.0 to −1.0	−13	—	—	—

As is evident from Figures 3.55 through 3.57, the transverse relaxation time values for signals 3 and 3′ considerably differ. This is in contradiction with a suggestion made by Kolychev et al. (1992) that signal 3 in liquid phase freezing experiments results exclusively from the compound filling adsorbent nanopores. In such a case, it could be expected that the relaxation time values for signal 3 (corresponding to completely filled pores) would be either similar to or larger than those for signal 3′. Besides, in the case of signal 3, there are benzene molecules bound to the surface as well as those situated at a distance of several molecular layers from the surface with larger T_2 values. However, the experimentally determined transverse relaxation times for signal 3 are less than the corresponding T_2 values for signal 3′ by a factor of ca 5 (Figure 3.57). Thus, with allowance for the data in Table 3.9, it can be concluded that signal 3 (Figure 3.55) is attributed to the compound located not only in the nanopores but also in larger pores, presumably supernanopores.

To explain the intensity redistribution pattern for signals 2 and 3 (Figure 3.55) it may be assumed that with a decrease in temperature, a transfer of molecules between surface areas with different shielding properties takes place. Such a reasoning would be justified for adsorption of gaseous substances when adsorbed molecules can easily diffuse in pores of an adsorbent. In our liquid phase

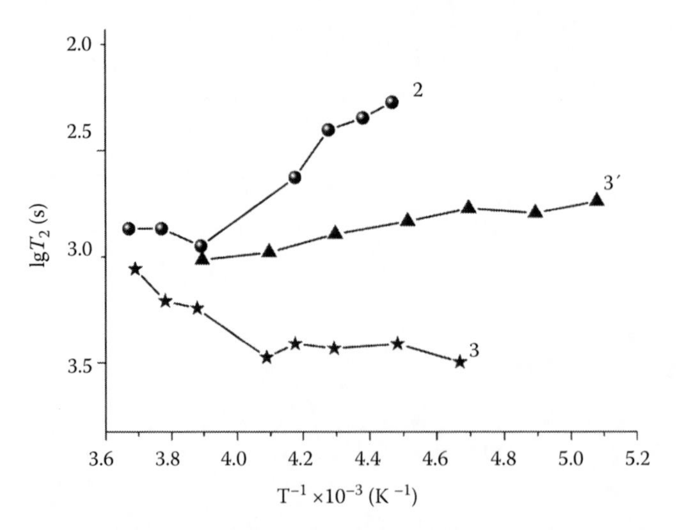

FIGURE 3.57 Dependence of the transverse relaxation time for benzene on the reciprocal of temperature for signals 2, 3, and 3′ recorded in ^1H NMR spectra. (According to Turov, V.V. et al., *Ukr. Khim. Zh.*, 58, 470, 1992.)

freezing experiments, however, all adsorbent pores are filled with a solid or adsorbed substance. Thus, to interpret the data in Figure 3.55, it is necessary to suggest that in SCA, there are adsorption sites unoccupied at high temperatures and binding of adsorbate molecules on such sites is an exothermal process facilitated by a decrease in temperature. Still, such an approach has some considerable disadvantages as it can be expected that instead of adsorbate molecules transferred to free adsorption sites additional molecules from a solid phase will come up to the reaction site. The presence of an unfrozen layer of the compound near the surface implies that the free energy of the molecules bound to the surface is less than that for solid benzene. Therefore, the intensity of signal 2 should vary only slightly.

To ascertain the nature of anomalous chemical shifts for molecules adsorbed on SCA, calculations of these parameters were performed for molecules in slit-shaped pores formed by graphitized surfaces (Turov et al. 1993). The geometrical position of molecules in a slit-shaped pore was determined using Scheraga's function (Momany et al. 1974) for interatomic potentials. Chemical shifts were computed with allowance for a screening effect of a graphitized surface by ring current method taking account of contributions of adsorbed molecule protons and anisotropy of a magnetic susceptibility tensor for a carbon material (Dubinin et al. 1988).

As the graphite-like structure is characteristic of only a certain portion of the carbon material, it is essential to know the extent of its graphitization at sites of adsorbed molecule localization. The extent of local graphitization can be taken into account by introducing a coefficient determined from a comparison between experimental and theoretical values of contributions of the material surface and bulk solid phase to magnetic shielding. The magnetic shielding value for an adsorbed molecule proton will be determined by a term accounting for the effect of ring currents and a term representing anisotropy of a magnetic susceptibility tensor caused by graphitization of a bulk carbon material:

$$\Delta\sigma = \Delta\langle\sigma\rangle_{RC} + \alpha\Delta\langle\sigma\rangle_{AMS} \quad (3.2)$$

where

- σ is the magnetic shielding of a proton in an adsorbed molecule
- σ_{RC} is the contribution of ring currents
- σ_{AMS} is the contribution of anisotropy of a magnetic susceptibility tensor for graphitized carbon atoms in the bulk of the material
- α is the extent of effective graphitization of the material

The α value was calculated for each adsorbed compound on condition that experimental and calculated chemical shifts are equal in value. The calculated values of chemical shifts and coefficient α are listed in Table 3.9. From Table 3.9 it follows that, in general, the theoretical values of chemical shifts agree satisfactorily with those determined experimentally with the exception of acetone and methylene dichloride. All the calculations were performed for a slit-shaped pore of the graphitized material with a small interplanar spacing of 0.58 nm. As can be seen, for different compounds (C_6H_6, CH_4, $C_2H_2Cl_2$, and CH_3CN) the α values are close with each other. This fact corroborates the validity of the assumption that the α value is determined solely by shielding properties of the studied carbon material.

3.4 CARBON NANOTUBES

CNTs can be SWCNT and MWCNT, the properties of which strongly differ, as well as the properties of SLG, MLG, EG, and graphite (Celzard et al. 2005, Harris 2009, Dreyer et al. 2010) because of the difference in carbon sheet–sheet interactions and conformations of the sheets (plane, flat, tubular, bent, etc.). In the case of CNT, chiral aspects in their structure play an important role. Additionally, the presence of O-containing groups at the periphery or inside the sheets can cause significant changes in the interfacial phenomena at a surface of carbons, especially with the participation of water or other polar compounds. The electronic properties of CNT play an important role in their other properties. According to band-structure calculations the basic electronic structure of SWCNT is expected to depend on the chiral wrapping vector (n, m), across the graphene plane, where tubes for which $(2n+m)/3=$integer are metallic, while all other tubes are semiconducting with a large ~1 eV gap (Singer et al. 2006, Harris 2009).

Kleinhammes et al. (2003) and Mao et al. (2006) used the ^{1}H NMR spectroscopy to measure methane, ethane, and water adsorption isotherms (Figure 3.58) for cut SWCNTs and AC at room temperature (289 K). Water adsorption inside SWCNTs could be similar to the adsorption in mesoporous ACs. However, the adsorption isotherms showed the lack of wetting inside cut-SWCNTs. Quantitative analysis and various effects of adsorbed D_2O upon gas adsorption inside cut-SWCNTs

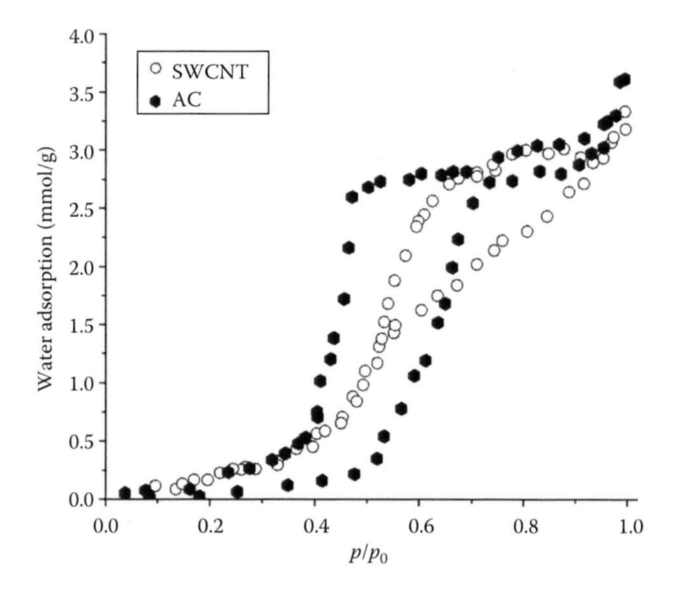

FIGURE 3.58 Water adsorption–desorption isotherms (289 K) for SWCNT and AC measured with ^{1}H NMR AC. (Taken from *Chem. Phys. Lett.*, 421, Mao, S., Kleinhammes, A., and Wu, Y., NMR study of water adsorption in single-walled carbon nanotubes, 513–517, 2006. Copyright 2006, with permission from Elsevier.)

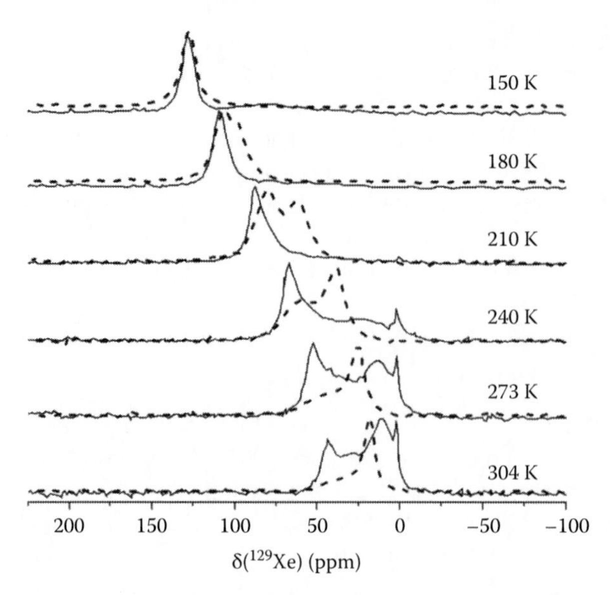

FIGURE 3.59 ^{129}Xe NMR spectra of open (solid lines) and closed (dashed lines) CNT at different temperatures. (Taken from *Solid State Nucl. Magn. Reson.*, 28, Romanenko, K.V., Fonseca, A., Dumonteil, S. et al., ^{129}Xe NMR study of Xe adsorption on multiwall carbon nanotubes, 135–141, 2005. Copyright 2005, with permission from Elsevier.)

indicate that adsorbed water molecules reside mostly near principal adsorption sites associated with tube ends and defects. Our calculations of the specific surface area (occupied by water molecules) from the water adsorption isotherms (Figure 3.58) give very low values ~100 and 20 m^2/g for SWCNT and AC, respectively, because of the hydrophobic nature of their surface. In other words, water does not cover the surface of both carbons at low pressures but at higher pressures capillary condensation occurs in mesopores. A broader hysteresis loop is observed for AC because of a more complex geometry of pores. The isotherm shape and capillary condensation of water occurred on SWCNT suggest that adsorbed water is SAW/WBW.

Romanenko et al. (2005) used ^{129}Xe NMR spectroscopy to study the adsorption of Xe on MWCNT closed and open by hydrogen exposure and of ball-milling. Significant changes observed in the ^{129}Xe NMR spectra after ball-milling of MWCNTs (Figure 3.59) were attributed to the destruction of the CNT aggregates and opening of the nanotubes that allows the Xe access inside them. However, these NMR results were not used to estimate the difference in the structural characteristics of closed and open CNT.

MWCNT can have a more complex texture, e.g., broad PSD, than SWCNT because of a non-ideal structure of the former (Tóth et al. 2012). MWCNTs produced by carbonization of methylene chloride in the channels (~50 nm in diameter) of a mineral matrix (alumina) are nonuniform materials with partially turbostrated structure of the walls and nonuniform surface (Gun'ko et al. 2009a). Catalytic synthesis (with nanosized catalyst particles, Figure 3.60b, nanoparticle inside a tube) of MWCNT gives nonideal structure of the walls (Figure 3.60).

A mosaic structure of the surface can provide enhanced clustering of water/organic mixture bound to MWCNTs. Mixtures with water and benzene were used to study this effect. The water amount (280 mg/g) corresponded to a quarter of the total pore volume of MWCNTs that a free part could be filled by benzene. All these MWCNTs were open; therefore, water adsorbed alone was SBW (frozen at $T < 230$ K and $-\Delta G < 2$ kJ/mol, Figure 3.61) because it was bound inside CNT channels. The $\Delta G(C_{uw})$ curves are typical for nano- or mesoporous materials as described earlier.

An increase in the water content from 70 to 280 mg/g the free surface energy value increases by a factor of 3.2 (Table 3.10, γ_S). Consequently, the area of contacts between water and MWCNT surface

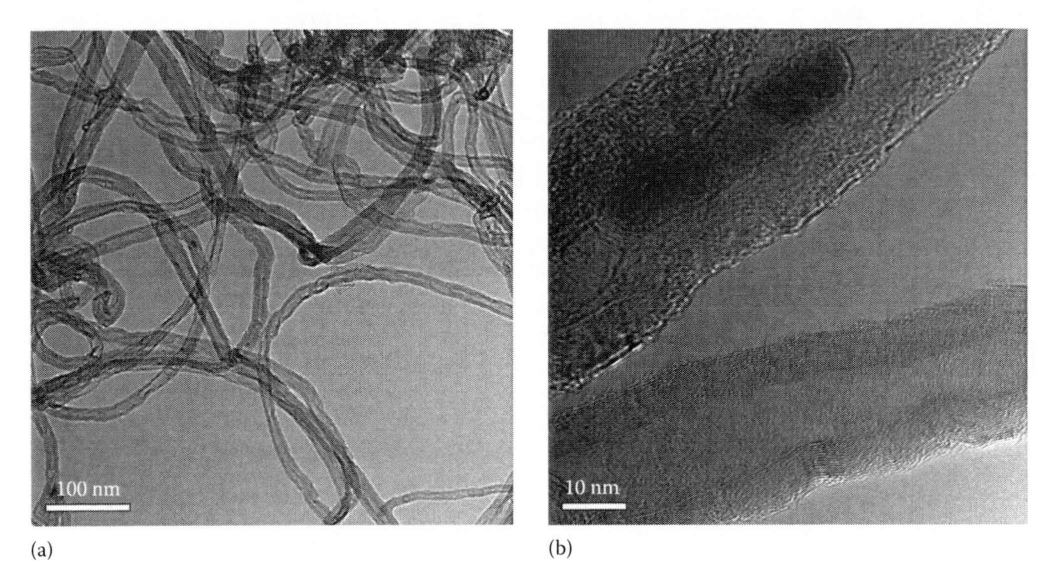

FIGURE 3.60 HRTEM images of MWCNT with bar (a) 100 nm and (b) 10 nm.

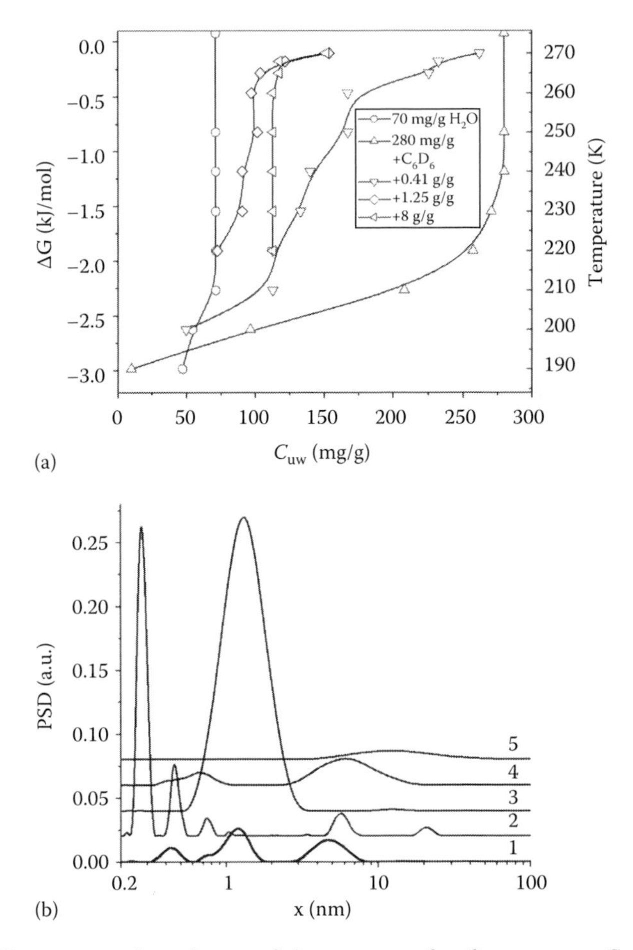

FIGURE 3.61 (a) Temperature dependences of the amounts of unfrozen water, C_{uw} bound to MWCNTs alone or with addition of C_6D_6 and relationships between changes in the Gibbs free energy and the C_{uw} values and (b) size distributions of pores filled by unfrozen water (IGT).

TABLE 3.10
Characteristics of Water Bound in MWCNT at Different Amounts of Added C_6D_6

h (g/g)	C_{C6D6} (g/g)	γ_S (J/g)	ΔG_s (kJ/mol)	ΔG_w (kJ/mol)	C^w_{uw} (mg/g)	C^s_{uw} (mg/g)	S_{uw} (m²/g)	S_{nano} (m²/g)	S_{meso} (m²/g)	⟨T⟩ (K)
0.07	0	11.8	−3.5	0	0	70	99	88	11	188.0
0.28	0	38.1	−2.97	0	0	280	165	0	165	205.3
0.28	0.41	22.2	−2.87	0.8	105	175	20	0	20	233.8
0.28	1.25	13.0	−2.94	0.6	125	100	11	0	11	236.6
0.28	8	16.9	−3.03	0.4	88	112	18.5	7	11	222.7

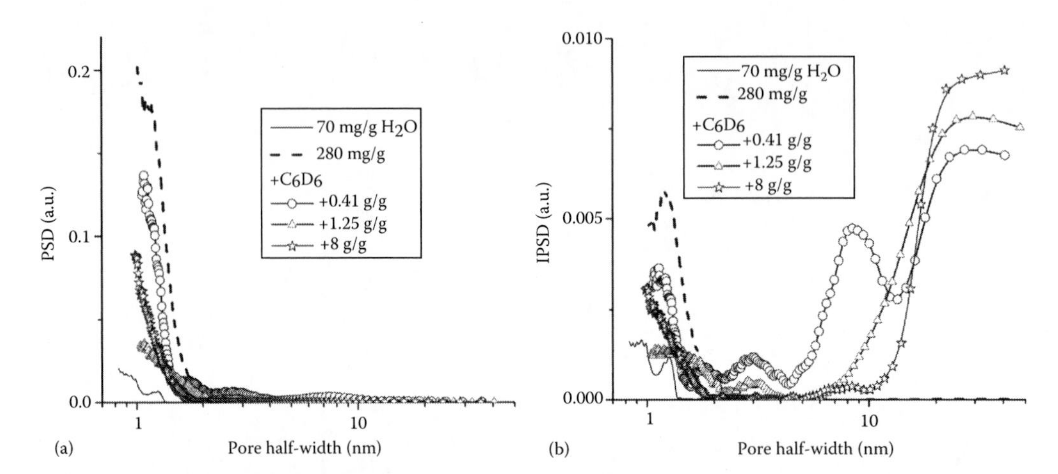

FIGURE 3.62 Size distributions of pores filled by unfrozen water: (a) differential and (b) incremental (GT, k_{GT}=85 K nm).

increases nearly linearly with increasing water content. According to calculations with the integral GT (IGT) equation, this area increases from 25 to 199 m²/g (IGT) or 99 and 165 m²/g (GT; close to the value determined from the Ar adsorption data) and the PSD of pores filled by unfrozen water changes (Figures 3.61 and 3.62). Water fills both narrow and broad pores.

Addition of C_6D_6 leads to changes in the characteristics of bound water (Table 3.10). A portion of water becomes WBW $\left(C^w_{uw}\right)$ and the γ_S value strongly decreases. Water fills mesopores (Figures 3.61 and 3.62), i.e., benzene displaces water from narrow pores. This process is enhanced with increasing content of added benzene.

Total dehydration of MWCNT at maximal C_6D_6 content corresponds to 29% (estimated as $(C^w_{uw}+C^s_{uw})/h$) but the content of SBW $\left(C^s_{uw}\right)$ decreases much strongly (Table 3.10) and the surface area in contact with water decreases. Water strongly interacts with MWCNT ($\Delta G_s \approx -3$ kJ/mol) that can be explained by the presence of O-containing surface sites. Therefore, benzene cannot totally remove water from the MWCNT surface even at large content (Table 3.10).

Thus, NMR spectroscopy and cryoporometry are effective methods to study the interfacial phenomena especially at a carbon surface because of an additional factor such as π-electron clouds at the carbon sheets affecting electron shielding of nuclei of adsorbed compounds depending on the pore size and structure of the pore walls.

4 Interfacial Phenomena at Carbon–Mineral Composites

Particular emphasis in transformation of organics at solid surfaces (e.g., catalysts) has been given to carbonaceous deposits (coke, pyrocarbon) inhibiting reactions on active surface sites due to their blocking (Delmon and Froment 1980, Hughes 1984). On the other hand, pyrocarbon on solid supports can be interesting as a catalyst (Lisovskii and Aharoni 1994) and an adsorbent (Hubbard 2002) per se or as an important component of hybrid adsorbents (Vasilieva et al. 1961, Leboda 1974, 1980a–d, 1981, 1987, 1992, 1993), which can be applied in chromatography (Colin and Guiochon 1976, Bebris and Kiselev 1978, Gierak and Leboda 1989), trace analysis (Rudzinski et al. 1995), catalyst supports (Vissers et al. 1988), etc. Variations in the nature of solid matrices, e.g., from catalysts with metals or metal oxides to practically inert silica gels or nanosilicas, can result in changes in the structure of pyrocarbon synthesized, however, under close conditions. The pyrolysis of organics is a complex multistage process. It includes dehydrogenation of organic substances $C_nH_mX_l \rightarrow C_nH_{m-z}+(H_2)_k+(HX)_i$ where X=OH, O, Cl, etc., $n \gg m-z$. The activation energy of these reactions can be mainly between 100 and 300 kJ/mol. They are accompanied by changes in the electronic configuration of carbon atoms from sp^3 or sp to sp^2 in polyaromatic structures. Carbon transport and chemisorption can occur on a matrix and then on formed pyrocarbon grains. There are many chemical transformations leading to formation of pyrocarbon (graphene) clusters or practically individual carbon particles with certain surface functionalities, especially O-containing ones (Buyanov 1983, Lisovskii and Aharoni 1994, Fenelonov 1995). However, there is an opinion based on analysis of intermediates upon coke formation on solid surfaces that the difference between catalytic and noncatalytic cokes is linked only with a thin structure of the reaction spectra but not overall coke morphology (Buyanov 1983). This can be explained by preferable growth of the carbon phase on carbon nuclei formed at the beginning of the process.

Carbonized silicas (carbosils [CS]) are appropriate materials to develop adsorbents capable of adsorbing equally well both polar and nonpolar compounds (Leboda 1992, 1993). Laboratory synthesis of CS was carried out in reactors of different types using nanosilica, silica gel, or other matrices and a variety of organics. At a temperature of 1000–1100 K, vapors of benzene (Bebris and Kiselev 1978), alcohols (Leboda 1987), methylene chloride (Leboda 1981), or other low- or high-molecular organic compounds (Leboda 1980c, Kamegawa and Yoshida 1995) were run through the silica layer with a controlled flow rate to form a carbonaceous coating. Clearly, the morphology and properties of the carbon deposits depend on the carbonization conditions (Tracz and Leboda 1985). Conducting the pyrolysis process at various temperatures makes it possible to study its separate stages (Leboda and Dabrowski 1996). However, when synthesizing adsorbents suitable for application in practice, the degree of chemical transformations in the carbonaceous layer should be as high as possible and characterized by a certain controlled texture and particle morphology.

The micro-/nanostructure of the carbon component of the CS surface seems to be close to that of carbon black particles since the conditions of their synthesis are similar in many aspects. However, as distinct from carbon black particles the size of carbonaceous deposits at a matrix surface is determined to a great extent by the nature and porosity of a substrate and sizes of its particles. Therefore, the morphology and the texture of carbons in CS may considerably differ from that of carbon blacks (Gun'ko and Leboda 2002, Gun'ko et al. 2002c).

CS as complex adsorbents include both accessible carbonaceous deposits and patches of a matrix. Therefore, the adsorption characteristics of CS depend on the ratio of areas of two types (or more phases) of the surfaces. During adsorption, various types of molecules can be predominantly adsorbed either on the matrix or on carbon components of the surfaces of hybrid particles (Gun'ko and Leboda 2002, Gun'ko et al. 2002c). Carbon deposits can have a structure similar to that of carbon black; therefore, the latter is analyzed later.

4.1 CARBON BLACKS

The characteristics of a set of compounds adsorbed to carbon black particles (P-245, specific surface area 110 m^2/g) were investigated applying the liquid phase freezing method with the ^1H NMR spectroscopy after the immersion of samples into water, acetonitrile, or benzene (Turov et al. 1994a, 1995b). The choice of the adsorbates was caused by the fact that all of them are capable to take part preferably in different intermolecular interactions with different surface structures. Hydroxyl groups of water molecules can interact with oxygen-containing surface functionalities by strong hydrogen bonds in which water molecules acts as a proton donor or a proton acceptor. Acetonitrile molecules contain CN groups having a great dipole moment but low electron-donor capacity. Therefore, the main mechanism of their binding to a surface is dipole–dipole interactions or the formation of weak hydrogen bonds. Benzene molecules can be adsorbed due to dispersion forces relatively strong because of the nature of both adsorbate and adsorbent. The carbon black studied has a specific feature consisting in that it belongs to weakly graphitized materials whose surface contains a substantial quantity of oxidized sites in the form of carbonyl, carboxyl, and other groups characteristic for carbon materials (Göritz et al. 1995, Brennan et al. 2001) that are predominantly situated on the boundaries of graphene planes and clusters.

The ^1H NMR spectra of benzene suspension with carbon black include two signals (Figure 4.1, $\delta_H = 6.3$ and 0.14 ppm) with up-field shifts relative to the signal of liquid benzene (7.2 ppm) due to the shielding effects of the carbon surface (see Chapter 3).

With varying temperature, the chemical shifts of signals and the ratios of intensities remained approximately constant. With decreasing temperature, the width of signal increases, which is caused by lowering of molecular mobility and is characteristic of the majority of substances in heterogeneous systems. As shown in Chapter 3, the up-field shift of NMR signals observed for substances adsorbed on carbon surfaces is attributed to the interaction of adsorbed molecules with π-electron systems of condensed aromatic rings forming the basis of carbonaceous particles. The carbon black surface has a considerable number of structural defects and it is characterized by low nanoporosity. Therefore, the shielding effect of the surface may be relatively weak.

It is possible to give the following assignment of the ^1H NMR signals for benzene adsorbed on the carbon black surface (Figure 4.1a). The signal at $\delta_H = 6.2$ ppm (up-field shift $\Delta\delta_H \approx -1$ ppm) can be assigned to the molecules adsorbed on basal graphite planes of the external surface of particles or in mesopores. The weak shielding effect of the surface is caused by a great number of structural defects partially destroying the ordered π-electron systems. There are broad slit-shaped or more complex pores formed by graphite planes and graphene clusters whose interplanar spacing substantially exceeds the interplanar distance in graphite or graphitized carbons. These pores as voids between nonplanar, turbostratic structures cannot provide strong shielding effects (since $\Delta\delta_H \approx -1$ ppm) in contrast to narrower pores with more ordered walls ($\Delta\delta_H \approx -7$ ppm). The oxygen-containing surface groups and adsorbed oxygen molecules do not make any appreciable contribution to the chemical shift value for adsorbed benzene molecules, which corroborates the inference about the physical nature of the benzene adsorption. In all probability, the adsorption of benzene on end faces and oxygen-containing sites can be neglected.

In the case of the interactions between the carbon black surface and water molecules, a somewhat different situation is observed (Figure 4.1b). As shown in this figure, at $T = 265$ K in the spectra there is only one signal whose shift is equal to 7 ppm, which corresponds to a down-field shift of 2.5 ppm

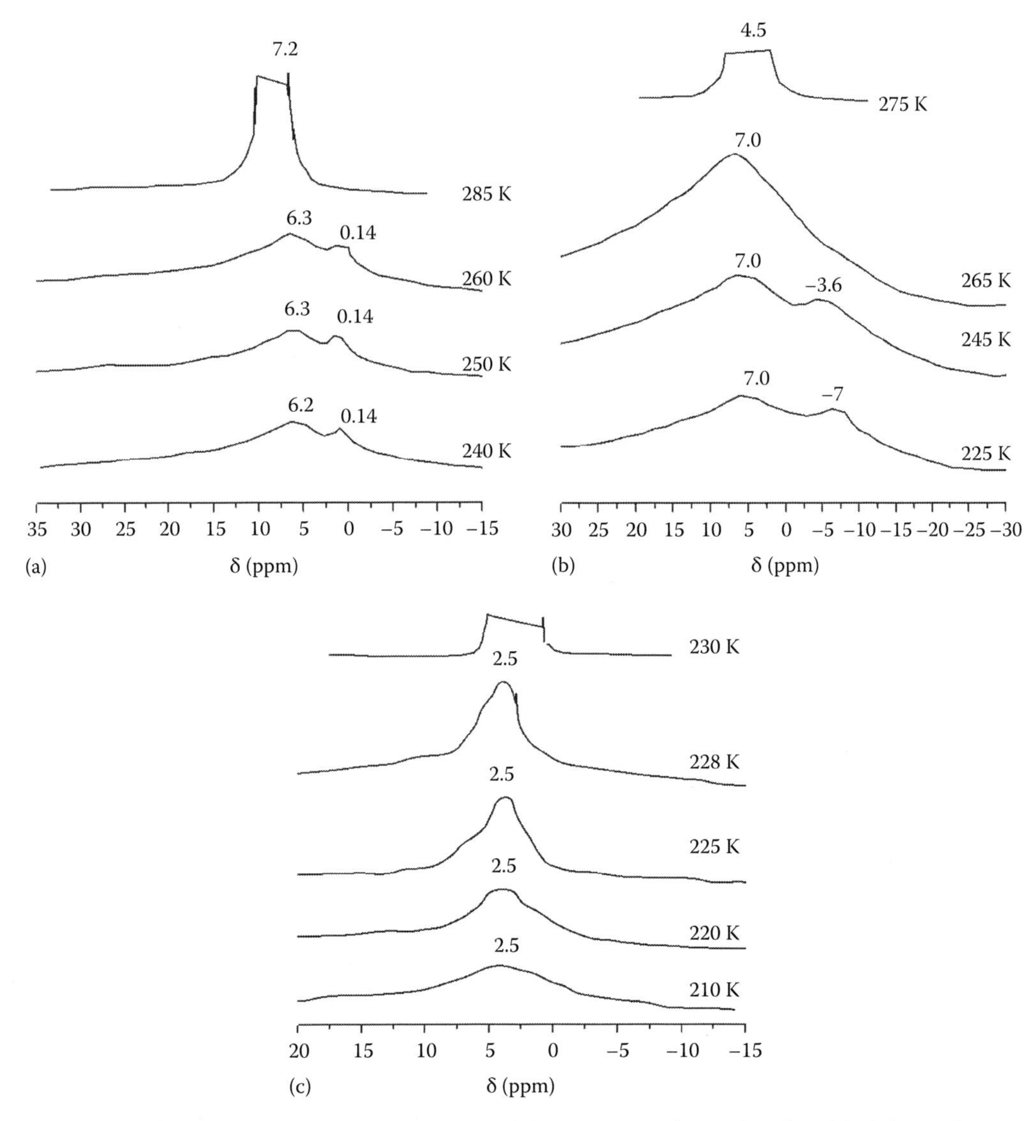

FIGURE 4.1 ^{1}H NMR spectra (static samples) of frozen suspensions of carbon black in (a) benzene, (b) water, and (c) acetonitrile. (According to Turov, V.V. et al., *Zh. Prikl. Spektros.*, 61, 106, 1994a; Turov, V.V. et al., *Biophysica*, 39, 988, 1994b, in Russian.)

in comparison to the signal of liquid water. With decreasing temperature, the chemical shift of this signal does not change, but at 245 K on its shoulder there appears a considerably less intensive signal at $\delta_H = -3.6$ ppm. As the temperature decreases, this signal demonstrates the up-field shift. Karpenko et al. (1990a,b) studying the interaction of water with the oxidized graphite surface observed a signal of water at $\delta_H \approx 7$ ppm ascribed to molecules interacting with carbonyl and carboxyl surface groups. On the other hand, for graphitized carbonaceous adsorbents and graphite, it has been shown (Turov et al. 1992) that the up-field shift for water adsorbed on basal graphite planes corresponded to $\Delta\delta = -5$ ppm. In the case of adsorption in slit-shaped nanopores formed by graphite planes, the up-field shift can be much greater $\Delta\delta = -6$ to -18 ppm. Then, a more intensive signal in Figure 4.1b may be related to water molecules interacting with the surface through the hydrogen bonds with polar O-containing surface sites. It is possible that a water layer with a disturbed network of the hydrogen bonds appears at the interfaces (i.e., clustered adsorption of water occurs). In this case, the deshielding contribution of hydrogen-bonded complexes of the C=O···HOH and C–OH···OH_2 types to the chemical shift is predominant. The low-intensity signal for bound water with the up-field shift ($\Delta\delta_H \approx -8$ to -11.5 ppm with respect to bulk water) may be assigned to water molecules located in narrow interplanar voids that can be assigned to WAW ($\Delta\delta_H \approx -5$ to -8.5 ppm with respect to typical WAW at $\delta_H \approx 1.5$ ppm). The strong shielding effect exerted by the surface on these water molecules in comparison to benzene molecules seems to be caused by penetration of the water molecules into very narrow nanopores at half-width $x < 0.4$ nm.

In the case of the interaction of acetonitrile with the carbon black surface, molecules bound to the surface are characterized by a single signal at $\delta_H \approx 2.5$ ppm (Figure 4.1c) whose chemical shift does not depend on temperature and is similar to the chemical shift of liquid acetonitrile. Consequently, acetonitrile molecules interact weakly with basal graphite planes of the surface and do not penetrate into narrow nanopores. The most probable mechanism of interactions in this case is that of dipole–dipole interactions of the CH_3CN molecules with polar groups that are predominantly situated on end faces and structural defects of basal planes of graphite in broad mesopores.

4.2 CARBONIZED SILICAS AND MIXED OXIDES

The adsorptive characteristics of hybrid pyrocarbon–mineral adsorbents depend on pyrolysis conditions (Gun'ko and Leboda 2002, Gun'ko et al. 2002b,c). With elevating temperature and pyrolysis time, enlargement of graphene clusters and their overlapping take place, and they form pregraphite structures, but if $T < 1000$ K, large basal planes (or their packs and stacks) are not formed, as a turbostratic structure is final (Fenelonov 1995). If the pyrolysis time is short, then disordered pyrocarbon deposits form on the support surfaces and the graphene size does not exceed several nanometers. The structure of relatively dense final pyrocarbon particles (specific density of 1.6–2.2 g/cm^3, size from several nanometers to 100–200 nm, Figure 4.2) can be described as turbostratic ($T < 1000$ K) with different fragments of "crumpled" graphite sheets types, but with elevating temperatures to 2000 K or higher, the pyrocarbon structure changes and becomes akin to graphite one. Typically, the internal porosity of pyrocarbon particles is very low (as the main contribution of the carbon phase to the porosity is provided by the outer surfaces of these particles and voids between them and the support surfaces), and it is observed a strong dependence of the accessible pore volume of hybrid adsorbents on the origin of precursors and the texture of oxide matrices (Figure 4.3 and Tables 4.1 through 4.5) (Gun'ko et al. 1999a, 2000a–h, 2001a–f, 2002b,c, 2003b,d, 2004b,c,d, Leboda et al. 2000b, 2001a–c, 2005, 2006, Seledets et al. 2003, 2005, 2006, Skubiszewska-Zięba et al. 2003, Gun'ko and Leboda 2005).

The difference in the structural characteristics of catalytic and noncatalytic pyrocarbons can also be linked to the formation of carbon deposits in confined space of pores of solid catalysts near active surface sites, but noncatalytic pyrocarbon can form mainly individual particles on the outer surfaces of the inert support. Therefore, the pore structure, active surface site distribution, and carbon transport from these sites to carbon grains impact the characteristics of the final carbon deposits. Preparation of pyrocarbon on various solid matrices using different precursors (individual organics or their mixtures) has been studied under different pressures (10^{-4}–10^4 Torr) and temperatures from 700–1000 K up to 2500 K. Despite the mentioned differences in synthetic techniques, pyrocarbon deposits possess some general properties such as low internal porosity of nanoscaled particles with the structures between turbostratic ($T < 1000$ K) and graphite-like ($T > 1450$ K), the absence of large basal planes, and disorder in sheet packs changing with elevating temperature.

Kinetic studies of carbonization of a methane/carbon tetrachloride mixture at low pressure (2 kPa) at $T < 800$°C allowed one to assume that pyrocarbon was formed from chlorinated intermediate species with apparent activation energy of 180 kJ/mol (Gun'ko and Leboda 2002). At higher temperatures, the apparent activation energy was lower (140 kJ/mol) and unsaturated hydrocarbons were the most probable carbon precursors (Feron et al. 1999). Chemical vapor infiltration of pyrocarbon was studied at 1100°C and methane or methane/hydrogen mixture pressures ranging from 5 to 100 kPa on porous alumina ceramic as a substrate (pore diameters from 1 to 36 μm)

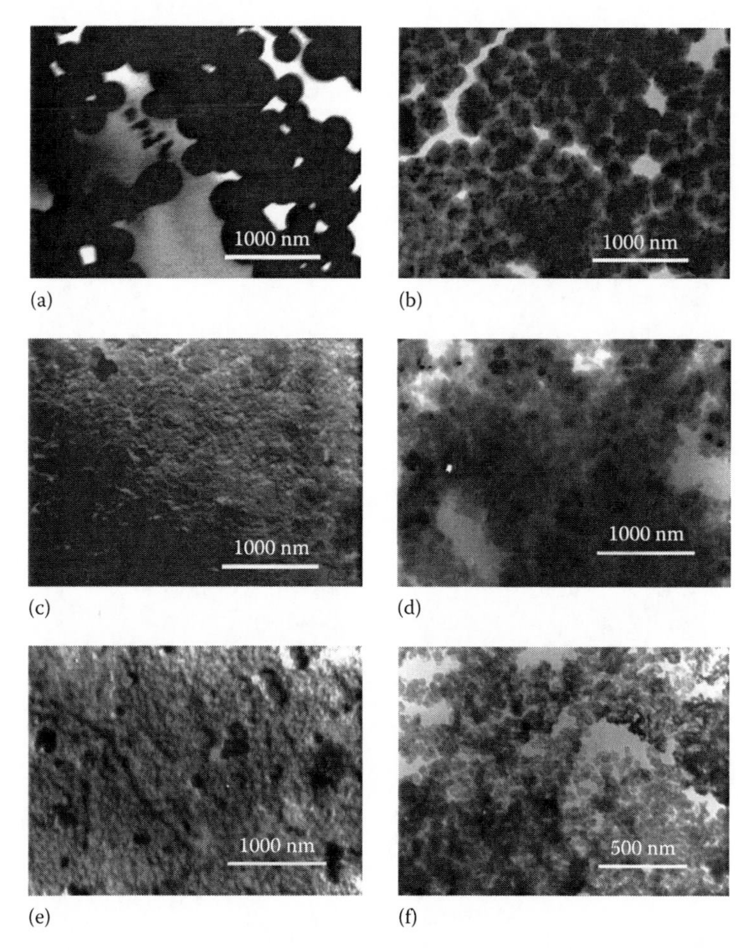

FIGURE 4.2 TEM micrographs of pyrocarbon/silica samples obtained on the basis of silica gel by pyrolysis of acetylacetone—AC-3 (a); glucose—Gl (b); acenaphthene—AN2 (c); nickel acetylacetonate—CS_{Ni} (d); cobalt acetylacetonate—CS_{Co} (e) and on the basis of pyrogenic silica by pyrolysis of CH_2Cl_2, C/A-300 ($C_C = 20\%$ w/w) (f). (Adapted with permission from Gun'ko, V.M., Leboda, R., Skubiszewska-Zięba, J., Turov, V.V., and Kowalczyk, P., Structure of silica gel Si-60 and pyrocarbon/silica gel adsorbents thermally and hydrothermally treated, *Langmuir* 17, 2001b, 3148. Copyright 2001 American Chemical Society.)

(Benzinger and Hüttinger 1999a,b). The degree of filling of pores increases with increasing methane pressure at a maximum at 20 kPa, but the overwhelming effect of hydrogen has to be ascribed to strongly lowered surface deposition rates.

Heterogeneous decomposition of allene C_3H_4 or carbon suboxide C_3O_2 on pyrocarbon at 1000°C–2000°C was performed using a continuous flow technique (low pressures 10^{-4}–10^{-2} Torr) that homogeneous processes become negligible (Wehrer et al. 1980). At 1800°C, a thin yellowish film deposits on the pyrex walls of the reactor: it is believed to result from polymerization of hydrocarbon free radicals. The allene decomposition reaction showed an unusual kinetic behavior, which was formerly observed in several carbon gasification reactions (with oxygen, water vapor, sulfur vapor, etc.) and in the heterogeneous decomposition of carbon suboxide: the reaction rate does not respond instantaneously to a temperature or a pressure change. The stationary rate determined by the new temperature or pressure conditions is attained through progressively changing transient rates. Such a behavior reveals that a carbon sample is not characterized by a unique surface state corresponding to a definite reactivity. Experiments showed that the surface state changed continuously with reaction temperature and allene pressure. The way the rate changes along the curves describing the

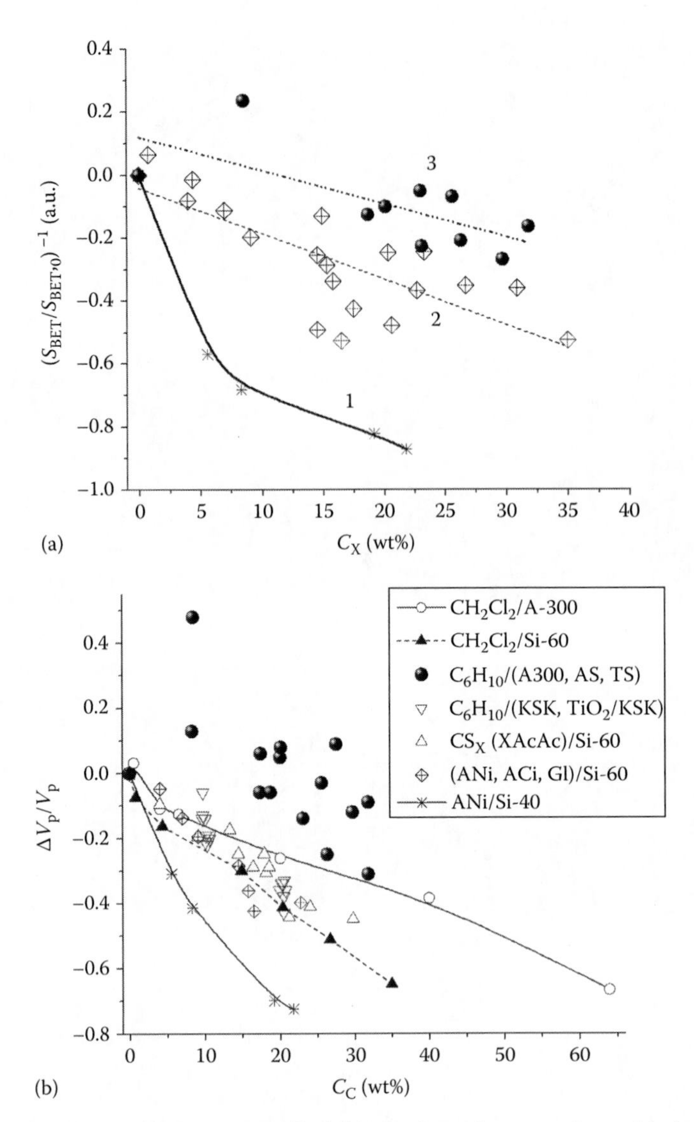

FIGURE 4.3 Relative changes in the texture of hybrid adsorbents in comparison with the matrices used for (a) specific surface area of (1) Si-40, (2) Si-60, and (3) nanooxides due to deposition of carbons or carbon/metal oxides and (b) the pore volume due to grafting of pyrocarbon on different silicas (silica gels Si-40, Si-60, and KSK, fumed silica A-300, Al_2O_3/SiO_2 [AS], and TiO_2/SiO_2 [TS]).

transient rates showed that the carbon sample became less reactive with increasing temperatures or decreasing allene pressure. This behavior allowed for an explanation of the unusual kinetics of the reaction where the activation energy of stationary rates could become negative and their reaction order greater than one. Such values were actually apparent values because stationary rates were obtained with different surface states and cannot then be directly compared (Wehrer et al. 1980).

The mentioned investigations showed a marked variety of pyrocarbon deposits. Now we will analyze pyrocarbon–silica (CS) using different methods to elucidate the relationships between the precursor composition and the parameters of the matrices, synthesis conditions, and the textural and adsorptive properties of final hybrid pyrocarbon–mineral materials affecting the interfacial phenomena with participation of water and organics adsorbed alone or co-adsorbed.

Different silicas, such as porous silica gels (Tables 4.1 through 4.4), fumed silicas with nonporous primary particles and mixed nanooxides (Table 4.5), natural minerals, and metals deposited on

TABLE 4.1
Structural Parameters of Silica Gel Si-60 (Merck) Initial and Modified by Pyrolysis of AN, AC, and Gl in an Autoclave at 500°C for 6 h

Samples	C_C (wt%)	S_{BET} (m²/g)	V_p (cm³/g)	R_p (nm)
Si-60		369	0.753	4.1
AN1	7	327	0.650	4.8
AN2	15.8	244	0.480	3.9
AN3	22.7	233	0.453	3.9
AC-1	4	339	0.717	4.2
AC-2	9.1	296	0.606	4.1
AC-3	14.5	275	0.566	4.1
Gl	16.5	174	0.433	5.0

AN, acenaphthene; AC, acetylacetone; Gl, glucose

TABLE 4.2
Parameters of Silica Gel Si-60 (Merck) Modified by Pyrolysis of Metal Acetylacetonates at 500°C for 6 h

Samples	C_C (wt%)	C_X (wt%)	X	S_{BET} (m²/g)	V_p (cm³/g)	R_p (nm)
CS_{Ti}	9.2	11.4	TiO_2	192	0.419	4.35
CS_{Cr}	12.0	11.3	Cr_2O_3	278	0.443	3.19
CS_{Co}	6.1	8.4	Co	187	0.524	5.56
CS_{Ni}	5.9	9.4	Ni, NiO	263	0.566	4.27
CS_{Zn}	6.5	11.0	Zn_2SiO_4	212	0.536	5.05
CS_{Zr}	13.8	17.1	ZrO_2	236	0.415	3.52

TABLE 4.3
Structural and Adsorptive Characteristics of Silica Gel Si-60 (Schuchardt München, Germany) Initial and Modified by Pyrolysis of CH_2Cl_2 at 550°C for 0.5, 1, 2, 3, 4, and 6 h

Samples	C_C (wt%)	S_{BET} (m²/g)	V_p (cm³/g)	R_p (nm)	$S_{BET,C}$ (m²/g)	d_C (nm)
Si-60	—	372	0.80	4.3	—	—
Si-60 (550°C)	—	344	0.75	4.4	—	—
CS-1	0.8	366	0.74	4.0	740	4.0
CS-2	4.4	339	0.67	4.0	213	14.1
CS-3	14.9	299	0.56	3.8	124	24.2
CS-4	20.3	259	0.47	3.6	96	31.3
CS-5	26.7	223	0.39	3.5	77	39.0
CS-6	35.0	163	0.28	3.4	56	53.6

Note: d_C is the average diameter of carbon particles.

TABLE 4.4
Structural Parameters of Silica Gel Si-40 (Merck) Initial and Modified by Pyrolysis of Acenaphthene at 500°C for 6 h

Samples	C_C (wt%)	S_{BET} (m²/g)	V_p (cm³/g)	R_p (nm)
Si-40	—	732	0.542	1.48
AN1	5.6	313	0.375	2.40
AN2	8.3	231	0.317	2.74
AN3	19.2	128	0.163	2.53
AN4	21.8	92	0.149	3.24

TABLE 4.5
Structural Parameters of Fumed Silica, SA, and ST Before and After Carbonization of Cyclohexene

Samples	C_X (wt%)	C_C (wt%)	S_{BET} (m²/g)	V_p (cm³/g)	R_p (nm)
A-300		—	312	0.65	3.8
C/A-300		8.5	385	0.96	4.3
SA1	1.3	—	207	0.42	3.7
C/SA1		31.8	173	0.38	3.9
SA23	23	—	353	0.80	4.0
C/SA23		25.6	328	0.77	4.1
AST	22	—	38	0.07	7.7
C/AST		23	36	0.07	9.0
ST2	1.7[a]	—	318	0.71	4.1
C/ST2		20.1	286	0.77	4.6
ST22	22[a]	—	251	0.63	4.4
C/ST22		18.7	219	0.60	4.8
ST33	33[a]	—	219	0.55	4.6
C/ST33		23.1	169	0.47	4.9
ST9	9[b]	—	238	0.57	4.3
C/ST9		26.3	188	0.43	4.1
ST36	36[b]	—	115	0.25	4.1
C/ST36		29.7	84	0.22	4.6

[a] CVD-TiO_2/fumed silica
[b] Fumed ST

oxides, were used as supports to prepare hybrid carbon–mineral adsorbents utilizing different organic precursors. Several general effects were observed upon the pyrolysis of organics on silicas. If organic molecules include oxygen atoms (especially OH groups, e.g., in glucose), then marked changes in the structure of the silica matrix occur upon the pyrolysis due to hydrolysis of the ≡Si–O–Si≡ bonds, and the average pore size increases (Table 4.1 and Figure 4.4).

The nitrogen adsorption isotherm shape changes slightly for silicas modified by pyrolysis ($T < 1000$ K) of different organics independent of the type of oxide matrices in comparison with that for pristine silicas (Figure 4.5).

However, the amounts of adsorbed nitrogen decrease due to filling of matrix pores by pyrocarbon or another grafted X phase (e.g., metal or metal oxide during pyrolysis of metal

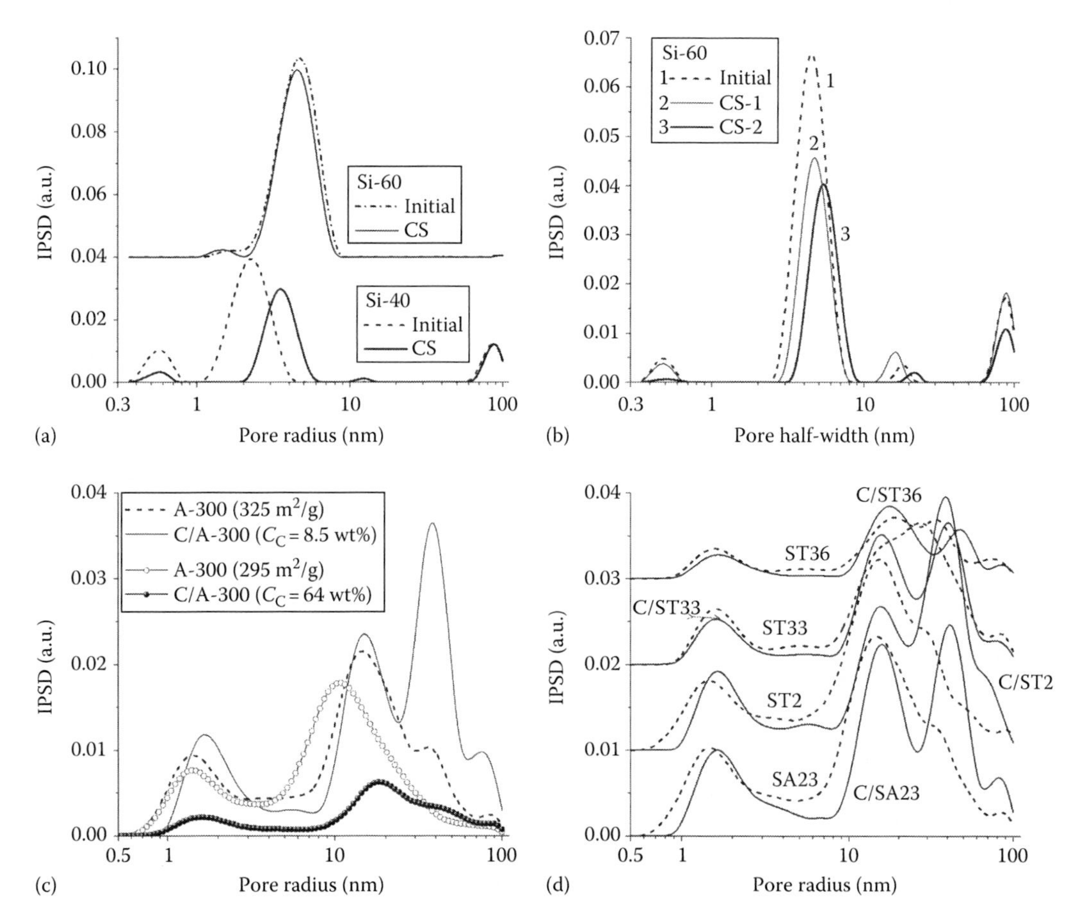

FIGURE 4.4 Incremental pore size distributions of (a) Si-40 pristine and modified by pyrolysis of acenaphthene (C_C=21.8 wt%) and Si-60 pristine and modified by pyrolysis of Zr acetylacetonate; (b) Si-60 pristine and modified by carbonization of acenaphthene (CS-1 at C_C=22.7 wt%) and glucose (CS-2 at C_C=16.5 wt%); (c) fumed silica A-300 modified by carbonization of CH_2Cl_2 (C_C=64 wt%) or cyclohexene (C_C=8.5 wt%), SiO_2/Al_2O_3 ($C_{Al_2O_3}$=23 wt%), SA23, and SiO_2/TiO_2 (C_{TiO_2}=33 wt% for CVD-TiO_2/SiO_2 and 2 or 36 wt% for fumed ST) initial and modified by pyrolysis of cyclohexene (SCR/SCV model with MND method, some PSDs in (a) and (d) are displaced along the *Y*-axis for better view).

acetylacetonates, Table 4.2). While the pyrocarbon structure is akin to that of carbon blacks, the nitrogen adsorption isotherms for pyrocarbon–silicas become close to that for carbon black with increasing carbon concentration (C_C) but substantially differ from that for nanoporous carbon Ajax (Figure 4.5). Filling of pores of silica gel by pure pyrocarbon (with no X phase in final material or O in a precursor) results in a decrease in the pore volume and the average pore size (Table 4.3) but own porosity of the carbon deposits does not practically appear in the PSDs whose shapes change slightly (Figure 4.4, CH_2Cl_2 as a precursor). For pyrocarbon at C_C<1 wt%, its own specific surface area can be great (S_{BET}>700 m²/g, d_C<3.9 nm) while carbon particles are very small at such a low C_C value (Table 4.3, d_C) and $S_{BET} \sim 1/d_C$. Notice that heating of Si-60 under conditions corresponding to pyrolysis leads to relatively small structural changes in the silica gel matrix (Table 4.3, Si-60, 550°C). Consequently, larger textural changes of CS (Tables 4.1 through 4.5) are due to the pyrolysis. For instance, the pyrolysis of acenaphthene $C_{12}H_{10}$ on silica gel Si-40 possessing very narrow mesopores (Table 4.4) leads to a marked displacement of the PSD toward larger pores (Figure 4.4a). A similar effect is also observed upon the glucose carbonization on silica gel Si-60 due to hydrolysis of the ≡Si–O–Si≡ bonds. However, the pyrolysis of acenaphthene on Si-60 does

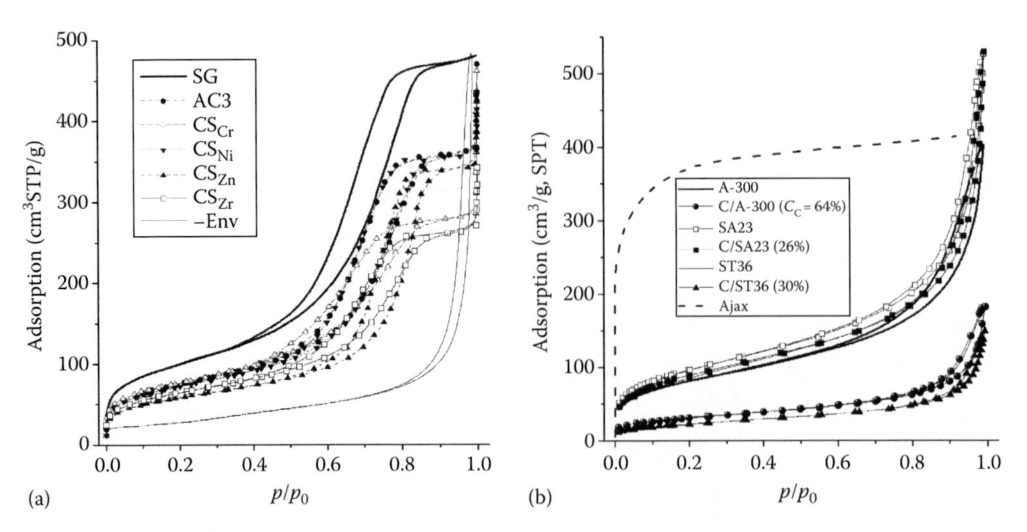

FIGURE 4.5 Nitrogen adsorption–desorption isotherms (77.4 K) on (a) silica gel Si-60 pristine and modified by pyrolysis of acetylacetone (CS) or acetylacetonates (AcAc) $Zr(AcAc)_4$, $TiO(AcAc)_2$, $Ni(AcAc)_2$, $Zn(AcAc)_2$, $Cr(AcAc)_3$, and $Co(AcAc)_2$; graphitized carbon black Envicarb (Supelco, particles of 40–60 μm, $S_{BET}=98$ m²/g) (Env) and (b) fumed silica A-300, SiO_2/Al_2O_3 (23 wt% Al_2O_3), and SiO_2/TiO_2 (36 wt% TiO_2) pristine and modified by pyrolysis of cyclohexene; nanoporous carbon Ajax ($S_{BET}=1345$ m²/g).

not give such a displacement (Figure 4.4 and Table 4.1). In the case of metal acetylacetonates, a new PSD peak of low intensity appears at R_p between 10 and 20 nm (Gun'ko et al. 2000c, 2001c), which can be linked to pores in the silica matrix transformed during pyrolysis or in C/X deposits or as voids between C/X particles and the silica gel surfaces. Additionally, the main peak of PSD for CS_X shifts markedly depending on the X nature (vide infra).

There is such a tendency as the higher the carbon concentration on silica gels, the lower the pore volume per gram of the adsorbent (Figure 4.3). This result is typical for all the studied pyrocarbon–mesoporous silica gel adsorbents (Tables 4.1 through 4.4) due to filling of mesopores by tiny carbon nanoparticles. Therefore, changes in the specific surface area per cm³ of the adsorbent are significantly lower than that per gram. In the case of the nitrogen adsorption isotherms normalized due to dividing by S_{BET}, a marked difference for carbosils CS-*i* (shown in Table 4.3) is observed only at low relative pressure $p/p_0<0.1$. This result suggests that changes in the nitrogen adsorption isotherms for hybrid CS adsorbents (CH_2Cl_2 as a precursor carbonized on Si-60 at 550°C) are related mainly to nanopores and narrow mesopores filled by nitrogen at low pressures, and the mesoporous character of C/Si-60 is saved (Figure 4.4). However, the volume and the specific surface area of mesopores significantly decrease (Figure 4.3 and Table 4.3), as own porosity of the carbon phase (as well as the specific surface area $S_{BET,C}$) is lower than that of silica gel at $C_C>1$ wt% (Table 4.3), and V_p strongly decreases with increasing C_C value.

In the case of metal acetylacetonates $M(AcAc)_n$ as precursors carbonized on silica gel Si-60, cobalt in carbosil CS_{Co} (Table 4.2) exists, according to the XRD data, as metallic phase (crystallite size 20–25 nm) due to its reduction during pyrolysis of $Co(AcAc)_2$. Nickel in CS_{Ni} is in both metallic (crystallite size ~25 nm) and oxide phases (~13 nm). Carbosil CS_{Cr} contains amorphous chromium oxide. During pyrolysis of zinc acetylacetonate (CS_{Zn}), zinc silicate is formed apart from carbon deposit due to reaction between zinc compounds and silica gel surfaces. Carbosils CS_{Zr} and CS_{Ti} include, respectively, zirconium dioxide (~4 nm) and titanium dioxide (anatase crystallites ~11 nm). Consequently, the sizes of X crystallites are typically larger than the average pore size of Si-60 (Table 4.2), i.e., a major portion of X deposits can form on the outer surfaces or in macropores of relatively large silica gel particles (0.1–0.2 mm). According to TEM micrographs, pyrocarbon deposits extracted from carbosils C/SiO_2 such as AC (Table 4.1) or CS-*i* consist of carbon globules (<100 nm) connected with adjacent ones. Changes in the adsorbent structure due to $M(AcAc)_n$ pyrolysis in comparison with that of C/SiO_2 are visible in micrographs of CS_X and features of X distribution reflect in the C/X structure as a whole. For example, more nonuniform and denser carbon deposit is observed for CS_{Ti} and the opposite result is for CS_{Zn}, which can be caused by the difference not only in the catalytic impact of the X phase on the pyrolysis but also in features of the X phase distribution. Average pore radius R_p increases for some CS_X samples but decreases for others; therefore, correlation between R_p and $C_C+C_X=C_\Sigma$ is poor (Table 4.2) that can be considered as evidence of a complicated structure of pores of $C/X/SiO_2$, as tiny C/X particles can be grafted on the pore walls, but larger C/X globules can be formed on the outer surfaces of silica gel particles with partial blocking of pores. In the case of formation of the deposit layer mainly on the pore walls at low C_C and C_Σ, the V and R_p values should decrease simultaneously, as it is seen for CS_{Zr} (small X particles ~4 nm) and CS_{Cr}. However, this effect is not observed for other CS_X samples, e.g., cobalt particles of 20–25 nm correspond to larger R_p, but smaller particles with NiO (~13 nm) give smaller average R_p (Table 4.2). Nevertheless, correlation between V and C_C or C_Σ is marked due to filling of pores by C/X deposits as the main effect on the pyrolysis. The C_X concentration of X phase has a larger influence on S_{BET} than C_C or C_Σ has due to changes in the structure not only of the X phase per se but also of pyrocarbon deposits because of the catalytic capability of the X phase in the pyrolysis. This can also impact small-scale nonuniformity and fractality of CS_X.

The structure of the titania phase in chemical vapor deposition (CVD)-TiO_2/SiO_2 (synthesized by CVD of $TiCl_4$ with subsequent hydrolysis of residual Ti–Cl bonds) controlled by means of CVD synthesis conditions can affect reactions of organics (coking, cracking, isomerization, disproportionation, etc.) occurring on the mixed oxide surfaces due to the catalytic ability of titania (anatase) and titania/silica in both acid–base and redox reactions. To elucidate this effect, the kinetics of the carbon deposition onto the surfaces of silica gel KSK and CVD-titania/silica gel was studied using cyclohexene pyrolysis (Leboda et al. 2000a, Gun'ko et al. 2000b).

Typically, the catalytic reaction rate strongly decreases during an induction period (τ^*, relaxation time) as follows (Kiperman 1979):

$$v = v_\infty + (v_0 - v_\infty)\times\exp\left(-\frac{t}{\tau^*}\right) \tag{4.1}$$

where

v_0 is the initial reaction rate

v_∞ is the steady-state reaction rate

while the effects are linked to the carbon deposit formation on the most active surface sites (Figure 4.6). In the case of heterogeneous surfaces involving several phases, the overall reaction rate v_∞ of the precursor carbonization can be represented as the sum

$$v_\infty = \sum_i k_{i0} f_i(\Theta_i) + k_C f(\Theta_C) \tag{4.2}$$

where

k_{i0} and k_C are the rate constants of the precursor carbonization on the i-th oxide and carbon phases, respectively

$f_i(\Theta_i)$ is a function of coverage of the i-th oxide phase (e.g., $f(\Theta)=(\Theta_{i0}-\Theta_i)^m$ where Θ_{i0} is an initial portion of the surface corresponding to the i-th oxide phase)

While anatase or anatase/silica can catalyze the pyrolysis, coking of the TiO_2/SiO_2 surfaces is due to interaction between organics and Brønsted (≡MO(H)M≡ at M = Ti, Si) or Lewis acid

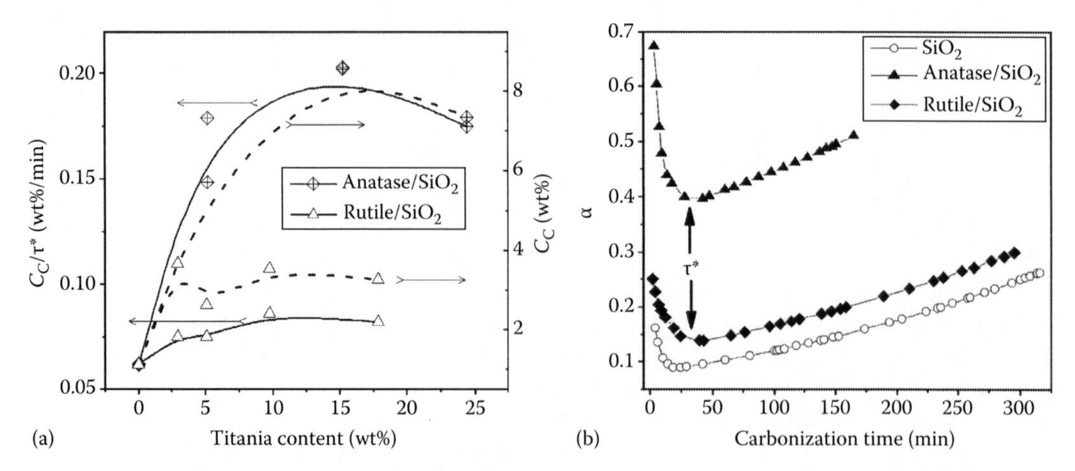

FIGURE 4.6 (a) Rate of the carbon deposit growth C_C/t during the relaxation time (τ^*) and changes in the carbon content during τ^* and (b) relative carbonization rate as a function of time for mesoporous silica gel KSK ($S_{BET}=377$ m^2/g, $V_p=0.98$ cm^3/g, $R_p=5.2$ nm), anatase/SiO_2 ($C_{TiO_2}=15.2$ wt%, $S_{BET}=281$ m^2/g, $V_p=0.69$ cm^3/g, $R_p=4.9$ nm), and rutile/SiO_2 ($C_{TiO_2}=17.9$ wt%, $S_{BET}=281$ m^2/g, $V_p=0.76$ cm^3/g, $R_p=5.4$ nm). (Adapted with permission from Gun'ko, V.M., Leboda, R., Marciniak, M. et al., CVD-titania/silica gel carbonized due to pyrolysis of cyclohexene, *Langmuir*, 16, 2000b, 3227. Copyright 2000 American Chemical Society.)

(incompletely O-coordinated Ti atoms) sites or radical sites (MO$^•$, M$^•$). The effectiveness of CVD-titania as a catalyst of the carbonization of organics can depend on the crystalline nature (e.g., catalytic capability of anatase is greater than that of rutile, Figure 4.6), the particle distribution in pores or on the outer surfaces of the support (which affects the accessible surface area of the titania and silica phases), and the particle size distribution (typically, the smaller the particles, the higher is the catalytic effect). Therefore, the reaction rate v of carbonization of organics on CVD-TiO_2/SiO_2 is higher than that on the pristine silica gel (Figure 4.6), as silica is practically inert in similar reactions. Additionally, the reactions on anatase have greater v values (depending also on the titania content in mixed oxides) than those for samples having a large portion of rutile. Rutile/silica provides the reaction rate close to that for pure silica depending slightly but nonlinearly on C_{TiO_2}. A linear increase in the v value with increasing reaction time at $t>\tau^*$ (Figure 4.6b) can be linked with a linear growth of the content of the carbon phase (increase in Θ_C in Equation 4.2), possessing a large number of active sites (e.g., nonclosed cycles) capable to interact with different gaseous reactants to form closed aromatic cycles through dehydrogenation. One can assume that the reaction mechanism does not change with increasing C_{TiO_2} and C_C at $t>\tau^*$, as the derivatives dv/dt for all the studied samples are closely related with consideration for a decrease in S_{BET} of $C/TiO_2/SiO_2$.

According to TEM micrographs (Figure 4.2, Gun'ko et al. 2000b), pyrocarbon particles formed on silica gel are relatively large (i.e., they were formed mainly on the outer surface of the substrate) and their distribution is nonuniform. In the case of the anatase/silica substrate (CVD-TiO_2 synthesized at 200°C), the pyrocarbon distribution is more uniform and its particles are smaller than those on silica gel. This effect can be caused by the distribution of CVD-titania (anatase) on the silica gel surfaces in the form of tiny particles (clusters) mainly in pores that result in formation of pyrocarbon in pores in a larger portion in comparison with C/SiO_2. In the case of titania (rutile/anatase blend) synthesized at higher temperature (600°C) with formation of dense and segregated particles, the pyrocarbon phase distribution is nonuniform and similar to that of C/SiO_2.

To estimate the deposit distribution on the substrate surface, the model developed to describe coke grafting (Fenelonov 1995) can be used. Coverage of the adsorbent surfaces by carbon deposits is given by

$$\Theta = 1 - \frac{S_{CS} - S_C x}{S_0(1-x)} \tag{4.3}$$

where

$S_{CS} = S_{BET}$ is the specific surface area of CS

S_C is the accessible surface area of the carbon phase (determined using probe molecules adsorbed mainly on carbon surfaces)

x is the relative content of carbon

S_0 is the specific surface area of the initial adsorbent

Relative change in the surface area per volume unit

$$\Psi = \frac{S_{CS}}{S_0(1-x)} \tag{4.4}$$

is linked to Θ as follows:

$$\Theta = \frac{1-\Psi}{1-\lambda} \tag{4.5}$$

where

$$\lambda = \frac{S_C x}{S_0 \Theta (1-x)} \tag{4.6}$$

In the case of $\lambda > 1$, formation of individual particles occurs on the outer surface of adsorbent, pore filling by carbon deposits corresponds to $\lambda < 1$, and the film covering the whole surface gives $\lambda \approx 1$. The Ψ values decrease with decreasing surface area per gram of adsorbent. Calculations of the Θ, λ, and Ψ values as functions of the amounts of carbon, titania, and titania + carbon deposits using the $S_{BET}(x)$ functions show that the titania phase possesses tighter contacts with the substrate than carbon particles have with silica or titania/silica ($\lambda_C > \lambda_{TiO_2}$) due to the differences in the nature and texture of these phases, the particle distributions in pores and on outer surfaces and deposit morphology as the whole. Thus, carbon particles can be formed in mesopores (or on the outer surfaces of silica gel particles) more loosely in comparison with titania. However, in the case of $C/TiO_2/SiO_2$, marked decrease in Ψ is observed with increasing total amounts of C/TiO_2 deposits. This effect can be explained by formation of a major portion of pyrocarbon (as well as CVD-titania) in mesopores, as titania grafted in mesopores can catalyze the pyrolysis and formation of pyrocarbon on titania is more probable than that on silica.

On the basis of the α_S plots for nitrogen adsorption on C/SiO_2 and $C/X/SiO_2$ (Gun'ko et al. 2001c), one can assume that larger changes in the pore structure can be for C/X/Si-60 prepared using carbonization of metal acetylacetonates in comparison with C/silica gels. The carbon phase distribution (in pores and on the outer surfaces of oxide particles) is nonuniform and depends on the nature of the matrix and the C_C values that can explain nonlinear dependencies of some structural parameters on the carbon concentration. In the case of pure silica gel, the carbon phase is distributed in the form of larger particles than that on titania/silica gels. This circumstance, as well as minor changes in the normalized isotherms, can be considered as a demonstration of relative uniformity (low contribution of nanopores) of the CS surfaces with respect to the nitrogen adsorption, which reflects in a slight dependence of the adsorption energy distribution $f(E)$ on C_C. One can assume that $f(E)$ for C/silica gel differs from that for pristine silica gel (Figure 4.7) due to the differences

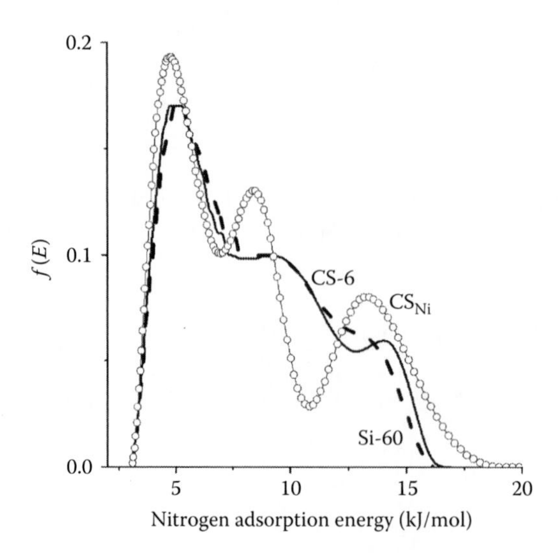

FIGURE 4.7 Nitrogen adsorption energy distribution for Si-60 (Schuchardt München, Germany) pristine and modified by carbonization of CH_2Cl_2 (C_C=35 wt%, CS-6) and $Ni(AcAc)_2$ (CS_{Ni}).

in dispersion interaction of nitrogen molecules with graphene sheets (greater) and silica (smaller) surfaces. Strong changes in the texture and chemistry of the adsorbent because of carbonization of $Ni(AcAc)_2$ result in significant changes in the nitrogen adsorption energy distribution in comparison with pristine Si-60 or CS-6. This is due to appearance of Ni and NiO nanoparticles and different structure of carbon deposits affected by these new phases.

It should be noted that the structure of C/silica and C/X/silica surfaces can be complex not only with respect to their topography but also to their chemical composition due to the availability of H, O, etc., atoms in surface groups, new noncarbon phases (oxide or metal as in the case of carbonization of acetylacetonates of Ni and Co), which can affect the adsorption energy of polar and nonpolar adsorbates.

The structural characteristics of carbon deposits formed at open surfaces of nanooxides (composed of nonporous primary nanoparticles) differ from deposits formed onto porous silicas (Gun'ko et al. 2001c). Conditions of pyrolysis of cyclohexene on nanooxides were the same for sample series (Gun'ko et al. 2000h), but the concentrations of pyrocarbon were different (Table 4.5) due to the catalytic effect of mixed oxides on this process, which can be accompanied by intensive coking of active sites. In contradiction with porous silica gels, the S_{BET} and V_p values increase after cyclohexene pyrolysis on the fumed silica surface (low C_C) (Figure 4.3). However, for $C/Al_2O_3/SiO_2$ or $C/TiO_2/SiO_2$ (large C_{TiO_2}), these values are typically smaller. Additionally, a reduction of the mesopore surface area is greater than that of nanopores for $C/X/SiO_2$. Consequently, formation of carbon deposits on fumed silica occurs mainly on the outer surface of aggregates (relatively dense as their apparent density $\rho \approx 0.7$ g/cm^3), but for $C/Al_2O_3/SiO_2$ and C/fumed TS, this process can take place not only on the outer surface of aggregates but also in the interparticle space inside aggregates on active surface sites. However, the average R_p values for C/SiO_2 and $C/X/SiO_2$ are slightly larger than that of initial oxide samples (Table 4.5) due to formation of pyrocarbon particles (larger than primary particles of nanooxides, i.e., average particle size increases) on the outer surface of aggregates and own low porosity of pyrocarbon deposits. The pore size distributions demonstrate changes in pores depending on the nature of X, the C_X and C_C values in X/SiO_2 and $C/X/SiO_2$ (Figure 4.4d). First, adsorption of nitrogen can occur near contacts between primary particles in aggregates corresponding to nanopores and narrow mesopores linked to the first PSD peak. Then adsorption (secondary filling of mesopores) occurs in channels (larger mesopores) between several neighboring particles in aggregates (i.e., on the accessible surfaces of particles packed in aggregates with pores at R_p>7 nm), and also between aggregates (which are not densely packed in agglomerates, which have the apparent density of only several percent of the true density), on their external surfaces at R_p>20 nm. Clearly, complete filling of this interparticle space (especially in agglomerates with very large empty volume of 15–25 cm^3/g) by nitrogen cannot be reached even at $p/p_0 \rightarrow 1$.

Pyrocarbon produced by carbonization of CH_2Cl_2 on nanooxides is denser than that synthesized using cyclohexene as a precursor (Figure 4.3). The increment in C_C due to the catalytic effect of mixed oxides on cyclohexene pyrolysis can be less than growth of carbon deposit on pure silica in the case of CH_2Cl_2. Clearly, the pyrolysis of molecules larger than CH_2Cl_2 under relatively soft conditions can result in formation of carbons with greater porosity and specific surface area due to the impact of their carbon skeleton (Figure 4.3 and Table 4.5). However, the use of such precursors as acenaphthene or acetylacetone gives larger reduction in S_{BET} of Si-60 in comparison with that on pyrolysis of such small molecules as CH_2Cl_2. Consequently, a precursor choice for preparation of hybrid C/SiO_2 or C/X/SiO_2 adsorbents allows one to change the texture of carbon deposits and adsorbents depending on pyrolysis conditions and oxide matrix structure.

The pyrocarbon surfaces include not only C (polyaromatics) but also H and O atoms and low amounts of other elements as impurities. Oxidized groups such as COH, COOH, C=O, C–O–C, etc., can possess acidic or basic properties. Pyrocarbon can contain during synthesis charged groups (C^+, CH^+, CO^-, etc.) and broken bonds ($C^\bullet$, $CH^\bullet$, $CO^\bullet$, etc.), which can play an important role on subsequent deposition of carbon (see Figure 4.6b, $t>\tau^*$) or on catalytic reactions with organic molecules. Additionally, at low C_C values, a portion of the oxide support surfaces (with different types of OH groups and incompletely O-coordinated metal atoms) is accessible and affects the adsorptive properties of CS as the whole.

Changes in the nature of disperse oxide surfaces during CVD of pyrocarbon lead to marked changes in the pH dependencies of electrokinetic (ζ) potential of hybrid particles in the aqueous suspensions of C/SiO_2 and C/X/SiO_2 (Figure 4.8). The ζ(pH, C_X, C_C) plots of hybrid adsorbents can lie at both over and below the ζ curves of initial X/SiO_2 oxides. These displacements are nonlinear with respect to pH, C_X, and C_C due to complexity of the surfaces of C/X/SiO_2 and dependence of the pyrocarbon structure on oxide matrix and precursor type.

Frequently, pyrocarbon grafting leads to an increase in the effective diameter D_{ef} (hydrodynamic diameter, i.e., the particle diameter plus the electrical double layer) of C/oxide particles in the aqueous suspensions (Figures 4.9 and 4.10).

Pyrocarbon deposits (graphene clusters and then carbon nanoparticles) form mainly around active acidic sites on the surfaces of Al_2O_3/SiO_2 and TiO_2/SiO_2 blocking these sites that result in changes in the acidic properties of oxides causing the displacement of the ζ(pH) plots toward larger pH and changes in the shape of these curves (Figure 4.8). The influence of pyrocarbon is minimal for C/CVD-TiO_2/fumed SiO_2, as CVD-titania is distributed in the form of relatively large particles with weak contacts to the silica matrix and pyrocarbon can mainly cover these TiO_2 particles, i.e., the silica surfaces remain more accessible for water. Additionally, this effect is stronger over the acidic range of pH due to blocking of the most acidic sites (such as M–O(H)–Si, M = Al, Ti) of the oxide surfaces by pyrocarbon deposits. Therefore, the shapes of ζ(pH) plots for C/X/SiO_2 differ less than those of X/SiO_2. Reduction of the acidic properties of C/nanooxide in comparison with pristine oxides can be caused by the polyaromatic character of graphene clusters and different basic sites on the carbon phase having oxidized groups, whose isoelectric point pH(IEP) can be close to 9. In the case of Al_2O_3/SiO_2, carbon deposits form first of all on the strongest acidic or basic sites, as the shape of the ζ(pH) curves for C/Al_2O_3/SiO_2 changes not only at low pH values (<5) (negative charge of particles is provided by acidic sites such as Si–O(H)–Al, Al–O(H)–Al, SiOH, etc.) but also at pH>6 (Figure 4.7). The pH(IEP) values of C/Al_2O_3/SiO_2 are about 3–4, but for pure Al_2O_3/SiO_2 samples, they are lower pH<2 due to acidic sites Si–O(H)–Al. Therefore, one can assume that pyrocarbon is distributed on Al_2O_3/SiO_2 nonuniformly mainly on the alumina/silica interfaces and alumina phase, as they possess the catalytic capability in the pyrolysis. Additionally, the ζ(pH) curves for C/Al_2O_3/SiO_2 are relatively close to that of C/SiO_2 (at pH between 3 and 5). Consequently, mainly silica

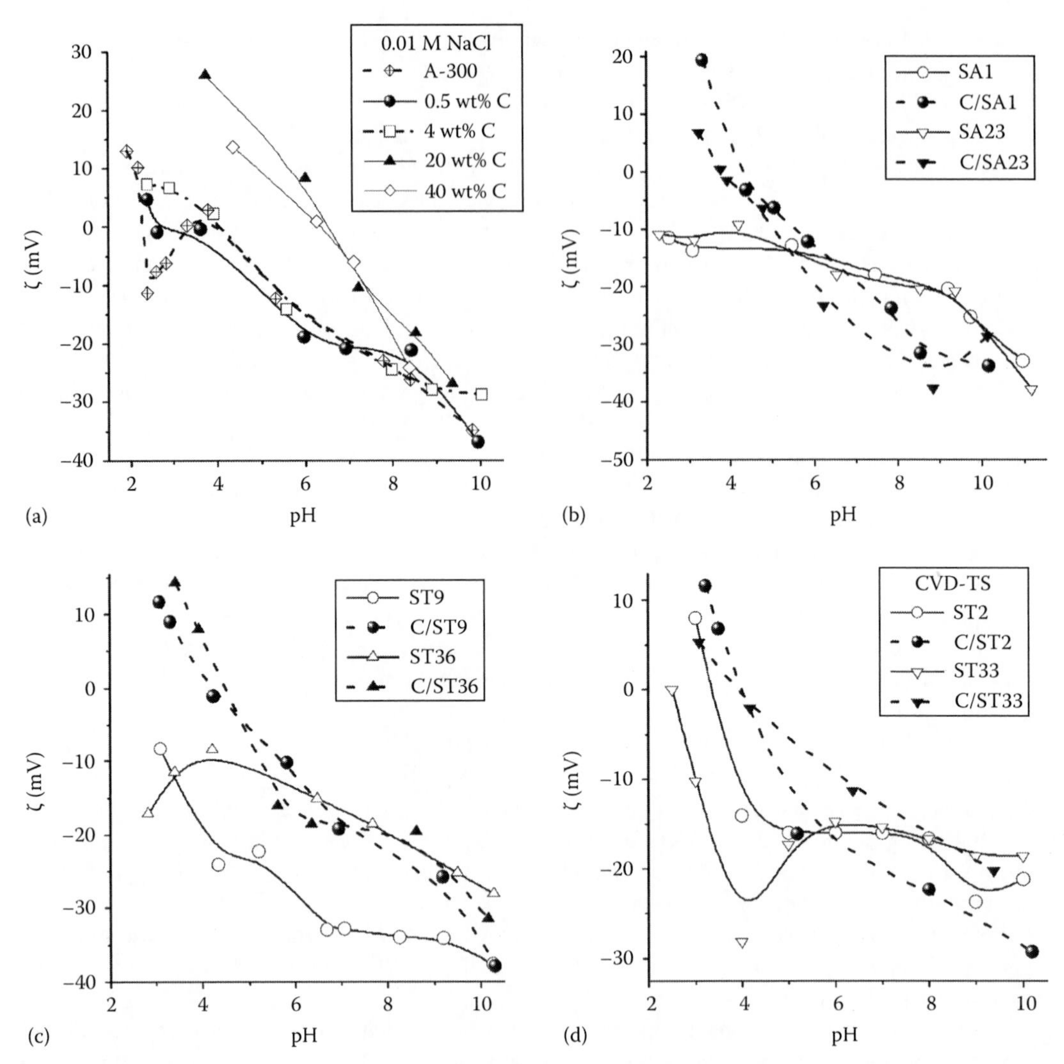

FIGURE 4.8 Electrokinetic potential of (a) fumed silica A-300 pristine and modified by pyrolysis of CH_2Cl_2; (b) Al_2O_3/SiO_2 and C/AS (cyclohexene as a precursor); (c) fumed TS and C/fumed TS (cyclohexene); and (d) CVD-TiO_2/SiO_2 and C/CVD-ST22 (cyclohexene) in aqueous suspensions (0.001 M NaCl) as a function of pH. (Adapted from *J. Colloid Interface Sci.*, 230, Gun'ko, V.M., Zarko, V.I., Leboda, R., Marciniak, M., Janusz, W., and Chibowski, S., Highly dispersed X/SiO_2 and $C/X/SiO_2$ (X=alumina, titania, alumina/titania) in the gas and liquid media, 396–409, 2000h, Copyright 2000, with permission from Elsevier.)

surface and carbon deposit in $C/X/SiO_2$ are accessible for liquid water. A similar effect is observed for $C/TiO_2/SiO_2$. There is tendency of an enhancement of pH(IEP) with increasing C_C in $C/Al_2O_3/SiO_2$ and C/fumed TiO_2/SiO_2. Additionally, the range of pH(IEP) for $C/X/SiO_2$ (especially for C/CVD-TiO_2/SiO_2) is more narrow (2.5–4.5) than that of X/SiO_2. These effects can be connected with formation of pyrocarbon on more active (in pyrolysis) patches of the X/SiO_2 surfaces such as the X/silica interfaces and X phase possessing Brønsted and Lewis acid sites. Therefore, the silica phase in $C/X/SiO_2$ is covered by carbon deposit to a lesser extent and pH(IEP) of $C/X/SiO_2$ can correspond to average value of silica (pH(IEP) ≈ 2.2) and carbon deposits.

Carbon deposits slightly influence the particle size distribution of C/A-300 in the aqueous suspension shifting it toward larger diameter (d_{PCS}) determined by using the PCS method, as the smallest particles have d_{PCS} = 65 nm (A-300) and 140 nm (C/A-300). However, the pyrocarbon

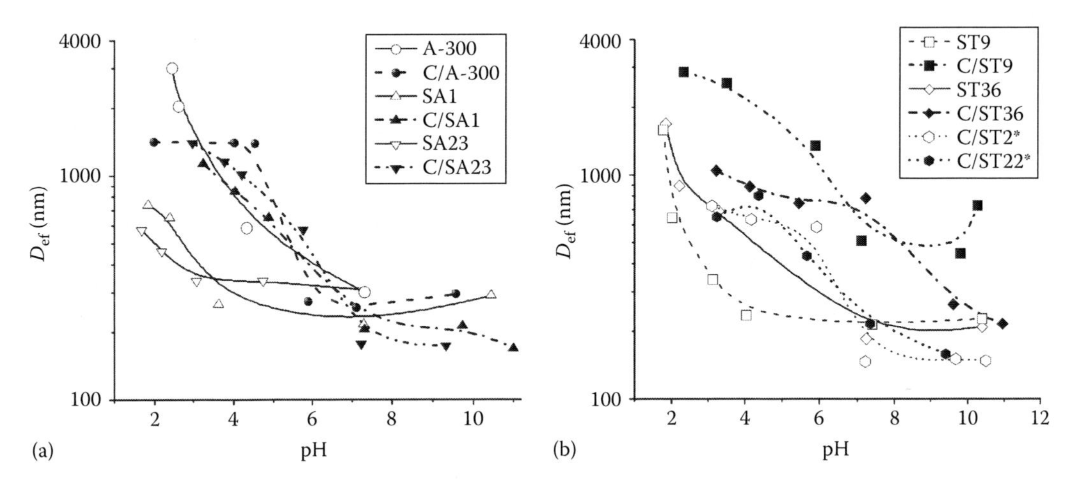

FIGURE 4.9 Effective diameter for (a) A-300 and Al_2O_3/SiO_2 and (b) fumed ST and CVD-TiO_2/SiO_2 (*) pristine and modified by pyrolysis of cyclohexene in aqueous suspensions (0.001 M NaCl) as a function of pH. (Adapted from *J. Colloid Interface Sci.*, 230, Gun'ko, V.M., Zarko, V.I., Leboda, R., Marciniak, M., Janusz, W., and Chibowski, S., Highly dispersed X/SiO$_2$ and C/X/SiO$_2$ (X = alumina, titania, alumina/titania) in the gas and liquid media, 396–409, 2000h, Copyright 2000, with permission from Elsevier.)

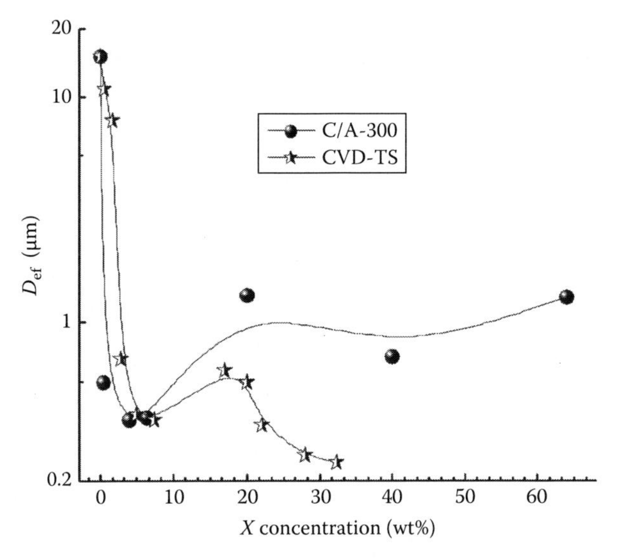

FIGURE 4.10 Effective diameter of C/A-300 and CVD-TiO_2/SiO_2 in the aqueous suspension (without addition of electrolytes) as a function of the deposit (C or TiO_2) concentration.

influence is stronger for C/Al_2O_3/SiO_2 and C/TS (Figures 4.9 and 4.10) due to the differences in the pyrocarbon distributions on silica and mixed oxides, as graphene clusters block active surface sites. Additionally, the lower the pH values, the larger are the aggregates and agglomerates. The last effect is linked to reduction of the electrostatic repulsive interactions between charged particles close to pH(IEP) (or PZC), as polar and dispersion interactions between them are attractive. For C/SA1 (SiO_2/Al_2O_3 at $C_{Al_2O_3} = 1$ wt%) and C/SA23 ($C_{Al_2O_3} = 23$ wt%), the particle size distribution is narrower and average diameter D_{ef} is larger than those of SA1 and SA23 at pH < 7 (Figure 4.9a), which can be due to changes in particle–particle interactions (especially between silica patches negatively charged and alumina fragments positively charged at pH between pH(IEP) for silica [≈2.2] and alumina [≈9.8]), as pyrocarbon blocks the most active (acidic or basic) sites on the mixed oxide surfaces. For fumed SiO_2/TiO_2 (ST36), the particle size distribution shifts toward larger particles and D_{ef} increases due to pyrocarbon in C/ST36 (Figure 4.9b). Additionally, the differences in the particle size distributions of $C/TiO_2/SiO_2$ and $C/Al_2O_3/TiO_2/SiO_2$ at different pH values are smaller than that for $C/Al_2O_3/SiO_2$ in comparison with those of initial oxides. For C/CVD-TiO_2/SiO_2, D_{ef}(pH) depends on C_{TiO_2} weaker than that for C/fumed ST36 due to features of the CVD-TiO_2 particle distribution and their segregation.

In the case of C/SiO_2, the pyrocarbon deposits synthesized at $T < 1000$ K represent mainly dense spherical globules (with turbostratic structures akin to that of carbon black), which can be relatively large (up to 100–200 nm) and possess low porosity, as changes in the pore size distributions correspond to reduction of the porosity of the supports practically without appearance of new pores, which can be assigned to the carbon phase per se. At $T > 1400$ K, the pyrocarbon structure becomes close to that of graphite, but its own porosity remains very low.

Structural features of titania–silica or alumina–silica adsorbents prepared using different silica matrices and various synthetic techniques cause significant differences in the distribution of pyrocarbon deposits on such supports, particle size distributions, and other structural and energetic properties of deposits and whole adsorbents. In general, pyrocarbon at concentration from several to dozens percents consists mainly of relatively large individual dense globules (up to 250–300 nm in size on silica gels) and possesses low porosity provided mainly by the voids between outer surfaces of the matrices and pyrocarbon globules. A smaller portion of pyrocarbon represents tiny dense graphene particles (clusters), which can fill mesopores of the supports or voids (channels) between primary particles of disperse mixed oxides. Pyrocarbon deposits form preferably near the titania particles (CVD-TiO_2) or TiO_2/SiO_2 and Al_2O_3/SiO_2 interfaces (carbon nuclei form at M–O(H)–Si groups and incompletely O-coordinated M atoms) possessing catalytic activity in the pyrolysis. In the case of metal–organics as precursors, pyrocarbon formation results in dramatic changes in the porosity of silica gel during simultaneous CVD processes for both pyrocarbon and titania phases during the Ti acetylacetonate pyrolysis. Similar but smaller effects are observed for C/X/Si-60 prepared by the pyrolysis of Zn and Zr acetylacetonates or for pyrolysis of cyclohexene on CVD-TiO_2/silica gel.

In the case of such mixed oxides as highly disperse Al_2O_3/SiO_2, TiO_2/SiO_2, and $Al_2O_3/TiO_2/SiO_2$, formation of carbon deposits due to pyrolysis of cyclohexene occurs mainly on the surface patches of the X/SiO_2 interfaces and X phase possessing catalytic activity, which leads to some similarity in the electrokinetic behavior of $C/X/SiO_2$ and C/SiO_2 particles, as the X phase is covered by the carbon deposits to a greater extent than silica, and the silica surface of C/SiO_2 is more accessible for water molecules than X of $C/X/SiO_2$. Variations in the structure of C/SiO_2 or $C/X/SiO_2$ allow one to change significantly the interaction of their surfaces with water (as well as other polar or nonpolar compounds) adsorbed from air or in the aqueous suspensions.

Thus, the choice of oxide matrices, organic precursors, and techniques of synthesis of both oxides and carbon deposits allow one to change dramatically the structural and adsorptive characteristics of hybrid carbon–mineral adsorbents as the whole. Notice that there is a possibility to change substantially the structural properties of pyrocarbon deposits by their additional activation similar to that applied to pure carbons to prepare nanoporous adsorbents. The availability of nanoscaled pyrocarbon particles (whose size can also be reduced on the activation) allows one to synthesize the adsorbents with substantially easier accessible nanopores in comparison with pure nanoporous carbons consisting with larger particles. Besides, carbon–mineral adsorbents can be synthesized using the technique applied to prepare exfoliated graphite for activation of pyrocarbon deposits. It is also of interest to synthesize pyrocarbon deposits on silicas in relatively soft conditions using polymers (with OH or Cl groups well removed in the pyrolysis) immobilized on oxide particles, as the structure of the carbon phase in this case can differ markedly from that for CVD-carbon prepared utilizing small molecules (such as CH_2Cl_2). These hybrid materials can possess interesting and easily varied adsorptive properties.

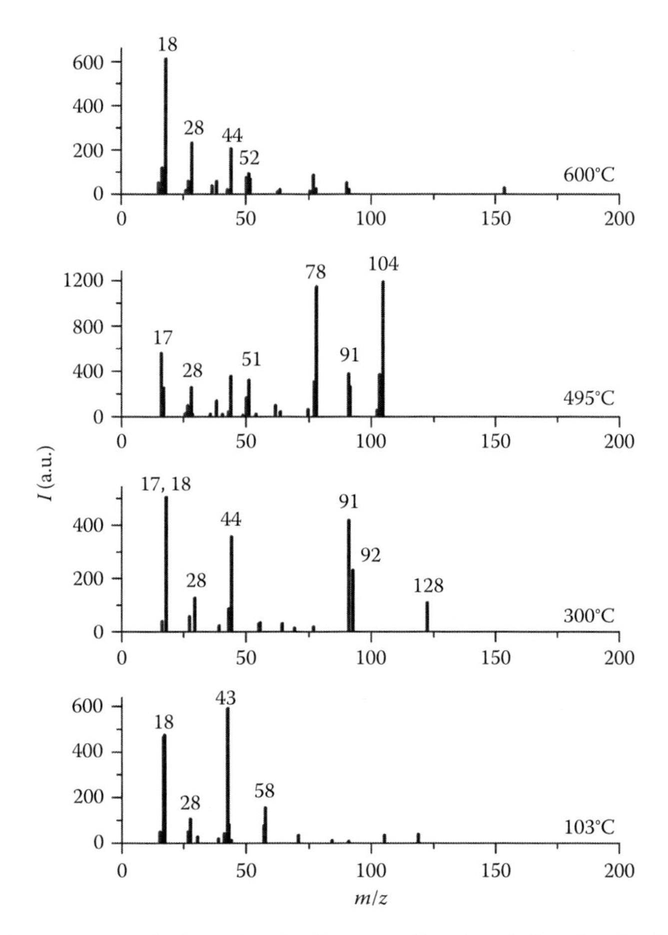

FIGURE 4.11 TPD mass spectra for heated carbosil prepared by phenylethanol carbonization at a silica gel surface. (Adapted from *Carbon*, 37, Pokrovskiy, V.A., Leboda, R., Turov, V.V., Charmas, B., and Ryczkowski, J., Temperature programmed desorption mass spectrometry of carbonized silica surface, 1039–1047, 1999, Copyright 1999, with permission from Elsevier.)

Pokrovskiy et al. (1999) studied the volatile products of thermal decomposition of the carbon-coated silica samples (silica gel coated by carbon deposits formed during carbonization of phenylethanol) using TPD mass spectrometry. The desorption mass spectra were recorded in the 10–220 Da range and the temperature step was about 10°C (Figure 4.11). The mass spectrum (representative one) corresponding to 103°C includes the lines at *m*/*z* 43 and 58 as the main components besides water (*m*/*z* 18 and 17). Their origin can be easily explained from the availability of adsorbed acetone used during sample pretreatment. Its TPD maximum lies at ca. 100°C. It should be noted that the lines corresponding to *N*,*N*-dimethylformamide used in parallel with acetone upon sample preparation were not observed in any spectrum.

The mass spectrum obtained at 300°C involves the lines at *m*/*z* 91, 92, and 122 caused by desorption of phenylethanol (this temperature corresponds to its maximal desorption). The main lines (*m*/*z* 51, 78, 92, 91, and 104) in the spectrum obtained at 495°C attest that phenylethylene and toluene desorb. The formation of phenylethylene is due to unimolecular decomposition of bound phenylethanol with H transfer from the CH_2 group (nearest to the aromatic ring) to O from a ≡SiOR group. The last mass spectrum, obtained at 606°C, includes the lines corresponding to benzene (*m*/*z* 78) and biphenyl (*m*/*z* 154). In as much as the existence of these molecules in the surface layer at such temperature is improbable, they can be formed due to migration of phenyl radicals along the

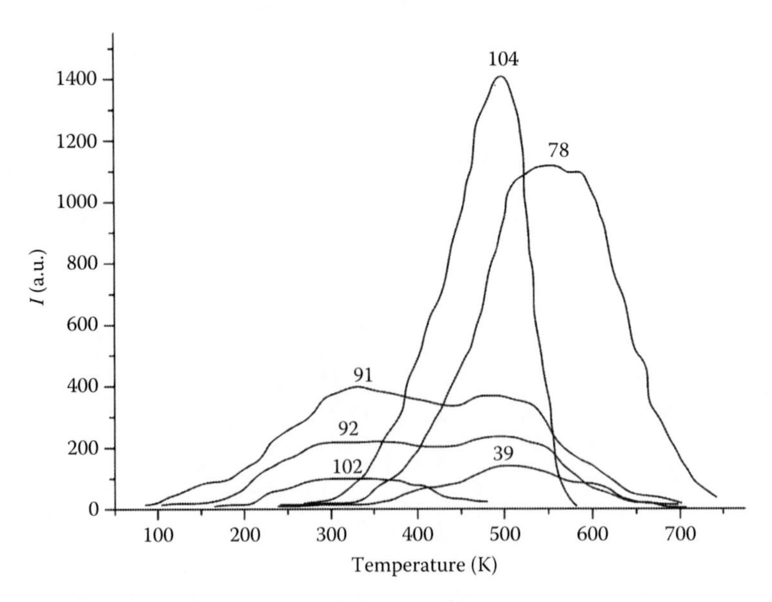

FIGURE 4.12 Thermograms of the deposition of organic substances at a heating of carbosils synthesized from phenylethanol. (Adapted from *Carbon*, 37, Pokrovskiy, V.A., Leboda, R., Turov, V.V., Charmas, B., and Ryczkowski, J., Temperature programmed desorption mass spectrometry of carbonized silica surface, 1039–1047, 1999, Copyright 1999, with permission from Elsevier.)

surface and their interaction with $H^{\bullet}$ or bound H (benzene formation) or bound C_6H_5 or free $C_6H_5^{\bullet}$ (biphenyl formation). At temperatures >600°C, the thermal transformations of bound organics have come to the end and only background lines are observed.

The temperature dependencies of the main components of the mass spectra observed are shown in Figure 4.12.

Phenylethanol desorption occurs in the 200°C–450°C range, and it is maximal at ca. 300°C. For phenylethylene, the temperature interval width of desorption is the same but it shifts to higher temperatures (300°C–550°C) with a maximum at 450°C. The thermogram of phenylethylene (*m*/*z* 104) desorption has an asymmetrical shape, which is typical for first-order reactions. The thermogram of *m*/*z* 91 has two maxima at 300°C and 470°C that can be explained from the origin of the fragment $C_6H_5CH_2^+$ (responsible for *m*/*z* 91) observed in the electron impact mass spectra for phenylethanol (*m*/*z* 122), phenylethylene (*m*/*z* 104), and toluene (*m*/*z* 92); therewith, the first maximum of *m*/*z* 91 corresponds to phenylethanol desorption. A benzene spectrum (*m*/*z* 78) is observed in the 300°C–700°C range and its shape indicates that $C_6H_5^+$ ion is formed through two mechanisms. At relatively low temperatures (300°C–500°C), this ion is generated as a fragment of phenylethylene or phenylethanol and, at high temperatures (500°C–700°C), it is obtained via ionization of benzene eliminated (Figure 4.12).

In the IR spectra of CS studied by Pokrovskiy et al. (1999), a narrow band at about 3727 cm^{-1} corresponds to free surface hydroxyl groups (isolated Si–OH) and broad band in the range 3700–3000 cm^{-1} (including resolved vibration at 3661, 3638, and 3624 cm^{-1}), as well as the band at about 1629 cm^{-1}, is due to surface hydroxyl groups bound with physically adsorbed water molecules. In turn, bands at about 2922 and 2951 cm^{-1} are usually attributed to asymmetric stretching vibration of C–H bonds in methyl (2920 cm^{-1}) and methylene (2955 cm^{-1}) groups. The two additional bands at 1989 and 1872 cm^{-1} can be attributed to C–H stretching of vinyl compounds or attributed benzenes. However, there are no bands at about 1600, 1650, and 3060 cm^{-1}, and this could be taken as evidence of absence of polycyclic aromatic compounds in the prepared sample. The band at 1456 cm^{-1} can be assigned to the asymmetric deformation mode of

CH_2 groups (Pokrovskiy et al. 1999). The earlier data support the conclusion that small amount of polymeric carbon chains may be presented on the surface of CS. This residual functionalization of the carbon deposits can affect the interactions with water and other adsorbates.

4.2.1 Structure of Adsorption Sites on CS Composed of Nonporous Nanoparticles

The surface structure of CS obtained as a result of methylene chloride pyrolysis at a surface of fumed silica (A-300, $S_{BET}=295$ m²/g) at 500°C was analyzed by Turov et al. (1995b) using benzene and water as adsorbates. In the process of carbonization, the specific surface area decreased to 260 m²/g. However, the carbon content in CS was not high (6.6 wt%). The ^{1}H NMR signals of benzene adsorbed onto a CS surface were with 1.5 ppm up-field shift and its intensity decreases with decreasing temperature from 280 to 230 K (Figure 4.13a). This up-field shift can be caused by interaction of benzene with the carbon deposits. The signal width increased with lowering temperature because of the decreased mobility of the adsorbed molecules.

For water adsorbed on a CS surface (Figure 4.13b) in the temperature range $268<T<273$ K, a single signal is observed at $\delta_H \approx 5.6$ ppm (i.e., the up-field shift observed for benzene is absent). At $T<268$ K, the signal splits into two signals of different intensity: signals 1 and 2 at $\delta_H=7$ and 4.5 ppm, respectively. When the temperature decreases the intensity of signal 1 decreases, but the signal 2 is practically unchanged. The chemical shift of signal 2 practically does not depend on the temperature and for the signal 1 only a slight down-field shift is observed.

NMR signal of benzene can be interpreted as follows: signal at $\delta_H=5.6$–7.1 ppm (Figure 4.13a) is caused by molecules adsorbed on basal graphite planes in broad pores and at accessible patches of the silica surface. A weak shielding effect of the carbon surface is determined by a large number of structural defects and the absence of nanopores in carbon particles. The average signal of molecules (located in broad pores) adjacent to the surface and distance from it is observed. Oxygen-containing surface groups and molecules of adsorbed oxygen do not contribute to the chemical shift of benzene molecules. Benzene adsorption on rough edges of the carbon sheet network is likely to be neglected for CS (as well as in the case of carbon blacks discussed earlier).

In the case of water, the main signal demonstrated rather down-field shift in comparison with signal of bulk water in the liquid state having $\delta_H \approx 4.5$ ppm. Approximate values of the chemical shifts ($\delta_H \approx 7$ ppm) were obtained for signal of water bonded with carboxylic and carbonyl groups of graphite oxide particles (Turov et al. 1991a–c). On the other hand, for carbon adsorbents the up-field

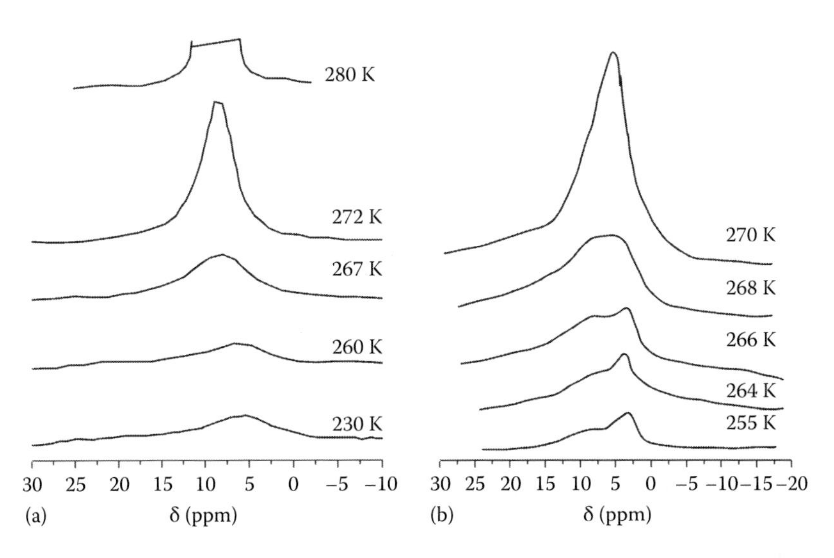

FIGURE 4.13 ^{1}H NMR spectra of benzene (a) and water (b) bound to carbosil based on A-300 ($C_C=6.6$ wt%).

shift of water molecules may be about −5 ppm or greater for molecules adsorbed in slit-shaped nanopores formed by ordered graphene planes. Thus, the main signal can be referred to water molecules interacting with the carbon surface through hydrogen-bonded complexes with electron-donor surface functionalities. Because of the large number of oxidized groups, a thin water layer built up the network of hydrogen bonds is formed on the interface boundaries.

For CS surface no water signal occurs with the up-down shift, but a relative narrower signal at $\delta_H = 4.5$ ppm is observed (Figure 4.13b). This value corresponds to water adsorbed onto nanosilica alone. The fact that widths of signals 1 and 2 essentially differ and that their overlapping is observed only close to the freezing temperature of water allow us to conclude that the adsorption centers responsible for these signals are spatially separated. With increasing temperature, the volume of unfrozen phase increases and the molecular exchange between various types of adsorption complexes becomes faster.

The earlier conclusions agree with the data from topographic studies of carbonized silica surface by microscopic methods (Gierak and Leboda 1989). The studies carried out succeeded in showing that at low carbonization of the surface, carbon occurs in the form of weakly developed carbon globules which form a mosaic structure on the silica surface.

Two types of CS with very low and large carbon contents were synthesized by pyrolysis of methylene chloride at a surface of nanosilica ($S_{BET} \approx 300$ m^2/g) at the surface concentration of carbon of 0.5 and 40 wt% for CS-1 and CS-2, respectively (Turov et al. 1997a). In order to elucidate the specificity of interactions of adsorbate molecules with silica and carbon deposits, the adsorption of water and benzene was studied (Figure 4.14). In the case of benzene, the spectra for both CS consist of two up-field shift signals, and for more intense signal for CS-1 $\Delta\delta_H = -2.7$ ppm, while for CS-2 $\Delta\delta_H = -2.3$ ppm. The second signal has similar values for both samples ($\Delta\delta_H = -8$ ppm). Clearly, the

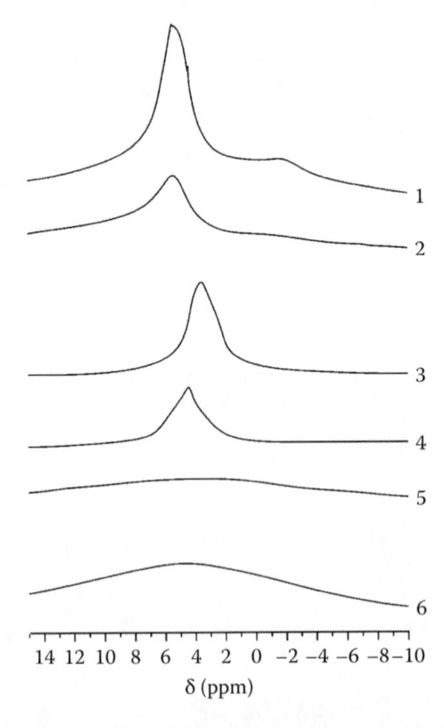

FIGURE 4.14 ^{1}H NMR spectra for benzene (1, 2) and water (3–6) adsorbed on the surface of nonporous carbosils (CS-1—1, 3, 6 and CS-2—2, 4, 5) in air (5, 6) or on $CDCl_3$ (3, 4). (From Turov, V.V., Leboda, R., Bogillo, V.I., and Skubiszewska-Zięba, J., Effect of carbon deposition on the surface of aerosil on the structure of adsorption sites and water and benzene interlayers – ^{1}H NMR spectroscopy study, *J. Chem. Soc. Faraday Trans.*, 93, 4047–4053, 1997a. Copyright 1997. Adapted by permission of Royal Society of Chemistry.)

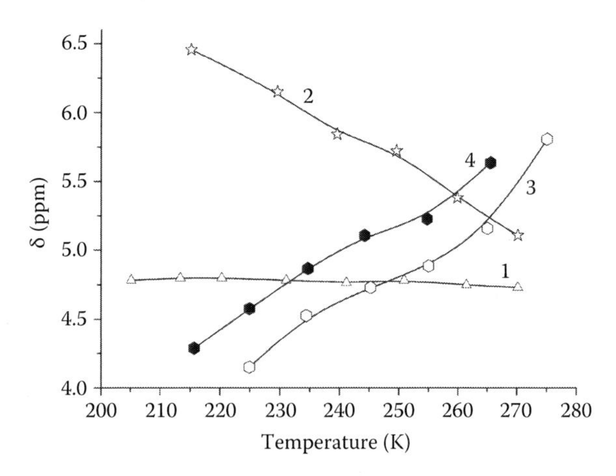

FIGURE 4.15 Temperature dependence of variations in the chemical shift for water (1, 2) and benzene (3, 4) adsorbed on the surface of CS-1 (1, 3) and CS-2 (2, 4). (From Turov, V.V., Leboda, R., Bogillo, V.I., and Skubiszewska-Zięba, J., Effect of carbon deposition on the surface of aerosil on the structure of adsorption sites and water and benzene interlayers–^{1}H NMR spectroscopy study, *J. Chem. Soc. Faraday Trans.*, 93, 4047–4053, 1997a. Copyright 1997. Adapted by permission of Royal Society of Chemistry.)

shape of the ^{1}H NMR spectra of water bound to the surface depends on conditions of experiments. The signal width (Figure 4.14) is larger by an order of magnitude as compared to that for water at the CS/chloroform interfaces. As the chemical shift of liquid water is 4.5 ppm, it may be concluded that values of $\Delta\delta_H$ differing from 0 are observed only in the case of water adsorbed on CS-2 (with large carbon content) in the presence of chloroform.

Figure 4.15 shows the temperature dependences of chemical shifts for benzene (3, 4) and water (1, 2) bound to CS-1 (1, 3) or CS-2 (2, 4). With increasing temperature, the chemical shift of adsorbed benzene molecules increases monotonously from 4.2–4.8 ppm at 220–230 K to 5.7 ppm at 270–279 K. The δ_H value for water bound to the CS-1 surface practically does not depend on the temperature, whereas for CS-2 a monotonous decrease in the δ_H value of bound water is observed as the temperature increases.

In accordance with the data presented in Figures 4.14 and 4.15, it is possible to draw a conclusion that irrespective of the carbon content the CS samples have sites of at least two types toward benzene adsorption. On the basis of the estimated $\Delta\delta$ values, they may be assigned to the basal graphite planes (where $\Delta\delta_H$ varies from −2 to −3 ppm) and slit-shaped pores between these planes ($\Delta\delta_H = -8$ ppm). With decreasing temperature, the $\Delta\delta_H$ value for both CS increases because the layer of unfrozen benzene becomes thinner. As the ^{1}H NMR signal for adsorbed benzene is an averaged one for the molecules contacting with the surface as well as for those remote from it at a distance of several molecular diameters, the $\Delta\delta_H$ value obtained in the experiments on bulk freezing is slightly reduced due to the shielding effects of the ring currents of the graphite planes having a maximum value for the first layer of adsorbed molecules. The variations in the thickness of the adsorption layer may be regarded to be responsible for the changes in the $\Delta\delta_H$ values of benzene molecules bound to the surface of CS-1 and CS-2 samples, i.e., the higher degree of the surface carbonization results in the thicker layers of bound benzene.

The difference in the character of the temperature dependences of chemical shifts for water adsorbed on the surface of CS-1 and CS-2 samples allows one to conclude that at different degrees of surface coverage with carbon water molecules interact with different oxygen-containing surface functionalities. At a low degree of coverage, the surface has extended carbon-free fragments, where aqueous structures (SAW) analogous to water structures bound to the pure nanosilica. Therefore, the chemical shift of water bound to the surface depends to a small extent on the temperature and is in agreement with that for liquid water (Figure 4.15, straight line 1). For the CS-2 sample (high degree of surface coverage by carbon), the chemical shift for adsorbed water is noticeably higher than that for water on pure silica or for liquid water. It gives evidence for the increase in the contribution of hydrogen-bonded complexes (where water molecules interact with stronger electron-donor sites [e.g., COOH] than those in water) to the total signal (Drago et al. 1970). In the case of carbon surfaces, such sites are usually oxidized carbon atoms (Karpenko et al. 1990a,b). Then, the increase in the chemical shift of adsorbed water on passing from CS-1 to CS-2 may be explained by the fact that water interacts predominantly with O-containing functionalities at the surface.

4.2.2 Ordered Layers of Water at a Surface of CS Composed of Nonporous Nanoparticles

Valuable information about processes at the water/adsorbent interfaces can be furnished by measurements of the thickness of a layer of bound water and the value of free surface energy of an adsorbent in an aqueous medium. As it was shown earlier, these characteristics can be obtained by measuring dependences of the signal intensity for unfrozen (melted) water on temperature. However, while for porous carbons such dependences are limited by the adsorbent pore size, in the case of nonporous materials it is possible to investigate the whole layer of water subjected to the disturbing action of the surface (in approximation of a smooth surface).

Direct experimental measurements and theoretical calculations show that at a distance of 5–6 molecular diameters the surface forces in the direction perpendicular to the flat surface show a tendency to change their sign (attraction to repulsion) (Israelachvili and Wennerstrom 1990, Zhu and Robinson 1991, Attard and Parker 1992, Delville 1993). This is true for both the hydrophobic and hydrophilic surfaces (Israelachvili 1994). As the surface of many industrial adsorbents is not flat and, additionally, it is influenced by the temperature changes, the averaged surface forces are repulsive forces decaying in accordance with the exponential law (Marrink et al. 1993). The general expression for the surface potential includes also the contribution due to the local polarization of the surface (Derjaguin et al. 1987). The simulations (Zhu and Robinson 1991, Marrink et al. 1993) showed that the surface strongly affects a liquid layer of thickness <3 nm. However, Derjaguin et al. (1987) showed variations in the properties of water at larger distances from the surface. At the same time, such parameters of liquids as density, viscosity, and dissolubility undergo variations near the liquid–solid interfaces (Chaplin 2011).

The thickness of the adsorbed water layer in nanometers or the number of statistical water monolayers can be easily derived from the ^{1}H NMR data (Figure 4.16) taking into account that a single monolayer corresponds to $C_{uw} = 18$ mmol/m^2 and its thickness is ≅0.3 nm. Two segments can be distinguished in the $\Delta G(d)$ dependence (Figure 4.16a and b, curve 3), namely the segment corresponding to a rapid decrease in the thickness of the layer of unfrozen water in a narrow range of ΔG changes (at temperatures close to 273 K) and the second corresponding to a relatively small decrease in d in a wide range of ΔG changes. The appearance of the former is attributed to a certain part of weakly bound water (frequently it is SAW). The second segment corresponds to strongly bound water located more closely to the surface (it can be both WAW and SAW depending on the surface structure and water content). The layer thickness values for each form of water (d_s and d_w for SBW and WBW, respectively) and the maximum values of the decrease in the free energy of water due to adsorption (ΔG_s and ΔG_w) are shown in Table 4.6.

From the data summarized in Figure 4.16 and Table 4.6, it can be concluded that for nonporous CS nanoparticles there is a simple correlation between the amounts of carbon deposits and the characteristics of bound water. Similar but nonlinear dependences were found for mixed oxides which were explained by nonlinear changes in the nonuniformity of the materials with active phase content which increases (γ_S grows) and then decreases (γ_S drops down). CS-1 is much more nonuniform (because of a great number of small carbon particles) than initial silica or CS-2 (a smaller number of large carbon particles deposited onto silica). Thus, in the case of the sample whose surface contains 0.5 wt% of carbon very large values for the thickness of bound water layers

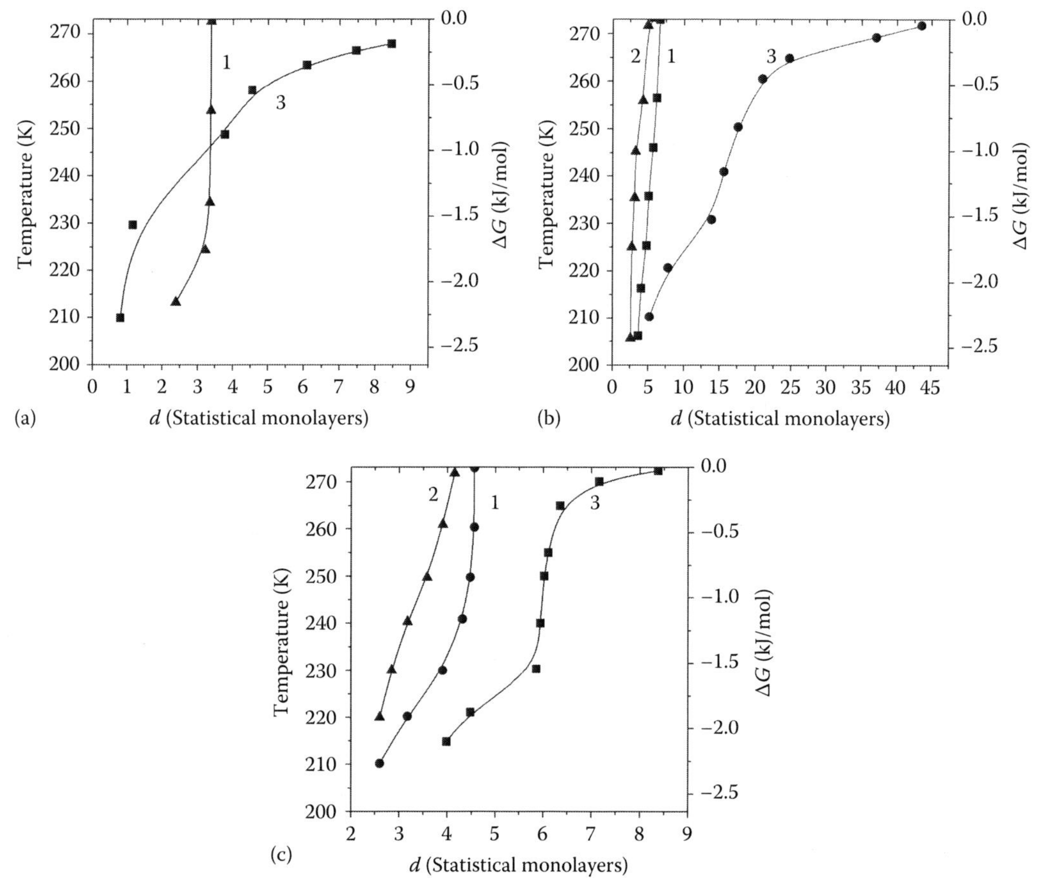

FIGURE 4.16 Variation in the free energy of water as a function of the thickness of nonfreezing water layers for frozen aqueous suspension of nanosilica A-300 (a), CS-1 (b), and CS-2 (c) in air (1); $CDCl_3$ (2); or water (3).

TABLE 4.6
Characteristics of Water Layers Adsorbed on the Surface of Carbosils and Pure Nanosilica A-300

Adsorbent	d_s	d_w	$-\Delta G_s$ (kJ/mol)	$-\Delta G_w$ (kJ/mol)	γ_S (mJ/m²)
CS-1	25	25	3.0	0.6	820
CS-2	7	3	3.0	0.8	184
Nanosilica	4	6	3.5	0.6	155

Note: d_s and d_w are the numbers of statistical monolayers of SBW and WBW, respectively.

and for free surface energy of adsorbents have been found. The structural features of CS can be estimated using NMR cryoporometry (Figure 4.17).

CS-2 seems more densely packed since contribution of narrow nanopores increases but for CS-1 contribution of mesopores is larger than that for CS-2. The distributions for CS-1 and CS-2 are broader than for silica alone that is due to more nonuniform texture and surface chemistry of the CS samples.

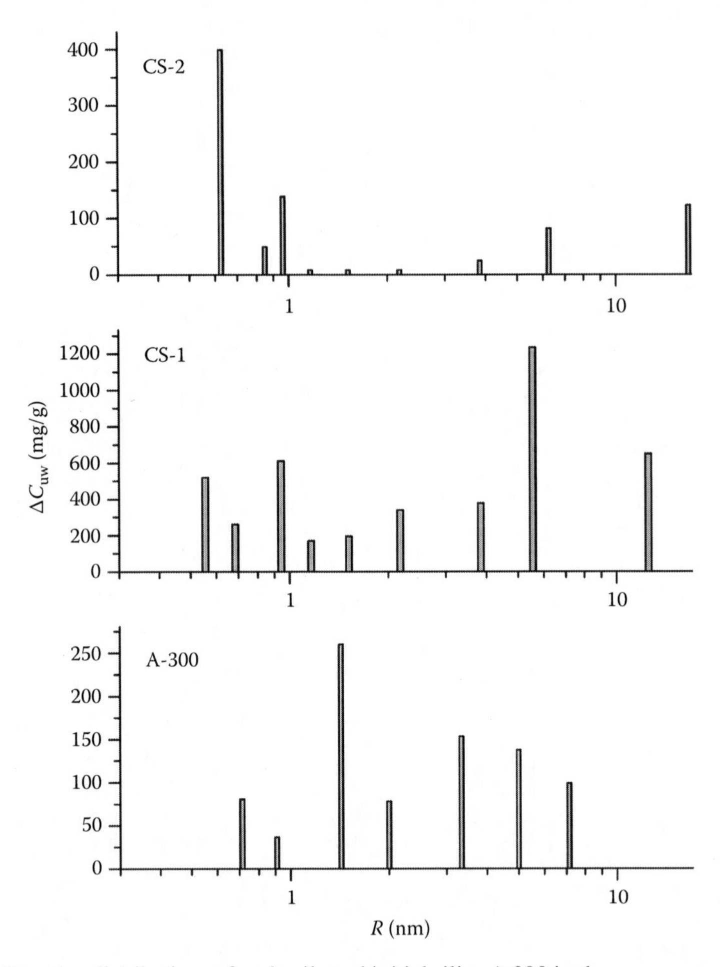

FIGURE 4.17 Pore size distributions of carbosils and initial silica A-300 in the aqueous suspensions.

Dukhin and Lyklema (1987) and Dukhin (1993) showed that if oppositely charged portions are on a surface of disperse particles, then a long-range component of the surface forces appears in the surrounding space due to the electrostatic field directed along the surface of the particles. This field affects the orientation of dipoles of water molecules around the disperse particles. The radius of action for the surface forces of this type is governed by the value of the electrical charges and distance between them. Since in an aqueous medium silica portions of the CS surface can acquire a negative charge (at the expense of a partial dissociation of surface hydroxyl groups) while carbon portions can acquire the positive charge (due to adsorption of solvated protons formed as a result of the dissociation of the electron-donating sites of the carbon portions of the surface or on the basal planes), it can be assumed that in this situation silica and carbon components of the surface carry charges of the opposite signs. The possibility of formation of a negative charge at the expense of adsorption of protons on basal planes of the carbon surface containing and not containing oxygen was considered in detail in Leon y Leon et al. (1992). This gives rise to polarization of adsorbent particles and appearance of a long-range component of the surface forces. The results achieved are in good agreement with the data gathered for other nanomaterials (Gun'ko et al. 2009a). In all the cases, only the surface polarization is a main factor that affects the thickness of adsorbed water layers. The increase in the carbon content to 40 wt% brings about a decrease in the free surface energy of the adsorbent, but in spite of this it remains significantly higher than that observed for the parent silica. It should be noted that in the case when the ΔG_S values are determined only by the hydrophilic properties of the surface, one can expect that the values of the free surface energy obtained for the CS should be lower than those obtained for the silica. In this connection in the case of nonporous adsorbents, the dominant role in the formation of the structurized (clustered) interfacial water layers relates to the tendency of the adsorbent surface to undergo polarization in the aqueous medium.

The surface of CS (C_C=5–13 wt%) was studied by the electron microscopy technique (Leboda 1987). In that case on the surface there were both portions of the parent silica and portions of silica covered by a carbon layer, but any uniform coverage of the starting surface by a carbon layer was not observed. This is typical picture for any silica substrates (Gun'ko and Leboda 2002). With increasing concentration of carbon, the area of the silica component of the surface noticeably decreases and at the carbon concentration equal to 13 wt% it became relatively small. It can be anticipated that at the carbon concentration of 40 wt% (CS-2) practically all the surface is covered by a carbon layer. So, the surface of the CS-2 sample may be more homogeneous than that of the CS-1 sample and it can be concluded that the tendency of the adsorbent surface to undergo polarization increases as a function of the surface heterogeneity of a given material.

4.2.3 Influence of the Degree of Carbonization on Hydration Parameters of Modified Porous Silica

Silica gel Si-60 (Schuchardt München, Germany), used as the initial material for CS production (Leboda et al. 2000b), was washed with an 18% HCl solution in a Soxhlet apparatus and then washed with distilled water to remove mineral impurities. The 0.15–0.25 mm particle size fraction was used. The silica gel samples were dried at 200°C for 24 h and cooled in a desiccator before the next stages of preparation. Carbon-coated silica gels were prepared as follows: 10 g of silica gel was placed in a quartz rotational reactor situated in an electric furnace. A portion of silica gel on the bottom of the reactor was conditioned at 550°C in a stream of deoxidized nitrogen for 1 h. Then the sample was subjected to the action of CH_2Cl_2 fed into the reactor (through a heated glass evaporator at 100°C) using a Masterflex (Cole Parmer) pump and pyrolyzed at 550°C. The rate of CH_2Cl_2 feeding was 0.6 cm^3/min for 0.5, 1, 2, 3, 4, and 6 h, corresponding to different synthesized carbosils CSG-*i* (Table 4.7).

The adsorption/desorption isotherms show a marked decrease in nitrogen adsorption due to grafted carbon deposits. As C_C increases the specific surface area, S_{BET}, mesopore surface, S_{meso}, pore volumes, V_p and V_{meso}, and pore radii, R_p (Table 4.7) decrease nearly linearly. Initial Si-60 and CSG are mesoporous materials (Figure 4.18 and Table 4.7). The deviations (Δw) of the pore model (cylindrical pores in silica gel and voids between spherical carbon or silica nanoparticles with the SCR procedure) are relatively small (4%–13%).

Figure 4.19 presents the temperature dependences of the shape of ^{1}H NMR spectra of water bound to CSG-1, CSG-5, and CSG-6 being in $CDCl_3$ medium. Figure 4.19d shows the analogous spectra

TABLE 4.7
Structural and Adsorption Characteristics of Carbosils (CSG) Based on Si-60

Adsorbent	C_C (wt%)	S_{BET} (m^2/g)	S_{meso} (m^2/g)	V_p (cm^3/g)	V_{meso} (cm^3/g)	V_{macro} (cm^3/g)	R_p (nm)	Δw
Si-60	0	376	376	0.798	0.798	0	4.3	0.038
CSG-1	0.8	373	372	0.751	0.737	0.014	4.0	−0.073
CSG-2	4.4	339	339	0.678	0.671	0.007	4.0	−0.086
CSG-3	14.9	299	297	0.566	0.565	0.001	3.8	−0.044
CSG-4	20.3	262	261	0.474	0.474	0.0	3.6	−0.061
CSG-5	26.7	222	219	0.392	0.390	0.001	3.5	−0.082
CSG-6	35.0	163	161	0.278	0.276	0.001	3.4	−0.133

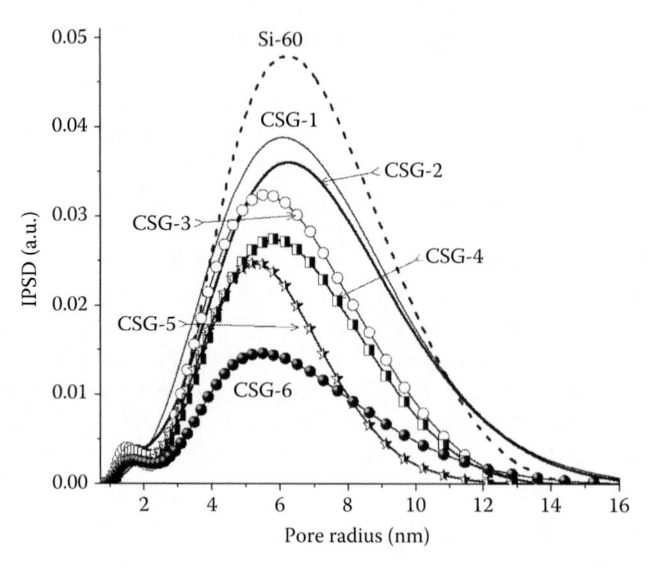

FIGURE 4.18 Pore size distributions of carbosils CSG and initial silica gel Si-60 (MND with CV/SCR model).

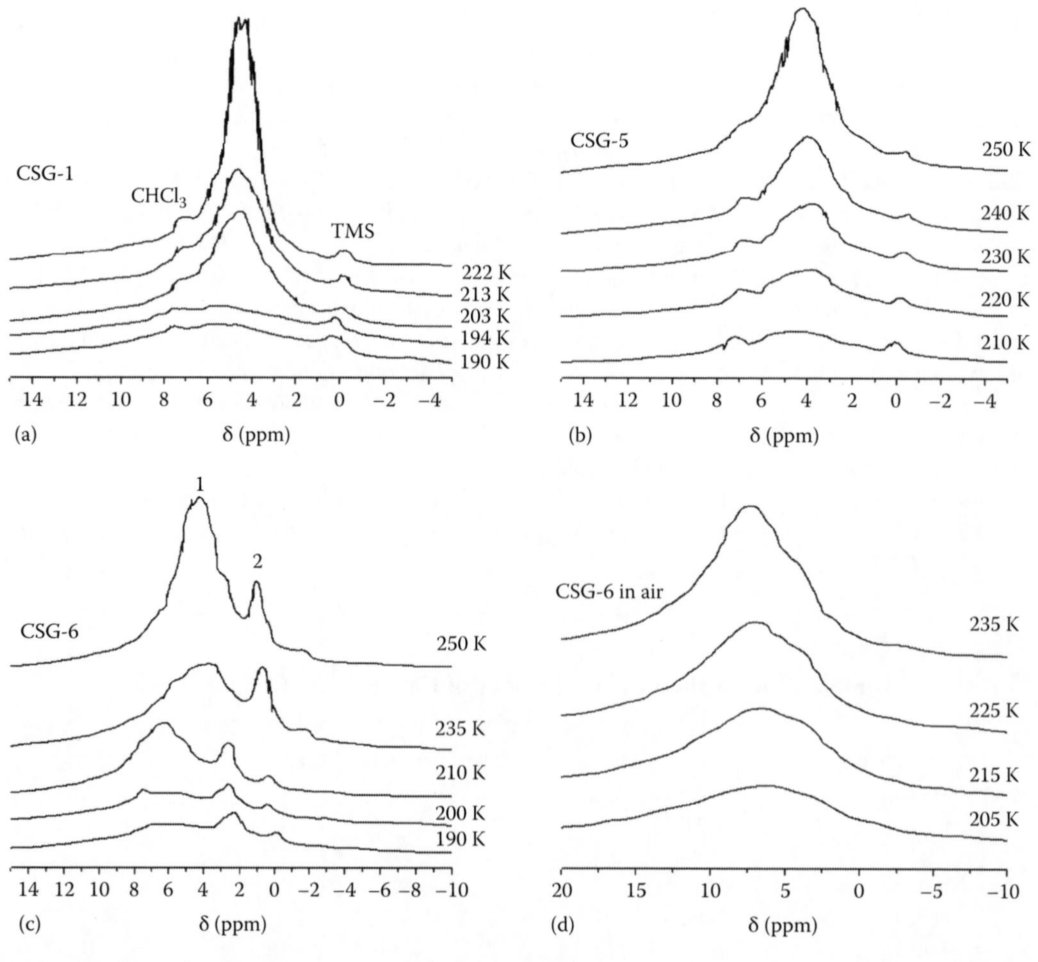

FIGURE 4.19 Temperature-dependent variations in the shape of ^1H NMR spectra for water adsorbed on the surface of mesoporous carbonized silicas (a) CSG-1, (b) CSG-5, and (c, d) CSG-6 in (a-c) chloroform-d and (d) air.

of water bound to CSG-6 in air. The concentration of the adsorbed water was 40–60 mg/g. In the case of CSG-2 and CSG-4, the spectra differ insignificantly from those presented in Figure 4.19b, and that is why they are not included here. In the spectra recorded for chloroform medium, there are not only signals for adsorbed water molecule protons but also weak signals of TMS and chloroform whose chemical shifts are 0 and 7.26 ppm, respectively. With decreasing temperature, the width of the adsorbed water signal increases and its intensity decreases that result from the decrease in the molecular mobility and partial freezing of the adsorbed water.

The chemical shift of the adsorbed water signal (signal 1) for different adsorbents falls within a range of $4.3 < \delta_H < 5.5$ ppm corresponding to SAW, so it is close to the chemical shift of liquid water. The spectrum for the CSG-6 sample contains the main signal of the adsorbed water (signal 1) as well as a weaker signal characterized by the chemical shift of 2.1 ppm (signal 2) (Figure 4.19). With decreasing temperature, the intensity of signal 2 decreases but to a smaller extent than that of the main signal, which provides evidence for strong adsorption interactions for water responsible for signal 2. Both these signals can be seen as separate only in a deuterochloroform medium. In the case of the air medium, the spectrum displays a variation in the shape of the adsorbed water signal, which points to the presence of another signal. However, recording of these signals as separate is not possible due to their great widths.

The presence of two signals of water adsorbed on the surface of the CSG-6 sample in the ^{1}H NMR spectra corroborates the hypothesis that there are two kinds of active sites on the surface, and in view of the timescale of NMR, the molecular exchange between these centers is a slow process. Then, a question concerning the character of active sites responsible for the appearance of these signals arises. As the carbon concentration in the sample reaches 35 wt%, it can be assumed that the whole surface of the initial silica is covered by carbon deposits and both signals correspond to the water adsorbed on the carbonized part of the surface. The up-field shift of the signal of the water adsorbed on the surface can be explained by two factors, namely by the influence of local magnetic anisotropy areas of the condensed aromatic systems (Chapter 3) or by the changes in the structure of the hydrogen-bonded complexes forming a hydrated shells of nanoparticles of the CSG-6 sample. The influence of the complex structure on the chemical shift is due to the fact that the chemical shift of the water is determined not only by the hydrogen bond strength but also by the average number (m) of hydrogen bonds per one water molecule.

To ascertain which of the earlier-mentioned mechanisms is responsible for the appearance of signal 2, we studied adsorption of benzene and methane molecules on the dehydrated surfaces of carbonized silicas (Figure 4.20). This figure also shows the spectra for benzene and methane adsorbed on the surface of the initial silica gel. It was assumed that these adsorbates in contrast to water molecules could adsorb on the surface fragments with the polyaromatic structure but they could not form the hydrogen-bonded complexes with electron-donor and proton-donor surface groups.

From Figure 4.20, it follows that for all the studied CS the chemical shift of both benzene and methane is close to the chemical shift of these adsorbates in the case of the initial silica gel.

But on the CS surfaces there are more adsorbed benzene molecules than on the silica surface, which is indicated by the differences in the signal intensity for benzene adsorbed on the Si-60 and CSG samples under the same conditions. However, methane adsorbs practically in the same way on both kinds of adsorbents. The signal intensity of the adsorbed methane is much smaller than the signal intensity of benzene adsorbed on the CS surfaces and is comparable with the signal intensity of benzene on the silica gel. The results obtained can be explained by the fact that methane can interact with the surface only through the dispersion mechanism, while benzene can be bound more specifically to the surface fragments formed by the condensed aromatic systems. The fact that the CSG-6 sample lacks signal 2 of both the adsorbates gives ground to conclude that the appearance of this signal in the case of adsorption of water molecules on the adsorbent CSG-6 is due to formation of small water clusters on this adsorbent. As carbon layer formation takes place in the silica gel pores, it can be assumed that signal 2 is attributed to water in the narrowest pores where water clusters of large sizes cannot be formed.

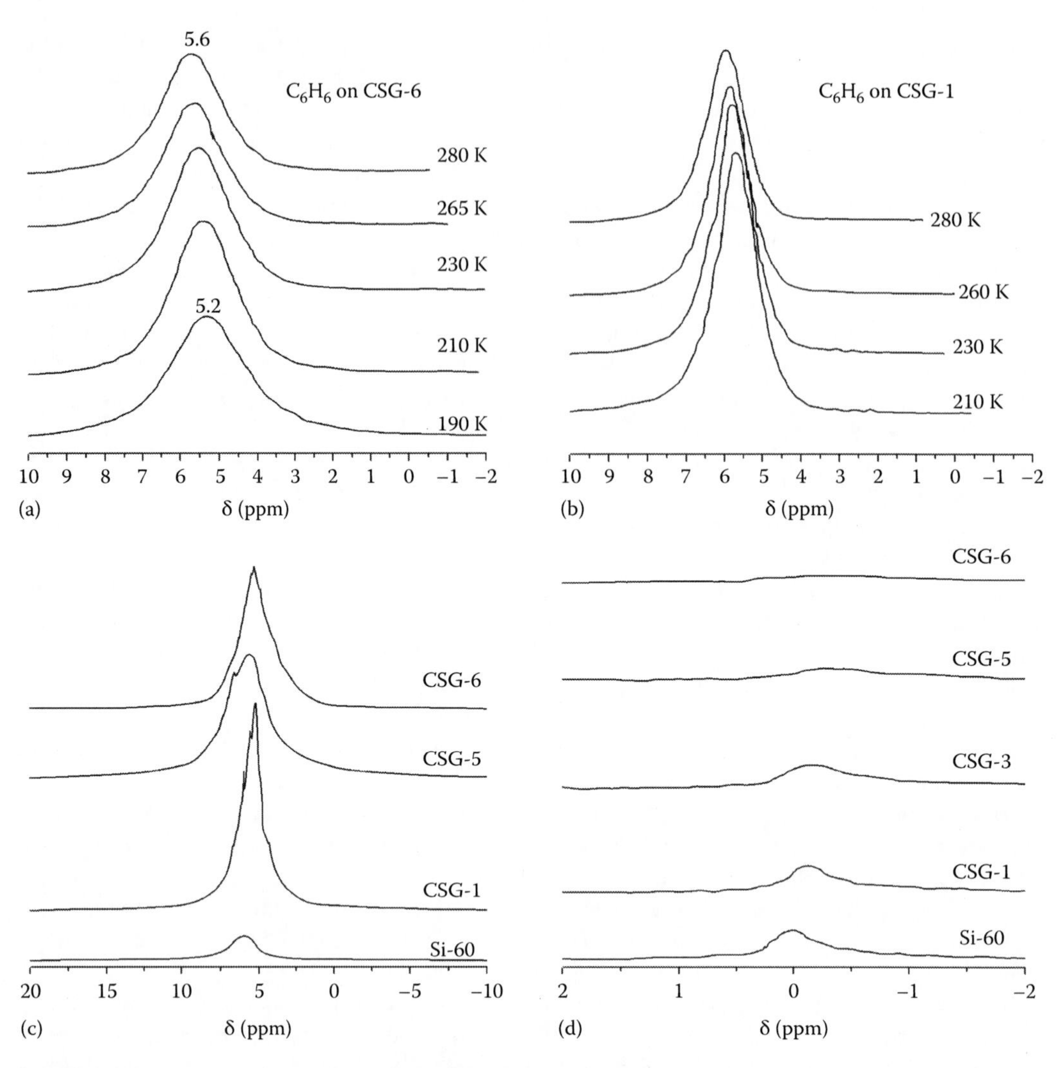

FIGURE 4.20 ^{1}H NMR spectra of (a–c) benzene and (d) methane adsorbed on the surface of the porous carbosils differing in their surface carbon contents at 293 K (c, d).

The $\Delta G(C_{uw})$ graphs (Figure 4.21) have two fragments, and the first one shows quick changes in the thickness of the unfrozen water layer with small changes in the ΔG value corresponding to weakly bound water. Another fragment is characterized by small changes in the thickness of the unfrozen water layer over a wide range of ΔG changes, and this fragment refers to strongly bound water. The thickness of the interfacial water layer (C_{uw}^{w} for weakly bound water and C_{uw}^{s} for strongly bound water) as well as the maximal changes in the free energy of water caused by adsorption (ΔG_w and ΔG_s, respectively) can be determined using extrapolation of the corresponding fragments of $\Delta G(C_{uw})$ to the axes. The characteristics of layers of water adsorbed on the CS surfaces and initial silica gel obtained from NMR data (Figures 4.20 and 4.21) are summarized in Table 4.8.

It should be noted that for porous materials the thickness of an adsorbed layer confined in pores cannot be greater than the radius of the adsorbent pores. Therefore, as the pyrolysis time increases, the total pore volume decreases, which results from carbon depositing in the pores, and so there is a tendency for the adsorbed water concentration to decrease when going from silica gel to CS with increasing C_C value. If the degree of carbonization is small, then C_{uw}^{w} and G can be determined sufficiently accurately, but in the case of other samples they can be estimated only approximately. The appearance of inflections

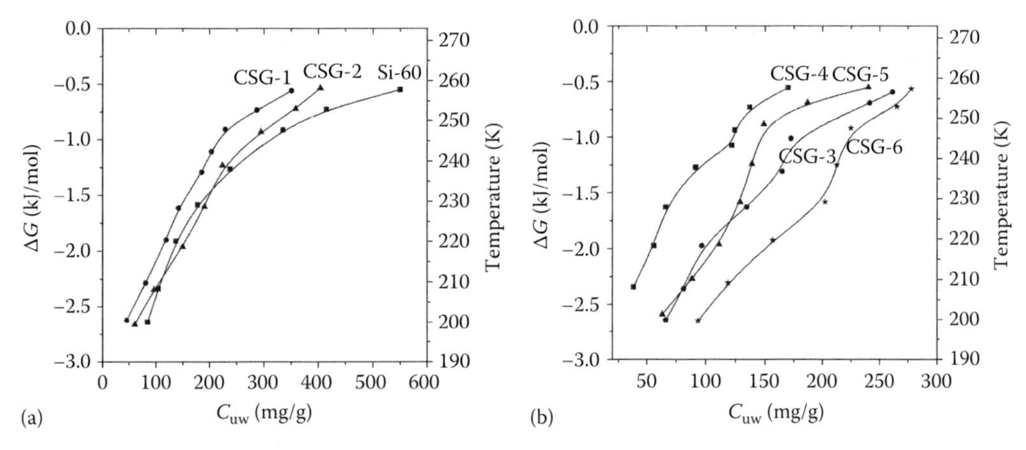

FIGURE 4.21 Relationships between changes in the Gibbs free energy (or melting temperature) of water bound to silica gel and CSG and the amounts of unfrozen water.

TABLE 4.8
^{1}H NMR Data about Layers of Water Adsorbed on the Surface of Mesoporous Carbosils

Adsorbent	$-\Delta G_s$ (kJ/mol)	C_{uw}^{s} (mg/g)	$-\Delta G_w$ (kJ/mol)	C_{uw}^{w} (mg/g)	γ_S (mJ/m^2)
Si-60	3.5	270	1.2	730	150
CSG-1	3.0	320	1.5	230	110
CSG-2	3.2	375	2.2	155	132
CSG-3	4.0	175	1.7	225	109
CSG-4	3.0	150	1.4	125	72
CSG-5	3.5	220	1.2	250	139
CSG-6	3.7	300	1.8	100	240

on the curve for the $\Delta G(C_{uw})$ dependence can be caused by the adsorbent surface heterogeneity and limited pore volume. In the case of heterogeneous surfaces, different kinds of radial functions describing variations in the free energy of adsorbed water correspond to different surface fragments. When considering the total surface of a sample, the function obtained for the corresponding adsorbent is averaged.

Hydration properties of the adsorbent surface at the adsorbent/water interfaces can be most completely expressed in terms of the free surface energy. From the data in Table 4.8, it follows that in the case when the carbonization time does not exceed 3 h (Table 4.7) the increases in the carbon content are accompanied with decreases in the free surface energy. The minimum value of γ_S was recorded for sample CSG-4 ($\gamma_S = 72$ mJ/m^2). It is in agreement with the statement that the hydration properties of the surface are determined by the concentration of primary adsorption sites that can form hydrogen-bonded complexes with water molecules.

On the amorphous carbon surface, such water adsorption sites are oxidized carbon atoms. Their concentration is usually much lower than that of hydroxyl groups on the silica surface, which should diminish the hydrophilic properties of the silica gel surface during carbonization. In contrast to it, however, there is an increase in the adsorbent hydrophilic properties for the CSG-5 and CSG-6 samples, which manifests itself in the increase of the γ_S and C_{uw}^{w} values. In the case of the CSG-6 sample, γ_S is even higher than for the initial silica gel. The regularities of variations in the hydrophilic properties of the surface observed after carbonization can be explained if it is assumed that at the initial stage of carbonization the carbon surface is formed through development of a disordered carbon lattice. The diminution of the hydrophilic properties of the surface is brought about by decreasing concentration of surface hydroxyl groups accessible for interaction with water molecules. With increasing carbonization time, the thickness of the carbon layer developed increases, and, in addition, its order increases too. Since in this case oxidized carbon atoms on the carbon surface are predominantly situated on the boundaries of mutually overlapped graphene clusters, they are easily accessible for formation of hydrogen-bonded complexes with water molecules. At the same time, the accessibility of basal graphite planes for water molecules decreases. As a result, the hydrophilic characteristics of the surface become more marked.

It is known (Kiselev and Lygin 1975, Dabrowski and Tertykh 1996, Legrand 1998, Gun'ko et al. 2004a,c, 2009d) that both thermal and hydrothermal treatments (HTT) of silica gels or nanosilica can change their texture. HTT reduces the specific surface area and enhances the pore size of porous silicas depending on the treatment temperature and time. Clearly, marked structural alterations in the silica matrix can occur during pyrolysis of organics at 700–1000 K depending on the nature of reactants, especially if water is one of the products of carbonization. Additionally, HTT of carbon–silica gel composites can weakly affect the texture in comparison with HTT of pure silica gels (Table 4.9).

Mesoporous silica gel Si-60 (Merck, grain fraction 0.2–0.5 mm) (Table 4.9) was studied before and after heating in the nitrogen stream at 823 K for 6 h (labeled Si-60T). These samples were compared with CS prepared by pyrolysis of CH_2Cl_2 (purity 99%, flow 0.6 cm^3/min) at Si-60 (10 g) under dynamic conditions in a flow rotary quartz reactor with the deoxidized nitrogen stream (100 cm^3/min) at 823 K for 0.5, 1, 2, 3, 4, and 6 h (labeled CS-*i* at $i = 1$–6, Table 4.9) (Gun'ko et al. 2004d). The carbon content (C_C) in CS-*i* was 1.5, 2.3, 5.9, 10, 13, and 19.5 wt%, respectively. HTT of the samples (additionally labeled HTT, Table 4.9) was performed in the steam phase at 473 K for 6 h.

TABLE 4.9
Structural Parameters of Initial, Heated, and Hydrothermally Carbosil Samples (Carbonization of CH_2Cl_2)

Sample	C_C (wt%)	S_{BET} (m^2/g)	S_C (m^2/g)	S_{meso} (m^2/g)	V_p (cm^3/g)	V_{meso} (cm^3/g)	D_p (nm)	Δw
Si-60	—	447		447	0.821	0.820	7.3	0.170
Si-60T	—	472		472	0.867	0.859	7.4	0.230
Si-60-HTT	—	61		38	0.264	0.062	17.2	1.084
Si-60T-HTT	—	42		33	0.126	0.084	12.0	0.839
CS-1	1.5	431	152	431	0.783	0.779	7.3	0.219
CS-1-HTT		403		403	0.829	0.829	8.2	0.215
CS-2	2.3	458	158	458	0.811	0.808	7.1	0.215
CS-2-HTT		370		370	0.808	0.808	8.7	0.214
CS-3	5.9	445	109	445	0.780	0.779	7.0	0.219
CS-3-HTT		361		361	0.733	0.731	8.1	0.217
CS-4	10.0	427	86	427	0.737	0.735	6.9	0.227
CS-4-HTT		360		360	0.676	0.671	7.5	0.229
CS-5	13.0	399	99	399	0.676	0.675	6.8	0.228
CS-5-HTT		324		324	0.622	0.622	7.7	0.220
CS-6	19.5	354	70	354	0.587	0.585	6.6	0.231
CS-6-HTT		292		291	0.535	0.535	7.4	0.224

Notes: C_C is the carbon content; S_C denotes the specific surface area of carbon deposits (determined using the PNP adsorption); D_p is the average pore diameter.

The surface topography after heating of pure silica gel was akin to that of the pristine oxide. Nonuniformity of pyrocarbon deposits and whole CS particles is affected by carbonization process and changes with increasing carbon content C_C. The formation of large pyrocarbon particles occurs at $C_C > 10$ wt% (Gun'ko et al. 2004d). At low $C_C < 10$ wt%, the CS surfaces was texturally more uniform. Notice that the textural, energetic, and hydrophilic/hydrophobic uniformities can depend differently on the C_C value. The porosity type of CS samples does not change in comparison with silica gel matrix (Figure 4.22).

The differences of CS-*i* and CS-*i*-HTT samples (Table 4.9) were significantly smaller than that of pure silica gel before and after HTT due to shielding of the oxide surfaces by pyrocarbon deposits. HTT of Si-60 and Si-60T results in practically total destroy of the texture since the S_{BET} value

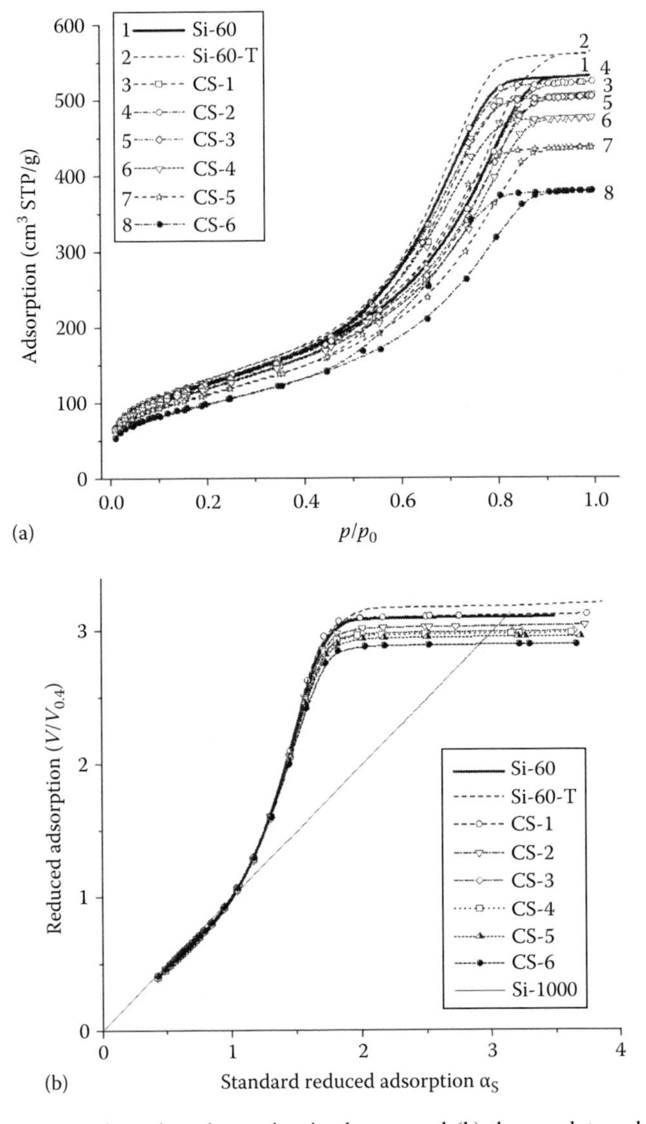

FIGURE 4.22 (a) Nitrogen adsorption–desorption isotherms and (b) the α_S plots reduced by dividing by the adsorption value at $p/p_0 = 0.4$ for Si-60, Si-60T, and CS-*i* (α_S is shown for LiChrospher Si-1000 silica used as a standard). (Adapted with kind permission from Springer Science+Business Media: *Adsorption*, Structural and adsorptive characteristics of pyrocabon/silica gel Si-60, 10, 2004d, 5–18, Gun'ko, V.M., Turov, V.V., Skubiszewska-Zięba, J., Charmas, B., and Leboda, R. Copyright 2004.)

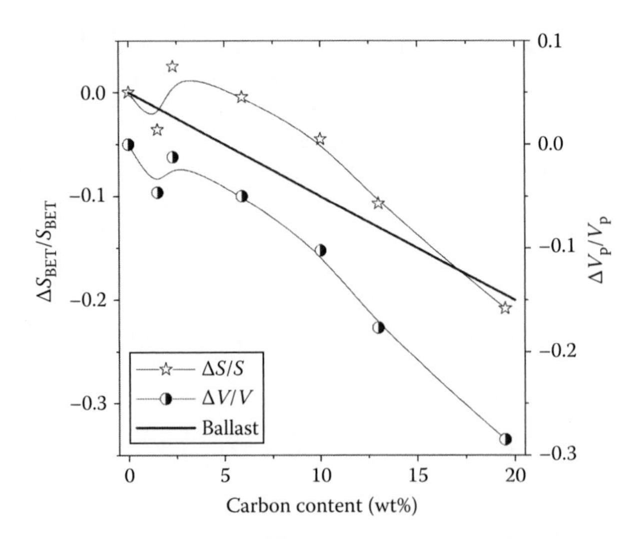

FIGURE 4.23 Changes in the pore volume and the specific surface area due to pyrocarbon deposits for carbosils based on Si-60. (Adapted with kind permission from Springer Science+Business Media: *Adsorption*, Structural and adsorptive characteristics of pyrocabon/silica gel Si-60, 10, 2004d, 5–18, Gun'ko, V.M., Turov, V.V., Skubiszewska-Zięba, J., Charmas, B., and Leboda, R. Copyright 2004.)

decreases by order of magnitude and the deviation of the pore shape from the cylindrical pore model (Table 4.9, Δw) is very large. For initial Si-60, Δw corresponds to 17% and 22%–23% for CS and CS-HTT samples but it is much higher for Si-60-HTT or Si-60T-HTT. The changes in the porosity of CS samples after HTT are much smaller than that for Si-60 or Si-60T after HTT.

Pyrocarbon deposits reduce the pore volume and the specific surface area of hybrid adsorbents nearly linearly with increasing C_C (Table 4.9 and Figure 4.23).

However, relative changes $\Delta V_p/V_p$ versus C_C decrease stronger at $C_C > 10$ wt% (due to blocking of Si-60 pores by carbon plugs) than that on addition of pure ballast (which does not contribute V_p or S_{BET}) in contrast to $\Delta S_{BET}/S_{BET}$ (decreasing smaller than $\Delta V_p/V_p(C_C)$ due to contribution of the outer surfaces of pyrocarbon particles possessing, however, small own porosity) (Figure 4.23 and Table 4.9). On the other hand at $C_C > 10$ wt%, the $\Delta V_p/V_p(C_C)$ graph is closer to the ballast line than $\Delta S_{BET}/S_{BET}$. These effects also reflect in the graphs of the nitrogen adsorption isotherms for CS-*i* samples (Figure 4.22a) lowering with C_C but without marked changes in the isotherm and hysteresis loop shapes. Additionally, the reduced (divided by $V_{0.4}$, i.e., amounts of nitrogen adsorbed at $p/p_0 = 0.4$) α_S plots (Figure 4.22b) are practically independent of the carbon content at $\alpha_S < 1.5$, i.e., changes in the shape of pores on pyrocarbon grafting are relatively small and CS samples remain mesoporous as $S_{BET} \approx S_{meso}$ and $V_p \approx V_{meso}$ (Table 4.9). Notice that the V_p values are slightly larger than those of V_{meso} due to contribution of macropores at $R_p > 25$ nm. Contribution of nanopores ($R_p < 1$ nm) to the surface area is very small for all samples (it is maximal for Si-60-HTT and Si-60T-HTT ~2 m^2/g, and ~0.5 m^2/g for CS-5 and CS-6, and 0–0.1 m^2/g for other samples) and $V_{nano} \approx 0$ for all samples. Macropores give a larger contribution to the pore volume after HTT, especially in the case of pure silica gel ($V_{macro} = 0.2$ and 0.08 cm^3/g for Si-60-HTT and Si-60T-HTT, respectively). Additionally, HTT leads to appearance of narrow mesopores at $R_p < 2$ nm (Figure 4.24) in consequence of pore wall "pitting." It should be noted that the c_{BET} values (from standard BET equation) for all the studied samples were between 80 (minimal for Si-60-HTT) and 111 (maximal for CS-6-HTT), which are close to that for the initial Si-60 ($c_{BET} = 91$) and CS-*i* (nearly constant between 87 and 88). The c_{BET} values are small due to a low contribution of nanopores to the total porosity for all the samples.

One can assume that strong inhibition of the hydrolysis of the silica gel surfaces during HTT by pyrocarbon deposits (even at low C_C) can be caused not only by blocking of pores by carbon

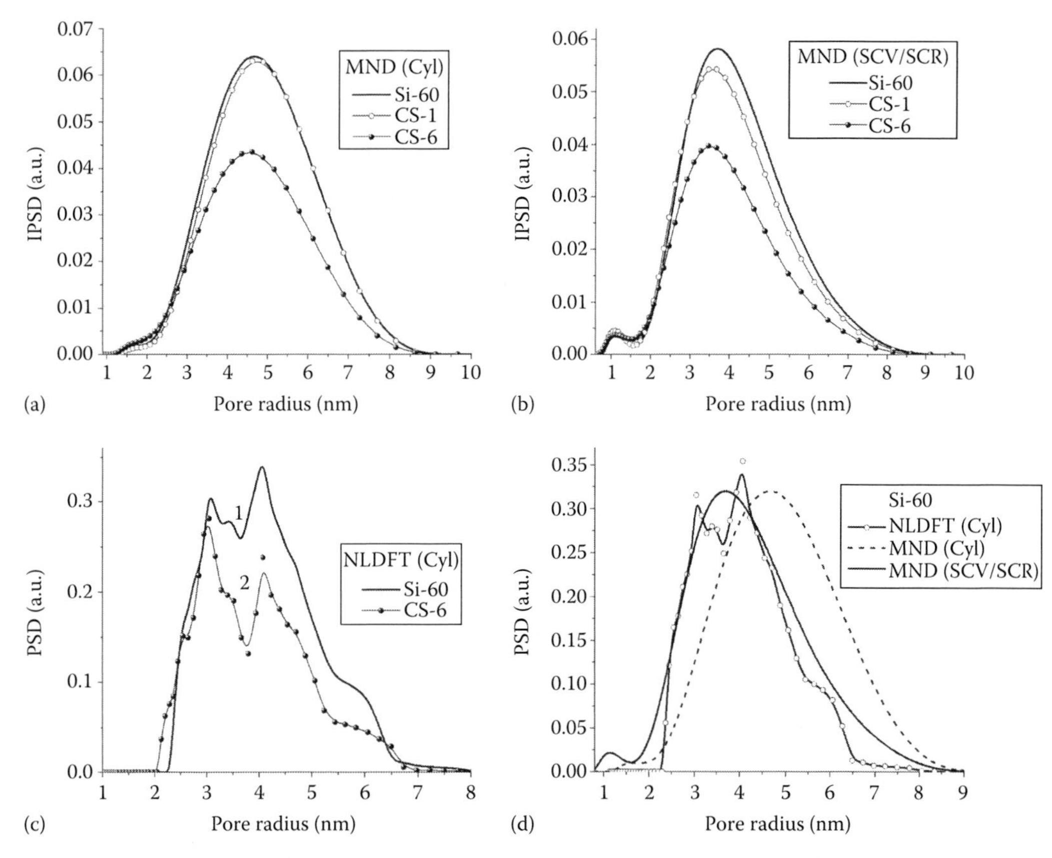

FIGURE 4.24 Pore size distributions for Si-60 and CS-*i* calculated with cylindrical (Cyl) and a mixture (slit-shaped and cylindrical pores and voids between spherical particles, SCV, with self-consistent regularization procedure, SCR) pore models and two methods (a, b, d) MND and (c, d) NLDFT.

plugs but also by blocking (or transforming) of the most active surface sites (e.g., islands of adjacent hydroxyl groups with formation of Si–O–C or Si–C bonds) in silica gel pores by small graphene clusters. Notice that the activation energy of hydrolysis of the Si–O–Si bonds reduces with increasing number of molecules in the water clusters near these bonds and if Si atoms already have OH groups (Gun'ko et al. 2009d). Pyrocarbon deposits can be noncontinuous at low C_C values but they cover active surface sites and can inhibit the formation of adsorption complexes of water that results in the reduction of the hydrolysis rate on HTT. Additionally, the lifetime (Adamson and Gast 1997) $\tau = \tau_0 \exp(E/RT)$ of water adsorption complexes decreases due to reduction of the energy of the water adsorption (E) on the pyrocarbon-covered surfaces possessing lowered hydrophilic properties (Kamegawa and Yoshida 1990, 1993, 1995, Gun'ko et al. 2002b,c). This effect can lead to lowering probability of hydrolysis of the siloxane bonds, i.e., to reduction of the reaction rate.

Relative changes in the pore volume and the specific surface area versus C_C of CS-*i*-HTT differ from those of CS-*i* (Table 4.9). One can assume that shielding effect of pyrocarbon slightly decreases with increasing C_C. This may be caused by, at least, two reasons. The first reason is the segregation of the carbon phase in the form of large particles formed on the outer surfaces of silica gel globules with increasing pyrolysis time (a similar effect was observed on CVD of a phase with a low compatibility with a substrate with increasing amounts of the CVD phase) (Gun'ko et al. 2009d). The second reason is the enhancement of the disorder and the reactivity of the silica surfaces with increasing time of carbonization, since HCl (as a product of CH_2Cl_2 pyrolysis) can react partially (as the carbonization occurs in the flow reactor and a major portion of HCl is

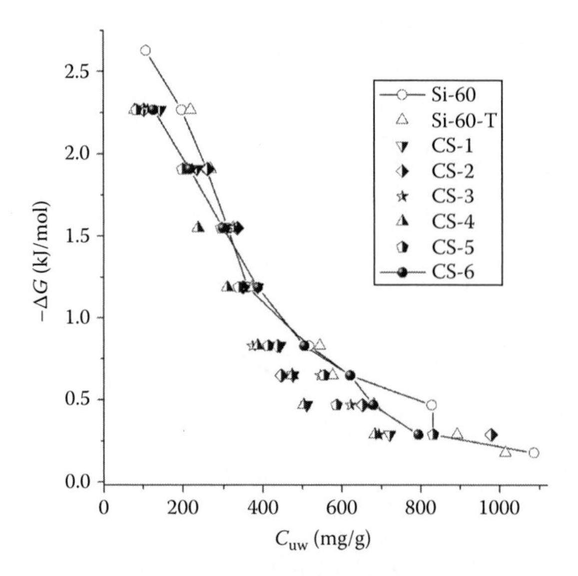

FIGURE 4.25 Changes in the free energy of the interfacial water in the aqueous suspension as a function of the amounts of unfrozen water. (Adapted with kind permission from Springer Science+Business Media: *Adsorption*, Structural and adsorptive characteristics of pyrocabon/silica gel Si-60, 10, 2004e, 5–18, Gun'ko, V.M., Turov, V.V., Skubiszewska-Zięba, J., Charmas, B., and Leboda, R. Copyright 2004.)

removed by the nitrogen flow) with silanols and siloxane bonds of the silica surface at 823 K to form ≡SiCl (because of substitution of OH in ≡SiOH for Cl which easily occurs at $T>623$ K or reaction with ≡Si–O–Si≡ occurring at $T>770$ K) (Tertykh and Belyakova 1991) and new ≡SiOH groups. However, the latter processes do not affect a thick silica layer (maybe due to shielding of the silica gel surfaces by pyrocarbon deposits forming Si–O–C and Si–C bonds with the silica surfaces) since carbonization of CH_2Cl_2 causes a small displacement of the main mesopore peak toward smaller R_p values (Figure 4.24) and average D_p values decreases (Table 4.9), i.e., disruption of the pore walls of silica gels due to the mentioned reactions is not great on the carbonization of CH_2Cl_2.

An increase in the pyrocarbon content leads to an enhancement of the CS surface roughness (Table 4.9, Δw). However, the specific surface area of pyrocarbon itself (S_C) decreases with C_C due to enlargement of pyrocarbon particles (observed in TEM micrographs). These structural features of CS samples (as well as shielding of active surface sites by pyrocarbon deposits) can impact their adsorptive properties of CS with respect to such a polar adsorbate as water.

The graph shapes of changes in the Gibbs free energy of the interfacial water versus the amount of unfrozen ($T<273$ K) water ($C_{uw} = C_{uw}^{S} + C_{uw}^{W}$) depend on the carbon content only slightly (Figures 4.25 and 4.26 and Table 4.10). However, the observed differences reveal an enhancement of the nonuniformity of CS-*i* samples with increasing C_C value. Notice that the energy of water adsorption on nonporous carbon black (similar to studied pyrocarbon) is about 20 kJ/mol, which is significantly lower than the heat of water condensation (≈45 kJ/mol) (Brennan et al. 2001). Changes in the free surface energy γ_S with increasing C_C depict that disturbance of the interfacial water in the confined space of pores and reduction of its Gibbs free energy depend nonlinearly on C_C. One can assume that the nonuniformity of pyrocarbon deposits increases with C_C (as well as γ_S at maximal C_C). However, for materials X/SiO_2 ($X=TiO_2$, Al_2O_3, GeO_2, and C), the maximal γ_S values were frequently observed at small C_X values (as shown in previous chapters).

4.2.4 CS Modified with Zinc, Titanium, and Zirconium Oxides

Silica gel Si-60 (Schuchardt München, Germany) was used to prepare pyrocarbon–silica gel samples using more complex precursors than CH_2Cl_2 (Table 4.11). Acetylacetone $C_5H_8O_2$ (L. Light Co.) in

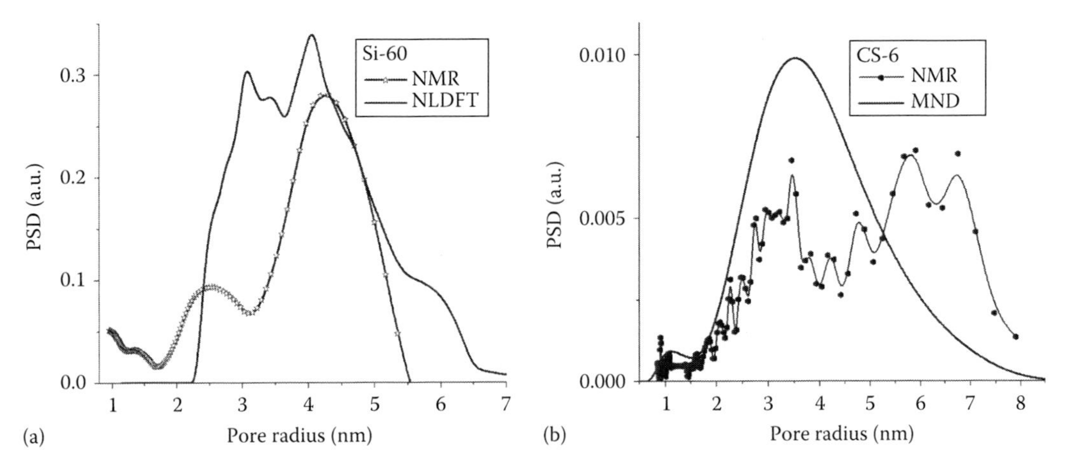

FIGURE 4.26 Pore size distributions of (a) Si-60 and (b) CS-6 according to NMR cryoporometry and nitrogen adsorption data.

TABLE 4.10
Characteristics of Interfacial Water in the Aqueous Suspension of Initial Silica Gel and Carbosils (Carbonization of CH_2Cl_2)

Sample	$-\Delta G^s$ (kJ/mol)	$-\Delta G^w$ (kJ/mol)	C^s_{uw} (mg/g)	γ_S (mJ/m²)
Si-60	3.4	0.8	500	154
Si-60T	4.0	1.1	500	119
CS-1	2.8	0.8	560	124
CS-2	2.4	0.8	480	127
CS-3	2.5	1.2	400	116
CS-4	2.5	1.2	550	109
CS-5	2.5	0.9	550	137
CS-6	2.7	1.3	630	164

Note: Superscripts w and s correspond to weakly and strongly bound waters, respectively, unfrozen at $T<273$ K.

the amounts of 0.02 M pyrolyzed on 5 g of Si-60 (Merck) at 773 K for 6 h gave the carbosils AC-1. Additionally, Si-60 (Merck) was utilized to prepare carbon/X/silica materials (carbosils CS_X) by pyrolysis of such metal acetylacetonates (AcAc) as $Zr(AcAc)_4$, $TiO(AcAc)_2$, and $Zn(AcAc)_2$ (Aldrich) in the autoclave at 773 K for 6 h (Gun'ko et al. 2001e). The utilization of the same amounts (0.01 M) of metal acetylacetonates results in deposition of equal amounts of metal moles but different amounts of moles of acetylacetonate groups, as the n values in $M(AcAc)_n$ compounds are different ($n=2$–4). These carbon–mineral adsorbents were labeled as CS(Ti), CS_{Zr1}, CS_{Zn}, and CS_{Zr}. Concentrations of X phases (grafted metal compounds, C_X) and pyrocarbon (C_C) are shown in Table 4.11 (Turov et al. 1997b, 1998b, Gun'ko et al. 2001c).

The nitrogen adsorption–desorption isotherms (Figure 4.27a) and the PSDs (Figure 4.27b) of modified silicas show that the texture of CS samples depends strongly on the type of the carbon precursors. In the case of $Zn(AcAc)_2$, these changes are maximal because of the reactions with the participation of silica since Zn_2SiO_4 is formed besides the carbon deposits (Table 4.11). Minimal changes in the silica texture are observed during pyrolysis of acetylacetone (AC-1).

TABLE 4.11
Concentrations of Metal Compounds and Grafted Carbon in CS Samples Based on Si-60 ($S_{BET}=356$ m²/g, $V_p=0.744$ cm³/g)

Sample	X phase	C_X (wt%)	C_C (wt%)	S_{BET} (m²/g)	V_p (cm³/g)	R_p (nm)
CS_{Ti}	TiO_2 (anatase)	11.37	9.2	192	0.419	4.35
CS_{Zn}	Zn_2SiO_4	10.98	6.5	212	0.536	5.05
CS_{Zr1}	ZrO_2	17.14	13.8	236	0.415	3.52
CS_{Zr2}	ZrO_2	10.19	7.4	281	0.537	3.82
AC-1	—	—	4.0	339	0.717	4.2
AC-2	—	—	9.1	288	0.559	3.9
AC-3	—	—	14.5	264	0.568	4.3

These CS samples differ in both texture and the surface chemistry that can affect the interfacial phenomena, especially interaction with water. Constitution of adsorption complexes of water confined in pores of adsorbents can be determined using the ^{1}H NMR spectral characteristics of bound water. Figure 4.28 displays the spectra of water adsorbed on the CS studied. The water bound to SG has a single signal at 4.5 ppm. This chemical shift value is close to that for liquid water and is typical for SAW.

Carbonization of acetylacetone at the Si-60 surface leads to significant changes in the ^{1}H NMR spectra of the adsorbed water (sample AC-1, Figures 4.28 and 4.29a). In this case, water gives two signals having chemical shifts of 2.8 and 0.6 ppm. The earlier signals have the up-field shift in contrast to the chemical shift of water on the pure silica surface. The presence of two up-field shifted signals gives strong evidence for the surface inhomogeneity and the influence of the carbon deposits. As the intensity of up-field shifted signals does not depend on temperature (Turov et al. 1997b, 1998b, Gun'ko et al. 2001c), the earlier signals may be attributed to the water molecules that are involved in the strongly bound surface complexes located in relatively narrow pores of CS.

Carbonization of organics used in this study was carried out at a sufficiently low temperature (~870 K). This condition leads to formation of the carbon layers whose graphene clusters have sizes that do not exceed 2 nm (Fenelonov 1995). Consequently, formation of partly graphitized areas on the carbon surface is hardly probable, and the up-field shifts of the ^{1}H NMR signal for the adsorbed water may be due to location of water molecules in narrow mesopores partially filled by carbon deposits because this displacement does not exceed 5 ppm (in slit-shaped nanopores this shift could be larger; see Chapter 3).

It is obvious that the signals of water adsorbed on the surface may be treated as averaged over several types of adsorption complexes. The main contribution to the chemical shift is made by the water molecules adsorbed on the carbon components of the CS surfaces which are formed by the system of graphene clusters and individual water molecules bound by hydrogen bonds with the surface hydroxyl groups. Therefore, the signal of water having a chemical shift at $\delta_H=0.6$ ppm should be attributed to the surface areas having a higher carbon content.

The spectra of CS_{Ti} and CS_{Zn} samples give two signals (Figure 4.29). The first signal being more intensive ($\delta_H=4.5$ ppm, SAW) has down-field shift as the temperature decreases. It is evident that this signal for both adsorbents is caused by the clusters of strongly associated water adsorbed on the surface that is covered with the CS_{Ti} or CS_{Zn} deposits formed in the course of oxidation of metal acetylacetonates. The data of the x-ray analysis as well as the correlation between spectral parameters of adsorbed water for the previous samples and water adsorbed on the surface of the oxides (Turov et al. 1998b) gave strong evidence in favor of the earlier conclusion. For both CS samples (Figure 4.29b and c), the chemical shift and the signal intensity

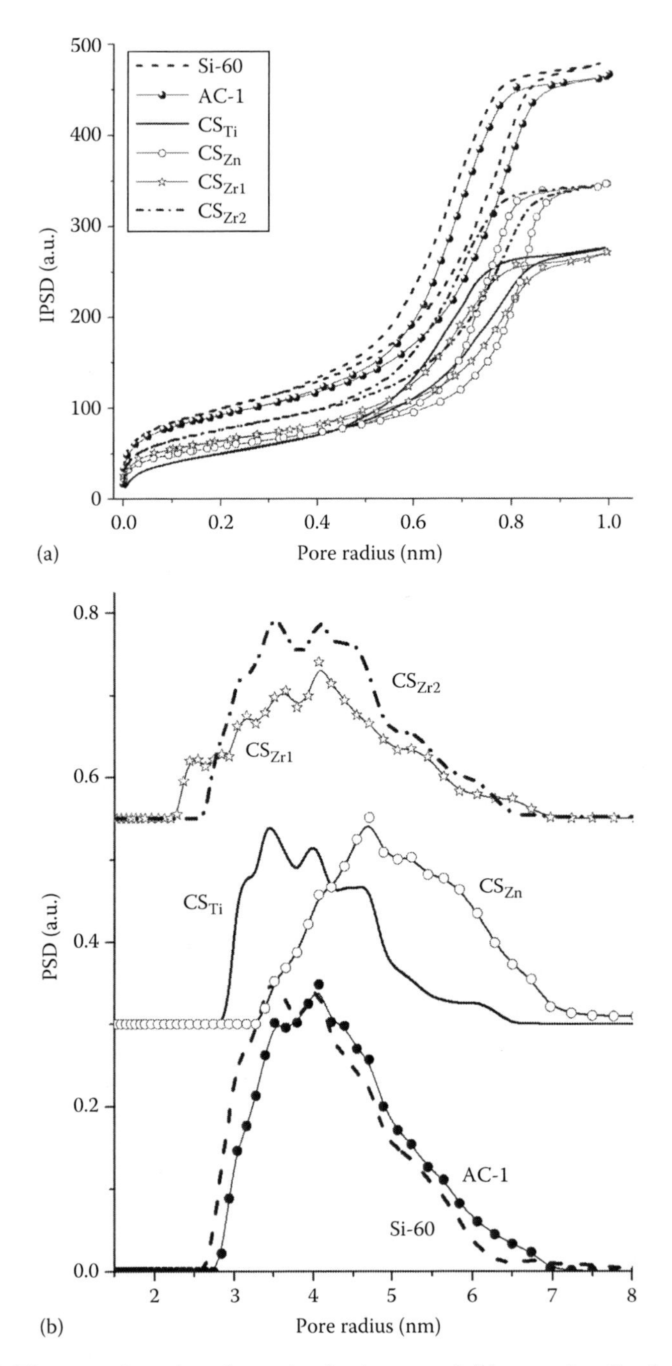

FIGURE 4.27 (a) Nitrogen adsorption–desorption isotherms and (b) pore size distributions of Si-60 and carbosils (NLDFT with the model of cylindrical pores, some PSD are displaced along the Y-axis for better view).

at $\delta_H \approx 1$ ppm depend weakly on temperature over a broad range. This water can be attributed to WAW bound to carbon deposit surfaces in narrow mesopores since nanopores are absent in CS (Figure 4.27b). The signal intensity of SAW decreases with lowering temperature much strongly than that of WAW (Figure 4.29).

The volume of unfrozen water bound to Si-60 (Figure 4.30) is close to the pore volume ($V_p = 0.744\,cm^3/g$). This can be explained by the absence of broad pores in Si-60 ($R < 7$ nm, Figure 4.27b).

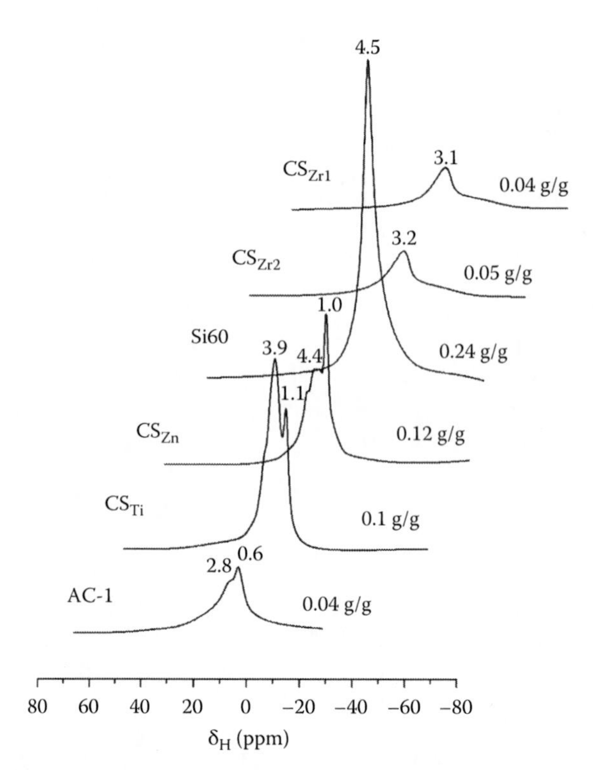

FIGURE 4.28 ^{1}H NMR spectra of water adsorbed on Si-60, AC-1, and C/X/Si-60 (Table 4.11) at room temperature (adsorbed water concentration is shown). (Adapted from *J. Colloid Interface Sci.*, 238, Gun'ko, V.M., Leboda, R., Turov, V.V. et al., Structural and energetic nonuniformities of pyrocarbon–mineral adsorbents, 340–356, 2001c, Copyright 2001, with permission from Elsevier.)

Therefore, it may be concluded that all the water confined in narrow mesopores of Si-60 is subjected to the perturbing action of the surface. The data given in Figure 4.30 make it possible to arrive at a conclusion that the hydrophilicity of the surface is minimal for AC-1 as a result of carbonization. It increases for CS samples prepared using metal acetylacetonates in comparison with AC-1 but it remains lower than that of Si-60. The amounts of WBW are minimal for Si-60.

Extrapolation of the $\Delta G(C_w)$ curves gives the maximum magnitude of $-\Delta G_{max}$. The ΔG_{max} value and the data on the free surface energy are listed in Table 4.12. The data given in Table 4.12 permit us to conclude that the free surface energy (in mJ/m^2) for the adsorbents studied increases in the following series:

$$(\gamma_S)_{AC\text{-}1} < (\gamma_S)_{CS_{Zr1}} < (\gamma_S)_{CS_{Zr2}} < (\gamma_S)_{Si\text{-}60} < (\gamma_S)_{CS_{Zn}} \approx (\gamma_S)_{CS_{Ti}}$$

It should be noted that the correlation between the hydrophilic properties of the surface and characteristics of its structure are not reflected by the revealed regularity in full measure. The cause of this is in the fact that, when going from sample AC-1 to samples CS$_{Zn}$ and CS$_{Ti}$, one can observe two trends, namely increasing in the concentration of surface hydroxyl groups (owing to formation of surface regions covered with titanium oxide or zinc silicate) and decreasing in the pore volume. In the case of both metal-containing samples, all the water in their pores becomes surface-bonded, which leads to leveling of differences in the values of the free surface energy. Sample CS$_{Zr1}$ is more hydrophobic than CS$_{Zr2}$ because the latter has larger pore volume and smaller content of carbon.

The water cluster sizes are larger for CS$_{Zn}$ than for Si-60 (Figure 4.31) because of the enlargement of pores during carbonization (Figure 4.27b). In the case of AC-1, small water domains ($R<2$ nm) observed for Si-60 are absent, and the PSD of AC-1 at $2<R<4$ nm slightly shifts toward larger pore

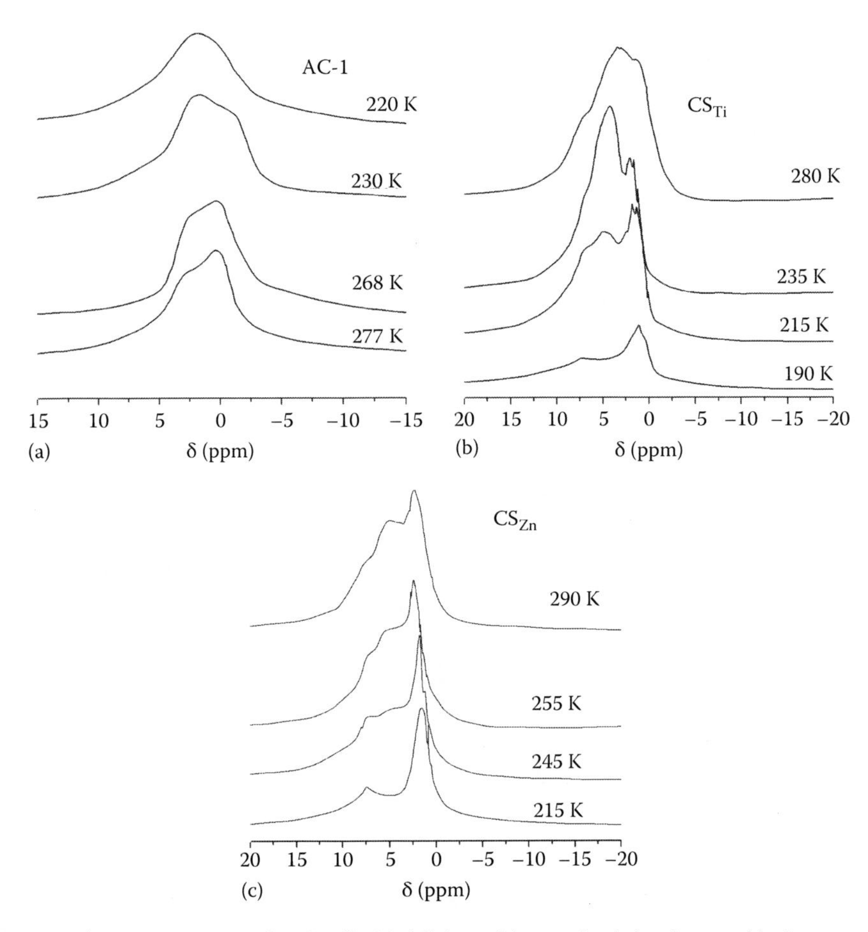

FIGURE 4.29 ^{1}H NMR spectra of carbosils (a) AC-1, or (b) contained titanium or (c) zinc compounds at different temperatures.

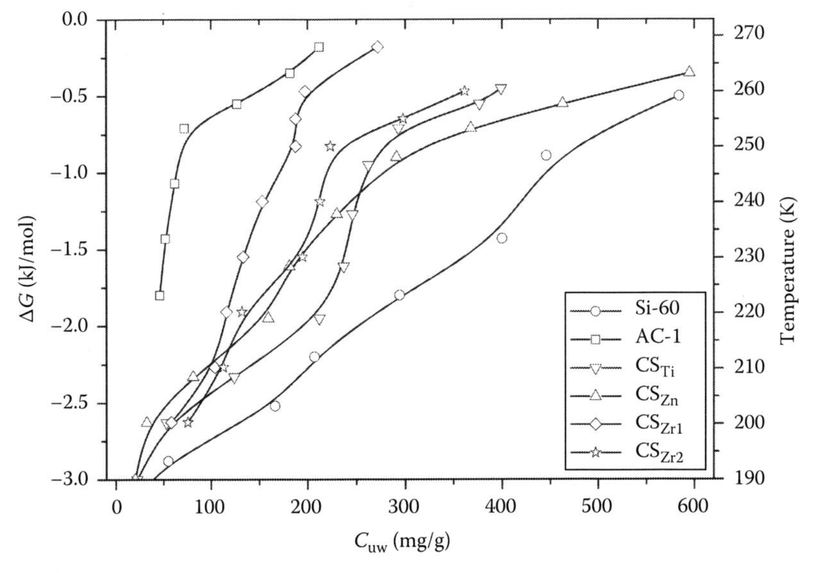

FIGURE 4.30 Relationships between the Gibbs free energy value (or melting temperature) and the amounts of unfrozen water bound to Si-60 or CS samples.

TABLE 4.12
Parameters of the Water Layers Adsorbed on Parent and Modified Si-60

Material	$-\Delta G_s$ (k/mol)	γ_S (J/g)	γ_S (mJ/m²)	C_{uw}^{max} (mg/g)	C_{uw}^{s} (mg/g)	S_{uw} (m²/g)
Si-60	3.16	62.8	176	700	100	396
AC-1	2.77	12.3	36	250	250	61
AC-2	2.39	33.2	115	908	380	58
AC-3	3.14	18.5	70	332	180	117
CS_{Zn}	3.30	45.0	209	535	340	210
CS_{Ti}	3.11	40.1	212	424	275	242
CS_{Zr1}	3.00	24.1	102	200	175	177
CS_{Zr2}	3.23	36.0	128	250	225	257

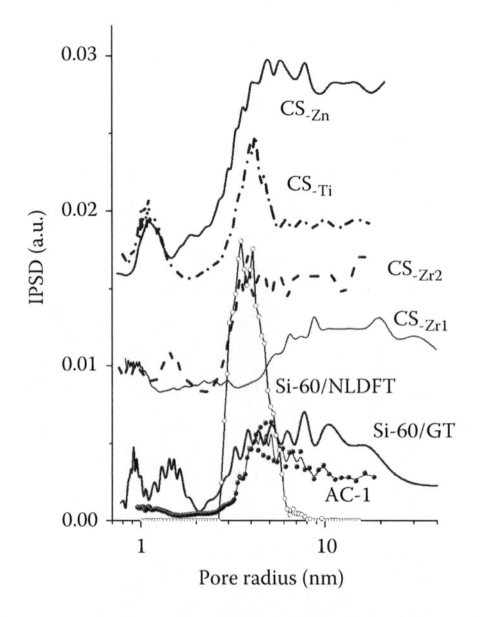

FIGURE 4.31 PSD (calculated with GT equation and corresponding to the size distributions of clusters of unfrozen bound water) for initial silica gel Si-60, AC-1, and carbosils with Ti, Zn, and Zr oxides (N_2 NLDFT PSD is shown for Si-60, some PSD are displaced along the *Y*-axis for better view).

sizes in comparison with Si-60 (Figure 4.31). This can be due to a partial filling of pores by carbon deposits despite that carbonization of O-containing acetylacetone results in certain increase in the pore sizes (Figure 4.27b) because of the HTT effects discussed earlier. Similar effects (Figure 4.31) are observed for CS_{Zn} with stronger enlarged pores. The difference between S_{uw} (the specific surface area being in contact with unfrozen water) and S_{BET} values can be positive for hydrophilic adsorbents (Si-60, CS-Ti) and negative for more hydrophobic adsorbents (especially AC-1). The GT PSDs are broader than N_2 NLDFT PSD because of both rearrangement of the adsorbent texture on suspending–freezing–thawing and the formation of unfrozen water structures bound at the external surface of modified silica gel.

Depending on the mode of carbonization of CS synthesized on the basis of a mesoporous silica gel, significant changes in the hydration properties of the surface are observed.

Thus, the thickness of the water layer subject to the distortive effect of the surface of the parent silica expands to the whole volume of pores. In a process of a high-temperature pyrolysis of acetylacetone in the silica gel pores, the hydrophilic properties of the surface suddenly disappear. Simultaneously, a carbon layer is formed on the adsorbent surface in which the carbonized fragments comprise predominantly individual hydroxyl groups. In the case of carbonization of metal acetylacetonates in the silica gel pores, the hydration properties of CS and of the parent silica are very similar. This is due to formation of new oxides on the silica gel surface. Probably, the carbon component of the surface is localized in pores and at the outer surface of the adsorbent particles. Carbonization of acetylacetone leads to certain increase in the pore size due to hydrothermal effects of water formed during decomposition of acetylacetone since the PSDs shift toward larger pore size (Figure 4.32c). Cryoporometry results for Si-60 with carbonized acetylacetone (Figure 4.32a) show that there are unfrozen water structures (formed during thawing process) of both smaller and larger sizes than the main PSD peak of mesopores of Si-60 initial or modified. A certain portion of water structures has the sizes corresponding to the main PSD peak at $R = 3–4$ nm. This is due to several factors: (i) an increase in size of pores during the ice formation; (ii) formation of ice crystallites and the presence of unfrozen water layer in the same broad pores; and (iii) formation of bound water structures at the outer surface of silica gel and carbon deposit particles.

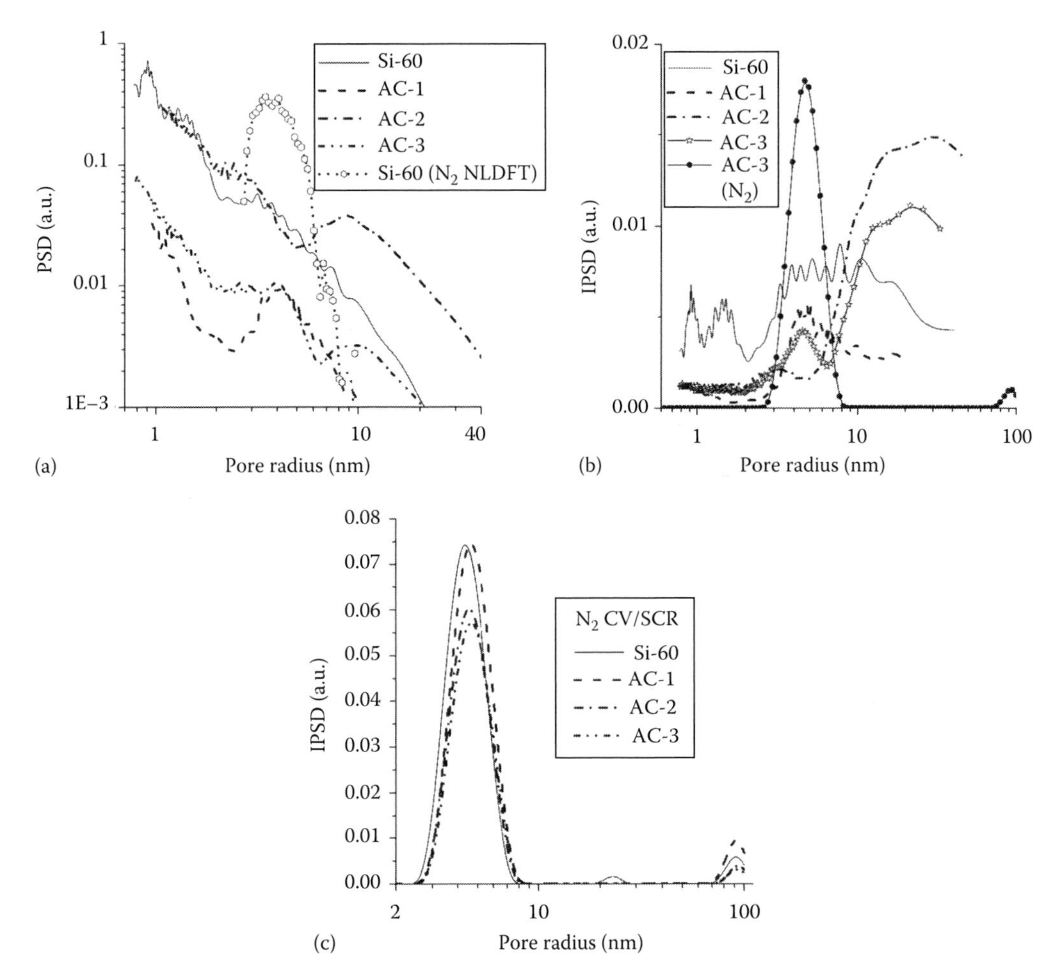

FIGURE 4.32 PSD calculated using (a, b) cryoporometry ((a) differential and (b) incremental PSD) and (c) nitrogen adsorption data for initial and modified Si-60 (IPSD).

Thus, investigations of carbon–mineral materials synthesized by carbonization of different organics at a surface of different substrates showed that the hydrophilicity of hybrid materials is lower than that of mineral matrices. This diminution is smaller if organic precursors are metal compounds (e.g., metal acetylacetonates). The presence of O-containing functionalities in the precursors can cause enlargement of pores of the matrices due to interaction of water (formed during carbonization in the inert atmosphere) with the pore walls that causes the HTT effects. However, the presence (or formation) of a carbon phase can provide shielding of the matrix surface and the HTT effects can be reduced in comparison with that observed for pure silica interacting with water.

In the case of a low content of carbon deposits ($C_C \approx 0.5$ wt%) formed at a surface of nanosilica, the amount of bound water increases but it decreases at large C_C values due to hydrophobic properties of carbon deposits. The properties of carbon deposits on silica are similar to that of carbon black but in the case of catalytically active substrate or the presence of metals in the precursors (forming a catalytically active phase) the texture of carbon can be different from carbon black. For water bound to carbon particles, a certain up-field shift can be observed due to the effects of the π-electrons current at the carbon sheets (small graphene clusters). However, this shift is not larger as in the case of nanoporous AC because carbon deposits do not have nanopores.

Thus, used methods of carbon deposits preparation at solid matrices typically have such an essential disadvantage as the absence of nanopores in carbon nanoparticles. To avoid this problem, carbon–mineral composites can be prepared using highly porous activated carbons (AC), which are composed of aggregated nanoporous nanoparticles (NPNP), and mineral nanoparticles (MNP) (e.g., nanosilica) using mechanochemical activation (MCA) with controlled mechanical loading without decomposition of nanoparticles (Gun'ko et al. 2011f). However, several questions arise: (i) how carbon NPNP (CNPNP) and MNP are distributed in composites; (ii) how structural properties of composites depend on sizes of MNP, CNPNP, and AC microparticles; (iii) what is optimal decomposition degree of microparticles and nanoparticles during MCA; and (iv) are chemical bonds formed between CNPNP and MNP during MCA? To answer the questions, the textural and electronic characteristics of composites should be studied (Gun'ko et al. 2011f).

Ultrasoft x-ray emission spectroscopy (USXES) is an informative and appropriate method to study the electronic structure of nanomaterials because penetration depth of electron beam can be varied in a nanometer range depending on the beam energy (Nordgren et al. 1991). Notice that x-ray photoelectron spectroscopy (XPS), which gives information on electronic states depending on chemical structure and phase composition of materials, was successfully applied to carbon and carbon–mineral composites (Liu et al. 2006, Wu et al. 2007, Peng et al. 2008, Wang and Stein 2008, Guo et al. 2009). Despite a certain difference in the spectra shapes obtained with XPS (electron-binding energy) and USXES (emitted photon energy), they give both complementary and similar information.

Changes in the particle sizes from macroscale to nanoscale can lead to changes in energetic states from local core electron levels of atoms to the valence band (with differently delocalized electrons) of the material. Valence band energy (E_b) is equal to

$$E_b = \int_{E_0}^{E_F} g(E) E \, dE \tag{4.7}$$

where

E_0 is the energy position of the valence band bottom
E_F is the Fermi energy
E is the binding energy of valence electrons
$g(E)$ is the distribution function of electron state density

Notice that a significant part (70%–80%) of the total energy of interatomic interactions is determined by the energy distribution of valence electrons. Experimental methods used to study the energy distributions of valence electrons undergoing different transfers (XPS, USXES, AES, etc.) can give detailed information on the particle size effects (Nordgren et al. 1991). X-ray emission and absorption spectroscopies reflect the energy distribution of valence electrons in the occupied bands and vacant electronic states in the conduction band of any symmetry as consistent with dipole rules. XPS reflecting the energy distribution of valence electrons depending on photoionization cross section of electrons of any symmetry allows the determination of the binding energy of core electrons in atoms. This makes it possible to superimpose emission bands (reflecting the energy distribution of electrons of defined symmetries) of atoms in a common energy scale. The USXE spectra reflect the energy distribution of valence electrons of surface and near-surface atoms that are on a distance of one to three tens of atomic planes deep in a specimen. A small width of core electron levels (photons are emitted due to transfer of electrons to these levels from the valence band) makes it possible to study peculiarities of the valence electron distribution with resolution of 0.2 eV. Practically, the same chemical structure of silica nanoparticles (fumed silica is very pure material with 99.5%–99.8% of SiO_2) of different sizes allows us to estimate the band energy changes due to the size effects.

Here the dependence of nanosilica/AC composite properties on size of MNP is analyzed using three nanosilicas A-50, A-300, and A-500 with different nanoparticle size distributions (Table 4.13). Amorphous nanosilicas A-50, A-300, and A-500 were heated at 650 K for several hours to remove adsorbed compounds. AC (MAST Carbon International Ltd., United Kingdom) in the form of beads (0.25–0.50 mm) was prepared from phenol–formaldehyde resin carbonized at 1073 K and CO_2 activated at 1183 K (burn-off degree 30%–47%). The textural and adsorption characteristics of similar materials were analyzed in Chapter 3. MCA of nanosilica/AC (40% burn-off) mixtures (2:1 and 1:1) was carried out in a stainless steel microbreaker at room temperature for 5 min. This short-term MCA was used to avoid too deep decomposition of AC particles. Average temperature of a mixture increased to 40°C during MCA. However, local heating of MCA-treated micro/nanoparticles can be higher that can lead to chemical reactions at the interfaces of carbon and mineral components.

USXE spectra were recorded using a RSM-500 (SCBXA, Burevestnik, St. Petersburg, Russia) spectrometer–monochromator with electron beam energy $E=5$ keV and intensity $I=6\times10^{17}$ electron/cm^2 (Gun'ko et al. 2011f). For comparative analysis of changes in the electronic structure of the valence band of MCA-treated silica/AC mixtures, emission bands SiL_α, OK_α, and CK_α (showing distributions of valence electrons localized on Si, C, and O atoms) of A-50, A-300, A-500, AC, and silica/AC samples were analyzed using the USXES method.

TABLE 4.13
Structural Characteristics (SCV/SCR Model)

Sample	S_{BET} (m²/g)	S_{nano} (m²/g)	S_{meso} (m²/g)	S_{macro} (m²/g)	V_p (cm³/g)	V_{nano} (cm³/g)	V_{meso} (cm³/g)	V_{macro} (cm³/g)	D_{FHH}	c_{slit}	c_{cyl}	c_{void}
A-50	52	5	44	3	0.126	0.003	0.075	0.048	2.596	0.285	0.371	0.344
A-300	337	60	270	7	0.714	0.033	0.573	0.108	2.556	0.483	0.239	0.278
A-500	492	85	402	6	0.874	0.044	0.746	0.084	2.631	0.317	0.343	0.340
A-300/AC	695	464	230	1	0.982	0.223	0.723	0.036	2.783	0.908	0.025	0.068
A-500/AC	859	543	314	2	1.159	0.262	0.854	0.043	2.769	0.889	0.036	0.076
AC	1520	1181	333	6	1.260	0.550	0.594	0.115	2.876	0.657	0.226	0.117

Notes: The weight ratio in silica/AC composites was 2:1. AC was with 40% burn-off activation.

The used excitation mode did not lead to formation of radiation defects in the material because they appear only at $I>5\times10^{20}$ electron/cm^2. However, the electron beam removed adsorbed water and other molecules from a surface of studied materials. To obtain stable spectra, the fourth–ninth records were used for analysis.

The distribution functions of photon energy in the CK_α and OK_α emission bands correspond to transition of valence C_{2p} and O_{2p} electrons on the vacant core C_{1s} and O_{1s} levels with emitted photon energy $h\nu = E_{C2p} - E_{C1s}$ and $E_{O2p} - E_{O1s}$, respectively. According to selection rules, the SiL_α (or labeled SiL_{23}) emission band corresponds to photon energy $h\nu = E_{Si3s3d} - E_{Si2p}$ due to transition of valence Si_{3s} (and Si_{3d}) electrons on the vacant Si_{2p} level. The vacancies on the core levels are formed due to interaction of atoms with the exciting electron beam. The USXES intensity ($I(\nu)$ or $I(E)$) depends on both the distribution of valence electron state density ($g_v(E)$) and the transition probability matrix elements. Therefore, the USXES band shape differs from the $g_v(E)$ shape; however, relative positions of the $I(E)$ and $g_v(E)$ peaks can be similar.

Diffraction grating used was with a period of 600 mm^{-1} and curvature radius of 6 m (OK_α and K_α) or 2 m (SiL_α). In the first case, a filter mirror was with gold, and in the second case, it was with polystyrene. Instrument distortion was less than 0.2 and 0.3 eV for the SiL_α and OK_α bandwidth, respectively, and the spectra recording accuracy was 0.1 eV. X-ray photons were registered using a secondary electronic multiplier with a primary photocathode with a deposited CsJ film.

XPS were recorded using an ES-2401 spectrometer (EPSIE, Chernogolovka, Russia) at sample heating at 200°C and residual pressure after evacuation $\sim10^{-9}$ Torr. To analyze the MCA effects on the texture of silica/AC composites, nitrogen adsorption isotherms for the mixtures (Figure 4.33, curves 3) were compared with the corresponding sum of the weighted isotherms of individual components (curves 4). There is a small difference in the MCA effects for A-300/AC and A-500/AC.

For the latter, curves 3 and 4 (Figure 4.33b) coincide at $p/p_0<0.4$ but at $p/p_0>0.4$ the adsorption on the mixture (curves 3) is higher. For A-300/AC (Figure 4.33a), the adsorption on the mixture is slightly lower at $p/p_0<0.8$ but at $p/p_0>0.85$ curves 3 and 4 practically coincide. These differences in the MCA effects on A-300/AC and A-500/AC can be due to certain variations in organization of the mixtures since primary nanoparticles of A-500 are smaller than those of A-300 but they are strongly aggregated. The small difference in the model (Figure 4.33, curves 4) and real (curves 3) isotherms for the mixtures at $p/p_0<0.4$ shows that AC structures responsible for the specific surface area (Table 4.13, S_{BET}) and the nanoporosity (V_{nano}, S_{nano}) of the materials remained practically

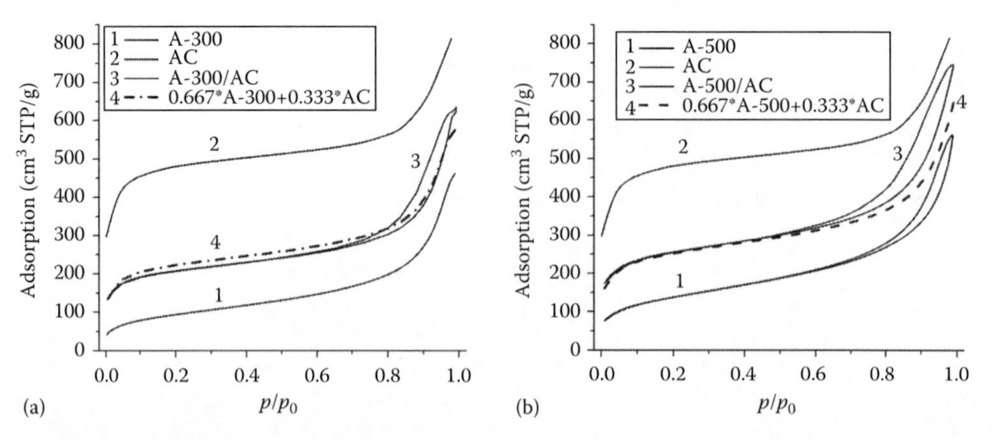

FIGURE 4.33 Nitrogen adsorption isotherms for nanosilica (curve 1), AC (curve 2), silica/AC (2:1) (curve 3) and a model isotherm for silica/AC estimated as the sum of the weighted terms of silica and AC (curve 4) for nanosilicas (a) A-300 and (b) A-500. (Adapted from *Appl. Surf. Sci.*, 258, Gun'ko, V.M., Zaulychnyy, Ya.V., Ilkiv, B.I. et al., Textural and electronic characteristics of mechanochemically activated composites with nanosilica and activated carbon, 1115–1125, 2011f, Copyright 2011, with permission from Elsevier.)

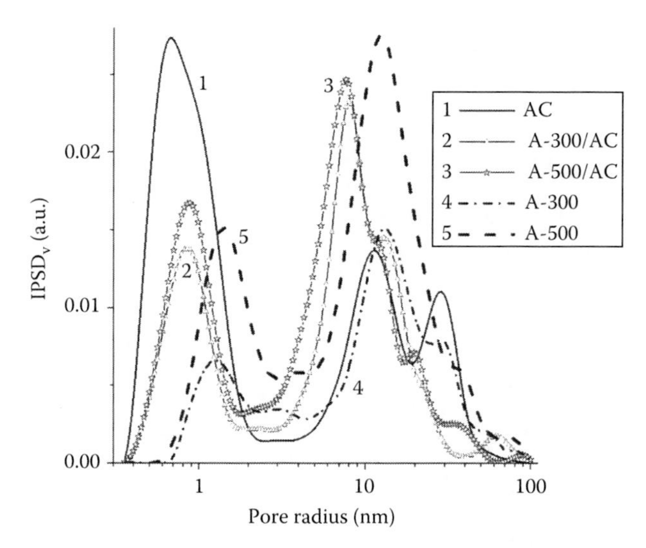

FIGURE 4.34 Incremental PSD_V for individual silicas and AC- and MCA-treated silica/AC composites calculated using the SCV/SCR method. (Adapted from *Appl. Surf. Sci.*, 258, Gun'ko, V.M., Zaulychnyy, Ya.V., Ilkiv, B.I. et al., Textural and electronic characteristics of mechanochemically activated composites with nanosilica and activated carbon, 1115–1125, 2011f, Copyright 2011, with permission from Elsevier.)

nondestructive. However, changes in the organization of aggregates of silica and AC particles (responsible for the textural porosity) cause an increase in the mesoporosity of the silica/AC mixtures (Table 4.13, V_{meso}) in comparison with individual materials. Additionally, there is a decrease in the macroporosity (V_{macro}, S_{macro}) of silica/AC. These textural changes can be explained by penetration of silica nanoparticles into broad mesopores and macropores of carbon microparticles during MCA. Therefore, the PSD intensity of macropores ($R > 25$ nm) of silica/AC decreases and the mesopore peak shifts toward smaller pore size (Figure 4.34).

In the case of A-500/AC, this effect is stronger than that for A-300/AC, i.e., smaller silica nanoparticles (Figure 4.35) can more easily penetrate into broad pores of AC. The porosity characteristics of the mixtures in the range of nanopores at $R < 1$ nm are determined mainly by AC component (Table 4.13, V_{nano}, S_{nano} and Figure 4.34) and the nanoporosity characteristics are nearly additive. However, contribution of nanopores as voids between nonporous silica nanoparticles is smaller by order of magnitude than that of inner nanopores in CNPNP. Nanoporosity constancy can be explained by (i) the absence of blocking of AC nanopores by silica nanoparticles and (ii) preservation of all AC nanoparticles despite a decrease of AC microparticles during MCA by a factor of 20–100 (Figure 4.35) in comparison with initial AC beads of 0.25–0.5 mm in diameter. Residual AC microparticles are smaller in the case of A-500/AC (Figure 4.35c) than A-50/AC (Figure 4.35a) or A-300/AC (Figure 4.35b).

Typically, the higher the specific surface area of nanosilicas, the narrower the nanoparticle size distribution (Figure 4.36). This affects the penetration of silica nanoparticles in broad pores of AC microparticles (Figures 4.33 and 4.34) and decomposition of AC microparticles (Figure 4.35) during MCA. A-50 particles are observed in the mixture (Figure 4.35a) in contrast to A-300/AC and A-500/AC.

Thus, the MCA treatment leads to significant diminution of AC microparticle sizes; however, the microparticles were not completely destroyed. Textural pores in AC microparticles are non-slit-shaped (see Chapter 3). Therefore, the observed decomposition of AC microparticles (Figure 4.35) results in an increase in contribution of slit-shaped pores (Table 4.13, c_{slit}) and a decrease in contribution of textural pores (cylindrical pores [c_{cyl}] and voids between spherical particles [c_{void}]) in silica/AC composites in comparison with initial AC.

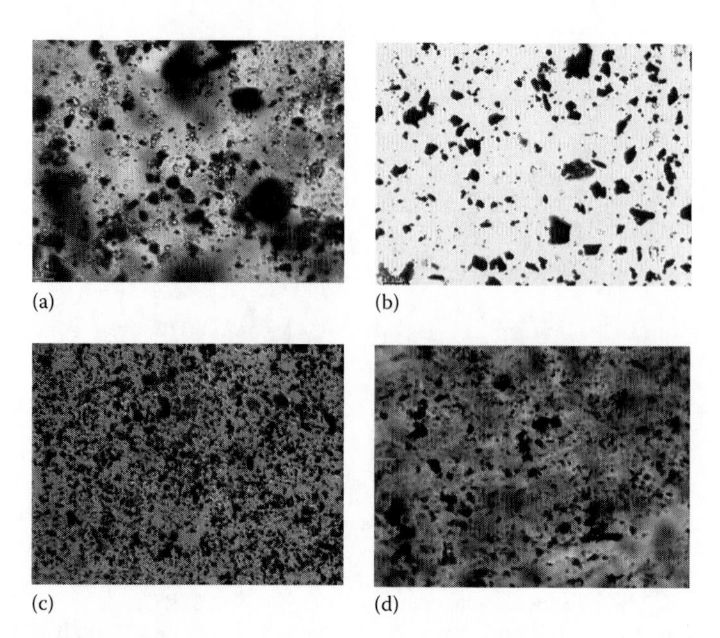

FIGURE 4.35 Microphotographs (Primo Star optical microscope, Carl Zeiss) of (a) A-50/AC; (b) A-300/AC; (c) A-500/AC; and (d) AC after 5 min MCA treatment (scale bar 10 μm). (Adapted from *Appl. Surf. Sci.*, 258, Gun'ko, V.M., Zaulychnyy, Ya.V., Ilkiv, B.I. et al., Textural and electronic characteristics of mechanochemically activated composites with nanosilica and activated carbon, 1115–1125, 2011f, Copyright 2011, with permission from Elsevier.)

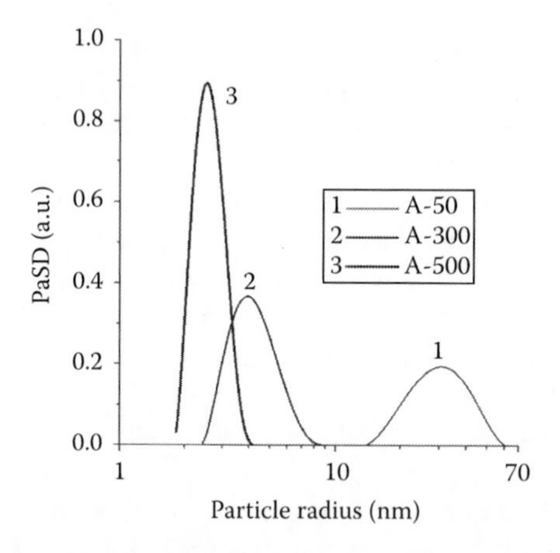

FIGURE 4.36 Primary nanoparticle size distributions of silicas estimated from the adsorption data using the SCR procedure with the model of voids between spherical nanoparticles. (Adapted from *Appl. Surf. Sci.*, 258, Gun'ko, V.M., Zaulychnyy, Ya.V., Ilkiv, B.I. et al., Textural and electronic characteristics of mechanochemically activated composites with nanosilica and activated carbon, 1115–1125, 2011f, Copyright 2011, with permission from Elsevier.)

Despite the increase in the slit-shaped pores contribution in the composites, the fractality of silica/AC (Table 4.13, D_{FHH}) is lower than that of AC. This is due to both contribution of smooth nonporous silica nanoparticles with lower fractality and partial destroy of AC microparticles. A weighted sum of the D_{FHH} values of A-300 and AC gives 2.663 and it is 2.713 for A-500+AC. These values are smaller than the D_{FHH} values of the composites because of nonadditive effects caused by the partial decomposition of AC microparticles and penetration of silica nanoparticles into broad pores of AC affecting the fractality of the mixtures.

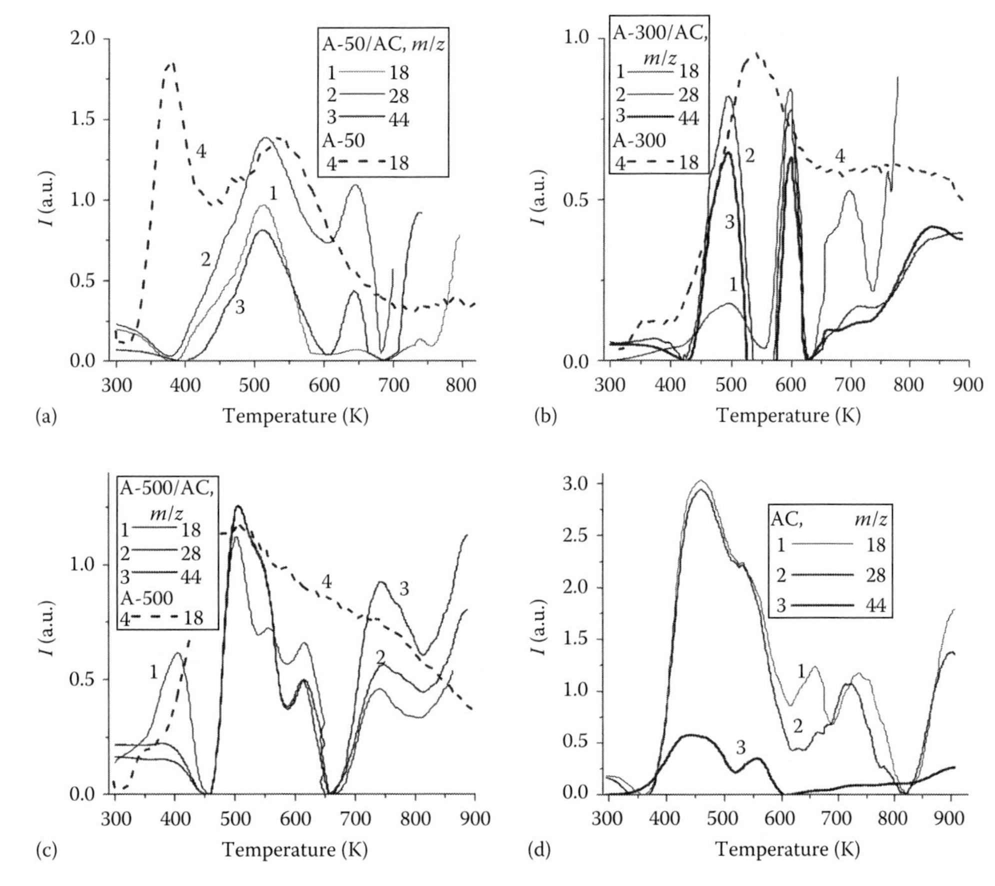

FIGURE 4.37 OPTPDMS thermograms of (a) A-50/AC, (b) A-300/AC, (c) A-500/AC, and (d) AC. (Adapted from *Appl. Surf. Sci.*, 258, Gun'ko, V.M., Zaulychnyy, Ya.V., Ilkiv, B.I. et al., Textural and electronic characteristics of mechanochemically activated composites with nanosilica and activated carbon, 1115–1125, 2011f, Copyright 2011, with permission from Elsevier.)

Changes in the OPTPDMS thermograms of desorption of CO (*m*/*z* 28), CO_2 (*m*/*z* 44), and H_2O (*m*/*z* 18) for silica/AC composites (Figure 4.37a through c) in comparison with that for individual AC (Figure 4.37d) allow one to assume that certain chemical reactions can occur between silica and carbon nanoparticles during MCA. These reactions can be between COH or COOH and SiOH groups with the formation of the Si–O–C bonds. Silica/AC composites, as well as silica alone or AC, contain both molecularly adsorbed water (desorbed at $T<450$ K from silica and at $T<375$ K from AC, Figure 4.37) and hydroxyl groups (appearing in the *m*/*z* 18 thermograms as several peaks at $T>400$–450 K). Concentration of the CO groups (*m*/*z* 28) (eliminated from ≡COH, ≡C–O–C≡, =C=O, –COOH) is greater than that of the C(O)O groups (*m*/*z* 44) (eliminated from –COOH) for both AC alone (Figure 4.37c) and A-50/AC and A-300/AC composites (Figure 4.37a and b). Similar shapes of the *m*/*z* 28 and *m*/*z* 44 lines (and certain similarity with the *m*/*z* 18 line) for composites are due to elimination of CO from COOH with parallel elimination of water from neighboring COOH and COH groups because of close location of O-containing functionalities at the edges of carbon sheets in AC. Partial decomposition of AC microparticles during MCA treatment in air (Figure 4.35) can result in an increase in the number of COOH groups (Figure 4.37a through c, *m*/*z* 44) in comparison with initial AC (Figure 4.37d). Changes in the chemical composition of silica/AC in comparison with individual components, as well as the nanoparticle size effects, can be analyzed using the USXES method.

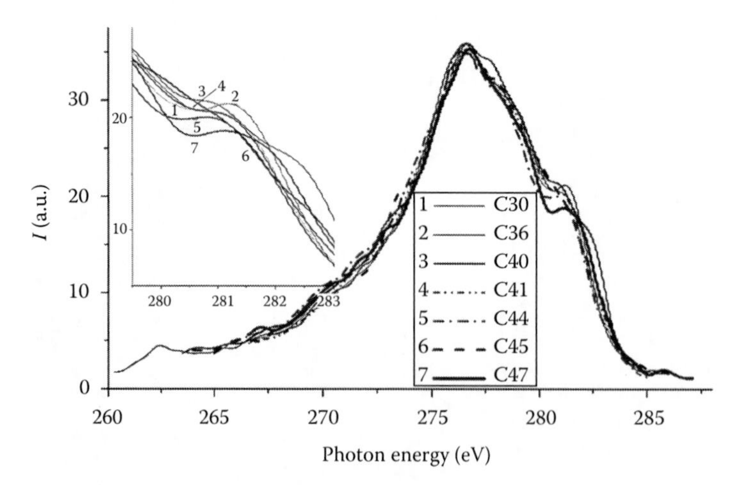

FIGURE 4.38 USXE spectra CK_α of AC (40% burn-off) samples with different burn-off degree (30%–47%). (Adapted from *Appl. Surf. Sci.*, 258, Gun'ko, V.M., Zaulychnyy, Ya.V., Ilkiv, B.I. et al., Textural and electronic characteristics of mechanochemically activated composites with nanosilica and activated carbon, 1115–1125, 2011f, Copyright 2011, with permission from Elsevier.)

Comparison of the emission CK_α bands of AC samples with different degree of burn-off activation (30%–47%) shows (Figure 4.38) that the main difference is due to variations in contribution of O-containing functionalities characterized by a band at 281 eV. The intensity of this band decreases with increasing burn-off degree. This result is in agreement with decreasing intensity of CO and CO_2 lines in the TPD MS spectra of AC with increasing burn-off degree. Notice that the C atoms directly bound to the O atoms have the positive charge, e.g., $q_C = 0.696$ (–C(O)OH), 0.549 (=C=O), 0.418 (≡COH), 0.315 (≡C–O–C≡) calculated using the HF/6-31G(d,p) basis set with a nanopore model with condensed 7+7 aromatic rings O-functionalized at the edges (Figure 4.39a).

However, the C atoms bound to the mentioned C atoms have the negative charges (–0.060, –0.183, –0.084, and –0.151, respectively). Therefore, there is the displacement of local density of electron states (LDES) of these C atoms toward the valence band top (Figure 4.39a). There is alternation of the positive and negative charges of the C atoms located near O-containing functionalities and this charge "wave" attenuates in the next coordination spheres. The CK_α peak at 281 eV can be attributed to the negatively charged C atoms from the second coordination sphere of the O-containing functionalities. The $I(E)$ related to the positively charged C atoms directly bound to the O atoms corresponds to a low-energy wing because the $g(E)$ peaks shift toward the valence band bottom if $q_C > 0$ (Figure 4.39a) and $h\nu = E_{C2p} - E_{C1s}$ decreases. According to the selection rules, the C_{2s} electrons do not contribute the CK_α line. However, the sp^2 and sp^3 hybridization of the C atoms in AC results in "smearing" of both C_{2s} and C_{2p} electrons over the total valence band.

Therefore, changes in the oxidation or reduction of the C atoms and inhomogeneity of AC particles result in the CK_α line practically without a fine structure (Figure 4.38) in contrast to model calculations (Figure 4.39).

The valence bandwidth is larger than that of the core levels; therefore, the width and peak positions in the USXE spectra (Table 4.14) are mainly determined by the characteristics of the valence electrons (with consideration of the selection rules). An increase in the electron density on a certain atom causes the displacement of its local $g(E)$ peaks toward the valence band top and the $I(E)$ peaks in the corresponding USXE spectrum shift toward greater energy. Notice that this effect is opposite to the shift of the XPS peaks because the electron-binding energy typically decreases with increasing electron density of the atoms.

A CK_α curve shape at lower energy $E < 280$ eV (related to C=C and C–C bonds with sp^2 and sp^3 hybridization, respectively, and C atoms in O-containing structures) is practically independent of

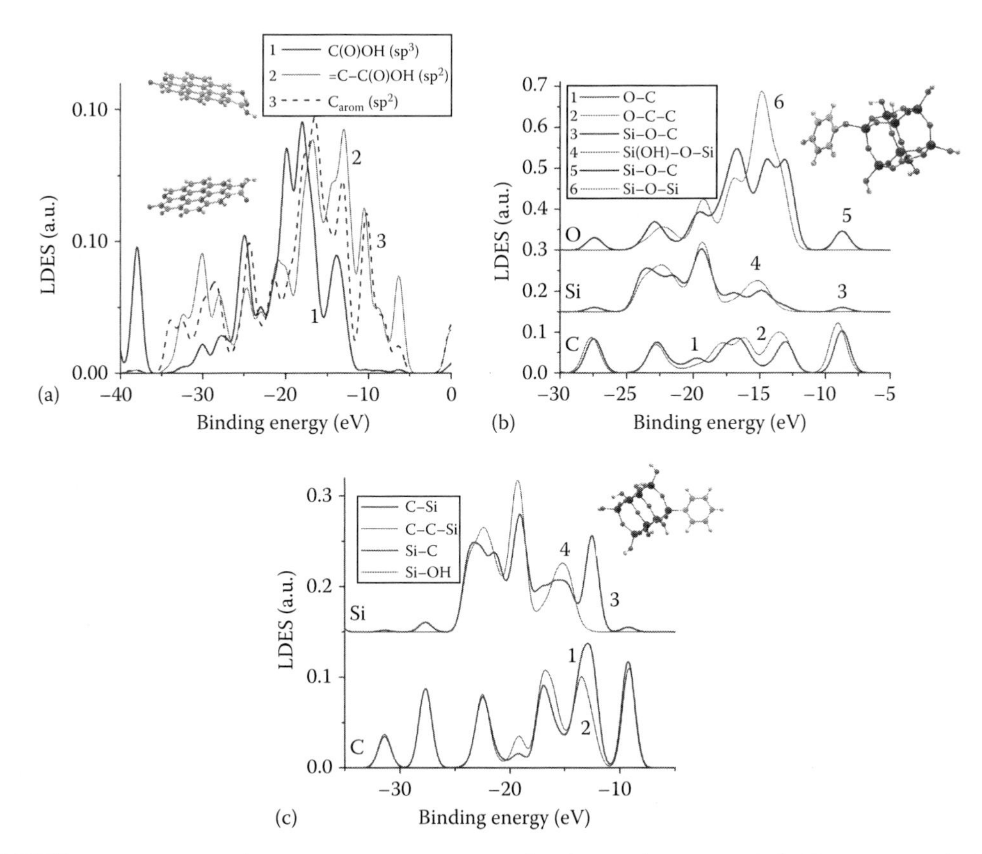

FIGURE 4.39 Local density of electron states (LDES) of (a) C atoms in groups –C(O)OH (curve 1), ≡C–C(O)OH (curve 2) and aromatic C in 7+7 aromatic ring cluster; (b) C, Si, and O atoms in cluster $Si_8O_{20}H_7C_6H_5$ with Si–O–C and (c) C and Si atoms in cluster $Si_8O_{19}H_7C_6H_5$ with Si–C bond (HF/6-31G(d, p) method). (Adapted from *Appl. Surf. Sci.*, 258, Gun'ko, V.M., Zaulychnyy, Ya.V., Ilkiv, B.I. et al., Textural and electronic characteristics of mechanochemically activated composites with nanosilica and activated carbon, 1115–1125, 2011f, Copyright 2011, with permission from Elsevier.)

the burn-off degree of AC samples. This similarity is due to the removal of more active, polar, and disordered fragments of AC during their burn-off activation (30%–47%) by CO_2 at high temperature, preservation of the main condensed aromatic skeleton of CNPNP at different burn-off degrees, and the inhomogeneity of whole AC nanoparticles.

There are differences in the USXE spectra not only of silica/AC and AC or silica (Figure 4.40) but also of silicas alone (Figure 4.41) depending on the size of nanoparticles (Table 4.14). A decrease in the size of silica nanoparticles from 52.4 nm (average diameter of A-50 primary particles) to 5.5 nm (A-500) results in a small shift of the main OK_α peak at 526 eV toward lower photon energy (Figure 4.41a) by 0.5 eV (Table 4.14), i.e., 2p-electron density on the O atoms in A-500 decreases. Notice that A-500 nanoparticles are being in the flame for shorter time during synthesis than A-300 or A-50 particles.

Additionally, a relative number of surface atoms is greater for A-500, as well as the Laplace pressure in the surface layers of smaller particles. All these effects result in changes of the electron density of both oxygen and silicon atoms in silicas appearing in their USXE spectra (Figure 4.40).

A decrease in the photon energy of the OK_α band in the composites in comparison with silicas alone suggests diminution of the charge density of the O atoms and a decrease in the bandwidth of smaller silica nanoparticles. In the AC/silica composites, the population of the 3s-Si levels increases in the low-energy range and the SiL_α band is broadened toward higher energy of photons

TABLE 4.14
Decomposition of USXES Bands: Peak Position, Width (FWHM), and Area

Parameter	A-50	A-50/AC	A-300	A-300/AC	A-500	A-500/AC	AC-40
CK_α (eV)		269.7				270.9	267.3
FWHM (eV)		6.7				6.6	10.5
Area (%)		11.2				16.6	6.6
CK_α (eV)		274.2		272.1		275.0	273.2
FWHM (eV)		4.7		6.8		4.4	6.9
Area (%)		21.4		23.5		21.1	25.7
CK_α (eV)		277.2		276.9		277.6	277.3
FWHM		3.5		4.0		3.7	4.8
Area (%)		40.4		48.4		40.0	51.2
CK_α (eV)		280.0		280.0		280.6	280.8
FWHM (eV)		2.3		2.4		2.3	2.7
Area (%)		18.2		17.2		15.5	11.7
CK_α (eV)		282.2		281.9, 283.3		282.5	282.5
FWHM (eV)		1.4		1.5, 0.8		1.4	1.6
Area (%)		8.7		8.4, 2.5		6.7	4.7
OK_α (eV)		519.7		520.6	514.5		
FWHM (eV)		1.2		1.0	3.1		
Area (%)		3.8		5.8	0.7		
OK_α (eV)	521.2	521.6	522.0	522.0	522.8	521.0	
FWHM (eV)	4.6	2.1	5.0	1.7	6.2	2.6	
Area (%)	23.8	15.6	32.3	11.5	45.5	17.5	
OK_α (eV)	525.8	525.8	524.5	525.5		525.2	
FWHM (eV)	4.2	4.0	1.7	3.5		4.0	
Area (%)	47.6	62.8	7.1	62.7		58.7	
OK_α (eV)	526.7	526.4	526.4	526.4	526.2	526.4	
FWHM (eV)	1.4	1.1	2.0	1.1	2.2	1.2	
Area (%)	15.9	14.8	41.4	20.0	43.7	20.0	
OK_α (eV)	530.6	530.4	528.4		529.9	529.8	
FWHM (eV)	8.4	2.4	8.1		4.0	6.1	
Area (%)	12.7	2.9	19.2		10.1	4.1	
SiL_α (eV)	89.1	89.1	89.2	89.3	89.0	89.2	
FWHM (eV)	3.5	2.8	3.9	3.3	4.3	3.2	
Area (%)	29.2	21.7	23.5	26.5	38.4	25.3	
SiL_α (eV)	91.6	91.6		91.7	91.6	91.7	
FWHM (eV)	1.8	2.2		1.8	1.7	1.6	
Area (%)	5.2	9.0		5.9	3.7	5.5	
SiL_α (eV)	94.6	93.8	94.4	94.7	94.2	94.8	
FWHM (eV)	3.4	12.8	3.6	16.7	2.9	3.4	
Area (%)	43.7	36.4	39.0	29.0	30.5	36.0	
SiL_α (eV)	95.2	94.9	97.2	94.8	95.5	95.0	
FWHM (eV)	10.9	3.3	19.0	3.7	5.7	15.2	
Area (%)	21.9	32.9	37.6	38.6	27.4	33.2	

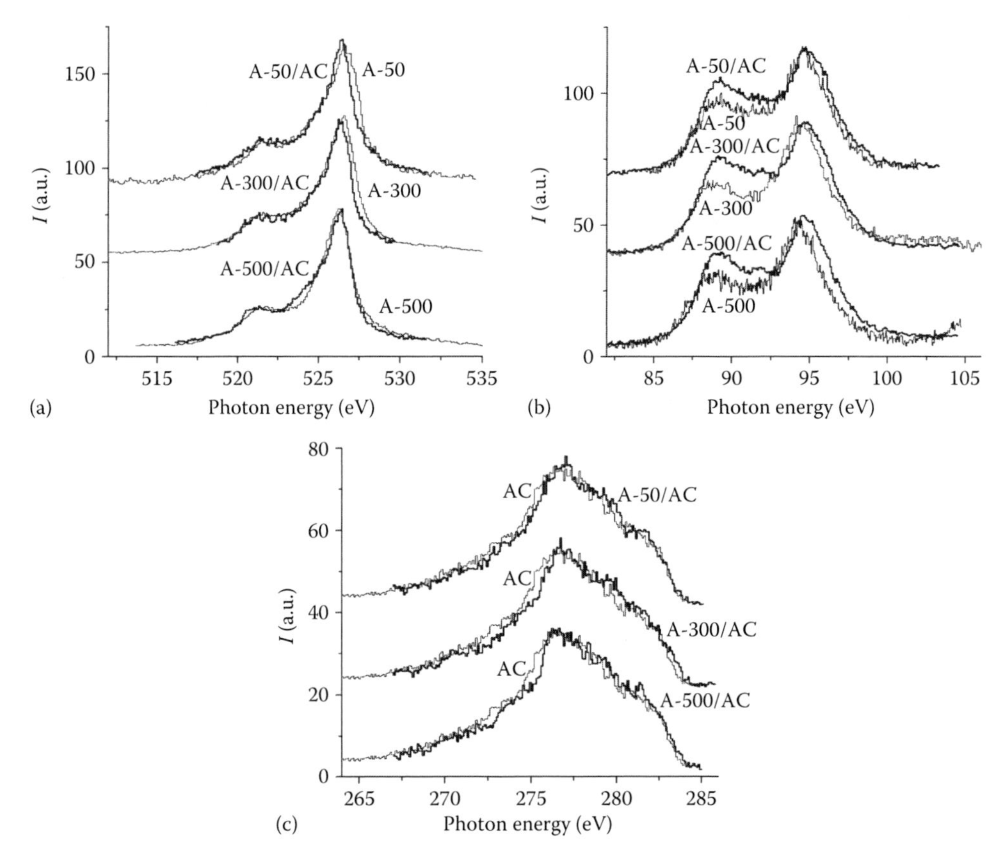

FIGURE 4.40 (a) OK_α, (b) SiL_α, and (c) CK_α spectra for silicas, AC (CK_α), and silica/AC composites. (Adapted from *Appl. Surf. Sci.*, 258, Gun'ko, V.M., Zaulychnyy, Ya.V., Ilkiv, B.I. et al., Textural and electronic characteristics of mechanochemically activated composites with nanosilica and activated carbon, 1115–1125, 2011f, Copyright 2011, with permission from Elsevier.)

(Figure 4.40b). This can be due to charge transfer of 2p-electrons from O and C atoms, overlapped with the sp^3-hybridized orbitals of Si atoms at the interfaces of silica and AC particles. An increase in this binding causes a decrease in the CK_α intensity in the low-energy range (Figure 4.40c). The displacement of the main OK_α peak at 526 eV of silica/AC decreases with decreasing silica nanoparticle size (Figure 4.41a).

Additional electron density appears on the O atoms located at the silica/AC interfaces (e.g., charge $q_O = -0.774$ (≡Si–O–C≡), −0.700 (≡SiOH), and −0.850 (≡Si–O–Si≡) calculated using HF/6-31G(d,p), Figure 4.39). The position of this peak for silica/AC samples is practically the same (Figure 4.41c) in contrast to that of pure silica (Figure 4.41a). However, relative intensity (normalized to the peak intensity) of OK_α at $E < 526$ eV increases for A-500/AC. This is due to changes in the electron density on a portion of the O atoms due to better contacts between AC and silica particles because smaller silica nanoparticles are better distributed in broad mesopores and macropores AC microparticles (Figure 4.34). Thus, the charge transfer between AC and silica nanoparticles depends on the size of silica nanoparticles. The maximal diminution of the electron density on the O atoms is for A-50/AC because the displacement of the main OK_α peak is greater than that for other composites (Figure 4.40a).

Changes in the SiL_α spectra (Figure 4.40b) in silica/AC are much greater than that of the OK_α (Figure 4.40a) and CK_α (Figure 4.40c) spectra, i.e., the electron density located on the Si atoms is more sensitive to the binding of silica nanoparticles to carbon microparticles. The relative intensity of a lower energy part of SiL_α at $E < 92$ eV increases for composites in comparison with that of

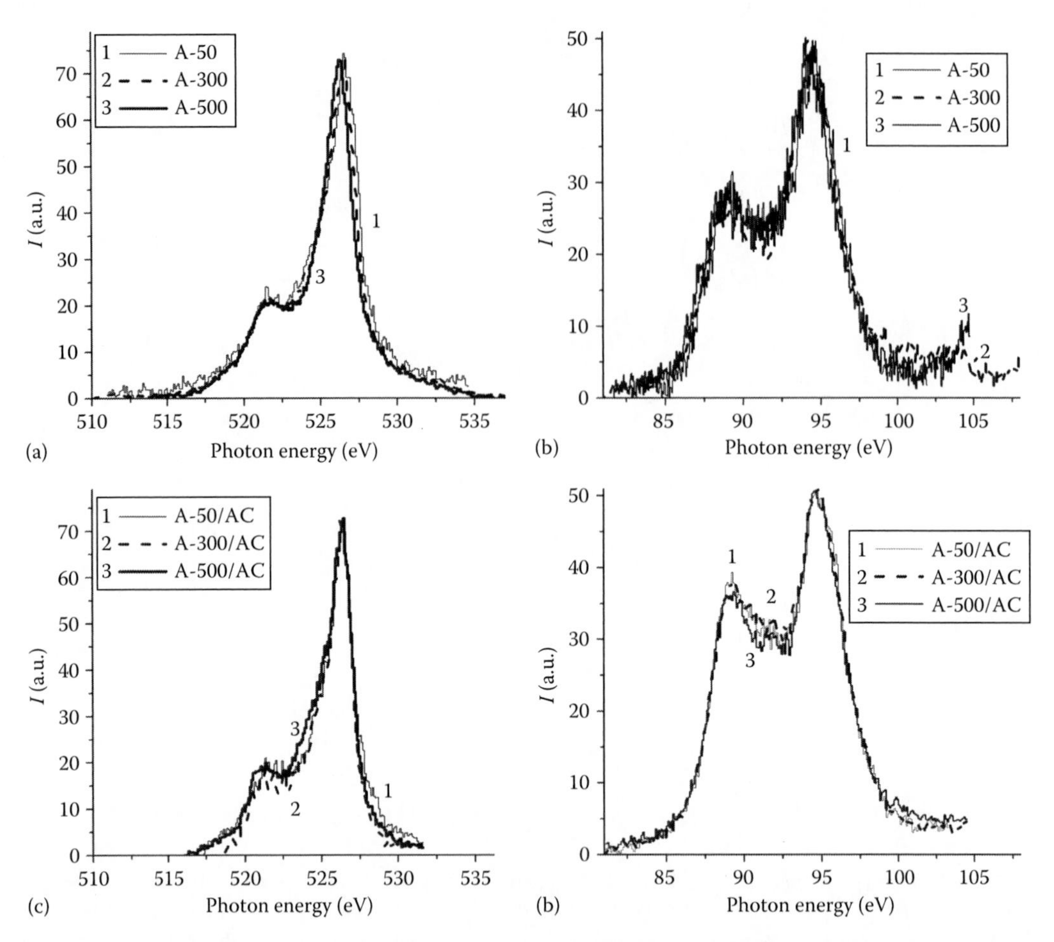

FIGURE 4.41 (a, c) OK_α and (b, d) SiL_α spectra of (a, b) silicas and (c, d) MCA-treated silica/AC composites. (Adapted from *Appl. Surf. Sci.*, 258, Gun'ko, V.M., Zaulychnyy, Ya.V., Ilkiv, B.I. et al., Textural and electronic characteristics of mechanochemically activated composites with nanosilica and activated carbon, 1115–1125, 2011f, Copyright 2011, with permission from Elsevier.)

silica alone. Additionally, the main peak at $E \approx 95$ eV shifts toward higher energy, as well as the peak at 89 eV, and this displacement increases for A-300/AC and A-500/AC in comparison with A-50/AC. Changes in relative intensity of the SiL_α main maxima in the nanocomposites show variations in the SiO_x clusters atomic composition. The silica/AC systems are more nonuniform than initial silicas and AC with respect to the electronic states of the O, C, and Si atoms due to changes in the chemical composition of the systems and possible chemical reactions (during MCA) at the interfaces resulting in the formation of the ≡Si–O–C≡ and ≡Si–OC(O)– (between silica and AC) and ≡Si–O–Si≡ (between adjacent silica nanoparticles) bonds. A SiL_α peak at 92 eV can be linked to the formation of the Si–Si bonds because of removal of surface O atoms interacting with C atoms of AC. However, the number of surface Si–Si (or Si–C) bonds is small.

During MCA treatment of the silica/AC mixtures, it is possible local heating that stimulates chemical reactions at the interfaces. The OPTPDMS thermograms (Figure 4.36) show elimination of water over the total temperature range from all samples. Molecularly adsorbed water can be desorbed from all pores at $T < 450$ K (first m/z 18 peak). At higher temperatures, water desorbs due to associative reactions of surface hydroxyls resulting in the formation of Si–O–C and Si–O–Si and other bonds. Contribution of C atoms with higher electron density ($q_C < 0$) increases in silica/AC composites in comparison with AC alone since the CK_α band slightly shifts toward higher energy

(Figure 4.40c). The oxidizing degree of the C atoms ($q_C > 0$) slightly decreases in silica/AC because the CK_α intensity in the lower energy range at $E < 277$ eV decreases (e.g., q_C is larger in COOH than COH or C–O–Si). However, changes in the CK_α band are smaller than that in the SiL_α band. Both mentioned effects suggest that the Si–C bonds were practically not formed (or their number is much smaller than that of the C–C, C=C, C–O, and Si–O–C bonds) during MCA treatment of silica/AC composites.

There are the hydrogen (electrostatic) and siloxane bonds between adjacent silica nanoparticles in aggregates (see Chapter 1). During MCA of silica/AC, the siloxane bonds between silica nanoparticles can be broken and Si–O–C (or Si–C and Si–Si) bonds can be formed. This reaction as well as the reaction ≡SiOH + ≡COH → ≡Si–O–C≡ + H_2O can cause both increase in lower energy SiL_α peak intensity (e.g., $q_{Si} = 1.640$ [Si–O–C] and 1.611 [Si–OH]) and a displacement of this peak toward higher energy (Figure 4.40b) (average electron density on all Si atoms emitting registered photons increases) depending on silica type. A small peak at $E \approx 92$ eV is also silica type-dependent and its intensity is maximal for A-500/AC when the contact area between silica and AC surfaces increases. This gives a stronger shift of both main peaks toward higher energy, i.e., electron density increases for all Si atoms.

Comparison of the normalized superposed OK_α and CK_α spectra (Figure 4.42) shows that relative intensity of the USXE spectra of the oxygen atoms decreases in the silica/AC mixtures with decreasing size of silica nanoparticles. This effect can be due to both changes in the structure of surface SiO_x clusters and more effective penetration of smaller silica nanoparticles of A-500 (Figure 4.36) into broad mesopores and macropores of AC microparticles (Figure 4.34), i.e., the number of the O atoms emitting registered photons decreases since AC includes a much smaller number of O atoms than silica. Despite a larger difference in the $\phi(a)$ functions of A-50 and A-300 than A-300 and A-500 (Figure 4.36), the difference in the OK_α spectra normalized to CK_α is larger for A-300 and A-500 than A-50 and A-300 (Figure 4.42).

Both textural and electronic characteristics of the nanosilica/AC mixtures show that short-term MCA (5 min) allows us to distribute small silica nanoparticles (especially at $a < 5$ nm of A-500) in broad mesopores and macropores of AC microparticles without blocking of nanopores in carbon NPNP. The XPS data on the valence band of silicas and composites (Figure 4.43) show that the presence of AC results in strong increase in relative intensity of the upper band (8–22 eV) and

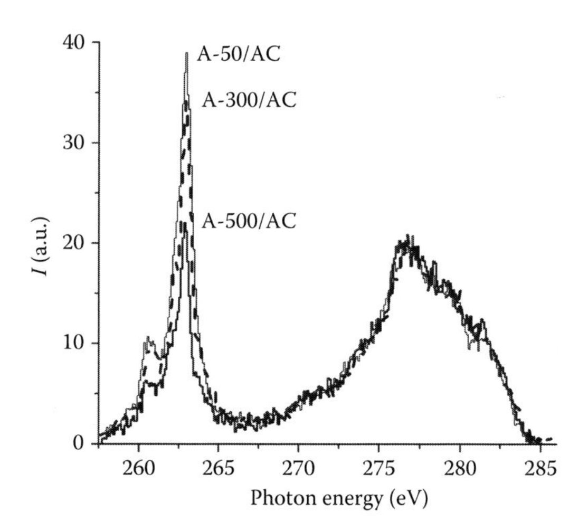

FIGURE 4.42 Emission bands OK_α (second order of diffraction $E^{2\ \text{order}}_{OK_\alpha} = 0.5E^{1\ \text{order}}_{OK_\alpha}$) and C$K_\alpha$ (first order of diffraction) of silica/AC composites normalized to the CK_α peak intensity. (Adapted from *Appl. Surf. Sci.*, 258, Gun'ko, V.M., Zaulychnyy, Ya.V., Ilkiv, B.I. et al., Textural and electronic characteristics of mechanochemically activated composites with nanosilica and activated carbon, 1115–1125, 2011f, Copyright 2011, with permission from Elsevier.)

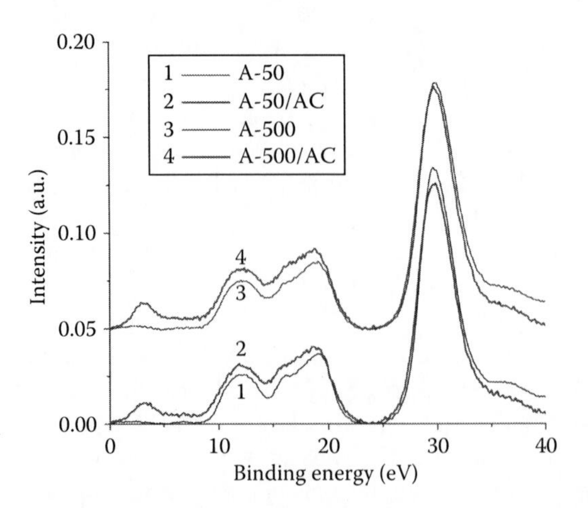

FIGURE 4.43 XPS valence band shape of silica and silica/AC samples (normalized to 1). (Adapted from *Appl. Surf. Sci.*, 258, Gun'ko, V.M., Zaulychnyy, Ya.V., Ilkiv, B.I. et al., Textural and electronic characteristics of mechanochemically activated composites with nanosilica and activated carbon, 1115–1125, 2011f, Copyright 2011, with permission from Elsevier.)

an additional peak appears at the forbidden zone of silica. This suggests that the composites can demonstrate good electrical conduction. Thus, the nanosilica/AC materials can be interesting for practical applications.

Thus, MCA treatment of nanosilica/AC mixtures can result in formation of the Si–O–C bonds between silica and carbon nanoparticles. Smaller silica nanoparticles (A-500 < A-300 < A-50) can more easily penetrate into broad mesopores and macropores of AC microparticles. Both processes are stronger for nanosilica A-500 composed of smaller nanoparticles (average diameter 5.5 nm). A decrease in the OK_{α} band intensity with diminishing size of silica nanoparticles is due to both changes in the surface structure of particles and penetration of a larger number of silica nanoparticles into broad pores of AC microparticles. AC nanoparticles remained nonbroken during MCA and nanopores in these particles are accessible for adsorbate molecules that suggest the effective practical applications of MCA nanosilica/AC composites.

5 Interfacial Phenomena at Polymer Surfaces

Natural and synthetic polymers are used in numerous applications, and the interfacial phenomena at polymer surfaces play an important role in the efficiency of their use, especially in the case of medical, food, and adsorption applications (Zhang et al. 2009, Jayakumar et al. 2010a,b, Long et al. 2011). Here we will analyze the interfacial phenomena with participation of different natural and synthetic polymers both hydrophilic and hydrophobic. In contrast to solid materials, polymers as soft materials can strongly change their texture during interaction with water or other solvents (a striking example with starch swelling and gelatinization). With changing temperature and hydration, polymer chains exhibit a complex hierarchy of dynamic processes, e.g., starting with very fast and local conformational rearrangements on the ps scale (γ relaxation at low temperatures), and extending into the range of seconds for slow, diffusive, and cooperative motions at $T>T_g$ (Doi and Edwards 1986, Kimmich and Fatkullin 2004, Saalwächter 2007). These processes are strongly affected by water or other solvents because of swelling, plasticizing, bonding, and other effects. As a whole, the behavior of polymers and bound water at the interfaces of solid and soft materials depends on several factors such as topology, topography, porosity, and chemistry of a surface, content and type of solvents and/or co-adsorbates, polymer length, cross-linking degree, temperature, etc. NMR techniques can give information on all the mentioned factors for polymers in native state, as well as differently treated (dried, heated, frozen, thawed, MCA, HTT, etc.). This is especially important for hydrogels, hydrated natural polymers, and materials, which can strongly change their characteristics during drying, wetting, or suspending. Therefore, the nature of the interactions between small guest molecules or ions (such as water, organics, salt ions) and large soft host structures (such as carbohydrate polymers, lipid membranes, proteins, swollen gels, cryogels) is of importance in many aspects. The properties of these structures can be intermediate between those of hard solids (crystalline, semicrystalline, amorphous) and liquids. Water bound to polymers can be divided into several types (strongly bound water [SBW] and weakly bound water [WBW], strongly associated water [SAW], and weakly associated water [WAW], nearly bulk water) as well as in the case of solid adsorbents discussed earlier.

5.1 NATURAL POLYMERS: CELLULOSE, STARCH, CHITOSAN, HYALURONIC ACID, AND OTHERS

Structure of swollen polymers and their hydrogels depends on their hydration degree. In the case of porous polymers, their swelling and subsequent freezing–thawing–drying can result in significant irreversible changes in the pore structure (Gun'ko et al. 2005a,d, 2009d). There are several methods that can be used for estimation of the PSD of native polymers and based on a specific behavior of confined liquids: (i) cryoporometry (Gallegos et al. 1987, Dullien 1992, Strange et al. 1993, 2003, Aksnes and Kimtys 2004, Gun'ko et al. 2005b, 2009d) based on the ^{1}H NMR measurements at $T<273$ K and the Gibbs–Thomson melting point depression (MPD) of confined liquids; (ii) relaxometry (Gallegos et al. 1987), which uses the enhanced relaxation of molecules at a pore surface depending on the pore size (Gallegos et al. 1987, Strange et al. 2003) and normally assumes rapid exchange between molecules at the surface and in the pores; (iii) thermoporometry based on the calorimetric (DSC) methods for characterizing pore structure from the melting or freezing point depression of a liquid confined in pores, by reason of the added contribution of surface curvature to the phase-transition free energy (Dullien 1992, Aksnes and Kimtys 2004); and (iv) dielectric techniques such as thermally stimulated depolarization current (TSDC) (Gun'ko et al. 2005d, 2007i, 2009a–f) and dielectric relaxation spectroscopy, DRS (Barbetta et al. 2012) describing relaxation processes affected by the interfacial orientational polarization of the aqueous phase at the polymer interface. Notice that in this book all these approaches are used.

The ^{1}H NMR cryoporometry (as well as TSDC, DRS, DSC, and some other methods) can be successfully applied to describe hydrogels and swelling powders of different carbohydrate polymers such as cellulose, starch (Figure 5.1), hyaluronic acid (HA), etc., in different states and different dispersion media. These nonstandard (for textural characterization) techniques allow one to study the characteristics of soft materials in their native hydrated states. This is of importance because polymers as soft materials can strongly change their texture during drying and degassing applied in the standard adsorption measurements of the textural characteristics. For instance, very significant structural changes of starches during gelatinization are well known (Galliard 1987, Atwell et al. 1988, Butler and Cameron 2000, Motwani et al. 2007). Similar processes should be considered during measurements of many of hydrophilic polymeric materials depending on the hydration degree.

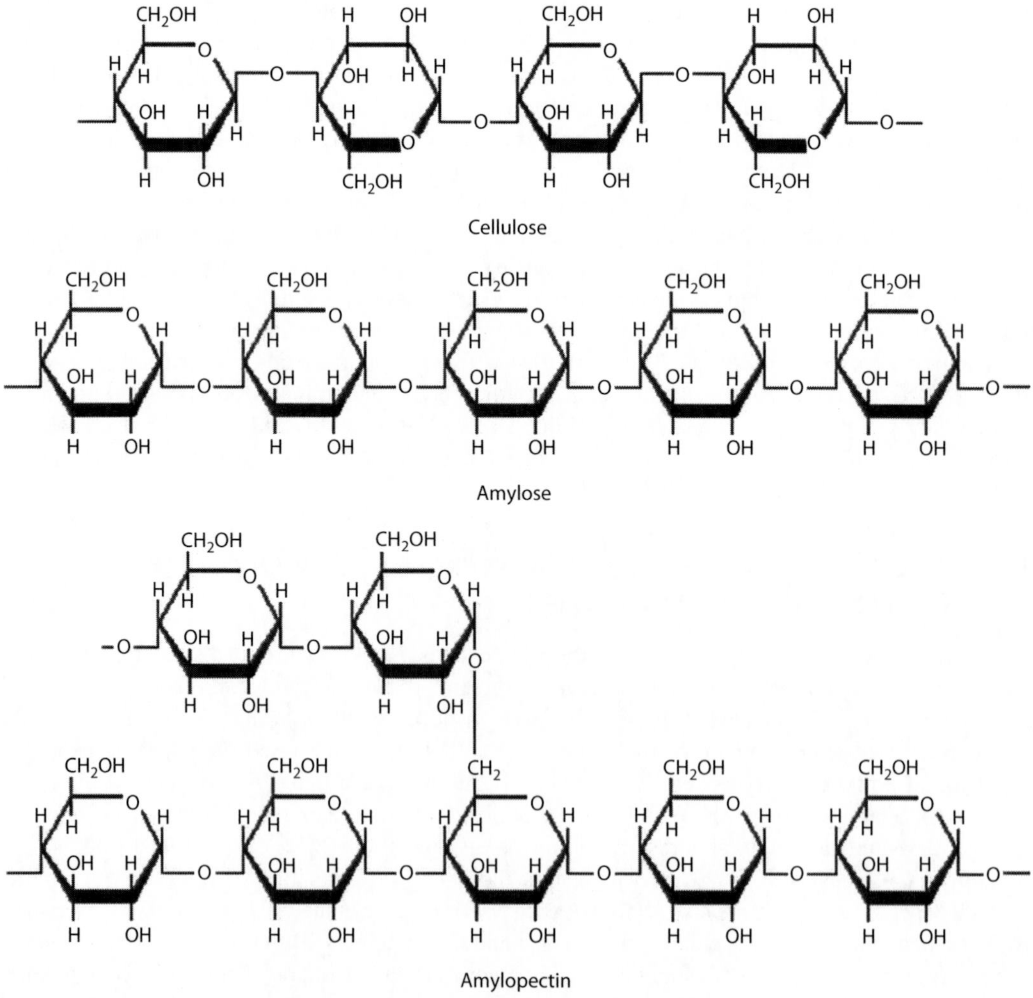

Cellulose

Amylose

Amylopectin

FIGURE 5.1 Structure of cellulose, amylose, and amylopectin (starch components).

Chemical structure of polysaccharides and their derivatives has a similar chain skeleton as shown in Figure 5.1 with the differences in the side groups and the ring connection places. Double-helix structure of starch consisting of essentially linear poly(1,4-α-D-glucan) (amylose, molecular weight $M_w \approx 0.1$–1.0 MDa) and branched amylopectin ($M_w \approx 0.01$–1.0 GDa) with 1,4-α-D-glucan chains connected through 1,6-α linkages (Figure 5.1) depends on biological origin, amounts of water, and temperature. In the case of potato starch (vide infra) having hexagonal unit cell ("B" type) of the space group $P6_1$, content of amylose is 17%–20% (apparent amylose ~23%) (Galliard 1987, Butler and Cameron 2000). Gelatinization of starch is accompanied by several processes such as swelling, leaching of amylose, loss of crystallinity, increase in viscosity, solubilization of macromolecules, etc. (Atwell et al. 1988, Motwani et al. 2007). Some of these effects lead to a decrease in the mobility of macromolecules and bound water molecules and influence the dielectric and other properties of the starch–water systems (Jaska 1971, Ryynänen et al. 1996, Gun'ko et al. 2008c) dependent on water content, starch gelatinization, and temperature (Bircan and Barringer 1998, Ndife et al. 1998a,b). Generally, the dielectric constant and the loss factor increase with increasing water content in the starch systems (Tsoubelli et al. 1995, Gun'ko et al. 2008c).

Different relaxation processes characteristic of modified and unmodified starches were studied by the DRS method over wide temperature and frequency ranges (Miller et al. 1991, Butler and Cameron 2000, Einfeldt et al. 2000, 2001, 2002, Majumder et al. 2006, Gun'ko et al. 2008c). Majumder et al. (2006) observed two α relaxations and two γ relaxations in starch depending on hydration, with excess of water able to unravel the double-helix structure. Subsequently, relaxation processes associated with collective helical motions depend on water concentration. Additionally, several kinds of phases, such as intercrystalline and liquid crystalline ones, were identified depending on water content and temperature. The total activation energy of relaxations increased with increasing water content. Einfeldt et al. (2000, 2001, 2002) observed three relaxations for starch/water: the β relaxation associated with the local chain motion, the δ relaxation with an unknown molecular origin, and the β_{wet} relaxation with motions of a water/starch mixing phase. They showed that the mixing of wet starches with vacuum grease did not affect the polymeric dynamics, and the type of starch had also no great influence on the polymeric dynamics. In addition, starches demonstrated a similar dynamic behavior like other pure polysaccharides, because the main point for the characterization of the segmental motion is the type of the glucosidic linkage and not its position and orientation at the repeating unit (Einfeldt et al. 2000, 2001, 2002). The water content has the most important influence on the position of the relaxations because of its impact as a plasticizer. According to Butler and Cameron (2000), two relaxations are observed in starch at ~153 K because of low motions of the chain backbone and rotation of methylol groups (with the latter process activated at lower temperatures than the former). Around ambient temperature a relaxation proposed to be the glass transition was observed and, in granular starch only, gelatinization (it is observed for potato starch at 332 K) caused a relaxation around 333–353 K. Thus structure of strongly swelling polymers such as starch in wetted powders and hydrogels depends on hydration, treatment temperature, and other conditions (Donovan 1979, Galliard 1987, Butler and Cameron 2000, Tananuwong and Reid 2004, Ma et al. 2007c, Gun'ko et al. 2008c, 2010a).

There are several methods utilized to estimate the structural characteristics (e.g., PSD) of wetted powders or hydrogels on the basis of a specific behavior of pore-confined liquids including not only NMR cryoporometry but also TSDC cryoporometry based on the dependence of dipolar relaxation of water affected by the size of confined space (Gun'ko et al. 2005d, 2007d,f,i, 2009d). Notice that in the case of rigid porous polymers, their swelling and subsequent freezing can result in significant irreversible changes in the pore structure (Gun'ko et al. 2005c). It was shown (Shulga et al. 2000, Gun'ko et al. 2005d, 2008c, 2009d) that interactions of polar polymers, proteins, DNA, etc., with nanosilica affect hydration and structure of macromolecules. A similar effect occurs during interactions of starch with nanosilica (Gun'ko et al. 2008c, 2010a).

A choice of starch (mixture with amylose and amylopectin), cellulose, and HA as probe natural polymers is caused by their high hydrophilicity which provides good contacts with water molecules, effective polymer swelling, and formation of hydrogels, which can be well mixed with aqueous suspension of nanosilica. Notice that the interactions of cellulosic materials with nanosilica can strongly affect the polymer hydration (Shulga et al. 2000). Cellulose fibrils with widths in the nanometer range are natural materials with unique and potentially useful features (Klemm et al. 2011). The fibrils are the basis of all the cellulosic materials. However, their properties can be strongly varied due to changes in the length and packing of fibrils and surroundings (as in the case of cotton and flax fibers compared later).

Water as an integral part of the structure in any biological porous materials strongly affects their textural characteristics because of the swelling effects. Removal of this water (e.g., during sample preparation for a study) results in a decreasing distance between supramolecular structural elements and a distortion of the structure. Therefore, methods such as NMR spectroscopy, which can be applied to native materials, should be rather nondestructive. Topgaard and Söderman (2002) studied water-swollen biological porous structures (wood pulp fibers and potato starch) using NMR to determine the amount and self-diffusion of water within the porous objects. The contribution of bulk water to the NMR signal was eliminated by performing experiments below the bulk freezing temperature. Further decrease of the temperature leads to a gradual freezing of water within the porous objects (Figure 5.2). The contribution of the freezing water fraction to the migration of water through the porous network was estimated. The results were rationalized in terms of the ultrastructure of the samples studied (wood pulp fibers and potato starch granules). The diffusion rate decreases with decreasing hydration degree and temperature (Figure 5.2). If the nonfreezing water diffuses locally with D_0 (bulk value), then D/D_0 equals the inverse tortuosity $1/\tau$, which is a purely geometric property of the pore space.

Ibbett et al. (2007) used complementary isotherm and NMR relaxation (the transverse T_2 relaxation time of the water protons in the saturated fiber samples) techniques to characterize a series of solvent-spun cellulosic (lyocell) fibers treated with a methylol urea type resin. Cross-linking reduced overall fiber swelling, caused a loss of both large and small pores, a reduction of the pore volume accessible (0.1–0.45 cm^3/g at water retention up to 1.2 cm^3/g) to large dye (Direct Blue 71, $M_w = 1030$) molecules, and a reduction in the free energy ($-\Delta G = 18$–29 kJ/mol) of the dye adsorption. Post-causticization of resin-treated fiber led to a dramatic improvement in dyeability, consistent with a recovery in total pore capacity and accessible volume. Structural rearrangement (total surface

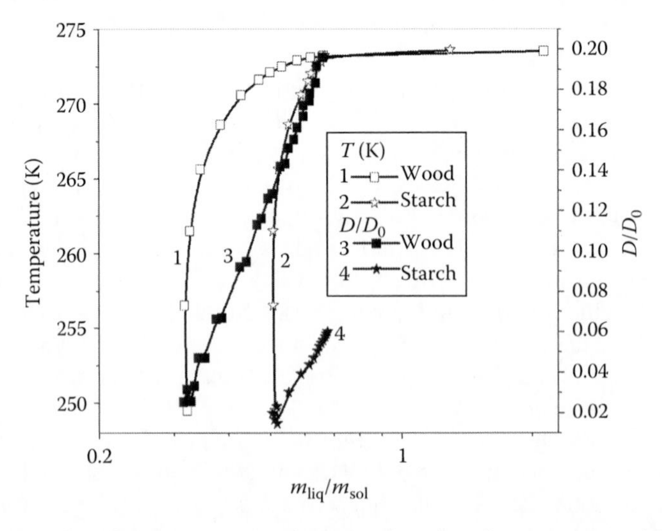

FIGURE 5.2 Amount of nonfreezing water (m_{liq}/m_{sol}) as a function of temperature and D/D_0 as a function of amount of nonfreezing water for wood pulp fibers and potato starch granules. (Adapted from *Biophys. J.*, 83, Topgaard, D. and Söderman, O., Self-diffusion of nonfreezing water in porous carbohydrate polymer systems studied with nuclear magnetic resonance, 3596–3606, 2002, Copyright 2002, with permission from Elsevier.)

area was between 343 and 126 m^2/g, pore size 4.4–9.4 nm) could be associated with relaxation of internal stresses at resin cross-linking and the reduction of pore aspect ratio. Contraction was driven by entropy effects, following the development of elastomeric behavior in the swelled state, creating new pore spaces that were free of cross-link molecules. Causticization caused the extraction of low-molecular-weight polymer with a higher proportion of negatively charged carboxylate groups, which also favored higher dye uptake (Ibbett et al. 2007).

Perkins and Batchelor (2012) measured water diffusion coefficients with PFG NMR applied to a variety of paper materials made from predominantly cellulose fiber and nanofibers, derived from wood, with different dimensions, internal porosity, and chemical composition. The moisture content ranged from 0.2 to 1.2 g of water per gram of dry fiber. Diffusion measurements were made both in the plane and through the thickness of the sheet. All data were generally well fitted by a simple two-component diffusion model. For moisture contents less than 0.55 and 0.85 g/g for measurements in the plane and through the thickness, respectively, it was found that both diffusion components increased approximately linearly with moisture content, with the faster diffusion coefficient being approximately five times larger than the smaller. The water appeared, within errors, to be evenly split between two components. The measured diffusion coefficients were not affected by fiber dimensions, internal structure, or chemical composition, but were consistently higher when measured in the plane. The relationship between signal attenuation and diffusion coefficient can be given by equation:

$$\ln\left(\frac{I}{I_0}\right) = -D\gamma^2\delta^2 g^2\left(\Delta - \frac{\delta}{3}\right) \tag{5.1}$$

where

I is the measured signal intensity
I_0 is the signal intensity if no diffusion occurred
D is the calculated diffusion coefficient
γ is the gyromagnetic ratio of the nuclei being observed
δ is the length
g is the strength of the applied gradients
Δ is the time between the two applied gradients

The experimental data were fitted with two-component equation (Perkins and Batchelor 2012):

$$I = I_0\left\{(1-p)\exp(-D_1\gamma^2\delta^2 g^2(\Delta - \delta/3)) + p\exp(-D_2\gamma^2\delta^2 g^2(\Delta - \delta/3)\right\} \tag{5.2}$$

where

p is the fraction of water in the slow component
D_1 and D_2 are the diffusion coefficients of the fast and slow components, respectively

The diffusion data was fitted by a two-component diffusion model and an alternate classification of the two components was discussed by Perkins and Batchelor (2012) in combination with traditional definitions of water types within such systems. At intermediate moisture contents, the diffusion coefficients of both components of water were observed to decrease with decreasing moisture content because SBW contribution increases. The relative mass fractions of water in each component were near equal and the ratio between the two diffusion coefficients remained near constant, implying that there was interaction between the two water components under these moisture conditions. Through plane and in-plane studies show an increase in the interaction effects and tortuosity of the diffusion motion for water through the sheet compared to that within the plane of the sheet. The fiber ultrastructure had no measured effect on the diffusion behavior (Perkins and Batchelor 2012).

Park et al. (2006) studied changes in pore size distribution during the drying of cellulose fibers using DSC with an isothermal step melting procedure.

Gellan is an anionic extracellular bacterial polysaccharide discovered in 1978 (Morris et al. 2012). Acyl groups present in the native polymer are removed by alkaline hydrolysis in normal commercial production, giving the charged tetrasaccharide repeating sequence: →3)-β-D-Glc*p*-(1→4)-β-D-Glc*p*A-(1→4)-β-D-Glc*p*-(1→4)-α-L-Rha*p*-(1→. Deacylated gellan converts on cooling from disordered coils to threefold double helices. The coil→helix transition temperature (T_m) is raised by salt in the way expected from polyelectrolyte theory: equivalent molar concentrations of different monovalent cations (group I and Me_4N^+) cause the same increase in T_m. There is also no selectivity between different divalent (group II) cations. However, divalent cations cause greater elevation of T_m than monovalent ones. Morris et al. (2012) analyzed the effects of temperature on ^{23}Na, ^{39}K, and ^{87}Rb NMR spectra for 20 mM NaCl, KCl, or RbCl in the presence of 1.5 wt% Na^+ gellan, and almost no thermal hysteresis was observed between the changes seen as temperature was raised or lowered. Group I cations decrease repulsion by binding to the helices in specific coordination sites around the carboxylate groups of the polymer. Strength of binding increases with increasing monovalent cation size ($Li^+ < Na^+ < K^+ < Rb^+ < Cs^+$). The extent of gellan aggregation and effectiveness in promoting gel formation increase in the same order (Morris et al. 2012).

The starch structure is well studied and documented (Whistler et al. 1984, Galliard 1987, Frazier et al. 1997, Hancock and Tarbet 2000, Li and Yeh 2001, Katopo et al. 2002, Szymonska and Krok 2003). However, detailed changes in its structure at different hydration degrees are complex, especially with consideration for the role of amylopectin and amylose structures forming granules (1–100 μm depending on the natural source of starch), which change (unfold) the structure during hydration/gelatinization or/and upon interaction with solid nanoparticles. Clearly, the spatial structure of composites with polymers (such as starch, cellulose, etc.) and solid nanoparticles (e.g., nanosilica) is much more complex than that of individual polymers or nanosilica. The concentration dependences of the structural and other parameters can be nonlinear functions. The properties of nanocomposites depend on features of interactions between nanoparticles and polymer molecules (Kasemo 2002) and individual components itself affected by the morphology, chemical structure, unfolding of polymer molecules, concentration and distribution of components in the system, temperature, pH, and salinity (aqueous media) and composition of dispersion organic media in certain cases (e.g., in the case of cryopreservation). Therefore, the structural and relaxational characteristics of starch and starch/nanosilica powders and hydrogels will be analyzed at different hydration degrees and temperatures using the low-temperature ^1H NMR spectroscopy (static samples), cryoporometry, TSDC, and DRS methods.

Potato starch (KhimLabReactive, Ukraine) was used as received to form hydrated powders (starch concentration $C_{str} = 50–71$ wt% without gelatinization) and hydrogels ($C_{str} = 4–38$ wt% with gelatinization of starch) (Gun'ko et al. 2008c). Gelatinized (temperature of gelatinization 332 K) starch hydrogels were also used with addition of modified A-300 ($S_{BET} \approx 320$ m^2/g) (stirred at 360 K) at the constant ratio of their content $C_{str}:C_{A\text{-}300} = 7.7:1$ but various hydration degrees at $C_{sp} = C_{str} + C_{A\text{-}300} = 5.6$, 8.3, 10.0, 17.5, and 29.9 wt% (Gun'ko et al. 2008c). Used nanosilica was previously modified by 3-propylaminodiethoxymethylsilane to enhance interactions with hydroxyl groups of starch. The reaction was carried out in toluene at 373 K for 10 h. Grafted 3-propylaminomethylsilyl functionalities substitute ~30% of free silanols that result in the formation of a mosaic surface with hydrophilic (≡SiOH and $-NH_2$) and hydrophobic ($-CH_3$ and $-(CH_2)_3$) functionalities. This structure of silica surface can provide both enhanced interactions with starch and appropriate conditions for appearance of such unusual interfacial water as WAW (Gun'ko et al. 2005d). The NH_2 groups can enhance interaction of starch with the silica surface because of the formation of strong hydrogen bonds $RCO–H\cdots(H_2)N(CH_2)_3(CH_3)(OH)Si–OSi\equiv$ (Figure 5.3).

An increase in the hydration degree ($h = (100 - C_{str})/C_{str}$) of gelatinized starch alone leads to strong diminution of the Gibbs free energy of the interfacial water since the γ_S value significantly increases (Table 5.1). However, the ΔG_s value, corresponding to diminution of the Gibbs free energy of a SBW layer which directly contacts with starch molecules, changes nonlinearly with C_{str} (Table 5.1, ΔG_s).

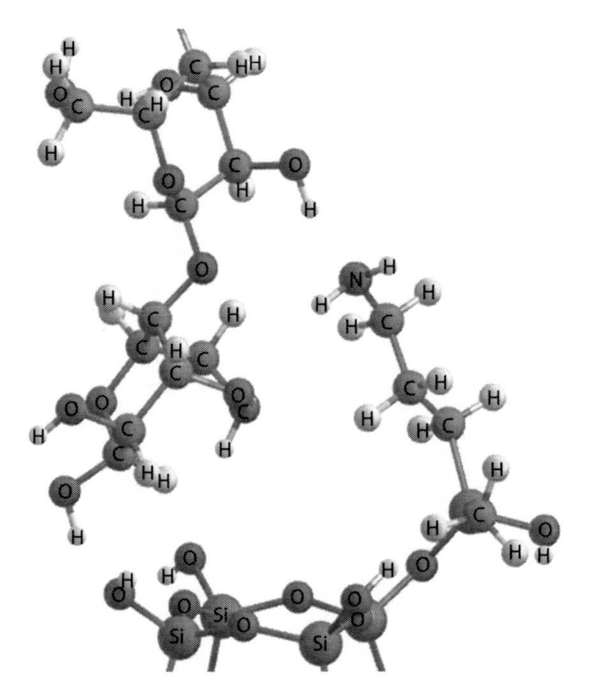

FIGURE 5.3 Hydrogen bonding of a starch fragment to a modified silica surface.

TABLE 5.1
Characteristics of Water Bound in Starch and Starch/Modified Nanosilica at Different Hydration Degrees

C_{sp} (wt%)	h (g/g)	γ_S (J/g)	$-\Delta G_s$ (kJ/mol)	C^s_{uw} (mg/g)	C^w_{uw} (mg/g)	Material
4.2	22.8	220	4.5	1383	776	Hydrogel
11.0	8.1	120	3.2	900	295	Hydrogel
25.4	2.9	80	3.5	549	208	Hydrogel
38.0	1.6	81	3.5	358	242	Hydrogel
50.0	1.0	61	3.2	434	53	Wetted powder
65.6	0.53	44	2.9	344	0	Wetted powder
71.0	0.41	38	3.2	296	0	Wetted powder
5.6	16.9	10	3.9	62	64	Starch/silica hydrogel
8.3	11.1	7	2.8	48	44	Starch/silica hydrogel
10.0	9.0	108	2.9	793	411	Starch/silica hydrogel
17.5	4.7	66	3.0	470	488	Starch/silica hydrogel
29.9	2.3	52	3.0	386	124	Starch/silica hydrogel

Notes: $C_{sp}=C_{str}+C_{A\text{-}300}$; h is the hydration; γ_S is the interfacial free energy; ΔG_s is the changes in the Gibbs free energy of strongly bound water; C^s_{uw} and C^w_{uw} are amounts of strongly and weakly bound waters.

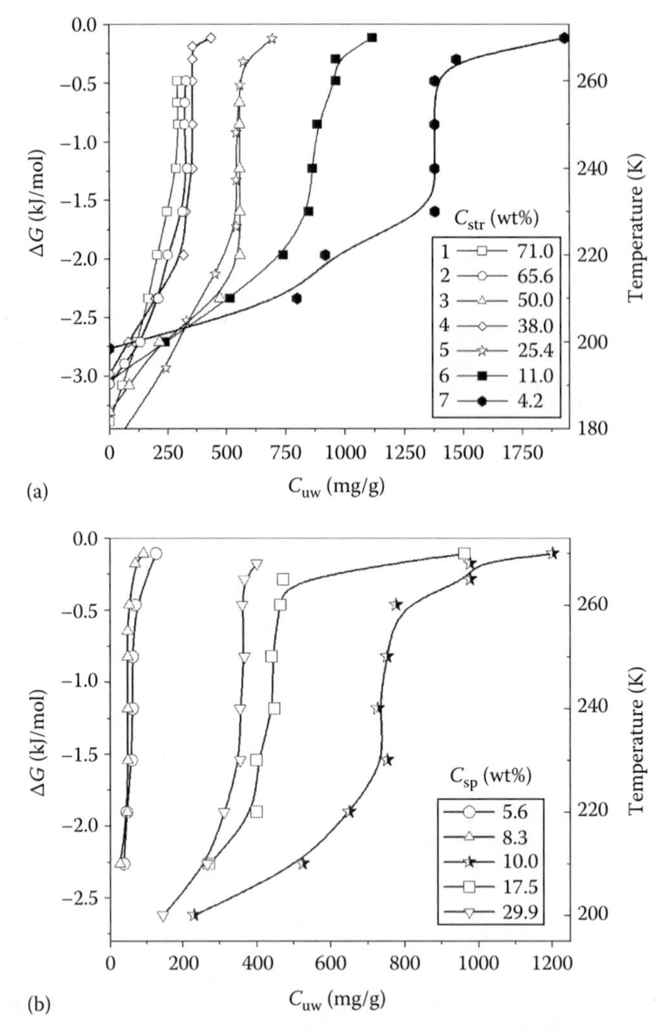

FIGURE 5.4 Temperature dependence of the amounts of unfrozen water (C_{uw}) bound in the starch powders and hydrogels (a) and starch/silica hydrogels (b) at different concentrations of starch (C_{str}) (a) or solid phase ($C_{sp}=C_{str}+C_{A\text{-}300}$ at C_{str}:$C_{A\text{-}300}$=7.7:1) (b) and the relationship between the amounts of unfrozen water and changes in the Gibbs free energy of the interfacial water layers. (Adapted from *Colloids Surf. A: Physicochem. Eng. Aspects* 320, Gun'ko, V.M., Pissis, P., Spanoudaki, A. et al. Interfacial phenomena in starch/fumed silica at varied hydration levels, 247–259, 2008a, Copyright 2008, with permission from Elsevier.)

In the case of the wetted starch powder (C_{str}>50 wt%), WBW is practically absent (Figure 5.4 and Table 5.1, C_{uw}^{W}) because all the water is strongly bound to the starch molecules, and their surface area in contact with structured water is high and increases with increasing hydration degree (Table 5.1, *S*). It should be noted that the surface area (S_{nano}), the volume (V_{nano}), and the radius (R_{nano}) of nanopores increase with increasing hydration but the surface area of mesopores (Table 5.2, S_{meso}) decreases with decreasing C_{str} value from 71 to 50 wt%.

These structural changes are due to increasing swelling of starch granules upon hydration with increasing nanoporosity stimulated by water molecules penetrating into intermolecular space in the starch granules and changing double-helix structure. In the case of the hydrogels (C_{str}=4.2–38 wt%), tendency of the increase in the nanoporosity with decreasing C_{str} value remains (Table 5.2). However, contribution of mesopores increases too in contrast to the wetted powders.

TABLE 5.2
Structural Characteristics of Starch and Starch/Modified Nanosilica Differently Hydrated

C_{sp} (wt%)	$S_{w,sum}$ (m²/g)	S_{nano} (m²/g)	S_{meso} (m²/g)	V_{sum} (cm³/g)	V_{nano} (cm³/g)	V_{meso} (cm³/g)	R_{av} (nm)	R_{nano} (nm)	R_{meso} (nm)
4.2[a]	2369	2219	150	1.84	1.05	0.79	10.01	0.94	10.05
11.0[a]	1964	1906	58	1.08	0.78	0.30	9.47	0.82	9.63
25.4[a]	1180	1145	35	0.67	0.49	0.18	9.62	0.85	9.73
38.0[a]	1184	1142	41	0.66	0.49	0.17	7.09	0.85	7.17
50.0[b]	1065	1052	13	0.47	0.40	0.07	8.74	0.77	9.23
65.6[b]	727	677	50	0.33	0.27	0.06	1.97	0.79	2.47
71.0[b]	641	552	89	0.29	0.22	0.07	1.41	0.78	1.74
5.6[c]	123	108	15	0.12	0.05	0.07	9.37	0.92	9.38
8.3[c]	116	111	5	0.08	0.05	0.03	10.47	0.87	10.50
10.0[c]	1775	1706	69	1.08	0.69	0.39	9.82	0.81	9.87
17.5[c]	1029	948	81	0.87	0.41	0.46	10.16	0.86	10.18
29.9[c]	834	814	20	0.46	0.33	0.13	10.43	0.80	10.48

Note: $S_{w,sum}=S_{nano}+S_{meso}+S_{macro}$; $\langle R\rangle=\int_{R_{min}}^{R_{max}} Rf(R)dR$ at $R_{min}=0.2$, 1, and 0.2 nm and $R_{max}=1$, 25, and 25 nm for R_{nano}, R_{meso}, and R_{av}, respectively.

[a] Hydrogel.

[b] Wetted powder of starch alone.

[c] Hydrogels with starch/modified silica.

The difference between the wetted powders and the hydrogels of starch is due to much stronger unfolding of molecules in the latter.

Vertical sections on the $C_{uw}(T)$ and $\Delta G(C_{uw})$ graphs (Figure 5.4) suggest the absence of freezing of water with significant lowering temperature because this water is strongly bound to the starch molecules or silica nanoparticles and has a strongly distorted hydrogen bond network. The position of this section on the graphs with respect to the point at $T \rightarrow 273$ K (close to glass transition temperature, T_g, of starch) corresponding to total amounts of bound water shows that a portion of SBW in the starch hydrogels is predominant and it is maximal (~75%) at $C_{str}=11$ wt% when the radius of nanopores (R_{nano}) increases. This is due to enhanced unfolding of starch molecules at a high hydration degree. In the case of the hydrogels, there is a small section on the $\Delta G(C_{uw})$ graphs to the right from the vertical section that corresponds to WBW characterized by small changes in the ΔG values ($\Delta G>-0.5$ kJ/mol). This water is not observed in the ^{1}H NMR spectra at $T<265$ K in contrast to SBW which is observed (as the corresponding signal in the NMR spectra) at much lower temperatures (Figure 5.4). Thus, the larger the total amounts of water, the higher the temperature of freezing of SBW because of swelling of granules and an increase in the pore size. A similar effect is observed for starch/nanosilica (Figure 5.4b).

Addition of modified nanosilica to the starch hydrogels can strongly change their structure, supramolecular interactions, and the corresponding thermodynamic parameters of the systems. Observed changes in the shape of the $C_{uw}(T)$ and $\Delta G(C_{uw})$ graphs depend on the ratio $C_{str}/C_{A\text{-}300}$ and the total content of solid phase ($C_{sp}=C_{str}+C_{A\text{-}300}$) in the aqueous medium (Figure 5.4b). At low C_{sp} values, interaction of starch with modified silica results in strong diminution of swelling, and the amount of bound water (Table 5.1, C_{uw} and Figure 5.4) decreases more than by an order of magnitude. An increase in concentration of starch/silica at $C_{sp}=10$ wt% gives a sharp increase in the C_{uw} and γ_S values. This can be caused by enhancement of interaction between polymer molecules that leads to changes in the structure of the interfacial layer, which becomes less dense than at low C_{sp} values. This assumption is confirmed by the changes in the values of the textural parameters (Table 5.2). For instance, the surface area in contact with structured water and the volume of this water are very low at $C_{sp} < 10$ wt% for starch/silica in contrast to starch alone. Strong swelling and unfolding of double-helix structure of starch is observed at $C_{sp} > 10$ wt% and the S and V values increase with respect to nanopores ($R < 1$ nm) and mesopores ($1 < R < 25$ nm). The PSDs (Figure 5.5) which have a bimodal shape in the presence of silica show that contribution of mesopores is maximal at $C_{sp} = 10$ wt%. However, the maximal contribution of nanopores is at the maximal C_{sp} value (Figure 5.5c) in contrast to starch alone (Figure 5.5b) for which this maximum is at minimal C_{str} value. An increase in contribution of nanopores at larger C_{sp} values is due to diminution in the particle–particle, polymer–particle, and polymer–polymer distances in more concentrated hydrogel. A similar effect is observed for the aqueous suspension of nanosilica alone (Figure 5.5a).

To study the effects of interaction of starch with silica, the broadband DRS method was applied to the starch/modified silica system at different hydration degrees. Several relaxations are observed for this system, and their temperature and frequency (i.e., relaxation time) depend on hydration of starch/silica (Figures 5.6 and 5.7). The relaxation at very low frequencies ($f < 1$ Hz) can be assigned to the Maxwell–Wagner–Sillars (MWS) mechanism associated with interfacial polarization and space charge polarization (which leads to diminution of $s < 1$ in Havriliak–Negami equation) or the δ relaxation, which can be faster because of the water effect (Figures 5.8 and 5.9).

Two β relaxations are observed over a middle frequency range ($f < 10^4$ Hz) at $T < T_g$ associated with the local chain motion and water/polymer mixing phase relaxation (Einfeldt et al. 2000, 2001, 2002, 2004, Majumder et al. 2004, Meißner and Einfeldt 2004). Dielectric loss spectra of gelatinized starch/modified silica (Figure 5.7) depend on hydration since intensity, frequency, and temperature characteristics of the observed relaxations change with increasing h value (Figures 5.6 through 5.10).

The γ relaxation is very weak (Figure 5.8b, $f \approx 1$ MHz) and masked by water in the studied system (comp. Figure 5.8b and d at $f \approx 1$ MHz). The β relaxation has a more complex spectral shape at lower h value since two β relaxations are clearly observed (Figure 5.6d). Comparison of normalized $\varepsilon''(f)$ graphs for starch and other polysaccharides (Einfeldt et al. 2001, Meißner and Einfeldt 2004) at 183 K (Figure 5.9a) shows that if practically all water is frozen in the system the β relaxations depend weakly on the hydration degree. However, at high frequencies the γ relaxation is more intensive at the lower h value. Certain mobility of structured water registered by the ^{1}H NMR method appears at $T > 180$–200 K (Figure 5.4).

Therefore, the β relaxation at 243 K (Figure 5.9b) has more complex shape at $h = 16.9$ g/g in contrast to the cases at lower temperatures ($T < 200$ K) when water is immobile (Figures 5.6 through 5.8).

Interaction of starch with silica nanoparticles results in slowing of the relaxations in comparison with starch alone (Figure 5.9) for both low- and high-frequency relaxations at the presence of mobile structured water (e.g., at $T = 213$ K) or on freezing-out of this water (at 173 K) (Figure 5.8). This result suggests that starch molecules strongly interact with modified silica nanoparticles, despite the large amount of water. The conductivity σ and ε'' values decrease with increasing amounts of water (Figures 5.6 through 5.8) because starch and silica are the main sources of mobile protons in the hydrogels. There are three relaxation processes (Figure 5.8b) such as δ, β, and γ, which can be assigned to the motions of different fragments of starch molecules with bound water molecules. Notice that high-frequency relaxation at $f > 10^5$ Hz is less visible at $h = 16.9$ g/g (Figure 5.8d) than at $h = 2.34$ g/g (Figure 5.8b).

The $\tau(T)$ graphs for the β relaxations (Figure 5.10a) demonstrate that there is deviation from the linear Arrhenius dependence $\ln\tau = a/T + b$ because of certain cooperative motions characteristic for starch having the double-helix structure and more complex supramolecular structures affected by water and silica nanoparticles.

This complex behavior of the starch/water/modified silica system is also confirmed by calculations of the distribution function of the activation energy of relaxations (Figure 5.10b) showing that the

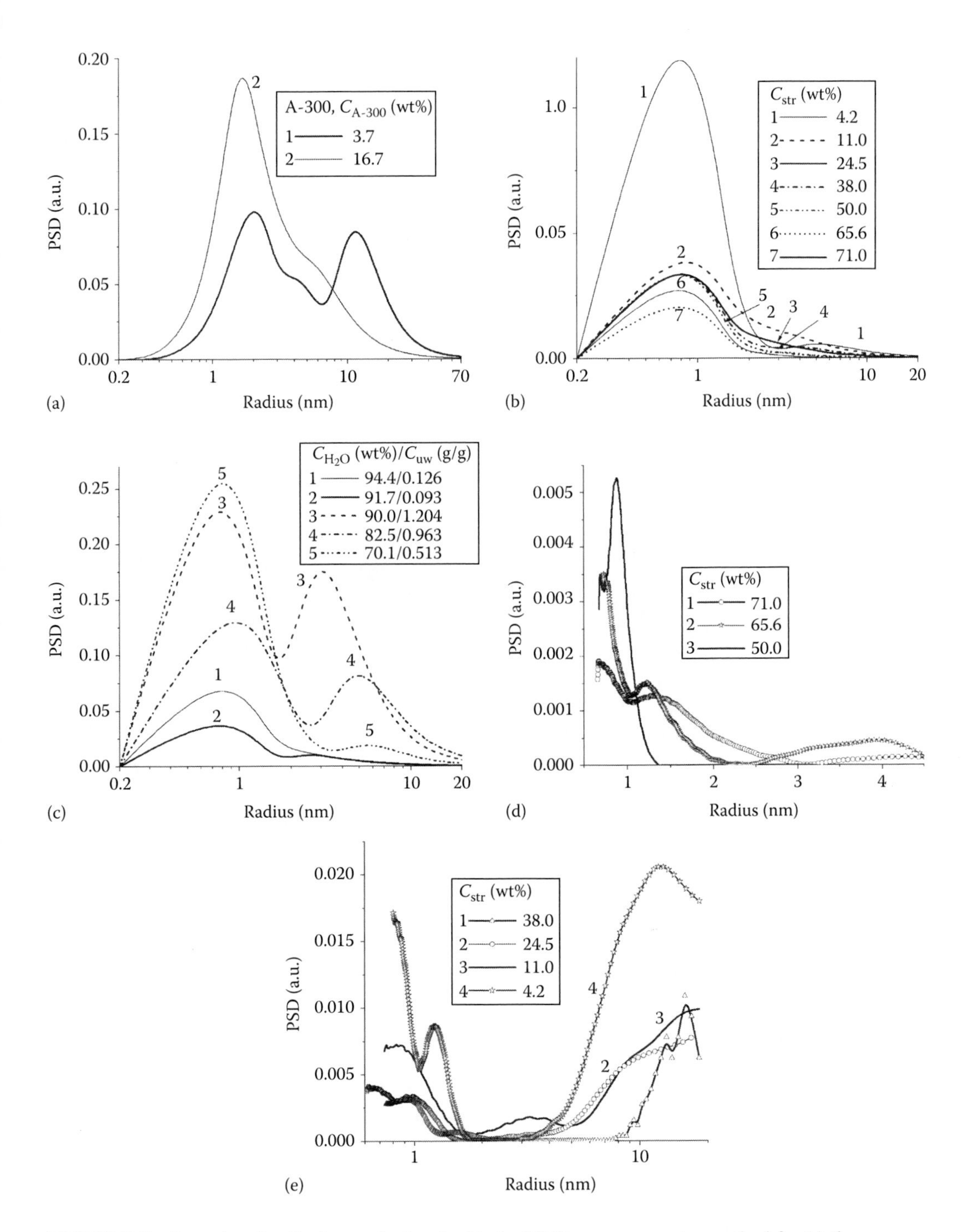

FIGURE 5.5 Pore size distributions calculated with the NMR cryoporometry method for (a) the aqueous suspensions of unmodified nanosilica A-300; (b) starch hydrogel (e) and wetted powder (d), and (c) starch/modified A-300 (7.7:1) at different amounts of water. PSDs were calculated according to (a–c) Aksnes and Kimtys (2004) or (d, e) Gun'ko et al. (2007f, 2008c). (a–c: Adapted from *Colloids Surf. A: Physicochem. Eng. Aspects* 320, Gun'ko, V.M., Pissis, P., Spanoudaki, A. et al. Interfacial phenomena in starch/fumed silica at varied hydration levels, 247–259, 2008a, Copyright 2008, with permission from Elsevier.)

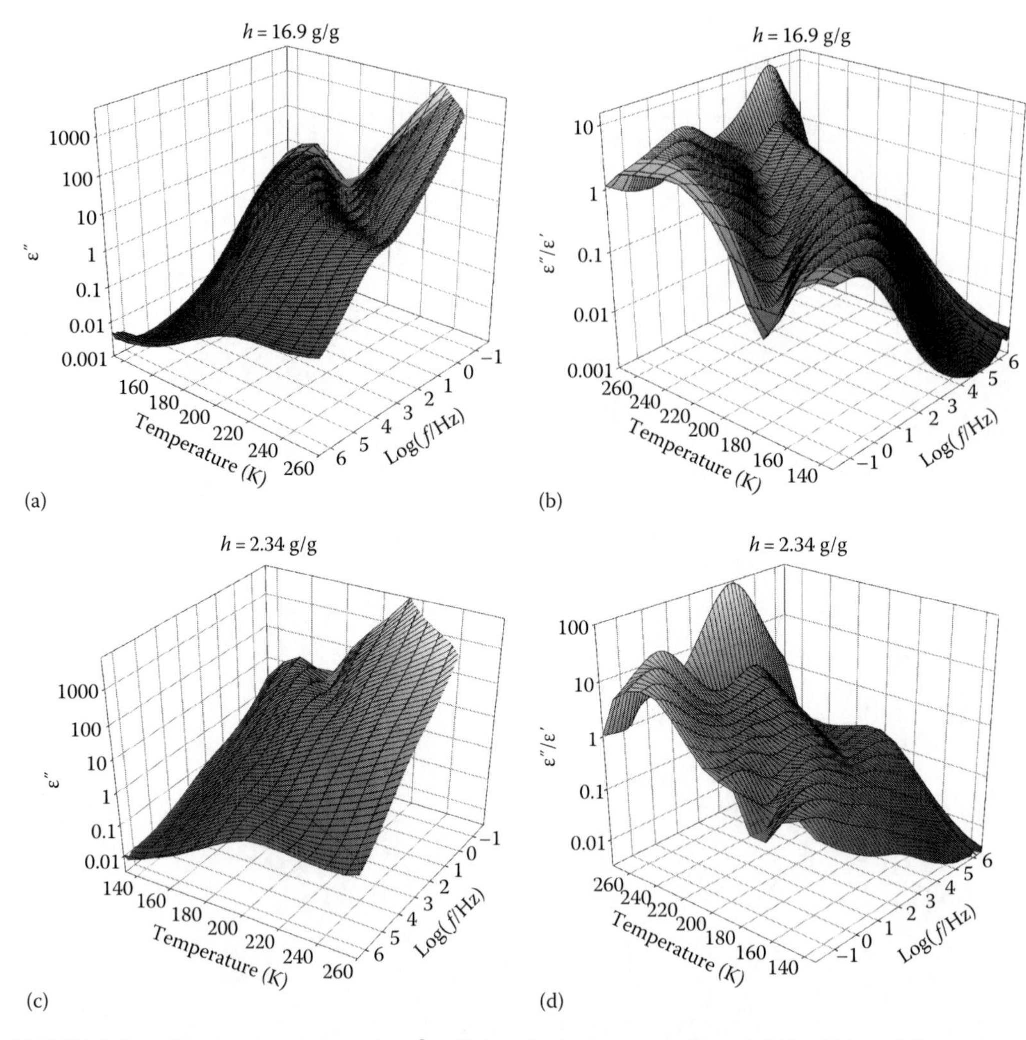

FIGURE 5.6 Dielectric loss (a, c) and $\tan\delta = \varepsilon''/\varepsilon'$ (b, d) for starch/modified A-300 (7.7:1) at different hydrations h = (a, b) 16.9 g/g and (c, d) 2.34 g/g. (Adapted from *Colloids Surf. A: Physicochem. Eng. Aspects* 320, Gun'ko, V.M., Pissis, P., Spanoudaki, A. et al. Interfacial phenomena in starch/fumed silica at varied hydration levels, 247–259, 2008a, Copyright 2008, with permission from Elsevier.)

β relaxation processes are characterized by a broad range of the E values and the $f(E)$ for the β_2 relaxation (higher frequency and temperature than those for the β_1 relaxation) at h = 2.34 g/g is close to the $f(E)$ for the β_1 relaxation (lower frequency and temperature) but at h = 16.9 g/g. Additionally, the β_2 relaxation at h = 2.34 g/g is characterized by a small peak at 140 kJ/mol which can be caused by strong interaction of the starch molecules with the functionalized silica surface.

Dry initial starch including low content of water demonstrates very low TSDC values at $T < 160$ K (Figure 5.11a, curve 1).

These temperatures correspond to relaxation of small water clusters or functional groups of polymers (Gun'ko et al. 2007i). At $T > 160$ K, the TSDC value increases but remains relatively low, and clear relaxation maxima are absent because of restricted mobility of dry polymer densely packed in starch granules. Hydrogel of gelatinized starch alone (Figure 5.11a, curve 2) demonstrates a typical TSDC thermogram observed for hydrogels of polar polymers (Gun'ko et al. 2007i) with two low- ($T < 160$ K, LT) and high-temperature ($T > 160$ K, HT) bands and dc relaxation at $T > 210$ K, which corresponds to appearance of mobile water (Figure 5.4). Concentrated solution of saccharose

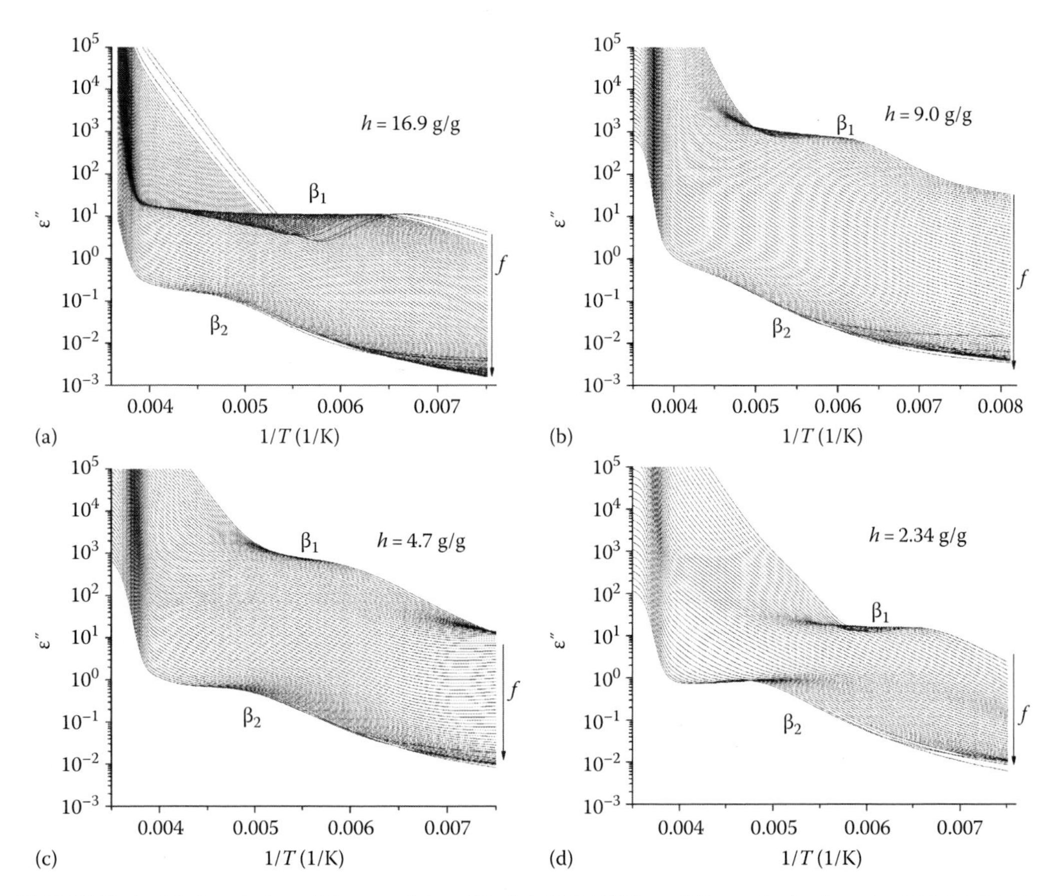

FIGURE 5.7 Dielectric loss ε″ for starch/modified A-300 (7.7:1) at different hydrations h = (a) 16.9, (b) 9.0, (c) 4.7, and (d) 2.34 g/g as function of $1/T$, arrows show increasing f values. (Adapted from *Colloids Surf. A: Physicochem. Eng. Aspects* 320, Gun'ko, V.M., Pissis, P., Spanoudaki, A. et al. Interfacial phenomena in starch/fumed silica at varied hydration levels, 247–259, 2008a, Copyright 2008, with permission from Elsevier.)

(Figure 5.11a, curve 4) and diluted hydrogel with gelatinized starch/nanosilica (curve 3) have close LT bands. However, more strongly diluted latter hydrogel has much higher TSDC values in the HT band due to higher contribution of relaxing large water structures. This hydrogel is also characterized by minimal value of the activation energy of the dc relaxation at $E_a = 56$ kJ/mol, since $E_a = 58$ kJ/mol for the hydrogel starch alone, 115 kJ/mol for saccharose solution, and 99 kJ/mol for starch silica at $h = 2.34$ g/g (TSDC thermogram is not shown here). The difference in the E_a values for the starch/silica hydrogel at $h = 2.34$ and 16.9 g/g can be due to appearance of a large number of the barriers of H^+ transferring (because starch molecules can attach protons) at the lower content of water.

Calculations of the distribution functions of sizes of relaxed structures (Figure 5.11b) show that the starch/silica hydrogel at $h = 2.34$ g/g is characterized by narrower molecular "pockets" than that at $h = 17.9$ g/g. The narrowest pockets (nanopores) are found for hydrogel with gelatinized starch alone. Notice that all distributions show complex structures of pockets (voids) between/inside starch molecules that are in agreement with complex structure of unfolding starch macromolecules.

Thus, the technique based on the ^1H NMR and TSDC cryoporometry is appropriate for calculations of the structural characteristics of wetted and swelling polymers and polymer hydrogels, and it allows the investigations of structural changes in swelling granules and unfolding molecules of gelatinized potato starch depending on the hydration degree and the presence of functionalized nanosilica in the hydrogels. Additionally, this approach allows the calculations

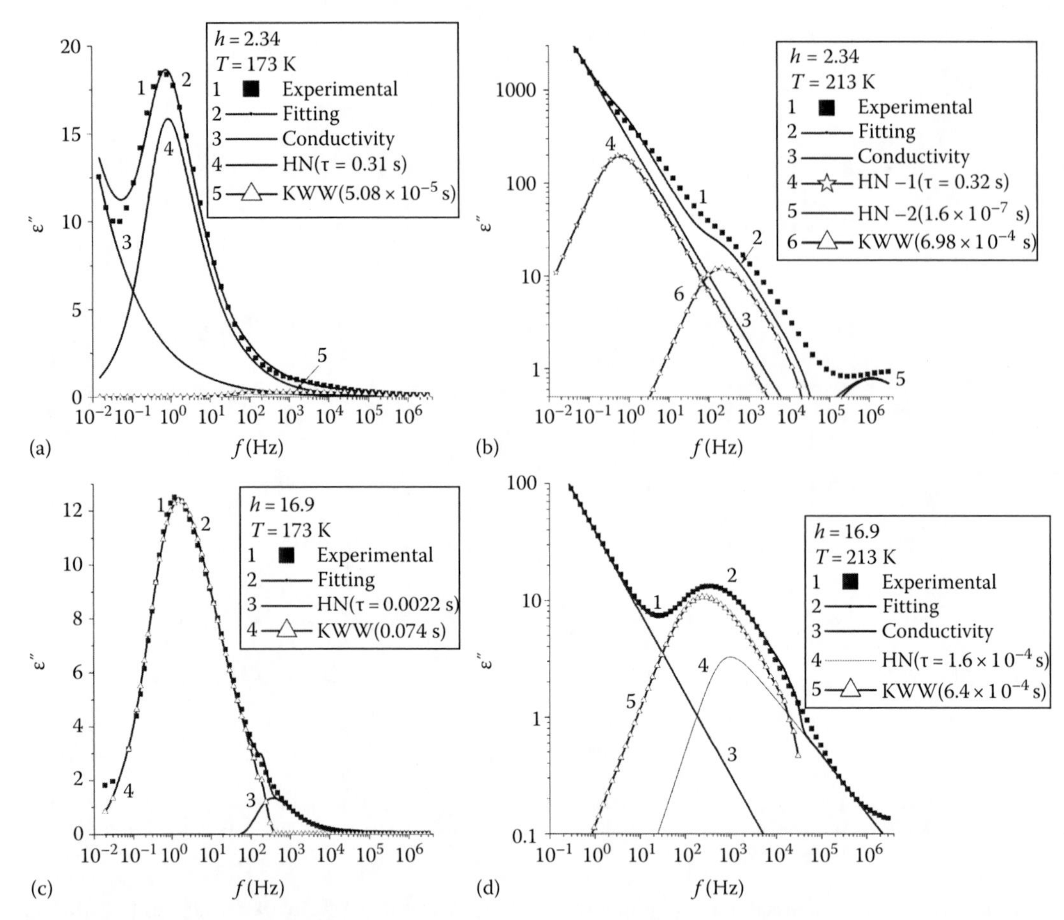

FIGURE 5.8 Deconvolution of the $\varepsilon''(f)$ curves with Equation 5.2 for starch/modified A-300 (7.7:1) at different hydrations h = (a, b) 2.34 and (c, d) 16.9 g/g and T = (a, c) 173 and (b, d) 213 K. (Adapted from *Colloids Surf. A: Physicochem. Eng. Aspects* 320, Gun'ko, V.M., Pissis, P., Spanoudaki, A. et al. Interfacial phenomena in starch/fumed silica at varied hydration levels, 247–259, 2008a, Copyright 2008, with permission from Elsevier.)

of the thermodynamic parameters of the water layers strongly and weakly bound to polymer molecules and solid nanoparticles.

The investigations of the starch/nanosilica hydrogel and the hydrated powder by the ^1H NMR spectroscopy with freezing-out of bulk and interfacial waters, TSDC, and broadband DRS show that the effects of a relatively low content of functionalized nanosilica at C_{str}:C_{silica} = 7.7:1 depend on the amounts of water, which can effectively unravel the double-helix structure. However, at low content of starch/nanosilica their supramolecular structure is denser than that at lower hydration and the critical content of the solid phase (starch + silica) is 10 wt% when significant changes in the properties of the hydrogel are observed.

HA (Figure 5.12) and its salt sodium hyaluronate (SH) are polymeric glycosaminoglycans (4–20 kDa) used in medicine as a bioactive compound for tissue regeneration (Hardingham and Fosang 1992, Toole 1997).

It is a component of the synovial fluid and different tissues (eye-bulb, mucous coat of stomach and intestine, connective tissue). HA and SH as well as other carbohydrates can bind great amounts of water and form colloid solutions or gel-like systems (Lehninger 1975). Freeze-dried HA (1.7 wt% water) wetted by controlled portions of water (h = 0.05, 0.15, and 2.15 g/g) was equilibrated

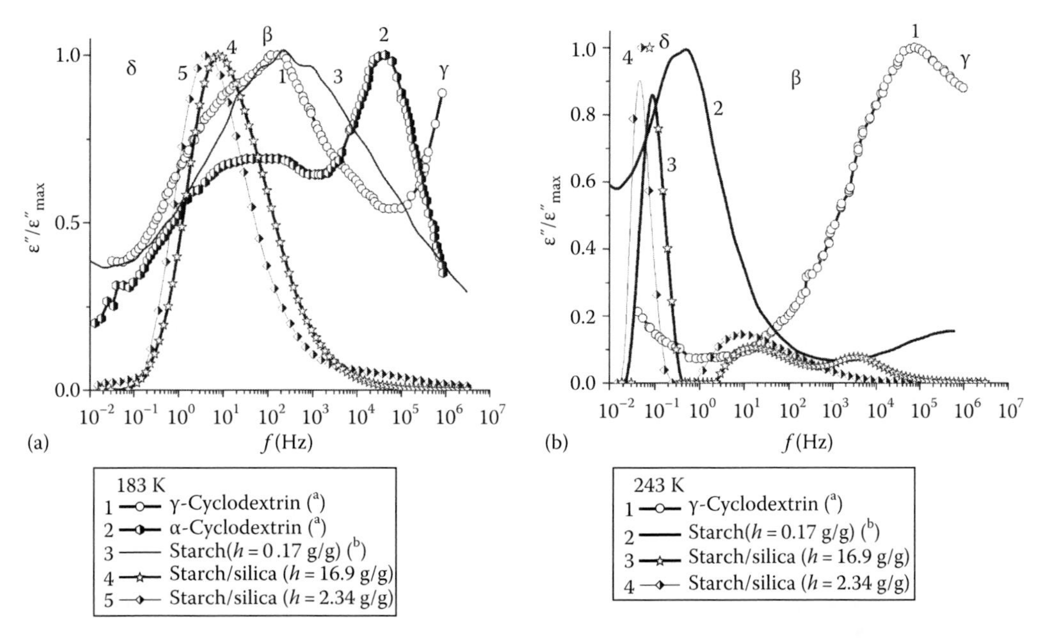

FIGURE 5.9 Normalized $\varepsilon''/\varepsilon''_{max}$ as a function of frequency for different polysaccharides differently hydrated at T = (a) 183 and (b) 243 K; [a](Meißner and Einfeldt 2004) and [b](Einfeldt et al. 2001). (Adapted from *Colloids Surf. A: Physicochem. Eng. Aspects* 320, Gun'ko, V.M., Pissis, P., Spanoudaki, A. et al. Interfacial phenomena in starch/fumed silica at varied hydration levels, 247–259, 2008a, Copyright 2008, with permission from Elsevier.)

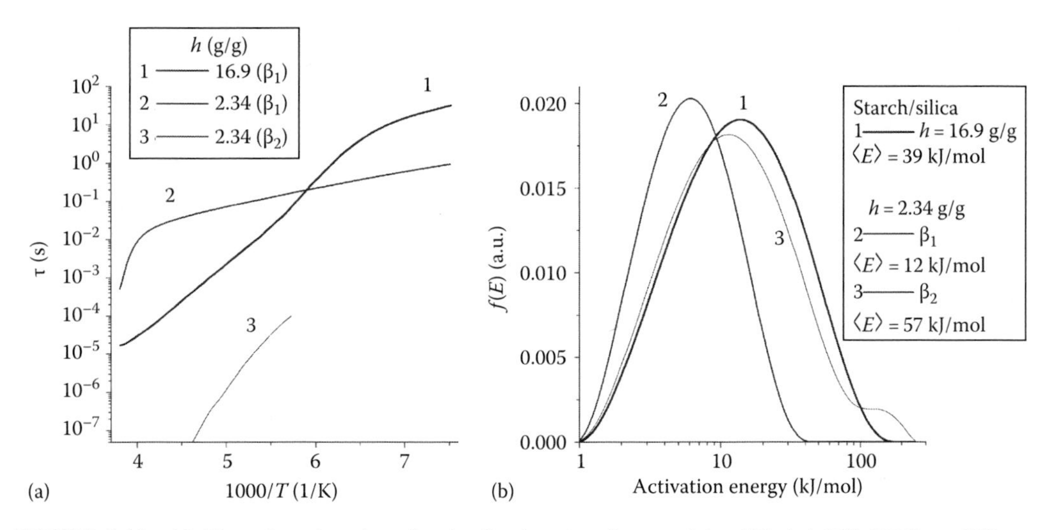

FIGURE 5.10 (a) The relaxation time for the β relaxation for starch/modified A-300 (7.7:1) at different hydrations h = 2.34 and 16.9 g/g as a function of reciprocal temperature and (b) the corresponding distribution functions of the activation energy of relaxation (average values <E> as the normalized first moments of the distributions are shown). (Adapted from *Colloids Surf. A: Physicochem. Eng. Aspects* 320, Gun'ko, V.M., Pissis, P., Spanoudaki, A. et al. Interfacial phenomena in starch/fumed silica at varied hydration levels, 247–259, 2008a, Copyright 2008, with permission from Elsevier.)

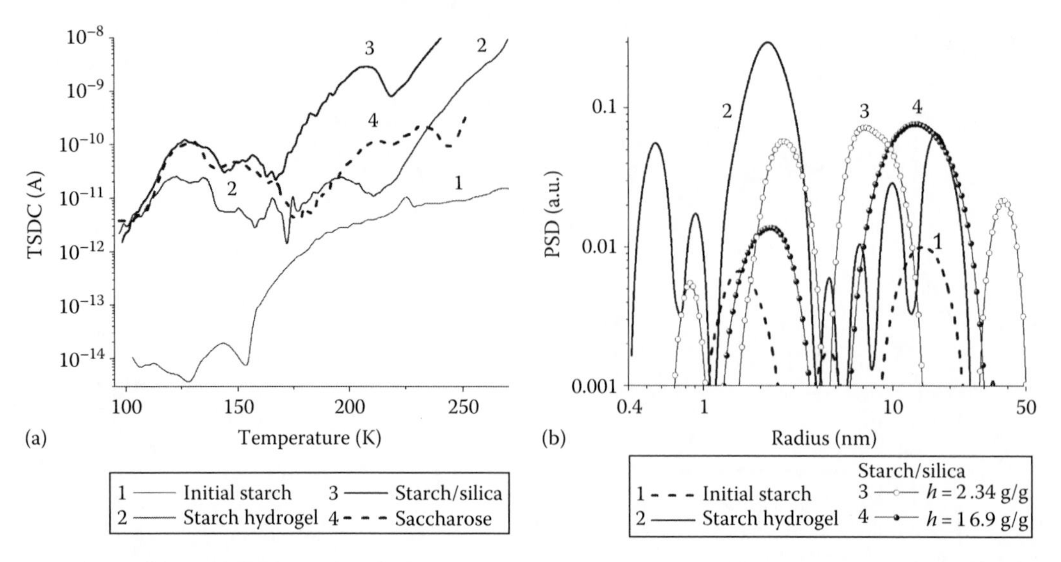

FIGURE 5.11 (a) TSDC thermograms of dry initial starch, hydrogel of gelatinized starch (14.2 wt%), gelatinized starch/modified nanosilica (7.7:1) at $h = 16.9$ g/g, and aqueous solution of saccharose (32 wt%) (all spectra were normalized to $F_p = 10^5$ V/m) and (b) size distributions of structures corresponding to the TSDC relaxations. (Adapted from *Colloids Surf. A: Physicochem. Eng. Aspects* 320, Gun'ko, V.M., Pissis, P., Spanoudaki, A. et al. Interfacial phenomena in starch/fumed silica at varied hydration levels, 247–259, 2008a, Copyright 2008, with permission from Elsevier.)

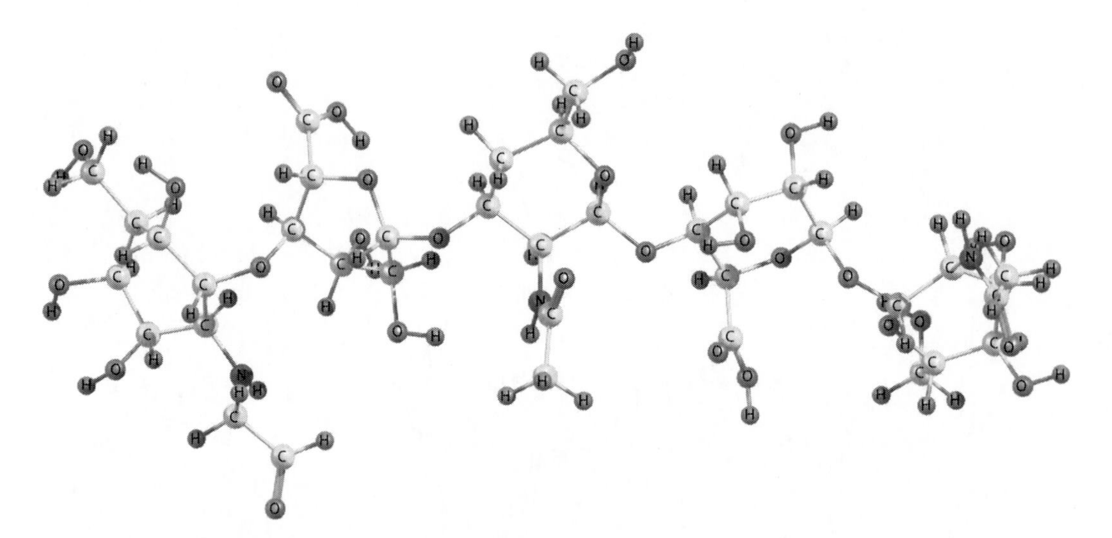

FIGURE 5.12 A fragment of hyaluronic acid (geometry was optimized by PM6 method).

for 1 h before NMR measurements. Initial HA is characterized by broad single ^{1}H NMR signal (Figure 5.13a) decreased with decreasing temperature due to partial freezing of bound water. The δ_H value increased from 5 (280 K) to 6.8 ppm (200 K) suggests that all bound water is SAW. This can be caused by a compacted structure of air-dried HA in which water forms continuous hydrogen bond network with O- and N-containing functionalities of HA. Even in more hydrated state, HA tends to form compacted structure calculated by PM6 method with total optimization of the geometry (Figure 5.14).

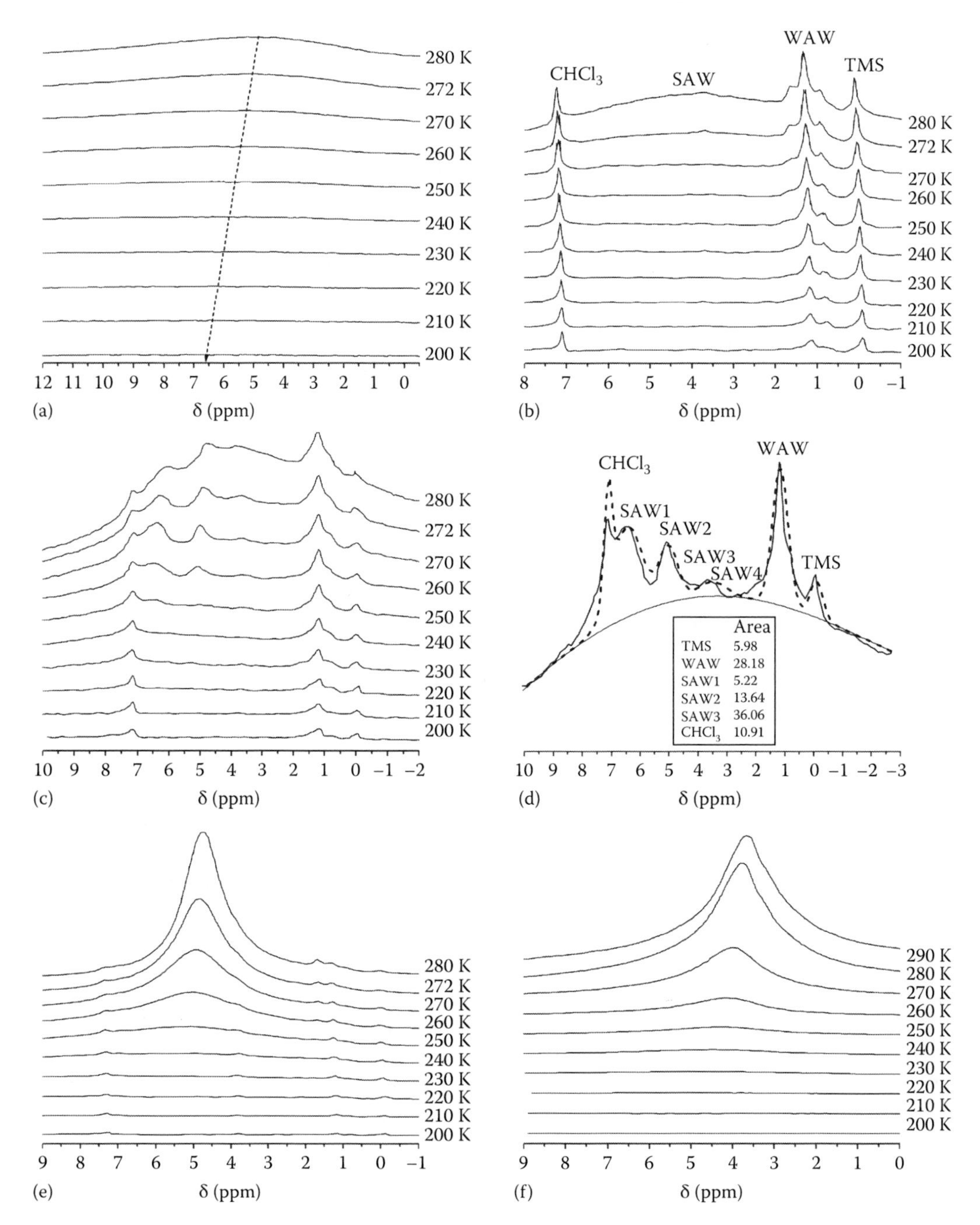

FIGURE 5.13 ^{1}H NMR spectra of water bound to HA differently hydrated at h=(a) 17 mg/g (in air), (b) 17, (c) 67, (d) 67 mg/g at 270 K, (e) 167, and (f) 2167 mg/g, (b–f) in $CDCl_3$ medium.

Weakly hydrated HA in $CDCL_3$ medium (Figure 5.13b) is characterized by a set of water structures corresponding to SAW ($\delta_H \approx 4$ ppm) and WAW ($\delta_H \approx 0.8$–1.5 ppm). The WAW amount increases with thawing of frozen water with increasing temperature. The spectra include signals of H-containing admixtures $CHCl_3$ (7.2 ppm) and TMS (0 ppm). The presence of several signals of WAW is due to its heterogeneity in structures bound to HA and dissolved in $CDCl_3$. The chloroform

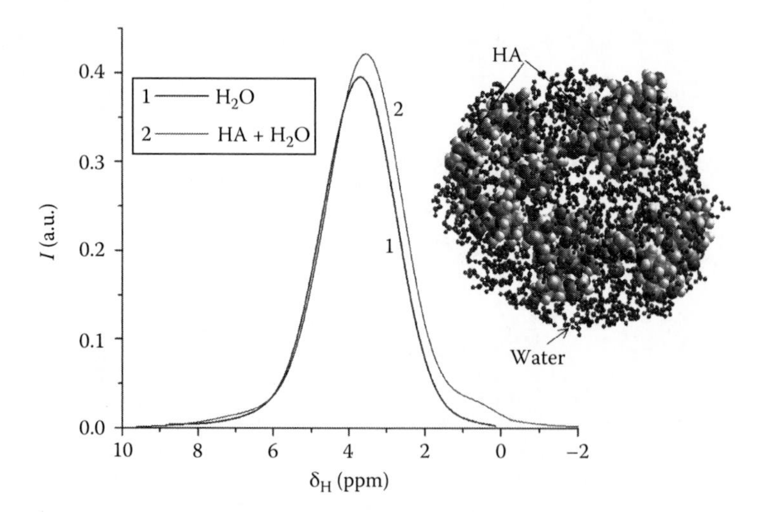

FIGURE 5.14 ^{1}H NMR spectra of only water bound to HA (1) and bound water+HA (2) (inset: cluster with water and HA with the optimized geometry by PM6).

medium prevents exchange between different structures. These clusters can be located in narrow pores in nearly dry HA. The maximal WAW content corresponds to 10% of the total amount of bound water.

An increase in HA hydration to 67 mg/g leads to changes in the spectra (Figure 5.13c). Several types of SAW are observed as narrow signals at $\delta_H=4$, 5, and 6.5 ppm with a broad background SAW at 4 ppm (Figure 5.13d). This can be interpreted as the formation of SAW clusters of different sizes with different association degrees of the molecules. With decreasing temperature the SAW signals become broader and have down-field shift because of decreased molecular mobility, increased order of the hydrogen bonds closer to that of hexagonal ice.

The relative amounts of WAW change weakly with addition of water to 67 mg/g. However, after subsequent addition of water to $h=167$ and 2167 mg/g relative contribution of WAW strongly decreases since practically single signal is observed at 4.5 and 3.5 ppm. These values are close to the calculated δ_H values of strongly hydrated HA (Figure 5.14). The disappearance of the fine structure of the signal is due to the formation of continuous water structure in more strongly wetted HA. In the aqueous solutions of HA (Figure 5.15), the signal intensity strongly decreases with lowering temperature because the main portion of water is bulk or WBW. At $C_{HA}=10$ mg/g, the ^{1}H NMR spectra include single symmetric signal of the Gaussian shape (Figure 5.15a and b). With increasing HA concentration, the signal splits into two to three signals which observed in the frozen and thawed solutions (Figure 5.15c–f). The HA solutions are characterized by larger δ_H values than wetted powders because water (both bulk and bound) is more strongly associated with the solutions.

The $C_{SAW}(T)$ curves are characterized by a bend at $T>260$ K (Figure 5.16a and b), which is a boundary between SBW ($-\Delta G>0.5$ kJ/mol) and WBW ($-\Delta G<0.5$ kJ/mol). The ΔG_s value (Table 5.3) corresponds to maximal changes in the first layer of bound water. At minimal hydration of HA being in the $CDCl_3$ medium, the $-\Delta G_s$ and γ_S values are lower than that in air. This is due to displacement of water from the HA surface and the formation of larger water clusters (Figure 5.17). The γ_S value increases with increasing hydration of HA being in $CDCl_3$. However, the $-\Delta G_s$ value is minimal at $C_{H2O}=167$ mg/g. In this case, the water clusters and nanodomains have maximal values (Figure 5.17) to reduce the surface area contact between water and hydrophobic solvent $CDCl_3$.

The C^s_{uw} and γ_S values change in concord (Table 5.3) because the SBW behavior determines the γ_S value course. In maximum hydrated HA powder ($C_{H_2O}=2167$ mg/g) being in $CDCl_3$, the main portion of water corresponds to bulk water, i.e., it does not interact with HA. However, in the HA solutions the amounts of SBW are lower than that in the maximum hydrated powder. This is due to

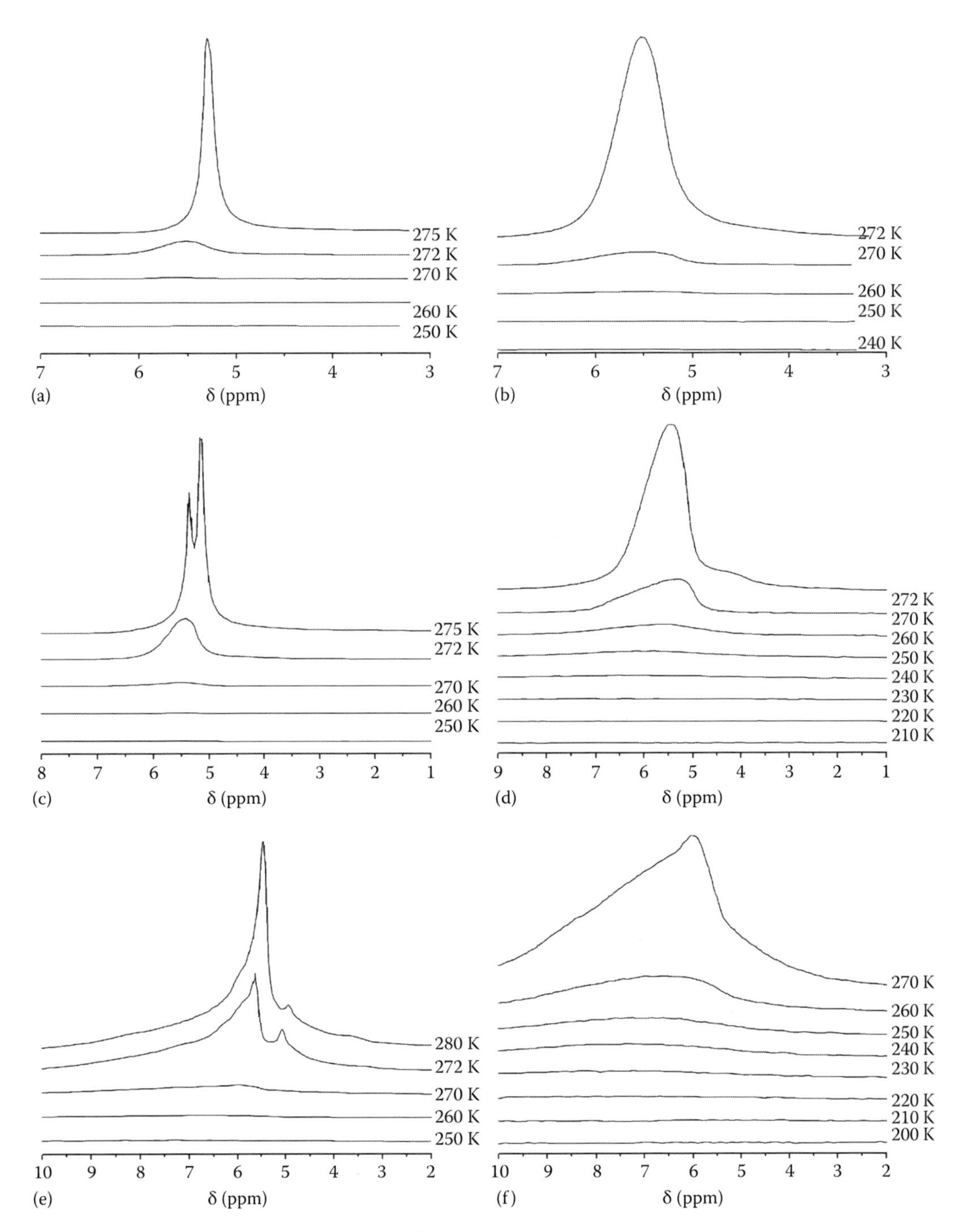

FIGURE 5.15 ^{1}H NMR spectra of water bulk and bound HA in solutions at C_{HA} = (a, b) 10 mg/g of water, (c, d) 20, and (e, f) 120 mg/g.

structural rearrangement of HA (Figure 5.17) reducing the confined space effects. This structural changes cause nonlinear changes in the γ_S value which is maximal at C_{HA} = 20 mg/g since it is less for both more and less concentrated solutions (Table 5.3).

NMR spectroscopy demonstrated that ketotifen fumarate in aqueous solution bound more strongly to tamarind seed polysaccharide (TSP) than to hydroxyethyl cellulose (HA) (Uccello-Barretta et al. 2008).

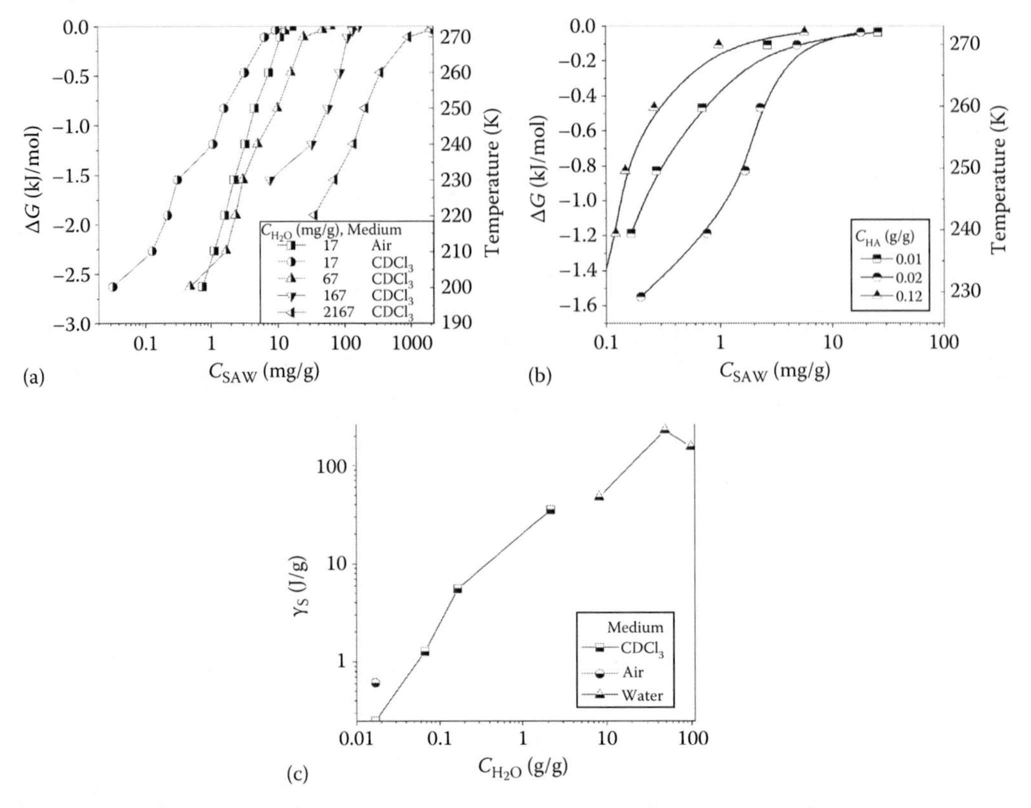

FIGURE 5.16 Temperature dependences of the amounts of unfrozen strongly associated water $C_{SAW}(T)$ and relationships between $C_{SAW}(T)$ and changes in the Gibbs free energy ΔG for (a) hydrated powder of HA and (b) aqueous solution of HA, and (c) relationship between free surface energy and total content of water in powders and solutions.

TABLE 5.3
Characteristics of SAW Bound to HA Powders or Solutions

C_{H_2O} (mg/g)	Medium	ΔG_s (kJ/mol)	ΔG_w (kJ/mol)	C^s_{uw} (mg/g)	C^w_{uw} (mg/g)	γ_S (J/g)
17	Air	−3.5	−0.4	5	12	0.6
17		−2.8	0.25	2	14	0.3
67	$CDCl_3$	−3.0	−0.25	25	42	1.3
167		−1.9	−0.3	120	47	5.6
2167		−2.3	−0.2	450	160	35.4
8333		−2.7	−0.12	300	590	47.7
5×10^4	Water	−1.8	−0.12	310	1500	228.0
1×10^5		−1.6	−0.1	250	2300	156.0

Despite high structural complexity of drug formulations involving polysaccharides as controlled release systems (that make spectroscopic investigations a challenging effort), NMR spectroscopy confirmed its versatility and efficiency in this field. Several fundamental aspects can be accurately ascertained by detecting NMR parameters, such as selective relaxation rates and diffusion coefficients, which are very sensitive probe of drug–drug and drug–polysaccharide interactions (Uccello-Barretta et al. 2008).

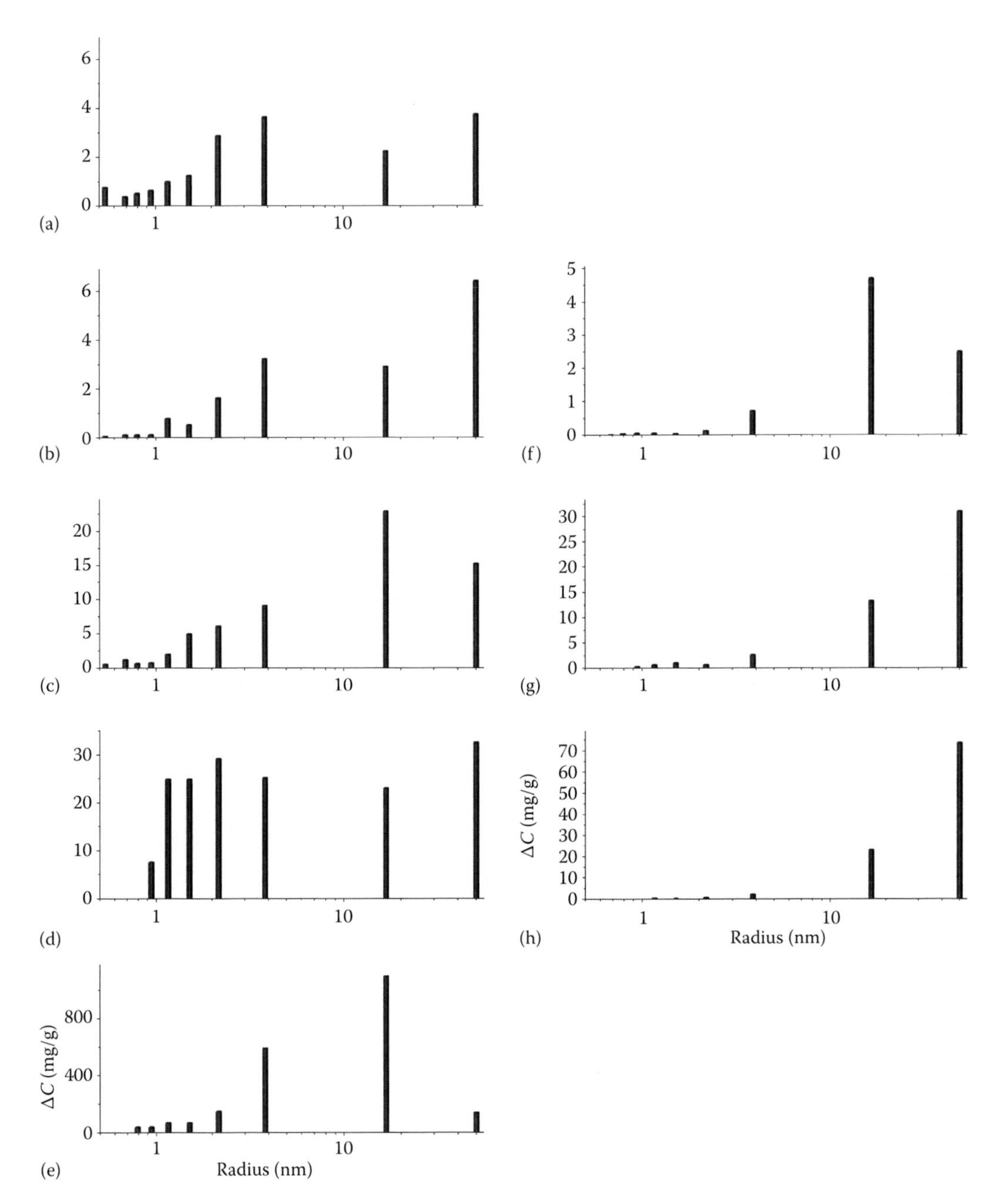

FIGURE 5.17 Distribution functions of cluster sizes of water bound to wetted HA powders at $h=(a, b)$ 17, (c) 67, (d) 167, (e) 2167 mg/g, and aqueous solutions of HA at $C_{HA}=$(f) 10, (g) 20, and (h) 120 mg/g.

Such samples as soils, sediments, plants, tissues, foods, and organisms, often containing liquid-, gel-, and solid-like polymeric phases, are very heterogeneous. There is the synergism between these phases in the materials that determine their environmental and biological properties. Studying each phase separately can perturb the sample, removing important structural information such as chemical interactions at the gel–solid interfaces, kinetics across boundaries, and conformation in the natural hydrated state. In order to overcome these limitations, a comprehensive multiphase-nuclear magnetic resonance (CMP-NMR) probe has been developed that permits all bonds in all phases to be studied and differentiated in whole unaltered natural samples (Courtier-Murias et al. 2012). Notice that mobile, single-sided NMR, which has its origin in inside-out NMR, where NMR spectrometers are lowered into bore holes for analysis of fluids in the surrounding rock matrix, can be used for measurement of product streams and product quality, and for medical diagnostics, i.e., to study native materials in native surroundings (Blümich et al. 2008). Additionally, NMR cryoporometry and relaxometry allow us to study the mentioned materials in native state over wide temperature range.

5.2 SYNTHETIC POLYMERS

NMR spectroscopy as a versatile tool to investigate structure and molecular mobility in soft materials (Scheler 2009) is a standard technique for structural characterization of polymers and polyelectrolytes and provides information on the effects of the surrounding medium and the counterions. High-resolution NMR spectroscopy enables the observation of structurizing of confined and bound liquids and counterion interactions. It gives information about spatial proximity of functional groups in polyelectrolyte complexes (PEC) and a whole structure of adsorption layer. Combination of pulsed field gradient (PFG) NMR and electrophoresis NMR permits the direct observation of counterion condensation. NMR cryoporometry shows different states of water interacting with polymers and polyelectrolytes in multilayers and allows one to determine the structural characteristics of porous media. Solid-state NMR can be applied to study both packing effects and local molecular dynamics in polymers and polyelectrolyte multilayers (PEM) (Scheler 2009).

McCormick et al. (2003) studied the structure and dynamics of adsorbed water and polymer components in PEM films and the bulk PEC using ^{1}H MAS NMR spectroscopy. The films (1–5 bilayers) with poly(diallyl dimethylammonium chloride) (PDADMAC) and poly(sodium-4-styrenesulfonate) (PSS) were adsorbed onto colloidal silica particles. Relaxation and line width measurements showed that the adsorbed water is less mobile in the films than in the analogous PEC. This result can be explained by compacting of the adsorption layer at a surface and enhancement of confined space effects for bound water. Relaxation measurements and ^{1}H double-quantum (DQ) NMR experiments revealed that polymer dynamics in the PEMs was strongly influenced by the layer number and water content (Figures 5.18 and 5.19). 2D spin

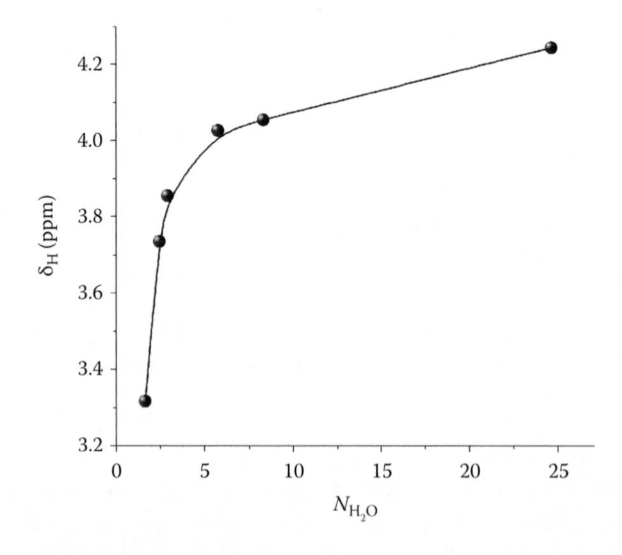

FIGURE 5.18 Water chemical shift as a function of water content per ion pair for a five bilayer film. (Adapted with permission from McCormick, M., Smith, R.N., Graf, R., Barrett, C.J., Reven, L., and Spiess, H.W., NMR studies of the effect of adsorbed water on polyelectrolyte multilayer films in the solid state, *Macromolecules* 36, 3616–3625, 2003. Copyright 2003 American Chemical Society.)

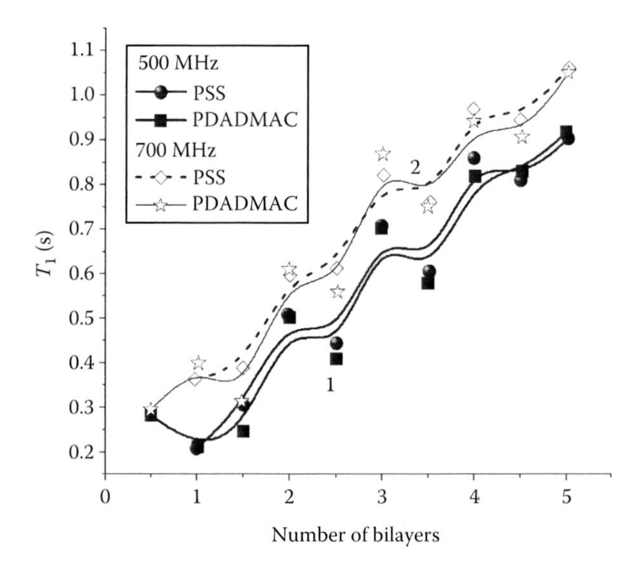

FIGURE 5.19 Spin–lattice relaxation time as a function of bilayer numbers for D_2O-saturated PEM films. (Adapted with permission from McCormick, M., Smith, R.N., Graf, R., Barrett, C.J., Reven, L., and Spiess, H.W., NMR studies of the effect of adsorbed water on polyelectrolyte multilayer films in the solid state, *Macromolecules* 36, 3616–3625, 2003. Copyright 2003 American Chemical Society.)

diffusion and DQ NMR were used to detect polymer–polymer and water–polymer association. The results supported the diffuse interpenetrating model of the different layers and a partitioning of the water to the PSS component and to the surface layer (McCormick et al. 2003).

The field dependence of the T_{1H} values (Figure 5.19) indicates that the motion of the polymers is slow compared to the Larmor frequency (McCormick et al. 2003).

Multilayers of cationic PDADMAC (200–350 kg/mol) and anionic PSS (70 kg/mol) were deposited onto silica (particles of 70–100 nm in size) colloids (Rodriguez et al. 2000). Adsorbed water peaks in PSS and PDADMAC were observed at $\delta_H = 3.66$ and 4.29 ppm, respectively (Figure 5.20). The water peak in the PSS/PDADMAC complex shifted to 3.85 ppm. For PEM on silica, two sharp peaks were observed 4.9 (protons associated with the silica surface) and 3.85 ppm (water adsorbed within the polymer film).

The peak at 3.85 ppm was observed to increase in intensity and shift slightly down-field (~0.3 ppm) when the sample was stored under ambient conditions, due to further uptake of water. The 3.85 ppm peaks and most of the peak at 5 ppm from the silica substrate disappeared when the complex and multilayer samples were dried under vacuum at 100°C. The bare silica has two strong signals at 5.0 and 4.0 ppm and a shoulder at 3.0 ppm. Peaks at 5.0, 4.1, 3.5, and 2.0 ppm can be assigned to hydrogen-bonded silanols, liquid-like water without any direct interaction with the silica surface, physisorbed water, and isolated silanols, respectively. The position of the 4.1 ppm water peak observed in humidified silica samples was proposed to be due to rapid proton exchange between liquid-like water (4.9 ppm) and the physisorbed water (3.5 ppm). When the first layer of PDADMAC was deposited, the peak at 4.0 ppm vanished and the peak at 5 ppm broadened, presumably due to displacement and/or rearrangement of bound water. A peak at 4.0 ppm reappeared when PSS was added as layer 2. This peak shifted slightly down-field when PDADMAC was added as layer 3 and up-field again when the last layer of PSS was added. Based on these observations, Rodriguez et al. (2000) assigned this peak to water associated with the polymer or water in rapid exchange with water associated with the polymer. The complexation was observed between two polymers in PEMs. Water associated with the polymer multilayers could be distinguished in the single-quantum spectra and suppressed in the DQF spectra. PSS/PDADMAC PEMs were structurally identical to the bulk PSS/PDADMAC PECs (Rodriguez et al. 2000).

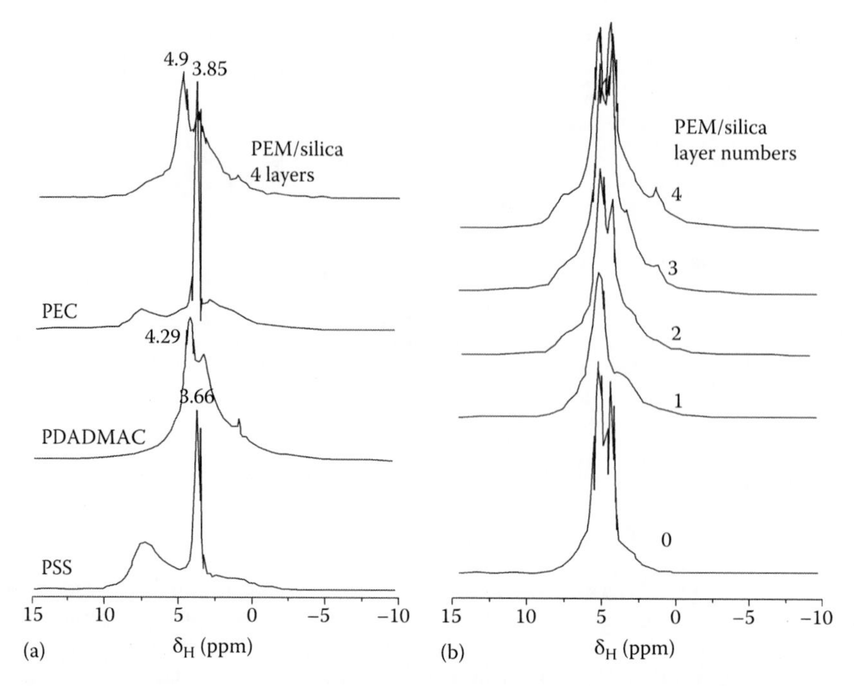

FIGURE 5.20 ^{1}H MAS NMR spectra, recorded with a spinning frequency of 30 kHz of (a) PSS, PDADMAC, the 1:1 PSS/PDADMAC PEC, and the PSS/PDADMAC multilayers on silica colloids (PEM/silica) and (b) the bare silica colloids (layer 0), silica colloids with adsorbed PDADMAC (layer 1), silica colloids with one layer each of adsorbed PDADMAC and PSS (layer 2), silica colloids with sequential adsorption of PDADMAC, PSS, and PDADMAC (layer 3), and silica colloids with sequential adsorption of PDADMAC, PSS, PDADMAC, and PSS (layer 4). (Adapted with permission from Rodriguez, L.N.J., De Paul, S.M., Barrett, C.J., Reven, L., and Spiess, H.W.: Fast magic-angle spinning and double-quantum ^{1}H solid-state NMR spectroscopy of polyelectrolyte multilayers. *Adv. Mater.* 2000. 12. 1934–1938. Copyright 2001 Wiley-VCH Verlag GmbH & Co. KGaA.)

Xu and Kirkpatrick (2006) studied NaCl-exchanged polyamide (PA) films, comparable to those of the active skin layer of many reverse osmosis (RO) membranes, using ^{23}Na NMR spectroscopy to analyze the influence of environment on the dynamical behavior of Na^+ in the films. The ^{23}Na NMR spectra for fresh polymer samples exchanged in 1 M NaCl solution showed significant relative humidity (RH) dependence. At near 0% RH, there were resonances for crystalline NaCl and rigidly held Na^+ in the PA. With increasing RH, a resonance for solution-like dynamically averaged Na^+ appeared and above 51% RH was the only signal observed. The slightly negative chemical shift of this resonance suggests a dominantly hydrous environment with some atomic scale coordination by atoms of the polymer. The greatly reduced ^{23}Na T_1 relaxation rates for this resonance relative to bulk solution and crystalline NaCl confirmed close association with the polymer. ^{23}Na NMR spectra for a sample equilibrated at 97% RH obtained from 193 to 293 K showed the presence of rigidly held Na^+ in a hydrated environment at low temperatures and replacement of this resonance by the dynamically averaged signal at $T > 253$ K. The results provided support for the solution–diffusion model for RO membranes transport and demonstrated the capabilities of multinuclear NMR methods to investigate molecular-scale structure and dynamics of the interactions between dissolved species and RO membranes (Xu and Kirkpatrick 2006).

Schonhöff et al. (2007) and Chávez and Schönhoff (2007) discussed different aspects of the hydration and internal properties of PEM formed by layer-by-layer assembly. Reflectivity techniques monitor the water content and swelling behavior, while spin relaxation monitors water mobility. Odd–even effects in dependence on the number of layers were discussed in terms of an influence of the terminating layer. X-ray microscopy and NMR cryoporometry were used to analyze the

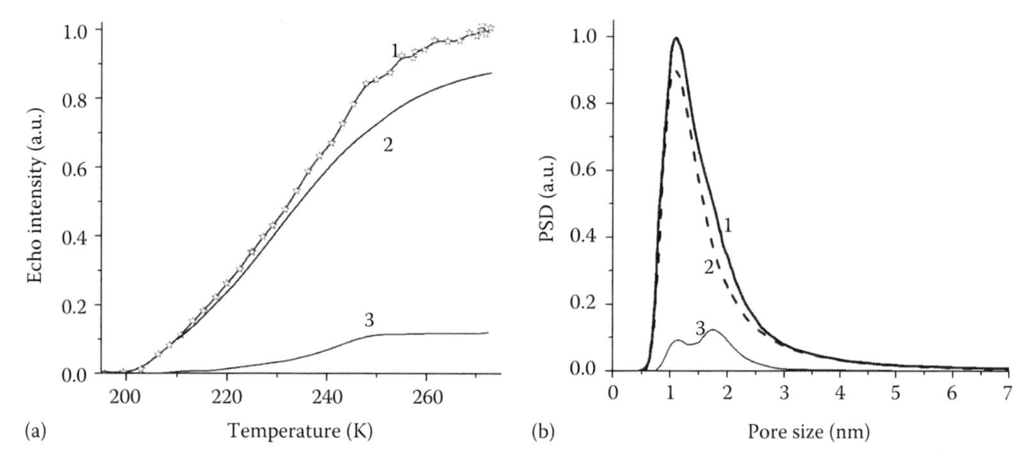

FIGURE 5.21 (a) Temperature dependence of the ^{1}H NMR echo intensity for an aqueous dispersion of colloidal silica coated by PDADMAC/(PSS/PAH)$_4$: (1) experimental data, contribution of water in (2) multilayer pores, and (3) silica pores and (b) corresponding PSDs (PDADMAC: poly(diallyl dimethylammonium chloride), PSS/PAH bilayer pair: poly(styrene sulfonate)/poly(allylamine hydrochloride). (Adapted with permission from (a) *Colloids Surf. A: Physicochem. Eng. Aspects*, 303, Schonhöff, M., Ball, V., Bausch, A.R. et al., Hydration and internal properties of polyelectrolyte multilayers, 14–29, 2007, Copyright 2007, with permission from Elsevier and (b) Chávez, V.F. and Schönhoff, M., Pore size distributions in polyelectrolyte multilayers determined by nuclear magnetic resonance cryoporometry, *J. Chem. Phys.*, 126, 104705-1–104705-7, 2007, Copyright 2007, American Institute of Physics.)

interfacial behavior of water in hollow capsules and pores of silica coated by five polymeric layers (Figure 5.21). Water is mainly bound to polyelectrolyte molecules. It is in the form of domains mainly <3 nm in size (Figure 5.21b). These sizes correspond to the sizes of silica pores but the volume of bound water is much larger because the lion's share of this water is bound to the polyelectrolyte molecules.

von Kraemer et al. (2008) studied sulfonated polysulfone (sPSU) membranes using NMR cryoporometry (with $k_{GT}=25$ K nm) in parallel with water uptake and DSC measurements. An sPSU membrane with high ion-exchange capacity (IEC 1.45 mequiv/g) possessed a significant amount of large pores after hydrothermal pretreatment at 80°C, which was related to its high hydrophilicity and low resistance toward swelling. A membrane with low IEC (0.95 mequiv/g) showed a significant fraction of narrow pores ($R \sim 1$ nm) after HTT. The NMR cryoporometry gave for Nafion membranes the PSD peak at $R \sim 1$ nm (Figure 5.22).

Above the lower critical solution temperature T_c (~34°C), poly(*N*-isopropylacrylamide) (PNIPA) hydrogels can be weakly hydrophobic and undergo microphase separation (László et al. 2010). Macroscopic deswelling was very slow, the out-of-equilibrium state of the gel being conserved for many days. László et al. (2010) studied the structure of the microphase-separated state at $T>T_c$ using SAXS and PFG NMR of water and showed that the gel included two microphases, separated by smooth interfaces. The cavities occupied by the water phase formed a connected network. The diffusion rate of the water molecules in this phase varied from one cavity to another and can be described by a Gaussian distribution function. Water molecules belonging to the polymer-rich phase were mobile, but their self-diffusion coefficient D was greatly diminished. Absence of compartmentalization of the water phase showed that the slow deswelling rate of the gel was not due to trapping of the water phase. PNIPA gels have a polymer content of 3%, collective diffusion coefficient $D_c \approx 4\times10^{-7}$ cm^2/s, and in bulk water at 40°C D is ~3.2×10^{-5} cm^2/s. Distribution of diffusion coefficients $g(D)$ for water in PNIPA hydrogel at 40°C has three maxima for deswollen gel at 3.2×10^{-5}, 0.22×10^{-5}, and 0.036×10^{-5} cm^2/s and two peaks at 2×10^{-5} and 0.25×10^{-5} cm^2/s for swollen gel. Lower D values correspond to water in the polymer-rich phase.

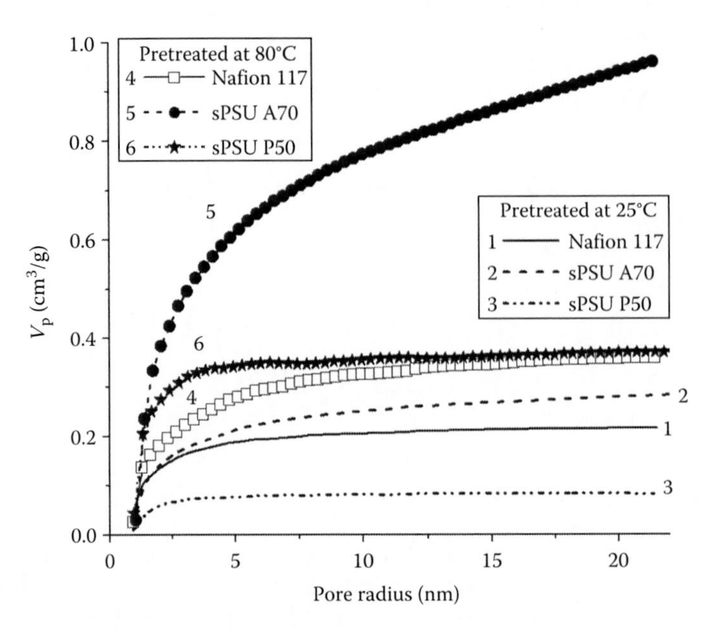

FIGURE 5.22 NMR cryoporometric melting curves for the specific pore volume related to the corresponding radius *R*. (Adapted from von Kraemer, S., Sagidullin, A.I., Lindbergh, G., Furó, I., Persson, E., and Jannasch, P.: Pore size distribution and water uptake in hydrocarbon and perfluorinated proton-exchange membranes as studied by NMR cryoporometry. *Fuel Cells*. 2008. 8. 262–269. Copyright 2008 Wiley-VCH Verlag GmbH & Co. KGaA.)

Alonso et al. (2002) used a combination of ^1H self-diffusion measurements and ^{13}C chemical shift analysis to study the solubilization of amphiphilic additives $C_nH_{2n+1}X$ (n=4, 6; X=OH, NH_2) in cetyltrimethylammonium bromide (CTAB) micelles that provided complementary data on structures of micelles and conformations of alkyl chains. All the additives studied behave as cosurfactants affecting micellar structures and the degree of their solubilization was determined solely by their alkyl chain length. For different additives, the effects depend on cosurfactant penetration into micelles.

Wu et al. (2011) studied the long-time behavior of the hydrolysis and condensation reaction of the tetraethoxysilane (TEOS) pre-solution at different pH values with and without addition of poly(ethylene glycol) (PEG) for various aging times using liquid ^1H, ^{13}C, and ^{29}Si NMR spectroscopy. Domján et al. (2010) analyzed proximity in a thermoresponsive host–guest gel system and the polymer–phenol distance using solid-state ^1H–^1H CRAMPS NMR spectroscopy and rate matrix analysis. The best agreement between calculated and measured values was found with methyl–phenol and methyne–phenol distances 0.5 and 0.55 nm, respectively.

Marcos et al. (2006) prepared porous substrates with poly(3-hydroxybutyrate-3-hydroxyvalerate) (PHBHV) by a particulate leaching method. After removing the salt by extraction in water, ^1H NMR relaxometry and imaging were performed on sets of PHBHV substrates immersed in PBS during 3 months at different time points. The results of NMR relaxometry showed two ^1H nuclei populations, well distinguishable on the FID, due to the different decay time constants, a factor of 100. Thus, it was possible to separate the two populations, giving separate distributions of T_1 relaxation times. One population could be associated with water protons in pores and other one was due to macromolecular protons. The distributions of T_1 and T_2 of the water proton-shifted to lower values with increasing immersion time to a constant value after 30 days. The results obtained by NMR imaging showed an initial increase in the apparent porosity, reaching a plateau after 25 days of immersion. This increase was attributed mainly to the absorption of water in nanopores as supported by the results of the relaxometry measurements and shown by SEM. The average porosity measured by NMR imaging at the plateau, 78% ± 3%, was slightly higher than that determined by optical microscopy, 73% ± 9%, which may be due to the fact that the latter method did not resolve the nanoporosity. Overall, the results suggested that at early stages after immersing the scaffolds in the aqueous medium, first 30 days approximately, NMR imaging could underestimate the porosity of the substrate (Marcos et al. 2006).

Vogel et al. (2000) studied the behavior of neat polystyrene and PS/benzene mixture using ^{2}H NMR spectroscopy. Below the glass transition, no further mobility on the timescale of 100 ms was detected. Benzene molecules were isotropically orientated on a timescale of microseconds at temperatures of about 80 K below glass transition temperature T_g. The dynamics of small benzene and large PS molecules decoupled while T_g was approached from higher temperatures. The distribution function of correlation time related to isotropic reorientation of the benzene molecules was much broader than in neat glass formers. The $\langle T_1 \rangle$ curve for the binary glass former shifted toward lower temperatures (Figure 5.23) due to the plasticizer effect caused by benzene ($T_g = 373$ K for PS and 278 K for the mixture).

For optimal design of polymer vehicles for controlled drug delivery, it is important to completely understand the nature and mechanisms of the structural evolution of the polymer matrix that ultimately controls drug release kinetics (Allison 2008, Perkins et al. 2010). NMR cryoporometry can be used to study drug release from polymer vehicles. Perkins et al. (2010) used the integrated cryoporometry and PFG NMR method (cryodiffusometry) for these purposes and showed that the true extent of the variability in structural evolution and transport properties between different batches of poly(D,L-lactide-*co*-glycolide) (PLGA) microspheres would be missed if these data were not available. Cryoporometry scanning loops were used to determine the overall network geometry. Cryoporometry freezing curves and PFG NMR were used to study the evolution in the pore-scale connectivity and the larger-scale interconnectedness of nanopores following immersion of microspheres in aqueous phase. The molecular weight of the polymer used and the presence of drug in the synthesis have a significant effect on structural changes in the polymer. It should be noted that Perkins et al. (2010) calculated only matrix pore neck size distributions at $d < 3$ nm in size.

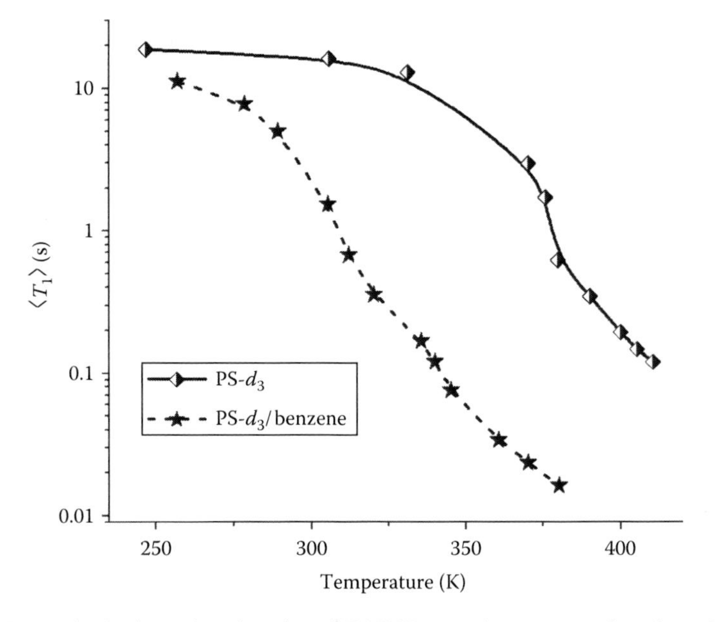

FIGURE 5.23 Mean spin–lattice relaxation time (^{2}H NMR experiments) as a function of temperature of neat polystyrene PS-d_3 and with addition of 13% of benzene. (Adapted from *J. Mol. Liquid.*, 86, Vogel, M., Medick, P., and Rössler, E., Slow molecular dynamics in binary organic glass formers, 103–108, 2000, Copyright 2000, with permission from Elsevier.)

However, SEM images of samples showed that these polymers are macroporous. Organic solvents could be used as adsorbates to study large pores.

Petrov et al. (2006) studied porous biodegradable polymer microparticles (with PLGA, 51–62 kDa, 21 samples) designed as devices for drug delivery in depot formulations using NMR cryoporometry (with k_{GT}=50 K nm) method. The main PSD peak for all samples was at 60–100 nm. There was no comparison of the NMR cryoporometry results with other methods. SEM images were of low resolution to analyze the porosity of polymer particles.

The PSD of an ultrafiltration (UF) membrane is of great importance as it governs liquid permeability, filtration ability, and the interaction between molecules and the membrane matrix (Jeon et al. 2008). They synthesized two series of UF membranes: commercial regenerated cellulose (PL) membranes and polyethersulfone (PB) membranes, each with molecular weight cutoff (MWCO) values of 5, 10, and 30 kDa. The PSDs of UF membranes were determined using ^{1}H NMR cryoporometry with the MPDs determined by analyzing the variation of the NMR signal intensity with temperature. From spin–echo intensity versus temperature curves, it was found that for each series of UF membranes the maximum MPDs of cyclohexane confined within the membrane pores were inversely proportional to the MWCO. For the same MWCO, the MPDs of the PL membrane were larger than that of the PB membrane, implying that the pores in the PL membranes were lower than those in the PB membranes (Figure 5.24), consistent with their water and solute flux performances. These findings indicated that NMR cryoporometry is an effective method to determine the PSDs of UF membranes with asymmetric pore structures.

Peinador et al. (2010) obtained good agreement of the structural characteristics of UF membranes measured using liquid–liquid displacement porosimetry with mean pores obtained from NMR characterization (Jeon et al. 2008).

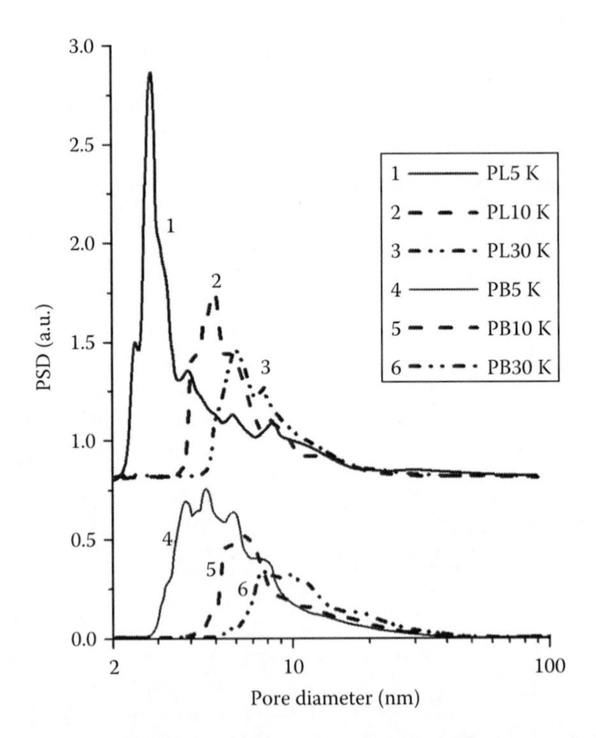

FIGURE 5.24 Pore size distribution curves of PL (curves 1–3) and PB (curves 4–6) series of membranes with 5, 10, and 30 kDa. (Adapted from *J. Membr. Sci.*, 309, Jeon, J.-D., Kim, S.J., and Kwak, S.-Y., ^{1}H nuclear magnetic resonance (NMR) cryoporometry as a tool to determine the pore size distribution of ultrafiltration membranes, 233–238, 2008, Copyright 2008, with permission from Elsevier.)

5.2.1 Structural Features of Polymer Adsorbent LiChrolut EN and Interfacial Behavior of Water and Water–Organic Mixtures

The behavior of organic adsorbates and water alone or in mixtures with organic co-adsorbates strongly differs in the bulk and pores of hydrophilic, hydrophobic, mosaic, or hybrid adsorbents because of the confined space effects and a strong influence of nonuniform surface fields on the adsorption phase, which can be strongly clustered. Hydration of organic molecules being in the surface force field differs from that of molecules in the bulk. The confined space effects in pores with mosaic hydrophilic/hydrophobic walls can cause structural and/or energetic differentiation of co-adsorbed liquids (up to their separation despite they can be miscible). These effects lead to the formation of self-organized structures of bound water co-adsorbed with organic liquids. Nonpolar or weakly polar adsorbates predominantly located at the pore walls, especially of low hydrophilicity, can displace a portion of water from narrow pores into larger pores or onto the outer surface of adsorbent, especially toward places with greater hydrophilicity. This occurs to reduce the contact area between two immiscible liquids in pores and to diminish the free energy of both liquid/liquid and liquid/solid interfaces. The effects and results of competitive adsorption of water and organics depend on structural characteristics of adsorbents and the type of organic adsorbates (polar or nonpolar, molecular and pore sizes), amounts and adsorption order of adsorbates, type and time of treatments, and external actions (e.g., electrostatic or electromagnetic fields, freezing, heating, sonication) (Brown et al. 1999, Fawcett 2004, Gun'ko et al. 2005d, 2007i, 2009d, Dominguez-Espinosa et al. 2006, Chaplin 2011).

The influence of structural features of polymer adsorbents poly(styrene divinylbenzene) (PSDVB), poly(ethylvinylbenzene divinylbenzene) (PEVBDVB), functionalized PSDVBs, etc., on the behavior of bound water alone or on co-adsorption with organics is of interest from both theoretical and practical points of view (Gun'ko et al. 2008e) because such PSDVB sorbents (e.g., LiChrolut EN, Merck) are effective solid-phase extraction (SPE) materials (Fiehn and Jekel 1996, Loos et al. 2000, López de Alda and Barceló 2001, Tomaszewski et al. 2005, López et al. 2007). For instance, LiChrolut EN provides the recovery rates >80% on the SPE of many organics (Fiehn and Jekel 1996). Therefore, the interfacial behavior of water adsorbed alone and in a mixture with polar, weakly polar, or nonpolar organics is of importance for deeper understanding of the observed phenomena and the SPE efficiency of the materials.

The pore walls of cross-linked PSDVB adsorbents include the aromatic rings as the main structural unit. Orientation of these rings in tangles of polymer chains is rather random in contrast to carbon adsorbents with the pore walls with the ordered condensed aromatic ring systems. Therefore, it is of interest to compare the confined space effects on the interfacial behavior of liquids co-adsorbed onto adsorbents of these two types (Gun'ko et al. 2008e).

TSDC and ^{1}H NMR spectroscopy measurements with layer-by-layer freezing-out of bulk liquids and the adsorption phase give detailed information on the structure of adsorbed liquids (especially on the molecular mobility and organization of adsorbed mixtures). This information can be obtained from the analysis of the TSDC thermograms (describing dipolar and direct current [dc] relaxations) and changes in the signal intensity and chemical shift of the proton resonance, δ_H, in the ^{1}H NMR spectra of adsorbed liquids depending on the amounts of water and/or organics, temperature, and other conditions. The δ_H value of water depends on the strength and the average number of the hydrogen bonds per water molecule, which differ in pores (and depends on pore size and pore wall structure) from that in the bulk. Many properties of bulk and interfacial water and water/organic mixtures (density, colligative properties, molecular mobility, etc.) depend on the characteristics of their hydrogen bond network, which changes in the confined space of pores.

LiChrolut EN adsorbent was selected from PSDVB and other polymer adsorbents studied (Gun'ko et al. 2005a, 2008e) due to its large specific surface area and pore volume (Table 5.4), stable pore structure (which was not practically changed in contrast to other polymer adsorbents on treatment in water or acetone for 24 h and then freezing in liquid nitrogen for 2 h before repeated nitrogen

TABLE 5.4
Structural Characteristics of Polymer and Carbon Adsorbents

Sample	S_{BET}	S_{nano}	S_{meso}	S_{macro}	V_p	V_{nano}	V_{meso}	V_{macro}	D_{N2}	D_{H2O}	Δw
LiChrolut	1512	1024	488	0.1	0.827	0.393	0.426	0.008	2.865	2.67	0.114
XAD-2	357	143	212	1.6	0.754	0.059	0.653	0.042	2.611		0.013
XAD-4	949	300	649	0.5	1.341	0.089	1.238	0.014	2.582		0.024
WVA	1710	1366	344	0.2	1.347	0.714	0.630	0.003	2.407	2.80	0.115

Notes: The specific surface area (in m^2/g) and the pore volumes (in cm^3/g) of nanopores (S_{nano} and V_{nano}) at radius for the model of cylindrical pores for LiChrolut EN (or half-width for carbon adsorbents) $R<1$ nm, mesopores (S_{meso} and V_{meso}) at $1<R<25$ nm and macropores (S_{macro} and V_{macro}) at $R>25$ nm were determined by integration of the $f_S(R)$ and $f_V(R)$ functions, respectively. The fractal dimension D_{N2} and D_{H2O} values were determined from the nitrogen and water adsorption isotherms. The Δw value determines the average error of the model of pores due to roughness of the pore walls.

FIGURE 5.25 (a) SEM (Superprobe-733, JEOL, Japan, accelerating voltage 25 kV) and (b) TEM (JEM-100CXII, accelerating voltage 100 kV) images of LiChrolut EN adsorbent particles. (Adapted from *J. Colloid Interface Sci*., 323, Gun'ko, V.M., Turov, V.V., Zarko, V.I. et al., Structural features of polymer adsorbent LiChrolut EN and interfacial behaviour of water and water-organic mixtures, 6–17, 2008e, Copyright 2008, with permission from Elsevier.)

adsorption–desorption measurements, Gun'ko et al. 2005a), the adsorbent stability over a wide pH range of the solution, and its high effectiveness as a SPE material. The textural and structural stability of LiChrolut EN adsorbent is of importance because the adsorbent with adsorbed water and water/organics was frozen at low temperatures during the TSDC and ^1H NMR measurements.

Commercial LiChrolut EN adsorbent (Merck) in the form of irregular particles (0.04–0.14 mm) (Figure 5.25a) was used (Gun'ko et al. 2008e) as received.

Notice that according to Merck's information (ChromBook 2006/2007) and some publications (López et al. 2007), LiChrolut EN adsorbent (orange) is composed of PEVBDVB; however, according to other publications (Hennion 1999, Ferreira et al. 2007), it is a PSDVB co-polymer, and there is no information on oxygen-containing functionalities and their origin in the adsorbent. According to these publications, the specific surface area of the adsorbent is 1200–1300 m^2/g and the pore volume is 0.75–0.8 cm^3/g. However, our measurements (Gun'ko et al. 2008e) gave slightly larger values (Table 5.4). Commercial PSDVB adsorbents Amberlite XAD-2 and Amberlite XAD-4 (Rohm and Haas, Philadelphia, PA) were used in comparative investigations using FTIR

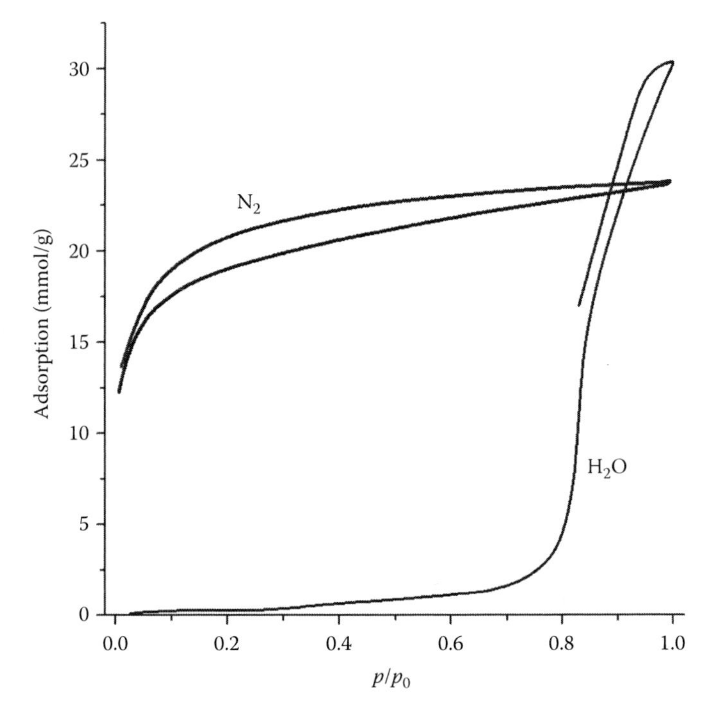

FIGURE 5.26 Nitrogen and water adsorption/desorption isotherms for LiChrolut EN.

spectroscopy and microcalorimetry. Wood-based activated carbon WVA (WVA-1100, Westvaco) described in detail (Gun'ko and Bandosz 2005g) was used to compare with LiChrolut EN with respect to their certain structural and adsorption properties.

According to the literature (ChromBook 2006/2007) and SEM images (Gun'ko et al. 2008e), LiChrolut EN adsorbent represents small particles (40–150 μm) of irregular shape. According to TEM image (Gun'ko et al. 2008e), these microparticles are composed of nanoparticles (40–150 nm). The outer surface of these nanoparticles can provide only a small portion (<100 m^2/g) of the specific surface area of the adsorbent (Table 5.4, S_{BET}). The porosity corresponding to voids between these nanoparticles can be assigned to mesoporosity and textural macroporosity (Figures 5.26 and 5.27, and Table 5.4). Polymer nanoparticles are relatively densely packed in microparticles because the textural porosity (macroporosity) weakly contributes the pore volume because an increase in the nitrogen adsorption at $p/p_0>0.95$ is very less (Figure 5.26) as well as contribution of macropores at $R>80$ nm (Figure 5.27). Contribution of mesopores to the pore volume (Table 5.4, V_{meso}) is predominant but the corresponding contribution to the specific surface area gives ~32%.

Consequently, LiChrolut EN is a nano/mesoporous adsorbent. This adsorbent is characterized by pores of a more complex shape than carbon adsorbents (even produced from natural raw materials) because cross-linked polymer chains form a disordered tangle structure. Therefore, the nitrogen adsorption/desorption isotherm for this adsorbent has a large and practically open hysteresis loop (Figure 5.26).

The PSD functions were calculated using the equation proposed by Nguyen and Do (1999) for carbon adsorbents and modified to study adsorbents with cylindrical and slit-shaped pores and voids between spherical nonporous particles or certain mixtures of these pores (Gun'ko et al. 2008e). The nitrogen desorption data were utilized to compute the PSD functions using the model of cylindrical pores for LiChrolut EN and other polymeric adsorbent and slit-shaped pores for carbon adsorbents.

This hysteresis loop differs strongly from that (much shorter) for carbon adsorbents (described earlier). This difference can be due to random orientation of aromatic rings and aliphatic chains in tangles with the cross-linked copolymers. This results in a complex shape of pores and retardation

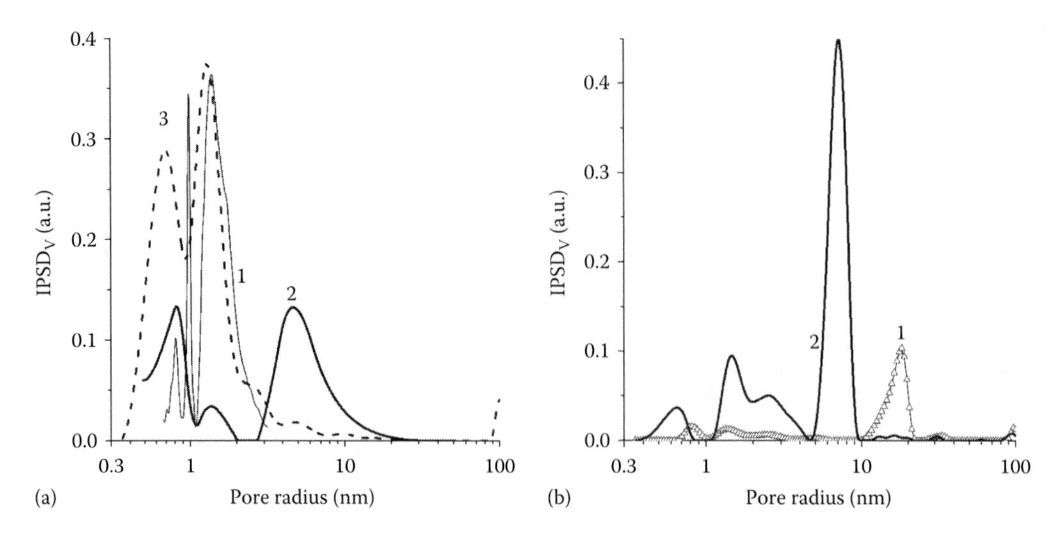

FIGURE 5.27 Incremental pore size distributions for polymers (a) LiChrolut EN and (b) Amberlites XAD-2 (curve 1) and XAD-4 (curve 2), calculated on the basis of the nitrogen desorption isotherm (a, curve 1, b), and the ^{1}H NMR (a, curve 2) and TSDC (a, curve 3) data. (Adapted from *J. Colloid Interface Sci.*, 323, Gun'ko, V.M., Turov, V.V., Zarko, V.I. et al., Structural features of polymer adsorbent LiChrolut EN and interfacial behaviour of water and water-organic mixtures, 6–17, 2008e, Copyright 2008, with permission from Elsevier.)

of the diffusion in nano- and mesopores (Figure 5.27) having many narrowing necks in contrast to more ordered basal planes in pores of carbon adsorbents without similar necks. Notice that these structural features manifest in very different values of the chemical shifts of the proton resonance of water and organic molecules adsorbed in pores of LiChrolut EN and C-47 adsorbents shown later. One can assume that some locally ordered wall structures can be only in narrow nanopores (0.3–0.5 nm) of LiChrolut EN adsorbent because of the π–π electron interactions of neighboring aromatic rings. However, according to the NMR investigations discussed later, this ordering effect causing the up-field shift of the signal is not observed for LiChrolut EN.

To analyze the adsorptive characteristics, water adsorption–desorption on LiChrolut EN adsorbent was studied by an adsorption apparatus with a McBain–Bark quartz scale at room temperature. LiChrolut EN particles have nanopores at $R < 1$ nm and narrow mesopores at $1 < R < 10$ nm (Figure 5.27) at large S_{BET} and V_p values. Therefore, both fractal values D_{N2} and D_{H2O} (Table 5.4) characterizing both pore and surface fractalities are significant. Broad mesopores at $10 < R < 25$ nm and macropores at $R > 80$ nm give a small contribution to the porosity and the specific surface area (Figure 5.27 and Table 5.4). Notice that the PSDs obtained with the NMR and TSDC cryoporometry differ from the PSD obtained on the basis of the nitrogen desorption isotherm (Figure 5.27) because of layer-by-layer freezing-out of water being in pores of different sizes that affect the PSD intensity based on the NMR and TSDC cryoporometry. Therefore, it is better to compare the position of the PSD peaks obtained by different methods than their intensity strongly affected by the layer-by-layer freezing processes (NMR) or the relaxation processes (TSDC). The positions of these peaks are relatively close (Figure 5.27). However, in the case of adsorbents with a simpler structure of pores (e.g., silica gels) the PSDs based on the adsorption and ^{1}H NMR cryoporometry methods are much close one to another. As a whole, three methods show the presence of, at least, three types of pores in LiChrolut EN: nanopores ($R < 1$ nm), narrow mesopores ($1 < R < 2$ nm), and broader mesopores ($R > 2$ nm). In these pores, the adsorption, the diffusion, and the relaxation of adsorbate molecules occur during different times and with different energetic characteristics.

The complexity of the pores and the heterogeneity of the chemical structure of the pore walls cause the adsorption energy distribution function $f(E)$ for LiChrolut EN strongly different from that for carbon adsorbents (see earlier), as well as the adsorption potential distribution functions $f(A)$ (Figure 5.28).

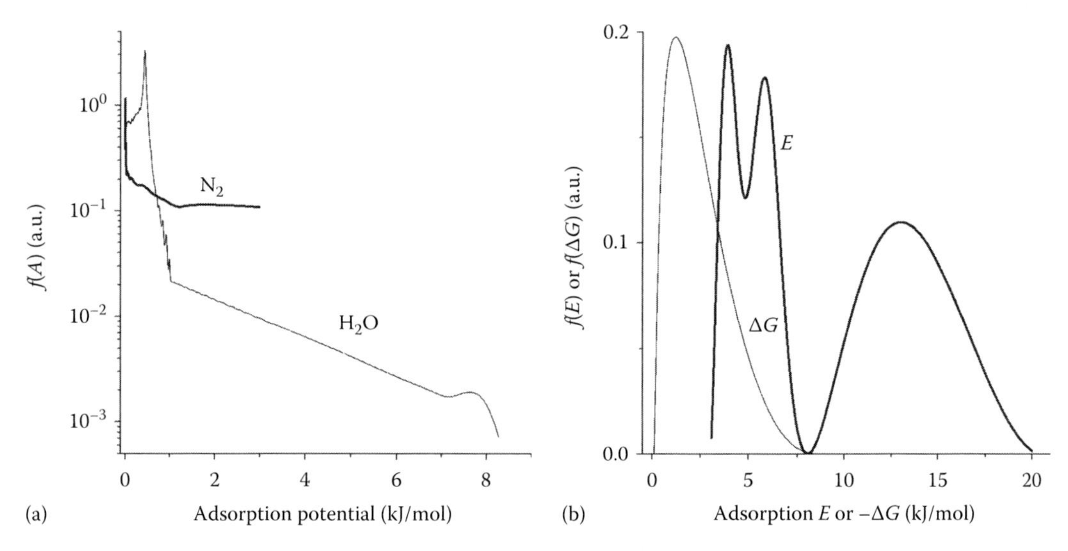

FIGURE 5.28 Distribution functions of (a) nitrogen and water adsorption potential and (b) nitrogen adsorption energy (E), and Gibbs free energy (ΔG) of water adsorption on LiChrolut EN. (Adapted from *J. Colloid Interface Sci.*, 323, Gun'ko, V.M., Turov, V.V., Zarko, V.I. et al., Structural features of polymer adsorbent LiChrolut EN and interfacial behaviour of water and water-organic mixtures, 6–17, 2008e, Copyright 2008, with permission from Elsevier.)

As a whole, the affinity of the LiChrolut EN surface to both nitrogen and water is lower than that for carbon adsorbents. The first effect is due to stronger dispersion interactions of the nitrogen molecules with more ordered carbon surface than with polymer adsorbent. Additionally, the hydrophilicity of LiChrolut EN is lower than that of the carbon adsorbents (obtained by carbonization of organic precursors with subsequent strong burn-off activation) due to lower amounts of oxygen-containing functionalities in LiChrolut EN.

On the PSDVB or PEVBDVB synthesis, peroxides are typically used. Therefore, oxygen-containing groups can be present in the polymer. According to chemical analysis of LiChrolut EN, the oxygen content is ~4 wt%. Therefore, similar polymers can interact with adsorbed water or other polar compounds due to not only dispersive forces but also the hydrogen bonding.

Desorption of water, CO and CO_2, and organics was studied by the one-pass temperature-programmed desorption (OPTPD) with mass spectrometry (MS) control (chamber pressure ~10^{-6} Torr, sample weight 5 mg, heating rate $\beta = 2$ K/s, with a short distance [~0.5 cm] between sample and a MS detector) with a MSC-3 ("Electron," Sumy, Ukraine) time-of-flight (ToF) mass spectrometer (sensitivity 2.2×10^{-5} A/Torr, accelerating voltage 0.5 kV, pulse frequency 3 kHz). TPD MS thermograms of water, CO and CO_2, and organic molecules and fragments were also recorded using a MX-7304A mass spectrometer ("Electron," Sumy, Ukraine) at a lower heating rate $\beta = 10$ K/min to 800°C. Volatile products are desorbed from the sample, ionized by the electron beam, separated by a mass analyzer, and come to a detector. Calculation of a distribution function of activation energy of desorption $f(E_a)$ was carried out using integral equation for a separated TPD peak or N overlapping peaks. The TPDMS thermograms (Figure 5.29) and the FTIR spectrum (Figure 5.30) reveal the presence of oxygen-containing groups in LiChrolut EN adsorbent.

Water molecularly adsorbed onto LiChrolut EN can be removed at relatively low temperatures (Figure 5.29a, curve 2, a shoulder at 105°C and a peak at 150°C). This water cannot be a sole source of desorbed water which intensively desorbs at higher temperatures $T = 200°C - 500°C$ (as well as CO and CO_2). Associative desorption of water at relatively high temperatures can be due to a set of chemical reactions at a surface. There are associative desorption of water (from –COOH, ≡COH, >C=O, and other groups) and oxidation of C–H groups by remaining molecular

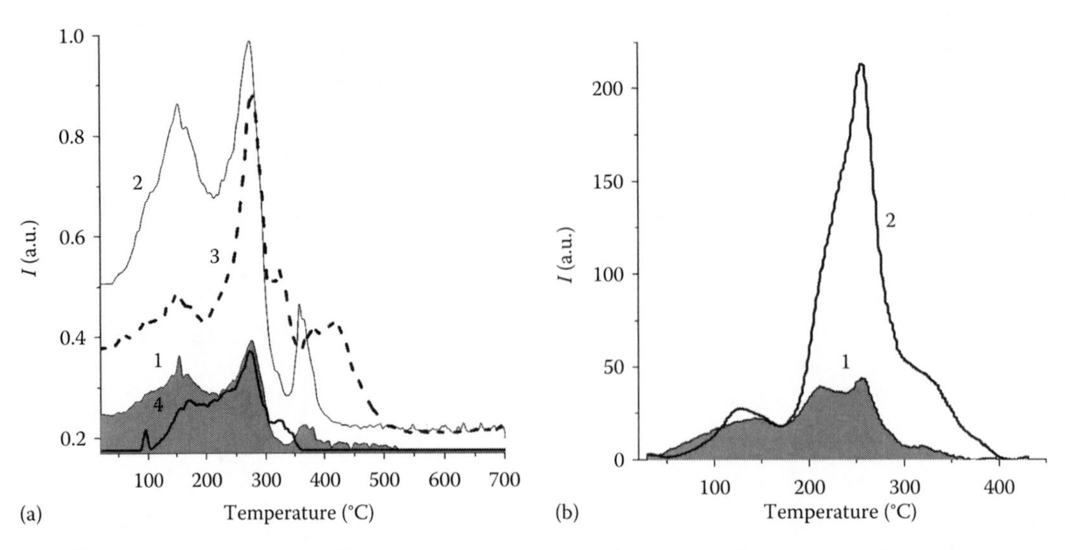

FIGURE 5.29 (a) TPDMS and (b) OPTPDMS thermograms for water (*m/z* 17 (a, curve 1) and 18 (a, curve 2 and b, curve 1), CO (*m/z* 28, a, curve 3 and b, curve 2), CO_2 (*m/z* 44, a, curve 4). (Adapted from *J. Colloid Interface Sci.*, 323, Gun'ko, V.M., Turov, V.V., Zarko, V.I. et al., Structural features of polymer adsorbent LiChrolut EN and interfacial behaviour of water and water-organic mixtures, 6–17, 2008e, Copyright 2008, with permission from Elsevier.)

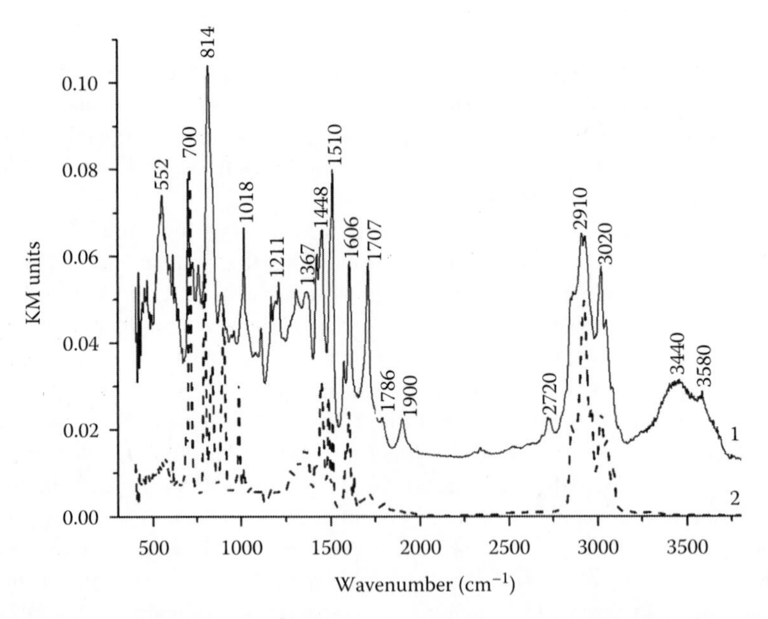

FIGURE 5.30 FTIR spectra (FTIR ThermoNicolet spectrophotometer with a diffusive reflectance mode) of (1) LiChrolut EN and (2) Amberlite XAD-4 adsorbents (mixture with KBr as 1:100).

oxygen forming >C=O groups, and other reactions. The first TPD MS peak of decomposition of the LiChrolut EN adsorbent at $T<300°C$ (the spectrum is not shown here) corresponds to the second peak of elimination of water, CO, and CO_2 (Figure 5.29a). The OPTPDMS thermograms (Figure 5.29b) recorded at a higher heating rate (β) are similar to the TPDMS thermograms (Figure 5.29a) recorded at a lower β value. However, the peak temperatures slightly shift due to this difference in the β values. Contributions of the DVB fragments are lower than that of the

EVB fragments, which affects the cross-linking of the polymer chains and, consequently, the pore shape. The presence of oxygen-containing groups in LiChrolut EN is also confirmed by the FTIR spectra (Figure 5.30).

According to the FTIR spectrum of LiChrolut EN adsorbent (Figure 5.30), there are the C–H-stretching vibrations at 3047, 3020 cm^{-1} (aromatic), saturated 2964 (asymmetric CH_3), 2929 (asymmetric CH_2), 2910 (symmetric CH_3), 2864 cm^{-1} (symmetric CH_2); deformation vibrations at 814, 760, and 702 cm^{-1} (out-of-plane bending of substituted benzenes); and deformation vibrations in aliphatic chains at 1367 and 1306 cm^{-1} (CH_2). The CC-stretching vibrations are observed at 1606, 1576, 1510, 1448, and 1421 cm^{-1} (aromatic). The CO-stretching vibrations are at 1786 and 1707 cm^{-1} (carbonyl groups). A broadband at $\nu_{OH}=3160$–3740 cm^{-1} can be attributed to adsorbed water and OH groups formed on synthesis of PSDVB-like polymers using the peroxide method. In the case of commercial PSDVB Amberlite XAD-4, the FTIR spectrum differs from that of LiChrolut EN adsorbent. Amberlite XAD-4 is more hydrophobic because the amount of adsorbed water is much lower than that adsorbed on LiChrolut EN (band at 3160–3740 cm^{-1}) at the same conditions. Additionally, a band at 2720 cm^{-1} is absent in the FTIR spectrum of XAD-4. Instead of an intensive band at 1900 cm^{-1} of LiChrolut EN adsorbent, there are several bands (1931, 1899, 1865, and 1811 cm^{-1}) of lower intensity for XAD-4 and the same effect is for the band at 1707 cm^{-1}. In other words, the number of oxygen-containing groups in LiChrolut EN, adsorbent is greater than in Amberlite XAD-4. In the latter according to chemical analysis, the oxygen content is less than 1.7 wt%, but it contains 0.14 wt% of nitrogen in contrast to LiChrolut EN which is with no nitrogen. However, this number is relatively low even in LiChrolut EN because the water adsorption isotherm (Figure 5.26) and the water adsorption potential (Figure 5.28) demonstrate rather low affinity of LiChrolut EN adsorbent to water. Additionally, water does not cover the total surface area of the adsorbent whereas the specific surface area determined from the water adsorption isotherm (at $0.05<p/p_0<0.25$) corresponds to 25 m^2/g, and $D_{H2O}<D_{N2}$. Consequently, water forms clusters and nanodomains in pores but not continuous thin film at the pore walls. For carbon adsorbents WVA and A2PS (possessing higher affinity to water than LiChrolut EN, Figure 5.28), the specific surface area estimated from the water adsorption isotherm is 511 and 621 m^2/g, respectively, i.e., 30% and 50% of the S_{BET} values. At $p/p_0\approx 1$, water fills ~66% of the pore volume of LiChrolut EN determined from the nitrogen adsorption. The water adsorption sharply increases only at $p/p_0>0.8$ due to capillary condensation and spreading of water in pores in the form of SAW nanodomains mounting the barriers at narrow hydrophobic necks in pores. Consequently, the LiChrolut EN adsorbent surface is rather hydrophobic than hydrophilic, and it is texturally, chemically, and energetically nonuniform. This leads to a complex temperature dependence of the behavior of adsorbed water and water/organics observed by the TSDC (Figure 5.31) and ^1H NMR (Figure 5.32) methods.

The tablets (diameter 30 mm, thickness ~1 mm) of frozen (at 265–260 K) adsorbent samples at water content $C_{H_2O}=70.2$ and 84.4 wt% were polarized by the electrostatic field at the intensity $F_p=0.14$–0.2 MV/m at 260 K and then cooled to 90 K with the field still applied and heated without the field at a heating rate $\beta=0.05$ K/s. The TSDC thermograms of hydrated LiChrolut EN adsorbent (Figure 5.31) are unusual because the depolarization or discharge current is observed at low temperatures $T<220$ K, and the direct current (dc) relaxation is absent at relatively high hydration $h=2.36$ and 5.4 g of water per gram of dry adsorbent.

Additionally, the depolarization current was transformed into the discharge current at low temperatures due to accumulation of charges and local thawing of frozen interfacial water in pores of LiChrolut EN with elevating temperature. Features of the TSDC thermograms can be caused by a mosaic hydrophobic/hydrophilic character of the pore walls. This influences the adsorption of water (Figures 5.26 and 5.30) and contributions of water molecules and polar surface functionalities into the TSDC spectra (Figure 5.31). However, hydrophobic functionalities (aromatic rings, CH_2 and CH_3 groups) give a major contribution. Therefore, despite the large specific surface area and the presence of narrow pores, the pore volume (Table 5.4) is incompletely filled by adsorbed water that affects the TSDC spectra. Various pores (Figure 5.27) of a complex shape (because of random

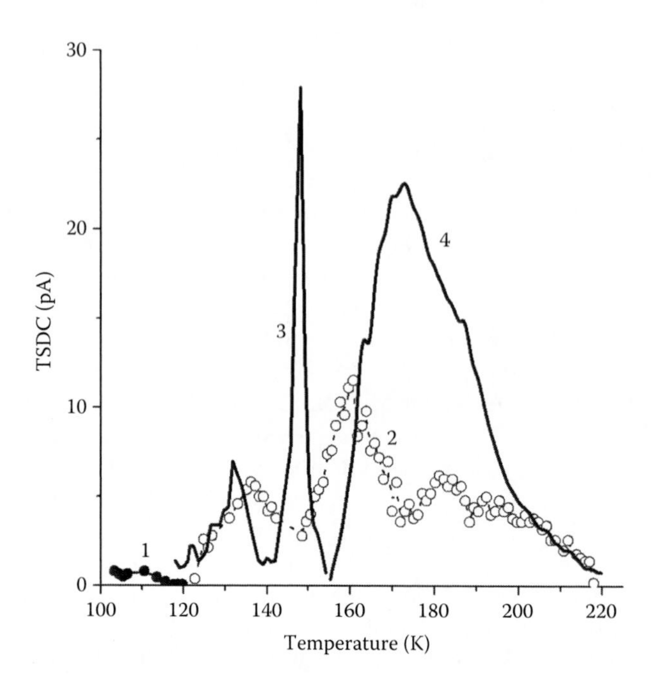

FIGURE 5.31 TSDC thermograms for differently hydrated LiChrolut adsorbent (depolarization [curves 1 and 3] and discharge [curves 2 and 4] currents are shown) at hydration $h = 2.36$ (curves 1 and 2) and 5.4 (curves 3 and 4) g/g. (Adapted from *J. Colloid Interface Sci.*, 323, Gun'ko, V.M., Turov, V.V., Zarko, V.I. et al., Structural features of polymer adsorbent LiChrolut EN and interfacial behaviour of water and water-organic mixtures, 6–17, 2008e, Copyright 2008, with permission from Elsevier.)

orientation of structural units in the pore walls, entangled and cross-linked polymer chains) can be differently filled by water on the TSDC measurements (Figure 5.27a, curve 1).

The ^1H NMR spectra (Figure 5.32) were recorded at 190–280 K using a Varian 400 Mercury spectrometer with 90° probe pulses with duration of 2 μs. During the NMR measurements, the predominantly hydrophobic character of the adsorbent also causes incomplete filling of pores by water at hydration $h = 2$ g/g. The amounts of bound water correspond to $V_w = 0.73$ cm^3/g (estimated from the amount of unfrozen water [C_{uw}] at 273 K), which is lower than $V_p = 0.83$ cm^3/g. However, this value is larger than that on the water vapor adsorption (30.44 mmol/g or 0.548 cm^3/g assuming that the density of adsorbed water $\rho_0 = 1$ g/cm^3) at $p/p_0 \approx 1$. The adsorbent is composed of main two structures with cross-linked [–CH–CH$_2$–Ph–CH–CH$_2$–]$_n$ (DVB) and linear [Ph(CH$_2$CH$_3$)–CH–CH$_2$–]$_m$ (EVB) chains. The intensity of the TPDMS peaks related to the EVB fragments is higher than that of the DVB fragments. It includes oxygen-containing groups (Figure 5.30) at the end of the chains because on free-radical polymerization of PSD, DVB, etc., activated by peroxides, growth of the chain occurs from the opposite end to peroxide.

Thus the number of polar active sites in LiChrolut EN is relatively low. Therefore, dispersive interactions of water with the pore surface are predominant similar to that for carbon adsorbents. The latter, however, are characterized by much higher ordering of condensed aromatic rings in basal planes than LiChrolut EN adsorbent. Therefore, the up-field shift of the spectra is not observed in contrast to carbon adsorbents (Figure 5.32).

The ^1H NMR spectra of water ($h = 2$ g/g) are characterized by two signals at $T > 273$ K ($\delta_H = 5$ and 4.5 ppm) and a single signal at $T < 273$ K, which has slight down-field shift with lowering temperature (Figure 5.32a).

Consequently, the shielding effect of π-electrons of aromatic rings (observed, e.g., in nanopores of carbon adsorbents as a significant up-field shift of the ^1H NMR signal, Figure 5.32b) on water

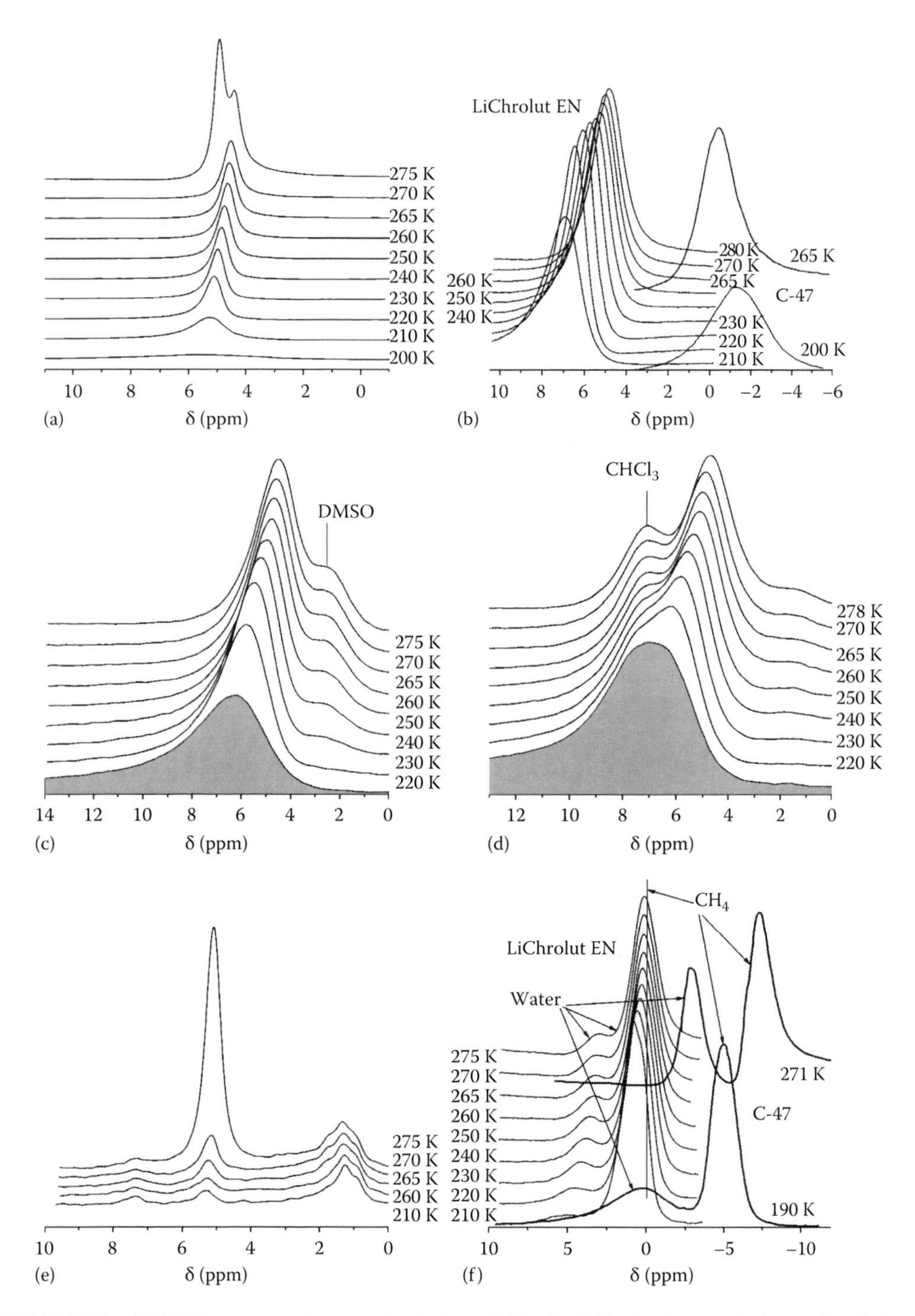

FIGURE 5.32 ^{1}H NMR spectra of water adsorbed on LiChrolut EN adsorbent at hydration h=(a) 2 and (b) 0.25 g/g and co-adsorption of water (pre-adsorbed at h=(c, d, f) 0.25 and (e) 0.175 g/g) and organics (c) 20 wt% of DMSO (half DMSO-d_6); (d) 30 wt% of $CHCl_3$; (e) in the $CDCl_3$ medium; and (f) 10 wt% of methane, hydrated activated carbon adsorbent C-47 at h=4.71 g/g (b) and $h \approx 0.1$ g/g and adsorbed CH_4 (f) recorded at different temperatures. (Adapted from *J. Colloid Interface Sci.*, 323, Gun'ko, V.M., Turov, V.V., Zarko, V.I. et al., Structural features of polymer adsorbent LiChrolut EN and interfacial behaviour of water and water-organic mixtures, 6–17, 2008e, Copyright 2008, with permission from Elsevier.)

adsorbed in narrow pores is very weak. This is due to orientation disordering of neighboring aromatic rings which do not form the condensed system as in carbon adsorbents.

The signal at $\delta_H=5$ ppm can be attributed to water located out of pores (the amount of water is greater than the pore volume) because it is frozen at 273 K. The second signal can be assigned to SAW, which includes strongly and weakly bound fractions. The first fraction is characterized by a larger diminution of the Gibbs free energy $\Delta G<-1$ kJ/mol than the second one at $\Delta G>-1$ kJ/mol (Figure 5.33d). The total Gibbs free surface energy $\gamma_S=89.4$ J/g (59 mJ/m^2) calculated from the NMR data is much larger than the heat of immersion of LiChrolut EN adsorbent in water

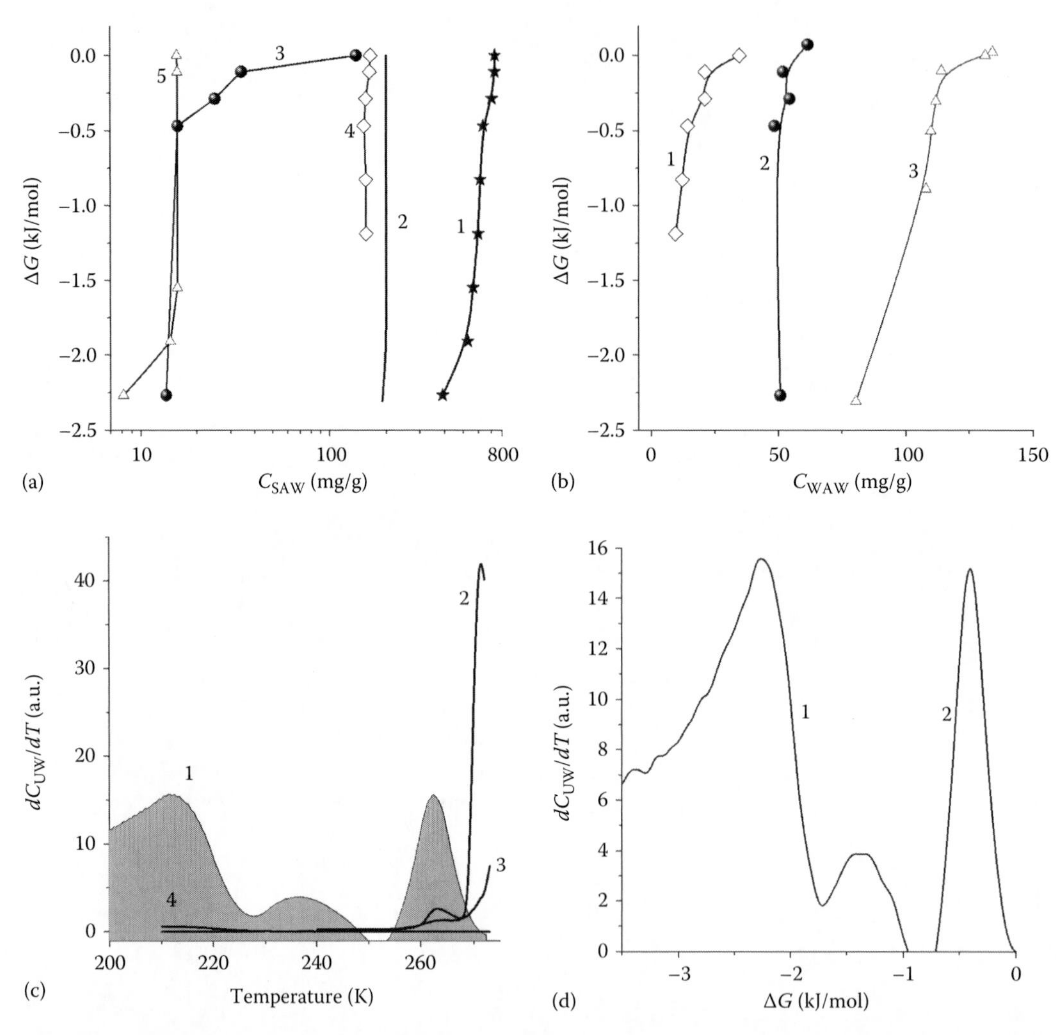

FIGURE 5.33 Relationships between changes in the Gibbs free energy (ΔG) and the amounts of (a) strongly (C_{SAW}) ($h=2$ (curve 1), and 0.25 (2–5) g/g) water alone (curves 1 and 2) and with addition of $CDCl_3$ (2 g/g) (curve 3), $CHCl_3$ (30 wt%) (curve 4), CH_4 (curve 5) and (b) weakly (C_{WAW}) associated water at $h=0.25$ g/g with addition of $CHCl_3$ (30 wt%) (curve 1), $CDCl_3$ (2 g/g) (curve 2), CH_4 (curve 3); derivatives of the total amounts of unfrozen water (dC_{uw}/dT) as a function of (c) temperature on adsorption of water alone at $h=2$ g/g (curve 1) and co-adsorption ($h=0.25$ g/g) with organics ($CDCl_3$ [curve 2], $CHCl_3$ [curve 3], CH_4 [curve 4]); and (d) changes in the Gibbs free energy of water adsorbed alone onto LiChrolut EN adsorbent at $h=2$ g/g (curves 1 and 2 correspond to strongly and weakly bound waters, respectively). (Adapted from *J. Colloid Interface Sci.*, 323, Gun'ko, V.M., Turov, V.V., Zarko, V.I. et al., Structural features of polymer adsorbent LiChrolut EN and interfacial behaviour of water and water–organic mixtures, 6–17, 2008e, Copyright 2008, with permission from Elsevier.)

at $\Delta H_{im}=34.7$ J/g (divided by S_{BET} is 23 mJ/m^2) calculated from the microcalorimetry data. This large difference in the γ_S and ΔH_{im} values can be due to different and incomplete filling of pores by water on the measurements. The adsorbent possesses relatively low hydrophilicity because in the case of hydrophilic adsorbents such as silica the γ_S and ΔH_{im} values are greater by a factor of 5–10. However, such PSDVB adsorbents as Amberlite XAD-2 (~5 wt% of oxygen and 0.14 wt% of nitrogen) and Amberlite XAD-4 are characterized by a lower hydrophilicity ($\Delta H_{im}=16$ mJ/m^2 for XAD-2 and 2.6 mJ/m^2 for XAD-4) than LiChrolut EN. This result is in agreement with the FTIR spectra of LiChrolut EN and Amberlite XAD-4 adsorbent (Figure 5.30) demonstrating lower amounts of water adsorbed on XAD-4 than on LiChrolut EN adsorbent.

The relationship between the dC_{uw}/dT and ΔG values (Figure 5.33d) calculated using the temperature dependence of the amounts of unfrozen water $C_{uw}(T)$ (determined from the ^{1}H NMR spectra recorded at different temperatures) demonstrates that there are, at least, two types of SBW at $\Delta G<-1$ kJ/mol. These types of water correspond to the adsorption in nanopores at $\Delta G<-2$ kJ/mol and in narrow mesopores at $-2<\Delta G<-1$ kJ/mol. WBW at $-1<\Delta G<0$ kJ/mol is more uniform than SBW. It can locate in broad pores giving smaller contributions to the porosity and the specific surface area than narrow pores (Figure 5.27 and Table 5.4).

Nonassociated or WAW is observed in the gaseous phase or in a mixture with nonpolar or weakly polar organic solvents (CCl_4, C_6H_6, $CHCl_3$, etc.). The chemical shift δ_H of such water is small (1–2 ppm) in contrast to bulk water at $\delta_H=3$–7 ppm. Water can be assigned to SAW if each molecule participates in more than two hydrogen bonds. The chemical shift of such water corresponds to $\delta_H=3$–7 ppm. Consequently, water adsorbed onto LiChrolut EN adsorbent is in the strongly associated state analogous to that of liquid water. Observation of two signals of SAW at $T>273$ K (Figure 5.32a) and the appearance of three peaks in the $dC_{uw}/dT(\Delta G)$ graph (Figure 5.33d) are due to the presence of bulk water and water bound in different nano-, meso-, and macropores of a complex shape (Figure 5.27). Notice that the position of the $dC_{uw}/dT(\Delta G)$ peak of SBW at $\Delta G\approx-2.5$ kJ/mol (Figure 5.33d) is in agreement with the position of the $f(\Delta G)$ peak (Figure 5.28b) calculated from the water adsorption isotherm (Figure 5.26). In the case of partial filling (~30% V_p) of pores, the water state changes (Figure 5.32b) in comparison with that on nearly completely filling (~90%) of pores (Figure 5.32a). An increase in the chemical shift δ_H is observed with decreasing temperature. Moreover at 210 K, the δ_H value is equal to 6.9 ppm close to the limiting value (7 ppm) for ice. This can be due to formation of ice crystallites in pores with hydrophobic polymer walls. Notice that all water on 30% filling of pores is SBW at $\Delta G<-2$ kJ/mol and located only in pores.

Added polar DMSO (20 wt% with respect to the weight of dry polymer sample) (Figure 5.32c) or weakly polar $CHCl_3$ (30 wt%) (Figure 5.32d) weakly influences the δ_H value of adsorbed water. This is due to location of water and organic phases in different pores with very low contact area between two liquids. For chloroform poorly water-soluble, this effect is typical and, therefore, can be expected. However, for DMSO well water-soluble, it would be possible to expect the formation of a homogeneous mixture with adsorbed water (the sample after addition of DMSO was shaken and equilibrated before the measurements) at an intermediate δ_H value between 3 ppm (water dissolved in DMSO) and 5 ppm (SAW alone). However, only SAW is observed for this mixture. Similar effects were previously observed for the water/organic mixtures adsorbed on carbon and oxide adsorbents. In the presence of $CHCl_3$ (Figure 5.32c), a portion of water transforms from SAW (Figure 5.33a) into WAW (Figure 5.33b). If the chloroform amount is larger than the pore volume that a significant portion of water becomes weakly associated (Figure 5.33), and certain amount of water corresponds to WBW because it is frozen at temperatures close to 273 K (Figure 5.32e). This occurs because chloroform can displace a portion of water from narrow pores. Similar effects were observed for other adsorbents because of several reasons: (i) diminution of contact area of practically immiscible liquids; (ii) weakly polar chloroform provides stronger dispersive interaction with hydrophobic pore walls than polar water molecules; (iii) small water clusters (as WAW) remain in pores at oxygen-containing sites. In the ^{1}H NMR spectra besides the signal of SAW ($\delta_H\approx5$ ppm) whose intensity rapidly decreases with decreasing temperature, the signal of WAW is observed at $\delta_H \approx 1.3$ ppm. The intensity of the latter depends weakly on temperature. A relatively complex shape of the ^{1}H NMR spectra of WAW can be due to different values of the magnetic susceptibility of adsorbates locating in pores of various sizes and the orientation effects of benzene rings (i.e., local currents of π-electrons) in these pores.

On co-adsorption of water and methane, the ^{1}H NMR spectra strongly change (Figure 5.32) because of contribution of methane and possible changes in the location and the state of adsorbed water. The signal intensity of SAW decreases and WAW appears (Figure 5.33). A residual quantity of SAW corresponds to not more than 25% of the initial one. The adsorption of methane is relatively high (up to 10 wt%) because of a significant contribution of nanopores (incompletely filled by water). Narrow pores have a complicated shape because the long-shape hysteresis loop is observed in the isotherm (Figure 5.26) due to random orientation of structural units in the pore walls. The signals of WAW and methane overlap but at $T < 220$ K a clearly visible shoulder related to the methane signal appears at $\delta_H \approx 0$ ppm. On co-adsorption of water and methane, transition of water from the strongly associated state into the weakly associated state is thermodynamically preferable because water and methane can form the hydrate system stabilized in narrow pores of the adsorbent. Probably, the main reasons responsible for the formation of such structures are the topological characteristics of polymer pores and the chemical structure of their walls with a low content of polar adsorption sites. The shielding effects of π-electrons of disordered benzene rings on adsorbed water/methane mixture are much smaller for LiChrolut EN than that of ordered condensed benzene rings in carbon adsorbent C-47 (Figure 5.32f) because for two adsorbents $|\Delta\delta_H| > 5$–7 ppm, which increases with temperature.

The associativity of water molecules depends on the presence of the second adsorbate. For instance, adsorbed water alone can be assigned to SAW (Figure 5.32a) because $\delta_H > 4$ ppm over the total temperature range. When the amount of adsorbed water corresponds to ~30% of the pore volume, co-adsorption of water and weakly polar chloroform causes the appearance of WAW (Figure 5.33b). The larger the amount of chloroform, the greater the contribution of WAW (Figure 5.33b) and the lower the amount of SAW (Figure 5.33a). This influence of organics leads to structural rearrangement of adsorbed water that causes changes in the temperature dependence of dC_{uw}/dT (Figure 5.33c) revealing changes in the amounts of unfrozen water with elevating temperature.

Relaxation phenomena (TSDC), molecular mobility (NMR, TPDMS), and chemical reactions (TPDMS of associative desorption of water) are observed for adsorbed water/LiChrolut EN adsorbent over a wide temperature range. These phenomena are characterized by very different activation energies from 10 kJ/mol (rotational mobility of hydroxyls in WAW molecules), 20–40 kJ/mol (rotational mobility of the molecules in SAW), 40–80 kJ/mol (rotational and translational mobility of the water molecules in pores of different sizes), and 60–200 kJ/mol (molecular and associative desorption of water) (Figure 5.34). As a whole, all the distribution functions of activation energy $f(E)$ obtained using different methods are well concordant. This is caused by the nature of activated processes whereas all the processes are caused by the molecular mobility of water dependent on the topological and chemical characteristics of confined space in nano- and mesopores in LiChrolut EN adsorbent.

Textural and structural features of nano/mesoporous LiChrolut EN adsorbent, its predominantly hydrophobic character at low content of oxygen-containing functionalities at the surface, cause the specific behavior of adsorbed water in a mixture with nonpolar or weakly polar organics. WAW ($\delta_H = 1$–2 ppm) appears for these mixtures. Adsorbed water can be divided into several types. There are weakly (frozen at T close to 273 K and $\Delta G > -0.8$ kJ/mol) and strongly (unfrozen at $T = 190$–250 K and $\Delta G < -0.8$ kJ/mol) bound, and WAW (small clusters or individual molecules at $\delta_H = 1$–2 ppm) and SAW (nanodomains and droplets at $\delta_H = 3$–5 ppm). Organic compounds can affect the amounts of different types of water and displace a portion of water from narrow pores of the adsorbent. This behavior of interfacial water and water/organics as well as the structural features of the adsorbent

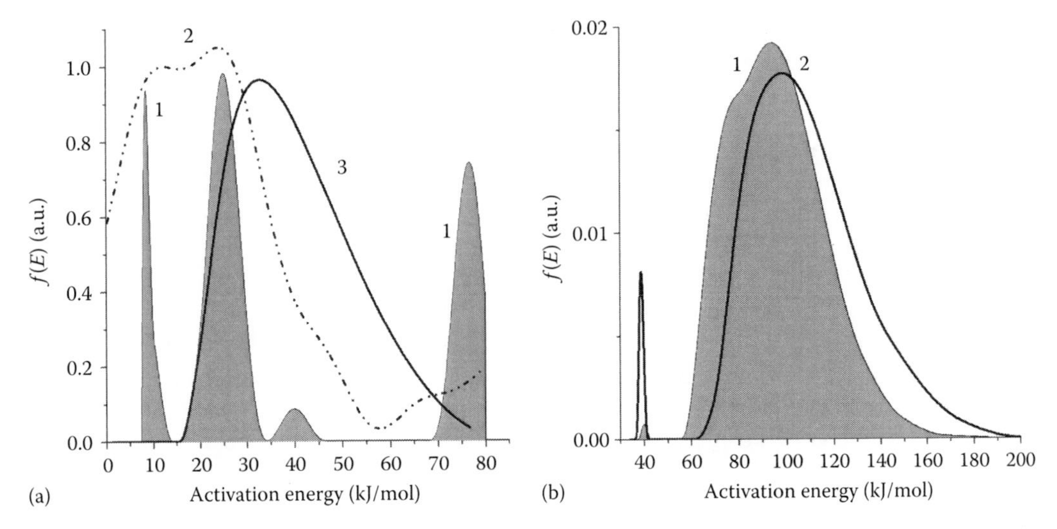

FIGURE 5.34 Distribution functions of the activation energy of the mobility and desorption of water molecules at the LiChrolut EN adsorbent surface calculated on the basis of (a) the TSDC (curves 1 [regularization] and 2 [piecewise linearization]) and ^{1}H NMR (curve 3) and (b) the TPDMS data (*m*/*z* 18) with OPTPDMS (curve 1) and TPDMS (curve 2). (Adapted from *J. Colloid Interface Sci.*, 323, Gun'ko, V.M., Turov, V.V., Zarko, V.I. et al., Structural features of polymer adsorbent LiChrolut EN and interfacial behaviour of water and water-organic mixtures, 6–17, 2008e, Copyright 2008, with permission from Elsevier.)

(large S_{BET} and V_p values, multimodal PSD at predominant contribution of nanopores and narrow mesopores) can be responsible for high effectiveness of LiChrolut EN adsorbent in the SPE processes.

Water can fill 70%–90% (dependent on filling conditions) of the total pore volume of LiChrolut EN adsorbent possessing nanopores and narrow mesopores of hydrophobic and partially hydrophilic characters. Adsorbed water is characterized by high associativity. It does not contact to the total surface area of the adsorbent. It is weakly affected by co-adsorbed polar DMSO. Weakly polar chloroform can displace a portion of adsorbed water from narrow pores into larger one or onto the outer surface of polymer particles. Methane can form the hydrate system with water adsorbed in narrow pores of LiChrolut EN adsorbent at low pressures.

Despite the similarity in the pore size distributions for polymer and carbon adsorbents (studied here), close values of the specific surface area and the pore volume, and the presence of aromatic rings in the pore walls for both types of the adsorbents, the confined effects on adsorbed water and water/organics differ strongly for them. This is because of the presence of the disordered (polymer) and ordered condensed (carbon) aromatic structures causing different values and cumulative effects of π-electron currents on adsorbates, especially located in narrow pores where strong up-field shifts can be observed for carbons but not for polymeric adsorbents. Therefore, the difference in the chemical shifts of bound water can reach 5–7 ppm for these adsorbents.

5.3 HYDROGELS AND CRYOGELS

Various polymer and protein hydrogels are used in numerous biomedical and pharmaceutical applications because of their high biocompatibility and rubbery nature that is similar to that of natural tissues (Peppas et al. 2000). Nano/macroporous hydrogels with a pore size $d > 1$ μm are widely used in tissue scaffold engineering, separation matrixes, bioreactors, and other applications (Mattiasson et al. 2009). To a large extent the application of hydrogels is defined by two factors, namely the state of water inside these highly hydrated polymer systems and the pore structure of the hydrogels allowing effective mass transport of solutes.

There are three states of the water in hydrogels: (i) free water, which does not interact with polymer and behaves as pure water; (ii) WBW weakly interacting with the polymer and freezing at subzero temperatures; and (iii) SBW, which is strongly bound to polymers through hydrogen bonding and unfrozen at $T<-15°C$. The amounts of water in different states effect interactions of solutes, biological substances, and cells with polymers within nanopores (in macropore walls), nano- and macropores. According to the life science (LSC) classification (ASTM International-Standards Worldwide 2010), pore sizes at diameter $d<0.1$ μm, $0.1<d<100$ μm, and $d>100$ μm correspond to nano-, micro-, and macropores, respectively. For more detailed classification of nanopores, they can be divided into three types such as narrow ($d<2$ nm), middle ($2<d<50$ nm), and broad ($50<d<100$ nm) nanopores (Gun'ko et al. 2010b), which correspond to nano-, meso-, and macropores of the IUPAC pore classification. The determination of the amounts of different types of water is of importance for understanding the nature of adsorption/desorption processes in hydrogels or characterization of their biocompatibility (Gun'ko et al. 2009d, Mattiasson et al. 2009, Savina et al. 2011). For these purposes, low-temperature ^{1}H NMR spectroscopy and DSC methods can be used.

Cryogelation (Lozinsky 2002) used here for synthesis of hydrogels at temperatures below the freezing point of a solvent has attracted considerable attention because it allows the production of micro/macroporous hydrogels from almost any gel precursor without using toxic organic solvents. When water as a solvent freezes out, the ice crystals (forming macropores of 1–250 μm in size) expel the gel precursors (monomers, polymer, cross-linker, and initiator), which concentrate into the unfrozen liquid phase remaining unfrozen even at −20°C to −30°C and forming macropore walls.

Techniques frequently used for structural characterization, e.g., gas/vapor adsorption, mercury porosimetry, pycnometry, and conventional scanning electron microscopy can be applied to dry materials. However, drying hydrogels can result in the shrinkage, deformation, and pore closure or collapse of the entire pore network significantly altering the natural pore structure. Therefore, nondestructive methods such as NMR, DSC, and confocal laser scanning microscopy (CLSM) were used for quantitative analysis of the pore structure of macroporous hydrogels (with poly(2-hydroxyethyl methacrylate-*co*-allyl glycidyl ether) [HEMA-AGE], gelatin [G], and gelatin–fibronectin [GF]) in hydrated state (Savina et al. 2011). Hydrogels with HEMA-AGE, gelatin, and GF were synthesized. The hydrogel samples native or freeze-dried were studied using SEM (JEOL JSM-6310), cryo-SEM (SEM coupled with a LT400 cryo-unit), CLSM (Leica TCS SP5), and multiphoton microscopy (MPM) (Zeiss LSM520 Meta NLO microscope). The images were analyzed using ImageJ (2007) and Fiji (2009) software. Differential scanning calorimetry (DSC) (Mettler Toledo) measurements were carried out using both hydrated and dried gel samples. The DSC cryoporometry was used for calculations of the pore size distributions. The ^{1}H NMR spectra were recorded at 200–280 K using a Varian 400 Mercury spectrometer of high resolution with 90° probe pulses with the duration of 2 μs. Hydrated fragments (up to 6000 atoms) of polymer and protein hydrogels were calculated using the MOPAC 2009 program suit with the PM6 method (Stewart 2008).

The morphology of HEMA-AGE gel (sample A) was studied using different imaging techniques: SEM, cryo-SEM, MPM, and CLSM. Images obtained by MPM and CLSM show the morphology of hydrated native gels. These images were used to estimate the interconnectivity and pore size distributions of 3D models reconstructed from 2D images obtained by nondestructive optical sample sectioning. Additionally, the image analysis gives the porosity, specific surface area, pore size, and wall thickness (Table 5.5). Both the techniques provide similar results with consideration of a certain limitation of image analysis with ImageJ and Fiji software (Savina et al. 2011).

The analysis of porous structure of hydrogels in freeze-dried state with the CLSM provides additional information on changes in the hydrogel structure during and/or after drying. Fluorescein isothiocyanate (FITC) stained HEMA-AGE gel was freeze-dried and CLSM image of the dried sample was compared with that obtained for hydrated hydrogel (Figure 5.35). This comparison shows that changes in the porous structure of the gel after freeze-drying were insignificant. Here the

TABLE 5.5
Characteristics of the Porous Structure of HEMA-AGE Gel Samples A, B, C, and D

Sample	A[a]	A[b]	B[a]	C[a]	D[a]
Porosity, %	91	88	91	82	68
Surface area (S), μm²/μm³	0.045	0.10	0.06	0.095	0.186
Wall thickness (t), μm	9.6±0.8	10.9±1.6	5.7±0.5	8±1	9.5±1.3
Pore size (d), μm	64±2	56±1.5	47±2.8	41.4±6	25.7±1

[a] CLSM image analysis.
[b] MPM image analysis.

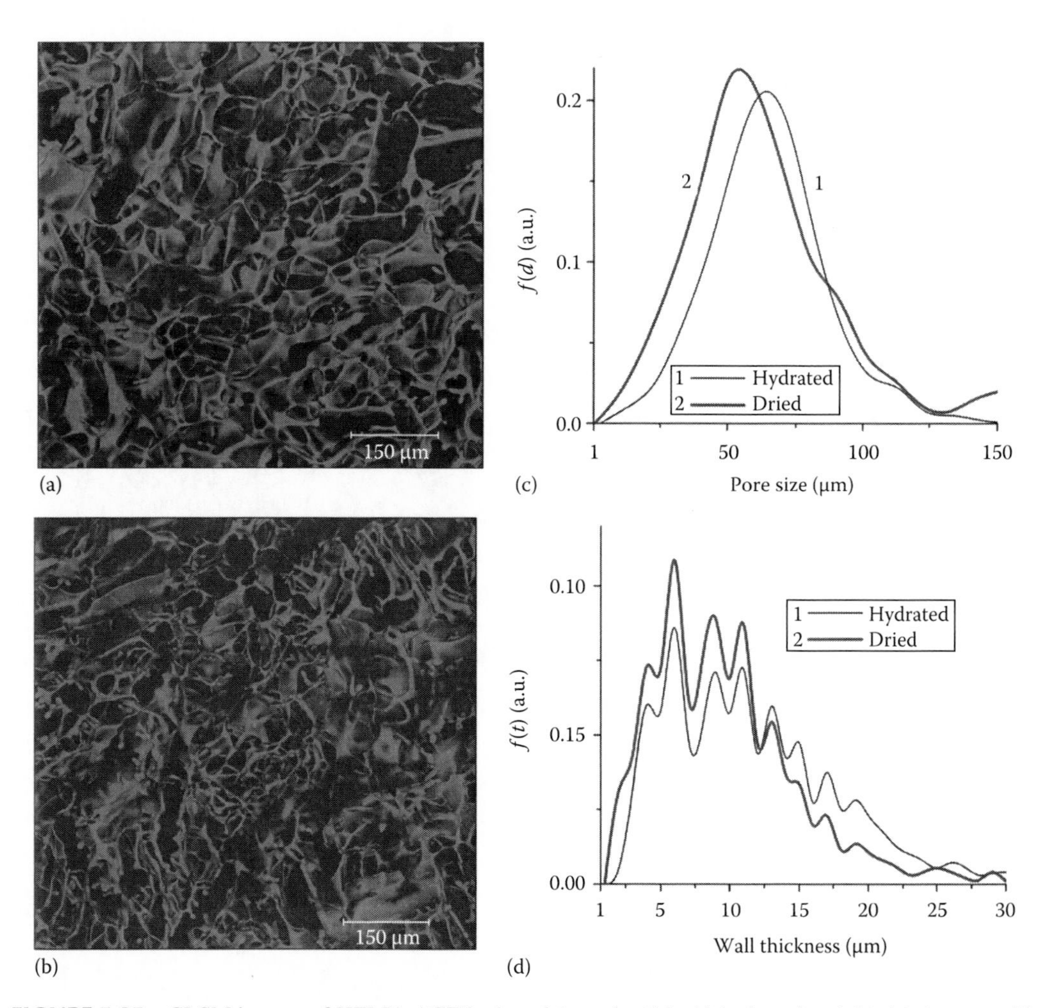

FIGURE 5.35 CLSM images of HEMA-AGE hydrogel (sample A) in (a) hydrated and (b) dried states with the pore (c) size and (d) wall thickness distributions. (Adapted from Savina, I.N., Gun'ko, V.M., Turov, V.V. et al., Noninvasive structural characterisation of macroporous cross-linked polymer and protein hydrogels, *Soft Matter*, 7, 4276–4283, 2011, Copyright 2011, with permission from The Royal Society of Chemistry.)

pore structure of dry and hydrated samples was compared quantitatively (Figure 5.35) that shows only a small diminution in the porosity, pore size, and wall thickness of HEMA-AGE gel after drying (Figure 5.35c and d).

The observed changes could be attributed to some shrinkage of the polymer walls rather than shrinking and collapsing of the whole 3D gel structure (Figure 5.35). These results confirm that the porous cryogels have relatively thick walls of macropores. During drying, the mean thickness of the walls decreased from 11.2 to 9.6 μm, which is within the errors of the analysis (Table 5.5). It can be concluded that most of the water in hydrated cryogels is located rather in macropores than in the swollen walls of the macropores. A relatively small thickness of the walls provides the hydrogel elasticity, spongelike morphology, and significant mechanical strength.

All HEMA-AGE gel samples had relatively uniform pore size and wall thickness distributions (Figure 5.36).

The pore sizes of samples A, B, and C were mainly in the range of 3–100 μm with mean pore sizes of 64, 47, and 41.4 μm, respectively (Table 5.5). Gel D displayed a narrower pore size distribution (3–83 μm) with a mean pore size of 25.7 μm. The wall thickness distribution of gel B calculated using ImageJ granulation plug-in was narrower (2–14 μm) while the gels A, C,

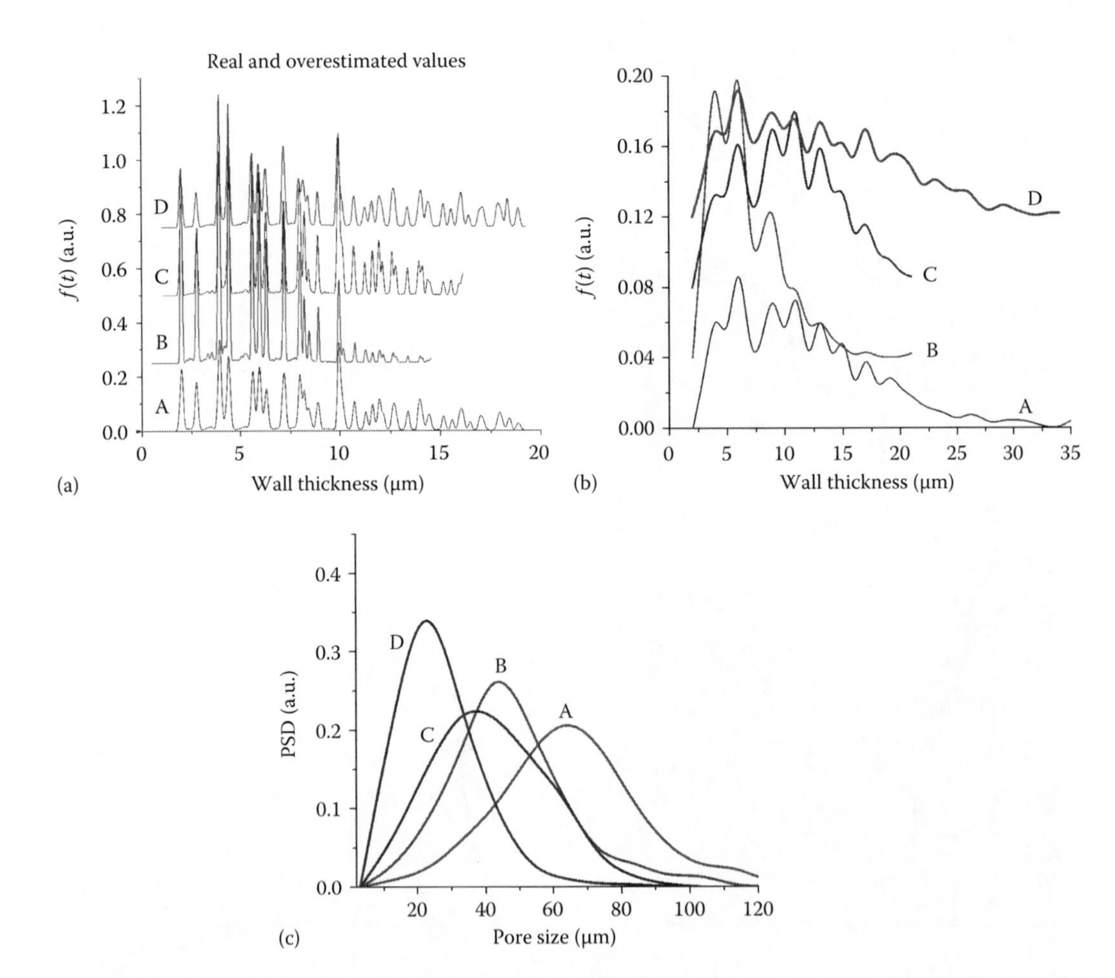

FIGURE 5.36 Wall thickness distributions with (a) Fiji, (b) ImageJ, and (c) pore size distributions for HEMA-AGE hydrogels A, B, C, and D (Table 5.5). (Adapted from Savina, I.N., Gun'ko, V.M., Turov, V.V. et al., Noninvasive structural characterisation of macroporous cross-linked polymer and protein hydrogels, *Soft Matter*, 7, 4276–4283, 2011, Copyright 2011, with permission from The Royal Society of Chemistry.)

and D had broader wall thickness distributions (2–32 μm, 2–23 μm, and 2–32 μm, respectively) (Figure 5.36b). Fiji software (local thickness plug-in with auto-threshold) gives systematically less average wall thickness (t) values (Figure 5.36a). The pore wall surface was assumed to be smooth. However, the DSC cryoporometry calculations of the specific surface area give $S \sim 80–90\ m^2/g$ for samples A–D because of the presence of nanopores in the macropore walls. Water (analyzed with DSC cryoporometry) is bound in these nanopores (1–30 nm in radius, Figure 5.37) located in the macropore walls of several micrometers in thickness (Figure 5.36).

The left peak or a shoulder of the main PSD peak (Figure 5.37a) corresponds to nanopores with the size close to that of icosahedral nanodomains (~3 nm in size) of water (Chaplin 2011). Other water structures are of larger sizes (Figure 5.37). The temperature of freezing of bound water is above −10°C. Water with icosahedral nanodomain or larger domain structure and frozen close to 0°C can be attributed to WBW. The main fraction of water in macropores corresponds to non-bound, practically bulk water, and bound water (<1% of total amount of water in initial hydrogel but ~100% of water in freeze-dried gels) is weakly bound by the polymer. The latter is due to the absence of narrow nanopores in the macropore walls (Figure 5.37). The weak binding of water in hydrogels is confirmed by the thermogravimetric measurements on heating at a constant rate.

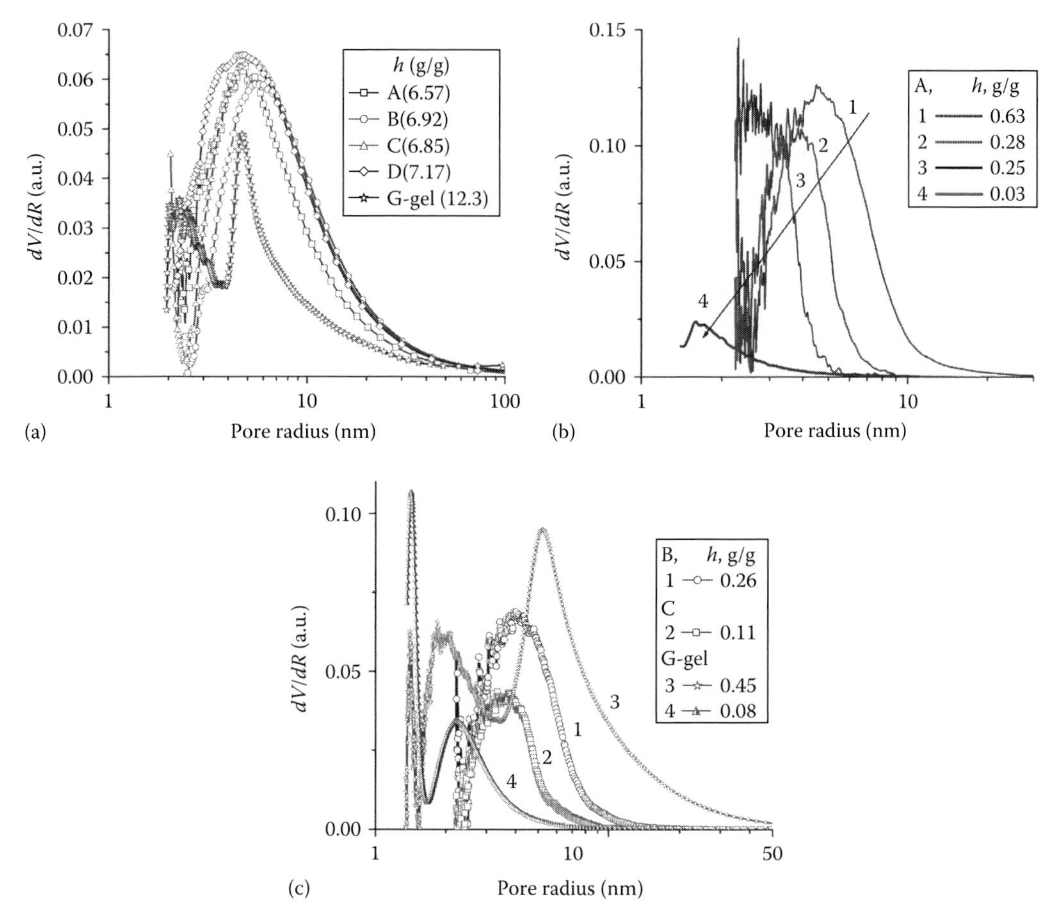

FIGURE 5.37 PSDs calculated from the DSC data for HEMA-AGE (A, B, C, and D samples) and G gels (hydration $h = m_w/m_d$ where m_w is the weight of water evaporated on DSC measurements to 160°C and m_d is the residual weight of heated sample) at (a) high and (b, c) low hydration. (Adapted from Savina, I.N., Gun'ko, V.M., Turov, V.V. et al., Noninvasive structural characterisation of macroporous cross-linked polymer and protein hydrogels, *Soft Matter*, 7, 4276–4283, 2011, Copyright 2011, with permission from The Royal Society of Chemistry.)

These measurements show that water desorbs gradually from 30°C and desorption is practically completed at 87°C–103°C depending on the hydrogel type. In the case of nanoporous materials, this process is completed at higher temperatures (~150°C) at the same heating rate. A decrease in hydration of HEMA-AGE or G gels leads to diminution of the PSD peaks of nanopores with the displacement toward lower *R* values (Figure 5.37b).

To study the PSD of G gel at different amounts of water, low-temperature ^{1}H NMR spectroscopy was applied to freeze-dried and then hydrated G gel samples. A freeze-dried G gel studied contained a low amount of water that causes a low intensity of signal of water (Figure 5.38a). The signals at the chemical shift of proton resonance $\delta_H = 7.2$ and 0 ppm (Figure 5.38a) are due to residual $CHCl_3$ in chloroform-*d* and tetramethylsilane added as a standard, respectively. Weak signals at $\delta_H = 4.8$ and 1.3 ppm can be attributed to SAW and WAW. SAW signal is not observed at $T < 250$ K, but the WAW signal, despite intensity decreasing with lowering temperature, is observed even at 210 K. Consequently, all strongly associated water is WBW but a portion of WAW corresponds to SBW. The addition of polar acetonitrile to weakly polar chloroform (Figure 5.38b) results in the appearance of the signal of methyl groups of CH_3CN (as an admixture in

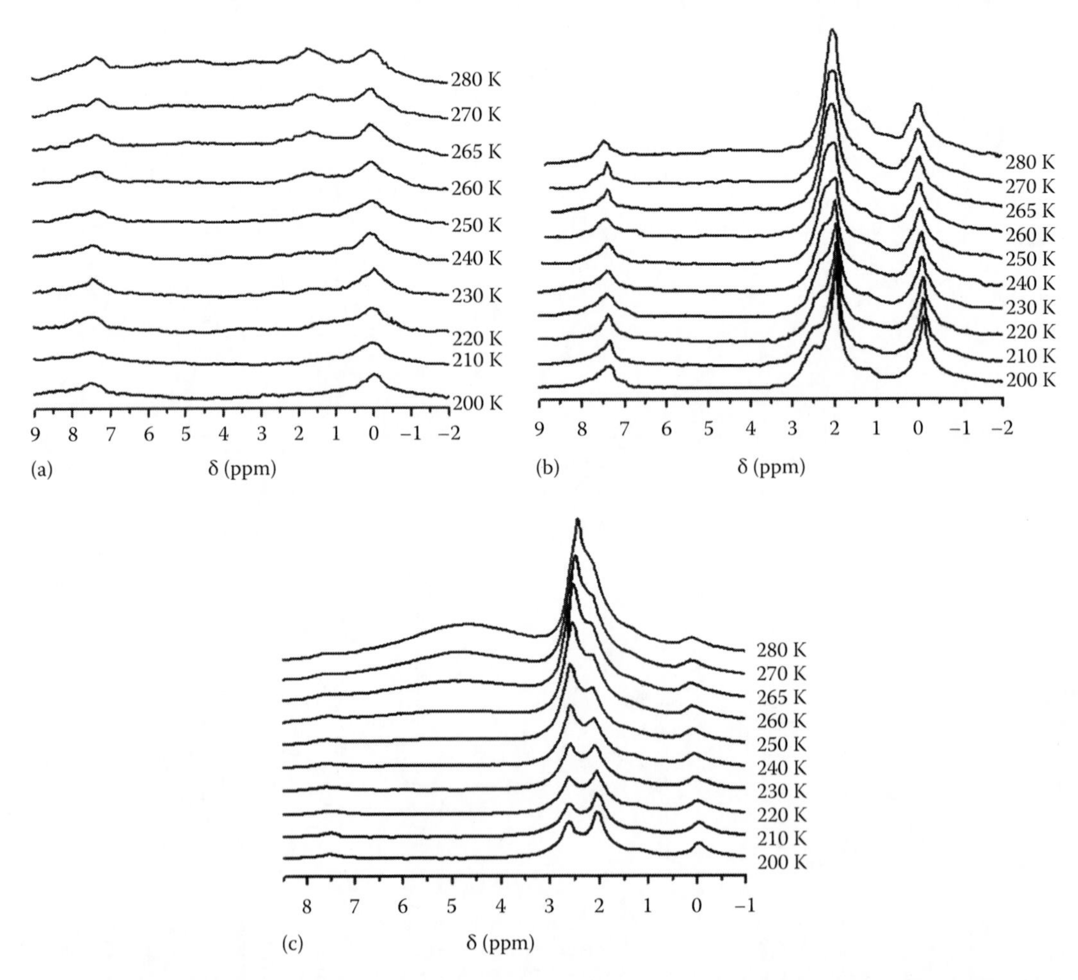

FIGURE 5.38 ^{1}H NMR spectra, recorded at different temperatures, of water adsorbed by G gel: (a) initial freeze-dried (0.3 wt% H_2O) in $CDCl_3$; and in a mixture $CDCl_3$:CD_3CN 3:1 at (b) 0.8 wt% and (c) 10 wt% of water. (Adapted from Savina, I.N., Gun'ko, V.M., Turov, V.V. et al., Noninvasive structural characterisation of macroporous cross-linked polymer and protein hydrogels, *Soft Matter*, 7, 4276–4283, 2011, Copyright 2011, with permission from The Royal Society of Chemistry.)

CD_3CN) at $\delta_H = 2$ ppm. This signal can overlap with a signal of water associated (ASW) with molecules of organic solvents. The main portion of this water is bound in complexes with acetonitrile HO–H···$NCCD_3$, ASW at $\delta_H = 2–2.5$ ppm. The signal of WAW is observed on the right wing of a signal of the CH_3 groups of acetonitrile. An increase in water concentration to 10 wt% (Figure 5.38c) leads to the appearance of a broad signal of SAW at $\delta_H = 4.8$ ppm. Its intensity sharply decreases with decreasing temperature but the ASW intensity increases. Notice that glutaraldehyde cross-linked gelatin (G gel) is characterized by diminished water bonding in comparison with unmodified gelatin (Gun'ko et al. 2007i, 2009d).

For more strongly hydrated G gel ($h = 1$ g/g provides about 10% of the total pore volume) being in air, a broad asymmetrical signal of SAW is observed (Figure 5.39a). This signal tends to split into two signals because of the presence of several forms of bound water and spatial heterogeneity of the sample with different magnetic susceptibility in different structures. A sample in the C_6D_6 medium (Figure 5.39b) is spatially more uniform (with respect to bound water) because of the filling of macropores by benzene and changes in water location in pores. Therefore, only one broad signal of SAW is observed. On the use of a mixture of solvents ($C_6D_6 + CD_3CN$, Figure 5.39c,

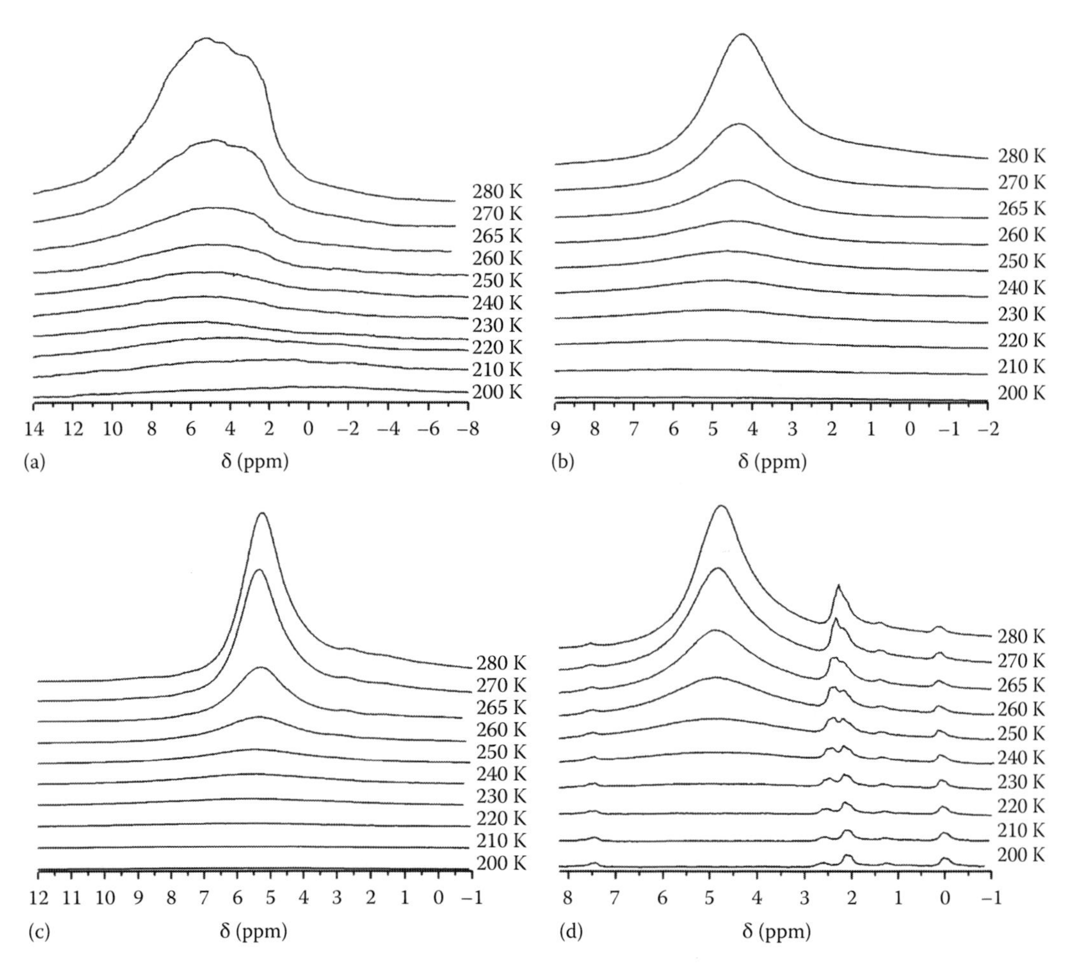

FIGURE 5.39 ^{1}H NMR spectra, recorded at different temperatures, of water bound in G gel at hydration $h = 1$ g/g of dried gelatin in different media: (a) air, (b) C_6D_6, (c) C_6D_6:$CD_3CN = 6:1$, and (d) $CDCl_3$:$CD_3CN = 3:1$. (Adapted from Savina, I.N., Gun'ko, V.M., Turov, V.V. et al., Noninvasive structural characterisation of macroporous cross-linked polymer and protein hydrogels, *Soft Matter*, 7, 4276–4283, 2011, Copyright 2011, with permission from The Royal Society of Chemistry.)

TABLE 5.6
Structural Characteristics of G Gel Based on NMR Cryoporometry

h (g/g)	Medium	S (m²/g)	S_{nn} (m²/g)	S_{mn} (m²/g)	S_{bn} (m²/g)	V_{nn} (cm³/g)	V_{mn} (cm³/g)	V_{bn} (cm³/g)	γ_S (J/g)
1	Air	465	423	43	0	0.157	0.431	0	34.4
1	C_6D_6	133	94	38	1	0.045	0.520	0.014	28.3
1	$C_6D_6+CD_3CN$	176	88	87	1	0.042	0.928	0.010	37.6
1	$CDCl_3+CD_3CN$	84	10	74	1	0.005	0.732	0.009	35.9
0.1	$CDCl_3+CD_3CN$	8	0	8	0	0	0.031	0	1.3

or $CDCl_3+CD_3CN$, Figure 5.39d), signals of WAW ($\delta_H=1.3$ ppm), ASW ($\delta_H=2$–2.5 ppm), and CH_3 groups of CH_3CN ($\delta_H=2$ ppm) are observed. Consequently, water structure and location depend on the presence of organic solvents, their polarity (hydrophilicity/hydrophobicity), and concentrations.

Water can be attributed to "weakly bound" if freezing temperature $T>250$ K that corresponds to changes in the Gibbs free energy $\Delta G>-0.5$ kJ/mol (due to water binding). Therefore, the main fraction of water bound in G gel is weakly bound, which is in agreement with the DSC data. The main reason of this effect is the macroporosity of the hydrogels. The amounts of water strongly bound in hydrated G gel at $h=1$ g/g correspond to the volume of 0.2–0.3 cm³/g (Table 5.6, V_{nn}). The main portion of this water can locate in a confined space between adjacent and cross-linked macromolecules. Organic solvents can displace this water that appears as diminution of the S_{nn} and V_{nn} values of narrow nanopores (Table 5.6) and changes in the size distribution of pores filled by bound water unfrozen at $T<273$ K (Figure 5.40). Water located in macropores of G gel can be attributed to bulk water. Therefore, the main portion of bound water ($V_{nn}+V_{mn}$) can be located in swollen pore walls or at their surfaces. Notice that the displacement of water from narrow nanopores by organic solvents can play an important role in practical applications of hydrogels. Notice that the PSD based on the NMR (Figure 5.40c) and DSC data for hydrated G gel is close to that for HEMA-AGE gel (Figure 5.37a). This similarity for different hydrated gels is due to great amounts of bulk water which can mask fine details of the effects related to bound water.

Regarding joint LSC and IUPAC classifications of pores, LSC nanopores ($d<100$ nm) can be divided into three types. The first one is narrow nanopores at $R<1$ nm (S_{nn}, V_{nn}, Table 5.6) corresponding to micropores in the IUPAC classification of pores. The second is nanopores of middle sizes at $1<R<25$ nm (S_{mn}, V_{mn}, Table 5.6) corresponding to IUPAC mesopores. The third is broad nanopores $25<R<100$ nm (S_{bn}, V_{bn}, Table 5.6) corresponding to IUPAC macropores.

Nanopores of 1–2 nm in radius are observed in the HEMA-AGE model (Figure 5.41, geometry was optimized using the PM6 method) in areas of both low and high hydration. The PM6/MOZYME calculations of the hydrated fragments of hydrogels (Figure 5.41) were carried out to estimate the hydration energy of polymers and proteins and the macromolecule–macromolecule interactions.

Hydrated two triple coils (i.e., six fragments) of collagen (Figure 5.41b), used as a model of gelatin, are characterized by hydration energy $E_h=-9.5$ kJ/mol per water molecule. This value is not great (it is 3–4 times lower than the energy of a strong hydrogen bond) because a significant portion of water molecules do not have direct contact with the protein molecules (Figure 5.41b). Protein–protein interaction energy in the triple coils is relatively high $E_{pp}=-19$ kJ/mol per each amino acid residue. With consideration of the solvation effect, this value is $E_{pp,h}=-106.7$ kJ/mol per each amino acid residue because water molecules can act as a bridge between protein functionalities of neighboring chains that enhance their interactions. In the case of the hydrated fibronectin (8-9FnI)–collagen complex (Figure 5.41c), the hydration energy $E_h=-28.6$ kJ/mol is greater than that for collagen because of more densely packed water layers. E_h value is minimal

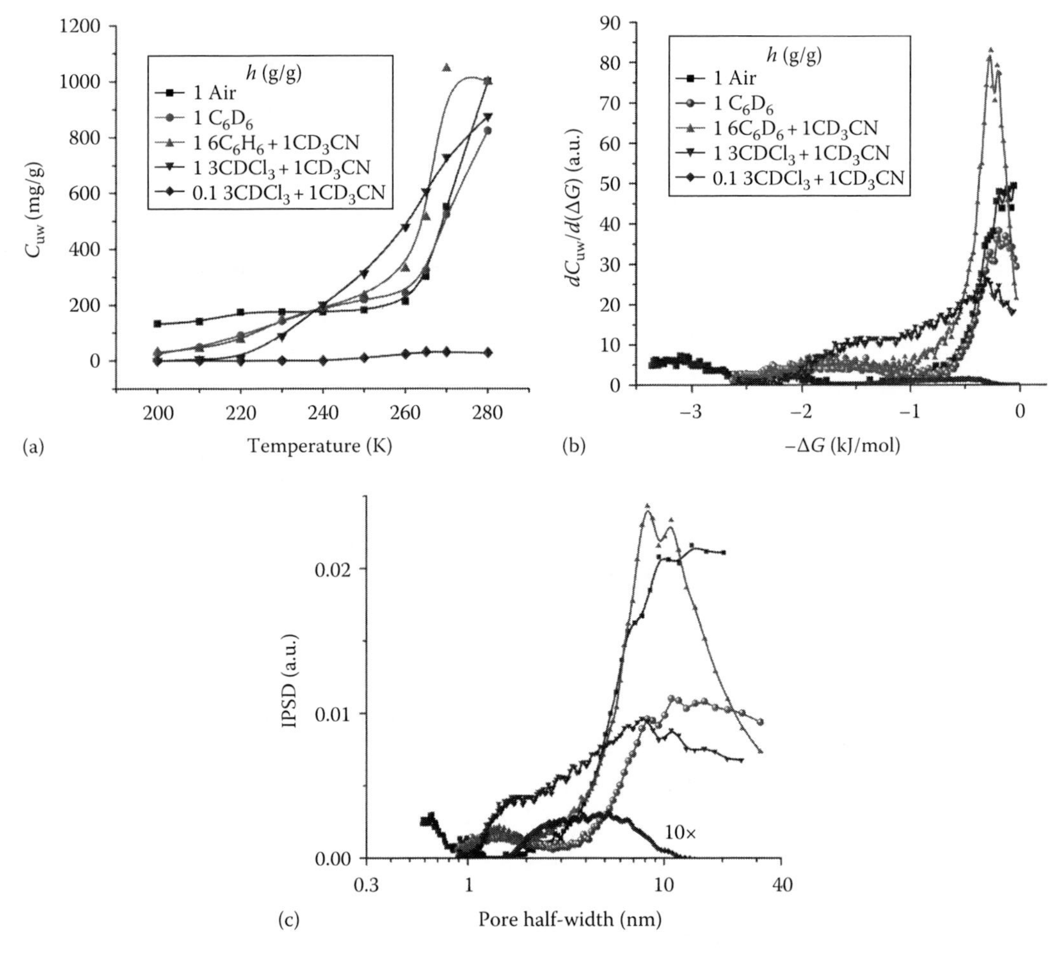

FIGURE 5.40 (a) Amounts of unfrozen water (C_{uw}) as a function of temperature; (b) derivative $dC_{uw}/d(\Delta G)$; and (c) pore size distribution (NMR cryoporometry) for G gel in different media. (Adapted from Savina, I.N., Gun'ko, V.M., Turov, V.V. et al., Noninvasive structural characterisation of macroporous cross-linked polymer and protein hydrogels, *Soft Matter*, 7, 4276–4283, 2011, Copyright 2011, with permission from The Royal Society of Chemistry.)

(−8.6 kJ/mol per water molecule) for HEMA-AGE gel (Figure 5.41a) because a significant portion of water molecules (larger than that for collagen model in Figure 5.41b) is not located in the first layer contacting with the polymer. This result is in agreement with the DSC and NMR data showing that water is weakly bound in the HEMA-AGE and G hydrogels. The NMR study gives changes in the free Gibbs energy of bound water $\Delta G > -3.5$ kJ/mol for the first layer (Figure 5.40b), i.e., $\Delta G > E_h$ ($E_h \approx \Delta H_f$) due to an entropy decrease in the bound water layer. The properties of bound water differ from that of bulk water, e.g., in the activity, mobility, diffusivity, etc. However, the relative amounts of water bound in the hydrogels are less (<10 wt% of bulk water) because macropores give the main contribution to the porosity of these materials in which water is practically nonbound (bulk). This factor is important for biocompatibility of hydrogels and normal functioning of cells in macropores since bulk water has much greater activity as a solvent than bound water has.

Thus, in the native hydrated state, polymer- and protein-based hydrogels produced by cryogelation are characterized by a large macropore volume but a small specific surface area of macropore walls, high pore interconnectivity, and high hydrophilicity. The main portion of water in macroporous hydrogels can be attributed to bulk water located in macropores. The amount of bound water located

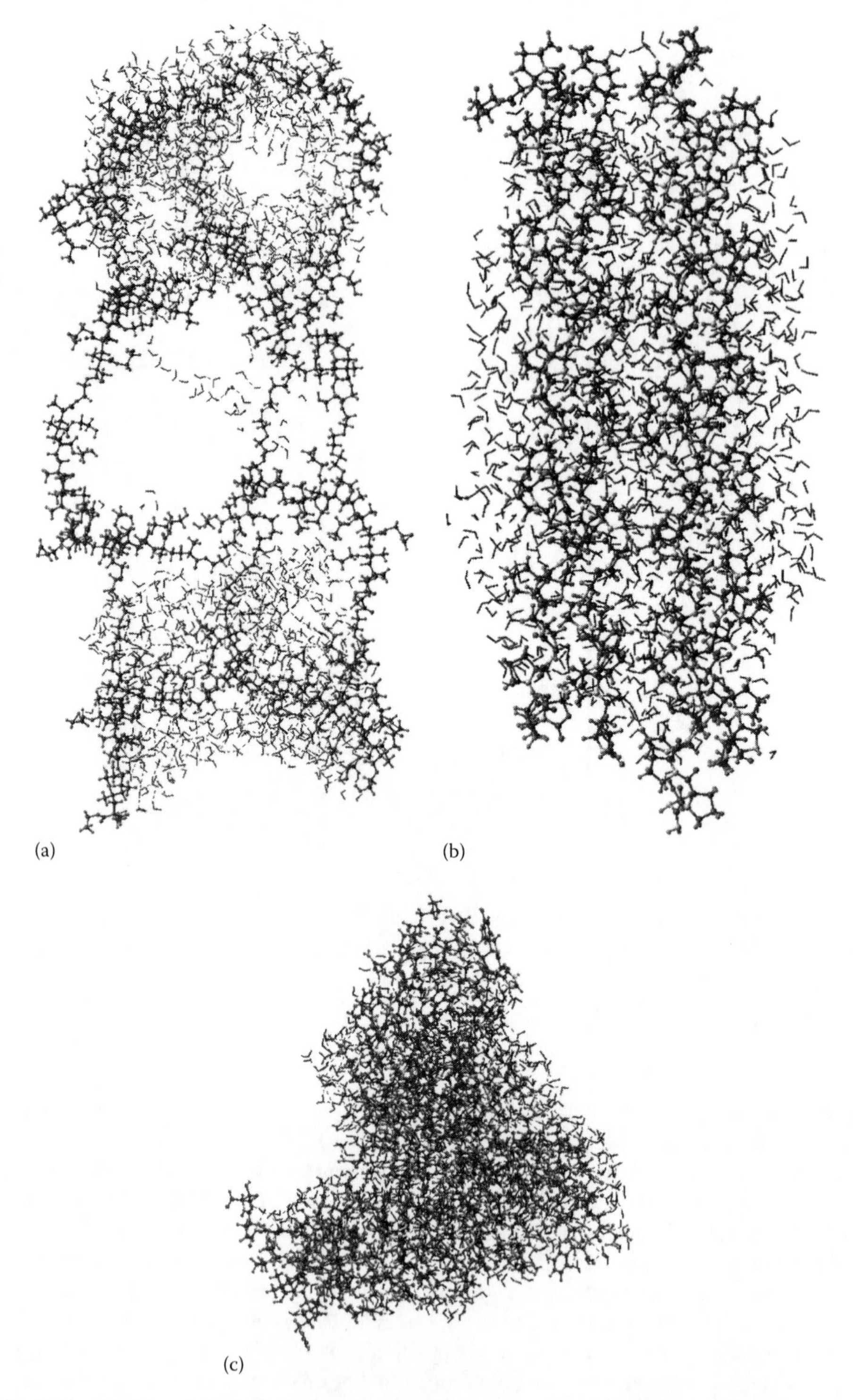

FIGURE 5.41 Models of partially hydrated gels with (a) cross-linked HEMA-AGE (2373 atoms) with $1192H_2O$; (b) collagen (two triple coils (1639 atoms) and $1032H_2O$); and (c) fibronectin (8-9FnI)–collagen (3200 atoms) with $827H_2O$ (PM6 geometry). (Adapted from Savina, I.N., Gun'ko, V.M., Turov, V.V. et al., Noninvasive structural characterisation of macroporous cross-linked polymer and protein hydrogels, *Soft Matter*, 7, 4276–4283, 2011, Copyright 2011, with permission from The Royal Society of Chemistry.)

in nanopores in the swollen macropore walls is less by an order of magnitude. This water is weakly bound that makes macroporous hydrogels an attractive system for biomedical applications as SBW is characterized by a low activity as a solvent and hence affecting adversely a variety of bioprocesses within the hydrogel matrix.

Collagen-based materials are widely used in tissue engineering (Nimni 1988, Rosenblatt et al. 1993, Vernon et al. 2005). Therapeutic properties of these preparations are determined by the structure of the collagen hydrogel, which allows transport of nutrients for tissue renewal and also removes necrotic products. Porous collagen–glycosaminoglycan (CG) hydrogels are used as artificial skin substitutes for treatment of thermal injuries (Yannas 2001) and as tissue scaffolds to provide physical support for cells and functional micro-architecture for cell growth (Breinan et al. 1997). The dermal layer of the CG-based artificial skin consists of cross-linked collagen and a glycosaminoglycan, chondroitin-6-sulfate. It is produced by a variety of methods such as precipitation of collagen–chondroitin sulfate, freeze-drying, and cross-linking of the network by chemical or physical treatment (Pek et al. 2004). The pore structure of soft biomaterials used in tissue engineering is an important parameter contributing to their performance. It strongly depends on the hydrogel interaction with water (Wallace and Rosenblatt 2003). The porous network in CG scaffolds with pore size ranging from 20 to 200 μm—micropores are suitable for cell accommodation, migration, and growth.

Characterization of the pore structure of collagen and other soft hydrogels presents a significant challenge, as there are very few experimental techniques that can examine the micro- and nanoporous structure of such materials in situ. Recently, an experimental method of micropore visualization and quantification in CG scaffolds based on digital image analysis and optical microscopy has been developed (Mattiasson et al. 2009). Data obtained using this method are broadly in agreement with those obtained using microscopic methods. However, detection of smaller pores in submicron and nanometer range—nanopores, which are important for supply of nutrients and oxygen and removal of wastes—presents an awesome problem. Investigating the behavior of water in hydrogels using ^{1}H NMR spectroscopy can provide useful information about their nanoporous structure because water properties and therefore the ^{1}H NMR signal depend on the local environment and interactions of water molecules with biomacromolecules and surfaces (Gun'ko et al. 2005d, 2007i, 2009d, Mitchell et al. 2008). NMR studies of the structural features of collagen and related materials such as gelatin, CG, and polypeptide models of collagen (Kozlov and Burdygina 1983, Melacini et al. 2000, Fantazzini et al. 2003, Slatter et al. 2003, Vernon et al. 2005) as well as visualization of these matrices (Mao and Kisaalita 2004, O'Brien et al. 2004) showed significant changes in their 3D structure upon hydration/dehydration. Investigations of collagen and gelatin, the product of denaturation, and structural degradation of collagen showed that collagen retains bound water more strongly than gelatin, with 0.12–0.47 g of water per gram of collagen and 0.05–0.37 g per gram of gelatin at a relative water pressure p/p_0 in the range from 0.25 to 0.90 (Kozlov and Burdygina 1983). This difference was attributed to differences in the molecular structure of these proteins.

Another technique that has been successfully used to study the behavior of water in emulsions and at surfaces is the TSDC, which measures dielectric properties of water related to relaxation of bound (dipoles of water molecules) and mobile (protons) charges, as well as the dynamics of polymers (Gun'ko et al. 2007f,i). Using both techniques, NMR and TPDC, it was shown that the freezing temperature of water is influenced by the mobility and state of its molecules in porous media. Unlike "bulk" water which freezes at 0°C, there is a fraction of water that remains unfrozen at temperatures below the normal freezing point. This water is known as "bound," "ice-like," or "structured water" (de Certaines 1989). It remains unfrozen at $T<273$ K because of interaction with a solid surface or polar solute molecules and ions which disturb the hydrogen bond network in interfacial water preventing ice formation. The NMR and TSDC spectra reflect the dynamic processes in the closest vicinity of a solid surface or biomacromolecules. Characteristics of the adsorbed water or water confined to pores can be explored using temperature dependence of the ^{1}H NMR and TSDC signal intensity of the unfrozen interfacial water at temperatures below 273 K.

Here, the nanoporous structure of CG hydrogels will be analyzed by measuring the amount of water bound by collagen matrix using ^{1}H NMR and TSDC techniques combined with freezing-out of bulk water at temperatures below its normal freezing point. An artificial skin with the dermal part comprising a copolymer of bovine collagen with chondroitin-6-sulfate (Integra Life Science Corp., Plainsboro, NJ) was used as the initial material. The "dermal" part, i.e., the protein hydrogel separated from the silicone membrane was thoroughly washed with distilled water and freeze-dried. The water and protein content calculated from the dried mass of the initial hydrogel was found to be 98.5% and 1.5% (w/w), respectively. A freeze-dried CG polymer (Figure 5.42) was further rehydrated with different amounts of water from 55.0 to 95.5 wt% and used (as well as the initial CG) for the ^{1}H NMR studies. The mass ratio of chondroitin-6-sulfate to collagen is low (between ~0.074 and 0.10); therefore, we can consider the material as being a collagen gel.

SEM and CLMS micrographs of CG (Figure 5.42) show the macroporous structure of the material. The macropores have relatively random spherical and cylindrical-like shapes. The SEM

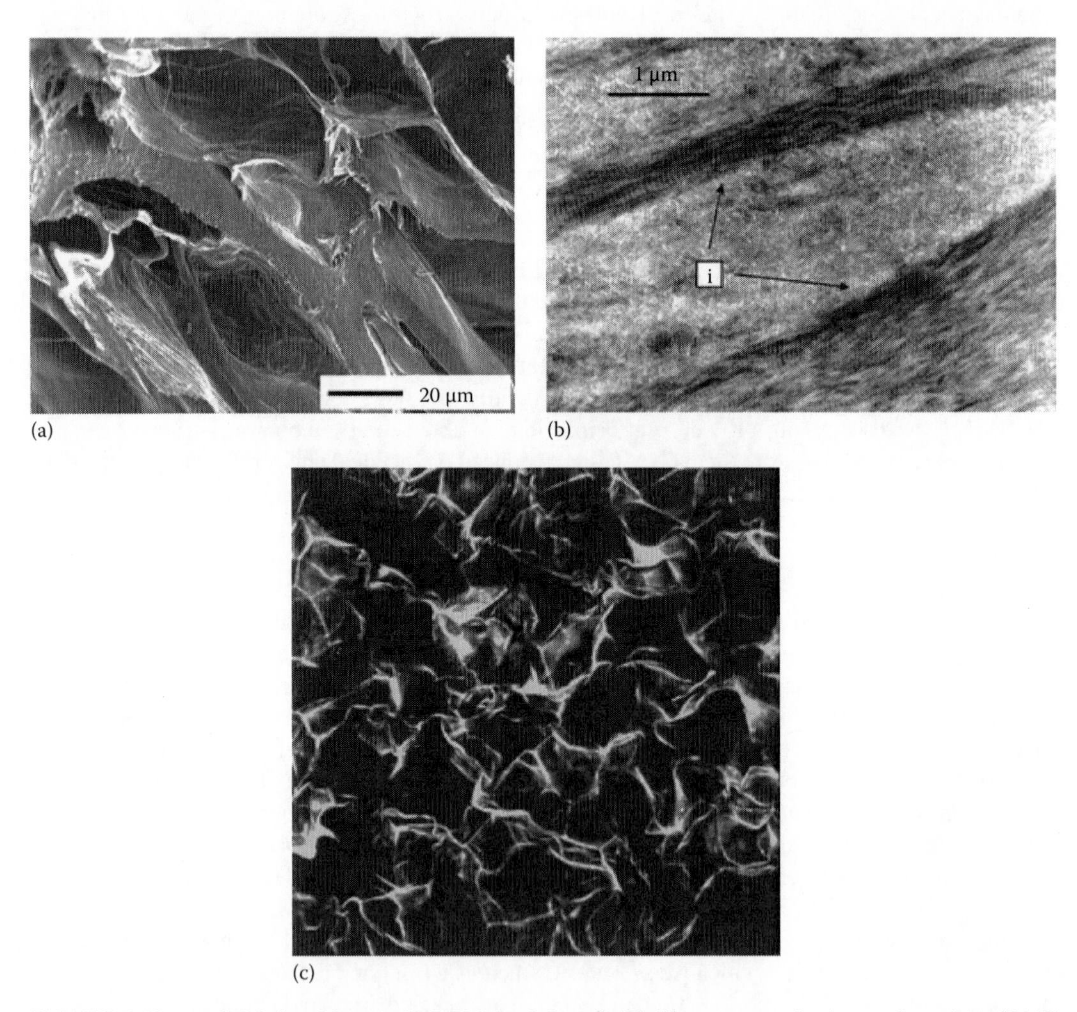

FIGURE 5.42 (a) SEM (bar 20 μm), (b) TEM (5 × 3.5 μm^2) (fibrillar structure is observed), and (c) CLMS (1 × 1 mm). (Images of CG gel adapted from (a, b) *Biomaterials*, 27, Mikhalovska, L.I., Gun'ko, V.M., Turov, V.V. et al., Characterisation of the nanoporous structure of collagen-glycosaminoglycan hydrogels by freezing-out of bulk and bound water, 3599–3607, 2006, Copyright 2006, with permission from Elsevier. and (c) Adapted from Gun'ko, V.M., Mikhalovska, L.I., Savina, I.N. et al., Characterisation and performance of hydrogel tissue scaffolds, *Soft Matter*, 6, 5351–8535, 2010b, Copyright 2010, with permission from The Royal Society of Chemistry.)

or CLMS resolution is insufficient to see nanopores. However, the micrographs provide useful information about the micro/macroporous structure over the 1 to 200 μm range. On the first sight, these pores have a relatively uniform size distribution and thin walls (Figure 5.42).

These structural features of the macropores in the CG hydrogel are important for cell accommodation, migration, and growth. Whether preparation of samples according to the procedure described earlier damages their intact porous structure is a topic for further discussion. However, in general, critical point drying is considered to be a gentle procedure preserving the structural integrity of the hydrogels studied here, and similar structures are also observed in CSLM images taken of native, hydrated gels (Figure 5.42c).

Figure 5.43 shows the temperature dependence of the concentration of unfrozen ($T<273$ K) water (C_{uw}) and the relationships between changes in the Gibbs free energy and this concentration (ΔG versus C_{uw}) caused by interaction of water with collagen molecules in the initial gel and the freeze-dried gel after rehydration. The amounts of water bound per gram of protein in rehydrated samples are significantly lower than in the initial CG hydrogel, with the lowest C_{uw}^{max} and C_{uw}^{w} being observed in the hydrogel having the highest collagen content (Table 5.7).

Thermodynamic parameters for layers of strongly ($\Delta G<-0.8$ kJ/mol) and weakly ($\Delta G>-0.8$ kJ/mol) bound water are calculated from the function ΔG versus C_{uw}. The WBW is a fraction of unfrozen water with free energy only slightly lowered by intermolecular interactions with the CG matrix (because this water is distant for CG functionalities); it freezes at temperatures close to 273 K. The SBW (located close to CG functionalities and forming the hydrogen bonds with them) may not freeze even upon significant cooling of the CG hydrogels. In the NMR experiments, some of the SBW remained unfrozen at 215 K. The thickness of the layers of each type of water (C_{uw}^{s} and C_{uw}^{w} for the SBW and WBW, respectively) and the maximum value of lowering of the free energy of water caused by interactions with collagen (ΔG_s and ΔG_w for the SBW and WBW, respectively) were estimated by linear extrapolation of appropriate sections of ΔG versus C_{uw} graphs to the corresponding axis.

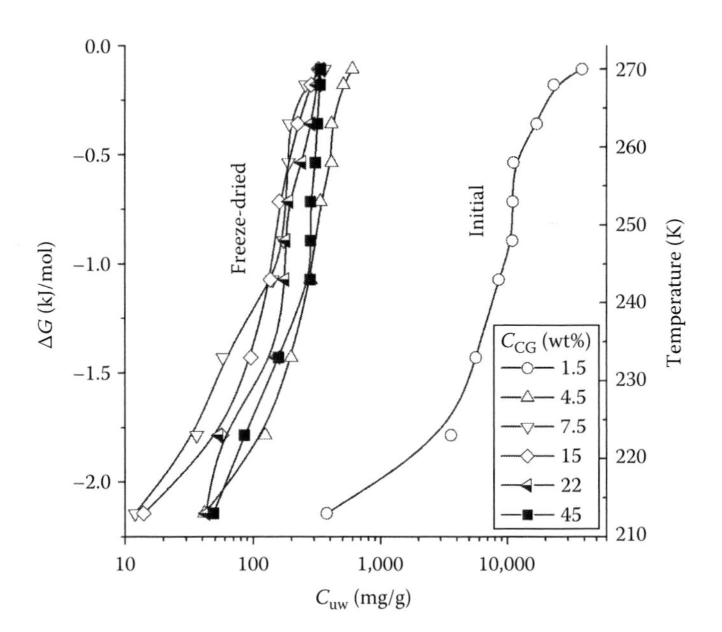

FIGURE 5.43 Amount of unfrozen water (C_{uw}) as a function of temperature, and changes in the Gibbs free energy of interfacial water versus C_{uw} at different concentrations of collagen in the hydrogel. (Adapted from *Biomaterials*, 27, Mikhalovska, L.I., Gun'ko, V.M., Turov, V.V. et al., Characterisation of the nanoporous structure of collagen-glycosaminoglycan hydrogels by freezing-out of bulk and bound water, 3599–3607, 2006, Copyright 2006, with permission from Elsevier.)

TABLE 5.7
Characteristics of Water Bound to CM1 (Samples 1–5) or CM2 (6) at Fixed Hydration ($h = 23$ mg/g) but Being in Different Media

Medium	Water Type	ΔG_s (kJ/mol)	C^s_{uw} (mg/g)	C^w_{uw} (mg/g)	γ_S (J/g)
Air	SAW	−3.5	15	8	1.1
$CDCl_3$	SAW	−2.8	17	6	0.94
CD_3CN	SAW+WAW	−2.7	20	3	1.61
$CD_3CN/CDCl_3$ 1:2.6	SAW	−2.2	14	0.75	0.62
	WAW	−10.0	4.5	5	1.1
$CD_3CN/CDCl_3$ 1:5	SAW	−2.7	12	11	0.81
	WAW	−7.0	0	11	1.1
$CD_3CN/CDCl_3$ 1:5	SAW	−2.6	6	6	0.45
	WAW1/WAW2	−3.0/−2.4	2/2	4.7/4.7	0.17/0.19

A negative correlation between collagen concentration (C_{CG}) and the amount of bound water C^w_{uw} (Table 5.7) is unusual, as this correlation is positive in aqueous solutions of proteins and aqueous suspensions of dispersed oxides (see Chapter 4). This effect can be linked to the rearrangement of the hydrogel structure upon rehydration of the collagen gel. Freeze-drying and rehydration of the CG gel reduces the contribution of the WBW to the total amount of the bound water from ~80 to 10 wt%. A plot showing the dependence of the interfacial Gibbs free energy (γ_S) on collagen concentration (C_{CG}) has a minimum at $C_{CG} \approx 10$ wt% (Figure 5.44 and Table 5.7). In the initial hydrogel, collagen molecules are strongly hydrated and the thickness of the interfacial water layer disturbed as a result of interactions with proteins could be >10 nm. By removing water

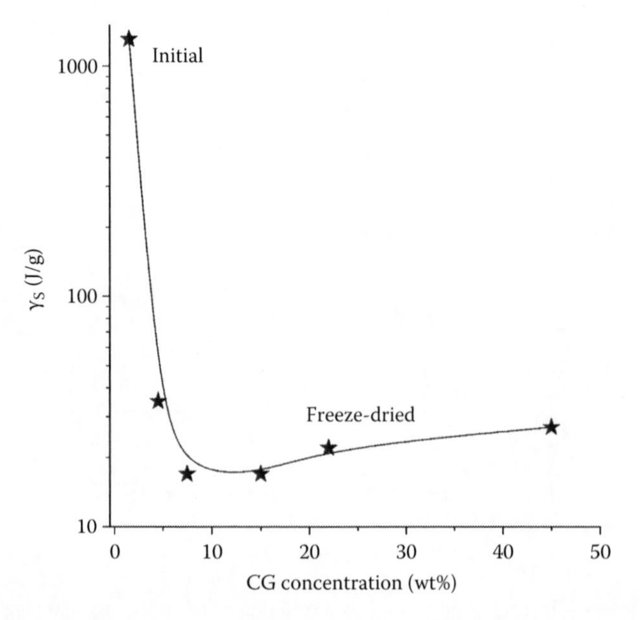

FIGURE 5.44 The free surface energy as a function of the collagen concentration in the CG hydrogel. (Adapted from *Biomaterials*, 27, Mikhalovska, L.I., Gun'ko, V.M., Turov, V.V. et al., Characterisation of the nanoporous structure of collagen-glycosaminoglycan hydrogels by freezing-out of bulk and bound water, 3599–3607, 2006, Copyright 2006, with permission from Elsevier.)

from the hydrogel, freeze-drying leads to the reduction of distance between protein molecules, thus reducing the amount of the bound water and the interfacial free energy of the protein–water interaction accordingly. A steep decrease in the γ_S value of the rehydrated CG gels in comparison with the initial hydrogel is probably caused by enhancement of interactions between collagen molecules and their aggregation upon drying; these interactions are not significantly weakened after rehydration (Figure 5.44). This process is not reversed upon rehydration due to a noticeable diminution of the interfacial area between water and collagen molecules. Similar results were observed upon drying–rehydration of gelatin and albumin–nanosilica system, where aggregates of 1 μm in size and larger were found (Gun'ko et al. 2009d).

The sensitivity of the TSDC method and information it provides is akin to that of ^{1}H NMR spectroscopy with freezing-out of bulk water in respect to the state of hydrogen-bonded network (i.e., the mobility of dipoles as bound charges) and the spatial confinement of interfacial water layers or clusters (i.e., the average number of hydrogen bonds per water molecule) to a surface. However, the temperature range used in the TSD current measurements is broader than that of ^{1}H NMR. Therefore, comparing the data obtained by these two methods provides a deeper insight of such complex systems as CG hydrogels.

The TSDC spectra (Figure 5.45a) and the corresponding distributions of the activation energy of relaxation of the initial CG hydrogel (Figure 5.45c) and pure (free, or bulk) water show that even a less amount of protein causes a noticeable change in the structure of the hydrogen-bonded network of water. Changes in the polarization field (500 and 350 V/cm) slightly affect the TSDC spectra (Figure 5.45a) that confirm a strong effect of the surroundings on the dipolar and dc relaxations of water molecules and ions of the CG hydrogel. Water in this system is strongly associated since the first $f(E)$ peak (Figure 5.45c) at low energy corresponding to molecules with the lowest number of the hydrogen bonds per water molecule shifts toward higher energies as well as the second $f(E)$ peak. This indicates formation of strong hydrogen bonds in the hydrogel in agreement with the ^{1}H NMR data, which also show relatively large values in the chemical shift of protons participating in hydrogen bonds. However, the high-energy $f(E)$ peak (Figure 5.45c) for water molecules that have approximately four hydrogen bonds per molecule is narrower and shifts slightly toward lower energies in comparison with the free water peak. This result may reflect a certain ordering action of the protein matrix on water in the initial CG hydrogel. It should be noted that according to the NMR data the amount of perturbed water corresponds to 90 wt% showing enhanced (mainly rotational) mobility of water molecules associated with protein molecules since the freezing temperature lowers. Consequently, results obtained by NMR and TSDC methods are in agreement as they both detect a very large amount of water perturbed by the CG matrix in the initial hydrogel.

Additionally, ^{1}H NMR cryoporometry in combination with the CLSM data gives the PSD of CG hydrogel over the total range of pores from nanopores to broad macropores (Figure 5.46). It shows that the volume of WBW (located in broad mesopores and frozen at temperature close to 273 K) is much greater (more than an order of magnitude) than the volume of SBW (located in nanopores and frozen at $T < 260$ K). However, macropores give the main contribution to the porosity of the CG hydrogels.

Estimation of the mesh size (ξ) (Canal and Peppas 1989, Wallace and Rosenblatt 2003), assuming uniform hydrogel structure at a protein volume fraction $v_{2,s} \approx 0.05$, gives $\xi \approx 20$–25 nm. This ξ value is in the nanopore (in LSC classification of pores) range and much lower than the average pore diameter d. The difference between the d and ξ values is due to structural nonuniformity of the CG hydrogel composed of thin pore walls ($v_{2,s} \approx 0.6$–0.8 and $\xi < 2$–3 nm) and broad micro/macropores filled by water ($v_{2,s} \approx 0$). Consequently, the pore walls can be impermeable for proteins and other macromolecules.

Three proteins different in size and shape: bovine serum albumin (BSA) (Sigma, mass average molar mass, m, 67 kDa), human serum fibrinogen, Fg (chromogenic, 340 kDa), and aprotinin (basic [bovine] pancreatic tripsin inhibitor [BPTI] 6.7 kDa) were used (as received from the manufacturer)

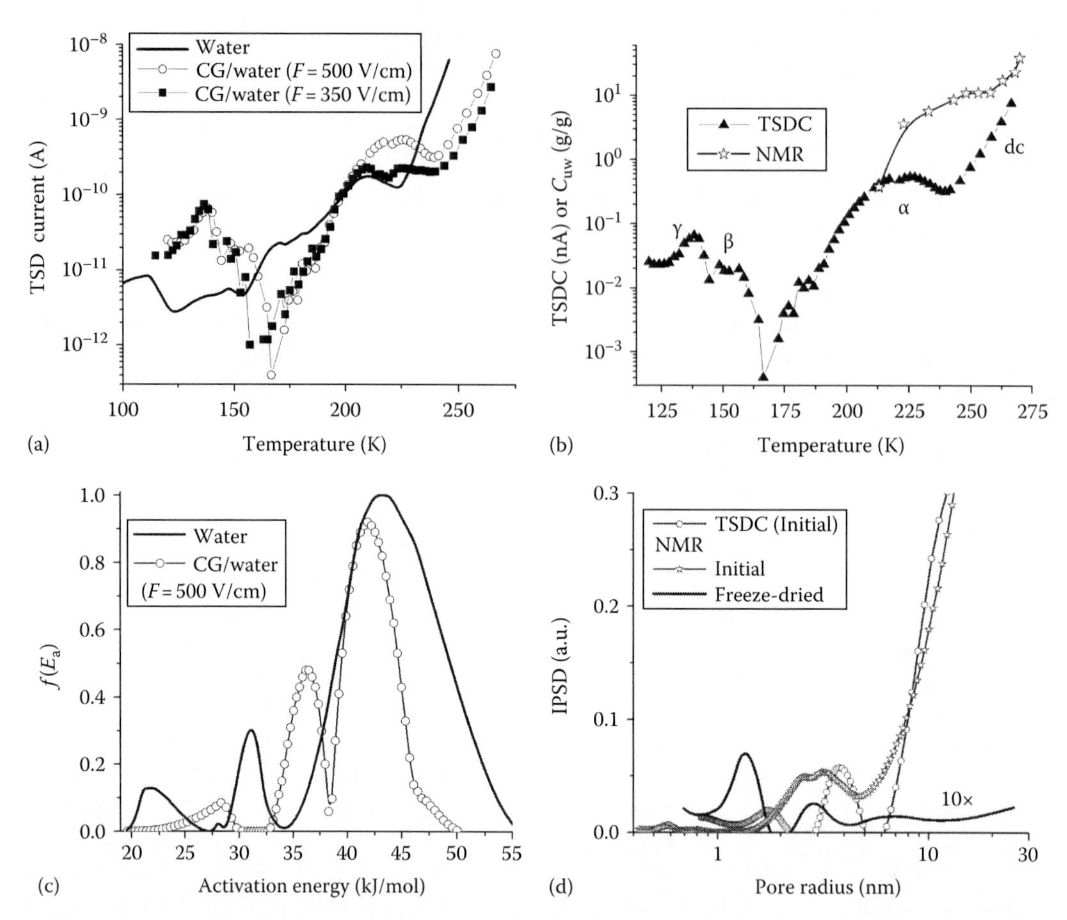

FIGURE 5.45 (a) Temperature dependence of the TSD current for the initial collagen hydrogel and "free" (bulk) water; (b) temperature dependences of the TSD current and the amounts of unfrozen water (C_{uw}) (NMR) for initial collagen hydrogel (98.5 wt% of water); (c) distribution function of the activation energy of relaxation in these systems; and (d) incremental pore size distributions for the initial CG hydrogel calculated on the basis of ^{1}H NMR and TSDC data. (a–c: Adapted from *Biomaterials*, 27, Mikhalovska, L.I., Gun'ko, V.M., Turov, V.V. et al., Characterisation of the nanoporous structure of collagen-glycosaminoglycan hydrogels by freezing-out of bulk and bound water, 3599–3607, 2006, Copyright 2006, with permission from Elsevier.)

in the diffusion experiments. Both BPTI and BSA are globular proteins, but of different sizes while Fg is a much larger rodlike protein. The CG hydrogel was characterized by the ability of these different molecular markers to pass through a thin CG membrane (~1 mm in thickness). Protein diffusion was studied using both fresh CG samples and hydrogels that had already been exposed to a first run through with BPTI, BSA, Fg, or Fg/BSA. A diffusion cell includes two glass compartments of 8 mL volume each separated by the CG membrane. The membrane was reinforced by two meshes and two rubber rings. The diffusion cell was agitated on a shaker at 300 rpm and room temperature. The protein solution was added into compartment 1 (feeder). Protein molecules diffused through the membrane into compartment 2. Aliquots of 0.1 mL were taken from compartment 2 at appropriate time intervals. The optical density (OD_{280}) of the protein solution was measured at 280 nm. To keep the volume of the diffusion system constant, 0.1 mL of PBS was added each time into compartment 2.

To gain more information from the diffusion kinetics data, the distribution function of the diffusion coefficient $f(D)$ was calculated using an integral equation:

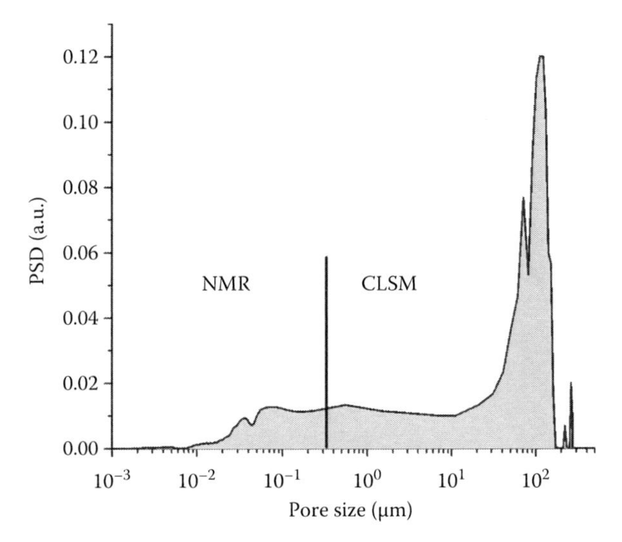

FIGURE 5.46 Pore size distribution calculated using ^{1}H NMR cryoporometry and CLSM methods. (Adapted from Gun'ko, V.M., Mikhalovska, L.I., Savina, I.N. et al., Characterisation and performance of hydrogel tissue scaffolds, *Soft Matter*, 6, 5351–8535, 2010b, Copyright 2010, with permission from The Royal Society of Chemistry.)

$$c(x,t) = \frac{c_0}{2\sqrt{\pi}} \int_{D_{min}}^{D_{max}} \frac{1}{\sqrt{Dt}} \exp\left(-\frac{x^2}{4Dt}\right) f(D)dD \tag{5.3}$$

where

- c_0 is the initial concentration in the feeder cell
- x is the CG membrane thickness
- D_{min} and D_{max} are the minimal and maximal values of the diffusion coefficient, respectively, used on integration
- t is the observation time

This equation was solved using a regularization procedure with unfixed regularization parameter.

The large macroporosity and interconnectivity of pores in the CG material (Figures 5.42, 5.45, and 5.46) can provide good conditions for macromolecular diffusion of, for example, proteins through a CG membrane. The results of measurements of the diffusion kinetics of BPTI (Figure 5.47a), BSA (Figure 5.47b), and Fg (Figure 5.47c) different in size and shape are shown in Figure 5.48.

Using the twin-chambered diffusion cell, with a protein solution on one side of a sample and PBS on the other, experiments have shown that within 60–65 min from the start of the diffusion of relatively small BPTI molecules, the system reaches equilibrium (Figure 5.48a). The BPTI concentration and pre-run of larger proteins affect the equilibrium time (Figure 5.48a). The larger BSA molecules need about 75 min to reach the equilibrium (Figure 5.48c), and Fg molecules equilibrate within 90 min (Figure 5.48e).

This pattern is characteristic of the relationship between the diffusion coefficient of individual molecules and their hydrodynamic radius: the larger the molecules, the slower their motion (diffusion) if their aggregation and interaction with the pore and cell walls could be ignored. For example, the diffusion coefficients D_0 of BPTI, BSA, and Fg molecules in the aqueous solution are equal to $(0.8–1.9) \times 10^{-6}$, $(0.6–1.0) \times 10^{-6}$, and $(1.9–3.1) \times 10^{-7}$ cm^2/s, (Young et al. 1980,

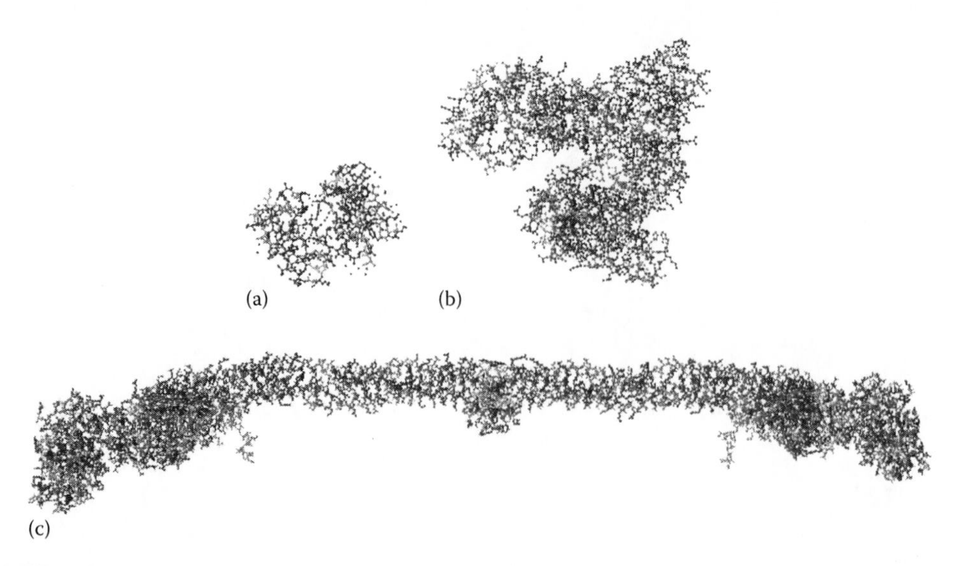

FIGURE 5.47 Molecular structures of (a) hydrated BPTI dimer (atomic coordinates from Islam et al. 2009), (b) HSA (Yang et al. 2007) (having a shape close to BSA), and (c) Fg (Kollman et al. 2009) (H atoms are not shown here).

Schmitz and Lu 1983, Hoffman et al. 2000, Gottschalk et al. 2003), respectively. The differences in the D_0 values are due to much smaller (by an order of magnitude) m_{BPTI} than m_{BSA}; however, the latter is five times lower than m_{Fg}.

The D_0 values also depend on the amount of dissolved compounds (influencing their aggregation), pH value, and salinity of the aqueous solutions (affecting the shape of protein molecules). These effects and protein interactions with the CG membrane can result in broadening of the D value range. This range corresponds to a certain $f(D)$ distribution that depends on individual protein types.

The left $f(D)$ peak (Figure 5.48d) or the left shoulder at $D<D_0$ of broad $f(D)$ curves in the case of monomodal distributions (Figure 5.48b, d, and f) indicates that the diffusion of a portion of protein molecules through the CG membrane is slower than in the free aqueous solution. This is due to their interactions with the pore walls and adsorption within the collagen matrix. The adsorption can reduce the membrane pore size that will have a negative impact on subsequent diffusion rates, especially within nanopores. Diffusion of BPTI through the CG membrane is slower after a pre-run of Fg, and particularly after a pre-run of BSA and then Fg (Figure 5.48b). Large protein molecules and their aggregates can affect membrane permeability more effectively than the smaller BPTI molecules because of more effective adsorption (Gun'ko et al. 2003b) and physical trapping of larger macromolecules from the previous run. Additionally, narrowing of pores on adsorption of large protein molecules can enhance the interactions of dissolved protein molecules with the molecules adsorbed during pre-run or this run. Therefore, a pre-run of small BPTI itself does not have this negative effect on the second BPTI run as the main $f(D)$ peak has remained in the same position. Additionally, a small $f(D)$ peak of BPTI appears at $D>10^{-5}$ cm^2/s due to diffusion of those protein molecules that remain in the membrane after the first run. However, retardation of the BSA diffusion is observed after the first run with the same protein (Figure 5.48d) or Fg (data not shown here). This could be due to the same effects of pre-adsorbed Fg and BSA for BPTI (Figure 5.48b).

The appearance of several peaks on the $f(D)$ curves for certain samples (Figure 5.48) can be explained by the mechanism of protein diffusion through the CG hydrogel (without or with interactions with the pore walls) and aggregation of proteins in the solution and at the pore walls. However, the position of the main $f(D)$ peaks for all the proteins studied reveals that the D values at the $f(D)$ maxima are close to the D_0 values shown earlier. Consequently, the diffusion of the

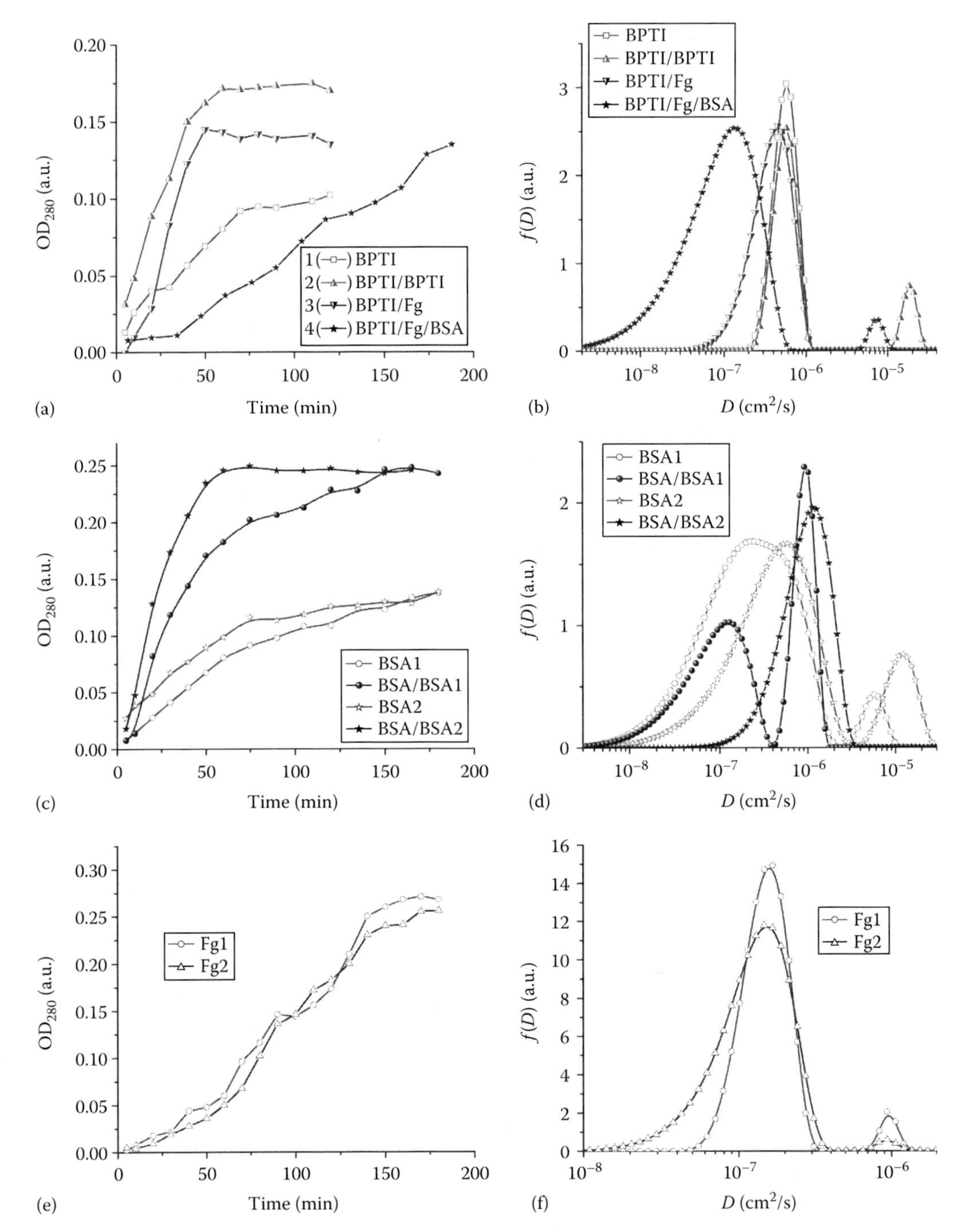

FIGURE 5.48 Diffusion kinetics through a collagen hydrogel membrane (~1 mm in thickness) for (a) BPTI. Curve 1 is for an initial concentration of 1.23 mg/mL in the feeder cell ($OD_{280}=0.09$); curves 2–4 are for an initial concentration of 2.46 mg/mL ($OD_{280}=0.18$); curve 2—BPTI run after the first BPTI run; curve 3—BPTI run after Fg; curve 4—BPTI run after Fg and BSA; (c) BSA and BSA (with twice concentration) after the first BSA run, (e) Fg (initial concentration 1.7 mg/mL); curves (b), (d), and (f) show the corresponding distribution functions of the diffusion coefficient $f(D)$ for (b) BPTI, (d) BSA, and (f) Fg. (Adapted from Gun'ko, V.M., Mikhalovska, L.I., Tomlins, P.E., and Mikhalovsky, S.V., Competitive adsorption of macromolecules and real-time dynamics of Vroman-like effects, *Phys. Chem. Chem. Phys.*, 13, 4476–4485, 2011b, Copyright 2011, with permission from The Royal Society of Chemistry.)

proteins studied that occurs through the CG membrane is similar to that of their free individual molecules in aqueous solution without the hydrogel, and without strong agglutination of individual macromolecules; in other words, $D_{por} \approx D_0$. This confirms that proteins at different molecular weight, shape, and size such as the small globular polypeptide BPTI (6.7 kDa, size 2.9×1.9 nm), the relatively large globular protein BSA (67 kDa, 14×6×4 nm), and glycoprotein Fg (340 kDa, 45×9×6 nm) can penetrate through the CG hydrogel membrane of 1 mm in thickness without significant retardation having D values close to D_0 values. This result is due to the macroporosity of the hydrogel (Figure 5.42). However, the broad shape of the $f(D)$ distributions shows that a certain portion of the proteins strongly interacts with the pore walls in the hydrogel and that the pore tortuosity factor affects the measured D values, as well as protein–protein interactions in the solution and at the pore walls.

There is an additional effect caused by the pre-adsorption of a protein which is related to the diminution of the contribution of very slow diffusion of BSA after its first run (Figure 5.48d). This effect can be explained by two phenomena. First, the number of strong adsorption sites at the CG surface is reduced (as well as the pore tortuosity factor) on the second run due to the adsorption of the protein molecules during the first run. Second, pre-adsorbed macromolecules can block pores (especially narrow necks and bends of pores) but relatively weakly. The first effect results in weakening of interactions of the proteins with the CG surfaces covered by pre-adsorbed protein molecules, i.e., the D value increases. Pore blocking gives the opposite but lower changes in the D values. Protein molecules can form aggregates (Gun'ko et al. 2003a). However, these structures must be smaller in size than the CG macropores because protein diffusion occurs despite the pre-runs, but at a lower rate (Figure 5.48). In other words, complete blocking of the CG pores by the proteins studied does not occur. The overall effect can play an important role on cell migration and proliferation within the CG hydrogel. This result is of great importance from a practical point of view since the possibility of nearly free diffusion of macromolecules is important for the normal functioning of cells within any hydrogel used as a dermal substitute.

The attachment of cells to CG hydrogel layered onto a surface of a quartz crystal sensor (QCM) reduces the resonant oscillation frequency by $\Delta f_1 = -75$ Hz after the first cell injection (Figure 5.49, curve 1 at $t < 600$ s). The second cell injection leads to the total $\Delta f_{12} \approx -105$ Hz, i.e., $|\Delta f_2| < |\Delta f_1|$. Both injections enhance the auto-gain controller voltage (Figure 5.49, curve 2) that indicates cells interactions with the CG matrix. These frequency shifts can be interpreted in terms of mass addition at the sensor surface according to the equation

$$\Delta f = -f_0^{1.5}\left(\frac{\eta_l \rho_l}{\pi \mu_q \rho_q}\right)^{0.5} - \frac{2 f_0^2 \rho_f h_f}{(\mu_q \rho_q)^{0.5}} \quad (5.4)$$

where

Δf is the measured frequency shift

f_0 is the resonant frequency of the unloaded crystal

ρ_l and η_l are the density and viscosity, respectively, of the liquid in contact with the coated crystal surface

ρ_q and μ_q refer to the specific density and the shear modulus of quartz

Treatment of the data shown in Figure 5.49 with Equation 5.4 gives $h_f \approx 5$ nm.

The use of this value and a minimal size of cells (~10 μm) allow us to estimate a part of the CG surface covered by cells. The cell covered area corresponds approximately to 0.05% of the total outer area of the CG matrix. This low value can be explained by the short time (~20 min) of the QCM observation, which is too short for cell migration deeper into the CG matrix. In other words, the observed low Δf value is due to cell attachment only to the outer surface of the CG matrix without cells penetration into deeper layers of the material.

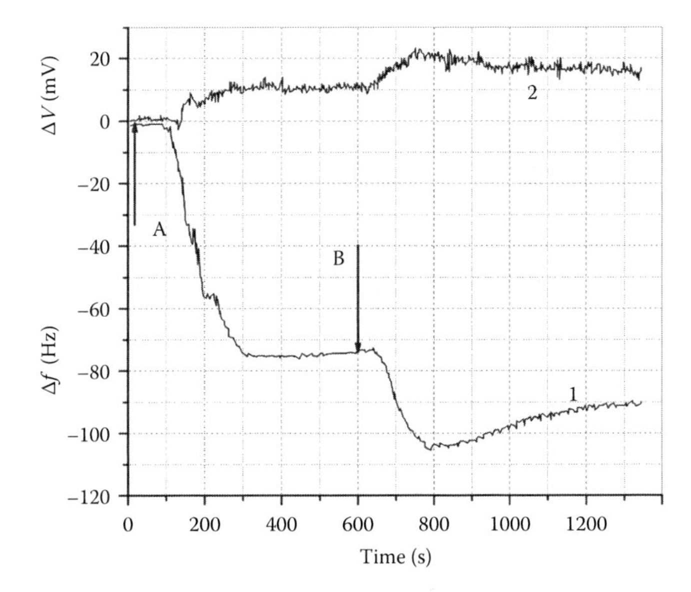

FIGURE 5.49 Changes in frequency (1) and auto-gain controller voltage (2) on two injections (A and B) of 0.1 mL aliquot of 3T3 fibroblast suspension (2.0×10^6 cell/mL) upon an unsupported section of CG hydrogel laid the surface of a 10 MHz gold-coated crystal. Flow injection rate 0.01 mL/min, 37°C±0.1°C, pH 7.2, PBS. (Adapted from Gun'ko, V.M., Mikhalovska, L.I., Tomlins, P.E., and Mikhalovsky, S.V., Competitive adsorption of macromolecules and real-time dynamics of Vroman-like effects, *Phys. Chem. Chem. Phys.*, 13, 4476–4485, 2011b, Copyright 2011, with permission from The Royal Society of Chemistry.)

The response of the QCM also reflects the dissipation of energy within the system and hence changes of viscoelastic properties of the CG matrix and cellular overlayers. The QCM study shows that interactions of fibroblast cells with the CG occur relatively quickly since the f minima are reached during 3–5 min. Subsequent diminution of the $|\Delta f|$ value (by 13%) after the second minimum is due to removal of a portion of cells because of several factors. The liquid flow can wash off the cells from the outer surface of the CG matrix, which can be enhanced by sensor longitudinal vibrations. The repulsive interaction of cells from the second portion with pre-adsorbed cells, attached after the first injection onto the outer surface without migration into deeper CG layers, can result in weakening of the cell attachment. The last effect is similar to the effect of a pre-adsorbed protein on a subsequently adsorbed protein observed on the diffusion study discussed earlier. However, it is very weak due to a small portion of the CG surface covered by attached cells.

Determination of the porous structure of intact soft hydrogels is a challenging task, which is of extreme importance for predicting their performance and optimizing their properties as tissue scaffolds. No single analytical method can provide all the necessary information; however, combining data obtained using modern experimental techniques such as NMR cryoporometry, CLSM, desorption, diffusion, etc., can help to solve this problem. All the results obtained by us using the ^{1}H NMR, cryoporometry, water desorption, CLSM, SEM, protein diffusion, cell attachment (QCM), and migration methods are in good agreement with one another. The investigations discussed earlier reveal the dominant role of pores at $d > 20$ μm contributing to the total porosity, channels for protein diffusion, places for cell attachment and for cell migration. The contribution of nanopores at $d < 50$ nm into the total porosity is lower than 1.7%. However, they provide the majority of the specific surface area of the material ~120 m^2/g. The main portion of water in the CG hydrogel corresponds to weakly bound (frozen at temperatures close to 0°C) and nonbound (frozen at $T = 0$°C and desorbed at $T \leq 100$°C) water located in micro- and macropores. This is of importance for dissolution and diffusion of nutrients, proteins, etc., as well as migration of cells, because SBW is characterized by very low activity as a solvent.

Thus, ^{1}H NMR spectroscopy, TSDC, diffusion, QCM, and microscopic methods show that initial CG hydrogel contains a large amount of bound water (90 wt%), most of which is lost upon freeze-drying of the gel. Rehydration of the CG gel did not restore the amount of bound water. This effect could be explained by structural changes in the 3D network of collagen molecules. Freeze-drying of the initial hydrogel causes the formation of additional collagen–collagen bonds, which are not destroyed in the rehydration step, and this leads to diminution of the amount of the bound water and a decrease in the absolute value of free surface energy measured by ^{1}H NMR. Analysis of the state and amounts of unfrozen water as functions of temperature below the normal freezing point of water allowed us to study the nanoporous structure of biopolymer gels. Estimation of the temperature dependence of the size and volume of nanopores filled with unfrozen water shows that short-range order in freeze-dried CG gels changes insignificantly in comparison with long-distance order. The proposed approach based on ^{1}H NMR method combined with freezing-out of bulk water and comparing the structural features of the unfrozen water with its thermodynamic properties provides important information on the nanoporous structure of biopolymer hydrogels and other soft biomaterials whose properties change upon hydration/dehydration and heating.

5.4 POLYMER–NANOOXIDE SYSTEMS

Competitive interactions of macromolecules with a surface, water as a solvent, and neighboring macromolecules cause not very large changes in the Gibbs free energy. For instance, in the case of PVA/A-300 $-\Delta G$ is in the range 5–16 kJ/mol (strong PVA–PVA interactions), 1–5 kJ/mol for fibrinogen (great desolvation energy), 5–25 kJ/mol for PVP (weak PVP–PVP interactions), and 5–35 kJ/mol for POE (very weak POE–POE interactions) (Gun'ko et al. 2005d, 2007i, 2008b, 2009d). These adsorption features appear in the measurement results obtained for dry or wetted composite powders or aqueous suspensions by DRS (Gun'ko et al. 2007b,e, 2008c, Klonos et al. 2010), TSDC (Gun'ko et al. 2007f,i, Klonos et al. 2010), DSC (Bershtein et al. 2009, 2010a–c), QELS (Gun'ko et al. 2001e, 2003a, 2006i, 2009d), and NMR (Gun'ko et al. 2005d, 2009d).

The liquid- and solid-state NMR spectroscopy techniques, including cryoporometry and relaxometry, are very informative in investigations of complex polymer and hybrid materials (Alexander et al. 2006, Iuliucci et al. 2006, Jagur-Grodzinski 2007, Utracki et al. 2007, Mizoshita et al. 2008, Rahman et al. 2008, Takahashi et al. 2009, Belon et al. 2010, Karakoti et al. 2011, Lavalle et al. 2011, Shapiro 2011), and especially for the analysis of the interfacial behavior of water in these systems (Gun'ko et al. 2005d, 2009d). Blum and Krisanangkura (2009) used NMR spectroscopy in parallel to temperature-modulated DSC (TMDSC) and FTIR measurements on samples with very low amounts of polymers (PMMA, PMA, PS, PVA) adsorbed on nanosilica particles. The hydrogen bonding of the polymers to surface silanols reduced the mobility of some of the adsorbed polymer segments and resulted in higher glass transition temperatures for those polymer segments. In contrast to bulk polymers, where the glass transition process can be considered spatially homogeneous, interfacial polymers can be considered heterogeneous on the scale of the interface. The results also suggested that the heterogeneous nature of the interfacial polymers and various sensitivities of the different techniques may result in apparently different conclusions about the surface effects (Blum and Krisanangkura 2009). Bershtein et al. (2009, 2010a–c) and Klonos et al. (2010) observed strong broadening of the temperature range of glass transition process for PVA, PEG, and PVP bound to nanosilica, nanoalumina, and mixed nanooxides that confirm enhanced heterogeneity of the interfacial layer of polymers. The temperature ranges and activation energies of different relaxations (from low-temperature γ relaxation to primary α relaxation, as well as β, MWS, dc, CR relaxations) observed for immobilized polymers depend strongly on the presence of water because of its plasticizing and polarization effects, as well as proton deattachment and proton conductivity. Gun'ko et al. (2009e) compared the temperature dependences of the ^{1}H NMR signal of bound water and TSDC relaxations for different systems and showed certain regularities in the data. For instance, the dc relaxation in TSDC caused by the throughout conductivity (disappearance of the percolation barrier with thawing of bound frozen water) appears at higher temperatures than ^{1}H NMR signal of unfrozen water. The latter can appear in local structures before the appearance of continuous pathways for dc relaxation because of strong heterogeneity of both texture and the surface chemistry of composites or adsorbents.

Shen et al. (2008), using PFG NMR spectroscopy, studied ligand exchange between poly(2-(*N*,*N*-dimethylamino)ethyl methacrylate) (PDMA) ($M_n = 12{,}000$, $M_w/M_n = 1.20$, $N_n = 78$) and trioctylphosphine oxide (TOPO) bound to the surface of CdSe/TOPO quantum dots (QDs). They showed that PFG ^{1}H NMR can be used to quantify the displacement of TOPO by PDMA because this technique can differentiate signals of TOPO bound to the QDs versus those from TOPO molecules free in solution.

Composites with starch or cellulose, drugs, and highly disperse adsorbents can be used for medicinal purposes. These nanocomposites are characterized by enhanced activity of adsorbed biopolymers or polymers and/or drugs because of their transformation into a "nanostate" during interactions with solid nanoparticles because each adsorbate molecule interacts with nanoparticles. TSDC, DRS, DSC, FTIR, and other methods allow to study structural features of similar nanocomposites. ^{1}H NMR spectroscopy with layer-by-layer freezing-out of bulk and bound water as a method sensitive to structure of interfacial water can be used to characterize both interfacial water and nanocomposite structure. Bound water can be divided into SBW (changes in the Gibbs free energy $\Delta G < -(0.5/0.8)$ kJ/mol) and WBW ($\Delta G > -(0.5/0.8)$ kJ/mol) whose hydrogen bond network can correspond to SAW (chemical shift of the proton resonance $\delta_H > 3$ ppm) and WAW ($\delta_H = 1$–2 ppm). Bound water is also characterized by certain freezing point depression; therefore, the temperature dependence of content of unfrozen water $C_{uw}(T)$ can be used to estimate the structural characteristics of nanocomposites and their interactions with organics, e.g., drugs, depending on a hydration degree. Therefore, hydrated starch (powder and gel) is studied here alone and in composition with nanosilica A-300 and quercetin (natural antioxidant as a model drug) and an influence of weakly polar ($CDCl_3$) and polar (DMSO) solvents on structure of bound water in these systems.

Interaction of hyaluronic acid (HA) with water was discussed in Section 5.1. Here, the interfacial behavior of water is analyzed for composites with HA/A-300 (A-300 with $S_{BET} = 280$ m^2/g). Two HA/A-300 composite materials (CM) were prepared using low (CM1) and high (CM2) amounts of water then dried. The FTIR spectra show that HA is poorly distributed on the silica surface in CM1 because the characteristic band intensity of HA is very low (Figure 5.50). The suspending, stirring, and drying of CM2 result in the appearance of the characteristic bands of HA. The calculations of the specific surface area from the ratio of the intensities of SiO–H stretching vibrations at 3750 cm^{-1} and Si–O overtone at 1860 cm^{-1} (McCool et al. 2006) give 261 (A-300 alone), 283 (CM1),

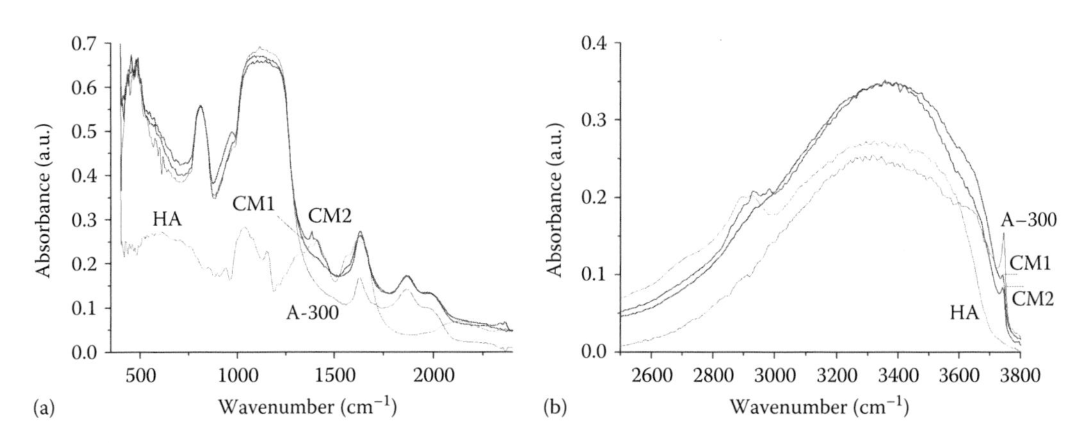

FIGURE 5.50 FTIR (ThermoNicolet Nexus) spectra of A-300, HA, and two composites.

and 228 (CM2) m²/g (~12% decrease for CM2). Consequently, HA (6 wt%) is well distributed at the silica surface in CM2 since monolayer coverage of A-300 by 1D polymers is in the range of 15–25 wt% (Gun'ko et al. 2009d).

The ^{1}H NMR spectra of water at low amount (2.3 wt%) bound to CM1 and CM2 being in different media (Figure 5.51) confirm that HA is much more effectively distributed in CM2

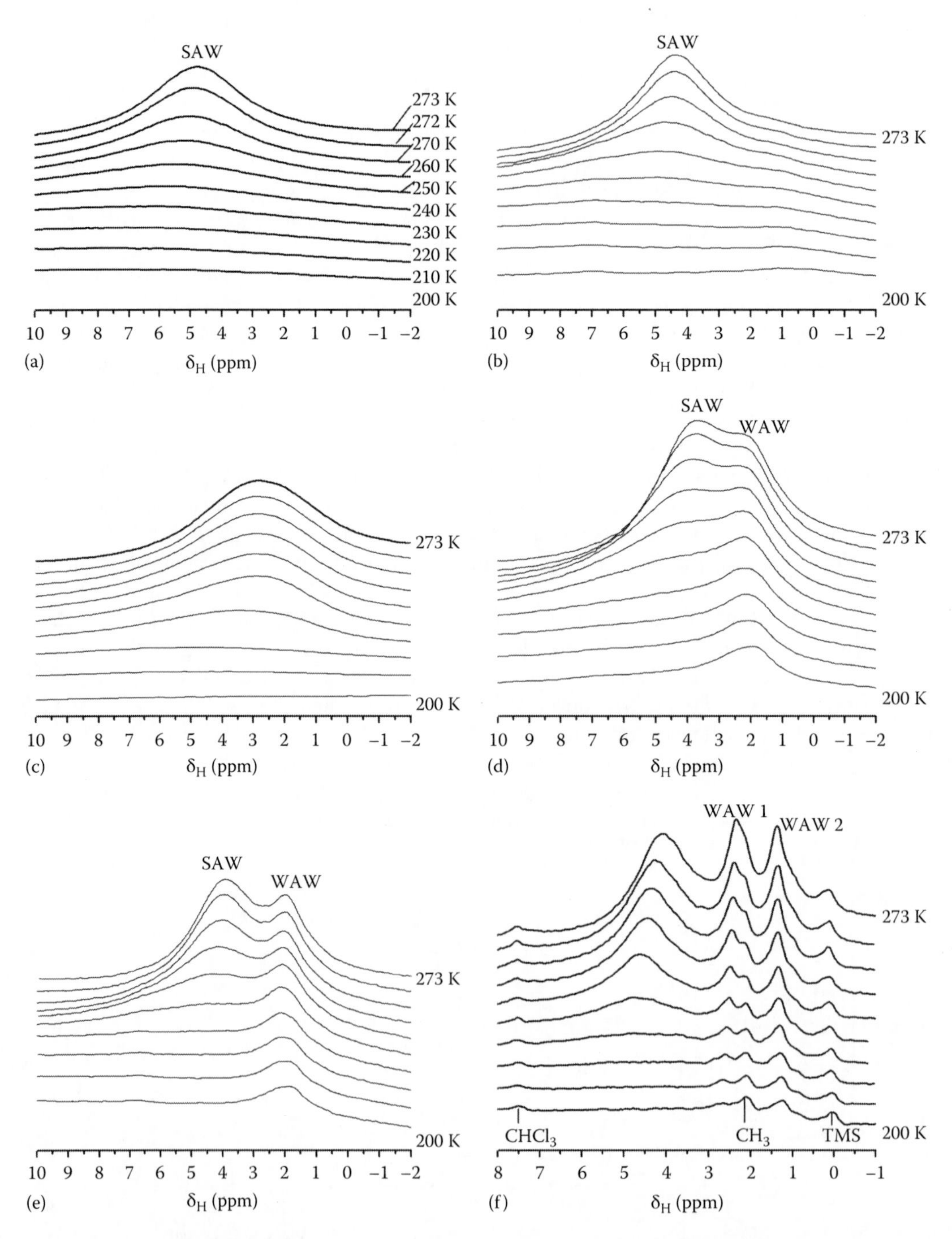

FIGURE 5.51 ^{1}H NMR spectra of water bound to (a–e) CM1 and (f) CM2 at fixed hydration (h = 23 mg/g) but being in different media: (a) air, (b) $CDCl_3$, (c) CD_3CN, (d) $CD_3CN:CDCl_3$ = 1:2.6, and (e, f) $CD_3CN:CDCl_3$ = 1:5.

than CM1 because in the former the spectra have a fine structure (Figure 5.51f). At a low water content (23 mg/g), WAW is observed in composites being in weakly polar $CDCl_3$ or a mixture with $CDCl_3$ and CD_3CN (Figure 5.51 and Table 5.7). However, for CM1 in air only SAW is observed (Figure 5.51a) which includes both SBW and WBW (Table 5.7, C^{s}_{uw} and C^{w}_{uw}, respectively). These types of water are observed in all samples. Only in CM1 being in the $CDCl_3/CD_3CN$ (5:1), WAW/SBW is absent (Table 5.7). In a greater number of samples, the SBW content is greater than WBW content that can be caused by the hydrophilic properties of both A-300 and HA and low content of water. In $CDCl_3$ (Figure 5.51b) or CD_3CN (Figure 5.51c), only single signal is observed and WAW appears as a shoulder (Figure 5.51b) or the spectra have up-field shift (Figure 5.51c). In the $CDCl_3$ and CD_3CN mixture, the signals of WAW and SAW split (Figure 5.51d–f).

The interfacial behavior of water bound in HA/nanosilica composites depends strongly on the type of dispersion medium (Figure 5.52). The amounts of SAW are maximal in CD_3CN and minimal in $CD_3CN/CDCl_3$ mixture for CM2. Additionally, contribution of SBW and WBW in SAW (Figure 5.52b) and WAW (Figure 5.52d) depends on medium and WBW content is greater for CM2.

Fast proton exchange between water and HCl results in great δ_H values (9–11 ppm) dependent on the HCl content. The ^1H NMR spectra of aqueous solution of HCl (18%) adsorbed onto CM1 is characterized by four signals (Figure 5.53a): WAW (1.3 ppm) and SAW (~4 [signal 1], 6–7 [signal 2], and 8–9 [signal 3] ppm). The amounts of HCl in these WAW and SAW structures differ: no HCl in

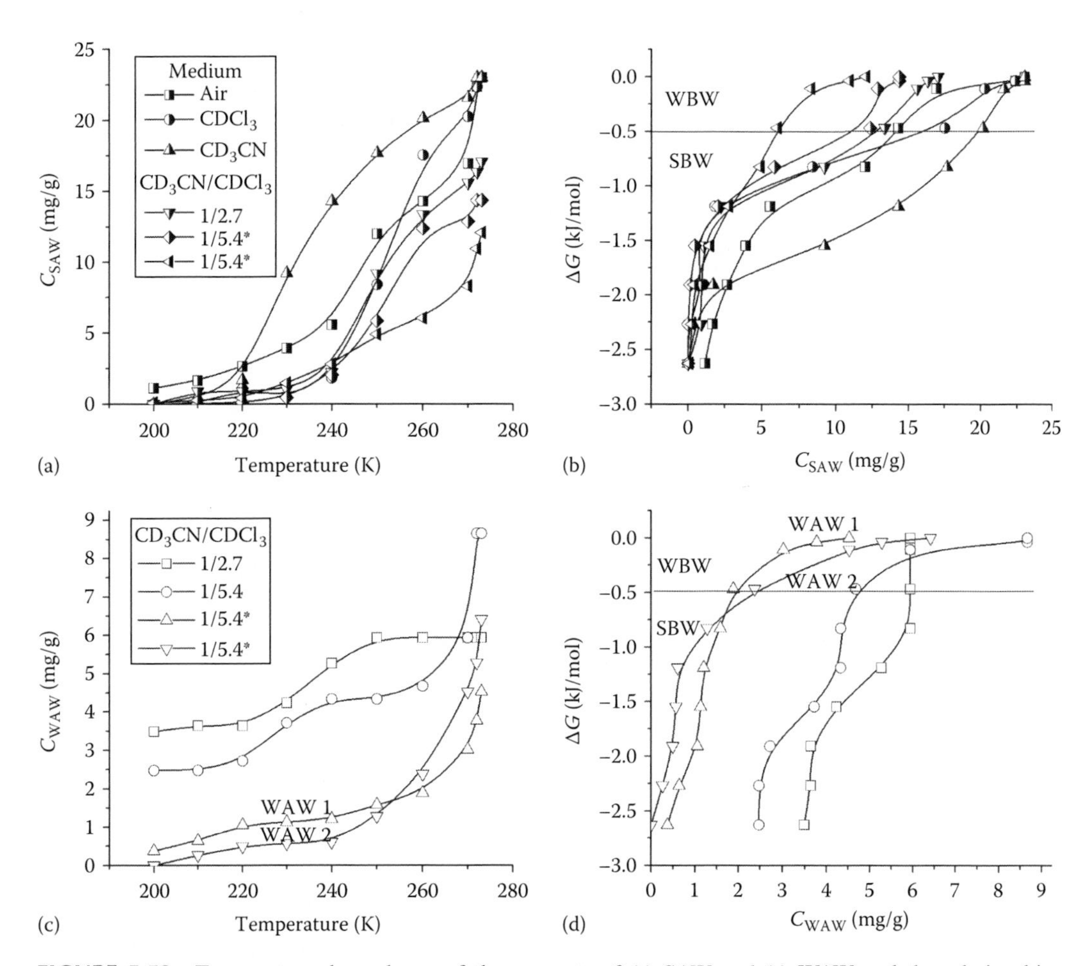

FIGURE 5.52 Temperature dependence of the amounts of (a) SAW and (c) WAW and the relationships between changes in the Gibbs free energy and the amounts of (b) SAW and (d) WAW for HA/A-300 composites CM1 and CM2(*).

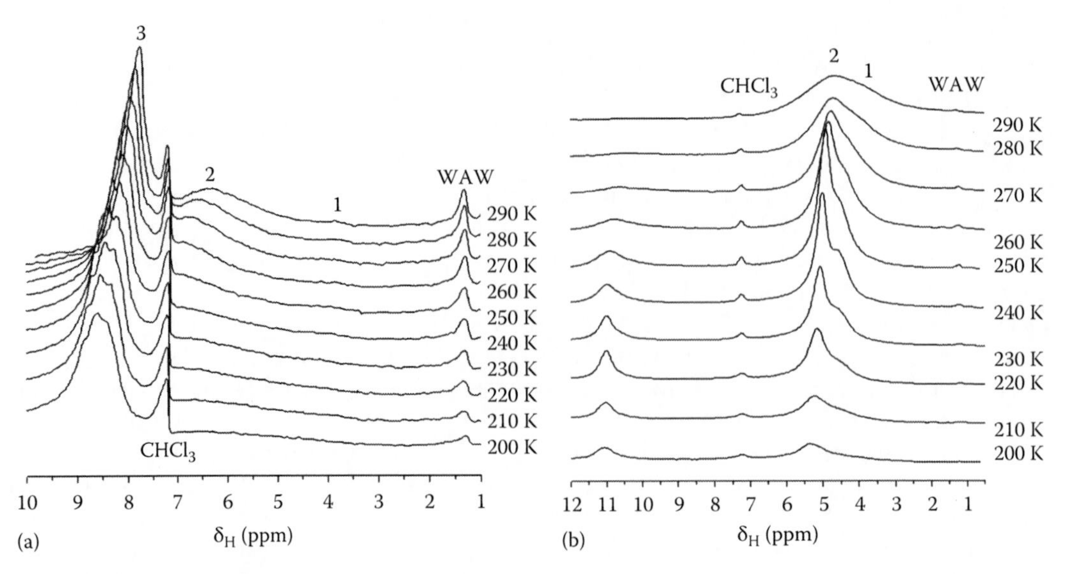

FIGURE 5.53 ^{1}H NMR spectra of CM1 with adsorbed aqueous solutions (150 mg/g) of (a) 18% HCl and (b) 16% H_2O_2 in $CDCl_3$ medium.

WAW and SAW with signal 1, low amount of HCl in SAW with signal 2, and more concentrated HCl solution in SAW with signal 3. The intensity of signals 1 and 2 decreases with lowering temperature but signal 3 does not decrease. A fine structure of signal 3 appears at low temperatures that can be caused by nonuniformity of HA/A-300 composite.

In the case of adsorption of the aqueous solution of H_2O_2, the ^{1}H NMR spectra (Figure 5.53b) demonstrate peculiarities different from those observed for the adsorbed HCl solution (Figure 5.53a). Signal of concentrated H_2O_2 (~11 ppm) is weak but SAW with relatively low content of H_2O_2 (4–6 ppm) has high signal intensity. Additionally, the WAW signal intensity is much lower than that in the HCl solution. Thus, the HCl solution tends to be in more strongly concentrated state than H_2O_2 at the interfaces of HA/A-300 composites.

Shulga et al. (2000) studied interaction of microcrystalline cellulose (MCC), used in pharmaceutical industry as a component of tablet formulation, with nanosilica A-300 or titania (100 nm in average size of particles) at content ratio 5.6:1 and 3:1, respectively, in aqueous suspensions and wetted powders using ^{1}H NMR spectroscopy (Tables 5.8 and 5.9, and Figure 5.54).

TABLE 5.8
Characteristics of Water Bound to MCC in Aqueous Suspensions and Hydrated Powders

C_{MCC} (wt%)	$-\Delta G_s$ (kJ/mol)	C^s_{uw} (mg/g)	S_{uw} (m^2/g)	S_{nano} (m^2/g)	S_{meso} (m^2/g)	V_{uw} (cm^3/g)	V_{nano} (cm^3/g)	γ_S (J/g)	γ_S (mJ/m^2)
6	3.08	230	338	288	50	0.567	0.126	37.5	111
11	3.08	200	363	339	23	0.349	0.147	31.6	87
20	3.01	160	235	218	17	0.155	0.098	21.1	132
23	2.90	75	111	102	9	0.089	0.045	10.0	90
43	3.17	125	230	210	19	0.131	0.092	18.6	81
44	3.02	280	426	402	24	0.281	0.177	39.4	92
90.4	3.01	95	133	115	18	0.097	0.051	12.8	96
94.5	2.37	52	9	0	9	0.054	0.0	4.9	

TABLE 5.9
Characteristics of Water Bound to MCC/A-300 (1–8 Samples) or MCC/TiO_2 (9–13 Samples) at Different Water Contents (*h*)

h (wt%)	$-\Delta G_s$ (kJ/mol)	S_{uw} (m²/g)	S_{nano} (m²/g)	S_{meso} (m²/g)	V_{uw} (cm³/g)	V_{nano} (cm³/g)	γ_S (J/g)	γ_S (mJ/m²)
6	3.14	55	48	7	0.060	0.021	6.3	115
15	2.80	164	131	33	0.150	0.061	18.7	125
21	2.83	198	144	53	0.201	0.066	22.0	111
62	3.25	293	287	6	0.150	0.121	22.7	77
71	2.88	162	153	9	0.137	0.070	16.2	100
81	3.19	153	148	5	0.107	0.064	13.2	86
89	2.78	84	79	5	0.072	0.037	9.3	111
94	3.25	176	168	8	0.143	0.071	17.6	100
71	2.78	138	111	27	0.115	0.052	14.8	107
78	2.88	65	60	5	0.072	0.027	8.6	132
83	2.94	66	62	4	0.065	0.028	7.5	113
86	2.96	116	107	9	0.088	0.048	11.9	103
90	3.31	120	115	5	0.088	0.049	11.2	93

There is a vertical part in all the $\Delta G(C_{uw})$ curves (Figure 5.54), which corresponds to the case of the constant concentration of unfrozen water in wide range of ΔG changes (i.e., water adsorbed on the surface does not freeze in a wide range of temperature variation). Typically, a similar feature is observed for adsorbents having nanopores and/or narrow mesopores (Gun'ko et al. 2005d, 2007i, 2009d). The presence of these pores is well seen for all samples studied (Figure 5.54b, d, and f). Additionally, contribution of nanopores into the specific surface area of solids being in contact with unfrozen water is predominant (Tables 5.8 and 5.9, S_{uw} and S_{nano}); however, relative contribution of nanopores into the pore volume is less since $(S_{uw} - S_{nano})/S_{uw} < (V_{uw} - V_{nano})/V_{uw}$. Relative changes in the γ_S values in mJ/m² are much lower than that of the γ_S values in J/g because the first water layer gives the main contribution to the γ_S values.

In the case of the MCC/oxide mixtures, strong interactions of cellulose with the silica or titania surfaces result in decrease in the amounts of bound water. Relative contributions of mesopores decrease (Tables 5.8 and 5.9, and Figure 5.54). This is typical result for soft/solid systems because of the formation core (solid)–shell (soft, nonrigid) structures with the displacement of a significant amount of water, which was bound to both soft and solid surfaces alone. This effect is much lower in the case of solid–solid particles interactions because of the rigid, fixed particle structure of solids.

Interaction of poly(vinyl pyrrolidone) (PVP), $(-CH_2CHR-)_n$, where $R = NC_4H_6O$, $n \approx 100$, molecular weight of 12,600 ± 2700, with nanosilica A-300 and some other nanooxides was studied using NMR, DSC, QELS, FTIR, AFM, TSDC, DRS, adsorption, and rheometry at different hydration of dry and wetted powders and aqueous suspensions (Gun'ko et al. 2001e,f, 2004e, 2005d, 2006g–i, 2007e,i, 2008b, Bershtein et al. 2009, 2010a). PVP has only electron-donor functionalities N–C=O. Therefore, it stronger interacts with oxide surfaces than PVP–PVP. At $C_{PVP}/C_{SiO_2} < 0.1$, the adsorption is practically irreversible.

Two specific regions in the graphs of ΔG versus the amounts of unfrozen water (C_{uw}) can be found for pure silica, PVP/silica, and ball-milled PVP/silica suspensions (Figure 1.175a) similar to previously studied samples. The first corresponds to a significant decline in the amount of the unfrozen water in a narrow range of ΔG at T near 273 K. For another region, the amount of unfrozen water goes down slightly but ΔG changes stronger. The interfacial water responsible for the first region corresponds to a thick layer of the water weakly bound to the surface and perturbed by long-range intermolecular forces (mainly electrostatic). The water strongly bound to the surface

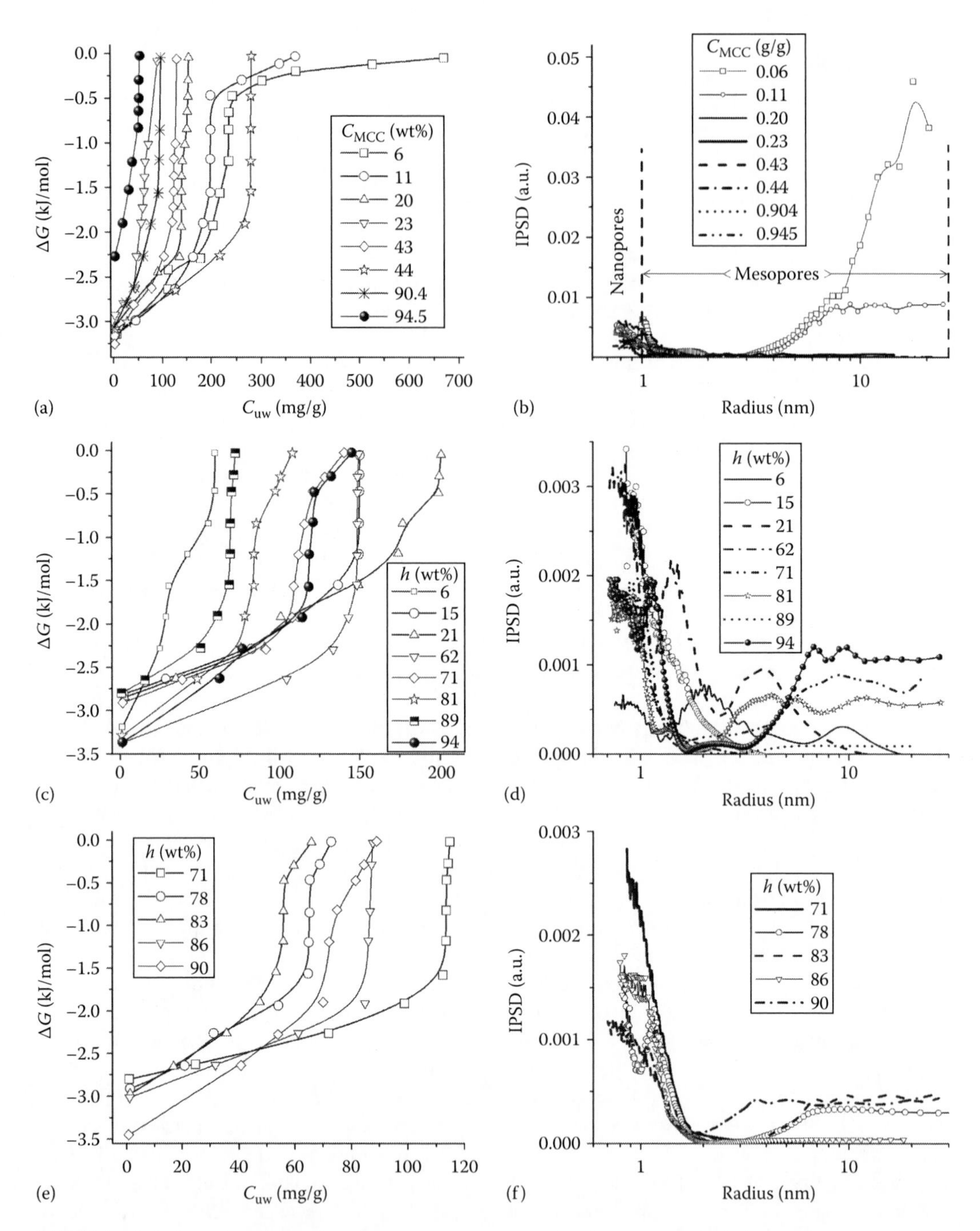

FIGURE 5.54 (a, c, e) Relationships between the amounts of unfrozen water and changes in the Gibbs free energy and (b, d, f) the corresponding PSD (NMR cryoporometry with GT equation at k_{GT} = 67 K nm) for (a, b) MCC, (c, d) MCC/A-300 (5.6:1), and MCC/TiO_2 (3:1).

responsible for the second region in the ΔG graph is linked to a thin layer adjacent to the silica surface. Its characteristics are affected by short-range forces (polar, Lifshitz–van der Waals, and electrostatic) between the surface and adsorbate molecules. The properties of the SBW and its interaction with the surface determine the immersion enthalpy and related characteristics of oxides. The interaction between the PVP molecules and the silica surface changes $-\Delta G$, C^{s}_{uw}, and C^{w}_{uw}.

TABLE 5.10
Parameters of Interfacial Water for Pure Silica and Silica/PVP Suspensions (C_{SiO_2} = 6 wt%)

System	$-\Delta G_s$ (kJ/mol)	$-\Delta G_w$ (kJ/mol)	C^s_{uw} (mg/g)	C^w_{uw} (mg/g)	γ_S (mJ/m²)	S_{uw} (m²/g)	S_{nano} (m²/g)
SiO_2	2.76	1.3	700	700	257	297	171
SiO_2 + 0.3 wt% PVP[a]	2.99	1.0	500	900	184	389	264
1 wt% PVP	1.64		300			30	0
SiO_2 + 1 wt% PVP	2.89	0.9	520	1600	195	347	158
SiO_2 + 5 wt% PVP	2.75	1.0	1100	1400	270	465	256

[a] Physiological buffer.

An enhancement in C^w_{uw} may be due to nonuniform electrostatic field in a relatively thick interfacial layer containing tails and segments of adsorbed polymer molecules. The amount of the WBW in this layer is larger than that around pure silica particles (Table 5.10), as well as the thickness of the interfacial layer perturbed by PVP + dissolved ions and silica particles in comparison with that for the initial silica suspension. This leads to an increase in the surface area in contact with unfrozen water (Table 5.10, S_{uw}). The polymer molecules set a significant portion (all PVP molecules are bound to the silica surface) in this layer. However, the sum $C^s_{uw} + C^w_{uw}$ is the same at 0.3 wt% PVP. The WBW is perturbed by 0.3 wt% of PVP in the physiological buffer solution weaker than by 1 wt% of pure PVP; however, changes in the Gibbs free energy of this layer (ΔG_w) are similar. An increase in C_{PVP} to 5 wt% leads to growth of the thickness of the unfrozen water layer (Figures 1.175 and 5.55), S_{uw}, S_{nano}, $C^s_{uw} + C^w_{uw}$, and γ_S (Table 5.10).

The PSDs (Figure 5.55) show that "pores" formed by silica nanoparticles and PVP macromolecules and filled by unfrozen structured water are of the mesoporous range that is typical for suspensions of nanosilica and other nanooxides.

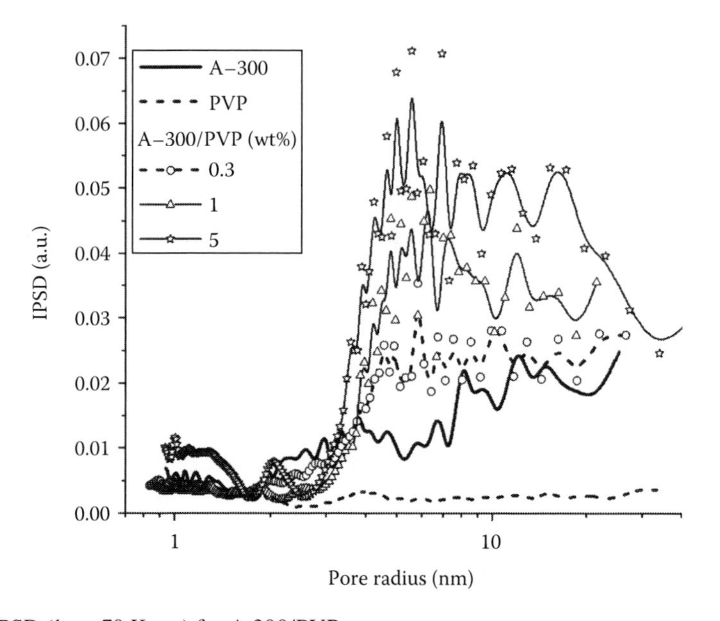

FIGURE 5.55 PSD (k_{GT} = 70 K nm) for A-300/PVP systems.

Biodegradable polymers can be mainly classified as agro-polymers (starch, protein, etc.) and biodegradable polyesters (polyhydroxyalkanoates, poly(lactic acid), etc.). These latter, also called biopolyesters, can be synthesized from fossil resources but main productions can be obtained from renewable resources (Bordes et al. 2009). However for certain applications, biopolyesters cannot be fully competitive with conventional thermoplastics since some of their properties are too weak. Therefore, to extend their applications, these biopolymers have been formulated and associated with nano-sized fillers, which could bring a large range of improved properties (stiffness, permeability, crystallinity, thermal stability). The resulting "nano-biocomposites" have been the subject of many recent publications. Bordes et al. (2009) analyzed this novel class of materials based on clays, which are nowadays the main nanofillers used in nanocomposite systems.

Other works attempted to better understand the mechanism of intercalation/exfoliation process either by melt intercalation or in situ ε-CL polymerization by molecular dynamics simulations (Calberg et al. 2004). During last decade, solid-state NMR has emerged as a tool to characterize clay/polymer nanocomposites in complement to data from classical methods (XRD, TEM) since it is a powerful technique for probing the molecular structure, conformation, and dynamics of species at interfaces. Therefore, solid-state NMR was used in PCL-based nanocomposites to investigate how the surfactant conformation and mobility are changed by the polymer adsorption and how the polymer motion is perturbed after intercalation in the nanocomposites (Calberg et al. 2004, Hrobarikova et al. 2004, Urbanczyk et al. 2006). Finally, Calberg et al. (2004) validated this characterization method to determine the structure of PCL-based nanocomposites since they demonstrated that there was a correlation between variations in the proton relaxation times $T_{1(H)}$ and the quality of clay dispersion.

The structure and dynamics of surfactant and polymer chains in intercalated poly(ε-caprolactone)/clay nanocomposites are characterized by ^{31}P magic-angle spinning (MAS) and ^{13}C cross-polarization MAS NMR techniques. To obtain hybrid materials with the low polymer content required for this study, in situ intercalative polymerization was performed by adapting a published procedure. After nanocomposite formation, the chain motion of the surfactant is enhanced in the saponite-based materials but reduced in the laponite ones. Compared to the initial clay, the *trans*-conformer population of the surfactant hydrocarbon chains in the nanocomposite decreases for the saponite systems. Mobility of the polymer chain is higher in the nanocomposites than in the bulk phase. The charge of the modified saponite does not significantly influence chain mobility in the nanocomposites.

VanderHart et al. (2001a–c) studied different clay nanocomposites measuring clay exfoliation by relaxation times of hydrogen that "sees" iron in the montmorillonite clay. They used Fe atoms in montmorillonite clay to determine clay dispersion in Nylon-6 matrix, and degraded alkyl ammoniums (from thermal processing above 200°C) were observed by NMR technique. Hou et al. (2002,2003) studied clay intercalation of poly(styrene–ethylene oxide)-*block*-copolymers using multinuclear solid-state NMR. Hrobarikova et al. (2004) prepared polycaprolactone with laponite or saponite nanocomposites by in situ polymerization and characterized by ^{13}CAP NMR to understand how surfactants at clay surface interacted with polymer matrix. Hrobarikova et al. (2004) used solid-state NMR to study intercalated species in poly(ε-caprolactone)/clay nanocomposites.

5.5 POLYMERS IN CONFINED SPACE OF PORES

Interactions of 1D polymers with porous solid particles depend on many factors: types (chemistry) of a polymer and a solid, textural, and morphological characteristics of solid particles (*S*, *V*, PSD, particle size), dispersion medium (solvent can be a strong competitor especially in narrow pores), concentration, temperature, and time of interactions. In this case, the confined space effects can play a much stronger role than in the case of nanooxides composed of nonporous nanoparticles. A polymer shell on adsorbent particles can form a strong barrier (thick cross-linked 3D strictures) or a semipermeable membrane (thin 2D structures at a lower degree of cross-linking). Core–shell particles can be used to protect target objects (e.g., native cells, RBC) from the adsorbent surfaces or the adsorbent surface from aggressive media. Clearly, water plays an important role in the interactions of polymers with porous particles in the aqueous media. Here, interactions of polymers with different solids (oxide, carbons, 3D polymers) studied using the NMR techniques are analyzed.

Fatkullin et al. (2004) used a spinodal demixing technique to prepare linear poly(ethylene oxide) (PEO) confined in nanoscopic strands, which in turn were embedded in a 3D quasi-solid methacrylate matrix (prepared as semi-interpenetrating networks from a variable mixture of hydroxyethyl-methacrylate, dodecyl-methacrylate, and 1,2-ethyleneglycol-dimethacrylate) impenetrable to PEO. Chain dynamics of the PEO in the molten state were analyzed using field-gradient NMR diffusometry and field-cycling NMR relaxometry. Reptation was the dominating mechanism for translational displacements of PEO in the nanoscopic strands. The timescale of the diffusion measurements was 10–300 ms but it was much shorter (10^{-9}–10^{-4} s) for field-cycling NMR relaxometry. A frequency dispersion of the spin–lattice relaxation time characteristic for reptation was clearly showed for all samples. In this case, an effective tube diameter of only 0.6 nm was found even when the pore diameter was larger than the radius of gyration of the PEO chain random coils. The finding that the tube diameter effective on the short timescale of field-cycling NMR relaxometry was much lower than the diameter of the confining structure was termed the "corset effect," and was traced back to the lack of local free-volume fluctuation capacity under nanoscale confinements. Fatkullin et al. (2004) analyzed application of three polymer dynamics theories: the Rouse model, the renormalized Rouse formalism, and the tube/reptation concept, which have specific NMR relaxometry signatures for the studied systems. Each of these predictions has been verified in experiments that comply with the theoretical prerequisites: bulk melts for $M_w < M_c$ (Rouse model), bulk melts for $M_w > M_c$ (renormalized Rouse formalism, where M_c is the critical molecular mass, an empirical constant which indicates the crossover to "entangled" chain dynamics), and polymer melts of any molecular mass confined in nanoscopic pores of a solid matrix (tube/reptation concept).

NMR spectroscopy data showed that the order of C_{30} alkyl chains (after reaction of C_{30} trichlorosilane with oxide surfaces) is higher on titania ($S_{BET} = 17$ m^2/g, $V_p = 0.12$ cm^3/g, pore size $d = 28.6$ nm) and zirconia (31 m^2/g, 0.23 cm^3/g, 29 nm) than on silica (77 and 109 m^2/g, 0.74 and 0.9 cm^3/g, 30 and 26 nm) microparticles (~3 μm in size for all) (Pursch et al. 2000). The C_{30} self-assembled monolayers (SAM) are thicker (4.05 nm) than that of C_{18} SAM (2.82 nm).

Mirau et al. (1999) used the ^{1}H MAS NMR to study a low T_g acrylate polymer in bulk and polymerized in a Vycor glass with 4 nm pores (250 m^2/g, porosity 0.28). The combination of chain motion and fast MAS averaged the proton–proton dipolar interactions such that high-resolution spectra were observed at ambient temperature. Multiple-quantum NMR showed that some of the chains were severely restricted relative to the bulk material due to confined effects in narrow mesopores of the glass. These restricted chains could be returned to their bulk-like state by surface treatment of the porous glass.

Bose et al. (2006) studied physiosorbed cyanopropyl–methyl–phenyl–methyl–siloxane polymer ($M_w = 8000$, 19 wt%) on a silica surface (340 m^2/g, average pore diameter 14 nm) using one- and two-dimensional solid-state NMR techniques including heteronuclear proton–silicon correlation (HETCOR) spectroscopy. Spin–lattice relaxations of protons of the siloxane polymer exhibited only small changes (thus, there is no strong bonding) upon anchoring to the silica surface. This indicated slightly changed molecular dynamics of proton moieties that contributed to the relaxation process. However, the same relaxation rates of the siloxane polymer's silica atoms were reduced due to restricted mobility of the polymer confined in pores. ^{1}H–^{29}Si HETCOR spectroscopy revealed strong correlations of silanol protons with both Q_3 ((≡SiO)$_3$SiOH) and Q_4 ((≡SiO)$_4$Si) sites of the silica surface. A correlation between methyl protons and the Q_3 site of the silica surface was observed when HETCOR experiments with very less mixing time (5 ms) were performed. The presence of these correlations was indicative of the coherent magnetization transfer mainly through dipolar mechanisms. Since magnetization transfer through the dipolar mechanism is

$1/r^3$-dependent, methyl protons must lie in close proximity to the silica surface. Hydrogen bonding of the silica surface's hydroxyl protons with the bridging oxygen of the siloxane polymer is most likely responsible for positioning the methyl protons closer to the surface. Additional correlations between ^{29}Si nuclei and methylene protons next to cyano group was also observed with mixing time indicating the closer proximity of these protons to the silica surface as well. This juxtaposition of methylene protons was most likely due to hydrogen bonding of the siloxane polymer through the cyano moiety. Furthermore, the hydrogen bonding through the cyano group was most likely to be in parallel orientation to the surface. The aromatic protons exhibited weak correlations only with Q_3 sites, indicating that these protons must also lie in close proximity to the silica surface.

Fischer et al. (2011) synthesized and characterized novel organic–inorganic hybrid nanotubes containing silica and ethane (EtSNT), ethylene (ESNT), and acetylene (ASNT) units. The unsaturated hydrocarbon linkers can be applied for a chemical functionalization of the surface of the silica nanotubes, e.g., for the grafting of organic/inorganic or organometallic molecules. These new materials were synthesized using a template method with V_3O_7 H_2O fibers and characterized by solid-state MAS NMR and other methods.

Jagadeesh et al. (2004) studied surface-induced order and anisotropic dynamics in ultrathin films (up to about four monomer thick, 0.25–3.6 mg/m^2) of a flexible polymer, polydimethylsiloxane (PDMS) at alumina macroporous membranes (average pore size 200 nm, thickness 60 μm, 7.6 m^2/g) using proton multiple-quantum NMR techniques. The ^{1}H–^{1}H residual dipolar coupling, and dynamic order parameter of the surface bound layer estimated at different PDMS coverages showed that the thin PDMS films exhibited discrete dynamic layers with stepwise increase in their motional freedom with increasing film thickness. Dynamic heterogeneity was more pronounced at one monomer thick film, which was absent for sub-monomer coverage. These findings were attributed to the possible conformational changes of PDMS backbone (Tuncer and Gubański 2000, Jagadeesh et al. 2004).

Many of properties of composites with filled polymers can depend significantly on intermolecular interactions of solid particles of fillers with media. Carbon blacks and nanooxides (e.g., silica, titania, alumina) are the most extensively studied as fillers. However, hybrid materials with pyrocarbon/nanooxides can be more appropriate as fillers for some systems. Such materials can be applied not only as fillers for polymers but also as an immobile phase in chromatography, adsorbents of polar and nonpolar organics in different media, etc. Changes in the surface properties of carbon–mineral adsorbents can occur due to immobilization of polar polymers (e.g., PEG). Colloidal dispersions with solid adsorbents, polymers, and low-molecular compounds are applied, e.g., in high-selective chromatographic columns. The characteristics of such dispersions depend on the state of the solvate shells (EDL) of solid particles and polymer molecules. Leboda et al. (2001c) studied the properties of the interfacial layers in water–PEG–pyrocarbon/nanosilica suspensions in comparison with those for the aqueous suspensions of individual solid adsorbents and pure PEG solutions using ^{1}H NMR and theoretical methods.

Pyrocarbon/nanosilica samples were synthesized by pyrolysis of CH_2Cl_2 on nanosilica A-300 ($S_{BET}=295$ m^2/g) at 773 K for 40 min (CS_4, $C_C=4$ wt%) or 823 K for 90 min (CS_{40}, $C_C=40$ wt%) (Leboda et al. 2001c). CS_4 and CS_{40} have the specific surface area of 280 and 160 m^2/g, respectively. The surfaces of CS_4 possess mosaic structures with pyrocarbon patches on oxide particles, individual carbon particles of 10–50 nm and free silica surfaces, but in the case of CS_{40}, practically entire surfaces of silica particles are covered by carbon deposits. PEG (Ferak) with molecular weight ~1250 was used as received.

For recording ^{1}H NMR spectra of water bound to the solid surfaces or polymer molecules, a high-resolution WP-100 SY (Bruker) NMR spectrometer with a bandwidth of 50 kHz was used.

The ^{1}H NMR signal intensity (Figure 5.56a) was normalized in respect to that for the solution at temperature higher than the melting temperature (T_m) of pure PEG (measured $T_m \approx 315$ K).

$T_m \approx 312$ K for PEG 1000 and $T_m \approx 318$ K for PEG 1500, i.e., one can assume that $M \approx 1250$ for studied PEG. The observed intensity for all the samples at $T < T_m$ is significantly lower than that at $T > T_m$. Consequently, one can conclude that the ^{1}H signal of solid PEG is not observed in the high-resolution NMR spectra due to the short time of cross-relaxation of protons of solids. The lowest

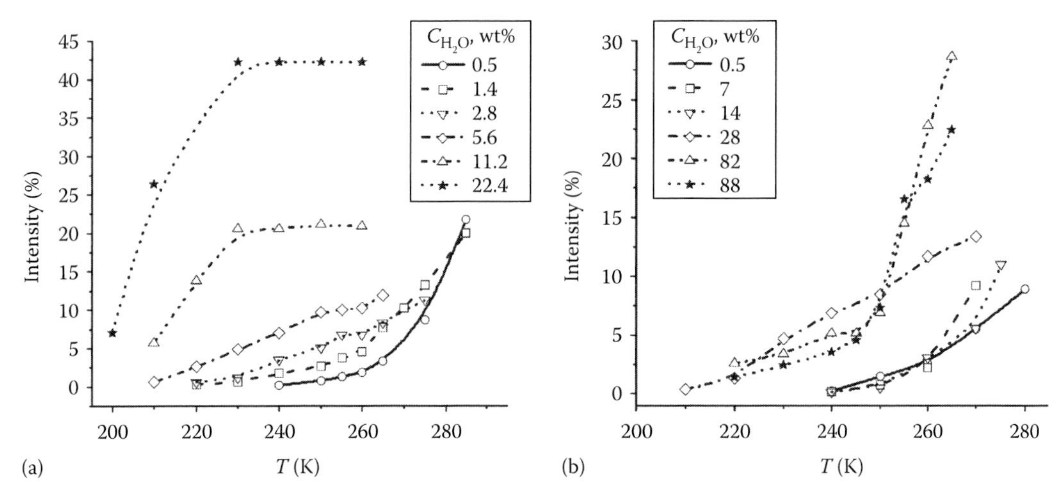

FIGURE 5.56 ^{1}H NMR signal intensity of unfrozen water bound to differently hydrated (a) PEG and (b) PEG/CS_4 at $C_{CS4} \approx 5$ wt%. (Adapted from *J. Colloid Interface Sci.*, 237, Leboda, R., Turov, V.V., Gun'ko, V.M., and Skubiszewska-Zieba, J., Characterization of aqueous suspensions of poly(ethylene glycol)/pyrocarbon/silica, 120–129, 2001c, Copyright 2001, with permission from Elsevier.)

curve in Figure 5.56a corresponds to the initial PEG containing ~0.5 wt% of water. The ^{1}H NMR signal intensity for this sample rises markedly with T elevating to T_m. This signal can be attributed to hydrogen atoms in CH_2 groups of PEG, as water content is very low. In the case of hydrated PEG, two signals with close widths (Figure 5.57a) are observed with increasing C_{H_2O}, which are linked to CH_2 groups of PEG and water molecules, and behavior of PEG molecules is akin to that of low-molecular organics.

On freezing, water forms pure ice and PEG molecules concentrate in the unfrozen phase, whose volume enhances with C_{PEG} and the freezing temperature of water lowers. This temperature depends on the molecular weight of polymers and the cryoscopic constant of water. The concentration of PEG in the unfrozen phase can be estimated from the relationship between the ^{1}H NMR signal intensity of water and PEG, as at $C_{PEG} = 50$ wt%, they have the signals with equal intensity and $\delta_{H,iso} \approx 5.0$ ppm (OH) and 4.2 ppm (CH_2) (Figure 5.57a). The last solution is frozen as eutectic mixture.

On temperature dependencies of the ^{1}H NMR signal intensity at different C_{uw}, there are portions with nearly constant intensity with lowering temperature (Figure 5.56a). Relative intensity of the signals corresponding to these curve portions is two times larger than expected from the concentration of water and the signal shape is the same as in the case of the PEG solution at $C_{PEG} = 50$ wt%. A fast reduction in the signal intensity observed with lowering temperature (Figures 5.56 and 5.57) is due to freezing of both PEG and water and formation of their eutectic mixture. Similar regularities are observed on addition of water to solid PEG, as the temperature of PEG freezing goes down according to its cryoscopic constant and PEG crystallizes in the form of a separated (segregated) phase.

Addition of CS_4 ($C_{CS4} = 5$ wt%) to the PEG solution results in dramatic changes in the shape of $I(T)$ (Figure 5.56b), as the signal intensity of mobile molecules strongly decreases and the curve portions linked to the bulk solution of PEG (non-bonded to solid particles) disappear. This effect (observed even at $C_{H_2O} = 88$ wt%) can be due to an increase in the amounts of the bulk water, as PEG molecules bonded rather to pyrocarbon can cover a part of the silica surfaces. Therefore, only a minor portion of water can be localized near accessible silica patches but the disturbing impact of PEG on the interfacial water is weaker than that of the silica or CS surfaces. Besides, the bulk solution of PEG nonbonded to solid particles does not form over a broad range of the amounts of components. A similar change in $I(T)$ is observed on addition of nanosilica or CS_{40} (Figure 5.57c) to the PEG solution. Consequently, the availability of highly dispersed particles of both silica and CS (i.e., independent of their nature) leads to strong alterations in the characteristics of the frozen PEG solution.

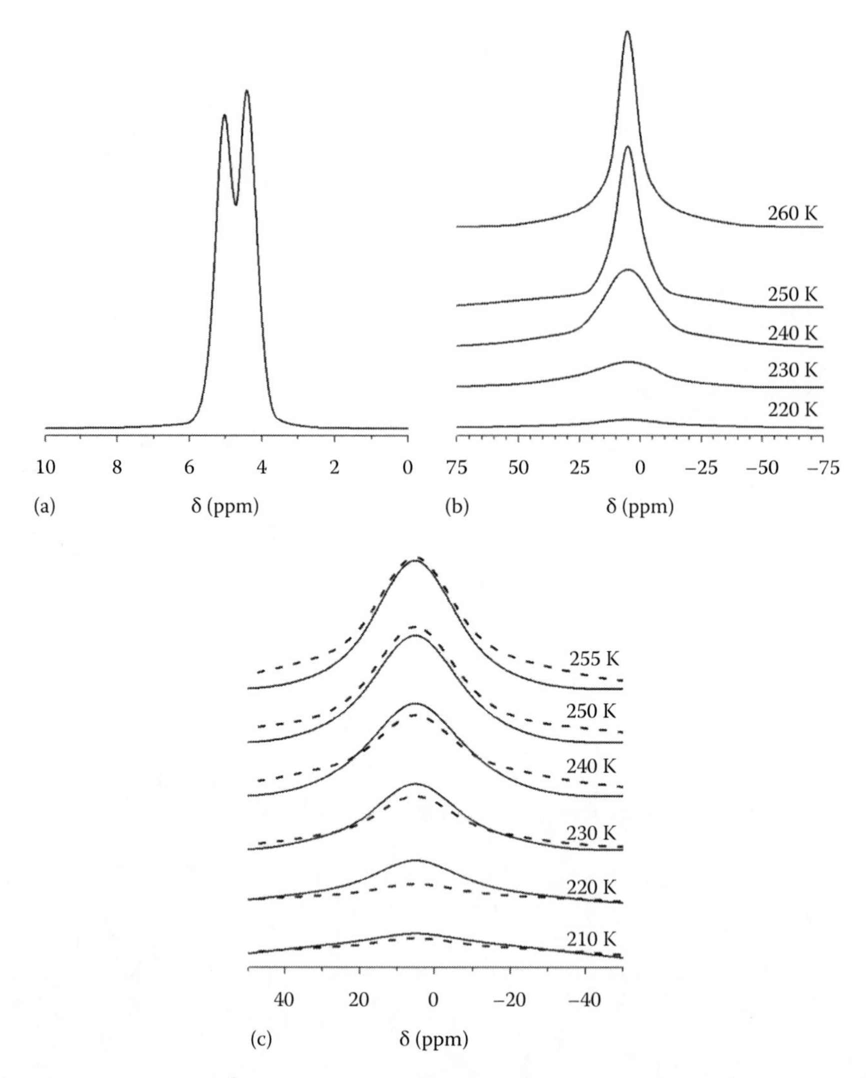

FIGURE 5.57 Chemical shift δ_H for aqueous solution of PEG (a) 50 wt% PEG at 260 K and (b) 2.8 wt% PEG; and for aqueous suspension of (c) 5 wt% CS_4 + 1 wt% PEG (solid lines) and 5 wt% CS_{40} + 1 wt% PEG (dashed lines) at different temperatures. (Adapted from *J. Colloid Interface Sci.*, 237, Leboda, R., Turov, V.V., Gun'ko, V.M., and Skubiszewska-Zieba, J., Characterization of aqueous suspensions of poly(ethylene glycol)/pyrocarbon/silica, 120–129, 2001c, Copyright 2001, with permission from Elsevier.)

Nanosilica A-300 ($S_{BET} \approx 300$ m²/g) consists of spherical primary particles (d = 5–15 nm), which form aggregates (100–500 nm, apparent density $\rho_{ap} \approx 0.7$ g/cm³, consisting of many thousands of primary particles randomly sticking together on synthesis in the flame or just after leaving it), looser agglomerates of aggregates (>1 μm, $\rho_{ap} \approx 0.07$–0.09 g/cm³, formed later than aggregates), and visible flocks ($\rho_{ap} \approx 0.04$–0.07 g/cm³). Pyrocarbon can form individual particles (up to 100–150 nm at large C_C) on the outer surfaces of silica aggregates, but smaller graphene clusters can be located in channels (gaps) between primary silica particles in swarms that results in a significant diminution of the accessible volume and surfaces of nanopores and mesopores of the oxide matrix at $R_p < 20$ nm for CS_{40} (Figure 5.58), but even in this case the surface area linked mainly to narrow pores (Figure 5.58b).

The structural characteristics of CS samples depend strongly on C_C, as own texture of pyrocarbon differs from that of nanosilica. The CS_{40} PSD has a maximum for mesopores at larger pore radii

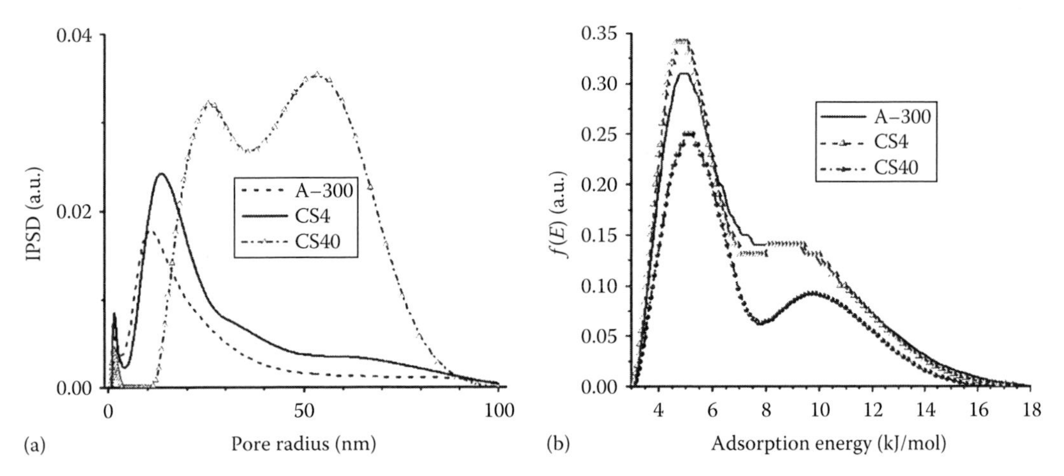

FIGURE 5.58 (a) Pore size distribution and (b) nitrogen adsorption energy distributions for nanosilica A-300, CS_4, and CS_{40}. (Adapted from *J. Colloid Interface Sci.*, 237, Leboda, R., Turov, V.V., Gun'ko, V.M., and Skubiszewska-Zieba, J., Characterization of aqueous suspensions of poly(ethylene glycol)/pyrocarbon/silica, 120–129, 2001c, Copyright 2001, with permission from Elsevier.)

than silica or CS_4 (Figure 5.58a). At $C_C = 40$ wt%, channels in oxide swarms can be blocked by pyrocarbon nearly entirely as the external surfaces of silica aggregates covered densely by carbon deposits, whose porosity (pore volume V_p) can, however, be close to that of initial silica. For instance, an estimation of the pore volume from the nitrogen adsorption at $p/p_0 \approx 0.99$ gives $V_{p,CS40} \approx 0.86$ cm^3/g and $V_{p,A300} \approx 0.62$ cm^3/g; however, at $p/p_0 \approx 0.985$, the corresponding volumes are $V_{p,CS40} \approx 0.42$ cm^3/g and $V_{p,A300} \approx 0.51$ cm^3/g, which are in agreement with the PSD shapes for these adsorbents (Figure 5.58a), while larger pores of CS_{40} are filled by nitrogen at higher $p/p_0 \geq 0.99$. These pore volumes are close to the volume of channels in aggregates, as the empty volume in agglomerates is substantially greater (up to 15 cm^3/g) and it is not entirely filled by nitrogen even at high p/p_0 and the plateau adsorption for both nanosilica and CS is absent even at $p/p_0 \to 1$. The specific surface area of CS_{40} is nearly twice smaller than that of pristine silica but the PSD of CS_{40} is much broader (Figure 5.58), while large pores in CS_{40} give a very small contribution to the specific surface area mainly linked with the interior surface area of narrower pores (voids between particles) and rough external surfaces of aggregates covered by pyrocarbon.

The enhancement of the mesopore sizes and the diminution of nanopore contribution for CS_{40} (Figure 5.58a), as pyrocarbon consists of mainly nonporous particles akin to those of carbon black, can also reflect in a decrease in the nitrogen adsorption energy (Figure 5.58b), especially in respect to the second $f(E)$ peak, as the narrower the pores, the greater the adsorption energy. Secondary filling of large pores corresponds to the adsorption energy (Figure 5.58b, first $f(E)$ peak) close to the liquefaction heat of nitrogen. The observed structural and energetic differences between pristine nanosilica and pyrocarbon/silica samples can influence not only the properties of the interfacial water layer but also the distribution of PEG molecules in this layer. For instance, the adsorption of such linear polymer molecules as PEG in channels of aggregates of pure nanosilica can be practically irreversible and causes decomposition of large swarms to smaller aggregates with PEG–SiO_2 of 50–80 nm, while >90% of the particle number in the aqueous suspensions of nanosilica at $C_{SiO_2} = 3$–6 wt% have the sizes of 20–60 nm. However, for the CS suspension in the presence of PEG, the swarm size distribution can be different especially for CS_{40} with larger particles, while pyrocarbon enhances the pH value of the isoelectric point with C_C (i.e., electrostatic repulsion between particles in the suspensions decreases with no addition of electrolytes), inhibits decomposition of swarms, and the effective diameter d_{ef} increases markedly (Gun'ko et al. 2000d). At $C_C \geq 20$ wt%, CS particles observed in the aqueous suspensions have $d_{ef} \geq 1$ μm, which is substantially larger than that for pristine nanosilica at the same pH and concentration values.

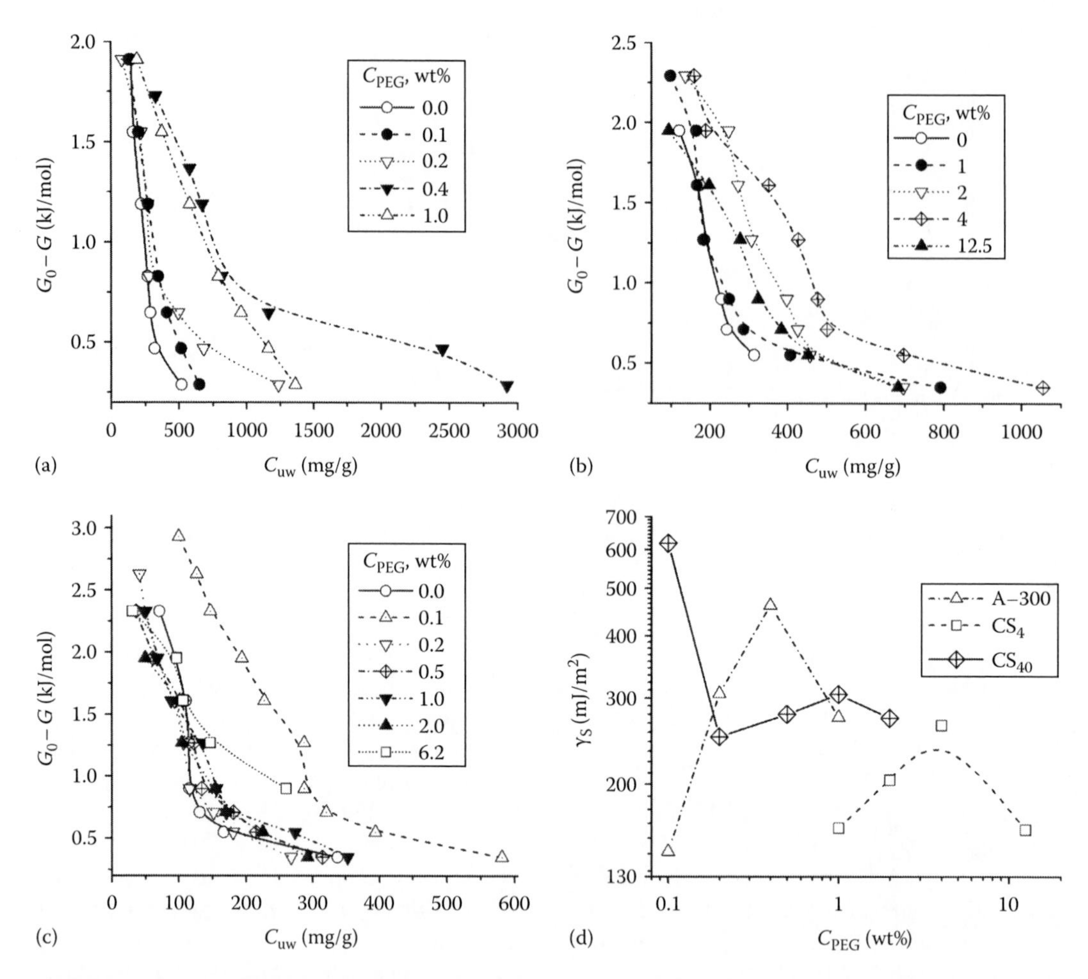

FIGURE 5.59 (a–c) Changes in the Gibbs free energy as a function of the amounts of unfrozen water (C_{uw}) and (d) free surface energy as a function of C_{PEG} for the aqueous suspensions of (a) nanosilica at $C_{SiO_2} \approx 5$ wt%, (b) CS_4 at $C_{CS} \approx 5$ wt%, and (c) CS_{40} at $C_{CS} \approx 5$ wt%; $\gamma_S(C_{PEG})$ curves for nanosilica A-300, CS_4, and CS_{40}. (Adapted from *J. Colloid Interface Sci.*, 237, Leboda, R., Turov, V.V., Gun'ko, V.M., and Skubiszewska-Zieba, J., Characterization of aqueous suspensions of poly(ethylene glycol)/pyrocarbon/silica, 120–129, 2001c, Copyright 2001, with permission from Elsevier.)

The graphs of the ^{1}H NMR signal intensity versus temperature for the PEG–SiO_2 suspensions (Figure 5.59a) are akin to those for oxides and pyrocarbon/mineral adsorbents shown earlier.

Notice that on analysis of the ^{1}H NMR spectra of the frozen suspensions of PEG solids, contribution of CH_2 groups of PEG was neglected, as C_{PEG} was low in comparison with C_{uw} and immobilized polymer molecules do not contribute to the observed ^{1}H NMR signal. Two portions of the plots $\Delta G(C_{uw})$ corresponding to fast and slow diminutions of ΔG can be attributed to SBW and WBW, whose parameters are shown in Table 5.11.

Strong adsorption interaction at a low concentration of adsorbate is characteristic for many low-molecular (water, nitrogen, alcohols, etc.) and polymer (proteins, PVP, etc.) compounds due to nonuniformity of the surfaces, i.e., the availability of strong and weak adsorption sites on them. In other words, the adsorption of low amounts of PEG results in enhancement of the interaction energy of solid particles with surrounding due to an increase in the nonuniformity of the interfaces. This process is best demonstrated by the $\gamma_S(C_{PEG})$ plot (Figure 5.59), as the free surface energy changes significantly in the presence of 0.4 wt% of PEG in the aqueous suspension of nanosilica (Table 5.11).

TABLE 5.11
Characteristics of Bound Water in Aqueous Suspensions of Nanosilica, CS_4, or CS_{40} (5 wt%) at Different Concentrations of PEG

C_{PEG} (wt%)	$-\Delta G_s$ (kJ/mol)	$-\Delta G_w$ (kJ/mol)	C^s_{uw} (mg/g)	C^w_{uw} (mg/g)	γ_S (mJ/m²)
(A-300)	2.8	0.6	350	650	120
0.1	2.6	0.7	500	500	150
0.2	2.2	0.7	550	1000	300
0.4	2.2	0.8	1300	2150	460
1	2.2	—	1200	700	270
(CS_4)	2.4	—	300	—	110
1	2.6	0.7	400	1100	160
2	2.6	0.8	550	1000	200
4	3.2	1.1	600	900	260
12.5	2.5	1.0	550	700	160
(CS_{40})	2.8	0.7	125	500	320
0.1	3.6	0.8	350	475	620
0.2	3.2	1.0	138	288	250
0.5	2.6	0.9	163	338	280
1	3.3	0.9	200	300	310
2	2.2	1.0	200	200	270
6.2	2.5	—	225	—	—

In the case of the aqueous suspensions of CS_4 (i.e., relatively low amount of pyrocarbon in CS), the $\Delta G(C_{uw})$ and $\gamma_S(C_{PEG})$ graphs (Figure 5.59) are akin to those for the initial nanosilica; however, maximal values of γ_S are observed at larger $C_{PEG}=4$ wt%. Another picture is observed for the aqueous suspensions of CS_{40} (Figure 5.59 and Table 5.11), while the characteristics of the interfacial water depend on C_{PEG} for the studied adsorbents differently, and maximal changes in γ_S are observed at a low C_{PEG} value.

Since the γ_S value determines the free energy of interaction between water and adsorbent surfaces, one can conclude that the adsorption of low amounts of PEG on CS_{40} enhances this energy (Figure 5.59) due to an increase in the nonuniformity of the interfaces. A similar effect was previously observed for nonporous CS and mixed oxides with low amount of the deposits (pyrocarbon or second metal oxide). The reasons of this phenomenon can be connected with changes in the surface charge distribution influencing long-range surface forces and enlargement of the thickness of the EDL, i.e., interfacial water disturbed by polar or charged solid surfaces. A higher energy of interaction of a water molecule with a solid particle than with other water molecules leads to reorientation of its dipole in the nonuniform electrostatic field (especially in a dense portion of the EDL). In the case of a large concentration of PEG, the interface nonuniformity reduces, as the solid surfaces are entirely covered by polymer molecules, and mainly PEG interacts (but weaker than silica surfaces) with water; therefore, the volume of bulk undisturbed water increases while the ^1H NMR signal intensity of the unfrozen water decreases. However, the adsorption of low amounts of PEG can promote disturbance of water in large pores of CS_{40} (Figure 5.58a), but increase in C_{PEG} reduces the water concentration in mesopores and γ_S decreases (Figure 5.59) and becomes close to that for the aqueous suspension of nanosilica possessing narrower channels in aggregates (blocked by pyrocarbon in CS_{40}).

An increase in the C_{PEG} range corresponding to a relative enhancement of γ_S for the suspension of CS_4 (however, γ_S for CS_{40} or A-300 is larger than that for CS_4, Figure 5.59d) in comparison with this range for initial silica can be due to preferential adsorption of PEG molecules on the pyrocarbon patches keeping a significant surface nonuniformity. In the case of CS_{40} (which possesses very small portion of accessible silica surfaces), enhancement of the surface nonuniformity is observed only due to adsorption of polymer molecules (in mesopores) at low C_{PEG}. Consequently, modification of silica surface by carbon deposits and immobilized polymer molecules at different concentrations allows one to control the interaction with the aqueous media. Similar regularities for CS were found on the investigations of the PEG impact on the separation ability of columns used in the gas chromatography, which was maximal at approximately monolayer coverage with PEG. In aqueous media at low amounts of PEG, deactivation of the adsorbent surfaces in respect to the adsorption of dissolved molecules can be due to an enhancement of the interaction with water molecules, as the greater the water adsorption energy the lower the probability of adsorption of dissolved compounds. At monolayer coverage by PEG, the surface becomes more homogeneous, interaction with water decreases (see, e.g., Figure 5.56), and the adsorption of dissolved compounds is determined by their interaction energy with adsorbed polymer molecules.

Computations of the nuclear magnetic shielding constants or chemical shifts for H atoms in ethylene glycol (EG) or its oligomers with the GIAO method at the B3LYP/6-31G(d,p) basis set give $\delta_{H,iso}=3.85$ ppm for CH_2 groups for EG (Figure 5.60a) and 3.89 and 3.54 ppm for dimer (PEG2, Figure 5.60b).

In the case of EG interacting with five water molecules (Figure 5.60c), average value of $\delta_{H,iso}(CH_2)=4.01$ and 4.26 ppm for H atoms from water molecules participating in the hydrogen bonds and $\delta_{H,iso}(OH)=6.57$ ppm in RCO–H...OH_2. The solvation effect for EG computed with SM5.42/6-31G(d) method with consideration for the geometry relaxation under solvation shows that H charges $q_{H(CH)}$ increase by ≈8% due to solvation, which corresponds to enhancement of

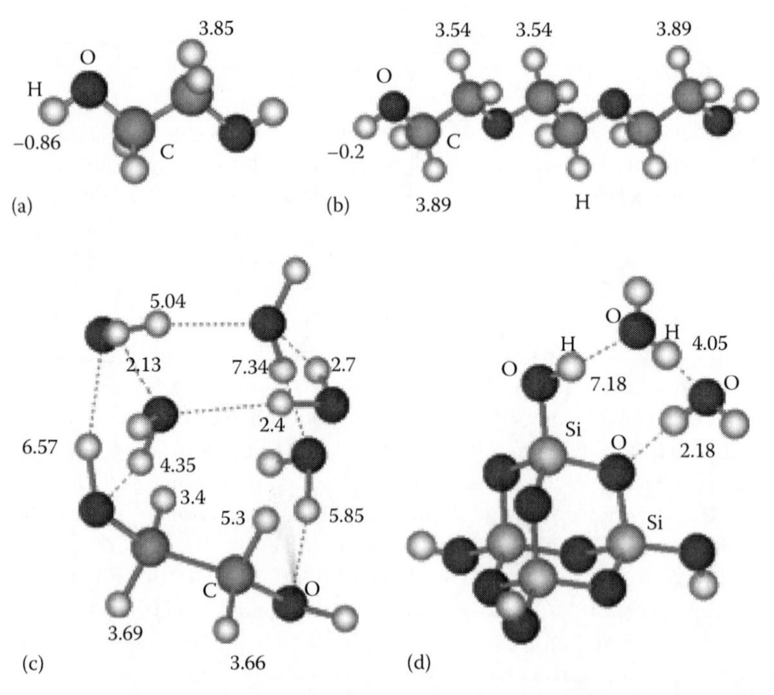

FIGURE 5.60 Chemical shift for ^1H in (a) ethylene glycol, (b) PEG2, (c) ethylene glycol and water molecules, and (d) silica cluster with adsorbed water molecules computed with GIAO method using the (a–c) B3LYP/6-31G(d, p) and B3LYP/6-31G(d ,p)//6-31G(d, p) basis sets. (Adapted from *J. Colloid Interface Sci.*, 237, Leboda, R., Turov, V.V., Gun'ko, V.M., and Skubiszewska-Zieba, J., Characterization of aqueous suspensions of poly(ethylene glycol)/pyrocarbon/silica, 120–129, 2001c, Copyright 2001, with permission from Elsevier.)

deshielding of protons, and $\delta_{H,iso}(CH_2)$ can be equal to 4.2–4.4 ppm estimated using a linear relationship between $\delta_{H,iso}$ and q_H, which is in agreement with experimental $\delta_{H,iso}(CH_2) \approx 4.2$ ppm. However, such an effect can also rise in the case of interaction between PEG2 and silica surface (cluster with 36 tetrahedrons $SiO_{4/2}$ and 24 OH) in the presence of water molecules ($n_{H2O} = 7$) and $\delta_{H,iso}(CH_2)$ increases up to 4.5 ppm estimated using a linear relationship between q_H and $\delta_{H,iso}$ and maximal $\delta_{H,iso}(OH) = 7.08$ ppm. At $n_{H2O} = 12$, maximal values of $\delta_{H,iso}(OH) = 7.15$ ppm and $\delta_{H,iso}(CH_2) = 4.7$ ppm. The strongest hydrogen bond forms between OH group of PEG4 and charged $\equiv SiO^-$ group ($r_{O...H} = 0.1702$ nm), which results in $\delta_{H,iso}(OH) = 10.2$ ppm (estimated with liner equation $\delta_{H,iso} = a + bq_H$). For ≡SiO(H)...H–OCR bond, $\delta_{H,iso}(OH)$ changes over the 4.4–7.2 ppm range for adsorption complexes of EG or PEG2–PEG4 on silica clusters with 12 or 36 tetrahedrons (with 12 and 24 OH groups, respectively) with the participation of water molecules ($7–18H_2O$) at $r_{O...H} = 0.1752–0.1823$ nm and $\angle O–H...O = 155°–171°$ for the hydrogen bonds. These values can be utilized to estimate the $\delta_{H,iso}$ range (which typically increases with n_{H2O} in adsorption complexes) using the corresponding linear relationships between structural characteristics of the hydrogen bonds and the chemical shifts.

Interaction in $\equiv SiO–H...O(CR)_2$ or $HO–H...O(CR)_2$ results in slightly weaker hydrogen bonds corresponding to $\delta_{H,iso}(OH) = 2.5–5.0$ ppm at $r_{O...H} = 0.18–0.2$ nm. Interaction of ≡Si–O–Si≡ bonds with CH_2 groups of PEG can be preferable than their interaction with water molecules, as the ≡Si–O–Si≡ bridges possess the reduced hydrophilic properties in comparison with ≡SiOH groups (see, e.g., Figure 5.60d, δ_H for H in the hydrogen bonds between H_2O and ≡SiOH or ≡Si–O–Si≡). A similar effect related to interaction between CH_2 groups of PEG with pyrocarbon on CS, which is more hydrophobic than the silica surfaces, can enhance changes in γ_S with C_{PEG} at low $C_C = 4$ wt% when only a portion of the silica surfaces is covered by carbon deposits. At $C_C = 40$ wt%, this effect is seen only at low amounts of PEG (Figure 5.59) due to the differences in the nature of the surfaces and the pore structure of adsorbents (Figure 5.58a). All these results show possible mechanisms of broadening of the ^1H NMR signals for the aqueous solutions of PEG or PEG–solid suspensions observed experimentally (Figure 5.57) in comparison with narrower ^1H NMR spectra for water adsorbed on silica from the gas phase. Additionally, on the basis of obtained results, one can assume that the tails of PEG molecules (i.e., COH groups) always interact directly with the adsorbent surfaces (e.g., with surface hydroxyls) and polymer molecules locate in the interface layer close to the surfaces of solid particles and do not form the bulk solution. This leads to an increase in the amounts of the bulk undisturbed water observed indirectly as a diminution of the ^1H NMR signal intensity of the unfrozen water for the aqueous suspensions of PEG–CS on freezing-out of the bulk water (Figure 5.56).

Behavior of PEG ($M \approx 1250$)–water on freezing is akin to that of solutions of low-molecular organics, as frozen PEG eliminates from solution as a solid phase containing low amounts of water. Addition of nanosilica or CS (~5 wt%) to PEG solutions at different C_{PEG} values causes an increase in the amount of bulk undisturbed water, i.e., the concentration of the unfrozen interfacial water decreases, while water interaction with PEG is weaker than with the solid surfaces. Dissolved PEG molecules adsorb effectively on the surfaces of both nanosilica and pyrocarbon/silica. On freezing of these suspensions, the bulk solution of PEG non-bonded to solid particles does not form over a large concentration range of PEG. For the aqueous suspensions of nanosilica or pyrocarbon/silica at $C_C = 4$ wt%, a linear enhancement of the amount of strongly bound unfrozen water, adhesion force, and free surface energy is observed with increasing C_{PEG} over its larger range than that for the suspension of CS_{40} possessing accessible pyrocarbon surfaces and nearly entirely shielded silica surfaces.

Theoretical calculations show that adsorption energy and chemical shifts in the ^1H NMR spectra of adsorbed PEG fragments depend on their length and amounts of adsorbed water and elucidate the mechanisms of broadening of ^1H NMR signals for aqueous suspensions of PEG–solids.

6 Interactions of Biomacromolecules with Water, Organic Compounds, and Oxides, Polymers, and Carbon Adsorbents

6.1 PROTEINS

Most systems with biomacromolecules (e.g., proteins, DNA, phospholipid/water dispersions), membranes, cells, tissues, etc. are strongly heterogeneous, being essentially constituted by a liquid component, mainly water, in contact with a solid-like proton-rich macromolecular matrix. This matrix can be considered as solid because of the slow reorientational motions (with correlation times $\tau > 10^{-9}$ s), but on a macroscopic scale it is soft, deformable (Calucci and Forte 2009). Water molecules fill the available spaces (pores, "pockets," voids) of different hydrophilic or hydrophobic or mosaic characters interacting with binding sites of the matrix (mainly polar or charged groups) and form clusters (WAW with SBW or WBW), bulk-like domains (SAW with smaller contribution of SBW and greater contribution of WBW) in large pores, and a full hydration shell of macromolecules. Other (outer) portion of water can be assigned to weakly bound or bulk water. Water molecules with different dynamic properties and differently interacting with macromolecules coexist and experience all the possible environments due to fast chemical exchange (lifetime ~10^{-11} s) and fast diffusion (diffusion coefficient for bulk water or WBW is about ~10^{-9} m^2/s) at ambient conditions. At low temperatures, picture strongly changes due to freezing-out of a water portion and cryoconcentration of dissolved compounds (e.g., salts). A quantitative description of water dynamics and water–macromolecule interactions is of fundamental importance for deeper insight into the behavior of biosystems in native state since for most biomacromolecules the biofunctions depend strongly on the properties of the macromolecule–water interfaces.

NMR technique is appropriate and informative to be applied to these systems, although the ^{1}H NMR spectra itself are not very informative. However, significantly different transverse relaxation times characterizing the macromolecular ($T_2 < 100$ μs) and mobile water ($T_2 > 10$ ms) protons allow detailed analysis of the behavior of bound water in these complex systems. Relatively narrow signals are observed for bound water with a linewidth typically on the order of tens of hertz. A very broad signal of the macromolecular protons (or ice at $T < 273$ K), characterized by a linewidth on the order of a few tens of kilohertz, is often hardly detectable. This effect can be enhanced in most systems due to the appreciably higher relative concentration of water protons. Information on the structural and dynamic properties of water and macromolecules in heterogeneous systems can be obtained by exploiting the longitudinal relaxation coupling between macromolecular and water protons (Hays and Fennema 1982, Otting 1997, Wider 1998, Calucci and Forte 2009). Ever since the first studies in the early 1960s, an enhancement in longitudinal relaxation of water protons was observed in hydrated macromolecular systems. This was initially interpreted in terms of the sole constraining effects of the macromolecular matrix on water dynamics. It was only in the late 1970s that the crucial importance of cross-relaxation between water and macromolecular protons was recognized by Edzes and Samulski (1977), although spin diffusion effects between macromolecular and water protons had already been suggested by other authors. Since then much work has been done, both experimentally and theoretically, in order to characterize the kinetics and mechanism of relaxation coupling and to exploit it to obtain information on the water status, on the dynamical and structural properties of the macromolecular systems, and on macromolecule–water interactions in many different fields of science (materials science, life sciences, and medicine) using both spectroscopic and imaging NMR techniques (Calucci and Forte 2009).

The determination of high-resolution structures of macromolecules relies on the availability of numerous structural constraints distributed throughout the system: accordingly, assignment of the spectra is a central objective in NMR-based structure determination (James 1975, Iino 1994, McDermott 2004, Opella and Marassi 2004, Hologne et al. 2006, Baldwin and Kay 2009, Mandell and Kortemme 2009, Mondal et al. 2011). Based on the assigned spectra, it is possible not only to solve 3D structures, but also to probe such issues as conformational dynamics and ligand binding, as well as to obtain valuable spectroscopic parameters unavailable from other methods, such as shielding anisotropies (McDermott 2004, Hunter et al. 2005). Additionally, NMR spectroscopy gives detailed information on the behavior of water bound in macromolecules alone or in adsorption complexes at a solid surface open (as in the case of nanooxides) or porous (Otting 1997, Wider 1998, Gun'ko et al. 2005d, 2009d, Mitchell et al. 2008, Petrov and Furó 2009, Simpson et al. 2011).

Proteins can strongly interact with different surfaces such as silica, titania, alumina, and mixed oxides (Gun'ko et al. 1997b, 2000f, 2002d, 2003c,g, 2006i, 2007i, 2009d), nanodiamond (Wu and Kure 2010), activated carbons (Melillo et al. 2004), and other materials (He et al. 2006, Gun'ko et al. 2011b) that can change inner protein structure, especially in the case of hydrophobic surfaces (Horbett and Brash 1995, Shpak and Gorbyk 2010). The temperature dependence of order parameters of proteins can be interpreted in terms of the conformational heat capacity ($C_{p,conf}$) of the protein (Jarymowycz and Stone 2006). Both protein unfolding and many binding events can be accompanied by significant changes in the heat capacity of the system. In general, the major contribution to these changes is the hydrophobic effect; burial of hydrophobic surface area releases ordered water from these surfaces, giving rise to negative heat capacity changes. However, changes in the intrinsic motional landscape of the protein could also contribute to heat capacity changes. Studying the temperature dependence of internal dynamics may provide important insights in this regard (Jarymowycz and Stone 2006, Tolman and Ruan 2006). The internal dynamics of macromolecules, which depends on their interaction with surroundings (Hologne et al. 2006, Tamm and Liang 2006), affects the bound water state that reflects in the temperature dependences of the signal intensity and chemical shift. Long-range "transmission" of perturbation of side chain dynamics correlated with changes in functional state of proteins was clearly observed in several systems (Igumenova et al. 2006). Kamerzell et al. (2011) analyzed the key mechanisms of protein–excipient interactions such as electrostatic and cation–π-electron interactions, preferential hydration, dispersive forces, and hydrogen bonding in the context of different physical states of the formulation such as frozen liquids, solutions, gels, freeze-dried solids, and interfacial phenomenon.

The understanding of the mechanisms involved in the interactions of biosystems with inorganic materials is of interest in both fundamental and applied disciplines. The adsorption of proteins modulates the formation of biofilms onto surfaces, a process important in infections associated to medical implants, in dental caries, in environmental technologies (Fenoglio et al. 2011). The interaction with biomacromolecules is crucial to determine the beneficial/adverse response of cells to foreign inorganic materials as implants, engineered or accidentally produced inorganic nanoparticles (NP). A detailed knowledge of the surface–biofluid interface processes is needed for the design of new biocompatible materials. Researchers involved in the different disciplines face up with similar difficulties in describing and predicting phenomena occurring at the interfaces between solid phases and biofluids. Fenoglio et al. (2011) analyzed the knowledge from different research areas by focusing on the search for determinants driving the interactions of inorganic surfaces with biomatter.

Chemical exchange processes arise due to conformational transitions or reactions on the chemical shift timescale that modify the magnetic environment of a given spin and consequently stochastically modulate its isotropic chemical shift. The effective field strengths typically employed in $R_{1\rho}$ relaxation experiments are of the order of 1–6 kHz, although weaker fields can be utilized to provide overlap with the CPMG experiment. Consequently, $R_{1\rho}$ experiments are most often used for faster microsecond timescale chemical exchange processes (Palmer and Massi 2006). By analogy to the laboratory-frame Bloch equations, $R_{1\rho}$ is the relaxation rate constant for magnetization locked along the direction of the effective field in the rotating frame, and $R_{2\rho}$ is the relaxation rate constant for magnetization orthogonal to the direction of the effective field in the rotating frame. $R_{1\rho}$ spin relaxation in the rotating frame is one of a set of NMR techniques that can be used to characterize chemical exchange processes occurring on the microsecond to millisecond timescale. In biomacromolecules, these processes often reflect time-dependent phenomena that are critical to functions such as protein folding, ligand binding, and catalysis (Palmer and Massi 2006) as well as adsorption of macromolecules to a solid surface. Analysis of $R_{1\rho}$ relaxation dispersion data allows determination of the kinetic and thermodynamic properties of the conformational transitions or chemical reactions that give rise to exchange broadening and evaluation of the changes in isotropic shifts between the different molecular conformations or chemical species. Clearly, water bound to macromolecules can affect these processes and this effect depends on the water content. $R_{1\rho}$ experiments have three principal advantages compared with other techniques, such as CPMG relaxation dispersion: (i) a wider range of effective fields in the rotating frame is accessible, allowing investigation of a large range of exchange rate constants, (ii) combined variation of both ω_1 (the amplitude of the spin-locking radio frequency [RF] field, assumed without loss of generality to be applied with *x*-phase) and ω_{rf} allows complete characterization of the exchange process using data recorded at single static magnetic field strengths, and (iii) the absolute sign of chemical shift changes can be determined if exchange is outside the fast exchange limit (Palmer and Massi 2006). Values of isotropic chemical shift differences, $\Delta\omega_{ij}$, obtained from dispersion measurements contain structural information about normally unobservable minor molecular or chemical species. However, chemical shifts depend on multiple conformational parameters, such as backbone and side chain dihedral angles, electrostatic interactions, hydrogen bonding, π-electron current effects, etc. Furthermore, a given conformational transition can affect the chemical shifts of different nuclei in different ways. These issues complicate the interpretation of chemical shift differences for any limited set of nuclear spins. Therefore, measurements of $R_{1\rho}$ dispersion data for multiple nuclei (e.g., ^{15}N, ^{1}HN, ^{13}CO, and ^{13}C$^{\alpha}$ for the protein backbone) provide a more detailed picture of the exchange process (Palmer and Massi 2006).

2D HETCOR experiments with amino acids or proteins have substantial advantages compared to 1D heteronuclear measurements when information about hydrogen bonding is desired (Gu et al. 1996, McDermott 2004). Although some information about hydrogen bonding can be derived from the HETCOR NMR spectra, the information available from the proton shifts is often independent from that obtained from the heteronuclear shifts. One specific example is carboxylic acids: proton shift depends on the interactions of the proton with any nearby hydrogen-bond acceptors, while σ_{22} of ^{13}C shift depends on the hydrogen-bonding interactions involving carboxyl oxygen and other nearby protons (Gu et al. 1996, McDermott 2004). In the case of the carboxylic anion, the effects of hydrogen bonding on the carbon spectra are pronounced only if the proton is in the so-called *syn* position; carbon shift is rather insensitive to the protons in the "*anti*" position. The shift of the nearby proton is sensitive to the interaction regardless of the geometry (McDermott 2004). Therefore, correlated carbon and proton shifts provide much more information about the geometry and strength of the hydrogen bonds than either proton or carbon shifts alone. HETCOR spectra are useful for assignment of carbon spectra and identification of the chemical nature of the hydrogen-bonding partner. The WIM transfer and rotor synchronized WIM sequence can transfer magnetization within C–H pairs separated by as much as 0.3 nm, which is long enough to cover the range of most hydrogen-bonding distances (Gu et al. 1996).

Proteins as bioligands are much more complicated than DNA. They can be either hydrophilic or hydrophobic (due to side group types), with either negative or positive charge, which makes it very challenging to suppress or avoid nonspecific interactions. Common strategy is post-coating and modifying the NP surface with different functional groups, including carboxylate, octadecyl, polyethylene glycol (PEG), or combinations of different functionalities (Knopp et al. 2009). Picard et al. (1998) studied the lateral diffusion (coefficient D_L) in the fluid phase of dipalmitoylphosphatidylcholine (DPPC) in the presence and absence of melittin using ^{31}P two-dimensional exchange solid-state NMR spectroscopy. They used spherical silica particles (radius 320 ± 20 nm) on which lipids and peptides were adsorbed together to analyze the 2D exchange patterns afforded by a narrow distribution of D_L centered at a value of $(8.8 \pm 0.5) \times 10^{-8}$ cm^2/s for the pure lipid system and between 1×10^{-8} and 10×10^{-8} cm^2/s for the lipids in the presence of melittin. The determination of D_L for non-supported DPPC multilamellar vesicles (MLVs) suggested that the support did not slow down the lipid diffusion and that the radii of the bilayers varied from 300 to 800 nm.

The NMR spectroscopy can give a complete picture of the reconstitution process of milk powders in aqueous media (Davenel et al. 2002) since it allows the quantification of water bound to powder particles, their reconstitution rate, and the detection of the presence of insoluble materials. The NMR transverse relaxation rate of the reconstituted product is proved to be an excellent indicator of the structural state of milk proteins. It was shown that the poor reconstitute ability of native cow phosphocasein micelles concentrated by tangential microfiltration and powdered by spray-drying can be significantly improved by addition of whey proteins, suitable polydextroses, or NaCl before spray-drying powder, without significantly affecting their micellar structure. The decrease of the relaxation rate showed an important modification of the casein structure when a citrate solution or, to a lesser extent, when a phosphate solution was added to the retentate. The addition of $CaCl_2$ strongly disturbed the micellar organization and led to the formation of insoluble structures during spray-drying (Davenel et al. 2002). They approximated the CPMG curves by a sum of two exponential curves with fast and slow decay components. The fast processes were attributed to water protons in fast exchange with exchangeable protons of non-dissolved powder particles. The slow decay component (rate R_{2S}; Figure 6.1) was attributed to water protons and exchangeable protons in the reconstituted phase.

The native phosphocaseinate suspension (NPCS) was prepared from skim milks of cow, ewe, and goat by microfiltration and diafiltration. Whey protein concentrate (WPC) was separated from previous cow milk microfiltrate. Combined cow casein powders (CCP) were obtained by two techniques: (i) WPC and Polydextrose Litesse (PDL) from Pfizer Co. (Orsay, France) were first mixed into cow NPCS retentate before freeze-drying to obtain 4%, 8%, and 12% PDL/total solid (TS) and 4%, 8%, and 12% WPC/TS powders and (ii) mineral salts were added to cow NPCS concentrate at 50°C for 30 min before spray co-drying. In the NMT measurements, the samples include 1 g of a powder per 20 mL of water.

Fast transverse relaxation of ^1H, ^{15}N, and ^{13}C by dipole–dipole (DD) coupling and chemical shift anisotropy (CSA) modulated by rotational molecular motions has a dominant impact on the size limit for biomacromolecular structures that can be studied by NMR spectroscopy in solution (Pervushin et al. 1997). Transverse relaxation-optimized spectroscopy, TROSY, is an approach for suppression of transverse relaxation in multidimensional NMR experiments, which is based on constructive use of interference between DD coupling and CSA. For example, a TROSY-type 2D ^1H–^{15}N correlation experiment with a uniformly ^{15}N-labeled protein in a DNA complex of molecular mass 17 kDa at a ^1H frequency of 750 MHz showed that ^{15}N relaxation during ^{15}N chemical shift evolution and $^1H^N$ relaxation during signal acquisition both were significantly reduced by mutual compensation of the DD and CSA interactions. The reduction of the linewidths when compared with a conventional two-dimensional ^1H–^{15}N correlation experiment was 60% and 40%, respectively, and the residual linewidths were 5 Hz for ^{15}N and 15 Hz for $^1H^N$ at 4°C. Because the ratio of the DD and CSA relaxation rates was nearly independent of the molecular size, a similar percentage wise reduction of the overall transverse relaxation rates was expected for larger proteins.

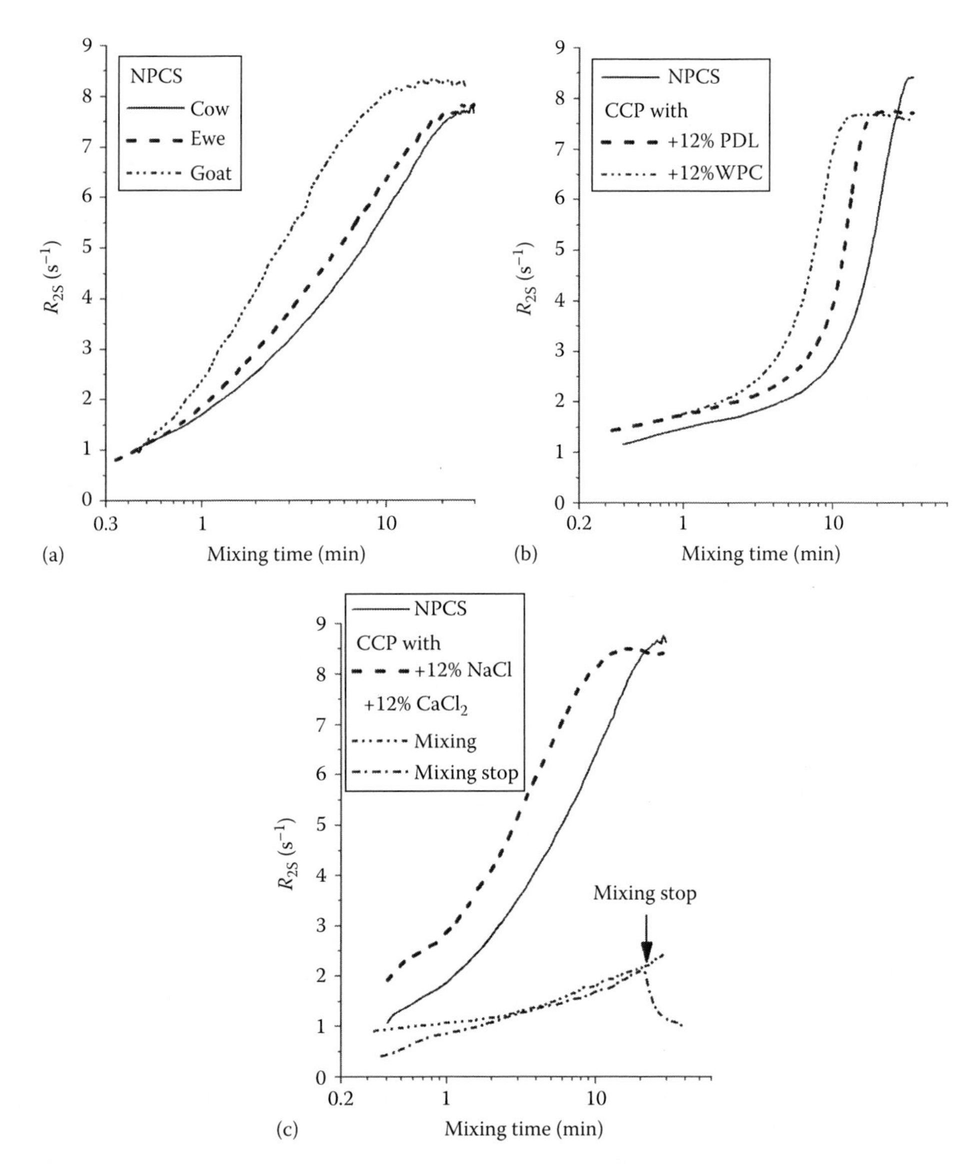

FIGURE 6.1 Relaxation rate R_{2s} (a) for pure NPCS, and after addition of (b) polydextrose PDL or whey protein WPC, or (c) salt (12% in respect to the TS). (According to Davenel, A. et al., *Lait*, 82, 465, 2002.)

For a ^{15}N-labeled protein of 150 kDa at 750 MHz and 20°C, one predicts residual linewidths of 10 Hz for ^{15}N and 45 Hz for $^{1}H^{N}$, and for the corresponding uniformly ^{15}N, ^{2}H-labeled protein the residual linewidths are predicted to be smaller than 5 and 15 Hz, respectively. The TROSY principle should benefit a variety of multidimensional solution NMR experiments, especially with future use of yet somewhat higher polarizing magnetic fields than are presently available, and thus largely eliminate one of the key factors that limit work with larger molecules (Pervushin et al. 1997).

Despite recent advances in tissue engineering to regenerate biofunction by combining cells with material supports, development is hindered by inadequate techniques for characterizing biomaterials in vivo (Mattiasson et al. 2009, Karfeld-Sulzer et al. 2011). Magnetic resonance imaging (MRI) has high temporal and spatial resolution and represents an excellent imaging modality for longitudinal noninvasive assessment of biomaterials in vivo. To distinguish biomaterials from surrounding tissues for MRI, protein polymer contrast agents (PPCAs) were developed and incorporated into hydrogels (Karfeld-Sulzer et al. 2011). In vitro and in vivo images of protein polymer hydrogels, with and without covalently incorporated PPCA, were acquired by MRI. The T_1 values of the labeled gels were consistently lower when PPCAs were included. Therefore, the PPCA hydrogels facilitated fate tracking, quantification of degradation, and detection of immune response in vivo. For the duration of the in vivo study, the PPCA-containing hydrogels could be distinguished from adjacent tissues and from the foreign body response surrounding the gels. The hydrogels containing PPCA have a contrast-to-noise ratio twofold greater than hydrogels without PPCA. In the absence of the PPCA, hydrogels cannot be distinguished by the end of the gel lifetime (Karfeld-Sulzer et al. 2011).

Zeeb et al. (2003) studied effects of different solvent viscosities analyzing the dynamics of the cold shock protein CspB from *Bacillus subtilis* from ^{15}N transverse and longitudinal relaxation rates and heteronuclear Overhauser effects. At a low relative viscosity of 2 (27% ethylene glycol, EG), the overall correlation time followed the linear Stokes–Einstein equation. At an increase in a relative viscosity to 6 (70% EG) the correlation time deviated from linearity by 30%, indicating that CspB tumbles at a higher rate as expected from the solvent viscosity probably due to a preferential binding of water molecules at the protein surface. The corresponding hydrodynamic radii, determined by NMR diffusion experiments, showed no variation with viscosity. The amplitudes of intramolecular motions on a sub-nanosecond timescale revealed by an extended Lipari–Szabo analysis were mainly independent of the solvent viscosity. The lower limit of the NMR "observation window" for the internal correlation time shifted above 0.5 ns at 70% EG, which is directly reflected in the experimentally derived internal correlation times. Chemical exchange contributions to the transverse relaxation rates derived from the Lipari–Szabo approach coincided with the experimentally determined values from the transverse ^{1}H–^{15}N dipolar/^{15}N CSA relaxation interference. These contributions originate from fast protein folding reactions on a millisecond timescale, which get retarded at increased solvent viscosities (Zeeb et al. 2003).

Babu et al. (2003) studied encapsulating proteins in reverse micelles and dissolving it in a low-viscosity solvent that can lower the rotational correlation time of the protein. They examined the applicability of several strategies for the preparation and characterization of encapsulated proteins dissolved in low-viscosity fluids that were suitable for high-performance NMR spectroscopy. Ubiquitin was used as a model system to explore various issues such as the homogeneity of the encapsulation, characterization of the hydrodynamic performance of reverse micelles containing protein molecules, and the effective pH of the water environment of the reverse micelle.

The structural analysis of protein–carbohydrate interactions is essential for the long-range aim to sort out entropic/enthalpic factors in the binding process (Siebert et al. 2003). They showed that it is possible to use exchangeable hydroxyl protons of carbohydrate ligands as conformational sensors to analyze their bound-state topology by measurements in dimethyl sulfoxide (DMSO)-d_6. However, the proteins are required to maintain binding capacity in the aprotic solvent. To define conditions to limit its harmful effect on sensitive protein structures while still being able to pick up solvent-exchangeable hydroxyl signals, binary solvent mixtures of DMSO and acetone (Ace) with water were studied. These solvent mixtures did not preclude to monitor hydroxyl protons of carbohydrate ligands even at temperatures well above 0°C. Hydrogen bonding of the two tested disaccharides (Galβ1–4Glcα/β and Galα1–3Galα/β or Galα1–3Galβ1–OCH_3), which are common lectin ligands, resembled the situation under physiological conditions. A refined topological description for hydroxyl positioning could be achieved for Galα1–3Gal. At least equally important, this approach worked to elucidate the mistletoe–lectin-bound topology of lactose in its *syn*-conformation with indication for formation of a characteristic inter-residual hydrogen bond. These measurements were performed in a binary DMSO-d_6:water mixture (6:4 ratio, v/v) at −12°C and encourage to pursue this line of investigation by monitoring in the course of stepwise temperature increases. The experiments revealed that binary mixtures have favorable properties for the conformational analysis of the free and bound-state topologies of bioactive ligands (Siebert et al. 2003).

Tompa et al. (2009) studied aqueous solutions and on lyophilized samples of human ubiquitin between −70°C and 45°C using wide-line ^{1}H-NMR and DSC methods. The measured properties

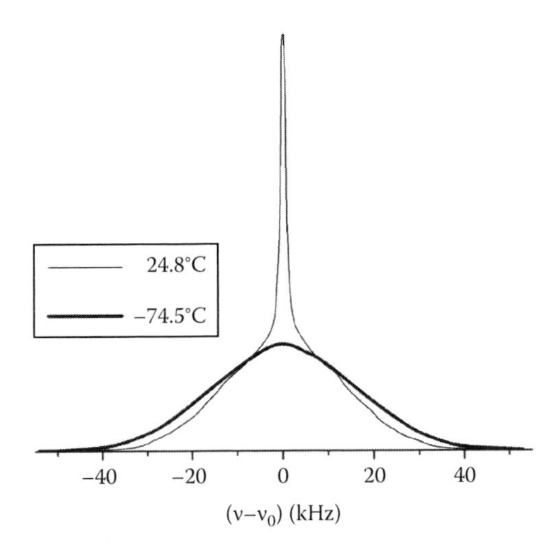

FIGURE 6.2 ^{1}H-NMR spectra for ubiquitin powder at different temperatures. (Adapted from *Biophys. J.*, 96, Tompa, K., Bánki, P., Bokor, M., Kamasa, P., Lasanda, G., and Tompa, P., Interfacial water at protein surfaces: wide-line NMR and DSC characterization of hydration in ubiquitin solutions, 2789–2798, 2009, Copyright 2009, with permission from Elsevier.)

(size, thermal evolution, and wide-line NMR spectra) of the protein–water interfaces were substantially different in the double-distilled and buffered water solutions of ubiquitin. The characteristic transition in water mobility was identified as the melting of the unfrozen hydrate water. Typical wide- and narrow-line signals are observed for the ubiquitin powder (Figure 6.2).

The amount of water in the low-temperature mobile fraction was 0.4 g/g protein for the pure water solution. The amount of mobile water was higher and its temperature dependence more pronounced for the buffered solution. The specific heat of the non-freezing/hydrate water was evaluated using combined DSC and NMR data. Considering the interfacial region as an independent phase, the values obtained were 5.0–5.8 J/gK that were higher than that of pure/bulk water (4.2 J/gK). This unexpected discrepancy can only be resolved in principle by assuming that hydrate water was in tight H-bond coupling with the protein matrix. The specific heat for the system composed of the protein molecule and its hydration water was 2.3 J/gK. It could be concluded that the protein ubiquitin and its hydrate layer behaved as a highly interconnected single phase in a thermodynamic sense. The actual size of the hydration shell of the protein molecule could be identified with the number of mobile water molecules for the 1 UBQ/water solution at $T<5$°C. It was found that the hydration shell around the ubiquitin molecule consists of $187 \pm 2H_2O$ (Tompa et al. 2009). This water can be assigned to SBW.

Lusceac et al. (2010) studied the temperature-dependent dynamics of water in the hydration shells of myoglobin, elastin, and collagen using ^{2}H NMR spin–lattice relaxation and lineshape analyses. The results showed that the dynamic behavior of the hydration water was similar for these proteins at comparable hydration degrees of $h=0.25$–0.43 g/g. Since water dynamics was characterized by strongly non-exponential correlation functions, they used a Cole–Cole spectral density for spin–lattice relaxation analysis, leading to correlation times, which were in agreement with results for the main dielectric relaxation process observed for various proteins. The temperature dependence can roughly be described by an Arrhenius law, with the possibility of a weak crossover in the vicinity of 220 K. Near ambient temperatures, the results substantially depended on the exact shape of the spectral density so that deviations from an Arrhenius behavior cannot be excluded in the high-temperature regime. However, for the studied proteins, the data gave no evidence for the existence of a sharp fragile-to-strong transition reported for lysozyme at about 220 K. Lineshape analysis revealed that the mechanism for the rotational motion of hydration waters changed in the vicinity of 220 K (notice that TSDC measurements showed the appearance of the dc relaxation at similar temperatures for hydrated proteins; Gun'ko et al. 2007i). For myoglobin, Lusceac et al. (2010) observed an isotropic motion at high temperatures and an anisotropic large-amplitude motion at low temperatures. Both mechanisms coexisted in the vicinity of 220 K. ^{13}C CP MAS spectra showed that hydration results in enhanced elastin dynamics at ambient temperatures, where the enhancement varies among different amino acids. Upon cooling, the enhanced mobility decreases. Comparison of ^{2}H and ^{13}C NMR data revealed that the observed protein dynamics was slower than the water dynamics.

The obtained results have also been confirmed by studies on water dynamics and water interaction in bovine serum albumin (BSA_{np}) suspension performed on W_{tot} ~300% swollen albumin NP by analysis of ^{1}H NMR relaxation curves and self-diffusion measurements that showed the presence of two well-separated relaxation rates (Bellotti et al. 2010). These rates have been accounted for a surface-limited relaxation regime of water molecules inside albumin NP (60 wt% of the total water content) and a diffusion-limited relaxation of water molecules into the meso-cavities between the packed NP (40 wt% of the total water content). At higher water content, the appearance of slowly relaxing components was assigned to free water molecules external to the NP, supernatant on the top of the sample. Therefore, the non-freezing and freezing water fractions up to $W_{tot}=300\%$, determined by DSC measurements, may be associated to the water molecules inside and between the packed NP, respectively. While, at higher water content free molecules external to the NP were also present (Bellotti et al. 2010).

Water mobility plays a crucial role in determining transport properties of small molecules in polymer matrices or in biosystems. In particular, in drug delivery systems, water state affects the pharmacokinetics, especially drug absorption, diffusion, and release. Martinelli et al. (2011) studied the state of water in an antibiotic-loaded composite consisting of albumin NP (BSA_{np}) dispersed into a carboxylated polyurethane (PEUA) and compared with that of the single drug-loaded components. The antibiotic cefamandole nafate was used as a model drug. DSC analysis, used to evaluate the freezing and non-freezing water fractions in the hydrated samples, showed that in BSA_{np} water can be bound both in the interparticles regions and inside the particles. With increasing total adsorbed water amount, the contribution of the freezing water fraction was higher than the non-freezing one. As for PEUA, the majority of water molecules absorbed is in a mobile freezing state (about 60% of the W_{tot}). As for the PEUA/BSA_{np} composite, the higher PEUA phase segregation induced by the NP as well as the higher non-freezing water fraction significantly enhanced drug uptake with respect to PEUA. Moreover, the greater non-freezing water fraction allowed the drug to penetrate within BSA NP and to give rise then to a controlled drug release. Indeed, the diffusion barrier exerted by NP and the matrix prolonged the antimicrobial activity from 4 to 9 days. Finally, the higher PEUA phase segregation also improved composite mechanical properties, as evidenced in stress–strain experiments and dynamic mechanical analysis (Martinelli et al. 2011). The amount of unfrozen water bound to BSA estimated as 170 wt% using DSC method was independent of the total water content.

Galán-Marín et al. (2010) studied the stabilization of soils with natural polymers and fibers to produce a composite, sustainable, nontoxic, and locally sourced building material. Mechanical tests have been conducted with a clay soil supplied by a Scottish brick manufacture. Alginate (a natural polymer from the cell walls of brown algae) was used as bonding in the composite. Sheep's wool was used as reinforcement. Tests done showed that the addition of alginate separately increases compression strength from 2.23 to 3.77 MPa and the addition of wool fiber increases compression strength 37%. The potential benefit of stabilization was found to depend on the combinations of both stabilizer and wool fiber. Adding alginate and reinforcing with wool fiber doubled the soil compression resistance. Better results were obtained with a lower quantity of wool. A review on the existing literature showed that most studies of natural fibers were focused on cellulose-based/vegetal fibers obtained from renewable plant resources. This is due to the fact that natural protein fibers have poor resistance to alkalis and cement is present nowadays in many building construction material. There are very few studies detailing composites made from protein fibers (animal hairs).

Barone and Schmidt (2005) reported on the use of keratin feather fiber as a short-fiber reinforcement in LDPE composites and showed that protein fibers have good resiliency and elastic recovery. Besides protein fibers have higher moisture regain and warmthness than natural cellulosic fibers properties, all related to its possible use in earth material. The keratin feather fiber for these tests was obtained from chicken feather waste generated by the U.S. poultry industry (Galán-Marín et al. 2010). Clearly, water can play a very important role in these materials that can be elucidated using NMR spectroscopy; however, this aspect was poorly studied.

Investigations of protein–surfactant interactions go back more than a century, and have been put to practical uses in everything from the estimation of protein molecular weights to efficient washing powder enzymes and products for personal hygiene (Otzen 2011). After a burst of activity in the late 1960s and early 1970s that established the general principles of how charged surfactants bind to and denature proteins, the field has kept a relatively low profile until the last decade. Within this period, there has been a maturation of techniques for more accurate and sophisticated analyses of protein–surfactant complexes such as DSC, NMR, SAXS, and some other techniques. Otzen (2011) analyzed different useful approaches to study these complexes and identify issues which define central concepts in the field: (i) Are proteins denatured by monomeric surfactant molecules, micelles, or both? (ii) How does unfolding of proteins in surfactant compare with "proper" unfolding in chemical denaturants? Recent work has highlighted the role of shared micelles, rather than monomers, below the critical micelle concentration (CMC) in promoting both protein denaturation and formation of higher order structures. Kinetic studies have extended the experimentally accessible range of surfactant concentrations to far above the CMC, revealing numerous different modes of denaturation by ionic surfactants below and above the CMC which reflected micellar properties as much as protein unfolding pathways. Uncharged surfactants follow a completely different denaturation strategy involving synergy between monomers and micelles. The high affinity of charged surfactants for proteins (charged too) means that unfolding pathways are generally different in surfactants versus chemical denaturants, although there are common traits. Other issues are as follows: (iii) Are there non-denaturing roles for sodium dodecyl sulfate (SDS)? (iv) How reversible is unfolding in SDS? (v) How do solvent conditions affect the way in which surfactants denature proteins? The last three issues compare SDS with "proper" membranes. (vi) Do anionic surfactants such as SDS mimic biomembranes? (vii) How do mixed micelles interact with globular proteins? (viii) How can mixed micelles be used to measure the stability of membrane proteins? The growing efforts to understand the unique features of membrane proteins have encouraged the development of mixed micelles to study the equilibria and kinetics of this class of proteins, and traits which unite globular and membrane proteins have also emerged. These issues emphasize the amazing power of surfactants to both extend the protein conformational landscape and at the same time provide convenient and reversible shortcuts between the native and denatured state for otherwise obdurate membrane proteins (Otzen 2011). It is possible that transformation SAW $\leftrightarrow$ WAW and WAW itself can play a specific role during interactions of surfactants with proteins. This changes the protein conformation with hidden hydrophobic functionalities in the aqueous medium toward unfolding and binding surfactant molecules by these hydrophobic structures that result in protein denaturation. Clear, a portion of water (SAW) bound to protein can be more strongly clustered (WAW) during the mentioned processes.

Malmendal et al. (2010) recently developed an NMR-based approach to directly compare the unfolding pathways of a protein under different denaturing conditions, namely Global Protein Folded State (GPS) NMR. ^{1}H-liquid state NMR spectra of the protein during titration with different denaturing agents (e.g., surfactants, chemical denaturants, organic solvents, extreme pH, temperature) were analyzed by Principal Component Analysis to identify principal components that can be used to describe the variation in these spectra. By mapping out how the values of the major two or three principal components vary with each other, they obtained a trajectory of protein unfolding in two- or three-dimensional space, where each linear segment represents one transition.

To deeper insight into the functions of proteins and their specific roles, it is important to establish efficient procedures for exploring the states that encapsulate their conformational space (Malmendal et al. 2010). GPS mapping by multivariate NMR (GPS NMR) is a powerful high-throughput method that provides such an overview. GPS NMR exploits the unique ability of NMR to simultaneously record signals from different H atoms in complex macromolecular systems and of multivariate analysis to describe spectral variations from these by a few variables for establishment of, and positioning in, protein folding state maps. The method is fast, sensitive, and robust, and it works without isotope labeling. The unique capabilities of GPS NMR to identify different folding states and to compare different unfolding processes were demonstrated by mapping of the equilibrium folding space of bovine α-lactalbumin in the presence of the anionic surfactant SDS and compared these with other surfactants, acid, denaturants, and heat effects (Malmendal et al. 2010).

NMR spectroscopy is useful to probe conformational changes through modification of dipolar interactions between distinct parts of the protein (Read and Burkett 2003, Lu et al. 2004a). Localized conformational information has been obtained for some adsorbed peptides by using solid-state NMR (Billsten et al. 1995). A recent study by Calzolai et al. (2010) reports that NMR measurements, chemical shift perturbation analysis, and dynamic light scattering (DLS) allowed to identify the interaction site of ubiquitin with gold NPs at amino acid scale in solution.

Magnetization transfer (MT) or cross-relaxation spectroscopy, also termed Z-spectroscopy by Grad and Bryant (1990), has been extensively used to obtain information on the spectral lineshape of protons in macromolecules as well as on relaxation in heterogeneous systems (Calucci and Forte 2009). A MT spectrum was obtained by performing saturation transfer (ST) experiments at many discrete frequencies and plotting the ratio of the steady-state water signal with and without RF saturation M_{Wz}/M_{W0} as a function of the saturation frequency offset. For spatially homogeneous samples, a broadband pulse sequence was proposed by Swanson allowing the entire MT spectrum to be acquired with only two acquisitions (i.e., with and without saturation) (Swanson 1991). In Figure 6.3, representative MT spectra are shown; the relatively narrow dip near offset ~0 Hz reflects the direct saturation of the water resonance, whereas the broad shape of the spectrum arises from the indirect water–macromolecular MT and is also a measure of the lineshape of the macromolecular proton population to which the water protons are coupled. The depth of the MT spectrum is a function of the magnitude of water–macromolecular spin coupling, relative proton populations, and relaxation parameters for each population, as well as of the RF irradiation power. As shown in the following,

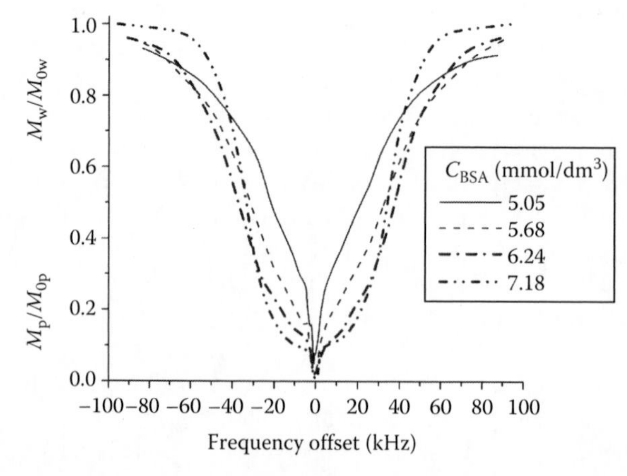

FIGURE 6.3 Magnetization transfer (MT) spectra of water (w) protons in BSA (p) solutions at different concentrations: 5.05, 5.68, 6.24, and 7.18 mmol of BSA per dm^3 of solution. (Iino, M.: Transition from Lorentzian to Gaussian lineshape of magnetization transfer spectrum in bovine serum albumin solutions. *Magn. Reson. Med.* 1994. 32. 459–463. Copyright 1994 Wiley-VCH Verlag GmbH & Co. KGaA. Adapted with permission, and according to Calucci, L. and Forte, C., *Prog. Nuclear Magn. Reson. Spectr.*, 55, 296, 2009.)

if the rate constant for MT between the two phases dominates the intrinsic longitudinal relaxation rate for water protons, then the MT spectrum faithfully reflects that of the solid-like phase (Calucci and Forte 2009).

Samouillan et al. (2011) studied the dielectric properties of elastin at different degrees of hydration and specifically at the limit of freezable water apparition. The quantification of freezable water was performed by DSC. Two dielectric techniques were used to explore the dipolar relaxations of hydrated elastin: dynamic dielectric spectroscopy (DDS), performed isothermally with the frequency varying from 10^{-2} to 3×10^{6} Hz, and the TSDC technique, an isochronal spectrometry running at variable temperature, analogous to a low-frequency spectroscopy (10^{-3} to 10^{-2} Hz). A complex relaxation map was evidenced by the two techniques. Assignments for the different processes can be proposed by the combination of DDS and TSDC experiments and the determination of the activation parameters of the relaxation times. As already observed for globular proteins, the concept of "solvent-slaved" protein motions was checked for the fibrillar hydrated elastin (Samouillan et al. 2011).

The use of intermolecular water–peptide nuclear Overhauser effects (NOEs) for the detection of solvent exposure was already proposed in 1974 (Otting 1997). With improved equipment, it is possible to obtain a much more complete picture of the hydration of biomolecules in aqueous solution using the NOE in high-resolution NMR spectroscopy to detect and localize the water molecules hydrating proteins, DNA, RNA, and other biomacromolecules (Otting 1997, Wider 1998). Intermolecular NOEs observed between the water and the solute protons allow the identification of individual hydration water molecules in the presence of a very large excess of bulk water appearing at the same chemical shift as the signals from the hydration water. This is possible because NOEs are effectively observed only for internuclear distances shorter than 0.4–0.5 nm. NOEs observed between the single, averaged water resonance and the solute thus report on direct interactions between the solute and the water molecules in the first shell of hydration. NOEs can be observed by the transfer of either longitudinal or transverse magnetization between spins. The latter is also referred to as ROE (rotating frame NOE). The distinction between NOE and ROE is made only in the terms σ_{NOE} and σ_{ROE} which describe the rates of MT between two protons by cross-relaxation in the respective frames of reference (Otting 1997).

Protein–water dynamics in mixtures of water and a globular protein, BSA, was studied over wide ranges of composition, in the form of solutions or hydrated solid pellets, using DSC, TSDC, and DRS (Panagopoulou et al. 2011). Additionally, water equilibrium sorption isotherm (ESI) measurements were performed at room temperature. The crystallization and melting events were studied by DSC and the amount of uncrystallized water was calculated by the enthalpy of melting during heating. The glass transition of the system was detected by DSC for water contents higher than the critical water content corresponding to the formation of the first sorption layer of water molecules directly bound to primary hydration sites, $h = 0.073$ g/g (in grams of water per gram of dry protein), estimated by ESI. A strong plasticization influencing the T_g value was observed by DSC for hydration degrees lower than those necessary for crystallization of water during cooling, i.e., lower than $h \approx 0.3$ g/g followed by a stabilization of T_g at ca. −80°C for higher water contents. The α relaxation associated with the glass transition was also observed using dielectric measurements. In TSDC, a microphase separation could be detected resulting in double T_g for some hydration degrees. A dielectric relaxation of small polar groups of the protein plasticized by water, overlapped by relaxations of uncrystallized water molecules, and a separate relaxation of water in the crystallized water phase (bulk ice crystals) were also observed (Panagopoulou et al. 2011).

DRS measurements provide the possibility to vary both frequency and temperature over wide ranges (10^{-1} to 10^{6} Hz and −150°C to 20°C, respectively) and so to analyze dynamics in more details than by TSDC (Panagopoulou et al. 2011). Isochronal (constant frequency) presentation of the DRS data is very convenient (Kremer and Schönhals 2002). The equivalent frequency of TSDC measurements is in the range from 10^{-4} to 10^{-2} Hz and therefore the lower equivalent frequency of TSDC shifts the peaks toward lower temperatures (Figure 6.4).

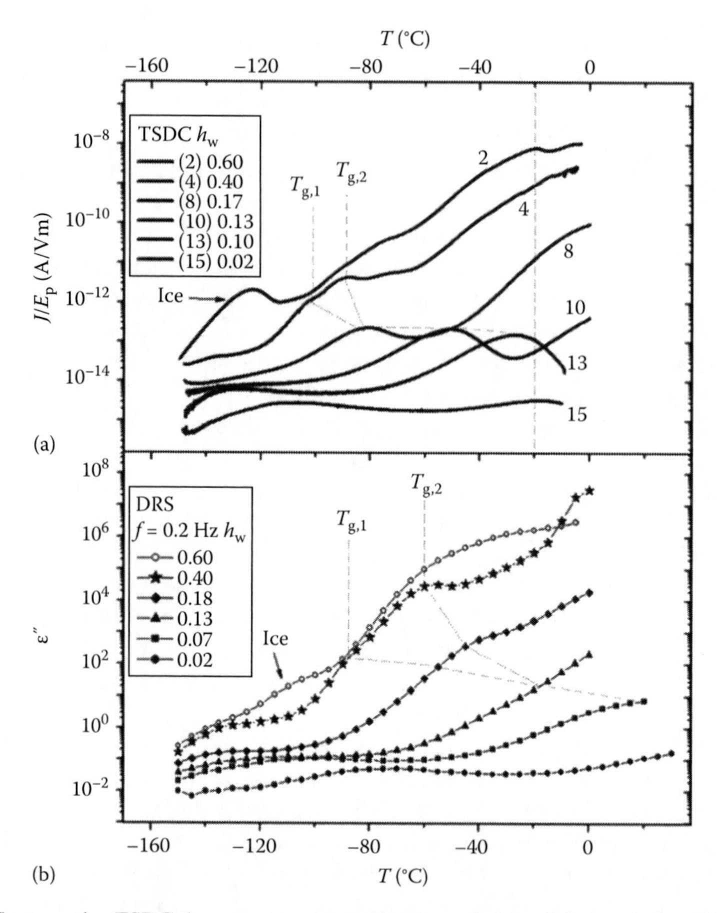

FIGURE 6.4 Comparative TSDC thermograms (a) and isochronal plots of dielectric loss (b), for compressed BSA pellets and solutions, at the water fractions h_w indicated on the plot: (a) normalized TSDC thermograms (density of depolarization current divided by polarizing electric field, J/E_p, versus temperature T) for the BSA–water solutions (samples 2 and 4) and the hydrated solid pellets (samples 8, 10, 13, and 15), of water fractions h_w. (b) Isochronal plots of dielectric loss (DRS) at the frequency of 0.2 Hz for the BSA–water samples of water fractions h_w indicated on the plot. (Adapted from *Biochim. Biophys. Acta. Protein. Proteom.*, 1814, Panagopoulou, A., Kyritsis, A., Sabater i Serra, R., Gómez Ribelles, J.L., Shinyashiki, N., and Pissis, P., Glass transition and dynamics in BSA–water mixtures over wide ranges of composition studied by thermal and dielectric techniques, 1984–1996, 2011, Copyright 2011, with permission from Elsevier.)

Starting with the sample at $h_w = 0.02$ g/g, the broad relaxation is attributed to a relaxation of side polar groups of the protein plasticized by water (TSDC, Figure 6.4a) and is observed by DRS at approximately −75°C (Figure 6.4b). For the next three solid pellets, in terms of increasing water content, namely DRS samples at $h_w = 0.07$, 0.13, and 0.18 g/g, the dielectric dispersions consist of two contributions in quite good accordance to TSDC data (Figure 6.4). At low temperatures, the relaxation of small polar side functionalities of the protein plasticized by water (at ca. −100°C for the sample at $h_w = 0.07$ g/g) was observed. At higher temperatures, the α peak was observed, although masked by conductivity at the high-temperature side, plasticized by water. At this point it is obvious that conductivity contributions in TSDC may be suppressed, presumably because the steps of applying the stimulus and of recording the response, i.e., polarization and depolarization steps are separated from each other. For more strongly hydrated sample at $h_w = 0.4$ g/g, three relaxation processes are observed instead of two in case of solid pellets. These are, with respect to the TSDC thermogram at the same water content, the relaxation of small polar groups that is plasticized by water at approximately −130°C and two α relaxations associated to the glass transition, which are denoted as $T_{g,1}$ at approximately −90°C and $T_{g,2}$ at ca. −60°C (Figure 6.4). Finally, regarding the sample at the highest water content in this diagram, $h_w = 0.6$ g/g, all three relaxations mentioned for the sample at $h_w = 0.4$ g/g were observed, though more vaguely for temperatures higher than approximately −70°C, due to high conductivity contributions, plus an additional peak at approximately −110°C due to bulk ice crystals. This good agreement between the data by DRS and TSDC suggested that the same mechanisms are at the origin of the peaks measured by the two techniques (Panagopoulou et al. 2011).

Solid-state NMR dipolar coupling measurements were used to obtain high-resolution structures for proteins in microcrystalline form, for proteins oriented in lipid bilayers and lipid vesicles and for proteins in fibrillar form (Goobes et al. 2007). They showed that the same solid-state dipolar recoupling techniques can be used to obtain the structure of proteins at biomineral surfaces, e.g., HAP (Figure 6.5). Solid-state NMR structural techniques can be extended to a broad class of biomaterial structural problems intractable to traditional structural biology techniques. The results of the NMR studies of hydroxyapatite (HAP)-bound statherin have numerous biological implications (Goobes et al. 2007).

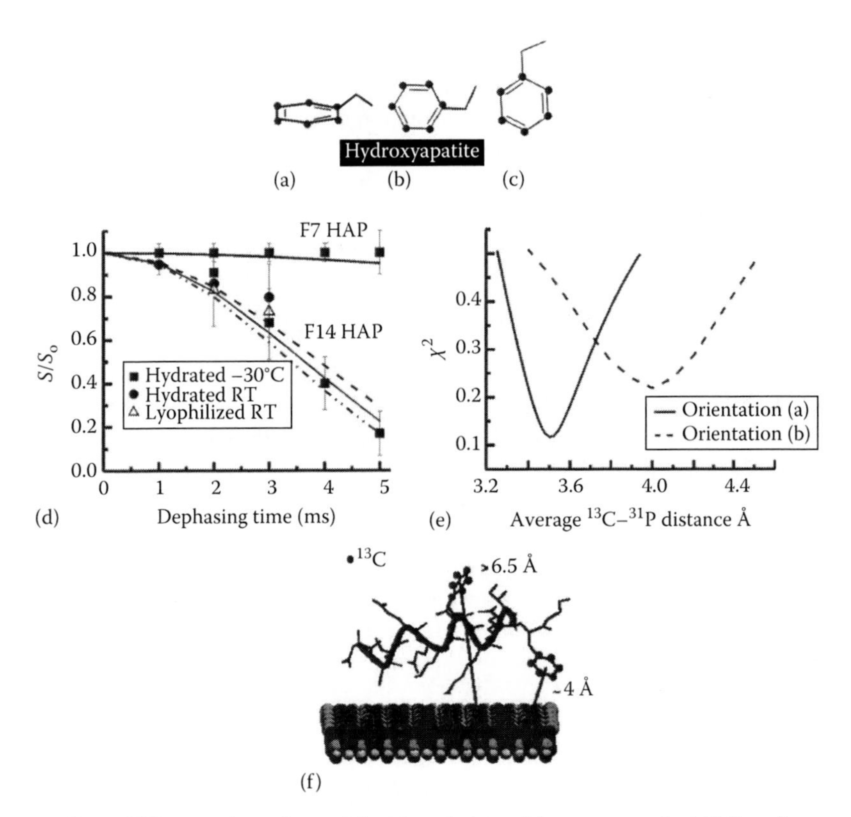

FIGURE 6.5 Three different orientations of the phenyl ring with respect to the HAP surface were considered: (a) the ring plane is parallel to the HAP surface, (b) a line perpendicular to the C_2–C_3 and C_5–C_6 bonds is also perpendicular to the HAP surface, and (c) a line from C_1 to C_4 is perpendicular to the HAP surface. (d) rotational-echo double resonance, REDOR analysis of the both samples bound to HAP is shown. The experimental data and best simulated fits are shown for both samples. (e) A χ^2 map is shown for the F14 (the side chain of the phenylalanine nearest to the C-terminus of the peptide) sample based on the orientations (a) and (b) depicted in (a). (f) A modified SN-15 (N-terminal 15 amino acids of salivary statherin) model with peptide bound to HAP is presented, in which the C-terminus loses its α-helicity, and both the K6 and F14 side chains are within close proximity to the HAP surface. (Adapted from *Prog. Nuclear Magn. Reson. Spectr.*, 50, Goobes, G., Stayton, P.S., and Drobny, G.P., Solid state NMR studies of molecular recognition at protein–mineral interfaces, 71–85, 2007, Copyright 2007, with permission from Elsevier.)

Statherin mediates the binding of several bacterial fimbriae via its C-terminal region only when it is adsorbed onto HAP. This recognition site is being exposed during the adsorption of statherin to the mineral surface. The α-helical conformation in the bacterial recognition site, shown by Goobes et al. (2007) to form upon adsorption, provides a clear avenue for the differential recognition of immobilized versus free statherin by bacteria. NMR measurements of statherin in 50% TFE solution report a 3_{10} helix in a C-terminus region that slightly overlaps with the region where the solid-state NMR measurements were taken. The C-terminus was helical on the surface, but the ^{13}C–^{15}N-REDOR measurements rule out the presence of a 3_{10} helix between residue 34 and residue 38. The C-terminus of statherin has been shown to be key to the lubricative action of statherin on tooth enamel, and the connection between helical secondary structure and viscoelastic function has been made with human serum albumin (HSA). The folding of statherin's C-terminus into an α-helical conformation thus serves as a structural context to understand both fimbrillin binding and consequently bacterial adhesion, as well as the important lubricative properties of bound statherin. The lubricative action of statherin in concert with other proteins in the pellicle should be greatly influenced by the structure the protein assumes when adsorbed. Particularly, if a cooperative mechanism of friction reduction exists, recognition between different proteins requires that statherin adopts a well-determined structure in the pellicle. In general, statherin is a paradigmatic system whose structural transition upon adsorption to mineral surfaces may explain the structural basis of the biological function of the protein complement of the extracellular matrix (ECM). Many ECM proteins like osteopontin when studied in solution have been found to be unstructured. In the case of statherin, the folding of the protein on the mineral surface may remove hydrophilic residues from solution, thus preventing further nucleation. In other words, the tertiary folding of statherin is the basis for its inhibition of HAP secondary crystallization, and this folding is naturally induced by contact between the HAP surface and the protein. Thus a protein that is unstructured in solution may nevertheless be functional on a surface where a folding transition occurs. This remarkable feature of statherin reverses the common paradigm observed in many proteins which unfold when surfaces, especially hydrophobic surfaces, are contacted (Goobes et al. 2007).

Nielsen et al. (2012) described a new and highly accurate, precise, and robust formulation for the prediction of NMR chemical shifts from protein structures using an approach, shAIC (shift prediction guided by Akaikes Information Criterion), which capitalizes on mathematical ideas and an information-theoretic principle, to represent the functional form of the relationship between structure and chemical shift as a parsimonious sum of smooth analytical potentials, which optimally takes into account short-, medium-, and long-range parameters in a nuclei-specific manner to capture potential chemical shift perturbations caused by distant nuclei. shAIC outperforms the state-of-the-art methods that use analytical formulations. Moreover, for structures derived by NMR or structures with novel folds, shAIC delivers better overall results (Figure 6.6), even when it is compared to sophisticated machine learning approaches. shAIC provides for a computationally lightweight implementation that is unimpeded by molecular size, making it an ideal for use as a force field (Nielsen et al. 2012).

Kitevski-LeBlanc and Prosser (2012) analyzed applications of ^{19}F NMR to study protein structure and dynamics since it provides a unique perspective of conformation, topology, and dynamics and the changes that ensue under biological conditions. In particular, ^{19}F NMR has provided insight into biologically significant events such as protein folding and unfolding (Bann et al. 2002, Boulègue et al. 2006, Li and Frieden 2007), enzymatic action (Thorson et al. 1998, Lee et al. 2000, Rozovsky et al. 2001), protein–protein (Quint et al. 2006), protein–lipid (Bouchard et al. 1998, Anderluh et al. 2005), or protein–ligand (Yu et al. 2006, Papeo et al. 2007) interactions as well as aggregation and fibrillation (Frieden 2007, Li et al. 2009a) in both soluble and membrane protein systems (Klein-Seetharaman et al. 1999, Oxenoid et al. 2002, Elvington et al. 2009, Kitevski-LeBlanc and Prosser 2012).

^{19}F NMR spectroscopy can answer specific questions which address structure and function as exemplified in Figure 6.7, where ^{19}F NMR shifts of the GPCR, β_2AR, in which Cys-265 has been

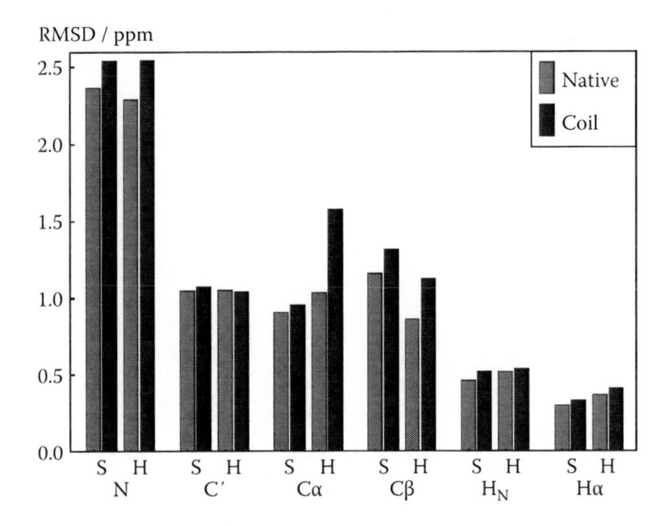

FIGURE 6.6 Root-mean-square deviations (RMSDs) between observed and predicted chemical shift for native and misclassified secondary structures. The bars show the RMSD in the evaluation set of 39 protein chains for the residues at the end of the helixes (H) and sheets (S) using the correctly assigned secondary structure (gray) and using a state misclassified to a coil residue (black), hence using the native and the coil shAIC parameters, respectively, to calculate the predicted shift. (Adapted from *Prog. Nuclear Magn. Reson. Spectr.*, 60, Nielsen, J.T., Eghbalnia, H.R., and Nielsen, N.Chr., Chemical shift prediction for protein structure calculation and quality assessment using an optimally parameterized force field, *Prog. Nuclear Magn. Reson. Spectr.*, 60, 1–28, 2012, Copyright 2012, with permission from Elsevier.)

tagged with BFTA, reveal an up-field shift as a function of activation (Kitevski-LeBlanc and Prosser 2012). β_2AR was currently prepared from an insect cell expression system, meaning that traditional labeling is prohibitive, while an NMR sample consisting of 50–100 μM (unlabeled) GPCR can be prepared and purified from 1 to 2 L. Thus 1D ^{19}F NMR is ideal, since relatively small amounts of protein are needed to acquire data. Addition of the inverse agonist, carazolol, is seen to shift the equilibrium toward the inactive state (less conformationally mobile) resulting in a more down-field resonance. Experiments were ongoing to improve labeling efficiency and attain information on millisecond dynamics (Kitevski-LeBlanc and Prosser 2012).

Disordered proteins pose a methodological challenge as they cannot be described by a single structure, but occupy an ensemble of interconverting conformations (Kjaergaard and Poulsen 2012). NMR spectroscopy is the only technique capable of describing these proteins at atomic resolution, which explains the large interest in developing NMR methods suitable for disordered proteins. The disordered proteins studied by NMR can roughly be divided into two groups depending on whether or not the protein is disordered in the native state. The proteins that are disordered in their native state are called intrinsically disordered proteins (IDPs) (Wright and Dyson 1999). The other group of disordered proteins contains proteins that are normally folded, but are denatured due to the external conditions (Kjaergaard and Poulsen 2012). One can assume that adsorbed proteins studied are in a partially disordered state.

HSA in native state (as well as a monomer form) is stabilized by bound water and certain amounts of salts. One can assume that changes in the dispersion media can affect the state and the distribution of water bound by HSA molecules. Water bound by HSA molecules ($h<0.2$ g/g) is characterized by two ^{1}H NMR signals (static samples) at $\delta_H=3$–5 (signal 1) and 1.1–1.5 (signal 2) ppm in the presence of different nonpolar or weakly polar solvents (Figure 6.8). The up-field signal at low δ_H values is predominant with lowering temperature, especially in the case of the chloroform medium. In the absence of these solvents, only typical signal of water (SAW) at $\delta_H=3$–5 ppm is observed. Acetonitrile as a more polar solvent (permittivity $\varepsilon=35.7$) than chloroform ($\varepsilon=4.7$) and benzene

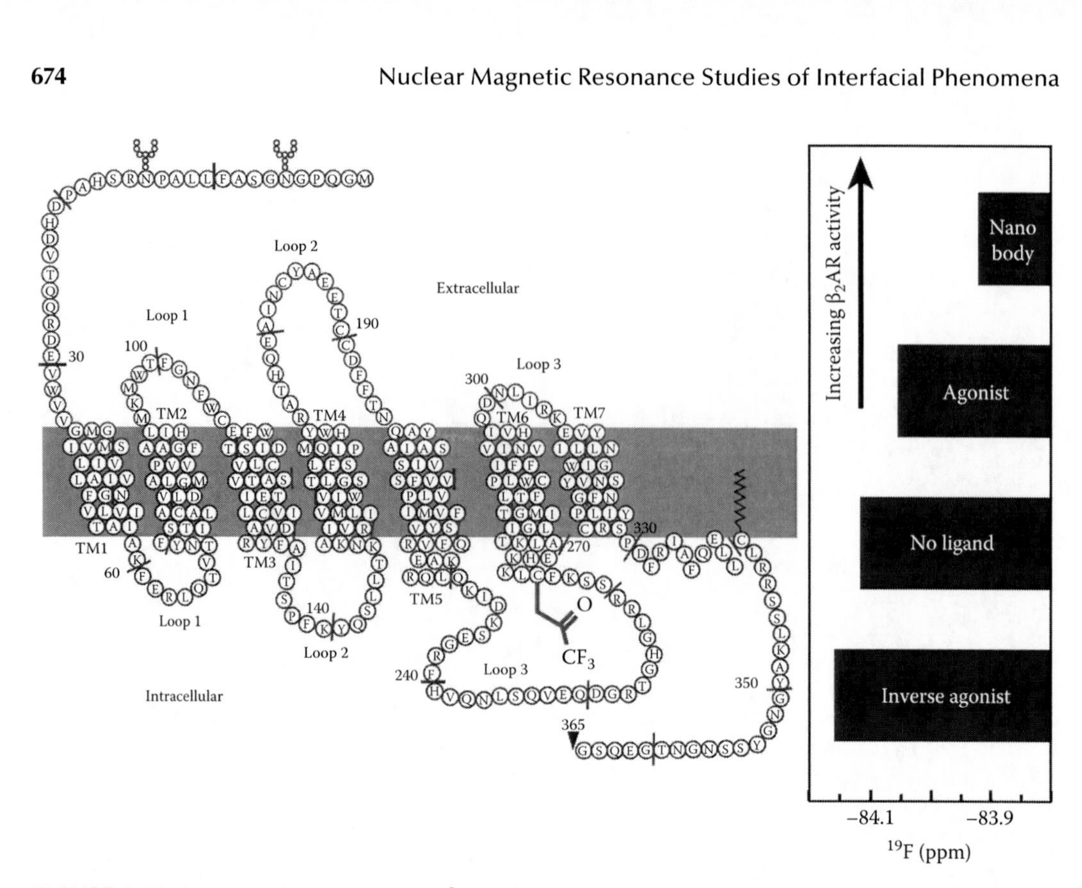

FIGURE 6.7 A schematic of the GPCR, β_2AR, illustrating the transmembrane residues and the location of the trifluoromethyl probe at C265. ^{19}F NMR chemical shift of CF_3-labeled β_2AR at 25°C, as a function of increasing activity from bottom to top, resulting from the addition of inverse agonist (Cz), no ligand (NL), agonist (Bi), or agonist plus activating nanobody (NB). Spectra were acquired at 564.3 MHz on an Agilent HCN/FCN cryogenic NMR probe, with 4000 transients (2 h), using a 90°-τ–180°-τ-acq pulse sequence with a 15 μs pulse length and a pulse spacing of τ=55 μs. Samples consisted of ~25% labeled 40 μM protein in a 275 μL volume (5 mm diameter Shigemi tube). (Adapted from *Prog. Nuclear Magn. Reson. Spectr.*, 62, Kitevski-LeBlanc, J.L. and Prosser, R.S., Current applications of ^{19}F NMR to studies of protein structure and dynamics, 1–33, 2012, Copyright 2012, with permission from Elsevier.)

(ε=2.3) demonstrates smaller effects because the intensity of up-field signal 2 is lower than that of down-field signal 1 even at T=220 K (Figure 6.8c).

HSA in aqueous medium can strongly bind about 400 mg of water per gram of protein (Figure 6.8d). Therefore, at a low hydration degree of the protein in air dispersion medium all water is SBW (Figure 6.8). The effects of addition of solvents of different polarities depend on their amounts (Figures 6.8 and 6.9 and Table 6.1). In the case of small amounts of solvents (benzene, acetonitrile), they displace water mainly into narrow nanopores (Table 6.1, S_{nano} and V_{nano} increase). However, chloroform (even at a low amount) displaces water into mesopores. Subsequent increase in the solvent amounts leads to displacement of water into larger pores (S, S_{nano}, and V_{nano} decrease). These effects can be explained in the case of weakly polar chloroform and benzene (immiscible with bulk water) by the tendency of diminution of the contact area between water and organic solvents.

In the case of polar acetonitrile, this effect is absent since the S, S_{nano}, and V_{nano} values (as well as other characteristics) depend weakly on the CD_3CN amount (Table 6.1). This is also seen from small changes in the pore size distributions (PSDs) (Figure 6.9g and h). Maximal changes in the PSDs are observed in suspensions of hydrated albumin in benzene (Figure 6.9e and f) and chloroform (Figure 6.9c and d) because water must contact with hydrophobic solvents but it tends to reduce the area of these contacts. Therefore, larger water structures form because their outer surface in

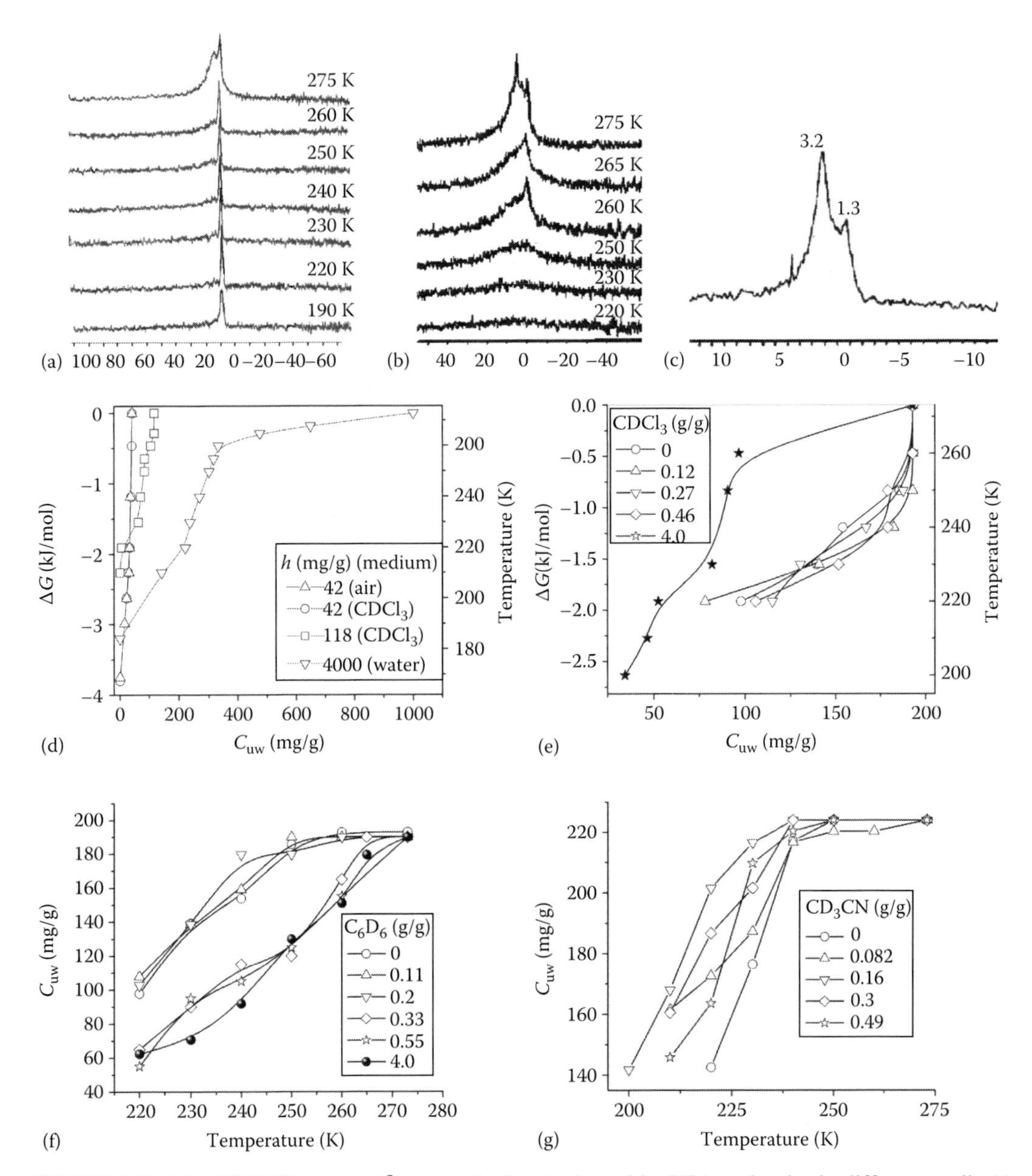

FIGURE 6.8 The ^{1}H NMR spectra (δ_H in ppm) of water bound by HSA molecules in different media (a) chloroform-*d*; (b) benzene-d_6, and (c) acetonitrile-d_3 (220 K) at h=(a) 193, (b) 190, and (c) 224 mg/g. (d–g) Content of unfrozen water bound by HSA as a function of temperature with addition of different solvents (d and e) chloroform-*d*, (f) benzene-d_6, and (g) acetonitrile-d_3; and (d and e) relationship between differential Gibbs free energy of unfrozen adsorbed water versus its content in the chloroform-*d* medium at h=(d) 42, 118, 4000, (e) 193, (f) 190, and (g) 224 mg/g.

contact with the organic solvents is minimal. However, complete displacement of water from polar structures of the albumin does not occur. The interfacial free energy of the system depends on the polarity of the solvents (Table 6.1) and minimal γ_S values are observed for weakly polar chloroform and benzene and it is maximal in the aqueous suspension. These solvents cause two effects: (i) a portion of bound water is displaced from the HSA molecules to form relatively large domains of SAW frozen at higher temperatures (therefore, the signal 1 is observed at temperatures close to 273 K)

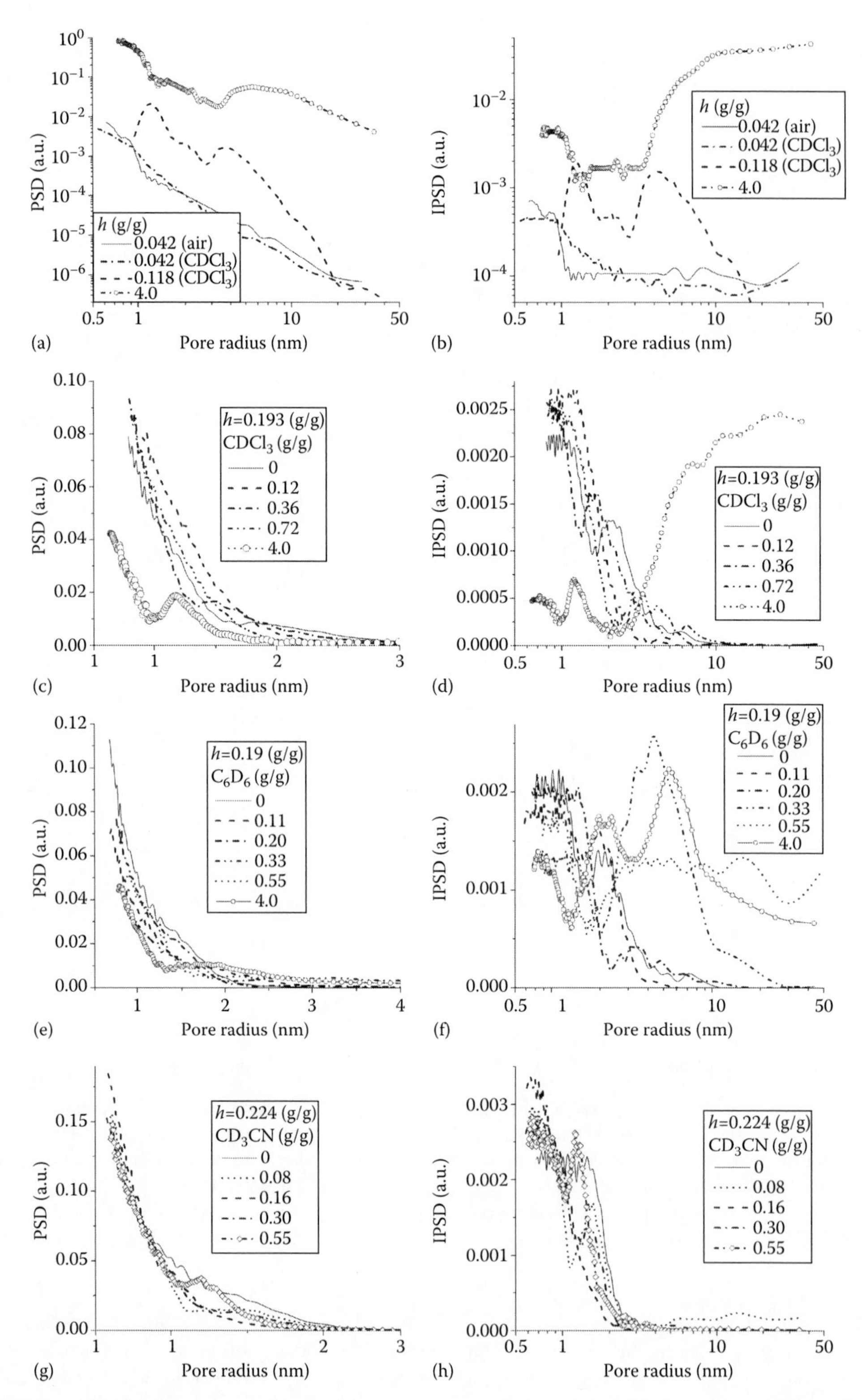

FIGURE 6.9 Distribution functions of HSA "pores" filled by unfrozen water as functions of the amounts of solvents: (a–d) chloroform-d, (e, f) benzene-d_6, and (g, h) acetonitrile-d_3 at h = (a, b) 42, 118, 4000, (c, d) 193, (e, f) 224, and (g, h) 190 mg/g; (a, c, e, g) differential and (b, d, f, h) incremental PSD.

TABLE 6.1
Characteristics of Water Bound to HSA in Different Media (k_{GT} = 60 K nm)

Medium	C_{sol} (g/g)	h (g/g)	S (m²/g)	S_{nano} (m²/g)	S_{meso} (m²/g)	V_{nano} (cm³/g)	V_{meso} (cm³/g)	R_{av} (nm)	γ_S (J/g)	$\langle T \rangle$ (K)	$-\Delta G_s$ (kJ/mol)
Air	—	0.042	82	81	1	0.032	0.010	20.0	5.7	203.7	3.35
Air	—	0.193	190	147	43	0.066	0.127	6.20	20.2	222.2	2.72
Water	—	4.0	428	368	57	0.161	0.728	26.37	44.4	249.3	2.85
$CDCl_3$	4.0	0.042	79	78	1	0.030	0.012	16.64	6.0	200.5	3.70
$CDCl_3$	4.0	0.118	33	3	30	0.001	0.117	9.02	8.8	237.1	2.31
$CDCl_3$	0.12	0.193	113	75	38	0.036	0.157	9.04	19.7	223.9	2.41
$CDCl_3$	0.36	0.193	203	166	37	0.075	0.118	6.55	20.7	220.8	2.70
$CDCl_3$	0.72	0.193	188	151	36	0.069	0.124	6.99	20.9	220.5	2.66
$CDCl_3$	4.0	0.193	132	119	13	0.048	0.132	21.92	13.2	237.7	3.27
C_6D_6	0.0	0.19	193	151	42	0.067	0.125	5.89	20.3	222.1	2.75
C_6D_6	0.11	0.19	246	200	45	0.084	0.106	4.62	21.8	216.9	3.09
C_6D_6	0.20	0.19	225	169	56	0.074	0.116	3.71	20.9	219.4	2.82
C_6D_6	0.33	0.19	147	105	43	0.046	0.144	6.68	15.8	233.1	2.81
C_6D_6	0.55	0.19	72	59	13	0.028	0.158	26.74	14.4	235.8	2.47
C_6D_6	4.0	0.19	111	95	15	0.043	0.146	20.34	14.5	236.0	2.75
CD_3CN	0	0.224	330	272	59	0.114	0.110	3.43	27.3	213.3	3.13
CD_3CN	0.08	0.224	413	405	8	0.159	0.064	15.6	31.1	203.1	3.43
CD_3CN	0.16	0.224	476	457	19	0.177	0.047	2.94	33.7	196.8	3.48
CD_3CN	0.30	0.224	442	420	22	0.164	0.060	3.54	32.5	200.0	3.54
CD_3CN	0.55	0.224	398	370	28	0.147	0.077	3.94	31.0	204.0	3.41

than more strongly bound water and (ii) another portion of water is distributed in the form of smaller clusters of WAW (characterized by the signal 2) than those formed without these solvents. Protein preparations can be components of most media used for cryopreservation of cells and tissue cultures (Tsuraeva 1983). It was experimentally established that the presence in the medium of 0.2%–1.0% protein substance substantially raises the survival rate of the preserved cells (Fedoteikov et al. 1963). However, in practice, the mechanism of the protective action of protein molecules needs additional investigations. A promising method to study the behavior of solutes in complex solutions containing biopolymers and cell cultures is low-temperature ^{1}H NMR spectroscopy (Bizunok et al. 1984, Gun'ko et al. 2005d, 2009d). The influence of a protein substance on the temperature of formation of eutectics in aqueous solutions of simple organic molecules differing in their ability to take part in the formation of hydrogen-bound associates was studied using HSA (Turov et al. 1994b).

One percent aqueous solution of HSA without addition of buffer was investigated. The concentration of the solutes did not exceed 0.5 vol%. The measurements were made with a Bruker WP-100 SY high-resolution NMR spectrometer in the temperature range of 240–270 K. Substances soluble in water at all concentrations and containing groups well identified in the ^{1}H NMR spectra were used including proton-donor acetic acid (AA) and 2-ethoxyethanol (ME) and electron-donor dimethylformamide (DMFA) and Ace.

The typical dependences of the form of the ^{1}H NMR spectra on temperature in conditions of freezing are presented in Figure 6.10a for the system H_2O+HSA+DMFA. Three widened signals are observed in the spectra. The signal with the chemical shift δ_H=5 ppm corresponds to the protons of the water molecules (SAW) and the signals with the chemical shifts δ_H=3 ppm and δ_H=8 ppm correspond to the CH_3 and CH groups of DMFA. With a fall in temperature, the intensity of the water signal decreases through freezing-out. The signal from the ice protons and HSA does not show up in the high-resolution spectra in view of extremely short transverse relaxation time.

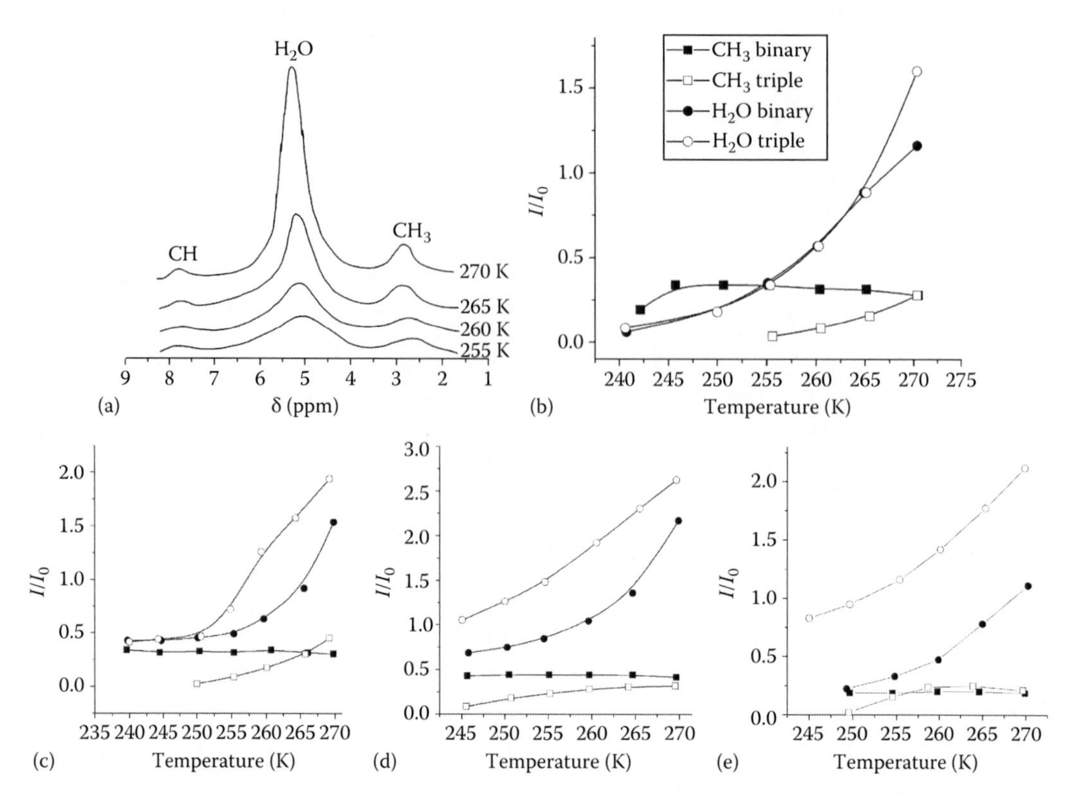

FIGURE 6.10 (a) Influence of temperature on ^1H NMR spectra of DMFA in frozen aqueous solution; and temperature dependence of intensity of ^1H NMR signal of water molecules and alkyl groups in frozen aqueous solutions: (b) CH_3COOH; (c) $CH_3O(CH_2)_2OH$; (d) $(CH_3)_2CHNO$; and (e) $(CH_3)_2CO$ (solid and open symbols correspond to binary and triple systems, respectively).

By measuring the intensities of the signals of unfrozen water and the solute, one may measure the concentration of the solution in the liquid phase. This concentration for binary solutions is determined by temperature and cryoscopic constant of water, but in the presence of biopolymers also depends on the amount of protein-bound water and the structure of the hydrate shell of the protein molecule (Gun'ko et al. 2005d, 2009d). If the concentration of the solute does not exceed 1 wt%, then upon freezing, in the case of binary solutions, it concentrates in the voids surrounded by ice and, in the presence of protein concentrated in its hydrate shells (Gun'ko et al. 2005d). As follows from ratio of intensity CH_3 and H_2O signals (Figure 6.10) even for dilute ($C_{DMFA}=0.5$ wt%) solutions, the concentration of organic additives in the non-freezing phase may amount to 10%–20% (it increases with decreasing temperature). At such high concentrations of proton-active molecules, denaturing of proteins is possible and, therefore, in the experiments with freezing one often deals with the denatured forms of biopolymers.

The process of formation of eutectics with a fall in temperature may be observed from the decrease in the intensity of the signals of the alkyl groups. For the substances studied, the dependences of the intensity of the signals of water and alkyl groups on temperature over the temperature range $240<T<270$ K are given in Figure 6.10b through e. All the intensities of the signals are normalized in relation to that of bound water in frozen aqueous solutions of HSA at $C_{HSA}=1$ wt% and 270 K. The temperature dependences for binary (H_2O+organic additive) and triple (H_2O+organic additive+HSA) systems were studied. As follows from Figure 6.10, for binary solutions the temperature of eutectics is reached only for solutions of AA (Figure 6.10b) where reduction in the intensity of the signal of the CH_3 groups is observed at $T<245$ K. A different situation applies in the case of triple systems. For the electron-donor molecules (DMFA, Ace), a fall in the intensity of the signals of the CH_3 groups is observed at $T<260$ K and for proton donors (AA, ME) already at $T<270$ K the intensity of the signal of the alkyl groups decreases with a fall in temperature. Thus, the presence of protein molecules raises the freezing temperature of the organic substances dissolved in water by several tens of degrees.

The precipitation of the eutectics from solutions upon cooling occurs when the free energy of the solute molecules in the crystalline state becomes less than that of the molecules in solution. Therefore, a change in the temperature of precipitation of the eutectics indicates a change in the chemical potentials of the components of the solutions in the presence of protein molecules. On concentration of the solution, this may occur in the hydrate shells of the protein molecules. Since for the organic solute molecules close to the macromolecules not only do the molecules of the solvent but also the structural elements of the protein molecules form the medium, it is most probable that the observed phenomenon of a rise in the temperature of freezing is based on the interaction of the molecules of the organic additive with the biopolymer.

The organic molecules forming part of the hydrate shell of the protein expel from it several water molecules, the number of which in the simplest case is determined by the ratio of the volumes of the organic and water molecules. In the presence of interaction of the organic molecule with protein, with rise in the concentration of the additive, the intensity of the SAW bound to the biopolymer will decrease to a greater degree the more effective the process of interaction. However, such an approach implies that all the substance of the additive is in solution. For a considerable number of substances in the temperature range $260<T<270$ K, this is true. When crystallization of the solute is observed, the method may be so modified that one measures not the concentration but the temperature dependence of the intensity of the SAW and solute for binary and triplet systems (Figure 6.10). As in the case of the concentration dependences, upon formation on solvate complexes with organic, water, and biopolymer molecules, the amount of bound water changes. Where the number of molecules in the hydrate shell in the presence of organic additive does not change, the intensities of the SAW for binary and triple systems differ by the value of the intensity of signal of bound water. In coordinates $I/I_0(T)$ (Figure 6.10), the difference in the intensities of the signal amounts to unity. This case is presented in Figure 6.10e, for the system H_2O+Ace+HSA. When there is rearrangement of the hydrate shell of the biomolecules, the intensity of the signal of unfrozen water is far lower (Figure 6.10b and c). Thus, from Figure 6.10, it may be concluded that both for proton donors (AA, ME) and electron donors (DMFA, Ace), the thickness of the solvate shell of HSA may change through incorporation into it of the molecules of the organic additive.

The different form of the temperature dependences for electron- and proton-donor additives apparently reflects the special features of the interactions of these substances with the protein molecules. Thus, the proton-donor molecules significantly reduce the amount of water bound to the biopolymer. Consequently, these molecules interact with the active centers, ensuring formation of the hydrate cover of the protein molecule. Such centers are the hydroxyl, carbonyl, and other polar side groups of the amino acids and also the imino and carbonyl groups of the polypeptide chains. Forming with them hydrogen-bound associates, the molecules of the proton donors impede the formation of the ordered structures of water close to the protein molecule. The effect grows with fall in temperature and at $T<250$ K the intensity of signal of unfrozen water becomes even less than the intensity of signal of bound water in binary solutions when protein-bound water is absent (Figure 6.10b and c). The electron-donor molecules of DMFA and Ace may form hydrogen-bound complexes with the proton-donor centers of the protein molecules to which the hydroxyl, amino, and sulfhydric groups of the amino acid residues primarily belong. The protein-bound molecules of the electron donor are in the direct vicinity of the hydrophobic regions of the protein molecules so that they weakly influence the thickness of the hydrate shell of the biopolymer (Figure 6.10d and e). Both the hydrophobic and hydrophilic regions of the protein molecules are capable of stabilizing the ordered structures of the solutes. Probably, the solvate shells of biopolymers are the scene of the formation of the microcrystallites of the solute in which the mobility of the molecules is far lower than that of the organic molecules associated with the active centers of the protein molecules.

The molecules forming part of such formations cease to be observed in the high-resolution spectra in view of the short transverse relaxation times of the protons, like the molecules in frozen liquids.

The results agree with the notion of the microcavity (voids, pockets) structure of protein molecules (Kraivyaryainen 1980) according to which the protein molecule may be represented as an entity of spatially separated hydrophobic, hydrophilic, and mosaic cavities. Organic substances concentrated in such cavities behave like a liquid in a confined volume for which in a number of cases a rise in the temperature of freezing of water has been recorded (Drost-Hansen and Etzler 1989).

6.2 PROTEINS IN ADSORBED STATE

6.2.1 Interactions of Proteins with Oxides

Interactions of proteins with oxide materials are of interest because of biomedical applications of these materials as medical adsorbents, components of medical preparations including drug delivery systems, implants, etc. In all these applications, the materials can contact proteins and other biomacromolecules, cells, or whole tissues. Clearly, during adsorption 3D structure of macromolecules and their biofunctions can change. Special interest is focused on interactions of NP with cells, membranes (lipids), and membrane-integrated proteins. For instance, a key factor in the development of type II diabetes is the loss of insulin producing pancreatic β-cells, and the amyloidogenic human islet amyloid polypeptide (hIAPP, human amylin) can play a crucial role in this process since hIAPP can form small aggregates (NP) that kill β-cells by disrupting the cellular membrane. Brender et al. (2007) studied membrane fragmentation by the hIAPP aggregates in solid-state NMR experiments using nanotube arrays of anodic Al_2O_3 (AAO) containing aligned phospholipid membranes. In a narrow concentration range of hIAPP, an isotropic ^{31}P chemical shift signal indicative of the peptide-induced membrane fragmentation was detected. Solid-state NMR results suggest that membrane fragmentation was related to peptide aggregation as the presence of Congo Red, an inhibitor of amyloid formation, prevented membrane fragmentation and the non-amyloidogenic rat-IAPP did not cause membrane fragmentation (Figures 6.11 and 6.12). The disappearance of membrane fragmentation at higher concentrations of hIAPP suggested an alternate kinetic pathway to fibril formation in which membrane fragmentation was inhibited (Brender et al. 2007). The isotropic peak at about 0 ppm, indicative of membrane fragmentation, is suppressed in the presence of Congo Red (b and c) (Figure 6.12).

Clay minerals are widely used materials in drug products both as excipients and active agents. When administered simultaneously, drug–clay interactions can be been observed and, therefore, studied, but until recently were not considered as a possible mechanism to modify drug release. During last decade, based on their high retention capacities as well as swelling and colloidal properties, clays have been proposed as very useful materials for modulating drug delivery. Aguzzi et al. (2007) analyzed the studies on drug–clay interactions focusing on the applications of natural clays and their semisynthetic or synthetic derivatives to carry out specific functions in new drug delivery systems. Clays can be used to delay and/or target drug release or even improve drug dissolution. New strategies were reported for increasing drug stability and simultaneously modifying drug delivery patterns through the use of clay minerals. The application properties of clay systems (Zhu et al. 2005, Aguzzi et al. 2007, Gopinath and Sugunan 2007, Betega de Paiva et al. 2008, Bae et al. 2009) are close to that of nanosilica and mixed nanooxides (Chuiko 2003, Blitz and Gun'ko 2006). Clear interactions of these materials with proteins can play an important role in their efficiency as drug delivery materials or adsorbents.

Gopinath and Sugunan (2007) immobilized α-amylase, glucoamylase, and invertase on acid-activated montmorillonite via adsorption and grafting. The NH_2 groups interact with the acid sites of clay mineral. Grafting between the enzyme and the glutaraldehyde spacers occurs via condensation of NH_2 groups of enzyme and CHO groups of glutaraldehyde. After immobilization of enzymes, the basal spacing of montmorillonite increased, but the whole enzyme molecules were not intercalated.

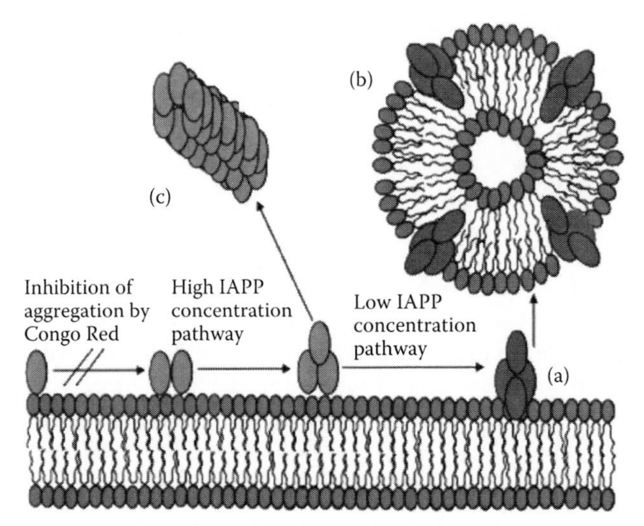

FIGURE 6.11 A schematic model for membrane fragmentation by hIAPP: (a) at lower concentrations, the peptide aggregates on the lipid bilayer surface to form an intermediate state that is capable of extracting phospholipid molecules from the bilayer and leading to peptide–lipid vesicles (b); at higher concentrations, fiber formation follows an alternate pathway bypassing the formation of this intermediate (c). (Adapted from *Biochim. Biophys. Acta Biomembr.*, 1768, Brender, J.R., Dürr, U.H.N., Heyl, D., Budarapu, M.B., and Ramamoorthy, A., Membrane fragmentation by an amyloidogenic fragment of human Islet Amyloid Polypeptide detected by solid-state NMR spectroscopy of membrane nanotubes, 2026–2029, 2007, Copyright 2007, with permission from Elsevier.)

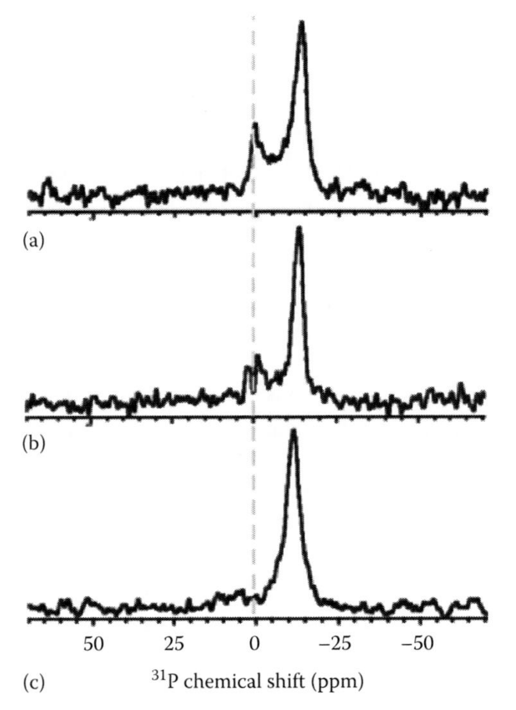

FIGURE 6.12 ^{31}P chemical shift spectra of DMPC bilayers containing 1 mol% of hIAPP aligned in AAO nanotubes with varying concentrations of Congo Red. (a) 0, (b) 2, and (c) 5 mol%. (Adapted from *Biochim. Biophys. Acta Biomembr.*, 1768, Brender, J.R., Dürr, U.H.N., Heyl, D., Budarapu, M.B., and Ramamoorthy, A., Membrane fragmentation by an amyloidogenic fragment of human Islet Amyloid Polypeptide detected by solid-state NMR spectroscopy of membrane nanotubes, 2026–2029, 2007, Copyright 2007, with permission from Elsevier.)

The protein backbone is situated at the periphery of the clay mineral particles while the side chains enter in the layers. Surface area and pore volume decreased. The morphology of the particles was changed. ^{27}Al NMR analysis showed that the tetrahedral Al species are involved in the adsorption process whereas the octahedral sites are influenced by grafting.

Betega de Paiva et al. (2008) analyzed properties, synthesis, and applications of organoclays (especially bentonite and polymer nanocomposites), modifications with several chemical compounds as quaternary alkylammonium salts and also biomolecules as enzymes that provide applicability in a variety of fields.

Clays are stable aluminosilicates with high cation exchange capacity, and exfoliated clay particles have a platelet shape with nanoscopic size (Murray 1992). Compared with organic polyelectrolytes, clay has the advantages of high chemical stability, good adsorption property due to its appreciable surface area, special structural feature, and unusual intercalation property (Bae et al. 2009). The interaction of proteins with clay has been studied extensively, and the direct electrochemistry of redox proteins such as hemoglobin (Hb), cytochrome *c* (Cyt *c*), horseradish peroxidase (HRP), and Mb in clay-related films has also been reported. For example, the electrochemical and electrocatalytic properties of heme proteins in cast protein–clay films and layer-by-layer $\{clay/proteins\}_n$ films modified on pyrolytic graphite (PG) electrodes were studied in detail (Chen et al. 1999a, Zhou et al. 2002, Liu et al. 2005).

In the case of contact of nanosilica, which can be used as a component of tourniquet preparations or an adsorbent on washing of organ cavities by a diluted suspension, with damaged tissues it can interact with blood (i.e., with blood plasma proteins). This interaction can strongly affect the state of blood because of coagulation of plasma proteins (human plasma fibrinogen [HPF, Fg], albumin, etc.) with silica NP, conformation changes of adsorbed macromolecules, hemolysis of red blood cells (RBCs), etc. Long HPF molecule (molecular weight $M_w \approx 340$ kDa) (Figure 6.13a) is composed of two identical molecular halves consisting of three nonidentical Aα-, Bβ-, and γ-chain subunits (compared in size with primary silica NP) held together by multiple disulfide bonds (Mosesson et al. 2004, Oenick 2004, Kollman et al. 2009).

HPF molecules have "loose ends" which are extremely mobile that can be functionally important, as well as they can play a specific role on the adsorption of HPF especially onto nonplanar surfaces as of nanosilica. Thrombin initiates the clotting of fibrinogen by proteolysis, thus splitting off fibrinopeptides from the fibrinogen molecule. On addition of thrombin to fibrinogen, initial polymers or protofibers were formed and after reaching certain lengths gel formation took place due to 3D interaction between the polymers. These phenomena can be affected by interaction of HPF with nanosilica resulting in certain changes in the blood clotting.

The effect of surface adsorption on protein structure is of importance because first adsorption of plasma proteins occurs on blood contact with a foreign material. Fibrinogen as an adhesive protein can strongly bind to physiological and pathophysiological surfaces (e.g., cytoplasmic membranes of activated platelets and macrophages, cytoplasmic membrane of tumor cells, circulatory prosthetics and other implantable devices, and atherosclerotic plaques) which are in contact with blood. Interaction of fibrinogen with solid NP or implants can affect fibrosis. However, fumed nanosilica as an amorphous material composed of spherical NP is characterized by weak similar negative effects (Chuiko 2003).

On the HPF adsorption on a smooth crystalline rutile surface, the mean length and the height of single molecules estimated by AFM were 46 ± 3 nm and 1.4 ± 0.2 nm, respectively. HPF molecules in the adsorbed state looks "thinner" due to interaction with the surface, which can also result in conformational changes of adsorbed and desorbed proteins (Yongli et al. 1999). AFM investigations of the HPF adsorption to a silica surface show that HPF molecules mainly adsorb through their D and E globules (Hemmerlè et al. 1999). Adsorption of proteins at a hydrophobic surface of silylated silica is irreversible. The hydropathic index for HPF indicates that the E and D domains are substantially more hydrophobic than the αC domains (Malmsten and Lassen 1994, Jung et al. 2003).

This could affect the structure of the adsorption complexes of HPF on both hydrophilic and hydrophobic surfaces (Tunc et al. 2005).

Interaction of HPF with nanosilica is of special interest because of the medical applications of this material, especially as a component of tourniquet preparations. Therefore, here we will analyze adsorption interactions of fibrinogen with nanosilica depending on concentration of components, salinity, and pH. Application of several methods such as (i) adsorption of HPF from the aqueous solutions; (ii) UV and FTIR spectroscopy of adsorbed layer with HPF; and (iii) TSDC and ^{1}H NMR spectroscopy with layer-by-layer freezing-out of bulk and bound water interacting with HPF and HPF/nanosilica allows us to deeply understand features of interaction of HPF with silica NP that is of importance for elucidation of the mechanism of interaction of nanosilica preparations with blood.

HPF was obtained from plasma of donor blood by using fractional salting out with sodium sulfate (Varetska et al. 1972). Initial HPF concentration (C_{HPF}) was 2.6 wt% in an aqueous solution at 0.15 mol NaCl. HPF was frozen and stored at 245 K. Before studies, frozen HPF was thawed in a water bath at 308–310 K for 15–20 min and maintained at room temperature for 10 min. To obtain the HPF solutions of given concentrations, a phosphate buffer solution was used for dilution of the initial solution directly before experiments.

Adsorption of HPF on nanosilica A-300 was studied at 293 ± 1 K using 0.2 g of the dry A-300 powder and 40 mL of the HPF solution at different concentrations. The HPF adsorption was studied at pH 7.4 because the pH value of blood is 7.35–7.45 (Taylor et al. 2000). Concentration of HPF before and after the adsorption was spectrophotometrically controlled by using a Specord M-40 (Carl Zeiss, Germany) UV/vis spectrophotometer at $\lambda = 280$ nm. For achievement of the adsorption equilibrium, the system was incubated for 1.5 h and then it was centrifuged at 8000 rpm for 15 min. Desorption of HPF was determined for samples with a known content of adsorbed HPF maintained in the phosphate buffer solution at pH 7.4 for 1.5 h, centrifuged for 15 min at 8000 rpm by the spectrophotometric measurements of the supernatant liquid.

UV spectra ($\lambda = 200$–400 nm) in a diffusive reflection mode of HPF alone and adsorbed onto nanosilica were recorded using a Specord M-40 (Carl Zeiss, Germany). The HPF/A-300 suspensions were centrifuged at 8000 rpm for 15 min and the solid residue was dried on filter paper in air at room temperature for 12 h.

The largest size of a HPF molecule (Figure 6.13) corresponds to the size of the smallest aggregates of silica NP composed of dozens of silica NP.

Therefore, very flexible HPF molecules could interact with several silica NP from the same aggregate or different aggregates simultaneously. This could cause a high nonuniformity of the adsorption layer because adsorbed HPF molecules could create barriers for other HPF molecules to be adsorbed and could inhibit decomposition of aggregates of primary silica particles. The adsorption isotherm at $C_{HPF} = 0.0175$–0.104 wt% has a shape of the Langmuir isotherm (Figure 6.14a) at the adsorption monolayer capacity of 156 mg of HPF per gram of silica. This adsorption corresponds to changes in the Gibbs free energy between −1 and −5 kJ/mol at a maximum at −2 kJ/mol (Figure 6.14c).

At high concentrations $C_{HPF} > 0.1$ wt% (or $C_{eq} > 0.3$ mg/mL), a sharp increase in the adsorption was observed. This was due to both polymolecular HPF adsorption and a decrease of the average area occupied by a HPF molecule at a higher concentration because of lateral interactions of the adsorbed protein molecules. At the high HPF concentration, the solution became opalescent that could be caused by coagulation of protein molecules themselves and by their aggregation with silica NP to form hybrid aggregates (this assumption could be confirmed by the ^{1}H NMR and TSDC investigations described later). An increase in the turbidity corresponds to a decrease in the optical density at 320 nm. This decrease is stronger ($I_{320} = 0.075$–0.201) (Figure 6.15) than typical one ($I_{320} = 0.014 \pm 0.006$) for HPF alone. This result suggests that stronger coagulation of the HPF molecules occurred in the presence of nanosilica than in the solution of HPF alone. The coagulation effects could be enhanced by interaction of HPF molecules with several primary silica particles

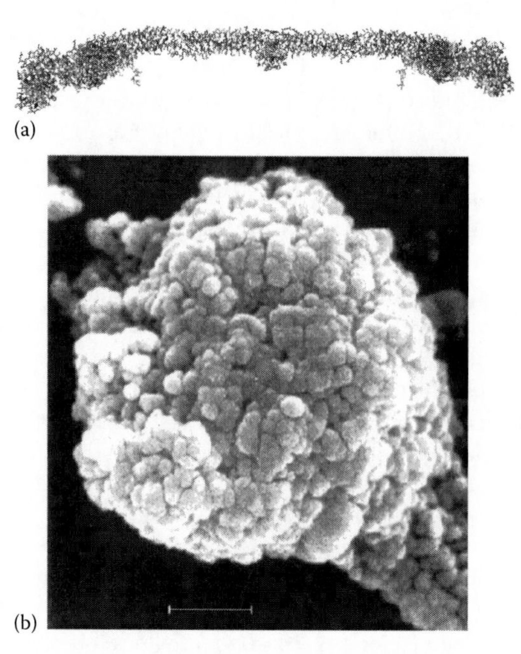

FIGURE 6.13 (a) Molecular structure of fibrinogen (length ~47.5 nm, H atoms were not shown) (atomic coordinates from Kollman, J.M. et al., *Biochemistry*, 48, 3877, 2009) and (b) SEM image of agglomerate (~5 μm in diameter) of aggregates (50–500 nm) of silica NP, and the largest HPF molecule size (~47 nm) corresponds to the size of the smallest aggregates composed of dozens of silica NP (bar 1 μm). ((b) Adapted with kind permission from Springer Science+Business Media: *Cent. Eur. J. Chem.*, Interaction of fibrinogen with nanosilica, 5, 2007, 32–54, Rugal, A.A., Gun'ko, V.M., Barvinchenko, V.N., Turov, V.V., Semeshkina, T.V., and Zarko, V.I.)

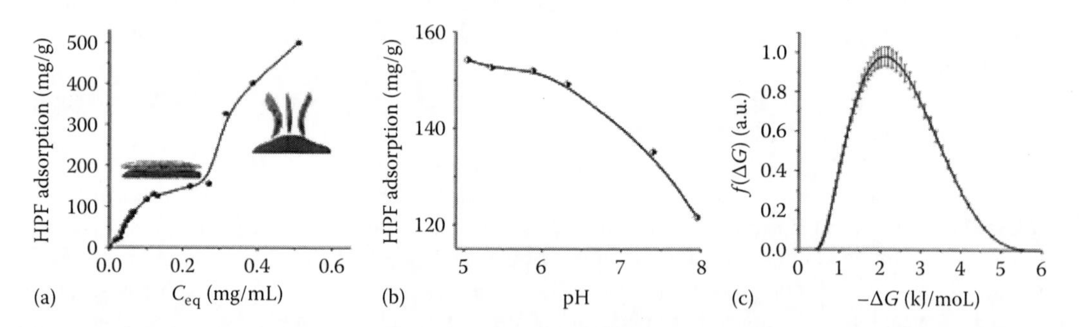

FIGURE 6.14 HPF adsorption as a function of (a) equilibrium C_{HPF} (PBS, pH 7.4); (b) pH (C_{HPF}=0.0795 wt%); and (c) distribution of Gibbs free energy of HPF adsorption at $A < 160$ mg/g. (Adapted with kind permission from Springer Science+Business Media: *Cent. Eur. J. Chem.*, Interaction of fibrinogen with nanosilica, 5, 2007, 32–54, Rugal, A.A., Gun'ko, V.M., Barvinchenko, V.N., Turov, V.V., Semeshkina, T.V., and Zarko, V.I.)

simultaneously from the same or different aggregates that lead to the formation of larger and denser hybrid HPF–silica aggregates and agglomerates.

The electronic reflection spectra of HPF adsorbed on silica (Figure 6.15) at moderate concentrations correspond to that of HPF alone. Broadening of the spectra and their displacement toward shorter wavelengths by 2–3 nm were observed with increasing HPF concentration to 500 mg/g of silica.

This behavior of adsorbed HPF could be caused by certain conformational changes because of the adsorption interactions, which were also observed in the FTIR spectra (see later). However, these

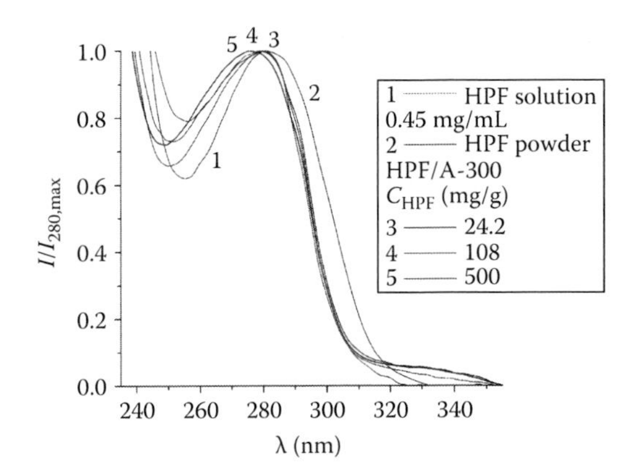

FIGURE 6.15 UV reflection spectra of powder and dissolved HPF (0.45 mg/mL, PBS) and HPF adsorbed onto silica normalized to the intensity value of a maximum at 280 nm. (Adapted with kind permission from Springer Science+Business Media: *Cent. Eur. J. Chem.*, Interaction of fibrinogen with nanosilica, 5, 2007, 32–54, Rugal, A.A., Gun'ko, V.M., Barvinchenko, V.N., Turov, V.V., Semeshkina, T.V., and Zarko, V.I.)

changes are not very strong. Consequently, the state of HPF adsorbed onto nanosilica is close to the native state. Consequently, nanosilica is characterized by certain biocompatibility with respect to proteins. Interaction of RBCs with pure nanosilica leads to intensive hemolysis. However, the hemolysis strongly decreased with the presence of polymers (e.g., poly(vinyl alcohol) [PVA], poly(vinyl pyrrolidone) [PVP]) or plasma proteins or in the case of interaction of whole blood with nanosilica.

The adsorption of HPF (initial $C_{HPF}=0.0795$ wt% in the solution) is maximal at pH 5–6 which is close to the pH value of the isoelectric point (IEP) of HPF. The adsorption slightly decreases (by ~20% at pH 8) at pH>6 (Figure 6.14b) because the adsorption of proteins is maximal at their IEP. Consequently, the state of HPF is stable at pH 5–8 that is of importance because HPF can easily denature in strongly acid or base solutions, as well as on the adsorption. A similar type of the pH dependence of the adsorption is characteristic for the systems, in which interaction of proteins with a silica surface occurs predominantly due to electrostatic forces and the hydrogen bonds (the main portion of which is due to electrostatic forces).

It is known that the adsorption and desorption isotherms of proteins do not coincide that is explained by multicenter binding of macromolecules to a surface, a higher activation energy of desorption (under strong condition of breaking of all the adsorption bonds simultaneously) than that of the adsorption (which can be non-activated). An increase in concentration of HPF adsorbed on silica led to an increase in HPF desorption. However, desorption did not exceed 0.002 wt%. Consequently, the adsorption of HPF (monolayer or smaller) on the nanosilica surface is practically irreversible under conditions used here. This could be also caused by the formation of hybrid HPF/silica aggregates and agglomerates in which the HPF molecules interacted with several silica NP simultaneously. Clearly, desorption of HPF from these complex aggregates is much more difficult than that from a flat surface.

Interaction of HPF with the silica surface could be analyzed for dried HPF/silica samples on the basis of changes in the FTIR spectra of the OH-stretching vibrations of free surface silanols at 3748 cm^{-1} with increasing surface concentration ($C_{HPF,ads}$) of protein (Figure 6.16).

The increase in the $C_{HPF,ads}$ value led to a decrease in the intensity of the mentioned band because of the formation of the hydrogen bonds between silanols and adsorbed molecules. This band was observed on the adsorption of 500 mg of HPF per gram of A-300 (Figure 1.13b), despite the monolayer capacity from the adsorption data was much smaller (156 mg/g). This could be caused by at least two reasons: (i) HPF molecules have a complex shape (Figure 6.13); therefore, an occupied area outline of an adsorbed HPF molecule has complex shape and a portion of undisturbed silanols closely located to this molecule could be inaccessible for other adsorbed HPF molecules

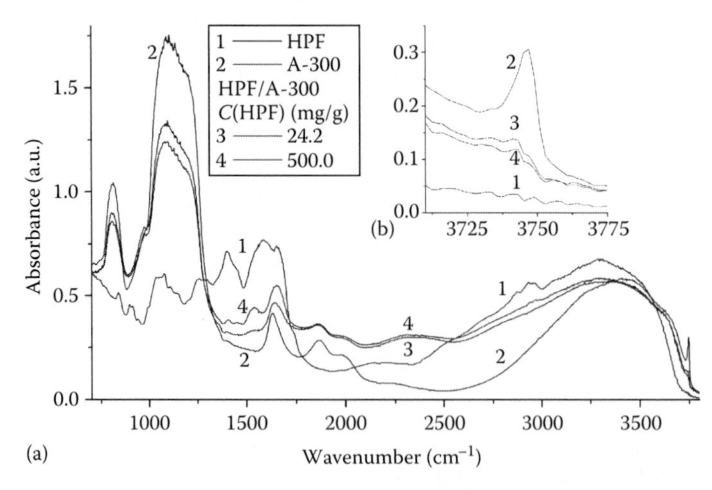

FIGURE 6.16 (a) FTIR spectra of powder HPF, A-300 individual and with adsorbed HPF (pH 7.4, PBS) and (b) OH-stretching vibrations of silanols at 3748 cm^{-1}. (Adapted with kind permission from Springer Science+Business Media: *Cent. Eur. J. Chem.*, Interaction of fibrinogen with nanosilica, 5, 2007, 32–54, Rugal, A.A., Gun'ko, V.M., Barvinchenko, V.N., Turov, V.V., Semeshkina, T.V., and Zarko, V.I.)

and (ii) not all silanols were accessible for HPF molecules because primary silica particles could form aggregates and a portion of which could remain after interaction with HPF. Consequently, there were certain voids and cavities at the interfaces of dried HPF/silica samples. The first assumption is confirmed by previous results showing that the adsorption of 400 mg BSA (which is much smaller and has a simpler shape than HPF) per gram of A-300 results in complete disappearance of the band of free silanols at 3748 cm^{-1}. The second assumption is confirmed by a large difference in the amounts of adsorbed globular BSA and linear polymers such as PVP or poly(ethylene oxide) to provide the complete disturbance of free silanols (400 and 180–200 mg/g, respectively).

Revealing of HPF functionalities responsible for the formation of the hydrogen bonds with the silanols is a complex task because HPF has many different O- and N-containing groups in the D and E globules and mobile "tails" (Figure 6.13a) which could take part in the HPF bonding to the silica surface. Additionally, the positively charged αC domains of HPF or other structures could form the $RN{-}H^{+}\cdots O(H)Si\equiv$ bonds. On the HPF adsorption, the content of α-helix slightly decreased. Consequently, peptide bonds could take part in the binding of HPF to the silica surface. A band of the C=O stretching vibrations of HPF at 1655 cm^{-1} slightly shifted toward lower wavenumbers (1640–1637 cm^{-1}) in comparison with alone HPF (Figure 6.16), as well as the δ_{CNH} bending vibrations at 1540 cm^{-1} shifted toward 1530 cm^{-1}. This displacement was higher at a lower C_{HPF} value (24.2 mg/g of silica) because at a higher C_{HPF} value (500 mg/g) a relative number of these groups interacting with silica was much smaller. A band of the C=O stretching vibrations has a shoulder at 1521 cm^{-1}, which is characteristic for alone and adsorbed HPF and corresponds to the α-helix of HPF. The FTIR bands at 2963, 2932, and 2872 cm^{-1} correspond to the CH stretching vibrations of HPF, and the bending vibrations of these groups are observed at 1450 cm^{-1}.

There are two dominant bands of HPF in the buffer solution, namely amide I around 1650 cm^{-1} and amide II band (Tunc et al. 2005) around 1550 cm^{-1} (Figure 6.16, curve 1). The bands at 1460 and 1400 cm^{-1} (curve 1) are assigned to the CH_2 deformation and to vibrations of the amino acid side chains, respectively. Band positions for different structures of HPF in PBS are 1689 ± 2 cm^{-1} and 1675 ± 2 cm^{-1} for β-turn (there is a shoulder on curve 1), 1658 ± 1 cm^{-1} (α-helix) (a peak is observed on curve 1) and 1633 ± 2 cm^{-1} (β-sheet) (this band is masked on curve 1). The FTIR spectra of HPF adsorbed on hydrophilic surfaces of oxidized silicon (i.e., similar to amorphous silica) show a maximum of the amide I band around 1658 cm^{-1} (1655 cm^{-1} on curve 1) which is typical of proteins with high α-helix content. On the adsorption, there is an increase of turn structure and a decrease of β-sheet structure. The secondary structure of adsorbed HPF on a hydrophilic surface turned out to be rather similar to that of the protein in solution phase with a major α-helix content. The general trend is that the amount of β-sheet decreases and less ordered structures, turn and random ones, increase on the HPF adsorption. β-Sheets of C-terminal domains of β- and γ-chain in fibrinogen are close to the domain centers, whereas α-helices are located at the outer parts of the protein. This indicates a more hydrophobic nature of side chains of β-sheets than of α-helices. Adsorption on hydrophobic surfaces induces extensive conformation changes (greater than that on the adsorption onto hydrophilic surfaces), which also disrupt hydrogen bonding within β-sheets. In addition, β-sheets are known to be not flexible; when they attach to a surface, they must soften by releasing their structure and go to more flexible structures. The structural changes of fibrinogen upon adsorption onto hydrophilic surfaces, e.g., titania, are quite small, as well as in the case of the HPF/nanosilica samples studied here (Figures 6.15 and 6.16).

Additional information related to changes in the state of adsorbed HPF and the structure of the interfaces could be obtained from a study of the HPF/silica systems by the ^{1}H NMR spectroscopy and TSDC with layer-by-layer freezing-out of bulk and bound water at $T<273$ K which allows one to distinct the water layers differently affected by protein molecules, silica surface, and dissolved ions of the PBS (which can be concentrated at the interfaces on freezing). Figure 6.17a shows the ^{1}H NMR spectra of unfrozen water in 0.15 mol NaCl solution at different temperatures.

Figure 6.17b gives a temperature dependence of concentration of unfrozen water per weight unit of salt. The concentration of unfrozen water rapidly decreased with lowering temperature. Nevertheless, certain amounts of water remained unfrozen even at 220 K and lower. There are two practically linear sections of $C_{uw}(T)$ with different inclination to the X-axis. Notice that similar linear sections were observed for many systems containing bound water locating in confined space of pores, voids, or pockets. The main reason for this behavior of pore water is the difference in the surface effects for nearby and distant water layers, cooperativity in structural changes of pore water over a narrow temperature range on the adsorption in narrow pores, and the absence of a similar effect for water adsorbed in broad pores. Concurrence of the lines is observed at 259 K (Figure 6.17b) close to temperature of precipitation of eutectic NaCl–H_2O (T_e). A section of the $C_{uw}(T)$ graph located at $T>T_e$ can be attributed to water unfrozen because of concentrating of the NaCl solution, but another section at $T<T_e$ corresponds to water adsorbed on a surface of fine NaCl crystals. It is possible to expect that appearance of strongly hydrated protein molecules and silica NP in the aqueous solution of NaCl would lead to additional increase in the amounts of unfrozen water. However, on freezing, salt ions can concentrate in the hydrated shells of solid NP or macromolecules instead of the formation of a bulk phase of the concentrated NaCl solution. To determine localization of components of the solutions on freezing, measurements of the signal width of the ^{1}H NMR spectra of unfrozen water can be used.

Figure 6.17c shows the ^{1}H NMR spectra of unfrozen water in the aqueous suspension with 9 wt% of silica and 0.15 mol NaCl recorded at different temperatures. The width of these spectra is several times larger than that for the alone NaCl solution. A large width of the signal can testify that the concentrating of solution components occurs in the hydrated shells of silica particles because the mobility of water molecules is substantially lower in the adsorbed state and in an immobile portion of the electrical double layer (EDL) than in the bulk.

The signal width of water bound by HPF in the frozen solution of HPF/PBS(NaCl) differs only slightly from that for the frozen solution of NaCl alone. To elucidate localization of salt ions in the HPF/PBS(NaCl)/H_2O system, the temperature dependences of relative intensity of the signal of unfrozen water were analyzed (Figure 6.17d) (intensity is referred to that before freezing). The presence of HPF caused a sharp increase in the signal intensity of unfrozen water, the total amount of which increased several times except for the sample with 0.6 wt% of HPF for which the I/I_0 values for both solutions were close at low temperatures. Thus, one can assume that the salts (PBS) only slightly contribute to the amounts of unfrozen water in the HPF solutions at $C_{HPF}=1.25$ and 2.5 wt%. Therefore, similar HPF concentrations were used in the investigations.

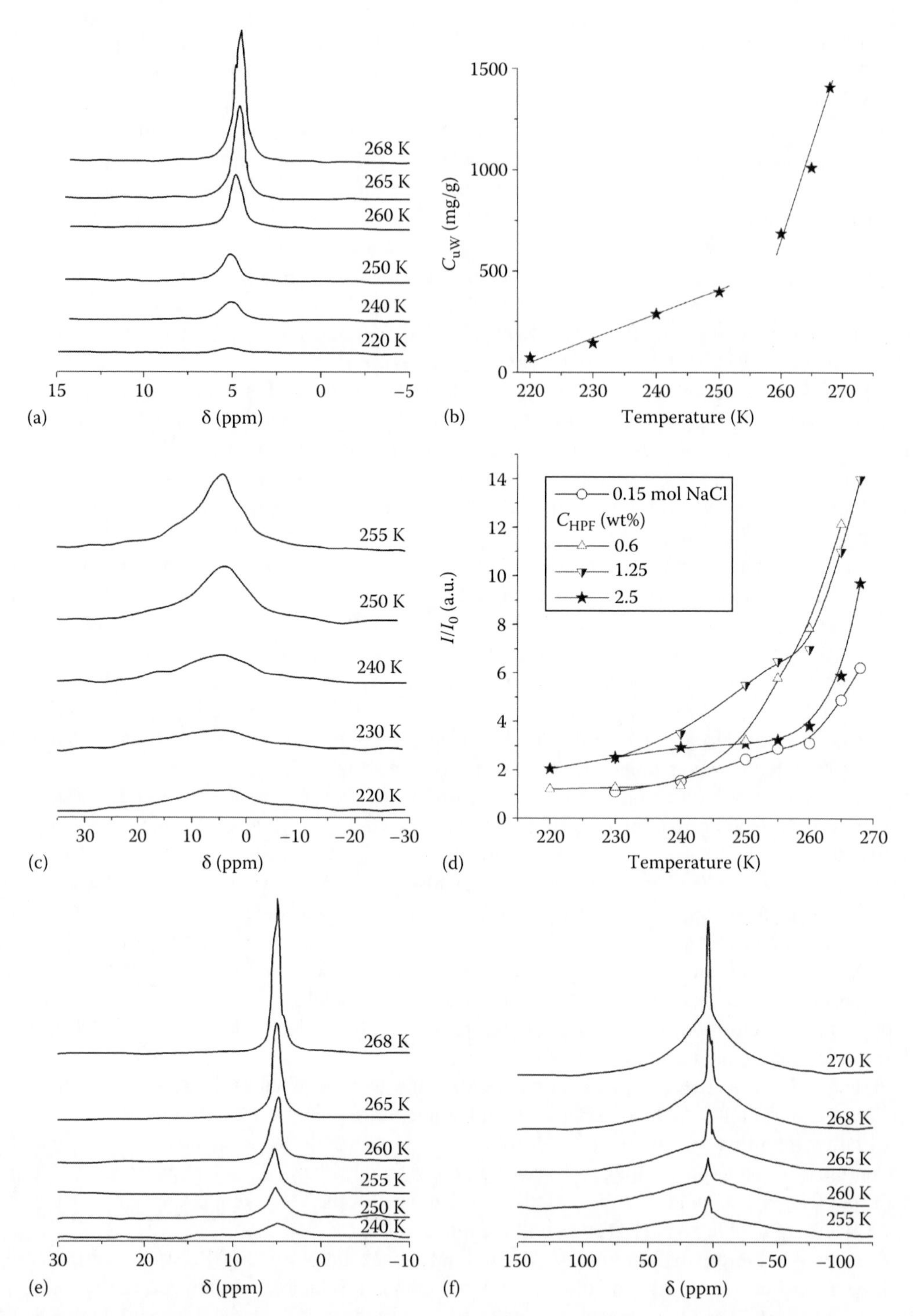

FIGURE 6.17 (a) ^{1}H NMR spectra of unfrozen water in the aqueous solution of 0.15 mol NaCl at different temperatures; (b) concentration of unfrozen water (mg) per gram of NaCl; (c) ^{1}H NMR spectra of unfrozen water in the aqueous suspension (0.15 mol NaCl) of silica (9 wt%) at different temperatures; (d) relative intensity of the signal of unfrozen water in the buffer solutions of HPF; and ^{1}H NMR spectra of frozen solutions of HPF/PBS at C_{HPF} = 1.25 wt% and C_{NaCl} = (e) 0.15 mol and (f) 0.075 mol. (Adapted with kind permission from Springer Science+Business Media: *Cent. Eur. J. Chem.*, Interaction of fibrinogen with nanosilica, 5, 2007, 32–54, Rugal, A.A., Gun'ko, V.M., Barvinchenko, V.N., Turov, V.V., Semeshkina, T.V., and Zarko, V.I.)

Figure 6.17e and f shows the ^{1}H NMR spectra of unfrozen water in the solutions at C_{HPF} = 1.25 wt% at 0.15 and 0.075 mol NaCl, respectively (NaCl is the main component in the PBS; therefore, only content of NaCl is shown later) recorded at different temperatures. A decrease in the C_{NaCl} value led to the appearance of a wide component in the signal at the width and the intensity larger by order than that of the narrow component. Notice that a similar signal shape as in Figure 6.17f was shown in Figure 6.2 for ubiquitin. At $T \to 273$, it became comparable with the SAW before freezing. The wide component can be attributed to amorphous ice in which the mobility of water molecules is much greater than that in bulk ice. This "mobile" ice was formed under action of strong surface forces of silica/protein disordering the ice structure, which resulted in a significant quantity of defects of the crystal lattice (amorphous ice). These defects enhanced the mobility of water molecules in comparison with bulk crystalline ice with much smaller amounts of defects.

Figure 6.18 shows the relationships between changes in the Gibbs free energy and concentration of unfrozen water for the solutions of (a) HPF/PBS at C_{HPF} = 1.25 and 2.5 wt%; (b) silica at C_{SiO_2} = 4.7 wt% without NaCl and 7.0 and 9.0 wt% at 0.15 mol NaCl; and (c) HPF/PBS and silica.

The characteristics of bound water are given in Table 6.2. Notice that the ΔG values (Table 6.2) are in agreement with the $f(\Delta G)$ function (Figure 6.14c) calculated from the HPF adsorption data

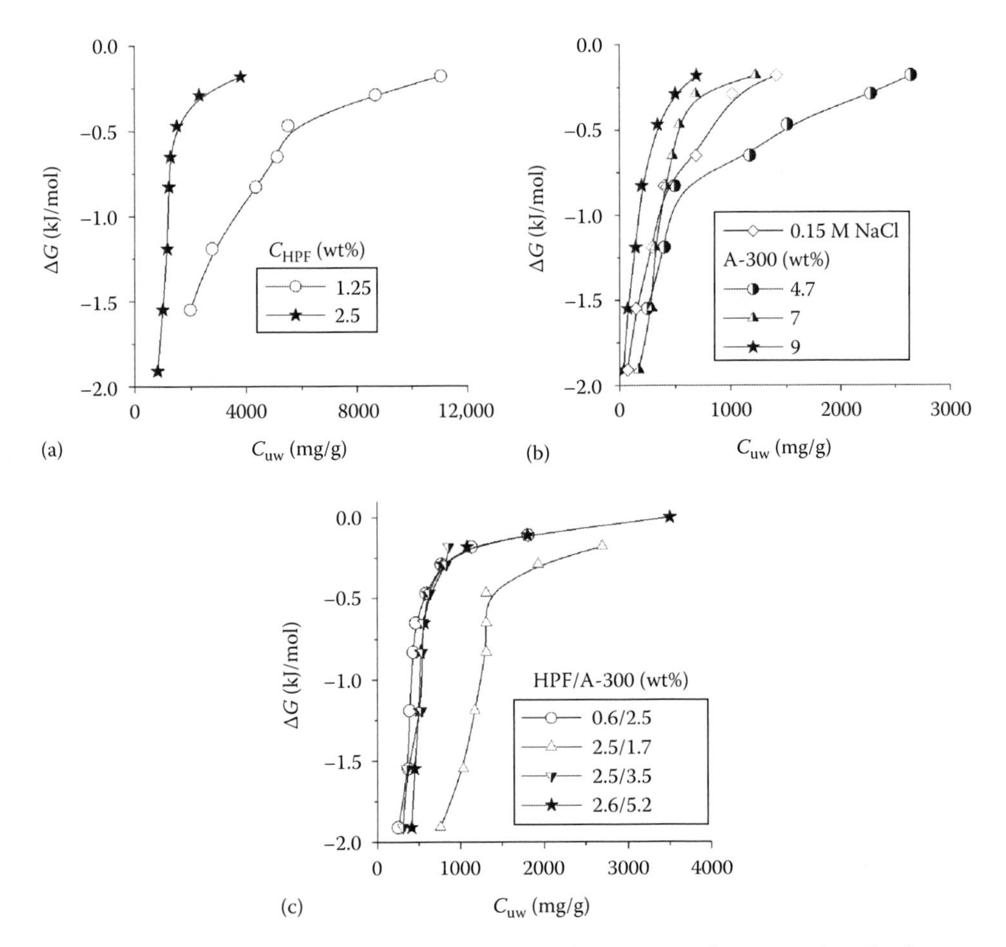

FIGURE 6.18 Relationships between changes in the Gibbs free energy and concentration of unfrozen water for (a) HPF solution at 1.25 and 2.5 wt%, (b) 0.15 M NaCl and A-300, and (c) HPF/A-300 at different concentrations of protein and silica. (Adapted with kind permission from Springer Science+Business Media: *Cent. Eur. J. Chem.*, Interaction of fibrinogen with nanosilica, 5, 2007, 32–54, Rugal, A.A., Gun'ko, V.M., Barvinchenko, V.N., Turov, V.V., Semeshkina, T.V., and Zarko, V.I.)

TABLE 6.2
Characteristics of Weakly and Strongly Bound Waters (in mg/g of a Solid Material) and Free Surface Energy γ_S (in J/g of Unfrozen Water) for Solution of 0.15 mol NaCl, Buffered HPF, Suspension of A-300, and HPF/A-300

Sample	C_{HPF} (wt%)	C_{SiO_2} (wt%)	γ_S (J/g)	C^s_{uw} (mg/g)	C^w_{uw} (mg/g)	$-\Delta G_s$ (kJ/mol)	$-\Delta G_w$ (kJ/mol)	S (m²/g)	S_{nano} (m²/g)	S_{meso} (m²/g)	S_{macro} (m²/g)	V_{nano} (cm³/g)	V_{meso} (cm³/g)	V_{macro} (cm³/g)
0.15 mol NaCl			60	500	1250	2.2	1.0	360[a]	22	315	24	0.011	1.150	0.589
A-300[b]		4.7	106	700	2500	2.4	1.0	179[a] (263[c])	0	115	64	0	1.203	1.997
A-300[d]		7.0	59	600	1600	2.6	0.5	187[a] (275[c])	2	160	26	0	1.537	0.662
A-300[d]		9.0	27	250	1000	2.2	0.6	172[a] (253[c])	1	169	2	0	1.191	0.058
HPF[e]	1.25		494	6000	8000	2.2	0.8	1331[c]	6	819	506	0.003	2.217	11.780
HPF[e]	2.5		204	1250	6750	3.0	0.5	783[c]	54	545	185	0.022	3.017	4.961
HPF+A-300[e]	0.6	2.5	72	350	3650	2.6	0.4	217[c]	1	71	145	0	0.164	3.836
HPF+A-300[e]	2.5	1.7	172	1250	1750	3.0	0.7	374[c]	8	323	43	0.003	2.310	0.687
HPF+A-300[e]	2.5	3.5	68	600	3100	3.0	1.0	200[c]	6	57	137	0.001	0.208	3.491
HPF+A-300[e]	2.5	5.2	103	400	400	4.5	0.4	117[c]	0	112	6	0	0.703	0.097

The models of [a]voids between spherical particles and [c]cylindrical pores.
[b] Aqueous suspension without addition of NaCl.
[d] Aqueous suspension with addition of 0.15 mol NaCl.
[e] Buffer solution.

(Figure 6.14a) and Langmuir equation as the kernel of integral equation solved using the regularization procedure. The graphs $\Delta G(C_{uw})$ for the majority of the studied systems have two portions (nearly linear sections) related to both strongly and weakly bound waters. The maximum amount of bound water ($C_{uw}^{s} + C_{uw}^{w} = 14$ g/g of dry HPF) is observed for the solution of HPF at $C_{HPF} = 1.25$ wt%. As a whole, a decrease in the concentration of a dispersion phase (HPF or HPF/A-300) was accompanied by an increase in the free surface energy (γ_S), i.e., the Gibbs free energy of the bound water reduced, and the layer thickness of unfrozen water increased. This was caused by a decrease of particle–particle interactions and diminution of the overlapping of their EDLs in the diluted colloidal systems. In other words, coagulation interaction of the HPF molecules occurred in a distant potential minimum. Addition of nanosilica led to a sharp decrease of the free surface energy (Table 6.2, γ_S) because of interaction of the HPF molecules with the silica surface (in the short distance potential minimum) and the corresponding diminution of the amounts of the structured water both strongly (Table 6.2, C_{uw}^{s}) and weakly (C_{uw}^{w}) bound ones. This could be also caused by conformational changes in the adsorbed HPF molecules.

The PSDs (Figure 6.19) were calculated on the basis of the $C_{uw}(T)$ graphs and Gibbs–Thomson relation converted into integral equation. The $f_V(R)$ and $f_S(R)$ functions were used to calculate the structural characteristics of the systems (Table 6.2). It should be noted that the use of the model of cylindrical pores gave the specific surface area (in contact with unfrozen water) (Table 6.2, S) close to the specific surface area ($S_{BET} = 292$ m^2/g) of dry silica calculated from the nitrogen adsorption isotherm. Figure 6.16b and the structural parameters (Table 6.2) show that certain conformational changes could occur in the solution of HPF with increasing C_{HPF} value from 1.25 to 2.5 wt%.

For instance, the PSD strongly decreases at $R = 0.8$–8.0 nm and $R > 16$ nm but it increases at $R = 8$–16 nm. The surface area (Table 6.2, S, S_{meso}, S_{macro}) in contact with structured water and the "macropore" volume V_{macro} strongly decrease. However, contributions of narrow pores (S_{nano}, V_{nano}) and the mesopore volume (V_{meso}) increase. Thus the HPF molecules tend to stronger coagulation in the more concentrated solution.

The structures with unfrozen water (domains and clusters) are smaller in the NaCl solution than that for the HPF solution (Figure 6.19 and Table 6.2). A similar diminution of the PSD, V_{macro}, and S_{macro} values was also observed for the suspensions of silica with increasing content of silica (Figure 6.19a and Table 6.2). Notice that the main peak of mesopores locates at $R = 10$ nm for both concentrated suspensions ($C_{SiO_2} = 7$ and 9 wt%, curves 2 and 3) and the dry silica powder (curve 4). The concentration $C_{SiO_2} = 7$ wt% corresponds to the bulk density of dry silica A-300 (60–70 g/dm^3). However, in the case of the smaller content of silica (4.7 wt%) this PSD peak shifts toward larger pore sizes, i.e., the distances between particles (between aggregates in agglomerates) increase similar to that for the diluted HPF solution. Therefore, one could assume that aggregates (responsible for the textural porosity of nanosilica) of primary silica particles are similar in the aqueous and gaseous media. The PSD value was much smaller for the suspension of silica than that for the HPF solution because proteins and its oligomers formed a looser structure than solid NP and their aggregates. The PSDs for the HPF/A-300 systems at $C_{SiO_2} = 1.7$–3.5 wt% locate between those for individual systems, which are, however, closer to those of nanosilica than that of HPF alone. At $C_{SiO_2} = 5.2$ wt%, the PSD value was lower than that for alone silica. The surface area of the HPF/silica samples at $C_{HPF} < C_{SiO_2}$ was lower than that of the silica samples (calculated using the model of cylindrical pores). Consequently, the space occupied by unfrozen water in the concentrated HPF/A-300 suspensions strongly decreased in comparison with the systems with HPF and silica alone due to (i) overlapping of the EDL of interacting particles and macromolecules on the HPF adsorption and HPF–HPF coagulation; (ii) compacting of the HPF molecules on the adsorption because adsorbed HPF molecules were flattened due to strong interaction with a solid surface; and (iii) formation of hybrid aggregates more densely packed than the aggregates of silica NP with the displacement of the bound water into the bulk.

For visualization of the mentioned regularities, the dependence of the free surface energy γ_S is shown as a function of the contents of HPF and A-300 (Figure 6.20).

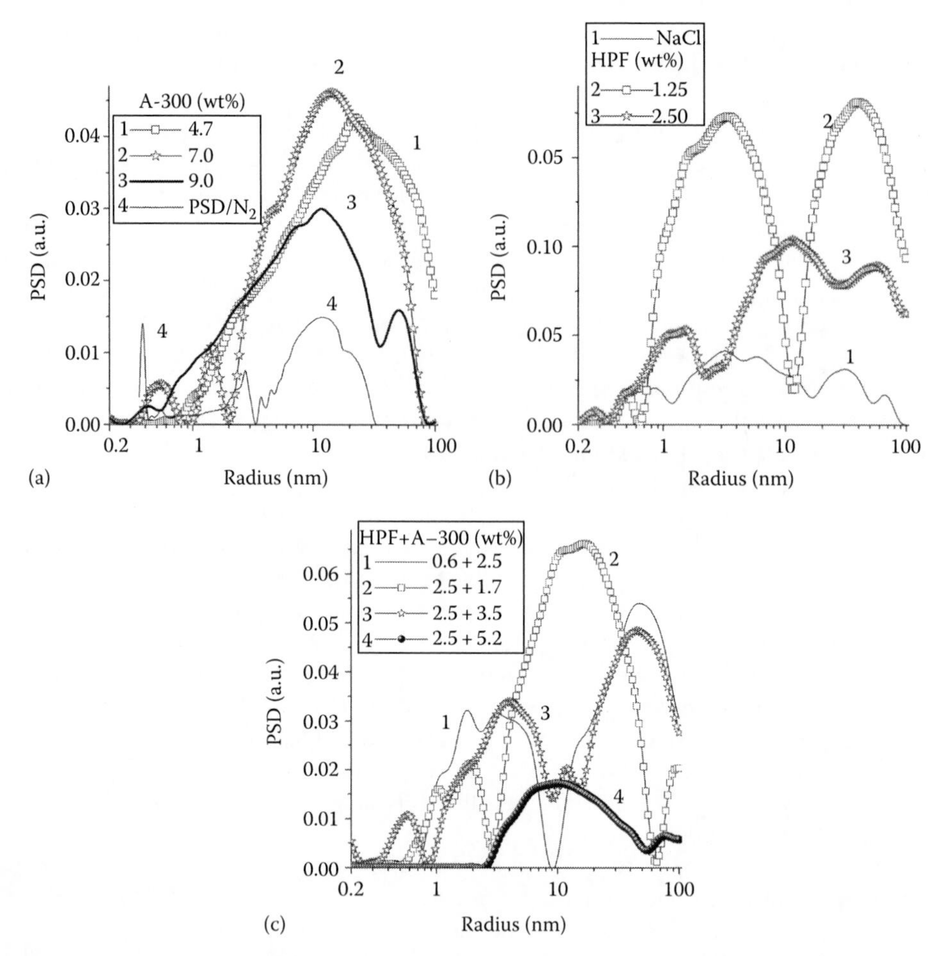

FIGURE 6.19 The size distributions of pores filled by unfrozen water for (a) aqueous suspension of A-300 and the PSD calculated on the basis of the nitrogen adsorption/desorption isotherm (with the model of voids between spherical particles); (b) 0.15 mol NaCl or HPF solution at 1.25 and 2.5 wt%; and (c) HPF/A-300 at different concentrations of protein and silica. (Adapted with kind permission from Springer Science+Business Media: *Cent. Eur. J. Chem.*, Interaction of fibrinogen with nanosilica, 5, 2007, 32–54, Rugal, A.A., Gun'ko, V.M., Barvinchenko, V.N., Turov, V.V., Semeshkina, T.V., and Zarko, V.I.)

There are several types of the interactions in the system: silica particle–particle, particle–HPF, and HPF–HPF molecules, as well as water–particle, water–HPF, and water–HPF–particle. Points of intersection of the dependences $\gamma_S(C_{HPF})|_{SiO_2}$ and $\gamma_S(C_{SiO_2})|_{HPF}$ extrapolated to the *ZX*-coordinate plane correspond to the free surface energy without one of the mentioned interactions (for silica and HPF). This value for the HPF solution corresponds to the Gibbs free energy of self-association of protein molecules or silica particle–particle themselves. Self-association of HPF molecules gave changes in the γ_S value by 600 J/g, and in the case of silica particle–particle interaction it was equal to 200 J/g. It should be noted that the γ_S value for the HPF/A-300/PBS system could be even less than that for the aqueous suspension of silica/NaCl at the same concentration. This strong dehydration of protein molecules is well seen in Figures 6.18 and 6.19 and Table 6.2. Notice that at the zero concentration of HPF, all points of the dependence $\gamma_S(C_{SiO_2})$ are on a straight line despite one of the suspension of A-300 was without NaCl. This fact confirms the absence of the bulk phase of the salt solution in the frozen aqueous suspensions of silica.

In the case of the adsorption of HPF (Figure 6.21), changes in the TSDC thermogram are much larger than that for A-300 or HPF solution (Figure 6.21a). The glass transition temperature for the

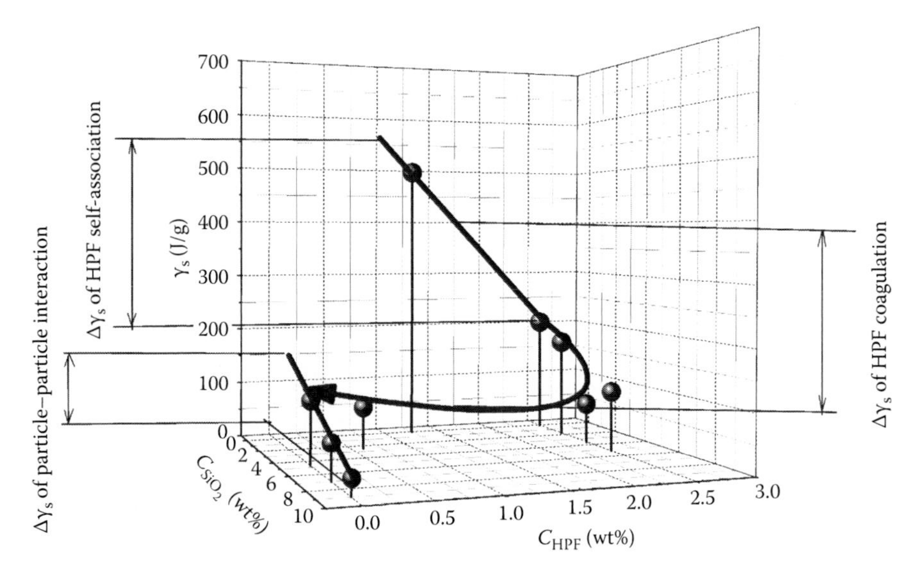

FIGURE 6.20 Free surface energy γ_S as a function of concentration of HPF and silica. (Adapted with kind permission from Springer Science+Business Media: *Cent. Eur. J. Chem.*, Interaction of fibrinogen with nanosilica, 5, 2007, 32–54, Rugal, A.A., Gun'ko, V.M., Barvinchenko, V.N., Turov, V.V., Semeshkina, T.V., and Zarko, V.I.)

HPF–silica system from the TSDC thermogram is ~202 K. This temperature corresponds to the appearance of mobile water molecules registered in the ^{1}H NMR spectra. The activation energy of dc relaxation (caused by the proton and ion conductivity with the through current between electrodes corresponding to ion percolation effect) is greater for HPF/A-300 ($E_{a,dc}$ = 168 kJ/mol) than that for A-300 (75–85 kJ/mol) or ossein/A-300 (90 kJ/mol) (Gun'ko et al. 2007i, 2009d). This suggests the formation of stronger complexes of HPF with silica particles than that for silica–silica or ossein–silica (because ossein is a small protein at $M_w \approx 20$ kDa). Calculations of the size distributions of the water clusters (which are similar to the PSDs shown in Figure 6.19c) using the TSDC cryoporometry (Figure 6.21c) showed that contribution of water superdomains at $R > 1.5$ nm increased for HPF–silica in comparison with the aqueous suspension of pure silica. The NMR cryoporometry gave similar results showing increased contribution of large superdomains of water in the HPF–silica systems in comparison with those for the silica suspensions (Figure 6.19). Certain dehydration of HPF was observed on the adsorption because the PSD of HPF–silica decreased at $R < 1.5$ nm (Figure 6.21c).

The local mobility of water molecules (NMR) and the dc relaxation (TSDC) appear at close temperatures (Figure 6.21b) maybe because of the presence of dissolved NaCl causing better conditions for the ion percolation on the dc relaxation.

Thus, interaction of fibrinogen molecules with water and nanosilica depends on their concentrations, as well as on pH and salinity of the solution. The adsorption of HPF on nanosilica is practically irreversible and the maximum value of the absorption is observed close to its IEP. The adsorption at C_{HPF} = 156 mg/g of silica corresponds to the monolayer coverage because subsequent increase in the HPF concentration C_{eq}>0.3 mg/mL leads to the polymolecular adsorption. However, even at C_{HPF} = 500 mg/g not all surface silanols take part in the hydrogen bonds with adsorbed HPF (in dried residue). Consequently, the HPF–silica interfaces include certain voids (in dried samples) which can be filled by water in the solutions. On freezing of the buffer solution of HPF at C_{NaCl} < 0.15 mol, protein molecules influence the structure of ice in such a way that the mobility of the molecules of water increases that leads to appearance of amorphous ice. This "mobile" ice is observed in the ^{1}H NMR spectra as a wide signal at intensity up to 30% of that of water before freezing.

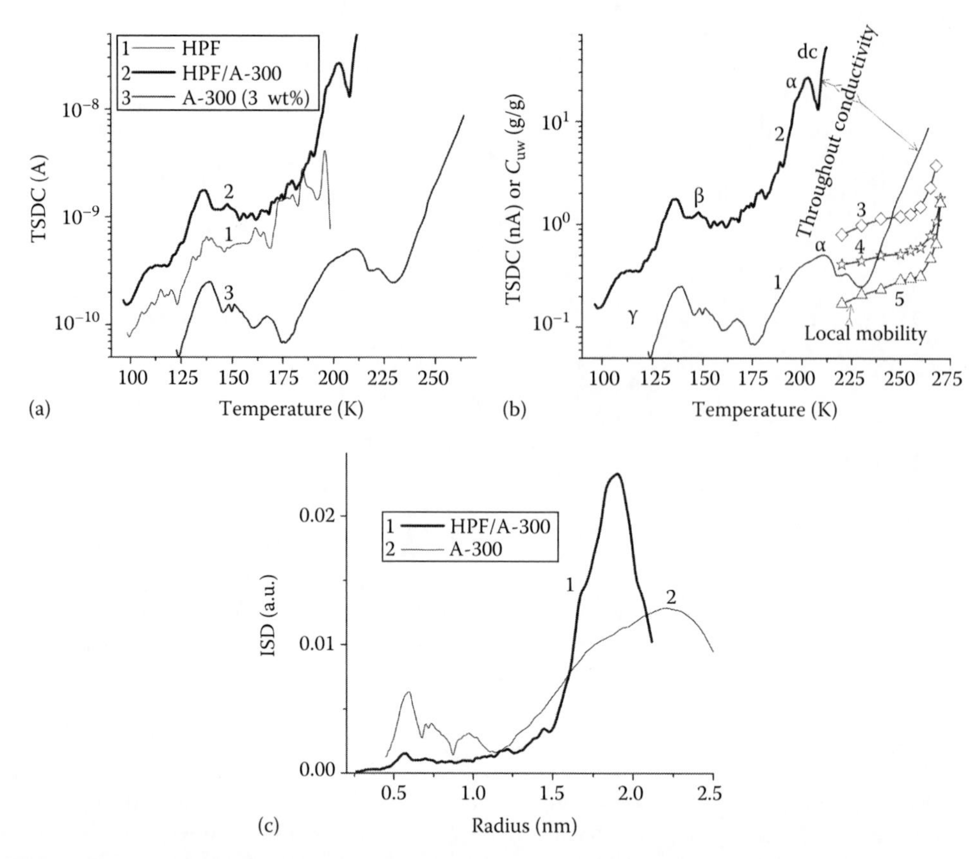

FIGURE 6.21 (a, b) TSDC thermograms of fibrinogen (2.5 wt% HPF in aqueous solution with 0.15 mol NaCl) and after addition of nanosilica A-300 (3.5 wt%); and aqueous suspension of A-300 (3 wt%); (b) TSDC (curves 1 and 2) and C_{uw} (NMR, curves 3, 4, and 5) as functions of temperature for A-300 alone (curves 1 and 5), solution of fibrinogen (2.5 wt% HPF and 0.15 M NaCl, curve 3) and after addition of nanosilica A-300 (3.5 wt%) (curves 2 and 4); and (c) corresponding incremental PSD for the HPF/A-300/water and A-300/water (based on the TSDC data). (Adapted with kind permission from Springer Science+Business Media: *Cent. Eur. J. Chem.*, Interaction of fibrinogen with nanosilica, 5, 2007, 32–54, Rugal, A.A., Gun'ko, V.M., Barvinchenko, V.N., Turov, V.V., Semeshkina, T.V., and Zarko, V.I.)

In the aqueous medium, HPF molecules can bind significant amounts of water which, however, strongly decreases with increasing concentration of HPF. The free surface energy on self-association of HPF molecules reaches 350 J/g. Addition of nanosilica to the solution of HPF leads to sharp reduction of the amounts of structured water and the free surface energy value. This testifies that more intensive coagulation of protein molecules occurs in the presence of nanosilica. The spectral data show that there are certain (but small) conformational changes of adsorbed HPF molecules which can suggest that the amount of α-helix diminishes. This is in agreement with the published results related to the adsorption of HPF onto hydrophilic surfaces. The absence of destructive changes and insignificant structural transformations of the HPF molecules adsorbed on nanosilica suggest that this material has appropriate biocompatibility with respect to plasma proteins.

The used techniques for calculations of the PSD on the basis of the ^{1}H NMR and TSDC spectra and modified regularization procedure applied to the integral equation based on the Gibbs–Thomson relation for the freezing point depression for confined liquids allowed us to reveal significant changes in the structure of bound and bound waters depending on the concentrations of HPF and nanosilica in their aqueous solutions.

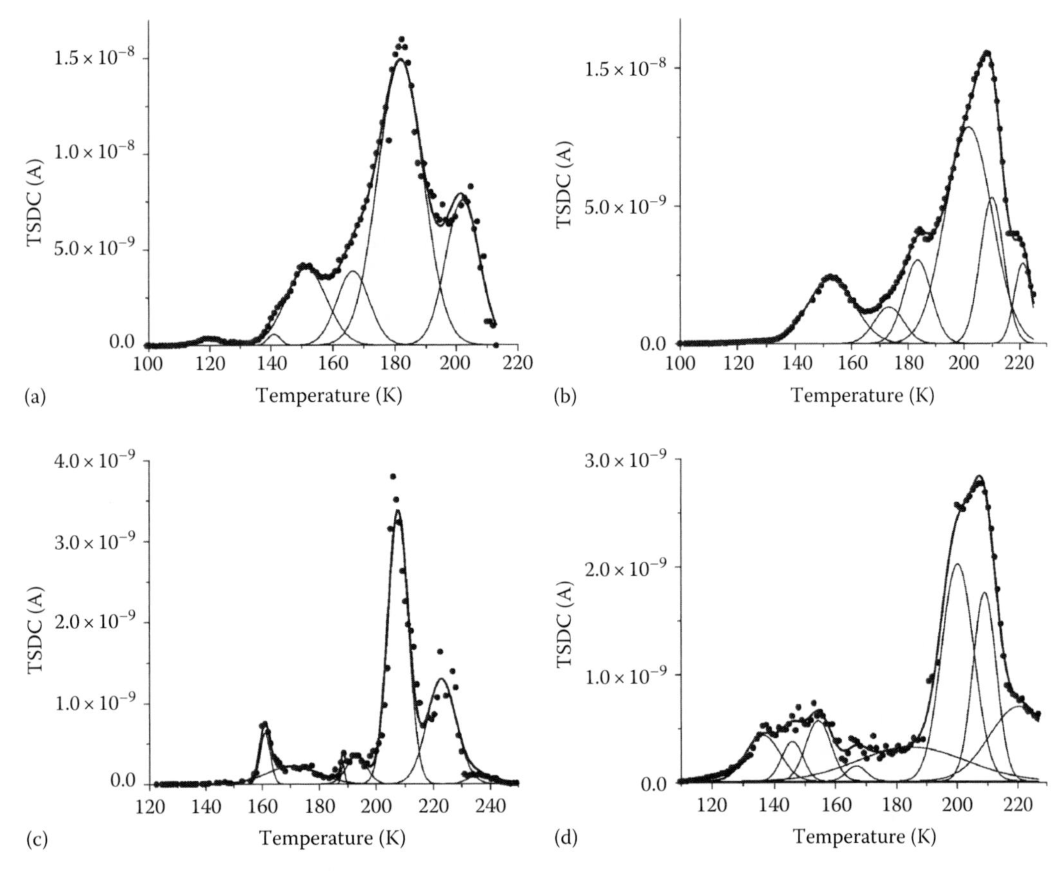

FIGURE 6.22 TSDC thermograms of (a) BSA (0.146 wt%), (b) BSA (0.162 wt%)–A-300, (c) BSA (0.146 wt%)–SA1, and (d) BSA (0.136% wt%)–ST20 at oxide content of 1 wt% at $T < T_{dc}$. (According to Gun'ko, V.M. et al., *Surface*, 16, 14, 2009f (in Russian).)

TSDC method was also used to study relaxation processes at a surface of nanooxides A-300 ($S_{BET} = 285$ m²/g), ST20 ($C_{TiO_2} = 20$ wt%, $S_{BET} = 84$ m²/g), and SA1 ($C_{Al_2O_3} = 1$ wt%, $S_{BET} = 203$ m²/g) alone and with adsorbed BSA (Fluka, fraction V) in aqueous medium (Gun'ko et al. 2009f). BSA adsorption was carried out at pH 4.8 at room temperature for 2 h at stirring, and then it was centrifuged at 4000 rpm. The adsorbed BSA content determined using a spectrophotometric method was 0.162, 0.146, and 0.136 g/g of A-300, SA1, and ST20, respectively. In TSDC measurements, 1 wt% suspensions of BSA–oxide and 0.146 wt% BSA solution were used. The TSDC thermograms (Figure 6.22) were recorded at heating of samples (cooled from 265 to 90 K and polarized using the electrostatic field at $F_p \approx 2 \times 10^5$ V/m) from 90 to 270 K at a heating rate of 3 K/min without the field.

TSDC thermogram of frozen aqueous solution of BSA (Figure 6.22a) is characterized by smaller current values in low-temperature, LT region at $T < 160$ K (where dipolar relaxation of OH groups, water molecules, polar bonds in the protein, and surface functionalities occurs) than in high-temperature, HT region at $T > 160$ K. Relatively low TSDC values in the LT region are due to bonding of a major portion of water to BSA (having large contact area $S > 1000$ m²/g with water in diluted solutions). This large S value can be seen from comparison of sizes of A-300 particle (providing $S_{BET} = 300$ m²/g) and HSA molecule (Figure 6.23).

The mobility of water molecules bound to BSA depends strongly on the mobility of macromolecule fragments which appears in the TSDC spectra at higher temperature ($T > 160$ K) than the mobility of water molecules alone at $T < 160$ K. Therefore, TSDC for BSA–water significantly

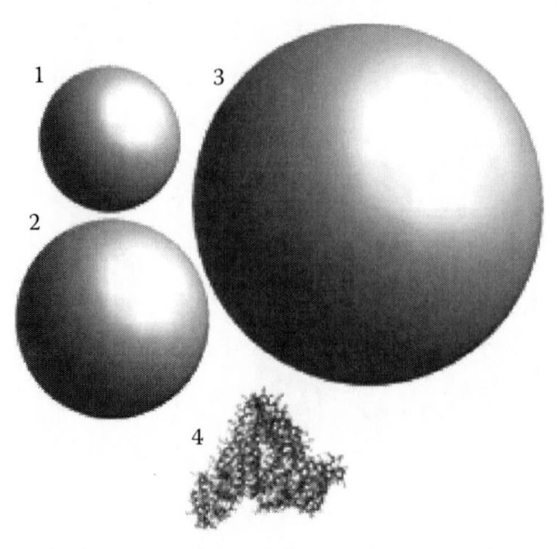

FIGURE 6.23 Relative sizes of primary particles of (1) A-300 (average diameter 9.6 nm), SA1 (2, 13.3 nm), ST20 (3, 27.7 nm), and HSA molecule (4). (According to Gun'ko, V.M. et al., *Surface*, 16, 14, 2009f (in Russian).)

increases in the HT region at $T > 160$ K where BSA fragments with bound water molecules participate in the relaxation. A similar relationship of the TSDC intensity in LT and HT regions is characteristic for BSA–oxide–water systems (Figure 6.22). However, the TSDC intensity for the latter is lower than that for the BSA solution because of the interactions of macromolecules with NP (of similar sizes, Figure 6.23) resulting in the formation of hybrid aggregates (50–500 nm in size) and a decrease in the amount of bound water there. This is due to adsorptional compaction of BSA with partial dehydration. In contrast to the systems with BSA–oxide–water, 1 wt% aqueous dispersion of oxides alone, as well as pure water, are characterized by the opposite ratio of the intensities in the LT region ($100 < T < 160$–170 K) and HT region (160–$170 < T < T_{dc}$, where T_{dc} is the temperature of the beginning of the dc relaxation). TSDC intensity is higher for A-300–water than for pure water (Figure 6.24a) because of the polarization of bound water by the electrostatic field of the oxide surface. For SA1 and ST20, this effect is lower because these oxides have smaller S_{BET} value.

Interaction of BSA with A-300 causes an increase in the temperature of dipolar relaxation in comparison with BSA alone, e.g., the main peak in the HT region shits by 26 K (Figure 6.22). This effect is accompanied by an increase in the free energy of activation of relaxation (Table 6.3, $\Delta G^{\neq}$) for both BSA fragments bound to silica and bound water, an increase in the effective permittivity (ε), relaxation time (τ), and activation energy of depolarization (Figure 6.25). One can assume that the sizes of relaxed structures increase because of aggregation of oxide NP with macromolecules.

Therefore, the mobility of macromolecules decreases but the $\Delta G^{\neq}$ and E values increase since the relaxation occurs at higher temperatures and the TSDC intensity decreases. The $f(E)$ peaks in the 20–45 kJ/mol (BSA) shift toward 30–55 kJ/mol (BSA–A-300) (Figure 6.25a). For BSA–ST20, a similar displacement is observed for the $f(E)$ peak at 40 → 55 kJ/mol; however, another peak shifts in the opposite direction (25 → 15 kJ/mol) (Figure 6.25c). For BSA–SA1, the $f(E)$ peak positions are practically the same (Figure 6.25b). However, the intensity of a low-energy peak (10 kJ/mol) decreases but for high-energy peak (70 kJ/mol) and a peak at 25 kJ/mol, it increases.

The latter can be explained by the presence of stronger acidic sites at SA1 surface in comparison with ST20 or A-300. The adsorption layer structure with BSA–water differs for different oxides that reflect in TSDC thermograms (Figures 6.22 and 6.24), $f(E)$ (Figure 6.25), and activation parameters (Table 6.3).

Notice that the ε value for BSA–oxide–water systems can be greater (Table 6.3, $\varepsilon = 100$–230) than that of bulk water ($\varepsilon = 78$).

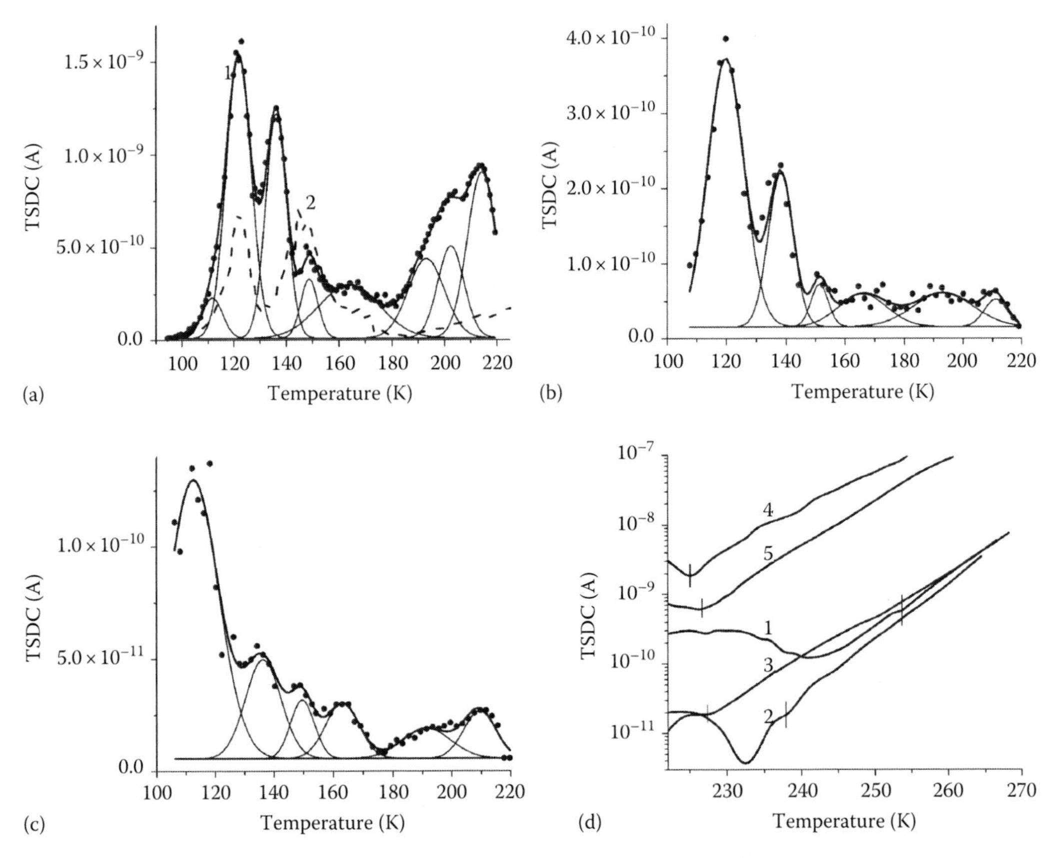

FIGURE 6.24 TSDC thermograms of 1 wt% aqueous suspensions of A-300 (curve 1) (curve 2 is pure water, TSDC is normalized to $F_p = 2 \times 10^5$ V/m) (a), SA1 (b), and ST20 (c); region of dc relaxation for water (curve 1), 1 wt% aqueous suspensions of A-300 (curve 2), ST20 (curve 3), BSA-A-300 (curve 4), and BSA-ST20 (curve 5). (According to Gun'ko, V.M. et al., *Surface*, 16, 14, 2009f (in Russian).)

However, for the oxide–water systems $\varepsilon < 23$ that correspond to the relaxation processes with the participation of water clusters but not continuous structures because oxide NP disorder the continuous structure of bulk water. The relaxation time of bound water (Table 6.3, $\tau = 300$–500 s) is much longer than the relaxation time of OH groups or individual water molecules. This is due to both sizes of relaxed structures and strong electrostatic fields at the oxide surface which hinder reorientation processes in the adsorption layer. At the presence of BSA, the relaxation time changes smaller (<50%) than the ε value which increases by order of magnitude (Table 6.3) because the number of bound charges (polar groups) strongly increases.

The activation energy of the dc relaxation (direct current with the through conductivity appearing due to charge percolation between electrodes) of pure water and 1 wt% A-300 suspension is similar ($E_{dc} = 99.9$ and 98.7 kJ/mol, respectively). However, it decreases for the ST20 suspension (72.9 kJ/mol) due to easier deprotonation of acidic bridges SiO(H)Ti in comparison with SiOH groups. For BSA–A-300–water, $E_{dc} = 62.1$ kJ/mol (i.e., the proton mobility increases), and for BSA–ST20–water, $E_{dc} = 74.7$ kJ/mol (the proton mobility decreases in comparison with the ST20 suspension), and T_{dc} for BSA–A-300–water decreases but for BSA–ST20–water it does not. This can be explained by effective interactions of BSA with surface acidic hydroxyls that lead to decrease in the mobility of protons of these groups.

The binding of low-molecular organics to proteins was investigated for HSA and BSA–water–sugar (glucose, fructose, or saccharose) systems and upon addition of nanosilica (Turov et al. 2010e).

TABLE 6.3
Time (τ) and Free Energy of Activation ($\Delta G^{\neq}$) of Dipolar Relaxation at Peak Temperature (T_i) and Permittivity (ε) Corresponding to Relaxed Structures

Sample	T_i (K)	τ (s)	ε	$\Delta G^{\neq}$ (kJ/mol)
A-300	122.9	342	1.3	11.6
A-300	136.3	300	22.6	12.9
A-300	147.7	480	11.6	14.6
A-300	213.3	318	8.1	21.0
SA1	119.8	288	3.7	11.1
SA1	138.1	313	5.0	13.1
SA1	150.4	428	3.2	14.8
SA1	172.4	468	2.6	17.3
SA1	207.0	206	2.3	19.6
ST20	112.2	302	3.6	10.4
ST20	134.3	376	2.4	12.9
ST20	165.1	469	1.8	16.5
ST20	211.6	500	1.3	21.6
BSA	119.4	126	1.0	10.3
BSA	150.5	95	4.3	12.9
BSA	182.2	518	118.9	18.5
BSA	204.6	596	228.4	21.2
BSA–A-300	152.4	351	1.1	14.7
BSA–A-300	184.1	633	40.0	19.0
BSA–A-300	208.0	614	139.6	21.6
BSA–A-300	216.8	172	151.6	20.3
BSA–SA1	161.0	587	1.0	16.3
BSA–SA1	188.6	439	3.5	18.9
BSA–SA1	206.0	99	8.7	18.2
BSA–SA1	222.3	284	28.4	21.8
BSA–ST20	137.3	177	1.3	12.4
BSA–ST20	152.8	324	14.1	14.7
BSA–ST20	165.0	130	7.3	14.7
BSA–ST20	207.2	730	101.5	21.9

Source: According to Gun'ko, V.M. et al., *Surface*, 16, 14, 2009f (in Russian).

BSA and HSA (Biopharma, Kiev, Ukraine), fructose, glucose, and saccharose (Merck, Germany) were used as received. The solutions of the proteins and sugars were prepared using the 0.9 wt% NaCl solution in distilled water. Immunoglobulin (Ig) solutions were prepared by dilution of 10% Ig solution (Biopharma, Kiev, Ukraine, 97% purity with 92% of monomers and dimers, 3% of fragments, and 5% of oligomers) by water or buffer solutions. Concentration of Ig was determined spectrophotometrically at $\lambda = 278$ and 550 nm using the Biuret method. The pH values of the solutions were measured using an EV-74 pH meter with a glass electrode. The pH value was adjusted using the buffer solutions, HCl or KOH. Adsorption of proteins on nanosilica was determined spectrophotometrically. Desorption of proteins was measured for samples incubated at room temperature for 1.5 h and centrifuged at 8000 rpm for 15 min, and protein concentration was

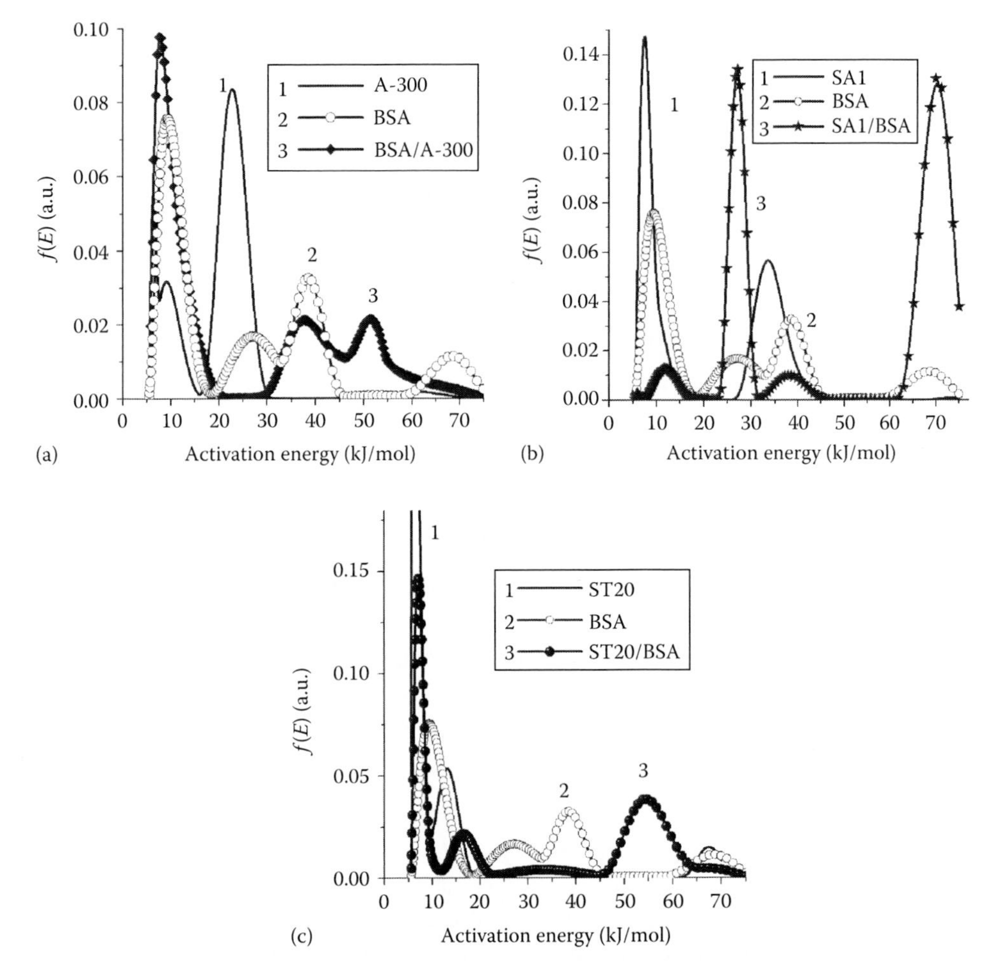

FIGURE 6.25 Distribution functions of activation energy of dipolar relaxation of frozen 1 wt% suspensions of oxides (curve 1), 0.146 wt% BSA solution (curve 2), and oxide–BSA suspensions (curve 3) for A-300 (a), SA1 (b), ST20 (c) at pH 4.8. (According to Gun'ko, V.M. et al., *Surface*, 16, 14, 2009f (in Russian).)

determined in the supernatant liquid. Highly disperse silicas A-175 (S_{BET} = 186 m^2/g) and A-300 (320 m^2/g) were heated at 700 K for several hours before they were used to remove adsorbed organics and residual HCl.

Serum albumins (HSA, BSA) having native globular structure include a significant quantity of bound water affecting their characteristics. Albumins fulfill a transport function in the organisms; therefore, they have labile and pH-dependent structures (Chager 1975), and organic compounds present in a solution and competing with water molecules for albumin adsorption sites can influence the protein characteristics. The difference in the properties of albumins in unloaded and loaded states is used by the receptor system for utilization of transported compounds.

Figure 6.26 shows the ^{1}H NMR spectra of unfrozen water bound to BSA and HSA in a frozen solution at $T < 273$ K. At a high resolution (400 MHz, Figure 6.26a), the spectra represent a broad asymmetrical signal whose width increases with lowering temperature because of a decrease in the mobility of water molecules (Mank and Lebovka 1988). The chemical shift of the proton resonance of this water is $\delta_H = 4–5$ ppm, i.e., this is SAW (Gun'ko et al. 2005d). At low temperatures, the asymmetry decreases, and the down-field shift of signal is observed. This can be interpreted as the bonding of the water molecules to most polar sites of BSA that result in an increase in the atomic charges and the proton shielding decreases. A more complex signal is observed for the HSA

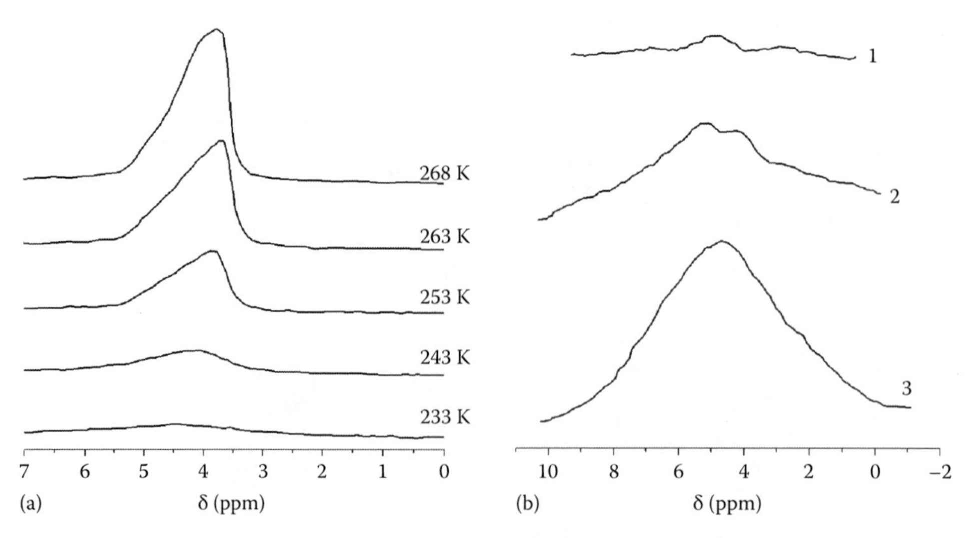

FIGURE 6.26 ^{1}H NMR spectra of unfrozen water: (a) BSA/water at $C_{BSA}=2$ wt% recorded at different temperatures and (b) HSA/saccharose/water at $C_{HSA}=1$ wt% and saccharose concentration $C_{Sc}=0$ (curve 1), 0.1 (curve 2), and 1 (curve 3) wt% recorded at 250 K.

solution with the presence of sugar (lower resolution at 100 MHz, Figure 6.26b) as its components differ by ~1 ppm.

These features of the ^{1}H NMR spectra of unfrozen water bound to protein–sugar reveal the existence of several forms of water characterized by fast or slow (in NMR timescale) molecular exchange. The structural differentiation of water bound to proteins can occur due to its interactions with hydrophobic and hydrophilic protein functionalities (Shrade et al. 2001, Gun'ko et al. 2009d). The water characteristics depend on the average number of the hydrogen bonds per molecule, n_H. A decrease in the n_H value is accompanied by the up-field shift of ^{1}H NMR signal. Consequently, signals at smaller δ_H values correspond to more strongly clustered (or more weakly associated) water characterized by a smaller average number of the hydrogen bonds per molecule.

There is a regularity in the albumin–sugar–water system corresponding to a sugar concentration range giving a decrease in unfrozen water content. This is clearly observed for the HSA solution with saccharose (Figure 6.26b) since addition of 1 wt% of saccharose leads to a decrease in the signal intensity of unfrozen water (at 250 K) more than five times. The obtained results testify about effective interaction of sugar with the protein molecules, i.e., the saccharose molecules can displace the major portion of water bound to protein.

Addition of very small quantities of sugars leads to a decrease in the amounts of unfrozen water. For fructose (Figure 6.27b), this effect is in a narrow concentration range. For glucose (Figure 6.27a), a considerable decrease in the C_{uw} value is observed only at $C_{Gl}<0.38$ wt% and T near 273 K, and a section of an increase in C_{uw} versus C_{Gl} is over a wide range of the concentrations and temperatures. The greatest changes in the amounts of unfrozen water are observed for the saccharose solution (Figure 6.27c), and diminution in C_{uw} is greater at lower temperatures. At 250 K, the signal of unfrozen water is not observed with increasing saccharose concentration at $C_{Sac}>1$ wt%. Thus, the aqueous solutions of glucose, fructose, and saccharose are characterized by different C_{uw} dependences on temperature and sugar content. Therefore, one can assume that the effects of these sugars on the aqueous solutions of proteins could be strongly different.

The surface energy γ_S diminishes with increasing concentration of BSA in the aqueous solution (Figure 6.28a).

Notice that similar dependences were observed for other heterogeneous systems. Protein molecules are strongly hydrated. Therefore, to provide tight contacts between them (on self-association), a certain quantity of bound water should be removed from intermolecular voids to the

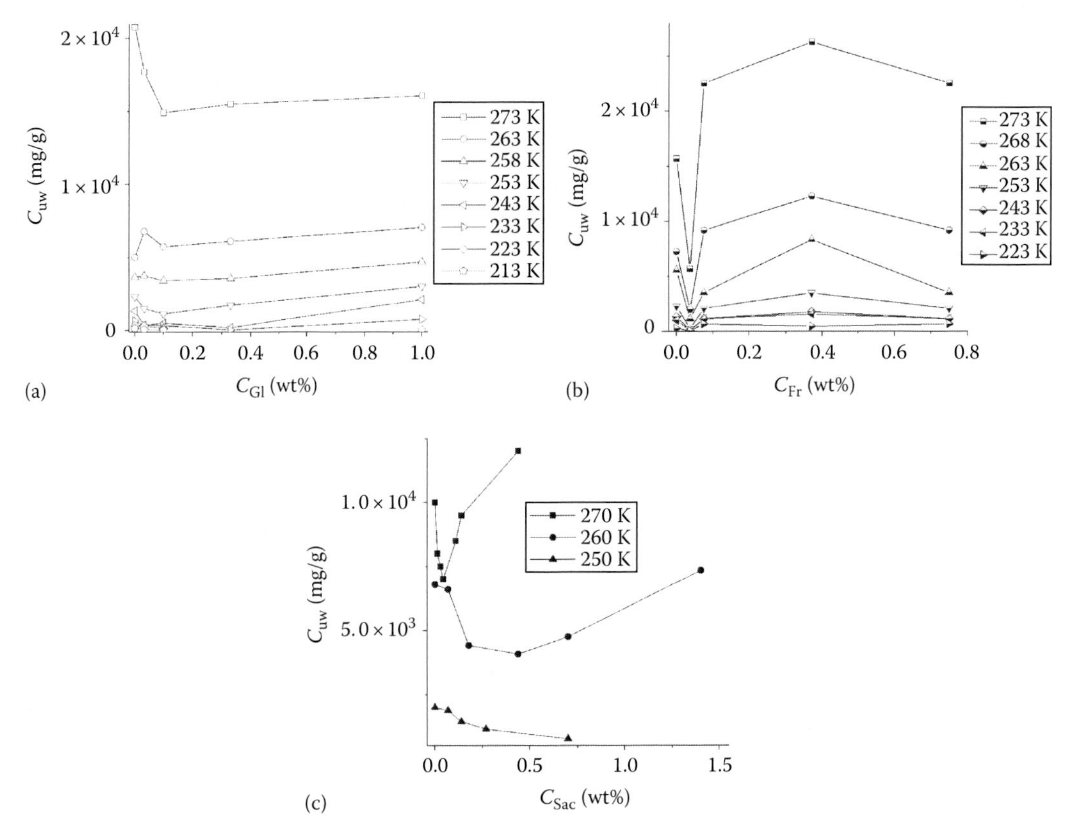

FIGURE 6.27 Dependence of C_{uw} on concentration of sugars in aqueous solutions at different temperatures: (a) glucose, (b) fructose, and (c) saccharose.

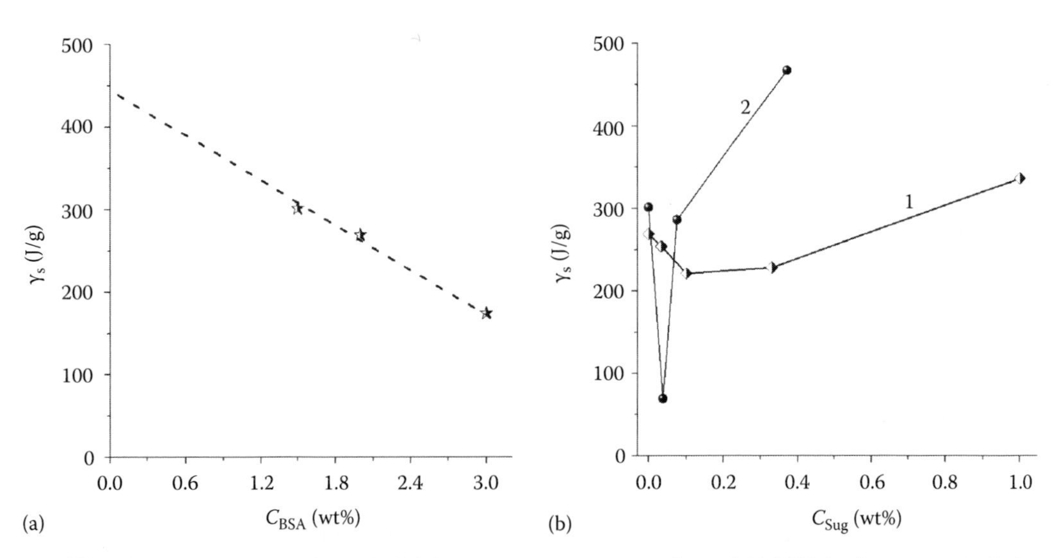

FIGURE 6.28 Dependence of interfacial free energy on concentrations of (a) BSA in the aqueous solution and (b) glucose at $C_{BSA} = 2$ wt% (curve 1) or fructose at $C_{BSA} = 1.5$ wt% (curve 2) in the ternary systems BSA/monosugar/water.

bulk that result in a decrease in the integral γ_S value. Thus, the amounts of bound water decrease as well as the γ_S value with increasing protein content in the solution. The difference in the γ_S values determines the energy of self-association of the protein molecules. Extrapolating $\gamma_S(C_{BSA})$ dependence to the zero concentration gives the surface energy of non-associated BSA molecules at $\gamma_S \approx 450$ J/g. The mentioned effects can be changed by addition of sugars to the aqueous solutions of proteins.

The $\gamma_S(C_{Sug})$ dependences (Figure 6.28b) have the shapes similar to the $C_{uw}(C_{Gl(Fr)})$ graphs (Figure 6.27). Simple estimations show that in the 2 wt% BSA solution, 8.8×10^5 H_2O are per protein molecule and $\sim 10^4$ H_2O from them are bound to macromolecules. Each sugar molecule can reduce a quantity of bound water by $\sim 10^3$ H_2O or even more since significant changes in the γ_S value are observed at $C_{Sug} < 0.1$ wt% (Figure 6.28b). It is not possible to explain the substantial dehydration of BSA by only replacement of bound water by sugar molecules. Therefore, one can assume that on the bonding of sugars to the protein molecules, certain changes occur in the conformation of macromolecules that result in strong diminution of the quantity of bound water. The effect of the significant dehydration of protein molecules in the presence of sugars (maximal at low temperatures) can be one of the factors responsible for the cryoprotective properties of saccharides used in the nature. Relatively small molecules of sugars can penetrate through cell membranes, cause certain dehydration of intracellular structures, and reduce the formation of intracellular ice crystallites. Additionally, sugars do not have a negative influence on living cells because they can be easily utilized in the cellular metabolism processes. The relationships between the amounts of unfrozen water and its energetic characteristics for the solutions with BSA alone (Figure 6.29a, curves 1–5) and on addition of nanosilica (curve 6) suggest that practically entire amounts of unfrozen water correspond to SBW. A section corresponding to WBW is observed only for the 0.25% BSA solution (Table 6.4). The γ_S values for the hydrated powder and aqueous solution of BSA are characterized by almost linear increase with increasing C_{uw} value (Figure 6.29b) dependent on protein hydration (total content of water C_{H_2O}) (Figure 6.27).

For wet BSA powders, the maximum changes in the γ_S value caused by the water bonding are equal to 70 J/g (Figure 6.29b). Addition of 1 wt% of nanosilica to the 10 wt% BSA solution leads to changes in the γ_S value decreased by 11 J/g. This value characterizes the interaction of silica NP with BSA causing the displacement of water bound to both macromolecules and NP. Notice that silica (A-300) NP and albumin molecules (BSA and HSA molecules have close sizes) have close sizes (Figure 6.29) and the A_m value (~500 mg/g) corresponds to the protein monolayer adsorption

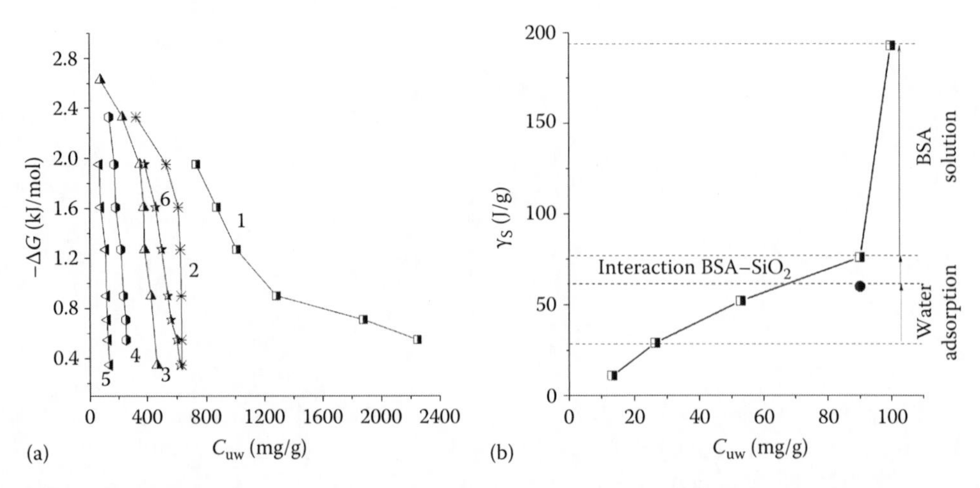

FIGURE 6.29 Relationships between changes in the Gibbs free energy and the amounts of unfrozen water for (a) BSA differently hydrated at $C_{H_2O} = 95.8$ (curve 1), 90 (curve 2), 52.8 (curve 3), 26.4 (curve 4), and 13.2 (curve 5) wt% for protein alone and on addition of 1 wt% of silica A-175 at $C_{H_2O} = 89$ wt% (curve 6) and (b) free surface energy as a function of C_{uw}.

TABLE 6.4
Characteristics of Bound Water in Hydrated BSA

No.	C_{BSA} (wt%)	γ_S (J/g)	C^s_{uw} (mg/g)	C^w_{uw} (mg/g)	$-\Delta G_s$ (kJ/mol)	$-\Delta G_w$ (kJ/mol)
1	0.25	193	1500	1700	3.4	1.6
2	10	76	620	—	2.8	—
3	47.2	52	528	—	2.8	—
4	73.6	29	264	—	2.8	—
5	86.8	11	132	—	3.0	—
6	10 (+1 wt% A-175)	65	700	—	2.8	—

Notes: Types of water: strongly (C^s_{uw}, SBW frozen at $T<250$ K and $\Delta G_s<-0.5$ kJ/mol) and weakly (C^w_{uw}, WBW frozen at $T>250$ K and $\Delta G_w>-0.5$ kJ/mol) bound waters.

at protein:silica ≈ 1:1, since their densities twice differ. Therefore, during the adsorption of albumins, macromolecules and silica NP form hybrid aggregates characterized by certain changes in the protein conformation affecting the adsorption of sugars.

An important task in immunology is development of new methods of immune activation using artificial antigens or vaccines, control of concentration and activity of antigens (antibodies such as Ig). Ig like other proteins can irreversibly adsorb on silica (Larsericsdotter et al. 2001). The Ig molecules (~160 kDa, length ~23 nm and average diameter ~5 nm) have a complex shape with two larger and two smaller polypeptide chains bonding by disulfide bridges. Ig adsorption (*A*) versus pH has a bell-shaped form at a maximum close to the PZC (or IEP) of Ig at pH 6.6 (Figure 6.30a). A similar *A*(pH) shape is typical for proteins as ionogenic macromolecules because an area occupied by a molecule is minimal at the PZC when it has the most compacted shape and the repulsive electrostatic interactions between adsorbed macromolecules are minimal. The appearance of negative charges on the molecules at $pH>pH_{PZC}$ leads to a decrease in the adsorption (in comparison with the adsorption at $pH<pH_{PZC}$) as a result of the electrostatic repulsion between them and negatively charged silica surface.

At $pH<pH_{PZC}$, adsorption decreases smaller because the electrostatic repulsion remains only between macromolecules but not between the molecules and silica having $pH_{PZC}\approx 3.5$ and low negative surface charge density at $pH_{PZC}<pH<7$ (see, e.g., Figure 2.61).

The Ig adsorption isotherms (Figure 6.30b) have the Langmuir shape and the corresponding monolayer adsorption capacity is $A_m = 105$ and 120 mg/g at pH 2.2 and 6.4, respectively. These A_m values correspond to a minimal thickness of the Ig adsorption layer due to its planar adsorption. Notice that for globular albumins $A_m = 300$–600 mg/g. For proteins, the adsorption and desorption isotherms do not coincide because of difficulties of macromolecule desorption requiring simultaneous breakage of all intermolecular bonds between macromolecules and a solid surface. Therefore, if desorption and adsorption ($A \leq A_m$) of Ig occurs at the same pH then the adsorption is practically irreversible. Desorption of Ig increases with increasing amount of adsorbed Ig but the quantity of desorbed protein is small (Figure 6.30c) even at significant changes in pH. The dependence of free surface energy as a function of concentrations of Ig and silica (Figure 6.31) includes three specific sectors. At $C_{SiO_2}=0$, $\gamma_S(C_{Ig})$ represents the modulus of the total changes in the Gibbs free energy of bound water changing due to self-association of macromolecules. At $C_{Ig}=0$, the $\gamma_S(C_{SiO_2})$ dependence describes the corresponding changes caused by interparticle interactions. If concentrations of both Ig and silica are not zero, then the $\gamma_S(C_{Ig}, C_{SiO_2})$ dependence is determined by both adsorption and coagulation processes. Taking into account the Langmuir shape of the Ig adsorption at $A<A_m$, one can assume that strong coagulation of macromolecules interacting with different silica particles does not occur at used concentrations. The Ig molecules are surrounded by a

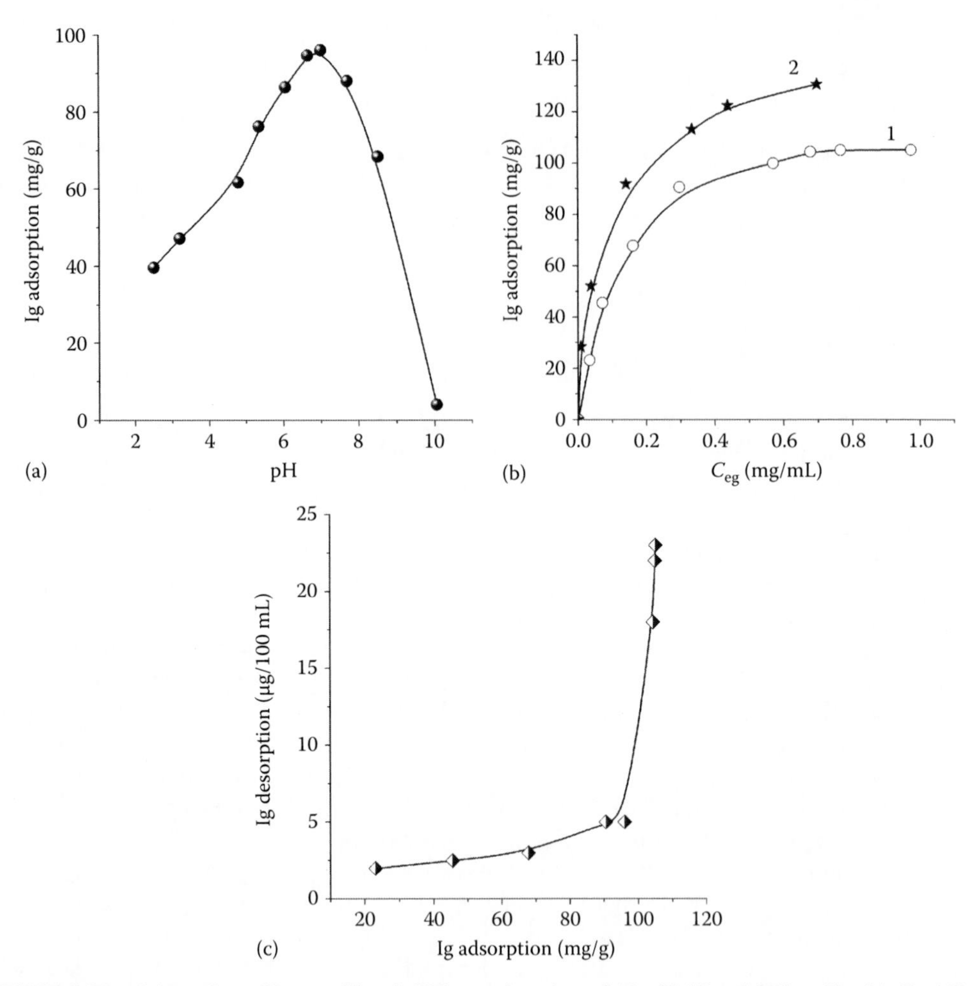

FIGURE 6.30 Adsorption of Ig on silica A-300 as a function of (a) pH (C_{Ig}=0.075 wt%), (b) C_{Ig} (pH 2.2 [curve 1] and 6.4 [curve 2]), and (c) desorption of Ig as a function of adsorbed amounts.

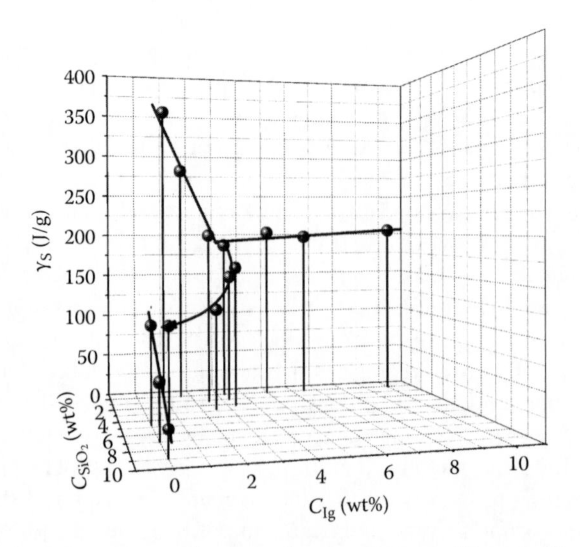

FIGURE 6.31 Free surface energy γ_S as a function of concentrations of Ig and silica A-300.

TABLE 6.5
Characteristics of Water Bound in Hydrated Ig and Suspensions of A-300 and Ig/A-300

C_{Ig}/C_{SiO_2} (wt%)	C_{solid} (wt%)	γ_S (J/g)	C^s_{uw} (g/g)	C^w_{uw} (g/g)	$-\Delta G_s$ (kJ/mol)	$-\Delta G_w$ (kJ/mol)
1/0	1	352	2.1	9.9	3	0.7
1.65/0	1.65	280	2.1	7.9	2.5	0.5
3.3/0	3.3	187	1.7	3.8	2.4	0.7
5/0	5	203	2.0	5.0	2.5	0.7
6.5/0	6.5	197	2.2	4.8	2.7	0.5
10/0	10	205	1.7	5.3	2.9	0.5
0/4.7	4.7	106	0.7	2.5	2.4	1
0/7	7	59	0.6	1.6	2.6	0.5
0/9	9	27	0.25	1.0	2.2	0.6
2.2/3.3	5.7	162	1.8	6.2	3	0.5
1/3.3	4.3	149	1.1	6.9	3	0.5
2.7/2.5	5.2	115	0.8	5.2	4	0.4
1.2/2.5	4.7	200	0.8	6.2	4.8	0.3
4.9/0.25	5.05	45	0.5	2.5	3	0.25
5.3/0.5	5.8	107	0.75	4.25	2.2	0.3

thick layer of bound water (Table 6.5) whose thickness increases in dilute solutions with decreasing concentration of proteins because in the dilute buffered solutions the probability of the formation of protein oligomers is low and undistorted hydrate shells of macromolecules have a large thickness.

Maximum diluted Ig solutions (Table 6.5) are characterized by minimal Gibbs free energy of bound water (the γ_S value is maximal ~400 J/g and $C_{uw} \approx 10$ g/g at $C_{Ig} \to 0$) (Figure 6.31). A small changes in the free surface energy at $C_{Ig}=3$–10 wt% can be explained by increased interaction between macromolecules leading to a decrease in the amounts of bound water per macromolecule. Interactions between macromolecules and silica NP are much stronger than between macromolecules; therefore, silica added to the Ig solution leads to diminution of the amounts of water bound by protein. The reduction of the γ_S value for the system Ig–A-300–water is maximal at small C_{Ig} values because relative adsorption is maximal at minimal C_{Ig} values (e.g., with increasing $C_{Ig,eq}$ from 0.1 to 1.0 mg/mL the adsorption increases only twice, Figure 6.30b). At $C_{Ig} > C_{SiO_2}$, changes in the γ_S values are relatively small due to weak interaction of Ig monolayer-coated silica NP with dissolved protein molecules. This is confirmed by the rheometry results showing an extreme dependence of the viscosity of the protein (polymer)–nanosilica suspensions on concentration of proteins (polymers).

For treatment of certain diseases (e.g., wound and purulent infections of internal cavities), the preparations based on nanosilica are successfully used (Chuiko 2003). In some of these cases, silica NP can contact blood. Blood as a multicomponent heterogeneous system contains many types of cells and macromolecules, and the aqueous solution of low-molecular organic and inorganic compounds plays a role of the dispersion medium. Therefore, investigations of hydrate shells of blood components, intermolecular interactions between them alone and upon contacts with solid NP are of importance for deeper understanding the mechanisms of actions of medicinal nanocomposites.

The supramolecular systems with nanosilica, proteins, and monosugars or disaccharides can possess a high bioactivity and be used in bionanocomposites perspective for biotechnology and medicine (Devis and Robinson 2002, Chuiko 2003). It was shown that composites on the basis of nanosilica, BSA, polyol (sorbitol, xylitol), or monosaccharides (fructose, glucose) in wide range of concentrations can stimulate activity and prolong lifetime of cells after their cryopreservation.

Since molecules of mono- and disaccharides poorly adsorb on the nanosilica surface, one can assume that they adsorb to BSA molecules immobilized on silica. As it was shown earlier, sugars

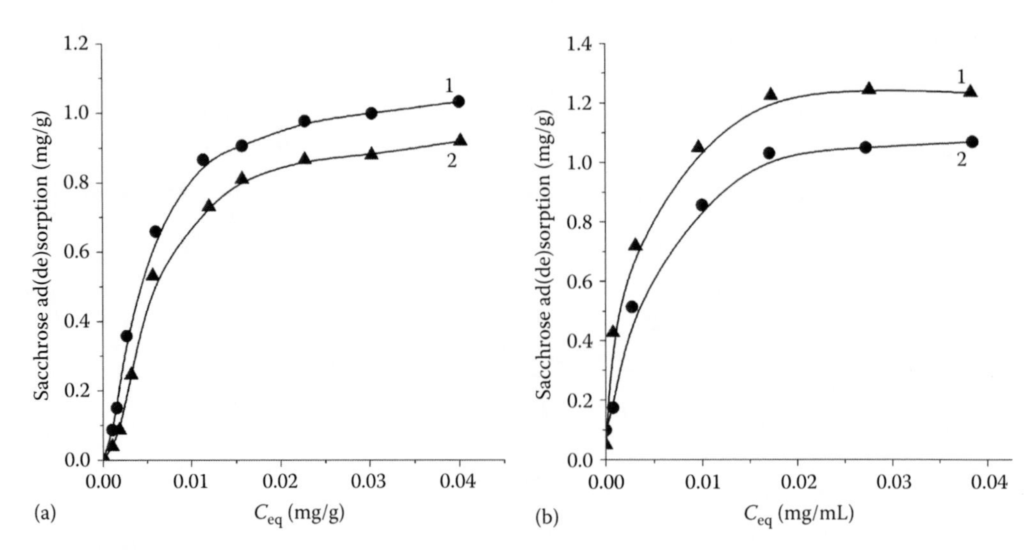

FIGURE 6.32 Adsorption (curve 1) and desorption (curve 2) isotherms of saccharose on surface of (a) nanosilica and (b) BSA/nanosilica at $C_{BSA} = 515$ mg/g.

can displace large amounts of water bound to albumins due to effective protein–sugar interactions. Especially, great effects were observed on the use of saccharose. One can assume that sugars bound in protein–silica nanocomposites and desorbed during interactions with cells are responsible for significant changes in the stimulating effects on cells.

Immobilization of BSA on silica leads to an increase in the adsorption of saccharose by ~20% (Figure 6.32). This difference is lower than could be expected from the data discussed earlier. However, it should be noted that conformational changes in adsorbed protein molecules are more difficult (because of the surface influence and confined space effects) than in dissolved macromolecules alone. Additionally, BSA–nanosilica can form compact hybrid aggregates (vide infra) with reduced accessible surface area of both protein molecules and silica surface for sugar molecules. These effects can be responsible for not high increase in the adsorption of saccharose on BSA–silica in comparison with silica alone.

The interactions of proteins with solid surfaces in biofluids (e.g., blood) are critically important for most applications of biomaterials (Williams 1987, Bamford et al. 1992, Horbett and Brash 1995, van Blaaderen 2003, King et al. 2003, Ma et al. 2007b, Vroman 2008). Body fluids are complex mixtures containing a range of proteins; hence the behavior of biomacromolecules in biofluids or in model systems, particularly the role of proteins acting as precursors for cell attachment, has been extensively studied (Vroman and Adams 1969, Malmsten 1998, Unsworth et al. 2005, Noh and Vogler 2007, Choi et al. 2008, Choi and Chae 2009). The composition of the protein layer adsorbed onto any surface that comes into contact with these fluids depends on the following factors: (i) the surface chemistry, i.e., the degree of hydrophobicity or hydrophilicity and the type and content of surface sites; (ii) the physical characteristics of a surface, i.e., its texture; (iii) protein mobility, conformation, and chemistry. The composition of the adsorbed layer can change within a short period of time after initial exposure to a protein-containing solution (Mandrusov et al. 1998, King et al. 2003, Ma et al. 2007b, Vroman 2008). The rate at which these compositional changes occur will depend on individual rates of molecular diffusion and the binding affinity that specific molecules have with the surface. Protein adsorption depends on the size of the molecule, the distribution of charge on it and the underlying substrate, any adsorption-induced conformational change, and both the chemical and morphological characteristics of the surface (Williams 1987, Horbett and Brash 1995, Siegel et al. 1997, Malmsten 1998). Clearly, the interfacial phenomena become more complicated for mixtures of different biomolecules and polymers which can be used as components of medicinal preparations. In this case, competitive adsorption depends on their mobility and binding affinity to a surface as well as on the factors (governing the interfacial layer structure) that affect protein adsorption. Smaller molecules adsorb more rapidly onto a surface as a result of having a faster diffusion coefficient than larger molecules; however, the latter can displace the former due to differences in binding affinity. This phenomenon is known as the Vroman effect and was discovered in a mixture containing fibrinogen competing with other plasma proteins (Vroman and Adams 1969, Bamford et al. 1992, Horbett and Brash 1995, Vroman 2008). It has been observed for blends of various proteins (Horbett and Brash 1995, Jung et al. 2003). The Vroman effect plays a key role in determining the biocompatibility of materials because it governs the adlayer structure formed on artificial surfaces in the body.

Real-time observation of the interfacial phenomena involving biomacromolecules and polymers is a difficult task (Gun'ko et al. 2011b). Microscopic visualization of the adsorbed layer (Hörber and Miles 2003) cannot provide qualitative data about the adlayer dynamics in liquids (Ta et al. 1998). Adsorption methods, which provide information on static equilibrium adsorption, require separation of liquid and solid components, e.g., by centrifugation, that can affect the structure of the adsorbed layer because of removal of the solvent. Real-time dynamics of protein adsorption can be followed by monitoring changes in the adsorbed mass of the protein layer using the quartz crystal microbalance (QCM) (Mikhalovsky et al. 2005b) or, for example, through changes in surface plasmon resonance. The sensitivity of the crystal sensors used in the QCM is significantly higher than that of most adsorption methods, up to 10^{-17} g/cm^2 with rupture event scanning (Kunze et al. 2009). Hence the QCM method is frequently used to study the interaction of complex compounds with different surfaces in liquid and gaseous media (Sultana et al. 2005, Dennes and Schwartz 2009). Here the real-time dynamics of the Vroman effect is analyzed using the adsorption of plasma proteins (fibrinogen, HSA) to various surfaces that have different topographies and chemical properties and can be used as biomaterials. Additionally, competitive adsorption effects have been studied for various polymers (PVP, PVA, and PEG) and proteins (BSA, ovalbumin, and gelatin) interacting with fumed nanosilica, used as an enterosorbent or a component of medicinal preparations (Gun'ko et al. 2001e, 2003a, 2008b, 2009d, Bershtein et al. 2010b, Klonos et al. 2010, Shpak and Gorbyk 2010) using quasi-elastic light scattering, QELS, and adsorption methods. Note that QELS can be used to noninvasively analyze the interaction of adsorption complexes, formed by proteins and polymers acting as co-adsorbates, with solid NP.

HPF, Fg (chromogenic, mass average molar mass, m, 340 kDa), HSA (Sigma, 67 kDa), BSA ("Allergen," Stavropol, Russia, 67 kDa), gelatin (Merck), ovalbumin (44 kDa), PVP ("Biopharma," Kiev, 12.6 ± 2.7 kDa), PVA ("Stirol," Severodonetsk, Ukraine; 43 ± 7 kDa), and PEG (Fluka, 2 and 20 kDa) were used as received. All these compounds are of pharmaceutical purity.

Nanosilica A-300, nanosilica as composed of NP of 5–15 nm in diameter (Pilot plant of the Institute of Surface Chemistry, Kalush, Ukraine; 99.8% purity, specific surface area $S_{BET} = 297$ m^2/g) was heated at 400°C for several hours to remove residual hydrogen chloride and other adsorbed compounds. The properties of aqueous suspensions of nanosilica depend on its structural characteristics, concentration, pH, salinity, temperature, sonication of the suspension, and the presence of dissolved organics and adsorbed polymers (Gun'ko et al. 2001e, 2003a).

Quartz crystal blanks (8.6 mm diameter, $f_0 = 10$ MHz, Spartan Europe, Portsmouth, United Kingdom or Crystal Works, OK) were furnished on both faces with a thin microcrystalline Au film (5 mm diameter). Then certain quartz crystals with the gold film were covered by a thin layer with (i) microcrystalline DLC with a graphite portion; (ii) Ti with a TiO_x surface layer formed due to interaction of the metal with oxygen from air; or (iii) TiN with TiN_x surface layer by means of a plasma vapor deposition (PVD) with an unbalanced magnetron sputtering technique (Teer Coatings Ltd., United Kingdom). These materials were studied here because they can be used for coating of cardiovascular stents to control their biocompatibility to reduce in-stent restenosis with subsequent target vessel revascularization (Gun'ko et al. 2011b).

Electrical connections of the sensors were made using gold-plated pins. The sensors were mounted in PEEK™ cartridges using a two-part low-viscosity elastomer (RS, Corby, United Kingdom), and 5 μm silver wires to yield a pressure-free cartridge crystal assembly prior to insertion into a custom-built flow injection system. The cartridges were in turn mounted within a Peltier temperature-controlled block, held at 25°C ± 0.1°C. Crystal oscillation was achieved using an oscillation circuit based upon auto-gain control technologies designed to give extremely high oscillation capabilities with low noise under high liquid loading (with consideration for the dissipation factor *D*). Real-time crystal oscillation data, auto-gain controller voltage, and temperature were recorded at 1 s intervals using a custom-written Labview™ application (National Instruments, Newbury, United Kingdom). Liquid flow through the crystal flow cell was achieved using a Shimadzu LC-9a dual piston HPLC pump with all experiments carried out at the flow rate of 5 μL/min. Introduction of a protein sample was through an all PEEK MBB Rheodyne™ pulseless six-way injection valve fitted with 20 μL loops using a Gilson 235 autoinjector. All tubing and fittings were of PEEK polymer to greatly reduce non-specific binding of biological materials within the flow system. Prepared surfaces are characterized by microscaled nonuniformity and this surface roughness can affect the protein adsorption and retention (Gun'ko et al. 2011b).

The adsorption kinetics of Fg and HSA onto various surfaces was measured using the QCM method with an injection of a single protein, e.g., fibrinogen or albumin solution followed by injection of a second protein to the liquid flow. Applying a coating to a quartz crystal sensor in a fluid environment lowers the resonant frequency by an amount, Δf according to equation (Kanazawa and Gordon 1985, Lucklum and Hauptmann 2000):

$$\Delta f = -f_0^{1.5}\left(\frac{\eta_l \rho_l}{\pi \mu_q \rho_q}\right)^{0.5} - \frac{2 f_0^2 \rho_f h_f}{\left(\mu_q \rho_q\right)^{0.5}} \tag{6.1}$$

where

Δf is the measured frequency shift
f_0 is the resonant frequency of the unloaded crystal
ρ_l and η_l are the density and viscosity, respectively, of the liquid in contact with the coated crystal surface
ρ_q and μ_q refer to the specific density and the shear modulus of quartz

The specific density of the adsorbed film is ρ_f, and its thickness is h_f. If this film is relatively rigid and the liquid is Newtonian, then the increased mass from the adsorbed film and liquid are approximately additive terms, and Equation 6.1 is valid (Lucklum and Hauptmann 2000). The f_0 value experimentally measured in air includes contributions from the quartz crystal coated with gold and other materials. Equation 6.1 may be used to estimate the thickness of coatings, from h_f values on the basis of results obtained using buffer solutions with and without proteins, assuming that all terms as additive (Shpak and Gorbyk 2010). The ρ_f value used (1.3 g/cm^3), estimated from experimental data, is in agreement with that reported in the literature (Lucklum and Hauptmann 2000). Note that variations in estimations of ρ_f with the accuracy of ±0.1 g/cm^3 have a minor effect on calculated h_f values, less than ±8%.

Protein adsorbing on to the surface of a quartz crystal can be monitored through the time dependence of changes in $\Delta f(t)$, curve portion corresponding to a sharp decrease in frequency because of the protein loading to the sensor surface can be given by equation (Kanazawa and Gordon 1985, Lucklum and Hauptmann 2000, Gun'ko et al. 2011b)

$$\ln\frac{\Delta f}{A} = -kt \tag{6.2}$$

where k is the adsorption rate constant of pseudo first order. The k value decreases with time because of increasing coverage, Θ of the sensor surface (assuming that only monomolecular coverage of a surface is stable). Additionally, the concentration of adsorbate decreases in the carrier fluid with time. Therefore, an exponential decay type function can be used to describe this time dependence.

The total coverage, Θ, depends on two processes: adsorption/desorption and unfolding of macromolecules $\Theta = \Theta_a + \Theta_u$, where Θ_a is the surface fraction covered by protein molecules in their native configuration and Θ_u is the surface fraction covered by partially unfolded protein. One can assume that concentration of proteins in the flow near a sensor surface ($C_f(t)$) is roughly a linear function of time from the initial interaction which occurs at some time interval after injecting the protein to the time at which adsorption reaches a plateau (C_p) (Gun'ko et al. 2011b). This is valid as the flow rate is both small and constant. The injected protein solution reaches the sensor in several minutes when the protein molecules are distributed in a certain volume of the liquid. The region of the $h_f(t)$ curves corresponding to a sharp increase in the h_f value is due to the formation of an adsorbed protein monolayer. The $h_f(t)$ functions can be recalculated to extract the adsorption per surface unit of the sensor ($C_s(t)$). On this basis, it is possible to write an overall equation (Gun'ko et al. 2011b):

$$\frac{C_s(t)}{C_p} = \int_{\Delta G_{min}}^{\Delta G_{max}} \frac{aC_f(t)e^{-(\Delta G+z)/R_gT}}{1+aC_f(t)e^{-(\Delta G+z)/R_gT}} f(\Delta G)d(\Delta G) \tag{6.3}$$

where

$f(\Delta G)$ is the distribution function of the free energy of protein interacting with a surface
a is an equation parameter
z depends on the presence of pre-adsorbed protein ($z=0$ for a pure surface)
R_g is the gas constant

The z value was estimated as $2R_gT$ corresponding to relatively weak lateral interactions between protein molecules (<5 kJ/mol). The kernel in Equation 6.3 corresponds to a kinetic analog of the Langmuir equation. Equation 6.3 can be solved using a regularization procedure based on the CONTIN algorithm (Provencher 1982). An integral equation similar to Equation 6.3 was used to calculate the $f(\Delta G)$ functions for the equilibrium adsorption of gelatin onto polymer-covered nanosilica (the isotherms, that show typical Langmuir behavior are not shown here).

An UBM laser profilometer was used to quantify the surface texture of features in excess of approximately ≥ 0.1 μm in terms of an average roughness, R_a. The hydrophilic/hydrophobic properties of the surfaces were assessed through measurements of dynamic contact angle (CAHN Dynamic Contact Angle analyzer) (Gun'ko et al. 2011b).

Adsorption of fibrinogen at a concentration, $C_{Fg} > 3 \times 10^{-4}$ g/cm^3 provides surface coverage, (Θ_a) with negligible molecular unfolding (Θ_u), i.e., $\Theta_u \ll \Theta_a$ (Seigel et al. 1997). In our experiments (Gun'ko et al. 2011b), C_{Fg} was more than an order of magnitude larger than the boundary value cited earlier and since $C_{Fg} \ll C_{HSA}$ (as in blood). Therefore, we expect a monolayer coverage of the sensor surface by proteins that have not unfolded. According to microscopic images, fibrinogen, which is "softer" than albumin, keeps its form after adsorption on to a very flat titania surface (Cacciafesta et al. 2000). Hence, it can be assumed that the shape of the protein molecules do not change significantly after being adsorbed on to the sensor surface. As a whole conformational changes of the proteins studied are not discussed here because the main aspect of the work is the competitive adsorption of proteins and polymers onto different surfaces. Adsorption-induced conformational changes of proteins are discussed in detail elsewhere (Williams 1987, Horbett and Brash 1995, Malmsten 1998).

The adsorption kinetics of individual proteins, reflected in changes in the frequency diminution $\Delta f(t)$ as a function of time (Figure 6.33), depends on the chemical nature of the surfaces studied and their roughness (Figures 6.34 through 6.36).

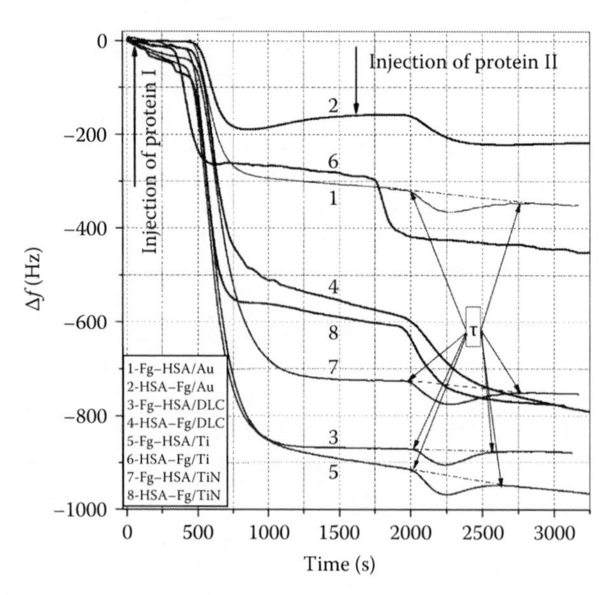

FIGURE 6.33 Changes in frequency shift, Δf due to adsorption of Fg_I (curves 1, 3, 5, and 7) (and then HSA_{II}) and HSA_I (curves 2, 4, 6, and 8) (and then Fg_{II}) onto surfaces of (curves 1 and 2) gold, (curves 3 and 4) DLC, (curves 5 and 6) Ti, and (curves 7 and 8) TiN; τ is the relaxation time on HSA adsorption on Fg-coated surfaces. (Adapted from Gun'ko, V.M., Mikhalovska, L.I., Tomlins, P.E., and Mikhalovsky, S.V., Competitive adsorption of macromolecules and real-time dynamics of Vroman-like effects, *Phys. Chem. Chem. Phys.*, 13, 4476–4485, 2011b, Copyright 2011, with permission from The Royal Society of Chemistry.)

Notice that the experimental conditions used were the same for all the adsorption measurements regardless of the surface type. Changes in the topography of bare and protein-coated surfaces are insignificant at the micrometer level (Figures 6.34 and 6.36) as the thickness of the protein layer is less than 40 nm (Figure 6.35). The thickness of the adsorbed protein layer as a function of time, $h_f(t)$, can be estimated from the $\Delta f(t)$ values using Equation 6.1. For this purpose, additional measurements for the blank buffer solution were also carried out to estimate contribution of the first term in Equation 6.1. The $h_f(t)$ value is between 6 and 25 nm after adsorption of the albumin is complete, and 14 to 39 nm for fibrinogen, which is in agreement with published results (Malmsten 1998) (Figure 6.35).

These values are close to the molecular sizes of the proteins, but much smaller than the surface micro-roughness (Figure 6.34) which occurs in the 0.1–0.5 μm range (Figures 6.34 and 6.36). The inequality $d_{min} < h_f < d_{max}$ (where $d_{min} \approx 6$ nm and $d_{max} \approx 45$ nm are minimal and maximal principal axes of Fg molecule, respectively) is true on the Fg adsorption completion. A similar relation has been found for albumin adsorbed onto gold and Ti surfaces, but $h_f > d_{max,HSA} \approx 14$ nm in the case of DLC and TiN.

This result can be caused by the formation of a dense layer that consists of "vertically" immobilized albumin molecules which forms as a result of the relatively high concentration of solution used. Such grafting of molecules on the rough surfaces (Figure 6.36) can provide the adsorption higher than a monolayer at a flat surface. Notice that a porous gold film deposited onto the QCM sensor surface strongly enhances its sensitivity due to adsorbate loading higher than a monolayer (Hieda et al. 2004). Thus, the amount of adsorbed protein depends, at least in part, on the available surface area; therefore, porous coatings can potentially support more adsorbed protein than their nonporous equivalents.

The surface roughness may enhance random adsorption of protein molecules with formation of more complex adlayer. Thus, the $h_f(t)$ values for individual proteins depend on the surface nano-roughness (R_{nr}). We may assume that a relative nano-roughness $r_{nr,X}$ at X = DLC, Ti, and TiN can be estimated with respect to the nano-roughness of the smoother gold surface ($R_{nr,Au}$) as follows: $r_{nr} = R_{nr,X}/R_{nr,Au} \sim h_f(X)/h_f(Au)$. For fibrinogen adsorption, r_{nr} is between 2.4 and 2.9, and

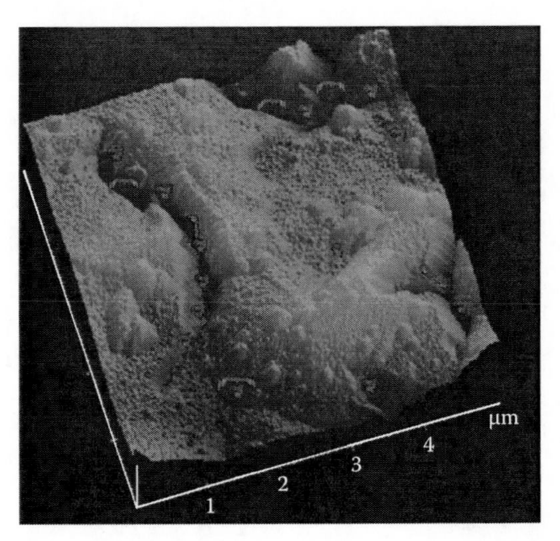

FIGURE 6.34 Microcrystal gold film deposited onto a quartz crystal was imaged (5×5×0.5 µm) using a Nanoscope IIIa microscope and scanning probe microscope controller in combination with Nanoscope software (Digital Instruments, Santa Barbara, CA). For better view, the size of the Fg and HSA molecules drawn in valley at the gold surface is enlarged by an order of magnitude. (Adapted from Gun'ko, V.M., Mikhalovska, L.I., Tomlins, P.E., and Mikhalovsky, S.V., Competitive adsorption of macromolecules and real-time dynamics of Vroman-like effects, *Phys. Chem. Chem. Phys.*, 13, 4476–4485, 2011b, Copyright 2011, with permission from The Royal Society of Chemistry.)

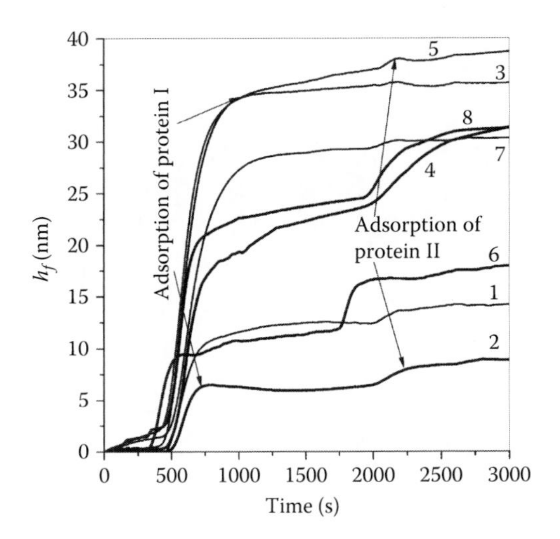

FIGURE 6.35 The protein layer thickness as a function of time of the adsorption of Fg_I–HSA_{II} (curves 1, 3, 5, and 7) and HSA_I–Fg_{II} (curves 2, 4, 6, and 8) on to gold (curves 1 and 2), DLC (curves 3 and 4), Ti (curves 5 and 6), and TiN (curves 7 and 8) coated surfaces. (Adapted from Gun'ko, V.M., Mikhalovska, L.I., Tomlins, P.E., and Mikhalovsky, S.V., Competitive adsorption of macromolecules and real-time dynamics of Vroman-like effects, *Phys. Chem. Chem. Phys.*, 13, 4476–4485, 2011b, Copyright 2011, with permission from The Royal Society of Chemistry.)

for albumin it is between 1.9 and 4.3. These r_{nr} ranges are larger than that of the micro-roughness (Table 6.6, R_a).

In other words, the studied surfaces more strongly differ at nano-lengthscale than at micro-lengthscale. The difference in the nanoscale roughness is of importance on the adsorption of protein molecules, which can penetrate into surface valleys (Figure 6.34) and pores. On the other hand,

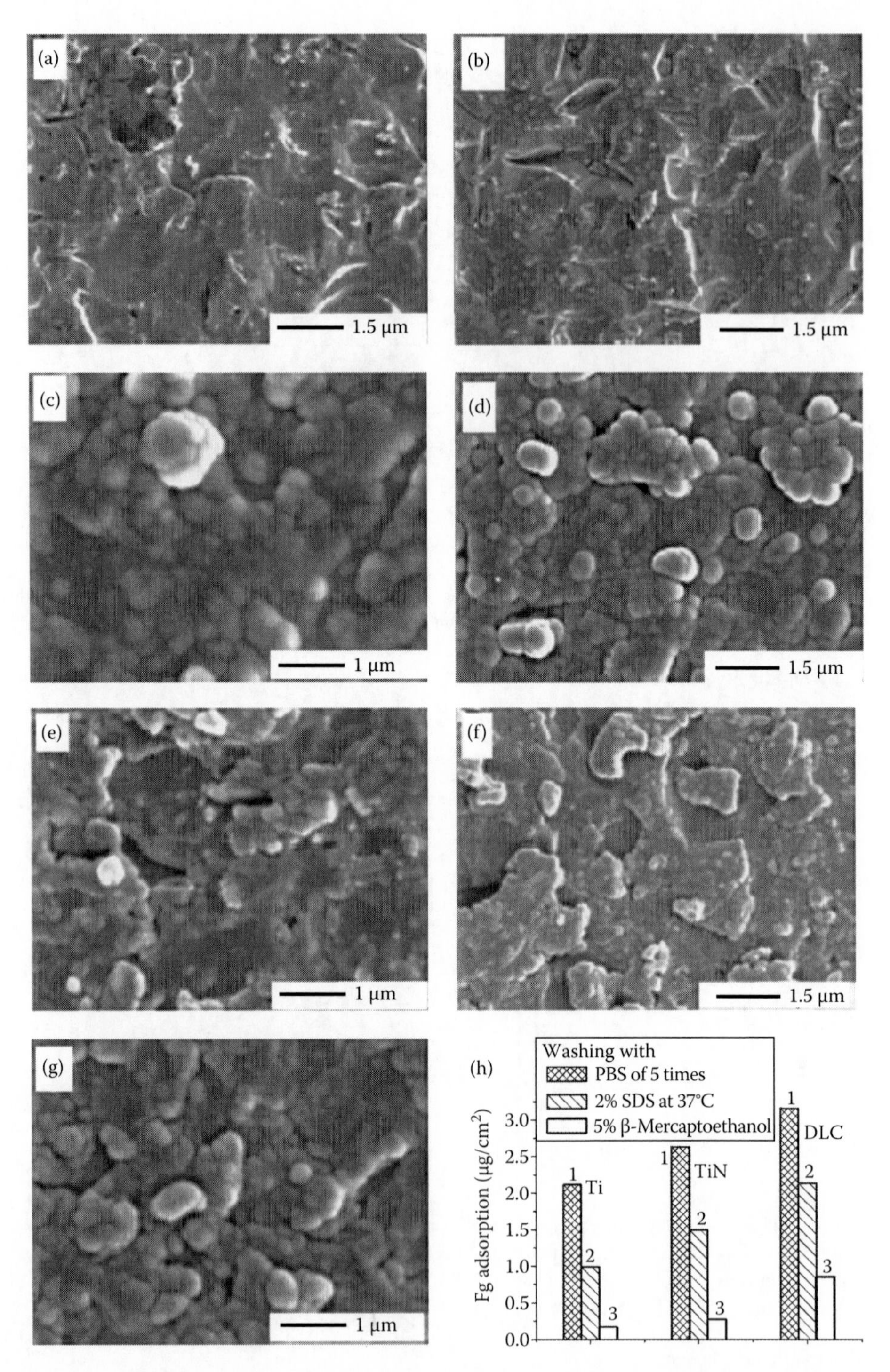

FIGURE 6.36 SEM (model JSM-6310, Japan Electron Optics Ltd.) images (20,000×) of (a, b) gold, (c, d) DLC, (e, f) Ti, and (g) TiN surfaces; (a, c, e, g) initial and (b, d, f) after Fg adsorption and drying; (h) Fg adsorption (in μg/cm^2) onto Ti, TiN, and DCL surfaces determined after washing with (1) PBS buffer (five times), (2) 2% sodium dodecyl sulfate, SDS, and (3) 5% β-mercaptoethanol at 37°C. (Adapted from Gun'ko, V.M., Mikhalovska, L.I., Tomlins, P.E., and Mikhalovsky, S.V., Competitive adsorption of macromolecules and real-time dynamics of Vroman-like effects, *Phys. Chem. Chem. Phys.*, 13, 4476–4485, 2011b, Copyright 2011, with permission from The Royal Society of Chemistry.)

TABLE 6.6
Relative Coverage (Θ_2) of Surfaces by Fg after Pre-Adsorption of HSA, Average Roughness, R_a, and Advancing Dynamic Contact Angle (θ)

Parameter	Gold	DLC	Ti	TiN
Θ_2	0.30	0.29	0.22	0.29
θ (degree)	47.9	76.7 ± 1.6	43.7 ± 3.0 −1 ± 0 (TiO_2)	27.1 ± 9
R_a (μm)	1.00005	1.060	1.094	1.094

Note: $\Theta_2 = (h_{HSA-Fg} - h_{HSA})/h_{Fg}$ after completion of the protein adsorption.

adsorbed macromolecules can be "thinner" (Cacciafesta et al. 2000, Bergna 2005). Therefore, the h_f value of the protein monolayer on a flat surface could be smaller than its molecular size in the solution. This occurs due to strong adsorption forces. Additionally, removal of a portion of water bonding by protein molecules occurs on the adsorption that was observed using the low-temperature ^{1}H NMR method. However, $h_f > d_{min}$ for HSA even on a gold surface, therefore $R_{nr,Au} > 1$.

The analysis of the adsorption kinetics (Figures 6.33 and 6.35) and micro- (R_a) and nano-roughness (r_{nr}) may suggest that the hydrophobic properties of the surfaces (Table 6.6, θ) affect the protein adsorption. For instance, the hydrophobicity is greatest for DLC that may cause the average maximal h_f value for this material (Figure 6.35), despite its R_a being smaller than that of Ti or TiN. The equilibrium adsorption of fibrinogen is maximal on DLC (Figure 6.36h). Additionally, the average $h_f(t)$ value (for both proteins) and the equilibrium adsorption (Figure 6.36h) is larger for TiN than that for Ti, since TiO_2 as a surface layer at Ti has a smaller θ value. However, these surfaces are characterized by the same R_a value and their SEM images (Figure 6.36) look similar. Consequently, the estimated r_{nr} and $h_f(t)$ values may be affected not only by topographic factors, mainly at nano-lengthscale, but also by chemical ones, i.e., hydrophobicity.

The lowest albumin adsorption is observed on the gold surface smoothest among the studied surfaces and $h_f(t)$ even decreases after a maximum (Figure 6.35). This $h_f(t)$ maximum corresponds to the final stages of HSA adsorption on to the sensor surface. The subsequent relaxation can be explained by desorption of a small fraction of albumin associated in the second adlayer (second term in Equation 6.1). Desorption of fibrinogen from the gold surface is not observed. This is due to its stronger interaction with the surface (Fg has many adsorption sites providing its high adhesive properties) and a lower concentration solution used than that of HSA. This gives $C_{Fg}/C_{HSA} \ll h_{Fg}/h_{HSA} < m_{Fg}/m_{HSA}$, which is valid for all the surfaces studied. This result can be explained by the higher affinity of fibrinogen to all the surfaces studied compared to albumin (Figure 6.33) that leads to a more planar adsorption of the Fg molecules. It is known that fibrinogen binds to the gold surface practically irreversibly, and its desorption is so slow that the adsorbed layer remains essentially constant (Dyr et al. 1998). However, fibrinogen adsorption on to the gold surface is significantly lower than on other surfaces because of lower roughness and low hydrophobicity (advancing $\theta \approx 48°$) of the Au surface (Table 6.6).

The $h_f(t)$ curves (Figure 6.35) describing the adsorption kinetics of the HSA_I–Fg_{II} and Fg_I–HSA_{II} pairs (see also adsorption rate constants in Figure 6.37) are essentially different in respect to the effect of the second protein. For the pair HSA_I–Fg_{II}, the relaxation effect is absent for the second protein. In contrast, for Fg_I–HSA_{II}, the $h_f(t)$ value increases then decreases to practically full relaxation, as the graphs show the same rate of decrease before and after the f minimum (Figure 6.33). The decrease in the $h_f(t)$ value is due to the albumin desorption but fibrinogen is not desorbed.

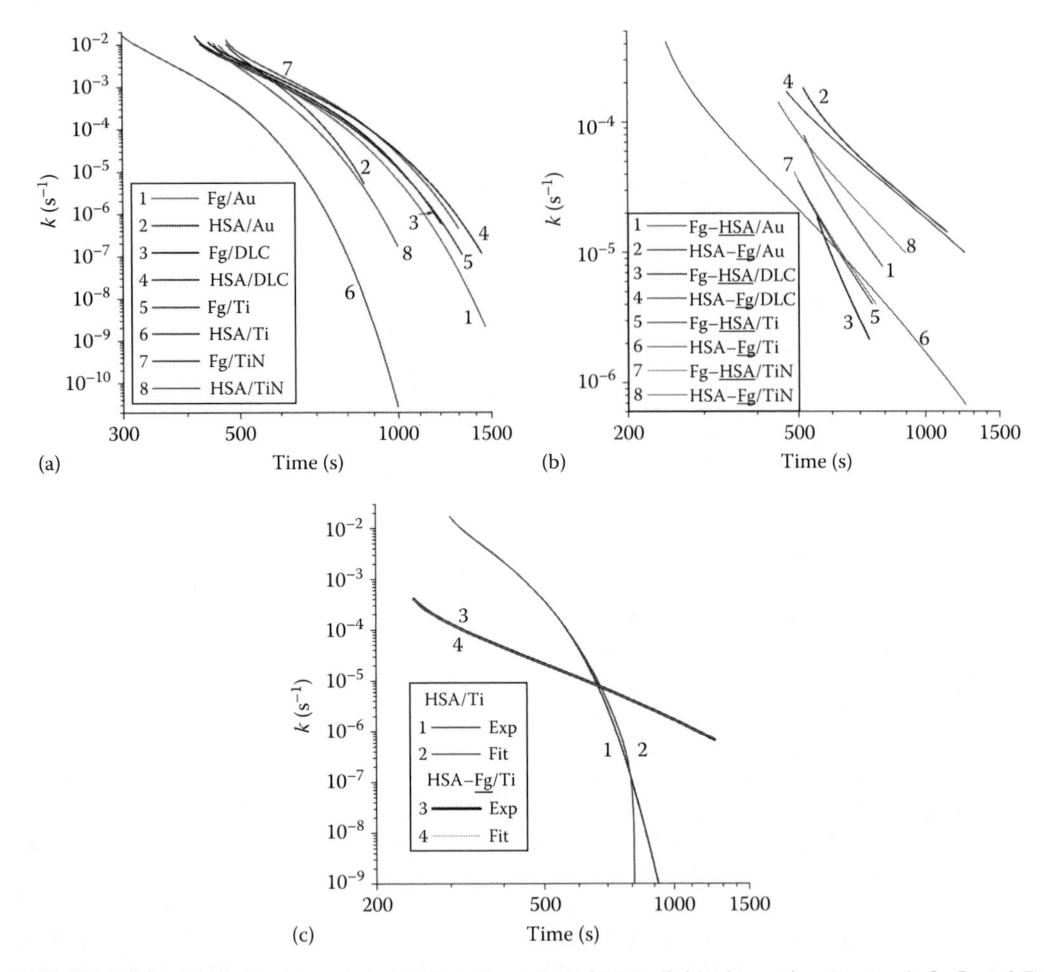

FIGURE 6.37 Adsorption rate constants for adsorption of (a) individual proteins (curves 1, 3, 5, and 7) Fg and (curves 2, 4, 6, and 8) HSA (first shift in Δf); (b) the second protein (curves 1, 3, 5, and 7) HSA after Fg and (curves 2, 4, 6, and 8) Fg after HSA (second shift in Δf, i.e., t corresponds to time after the second protein injection) on (curves 1 and 2) Au, (curves 3 and 4) DLC, (curves 5 and 6) Ti, and (curves 7 and 8) TiN; (c) fitting of $k(t)$ for two samples (curve 6 in (a) and (b)) using Equation 6.4. (Adapted from Gun'ko, V.M., Mikhalovska, L.I., Tomlins, P.E., and Mikhalovsky, S.V., Competitive adsorption of macromolecules and real-time dynamics of Vroman-like effects, *Phys. Chem. Chem. Phys.*, 13, 4476–4485, 2011b, Copyright 2011, with permission from The Royal Society of Chemistry.)

Relaxation time (τ) of the adlayer corresponds to time between beginning and completion of changes in QCM crystal frequency due to protein adsorption (Figure 6.33). The τ value is minimal on the albumin adsorption on Fg/DLC and maximal for Fg/Au. Rearrangement of the protein layer on the smoother gold surface (Table 6.6) pre-adsorbing a smaller amount of fibrinogen can be greater (the τ value is larger) than that on other rougher surfaces (the τ value is smaller). However, albumin cannot displace fibrinogen pre-adsorbed on all the surfaces studied.

The behavior of $k(t)$ (Figure 6.37) with time can be described by an exponential decay function, following

$$k(t) = b_0 + \sum_{i=1}^{3} b_i \exp\left(-\frac{(t-t_i)^{\alpha}}{c_i}\right) \tag{6.4}$$

where b_0, b_i, t_i, c_i, and α are equation parameters (Figure 6.37c). Notice that simpler exponential decay functions in Equation 6.4 provide poor fits to the experimental data. This finding shows that changes in $k(t)$ are complex as a result of several factors changing simultaneously during the observation time. Additionally, the shape of the $k(t)$ curves differ for the first and second runs (Figure 6.37). The initial $k(t)$ values are larger and change more slowly with time on the first run than on the second one. This difference can be attributed to the stronger interactions that a protein (A) has with a bare surface (great k_{1A} at $t<600$ s) compared to that of a protein-coated surface.

Then the amounts of non-adsorbed protein strongly decrease (low k_{2A} at $t<700–1000$ s) due to the adsorption and removal of non-adsorbed protein A by the liquid flow. On the second run, the interaction of the second protein (B) with a surface coated by protein A is weaker than with a bare surface ($k_{1B}<k_{1A}$). However, a smaller amount of the adsorbed protein B leads to a slower decrease in $k(t)$ with time and $k_{2B}>k_{2A}$ at $t>800$ s (Figure 6.37), where t is the observation time after the protein injection for both runs. Fibrinogen molecules, which are larger (Figure 6.38) and have much greater binding affinity (as an adhesive protein) to the surfaces studied than albumin (as a transport protein), can displace adsorbed HSA according to the Vroman effect (Vroman and Adams 1969, Bamford et al. 1992, Horbett and Brash 1995, Vroman 2008).

Comparison of the h_f values (e.g., $h_{HSA-Fg}-h_{HSA}>h_{Fg-HSA}-h_{Fg}$), nano- ($r_{nr}$) and micro-roughness (Table 6.6, R_a) reveals that the larger the roughness and surface hydrophobicity, the greater is the absolute value of the Vroman effect (comp. curves 2, 4, 6, and 8 in Figure 6.35). However, the Θ_2 value ($\Theta_2=(h_{HSA-Fg}-h_{HSA})/h_{Fg}$ after protein adsorption has completed), which can be used as a relative measure of the Vroman effect, is greatest for the gold surface. This value is less than unity as a result of several factors. First, the fibrinogen injection provides a relatively short time for interactions to occur with pre-adsorbed albumin. Second, the rate of replacement is slower than the rate at which protein adsorbs on to the bare surface (Figure 6.37, initial part of curves). Third, replacement of HSA by fibrinogen on a rougher surface can be more difficult because of their differences in their molecular size, and enhancement of the adsorption potential for smaller albumin molecules located in "pores" in the rough surface. According to published data (Malmsten 1998, Vogler 1998), proteins strongly adsorb on to hydrophobic surfaces. Therefore, their displacement from more hydrophobic surfaces can be more difficult than from hydrophilic ones.

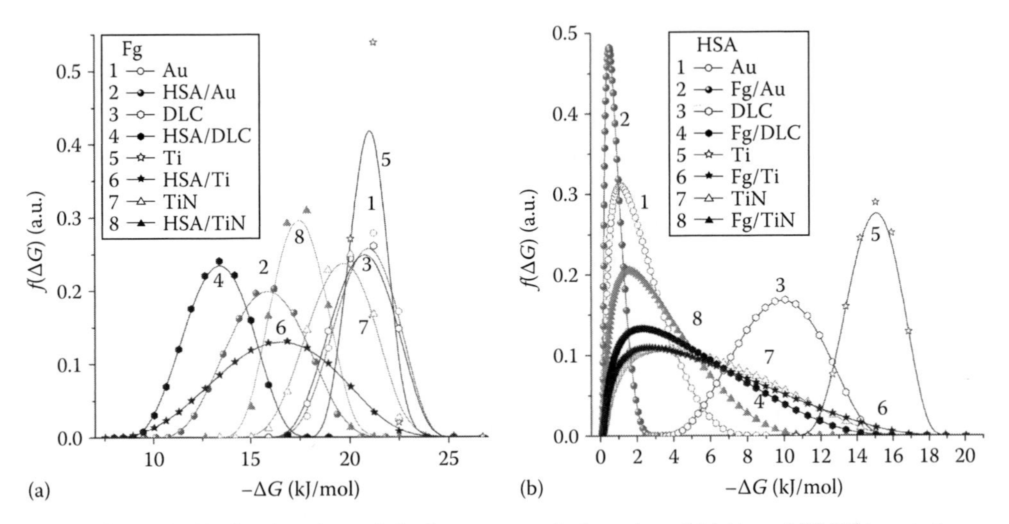

FIGURE 6.38 Distribution functions of the free energy of adsorption of (a) Fg and (b) HSA on to bare and protein (HSA (a) and Fg (b))-coated surfaces. (Adapted from Gun'ko, V.M., Mikhalovska, L.I., Tomlins, P.E., and Mikhalovsky, S.V., Competitive adsorption of macromolecules and real-time dynamics of Vroman-like effects, *Phys. Chem. Chem. Phys.*, 13, 4476–4485, 2011b, Copyright 2011, with permission from The Royal Society of Chemistry.)

The replacement of albumin by fibrinogen depends on the surface type. For instance, the absolute value of the amounts of the second protein is lower on gold than on DLC (Figure 6.35). The corresponding changes in $h_f(t)$ are largest for HSA_I–Fg_{II}/DLC and least for Fg_I–HSA_{II}/DLC because of the differences in hydrophobicity and noticeable roughness of the DLC surface. In the later pair, albumin senses mainly the fibrinogen layer densely covering the DLC surface. Therefore, the Θ_2 value (Table 6.6) is maximal for the gold surface covered to a smaller extent because the displacement of albumin by fibrinogen can be easier on a smoother and relatively hydrophilic surface (Table 6.6). The Θ_2 value is minimal for the rough Ti surface (Figure 6.36) due to difficulties in displacement of albumin molecules located in surface valleys and pores by fibrinogen molecules.

The values of changes in the Gibbs free energy of interaction of Fg with bare and HSA-coated surfaces show (Figure 6.38a) that this interaction is stronger than HSA interacts with the bare and Fg-coated surfaces (Figure 6.38b). The $f(\Delta G)$ peaks are observed at lower ΔG values for Fg adsorbed as the first and second co-adsorbate than that for HSA. This result can be explained by the shape of the proteins. The contact area of long non-globular Fg molecules, possessing greater molecular weight and a larger number of active adsorption sites than HSA (Williams 1987, Horbett and Brash 1995) with a sensor surface can be much larger than that of globular HSA molecules. Interactions of HSA with Fg-coated surfaces are weak (Figure 6.38b) in contrast to interactions of Fg with HSA-coated surface (Figure 6.38a). Changes in the $f(\Delta G)$ distribution functions (position and half-width of peaks) are smaller on interaction of Fg with bare surfaces than on interactions with pre-coated surfaces (Figure 6.38a). For HSA, the $f(\Delta G)$ distribution functions more strongly change (Figure 6.38b) than that for Fg (Figure 6.38a). As whole HSA weaker interacts with both bare and pre-coated surfaces than Fg. Fibrinogen adsorption onto the HSA-coated surface is weaker than on the bare surfaces. However, the corresponding ΔG range suggests that Fg interacts with HSA-coated surfaces stronger than HSA interacts with the bare surfaces with the one exception of the Ti surface (Figure 6.38). These effects are characteristic for both hydrophobic and hydrophilic surfaces (Table 6.6). Similar effects of a reduced adsorption of gelatin and ovalbumin (Figures 6.39 and 6.40) or BSA (Figure 6.41) on polymer-coated silica surfaces were observed on equilibrium adsorption. The protein adsorption decreases practically linearly with increasing amounts of pre-adsorbed PEG or PVP having only electron-donor sites. However, pre-adsorption of PVA having both electron-donor and proton-donor OH groups leads to smaller diminution of the gelatin adsorption (Gun'ko et al. 2009d).

The appearance of the Vroman effect and related phenomena can be inhibited by the roughness and the porosity of a surface or due to the formation of hybrid aggregates with macromolecules and oxide NP (Figure 6.41). All these structural features of the interfaces cause reduction of accessibility of pre-adsorbed macromolecules for co-adsorbate molecules as well as for solvent molecules. The confinement effects in restricted space of pores or surface roughness (valleys, Figure 6.34) diminish the mobility of the adsorbed molecules. Therefore, the possibility of the displacement of these molecules by other molecules (even of a larger size) decreases.

Thus, the surface properties enhancing the adsorption of the first protein can cause diminution of its fraction displaced by the larger second protein subsequently adsorbed. The real-time dynamics of the Vroman effect for HSA and Fg subsequently adsorbed on various (metal, oxide, nonmetal) surfaces monitored by the QCM technique demonstrates the influence of the nature and nano-relief of the surfaces onto the competitive adsorption of proteins. Obtained results suggest that the dynamics of the competitive adsorption of any polymers, biomacromolecules, cells, and microorganisms in aqueous media may be studied in real timescale using the QCM method in parallel with observation by the QELS method.

The QELS and adsorption studies of competitive adsorption and aggregation of macromolecules/NP for polymers (PVP, PEG, PVA), gelatin, ovalbumin or BSA, and nanosilica are in agreement with the QCM data with respect to the effects of surface roughness, confinement space in pores, and the characteristics of macromolecules and their interactions in pre-adsorbed macromolecules and with partially coated solid surfaces. Proteins can more effectively displace smaller polymer molecules if the latter have only electron-donor groups (PEG, PVP). In the case of a pre-adsorbed

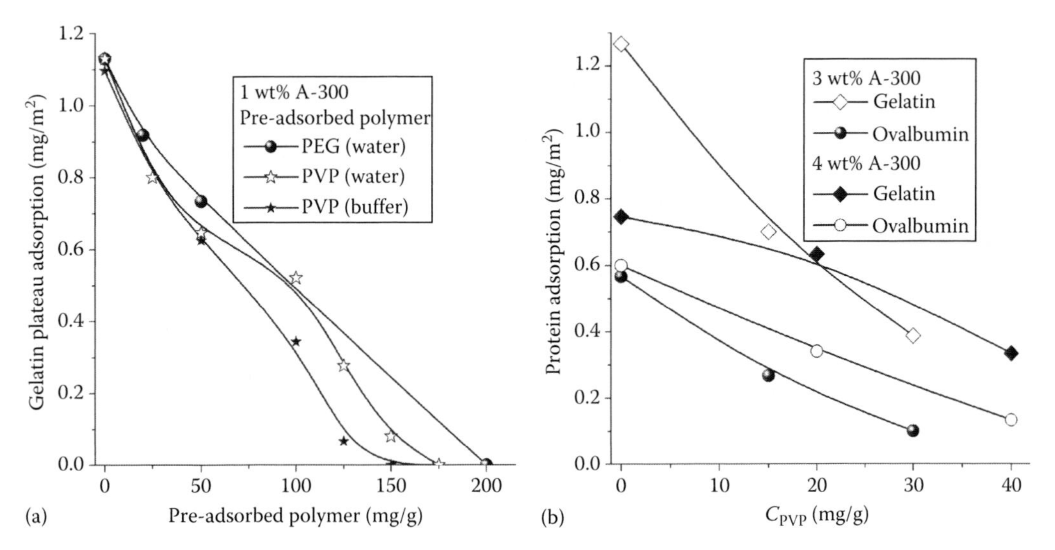

FIGURE 6.39 Plateau adsorption of (a, b) gelatin or (b) ovalbumin onto A-300 at C_{SiO_2} = (a) 1 wt% and (b) 3 or 4 wt% as a function of the amounts of (a) pre-adsorbed PEG (20 kDa, aqueous solution) and PVP (aqueous and phosphate buffer solutions) and (b) PVP. (Adapted from Gun'ko, V.M., Mikhalovska, L.I., Tomlins, P.E., and Mikhalovsky, S.V., Competitive adsorption of macromolecules and real-time dynamics of Vroman-like effects, *Phys. Chem. Chem. Phys.*, 13, 4476–4485, 2011b, Copyright 2011, with permission from The Royal Society of Chemistry.)

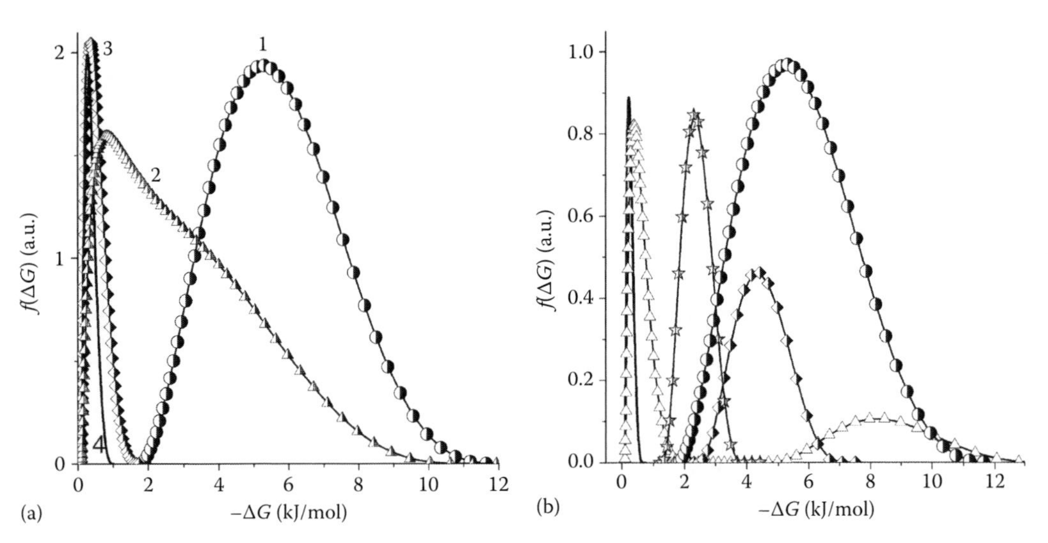

FIGURE 6.40 Distribution functions of changes in Gibbs free energy on the adsorption of gelatin on nanosilica A-300 (1 wt%) with pre-adsorbed (a) PEG (20 kDa) in amounts (curve 1) 0, (curve 2) 20, (curve 3) 50, and (curve 4) 200 mg/g of silica and (b) PVP in amounts (curve 1) 0, (curve 2) 25, (curve 3) 50, (curve 4) 100, and (curve 5) 125 mg/g of silica (last PEG and PVP coverage corresponds to monolayer). (Adapted from Gun'ko, V.M., Mikhalovska, L.I., Tomlins, P.E., and Mikhalovsky, S.V., Competitive adsorption of macromolecules and real-time dynamics of Vroman-like effects, *Phys. Chem. Chem. Phys.*, 13, 4476–4485, 2011b, Copyright 2011, with permission from The Royal Society of Chemistry.)

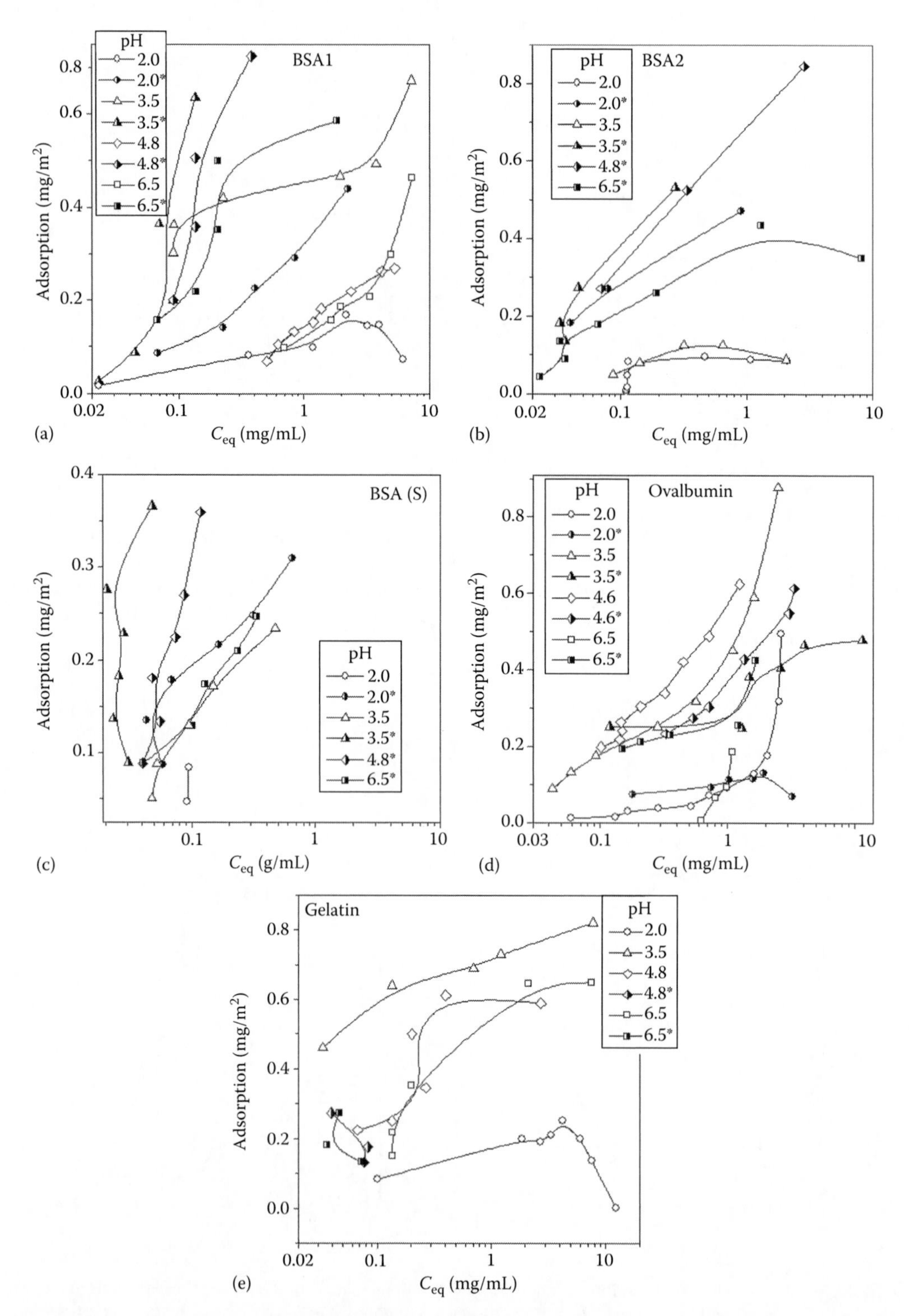

FIGURE 6.41 Adsorption isotherms of different proteins: (a) BSA1, (b) BSA2, (c) BSA(S), (d) ovalbumin, and (e) gelatin on nanosilica A-300 (3.6 wt%) at different pH values, which with asterisk show the systems with addition of 0.9 wt% NaCl. (Adapted from *J. Colloid Interface Sci.*, 260, Gun'ko, V.M., Mikhailova, I.V., Zarko, V.I. et al., Study of interaction of proteins with fumed silica in aqueous suspensions by adsorption and photon correlation spectroscopy methods, 56–69, 2003c, Copyright 2003, with permission from Elsevier.)

polymer with OH groups and larger molecular weight (PVA), aggregation of BSA with PVA/silica structures is predominant. The results obtained on the competitive adsorption, displacement, and aggregation of different macromolecules with NP or at solid surfaces allow one to insight into the interfacial behavior of biomaterials in aqueous media and biofluids.

Bovine, BSA, HSA, ovalbumin, OA, human Hb (HHb), and gelatin were used in comparative investigations to analyze features in the protein interactions with different solid surfaces (Gun'ko et al. 2003a,c). Albumins (as well as HHb) are soft proteins which can easily change their conformation with pH. For instance, BSA and HSA have several relatively stable conformations such as native "N" at pH ≈ 7, basic "B" at pH > 8, "F" (pH < 4), "E" (pH < 3.5), and "A" for aged proteins at pH > 8. At neutral pH ≈ 7, HSA has a total charge $q_{HSA} = -14$ and $q_{BSA} = -17$. The free energy of their adsorption onto the silica surface as well as onto other solids can be several kT due to a great entropy factor destabilizing adsorption complexes. However, formation of many adsorption bonds, e.g., hydrogen bonds (as a cross-sectional area of protein molecules upon adsorption onto oxides can reach several nm^2) causes irreversible adsorption of proteins due to a small probability of simultaneous breaking all these bonds. Changes in the Gibbs free energy of the bound water in the systems with protein (1 wt% of BSA hydrolyzate)/nanosilica (3 wt%) are up to −5 kJ/mol (Gun'ko et al. 1997b, 2003a,c).

Solved proteins can represent monomers, dimers, trimers, or polymers, which have different rate constants of adsorption, desorption, and diffusion. However, plateau adsorption is typically observed at incubation time $t_i = 1$–2 h for different proteins (on different nanooxides), and their maximal adsorption/flocculation is found at pH close to their IEP, as the molecules have the smallest cross-sectional area on the surface and repulsive electrostatic lateral interaction between them is minimal. Lipids and fatty acids (having large hydrophobic CH chains) present in different amounts in protein preparations can also affect protein adsorption due to hydrophobic interactions and ordering of bound water leading to formation of micelles, as adsorption of the latter enhances the amounts of adsorbed proteins shielding hydrophobic compounds or their tails. It should be noted that adsorption of simple amino acids (not to mention proteins) onto the oxide surfaces has a complicated mechanism and strongly depends on their structure, the origin and morphology of oxide (porosity, particle size distribution, types of active sites, etc.), and pH. Parallel investigations of the influence of the protein type, its preparation technique, pH, salinity, and nanosilica concentration on protein adsorption was not performed previously by adsorption and QELS methods. Therefore, here features of interactions of several types of albumins, HHb, and gelatin with nanosilica are analyzed depending on pH, salinity, and concentrations of components by adsorption and QELS methods. Ovalbumin (Biom, Omutninsk, Russia, molecular weight $M \approx 4.4 \times 10^4$ Da, pH(IEP) ≈ 4.6), gelatin (Moskhimfarmpreparaty, Moscow, Russia, $M \approx 3.5 \times 10^5$ Da), BSA1 (TTM, Moscow, Russia, prepared by salting-out method, $M \approx 6.7 \times 10^4$ Da, pH[IEP] ≈ 4.8, total concentration of lipids and fatty acids $C_{lfa} \approx 1$ wt%), BSA2 (Allergen, Stavropol, Russia, prepared by ethanol fractionating), BSA (Sigma, $C_{lfa} < 0.01$ wt%) (BSA(S)), BSA (Olaine, Latvia, $C_{lfa} \approx 2$ wt%) (BSA(O)), and human hemoglobin (VMU, Ukraine, HHb, $M \approx 6.5 \times 10^4$ Da, pH[IEP] ≈ 6.8) were used as received (Gun'ko et al. 2003a,c).

Comparison of equilibrium adsorption (Figure 6.41) and minute protein adsorption/flocculation (changes in the optical density in minute of adsorption, ΔD_1 as a function of the total protein concentration, C_p are shown in Figure 6.42) on nanosilica demonstrates strong but different effects of pH and salinity (Gun'ko et al. 2003a,c). The equilibrium adsorption of proteins is up to $\Gamma \approx 1$ mg/m^2 (or ≈300 mg/g) at pH = 3.5 (i.e., between pH[IEP] of silica and proteins) for BSA with 0.9 wt% NaCl and gelatin without NaCl or at pH[IEP] of protein for ovalbumin without NaCl. The lowest equilibrium adsorption $\Gamma \approx 0.1$–0.2 mg/m^2 is typically observed at pH = 2, which is close to pH(IEP_{SiO_2}) ≈ 2.2, and without NaCl (Figure 6.41). It should be noted that the maximal Γ_{max} values for BSA preparations with marked amounts of lipids and fatty acids can be up to 600 mg/g (using concentrated A-300 suspensions at $C_{SiO_2} = 3$–5 wt%) and significantly lower (280–300 mg/g) for pure BSA (such as BSA(S)) that can be caused by the impact of hydrophobic admixtures on the structure of adsorbed complexes, as polar and charged proteins should shield hydrophobic compounds from

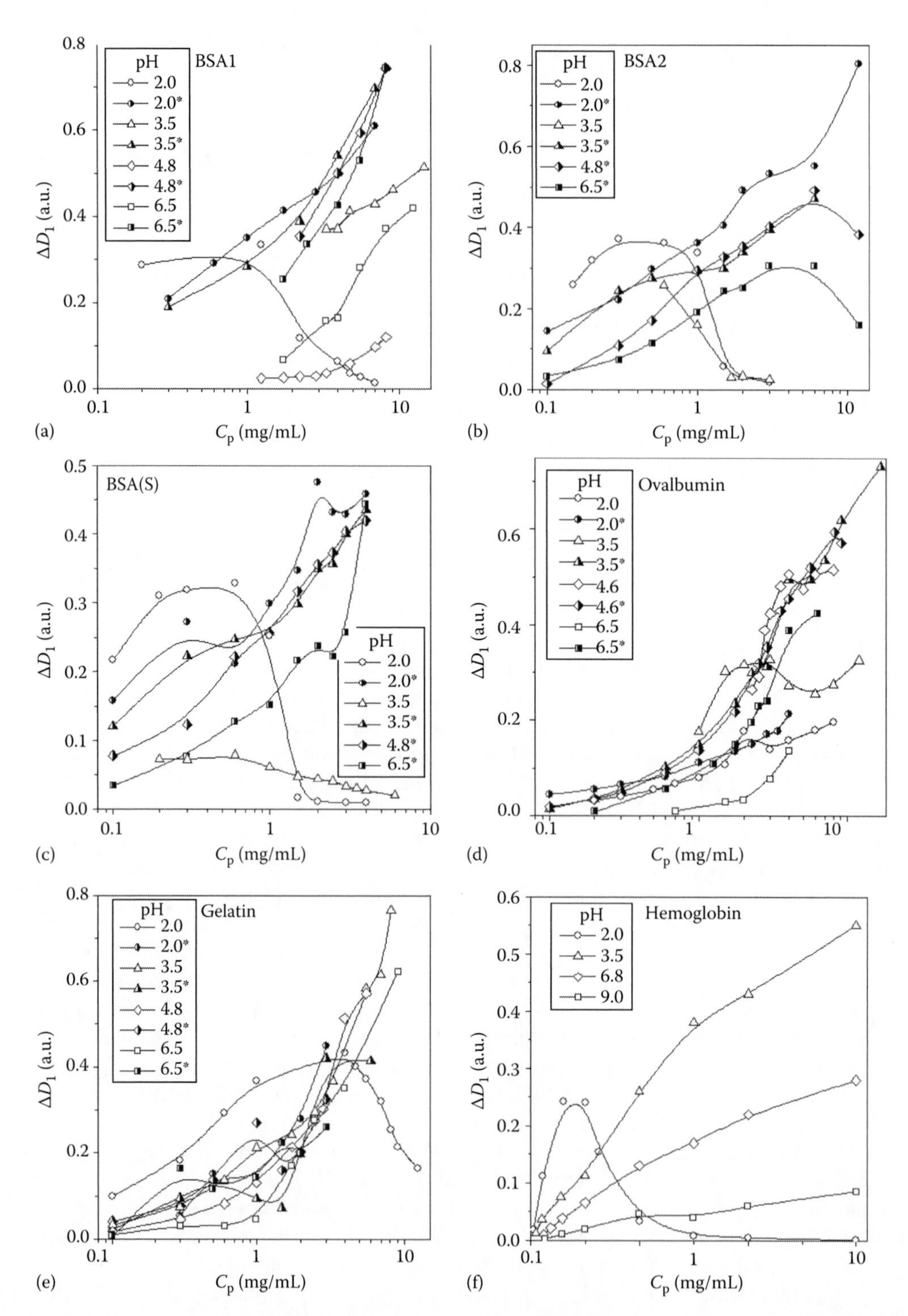

FIGURE 6.42 Changes in the optical density (ΔD_1) in 1 min after addition of different proteins: (a) BSA1, (b) BSA2, (c) BSA(S), (d) ovalbumin, (e) gelatin, and (f) hemoglobin to nanosilica A-300 suspension (3.6 wt%) at different pH values, which with asterisk show the systems with addition of 0.9 wt% NaCl. (Adapted from *J. Colloid Interface Sci.*, 260, Gun'ko, V.M., Mikhailova, I.V., Zarko, V.I. et al., Study of interaction of proteins with fumed silica in aqueous suspensions by adsorption and photon correlation spectroscopy methods, 56–69, 2003c, Copyright 2003, with permission from Elsevier.)

the bound water. In the case of diluted suspensions (C_{SiO_2}=0.1–0.2 wt%), Γ_{max} can increase up to 1500–4500 mg/g due to very intensive flocculation of proteins with silica aggregates and agglomerates (typically observed in diluted or weakly treated suspensions) because of formation of protein bridges between them, which is also affected by lipids and fatty acids present in the BSA preparations (Gun'ko et al. 2003a,c).

In the case of BSA (Figure 6.41a through c), addition of 0.15 M NaCl leads to enhancement of adsorption/flocculation. However, in the case of ovalbumin, the opposite effect is observed (Figures 6.41d and 6.42d). One can see an increase in minute flocculation at different pH values with C_p except pH=2 (for all the samples without NaCl) and 3.5 (BSA2 and BSA(S)) with a $\Delta D_1(C_p)$ maximum (Figure 6.42). At low concentrations and pH=2 or 3.5, $\Delta D_1(C_p)$ (connected with the adsorption/flocculation rate) is higher than that at greater pH values. Thus, fast (t_i=1 min) flocculation gives a complex picture depending on pH with a $\Delta D_1(C_p)$ maximum marked at pH=2. The position of this maximum depends on the type of proteins (Figure 6.42), as it is at C_p<1 mg/mL for BSA and HHb but for gelatin it is at C_p>1 mg/mL. For ovalbumin, maximal ΔD_1 values are observed at C_p>0.5 mg/mL and pH=2, 3.5, and 4.6. A similar maximum at pH=2 is also observed on the adsorption isotherms but at higher protein concentration (Figure 6.41) (Gun'ko et al. 2003a,c).

The S-like shape of the BSA adsorption isotherms (Figure 6.41a) can be explained by ordering of protein due to interaction between non-adsorbed and adsorbed segments resulting in enhancement of adsorption and changes in the protein conformation approaching to native one. Additionally, this type of the isotherm can be explained by self-association of protein molecules leading to increase in the effective molecular weight of protein. The steps on the isotherms can be due to reorientation of adsorbed ellipsoid-like molecules from the horizontal position to vertical one (Gun'ko et al. 2003a,c).

Addition of BSA1 (pH[IEP]=4.8, prepared by the salting-out method) to the silica suspension gives a minimal stability (maximal ΔD_1) at low pH values (2 and 3.5) (Figure 6.42a). However, the stability increases with C_p>1 mg/mL at pH=2 as ΔD_1 decreases sharply. Minute flocculation of BSA1 molecules with silica particles increases with C_p at pH=6.5 greater than that at pH=4.8 (without NaCl). However, the isotherms at these pH values are close (Figure 6.41a). Thus, BSA can adsorb on hydrophilic surfaces at high pH values when the surfaces and molecules were charged negatively, since electrostatic repulsive interaction could be weaker than attractive polar interaction and hydrophobic effects related to ordering of the bound water. Additional attractive effect can be caused by lipids and fatty acids present in BSA1 and ordering bound water stronger (as more hydrophobic) than more polar (and charged) protein molecules (ions). Larger ΔD_1 values at pH=3.5 and 6.5 than at pH(IEP)=4.8 (in contrast to the equilibrium adsorption, Figure 6.41a) can correspond to adsorption of protein molecules in the unfolded state. Addition of NaCl leads to decrease in the difference in $\Delta D_1(C_p)$ with changes in pH values. However, these $\Delta D_1(C_p)$ curves lie over those for the system without NaCl (Figure 6.42a) and drop with pH in contrast to ovalbumin (Figure 6.42d) or gelatin (Figure 6.42e) due to different contributions of electrostatic, polar, and hydrophobic interactions of these proteins (and present admixtures) with the silica surfaces and the bound water (Gun'ko et al. 2003a,c).

Minute flocculation of BSA2 (prepared by ethanol fractionating and containing lower amounts of lipids and fatty acids than BSA1) with nanosilica particles (Figure 6.42b), as well as equilibrium adsorption (Figure 6.41b), differs from that for BSA1, especially at pH=3.5 (Gun'ko et al. 2003a,c). The suspension is less stable at low pH=2 and low C_p. Increase in C_p leads to enhancement of the suspension stability, as ΔD_1 decreases with C_p. At pH=4.8 and 6.5, the ΔD_1 values (as well as equilibrium adsorption) were very small and could not be measured. However, addition of NaCl leads to significant adsorption/flocculation of BSA2 (Figures 6.41b and 6.42b), which is minimal at pH=6.5. Notice that adsorption of BSA2 can differ from that of BSA1 due to the difference in the methods of their preparation, and BSA1 includes larger amounts of lipids and fatty acids than BSA2 that can affect their interaction with nanosilica in the aqueous medium (Gun'ko et al. 2003a,c).

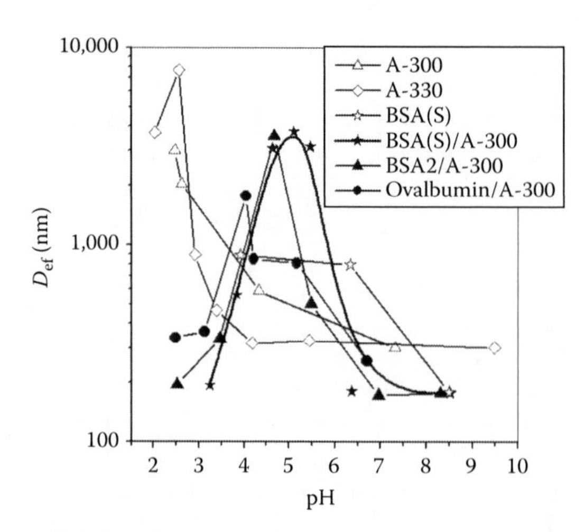

FIGURE 6.43 Effective diameter of particles as a function of pH in the aqueous suspensions of nanosilica (A-300 and A-330) and protein/A-300. (Adapted from *J. Colloid Interface Sci.*, 260, Gun'ko, V.M., Mikhailova, I.V., Zarko, V.I. et al., Study of interaction of proteins with fumed silica in aqueous suspensions by adsorption and photon correlation spectroscopy methods, 56–69, 2003c, Copyright 2003, with permission from Elsevier.)

For BSA(S) (containing very low amounts of lipids and fatty acids), a ΔD_l maximum is observed at pH = 2 too (Figure 6.42c) due to aggregation of protein molecules with large silica agglomerates (formed at pH close to pH[IEP_{SiO_2}], Figure 6.43) at low C_p. Increase in C_p leads to decomposition of these large silica agglomerates due to interaction with protein molecules and ΔD_l decreases at pH = 2 and 3.5 similar to BSA2. Enhancement of pH at low C_p values is responsible for increase in the stability of the suspension (as ΔD_l decreases) (Figure 6.42c) similar to other proteins. Notice that ΔD_l could not be measured at pH = 4.8 and 6.5 without addition of NaCl. As a whole, adsorption BSA(S) is akin to that of BSA2, as both preparations are characterized by low amounts of lipids and fatty acids (Gun'ko et al. 2003a,c).

For ovalbumin (pH[IEP] = 4.6) at low C_p, the suspension stability is minimal at pH = 2 when the surface charge density on silica particles is zero and large agglomerates are observed (Figure 6.43); however, a $\Delta D_l(C_p)$ maximum similar to that for BSA is absent (Figure 6.42d) (Gun'ko et al. 2003a,c). In addition to flocculation effect of protein, aggregation of nanosilica particles at pH close to pH (IEP_{SiO_2}) = 2.2 can give a marked contribution, as the effective diameter D_{ef} increases more than by order in comparison with that at pH > 5 (Figure 6.43). The ΔD_l values become nearly constant with increasing C_p of ovalbumin, especially at pH = 2 and 3.5 (Figure 6.42d), as adsorption/flocculation of positively charged protein molecules is restricted due to their electrostatic lateral interactions. At pH = 3.5 and low C_p (between 0.7 and 2.6 mg/mL), ΔD_l is maximal (Figure 6.42d) due to different charges of ovalbumin molecules and silica surface, but ΔD_l changes slower at pH = 4.6 (IEP of ovalbumin) with C_p, as its long-range electrostatic interaction with the silica surfaces reduces. Negative charges of the silica surfaces and ovalbumin molecules at pH = 6.5 provide more stable suspensions (without NaCl), as $\Delta D_l(C_p)$ minimal, especially at low C_p. However, increase in C_p to 4 mg/mL leads to enhancement of the aggregation rate (Figure 6.42d) as well as the equilibrium adsorption (Figure 6.41d). Addition of NaCl (0.9 wt%) provides a lower effect than that for BSA (Figure 6.41) and does not give enhancement of ovalbumin adsorption at pH = 2, 3.5, and 4.6. A similar picture is observed for $\Delta D_l(C_p)$ (Figure 6.42). From comparison of adsorption of BSA and ovalbumin, one can assume that BSA molecules are softer than ovalbumin ones (Gun'ko et al. 2003a,c).

The ΔD_l values for gelatin increase with C_p (except pH = 2) (Figure 6.42e). Since gelatin is a product of collagen hydrolysis and does not possess a native conformation, its adsorptive and aggregative capabilities depend on pH weaker than that of BSA. Adsorption of gelatin occurs mainly due to formation of the hydrogen bonds with silica particles, and the number of these bonds can decrease with pH > 2. The ΔD_l values and equilibrium adsorption decrease with pH from 3.5 to 6.5. The suspension is less stable at pH = 2 and $C_p < 1$ mg/mL (as well as for other proteins) (Figure 6.42e); however, equilibrium adsorption is minimal at this pH value as well as for BSA (Figure 6.41). Addition of NaCl gives a weaker effect for minute flocculation of gelatin with A-300 than that for BSA/A-300 but it is akin to that of ovalbumin/A-300 (Figure 6.42) (Gun'ko et al. 2003a,c).

The impact of HHb (pH([EP] ≈ 6.8, plateau adsorption onto fine silica with $S_{BET} \approx 200$ m^2/g is ~2 h [Gun'ko et al. 2009a]) on the stability of the aqueous suspension of nanosilica is the most typical among the studied proteins, as the ΔD_l values decreases with pH (except pH = 2 with a maximum of $\Delta D_l(C_p)$) (Figure 6.42f) that corresponds to enhancement of the stability of the HHb/A-300 suspension with increasing pH values and shows the importance of electrostatic contribution to repulsive interaction between protein molecules and silica surface both negatively charged (Gun'ko et al. 2003a,c).

Thus, despite the difference in the aggregative effects of the studied proteins, an important contribution to their interaction with the silica surfaces is connected with the electrostatic interaction increasing with pH values that results in the enhancement of the aggregative stability of the suspensions. At pH = 2 close to pH(IEP_{SiO_2}) = 2.2, a low amount of protein leads to marked flocculation of particles (Figure 6.43) such as large silica agglomerates, which are not decomposed (rearranged) at low protein concentration and protein molecules form bridges between them. At higher pH values (or at longer incubation time on the equilibrium adsorption, as similar maximum is observed at greater protein concentration, Figure 6.41), greater amounts of proteins are needed for neutralization of the negative surface charges and for fast flocculation (Figure 6.42), as decomposition (corresponding to diminution of ΔD_l) of large silica agglomerates can occur due to interaction with protein molecules with increasing electrostatic interaction between silica surfaces and protein molecules. For soft proteins such as BSA and HHb, the aggregation rate can depend not only on adsorption values but also on the conformation state of protein molecules dependent on pH, which can also change due to adsorption and lateral interaction (Gun'ko et al. 2003a,c).

To elucidate some of these effects, the PCS measurements of the protein/A-300 systems were performed (Gun'ko et al. 2003a,c). The D_{ef}(pH) graphs have a maximum close to pH(IEP) of proteins (Figure 6.43). However, at pH far from pH(IEP) of proteins the PSDs of protein/A-300 are akin to those for the pure silica suspension but D_{ef} is smaller than that for pure silica due to decomposition of silica agglomerates and aggregates under action of protein molecules. This effect is independent of the protein type, as interaction of protein molecules with silica particles can be stronger than that between silica particles per se especially in loose agglomerates (Gun'ko et al. 2001a, 2009a). Therefore, the Γ and ΔD_l maxima are observed for all the studied systems at pH = 2 (Figures 6.41 and 6.42) (Gun'ko et al. 2003a,c).

The PSDs of A-300 in the aqueous suspension at C_{SiO_2} from 0.1 to 3 wt% (without NaCl) and far from IEP of silica are typically bimodal and correspond to small aggregates of 20–60 nm (including from several to dozens of primary particles with the size of 5–12 nm) and larger ones with the size up to 500 nm (up to dozens of thousands of primary particles) (Figure 6.44). Agglomerates (which are typical for untreated or weakly treated suspensions [Gun'ko et al. 2001a]) with the size $d_{PCS} > 1$ μm are not observed in the strongly sonicated suspensions at pH > 5. Addition of 0.15 M NaCl to 3% suspension (Figure 6.44f) results in disappearance of the first PSD peak at d_{PCS} between 10 and 30 nm observed without NaCl (Figure 6.44e). As a whole the bimodal PSD (Figure 6.44f) slightly shifts toward larger d_{PCS} in comparison with other PSDs shown in Figure 6.44 (Gun'ko et al. 2003a,c).

The PSDs of BSA(S) (Figure 6.45) depict large swarms of protein molecules with the main peak over the 60–200 nm range, since special conditions (T, pH, C_p, salinity) to decompose these protein molecule swarms were not applied. Notice that for BSA2, $D_{ef} > 1$ μm at different pH values (3.39, 5.19, 6.96, 8.08) and $C_p = 0.045$ wt%; and for ovalbumin, $D_{ef} > 3$ μm at pH = 6.69 and $C_p = 0.1$ and 1 wt% (Gun'ko et al. 2003a,c).

Three types of BSA (Figures 6.46 through 6.48) and ovalbumin (Figure 6.49) interacting with secondary silica particles typically give the PSDs shifted toward larger d_{PCS} values near or above 100 nm (except BSA(S)/A-300 at pH = 8.53, Figure 6.46d).

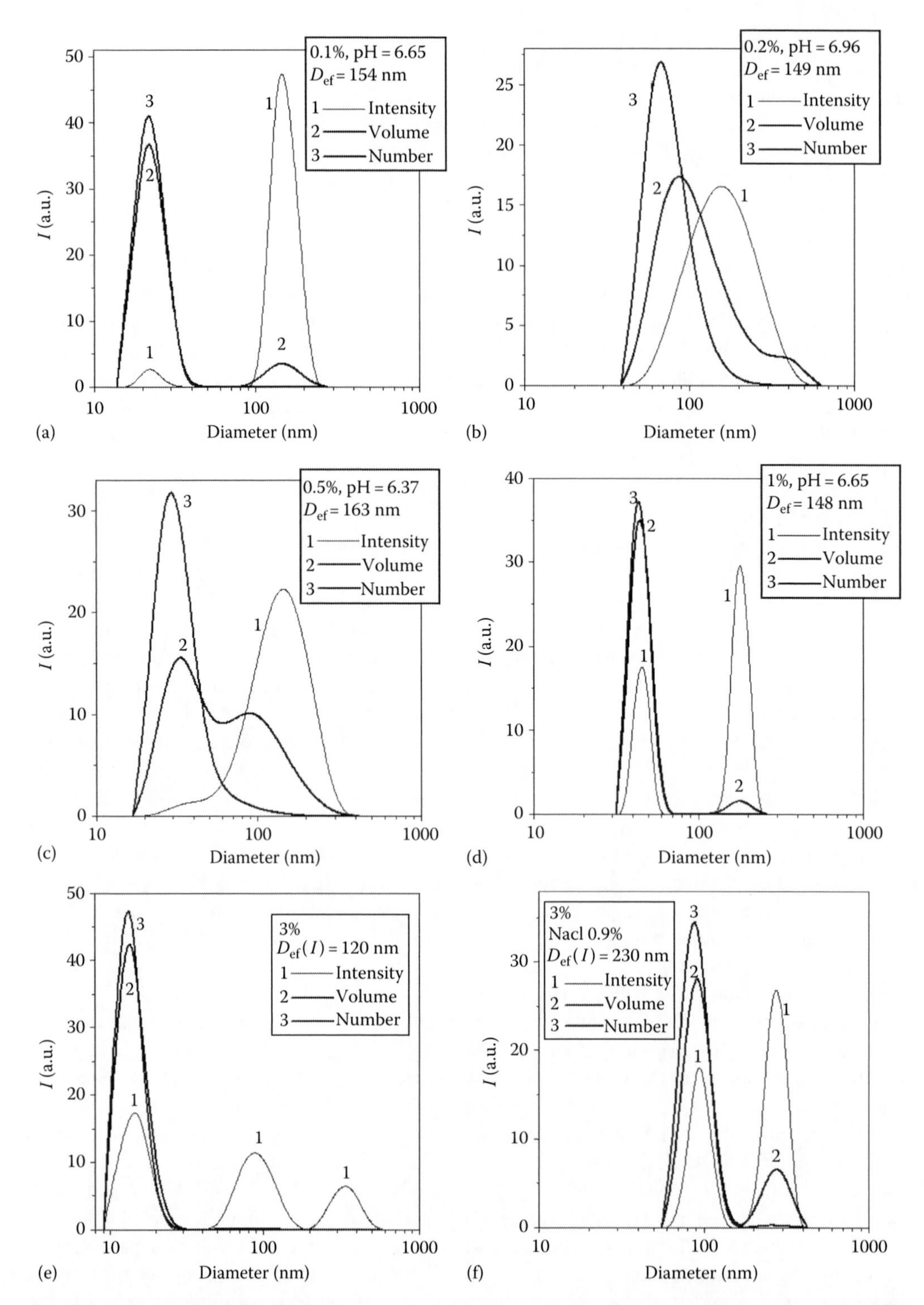

FIGURE 6.44 Particle size distributions in the aqueous suspensions of nanosilica A-300 at different pH values and concentrations: (a) 0.1, (b) 0.2, (c) 0.5, (d) 1, and (e, f) 3 wt% with respect to the light scattering intensity, particle volume and number; (f) with 0.9 wt% NaCl. (Adapted from *J. Colloid Interface Sci.*, 260, Gun'ko, V.M., Mikhailova, I.V., Zarko, V.I. et al., Study of interaction of proteins with fumed silica in aqueous suspensions by adsorption and photon correlation spectroscopy methods, 56–69, 2003c, Copyright 2003, with permission from Elsevier.)

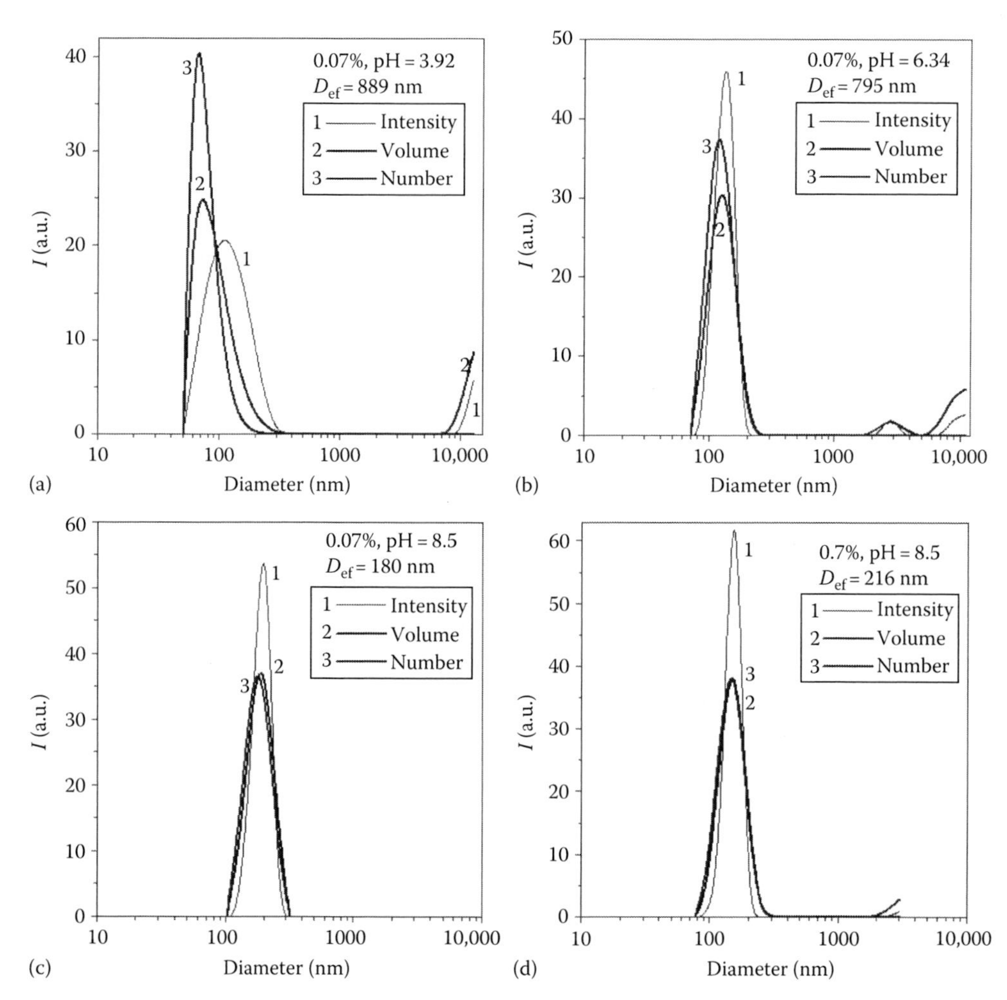

FIGURE 6.45 PSDs of BSA(S) oligomers at pH = (a) 3.92, (b) 6.37, and (c, d) 8.5 and concentration C_p = (a, b, c) 0.07 wt% and (d) 0.7 wt%. (Adapted from *J. Colloid Interface Sci.*, 260, Gun'ko, V.M., Mikhailova, I.V., Zarko, V.I. et al., Study of interaction of proteins with fumed silica in aqueous suspensions by adsorption and photon correlation spectroscopy methods, 56–69, 2003c, Copyright 2003, with permission from Elsevier.)

However, only a minor number of the secondary particles is larger than 1 μm (Figures 6.46 through 6.49, curves 3), especially at pH far from pH(IEP) of proteins. Notice that larger swarms give greater contributions to the PSDs with respect to the light scattering (curves 1) and particle volume (curves 2) than that of the PSDs connected to the particle number (curves 3). Similar relationships are characteristic for the aqueous suspensions of fumed oxides due to features of the PCS method (relatively long wavelength of the used lasers) and the type of the PSDs of nanosilicas including different types of secondary particles. At pH values close to the IEP of proteins, their aggregation with nanosilica particles is more effective and the D_{ef} value becomes larger than 1 μm (Figure 6.43) and the PSDs depict only large swarms (therefore these PSDs are not shown in Figures 6.44–6.49). Notice that pH(IEP) of the studied proteins slightly differs that can give a small displacement of the D_{ef}(pH) curves. The displacement toward higher or lower pH values gives diminution of D_{ef} (Figure 6.43) and aggregates even smaller than 100 nm appear in the PSDs (Figures 6.46 through 6.49), and the higher or the lower the pH value than pH(IEP) of proteins, the smaller the aggregates and the D_{ef} value. For ovalbumin/A-300 at C_p = 1 mg/mL (Figure 6.49), changes in the PSDs with pH are in agreement with analogous changes in ΔD_1 (Figure 6.42d). For example, a minimal D_{ef}

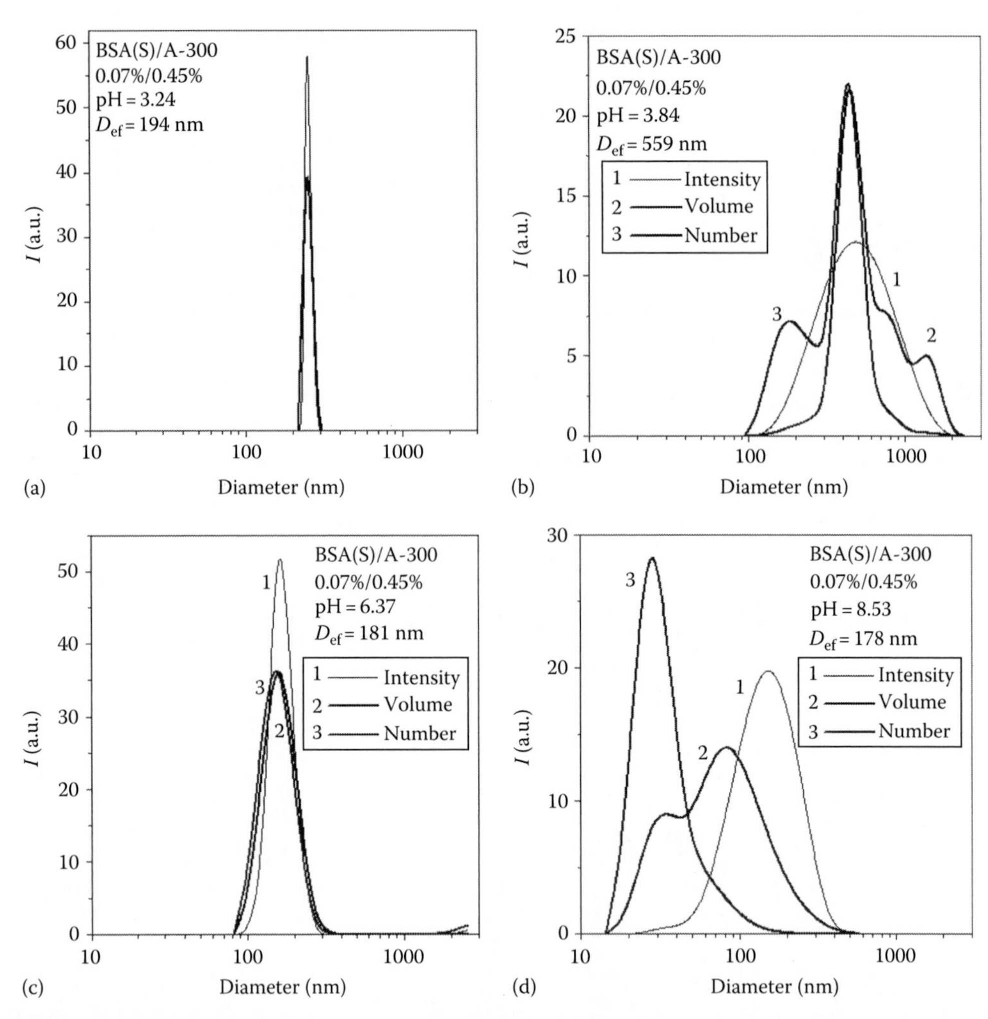

FIGURE 6.46 Particle size distributions in the aqueous suspensions of nanosilica A-300/BSA(S) at the concentrations of 0.45 wt% A-300 and 0.07 wt% BSA and different pH values: (a) 3.24, (b) 3.84, (c) 6.40, and (d) 8.53 with respect to the light scattering intensity, particle volume, and number. (Adapted from *J. Colloid Interface Sci.*, 260, Gun'ko, V.M., Mikhailova, I.V., Zarko, V.I. et al., Study of interaction of proteins with fumed silica in aqueous suspensions by adsorption and photon correlation spectroscopy methods, 56–69, 2003c, Copyright 2003, with permission from Elsevier.)

is observed at pH = 6.71 (Figure 6.49f), the next one is at pH = 2.48 (Figure 6.49a), and large D_{ef} is observed at pH = 4.04 (Figure 6.49c), 4.21 (Figure 6.49d), and 5.17 (Figure 6.49e) and this order is akin to that for ΔD_1 at similar C_p values (Figure 6.42d). For BSA(S) and BSA2, changes in the PSDs (Figures 6.46 and 6.47) and ΔD_1 (Figure 6.42) at low pH are not so compatible as that for ovalbumin maybe due to lower C_{SiO_2} values. However, at high pH values, e.g., at pH = 6.96 (Figure 6.47d), the PSDs for BSA2/A-300 and A-300 (Figure 6.44b) are very similar that is in agreement with very low ΔD_1 values (not measured) for BSA2/A-300 at pH = 6.5. The PSD for BSA(S)/A-300 at high pH (Figure 6.46d) is similar to that for A-300 at pH = 6.40 (Figure 6.44c), and ΔD_1 values for BSA(S)/A-300 were very low at pH = 6.5 (Figure 6.42c). Notice that the PSDs for more concentrated suspensions with protein/A-300 (C_{SiO_2} = 3.6 wt%) were not measured by the PCS method due to strong scattering effects (Gun'ko et al. 2003a,c). Similarity in the PSDs (Figures 6.44 through 6.49) corresponds to similarity in the electrophoretic mobility U_e of protein/A-300 aggregates (Figure 6.50) (Gun'ko et al. 2003a,c).

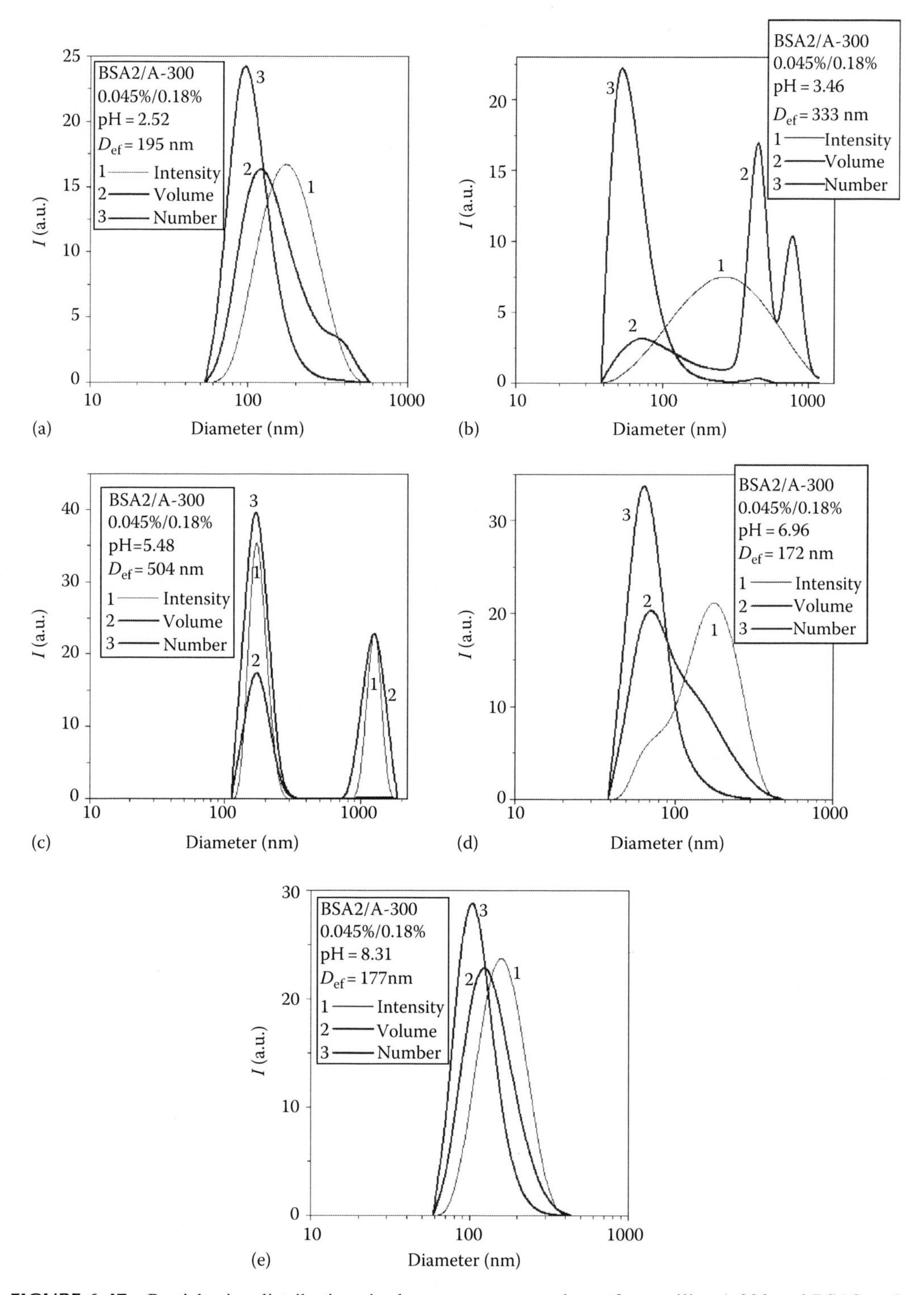

FIGURE 6.47 Particle size distributions in the aqueous suspensions of nanosilica A-300 and BSA2 at the concentrations of 0.18 wt% A-300 and 0.045 wt% BSA and different pH values: (a) 2.52, (b) 3.46, (c) 5.48, (d) 6.96, and (e) 8.31 with respect to the light scattering intensity, particle volume, and number. (Adapted from *J. Colloid Interface Sci.*, 260, Gun'ko, V.M., Mikhailova, I.V., Zarko, V.I. et al., Study of interaction of proteins with fumed silica in aqueous suspensions by adsorption and photon correlation spectroscopy methods, 56–69, 2003c, Copyright 2003, with permission from Elsevier.)

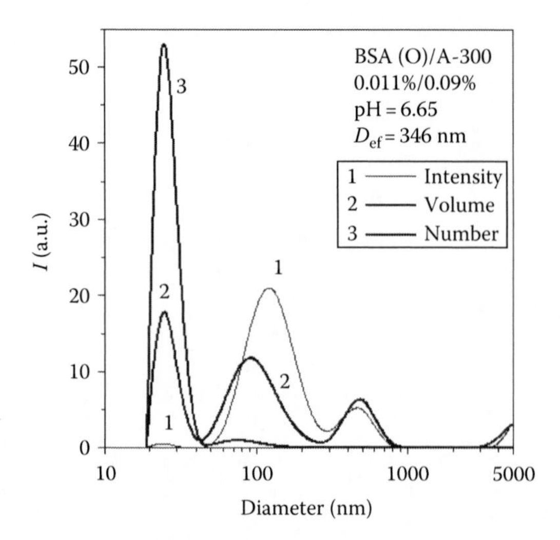

FIGURE 6.48 Particle size distributions in the aqueous suspensions of nanosilica A-300 and BSA(O) at the concentration of 0.09 wt% A-300 and 0.011 wt% BSA and pH = 6.65 with respect to the light scattering intensity, particle volume, and number. (Adapted from *J. Colloid Interface Sci.*, 260, Gun'ko, V.M., Mikhailova, I.V., Zarko, V.I. et al., Study of interaction of proteins with fumed silica in aqueous suspensions by adsorption and photon correlation spectroscopy methods, 56–69, 2003c, Copyright 2003, with permission from Elsevier.)

Additionally, the U_e(pH) curve order is akin to that of D_{ef}(pH) for protein/A-300 (Figure 6.43). The mobility of swarms of protein/silica particles (especially for BSA(S)/A-300) is greater than that of A-300 or the corresponding proteins far from pH(IEP) of proteins, as their mobility is close to zero at this pH. Consequently, the U_e values of protein/silica swarms depend mainly on the protein coverage of silica particles. Strong interaction between protein molecules and silica particles does not give significant changes in the PSDs far from pH(IEP) of proteins (Figures 6.46 through 6.49) but D_{ef} of protein/A-300 is smaller than that of silica (Figure 6.43), and their surface charge density σ_0 can be higher (e.g., due to orientation of $-COO^-$ groups of adsorbed protein molecules outside) than that for pure silica, which is relatively low at $pH < 7$ (Figure 6.51).

Additionally, according to the infrared spectroscopy data (see earlier), interaction between protein molecules and nanosilica particles leads to disturbance of a significant portion of ≡SiOH groups at a low amount of protein and to their entire disturbance at $C_{BSA}/C_{SiO_2} \approx 0.4$. This is possible if a major portion of oligomers and polymers of BSA molecules decomposes on interaction with nanosilica, whose aggregates and agglomerates are decomposed too. Thus, the enhancement of U_e of protein/silica comparing with that of individual protein molecules or silica particles can be explained by changes in the structure of adsorbed protein molecules leading to augmentation of the charge density at the shear plane in the EDL, as well as by partial decomposition of silica aggregates and protein swarms due to their interaction (Gun'ko et al. 2003a,c).

Adsorption of gelatin at pH = 3.5 corresponds to maximal changes in the Gibbs free energy of adsorption, ΔG (Figure 6.52e) calculated as follows:

$$\Delta G = -R_g T \ln \frac{x}{1-x} \tag{6.5}$$

where

$x = C_{ads}/C_p$, where C_{ads} is the adsorbed protein concentration
R_g is the gas constant

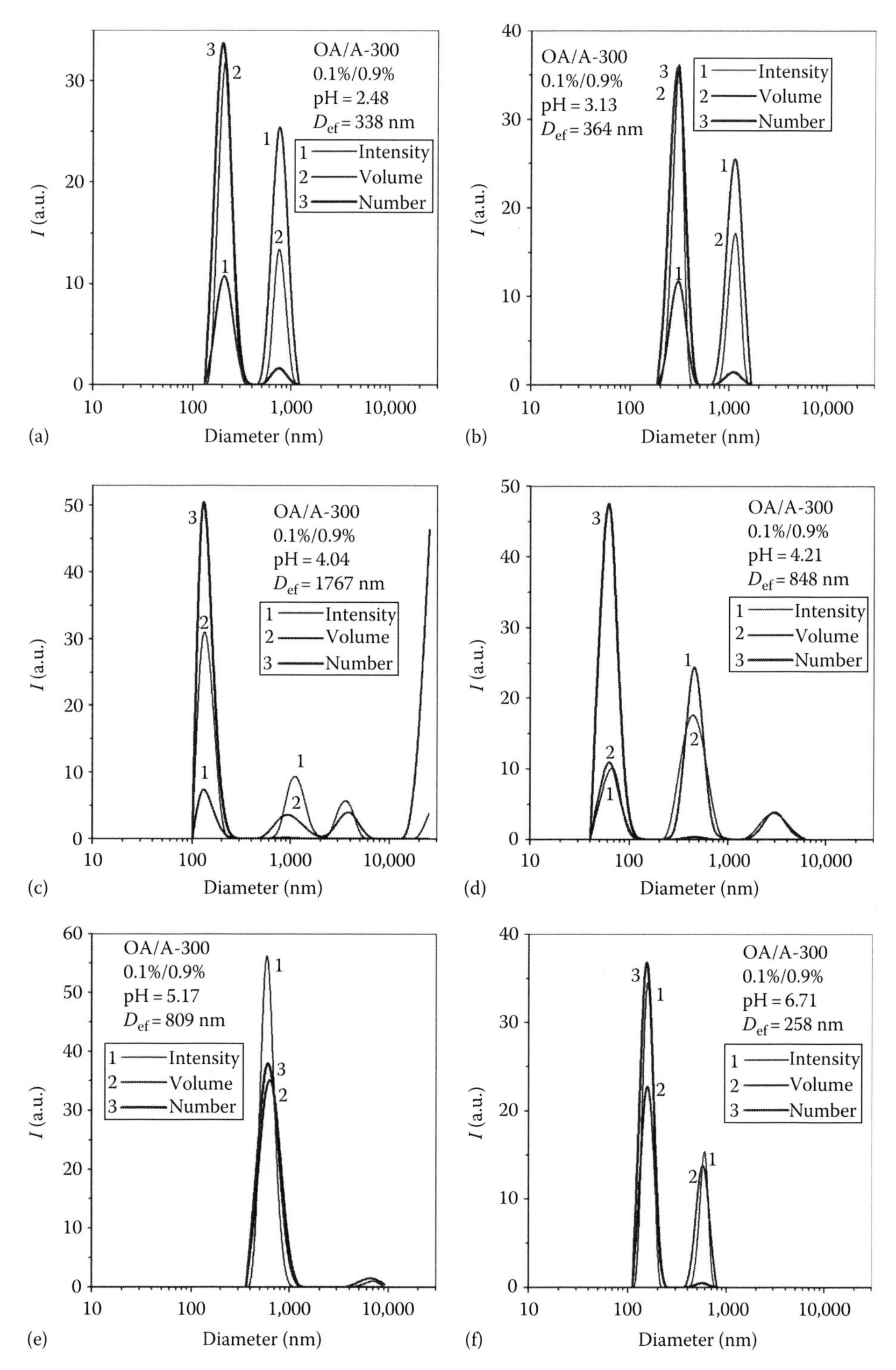

FIGURE 6.49 Particle size distributions in the aqueous suspensions of nanosilica A-300 and ovalbumin (OA) at the concentrations of 0.9 wt% A-300 and 0.1 wt% OA and different pH values: (a) 2.48, (b) 3.13, (c) 4.04, (d) 4.21, (e) 5.17, and (f) 6.71 with respect to the light scattering intensity, particle volume, and number. (Adapted from *J. Colloid Interface Sci.*, 260, Gun'ko, V.M., Mikhailova, I.V., Zarko, V.I. et al., Study of interaction of proteins with fumed silica in aqueous suspensions by adsorption and photon correlation spectroscopy methods, 56–69, 2003c, Copyright 2003, with permission from Elsevier.)

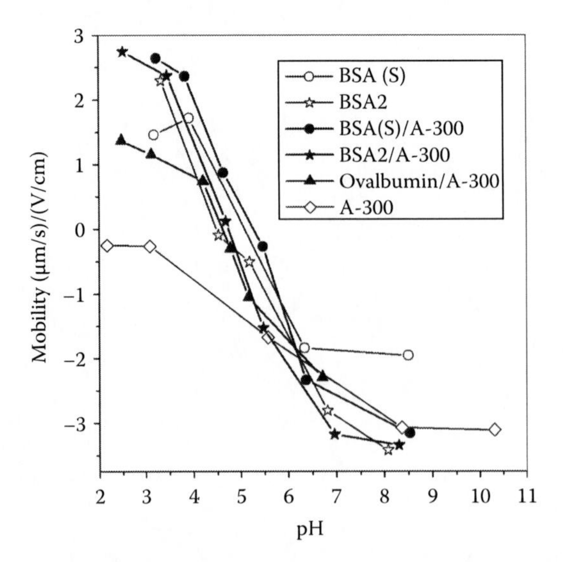

FIGURE 6.50 Electrophoretic mobility as a function of pH of particles of pure A-300, BSA(S), BSA2, and protein/A-300. (Adapted from *J. Colloid Interface Sci.*, 260, Gun'ko, V.M., Mikhailova, I.V., Zarko, V.I. et al., Study of interaction of proteins with fumed silica in aqueous suspensions by adsorption and photon correlation spectroscopy methods, 56–69, 2003c, Copyright 2003, with permission from Elsevier.)

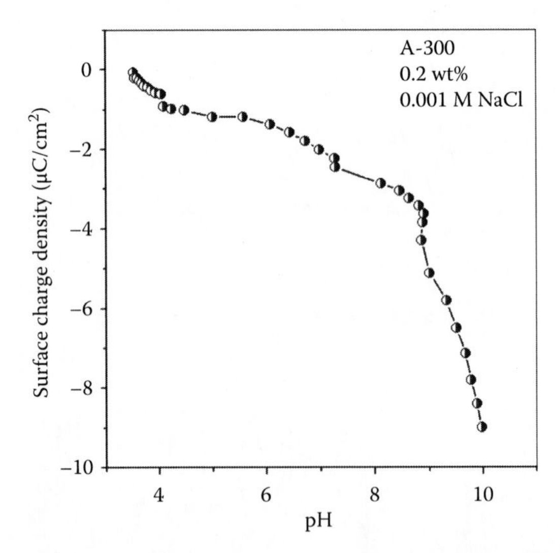

FIGURE 6.51 Surface charge density for A-300 in the aqueous suspension (C_{SiO_2}=0.2 wt%) at 0.001 M NaCl as a function of pH (measured by titration method). (Adapted from *J. Colloid Interface Sci.*, 260, Gun'ko, V.M., Mikhailova, I.V., Zarko, V.I. et al., Study of interaction of proteins with fumed silica in aqueous suspensions by adsorption and photon correlation spectroscopy methods, 56–69, 2003c, Copyright 2003, with permission from Elsevier.)

Similar ΔG values (relatively large) are characteristic for BSA/silica at the same pH but with addition of NaCl (Figure 6.52a–c), as this pH value corresponds to different sign of the charges of silica particles and protein molecules giving their attractive electrostatic interaction. For ovalbumin/silica, the maximal $-\Delta G$ values are smaller (Figure 6.52d) maybe due to lower contribution of this interaction. Notice that maximal values of the ΔG modulus close to 8–10 kJ/mol are typical for adsorption of proteins onto oxide surfaces. The presence of lipids and fatty acids in the BSA preparations can give a slight diminution of $-\Delta G$ (comparing BSA1 [$C_{lfa} \approx 2$ wt%] [Figure 6.52a]

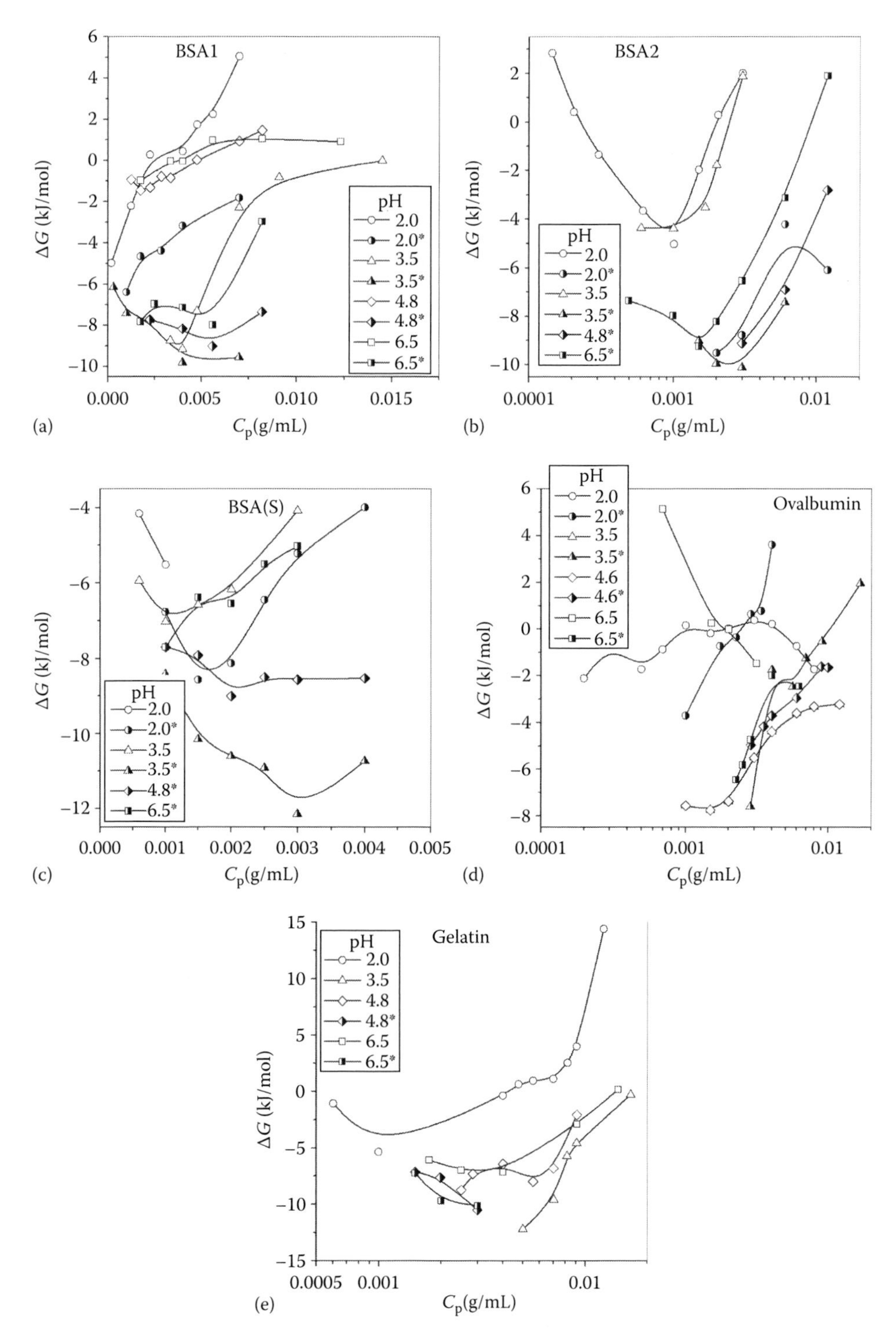

FIGURE 6.52 Changes in the Gibbs free energy (ΔG) due to adsorption of different proteins: (a) BSA1, (b) BSA2, (c) BSA(S), (d) ovalbumin, and (e) gelatin on nanosilica A-300 (3.6 wt%) at different pH values, which with asterisk show the systems with addition of 0.9 wt% NaCl. (Adapted from *J. Colloid Interface Sci.*, 260, Gun'ko, V.M., Mikhailova, I.V., Zarko, V.I. et al., Study of interaction of proteins with fumed silica in aqueous suspensions by adsorption and photon correlation spectroscopy methods, 56–69, 2003c, Copyright 2003, with permission from Elsevier.)

and BSA(S) [C_{lfa}<0.01 wt%] [Figure 6.52c] at pH=3.5 with NaCl) due to the differences in the adsorption complex structures caused by the necessity to reduce the ordering action of hydrophobic admixtures on the bound water (Gun'ko et al. 2003a,c).

To compute the distributions of changes in the Gibbs free energy ($f(\Delta G)$) on protein adsorption, the Langmuir equation was used as the kernel of the adsorption isotherm equation in the form of Fredholm integral equation of the first kind (see Chapter 10).

The graphs of $f(\Delta G)$ (Figure 6.53) as well as $\Delta G(C_p)$ (Figure 6.52) show that the $-\Delta G$ values are typically greater on addition of 0.15 M NaCl.

This effect can be due to several reasons such as changes in (i) the EDL of protein molecules (ions) and charged silica particles and electrostatic interaction between them, (ii) the structure of protein molecules, (iii) aggregation of these molecules per se or with silica particles, and (iv) PSDs of silica swarms. The $f(\Delta G)$ distribution for BSA(S)/silica (with NaCl) shifts toward large $-\Delta G$

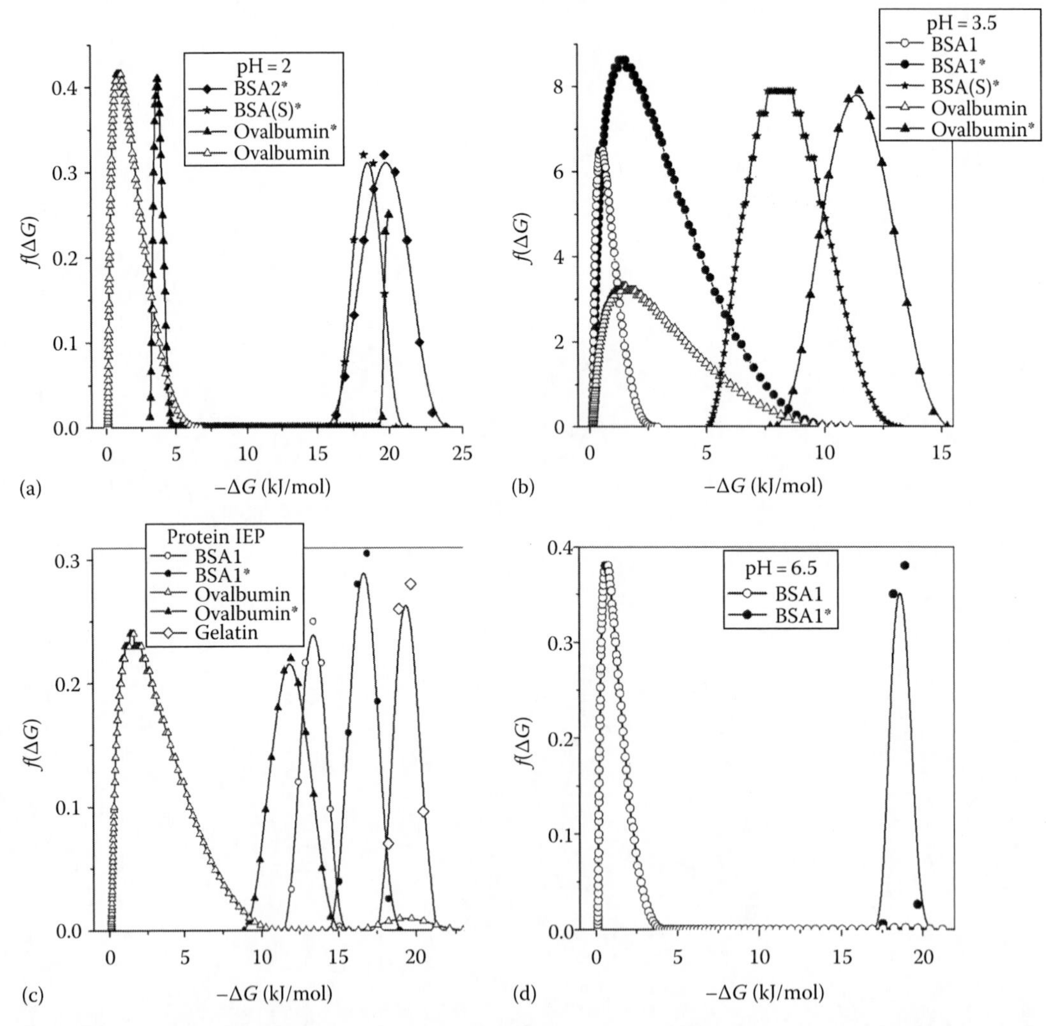

FIGURE 6.53 The distributions of changes in the Gibbs free energy ($f(\Delta G)$) due to adsorption of proteins on nanosilica A-300 (3.6 wt%) at pH=(a) 2, (b) 3.5, (c) pH(IEP) of proteins (4.8 or 4.6), and (d) 6.5; asterisk (and solid symbols) shows the systems with addition of 0.9 wt% NaCl. (Adapted from *J. Colloid Interface Sci.*, 260, Gun'ko, V.M., Mikhailova, I.V., Zarko, V.I. et al., Study of interaction of proteins with fumed silica in aqueous suspensions by adsorption and photon correlation spectroscopy methods, 56–69, 2003c, Copyright 2003, with permission from Elsevier.)

values comparing with that for BSA1 (Figure 6.53b) (due to the mentioned effect of hydrophobic admixtures), as the BSA(S) adsorption isotherm is steeper (Figure 6.41a and c). However, $f(\Delta G)$ for ovalbumin/silica (with NaCl) lies at higher $-\Delta G$ than that for adsorbed BSA(S) due to elevating adsorption with decreasing $C_{eq} < 1$ mg/mL (Figure 6.41d). Similar results are seen at pH = 2 for a high-energy peak at $-\Delta G > 15$ kJ/mol (Figure 6.53a); however, for ovalbumin, a low-energy peak at $-\Delta G < 5$ kJ/mol is also observed. Opposite order of the $f(\Delta G)$ curves for these proteins is seen at their IEP (Figure 6.53c). Notice that some difference in the positions of the high-energy $f(\Delta G)$ peaks (Figure 6.53) and the ΔG minima (Figure 6.52) can be caused not only by approximate calculations of these values using simplified equations but also by experimental errors for the $\Gamma(C_{eq})$ graphs. Consequently, changes in the free energy on protein adsorption onto nanosilica depend strongly on pH, salinity, presence of hydrophobic admixtures, and protein concentration.

Thus, according to measurements of flocculation of protein molecules with nanosilica particles at short incubation time of 1 min, low amounts of proteins (taking part in two main processes: flocculation with silica particles and decomposition of silica swarms with weak inter-aggregate binding in agglomerates and interparticle binding in aggregates) do not provide significant decomposition of large silica agglomerates at pH = 2 (close to IEP of nanosilica A-300) and protein molecules can form several bonds with particles from neighboring aggregates that results in appearance of the $\Delta D_1(C_p)$ maximum corresponding to maximal flocculation. A similar maximum is observed upon equilibrium adsorption at pH = 2 but at greater protein concentration due to changes in different swarm rearrangement during longer incubation time of 1 h. Increase in pH from 3.5 to 6.5 gives decrease in the equilibrium and minute adsorption/flocculation of proteins on A-300. Increase in the protein concentration and addition of 0.15 M NaCl differently affect adsorption/flocculation at the incubation time of 1 min and 1 h. The PCS measurements show that intensive rearrangement of secondary particles of nanosilica occurs at pH close to IEP of silica under action of protein molecules at high concentrations, which is in agreement with the adsorption data. Maximal flocculation of proteins with silica is observed at pH close to IEP of proteins, as effective diameter of particle swarms increases by order. Electrokinetic mobility of protein/silica particles is greater than that of individual protein molecules or silica particles at pH far from IEP of proteins. Changes in the Gibbs free energy on protein adsorption depend on (i) pH: $-\Delta G$ is minimal at pH = 2 close to IEP of silica and maximal at pH between IEP of protein and silica; (ii) protein concentration: $-\Delta G$ is maximal at C_p between 1 and 6 mg/mL; (iii) type of proteins: $-\Delta G$ is smaller for ovalbumin than BSA or gelatin; and (iv) protein preparation technique: $-\Delta G$ is greater for the BSA preparations with lower concentrations of lipids and fatty acids (Gun'ko et al. 2003a,c).

6.2.2 Interactions of Proteins with Carbon Materials

As mentioned earlier, NMR studies of water in solutions of proteins (Kuntz et al. 1969, Gallier et al. 1987) and other biostructures have demonstrated that certain amount of water does not freeze at $T < 273$ K. This non-freezing water, also called "bound," "ice-like," or "structured water" is characterized by significantly longer rotational correlation times than bulk water (Zipp et al. 1976, Hays and Fennema 1982, de Certaines 1989). Analysis of the temperature-dependent relaxation rates of protein solutions allows one to separate bulk and structured water on the basis of their different thermodynamic properties (de Certaines 1989, Pouliquen and Gallois 2001). Proton relaxation times, both longitudinal T_1 and transverse T_2, respond to changes in the molecular environment and thus are affected by a solid surface or solute molecules and ions (Mank and Lebovka 1988, Wüthrich 1995). The relaxation times, as well as the NMR spectra as a whole, reflect the dynamic processes in the closest vicinity of a solid surface or biomacromolecules such as proteins or DNA. They also provide information on the structure of the interfacial layer and the nature of an adsorbate–adsorbent interactions. The ^{1}H NMR spectroscopy applied to molecules adsorbed on a carbon surface allows one to distinguish between different types of adsorption complexes by comparing the chemical shifts in the ^1H NMR spectra of adsorbed and unbound compounds. Characteristics of the adsorbed water disturbed by the surface can be explored using temperature dependence of the ^1H NMR signal intensity of the unfrozen bound water in the adsorbent suspensions frozen at temperatures below 273 K.

The protein adsorption on active carbons is an important process contributing to their biocompatibility and the therapeutic effect of hemoperfusion (Arshady 1999). Covalent binding of proteins to a carbon surface allows one to synthesize adsorbents with high selectivity of action similar to that achieved in affinity chromatography. Despite numerous studies of carbon adsorbent interactions with biomolecules in aqueous media, the influence of the adsorbent structure on the characteristics of the interfacial layer of water in the presence of immobilized proteins remains unclear. Here the relationships between the structural characteristics of activated carbon SCN and the state of the bound water in the presence of BSA and mouse γ-globulin (IgG) physically adsorbed or covalently attached to the carbon surface via carbodiimide coupling are analyzed using adsorption and ^1H NMR methods.

Activated carbon SCN (the abbreviation stands for *s*pherical *c*arbon with *n*itrogen) was produced (Institute of Sorption and Problems of Endoecology, Kiev, Ukraine) by a step pyrolysis of a mesoporous vinylpyridine–divinylbenzene (VP-DVB) copolymer containing 10% of DVB at 350°C–900°C followed by steam activation at 900°C with burn-off up to 85%, bulk density 0.2 g/cm^3 (Table 6.7). In these conditions, carbon with the turbostratic structure is formed (Figueiredo and Mouliin 1986).

It is characterized by a large specific surface area (S_{BET} = 1118 m^2/g), a significant total pore volume (V_p = 0.7 cm^3/g), and a broad PSD (Figure 6.54c) with a significant deviation of the pore shape from the slit-shaped model Δw_{slit} = 0.212. SCN contains ~1% structural nitrogen, which could form polar groups at the surface and contribute to the adsorption of water and other polar compounds. SCN is a granulated carbon with spherical beads of 0.2–0.5 mm in diameter.

BSA (molecular weight $M_W \approx 68$ kDa, Cohn fraction V, Sigma) was recrystallized to obtain an essentially globulin-free protein. HPLC-purified mouse γ-globulin (IgG, $M_W \approx 158$ kDa, Palladin Institute of Biochemistry, Kiev, Ukraine) and water-soluble carbodiimide (*N*-cyclohexyl-*N*′-(2-morpholinoethyl) carbodiimide methyl-*p*-toluenesulfonate, CMC, Sigma) were used as received (Alexeeva et al. 2004). Adsorption experiments were carried out by shaking 0.2% aqueous solutions of individual solutes or proteins with CMC and a carbon sample at room temperature for 1 h. The adsorption of proteins was about 0.2 g/g. In these experimental conditions, unreacted CMC is hydrolyzed and washed out of the system (Alexeeva et al. 2004). Analysis of porous structure of SCN was done using nitrogen adsorption–desorption isotherm recorded at 77.4 K (Alexeeva et al. 2004). The ^1H NMR spectra were recorded by means of a high-resolution WP-100 SY (Bruker) NMR spectrometer at the bandwidth of

TABLE 6.7
Structural Characteristics of Unmodified and Modified SCN (k_{GT} = 60 K nm)

Sample	S_{uw} (m^2/g)	S_{nano} (m^2/g)	S_{meso} (m^2/g)	V_{nano} (cm^3/g)	V_{meso} (cm^3/g)	γ_S (J/g)	$-\Delta G_s$ (kJ/mol)	$\langle T \rangle$ (K)	C_{max} (g/g)
SCN (N_2)	1096[a]	1072	44	0.51	0.13				
SCN (H_2O)	1237	1232	4	0.45	0.01	74.9	3.35	188.8	0.455
SCN/BSA	1307	1306	1	0.47	0.001	78.0	3.32	188.4	0.471
+CMC	786	785	1	0.29	0.001	47.9	3.43	190.5	0.296
SCN/IgG	1122	1122	1	0.41	0.001	67.5	3.31	187.8	0.406
+CMC	798	797	1	0.30	0.002	48.8	3.41	190.2	0.301
SCN/CMC	877	876	1	0.33	0.002	53.4	3.37	190.7	0.331

[a] S_{BET}

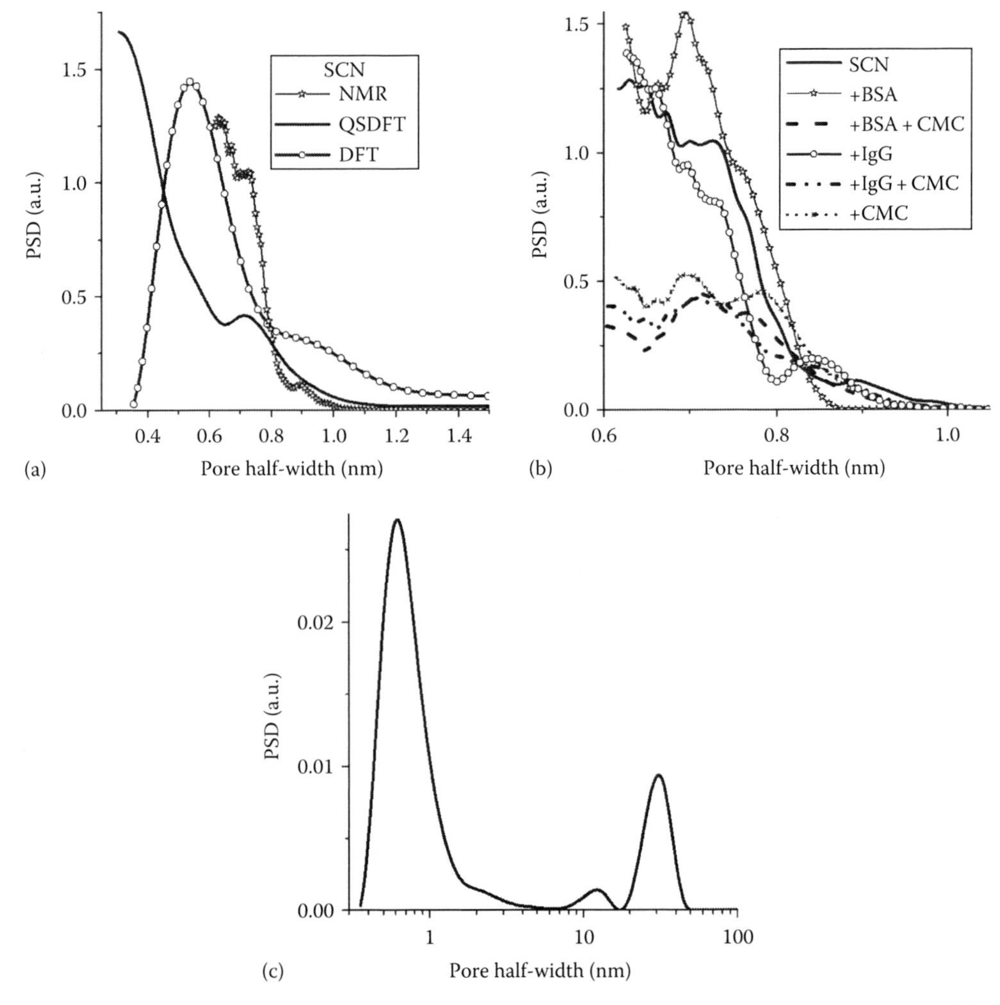

FIGURE 6.54 Pore size distributions of (a) SCN calculated using N_2 adsorption isotherm and different methods, (b) SCN with adsorbed proteins calculated with NMR cryoporometry, and (c) incremental pore size distribution of SCN related to the pore volume using N_2 adsorption (MND method).

50 or 15 kHz (Alexeeva et al. 2004). A standard program of spin–echo pulse sequence was used for measuring transverse proton relaxation time T_2 at $T < 250$ K. To decrease contribution of bulk water to the transverse proton relaxation, the freezing-out technique was utilized. For this purpose, relaxation measurements were performed at temperatures 180–250 K, which are below the freezing point of pure bulk water and water adsorbed in macropores and distant from the pore walls (this weakly bound water is frozen at T close to 273 K). The temperature was controlled by a Bruker VT-1000 device with an accuracy of ±0.5 K.

Under the experimental conditions and due to the short time of transverse relaxation of proton in solids, the contribution of protons from ice and surface hydroxyls to the recorded ^{1}H NMR signal can be neglected. Only the signal from mobile water molecules remaining unfrozen (due to the interaction with the carbon surface or solute molecules and ions) at temperature below 273 K was measured. The amount of unfrozen water bound to carbon surface and protein molecules is temperature-dependent. It was determined from the ratio of signal intensities before and after freezing using a special calibrated function. The total amount of water in each sample was constant (2 mL).

Assuming "unimolecular" mechanism of such an exchange, the Arrhenius type equation can be applied:

$$(T_2)^{-1} = (T_2)_0^{-1} \exp\left(-\frac{\Delta G_a}{RT}\right) \quad (6.6)$$

where

$(T_2)_0^{-1}$ is the pre-exponential factor dependent on temperature and at high temperatures $T_2 \to T_{2,0}$

Rearranging Equation 6.6 and taking into account that the Gibbs free energy of activation is

$$\Delta G_a = \Delta H^{\neq} - T\Delta S^{\neq} \quad (6.7)$$

the thermodynamic parameters of the exchange processes can be estimated.

The distribution function of the activation free energy of relaxation $f(\Delta G_a)$ can be calculated using Equation 6.6 as the kernel of the Fredholm integral equation of the first kind:

$$k(T) = \int_0^{\Delta G_{max}} k'(T, \Delta G_a) f(\Delta G_a) d(\Delta G_a) \quad (6.8)$$

which is solved using the regularization method under condition of non-negativity ($f(\Delta G_a) \geq 0$ at any ΔG_a) at a fixed regularization parameter ($\alpha = 0.01$).

It is well known that water can remain unfrozen at $T < 273$ K because of interaction with a solid surface or polar solute molecules and ions which disturb the hydrogen-bond network in bound water preventing ice formation. The bound water it can be assumed that (i) non-freezing water is either located in adsorbent pores or structured due to interaction with solutes; (ii) the stronger the interaction (adsorption potential), the lower the freezing temperature of the bound water; (iii) since the adsorption potential is stronger in narrow pores than in broad pores, the water adsorbed in narrow pores freezes at lower temperature than the water adsorbed in broad pores; (iv) transverse relaxation time of protons in the bound unfrozen water differs significantly from that for ice; (v) ^{1}H NMR signal of solids (e.g., OH groups in ice, surface O-containing functionalities, etc.) does not contribute to the recorded signals in the conditions of the experiment. It should be noted that although proteins and other biomacromolecules adsorb mainly onto the outer surface of nanoporous carbon particles, they can accommodate in broad meso- and macropores (Melillo et al. 2004, Gun'ko et al. 2006a) akin to those observed for SCN at the pore half-width $x > 10$ nm (Figure 6.54c) subject to the pore accessibility.

The amount of unfrozen water C_{uw} (Figure 6.55) depends not only on the pore volume but also on the specific surface area and the PSD, since the water contained in large pores at a distance from the pore walls would behave like bulk water and freeze at T close to 273 K. Assuming that water is adsorbed in nano-, meso-, and macropores, one can examine the relationships between (i) amounts of unfrozen water and changes in its Gibbs free energy (Figure 6.55); (ii) the PSD of pores filled by unfrozen water (Figure 6.54a and b); and (iii) the corresponding characteristics (Table 6.7). Figure 6.54a and b show that the major fraction of unfrozen water is located in nanopores with $x < 1.0$ nm. The amount of water remaining unfrozen at $T < 220$ K is 0.455 g/g of carbon. This value is close to the nanopore volume computed from the differential $f_V(x)$ distribution ($V_{nano} = 0.51$ cm^3/g). Thus, according to the shape of the $-\Delta G(C_{uw})$ graphs one can conclude that the water unfrozen at $T < 220$ K is strongly bound to the carbon surface and it is located in narrow pores, which is typical of nanoporous adsorbents. Therefore, the relaxation processes at $T < 250$ K in the interfacial unfrozen water at the SCN surface occur mainly in narrow pores at $x < 1.0$ nm (Alexeeva et al. 2004).

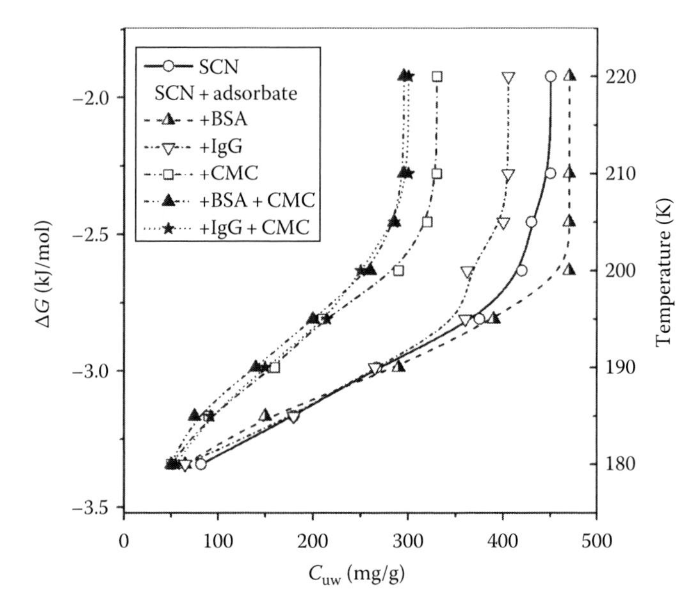

FIGURE 6.55 Concentration of unfrozen water (strongly bound water) as a function of the temperature in the presence of organics; and relationships between changes in the Gibbs free energy of bound water and the amount of unfrozen water C_{uw}. (Adapted from *J. Colloid Interface Sci.*, 278, Alexeeva, T.A., Lebovka, N.I., Gun'ko, V.M., Strashko, V.V., and Mikhalovsky, S.V., Characteristics of interfacial water affected by proteins adsorbed on activated carbon, 333–341, 2004, Copyright 2004, with permission from Elsevier.)

In the absence of CMC, physically adsorbed protein molecules have a weak influence on water adsorbed in the narrowest pores with $x < 1$ nm (Figure 6.54b) because of a large size of protein molecules which cannot penetrate narrow pores. These pores, however, are accessible to small CMC molecules, which have a stronger effect on the bound water confined to these pores, e.g., CMC molecules can displace water molecules from narrow pores to larger ones. Most interestingly, protein molecules covalently attached to the surface via the coupling reaction with CMC, have the highest impact on the water contained in narrow pores. It is likely that the immobilized protein molecules partly block access for water molecules to the pores and residual CMC molecules displace a portion of water from narrow pores. Additionally, ions from salts typically accompanying protein preparations diffuse in liquid unfrozen water and can strongly affect the state of the bound water with lowering temperature. A difference between physically adsorbed and covalently attached to the surface protein molecules may reflect the loss of conformational flexibility by CMC-linked proteins, which influences the bound water. Carbodiimide attaches protein to the surface by peptide-like bonds between surface carboxylic groups and amine groups of the protein (Bodanszky and Bodanszky 1984). It could also cross-link carboxylic and amine groups of the protein itself, further increasing the rigidity of its structure. Clearly, these structural changes of adsorbed proteins can affect the structured water (de Certaines 1989).

The T_2 value for water adsorbed on SCN (Figure 6.56a) is low (< 0.003 s) but much longer compared to T_2 in ice in the studied temperature range between 180 and 250 K.

Under these conditions, T_2 arises mainly from mobile unfrozen water molecules. Spatial structure of water clusters inside nanopores and narrow mesopores of carbons (Brennan et al. 2001, Gun'ko et al. 2005d, 2009d) does not correspond to the ice lattice and water exists in a pseudo-liquid mobile state. The plot of T_2 versus temperature in the Arrhenius coordinates, $\ln T_2$ versus T^{-1}, is nearly linear without breaking points in the studied temperature range for the systems "SCN–BSA" and "SCN–CMC–IgG" (Figure 6.56). For the other systems, however, two to three different regions are clearly seen (Figure 6.56b). Surface heterogeneity affects proton exchange in water and different exchange patterns suggest that several types of water clusters and nanodroplets can be formed. Their structure is

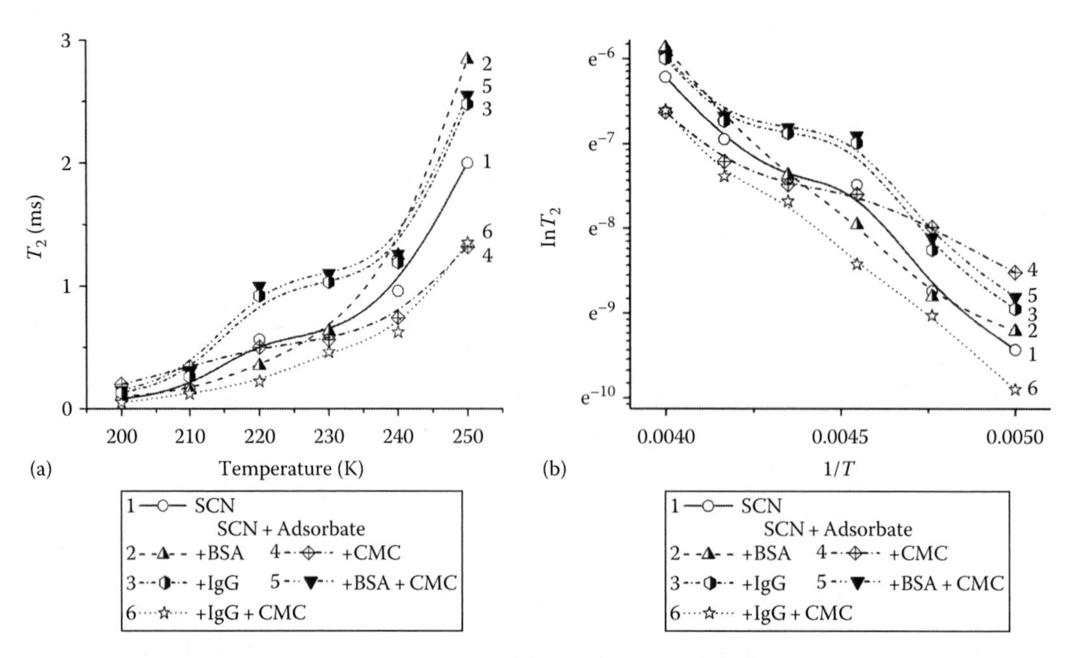

FIGURE 6.56 The transverse relaxation time T_2 of protons in the bound water as a function of the temperature in (a) linear scale and (b) $\ln T_2$ versus T^{-1}; curves: (curve 1) SCN, and SCN + adsorbate: (curve 2) BSA, (curve 3) IgG, (curve 4) CMC, (curve 5) BSA + CMC, and (curve 6) IgG + CMC. (Adapted from *J. Colloid Interface Sci.*, 278, Alexeeva, T.A., Lebovka, N.I., Gun'ko, V.M., Strashko, V.V., and Mikhalovsky, S.V., Characteristics of interfacial water affected by proteins adsorbed on activated carbon, 333–341, 2004, Copyright 2004, with permission from Elsevier.)

strongly influenced by the texture of the porous matrix and the structure of hydrated protein molecules as well as dissolved ions (Na^+, Cl^-, etc.) which can concentrate in liquid water in narrow pores as the temperature is lowered. It is possible that the adsorbed BSA or covalently attached IgG replace lesser amounts of water from pores because these molecules are retained mainly on the external surface of activated carbon or in large pores at $x > 10$ nm (Figure 6.54). Different behavior of BSA and IgG can be explained in terms of their conformational features. Native BSA is a flexible molecule and its adsorption on porous activated carbon is irreversible because of the multi-site character of its binding to the surface. Due to the conformational flexibility of BSA, its amino acid fragments can penetrate the pores that are much smaller than the size of the protein molecule irreversibly immobilizing it on the surface. In "SCN–CMC–protein" systems, carbodiimide binds protein molecules to the surface via amino groups reacting with activated carboxylic groups on the carbon surface. Additionally, CMC can cross-link protein molecules via their own amino and carboxylic groups. Covalent binding to the surface and cross-linking reduce the flexibility of BSA molecules and their ability to displace adsorbed water molecules from narrow pores. Contrary to BSA, native IgG has more rigid structure which is not affected significantly by the carbon surface. Besides, IgG has a larger molecular size than BSA; therefore, it can accommodate only in broader pores. Hence, physically adsorbed IgG cannot displace the adsorbed water from nano- and narrow mesopores resembling BSA treated with CMC (Figure 6.55). However, the molecular weight of IgG is significantly higher than that of BSA, respectively, the specific surface area of BSA molecules is larger than that of IgG by the factor of $(M_{W,IgG}/M_{W,BSA})^{1/3} \approx 1.3$. Consequently, the amount of bound water (per volume unit) interacting with the IgG molecules is lower than that with BSA. Therefore, $C_{uw}(T)$ curve for SCN–IgG system lies below $C_{uw}(T)$ curve for SCN–BSA (Figure 6.55). Covalent binding results in a multi-site attachment of IgG to the carbon surface via CMC-activated surface groups, which is more significant than cross-linking of the protein in the solution. It leads to the displacement of a fraction of adsorbed water by IgG (and reduction of amount of water bound to the protein molecules). Thus, the nonlinear relationship between $\ln T_2$ and T^{-1} (Figure 6.56b) may reflect existence of several types of structured water in the interfacial layer (in the pores, at the outer surface of carbon particles and water bound by macromolecules), which are disturbed differently by the pore walls and the protein molecules (Alexeeva et al. 2004).

Addition of carbodiimide reduces amount of the bound water by 20%–25% (Figure 6.55) and elevates its freezing point by 10–15 K compared to the SCN–protein system. The difference between the $C_{uw}(T)$ curves for SCN–CMC and "SCN–CMC–protein" is smaller than the difference between them and CMC-free systems (Figure 6.55). The role of carbodiimide in SCN–CMC system is not easy to explain; it is possible that it creates certain chemical and spatial structures in narrow pores by cross-linking surface functional groups, for example, acidic and basic sites or even acidic and acidic sites. Such intra-porous bonding may reduce water adsorption on these polar sites which is reflected in the position of C_{uw} (Figure 6.55).

The distribution function $f(\Delta G_a)$ characterizing the proton relaxation processes can be used to estimate the impact of CMC and proteins on the adsorbed non-freezing water (Figure 6.57) (Alexeeva et al. 2004).

The maximal ΔG_a value is observed for the aqueous suspension of SCN with BSA and the minimal ΔG_a value is found for the SCN–CMC system. It should be noted that similar values of the activation energy (12–26 kJ/mol) for the proton relaxation (T_1) in structured water were found using the ^{1}H NMR probing in normal and dehydrated rabbit lenses (Bodurka et al. 1996). The distortion action of the adsorbates and the carbon surface on the hydrogen-bond network in the bound water decreases in suspensions with relatively small CMC molecules (Figures 6.55 and 6.57). This effect can be caused by CMC penetration into pores with $x>0.4$ nm and partial displacement of the adsorbed water into broader pores. This leads to diminution of the $C_{uw}(T)$ (Figure 6.55) and the activation energy of relaxation, since the concentration of unfrozen water in narrow pores decreases. The CMC impact is larger for SCN–BSA than that for SCN–IgG due to structural features of these proteins. However, the corresponding curves of $C_{uw}(T)$ for both "SCN–CMC–protein" systems are close to each other because of the mentioned effect of CMC on the bound water and proteins (Figure 6.55).

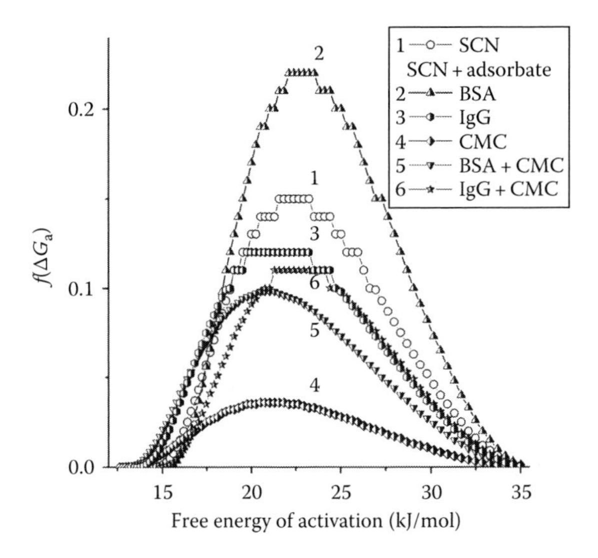

FIGURE 6.57 The distribution functions of the Gibbs free energy of activation of the transverse proton relaxation for different adsorbates (calculation with fixed regularization parameter $\alpha=0.01$); curves: (1) SCN, and SCN+adsorbate: (2) BSA, (3) IgG, (4) CMC, (5) BSA+CMC, and (6) IgG+CMC. (Adapted from *J. Colloid Interface Sci.*, 278, Alexeeva, T.A., Lebovka, N.I., Gun'ko, V.M., Strashko, V.V., and Mikhalovsky, S.V., Characteristics of interfacial water affected by proteins adsorbed on activated carbon, 333–341, 2004, Copyright 2004, with permission from Elsevier.)

Thus, ^{1}H NMR measurement of the transverse relaxation time at temperatures below 273 K has been used to study water adsorbed in the pores of activated carbon and water bound to protein molecules. Application of the technique of freezing-out of bulk water allows studying of the effect of BSA and mouse γ-globulin adsorbed from aqueous solutions on the properties of the bound water at the surface of activated carbon SCN. Water remaining unfrozen at $T < 273$ K is located predominantly in nano- and narrow mesopores (with the pore half-width <2 nm) of the carbon. Presence of physically adsorbed BSA enhances the amount of unfrozen water in the system in comparison with that in the frozen "SCN–water" or "SCN–IgG–water" suspensions. Addition of water-soluble CMC significantly reduces the amount of unfrozen water and the activation energy of transverse proton relaxation due to covalent binding of protein molecules to the carbon surface and cross-linking of protein molecules. CMC displaces the bound water from narrow pores thus reducing the amount of unfrozen water (Alexeeva et al. 2004).

6.2.3 Interactions of Proteins with Polymers

Controlling interactions of synthetic polymers with proteins are of practical interest for biomedical and other applications (Slater et al. 2006, Börner 2009, Kumar et al. 2009, Jaeger et al. 2010, Gokmen and Du Prez 2012). Advanced structural and functional control can lead to rational design of integrated nano- and microstructures with polymers/peptides. To achieve this, properties of monomer sequence-defined peptides were exploited. Through their incorporation as monodisperse segments into synthetic polymers, it was learned in recent years how to program the structure formation of polymers, to adjust and exploit interactions in such polymers, to control inorganic–organic interfaces in fiber composites, and to induce structure in biomacromolecules like DNA for biomedical applications (Börner 2009).

The polymer–NP/nanocomposites have been the exponentially growing field of research for developing the materials in last few decades and have been mainly focusing on the structure–property relationships and their development (Kumar et al. 2009). Since the polymer–nanocomposites have been the staple of modern polymer industry, their durability under various environmental conditions and degradability after their service life are also essential fields of research. Among various nanoparticulates, clay minerals and carbon nanotubes are more often used in enhancing physical, mechanical, and thermal properties of polymers. In very few systems, the nanoparticulates have been incorporated into polymer as "nano-additives" for both purposes: degradation and stabilization of polymers. The degradation and durability of polymers in the presence of NP/nanocomposites under different environmental conditions, as well as NP-induced biodegradation of polymers, is of importance (Kumar et al. 2009). Therefore, carbohydrate polymers such as cellulose, starch, chitin, hyaluronic acid are most appropriate for these applications. Chitin whiskers prepared by partial hydrolysis, and incorporated polymers such as poly(styrene-*co*-butylacrylate) (Paillet and Dufresne 2001), polycaprolactone (Morin and Dufresne 2002), natural rubber (Nair and Dufresne 2003), and soy protein isolate (Lu et al. 2004b) are attractive for these applications. Both the addition of chitin whiskers and heat treatment to PVA helped improve water resistance, leading to decreased percentage degree of swelling, of the nanocomposite films. Moreover, the thermal stability of the nanocomposites was found to be more than the pristine PVA film, and it increases with the CS whiskers content (Sriupayo et al. 2005). Motornov et al. (2010) analyzed the synthesis and applications of stimuli-responsive polymeric and hybrid nanostructured particles in a range of sizes from nanometers to a few micrometers: nano- and microgels, core–shell structures, polymerosomes, block-copolymer micelles, and more complex architectures in colloidal dispersions, thin films, delivery devices, and sensors.

Reception of chemical and biochemical signals can be based on a physical interaction or chemical reaction between functional groups in the polymer and signaling molecules (Motornov et al. 2010). There are many examples of specific complex formation between synthetic polymer materials and ligands, e.g., glucose-responsive polymers with phenylboronic side groups (Lapeyre et al. 2008).

The selective interaction of the responsive polymers with signaling molecules relies on selective molecular recognition phenomena using the conjugation of responsive polymers with biological molecules such as DNA (Li et al. 2004, Yurke et al. 2006), enzymes (Miyata et al. 2002, Ulijn 2006), antibodies (Miyata et al. 1999), and other proteins (Stayton et al. 1995, Hoffman 2000, Cutler and Garcia 2003, Rosso et al. 2005). The interactions that involve DNA and proteins result in formation or cleavage of junction points between the functional polymers and signaling molecules. Applications of enzymes refer to two possible situations: (i) polymers with immobilized enzymes (Hassan et al. 1997) and (ii) materials with fragments that are substrates for enzymes. In the first case, a substrate diffusing to the polymer from the surrounding aqueous medium can be biocatalytically converted into products which interact with the responsive polymer and cause chemical changes in the polymer. For example, catalytic oxidation of glucose by GO_x yields gluconic acid and, consequently, results in a decrease of pH in the local environment. These changes are then used to trigger response of the NPs (Patel et al. 2008). In the second case, the enzyme is used as an "external stimulus" that cleaves the chemical bonds in the material.

Hatakeyama et al. (2009) studied several simple, lightly cross-linked quaternary phosphonium- and ammonium-based polymer coatings which can effectively resist the nonspecific adsorption of proteins (i.e., BSA and fibrinogen [Fg]) from aqueous solution under both static exposure and dynamic membrane fouling conditions. In some cases, their protein-resistance performance was comparable to, or even better than, cross-linked PEG-based polymers, which are considered benchmark protein-resistant coating materials. Similarly, these quaternary phosphonium and ammonium polymers exhibited comparable or better resistance to protein adsorption compared to polymeric analogs of some of the best organic functional groups identified in prior self-assembled monolayer-based protein-resistance studies. In particular, initial results of dynamic membrane fouling experiments showed that lightly cross-linked poly[trimethyl-(4-vinyl-benzyl)-phosphonium bromide] had exceptional protein fouling resistance and better water transport properties than a representative PEG-based polymer coating. In addition to surface functional group chemistry, it was also found that the sub-surface chemistry, the nature of the substrate that the coating is on (i.e., substrate type and morphology), and the protein exposure conditions (i.e., static adsorption versus dynamic filtration testing) can greatly affect the overall protein adsorption resistance behavior of the coating. The presence of a regular nanostructure on the polymeric coating surface can lead to enhancement of protein resistance under static exposure conditions even with the same functional groups present, similar to what has been observed with inorganic surfaces (Hatakeyama et al. 2009).

If the most important parameters affecting protein–polysaccharide interactions have been well documented, as well as the structure-building kinetics, thermodynamics, and structure of mixtures (Turgeon et al. 2003). As far as complex coacervation is concerned, experimental tools such as light scattering techniques, confocal and electron microscopy, and high sensitivity or isothermal titration calorimetry have brought insights on the molecular structure of the systems and gave access to thermodynamic parameters. These experimental data, combined with intensive numerical simulations on polyelectrolyte systems, were used to build up various models of protein–polysaccharide interaction taking into account both intrinsic and extrinsic parameters. In the case of thermodynamic incompatibility, probing interfaces of phase-separated systems was bringing some insights on the spatial and temporal behavior of such systems. The use of w/w emulsion-like approach successfully allowed better control of the morphology of segregative phase-separated systems (Turgeon et al. 2003).

Investigations of interaction of soluble polymers of different molecular weights with a solid surface in aqueous media show that smaller molecules adsorb faster but larger molecules adsorbed more slowly can displace smaller molecules from the surface. Composition of the adsorbed layer containing macromolecules can change during a short period of time after interaction with solution of the second polymer or protein (Horbett and Brash 1995). The rate at which these changes occur depends on the diffusion rate and the binding affinity that the molecules have with the surface. The adsorption of polymers depends on the molecular size, the charge distribution, polymer solubility, intermolecular interactions between polymer molecules, solvation–desolvation effects on adsorption, adsorption-induced conformational changes, and both the chemical and morphological characteristics of solids, as well as on the medium characteristics (pH, salinity, concentrations of polymers and solid particles, and temperature) (Horbett and Brash 1995, Malmsten 1998). As a whole, the interfacial phenomena become more complex for mixtures of adsorbates, especially macromolecules, with strongly different spatial and chemical structures. In this case, results of their competitive adsorption depend on their mobility and affinity to a surface as well as on the mentioned factors and applied external electrical or mechanical forces governing the interfacial layer structure with time. Smaller molecules adsorbed faster (because of limitation of this process by the diffusion rate) than larger but slower ones can be displaced by the latter if pores are much larger than molecules. This phenomenon known as the Vroman effect related to the adsorption of proteins and discovered for fibrinogen competing with other plasma proteins has been observed for blends of various proteins (Horbett and Brash 1995). The Vroman effect has a substantial significance for the biocompatibility of biomaterials because it governs the adlayer structure with time at the artificial surfaces (e.g., endovascular stents) in the body or the structure of the interfaces of administrated enterosorbents such as nanosilica, activated carbons, etc. Real-time observation of the interfacial phenomena involving macromolecules is a difficult task. Microscopic visualization of the adsorbed layer cannot give the qualitative data about the interfacial dynamics in liquids. Many investigations of interaction of mixtures of macromolecules with solids were carried out for large solid particles or continuous solid surfaces. However, NP, which can be, e.g., effective carriers for therapeutics, demonstrate a very specific behavior in the liquid media. For instance, interaction of solid NP (forming true colloidal systems with unusually high stability) with polymer mixtures can strongly differ from that of microparticles (which strongly precipitate and do not form fine dispersions). Therefore, we analyze the kinetics of successive interaction of pairs of organics (BSA, linear polymers PVA, PVP, PEG) with NP of nanosilica A-300 and A-50 characterized by different size distributions of primary particles (5–15 nm and 15–100 nm, respectively) and the aggregative capability using DLS method and to compare these results with the equilibrium successive adsorption of the pairs using the adsorption and FTIR spectroscopy methods. It should be noted that the measurement techniques on the DLS and adsorption (and FTIR) investigations strongly differ because the DLS measurements were carried out without strong mechanical action on the suspensions with time (only the suspensions with individual silica was sonicated but after addition of the polymer or protein solutions the suspensions were only shaken up to 30 s) that allow us to study much weaker interactions in the systems with organics/silica than that on the adsorption and FTIR measurements carried out for liquid and solid residua centrifuged at 6000 rpm.

Nanosilica A-300 and A-50 ($S_{BET}=232$ or 297 and 52 m^2/g, respectively) was heated at 673 K for several hours to remove residual HCl and other adsorbed compounds (Gun'ko et al. 2006i). Proteins such as BSA ("Allergen," Stavropol, Russia, molecular weight $W_M \approx 67$ kDa), gelatin (Merck, powder food grade), ovalbumin ($W_M \approx 44$ kDa), and ossein ("Indar," Kiev, Ukraine, 24.5 ± 4.5 kDa); polymers such as PVP ("Biopharma," Kiev, 12.6 ± 2.7 kDa), PVA ("Stirol," Severodonetsk, Ukraine; 43 ± 7 kDa), and PEG (Fluka, $W_M \approx 2$, 20 and 35 kDa) were used in QELS experiments to analyze the kinetic interactions of co-adsorbates (Gun'ko et al. 2006i).

Before adsorption of polymers, nanosilica ($S_{BET}=232$ m^2/g) was heated in air at 450°C for several hours to remove adsorbed compounds. Certain amounts of the nanosilica powder and a polymer (PVP or PEG) solution were loaded into a glass reactor (2 dm^3) with a mixer (>500 rpm) providing a pseudo-liquid state (PLSR) of a powder and agitated at room temperature for 0.5 h. Then certain amount of ethanol was dripped (one or two drops per second) in the reactor at room temperature and the mixture was agitated for 1–6 h, and then it was dried at room temperature (Gun'ko et al. 2006i).

DLS investigations were performed using a Zetasizer 3000 (Malvern Instruments) apparatus ($\lambda=633$ nm, $\Theta=90°$) at 298 K (Gun'ko et al. 2006i). Deionized distilled water was used for preparation of nanosilica suspension ($C_{SiO_2}=0.1$ and 0.5 wt% for the systems based on A-300 and

BSA/PVP/A-50 and 1.0 wt% for other systems with A-50) sonicated for 15 min using an ultrasonic disperser (Sonicator Misonix Inc.) (22 kHz and 500 W). Low-concentrated suspensions ($C_{SiO_2}=0.1$ wt%) were used to avoid strong aggregation caused by silica particle–organic compound I (OC-I) interaction in more concentrated suspensions of A-300. The concentrated solutions of OC-I and then organic compound II (OC-II) were added to the treated aqueous suspension of nanosilica A-300 or A-50 in a given time without additional treatment, and the PSDs were measured several times during 1 h without motion of the measuring cell. The pH value measured by a precision digital pH meter was 5.34–8.02, and the salinity was 0.9 wt% NaCl. To compute the PSD with respect to the intensity of light scattering (PSD_I) and particle volume (PSD_V) and number (PSD_N), the Malvern Instruments software (version 1.3) was utilized, assuming that particles had a roughly spherical shape. The concentrations of organics were $C_{BSA}=0.4$ g/g of silica (A-300 or A-50), 0.15 g of PVP, PEG (2 kDa), or PVA per gram of silica (A-300 or A-50).

PVP and PEG molecules have polar basic groups N–C=O and C–O–C, respectively, possessing the electron-donor properties in contrast to PVA (C–O–H), and proteins (BSA, gelatin, ovalbumin, or ossein with a large set of side functionalities characteristic for amino acids) having both electron-donor (basic) and proton-donor (acidic) groups, as well as silica with amphoteric silanols ≡Si–O–H. Intermolecular interaction between PVP or PEG molecules (that is provided by dispersive forces and very weak hydrogen bonding C–H…O) is weaker than that for PVA, and proteins because for the last molecules the binding can be caused by electrostatic forces (e.g., $COO^-…H^+N$) and strong hydrogen bonds (e.g., O–H…O and O–H…N) as well as by significant dispersive interaction. The strongest intermolecular interaction in the adlayer with PVP or PEG at the silica surface is linked to the hydrogen bonds between the basic groups of adsorbates and surface silanols ≡SiO–H…OR. Features of the adsorption and intermolecular interactions cause different dependences of the perturbation degree of free silanols (estimated from the FTIR spectra of solid residua with A-300/polymer) $\Phi=1-D/D_0$ (where D and D_0 are the optical density of the IR band of free silanols at 3748 cm^{-1}) on the adsorption. The Φ values are much lower for PVA than that for PEG at the same loading because of stronger intermolecular interaction between PVA molecules than that between PEG molecules. In the case of the adsorption of PVA and proteins, the lateral interactions between adsorbed molecules are characterized by the energy comparable to that of their binding to the silica surface. Additionally, the effects of water as a solvent (i.e., solvation–desolvation effects on adsorption) are stronger in the case of PVA and especially proteins than that on the adsorption of PEG or PVP. These circumstances provide lower Gibbs free energy of the adsorption (ΔG_{ads}) of PEG or PVP ($\Delta G_{ads}<-15$ kJ/mol) onto a silica surface in comparison with BSA ($\Delta G_{ads}\approx-10$ kJ/mol) or PVA ($\Delta G_{ads}>-15$ kJ/mol) from the aqueous media. On the adsorption of protein (gelatin) on silica A-300 with pre-adsorbed PEG (Figure 6.58a), the ΔG_{ads} value depends on the amounts of pre-adsorbed polymer (Figure 6.58b) and at the PEG coverage >50 mg/g, $-\Delta G_{ads}<1$ kJ/mol, i.e., the adsorption of gelatin is weak.

Phenomena related to the adsorption of macromolecules to a solid surface or coagulation of macromolecules and NP can be studied using noninvasive analysis with, e.g., DLS or using techniques including stages with relatively strong actions on the samples, e.g., centrifugation, sonication, high-frequency vibration, drying–heating, etc. Therefore, observations based on the mentioned methods can give different (even opposite) information on the interfacial phenomena depending on balance of intramolecular, intermolecular, and external forces. The adsorption of proteins on unmodified and modified (by pre-adsorbed polymers) silicas measured using solid and liquid components separated by centrifugation of the suspensions is caused by the strong binding of adsorbed molecules because centrifugation work (6000 rpm, radius of rotation 3–5 cm) is close or even higher than the hydrogen-bonding energy. Therefore, significant desolvation of solid residue is observed on centrifugation, and relative weak complexes formed in the second or next layers with protein molecules can be completely destroyed if pre-adsorbed polymer forms a monolayer totally covering the silica surface, i.e., direct access of protein molecules cannot occur. These effects lead to strong diminution of the amounts of adsorbed gelatin with increasing pre-adsorption of PEG (Figure 6.58a) or PVP

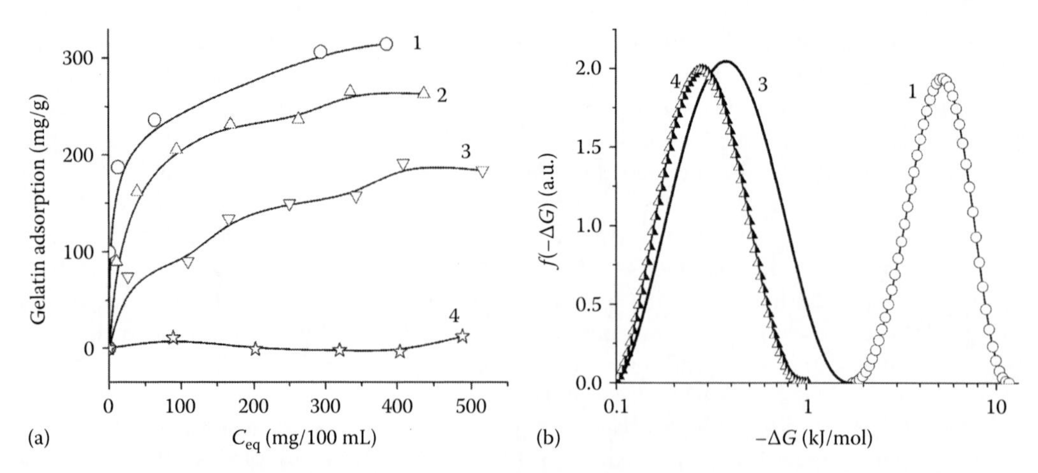

FIGURE 6.58 (a) Adsorption isotherms of gelatin from aqueous solution on nanosilica A-300 (1 wt%) with preadsorbed PEG (20 kDa) in amounts (1) 0, (2) 20, (3) 50, and (4) 200 mg per gram of silica, (b) distribution functions of changes in Gibbs free energy on the adsorption of gelatin corresponding to adsorption curves shown in (a). (Adapted from *J. Colloid Interface Sci.*, 300, Gun'ko, V.M., Zarko, V.I., Voronin, E.F. et al., Successive interaction of pairs of soluble organics with nanosilica in aqueous media, 20–32, 2006f, Copyright 2006, with permission from Elsevier.)

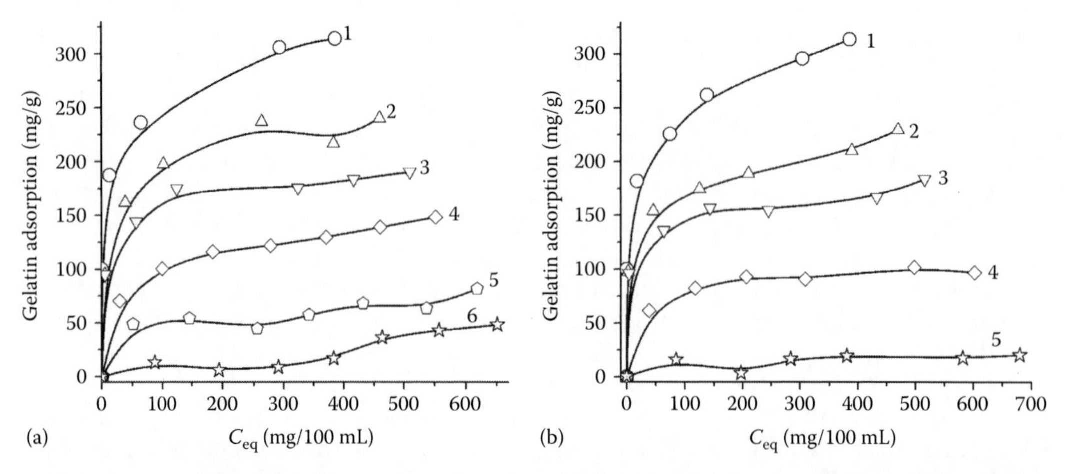

FIGURE 6.59 Adsorption isotherms of gelatin from (a) aqueous solution and (b) phosphate buffer solution on A-300 (1 wt%) with pre-adsorbed PVP in amounts (1) 0, (2) 25, (3) 50, (4) 100, (5) 125, and (6) 200 mg per gram of silica. (Adapted from *J. Colloid Interface Sci.*, 300, Gun'ko, V.M., Zarko, V.I., Voronin, E.F. et al., Successive interaction of pairs of soluble organics with nanosilica in aqueous media, 20–32, 2006f, Copyright 2006, with permission from Elsevier.)

(Figure 6.59) and changes in the Gibbs free energy on the adsorption of gelatin become very small (Figure 6.58b). Similar effects were observed using QCM method applied to protein pairs (fibrinogen and HSA) adsorbed onto different surfaces when the second adlayer with protein molecules (especially for HSA) can be easily and totally removed by the liquid flow from vibrating (10 MHz) solid surface (Mikhalovsky et al. 2005a,b, 2006, Gun'ko et al. 2011b). The FTIR spectra of gelatin/polymer/silica demonstrate complete disturbance of free silanols because a band at 3750 cm^{-1} is absent for both polymer–silica and protein–polymer–silica samples. At the same time, after the adsorption of gelatin, a new band appears at 1570–1500 cm^{-1} characteristic for this protein.

Pre-adsorbed PVP and PEG give a close dependence of the plateau adsorption of gelatin as a function of the pre-adsorbed amounts of the polymer (Figure 6.60). Consequently, the strong adsorption

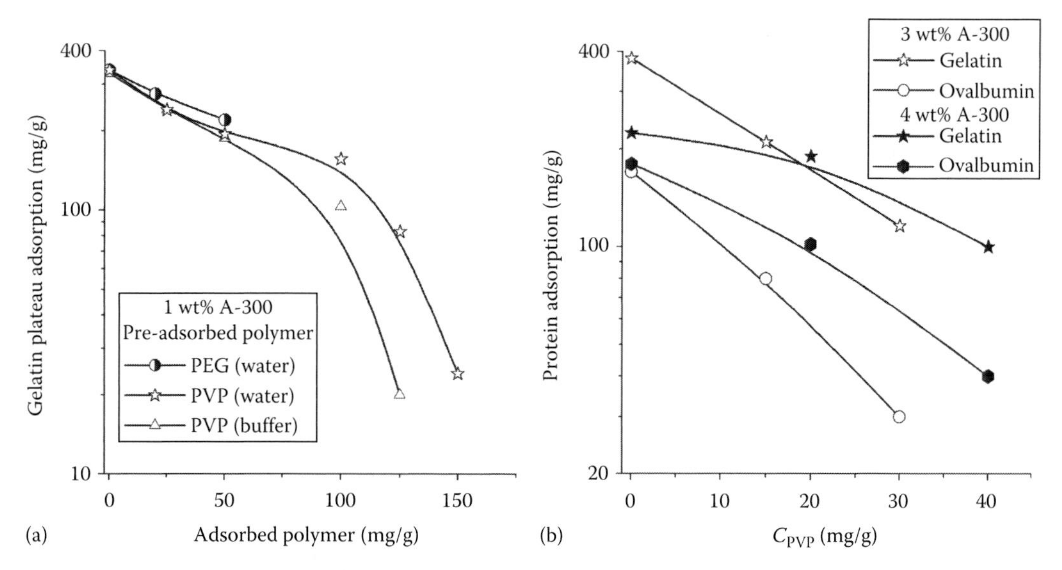

FIGURE 6.60 Plateau adsorption of (a, b) gelatin or (b) ovalbumin onto A-300 (a) 1 wt% and (b) 3 or 4 wt% as a function of the amounts of (a) pre-adsorbed PEG (20 kDa, aqueous solution) and PVP (aqueous and phosphate buffer solutions) and (b) PVP. (Adapted from *J. Colloid Interface Sci.*, 300, Gun'ko, V.M., Zarko, V.I., Voronin, E.F. et al., Successive interaction of pairs of soluble organics with nanosilica in aqueous media, 20–32, 2006f, Copyright 2006, with permission from Elsevier.)

of gelatin onto polymer/silica is observed when a portion of the surface is uncovered by the polymer, and protein molecules can form the hydrogen or other kind of bonds directly with the silica surface. In the case of the formation of intermolecular bonds between protein and polymer molecules, centrifugation leads to decomposition of these adsorption complexes that depict as diminution of the amounts of adsorbed protein. The formation of these complexes will be studied later using the QELS method. One can expect that investigations of the interaction of proteins with nanosilica modified by adsorbed polymers using noninvasive analysis with the DLS method (without sonication of the suspensions with OC-II/OC-I/silica) reveal additional details of the studied interfacial phenomena.

It should be noted that the polymers used on these investigations form the true aqueous solutions as their oligomers are not practically observed in the used solution. However, pair mixtures (at 1 wt% of each organic compound and the constant salinity 0.9 wt% NaCl) with ossein+PVA (Figure 6.61) demonstrate strong coagulation of organics that is characterized by multimodal particle size distributions. Nanosilicas are characterized by a stepwise structural hierarchy with primary particles of 5–15 nm in diameter for A-300 and 15–100 nm for A-50, which form aggregates (<1 μm in size) and agglomerates of aggregates (>1 μm). Sonication of aqueous suspension of nanosilica or the interaction with polymer molecules lead to decomposition of a portion of secondary particles to primary ones and formation of new secondary particles as hybrid aggregates due to coagulation of macromolecules and solid NP. Sonication of the concentrated aqueous suspensions of nanosilicas at $C_{SiO_2} \geq 3$ wt%, which can be substantially stable than diluted ones, can give the dispersion composed mainly of primary particles up to 99.9% of the total number of particles corresponding to 96%–98% of the particle volume. However, fast coagulation can be observed in the suspensions after addition of the polymers or proteins. Therefore, we used non-concentrated suspensions at $C_{SiO_2}=0.1$ and 0.5 wt% (A-300 and A-50) and 1 wt% (A-50). Successive interaction of OC-II with pre-adsorbed OC-I/nanosilica is characterized by time-dependent rearrangement of secondary particles, i.e., the displacement of the PSD peaks toward larger or smaller size depends on time. For certain systems, this displacement is small but for others very strong coagulation occurs, e.g., on addition of gelatin or ossein (proteins of close origin) because these proteins have an extended

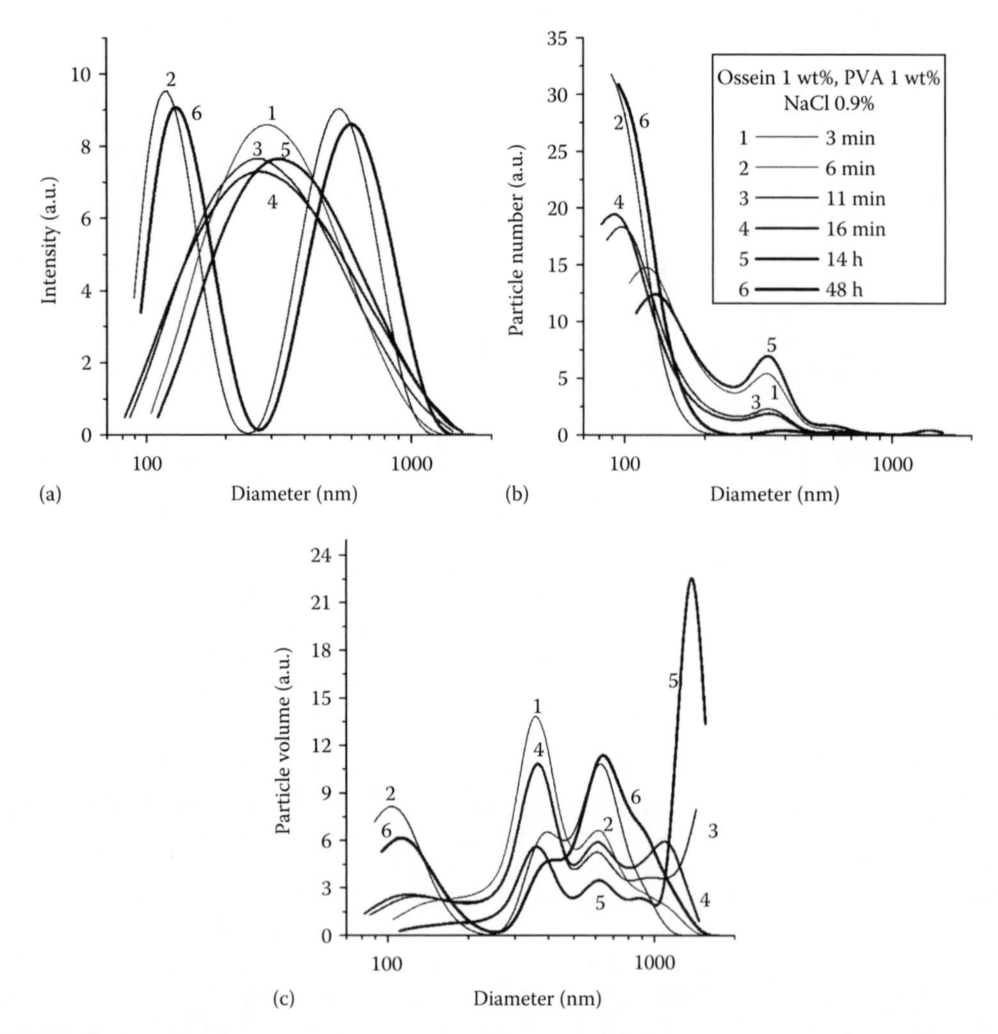

FIGURE 6.61 Oligomer size distribution with respect to (a) light scattering, (b) particle number, and (c) particle volume for a mixture of ossein (1 wt%) and PVA (1 wt%) solutions (salinity 0.9 wt% NaCl) as a function of observation time from 3 min to 48 h. (Adapted from *J. Colloid Interface Sci.*, 300, Gun'ko, V.M., Zarko, V.I., Voronin, E.F. et al., Successive interaction of pairs of soluble organics with nanosilica in aqueous media, 20–32, 2006f, Copyright 2006, with permission from Elsevier.)

form with much larger ratio between the largest and smallest sizes than globular BSA molecules. Therefore, gelatin and ossein molecules can interact with larger number of silica particles than BSA that leads to stronger aggregation.

Notice that several time-dependent effects can be responsible for the secondary particle rearrangement: (i) decomposition of residual secondary particles of silica upon interaction with OC-I and then OC-II because of stronger binding of organics to silica surface than silica to silica particles; (ii) coagulation of silica NP with polymer or protein molecules to form new secondary hybrid particles due to electrostatic, polar, and hydrogen bonding; (iii) rearrangement of adsorbed OC-I/silica aggregates on interaction with OC-II because of the displacement of adsorbed smaller molecules by larger ones (i.e., due to the Vroman effect); and (iv) rearrangement of aggregates with adsorbed OC-II/OC-I/silica with time because of perikinetic and orthokinetic aggregation and differential sedimentation of organics/solid particles. Notice that the Brownian kernel predictions were found to describe aggregation kinetics correctly only at high electrolyte concentrations. It should be noted that the adsorption of organics from the aqueous media or gases (e.g., N_2, Ar, etc.) from the gas phase onto silica surface modified by adsorbed polymers or grafted organic or organosilicon groups reduces in comparison with the adsorption onto unmodified silica. Consequently, diminution of the adsorption potential of OC-I/silica in comparison with silica can reduce this effect for OC-II. Certain aspects of the behavior of the adsorbed OC-II/OC-I/nanosilica/water systems with time can be elucidated on the basis of the DLS investigations of the corresponding aqueous suspensions during relatively short period (from several minutes to 1 h) because the adsorption of polymers or proteins onto nonporous primary particles of nanosilica is enough fast. For instance, the adsorption of BSA onto nanosilica A-300 during a minute gives about 50% of the plateau adsorption value (A_{eq}), and the adsorption for 10 min and 1 h gives about 85% and ≈100% of the A_{eq} value, respectively. Notice that the sediment was not formed during a longer period than the observation time for all the systems shown here. Additionally, investigations of the turbidity of aqueous suspensions of silica (particles 0.5–1.5 μm) on the dynamic turbulent aggregation show that the main changes in the turbidity because of coagulation occur during 10 min. Therefore, the kinetics of coagulation of OC-I with nanosilica particles was studied during up to 34 min, and the kinetics of successive coagulation of OC-II with adsorbed OC-I/silica was investigated for additional 20–30 min.

Figure 6.62 depicts the time-dependent interaction of PEG(I)–BSA(II) and PVA(I)–BSA(II) with A-300 (C_{SiO_2}=0.1 wt% and salinity 0.9 wt% of NaCl) at polymer concentrations close to their monolayer coverage. For the first pair, rather smaller particles appear after addition of the BSA solution to the PEG/A-300 suspension. However, certain amount of large particles at the size d>600 nm are observed (Figure 6.62a and c). The behavior of the second pair strongly differs

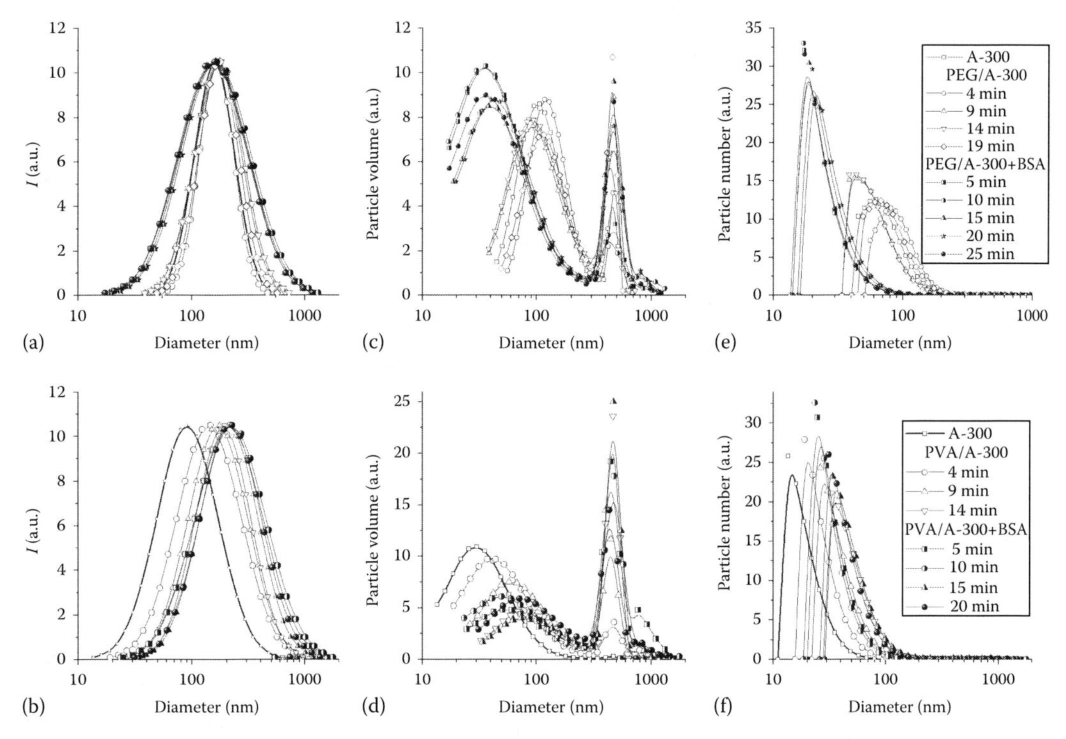

FIGURE 6.62 Particle size distributions with respect to (a, b) light scattering (PSD_I), (c, d) particle volume (PSD_V), and (e, f) particle number (PSD_N) in the aqueous suspension of (a, c, e) A-300, PEG (2 kDa)/A-300, and BSA(II)/PEG(I)/A-300 and (b, d, f) A-300, PVA/A-300, and BSA(II)/PVA(I)/A-300. (Adapted from *J. Colloid Interface Sci.*, 300, Gun'ko, V.M., Zarko, V.I., Voronin, E.F. et al., Successive interaction of pairs of soluble organics with nanosilica in aqueous media, 20–32, 2006i, Copyright 2006, with permission from Elsevier.)

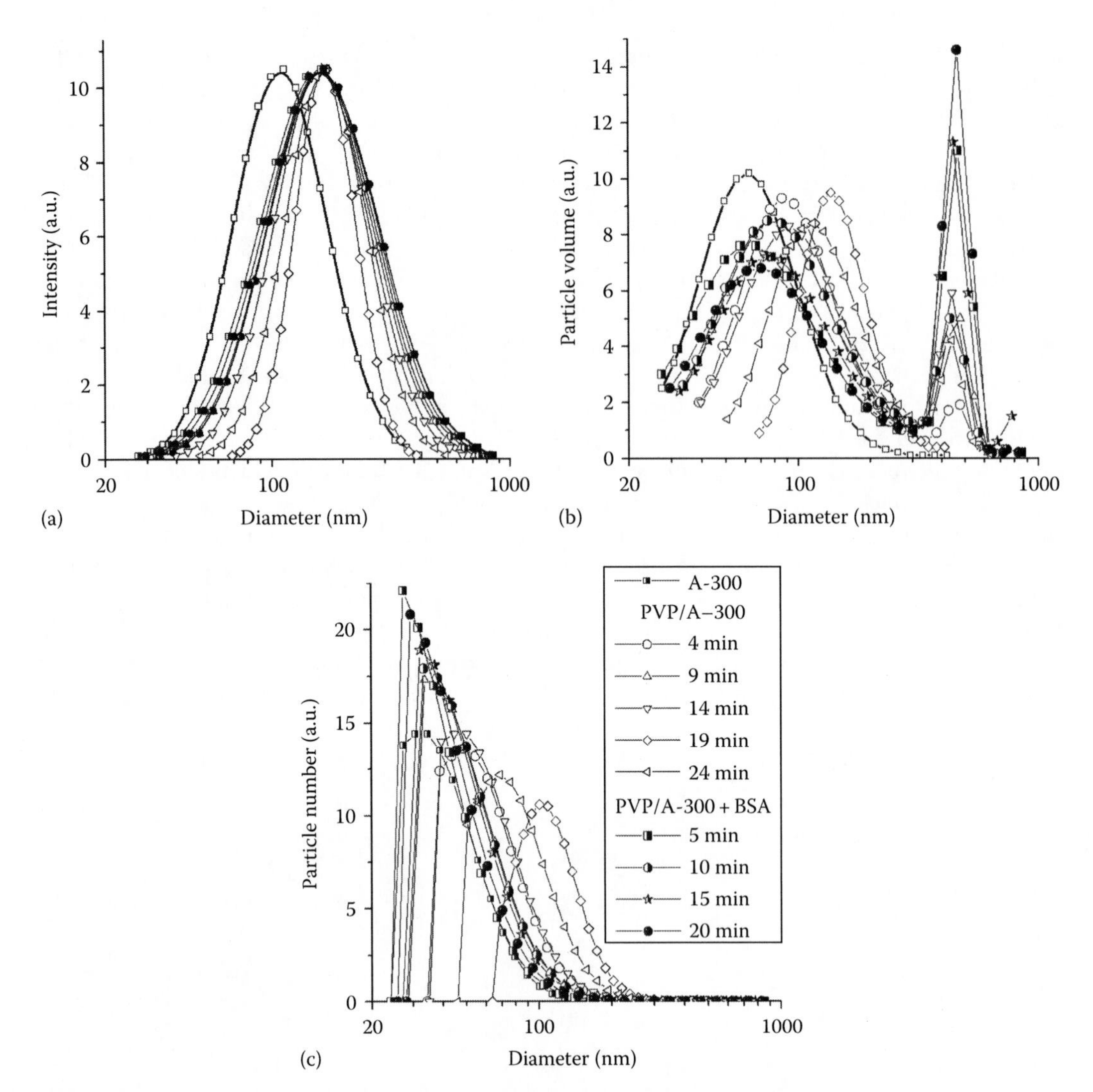

FIGURE 6.63 Particle size distributions with respect to (a) light scattering, (b) particle volume, and (c) particle number in the aqueous suspension of A-300, PVP/A-300, and BSA(II)/PVP(I)/A-300. (Adapted from *J. Colloid Interface Sci.*, 300, Gun'ko, V.M., Zarko, V.I., Voronin, E.F. et al., Successive interaction of pairs of soluble organics with nanosilica in aqueous media, 20–32, 2006i, Copyright 2006, with permission from Elsevier.)

because rather large particles form after addition of the BSA solution to the PVA/A-300 suspension. This enlargement of aggregates increases with time (Figure 6.62b, d, and f).

The behavior of the pair PVP(I)–BSA(II) (Figure 6.63) is akin to that of PEG–BSA (because polymers have only electron-donor groups). However, certain differences in the PSDs are observed with time because the molecular weight of PVP (12.6 kDa) is higher than that of PEG (2 kDa).

Similar effects are observed in the case of A-50 (Figure 6.64), despite it is composed of larger primary particles (15–100 nm) than A-300 (5–15 nm). The time-dependent effects for the BSA(II)/PVP(I)/A-50 (Figure 6.64) and BSA/PVP/A-300 (Figure 6.63) or BSA/PVA/A-50 (Figure 6.64) and BSA/PVA/A-300 (Figure 6.62) systems are close, despite the differences in the concentrations of oxides and polymers and in the size of primary particles of A-300 and A-50.

Large BSA molecules can be more difficult to be removed from the silica surface than smaller polymers because of the formation of a larger number of bonds (e.g., hydrogen bonds) between BSA

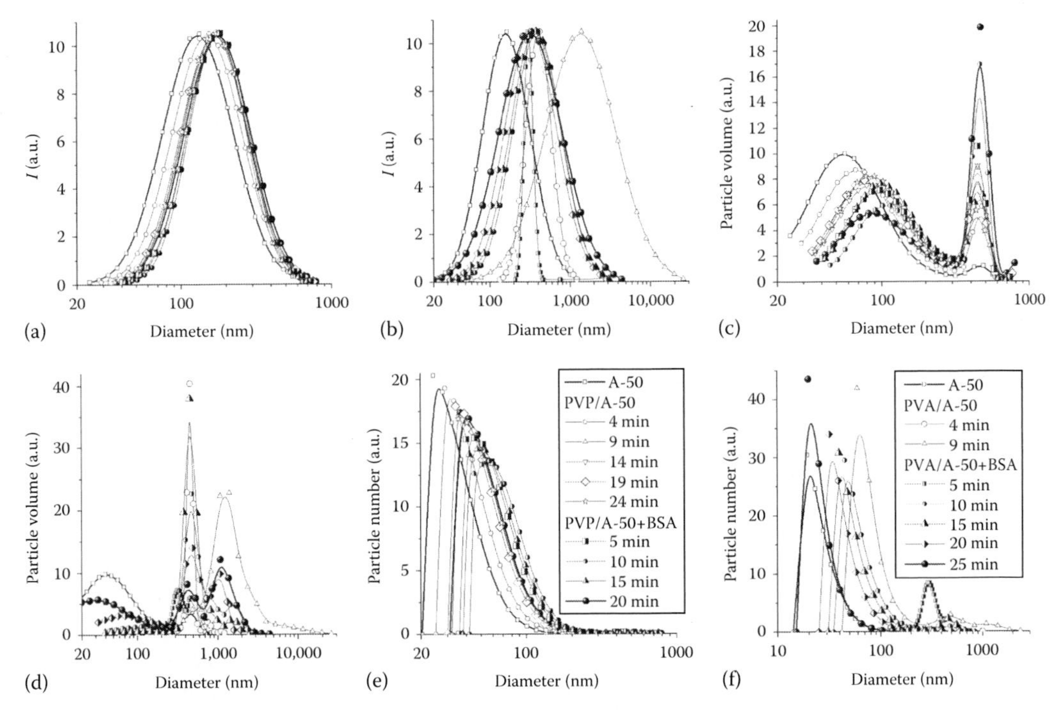

FIGURE 6.64 Particle size distributions with respect to (a, b) light scattering, (c, d) particle volume, and (e, f) particle number in the aqueous suspension of (a, c, e) A-50, PVP/A-50, and BSA(II)/PVP(I)/A-50 and (b, d, f) A-50, PVA/A-50, and BSA(II)/PVA(I)/A-50. (Adapted from *J. Colloid Interface Sci.*, 300, Gun'ko, V.M., Zarko, V.I., Voronin, E.F. et al., Successive interaction of pairs of soluble organics with nanosilica in aqueous media, 20–32, 2006i, Copyright 2006, with permission from Elsevier.)

molecules and the surface in comparison with smaller polymers. On the other hand, BSA molecules can displace smaller PEG or PVP molecules and decompose a portion of hybrid aggregates that reflect in the time dependence of the PSDs. For instance, larger and smaller particles compared to those of OC-I/silica simultaneously appear. In the case of PVA possessing the electron-donor and proton-donor properties and the W_M value higher than that of PEG or PVP, the effect of displacement of the molecules and decomposition of hybrid aggregates by BSA is weak because only increased aggregation in the BSA/PVA/nanosilica system is observed with time.

BSA molecules can destroy hybrid aggregates composed of primary silica particles and relatively small PEG molecules, which can weakly block the silica surface against much larger BSA molecules. Therefore, the PSD_V and PSD_N peaks for PEG/A-300 at $d \approx 100$ nm disappear but the peaks at 20–40 nm appear after interaction with BSA (Figure 6.62c and e). The second PSD_V peak at 400–500 nm is also time-dependent and affected by adsorbed BSA (Figure 6.62c) as well as PSD_I (Figure 6.62a). In the case of pre-adsorbed PVA, decomposition of aggregates of PVA/A-300 by BSA molecules is not practically observed because aggregates of electron-donor/proton-donor PVA/silica can strongly interact with BSA practically without decomposition. Additionally, PVA molecules can more strongly shield the silica surface against BSA molecules than smaller PEG or PVP molecules. The PVP molecules at $W_M \approx 12.6$ kDa, which is between W_M of PEG (2 kDa) and PVA (43 kDa) but closer to that of PEG, give the effects closer to those of PEG because both polymers (i.e., PVP and PEG) have only electron-donor groups, which can more effectively interact with ≡SiOH groups than with other PVP (or PEG) molecules. Therefore, the interaction between PVP–PVP (or PEG–PEG) molecules can be weaker than that between PVP (or PEG) and BSA molecules. The system PVA/PEG/A-300 is characterized by a stronger effect of the second polymer (PVA) than that for PVA/PVP/A-300 because of $W_{M,PEG} \ll W_{M,PVP}$, and the displacement of the PVP molecules and decomposition of hybrid aggregates with PVP/silica by the PVA molecules are more difficult than that for the PEG/silica system. Additionally, linear PVA and PEG molecules are structurally closer than PVA and PVP that can provide more effective interaction of PVA with pre-adsorbed PEG due to formation of the hydrogen bonds CO–H···OCC. The formation of these bonds can assist in removal of smaller PEG molecules from silica surface by larger PVA molecules.

The BSA/PVP/A-300 (Figure 6.63) and BSA/PVP/A-50 (Figure 6.64) systems (with the same concentrations of the components) have very close shapes of the PSD_I, PSD_V, and PSD_N graphs. However, the PSD_N curves for the suspension with A-50 shift toward smaller d values because A-50 is characterized by smaller aggregation of primary particles than A-300. Therefore, the effects of rearrangement of secondary particles with A-50 can be slightly different (Figures 6.63b and 6.64c or Figures 6.63c or 6.64e). However, as a whole the rearrangement results for hybrid aggregates are close for A-300 and A-50 because of close morphological properties of nanosilicas of different specific surface areas. This is caused by the formation of roughly spherical particles in the flame upon the synthesis of nanosilica with their subsequent aggregation into aggregates and agglomerates despite the difference in the size of primary particles of A-300 and A-50. Larger differences in the behavior of the BSA/PVA/A-300 (Figure 6.62) and BSA/PVA/A-50 (Figure 6.64) systems can be caused by the difference in the concentrations of oxides (0.1 and 1.0 wt%, respectively) and polymers (C_{pol} is linked to C_{SiO_2} as $C_{BSA}/C_{SiO_2}=0.40$ and 0.15 for other polymers). Therefore, aggregation in more concentrated suspension (based on A-50) gives more broadened PSDs with respect to both light scattering and particle volume and number. Clearly, these features in the interfacial behavior of macromolecules interacting with nanosilica particles and bound water are reflected in results of the investigations of the interfacial phenomena in similar systems using low-temperature NMR spectroscopy, TSDC, DRS, and DSC methods.

6.3 DNA

NMR spectroscopy is widely used in investigations of nucleic acids and related systems (Zídek et al. 2001). The assignment of DNA fragments requires high sensitivity of the NMR experiments used to correlate the nuclei of the studied molecule. The detrimental influence of the amplified dipolar relaxation in isotopically labeled oligonucleotides has been known since the introduction of ^{13}C and ^{15}N labeling of nucleic acids. As different spin orders vary in their relaxation properties, the experimental design preferring the spin system evolution in the state that exhibits the slowest relaxation rate can reduce the sensitivity loss due to the relaxation processes during an experiment. One of the possibilities is to replace the evolution of the single quantum (SQ) states by the multiple-quantum (MQ) ones (Zídek et al. 2001). In DNA research, the correlation via hydrogen bonds can help to assign the nuclei of bases by extending the scalar coupling network to the hydrogen-bonded partners. The information about hydrogen bonds also provides very useful additional restraints for structure determination, especially in the case of noncanonical motifs. Understanding at a molecular level, the mechanisms for the control of genetic information and its replication, packaging, and repair necessitate the elucidation of the detailed interactions between proteins and DNA (Jamin and Toma 2001). Double-stranded RNA-dependent protein kinase (PKR) is an integral part of the cellular response to viral infection (Lindhout et al. 2007). They showed a role of NMR as a structural tool in elucidating the mechanism of autophosphorylation within the 66 kDa enzyme PKR. The modularity of PKR coupled with the favorable spectral dispersion and cross-peak signal to noise of the domains involved within the individual states of the activation mechanism have been shown to be an excellent proof of concept argument in the pursuit of studying high-molecular-weight complexes in excess of 80 kDa using NMR as both a structural and diagnostic tool. Indeed, the in vitro use of NMR has been a cornerstone in the growing understanding regarding the mechanism of activation of PKR in a physiological, in vivo setting (Lindhout et al. 2007). Clearly, NP penetrating into cell can affect the mentioned processes. The importance of the knowledge of hydrogen bonds is even greater for intermolecular complexes of DNA with proteins, small ligands, drugs (vide infra), and other compounds. Scalar coupling constants due to hydrogen bonds in Watson–Crick pairs ($^{1h}J_{HN}$ and $^{2h}J_{NN}$) were first measured by Pervushin et al. (1998) using a ^{15}N–^{1}H TROSY E.COSY experiment and by Dingley and Grzesiek (1998) using an HNN-COSY experiment (Figure 6.65).

Historically, the first NMR structure of a peptide–ribonucleic acid (RNA) complex was determined in 1995 and consisted of the structure of a small peptide (14 amino acids) bound to a 26-nucleotide (26-nt) RNA stem loop (Puglisi et al. 1995, Ye et al. 1995, Dominguez et al. 2011). In 1996, the first structure of a protein–RNA complex was solved by NMR that corresponded to the N-terminal RNA recognition motif (RRM) of the U1A protein (100 amino acids) in complex with a 30-nt stem-loop RNA (Allain et al. 1996). Complex formation can easily be monitored by NMR spectroscopy using the so-called chemical shift perturbation mapping or titration experiments (Dominguez et al. 2011) because the chemical shift of a nucleus is highly dependent on its chemical environment and can be accurately measured. Changes in the environment of the nucleus results in changes of its chemical shift and these changes can be measured to identify the interface of a macromolecular complex (Zuiderweg 2002, Marintchev et al. 2007, Stefl et al. 2010, Dominguez et al. 2011). There are many possible ways of following the complex formation of a protein–RNA complex by NMR. The complex can be formed by adding the protein into the RNA or vice versa.

Furthermore, chemical shifts of different nuclei can be used. Chemical shift perturbations of the protein can be followed by measuring ^{15}N–^{1}H HSQC (heteronuclear SQ correlation) spectra in absence or in presence of increasing amounts of unlabeled RNA. Similarly, RNA chemical shift perturbations can be followed by adding increasing amount of protein into the RNA. If the RNA is unlabeled, ^{1}H chemical shift perturbations of the H5–H6 cross-peaks of pyrimidines can be monitored using 2D ^{1}H–^{1}H TOCSY (total correlation spectroscopy) spectra (Dominguez et al. 2011). Alternatively, chemical shift perturbations of imino protons can be monitored by 1D NMR if they are visible. If ^{15}N- and/or ^{13}C-labeled RNA samples are available, chemical shift perturbations of imino or non-exchangeable protons can be followed using ^{15}N–^{1}H HSQC or ^{13}C–^{1}H HSQC spectra, respectively. In the slow exchange regime, when a component (e.g., RNA) is gradually added to the other component (the protein), two sets of signals are observed, one corresponding to the protein free state and the other one corresponding to the protein-bound state, as was observed in the case of the Fox-1–RNA complex (Auweter et al. 2006) (Figure 6.66). The integral of each signal is linearly dependent on the population of the two states and is directly correlated to the molar ratio of both components. Therefore, while gradually adding the RNA, the signal of the protein corresponding to the free state decreases and the signal corresponding to the bound state increases. Slow exchange regimes were reported for protein–RNA complexes with high affinity corresponding to dissociation constants ranging from 0.5 (Fox-1–UGCAUGU) (Auweter et al. 2006) to 250 nM (protein NC–AACAGU) (Dey et al. 2005, Dominguez et al. 2011).

Silica-based NP/nanobeads are very useful in bioanalysis once conjugated with bio-entities (such as DNA or antibodies) for analyte recognition and/or signal generation (Knopp et al. 2009). Interactions of DNA with such nanomaterials as graphenes (Varghese et al. 2009, Tang et al. 2010), CNT (Katz and Willner 2004), fullerenes, or nanooxides (Turov et al. 2010a) can result not only in structural changes of the macromolecules but also in changes in their biological properties as well as the properties of whole modified solids. For instance, the single-stranded DNA constrained on functionalized graphene can be effectively protected from enzymatic cleavage, and the constraint of DNA on the graphene improves the specificity of its response to complementary DNA (Tang et al. 2010). Theoretically calculated values of interaction energies of nucleobases with graphene are large (between −33 and −63 kJ/mol, Varghese et al. 2009). However, the free energy of adsorption of polyions such as proteins and DNA is smaller (about −10 kJ/mol) than these values because of desolvation effects and destabilizing entropy contribution. According to NMR data for the first water layer bound to biomacromolecules, the ΔG_s value is close to −3 kJ/mol (Gun'ko et al. 2005d, 2009d), i.e., it is ~30% of the total free energy of the interactions of macromolecules with the solid surface.

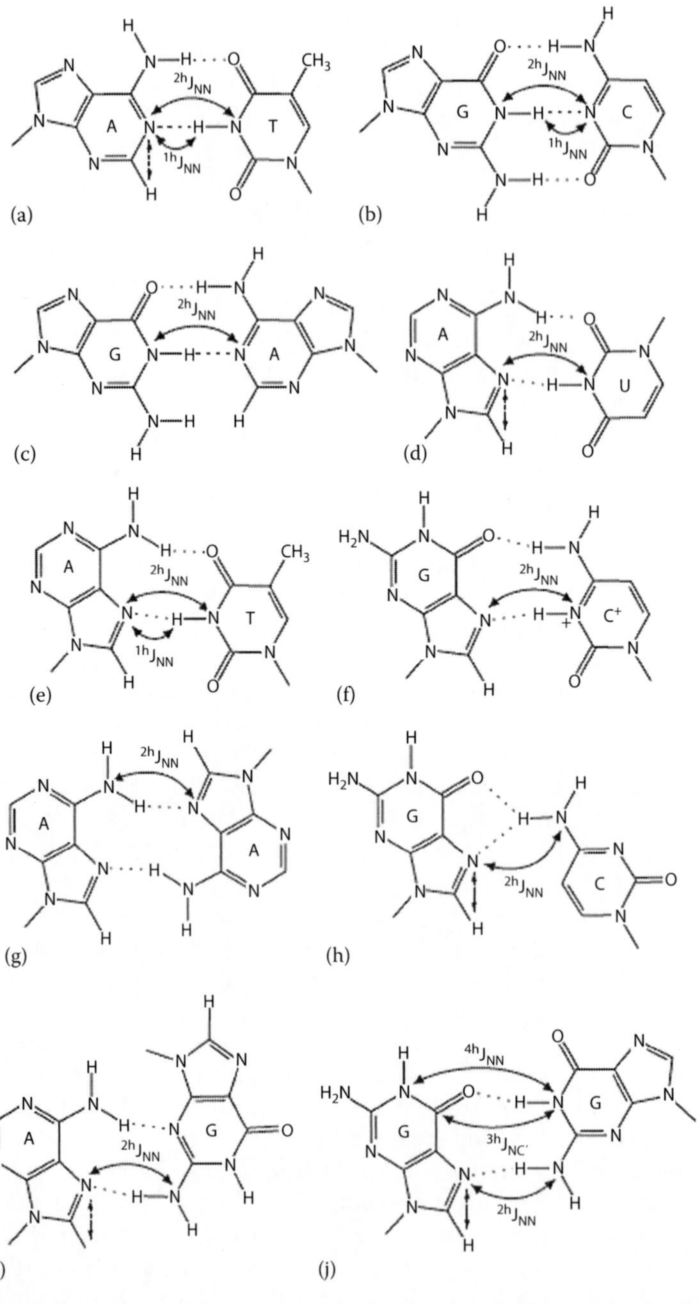

FIGURE 6.65 Hydrogen bonds between purine and pyrimidine bases, with indicated *trans*-hydrogen-bond scalar interactions and related coupling constants, which can be measured using NMR. In addition to correlations between exchangeable protons and nitrogens, a relayed transfer to non-exchangeable aromatic protons, shown by a dashed arrow, can also be employed. The following base pairs are depicted: (a, b) Watson–Crick pairs (Pervushin et al. 1998, Dingley and Grzesiek 1998, Dingley et al. 1999); (c) imino-hydrogen-bonded G•A pairs (Wöhnert et al. 1999); (d) reversed Hoogsteen pairs (Wöhnert et al. 1999); (e, f) Hoogsteen pairs (Dingley et al. 1999); (g) A•A mismatch pairs (Majumdar et al. 1999a); (h) C•G in the G•G•G•C tetrad (Majumdar et al. 1999b); (i) A•G in the A•(G•G•G•G)•A hexad (Majumdar et al. 1999b); and (j) G•G in the G•G•G•G tetrad (Liu et al. 2000, Dingley et al. 2000) and in the A•(G•G•G•G)•A hexad (Majumdar et al. 1999b). (Adapted from *Curr. Opin. Struct. Biol.*, 11, Zídek, L., Štefl, R., and Sklenár, V., NMR methodology for the study of nucleic acids, 275–281, 2001, Copyright 2001, with permission from Elsevier.)

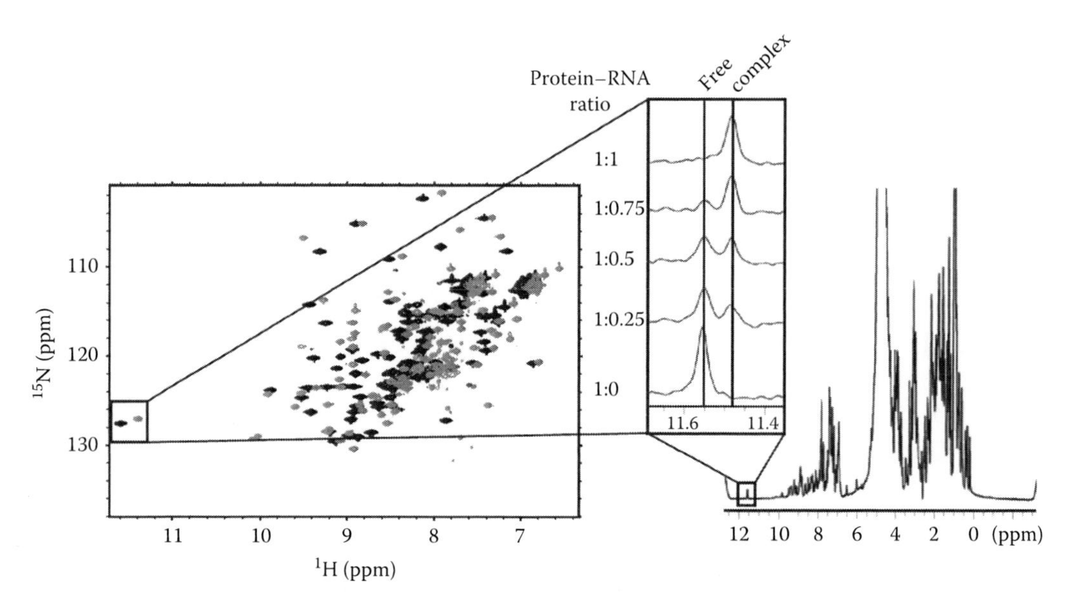

FIGURE 6.66 Example of a protein–RNA complex in the slow exchange regime. NMR spectra at 900 MHz of the Fox-1–RNA complex. Left: ^{15}N–^{1}H HSQC spectra of Fox-1 free (black) or in complex with a UGCAUGU RNA at a 1:1 molar ratio (light). Right: 1D spectrum of Fox-1 free. The signal at 11.5 ppm is shown as a function of the protein:RNA molar ratio. The complex formation is in slow exchange on the NMR timescale. (Adapted from *Prog. Nucl. Magn. Reson. Spectrosc.*, 58, Dominguez, C., Schubert, M., Duss, O., Ravindranathan, S., and Allain, F.H.-T., Structure determination and dynamics of protein–RNA complexes by NMR spectroscopy, 1–61, 2011, Copyright 2011, with permission from Elsevier.)

Composites prepared with DNA-modified silica are promising materials in development of new drug delivery systems (Turov et al. 2010a). DNA is well soluble in water, forms gel-like or micro gel-like structure, and binds a large amount of water. Adsorption/distribution of DNA on a silica surface can be performed using solution, impregnation, and mechanochemical activation, MCA of macromolecules and silica NP. Notice that MCA in the presence of some amount of water or an organic solvent allows better distribution of macromolecules on the surface of solid particles (Gun'ko et al. 2009d). During this process, mechanical loading allows equal distribution of macromolecules on the surface of silica matrix that creates a large contact area for substrate and external media.

Water in solid DNA investigated by the ^{1}H NMR spectroscopy is in a clustered state and several types of water clusters with different degree of association of the molecules could be identified. The bound water behavior is affected by temperature and co-adsorbates or the presence of organic media. Water bound to DNA includes both WAW and SAW components as well as molecules dissolved in the organic media (ASW). Composites DNA/nanosilica could be used for selective binding of biologically active compounds in diluted solutions and biological media as well as they could be used as carriers for drug delivery, DNA transport into cells, etc.

Figure 6.67a shows ^{1}H NMR spectra of water bound in the crystalline DNA powder at different temperatures. Bound water is observed as a broad, Gaussian shape single signal of SAW (including both WBW and SBW components) of intensity decreasing with lowering temperature. The signal is not observed at $T \leq 220$ K.

This signal shape is explained by low mobility of bound water molecules at low temperatures as well as by non-uniform broadening of the NMR spectra caused by significant differences in the diamagnetic susceptibility of DNA and air. In the aqueous media (25% gel of DNA), the signal width decreases significantly (Figure 6.67b) and several signals of unfrozen water appear, i.e., bound water differentiation occurs. Signals at $\delta_H > 4$ ppm are linked to SAW; signals at $\delta_H < 2$ ppm are

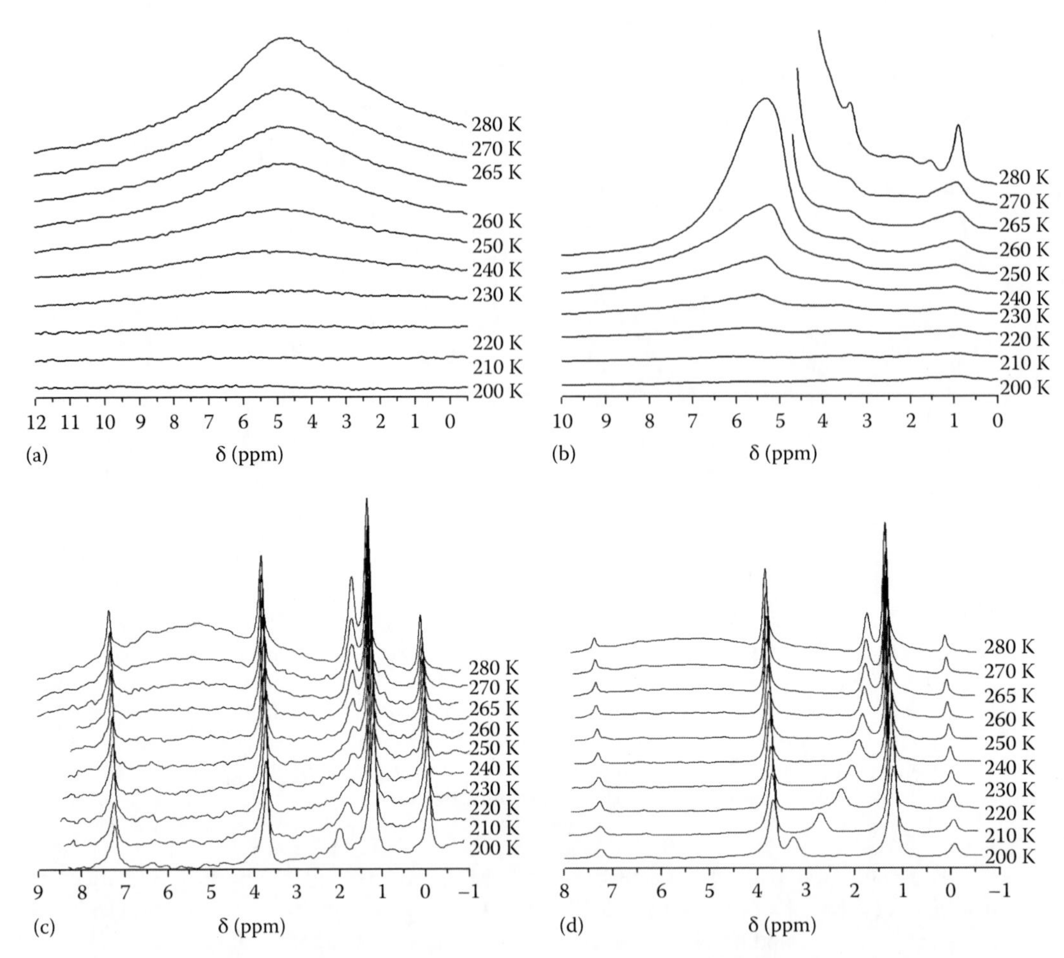

FIGURE 6.67 ^{1}H NMR spectra of water bound to DNA at different temperatures: (a) 50 mg/g H_2O on DNA in air; (b) 25% gel of DNA in water; 0.1 g/g H_2O on DNA in $CDCl_3$ media in (c) 10 min or (d) 8 days after preparation.

attributed to WAW; and at $\delta_H = 3–4$ ppm, it corresponds to water hydrogen-bonded to electron-donor centers of DNA (Anglin et al. 2008). Notice that in 2D cyclic clusters (in hydrophobic surroundings) half of the H atoms do not form the hydrogen bonds and, therefore, have small δ_H values (~1 ppm). These structures are characterized by the chemical shift with averaged signals from the H atoms without and with hydrogen bonds.

The distribution functions of the Gibbs free energy changes of unfrozen water (Figure 6.68a) and the size distributions of unfrozen water clusters, UWCSD (Figure 6.68b) show that a significant part of SAW corresponds to structures larger in size than the width of the DNA double helix (2.4 nm). Consequently, not only SAW is intra-structural water in DNA, but also it forms the clusters at the outer surface of the macromolecules. WAW and ASW structures observed in the DNA gel can have smaller size and freeze only at very low temperatures, i.e., they correspond to SBW. Therefore, these two types of water can be attributed to intra-structural water in DNA. The value of free surface energy, γ_S for SAW (Figure 6.68a) depends not only on the interaction features at the interfaces but also on the amount of SAW. SAW, in contrast to WAW and ASW, is characterized by significant contribution of WBW at $-\Delta G < 0.5$ kJ/mol. It has mainly an extra-structural character, i.e., it is bound to the outer surface of the DNA helix (Turov et al. 2010a–c, 2011d,e, Turov and Gun'ko 2011).

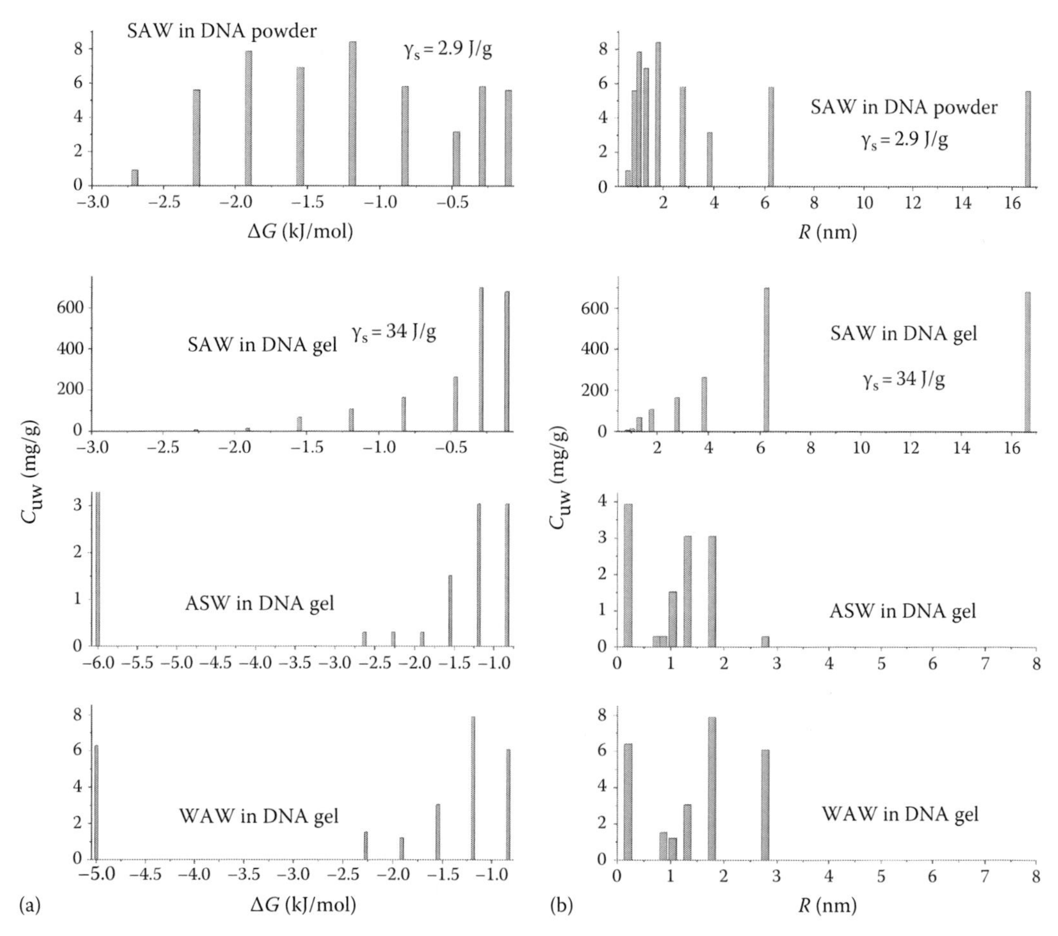

FIGURE 6.68 Distributions of (a) non-frozen water amounts versus changes in the Gibbs free energy and (b) cluster sizes (water-filled cavities) in the DNA powder containing 50 mg/g of adsorbed water and aqueous 25% DNA gel.

It is probably that water signal at $\delta_H = 3.5$ ppm is caused by crystallization water bound to electron-donor groups of nucleotides and localized in the inner space of DNA molecules. WAW ($\delta_H = 1$–1.5 ppm) is probably localized in the voids between structural DNA elements. This type of water is strongly bound (does not freeze at low temperatures) and can be characterized by low average number of the hydrogen bonds per water molecule (single molecules and 1D and 2D clusters) with significant distortion of the hydrogen bonds compared to bulk water. Formation of small water clusters inside of the DNA helix caused by alternation of charged, polar, and relatively hydrophobic structures in the slit cavities between neighboring nucleotides. According to data presented in Figure 6.68a, C_{WAW} is almost twice as high as C_{ASW}.

As temperature decreases ASW and WAW signal intensity gets somewhat smaller, which suggests that these types of water can be partially frozen. When ice formation process takes place, coordination number of water molecules inside of the crystallites is equal to four. Therefore for the nucleation process of ASW and WAW, the formation of 3D water clusters is necessary. The formation of 3D crystallites in the narrow slits is hindered by steric factors. Before freezing ASW and WAW molecules possess certain mobility, which assists formation of ordered structures of bound water from amorphous to crystalline as a result of rotational and translational diffusion of molecules. It can be suggested that ASW- and WAW-formed ice crystals have small size and slightly distorted structure. Hence they may be considered as cluster water structures, which form small 3D structures as well as low-sized 1D and 2D ice crystallites.

Figure 6.67c and d shows ^{1}H NMR spectra of DNA (being in weakly polar $CDCl_3$) with the hydration degree $h = 100$ mg/g. As it can be observed, besides water signal the spectra contain signals of undeuterated fraction of chloroform (7.3 ppm) and tetramethylsilane (0.0 ppm) used as an internal standard. Signals of ASW and WAW are significantly narrower than that in DNA hydrogel (Figure 6.67b). There are two forms of WAW observed. Signal intensity at $\delta_H = 1.2$ ppm does not change significantly over the total temperature range used. At the same time, the δ_H value of WAW (1.6 ppm at 280 K) increases with decreasing temperature. This effect strongly depends on observation time. Thus, in 10 min after sample preparation this shift does not exceed 0.4 ppm. However, if a spectrum is recorded in 8 days for the same sample the shift reaches 1.8 ppm. Probably, water molecules, which are responsible for this signal, are sensitive to structural changes of DNA crystals, which may occur after prolong contact of sample with chloroform. As a result, water molecules may increase their coordination number with decreasing temperature. In accordance with data presented in Figure 6.67a and c, chloroform stabilizes weakly associated forms of water and amount of WAW becomes to be equal to ASW. In chloroform media, both forms of water do not freeze. It is probably due to hindrance of diffusion of WAW molecules to SAW by weakly polar chloroform, which may also decrease probability of formation of low-sized nanocrystallites of ice from WAW. Addition of weak electron-donor acetonitrile leads to a higher degree of stabilization of weakly associated forms of water (Figure 6.69). In pure acetonitrile, signal for water bound to DNA molecules is observed at $T > 220$ K. At lower temperatures, acetonitrile freezes and significantly decreases mobility of adsorbed water molecules. It may occur due to steric hindrances which affect translational and rotational motions of water molecules, which are localized in the slit-shaped space between nucleotides of biopolymer and between DNA and solid acetonitrile. Acetonitrile not only can be located on the outer surface of DNA molecules, but also fills significant part of the inner space in the DNA helixes, since crystallized and weakly associated waters are located there.

WAW is observed as two signals, one of which has down-field shift as temperature decreases. ASW signal appears to be significantly broader than that observed in $CDCl_3$ media. It is probably due to acceleration of molecular (or proton) exchange with other forms of bound water. Addition of equal amount of $CDCl_3$ allows one to perform the measurements at 200 K. However, the total amount of WAW decreases. Calculations of distributions of different forms of bound water as a

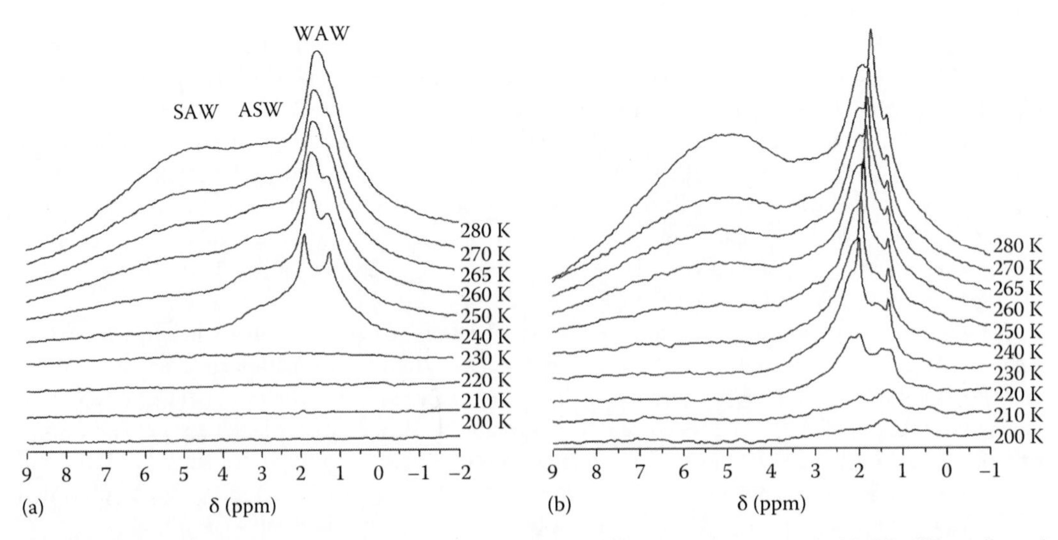

FIGURE 6.69 ^{1}H NMR spectra of water bound to DNA at different temperatures in (a) CD_3CN media and (b) mixture $CDCl_3 + CD_3CN$.

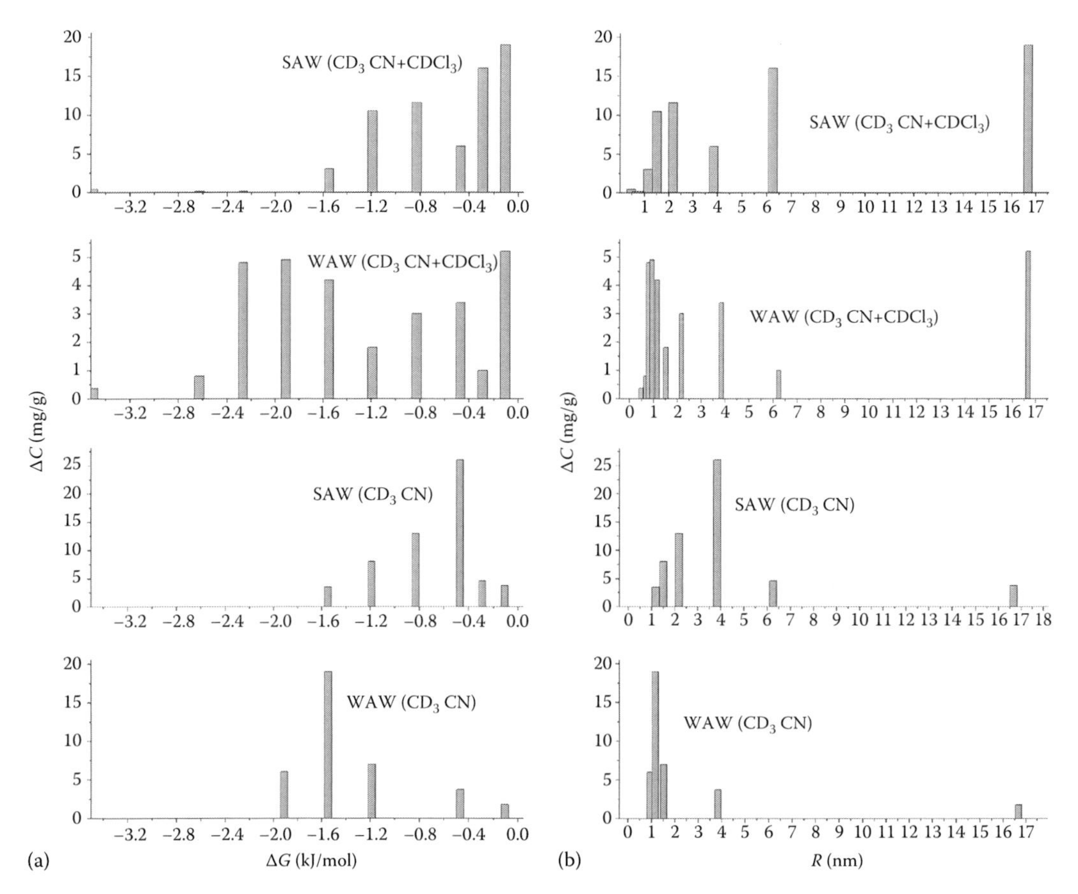

FIGURE 6.70 Distributions of (a) Gibbs free energy changes and (b) cluster sizes of bound water in CD_3CN media or CD_3CN and $CDCl_3$ mixture.

function of Gibbs free energy change and radii of clusters of bound (Figure 6.70) show that in acetonitrile a maximum of the cluster size distribution of SAW is at $R=4$ nm, and for WAW it is at 1.5 nm. Addition of chloroform to the system leads to an increase in the quantities of large and very small clusters of bound water of both SAW and WAW.

Upon addition of small quantities of DMSO (~8 wt%) to $CDCl_3$, the spectra change (Figure 6.71a). Signal of WAW disappears or masked by signal of CH_3 groups of DMSO ($\delta_H=2.5$ ppm). Instead, a new ASW signal at $\delta_H=3.5$–4 ppm appears. The appearance of this signal is caused by formation of the hydrogen-bonded complexes between water molecules and DMSO ($HO–H\cdots O=S(CH)_3)_2$, dissolved in the organic phase. At the same time, SAW signal disappears. When temperature reaches 220–230 K, signal of SAW dissolved in the mixed organic solvents appears as two signals, one is broad and another one is narrow.

Complexes of DNA with a drug doxorubicin (also known as adriamycin or hydroxydaunorubicin), Dox, used in cancer chemotherapy were studied using the ^{1}H NMR spectroscopy. Dox is an anthracycline antibiotic closely related to natural daunomycin and like all anthracyclines, it works by intercalating DNA. The intercalation of DNA by Dox (Figures 6.71b, 6.72, and 6.73) affects the behavior of bound water. The used concentration of Dox provides interaction of a large number of Dox molecules with each DNA macromolecule. Dox molecules are located in the places occupied by WAW in the slits between neighboring nucleotide planes.

Therefore, the fact of appearance of significant amount of DNA-bound SAW is an evidence of the displacement of WAW by Dox from that space. Upon intercalation, a glycoside part of Dox

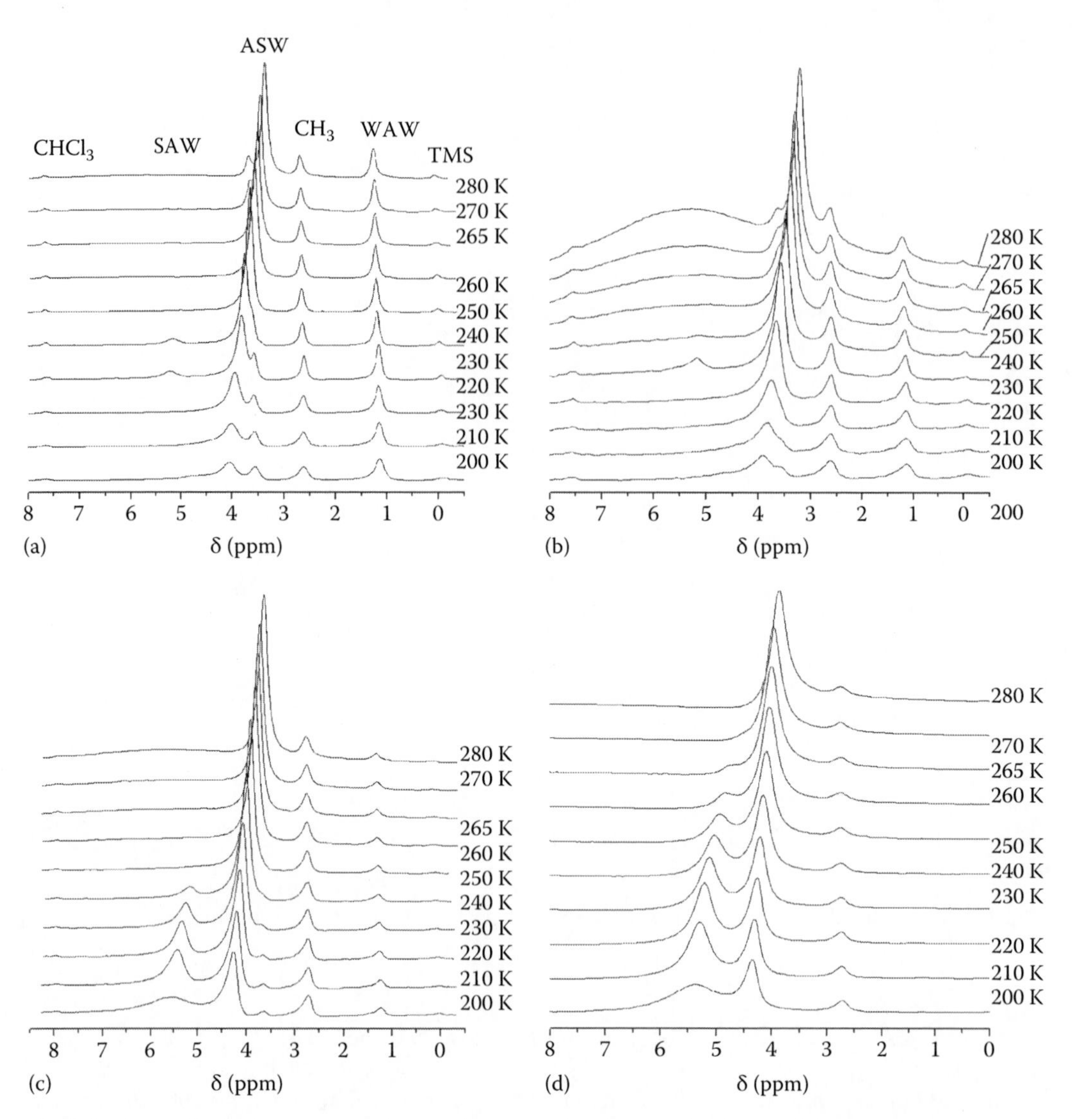

FIGURE 6.71 ^{1}H NMR spectra of water bound to DNA (h=0.1 g/g) at different temperatures: (a) $CDCl_3$:DMSO=12:1; (b) $CDCl_3$:DMSO=9:1 and 0.75 wt% Dox; (c) $CDCl_3$:DMSO=9:2 and 1.5 wt% Dox; and (d) $CDCl_3$:DMSO=9:3 and 1.5 wt% Dox.

molecule is located outer of the nucleotide planes but a hydrophobic part is located between the nucleotide planes (Figure 6.72). Therefore, water molecules can locate around a hydrophilic part of Dox to form the SAW structures, and contribution of WAW diminishes. As concentration of DMSO increases, SAW clusters located in the liquid organic phase (signal at δ_H=5–5.7 ppm) are stabilized. As temperature decreases, this form of water freezes whereas at high temperatures it transforms to ASW due to its dissolution in the organic phase.

That is why at low concentration of DMSO, SAW signal is observed only in the narrow temperature range, which significantly widens with increasing concentration of DMSO.

If concentration of water in the model system is relatively low, an increase in WAW contribution occurs when DNA is intercalated by Dox (Figure 6.73). This effect can be explained by swelling of DNA during intercalation. Since system has relatively low concentration of water to increase SAW structures, water molecules are associated to a smaller degree and therefore SAW contribution decreases. That is why the $f(\delta_H)$ peak shifts toward smaller δ_H values. This effect may also

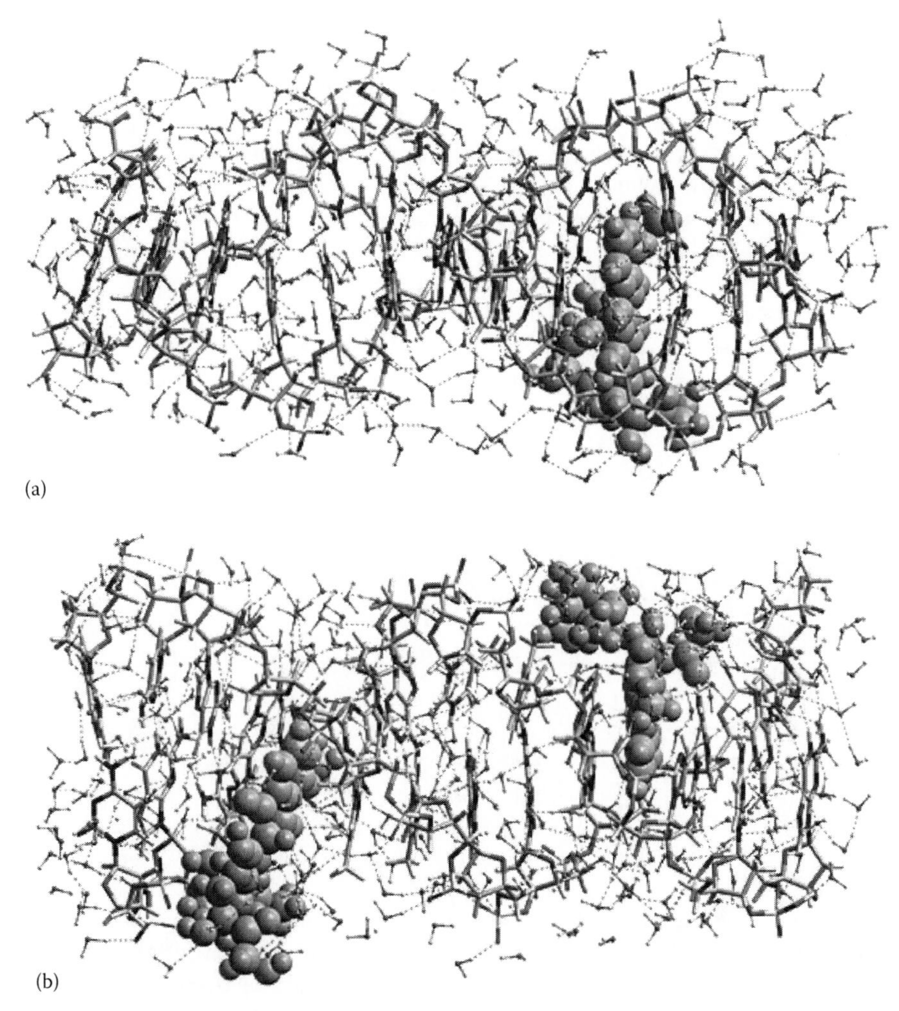

FIGURE 6.72 Intercalation of hydrated fragment of DNA molecule by one (a) and two (b) molecules of doxorubicin (geometry optimized by PM6/MOZYME method).

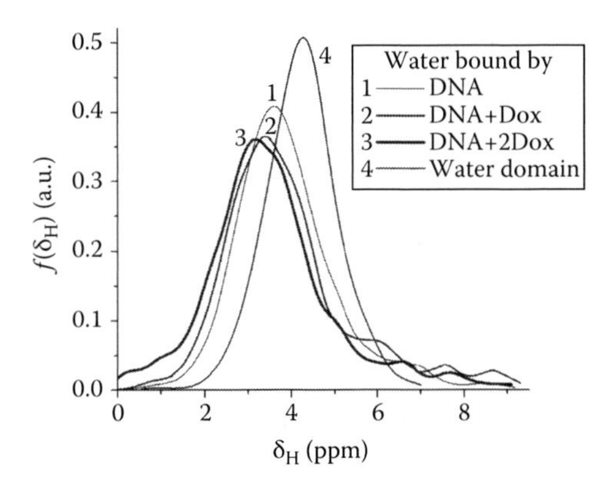

FIGURE 6.73 Theoretically (PM6 with a correlation function) calculated ^{1}H NMR spectra of water bound by the fragment of DNA molecule in the absence (curve 1) and in the presence of one (curve 2) and two (curve 3) molecules of Dox and for pure water domain which contains 2000 molecules (curve 4).

testify about dependence of bioactivity of Dox on hydration degree of cells. Biological activity may increase as hydration increases since probability of de-intercalation of Dox from DNA decreases due to weak interaction of hydrophobic fragments of Dox with bulk water.

Figure 6.74 shows temperature changes in the ^{1}H NMR SAW bound to DNA in the presence of basic pyridine (Py) which was chosen due to its electron-donor properties which are similar to that of double helix–forming nucleotides of DNA. As it was observed earlier for DMSO, several signals of water are observed. Signal at $\delta_H = 1.4$ ppm is due to WAW and signal at $\delta_H = 3.8$ ppm is due to ASW. The intensity of these peaks is temperature-independent. The other signals are of DNA-bound SAW at $\delta_H = 6–6.5$ ppm, ASW, and complex signals. As temperature decreases SAW freezes easily, which is due to large amount of clusters and nanodomains with this water. However, ASW signal intensity increases. This process occurs in parallel and it is due to partial dissolution of SAW in the organic phase. This process is stabilized by temperature increase and increasing Py concentration (Figure 6.74a through c). Signals of undeuterated Py ($\delta_H = 7.2–7.9$ ppm) and chloroform ($\delta_H = 8.7$ ppm) as well as TMS ($\delta_H = 0.0$ ppm) are also observed.

Slightly greater values of chemical shift of $CDCl_3$ signal are due to formation of H-bonded complexes $CCl_3–H\cdots NC_5H_5$. In contrast to DMSO, SAW is not observed at low temperatures in liquid mixtures of Py–$CDCl_3$. An increase in the chemical shift of SAW (compared to that in pure chloroform) could be due to formation of hydrogen bonds between water molecules and Py. Dissolution of Py in SAW is energetically unfavorable process since a large number of hydrogen bonds need to be broken in the SAW clusters.

DNA-bound water can be characterized by several signals in ^{1}H NMR spectra. These signals can be attributed to weakly associated (1D and 2D) and strongly associated (3D) structures. These structures can be characterized by different energy of interactions between DNA and water molecules as well as different association of water molecules and ability of bound water to be frozen at low temperatures. Several signals are due to strongly associated water (5 ppm), water which forms strong hydrogen-bonded complexes with electron-donor centers ($\delta_H = 3.5$ ppm) and two types of weakly associated water ($\delta_H = 1.2–1.8$ ppm). All these types of water may play an important role in conformational changes of DNA and have an effect on efficacy of drug molecules, which interact with DNA molecules. For example, it was shown in DNA intercalation by Dox. Weakly polar media of chloroform stabilizes weakly associated forms of water. Furthermore, WAW signal may have down-field shift more strongly with decreasing temperature. Thus, the degree of association of this type of water increases. In the presence of DMSO, this signal disappears.

Thus, during intercalation of DNA by Dox the amount of weakly associated water decreases. This fact suggests that WAW is located in narrow cavities between neighboring pairs of nucleotides and its substitution by Dox molecules. At the same time, the amount of SAW increases due to displacement of WAW by Dox, its association with SAW, and interaction with hydrophilic glycoside fragments of Dox located on the outer side of DNA helix. Electron-donor organic solvents DMSO and Py at high concentrations may displace water clusters from inner space of DNA molecules. This testifies about high energetic stability of DNA–water adducts which can be presented as extremely stable to external influence composite system.

Highly disperse silica A-300 was chosen as a mineral component of nanocomposite with DNA. Nanocomposite was prepared by mixing lyophilized DNA from salmon sperm (Sigma) with nanosilica upon addition of fixed, known amounts of water, followed by mechanical agitation for 20 min. Prepared composite dried at 60°C for 48 h. Deuterated solvents ($CDCl_3$, C_6D_6, CD_3CN, and $CDCl_3$ and C_6D_6 mixtures) were used. Signal of SiOH groups, which do not participate in the hydrogen bonds (with water molecules or others), is observed as narrow characteristic IR band at 3750 cm^{-1}. Intensity of this band in the nanocomposite is significantly lower than in the initial silica because of interaction of DNA with the silica surface.

Figure 6.75 shows ^{1}H NMR spectra of nanocomposite bound water recorded at different temperatures. The amount of water in the nanocomposite alone was 0.5 g/g (a) or 0.03 g/g in $CDCl_3$

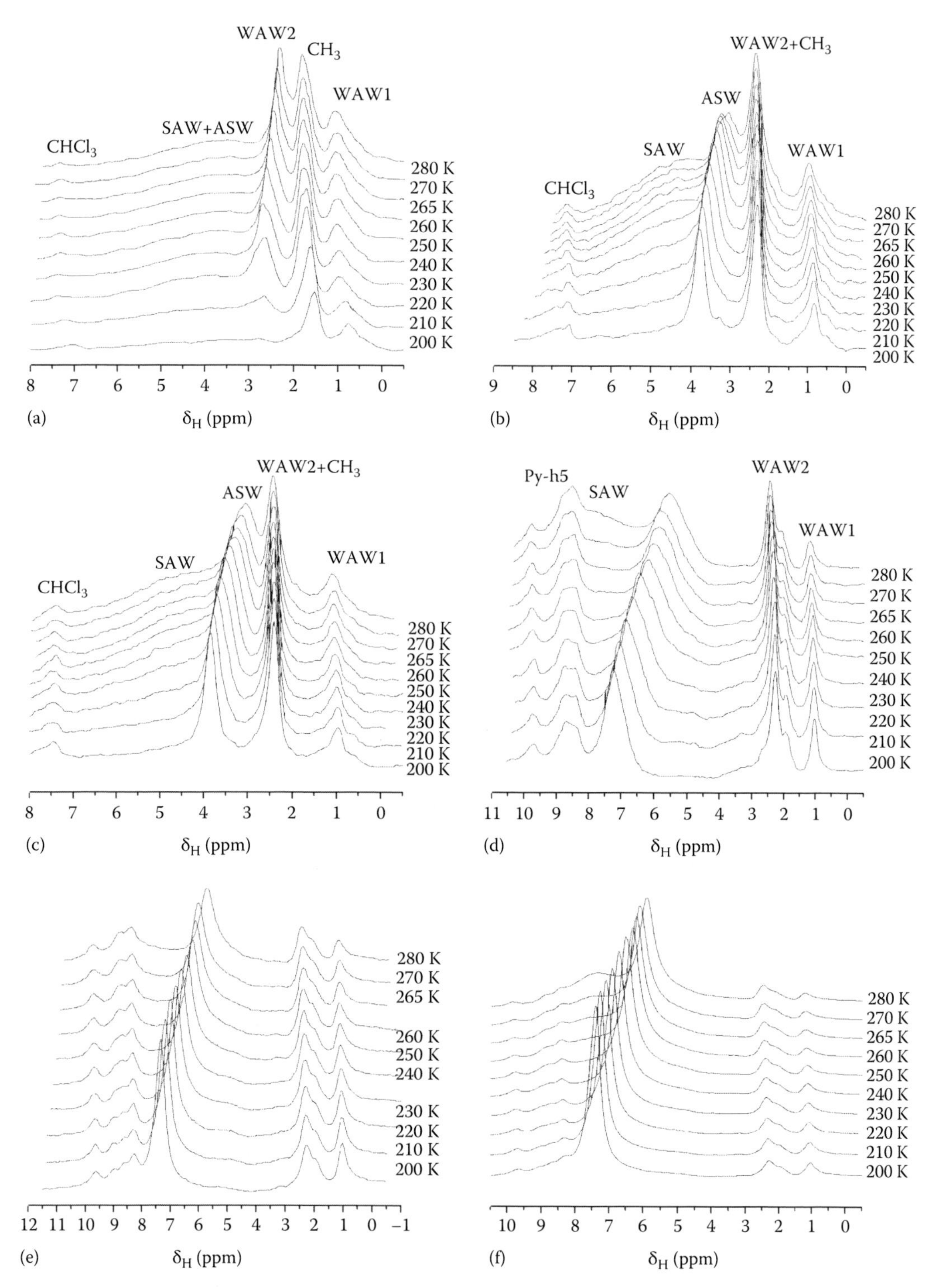

FIGURE 6.74 ^{1}H NMR spectra of water bound to A-300–DNA at hydration h=(a–e) 30 mg/g and (f) 60 mg/g in $CDCl_3$ with addition of (a) CD_3CN (2:1), (b, c) DMSO-d_6 (8:1 and 8:2), (d–f) Py-d_5, (d) 8:1 and (e, f) 8:2 recorded at different temperatures.

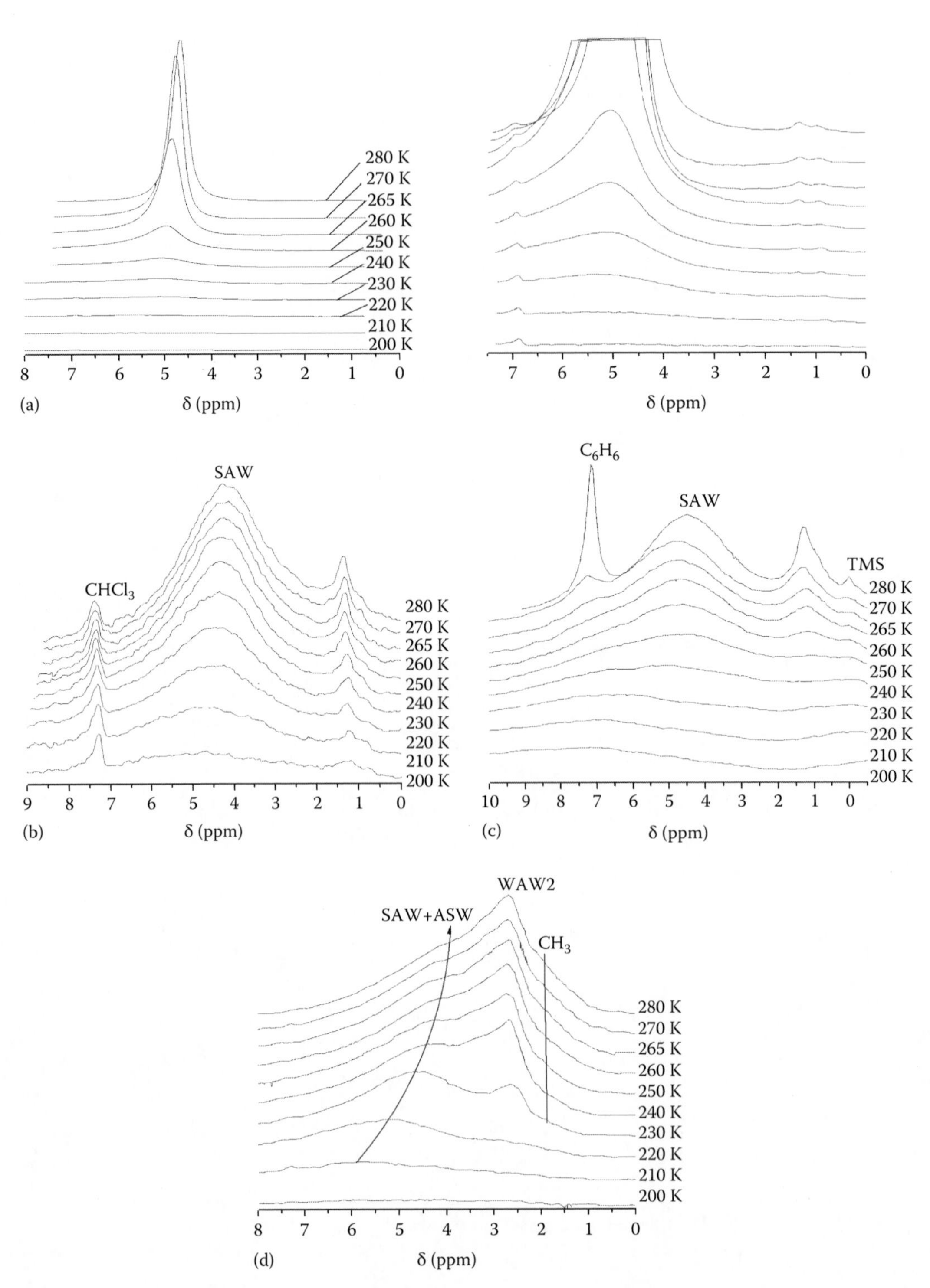

FIGURE 6.75 ^{1}H NMR spectra of water bound to A-300–DNA at hydration $h=0.5$ g/g (a), and in $CDCl_3$ (b), C_6D_6 (c), and CD_3CN (d) media at $h=0.03$ g/g recorded at different temperatures.

(b), C_6D_6 (c), and CD_3CN (d) media. At 0.5 g/g of water (Figure 6.75a), the chemical shift for main signal was 4.5–5 ppm, which corresponds to the value of chemical shift found for liquid water and is due to strongly associated water. At higher magnification (right Figure 6.75a), signals of a weak intensity are observed at 1.2–1.7 ppm and 7 ppm. The signals at 1.2–1.7 ppm are due to the presence of small amount of weakly associated water, whereas the signals at 7 ppm are probably due to water molecules which interact with strong electron-donor centers capable to form proton-transfer complexes. Exact measurement of intensity of such weak signals on the background of strong SAW signal is complicated. Estimation shows that their intensity does not exceed 0.2% of intensity of the SAW signal.

For nanocomposite at hydration $h=0.03$ g/g in weakly polar chloroform medium, ^1H NMR spectra show signals of both SAW and WAW. As temperature decreases, intensity of these peaks decreases due to partial freezing of water. It can be noticed that signal intensity of SAW decreases faster than intensity of WAW (Figure 6.75). Free energy of WAW is decreased by adsorption interactions with composite surface much stronger than for SAW. Signal of nondeuterated chloroform is also observed at 7.2 ppm.

In benzene media at 280 K (Figure 6.75c), the amount of WAW increases significantly (approximately two times) compared to that in $CDCl_3$ media. However, as temperature decreases quantity of WAW decreases fast (Figure 6.76). At the same time, the intensity of relatively narrow signal of undeuterated liquid benzene (7.2 ppm) decreases and disappears. Solid benzene exists as molecular crystal. Its signal can be observed as very broad band at 7.2 ppm. Possibility of existence of water in weakly associated state is dictated by the presence of the liquid phase of nonpolar organic solvent (Figure 6.75).

For composite in acetonitrile (whose polarity is significantly higher than that of benzene), the NMR spectra of bound water change significantly. Concentration of WAW significantly increases and becomes larger than SAW until freezing temperature of acetonitrile (233 K). At lower temperature, WAW signal intensity decreases quickly with decreasing temperature and disappears at $T<220$ K (Figure 6.75d). Chemical shift of WAW in acetonitrile media is somewhat higher (2.5 ppm) than that in weakly polar solvents. Therefore, it can be suggested that in SiO_2–DNA nanocomposite water exists as WAW1 and WAW2. Probably, WAW2 corresponds to somewhat more ordered water clusters which are stabilized by polar media. Chemical shift of SAW decreases from 6 ppm at 200 K to 4 ppm at 280 K. This $\delta_H(T)$ dependence may be explained by existence of water as clusters of SAW and in the form of hydrogen-bonded complexes HOH···$NCCD_3$ (ASW), which

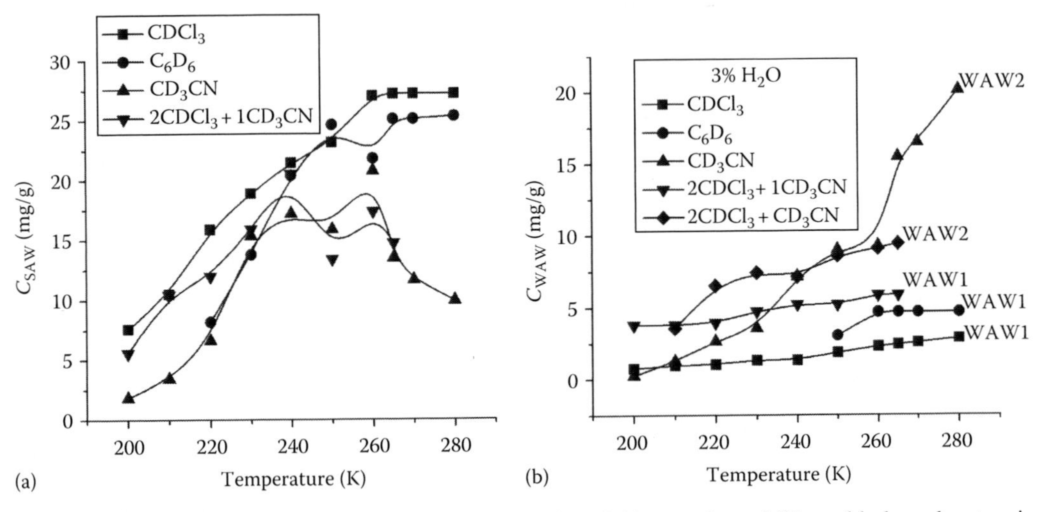

FIGURE 6.76 Temperature dependencies of concentration of (a) strongly and (b) weakly bound waters in nanocomposite A-300–DNA at hydration $h=0.03$ g/g being in organic solvents.

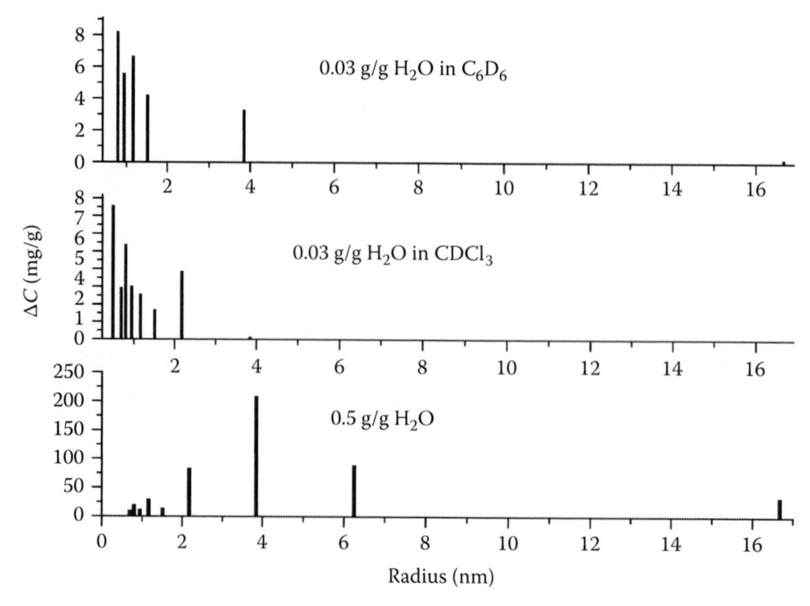

FIGURE 6.77 Size distribution of SAW clusters in SiO_2–DNA nanocomposites in different media.

are localized in the liquid acetonitrile media. Since only one signal is present in the spectra, it can be suggested that fast molecular exchange between these states takes place. As acetonitrile freezes, SAW signal becomes noticeably larger than that of liquid water. It could be due to formation of hydrogen bonds between water molecules located on the outside of SAW clusters and acetonitrile and as a result of this coordination number of water increases. Weak intensity signal observed at 2 ppm is due to methyl groups of nondeuterated acetonitrile. When water content in the nanocomposite was 0.5 g/g, the average size of the cluster is about 4 nm. Whereas, when nanocomposite contained 0.03 g/g of adsorbed water in weakly polar solvent media the size of SAW clusters was found to be 1 nm (Figure 6.77).

Figures 6.74 and 6.78 show ^{1}H NMR spectra of water adsorbed on SiO_2–DNA nanocomposite in the mixtures of chloroform or benzene with such electron-donor solvents as acetonitrile, DMSO, and pyridine at $C_{H_2O}=0.03$ or 0.06 g/g. In 2:1 binary mixture of $CDCl_3/CD_3CN$ (Figure 6.74a), signals due to CH_3 and CH groups of acetonitrile and chloroform are observed together with SAW (SAW+ASW), WAW1, and WAW2. WAW1 signal does not strongly change with decreasing temperature.

At the same time, intensity of WAW2 signal may either decrease due to freezing of water at $T<230$ K or increase due to partial transform of water into weakly associated state identified as WAW + ASW.

In $CDCl_3$/DMSO-d_6 mixtures (Figure 6.74b and c), signal for methyl groups of DMSO exactly coincides with chemical shift of WAW2. The presence of both WAW1 and WAW2 is described by temperature dependence of signal intensity at 2.5 ppm. However, in contrast to $CDCl_3/CD_3CN$ mixture, increase in WAW2 signal intensity takes place with decreasing temperature. As temperature increases, a part of WAW2 is transformed into ASW registered separately from SAW (Figure 6.74b and c). SAW has a tendency to decrease due to transformation of water into a state with hydrogen-bonded complexes HO–H···O=S$(CD_3)_2$ (ASW). As concentration of DMSO increases (Figure 6.74c) concentration of SAW decreases, whereas concentration of ASW increases. Chemical shift of SAW is almost the same as chemical shift of liquid water.

Addition of pyridine (Figure 6.74) results in separated WAW1 and WAW2 signals. Separate registration of SAW and ASW signals depends on temperature and relative concentrations of the components. That is why it is registered at relatively high temperature. Both ASW and SAW have large values of chemical shift, which can reach 7.5 ppm, which is higher than it was found

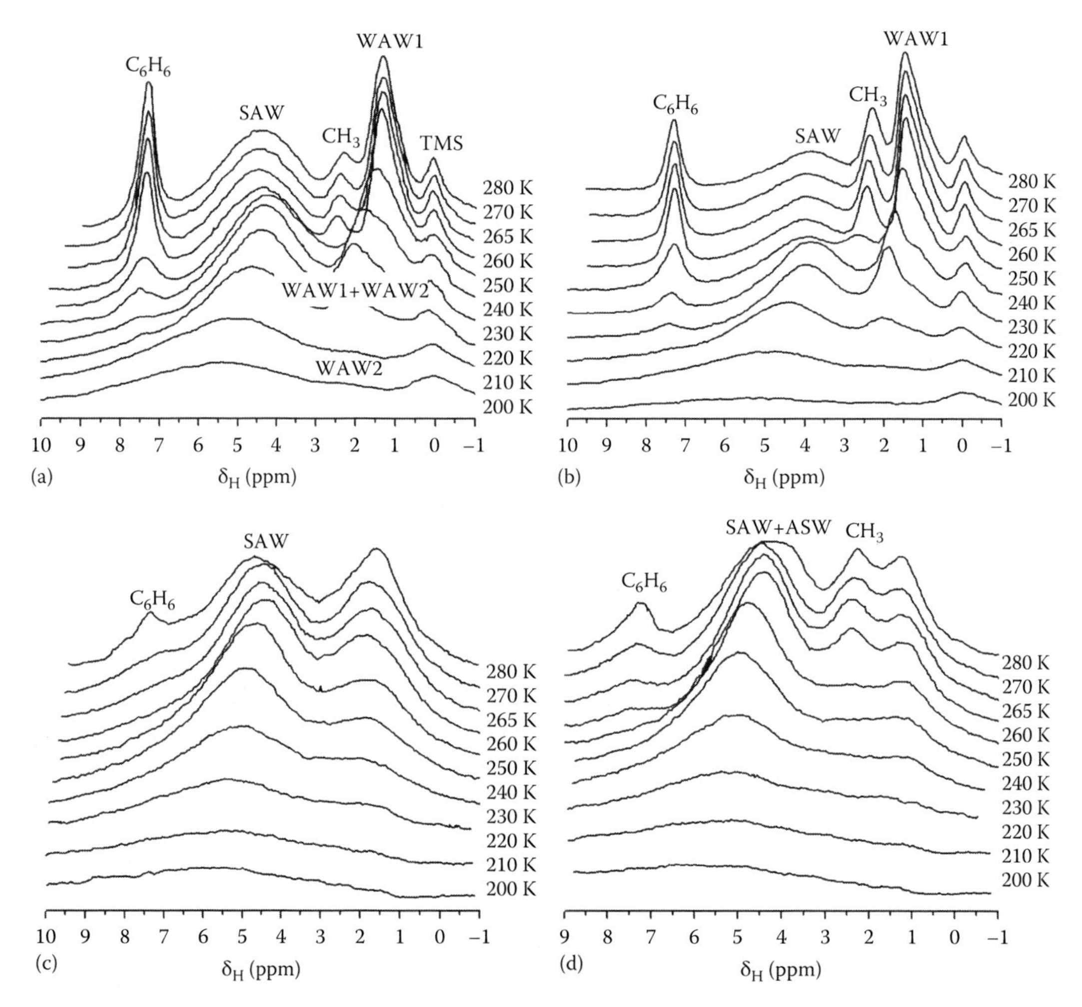

FIGURE 6.78 ^1H NMR spectra of water adsorbed on SiO_2–DNA nanocomposite recorded at different temperatures in suspensions of C_6D_6, which contain electron-donor solvents: CD_3CN (a, b) and DMSO (c, d) at concentration of adsorbed water of 0.03 g/g.

for tetra-coordinated water-forming hexagonal ice crystallites. It can be explained by ability of pyridine molecules to attach protons. Decrease in temperature stabilizes complexes with transferred protons. Therefore, ASW signal has a down-field shift. Since at $T<280$ K benzene freezes easily, in mixtures with benzene the state of bound water depends not only on the ratio of the components, but also on phase state of the dispersion media, which can be analyzed using signals of a nondeuterated fraction of organic solvents. When nanocomposite is placed in C_6D_6/CD_3CN (3:3) mixture, signal of CH groups of benzene is observed at $T>230$ K and signal of CH_3 groups of acetonitrile is observed at $T>250$ K (Figure 6.78a). Dispersion phase freezes as a single phase after the beginning of crystallization of benzene. At 280 K, both SAW and WAW1 are registered separately. Concentration of WAW is approximately two times smaller than that of SAW. Freezing of organic phase is accompanied by decrease of WAW signal intensity and its down-field shift, and at $T<240$ K, WAW1–WAW2 transformation takes place. As concentration of acetonitrile increases (Figure 6.78b), the amount of WAW increases but the SAW concentration decreases.

In C_6D_6/DMSO mixtures (Figure 6.78c and d), dispersion media is present in the frozen or partially frozen state at $T<270$ K. This leads to significant broadening of the signals of bound water.

When C_6D_6/DMSO ratio is equal to 5:1, both SAW and WAW1 are registered separately in the spectra and intensity of their signals are similar (Figure 6.78c). As concentration of DMSO increases, its solubility in solid benzene decreases. Redistribution of signal intensities between SAW and WAW1 takes place. At 280 K, SAW signal broadens due to existence of ASW.

In the media of weakly polar organic solvent ($CDCl_3$, C_6D_6), the process (I) freezing of strongly associated water is observed.

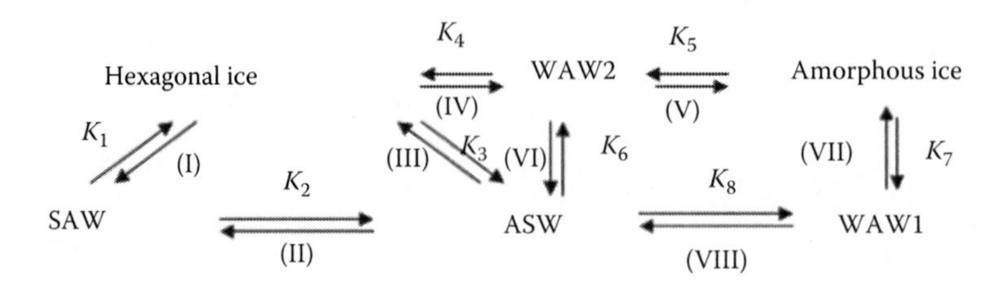

X-ray methods show that when aqueous–organic systems freeze at $T > 200$ K water undergoes transformation to hexagonal ice. Freezing of weakly associated water (WAW1) occurs only at very low temperatures, which allows us to suggest that the equilibrium VII exists. When electron-donor compounds ($CDCl_3$, $(CD_3)_2SO$, C_5H_5N) are present in dispersion media at different concentration and different temperatures, equilibria (II)–(VI) can be realized. Temperature dependencies of SAW, ASW, WAW1, and WAW2 for hydrated powder of SiO_2–DNA in chloroform (Figures 6.78 and 6.79) and benzene (Figure 6.80) show that in benzene suspensions freezing of dispersion media occurred over a broad temperature range. Therefore, it was possible to compare the effect of liquid and solid phases with regulated hydrophobic/hydrophilic characteristics on bound water.

In the chloroform mixtures with acetonitrile (weak electron-donor solvent), decrease in signal intensity for SAW+ASW, WAW1, and WAW2 is observed with decreasing temperature (Figures 6.74a and 6.76). Since the change in signal intensity of WAW is insignificant, one can conclude that processes (V) and (VII) take place. Chemical shift of SAW does not exceed 4 ppm, which allows us to conclude that SAW and ASW signals are registered jointly and are characterized by fast molecular exchange. Therefore, the equilibria (I), (II), and (III) take place at the interfaces of nanocomposite. In the presence of DMSO (Figures 6.74b and c and 6.79a and b), molecular exchange between SAW and ASW slows down, so signals of these water types are registered separately. As temperature decreases, the peak intensity of SAW significantly decreases due to the process (I). A large scatter of points for $C_{uw}(T)$ is due to low precision in integration of NMR peaks when several close signals with different widths are analyzed. Data (Figure 6.78) suggest that concentration of WAW (WAW1 + WAW2) reaches 10 mg/g that corresponds to one-third of all bound water. Therefore, unlike the initial silica SiO_2–DNA nanocomposite it is capable to stabilize weakly associated forms of water.

In the presence of Py (Figure 6.79c and d) at $T > 250$ K, an increase in the amount of ASW occurs simultaneously with a decrease in SAW concentration, which suggests about the presence of equilibrium (II). Besides, as temperature decreases a significant increase in WAW1 concentration occurs, which may take place due to partial transformation of unfrozen water to weakly associated state corresponding to equilibrium (VIII). More complicated transformations are also possible; however, it is not possible to define them due to low signal intensity of different clusters of bound water. Upon addition of $CDCl_3$ (Figure 6.77), DNSO, or Py, the size distribution of SAW clusters bound to SiO_2–DNA nanocomposite changes significantly (Figure 6.79e). As concentration of polar additive or its electron-donor ability increases, the size of SAW cluster increases. In the media of pure $CDCl_3$ the size of SAW clusters did not exceed 2 nm (Figure 6.77) but it increases in DMSO or Py (Figure 6.79e). However, twofold increase in concentration of water in the presence of Py led to increase of the 1.5–4 nm clusters.

When main component of dispersion media is benzene, hydrated SiO_2–DNA nanocomposite may exist either in liquid or solid state depending on the temperature. Appearance of signals of

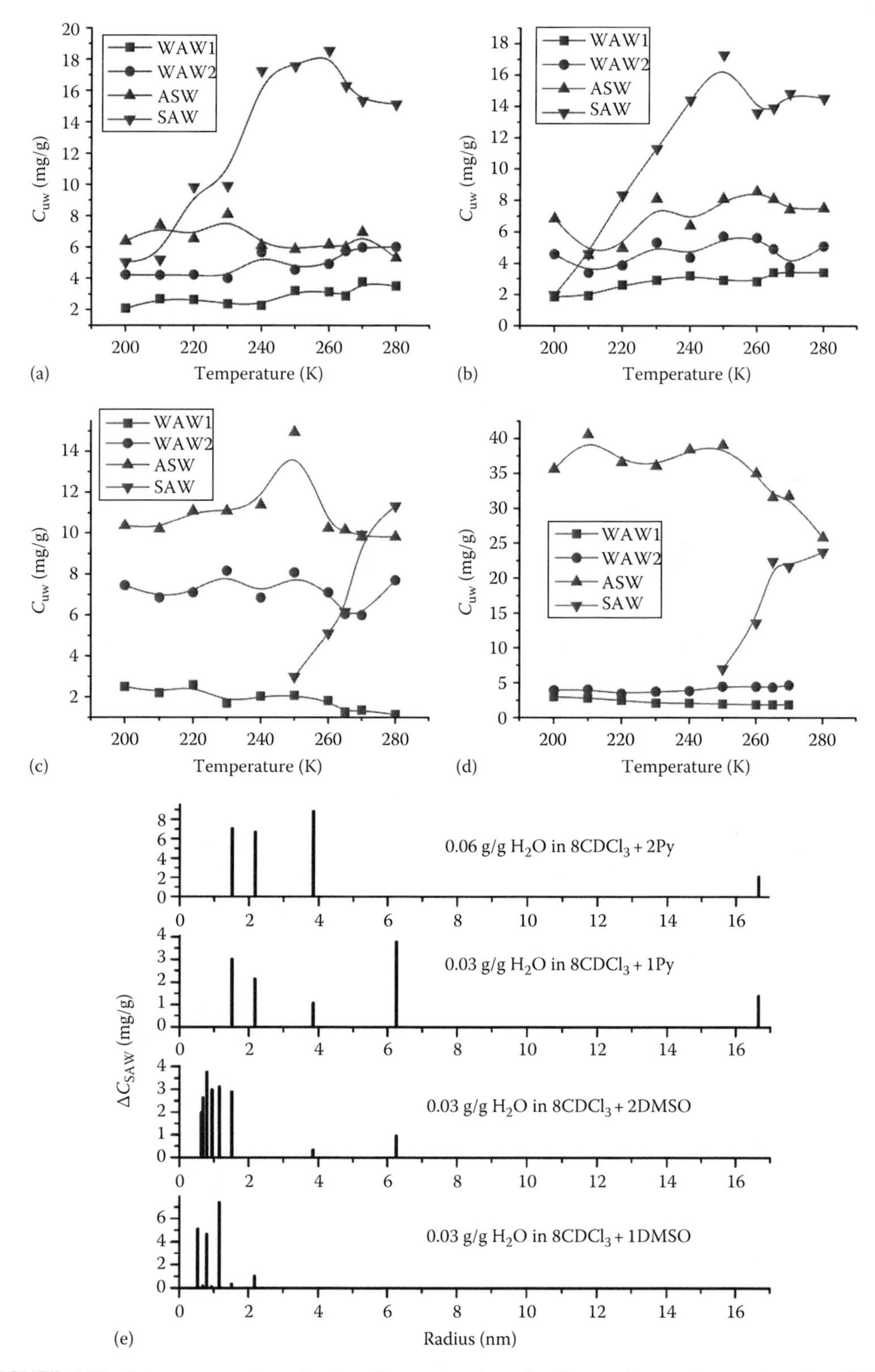

FIGURE 6.79 Temperature dependencies of concentration of different forms of water bound to SiO_2–DNA nanocomposite in mixed solvents $CDCl_3$+DMSO (a, b) and $CDCl_3$+Py (c, d), and SAW cluster size distributions (e).

CH_3 groups (from electron-donor admixture) is due to thawing of mixture of organic solvents (C_6D_6+CD_3CN or C_6D_6+DMSO). Temperature dependencies of concentration of different forms of water in benzene dispersion media with addition of CD_3CN (Figure 6.80a and b) and DMSO (Figure 6.80c and d) show if CD_3CN is present that the $C_{SAW}(T)$ dependence is complicated.

At $T>240$ K, an increase in concentration of SAW is observed with decreasing temperature. WAW1 signal is observed, and after freezing of dispersion media, WAW2 signal is registered.

As temperature decreases, the C_{SAW} and C_{WAW2} values decrease due to freezing of water according to the shift of the equilibria (I) and (IV). The WAW1–WAW2 transformation could occur due to occurrence of (VI) and (VIII) processes. The reason for decrease in C_{SAW} with increasing temperature could be the shift of equilibrium (II).

When acetonitrile is substituted with DMSO (higher freezing point), only wide signals of WAW1 and SAW (or SAW+ASW) can be registered (Figure 6.74). $C_{SAW}(T)$ and $C_{WAW1}(T)$ dependencies change similarly. It allows one to suggest the presence of the dependence between C_{WAW1} and the amount of unfrozen substance (SAW, ASW, Py, DMSO) localized at a surface of nanocomposite. This peculiarity allows us to calculate clusters size distributions not only for SAW but also for WAW1 (Figure 6.80e). Fairly large radii were determined for WAW clusters. Probably, these data characterize aqueous–organic structures where WAW1 exists in liquid state. After their freezing, signal of WAW disappears similarly to signals of methyl groups of solid DMSO or acetonitrile.

Fullerenes and their derivatives are able to penetrate through cell membranes, have strong antioxidant, antiviral, and antimicrobial properties, and can be used in medicine (Emerich and Thanos 2003, Bakry et al. 2007, Ryan et al. 2007). The earlier properties of fullerenes suggest that they can significantly affect the properties of DNA molecules even at very low concentrations. On the basis of immobilized DNA molecules, different types of sensing devices have been already successfully created. Most of them were sensitive to the presence of low concentrations of pathogens in biological environments. A considerable interest was given to the study of the effect of small addition of fullerenes on the hydrated properties of DNA molecules that can change the mentioned properties important for biosensors.

Highly stable water colloid solution of C_{60} fullerene (maximum concentration 2 mg/mL) was prepared by transfer of C_{60} (purity >99.5%) from toluene to water using ultrasound sonication. Analysis of the UV/vis, Raman, FTIR, and SANS spectroscopic data showed that the water colloid solution of C_{60} fullerenes contains single C_{60} molecules in the hydrated state. However, their clusters and solids (nanocrystallites) have different sizes (1–36 nm in dependence of C_{60} concentration in water). The aggregation kinetics indicated that the structure of C_{60} aggregates could be described as a fractal with a dimension of 1.8–2.0 nm.

The A-300/DNA nanocomposite was used to prepare two samples under the same conditions. The first sample was with 5 wt% of distilled water, and the second sample was with 5 wt% of an aqueous solution of 0.1 mg/mL (0.01 wt%) C_{60}. A mixture of 2:1 deuterochloroform and deuteroacetonitrile was used as a solvent.

Figure 6.81 shows the ^1H NMR spectra of water in aqueous solution of C_{60} recorded at different temperatures. A single signal observed at 4.5 ppm is characteristic for bulk water or SAW. With decreasing temperature, the intensity of the signal decreases sharply and it completely disappears at $T<265$ K. The concentration of bound water does not exceed 0.1–0.3 wt% and such small concentrations do not permit the determination of the thermodynamic parameters of water bound to C_{60} aggregates.

Figure 6.82 shows the ^1H NMR spectra of water adsorbed to the DNA–SiO_2 nanocomposite (hydration $h=0.08$ g/g). The measurements were carried out in a mixture of $CDCl_3/CD_3CN$ (2:1). The water is characterized by three signals at $\delta_H=4.5$, 1.8, and 1 ppm. The first of the signals is due to SAW, and the other two correspond to WAW. With decreasing temperature, the intensity of the SAW signal is reduced in accordance with the partial freezing of bound water. For WAW, the freezing process is observed only at temperatures near to 200 K. Therefore, all WAW can be attributed to

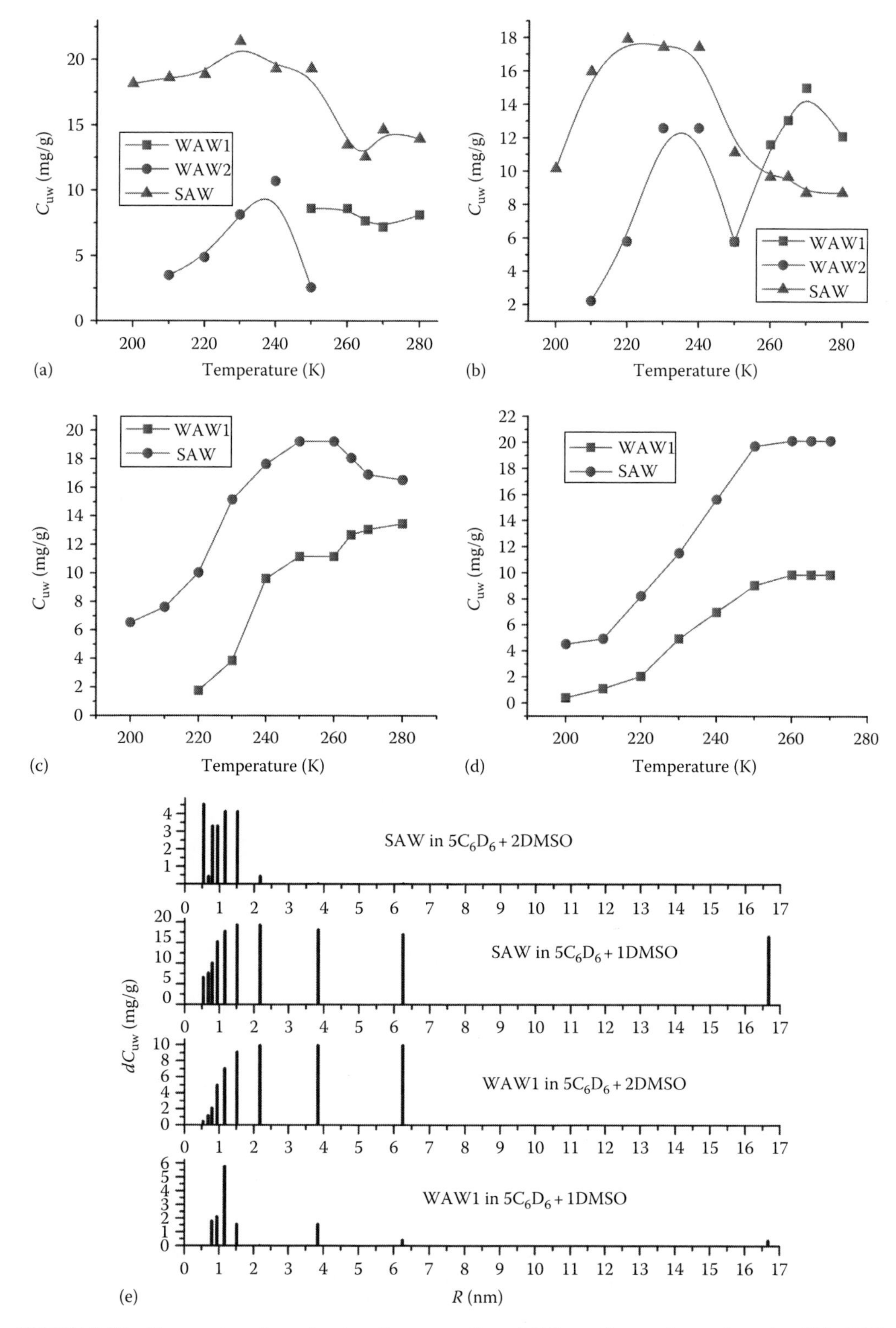

FIGURE 6.80 Temperature dependencies of concentration of different forms of water bound to SiO_2–DNA nanocomposite in mixed solvents $C_6D_6+CD_3CN$ (a, b) and $CDCl_3$+DMSO (c, d), and SAW cluster size distributions (e).

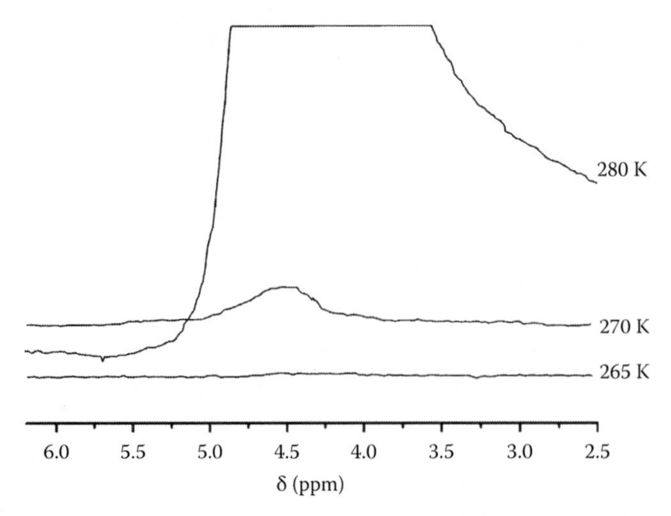

FIGURE 6.81 ^{1}H NMR spectra of water suspension of fullerene. (Adapted from *Chem. Phys. Lett.*, 496, Turov, V.V., Chehun, V.F., Krupskaya, T.V. et al., Effect of small addition of C_{60} fullerenes on the hydrated properties of nanocomposites based on highly dispersed silica and DNA, 152–156, 2010d, Copyright 2010, with permission from Elsevier.)

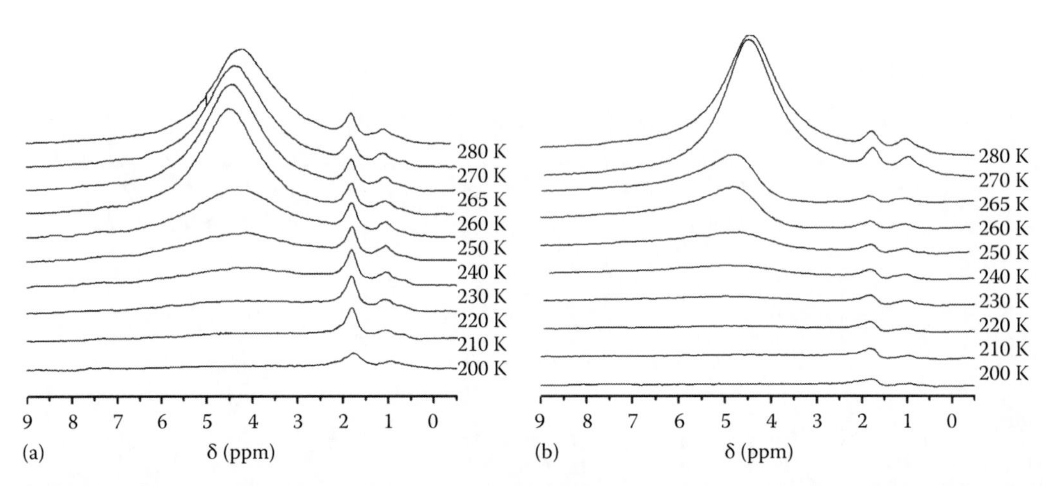

FIGURE 6.82 ^{1}H NMR spectra of water in nanocomposites SiO_2–DNA at hydration $h=0.08$ g/g, at different temperatures: (a) initial and (b) with 0.008% of fullerene C_{60}. (Adapted from *Chem. Phys. Lett.*, 496, Turov, V.V., Chehun, V.F., Krupskaya, T.V. et al., Effect of small addition of C_{60} fullerenes on the hydrated properties of nanocomposites based on highly dispersed silica and DNA, 152–156, 2010a, Copyright 2010, with permission from Elsevier.)

strongly bound water. In the presence of C_{60} fullerenes (0.008 wt%), the WAW spectra vary slightly (Figure 6.82b).

As it can be seen from Figure 6.82b, addition of a small amount of C_{60} fullerenes in the nanocomposite–water system led to significant changes in the characteristics of the SAW signal. Thus, in the initial sample (Figure 6.82a) the thawing process of SAW ended at 260 K, while in the presence of C_{60} it ended at 270 K.

Figure 6.83 shows the concentration of SAW as a function of the temperature (a), $\Delta G(C_{SAW})$ dependencies (b), and the distribution of the SAW clusters on the radii (c). The physical characteristics of the layers of bound water in the investigated samples are given in Table 6.8.

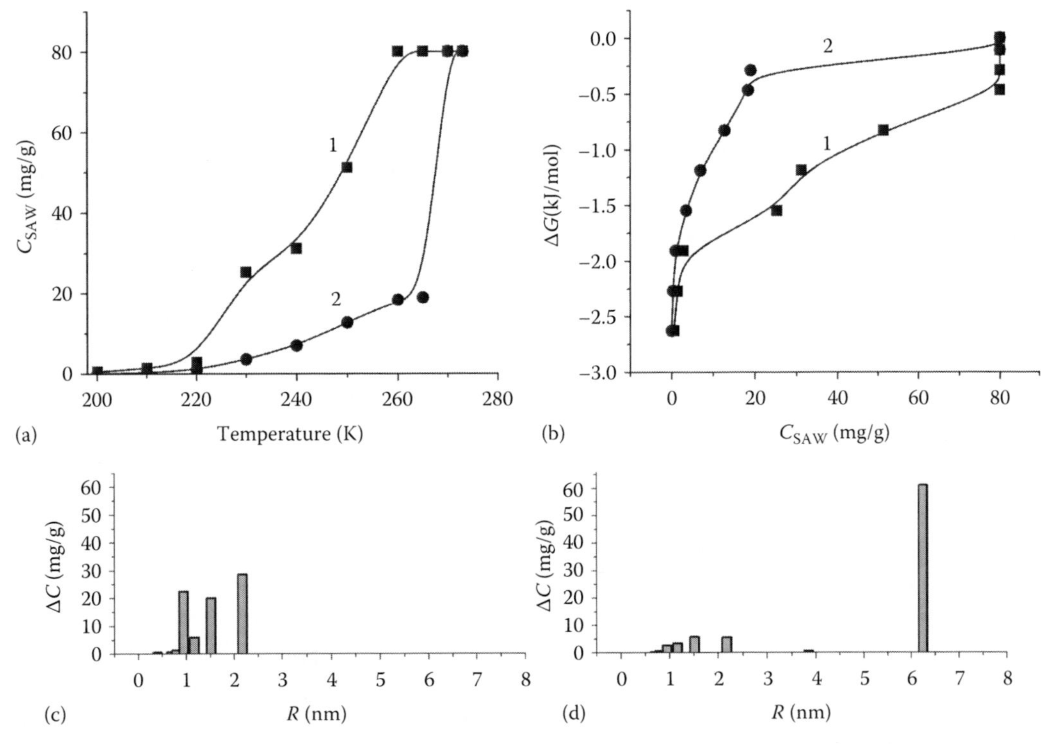

FIGURE 6.83 Dependences from temperature of SAW concentration in initial sample of SiO_2–DNA (a) and dependences of changes in the Gibbs free energy on C_{SAW} (b): 10–80 mg/g H_2O and 20–80 mg/g H_2O+C_{60} (0.008%); and (c, d) the corresponding ΔC (mg/g) size distributions of water structures. (Adapted from *Chem. Phys. Lett.*, 496, Turov, V.V., Chehun, V.F., Krupskaya, T.V. et al., Effect of small addition of C_{60} fullerenes on the hydrated properties of nanocomposites based on highly dispersed silica and DNA, 152–156, 2010a, Copyright 2010, with permission from Elsevier.)

TABLE 6.8
Parameters of Water Layers on Initial Composites of SiO_2–DNA and Composites Contained Fullerene

Sample	C^s_{uw} (mg/g)	C^w_{uw} (mg/g)	ΔG_s (kJ/mol)	ΔG_w (kJ/mol)	γ_S (J/g)
Initial	80	0	−3	—	5
With fullerene C_{60} (0.008%)	18	62	−3	−0.5	1.8

The $C_{SAW}(T)$ dependencies have a complicated shape which shows the heterogeneity of SAW clusters. They contain two regions with different C^s_{uw} values. Table 6.8 shows the C^s_{uw} value corresponding to the low-temperature region of the $C_{SAW}(T)$ dependence. The presence of C_{60} leads to a significant decrease in the concentration of strongly bound water and the appearance of weakly bound water. SAW clusters, bound by the DNA–A-300 nanocomposite, have radii in the range of 0.6–2.2 nm. As a consequence, more than 80% water is concentrated in the clusters with radii of 0.9, 1.5, and 2.2 nm. In the presence of C_{60} fullerenes, most of the bound water is concentrated in the clusters with a radius of ~6 nm. Consequently, as it can be seen in Figure 6.83, the contribution of clusters at radius <2.5 nm decreases. Finally, the surface free energy is reduced to the half in the presence of C_{60}.

Hydrated properties silica–DNA nanocomposite are sensitive to the presence of small amounts of C_{60} fullerenes which cause an increase in the size of bound water clusters and decrease in the interfacial free energy. The DNA molecules immobilized on the A-300 surface and hydrophobic C_{60} aggregates significantly affect the conditions of the formation of SAW clusters. These clusters are displaced to the outer surface of the DNA–A-300–C_{60} composite and transferred from strongly to weakly bound state characterized by larger cluster sizes. Thus, the ratio between the weakly and strongly bound water clusters, adsorbed by DNA molecules, can be controlled by a small amount of C_{60} fullerenes and consequently C_{60} has a strong influence on the DNA function in the biological systems.

6.4 LIPIDS

Wattraint et al. (2005a,b) and Wattraint and Sarazin (2005) investigated the mobility of water, ubiquinone, and tethered phospholipids, components of a biomimetic model membrane using stimulated echo pulsed-field gradient ^{1}H and ^{2}H MAS NMR method. The diffusion constant of water corresponded to an isotropic motion in a cylinder. When the lipid bilayer was obtained after the fusion of small unilamellar vesicles (SUVs), the extracted value of lipid diffusion indicated unrestricted motion. The cylindrical arrangement of the lipids permits a simplification of data analysis since the normal bilayer was perpendicular to the gradient axis. This feature led to a linear relation between the logarithm of the attenuation of the signal intensity and a factor depending on the gradient strength, for lipids covering the inner wall of anodic Al_2O_3 (AAO) pores (200 nm in diameter) as well as for lipids adsorbed on a polymer sheet rolled into a cylinder. The effect of the bilayer formation on water diffusion was observed (Figure 6.84). The lateral diffusion coefficient of ubiquinone was in the same order of magnitude as the lipid lateral diffusion coefficient, in agreement with its localization within the bilayer. The corresponding slope gave the apparent diffusion coefficient for lipids. A steady-state value of $(1.7 \pm 0.2) \times 10^{-12}$ m^2/s was reached for the lateral diffusion coefficient after a diffusion time of 250 ms. This value agreed with the lateral diffusion coefficient of phospholipids $(2.8 \pm 0.4) \times 10^{-12}$ m^2/s obtained by ^{31}P NMR spin–spin relaxation time with a spherical geometry (Wattraint and Sarazin 2005).

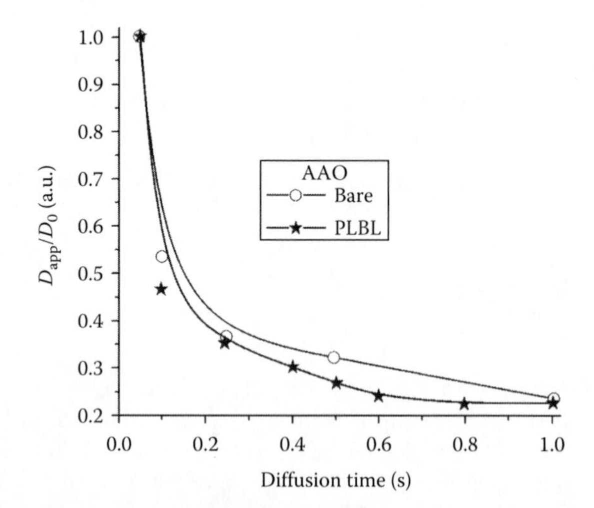

FIGURE 6.84 Plot of normalized D_{app}/D_0 (D_0 is the diffusion constant of bulk water) as a function of diffusion time for deuterated water in the bare AAO support and in the tethered phospholipid bilayers (PLBL) inside the AAO support. (Adapted from *Biochim. Biophys. Acta—Biomembranes*, 1713, Wattraint, O. and Sarazin, C, Diffusion measurements of water, ubiquinone and lipid bilayer inside a cylindrical nanoporous support: A stimulated echo pulsed-field gradient MAS-NMR investigation, 65–72, 2005, Copyright 2005, with permission from Elsevier.)

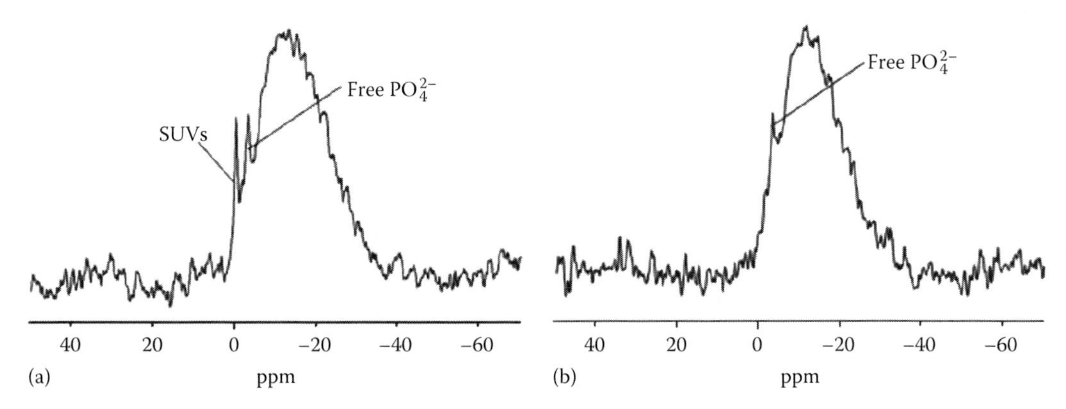

FIGURE 6.85 MAS (10 kHz) ^{31}P NMR spectra of phospholipids inside AAO pores (a) before PEG treatment and (b) after PEG treatment. (Adapted from *Analyt. Biochem.*, 336, Wattraint, O., Arnold, A., Auger, M., Bourdillon, C., and Sarazin, C., Lipid bilayer tethered inside a nanoporous support: A solid-state nuclear magnetic resonance investigation, 253–261, 2005a, Copyright 2005, with permission from Elsevier.)

SUVs composed of egg yolk 1-*a*-phosphatidylcholine and 1,2-dioleoyl-*sn*-glycero-3-phosphatidylethanolamine, in a molar ratio of 64.5%/35% and containing 0.5% of 1,2-dipalmitoyl-*sn*-glycero-3-phosphatidylethanolamine-*N*-biotinyl were obtained by sonication of the mixture at a concentration of 5 mM (Wattraint et al. 2005a). Fusion was promoted by a treatment with 30% (w/v) PEG8000 solution (Figure 6.85). Yuan et al. (1995) studying interaction of DPPC liposomes with silica particles showed that the silica surface binds to the phosphate moiety of the polar headgroups of DPPC bilayers through the hydrogen bonds between SiO–H and O–P groups. However, they have no significant effect on the conformation of the choline group of the polar headgroups. The addition of silica particles decreases slightly the mobility of the hydrocarbon chains. Proton T_1 data indicated that aluminum citrate reduced the effect of silica.

Leftin and Brown (2011) reporting a database of NMR results for membrane phospholipids summarized experimental ^{13}C–^{1}H and ^{2}H NMR segmental order parameters (SCH or SCD) and spin–lattice (Zeeman) relaxation times (T_{1Z}) in convenient tabular form for various saturated, unsaturated, and biological membrane phospholipids. Segmental order parameters give direct information about bilayer structural properties, including the area per lipid and volumetric hydrocarbon thickness, and relaxation rates provide complementary information about molecular dynamics. Particular attention was paid to the magnetic field dependence (frequency dispersion) of the NMR relaxation rates in terms of various simplified power laws. Model-free reduction of the T_{1Z} studies in terms of a power-law formalism showed that the relaxation rates for saturated phosphatidylcholines followed a single frequency-dispersive trend within the MHz regime. They showed how analytical models can guide the continued development of atomistic and coarse-grained force fields. The interpretation suggested that lipid diffusion and collective order fluctuations were implicitly governed by the viscoelastic nature of the liquid–crystalline ensemble. Collective bilayer excitations were emergent over mesoscopic length scales that fall between the molecular and bilayer dimensions and are important for lipid organization and lipid–protein interactions (Leftin and Brown 2011).

According to the hierarchy (Figure 6.86) arranged from smallest to largest timescale, one can first focus on the individual C–H bonds (Leftin and Brown 2011). These segmental sites fluctuate with correlation times in the fs–ns range due to vibrational motion, trans–gauche isomerizations, and restricted segmental reorientation. One step up in the temporal hierarchy, one may consider the orientational fluctuations of the lipid molecules. Anisotropic motion with correlation times in the ns–ms range lends the molecular assembly properties of different liquid crystals (Luzzati and Husson 1962, Charvolin and Tardieu 1978). Lastly, the macroscopic organization and order of lipid molecules correspond to continuum thermal viscoelastic deformations with correlation times in the ms–s range (de Gennes 1969, Helfrich 1973). These structural and dynamic features

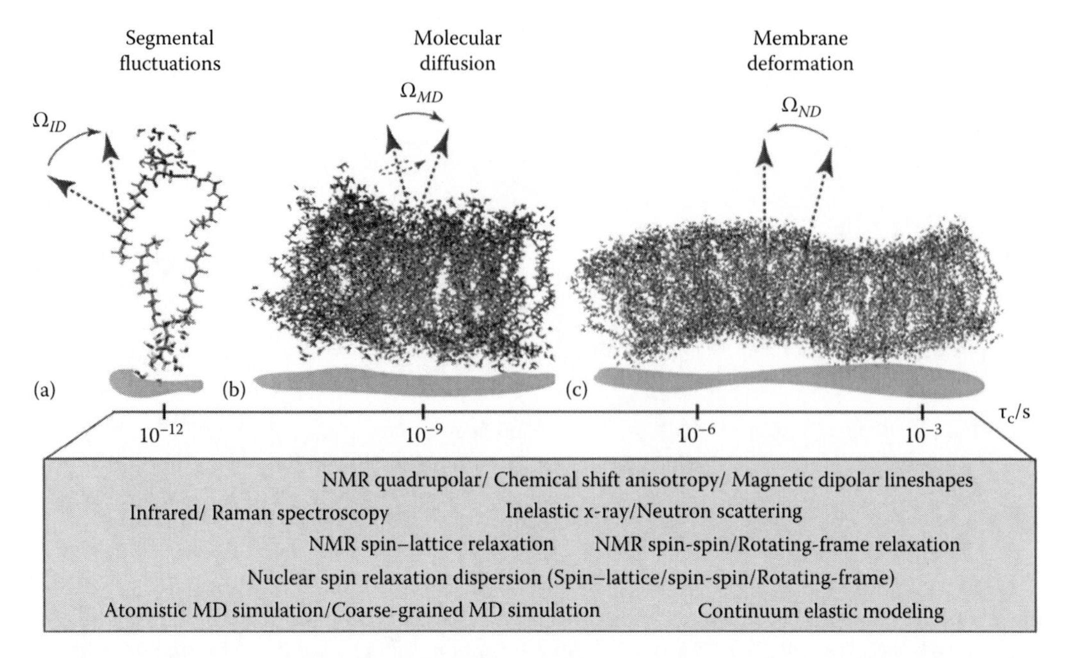

FIGURE 6.86 NMR spectroscopy reveals phospholipid membrane dynamics and structure over a range of timescales. The energy landscape of phospholipid mobility is characterized by segmental fluctuations, molecular diffusion, and viscoelastic membrane deformation. Orientational fluctuations correspond to geometry of interactions via Euler angles Ω and by correlation times τ_c of the motions. (a) Principal axis system of ^{13}C–^{1}H or C–^{2}H bonds fluctuates due to motions of internal segmental frame (*I*) with respect to the membrane director axis (*D*). (b) Diffusive phospholipid motions are described by anisotropic reorientation of molecule-fixed frame (*M*) with respect to the membrane director axis (*D*). (c) Liquid–crystalline bilayer lends itself to propagation of thermally excited quasi-periodic fluctuations in membrane curvature expressed by motion of the local membrane normal (*N*) relative to membrane director axis (*D*). The appropriate range of timescales of various complementary biophysical methods is indicated at the bottom of the figure. (Adapted from *Biochim. Biophys. Acta—Biomembranes*, 1808, Leftin, A. and Brown, M.F., An NMR database for simulations of membrane dynamics, 818–839, 2011, Copyright 2011, with permission from Elsevier.)

are characterized by various biophysical techniques, in particular SAXS (Nagle and Tristram-Nagle 2000), molecular dynamics simulations (Pastor et al. 2002), and NMR spectroscopy (Brown 1996, Leftin and Brown 2011).

Motions much faster than the static coupling result in isotopic frequency spectra, such as the ^{13}C chemical shift spectrum of 1,2-dilauroyl-*sn*-glycero-3-phosphocholine (DLPC) obtained using MAS shown in Figure 6.87 (Leftin and Brown 2011).

These high-resolution experiments report on the acyl chain, backbone, and headgroup regions simultaneously at multiple nuclear sites at natural isotopic abundance. The ^{13}C isotropic chemical shifts in the MAS experiment are similar to those recorded under comparable solution phase conditions (Chan et al. 1971, Lee et al. 1972, Levine et al. 1972, Feigenson and Chan 1974, Lyerla and Levy 1974, Lyerla et al. 1974, Kainosho et al. 1978, Oldfield et al. 1987). Furthermore, the spin–lattice relaxation rates (R_{1Z}) of multilamellar lipids measured with solid-state NMR methods and SUVs measured with solution-state NMR are nearly identical (Figure 6.87, inset). This similarity has been reproduced in recent studies where the NMR relaxation times of small vesicles and vesicle dispersions have been compared through Brownian dynamics simulations that demonstrate the influence of torsional potentials on the dynamics of the bilayer lipids (Klauda et al. 2008). The extended data sets presented in the database offer an opportunity to compare the magnetic field (Larmor frequency) dispersions of ^{13}C NMR relaxation rates for multiple saturated and unsaturated

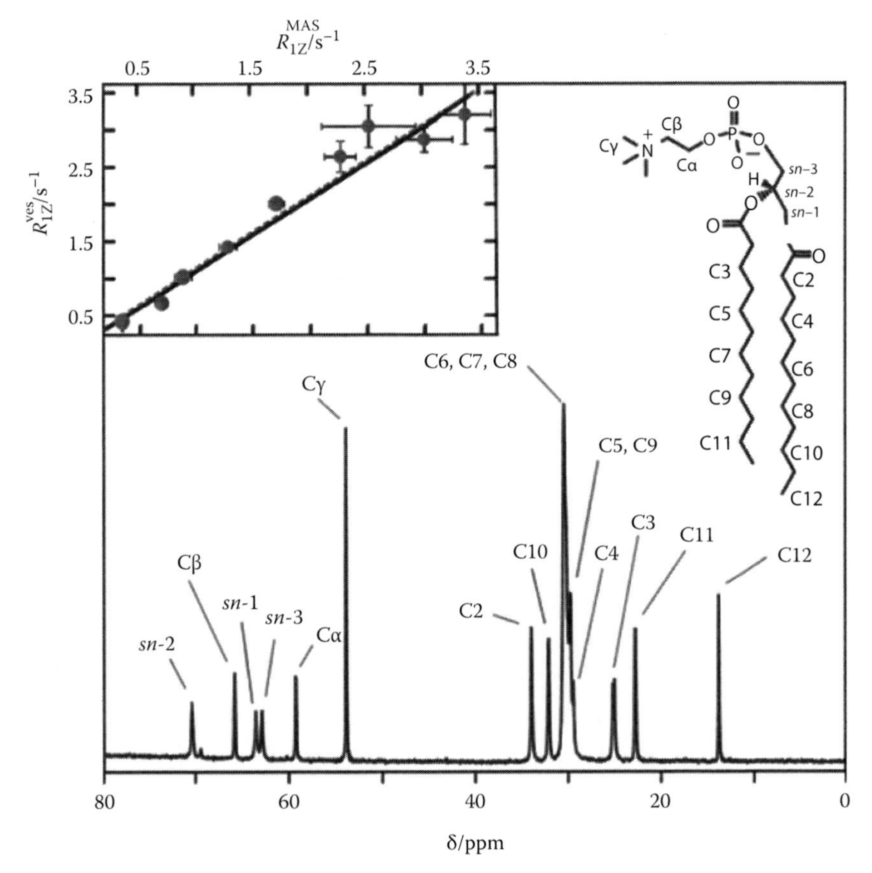

FIGURE 6.87 High-resolution solid-state ^{13}C NMR spectroscopy of membrane lipids reveals chemical shifts and nuclear spin–lattice (R_{1Z}) relaxation rates. Isotropic ^{13}C NMR chemical shift (δ) frequency spectrum obtained under MAS conditions (6 kHz ± 2 Hz, SPINAL-64 ^1H-decoupling field strength of 50 kHz) for multilamellar DLPC vesicle dispersion at 30°C. Inset depicts nuclear spin–lattice relaxation rates (R_{1Z}) for $(CH_2)_n$ carbons (C_4–C_{11}) obtained for small unilamellar vesicles (ves) under stationary high-resolution solution NMR conditions versus those for multilamellar dispersions using solid-state MAS techniques. The molecular structure of DLPC and assignments corresponding to the chemical shift assignments in the ^{13}C NMR spectra are shown. (Adapted from *Biochim. Biophys. Acta—Biomembranes*, 1808, Leftin, A. and Brown, M.F., An NMR database for simulations of membrane dynamics, 818–839, 2011, Copyright 2011, with permission from Elsevier.)

lipids, thus allowing simulation comparisons invoking multiple timescales (Leftin and Brown 2011). A comparison of a phospholipid liquid–crystalline system to an isotropic alkane in the liquid state is illustrated in Figure 6.88.

Here the R_{1Z} values provided in the Supplementary Content (Leftin and Brown 2011) for the $(CH_2)_n$ segments of DLPC and the liquid hydrocarbon *n*-dodecane are compared. Whereas scaling of the relaxation rates for the lipid depends on segmental motion, anisotropic molecular motion, and collective membrane motion, the relaxation rate of the alkane depends on isotropic, fast segmental, and molecular motions only. The frequency dispersion for the liquid is linear with a slope nearly equal to zero at all frequencies. However, for DLPC the slopes of the dispersion depend significantly on temperature. This shows that the phase behavior of the membrane contributes to the structural dynamics observed, and that the rate of the acyl chain motion becomes more like the isotropic alkane with increasing temperature, thereby highlighting the contribution of order fluctuations to

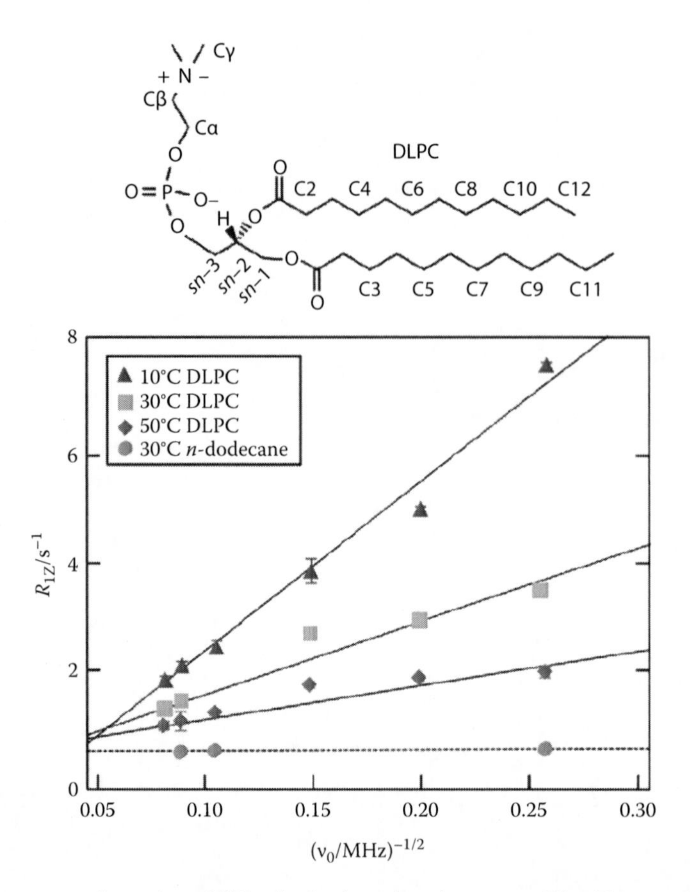

FIGURE 6.88 Frequency dispersion of ^{13}C spin–lattice relaxation rates of liquid–crystalline DLPC membrane bilayers compared to liquid hydrocarbon. Relaxation rates obtained for the $(CH_2)_n$ segments of DLPC are shown at (▲) 10°C, (■) 30°C, and (◆) 50°C. The slope of the power-law relaxation dispersion ($\nu_0 = \nu_C$) decreases with an increase in temperature versus corresponding data (•) for $(CH_2)_n$ carbons of *n*-dodecane at 30°C. Extrapolating DPLC relaxation rates to high frequency or high temperature approximately matches results for *n*-dodecane in the isotropic liquid state. The center of the phospholipid membrane at high temperature resembles isotropic motion of the hydrocarbon liquid. The molecular structures of *n*-dodecane and DLPC are shown. (Adapted from *Biochim. Biophys. Acta—Biomembranes*, 1808, Leftin, A. and Brown, M.F., An NMR database for simulations of membrane dynamics, 818–839, 2011, Copyright 2011, with permission from Elsevier.)

the frequency dispersion. Interestingly, the ordinate intercept of the phospholipid rate dispersion in the inverse square-root presentation of the dispersion is non-zero, and is nearly the same for the $(CH_2)_n$ and liquid hydrocarbon segments. In this high-frequency limit, the relaxation rates are sensitive to the fast segmental motions and molecular reorientation of the acyl chain only. Thus, fundamentally the motions of the lipid acyl chains are similar to dynamics observed for liquid hydrocarbons, but are distinguished by molecular organization in the membrane, which allows for the possibility of lower-frequency motional contributions that arise from quasinematic lipid order fluctuations (Brown 1982, 1984).

On the basis of shown results, one can conclude that NMR is one of the most powerful tools to study different aspects of the interfacial phenomena with participation of biomacromolecules. This is true with respect to more complex biosystems, as shown in Chapter 7.

7 Water Associated with Bio-Objects
Cells and Tissues

High-resolution nuclear magnetic resonance (NMR) techniques are powerful tools to study biosystems both in vivo and ex vivo using static samples or rotated ones at appropriate speed of magic angle spinning (Schick 1996, García-Martín et al. 2001, Lindon et al. 2001, 2009, Gun'ko et al. 2005d, 2009d, Barker and Lin 2006, Wind and Hu 2006, Grivet and Delort 2009, Gröger et al. 2009, Michaelis et al. 2009, Muja and Bulte 2009, Courtier-Murias et al. 2012, Duarte and Gil 2012). Water as an essential component of bio-objects plays a very important role in intracellular (Figure 7.1) and extracellular processes which can be studied using the NMR techniques (Finch and Schneider 1975, García-Martín et al. 2001, Turov and Gorbyk 2003, Gun'ko et al. 2005d, 2009d, Tompa et al. 2009, 2010). In bio-objects, many metabolic compounds are mobile enough to be observed with standard high-resolution NMR, and although the metabolite concentrations are low (typically in the micromolar range) many metabolites can be detected by the NMR methods. Therefore in addition to MRI, non- or minimally invasive magnetic resonance spectroscopy (MRS) and spectroscopic or chemical shift imaging (CSI) are increasingly used for biochemical and biomedical research. With MRS that has been employed both in vivo, ex vivo (i.e., using intact excised lesions and tissues), and in vitro (using, e.g., urine, plasma, and tissue and cell extracts), the resonance lines of several key metabolites are measured, and several metabolites have been used as biomarkers as their presence and concentrations could be linked to, e.g., drug response, toxic effects, specific metabolic pathways, tumor phenotype, tumorigenesis, tumor size, tumor diagnosis, increased cell proliferation, cell apoptosis and necrosis, and brain disorders. Hence, MRS provides important complementary information to MRI, and it has been shown that in several cases the combination of MRI and MRS improves the diagnosis of diseases and lesions such as brain, breast, and prostate lesions, sometimes even to the extent that biopsies may no longer be necessary (Wind and Hu 2006).

Figure 7.2 demonstrates the effects of different spinning speeds in ^1H NMR (water suppressed) metabolite spectra of excised rat liver, obtained immediately after excision on a stationary sample (a) and with MAS (b–d), illustrating the appearance of spinning side bands (SSBs) at low spinning speeds. In fact, at 1 Hz spinning the side bands are so dense that the spectrum is virtually the same as the static one. The spectrum obtained at a 4 kHz MAS speed is free of SSBs, and in practice usually similar spinning rates have been used to obtain high-resolution ^1H MAS metabolite spectra in cells and excised tissues (Wind and Hu 2006).

Chemical structural aspects of bio-objects will not be analyzed here because the main attention in the book is focused onto the interfacial phenomena and the textural characteristics of cells, tissues, seeds, etc., analyzed using NMR, cryoporometry, and other methods.

7.1 YEAST *SACCHAROMYCES CEREVISIAE* CELLS

Water as the unique natural medium for native cells contributes a major portion of their weight. Water (with dissolved inorganic and organic compounds) can penetrate into or from cells through membrane channels or by transport protein molecules (e.g., aquaporins) (Finkelstein 1987, Murata et al. 2000, Voríšek 2000, Fujiyoshi et al. 2002). The length of transmembrane channels is about several nanometers but their diameter is significantly smaller. Despite this, their flow capacity is very high up to 10^9 mol/s (Fujiyoshi et al. 2002). Water can be strongly associated in the channels because of the formation of hydrogen-bonded complexes with polar functionalities of membranes and transport proteins or themselves since a clustered structure is typical for water (Wiggins 2008, Chaplin 2011). Formation of hydrogen-bonded complexes of water molecules in narrow channels could reduce the diffusion rate, i.e., water transport as a whole. Clearly, intermolecular interactions of water molecules in a confined space of channels or small cavities of aquaporins can strongly differ from those in bulk water. Additionally, hydrophobic fragments of the supramolecular cellular structures can affect the structure of intracellular water clusters and nanodomains. For instance, water molecules form small clusters from dimers to pentamers in nanopores of hydrophobic materials and the heat of water adsorption there can be relatively low in the 15–30 kJ/mol range (Brennan et al. 2001). This may enhance the influence of organic molecules (especially having both polar and nonpolar fragments) on the behavior of water in partially hydrophobic environment.

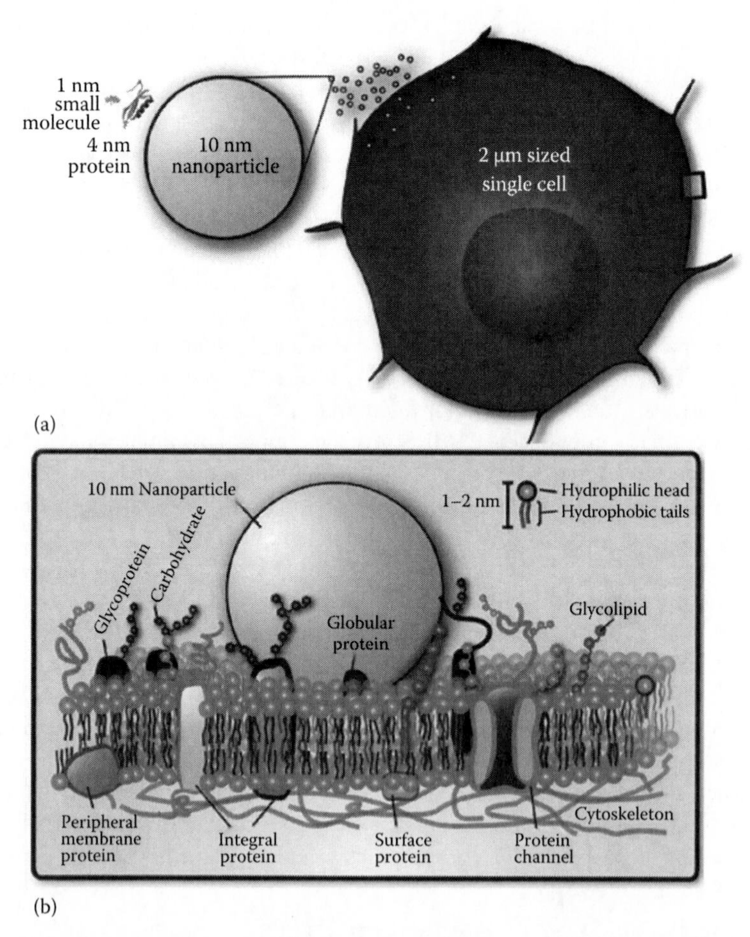

FIGURE 7.1 Size matters. (a) Compared to a 10 nm nanoparticle (e.g., nanosilica A-300), proteins, e.g., APP (amyloid precursor protein), ID 2FKL, and small molecules (e.g., small drug molecule of dehydroevodiamine hydrochloride) are small in size and volume. A mammalian cell, which is made up of proteins, nucleic acids, and other small to large molecules, is thousand times larger in volume and size compared to a 10 nm nanoparticle. (b) Cell membrane incorporating various proteins and a single 10 nm nanoparticle. (Adapted from *Prog. Neurobiol.*, 87, Suh, W.H., Suslick, K.S., Stucky, G.D., and Suh, Y.-H., Nanotechnology, nanotoxicology, and neuroscience, 133–170, 2009, Copyright 2009, with permission from Elsevier.)

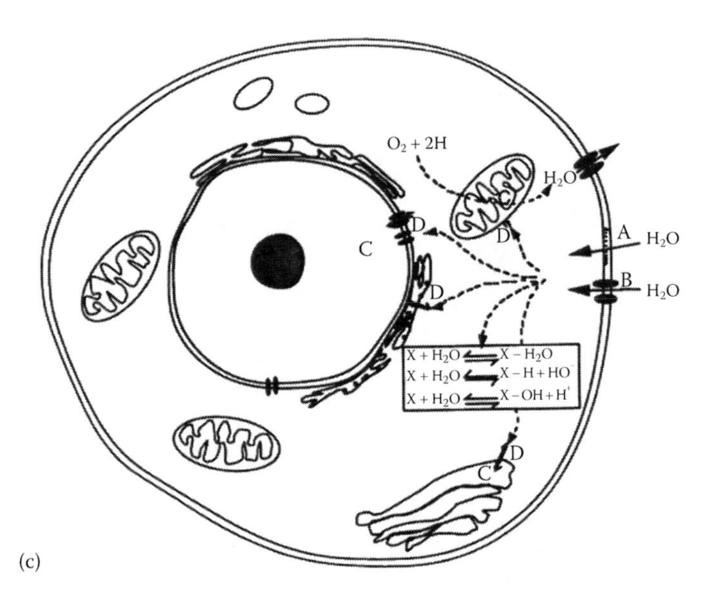

FIGURE 7.1 (continued) Size matters. (c) Overview of water metabolism in cells. Transport of water across the plasma membrane occurs through the phospholipid bilayer (A) or through water channels (B). Cytoplasmic water diffuses translationally, being exchanged with water from other subcellular organelles (D). There it provides the solvent for biochemical transformations and experiences a variety of water, hydrogen, or hydroxyl exchange reactions (C). Eventually water molecules abandon the organelle and the cell. (Adapted from *Prog. Nuclear Magn. Reson. Spectr.*, 39, García-Martín, M., Ballesteros, P., and Cerdán, S., The metabolism of water in cells and tissues as detected by NMR methods, 41–77, 2001, Copyright 2001, with permission from Elsevier.)

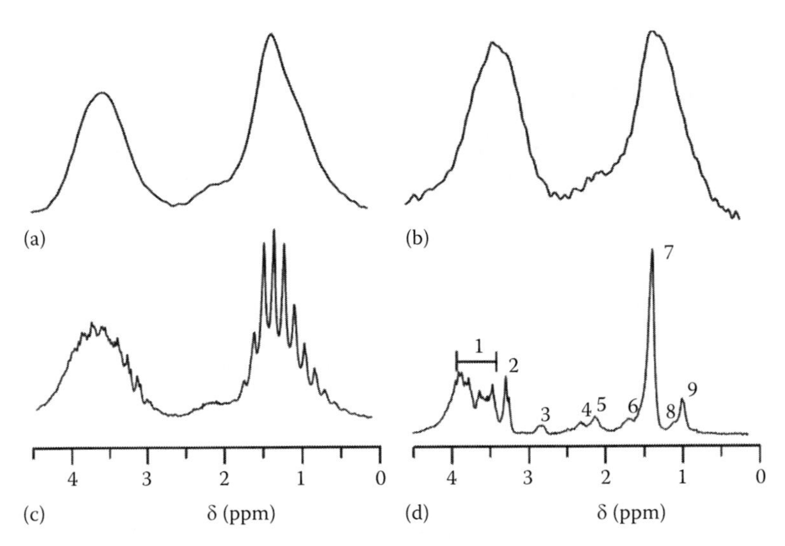

FIGURE 7.2 300 MHz water-suppressed ^{1}H NMR spectra of a freshly excised liver sample obtained from a male Fisher 344 rat with SP on (a) a static sample and with SP-MAS using spinning speeds of (b) 1 Hz, (c) 40 Hz, and (d) 4 kHz: (1) glucose, glycogen, aliphatic amino acids; (2) choline methyl, phosphocholine methyl and β-glucose, and trimethylamine-*N*-oxide methyl; (3) triglycerides $CH{=}CH{-}CH_2{-}CH{=}CH$; (4), triglycerides $CH_2{-}CH_2{-}COO$; (5) triglycerides $-CH{=}CH{-}CH_2{-}CH_2$; (6) triglycerides $-CH_2{-}CH_2{-}COO$; (7) triglycerides $-(CH_2)_n-$; (8) valine, leucine, isoleucine methyl; and (9) triglycerides CH_3 terminal, or neutral amino acid methyl. (Adapted from *Prog. Nuclear Magn. Reson. Spectr.*, 49, Wind, R.A. and Hu, J.Z., In vivo and ex vivo high-resolution ^{1}H NMR in biological systems using low-speed magic angle spinning, 207–259, 2006, Copyright 2006, with permission from Elsevier.)

(a)

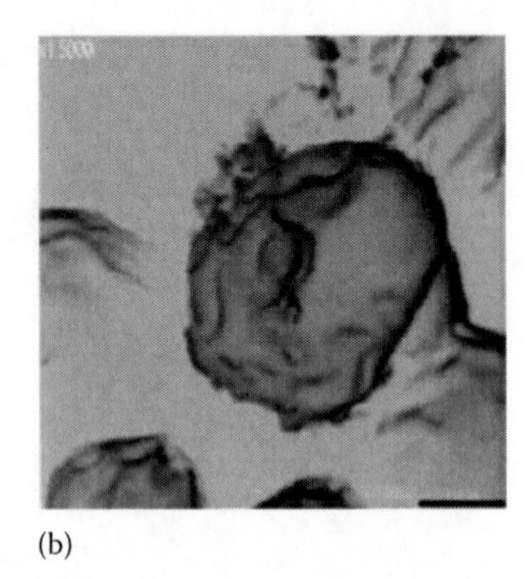

(b)

FIGURE 7.3 SEM images of dry yeast *S. cerevisiae* cells (5–10 μm in diameter in native state) at different magnification: (a) 4,500× (bar 10 μm) and (b) 15,000× (bar 1 μm). (Adapted from *J. Colloid Interface Sci.*, 283, Turov, V.V., Gun'ko, V.M., Bogatyrev, V.M. et al., Structured water in partially dehydrated yeast cells and at partially hydrophobized fumed silica surface, 329–343, 2005, Copyright 2005, with permission from Elsevier.)

Intramembrane water corresponds to a small portion of intracellular water for native cells. To study the state of water bound in membranes by the ^{1}H NMR method, the corresponding relative portion of intramembrane water should be enlarged, e.g., by partial drying. For dry cells, very slow vital processes are characteristic. However, after rehydration, a major portion of them can renew their vitality. This effect is widely used for yeast *S. cerevisiae* cells (Figure 7.3) which are, therefore, appropriate to study the drying–hydration effects using low-temperature ^{1}H NMR spectroscopy (Turov et al. 2005). Notice that "*Saccharomyces*" derives from Latinized Greek and means "sugar mold" or "sugar fungus," saccharo being the combining form "sugar" and myces being "fungus," and *cerevisiae* comes from Latin and means "of beer."

7.1.1 Water Bound in Weakly and Strongly Hydrated Yeast *S. cerevisiae* Cells

Almost dry and rehydrated yeast *S. cerevisiae* cells were chosen because of their high viability during strong drying, rehydration, and freezing (Tsuraeva 1983, Turov et al. 2005). For low-temperature ^{1}H NMR investigations, dry yeast *S. cerevisiae* cells (100–200 mg) were loaded into a measuring NMR ampoule and a certain amount (10–200 mg) of distilled water was added, and then to equilibrate the sample was shaken for 30 min. The static ^{1}H NMR spectra were recorded by means of a Bruker WP-100 SY spectrometer of high resolution with a bandwidth of 50 kHz or a Varian 400 Mercury spectrometer. To prevent supercooling of the studied systems, measurements of amounts of unfrozen water were carried out on heating them preliminary cooled to 200 K (Gun'ko et al. 2005d, 2009d). The signals of water molecules from ice (as well as of macromolecules) were not detected here because of features of the measurement technique with static samples and the short time of cross-relaxation of protons in solids and macromolecules.

Measurements of chemical shift of water bound in cells were carried out in air as a dispersion medium using the external standard ($CHCl_3$, $\delta_H = 7.26$ ppm) or in $CDCl_3$ medium. Since chloroform may dissolve a small amount of water (<0.5 wt%), it was supposed that during freezing of partially hydrated cells water dissolved in chloroform does not strongly contribute the total ^{1}H NMR signal linked mainly to water located in cells and bound to intracellular functionalities. Application of inert solvent medium allows us to determine the chemical shift of bound water, to diminish non-uniform broadening of the ^{1}H NMR signals, and to reduce exchange of water molecules bonded to different surface sites (Gun'ko et al. 2009d).

Figure 7.4a shows the ^{1}H NMR spectrum of water in maximum dry yeast cells containing ~70 mg of water per gram of dry matter, and Figure 7.4b shows the ^{1}H NMR spectra of rehydrated cells

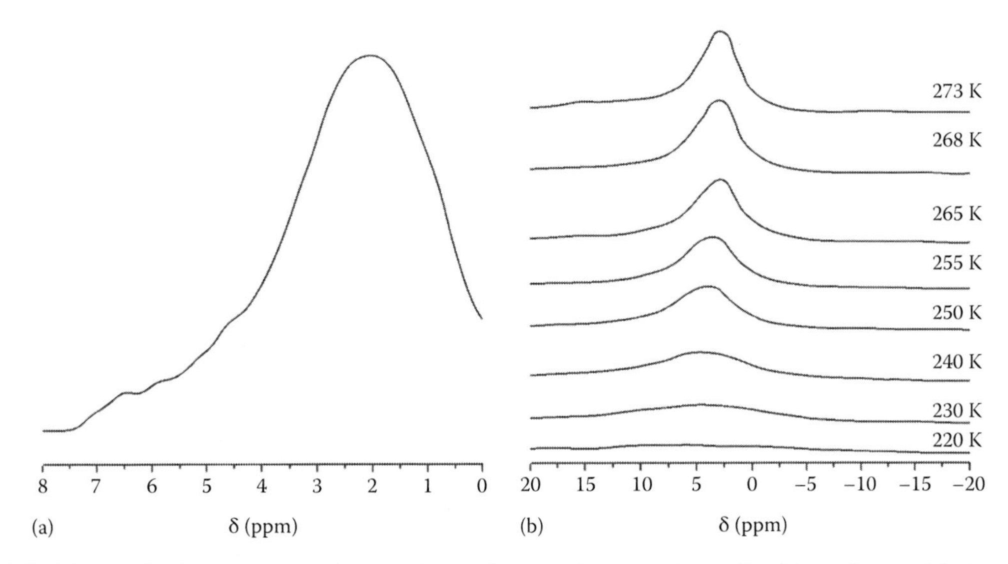

FIGURE 7.4 ^{1}H NMR spectra of water bound in yeast *S. cerevisiae* cells: (a) maximum dehydrated (C_{H_2O} = 70 mg/g) at 293 K and (b) at C_{H_2O} = 320 mg/g and different temperatures (in air). (Adapted from *J. Colloid Interface Sci.*, 283, Turov, V.V., Gun'ko, V.M., Bogatyrev, V.M. et al., Structured water in partially dehydrated yeast cells and at partially hydrophobized fumed silica surface, 329–343, 2005, Copyright 2005, with permission from Elsevier.)

(C_{H_2O} = 320 mg/g) at different temperatures. The spectra of water bound in cells represent a broad single signal with a maximum of the chemical shift at $\delta_H \approx 2$ ppm (i.e., mainly WAW) at a low water content and 3–5 ppm (i.e., mainly SAW) at the higher concentration. The signal intensity decreases and the δ_H value increases with lowering temperature.

In the chloroform medium, the ^{1}H NMR signal width of water bound in cells strongly decreases (Figure 7.5a through d) that allows us to observe changes in the chemical shift and the shape of signal of bound water depending on C_{H_2O}. If concentration of water in cells does not exceed 270 mg/g, then unfrozen water is observed as a signal at $\delta_H \approx 1.1$ ppm but its intensity decreases with increasing water amount in cells. With lowering temperature, appreciable reduction of the signal intensity caused by freezing of water is observed only upon cooling of samples to $T < 210$ K. The signal width changes slightly over total temperature range used that testifies relatively high mobility of the bound water molecules. The δ_H value of water in partially dehydrated cells is close to that of water molecules which do not take part in the formation of hydrogen-bonded complexes (as a proton donor) or forming the hydrogen bonds only through alone electron pairs of oxygen atoms (Turov and Leboda 1999, Gun'ko et al. 2005d, 2009d).

The signal intensity of water at $\delta_H \approx 1.1$ ppm (first signal, WAW) appreciably decreases with increasing water amounts in cells to 400 mg/g, but simultaneously a broad signal appears at $\delta_H = 5$ ppm (second signal, SAW). The intensity of this signal decreases with lowering temperature and it is not observed at $T < 230$ K (Figure 7.5c and d). Increase in C_{H_2O} leads to reduction of the width of the second signal. Thus, in rather narrow range of changes in hydration of yeast cells ($270 < C_{H_2O} < 400$ mg/g), the state of the bound water undergoes significant changes, as a result of which water passes from a weakly associated state ($\delta_H \approx 1.1$ ppm) to strongly associated one ($\delta_H \approx 5$ ppm).

The δ_H value of water dissolved in chloroform coincides with that of the first signal (Figure 7.5a through d). Notice that the ^{1}H NMR spectrum of the dispersion medium is separately shown in Figure 7.5c (the bottom spectrum at 220 K). Apparently from Figure 7.5c, the signal intensity of the water dissolved in chloroform is much lower than that of the intracellular water. Taking into account that the volume of the dispersion phase in the studied cellular systems is more than that

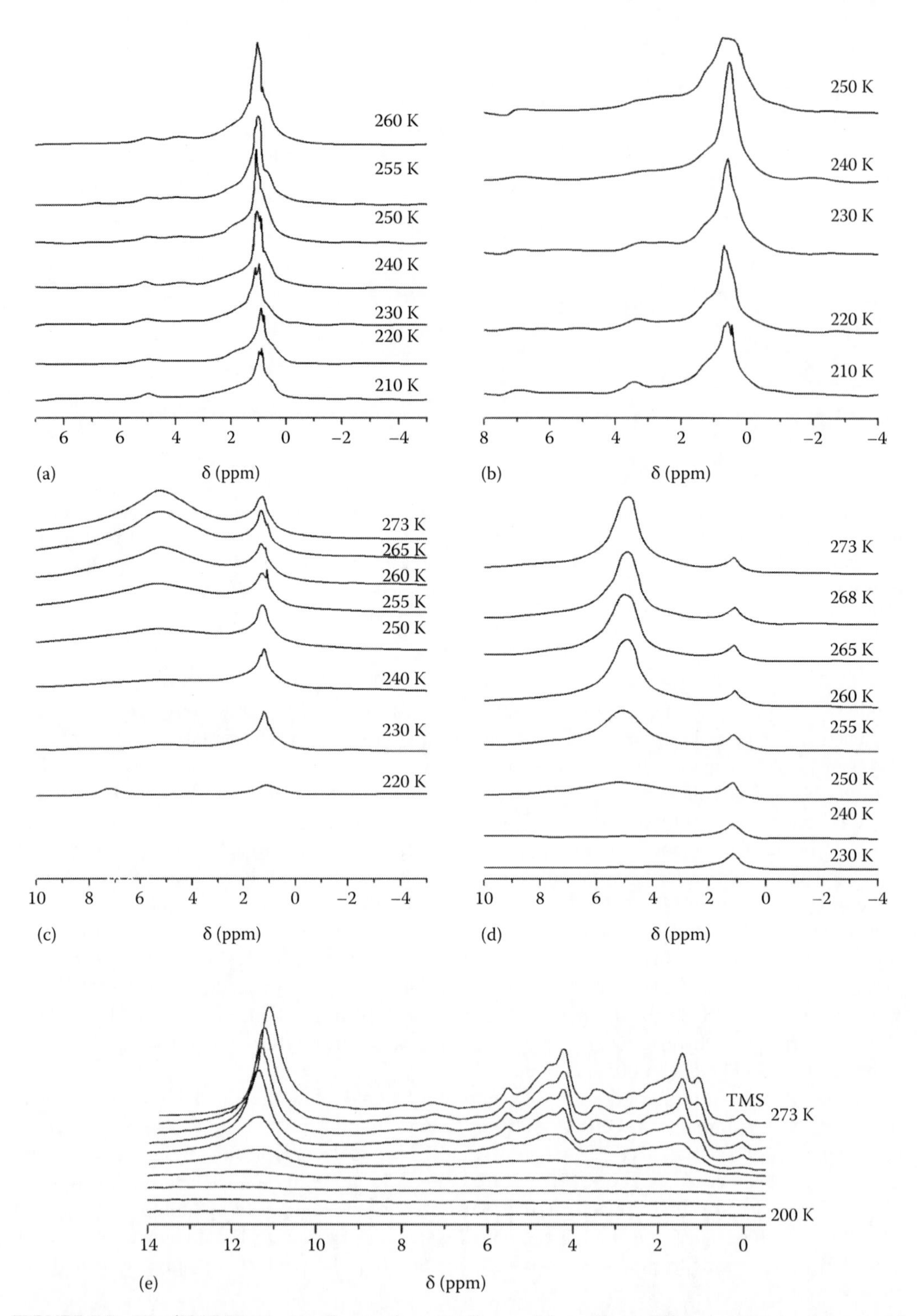

FIGURE 7.5 The ^{1}H NMR spectra of water in yeast *S. cerevisiae* cells at different amounts of water and temperatures in (a–d) chloroform-*d* (Bruker WP-100 SY) (Adapted from *J. Colloid Interface Sci.*, 283, Turov, V.V., Gun'ko, V.M., Bogatyrev, V.M. et al., Structured water in partially dehydrated yeast cells and at partially hydrophobized fumed silica surface, 329–343, 2005, Copyright 2005, with permission from Elsevier), and (e) CF_3COOD at $C_{H_2O} = 10$ mg/g (Varian Mercury 400).

of the dispersion medium, actually the difference between the signal intensities of the dissolved and intracellular waters in the spectra shown in Figure 7.5c is even more. Therefore, one could assume that the water dissolved in chloroform does not essentially contribute the intensity of the first signal.

The CH_2 groups of the aliphatic chains of phospholipids forming the cell membranes can be characterized by the ^1H signals in the range characteristic for WAW (~1–2 ppm). To analyze the WAW signal on the background signal of the CH_2 groups, we can use a strong acid because water can easily participate in the proton exchange reactions with similar acids in contrast to CH_2 groups.

Figure 7.5e shows the ^1H NMR spectra of dry yeast cells (with about 10 wt% of water) stirred with anhydrous trifluoroacetic acid CF_3COOD, then heated at 300 K for 10 min to enhance the exchange reaction. The deuterons of CF_3COOD were partially replaced by protons from water or OH and NH groups of organic components of the cells. According to the spectra (Figure 7.5), the signals of protons registered in the 1–2 ppm range for the initial cells were transformed after addition of CF_3COOD into four bands at $\delta_H = 11$–12, 6–8, 3–6, and 1–2 ppm. Consequently, in the acidic medium, the main portion of WAW (~70%) was transformed into structures related to SAW. These structures can differently dissolve the acid. The signal at 11–12 ppm can be attributed to structures with hydrated CF_3COOH ($\delta_H \approx 13$ ppm is for pure acid) and the water amount in these structures is about 10%. If we assume that the signal (in the acidic medium) at 1–2 ppm is linked only to phospholipids, then their signal intensity in the initial cells is less than 20% in this spectral range.

Dependences of $\Delta G(C_{uw})$ and $\gamma_S(C_{H_2O})$ for rehydrated cells at variation of the water amounts in samples are given in Figure 7.6. The first two dependences in Figure 7.6a and the first two points in Figure 7.6b concern to WAW ($\delta_H \approx 1.1$ ppm) with the corresponding spectra given in Figure 7.5a and b. A sharp reduction of interface energy is observed in the $\gamma_S(C_{H_2O})$ graph at $C_{H_2O} = 320$ mg/g when a signal of SAW ($\delta_H = 5$ ppm) is observed in the ^1H NMR spectra of water bound in cells. Change in the distribution of free energy of interfacial water upon rehydration of cells is shown in Figure 7.7. Apparently from this figure, the share of WAW ($-\Delta G < 0.8$ kJ/mol) does not exceed 20% of the total amount of the bound water. The concentration of the weakly bound water (WBW) increases with increasing C_{H_2O} to 520 mg/g that 80% of water in cells becomes weakly bound.

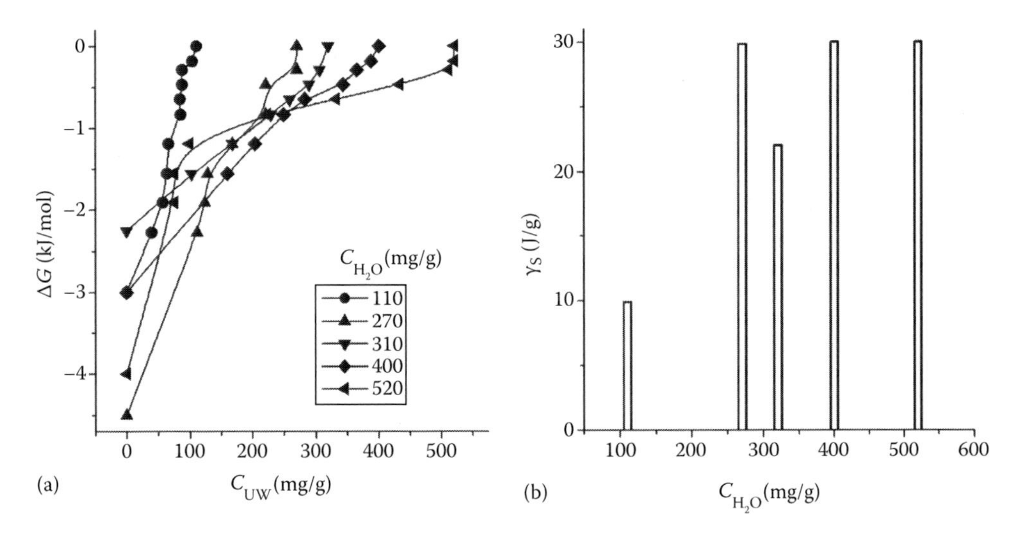

FIGURE 7.6 Dependence of changes in (a) Gibbs free energy of water in yeast *S. cerevisiae* cells on amounts of unfrozen water and (b) free surface energy of cell–water on concentration of total amount of water (in chloroform-*d*). (Adapted from *J. Colloid Interface Sci.*, 283, Turov, V.V., Gun'ko, V.M., Bogatyrev, V.M. et al., Structured water in partially dehydrated yeast cells and at partially hydrophobized fumed silica surface, 329–343, 2005, Copyright 2005, with permission from Elsevier.)

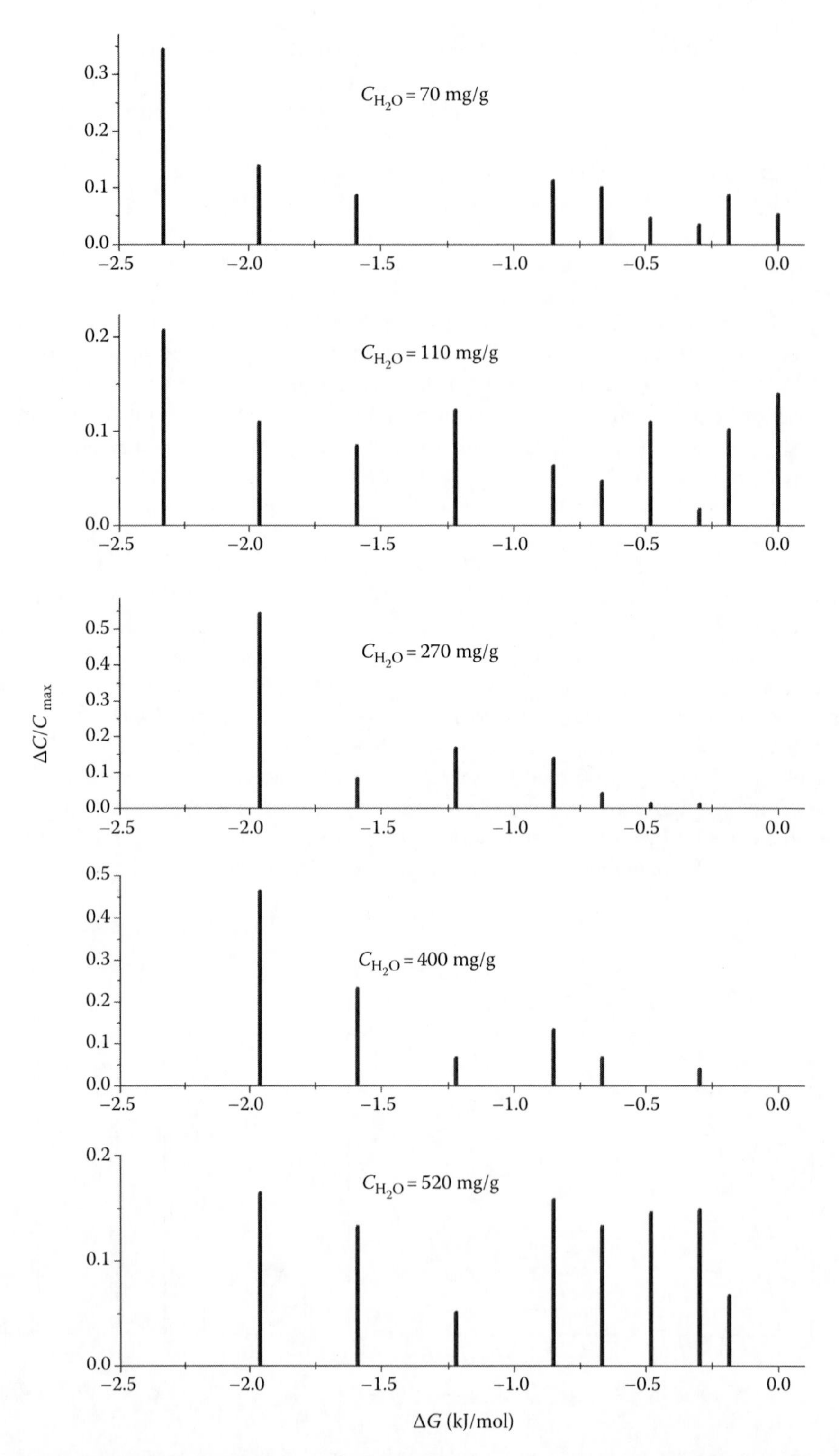

FIGURE 7.7 Influence of cell hydration on the distribution of unfrozen water versus changes in Gibbs free energy of this water (in chloroform-*d*). (Adapted from *J. Colloid Interface Sci.*, 283, Turov, V.V., Gun'ko, V.M., Bogatyrev, V.M. et al., Structured water in partially dehydrated yeast cells and at partially hydrophobized fumed silica surface, 329–343, 2005, Copyright 2005, with permission from Elsevier.)

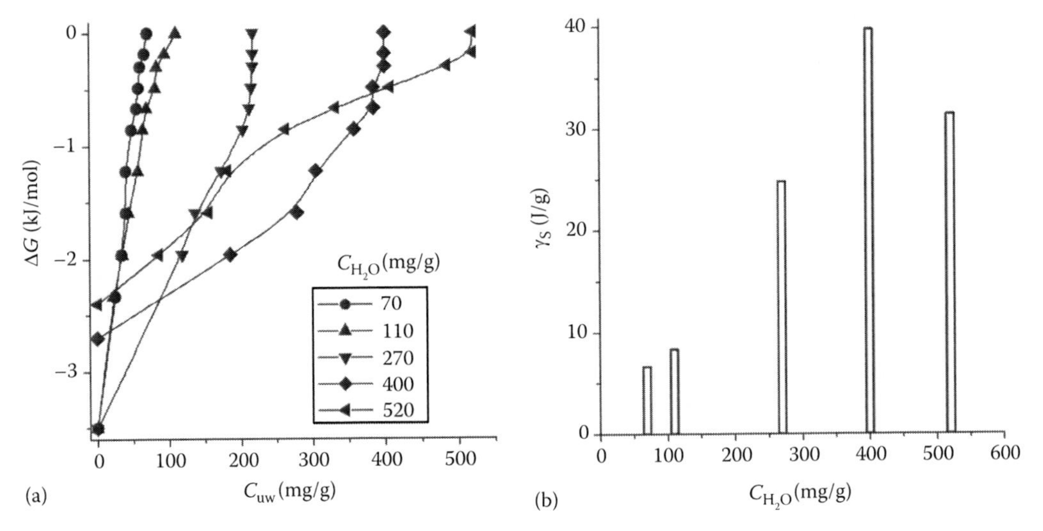

FIGURE 7.8 Dependence of changes in (a) Gibbs free energy of water in yeast *S. cerevisiae* cells on the amounts of unfrozen water and (b) interfacial free energy of cell–water on concentration of total amount of water in cells (in air). (Adapted from *J. Colloid Interface Sci.*, 283, Turov, V.V., Gun'ko, V.M., Bogatyrev, V.M. et al., Structured water in partially dehydrated yeast cells and at partially hydrophobized fumed silica surface, 329–343, 2005, Copyright 2005, with permission from Elsevier.)

Figure 7.8 shows $\Delta G(C_{uw})$ and $\gamma_S(C_{H_2O})$ for cells rehydrated in air. General tendencies of changes in the characteristics of the bound water in the air medium are the same as in chloroform. Distinctions are that a minimum of $\gamma_S(C_{H_2O})$ dependence shifts toward larger C_{H_2O} values and at $270 < C_{H_2O} < 400$ mg/g concentration of WBW remains negligibly low and increases only at $C_{H_2O} = 520$ mg/g.

It is possible to explain the described regularities in changes of the spectral characteristics of water on rehydration of cells as follows. If the amounts of water in cells do not exceed a certain critical value, then water in dehydrated cells is localized mainly in cellular membranes (around protein molecules and other charged functionalities) and hydration shells of organelles and biomacromolecules. Such water (a portion of protons from this water) has a low value of the chemical shift characteristic for water molecules not participating in the hydrogen bonds as proton donors.

From the experimental data and quantum chemical calculations of water clusters in the gas phase and with consideration for solvation effects (Tables 1.11 and 1.12), one can conclude that there are several reasons of appearance of the ^{1}H NMR signal at 1.1 ppm (Figure 7.5). First, in the case of air medium, the signal strongly differs from that in chloroform-*d* (Figures 7.4 and 7.5) and the mentioned signal appears only in the second medium. Second, in small water clusters (e.g., 2D cycles with 4–6 molecules akin to that found for water adsorbed in carbons; see Chapter 3) at least half of hydrogen atoms do not form the hydrogen bonds and are in contact with $CDCl_3$ molecules and nonpolar or weakly polar cellular functionalities. According to ab initio calculations of the chemical shifts in the gas phase and consideration for their correction (using linear approximation $\delta_H = a + bq_H$ where a and b are constants, and q_H is the atom charge) due to the solvent influence, δ_H of "free" hydrogen atoms may be between 0.5 and 2.5 depending on polarity of the surrounding and the number of water molecules in a cluster.

Notice that the free energy of solvation of water clusters in chloroform decreases with increasing their size (Tables 1.11 and 1.12). Consequently, there are two tendencies on interaction of water with cellular structures in the chloroform-*d* medium. The first is the formation of larger and larger water droplets with increasing hydration degree of cells (compare the signal intensity at 5 and 1 ppm in Figure 7.5). The second is the water distribution in the form of small clusters, dimers, and alone molecules in the cellular/chloroform-*d* medium, and the size of these clusters decreases with decreasing hydration degree of cells, since the signal at 5 ppm is significantly lower than that at $\delta_H \approx 1.1$ ppm at a low hydration (Figure 7.5a and b). The first tendency becomes dominant at greater hydration of cells and especially in the air medium when the signal at 2 ppm is observed only at minimal hydration ($C_{H_2O} = 70$ mg/g) of cells (Figures 7.4 and 7.5). Hence, chloroform provides decomposition of water into small clusters and the formation of such "emulsion" is accompanied by appearance of the intensive ^{1}H NMR signal at $\delta_H \approx 1.1$ ppm, but other portion of water can form domain structures with minimal surface area between chloroform and water.

Thus, investigations of states of the interfacial water in dry and rehydrated yeast *S. cerevisiae* cells in the air and chloroform-*d* media by means of the ^{1}H NMR spectroscopy with freezing-out layer-by-layer of the bulk and interfacial (weakly and strongly bound, SAW and WAW) waters show that the interfacial water can be in two main states characterized by the chemical shift $\delta_H \approx 1$ (WAW) and 5 (SAW) ppm. Relative contribution of these states depends on the hydration degree, since the first signal reduces with increasing water amounts and the opposite effect is observed for the second signal. Quantum chemical calculations of different water clusters in the gas and liquid phases and estimations of the changes in the chemical shifts because of the solvent influence show that distribution of dimer and alone water molecules in chloroform is more energetically favorable than larger water clusters. Hydrogen atoms of water molecules, which do not take part in the formation of the hydrogen bonds with adjacent molecules (in chloroform), have the δ_H values slightly greater than that of the first signal ($\delta_H \approx 1.1$ ppm) observed experimentally. However, in the case of modeling of the surrounding as nonpolar cyclohexene, this value reduces to $\delta_H = 0.8$ ppm. Consequently, simultaneous influence of nonpolar cellular functionalities and chloroform molecules could lead to the appearance of the intensive signal of water at δ_H close to 1 ppm and a similar effect is observed during the adsorption of water on partially hydrophobic fumed silica (see Chapter 1).

In the case of yeast cell suspensions at a great content of water, the behavior of bound water can differ from that in dry and partially hydrated cells (Turov et al. 2002a,b, 2005). Yeast *S. cerevisiae* cells are cryostable (Tsuraeva 1983); however, this stability can depend on the water content. To study the cell viability by cytochemical method (Lee et al. 1981), the viability degree $W = [(N_0 - N)/N_0] \times 100\%$ (where N_0 is the total number of cells and N is the number of death cells) was determined. At room temperature, the W value for yeast cells is close to 100%. In the case of freezing of dry initial cells by liquid nitrogen for 10 min and then heating to room temperature $W \approx 90\%$ but the same treatment of the suspension gives a smaller value $W \approx 70\%$. This is a relatively high degree for strongly frozen cells with great amount of water.

Two signals of water bound in yeast cells strongly hydrated at $C_{H_2O} = 865$ mg/g are observed in the ^{1}H NMR spectra at $\delta_H \approx 5$ (signal 1) and 1 (signal 2) ppm (Figure 7.9). The intensity of signal 1 decreases with lowering temperature much stronger than that of signal 2, and at 250 K their intensities become close.

The behavior of intracellular water (especially its colligative properties) is affected by interactions of water with biomacromolecules and low-molecular compounds dissolved in water due to the Raoult law (Frolov 1982, Chaplin 2011). This influence can cover both signals. Signal 2 at $\delta_H \approx 1$ ppm can be linked to intracellular WAW located in small voids at the boundary of hydrophilic and hydrophobic intracellular (mosaic) structures. Typically, low-molecular compounds cannot be dissolved in WAW but these compounds can be at the boundary of WAW clusters and prevent the formation of SAW structures. Relative contribution of WAW decreases with increasing water content because conditions become more favorable for the formation of SAW structures. With increasing size of SAW structures, the water activity as a solvent increases and it can dissolve low-molecular compounds in contrast to WAW.

In the case of the cellular suspensions, the shape of the temperature dependences of the ^{1}H NMR signal intensity of unfrozen water (Figure 7.10a) is similar to that observed for aqueous

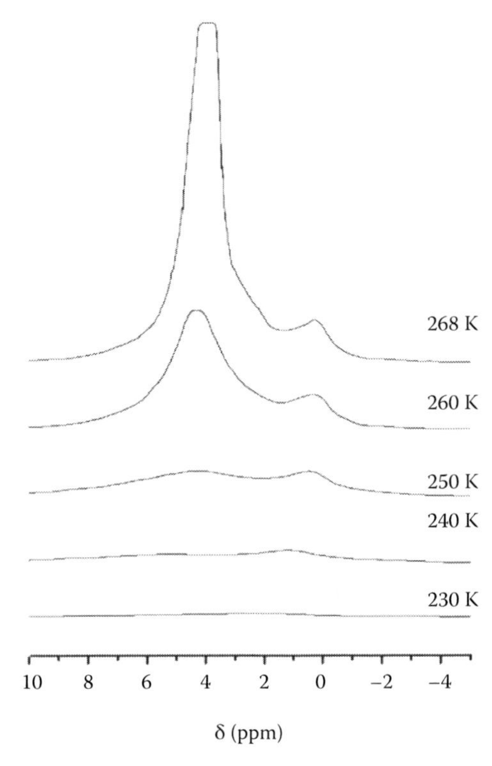

FIGURE 7.9 ^{1}H NMR spectra of water in strongly hydrated yeast *S. cerevisiae* cells at 86.5 wt% of water ($C_{cell}+C_{ICW}=50$ wt%, ICW is the intracellular water).

suspensions of macromolecules and nanooxides (Gun'ko et al. 2005d, 2009d) (see Chapters 1, 2, 5 and 6). A maximal intensity of bound unfrozen water is observed at a minimal amount of water (73 wt%, practically all water is intracellular) because in this case a relative content of bulk water (~40%) is minimal.

Notice that the C_{uw} values are attributed to both intracellular and extracellular waters. Therefore, the γ_S values (Table 7.1) could be calculated per gram of dry cells. On the basis of the data shown in Table 7.1, one can conclude that there is not a simple correlation between the amount of water and the characteristics of bound water. An increase in the cell content in the suspension leads to a decrease in the amounts of bound water and the γ_S value; however, these changes are nonlinear.

The most extremum behavior of all the characteristics is observed at low content of cells in the suspension (Figure 7.11). A minimum of γ_S and C^s_{uw} and a maximum of C^w_{uw} are at C_{H_2O} = 98.4 wt% or $C_Y=C_{cell}+C_{ICW}=6.1$ wt%. Notice that there is the extreme dependence of the γ_S value on the total concentration of water in the aqueous suspensions of nanooxides (see Section 1.1.6) at a minimum at $C_{H_2O} \approx 93$ wt%. This boundary concentration corresponds to transition from diluted suspensions to concentrated ones characterized by different particle–particle interactions. In the diluted suspensions, the systems can separate into a gel-like layer and upper layer with bulk, almost pure water. In the concentrated suspensions, the systems represent a continuous gel-like structure without separation of bulk water. With increasing size of particles, the critical concentration (C_c) should increase. Therefore, one could expect a larger C_c value for yeast *S. cerevisiae* cells (5–10 μm) than for nanosilica (primary particles ~10 nm). However, the C_c values for yeast cells and nanosilica are relatively close due to the formation of silica nanoparticles aggregates ~0.5–1 μm and agglomerates >l μm, which have sizes close to sizes of cells. Therefore, at $C_Y<10$ wt% ($C_Y<C_c$) the systems represent colloidal dispersion with relatively weak intercell interactions. At these C_Y values, the adhesion

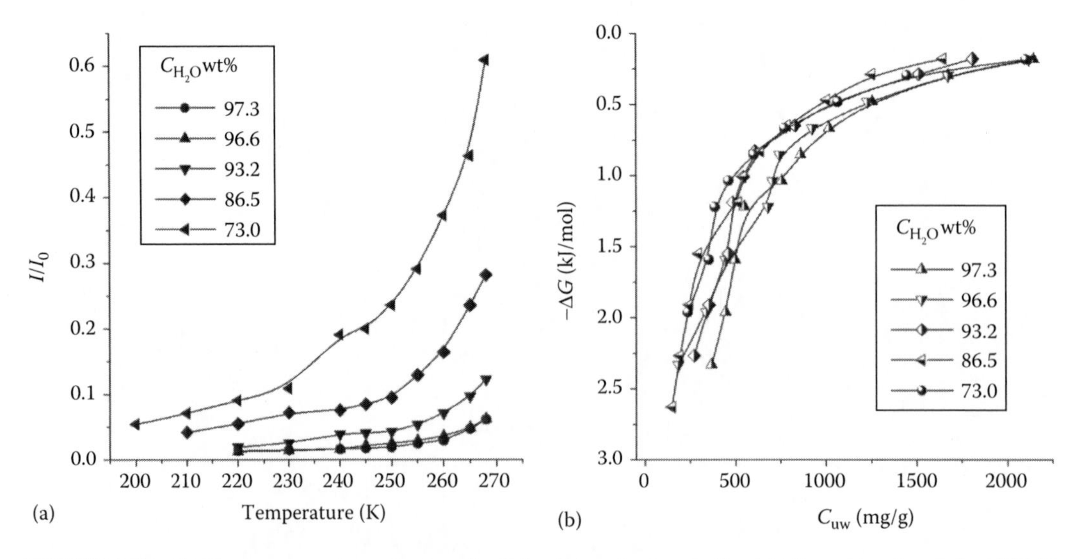

FIGURE 7.10 (a) Temperature dependences of relative intensity (normalized to the water signal, I_0 at 275 K before sample freezing) of the ^{1}H NMR signal of unfrozen water for frozen aqueous suspensions of yeast cells at different total content of water and (b) the corresponding relationships between the amounts of unfrozen water (C_{uw}) and changes in the Gibbs free energy of bound water.

TABLE 7.1
Characteristics of Unfrozen Water Bound in Yeast *S. cerevisiae* Cells being in the Aqueous Suspensions of Different Concentrations

	$C_{cell} + C_{ICW}$ (wt%)							
Parameter	**3.0**	**4.5**	**6.1**	**10.0**	**12.5**	**25.0**	**50.0**	**100.0**
C_{H_2O}(wt%)	99.2	98.8	98.4	97.3	96.6	93.2	86.5	73.0
$-\Delta G_s$ (kJ/mol)	3.19	3.99	3.34	3.80	3.89	2.73	3.19	3.35
$-\Delta G_w$ (kJ/mol)	0.9	0.8	0.9	0.7	0.8	0.8	0.8	0.7
C_{uw}^{s}(mg/g)	1500	950	820	600	750	900	750	700
C_{uw}^{w}(mg/g)	2888	2983	1878	3105	2152	1997	1671	1923
C_{uw}^{max} (mg/g)	4388	3933	2698	2705	2902	2897	2421	2623
γ_S (J/g)	208.3	168.9	123.3	137.7	139.9	119.2	106.8	97.4
$\langle T \rangle$ (K)	247.9	245.3	248.8	245.6	246.6	252.0	249.9	252.8
S_{uw} (m^2/g)	1551	1641	840	1181	1182	593	769	643
S_{nano} (m^2/g)	1217	1279	623	916	941	379	577	439
S_{meso} (m^2/g)	323	362	208	265	241	209	185	193
S_{macro} (m^2/g)	11	0	8	0	0	5	6	12
V_{nano} (cm^3/g)	0.519	0.488	0.260	0.360	0.366	0.175	0.246	0.182
V_{meso} (cm^3/g)	3.490	2.492	2.190	2.050	2.212	2.543	1.984	2.105
V_{macro} (cm^3/g)	0.151	0	0.106	0	0	0.081	0.082	0.147

Notes: C_{H_2O} is the total content of intra- and extracellular water; C_{ICW} is the content of intracellular water (ICW); and C_{cell} is the weight of dry cells.

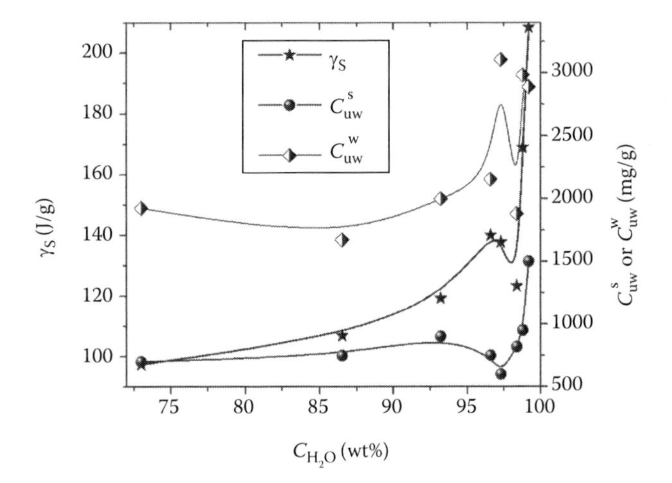

FIGURE 7.11 Relationships between total content of water and changes in the free surface energy and the amounts of strongly and weakly bound waters in yeast cells suspensions.

forces for bound water can be calculated. The distance from the cell surface (x) can be calculated in approximation of a plane interface:

$$E(\text{nm}) = 0.9\frac{C_{uw}^{ext}}{S} \quad (7.1)$$

where

S is the external specific surface area of cells (estimated as 1 m^2/g for a smooth cell surface)
C_{uw}^{ext} is the amount of extracellular water (Figure 7.12)

In this approach, the adhesion forces for the cell suspensions are much stronger than that for nano-oxides (Gun'ko et al. 1999a, 2005a, 2009a). This difference can be explained by the use of too small S value in Equation 7.1. Real specific surface area of the cells is much larger because of a significant nanoscaled roughness of the cell surface (Figure 7.3) than that estimated for the smooth

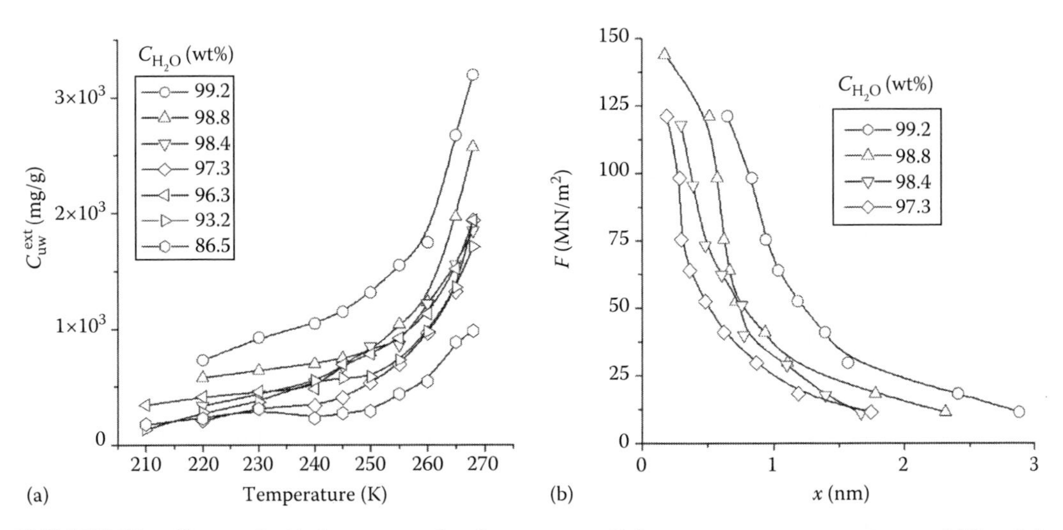

FIGURE 7.12 Changes in (a) the content of unfrozen extracellular water versus temperature and (b) radial forces of adhesion of water layers for more diluted systems.

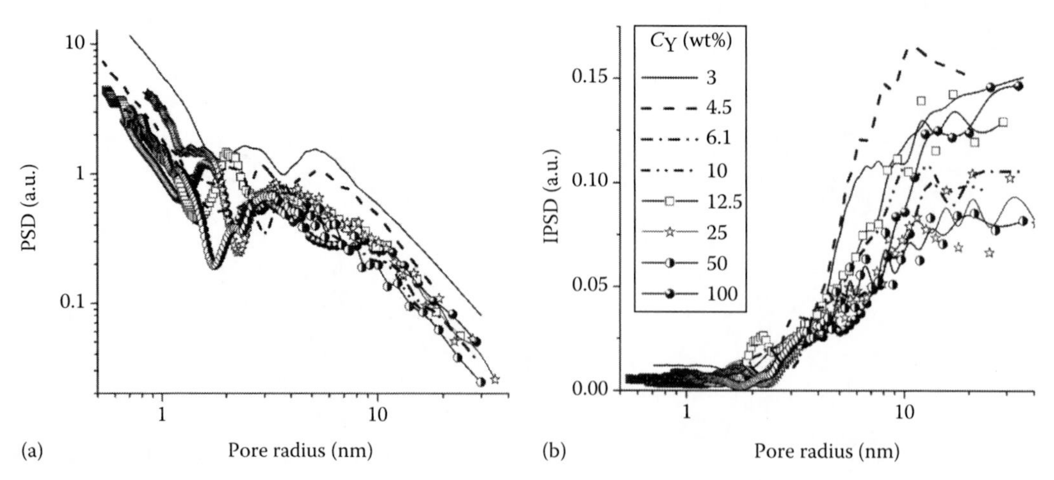

FIGURE 7.13 Size distributions: (a) differential and (b) incremental of pores filled by unfrozen water (cylindrical pore model) bound in strongly hydrated yeast *S. cerevisiae* cells ($C_Y = C_{cell} + C_{ICW}$).

cell surface. The estimation of the S value using NMR cryoporometry as the surface area in contact with unfrozen water (Table 7.1, S_{uw}) gives very large values, which, however, significantly change with changing hydration degree of cells.

The adhesion forces affect a relatively thick water layer (~3 nm) (Figure 7.12b). This is due to the cell surface structure including polar integral proteins and saccharides in the glycocalyx layer of 20–50 nm in thickness (Chou and Fasman 1978, Davis 2002, Beardsley et al. 2003). However, the $\langle T \rangle$ values (Table 7.1) are relatively large that suggest the absence of very strong confined space effects characteristic for nanopores in activated carbons with $S_{BET} > 1000$ m^2/g. Thus, the main effects of the cell structures are due to polar and charged functionalities in "shaggy" macromolecular cell surface, which provide confined space effects in pores at $R > 0.5$ nm (Figure 7.13).

Incremental PSD (Figure 7.13b) shows that mesopores at $2 < R < 25$ nm and macropores at $R = 25$–40 nm provide the main contribution to the pore volume. However, narrow pores more strongly contribute the specific surface area of cellular structures (Table 7.1, S_{uw} and S_{nano}). Notice that these structural characteristics are related to both intra- and extracellular waters. Notice that the ion percolation effects (i.e., throughout conductivity) are practically absent for the initial dry cells (Figure 7.14).

After swelling of cells in water at room temperature (content of water increases by order of magnitude), the direct current (dc) relaxation is observed at $T > 217$ K. However, the activation energy of dc is relatively high (115 kJ/mol) even at high hydration of cells because of the negative influence of intracellular structures (as traps for charged particles) on the mobility of protons and other ions bound to water molecules.

7.1.2 Structural and Energetic Differentiation of Water in Partially Dried Yeast Cells Affected by Organic Solvents

Microorganisms in anabiosis can be subjected to cold stress and freezing to temperature of liquid nitrogen without damage of them (Crowe et al. 1992, Armitage and Juss 1996, Tunnacliffe et al. 2001). Renewal of the active state after anabiosis occurs on changes in conditions (e.g., appearance of a nutrient medium at appropriate temperature) corresponding to optimum parameters for vital activity and cells growth. Rehydration of cells is accomplished sufficiently rapidly that indicates the accessibility of the internal space of cells for water molecules and dissolved substances (Karlsson 2001). The anabiosis state is widely used on cryopreservation and long storage of cells and tissues at temperature of liquid nitrogen (Wang 2000, Hubálek 2003).

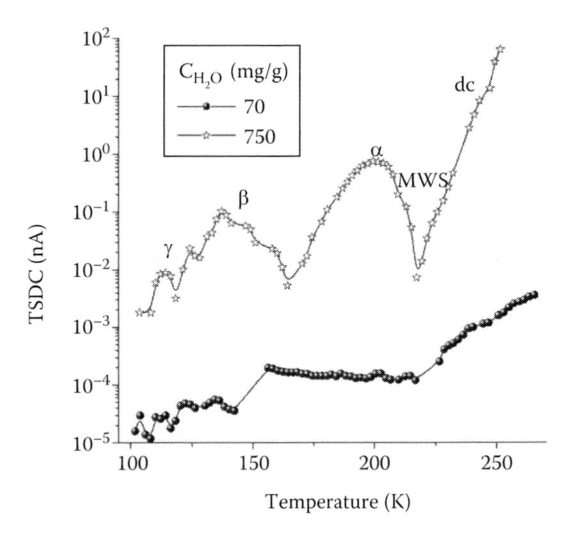

FIGURE 7.14 TSDC spectra of yeast *S. cerevisiae* cells initial (70 mg/g of water) and wetted (750 mg/g of water). (Adapted from *J. Colloid Interface Sci.*, 283, Turov, V.V., Gun'ko, V.M., Bogatyrev, V.M. et al., Structured water in partially dehydrated yeast cells and at partially hydrophobized fumed silica surface, 329–343, 2005, Copyright 2005, with permission from Elsevier.)

Strong dehydration of cells occurs on cryopreservation (utilizing low-molecular cryopreservatives) similar to that under natural conditions of anabiosis of cells. This dehydration is caused by the displacement of a significant portion of intracellular water by organic compounds and formation of the concentrated solutions of cryopreservatives in the cells. Dehydrated cells can be considered as complex systems with connected nanodimensional structures formed by organelles, biomacromolecules, cell membranes, etc., and only a minor portion of "pockets" there (i.e., internal areas in biostructures) is filled by structured water with dissolved low-molecular organic or inorganic compounds. Changes in the environment, e.g., on hydration of cells, lead to appearance of concentration gradients and strong osmotic pressure at the cellular interfaces (membranes and inner structures) that result in fast rehydration of cells. Here, we will analyze the effect of nonpolar (C_6H_6, CCl_4) and polar (CD_3CN, $(CH_3)_2CO$, $(CD_3)_2CO$, $(CH_3)_2SO$, $(CD_3)_2SO$) organic solvents on the structure of intracellular water in yeast cells and the Gibbs free energy of interaction of water with intracellular structures.

Commercial dry yeast *S. cerevisiae* cells (containing <7 wt% of water) were used as the initial material and then wetted by a given amount of water and organic solvents. Before NMR experiments, the same amounts of water (100 mg/g) were added to cell granules (~0.25 mm in diameter) maintained for 7 days at room temperature. Proton-containing and deuterated organic solvents (Aldrich) were added to partially hydrated cell samples (300–400 mg) directly into the NMR ampoules shaken for 10 min and equilibrated for 1 h. Then ampoules were placed into the NMR spectrometer and temperature was decreased to 200–210 K for 15 min, and then it was elevated by a step of 5–10 K and samples were equilibrated at each temperature for 5 min under thermostatic control.

According to the TG/DTA (Figure 7.15) and FTIR (Figure 7.16) data, addition of water (100 mg/g) and then benzene (0.1 or 1.7 g/g) leads to changes in the organization and the temperature behavior of intracellular water.

Added water enhances the endothermic effect on desorption (Figure 7.15c, DTA minimum at 144°C). Despite a DTG minimum (related to a maximal rate of water desorption) shifts toward lower temperatures (by ~5°C), addition of water does not affect a maximum of cell oxidation at 305°C. Benzene desorbs from cells more easily than water (Figure 7.15, sample 4). Benzene slightly

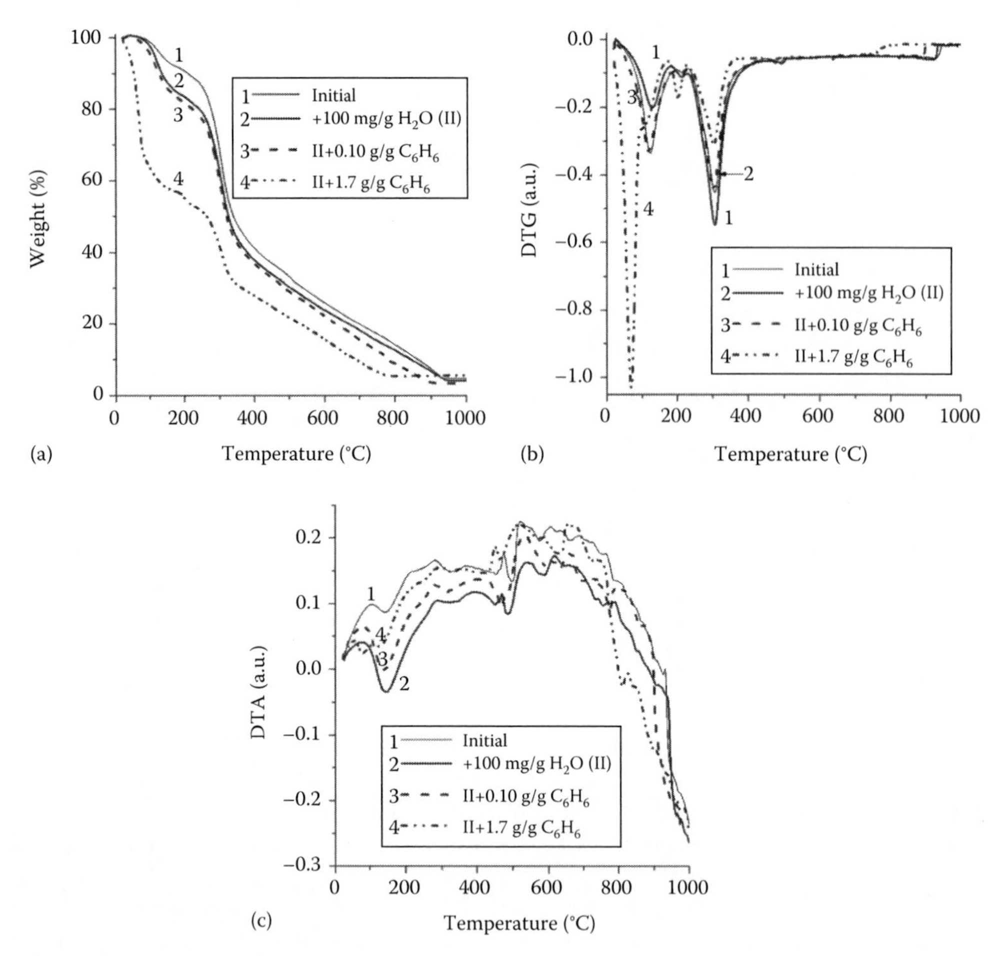

FIGURE 7.15 (a) TG, (b) differential TG (DTG), and (c) DTA curves for initial yeast *S. cerevisiae* cells (curve 1), after addition of 100 mg/g of distilled water (curve 2), and subsequent addition of benzene of 0.1 (curve 3) or 1.7 g/g (curve 4).

shifts the oxidation maximum (by 2°C–3°C) toward lower temperatures. Deconvolution of the FTIR spectra of cells was done over the range of the O–H stretching vibrations at 2200–3800 cm^{-1} (Figure 7.16) into five bands at 3600–3630 cm^{-1} (peak 1, P1), 3500–3550 (P2), 3280–3370 (P3), 2850–2950 (P4), and 2500–2540 cm^{-1} (P5) (in this range, the C–H stretching vibrations were factorized into two to three bands but not analyzed here).

These O–H bands can be assigned to more strongly disturbed OH groups (i.e., more strongly bound and strongly associated waters [SAWs]) with decreasing ν_{OH} values. In other words, the band at 2500 cm^{-1} is connected to the most strongly associated and strongly bound water (SBW, e.g., water interacting with charged functionalities, structures with attached protons, disturbed acidic COOH groups, etc.), but the band at 3600–3630 cm^{-1} is linked to WAW. Several effects are observed on addition of water (sample 2) and benzene (samples 3 and 4) (Figure 7.16). Addition of water diminishes the amounts of WAW because contribution of peaks 1–3 decreases but the amounts of SBW and SAW increase (peaks 4 and 5). Addition of a low amount of benzene affects intracellular water and contribution of the most SBW grows. The increase in benzene content leads to a small decrease in contribution of SBW and WAW. Notice that addition of different organic solvents to partially hydrated cells does not lead to significant injury of cells since their integrity remains. However, in the case of acetone some cells were damaged.

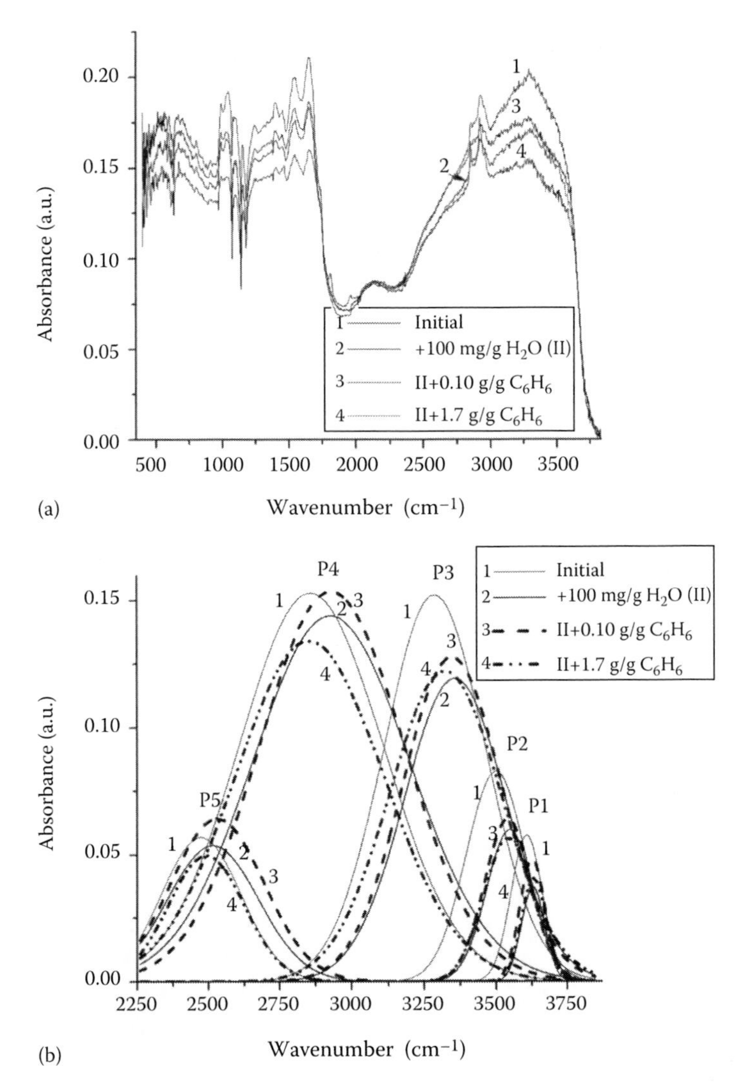

FIGURE 7.16 (a) FTIR spectra of initial yeast *S. cerevisiae* cells (curve 1), after addition of 100 mg/g of distilled water (curve 2), and subsequent addition of benzene of 0.1 (curve 3) or 1.7 g/g (curve 4) and (b) deconvolution of the spectra to five bands over the O–H stretching vibrations.

Figure 7.17d shows the ^{1}H NMR spectra of water in initial yeast *S. cerevisiae* cells (with 70 mg/g of water) recorded at different temperatures. The spectra include a single signal with the chemical shift at 1.5–3 ppm, i.e., a major portion of this water is weakly associated and freezes at $T>250$ K. In decrease in the Gibbs free energy of this portion of structured water is smaller than 1 kJ/mol; consequently, this water can be assigned to WBW. A similar shape and the temperature behavior of the ^{1}H NMR spectra are observed for cell samples containing 170 mg/g of intracellular water.

Yeast *S. cerevisiae* cells are capable to absorb significant amounts not only of organic solvents of various polarity but also of gaseous methane. Figure 7.17a shows the ^{1}H NMR spectra of yeast cells containing 170 mg/g of water after being in methane flow (rate 50 mL/min) for 30 min.

The ^{1}H NMR signal of methane (0.5 wt% with respect to cell weight) at $\delta_H=0$ ppm is well observed in the spectra at $T<250$ K when the signal of water becomes negligible. However, at higher temperatures both signals overlap.

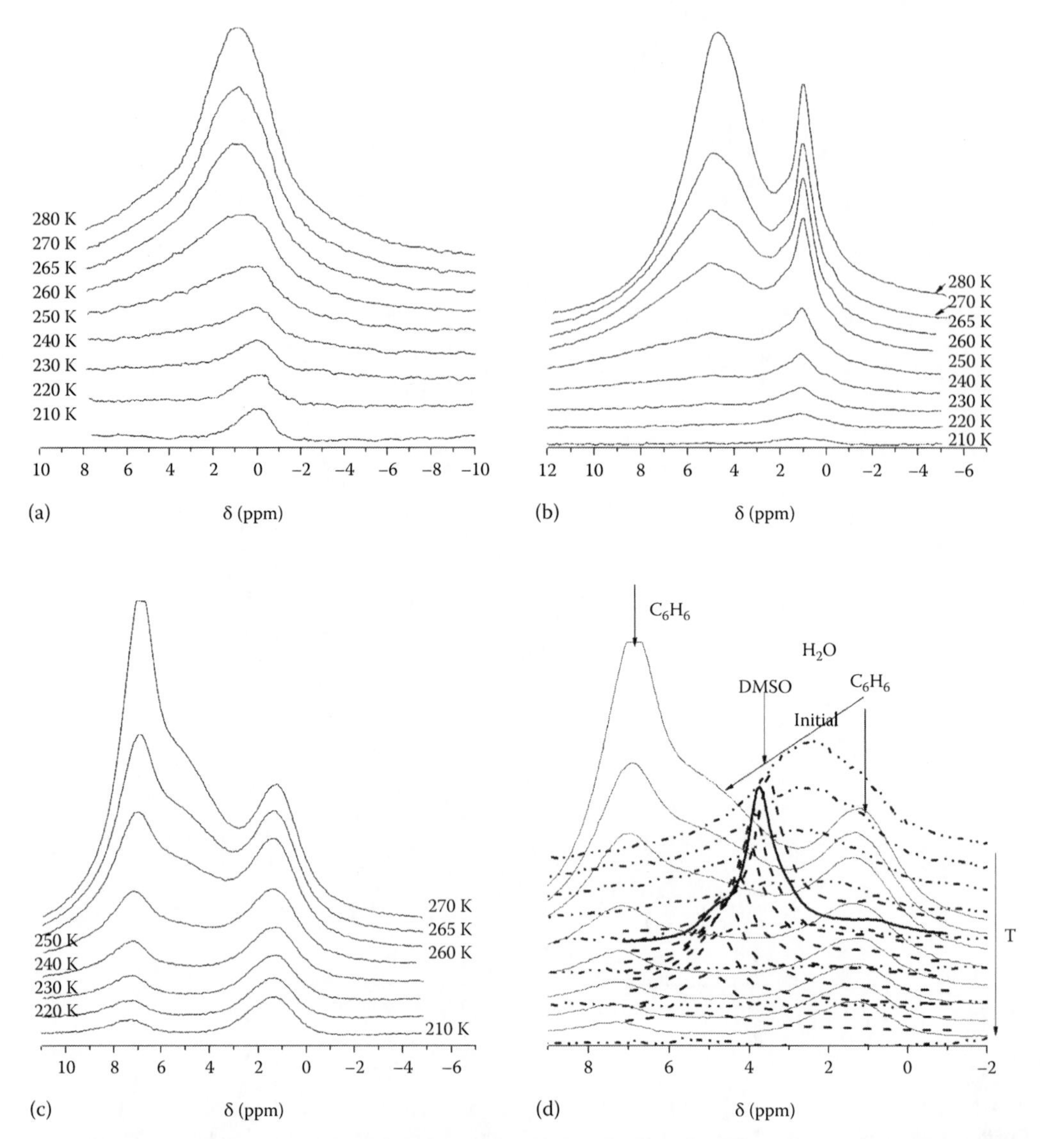

FIGURE 7.17 ^{1}H NMR spectra of intracellular water (170 mg/g) and (a) methane, and for yeast *S. cerevisiae* cells (170 mg/g of water) being in (b) CCl_4; (c) C_6H_6 media at different temperatures; and (d) comparison of the spectra of initial cells and cells in C_6H_6 and DMSO-d_6 media.

Nonpolar solvents (CCl_4 or C_6H_6) can penetrate into yeast *S. cerevisiae* cells. Figures 7.17b and 7.17c show the ^{1}H NMR spectra of water adsorbed in yeast cells placed into CCl_4 and benzene media, respectively (signal of benzene is observed at $\delta_H = 7.2$ ppm). The presence of CCl_4 or benzene in the cells leads to significant changes in the spectra of adsorbed water, i.e., its organization changes. A substantial fraction of intracellular water affected by these solvents becomes strongly associated and characterized by signal at $\delta_H = 5$ ppm. Another portion of water remains weakly associated and characterized by signal at $\delta_H = 1.2$ ppm. In the case of benzene, the signal of SAW is observed as a shoulder on the right wing of the benzene signal. The observed effect of the solvents can be considered as the structured differentiation of intracellular water.

Figure 7.18 depicts the temperature dependences of concentration of weakly associated water (WAW) (Figure 7.18a) and the dependences of changes in the Gibbs free energy of weakly associated (Figure 7.18b) and strongly associated (Figure 7.18c) waters. The replacement of air by the

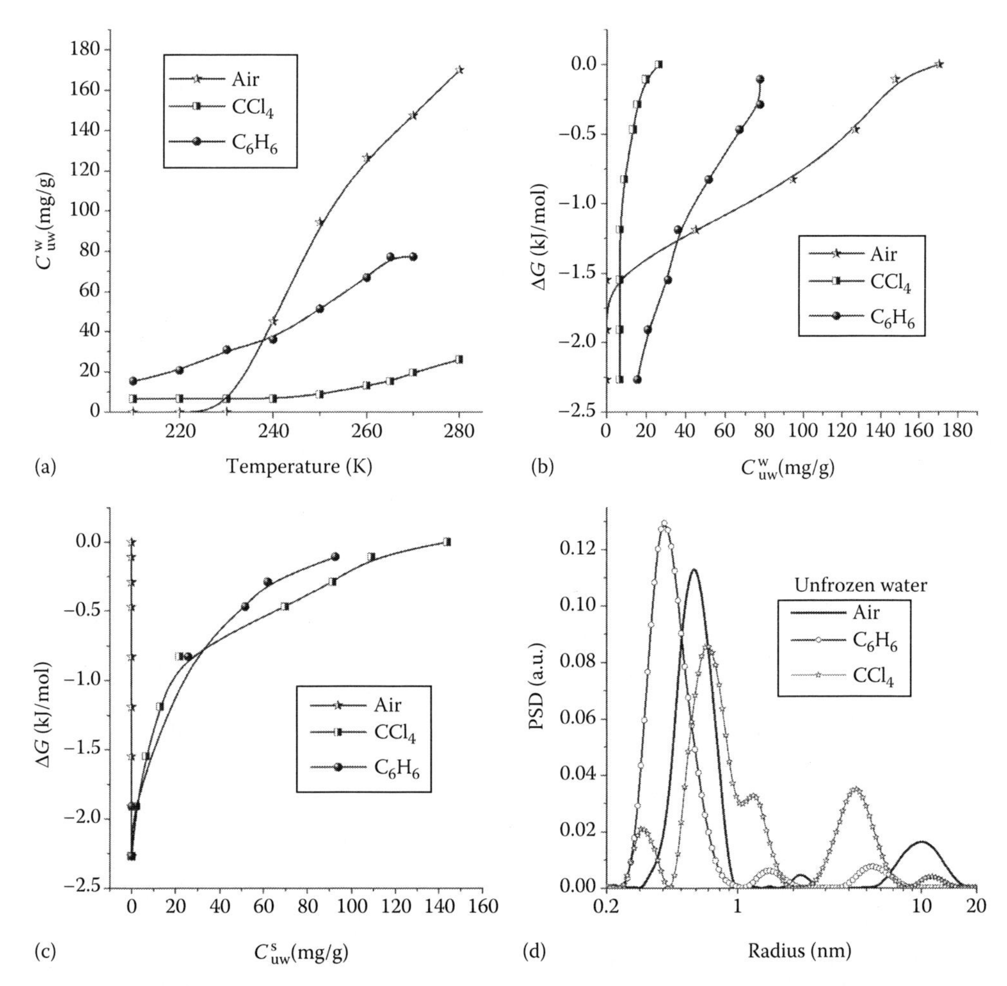

FIGURE 7.18 (a) The temperature dependences of concentration of weakly associated water, and changes in the Gibbs free energy of (b) weakly associated water, (c) strongly associated water in the yeast *S. cerevisiae* cells containing 170 mg/g of water in air, CCl_4, and benzene, and (d) pore size distribution functions calculated using Gibbs–Thomson equation.

nonpolar organic solvents leads to a considerable decrease in the amounts of WAW (Figure 7.18a) (by two times for benzene and five times for CCl_4). This occurs due to the transformation of a fraction of WAW into strongly associated one. The transformation of a portion of WAW occurs from weakly bound to strongly bound one which does not freeze even at 210 K. A significant portion of SAW is weakly bound ($-\Delta G < 1$ kJ/mol).

Thus, energetic differentiation of intracellular water affected by nonpolar organic solvents occurs together with its structural differentiation (Figure 7.18). The observed phenomenon is analogous to that observed during self-organization of the water–organic systems in confined space of pores with hydrophilic/hydrophobic walls in nanoporous activated carbons (Turov et al. 2002c, Gun'ko et al. 2005e), mesoporous silicas (Gun'ko et al. 2007g), and nanosilicas (Turov et al. 2007) (see Chapters 1 and 2). These rearrangements occur under the principle of minimization of the free energy of the systems with consideration for interactions of cell structures with water and organic solvent, as well as water–organic, water–water, and organic–organic interactions. In the case of a nonpolar organic solvent, a minimal surface area of its contact with water is predominant. To reach this, the organic solvent can displace water to larger pores (to form weakly bound but SAW structures such as large

TABLE 7.2
Characteristics of Cell Structures in Contact with Bound Intracellular Water Unfrozen at $T<273$ K

Medium	S_w (m²/g)	$S_{w,nano}$ (m²/g)	$S_{w,meso}$ (m²/g)	$S_{w,macro}$ (m²/g)	$V_{w,nano}$ (cm³/g)	$V_{w,meso}$ (cm³/g)	$V_{w,macro}$ (cm³/g)	γ_S (J/g)
Air	137	114	22	1	0.035	0.107	0.28	8.3
C_6H_6	462	428	33	0	0.100	0.070	0	10.0
CCl_4	162	98	63	1	0.034	0.117	0.019	5.9
$(CD_3)_2CO$[a]	48	6	37	5	0.001	0.047	0.122	19.0
CD_3CN[a]	88	11	77	0	0.004	0.160	0.005	5.1
$(CD_3)_2SO$[a]	79	3	76	0	0.001	0.168	0.001	7.0

Note: C_{H_2O} = 170 mg/g in all samples.
[a] 100 mg/g of organic solvents.

clusters and nanodomains) or to very narrow pores with hydrophilic pore walls (to form strongly bound and WAW in the form of small clusters or even individual molecules).

Figure 7.17c shows that signals of weakly associated but SBW and strongly bound benzene are observed in the cells at low temperatures. The PSD functions (Figure 7.18d) and the structural characteristics (Table 7.2) reveal that the effects of benzene and CCl_4 on intracellular water strongly differ. For instance, benzene and CCl_4 give the displacement of the nanopore peak toward opposite directions and the PSD functions for mesopores differ too since CCl_4 enhances the mesoporosity but benzene increases the nanoporosity (Figure 7.18d and Table 7.2).

Figure 7.19 shows the ^{1}H NMR spectra of water and methyl groups in acetone and DMSO (for the proton-containing solvents) and only water in the case of the deuterated solvents. In contrast to nonpolar solvents, these organic compounds possess high electron-donor ability and can form mixtures with water at any concentration in the bulk. Water dissolved in these organics at excess of them is characterized by the chemical shifts at $\delta_H = 2.5–3.0$ ppm.

Water in yeast *S. cerevisiae* cells in the acetonitrile medium (Figure 7.19a) gives two signals at $\delta_H = 1$ and 3 ppm. These water structures can be assigned to WAW in contact with cell structures and water dissolved in acetonitrile, ASW. The intensity of the latter changes weakly with decreasing temperature from 280 to 250 K and it vanishes at lower temperatures because of freezing-out of the eutectic mixture with water and acetonitrile. The signals of WAW and water dissolved in acetonitrile are observed separately, i.e., the molecular exchange between these water structures occurs slowly in the NMR timescale (Pople et al. 1959). Consequently, these structures containing different types of water are spatially separated. Since the content of acetonitrile in the sample was higher by order than water concentration and the signal of water dissolved in acetonitrile has a large width, one can conclude that water does not pass to the bulk solvent out of cells, i.e., water remains in the intracellular space. As in the case of nonpolar solvents, a fraction of WAW transforms from weakly bound into strongly bound one and does not freeze to 210 K. However, SAW is absent in yeast cells immersed in liquid acetonitrile.

Addition of 100 mg/g of acetone or DMSO (possessing stronger electron-donor properties than acetonitrile) to the cells (Figure 7.19) leads to appearance of the signals of strongly and WAW (signal of methyl groups is also observed in the case of the solvents in the H form). The ^{1}H NMR spectra of water upon the use of deuterated acetone and DMSO are very close at high temperatures (Figure 7.19), some difference is observed at $T<260$ K. At low temperatures WAW is stabilized in the cells containing acetone, but SAW is observed in cells with DMSO.

Water dissolved in DMSO is observed only in the case of great excess of the solvent (Figure 7.19f). At $T=280$ K (lower than the freezing point of liquid DMSO at 291.5 K), there are three signals

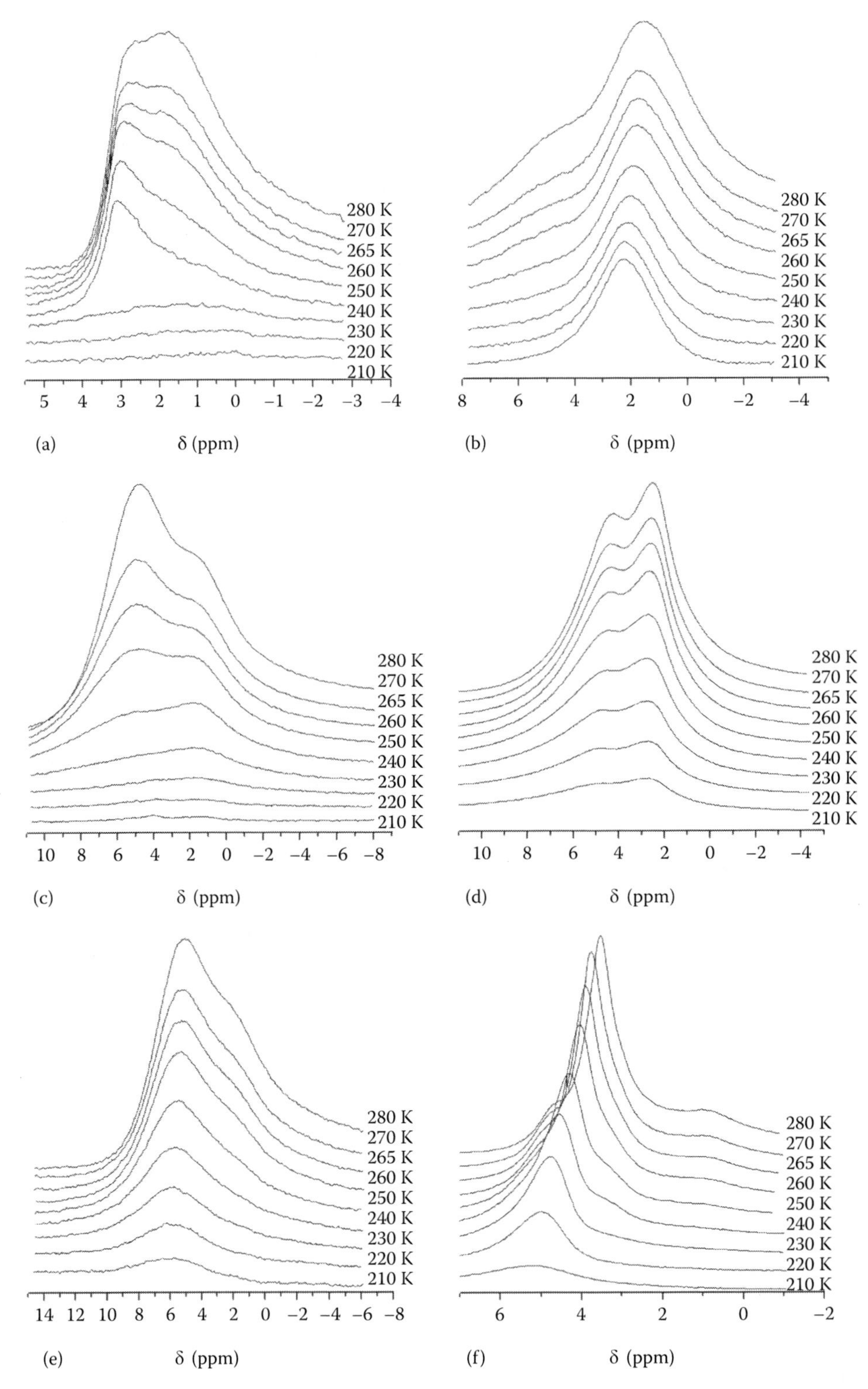

FIGURE 7.19 ^{1}H NMR spectra of water (170 mg/g) in yeast *S. cerevisiae* cells (a) in CD_3CN medium, and on addition of 100 mg/g of (b) $(CH_3)_2CO$, (c) $(CD_3)_2CO$, (d) $(CH_3)_2SO$, (e) $(CD_3)_2SO$, and (f) in $(CD_3)_2SO$ medium at different temperatures.

of water at $\delta_H = 1$, 3, and 5 ppm corresponding to WAW, water dissolved in DMSO, and SAW, respectively. Water dissolved in DMSO transforms into SAW with lowering temperature, probably, because of the DMSO crystallization and pressing-out of dissolved water on the outer surface of crystallites. The chemical shift of this water increases from $\delta_H = 3$ to 5 ppm that indicates the existence of fast molecular exchange between both forms of water. However, a weak signal of SAW ($\delta_H = 5$ ppm) is observed at $260 < T < 280$ K (Figure 7.19f) and its intensity changes weakly in this temperature range. This behavior of the ^{1}H NMR spectra can be explained by the fact that strongly associated intracellular water locates in "pores" or "pockets" of different sizes with the walls of different chemical structures. These pores are spatially separated; therefore, the molecular exchange of water between them does not occur.

Figure 7.20 shows the dependences $\Delta G(C_{uw})$ for WAW and SAW in yeast *S. cerevisiae* cells containing 170 mg/g of water and 100 mg/g of organics and Figure 7.21 depicts the corresponding PSD functions calculated using the Gibbs–Thomson equation and the total content of unfrozen water as a function of temperature $C_{uw}(T)$.

The effects of acetonitrile and acetone lead to transformation of WAW into weakly bound one. However, with the presence of DMSO-d_6 WBW gives ~60% of WAW. The dependences $\Delta G\left(C_{uw}^{w}\right)$ for acetone and acetonitrile coincide but in the case of DMSO-d_6 the corresponding dependence displaced by −0.7 kJ/mol toward the region of lower Gibbs free energy. Approximately half of SAW corresponds to SBW with the presence of acetone and DMSO-d_6 in the cells, and in the case of acetonitrile 90% of SAW is weakly bound. All polar solvents in contrast to nonpolar solvents (Figure 7.18d) displace bound water into larger mesopores (Figure 7.21 and Table 7.2). This difference can be caused by the possibility for these solvents to form mixtures with water at any concentration in the bulk. However, the capability of interfacial water to dissolve organics diminishes (Chaplin 2011). Therefore, the formation of the separated structures with water and organic solvents is possible in the intracellular space. We may conclude that water in the weakly hydrated yeast *S. cerevisiae* cells is in unusual weakly associated state characterized by the chemical shift of proton resonance at $\delta_H = 1$–2 ppm. A significant fraction of this water is weakly bound and it is frozen at $T > 250$ K. The behavior of structured water bound in yeast *S. cerevisiae* cells affected by organic solvents is similar to that in pores with weakly hydrophilic walls of activated carbons or mesoporous silicas. The structural and energetic differentiation of intracellular water occurs under the effects of organic solvents. In the presence of nonpolar CCl_4 and benzene, a substantial part of WAW transforms into

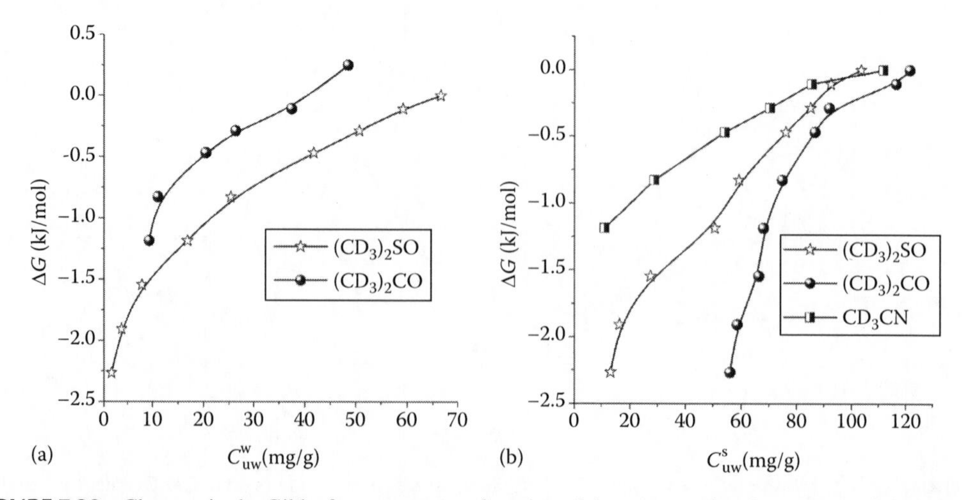

FIGURE 7.20 Changes in the Gibbs free energy as a function of the content of unfrozen water for (a) weakly associated water and (b) strongly associated water in yeast *S. cerevisiae* cells containing 170 mg/g of water and 100 mg/g of $(CD_3)_2SO$, $(CD_3)_2CO$, and CD_3CN.

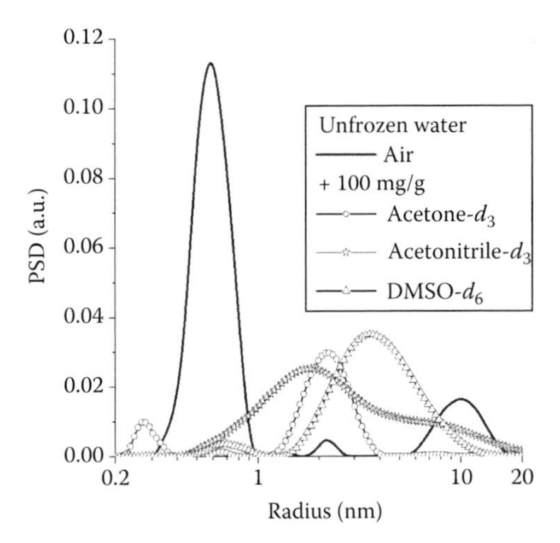

FIGURE 7.21 PSD functions calculated using the temperature dependences of the total content of unfrozen water in yeast *S. cerevisiae* cells containing 170 mg/g of water and 100 mg/g of $(CD_3)_2SO$, $(CD_3)_2CO$, and CD_3CN.

strongly associated one, and another portion of water becomes strongly bound. This occurs to reduce the Gibbs free energy of the system to diminish the contact area between water structures and nonpolar solvents immiscible with water.

Polar organic solvents (acetonitrile, acetone, DMSO) capable to be dissolved in bulk water at any concentrations are characterized by lower solubility in the interfacial or intracellular water. The structural and energetic differentiation of intracellular water occurs in yeast *S. cerevisiae* cells with the presence of these polar solvents. A significant portion of intracellular water transforms into strongly associated state in which water has a minimal contact area with the organic solvent. Consequently, conditions of the formation of mixtures of organic solvents with intracellular water strongly differ from the bulk and over a wide range of concentrations the aqueous and organic solvent phases can exist separately in the intracellular space. The formation of mixtures occurs only at great excess of organic solvents. However, in this case, this mixture exists separately from WAW structures remaining in the cells.

7.1.3 Effect of Disperse Silica on Bound Water in Frozen Cellular Suspensions

During many years, disperse silica has found various applications in pharmaceutics and medicine when creating medicines possessing antimicrobial activity, preparations of prolonged effect, as well as various pastes and ointments used for treating purulent infections (Chuiko 1993, 2001, 2003, Cruz et al. 2009). Being entered into an organism, the particles of nanosilica contact tissues, separate cells, and can penetrate into cells (Nabeshi et al. 2011). Therefore, the investigation of the process of interaction of living cells with silica nanoparticles is of a significant interest. Cellular suspensions, containing colloid silica nanoparticles, can be studied as a model system. The behavior of such systems is frequently subjected to the same regulations as simple colloids (Hermansson 1999). The number of nanomaterials used in medicine and biotechnology is great (Figure 7.22). However, here we will analyze interactions of nanoparticles of a simple type (composed with silica, alumina, and titania alone or related mixed oxides, i.e., particles of type 1 shown in Figure 7.22) with cells, tissues, and seeds.

Here we will analyze the processes of hydration and rehydration, occurring in frozen aqueous suspensions of living cells, containing the additives of disperse silica, and changes in the value of free surface energy due to cell interactions with nanosilica (Turov et al. 2002b,c).

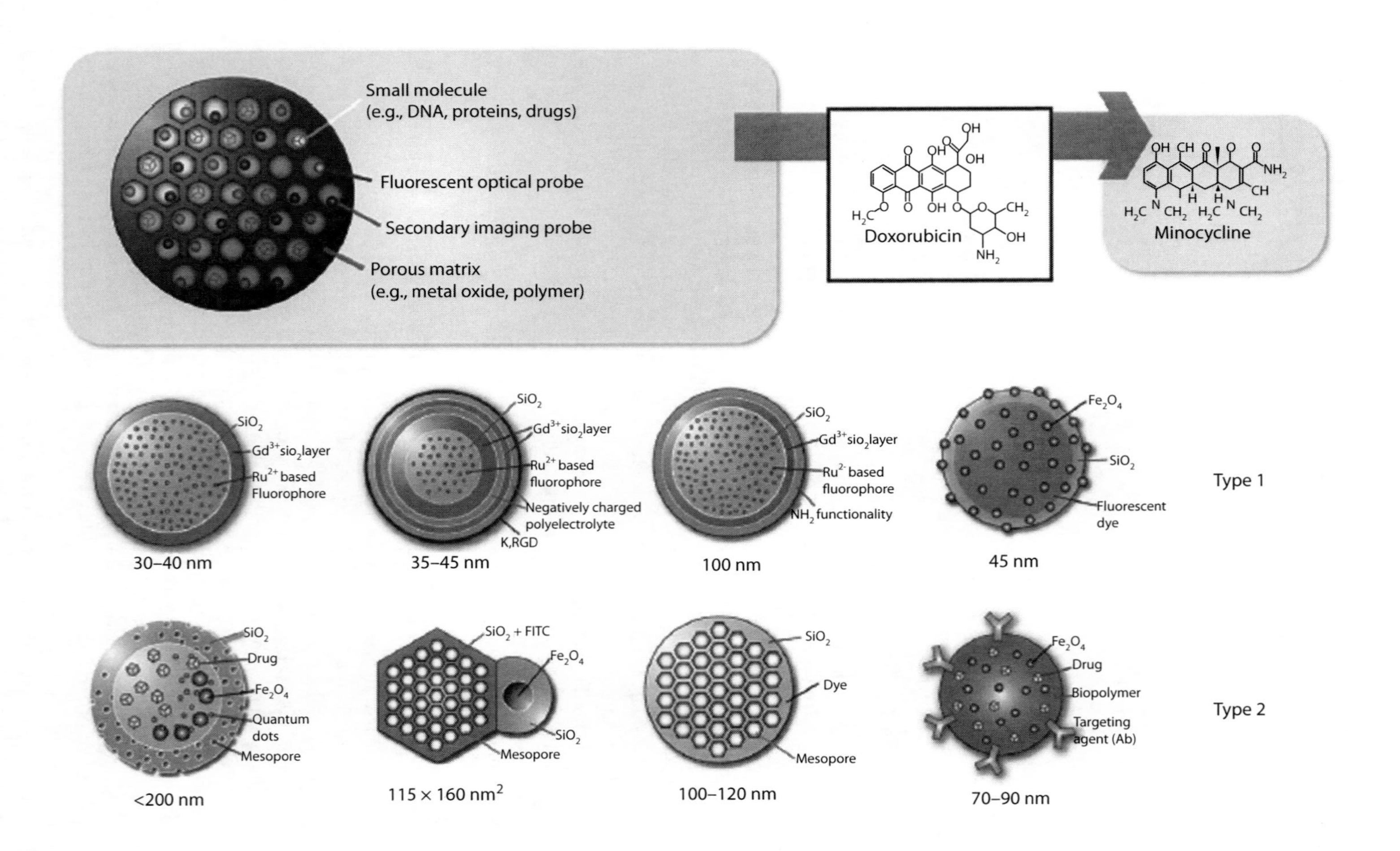

Small molecule
(e.g., DNA, proteins, drugs)
Fluorescent optical probe
Secondary imaging probe
Porous matrix
(e.g., metal oxide, polymer)
Doxorubicin
Minocycline
SiO2
Gd3+ sio2 layer
Ru2+ based
Fluorophore
30–40 nm
SiO2
Gd3+ sio2 layer
Ru2+ based
fluorophore
Negatively charged
polyelectrolyte
K,RGD
35–45 nm
SiO2
Gd3+ sio2 layer
Ru2- based
fluorophore
NH2 functionality
100 nm
Fe2O4
SiO2
Fluorescent
dye
45 nm
Type 1
SiO2
Drug
Fe2O4
Quantum
dots
Mesopore
<200 nm
SiO2 + FITC
Fe2O4
SiO2
Mesopore
115 × 160 nm2
SiO2
Dye
Mesopore
100–120 nm
Fe2O4
Drug
Biopolymer
Targeting
agent (Ab)
70–90 nm
Type 2

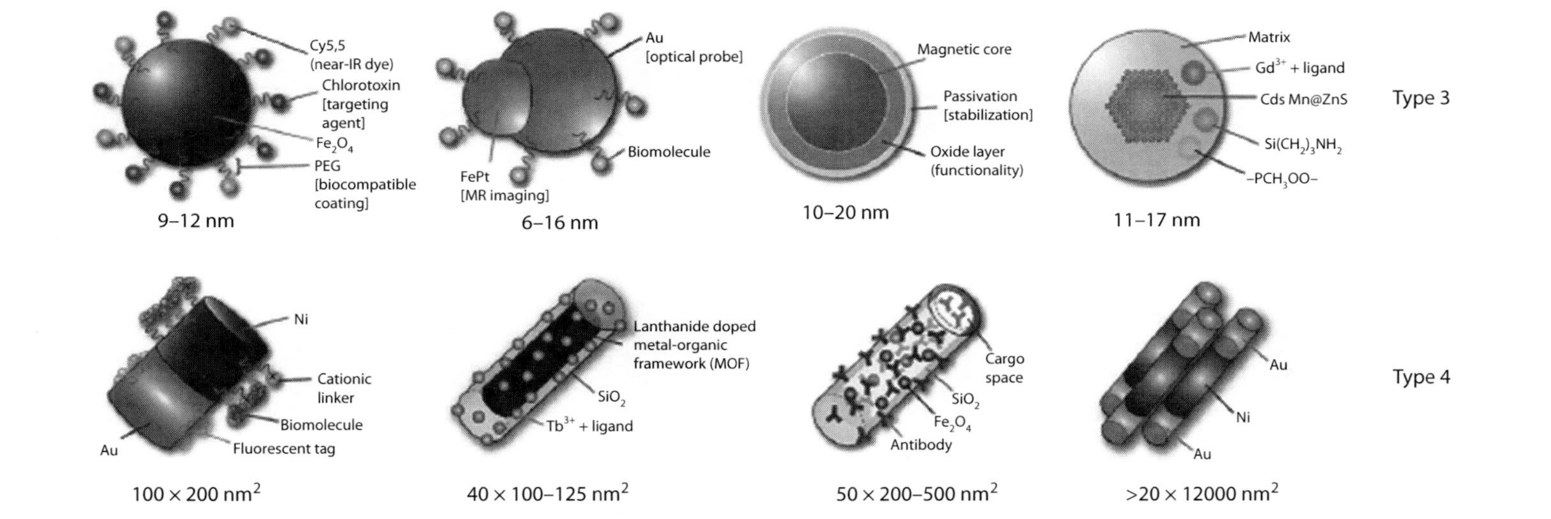

FIGURE 7.22 Multifunctional nanoparticle systems (MFNPS) for biomedical applications. MFNPS can be divided into four distinctive types. Type 1 is non-porous but spherical SiO_2 based sub-100 nm nanoparticles with two or more components. Type 2 is sub-200 nm spherical nanoparticles that is either porous or can incorporate and, in time, release small molecules such as drug molecules. Type 3 is sub-20 nm nanoparticles with functionalizable ligands or biomolecules stabilized (passivated) onto the nanoparticles and are, in most cases, first synthesized in organic conditions which offer good size control and then phase-exchanged to become dispersable in aqueous media. Finally, type 4 is nonspherical nanoparticle systems that have multiple components such as fluorescent tags and antibodies. (Adapted from *Prog. Neurobiol.*, 87, Suh, W.H., Suslick, K.S., Stucky, G.D., and Suh, Y.-H., Nanotechnology, nanotoxicology, and neuroscience, 133–170, 2009, Copyright 2009, with permission from Elsevier.)

Because of the fact that many cells in the process of freeze–thawing die (Lee et al. 1981) for the experiments, yeast *S. cerevisiae* cells were used because they possess a high cryoresistance as mentioned earlier.

It should be noted that in frozen cellular suspensions similar to suspensions of disperse adsorbents and biopolymers, a part of water is unfrozen at $T<273$ K. There is concentrating of the solution of low-molecular substances, presenting in intracellular liquid. Experimentally, it is difficult to divide intra- and extracellular types of unfrozen water and water affected by dissolved low-molecular compounds (e.g., saccharides). Therefore, the value of free surface energy contains some contribution due to changes in water free energy, stipulated by the presence in intracellular liquid of low-molecular substances. It could be expected that on the dependence $\Delta G(C_{uw})$ such a "volumetric" unfrozen water will manifest predominantly as WBW.

The dependence of the concentration of unfrozen water on temperature and related dependence of the change in the free energy on concentration of yeast *S. cerevisiae* cells (C_Y) within the range of $0 \le C_Y \le 25$ wt% at a constant concentration of silica in the suspension (C_{SiO_2} = 5 wt%) and wetted cells (100 wt%) without silica are shown in Figure 7.23.

If at certain fixed temperature ($T < 273$K) the concentration of non-freezing water in aqueous suspension of silica C_{uw,SiO_2} and the concentration of non-freezing water/stipulated by yeast cells $C_{uw,Y=100\%} \cdot C_Y/100$, then change in concentration of non-freezing water, connected to silica presence, is determined by the ratio:

$$\Delta C_{uw} = C_{uw} - C_{uw,SiO_2} - C_{uw,Y=100\%} \cdot C_Y/100 \tag{7.2}$$

The $\Delta C_{uw}(T)$ and $\Delta C_{uw}(C_Y)$ dependencies are presented in Figure 7.24a and b, respectively. Figure 7.24 shows that these dependencies are not monotonous, although in a whole, the silica presence in cellular suspension resulted in a reduction of concentration of non-freezing water in the system.

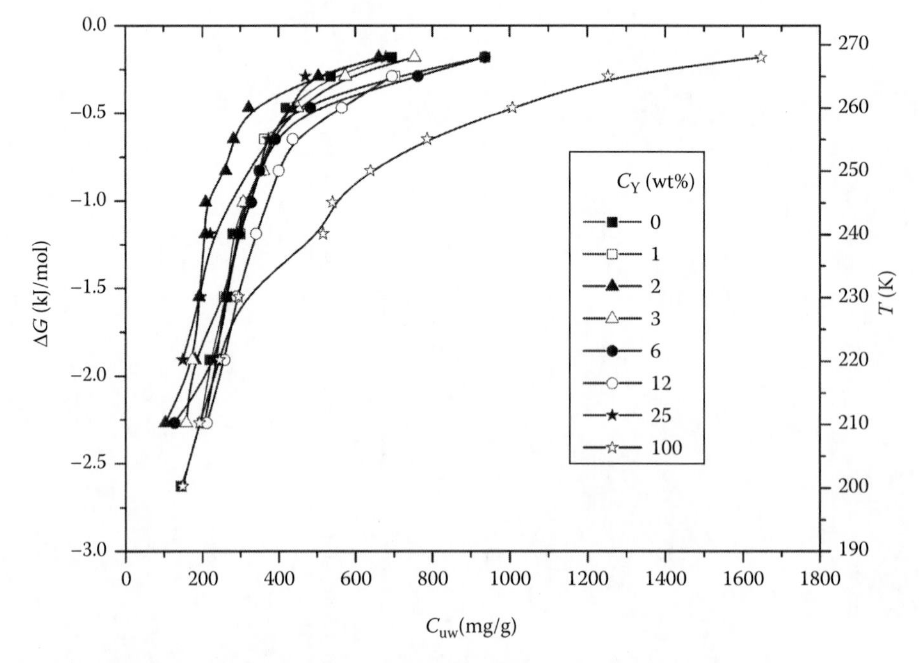

FIGURE 7.23 Dependency of changes in free energy on the concentration of non-freezing water of concentration of non-freezing water for aqueous suspensions of yeast, containing 5 wt% of nanosilica (C_Y = 100 wt% corresponds to wetted cells without silica).

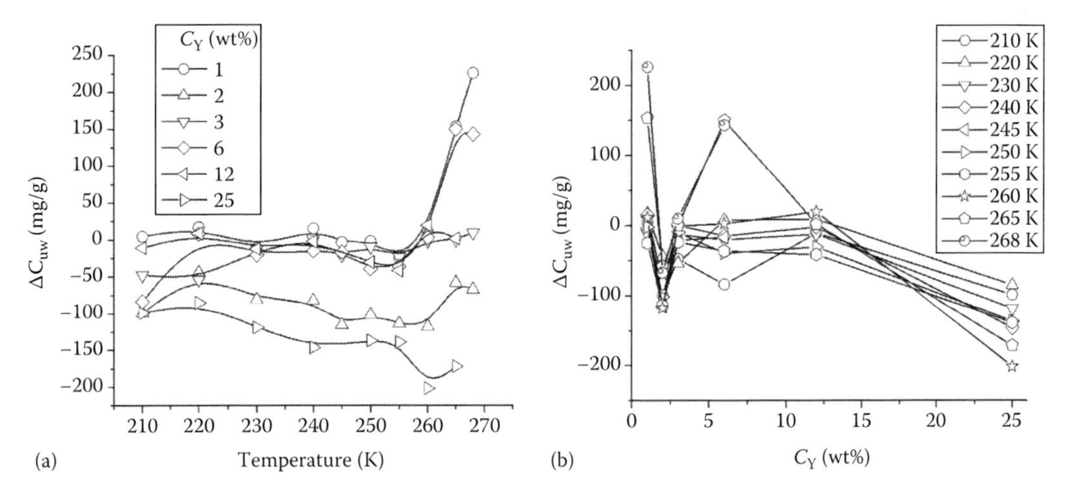

FIGURE 7.24 Effect of temperature (a) and yeast cell concentration (b) on hydration of aqueous suspensions of yeast cells containing 5 wt% of SiO_2.

TABLE 7.3
Characteristics of the Layers of Bound Water in Aqueous Solutions of Baker's Yeast, Containing 5% Mass of Silica (A-400)

C_Y (wt%)	$-\Delta G_s$ (kJ/mol)	$-\Delta G_w$ (kJ/mol)	C^s_{uw} (mg/g)	$C^w_{H_2O}$ (mg/g)	γ_S (J/g)
0	3.75	0.4	420	980	78
1	4.2	0.6	400	1100	83
2	2.7	0.6	250	850	61
3	3.25	0.55	400	900	72
6	2.6	0.75	400	900	72
12.5	3.5	1.2	440	560	55
25	2.8	0.6	550	550	61

In the range of low temperatures with the rise in concentration of cells and temperature increase, there is a tendency to the decrease in the ΔC_{uw} value. The highest dehydration of samples is observed at $C_Y=2$ and 25 wt%. At $T \to 273$ K for all studied samples, there is a tendency of the growth of the ΔC_{uw} value, in addition for the samples, containing 1% and 6% of cellular mass, ΔC_{uw} gets positive.

As Table 7.3 data show, with the growth of C_Y the value C^s_{uw} first reduces and then increases ~1.5 times in respect of initial value, which is the suspension of silica at the absence of yeast. The concentration of WBW with the growth of C_Y from 0 to 25 wt% decreases twice. The concentration of both types of bound water coincides at $C_Y=25$ wt%. If do not take into account an abrupt decrease in the γ_S value at $C_Y=2$ wt%, on the base of the $\gamma_S(C_Y)$ dependence we can conclude that the value of free surface energy with the growth in C_Y reduces, if $C_Y<10$ wt%, and further growth of cell concentration does not result in a significant change in the γ_S value.

In order to determine how the surface of silica affects the value of the cell–aqueous medium free surface energy, one can compare the data, presented in Table 7.3, with the corresponding dependence for binary cellular suspension, mentioned in Turov and Barvinchenko (1997) (Figure 7.25a). The differences in the γ_S values under fixed concentration of cellular mass in the suspension determine the changes in free energy in the result of the cell–surface interaction (Figure 7.25b). The mentioned

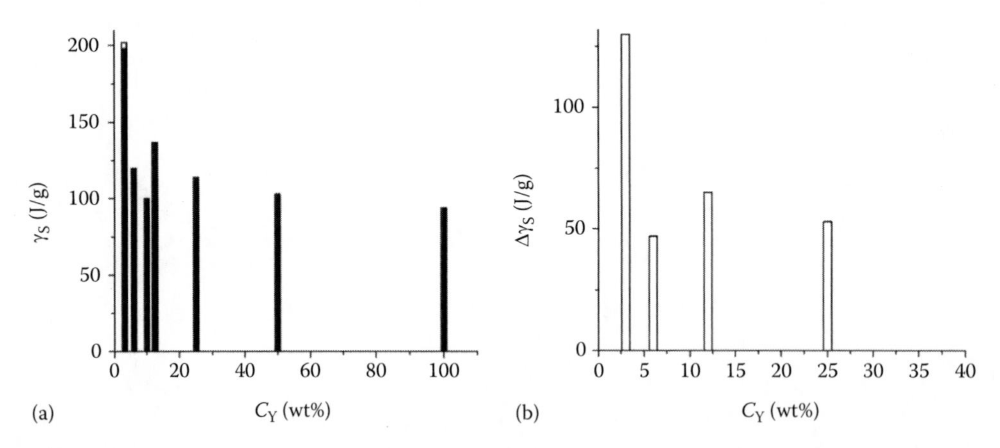

FIGURE 7.25 Effect of concentration of cellular mass in the aqueous suspension on (a) surface free energy in the system containing water and cellular mass (without silica) and (b) changes in free surface energy caused by the presence of highly disperse SiO_2 for the suspension containing 5 wt% of SiO_2.

data demonstrate that the presence of silica sharply decreases free surface energy at the boundary of cell–water phase. This testifies to an intensive binding of cells with silica surface. Especially, large difference is observed in the γ_S values of binary and triple suspensions under low concentrations of cellular mass. The site of $\gamma_S(C_Y)$ to the left from the break (Figure 7.25a) corresponds to a weak intercellular interaction (gel-like phase does not form). The presence of disperse silica particles in suspension results in the decrease in γ_S value by 130 J/g. Then this value determines a maximum energy of cell interaction with silica surface.

Under high concentration due to strong intercellular interactions, cellular suspension forms gel-like phase. The introduction of 5 wt% of disperse silica into such a suspension results in small changes in the free surface energy. Probably within this concentration interval, a major part of the cells participates in the formation of the cell–cell bonds and does not participate in the formation of the cell–surface complexes. The break on the $\gamma_S(C_Y)$ dependence at $C_Y = 2$ wt% in triple suspensions (Table 7.3) and in binary suspensions as well (Figure 7.25a) is rather caused with a phase transition occurring in cellular suspension with the rise in concentration.

For more detailed study of the processes of silica interaction with concentrated cellular suspension, we have investigated the changes in characteristics of the layers of bound water in triple systems containing fixed concentration of cellular mass ($C_Y = 12.5\%$ w/w) and being varied by silica concentration within the range of 0–5 wt%. Figure 7.26 shows the diagrams of dependencies of the changes in free energy on the concentration of non-freezing water and calculated on their base the dependence of the change in the free surface energy on concentration of SiO_2 in the suspension. The characteristics of layers of bound water are summarized in Table 7.4.

The mentioned results demonstrate that with the rise in C_{SiO_2} free surface energy monotonously reduces if $C_{SiO_2} \leq 1$ w/w%. Under following concentration of the suspension, the value γ_S remains constant within a wide range of C_{SiO_2} change. In contrast to the data presented in Figure 7.26a, the decrease in free surface energy is stipulated with the reduction of both WBW and SBW (Table 7.4).

Obtained results testify to the fact that silica is capable to actively affect the phase state of concentrated cellular suspensions. The dependence in $\gamma_S\left(C_{SiO_2}\right)$ in Figure 7.26b can be explained from the method of formation in a triple system containing the particles of highly disperse SiO_2 and cells of yeast of new phase state in which the particles of silica contribute to intercellular interactions. This state is characterized with the 12:1 ratio of cell concentrations and SiO_2. Cell hydration rate for it is observed as twice less when comparing with binary system. The growth silica concentration higher than the equal ratio does not affect the suspension state.

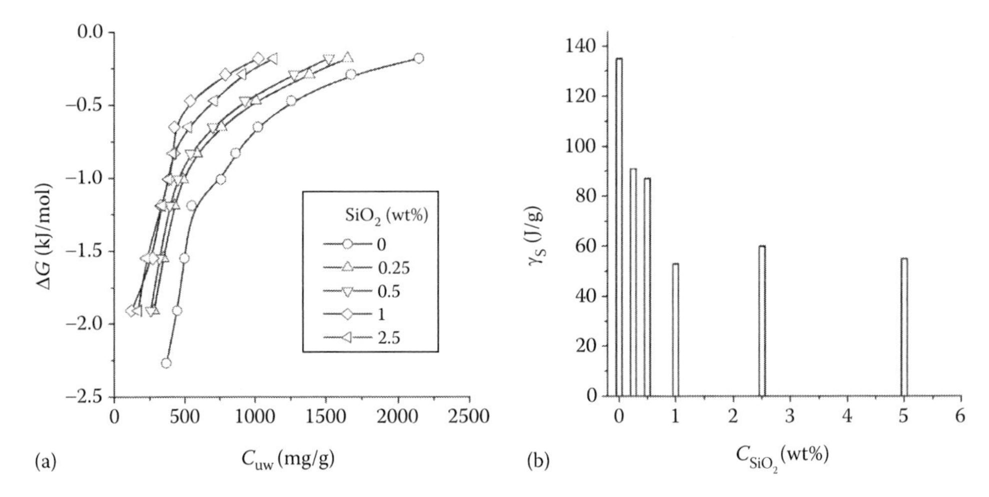

FIGURE 7.26 Dependence in the change of free energy on the concentration of non-freezing water (a) and building on its base the dependence of the change in free surface energy on the concentration of silica (b) for aqueous suspensions containing 12.5 wt% of cellular mass.

TABLE 7.4
Characteristics of the Layers of Bound Water in Colloid System Containing Disperse Silica and 12.5 wt% of Yeast *S. cerevisiae* Cells

C_{SiO_2} (w/w%)	$-\Delta G_s$ (kJ/mol)	$-\Delta G_w$ (kJ/mol)	C^s_{uw} (mg/g)	C^w_{uw} (mg/g)	γ_S (J/g)
0	4.25	0.7	500	2300	135
0.25	3.4	0.9	700	1500	91
0.5	3.75	0.9	600	1600	87
1	2.2	0.5	450	1250	53
2.5	2.8	0.85	470	1030	60

7.1.4 Effects of Nanosilica on Phase Equilibrium in Aqueous Suspensions of Yeast Cells and Gelatin

Numerous investigations of interactions of nanosilica with different microorganisms showed that it can effectively bind them (Hermansson 1999, Chuiko 2003). Microorganisms in this bound state can remain alive. However, their metabolism becomes slower and they can be removed without proliferation. At low content (0.03–0.1 wt%), nanosilica can stimulate cell activity and increase their survivability during cryopreservation but at higher concentrations the opposite influence can be observed. One of the possible mechanisms of negative action of nanosilica on cells is its effective interaction, blocking, and removal of integral proteins from the cell membranes. However, immobilization of proteins or other macromolecules on the silica surface from the solution (before interactions with cells) can reduce the negative effects (Chuiko 2003, Blitz and Gun'ko 2006). Therefore, here the behavior of yeast *S. cerevisiae* cells affected by nanosilica is studied in the presence of gelatin (Gorbyk et al. 2003, Turov et al. 2003). Changes in the phase composition of a complex colloidal system can be analyzed using the concentration dependences of free surface energy (Gun'ko et al. 2005d, 2009d). Typically, if the interaction between disperse particles (solid particles, cells, macromolecules) is weak or practically absent then maximum possible values of γ_S are observed.

TABLE 7.5
Characteristics of Bound Water in the System Containing Yeast *S. cerevisiae* Cells, Gelatin and Water

C_Y (wt%)	C_G (wt%)	$-\Delta G_s$ (kJ/mol)	$-\Delta G_w$ (kJ/mol)	C^s_{uw} (mg/g)	C^w_{uw} (mg/g)	γ_S (J/g)
5	0	3.2	0.55	400	1600	56
5	0.4	2	0.65	750	2000	75
5	0.625	2.4	0.60	600	2900	90
5	1.25	2.5	0.5	650	2350	79
5	5	2.8	0.6	650	1150	67
0.62	0.62	2.4	0.65	900	1900	87
1.25	1.25	2.7	0.8	600	1000	60
2.5	2.5	2.6	0.5	500	1800	68
0	5	3.25	0.75	380	820	60
0.5	5	3.5	0.6	600	900	75
2.5	5	3.25	0.55	500	1200	75
3	0	3.4	0.9	1500	3000	202
4.5	0	5.0	0.8	950	3050	178
6.1	0	3.5	0.9	820	1780	120
10	0	3.0	0.65	600	3250	100

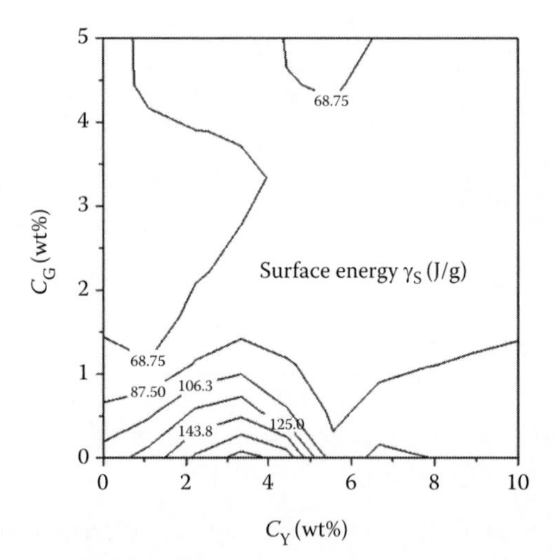

FIGURE 7.27 Dependence of the free surface energy on the amounts of yeast *S. cerevisiae* cells and gelatin in the aqueous suspension.

Table 7.5 and Figure 7.27 show the characteristics of bound water in the system with cells–gelatin–nanosilica.

In a major part of the used concentrations, the system is in gel-like state since small γ_S values are observed. The gel is decomposed with decreasing content of gelatin that appears as an increase in the WBW amounts and the γ_S value. However, the complete decomposition of the gel at low gelatin content does not occur and the γ_S value remains smaller the maximal value (202 J/g) observed at $C_Y=3$ wt% and $C_G=0$ (Table 7.5).

To study the effects of nanosilica, the system at $C_Y=10$ and 2.5 wt% of gelatin with partially decomposed gel structure (Table 7.6 and Figure 7.28). At low content of nanosilica, the γ_S value

TABLE 7.6
Characteristics of Bound Water in the Aqueous Suspensions with 10 wt% of Yeast *S. cerevisiae* Cells and 2.5 wt% of Gelatin at Varied Content of Nanosilica

C_{SiO_2} (wt%)	$-\Delta G_s$ (kJ/mol)	$-\Delta G_w$ (kJ/mol)	C^s_{uw} (mg/g)	C^w_{uw} (mg/g)	γ_S (J/g)
0	3	0.7	1000	2600	140
0.5	3	0.7	1100	1900	114
1	3	0.6	1250	1750	111
1.5	3	0.8	600	800	56
2	2.4	0.8	800	1600	76
2.5	3	0.8	700	1000	66

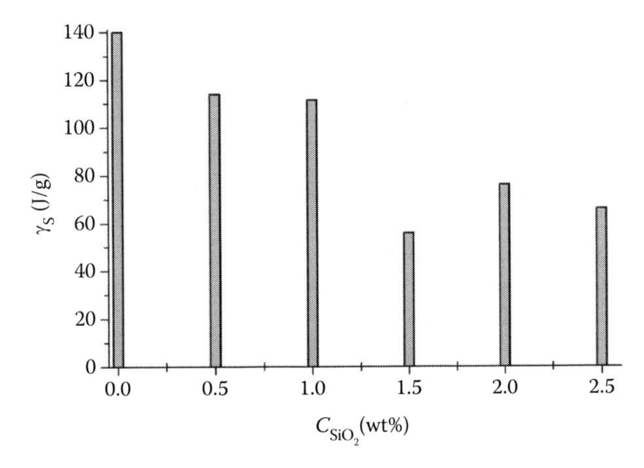

FIGURE 7.28 The effect of nanosilica (A-400) on the free surface energy of yeast cells/gelatin (10 wt%/2.5 wt%) in the aqueous suspension.

decreases because silica enhanced strength of the gel. When the silica content is equal to 1.5 wt%, the surface energy strongly decreases. Subsequent increase in the C_{SiO_2} value results in certain increase (15%–20%) in the γ_S value (Table 7.6).

7.1.5 Influence of Highly Disperse Materials on Physiological Activity of Yeast Cells

Recently, a great attention was paid to investigations of interactions of microorganisms with highly disperse materials capable to essentially influence physiological activity of microbe populations via changes in intensity and direction of biochemical processes, proliferation rate of microorganisms, and biomass growth. In some cases, a decrease in the metabolic activity of microorganisms due to their interaction with nanocomposites has an important biotechnological significance since it favors their survival in and after the anabiosis state (Kurdish et al. 1991, Kurdish and Kigel 1997, Chuiko 2003). However, it is known that the physiological activity of microorganisms can be enhanced due to their contact with solid materials (Zviyginzev 1987).

Krupska et al. (2009) studied the influence of modified silicas and composites with starch hydrogel and modified nanosilicas on the vital activity of yeast *S. cerevisiae* cells. Modified silicas were nanosilica A-300 (S_{BET} = 300 m^2/g) and macroporous silica Silochrome (S_{BET} = 120 m^2/g) with grafted aminopropylmethylsilyl groups substituted approximately one-third of surface hydroxyls. APMS-modified macroporous Silochrome (APMS-MS) and APMS-modified nanosilica (APMS-NS) have a mosaic surface with hydrophilic and hydrophobic functionalities (Turova et al. 2007). Additionally, one of the protons in amino groups of APMS-NS was substituted by phosphoric acid residue (P-APMS-NS). The concentration of surface functionalities was controlled by element analysis (Klimova 1975).

Potato starch, used as additive to affect the yeast cell behavior, as a hydrogel (4–50 wt% of starch) was mixed with nanosilica at a starch/silica ratio of 9:2, 7.8:1, 1.4:1, 1:1.4, and 1:3. The hydrogels were prepared with careful stirring powderlike components and certain amounts of water. Then the mixture was heated to 87°C with intensive stirring to gelatinize starch.

Yeast *S. cerevisiae* cells ("Enzyme," Ukraine) were cultivated in 8% glucose solution (solution for injections, "Farmak," Ukraine) at 28°C. To analyze yeast cells growth, the number of cells was controlled. The solution turbidity and CO_2 elimination measured depend on the number of cells that can be estimated according to a known procedure (Egorova 1976). The aqueous solution (0.1 L) containing 8% of glucose was put in a 0.25 L flask, 1 g of yeast cells was added, and then the flask was blanked with a stopper having a built-in bib. As a result of fermentation, carbon dioxide was formed in the flask and evaporated freely. Therefore, the mass of the reaction vessel decreased that can be determined quantitatively. First, the flask with a sample was weighted with the accuracy of ±0.01g. The reaction vessel was weighted periodically to find the weight of CO_2 eliminated. The number of cells was counted by means of a microscope in a Goryaev chamber. The suspension turbidity was measured from the optical density using a KFK photocolorimeter at $\lambda=670$ nm and the optical length of 0.1 cm using a red filter with the band maximum at 670 nm.

The bionanocomposites have a stimulating impact on the cell fission in contrast to the initial oxides (Table 7.7). The maximum biomass growth (29%–33%) is observed for APMS-NS (Table 7.7, samples 14 and 15). Changes in the concentration ratio of mineral and biopolymeric components in the nanocomposites do not result in a systematic varying their bioactivity. However, it was shown (Turov et al. 2005) that this ratio can strongly influence the conformational state of starch macromolecules.

TABLE 7.7
Influence of Modified Silicas and Their Composites with Starch on Yeast Cell Fission

Sample	Composition	Cell Number (%) to Control[a]
1	Control	100
2	P-APMS-NS	90.60
3	APMS-MS	96.02
4	APMS-NS	83.11
5	Starch (7.8)[b] :NS (1)[b] :water (138.8)[b]	125.90
6	Starch (7.8)[b]:NS (1)[b]:water (72)[b]	131.2
7	Starch (7.8)[b]:NS (1)[b]:water (12.2)[b]	116.40
8	Starch (1)[b]:NS (3)[b]:water (30)[b]	120.40
9	Starch (1)[b] :NS (1.4)[b]:water (20)[b]	122.50
10	Starch (1.4)[b]:NS (1)[b]:water (14)[b]	108.80
11	Starch (4.5)[b]:NS (1)[b]:water (16)[b]	121.50
12	Starch (1)[b]:APMS-NS (3)[b]:water (30)[b]	123.80
13	Starch (1)[b]:APMS-NS (1.4)[b]:water (20)[b]	118.20
14	Starch (1.4)[b]:APMS-NS (1)[b]:water (14)[b]	133.50
15	Starch (4.5)[b]:APMS-NS (1)[b]:water (16)[b]	129.30

[a] Control sample corresponds to the suspension of cells in the 8% solution of glucose.
[b] Weight ratio of components.

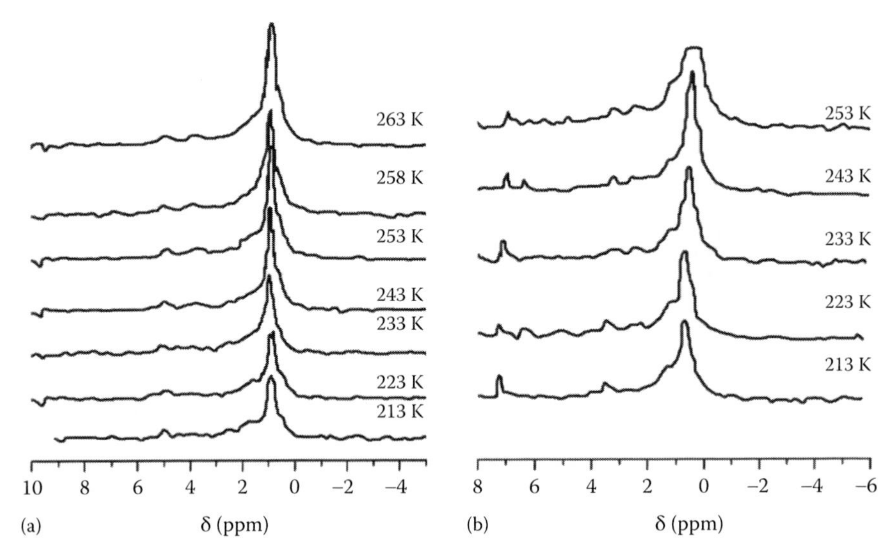

FIGURE 7.29 ^{1}H NMR spectra of water bound in partially hydrated yeast *S. cerevisiae* cells containing 110 (a) and 270 (b) mg/g of water.

The ^{1}H NMR spectra of water in partially hydrated yeast *S. cerevisiae* cells (Figure 7.29), silica powder (prepared by drying of the aqueous suspension) containing 4 wt% of bound water (Figure 7.30a), and the composites consisting of a dried mixture (2:1) of yeast cells and silica including various quantities of bound water (Figure 7.30b through d) are shown. The measurements were carried out in weakly polar chloroform medium which considerably reduced the NMR signal width and allowed us to determine the chemical shifts of protons in bound water, as well as stabilized the water state both in yeast cells and at the interface of cell–highly disperse silica particles.

The water signals in yeast *S. cerevisiae* cells (Figure 7.29) containing 110 or 270 mg/g of water are observed as single signals at $\delta_H \approx 1.3$ ppm. All intracellular water is strongly clustered, attributed to WAW but SBW, i.e., it undergoes a strong disturbing effect of the biosurroundings.

A few signals are observed in the ^{1}H NMR spectra of the silica powder (prepared by drying of the aqueous suspension) containing about 40 mg/g of water and placed in $CDCl_3$ medium. The most intensive signal at $\delta_H = 4.2$–4.8 ppm is due to water adsorbed onto the silica surface. Despite relatively low water content, this signal corresponds to SAW. This can be explained by location of water molecules in the contact zones in voids between adjacent primary particles, i.e., water does not form a continuous layer at the silica surface. Other signals at $\delta_H = 7.26$, 5, and 1.3 ppm are linked to the protons in CH groups of non-deuterated chloroform present as an admixture, water in nanodrops dispersed in the chloroform phase, and bound WAW, respectively.

For the composite system containing cells and silica, dehydrated under the same conditions as silica suspension, at $C_{H_2O} = 120$ mg/g (Figure 7.30b), more than 90% of the total quantity of bound water can be attributed to WAW. Nevertheless, unlike the sample of pure cells (Figure 7.29), an intensive signal of SAW (but much weaker than WAW) is registered at $C_{H_2O} = 120$ mg/g (Figure 7.29b). An increase in the water content results in significant increase in the signal of SAW (Figure 7.30c and d), and these systems include about 30% and 20% of WAW, respectively.

Thus, both initial and modified nanosilica strongly influence the structure of interfacial water layers in partially hydrated cells. When silica nanoparticles contact with cell surface, not only a decrease in concentration of bound water takes place caused by cell–surface interactions (Gun'ko et al. 2006d), the displacement of boundary water, and dehydration of both surfaces but

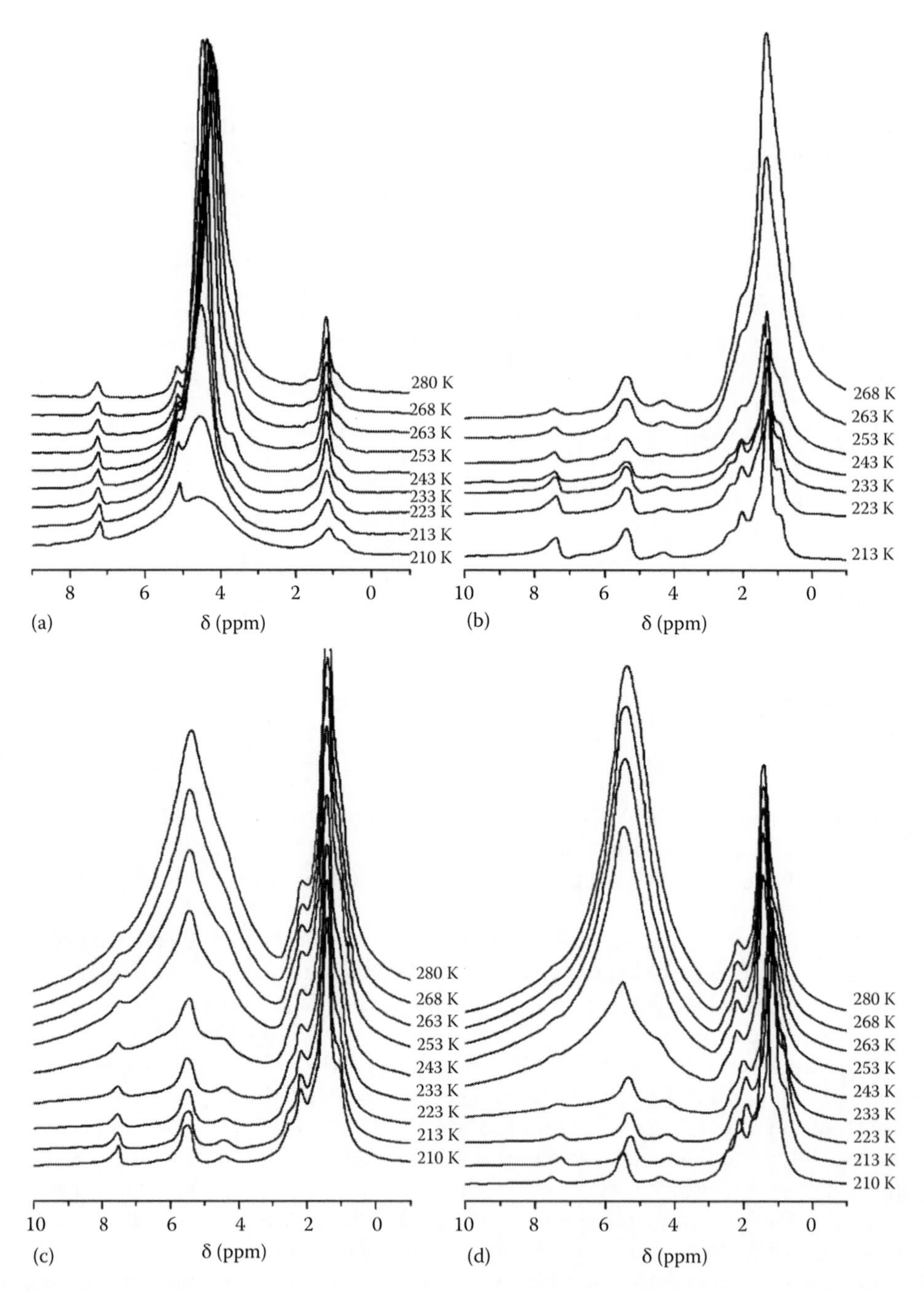

FIGURE 7.30 ^{1}H NMR spectra of water bound in (a) nanosilica powder prepared by drying of the aqueous suspension (residual water of 40 mg/g), and dehydrated composites at cells:silica = 2:1 at different C_{H_2O} = (b) 120, (c) 220, and (d) 320 mg/g.

also changes in the structure of interfacial water occur. The most part of water localized at the cell–silica interfaces transforms from SAW to WAW. Therefore, silica particles locating at the cell surfaces can strongly influence the penetrability of the cellular membranes since WAW presents a weak barrier for both polar and nonpolar substances and promotes the intensification of the cellular metabolism processes. This activates the vital functions of the cells and results in strengthening gas release and biomass growth that is observed experimentally.

Reasoning from the results obtained, one can draw a conclusion on the considerable stimulating influence of modified silicas and related nanocomposites on the vital activity of yeast *S. cerevisiae* cells. The mechanism of this effect is enough complicated. It is conditioned, to all appearance, by the transition of the water–organic systems located at the cell–silica interfaces into a clustered state characterized by the presence of WAW. This is a peculiar state of matter where interfacial water becomes easily permeable for both polar and nonpolar organic substances that promotes the cellular metabolism processes. Nevertheless, despite the vital activity (CO_2 elimination) of cells increases in case of the use of both nanosilica and related nanocomposites, the reproductive performance of cells increases only on the use of starch–silica nanocomposites.

Thus, application of bio-nanostimulators gives an opportunity to control different aspects of cellular metabolism directed to both increase of cell biomass and intensification of the process of fermentation. These properties of nanosized biostimulators can be utilized in various biotechnological processes.

7.2 INTRACELLULAR WATER IN PARTIALLY DEHYDRATED BONE MARROW CELLS

Before cryopreservation of cells, significant dehydration of them is utilized whereas concentration of intracellular water should be decreased by several times to reduce damage of cells by intracellular ice crystallites (Karlsson and Toner 1996, Hubálek 2003, Yoon et al. 2003, Han et al. 2005). Optimal dehydration of cells performed without damage of intracellular structures provides fast penetration of water into intracellular space upon renewal of vital activity of the cells. Many microorganisms can change their state from anabiosis to normal state, rapidly restoring their vital activity (Feofilova 2003). However, transformation of functionally differentiated cells into anabiosis state is more difficult because of their poor adaptability to deep dehydration and the absence of the specific dormancy mechanisms characteristics for microorganisms, some plants, etc. Therefore, development of effective methods for storage of functionally differentiated cells is of practical interest. The problems appearing during these treatments of cells can be partially solved by the use of some substance mediators or cryopreservatives to reduce the negative effects of both dehydration of cells and the formation of ice crystallites. Cryopreservatives can (i) retain and/or displace a certain quantity of intracellular water; (ii) affect structure of water; (iii) change the colligative properties (Chaplin 2011) of bound water; (iv) affect formation of ice crystallites in confined intracellular space; and thus (v) prevent too deep dehydration and damage of cells during cryopreservation. Therefore, development of methods to control changes in the state of intracellular water upon dehydration/rehydration and addition of cryopreservatives is of interest.

It should be noted that the physicochemical properties of water being in confined space can strongly differ from that of bulk water because of the effects of strong electrostatic fields produced by surface (macromolecules) atoms changing interfacial water structure (i.e., hydrogen bond network) (Derjaguin et al. 1987, 1989, Israelachvili and Wennerstrom 1990, Israelachvili 1994, Gun'ko et al. 2005d). Addition of nonpolar, weakly polar, or polar organic liquids to partially dehydrated cells causes different changes in the structures of both macromolecules and intracellular water because of the difference in the interaction of water and organics with hydrophilic and hydrophobic intracellular functionalities (Gun'ko et al. 2005d, 2007i). In the bulk, water nanodomains have the size close to 3 nm (Chaplin 2011) but in narrow nanopores these structures undergo transformations into smaller clusters, the characteristics of which depend on concentration of water, the nature of pore walls, surface functionalities, the type and content of other liquids and dissolved compounds (e.g., salts), and other conditions (Gun'ko et al. 2005d, 2007i, 2009d, Turov and Gun'ko 2011). The associativity of intracellular water (i.e., the average number of the hydrogen bonds per molecule) decreases with decreasing its amount during dehydration of cells as well as on addition of organic liquids (Gun'ko et al. 2005d).

Investigations of functionalized nanooxides, mesoporous and nanoporous silicas, carbons, and some bio-objects showed (Gun'ko et al. 2005d, 2009d; see Chapters 1–3 in this volume) that a significant fraction of bound water can be transformed into WAW characterized by $\delta_H = 1.2–1.7$ ppm (characteristic for alone molecules, small water clusters [or any OH groups] surrounded by nonpolar or weakly polar functionalities) depending on hydrophobic–hydrophilic balance of surface or macromolecular functionalities and the pore size distribution. Such water was found in partially dehydrated yeast cells, seeds, porous component of bone tissues, and other bio-objects described in this book. There are some questions related to the influence of organic compounds (e.g., used as cryopreservatives) on the characteristics of intracellular water. Turov et al. (2009) analyzed features of intracellular water bound in partially dehydrated bone marrow cells of broiler chickens depending on addition of weakly polar or polar organic solvents to model the effects observed during cryopreservation of cells.

Bone marrow cells (Muller-Sieburg et al. 2002, Ulumbekov and Chelyshev 2002, Kondo et al. 2003, Theml et al. 2004, Rubin and Strayer 2007) released from tubular bones (stored for 12–14 h at 277 K) of broiler chickens are of several types (Figure 7.29). Bone marrow cells were received by washout of marrow from tubular bones by the physiological buffer solution (0.01 M phosphate buffer with NaCl solution, pH 7.2–7.4, PBS) with additions of heparin and gentamicin. The suspension of cells was centrifuged (10 min, 1000 rpm) at room temperature, supernatant liquid was removed, and cell residue was resuspended in the PBS and then centrifuged for 10 min (1000 rpm). The residue with cells was carefully dried using a jet pump at 295 K and used as the initial material (corresponding to 17.4 wt% of initial weight) containing ~20 wt% of intracellular water.

Figure 7.31 shows a micrograph recorded using a phase contrast microscope (MIKMED 2, LOMO, Russia, magnification 1000×) of bone marrow cells of different morphology. The NMR measurements were carried out without separation of these cells. The initial, partially dehydrated cells (100–200 mg) were placed in a NMR ampoule and required quantity of water and an organic solvent was added and equilibrated for 1–2 h.

The ^{1}H NMR spectra (recorded at different temperatures) of water bound in partially dehydrated bone marrow cells initial (Figure 7.32a) and after addition of certain amounts of chloroform-*d* (Figure 7.32b and c) strongly differ. Intracellular water bound in the initial cells appears as a singlet at $\delta_H \approx 5$ ppm which is close to the signal of liquid bulk water. Therefore, it can be assigned to SAW

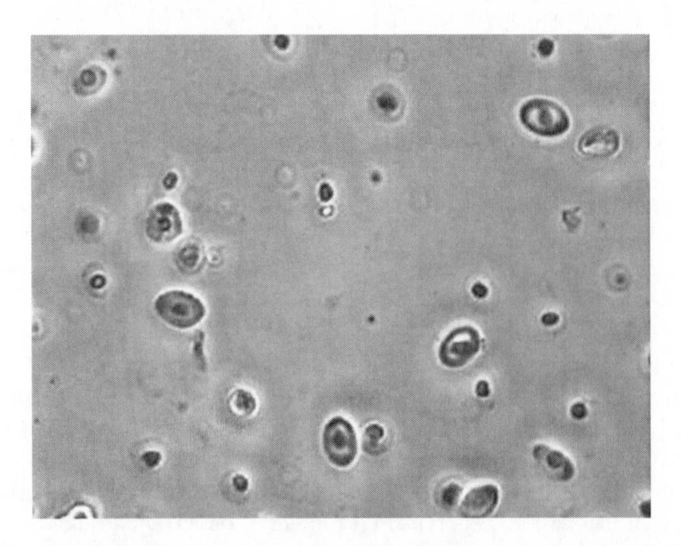

FIGURE 7.31 Microphotography of bone marrow cells of different morphology (1000×). (Adapted from *Cryobiology*, 59, Turov, V.V., Kerus, S.V., and Gun'ko, V.M., Behaviour of water bound in bone marrow cells affected by organic solvents of different polarity, 102–112, 2009, Copyright 2009, with permission from Elsevier.)

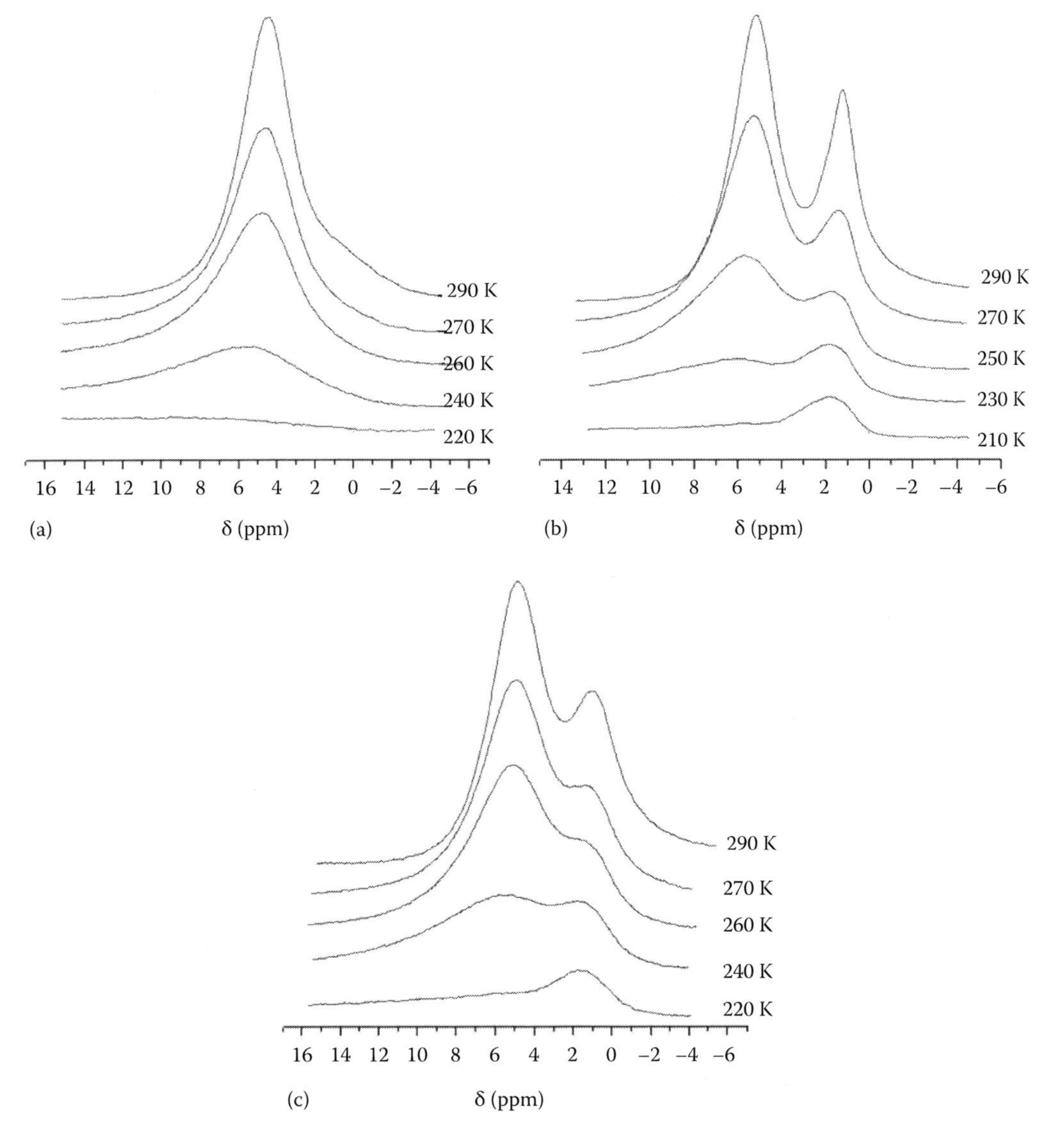

FIGURE 7.32 ^{1}H NMR spectra of unfrozen water in partially dehydrated bone marrow cells containing 20 wt% of residual water (a) initial and after addition of (b) 1 g/g and (c) 2 g/g of $CDCl_3$. (Adapted from *Cryobiology*, 59, Turov, V.V., Kerus, S.V., and Gun'ko, V.M., Behaviour of water bound in bone marrow cells affected by organic solvents of different polarity, 102–112, 2009, Copyright 2009, with permission from Elsevier.)

(Gun'ko et al. 2005d). However, this signal can be divided by deconvolution of the spectra into narrower and broader components at practically the same δ_H values ($\Delta\delta_H \approx 0.3$ ppm) observed at $T > 250$ ppm (Figure 7.32a, deconvolution was shown at $T = 290$ K and 250 K, Turov et al. 2009) because of different confined space effects in cells. The signal intensity decreases but the signal width increases with lowering temperature because of decreasing molecular mobility of bound water (Abragam 1961), and it is not observed at $T < 230$ K. Consequently, SAW includes two fractions with SBW and WBW. Additionally, on the right wing of the signal of SAW a low signal of WAW appears at $\delta_H \approx 1$ ppm, whose intensity rapidly diminishes with lowering temperature and it is not observed at $T < 250$ K, i.e., this WAW can be attributed to WBW since $\Delta G > -0.8$ kJ/mol.

After the addition of weakly polar $CDCl_3$, the ^1H NMR spectra of intracellular water substantially change because a fraction (~30%) of SAW transforms into WAW characterized by $\delta_H \approx 1.0$–1.7 ppm. Contribution of this water increases with increasing C_{CDCl_3} value. A certain fraction of this water does not freeze on cooling to 210 K; therefore, it can be identified as SBW at $\Delta G < -0.8$ kJ/mol (Gun'ko et al. 2005d). However, another portion of WAW is frozen at $T < 250$ K, i.e., it can be assigned to WBW ($\Delta G > -0.8$ kJ/mol). Similar to WAW, SAW ($\delta_H = 4$–5 ppm) can be divided into SBW (unfrozen at $T < 250$ K) and WBW (frozen at $T > 250$ K) (Figure 7.32). However, a fraction of SBW in WAW is larger than that in SAW, and only WAW is unfrozen at 210 K (Figure 7.32) at $\Delta G < -2$ kJ/mol. At $C_{CDCl_3} = 2$ g/g, the signal width is smaller than that for the initial cells at $C_{CDCl_3} = 1$ g/g, and the singlet at $\delta_H \approx 5$ ppm does not have broader and narrower components in contrast to the initial system or after addition of acetonitrile-d_3 or DMSO-d_6. This is due to the difference in the hydrophilicity of the solvents since chloroform as practically immiscible with water (in contrast to acetonitrile and DMSO) more strongly prevents the exchange processes between neighboring water clusters in cells.

Addition of acetonitrile-d_3 as a stronger electron donor than chloroform (donor number $DN_{CH_3CN} = 59$ kJ/mol and the permittivity $\varepsilon = 35.7$; DN_{CHCl_3} is close to zero but $\varepsilon = 38.4$, and $DN_{H_2O} = 75$ kJ/mol at $\varepsilon = 78.4$; Gutmann 1978, Tabata et al. 1994, Onjia et al. 2001) to cells leads to the effects (Figure 7.33) similar to those observed on the addition of chloroform-d. However, acetonitrile in contrast to chloroform (immiscible with water as less than 0.7 wt% of water can be dissolved in chloroform at $T < 293$ K) can be mixed with water in the bulk at any proportion.

For samples shown in Figure 7.33, entire amounts of acetonitrile were absorbed by cells but at $C_{CD_3CN} = 0.8$ g per gram of dry cells its certain amount remained out of the cells. The signal of WAW increases with increasing C_{CD_3CN}, and its relative intensity increases with lowering temperature. Consequently, a significant fraction of this water represents SBW. Contribution of SAW into SBW is smaller than that for the initial cells or after addition of chloroform-d (Figure 7.32), i.e., acetonitrile reduces the interaction of water with intracellular functionalities stronger than chloroform-d (Table 7.8, γ_S). However, on addition of acetonitrile-d_3 two components at close δ_H values ($\Delta\delta_H < 1.5$ ppm) but at different widths can be found in the signal of SAW at $T > 250$ K.

A different picture is observed on addition of DMSO-d_6 (which is the strongest electron donor at $DN = 124$ kJ/mol and $\varepsilon = 46.8$ [Gutmann 1978, Tabata et al. 1994, Onjia et al. 2001] among the studied solvents) to cells (Figure 7.34). Notice that DMSO can be used as an intracellular cryopreservative (Tsuraeva 1983). Although the signal of WAW appears in the spectra, its intensity is low (<15% of the total intensity), and in contrast to the effects of chloroform-d and acetonitrile-d_3 (Figures 7.32 and 7.33), a major fraction of WAW corresponds to WBW frozen at $T < 250$ K ($\Delta G > -0.8$ kJ/mol). At the same time, a major fraction of SAW becomes very strongly bound since it freezes at lower temperatures and a significant portion of this water is unfrozen at 210 K ($\Delta G < -2$ kJ/mol) even at $C_{DMSO} = 0.2$ g/g (Figure 7.34). One can expect that addition of DMSO destroys the hydrogen bond network in intracellular SAW because of easy mixing of DMSO with water in the bulk (notice that $\delta_H = 3.5$ ppm is characteristic for water dissolved in DMSO). However, despite high concentration of DMSO, the δ_H value of SAW slightly increases (from 5 to 5.5 ppm) with lowering temperature (Figure 7.32). Notice that poor mixing of DMSO with confined water was also observed on their competitive adsorption on porous solid adsorbents (Gun'ko et al. 2005a). Notice that narrow and broader components ($\Delta\delta_H < 1.5$ ppm) are observed in the signal of SAW at $T > 250$ K.

The chemical shift of WAW (Figure 7.32) is close to that of aliphatic groups in phospholipids at $\delta_H = 1$–2 ppm. Therefore to analyze possible contribution of phospholipids (or other low-molecular intracellular organics) to the signals of WAW, the ^1H NMR spectra were recorded for molecular crystals of lecithin (as a representative of lipids) with adsorbed water from air at 293 K and the solution (3 wt%) of hydrated lecithin ($h = 3$ g/g) in deuterochloroform (Figure 7.35). In a dehydrated cellular material, phospholipids are in low-mobile state similar to that in the crystalline state. However, on destruction of cellular membranes a fraction of phospholipids can be in the solution.

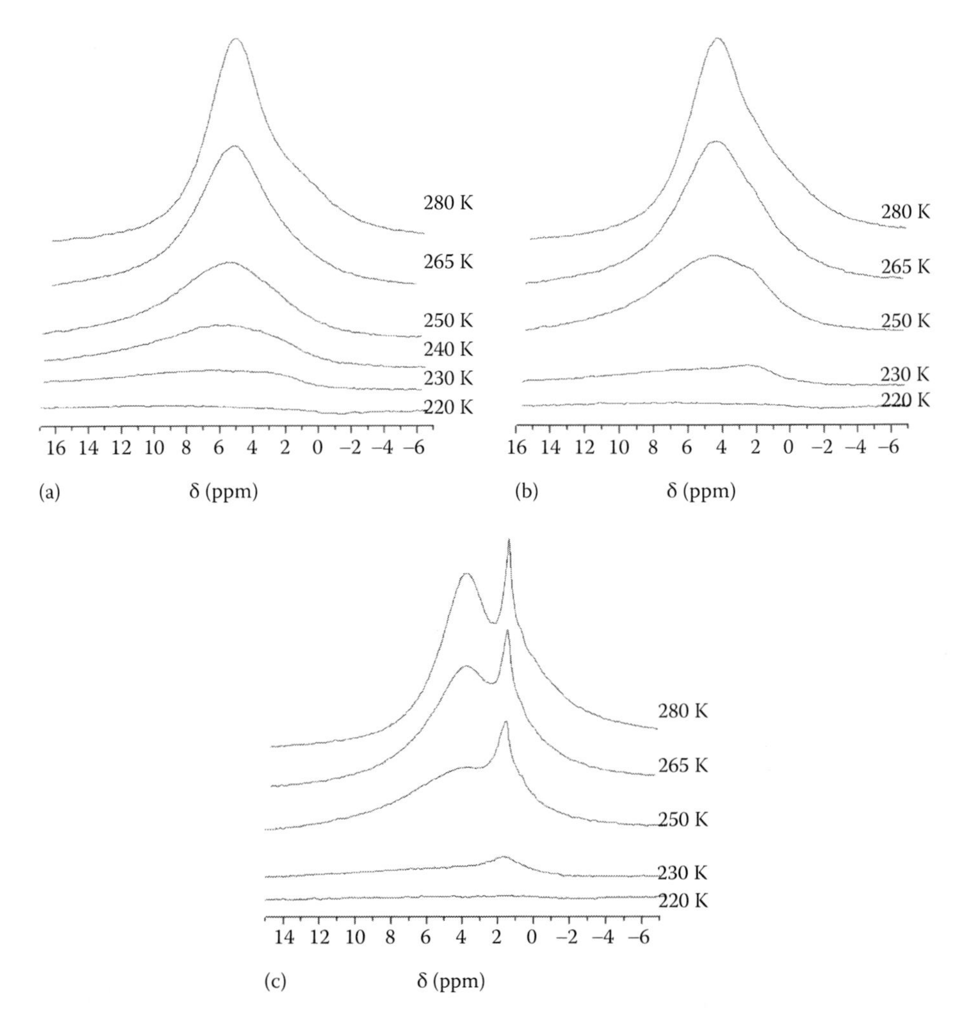

FIGURE 7.33 ^{1}H NMR spectra of unfrozen water in partially dehydrated bone marrow cells containing 20 wt% of residual water on addition of acetonitrile-d_3: (a) 0.2, (b) 0.4, and (c) 0.8 g/g. (Adapted from *Cryobiology*, 59, Turov, V.V., Kerus, S.V., and Gun'ko, V.M., Behaviour of water bound in bone marrow cells affected by organic solvents of different polarity, 102–112, 2009, Copyright 2009, with permission from Elsevier.)

In the case of solid lecithin, a very wide signal is observed at $\delta_H \approx 1$ ppm (Figure 7.35). This signal can weakly contribute the baseline of the spectra of mobile water recoded here using the bandwidth of 10 kHz. The signal width changes weakly with lowering temperature in contrast to that of WAW (Figure 7.32). Hydrated lecithin dissolved in $CDCl_3$ is characterized by a set of signals. Signal of CH_2 groups at $\delta_H \approx 1$ ppm has maximal intensity but changes in its intensity with temperature are small (in contrast to WAW). A significant decrease in the intensity is observed with lowering temperature only for SAW at $\delta_H \approx 5$ ppm.

According to microscopic investigations, addition of organic solvents to cells does lead to their destruction; however, solvents can penetrate into cells. The difference in the temperature behavior of the ^{1}H NMR signals of intracellular water and hydrated lecithin dissolved in chloroform suggests that interpretation of the data given earlier related to changes in contributions of intracellular SAW and WAW due to organic solvent effects on water bound in cells is correct. Additionally, these results are in agreement with the data related to redistribution of SAW and WAW in partially dehydrated yeast cells (Tsuraeva 1983).

TABLE 7.8
Characteristics of Intracellular Water (20 wt%) Bound in Marrow Cells with the Presence of Organic Solvents

		SAW				WAW			
Solvent	**C_{sol} (g/g)**	**C^{s}_{uw} (mg/g)**	**C^{w}_{uw} (mg/g)**	**ΔG_{max} (kJ/mol)**	**γ_S (J/g)**	**C^{s}_{uw} (mg/g)**	**C^{w}_{uw} (mg/g)**	**ΔG_{max} (kJ/mol)**	**γ_S (J/g)**
—	—	170	—	−2.0	9.2	—	30	−2	—
$CDCl_3$	1	130	—	−2.2	6.9	55	15	−3.5	5.8
$CDCl_3$	2	125	—	−2.5	6.7	60	15	−3.5	5.9
CD_3CN	0.2	160	—	−2.0	8.4	40	—	—	—
CD_3CN	0.4	110	—	−2.0	4.6	60	30	−2	3.1
CD_3CN	0.8	70	—	−2.0	2.8	80	50	−2	4.5
DMSO-d_6	0.2	120	20	−2.3	11.8	—	60	−1	1.2
DMSO-d_6	0.4	130	15	−3.0	16.1	—	55	−1	1.4
DMSO-d_6	0.8	145	—	−3.0	16.4	10	45	−1	1.3

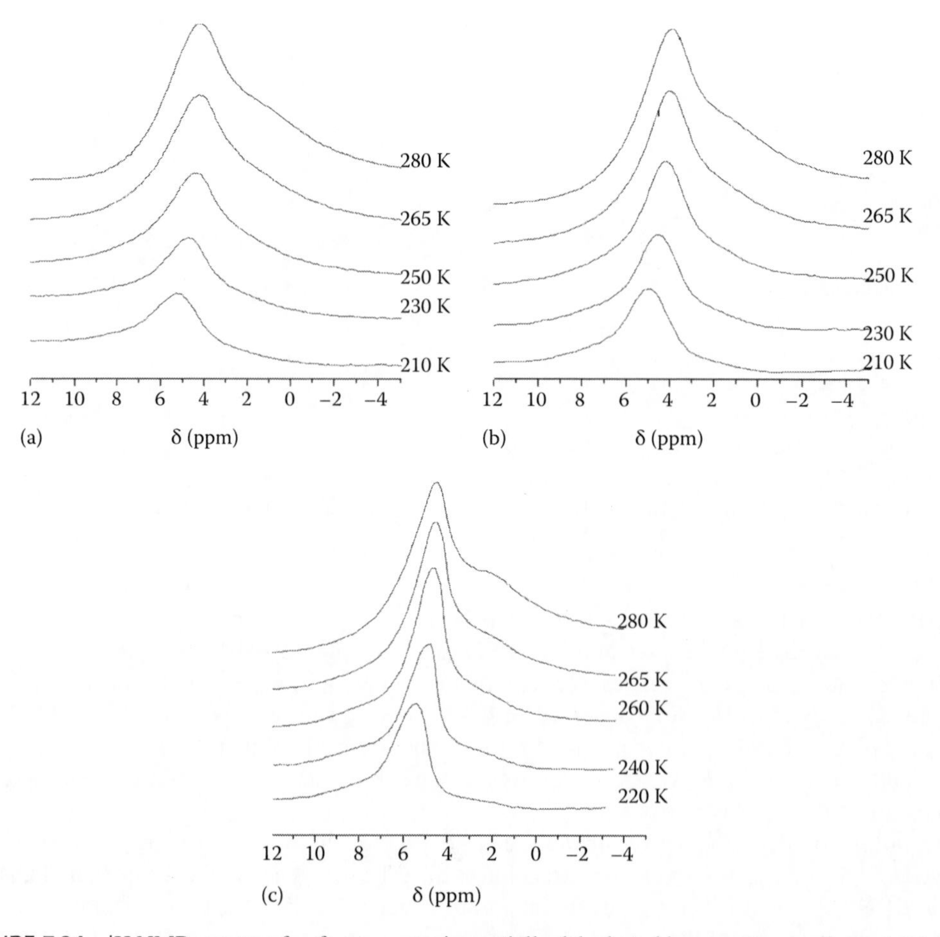

FIGURE 7.34 ^{1}H NMR spectra of unfrozen water in partially dehydrated bone marrow cells containing 20 wt% of residual water on addition of DMSO-d_6: (a) 0.2, (b) 0.4, and (c) 0.8 g/g. (Adapted from *Cryobiology*, 59, Turov, V.V., Kerus, S.V., and Gun'ko, V.M., Behaviour of water bound in bone marrow cells affected by organic solvents of different polarity, 102–112, 2009, Copyright 2009, with permission from Elsevier.)

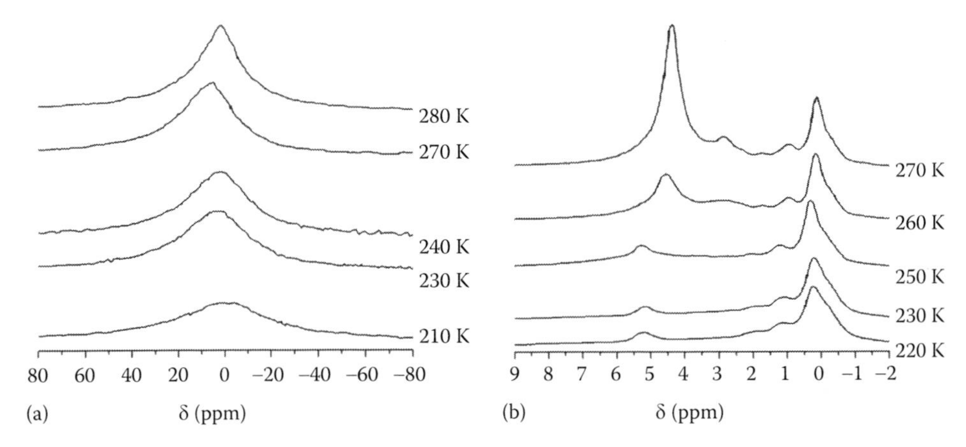

FIGURE 7.35 ^{1}H NMR spectra of crystalline lecithin with adsorbed water from (a) air and (b) the hydrated lecithin dissolved in deuterochloroform. (Adapted from *Cryobiology*, 59, Turov, V.V., Kerus, S.V., and Gun'ko, V.M., Behaviour of water bound in bone marrow cells affected by organic solvents of different polarity, 102–112, 2009, Copyright 2009, with permission from Elsevier.)

Figures 7.36 and 7.37 show the relationships between the amounts C_{uw} and $dC_{uw}(T)/dT$, the δ_H values, and changes in the Gibbs free energy (ΔG) of unfrozen SAW and WAW fractions bound in cells on addition of organic solvents. The characteristics of these bound waters calculated on the basis of the $\Delta G(C_{uw})$ graphs for the SAW and WAW fractions are given in Table 7.8. For the used concentrations of chloroform-*d*, 30% of the total amount of water (in the initial cells) transforms

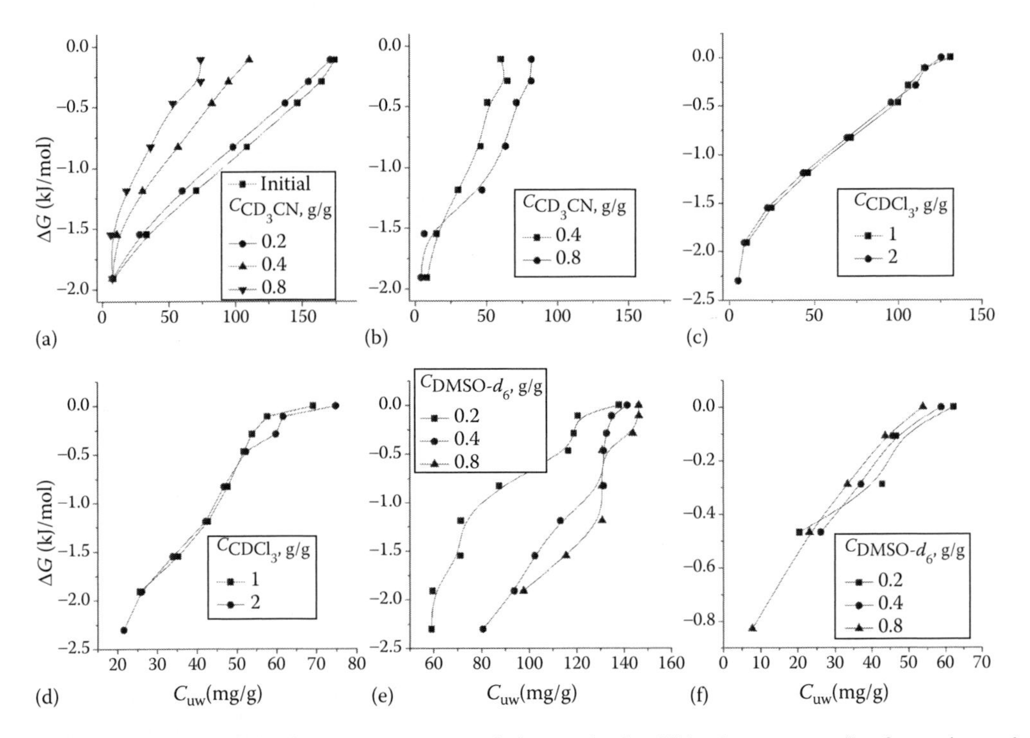

FIGURE 7.36 Relationships between amounts and changes in the Gibbs free energy of unfrozen intracellular water bound in cells on addition of (a, b) CD_3CN, (c, d) $CDCl_3$, and (e, f) $(CD_3)_2SO$ for (a, c, e) SAW and (b, d, f) WAW. (Adapted from *Cryobiology*, 59, Turov, V.V., Kerus, S.V., and Gun'ko, V.M., Behaviour of water bound in bone marrow cells affected by organic solvents of different polarity, 102–112, 2009, Copyright 2009, with permission from Elsevier.)

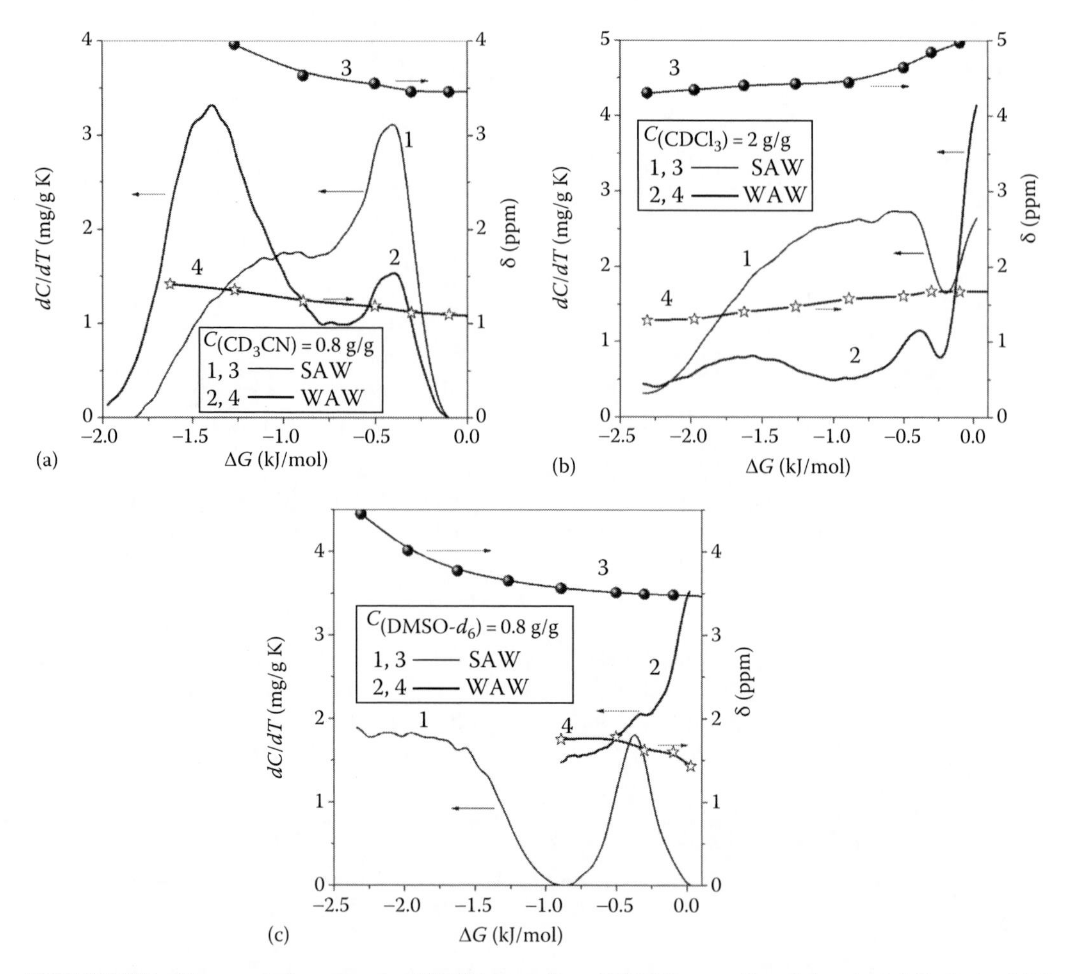

FIGURE 7.37 Changes in the amounts of SAW (curves 1) and WAW (curves 2) and chemical shifts δ_H for SAW (curves 3) and WAW (curves 4) versus changes in their Gibbs free energy at the maximal content of organic solvents: (a) CD_3CN (0.8 g/g), (b) $CDCl_3$ (2 g/g), and (c) Me_2SO-d_6 (0.8 g/g). (Adapted from *Cryobiology*, 59, Turov, V.V., Kerus, S.V., and Gun'ko, V.M., Behaviour of water bound in bone marrow cells affected by organic solvents of different polarity, 102–112, 2009, Copyright 2009, with permission from Elsevier.)

from SAW into WAW and the Gibbs free energy of the interfaces grows, i.e., the γ_S value becomes smaller (Table 7.8).

Substantial portions of SAW and WAW (larger for SAW) correspond to SBW (Table 7.8, C^s_{uw} and Figure 7.37b). The ΔG_{max} value (Table 7.8) characterizing changes in the Gibbs free energy of the first layer of bound water suggests that WAW is more strongly bound than SAW that is well seen from the ^1H NMR spectra recorded at different temperatures (Figure 7.32) and the relationships between dC_{uw}/dT and ΔG (Figure 7.37).

On addition of acetonitrile-d_3 the γ_S value of SAW decreases with increasing C_{CD_3CN} value, and contribution of SBW to SAW decreases but the opposite result is for WAW (Table 7.8, and Figures 7.33, 7.36, and 7.37). However, the ΔG_{max} value for both types of bound water remains practically the same as that for the initial cells (Table 7.8). On addition of DMSO, the γ_S value of SAW substantially increases but it is low for WAW (Table 7.8, and Figures 7.34, 7.36, and 7.37). A fraction of SBW in SAW grows but it decreases in WAW with increasing C_{DMSO}. The $\Delta G(C_{uw})$ graphs demonstrate a nonlinear shape with some bends (Figure 7.36) that reveal energetic heterogeneity of SAW, especially on addition of DMSO-d_6, and the presence of at least two types of water in both SAW and WAW (Figure 7.37).

Thus, water–organic solvent mixtures bound in partially dehydrated bone marrow cells form certain heterogeneous clustered or nanodomain structures, features of which depend on the types and the amounts of organic solvents which affect not only bound water but also intracellular macromolecular structures. SAW and WAW are observed as separated fractions of bound water (that is well seen in Figure 7.37) since the molecular exchange between them is slow (in the NMR timescale). Previously, it was shown that organic solvents can reduce the interaction of water (in water–organic mixtures) with both hydrophobic and hydrophilic surfaces (Gun'ko et al. 2005d, 2008d, 2009d, Turov and Gun'ko 2011). This spatial self-organizing of the systems occurring in confined space to reduce the free energy leads to the displacement of water from narrow pores into larger ones or out of them. However, a portion of water can remain in very narrow nanopores or it can be displaced into these pores. Similar effects can appear in the studied cells as in the presence of DMSO-d_6 (Figure 7.38c). However, self-organizing of water–organic solvent mixtures is more complicated in cells because changes in confined intracellular space formed by macromolecular functionalities (nonpolar, polar, and charged) can be considered as soft pores and pockets.

The structure of these soft pores can be differently affected by nonpolar and polar organic solvents and intracellular water. In comparison with chloroform, acetonitrile more strongly reduces the interaction of water with intracellular functionalities (Table 7.8, γ_S). Therefore, one can assume that the acetonitrile molecules (in water/acetonitrile clusters and domains) predominantly contact the intracellular functionalities. A homogeneous mixture of water with acetonitrile-d_3 or DMSO-d_6 is not formed in cells because the δ_H value of water depends weakly on the presence of organics over a wide temperature range (Figures 7.34 and 7.35). An increase in the γ_S value on addition of DMSO-d_6 can be caused by reorganization of intracellular macromolecules and water structures with enhancement of filling of narrow intracellular "pores" (having polar walls, i.e., formed by polar macromolecular functionalities) by SBW (Figure 7.38).

Water structures (clusters and domains) in SAW have larger sizes, i.e., broader distribution function $f(R)$ (Figure 7.38), than that in WAW. The shape of the $f(R)$ functions depends on the type and content of organic solvent in cells. Broader $f(R)$ functions are observed on addition of weakly polar $CDCl_3$ (Figure 7.38b) and smaller structures appear on addition of polar DMSO (Figure 7.38c). This leads to an increase in the specific surface area of cell functionalities in contact with unfrozen water from $S=36$ m^2/g (SAW with a domain structure in cells without addition of organic solvents) to ~60 m^2/g (SAW+WAW in cells with CD_3CN), ~80 m^2/g (SAW+WAW with $CDCl_3$), and ~200 m^2/g (SAW+WAW with DMSO-d_6). Consequently, maximum clustered water is in cells on addition of DMSO.

We may conclude that in complex bio-objects such as bone marrow cells, four types of intracellular water can be registered using the ^{1}H NMR spectroscopy: WBW ($\Delta G > -0.8$ kJ/mol) and SBW ($\Delta G < -0.8$ kJ/mol), and WAW ($\delta_H = 1.2$–1.7 ppm) and SAW ($\delta_H = 4$–5 ppm). The formation of small structures with intracellular water or water–organic mixtures in cells is enhanced due to the confined space effects and the presence of macromolecular functionalities with various hydrophilic/hydrophobic properties differently interacting with water and organic solvent molecules. Therefore, addition of certain amounts of an organic solvent leads to structural (reflecting in changes in δ_H) and energetic (ΔG) differentiations of intracellular water. For partially dehydrated bone marrow cells, additions of weakly polar, weak electron-donor solvents (chloroform, acetonitrile) or polar, strong electron-donor DMSO more strongly stabilize WAW than SAW. Under the effects of the organic solvents spatial characteristics of intracellular structures change, and these changes depend on the type of a solvent and reflect in changes of the properties of intracellular water. More strongly clustered water is observed with the presence of DMSO, and acetonitrile provides minimum clustering of intracellular water. Therefore, DMSO can be used as an effective cryopreservative reducing the negative effects caused by the formation of large ice crystallites in cells. Similar NMR investigations of differently hydrated cells with addition of various cryopreservatives can allow deeper insight into their effects important for better understanding the mechanism of cryopreservation of functionally differentiated cells.

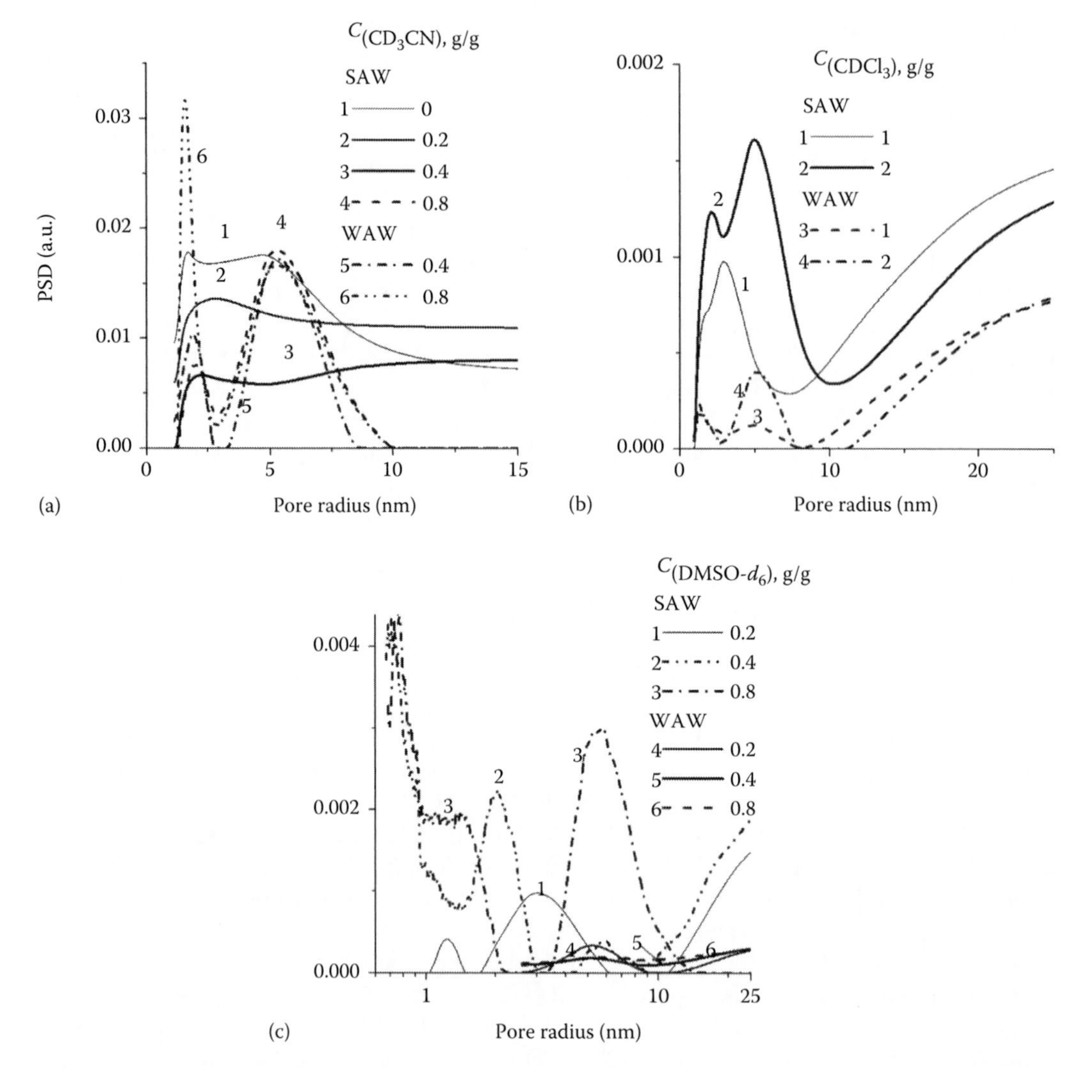

FIGURE 7.38 Size distributions of water structures in SAW and WAW bound in cells without addition of organics (curve 1, a) and on addition of different amounts of (a) CD_3CN, (b) $CDCl_3$, and (c) DMSO-d_6. (Adapted from *Cryobiology*, 59, Turov, V.V., Kerus, S.V., and Gun'ko, V.M., Behaviour of water bound in bone marrow cells affected by organic solvents of different polarity, 102–112, 2009, Copyright 2009, with permission from Elsevier.)

7.3 FREEZE-DRIED BOVINE GAMETES WITH ORGANIC ADDITIVES

To storage reproductive cells, different freeze-dry and cryopreservation techniques are used (Karlsson and Toner 1996, Feofilova 2003, Hubálek 2003, Yoon et al. 2003, Han et al. 2005). Cryopreservation uses special media that prevent destruction of the cell membranes by ice crystallites and provide vital functions of the cells after thawing (Feofilova 2003, Han et al. 2005). During cryopreservation, strong dehydration of cells is accompanied by changes in composition of intracellular liquid. After dehydration, cells include a certain small content of water which is typically strongly clustered. This clustered state of residual intracellular water is necessary to prevent the formation of large ice crystallites which can destroy the cells. Spatial and chemical complexity of intracellular structures causes a complex structure of clustered intracellular water. State of intracellular water bound in dehydrated bovine gametes and the effects of strongly polar DMSO (used as a cryopreservative) and weakly polar chloroform were studied using low-temperature ^{1}H NMR spectroscopy with a Varian Mercury 400 spectrometer (Turov et al. 2008b).

Native bovine gametes were centrifuged (3000 rpm) for 10 min, residual cells were four times washed by 2.9% sodium citrate with control of the protein content in the supernatant (spectrophotometrically at $\lambda = 280$ nm). Cells (with residual amount of sodium citrate) freeze-dried (the bulk density of 0.12 g/cm^3) included about 2 wt% of residual water. The same cell sample (equilibrated after each addition for 10 min by shaking) was studied with subsequent addition of water, $CDCl_3$, and DMSO (which can be solved in chloroform). Extracellular chloroform was evaporated from the samples for 2 weeks and then an additional portion of DMSO (up to 5 g/g) was added and the system was equilibrated for 2 h (Turov et al. 2008a).

The initial cell sample hydrated at $h = 0.1$ g/g is characterized by a single signal of SAW at $\delta_H = 5$ ppm (Figure 7.39a) similar to that of bulk water. Its intensity decreases with lowering temperature due to partial freezing of intracellular water.

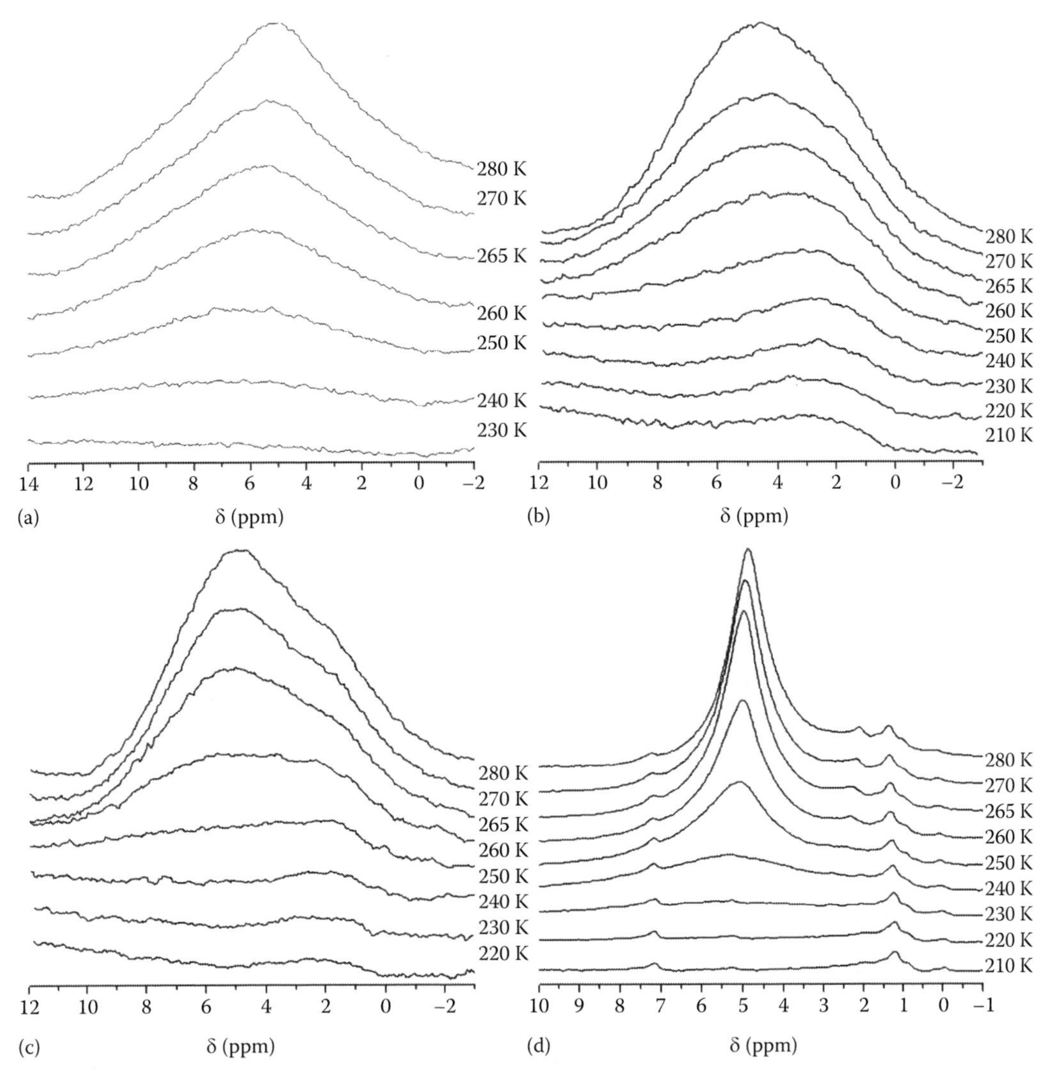

FIGURE 7.39 ^{1}H NMR spectra of intracellular water bound in bovine gametes containing (a–c) 100 and (d) 300 mg/g of water at different contents of $CDCl_3$: (a) 0, (b) 1, (c) 3, and (d) 5 g/g.

The SAW signal is not observed at $T < 230$ K. Addition of chloroform (Figure 7.39b and c) leads to the appearance of a signal of WAW (besides SAW) $\delta_H \approx 2$ ppm. Its intensity is not higher than 12% of the SAW signal intensity and weakly depends on temperature till 210 K. Therefore, practically all WAW can be attributed to SBW (Gun'ko et al. 2009d).

The signal width decreases with increasing content of water (Figure 7.39d) that can be explained by increased mobility of water molecules (Emsley et al. 1965). Additionally, WAW appears with two signals at $\delta_H = 1.3$ and 2 ppm. There are weak signals at $\delta_H = 7.16$ ppm (admixture $CHCl_3$ in $CDCl_3$) and $\delta_H = 0$ ppm (TMS was added as a standard compound). Consequently, chloroform can easily penetrate into cells and affect the behavior of intracellular water.

Addition of DMSO to the cell sample (Figure 7.40) leads to the appearance of (besides signals of WAW and SAW) a signal at $\delta_H = 3.5$ ppm which can be assigned to associates $(CD_3)_2SO \cdots HOH$. Its intensity increases with increasing DMSO content. DMSO can be easily dissolved in chloroform.

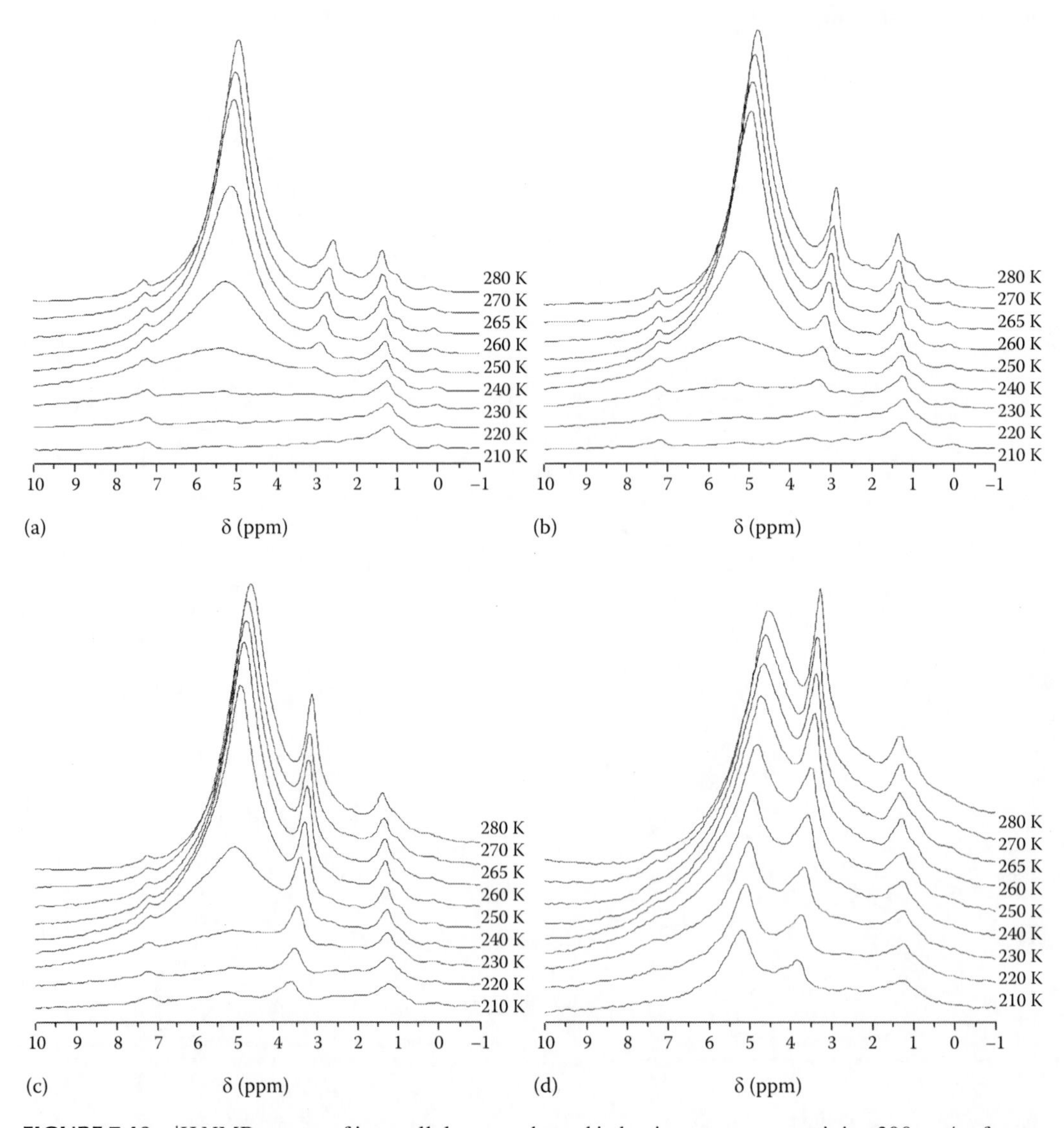

FIGURE 7.40 ^{1}H NMR spectra of intracellular water bound in bovine gametes containing 300 mg/g of water and 5 g/g of $CDCl_3$ at different contents of $(CD_3)_2SO$: (a) 0.1, (b) 0.3, and (c, d) 0.65 g/g, (d) after slow evaporation of $CDCl_3$ (to ~0.6 g/g).

Therefore, its penetration into cells depends on the changes in the free energy of complexes with cell–DMSO and chloroform–DMSO. DMSO penetrating into cells can interact with intracellular water and a portion of which transfers in complexes $(CD_3)_2SO\cdots HOH$ characterized by signal at $\delta_H = 3.5$ ppm (Figure 7.40).

After evaporation of bulk phase of chloroform both organic solvents are mainly located inside cells that results in changes in the spectral characteristics (Figure 7.40d). Freezing temperature of SAW decreases because a portion of DMSO dissolves in SAW. The chemical shift value of SAW increases with lowering temperature that can be explained by surrounding of SAW nanodomains by DMSO molecules $(HOH\cdots OS(CD_3)_2{}^*CDCl_3)$ oriented by CH_3 groups toward hydrophobic cellular structures or $CDCl_3$. The amount of intracellular DMSO is large because of a great intensity of the signal at $\delta_H = 3.5$ ppm. The intensity of WAW doubles after chloroform evaporation.

Water in associates $(CD_3)_2SO\cdots HOH$ is mainly SBW since the corresponding signal is observed even at 210 K (Figure 7.40). WAW is SBW in all the systems studied but its amount and freezing behavior are complex functions of component concentrations that reflect the capability of cells of changes in the inner topology depending on the medium characteristics and composition of intracellular liquids.

A fraction of SBW in SAW decreases with increasing content of chloroform that leads to a decrease in the γ_S value (Table 7.9). Similar phenomena were earlier described for nanostructured mineral or organic materials. The γ_S value increases by order of magnitude with increasing hydration from 0.1 to 0.3 g/g, and a contribution of SBW increases. DMSO weakly affects the γ_S value which slightly increases. This is due to easily mixing DMSO with both SAW and WAW and formation of interfacial layer with DMSO molecules at the boundary of SAW domains or WAW clusters that result in decreasing contacts of water with hydrophobic structures. At a large content of DMSO (5 g/g), the spectra of SAW become more complex including 4 signals of SAW and 1–2 signals of WAW (Figure 7.41). One can assume that a significant portion of intracellular water (SAW) could transfer into DMSO (signal at $\delta_H = 3.5$ ppm). However, a signal at 4 ppm (close to $\delta_H = 3.5$ ppm) is not maximum intensive. Other SAW signals (Figure 7.41, signals 1–3) can be attributed to intracellular water. Their intensity depends differently on temperature. Therefore, they can be attributed to different intracellular structures with the participation of DMSO molecules.

TABLE 7.9
Characteristics of Intracellular Water Bound in Bovine Gametes Being in Different Dispersion Media

	C_{sol} (g/g)						
Water Type	**$CDCl_3$**	**DMSO**	**C_{H_2O} (g/g)**	**C^s_{uw} (mg/g)**	**C^w_{uw} (mg/g)**	**ΔG^{max} (kJ/mol)**	**γ_S (J/g)**
SAW	0	0	0.1	55	45	−1.6	3.2
	1	0	0.1	40	60	−1.2	1.9
	3	0	0.1	15	85	−1.2	1.8
	5	0	0.3	280	20	−2.0	17.0
	5	0.1	0.3	300	0	−2.3	17.4
	5	0.3	0.3	300	0	−2.3	18.3
	5	0.6	0.3	300	0	−2.6	20.1
	≈0.6	0.6	0.3	300	0	−3.5	22.7
ASW	5	0.1	0.3	18	7	−1.3	1.0
	5	0.3	0.3	23	14	−2.6	2.1
	5	0.6	0.3	30	0	−2.6	2.5
	≈0.6	0.6	0.3	38	22	−3.0	4.0

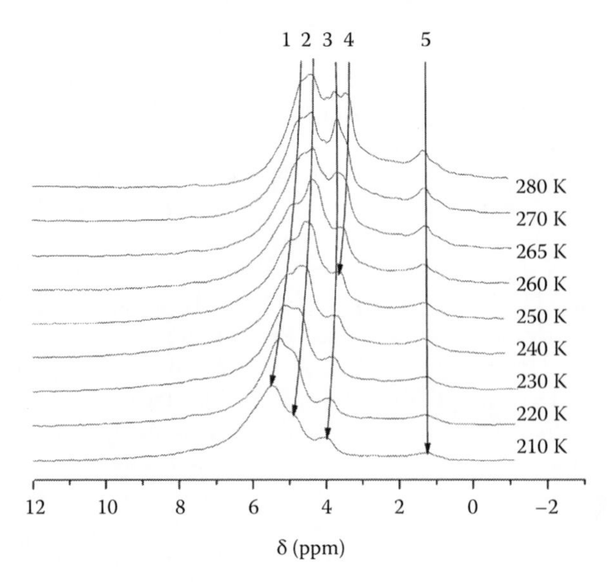

FIGURE 7.41 ^1H NMR spectra of intracellular water bound in bovine gametes containing 0.3 g/g of water and 0.3 g/g of $CDCl_3$ at 5 g/g of $(CD_3)_2SO$.

Thus, strong dehydration of bovine gametes (0.1 g/g) and subsequent partial hydration (0.3 g/g) result in an increase in the content of SBW unfrozen at low temperature (i.e., less dangerous for cells due to the formation of ice crystallites). Addition of weakly polar $CDCl_3$ or strongly polar DMSO to partially hydrated cells allows controlling the structure of intracellular water, i.e., amounts of WBW and SBW and WAW and SAW.

7.4 RED BLOOD CELLS

Red blood cells (RBCs) in the normal state (i.e., normal discocytes) have a shape of a convexoconvex disk, and these cells contribute ~85% of their total number. RBCs are characterized by a high flexibility of the membrane, which provides significant reversible changes in their forms on penetration through thin capillaries. RBCs leaving the capillaries can quickly transform into the normal discocyte state (Kozlov and Markin 1986). This transformation occurs through an echinocyte state (mottled discocyte) characterized by formation of 30–50 spicules on each cell. Echinocytes can be divided into four (I–IV) groups. Irreversible transformation of normal discocytes into echinocytes and then into spherocytes occurs through several stages: normal and mottled discocytes, spinous discoids and spinous oblate ellipsoids, and the end shape corresponding to spherocytes (Krymsky et al. 1976). If a RBC is in the echinocyte state during a long period (because of a low flexibility of the membrane on a diminution of the ATP content in the cell or due to certain external factors), then it can lose a portion of the spicules (i.e., a portion of the membrane) and the cell transformation goes to a spherical shape which can be irreversible (Muromtsev and Kidalov 1998, Rudenko et al. 1998). Structural features of RBCs can play an important role on their interaction with solid nanoparticles (Harley and Margolis 1961, Lindqvist et al. 1974, Lysenko et al. 1977, Chuiko 2003) as a component of tourniquet preparations or other medicines.

Nanosilica composed of bioactive nanoparticles can affect the RBC state up to strong hemolysis (erythrolysis) at relatively low concentration of silica $C>0.1$ wt% (Diociaiuti et al. 1999, Gerashchenko et al. 2002, Blitz and Gun'ko 2006). The hemolytic effect of nanosilica can be reduced due to its modification by immobilized polymers (e.g., poly(vinyl pyrrolidone), poly(vinyl alcohol)) or on the use of binary nanooxides such as silica/alumina and silica/titania (Blitz and Gun'ko 2006).

Notice that the greater the content of the second oxide or adsorbed polymer, the lower the hemolytic effect of nanooxides. Changes in the whole RBCs and the RBC membranes interacting with solid nanoparticles can be complex and dependent on component concentrations and conditions (Chuiko 2003, Blitz and Gun'ko 2006).

Nanosilica A-300 ($S_{BET}=285$ m^2/g) was used as the initial material. Nanosilica was modified by hexamethyldisilazane to substitute ~50% of free surface silanols for trimethylsilyl (TMS) groups. This modification reduces the specific surface area by ~15% but the hydrophilic properties practically remain since the heat of immersion in water and Gibbs free energy of the interfacial water layer are close to the corresponding values for the initial silica A-300.

Aqueous suspensions of unmodified and modified nanosilicas were prepared by mixing of the dry powder with 3.8 wt% solution of trisodium citrate (TSC). The concentration of silica was between 0.0001 and 1.0 wt%. To produce more homogeneous suspensions with partially decomposed aggregates of primary particles and strongly decomposed agglomerates of aggregates and to enhance the bioactivity of nanosilica, the suspensions were sonicated for 40 min with an ultrasonic dispenser (35 kHz). Human donor RBCs (blood group A, Kiev Hemotransfusion Center) placed initially in a standard buffer with sodium hydrocitrate (2%) and glucose (3%) (SCG) (8×10^7 cell/mL) were added to the suspension of silica in the TSC buffer to obtain the RBC concentration of about 1×10^7 cell/mL (incubation time 20 min).

It was found (Gun'ko et al. 2007c) that the shape of RBCs changes during interaction with unmodified and modified silicas from discocytes to spherocytes through echinocytes and then to shadow corpuscles (or hemoglobin-free ghost of RBC, Dodge et al. 1963) depending on silica concentration in the suspension (Table 7.10). All the stages of RBC transformation due to interaction with nanosilica were well identified (Figure 7.42). The transformation of discocytes (Figure 7.42, form 1) to echinocytes (forms 2–5) begins from distortion to a convexoconvex contour of normal RBCs.

Rough spicules appear first on the edge of the disk (form 3) and then on the whole RBC surface (forms 4 and 5).

The echinocyte spicules gradually become thinner and more uniformly distributed on a cellular surface. Then cells become of a spherical shape (form 6). On the final transformation, cells lose a part of the spicules and the transformation into the spherocyte shape becomes irreversible. A strong distortion of the membrane (e.g., on interaction with silica nanoparticles) leads to loss of its flexibility and resiliency. RBCs swell and increase in size in comparison with the spherocyte that leads to membrane break. Eliminated hemoglobin can be detected, i.e., hemolysis of RBCs occurs. However, the perforated cellular membrane remains as unique whole and forms so-called shadow corpuscle (Figure 7.42, forms 7 and 8).

In the case of the RBC/A-300 systems, contribution of the different RBC forms changes due to interaction with nanosilica (Table 7.10). In the control sample, a major portion of RBCs corresponds to discocytes (Table 7.10 and Figure 7.43a). On interaction of RBCs with nanosilica at $C_{A\text{-}300}=0.0001$–0.01 wt%, there is a monotonic diminution of discocytes and echinocytes contributions. At the same time, the amounts of spherocytes increase, as well as contribution of deformed cells. At these silica concentrations, the hemolysis of the RBCs is not intensive and the content of shadow corpuscles is small.

An increase in concentration of silica to 0.1 wt% leads to break of the membranes of many cells and shadow corpuscles give the major portion of RBCs (Table 7.10). The spherocytes and deformed cells give much smaller contribution. Agglutination of cells is observed for this system (Figures 7.43b and 7.44).

It is possible that nanosilica aggregates promote this process as it is shown in Figure 7.44. The agglutination of RBCs can be enhanced due to interaction with aggregates of silica nanoparticles which prevent the electrostatic repulsive interaction of negatively charged cells. This is due to strong interaction of silica nanoparticles with proteins integrated into the RBC membranes (Chuiko 2003). Changes in the interaction between RBCs are observed after addition of nanosilica since microscopic images demonstrate the absence of agglutination for RBCs alone (Figure 7.43a) and

TABLE 7.10
Contributions of Different RBC Forms on Interaction with Unmodified and Silylated Nanosilica at Different Silica Contents

		Silica Content (wt%)									
		0.0001		0.001		0.01		0.1		1.0	
Form	**Control**	**A-300**	**TMS/A-300**	**A-300**	**TMS/A-300**	**A-300**	**TMS/A-300**	**A-300**	**TMS/A-300**	**A-300**	**TMS/A-300**
Discocytes	48.1 ± 27.8	37.1 ± 26.2	34.0 ± 19.6	32.2 ± 18.5	30.4 ± 21.4	26.4 ± 15.2	43.5 ± 30.7	0	0	0	0
Echinocytes	23.5 ± 13.6	23.5 ± 13.6	21.0 ± 12.1	16.8 ± 9.6	25.7 ± 18.2	14.8 ± 8.5	20.7 ± 14.6	0	0	0	0
Spherocytes	20.8 ± 12.0	37.0 ± 26.1	41.6 ± 24.0	39.5 ± 22.8	41.1 ± 29.1	46.7 ± 27.0	31.3 ± 22.1	6.4 ± 4.5	59.0 ± 34.1	0	0
Deformed cells	7.6 ± 4.3	2.6 ± 1.8	3.4 ± 2.0	10.8 ± 6.2	2.8 ± 1.9	11.9 ± 6.9	4.5 ± 3.2	2.6 ± 1.8	0.9 ± 0.5	0	16.7 ± 9.6
Shadow corpuscles	0	0.4 ± 0.2	0	0.7 ± 0.4	0	0.2 ± 0.1	0	91.0 ± 64.3	40.1 ± 23.2	100	83.3 ± 48.1

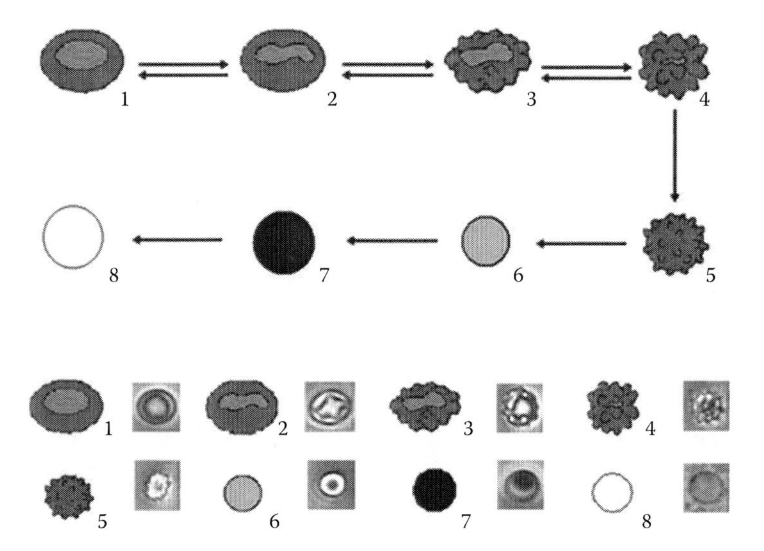

FIGURE 7.42 Scheme of RBC transformation from a discocyte to a shadow corpuscle on interaction with nanosilica: discocyte (1), echinocyte I–IV (2–5), spherocyte (6), and shadow corpuscle (7, 8) shown with consideration for their relative sizes. (Adapted with kind permission from Springer Science+Business Media: *Centr. Eur. J. Chem.*, Interaction of unmodified and partially silylated nanosilica with red blood cells, 5, 2007i, 951–969, Gun'ko, V.M., Galagan, N.P., Grytsenko, I.V. et al.)

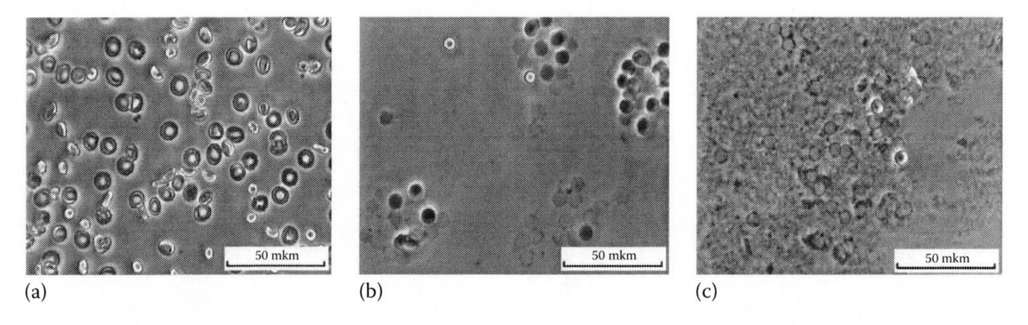

FIGURE 7.43 Microphotographs of RBCs in the buffer suspension of (a) control sample, and with the presence of (b) 0.1 wt% and (c) 1.0 wt% of unmodified silica A-300. (Adapted with kind permission from Springer Science+Business Media: *Centr. Eur. J. Chem.*, Interaction of unmodified and partially silylated nanosilica with red blood cells, 5, 2007c, 951–969, Gun'ko, V.M., Galagan, N.P., Grytsenko, I.V. et al.)

their agglutination for the RBCs/silica system (Figure 7.43b). The hemolytic activity of 1% suspension of A-300 provides 100% hemolysis of RBCs (Table 7.10) and only the shadow corpuscles are observed (Figure 7.43c).

Nanosilica with approximately half free silanols substituted for TMS groups (TMS/A-300) possesses a lower hemolytic activity in comparison with unmodified silica of the same concentration in the suspension (Table 7.10). This effect can be caused by several reasons discussed later. At concentration of modified silica from 0.0001 to 0.001 wt%, a portion of discocytes decreases but a fraction of spherocytes grows. On the use of 0.01 wt% suspension of modified silica, the number of discocytes is smaller than 50% but larger than that of spherocytes. The amounts of echinocytes in these systems are close to that for the control samples but the amounts of deformed cells are lower and the shadow corpuscles are not observed.

Discocytes and echinocytes are absent at TMS/A-300 concentration of 0.1 wt% (Table 7.10). Spherocytes gives the main fraction and the second fraction with shadow corpuscles is approximately two times less than that on interaction of RBCs with unmodified nanosilica. Agglutination

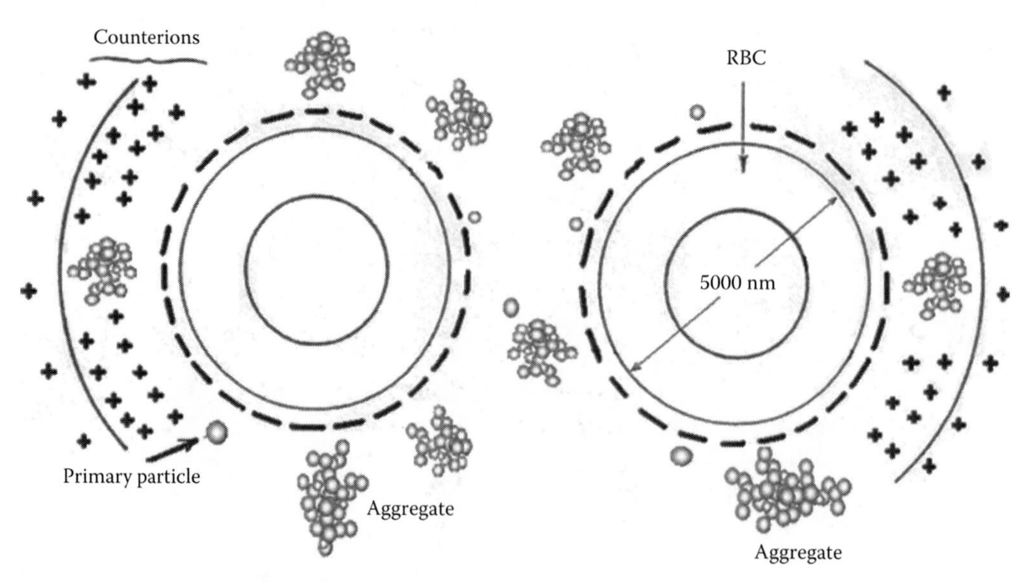

FIGURE 7.44 Model of agglutination of RBCs with the participation of aggregates with primary silica nanoparticles. (Adapted with kind permission from Springer Science+Business Media: *Centr. Eur. J. Chem.*, Interaction of unmodified and partially silylated nanosilica with red blood cells, 5, 2007c, 951–969, Gun'ko, V.M., Galagan, N.P., Grytsenko, I.V. et al.)

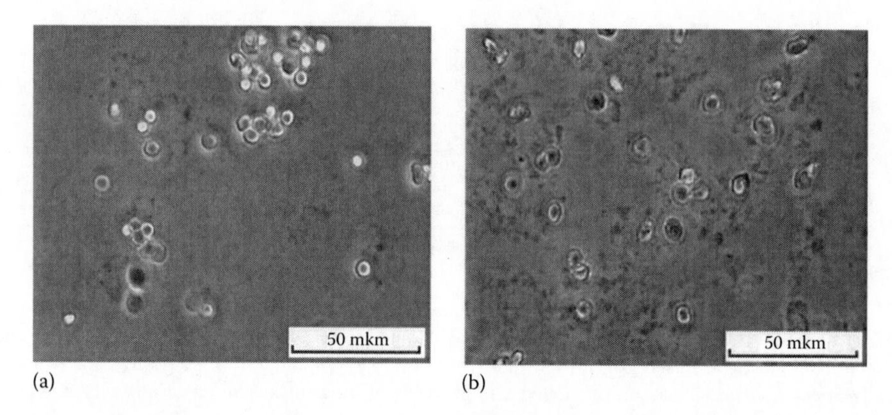

FIGURE 7.45 Microphotographs of RBCs in the buffer suspension with (a) 0.1 wt% and (b) 1.0 wt% modified silica TMS/A-300. (Adapted with kind permission from Springer Science+Business Media: *Centr. Eur. J. Chem.*, Interaction of unmodified and partially silylated nanosilica with red blood cells, 5, 2007c, 951–969, Gun'ko, V.M., Galagan, N.P., Grytsenko, I.V. et al.)

of cells is observed (Figure 7.45). If $C_{TMS/A\text{-}300} = 1$ wt% intensive formation of shadow corpuscles occurs, then it is accompanied by strong hemolysis of RBCs. However, this process is less intensive than that on RBC interaction with unmodified nanosilica because deformed cells (16.7%) remain in the suspension with TMS/A-300 (Figure 7.45b).

There are, at least, three reasons responsible for diminution of the hemolytic effect on interaction of RBCs with TMS/A-300: (i) strong aggregation of partially hydrophobic silica particles and diminution of the amounts of individual primary silica nanoparticles, which are maximum bioactive, that causes diminution of the total contact area between RBCs and silica (per gram of silica) and a decrease of local interaction of solid particles with membrane proteins, (ii) changes in interaction energy between modified silica surface and the membrane structures due to a decrease in the adsorption potential of modified silica, and (iii) changes in the structure of the interfacial water layer (Gun'ko et al. 2005d, Blitz and Gun'ko 2006). In the case of TMS/A-300, the effective diameter of particles (measured by DLS method with a Zetasizer 3000 [Malvern Instruments]) increases in comparison with unmodified silica by 5–10 times (up to 0.9–3.6 μm) depending on sonication time (Gun'ko et al. 2005d, 2007c, Blitz and Gun'ko 2006). We can assume that the first reason from the mentioned ones is predominant; however, two other reasons can play an important role too because changes in hydration/dehydration and swelling of RBCs (Figure 7.42) affect the state of membrane (Bennekou et al. 2006, Sauviat et al. 2006).

XRD patterns (Figure 7.46) depict that air-dried RBCs have two broad peaks at $2\theta = 10.7°$ and 24.5° corresponding to average cleavage spacing of 0.940 and 0.422 nm. Dry nanosilica has an asymmetrical peak $2\theta = 25.3°$ (spacing 0.404 nm corresponding to the distance between oxygen atoms in ring structures with 6–5 silica tetrahedrons in amorphous silica). The initial RBCs and RBCs/A-300 (1 wt%) are similar. However, besides the peak at 10.7° they depict a broad peak at $2\theta = 32°$ and a shoulder at $2\theta = 47°$, which are linked to water (because hexagonal ice has the lines in these ranges, Goto et al. 1990). These peaks decrease with increasing content of silica because of strong distortion of the structure of interfacial water in comparison with bulk water (Gun'ko et al. 2005d, 2007i). Content-dependent contributions of the bands of silica ($2\theta = 25.3°$), RBCs ($2\theta = 24.5°$), and water ($2\theta = 32°$) cause the observed displacement of the total broadband toward smaller 2θ values with increasing silica content. The intensity of the XRD peak of RBCs at $2\theta = 10.7°$ decreases with decreasing content of RBCs. The intensity of the water band decreases more strongly with increasing silica content from 1 to 8 wt% than that on a subsequent increase of the $C_{A\text{-}300}$ value to 30 wt%. The XRD data (Figure 7.46) reveal that nanosilica interacting with RBCs strongly affect the structure of extra- and intracellular water that can result in certain dehydration of cells at $C_{A\text{-}300} = 1$ wt% when the hemolysis of all the cells is observed and only shadow corpuscles can be found. At higher silica content, i.e., on a decrease in contribution of the cells, the amounts of interfacial water strongly increase.

In the case of pure water and the TSC or SCG buffers, there is a TSDC peak at 115–120 K (Figure 7.47b) characteristic for bulk water (Gun'ko et al. 2007i). For 0.9 wt% NaCl solution, this peak shifts toward higher temperatures because of distortion of the hydrogen bond network by the Na^+ and Cl^- ions (Gun'ko et al. 2005d, 2007c). The dc relaxation (observed at $T > 210$ K) depends

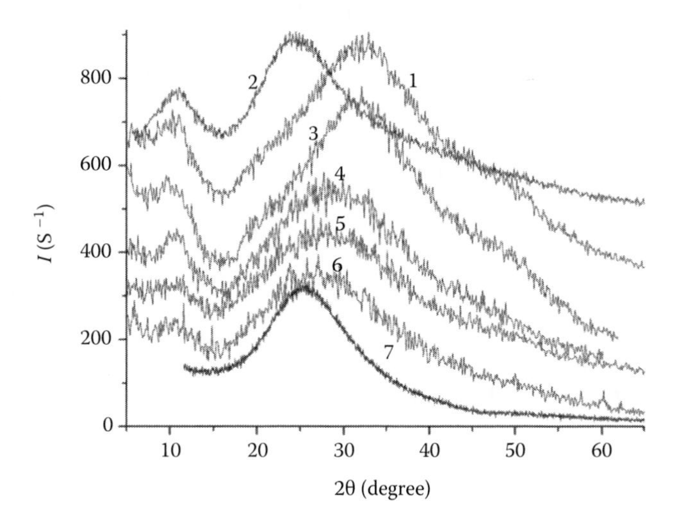

FIGURE 7.46 XRD patterns of initial (curve 1) and air-dried RBCs (curve 2), RBCs/A-300 at $C_{A\text{-}300} = 1$ (curve 3), 8 (curve 4), 20 (curve 5), and 30 wt% (curve 6), and dry A-300 (curve 7). (Adapted with kind permission from Springer Science+Business Media: *Centr. Eur. J. Chem.*, Interaction of unmodified and partially silylated nanosilica with red blood cells, 5, 2007i, 951–969, Gun'ko, V.M., Galagan, N.P., Grytsenko, I.V. et al.)

more strongly on the presence of dissolved salts than the TSDC shape at $T < 150$ K because salt anions and cations (in addition to protons) strongly contribute the conductivity (Table 7.11).

The difference in the TSC and SCG buffers causes certain differences in their TSDC thermograms (Figure 7.47b) and the corresponding relaxation parameters. This suggests changes in the structure of water in the buffers in comparison with distilled water.

Addition of 0.01 wt% of A-300 does not affect the TSDC spectrum of buffered RBCs (Figure 7.47a) that is in agreement with the microscopic data and the absence of the hemolysis. However, addition of a larger amount of A-300 (1 wt%) causes significant changes in the TSDC spectrum shape at $T < 130$ K (relaxing of bulk water structures) and $170 < T < 230$ K (relaxing bound water and biopolymers). For instance, TSDC relaxation intensity of bulk water strongly decreases with increasing $C_{A\text{-}300}$ value that is in agreement with the XRD data. These changes in the TSDC spectra correspond to 100% hemolysis of RBCs. However, at $130 < T < 170$ K the TSDC thermograms have the same shape (Figure 7.47a). Therefore, we can assume that the corresponding relaxing structures are connected only with RBC/TSC. However, only shadow corpuscles remain after interaction of RBCs with nanosilica in the aqueous suspension at $C_{A\text{-}300} = 1$ wt%.

Consequently, the relaxing structures can be linked to water/RBC membrane and biomolecules eliminated by silica nanoparticles from RBCs because TSC, SCG, and water weakly contribute to the TSDC spectra at these temperatures. Diminution of the TSDC spectra intensity at $T < 130$ K suggests that the water structure (i.e., the hydrogen bond network) undergoes significant changes because of the interaction with silica nanoparticles (similar results were observed by XRD). Additionally, silica particles ($C_{A\text{-}300} = 1$ wt%) strongly bind macromolecules (e.g., hemoglobin, other proteins eliminated from membranes, etc.) (Chuiko 2003, Blitz and Gun'ko 2006, Gun'ko et al. 2007i, Rugal et al. 2007) and ions because the TSDC intensity decreases at $T > 170$ K but the E_{dc} value increases (Figure 7.47d).

The $f(E)$ distribution functions of activation energy of depolarization (Figure 7.48) reveal an increase in the activation energy of depolarization on addition of A-300 (1 wt%) to buffered RBCs. A patch-wise linearization according to equation $\ln I = -E/k_BT + B$ (where B is a constant) (Figure 7.47d) shows that the activation energy of depolarization is higher for RBC/A-300 at $T < 120$ K and $T > 170$ K. These results suggest that the structure of bulk water and bound water strongly change on addition of nanosilica to buffered RBCs.

Turov et al. (2006a) studied hydration of the RBC membranes (shadow corpuscles, RBCSC) affected by nanosilica using low-temperature ^{1}H NMR method. Shadow corpuscles were prepared using a method described by Dodge et al. (1963). RBCs were triply washed (by 0.15M NaCl/PBS) and centrifuged. Then RBCs in 0.05M PBS (1: 20) were centrifuged at 16,000 rpm and 4°C for 20 min that provided hemolysis of the cells, and the residue (shadow corpuscles) was washed six times. Shadow corpuscles (1 mL includes 7.5 mg of dry matter) were stored at −15°C.

Water bound to shadow corpuscles gives a singlet ^{1}H NMR signal (Figure 7.49). Its width strongly increases (by order of magnitude) with decreasing temperature. The average δ_H value is equal to 4.5 ppm, which corresponds to that of bulk water. The suspension with shadow corpuscles included certain amounts of dissolved compounds (NaCl, etc.). Therefore, the appearance of unfrozen water can be caused by its interaction with both shadow corpuscles (Gibbs–Thomson relation for the freezing point depression) and dissolved compounds (Raoult law). During freezing of biomacromolecules, salts are concentrated in a liquid fraction of water because ice crystallites tend to be free of impurities. Therefore, water responsible for the observed signals (Figure 7.49) corresponds to water (with dissolved ions) bound to RBC membranes. This is also confirmed by a large width of the signal.

The $\Delta G(C_{uw})$ graphs (Figure 7.50) have two characteristic sections caused by SBW (strong changes in the Gibbs free energy at $\Delta G < -0.5$ kJ/mol but small changes in C_{uw}) and WBW (small changes in ΔG at $\Delta G > -0.5$ kJ/mol but large changes in C_{uw}).

Results for RBCSC (Figures 7.50 and 7.51, and Table 7.12) are in agreement with results for yeast cells described in Chapter 7, despite yeast cells were native but RBCSC represent mainly the cellular membranes. This suggests that the behavior of intracellular water depends strongly on the

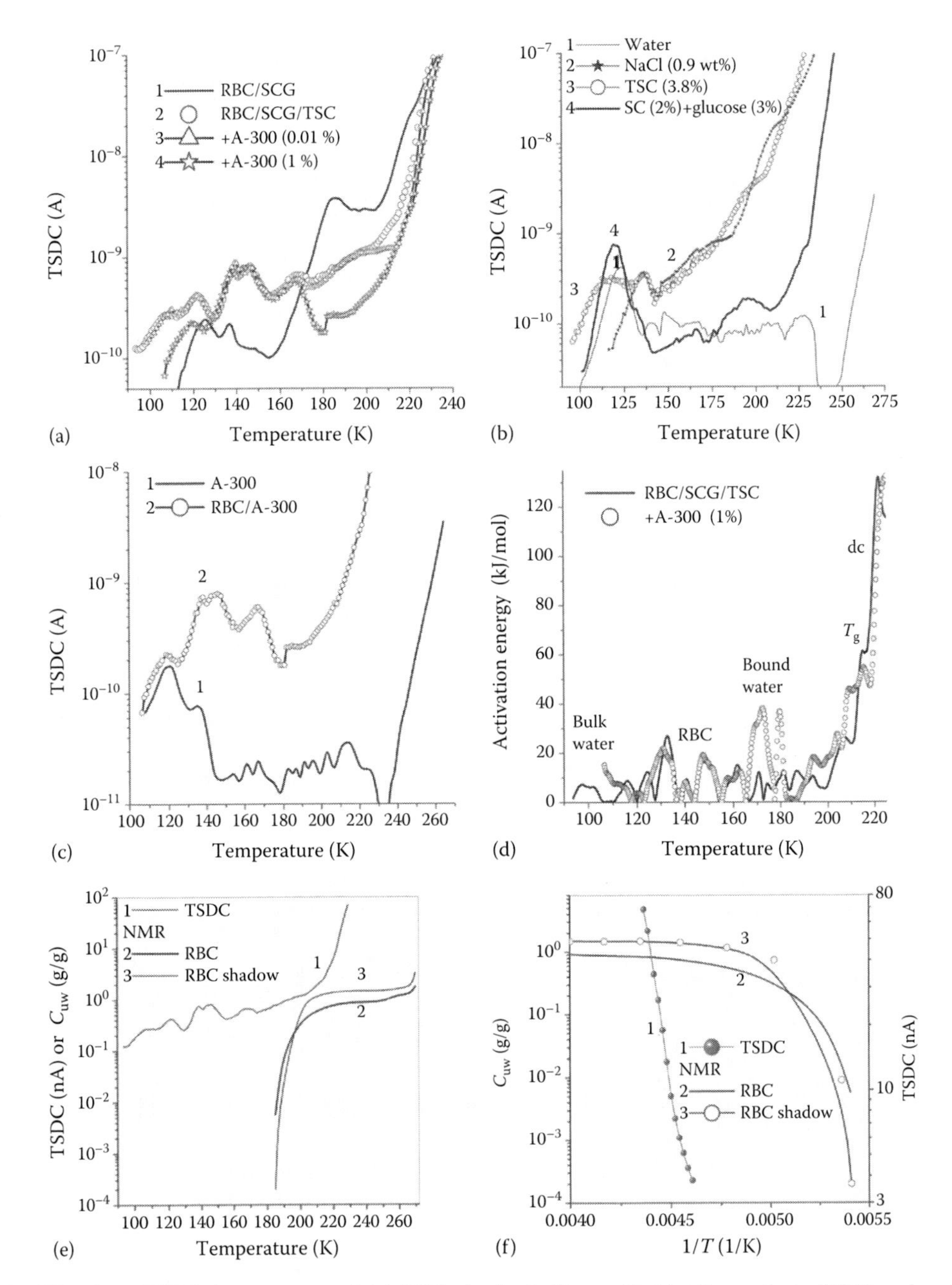

FIGURE 7.47 TSDC thermograms of (a) RBCs in the buffers with trisodium citrate (TSC) and sodium hydrocitrate with glucose (SCG) and on interaction with nanosilica A-300 at $C_{A\text{-}300} = 0.01$ and (a) 1.0 wt%, (b) pure water and different buffer solutions, (c) A-300 (in water) and RBC/A-300 (in the buffer) at $C_{A\text{-}300} = 1.0$ wt%, (d) activation energy of relaxation calculated from patch-wise linearization for $\ln I = E/k_BT + B$ (B is a constant), the spectra were normalized to $F_p = 100$ kV/m using a linear scale for $I(F_p)$; temperature dependences of (e) C_{uw} (NMR) and TSD current versus T and (f) dc relaxation (TSDC) and C_{uw} value versus $1/T$ for RBC or RBC shadow corpuscles. (a–d: Adapted with kind permission from Springer Science+Business Media: *Centr. Eur. J. Chem.*, Interaction of unmodified and partially silylated nanosilica with red blood cells, 5, 2007i, 951–969, Gun'ko, V.M., Galagan, N.P., Grytsenko, I.V. et al.; and e, f: *Colloids Surf. A: Physicochem. Eng. Aspects*, 336, Gun'ko, V.M., Turov, V.V., Zarko, V.I., Goncharuk, E.V., and Turova, A.A., Regularities in the behaviour of water confined in adsorbents and bioobjects studied by ^{1}H NMR spectroscopy and TSDC methods at low temperatures, 147–158, 2009e, Copyright 2009, with permission from Elsevier.)

TABLE 7.11
Activation Energy (kJ/mol) of the dc Relaxation

System	$E_{a,dc}$
Distilled water	125
NaCl (0.9 wt%)	42
SCG	155
TSC	74
RBC/SCG	47
RBS/SCG/TSC	122
RBC/A-300/SCG/TSC (0.01 wt%)	149
RBC/A-300/SCG/TSC (1.0 wt%)	132

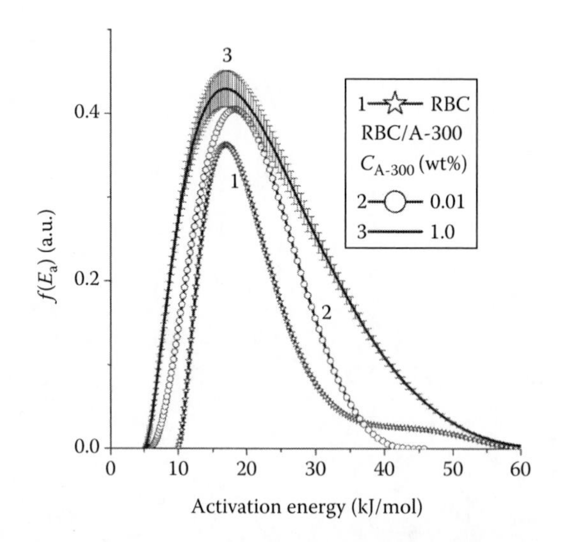

FIGURE 7.48 Distribution function of activation energy of depolarization for the RBCs in the trisodium citrate buffer and in the A-300 suspension (in TSC) at CA-300 = 0.01 and 1.0 wt%. (Adapted with kind permission from Springer Science+Business Media: *Centr. Eur. J. Chem.*, Interaction of unmodified and partially silylated nanosilica with red blood cells, 5, 2007c, 951–969, Gun'ko, V.M., Galagan, N.P., Grytsenko, I.V. et al.)

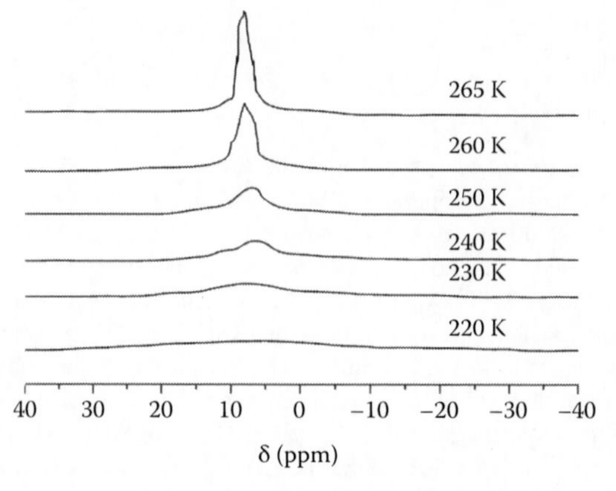

FIGURE 7.49 ^{1}H NMR spectra of water bound to RBC membranes (shadow corpuscles) in the 1.5 wt% aqueous suspension.

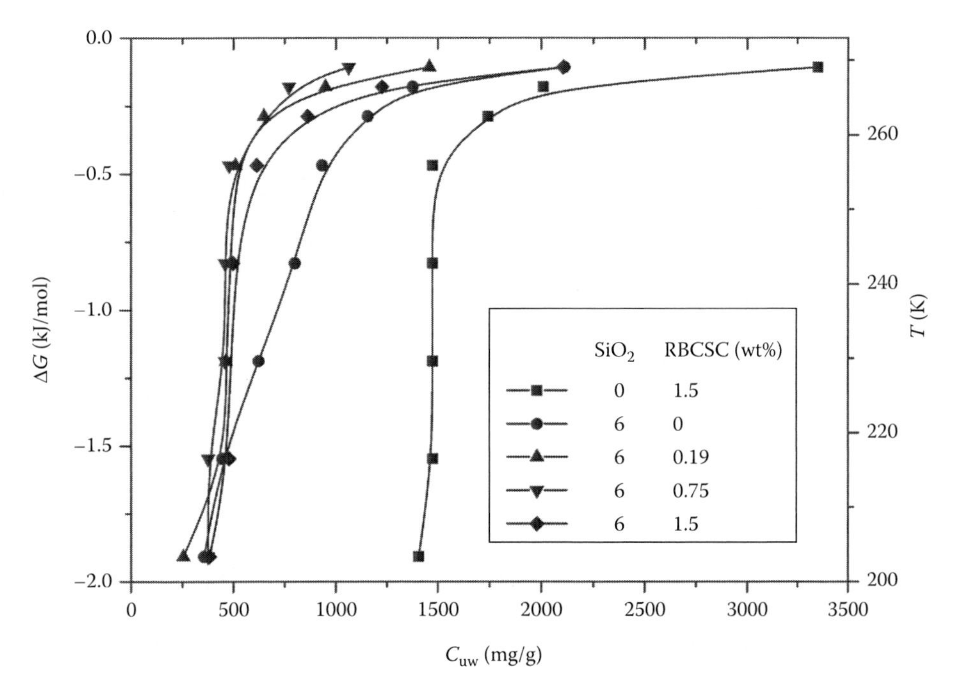

FIGURE 7.50 Relationships between changes in the Gibbs free energy of unfrozen water and its amounts and the corresponding thawing temperature (measurements were carried out on heating of the samples) of aqueous suspensions of RBCSC (0.0–1.5 wt%) with added nanosilica (6 wt%).

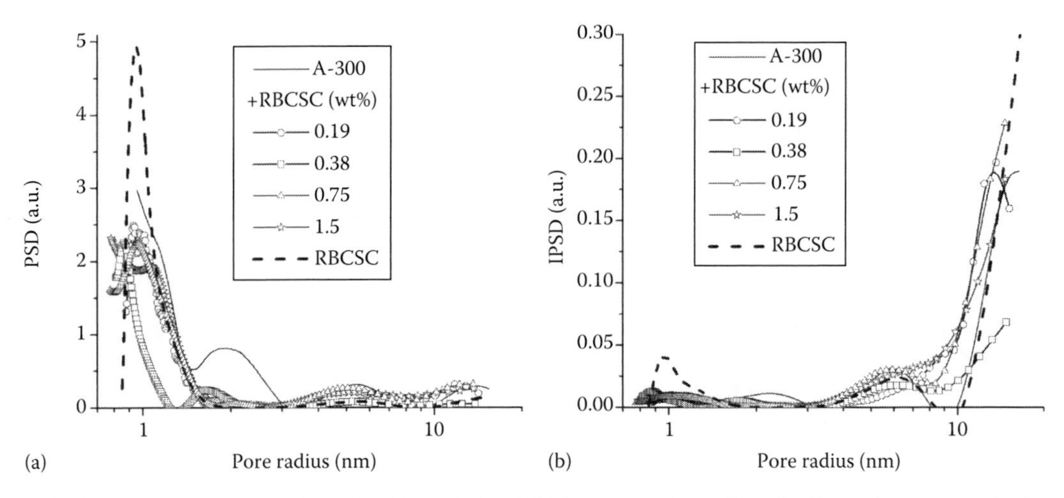

FIGURE 7.51 Size distributions (a) differential and (b) incremental of voids filled by unfrozen water in the aqueous suspensions of A-300 alone or with RBCSC.

properties of the cellular membranes because the contact area of unfrozen water with the native yeast cells and RBCSC is similar (~1500 m^2/g).

However, there is a certain difference in the PSD of yeast cells (Figure 7.13) and RBCSC (Figure 7.51) because hemolysis of RBC and removal of intracellular organics cause the formation of larger water structures in the shadow corpuscles. RBCSC bind a large amount of water (Table 7.12) with much greater volume of nanopores (V_{nano}) than in the mixtures with nanosilica. The interaction of RBCSC with silica nanoparticles leads to compaction of the membranes and both nanopores and

TABLE 7.12
Characteristics of Water Bound to RBCSC Alone or with the Presence of A-300 in the Aqueous Media

C_{RBCSC} (wt%)	C_w (mg/g)	C_s (mg/g)	$-\Delta G_w$ (kJ/mol)	$-\Delta G_s$ (kJ/mol)	γ_S (J/g)	S_{uw} (m²/g)	S_{nano} (m²/g)	S_{meso} (m²/g)	V_{nano} (cm³/g)	V_{meso} (cm³/g)	$\langle T \rangle$ (K)
1.5[a]	3475	1480	0.3	3.07	238	1512	1307	205	0.614	1.518	223.7
0.0[b]	2028	1000	0.3	2.67	101	323	106	217	0.052	1.560	246.1
0.19[b]	2074	500	0.4	2.65	77	486	293	193	0.138	1.264	249.0
0.38[b]	913	500	0.3	2.85	69	732	648	84	0.291	0.528	234.9
0.75[b]	2141	500	0.3	2.94	86	612	429	183	0.193	1.167	246.7
1.5[b]	2685	500	0.4	2.91	88	619	438	181	0.196	1.145	245.8

$C_{A\text{-}300}$ = [a]0 or [b]6 wt%.

mesopores (V_{meso}) decrease. However, S_{meso} decreases much smaller than V_{meso} in contrast to S_{nano}. This can be explained by a partial collapse of nanopores and compaction of mesopores without their collapse (Table 7.12 and Figure 7.51). These conformational changes are accompanied by partial dehydration of the membranes due to interaction with silica because the sum $C_w + C_s$ (as well as γ_S) is maximal for RBCSC alone and its value is smaller for the mixtures than that for silica alone. Additionally, conformational changes of RBCSC interacting with nanosilica lead to an increase in the average freezing temperature (Table 7.12, $\langle T \rangle$). Notice that the compaction of mesopores and diminution of the γ_S value are maximal at $C_{RBCSC} = 0.38$ wt%. This system can be characterized by relatively maximum interactions between silica and RBCSC.

The size of free RBCSC can be equal to 5–7 μm (Kozlov and Markin 1986). Therefore, a large number of primary nanoparticles can interact with the membrane (Figures 7.1 and 7.44). This interaction results in diminution of the amounts of SBW (Table 7.12).

Calculations of the SDWC distributions with the TSDC cryporometry and the size distributions of unfrozen water structures with the NMR cryoporometry (Figure 7.52) confirm changes in the water structures because of changes in the buffer composition, RBC content, and on addition of silica. Calculations of the specific surface area of structures being in contact with unfrozen bound water on the basis of the NMR cryoporometry (Gun'ko et al. 2007f,g) give 248 m²/g for silica ($S_{BET} = 285$ m²/g) in the aqueous suspension (6 wt%), 289 m²/g for shadow corpuscles (1.5 wt%), and 215 m²/g for their mixture. Consequently, the interaction of silica nanoparticles with the RBC membranes (as shadow corpuscles) leads to displacement of the interfacial water. Similar effects were observed on interaction of nanosilica with proteins (Gun'ko et al. 2005d, 2007i, Blitz and Gun'ko 2006). In the RBC/SCG system, much higher intensity of the SDWC distribution at $R > 1$ nm is due to large concentration of RBCs in comparison with other systems. Notice that besides water molecules polar macromolecules (e.g., intracellular and membrane proteins [Haris et al. 1995], phospholipids, etc.) can contribute the TSDC (Gun'ko et al. 2007i), as well as salt ions on the dc relaxation. According to the TSDC data, mobile water molecules and mobile ions appear in the buffered RBCs at ~210 K for all studied systems. Typically, at similar temperatures mobile interfacial water is observed (e.g., by ^{1}H NMR spectroscopy; Turov et al. 2006a) in hydrated proteins, cells, and other bio-objects (Gun'ko et al. 2005, 2007i). Decomposition of large water structures in the RBC/A-300 system at $C_{A\text{-}300} = 0.01$ and 1 wt% is also confirmed by the calculations of the dielectric constant (Gun'ko et al. 2007c) (for unfrozen bound water/RBCs/silica structures) which decreases from 7.8–10 (buffered individual RBCs) to 3.7–4.4 (RBC/A-300 at $C_{A\text{-}300} = 0.01$ wt%) and 2.2–2.7 ($C_{A\text{-}300} = 1$ wt%) calculated from the TSDC data at $T < 210$ K.

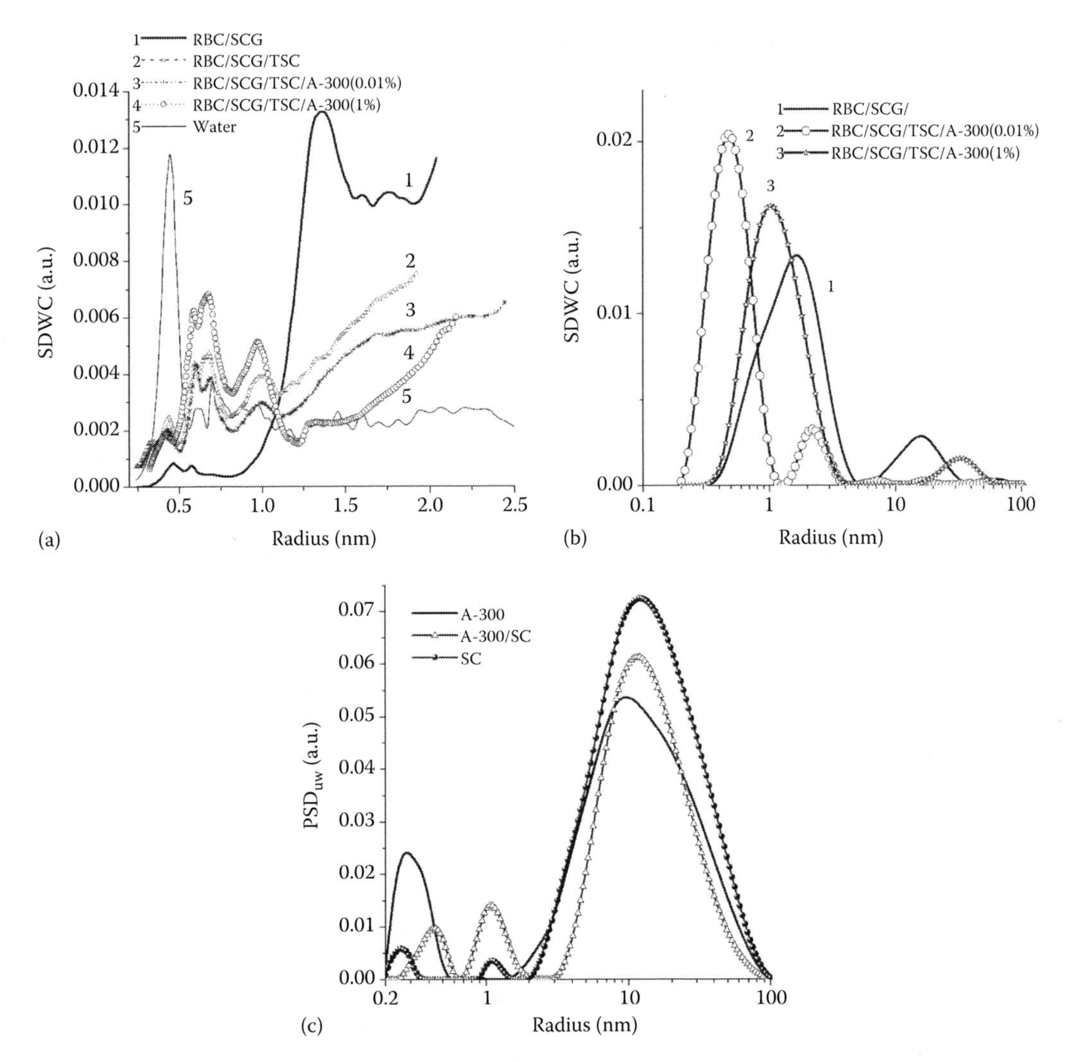

FIGURE 7.52 Distribution functions of sizes of water clusters relaxing during TSDC measurements in different systems with (a) RBCs (in two buffers), RBC/A-300 (two concentrations of A-300) and pure water (GT equation), (b) IGT equation, and (c) distribution functions of sizes of unfrozen water structures in aqueous suspensions of A-300 (6 wt%), shadow corpuscles (CS, 1.5 wt%), and a mixture of A-300 (6 wt%) and CS (1.5 wt%) IGT calculated on the basis of ^{1}H NMR spectra of unfrozen water at $200<T<273$ K. (Adapted with kind permission from Springer Science+Business Media: *Centr. Eur. J. Chem.*, Interaction of unmodified and partially silylated nanosilica with red blood cells, 5, 2007c, 951–969, Gun'ko, V.M., Galagan, N.P., Grytsenko, I.V. et al.)

Typically, the smaller the water clusters (water with bio-organics here) the lower the dielectric constant. For the TSDC peak which is close to the mentioned temperature (210 K), the E_a value (as well as the free energy of activation) (Gun'ko et al. 2007i) slightly increases due to the A-300 effects. However, large differences in the activation energy of the dc relaxation ($E_{a,dc}$) are observed for the studied systems (Table 7.11). Relatively high $E_{a,dc}$ values for the buffered systems (containing large number of ions) can be caused by much higher activation energy of the motion of citrate anions than that of protons and the trap effects of anions (and glucose molecules) for protons. Notice that the intracellular conductivity in RBCs is 0.5–0.8 S/m (Fomekong et al. 1998), and the RBC cytoplasm is not an ideal suspension, since the conductivity for the given ionic strength should be as twice as high, i.e., about 1.2–1.4 S/m (Fomekong et al. 1998) that can be explained by the interactions of ions with the large hemoglobin and other macromolecules, which decrease the mobility of ions.

The effects of silica on the $E_{a,dc}$ value is due to changes in the structure of the hydrogen bond network in the interfacial water, induced conformation changes in biomacromolecules and RBCs, and an increase in concentration of mobile protons on dissociation of silanols. In the case of changes in the medium (with RBCs), larger changes are observed for the high-temperature band of the TSDC spectra than that for the low-temperature band (Figures 7.47 and 7.48). Consequently, the structure of large water domains (Gun'ko et al. 2005d) changes strongly in comparison with that of the first interfacial water layers because structured water is strongly bound in the biomacromolecules (Gun'ko et al. 2005d, 2007i) and other structures composing RBCs.

Certain similarity in the interfacial behavior of water studied by NMR and TSDC methods is observed for RBC since the local molecular mobility in intracellular water (in RBC or shadow corpuscles obtained on 100% hemolysis of RBC) appears at ~185 K (NMR) but the dc relaxation (TSDC) of water bound in RBC begins at 217 K (Figure 7.47).

The summarized microscopic, XRD, TSDC, and cryoporometry data testify the presence of correlation between concentration of silica in the buffered aqueous suspensions and morphological changes of RBCs with appearance of the cells with irreversible destruction of membranes and formation of shadow corpuscles. The effects of modified nanosilica with half silanols substituted by TMS groups are weaker in comparison with those of unmodified A-300 because of aggregation of silica particles and the absence of very bioactive individual primary particles which are observed in the aqueous suspension of unmodified nanosilica. Silica nanoparticles strongly interacting with RBCs change the structure of not only the interfacial water bound by cells (e.g., their membranes) but also large water structures in the bulk.

7.5 BONE TISSUE

Bone tissue includes hydroxyapatite (HAP), collagen, non-collagenous proteins, proteoglycans, and water as the main components (Katz 1969, Burr 2002). The structure of bone tissues was studied in detail by FTIR and NMR spectroscopies (Wu et al. 2002, Wilson at al. 2005a,b, Gröger et al. 2009, Sigmund et al. 2009), TEM, SEM, AFM (Robinson 1952, Eppell et al. 2001), and other methods. The HAP crystallites in size of ~2 nm can have about 98% of the mineral matrix (Eppell et al. 2001). Bone tissue is not inert and replacement of 5%–10% of the tissue occurs in renovation processes during a year (Malcolm 2002). The metabolic disorder (resulting in, e.g., osteoporosis) can lead to changes in the mineral matrix of bone tissue (Flavus 1999) with reduction of the density in the HAP content. HAP is not practically dissolved in water; therefore, specific conditions are necessary for its synthesis and accumulation in bones. The presence of water bound in bone tissue and having a specific state (low or high density waters; Chaplin 2011, Wiggins 2008) can be one of the factors influencing the HAP formation. An ordered water layer sandwiched between the mineral and the organic collagen fibers may affect the biomechanical properties of bone tissue as a complex composite material (Wilson et al. 2005b).

7.5.1 WAW and SAW Bound in Bone Tissue

Turov et al. (2006a) studied tail bones of rats using low-temperature NMR spectroscopy. The bone tissue is very heterogeneous. Therefore, to obtain average characteristics of bound water three rat tail bones were placed simultaneously in 5 mm measuring NMR ampoule using nine bones from nine rats. Sample I (total weight 150 mg) corresponds to the initial bones. Then it was saturated by 160 mg of water (sample II). Sample III (132 mg) corresponds to sample I after 35 day storage at 253 K. Sample IV corresponds to sample III after heating at 393 K for 1 h. Then this sample was wetted for 1 h (sample V).

Figure 7.53 shows the ^1H NMR spectra of water in sample I in air and deuterochloroform. The spectra include two signals, one of which (signal 1) has chemical shift $\delta_H = 4$–5 ppm, and the second (signal 2) $\delta_H = 1.4$–1.7 ppm. The intensity of signal 1 more strongly decreases because of partial

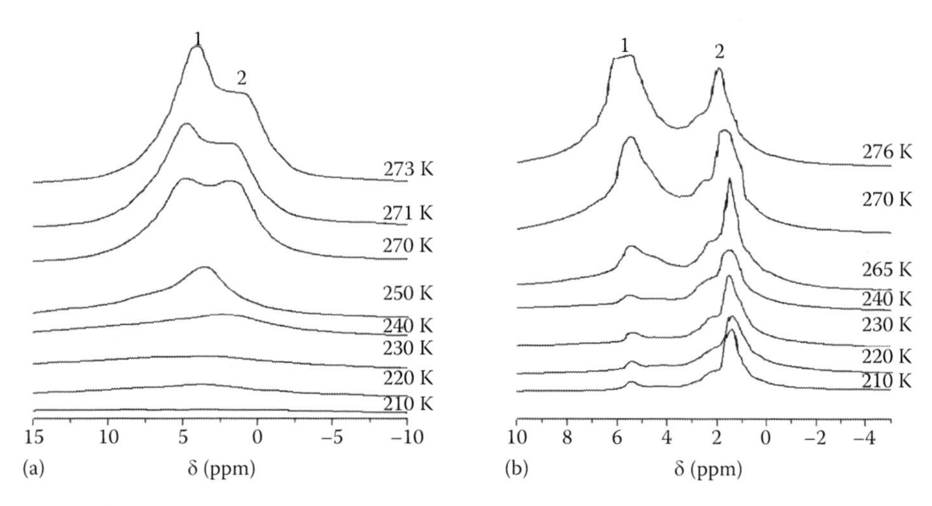

FIGURE 7.53 ^{1}H NMR spectra of water in partially hydrated bones at different temperatures in (a) air and (b) $CDCl_3$. (Adapted from *Colloids Surf. B: Biointerfaces*, 48, Turov, V.V., Gun'ko, V.M., Zarko, V.I. et al., Weakly and strongly associated nonfreezable water bound in bones, 167–175, 2006a, Copyright 2006, with permission from Elsevier.)

freezing of bound water, while signal 2 remains practically without the changes at $T>250$ K. The freezing of water, critical for signal 2, occurs at $T<250$ K. The replacement of air by chloroform (Figure 7.53b) leads to a significant decrease of the width of both signals. This occurs because of decrease of heterogeneous broadening and retarding of the molecular exchange between water molecules linked to separate sites at the interfaces (Gun'ko et al. 2009d). However, the character of changes in the intensity of signal 1 with temperature remains nearly the same as in air. However, the intensity of signal 2 decreases in the $CDCl_3$ medium less than in air.

The chemical shift of signal 1 coincides with that of liquid bulk water or water adsorbed on nonporous nanoparticles or porous solids or in hydrated shells of biomacromolecules (Turov and Leboda 1999, Gun'ko et al. 2005d). It is determined by the strength and the average number of the hydrogen bonds per water molecule (Gun'ko and Turov 1999). Experimental studies (Kinney et al. 1993, Chuang and Maciel 1996, Gun'ko et al. 2009d) and quantum chemical calculations (Gun'ko and Turov 1999, Gun'ko et al. 2005d) show that the maximum δ_H value of water molecules participating in four hydrogen bonds (of optimal structure corresponding to hexagonal ice) with neighboring molecules is 7 ppm. If a water molecule participates in a hydrogen bond as an electron donor (i.e., $H_2O\cdots H{-}X$), the δ_H is 1.2–1.7 ppm for the protons of this molecule (e.g., water in nonpolar or weakly polar media). This δ_H value is close to that of water dissolved in chloroform ($\delta_H = 1.7$ ppm).

The δ_H value of bound water depends on the structure of associates in the hydrated shells of bio-objects or solid particles. Water with signal 2 (Figures 7.53 and 7.54) corresponds to SBW because it freezes at temperatures considerably lower than the freezing temperature characteristic for water with signal 1. Thus, signal 2 is linked to water strongly bound but characterized by a small number of the hydrogen bonds per molecule, i.e., this water is weakly associated. According to theoretical calculations, signal 2 can be attributed to small clusters with two to five molecules, in which a substantial part of protons does not participate in the hydrogen bonds but can interact with nonpolar functionalities, e.g., of biomacromolecules (as well as with weakly polar chloroform molecules).

Quantum chemical calculations (with consideration of the effects of a solvent) of water clusters $(H_2O)_n$ at $n = 1, 2, 4, 6$ in cyclohexene (as a model nonpolar medium) show that the free energy of solvation is minimal for a dimmer in which 75% of protons do not participate in the hydrogen bonds and $\delta_H = 0.8$–1.9 ppm corresponding to signal 2. These calculations show a possible reason for the appearance of the aforementioned signal for bone tissue.

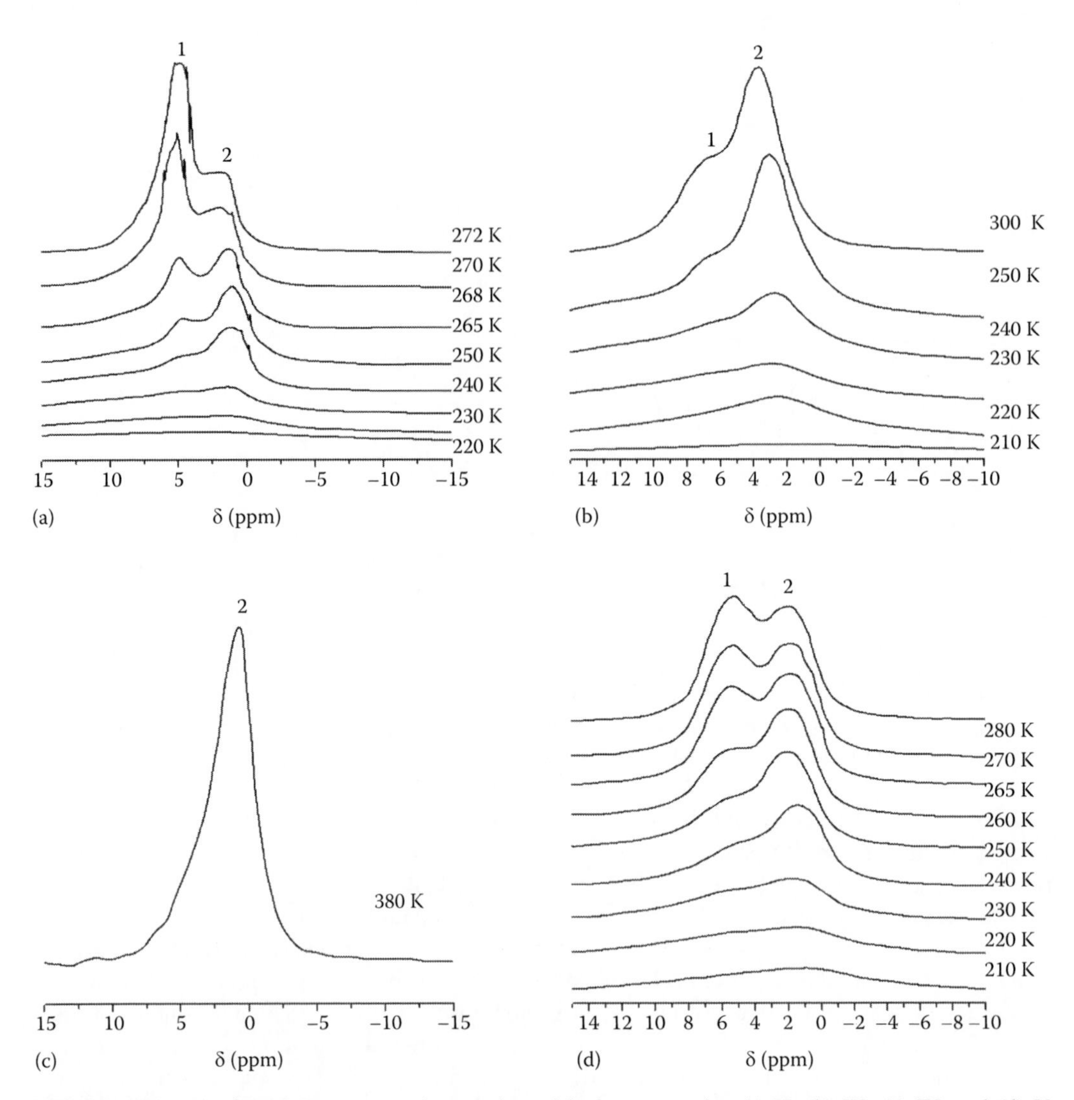

FIGURE 7.54 The ^{1}H NMR spectra of water bound in bone samples (a) II, (b) III, (c) IV, and (d) V. (Adapted from *Colloids Surf. B: Biointerfaces*, 48, Turov, V.V., Gun'ko, V.M., Zarko, V.I. et al., Weakly and strongly associated nonfreezable water bound in bones, 167–175, 2006b, Copyright 2006, with permission from Elsevier.)

Figure 7.54 shows changes in the intensities of signals 1 and 2 of water bound in bones of samples I–V. Saturation of the sample I by water (sample II) leads to a relative increase in the intensity of signal 1 (Figure 7.54a). Prolonged storage of a sample at a low temperature (sample III) leads to its partial dehydration (Figure 7.54b) which occurs predominantly because of a decrease in the amount of water characterized by signal 1. Heating of the sample at 393 K for 1 h (Figure 7.54c) leads to complete disappearance of water with signal 1. This signal is restored after wetting of the sample (Figure 7.54d). However, substantial changes in the intensity of signal 2 are not observed.

Freezing of water in sample III is observed only at $T<250$ K; consequently, this water is strongly bound. For the samples I and II, there is a section on the $\Delta G(C_{uw})$ graphs (Figure 7.55), which corresponds to WBW ($\Delta G>-0.5$ kJ/mol) freezing near 273 K. The data shown in Table 7.13 and Figures 7.53 and 7.54 allow us to conclude that in the case of sample I the relationship of the intensities of signals 1 and 2 is close to that of the amounts of SBW and WBW, respectively. For the samples II and III, a portion of SBW is linked to signal 2. The ΔG_s value for all the samples is identical and equal

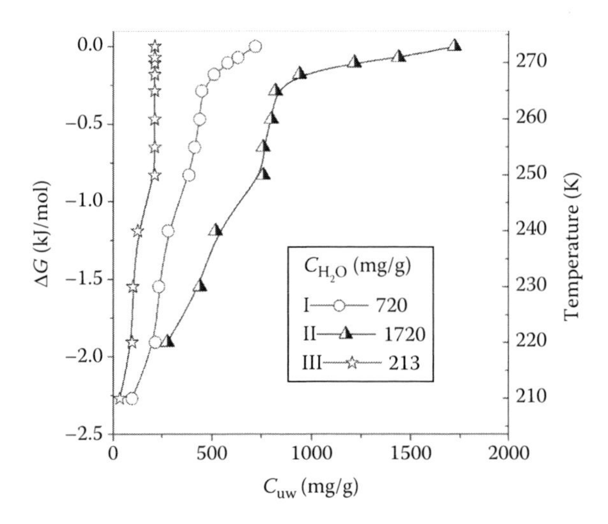

FIGURE 7.55 Amounts of unfrozen water C_{uw} is a function of temperature and ΔG versus C_{uw} for bone samples I–III. (Adapted from *Colloids Surf. B: Biointerfaces*, 48, Turov, V.V., Gun'ko, V.M., Zarko, V.I. et al., Weakly and strongly associated nonfreezable water bound in bones, 167–175, 2006b, Copyright 2006, with permission from Elsevier.)

TABLE 7.13
Characteristics of Bound Water in Bone Samples I–III

Sample	C_{H_2O} (mg/g)	$-\Delta G_s$ (kJ/mol)	$-\Delta G_w$ (kJ/mol)	γ_S (J/g)	γ_S* (J/g)	C_{uw}^{w} (mg/g)	C_{uw}^{s} (mg/g)
I	720	2.5	0.6	42	58	270	450
II	1720	2.5	0.4	76	44	920	800
III	213	2.5	0	19	89	0	213

Note: $\gamma_S^* = \gamma_s/C_{H_2O}$ (at C_{H_2O} in g/g).

to −2.5 kJ/mol, which is smaller by the modulus than that for the majority of biological subjects and disperse or porous solid particles. This value determines the maximum reduction of the Gibbs free energy in the first water monolayer at the interfaces. Typically, this reduction is greater if the hydrogen bonds of the water molecules with surface sites are stronger and their number is higher. Hence, it is possible to conclude that water bound in bone tissues relatively weakly interacts with the mineral matrix and biomolecules. A significant intensity of signal 2 confirms this conclusion because this signal is linked with the H atoms which do not participate in the strong hydrogen bonds. The γ_S value changes proportionally to the water concentration in samples. However, the $\gamma^* = \gamma_S/C_{H_2O}$ value (Table 7.13) determining the interfacial free energy averaged per unit of the mass of the total amount of bound water demonstrates opposite dependence because of stronger changes in C_{uw}^{w} than C_{uw}^{s} (Table 7.13).

Signals 1 and 2 (Figures 7.53 and 7.54) are observed as separated ones. Therefore, it is possible to conclude that the molecular exchange between the different forms of water is slow in bone tissue. This is possible if there are spatial or energetic limitations of these exchange processes. The mineral matrix or biomacromolecules located in the structural voids of the bone can create certain barriers for this exchange. The specific surface area of the mineral component of the bone tissue is relatively small. If one assumes that small water clusters are placed along the entire surface of the mineral matrix, then the volume of bound water gives a small portion of the volume of solid. From other side, proteins and other macromolecules forming the three-dimensional network in the volume

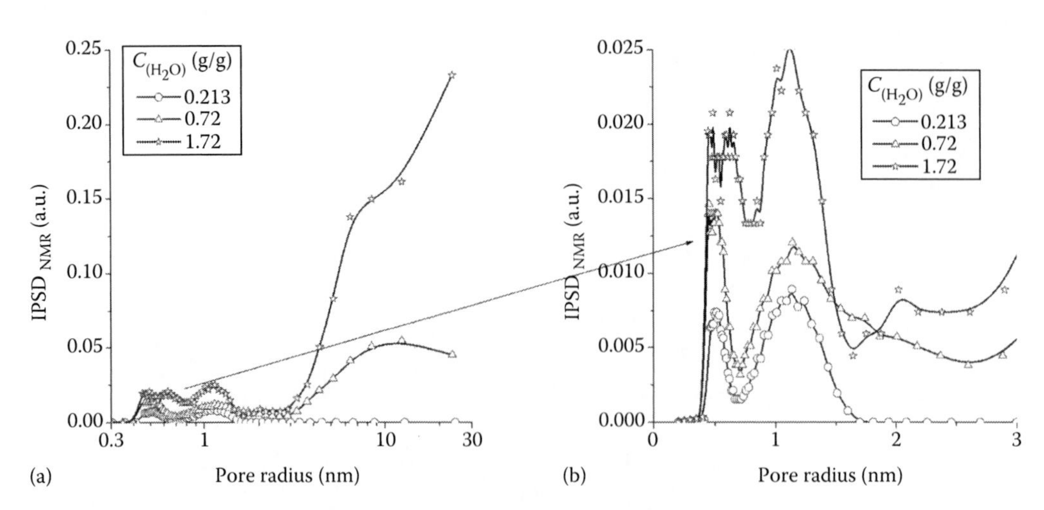

FIGURE 7.56 Pore size distribution in bone tissue calculated on the basis of the $C_{uw}(T)$ graphs. (Adapted from *Colloids Surf. B: Biointerfaces*, 48, Turov, V.V., Gun'ko, V.M., Zarko, V.I. et al., Weakly and strongly associated nonfreezable water bound in bones, 167–175, 2006b, Copyright 2006, with permission from Elsevier.)

of the structural voids of the bone are submerged in the aqueous medium. The "surface" area of biomacromolecules can reach a very high value up to 1000 m^2/g because the size of these macromolecules corresponds to several nanometers and their "surface" is not smooth. These molecules form numerous pockets with both hydrophilic and hydrophobic walls. These structural features of bone tissue set conditions for the formation of small water clusters located in these pockets. If the total amount of water in bone tissue is not high, then these clusters are segregated that results in the maximum intensity of signal 2. An increase in the amount of water in bone tissue causes association of small water clusters into larger structures, and the intensity of signal 1 increases.

The use of Gibbs–Thomson equation allows us to calculate the IPSD for pores filled by unfrozen water (Figure 7.56). In the case of minimal amount of water, it fills only narrow pores at the pore radius $R<1.5$ nm. These IPSD peaks corresponds to water stronger ($0.7<R<1.5$ nm) and weaker ($R<0.7$ nm) associated in clusters. Additionally, at $C_{H_2O} = 213$ mg/g all water is strongly bound as it is located only in narrow pores (Figure 7.56a) and this is in agreement with the shape of $\Delta G(C_{uw})$ (Figure 7.55). The amount of SAW increases much faster than that of WAW with increasing C_{H_2O} value. This leads to appearance of water in pores of larger size (Figure 7.56) in the form of larger clusters in which the associativity of water molecules increases. Therefore, the intensity of signal 1 increases (Figures 7.53 and 7.54).

The specific spatial distribution of hydrophobic structures (cavities, pockets) in bone tissue occupied by water can serve as a stabilizing factor for WAW. For instance, hydrophobic functionalities at a surface of partially modified silicas cause the appearance of WAW in the aqueous suspensions (Pople et al. 1959, Chuang and Maciel 1996). It is possible to assume that WAW is an essential component of living organism tissues and cells because it plays an important role in stabilizing the structure of biomacromolecules, in transport and metabolic processes. For instance, in the case of strongly braked processes in dry seeds and yeast cells, the main portion of intracellular water is in the weakly associated state. Acceleration of metabolic processes is observed after substantial increase in the amount of intracellular water when a portion of SAW increases creating a continuous medium for transport processes.

Rearrangement of water bound in bone tissue with increasing C_{H_2O} value affects the relaxation processes which appear in the TSDC spectra (Figure 7.57). A low amount of water in the studied bones causes a very low intensity of the TSDC spectrum (Figure 7.57, initial bone tissue). This spectrum depicts the presence of several types of water such as (i) WAW relaxing at $T<120$ K, (ii) small water clusters with 2–3 hydrogen bonds per molecule ($120<T<160$ K), and (iii) water microdomains at 3–4 hydrogen bonds per molecule ($T>160$ K). The TSDC spectrum strongly changes

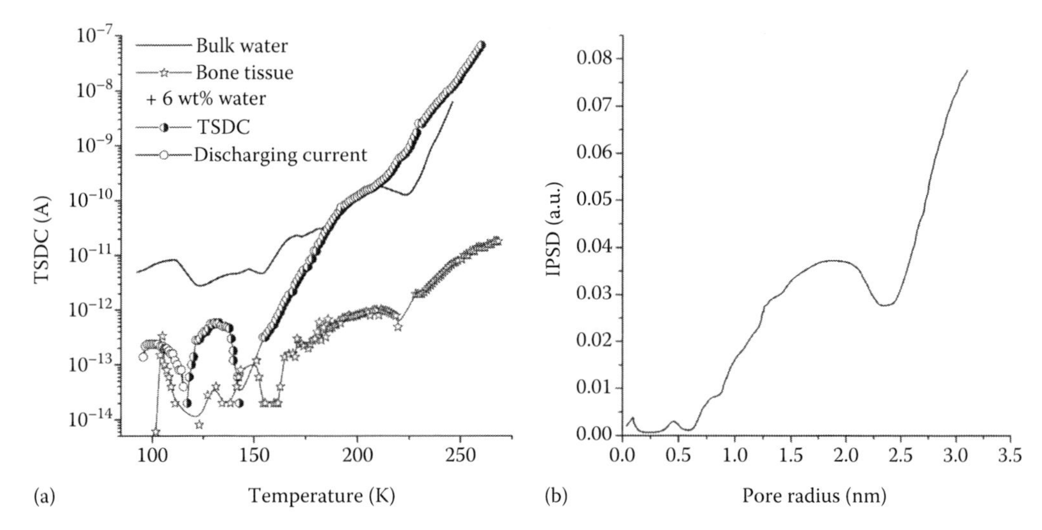

FIGURE 7.57 (a) TSDC spectra of individual water and water located in rat tail vertebrae initially and after additional adsorption of water vapor for 24 h (+6 wt%) (discharging current has the opposite sign to that of the TSD current) and (b) "pore" size distribution calculated on the basis of the TSDC for initial bone tissue (sample I) for dipolar relaxation at $T<225$ K. (Adapted from *Colloids Surf. B: Biointerfaces*, 48, Turov, V.V., Gun'ko, V.M., Zarko, V.I. et al., Weakly and strongly associated nonfreezable water bound in bones, 167–175, 2006a, Copyright 2006, with permission from Elsevier.)

after additional adsorption of ~6 wt% of water (adsorption of saturated water vapor for 24 h), especially at $T>160$ K because of the formation of larger continuous water structures. The TSDC spectra revealing changes in relaxation processes in the interfacial water with increasing its content are in agreement with the NMR data.

Thus, a substantial part of bound water in bone tissue is in the weakly associated state characterized by the chemical shift $\delta_H=1.4$–1.7 ppm close to that of individual water molecules in hydrophobic environment. A mass fraction of this anomalous water is compared with a mass fraction of mineral component. This water is strongly bound in bone tissue, fills very narrow pores, and freezes at $T<250$ K. It is moved away from the bones only on strong heating.

7.5.2 Characteristics of Water Bound in Human Bone Tissues Healthy and Affected by Osteoporosis

The influence of the medium and temperature on a state of water bound in a spongy component of human bones healthy and affected by osteoporosis was analyzed using NMR and other methods (Gun'ko et al. 2006i). Two samples of human vertebral bone tissue were from healthy bone (man of 37 year old, sample M37) and bone affected by osteoporosis (man of 61 year old, sample M61). The bone mineral density (as a planimetric "x-ray shadow" density estimated from the XRD data obtained with a Hologic QDK-4500A apparatus) of a central part of the vertebral body in the vertical projection of the M37 sample is higher by 46% than that of M61.

The ^{1}H NMR spectra of water bound in both bone samples being in air and aqueous media (equilibrating for 2 h) are shown in Figure 7.58. In the case of sample M37 in air, two signals at $\delta_H\approx5$ ppm (signal 1) and 1.4 ppm (signal 2) are observed at $T>260$ K (Figure 7.58a). At $T\approx270$ K, the intensities of both signals are close. However, at higher temperatures signal 1 is more intensive but at $T<270$ K the opposite relationship is observed and only signal 2 is observed at $T<260$ K. In the case of sample M61 with bone tissue affected by osteoporosis, the intensity of signal 1 is predominant since signal 2 is low (Figure 7.58b). In the aqueous medium, the width of both signals strongly decreases (probably, because of a decrease in heterogeneous broadening). The relationship

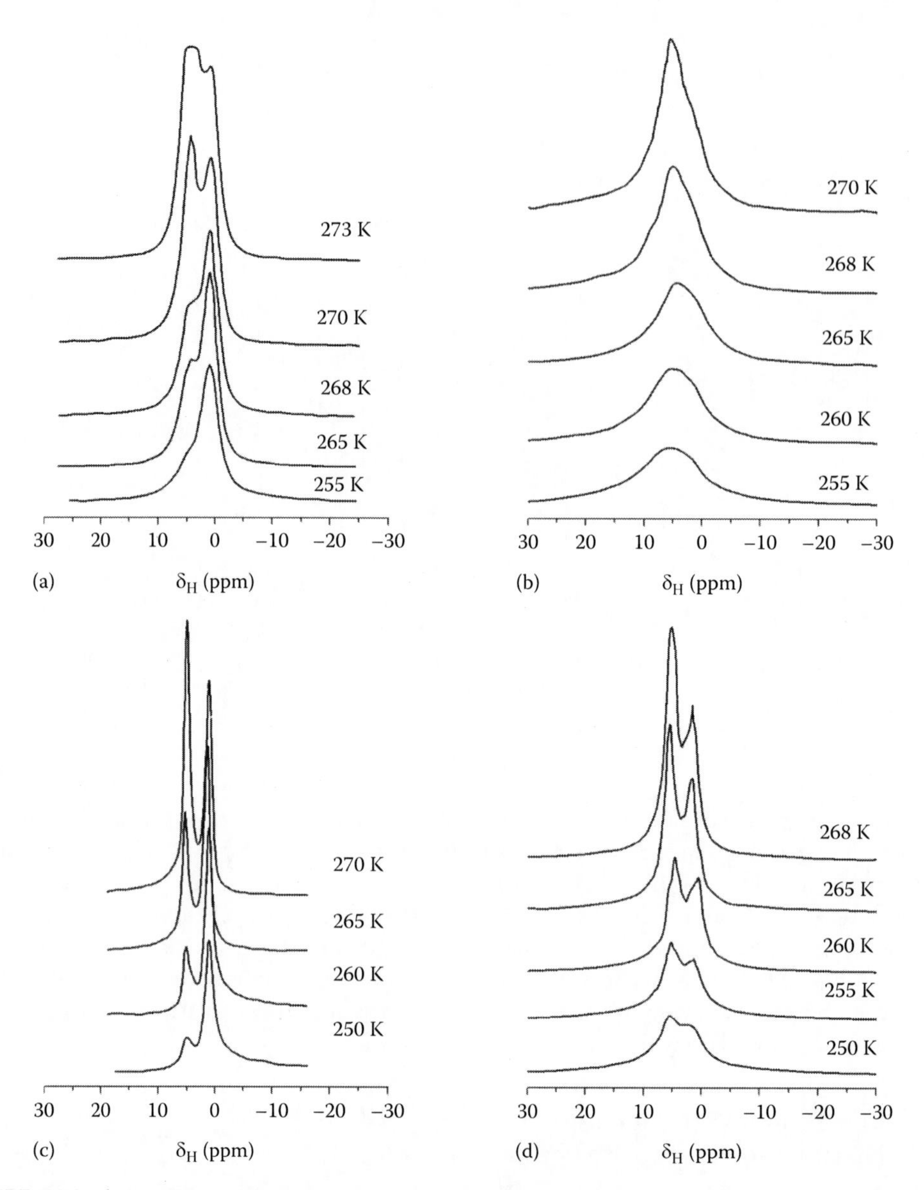

FIGURE 7.58 ^{1}H NMR spectra for samples (a, c) M37 and (b, d) M61 in (a, b) air (h=0.5 and 0.55 g/g, respectively) and (c, d) water (h=5.7 and 7.1 g/g, respectively) (a, c, d) the left and right signals correspond to signals 1 and 2, respectively. (Adapted from *Colloid. Surf. B: Biointerfaces*, 53, Gun'ko, V.M., Turov, V.V., Shpilko, A.P. et al., Relationships between characteristics of interfacial water and human bone tissues, 29–36, 2006f, Copyright 2006, with permission from Elsevier.)

of intensities of signals 1 and 2 remains nearly the same for M37 (Figure 7.58c). Signal 2 in M61 is more intensive (Figure 7.58d) than that in the air medium (Figure 7.58b); however, it is lower than that of M37. The intensity of signal 1 rapidly decreases for M37 because of freezing of water with lowering temperature. The freezing point depression for water with signal 1 is more significant for M61 than that for M37. However, the opposite result is observed for water responsible for signal 2.

Figure 7.59 shows dependence of the amounts of unfrozen water $C_{uw}(T)$ as a function of temperature and changes in the Gibbs free energy ΔG of structured water caused by interaction with the surrounding versus C_{uw}. Sample M37 was investigated in air, chloroform-d, water, and aqueous suspension of fumed silica A-300 (specific surface area ~300 m^2/g) at concentration

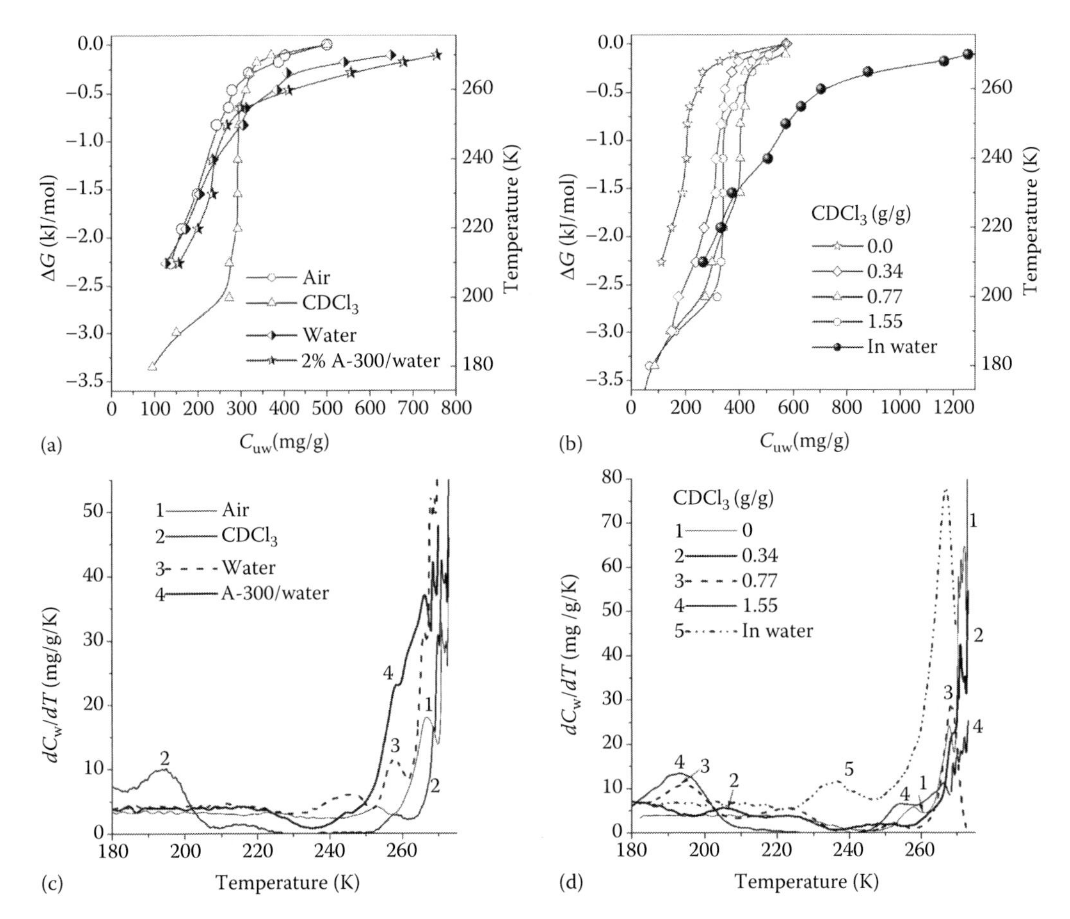

FIGURE 7.59 Amounts of bound unfrozen water C_{uw} as a function of temperature for samples (a) M37 and (b) M61, changes in the Gibbs free energy of bound water versus C_{uw} for samples (a) M37 and (b) M61, and derivatives dC_{uw}/dT for samples (c) M37 and (d) M61. (Adapted from *Colloid. Surf. B: Biointerfaces*, 53, Gun'ko, V.M., Turov, V.V., Shpilko, A.P. et al., Relationships between characteristics of interfacial water and human bone tissues, 29–36, 2006f, Copyright 2006, with permission from Elsevier.)

$C_{A\text{-}300} = 2$ wt%. Sample M61 was studied in air, water, and in the deuterochloroform medium at 0.34–1.55 g of chloroform-*d* per gram of bone tissue. There are two or three fragments on the $\Delta G(C_{uw})$ graphs corresponding to fast and slow changes in ΔG versus C_{uw}. The first section of the slow changes at $\Delta G > -0.5$ kJ/mol (or slightly lower) (Figure 7.59a and b) corresponds to WBW which is mainly strongly associated (Figure 7.60) (Dullien 1992, Gun'ko et al. 2005d), i.e., it is responsible for signal 1.

This section corresponds to significant changes in $C_{uw}(T)$ at maximal values of the derivative $dC_{uw}(T)/dT$ (Figure 7.59c and d). The second portion of the $\Delta G(C_{uw})$ graphs is nearly vertical, i.e., significant changes in ΔG correspond to small changes in the amounts of unfrozen water (small values of the derivative $dC_{uw}(T)/dT$). This water is strongly bound and it is contributed by both WAW and SAW because both signals 1 and 2 are observed in the ^{1}H NMR spectra over the corresponding temperature range (Figure 7.58). The third section of slow changes in ΔG at significant changes in C_{uw} is observed only for samples in the $CDCl_3$ medium (Figure 7.59). This water can be assigned to weakly associated one (comp. Figures 7.59 and 7.60) because of the small δ_H value (signal 2) and it can freeze only with significant lowering temperature $T < 220$ K for M37 and $T < 240$ K for M61.

Calculations by using the cryoporometry give complex IPSDs for both samples (Figure 7.61). The calculated IPSDs are affected by the media because of rearrangement of the pore water at

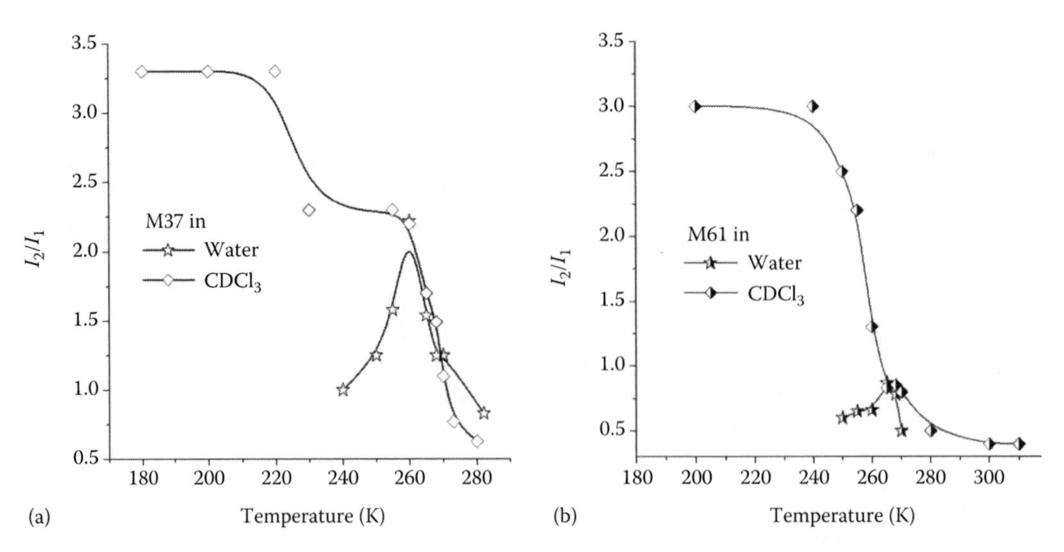

FIGURE 7.60 Relationship between the ^{1}H NMR signals of weakly (I_2) and strongly (I_1) associated waters as a function of temperature for samples (a) M37 (in chloroform at C_{CDCl_3} = 1.21 g/g and h=0.5 g/g and in water at h=5.7 g/g) and (b) M61 (in chloroform at C_{CDCl_3} = 1.55 g/g and h=0.55 g/g and in water at h=7.1 g/g). (Adapted from *Colloid. Surf. B: Biointerfaces*, 53, Gun'ko, V.M., Turov, V.V., Shpilko, A.P. et al., Relationships between characteristics of interfacial water and human bone tissues, 29–36, 2006f, Copyright 2006, with permission from Elsevier.)

different interfaces with air, chloroform-*d*, or aqueous suspension of fumed silica A-300. This leads to diminution in the surface area in contact with unfrozen water (Table 7.14, S_w) in the case of the $CDCl_3$ medium. This result shows that chloroform can change the organization of the pore water. As a whole, the IPSDs for M37 correspond to narrower mesopores (cavities filled by water frozen at T<273 K) than that for M61. However, enlargement of mesopores in M61 is accompanied by increased contribution of nanopores at R<1 nm. Consequently, two tendencies are observed on osteoporosis: (i) broadening of mesopores and (ii) an increase in contribution of nanopores. This causes much larger S_w value for M61 than that for M37 in the case of the aqueous medium at maximum contact of water with biostructures in bone tissue. These results are in agreement with the morphological changes in bone tissue affected by osteoporosis because it is characterized by much lower density of the mineral matrix. The chloroform-*d* medium is responsible for two effects: (i) a fraction of water bound in narrow pores (mainly WAW) grows and (ii) occupation of maximum broad pores by structured water (SAW) increases because of the displacement by chloroform. This displacement of water into larger pores results in smaller values of the surface area S_w in contact with unfrozen water. Gaps between primary silica particles in their aggregates contribute narrow mesopores (Figure 7.61b, curve 3) that is typical for the aqueous suspensions of fumed oxides (Moelbert et al. 2004).

The characteristics of structured water bound in spongy component of bone tissue (healthy and affected by osteoporosis) being in the different media are given in Table 7.14. These characteristics of samples M37 and M61 being in air are relatively close; however, they strongly differ from the samples being in the aqueous and chloroform-*d* media. In the case of M37, the replacement of air by $CDCl_3$ does not lead to the redistribution of SBW and WBW (Table 7.14, C^{S}_{uw}, C^{W}_{uw}, and S_w). However, the value of free surface energy γ_S increases by 38%. This is caused by changes in the form of the $\Delta G(C_{uw})$ curve because the freezing temperature depression of SBW changes (Figures 7.58 and 7.59). In the case of M37 in the aqueous medium, the total amount of structured water increases by 80% and the surface area S_w in contact with unfrozen water grows. However, the ΔG_s value is the same and the γ_S value is between that for the air and chloroform media.

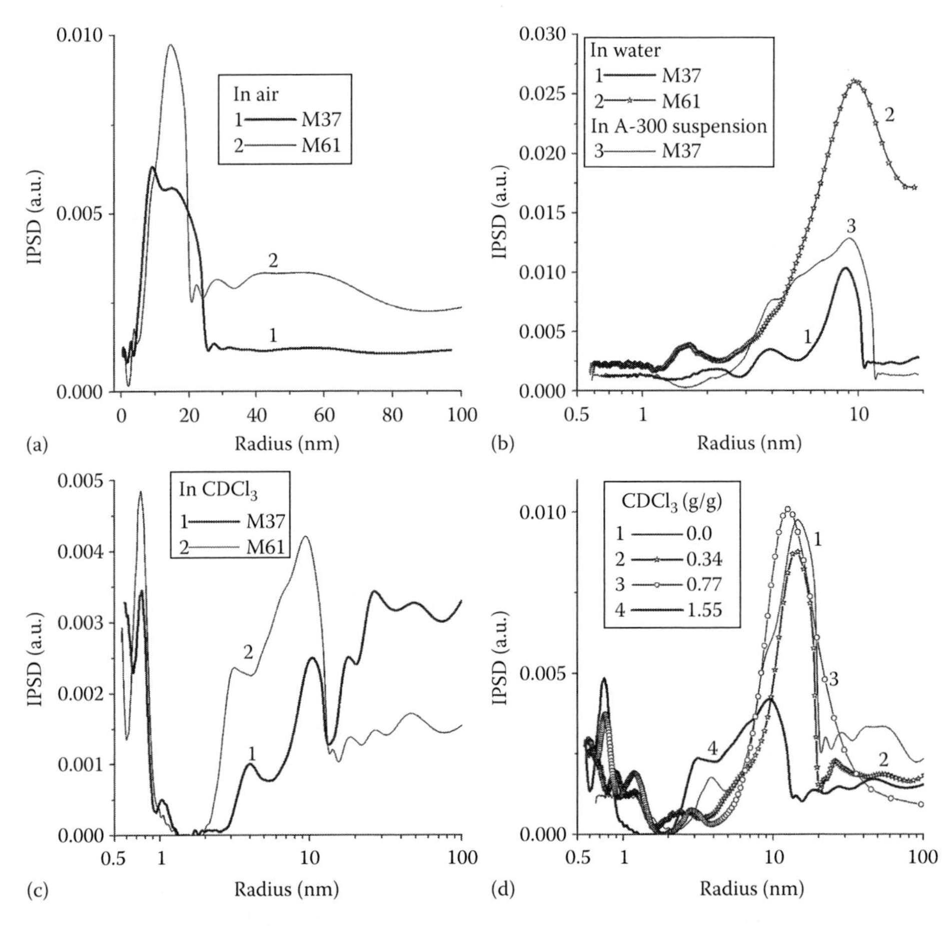

FIGURE 7.61 Incremental pore size distributions calculated using the cryoporometry method (IGT) for M37 and M61 in (a) air, (b) aqueous medium, and (c, d) chloroform-*d*. (Adapted from *Colloid. Surf. B: Biointerfaces*, 53, Gun'ko, V.M., Turov, V.V., Shpilko, A.P. et al., Relationships between characteristics of interfacial water and human bone tissues, 29–36, 2006f, Copyright 2006, with permission from Elsevier.)

TABLE 7.14
Characteristics of Water Bound in Bone Tissue of Samples M37 and M61

Sample	Medium	h (g/g)	C_{uw} (mg/g)	C^{s}_{uw} (mg/g)	C^{w}_{uw} (mg/g)	γ_S (J/g)	$-\Delta G_s$ (kJ/mol)	S_w (m²/g)
M37	Air	0.5	500	300	200	37	3.8	88
	$CDCl_3$ 1.21 g/g	0.5	500	300	200	51	3.8	77
	Water	5.8	900	300	600	44	3.8	145
	Water + 2% A-300	5.7	900	300	600	48	3.7	183
M61	Air	0.55	550	200	350	31	3.3	70
	$CDCl_3$ 0.34 g/g	0.55	550	320	230	52	3.9	81
	$CDCl_3$ 0.77 g/g	0.55	550	400	150	63	3.7	101
	$CDCl_3$ 1.55 g/g	0.55	550	330	220	61	3.9	122
	Water	7.1	1300	600	700	84	3.6	263

h is the hydration in gram of water per gram of dry material; C^{s}_{uw} and C^{w}_{uw} are concentrations of strongly and weakly bound waters, respectively; C_{uw} is the total amount of unfrozen water at $T \to 273$ K; ΔG_s is the maximum change in the Gibbs free energy of strongly bound water in the first interfacial monolayer; and γ_S is the value of free surface energy equal to summary reduction in the Gibbs free energy of bound water.

A larger increase in the free surface energy is observed for sample M61 comparing the results for the air and aqueous media because in the last case the sample is characterized by significantly increased amounts of WBW and SBW (Table 7.14), as well as greater values of S_w and γ_S. The effect of $CDCl_3$ (at its maximal amount) in comparison with air is stronger for M61 than that for M37 that is caused by an increase in concentration of SBW (Table 7.14, C^s_{uw}) and simultaneous decrease in the amounts of WBW (C^w_{uw}). The surface area in contact with unfrozen water is greater for M61 than that for M37 being in the $CDCl_3$ medium. For M61 in comparison with M37, weakly polar chloroform-*d* medium slightly strongly influences the WBW and its fraction transforms into SBW because of changes in its distribution in cavities of different sizes over ranges of nanopores at $R < 1$ nm and mesopores at $R > 1$ nm (Figure 7.61). In other words, chloroform displaces a fraction of water into both nanopores at $R < 1$ nm and mesopores at $R > 20$ nm. A whole bone tissue affected by osteoporosis is more loosely packed with HAP structures than the M37 sample over the total range of the pore sizes.

The maximum decrease in the Gibbs free energy in the first layer of SBW (ΔG_s) changes between −3.9 and −3.3 kJ/mol and weakly depends on the medium. This weak effect is caused by the fact that the ΔG_s value relates to weakly associated and SBW maximum disturbed by bone tissue functionalities and the medium replacement can only slightly influence this first layer because exchange between molecules of an added liquid (water, $CDCl_3$) and adsorbed water molecules is very slow especially in narrow pores at $R < 1$ nm (Figure 7.61) filled by WAW.

The effects of the aqueous suspension of fumed silica A-300 onto water bound in bone tissue of the M37 sample are close to that of pure aqueous medium (Table 7.14). Silica surface is strongly hydrated (Strange et al. 2003, Gun'ko et al. 2005d); however, the total amount of water bound to this surface is smaller than that bound in bone tissue. Perhaps, penetration of silica nanoparticles into narrow pores of the studied spongy component of bone tissue (where major amounts of structured water locate) does not occur; therefore, silica particles being out (or in macropores) of bone tissue are not capable to affect water locating in bone tissue. An increase in the S_w value by 27% with the presence of silica is due to its high specific surface area.

Figure 7.60 shows the temperature dependences of the relationship between the amounts of WAW and SAW as relationship between intensities of signals 2 and 1 (I_2/I_1) in the aqueous and chloroform-*d* media for samples M37 (Figure 7.60a) and M61 (Figure 7.60b). The shape of these dependences differs. For M61 in chloroform-*d*, the maximum contribution of signal 2 to the total intensity is reached at 240 K but for M37 this is observed at 220 K because the narrower the pores, the lower is the freezing temperature of water locating there. A large difference is observed for the aqueous medium because contribution of WAW in sample M37 is much larger than that for M61 since water can form larger structures in the M61 sample (Figure 7.61b).

The main difference in the properties of water bound in spongy component of healthy bone and bone affected by osteoporosis is linked to the concentrations of WAW and SAW (Figure 7.58) because of the structural differences in the mineral matrix of these samples (Figure 7.61). In the aqueous medium, the maximum amount of bound water is equal to 900 mg/g for M37 and 1300 mg/g for M61. The greater amounts of water adsorbed onto bone tissues affected by osteoporosis are due to the formation of larger pores (Figure 7.61) and lower density of this bone tissue. The increase in the total amounts of bound water in the aqueous medium causes an increase in concentration of the SAW and contribution of unusual (interstitial; Wilson et al. 2005b) water with the chemical shift at $\delta_H = 1.3$ ppm decreases.

The present investigations show that the ^1H NMR spectroscopy with layer-by-layer freezing-out of bulk and interfacial waters can be used as the cryoporometry to estimate the pore size distribution in the bone tissue. This approach allows the characterization of different bone tissues (healthy and affected by osteoporosis) in terms of amounts of WBW and SBW and WAW and SAW affected differently by the bio-matrix and the media (air, water, and chloroform-*d*). A significant portion of the bound water is in the weakly associated state which is more characteristic for healthy bone tissue than that for one affected by osteoporosis. Weakly polar chloroform-*d* medium more strongly affect the organization of water bound by spongy component of bone tissue affected by osteoporosis. This is caused by easier penetration of chloroform-*d* molecules into larger pores of bone tissue affected by osteoporosis that leads to the displacement of a portion of water into larger pores and changes conditions of the formation of small water clusters in nanopores and narrow mesopores. The pore size distribution in healthy bone tissue differs from that for bone affected by osteoporosis over the wide range of the pore sizes and explains the difference in the density of the studied samples.

7.5.3 Self-Organization of Water–Organic Systems in Bone Tissue and Products of Its Chemical Degradation

A sample of human vertebral bone tissue (Medical University, Lublin) was from healthy bone. Two samples (BB1 and BB2) of degradation products of bovine tubular bone tissue were used. Bovine bone sample was boiled in water in a reactor for 0.5 h and treated at 70°C for 5 h, then hydrolyzed in 0.1 M HCl for several hours, washed off by water, and dried at 150°C in a drying chamber. For degreasing, cutting bone chips (5–10 mm length, 0.1–0.2 mm thickness) were washed off by 0.5% solution of sodium carbonate on stirring for 2 h, decanted, and second washed off for 2 h. After removal of the alkaline solution, the sample was washed off by distilled water. For removal of mineral component, the sample was treated by 1.35 g equiv/L solution of HCl on stirring for 15 min, decanted, and washed off by water to neutral pH controlled using an indicator. After drying at 105°C–110°C, the sample (BB1) as a white powder at the specific surface area ~50 m^2/g included predominantly protein fraction of bone tissue. The decanted solution included a dissolved mineral fraction of bone. For precipitation, potassium phosphate (1.6 g) was added to 100 mL of the solution and neutralized to pH 7.0–7.3 by caustic soda on stirring. The precipitate of white color was washed off by water and dried at 105°C–110°C. The obtained powder sample (BB2) corresponds to recrystallized HAP.

The ^1H NMR spectra of human bone tissue containing 0.3 g/g (initial sample) and 0.6 g/g of water are shown in Figure 7.62.

The temperature dependences of the amounts of unfrozen water (C_{uw}) and the relationships between the C_{uw} and ΔG values calculated as described previously (Turov and Leboda 1999, Gun'ko et al. 2005d) for SAW and WAW are given in Figure 7.63. The characteristics of the bound water layers are shown in Table 7.15.

The ^1H NMR spectra of porous component of bone tissue (Figure 7.62a) were recorded at T=210–280 K. The spectra include two overlapping signals at $\delta_H \approx 1.5$ (more intense signal) and 5 ppm. The intensity of both signals decreases with lowering temperature and they are not observed at 210 K. These signals can be attributed to different forms of water: WAW ($\delta_H \approx 1.5$ ppm) and SAW ($\delta_H \approx 5$ ppm) (Gun'ko et al. 2005d). Water addition (+30 wt%) to the initial sample (Figure 7.62b) leads to redistribution of the signal intensity, since in the initial sample, the amounts of SAW and WAW were C_{uw}^{sa} = 75 mg/g and C_{uw}^{wa} = 220 mg/g and after addition of water they were 350 and 240 mg/g, respectively. From the data (Figure 7.62 and Table 7.15), an increase in the amounts of SAW in bone tissue is accompanied by a predominate increase in the amounts of WBW. An increase in a WBW fraction in WAW occurs due to a decrease in the amounts of SBW. The total amounts of SBW increase by ~30%. Bone tissue as other biotissues is a soft material, internal structure of which depends on a quantity of adsorbate (e.g., water). Therefore, the increase in the amounts of SBW can be explained by an increase in the volume of soft cavities accessible for water and characterized by a high adsorption potential.

Addition of a certain quantity of nonpolar organic solvent (benzene) to the bone tissue sample leads to significant changes in the shape of the ^1H NMR spectra (Figure 7.62c). The signal width of bound water sharply decreases, and signals of SAW and WAW become easily resolved. Notice that besides signals of water (Figure 7.62) signal of CH groups of benzene and chloroform is observed at δ_H=7.2–7.3 ppm. Redistribution of the signal intensities of SAW and WAW and the amounts of

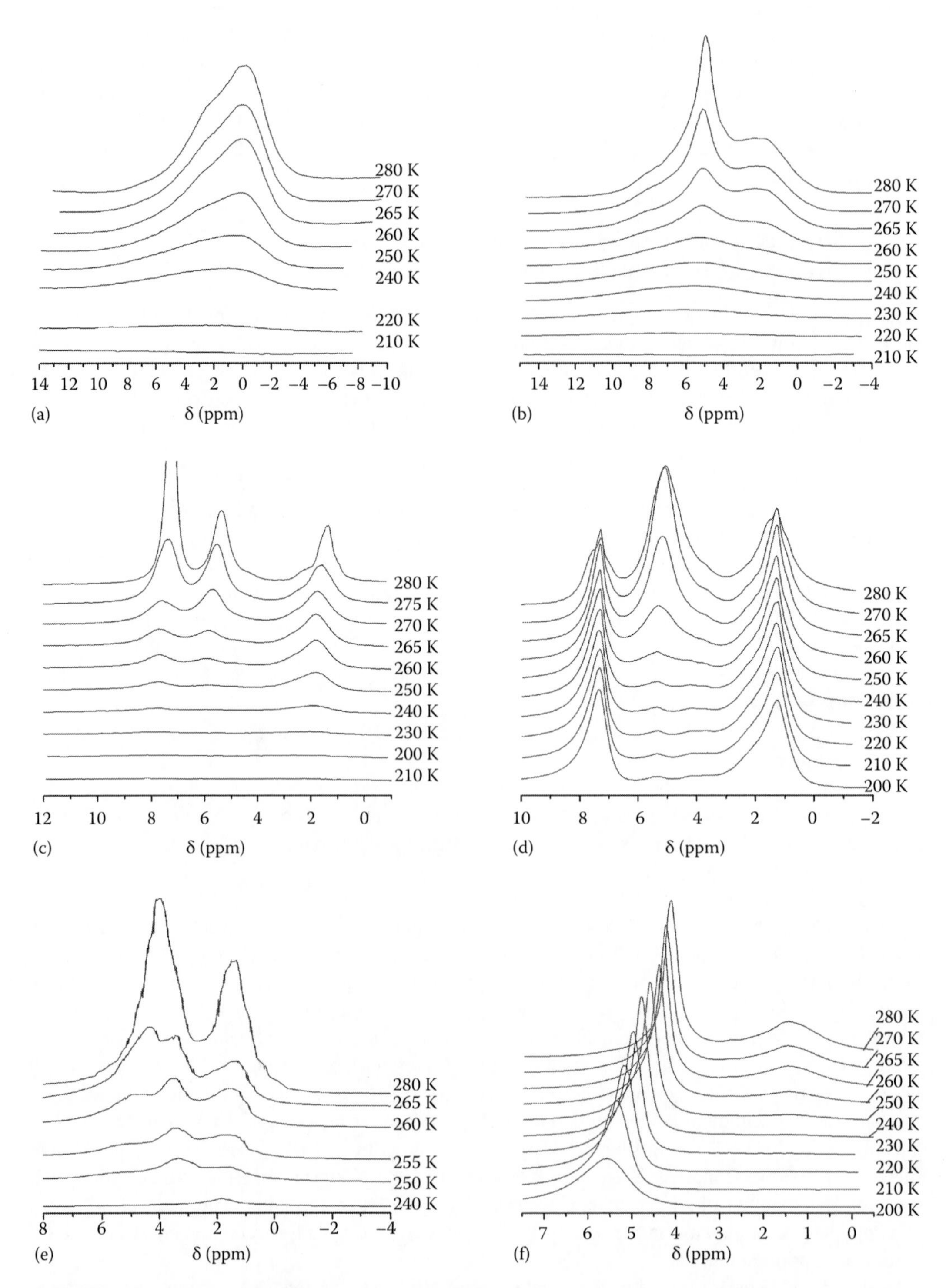

FIGURE 7.62 Influence of solvents on water state in human bone tissue: (a) initial sample, (b) after addition of 0.3 g/g of water, initial sample in (c) benzene, (d) chloroform (1 g/g), and (e) CD_3CN (0.8 g/g), and (f) sample with added 0.3 g/g of water in DMSO-d_6.

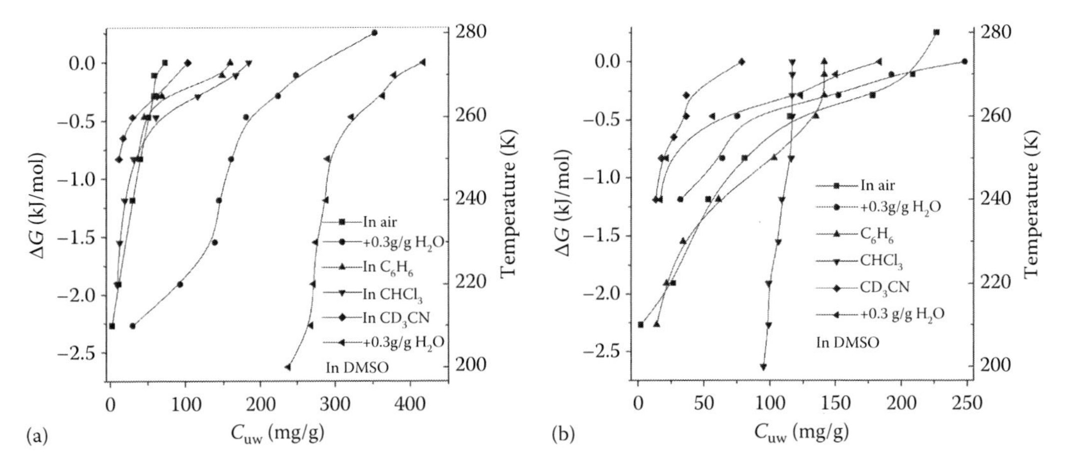

FIGURE 7.63 Influence of solvents on water state in human bone tissue for (a) SAW and (b) WAW for initial sample, after addition of 0.3 g/g of water or in benzene, chloroform, and acetonitrile, and in DMSO-d_6 (after addition of 0.3 g/g of water).

TABLE 7.15
Characteristics of Water Bound in Human Bone Tissue with Addition of Organic Solvents

Water Type	Solvent	Water Content (g/g)	C^s_{uw} (mg/g)	C^w_{uw} (mg/g)	$-\Delta G_s$ (kJ/mol)	γ_S (J/g)	S_{uw} (m²/g)	V_{uw} (cm³/g)
SAW	—	0.3	65	10	2.40	4.2	4.4	0.061
	—	0.6	200	150	2.46	19.5	15.8	0.269
	C_6H_6	0.3	40	120	2.54	5.3	22.0	0.158
	$CHCl_3$	0.3	40	140	2.95	5.8	24.8	0.169
	CD_3CN	0.3	20	85	2.50	2.2	15.3	0.076
	DMSO-d_6	0.6	305	115	5.16	60.0	551.0	0.385
WAW	—	0.3	120	100	2.34	10.3	21.0	0.214
	—	0.6	90	150	2.50	7.1	31.5	0.204
	C_6H_6	0.3	140	—	2.93	10.2	71.2	0.141
	$CHCl_3$	0.3	115	—	2.50	36.2	218.4	0.117
	CD_3CN	0.3	38	42	2.37	2.4	5.7	0.050
	DMSO-d_6	0.6	28	152	2.52	5.2	26.9	0.157

SBW and WBW (Table 7.15) occurs simultaneously with decreasing signal width. The amount of SAW increases by a factor of 1.5, and the corresponding decrease in the concentration of WAW is observed. These changes occur due to a decrease in the amounts of WBW which completely vanishes in the WAW structures. In the SAW structures, a quantity of SBW decreases by a factor of 1.5. Similar changes in the spectra are observed with the presence of weakly polar deuterochloroform (Figure 7.62d); however, in this case the decrease in the WAW fraction is larger. Features of the water–chloroform mixture are caused by changes in WAW becoming much strongly bound that it does not practically freeze over the used temperature range.

In the presence of acetonitrile as well as in the case of nonpolar or weakly polar solvents, an increase in the SAW fraction is observed (Table 7.15); however, this effect is weaker. At $T<280$ K, signal of SAW splits into two signals at $\delta_H \approx 4$–5 and 3 ppm. The SAW signal with up-field shift corresponds to water dissolved in acetonitrile. This mixture can be formed in pores of bone tissue or out of them. Since bone tissue has a high affinity to water and a quantity of acetonitrile is not large (80 wt% of dry bone material), it is possible to consider that a certain portion of water–acetonitrile complexes forms within pores.

An increase in the signal intensity and appearance of a strong temperature dependence of the chemical shift value for SAW (δ_H decreases from 5.5 to 4 ppm with elevating temperature) are observed for hydrated bone tissue placed in DMSO. In this case, a major fraction of WAW becomes WBW (Table 7.15). The strong temperature dependence of the chemical shift testifies about the existence of several types of water–DMSO structures (frozen at different temperatures) such as small water clusters or individual water molecules surrounded by DMSO molecules, confined water nanodomains with small amounts of dissolved DMSO molecules, and water structures located out of pores.

The obtained results (Table 7.15, and Figures 7.62 and 7.63) suggest that in the case of samples being in solvent media (chloroform, benzene, and acetonitrile), an increase in the quantity of WBW is characteristic for the SAW structures. However, in contrast to inorganic adsorbents (Gun'ko et al. 2005d, 2009d), this growth occurs due to a decrease in a quantity of WAW. As a result, the value of interfacial free energy (with exception of water–acetonitrile) has a tendency to increasing. For DMSO, as the most polar solvent among studied ones, a substantial fraction of water is dissolved in the solvent. This leads to a sharp increase in the γ_S value (Table 7.15), which reflects interaction of water rather with DMSO than with the pore walls of bone tissue.

Thus, water in bone tissue is present in two forms of spatially divided clustered structures as SAW characterized at $\delta_H = 4.5–6$ ppm and WAW at $\delta_H = 1–2$ ppm. WAW is more strongly bound and freezes at temperatures considerably lower than the freezing point of the major fraction of SAW. The chloroform medium stabilizes WAW and decreases the interaction of water with the surfaces of the bovine bone materials. The same tendency is outlined for protein and mineral components of the bone material. The presence of such electron-donor solvents as acetonitrile and DMSO leads to the formation of clustered structures (with the electron-donor molecules and SAW or the solution of water in the solvents) in the hydrate shells of structural elements of the bone material. The ^{1}H NMR signals of these structures are observed in the form of separate signals. A small fraction of WAW is characteristic for protein and mineral components in contrast to the native bone tissue. For protein component in chloroform medium, contribution of this fraction increases, while for the mineral component, it is present only in the form of a broad and weak signal. There are several factors in the native bone material, which are appropriate for the formation of WAW structures, such as hydrophilic HAP nanoparticles in hydrophilic/hydrophobic collagen matrix and hydrophobic grease functionalities as weakly polar or nonpolar components creating mosaic hydrophilic/hydrophobic structures.

Figure 7.64 shows temperature changes in the ^{1}H NMR spectra of BB1. The initial powder does not practically contain water (Figure 7.64a) since a small quantity of residual water appears in the spectra in the form of very broad signal with a maximum at $\delta_H \approx 5$ ppm. Estimation of the signal intensity shows that the amount of water in this sample is smaller than 1 wt%. Addition of 10 wt% of water to this sample leads to a certain decrease in the signal width (Figure 7.64b) and its chemical shift is displaced from 5 to 5.8 ppm with lowering temperature. The replacement of air medium by such weakly polar solvent as deuterochloroform leads to the appearance of WAW, whose quantity is equal to 2 wt% at 280 K (Figure 7.64c). In contrast to SAW, practically total amount of WAW corresponds to SBW. In contrast to the native bone material (Figure 7.62), addition of electron-donor organic solvents (acetonitrile-d_3 and DMSO-d_6) to BB1 did not lead to stabilization of WAW (Figure 7.64 and Table 7.16).

In the presence of acetonitrile besides signal of SAW, a weak signal of CHD_2 groups of non-deuterated acetonitrile is observed at $\delta_H \approx 2$ ppm (Figure 7.64d). In the case of a relatively small concentration of DMSO (0.3 g/g), two close signals of SAW and the solution of water in DMSO (signal with up-field shift) are observed in the spectra (Figure 7.64e). In the DMSO medium (Figure 7.64f),

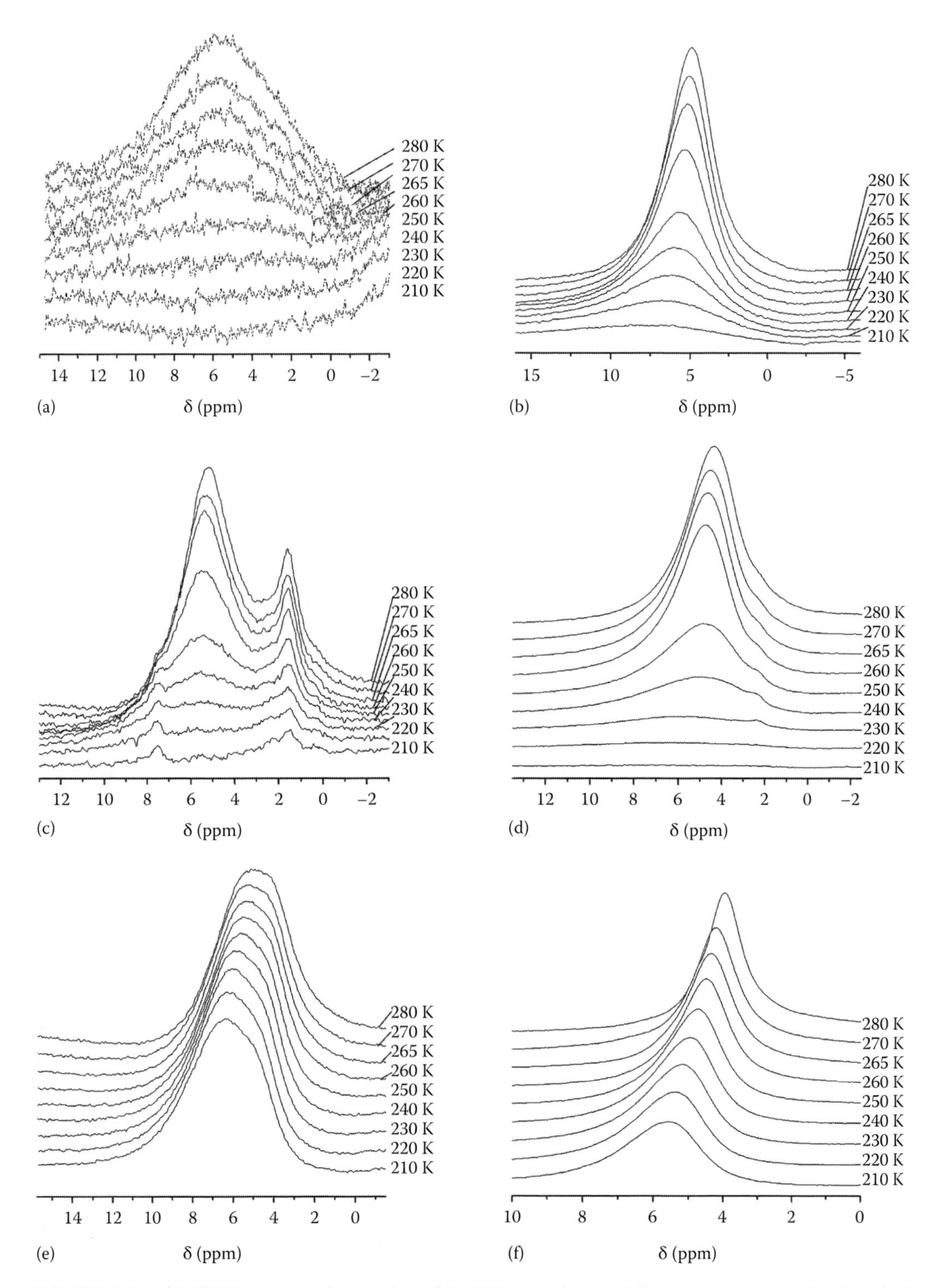

FIGURE 7.64 ^{1}H NMR spectra of water bound in BB1 recorded at different temperatures for (a) initial sample, (b) after addition of 0.1 g/g of water, and then addition of (c) 0.6 g/g of $CDCl_3$, (d) 0.8 g/g of CD_3CN, (e) 0.3 g/g, and (f) 0.8 g/g of DMSO-d_6.

TABLE 7.16
Characteristics of Water Bound in BB1 at 10 wt% of Water and after Addition of Solvents

Solvent	Solvent Content (g/g)	C^{s}_{uw} (mg/g)	C^{w}_{uw} (mg/g)	ΔG^{s} (kJ/mol)	γ_S (J/g)	S_{uw} (m^2/g)	V_{uw} (cm^3/g)
—	—	80	20	−2.5	7.1	7.3	0.094
$CDCl_3$ (WAW)	1	10	7	−2.3	0.68	2.6	0.017
$CDCl_3$ (SAW)	1	50	33	−2.5	4.5	6.1	0.077
CD_3CN	0.1	100	0	−2.5	6.2	53.1	0.1
CD_3CN	0.3	96	4	−2.5	5.1	49.1	0.1
CD_3CN	1	90	10	−2.5	5.8	38	0.1
$(CD_3)_2SO$	0.1	90	10	−6	16.3	24.7	0.1
$(CD_3)_2SO$	0.3	100	0	−15	41.6	93.7	0.1
$(CD_3)_2SO$	1	100	0	−6	19.1	65.9	0.1

there is dynamic equilibrium between these forms of water, and the signal shifts from 6 ppm at 210 K to 3.5 ppm at 280 K.

Since the δ_H value at low temperatures is slightly larger than that of liquid water, one can assume that SAW and DMSO form clustered structures in which the DMSO molecules surround water clusters. Sample BB1 includes mainly protein component of bone tissue; therefore, one can assume that WAW localizes in cavities with mainly hydrophobic walls (hydrophobic side groups of certain amino acids). A similar effect was previously observed for frozen-dried serum albumin (Gun'ko et al. 2005d). Figure 7.65 shows the temperature dependences of the amounts of unfrozen water $C_{uw}(T)$ and the corresponding relationships between the ΔG and C_{uw} values, and the characteristics of bound water layers are shown in Table 7.16.

These results suggest that the amount of SBW in SAW is significantly decreased only in the presence of chloroform. In the case of added electron-donor solvents, entire water becomes SBW

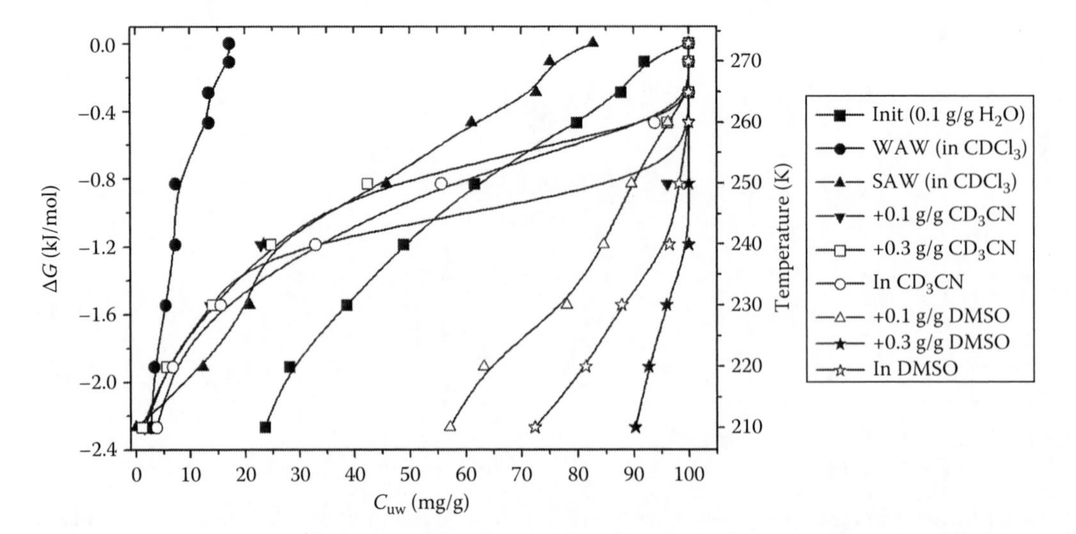

FIGURE 7.65 Influence of solvents on temperature dependence of the amounts of unfrozen water and relationships between the C_{uw} and ΔG values for BB1 with added 0.1 g/g of water and different amounts of solvents.

TABLE 7.17
Characteristics of Water Bound in BB2 with Added 10 wt% of Water and Solvents (30 or 100 wt%)

Solvent	Solvent Content (g/g)	C^{s}_{uw} (mg/g)	C^{w}_{uw} (mg/g)	ΔG_s (kJ/mol)	γ_S (J/g)	S_{uw} (m²/g)	V_{uw} (cm³/g)
—	—	60	40	−2.5	4.1	23.2	0.1
$CDCl_3$	0.3	60	40	−2.5	4.3	21.6	0.099
$CDCl_3$	1	55	35	−2.3	4.0	15.7	0.095
CD_3CN	0.3	100	0	−2.5	5.6	41.1	0.1
CD3CN	1	100	0	−2.5	6.3	49.3	0.1
$(CD_3)_2SO$	0.3	100	0	−6	20.1	19.5	0.1
$(CD_3)_2SO$	1	100	0	−6	20.1	74.4	0.1

and partially freezes out only at low temperatures. The proximity of the ΔG^s values (characterizing changes in the Gibbs free energy in the first adsorbed water layer; Turov and Leboda 1999, Gun'ko et al. 2005d) for practically all samples (with exception of samples with DMSO) to that for the initial BB1 without solvents suggests that water is concentrated in the solvation shells of collagen structures. A significant increase in the ΔG^s and γ_S values is observed for samples containing DMSO that can be caused by a great energy of hydration of DMSO molecules and dissolution of water molecules in DMSO. The maximum γ_S value for sample containing 0.3 g/g of DMSO can be caused by the maximal clusterization of the DMSO/water structures. The presence of several water signals at $\delta_H = 6$ and 4–4.5 ppm is due to heterogeneity of these structures. Stabilization of a considerable quantity of WAW by chloroform does not occur for sample BB2 (mineral component of bone tissue) (Figure 7.66).

In the presence of deuterochloroform, WAW is observed as a weak signal whose intensity is difficult to be measured against the background of signal of SAW. In other respects, features of changes in the spectral characteristics of BB2 with organic solvents are close to that observed for BB1. Chloroform slightly reduces a quantity of SAW (Table 7.17 and Figure 7.67); however, this decrease is considerably less than for BB1. In the presence of acetonitrile, signals of SAW and a solution of water in acetonitrile are distinguished but a fraction of the solution is higher than that for BB1. In the case of DMSO, this equilibrium shifts toward the solution of water in DMSO. Probably, this is due to a smaller value of the specific surface area of the HAP crystallites/water interfaces because a smaller γ_S value is for the initial BB2 in comparison with the initial BB1. The characteristics of bound water layers in the presence of 30 and 100 wt% of DMSO for BB2 are close to that for BB1 with the addition of 1 g/g of DMSO.

In the presence of DMSO, the ΔG^s values are equal to 5–6 kJ/mol for the majority of samples of native human bone tissue and protein or mineral components of water in DMSO in the interfacial layers under action of the surface forces of the materials. Taking into account that the heat effect of water crystallization is 4 kJ/mol, one can conclude that the solvation energy of water in DMSO ($T < 273$ K) is higher than the interaction energy of water with functionalities of both protein and mineral components of bone tissue.

For the majority of samples, addition of solvents leads to an increase in the surface area of bone structures being in contact with unfrozen water (Tables 7.15–7.17, S_{uw}). This effect is strongest on addition of DMSO because contribution of narrow cavities filled by water (or water/DMSO mixture) increases (Figure 7.68). In other words, DMSO/water mixture can penetrate into narrower cavities than water alone. The observed results are in agreement with well-known properties of DMSO as a carrier of other compounds and a cryoprotectant (Tsuraeva 1983, Hubálek 2003). The effect of

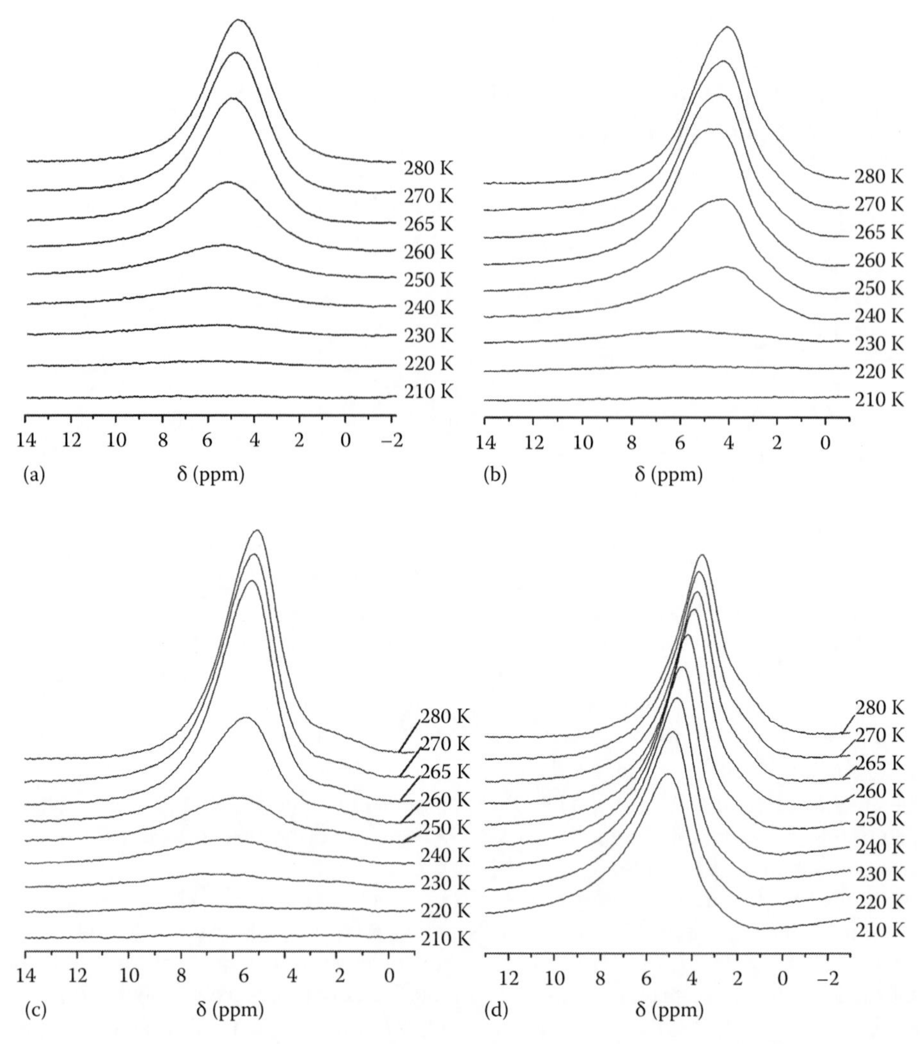

FIGURE 7.66 ^1H NMR spectra of water bound in BB2 recorded at different temperatures after addition of (a) 0.1 g/g of water and then 0.8–1.0 g/g of solvents (b) $CDCl_3$, (c) CD_3CN, and (d) DMSO-d_6.

DMSO is stronger for SAW than WAW (Table 7.15 and Figure 7.68) because WAW as water bound in the form of small interfacial clusters possesses a lower bovine bone tissue.

Probably, this value is affected by the free energy of salvation ability to form mixtures with DMSO (or other polar solvents) than SAW present in the form of nanodomains. Therefore, the effects of nonpolar (benzene) or weakly polar (chloroform) solvents (which can more strongly interact with hydrophobic functionalities than water) on WAW is stronger than that of DMSO (Tables 7.15–7.17 S_{uw} and γ_S). Notice that the effects of CD_3CN (as more polar than C_6H_6 but less polar than DMSO) for both SAW and WAW are minimal.

In the case of protein or mineral components of bovine bone tissue, the influence of solvents on bound water differs from that for human bone tissue (Tables 7.15 and 7.17 and Figure 7.68). Addition of less polar solvents gives smaller changes in the S_{uw}, γ_S, and ΔG^s values. These results can be explained by a lower heterogeneity of BB1 and BB2 samples than native one, i.e., model samples BB1 and BB2 have less mosaic (hydrophilic/hydrophobic) surfaces than native bone tissue. WAW is more strongly bound and freezes at temperatures considerably lower than the freezing point of the major fraction of SAW.

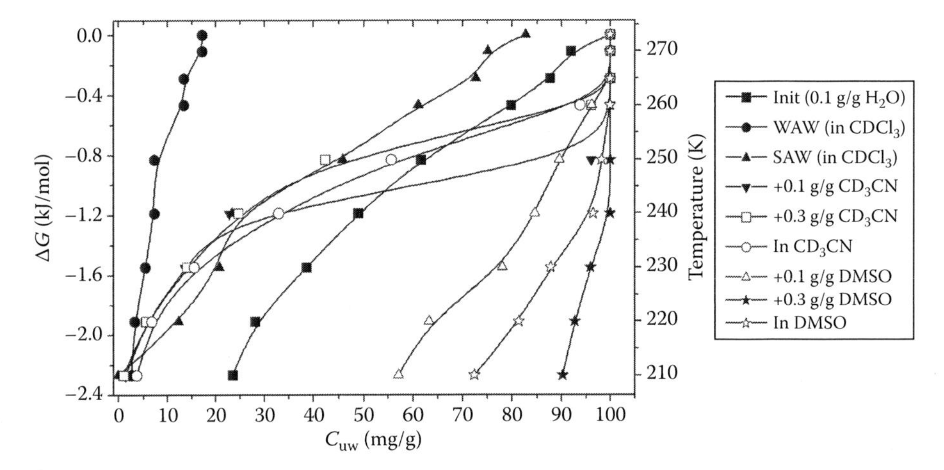

FIGURE 7.67 Influence of solvents on temperature dependence of the amounts of unfrozen water and relationships between the C_{uw} and ΔG values for BB2 with 0.1 g/g of water and different amounts of solvents.

We may conclude that chloroform medium stabilizes WAW and decreases the interaction of water with the surfaces of the bovine bone materials. The same tendency is outlined for protein and mineral components of the bone material. The presence of such electron-donor solvents as acetonitrile and DMSO leads to the formation of clustered structures (with the electron-donor molecules and SAW or the solution of water in the solvents) in the hydrate shells of structural elements of the bone material. The ^{1}H NMR signals of these structures are observed in the form of separate signals.

A small fraction of WAW is characteristic for protein and mineral components in contrast to the native bone tissue. For protein component in chloroform medium, contribution of this fraction increases, while for the mineral component, it is present only in the form of a broad and weak signal. There are several factors in the native bone material, which are appropriate for the formation of WAW structures, such as hydrophilic HAP nanoparticles in hydrophilic/hydrophobic collagen matrix and hydrophobic grease functionalities as weakly polar or nonpolar components creating mosaic hydrophilic/hydrophobic structures.

7.5.4 Bone Tissue Models

Hydrated composites with nanostructured HAP and proteins (gelatin, BSA), HGA, were studied as models of bone tissue (Golovan et al. 2010). BSA was used because albumins are the main non-collagen proteins in bone tissue (Krasavina and Torbenko 1979), and gelatin was used instead of collagen.

Nanostructured hydroxyapatite $Ca_{10}(PO_4)_6(OH)_2$ (HAP) was synthesized using aqueous solutions of the corresponding salts of phosphorus ($(NH_4)_2HPO_4$) and nitric ($Ca(NO_3)*2H_2O$) acids (Vysotskaja et al. 2002):

$$10Ca(NO_3)_2 + 6(NH_4)_2HPO_4 + 8NH_4OH \rightarrow Ca_{10}(PO_4)_6(OH)_2 + 20NH_4NO_3 + H_2O \quad (7.3)$$

Composites with HAP, gelatin, and albumin (HGA) were prepared by precipitation of a mixture of the mentioned salts and proteins in the 25% ammoniac solution (Weiner et al. 1999) then dried at 120°C. HGA (S_{BET} = 118 m^2/g) represents a white powder (Figure 7.69).

Infrared spectra of components alone and HGA (Figure 7.70) show the bands characteristic for main structural components of HAP (PO_4^{3-} tetrahedral, CO_3^{2-}, OH, adsorbed water)

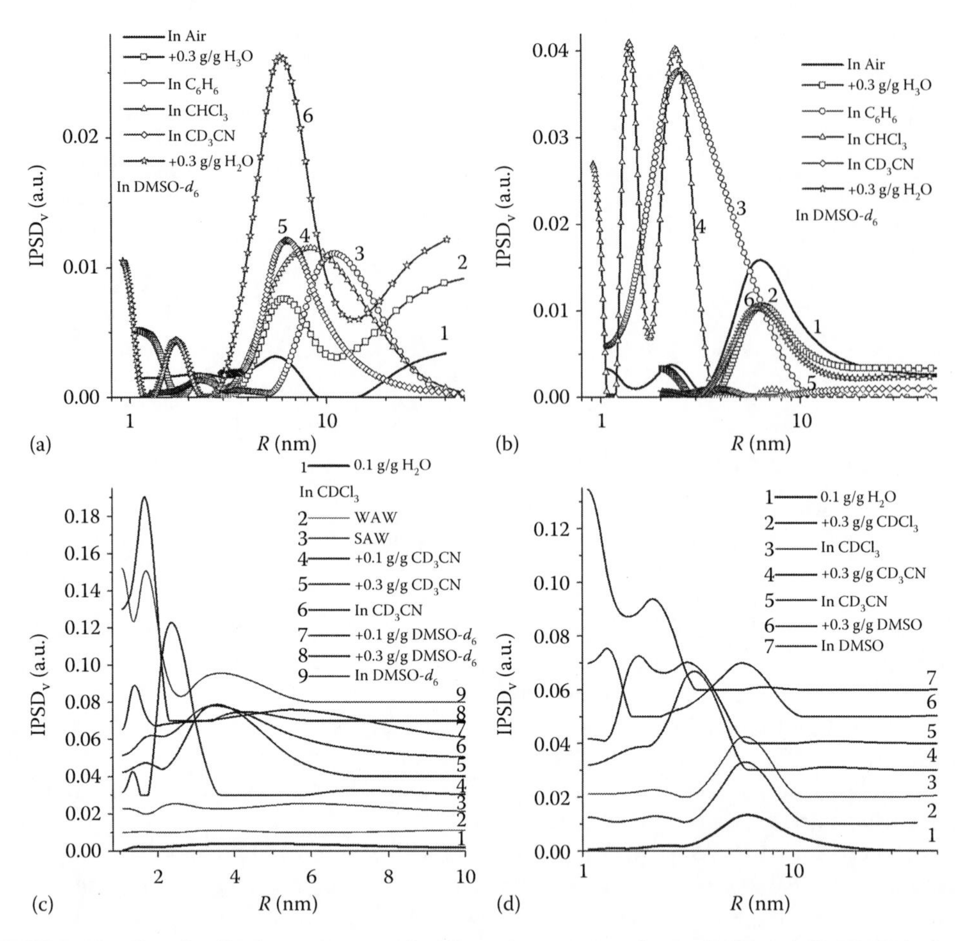

FIGURE 7.68 Size distributions of pores filled by unfrozen water for (a) SAW and (b) WAW in human bone tissue initially and after addition of water and solvents, (c) BB1 and (d) BB2 after addition of water and solvents.

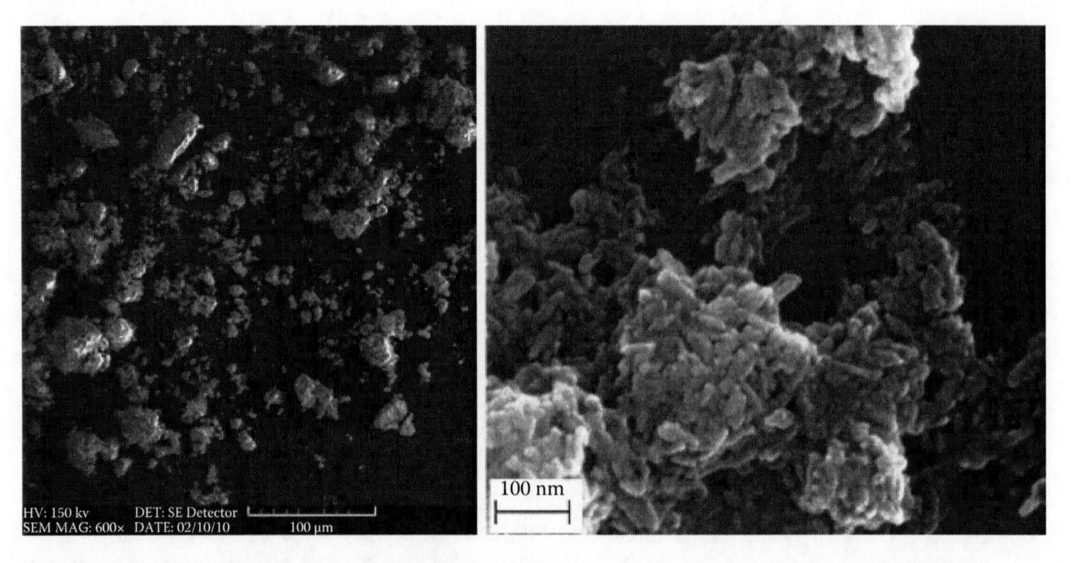

FIGURE 7.69 SEM images of HGA composite at different magnification (bar 100 μm and 100 nm).

(Gun'ko et al. 2006d, Shu et al. 2007) and proteins. Bands at 1095, 1050, 960, and 606 cm^{-1} can be attributed to valence and deformation vibrations of phosphate groups, as well as the vibrations at 1490–1410 and 876 cm^{-1} related to carbonate groups. Consequently, a part of phosphate groups in HAP was replaced by carbonate groups (Danilchenko et al. 2005). It is known that biogenic HAP includes 7.4 wt% of carbonates (Kikuchi et al. 2004).

The IR spectra of HGA include the bands at 1645–1649 and 1548 cm^{-1} (amides I and II) of proteins. The band of free hydroxyls of HAP in HGA shifts toward 3569 cm^{-1} from 3572 cm^{-1} characteristic for pure HAP. A relatively large specific surface area of hydrophilic HGA and HAP causes the adsorption of significant amounts of water (bands at 3500–3350 cm^{-1} of the OH stretching vibrations and 1630 cm^{-1} of the deformation vibrations).

Thermogravimetric study of HGA and component alone (Figure 7.71) shows that proteins correspond to 12.7 wt% in HGA. The weight loss for HAP alone is 7.7 wt% at 1000°C. Water desorbed at $T < 150$°C corresponds to 6.5 wt% in HGA and ~10 wt% in proteins. Proteins bound in HGA (transformed into nanostate) are less stable because the second extremum of the DTG curve of HGA shifts toward lower temperatures (~30°C–40°C) in comparison with that of pure proteins.

The absence of a very narrow crystallization exotherm (characteristic to bulk water) and relatively low temperature of observed exotherms of a complex shape in the DSC curves of HGA freezing (Figure 7.72a) suggest that all water is bound in the system.

This water includes both WBW and SBW frozen over a broad temperature range. Calculations of the PSD using DCS thermoporometry (Figure 7.72b) with both freezing and heating curves show that water fills different voids mainly in the mesopore range. This result is in agreement with SEM image of the material (Figure 7.69).

The particle size distribution (PaSD) of HAP (Figure 7.73) depends on pH because a decrease in the surface charge density results in enhanced aggregation of nanoparticles. This is typical for all nanooxides (Gun'ko et al. 2001e). The QESL results for HGA (Figure 7.73) are in agreement with SEM images (Figure 7.69) and show the presence of both primary particles (~50 nm) and their small aggregates (100–600 nm). HGA is composed of smaller particles than HAP (Figure 7.73).

At relatively low content of water (0.07–0.085 g/g) in the HGA materials, the ^{1}H NMR spectra represent broad single signal at $\delta_H \approx 5$ ppm (Figure 7.74a and b) that is close to signal of bulk water.

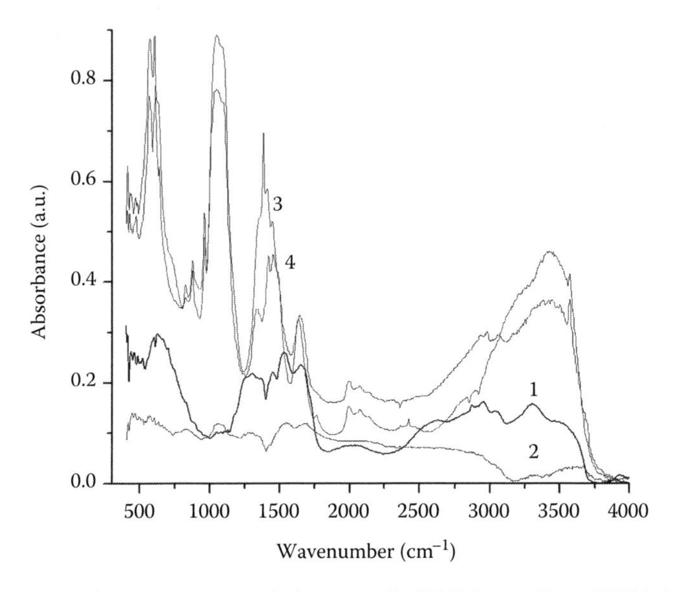

FIGURE 7.70 IR spectra of BSA (curve 1), gelatin (curve 2), HAP (curve 3), and HGA (curve 4).

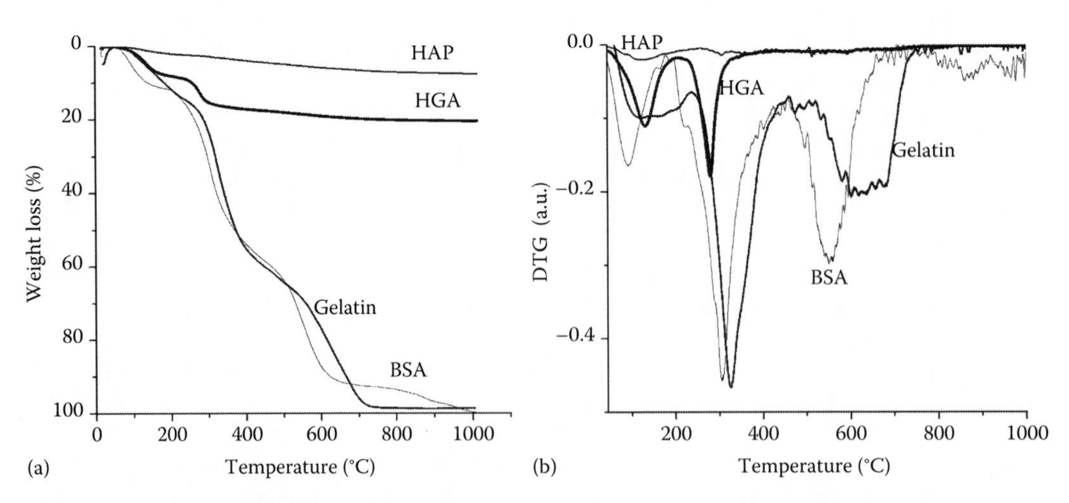

FIGURE 7.71 (a) Weight loss and (b) differential thermogravimetric curves for HGA and individual BSA, gelatin, and HAP.

This water can be assigned to SAW that is in agreement with the DSC PSD (Figure 7.72b) showing water-filled voids in the range 1.5–20 nm in radius. The signal intensity decreases with decreasing temperature and it is not observed at $T<230$ K. This range shows that both WBW and SBW can be found in hydrated HGA. In the case of dispersion medium with weakly polar $CDCl_3$ (Figure 7.74c) or weak electron-donor CD_3CN (Figure 7.74d), the signal is with maximum up-field shift at $\delta_H \approx 2.5$ ppm ($CDCl_3$) or 3 ppm (CD_3CN).

This suggests enhanced clustering of water with stabilization of WAW. Only one signal of mobile water (other protons are not observed in the spectra) can be caused by fast (in the NMR timescale) exchange of water molecules from different structures (Emsley et al. 1965) because the amounts of water are enough to form multilayer coverage of the HGA particles, and the HGA powder is characterized by the textural porosity.

At $C_{H_2O} = 17$ wt% (Figure 7.74e through h), the spectra include several signals. In the chloroform medium, the signals of WAW and SAW are observed at $\delta_H = 2$ and 5 ppm, respectively (Figure 7.74e and f). The WAW signal (Figure 7.74e) is not observed at $T<240$ K and a significant portion of this water is WBW. Addition of a small amount of acetonitrile to chloroform (1:11) (Figure 7.74f) weakly affect the spectra. An increase in the acetonitrile amount (Figure 7.74g) results in the appearance of signal at $\delta_H = 3.5$ ppm corresponding to water (ASW as $HOH \cdots N{\equiv}CCD_3$) dissolved in the organic solvents mixture. The WAW signal decreases and shifts toward $\delta_H = 1$ ppm, i.e., the associativity of WAW decreases. Temperature lowering weakly affects its intensity but it splits into two signals at low temperatures (Figure 7.74g). Thus, WAW is mainly SBW and nonuniform because of the complex surrounding effects.

Addition of a stronger electron-donor DMSO (Figure 7.74h) than CD_3CN changes the ^1H NMR spectra. The WAW and SAW intensities decrease. The latter is with down-field shift toward $\delta_H = 6.2$ ppm. The ASW signal splits into two signals differently dependent on temperature. Thus, the $(CD_3)SO–CDCl_3$ mixture displaces water from pores much stronger than the mixture acetonitrile–chloroform. The DMSO molecules can increase the average number of the hydrogen bonds of water molecules at the periphery of SAW domains. This also increases contribution of SBW in SAW.

In more strongly hydrated systems at 0.37 g/g H_2O (Figure 7.74i through k), the SAW intensity significantly increases. Bound water is mainly WBW in the CD_3CN medium (Figure 7.74k); therefore, it can be easily transferred into the acetonitrile medium. In the weakly polar benzene (Figure 7.74j) and more polar acetonitrile (Figure 7.74k), a certain amount of WAW is observed. This water is structurally nonuniform since two signals are observed.

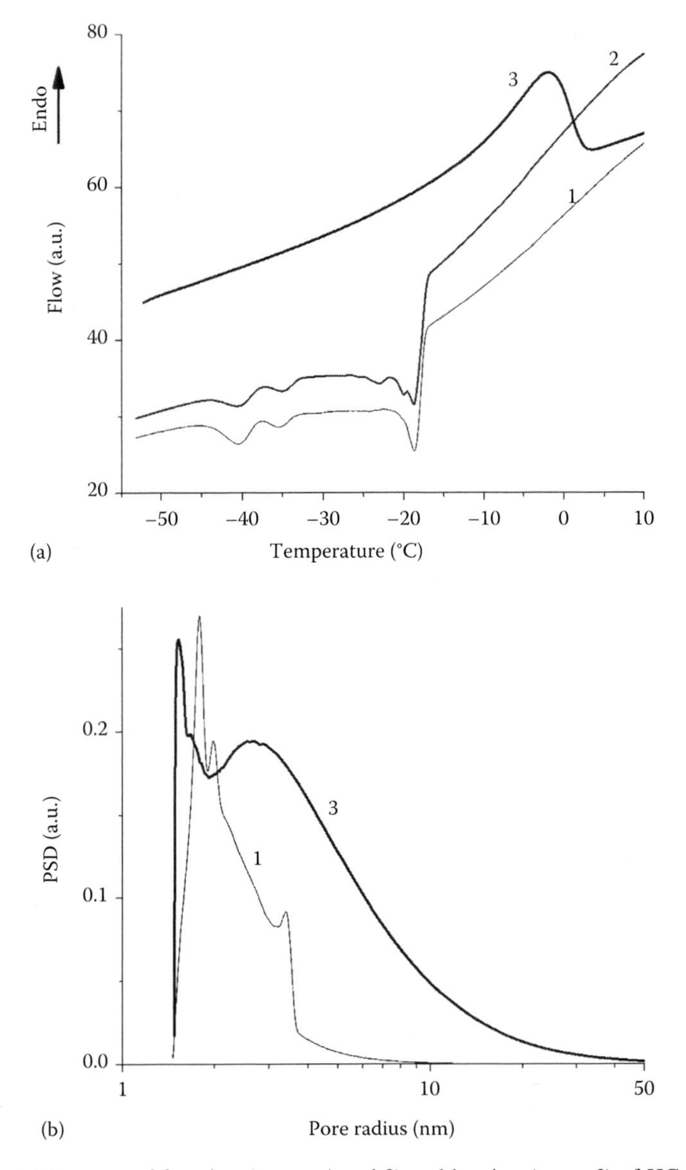

FIGURE 7.72 (a) DSC curves of freezing (curves 1 and 2) and heating (curve 3) of HGA with addition of water (0.1 g/g) without (curves 1 and 3) and with (curve 2) addition of 0.2 g/g of chloroform, and (b) the pore size distributions calculated using curves 1 and 3.

In the case of a similar hydration (0.07 and 0.085 g/g), the $C_{uw}(T)$ and $\Delta G(C_{uw})$ curves significantly differ (Figure 7.75a).

At C_{H_2O} = 0.085 g/g, a significant fraction of bound water becomes WBW at $-\Delta G<0.5$ kJ/mol (Table 7.18). This suggests changes in the structure of water clusters and domains (Figure 7.76a). At C_{H_2O} = 0.07 g/g, a significant portion of water represents clusters of about 1 nm in radius but at C_{H_2O} = 0.085 g/g contribution of larger domains (1.5–6 nm in radius) increases. Changes in the distribution and structure of bound water lead to reduction of the γ_S value (Table 7.18). The chloroform medium decreases the γ_S value. At low hydration, a major portion of SAW and WAW is SBW (Figure 7.75 and Table 7.18) but this portion decreases with increasing hydration. A minimal SAW content is observed in HGA in the chloroform medium stabilizing WAW. At C_{H_2O} = 0.37 g/g (Figure 7.75d),

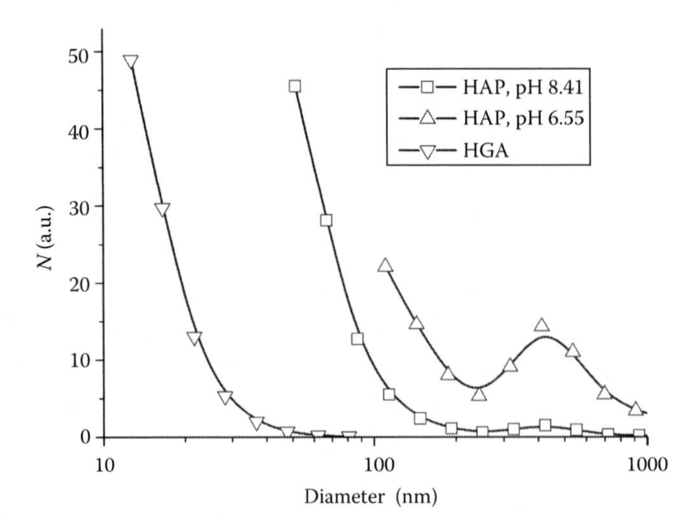

FIGURE 7.73 Particle size distributions of HAP and HGA in 0.25 wt% aqueous suspension (salinity 0.01 M NaCl).

the $\Delta G(C_{uw})$ curves include large sections linked to WBW and $C_{WBW} = 0.22$ g/g (in HGA in air) and 0.27 g/g (in C_6D_6) and the SAW domain sizes increase (Figure 7.76d). However, in acetonitrile large domains are not observed because water can be dissolved in the organic solvent.

7.6 MUSCULAR TISSUES

Muscle is a highly organized tissue, composed of individual cells known as fibers, which are structured by connective tissue (Pearce et al. 2011). A schematic structure of muscle tissue is shown in Figure 7.77 (Baechle and Earle 2008). The order of component breakdown is (1) muscle, (2) muscle fasciculus, (3) muscle fiber/muscle cell, (4) myofibril, and (5) myofilaments. Each muscle fiber consists of a great number of single strands or organelles called myofibrils and these are comprised of myofilaments, which are based on thin and thick filaments predominantly made up of actin and myosin. Myosin is a large protein made up of two heavy chains ($M_w = 220$ kDa) and two pairs of small subunits referred to as light chains ($M_w = 130$ kDa) (Bandman 1999).

Lean muscle contains ~75% water. The other main components include proteins (~20%), lipids (~5%), carbohydrates (~1%), and vitamins and minerals (often analyzed as ash, ~1%) (Offer and Knight 1988).

7.6.1 Chicken Muscular Tissue

Figure 7.78 shows the ^{1}H NMR spectra (recorded using a Bruker WP-100SY spectrometer at the working frequency of 100 MHz) of water bound in chicken muscular tissue partially dried in air at 330 K. Bound water is observed at $\delta_H \approx 5$ ppm (half-width of 0.25–3 ppm) that correspond to SAW. At minimal hydration, all the water can be attributed to SBW (Figure 7.79 and Table 7.19) forming much smaller clusters and domains than in the case of more strongly hydrated (more weakly dried) samples (Figure 7.80).

In the case of the maximum hydrated sample, a major portion (65%) of bound water corresponds to WBW (Table 7.19) and water domains have larger sizes (Figure 7.80). At hydration $C_{H_2O} = 0.18$ g/g, a fraction of WBW is of 22% and water structures have smaller sizes. The surface free energy γ_S depends weakly on the hydration degree and has a minimal value at minimal hydration. The average freezing temperature (Table 7.19, $\langle T \rangle$) decreases with decreasing hydration because contribution of SAW/WBW decreases and WBW is absent at minimal hydration. Contribution of nanostructures

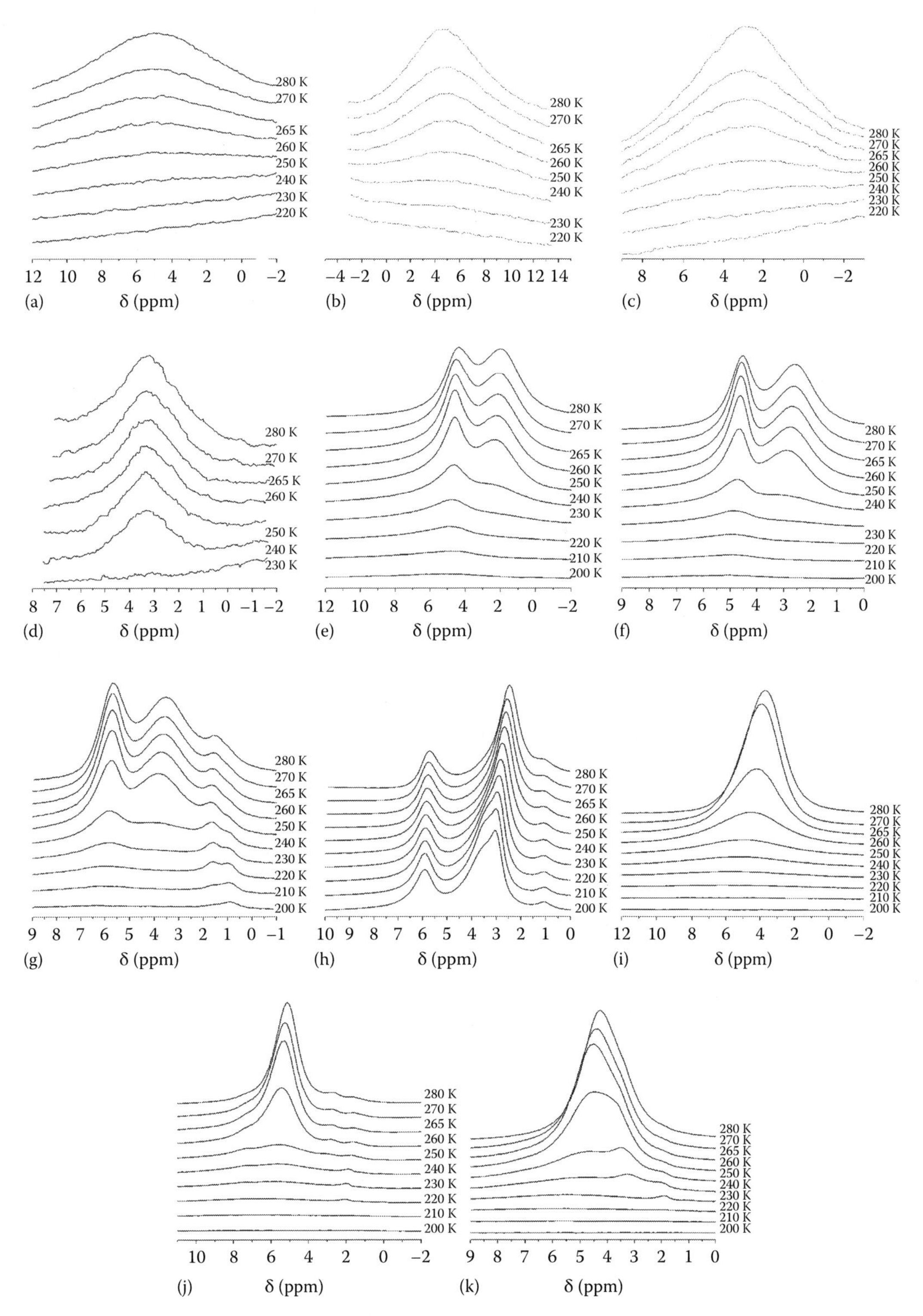

FIGURE 7.74 ^{1}H spectra of water in hydrated HGA: (a) 0.07, (b–d) 0.085, (e–h) 0.17, and (i–k) 0.37 g/g in different dispersion media (a, b, i) air, (c, e) $CDCl_3$, (d) CD_3CN, $CDCl_3$+ CD_3CN (f) 11:1, (g) 2:1, (h) $CDCl_3$+ DMSO-d_6 (1.5:1), (j) C_6D_6, and (k) CD_3CN.

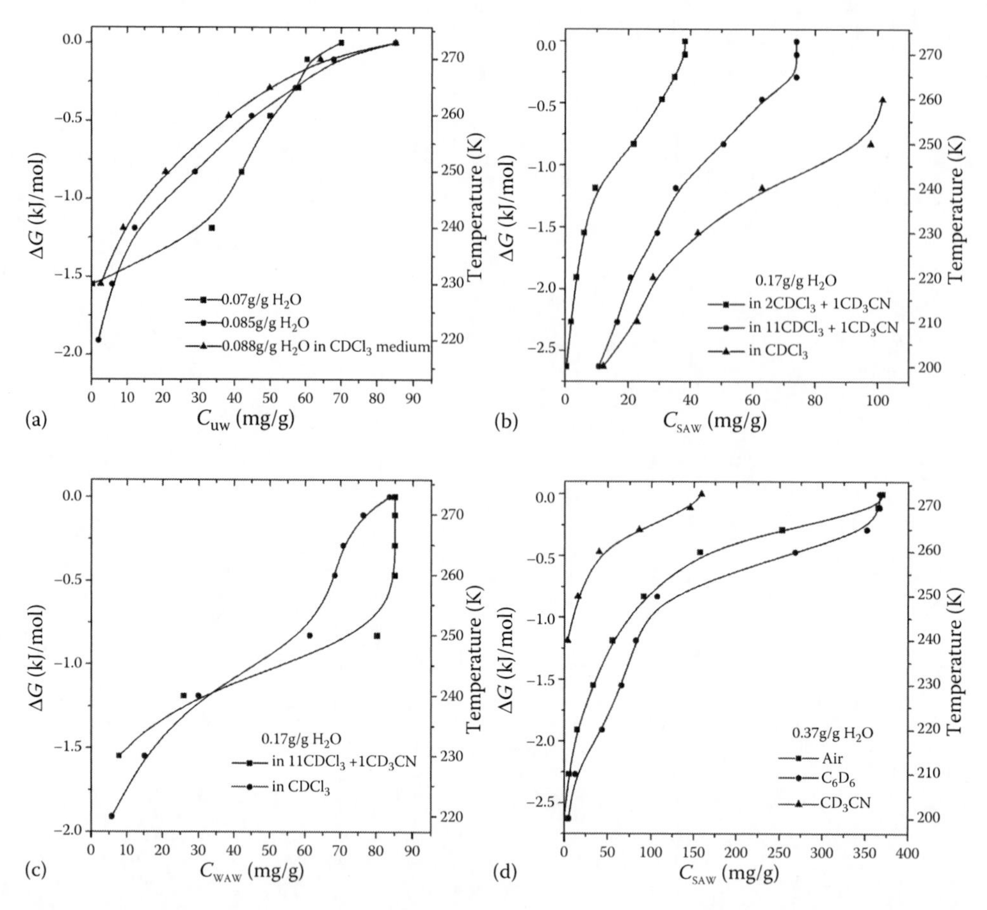

FIGURE 7.75 Temperature dependences of the amounts of unfrozen water (a) total, (b, d) SAW, (c) WAW and related changes in the Gibbs free energy for differently hydrated HGA being in different media at C_{H_2O} = (a) 0.07 and 0.085, (b, c) 0.17, and (d) 0.37 g/g.

TABLE 7.18
Characteristics of Water Bound in HGA Being in Different Media

C_{H_2O} (g/g)	Medium	C^s_{uw} (mg/g)	C^w_{uw} (mg/g)	ΔG^s (kJ/mol)	γ_S (J/g)
0.07	Air	50	20	−1.5	3.5
0.085	Air	45	40	−2.2	2.8
	$CDCl_3$	35	55	−1.8	2.2
0.17	$CDCl_3$	30 (SAW)	10	−3	2.1
	$11CDCl_3 + 1CD_3CN$	65 (SAW)	10	−3	5.7
	$2CDCl_3 + 1CD_3CN$	100 (SAW	—	−3	8.9
0.37	Air	150 (SAW)	220	−3	12.9
	C_6D_6	100 (SAW)	270	−3	17.1
	CD_3CN	25 (SAW)	125	−1.5	8.9

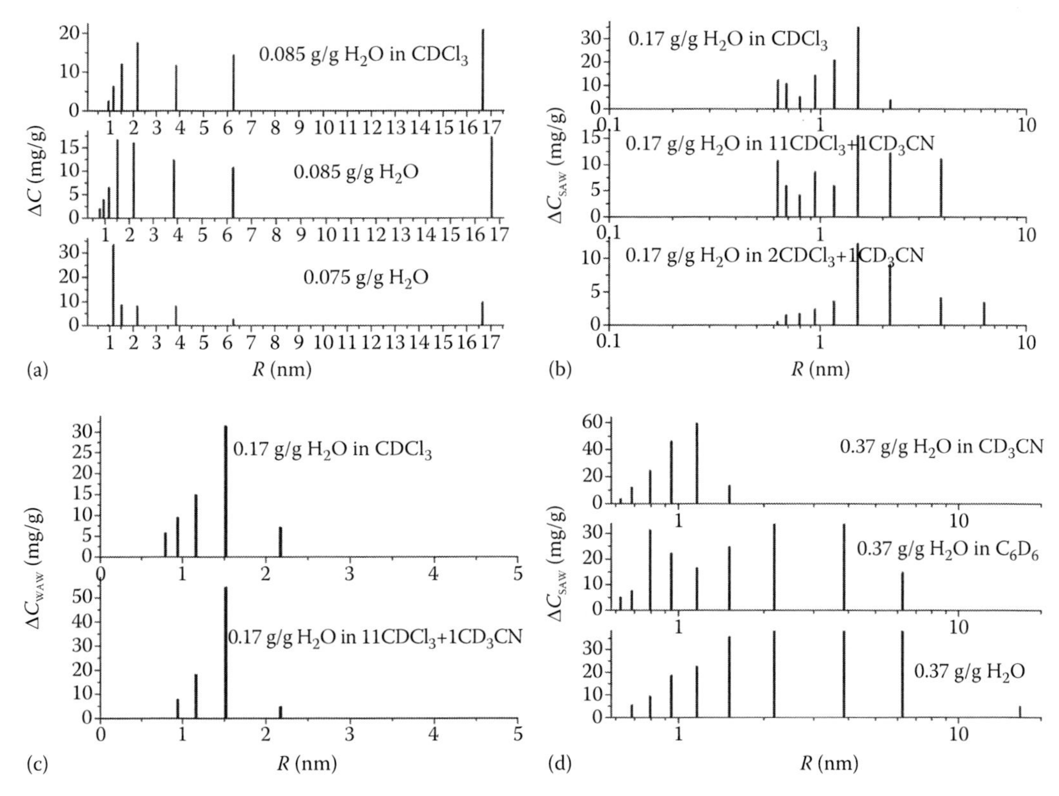

FIGURE 7.76 Size distributions of water structures in hydrated HGA in different media at C_{H_2O} = (a) 0.07 and 0.085, (b, c) 0.17, and (d) 0.37 g/g.

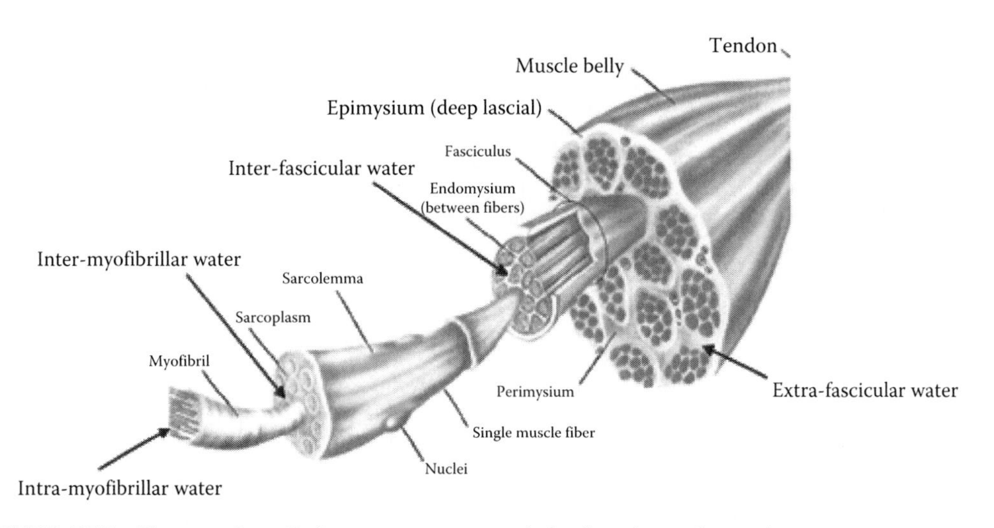

FIGURE 7.77 The muscle split into components and the locations of muscle water compartments: the intra-myofibrillar component and the extra-myofibrillar component which is composed of the inter-myofibrillar, inter-fascicular, and extra-fascicular water populations. (According to Baechle, T.R. and Earle, R.W., *Essentials of Strength Training and Conditioning*, Second edition, Human Kinetics, Champaign, IL, 2008.)

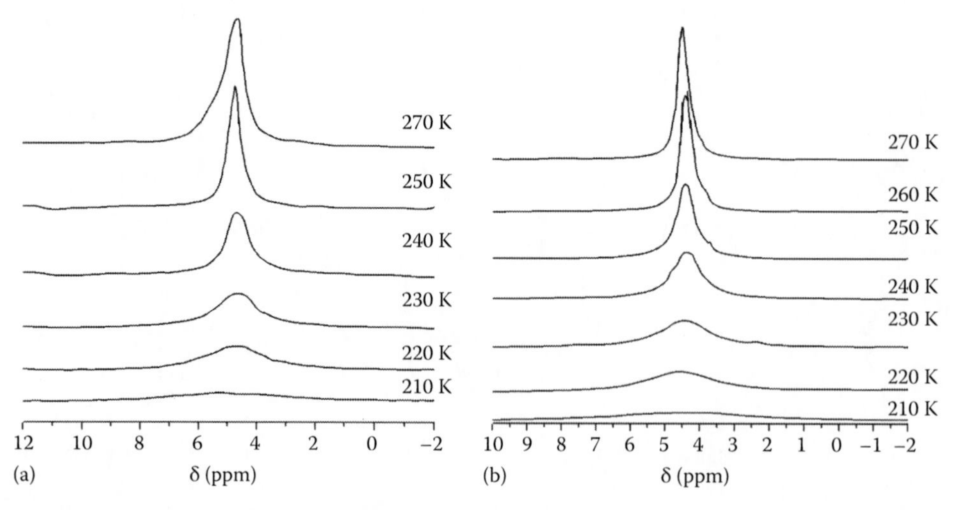

FIGURE 7.78 ^{1}H NMR spectra of water bound in chicken muscular tissue at different hydrations C_{H_2O} = (a) 0.4 and (b) 0.105 g/g.

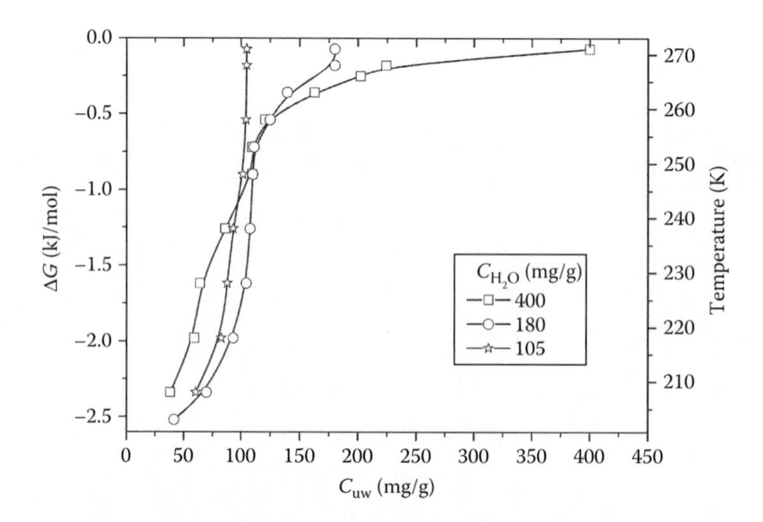

FIGURE 7.79 Temperature dependences of the amounts of unfrozen water bound in chicken muscular tissue at different hydrations C_{H_2O} = 0.4, 0.18, and 0.105 g/g and related changes in the Gibbs free energy.

TABLE 7.19
Characteristics of Water Bound in Chicken Muscular Tissue at Different Hydrations

C_{H_2O} (g/g)	C^{s}_{uw} (mg/g)	C^{w}_{uw} (mg/g)	ΔG_s (kJ/mol)	ΔG_w (kJ/mol)	γ_S (J/g)	S_{uw} (m²/g)	S_{nano} (m²/g)	S_{meso} (m²/g)	V_{nano} (cm³/g)	V_{meso} (cm³/g)	$\langle T \rangle$ (K)
0.4	140	260	−2.91	−0.5	15.9	116	81	34	0.037	0.285	250.9
0.18	140	40	−2.78	−0.5	16.1	165	141	24	0.065	0.107	228.7
0.105	105	0	−3.21	—	13.8	167	139	28	0.060	0.045	208.2

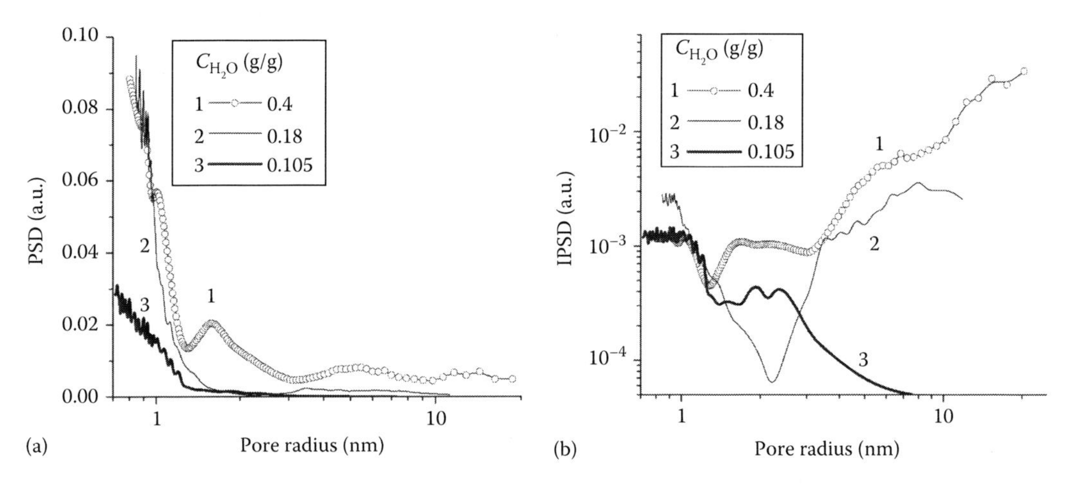

FIGURE 7.80 Size distributions of water structures (i.e., sizes of pores filled by structured water) bound in chicken muscular tissue at different hydrations $C_{H_2O} = 0.4$, 0.18, and 0.105 g/g: (a) differential and (b) incremental PSD.

(S_{nano}, V_{nano}) increases, as well as total area in contact with bound water (S_{uw}), but contribution of mesostructures (S_{meso}, V_{meso}) decreases with drying. Notice that the WAW signal (1–2 ppm) is not observed in all the samples due to relatively large content of water and absence of very narrow voids with mosaic hydrophobic/hydrophilic walls.

7.6.2 Effect of Dispersed Silica on Hydration of Forcemeat

The amount of water in meat is one of the main characteristics, which defines technological and gustatory properties of many beef products (Gorbatov 1982, Zhuravskaya et al. 1985, Decareau 1985). Dehydration of meat raw products during their processing and storage reduces the yield of readymade products. A promising trend in improving the properties of beef products is incorporation into them of biologically inert mineral additives capable to increase the concentration of bound water and prevent its removal in the process of their storage and production. One of these mineral additives may be nanosilica, which is widely used as an adsorbent and filling agent in medicinal preparations (Chuiko 2003).

The developed surface of nanosilica with a large amount of structural hydroxyl groups provides a high affinity to both the water molecules and biopolymer molecules (Gun'ko et al. 2005d, 2009d), while its biological passivity in the range of pH characteristic of the gastrointestinal tract results in its complete elimination from the organism in the process of vital activities. Biomedical trials demonstrated that silica does not have any toxic properties and inflict any harm to the organism even when applied in high doses and for a long time (Chuiko 1993, 2003).

The samples of forcemeat (beef) were minced with a whirlabout with the diameters of holes in the grid of 2–3 mm, and processed with a cutter. Comparative characteristics of the original forcemeat and the minced meat, containing 1% of nanosilica (A-300, $S_{BET} \approx 300$ m²/g), were studied prior to and after temperature treatment (water bath at 373 K for 1.5 h). The content of moisture in the forcemeat, as calculated with respect to the dry substance, was determined routinely (Zhuravskaya et al. 1985), and it comprised 69%.

The ^{1}H NMR spectra of forcemeat in Sych et al. (2000) were registered with a Bruker WP-100 SY NMR spectrometer of high resolution (operating frequency of 100 MHz and a bandwidth of 50 kHz).

The ^{1}H NMR spectra of forcemeat at $T < 255$ K represent a single broad signal at $\delta_H = 5$ ppm (Figure 7.81). At $T \geq 255$ K, a weaker signal is observed at 1.5 ppm. This signal can be linked to both WAW and low-molecular organics.

Figure 7.82 shows the dependencies of concentration of the unfrozen water in the samples on temperature, and the dependencies of changes in free energy of the bound water on its concentration. The calculations were done by the method described in Gun'ko et al. (2005d, 2009d). The submitted dependencies characterize changes in free energy of water molecules on the interphase of food product/water. The obtained dependencies $\Delta G(C_{uw})$ differ from the dependence for aqueous suspension of nanosilica (Chapter 1). Midle parts of the dependencies demonstrate inflections, characteristic of the samples with heterogeneous interfacial layers (Gun'ko et al. 2005d).

Concentrations of strongly ($\Delta G<-0.5$ kJ/mol) and weakly ($\Delta G>-0.5$ kJ/mol) bound waters (Table 7.20) were determined from the dependencies $\Delta G(C_{uw})$ (Figure 7.82).

Addition of nanosilica into the original sample of the beef forcemeat results in certain changes in the characteristics of the bound water (Table 7.20 and Figures 7.81 through 7.83). The amounts of WBW and SBW increase and contribution of mesopores at $R=2$–6 nm increases (Figure 7.83b). However, the S_{meso} value remained the same but S_{uw} and S_{nano} increased (Table 7.20), as well as V_{meso} increased from 1.494 to 1.556 cm^3/g, but the V_{macro} value decreased from 0.191 to 0.189 cm^3/g. The average freezing temperature of water is practically the same $\langle T\rangle=261.3$ and 261.2 K, respectively; however, the γ_S value slightly increases (~5%). In the case of thermal treatment of the forcemeat, addition of nanosilica gives the opposite effects (Table 7.20) since the S_{uw}, S_{nano}, S_{meso}, γ_S values

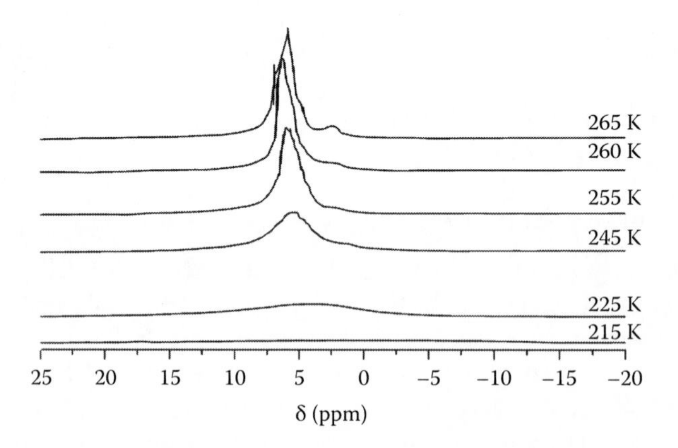

FIGURE 7.81 ^{1}H NMR spectra of initial beef forcemeat at different temperatures.

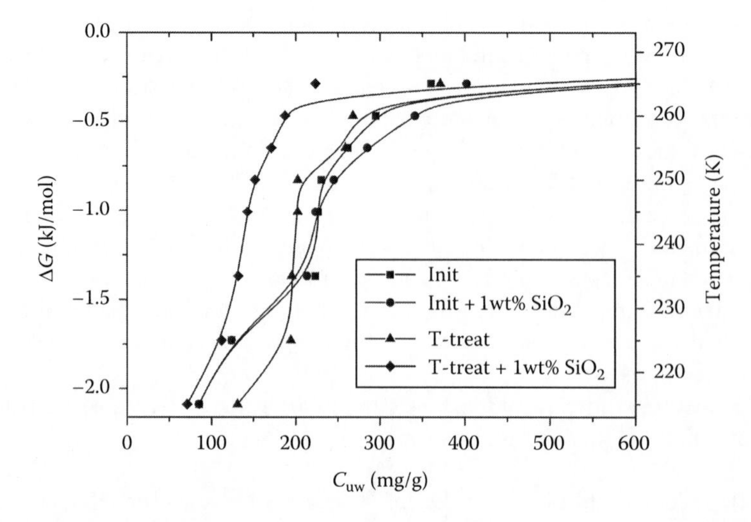

FIGURE 7.82 Temperature dependences of the amounts of unfrozen water bound in forcemeat (beef) initial and treated without and with addition of 1 wt% of nanosilica and related changes in the Gibbs free energy.

TABLE 7.20
Action of the Dispersed Silica Additives and Thermal Treatment on the Characteristics of the Layers of Bound Water in Beef Forcemeat (k_{GT} = 65 K nm)

Sample	C^{w}_{uw} (mg/g)	C^{s}_{uw} (mg/g)	$-\Delta G_w$ (kJ/mol)	$-\Delta G_s$ (kJ/mol)	γ_S (J/g)	S_{uw} (m²/g)	S_{nano} (m²/g)	S_{meso} (m²/g)
Initial forcemeat	240	260	0.34	2.69	38.8	165	51	101
+1% SiO_2	260	300	0.37	2.74	40.7	171	58	101
Thermal treatment	390	210	0.33	2.76	40.2	178	66	100
Treated forcemeat + 1% SiO_2	200	180	0.31	2.65	29.0	140	38	86

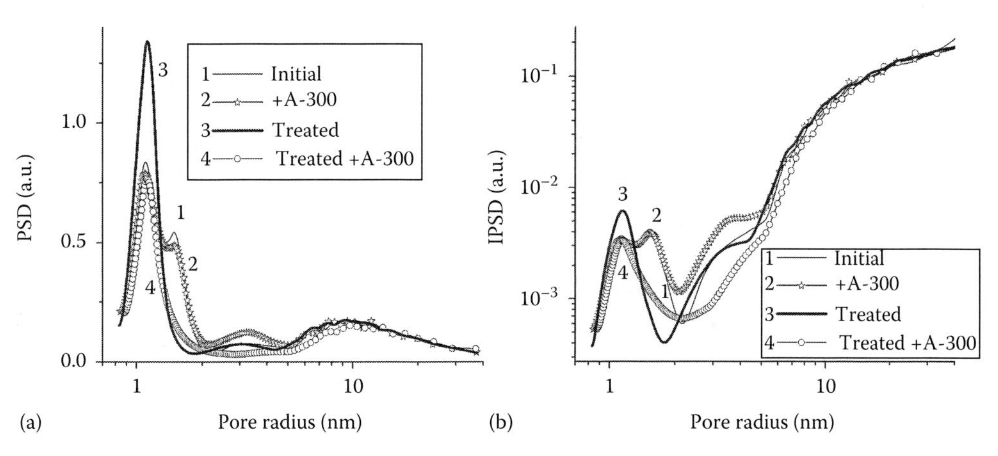

FIGURE 7.83 Pore size distributions in initial beef forcemeat (curve 1), with addition of A-300 (curve 2), treated (curve 3), and treated with addition of A-300 (curve 4): (a) differential and (b) incremental PSD (k_{GT} = 65 K nm).

decrease, as well as V_{nano} (0.031 and 0.018 cm³/g) and V_{meso} (1.542 and 1.369 cm³/g) and the corresponding diminution of the PSD (Figure 7.83) but V_{macro} increased from 0.186 to 0.252 cm³/g that caused an increase in the $\langle T \rangle$ value from 261.2 to 263.3 K and a decrease in the γ_S value (Table 7.20).

Thus, small additions of nanosilica into the beef forcemeat fresh and heated cause rather opposite effects with respect to the amounts of bound water that can be used to control the food characteristics during storage and preparation. Rapid reduction of the concentration of bound water in the readymade product may shorten the duration of thermal treatment, which is required for production of beef foodstuff, capable of long-term storage.

7.6.3 Pork Meat and Fat

Pork in contrast to beef and chicken meat includes a higher content of fat with different fatty acids (C_{15}–C_{17}) and glycerin esters giving ^{1}H NMR signals at 1–2 ppm. The ^{1}H NMR spectra of pork meat includes signals of SAW at 5 ppm and WAW with organics (fatty acids and esters) at 1.3 and 2 ppm (Figure 7.84a).

For fat (Figure 7.84b), the intensity of signals at 1–2 ppm is much higher than that for meat (Figure 7.84a). The fat spectrum includes two signals of SAW at 4.5 and 5.5 ppm with contribution of OH groups from organics. Water content in the fat tissue studied is about 0.2 g/g which is lower than that in the pork meat (2.2 g/g) (Table 7.21 and Figure 7.85).

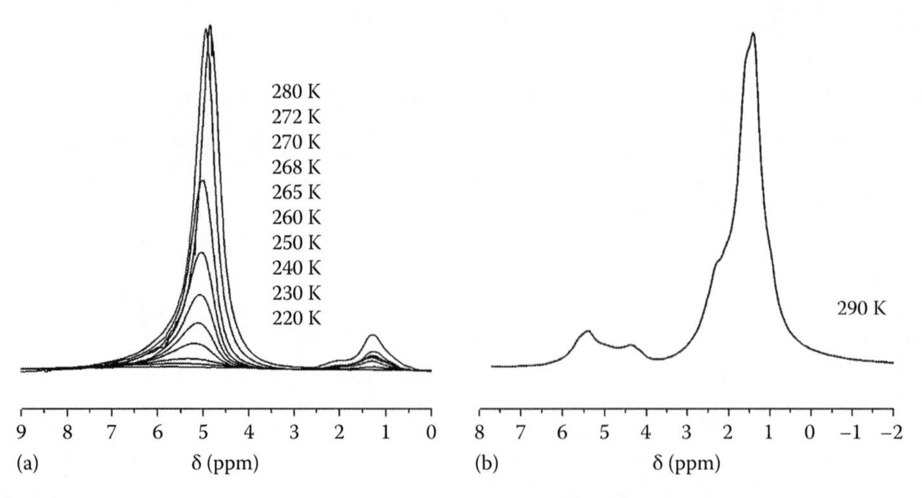

FIGURE 7.84 The ^{1}H NMR spectra of pork (a) meat at different temperatures and (b) fat at 290 K (Varian Mercury 400).

TABLE 7.21
Characteristics of Water Bound in Pork Muscular Tissue

C_{H_2O} (g/g)	C^{s}_{uw} (mg/g)	C^{w}_{uw} (mg/g)	ΔG_s (kJ/mol)	ΔG_w (kJ/mol)	γ_S (J/g)	S_{uw} (m²/g)	S_{nano} (m²/g)	S_{meso} (m²/g)	V_{nano} (cm³/g)	V_{meso} (cm³/g)	$\langle T \rangle$ (K)
2.21	700	1500	−2.29	−0.5	59.6	121	0	115	0.0	2.033	259.1

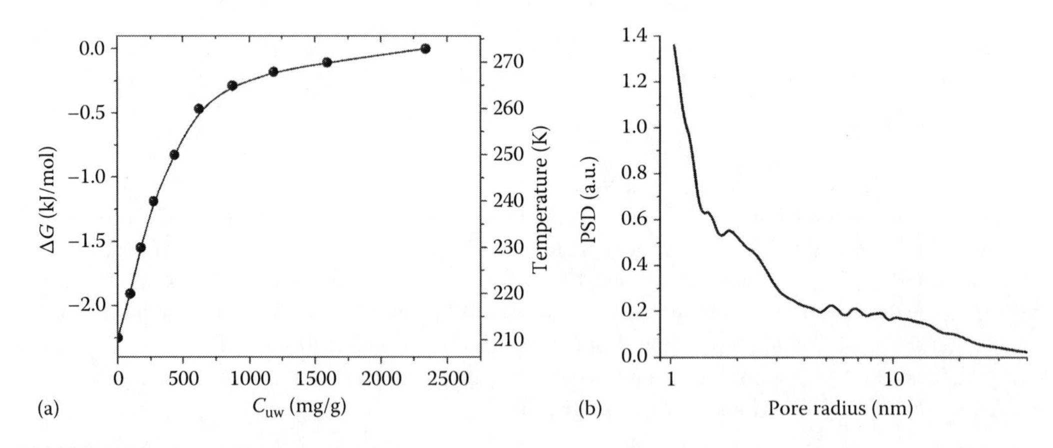

FIGURE 7.85 (a) The amount of unfrozen water versus T and the corresponding changes in the Gibbs free energy and (b) differential PSD (k_{GT}=65 K nm).

The average freezing temperature $\langle T \rangle$ value is lower for pork meat (Table 7.21) than that for beef forcemeat (261 K) but higher than that for chicken muscular tissue (251 K). The structural characteristics of the meats studied differ with respect to bound water, as well as the amounts of this water. Voids filled by structured water are mainly in the mesopore range with significant contribution of SAW (Figure 7.85). One of the consequences from this structure is the possibility of formation of relatively large ice crystallites which can destroy the meat cells; therefore, the second freezing of the meat is not strongly recommended.

7.7 INTRACELLULAR WATER AND CRYOPRESERVATION

7.7.1 Water–DMSO Mixtures

To elucidate the influence of cellular structures on the characteristics of DMSO/water, this mixture was studied alone. For the binary solutions, the freezing point depression is caused by the solvation effects and the colligative properties of solutions (Gun'ko et al. 2005d, Chaplin 2011). In the ^{1}H NMR spectra of the water/DMSO mixtures (without cells), two signals of H_2O and CH_3 (from DMSO) are observed at $\delta_H \approx 5$ and 2.5 ppm, respectively (Figure 7.86). The δ_H value of methyl groups does not practically depend on temperature and the DMSO signal intensity remains nearly constant over the total temperature range and used concentrations of components. This can be due to non-freezing-out of DMSO in the mixture with water and the formation of various clustered structures (Figure 7.87). For dilute solutions, the $\Delta I/I_0(T)$ graphs (Figure 7.88a), which characterize a decrease in the signal intensity of structured water referenced to its intensity before freezing, have

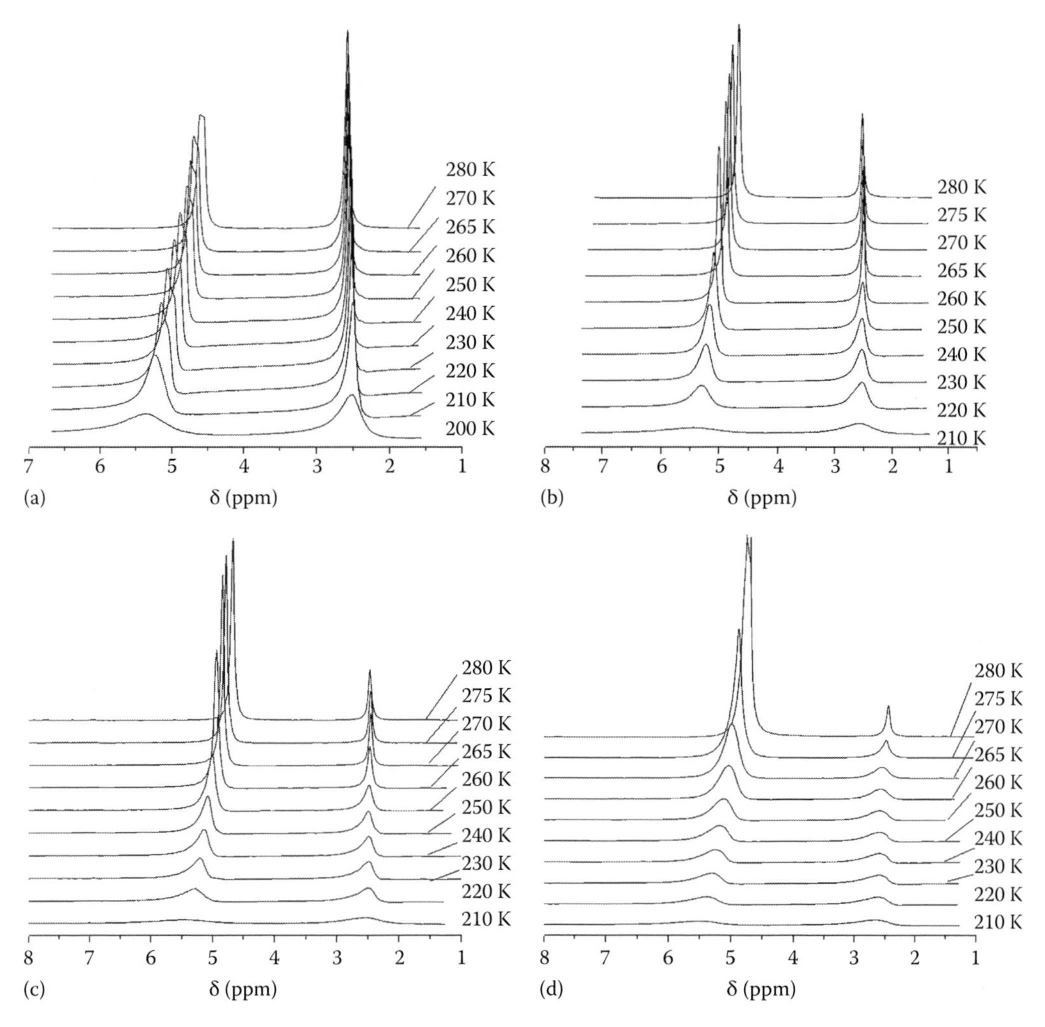

FIGURE 7.86 ^{1}H NMR spectra of aqueous solution of DMSO at different ratios of the water and DMSO contents as (a) 1:2, (b) 1:1, (c) 2.25:1, and (d) 9:1 and different temperatures. (Adapted from *Cryobiology*, 59, Turov, V.V., Kerus, S.V., and Gun'ko, V.M., Behaviour of water bound in bone marrow cells affected by organic solvents of different polarity, 102–112, 2009, Copyright 2009, with permission from Elsevier.)

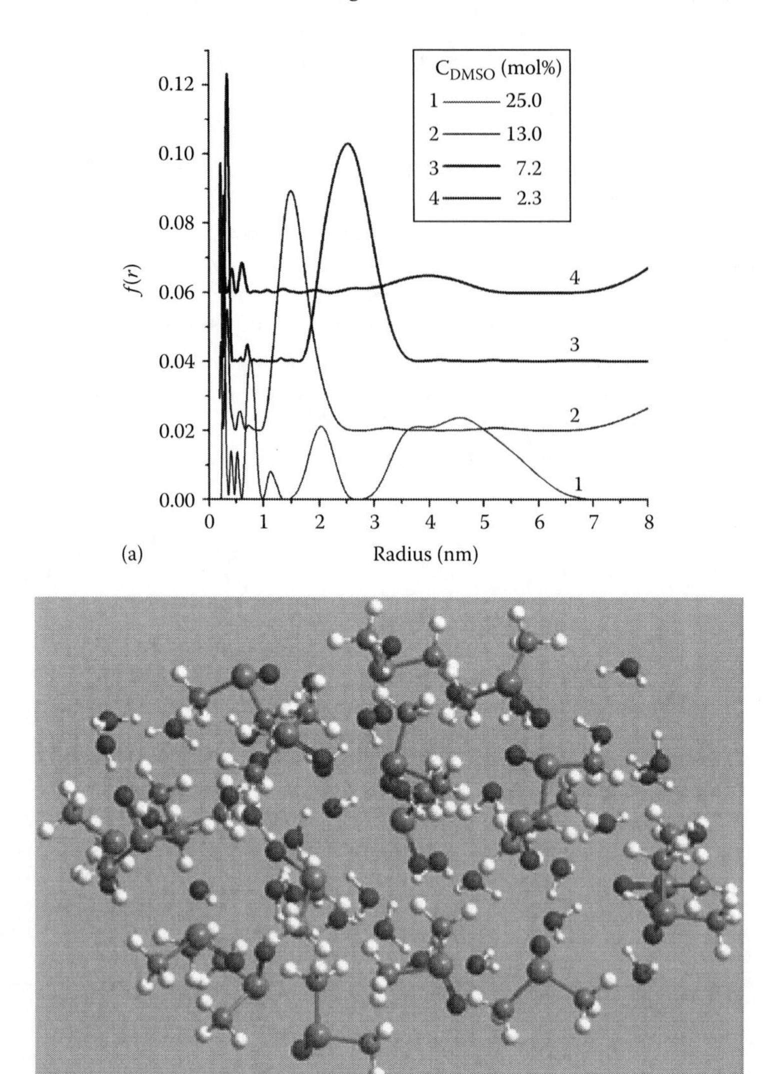

FIGURE 7.87 (a) Size distributions of water structures in water–DMSO mixtures of different compositions and (b) a cluster (2.4 nm × 1.5 nm in size) with $19(CH_3)_2SO$ and $37H_2O$. (Adapted from *Cryobiology*, 59, Turov, V.V., Kerus, S.V., and Gun'ko, V.M., Behaviour of water bound in bone marrow cells affected by organic solvents of different polarity, 102–112, 2009, Copyright 2009, with permission from Elsevier.)

the shape close to that for frozen aqueous suspensions of highly disperse and porous oxides (Gun'ko et al. 2005d, 2009d). There is a section of a relatively sharp decrease in the signal intensity over a relatively narrow temperature range (similar to that for WBW) and a section of small changes in the signal over a wide temperature range (similar to that for SBW). An increase in concentration of DMSO and elevating temperature cause a decrease in the δ_H value of water (Figure 7.88b) because of a decrease in the average coordination number of H_2O molecules. This value approaches that of water dissolved in DMSO and corresponding to $\delta_H \approx 3.5$ ppm (Emsley et al. 1965).

However, these dependences are slightly nonlinear and have different inclination at different temperatures and concentrations of DMSO because of the formation of a variety of water–DMSO complexes, clusters, and nanodomains (Figure 7.87). The structure of the mixtures (especially small clusters at interfacial surface area $S > 1000$ m^2/g) depends on C_{DMSO} since more strongly

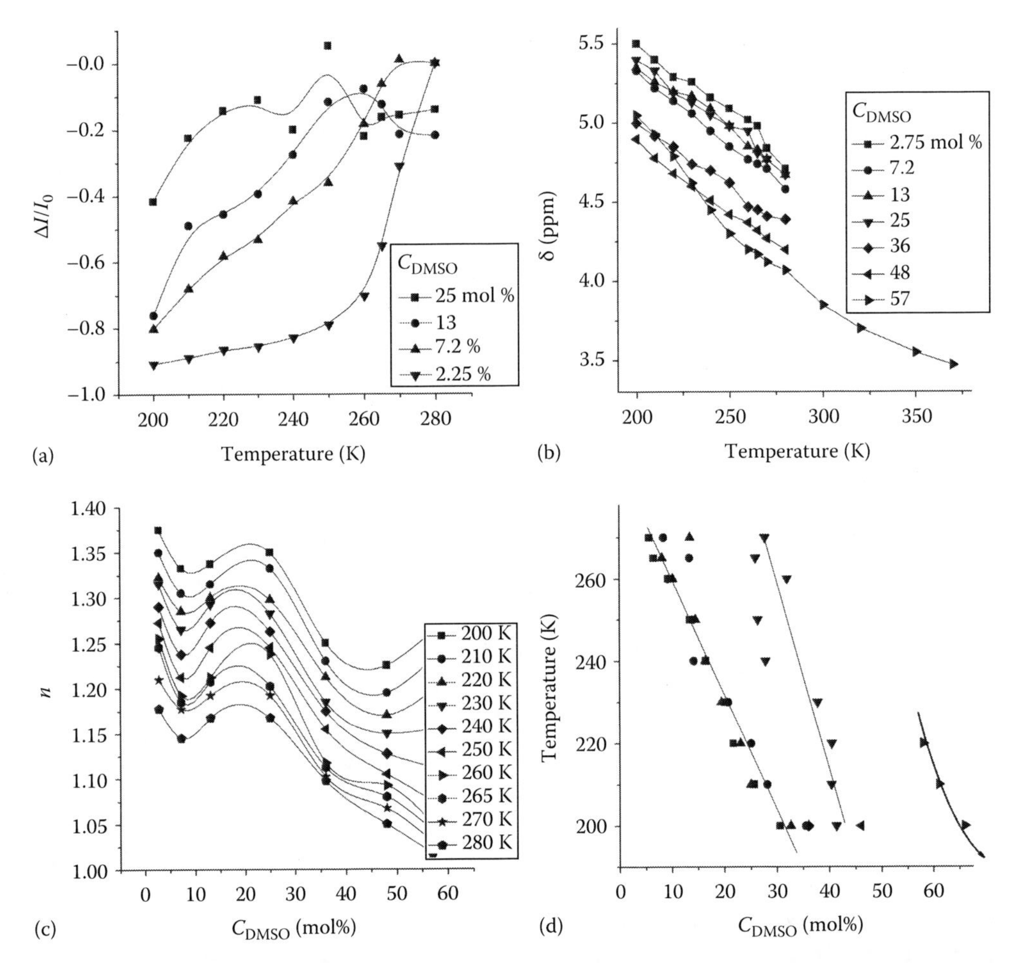

FIGURE 7.88 Water–DMSO system: (a) temperature dependences of a relative decrease in the intensity of the water signal on freezing; (b) temperature dependences of the δ_H value of water at different concentrations of DMSO; (c) change in the average number (n) of protons of water molecules participating in the hydrogen bonds as a function of concentration of DMSO; and (d) fluid–solid phase diagram. (Adapted from *Cryobiology*, 59, Turov, V.V., Kerus, S.V., and Gun'ko, V.M., Behaviour of water bound in bone marrow cells affected by organic solvents of different polarity, 102–112, 2009, Copyright 2009, with permission from Elsevier.)

clustered water is observed at larger amounts of DMSO. For simple complexes, HO–**H**···O=S$(CH_3)_2$ and HO–**H**···OH_2 (calculated using GIAO/B3LYP/6-31G(d,p) as described previously; Gun'ko and Turov 1999) $\delta_{H,iso} = 3.2$ and 3.1 ppm, respectively, and $\delta_{H,iso} = 3.3$ ppm for $(CH_3)_2$S=O···**H**–O–**H**···O=S$(CH_3)_2$. In larger water domains (for water alone), the δ_H value grows because the number of molecules having four hydrogen bonds (per a molecule) increases in dense structures similar to those formed in ice (Gun'ko et al. 2005d, Chaplin 2011). Therefore, the formation of larger water clusters and domains in the DMSO media with decreasing C_{DMSO} leads to an increase in the δ_H value to 5–5.5 ppm.

A relative small water/DMSO droplet with the geometry optimized using the PM3 method (Figure 7.87b) is characterized by a smaller average value $\delta_H = 2.1$ ppm (the δ_H values change over the 0.5–5.5 ppm range) for water molecules because some H atoms of them do not participate in the hydrogen bonds and have $\delta_H < 2$ ppm. The average value $\delta_H = 3.6$ ppm is for the H atoms participating in the hydrogen bonds. These δ_H values were calculated on the basis of the atom

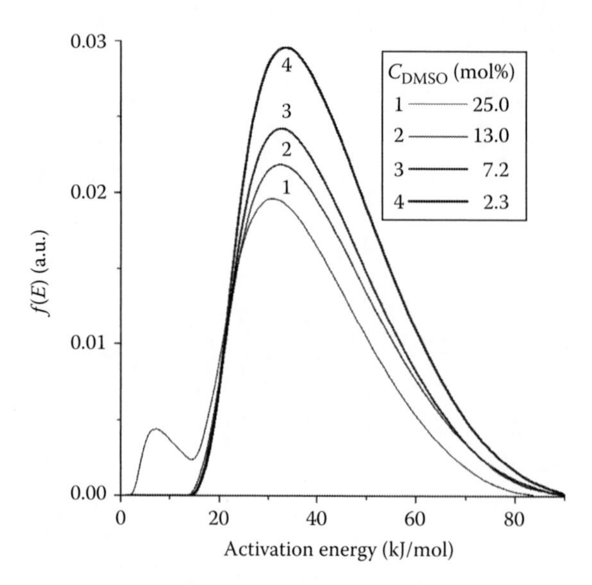

FIGURE 7.89 Distribution functions of the activation energy of molecular mobility of water in water–DMSO mixtures of different compositions. (Adapted from *Cryobiology*, 59, Turov, V.V., Kerus, S.V., and Gun'ko, V.M., Behaviour of water bound in bone marrow cells affected by organic solvents of different polarity, 102–112, 2009, Copyright 2009, with permission from Elsevier.)

charges q_H (PM3) in the DMSO/water nanodroplet (Figure 7.87b) and a linear approximation $\delta_{H(H_2O)} = -11.72921 + q_{H(H_2O)} \times 62.30708$ obtained using B3LYP/6-31G(d,p) (to calculate δ_H) and PM3 (q_H) calculations of small complexes with one or two DMSO molecules and H_2O (Gun'ko et al. 2005d, 2009d).

The freezing point depression is directly proportional to mole concentration of DMSO (because of Raul's law) at $C_{DMSO} < 30$ mol% (Figure 7.88d). At higher concentrations, the freezing depends nonlinearly on C_{DMSO} because of the existence of a variety DMSO–water structures. For the used range of concentrations, the point of eutectic was not achieved because of the absence of entire freezing-out of DMSO. In the DMSO/water mixtures, the activation energy of the molecular mobility of water depends on C_{DMSO} (Figure 7.89) because of enhancement of clusterization of water that leads to a slight decrease in the average number of the hydrogen bonds per molecule, and therefore the E value decreases.

Addition of organic solvents to cells causes diminution of the activation energy (E) of the molecular mobility of intracellular water since the peak value of $f(E)$ shifts toward smaller E values (Figure 7.90). This can be caused by diminution of the average number of the hydrogen bonds per water molecule; however, these changes are nonlinear even for DMSO/water without cells (Figure 7.88c). This effect is stronger for intracellular water/organics than that for the DMSO/water mixtures without cells (Figure 7.89) because of stronger clusterization of intracellular water. This effect is of importance on cryopreservation of cells as it provides diminution of the negative effects (damage of cells) caused by the formation of large ice crystallites based on a fraction of intracellular SAW.

7.7.2 Clusterization of Water in Ternary Systems with DMSO or Acetonitrile and Chloroform

Binary and ternary solvent systems with water and DMSO (as a cryopreservation compound) with polar (acetonitrile) and weakly polar (chloroform) cosolvents were used during investigations of

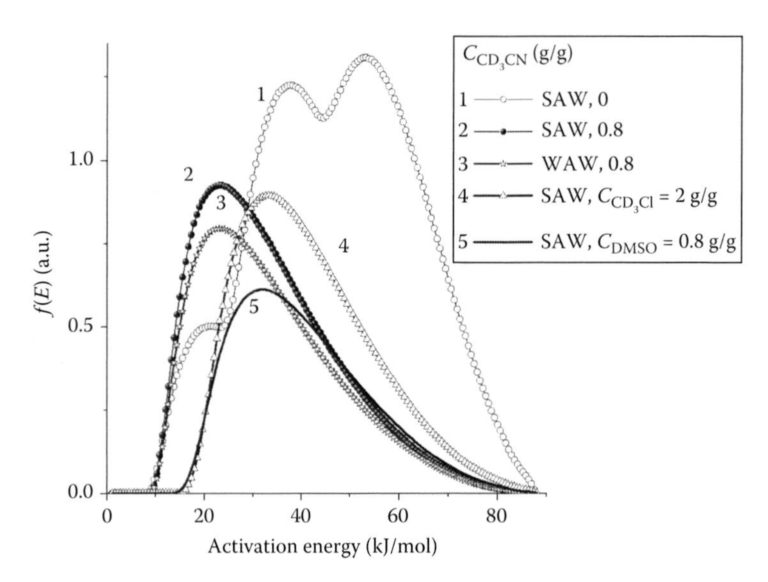

FIGURE 7.90 Distribution functions of the activation energy of molecular mobility of water (in SAW and WAW) bound in cells without or with addition of organics. (Adapted from *Cryobiology*, 59, Turov, V.V., Kerus, S.V., and Gun'ko, V.M., Behaviour of water bound in bone marrow cells affected by organic solvents of different polarity, 102–112, 2009, Copyright 2009, with permission from Elsevier.)

bio-objects. Therefore, it is of interest to analyze these solvent systems alone. The phase state of binary and more complex systems is determined by relationships of the thermodynamic parameters of a homogeneous solution and a system with two or more phases enriched by one of the components (Shah and Roberts 2008). At a small difference in the free energies of homogeneous and heterogeneous systems, the separation of mixtures can occur due to changes in environmental conditions (pressure, temperature) or with time because for the formation of one phase in another phase a value of fluctuations of component concentration should be larger than a certain critical value. The solution stability with polar and nonpolar or weakly polar components (e.g., water and organics) is of importance for liquid fuels, in which the formation of a water phase leads to negative results (Gani and Brignole 1983, Pereda et al. 2009). Binary systems are widely used in analytical chemistry for extraction of poorly soluble compounds (Barwick 1997, Kholkin et al. 1999), or as additives to cryopreservants (Hubálek 2003, Varisli et al. 2009).

Many mixtures of organic solvents, which are very different in their hydrophobic/hydrophilic properties, can form true solutions in a wide range of their concentrations and these mixtures are not separate even with decreasing temperature lower than freezing temperatures of both components. Despite a polar component can dissolve water in any proportion, addition of a small amount of water can lead to separation of the system (Robbins et al. 2007). Therefore, clusterization of DMSO (or acetonitrile)–chloroform was studied with addition of small amounts of water (Turov and Gun'ko 2011).

The $CDCl_3$/DMSO-d_6 and $CDCl_3/CD_3CN$ mixtures were prepared in NMR ampoules (shaken and equilibrated) and then a certain water amount was added and the mixtures were shaken. If water content was <0.025 g/g, then the mixtures were transparent but at larger content of water they became opalescent due to phase separation. $CDCl_3$ has a greater density (1.5 g/cm^3) than DMSO-d_6 (1.19 g/cm^3) or CD_3CN (0.844 g/cm^3). Therefore, a phase enriched by chloroform is located in the bottom part of the ampoule in the working area of the NMR censor.

The $H_2O/CDCl_3$/DMSO-d_6 systems (Figure 7.91a through c) are characterized by three main signals of protons at $\delta_H \approx 2$, 2.8–3.8, and 4.5 ppm (signals 1, 2, and 3, respectively). The characteristics of signal 1, which can be assigned to CHD_2 groups in DMSO, are practically independent of the conditions. A similar signal of CHD_2 is observed for acetonitrile at $\delta_H \approx 2$ ppm (Figure 7.91d through f).

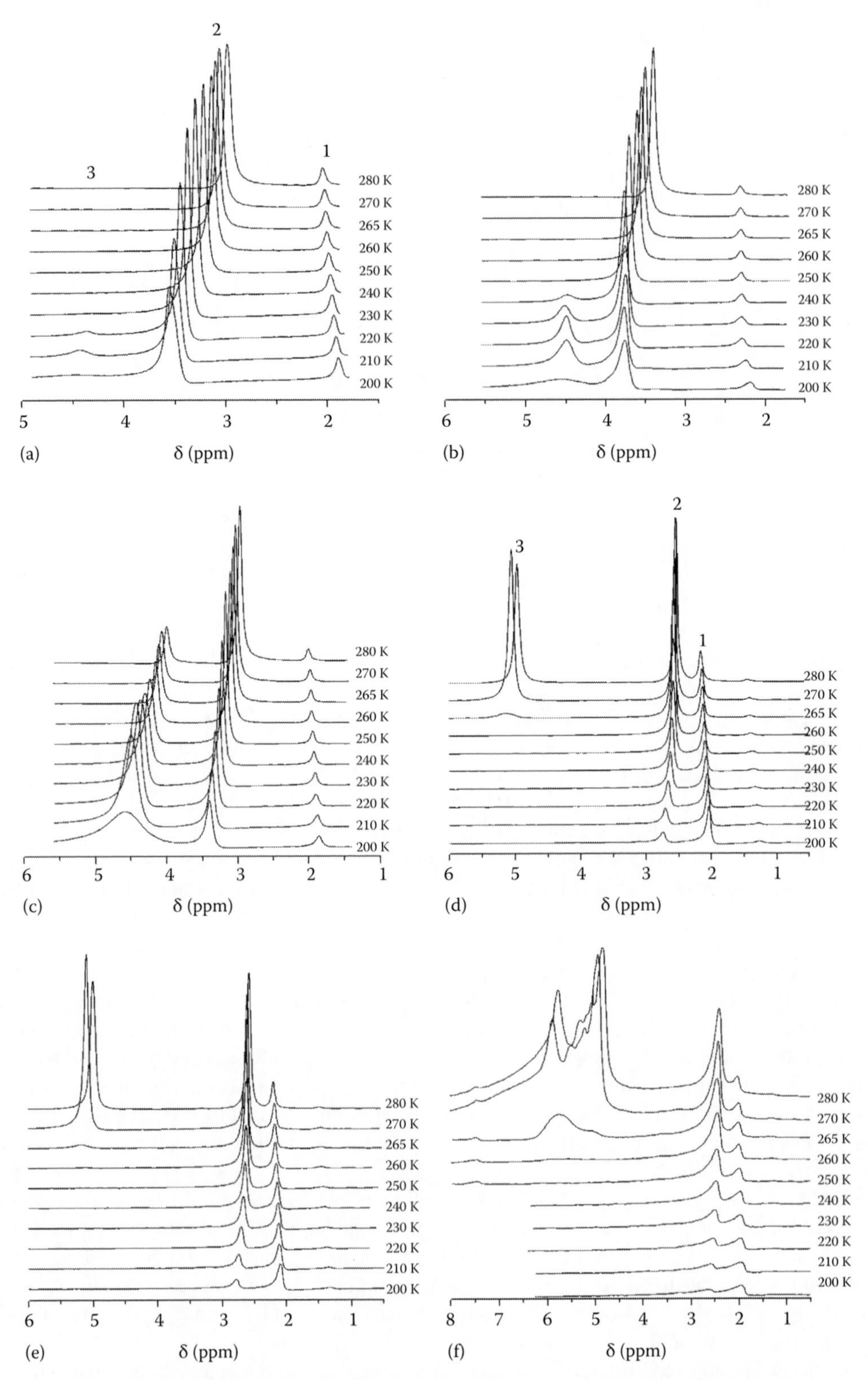

FIGURE 7.91 ^{1}H NMR spectra of water added to a mixture with (a–c) $CDCl_3$/DMSO-d_6 (2.7:1) and (d–f) $CDCl_3$/CD_3CN (3.8:1) at water content (a) 0.011, (b) 0.025, (c) 0.062, (d) 0.013, (e) 0.02, and (f) 0.04 g of water per gram of organic solution.

Signal 2 can be assigned to water dissolved in an electron-donor solvent and forming such complexes (associated water, ASW) as HOH···OS$(CD_3)_2$ or HOH···$NCCD_3$ (Gun'ko et al. 2005a). The spectral characteristics of signal 2 depend on temperature differently for the system studied. In the case of $H_2O/CDCl_3/CD_3CN$, the δ_H value of signal 2 depends weakly on temperature but for $H_2O/CDCl_3$/DMSO-d_6, it shifts toward stronger magnetic field with increasing temperature.

Signal 3 can be assigned to SAW and its intensity increases with increasing water content in the mixtures. Its temperature behavior differs for the systems with DMSO and acetonitrile (Figures 7.91 and 7.92). For the system with DMSO, the intensity of signal 3 increases with decreasing temperature to a certain maximum and then it decreases due to freezing-out of SAW at $T \leq 210$ K. However, for the system with acetonitrile, the intensity of signal 3 decreases with decreasing temperature at $T < 273$ K.

A decrease in the intensity of signal of ASW with lowering temperature can be due to freezing-out of water because the protons of ice are not observed in the high-resolution ^{1}H NMR

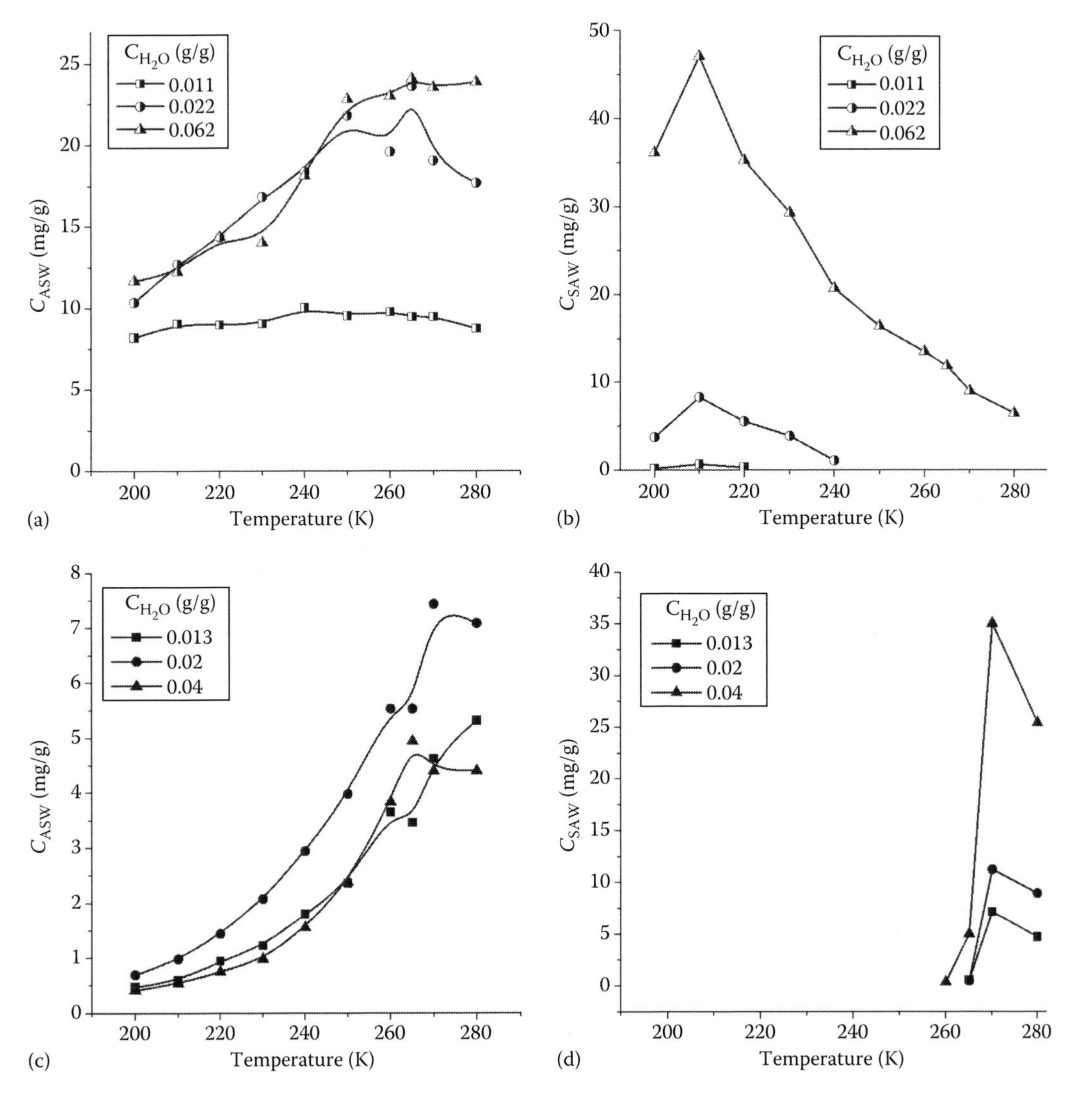

FIGURE 7.92 Temperature dependences of the amounts of water in complexes (ASW) with (a) DMSO or (c) CD_3CN and SAW in the mixtures with (b) $CDCl_3$/DMSO-d_6 (2.7:1) and (d) $CDCl_3/CD_3CN$ (3.8:1).

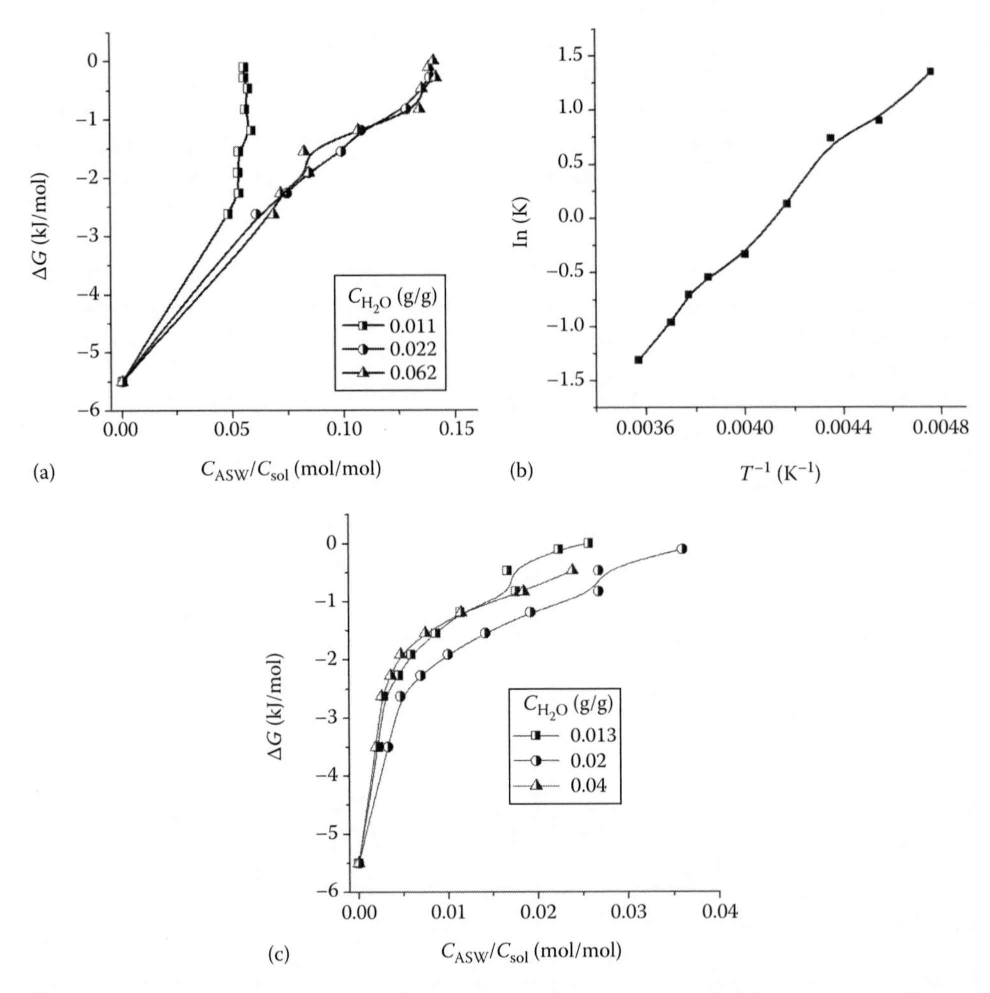

FIGURE 7.93 (a) Changes in the Gibbs free energy of water (ASW) versus relative content of ASW in (a) $H_2O/CDCl_3/DMSO\text{-}d_6$ or (c) $H_2O/CDCl_3/CD_3CN$ and (b) equilibrium constant versus temperature of the process $HOH \cdots OS(CD_3)_2 \leftrightarrow$ SAW ($C_{H_2O} = 0.062$ g/g); point at $\Delta G = -5.5$ kJ/mol corresponds to extrapolation.

spectra of static samples due to small values of transverse relaxation time (Emsley et al. 1965, Kinney et al. 1993).

From the signal intensity (Figures 7.92 and 7.93), the equilibrium constant of the formation of SAW from ASW could be estimated (Figure 7.93b). Changes in the Gibbs free energy of this process correspond to −18.4 kJ/mol that is close to the heat of ice formation. Figure 7.93 demonstrates the graphs of changes in the Gibbs free energy versus relative content of ASW.

Thus, water in mixtures with polar (DMSO, acetonitrile) and weakly (chloroform) polar organic solvents can form both complexes with electron-donor molecules and large clusters of SAW (as dispersion phase). In the mixtures with DMSO, the amount of ASW is greater than in the case of the mixtures with acetonitrile. The behavior of SAW in the mixtures with DMSO and acetonitrile differs and this difference explains good properties of DMSO as a cryopreservative preventing the formation of large SAW structures (i.e., ice crystallites).

8 Interaction of Seeds, Herbs, and Related Materials with Water and Nanooxides

Although the application of NMR techniques to seeds, herbs, plant tissues, etc., is a minority interest among spectroscopists, NMR provides a range of problem-solving tools that can be applied in many areas of plant biochemistry and plant physiology (Pouliquen et al. 1997, Ishida et al. 2000, Ratcliffe et al. 2001, Glidewell 2006, Garnczarska et al. 2007, 2008) as well as related foods (Todt et al. 2006, Spyros and Dais 2009). These methods have been employed productively for over 35 years, and the combination of the continuing technological development of NMR itself, and the rapid strides that are taking place in the genetic manipulation of plants should ensure that application of plant NMR spectroscopy is very informative in structural analysis of macromolecules and whole biosystems (cells, tissues, seeds, etc.). In addition to different spectral techniques, imaging can provide very clear pictorial information on complex biosystems. For instance, an NMR microscope can nondestructively detect free water in tissues and create anatomical images of the plant tissues. Since the quantity and mobility of cell-associated water is closely related to the condition of the cells, ^{1}H NMR images represent physiological maps of the tissue (Ishida et al. 2000). In addition, the technique locates soluble organic compounds accumulated in the tissues, such as sugars in vacuoles or fatty acids stored as oil droplets in vesicles. ^{23}Na NMR imaging is suitable for studying the physiology of salt-tolerant plants. Diffusion measurements provide information about the transport of substances and ions accompanied by water movement. The developed techniques of 3D imaging, flow-encoded imaging, and spectroscopic imaging open up new opportunities for plant biologists. NMR microscope is thus a unique and promising tool for the study of living plant systems in relation to morphology, the true features of which are often lost during preparation for more conventional tissue analysis (Ishida et al. 2000). In this chapter, the behavior of water bound in different plants, seeds, and the corresponding products and related interfacial phenomena are briefly analyzed.

Gödecke et al. (2012) analyzed active constituents of *Angelica sinensis* (AS) extracts and fractions that demonstrate activity in a panel of in vitro bioassays. They demonstrated how NMR can satisfy the requirement for simultaneous, multitarget quantification and qualitative identification of bioactive compounds. First, the AS activity was concentrated into a single fraction by reversed-phase (RP) solid-phase extraction (SPE), as confirmed by an alkaline phosphatase, (anti-)estrogenicity, and cytotoxicity assay. Second, a quantitative ^{1}H NMR (qHNMR) method was established and validated using standard compounds and comparing processing methods. Subsequent 1D/2D NMR and qHNMR analysis led to the identification and quantification of ligustilide and other minor components in the active fraction, and to the development of quality criteria for authentic AS preparations. The absolute and relative quantities of ligustilide, six minor alkyl phthalides, and groups of phenylpropanoids, polyynes, and polyunsaturated fatty acids were measured by a combination of qHNMR and 2D COSY techniques. The qNMR approach enables multitarget quality control of the bioactive fraction, and enables the integrated biological and chemical standardization of AS botanicals. This methodology can potentially be transferred to other botanicals with active principles that act synergistically, or that contain closely related and/or constituents, which have not been conclusively identified as the active principles (Gödecke et al. 2012).

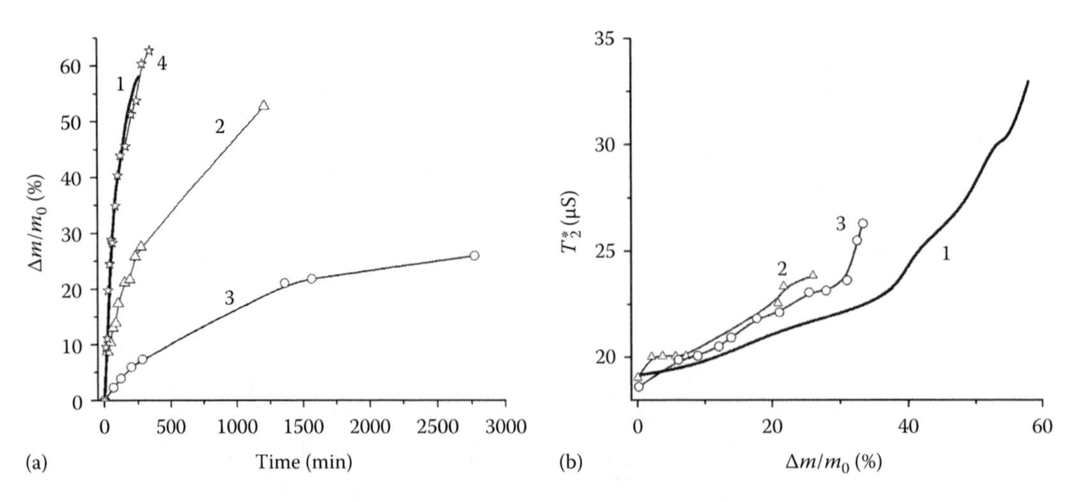

FIGURE 8.1 (a) Wheat seed hydration expressed as relative mass increase ($\Delta m/m_0$) versus time t and (b) the hydration ($\Delta m/m_0$) dependence of the proton spin–spin relaxation time T_2* for the solid (Gaussian) signal component of wheat seed: (curve 1) Seed soaked between two layers of blotting paper; (curve 2) seed completely immersed in water; (curve 3) Seed hydrated over the water surface (relative humidity $p/p_0 = 100\%$); (curve 4) seed soaked in D_2O (procedure as curve 1). (Adapted from *Colloid. Surf. A: Physicochem. Eng. Aspects*, 115, Harańczyk, H., Strzałka, K., Jasiński, G., and Mosna-Bojarska, K., The initial stages of wheat (*Triticum aestivum, L.*) seed imbibition as observed by proton nuclear magnetic relaxation, 47–54, 1996, Copyright 1996, with permission from Elsevier.)

Harańczyk et al. (1996) studied water imbibition by wheat seeds using NMR spectroscopy analyzing the total proton NMR signal as a function of H_2O or D_2O mass uptake. The analysis of the signal components coming from the solid matrix of the seed (S) and from the water absorbed showed that the first one was not much affected by seed hydration, but its magnitude decreased as the starchy endosperm of the wheat seed was decomposed. The decrease was linear as a function of the water mass absorbed, causing the nonlinear increase of the total liquid proton signal (L). Models describing L/S as a function of H_2O mass absorbed and $S/(L + S)$ as a function of D_2O absorbed were analyzed. The seed hydration dependence (Figure 8.1a) may be described by a single exponential function with sufficient accuracy, strongly suggesting first-order kinetics.

However, one cannot exclude the very fast component which may cause an initial rapid mass increase of about $\Delta m/m_0 \approx 5\%$, as two-exponential fits of the mass increase versus time dependence diverged. Figure 8.1b shows the hydration ($\Delta m/m_0$) dependence of T_2* for the solid signal component of a wheat seed. For all types of hydration, the value of T_2* changes only weakly with mass increase, which means that the solid matrix of the seed (mostly starch body) is little affected by water added in regions that are not digested or dissolved (Harañczyk et al. 1996).

Rascio et al. (1999, 2005) analyzed pressure–volume (PV) and adsorption isotherm (AI) curves for fresh and dry tissues of durum wheat to assess the relation between the water status of living leaves and the properties of water that is bound (BW) with different strength to ionic, polar, or hydrophobic sites of macromolecules. The leaves were collected from six genotypes grown in field, in 2 years. The amounts of the nonosmotic BW fraction, free water, and osmotic potential at full turgor were determined by PV curves. Three parameters that relate to the amounts of the weakly and strongly bound water (quantitative BW properties) and five parameters related to tissue-binding strength for the same water fractions (qualitative BW properties) were calculated using AI curves. The nonosmotic volume of PV curves did not correspond to the water fraction bound to the charged or polar sites of macromolecules as determined by AI curves. The qualitative parameters, in contrast to quantitative parameters, may be affected by common physical–chemical factors: the changes in the amount of strongly bound water in tissues were independent from

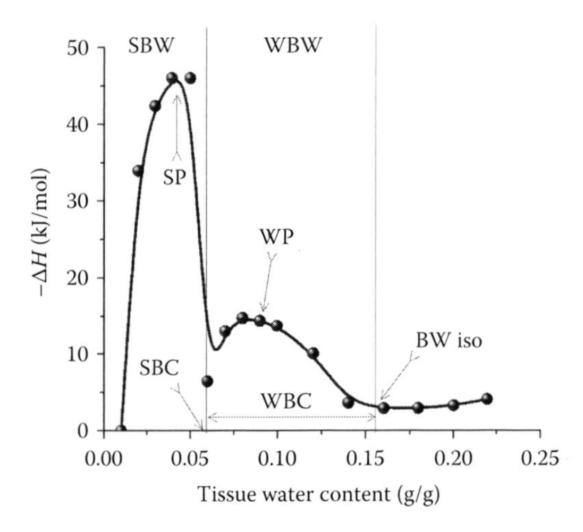

FIGURE 8.2 The differential heat of water sorption calculated using the Clausius–Clapeyron equation. Arrows depict the most negative enthalpy value in the region of strongly (SP) and weakly (WP) bound water, the limit of strongly bound water region (SBC), the moisture content where bound water first appeared (BW iso) and the tissue moisture range corresponding to weakly bound water (WBC). (Adapted from *Plant Sci*., 169, Rascio, A., Nicastro, G., Carlino, E., and Di Fonzo, N., Differences for bound water content as estimated by pressure–volume and adsorption isotherm curves, 395–401, 2005, Copyright 2005, with permission from Elsevier.)

those in the weakly bound fraction. In contrast, an increase in tissue affinity for strongly bound water implied a simultaneous increase in the affinity for weakly bound water. The qualitative properties of bound water may be particularly important for drought adaptation in durum wheat, being associated with the solute potential and with the succulence degree of the leaves. Plots of differential energies of water sorption at increasing moisture content were constructed for each sample (Figure 8.2). The quantities of BW fractions were determined graphically by SBC (tissue moisture content where enthalpies reached about zero from their very negative value, indicating the moisture content below which all the water is presumably chemisorbed and, therefore, tightly bound), BWiso (tissue moisture where free water first appeared, that indicates the end point of the weakly plus strongly bound regions of the AI), and WBC (weakly bound tissue moisture calculated as difference between BWiso and SBC). The following qualitative properties of BW fractions were estimated: SP (the most negative values of enthalpy observed in the strongly bound water region), ΔH_{strong} (the strength of water binding for strongly bound water, calculated as the average enthalpy in the first isotherm region), WP (the most negative value of enthalpy observed in the region of weakly bound water), ΔH_{weak} (the strength of water binding for weakly bound water, calculated as the average enthalpy for the second isotherm region), and ΔH_{tot} (the average strength of water binding of both water fractions calculated as the average enthalpy for the first and the second isotherm regions) (Rascio et al. 1999, 2005) (Figure 8.2).

Pouliquen et al. (1997) used NMR technique to study water and lipid reserves in seeds. The temperature dependence of T_1 relaxation time was used to identify differences in the thermodynamic properties of water between dry seeds and during germination (Figure 8.3).

The T_1 minimum position depends on both the seed type and the seed state. Among the species studied, T_2 measurements distinguished two categories of seeds: pea ($T_2 = 1.1$ and 45 ms), maize (1.2 and 60 ms), and wheat (1.3 and 30 ms) for which two components of T_2 were observed, and lettuce (103 ms), tomato (98 ms), and radish (85 ms), which presented one single component (Pouliquen et al. 1997). The main short component was attributed to water (T_1:T_2 ratio is very high, which suggests that these protons are in a near-solid state at $T = 39°C$, i.e., this water is SBW) whereas the long one was attributed to lipids from oil bodies (T_1:T_2 is close to 1, which means that the protons

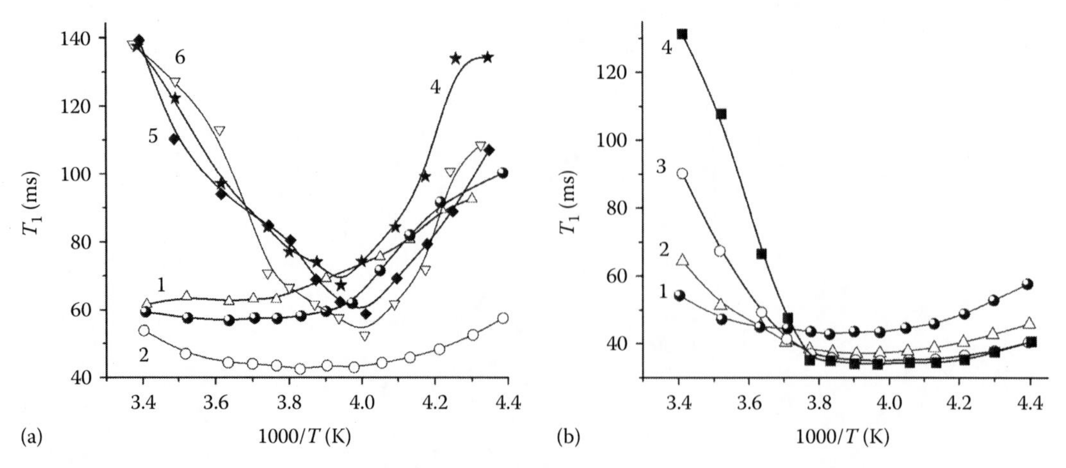

FIGURE 8.3 Temperature dependence of T_1 relaxation time for (a) dry seeds (curve 1) pea, (curve 2) maize, (curve 3) wheat (20 MHz), and radish seeds during germination t=(curve 4) 0, (curve 5) 3, and (curve 6) 24 h (40 MHz) and (b) wheat seeds during germination t=(curve 1) 0, (curve 2) 4 (curve 3) 24, and (curve 4) 70 h (20 MHz). (Adapted from *C. R. Acad. Sci. Paris Life Sci.*, 320, Pouliquen, D., Gross, D., Lehmann, V., Ducoufwau, S., Demilly, D., and Léchappé, J., Study of water and oil bodies in seeds by nuclear magnetic resonance, 131–138, 1997, Copyright 1997, with permission from Elsevier.)

generating the NMR signal belong to molecules that are in the liquid state). These observations are in good correlation with the hydration degree of the seeds. For instance, lettuce and radish seeds, which have the lowest water contents (7.3% and 5.6%, respectively), present the highest magnetization, but maize and pea seeds, which have much higher water contents (12.2% and 9.0%, respectively), show the lowest magnetization. Images of two dry seeds, one of pea and the other of radish, showed marked differences in the distribution of NMR signal intensity, which suggested a deferent distribution of oil bodies (Pouliquen et al. 1997).

Inter-seed processes and the behavior of bound water are the most interesting during seed germinating. Therefore, the effects of wheat seed germinating for 1 and 2 days (at 22°C–25°C in drinking water) and addition of nanosilica A-300 (1 wt%) were studied using ^{1}H NMR spectroscopy. The water spectra of initial dry wheat seeds (~7 wt% of water) at T>260 K include two signals at 5 ppm, signal 1 (SAW), and 1.4 ppm, signal 2 (WAW) (Figure 8.4a). The intensity of signal at 5 ppm strongly decreases with decreasing temperature because this is SAW with a certain contribution of WBW frozen at subzero temperatures. Water responsible for the signal at 1.4 ppm (WAW) includes a significant portion of SBW since it is not frozen at 220 K. After seed germinating, the intensity of signal 2 strongly decreases because seed swelling (Figure 8.5 and Table 8.1) increases contribution of large voids (Figure 8.6) filled by SAW/WBW. In the presence of A-300, the seed swelling is smaller and contribution of large voids filled by SAW/WBW decreases (Figures 8.5 and 8.6, and Table 8.1).

The effects of nanosilica A-300 on seed germinating can be caused by several factors. First, the viscosity of the liquid medium increases, that is, negative effect for the diffusion of water molecules into seeds. Interaction of nanoparticles with a seed shell and especially with a shoot bud inhibits the germinating process. Dissolution silica products penetrating into seeds can cause certain influence on the vital functions of the seeds. The first factor is not strong due to a low content of silica (1 wt%) and the viscosity of the dispersion is close to that of pure water. It is known that low content of silica (<0.1 wt%) can stimulate the vital function of cells but larger content (>1 wt%) can provide the opposite effects (Chuiko 2003, Blitz and Gun'ko 2006).

To change influence of nanooxides on the water state in seeds, hydrophobic nanosilica AM-1 prepared by silylation of A-300 by trimethylchlorosilane was used as a component of complex preparation. This preparation included AM-1 with protective-stimulating means (PSM) used to affect

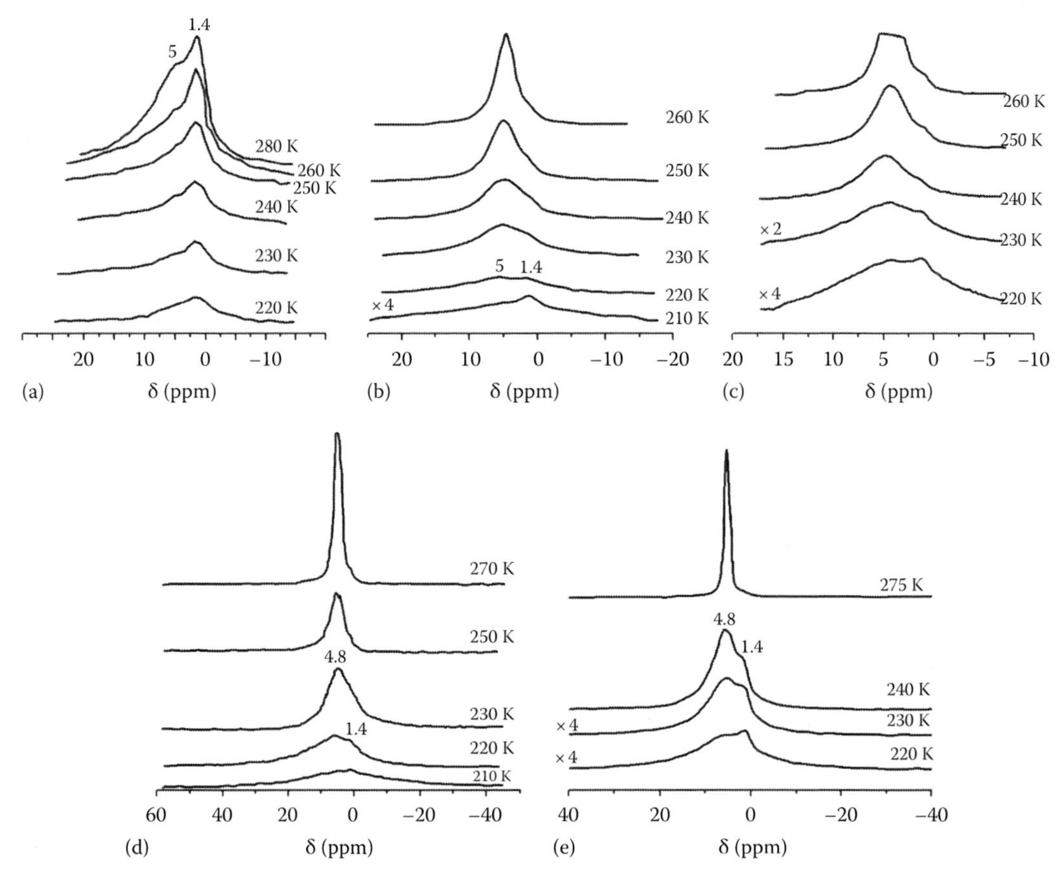

FIGURE 8.4 ^1H NMR spectra of water bound in wheat seeds (a) initial and after (b, d) 1 and (c, e) 2 days of germinating in (b, c) pure water or (d, e) with addition of nanosilica A-300.

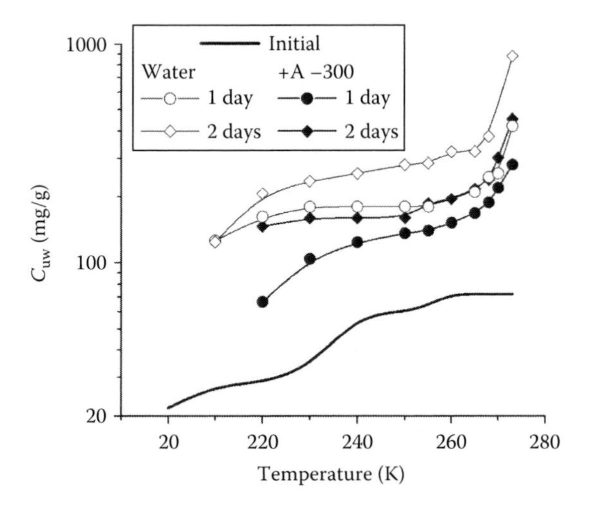

FIGURE 8.5 Temperature dependences of the amounts of water bound in wheat seeds initial and after 1 and 2 days of germinating in pure water or with addition of nanosilica A-300.

TABLE 8.1
Characteristics of Water Bound in Wheat Seeds Germinating for 1 or 2 Days

Sample	S_{uw} (m²/g)	S_{nano} (m²/g)	S_{meso} (m²/g)	V_{uw} (cm³/g)	V_{nano} (cm³/g)	V_{meso} (cm³/g)	γ_S (J/g)	$-\Delta G_s$ (kJ/mol)
Initial	88	71	17	0.072	0.027	0.045	7.6	3.36
1 day	348	332	16	0.344	0.135	0.188	27.6	3.37
2 days	378	332	42	0.775	0.147	0.576	38.0	2.77
1 day[a]	107	92	15	0.261	0.042	0.209	16.2	2.61
2 days[a]	262	242	18	0.403	0.108	0.272	23.8	2.66

[a] With addition of 1 wt% of A-300.

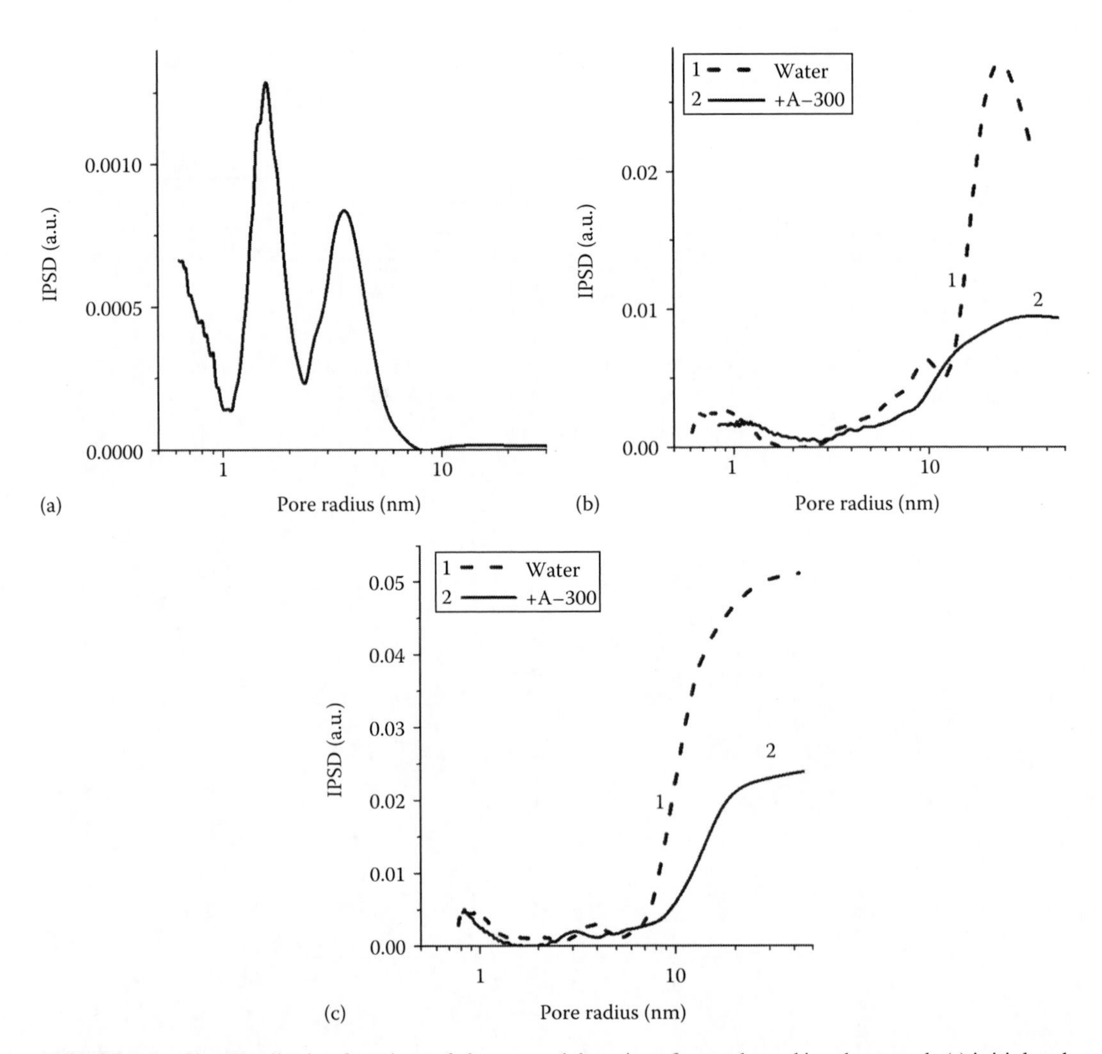

FIGURE 8.6 Size distribution functions of clusters and domains of water bound in wheat seeds (a) initial and after (b) 1 and (c) 2 days of germinating in (curve 1) pure water or (curve 2) with addition of nanosilica A-300.

TABLE 8.2
Biometric Parameters of Overgrown Wheat Seeds Affected by Nanosilica, or PSM

Parameter	Control Sample	1% SiO_2	PSM
Germinability (%)	74	81 (+7%)	95 (+21%)
Length of sprouts (cm)	7.5	10.7 (+42%)	11.3 (+50%)
Weight of 100 sprouts (g)	1.54	2.21 (+43%)	1.9 (+23%)
Length of roots (cm)	4.5	3.4 (−24%)	8.4 (+86%)
Root number per seed	2.6	3.7 (+42%)	4.2 (+61%)

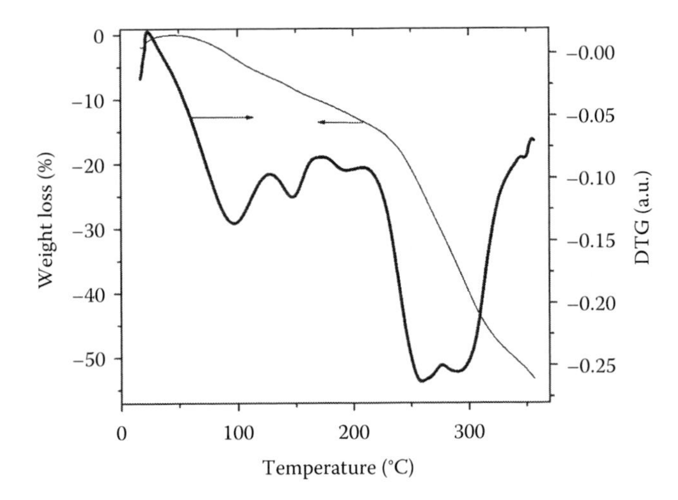

FIGURE 8.7 Weight loss and DTG data for wheat seed roots.

germination of wheat seeds. It was prepared by MCA of AM-1 with a mixture of potash, phosphate, and nitrogen (1:1:1) fertilizers. Wheat seeds were couched in pure water (control sample 1) or the same volume of water with addition of initial A-300 (sample 2) or PSM (sample 3) at 22°C–25°C in Petri dishes with filter paper at the bottom and controlled degree of hydration. Measurements were performed for seed roots in 3 and 7 days of sprouting. Table 8.2 shows that both A-300 and PSM can strongly affect germinating seeds; however, the PSM effects were stronger.

Weakly and strongly bound waters (~10 wt%) give negative DTG peaks at 90°C and 130°C, respectively (Figure 8.7). At higher temperatures, associative desorption of water and other decomposition reactions of organic components in the seed roots occur and at 360°C the weight loss corresponds to 55%. Thus, several types of water are present in the wheat seed roots that can be studied using low-temperature NMR spectroscopy.

The ^{1}H NMR spectra of water in partially dried roots of germinating seed samples (for 3 or 7 days) (control and affected by A-300 and PSM) were recorded in air (Figure 8.8) or $CDCl_3$ (Figure 8.9). Water (about 7 wt%) bound in seed roots being in air is observed as two broad signals at 1 (WAW) and 5 (SAW) ppm. In weakly polar $CDCl_3$ medium, the spectral width decreases and SAW gives two signals. The SAW intensity decreases more strongly than that of WAW, i.e., a significant portion of SAW corresponds to WBW (frozen at 250–270 K), but a significant portion of WAW corresponds to SBW (unfrozen at $T<250$ K).

Weakly polar $CDCl_3$ medium reduces the freezing temperature of WAW (Figure 8.8c and d). The intensity ratio of WAW and SAW changes for the wheat seed roots with increasing germinating time.

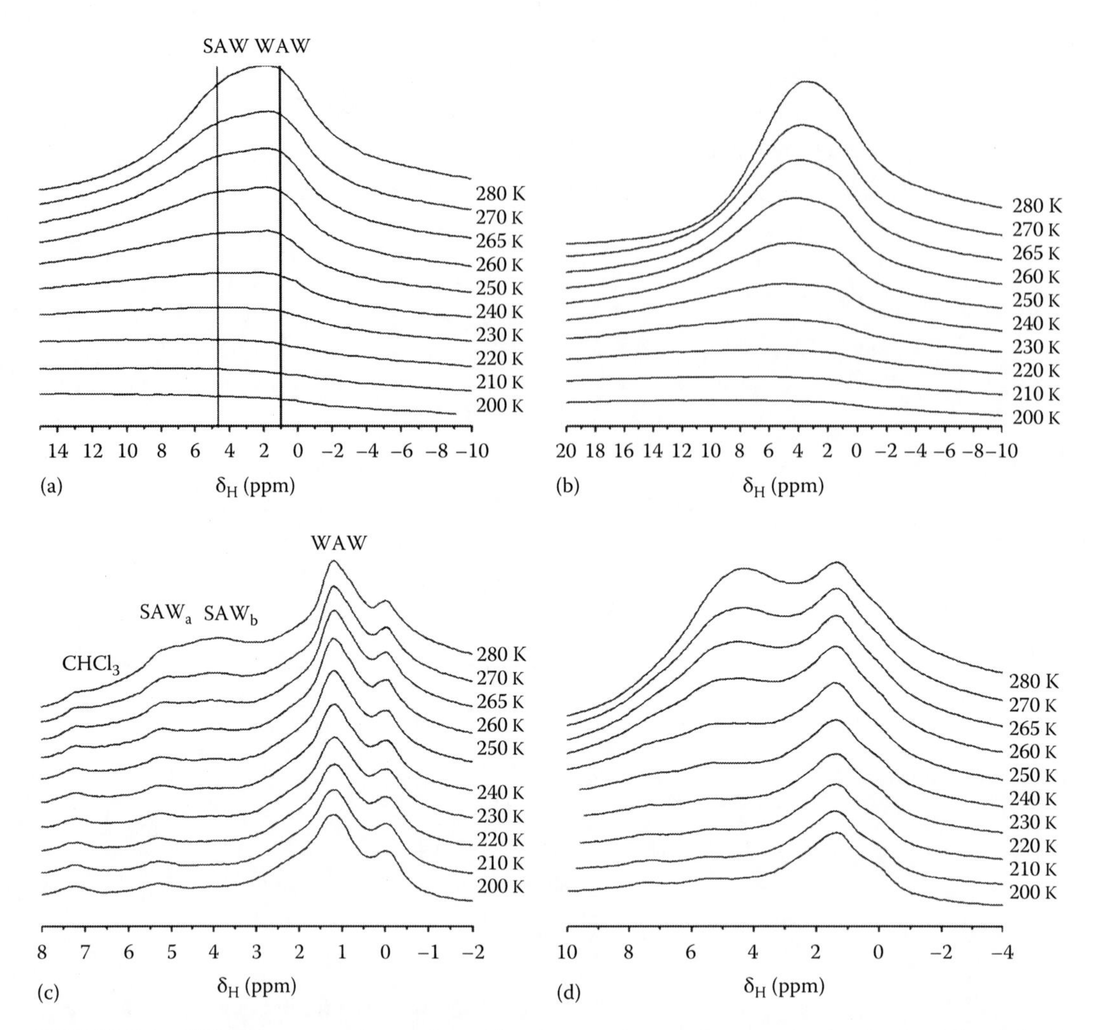

FIGURE 8.8 ^{1}H NMR spectra of water bound in wheat seed roots after germinating for (a, c) 3 and (b, d) 7 days in pure water recorded in air (a, b) or $CDCl_3$ (c, d) media.

Addition of A-300 (Figure 8.9) or PSM (Figure 8.10) to water during seed germinating changes the characteristics of bound water in the roots (Table 8.3 and Figures 8.11 through 8.13). These changes depend on the type of additive. The presence of A-300 nanoparticles leads to an increase in contribution of nanopores in the roots (mainly in cellulosic fibers as voids between fibrils) (Table 8.3, and Figure 8.13a and b). Addition of PSM affects rather mesopores than nanopores filled by water unfrozen at $T<273$ K. In the $CDCl_3$ medium, water organization in the roots undergoes significant changes because chloroform displaces a portion of bound water from the roots (especially from mesopores) (Figure 8.13c and d). The free surface energy (Table 8.3, γ_s) increases in the $CDCl_3$ medium for control sample and affected by PSM (germinating for 3 days) but it decreases for these samples after 7 days germinating. For both samples affected A-300, it decreases in the $CDCl_3$ medium. These effects are due to changes in the root structure during germinating differently affected by A-300 and PSM and the displacement of bound water by $CDCl_3$ dependent on both topology and chemistry of the surface of root supramolecular structures.

The NMR spectra (Figures 8.8 through 8.10) demonstrate that water bound in the wheat seed roots includes both SAW and WAW but contribution of SBW in WAW is much greater than in SAW because a significant portion of SAW is frozen at $250<T<273$ K. However, intensity of WAW

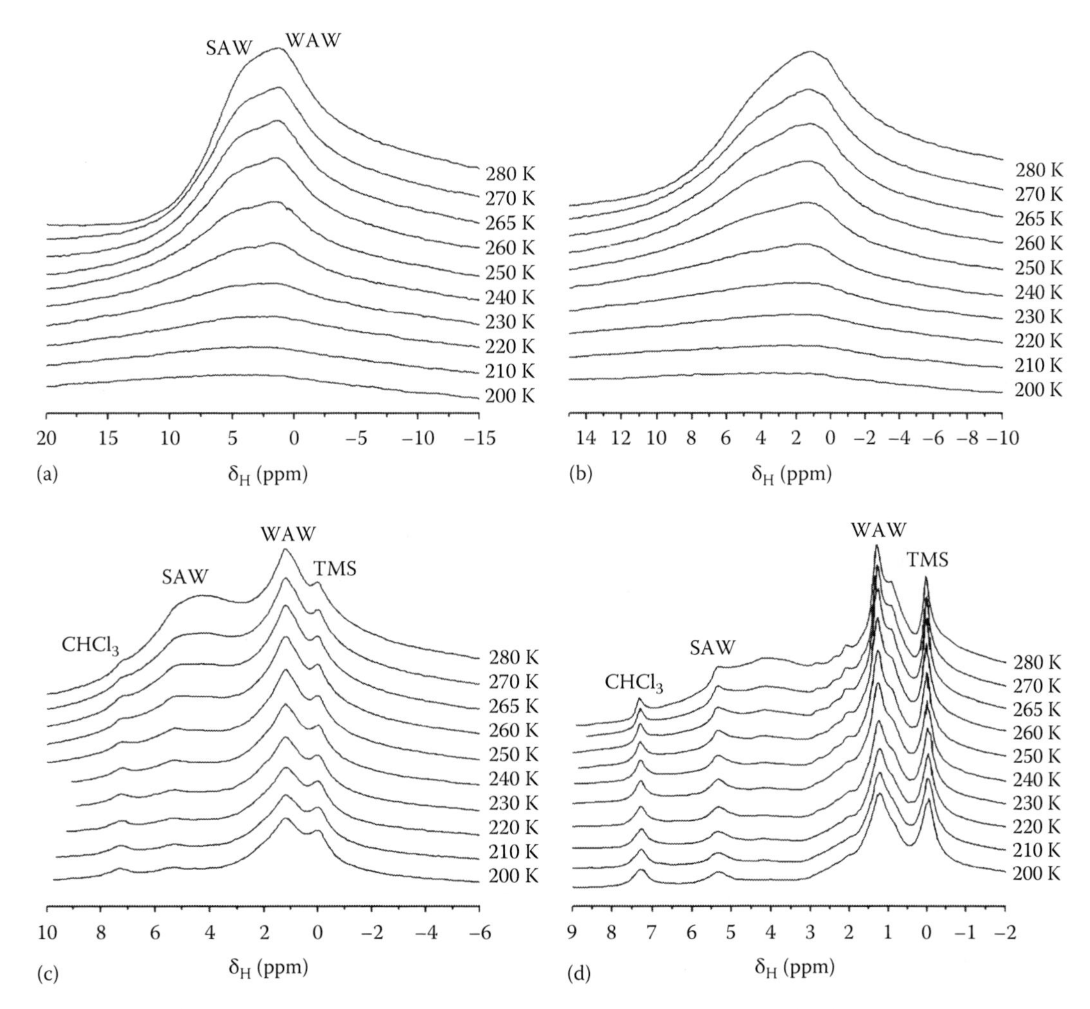

FIGURE 8.9 ^{1}H NMR spectra of water bound in wheat seed roots after germinating for (a, c) 3 and (b, d) 7 days in water with addition of A-300; spectra were recorded in air (a, b) or $CDCl_3$ (c, d) media.

weakly depends on temperature till 200 K. This can be explained by specific surroundings of this WAW confined in root cells and the influence of low-molecular organic and inorganic compounds present in the roots.

The nonuniformity of both SAW and WAW bound in the roots appears in the complex shape of the lines with the presence of several signals. However, the differentiation of these signals occurs only in the $CDCl_3$ medium since in the air dispersion medium only one broad signal is observed independently of the presence of additives with A-300 or PSM (Figures 8.8 through 8.10). This $CDCl_3$ effect is due to its penetration into the roots and blocking fast exchange processes between different water structures. This water clusterization reveals the nonuniformity of the surroundings for both SAW and WAW structures. These effects are also well seen in the temperature dependences of the total amounts of bound water (Figure 8.11a and b) depending on the germinating period of 3 or 7 days, as well as in the behavior of SAW (Figure 8.11c and d) and WAW (Figure 8.11e and f).

Figures 8.11 and 8.12 show that the amounts of differently structured water (i.e., SAW and WAW) bound in the roots (<10 wt%) depend on the presence of A-300 or PSM and these dependences change during germination of seeds. These changes are well observed in the changes in the PSD functions (Figure 8.13).

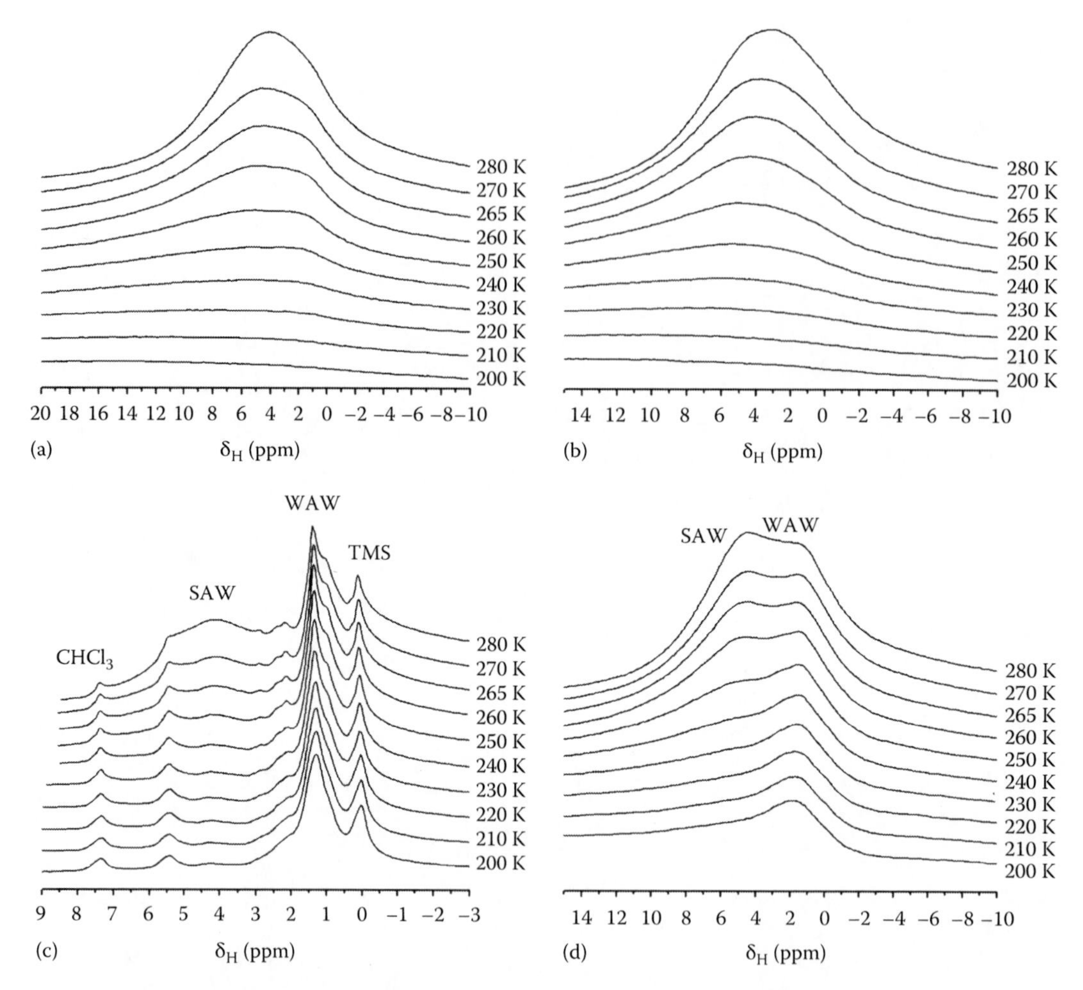

FIGURE 8.10 ^{1}H NMR spectra of water bound in wheat seed roots after germinating for (a, c) 3 and (b, d) 7 days in water with addition of PSM; spectra were recorded in air (a, b) or $CDCl_3$ (c, d) media.

These changes occur over all range of voids in the roots from nanopores at $R < 1$ nm up to broad mesopores of 10–25 nm in radius. Notice that the influence of nanosilica A-300 on the PSD shape is stronger than that of PSM in comparison with the control sample (Figure 8.13). The PSD function shapes demonstrate the influence of $CDCl_3$ on the organization of water in the roots. This effect is stronger for samples after 3 days of germination (Figure 8.13a and c) than after 7 days of germination (Figure 8.13b and d).

During germinating certain changes occur with seeds, and the first effect is their swelling with water excess. The TSDC thermograms for dry and wetted wheat seeds very differ with respect to the ratio of the intensities of the LT and HT bands (Figure 8.14) since the HT band intensity is larger by order in magnitude for wetted seeds (due to plasticizing effects of water on intracellular macromolecules and their polar groups plus the effects of water itself with increasing its amount). The ion percolation is absent for dry seeds (as well as in yeast cells described earlier) because water forms only local intracellular structures (mainly WAW). It appears after wetting of seeds because of the formation of continuous structures with water intracellular or located at the outer surface (seed shells or cell membranes) of seeds or cells. The local mobility of water molecules (observed by the NMR method) appears in wetted seeds (or cells) at slightly lower temperatures than the dc relaxation (observed in TSDC thermograms) begins. Consequently, continuous structures with mobile

TABLE 8.3
Characteristics of Water Bound in Roots of Wheat Seeds Germinating for 3 and 7 Days

Sample	S_{uw} (m^2/g)	S_{nano} (m^2/g)	S_{meso} (m^2/g)	V_{uw} (cm^3/g)	V_{nano} (cm^3/g)	V_{meso} (cm^3/g)	γ_S (J/g)	$-\Delta G_s$ (kJ/mol)
Control[a]	26	18	8	0.063	0.007	0.056	4.1	3.16
Control[b]	88	86	1	0.042	0.030	0.012	5.8	3.43
Control[c]	44	35	9	0.066	0.014	0.052	5.4	3.39
Control[d]	41	38	3	0.033	0.014	0.019	3.2	3.38
+A-300[a]	72	66	6	0.063	0.027	0.036	6.3	3.13
+A-300[b]	56	54	2	0.035	0.019	0.016	4.3	3.33
+A-300[c]	82	65	17	0.070	0.026	0.044	7.1	3.16
+A-300[d]	61	58	3	0.040	0.022	0.017	4.7	3.14
+PSM[a]	29	22	7	0.060	0.009	0.050	4.0	3.16
+PSM[b]	72	70	2	0.037	0.025	0.012	4.8	3.37
+PSM[c]	35	28	7	0.063	0.012	0.051	4.6	3.06
+PSM[d]	32	29	3	0.031	0.011	0.020	2.8	3.42

Note: Wheat seed sprouting for [a,b]3 or [c,d]7 days (NMR measurements of roots in [a,c]air and [b,d]$CDCl_3$).

molecules as carries of charges (e.g., H_3O^+) can appear in bio-objects at temperatures higher by 10–30 K than the local molecular mobility appears in water bound by polar intracellular functionalities. Notice that the local mobility of water molecules is characteristic for dry seeds (Figure 8.14).

The PSDs for wetted rape seeds (Figure 8.15b) demonstrate a significant contribution of narrow pores in swelling seeds (wetted for 3–4 days to appearance of small shoots). The specific surface area of intracellular structures being in contact with bound unfrozen water is 550 m^2/g (estimated from the ^1H NMR data). The PSDs based on the NMR and TSDC data are close. However, the PSD_{NMR} graphs demonstrate appearance of broad pores at $R>20$ nm because of longer wetting (4 days) than on the TSDC measurements (3 days). The dc relaxation appears at slightly higher temperatures and at a larger activation energy ($E_{dc}=53$ kJ/mol) than the local molecular mobility starts ($E_{LMM}=36$ kJ/mol) (estimated from data shown in Figure 8.15). Deviations from the Arrhenius-type temperature dependence (linear) are relatively small for both processes (i.e., cooperative effects do not practically contribute these processes) observed by the NMR and TSDC methods. There are similar results for wetted wheat seeds (Figure 8.14).

The mentioned effects were observed for such living bio-objects as germinating seeds. It is of importance to analyze similar effects for such related but nonliving systems as flour prepared from dry seeds. Properties of wheat flour as the most important ingredient in baked food products depend not only on flour quality (i.e., wheat seeds) but also on water amount and its interaction with flour. The behavior of water bound in dry or wetted flours (as well as seeds) can be studied using NMR, TSDC, DRS, and DSC techniques (Henry et al. 2003, Calucci et al. 2004, Gun'ko et al. 2007i, 2009d, Zarko et al. 2008).

Calucci et al. (2004) studied the effects of accelerated aging of wheat seeds on structural and dynamic properties of dry and hydrated (ca. 10 wt% H_2O) flour at a molecular level using high- and low-resolution solid-state NMR techniques (Figure 8.16).

Identification and characterization of domains with different mobilities were performed by ^{13}C direct excitation (DE) and CP MAS, as well as by ^1H static and MAS experiments. ^1H spin–lattice relaxation time (within laboratory, T_1 and rotating, $T_{1\rho}$ frames) measurements were carried out (Figure 8.16) to investigate molecular motions in different frequency ranges. Experimental data show that the main components of flour (starch and gluten proteins) are in a glassy phase, whereas

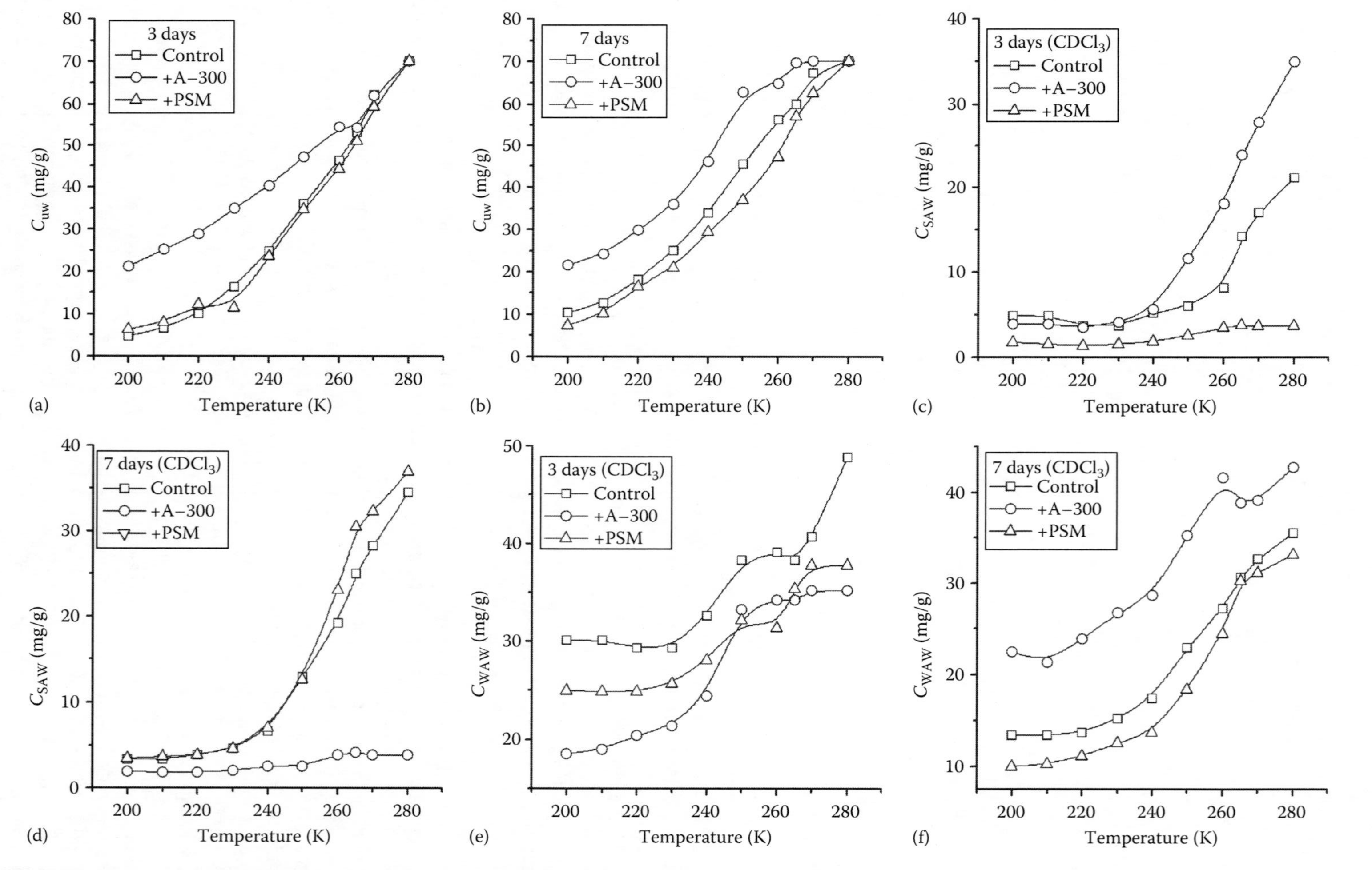

FIGURE 8.11 Temperature dependences of the amounts of unfrozen water (a, b), SAW (c, d) and WAW (e, f) contributions in wheat seed roots after germinating for 3 (a, c, e) or 7 (b, d, f) days in different media (pure water [control sample]) and with addition of nanosilica or PSM (spectra were recorded in air or $CDCl_3$ medium).

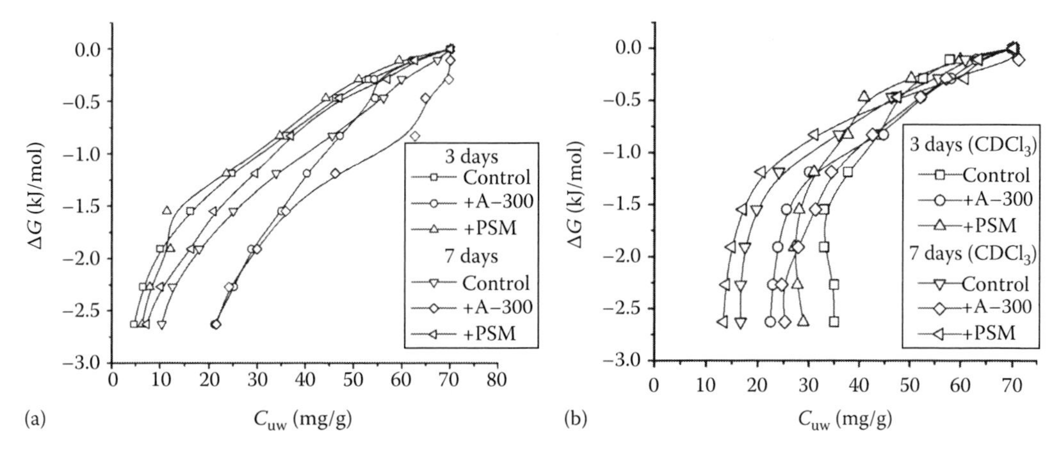

FIGURE 8.12 Relationships between the amounts of unfrozen water and changes in the Gibbs free energy calculated from the ^{1}H NMR spectra recorded in (a) air or (b) $CDCl_3$ for wheat seed roots after germinating for 3 or 7 days in different media.

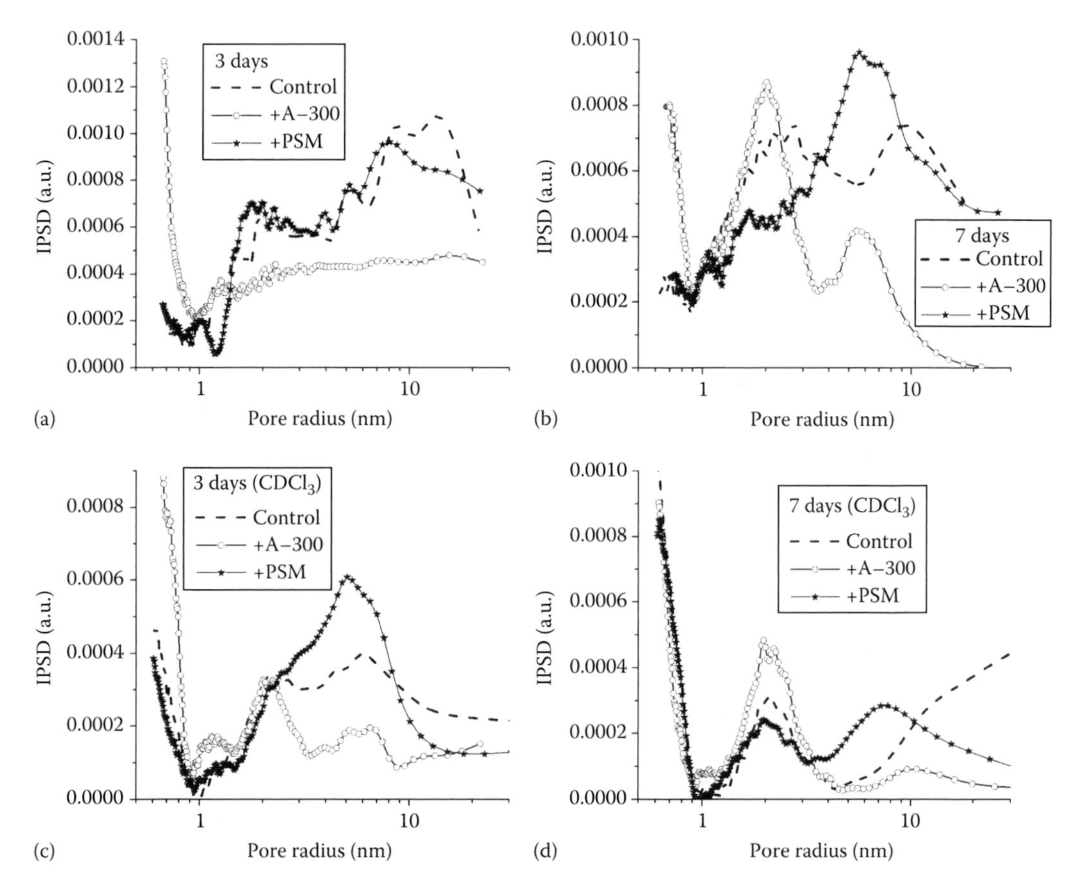

FIGURE 8.13 Incremental size distribution functions of unfrozen water clusters and domains bound in the wheat seed roots after germinating for 3 (a, c) or 7 (b, d) days in different media.

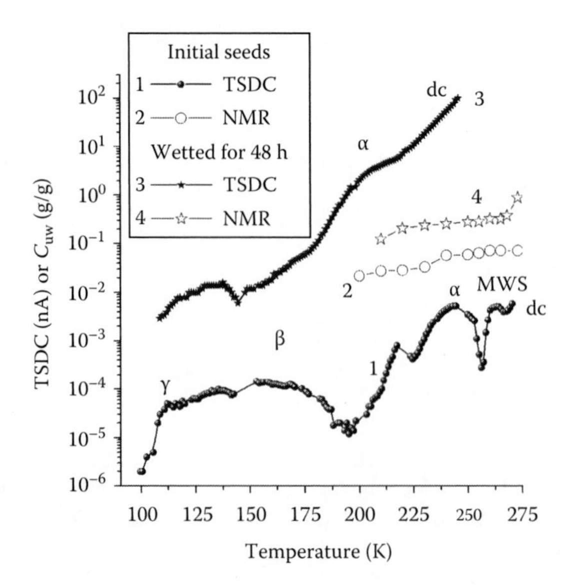

FIGURE 8.14 TSDC thermograms and amounts of unfrozen water (^{1}H NMR) for initial dry wheat seeds and after their wetting for 48 h. (Adapted from *Colloids Surf. A: Physicochem. Eng. Aspects*, 336, Gun'ko, V.M., Turov, V.V., Zarko, V.I., Goncharuk, E.V., and Turova, A.A., Regularities in the behaviour of water confined in adsorbents and bioobjects studied by ^{1}H NMR spectroscopy and TSDC methods at low temperatures, 147–158, 2009e, Copyright 2009, with permission from Elsevier.)

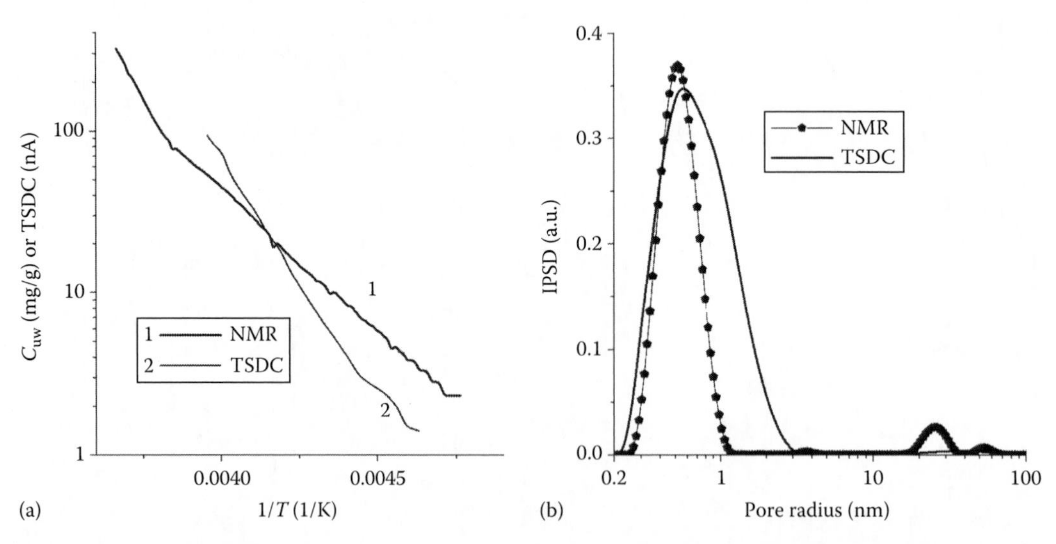

FIGURE 8.15 (a) Amounts of unfrozen water (NMR, curve 1) and dc (TSDC, curve 2) as functions of reciprocal temperature and (b) pore size distributions calculated using integral Gibbs–Thomson equation for $C_{uw}(T,R)$ and I_{TSDC} for rape seeds wetted for 3 (TSDC) and 4 (NMR) days. (Adapted from *Colloids Surf. A: Physicochem. Eng. Aspects*, 336, Gun'ko, V.M., Turov, V.V., Zarko, V.I., Goncharuk, E.V., and Turova, A.A., Regularities in the behaviour of water confined in adsorbents and bioobjects studied by ^{1}H NMR spectroscopy and TSDC methods at low temperatures, 147–158, 2009e, Copyright 2009, with permission from Elsevier.)

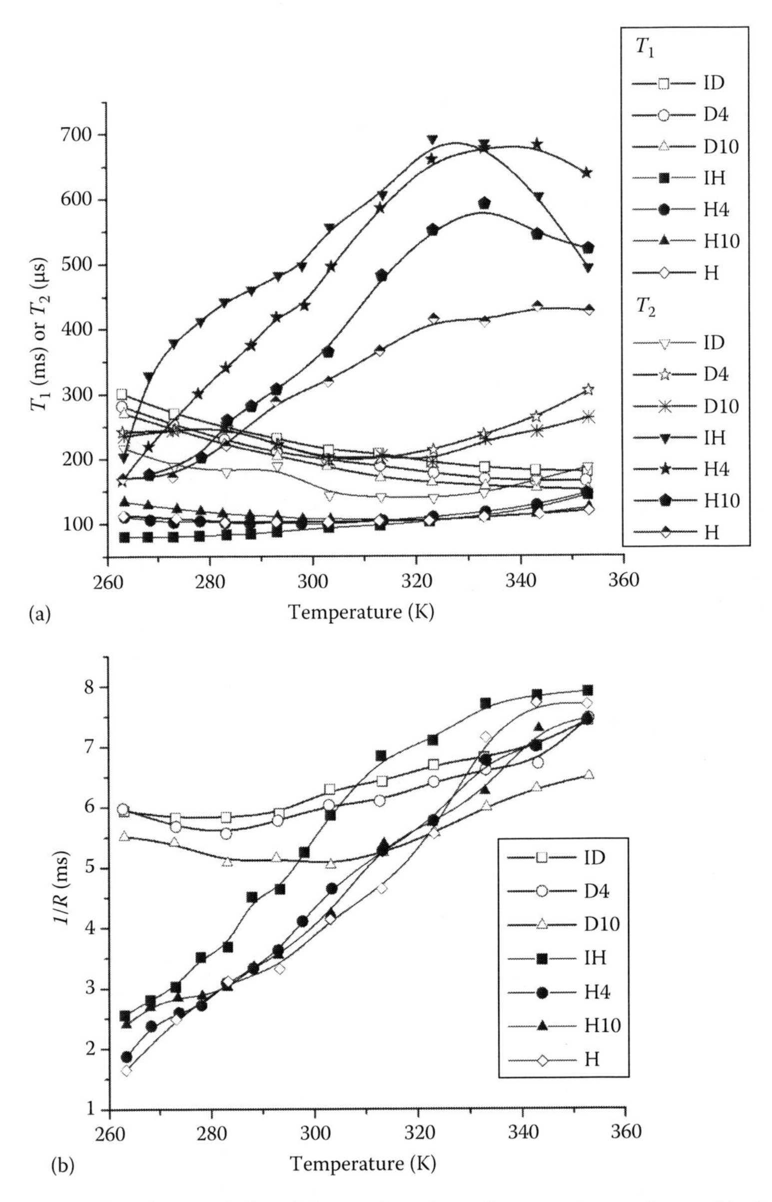

FIGURE 8.16 Relaxation characteristics of flour as functions of temperature, aging, and hydration: (a) ^{1}H T_1 and T_2 and (b) ^{1}H $T_{1\rho}$ $1/R$ ($R=(\Sigma w_i/T_{1\rho,i})/\Sigma w_i$ is the average relaxation rate); open and solid symbols represent dry and hydrated samples, respectively, to flour from seeds unaged (ID or IH) and aged for 4 (D4 or H4) and 10 (D10 or H10) days, respectively, and H refers to the artificially hydrated sample. (Adapted with permission from Calucci, L., Galleschi, L., Geppi, M., and Mollica, G., Structure and dynamics of flour by solid state NMR: effects of hydration and wheat aging, *Biomacromolecules* 5, 1536–1544, 2004. Copyright 2004 American Chemical Society.)

the mobile fraction is constituted by lipids and, in hydrated samples, absorbed water. A lower proportion of rigid domains, as well as an increased dynamics of all flour components is observed after both seeds aging and flour hydration. Linear average dimensions between 2 and 20 nm are estimated for water domains in hydrated samples. Wheat flour includes a greater amount of starch than oat flour (Table 8.4) that can affect the amounts and organization of bound water.

TABLE 8.4
Composition of Wheat and Oat Flours (wt%)

Flour	Water	Proteins	Carbohydrates	Fat	Ash
Wheat	6.0	15.7	71.8	2.5	4.0
Oat	6.0	14.8	65.0	10.7	3.5

Treatment temperatures (40°C and 85°C) of aqueous dispersions with flours are 20°C above and below of the gelatinization temperature of starches (T_{gel}=60°C–66°C) as the main components of flour (Thomas and Atwell 1999).

Oat flour dispersion includes larger particles than wheat flour studied by QELS method (Figure 8.17a). However, the swelling degree of wheat flour is larger (Figure 8.17b) due to a larger content of carbohydrates (Table 8.4). Content of bound water increases after gelatinization (Table 8.5 and Figure 8.17b). Viscosity of water flour increases with increasing flour content (Figure 8.18) and it is higher for oat flour because of the gum presence. The temperature dependence of the viscosity at $T < T_{gel}$ is much weaker because the starch molecules are not completely unfolded.

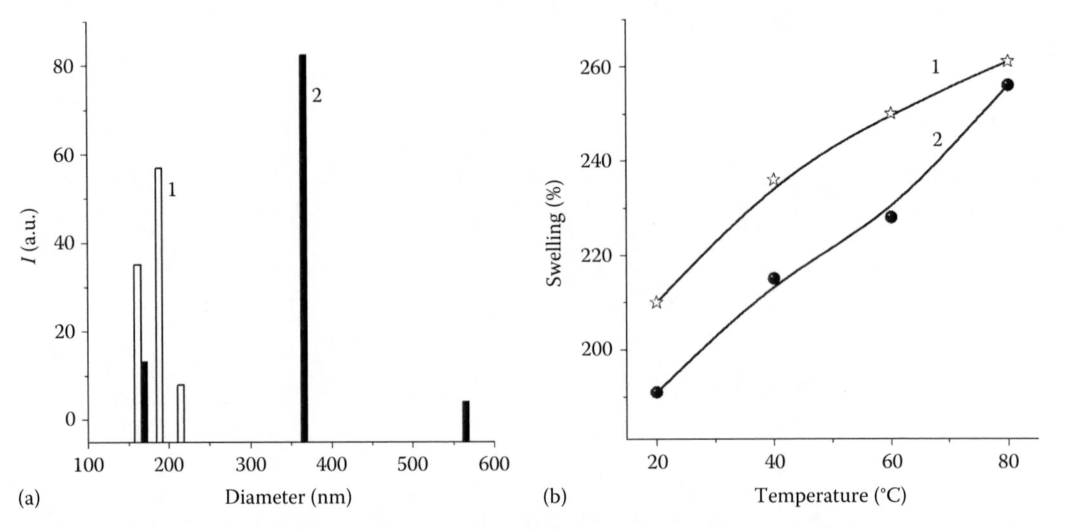

FIGURE 8.17 (a) Particle size distribution for aqueous dispersion of wheat (1) and oat (2) flours at 20°C and (b) swelling degree of wheat (curve 1) and oat (curve 2) flour as a function of temperature.

TABLE 8.5
Content of Free and Bound Water before (40°C) and after (85°C) Flour Gelatinization

	Wheat Flour		Oat Flour	
Water Type	**40°C**	**85°C**	**40°C**	**85°C**
Hydration (g/g)	4.8	7.0	6.2	6.8
Content of free water (wt%)	74.3	66.0	70.1	62.4
Content of bound water (wt%)	25.7	34.0	29.9	37.6
Relative content of bound water (g_{water}/g_{flour})	1.49	2.72	2.15	2.94
Relative content of bound water ($g_{water}/g_{carbohydrate}$)	2.07	3.79	3.31	4.52

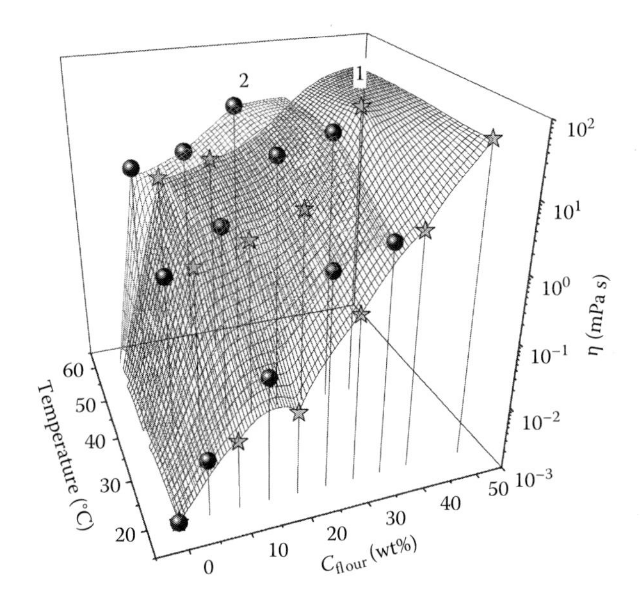

FIGURE 8.18 Viscosity as a function of temperature and content of flour of wheat (1) and oat (2).

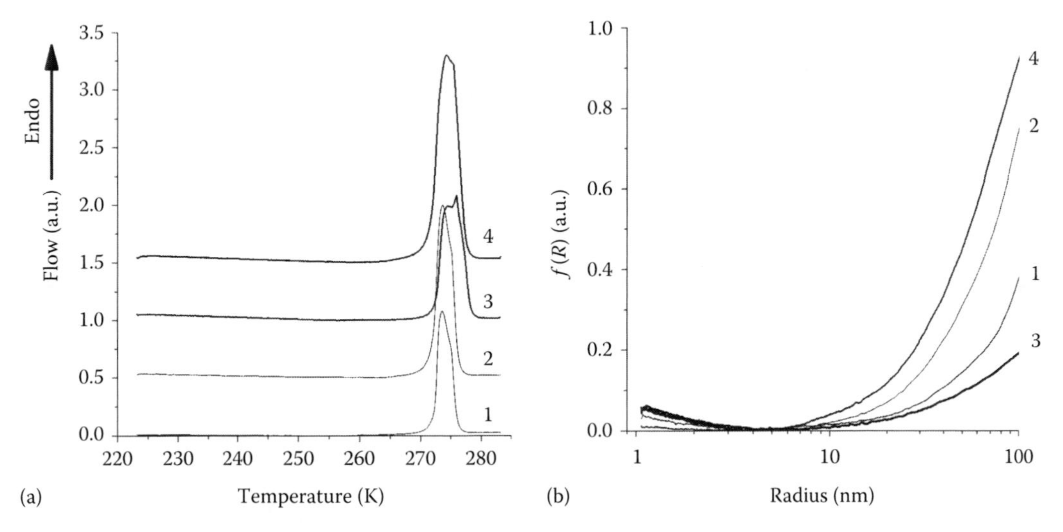

FIGURE 8.19 (a) DSC curves for hydrated oat (curves 1 and 2) and wheat (curves 3 and 4) flours prepared at 40 (curves 1 and 3) and 85°C (curves 2 and 4) and (b) size distributions of clusters and domains of bound water.

Calculations of the size distributions of clusters and domains of bound water (Figure 8.19b) based on the DSC data (Figure 8.19a) show increasing content of bound water after gelatinization with increasing contribution of broad mesopores ($10 < R < 25$ nm) in supramolecular structures that correspond to swelling and unfolding of macromolecules.

Detailed information on these processes can be obtained using TSDC, DSC, and DRS methods (Ratcliffe et al. 2001, Tananuwong and Reid 2004, Gun'ko et al. 2007i, 2008c, 2009d, Zarko et al. 2008).

The displacement of the TSDC peaks in the LT range ($100 < T < 170$ K) toward higher temperatures is observed for hydrated wheat flour at $T_{pr} > T_{gel}$ and the opposite displacement is for the peaks in the HT ($170 < T < 260$ K) (Figure 8.20a). The effects in the HT range are caused by diminution of the activation energy of dipolar relaxation of macromolecular fragments with bound water due to

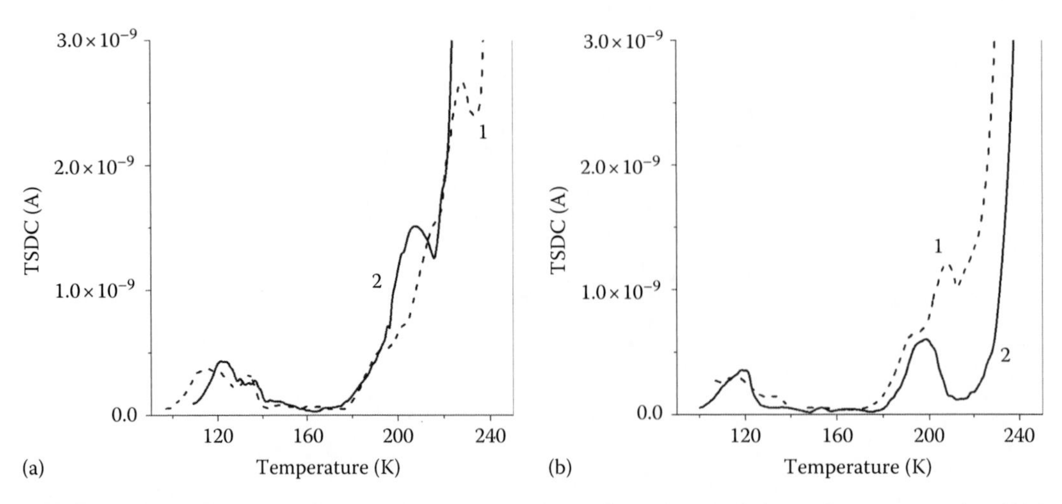

FIGURE 8.20 TSDC thermograms of frozen suspensions of (a) wheat and (b) oat flours prepared at 40°C (curve 1) and 85°C (curve 2). (According to Zarko, V.I. et al., *Chem. Phys. Technol. Surf.*, 14, 467, 2008 [in Russian].)

plasticizing effects of water and additional water clustering during unfolding of macromolecules. The effects in the LT range are caused by an increase in the number of strong hydrogen bonds per a water molecule that cause an increase in the energetic barriers for rotational motion of the water molecules and polar groups of macromolecules. The displacement of the beginning of dc relaxation for wheat and oat flours is opposite for samples prepared at 85°C in comparison with that at 40°C (Figure 8.20).

Carbohydrates play the main role in structural changes during flour gelatinization. Therefore, pure potato starch was studied after gelatinization at $T_{pr} > T_{gel}$ (Figures 8.21 and 8.22). Obtained results (Figures 8.21 and 8.22 and Tables 8.6 and 8.7) show that the TSDC peak in LT at 104 K is observed for pure starch and wheat flour at $T_{pr} < T_{gel}$, but third, fourth, and fifth peaks are observed for both preparation temperatures for all samples.

Integral peak intensity at 111–114 K for dispersion of oat flour remains practically the same, and for wheat flour it is not observed at $T_{pr} > T_{gel}$. Integral peak intensity at $T_{max} = 120–122$ K for dispersion of wheat flour increases at $T_{pr} > T_{gel}$, but for oat it decreases. Integral peak intensity at $T_{max} = 133–136$ K for wheat flour remains the same, but for oat it significantly decreases at $T_{pr} = 85$°C. Integral peak intensity at $T_{max} = 147–153$ K for both wheat and oat flour does not change, as well as for a peak at $T_{max} = 157–165$ K, but the latter disappears for wheat flour at $T_{pr} = 85$°C.

In HT range, gelatinization weakly affects a peak at $T_{max} = 191–195$ K. The presence and location of subsequent peaks depend on treatment temperature and a flour type because they depend more strongly on the relaxation of dipolar groups and fragments of macromolecules than the peaks in the LT range. The TSDC peak temperatures of gelatinized starch are in agreement with those of oat and wheat flours (Figure 8.22). This confirms that changes in the behavior of the system studied during phase transition caused by flour gelatinization are mainly due to rearrangement of supramolecular structures of starch component.

The size distribution functions ($f(R)$) of clusters and domains of bound water (Figure 8.23) calculated using TSDC cryoporometry (IGT equation) give more detailed information than PSD based on the DSC data (Figure 8.19b).

Integration of the $f(R)$ functions shows that contribution of small clusters of bound water located in nanopores (voids in macromolecules) at $R < 1$ nm decreases after gelatinization more than by order of magnitude for oat flour and approximately six times for wheat flour. Contribution of narrow mesopores at $R < 3$ decreases, and contributions of broad mesopores and macropores are predominant

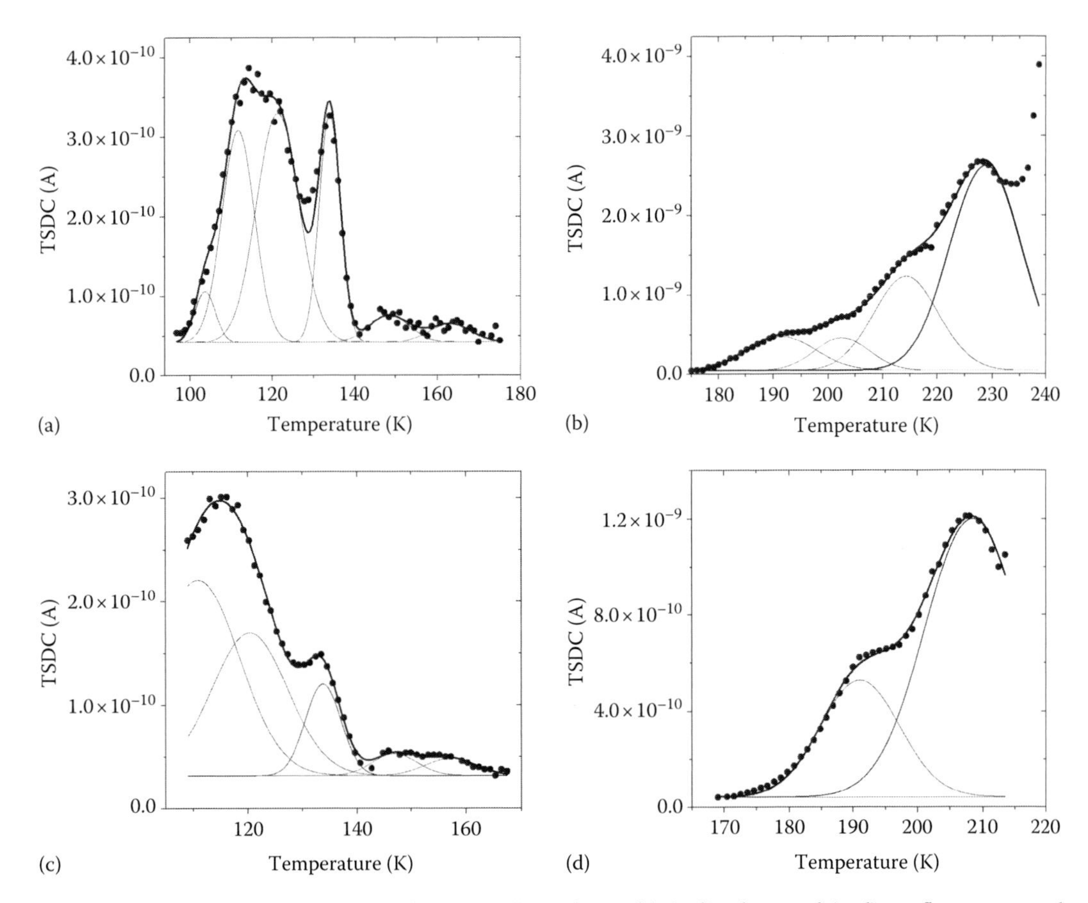

FIGURE 8.21 TSDC thermograms of aqueous dispersions with (a, b) wheat and (c, d) oat flours prepared at 40°C in the (a, c) LT and (b, d) HT ranges. (According to Zarko, V.I. et al., *Chem. Phys. Technol. Surf.*, 14, 467, 2008 (in Russian).)

(Figure 8.23). This is in agreement with the data about swelling of microparticles and unfolding of starch macromolecules during gelatinization.

Distribution functions of activation energy of dipolar relaxation of frozen aqueous suspensions of oat and wheat flours (Figure 8.24) show that gelatinization ($T_{pr}=85$°C) leads to a significant diminution of the E_a values in comparison with the systems prepared at 40°C. This result is due to the plasticizing effects caused by water after swelling of flour microparticles and unfolding of the macromolecules during gelatinization, since in compacted folding structures intermolecular interactions of neighboring glycoside rings of neighboring macromolecules cause higher energetic barriers of the dipolar relaxation (both α and β relaxations) of these fragments and whole molecules. The barriers decrease by 20–100 kJ/mol after breaking the hydrogen bonds between neighboring glycoside structures and an increasing water layer thickness between them because water molecules much weakly hinder the dipolar relaxation of the polymer fragments. Additionally, a greater amount of water in swollen samples takes part in this relaxation. Notice that the activation energy of the β relaxation at $T<220$ K in the system with gelatinized starch/water/nanosilica is in the range 10–100 kJ/mol (on the basis of the DRS data) (Gun'ko et al. 2009d) but for pure water $E_a=10$–45 kJ/mol. After flour gelatinization, the dipolar relaxation is in the same E_a range (Figure 8.24) as pure water.

The activation energy of dc relaxation (due to through current appearing when the percolation channels appear) E_{dc} (caused by mobile protons and other small ions) corresponds to a linear

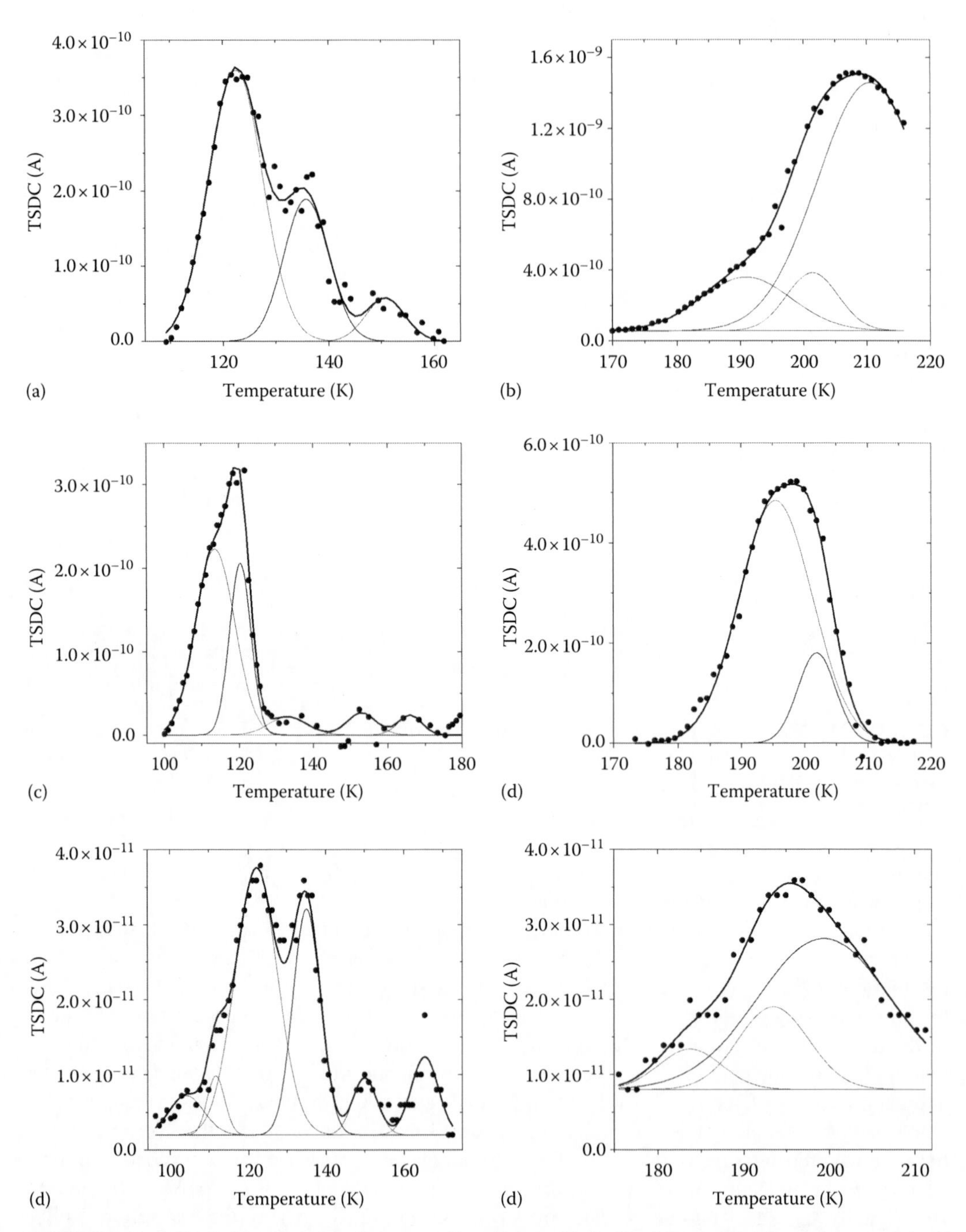

FIGURE 8.22 TSDC thermograms of aqueous dispersions with (a, b) wheat and (c, d) oat flours and (e, f) starch prepared at 85°C in the (a, c, e) LT and (b, d, f) HT ranges. (According to Zarko, V.I. et al., *Chem. Phys. Technol. Surf.*, 14, 467, 2008 (in Russian).)

TABLE 8.6
Peak Temperatures T_{max} (K) and Integral Intensities I ($nA \times K$) of TSDC Maxima for Frozen Aqueous Dispersion (LT Range)

	TSDC Maximum											
	1		2		3		4		5		6	
Sample	T_{max}	I	T_{max}	I	T_{max}	I	T_{max}	I	T_{max}	I	T_{max}	I
Wheat[a]	104	0.39	112	2.7	121	3.8	134	1.8	148	0.42	163	2.9
Wheat[b]	—	—	—	—	122	4.6	136	2.0	151	0.53	—	—
Oat[a]	—	—	111	2.1	120	2.3	134	0.72	147	0.22	157	0.19
Oat[b]	—	—	114	2.9	120	1.5	133	0.28	153	0.27	166	0.19
Starch[b]	104	0.057	112	0.044	122	0.49	135	0.27	150	0.06	165	0.089

[a] $T_{pr}=40°C$.
[b] $T_{pr}=85°C$.

TABLE 8.7
Peak Temperatures T_{max} (K) and Integral Intensities I ($nA \times K$) of TSDC Maxima for Frozen Aqueous Dispersion (HT Range)

	TSDC Maxima									
	7		8		9		10		11	
Sample	T_{max}	I	T_{max}	I	T_{max}	I	T_{max}	I	T_{max}	I
Wheat flour, 40°C	—	—	191	6.8	203	5.2	214	18	229	39
Wheat flour, 85°C	—	—	191	5.5	201	3.4	211	22	—	—
Oat flour, 40°C	—	—	191	7.5	—	—	208	16	—	—
Oat flour, 85°C	—	—	195	7.7	202	1.4	—	—	—	—
Starch, 85°C	184	0.049	193	0.11	199	0.37	—	—	—	—

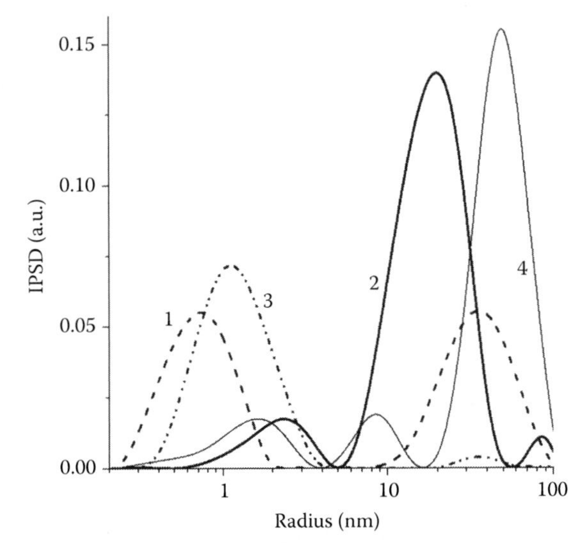

FIGURE 8.23 Size distributions of clusters and domains of bound water in the frozen dispersions of oat (curves 1 and 2) and wheat (curve 3 and 4) flours prepared at 40°C (curve 1 and 3) and 85°C (curves 2 and 4).

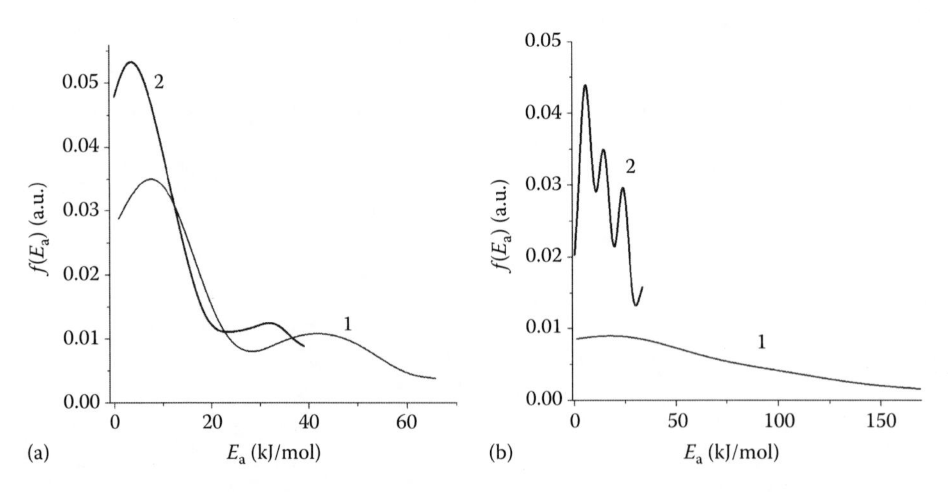

FIGURE 8.24 Distribution functions of activation energy of dipolar relaxation of frozen aqueous suspensions of oat (a) and wheat (b) flours at preparation temperatures of (curve 1) 40°C and (curve 2) 85°C.

increase in the TSDC logarithm as a function of temperature is equal to from 58 kJ/mol (oat flour at $T_{pr}=40°C$) to 79 kJ/mol (wheat flour at $T_{pr}=40°C$) and 65–67 kJ/mol for the systems gelatinized at 85°C. These values are lower than that for pure water (91–125 kJ/mol depending on residual content of salts and other experimental conditions) because of the presence of salts in flour and other sources of mobile charges. For pure gelatinized starch, E_{dc} is low (≈50 kJ/mol) that suggests abound good conditions for percolation effects for mobile ions despite gelatinization of starch and strong water structuring. Percolation effect in the TSDC spectra (dc relaxation) appears at higher temperatures (by 10–30 K) than local molecular mobility of water observed in the low-temperature NMR spectra. Both effects are observed at $T<273$ K. Therefore, they occur in water bound at the interfaces of solids or macromolecules. During flour gelatinization, the surface area of the interfaces increases that result in an increase in the number of additional percolation channels for the dc relaxation. This effect is well observed for the dispersion of wheat flour. In the case of oat flour, the presence of gum component causes the opposite effect and the dc relaxation appears at higher temperatures after gelatinization at 85°C.

Such another type of botanics as natural plant fibers represents an appropriate object for the analysis of the interfacial phenomena with the participation of bound water using NMR and DSC techniques. Cotton and flax as natural plant fibers are typical soft materials widely used in textile industry, medical, and technical applications (Satyanarayana et al. 2009). Of particular importance is understanding of their interactions with water, as water strongly influences the properties and performance of fibers and fiber-based materials, which can undergo substantial swelling. Cellulose fibrils aligned along the fiber length provide tensile and flexural strength and rigidity of fibers. The textural and structural organization of natural fibers creates certain inner porosity as voids between adjacent cellulosic fibrils or in amorphous fragments (Rowell et al. 1997, Zugenmaier 2008). This porosity can be changed due to interaction with water, especially during fiber swelling. As a result, the adsorption capacity of fibers can be very different in dry and hydrated states of the materials (Mikhalovska et al. 2012, Mikhalovsky et al. 2012).

Cotton and flax fibers can absorb relatively large amount of moisture that can lead to microbial attack under certain conditions of humidity and temperature. Cotton and flax may act as a nutrient, becoming suitable medium for bacterial and fungal growth. Therefore, cotton and flax fibers are treated with numerous chemicals to get better antimicrobial textiles. Additionally, many of physicochemical properties of natural fibers and composites based on fibers depend on the amounts of bound water, adsorption capacity in respect to water, and possible swelling degree. Thus, from practical and theoretical points of view, it is important to study the behavior of water in air-dry and swollen flax and cotton fibers.

Textural, adsorption, and other characteristics of crude and bleached cottonized flax fibers prepared from the side product of flax scutching (Makarov Lenzavod, Kiev region, Ukraine), i.e., short fibers were carded in order to divide the technical fibers still linked by their tricellular junctions into elementary fibers, were compared with those of commercial (Fisher) carded cotton fibers (Mikhalovska et al. 2012, Mikhalovsky et al. 2012). Here crude and bleached flax fibers are designated as flax and bleached flax, respectively.

SEM images of flax and cotton fibers (Figure 8.25a and b) demonstrate that both flax and cotton wool have fibers of comparable diameter values (10–30 μm). However, their surface roughness is different reflecting different origin of fibers (Figure 8.25c and d) and preparation techniques. TEM images reveal the presence of some inner slit-shaped macropores (Figure 8.25e and f); however, the fiber cell body is only weakly porous.

Information on interaction of fibers with water absorbed from air has been obtained from TG measurements. Weight loss versus temperature is shown in Figure 8.26. Cotton fibers absorbed 4.2 wt% water, whereas both types of flax fibers absorbed 5.7 wt% of water from air at 40% RH. The absorbed water is mainly lost upon heating below 373 K because the fibers are not nanoporous materials. The TG data show no considerable difference in the thermal behavior of all three materials. Thermal destruction of the fibers started in the same temperature range of 500–510 K, with the main weight loss observed at 570–600 K (Figure 8.26) due to decomposition/oxidation of cellulose and other organic components. More detailed information about the state of water retained by fibers has been obtained with DSC. This analysis is based on measuring temperature, magnitude, and shape of the melting endotherms (unfreezing of mobility) or the exotherms of ordering (crystallization or quasi-crystallization) of adsorbed water. The endotherms of polymer dehydration have also been measured. The phase transitions in adsorbed water have been observed in hydrophobic polymers, where the water formed a separate phase as droplets or clusters of micro- or submicrovolumes in cavities.

The state of adsorbed water in hydrophilic polymers is more complex. It has been found to exist in three states. Water I is the same as "normal" bulk water, and its phase transitions are characterized by relatively narrow endothermic (melting) and exothermic (crystallization) peaks on DSC curves (Figure 8.27); the temperature of these phase transitions is typically close to 273 K for ice–bulk water. It can be shifted at a high rate of cooling or heating. Notice that the ^{1}H NMR signals of this water are not observed at $T<273$ K since only mobile water is registered. The NMR cryoporometry method based on measuring the proton magnetic resonance signal of unfrozen water inside pores cannot detect water I at $T<273$ K. Water II is weakly bound but still capable of freezing by undergoing some ordering, structural reorganization, and quasi-crystallization upon cooling, and disordering, or "melting" upon thawing at temperatures below 273 K.

Water III is strongly bound to polar groups of polymers. It does not show crystallization or melting peaks on DSC curves. Water III forms a monomolecular water layer or a few monolayers (with a clustered structure) attached to macromolecules. The relative contributions of the three types of adsorbed water depend on a number of factors such as the size of submicrocavities in the studied materials (inner pores in fibers) and molecular packing (free volume as voids between fibers). This is of importance in particular for polymer membranes, permeability of which is probably controlled mainly by water I. The thermal behavior of waters II and III is the base of both DSC and NMR cryoporometry. However, the quantitative assignment of bound water to the mentioned three types on the basis of DSC and NMR methods can give slightly different contributions because of different sensitivities and different physical phenomena used in these methods. Water III content may characterize to a certain extent the value of water/macromolecule contact surface. All three types of water are removed during dehydration of heated samples. Combined DSC, TG, and NMR cryoporometry approach allowed us to reveal the differences in the state of water adsorbed by

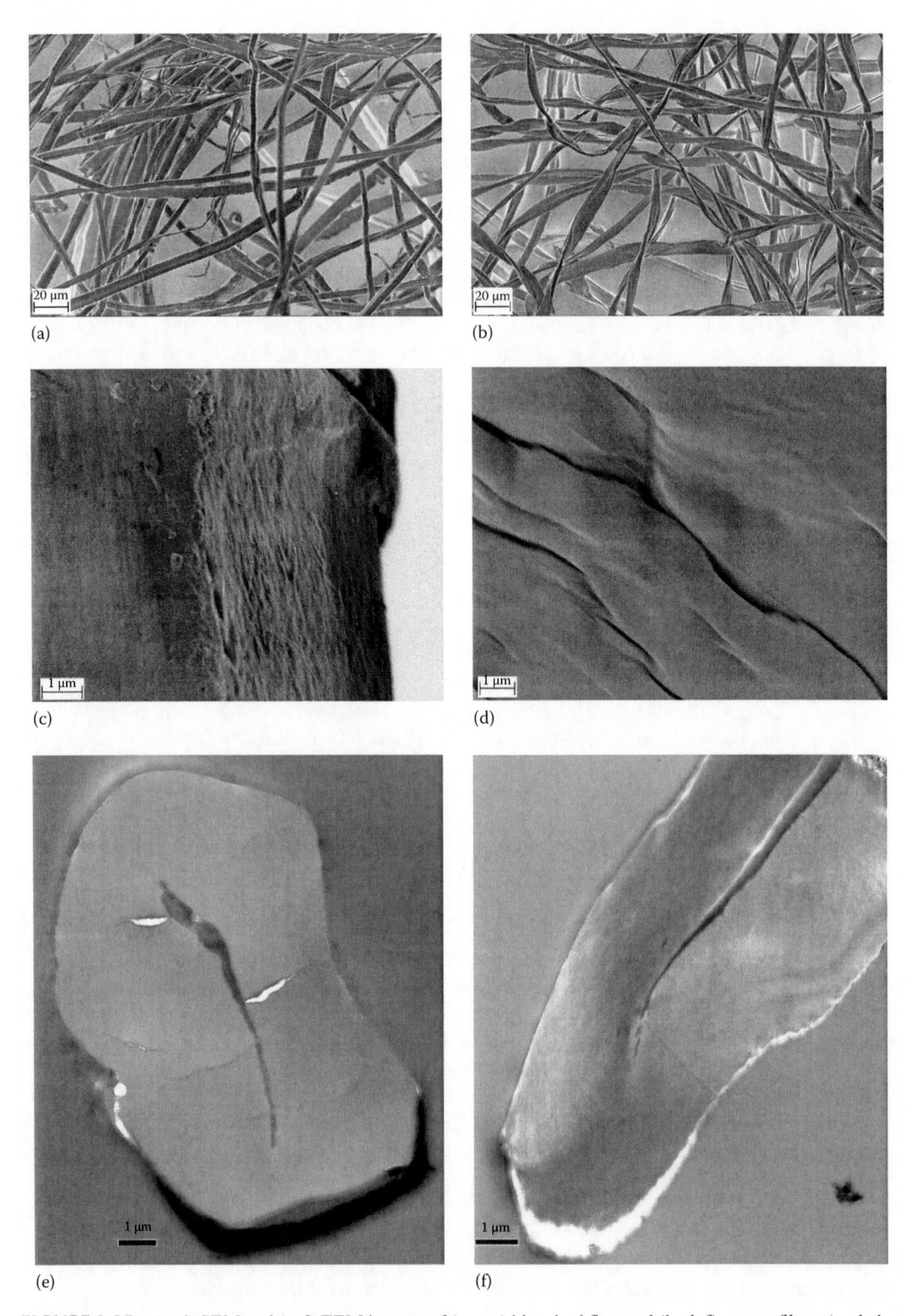

FIGURE 8.25 (a–d) SEM and (e, f) TEM images of (a, c, e) bleached flax and (b, d, f) cotton fibers (scale bar (a, b) 20 μm and (c–f) 1 μm). (Mikhalovsky, S.V., Gun'ko, V.M., Bershtein, V.A. et al., A comparative study of air-dry and water swollen flax and cotton fibres, *RSC Adv.*, 2, 2868–2874, 2012. Adapted with permission from The Royal Society of Chemistry.)

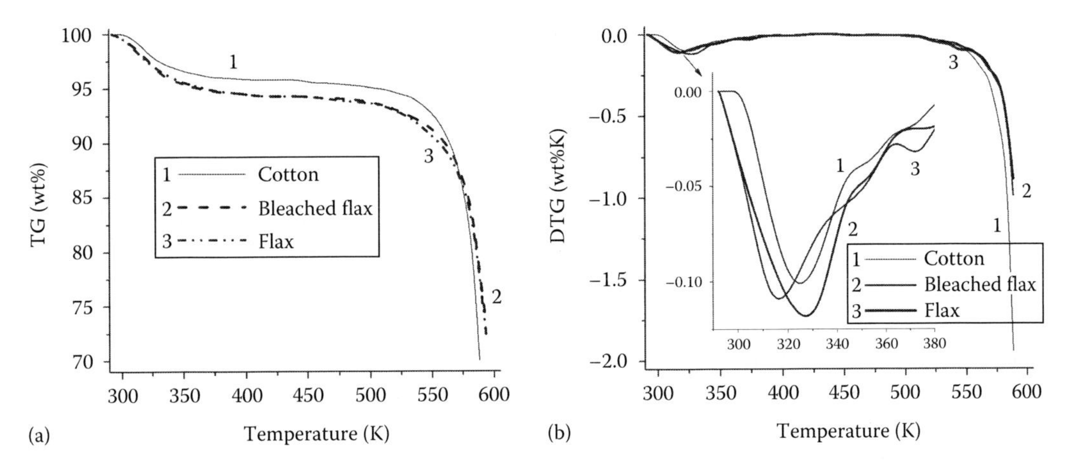

FIGURE 8.26 Thermogravimetric analysis of fiber samples (storage in air at RH = 40%): (a) TG, (b) DTG, and inset in (b) enlarged fragment of DTG curves in the range of 300–380 K. (Mikhalovsky, S.V., Gun'ko, V.M., Bershtein, V.A. et al., A comparative study of air-dry and water swollen flax and cotton fibres, *RSC Adv.*, 2, 2868–2874, 2012. Adapted with permission from The Royal Society of Chemistry.)

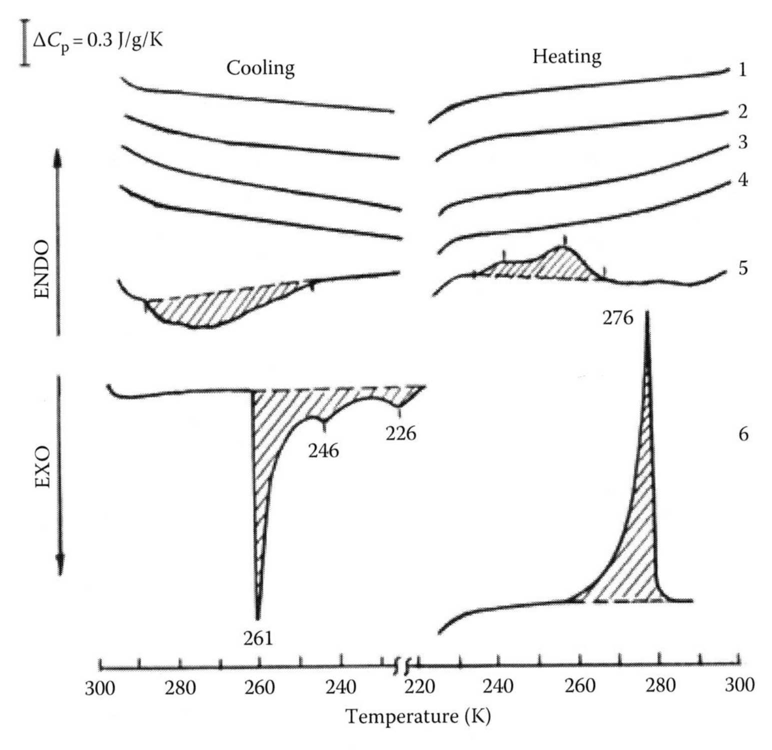

FIGURE 8.27 DSC curves of fiber samples obtained below room temperature at cooling (on the left) or heating (on the right) with the rate of 5 K/min. Curves 1–3 are for air-dry samples stored in air at 40% RH and curves 4–6 are for swollen samples after incubation for 24 h in distilled water. Curves 1 and 4—Bleached flax; curves 2 and 5—flax; curves 3 and 6—cotton. (Mikhalovsky, S.V., Gun'ko, V.M., Bershtein, V.A. et al., A comparative study of air-dry and water swollen flax and cotton fibres, *RSC Adv.*, 2, 2868–2874, 2012. Adapted with permission from The Royal Society of Chemistry.)

TABLE 8.8
Water Content in Fibers and Enthalpic Changes Upon Heating of Water-Containing Fibers from 223 to 423 K[a]

Sample	State	C_w (wt%)	ΔH_d (kJ/g Water)	$\Delta H_d/\Delta H_{vap}$	ΔH_m (J/g Ice)	$\Delta H_m/\Delta H_m^0$(%)
Cotton	Air-dry, 40% RH	4.2	11.74	5.20	n.d.[b]	0
	Swollen for 24 h	26	6.15	2.72	51.1	15.3
Bleached flax	Air-dry, 40% RH	5.7	6.30	2.78	n.d.	0
	Swollen for 24 h	10	5.54	2.45	n.d.	0
Flax	Air-dry, 40% RH	5.7	3.54	1.57	n.d.	0
	Swollen for 24 h	17	5.32	2.35	22.9	6.8

[a] ΔH^0_{vap} of water = 2.257 kJ/g; ΔH^0_{mice} = 334 J/g.
[b] n.d., Not detected.

fibers studied and to estimate the strength of water–fiber interactions and structural features of bound water and fibers. The dehydration enthalpy ΔH_d values (Table 8.8) were determined from the area of fiber dehydration endotherms in the DSC curves. These values were compared with the standard value of specific enthalpy of "free" water evaporation, ΔH^0_{vap}. The ΔH_d value is a complex parameter, which includes enthalpy required for breaking down water–fiber hydrogen bond interactions, enthalpy of water diffusion within a fiber, and ΔH_{vap} contribution. The $\Delta H_d/\Delta H_{vap}$ ratio characterizes to a certain extent the relative strength of water–fiber interactions. The exothermic peaks of crystallization and endothermic melting peaks of adsorbed water were identified in two cases (Figure 8.27, hatched areas, swollen flax, and swollen cotton). The melting enthalpy ΔH_m values were determined and compared with ice melting enthalpy ΔH^0_m. The $\Delta H_m/\Delta H^0_m$ ratio characterizes the fraction of water which converted into ice crystals. In swollen cotton and swollen flax, only a small part of adsorbed water (15.3% and 6.8%) is ice crystals at low temperatures (Table 8.8, $\Delta H_m/\Delta H^0_m$). The other samples do not contain ice crystals at all. The $\Delta H_m/\Delta H^0_m$ ratio characterizes the fraction of adsorbed water I and partly water II because water III does not have a crystallization peak in the thermograms (Figure 8.27).

DSC measurements in a high-sensitivity mode recorded a big difference in heat capacities ΔC_p between swollen samples (Figure 8.27, curves 4–6) in the low-temperature region. The DSC curve showed no peaks for swollen bleached flax at a relatively low content of water (10 wt%). For two other swollen samples, the DSC curves revealed distinctive endo- and exothermic effects (Figure 8.27, curves 5 and 6). For swollen flax (C_w = 17 wt%), a small exothermic effect at 250–290 K upon cooling and the same endothermic effect at 235–268 K upon heating, with maxima at ~240 and 260 K, were observed. It means that not only water III but also weakly bound water II is present in this sample accounting for about 6.8% of the total quantity of adsorbed water. In the swollen cotton (C_w = 26 wt%), all three types of adsorbed water were present: sharp peaks of water crystallization (T_{max} = 261 K) and melting (T_{max} = 276 K) indicate the presence of clusters of mobile water I; small exothermic peaks at 226 and 246 K are related to water II (Figure 8.27). Waters I and II account for 15.3% of their total amount in the swollen cotton (Table 8.8). The rest of water in this material corresponds to unfrozen water III, i.e., its structure is strongly distorted due to interactions with fibers. No such effects were found in the low-temperature DSC curves for air-dry samples (Figure 8.27, curves 1–3). It means that practically all water (~4–6 wt%) absorbed by air-dry fibers was strongly bound water of type III.

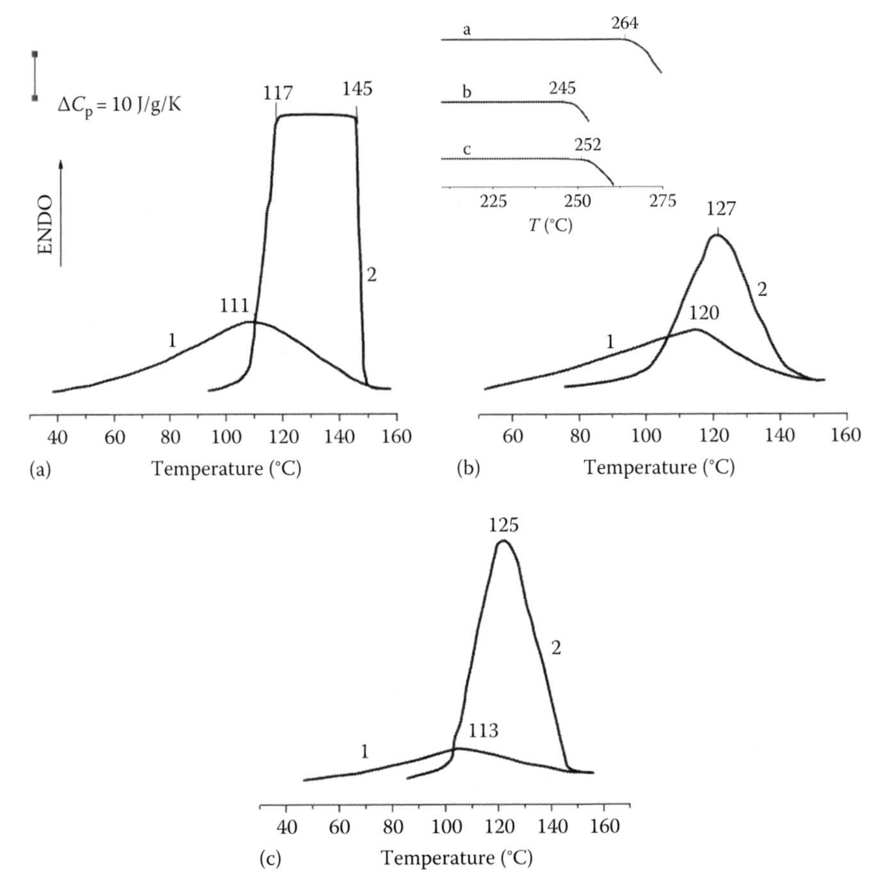

FIGURE 8.28 DSC dehydration endothermic curves of fiber samples obtained at elevated temperatures. (Curve 1) Air-dry samples and (curve 2) water swollen samples; heating rate was 20 K/min. Inset: the exothermic bends in the DSC curves corresponding to the onset of cellulose destruction process in (a) cotton, (b) bleached flax, and (c) flax. (Mikhalovsky, S.V., Gun'ko, V.M., Bershtein, V.A. et al., A comparative study of air-dry and water swollen flax and cotton fibres, *RSC Adv.*, 2, 2868–2874, 2012. Adapted with permission from The Royal Society of Chemistry.)

Figure 8.28 shows the DSC curves of the fibers analyzed at elevated temperatures. The endothermic effects characterize the dehydration process. It starts at lower temperatures and T_{max} values are lower for air-dry samples than those for swollen samples, suggesting that it is more difficult to dehydrate the swollen fibers than air-dry samples. Second, the unusual plateau in the endotherm was reproducibly observed for swollen cotton at 390–418 K (Figure 8.28a), which corresponds to a practically constant dehydration rate in this temperature region. This may be associated with a compensating effect of decreasing amount of adsorbed water versus temperature rise.

At elevated temperatures all air-dry samples have very different ΔH_d values, which are significantly larger than ΔH_{vap} values (Table 8.8). The $\Delta H_d/\Delta H_{vap}$ ratios indicate that the "strength" of water III–fiber interactions in air-dry materials decreases in a row:cotton > bleached flax > flax. The differences in ΔH_d and $\Delta H_d/\Delta H_{vap}$ for swollen samples were much smaller. The inset in Figure 8.28 shows the temperatures at which thermal destruction of fiber samples started. These temperatures are a little higher than values obtained in TG measurements, perhaps due to a higher heating rate in DSC experiments.

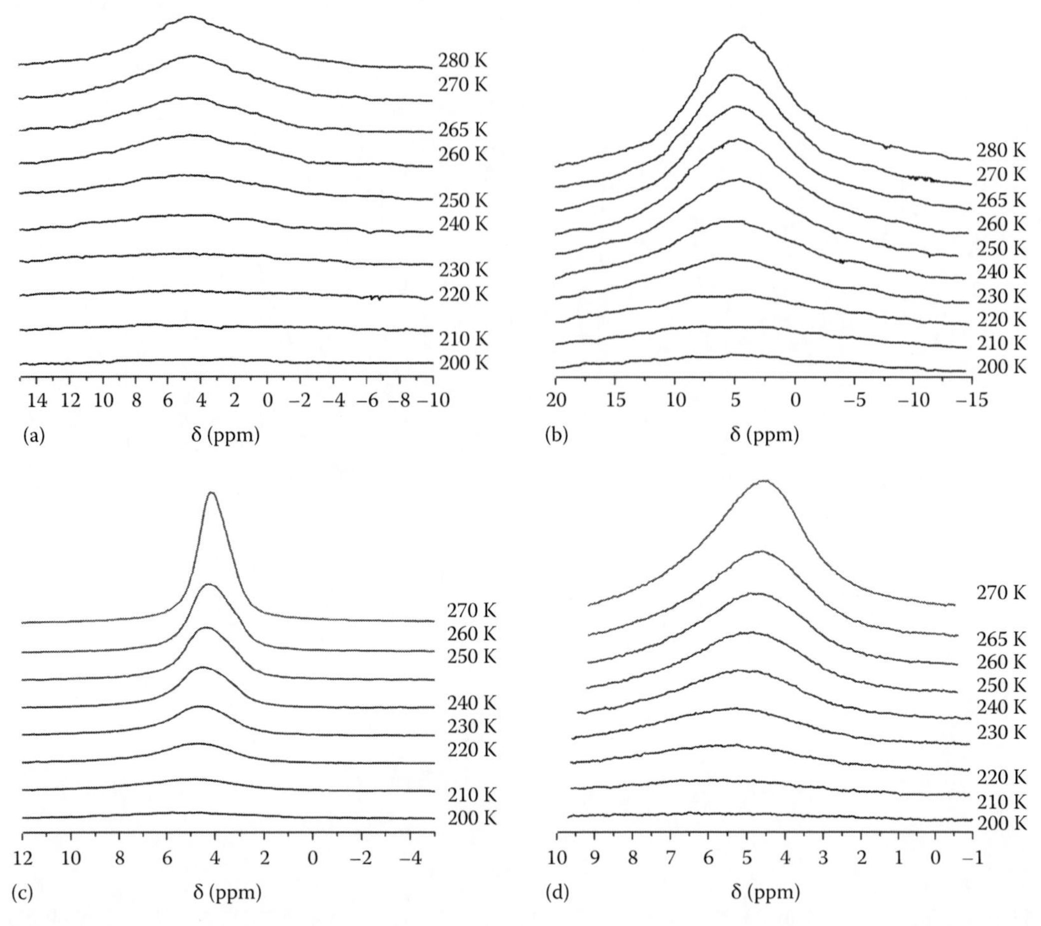

FIGURE 8.29 ^{1}H NMR spectra of air-dry (a, b) and 1 h incubated with water (c, d) fibers recorded at different temperatures. (a, c) Bleached flax and (b, d) cotton. (Mikhalovsky, S.V., Gun'ko, V.M., Bershtein, V.A. et al., A comparative study of air-dry and water swollen flax and cotton fibres, *RSC Adv.*, 2, 2868–2874, 2012. Adapted with permission from The Royal Society of Chemistry.)

Additional information on the state of water in air-dry and swollen fibers was obtained using the low-temperature ^{1}H NMR spectroscopy (Figures 8.29 and 8.30) and cryoporometry (Figures 8.31 and 8.32).

The ^{1}H NMR spectra (Figure 8.29), amount of unfrozen water as a function of temperature ($C_{uw}(T)$) and the corresponding relationships between the $C_{uw}(T)$ values and changes in the Gibbs free energy of water bound to fibers (Figure 8.30), show that swollen cotton fibers contain a larger amount of both strongly bound water of type III (C_{SBW}) and weakly bound water of type II (C_{WBW}) than swollen bleached flax fibers (Table 8.9).

This conclusion is in agreement with the DSC data despite the difference in swelling time. The air-dry bleached flax fibers contain larger amount of weakly bound water than air-dry cotton fibers. NMR cryoporometry was used to calculate the pore size distribution, PSD (Figure 8.31) using the $C_{uw}(T)$ functions (Figure 8.30).

Integration of PSD functions gives the textural characteristics of the samples (Table 8.9). The specific surface area ($S_{uw}=51.7$ m^2/g, Table 8.9) of bleached flax fibers in contact with unfrozen bound water at 273 K is close to the specific surface area determined from the water sorption isotherm ($S_{BET,w}=46$ m^2/g, Table 8.9). However, $S_{uw} \gg S_{BET,w}$ for 1 h incubated cotton. This difference in the bleached flax and cotton fiber properties can be explained by features in the fibers

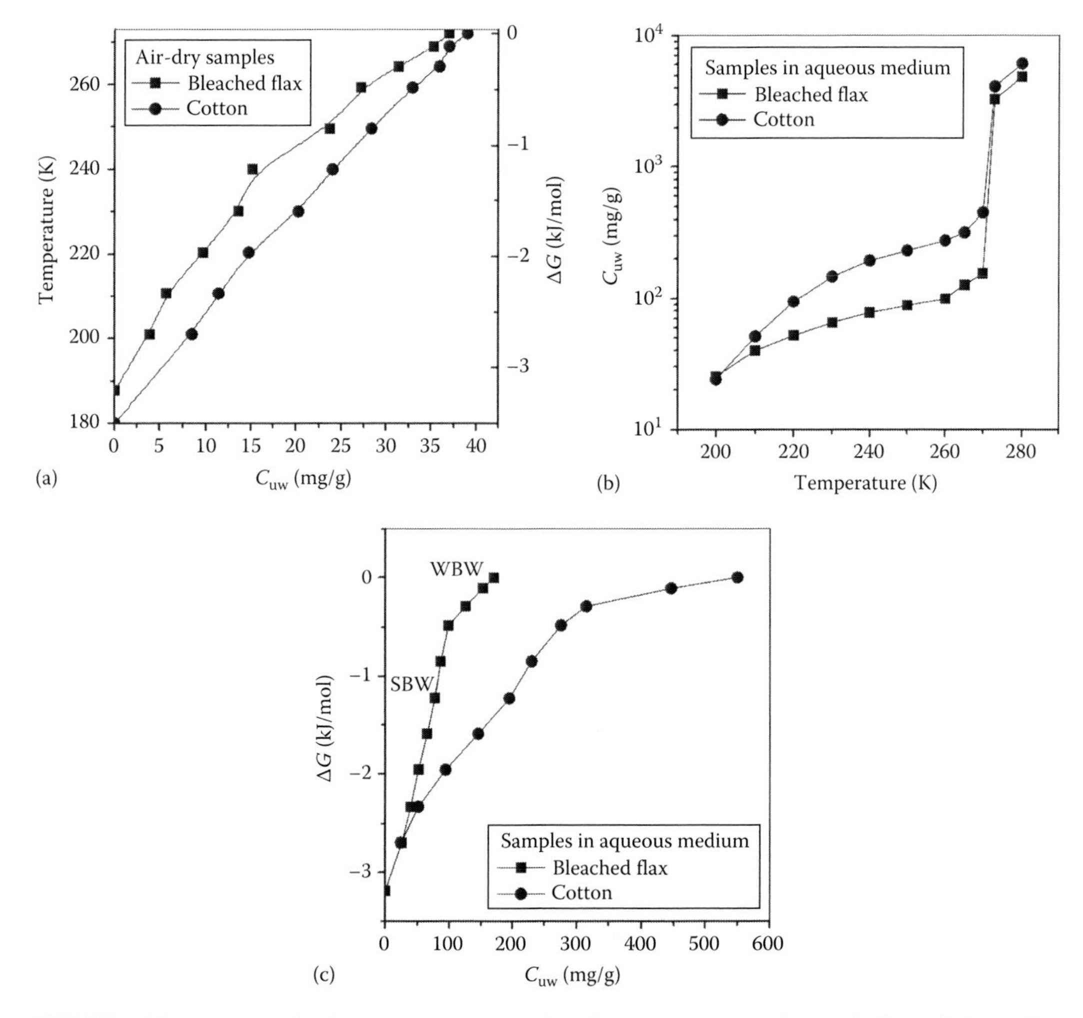

FIGURE 8.30 Amount of unfrozen water C_{uw} as a function of temperature for (a) air-dry and (b) swollen samples, and (a, c) corresponding relationships between C_{uw} and changes in the Gibbs free energy of bound water, ΔG. (Mikhalovsky, S.V., Gun'ko, V.M., Bershtein, V.A. et al., A comparative study of air-dry and water swollen flax and cotton fibres, *RSC Adv.*, 2, 2868–2874, 2012. Adapted with permission from The Royal Society of Chemistry.)

interactions with water (see earlier DSC results) and smaller amounts of water adsorbed from the water vapor (0.15–0.2 g H_2O per gram of fibers but close to that used on the DSC measurement of swollen fibers) during measurements of the water sorption isotherms than during the NMR measurements (Figure 8.30b).

The PSD (Figure 8.31) and the textural characteristics (Table 8.9) of swollen fibers show that contribution of mesopores (pore radius $1<R<25$ nm) to the total surface area and especially to the total pore volume are significantly larger than that of nanopores ($R<1$ nm). Compared to the air-dry samples, both S_{meso} and V_{meso} increased substantially whereas changes in S_{nano} and V_{nano} were minor (because fibers studied are not nanoporous materials). Increase of the S and V parameters is due to swelling since bound water enlarges the porosity of fibers, which is evident by comparing PSD curves for air-dry (Figure 8.31a) and swollen (Figure 8.31b) fibers, and their V_{meso} values (Table 8.9). Notice that water was not removed from swollen samples used in the NMR measurements ($C_{uw}=C_{SBW}+C_{WBW}<0.55$ g/g) (Table 8.9) in contrast to the DSC measurements ($C_w \leq 0.26$ g/g).

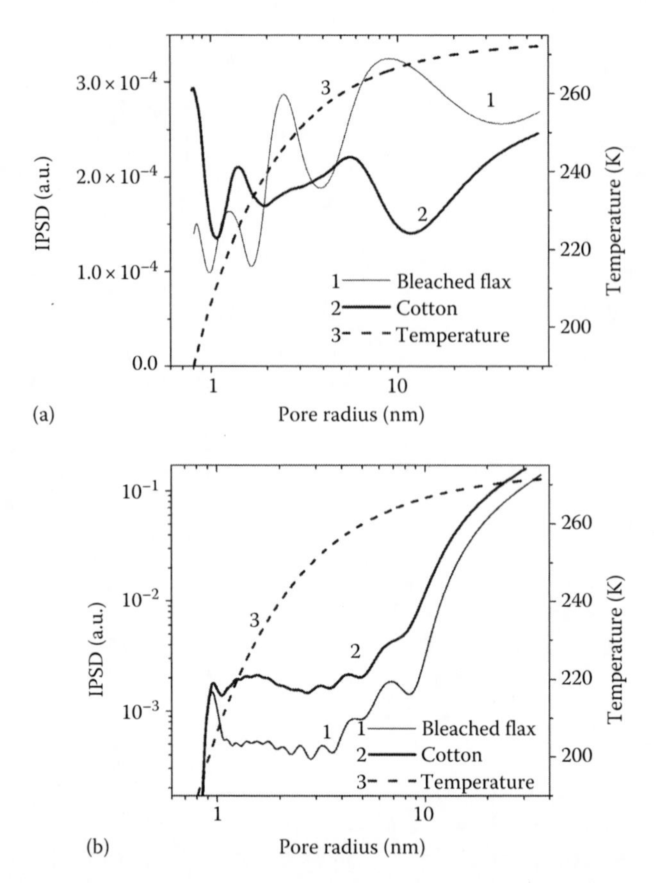

FIGURE 8.31 Integral pore size distributions in (a) air-dry fibers and (b) fibers incubated with water for 1 h. (Mikhalovsky, S.V., Gun'ko, V.M., Bershtein, V.A. et al., A comparative study of air-dry and water swollen flax and cotton fibres, *RSC Adv.*, 2, 2868–2874, 2012. Adapted with permission from The Royal Society of Chemistry.)

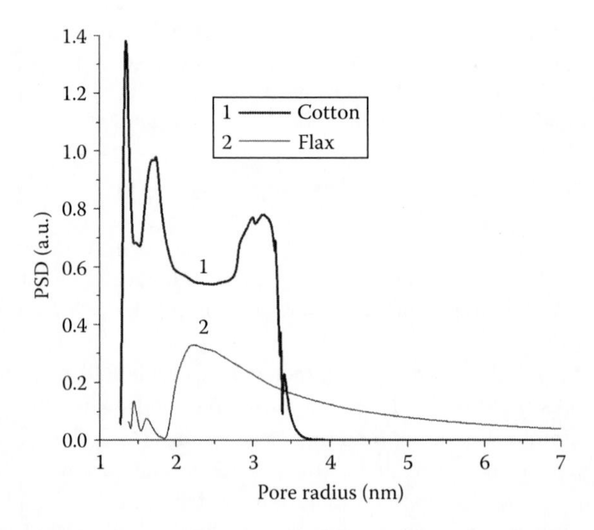

FIGURE 8.32 Pore size distributions for flax and cotton fibers calculated using DSC data (Figure 8.27) and DSC cryoporometry method. (Mikhalovsky, S.V., Gun'ko, V.M., Bershtein, V.A. et al., A comparative study of air-dry and water swollen flax and cotton fibres, *RSC Adv.*, 2, 2868–2874, 2012. Adapted with permission from The Royal Society of Chemistry.)

TABLE 8.9
Characteristics of Bound Water and Structural Characteristics of Air-Dry Fibers and Swollen Fibers Incubated for 1 h with Water

Sample	C_{SBW} (mg/g)	C_{WBW} (mg/g)	ΔG_s (kJ/mol)	S_{uw} (m²/g)	S_{nano} (m²/g)	V_{nano} (cm³/g)	S_{meso} (m²/g)	V_{meso} (cm³/g)	$S_{BET,w}$ (m²/g)
Air-dry bleached flax	28	10	−2.94	10.9	6.0	0.003	4.8	0.032	
Air-dry cotton	33	6	−3.00	19.2	14.0	0.007	5.1	0.031	
Swollen bleached flax	100	70	−2.97	51.7	10.4	0.005	31.7	0.533	46
Swollen cotton	270	280	−2.94	88.5	9.3	0.005	67.7	0.946	34

The melting temperature varies in a broad range for strongly bound water III, which filled nanopores with $R < 1$ nm (Figure 8.31, curves 3). Weakly bound water II is located in mesopores with $R > 1$ nm. Its melting occurs in a narrow temperature range close to 273 K (Figure 8.30b). The major amount of water II in swollen fibers (Figure 8.31b) is located in pores with $R > 10$ nm, where the influence of the surface is low. Therefore, T_m for this type of water is close to 273 K. A portion of water located in mesopores ($R < 25$ nm) can be assigned to water I because the V_{meso} value is larger than the volume occupied by SBW and WBW for both swollen cotton and swollen bleached flax (Table 8.9). Calculations of the PSD for hydrated flax and cotton fibers using the DSC cryoporometry data (Figure 8.27) give a picture (Figure 8.32) which is in general agreement with the NMR cryoporometry results (Figure 8.31).

However, certain details of the NMR and DSC PSD differ because of different conditions of hydration (1 and 24 h), different amounts of water in the samples, and different nature of these cryoporometry methods. The narrowest DSC PSD peaks at $R \approx 1.4$ nm (Figure 8.32) are shifted toward larger R values in comparison with the NMR PSD at $R = 1$ nm (Figure 8.31b) because of longer time of swelling before the DSC measurements that can result in larger nanopore size. Intensity of the DSC PSD is lower for flax than cotton (Figure 8.32) because of a lower content of water (17 and 26 wt%, respectively). The DSC measurements for bleached flax did not produce curves (Figure 8.27) suitable for calculations of the PSD due to a low content of bound water (10 wt%).

Thus, flax and cotton fibers have comparable thermal stability, high hydrophilicity, and swelling ability but different distributions of states of water adsorbed on air-dry and wetted samples. The air-dry fiber samples of both cotton and bleached flax contain mainly strongly bound water (type III) located in nanopores with radius $R < 1$–2 nm. In addition to water type III, the swollen cotton and swollen bleached flax fibers contain a significant fraction of weakly bound water (type II), which is located in mesopores with $1 < R < 10$ nm, and practically unbound bulk water in large mesopores at $R > 10$ nm. According to the NMR cryoporometry data, even short swelling of fibers in water significantly changes the pore size distribution in bleached flax and cotton fibers substantially increasing mesopore volume and surface area. The results obtained provide a deeper insight into features of sorption of different compounds on air-dry and swollen natural fibers such as cotton and flax.

This brief analysis of applications of the NMR techniques to different botanics and related materials to study the behavior of bound water affected by biosurroundings shows that the parallel use of TSDC, DSC, and other methods allows deeper insight into the problems related to drying, wetting, swelling, germinating, etc., of these complex bio-objects.

9 Recurring Trends in Adsorption, Spectroscopy, and Other Interfacial Experiments

There are top NMR techniques to study organics in solution and at the interfaces such as ^{1}H and ^{13}C NMR, diffusion-edited ^{1}H NMR spectroscopy (DE), correlation spectroscopy (COSY, TOCSY), heteronuclear single quantum coherence or heteronuclear multiple quantum coherence (HSQC, HMQC), heteronuclear multiple bond correlation (HMBC), nuclear overhauser effect spectroscopy (NOESY, ROESY), and heteronuclear single quantum coherence–total correlation spectroscopy (HSQC–TOCSY) (Kumar et al. 2000, Kolodziejski and Klinowski 2002, Webb 2003, 2009, Hennel and Klinowski 2004, Kimmich and Anoardo 2004, Igumenova et al. 2006, Jarymowycz and Stone 2006, Palmer and Massi 2006, Tolman and Ruan 2006, Calucci and Forte 2009, Mao et al. 2009, 2010, Mao and Pruski 2009, 2010, Simpson et al. 2011, Volovenko et al. 2011, Brown 2012). These methods are mainly oriented toward obtaining structural information on complex compounds affected by surroundings. A portion of these measurements can give very useful textural information on complex objects but mainly indirect. Such methods as MRI and NMR microscopy (NMRM) can give direct textural information, which, however, is rather qualitative information than quantitative textural one. To obtain quantitative textural information, the MRI and NMRM images (as in the case of images obtained by SEM, TEM, AFM, CLSM, and other microscopic methods) should be treated using specific software as it was demonstrated for different samples in this book using ImageJ and Fiji software applied to images obtained with HRTEM, HRSEM, CLSM, etc. This trend in the development of NMR spectroscopy (MRI, NMR cryoporometry, and relaxometry studies of soft and solid materials, especially macroporous ones) is obvious for near future (Mitchell et al. 2008, Ono et al. 2009, Webber 2010, Shapiro 2011, Rouquerol et al. 2012, Yao and Liu 2012). For MRI study of porous materials (Koptyug 2011), ^{1}H NMR is carried out with a wetting probe fluid in pores (water, hydrocarbon). Pores of a width above 10 μm are directly visualized by MRI. Smaller pores can be assessed but requires chemically homogeneous samples (pure silica, pure alumina, etc.) and NMR contrast methods (relaxation time or pulsed-field gradient techniques). According to Rouquerol et al. (2012), one can stress the importance of developing alternative methods to mercury porosimetry to study the texture of macroporous materials and of extending the range of application, ease of handling, and experimental rapidity of the methods already existing today. They wrote that at least for the time-being, there is still no well-established alternative to mercury porosimetry for the characterization of macropores up to 400 μm width. This is partially incorrect because confocal laser scanning microscopy (CLSM), multiphoton microscopy (SEM), and cryo-SEM provide quantitative textural characterization of macroporous materials in both native hydrated and freeze-dried states (Mattiasson et al. 2009, Savina et al. 2011). MRI and NMR cryoporometry can be appropriate tools for the textural characterization of macroporous materials. We could expect further progress in the development of NMR cryoporometry and relaxometry, as well as MRI, for macroporous materials and to study a variety of the interfacial phenomena in confined space of pores, voids, and pockets in solid and soft materials and bio-objects.

There are several general problems related to the influence degree of pretreatment (drying, degassing, heating, wetting, freezing, swelling, pressing, etc.), measurement conditions, and measurement itself on studied objects because this influence can be too strong for appropriately correct

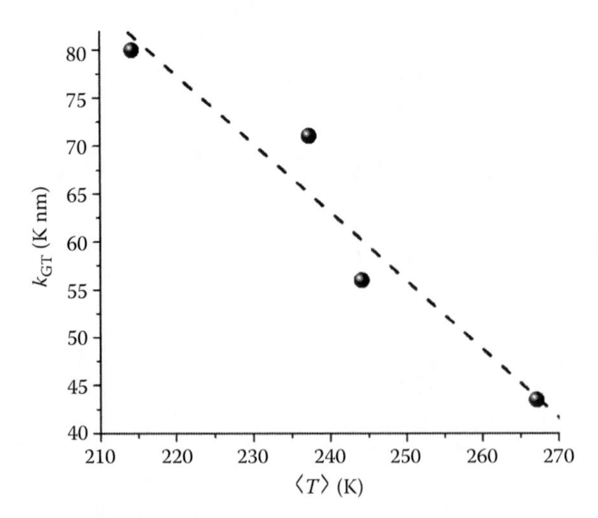

FIGURE 9.1 Relationship between the k_{GT} value (determined at condition $S_{uw}=S_{BET}$) and $\langle T \rangle$ for silica gels (Si-40, Si-60, Si-100) and nanosilica A-300 (linear approximation corresponds to $k_{GT}=234.603-0.71462\langle T \rangle$).

description of the systems studied. Therefore, the development of nondestructive, noninvasive methods that can be applied to objects in their native state is of importance, and the NMR spectroscopy as a nondestructive method is an appropriate candidate. However, there is a problem of the analysis and interpretation of indirect information obtained by the NMR spectroscopy, cryoporometry, and relaxometry. The use of calibration functions (e.g., transformation of the NMR signal intensity into the amount of unfrozen water $I(T) \rightarrow C_{uw}(T)$ or other confined liquids) or constants (e.g., k_{GT} in the Gibbs–Thomson equation; Figure 9.1), which were obtained for certain systems but applied for other nontested systems, can give both accurate and inaccurate results that depend on the similarity of the studied and reference systems. The situation is much simpler for simple materials, e.g., oxide or carbon adsorbents, than for bio-objects (cells, tissues, etc.). For instance, there is a nearly linear relationship between the k_{GT} value (determined using condition $S_{uw}=S_{BET}$) and the average freezing temperature $\langle T \rangle$ (characterizing averaged interactions between adsorbate and adsorbent) for different silicas (Figure 9.1):

$$\langle T \rangle = \frac{\int_{T_{min}}^{T_{max}} T(dC_{uw}/dT)dT}{\int_{T_{min}}^{T_{max}} (dC_{uw}/dT)dT} \tag{9.1}$$

The k_{GT} value decreases with decreasing specific surface area of silicas. However, despite apparent similarity of silica gels and fumed silicas in respect to the surface chemistry (i.e., interactions of water or other adsorbates with the surfaces), the differences in pretreatment history of the reference and studied samples, admixtures, residual surface functionalities, etc. can give a significant scatter in the characteristics in comparison with those obtained by other independent methods. For instance, the k_{GT} value is not constant for silica gels and nanosilica (Figure 9.1) because of different confined space effects for bound water. There is a similar situation in the calculations of the BET-specific surface area of silicas since the area occupied by N_2 molecule decreases with decreasing pore size of mesoporous ordered silicas (Gun'ko et al. 2007g). Additionally, for MCM-41 ($S_{BET}=1196$ m^2/g) with a certain residual amount of the used organic template (Gun'ko et al. 2007g), $k_{GT}=50.5$ K nm at very low $\langle T \rangle=204.3$ K. Consequently, the presence of the surface organic functionalities in MCM-41 results in diminution of the k_{GT} value and a strong deviation from the dependence observed for pure silicas (Figure 9.1). However, for oxidized AC at the S_{BET} value (1162 m^2/g) close to that of MCM-41, $k_{GT}=82.4$ K nm at $\langle T\rangle=206.5$ K, and the ratio between the k_{GT} and $\langle T\rangle$ values corresponds to the dependence shown in Figure 9.1 for pure hydrophilic silicas. These results show that besides pore sizes and the specific surface area, the wettability (hydrophilicity) of a surface is an important property that can affect the $k_{GT}/\langle T\rangle$ ratio. The variability of the k_{GT} value even for the same material (Figure 9.1) is the main disadvantage of the NMR cryoporometry with water as a probe compound, and it demands independent measurements on the system studied to check the correctness of the NMR cryoporometry results. The use of organic or inorganic compounds as a probe in the NMR cryoporometry and relaxometry is therefore necessary. On the other hand, the NMR cryoporometry or relaxometry, as well as DSC thermoporometry, gives not only textural information but also thermodynamic information on the interfacial phenomena including layer-by-layer freezing bulk and bound water of other liquids, structuring interfacial layers, clustering of water, quantitative results on weakly and strongly bound and strongly and weakly associated waters.

There are several effects responsible for the changes in the k_{GT} value for both the same and different materials: (i) an increase in average polarity of water molecules adsorbed in narrower pores of a polar adsorbent; (ii) stronger compaction of the adsorbed liquid (fluid) in narrower pores; (iii) changes in the geometry of the hydrogen bond network and clustered structure of water in narrow pores; (iv) a stronger influence of the opposite pore wall in narrower pores; and (v) an influence of the chemical structure of the pore wall surface on the interfacial layer structure. Therefore, the NMR cryoporometry with water as a probe liquid can give worse results (especially if the k_{GT} value poorly corresponds to a studied material) than that with other probe compounds (e.g., ^{129}Xe) less sensitive to the mentioned factors. Therefore, it will be expected trend in the NMR cryoporometry and relaxometry (and MRI; Koptyug 2011) toward the use of more appropriate probes than water with appropriate k_{GT} values (Petrov and Furó 2009) determined with certain corrections (Webber 2010) for textural characterization of soft and solid materials, especially macroporous (Rouquerol et al. 2012). Clearly, these aspects of the mentioned NMR techniques, as well as the related interfacial phenomena, need additional detailed investigations.

A maximum comprehensive study of any object is the best way to obtain maximum reliable information. For instance, a combination of NMR cryoporometry or relaxometry (or DSC thermoporometry) with nitrogen (or argon) and water (or another compound used as a cryoporometry/relaxometry probe) adsorption, SAXS (Gun'ko et al. 2011a), FTIR (Shen and Ostroverkhov 2006, Gun'ko et al. 2009d), DRS (Buchner 2008), TSDC (Gun'ko et al. 2007i), and CLSM (Savina et al. 2011) can give more reliable information on the structure, texture, and morphology than any pair of the mentioned methods can give (Brown et al. 1999, Hindmarsh et al. 2004, Petrov and Furó 2009, Webber 2010, Gribble et al. 2011, Hitchcock et al. 2011, Yao and Liu 2012). Even several NMR techniques (low and high temperature, static and MAS, 1D and 2D, HSQC and HMQC, etc.) applied to the same object can give a clearer picture of a phenomenon or material properties.

The next feature of low-temperature NMR investigations (structural/textural) with water as a probe compound is due to the fact that textural information can be affected by the "memory" effects of studied materials undergoing drying–wetting–suspending–freezing–thawing since the ice volume is larger than that of liquid water, and swelling can change the texture of soft materials in a solvent. In other words, a material can undergo irreversible changes during cycling pretreatment and the hysteresis effects can depend on many factors. For instance, commercial porous polymers in the form of beads such as Amberlite XAD-7 (Fluka) (with acrylic ester polymer, $S_{BET}=341$ [initial], 462 [frozen with acetone by liquid nitrogen], 488 [frozen with water] m^2/g) and LiChrolut EN (Merck) ($S_{BET}=1512$ [initial], 1479 [frozen with acetone], 1521 [frozen with water] m^2/g), and activated carbon PS3 ($S_{BET}=1873$ m^2/g, $V_p=1.037$ cm^3/g, and $R_p=2.21$ nm for initial sample and $S_{BET}=2233$ m^2/g, $V_p=1.255$ cm^3/g, and $R_p=2.25$ nm after suspending/freezing) were used for comparison of the freezing effect on the porous structure (Gun'ko et al. 2005a). Some of the polymers (e.g., XAD-7; Figure 9.2a) demonstrated significant structural changes after suspending–freezing–thawing–degassing.

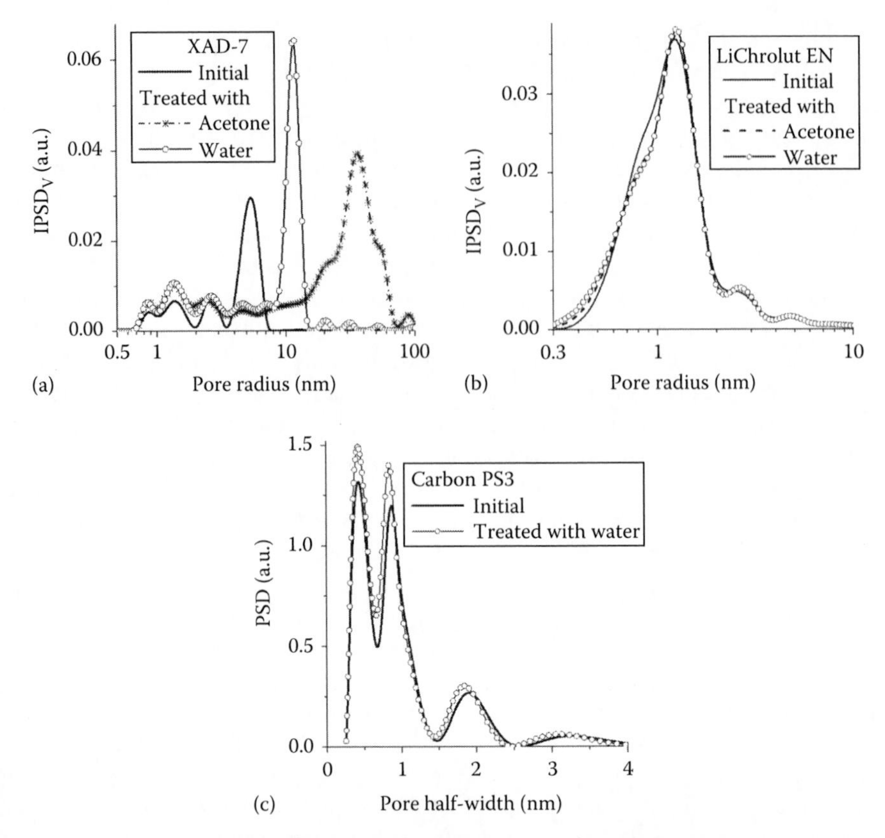

FIGURE 9.2 Pore size distributions of (a, b) porous polymers initial and frozen with water or acetone (stored for 24 h with a liquid, frozen by liquid nitrogen for 2 h, heated to room temperature, and then degassed at 353 K for 2 h before the second nitrogen adsorption measurement) or activated carbon initial or frozen with water; pore models: (a, b) cylindrical and (c) slit-shaped. (Adapted from *Appl. Surf. Sci.*, 252, Gun'ko, V.M., Leboda, R., Skubiszewska-Zięba, J., Gawdzik, B., and Charmas, B., Structural characteristics of porous polymers treated by freezing with water or acetone, 612–618, 2005a. Copyright (2005), with permission from Elsevier.)

However, other polymers (e.g., LiChrolut EN; Figure 9.2b) or AC (Figure 9.2c) demonstrated small changes in the textural characteristics (Gun'ko et al. 2005a). Notice that the applied treatments were used to model possible effects during low-temperature measurements using the NMR cryoporometry (with water or organic solvents as a probe compound). Obtained results showed in general that it is difficult to predict possible structural (textural) changes of different materials (especially complex soft materials or bio-objects) undergoing drying–wetting–suspending–freezing–thawing. Therefore, adsorption, SAXS (Gun'ko et al. 2011d), neutron diffraction cryoporometry (Dore et al. 2004, Webber and Dore 2008), CLSM, and cryo-SEM measurements (Savina et al. 2011) could be carried out in parallel to NMR measurements to elucidate different aspects of changes in the texture, morphology, and structure of the materials as well as features of the interfacial phenomena itself.

Investigations of textural changes as well as the relaxation (of both soft adsorbents and adsorbates, especially high-molecular ones) and diffusion (of low-molecular adsorbates or macromolecules) processes are of interest for deeper insight into the interfacial phenomena occurring in strongly hydrated, wetted, and air-dry systems. This is especially interesting when strong hydration starts up such specific processes as germination of seeds or when drying of hydrogels results in significant textural rearrangement of the materials. In this case, the NMR investigations could be accompanied by DRS, TSDC, DSC, and diffusion measurements to elucidate structural, morphological, and textural changes in the materials or bio-objects. For instance, local mobility of water in

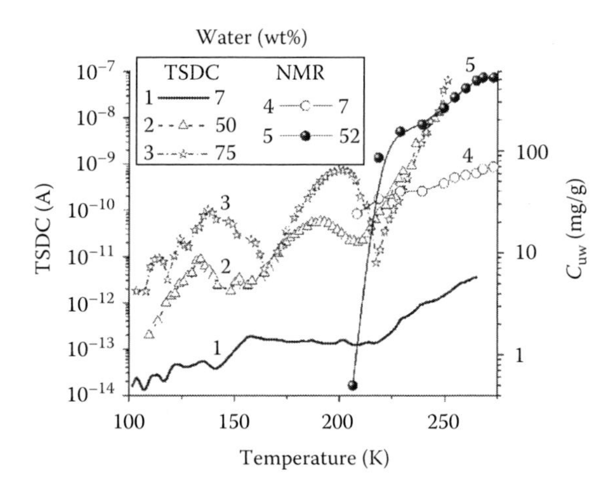

FIGURE 9.3 Temperature dependences of TSD current and amount of unfrozen water, C_{uw} (^{1}H NMR) for yeast cells initial dry containing 7 wt% of water (curves 1 and 4) and hydrated at 50 (curve 2), 52 (curve 5), and 75 (curve 3) wt% of bound water.

dry and hydrated yeast cells, registered by the ^{1}H NMR spectroscopy, appears at lower temperatures (~200 K) than the through conductivity (ion percolation with dc relaxation) observed by TSDC (210–220 K) (Figure 9.3).

Besides the dc relaxation, the TSDC measurements allow one to analyze the dipolar relaxations observed at $T < 200$ K and related to γ (polar bonds), β (small fragments of polar macromolecules with bound water), and α (water clusters and polar organic molecules) relaxations. Typically, the activation energy of the through conductivity (TSDC) is higher than the activation of the local molecular mobility (NMR). CLSM, cryo-SEM, SEM, and AFM studies of dry and hydrated cells can give additional information about rearrangement of cellular structures during hydration or drying of cells. Thus, the textural characteristics of cells and other soft matters undergoing drying, wetting, suspending, swelling, etc. should be analyzed using a set of methods because these changes are complex and affected by both changes in the hydration degree and temperature regime. Despite NMR techniques are powerful tools for structural, textural, and diffusion characterization of solid and soft materials, liquids, etc., the application of other informative methods, especially noninvasive ones in the case of bio-objects or soft materials, to the same objects at the same conditions is very necessary for deeper insight into the interfacial phenomena and the properties of whole complex objects. This can be the main trend in the investigations of the interfacial phenomena.

10 Methods

10.1 LOW-TEMPERATURE ^{1}H NMR SPECTROSCOPY

The NMR spectroscopy has been described in detail by many authors (Pople 1956, 1957, Pople et al. 1959, Abragam 1961, Stejskal and Tanner 1965, Tanner and Stejskal 1968, Wüthrich 1986, 1995, Engelhardt and Michel 1987, Abraham et al. 1988, Martin and Zekter 1988, Facelli and de Dois 1999, Akitt and Mann 2000, Hennel and Klinowski 2004, Webb 2003, 2009, Keeler 2005, *Encyclopedia of Magnetic Resonance* 2007, 2009, 2011, Schilf et al. 2011). In this book, we discussed the applications of NMR techniques to the interfacial phenomena; therefore, description of the NMR methods is given only briefly.

Phenomenon of absorption of the radiofrequency energy by the nuclei possessing own magnetic moment is the basis of the NMR method. The nuclei having nonzero spin S are oriented in external magnetic field B_0 along quantified directions that the spin projection (m) on the field direction takes the $2S+1$ values and $m=S, S-1, \ldots, -S$. This effect results in splitting of the corresponding energetic level into $2S+1$ sublevels with an energetic gap between them:

$$E_{\rm m} = -\hbar m \gamma B_0 \quad (10.1)$$

where

$\hbar$ is the Planck's constant

γ is the gyromagnetic ratio of the proton

The imposition of an additional field $B_{\rm r}$ with a high frequency ω allows one to study transitions between these sublevels. The resonance absorption of the energy occurs under condition

$$\omega_0 = \gamma B_0 \quad (10.2)$$

The last equation describes general condition of the nuclear magnetic resonance. However, local magnetic field acting on nuclei with the S spin differs from B_0 generated by a magnet of a NMR apparatus due to the magnetic shielding of nuclei by their electron shells and local magnetic fields produced by adjacent atoms and molecules. This difference gives the magnetic shielding constant, σ. The last constant represents an asymmetric tensor with nine independent components:

$$\sigma_{\alpha\beta}^{\rm N} = \frac{\partial^2 E}{\partial B_\alpha \partial \mu_{{\rm N}\beta}} \quad (10.3)$$

where α, $\beta = x, y, z$ denote components of the external magnetic field $\mathbf{B}_0$ and the nuclear magnetic moment $\boldsymbol{\mu}_{\rm N}$. Typically, an isotropic value $\sigma = (\sigma_{11} + \sigma_{22} + \sigma_{33})/3$ is utilized. The shielding tensor is an antisymmetric tensor because exchanging the subindices α and β in Equation 10.3 leads to a different quantity (Facelli 2011). This property of the tensor should not to be confused with the fact that the first-order response function to the shielding interaction, i.e., the NMR spectrum, is sensitive only to the symmetric part of the tensor and that the antisymmetric components contribute only to the second-order response, i.e., the relaxation of the spin magnetization.

In NMR, magnetic nuclei placed in external magnetic field absorb and reemit electromagnetic radiation at a specific resonance frequency depending on the parameters of the field applied and the quantum properties of the atoms studied. All atoms (isotopes) containing odd numbers of protons and/or neutrons are characterized by an intrinsic magnetic moment and angular moment (i.e., non-zero spin state). There are two isotopes ^{1}H and ^{13}C mainly studied by the NMR, especially in organic and bioorganic chemistry. Other isotopes such as ^{2}H, ^{6}Li, ^{10}B, ^{11}B, ^{14}N, ^{15}N, ^{17}O, ^{19}F, ^{23}Na, ^{25}Mg, ^{27}Al, ^{29}Si, ^{31}P, ^{33}S, ^{35}Cl, ^{39}K, ^{41}K, ^{47}Ti, ^{49}Ti, ^{113}Cd, ^{127}I, ^{129}Xe, ^{133}Cs, ^{195}Pt, ^{199}Hg, and some others are used in NMR depending on studied materials. For instance, many important results in chemistry of silica and other natural and synthetic minerals were received using ^{29}Si and ^{27}Al NMR spectroscopy.

The NMR spectrum can also reveal information on the molecular dynamics but for ^{1}H-rich materials in the solid state, the multitude of strong homonuclear ^{1}H–^{1}H dipolar couplings invariably leads to broad, relatively featureless ^{1}H NMR absorption lines (Horsewill 2008). In this case, only low-resolution information on the second moment of the absorption line and its temperature dependence can be extracted. Such data reveals insight into the partial averaging of the dipolar couplings due to molecular motion but provides little additional information and is only supplementary to that obtained by spin–lattice relaxation. However, in a few ideal cases, intramolecular dipolar couplings involving the hydrogen-bond protons dominate over all other intermolecular terms. This leads to the opportunity to resolve one or two dipolar splitting in the ^{1}H NMR spectrum. These can then be studied as a function of temperature (Horsewill 2008).

For characterization of the magnitude of magnetic shielding of a given nuclei, the value of chemical shift (δ) is used, which corresponds to the difference between nuclear magnetic shielding of standard (σ_1) and tested (σ) compounds:

$$\delta = (\sigma_1 - \sigma) \tag{10.4}$$

and is measured in million proportions (ppm) of the magnitude of the magnetic field frequency of a NMR spectrometer. The nuclear magnetic shielding constant is determined by two contributions such as diamagnetic (σ_d) and paramagnetic (σ_p) $\sigma = \sigma_d + \sigma_p$. For alone atom in the 1s state, paramagnetic contribution equals zero, but diamagnetic one is

$$\sigma_d = \frac{\mu_0 e^2}{12\pi m_e c^2} \int r^{-1} \rho(r) d^3 r \tag{10.5}$$

where

μ_0 is the magnetic constant
m_e is the electron mass
c is the light velocity
e is the electron charge

The calculations show that changes in the charge in $1e$ result in changes in σ value by approximately 20 ppm.

Tetramethylsilane (TMS) is frequently used as a standard compound to determine the chemical shift values for the protons of probe compounds. Relatively low charges of H atoms in CH_3 groups, sp^3 hybridization of atomic orbitals of C atoms, and the absence of electron-donor groups in TMS result in the down-field (lower resonance frequency) shifts of ^{1}H of organic compounds in comparison with TMS. The values of chemical shifts depend on the characteristics of intermolecular interactions of a given molecule with surroundings because of changes in nuclear magnetic shielding of its nuclei. The hydrogen bonding of molecules strongly affects the δ_H values. Formation of the hydrogen bonds leads to a decrease in shielding of the protons of the H atoms, i.e., the down-field shift occurs. The σ value for alone molecule is practically independent of pressure and temperature, but for hydrogen-bonded complexes the opposite effect is observed. This is due to destruction of a portion of the hydrogen bonds with elevating temperature resulting in up-field shift of the proton resonance (or a shift of the resonance to a higher frequency). For explanation of the dependence of the δ value on temperature, a model of two states (Haggis et al. 1952, 1953) is utilized, according to which, the measured chemical shift can be written as follows:

$$\delta_H(T) = (1 - P_F)\delta_{HB} + P_F\delta_F \quad (10.6)$$

where

P_F is the portion of broken hydrogen bonds

δ_{HB} and δ_F are the chemical shifts of ^1H, which take part or do not participate in the hydrogen bonds, respectively

Recording of only one average signal is conditioned by a very short lifetime of each state. On adsorption interactions of non-hydrogen-bonding type or limitation of the molecular exchange rate by a spatial factor in adsorbent pores, the lifetime of the adsorption complexes can change over a broad range. According to Gutowsky–Holm theory of exchange processes (Gutowsky and Saika 1953, Gutowsky and Holm 1956), the intensity of the NMR signal as a function of a frequency ν of two states A and B, possessing the lifetimes τ_A and τ_B and resonance frequencies ν_A and ν_B, can be given by

$$g(\nu) = \frac{K\tau(\nu_A - \nu_B)^2}{[(\nu_A - \nu_B)/2 - \nu]^2 + 4\pi\tau^2(\nu_A - \nu)^2(\nu_B - \nu)^2} \quad (10.7)$$

where

$\tau = \tau_A\tau_B/(\tau_A + \tau_B)$

K is a normalizing constant

If the exchange between the A and B states occurs slowly in NMR scale, i.e., a relationship $\tau_A(\tau_B) \gg 1/(\nu_A - \nu_B)$ is valid, then the signals corresponding to these states are observed separately. On fast exchange when $\tau_A(\tau_B) \ll 1/(\nu_A - \nu_B)$, only one average signal is observed and its frequency is determined by a statistical portion of each state:

$$\nu = P_A\nu_A - P_B\nu_B \quad (10.8)$$

where

$P_A = \tau_A/(\tau_A - \tau_B)$

$P_B = \tau_A/(\tau_A + \tau_B)$

For interim exchange rates, the shape of resonance curves can be determined using Equation 10.7. The velocity of molecule motions decreases with lowering temperature; therefore on sample cooling, the spectrum shape changes. According to Equation 10.7, for separated recording of the signals related to different types of adsorption complexes in the case of slow exchange, the difference in the chemical shifts of the corresponding states should be larger than the bandwidth of the NMR signals of these states. The range of the chemical shifts of the proton resonance is relatively small (10–15 ppm), but on the other hand, the bandwidth of the signals of adsorbed molecules is larger than that in pure liquids because of reduced mobility of adsorbed molecules. Therefore, conditions of separated observation of the signals of the different adsorption complexes are difficult to be achieved. This circumstance hampers the use of the ^1H NMR spectroscopy to study the structure of active surface sites of many adsorbents.

In the case of proton transfer in the hydrogen bond, the de Broglie wavelength may have similar magnitude to the distance displaced by the hydrogen atom; therefore, the influence of quantum tunneling can be profound (Horsewill 2008). This has important implications for many chemical and biological processes, for example, enzyme catalysis, transport through cell membranes, the physics of ice, protein conformations, hydration around proteins, etc. The clearest signatures of tunneling come at low temperature and in this review we have considered systems in the solid state where strong deviations from Arrhenius behavior reveal the underlying quantum effects. As an experimental technique, solid-state NMR has provided some of the deepest insight into the quantum motion of the hydrogen bond. It has both good resolving power and wide dynamic range. For slow motions, the evolution of the NMR spectrum with temperature provides accurate data on the proton transfer exchange rates. Faster motions can be well characterized by spin–lattice relaxometry. In particular, field-cycling NMR relaxometry facilitates a direct mapping of the spectral density providing some of the most accurate measurements. Additionally, with both ^{1}H and ^{2}H receptive to NMR, our deeper understanding of quantum tunneling is facilitated through isotope effect measurements (Horsewill 2008).

A major portion of the ^{1}H NMR spectra of bound water discussed in this book was recorded for static samples (to suppress signals of immobile solid or macromolecular parts of samples) using a Varian 400 Mercury spectrometer or a Bruker WP-100 SY spectrometer of high resolution with the probing 90° or 40° pulses at duration of 2–5 μs. The temperature (170–350 K, frequently 200–280 K) was controlled by means of a Bruker VT-1000 device. Relative mean errors were $<\pm 10\%$ for ^{1}H NMR signal intensity for overlapping signals and $<\pm 5\%$ for single signals and ±1 K for temperature. To prevent supercooling of the studied systems, the measurements of the amounts of unfrozen water were carried out upon heating of samples preliminarily cooled to 170–200 K and equilibrated during 5–15 min for each temperature. The ^{1}H NMR spectra recorded include the signals only of unfrozen mobile molecules of water or other small molecules. The signals of water molecules from ice as well as protons form solid materials or macromolecules do not contribute the ^{1}H NMR spectra because of features of the measurement technique (narrow bandwidth of 10–50 kHz) and the short duration of transverse relaxation of protons in immobile structures, which is shorter by several orders than that of mobile water.

The transverse relaxation times were measured with standard Carr–Purcell–Meiboom–Gill (CPMG) pulse sequence: $90x$-(τ-$180y$-τ-echo)$_n$ with 90° pulse of 3 μs and total echo time of 500 μs. A total of eight scans were used for each measurement. The relaxation time between scans was 0.5 s. Equilibration time of each temperature was 5 min. To calculate a distribution function of transverse relaxation time $f(T_2)$, CPMG echo decay envelopes ($I(t)$) were used as the left term of integral equation:

$$I(t) = A \int_{T_{2\min}}^{T_{2\max}} \exp\left(-\frac{t}{T_2}\right) f(T_2)\,dT_2 \tag{10.9}$$

where

t is the time

$T_{2\min}$ and $T_{2\max}$ are the minimal (10^{-6} s) and maximal (1 s) T_2 values upon integration, respectively

A is a constant

Equation 10.9 was solved using a regularization procedure based on the CONTIN algorithm (Provencher 1982) under non-negativity condition ($f(T_2) \geq 0$ at any T_2) and an unfixed value of the regularization parameter (α) determined on the basis of the F-test and confidence regions using the parsimony principle. It should be noted that the solution of Equation 10.9 is insensitive to the A value, which can affect only relative intensity of $f(T_2)$. A similar procedure applied the NMR data to calculate $f(T_2)$ was described by Tang et al. (2000) and Choi and Kerr (2003).

One of the first methods to explore interaction between adsorbed molecules and solid surfaces was based on the measurements of the temperature dependencies of longitudinal and cross relaxation of nuclear spins of adsorbed molecules. The mobility of adsorbed molecules decreases with strengthening adsorption interactions, which results in a reduction of the relaxation time. Intermolecular (as well as adsorption) interactions can be described more entirely using the self-diffusion coefficient, which can be measured with pulse gradient of the magnetic field. These measurements demonstrate that the interaction between adsorbate and adsorbent is propagated through a relatively thick layer of compound adsorbed on nanoporous and layered matters, which can reach 2–3 nm or larger. In the aqueous suspensions of oxides, interaction of water with the solid surfaces can be detected even at 100 nm distance due to long-range component of intermolecular interaction, formation of the electrical double layer (EDL), and electrostatic and polarization effects. One of the main parameters characterizing interaction between solids and liquids is the Gibbs free surface energy (γ_S), which equals to change in the Gibbs free energy with increasing surface area by unit. Notice that the corresponding exact measurements are very difficult. One of the well-known methods to determine the surface energy of solids is a method of contact angles. It is assumed in this method that the free surface energy equals to the sum of two terms, and the first of them is conditioned by Lifshits–van der Waals forces, and the second is related to acid–base interactions $\gamma_S = \gamma_S^{lw} + \gamma_S^{ab}$. The acid–base term consists of electron-acceptor (γ^+) and electron-donor (γ^-) components $\gamma_S^{ab} = 2\left(\gamma_S^+\gamma_S^-\right)^{1/2}$. Then the contact angle value for the system with vapor (V), liquid (L), and solids (S) can be determined as follows: $\cos\theta = -1 + 2\left[\left(\gamma_S^{LV}\gamma_L^{LW}\right)^{1/2}/\gamma_{LW}\right] + 2\left[\left(\gamma_S^+\gamma_L^-\right)^{1/2}/\gamma_{LV}\right] + 2\left[\left(\gamma_S^-\gamma_L^+\right)^{1/2}/\gamma_{LV}\right]$. Components of the free surface energy can be estimated on the basis of the measurements of the contact angle for liquids with different types of intermolecular interactions. Here, we use the low-temperature ^{1}H NMR spectroscopy to estimate the values of changes in the Gibbs free energy. The condition of freezing of pore water (or benzene or other liquid used as a probe) corresponds to the equality of the Gibbs free energies of molecules of water and ice. Lowering the temperature of freezing of structured pore water ($T_f < 273.15$ K) is defined by diminution of its Gibbs free energy caused by strong intermolecular interactions ($\Delta G = G - G_0 < 0$, where G_0 is the free energy of ice at 273.15 K) that disrupt the hydrogen-bond network characteristic of the bulk water. The temperature dependence of the Gibbs free energy of ice (Glushko 1978) can be approximated as follows:

$$\Delta G_{ice} = 0.0295 - 0.0413\Delta T + 6.64369\times10^{-5}(\Delta T)^2 + 2.27708\times10^{-8}(\Delta T)^3 \text{(kJ/mol)} \quad (10.10)$$

where $\Delta T = 273.16 - T$ at $T \le 273.15$ K, or by a simpler equation $\Delta G_{ice} = -0.036\Delta T$. The fact that pore water may remain unfrozen at $T < 273$ K suggests that its Gibbs free energy remains lower than that of bulk water or bulk ice $G_w^i < G_{ice}$ because of interaction with a solid surface. Further lowering of temperature reduces this inequality until the point T_c, at which a certain amount of bound water becomes frozen. At T_c, the relationship $\Delta G_w = \Delta G_{ice}$ pertains, where $\Delta G_w^i = G_w^i(T) - G_w^0$ (G_w^0 denotes the Gibbs free energy of unperturbed bulk water at 273.15 K and superscript i stands for interfaces). It is assumed that neither G_{ice}, nor ΔG_{ice}, are influenced by the solid surface. The area under the $\Delta G(C_{uw})$ curve (temperature dependences $\Delta G(T)$ and amounts of unfrozen pore water $C_{uw}(T)$ can be simply transformed into the relationship $\Delta G(C_{uw})$) determines overall changes in the Gibbs free energy of interfacial water:

$$\gamma_S = A\int_0^{C_{uw}^{max}} \Delta G dC_{uw} \quad (10.11)$$

where

C_{uw}^{max} is the total amount of unfrozen water at $T \rightarrow 273$ K
A is a constant dependent on the type of units used in this equation

Weakly bound water corresponding to a portion of unfrozen water with the free energy slightly lowered by intermolecular interactions with a solid surface freezes at temperature close to 273 K. Strongly bound water can be unfrozen even on significant cooling of the system and corresponds to the maximum disturbed water layer at the interfaces. The thickness of the layers of each type of water or water + organic solvent (e.g., benzene) (C_u^s and C_u^w denoting amounts of strongly and weakly bound liquids) and the maximum values for lowering of the Gibbs free energy of these liquids (ΔG_s and ΔG_w) can be estimated from a linear extrapolation of appropriate linear portions of the $\Delta G(C_u)$ graphs to the corresponding axis. Notice that there are two types of water (parameter subscript "w") and benzene (subscript "b"), chloroform-*d*, or another organics amounts: (i) unfrozen fraction ($C_u = C_{uw} + C_{ub}$) and (ii) total amounts of water (*h*), benzene (*b*), and chloroform-*d* (*c*) in milliliters per gram of dry silica. Derivative dC_u/dT (calculated using spline approximation of $C_u(T)$) as a function of temperature can be used to analyze characteristic changes in the amounts of structured unfrozen water and water–benzene dependent on temperature. A portion of unfrozen water or water–benzene corresponding to $\Delta G > -0.5$ kJ/mol may be considered as a liquid weakly bound in pores.

In this book, solvents (water and different organics) are described as probe compounds in the H form to analyze the interfacial phenomena or in the D form to avoid their signals in ^{1}H NMR spectra. However, there is a problem of solvent signal suppression when pure information on a material is needed but it is masked by the solvent signals. Zheng and Price (2010) analyzed solvent signal suppression techniques in NMR that attempt to meet new and increasingly stringent requirements (e.g., lower solute concentration, multiple signals, and greater selectivity). Of all the pulse sequence suppression methods, the frequency differentiation-based methods are the most versatile; and future developments will likely include the invention of novel frequency-selective RF pulses with large excitation bandwidths, sharper null regions, and immunity to nonideal conditions such as B_0, B_1, and T_1 inhomogeneity as well as to the effects of radiation damping. The performance of suppression methods is often illustrated on (biomolecular) samples with relatively high concentrations (e.g., mM range) under ideal conditions (e.g., starting from thermal equilibrium) while in reality the intended application is for samples at super low concentrations (e.g., $<\mu$M) as is commonly encountered, for example, ex vivo (e.g., NMR diffusion studies on peptide–membrane interactions); therefore, testing suppression performance on super low concentration samples under realistic conditions should become standard during the development of new suppression methods. As the availability of higher field clinical MRI equipment becomes more widespread in conjunction with the development of digitizers with higher resolution and more efficient methods for the suppression of sideband modulation; in vivo NMR experiments without water suppression will find more and more applications in medicine. This will also spur greater developments in post-acquisition data processing (Zheng and Price 2010). However, here the interfacial behavior of solvents is one of the main objects under investigations.

Water or other liquids can be frozen in narrower pores at lower temperatures that can be described by the Gibbs–Thomson relation for the freezing point depression:

$$\Delta T_m = T_m(R) - T_{m,\infty} = \frac{2\sigma_{sl} T_{m,\infty}}{\Delta H_f \rho R} = -\frac{k_{GT}}{R} \tag{10.12}$$

where

$T_m(R)$ is the melting temperature of a frozen liquid in pores of radius R
$T_{m,\infty}$ is the bulk melting temperature
ρ is the density of the solid
σ_{sl} is the energy of solid–liquid interaction
ΔH_f is the bulk enthalpy of fusion
k is a constant (Strange et al. 1993, 1996, 2003)

According to Aksnes and Kimtys (2004), $k_b = 44$ K·nm for benzene freezing in porous silicas. For pure water adsorbed in pores of silica, $k_w = 67$ K·nm. Notice that the k_{GT} values (in the 40–90 K·nm range) for certain systems were determined using condition $S_{uw} = S_{BET}$. Equation 10.12 can be utilized to calculate the incremental pore size distribution (IPSD) on the basis of the temperature dependence $C_u(T)$ estimated from the ^{1}H NMR spectra recorded at $T < T_{m,\infty}$ for individually adsorbed liquids. As shown in Chapters 1–9, this approach was employed for different materials and gave the IPSDs close to those calculated on the basis of nitrogen adsorption–desorption isotherms. In the case of a mixture of immiscible liquids (e.g., water and benzene or chloroform-*d*), Equation 10.12 can be used for both liquids with consideration for their amounts in a mixture and their temperature-dependent signal intensity (obtained on deconvolution of the total signal):

$$\frac{dV_{u,i}}{dR} = \frac{A_i}{k_{GTi}}(T - T_{m,\infty,i})^2 \frac{dC_{u,i}}{dT} \tag{10.13}$$

where

i denotes an adsorbate number

$C_{u,i}(T)$ is the integral intensity of a δ_H band for the *i*th adsorbate as a function of temperature

A_i is a weight constant dependent on the molecular volume (v_i), the number of protons (n_i) in a molecule of the *i*th adsorbate, and the used units

Notice that $v_w/n_w \approx v_b/n_b$ because of $v_b/v_w \approx 2.83$ and $n_b/n_w = 3$. In the case of the water–benzene mixtures, the $\sigma_{sl,w}$ value determined for the water–silica system was multiplied by $\Delta G_{s,h,b}/\Delta G_{s,h}$ (where $\Delta G_{s,h}$ and $\Delta G_{s,h,b}$ are changes in the free energy of strongly bound water without and with the presence of benzene, respectively) for consideration for the effect of benzene. Clearly, this effect leads to a certain diminution of the k_w value varied between 67 and 40 K·nm for different water–benzene mixtures. For simplicity, the k_b value was assumed as a constant (44 K·nm) because benzene is being in contact with the pore walls since it can displace water from pores and the pore walls. Equations 10.12 and 10.13 allow us to calculate the distribution functions of the sizes of pores filled by an individual adsorbate (e.g., water) and an incremental size distribution (ISD) of structures in a liquid mixture (e.g., water–benzene)–filling pores. In the last case, the pore size can correspond to a sum of the sizes (thickness) of the pore liquid layers if liquids simultaneously locate in the same pores. In the case of a mixture of water with chloroform-*d*, the latter does not contribute the ^{1}H NMR spectra. However, chloroform-*d* can fill a portion of pores and affect the occupation of them by water, i.e., change the $C_{uw}(T)$ shape. Equations 10.12 and 10.13 can be transformed into integral equation (IGT), replacing dV/dR by $f(R)$, converting dC/dT to dC/dR with Equation 10.12 assuming that $dV_{uw}(R)/dR = dC_{uw}/dR$ and using Equation 10.12 $dT_m(R)/dR = k_{GT}/R^2 = (T_m(R) - T_{m,\infty})^2/k_{GT}$ that give $dV_{uw}(R)/dR = (A/k_{GT})(T_m(R) - T_{m,\infty})^2 dC_{uw}(T)/dT$, where $V_{uw}(R)$ is the volume of unfrozen water in pores of radius R, C_{uw} the amount of unfrozen water per gram of adsorbent as a function of temperature, and A is a constant. This equation can be transformed into integral equation on replacing $dV_{uw}(R)/dR$ by $f_V(R)$, converting dC_{uw}/dT to dC_{uw}/dR as $dT = dT/dR\ dR$ and integrating by R,

$$C_{u,i}(T) = A \int_{R_{min}}^{R_{max}} \left(\frac{k_{GTi}}{(T_{m,\infty,i} - T_{m,i}(R))R} \right)^2 f_i(R) dR \tag{10.14}$$

where

R_{max} and R_{min} are the maximal and minimal pore radii (or sizes of unfrozen liquid structures) respectively

i is the index corresponding to *i*th pore liquid

A is a normalization factor

In the case of the use of the relaxometry method for estimation of the sizes of the water and benzene structures (layers, clusters, etc.) filling pores, Equation 10.9 can be transformed for consideration for the dependence of the CPMG echo decay envelopes (i.e., transverse relaxation time) on the pore size:

$$I_i(t) = B\int_{R_{\min}}^{R_{\max}} \exp\left(-\frac{t}{T_{2i,\mathrm{m}}}\right)\frac{(T_{\mathrm{m},\infty,i} - T_{\mathrm{m},i}(R))}{k_{\mathrm{GT}i}} f_i(R)dR \tag{10.15}$$

where

B is a normalization factor

$k_{\mathrm{GT}i}$ is a constant analogous to that in Equations 10.12 and 10.13

Equations 10.14 and 10.15 can be solved using the regularization procedure similar to that applied to Equation 10.9.

To more accurately calculate the distribution function on the basis of the IGT equation, an additional regularizer was derived using the maximum entropy principle (Muniz et al. 2000) applied to $f_V(R)$ written as N-dimension vector (N is the number of the grid points for f):

$$\mathrm{VAR} + \alpha^2\left(1 - \frac{S(\vec{p}(\vec{f}))}{S_{\max}}\right) \to \min \tag{10.16}$$

where

VAR is the regularizer

α the regularization parameter

S is the entropy

$$\vec{p}^{\,0}(\vec{f}) = \vec{f},\; p_i^1(\vec{f}) = f_{i+1} - f_i + (f_{\max} - f_{\min}); i = 1,\ldots,N-1$$

$$p_i^2(\vec{f}) = f_{i+1} - 2f_i + f_{i-1} + 2(f_{\max} - f_{\min}),\; i = 2,\ldots,N-1$$

$$S(\vec{f}) = -\sum_{k=1}^{N} s_k \ln(s_k),\; s_k = \frac{f_k}{\sum_{k=1}^{N} f_k}, \quad \text{and} \quad S_{\max} = -\ln\left(\frac{1}{N}\right)$$

The $\vec{p}^{\,j}(\vec{f})$ vector corresponds to the maximum entropy principle of the j-order (Muniz et al. 2000). This procedure was used to modify the CONTIN algorithm (CONTIN/MEM-j where j denotes the order of $\vec{p}^{\,j}(\vec{f})$). A self-consistent regularization procedure (starting calculations were done without application of MEM) with an unfixed regularization parameter (for better fitting) was used on CONTIN/MEM-j calculations. This procedure was also applied to the nitrogen adsorption data with the overall adsorption equations described later.

Equation 10.14 (and 10.15) can be used to calculate the PSD $f(R)$ from the amount of unfrozen water as a function of temperature $C_{\mathrm{uw}}(T)$. The $f_V(R)$ function can be converted into the distribution function $f_S(R)$ with respect to the specific surface area in contact with unfrozen water or other liquids:

$$f_S(R) = \frac{w}{R}\left(f_V(R) - \frac{V(R)}{R}\right) \tag{10.17}$$

where $w = 1$, 2, and 1.36 for slit-shaped, cylindrical pores and voids between spherical particles packed in the cubic lattice, respectively. Integration of the $f_S(R)$ function determined with Equation 10.17, on the basis of the IGT equation, gives the specific surface area (S_{IGT}) of the studied materials in contact with structured water. Integration of the $f_V(R)$ and $f_S(R)$ functions at $R < 1$ nm, $1 < R < 25$ nm, and $R > 25$ nm gives the volume and the specific surface area of nanopores, mesopores, and macropores. The specific surface area (S_{uw}) of adsorbents in contact with bound water (assuming for simplicity that the density of unfrozen bound water $\rho_{uw} = 1$ g/cm^3) can be determined from the amount of this water C_{uw}^{max} (estimating pore volume as $V_{uw} = C_{uw}^{max}/\rho_{uw}$) at $T = 273.15$ K and PSD $f(R)$ (used to estimate the average pore radius R_{av}) with a model of cylindrical pores:

$$S_{w,sum} = \frac{V_{uw}}{2R_{av}} = \frac{2C_{uw}^{max}}{\rho_{uw}} \frac{\int_{R_{min}}^{R_{max}} f(R)dR}{\int_{R_{min}}^{R_{max}} f(R)RdR} \tag{10.18}$$

where R_{min} and R_{max} are the minimal and maximal radii of pores filled by unfrozen water, respectively. In the case of calculations of the structural characteristics of nanopores ($R < 1$ nm), mesopores ($1 < R < 25$ nm), and macropores ($R > 25$ nm), the R_{min} and R_{max} values are the boundary R values for the corresponding pore types (including $R_{min} = 0.2$ nm for nanopores), and the C_{uw}^{max}/ρ_{uw} value should be replaced by the corresponding values of the volumes of nanopores, mesopores, or macropores. A limitation of the use of Equation 10.18 is due to possible incomplete filling of pores by water because air bubbles can remain in pores. However, this effect can be estimated from the difference between expected and determined densities of a material with empty and filled pores.

Aksnes and Kimtys (2004) proposed to calculate the distribution function of the pore size $f(R)$ as follows:

$$f(R) = \frac{10^3 k}{\sqrt{2\pi}(RT_{m,\infty} - k)^2} \sum_{i=1}^{N} \frac{I_{0,i}}{\sigma_i} \exp\left[-\left(\frac{10^3 R - X_{ci}(RT_{m,\infty} - k)}{\sqrt{2}\sigma_i(RT_{m,\infty} - k)}\right)^2\right] \tag{10.19}$$

where

$X = 10^3/T$

X_{ci} is the normalized inverse transition temperature of phase i

$I_{0,i}$ and σ_i are the intensity and the width of the temperature distribution curve of phase i

Equation 10.19 was used to calculate the $f(R)$ distribution functions for the aqueous suspensions of fumed silica, hydrated powders, and hydrogels with starch or starch/functionalized nanosilica.

Relaxation measurements of adsorbates bound to carbons were performed at temperatures 180–250 K (lower than in the case of oxide adsorbents because of a larger contribution of nanopores in carbons), which are below the freezing point of pure bulk water and water adsorbed in macropores and distant from the pore walls (this weakly bound water is frozen at T close to 273 K). Changes in the Gibbs free energy (ΔG) of unfrozen interfacial water were calculated (with relative mean error ±15%) using the known dependence of change in the Gibbs free energy of ice ΔG_{ice} on temperature T (Equation 10.10). The fact that at temperatures below 273 K surface bound water may remain unfrozen suggests that the Gibbs free energy of the interfacial water remains lower than that of bulk water and bulk ice due to the interaction with the solid surface or biomacromolecules:

$$G_w^i < G_{ice} \tag{10.20}$$

Further lowering of the temperature reduces this inequality until the point T_c, at which certain amount of bound water becomes frozen. At T_c,

$$\Delta G_w = \Delta G_{ice} \tag{10.21}$$

where

$$\Delta G_w^i = G_w^i(T) - G_w^0 \tag{10.22}$$

G_w^0 denotes the Gibbs free energy of unperturbed bulk water at 273 K and superscript i stands for interface, and

$$\Delta G_{ice} = G_{ice}(T) - G_{ice}^0 \tag{10.23}$$

G_{ice}^0 is the Gibbs free energy of ice at 273 K. It is assumed that neither G_{ice} nor ΔG_{ice} is influenced by the solid surface.

The water molecules inside pores can form clusters, in which the average number of hydrogen bonds per molecule varies between 0.5 (for dimers adsorbed on nonpolar functionalities due to dispersion interaction) and approximately 3 (tiny droplets in pores). Mobility of water molecules, particularly their rotational characteristics in liquid water at interfaces, is influenced by the number of hydrogen bonds. The proton exchange between these clusters can be represented by the following scheme:

$$(H_2O)_A \leftrightarrow (H_2O)^* \leftrightarrow (H_2O)_B \tag{10.24}$$

where a proton exchanges between cluster A and cluster B at a surface through the transition state $(H_2O)^*$. Assuming "unimolecular" mechanism of such an exchange, the Arrhenius-type equation can be applied:

$$(T_2)^{-1} = (T_2)_0^{-1}\exp\left(-\frac{\Delta G_a}{RT}\right) \tag{10.25}$$

where $(T_2)_0^{-1}$ is the pre-exponential factor dependent on temperature and at high temperatures $T_2 \rightarrow T_{2,0}$. Rearranging Equation 10.25 and taking into account that the Gibbs free energy of activation is

$$\Delta G_a = \Delta H^{\neq} - T\Delta S^{\neq} \tag{10.26}$$

the thermodynamic parameters of the exchange processes can be estimated.

The distribution function of the activation free energy of relaxation $f(\Delta G_a)$ can be calculated using Equation 10.25 as the kernel of Fredholm integral equation of the first kind:

$$k(T) = \int_0^{\Delta G_{max}} k'(T, \Delta G_a) f(\Delta G_a) d(\Delta G_a) \tag{10.27}$$

which is solved using the regularization method under condition of non-negativity ($f(\Delta G_a) \geq 0$ at any ΔG_a) at fixed regularization parameter ($\alpha = 0.01$).

According to Wiench et al. (2008), a remarkable enhancement of sensitivity can be achieved in ^{29}Si solid-state NMR by applying the Carr–Purcell–Meiboom–Gill (CPMG) train of rotor-synchronized π pulses during the detection of Si magnetization. They used several one- and two-dimensional (1D and 2D) techniques to demonstrate the capabilities of this approach. Examples include 1D ^{29}Si{X} CPMAS spectra and 2D ^{29}Si{X} HETCOR spectra of mesoporous silicas, zeolites, and minerals, where X = ^{1}H or ^{27}Al. Wiench et al. (2008) discussed data processing methods, experimental strategies, and sensitivity limits and illustrated by experiments. The mechanisms of transverse dephasing of ^{29}Si nuclei in solids were analyzed. Fast magic angle spinning, at rates between 25 and 40 kHz, was instrumental in achieving the highest sensitivity gain in some of these experiments. In the case of ^{29}Si–^{29}Si double-quantum (DQ) techniques, CPMG detection can be exploited to measure homonuclear J-couplings (Wiench et al. 2008).

^{29}Si is one of the most widely studied nuclei in solid-state NMR spectroscopy applied to silicates (Engelhardt and Michel 1987, Wiench et al. 2008), silica surfaces (Fitzgerald 1999), silicon alloys and ceramics (Grant and Harris 1996), glasses (Eckert 1992), and other crystalline and amorphous materials (Takeuchi and Takayama 1998, Kuroki et al. 2004). 1D measurements with direct ^{29}Si polarization (DP) or ^{29}Si{^{1}H} cross polarization (CP), magic angle spinning (MAS), and ^{1}H decoupling using a radiofrequency (RF) magnetic field were predominated. However, 2D methods have also been employed, e.g., in CP-based ^{29}Si{^{1}H} heteronuclear (HETCOR) NMR studies of organic–inorganic nanocomposites (Melosh et al. 1999, Christiansen et al. 2001, Brus et al. 2004, Baccile et al. 2007), zeolites (Vega 1988), silica (Fyfe et al. 1992), and mesoporous aluminosilicates (Janicke et al. 1998). 2D methods were used to obtain ^{29}Si–^{29}Si correlation spectra that exploit DQ through-bond (scalar) or through-space (dipolar) interactions (Fyfe et al. 1990b, Hedin et al. 2004, Brouwer et al. 2005, Wiench et al. 2008).

However, wider exploitation of multidimensional solid-state NMR methods is hindered by low ^{29}Si sensitivity, which results from low natural abundance of spin-1/2 isotope (4.7%), small gyromagnetic ratio, and unusually slow longitudinal relaxation. Fortunately, the presence of a long relaxation time T_1 in inorganic solids is often accompanied by a long transverse dephasing (decoherence) time T_2^{CPMG}, defined here as the decay time of the echo train due to time-dependent interactions that are non-refocusable by the CPMG sequence (Meiboom and Gill 1958) of π pulses combined with MAS and/or RF decoupling schemes (Wiench et al. 2008). This makes it possible to detect the signal multiple times at intervals that are longer than the free induction decay (FID) observed following the single-pulse excitation (often denoted as T_2^*) but much shorter than T_1. The multiple pulse CPMG sequence, which refocuses inhomogeneous line broadening and reduces homonuclear dipolar broadening in solids, was first introduced to measure the decay of transverse nuclear magnetization (Ostroff and Waugh 1966, Garroway 1977). It later found applications in various areas of magnetic resonance spectroscopy, including experiments with field gradients or strongly inhomogeneous static fields (e.g., imaging [Maudsley 1986, Perlo et al. 2004, Ahola et al. 2006] and diffusion measurements [Pavesi and Balzarini 1996, Zielinski and Huerlimann 2005, Rata et al. 2006]), high-resolution liquid-state NMR (solvent peak suppression) (Bryant and Eads 1985, Magnuson and Fung 1992), HETCOR experiments (Luy and Marino 2001, Koskela et al. 2003, Koever et al. 2006), and exchange studies (Frahm 1982), homonuclear distance measurements in solids (Molitor et al. 1989, Nordon et al. 1998), and electron spin resonance (ESR) (Harbridge et al. 2003, Witzel and Das Sarma 2007). In solid-state NMR applications, the most appealing advantage offered by the CPMG pulse sequence is the sensitivity gain resulting from refocusing the magnetization multiple times. Indeed, enhancement of the signal-to-noise (S/N) ratio has been demonstrated in various studies where low natural abundance, a low gyromagnetic ratio (γ), long relaxation time T_1, large line width, or any combination of these factors rendered "standard" averaging of FIDs impractical. These studies used static (Cheng and Ellis 1989, Bryce et al. 2001, Lipton et al. 2001, Schurko et al. 2003, Hung et al. 2004, Siegel et al. 2004), MAS (Larsen et al. 1998), MQMAS (Vosegaard et al. 1997, Lefort et al. 2002), and PHORMAT (Hu and Wind 2003) experiments with spin-1/2 and quadrupolar nuclei. In spite of long decoherence times of silicon nuclei, the CPMG technique has been used only rarely in ^{29}Si NMR. The strategy was applied in a study of silicates (α-crystobalite and Zircon), where the sensitivity of ^{29}Si MAS spectra was increased by an order of magnitude using a rotor-synchronized CPMG sequence (Larsen and Farnan 2002) and heteronuclear correlation (HETCOR) experiments on clay minerals, where the ^{29}Si signal was stretched by means of multiple echoes at the expense of the chemical shift information (Hou et al. 2002). Highly resolved ^{29}Si{^{1}H} HETCOR

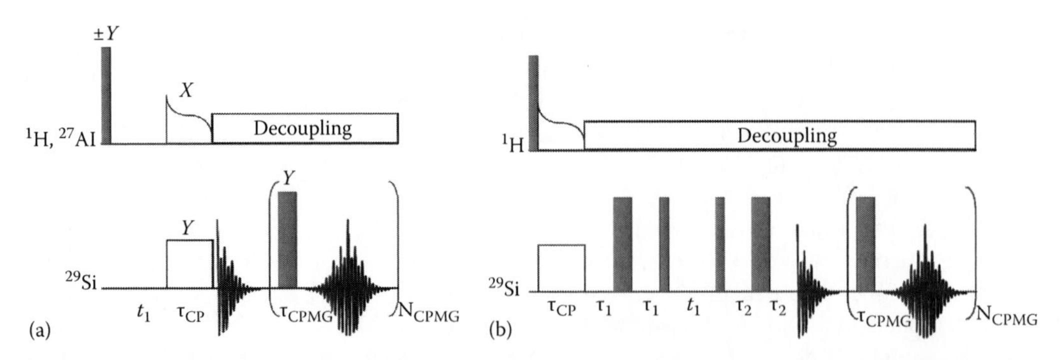

FIGURE 10.1 Pulse sequences for 2D ^{29}Si NMR experiments with CPMG detection: (a) ^{29}Si{^{1}H} and ^{29}Si{^{27}Al} HETCOR, (b) ^{29}Si–^{29}Si{^{1}H} refocused INADEQUATE. Solid rectangles represent p and $p/2$ pulses, whereas open pulses represent cross polarization and decoupling. In sequence (a), the phases are the same as in Trebosc et al. (2005b), with the receiver phase being inverted in concert with the phase of $\pi/2$ pulse in the ^{1}H channel. In sequence (b), the phase cycling during INADEQUATE is the same as used by Lesage et al. (1999). The phases of π pulses used in the CPMG sequence followed the phase of the last (refocusing) pulse in INADEQUATE. In addition, the hypercomplex method was used to achieve quadrature detection in the ν_1 dimension in all 2D experiments. (Adapted from *J. Magn. Reson.*, 193, Wiench, J.W., Lin, V.S.-Y., and Pruski, M., ^{29}Si NMR in solid state with CPMG acquisition under MAS, 233–242, Copyright (2008), with permission from Elsevier.)

NMR spectra of silica-based catalytic systems were obtained with excellent sensitivity using multiple pulse CPMG refocusing and appropriate data processing (Trebosc et al. 2005a,b, Wiench et al. 2007a,b). These experiments were made possible by the use of fast MAS, at rates of 25–40 kHz, which provided adequate ^{1}H–^{1}H decoupling. More importantly, fast MAS allowed the heteronuclear decoupling to be performed using low RF power, which was essential in these experiments due to unusually long acquisition periods (up to several seconds). Herein, we analyze the utility of CPMG refocusing in a number ^{29}Si NMR techniques, and discuss the experimental strategies that maximize sensitivity, as well as the future applications of these methods. Examples of 1D ^{29}Si{X} CPMAS spectra, 2D ^{29}Si{X} HETCOR (X = ^{1}H and ^{27}Al) spectra and ^{29}Si–^{29}Si DQMAS homonuclear correlation spectra are shown. This effort was largely motivated by interest in studying the mesoporous catalytic materials, which is reflected in the choice of samples for this study (Wiench et al. 2008). Figure 10.1 describes the pulse sequence used in 2D ^{29}Si{X} experiments (Wiench et al. 2008).

This sequence was used in 1D CPMAS experiments by setting $t_1 = 0$. The DQMAS homonuclear correlation experiments (Figure 10.1b) utilized the ^{29}Si{^{1}H} CP (Figure 10.2) followed by the INADEQUATE scheme for excitation and reconversion of DQ coherences (Bax et al. 1980, Fyfe et al. 1990a, Lesage et al. 1999). The π pulses were always applied synchronously with the rotor (Wiench et al. 2008).

In NMR, the resonance frequency of a certain isotope is directly proportional to the strength of the applied magnetic field. Therefore, the chemical shift of this resonance caused by local surrounding affecting electron shielding of nuclei and measured in parts per million (ppm) is field-independent parameter. For instance, the chemical shift of the proton resonance, δ_H of liquid water and ice is equal to approximately 5 and 7 ppm, respectively, measured using both a Bruker 100 (100 MHz) and a Bruker 1000 (1 GHz) spectrometer. Of course, new equipment such as 1 GHz NMR Bruker Avance 1000 Spectrometer (23.5 T) and 900 MHz Varian (21.2 T) is characterized by much higher resolution, which is very important for identification of NMR spectra to describe chemical structure of very complex biomacromolecules. The mentioned feature of NMR is used in MRI with a nonuniform magnetic field. The NMR efficiency can be improved using hyperpolarization, two-, tri-, or higher-dimensional multifrequency techniques. Some details of NMR techniques are described in Chapters 1–8 where applications of these methods are shown.

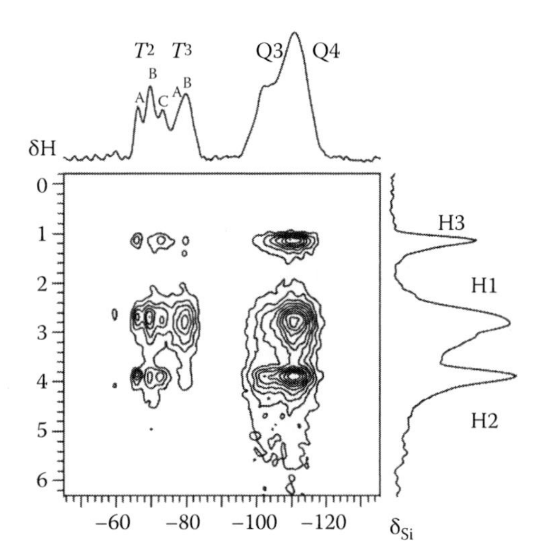

FIGURE 10.2 $^{29}Si\{^{1}H\}$ experiments on CMTES-MSN. $^{29}Si\{^{1}H\}$ HETCOR: $B_0 = 14.1$ T, $\nu_R = 25$ kHz, $\tau_{CP} = 8$ ms, $\nu_{RF}^{H} = 104$ kHz during excitation, $\nu_{RF}^{H} = 30 \pm 3$ kHz during CP (ramped), $\nu_{RF}^{H} = 6$ kHz during TPPM decoupling, $\nu_{RF}^{H} = 55$ kHz during CP and CPMG pulses, $N_{CPMG} = 160$, $\tau_{CPMG} = 6$ ms, $\tau_{RD} = 1$ s, and NS = 80; and 100 increments of 80 μs were used in the t_1 dimension. The total experimental time was 9 h. (Adapted from *J. Magn. Reson.*, 193, Wiench, J.W., Lin, V.S.-Y., and Pruski, M., ^{29}Si NMR in solid state with CPMG acquisition under MAS, 233–242, Copyright (2008), with permission from Elsevier.)

10.2 LOW-TEMPERATURE NITROGEN ADSORPTION

To calculate the textural characteristics of materials, the nitrogen adsorption–desorption isotherms were recorded at 77.4 K using Micromeritics ASAP 2010 or 2405N or Quantachrome Autosorb adsorption analyzers. The specific surface area (S_{BET}) was calculated according to the standard BET method (Gregg and Sing 1982). The total pore volume V_p was evaluated from the nitrogen adsorption at $p/p_0 = 0.98$–0.99 (p and p_0 denote the equilibrium pressure and the saturation pressure of nitrogen at 77.4 K, respectively). The nitrogen desorption data (Nguyen and Do 1999) were used to compute PSDs (differential $f_V(R_p) \sim dV_p/dR_p$ and $f_S(R_p) \sim dS/dR_p$) using the overall isotherm equation:

$$a = \int_{r_{min}}^{r_k(p)} f(R_p)dR_p + \int_{r_k(p)}^{r_{max}} \frac{w}{R_p} t(p, R_p) f(R_p) dR_p \tag{10.28}$$

where

r_{min} and r_{max} are the minimal and maximal half-width or pore radius, respectively
$w = 1$ (slit-like pores), 2 (cylindrical pores), and 1.36 (voids between spherical particles packed in a cubic lattice)

$$r_k(p) = t(p, R_p) + \frac{2\gamma v_m \cos\theta}{R_g T \ln(p_0/p)} \tag{10.29}$$

$$t(p, R_p) = \frac{a_m}{S_{BET}} \frac{cz}{(1-z)} \frac{[1 + (nb/2 - n/2)z^{n-1} - (nb+1)z^n + (nb/2 + n/2)z^{n+1}]}{[1 + (c-1)z + (cb/2 - c/2)z^n - (cb/2 + c/2)z^{n+1}]} \tag{10.30}$$

where

$b = \exp(\Delta\varepsilon/R_gT)$

$\Delta\varepsilon$ is the excess of the evaporation heat due to the interference of the layering on the opposite wall of pores (determined as a varied parameter using local isotherm approximation [LIA])

$t(p, R_p)$ is the statistical thickness of adsorbed layer

a_m is the BET monolayer capacity

$c = c_s \exp((Q_p - Q_s)/R_gT)$

c_s is the BET coefficient for adsorption on flat surface (calculated using LIA)

Q_s and Q_p are the adsorption heat on flat surface and in pores, respectively

$z = p/p_0$

$n = R_p/t_m$ is the number (noninteger) of statistical monolayers of adsorbate molecules

Desorption data were utilized to compute the $f(R_p)$ distribution with Equation 10.28 using a regularization procedure (Provencher 1982) under non-negativity condition ($f_V(R_p) \geq 0$ at any R_p), at a fixed regularization parameter $\alpha = 0.01$. A model of a pore mixture with slit-shaped and cylindrical pores and voids between spherical nanoparticles packed in random aggregates (SCV model, MND method) was developed (Gun'ko 2000b) because the standard pore models with slit-shaped or cylindrical pores are inappropriate to describe the textural porosity of nanosilicas. The self-consistent regularization (SCR) procedure was used to calculate the PSDs and to determine contributions of different types of pores to the pore volume (Gun'ko et al. 2009c,d). The $f_V(R_p)$ functions linked to pore volume can be transformed to the $f_S(R_p)$ distribution functions with respect to surface area using the corresponding pore models:

$$f_S(R_p) = \frac{w}{R_p}\left(f_V(R_p) - \frac{V_p}{R_p}\right) \tag{10.31}$$

where $w = 1$, 2, 3, and 1.36 for slit-like, cylindrical, spherical pores, and gaps between spherical particles packed in the cubic lattice. The $f_V(R_p)$ and $f_S(R_p)$ functions were used to calculate contributions of nanopores (V_{nano} and S_{nano} at the pore radius $R_p < 1$ nm), mesopores (V_{meso} and S_{meso} at $1 \text{ nm} \leq R_p \leq 25$ nm), and macropores (V_{macro} and S_{macro} at $R_p > 25$ nm) to the total pore volume and the specific surface area. The values of S_{nano}, S_{meso}, and S_{macro} were corrected that $S_{nano} + S_{meso} + S_{macro} = S_{BET}$. For estimation of deviation of the pore shape from the model, a criterion $\Delta w = S_{BET}/S_{sum} - 1$ (Gun'ko and Mikhalovsky 2004) where

$$S_{sum} = \sum_j c_j \int_{R_{min}}^{R_{max}} f_{S,j}(R_p)dR_p = \sum_j c_j \int_{R_{min}}^{R_{max}} \frac{w_j}{R_p}\left(f_{V,j}(R_p) - \frac{V_{p,j}}{R_p}\right)dR_p \tag{10.32}$$

where

c_j is the weight coefficient

$R_{min} = 0.3$ nm

$R_{max} = 100$ nm

j denotes a pore model

$f_{S,j}(R_p)$ is the differential distribution function with respect to the specific surface area of pores of the jth type

The c_j values were determined for the best fitting of the experimental isotherms using self-consistent regularization procedure with several steps in solution of the sum of integral equations:

$$a_\Sigma = \sum_j c_j a_j = \sum_j c_j \left[\int_{r_{min}}^{r_{k,j}(p)} f_{V,j}(R_p)dR_p + \int_{r_{k,j}(p)}^{r_{max,j}} \frac{w_j}{R_p - \sigma_{sf}/2} t_j(p, R_p) f_{V,j}(R_p)dR_p\right] \tag{10.33}$$

where

a_Σ and a_j are the total adsorption and adsorption in pores of a certain shape

$\sigma_{sf}=(\sigma_s+\sigma_f)/2$ is the average collision diameter of surface and fluid atoms

$r_{k,j}$ is determined through modified Kelvin equation including the thickness of the adsorbate layer t_j

Lennard–Jones (LD) potential was used for cylindrical and slitshaped pores and gaps between spherical particles. This method gives better results than that used previously for complex adsorbents with one distribution function for all the pore shape models because of the use of independent $f_{V,j}(R_p)$ functions for each pore type. For a pictorial presentation of the PSD, the $f_V(R_p)$ functions were recalculated to incremental PSDs (IPSDs):

$$\Phi_V(R_{p,i}) = 0.5(f_V(R_{p,i}) + f_V(R_{p,i-1}))\,(R_{p,i} - R_{p,i-1}) \tag{10.34}$$

Additionally, $f_S(R)$ was used to estimate the deviation (Δw) of the pore shape from the model using a self-consistent (to better fit the nitrogen adsorption isotherms) regularization in the case of the use of a complex model of the pore shape (Gun'ko and Mikhalovsky 2004):

$$\Delta w = \frac{S_{BET}}{\int_{R_{min}}^{R_{max}} f_S(R)dR} - 1 \tag{10.35}$$

where R_{max} and R_{min} are the maximal and minimal pore radii respectively. The S^*_{nano}, S^*_{meso}, and S^*_{macro} values were corrected by multiplication by ($\Delta w+1$) that gives $S^*(\Delta w+1)=S_{sum}=S_{nano}+S_{meso}+S_{macro}=S_{BET}$. The effective w value (w_{ef}) can be estimated with equation:

$$w_{ef} = \frac{S_{BET}}{V_p}\frac{\int_{R_{min}}^{R_{max}} Rf_V(R)dR}{\int_{R_{min}}^{R_{max}} f_V(R)dR} \tag{10.36}$$

The specific surface area (S_ϕ) of materials composed of spherical nanoparticles characterized by the particle size distribution $\phi(a)$ (calculated using the self-consistent regularization for $f_V(R)$ and $\phi(a)$ [normalized to 1] with the model of voids between spherical particles) can be calculated with equation:

$$S_\phi = \int_{a_{min}}^{a_{max}} \frac{3}{2a^3\rho}\left[2(a+t)^2 + Nr_m \arcsin\left(\frac{a}{A}\right)\sqrt{A^2-a^2} - N(a+t)\left(\frac{ar_m}{A}+t\right)\right]\phi(a)da \tag{10.37}$$

where

$A=a+t+r_m$

a is the particle radius

ρ is the density of material

N is the average coordination number of nanoparticles in aggregates

t is the thickness of an adsorbed nitrogen layer

r_m is the meniscus radius determined at $0.05<p/p_0<0.2$ corresponding to the effective radius R of voids between spherical particles

Condition $S_\phi = S_{BET}$ can be used to estimate the N value. An additional criterion $|\langle S_\phi \rangle - S_{BET}| < 1$ m^2/g was used to determine the a_{min} and a_{max} values for the $\phi(a)$ distributions calculated at $p/p_0 < 0.5$ (i.e., before capillary condensation starts) with

$$\langle S_\phi \rangle = \frac{\iint S_\phi(r_m, t)\, dt dr_m}{\iint dt dr_m} \tag{10.38}$$

The PSDs were also calculated using overall equation (with the framework of density functional theory, DFT) (Do et al. 2001, Gun'ko et al. 2007f):

$$W(p) = v_M \left[\int_{\sigma_{ss}/2}^{r_k(p)} \rho_f(R) f(R) dR + \int_{r_k(p)}^{R_{max}} \frac{t}{R - \sigma_{ss}/2} \rho_m(R) f(R) dR \right] \tag{10.39}$$

where

W is the adsorption

v_M is the liquid molar volume

ρ_f is the fluid density in occupied pores

ρ_m is the density of the multilayered adsorbate in pores

r_k is the radius of pores occupied at the pressure p

σ_{ss} is the collision diameter of the surface atoms

To calculate the density of a gaseous adsorbate (nitrogen) at a given pressure p, Bender equation (Platzer and Maurer 1993) was used:

$$p = \rho T[R_g + B\rho + C\rho^2 + D\rho^3 + E\rho^4 + F\rho^5 + (G + H\rho^2)\rho^2 \exp(-a_{20}\rho^2)] \tag{10.40}$$

where

$B = a_1 - a_2/T - a_3/T^2 - a_4/T^3 - a_5/T^4$; $C = a_6 + a_7/T + a_8/T^2$; $D = a_9 + a_{10}/T$; $E = a_{11} + a_{12}/T$; $F = a_{13}/T$; $G = a_{14}/T^3 + a_{15}/T^4 + a_{16}/T^5$; $H = a_{17}/T^3 + a_{18}/T^4 + a_{19}/T^5$; $a_{20} = \rho_c^{-2}$; a_i are constants

R_g is the gas constant

Transition from gas (subscript g) to liquid (l) or fluid in the form of multilayered adsorbate in pores (m) can be linked to the corresponding fugacity f:

$$\ln \frac{f(T,\rho)}{R_g T \rho} = \frac{p(T,\rho)}{R_g T \rho} - 1 + \frac{1}{R_g T} \int_0^\rho [p(T,\rho) - R_g T \rho] \frac{d\rho}{\rho^2} \tag{10.41}$$

and

$$f_{l,m} = f_g \exp\left(\frac{E_{i,m}}{RT} \right) \tag{10.42}$$

where E is the interaction energy of an adsorbate molecule with the pore walls and neighboring molecules calculated with the LD potentials. The CONTIN/MEM procedure was applied to solve Equation 10.39. There are certain limitations of the use of Equation 10.40 described by Platzer and Maurer (1993). Therefore, the minimal pressure used with Equation 10.40 should correspond to condition $p/p_0 > 0.01$. Certain PSDs were calculated using nonlocal DFT, NLDFT method (with the model of slit-shaper, silt-shaped/cylindrical, or cylindrical pores in carbons or silicas) and quenched solid DFT, QSDFT method (slit-shaped pore model for carbons) (Quantachrome software).

To calculate the adsorption energy distribution functions, the Fowler–Guggenheim (FG) equation was used to describe localized monolayer adsorption with lateral interactions:

$$\theta_i(p,E) = \frac{Kp\exp(zw\Theta)/k_BT}{1+Kp\exp(zw\Theta)/k_BT} \tag{10.43}$$

where

$K = K_0(T)\exp(E/k_BT)$ is the Langmuir constant for adsorption on energetically uniformed sites and the pre-exponential factor $K_0(T)$ is expressed in terms of the partition functions for isolated gas and surface phases
z is the number of nearest neighbors of an adsorbate molecule (assuming $z=4$)
w is the interaction energy between a pair of nearest neighbors
k_B is the Boltzmann constant, e.g., $zw/k_B = 380\,K$ for nitrogen (Jaroniec and Madey 1988)

The right term of Equation 10.43 was used as the kernel in the overall adsorption isotherm equation to calculate the distribution function $f(E)$ of the nitrogen adsorption energy.

Calculation of the fractal dimension (D_{AJ}) was performed on the basis of the nitrogen adsorption data using equation

$$\ln(\Theta) = \text{const} + (D_{AJ} - 3)\left[\ln\ln\left(\frac{p_0}{p}\right)\right] \tag{10.44}$$

where Θ denotes the relative adsorption a/a_m (a_m is the volume of adsorbed gas for the monolayer coverage calculated with the BET method), at $p/p_0 \le 0.85$. Additionally, the adsorption isotherm as a fractal analog of the Dubinin–Astakhov equation

$$\Theta = \frac{\rho}{n}\left[\gamma\left(\frac{3-D}{n}, x_{max}^n\,\mu A^n\right) - \gamma\left(\frac{3-D}{n}, x_{min}^n\,\mu A^n\right)\right]\mu^{(D-3)/n}A^{D-3} \tag{10.45}$$

where

$\rho = (3-D)/\left(x_{max}^{3-D} - x_{min}^{3-D}\right)$
$\mu = (k\beta)^{-n}$
n is the varied equation parameter
x_{max} and x_{min} are the maximal and minimal half-widths of pores
$A = R_gT\ln(p_0/p)$ is the differential molar work equal (with inverse sign) to the variation in the Gibbs free energy
k is the constant
γ denotes the incomplete gamma function

was utilized to estimate fractal dimension D_{FRDA} at $p/p_0 \le 0.1$, $x_{min} = 0.2\,\text{nm}$ and $x_{max} = 4.0\,\text{nm}$.

The adsorption potential (U_0) and free energy changes (ΔG) upon water adsorption onto the silica surface from air can be calculated using the Langmuir equation:

$$\Theta = \frac{bC}{1+bC} \tag{10.46}$$

where $b = \gamma_L e^{-\Delta G/R_g T}$ or $K_L e^{U_0/R_g T}$, γ_L and K_L are the constants, using the minimization of the discrepancy functional.

Instead of the average values determined with Equation 10.46, the distribution functions of the energetic parameters can be calculated using Fredholm integral equation of the first kind with the kernel Θ_l similar to the right term in Equation 10.46:

$$\Theta(T,p) = \int_{x_{\min}}^{x_{\max}} \Theta_l(T,p,x) f(x)dx \tag{10.47}$$

where $f(x)$ is the unknown distribution function of a given parameter x. To calculate the $f(x)$ function, the regularization method can be used, as solution of Equation 10.47 is well-known ill-posed problem due to a strong influence of noise components on experimental data, which do not allow one to effectively utilize exact inversion formulas or iterative algorithms (Provencher 1982). Additionally, to calculate the water adsorption energy distributions $f(E)$, modified Equation 10.30 can be applied at $n \approx 2$–3. Equation 10.43 can also be applied to compute $f(E)$ at $a < a_m$ assuming that lateral interactions at coverage less than monolayer can be smaller than the latent heat of bulk water condensation (45 kJ/mol; Brennan et al. 2001). Another approach based on the model of clustered adsorption of water can be used to calculate the $f(E)$ function.

10.3 ADSORPTION OF WATER AND ORGANICS

Water adsorption onto a set of samples was studied using an adsorption apparatus with a McBain–Bark quartz scale. After evacuation to 10^{-3} Torr for 1–2 h, samples were heated at 613 K for 3–4 h to a constant weight, then cooled to 293 ± 0.2 K, and adsorption of water vapor was studied at pressure (p) varied in the 0.06–0.999 p/p_0 range. The measurement accuracy was 1×10^{-3} mg with relative mean error $<\pm 5\%$.

The water adsorption a (in g/g) can be converted to the Gibbs adsorption Γ (in g/m^2):

$$\Gamma = \frac{a - \rho V_a}{S} = -\frac{\rho}{R_g T}\left(\frac{\partial \sigma}{\partial \rho}\right) \tag{10.48}$$

where

V_a is the volume of the adsorption layer
S is the specific surface area
ρ is the density of water vapor as a function of the relative pressure p/p_0
σ is the surface tension
R_g is the gas constant

Integrating this equation could be done with respect to σ assuming that changes in the surface tension $\Delta\sigma$ upon the water adsorption correspond to changes in the Gibbs free surface energy (γ_S) or changes in the Gibbs free energy upon the adsorption (ΔG_{ads} in mJ/m^2) as follows:

$$\Delta G_{ads} = \Delta\sigma + \Gamma \Delta\mu_s \tag{10.49}$$

where $\Delta\mu_s$ is the change in the chemical potential of adsorbate due to the adsorption. The term $\Gamma\Delta\mu_s$ determines the compaction of the adsorption layer.

Equation 10.46 was modified to describe clustered adsorption of water (Brennan et al. 2001, Verdaguer et al. 2006) onto carbon materials (Tarasevich et al. 2007):

$$\frac{a}{a_m} = \frac{bx\left[1+\sum_{i=2}^{m} i\varepsilon_i (bx)^{i-1}\right]}{1+bx\left[1+\sum_{i=2}^{m} \varepsilon_i (bx)^{i-1}\right]} \quad (10.50)$$

where

a_m is the monolayer capacity

$b = z_0 q_1/(q_0 q_a)$, where $z_0 = \exp(\mu_0/R_gT)$ is the absolute activity, q_i is the statistical sum of the adsorption complex with i molecules

$\varepsilon_i = \exp(-i(\Delta E_i - \Delta E_0)/R_gT)$, where ΔE_i is the energy of adsorption of the ith molecule, $x = p/p_s$

m is the molecule number at the adsorption site

Equation 10.50 was transformed into the integral equation (Gun'ko et al. 2009d):

$$\frac{a}{a_m} = \int_{E_{min}}^{E_{max}} f(E) \frac{bx\left[1+\sum_{i=2}^{m} i\exp(-E/R_gT)(bx)^{i-1}\right]}{1+bx\left[1+\sum_{i=2}^{m} \exp(-E/R_gT)(bx)^{i-1}\right]} dE \quad (10.51)$$

where $f(E)$ is the distribution function of the adsorption energy of water molecules forming clusters at the adsorption sites (normalized to one hydrogen bond per one molecule).

To analyze the adsorption characteristics of nanooxides, adsorption of nonpolar hexane ("hex" was used as a subscript), weakly polar acetonitrile (ac), and polar diethylamine (DEA), triethylamine (TEA), and water (w) was studied using an adsorption apparatus with a McBain–Bark quartz scale at 293 K. Oxide samples were evacuated at 10^{-3} Torr and 473 K for several hours to a constant weight, then cooled to 293 ± 0.2 K. The measurement accuracy was 1 ± 10^{-3} mg with a relative mean error of ±5%.

The surface area occupied by a molecule of acetonitrile ($\sigma = 0.438\,nm^2$), hexane ($0.744\,nm^2$), DEA ($0.724\,nm^2$), and TEA ($0.760\,nm^2$) was estimated using quantum chemical calculations. These σ values were used to calculate the BET surface area. The validity of the standard BET method to the adsorption of different adsorbates onto heterogeneous surfaces was discussed and criticized by many authors, and some improved models were developed (Fagerlund 1973, Gregg and Sing 1982, Adamson and Gast 1997, Gelb and Gubbins 1998, Jonquières and Fane 1998, Choma et al. 2003). However, there are several reasons to use the standard BET method here. First, the particle morphology as well as the textural porosity of all the nanooxides studied is of the same nature because of the flame synthesis. Therefore, the systematic errors in determination of the S_{BET} values could be similar for all the samples. Second, primary nanoparticles ($d_{av} = 6.9$–52.5 nm) are spherical and nonporous, i.e., the porosity in the powders is of a textural character as voids between nanoparticles in aggregates, and nanopores are practically absent there. Therefore, the steric effects are relatively low. Third, the BET equation gives good fitting for all the adsorbates studied, and $S_{BET,X}$ values were used mainly to compare the surface composition effects. Notice that calculations of the specific surface area of heterogeneous adsorbents with, e.g., the DFT method require additional information on interaction potentials dependent on the pore types and the nature of an adsorbate and an adsorbent. Additionally, it is difficult to estimate the potentials for binary and ternary oxides because their surface composition can differ from the total composition and the surface can be mosaiclike with different oxide patches.

10.4 POLYMER AND PROTEIN ADSORPTION

Commercial poly(vinyl pyrrolidone) (PVP) (Biopharma, Kiev, pharmaceutical purity, molecular weight 12.6±2.7 kDa), poly(vinyl alcohol) (PVA) (PO "Stirol," Severodonetsk, Ukraine; 43±7 kDa), poly(ethylene glycol) (PEG) (Fluka, 35 kDa), poly(ethylene oxide), (PEO) (Sigma, 600 kDa), polydimethylsiloxane (PDMS) ("Kremniypolymer," Zaporozhye, Ukraine, 7.96 kDa), bovine serum albumin (BSA) (Allergen, Stavropol, Russia, 67 kDa), gelatine (Merck, powder food grade), ossein (Indar, Kiev, 20–29 kDa), and some other polymers were used as received. Several preparations of the oxide–polymer powders were used: (i) ball milling of dry mixtures of polymer (protein) and oxides powders for 1 h; (ii) treatment of similar mixtures in saturated vapors of water or ethanol for 24 h followed by ball milling for 1 h and drying; (iii) addition of 30 wt% of water or ethanol to the treated dry oxide–polymer (protein) powders; (iv) adsorption of polymers (proteins) from aqueous solution, centrifugation, and drying of the residual solid. Similar treatments were used for the preparation of oxide materials alone. Before the nitrogen adsorption–desorption isotherm measurements oxides were heated at 200°C, oxide–polymer materials were heated at 80°C for 2 h.

The PVA adsorption onto fumed silica was carried out from distilled water and measured using a viscometric method, which is frequently applied to study polymer solutions, adsorption, molecular weight, etc. The adsorption was determined using a calibrating graph from the difference in the flow velocity of the PVA solution in a glass capillary before and after the adsorption (solid residue with adsorbed PVA was removed by centrifugation) at pH 6.3±0.2 and $T=298\pm2$ K. Three measurements were performed for each adsorption point, and the average errors were smaller than ±5%.

The PVA–silica samples were used after drying in the IR spectroscopy measurements. To calculate a distribution function of Gibbs free energy of PVA adsorption, the modified Langmuir equation with consideration for lateral interaction (Adamson and Gast 1997) was used as the kernel of the integral adsorption equation (in the form of Fredholm equation of the first kind) solved using a regularization procedure. The Langmuir equation was used to estimate the monolayer coverage of silica by PVA.

10.5 INFRARED SPECTROSCOPY

Variable temperature diffuse reflectance FTIR spectra of TiO_2, Al_2O_3, SiO_2, SA, ST, and AST samples were acquired at 4 cm^{-1} nominal resolution by co-addition of 64 scans using a Digilab Excalibur FTIR spectrometer equipped with a liquid nitrogen cooled MCT detector. The spectra were subjected to a 9-point quartic smoothing routine to minimize problems from incomplete rationing of vapor phase water in the spectrometer. Both the optical accessory (DRA-2CN) and variable temperature cell (HVC-DRP), obtained from Harrick Scientific (Ossining, NY, USA), were used with minor modifications to permit more accurate sample height adjustments as a function of temperature. The sample temperature was monitored and controlled with a programmable temperature controller from Omega Engineering. A 5% (w/w) dispersion of fumed oxide was dispersed in ground and dried KCl (particle size ~5 ìm). The dispersion was heated to 200°C or 300°C for a minimum of 15 min prior to spectral data acquisition, and ratioed to a spectrum of pure KCl acquired under the same conditions. The percentage reflectance data was converted to log(1/*R*) values to approximate absorbance spectra.

For spectral analysis below 2000 cm^{-1}, FTIR transmission spectra were obtained. A small amount of sample (approximately 10–20 mg) was placed between two NaCl plates, which were then rotated to disperse the particles. Spectra were obtained at 4 cm^{-1} nominal resolution with a DTGS detector by co-addition of 16 scans.

The FTIR spectra of certain oxide samples were recorded over the 4000–400 cm^{-1} range by means of a FTIR 1725× (Perkin-Elmer) or a Specord M-80 (Carl Zeiss) spectrophotometer. Before FTIR measurements, samples were dried at 200°C for 2 h. They (0.33 wt%) were stirred with KBr (Merck, spectroscopy grade) and then pressed to form appropriate tablets.

The perturbation degree $\Phi = 1 - (I/I_0)$, where I_0 and I are intensities of a band of free silanols at 3750 cm^{-1} of initial silica and silica–polymer samples, respectively. Φ was used as a measure of interaction of polymers with the silica surface, because free silanols are the main adsorption sites for polar compounds adsorbed on the silica surface.

10.6 THERMOGRAVIMETRY

Thermo-oxidizing destruction of samples (250–300 mg) with recoding thermogravimetric (TG), differential TG (DTG), and differential thermal analysis (DTA) curves over the 20°C–1000°C range at a heating rate of 10°C/min was studied using a Q-1500D Derivatograph (MOM, Budapest).

10.7 DIFFERENTIAL SCANNING CALORIMETRY

Microcalorimetric studies of nanooxides, carbons, composites, etc. were carried out by means of a DAC 1.1A (EPSE, Chernogolovka, Russia) differential automatic calorimeter. Before the measurements of the heat of immersion in water (Q_w) or decane (Q_d), solid samples (50 mg) were degassed at 473 K and 0.01 Pa for 2 h and then used without contact with air. The amount of a sample used was 50 mg per 3 mL of distilled water or decane. The Q values were measured during exposure of the mixture for several hours. The average errors of the Q measurements repeated several times were ±5%. The Q values were calculated per surface area unit (i.e., Q measured in J/g was divided by S_{BET,N_2}) of samples because the studied oxides differ strongly in the S_{BET,N_2} values. To estimate the hydrophilicity of nanooxides, the hydrophilicity coefficient $K = Q_w/Q_d$ was calculated.

A Perkin-Elmer DSC-2 apparatus was used to determine the polymer glass transition temperatures, T_g at the half-height of a heat capacity step ΔC_p, and the transition breadth $\Delta T_g = T_g'' - T_g'$ where T_g' and T_g'' are the temperatures of the transition onset and completion, respectively. The second scans were taken for all compositions to exclude the side endothermic effect of water desorption. The measurements were performed under nitrogen atmosphere at a heating rate of 20°C/min, over the temperature range from −20°C to 197°C, after cooling from 197°C to −20°C with the rate of 320°C/min. Amorphous quartz was used as the reference sample.

10.8 AUGER ELECTRON SPECTROSCOPY

The surface content of aluminum in SA and AST and titanium in ST and AST was determined by Auger electron spectra (AES) recorded by a JAMP-10S (JEOL) spectrometer. The nanooxide powders were prepared with an indium–carbon matrix to avoid charge buildup on these dielectric samples. Spectra with minimum indium and carbon intensity were selected for subsequent analysis. Sample regions most characteristic and optimum for the AES studies were determined by scanning electron microscopy (SEM) (electron beam energy = 5 keV, beam current = 2×10^{-10} A, beam diameter = 0.05–0.1 μm). Differential AES $E \times dN(E)/dE$ were recorded with an energy analyzer such as a "cylindrical mirror" with an energy resolution of $\Delta E/E = 0.007$, a step of 1 eV, modulation amplitude of voltage on the energy analyzer = 4 V, circuit voltage = 2.5 kV, and an amplification circuit time constant of 1 s. The surface Si and Al content was determined by analysis of the LVV lines, and the surface titanium content was estimated using the LMM line (Briggs and Seach 1983).

10.9 TEMPERATURE-PROGRAMMED DESORPTION WITH MASS-SPECTROMETRY CONTROL

Water and CO desorption from oxides, carbons, or polymers was studied at 300–900 K by the one-pass temperature-programmed desorption (TPD) with mass-spectrometry control (OPTPD-MS) method (chamber pressure ~10^{-7} Torr, sample weight 5–7 mg, heating rate 2 K/s, with a short distance [~0.5 cm] between sample and MS detector) with a MSC-3 ("Electron," Sumy, Ukraine) time-of-flight mass spectrometer (sensitivity 2.2×10^{-5} A/Torr, accelerating voltage 0.5 kV, pulse frequency 3 kHz).

TPD of water was studied with mass-spectrometric control (MX-7304A mass spectrometer, "Electron," Sumy, Ukraine). Sample (~4 mg) placed in a quartz–molybdenum cell (diameter 11 mm) was degassed at room temperature and pressure 7×10^{-5} Pa and then heated to 800°C at a linear heating rate of 10 K/min. Volatile products are desorbed from the sample, ionized by the electron beam, separated by a mass analyzer, and come to a detector. Calculation of a distribution function of activation energy of water desorption $f(E_a)$ was carried out using integral equation for a separated TPD peak:

$$A(T) = \frac{c_1 T[\Theta(T)]^n}{\sqrt{2\pi\sigma^2}} \exp\left(-\frac{(T - T_{max})^2}{2\sigma^2}\right) \int_{E_{min}}^{E_{max}} f(E_a) \exp\left(-\frac{E_a}{R_g T}\right) dE_a \tag{10.52}$$

where

c_1 is a constant
σ is the half-width of a desorption peak
T_{max} is the temperature of the maximal water desorption
$\Theta(T)$ is the temperature dependence of the surface coverage by the hydroxyl groups
n is the reaction order ($n = 1$ for molecularly adsorbed–desorbed water and $n = 2$ for associatively desorbed water)
E_{min} and E_{max} are the minimal and maximal values of the activation energy E_a

Equation 10.52 was solved using a regularization procedure assuming that normalized desorption and residual coverage are bonded by equation:

$$\int_{T_{min}}^{T_{max}} \left[\frac{A(T)}{\int A(T')dT'} + \Theta(T) \right] dT = 1 \tag{10.53}$$

The physical model of the TPD experiments used as a base of the kernel in Equation 10.52 was described by Russell and Ekerdt (1996). In the case of a complex shape of a TPD spectrum including N peaks, Equation 10.52 was transformed into the sum of the integral equations:

$$A_\Sigma(T) = \sum_{i=1}^{N} A_i(T) = \sum_{i=1}^{N} \frac{c_i T[\Theta_i(T)]^{n_i}}{\sqrt{2\pi\sigma_i^2}} \exp\left(-\frac{(T - T_{max,i})^2}{2\sigma_i^2}\right) \int_{E_{min,i}}^{E_{max,i}} f_i(E_{a,i}) \exp\left(-\frac{E_{a,i}}{R_g T}\right) dE_{a,i} \tag{10.54}$$

solved by using SCR procedure assuming

$$\int \sum_{i=1}^{N} \left(\frac{A_i(T)}{A_I} + \Theta_i(T) \right) dT = 1 \tag{10.55}$$

where A_I is the total amounts of desorbed water.

10.10 THERMALLY STIMULATED DEPOLARIZATION CURRENT

The tablets (diameter 30 mm, thickness ~1 mm) with frozen studied materials differently hydrated (hydration h=0.03–99 g of water per gram of dry material) and bio-objects were polarized by the electrostatic field at the intensity F_p=0.1–0.5 MV/m at 260–265 K then cooled to 90 K with the field still applied and heated without the field to 265–270 K at a heating rate β=0.05 K/s. The current evolving due to sample depolarization (after polarization at a certain temperature) was recorded by an electrometer over the 10^{-15}–10^{-5} A range. Relative mean errors for measured TSDC were δ_I=±5%, δ_T=±2 K for temperature, and δ_b=±5% for the temperature change rate. Activation energy of depolarization (i.e., dipolar relaxation of bound charges) was calculated using the following equation for the TSD current (Gun'ko et al. 2007i):

$$I(T) = \frac{S_{el}\Pi_0}{\tau_0}\exp\left(-\frac{E}{k_BT}\right)\exp\left(-\frac{1}{b\tau_0}\int_{T_0}^{T}\exp\left(-\frac{E}{k_BT}\right)dT\right) \quad (10.56)$$

and overall equation for a TSDC spectrum with several peaks

$$I(T) = S_{el}\Pi_0\int_{E_{min}}^{E_{max}}\sum_{i=1}^{N} w_0^i\exp\left\{-\frac{E}{k_BT}-\frac{w_0^ik_BT^3}{Eb}\exp\left(-\frac{E}{k_BT}\right)+\frac{w_0^ik_BT_0^3}{Eb}\exp\left(-\frac{E}{k_BT_0}\right)\right\}f(E)dE \quad (10.57)$$

where

$$w_0^i = \frac{b}{k_BT_i^3}(E+k_BT_i)\exp\left(\frac{E}{k_BT_i}\right) \quad (10.58)$$

S_{el} is the surface area of the electrodes
Π_0 is the frozen polarization (relationship between Π_0 and F_p is shown later)
k_B is the Boltzmann constant
$1/\tau_0$ is the natural frequency
T_0 is the initial temperature of depolarization
T_i is the temperature of the ith TSDC maximum

Transition probability of dipoles is

$$w_1 = \frac{w_0}{4\pi}T\exp\left(-\frac{E}{k_BT}\right) \quad (10.59)$$

the relaxation time is

$$\tau = \frac{1}{4\pi w_1} \quad (10.60)$$

and susceptibility

$$\chi = \frac{mA_i}{S_{el}F_p} \quad (10.61)$$

where
m is the normalization factor
A_i is the i-peak area, the dielectric constant

$$\varepsilon = 1 + 4\pi\chi \tag{10.62}$$

effective dipole moment

$$\mu_{\text{eff}} = \sqrt{\frac{2mA_i k_B T'}{N_0 F_p S_{\text{el}}}} \tag{10.63}$$

where

N_0 is the number of molecules/cm^3

T' is the temperature for which the relaxation time is still sufficiently short to let the polarization corresponding to the temperature T' adjust during the time of the applied electric field

T' can be computed from Equations 10.58 and 10.60. The effective dipole moment μ_{eff} deviates from the dipole moment (μ) of the free molecule:

$$\mu = \frac{3\mu_{\text{eff}}}{(n^2 + 2)\sqrt{1 + z\langle\cos\vartheta\rangle}} \tag{10.64}$$

where

n is the refractive index of the substance

z is the number of nearest neighbors to the molecule

$\langle\cos\vartheta\rangle$ is the spatial average of the cosine of the angle between two adjacent dipoles

Dipole–dipole interaction energy between two adjacent dipoles is given by

$$E = \frac{\mu_{\text{eff}}^4}{3k_B T R^6} \tag{10.65}$$

where R is the mean distance between the dipoles.

To calculate a distribution function $f(E)$ with integral equation 10.57, a constrained regularization method can be used because solution of similar equations is well-known ill-posed problem due to the impact of noise on the measured TSDC data, which does not allow the utilization of exact inversion formulas or iterative algorithms. For this purpose, the constrained regularization procedure CONTIN (Provencher 1982) was modified to apply Equation 10.57 with non-negativity condition ($f(E) \geq 0$ at any E) and an unfixed value (determined on the basis of the F-test and confidence regions using the parsimony principle) or a fixed value ($\alpha = 0.01$–0.001) of the regularization parameter.

For TSDC measurements, samples of a disk shape with diameter 10–30 mm and thickness 0.3–2.0 mm were polarized by an electrostatic field of intensity $F = (3.5 - 5) \times 10^5$ V/m at 263–265 K for 2 min, cooled to 90–100 K *with* the field still applied and subsequently heated *without* the field at a rate $h = 0.05$ K/s to 265 K (Gun'ko et al. 2007i). The TSDC method can be effectively used to analyze the associativity and the mobility of the interfacial water because the measured temperature dependence of depolarization characterizes environment of water molecules: (i) how these molecules interacts with neighbors; (ii) how many hydrogen bonds they form per molecule; (iii) what strength of these bonds; (iv) what mobility of bound charges (water dipoles) and free charges (protons); and (v) what the size of water clusters (Gun'ko et al. 2007i).

10.11 DIELECTRIC RELAXATION SPECTROSCOPY

On DRS measurements (Jonscher 1983), the complex dielectric permittivity

$$\varepsilon^*(f) = \varepsilon'(f) - j\varepsilon''(f) \tag{10.66}$$

was determined as a function of frequency ($f = 10^{-1}$–3×10^6 Hz) in the temperature range from 133 to 423 K (controlled better than ±0.1 K). An alpha dielectric analyzer with a Quatro Cryosystem for the temperature control was employed (both from Novocontrol GmbH, Germany) to sample pellets (pressed at 5 tons) of 13 mm in diameter and of 1 mm in thickness. The complex dielectric permittivity can be written as a function of angular frequency ($\omega = 2\pi f$) in accordance with several possible relaxation processes caused by the mobility of different dipoles and charges in the systems with polymer–water–solid nanoparticles:

$$\varepsilon^*(\omega) = \varepsilon_\infty + \sum_{i=1}^{N} \frac{\Delta\varepsilon_i}{[1+(j\omega\tau_i)^{\alpha_i}]^{\beta_i}} + \sum_{l=1}^{M} \Delta\varepsilon_l \int_0^\infty \left\{ -\frac{d}{dt}\left[\exp\left(-\frac{t}{\tau_l}\right)^{\beta_l}\right]\right\} \exp(-j\omega t)dt - \left(\frac{\sigma}{j\varepsilon_0\omega}\right)^s \tag{10.67}$$

where

$\varepsilon_\infty = \varepsilon'(\omega \to \infty)$, ε_0 is the vacuum dielectric permittivity, $\Delta\varepsilon_i$ is the relaxation strength of the ith relaxation process, τ_i is the relaxation time of the ith process described by the Havriliak–Negami (HN) parameters α_i and β_i ($0 < \alpha_i$, $\beta_i \le 1$) related to symmetric and asymmetric broadening of the ith relaxation peak (Havriliak and Negami 1966, 1967), the sum of integrals corresponds to the processes described by parameters τ_l and β_l ($0 < \beta_l \le 1$) according to Kohlrauch–Williams–Watts (KWW) for a broad and asymmetrical relaxation peak (the response function $\exp(-t/\tau_l)^{\beta_l}$ is sensitive to segmental motion in a polymer), and the last term dependent on frequency is due to the conductivity σ (dc + hopping conductivity) at $s < 1$ due to interfacial polarization and space charging, when $s = 1$, the conduction term reduces to dc one. Notice that the relaxations suggested by Debye, Cole–Cole (Cole and Cole 1941), and Davidson–Cole are described as special cases of the HN and KWW expressions shown as the sums in Equation 10.67. The number of the HN and/or KWW relaxation processes was determined as a number of peaks and shoulders according to the $\varepsilon''(\omega)$ function shape to best fit this function on minimization of the functional:

$$\Phi(x) = \int_{\omega_{min}}^{\omega_{max}} \left(\varepsilon''_{exp} - \varepsilon''_{theor}\right)^2 d\log\omega \tag{10.68}$$

where

x is the set of the equation parameters

ω_{min} and ω_{max} are the minimal and maximal angular frequencies on the measurements respectively

ε''_{exp} and ε''_{theor} are the experimental and theoretical (calculated with Equation 10.67) dielectric loss, respectively

Here the number of HN terms $N = 2$ and the number of KWW terms $M = 0$ or 1 depending of the spectrum shape. The presence of the conductivity term in Equation 10.67 was determined from the $\varepsilon''(\omega)$ function shape at low frequencies as the corresponding decay.

The dielectric loss can be written using the distribution function of relaxation time $f(\log\tau)$ (or $f(\tau)$) assuming the Debye-type relaxations for simplicity of the kernel:

$$\varepsilon''(\omega) = \left(\frac{\sigma}{\varepsilon_0\omega}\right)^s + \int_{\tau_{min}}^{\tau_{max}} f(\log\tau)\frac{\omega\tau}{1+(\omega\tau)^2}d(\log\tau) \qquad (10.69)$$

The experimental $\varepsilon''(\omega,T)$ functions have an incomplete maximum at high ω values because in our experiments the maximal ω value was 1.88×10^7 Hz (i.e., boundary $\tau_b=5.32\times10^{-8}$ s). However, the used regularization procedure can retrieve a complete shape of a boundary $\varepsilon''(\omega,T)$ maximum on the basis of the shape of the corresponding boundary tail of this function. Therefore, we used in Equation 10.60 $\tau_{min}=1.0\times10^{-9}$ s$<\tau_b$ that give better fitting than that at $\tau_{min}=\tau_b$. However, the calculations with Equation 10.69 gave only a couple of the $f(\tau)$ distributions nonzero at $\tau_{min}<\tau<\tau_b$.

Assuming that each relaxation process obeys the Arrhenius law: $1/\tau=(1/\tau_0)\exp(-E/R_gT)$, the dependence of relaxation on temperature can be written using the distribution function of the activation energy $f(E)$:

$$\tau(T) = \tau_0 \int_{E_{min}}^{E_{max}} \exp\left(\frac{E}{R_gT}\right) f(E)dE \qquad (10.70)$$

where

E is the activation energy of relaxation
R_g is the gas constant

Equations 10.69 and 10.70 as Fredholm integral equations of the first kind were solved using a regularization procedure based on the CONTIN algorithm. The use of Equation 10.70 allows us to describe the sum relaxation, which can demonstrate certain deviation from the Arrhenius law because of, for example, the cooperative effects characteristic for such supramolecular systems as PVA or PVA/nanosilica.

Water adsorption was performed at 300 K with mean errors near ±0.01 g of water/g of oxide. The dielectric characteristics (dielectric permittivity, ε', and dielectric loss, ε'') were measured by a Q-meter VM-560 ("Tesla") at $f=0.15$, 0.25, 0.45, 0.8, 1.3, 3.0, 8.0, and 9.0 MHz at $T=100$–300 K. The measurements of $\varepsilon'(T)$ and $\varepsilon''(T)$ were taken by a thermochamber with programmed temperature changes. The heating rate, β, was equal to 0.05 K/s with relative mean errors $\delta\beta=\pm5\%$. The $tg\delta(f)=\varepsilon''/\varepsilon'$ function was detected at 300 K in the 10^4–10^7 Hz range.

10.12 ULTRAVIOLET–VISIBLE SPECTROSCOPY

(Dimethylamino)azobenzene (n-DMAAB, p$K=3.3$) was chosen as an Hammet color indicator, and its diffuse reflection spectra have been recorded by a SF-18 (LOMO, Russia) spectrophotometer upon indicator adsorption on oxides, then they were recalculated to the absorption spectra according to the Kubelka–Munk formula. n-DMAAB adsorption on the samples previously evacuated and heated at 473 K for 1 h was performed from the gas phase at 338 ± 5 K for 2–4 h. Assignment of the n-DMAAB absorption bands has been done on the analogy of the spectra for this substance in neutral and acidic solutions. Four absorption bands of DMAAB adsorbed on mixed oxides can be detected: (1) 430–460 nm, d-DMAAB (physically adsorbed n-DMAAB via dispersive, nonspecific interaction); (2) 480–490 nm, H-DMAAB (hydrogen-bonded complex); (3) 520–545 nm, H^+-DMAAB (complex with H^+ transfer from the B-sites to DMAAB); and (4) 555–560 nm, L-DMAAB (complex with the Lewis acid sites, L-sites) (Gun'ko et al. 1999a,b).

10.13 RHEOMETRY

Rheological investigations of aqueous suspensions were performed using a Rheotest 2.1 (VEB MLW Prufgerate-Werk Medingen Sitz Ftreital, Germany) rotary viscometer with a thermostated cylinder–cylinder system at the clearance between the cylinders of 0.4 mm and the suspension volume of 10 mL. The effective viscosity (η) was determined at the shear rate $\dot{\gamma}=5.4$–$1312\,s^{-1}$ (increased and then decreased) at 298 K. Additionally, the temperature dependence of the viscosity was measured at a fixed shear rate $\dot{\gamma}=1312\,s^{-1}$.

Models of Ree and Eyring (Eirich 1958) and Baxter-Drayton and Brady (1996) give the dependence of the viscosity η on a shear rate assuming that the motion of particles can be described as activated jumps from the cage formed by the nearest neighbors to holes into the adjoining cages (Quemada and Berli 2002):

$$\eta = (\eta_0 - \eta_\infty)\frac{\sin h^{-1}(\beta\dot{\gamma})}{\beta\dot{\gamma}} + \eta_\infty \tag{10.71}$$

where

$$\beta = c\frac{a^2}{D_0}\frac{k_B T}{E}\exp\left(\frac{E}{k_B T}\right) \tag{10.72}$$

η_0 and η_∞ are the viscosity at the shear rate $\dot{\gamma} \to 0$ and $\dot{\gamma} \to \infty$, respectively
a is the particle radius
D_0 is the diffusion coefficient
k_B is the Boltzmann constant
T is the temperature
E is the activation energy of the particle (molecules) jumps

Any colloidal system is nonuniform with respect to conditions of particle–molecule motions. Therefore, the activation energy should be determined as a distribution function $f(E)$. Equation 10.71 can be easily transformed into an integral equation (Gun'ko et al. 2006b):

$$\eta = \int_{E_0}^{E_{max}} [(\eta_0 - \eta_\infty)\left(\frac{1}{z}\int_0^z \frac{dx}{\sqrt{1+x^2}} + \eta_\infty\right) f(E)dE \tag{10.73}$$

where

$$z = c\dot{\gamma}\frac{a^2}{D_0}\frac{k_B T}{E}\exp\left(\frac{E}{k_B T}\right) = A\dot{\gamma}\frac{k_B T}{E}\exp\left(\frac{E}{k_B T}\right) \tag{10.74}$$

Equation 10.73 can be solved using a regularization procedure under conditions of non-negativity $f(E) \geq 0$ at any E and fixed or unfixed regularization parameter determined by statistical analysis of the experimental data and (Quemada and Berli 2002):

$$\frac{\eta - \eta_\infty}{\eta_0 - \eta_\infty} = \frac{1}{z}\int_0^z \frac{dx}{\sqrt{1+x^2}} \tag{10.75}$$

Equation 10.75 allows us to determine A in Equation 10.74 at different $\dot{\gamma}$ values. There is a noticeable advantage of Equation 10.73 related to the possibility to use it for analysis of both measured $\eta(\dot{\gamma})$ (at a constant T value) and $\eta(T)$ (at a constant $\dot{\gamma}$ value) functions or even in the case of simultaneously varied T and $\dot{\gamma}$ values.

10.14 POTENTIOMETRIC TITRATION

To evaluate the surface charge density (σ_0), potentiometric titrations were performed using a thermostated Teflon vessel in nitrogen atmosphere free from CO_2 at 25°C ± 0.2°C. The solution pH was measured using a PHM240 Research pH meter (G202C and K401 electrodes) coupled with an REC-61 recorder. The surface charge density was calculated using the potentiometric titration data for a blank electrolyte solution and oxide suspensions ($C_{ox}=0.2$ wt% for all oxides), at a constant salinity of 10^{-3} M NaCl. From the difference of acid or base volume utilized to obtain the same pH value as that for the background electrolyte of the same ionic strength according to the following equation, one can calculate $\sigma_0=\Delta VcF/mS_{BET}$, where $\Delta V=V_s-V_e$ is the difference between the base (acid) volumes added to the electrolyte solution V_e and suspension V_s to achieve the same pH; F is the Faraday constant, c is the concentration of base (acid), and m is the weight of the oxide.

10.15 PHOTON CORRELATION SPECTROSCOPY

Electrophoretic and secondary particle size distribution (SPSD) investigations were performed using a Zetasizer 3000 (Malvern Instruments) apparatus based on the photon correlation spectroscopy (PCS) ($\lambda=633$ nm, $\Theta=90°$, software version 1.3). Deionized distilled water and oxide samples (0.4 g/dm^3 of the water) were utilized to prepare suspensions ultrasonicated for 5 min using an ultrasonic disperser (Sonicator Misonix, power 500 W and frequency 22 kHz). The pH values were adjusted by addition of 0.1 M HCl or NaOH solutions, and the salinity was constant (0.001 M NaCl). The electrophoretic behavior and the SPSDs with respect to the intensity of scattered light ($SPDS_I$) and the particle number ($SPDS_N$) of fumed oxides in the aqueous suspensions studied by using PCS method were described in detail elsewhere (Zarko et al. 1997, Gun'ko et al. 2001e, 2003g).

According to the Smoluchowski theory (Hunter 1981, 1993), there is a linear relationship between the electrophoretic mobility U_e and the ζ potential: $U_e=A\zeta$, where A is a constant for a thin EDL at $\kappa a \gg 1$ (where a denotes the particle radius and κ is the Debye–Huckel parameter). For a thick EDL ($\kappa a<1$), e.g., at pH close to the isoelectric point (IEP), the equation with the Henry correction factor is more appropriate:

$$U_e = \frac{2\varepsilon\cdot\zeta}{3\eta} \quad (10.76)$$

where

ε is the dielectric permittivity
η is the viscosity

However, the relationship between the ζ potential and the mobility for aggregates of primary particles should be corrected considering the particle volume fraction (ϕ) and the shear-plane potential (Ψ_a) at the particle–fluid interfaces within the porous aggregates (Miller et al. 1992, Miller and Berg 1993):

$$U_e = \frac{\varepsilon}{3\eta}\left(1+\frac{\varphi}{2}\right)(2\zeta+\Psi_a F) \quad (10.77)$$

where

F is the electroosmotic flow factor

$$\Psi_a = \frac{\zeta(1+\kappa a)}{B} \tag{10.78}$$

$$B = \frac{(1-\kappa^2 ab)\sin h(\kappa b - \kappa a) - (\kappa b - \kappa a)\cos h(\kappa b - \kappa a)}{\sin h(\kappa b - \kappa a) - \kappa b \cos h(\kappa b - \kappa a)} \tag{10.79}$$

b is the outer radius of a unit cell representative, in an average sense, of the aggregate interior defined by $\phi = a^3/b^3$ (e.g., for aggregates of fumed silicas $\phi \approx 0.3$–0.4 estimated from the pore volume of aggregates determined on the basis of the nitrogen adsorption).

Differences in the ζ potentials calculated using Equations 10.76 and 10.77 can reach up to 25% (Miller et al. 1992, Miller and Berg 1993). One can estimate the κ values according to Xu (1998) and the average size of primary particles from the specific surface area (or from direct observation of primary particles using TEM, AFM, or SEM methods) or the primary particle size distribution $f(a)$ with minimal (a_{min}) and maximal (a_{max}) radii. Equation 10.77 for the electrophoretic mobility can be written as follows:

$$U_e = \frac{\varepsilon}{3\eta}\int_{a_{min}}^{a_{max}}\left(1+\frac{\varphi}{2}\right)(2\zeta + \Psi_a F)f(a)da \tag{10.80}$$

Calculations of the Debye screening length (κ^{-1}) were carried out using Loeb equation (Hunter 1981, Chan and Lewis 2005, Gun'ko et al. 2007d):

$$q = 4\pi\varepsilon\varepsilon_0 \frac{k_B T}{ze}\kappa a^2\left[2\sin h\left(\frac{ze\zeta}{2k_B T}\right) + \frac{4}{\kappa a}\tan h\left(\frac{ze\zeta}{4k_B T}\right)\right] \tag{10.81}$$

which links the surface charge (q) of a particle with its ζ potential (where z is the charge of the ions in solution, e is the electron charge, $\varepsilon\varepsilon_0$ is the dielectric constant of the solution, and a is the particle radius). This equation gives the surface charge density $\sigma = q/(4\pi a^2)$ as a function of ζ. If the σ, ζ, and a values are measured experimentally under the same conditions, then κ can be determined as

$$\kappa = \frac{(\sigma/A - C)}{B} \tag{10.82}$$

where

$$A = \varepsilon\varepsilon_0 \frac{k_B T}{ze}$$

$$B = 2\sin h\left(\frac{ze\zeta}{2k_B T}\right)$$

$C = (4/a)\tan h(ze\zeta/4k_B T)$ with consideration for the porosity of aggregates (characterized by the ζ potential and the effective diameter D_{ef}), i.e., a volume fraction (ϕ) of nanoparticles in porous aggregates

For the studied systems, the ϕ value changes over the 0.012–0.088 range which is in agreement with the bulk density of the fumed oxide powder because the structures of aggregates in the gaseous and aqueous media can be close.

DLS investigations were also performed using a Malvern Zetasizer Nano S apparatus ($\lambda = 632$ nm, 5 mW, $\Theta = 173°$) at 298 K. Deionized distilled water was used for preparation of nanooxide suspensions ($C = 3.0$ wt%, pH 4.5–5.5), and sonicated for 3 min using an ultrasonic disperser (22 kHz and 500 W) prior to analysis. To compute the particle size distribution with respect to the intensity of light scattering (I_S), particle volume (I_V), and particle number (I_N), the Malvern Instruments software was used, assuming that particles had a roughly spherical shape.

10.16 ADSORPTION OF METAL IONS

The Pb(II) adsorption on oxide surfaces ($C_{ox} = 0.2$ wt%) was performed from the aqueous solution of $Pb(ClO_4)_2$ at initial Pb(II) concentration of 10^{-5}, 10^{-4}, or 10^{-3} M (concentration of radioactive species ^{210}Pb(II) was equal to 10^{-6} M) with addition of a neutral electrolyte (10^{-3} M $NaClO_4$) using a Teflon cell (50 cm^3) temperature-controlled at $T = 25°C \pm 0.2°C$. The pH value was varied by addition of 0.1 M HCl or NaOH solutions. The gamma radioactivity of the solution containing ^{210}Pb was determined using a Beckman Gamma 5500B counter. The Sr(II) adsorption (with ^{90}Sr and 10^{-3} M NaCl) and Cs(I) (with radioisotope ^{137}Cs and without a neutral electrolyte) was carried out at metal ion concentrations of 10^{-5}, 10^{-4}, and 10^{-3} M. In the case of the Sr(II) adsorption, the measurements were performed using two channels to eliminate ^{90}Y contribution.

To calculate a distribution function $f(pK_n)$ of a normalized constant (K_n) (relative binding strength) of the adsorption of metal ions, integral equation (Gun'ko et al. 2004f)

$$\theta(pH) = \int_0^\infty \frac{K_n z}{1 + K_n z} f(pK_n) dK_n \tag{10.83}$$

where θ is the normalized cation uptake $C_{cat}/C_{cat,0}$ ($C_{cat,0}$ is the initial concentration of metal ions; $\log z = -\text{pH}$ and $pK_n = -\log K_n$) was solved using a modified regularization procedure CONTIN at an unfixed regularization parameter (automatically determined on the basis of F-test and confidence regions) and non-negativity condition for $f(pK_n)$ ($f(pK_n) \geq 0$ at any K_n.

10.17 X-RAY DIFFRACTION

X-ray diffraction (XRD) patterns were recorded over $2\theta = 10°$–$70°$ range using a DRON-4-07 (Burevestnik, St. Petersburg) diffractometer with Cu K_α ($\lambda = 0.15418$ nm) radiation and a Ni filter. The average size of γ-alumina crystallites (d_{cr}) was estimated according to the Scherrer equation. Analysis of the crystalline structure of alumina alone and in composite materials was carried out using the JCPDS Database (International Center for Diffraction Data, PA, 2001).

The PSD $f(r)$ based on the small-angle x-ray scattering (SAXS) data was calculated using integral equation for the scattering intensity $I(q)$ (Pujari et al. 2007):

$$I(q) = C \int_{R_{min}}^{R_{max}} \frac{(\sin qr - qr\cos qr)^2}{(qr)^2} v(r) f(r) dr \tag{10.84}$$

where

- $q = 4\pi\sin(\theta)/\lambda$ the scattering vector value, 2θ is the scattering angle, λ is the wavelength of incident x-ray
- $v(r)$ is the volume of a pore with radius r (proportional to r^3)
- $f(r)dr$ represents the probability of having pores with radius r to $r + dr$
- R_{min} ($= \pi/q_{max}$) and R_{max} ($= \pi/q_{min}$) correspond to lower and upper limits of the resolvable real space due to instrument resolution

The chord size distribution, $G(r)$ as a geometrical statistic description of a multiphase medium, was calculated from the SAXS data (Brumberger 1965, Dieudonné et al. 2000):

$$G(r) = C\int_0^\infty \left[K - q^4 I(q)\right]\frac{d^2}{dr^2}\left(-4\frac{\sin qr}{qr}\right)dq \tag{10.85}$$

where K is the Porod constant (scattering intensity $I(q) \sim Kq^{-4}$ in the Porod range).

The specific surface area from the SAXS data was calculated (in m^2/g) using equation:

$$S_{SAXS} = 10^4\pi\phi(1-\phi)\frac{K}{Q\rho_a} \tag{10.86}$$

where

$\phi = \rho_a/\rho_0$ is the solid fraction of adsorbent

Q is the invariant

$$Q = \int_0^\infty q^2 I(q)dq \tag{10.87}$$

The Q value is sensitive to the range used on integration of Equation 10.78 (since experimental q values are measured between the q_{min} and q_{max} values different from 0 and ∞). Therefore, the invariant value Q was calculated using equation (Fairén-Jiménez et al. 2006)

$$Q = \sum_{q_{min}}^{q_{max}} \frac{(I(q_i) - b)q_i^2\Delta q_i + K}{q_{max}} \tag{10.88}$$

where b is a constant determined using equation

$$I(q)q^4 = K + bq^4 \tag{10.89}$$

valid in the Porod range.

10.18 RAMAN SPECTROSCOPY

Raman spectra of certain nanooxides were recorded over the range 150–3200 cm^{-1} using an inVia Reflex (Renishaw, United Kingdom) microscope. Only the most interesting spectra of titania-containing samples are shown here.

10.19 AFM, SEM, AND TEM

Atomic force microscopic (AFM) images were obtained by means of a NanoScope III (Digital Instruments, United States) apparatus using the Tapping Mode AFM measurement technique. Before AFM scanning, powder samples were gently smoothed by handpressing using a glass plate, which does not affect the structure of primary and secondary particles of these materials.

For transmission electron microscopy (TEM), gel samples were dehydrated in acetone, stained with osmium tetroxide, and embedded in epoxy resin. The embedded samples were cut into slices of 50–100 nm thickness with an ultramicrotome LKB-V (LKB, Sweden) and examined using TEM (Hitachi H-600, Japan).

For SEM, gel samples were dried at critical point, sputter-coated with gold, and examined using SEM JSM-6310 (JEOL, Japan).

10.20 QUANTUM CHEMISTRY

Quantum chemical calculations of oxide, carbon, polymer, or complex materials were carried out by using ab initio, DFT (with different basis sets) (GAMESS 7, WinGAMESS 10 and 11 (Schmidt et al. 1993), GAMESS Firefly (Granovsky 2009), Gaussian 03 (Frisch et al. 2004)), and semiempirical PM3 or PM6 (MOPAC 2009, Stewart 2008) methods. The solid clusters up to 220 atoms were calculated using ab initio or DFT methods, and larger systems (up to 15,000 atoms) were calculated using semiempirical PM3 or PM6 methods.

The magnetic shielding is the tensor describing relative changes in the local magnetic field at the nucleus position relative to the external magnetic field (Facelli 2011). This change in the local magnetic field, which originates in the interaction of the electron cloud with the external magnetic field, can produce shielding or deshielding of the nucleus. In the first case the local magnetic field is increased with respect to the external field, while in the second case the local field is decreased. In general, shielding effects are associated with diamagnetic effects from spherical charge distributions, while deshielding effects are associated with a nonspherical charge distribution originating from p or higher angular momentum electrons. When experiments are performed at a constant magnetic field, as it is normally done in modern NMR spectrometers, a shielding effect results in a shift of the resonance to a higher frequency, while a deshielding effect will result in a lower resonance frequency. For historic reasons associated with the early use of fix-frequency/variable field spectrometers, it is still common practice today to associate shielding with the term "up-field shift" and deshielding with "down-field shift" (Harris et al. 2008, Facelli 2011).

The magnitudes of NMR shielding σ_{iso} (or chemical shift δ_{iso} referenced to TMS using different basis sets) and the shielding anisotropy $\Delta\sigma_{aniso}=\sigma_{11}-(\sigma_{22}+\sigma_{33})$ were computed with the continuous set of gauge transformations (CSGT) or the gauge-independent atomic orbital (GIAO) (Hadzi 1997, Frisch et al. 2004) methods using the Hartree–Fock (and DFT was used with the combined B3LYP exchange and correlation functional) methods. The nuclear magnetic shielding constant of a nucleus represents an asymmetric tensor with nine independent components $\sigma^{N}_{\alpha\beta}=\partial^2 E/\partial B_\alpha \partial\mu_{N\beta}$, where α, $\beta = x, y, z$ denote components of the external magnetic field $\mathbf{B}$ and the nuclear magnetic moment $\boldsymbol{\mu}_N$. In a finite basis set, the magnetic shielding tensor of a nuclei N can be written in general form as

$$\sigma^{N}=\mathrm{Tr}\left[\mathbf{D}^{(0)}\mathbf{H}_{N}^{(11)}\right]+\mathrm{Tr}\left[\mathbf{D}^{(10)}\mathbf{H}_{N}^{(01)}\right] \tag{10.90}$$

where

Tr stands for the trace of a matrix

The superscripts represent the order of the perturbations in respect to the external magnetic field and the nuclear magnetic moment

$\mathbf{D}^{(0)}$ and $\mathbf{D}^{(10)}$ are the unperturbed and first-order perturbed density matrices

$\mathbf{H}$ is the one-electron Hamiltonian integral

Equation 10.90 is valid at various levels of theory (HF SCF, MP-n, CC, DFT) with the only difference being in the definition of the zero-order and first-order density matrices $\mathbf{D}^{(0)}$ and $\mathbf{D}^{(10)}$; also, this equation holds for both field-independent ϕ_μ and field-dependent $\phi_\mu(\mathbf{B})$ atomic orbital (AO) basis sets. In the GIAO method, the last functions are defined as

$$\phi_\mu(\mathbf{r}, B) = \exp\left\{\left(\frac{-i}{2c}\right)[\mathbf{B}\times(\mathbf{R}_\mu - \mathbf{R}_G)]\mathbf{r}\right\}\phi_\mu(\mathbf{r}, 0) \quad (10.91)$$

where
- c is the velocity of light
- $\phi_\mu(\mathbf{r},0)$ denotes the usual field-independent basis function centered at $\mathbf{R}_\mu$

The GIAO method provides the gauge-invariant results for magnetic shielding tensor. Recently, Facelli (2011) analyzed different methods of the calculations of the chemical shifts. Notice that errors for δ_H are much lower than for more heavy atoms. Therefore, in this book theoretical results are shown only for the δ_H values.

The δ_H values were calculated as the difference between isotropic values (average value of diagonal values of the magnetic shielding tensor of protons, $\sigma_{H,iso}$) of TMS (e.g., $\sigma_{H,iso} = 31.76$ ppm by GIAO/B3LYP/6-31G(d,p)) and studied compounds $\delta_H = (1/3)\mathrm{Tr}\sigma_{TMS} - (1/3)\mathrm{Tr}\sigma_H$. The distribution functions of the δ_H values were calculated using a simple equation:

$$f(\delta_H) = (2\pi\sigma^2)^{-0.5}\sum_j \exp\left[-\frac{(\delta_j - \delta_H)^2}{2\sigma^2}\right] \quad (10.92)$$

where
- j is the number of H atoms in the system
- σ^2 is the distribution dispersion
- δ_j is the calculated value of the chemical shift of the jth H atom

Large models were calculated using the PM3 or PM6 methods (MOPAC 2009). To calculate the $f(\delta_H)$ functions using the PM6 results, calibration functions for H_2O, H_2O_2, HCl, CH_4, etc. were used to describe the dependencies between atomic charges q_H (PM6) and the δ_H values (GIAO/B3LYP/6-31G(d,p)) for certain systems.

For simultaneous analysis of the Hamiltonian eigenvalues and eigenvectors (related to the valence zone, as the most interesting for chemical purposes), the local density of the electronic states (LDES)

$$\rho_i(E) = \left(2\pi\sigma_k^2\right)^{-0.5}\sum_{n,l,j}|C_{nl}^{ij}|^2 \exp\left[-\frac{(E_j - E)^2}{2\sigma_k^2}\right] \quad (10.93)$$

where
- i denotes the atom number
- n and l are the main and orbital quantum numbers
- σ^2 is the dispersion
- C is the MO coefficient
- E_j is the j-orbital energy

has been calculated for certain atoms having different neighbors in oxide clusters. The σ_k^2 dispersion value could be estimated from the half-width of the corresponding XPS band in the valence zone or in other zones of oxides, but we assumed $\sigma = 0.4$–0.5 eV in order to well resolve the band structure.

10.21 CONCLUSIONS

Interfacial phenomena such as adsorption–desorption, clusterization, separation, structurization, relaxation, diffusion, etc. have certain features at open surfaces (e.g., nanooxides) or in pores of solid and soft materials, biomaterials, bio-objects that depend on textural, morphological, and chemical characteristics of both adsorbates (guest) and adsorbents (host). The adsorption, relaxation, and energetic parameters of nanopowders, porous materials, composite, and hybrid carbon–mineral or polymer–solid materials correlate with their surface composition and textural characteristics. Comparative investigations of water, organics, or their mixtures confined in pores of solid materials and bio-objects using low-temperature NMR, TSDC, calorimetry, and other methods show that the difference in temperature of transition from the local molecular mobility of water and organics bound at the pore walls or intracellular functionalities to molecular (ion) percolation effects (throughout conductivity) can be 10–40 K. This value depends on several factors such as the morphology of particles, textural and structural characteristics of adsorbents, chemical structure of the surfaces, concentrations of components, the presence of dissolved salts (e.g., NaCl) and other compounds. This result confirms a clustered structure of the adsorption layer.

Besides bulk free water, there are four types of water bound to adsorbents or bio-objects such as strongly (changes in the Gibbs free energy $-\Delta G > 0.5$–1 kJ/mol) and weakly ($-\Delta G < 0.5$–1 kJ/mol) bound and strongly ($\delta_H = 4$–5 ppm) and weakly ($\delta_H = 1$–2 ppm) associated waters. This boundary value $-\Delta G = 0.5$–1 kJ/mol depends on system composition and the textural characteristics of the materials. At low amounts of adsorbed water (much smaller than the pore volume), typically all water is strongly bound. A portion of this water can be weakly associated if the pore walls are mosaic (partially modified oxides or hybrid carbon–mineral or polymer–mineral systems) and composed with hydrophilic and hydrophobic patches such as partially silylated silica surface. Adsorbed water clusterization depends also on the presence of co-adsorbates such as chloroform, acetonitrile, DMSO, benzene, etc.

The local molecular mobility of bound water, organics, or water–organic mixtures depicts a non-Arrhenius character or includes several Arrhenius-type processes characterized by different activation energies because the corresponding structures are differently clustered and bound to different surface functionalities at the interfaces. Therefore, the dynamic characteristics of these types of bound molecules can significantly differ. The temperature dependence of ion percolation in bound water has the Arrhenius character for relatively uniform systems such as the aqueous suspensions of nanosilica or mesoporous silicas. However, in the case of complex materials (e.g., mixed nanooxides, cells, etc.) or adsorbents possessing both broad and very narrow pores (e.g., activated carbons) certain deviations from the Arrhenius behavior are observed on the dc relaxation or α relaxation and glass transition observed by DRS, DSC, or other methods. The behavior of interfacial water depends strongly on the presence of nonpolar or polar organic co-adsorbates, which can displace a portion of water from narrow pores into larger or smaller ones, change the associativity of bound water and its interaction energy with solid surfaces or intracellular functionalities. However, confined water and polar organics are much poorly mixed in narrow pores than in the bulk that leads to enhanced clusterization of water–organics bound to the mosaic surfaces.

Comparison of the NMR and TSDC data (the latter technique is appropriate to study relaxation phenomena in any materials because electron clouds in atoms, molecules, solids, and other materials sensitive to an external electrostatic field, and corresponding reorientation of dipoles or location of space charge can be frozen) for thawing of the bound water and other adsorbates at the interfaces gives information on different processes at the interfaces such as local, segmental, cluster and full mobility of a liquid phase, depolarization, dipolar, dc and space charge relaxation origin, etc. It is possible to estimate the activation energy of relaxation, relaxation time, glass transition temperature, contribution of crystalline or amorphous phases, diffusion, denaturation of proteins, inactivation of cells, and many other interfacial processes. The NMR spectra, TSDC, and DSC thermograms demonstrate the dependence of the characteristic structural and energetic parameters of polymers, powders, cells, seeds, and other objects on their hydration–dehydration.

The NMR results are in agreement with results from DRS, TSDC, DSC, XRD, and other methods with respect to clusterization of the adsorption layers. It is possible to estimate the size distribution functions of water clusters that become (melting) or remain (freezing) liquid at $T<273$ K (NMR) or relaxing (TSDC, DRS, DSC) at different temperatures. TSDC and NMR cryoporometry give results for solid adsorbents in agreement with the structural characteristics estimated from nitrogen adsorption isotherms. However, NMR and TSDC cryoporometry can be applied to systems (e.g., seeds, cells, tissues) which cannot be studied by conventional adsorption methods to determine their structural and textural characteristics.

The low-temperature ^{1}H NMR spectroscopy used to determine the interfacial energy of biomacromolecules and related bionanocomposites has some advantages in comparison with other methods measuring similar characteristics. In contrast to calorimetric method determining the adhesion energy, it does not require long time to reach equilibrium and it is more sensitive at low concentrations of solid phase. Additionally, it allows the determination of the radial dependences of adhesive forces in aqueous media, size distributions of cavities (pores, voids) filled by unfrozen bound water in any materials, and the thickness of bound water layers that is impossible or more difficult by using other methods. The concentration dependences of interfacial energy can be used to determine the energy of intermolecular interaction of low- or high-molecular compounds, energy of self-association, and energy of swelling or destroying of gel or gel-like structures formed in the suspensions of nanoparticles or supramolecular systems. For aqueous solutions of biopolymers, solid nanoparticles, and low-molecular organics (e.g., sugars), the dependence of interfacial energy on concentrations of dispersion components are quite informative since it is possible to trace the processes of adsorption, gel formation, coagulation, etc. The obtained results for supramolecular structures with nanopowders, proteins, and mono- and disugars allow us to explain certain features of the influence of the bionanocomposites on living cells, in particular the effects of saccharose on the activity of RBCs and their transformations from normal discocytes to echinocytes, spherocytes to shadow corpuscles on interaction with bionanocomposites.

Water bound at the hydrophilic–hydrophobic interfaces can be assigned to several structural types which can also be analyzed in terms of high-density (HDW) and low-density (LDW) waters with collapsed (CS) or expanded (ES) structures. The molecular mobility of weakly associated interfacial water (unusual water because atypical value $\delta_H = 1$–2 ppm) depends weaker on temperature, and it is higher than that of strongly associated water because of smaller number of the hydrogen bonds per molecule. Additionally, this water can be characterized by considerable bending of the hydrogen bonds or it corresponds to interstitial water in the interfacial HDW layer. These differences in typical (bulk or weakly clustered) and unusual (strongly clustered) interfacial waters result in the different temperature dependences of their ^{1}H NMR spectra at $T<273$ K. Immersion of the mosaic hydrophobic–hydrophilic systems (e.g., partially silylated fumed silica or silica gel, proteins, yeast cells, wheat seeds, and bone or muscular tissues containing certain amounts of adsorbed water) in weakly polar or nonpolar solvents ($CDCl_3$, benzene, CCl_4) allows increasing contribution of weakly associated bound water, i.e., more clustered water. This water gives a peak in the ^{1}H NMR spectra at $\delta_H = 1$–2 ppm, which demonstrates the atypical behavior with lowering temperature.

The unusual properties of the weakly associated interfacial water can be explained by its augmented mixing with a hydrophobic weakly polar solvent at the mosaic hydrophobic–hydrophilic interfaces in comparison with its solubility in the bulk of this solvent because of very local chaotropic and kosmotropic effects of adjacent different surface and macromolecular hydrophobic–hydrophilic functionalities in small confined space of the voids (pores, pockets) filled by water or water–organic solvent. The weakly polar solvent (chloroform) or its mixture with a polar solvent (DMSO, acetonitrile) can provide shielding of nonpolar surface functionalities from interfacial water. The Gibbs free energy of water solved in $CHCl_3$ is negative (experimental $\Delta G_{s,exp} \approx -5$ kJ/mol) in contrast to its solution in nonpolar solvents (e.g., $\Delta G_{s,exp} \approx 11$ kJ/mol for water solved in heptane, and ΔG_s is positive for water solved other aliphatic solvents and vice versa; however, it is negative for aromatic solvents, e.g., $\Delta G_{s,exp} \approx -4$ kJ/mol for water solved in benzene or toluene), which can be considered for explanation of separation of water in tiny clusters in vicinity to both polar silanols or other polar groups and nonpolar CH functionalities of hydrophobized solid surfaces or intracellular biostructures. The polarity both of water and weakly polar solvent molecules can increase near polar patches of the surface that can result in increase in their mixing at the mosaic interfaces, despite decrease activity of the adsorbed water, i.e., dissolution of nonpolar or weakly polar solvents in the interfacial water (as well in the bulk water) is lower than that for water in these solvents. The use of solvents of lower polarity other than chloroform cannot provide the appearance of unusual interfacial water at $\delta_H = 1–2$ ppm if $\Delta G_s \rightarrow 0$ or $\Delta G_s > 0$. However, in the case of a low content of the interfacial water, similar effects could be observed even without a weakly polar solvent because water is distributed in the form of tiny clusters with strong bending of the hydrogen bonds and individual molecules interacting with nonpolar or weakly polar organic functionalities. This results in the appearance of the ^{1}H NMR signal at $\delta_H = 1.2–1.7$ ppm, as in the case of minimum hydrated yeast cells, bone, and muscular tissues.

Thus, soft tissues in the body can contain certain quantity of weakly associated water. One can assume that this unusual water is typical for hydrophilic–hydrophobic biosystems of different origin such as muscular and bone tissues and cells. This makes it possible to assume that weakly associated water can fulfill an important function in the cells and multicellular biosystems. In particular, it can stabilize the conformation of biomacromolecules, participate in transmembrane transfer of hydrophobic and high-molecular substances, for which the penetration through the layer containing tiny clusters of water requires smaller reconstruction of the intermolecular bonds in comparison with that in typically clustering water with the microdomain structure. Since the weakly associated water does not freeze during cooling of biological objects up to 190 K, it can contribute to the safety of living organisms on cryopreservation. Possibly, new methods of cryopreservation can be developed by creating conditions of passing of certain amounts of intracellular water into the weakly associated state. Thus, the present investigations of the properties of unusual interfacial water give additional information to the conception of high-density and low-density waters with respect to the interfacial phenomena at mosaic hydrophobic–hydrophilic surface structures or endocellular functionalities.

ACKNOWLEDGMENTS

These investigations were partially supported by the Seventh Framework Programme (FP7/2007–2013), Marie Curie International Research Staff Exchange Scheme (grant no. 230790), and the STCU (Ukraine) (grants no. 1946, 3832, and 4481). The authors thank Prof. A.V. Turov (Taras Shevchenko University, Kiev, Ukraine), Dr. V.I. Zarko, Dr. E.F. Voronin, Dr. V.M. Bogatyrev, Dr. E.M. Pakhlov, Dr. E.V. Goncharuk, Dr. O.I. Oranska, Dr. T.V. Kulyk, Dr. V.N. Barvinchenko, Dr. A.A. Turova, Dr. T.V. Krupska, Dr. A.P. Golovan, Dr. G.R. Yurchenko, Dr. A.A. Rugal, Dr. M.V. Borysenko, and Dr. G.P. Prykhodko (Chuiko Institute of Surface Chemistry, Kiev, Ukraine), Prof. J.P. Blitz (Eastern Illinois University, Charleston, Illinois), Prof. R. Leboda, Prof. W. Janusz, and Dr. J. Skubiszewska-Zięba (Maria Curie-Sklodowska University, Poland), Prof. S.V. Mikhalovsky, Dr. L.I. Mikhalovska, Dr. R.L.D. Whitby, and Dr. I.N. Savina (University of Brighton, United Kingdom), Dr. O.P. Kozynchenko (MAST Carbon International Ltd, United Kingdom), Prof. V.A. Bershtein (Ioffe Physical-Technical Institute, St. Petersburg, Russia), Prof. P. Pissis (National Technical University of Athens, Greece), Prof. I.F. Myronyuk (Stefanyk Precarpathion National University, Ivano-Frankivsk, Ukraine), and Prof. Y.G. Ptushinskii (Institute of Physics, Kiev, Ukraine) for their help in synthesis and characterization of materials and results discussion.

References

Abe, I., Sato, K., Abe, H., and Naito, M. 2008. Formation of porous fumed silica coating on the surface of glass fibers by a dry mechanical processing technique. *Adv. Powder Technol.* 19:311–320.

Abragam, A. 1961. *The Principles of Nuclear Magnetism.* London, U.K.: Oxford University Press.

Abragam, A. 1989. *Principles of Nuclear Magnetic Resonance.* Oxford, U.K.: Oxford Science Publications.

Abraham, R.J., Fisher, J., and Loftus, P. 1988. *Introduction to Spectroscopy.* New York: John Wiley & Sons.

Abrahamson, J. 1973. The surface energies of graphite. *Carbon* 11:337–362.

Adamson, A.W. and Gast, A.P. 1997. *Physical Chemistry of Surface*, 6th edn. New York: Wiley.

Agoudjila, N., Kermadia, S., and Larbot, A. 2008. Synthesis of inorganic membrane by sol–gel process. *Desalination* 223:417–424.

Aguzzi, C., Cerezo, P., Viseras, C., and Caramella, C. 2007. Use of clays as drug delivery systems: Possibilities and limitations. *Appl. Clay Sci.* 36:22–36.

Ahola, S., Perlo, J., Casanova, F., Stapf, S., and Bluemich, B. 2006. Multiecho sequence for velocity imaging in inhomogeneous RF fields. *J. Magn. Reson.* 182:143–151.

Aihara, Y., Sonai, A., Hattori, M., and Hayamizu, K. 2006. Ion conduction mechanisms and thermal properties of hydrated and anhydrous phosphoric acids studied with ^{1}H, ^{2}H, and ^{31}P NMR. *J. Phys. Chem. B* 110:24999–25006.

Akitt, J.W. and Mann, B.E. 2000. *NMR and Chemistry.* Cheltenham, U.K.: Stanley Thornes.

Akporiaye, D., Hansen, E.W., Scmidt, R., and Stöcker, M. 1994. Water-saturated mesoporous MCM-41 systems characterized by ^{1}H NMR. *J. Phys. Chem.* 98:1926–1928.

Aksnes, D.W. and Kimtys, L. 2004. ^{1}H and ^{2}H NMR studies of benzene confined in porous solids: Melting point depression and pore size distribution. *Solid State Nucl. Magn. Reson.* 25:146–152.

Al-27. 2011. Solid-state aluminum NMR references. http://www.pascal-man.com/periodic-table/aluminum.shtml (accessed April 02, 2011).

Al-Abadleh, H.A. and Grassian, V.H. 2003. Oxide surfaces as environmental interfaces. *Surf. Sci. Rep.* 52:63–161.

Alberius, P., Frindell, K., Hayward, R., Kramer, E., Stucky, G., and Chmelka, B. 2002. General predictive synthesis of cubic, hexagonal and lamellar silica and titania mesostructured thin films. *Chem. Mater.* 14:3284–3294.

Alexander, C., Andersson, H.S., Andersson, L.I. et al. 2006. Molecular imprinting science and technology: A survey of the literature for the years up to and including 2003. *J. Mol. Recogn.* 19:106–180.

Alexeeva, T.A., Lebovka, N.I., Gun'ko, V.M., Strashko, V.V., and Mikhalovsky, S.V. 2004. Characteristics of interfacial water affected by proteins adsorbed on activated carbon. *J. Colloid Interface Sci.* 278:333–341.

Allain, F.H., Gubser, C.C., Howe, P.W., Nagai, K., Neuhaus, D., and Varani, G. 1996. Specificity of ribonucleoprotein interaction determined by RNA folding during complex formulation. *Nature* 380:646–650.

Allen, S.G., Mallett, M.J.D., and Strange, J.H. 2001. Morphology of porous media studied by nuclear magnetic resonance line shapes and spin-echo decays. *J. Chem. Phys.* 114:3258–3264.

Allen, S.G., Stephenson, P.C.L., and Strange, J.H. 1998. Internal surfaces of porous media studied by nuclear magnetic resonance cryoporometry. *J. Chem. Phys.* 108:8195–8198.

Allison, S.D. 2008. Effect of structural relaxation on the preparation and drug release behaviour of poly(lactic-co-glycolic) acid microparticle drug delivery systems. *J. Pharmacol. Sci.* 97:2022–2035.

Alnaimi, S.M., Mitchell, J., Strange, J.H., and Webber, J.B.W. 2004. Binary liquid mixtures in porous solids. *J. Chem. Phys.* 120:2075–2077.

Alonso, B., Harris, R.K., and Kenwright, A.M. 2002. Micellar solubilization: Structural and conformational changes investigated by ^{1}H and ^{13}C liquid-state NMR. *J. Colloid Interface Sci.* 251:366–375.

Alonso, B. and Massiot, D. 2003. Multi-scale NMR characterisation of mesostructured materials using 1H→13C through-bond polarisation transfer, fast MAS, and 1H spin diffusion. *J. Magn. Res.* 163:347–352.

Anderluh, G., Razpotnik, A., Podlesek, Z., Macek, P., Separovic, F., and Norton, R.S. 2005. Interaction of the eukaryotic pore-forming cytolysin equinatoxin II with model membranes: F-19 NMR studies. *J. Mol. Biol.* 347:27–39.

Andersen, M.D., Jakobsen, H.J., and Skibsted, J. 2004. Characterization of white Portland cement hydration and the C-S-H structure in the presence of sodium aluminate by ^{27}Al and ^{29}Si MAS NMR spectroscopy. *Cem. Concr. Res.* 34:857–868.

Anderson, M.W., Rocha, J., Lin, Z., Philippou, A., Orion, I., and Ferreira, A. 1996. Isomorphous substitution in the microporous titanosilicate ETS-10. *Micropor. Mater.* 6:195–204.

Anglin, E.J., Cheng, L., Freeman, W.R., and Sailor, M.J. 2008. Porous silicon in drug delivery devices and materials. *Adv. Drug Deliv. Rev.* 60:1266–1277.

Arce, V.B., Bertolotti, S.G., Oliveira, F.J.V.E. et al. 2009. The use of molecular probes for the characterization of dispersions of functionalized silica nanoparticles. *Spectrochim. Acta A* 73:54–60.

Argauer, R.J. and Landolt, G.R. 1972. US Pat. 3 702 886.

Armitage, W.J. and Juss, B.K. 1996. The influence of cooling rate on survival of frozen cells differs in monolayers and in suspensions. *Cryo Lett.* 17:213–218.

Arruebo, M. 2012. Drug delivery from structured porous inorganic materials. *WIDRs: Nanomed. Nanobiotechnol.* 4:16–30.

Arshady, R. (Ed.) 1999. *Medical & Biotechnology Applications*. London, U.K.: Citus Books.

ASTM International-Standards Worldwide. Standard guide for assessing microstructure of polymeric scaffolds for use in tissue-engineered medical products. http://www.astm.org/Standards/F2450.htm (April 2010). doi: 10.1520/F2450-10.

Atkins, D., Kékicheff, P., and Spalla, O. 1997. Adhesion between colloidal silica as seen with direct force measurement. *J. Colloid Interface Sci.* 188:234–237.

Attard, P. and Parker, J.L. 1992. Oscillatory solvation forces: A comparison of theory and experiment. *J. Phys. Chem.* 96:5086–5093.

Atwell, A.W., Hood, P.L., Lineback, R.D., Varriano-Martson, E., and Zobel, F.H. 1988. The terminology and methodology associated with basic starch phenomena. *Cereal Foods World* 33:306–311.

Auner, N. and Weis, J. (Eds.) 1996. *Organosilicon Chemistry II. From Molecules to Materials*. Weinheim, Germany: VCH.

Auner, N. and Weis, J. (Eds.) 2003. *Organosilicon Chemistry V. From Molecules to Materials*. Weinheim, Germany: Wiley.

Auner, N. and Weis, J. (Eds.) 2005. *Organosilicon Chemistry VI. From Molecules to Materials*. Weinheim, Germany: Wiley.

Auweter, S.D., Fasan, R., Reymond, L. et al. 2006. Molecular basis of RNA recognition by the human alternative splicing factor Fox-1. *EMBO J.* 25:163–173.

Babonneau, F. and Maquet, J. 2000. Nuclear magnetic resonance techniques for the structural characterization of siloxane–oxide hybrid materials. *Polyhedron* 19:315–322.

Babu, C.R., Flynn, P.F., and Wand, A.J. 2003. Preparation, characterization, and NMR spectroscopy of encapsulated proteins dissolved in low viscosity fluids. *J. Biomol. NMR* 25:313–323.

Baccile, N. and Babonneau, F. 2008. Organo-modified mesoporous silicas for organic pollutant removal in water: Solid-state NMR study of the organic/silica interactions. *Micropor. Mesopor. Mater.* 110:534–542.

Baccile, N., Laurent, G., Bonhomme, C., Innocenzi, P., and Babonneau, F. 2007. Solid-state NMR characterization of the surfactant–silica interface in templated silicas: Acidic versus basic conditions. *Chem. Mater.* 19:1343–1354.

Bae, H.J., Darby, D.O., Kimmel, R.M., Park, H.J., and Whiteside, W.S. 2009. Effects of transglutaminase-induced cross-linking on properties of fish gelatin-nanoclay composite film. *Food Chem.* 114:180–189.

Baechle, T.R. and Earle, R.W. 2008. *Essentials of Strength Training and Conditioning*, 2nd edn. Champaign, IL: Human Kinetics.

Bakry, R., Vallant, R.M. Najam-ul-Haq, M. et al. 2007. Medicinal applications of fullerenes. *Int. J. Nanomed.* 2:639–649.

Balard, H., Donnet, J.-B., Oulanti, H., Gottschalk-Gaudig, T., and Barthel, H. 2011. Study of aging of pyrogenic silicas by gravimetry and microcalorimetry. *Colloids Surf. A: Physicochem. Eng. Aspects* 378:38–43.

Baldwin, A.J. and Kay, L.E. 2009. NMR spectroscopy brings invisible protein states into focus. *Nat. Chem. Biol.* 5:808–814.

Ballard, L. and Sloan, Jr., E.D. 2004. The next generation of hydrate prediction IV A comparison of available hydrate prediction programs. *Fluid Phase Equilib.* 216:257–270.

Bamford, C.H., Cooper, S.L., and Tsuruta, T. (Eds.) 1992. *The Vroman Effect*. Utrecht, the Netherlands: VSP.

Bandman, E. 1999. Solubility of myosin and the binding quality of meat products. *Proceedings of 45th International Congress of Meat Science & Technology*, Yokohama, Japan, pp. 236–244.

Bandosz, T.J. (Ed.) 2006. *Activated Carbon Surfaces in Environmental Remediation*. Oxford, U.K.: Elsevier.

Bann, J.G., Pinkner, J., Hultgren, S.J., and Frieden, C. 2002. Real-time and equilibrium F-19-NMR studies reveal the role of domain–domain interactions in the folding of the chaperone PapD. *Proc. Natl Acad. Sci. USA* 99:709–714.

Bansal, R.C., Donnet, J.B., and Stoeckli, F. 1988. *Active Carbon*. New York: Marcel Dekker.

Bansal, R.C. and Goyal, M. 2005. *Activated Carbon Adsorption*. Boca Raton, FL: Taylor & Francis Group.

Barbetta, A., Cametti, C., Rizzitelli, G., and Dentini, M. 2012. Geometrical characterization of polymeric matrices by means of dielectric spectroscopy measurements. *Soft Matter* 8:1120–1129.

Barker, P.B. and Lin, D.D.M. 2006. In vivo proton MR spectroscopy of the human brain. *Prog. Nucl. Magn. Reson. Spectrosc.* 49:99–128.

Barone, J.R. and Schmidt, W.F. 2005. Polyethylene reinforced with keratin fibres obtained from chicken feathers. *Compos. Sci. Technol.* 65:173–181.

Barrett, E.P., Joyner, L.G., and Halenda, P.C. 1951. The determination of pore volume and area distributions in porous substance. I. Computations from nitrogen isotherm. *J. Am. Chem. Soc.* 73:373–380.

Barthel, H. 1995. Surface interactions of dimethylsiloxy group-modified fumed silica. *Colloids Surf. A: Physicochem. Eng. Aspects* 101:217–226.

Barthel, H., Heinemann, M., Stintz, M., and Wessely, B. 1998a. Particle sizes of fumed silica. *An International Conference on Silica Science and Technology "Silica 98"*, Mulhouse, France, September 1–4, 1998, pp. 323–326.

Barthel, H., Heinemann, M., Stintz, M., and Wessely, B. 1998b. Particle sizes of fumed silica. *Chem. Eng. Technol.* 21:745–752.

Barwick, V.J. 1997. Strategies for solvent selection. *Trends Anal. Chem.* 16:293–310.

Bastow, T.J., Moodie, A.F., Smith, M.E., and Whitfield, H.J. 1993. Characterisation of titania gels by ^{17}O nuclear magnetic resonance and electron diffraction. *J. Mater. Chem.* 3:697–702.

Bäumer, M. and Freund, H.-J. 1999. Metal deposits on well-ordered oxide films. *Prog. Surf. Sci.* 61:127–198.

Bax, A., Freeman, R., and Kempsell, S.P. 1980. Natural abundance carbon-13-carbon-13 coupling observed via double-quantum coherence. *J. Am. Chem. Soc.* 102:4849–4851.

Baxter-Drayton, Y. and Brady, J.F. 1996. Brownian electrorheological fluids as a model for flocculated dispersions. *J. Rheol.* 40:1027–1056.

Beardsley, D.J.S., Tang, C., Chen, B.-G., Lamborn, C., Gomes, E., and Srimatkandada, V. 2003. The disulfide-rich region of platelet glycoprotein (GP) IIIa contains hydrophilic peptide sequences that bind anti-GPIIIa autoantibodies from patients with immune thrombocytopenic purpura (ITP). *Biophys. Chem.* 105:503–515.

Bebris, N.K. and Kiselev, A.V. 1978. Liquid chromatography on carbosilochrom and carbosilica gel–silica adsorbents with a modified carbon surface. *Chromatographia* 11:206–211.

Beck, J.S., Vartuli, J.C., Roth, W.J. et al. 1992. A new family of mesoporous molecular sieves prepared with liquid crystal templates. *J. Am. Chem. Soc.* 114:10834–10843.

Bell, A.T. and Pines, A. (Eds.) 1994. *NMR Techniques in Catalysis*. New York: Marcel Dekker.

Bellotti, M., Martinelli, A., Gianferri, R., and Brosio, E. 2010. A proton NMR relaxation study of water dynamics in bovine serum albumin nanoparticles. *Phys. Chem. Chem. Phys.* 12:516–522.

Belon, C., Chemtob, A., Croutxé-Barghorn, C. et al. 2010. Nanocomposite coatings via simultaneous organic–inorganic photo-induced polymerization: Synthesis, structural investigation and mechanical characterization. *Polym. Int.* 59:1175–1186.

Bennekou, P., Barkmann, T.L., Christophersen, P., and Kristensen, B.I. 2006. The human red cell voltage-dependent cation channel. Part III: Distribution homogeneity and pH dependence. *Blood Cell. Mol. Dis.* 36:10–14.

Benzinger, W. and Hüttinger, K.J. 1999a. Chemical vapor infiltration of pyrocarbon—III: The influence of increasing methane partial pressure at increasing total pressure on infiltration rate and degree of pore filling. *Carbon* 37:181–193.

Benzinger, W. and Hüttinger, K.J. 1999b. Chemistry and kinetics of chemical vapor infiltration of pyrocarbon-IV. Investigation of methane/hydrogen mixtures. *Carbon* 37:931–940.

Bergna, H.E. (Ed.) 1994. *The Colloid Chemistry of Silica*. Adv. Chem. Ser., Vol. 234. Washington, DC: American Chemical Society.

Bergna, H.E. (Ed.) 2005. *Colloidal Silica: Fundamentals and Applications*. Salisbury, MD: Taylor & Francis Group.

Bernstein, H.J., Schneider, W.G., and Pople, J.A. 1956. The proton magnetic resonance spectra of conjugated aromatic hydrocarbons. *Proc. R. Soc. A* 236:515–520.

Bershtein, V.A. and Egorov, V.M. 1994. *Differential Scanning Calorimetry of Polymers. Physics, Chemistry, Analysis, Technology*. New York: Ellis Horwood.

Bershtein, V., Gun'ko, V., Egorova, L. et al. 2009. Well-defined silica core-poly(vinyl pyrrolidone) shell nanoparticles: Interactions and multi-modal glass transition dynamics at interfaces. *Polymer* 50:860–871.

Bershtein, V., Gun'ko, V., Egorova, L. et al. 2010a. Well-defined oxide core-poly(vinyl pyrrolidone) shell nanoparticles: Interactions and multi-modal glass transition dynamics at interfaces. *Macromol. Symp.* 296:541–549.

Bershtein, V.A., Gun'ko, V.M., Egorova, L.M. et al. 2010b. Well-defined oxide core-polymer shell nanoparticles: Interfacial interactions, peculiar dynamics and transitions in polymer nanolayers. *Langmuir* 26:10968–10979.

Bershtein, V.A., Gun'ko, V.M., Karabanova, L.V. et al. 2010c. Hybrid polyurethane-poly(2-hydroxyethyl methacrylate) semi-IPN-silica nanocomposites: Interfacial interactions and glass transition dynamics. *J. Macromol. Sci. B: Phys.* 49:18–32.

Berthod, A., Rollet, M., and Farah, N. 1988. Dry adsorbed emulsions: An oral sustained drug delivery system. *J. Pharmacol. Sci.* 77:216–221.

Betega de Paiva, L., Morales, A.R., and Díaz, F.R.V. 2008. Organoclays: Properties, preparation and applications. *Appl. Clay Sci.* 42:8–24.

Beutel, T., Peltre, M.-J., and Su, B.L. 2001. Interaction of phenol with NaX zeolite as studied by ^{1}H MAS NMR, ^{29}Si MAS NMR and ^{29}Si CP MAS NMR spectroscopy. *Colloids Surf. A: Physicochem. Eng. Aspects* 187–188:319–325.

Beysens, D., Boccara, N., and Forgacs, G. (Eds.) 1993. *Dynamical Phenomena at Interfaces, Surfaces and Membranes*. New York: Nova Science Publishers.

Billsten, P., Wahlgren, M., Arnebrant, T., McGuire, J., and Elwing, H. 1995. Structural-changes of T4 lysozyme upon adsorption to silica nanoparticles measured by circular dichroism. *J. Colloid Interface Sci.* 175:77–82.

Binks, B.P. and Murakami, R. 2006. Phase inversion of particle-stabilized materials from foams to dry water. *Nat. Mater.* 5:865–869.

Bircan, C. and Barringer, S.A. 1998. Salt-starch interactions as evidenced by viscosity and dielectric property measurements. *J. Food Sci.* 6:983–986.

Birdi, K.S. (Ed.) 2009. *Handbook of Surface and Colloid Chemistry*, 3rd edn. Boca Raton, FL: Taylor & Francis Group.

Biscoc, J. and Warren, B.E. 1942. Structure of carbon black surface. *J. Appl. Chem.* 13:364–371.

Bizunok, S.N., Popov, V.G., and Sventitskij, E.N. 1984. Study of low-temperature crystallization in *E. coli* cells by NMR method. *Biophysica* 29:268–271 (in Russian).

Bladgen, N., de Matas, M., Gavan, P., and York, P. 2007. Crystal engineering of active pharmaceutical ingredients to improve solubility and dissolution rates. *Adv. Drug Deliv. Rev.* 59:617–630.

Bland, P.A., Jackson, M.D., Coker, R.F. et al. 2009. Why aqueous alteration in asteroids was isochemical: High porosity≠high permeability. *Earth Planet. Sci. Lett.* 287:559–568.

Blitz, J.P. and Gun'ko, V.M. (Eds.) 2006. *Surface Chemistry in Biomedical and Environmental Science*. NATO Science Series II: Mathematics, Physics and Chemistry, Vol. 228. Dordrecht, the Netherlands: Springer.

Blum, F.D. and Krisanangkura, P. 2009. Comparison of differential scanning calorimetry, FTIR, and NMR to measurements of adsorbed polymers. *Thermochim. Acta* 492:55–60.

Blümich, B., Perlo, J., and Casanova, F. 2008. Mobile single-sided NMR. *Prog. Nucl. Magn. Reson. Spectrosc.* 52:197–269.

Bockel, C. and Thomy, A. 1981. Adsorption properties of exfoliated graphite's prepared by dissociation of different intercalation compounds. *Carbon* 19:142.

Bodanszky, M. and Bodanszky, A. 1984. *The Practice of Peptide Synthesis*. Berlin, Germany: Springer-Verlag.

Boddenberg, B. and Grosse, R. 1988. ^{2}H NMR pattern of methylaromatics adsorbed on graphite. *Z. Naturforsch. A* 43:497–504.

Boddenberg, B. and Moreno, J.A. 1977. NMR studies of structure and dynamics of physically adsorbed layers on uniform solid surfaces. *J. Phys.* 38:C4-52–C4-55.

Boddenberg, B. and Neue, G. 1989. An NMR study of am n-butane film on graphite. I. ^{2}H relaxation of the 2D fluid state. *Mol. Phys.* 67:385–398.

Bodurka, J., Buntkowsky, G., Olechnowicz, R., Gutsze, A., and Limbach, H.-H. 1996. Investigations of water in normal and dehydrated rabbit lenses with ^{1}H NMR and calorimetric measurements. *Colloids Surf. A: Physicochem. Eng. Aspects* 115:55–62.

Boehm, H.P. 1966. Chemical identification of surface groups. *Adv. Catal.* 16:179–274.

Bogdan, A., Kulmala, M., Gorbunov, B., and Kruppa, A.I. 1996. NMR study of phase transitions in pure water and binary H_2O/HNO_3 films adsorbed on surface of pyrogenic silica. *J. Colloid Interface Sci.* 177:79–87.

Boissiere, C., Grosso, D., Chaumonnot, A., Nicole, L., and Sanchez, C. 2011. Aerosol route to functional nanostructured inorganic and hybrid porous materials. *Adv. Mater.* 23:599–623.

Bonneviot, L., Belard, F., Danumah, C., Giasson, S., and Kaliaguine, S. (Eds.) 1998. *Mesoporous Molecular Sieves, Studies in Surface Science and Catalysis*, Vol. 117. Amsterdam, the Netherlands: Elsevier.

Bordes, P., Pollet, E., and Avérous, L. 2009. Nano-biocomposites: Biodegradable polyester/nanoclay systems. *Prog. Polym. Sci.* 34:125–155.

Börner, H.G. 2009. Strategies exploiting functions and self-assembly properties of bioconjugates for polymer and materials sciences. *Prog. Polym. Sci.* 34:811–851.

Borsacchi, S., Geppi, M., Veracini, C.A., Lazzeri, A., Di Cuia, F., and Geloni, C. 2008. A multinuclear solid-state magnetic resonance study of the interactions between the inorganic and organic coatings of $BaSO_4$ submicronic particles. *Magn. Reson. Chem.* 46:52–57.

Borysenko, M.V., Gun'ko, V.M., Dyachenko, A.G., Sulim, I.Y., Leboda, R., and Skubiszewska-Zięba, J. 2005. CVD-zirconia on fumed silica and silica gel. *Appl. Surf. Sci.* 242:1–12.

Bose, A.B., Gangoda, M., Jaroniec, M., Gilpin, R.K., and Bose, R.N. 2006. Two-dimensional solid state NMR characterization of physisorbed siloxane polymer (OV-225) on silica. *Surf. Sci.* 600:143–154.

Bothner-By, A.A. and Glick, R.E. 1956. Specific medium effects in nuclear magnetic resonance spectra of liquids. *J. Am. Chem. Soc.* 78:1071–1072.

Bouchard, M., Pare, C., Dutasta, J., Chauvet, J., Gicquaud, C., and Auger, M. 1998. Interaction between G-actin and various types of liposomes: A F-19, P-31 and H-2 nuclear magnetic resonance study. *Biochemistry* 37:3149–3155.

Boulègue, C., Milbradt, A.G., Renner, C., and Moroder, L. 2006. Single proline residues can dictate the oxidative folding pathways of cysteine-rich peptides. *J. Mol. Biol.* 358:846–856.

Bourrat, X. 1993. Electrically conductive grades of carbon black: Structure and properties. *Carbon* 31:287–302.

Bradley, R.H. and Rand, B.J. 1995. On the physical adsorption of vapors by microporous carbons. *J. Colloid Interface Sci.* 169:168–176.

Braga, P.R.S., Costa, A.A., de Macedo, J.L. et al. 2011. Liquid phase calorimetric-adsorption analysis of Si-MCM-41: Evidence of strong hydrogen-bonding sites. *Micropor. Mesopor. Mater.* 139:74–80.

Brar, S.K., Verma, M., Tyagi, R.D., and Surampalli, R.Y. 2010. Engineered nanoparticles in wastewater and wastewater sludge—Evidence and impacts. *Waste Manage.* 30:504–520.

Bräuniger, T., Madhu, P.K., Pampel, A., and Reichert, D. 2004. Application of fast amplitude-modulated pulse trains for signal enhancement in static and magic-angle-spinning ^{47,49}Ti-NMR spectra. *Solid State Nucl. Magn. Reson.* 26:114–120.

Bredereck, K., Effenberger, F., and Tretter, M. 2011. Preparation and characterization of silica aquasols. *J. Colloid Interface Sci.* 360:408–414.

Breinan, H.A., Minas, T., Hsu, H.P., Nehrer, S., Sledge, C.B., and Spector, M. 1997. Effect of cultured autologous chondrocytes on repair of chondral defects in a canine model. *J. Bone Joint Surg. Am.* 79A:1439–1451.

Brender, J.R., Dürr, U.H.N., Heyl, D., Budarapu, M.B., and Ramamoorthy, A. 2007. Membrane fragmentation by an amyloidogenic fragment of human Islet Amyloid Polypeptide detected by solid-state NMR spectroscopy of membrane nanotubes. *Biochim. Biophys. Acta Biomembr.* 1768:2026–2029.

Brennan, J.K., Bandosz, T.J., Thomson, K.T., and Gubbins, K.E. 2001. Water in porous carbons. *Colloids Surf. A: Physicochem. Eng. Aspects* 187–188:539–568.

Briggs, D. and Seach, M.P. (Eds.) 1983. *Practical Surface Analysis by Auger and X-Ray Photoelectron Spectroscopy*. Chichester, U.K.: John Wiley & Sons.

Brinker, C.J. and Scherer, G.W. 1990. *Sol-Gel Science, the Physics and Chemistry of Sol-Gel Processing*. New York: Academic Press.

Bronnimann, C.E., Ridenour, C.F., Kinney, D.R., and Maciel, G.E. 1992. 2D ^{1}H-^{13}C heteronuclear correlation spectra of representative organic solids. *J. Magn. Reson.* 97:522–534.

Brouwer, D.H., Kristiansen, P.E., Fyfe, C.A., and Levitt, M.H. 2005. Symmetry-based ^{29}Si dipolar recoupling magic angle spinning NMR spectroscopy: A new method for investigating three-dimensional structures of zeolite frameworks. *J. Am. Chem. Soc.* 127:542–543.

Brovchenko, I. and Oleinikova, A. 2008. *Interfacial and Confined Water*. Amsterdam, the Netherlands: Elsevier.

Brown, M.F. 1982. Theory of spin-lattice relaxation in lipid bilayers and biological membranes. ^{2}H and ^{14}N quadrupolar relaxation. *J. Chem. Phys.* 77:1576–1599.

Brown, M.F. 1984. Theory of spin-lattice relaxation in lipid bilayers and biological membranes. Dipolar relaxation. *J. Chem. Phys.* 80:2808–2831.

Brown, W. 1993. *Dynamic Light Scattering. The Method and Some Applications*. Oxford, U.K.: Clarendon Press.

Brown, M.F. 1996. Membrane structure and dynamics studied with NMR spectroscopy. In: Merz, K., Jr. and Roux, B. (Eds.) *Biological Membranes. A Molecular Perspective from Computation and Experiment*. Basel, Switzerland: Birkhäuser, pp. 175–252.

Brown, S.P. 2012. Applications of high-resolution ^{1}H solid-state NMR. *Solid State Nucl. Magn. Reson.* 41:1–27.
Brown, R.J.S. and Fatt, I. 1956. Measurements of fractional wettability of oil-field rocks by the nuclear magnetic relaxation method. *Pet. Trans. AIME* 207:262–270.
Brown Jr., G.E., Henrich, V.E., Casey, W.H. et al. 1999. Metal oxide surfaces and their interactions with aqueous solutions and microbial organisms. *Chem. Rev.* 99:77–174.
Brownstein, K.R. and Tarr, C.E. 1979. Importance of classical diffusion in NMR studies of water in biological cells. *Phys. Rev. A* 19:2446–2453.
Brumberger, H. (Ed.) 1965. *Small Angle X-Ray Scattering*. Syracuse, NY: Gordon and Breach.
Brunet, F., Charpentier, T., Le Caër, S., and Renault, J.-P. 2008. Solid-state NMR characterization of a controlled-pore glass and of the effects of electron irradiation. *Solid State Nucl. Magn. Reson.* 33:1–11.
Brunner, E. and Pfeifer, H. 2008. NMR spectroscopic techniques for determining acidity and basicity. In: Karge, H.G. and Weitkamp, J. (Eds.) *Acidity and Basicity*, Molecular Sieves, Vol. 6. Springer, Berlin, Germany, pp. 1–43.
Brus, J., Spirkova, M., Hlavata, D., and Strachota, A. 2004. Self-organization, structure, dynamic properties, and surface morphology of silica/epoxy films as seen by solid-state NMR, SAXS, and AFM. *Macromolecules* 37:1346–1357.
Bryant, R.G. and Eads, T.M. 1985. Solvent peak suppression in high resolution NMR. *J. Magn. Reson.* 64:312–315.
Bryce, D.L., Gee, M., and Wasylishen, R.E. 2001. High-field chlorine NMR spectroscopy of solid organic hydrochloride salts: A sensitive probe of hydrogen bonding environment. *J. Phys. Chem. A* 105:10413–10421.
Buchholz, A., Wang, W., Arnold, A., Xu, M., and Hunger, M. 2003. Successive steps of hydration and dehydration of silicoaluminophosphates H-SAPO-34 and H-SAPO-37 investigated by in situ CF MAS NMR spectroscopy. *Micropor. Mesopor. Mater.* 57:157–168.
Buchholz, A., Wang, W., Xu, M., Arnold, A., and Hunger, M. 2002. Thermal stability and dehydroxylation of Brønsted acid sites in silicoaluminophosphates H-SAPO-11, H-SAPO-18, H-SAPO-31, and H-SAPO-34 investigated by multi-nuclear solid-state NMR spectroscopy. *Micropor. Mesopor. Mater.* 56:267–278.
Buchner, R. 2008. What can be learnt from dielectric relaxation spectroscopy about ion solvation and association? *Pure Appl. Chem.* 80:1239–1252.
Buckingham, A.D. 1958. Hydrogen bonds force and charge density. *Can. J. Chem.* 38:300–307.
Buesser, B. and Pratsinis, S.E. 2010. Design of aerosol particle coating: Thickness, texture and efficiency. *Chem. Eng. Sci.* 65:5471–5481.
Burr, D.B. 2002. The contribution of the organic matrix to bone's material properties. *Bone* 31:8–11.
Burton, A.W., Zones, S.I., Elomari, S. et al. 2007. Comparative study of three closely related unsolved zeolite structures. In: Xu, R., Gao, Z., Chen, J., and Yan, W. (Eds.) *Zeolites to Porous MOF Materials—The 40th Anniversary of International Zeolite Conference*. Elsevier.
Burum, D.P. and Rhim, W.K. 1979. Analysis of multiple pulse NMR in solids. *J. Chem. Phys.* 71:944–956.
Busca, G.1998. Spectroscopic characterization of the acid properties of metal oxide catalysts. *Catal. Today* 41:191–206.
Butler, M.F. and Cameron, R.E. 2000. A study of molecular relaxations in solid starch using dielectric spectroscopy. *Polymer* 41:2249–2263.
Buyanov, R.A. 1983. *Catalyst Coking*. Novosibirsk, Russia: Nauka.
Buyanov, R.A., Molchanov, V.V., and Boldyrev, V.V. 2009. Mechanochemical activation as a tool of increasing catalytic activity. *Catal. Today* 144:212–218.
Buzzoni, R., Bordiga, S., Ricchiardi, G., Spoto, G. and Zecchina, A. 1995. Interaction of H_2O, CH_3OH, $(CH_3)_2O$, CH_3CN, and pyridine with the superacid perfluorosulfonic membrane Naflon: An IR and Raman study. *J. Phys. Chem.* 99:11937–11951.
Cabot Corporation 2011. Fumed silica and fumed alumina in coatings applications. http://www.cabot-corp.com/Silicas-And-Aluminas (accessed October 27, 2006).
Cacciafesta, P., Humphris, A.D.L., Jandt, K.D., and Miles, M.J. 2000. Human plasma fibrinogen adsorption on ultraflat titanium oxide surfaces studied with atomic force microscopy. *Langmuir* 16:8167–8175.
Calberg, C., Jerome, R., and Grandjean, J. 2004. Solid-state NMR study of poly(ε-caprolactone)/clay nanocomposites. *Langmuir* 20:2039–2041.
Callahan, P.T., Gros, M.A.L., and Pinder, D.N. 1983. The measurement of diffusion using deuterium pulsed field gradient nuclear magnetic resonance. *J. Chem. Phys.* 79:6372–6381.
Calucci, L. and Forte, C. 2009. Proton longitudinal relaxation coupling in dynamically heterogeneous soft systems. *Prog. Nucl. Magn. Reson. Spectrosc.* 55:296–323.

Calucci, L., Galleschi, L., Geppi, M., and Mollica, G. 2004. Structure and dynamics of flour by solid state NMR: Effects of hydration and wheat aging. *Biomacromolecules* 5:1536–1544.

Calzolai, L., Franchini, F., Gilliland, D., and Rossi, F. 2010. Protein–nanoparticle interaction: Identification of the ubiquitin–gold nanoparticle interaction site. *Nano Lett.* 10:3101–3105.

Canal, T. and Peppas, N.A. 1989. Correlation between mesh size and equilibrium degree of swelling of polymeric networks. *J. Biomed. Mater. Res.* 23:1183–1193.

Cao, H., Gan, J., Wang, S. et al. 2008. Novel silica-coated iron–carbon composite particles and their targeting effect as a drug carrier. *J. Biomed. Mater. Res. A* 86:671–677.

Capel-Sanchez, M.C., Barrio, L., Campos-Martin, J.M., and Fierro, J.L.G. 2004. Silylation and surface properties of chemically grafted hydrophobic silica. *J. Colloid Interface Sci.* 277:146–153.

Cardoso, M.B., Luckarift, H.R., Urban, V.S., O'Neill, H., and Johnson, G.R. 2010. Protein localization in silica nanospheres derived via biomimetic mineralization. *Adv. Funct. Mater.* 20:3031–3038.

Carroll, S.A., Maxwell, R.S., Bourcier, W., Martin, S., and Hulsey, S. 2002. Evaluation of silica-water surface chemistry using NMR spectroscopy. *Geochim. Cosmochim. Acta* 66:913–926.

Cataldo, F., Capitani, D., Proietti, N., and Ragni, P. 2008. γ radiolyzed amorphous silica: A study with ^{29}Si CP-MAS NMR spectroscopy. *Radiat. Phys. Chem.* 77:267–272.

Čejka, J. 2003. Organized mesoporous alumina: Synthesis, structure and potential in catalysis. *Appl. Catal. A: Gen* 254:327–338.

Celzard, A. and Marêché, J.F. 2006. Optimal wetting of active carbons for methane hydrate formation. *Fuel* 85:957–966.

Celzard, A., Marêché, J.F., and Furdin, G. 2005. Modelling of exfoliated graphite. *Prog. Mater. Sci.* 50:93–179.

Celzard, A., Perrin, A., Albiniak, A., Broniek, E., and Marêché, J.F. 2007. The effect of wetting on pore texture and methane storage ability of NaOH activated anthracite. *Fuel* 86:287–293.

Chadwick, A.V., Mountjoy, G., Nield, V.M. et al. 2001. Solid state NMR and X-ray structural studies of the structural evolution of nanocrystalline zirconia. *Chem. Mater.* 13:1219–1229.

Chager, S.I. 1975. *Transport Function of Serum Albumin*. Bucharest, Romania: Academy of Sciences of SRR.

Chakrabarti, H. 1996. Anomalies in the ion transport of phosphoric acid in water and heavy-water environments. *J. Phys.: Condens. Matter* 8:7019–7029.

Chalyi, O.V. 2000. *Synergetic Principles in Education and Science*. Kyiv, Ukraine: VIPOL (in Russian).

Chan, S.I., Feigenson, G.W., and Seiter, C.H.A. 1971. Nuclear relaxation studies of lecithin bilayers. *Nature* 231:110–112.

Chan, A.T. and Lewis, J.A. 2005. Size ratio effects on interparticle interactions and phase behavior. *Langmuir* 21:8576–8579.

Chang, W.K., Liao, M.Y., and Gleason, K.K. 1996. Characterization of porous silicon by solid-state nuclear magnetic resonance. *J. Phys. Chem. B* 100:19653–19658.

Chaplin, M. 2011. Water structure and behavior. http://www.lsbu.ac.uk/water/ (accessed December 05, 2011).

Charvolin, J. and Tardieu, A. 1978. Lyotropic liquid crystals: Structure and molecular motion. In: Liebert, L. (Ed.) *Solid-State Physics*, Supp. 14. New York: Academic Press, pp. 209–256.

Chary, K.V.R., Rao, P.V.R., and Rao, V.V. 2008. Catalytic functionalities of nickel supported on different polymorphs of alumina. *Catal. Commun.* 9:886–893.

Chávez, V.F. and Schönhoff, M. 2007. Pore size distributions in polyelectrolyte multilayers determined by nuclear magnetic resonance cryoporometry. *J. Chem. Phys.* 126:104705-1–104705-7.

Chekmenev, E.Y., Hu, J., Gor'kov, P.L. et al. 2005. ^{15}N and ^{31}P solid-state NMR study of transmembrane domain alignment of M2 protein of influenza A virus in hydrated cylindrical lipid bilayers confined to anodic aluminum oxide nanopores. *J. Magn. Reson.* 173:322–327.

Chen, S.-H., Chu, B., and Nossal, R. (Eds.) 1981. *Scattering Techniques Applied to Supramolecular and Nonequilibrium Systems*. New York: Plenum Press.

Chen, X., Hu, N., Zeng, Y., Rusling, J.F., and Yang, J. 1999a. Ordered electrochemically active films of hemoglobin, didodecyldimethylammonium ions, and clay. *Langmuir* 15:7022–7030.

Chen, H.-T., Huh, S., Wiench, J.W., Pruski, M., and Lin, V.S.-Y. 2005. Dialkylaminopyridinefunctionalized mesoporous silica nanosphere as an efficient and highly stable heterogeneous nucleophilic catalyst. *J. Am. Chem. Soc.* 127:13305–13311.

Chen, X. and Mao, S.S. 2007. Titanium dioxide nanomaterials: Synthesis, properties, modifications, and applications. *Chem. Rev.* 107:2891–2959.

Chen, L.F., Wang, J.A., Noreña, L.E. et al. 2007. Synthesis and physicochemical properties of Zr-MCM-41 mesoporous molecular sieves and Pt/$H_3PW_{12}O_{40}$/Zr-MCM-41 catalysts. *J. Solid State Chem.* 180:2958–2972.

Chen, T.-H., Wouters, B.H., and Grobet, P.J. 1999b. Enhanced resolution of Al sites in the molecular sieve SAPO-37 by ^{27}Al MQ MAS NMR. *Colloids Surf. A: Physicochem. Eng. Aspects* 158:145–149.

Cheng, J.T. and Ellis, P.D. 1989. Adsorption of Rb^+ to γ-alumina as followed by solid-state ^{87}Rb NMR spectroscopy. *J. Phys. Chem.* 93:2549–2555.

Cherry, B.R., Nyman, M., and Alam, T.M. 2004. Investigation of cation environment and framework changes in silicotitanate exchange materials using solid-state ^{23}Na, ^{29}Si, and ^{133}Cs MAS NMR. *J. Solid State Chem.* 177:2079–2093.

Chodorowski, S., Leboda, R., Brei, V.V., Khomenko, K.N. and Turov, V.V. 1998. The water adsorption on silicate surface studied by ^{1}H NMR spectroscopy method. In: Rozwadowski, M. (Ed.) *Proceedings of the Third Polish-German Zeolite Colloquium*. Nicolaus Copernicus University Press. Torun, pp. 63–73.

Choi, S. and Chae, J. 2009. A microfluidic biosensor based on competitive protein adsorption for thyroglobulin detection. *Biosens. Bioelectron.* 25:118–123.

Choi, S.-G. and Kerr, W.L. 2003. Effects of chemical modification of wheat starch on molecular mobility as studied by pulsed ^{1}H NMR. *Lebensm. Wiss. Univ. Technol.* 36:105–112.

Choi, S., Yang, Y., and Chae, J. 2008. Surface plasmon resonance protein sensor using Vroman effect. *Biosens. Bioelectron.* 24:893–899.

Choma, J., Kloske, M., and Jaroniec, M. 2003. An improved methodology for adsorption characterization of unmodified and modified silica gels. *J. Colloid Interface Sci.* 266:168–174.

Chou, P.Y. and Fasman, G.D. 1978. Prediction of the secondary structure of proteins from their amino acid sequence. *Adv. Enzymol. Related Areas Mol. Biol.* 47:45–148.

ChromBook 2006/2007, http://www.chromatography.merck.de

Christiansen, S.C., Zhao, D., Janicke, M.T., Landry, C.C., Stucky, G.D. and Chmelka, B.F. 2001. Molecularly ordered inorganic frameworks in layered silicate surfactant mesophases. *J. Am. Chem. Soc.* 123:4519–4529.

Chruszcz, K., Baranska, M., Czarniecki, K., and Proniewicz, L.M. 2003. Experimental and calculated ^{1}H, ^{13}C and ^{31}P NMR spectra of (hydroxypyridin-3-yl-methyl)phosphonic acid. *J. Mol. Struct.* 651–653:729–737.

Chu, B. 1991. *Laser Light Scattering: Basic Principles and Practice*. New York: Academic Press.

Chuang, I.-S., Kinney, D.R., Bronnimann, C.E., Zeigler, R.C., and Maciel, G.E. 1992. Effects of proton-proton spin exchange in the silicon-29 CP-MAS NMR spectra of the silica surface. *J. Phys. Chem.* 96:4027–4034.

Chuang, I.-S., Kinney, D.R., and Maciel, G.E. 1993. Interior hydroxyls of the silica gel system as studied by ^{29}Si CP-MAS NMR spectroscopy. *J. Am. Chem. Sci.* 115:8695–8705.

Chuang, I.-S. and Maciel, G.E. 1996. Probing hydrogen bonding and the local environment of silanols on silica surfaces via nuclear spin cross polarization dynamics. *J. Am. Chem. Soc.* 118:401–406.

Chuiko, A.A. (Ed.) 1993. *Silicas in Medicine and Biology*. Stavropol, Russia: SMI (in Russian).

Chuiko, A.A. (Ed.) 2001. *Chemistry of Silica Surface*. Kiev, Ukraine: UkrINTEI (in Russian).

Chuiko, A.A. (Ed.) 2003. *Medical Chemistry and Clinical Application of Silica*. Kiev, Ukraine: Naukova Dumka (in Russian).

Chung, S.H., Bajue, S., and Greenbaum, S.G. 2000. Mass transport of phosphoric acid in water: AH-1andP-31pulsed gradient spin-echonuclear magnetic resonance study. *J. Chem. Phys.* 112:8515–8521.

Ciesla, U. and Schüth, F. 1999. Ordered mesoporous materials. *Micropor. Mesopor. Mater.* 27:131–149.

Claesson, P.M., Christenson, H.K., Berg, J.M., and Neuman, R.D. 1995. Interaction between mica surface in the presence of carbohydrates. *J. Colloid Interface Sci.* 172:415–424.

Coetzee, J.F. and Chang, T.H. 1986. Recommended methods for the purification of solvents end tests for impurities acetone. *Pure Appl. Chem.* 58:1535–1540.

Cole, K.S. and Cole, R.H. 1941. Dispersion and absorption in dielectrics: 1. Alternating current characteristics. *J. Chem. Phys.* 9:341–351.

Colin, H. and Guiochon, G. 1976. Development and use of carbon adsorbents in high–performance liquid–solid chromatography. *J. Chromatogr.* 126:43–62.

Collins, K.E., Dimiras, A.B., de Camargo, V.R., and Collins, C.H. 2006. Use of kinetic H_2O-adsorption isotherms for the determination of specific surface areas of fully hydroxylated mesoporous silicas. *Micropor. Mesopor. Mater.* 89:246–250.

Compernolle, S., Chibotaru, L.F., and Ceulemans, A. 2006. Novel type of magnetic response in carbon nanomaterials. *Chem. Phys. Lett.* 428:119–124.

Corcoran, C.T. and Hirschfelder, J.O. 1980. The magnetic susceptibility of BH. *J. Chem. Phys.* 72:1524–1528.

Coulson, C.A., Gomes, J.A.N.F., and Mallion, R.B. 1975. Ring magnetic susceptibilities in conjugated hydrocarbons. *Mol. Phys.* 30:713–732.

Courtier-Murias, D., Farooq, H., Masoom, H. et al. 2012. Comprehensive multiphase NMR spectroscopy: Basic experimental approaches to differentiate phases in heterogeneous samples. *J. Magn. Reson.* doi:10.1016/j.jmr.2012.02.009.

Crowe, J.H., Hoekstra, F.A., and Crowe, L.M. 1992. Anhydrobiosis. *Annu. Rev. Physiol.* 54:579–599.

Cruz, J.C., Pfromm, P.H., and Rezac, M.E. 2009. Immobilization of Candida antarctica Lipase B on fumed silica. *Process Biochem.* 44:62–69.

Cui, Y., Qian, G., Chen, L., Wang, Z., and Wang, M. 2008. Second-order nonlinear optical properties of cross-linked silica films prepared through sol–gel process. *Thin Solid Films* 516:5483–5487.

Cundy, C.S. and Cox, P.A. 2005. The hydrothermal synthesis of zeolites: Precursors, intermediates and reaction mechanism. *Micropor. Mesopor. Mater.* 82:1–78.

Cutler, S.M. and Garcia, A.J. 2003. Engineering cell adhesive surfaces that direct integrin alpha(5)beta(1) binding using a recombinant fragment of fibronectin. *Biomaterials* 24:1759–1770.

Dabrowski, A. and Tertykh, V.A. (Eds.) 1996. *Adsorption on New and Modified Inorganic Sorbents, Studies in Surface Science and Catalysis*, Vol. 99. Amsterdam, the Netherlands: Elsevier.

Dahneke, B. (Ed.) 1983. *Measurements of Suspended Particles by Quasi-Elastic Light Scattering*. New York: Wiley.

D'Alessandro, D.M., Smit, B., and Long, J.R. 2010, Carbon dioxide capture: Prospects for new materials. *Angew. Chem. Int. Ed.* 49:6058–6082.

Dalitz, F., Cudaj, M., Maiwald, M., and Guthausen, G. 2012. Process and reaction monitoring by low-field NMR spectroscopy. *Prog. Nucl. Magn. Reson. Spectrosc.* 60:52–70.

Dampeirou, C. 2005. Hydrophobic silica-based water powder. Patent WO2005034917.

Danilchenko, S.N., Pokrovskiy, V.A., Bogatyrov, V.M., Sukhodub, L.F., and Sulkio-Cleff, B. 2005. Carbonate location in bone tissue mineral by X-ray diffraction and temperature-programmed desorption mass spectrometry. *Cryst. Res. Technol.* 40:692–697.

Darmstadt, H., Roy, C., and Kaliaguine, S. 1994. ESCA characterization of commercial carbon blacks and of carbon blacks from vacuum pyrolysis of used tires. *Carbon* 32:1399–1406.

Datka, J., Gil, B., and Baran, P. 2003. IR study of heterogeneity of OH groups in zeolite HY-splitting of OH and OD bands. *J. Mol. Struct.* 645:45–49.

Davenel, A., Schuck, P., Mariette, F., and Brulé, G. 2002. NMR relaxometry as a non-invasive tool to characterize milk powders. *Lait* 82:465–473.

Davidov, D. and Selig, H. 1986. Recent developments in magnetic resonance studies of graphite intercalated compounds. In *Intercalate Layer Matter. Proc. 10th* Course Erisi Summer, New York, pp. 433–456.

Davidson, A. 2002. Modifying the walls of mesoporous silicas prepared by supramolecular-templating. *Curr. Opin. Colloid Interface Sci.* 7:92–106.

Davis, B.G. 2002. Synthesis of glycoproteins. *Chem. Rev.* 102:579–601.

Decareau, R.V. (Ed.) 1985. *Microwaves in Food Processing Industry*. New York: Academic Press.

de Certaines, J.D. (Ed.) 1989. *Magnetic Resonance Spectroscopy of Biofluids, a New Tool in Clinical Biology*. Singapore: World Scientific.

de Gennes, P.G. 1969. Dynamics of fluctuations in nematic liquid crystals. *J. Chem. Phys.* 51:816–822.

de Gennes, P.-G., Brochard-Wyart, F., and Quéré, D. 2004. *Capillarity and Wetting Phenomena: Drops, Bubbles, Pearls, Waves*. Springer, Berlin, Germany.

Degussa AG. 1996. *Technical Information*. Frankfurt, Germany: Degussa AG, TI 1176.

Degussa AG. 1997. *Basic Characteristics of Aerosil*. Technical Bulletin Pigments, No 11. Hanau: Degussa AG.

Delattre, L. and Babonneau, F. 1997. ^{17}O Solution NMR characterization of the preparation of sol–gel derived SiO_2/TiO_2 and SiO_2/ZrO_2 glasses. *Chem. Mater.* 9:2385–2394.

Deleplanque, J., Hubaut, R., Bodart, P., Fournier, M., and Rives, A. 2009. ^{1}H and ^{31}P solid-state NMR of trimethylphosphine adsorbed on heteropolytungstate supported on silica. *Appl. Surf. Sci.* 255:4897–4901.

Delmon, B. and Froment, G.F. (Eds.) 1980. *Catalyst Deactivation*. Amsterdam, the Netherlands: Elsevier.

Delville, A. 1993. Structure and properties of confined liquids: A molecular model of clay-water interface. *J. Phys. Chem.* 97:9703–9712.

DeMott, P.J. 1995. Quantitative descriptions of ice formation mechanisms of silver iodide-type aerosols. *Atmos. Res.* 38:63–99.

Deng, D., Li, H., Yao, J., and Han, S. 2003. Simple local composition model for ^{1}H NMR chemical shift of mixtures. *Chem. Phys. Lett.* 376:125–129.

Dennes, T.J. and Schwartz, J. A nanoscale adhesion layer to promote cell attachment on PEEK. 2009. *J. Am. Chem. Soc.* 131:3456–3457.

Derjaguin, B.V., Churaev, N.V., and Muller, V.M. 1987. *Surface Forces*. New York: Consultants Bureau.

Derjaguin, B.V., Ovcharenko, F.D., and Churaev, N.V. (Eds.) 1989. *Water in Disperse Systems*. Moscow, Russia: Khimiya (in Russian).

d'Espinose de la Caillerie, J.-B., Aimeur, M.R., El Kortobi, Y., and Legrand, A.P. 1997. Water adsorption on pyrogenic silica followed by ^{1}H MAS NMR. *J. Colloid Interface Sci.* 194:434–439.

Devis, B.G. and Robinson, M.A. 2002. Drug delivery systems based on sugar—Macromolecule conjugates. *Curr. Opin. Drug Discov. Dev.* 5:279–288.

Dey, A., York, D., Smalls-Mantey, A., and Summers, M.F. 2005. Composition and sequence-dependent binding of RNA to the nucleocapsid protein of Moloney murine leukemia virus. *Biochemistry* 44:3735–3744.

Diebold, U. 2003. The surface science of titanium dioxide. *Surf. Sci. Rep.* 48:53–229.

Dieudonné, Ph., Hafidi Alaoui, A., Delord, P., and Phalippou, J. 2000. Transformation of nanostructure of silica gels during drying. *J. Non-Crystal. Solid.* 262:155–161.

Dill, K.A. and Bromberg, S. 2003. *Molecular Driving Forces*. New York: Garland Science.

Ding, L.P. 2002. *Multicomponent Adsorption in Heterogeneous Microporous Solids*. St. Lucia, Queensland, Australia.

Ding, P., Orwa, M.G., and Pacek, A.W. 2009. De-agglomeration of hydrophobic and hydrophilic silica nano-powders in a high shear mixer. *Powder Technol.* 195:221–226.

Dingley, A.J. and Grzesiek, S. 1998. Direct observation of hydrogen bonds in nucleic acid base pairs by internucleotide $^2J_{NN}$ couplings. *J. Am. Chem. Soc.* 120:8293–8297.

Dingley, A.J., Masse, J.E., Feigon, J., and Grzesiek, S. 2000. Characterization of the hydrogen bond network in guanosine quartets by internucleotide $^{3h}J_{NC'}$ and $^{2h}J_{NN}$ scalar couplings. *J. Biomol. NMR* 16:279–289.

Dingley, A.J., Masse, J.E., Peterson, R.D., Barfield, M., Feigon, J., and Grzesiek, S. 1999. Internucleotide scalar couplings across hydrogen bonds in Watson-Crick and Hoogsteen base pairs of a DNA triplex. *J. Am. Chem. Soc.* 121:6019–6027.

Diociaiuti, M., Bordi, F., Gataleta, L., Baldo, G., Crateri, P., and Paoletti, L. 1999. Morphological and functional alterations of human erythrocytes induced by SiO_2 particles: An electron microscopy and dielectric spectroscopy study. *Environ. Res. A.* 80:197–207.

Dippel, Th. and Kreuer, K.D. 1991. Proton transport mechanism in concentrated aqueous solutions and solid hydrates of acids. *Solid State Ionics* 46:3–9.

Dippel, Th., Kreuer, K.D., Lassègues, J.C., and Rodriguez, D. 1993. Proton conductivity in fused phosphoric acid; A^{1}H/^{31}P PFG-NMR and QNS-study. *Solid State Ionics* 61:41–46.

Do, D.D. 1998. *Adsorption Analysis: Equilibria and Kinetics*. London, U.K.: Imperial College Press.

Do, D.D., Nguyen, C., and Do, H.D. 2001. Characterization of micro-mesoporous carbon media. *Colloids Surf. A: Physicochem. Eng. Aspects* 187–188:51–71.

Dodge, J., Mitchell, C., and Hanahan, D. 1963. The preparation and chemical characteristics of hemoglobin-free ghosts of human erythrocytes. *Arch. Biochem. Biophys.* 100:119–130.

Doi, M. and Edwards, S.F. 1986. *The Theory of Polymer Dynamics*. Oxford, U.K.: Clarendon Press.

Dominguez, C., Schubert, M., Duss, O., Ravindranathan, S., and Allain, F.H.-T. 2011. Structure determination and dynamics of protein–RNA complexes by NMR spectroscopy. *Prog. Nucl. Magn. Reson. Spectrosc.* 58:1–61.

Dominguez-Espinosa, G., Díaz-Calleja, R., Riande, E., Gargallo, L., and Radic, D. 2006. Influence of the fine structure on the response of polymer chains to perturbation fields. *Macromolecules* 39:3071–3080.

Domján, A., Geissler, E., and László, K. 2010. Phenol–polymer proximity in a thermoresponsive gel determined by solid-state ^{1}H–^{1}H CRAMPS NMR spectroscopy. *Soft Matter* 6:247–249.

Donnet, J.B. 1994. Fifty years of research and progress on carbon black. *Carbon* 32:1305–1310.

Donnet, J.B. and Custodero, E. 1992. Ordered structure observed by scanning tunneling microscopy at atomic scale on carbon black surface. *Carbon* 30:813–817.

Donnet, J.B. and Voet, A. 1976. *Carbon Black*. New York: Marcel Dekker.

Donnet, J.B., Wang, T.K., and Custodero, E. 1995. STM and AFM study of carbon black structure. In: *Proc. International Conference Eurofillers 95*, Mulhouse, France, September 11–14, pp. 187–190.

Donovan, J.W. 1979. Phase transition of the starch-water system. *Biopolymers* 18:263–275.

Dore, J.C., Webber, J.B.W., and Strange, J.H. 2004. Characterisation of porous solids using small-angle scattering and NMR cryoporometry. *Colloids Surf. A: Physicochem. Eng. Aspects* 241:191–200.

Dorémieux-Morin, C., Batamack, P., Martin, C., Brégeault, J.-M., and Fraissard, J. 1991. Comparison of amorphous silica-alumina and highly dealuminated HY zeolite by ^{1}H high resolution MAS-NMR of solids. *Catal. Lett.* 9:403–410.

Dourdain, S., Mehdi, A., Bardeau, J.F., and Gibaud, A. 2006. Determination of porosity of mesoporous silica thin films by quantitative X-ray reflectivity analysis and GISAXS. *Thin Solid Films* 495:205–209.

Drago, R.S. 1977. *Physical Methods in Chemistry*. Philadelphia, PA: W.B. Saunders Co.

Drago, R.S., O'Bryan, N., and Vogel, G.C. 1970. A frequency shift enthalpy correlation for a given donor with various hydrogen bonding acids. *J. Am. Chem. Soc.* 92:3924–3929.

Dreyer, D.R., Park, S., Bielawski, C.W., and Ruoff, R.S. 2010. The chemistry of graphene oxide. *Chem. Soc. Rev.* 39:228–240.

Drost-Hansen, W. and Etzler, F.M. 1989. Melting of ice in silica pores. *Langmuir* 5:1439–1441.

Duarte, I.F. and Gil, A.M. 2012. Metabolic signatures of cancer unveiled by NMR spectroscopy of human biofluids. *Prog. Nucl. Magn. Reson. Spectrosc.* 62:51–74.

Dubin, C.H. 2006. Formulation strategies for poorly soluble drugs. *Drug Deliv. Technol.* 6:34–38.

Dubinin, M.M. 1960. The potential theory of adsorption of gases and vapors for adsorbents with energetically nonuniform surfaces. *Chem. Rev.* 60:235–241.

Dubinin, M.M. 1983. Microporous structure and adsorption properties of carbonaceous adsorbents. *Carbon* 21:359–366.

Dubinin, M.M., Vartapetian, R.S., Voloshchuk, A.M., Kärger, J., and Pfeifer, H. 1988. NMR study of translational mobility of molecules adsorbed on active carbons. *Carbon* 26:515–520.

Dukhin, S.S. 1993. Non-equilibrium electric surface phenomena. *Adv. Colloid Interface Sci.* 44:1–134.

Dukhin, S.S. and Lyklema, J. 1987. Dynamics of colloid particle interaction. *Langmuir* 3:94–98.

Dullien, F.A.L. 1992. *Porous Media Fluid Transport and Pore Structure*, 2nd edn. San Diego, CA: Academic Press.

Dvoyashkin, M., Khokhlov, A., Naumov, S., and Valiullin, R. 2009. Pulsed field gradient NMR study of surface diffusion in mesoporous adsorbents. *Micropor. Mesopor. Mater.* 125:58–62.

Dyr, J.E., Tichý, I., Jiroušková, M. et al. 1998. Molecular arrangement of adsorbed fibrinogen molecules characterized by specific monoclonal antibodies and a surface plasmon resonance sensor. *Sens. Actuat. B* 51:268–272.

Eberhardt, W. 2002. Clusters as new materials. *Surf. Sci.* 500:242–270.

Eckert, H. 1992. Structural characterization of noncrystalline solids and glasses using solid state NMR. *Prog. Nucl. Magn. Reson. Spectrosc.* 24:159–293.

Edzes, H.T. and Samulski, E.T. 1977. Spin diffusion and cross relaxation in the proton NMR of hydrated collagen. *Nature* 265:521–522.

Egorova, N.S. (Ed.) 1976. *Microbiology Workshop*. Moscow, Russia: Moscow State University Press.

Einfeldt, J., Einfeldt, L., Dicke, R., Klemm, D., and Kwasniewski, A. 2002. The influence of the type of substituents on the polymer dynamics of different regioselective 2-O-esters of starch using dielectric spectroscopy. *Polymers* 43:1391–1397.

Einfeldt, J., Kwasniewski, A., Klemm, D., Dicke, R., and Einfeldt, L. 2000. Analysis of side group motion in O-acetyl-starch using regioselective 2-O-acetyl-starches by means of dielectric spectroscopy. *Polymers* 41:9273–9281.

Einfeldt, J., Meißner, D., and Kwasniewski, A. 2004. Molecular interpretation of the main relaxations found in dielectric spectra of cellulose—Experimental arguments. *Cellulose* 11:137–150.

Einfeldt, J., Meißner, D., Kwasniewski, A., and Einfeldt, L. 2001. Dielectric spectroscopy analysis of wet and well dried starches in comparison with other polysaccharides. *Polymers* 42:7049–7062.

Eirich, F. (Ed.) 1958. *Rheology*. London, U.K.: Academic Press.

Ek, S., Root, A., Peussa, M., and Niinistö, L. 2001. Determination of hydroxyl group content in silica by thermogravimetry and a comparison with ^{1}H MAS NMR results. *Thermochim. Acta* 379:201–212.

El Horr, N., Bourgerette, C., and Oberlin, A. 1994. Mesophase powders (carbonization and graphitization). *Carbon* 32:1035–1044.

Eliasson, A.-Ch. 2004. *Starch in Food: Structure, Function and Applications*. Cambridge, U.K.: Woodhead Publishing.

Elvington, S.M., Liu, C.W., and Maduke, M.C. 2009. Substrate-driven conformational changes in CIC-ec1 observed by fluorine NMR. *EMBO J.* 28:3090–3102.

Elyashberg, M.E., Williams, A.J., and Martin, G.E. 2008. Computer-assisted structure verification and elucidation tools in NMR-based structure elucidation. *Prog. Nucl. Magn. Reson. Spectrosc.* 53:1–104.

Emerich, D.F. and Thanos, C.G. 2003. Nanotechnology and medicine. *Expert Opin. Biol. Ther.* 3:655–663.

Emsley, J.W., Feenej, J., and Sutcliffe, L.H. 1965. High *Resolution Nuclear Magnetic Resonance Spectroscopy*. Oxford, U.K.: Pergamon Press.

Encyclopedia of Magnetic Resonance. 2007, 2009, 2011. New York: John Wiley & Sons.

Engelhardt, G. and Michel, D. 1987. *High-Resolution Solid-State NMR of Silicates and Zeolites*. Chichester, U.K.: John Wiley & Sons.

Epifani, M., Giannini, C., Tapfer, L., and Vasanelli, L. 2000. Sol-gel synthesis and characterization of Ag and Au nanoparticles in SiO_2, TiO_2 and ZrO_2 thin films. *J. Am. Ceram. Soc.* 83:2385–2393.

Eppell, S.J., Tong, W., Katz, J.L., Kuhn, L., and Glimcher, M.J. 2001. Shape and size of isolated bone mineralites measured using atomic force microscopy. *J. Orthop. Res.* 19:1027–1034.

Epping, J.D. and Chmelka, B.F. 2006. Nucleation and growth of zeolites and inorganic mesoporous solids: Molecular insights from magnetic resonance spectroscopy. *Curr. Opin. Colloid Interface Sci.* 11:81–117.

Erickson, K., Erni, R., Lee, Z., Alem, N., Gannett, W., and Zettl, A. 2010. Determination of the local chemical structure of graphene oxide and reduced graphene oxide. *Adv. Mater.* 22:4467–4472.

Ernst, R.R., Bodenhausen, G., and Wokaun, A. 1987. *Principles of Nuclear Magnetic Resonance in One and Two Dimensions*. Oxford, U.K.: Oxford University Press.

Ertl, G., Knozinger, H., and Weitkamp, J. (Eds.) 1997. *Handbook of Heterogeneous Catalysis*. Weinheim, Germany: VCH.

Eshtiaghi, N., Arhatari, B., and Hapgood, K.P. 2009. Producing hollow granules from hydrophobic powders in high-shear mixer granulators. *Adv. Powder Technol.* 20:558–566.

Estrade-Szwarckopf, H. 1985. Magnetic resonance in graphite intercalation compounds. *Helv. Phys. Acta* 58:139–161.

Evanoff, Jr., D.D. and Chumanov, G. 2005. Synthesis and optical properties of silver nanoparticles and arrays. *ChemPhysChem* 6:1221–1231.

Faßbender, B. 2009. Solid-state NMR investigation of phosphonic acid based proton conducting materials. PhD thesis. Johannes Gutenberg-Universität Mainz.

Facelli, J.C. 2011. Chemical shift tensors: Theory and application to molecular structural problems. *Prog. Nucl. Magn. Reson. Spectrosc.* 58:176–201.

Facelli, J.C. and de Dios, A.C. 1999. *Modeling NMR Chemical Shifts. Gaining Insights into Structure and Environment*. ACS Symposium Series 732. Washington, DC: American Chemical Society.

Fackler, J.P., Jr. (Ed.) 1990. *Metal-Metal Bonds and Clusters in Chemistry and Catalysis*. New York: Plenum Press.

Fagerlund, G. 1973. Determination of specific surface by the BET method. *Mater. Struct.* 6:239–245.

Fairén-Jiménez, D., Carrasco-Marín, F., Djurado, D., Bley, F., Ehrburger-Dolle, F., and Moreno-Castilla, C. 2006. Surface area and microporosity of carbon aerogels from gas adsorption and small- and wide-angle X-ray scattering measurements. *J. Phys. Chem. B* 110:8681–8688.

Fanning, P.E. and Vannice, M.A. 1993. A DRIFTS study of the formation of surface groups on carbon by oxidation. *Carbon* 31:721–730.

Fantazzini, P., Brown, R.J.S., and Borgia, G.C. 2003. Bone tissue and porous media: Common features and differences studied by NMR relaxation. *Magn. Reson. Imag.* 21:227–234.

Fatkullin, N., Fischer, E., Mattea, C., Beginn, U., and Kimmich, R. 2004. Polymer dynamics under nanoscopic constraints: The "corset effect" as revealed by NMR relaxometry and diffusometry. *ChemPhysChem* 5:884–894.

Fawcett, W.R. 2004. *Liquids, Solutions, and Interfaces: From Classical Macroscopic Descriptions to Modern Microscopic Details*. Oxford, U.K.: Oxford University Press.

Fedoteikov, A.G., Shisgkina, I.D., and Levitskaja, L.A. 1963. Freezing of bone morrow at its preservation at low temperature. *Probl. Gematol. Pereliv. Krovi.* 8:16–22.

Feigenson, G.W. and Chan, S.I. 1974. Nuclear magnetic relaxation behavior of lecithin multilayers. *J. Am. Chem. Soc.* 96:1312–1319.

Fenelonov, V.B. 1995. *Porous Carbon*. Novosibirsk, Russia: Institute of Catalysis (in Russian).

Fenoglio, I., Fubini, B., Ghibaudi, E.M., and Turci, F. 2011. Multiple aspects of the interaction of biomacromolecules with inorganic surfaces. *Adv. Drug Deliv. Rev.* 63:1186–1209.

Feofilova, E.P. 2003. Deceleration of vital activity as a universal biochemical mechanism ensuring adaptation of microorganisms to stress factors. *Appl. Biochem. Microbiol.* 39:5–24.

Feron, O., Langlais, F., and Naslain, R. 1999. In-situ analysis of gas phase decomposition and kinetic study during carbon deposition from mixtures of carbon tetrachloride and methane. *Carbon* 37:1355–1361.

Ferreira, V., Ortín, N., and Cacho, J.F. 2007. Optimization of a procedure for the selective isolation of some powerful aroma thiols. Development and validation of a quantitative method for their determination in wine. *J. Chromatogr. A* 1143:190–198.

Fiehn, O. and Jekel, M. 1996. Comparison of different sorbents for on-line solid-phase extraction of pesticides and phenolic compounds from natural water followed by liquid chromatography. *Anal. Chem.* 68:3083–3089.

FieldView 2.0.2. 2011. http://www.cresset-group.com

Figueiredo, J.L. and Mouliin, J.A. (Eds.) 1986. *Carbon and Coal Gasification*. NATO Series, Series E, Applied Sciences No. 105. Dordrecht, the Netherlands: Martinus Nijhoff Publishers.

Fiji. 2009. Fiji is just ImageJ. http://pacific.mpi-cbg.de/wiki/index.php (accessed November 13, 2009).

Finch, E.D. and Schneider, A.S. 1975. Mobility of water bound to biological membranes a proton NMR relaxation study. *Biochim. Biophys. Acta* 406:146–154.

Findenegg, G.H., Jähnert, S., Akcakayiran, D., and Schreiber, A. 2008. Freezing and melting of water confined in silica nanopores. *ChemPhysChem* 9:2651–2659.

Finkelstein, A. 1987. *Water Movements through Lipid Bilayers, Pores, and Plasma Membranes: Theory and Reality*. New York: John Wiley & Sons.

Fischer, C.E., Raith, A., Mink, J., Raudaschl-Sieber, G., Cokoja, M., and Kühn, F.E. 2011. Organic-inorganic nanotube hybrids: Organosilica-nanotubes containing ethane, ethylene and acetylene groups. *J. Organometal. Chem.* 696:2910–2917.

Fitzgerald, J.J. (Ed.) 1999. *Solid-State NMR Spectroscopy of Inorganic Materials*. ACS Symposium Series 717. Washington, DC: American Chemical Society.

Fitzgerald, J.J., Hamza, A.I., Bronnimann, C.E., and Dec, S.F. 1997. Studies of the solid/solution "interfacial" dealumination of kaolinite in HCl(aq) using solid-state ^{1}H CRAMPS and SP/MAS ^{29}Si NMR spectroscopy. *J. Am. Chem. Soc.* 119:7105–7113.

Fitzgerald, J.J., Hamza, A.I., Dec, S.F., and Bronnimann, C.E. 1996. Solid-state ^{27}Al and ^{29}Si NMR and ^{1}H CRAMPS studies of the thermal transformations of the 2:1 phyllosilicate pyrophyllite. *J. Phys. Chem.* 100:17351–17360.

Flavus, M.J. (Ed.) 1999. *Primer of Metabolic Bone Diseases and Disorders of Mineral Metabolism*. Philadelphia, PA: Lippincott-Williams & Wilkins.

Flesch, J., Kerner, D., Riemenschneider, H., and Reimert, R. 2008. Experiments and modeling on the deacidification of agglomerates of nanoparticles in a fluidized bed. *Powder Technol.* 183:467–479.

Fomekong, R.D., Pliquett, U., and Pliquett, F. 1998, Passive electrical properties of RBC suspensions: Changes due to distribution of relaxation times in dependence on the cell volume fraction and medium conductivity. *Bioelectrochem. Bioenerg.* 47:81–88.

Fomkin, A.A. 2005. Adsorption of gases, vapors and liquids by microporous adsorbents. *Adsorption* 11:425–436.

Forny, L., Pezron, I., Saleh, K., Guigon, P., and Komunjer, L. 2007. Storing water in powder form by self-assembling hydrophobic silica nanoparticles. *Powder Technol.* 171:15–24.

Forny, L., Saleh, K., Pezron, I., Komunjer, L., and Guigon, P. 2009. Influence of mixing characteristics for water encapsulation by self-assembling hydrophobic silica nanoparticles. *Powder Technol.* 189:263–269.

Frahm, J. 1982. Multiple-pulse FT-NMR experiments for kinetic applications. *J. Magn. Reson.* 47:209–226.

Franks, F. (Ed.) 1982. *Water, a Comprehensive Treatise*, Vol. 7. New York: Plenum Press.

Franks, F. 2000. *Water: A Matrix of Life*, 2nd edn. Cambridge, MA: Royal Society of Chemistry.

Frazier, P.J., Donald, A.M., and Richmond, P. (Eds.) 1997. *Starch, Structure and Functionality*. Cambridge, MA: The Royal Society of Chemistry.

Freitas, J.C.C., Bonagamba, T.J., and Emmerich, F.G. 2001. Investigation of biomass- and polymer-based carbon materials using ^{13}C high-resolution solid-state NMR. *Carbon* 39:535–545.

Freude, D., Loeser, T., Michel, D., Pingel, U., and Prochnow, D. 2001. ^{17}O NMR studies of low silicate zeolites. *Solid State Nucl. Magn. Reson.* 20:46–60.

Frieden, C. 2007. Protein aggregation processes: In search of the mechanism. *Protein Sci.* 16:2334–2344.

Frisch, M.J., Trucks, G.W., Schlegel, H.B. et al. 2004. *Gaussian 03*, Revision E.01. Wallingford, CT: Gaussian, Inc.

Frolov, Yu.G. 1982. *Course of Colloid Chemistry*. Moscow, Russia: Khimia (in Russian).

Fry, R., Pantano, C.G., and Mueller, K.T. 2003a. Effect of boronoxide on surface hydroxyl coverage of aluminoborosilicate glass fibers: A ^{19}F solid-state NMR study. *Phys. Chem. Glasses* 44:64–68.

Fry, R.A., Tsomaia, N., Pantano, C.G., and Mueller, K.T. 2003b. ^{19}F MAS NMR quantification of accessible hydroxyl sites on fiberglass surfaces. *J. Am. Chem. Soc.* 125:2378–2379.

Frydman, L. and Harwood, J.S. 1995. ^{59}Co solid state NMR of hexaaminecobalt(III): A new probe for elucidating metal binding in polynucleotides. *J. Am. Chem. Soc.* 117:5367–5368.

Fujii, S. and Enoki, T. 2010. Cutting of oxidized graphene into nanosized pieces. *J. Am. Chem. Soc.* 132:10034–10041.

Fujiyoshi, Y., Mitsuoka, K., Groot, B.L. et al. 2002. Structure and function of water channels. *Curr. Opin. Struct. Biol.* 12:509–515.

Fukahori, S., Fujiwara, T., Ito, R., and Funamizu, N. 2011. pH-Dependent adsorption of sulfa drugs on high silica zeolite: Modeling and kinetic study. *Desalination* 275:237–242.

Furmaniak, S., Terzyk, A.P., Gauden, P.A., Marks, N.A., Powles, R.C., and Kowalczyk, P. 2011. Simulating the changes in carbon structure during the burn-off process. *J. Colloid Interface Sci.* 360:211–219.

Fyfe, C.A., Feng, Y., Gies, H., Grondey, H., and Kokotailo, G.T. 1990a. Natural-abundance twodimensional solid-state ^{29}Si NMR investigations of three-dimensional lattice connectivities in zeolite structures. *J. Am. Chem. Soc.* 112:3264–3270.

Fyfe, C.A., Grondey, H., Feng, Y., Kokotailo, G.T. 1990b. Natural-abundance twodimensional silicon-29 MAS NMR investigation of the three-dimensional bonding connectivities in the zeolite catalyst ZSM-5. *J. Am. Chem. Soc.* 112:8812–8820.

Fyfe, C.A., Wong-Moon, K.C., Huang, Y., and Grondey, H. 1995. Structural investigations of SAPO-37 molecular sieve by coherence-transfer and dipolar-dephasing solid-state nuclear magnetic resonance experiments. *Micropor. Mater.* 5:29–37.

Fyfe, C.A., Zhang, Y., and Aroca, P. 1992. An alternative preparation of organofunctionalized silica gels and their characterization by twodimensional high-resolution solid-state heteronuclear NMR correlation spectroscopy. *J. Am. Chem. Soc.* 114:3252–3255.

Gaede, H.C., Luckett, K.M., Polozov, I.V., and Gawrisch, K. 2004. Multinuclear NMR studies of single lipid bilayers supported in cylindrical aluminum oxide nanopores. *Langmuir* 20:7711–7719.

Galán-Marín, C., Rivera-Gómez, C., and Petric, *J.* 2010. Clay-based composite stabilized with natural polymer and fibre. *Construct. Build. Mater.* 24:1462–1468.

Gallegos, D.P., Munn, K., Smith, D.M., and Stermer, D.L. 1987. A NMR technique for the analysis of pore structure: Application to materials with well-defined pore structure. *J. Colloid Interface Sci.* 119:127–140.

Galliard, T. 1987. *Starch: Properties and Potential. Critical Reports on Applied Chemistry*, Vol. 13. Chichester, U.K.: Wiley.

Gallier, J., Rivet, P., and de Certaines, J.D. 1987. ^{1}H- and ^{2}H-NMR study of bovine serum albumin solutions. *Biochim. Biophys. Acta* 915:1–18.

Ganapathy, S., Gore, K.U., Kumar, R., and Amoureux, J.-P. 2003. Multinuclear (^{27}Al, ^{29}Si, ^{47,49}Ti) solid-state NMR of titanium substituted zeolite USY. *Solid State Nucl. Magn. Reson.* 24:184–195.

Ganapathy, S., Kumar, R., Montouillout, V., Fernandez, C., and Amoureux, J.P. 2004. Identification of tetrahedrally ordered Si-O-Al environments in molecular sieves by {^{27}Al}-^{29}Si REAPDOR NMR. *Chem. Phys. Lett.* 390:79–83.

Gani, R. and Brignole, E.A. 1983. Molecular design of solvents for liquid extraction based on UNIFAC. *Fluid Phase Equilib.* 13:331–340.

García-Martín, M., Ballesteros, P., and Cerdán, S. 2001. The metabolism of water in cells and tissues as detected by NMR methods. *Prog. Nucl. Magn. Reson. Spectrosc.* 39:41–77.

Garnczarska, M., Zalewski, T., and Kempka, M. 2007. Changes in water status and water distribution in maturing lupin seeds studied by MR imaging and NMR spectroscopy. *J. Exp. Bot.* 58:3961–3969.

Garnczarska, M., Zalewski, T., and Wojtyla, Ł. 2008. A comparative study of water distribution and dehydrin protein localization in maturing pea seeds. *J. Plant Physiol.* 165:1940–1946.

Garroway, A.N. 1977. Homogeneous and inhomogeneous nuclear spin echoes in organic solids: Adamantine. *J. Magn. Reson.* 28:365–371.

Gelb, L.D. and Gubbins, K.E. 1998. Characterization of porous glasses: Simulation models, adsorption isotherms, and the Brunauer-Emmett-Teller analysis method. *Langmuir* 14:2097–2111.

Gerashchenko, B.I., Gun'ko, V.M., Gerashchenko, I.I., Leboda, R., Hosoya, H., and Mironyuk, I.F. 2002. Probing the silica surfaces by red blood cells. *Cytometry* 49:56–61.

Gerstein, B.C., Pembleton, R.G., Wilson, R.C., and Ryan, L.M. 1977. High resolution NMR in randomly oriented solids with homonuclear dipolar broadening: Combined multiple pulse NMR and magic angle spinning. *J. Chem. Phys.* 66:361–362.

Gerstein, B.C., Pruski, M., and Hwang, S.-J. 1993. Determination of proton densities on silica gel catalyst supports by n-quantum coherence in NMR. *Anal. Chim. Acta* 283:1059–1079.

Gervais, C., Babonneau, F., and Smith, M.E. 2001. Detection, quantification, and magnetic field dependence of solid-state ^{17}O NMR of X-O-Y (XY = Si, Ti) linkages: Implications for characterising amorphous titania–silica based materials. *J. Phys. Chem. B* 105:1971–1977.

Gierak, A. and Leboda, R. 1989. Preparation of carbonaceous adsorbents by catalytic decomposition of hydrocarbons for chromatographic applications. *J. Chromatogr.* 483:197–207.

Glidewell, S.M. 2006. NMR imaging of developing barley grains. *J. Cereal Sci.* 43:70–78.

Glushko, V.P. (Ed.) 1978. *Handbook of Thermodynamic Properties of Individual Substances.* Moscow, Russia: Nauka (in Russian).

Gocmez, H. and Özcan, O. 2008. Low temperature synthesis of nanocrystalline α-Al_2O_3 by a tartaric acid gel method. *Mater. Sci. Eng.* A 475:20–22.

Gödecke, T., Yao, P., Napolitano, J.G. et al. 2012. Integrated standardization concept for Angelica botanicals using quantitative NMR. *Fitoterapia* 83:18–32.

Gokmen, M.T. and Du Prez, F.E. 2012. Porous polymer particles - A comprehensive guide to synthesis, characterization, functionalization and applications. *Progr. Polym. Sci.* 37:365–405.

Golovan, A.P., Rugal, A.A., Gun'ko, V.M. et al. 2010. Modeling of bone tissue by nanocomposites on the base of albumin, zhelatine and hydroxyapatite and its properties. *Surface* 2:244–265 (in Russian).

Gomes, J.A.N.F. and Mallion, R.B. 2001. Aromaticity and ring currents. *Chem. Rev.* 101:1349–1383.

Gómez-de-Salazar, C., Sepúlveda-Escribano, A., and Rodrígues-Reinozo, F. 2000. Preparation of carbon molecular sieves by controlled oxidation treatments. *Carbon* 38:1879–1902.

Goobes, G., Stayton, P.S., and Drobny, G.P. 2007. Solid state NMR studies of molecular recognition at protein–mineral interfaces. *Prog. Nucl. Magn. Reson. Spectrosc.* 50:71–85.

Gopinath, S. and Sugunan, S. 2007. Enzymes immobilized on montmorillonite K10: Effect of adsorption and grafting on the surface properties and the enzyme activity. *Appl. Clay Sci.* 35:67–75.

Gorbatov, A.V. (Ed.) 1982. *Structure-Mechanical Characteristics of Foodstuff*. Moscow, Russia: Light and Heavy Industry.

Gorbunov, B., Baklanov, A., Kakutkina, N., Windsor, H.L., and Toumi, R. 2001. Ice nucleation on soot particles. *J. Aerosol Sci.* 32:199–215.

Gorbyk, S.P., Turov, V.V., and Chuiko, A.A. Study of interface interactions in the system cells/water/silica by ^{1}H NMR spectroscopy method. 2003. *Phys. Chem. Technol. Surf.* 9:123–127 (in Russian).

Göritz, D., Neidermeier, W., and Raab, H. 1995. Structure of carbon black fillers. In: *Proc. Eurofillers 95*, Mulhouse, France, September 11–14, 1995, pp. 183–186.

Goto, A., Hondoh, T., and Mae, S. 1990. The electron density distribution in ice Ih determined by single-crystal X-ray diffractometry. *J. Chem. Phys.* 93:1412–1417.

Gottschalk, M., Venu, K., and Halle, B. 2003. Protein self-association in solution: The bovine pancreatic trypsin inhibitor decamer. *Biophys. J.* 84:3941–3958.

Grad, J. and Bryant, R.G. 1990. Nuclear magnetic cross-relaxation spectroscopy. *J. Magn. Reson.* 90:1–8.

Gradstajn, S., Conard, J., and Benoit, H. 1970. Etude de la surface de solides par RMN de liquides adsorbe I. Origine du deplacement et de la largeur des raies dans le cas des carbones. *J. Phys. Chem. Solids* 31:1121–1135.

Gran, H.Chr. and Hansen, E.W. 1997. *Cem. Concr. Res.* 27:1319–1331.

Grandjean, J. and László, P. 1989. Multinuclear and pulsed gradient magnetic resonance studies of sodium cations and of water reorientation at the interface of a clay. *J. Magn. Reson.* 83:128–137.

Granovsky, A.A. 2009. Firefly version 7.1G. http://classic.chem.msu.su/gran/gamess/index.html (accessed December 04, 2009).

Grant, D.M. and Harris, R.K. (Eds.) 1996. *Encyclopedia of Nuclear Magnetic Resonance*. Chichester, U.K.: John Wiley & Sons.

Gregg, S.J. and Sing, K.S.W. 1982. *Adsorption, Surface Area and Porosity*, 2nd edn. London, U.K.: Academic Press.

Gregg, S.J., Sing, K.S.W., and Stoeckli, H.F. (Eds.) 1979. *The Characterisation of Porous Solids*. London, U.K.: The Society of Chemistry and Industry.

Gribble, C.M., Matthews, G.P., Laudone, G.M. et al. 2011. Porometry, porosimetry, image analysis and void network modelling in the study of the pore-level properties of filters. *Chem. Eng. Sci.* 66:3701–3709.

Grivet, J.-P. and Delort, A.-M. 2009. NMR for microbiology: In vivo and in situ applications. *Prog. Nucl. Magn. Reson. Spectrosc.* 54:1–53.

Gröger, C., Lutz, K., and Brunner, E. 2009. NMR studies of biomineralisation. *Prog. Nucl. Magn. Reson. Spectrosc.* 54:54–68.

Grose, R.W. and Flanigen, E.M. 1977. US Pat. 4 061 724.

Grosse, R., and Boddenberg, B. 1987. ^{2}H NMR studies on the dynamics of benzene on graphytized carbon and boron nitride. *Z. Phys. Chem. Neue Folge* 152:259–270.

Grünberg, B., Emmler, Th., Gedat, E. et al. 2004. Hydrogen bonding of water confined in mesoporous silica MCM-41 and SBA-15 studied by ^{1}H solid-state NMR. *Chem. Eur. J.* 10:5689–5696.

Grunenwald, A., Ayral, A., Albouy, P.-A. et al. 2012. Hydrophobic mesostructured organosilica-based thin films with tunable mesopore ordering. *Micropor. Mesopor. Mater.* 150:64–75.

Gu, Z., Ridenour, C.F., Bronnimann, C.E., Iwashita, T., and McDermott, A. 1996. Hydrogen bonding and distance studies of amino acids and peptides using solid state 2D ^{1}H-^{13}C heteronuclear correlation spectra. *J. Am. Chem. Soc.* 118:822–829.

Guliants, V.V., Carreon, M.A., and Lin, Y.S. 2004. Ordered mesoporous and macroporous inorganic films and membranes. *J. Membr. Sci.* 235:53–72.

Gunawidjaja, P.N., Holland, M.A., Mountjoy, G., Pickup, D.M., Newport, R.J., and Smith, M.E. 2003. The effects of different heat treatment and atmospheres on the NMR signal and structure of TiO_2–ZrO_2–SiO_2 sol–gel materials. *Solid State Nucl. Magn. Reson.* 23:88–106.

Gun'ko, V.M. 2000a. The effect of the nature and the state of the surface of highly dispersed silicon, aluminum and titanium oxides on their sorption characteristics. *Theoret. Exp. Chem.* 36:1–30.

Gun'ko, V.M. 2000b. Consideration of the multicomponent nature of adsorbents during analysis of their structural and energy parameters. *Theoret. Exp. Chem.* 36:319–324.

Gun'ko, V.M. 2007. Competitive adsorption. *Theor. Exp. Chem.* 43:139–183.

Gun'ko, V.M. and Bandosz, T.J. 2005. Heterogeneity of adsorption energy of water, methanol and diethyl ether on activated carbons: Effect of porosity and surface chemistry. *Adsorption* 11:97–102.

Gun'ko, V.M., Betz, W.R., Patel, S., Murphy, M.C., and Mikhalovsky, S.V. 2006a. Adsorption of lipopolysaccharide on carbon sieves. *Carbon* 44:1258–1262.

Gun'ko, V.M., Blitz, J.P., Gude, K. et al. 2007a. Surface structure and properties of mixed fumed oxides. *J. Colloid Interface Sci.* 314:119–130.

Gun'ko, V.M., Blitz, J.P., Zarko, V.I. et al. 2009a. Structural and adsorption characteristics and catalytic activity of titania and titania-containing nanomaterials. *J. Colloid Interface Sci.* 330:125–137.

Gun'ko, V.M., Bogatyrev, V.M., Borysenko, M.V. et al. 2010a. Morphological, structural and adsorptional features of oxide composites of different origin. *Appl. Surf. Sci.* 256:5263–5269.

Gun'ko, V.M., Bogatyrev, V.M., Leboda, R. et al. 2009b. Titania deposits on nanosilicas. *Ann. Univ. Marie Curie-Sklodowska Sect. Chem.* LXIV:21–48.

Gun'ko, V.M., Borysenko, M.V., Pissis, P. et al. 2007b. Polydimethylsiloxane at the interfaces of fumed silica and zirconia/fumed silica. *Appl. Surf. Sci.* 253:7143–7156.

Gun'ko, V.M., Dyachenko, A.G., Borysenko, M.V., Skubiszewska-Zięba, J., and Leboda, R. 2002a. CVD-titania on mesoporous silica gels. *Adsorption* 8:59–70.

Gun'ko, V.M., Galagan, N.P., Grytsenko, I.V. et al. 2007c. Interaction of unmodified and partially silylated nanosilica with red blood cells. *Central Eur. J. Chem.* 5:951–969.

Gun'ko, V.M., Goncharuk, E.V., O.V. Nechypor, Pakhovchishin, S.V., and Turov, V.V. 2006b. Integral equation for calculation of distribution function of activation energy of shear viscosity. *J. Colloid Interface Sci.* 304:239–245.

Gun'ko, V.M., Klyueva, A.V., Levchuk, Yu.N., and Leboda, R. 2003a. Photon correlation spectroscopy investigations of proteins. *Adv. Colloid Interface Sci.* 105:201–328.

Gun'ko, V.M., Kozynchenko, O.P., Tennison, S.R., Leboda, R., Skubiszewska-Zięba, J., and Mikhalovsky, S.V. 2012a. Comparative study of nanopores in activated carbons by HRTEM and adsorption methods. *Carbon* 50:3146–3153.

Gun'ko, V.M., Kozynchenko, O.P., Turov, V.V. et al. 2008a. Structural and adsorption characteristics of activated carbons possessing significant textural porosity. *Colloids Surf. A: Physicochem. Eng. Aspects* 317:377–387.

Gun'ko, V.M. and Leboda, R. 2002. Carbon-silica gel adsorbents. In: *Encyclopedia of Surface and Colloid Science*, Hubbard, A.T. (Ed.), New York: Marcel Dekker, pp. 864–878.

Gun'ko, V.M. and Leboda, R. 2005. Pore size distributions of complex systems. *Ann. Univ. Marie Curie-Sklodowska Sect. Chem.* LX:246–255.

Gun'ko, V.M., Leboda, R., Charmas, B., and Villieras, F. 2000a. Characterization of spatial and energetic structures of carbon-silica gels. *Colloids Surf. A: Physicochem. Eng. Aspects* 173:159–169.

Gun'ko, V.M., Leboda, R., Marciniak, M. et al. 2000b. CVD-titania/silica gel carbonized due to pyrolysis of cyclohexene. *Langmuir* 16:3227–3243.

Gun'ko, V.M., Leboda, R., and Pokrovskiy, V.A. 1999a. Pathways for decomposition of phenylethanol bound to silica surface. *Polish J. Chem.* 73:1345–1356.

Gun'ko, V.M., Leboda, R., Pokrovskiy, V.A., Charmas, B., Turov, V.V., and Ryczkowski, J. 2001a. A study of the organic carbon content of silica gel carbonised by pyrolysis of alcohols. *J. Anal. Appl. Pyrol.* 60:233–247.

Gun'ko, V.M., Leboda, R., Skubiszewska-Zięba, J. et al. 2008b. Influence of different treatments on characteristics of nanooxide powders alone or with adsorbed polar polymers or proteins. *Powder Technol.* 187:146–158.

Gun'ko, V.M., Leboda, R., and Skubiszewska-Zięba, J. 2009c. Heating effects on morphological and textural characteristics of individual and composite nanooxides. *Adsorption* 15:89–98.

Gun'ko, V.M., Leboda, R., Skubiszewska-Zięba, J., Gawdzik, B., and Charmas, B., 2005a. Structural characteristics of porous polymers treated by freezing with water or acetone. *Appl. Surf. Sci.* 252:612–618.

Gun'ko, V.M., Leboda, R., Skubiszewska-Zięba, J., and Rynkowski, J. 2000c. Silica gel modified due to pyrolysis of acetylacetone and metal (Ti, Cr, Co, Ni, Zn, Zr) acetylacetonates. *J. Colloid Interface Sci.* 231:13–25.

Gun'ko, V.M., Leboda, R., Skubiszewska-Zięba, J., Turov, V.V., and Kowalczyk, P. 2001b. Structure of silica gel Si-60 and pyrocarbon/silica gel adsorbents thermally and hydrothermally treated. *Langmuir* 17:3148–3161.

Gun'ko, V.M., Leboda, R., Turov, V.V. et al. 2001c. Structural and energetic nonuniformities of pyrocarbon–mineral adsorbents, *J. Colloid Interface Sci.* 238:340–356.
Gun'ko, V.M., Leboda, R., Turov, V.V., Charmas, B., and Skubiszewska-Zięba, J. 2002b. Structural and energetic heterogeneities of hybrid carbon-mineral adsorbents. *Appl. Surf. Sci.* 191:286–299.
Gun'ko, V.M., Leboda, R., Zarko, V.I. et al. 2003b. Fumed oxides modified due to pyrolysis of cyclohexene. *Colloids Surf. A: Physicochem. Eng. Aspects* 218:103–124.
Gun'ko, V.M., Meikle, S.T., Kozynchenko, O.P. et al. 2011a. Comparative characterisation of carbon and polymer adsorbents by SAXS and nitrogen adsorption methods. *J. Phys. Chem. C* 115:10727–10735.
Gun'ko, V.M., Mikhailova, I.V., Zarko, V.I. et al. 2003c. Study of interaction of proteins with fumed silica in aqueous suspensions by adsorption and photon correlation spectroscopy methods. *J. Colloid Interface Sci.* 260:56–69.
Gun'ko, V.M., Mikhalovska, L.I., Savina, I.N. et al. 2010b. Characterisation and performance of hydrogel tissue scaffolds. *Soft Matter* 6:5351–5358.
Gun'ko, V.M., Mikhalovska, L.I., Tomlins, P.E., and Mikhalovsky, S.V. 2011b. Competitive adsorption of macromolecules and real-time dynamics of Vroman-like effects. *Phys. Chem. Chem. Phys.* 13:4476–4485.
Gun'ko, V.M. and Mikhalovsky, S.V. 2004. Evaluation of slitlike porosity of carbon adsorbents. *Carbon* 42:843–849.
Gun'ko, V.M., Mironyuk, I.F., Zarko, V.I. et al. 2001d. Fumed silicas possessing different morphology and hydrophilicity. *J. Colloid Interface Sci.* 242:90–103.
Gun'ko, V.M., Mironyuk, I.F., Zarko, V.I. et al. 2005b. Morphology and surface properties of fumed silicas. *J. Colloid Interface Sci.* 289:427–445.
Gun'ko, V.M., Morozova, L.P., Turova, A.A. et al. 2012b. Hydrated phosphorus oxyacids alone and adsorbed on nanosilica. *J. Colloid Interface Sci.* 368:263–272.
Gun'ko, V.M., Nychiporuk, Yu.M., Zarko, V.I. et al. 2007d. Relationships between surface compositions and properties of surfaces of mixed fumed oxides. *Appl. Surf. Sci.* 253:3215–3230.
Gun'ko, V.M., Palijczuk, D., Leboda, R., Skubiszewska-Zięba, J., and Ziętek, S. 2006c. Influence of pore structure and pretreatments of activated carbons and water effects on breakthrough dynamics of tert-butylbenzene. *J. Colloid Interface Sci.* 294:53–68.
Gun'ko, V.M., Pissis, P., Spanoudaki, A. et al. 2007e. Relaxation phenomena in poly(vinyl alcohol)/fumed silica affected by interfacial water. *J. Colloid Interface Sci.* 312:201–213.
Gun'ko, V.M., Pissis, P., Spanoudaki, A. et al. 2008c. Interfacial phenomena in starch/fumed silica at varied hydration levels. *Colloids Surf. A: Physicochem. Eng. Aspects* 320:247–259.
Gun'ko, V.M., Seledets, O., Skubiszewska-Zięba, J. et al. 2005c. Phosphorus-containing pyrocarbon deposits on silica gel Si-100. *Micropor. Mesopor. Mater.* 87:133–145.
Gun'ko, V.M., Seledets, O., Skubiszewska-Zięba, J., Leboda, R., and Pasieczna, S. 2003d. Structural characteristics of pyrocarbon-fumed silica. *Colloids Surf. A: Physicochem. Eng. Aspects* 220:69–81.
Gun'ko, V.M., Skubiszewska-Zięba, J., Leboda, R. et al. 2004a. Influence of morphology and composition of fumed oxides on changes in their structural and adsorptive characteristics on hydrothermal treatment at different temperatures. *J. Colloid Interface Sci.* 269:403–424.
Gun'ko, V.M., Skubiszewska-Zięba, J., Leboda, R. et al. 2004b. Pyrocarbons prepared by carbonisation of polymers adsorbed or synthesised on a surface of silica and mixed oxides. *Appl. Surf. Sci.* 227:219–243.
Gun'ko, V.M., Skubiszewska- Zięba, J., Leboda, R., and Turov, V.V. 2004c. Impact of thermal and hydrothermal treatments on structural characteristics of silica gel Si-40 and carbon/silica gel adsorbents. *Colloids Surf. A: Physicochem. Eng. Aspects* 235:101–111.
Gun'ko, V.M., Skubiszewska-Zięba, J., Leboda, R., and Zarko, V.I. 2000d. Fumed silica carbonized due to pyrolysis of methylene chloride. *Langmuir* 16:374–382.
Gun'ko, V.M., Turanskaya, S.P., Nechipor, O.V. et al. 2006d. *Chem. Phys. Technol. Surf.* Kyiv: Naukova Dumka, 11:397–430 (in Russian).
Gun'ko, V.M. and Turov, V.V. 1999. Structure of hydrogen bonds and ^{1}H NMR spectra of water at the interface of oxides. *Langmuir* 15:6405–6415.
Gun'ko, V.M., Turov, V.V., Barvinchenko, V.N. et al. 2006e. Characteristics of interfacial water at nanosilica surface with adsorbed 1,3,5-trihydroxybenzene over wide temperature range. *Colloids Surf. A: Physicochem. Eng. Aspects* 278:106–122.
Gun'ko, V.M., Turov, V.V., Barvinchenko, V.N. et al. 2010c. Nonuniformity of starch/nanosilica composites and interfacial behaviour of water and organic compounds. *Appl. Surf. Sci.* 256:5275–5280.
Gun'ko, V.M., Turov, V.V., Bogatyrev, V.M. et al. 2003e. Influence of partial hydrophobization of fumed silica by hexamethyldisilazane on interaction with water. *Langmuir* 19:10816–10828.

Gun'ko, V.M., Turov, V.V., Bogatyrev, V.M. et al. 2005d. Unusual properties of water at hydrophilic/hydrophobic interfaces. *Adv. Colloid Interface Sci.* 118:125–172.

Gun'ko, V.M., Turov, V.V., Bogatyrev, V.M. et al. 2011c. The influence of pre-adsorbed water on adsorption of methane on fumed and nanoporous silicas. *Appl. Surf. Sci.* 258:1306–1316.

Gun'ko, V.M., Turov, V.V., and Gorbik, P.P. 2009d. *Water at the Interfaces*. Kiev, Ukraine: Naukova Dumka (in Russian).

Gun'ko, V.M., Turov, V.V., Kozynchenko, O.P. et al. 2008d. Characteristics of adsorption phase with water/organic mixtures at a surface of activated carbons possessing intraparticle and textural porosities. *Appl. Surf. Sci.* 254:3220–3231.

Gun'ko, V.M., Turov, V.V., Kozynchenko, O.P. et al. 2011d. Activation and structural and adsorption features of activated carbons with highly developed micro-, meso- and macroporosity. *Adsorption* 17:453–460.

Gun'ko, V.M., Turov, V.V., and Leboda, R. 2002c. Structure-adsorption characteristics of carbon-oxide materials. *Theoret. Exp. Chem.* 38:199–228.

Gun'ko, V.M., Turov, V.V., Leboda, R., Skubiszewska-Zięba, J., Tsapko, M.D., and Palijczuk, D. 2005e. Influence of organics on structure of water adsorbed on activated carbons. *Adsorption* 11:163–168.

Gun'ko, V.M., Turov, V.V., Leboda, R., Zarko, V.I., Skubiszewska-Zięba, J., and Charmas, B. 2007f. Adsorption, NMR and thermally stimulated depolarization current methods for comparative analysis of heterogeneous solid and soft materials. *Langmuir* 23:3184–3192.

Gun'ko, V.M., Turov, V.V., Shpilko, A.P. et al. 2006f. Relationships between characteristics of interfacial water and human bone tissues. *Colloid. Surf. B: Biointerfaces* 53:29–36.

Gun'ko, V.M., Turov, V.V., Skubiszewska-Zięba, J., Charmas, B., and Leboda, R. 2004d. Structural and adsorptive characteristics of pyrocabon/silica gel Si-60. *Adsorption* 10:5–18.

Gun'ko, V.M., Turov, V.V., Skubiszewska-Zięba, J., Leboda, R., Tsapko, M.D., and Palijczuk, D. 2003f. Structural characteristics of a carbon adsorbent and influence of organic solvents on interfacial water. *Appl. Surf. Sci.* 214:178–189.

Gun'ko, V.M., Turov, V.V., Turov, A.V. et al. 2007g. Behaviour of pure water and water mixture with benzene or chloroform adsorbed onto ordered mesoporous silicas. *Central Eur. J. Chem.* 5:420–454.

Gun'ko, V.M., Turov, V.V., Zarko, V.I. et al. 1997a. Active site nature of pyrogenic alumina/silica and water bound to surfaces. *Langmuir* 13:1529–1544.

Gun'ko, V.M., Turov, V.V., Zarko, V.I. et al. 1997b. Features of aqueous suspensions of fumed silica and interaction with proteins. *J. Colloid Interface Sci.* 192:166–178.

Gun'ko, V.M., Turov, V.V., Zarko, V.I. et al. 2007h. Comparative characterization of polymethylsiloxane hydrogel and silylated fumed silica and silica gel. *J. Colloid Interface Sci.* 308:142–156.

Gun'ko, V.M., Turov, V.V., Zarko, V.I. et al. 2008e. Structural features of polymer adsorbent LiChrolut EN and interfacial behaviour of water and water-organic mixtures. *J. Colloid Interface Sci.* 323:6–17.

Gun'ko, V.M., Turov, V.V., Zarko, V.I., Goncharuk, E.V., and Turova, A.A. 2009e. Regularities in the behaviour of water confined in adsorbents and bioobjects studied by ^{1}H NMR spectroscopy and TSDC methods at low temperatures. *Colloids Surf. A: Physicochem. Eng. Aspects* 336:147–158.

Gun'ko, V.M., Villiéras, F., Leboda, R., Marciniak, M., Charmas, B., and Skubiszewska-Zięba, J. 2000e. Characterization of CVD-titania/silica gel by means of low pressure nitrogen adsorption. *J. Colloid Interface Sci.* 230:320–327.

Gun'ko, V.M., Vlasova, N.N., Golovkova, L.P. et al. 2000f. Interaction of proteins and substituted aromatic drugs with highly disperse oxides in aqueous suspensions. *Colloids Surf. A: Physicochem. Eng. Aspects* 167:229–243.

Gun'ko, V.M., Voronin, E.F., Mironyuk, I.F. et al. 2003g. The effect of heat, adsorption and mechanochemical treatments on stuck structure and adsorption properties of fumed silicas. *Colloids Surf. A: Physicochem. Eng. Aspects* 218:125–135.

Gun'ko, V.M., Voronin, E.F., Nosach, L.V. et al. 2006g. Adsorption and migration of poly(vinyl pyrrolidone) at a surface of fumed silica, *Adsorp. Sci. Technol.* 24:143–157.

Gun'ko, V.M., Voronin, E.F., Nosach, L.V. et al. 2006h. Nanocomposites with fumed silica/poly(vinyl pyrrolidone) prepared at a low content of solvents. *Appl. Surf. Sci.* 253:2801–2811.

Gun'ko, V.M., Voronin, E.F., Nosach, L.V. et al. 2011e. Structural, textural and adsorption characteristics of nanosilica mechanochemically activated in different media. *J. Colloid Interface Sci.* 355:300–311.

Gun'ko, V.M., Voronin, E.F., Pakhlov, E.M. et al. 2000g. Features of fumed silica coverage with silanes having three or two groups reacting with the surface. *Colloids Surf. A: Physicochem. Eng. Aspects* 166:187–201.

Gun'ko, V.M., Voronin, E.F., Zarko, V.I. et al. 2004e. Interaction of poly(vinyl pyrrolidone) with fumed silica in dry and wet powders and aqueous suspensions. *Colloids Surf. A: Physicochem. Eng. Aspects* 233, 63–78.

Gun'ko, V.M., Voronin, E.F., Zarko, V.I., Pakhlov, E.M., and Chuiko, A.A. 1997c. Modification of some oxides by organic and organosilicon compounds. *J. Adhes. Sci. Technol.* 11:627–653.

Gun'ko, V.M., Yurchenko, G.R, Turov, V.V. et al. 2010d. Adsorption of polar and nonpolar compounds onto complex nanooxides with silica, alumina, and titania. *J. Colloid Interface Sci.* 348:546–558.

Gun'ko, V.M., Zarko, V.I., Chibowski, E., Dudnik, V.V., Leboda, R., and Zaets, V.A. 1997d. Structure of fumed titania and silica/titania and influence of the nature of surface sites on interaction with water. *J. Colloid Interface Sci.* 188:39–57.

Gun'ko, V.M., Zarko, V.I., Chuikov, B.A. et al. 1998a. Temperature-programmed desorption of water from fumed silica, titania, silica/titania, and silica/alumina. *Int. J. Mass Spectrom. Ion Process.* 172:161–179.

Gun'ko, V.M., Zarko, V.I., Goncharuk, E.V. et al. 2007i. TSDC spectroscopy of relaxational and interfacial phenomena. *Adv. Colloid Interface Sci.* 131:1–89.

Gun'ko, V.M., Zarko, V.I., Goncharuk, E.V., Klimenko, N.Yu., and Galagan, N.P. 2009f. Processes of low-temperature dipolar relaxation in BSA-nanooxide-water systems. *Surface* 16:14–25 (in Russian).

Gun'ko, V.M., Zarko, V.I., Leboda, R., and Chibowski, E. 2001e. Aqueous suspensions of fumed oxides: Particle size distribution and zeta potential. *Adv. Colloid Interface Sci.* 91:1–112.

Gun'ko, V.M., Zarko, V.I., Leboda, R., Marciniak, M., Janusz, W., and Chibowski, S. 2000h. Highly dispersed X/SiO_2 and $C/X/SiO_2$ (X = alumina, titania, alumina/titania) in the gas and liquid media. *J. Colloid Interface Sci.* 230:396–409.

Gun'ko, V.M., Zarko, V.I., Mironyuk, I.F. et al. 2004f. Surface electric and titration behaviour of fumed oxides. *Colloids Surf. A: Physicochem. Eng. Aspects* 240:9–25.

Gun'ko, V.M., Turov, V.V., and Turov, A.V. 2012c. Hydrogen peroxide - water mixture bound to nanostructured silica. *Chem. Phys. Lett.* 531:132–137.

Gun'ko, V.M., Zarko, V.I., Turov, V.V. et al. 1999b. Characterization of fumed alumina/silica/titania in the gas phase and aqueous suspension. *J. Colloid Interface Sci.* 220:302–323.

Gun'ko, V.M., Zarko, V.I., Turov, V.V. et al. 1998b. CVD-titania on fumed silica substrate. *J. Colloid Interface Sci.* 198:141–156.

Gun'ko, V.M., Zarko, V.I., Turov, V.V. et al. 2009g. Morphological and structural features of individual and composite nanooxides with alumina, silica, and titania in powders and aqueous suspensions. *Powder Technol.* 195:245–258.

Gun'ko, V.M., Zarko, V.I., Turov, V.V. et al. 2010e. Regularities in the behaviour of nanooxides in different media affected by surface structure and morphology of particles. In: Shpak, A.P. and Gorbyk, P.P. (Eds.) *Nanomaterials and Supramolecular Structures*. Dordrecht, the Netherlands: Springer, pp. 93–118.

Gun'ko, V.M., Zarko, V.I., Turov, V.V., Goncharuk, E.V., Voronin, E.F., and Kazakova, O.A. 2001f. Hydrogen bonds at silica-polyvinylpyrrolidone-water interfaces. *Theoret. Exp. Chem.* 37:75–79.

Gun'ko, V.M., Zarko, V.I., Turov, V.V., Leboda, R., and Chibowski, E. 1999c. The effect of second phase distribution in disperse X/silica (X = Al_2O_3, TiO_2, and GeO_2) on its surface properties. *Langmuir* 15:5694–5702.

Gun'ko, V.M., Zarko, V.I., Turov, V.V., Leboda, R., Chibowski, E., and Gun'ko, V.V. 1998c. Aqueous suspensions of highly disperse silica and germania/silica. *J. Colloid Interface Sci.* 205:106–120.

Gun'ko, V.M., Zarko, V.I., Turov, V.V., Voronin, E.F., Tischenko, V.A., and Chuiko, A.A. 1995. Dielectric properties and dynamic simulation of water bound to titania/silica surfaces. *Langmuir* 11:2115–2122.

Gun'ko, V.M., Zarko, V.I., Voronin, E.F. et al. 2002d. Impact of some organics on structural and adsorptive characteristics of fumed silica in different media. *Langmuir* 18:581–596.

Gun'ko, V.M., Zarko, V.I., Voronin, E.F. et al. 2006i. Successive interaction of pairs of soluble organics with nanosilica in aqueous media. *J. Colloid Interface Sci.* 300:20–32.

Gun'ko, V.M., Zaulychnyy, Ya.V., Ilkiv, B.I. et al. 2011f. Textural and electronic characteristics of mechanochemically activated composites with nanosilica and activated carbon. *Appl. Surf. Sci.* 258:1115–1125.

Guo, X., Liu, X., Xu, B., and Dou, T. 2009. Synthesis and characterization of carbon sphere-silica core–shell structure and hollow silica spheres. *Colloids Surf. A: Physicochem. Eng. Aspects* 345:141–146.

Gutmann, V. 1978. *The Donor-Acceptor Approach to Molecular Interactions*. New York: Plenum Press.

Gutowsky, H.S. and Holm, C.H. 1956. Rate processes and nuclear magnetic resonance spectra. II. Hindered internal rotation of amides. *J. Chem. Phys.* 25:1228–1234.

Gutowsky, H.S. and Saika, A. 1953. Dissociation, chemical exchange, and the proton magnetic resonance in some aqueous electrolytes. *J. Chem. Phys.* 21:1688–1694.

Gutsch, A., Krämer, M., Michael, G., Mühlenweg, H., Pridöhl, M., and Zimmermann, G. 2002. Gas-phase production of nanoparticles. *KONA* No.20:24–37.

Haacke, E.M., Brown, R.W., Thompson, M.R., and Venkatesan, R. 1999. *Magnetic Resonance Imaging—Physical Principles and Sequence Design*. New York: John Wiley & Sons.

Haddad, E., Nossov, A., Guenneau, F. et al. 2004. Exploring the internal structure of mesoporous powders and thin films by continuous flow laser-enhanced 129Xe NMR. *Stud. Surf. Sci. Catal.* 154B:1464–1470.

Hadzi, D. (Ed.) 1997. *Theoretical Treatments of Hydrogen Bonding*. New York: John Wiley & Sons.

Hagaman, E.W., Chen, B., Jiao, J., and Parsons, W. 2012. Solid-state ^{17}O NMR study of benzoic acid adsorption on metal oxide surfaces. *Solid State Nucl. Magn. Reson.* 41:60–67.

Haggis, G.H., Hasted, J.B., and Buchanan, T.J. 1952. The dielectric properties of water in solutions. *J. Chem. Phys.* 20:1452–1465.

Haggis, G.H., Hasted, J.B., and Buchanan, J.T. 1953. Temperature behavior of chemical shift caused by hydrogen bonds. *J. Chem. Phys.* 21:1688–1693.

Hakim, L.F., Portman, J.L., Casper, M.D., and Weimer, A.W. 2005. Aggregation behavior of nanoparticles in fluidized beds. *Powder Technol.* 160:149–160.

Hall, C.R. and Holmes, R.J. 1992. The preparation and properties of some activated carbons modified by treatment with phosgene or chlorine. *Carbon* 30:173–176.

Halperin, W.P., D'Orazio, F., Bhattacharya, S., and Tarczon, T.C. 1989. *Molecular Dynamics in Restricted Geometries*. New York: John Wiley & Sons, Chapter 11.

Hamer, M.A., Hendrickson, W.A., and Marti, J.J. 2001. Particulate encapsulation of liquid beads. Patent WO0185138. Aveka Inc.

Han, Y., Quan, G.B., Liu, X.Z. et al. 2005. Improved preservation of human red blood cells by lyophilization. *Cryobiology* 51:152–164.

Hancock, R.D. and Tarbet, B.J. 2000. The other double helix - the fascinating chemistry of starch. *J. Chem. Educ.* 77:988–992.

Hansen, T.C. 1986. Physical structure of hardened cement paste. a classical approach. *Mater. Struct.* 19:423–436.

Hansen, E.W., Stöcker, M., and Schmidt, R. 1996. Low-temperature phase transition of water confined in mesopores probed by NMR. Influence on pore size distribution. *J. Phys. Chem.* 100:2195–2200.

Harañczyk, H., Strzałka, K., Jasiñski, G., and Mosna-Bojarska, K. 1996. The initial stages of wheat (*Triticum aestivum, L.*) seed imbibition as observed by proton nuclear magnetic relaxation. *Colloid. Surf. A: Physicochem. Eng. Aspects* 115:47–54.

Harbridge, J.R., Eaton, S.S., and Eaton, G.R. 2003. Comparison of electron spin relaxation times measured by Carr–Purcell–Meiboom–Gill and two-pulse spin–echo sequences. *J. Magn. Reson.* 164:44–53.

Hardingham, T.E. and Fosang, A.J. 1992. Proteoglycans: Many forms and functions. *FASEB J.* 6:861–870.

Haris, P.I., Chapman, D., and Benga, G. 1995. A Fourier-transform infrared spectroscopic investigation of the hydrogen-deuterium exchange and secondary structure of the 28-kDa channel-forming integral membrane protein (CHIP28). *Eur. J. Biochem.* 233:659–664.

Harley, J.D. and Margolis, J. 1961. Haemolytic activity of colloidal silica. *Nature* 189:1010–1011.

Harris, P.J.F. 2009. *Carbon Nanotube Science. Synthesis, Properties and Applications*. Cambridge, MA: Cambridge University Press.

Harris, R.K., Becker, E.D., Cabral de Mendez, S.M., Granger, P., Hoffman, R.E., and Zilm, K.W. 2008. Further conventions for NMR shielding and chemical shifts (IUPAC Recommendation 2008). *Pure Appl. Chem.* 80:59–84.

Harris, R.K., Thompson, T.V., Forshaw, P. et al. 1996. A magic angle spinning NMR study into the adsorption of deuterated water by activated carbon. *Carbon* 34:1275–1279.

Harris, R.K., Thompson, T.V., Norman, P.R., Pottage, C., and Trethewey, A.N. 1995. *J. Chem. Soc. Faraday Trans.* 91:1795–1799.

Harrison, R.M. and van Grieken, R.E. (Eds.) 1998. *Atmospheric Particles*. New York: John Wiley & Sons.

Hasenzahl, S., Gray, A., Walzer, E., and Braunagel, A. 2005. Dry water for the skin, *SOÉFW J.* 131:1–8.

Hass, K.C., Schneider, W.F., Curioni, A., and Andreoni, W. 1998. The Chemistry of water on alumina surfaces: Reaction dynamics from first principles. *Science* 282:265–268.

Hassan, C.M., Doyle, F.J., and Peppas, N.A. 1997. Dynamic behavior of glucoseresponsive poly(methacrylic acid-g-ethylene glycol) hydrogels. *Macromolecules* 30:6166–6173.

Hatakeyama, E.S., Ju, H., Gabriel, C.J. et al. 2009. New protein-resistant coatings for water filtration membranes based on quaternary ammonium and phosphonium polymers. *J. Membr. Sci.* 330:104–116.

Havriliak, S. and Negami, S. 1966. A complex plane analysis of alpha-dispersion of some polymeric systems. *J. Polym. Sci. C* 14:99–117.

Havriliak, S. and Negami, S. 1967. A complex plane representation of dielectric and mechanical relaxation processes in some polymers. *Polymers* 8:161–210.

Hays, D.L. and Fennema, O., 1982. Methodology for determining unfreezable water in protein suspensions by low-temperature NMR. *Arch. Biochem. Biophys.* 213:1–6.

He, L., Dexter, A.F., and Middelberg, A.P.J. 2006. Biomolecular engineering at interfaces. *Chem. Eng. Sci.* 61:989–1003.

He, H., Guo, J., Zhu, J., Yuan, P., and Hu, C. 2004. ^{29}Si and ^{27}Al MAS NMR spectra of mullites from different kaolinites. *Spectrochim. Acta A* 60:1061–1064.

He, Q. and Shi, J. 2011. Mesoporous silica nanoparticle based nano drug delivery systems: Synthesis, controlled drug release and delivery, pharmacokinetics and biocompatibility. *J. Mater. Chem.* 21:5845–5855.

Hedin, N., Graf, R., Christiansen, S.C. et al. 2004. Structure of a surfactant-templated silicate framework in the absence of 3D crystallinity. *J. Am. Chem. Soc.* 126:9425–9432.

Heeribout, L., d'Espinose de la Caillerie, J.B., Legrand, A.P., and Mignani, G. 1999. A new straightforward approach to generate Si–H groups on silica. *J. Colloid Interface Sci.* 215:296–299.

Heimink, J., Sieger, P., and Koller, H. 2011. Two-dimensional ph mapping of release kinetics of silica-encapsulated drugs. *J. Pharmacol. Sci.* 100:4401–4412.

Heink, W., Kärger, J., Pfeifer, H., Vartapetian, R.S., Voloshchuk, A.M. 1993. PFG NMR study of multicomponent self-diffusion on active carbons. *Carbon* 31:1083–1087.

Helfrich, W. 1973. Elastic properties of lipid bilayers. Theory and possible experiments. *Z. Naturforsch.* 28c:693–703.

Hemmerlè, J., Altmann, S.M., Maaloum, M. et al. 1999. Direct observation of the anchoring process during the adsorption of fibrinogen on a solid surface by force-spectroscopy mode atomic force microscopy. *Proc. Natl Acad. Sci. USA* 96:6705–6710.

Henderson, M.A. 2002. Interaction of water with solid surfaces: Fundamental aspects revisited. *Surf. Sci. Rep.* 46:1–308.

Hennel, J.W. and Klinowski, J. 2004. Magic-angle spinning: A historical perspective. *Topic. Curr. Chem.* 246:1–14.

Hennig, G. 1959. Interstitial compounds of graphite. *Prog. Anorg. Chem.* 1:125–205.

Hennion, M.-C. 1999. Solid-phase extraction: Method development, sorbents, and coupling with liquid chromatography. *J. Chromatogr. A* 856:3–54.

Henry, F., Gaudillat, M., Costa, L.C., and Lakkis, F. 2003. Free and/or bound water by dielectric measurements. *Food Chem.* 82:29–34.

Hensen, E.J.M., Poduval, D.G., Magusin, P.C.M.M., Coumans, A.E., and van Veen, J.A.R. 2010. Formation of acid sites in amorphous silica-alumina. *J. Catal.* 269:201–218.

Hermansson, M. 1999. The DLVO theory in microbial adhesion. *Colloid. Surf. B: Biointerfaces* 14:105–119.

Hertrich, M. 2008. Imaging of groundwater with nuclear magnetic resonance. *Prog. Nucl. Magn. Reson. Spectrosc.* 53:227–248.

Hess, W.T. 1995. Hydrogen peroxide, In *Kirk-Othmer Encyclopedia of Chemical Technology*, Vol. 13, 4th edn. New York: Wiley, pp. 961–995.

Hieda, M., Garcia, R., Dixon, M., Daniel, T., Allara, D., and Chan, M.H.W. 2004. Ultrasensitive quartz crystal microbalance with porous gold electrodes. *Appl. Phys. Lett.* 84:628–630.

Hindman, J.S. 1966. Proton resonance shift of water in the gas and liquid states. *J. Chem. Phys.* 44:4582–4592.

Hindmarsh, J.P., Buckley, C., Russell, A.B. et al. 2004. Imaging droplet freezing using MRI. *Chem. Eng. Sci.* 59:2113–2122.

Hitchcock, I., Holt, E.M., Lowe, J.P., and Rigby, S.P. 2011. Studies of freezing–melting hysteresis in cryoporometry scanning loop experiments using NMR diffusometry and relaxometry. *Chem. Eng. Sci.* 66:582–592.

Hoffman, A.S. 2000. Bioconjugates of intelligent polymers and recognition proteins for use in diagnostics and affinity separations. *Clin. Chem.* 46:1478–1486.

Hoffman, R., Benz, E.J., Shattil, S.J. et al. 2000. *Hematology: Basic Principles and Practice*, 3rd edn. Philadelphia, PA: Churchill Livingstone.

Hoffmann, P. and Knözinger, E. 1987. Novel aspects of mid and far IR Fourier spectroscopy applied to surface and adsorption studies on SiO_2. *Surf. Sci.* 188:181–198.

Holland, G.P., Sharma, R., Agola, J.O. et al. 2007. NMR characterization of phosphonic acid capped SnO_2 nanoparticles. *Chem. Mater.* 19:2519–2526.

Hologne, M., Chevelkov, V., and Reif, B. 2006. Deuterated peptides and proteins in MAS solid-state NMR. *Progr. Nucl. Magn. Reson. Spectrosc.* 48:211–232.

Hörber, J.K.H. and Miles, M.J. 2003. Scanning probe evolution in biology. *Science* 302:1002–1005.

Horbett, T.A. and Brash, J.L. (Eds.) 1995. *Proteins at Interfaces II, Fundamentals and Applications*, ACS Symposium Series 602. Washington, DC: American Chemical Society.

Horsewill, A.J. 2008. Quantum tunnelling in the hydrogen bond. *Prog. Nucl. Magn. Reson. Spectrosc.* 52:170–196.

Hou, S.S., Beyer, F.L., and Schmidt-Rohr, K. 2002. High-sensitivity multinuclear NMR spectroscopy of a smectite clay and of clay-intercalated polymer. *Solid State Nucl. Magn. Reson.* 22:110–127.

Hou, S.S., Bonagamba, T.J., Beyer, F.L., Madison, P.H., and Schmidt-Rohr, K. 2003. Clay intercalation of poly(styrene–ethylene oxide) block copolymers studied by multinuclear solid-state NMR. *Macromolecules* 36:2769–2776.

Hrobarikova, J., Robert, J.-L., Calberg, C., Jerome, R., and Grandjean, J. 2004. Solid-state NMR study of intercalated species in poly(ε-caprolactone)/clay nanocomposites. *Langmuir* 20:9828–9833.

Hu, Y.J., Jin, J.A., Zhang, H., Wu, P., and Cai, C.X. 2010. Graphene: Synthesis, functionalization and applications in chemistry. *Acta Phys. Chim. Sin.* 26:2073–2086.

Hu, J.Z., Kwak, J.H., Herrera, J.E., Wang, Y., and Peden, Ch.H.F. 2005. Line narrowing in ^1H MAS spectrum of mesoporous silica by removing adsorbed H_2O using N_2. *Solid State Nucl. Magn. Reson.* 27:200–205.

Hu, J.Z. and Wind, R.A. 2003. Sensitivity-enhanced phase-corrected ultra-slow magic angle turning using multiple-echo data acquisition. *J. Magn. Reson.* 163:149–162.

Huang, K.-y., He, Z.-p., and Chao, K.-j. 2006. Mesoporous silica films- characterization and reduction of their water uptake. *Thin Solid Films* 495:197–204.

Hubálek, Z. 2003. Protectants used in the cryopreservation of microorganisms. *Cryobiology* 46:205–229.

Hubbard, A.T. (Ed.) 2002. *Encyclopedia of Surface and Colloid Science.* New York: Marcel Dekker.

Hughes, R. 1984. *Deactivation of Catalysts.* New York: Academic Press.

Huh, S., Chen, H.-T., Wiench, J.W., Pruski, M., and Lin, V.S.-Y. 2004. Controlling the selectivity of competitive nitroaldol condensation by using a bifunctionalized mesoporous silica nanosphere-based catalytic system. *J. Am. Chem. Soc.* 126:1010–1011.

Huh, S., Chen, H.-T., Wiench, J.W., Pruski, M., and Lin, V.S.-Y. 2005. Cooperative catalysis by general acid and base bifunctionalized mesoporous silica nanospheres. *Angew. Chem. Int. Ed.* 44:1826–1830.

Huh, S., Wiench, J.W., Yoo, J.-C., Pruski, M., and Lin, V.S.-Y. 2003. Organic functionalization and morphology control of mesoporous silicas via a co-condensation synthesis method. *Chem. Mater.* 15:4247–4256.

Huittinen, N., Sarv, P., and Lehto, J. 2011. A proton NMR study on the specific sorption of yttrium(III) and europium(III) on gamma-alumina [γ-Al_2O_3]. *J. Colloid Interface Sci.* 361:252–258.

Humbert, H. 1995. Estimation of hydroxyl density at the surface of pyrogenic silicas by complementary NMR and Raman experiments. *J. Non-Cryst. Solids* 191:29–37.

Hung, I., Rossini, A.J., and Schurko, R.W. 2004. Application of the Carr–Purcell–Meiboom– Gill pulse sequence for the acquisition of solid-state NMR spectra of spin-1/2 nuclei. *J. Phys. Chem. A* 108:7112–7120.

Hunger, M. 1996. Multinuclear solid-state NMR studies of acidic and non-acidic hydroxyl protons in zeolites. *Solid State Nucl. Magn. Reson.* 6:1–29.

Hunger, M., Karger, J., Pfeifer, H. et al. 1987. Investigation of internal silanol groups as structural defects in ZSM-5-type zeolites. *J. Chem. Soc. Faraday Trans. I.* 83:3459–3468.

Hunt, H.K., Lew, Ch.M., Sun, M., Yan, Y., and Davis, M.E. 2010. Pure-silica LTA, CHA, STT, ITW, and -SVR thin films and powders for low-k applications. *Micropor. Mesopor. Mater.* 130:49–55.

Hunter, R.J. 1981. *Zeta Potential in Colloid Science: Principles and Applications.* New York: Academic Press Inc.

Hunter, R.J. 1993. *Introduction to Modern Colloid Science.* London, U.K.: Oxford University Press.

Hunter, C.A., Packer, M.J., Zonta, C. 2005. From structure to chemical shift and vice-versa. *Prog. Nucl. Magn. Reson. Spectrosc.* 47:27–39.

Huo, Q., Margolese, D.I., and Stucky, G.D. 1996. Surfactant control of phases in the synthesis of mesoporous silica-based materials. *Chem. Mater.* 8:1147–1160.

Hüsing, N. and Schubert, U. 1998. Aerogels—Airy materials: Chemistry, structure, and properties. *Angew. Chem. Int. Ed.* 37:22–45.

Huttepain, M. and Oberlin, A., 1990. Microtexture of nongraphitizing carbons and TEM studies of some activated samples. *Carbon* 28:103–111.

Hyeon-Lee, J., Beaucage, G., Pratsinis, S.E., and Vemury, S. 1998. Fractal analysis of flame-synthesized nanostructured silica and titania powders using small-angle X-ray scattering. *Langmuir* 14:5751–5756.

Ibbett, R.N., Phillips, D.A.S., and Kaenthong, S. 2007. A dye-adsorption and water NMR-relaxation study of the effect of resin cross-linking on the porosity characteristics of lyocell solvent-spun cellulosic fibre. *Dye. Pigment.* 75:624–632.

Igumenova, T.I., Frederick, K.K., and Wand, A.J. 2006. Characterization of the fast dynamics of protein amino acid side chains using NMR relaxation in solution. *Chem. Rev.* 106:1672–1699.

Iino, M. 1994. Transition from Lorentzian to Gaussian lineshape of magnetization transfer spectrum in bovine serum albumin solutions. *Magn. Reson. Med.* 32:459–463.

Iler, R.K. 1979. *The Chemistry of Silica*. Chichester, U.K.: Wiley.

ImageJ. 2007. Image processing and analysis in Java. http://rsbweb.nih.gov/ij/download.html, http://rsb.info.nih.gov/ij/plugins/voxel-counter.html, http://rsb.info.nih.gov/ij/plugins/granulometry.html (accessed July, 2007).

Ishida, N., Koizumi, M., and Kano, H. 2000. The NMR microscope: A unique and promising tool for plant science. *Ann. Bot.* 86:259–278.

Islam, M.M., Sohya, S., Noguchi, K., Kidokoro, S., Yohda, M., and Kuroda, Y. 2009. Thermodynamic and structural analysis of highly stabilized BPTIs by single and double mutations. *Proteins* 77:962–970.

Ismail, A.A. and Bahnemann, D.W. 2011. Mesoporous titania photocatalysts: Preparation, characterization and reaction mechanisms. *J. Mater. Chem.* 21:11686–11707.

Isobe, T., Watanabe, T., d'Espinose de la Caillerie, J.B., Legrand, A.P., Massiot, D. 2003. Solid-state ^{1}H and ^{27}Al NMR studies of amorphous aluminum hydroxides. *J. Colloid Interface Sci.* 261:320–324.

Israelachvili, J. 1994. *Intermolecular and Surface Forces*, 2nd edn. New York: Academic Press.

Israelachvili, J.N. and Wennerstrom, H. 1990. Hydration or steric forces between amphiphilic surfaces? *Langmuir* 6:873–876.

Ito, O., Seki, H., and Iino, M. 1988. Diffuse reflectance spectra in near i.r. region of coals; a new index for degrees of coalification and carbonization. *Fuel* 67:573–579.

Iuliucci, R., Taylor, C., and Hollis, W.K. 2006. ^{1}H/^{29}Si cross-polarization NMR experiments of silica-reinforced polydimethylsiloxane elastomers: Probing the polymer–filler interface. *Magn. Reson. Chem.* 44:375–384.

Jackson, C.L. and McKenna, G.B. 1990. The melting behavior of organic materials confined in porous solids. *J. Chem. Phys.* 93:9002–9011.

Jacobsen, C.J.H., Madsen, C., Janssens, T.V.W., Jakobsen, H.J., and Skibsted, J. 2000. Zeolites by conned space synthesis characterization of the acid sites in nanosized ZSM-5 by ammonia desorption and ^{27}Al/^{29}Si-MAS NMR spectroscopy. *Micropor. Mesopor. Mater.* 39:393–401.

Jaeger, W., Bohrisch, J., Laschewsky, A. 2010. Synthetic polymers with quaternary nitrogen atoms—Synthesis and structure of the most used type of cationic polyelectrolytes. *Progr. Polym. Sci.* 35:511–577.

Jagadeesh, B., Demco, D.E., and Blümich, B. 2004. Surface induced order and dynamic heterogeneity in ultra thin polymer films: A ^{1}H multiple-quantum NMR study. *Chem. Phys. Lett.* 393:416–420.

Jagur-Grodzinski, J. 2007. Polymeric materials for fuel cells: Concise review of recent studies. *Polym. Adv. Technol.* 18:785–799.

Jähnert, S., Vaca Chávez, F., Schaumann, G.E., Schreiber, A., Schönhoff, M., and Findenegg, G.H. 2008. Melting and freezing of water in cylindrical silica nanopores. *Phys. Chem. Chem. Phys.* 10:6039–6051.

James, T.L. 1975. *Nuclear Magnetic Resonance in Biochemistry*. New York: Academic Press.

Jamin, N. and Toma, F. 2001. NMR studies of protein-DNA interactions. *Prog. Nucl. Magn. Reson. Spectrosc.* 38:83–114.

Janicke, M.T., Landry, C.C., Christiansen, S.C., Kumar, D., Stucky, G.D., and Chmelka, B.F. 1998. Aluminum incorporation and interfacial structures in MCM-41 mesoporous molecular sieves. *J. Am. Chem. Soc.* 120:6940–6951.

Jannin, V., Musakhanian, J. and Marchaud, D. 2008. Approaches for the development of solid and semi-solid lipid-based formulations. *Adv. Drug Deliv. Rev.* 60:734–746.

Jaroniec, M., Kruk, M., and Olivier, J.P. 1999. Standard nitrogen adsorption data for characterization of nanoporous silicas. *Langmuir* 15:5410–5413.

Jaroniec, M., Kruk, M., Shin, H.J., Ryoo, R., Sakamoto, Y., and Terasaki, O. 2001. Comprehensive characterization of highly ordered MCM-41 silicas using nitrogen adsorption, thermogravimetry, X-ray diffraction and transmission electron microscopy. *Micropor. Mesopor. Mater.* 48:127–134.

Jaroniec, M. and Madey, R. 1988. *Physical Adsorption on Heterogeneous Solids*. Amsterdam, the Netherlands: Elsevier.

Jarymowycz, V.A. and Stone, M.J. 2006. Fast time scale dynamics of protein backbones: NMR relaxation methods, applications, and functional consequences. *Chem. Rev.* 106:1624–1671.

Jaska, E. 1971. Starch gelatinization as detected by proton magnetic resonance. *Cereal Chem.* 48:437–444.

Jayakumar, R., Menon, D., Manzoor, K., Nair, S.V., and Tamura, H. 2010a. Biomedical applications of chitin and chitosan based nanomaterials—A short review. *Carbohydr. Polym.* 82:227–232.

Jayakumar, R., Prabaharan, M., Nair, S.V., and Tamura, H. 2010b. Novel chitin and chitosan nanofibers in biomedical applications. *Biotechnol. Adv.* 28:142–150.

JCPDS Database. 2001. International Center for Diffraction Data, PA.

Jeon, J.-D., Kim, S.J., and Kwak, S.-Y. 2008. ^{1}H nuclear magnetic resonance (NMR) cryoporometry as a tool to determine the pore size distribution of ultrafiltration membranes. *J. Membr. Sci.* 309:233–238.

Jerman, I., Orel, B., Vuk, A.Š., Koželj, M., and Kovač, J. 2010. A structural and corrosion study of triethoxysilyl and perfluorooctyl functionalized polyhedral silsesquioxane nanocomposite films on AA 2024 alloy. *Thin Solid Films* 518:2710–2721.

Jesionowski, T., Żurawska, J., Krysztafkiewicz, A., Pokora, M., Waszak, D., and Tylus, W. 2003. Physicochemical and morphological properties of hydrated silicas precipitated following alkoxysilane surface modification. *Appl. Surf. Sci.* 205:212–224.

Jieli, M. and Aida, M. 2009. Classification of OH bonds and infrared spectra of the topology-distinct protonated water clusters $H_3O^+(H_2O)_{n-1}$ ($n \leq 7$). *J. Phys. Chem. A* 113:1586–1594.

Jmol, 2011. An open-source Java viewer for chemical structures in 3D. http://jmol.sourceforge.net/ (accessed November, 2011).

Jobic, H., Bée, M. Kärger, J., Vartapetian, R.S., Balzer, C., and Julbe, A. 1995. Mobility of cyclohexane in a microporous silica sample: A quasielastic neutron scattering and NMR pulsed-field gradient technique study. *J. Membr. Sci.* 108:71–78.

Johari, G.P. 2009. Thermal relaxation of water due to interfacial processes and phase equilibria in 1.8 nm pores of MCM-41. *Thermochim. Acta* 492:29–36.

Johari, G.P. and Goldstein, M. 1970. Molecular mobility in simple glasses. *J. Phys. Chem.* 74:2034–2035.

Jones, C.W. 1999. *Applications of Hydrogen Peroxide and Derivatives*. Cambridge, MA: Royal Society of Chemistry.

Jonquières, A. and Fane, A. 1998. Modified BET models for modeling water vapor sorption in hydrophilic glassy polymers and systems deviating strongly from ideality. *J. Appl. Polym. Sci.* 67:1415–1430.

Jonscher, A.K. 1983. *Dielectric Relaxation in Solids*. London, U.K.: Chelsea Dielectric Press.

Jung, M. 2001. NMR characterization on the preparation of sol–gel derived mixed oxide materials. *Int. J. Inorg. Mater.* 3:471–478.

Jung, S.-Y., Lim, S.-M., Albertorio, F. et al. 2003. The Vroman effect: A molecular level description of fibrinogen displacement. *J. Am. Chem. Soc.* 125:12782–12786.

Jung, J.K., Ryu, K.-S., Kim, Y.-I., and Tang, C. 2004. NMR study of boron nitride nanotubes. *Solid State Commun.* 130:45–48.

Juvaste, H., Iiskola, E.I., and Pakkanen, T.T. 1999. Aminosilane as a coupling agent for cyclopentadienyl ligands on silica. *J. Organometal. Chem.* 587:38–45.

Kainosho, M., Kroon, P.A., Lawaczek, R., Petersen, N.O., and Chan, S.I. 1978. Chain length dependence of the ^{1}H NMR relaxation rates in bilayer vesicles. *Chem. Phys. Lipids* 21:59–68.

Kamegawa, K. and Yoshida, H. 1990. A method for measuring surface area of carbon of carbon-coated silica gel. *Bull. Chem. Soc. Jpn.* 63:3683–3685.

Kamegawa, K. and Yoshida, H. 1993. Carbon coated of silica surface I. Pyrolysis of silica gels esterified with alcohols. *J. Colloid Interface Sci.* 159:324–327.

Kamegawa, K. and Yoshida, H. 1995. Carbon coating of silica surface. II. Pyrolysis of silica gels esterified with phenols. *J. Colloid Interface Sci.* 172:94–97.

Kamerzell, T.J., Esfandiary, R., Joshi, S.B., Middaugh, C.R., and Volkin, D.B. 2011. Protein–excipient interactions: Mechanisms and biophysical characterization applied to protein formulation development. *Adv. Drug Deliv. Rev.* 63:1118–1159.

Kammler, H., Beaucage, G., Mueller, R., and Pratsinis, S. 2004. Structure of flamemade silica nanoparticles by ultra-small-angle X-ray scattering. *Langmuir* 20:1915–1921.

Kammler, H.K. and Pratsinis, S.E. 2000. Electrically-assisted flame aerosol synthesis of fumed silica at high production rates. *Chem. Eng. Process.* 39:219–227.

Kanan, S.M., El-Kadri, O.M., Abu-Yousef, I.A., and Kanan, M.C. 2009. Semiconducting metal oxide based sensors for selective gas pollutant detection. *Sensors* 9:8158–8196.

Kanazawa, K.K. and Gordon, J. II. 1985. Frequency of a quartz microbalance in contact with liquid. *Anal Chem.* 57:1770–1771.

Kaneko, K., Ishii, C., Ruike, M., and Kuwabara, H. 1992. Origin of super high surface area and microcrystalline graphitic structures of activated carbons. *Carbon* 30:1075–1088.

Kang, S.A, Li, W., Lee, H.E., Phillips, B.L., and Lee, Y.J. 2011. Phosphate uptake by TiO_2: Batch studies and NMR spectroscopic evidence for multisite adsorption. *J. Colloid Interface Sci.* 364:455–461.

Karakoti, A.S., Das, S., Thevuthasan, S., and Seal, S. 2011. PEGylated inorganic nanoparticles. *Angew. Chem. Int. Ed.* 50:1980–1994.

Karfeld-Sulzer, L.S., Waters, E.A., Kohlmeir, E.K. et al. 2011. Protein polymer MRI contrast agents: Longitudinal analysis of biomaterials in vivo. *Magn. Reson. Med.* 65:220–228.

Karger, J. and Ruthven, D.M. 1992. *Diffusion in Zeolites and Other Microporous Solids*. New York: John Willey & Sons.

Karlsson, J.O.M. 2001. A theoretical model of intracellular devitrification. *Cryobiology* 42:154–169.

Karlsson, J.O.M. and Toner, M. 1996. Long-term storage by cryopreservation: Critical issues. *Biomaterials* 17:243–256.

Karp, E.S., Newstadt, J.P., Chu, S., and Lorigan, G.A. 2007. Characterization of lipid bilayer formation in aligned nanoporous aluminum oxide nanotube arrays. *J. Magn. Reson.* 187:112–119.

Karpenko, G.A., Turov, V.V., and Chuiko, A.A. 1990a. Bound water molecules in graphite oxide structure. In: *Extended Abstracts of International Carbon 90 Conference*, pp. 676–677.

Karpenko, G.A., Turov, V.V., Kovtyukhova, N.I., Bakay, E.A, and Chuiko, A.A. 1990b. Structural peculiarities of graphite oxide its adsorption ability to water molecules. *Theoret. Eksp. Khim.* 26:102–106.

Kasemo, B. 2002. Biological surface science. *Surf. Sci.* 500:656–677.

Katopo, H., Song, Y., and Jane, J. 2002. Effect and mechanism of ultrahigh hydrostatic pressure on the structure and properties of starch. *Carbohydr. Polym.* 47:233–244.

Katz, E.P. 1969. The kinetics of mineralization in vitro. *Biochem. Biophys. Acta* 194:121–130.

Katz, E. and Willner, I. 2004. Biomolecule-functionalized carbon nanotubes: Applications in nanobioelectronics. *ChemPhysChem* 5:1084–1104.

Kawai, T. and Tsutsumi, K. 1998. Reactivity of silanol groups on zeolite surfaces. *Colloid Polym. Sci.* 276:992–998.

Kazakova, O.A., Gun'ko, V.M., Lipkovskaya, N.A., Voronin, E.F., and Pogorelyi, V.K. 2002. Interaction of quercetin with highly dispersed silica in aqueous suspensions. *Colloid J.* 64:412–418.

Kazansky, V.B. 2003. Localization of bivalent transition metal ions in high-silica zeolites with the very broad range of Si/Al ratios in the framework probed by low-temperature H_2 adsorption. *J. Catal.* 216:192–202.

Keeler, J. 2005. *Understanding NMR Spectroscopy*. New York: John Wiley & Sons.

Kennedy, G.J., Afeworki, M., and Hong, S.B. 2002. Probing the non-random aluminum distribution in zeolite merlinoite with ultra-high-field (18.8 T) ^{27}Al and ^{29}Si MAS NMR. *Micropor. Mesopor. Mater.* 52:55–59.

Kholkin, A.I., Belova, V.V., Pashkov, G.S., Fleitlikh, I.Yu., and Sergeev, V.V. 1999. Solvent binary extraction. *J. Molec. Liquids* 82:131–146.

Kikuchi, M., Ikoma, T., Itoh, S. et al. 2004. Biomimetic synthesis of bone like nanocomposites using the self organization mechanism of hydroxyapatite and collagen. *Compos. Sci. Technol.* 64:819–825.

Kimmich, R. 1997. *NMR: Tomography, Diffusometry, Relaxometry*. New York: Springer.

Kimmich, R. and Anoardo, E. 2004. Field-cycling NMR relaxometry. *Progr. Nucl. Magn. Reson. Spectrosc.* 44:257–320.

Kimmich, R. and Fatkullin, N. 2004. Polymer chain dynamics and NMR. *Adv. Polym. Sci.* 170:1–113.

King, N., Hittinger, C.T., and Carroll, S.B. 2003. Evolution of key cell signaling and adhesion protein families predates animal origins. *Science* 301:361–363.

Kinge, S., Crego-Calama, M., and Reinhoudt, D.N. 2008. Self-assembling nanoparticles at surfaces and interfaces. *ChemPhysChem* 9:20–42.

Kinney, D.R., Chaung, I.-S., and Maciel, G.E. 1993. Water and the silica surface as studied by variable temperature high resolution ^{1}H NMR. *J. Am. Chem. Soc.* 115:6786–6794.

Kiperman, S.L. 1979. *Fundamentals of Chemical Kinetics in Heterogeneous Catalysis*. Moscow, Russia: Khimiya (in Russian).

Kiselev, A.V. and Lygin, V.I. 1972. *IR Spectra of Surface Compounds and Adsorbed Substances*. Moscow, Russia: Nauka (in Russian).

Kiselev, A.V. and Lygin, V.I. 1975. *Infrared Spectra of Surface Compounds*. New York: Wiley.

Kitevski-LeBlanc, J.L., and Prosser, R.S. 2012. Current applications of ^{19}F NMR to studies of protein structure and dynamics. *Prog. Nucl. Magn. Reson. Spectrosc.* 62:1–33.

Kjaergaard, M. and Poulsen, F.M. 2012. Disordered proteins studied by chemical shifts. *Prog. Nucl. Magn. Reson. Spectrosc.* 60:42–51.

Klauda, J.B., Eldho, N.V., Gawrisch, K., Brooks, B.R., and Pastor, R.W. 2008. Collective and noncollective models of NMR relaxation in lipid vesicles and multilayers. *J. Phys. Chem. B* 112:5924–5929.

Kleinhammes, A., Mao, S.-H., Yang, X.-J. et al. 2003. Gas adsorption in single-walled carbon nanotubes studied by NMR. *Phys. Rev. B* 68:075418(1–6).

Klein-Seetharaman, J., Getmanova, E., Loewen, M., Reeves, P., and Khorana, H. 1999. NMR spectroscopy in studies of light-induced structural changes in mammalian rhodopsin: Applicability of solution ^{19}F NMR. *Proc. Natl. Acad. Sci. USA* 96:13744–13749.

Klemm, D., Kramer, F., Moritz, S. et al. 2011. Nanocelluloses: A new family of nature-based materials. *Angew. Chem. Int. Ed.* 50:5438–5466.

Klimova, V.A. 1975. *Fundamental Micromethods of Analysis of Organic Compounds*. Moscow, Russia: Chemistry (in Russian).

Klingshirn, C. 2007. ZnO: Material, physics and applications. *ChemPhysChem* 8:782–803.

Klonos, P., Pissis, P., Gun'ko, V.M. et al. 2010. Interfacial and relaxational phenomena in poly(ethylene glycol)—Nanooxide systems. *Colloids Surf. A: Physicochem. Eng. Aspects* 360:220–231.

Kneller, J.M., Pietraß, T., Ott, K.C., and Labouriau, A. 2003. Synthesis of dealuminated zeolites NaY and MOR and characterization by diverse methodologies: ^{27}Al and ^{29}Si MAS NMR, XRD, and temperature dependent ^{129}Xe NMR. *Micropor. Mesopor. Mater.* 62:121–131.

Knopp, D., Tang, D., and Niessner, R. 2009. Review: Bioanalytical applications of biomolecule-functionalized nanometer-sized doped silica particles. *Anal. Chim. Acta* 647:14–30.

Ko, K.R., Kim, C.G., and Ryu, S.K. 1977. Functional groups of carbon fibers and activated carbon fibers treated with ozone. In: *Proc. 23rd Biennial Conference on Carbon (Carbon 97)*, July 18–23, Penn. State, pp. 264–265.

Kobayashi, T., Mao, K., Wang, S.G., Lin, V.S., and Pruski, M. 2011. Molecular ordering of mixed surfactants in mesoporous silicas: A solid-state NMR study. *Solid State Nucl. Magn. Reson.* 39:65–71.

Koever, K.E., Batta, G., and Feher, K. 2006. Accurate measurement of long-range heteronuclear coupling constants from undistorted multiplets of an enhanced CPMG-HSQMBC experiment. *J. Magn. Reson.* 181:89–97.

Koga, N., Takasu, E., and Yanaki, T. 2004. Process for producing dry water. Patent EP1386599. Shiseido.

Kollman, J.M., Pandi, L., Sawaya, M.R., Riley, M., and Doolittle, R.F. 2009. Crystal structure of human fibrinogen. *Biochemistry* 48:3877–3886.

Kolodziejski, W. and Klinowski, J. 2002. Kinetics of cross-polarization in solid-state NMR: A guide for chemists. *Chem. Rev.* 102:613–628.

Kolychev, V.I., Turov, V.V., and Burushkina, T.N. 1992. Abnormal chemical shifts of benzene adsorbed on spherical carbon adsorbents. *Theoret. Eksp. Khim.* 28:80–84.

Kondo, M., Wagers, A.J., Manz, M.G. et al. 2003. Biology of hematopoietic stem cells and progenitors: Implications for clinical application. *Annu. Rev. Immunol.* 21:759–806.

König, R., Scholz, G., and Kemnitz, E. 2007. New inserts and low temperature - Two strategies to overcome the bottleneck in MAS NMR on wet gels. *Solid State Nucl. Magn.* Res. 32:78–88.

Koptyug, I.V. 2011. MRI of mass transport in porous media: Drying and sorption processes. *Prog. Nucl. Magn. Reson. Spectrosc.* 2011. doi:10.1016/j.pnmrs.2011.12.001.

Korb, J.-P., Monteilhet, L., McDonald, P.J., and Mitchell, J. 2007. Microstructure and texture of hydrated cement-based materials: A proton field cycling relaxometry approach. *Cem. Concr. Res.* 37:295–302.

Koskela, H., Kilpelainen, I., Heikkinen, S., and Cahsqc, L.R. 2003. An application of a Carr–Purcell–Meiboom–Gill-type sequence to heteronuclear multiple bond correlation spectroscopy. *J. Magn. Reson.* 164:228–232.

Kozlov, P.V. and Burdygina, G.I. 1983. The structure and properties of solid gelatin and principles of their modification. *Polymers* 24:651–666.

Kozlov, M.M. and Markin, V.C. 1986. Membran skeleton of erythrocytes. Theoretical model. *Biol. Membr.* 3:404–421.

Kraivyaryainen, A.I. 1980. *Dynamic Behaviour of Proteins in an Aqueous Medium and Their Function.* Leningrad: Nauka (in Russian).

Krasavina, B.C. and Torbenko, V.P. 1979. *Living of Bone Tissue*. Moscow, Russia: Nauka.

Kremer, F. and Schönhals, A. (Eds.) 2002. *Broadband Dielectric Spectroscopy*. Berlin, Germany: Springer.

Kreuer, K.D. 1996. Proton conductivity: Materials and applications. *Chem. Mater.* 8:610–641.

Kruk, M., Jaroniec, M., and Gadkaree, K.P. 1999. Determination of the specific surface area and the pore size of microporous carbons from adsorption potential distributions. *Langmuir* 15:1442–1448.

Krupska, T.V., Barvinchenko, V.M., Kaspersky, V.O., Turov, V.V., and Chuiko, O.O. 2006. Imobilization of chloramphenicol on a surface of highly disperse silica. *Pharmacol. J.* 2:59–64 (in Russian).

Krupska, T.V., Turova, A.A., Gun'ko, V.M., and Turov, V.V. 2009. Influence of highly disperse materials on physiological activity of yeast cells. *Biopolym. Cells* 25:290–297.

Krutyeva, M., Grinberg, F., Furtado, F. et al. 2009. Characterization of carbon materials with the help of NMR methods. *Micropor. Mesopor. Mater.* 120:91–97.

Krylova, G.V., Gnatyuk, Yu.I., Smirnova, N.P., Eremenko, A.M., and Gun'ko, V.M. 2009. Ag nanoparticles deposited onto silica, titania and zirconia mesoporous films synthesized by sol-gel template method. *J. Sol-Gel Sci. Technol.* 50:216–228.

Krymsky, L.D., Nestaiko, G.V., and Rybakov, A.G. 1976. *Raster Electronic Microscopy of Vessels and Blood.* Moscow, Russia: Medicine (in Russian).

Kumar, R., Chen, H.T., Escoto, J.L.V., Lin, V.S.-Y., and Pruski, M. 2006. Template removal and thermal stability of organically functionalized mesoporous silica nanoparticles. *Chem. Mater.* 18:4319–4327.

Kumar, A.P., Depan, D., Tomer, N.S., and Singh, R.P. 2009. Nanoscale particles for polymer degradation and stabilization—Trends and future perspectives. *Progr. Polym. Sci.* 34:479–515.

Kumar, A., Grace, R.C.R., and Madhu, P.K. 2000. Cross-correlations in NMR. *Progr. Nucl. Magn. Reson. Spectrosc.* 37:191–319.

Kumar, D., Shumacher, K., du Fresne von Hohenesche, C., Grün, M., and Unger, K.K. 2001. MCM-41, MCM-48 and related mesoporous adsorbents: Their synthesis and characterisation. *Colloids Surf. A: Physicochem. Eng. Aspects* 187–188:109–116.

Kuntz, I.D. Jr., Brassfield, T.S., Law, G.D., and Purcell, G.V. 1969. Hydration of macromolecules. *Science* 163:1329–1331.

Kunze, A., Sjövall, P., Kasemo, B., and Svedhem, S. 2009. *In situ* preparation and modification of supported lipid layers by lipid transfer from vesicles studied by QCM-D and TOF-SIMS. *J. Am. Chem. Soc.* 131:2450–2451.

Kuo, I.-F.W. and Mundy, C.J. 2004. An ab initio molecular dynamics study of the aqueous liquid-vapor interface. *Science* 303:658–660.

Kuochih, H. 2001. The development of hydrogen storage alloys and the progress of nickel hydride batteries. *J. Alloys Comp.* 321:307–313.

Kurdish, I.K., Bikhunov, V.L., Tsimberg, E.A., El'chits, S.V., Vygovskaia, E.L., and Chuiko, A.A. 1991. Effect of disperse silicon dioxide—Aerosil A-300 on the growth of *Saccharomyces cerevisiae* yeast cells. *J. Microbiol.* 53:41–44.

Kurdish, I.K. and Kigel, N.F. 1997. Effect of high-disperse materials on physiological activity of metanotrophic bacteria's. *J. Microbiol.* 59:29–36.

Kuroki, S., Kimura, H., and Ando, I. 2004. Structural characterization of Si-based polymer materials by solid-state NMR spectroscopy. *Ann. Rep. NMR Spectrosc.* 52:201–243.

Labouriau, A. and Earl, W.L. 1997. Titanium solid-state NMR in anatase, brookite and rutile. *Chem. Phys. Lett.* 270:278–284.

Lahanas, K.M., Vrabie, N., Santos, E., and Miklean, S. 2001. Powder to liquid compositions. Patent US6290941 B1.

Landry, M.R. 2005. Thermoporometry by differential scanning calorimetry: Experimental considerations and applications. *Thermochim. Acta* 433:27–50.

Lapeyre, V., Ancla, C., Catargi, B., and Ravaine, V. 2008. Glucose-responsive microgels with a core–shell structure. *J. Colloid Interface Sci.* 327:316–323.

Larsen, F.H. and Farnan, I. 2002. ^{29}Si and ^{17}O (Q)CPMG-MAS solid-state NMR experiments as an optimum approach for half-integer nuclei having long T_1 relaxation times. *Chem. Phys. Lett.* 357:403–408.

Larsen, F.H., Jakobsen, H.J., Ellis, P.D., and Nielsen, N.C. 1998. High-field QCPMG-MAS NMR of half-integer quadrupolar nuclei with large quadrupole couplings. *Mol. Phys.* 95:1185–1195.

Larsericsdotter, H., Oscarsson, S., and Buijs, J. 2001. Thermodynamic analysis of proteins adsorbed on silica particles: Electrostatic effects. *J. Colloid Interface Sci.* 237:98–103.

László, K., Guillermo, A., Fluerasu, A. Moussaïd, A., and Geissler, E. 2010. Microphase structure of poly-(n-isopropylacrylamide) hydrogels as seen by small- and wide-angle X-ray scattering and pulsed field gradient NMR. *Langmuir* 26:4415–4420.

Lavalle, P., Voegel, J.-C., Vautier, D., Senger, B., Schaaf, P., and Ball, V. 2011. Dynamic aspects of films prepared by a sequential deposition of species: Perspectives for smart and responsive materials. *Adv. Mater.* 23:1191–1221.

Lazzeretti, P. 2000. Ring currents. *Prog. Nucl. Magn. Reson. Spectrosc.* 36:1–88.

Lazzeretti, P. and Zanasi, R. 1983. The validity of Musher's model for the magnetic properties of aromatic molecules. *Chem. Phys. Lett.* 100:67–69.

Leboda, R. 1974. Modification of surface properties of silica gels from the point of view of their application in chromatography. PhD thesis. Lublin: Maria Curie-Skłodowska University.

Leboda, R. 1980a. Preparation of complex carbon–silica adsorbents with different properties. *Chromatographia* 9:549–554.

Leboda, R. 1980b. Modification of surface properties of silica gels in aspect of their application in chromatography. XIX. The effect of the condition of preparation of complex carbon–silica adsorbents on their surface properties. *Chem. Anal.* 25:979–991.

Leboda, R. 1980c. The chemical nature of adsorption centres in modified carbon–silica adsorbents prepared by the pyrolysis of alcohols. *Chromatographia* 11:703–708.

Leboda, R. 1980d. Modification of surface properties of complex silica–carbon adsorbents (carbosils). *Polish J. Chem.* 11:2305–2308.

Leboda, R. 1981. Preparation and modification of complex pyrolytic carbon-silica adsorbents. *Chromatographia* 9:524–528.

Leboda, R. 1987. Thermal behaviour of complex carbon–silica adsorbents (carbosils) J. *Thermal Anal. Calorim.* 32:1435–1448.

Leboda, R. 1992. Carbon–mineral adsorbents—New type of sorbents. Part I. The methods of preparation. *Mater. Chem. Phys.* 31:243–255.

Leboda, R. 1993. Carbon–mineral adsorbents—New type of sorbents. Part II. Surface properties and methods of their modification. *Mater. Chem. Phys.* 34:123–141.

Leboda, R., Charmas, B., Chodorowski, S., Skubiszewska-Zięba, J., and Gun'ko, V.M. 2006. Improved carbon–mineral adsorbents derived from cross-linking carbon-bearing residues in spent palygorskite. *Micropor. Mesopor. Mater.* 87:207–216.

Leboda, R., Charmas, B., Skubiszewska-Zięba, J. et al. 2005. Carbon-mineral adsorbents prepared by pyrolysis of waste materials in the presence of tetrachloromethane. *J. Colloid Interface Sci.* 284:39–47.

Leboda, R. and Dabrowski, A. 1996. Complex carbon mineral adsorbants: Preparation, surface properties and their modification. In: Dabrowski, A., and Tertykh, V.A. (Eds.) *Adsorption on New and Modified Inorganic Sorbents*, Studies in Surface Science and Catalysis, Vol. 99. Amsterdam, the Netherlands: Elsevier, pp. 115–146.

Leboda, R., Grzegorczyk, W., Łodyga, A., and Gierak, A. 1997. Carbon adsorbents as materials for chromatography. I. Gas chromatography. *Mater. Chem. Phys.* 51:216–232.

Leboda, R., Grzegorczyk, W., Łodyga, A., and Gierak, A. 1998a. Carbon adsorbents as materials for chromatography. II. Liquid chromatography. *Mater. Chem. Phys.* 55:1–29.

Leboda, R., Gun'ko, V.M., Marciniak, M. et al. 1999a. Structure of CVD-titania/silica gel. *J. Colloid Interface Sci.* 218:23–39.

Leboda, R., Gun'ko, V.M., Marciniak, M., Grzegorczyk, W., and Skubiszewska-Zięba, J. 2001a. Carbon deposit effect on the structure of hybrid adsorbents $C/TiO_2/SiO_2$. *Adsorp. Sci. Technol.* 19:385–396.

Leboda, R., Gun'ko, V.M., Skubiszewska-Zięba, J., Gierak, A., and Oleszczuk, P. 2001b. Properties of thin polyethylene glycol layers on the surface of silica gel and pyrocarbon/silica gel. Effect of topography and morphology of carbon deposit. *Mater. Chem. Phys.* 70:25–37.

Leboda, R., Marciniak, M., Gun'ko, V.M., Grzegorczyk, W., Malygin, A.A., and Malkov, A.A. 2000a. Structure of carbonized mesoporous silica gel/CVD-titania. *Colloids Surf. A: Physicochem. Eng. Aspects* 167:275–285.

Leboda, R., Mendyk, E., Gierak, A., and Tertykh, V.A. 1995. Hydrothermal modification of silica gels (xerogels) 2. Effect of the duration of treatment on their porous structure. *Colloids Surf. A: Physicochem. Eng. Aspects* 105:191–197.

Leboda, R., Tertykh, V.A., Sidorchuk, V.V., and Skubiszewska-Zieba, J. 1998b. Peculiarities of steam-phase hydrothermal modification of pyrogenic silicium dioxide. *Colloids Surf. A: Physicochem. Eng. Aspects* 135:253–265.

Leboda, R., Turov, V.V., Charmas, B., Skubiszewska-Zieba, J., and Gun'ko, V.M. 2000b. Surface properties of mesoporous carbon-silica gel adsorbents. *J. Colloid Interface Sci.* 223:112–125.

Leboda, R., Turov, V.V., Gun'ko, V.M., and Skubiszewska-Zieba, J. 2001c. Characterization of aqueous suspensions of poly(ethylene glycol)/pyrocarbon/silica. *J. Colloid Interface Sci.* 237:120–129.

Leboda, R., Turov, V.V., Marciniak, M., Malygin, A.A., and Malkov, A.A. 1999b. Characteristics of the hydration layer structure in porous titania-silica obtained by the chemical vapor deposition method. *Langmuir* 15:8441–8446.

Lee, A.G., Birdsall, N.J.M., Levine, Y.K., and Metcalfe, J.C. 1972. High resolution proton relaxation studies of lecithin. *Biochim. Biophys. Acta* 255:43–56.

Lee, T., Park, S.S., Jung, Y. et al. 2009. Preparation and characterization of polyimide/mesoporous silica hybrid nanocomposites based on water-soluble poly(amic acid) ammonium salt. *Eur. Polym. J.* 45:19–29.

Lee, S.S., Robinson, F.M., and Wang, H.G. 1981. Rapid determination of yeast viability. *Biotechn. Bioeng. Symp.* 11:641–649.

Lee, H., Sohn, J., Yeh, B., Choi, J., Jung, S., and Kim, K. 2000. F-19 NMR investigation of F-1-ATPase of Escherichia coli using fluorotryptophan labeling. *J. Biochem.* 127:1053–1056.

Le Feunteun, S., Diat, O., Guillermo, A., Poulesquen, A., and Podor, R. 2011. NMR 1D-imaging of water infiltration into mesoporous matrices. *Magn. Reson. Imag.* 29:443–455.

Lefort, R., Wiench, J.W., Pruski, M., and Amoureux, J.P. 2002. Optimization of data acquisition and processing in Carr–Purcell–Meiboom–Gill multiple quantum magic angle spinning nuclear magnetic resonance. *J. Chem. Phys.* 116:2493–2501.

Leftin, A. and Brown, M.F. 2011. An NMR database for simulations of membrane dynamics. *Biochim. Biophys. Acta Biomembranes* 1808:818–839.

Legrand, A.P. (Ed.) 1998. *The Surface Properties of Silicas*. New York: Wiley.

Legrand, A.P., Hommel, H., and d'Espinose de la Caillerie, J.B. 1999. On the silica edge, an NMR point of view. *Colloids Surf. A: Physicochem. Eng. Aspects* 158:157–163.

Legrand, A.P., Hommel, H., Tuel, A. et al. 1990. Hydroxyles of silica powders. *Adv. Colloid Interface Sci.* 33:91–330.

Legrand, A.P., Taibi, H., Hommel, H.H., Tougne, P., and Leonardelli, S. 1993. Silicon functionality distribution on the surface of amorphous silicas by ^{29}Si solid state NMR. *J. Non-Cryst. Solids* 155:122–130.

Lehninger, A.L. 1975. *Biochemistry. The Molecular Basis of Cell Structure and Function*, 2nd edn. New York: Worth Publ. Inc.

Leirose, G.D.S. and Cardoso, M.B. 2011. Silica–maltose composites: Obtaining drug carrier systems through tailored ultrastructural nanoparticles. *J. Pharmacol. Sci.* 100:2826–2834.

Leninsky, M.I., Mazanko, A.F., and Novikov, I.N. 1985. *Hydrogen Chloride and Hydrochloric Acid*. Moscow, Russia: Khimiya (in Russian).

Leon y Leon, C.A., Solar, J., Calemma, V., and Radovic, L.R. 1992. Evidence for the protonation of basal plane sites on carbon. *Carbon* 30:797–811.

Lesage, A., Bardet, M. and Emsley, L. 1999. Through-bond carbon–carbon connectivities in disordered solids by NMR. *J. Am. Chem. Soc.* 121:10987–10993.

LeVan, M.D. (Ed.) 1996. *Fundamentals of Adsorption*. Boston, MA: Kluwer Academic Publishers.

Levine, Y.K., Birdsall, N.J.M., Lee, A.G., and Metcalfe, J.C. 1972. ^{13}C nuclear magnetic resonance relaxation measurements of synthetic lecithins and the effect of spin-labeled lipids. *Biochemistry* 11:1416–1421.

Levitt, M.H. 2008. *Spin Dynamics*. New York: John Wiley & Sons.

Leyden, D.E. 1985. *Silanes, Surfaces and Interfaces*. New York: Gordon and Breach.

Leyden, D.E. and Collins, W.T. 1989. *Chemically Modified Oxide Surfaces*. New York: Gordon and Breach.

Li, Y.-S. and Ba, A. 2008. Spectroscopic studies of triethoxysilane sol–gel and coating process. *Spectrochim. Acta A* 70:1013–1019.

Li, H.L. and Frieden, C. 2007. Observation of sequential steps in the folding of intestinal fatty acid binding protein using a slow folding mutant and F-19 NMR. *Proc. Natl Acad. Sci. USA* 104:11993–11998.

Li, C., Lutz, E.A., Slade, K.M., Ruf, R.A.S., Wang, G.-F., and Pielak, G.J. 2009a. ^{19}F NMR studies of alpha-synuclein conformation and fibrillation. *Biochemistry* 48:8578–8584.

Li, X., Yang, T., Gao, Q., Yuan, J., and Cheng, S. 2009b. Biomimetic synthesis of copolymer–silica nanoparticles with tunable compositions and surface property. *J. Colloid Interface Sci.* 338:99–104.

Li, J.-Y. and Yeh, A.-I. 2001. Relationships between thermal, rheological characteristics and swelling power for various starches. *J. Food Eng.* 50:141–148.

Li, Z., Zhang, Y., Fullhart, P., and Mirkin, C.A. 2004. Reversible and chemically programmable micelle assembly with DNA block-copolymer amphiphiles. *Nano Lett.* 4:1055–1058.

Limnell, T., Santos, H.A., Mäkilä, E. et al. 2011. Drug delivery formulations of ordered and nonordered mesoporous silica: Comparison of three drug loading methods. *J. Pharmacol. Sci.* 100:3294–3306.

Lind, A., du Fresne von Hohenesche, C., Smått, J.-H., Lindén, M., and Unger, K.K. 2003. Spherical silica agglomerates possessing hierarchical porosity prepared by spray drying of MCM-41 and MCM-48 nanospheres. *Micropor. Mesopor. Mater.* 66:219–227.

Lindhout, D.A., McKenna, S.A., Aitken, C.E., Liu, C.W., and Puglisi, J.D. 2007. PKR: A NMR perspective. *Prog. Nucl. Magn. Reson. Spectrosc.* 51:199–215.

Lindon, J.C., Beckonert, O.P., Holmes, E, and Nicholson, J.K. 2009. High-resolution magic angle spinning NMR spectroscopy: Application to biomedical studies. *Prog. Nucl. Magn. Reson. Spectrosc.* 55:79–100.

Lindon, J.C., Holmes, E, and Nicholson, J.K. 2001. Pattern recognition methods and applications in biomedical magnetic resonance. *Prog. Nucl. Magn. Reson. Spectrosc.* 39:1–40.

Lindqvist, I., Nilsson, O., and Ronquist, G. 1974. Ultrastructural changes of erythrocyte membranes isolated with colloidal silica solution (ludox). *Ups. J. Med. Sci.* 79:1–6.

Lipton, A.S., Sears, J.A., and Ellis, P.D. 2001. A general strategy for the NMR observation of half-integer quadrupolar nuclei in dilute environments. *J. Magn. Reson.* 151:48–59.

Lisovskii, A.E. and Aharoni, C. 1994. Carbonaceous deposits as catalysts for oxydehydrogenation of alkylbenzenes. *Catal. Rev. Sci. Eng.* 36:25–74.

Liu, Y., Liu, H., and Hu, N. 2005. Core–shell nanocluster films of hemoglobin and clay nanoparticle: Direct electrochemistry and electrocatalysis. *Biophys. Chem.* 117:27–37.

Liu, C.C. and Maciel, G.E. 1996. The fumed silica surface: A study by NMR. *J. Am. Chem. Soc.* 118:5103–5119.

Liu, A.Z., Majumdar, A., Hu, W.D., Kettani, A., Skripkin, E., and Patel, D.J. 2000. NMR detection of N-H…O = C hydrogen bonds in ^{13}C, ^{15}N-labeled nucleic acids. *J. Am. Chem. Soc.* 122:3206–3210.

Liu, L., Ryu, S., Tomasik, M.R., Stolyarova, E. et al. 2008. Graphene oxidation: Thickness-dependent etching and strong chemical doping. *Nano Lett.* 8:1965–1970.

Liu, R., Shi, Y., Wan, Y. et al. 2006. Triconstituent co-assembly to ordered mesostructured polymer-silica and carbon-silica nanocomposites and large-pore mesoporous carbons with high surface areas. *J. Am. Chem. Soc.* 128:11652–11662.

Llewellyn, P.L. and Maurin, G. 2005. Gas adsorption microcalorimetry and modelling to characterize zeolites and related materials. *C. R. Chim.* 8:283–302.

Long, Y.-Z., Li, M.-M., Gu, C. et al. 2011. Recent advances in synthesis, physical properties and applications of conducting polymer nanotubes and nanofibers. *Prog. Polym. Sci.* 36:1415–1442.

Lonsdale, K. 1937. Magnetic anisotropy and electronic structure of aromatic molecules. *Proc. R. Soc. A* 159:149–161.

Loos, R., Alonso, M.C., and Barceló, D. 2000. Solid-phase extraction of polar hydrophilic aromatic sulfonates followed by capillary zone electrophoresis–UV absorbance detection and ion-pair liquid chromatography–diode array UV detection and electrospray mass spectrometry. *J. Chromatogr. A* 890:225–237.

López, P., Batlle, R., Nerín, C., Cacho, J., and Ferreira, V. 2007. Use of the new generation of poly(styrene-divinylbenzene)resins for gas trapping-thermal desorption. Application to the retention of eight volatile organic compounds. *J. Chromatogr. A* 1139:36–44.

López de Alda, M.J. and Barceló, D. 2001. Use of solid-phase extraction in various of its modalities for sample preparation in the determination of estrogens and progestogens in sediment and water. *J. Chromatogr. A* 938:145–153.

López-González, J. de D., Martínez-Vilchez, F. and Rodríguez-Reinoso, F. 1980. Preparation and characterization of active carbons from olive stones, *Carbon* 18:413–418.

Lorigan, G.A., Dave, P.C., Tiburu, E.K. et al. 2004. Solid-state NMR spectroscopic studies of an integral membrane protein inserted into aligned phospholipid bilayer nanotube arrays. *J. Am. Chem. Soc.* 126:9504–9505.

Lozano-Castello, D., Cazorla-Amoros, D., Linares-Solano, A., and Quinn, D.F. 2002. Activated carbon monoliths for methane storage: Influence of binder. *Carbon* 40:2817–2825.

Lozinsky, V.I. 2002. Cryogels on the basis of natural and synthetic polymers: Preparation, properties and areas of implementation. *Russ. Chem. Rev.* 71:489–511.

Lu, J.G., Chang, P., and Fan, Z. 2006. Quasi-one-dimensional metal oxide materials— Synthesis, properties and applications. *Mater. Sci. Eng. R* 52:49–91.

Lu, J.R., Perumal, S., Hopkinson, I. et al. 2004a. Interfacial nano-structuring of designed peptides regulated by solution pH. *J. Am. Chem. Soc.* 126:8940–8947.

Lu, Y., Weng, L., and Zhang, L. 2004b. Morphology and properties of soy protein isolate thermoplastic reinforced with chitin whiskers. *Biomacromolecules* 5:1046–1051.

Lucklum, R. and Hauptmann, P. 2000. The Δf-ΔR QCM technique: An approach to an advanced sensor signal interpretation. *Electrochim. Acta* 45:3907–3916.

Luhmer, M., d'Espinose, J.B., Hommel, H., and Legrand, A.P. 1996. High-resolution ^{29}Si solid-state NMR study of silicon functionality distribution on the surface of silicas. *Magnet. Reson. Imag.* 14:911–913.

Lusceac, S.A., Vogel, M.R., Herbers, C.R. 2010. ^{2}H and ^{13}C NMR studies on the temperature-dependent water and protein dynamics in hydrated elastin, myoglobin and collagen. *Biochim. Biophys. Acta* 1804:41–48.

Lussier, M.G., Shull, J.C., and Miller, D.J. 1994. Activated carbon from cherry stones. *Carbon* 32:1493–1498.

Luy, B. and Marino, J.P. 2001. ^{1}H-^{31}P CPMG-correlated experiments for the assignment of nucleic acids. *J. Am. Chem. Soc.* 123:11306–11307.

Luzzati, V. and Husson, F. 1962. The structure of the liquid crystalline phase of lipid–water systems. *J. Cell Biol.* 12:207–219.

Lyerla, J.R., Jr. and Levy, G.C. 1974. Carbon-13 nuclear spin relaxation. In: Levy, G.C. (Ed.) *Topics in Carbon-13 NMR Spectroscopy*, Vol. 1, New York: John Wiley & Sons, pp. 79–148.

Lyerla, J.R., Jr., McIntyre, H.M., and Torchia, D.A. 1974. A ^{13}C nuclear magnetic resonance study of alkane motion. *Macromolecules* 7:11–14.

Lyklema, J. 1991. *Fundamentals of Interface and Colloid Science. Fundamentals*, Vol. 1. London, U.K.: Academic Press.

Lyklema, J. 1995. *Fundamentals of Interface and Colloid Science. Solid-Liquid Interfaces*, Vol. 2. London, U.K.: Academic Press.

Lysenko, L.V., Chueshov, V.I., and Lavrushina, T.T. 1977. Comparative toxicity of modified samples of aerosil. *Pharmacology* 26:56–58.

Ma, Ch., Chang, Ya., Ye, W., Shang, W., and Wang, Ch. 2008. Supercritical preparation of hexagonal γ-alumina nanosheets and its electrocatalytic properties. *J. Colloid Interface Sci.* 317:148–154.

Ma, Z., Dunn, B.C., Turpin, G.C., Eyring, E.M., Ernst, R.D., and Pugmire, R.J. 2007a. Solid state NMR investigation of silica aerogel supported Fischer–Tropsch catalysts. *Fuel Process. Technol.* 88:29–33.

Ma, W.J., Ruys, A.J., Mason, R.S. et al. 2007b. DLC coatings: Effects of physical and chemical properties on biological response. *Biomaterials* 28:1620–1628.

Ma, X.F., Yu, J.G., and Wang, N. 2007c. Fly ash-reinforced thermoplastic starch composites. *Carbohydr. Polym.* 67:32–39.

Ma, Z. and Zaera, F. 2006. Organic chemistry on solid surfaces. *Surf. Sci. Rep.* 61:229–281.

Maciel, G.E. 1984. High-resolution nuclear magnetic resonance of solids. *Science* 226:282–288.

MacKenzie, K.J.D. 2004. Solid state multinuclear NMR: A versatile tool for studying there activity of solid systems. *Solid State Ionics* 172:383–388.

Macomber, R.S. 1998. *A Complete Introduction to Modern NMR Spectroscopy*. New York: John Willey & Sons.

Magnuson, M.L. and Fung, B.M. 1992. Solvent suppression in proton NMR by the use of oxygen-17-enriched water. *J. Magn. Reson.* 99:301–307.

Majumdar, A., Kettani, A., and Skripkin, E. 1999a. Observation and measurement of internucleotide $^{2}J_{NN}$ coupling constants between ^{15}N nuclei with widely separated chemical shifts. *J. Biomol. NMR* 14:67–70.

Majumdar, A., Kettani, A., Skripkin, E., and Patel, D.J. 1999b. Observation of internucleotide NH…N hydrogen bonds in the absence of directly detectable protons. *J. Biomol. NMR* 15:207–211.

Majumder, T.P., Meißner, D., and Schick, C. 2004. Dielectric processes of wet and well-dried wheat starch. *Carbohydr. Polym.* 56:361–366.

Majumder, T.P., Meißner, D., Schick, C., and Roy, S.K. 2006. Phase transition phenomena and the corresponding relaxation process of wheat starch–water polymer matrix studied by dielectric spectroscopic method. *Carbohydr. Polym.* 65:129–133.

Malcolm, A.J. 2002. Metabolic bone disease. *Curr. Diagn. Pathol.* 8:19–25.

Mallion, R.B. 2008. Topological ring–currents in condensed benzenoid hydrocarbons. *Croat. Chem. Acta* 81:227–246.

Malmendal, A., Underhaug, J., Otzen, D.E., and Nielsen, N.C. 2010. Fast mapping of global protein folding states by multivariate NMR: A GPS for proteins. *PLoS ONE* 5:e10262,1–6.

Malmsten, M. (Ed.) 1998. *Biopolymers at Interfaces*. New York: Marcel Dekker.

Malmsten, M. and Lassen, B. 1994. Competitive adsorption at hydrophobic surfaces from binary protein systems. *J. Colloid Interface Sci.* 166:490–498.

Mandell, D.J. and Kortemme, T. 2009. Computer-aided design of functional protein interactions. *Nat. Chem. Biol.* 5:797–807.

Mandrusov, E., Yang, J.D., Pfeiffer, N., Vroman, L., Puszkin, E., and Leonard, E.F. 1998. Kinetics of protein deposition and replacement from a shear flow. *AIChE J. Fluid Mech. Transp. Phenom.* 44:233–244.

Maniwa, Y., Sato, M., Kume, K., Kozlov, M.E., and Tokumoto, M. 1996. Comparative NMR study of new carbon forms. *Carbon* 34:1287–1291.

Mank, V.V. and Lebovka, N.I. 1988. *NMR Spectroscopy of Water in Heterogeneous Systems*. Kiev, Ukraine: Naukova Dumka (in Russian).

Mao, C. and Kisaalita, W.S. 2004. Characterization of 3-D collagen hydrogels for functional cell-based biosensing. *Biosens. Bioelectron.* 19:1075–1088.

Mao, S., Kleinhammes, A., and Wu, Y. 2006. NMR study of water adsorption in single-walled carbon nanotubes. *Chem. Phys. Lett.* 421:513–517.

Mao, K., Kobayashi, T., Wiench, J.W. et al. 2010. Conformations of silica-bound (pentafluorophenyl)propyl groups determined by solid-state NMR spectroscopy and theoretical calculations. *J. Am. Chem. Soc.* 132:12452–12457.

Mao, K. and Pruski, M. 2009. Directly and indirectly detected through-bond heteronuclear correlation solid-stat NMR spectroscopy under fast MAS. *J. Magn. Reson.* 201:165–174.

Mao, K. and Pruski, M. 2010. Homonuclear dipolar decoupling under fast MAS: Resolution patterns and simple optimization strategy. *J. Magn. Reson.* 203:144–149.

Mao, K., Wiench, J.W., Lin, V.S.-Y., and Pruski, M. 2009. Indirectly detected through-bond chemical shift correlation NMR spectroscopy under fast MAS: Studies of organic-inorganic hybrid materials. *J. Magn. Reson.* 196:92–95.

March, J. 1992. *Advanced Organic Chemistry*, 4th edn. New York: Wiley.

Marcos, M., Cano, P., Fantazzini, P., Garavaglia, C., Gomez, S., and Garrido, L. 2006. NMR relaxometry and imaging of water absorbed in biodegradable polymer scaffolds. *Magn. Reson. Imag.* 24:89–95.

Marintchev, A., Frueh, D., and Wagner, G. 2007. NMR methods for studying protein-protein interactions involved in translation initiation. *Methods Enzymol.* 430:283–331.

Marrink, S.J., Berkowitz, M., and Berendsen, H.J.C. 1993. Molecular dynamic simulation of membrane/water interface: The ordering of water and its relation to the hydration forces. *Langmuir* 9:3122–3131.

Marsh, H., Heintz, E.A., Rodríguez-Reinoso, F. (Eds.) 1997. *Introduction to Carbon Technologies.* Alicante, Spain: University of Alicante.

Marsh, H., Iley, M., Berger, J., and Siemieniewska, T. 1975. The adsorptive properties of activated plum stone chars. *Carbon* 13:103–109.

Marsh, H. and Rodríguez-Reinoso, F. 2006. *Activated Carbon.* London, U.K.: Elsevier.

Martin, G.E. and Zekter, A.S. 1988. *Two-Dimensional NMR Methods for Establishing Molecular Connectivity.* New York: VCH Publishers, Inc.

Martinelli, A., D'Ilario, L., Francolini, I., and Piozzi, A. 2011. Water state effect on drug release from an antibiotic loaded polyurethane matrix containing albumin nanoparticles. *Int. J. Pharmacol.* 407:197–206.

Marzouk, S., Rachdi, F., Fourati, M., and Bouaziz, J. 2004. Synthesis and grafting of silica aerogels. *Colloids Surf. A: Physicochem. Eng. Aspects* 234:109–116.

Massiot, D., Fayon, F., Capron, M. et al. 2002. Modelling one- and two-dimensional solid state NMR spectra. *Magn. Reson. Chem.* 40:70–76.

Masuda, S. and Takahashi, K. (Eds.) 1990. *Aerosols. Science, Industry, Health and Environment.* Oxford, U.K.: Pergamon Press.

Matsumoto, A., Zhao, J-X., and Tsutsumi, K. 1997. Adsorption behavior on slit-shaped micropores. *Langmuir* 13:496–501.

Mattiasson, B., Kumar, A., and Galaev, I.Yu. (Eds.) 2009. *Macroporous Polymers: Production, Properties and Biotechnological/Biomedical Applications.* New York: Taylor & Francis Group.

Maudsley, A.A. 1986. Modified Carr–Purcell–Meiboom–Gill sequence for NMR Fourier imaging applications. *J. Magn. Reson.* 69:488–491.

McCool, B., Murphy, L., and Tripp, C.P. 2006. A simple FTIR technique for estimating the surface area of silica powders and films. *J. Colloid Interface Sci.* 295:294–298.

McCormick, M., Smith, R.N., Graf, R., Barrett, C.J., Reven, L., and Spiess, H.W. 2003. NMR studies of the effect of adsorbed water on polyelectrolyte multilayer films in the solid state. *Macromolecules* 36:3616–3625.

McDermott, A.E. 2004. Structural and dynamic studies of proteins by solid-state NMR spectroscopy: Rapid movement forward. *Curr. Opin. Struct. Biol.* 14:554–561.

McGuire, M.J. and Suffet, I.H. (Eds.) 1983. *Treatment of Water by Granular Activated Carbon*, Advances in Chemistry Series N202. Oxford, U.K.: Oxford University Press.

Medek, A., Harwood, J.S., and Frydman, L. 1995. Multiple-quantum magic-angle spinning NMR: A new method for the study of quadrupolar nuclei in solids. *J. Am. Chem. Soc.* 117:12779–12787.

Meißner, D. and Einfeldt, J. 2004. Dielectric relaxation analysis of starch oligomers and polymers with respect to their chain length. *J. Polym. Sci. B: Polym. Phys.* 42:188–197.

Meiboom, S. and Gill, D. 1958. Modified spin–echo method for measuring nuclear relaxation times. *Rev. Sci. Instrum.* 29:688–691.

Melacini, G., Bonvin, A.M.J.J., Goodman, M., Boelens, R., and Kaptein, R. 2000. Hydration dynamics of the collagen triple helix by NMR. *J. Mol. Biol.* 300:1041–1048.

Meldrum, B.J. and Rochester, C.H. 1990. The study in situ of oxidation of the surface of active carbon in oxygen and carbon oxide by IR spectroscopy. *J. Chem. Soc. Faraday Trans.* 86:861–865.

Melillo, M., Gun'ko, V.M., Mikhalovska, L.I. et al. 2004. Structural characteristics of activated carbons and ibuprofen adsorption affected by bovine serum albumin. *Langmuir* 20:2837–2851.

Melosh, N.A., Lipic, P., Bates, F.S. et al. 1999. Molecular and mesoscopic structures of transparent block copolymer–silica monoliths. *Macromolecules* 32:4332–4242.

Metroke, T.L., Kachurina, O., and Knobbe, E.T. 2002. Spectroscopic and corrosion resistance characterization of GLYMO–TEOS Ormosil coatings for aluminium alloy corrosion inhibition. *Prog. Organ. Coating* 44:295–305.

Meynen, V., Cool, P., and Vansant, E.F. 2009. Verified syntheses of mesoporous materials. *Micropor. Mesopor. Mater.* 125:170–223.

Meziani, M.J., Zajac, J., Douillard, J.-M., Jones, D.J., Partyka, S., and Rozière, J. 2001. Evaluation of surface enthalpy of porous aluminosilicates of the MCM-41 type using immersional calorimetry: Effect of the pore size and framework Si:Al ratio. *J. Colloid Interface Sci.* 233:219–226.

Michaelis, T., Boretius, S., and Frahm, J. 2009. Localized proton MRS of animal brain in vivo: Models of human disorders. *Prog. Nucl. Magn. Reson. Spectrosc.* 55:1–34.

Mijatovic, J., Binder, W.H., and Gruber, H. 2000. Characterization of surface modified silica nanoparticles by ^{29}Si solid state NMR spectroscopy. *Mikrochim. Acta* 133:175–181.

Mikhalovska, L.I., Gun'ko, V.M., Rugal, A.A. et al. 2012. Cottonised flax vs cotton fibers: Texture and adsorption capacity of dry, wetted and swollen samples. *RSC Adv.* 2:2032–2042.

Mikhalovska, L.I., Gun'ko, V.M., Turov, V.V. et al. 2006. Characterisation of the nanoporous structure of collagen-glycosaminoglycan hydrogels by freezing-out of bulk and bound water. *Biomaterials* 27:3599–3607.

Mikhalovsky, S.V., Gun'ko, V.M., Bershtein, V.A. et al. 2012. A comparative study of air-dry and water swollen flax and cotton fibres. *RSC Adv.* 2:2868–2874.

Mikhalovsky, S.V., Gun'ko, V.M., Pavey, K.D., Tomlins, P., and James, S.L., 2005a. Microgravimetry. In: Vadgama, P. (Ed.) *Surfaces and Interfaces for Biomaterials*. Cambridge, MA: CRC Press and Woodhead Publish. Ltd., pp. 322–385.

Mikhalovsky, S.V., Gun'ko, V.M., Turov, V.V., Leboda, R. and Betz, W.R. 2005b. Investigation of structural and adsorptive characteristics of various carbons. *Adsorption* 11:657–662.

Mikhalovsky, S.V., Mikhalovska, L.I., James, S.L. et al. 2006. Characterization of hard and soft porous materials and tissue scaffolds. In: Loureiro, J.M. and Kartel, M.T. (Eds.) *Combined and Hybrid Adsorbents, Fundamentals and Applications*, NATO Security through Science Series C: Environmental Security. Dordrecht, the Netherlands: Springer, pp. 309–320.

Milette, J., Yim, C.T., and Reven, L. 2008. DNMR study of hydrophilic and hydrophobic silica dispersions in EBBA liquid crystals. *J. Phys. Chem. B* 112:3322–3327.

Miller, N.P. and Berg, J.C. 1993. Experiments on the electrophoresis of porous aggregates. *J. Colloid Interface Sci.* 159:253–254.

Miller, N.P., Berg, J.C., and O'Brien, R.W. 1992. The electrophoretic mobility of a porous aggregate. *J. Colloid Interface Sci.* 153:237–243.

Miller, L.A., Gordon, J., and Davis, E.A. 1991. Dielectric and thermal transition properties of chemically modified starched during heating. *Cereal Chem.* 68:441–448.

Minina, S.A. and Kaukhova, I.E. 2004. *Chemistry and Technology of Phytopreparations*. Moscow, Russia: Khimiya (in Russian).

Mirau, P.A., Heffner, S.A., and Schilling, M. 1999. Controlling the chain dynamics of polymers at interfaces. *Chem. Phys. Lett.* 313:139–144.

Mironyuk, I.F., Chuiko, O.O., Ognenko, V.M., Belyakov, V.S., Zagoruiko, V.O., and Valuiko, G.G. 1994. Adsorbent "Flotosorb-1" for clarification and stabilization of grape beverages. Ukrainian patent 3523/C1/.

Mironyuk, I.F., Gun'ko, V.M., Turov, V.V., Zarko, V.I., Leboda, R., and Skubiszewska-Zięba, J. 2001. Characterization of fumed silicas and their interaction with water and dissolved proteins. *Colloids Surf. A: Physicochem. Eng. Aspects* 180:87–101.

Mironyuk, I.F., Gun'ko, V.M., and Zarko, V.I. 1999. System with hydrophilic and hydrophobic silicas—Water. *Rep. NAS Ukr.* N3:149–154 (in Russian).

Mitchell, J., Chandrasekera, T.C., and Gladden, L.F. 2012. Numerical estimation of relaxation and diffusion distributions in two dimensions. *Prog. Nucl. Magn. Reson. Spectrosc.* 62:34–50.

Mitchell, J., Stark, S.C., and Strange, J.H. 2005. Probing surface interactions by combining NMR cryoporometry and NMR relaxometry. *J. Phys. D Appl. Phys.* 38:1950–1958.

Mitchell, J. and Strange, J.H. 2004. An NMR investigation of naphthalene nanostructures. *Mol. Phys.* 102:1997–2005.

Mitchell, J., Webber, J.B.W., and Strange, J.H. 2008. Nuclear magnetic resonance cryoporometry. *Phys. Rep.* 461:1–36.

Mitra, P.P., Sen, P.N., and Schwartz, L.M., 1993. Short time behaviour of the diffusion coefficient as a geometrical probe of porous media. *Phys. Rev. B* 47:8565–8574.

Miyata, T., Asami, N., and Uragami, T. 1999. A reversibly antigen-responsive hydrogel. *Nature* 399:766–769.

Miyata, T., Uragami, T., and Nakamae, K. 2002. Biomolecule-sensitive hydrogels. *Adv. Drug. Deliv. Rev.* 54:79–98.

Mizoshita, N., Goto, Y., Tani, T., and Inagaki, S. 2008. Highly fluorescent mesostructured films that consist of oligo(phenylenevinylene)–silica hybrid frameworks. *Adv. Funct. Mater.* 18:3699–3705.

Moelbert, S, Normand, B., and DeLos Rios, P. 2004. Kosmotropes and chaotropes: Modeling preferential exclusion, binding and aggregate stability. *Biophys. Chem.* 112:45–57.

Molitor, P.F., Shoemaker, R.K., and Apple, T.M. 1989. Detection and structural characterization of rhodium dicarbonyls adsorbed in Y zeolites. *J. Phys. Chem.* 93:2891–2893.

Momany, F.A., Carruthers, L.M., McGuire, R.F., and Scheraga, H.A. 1974. Intermolecular potential from crystal data. *J. Phys. Chem.* 78:1595–1620.

Mondal, S., Khelashvili, G., Shan, J., Andersen, O.S., and Weinstein, H. 2011. Quantitative modeling of membrane deformations by multihelical membrane proteins: Application to G-protein coupled receptors. *Biophys. J.* 101:2092–2101.

Morel, B., Autissier, L., Autissier, D., Lemordant, D., Yrieix, B., and Quenard, D. 2009. Pyrogenic silica ageing under humid atmosphere. *Powder Technol.* 190:225–229.

Morimoto, T. and Miura, K. 1985. Adsorption sites for water on graphite. 1. Effect of high temperature treatment of sample. *Langmuir* 1:658–662.

Morin, A. and Dufresne, A. 2002. Nanocomposites of chitin whiskers from Riftia tubes and polycaprolactone. *Macromolecules* 35:2190–2199.

Morishige, K. and Iwasaki, H. 2003. X-ray study of freezing and melting of water confined within SBA-15. *Langmuir* 19:2808–2811.

Morris, E.R., Nishinari, K., and Rinaudo, M. 2012. Gelation of gellan—A review. *Food Hydrocolloids* 28:373–411.

Morris, R. and Wheatley, P. 2008. Gas storage in nanoporous materials. *Angew. Chem. Int. Ed* 47:4966–4981.

Morterra, C. and Low, M.J.D. 1983. IR Studies of carbons. II. The vacuum pyrolysis of cellulose. *Carbon* 21:283–288.

Mosesson, M.W., DiOrio, J.P., Hernandez, I., Hainfeld, J.F., Wall, J.S., and Grieninger. G. 2004. The ultrastructure of fibrinogen-420 and the fibrin-420 clot. *Biophys. Chem.* 112:209–214.

Motornov, M., Roiter, Y., Tokarev, I., and Minko, S. 2010. Stimuli-responsive nanoparticles, nanogels and capsules for integrated multifunctional intelligent systems. *Progr. Polym. Sci.* 35:174–211.

Motwani, T., Seetharaman, K., and Anantheswaran, R.C. 2007. Dielectric properties of starch slurries as influenced by starch concentration and gelatinization. *Carbohydr. Polym.* 67:73–79.

Mueller, K.T., Chingas, G.C., and Pines, A. 1991a. A dynamic-angle spinning NMR probe. *Rev. Sci. Instrum.* 62:1445–1452.

Mueller, R., Kammler, H.K., Pratsinis, S.E., Vital, A., Beaucage, G., and Burtscher, P. 2004. Non-agglomerated dry silica nanoparticles. *Powder Technol.* 140:40–48.

Mueller, K.T., Sun, B.Q., Chingas, G.C., Zwanziger, J.W., Terao, T., and Pines, A. 1990. Dynamic-angle spinning of quadrupolar nuclei. *J. Magn. Reson.* 86:470–487.

Mueller, K.T., Wooten, E.W., and Pines, A. 1991b. Pure-absorption-phase dynamic-angle spinning. *J. Magn. Reson.* 92:620–627.

Muja, N. and Bulte, J.W.M. 2009. Magnetic resonance imaging of cells in experimental disease models. *Prog. Nucl. Magn. Reson. Spectrosc.* 55:61–77.

Müller, M., Harvey, G., and Prins, R. 2000a. Comparison of the dealumination of zeolites beta, mordenite, ZSM-5 and ferrierite by thermal treatment, leaching with oxalic acid and treatment with $SiCl_4$ by ^1H, ^{29}Si and ^{27}Al MAS NMR. *Micropor. Mesopor. Mater.* 34:135–147.

Müller, M., Harvey, G., and Prins, R. 2000b. Quantitative multinuclear MAS NMR studies of zeolites. *Micropor. Mesopor. Mater.* 34:281–290.

Muller-Sieburg, C.E., Cho, R.H., Thoman, M., Adkins, B., and Sieburg, H.B. 2002. Deterministic regulation of hematopoietic stem cell self-renewal and differentiation. *Blood* 100:1302–1309.

Muniz, W.B., Ramos, F.M., and de Campos Velho, H.F. 2000. Entropy- and Tikhonov-based regularization techniques applied to the backwards heat equation. *Comput. Math. Appl.* 40:1071–1084.

Murakami, H., Kushida, T., and Tashiro, H. 1998. Fast structural relaxation of polyvinyl alcohol below the glass-transition temperature. *J. Chem. Phys.* 108:10309–10318.

Murakhtina, T., Heuft, J., Meijer, E.J., and Sebastiani, D. 2006. First principles and experimental ^1H NMR signatures of solvated ions: The case of HCl(aq). *ChemPhysChem* 7:2578–2584.

Murata, K., Mitsuoka, K., Hirai, T. et al. 2000. Structural determination of water permeation through aquaporin-1. *Nature* 407:599–605.

Muromtsev, V.A. and Kidalov, V.N. 1998. *Medicine in 21 Century*, St. Petersburg, FL: INTAN.

Murray, R.W. (Ed.) 1992. *Molecular Design of Electrode Surface*. New York: Wiley.

Murray, D.K. 2010. Differentiating and characterizing geminal silanols in silicas by ^{29}Si NMR spectroscopy. *J. Colloid Interface Sci.* 352:163–170.

Myers, A.L. and Siperstein, F. 2001. Characterization of adsorbents by energy profile of adsorbed molecules. *Colloids Surf. A: Physicochem. Eng. Aspects* 187–188:73–81.

Myronyuk, I.F., Chelyadyn, V.L., Yakubovskyi, R.R., and Kotsyubynsky, V.O. 2010. Atomic structure and morphology of fumed silica nanoparticles. *Phys. Chem. Solid State* 11:409–418.

Nabeshi, H., Yoshikawa, T., and Matsuyama, K. et al. 2011. Systemic distribution, nuclear entry and cytotoxicity of amorphous nanosilica following topical application. *Biomaterials* 32:2713–2724.

Nagle, J.F. and Tristram-Nagle, S. 2000. Lipid bilayer structure. *Curr. Opin. Struct. Biol.* 10:474–480.

Nah, Y.-C., Paramasivam, I., and Schmuki, P. 2010. Doped TiO_2 and TiO_2 nanotubes: Synthesis and applications. *ChemPhysChem* 11:2698–2713.

Nair, K.G. and Dufresne, A. 2003. Crab shell chitin whiskers reinforced natural rubber nanocomposites. 1. Processing and swelling behaviour. *Biomacromolecules* 4:657–665.

Nakata, S., Tanaka, Y., Asaoka, S., and Nakamura, M. 1998. Recent advances in applications of multinuclear solid-state NMR to heterogeneous catalysis and inorganic materials. *J. Mol. Struct.* 441:267–281.

Nakayama, H., Kataoka, H., Taketani, Y., Sumita, C., and Tsuhako, M. 2002. Synthesis of a new template-free microporous silicoaluminophosphate ($Si_2AlP_3O_{13}$) and its characterization by solid-state NMR. *Micropor. Mesopor. Mater.* 51:7–15.

Nakayama, Y., Soeda, F., and Ishitani, A. 1990. XPS Study of the carbon fiber matrix interface. *Carbon* 28:21–26.

Narayan, R.L. and King, T.S. 1998. Hydrogen adsorption states on silica-supported Ru-Ag and Ru-Cu bimetallic catalysts investigated via microcalorimetry. *Thermochim. Acta* 312:105–114.

Narayanan, K.B. and Sakthivel, N. 2011. Synthesis and characterization of nano-gold composite using *Cylindrocladium floridanum* and its heterogeneous catalysis in the degradation of 4-nitrophenol. *J. Hazard. Mater.* 189:519–525.

Navrotsky, A. 2007. Calorimetry of nanoparticles, surfaces, interfaces, thin films, and multilayers. *J. Chem. Thermodyn.* 39:2–9.

Ndife, M.K., Şumnu, G., and Bayindirli, L. 1998a. Dielectric properties of six different species of starch at 2450 MHz. *Food Res. Int.* 3:43–52.

Ndife, M.K., Şumnu, G., and Bayindirli, L. 1998b. Differential scanning calorimetry determination of gelatinization rates in different starches due to microwave heating. *LWT Food Sci. Technol.* 31:484–488.

Nekrasov, B.V. 1973. *Fundamentals of General Chemistry*. Moscow, Russia: Khimia.

Ngai, K.L., Riande, E., and Wright, G.B. (Eds.) 1994. Proceedings of the International discussion meeting on relaxations in complex systems. *J. Non-Cryst. Solids* 172–174.

Nguyen, C. and Do, D.D. 1999. A new method for the characterization of porous materials. *Langmuir* 15:3608–3615.

Nicholson, D. and Cracknell, R.F. 1996. Self diffusion and transport in slite shiped pores. In: LeVan, M.D. (Ed.) *Proc. Fundamentals of Adsorption*. Boston, MA: Kluwer Academic Publishers, pp. 667–670.

Nielsen, J.T., Eghbalnia, H.R., and Nielsen, N.Chr. 2012. Chemical shift prediction for protein structure calculation and quality assessment using an optimally parameterized force field. *Prog. Nucl. Magn. Reson. Spectrosc.* 60:1–28.

Nimni, M. (Ed.) 1988. Collagen, Biotechnology. Vol. 3, Boca Raton, FL: CRC Press.

Noh, H. and Vogler, E.A. 2007. Volumetric interpretation of protein adsorption: Competition from mixtures and the Vroman effect. *Biomaterials* 28:405–422.

Nordgren, J., Glans, P., and Wassdahl, N. 1991. Progress in ultra-soft X-ray emission spectroscopy. *Phys. Scr.* 34:100–107.

Nordon, A., Hughes, E., Harris, R.K., Yeo, L., and Harris, K.D.M. 1998. Direct measurement of the distance between adjacent guest molecules in a disordered solid inclusion compound using solid-state ^{19}F–^{1}H NMR spectroscopy. *Chem. Phys. Lett.* 289:25–29.

Novakov, T. and Penner, J.E. 1993. Large contribution of organic aerosols to cloud-condensation-nuclei concentrations. *Nature* 365:823–826.

Novoselov, K.S., Geim, A.K., Morozov, S.V. et al. 2004. Electric field effect in atomically thin carbon films. *Science* 306:666–669.

O'Brien, F.J., Harley, B.A., Yannas, I.V., and Gibson, L. 2004. Influence of freezing rate on pore structure in freeze-dried collagen-GAG scaffolds. *Biomaterials* 25:1077–1086.

O'Donnell, A., Yach, K., and Reven, L. 2008. Particle—particle interactions and chain dynamics of fluorocarbon and hydrocarbon functionalized ZrO_2 nanoparticles. *Langmuir* 24:2465–2471.

Oenick, M.D.B. 2004. Studies on fibrin polymerization and fibrin structure - a retrospective. *Biophys. Chem.* 112:187–192.

Oepen, S.B. and Günther, H. 1996. ^{29}Si MAS solid state and ^{129}Xe NMR of porous silica. *Magn. Reson. Imag.* 14:993–994.

Offer, G. and Knight, P. 1988. *The Structural Basis of Water-Holding in Meat.* In: Lawrie, R.A. (Ed.) *Developments in Meat Science*. London, U.K.: Elsevier.

Ohring, M. 1992. *The Materials Science of Thin Films*. London, U.K.: Academic Press.

Oldfield, E., Bowers, J.L., and Forbes, J. 1987. High-resolution proton and carbon-13 NMR of membranes: Why sonicate? *Biochemistry* 26:6919–6923.

Oliver, M.S., Dubois, G., Sherwood, M., Gage, D.M., and Dauskardt, R.H. 2010. Molecular origins of the mechanical behavior of hybrid glasses. *Adv. Funct. Mater.* 20:2884–2892.

Olson, D.H., Haag, W.O., and Lago, R.W. 1980. Chemical and physical properties of the ZSM-5 substitutional series. *J. Catal.* 61:390–396.

Onjia, A.E., Milonjić, S.K., and Rajaković, Lj.V. 2001. Inverse gas chromatography of chromia. Part I. Zero surface coverage. *J. Serb. Chem. Soc.* 66:259–271.

Ono, Y., Mayama, H., Furó, I. et al. 2009. Characterization and structural investigation of fractal porous-silica over an extremely wide scale range of pore size. *J. Colloid Interface Sci.* 336:215–225.

Opella, S.J. and Marassi, F.M. 2004. Structure determination of membrane proteins by NMR spectroscopy. *Chem. Rev.* 104:3587–3606.

O'Reilly, D.E. 1974. Temperature and pressure dependence of the proton chemical shift in water. *J. Chem. Phys.* 61:1592–1593.

Oriňáková, R. and Oriňák, A. 2011. Recent applications of carbon nanotubes in hydrogen production and storage. *Fuel* 90:3123–3140.

Østergaard, K.K., Anderson, R., Llamedo, M., and Tohidi, B. 2002. Hydrate phase equilibria in porous media: Effect of pore size and salinity. *Terra Nova* 14:307–312.

Ostroff, E.D. and Waugh, J.S. 1966. Multiple spin echoes and spin blocking in solids. *Phys. Rev. Lett.* 16:1097–1098.

Otting, G. 1997. NMR studies of water bound to biological molecules. *Prog. Nucl. Magn. Reson. Spectrosc.* 31:259–285.

Otzen, D. 2011. Protein–surfactant interactions: A tale of many states. *Biochim. Biophys. Acta* 1814:562–591.

Ou, H.-H. and Lo, S.-L. 2007. Review of titania nanotubes synthesized via the hydrothermal treatment: Fabrication, modification, and application. *Sep. Purif. Technol.* 58:179–191.

Overloop, K. and Van Gerven, L. 1993. Freezing phenomena in adsorbed water as studied by NMR. *J. Magn. Reson. A* 101:179–187.

Oxenoid, K., Sonnichsen, F.D., and Sanders, C.R. 2002. Topology and secondary structure of the N-terminal domain of diacylglycerol kinase. *Biochemistry* 41:12876–12882.

Paillet, M. and Dufresne, A. 2001. Chitin whisker reinforced thermoplastic nanocomposites. *Macromolecules* 34:6527–6530.

Palmer, J.C. and Gubbins, K.E. 2012. Atomistic models for disordered nanoporous carbons using reactive force fields. *Micropor. Mesopor. Mater.* 154:24–37.

Palmer, A.G., III and Massi, F. 2006. Characterization of the dynamics of biomacromolecules using rotating-frame spin relaxation NMR spectroscopy. *Chem. Rev.* 106:1700–1719.

Panagopoulou, A., Kyritsis, A., Sabater i Serra, R., Gómez Ribelles, J.L., Shinyashiki, N., and Pissis, P. 2011. Glass transition and dynamics in BSA–water mixtures over wide ranges of composition studied by thermal and dielectric techniques. *Biochim. Biophys. Acta. Protein. Proteom.* 1814:1984–1996.

Papeo, G., Giordano, P., Brasca, M.G. et al. 2007. Polyfluorinated amino acids for sensitive F-19 NMR-based screening and kinetic measurements. *J. Am. Chem. Soc.* 129:5665–5672.

Parida, S.K., Dash, S., Patel, S., and Mishra, B.K. 2006. Adsorption of organic molecules on silica surface. *Adv. Colloid Interface Sci.* 121:77–110.

Park, S., Venditti, R.A., Jameel, H., and Pawlak, J.J. 2006. Changes in pore size distribution during the drying of cellulose fibers as measured by differential scanning calorimetry. *Carbohydr. Polym.* 66:97–103.

Parneix, C., Persello, J., Schweins, R., and Cabane, B. 2009. How do colloidal aggregates yield to compressive stress? *Langmuir* 25:4692–4707.

Pastor, R.W., Venable, R.M., and Feller, S.E. 2002. Lipid bilayers, NMR relaxation, and computer simulations. *Acc. Chem. Res.* 35:438–446.

Patel, K., Angelos, S., Dichtel, W.R. et al. 2008. Enzyme-responsive snap-top covered silica nanocontainers. *J. Am. Chem. Soc.*130:2382–2383.

Pavesi, L. and Balzarini, M. 1996. NMR study of the diffusion processes in gels. *Magn. Reson. Imag.* 14:985–987.

Pazé, C., Bordiga, S., Lamberti, C., Salvalaggio, M., Zecchina, A., and Bellussi, G. 1997. Acidic properties of H—β Zeolite as probed by bases with proton affinity in the 118–204 kcal mol^{-1} range: a FTIR investigation. *J. Phys. Chem. B.* 101:4740–4751.

Pearce, K.L., Rosenvold, K., Andersen, H.J., and Hopkins, D.L. 2011. Water distribution and mobility in meat during the conversion of muscle to meat and ageing and the impacts on fresh meat quality attributes—A review. *Meat Sci.* 89:111–124.

Pecora, R. 1985. *Dynamic light scattering*. New York: Plenum Press.

Pedretti, A.L., Villa, G., and Vistoli, G. 2004. VEGA—An open platform to develop chemo-bio-informatics applications, using plug-in architecture and script programming. *J. Comput. Aided Mol. Des.* 18:167–173.

Peeters, M.P.J. and Kentgens, A.P.M. 1997. A ^{27}Al MAS, MQMAS and off-resonance nutation NMR study of aluminium containing silica-based sol-gel materials. *Solid State Nucl. Magn. Reson.* 9:203–217.

Peinador, R.I., Calvo, J.I., Prádanos, P., Palacio, L., and Hernández, A. 2010. Characterisation of polymeric UF membranes by liquid–liquid displacement porosimetry. *J. Membr. Sci.* 348:238–244.

Pek, Y.S., Spector, M., Yannas, I.V., and Gibson, L.J. 2004. Degradation of a collagen-chondroitin-6-sulfate matrix by collagenase and chondroitinase. *Biomaterials* 25:473–482.

Peng, L.J. and Morris, J.R. 2012. Structure and hydrogen adsorption properties of low density nanoporous carbons from simulations. *Carbon* 50:1394–1406.

Peng, H., Zhu, Y., Peterson, D.E., and Lu, Y. 2008. Nanolayered carbon/silica superstructures via organosilane assembly. *Adv. Mater.* 20:1199–1204.

Peppas, N.A., Bures, P., Leobandung, W., and Ichikawa, H. 2000. Hydrogels in pharmaceutical formulations. *Eur. J. Pharm. Biopharm.* 50:27–46.

Pereda, S., Awan, J.A., Mohammadi, A.H. et al. 2009. Solubility of hydrocarbons in water: Experimental measurements and modeling using a group contribution with association equation of state (GCA-EoS). *Fluid Phase Equilib.* 275:52–59.

Perkins, E.L. and Batchelor, W.J. 2012. Water interaction in paper cellulose fibres as investigated by NMR pulsed field gradient. *Carbohydr. Polym.* 87:361–367.

Perkins, E.L., Lowe, J.P., Edler, K.J., and Rigby, S.P. 2010. Studies of structure–transport relationships in biodegradable polymer microspheres for drug delivery using NMR cryodiffusometry. *Chem. Eng. Sci.* 65:611–625.

Perkins, E.L., Lowe, J.P., Edler, K.J., Tanko, N., and Rigby, S.P. 2008. Determination of the percolation properties and pore connectivity for mesoporous solids using NMR cryodiffusometry. *Chem. Eng. Sci.* 63:1929–1940.

Perlo, J., Casanova, F., and Blumich, B. 2004. 3D imaging with a single-sided sensor: An open tomograph. *J. Magn. Reson.* 166:228–235.

Pervushin, K., Ono, A., Fernández, C., Szyperski, T., Kainosho, M., and Wüthrich, K. 1998. NMR scalar couplings across Watson-Crick base pair hydrogen bonds in DNA observed by transverse relaxation optimized spectroscopy. *Proc. Nat. Acad. Sci. USA* 95:14147–14151.

Pervushin, K., Riek, R., Wider, G., and Wüthrich, K. 1997. Attenuated T_2 relaxation by mutual cancellation of dipole–dipole coupling and chemical shift anisotropy indicates an avenue to NMR structures of very large biological macromolecules in solution. *Proc. Natl. Acad. Sci. USA* 94:12366–12371.

Petin, A.Yu., Gun'ko, V.M., Skubiszewska-Zięba, J., Leboda, R., and Turov, V.V. 2010. Co-adsorption of methane and nonpolar organics in mesoporous silica gel with the presence of water. *Chem. Phys. Technol. Surf.* 1:138–147 (in Russian).

Petrov, O.V. and Furó, I. 2006. Curvature-dependent metastability of the solid phase and the freezing-melting hysteresis in pores. *Phys. Rev. E* 73:011608.

Petrov, O.V. and Furó, I. 2009. NMR cryoporometry: Principles, applications and potential. *Prog. Nucl. Magn. Reson. Spectrosc.* 54:97–122.

Petrov, O.V. and Furó, I. 2010. A joint use of melting and freezing data in NMR cryoporometry. *Micropor. Mesopor. Mater.* 136:83–91.

Petrov, O.V. and Furó, I. 2011. A study of freezing–melting hysteresis of water in different porous materials. Part I: Porous silica glasses *Micropor. Mesopor. Mater.* 138:221–227.

Petrov, O., Furó, I., Schuleit, M., Domanig, R., Plunkett, M., and Daicic, J. 2006. Pore size distributions of biodegradable polymer microparticles in aqueous environments measured by NMR cryoporometry. *Int. J. Pharmacol.* 309:157–162.

Petrov, O.V., Vargas-Florencia, D., and Furó, I. 2007. Surface melting of octamethylcyclotetrasiloxane confined in controlled pore glasses: Curvature effects observed by ^{1}H NMR. *J. Phys. Chem. B* 111:1574–1581.

Pevzner, S., Regev, O., and Yerushalmi-Rozen, R. 2000. Thin films of mesoporous silica: Preparation and characterization. *Curr. Opin. Colloid Interface Sci.* 4:420–427.

Picard, F., Paquet, M.-J., Dufourc, E.J., and Auger, M. 1998. Measurement of the lateral diffusion of dipalmitoylphosphatidylcholine adsorbed on silica beads in the absence and presence of melittin: A ^{31}P two-dimensional exchange solid-state NMR study. *Biophys. J.* 74:857–868.

Pilkenton, S. and Raftery, D. 2003. Solid-state NMR studies of the adsorption and photooxidation of ethanol on mixed TiO_2–SnO_2 photocatalysts. *Solid State Nucl. Magn. Reson.* 24:236–253.

Pines, A. and Bell, A. (Eds.) 1994. *NMR Techniques in Catalysis*. New York: Marcel Dekker.

Piyasena, P., Ramaswamy, H.S., Awuah, G.B., and Defelice, C. 2003. Dielectric properties of starch solutions as influenced by temperature, concentration, frequency and salt. *J. Food Process Eng.* 26:93–119.

Platzer, B. and Maurer, G. 1993. Application of a generalized Bender equation of state to the description of vapour-liquid in binary systems. *Fluid Phase Equilib.* 84:79–110.

Pogorelyi, K.V., Turov, V.V., Mironova, L.A., and Chuiko, A.A. 1991. Mathematical modeling of PMR spectra of the molecules adsorbed on graphite surface. *Dokl. AN USSR* 9:128–131.

Pokrovskiy, V.A., Leboda, R., Turov, V.V., Charmas, B., and Ryczkowski, J. 1999. Temperature programmed desorption mass spectrometry of carbonized silica surface. *Carbon* 37:1039–1047.

Pool Jr., C.P. and Owens, F.J. 2003. *Introduction to Nanotechnology*. New York: Wiley.

Pople, J.A. 1956. Proton magnetic resonance of hydrocarbons. *J. Chem. Phys.* 24:1111–1115.

Pople, J.A. 1957. Proton resonance sifts in molecules with hydrogen bonds. *Proc. R. Soc.* 239:541–550.

Pople, J.A., Schneider, W.G., and Bernstein, H.J. 1959. *High-Resolution Nuclear Magnetic Resonance*. New York: McGraw-Hill Book Company.

Portsmouth, R.L., Duer, M.J., and Gladden, L.F. 1995. ^{2}H NMR studies of binary adsorption in silicalite. *J. Chem. Soc. Faraday Trans.* 91:963–969.

Pouliquen, D. and Gallois, Y. 2001. Physicochemical properties of structured water in human albumin and gammaglobulin solutions. *Biochimie* 83:891–898.

Pouliquen, D., Gross, D., Lehmann, V., Ducoufwau, S., Demilly, D., and Léchappé, J. 1997. Study of water and oil bodies in seeds by nuclear magnetic resonance. *C. R. Acad. Sci. Paris Life Sci.* 320:131–138.

Prasanth, K.P., Pillai, R.S., Bajaj, H.C. et al. 2008. Adsorption of hydrogen in nickel and rhodium exchanged zeolite X. *Int. J. Hydrogen Energy* 33:735–745.

Provencher, S.W. 1982. A constrained regularization method for inverting data represented by linear algebraic or integral equations. *Comp. Phys. Comm.* 27:213–227.

Puglisi, J.D., Chen, L., Blanchard, S., and Frankel, A.D. 1995. Solution structure of a bovine immunodeficiency virus Tat-TAR peptide-RNA complex. *Science* 270:1200–1203.

Pujari, P.K., Sen, D., Amarendra, G. et al. 2007. Study of pore structure in grafted polymer membranes using slow positron beam and small-angle X-ray scattering techniques. *Nucl. Instrum. Method Phys. Res. B* 254:278–282.

Pursch, M., Vanderhart, D.L., Sander, L.C. et al. 2000. C_{30} self-assembled monolayers on silica, titania, and zirconia: HPLC performance, atomic force microscopy, ellipsometry, and NMR studies of molecular dynamics and uniformity of coverage. *J. Am. Chem. Soc.* 122:6997–7011.

Qian, L. and Zhang, H. 2011. Controlled freezing and freeze drying: A versatile route for porous and micro-/nano-structured materials. *J. Chem. Technol. Biotechnol.* 86:172–184.

Quemada, D. and Berli, C. 2002. Energy of interaction in colloids and its implications in rheological modeling. *Adv. Colloid Interface Sci.* 98:51–85.

Quint, P., Ayala, I., Busby, S.A. et al. 2006. Structural mobility in human manganese superoxide dismutase. *Biochemistry* 45:8209–8215.

Quinton, M.F., Beguin, F., and Legrand, A.P. 1986. NMR characterization of heavy alkali-organic molecule-graphite compounds prepared from highly oriented pyrolytic graphite. In: *Proc. International Carbon Conference. Carbon-86*, pp. 427–429.

Quinton, M.F., Faccini, L., Beguin, F., and Legrand, A.P. 1982. Mouvments moleculaires du tetrahydrofuranne dans les composes d'insertion graphite-potassium-tetrahydrofuranne etudirs par R.M.N. du proton. *Rev. Chem. Miner.* 19:407–419.

Quinton, M.F., Legrand, A.P., and Faccini, L. 1988. ^{13}C NMR study of ternary intercalated compounds heavy alkali-tetrohydrofuran-graphite. *Synth. Metal* 23:271–276.

Rahman, M.M., Czaun, M., Takafuji, M., and Ihara, H. 2008. Synthesis, self-assembling properties, and atom transfer radical polymerization of an alkylated l-phenylalanine-derived monomeric organogel from silica: A new approach to prepare packing materials for high-performance liquid chromatography. *Chem. Eur. J.* 14:1312–1321.

Rai, D., Hod, O., and Nitzan, A. 2010. Circular currents in molecular wires. *J. Phys. Chem. C.* 114:20583–20594.

Ralston, J., Dukhin, S.S., and Mishchuk, N.A. 2002. Wetting film stability and flotation kinetics. *Adv. Colloid Interface Sci.* 95:145–236.

Ramseir, R.O. 1967. Self-diffusion of tritium in natural and synthetic ice monocrystals. *J. Appl. Phys.* 38:2553–2556.

Ramsey, N.F. 1952. Chemical effects in nuclear magnetic resonance and in diamagnetic susceptibility. *Phys. Rev.* 86:243–246.

Rannou, I., Bayot, V., and Lalaurain, M. 1994. Structural characterization of graphitization process in pyrocarbons. *Carbon* 32:833–843.

Rascio, A., Nicastro, G., Carlino, E., and Di Fonzo, N. 2005. Differences for bound water content as estimated by pressure–volume and adsorption isotherm curves. *Plant Sci.* 169:395–401.

Rascio, A., Russo, M., Platani, C., Ronga, G., and Di Fonzo, N. 1999. Mutants of durum wheat with alterations in tissue affinity for strongly bound water. *Plant Sci.* 144:29–34.

Rata, D.G., Casanova, F., Perlo, J., Demco, D.E., and Bluemich, B. 2006. Self-diffusion measurements by a mobile single-sided NMR sensor with improved magnetic field gradient. *J. Magn. Reson.* 180:229–235.

Ratcliffe, R.G., Roscher, A., and Shachar-Hill, Y. 2001. Plant NMR spectroscopy. *Progr. Nucl. Magn. Reson. Spectrosc.* 39:267–300.

Raviv, U. and Klein, J. 2002. Fluidity of bound hydration layers. *Science* 297:1540–1543.

Ray, G.J. and Samoson, A. 1993. Double rotation and variable field ^{27}Al n.m.r. study of dealuminated Y zeolites. *Zeolites* 13:410–413.

Read, M.J. and Burkett, S.L. 2003. Asymmetric alpha-helicity loss within a peptide adsorbed onto charged colloidal substrates. *J. Colloid Interface Sci.* 261:255–263.

Reid, G.C. 1997. The nucleation and growth of ice particles in the upper mesosphere. *Adv. Space Res.* 20:1285–1291.

Resing, H.A. and Davidson, D.W. 1975. Commentary on the NMR apparent phase transition effect in natrolite, fluor-montmorillionite and other systems: Generalizations when the Arrhenius law applies. *Can. J. Phys.* 54:295–300.

Ribeiro Carrott, M.M.L., Candeias, A.J.E., Carrott, P.J.M., Ravikovitch, P.I., Neimark, A.V., and Sequeira, A.D. 2001. Adsorption of nitrogen, neopentane, *n*-hexane, benzene and methanol for evaluation of pore size in silica grades of MCM-41. *Micropor. Mesopor. Mater.* 47:323–337.

Richet, P. and Polian, A. 1998. Water as a dense icelike component in silicate glasses. *Science* 281:396–398.

Rigby, S.P., Fairhead, M, and van der Walle, C.F. 2008. Engineering silica particles as oral drug delivery vehicles. *Curr. Pharmacol. Des.* 14:1821–1831.

Robbins, G.P., Hallett, J.P., Bush, D., and Eckert, C.A. 2007. Liquid–liquid equilibria and partitioning in organic–aqueous systems. *Fluid Phase Equilib.* 253:48–53.

Robinson, R.A. 1952. An electron microscopic study of the crystalline inorganic components of bone and its relationship to the organic matrix. *J. Bone Joint Surg.* 34A:389–434.

Rocha, J. and Anderson, M.W. 2000. Microporous titanosilicates and other novel mixed octahedral-tetrahedral framework oxides. *Eur. J. Inorg. Chem.* 801–818.

Rodriguez, L.N.J., De Paul, S.M., Barrett, C.J., Reven, L., and Spiess, H.W. 2000. Fast magic-angle spinning and double-quantum ^{1}H solid-state NMR spectroscopy of polyelectrolyte multilayers. *Adv. Mater.* 12:1934–1938.

Romanenko, K.V., Fonseca, A., Dumonteil, S. et al. 2005. ^{129}Xe NMR study of Xe adsorption on multiwall carbon nanotubes. *Solid State Nucl. Magn. Reson.* 28:135–141.

Roose, P., Van Craen, J., Andriessens, G., and Eisendrath, H. 1996. NMR study of spin–lattice relaxation of water protons by Mn^{2+} adsorbed onto colloidal silica. *J. Magn. Reson. A* 120:206–213.

Rosenblatt, J., Devereux, B., and Wallace, D. 1993. Dynamic rheological studies of hydrophobic interactions in injectable collagen biomaterials. *J. Appl. Polym. Sci.* 50:953–963.

Rosso, F., Marino, G., Giordano, A., Barbarisi, M., Parmeggiani, D., and Barbarisi, A. 2005. Smart materials as scaffolds for tissue engineering. *J. Cell Physiol.* 203:465–470.

Rouquerol, J., Baron, G.V., Denoyel, R. et al. 2012. The characterization of macroporous solids: An overview of the methodology. *Micropor. Mesopor. Mater.* 154:2–6.

Rowell, M.R., Young, R.A., and Rowell, J.K. (Eds.) 1997. *Paper and Composites from Agro-Based Resources.* New York: CRC Lewis Publishers.

Rozovsky, S., Jogl, G., Tong, L., and McDermott, A.E. 2001. Solution-state NMR investigations of triosephosphate isomerase active site loop motion: Ligand release in relation to active site loop dynamics. *J. Mol. Biol.* 310:271–280.

Rubin, R. and Strayer, D.S. 2007. *Rubin's Pathology: Clinicopathologic Foundations of Medicine.* Philadelphia, PA: Lippincott Williams & Wilkins.

Rudenko, S.V., Crowe, J.H., and Tablin, F. 1998. Change in the erythrocyte shape depending on the term. *Biochemistry* 63:1630–1639.

Rudzinski, W., Gierak, A., Leboda, R., and D⊠browski, A. 1995. Studies of properties of complex carbon–silica adsorbents used in sorption–desorption processes (solid–phase extraction). *Fresenius J. Anal. Chem.* 352:667–672.

Rugal, A.A., Gun'ko, V.M., Barvinchenko, V.N., Turov, V.V., Semeshkina, T.V., and Zarko, V.I. 2007. Interaction of fibrinogen with nanosilica. *Central Eur. J. Chem.* 5:32–54.

Ruiz-Bevia, F., Fernandez-Sempere, J., and Boluda-Botella, N. 1995. Variation of phosphoric acid diffusion coefficient with concentration. *AIChE J.* 41:185–189.

Ruland, W. 1965. X-ray studies on the carbonization and graphitization of acenaphthylene and bifluorenyl. *Carbon* 2:365–375.

Russell, N.M. and Ekerdt, J.G. 1996. Nonlinear parameter estimation technique for kinetic analysis of thermal desorption data. *Surf. Sci.* 364:199–218.

Ryan, J.J., Bateman, H.R., Stover, A. et al. 2007. Fullerene nanomaterials inhibit the allergic response. *J. Immunol.* 179:665–672.

Ryu, S.-Y., Kim, D.S., Jeon, J.-D., and Kwak, S.-Y. 2010. Pore size distribution analysis of mesoporous TiO_2 spheres by ^1H nuclear magnetic resonance (NMR) cryoporometry. *J. Phys. Chem. C* 114:17440–17445.

Ryynänen, S., Risman, P.O., and Ohlsson, T. 1996. The dielectric properties of native starch solutions—A research note. *J. Microwave Power Electromagn. Energy* 31:50–53.

Saalwächter, K. 2007. Proton multiple-quantum NMR for the study of chain dynamics and structural constraints in polymeric soft materials. *Prog. Nucl. Magn. Reson. Spectrosc.* 51:1–35.

Salame, I.I. and Bandosz, T.J. 1999a. Study of water adsorption on activated carbons with different degrees of surface oxidation. *J. Colloid Interface Sci.* 210:367–374.

Salame, I.I. and Bandosz, T.J. 1999b. Experimental study of water adsorption on activated carbons. *Langmuir* 15:587–593.

Saleh, K., Forny, L., Guigon, P., and Pezron, I. 2011. Dry water: From physico-chemical aspects to process-related parameters. *Chem. Eng. Res. Des.* 89:537–544.

Samouillan, V., Tintar, D., and Lacabanne, C. 2011. Hydrated elastin: Dynamics of water and protein followed by dielectric spectroscopies. *Chem. Phys.* 385:19–26.

Satyanarayana, K.G., Arizaga, G.G.C., and Wypych, F. 2009. Biodegradable composites based on lignocellulosic fibers - An overview. *Prog. Polym. Sci.* 34:982–1021.

Sauviat, M.-P., Boydron-Le Garrec, R., Masson, J.-B. et al. 2006. Mechanisms involved in the swelling of erythrocytes caused by Pacific and Caribbean ciguatoxins. *Blood Cell. Mol. Dis.* 36:1–9.

Savina, I.N., Gun'ko, V.M., Turov, V.V. et al. 2011. Noninvasive structural characterisation of macroporous cross-linked polymer and protein hydrogels. *Soft Matter* 7:4276–4283.

Scatena, L.F., Brown, M.G., and Richmond, G.L. 2001. Water at hydrophobic surfaces: Weak hydrogen bonding and strong orientation effects. *Science* 292:908–912.

Scheler, U. 2009. NMR on polyelectrolytes. *Curr. Opin. Colloid Interface Sci.* 14:212–215.

Schick, F. 1996. Bone marrow NMR *in vivo*. *Prog. Nucl. Magn. Reson. Spectrosc.* 29:169–227.

Schilf, W., Jameson, C.J., and Kuroki, S. 2011. *Nuclear Magnetic Resonance*, Vol. 40. London, U.K.: Royal Society of Chemistry.

Schmidt, M.W., Baldridge, K.K., Boatz, J.A. et al. 1993. General atomic and molecular electronic structure system. *J. Comput. Chem.* 14:1347–1363.

Schmitz, K.S. and Lu, M. 1983. Effect of titration charge on the diffusion of bovine serum albumin. *Proc. Natl. Acad. Sci. USA Biophys.* 80:425–429.

Schmücker, M., MacKenzie, K.J.D., Schneider, H., and Meinhold, R. 1997. NMR studies on rapidly solidified SiO_2-Al_2O_3 and SiO_2-Al_2O_3-Na_2O-glasses. *J. Non-Cryst. Solid.* 217:99–105.

Scholten, A.B., de Haan, J.W., Claessens, H.A., van der Ven, L.J.M., and Cramers, C.A. 1994. 29-Silicon NMR evidence for the improved chromatographic siloxane bond stability of bulk alkylsilane ligands on a silica surface. *J. Chromatogr. A* 688:25–29.

Schönfelder, W., Dietrich, J., Märten, A., Kopinga, K., and Stallmach, F. 2007. NMR studies of pore formation and water diffusion in self-hardening cut-off wall materials. *Cem. Concr. Res.* 37:902–908.

Schonhöff, M., Ball, V., Bausch, A.R. et al. 2007. Hydration and internal properties of polyelectrolyte multilayers. *Colloids Surf. A: Physicochem. Eng. Aspects* 303:14–29.

Schreiber, A., Ketelsen, I., and Findenegg, G.H. 2001. Melting and freezing of water in ordered mesoporous silica materials. *Phys. Chem. Chem. Phys.* 3:1185–1195.

Schumacher, K., Ravikovitch, P.I., Du Chesne, A., Neimark, A.V., and Unger, K.K. 2000. Characterization of MCM-48 materials. *Langmuir* 16:4648–4654.

Schurko, R.W., Hung, I., and Widdifield, C.M. 2003. Signal enhancement in NMR spectra of half-integer quadrupolar nuclei via DFS-QCPMG and RAPT-QCPMG pulse sequences. *Chem. Phys. Lett.* 379:1–10.

Schuster, M., Kreuer, K.-D., Steininger, H., and Maier, J. 2008. Proton conductivity and diffusion study of molten phosphonic acid H_3PO_3. *Solid State Ionics* 179:523–528.

Schwarz, J.A. and Contescu, C.I. 1999. *Surfaces of Nanoparticles and Porous Materials*. New York: Marcel Dekker.

Seayad, A.M. and Antonelli, D.M. 2004. Recent advances in hydrogen storage in metal-containing inorganic nanostructures and related materials. *Adv. Mater.* 16:765–777.

Seledets, O., Gun'ko, V.M., Skubiszewska- Zięba, J. et al. 2005. Structural and energetic heterogeneities of pyrocarbon/silica gel systems and their adsorption properties. *Appl. Surf. Sci.* 240:222–235.

Seledets, O., Leboda, R., Grzegorczyk, W. et al. 2006. Effect of sulphur on surface properties of complex carbon-silica adsorbents. *Micropor. Mesopor. Mater.* 93:90–100.

Seledets, O., Skubiszewska-Zięba, J., Leboda, R., and Gun'ko, V.M. 2003. On the surface properties of carbon-silica adsorbents (carboaerosils) prepared by pyrolysis of methylene chloride on the surface of fumed silica (aerosil). *Mater. Chem. Phys.* 82:199–205.

Shah, P.P. and Roberts, C.J. 2008. Solvation in mixed aqueous solvents from a thermodynamic cycle approach. *J. Phys. Chem. B* 112:1049–1052.

Shapiro, Yu.E. 2011. Structure and dynamics of hydrogels and organogels: An NMR spectroscopy approach. *Prog. Polym. Sci.* 36:1184–1253.

Shegokar, R. and Müller, R.H. 2010. Nanocrystals: Industrially feasible multifunctional formulation technology for poorly soluble actives. *Int. J. Pharmacol.* 399:129–139.

Shen, Y.R. and Ostroverkhov, V. 2006. Sum-frequency vibrational spectroscopy on water interfaces: Polar orientation of water molecules at interfaces. *Chem. Rev.* 106:1140–1154.

Shen, L., Soong, R., Wang, M. et al. 2008. Pulsed field gradient NMR studies of polymer adsorption on colloidal CdSe quantum dots. *J. Phys. Chem. B* 112:1626–1633.

Shim, W.G., Moon, H., and Lee, J.W. 2006. Performance evaluation of wash-coated MCM-48 monolith for adsorption of volatile organic compounds and water vapors. *Micropor. Mesopor. Mat.* 94:15–28.

Shipway, A.N., Katz, E., and Willner, I. 2000. Nanoparticle arrays on surfaces for electronic, optical, and sensor applications. *ChemPhysChem* 1:18–52.

Shkolin, A.V. and Fomkin, A.A. 2009. Deformation of AUK microporous carbon adsorbent induced by methane adsorption. *Colloid. J.* 71:119–124.

Shouro, D., Moriya, Y., Nakajima, T., and Mishima, S. 2000. Mesoporous silica FSM-16 catalysts modified with various oxides for the vapor-phase Beckmann rearrangement of cyclohexanone oxime. *Appl. Catal. A: Gen.* 198:275–282.

Shpak, A.P. and Gorbyk, P.P. (Eds.) 2010. *Nanomaterials and Supramolecular Structures*. Dordrecht, the Netherlands: Springer.

Shrade, P., Klein, H., Egry, I. et al. 2001. Hydrophobic volume effects in albumin solutions. *J. Colloid Interface Sci.* 234:445–447.

Shu, C., Xianzhu, Y., Zhangyin, X., Guohua, X., Hong, L., and Kangde, Y. 2007. Synthesis and sintering of nanocrystalline hydroxyapatite powders by gelatin–based precipitation method. *Ceram. Int.* 33:193–196.

Shulga, O.V., Dollimor, D., and Turov, V.V. 2000. Peculiarities in the hydration of cellulose in air and water suspensions as studied by ^{1}H NMR spectroscopy. *Adsorp. Sci. Technol.* 18:857–863.

Siebert, H.-C., André, S., Vliegenthart, J.F.G., Gabius, H.-J., and Minch, M.J. 2003. Suitability of binary mixtures of water with aprotic solvents to turn hydroxyl protons of carbohydrate ligands into conformational sensors in NOE and transferred NOE experiments. *J. Biomolec. NMR* 25:197–215.

Siegel, R.R., Harder, P., Dahint, R. et al. 1997. On-line detection of nonspecific protein adsorption at artificial surfaces. *Anal. Chem.* 69:3321–3328.

Siegel, R., Nakashima, T.T., and Wasylishen, R.E. 2004. Application of multiple-pulse experiments to characterize broad NMR chemical-shift powder patterns from spin-1/2 nuclei in the solid state. *J. Phys. Chem. B* 108:2218–2226.

Sigmund, E.E., Cho, H., and Song, Y.-Q. 2009. High-resolution MRI of internal field diffusion-weighting in trabecular bone. *NMR Biomed.* 22:436–448.

Simms, J.R. and Yang, C.Q. 1994. Infrared spectroscopy studies of the petroleum pitch carbon fiber: II. The distribution of the oxidation products between the surface and the bulk. *Carbon* 32:621–626.

Simpson, A.J., McNally, D.J., and Simpson, M.J. 2011. NMR spectroscopy in environmental research: From molecular interactions to global processes. *Progr. Nucl. Magn. Reson. Spectrosc.* 58:97–175.

Singer, P.M., Wzietek, P., Alloul, H., Simon, F., and Kuzmany, H. 2006. NMR study of spin excitations in carbon nanotubes. *Phys. Stat. Sol. B* 243:3111–31116.

Siperstein, F.R. and Myers, A.L. 2001. Mixed-gas adsorption. *AIChE J.* 47:1141–1159.

Sklari, S., Rahiala, H., Stathopoulos, V., Rosenhjlm, J., and Pomonis, P. 2001. The influence of surface acid density on the freezing behavior of water confined in mesoporous MCM-41 solids. *Micropor. Mesopor. Mater.* 49:1–13.

Skubiszewska-Zięba, J., Leboda, R., Seledets, O., Gun'ko, V.M. 2003. Effect of preparation conditions of carbon-silica adsorbents based on mesoporous silica gel Si-100 and carbonised glucose on their pore structure. *Colloids Surf. A: Physicochem. Eng. Aspects* 231:39–49.

Slater, M., Snauko, M., Svec, F., and Frechet, J.M.J. 2006. "Click chemistry" in the preparation of porous polymer-based particulate stationary phases for μ-HPLC separation of peptides and proteins. *Anal. Chem.* 78:4969–4975.

Slatter, D.A., Miles, C.A., and Bailey, A.J. 2003. Asymmetry in the triple helix of collagen-like heterotrimers confirms that external bonds stabilize collagen structure. *J. Mol. Biol.* 329:175–183.

Sliwinska-Bartkowiak, M., Dudzoak, G., Gras, R., Sikorski, Radhakrishnan, R., and Gubbins, K.E. 2001. Freezing behavior in porous glasses and MCM-41. *Colloids Surf. A: Physicochem. Eng. Aspects* 187–188:523–529.

Smirnova, I., Mamic, J., and Arlt, W. 2003. Adsorption of drugs on silica aerogels. *Langmuir* 19:8521–8525.

Smisek, M. and Cerny, S. 1970. *Active Carbon*. Amsterdam, the Netherlands: Elsevier.

Smith, R.K., Lewis, P.A., and Weiss, P.S. 2004. Patterning self-assembled monolayers. *Prog. Surf. Sci.* 75:1–68.

Socha, R.P. and Fransaer, J. 2005. Mechanism of formation of silica–silicate thin films on zinc. *Thin Solid Film.* 488:45–55.

Soldano, C., Mahmood, A., and Dujardin, E. 2010. Production, properties and potential of graphene. *Carbon* 48:2127–2150.

Spalla, O. and Kékicheff, P. 1997. Adhesion between oxide nanoparticles: Influence of surface complexation. *J. Colloid Interface Sci.* 192:43–65.

Spyros, A. and Dais, P. 2009. ^{31}P NMR spectroscopy in food analysis. *Prog. Nucl. Magn. Reson. Spectrosc.* 54:195–207.

Sriupayo, J., Supaphol, P., Blackwell, J., and Rujiravanit, R. 2005. Preparation and characterization of chitin whisker-reinforced poly(vinyl alcohol) nanocomposite films with or without heat treatment. *Polymers* 46:5637–5644.

Stallmach, F. and Kärger, J. 1999. The potentials of pulsed field gradient NMR for investigation of porous media. *Adsorption* 5:117–133.

Stark, J. 2011. Recent advances in the field of cement hydration and microstructure analysis. *Cem. Concr. Res.* 41:666–678.

Stayton, P.S., Shimoboji, T., Long, C. et al. 1995. Control of protein–ligand recognition using a stimuli-responsive polymer. *Nature* 378:472–474.

Stefl, R., Oberstrass, F.C., Hood, J.L. et al. 2010. The solution structure of the ADAR2 dsRBM-RNA complex reveals a sequence-specific readout of the minor groove. *Cell* 143:225–237.

Stein, A., Wang, Z., and Fierke, M.A. 2009. Functionalization of porous carbon materials with designed pore architecture. *Adv. Mater.* 21:265–293.

Stejskal, E.O. and Tanner, J.E. 1965. Spin diffusion measurements: Spin echoes in the presence of a time-dependent field gradient. *J. Chem. Phys.* 42:288–292.

Stenius, P. 2000. *Forest Products Chemistry*. Papermaking Science and Technology, Fapet OYp, Finland.

Stephenson, N.A. and Bell, A.T. 2005. Quantitative analysis of hydrogen peroxide by ^{1}H NMR spectroscopy. *Anal. Bioanal. Chem.* 381:1289–1293.

Stewart, J.J.P. 2008. *MOPAC2009*. Stewart Computational Chemistry, Colorado Springs, CO, http://openmopac.net/ (accessed December, 2008).

Stoeckli, H.F. 1990. Microporous carbons and their characterization: The present state of the art. *Carbon* 28:1–6.

Strange, J.H., Allen, S.G., Stephenson, P.C.L., and Matveeva, N.P. 1996. Phase equilibria of adsorbed liquids and the structure of porous media. *Magn. Reson. Imag.* 14:963–965.

Strange, J.H., Mitchell, J., and Webber, J.B.W. 2003. Pore surface exploration by NMR. *Magn. Reson. Imag.* 21:221–226.

Strange, J.H., Rahman, M., and Smith, E.G. 1993. Characterization of porous solids by NMR. *Phys. Rev. Lett.* 71:3589–3591.

Su, Y.-Z., Fu, Y.-C., Wei, Y.-M., Yan, J.-W., and Mao, B.-W. 2010. The electrode/ionic liquid interface: Electric double layer and metal electrodeposition. *ChemPhysChem* 11:2764–2778.

Subrahmanyam, K.S., Vivekchand, S.R.C., Govindaraj, A., and Rao, C.N.R. 2008. A study of graphenes prepared by different methods: Characterization, properties and solubilization. *J. Mater. Chem.* 18:1517–1523.

Subramanian, S. and Sloan Jr., E.D. 1999. Molecular measurements of methane hydrate formation. *Fluid Phase Equilib.* 158–160:813–820.

Suh, W.H., Suslick, K.S., Stucky, G.D., and Suh, Y.-H. 2009. Nanotechnology, nanotoxicology, and neuroscience. *Prog. Neurobiol.* 87:133–170.

Sultana, N., Schenkman, J.B., and Rusling, J.F. 2005. Protein film electro chemistry of microsomes genetically enriched in human cytochrome P450 mono oxygenases. *J. Am. Chem. Soc.* 127:13460–13461.

Sulym, I.Y., Borysenko, M.V., Goncharuk, O.V. et al. 2011. Structural and hydrophobic-hydrophilic properties of nanosilica/zirconia alone and with adsorbed PDMS. *Appl. Surf. Sci.* 258:270–2777.

Sun, Z. and Scherer, G.W. 2010. Pore size and shape in mortar by thermoporometry. *Cem. Concr. Res.* 40:740–751.

Suyetin, M.V. and Vakhrushev, A.V. 2009. Nanocapsule for safe and effective methane storage. *Nanoscale Res. Lett.* 4:1267–1270.

Swanson, S.D. 1991. Broadband excitation and detection of cross-relaxation NMR spectra. *J. Magn. Reson.* 95:615–618.

Sych, L.A., Barvinchenko, V.N., Pogorely, V.K., and Turov, V.V. 2000. Effect of dispersed silica on hydration of force-meat. *Probl. Cryobiol.* 4:3–7 (in Russian).

Sydorchuk, V., Khalameida, S., Zazhigalov, V., Skubiszewska-Zięba, J., Leboda, R., and Wieczorek-Ciurowa, K. 2010. Influence of mechanochemical activation in various media on structure of porous and non-porous silicas. *Appl. Surf. Sci.* 257:451–457.

Szekeres, M., Kamalin, O., Grobet, P.G. et al. 2003. Two-dimensional ordering of Stöber silica particles at the air/water interface. *Colloids Surf. A: Physicochem. Eng. Aspects* 227:77–83.

Szymonska, J. and Krok, F. 2003. Potato starch granule nanostructure studied by high resolution non-contact AFM. *Int. J. Biol. Macromol.* 33:1–7.

Ta, T.C., Sykes, M.T., and McDermott, M.T. 1998. Real-time observation of plasma protein film formation on well-defined surfaces with scanning force microscopy. *Langmuir* 14:2435–2443.

Tabata, M., Kumamoto, M., and Nishimoto, J. 1994. Chemical properties of water-miscible solvents separated by salting-out and their application to solvent extraction. *Anal. Sci.* 10:383–388.

Tabony, J. 1980. Nuclear magnetic resonance studies of molecules physisorbed on homogeneous surface. *Progr. Nucl. Magn. Reson. Spectrosc.* 14:1–26.

Tabony, J. White, J.W., Delachaume, J.C., and Coulon, M. 1980. Nuclear magnetic resonance studies of the magnetic and orientation of benzene adsorption upon graphite. *Surf. Sci.* 95:L282–L288.

Tajouri, T. and Bouchriha, H. 2002. Study by ^{1}H NMR of the concentration of monomer-units on interface polymer grafted on silica using magic angle spinning and relaxation techniques. *J. Colloid Interface Sci.* 252:450–455.

Takahashi, M., Figus, C., Kichob, T. et al. 2009. Self-organized nanocrystalline organosilicates in organic-inorganic hybrid films. *Adv. Mater.* 21:1732–1736.

Takeuchi, Y. and Takayama, T. 1998. ^{29}Si NMR spectroscopy of organosilicon compounds. *Chem. Org. Silicon Comp.* 2:267–354.

Tamm, L.K. and Liang, B. 2006. NMR of membrane proteins in solution. *Progr. Nucl. Magn. Reson. Spectrosc.* 48:201–210.

Tamman, G. and Hesse, W. 1926. The dependence of viscosity upon the temperature of supercooled liquids. *Z. Anorg. Allgem. Chem.* 156:645–657.

Tanabe, K. 1970. *Solid Acids and Bases*. Tokyo, Japan: Kodansha.

Tanahashi, I., Yoshida, A., and Nishino, A. 1990. Activated carbon fiber sheets as polarizable electrodes of electric double layer capacitors. *Carbon* 28:477–482.

Tananuwong, K. and Reid, D.S. 2004. DSC and NMR relaxation studies of starch-water interactions during gelatinization. *Carbohydr. Polym.* 58:345–358.

Tanev, P.T. and Pinnavaia, T.J. 1996. Mesoporous silica molecular sieves prepared by ionic and neutral surfactant templating: A comparison of physical properties. *Chem. Mater.* 8:2068–2079.

Tang, H.-R., Godward, J., and Hills, B. 2000. The distribution of water in native starch granules—A multinuclear NMR study. *Carbohydr. Polym.* 43:375–387.

Tang, Z., Wu, H., Cort, J.R. et al. 2010. Constraint of DNA on functionalized graphene improves its biostability and specificity. *Small* 6:1205–1209.

Tanner, J.E. and Stejskal, E.O. 1968. Restricted self-diffusion of protons in colloidal systems by the pulsed gradient, spin-echo method. *J. Chem. Phys.* 49:1768–1777.

Tao, Y., Kanoh, H., Abrams, L., and Kaneko, K. 2006. Mesopore-modified zeolites: Preparation, characterization, and applications. *Chem. Rev.* 106:896–910.

Tarasevich, Yu.I. 2006. Surface energy of oxides and silicates. *Theoret. Exp. Chem.* 42:133–149.

Tarasevich, Yu.I., Aksenenko, E.V., Bondarenko, S.V., and Zhukova, A.I. 2007. Complex investigation of clustered adsorption of water molecules onto hydrophilic sites of graphite and graphitized carbon black. *Theoret. Exp. Chem.* 43:176–182.

Tasis, D., Tagmatarchis, N., Bianco, A., and Prato, M. 2006. Chemistry of carbon nanotubes. *Chem. Rev.* 106:1105–1136.

Taylor, D.J., Fleig, P.F., and Hietala, S.L. 1998. Technique for characterization of thin film porosity. *Thin Solid Films* 332:257–261.

Taylor, D.J., Green, N.P.O., and Stout, G.W. 2000. *Biological Sciences*, Vols. 1 and 2, 3rd edn. Cambridge, MA: Cambridge University Press.

Teleki, A., Wengeler, R., Wengeler, L., Nirschl, H., and Pratsinis, S.E. 2008. Distinguishing between aggregates and agglomerates of flame-made TiO_2 by high-pressure dispersion. *Powder Technol.* 181:292–300.

Telkki, V.-V., Lounila, J., and Jokisaari, J. 2005a. Behavior of acetonitrile confined to mesoporous silica gels as studied by ^{129}Xe NMR: A novel method for determining the pore sizes. *J. Phys. Chem. B* 109:757–763.

Telkki, V.-V., Lounila, J., and Jokisaari, J. 2005b. Determination of pore sizes and volumes of porous materials by ^{129}Xe NMR of xenon gas dissolved in a medium. *J. Phys. Chem. B* 109:24343–24351.

Telkki, V.-V., Lounila, J., and Jokisaari, J. 2006a. Xenon porometry at room temperature. *J. Chem. Phys.* 124:034711(1–8).

Telkki, V.-V., Lounila, J., and Jokisaari, J. 2006b. Influence of diffusion on pore size distributions determined by xenon porometry. *Phys. Chem. Chem. Phys.* 8:2072–2076.

Tennison, S.R., 1998. Phenolic resin derived activated carbons. *Appl. Catal. A.* 173:289–311.

Tennison, S.R., Kozynchenko, O.P., Strelko, V.V., and Blackburn, A.J. 2004. US Patent 2004024074A1 (publ. 05 February, 2004).

Tertykh, V.A. and Belyakova, L.A., 1991. *Chemical Reactions Involving Silica Surface*. Kiev, Ukraine: Naukova Dumka (in Russian).

Theml, H., Diem, H., and Haferlach, T. 2004. *Color Atlas of Hematology*, 2nd edn. Stuttgart, Germany: Thieme.

Thomas, D.J. and Atwell, W.A. (Eds.) 1999. *Starches*. New York: Eagan Press.

Thommes, M., Köhn, R., and Fröba, M. 2002. Sorption and pore condensation behavior of pure fluids in mesoporous MCM-48 silica, MCM-41 silica, SBA-15 silica and controlled-pore glass at temperatures above and below the bulk triple point. *Appl. Surf. Sci.* 196:239–249.

Thorson, J., Shin, I., Chapman, E., Stenberg, G., Mannervik, B., and Schultz, P. 1998. Analysis of the role of the active site tyrosine in human glutathione transferase A1-1 by unnatural amino acid mutagenesis. *J. Am. Chem. Soc.* 120:451–452.

Ti-47,49. 2011. Solid-state titanium NMR references. http://www.pascal-man.com/periodic-table/titanium.shtml (accessed December, 2011).

Timonen, J.T. and Pakkanen, T.T. 1999. A qualitative ^{1}H NMR study of $CHCl_3$ adsorption on conjugated acid–base pairs in cation exchanged Y-zeolites. *Micropor. Mesopor. Mater.* 30:327–333.

Tischenko, V.A. and Gun'ko, V.M. 1995. Water electret relaxation at dispersed silica surface. *Colloids Surf. A: Physicochem. Eng. Aspects* 101:287–294.

Tkalcec, E., Kurajica, S., and Schmauch, J. 2007. Crystallization of amorphous Al_2O_3–SiO_2 precursors doped with nickel. *J. Non-Cryst. Solid.* 353:2837–2844.

Todt, H., Guthausen, G., Burk, W., Schmalbein, D., Kamlowski, A. 2006. Water/moisture and fat analysis by time-domain NMR. *Food Chem.* 96:436–440.

Tolman, J.R. and Ruan, K. 2006. NMR residual dipolar couplings as probes of biomolecular dynamics. *Chem. Rev.* 106:1720–1736.

Tolstoy, P.M., Smirnov, S.N., Shenderovich, I.G., Golubev, N.S., Denisov, G.S., and Limbacha, H.-H. 2004. NMR studies of solid state-solvent and H/D isotope effects on hydrogen bond geometries of 1:1 complexes of collidine with carboxylic acids. *J. Mol. Struct.* 700:19–27.

Tomaszewski, W., Gun'ko, V.M., Leboda, R., and Skubiszewska-Zięba, J., 2005. Interaction of amphetamines with micro- and mesoporous adsorbents in polar liquids. *J Colloid Interface Sci.* 282:261–269.

Tombari, E., Ferrari, C., Salvetti, G., and Johari, G.P. 2009. Dynamic and apparent specific heats during transformation of water in partly filled nanopores during slow cooling to 110K and heating. *Thermochim. Acta* 492:37–44.

Tompa, K., Bánki, P., Bokor, M., Kamasa, P., Lasanda, G., and Tompa, P. 2009. Interfacial water at protein surfaces: Wide-line NMR and DSC characterization of hydration in ubiquitin solutions. *Biophys. J.* 96:2789–2798.

Tompa, K., Bánki, P., Bokor, M., Kamasa, P., Rácz, P., and Tompa, P. 2010. Hydration water/interfacial water in crystalline lens. *Exp. Eye Res.* 91:76–84.

Toole, B.P. 1997. Hyaluronan in morphogenesis. *J. Intern. Med.* 42:35–40.

Topgaard, D. and Söderman, O. 2002. Self-diffusion of nonfreezing water in porous carbohydrate polymer systems studied with nuclear magnetic resonance. *Biophys. J.* 83:3596–3606.

Toriya, S., Takei, T., Fuji, M., and Chikazawa, M. 2003. Characterization of silica-pillared derivatives from aluminum-containing kanemite. *J. Colloid Interface Sci.* 268:435–440.

Tóth, A., Voitko, K.V., Bakalinska, O.N. et al. 2012. Morphology and adsorption activity of chemically modified MWCNT probed by nitrogen, n-propane and water vapor. *Carbon* 50:577–585.

Totland, C., Lewis, R.T., and Nerdal, W. 2011. ^{1}H NMR relaxation of water: A probe for surfactant adsorption on kaolin. *J. Colloid Interface Sci.* 363:362–370.

Tovbin, Yu.K., Petukhov, A.G., and Eremich, D.V. 2006. Capillary condensation of an adsorbate in the cylindrical channel so fMSM-41 mesoporous adsorbent. *Russ. J. Phys. Chem. A Focus Chem.* 80:406–412.

Tracz, E. and Leboda, R. 1985. A microscopic investigation of the surface of carbon–silica adsorbents. II. Relationships between the type of information obtainable about the surface and the microscopic techniques used for its examination. *J. Chromatogr.* 364:364–368.

Trebosc, J., Wiench, J.W., Huh, S., Lin, V.S.Y., and Pruski, M. 2005a. Solid state NMR study of MCM-41-type mesoporous silica nanoparticles. *J. Am. Chem. Soc.* 127:3057–3068.

Trebosc, J., Wiench, J.W., Huh, S., Lin, V.S.Y., and Pruski, M. 2005b. Studies of organically functionalized mesoporous silicas using heteronuclear solid-state correlation NMR spectroscopy under fast magic angle spinning. *J. Am. Chem. Soc.* 127:7587–7593.

Troyer, W.E., Holly, R., Peemoeller, H., and Pintar, M.M. 2005. Proton spin–spin relaxation study of hydration of a model nanopores. *Solid State Nucl. Magn. Reson.* 28:238–243.

Tsang, T. and Resing, H.A. 1985. ^{13}C Nuclear magnetic resonance in graphite intercalation compounds. *Solid State Commun.* 53:39–43.

Tsantilis, S., Kammler, H.K., and Pratsinis, S.E. 2002. Population balance modeling of flame synthesis of titania nanoparticles. *Chem. Eng. Sci.* 57:2139–2156.

Tsoubelli, M.N., Davis, E.A., and Gordon, J. 1995. Dielectric properties and water mobility for heated mixtures of starch, milk, protein and water. *Cereal Chem.* 7:64–69.

Tsuraeva, A.A. (Ed.) 1983. *Cryopreservation of Cells Suspensions.* Kyiv, Ukraine: Naukova Dumka (in Russian).

Tsutsumi, K. and Takahashi, H. 1985. Studies of surface modification of solids. *Colloid Polym. Sci.* 263:506–511.

Tunc, S., Maitz, M.F., Steiner, G., Vázquez, L., Pham, M.T., and Salzer, R. 2005. In situ conformational analysis of fibrinogen adsorbed on Si surfaces. *Colloids Surf. B: Biointerfaces* 42:219–225.

Tuncer, E. and Gubañski, S.M. 2000. Electrical properties of filled silicone rubber. *J. Phys.: Condens. Matter* 12:1873–1897.

Tunnacliffe, A., de Castro, A.G., and Manzanera, M. 2001. Anhydrobiotic engineering of bacterial and mammalian cells: Is intracellular trehalose sufficient? *Cryobiology* 43:124–132.

Turgeon, S.L., Beaulieu, M., Schmitt, C., Sanchez, C. 2003. Protein–polysaccharide interactions: Phase-ordering kinetics, thermodynamic and structural aspects. *Curr. Opin. Colloid Interface Sci.* 8:401–414.

Turova, A.A., Gun'ko, V.M., Turov, V.V., and Gorbyk, P.P. 2007. Influence of structural and chemical modification of silica on its surface hydration. *Adsorb. Sci. Technol.* 25:65–9.

Turov, V.V. and Barvinchenko, V.N. 1997. Structurally ordered surface layers of water at the SiO_2/ice interface and influence of adsorbed molecules of protein hydrolysate on them. *Colloids Surf. B: Biointerfaces* 8:125–132.

Turov, V.V., Barvinchenko, V.N., Krupska, T.V., Gun'ko, V.M., and Chekhun, V.F. 2011a. Hydration properties of composite material based on highly disperse silica and DNA. *Biotechnology* 4:34–49 (in Ukrainian).

Turov, V.V., Bogillo, V.I., and Utlenko, E.V. 1994a. Study of water, benzene and acetonitrile interaction with carbon black by NMR spectroscopy method. *Zh. Prikl. Spektros.* 61:106–113.

Turov, V.V., Brei, V.V., Khomenko, K.N., and Leboda, R. 1998a. ^{1}H NMR studies of the adsorption of water on silicalite. *Micropor. Mesopor. Mater.* 23:189–196.

Turov, V.V., Brei, V.V., Khomenko, K.N., and Leboda, R. 2000. The influence of aluminium concentration on the hydration of ZSM-5-type zeolites in aqueous suspensions. *Adsorp. Sci. Technol.* 18:75–82.

Turov, V.V., Chehun, V.F., Krupskaya, T.V. et al. 2010a. Effect of small addition of C_{60} fullerenes on the hydrated properties of nanocomposites based on highly dispersed silica and DNA. *Chem. Phys. Lett.* 496:152–156.

Turov, V.V., Chekhun, V.F., Gun'ko, V.M., Barvinchenko, V.N., Chekhun, S.V., and Turov, A.V. 2010b. Influence of organic solvents and doxorubicin on clusterization of water bound to DNA. *Chem. Phys. Technol. Surf.* 1:465–472 (in Russian).

Turov, V.V., Chodorowski, S., Leboda, R., Skubiszewska- Zięba, J., and Brei, V.V. 1999. Thermogravimetric and ^{1}H NMR spectroscopy studies of water on silicalites. *Colloids Surf. A: Physicochem. Eng. Aspects* 158:363–373.

Turov, V.V. Galagan, N.P., Grytsenko, I.V., and Chuiko, A.A. 2006a. Hydration of erythrocyte membranes and their interaction with fumed silica. *Biopolym. Cell.* 22:56–62.

Turov, V.V., Galagan, N.P., Gun'ko, V.M. et al. 2008a. Clusterization of water bound in freeze-dried bovine gametes containing organic additives. *Chem. Phys. Technol. Surf.* 14:494–501 (in Russian).

Turov, V.V. and Gorbyk, S.P. 2003. Determination of adhesion forces at interface of cell/water by ^{1}H NMR spectroscopy. *Ukr. Khim. Zh.* 69:80–86.

Turov, V.V., Gorbyk, S.P., and Chuiko, A.A. 2002a. Application of ^{1}H NMR spectroscopy for determination of energy for interaction of cells with aqueous medium. *Dokl. NAN Ukr.* 3:141–145.

Turov, V.V., Gorbyk, S.P., and Chuiko, A.A. 2002b. Effect of disperse silica on bound water in frozen cellular suspensions. *Probl. Cryobiol.* 3:16–23 (in Russian).

Turov, V.V., Gorbik, P.P., Laguta, I.V., and Ogenko, V.M. 1995a. Investigation of hydrated powders and suspensions of VO_2 by the ^{1}H NMR method of adsorbed molecules. *Colloids Surf. A: Physicochem. Eng. Aspects* 103:41–45.

Turov, V.V., Gorbik, P.P., Ogenko, V.M., Shulga, O.V., and Chuiko, A.A. 2001. Characteristic properties of metal–semiconductor phase transitions of vanadium dioxide in a polyethylene glycol medium containing tetraethylammonium bromide. *Colloids Surf. A: Physicochem. Eng. Aspects* 178:105–112.

Turov, V.V. and Gun'ko, V.M. 2011. *Clustered Water and Ways of Its Applications.* Kiev, Ukraine: Naukova Dumka (in Russian).

Turov, V.V., Gun'ko, V.M., Barvinchenko, V.N., Kerus, S.V., Buryak, O.A., and Chekhun, V.F., 2010c. Nanochemistry in development of novel remedies of transdermal insertion of medicinal preparations. *Dokl. NAS Ukraine* 7:180–188 (in Russian).

Turov, V.V., Gun'ko, V.M., Bogatyrev, V.M. et al. 2005. Structured water in partially dehydrated yeast cells and at partially hydrophobized fumed silica surface. *J. Colloid Interface Sci.* 283:329–343.

Turov, V.V., Gun'ko, V.M., and Chekhun, V.F. 2011b. Hydration properties of DNA and related nanocomposites. In: Shpak, A.P. and Chekhun, V.F. (Eds.) *Nanomaterials and Nanocomposites in Medicine, Biology, Ecology*. Kiev, Ukraine: Naukova Dumka, pp. 10–46 (in Russian).

Turov, V.V., Gun'ko, V.M., Gaishun, V.E., Kosenok, Y.A., and Golovan, A.P. 2010d. Application of NMR spectroscopy to determine the thermodynamic characteristics of water bound to OX-50 nanosilica. *J. Appl. Spectrosc.* 77:588–594.

Turov, V.V., Gun'ko, V.M., Galagan, N.P., Rugal, A.A., Barvinchenko, V.N., and Gorbyk, P.P. 2010e. Supramolecular structures with blood plasma proteins and nanosilica. In: Shpak, A.P. and Gorbyk, P.P. (Eds.) *Nanomaterials and Supramolecular Structures*. Dordrecht, the Netherlands: Springer, pp. 303–328.

Turov, V.V., Gun'ko, V.M., Gorbyk, P.P., Tsapko, M.D., Leboda, R., and Golovan, A.P. 2007. Processes of self-organization in water-organic systems in nanosized space of solids and biological objects. In: Shpak, A.P., and Gorbyk, P.P. (Eds.) *Physicochemistry of Nanomaterials and Supramolecular Structures*. Kiev, Ukraine: Naukova Dumka, pp. 91–156 (in Russian).

Turov, V.V., Gun'ko, V.M., Gorgyk, S.P. and Chuiko, A.A. 2003. Effect of fumed silica on phase equilibrium in water suspensions containing cells and proteins. *Dokl. NAN Ukr.* 9:150–156.

Turov, V.V., Gun'ko, V.M., Khomenko, K.N., Petin, A.Yu., Turov, A.V., and Gorbik, P.P. 2010f. Hydrogen adsorption on silicalite in the presence of benzene. *Russ. J. Phys. Chem.* 84:76–81.

Turov, V.V., Gun'ko, V.M., Kozynchenko, O.P., Tennison, S.P., and Mikhalovski, S.V. 2011d. Localization of hydrogen in porous space of activated carbon. *Chem. Phys. Technol. Surf.* 2:23–33 (in Russian).

Turov, V.V., Gun'ko, V.M., Kozynchenko, O.P., Tennison, S.R., and Mikhalovsky, S.V. 2011c. Effect of temperature and a weakly polar organic medium on water localization in slite-like pores of various sizes in microporous activated carbon. *Russ. J. Phys. Chem.* 85:2094–2099.

Turov, V.V., Gun'ko, V.M., Leboda, R. et al. 2002c. Influence of organics on structure of water adsorbed on activated carbons. *J. Colloid. Interface Sci.* 253:23–34.

Turov, V.V., Gun'ko, V.M., Petin, A.Yu. et al. 2010g. Co-adsorption of hydrogen and water in nanostrutured materials by the NMR-spectroscopy. *Nanosyst. Nanomater. Nanotechnol.* 8:153–175 (in Russian).

Turov, V.V., Gun'ko, V.M., Tsapko, M.D. et al. 2004. Influence of organic solvents on interfacial water at surface of silica gel and partly silylated fumed silica. *Appl. Surf. Sci.* 229:197–213.

Turov, V.V., Gun'ko, V.M., Turova, A.A., Morozova, L., and Voronin, E.F. 2011e. Interfacial behavior of concentrated HCl solution and water clustered at a surface of nanosilica in weakly polar solvents media, *Colloids Surf. A: Physicochem. Eng. Aspects* 390:48–55.

Turov, V.V., Gun'ko, V.M., Zarko, V.I., Bogatyr'ov, V.M., Dudnik, V.V., and Chuiko, A.A. 1996a. Water adsorption at pyrogenic silica surfaces modified by phosphorus compounds. *Langmuir* 12:3503–3510.

Turov, V.V., Gun'ko, V.M., Zarko, V.I. et al. 2006b. Weakly and strongly associated nonfreezable water bound in bones. *Colloids Surf. B: Biointerfaces* 48:167–175.

Turov, V.V., Karpenko, G.A., Bakai, E.A., and Chuiko, 1991a. Influence of electron–donor molecules on the properties of graphite–oxide dispersion. *Theoret. Eksp. Khim.* 27:201–205.

Turov, V.V., Kerus, S.V., and Gun'ko, V.M. 2009. Behaviour of water bound in bone marrow cells affected by organic solvents of different polarity. *Cryobiology* 59:102–112.

Turov, V.V., Kolychev, V.I., Bakay, E.A., Burushkina, T.N., and Chuiko, A.A. 1990. Study of water interaction with spherical carbon adsorbents by NMR spectroscopy with freezing of liquid phase. *Theoret. Eksp. Khim.* 26:111–115.

Turov, V.V. and Leboda, R. 1999. Application of ^1H NMR spectroscopy method for determination of characteristics of thin layers of water adsorbed on the surface of dispersed and porous adsorbents. *Adv. Colloid Interface Sci.* 79:173–211.

Turov, V.V. and Leboda, R. 2000. ^1H NMR spectroscopy of adsorbed molecules and free surface energy of carbon adsorbents. In: L.R. Radovic (Ed.) *Physical Chemistry of Carbons*, Vol. 27. Boca Raton, FL: CRC Press, pp. 67–124.

Turov, V.V., Leboda, R, Bogillo, V.I., and Skuboszewska-Zieba, J. 1995b. Structure of adsorption centers on a carbosil surface deduced from ^1H nuclear magnetic resonance spectroscopy data of adsorbed benzene and water molecules. *Langmuir* 11:931–935.

Turov, V.V., Leboda, R., Bogillo, V.I., and Skubiszewska-Zięba, J. 1996b. Surface properties of carbonaceous solids as examined by ^1H NMR spectroscopy using the bulk freezing procedure. *Adsorp. Sci. Technol.* 14:319–330.

Turov, V.V., Leboda, R., Bogillo, V.I., and Skubiszewska-Zięba, J. 1997a. Effect of carbon deposition on the surface of aerosil on the structure of adsorption sites and water and benzene interlayers—^1H NMR spectroscopy study. *J. Chem. Soc. Faraday Trans.* 93:4047–4053.

Turov, V.V., Leboda, R., Bogillo, V.I., and Skubiszewska-Zięba, J. 1997b. Study of hydrated structures on the surface of mesoporous silicas and carbosils by ^1H NMR spectroscopy of adsorbed water. *Langmuir* 13:1237–1244.

Turov, V.V., Leboda, R., and Skubiszewska- Zięba, J. 1998b. Changes in hydration properties of silica gel in a process of its carbonization by pyrolysis of acetylacetone Zn (Ti) acetylacetonates. *J. Colloid Interface Sci.* 206:58–65.

Turov, V.V. and Mironyuk, I.F. 1998. Adsorption layers of water on the surface of hydrophilic, hydrophobic and mixed silicas. *Colloids Surf. A: Physicochem. Eng. Aspects* 134:257–263.

Turov, V.V., Pogoreliy, K.V., Kolychev, V.I. and Burushkina, T.N. 1992. Proton magnetic relaxation of benzene in the pores of partly graphitized carbon adsorbents. *Ukr. Khim. Zh.* 58:470–475.

Turov, V.V., Pogorely, K.V., and Buruschkina, T.N. 1993. Anomalous chemical shifts of water and organic molecules adsorbed on graphitized surface of various types. *React. Kinet. Catal. Lett.* 50:279–284.

Turov, V.V., Pogorelyi, K.V., Mironova, L.A., and Chuiko, A.A. 1991b. Study of adsorption of simple molecules on thermoexfoliated graphite by ^1H NMR-spectroscopy. *Dokl. AN USSR* 128–131.

Turov, V.V., Pogorelyi, K.V., Mironova, L.A., and Chuiko, A.A. 1991c. Study of water adsorption on thermoexfoliated graphite. *Russ. J. Phys. Chem.* 65:170–174.

Turov, V.V., Pokrovskii, V.A., and Chuiko, A.A. 1994b. Effect of serum albumin on the temperature of formation of eutectics in binary solutions of organic compounds. *Biophysica* 39:988–992 (in Russian).

Turov, V.V., Turov, A.V., and Arkharov, A.V. 1998a. Dependence of ice formation of dioxynaphtalines on structure of hydrate shells of their crystals. *J. Appl. Chem.* 65:557–561 (in Russian).

Turov, V.V., Turova, A.A., Goncharuk, E.V., and Gun'ko, V.M. 2008b. Adsorption of methane with the presence of water on oxide, polymer, carbon adsorbents studied using ^1H NMR spectroscopy at low temperatures. *Appl. Surf. Sci.* 255:3310–3317.

Uccello-Barretta, G., Nazzi, S., Balzano, F. et al. 2008. Enhanced affinity of ketotifen towards tamarind-seed polysaccharide in comparison with hydroxyethylcellulose and hyaluronic acid: A nuclear magnetic resonance investigation. *Bioorg. Med. Chem.* 16:7371–7376.

Ulijn, R.V. 2006. Enzyme-responsive materials: A new class of smart biomaterials. *J. Mater. Chem.* 16:2217–2225.

Ulrich, G.D. 1984. Flame synthesis of fine particles. *Chem. Eng. News* 6:22–29.

Ulumbekov, E.G. and Chelyshev, Yu.A. (Eds.) 2002. *Histology*, 2nd edn. Moscow, Russia: GEOTAR.

Unsworth, L.D., Sheardown, H., and Brash, J.L. 2005. Polyethylene oxide surfaces of variable chain density by chemisorption of PEO-thiol on gold: Adsorption of proteins from plasma studied by radiolabelling and immunoblotting. *Biomaterials* 26:5927–5933.

Urbanczyk, L., Hrobarikova, J., Calberg, C., Jerome, R., and Grandjean, J. 2006. Motional heterogeneity of intercalated species in modified clays and poly(ε-caprolactone)/clay nanocomposites. *Langmuir* 22:4818–4824.

Utracki, L.A., Sepehr, M., and Boccaleri, E. 2007. Synthetic, layered nanoparticles for polymeric nanocomposites (PNCs). *Polym. Adv. Technol.* 18:1–37.

Vakhrushev, A. and Suyetin, M. 2009. Methane storage in bottle-like nanocapsules. *Nanotechnology* 20:125602–125605.

Valckenborg, R.M.E. 2001. NMR on technological porous materials. PhD thesis. Eindhoven, the Netherlands: Eindhoven University of Technology.

Valckenborg, R., Pel, L., and Kopinga, K. 2000. Cryoporometry and relaxation of water in porous materials. *Proceedings of the 15th European Experimental NMR Conference (EENC 2000)*, June 12–17, 2000, University of Leipzig, Germany.

Valckenborg, R., Pel, L., and Kopinga, K. 2001. Cryoporometry and relaxometry of water in silica-gels. *Magn. Reson. Imag.* 19 (2001) 489–491.

Valiullin, R. and Furó, I. 2002a. Low-temperature phase separation of a binary liquid mixture in porous materials studied by cryoporometry and pulsed-field-gradient NMR. *Phys Rev E* 66:031508.

Valiullin, R. and Furó, I. 2002b. The morphology of coexisting liquid and frozen phases in porous materials as revealed by exchange of nuclear spin magnetization followed by H-1 nuclear magnetic resonance. *J. Chem. Phys.* 117:2307–2316.

Valiullin, R. and Furó, I. 2002c. Phase separation of a binary liquid mixture in porous media studied by nuclear magnetic resonance cryoporometry. *J. Chem. Phys.* 116:1072–1076.

Valiullin, R., Kortunov, P., Kärger, J., and Timoshenko, V. 2005. Concentration-dependent self-diffusion of adsorbates in mesoporous materials. *Magn. Reson. Imag.* 23:209–214.

Valiullin, R., Vargas-Kruså, D., and Furó, I. 2004. Liquid–liquid phase separation in micropores. *Curr. Appl. Phys.* 4:370–372.

Vallet-Regi, M. 2010. Nanostructured mesoporous silica matrices in nanomedicine. *J. Int. Medicine* 267:22–43.

van Blaaderen, A. 2003. Chemistry: Colloidal molecules and beyond. *Science* 301:470–471.

VanderHart, D.L., Asano, A., and Gilman, J.W. 2001a. NMR measurements related to clay-dispersion quality and organic-modifier stability in nylon-6/clay nanocomposites. *Macromolecules* 34:3819–3822.

VanderHart, D.L., Asano, A., and Gilman, J.W. 2001b. Solid-state NMR investigation of paramagnetic Nylon-6 clay nanocomposites. 1. Crystallinity, morphology, and the direct influence of Fe^{3+} on nuclear spins. *Chem. Mater.* 13:3781–3795.

VanderHart, D.L., Asano, A., and Gilman, J.W. 2001c. Solid-state NMR investigation of paramagnetic Nylon-6 clay nanocomposites. 2. Measurement of clay dispersion, crystal stratification, and stability of organic modifiers. *Chem. Mater.* 13:3796–3809.

vanderHeijden, G.H.A., Huinink, H.P., Pel, L., and Kopinga, K. 1997. One-dimensional scanning of moisture in heated porous building materials with NMR. *J. Magn. Reson.* 208:235–242.

VanderWiel, D.P., Pruski, M., and King, T.S. 1999. A kinetic study on the adsorption and reaction of hydrogen over silica-supported ruthenium and silver–ruthenium catalysts during the hydrogenation of carbon monoxide. *J. Catal.* 188:186–202.

van Eck, E.R.H., Smith, M.E., and Kohn, S.C. 1999. Observation of hydroxyl groups by ^{17}O solid state multiple quantum MAS NMR in sol–gel-produced silica. *Solid State Nucl. Magn. Reson.* 15:181–188.

Vansant, E.F., Van Der Voort, P., and Vrancken, K.C. 1995. *Characterization and Chemical Modification of the Silica Surface*. Amsterdam, the Netherlands: Elsevier.

van Vleck, J.H. 1932. *The Theory of Electric and Magnetic Susceptibilities*. Oxford, U.K.: Oxford University Press.

Varetska, T.V., Tsinkalovska, S.M., and Demchenko, O.P. 1972. Physicochemical properties of tryptic fragment of fibrinogen—Inhibitor of fibrin polymerization. *Ukr. Biochem. J.* 44:418–422.

Vargas-Florencia, D., Petrov, O.V., and Furó, I. 2006. Inorganic salt hydrates as cryoporometric probe materials to obtain pore size distribution. *J. Phys. Chem. B* 110:3867–3870.

Vargas-Florencia, D., Petrov, O.V., and Furó, I. 2007. NMR cryoporometry with octamethylcyclotetrasiloxane as a probe liquid. Accessing large pores. *J. Colloid Interface Sci.* 305:280–285.

Varghese, N., Mogera, U., Govindaraj, A. et al. 2009. Binding of DNA nucleobases and nucleosides with graphene. *ChemPhysChem* 10:206–210.

Varisli, O., Uguz, C., Agca, C., and Agca, Y. 2009. Motility and acrosomal integrity comparisons between electro-ejaculated and epididymal ram sperm after exposure to a range of anisosmotic solutions, cryoprotective agents and low temperatures. *Anim. Reproduc. Sci.* 110:256–268.

Vartapetyan, R. Sh., Voloshchuk, A.M., Dubinin, M.M. and Moskovskaya, T.A. 1989. Adsorption of water vapors and micropore structures of carbon adsorbents. 16. Comparative method of analysis of the structure of carbon adsorbents. *Russ. Chem. Bull.* 37:1751–1755.

Vartzouma, Ch., Louloudi, M., Prodromidis, M., and Hadjiliadis, N. 2004. Synthesis and characterization of NAD$^+$-modified silica: A convenient immobilization of biomolecule via its phosphate group. *Mater. Sci. Eng. C* 24:473–477.

Vasilieva, V.S., Kiselev, A.V., Nikitin, J.S., Petrova, R.S., and Scherbakova, K.D. 1961. Graphitized carbon black as adsorbent in gas chromatography. *Zh. Fiz. Khim.* 35:1889–1891.

Vassiliou, A.A., Bikiaris, D., El Mabrouk, Kh., and Kontopoulou, M. 2011. Effect of evolved interactions in poly(butylene succinate)/fumed silica biodegradable in situ prepared nanocomposites on molecular weight, material properties, and biodegradability. *J. Appl. Polym. Sci.* 119:2010–2024.

Vega, A.J. 1988. Heteronuclear chemical-shift correlations of silanol groups studied by two-dimensional cross-polarization magic angle spinning NMR. *J. Am. Chem. Soc.* 110:1049–1054.

Velikov, V., Borick, S., and Angell, C.A. 2001. The glass transition of water, based on hyperquenching experiments. *Science* 294:2335–2338.

Verdaguer, A., Sacha, G.M., Bluhm, H., and Salmeron, M. 2006. Molecular structure of water at interfaces: Wetting at the nanometer scale. *Chem. Rev.* 106:1478–1510.

Vernon, R.B., Gooden, M.D., Lara, S.L., and Wight, T.N. 2005. Native fibrillar collagen membranes of micron-scale and submicron thickness for cell support and perfusion. *Biomaterials* 26:1109–1117.

Vissers, J.P.R., Merex, F.P.M., Bouwens, S.M.A.M., de Beer, V.H.J. and Prins, R. 1988. Carbon-covered alumina as support for sulfide catalysis. *J. Catal.* 114:291–302.

Vogel, M., Medick, P., and Rössler, E. 2000. Slow molecular dynamics in binary organic glass formers. *J. Mol. Liquid.* 86:103–108.

Vogler, E.A. 1998. Structure and reactivity of water at biomaterial surfaces. *Adv. Colloid Interface Sci.* 74:69–117.

Voitko, K.V., Whitby, R.L.D., Gun'ko, V.M. et al. 2011. Morphological and structural features of nano- and macroscale carbons affecting hydrogen peroxide decomposition in aqueous medium. *J. Colloid Interface Sci.* 361:129–136.

Voloshchuk, A.M., Vartapetian, R. Sh., Dubinin, M.M., Karger, J. and Pfeifer, H. 1986. Mobility of molecules adsorbed in active carbons. *Proc. Fourth International Carbon Conference—Carbon 86*. Baden-Baden, June 4., Proc. Baden-Baden, pp. 309–311.

Volovenko, Yu.M., Kartzev, V.G., Komarov, I.V., Turov, A.V., and Khilya, V.P. 2011. *Nuclear Magnetic Resonance Spectroscopy for Chemists*. Moscow, Russia: ICSPF.

Von Hippel, A.R. (Ed.) 1954. *Dielectrics and Waves*. New York: Wiley.

von Kraemer, S., Sagidullin, A.I., Lindbergh, G., Furó, I., Persson, E., and Jannasch, P. 2008. Pore size distribution and water uptake in hydrocarbon and perfluorinated proton-exchange membranes as studied by NMR cryoporometry. *Fuel Cells* 8:262–269.

Voríšek, J. 2000. Functional morphology of the secretary pathway organelles in yeast. *Microsc. Res. Tech.* 51:530–546.

Vosegaard, T., Larsen, F.H., Jakobsen, H.J., Ellis, P.D., and Nielsen, N.C. 1997. Sensitivity enhanced multiple-quantum MAS NMR of half-integer quadrupolar nuclei. *J. Am. Chem. Soc.* 119:9055–9056.

Vroman, L. 2008. Biointerface perspective. Finding seconds count after contact with blood (and that is all I did). *Colloids and Surfaces B: Biointerfaces* 62:1–4.

Vroman, L. and Adams, A.L. 1969. Findings with the recording ellipsometer suggesting rapid exchange of specific plasma proteins at liquid/solid interfaces. *Surf. Sci.* 16:438–446.

Vyalikh, A., Emmeler, Th., Grünberg, B. et al. 2007. Hydrogen bonding of water confined in controlled-pore glass 10–75 studied by ^{1}H-solid state NMR. *Z. Phys. Chem.* 221:155–168.

Vysotskaja, E.V., Tarasevich, Ju.I., Klimiva, L.N., and Kuzmenko, L.N. 2002. Synthesis of hydroxyapatite and the application of obtained materials for the extraction of heavy metal ions from aqueous solutions. *Chem. Technol. Water* 24:535–546.

Wagner, V., Dullaart, A., Bock, A.K., and Zweck, A. 2006. The emerging nanomedicine landscape. *Nat. Biotechnol.* 24:1211–1217.

Wagner, P. and Vali, G. (Eds.) 1988. *Atmospheric Aerosols and Nucleation*. Berlin, Germany: Springer.

Walaszek, B., Yeping, X., Adamczyk, A. et al. 2009. ^{2}H-solid-state-NMR study of hydrogen adsorbed on catalytically active ruthenium coated mesoporous silica materials. *Solid State Nucl. Magn. Reson.* 35:164–171.

Walker, P.L. Jr. 1990. Carbon an old but new material revisited. *Carbon* 28:261–279.

Wallace, D.G. and Rosenblatt, J. 2003. Collagen gel systems for sustained delivery and tissue engineering. *Adv. Drug Deliv. Rev.* 55:1631–1649.

Wang, J.-H. 2000. A comprehensive evaluation of the effects and mechanisms of antifreeze proteins during low-temperature preservation. *Cryobiology* 41:1–9.

Wang, K. and Do, D.D. 1999. Sorption equilibria and kinetics of hydrocarbons onto activated carbon samples having different micropore size distributions. *Adsorption* 5:25–37.

Wang, A. and Kabe, T. 1999. Fine-tuning of pore size of MCM-41 by adjusting the initial pH of the synthesis mixture. *Chem. Commun.* No 20:2067–2068.

Wang, K., Qiao, S., and Hu, X. 2000. On the Performance of HIAST and IAST in the prediction of multicomponent adsorption equilibrium. *Sep. Purif. Technol.* 20:243–249.

Wang, S.-D. and Scrivener, K.L. 2003. ^{29}Si and ^{27}Al NMR study of alkali-activated slag. *Cem. Concr. Res.* 33:769–774.

Wang, Zh. and Stein, A. 2008. Morphology control of carbon, silica, and carbon/silica nanocomposites: From 3D ordered macro-/mesoporous monoliths to shaped mesoporous particles. *Chem. Mater.* 20:1029–1040.

Warren, B.E. 1934. X-ray diffractions study of carbon black. *J. Chem. Phys.* 2:551–555.

Warren, W.S. (Ed.) 1990. *Advances in Magnetic Resonance: The Waugh Symposium*, Vol. 14. San Diego, CA: Academic Press.

Washton, N.M. 2007. Determination of oxide surface reactivity via solid-state nuclear magnetic resonance and ab initio methods. PhD thesis. Penn State University.

Washton, N.M., Brantley, S.L., and Mueller, K.T. 2008. Probing the molecular-level control of aluminosilicate dissolution: A sensitive solid-state NMR proxy for reactive surface area. *Geochim. Cosmochim. Acta* 72:5949–5961.

Watson, A.T. and Chang, C.T.P. 1997. Characterizing porous media with NMR methods. *Prog. Nucl. Magn. Reson. Spectrosc.* 31:343–386.

Wattraint, O., Arnold, A., Auger, M., Bourdillon, C., and Sarazin, C. 2005a. Lipid bilayer tethered inside a nanoporous support: A solid-state nuclear magnetic resonance investigation. *Anal. Biochem.* 336:253–261.

Wattraint, O. and Sarazin, C. 2005. Diffusion measurements of water, ubiquinone and lipid bilayer inside a cylindrical nanoporous support: A stimulated echo pulsed-field gradient MAS-NMR investigation. *Biochim. Biophys. Acta Biomembranes* 1713:65–72.

Wattraint, O., Warschawski, D.E., and Sarazin, C. 2005b. Tethered or adsorbed supported lipid bilayers in nanotubes characterized by deuterium magic angle spinning NMR spectroscopy. *Langmuir* 21:3226–3228.

Waugh, J.S., Huber, L.M., and Haeberlen, U. 1968. Approach to high-resolution NMR in solids. *Phys. Rev. Lett.* 20:180–182.

Webb, G.A. (Ed.) 2003. *Nuclear Magnetic Resonance*, Vol. 32. London, U.K.: Royal Society of Chemistry.

Webb, G.A. (Ed.) 2009. *Nuclear Magnetic Resonance*, Vol. 38. London, U.K.: Royal Society of Chemistry.

Webber, J.B.W. 2010. Studies of nano-structured liquids in confined geometries and at surfaces. *Prog. Nucl. Magn. Res. Spectrosc.* 56:78–93.

Webber, J.B.W., Anderson, R., Strange, J.H., and Tohidi, B., 2007a. Clathrate formation and dissociation in vapour/water/ice/hydrate systems in SBA-15, Sol-Gel and CPG porous media, as probed by NMR relaxation, novel protocol NMR cryoporometry, neutron scattering and ab-initio quantum-mechanical molecular dynamics simulation. *Magn. Reson. Imag.* 25:533–536.

Webber, J.B.W. and Dore, J.C. 2008. Neutron diffraction cryoporometry—A measurement technique for studying mesoporous materials and the phases of contained liquids and their crystalline forms. *Nucl. Instrum. Methods Phys. Res. A* 586:356–366.

Webber, J.B.W., Dore, J.C., Strange, J.H., Anderson, R., and Tohidi, B. 2007b. Plastic ice in confined geometry: The evidence from neutron diffraction and NMR relaxation. *J. Phys.: Condens. Matter* 19:415117 (1–12).

Webber, J.B.W., Strange, J.H., and Dore, J.C. 2001. An evaluation of NMR cryoporometry, density measurement and neutron scattering methods of pore characterisation. *Magn. Reson. Imag.* 19:395–399.

Wehrer, A., Wehrer, P., and Duval, X. 1980. Cinetique de la decomposition de l'allene sur le carbone porte a de hautes temperatures. *Bull. Soc. Chim. Fr.* 11–12:434–440.

Weiner, S., Traub, W., and Wagner, H.D. 1999. Lamellar bone: Structure–function relations. *J. Struct. Biol.* 126:241–255.

Wergzyn, J. and Gurevich, M. 1996. Adsorbent storage of natural gas. *Appl. Energy* 55:71–83.

Whistler, R.L., BeMiller, J.N., and Paschall, E.F. 1984. *Starch, Chemistry and Technology*. London, U.K.: Academic Press.

Whitby, R.L.D., Gun'ko, V.M., Korobeinyk, A. et al. 2012. Driving forces of conformational changes in single-layer graphene oxide. *ACS Nano* 10.1021/nn3002278.

Whitby, R.L.D., Korobeinyk, A., Gun'ko, V.M. et al. 2011. pH driven-physicochemical conformational changes of single-layer graphene oxide. *Chem. Comm.* 47:9645–9647.

Wickens, D.A. 1990. A generalization of the slit-model of pores in microporous carbons. *Carbon* 28:97–101.

Wider, G. 1998. Technical aspects of NMR spectroscopy with biological macromolecules and studies of hydration in solution. *Prog. Nucl. Magn. Reson. Spectrosc.* 32:193–275.

Wiench, J.W., Avadhut, Y.S., Maity, N. et al. 2007a. Characterization of covalent linkages in organically functionalized MCM-41 mesoporous materials by solid-state NMR and theoretical calculations. *J. Phys. Chem. B* 111:3877–3885.

Wiench, J.W., Bronniman, C.E., Lin, V.S.-Y., and Pruski, M. 2007b. Chemical shift correlation NMR spectroscopy with indirect detection in fast rotating solids: Studies of organically functionalized mesoporous silicas. *J. Am. Chem. Soc.* 129:12076–12077.

Wiench, J.W., Lin, V.S.-Y., and Pruski, M. 2008. ^{29}Si NMR in solid state with CPMG acquisition under MAS. *J. Magn. Reson.* 193:233–242.

Wiggins, P. 2008. Life depends upon two kinds of water. *PLoS ONE* 3:e1406(1–16). doi:10.1371/journal.pone.0001406.

Wikberg, E., Sparrman, T., Viklund, C., Jonsson, T., and Irgum, K. 2011. A ^{2}H nuclear magnetic resonance study of the state of water in neat silica and zwitterionic stationary phases and its influence on the chromatographic retention characteristics in hydrophilic interaction high-performance liquid chromatography. *J. Chromatogr. A* 1218:6630–6638.

Williams, D.F. (Ed.) 1987. *Blood Compatibility*, Vol. 1. Boca Raton, FL: CRC Press.

Wilson, E.E., Awonusi, A., Morris, M.D., Kohn, D.H., Techlenburg, M.M.J., and Beck, L.W. 2005b. Highly ordered interstitial water observed in bone by nuclear magnetic resonance. *J. Bone Mineral Res.* 20:625–634.

Wilson, R.M., Elliot, J.C., Dowker, S.E.P., and Rodrigues-Lorenco, L.M. 2005a. Rietveld refinements and spectroscopic studies of the structure of Ca-deficient apatite. *Biomaterials* 26:1317–1327.

Wind, R.A. and Hu, J.Z. 2006. In vivo and ex vivo high-resolution ^{1}H NMR in biological systems using low-speed magic angle spinning. *Prog. Nucl. Magn. Reson. Spectrosc.* 49:207–259.

Witzel, W.M. and Das Sarma, S. 2007. Multiple-pulse coherence enhancement of solid state spin qubits. *Phys. Rev. Lett.* 98:077601-1–077601-4.

Wöhnert, J., Dingley, A.J., Stoldt, M., Görlach, M., Grzesiek, S., and Brown, L.R. 1999. Direct identification of NH…N hydrogen bonds in non-canonical base pairs of RNA by NMR spectroscopy. *Nucl. Acids Res.* 27:3104–3110.

Wooten, E.W., Mueller, K.T., and Pines, A. 1992. New angles in NMR sample spinning. *Acc. Chem. Res.* 25:209–215.

Wouters, B.H., Chen, T., Dewilde, M., and Grobet, P.J. 2001. Reactivity of the surface hydroxyl groups of MCM-41 towards silylation with trimethylchlorosilane. *Micropor. Mesopor. Mater.* 44–45:453–457.

Wright, P.E. and Dyson, H.J. 1999. Intrinsically unstructured proteins: Re-assessing the protein structure–function paradigm. *J. Mol. Biol.* 293:321–331.

Wright, J.D. and Sommerdijk, N.A.J.M. 2001. *Sol–Gel Materials Chemistry and Applications*. London, U.K.: Taylor & Francis Group.

Wu, Y., Ackerman, J.L., Kim, H.M., Rey, C., Barroug, A., and Glimcher, M.J. 2002. Nuclear magnetic resonance spin-spin relaxation of the crystals of bone, dental enamel, and synthetic hydroxyapatite. *J. Bone Mineral Res.* 17:472–480.

Wu, C.-H., Jeng, J.-S., Chia, J.-L., and Ding, S. 2011. Multi-nuclear liquid state NMR investigation of the effects of pH and addition of polyethyleneglycol on the long-term hydrolysis and condensation of tetraethoxysilane. *J. Colloid Interface Sci.* 353:124–130.

Wu, Sh., Ju, H., and Liu, Y. 2007. Conductive mesocellular silica–carbon nanocomposite foams for immobilization, direct electrochemistry, and biosensing of proteins. *Adv. Funct. Mater.* 17:585–592.

Wu, V.W.-K. and Kure(Ko), F. 2010. Preparation for optimal conformation of lysozyme with nanodiamond and nanosilica as carriers. *Chin. J. Chem.* 28:2520–2526.

Wüthrich, K. 1986. *NMR of Proteins and Nucleic Acids*. New York: Wiley-Interscience.

Wüthrich, K. (Ed.) 1995. *NMR in Structural Biology*. Singapore: World Scientific.

Xia, Y., Dosseh, G., Morineau, D., and Alba-Simionesco, C. 2006. Phase diagram and glass transition of confined benzene. *J. Phys. Chem. B* 110:19735–19744.

Xidos, J.D., Li, J., Zhu, T. et al. 2002. *GAMESOL–Version 3.1*. Minneapolis, MN: University of Minnesota.

Xu, R. 1998. Shear Plane and hydrodynamic diameter of microspheres in suspension. *Langmuir* 14:2593–2597.

Xu, X. and Kirkpatrick, R.J. 2006. NaCl interaction with interfacially polymerized polyamide films of reverse osmosis membranes: A solid-state ^{23}Na NMR study. *J. Membr. Sci.* 280:226–233.

Xue, J.M., Tan, C.H., and Lukito, D. 2006. Biodegradable polymer–silica xerogel composite microspheres for controlled release of gentamicin. *J. Biomed. Mater. Res. B Appl. Biomater.* 78B:417–422.

Xuecheng, C. and Weimin, W. 1996. Nucleation and characteristics of liquid nitrogen. *J. Appl. Meteorol.* 35:1582–1586.

Yang, F., Bian, C., Zhu, L., Zhao, G., Huang, Z., and Huang, M. 2007. Effect of human serum albumin on drug metabolism: Structural evidence of esterase activity of human serum albumin. *J. Struct. Biol.* 157:348–355.

Yang, H., Coombs, N., Sokolov, I., and Ozin, G.A. 1997. Registered growth of mesoporous silica films on graphite. *J. Mater. Chem.* 7:1285–1290.

Yannas, I.V. 2001. *Tissue and Organ Regeneration in Adults*. New York: Springer.

Yao, Y. and Liu, D. 2012. Comparison of low-field NMR and mercury intrusion porosimetry in characterizing pore size distributions of coals. *Fuel* 95:152–158.

Ye, X., Kumar, R.A., and Patel, D.J. 1995. Molecular recognition in the bovine immunodeficiency virus Tat peptide-TAR RNA complex. *Chem. Biol.* 2:827–840.

Yoichiro, T., Yuriko, T., and Shinji, K. 2002. Water-containing powder composition, process for producing the same, and cosmetic preparation containing the powder composition. Patent EP1206928.

Yongli, C., Xiufang, Z., Yandao, G., Nanming, Z., Tingying, Z., and Xinqi, S. 1999. Conformational changes of fibrinogen adsorption onto hydroxyapatite and titanium oxide nanoparticles. *J. Colloid Interface Sci.* 214:38–45.

Yoon, Y., Pope, J., and Wolfe, J. 2003. Freezing stresses and hydration of isolated cell walls. *Cryobiology* 46:271–276.

Young, R. 1986. *Cellulose Structure Modification and Hydrolysis*. New York: Wiley.

Young, M.E., Carroad, P.A., and Bell, R.L. 1980. Estimation of diffusion coefficients of proteins. *Biotechnol. Bioeng.* 22:947–955.

Youngman, R.E. and Sen, S. 2004. Structural role of fluorine in amorphous silica. *J. Non-Cryst. Solid.* 349:10–15.

Yu, Y.-Y., Chen, C.-Y., and Chen, W.-C. 2003. Synthesis and characterization of organic–inorganic hybrid thin films from poly(acrylic) and monodispersed colloidal silica. *Polymers* 44:593–601.

Yu, L.P., Hajduk, P.J., Mack, J., and Olejniczak, E.T. 2006. Structural studies of Bcl-xL/ligand complexes using F-19 NMR. *J. Biomol. NMR* 34:221–227.

Yuan, C.B., Zhao, D.Q., Liu, A.Z., and Ni, J.Z. 1995. A NMR study of the interaction of silica with dipalmitoylphosphatidylcholine liposomes. *J. Colloid Interface Sci.* 172:536–538.

Yurke, B., Lin, D.C., and Langrana, N.A. 2006. Use of DNA nanodevices in modulating the mechanical properties of polyacrylamide gels. In: Carbone, A., and Pierce, N.A. (Eds.) *Lecture Notes in Computer Science*, Vol. 3892. Berlin, Germany: Springer-Verlag; p. 417–426.

Zaporozhets, O.A., Shulga, O.V., Nadzhafova, O.Yu., Turov, V.V., and Sukhan, V.V. 2000. The nature of the binding of high-molecular weight aminoammonium and quaternary ammonium salts with the amorphous silica surface. *Colloids Surf. A: Physicochem. Eng. Aspects* 168:103–108.

Zarko, V.I., Gun'ko, V.M., Chibowski, E., Dudnik, V.V., and Leboda, R. 1997. Study of some surface properties of pyrogenic alumina/silica materials. *Colloids Surf. A: Physicochem. Eng. Aspects* 127:11–18.

Zarko, V.I., Polischuk, G.E., Rybak, O.N. et al. 2008. Supramolecular effects in hydrated flour. *Chem. Phys. Technol. Surf.* 14:467–477 (in Russian).

Zatsepina, G.N. 1998. *Physical Properties and Structure of Water*, 3rd edn. Moscow, Russia: Moscow State University (in Russian).

Zawadzki, J. 1988. Infrared spectroscopy in surface chemistry of carbons. *Chem. Phys. Carbon* 21:147–380.

Zecchina, A., Spoto, G., and Bordiga, S. 2005. Probing the acid sites in confined spaces of microporous materials by vibrational spectroscopy. *Phys. Chem. Chem. Phys.* 7:1627–1642.

Zeeb, M., Jacob, M.H., Schindler, T., and Balbach, J. 2003. ^{15}N relaxation study of the cold shock protein CspB at various solvent viscosities. *J. Biomol. NMR* 27:221–234.

Zhang, S. 2007. Direct detection of hydrogen peroxide via ^{1}H CEST-MRI. *Proc. Int. Soc. Magn. Reson. Med.* 15:3461.

Zhang, L., Gu, F.X., Chan, J.M., Wang, A.Z., Langer, R.S., and Farokhzad, O.C. 2007a. Nanoparticles in medicine: Therapeutic applications and developments. *Clin. Pharmacol. Ther.* 83:761–769.

Zhang, W., Han, X., Liu, X., and Bao, X. 2001. The stability of nanosized HZSM-5 zeolite: A high-resolution solid-state NMR study. *Micropor. Mesopor. Mater.* 50:13–23.

Zhang, X., Wu, W., Wang, J., and Liu, C. 2007b. Effects of sol aging on mesoporous silica thin films organization. *Thin Solid Films* 515:8376–8380.

Zhang, X., Wu, W., Wang, J., and Tian, X. 2008. Direct synthesis and characterization of highly ordered functional mesoporous silica thin films with high amino-groups content. *Appl. Surf. Sci.* 254:2893–2899.

Zhang, Y.-f., Yin, P., Zhao, X.-q. et al. 2009. O-Carboxymethyl-chitosan/organosilica hybrid nanoparticles as non-viral vectors for gene delivery. *Mater. Sci. Eng.* C 29:2045–2049.

Zhao, T.S., Kreuer, K.D., and Van Nguyen, T. (Eds.) 2007. *Advances in Fuel Cells*, Vol. 1. Amsterdam, the Netherlands: Elsevier.

Zhao, J., Li, B., Onda, K., Feng, M., and Petek, H. 2006. Solvated electrons on metal oxide surfaces. *Chem. Rev.* 106:4402–4427.

Zhdanov, V.P. 1991. *Elementary Physicochemical Processes on Solid Surfaces*. New York: Plenum Press.

Zheng, G. and Price, W.S. 2010. Solvent signal suppression in NMR. *Progr. Nucl. Magn. Reson. Spectrosc.* 56:267–288.

Zhou, J. and Fedkiw, P.S. 2004. Ionic conductivity of composite electrolytes based on oligo(ethylene oxide) and fumed oxides. *Solid State Ionics* 166:275–293.

Zhou, W. and Klinowski, J. 1998. The mechanism of channel formation in the mesoporous molecular sieve MCM-41. *Chem. Phys. Lett.* 292:207–212.

Zhou, Y., Li, Z., Hu, N., Zeng, Y., and Rusling, J.F. 2002. Layer-by-layer assembly of ultrathin films of hemoglobin and clay nanoparticles with electrochemical and catalytic activity. *Langmuir* 18:8573–8579.

Zhou, L., Liu, X., Li, J., Sun, Y., and Zhou, Y. 2006. Sorption/desorption equilibrium of methane in silica gel with pre-adsorption of water. *Colloids Surf. A: Physicochem. Eng. Aspects* 273:117–120.

Zhu, J., He, H., Zhu, L., Wen, X., and Deng, F. 2005. Characterization of organic phases in the interlayer of montmorillonite using FTIR and ^{13}C NMR. *J. Colloid Interface Sci.* 286:239–244.

Zhu, Y.W., Murali, S., Cai, W.W. et al. 2010. Synthesis and characterization of large-area graphene and graphite films on commercial Cu-Ni alloy foils. *Adv. Mater.* 22:3906–3924.

Zhu, S.B. and Robinson, G.W. 1991. Structure and dynamics of liquid water between plates. *J. Chem. Phys.* 94:1403–1410.

Zhuravlev, L.T. 2000. The surface chemistry of amorphous silica. Zhuravlev model. *Colloids Surf. A: Physicochem. Eng. Aspects* 173:1–38.

Zhuravskaya, N.K., Alekhina, L.T., and Otryashenkova, L.M. 1985. *Study and Quality Control of Meat and Beef Foodstuff*. Moscow, Russia: Agropromizdat (in Russian).

Zhurko, G.A. and Zhurko, D.A. 2011. *Chemcraft* (version 1.6, build 350). http://www.chemcraftprog.com (accessed December, 2011).

Zídek, L., Štefl, R., and Sklenár, V. 2001. NMR methodology for the study of nucleic acids. *Curr. Opin. Struct. Biol.* 11:275–281.

Zielinski, L.J. and Huerlimann, M.D. 2005. Probing short length scales with restricted diffusion in a static gradient using the CPMG sequence. *J. Magn. Reson.* 172:161–167.

Zimmerman, J.R., Holmes, B.G., and Lasater, J.A. 1956. A study of adsorbed water on silica gel by nuclear resonance techniques. *J. Phys. Chem.* 60:1157–1161.

Zipp, A., Kuntz, I.D., and James, T.L. 1976. An investigation of bound water in frozen erythrocytes by proton magnetic resonance spin-lattice. *J. Magn. Reson.* 24:411–424.

Zugenmaier, P. 2008. *Crystalline Cellulose and Cellulose Derivatives. Characterization and Structures*. Berlin, Germany: Springer-Verlag.

Zuiderweg, E.R.P. 2002. Mapping protein-protein interactions in solution by NMR spectroscopy. *Biochemistry* 41:1–7.

Zviyginzev, D.G. 1987. *Soil and Microorganisms*. Moscow, Russia: Moscow State University Press (in Russian).

Zyss, J. 1994. *Molecular Nonlinear Optics Materials Physics and Devices*. Orlando, FL: Academic Press.

Index

A

B

C

L

M

N

P

R

S